Pump Handbook

OTHER McGRAW-HILL HANDBOOKS OF INTEREST

American Institute of Physics · American Institute of Physics Handbook
American Society of Mechanical Engineers · ASME Handbooks
  Engineering Tables Metals Engineering—Processes
  Metals Engineering—Design Metals Properties
Baumeister and Marks · Standard Handbook for Mechanical Engineers
Berry, Bollay, and Beers · Handbook of Meteorology
Blatz · Radiation Hygiene Handbook
Brady · Materials Handbook
Burington · Handbook of Mathematical Tables and Formulas
Burington and May · Handbook of Probability and Statistics with Tables
Callender · Time-Saver Standards for Architectural Design Data
Chow · Handbook of Applied Hydrology
Condon and Odishaw · Handbook of Physics
Considine · Process Instruments and Controls Handbook
Considine and Ross · Handbook of Applied Instrumentation
Croft, Carr, and Watt · American Electricians' Handbook
Dean · Lange's Handbook of Chemistry
Etherington · Nuclear Engineering Handbook
Fink and Carroll · Standard Handbook for Electrical Engineers
Flügge · Handbook of Engineering Mechanics
Grant · Hackh's Chemical Dictionary
Hamsher · Communication System Engineering Handbook
Harris and Crede · Shock and Vibration Handbook
Henney · Radio Engineering Handbook
Hicks · Standard Handbook of Engineering Calculations
Hunter · Handbook of Semiconductor Electronics
Huskey and Korn · Computer Handbook
Ireson · Reliability Handbook
Juran · Quality Control Handbook
Kaelble · Handbook of X-rays
Kallen · Handbook of Instrumentation and Controls
King and Brater · Handbook of Hydraulics
Klerer and Korn · Digital Computer User's Handbook
Koelle · Handbook of Astronautical Engineering
Korn and Korn · Mathematical Handbook for Scientists and Engineers
Landee, Davis, and Albrecht · Electronic Designer's Handbook
Machol · System Engineering Handbook
Mantell · Engineering Materials Handbook
Markus · Electronics and Nucleonics Dictionary
Meites · Handbook of Analytical Chemistry
Merritt · Standard Handbook for Civil Engineers
Perry · Engineering Manual
Perry, Chilton, and Kirkpatrick · Chemical Engineers' Handbook
Richey · Agricultural Engineers' Handbook
Rothbart · Mechanical Design and Systems Handbook
Streeter · Handbook of Fluid Dynamics
Terman · Radio Engineers' Handbook
Truxal · Control Engineers' Handbook
Tuma · Engineering Mathematics Handbook
Tuma · Technology Mathematics Handbook
Urquhart · Civil Engineering Handbook
Watt and Summers · NFPA Handbook of the National Electrical Code

Pump Handbook

EDITED BY

IGOR J. KARASSIK
WILLIAM C. KRUTZSCH
WARREN H. FRASER

Worthington Pump Inc.

and

JOSEPH P. MESSINA

Public Service Electric and Gas Company
New Jersey Institute of Technology

McGRAW-HILL BOOK COMPANY

New York St. Louis San Francisco Auckland Bogotá
Dusseldorf Johannesburg London Madrid Mexico
Montreal New Delhi Panama Paris Sao Paulo
Singapore Sydney Tokyo Toronto

Library of Congress Cataloging in Publication Data

Main entry under title:

Pump handbook.

1. Pumping machinery—Handbooks, manuals, etc.
I. Karassik, Igor J., date.
TJ900.P79 621.6 75-22343
ISBN 0-07-033301-7

121314151617181920 VBVB 8987654

The editors for this book were Harold B. Crawford and Ross Kepler, the designer was Naomi Auerbach, and the production supervisor was Teresa F. Leaden. It was set in Linotype 21 by Bi-Comp, Inc.

Contents

Contributors

ARNOLD, CONRAD L., B.S. (E.E.) SECTION 6.2.3. FLUID COUPLINGS.
Director of Engineering—American Standard Industrial Division.

BECK, WESLEY W., B.S. (C.E.), P.E. SECTION 14. PUMP TESTING.
Hydraulic Consulting Engineer. Formerly with the Chief Engineers Office of the U.S. Bureau of Reclamation.

BENJES, H. H., B.S. (C.E), P.E. SECTION 10.2. SEWAGE.
Partner, Black & Veatch, Consulting Engineers.

BERGERON, WALLACE L., B.S. (E.E.) SECTION 6.1.2. STEAM TURBINES.
Senior Market Engineer, Elliott Company, A Division of Carrier Corp.

BIHELLER, H. JOSEPH, B.S. (M.E.), M.S. (M.E.) SECTION 2.5. SPECIAL VARIANTS OF DYNAMIC PUMPS.
Pump Specialist, American Electric Power Service Corporation.

BIRGEL, W. J., B.S. (E.E.) SECTION 6.2.1. EDDY-CURRENT COUPLINGS.
Manager, Systems and Control Marketing Department, Electric Machinery Manufacturing Company.

BIRK, JOHN R., B.S. (M.E.), P.E. SECTION 10.6. CHEMICAL INDUSTRY.
Vice President, The Duriron Company, Inc.

BUSE, FRED, BACHELOR OF MARINE ENGINEERING SECTION 3.1. POWER PUMPS. PUMPS.
Chief Engineer, Standard Pump—Aldrich Division, Ingersoll-Rand Company.

CHAPUS, EDMOND E., M.S. (Mining), M.S. (C.E.) SECTION 10.21.1. HYDRAULIC TRANSPORT OF SOLIDS.
President, SOGREAH, Inc.

CLOPTON, D. E., B.S. (C.E.), P.E. SECTION 10.1. WATER SUPPLY.
Assistant Project Manager, Water Quality Division, URS/Forrest and Cotton, Inc., Consulting Engineers.

CONDOLIOS, ELIE, B.S. (M.E.) SECTION 10.21.1. HYDRAULIC TRANSPORT OF SOLIDS.
Head, Hydraulic Transportation Division, Techniques des Fluides, Grenoble, France.

CORONEOS, J. N., B.S. (M.E.) SECTION 10.15. HYDRAULIC SERVOSYSTEMS.
Engineering Manager, Fluid Power Drives, The Falk Corporation.

COSTIGAN, JAMES L., B.S. (Chem) SECTION 10.9. FOOD AND BEVERAGE PUMPING.
Sales Manager, Tri-Clover Division, Ladish Company.

COZZARIN, E., B.S. (M.E.), B.S. (I.E.) SECTION 6.2.6. CLUTCHES.
Product Manager, Clutches, Morse Chain, Division of Borg-Warner Corporation.

CZARNECKI, G. J. B.Sc., M.Sc. (Tech.) SECTION 3.3. SCREW PUMPS.
Chief Engineer, DeLaval IMO Pump Division.

DiVONA, A. A., B.S. (M.E.) SECTION 6.1.1. ELECTRIC MOTORS AND MOTOR CONTROLS.
Special Representative, Westinghouse Electric Corporation.

DOLAN, A. J., B.S. (E.E.), M.S. (E.E.), P.E. SECTION 6.1.1. ELECTRIC MOTORS AND MOTOR CONTROLS.
Fellow Engineer, Westinghouse Electric Corporation.

DORNAUS, WILSON, L., B.S. (C.E.), P.E. SECTION 11. INTAKES AND SUCTION PIPING.
Engineering Specialist, Bechtel Power Corporation.

ELVITSKY, A. W., B.S. (M.E.), M.S. (M.E.), P.E. SECTION 10.7. PETROLEUM INDUSTRY.
Vice President and Chief Engineer, United Centrifugal Pumps.

FOSTER, W. E., B.S. (C.E.), P.E. SECTION 10.2. SEWAGE.
Project Manager, Black & Veatch, Consulting Engineers.

FRASER, WARREN H., B.M.E. SECTION 5. MATERIALS OF CONSTRUCTION.
Chief Design Engineer, Research and Development, Engineered Products, Worthington Pump Inc.

FREEBOROUGH, ROBERT M., B.S. (Min.E.) SECTION 3.2. STEAM PUMPS.
Manager, Parts Marketing, Worthington Pump Corporation (U.S.A.).

GIDDINGS, J. F., DIPLOMA, MECHANICAL, ELECTRICAL, AND CIVIL ENGINEERING SECTION 10.8. PULP AND PAPER MILL SERVICES.
Development Manager, Parsons & Whittemore, Lyddon, Ltd. England.

GLANVILLE, ROBERT H., M.E. SECTION 10.20. METERING.
Vice President Engineering, BIF, A Unit of General Signal.

GRIMES, A. S., B.S. (M.E.) SECTION 8. SUPERVISORY AND MONITORING INSTRUMENTATION.
Assistant Head, Mechanical Engineering Division, American Electric Power Service Corporation.

GUNTHER, F. J., B.S. (M.E.), M.S. (M.E.) SECTION 6.1.3. ENGINES.
Sales Engineer, Waukesha Motor Company.

HAENTJENS, W. D., B.M.E., M.S. (M.E.), P.E. SECTION 10.10. MINING SERVICES.
President and Chief Engineer, Barrett, Haentjens & Company.

HARRIS, ROWLAND A., B.S. Engineering SECTION 10.4. FIRE PUMPS.
Manager, Johnston Pump Company Service Center.

HEISLER, S. I., B.S. (M.E.), P.E. SECTION 12. SELECTING AND PURCHASING PUMPS.
Project Manager, Bechtel Power Corporation.

HONEYCUTT, F. G., Jr., B.S. (C.E.), P.E. SECTION 10.1. WATER SUPPLY.
Assistant Vice President and Head Water Quality Division, URS/Forrest and Cotton, Inc., Consulting Engineers.

JEKAT, WALTER K., Dipl.-Ing SECTION 2.1. CENTRIFUGAL PUMP THEORY.
Designer of Turbomachinery, Munich, Germany. Pump Consultant to KSB, Germany.

JUMPETER, ALEX M., B.S. (Ch.E.) SECTION 4. JET PUMPS.
Product Manager, Krauss-Maffei Processes, Ametek Process Systems Division.

KARASSIK, IGOR J., B.S. (M.E.), M.S. (M.E.), P.E. SECTIONS 2.2. CENTRIFUGAL PUMP CONSTRUCTION; 10.5. STEAM POWER PLANT PUMPING SERVICES; 13. INSTALLATION, OPERATION, AND MAINTENANCE; APPENDIX—TECHNICAL DATA.
Vice President and Chief Consulting Engineer, Worthington Pump Inc.

KENT, G. R., B.S. (C.E.), P.E. SECTION 9.3. ECONOMICS OF PUMPING SYSTEMS.
Senior Engineer, United Engineers and Constructors, Inc. Power Division. Formerly with Stone and Webster Engineering Corporation.

KITTREDGE, C. P., B.S. (C.E.), DOCTOR OF TECHNICAL SCIENCE (M.E.) SECTION 2.3. CENTRIFUGAL PUMP PERFORMANCE.
Consulting Engineer.

KRON, H. O., B.S. (M.E.), P.E. SECTION 6.2.4. GEARS.
Executive Vice President, Philadelphia Gear Corporation.

KRUTZSCH, W. C., B.S. (M.E.), P.E. SECTION 1. INTRODUCTION AND CLASSIFICATION OF PUMPS. SI UNITS—A COMMENTARY.
Director, Research and Development, Engineered Products, Worthington Pump Inc.

LANDON, FRED K., B.S. (Aero. E.), P.E. SECTION 6.3. PUMP COUPLINGS AND INTERMEDIATE SHAFTING.
Manager, Engineering, Rexnord Inc., Coupling Division.

LIPPINCOTT, J. K., B.S.M.E. SECTION 3.3. SCREW PUMPS.
Vice President and General Manager, DeLaval IMO Pump Division.

LITTLE, C. W., Jr., B.E. (E.E.), D.Eng. SECTION 3.4. ROTARY PUMPS.
Vice President, General Manager, Manufactured Products Division, Waukesha Foundry Company.

LORD, J. ARTHUR, SECTION 10.16. PUMPS FOR MACHINE-TOOL SERVICES.
Retired Director of Sales and Chief Engineer, Pump Department, Brown and Sharpe Manufacturing Company.

McFARLIN, STANLEY, B.S. (M.E.) SECTION 10.11. PUMPS FOR CONSTRUCTION SERVICES.
Director of Engineering, The Warren Rupp Company.

MESSINA, J. P., B.S. (M.E.), M.S. (C.E.), P.E. SECTIONS 9.1. GENERAL CHARACTERISTICS OF PUMPING SYSTEMS AND SYSTEM HEAD CURVES; 9.2. BRANCH-LINE PUMPING SYSTEMS; APPENDIX—TECHNICAL DATA.
Senior Staff Engineer, Public Service Electric and Gas Company; Instructor of Mechanics and Hydraulics, New Jersey Institute of Technology (N.C.E.). Formerly Manager, Pump Application Engineering, Worthington Pump Inc.

O'KEEFE, W., A.B., P.E. SECTION 7. PUMP CONTROLS AND VALVES.
Associate Editor, Power Magazine, McGraw-Hill, Inc.

OLSON, RICHARD G., ME., M.S., P.E. SECTION 6.1.5. GAS TURBINES.
Marketing Supervisor, International Turbine Systems, Turbodyne Corporation.

O'NEEL, WILLIAM E., B.S. (E.E.), M.B.A. SECTION 10.16. PUMPS FOR MACHINE-TOOL SERVICES.
Manager, Design Engineering, Machine Tool Division, Brown and Sharpe Manufacturing Company.

PARMAKIAN, JOHN, B.S. (M.E.), M.S. (C.E.), P.E. SECTION 9.4. WATER HAMMER.
Consulting Engineer.

PEACOCK, JAMES H., B.S. (Met.E.) SECTION 10.6. CHEMICAL INDUSTRY.
Manager, Machine Division, The Duriron Company, Inc.

POTTHOFF, E. O., B.S. (E.E.), P.E. SECTION 6.2.2. SINGLE-UNIT ADJUSTABLE-SPEED ELECTRIC DRIVES.
Industrial Engineer, Industrial Sales Division, General Electric Company.

PRITCHETT, E. R., B.E. (M.E.), B.E. (E.E.), P.E. SECTION 10.13. STEEL MILLS.
Project Engineer, Bethlehem Steel Corporation.

RAMSEY, MELVIN A., M.E., P.E. SECTION 10.17. REFRIGERATION, HEATING, AND AIR CONDITIONING.
Consulting Engineer. Instructor, Stevens Institute of Technology.

RICH, GEORGE R., B.S. (C.E.), C.E., D.Eng., P.E. SECTION 10.18. PUMPED STORAGE.
Director, Senior Vice President–Chief Engineer, Chas. T. Main, Inc.

ROBERTSON, JOHN S., B.S. (C.E.), P.E. SECTION 10.3. DRAINAGE AND IRRIGATION PUMPS.
Assistant Branch Chief, Electrical and Mechanical Branch, Civil Works, Office Chief of Engineers, U.S. Army Corps of Engineers.

SNYDER, MILTON B., B.S. (B.A.) SECTION 6.2.5. ADJUSTABLE-SPEED BELT DRIVES,
Sales Engineer, Master-Reeves Division, Reliance Electric Company.

SOETE, G. W., B.S. (M.E.), M.S. (M.E.), P.E. SECTION 10.12. MARINE PUMPING.
Engineering Supervisor, DeLaval Turbine Inc., Turbine Division.

SZENASI, F. R., B.S. (M.E.), M.S. (M.E.), P.E. SECTION 9.5. VIBRATION AND NOISE IN PUMPS.
Senior Research Engineer, Applied Physics Division, Southwest Research Institute

TULLO, C. J., P.E. SECTION 2.4. CENTRIFUGAL PUMP PRIMING.
Retired Chief Engineer, Centrifugal Pump Engineering, Worthington Pump Inc.

WACHEL, J. C., B.S. (M.E.), M.S. (M.E.). SECTION 9.5. VIBRATION AND NOISE IN PUMPS.
Senior Research Engineer, Manager, Applied Mechanics, Applied Physics Division, Southwest Research Institute.

WEPFER, W. M., B.S. (M.E.), P.E. SECTION 10.19. NUCLEAR SERVICES.
Supervisor, Commercial Engineering, Electromechanical Division, Westinghouse Electric Corporation.

WHIPPEN, WARREN G., B.M.E., P.E. SECTION 6.1.4. HYDRAULIC TURBINES.
Manager, Product Development, Hydro-Turbine Division, Allis Chalmers Corporation.

WILSON, G., HNC in Hydromechanics, P.E. SECTION 10.21.2. CONSTRUCTION OF SOLIDS-HANDLING PUMPS.
Manager Engineering, Research and Development, Worthington (Canada) Ltd.

ZEITLIN, A. B., M.S. (M.E.), Dr.-Eng. (E.E.), P.E. SECTION 10.14. HYDRAULIC PRESSES.
Technical Director, Satra Corporation.

Preface

Considering that I had written the prefaces of the three books published so far under my name, my colleagues thought it both polite and expedient to suggest that I prepare the preface to this handbook, coedited by the four of us. Except for the writing of the opening paragraph of an article, a preface is the most difficult assignment that I know. Certainly the preface to a handbook should do more than describe minutely and in proper order the material which is contained therein.

Yet I submit that the saying "A book should not be judged by its cover" should be expanded by adding: "and not by its preface." If the reader will accept this disclaimer, I can proceed.

As will be stated in Section 1, "Introduction and Classification of Pumps," it can rightly be claimed that no machine and very few tools have had as long a history in the service of man as the pump, or have filled as broad a need in his life. Every process which underlies our modern civilization involves the transfer of liquids from one level of pressure or static energy to another. Thus pumps have played an essential role in our life ever since the dawn of civilization.

Thus it is that a constantly growing population of technical personnel is in need of information that will help it in either designing, selecting, operating, or maintaining pumping equipment. There has never been a dearth of excellent books and articles on the subject of pumps. But the editors and the publisher felt that a need existed for a handbook on pumps which would present this information in a compact and authoritative form. The format of a handbook permits a selection of the most versatile group of contributors, each an expert on his particular subject, each with a background of experience which makes him particularly knowledgeable in the area assigned to him.

This handbook deals first with the theory, construction details, and performance characteristics of all the major types of pumps—centrifugal pumps, power pumps, steam pumps, screw and rotary pumps, jet pumps, and many of their variants. It deals with prime movers, couplings, controls, valves, and the instruments used in pumping systems. It treats in detail the systems in which pumps operate and the characteristics of these systems. And because of the many services in which pumps have to be applied, a total of twenty-one different services—ranging from water supply, through steam power plants, construction, marine applications, and refrigeration to metering and solids pumping—are examined and described in detail, again by a specialist in each case.

Finally the handbook provides information on the selection, purchasing, installation, operation, testing, and maintenance of pumps. An appendix provides a variety of technical data useful to anyone dealing with pumping equipment.

We are greatly indebted to the men who supplied the individual sections which make up this handbook. We hope that our common task will have produced a handbook which will help its user to make a better and more economical pump installation than he would have done without it; to install equipment that will perform more satisfactorily and for longer uninterrupted periods; and when trouble occurs, to diagnose it quickly and accurately. If this handbook does all this, the contributors, its editors, and its publisher will be pleased and satisfied.

No doubt a few readers will look for subject matter that they will not find in this handbook. Into the making of decisions on what to include and what to leave out must always enter an element of personal opinion; therefore we will feel some responsibility for their disappointment. But we submit that it was quite impossible to include even everything we had wanted to cover. As to our possible sins of commission, they are obviously unknown to us at this writing. We can only promise that we shall correct them if an opportunity is afforded us.

Igor J. Karassik

SI Units—a Commentary

W. C. KRUTZSCH

REFERENCES

1. "Metric Practice Guide," American Society for Testing and Materials, ASTM Standard E 380-74, 1974.
2. "SI Units and Recommendations for the Use of Their Multiples and of Certain Other Units," International Organization for Standardization, ISO Standard 1000-1973.

This handbook uses contemporary U.S. engineering units of measurement throughout the entire volume. The decision to employ these units was made in recognition of the fact that they still constitute the only system of practical importance in the United States, and would thus be easier to work with for the majority of both readers and authors. This system is still well-known around the world, even though its utilization is declining, and it has the added advantage of being easily recognized and readily converted into other systems.

Increasingly, however, the world has recognized that it needs a single, unified system of measurement, eliminating the need for such conversions and the inevitable opportunities for error which they create. As business has become more and more international in its conduct, the world has been responding to that need, and has been focusing its attention on metric units. This trend is acknowledged in the handbook in C. P. Kittredge's section on Centrifugal Pump Performance (Sec. 2.3), where SI units have been used in addition to U.S. engineering units.

Unfortunately the term "metric system" as such cannot fully define a system of measurement, since there are differing systems employing metric units. Historically, scientists have tended to work in the "cgs" (centimetre-gram-second) system, while engineers in metric countries have worked largely in metres, kilograms force, and seconds (the "mks" system). Both of these systems will probably be superseded ultimately by the SI system.

The designation SI is the official abbreviation, for any language, of the French title "Le Système International d'Unités," given by the eleventh General Conference on Weights and Measures (sponsored by the International Bureau of Weights and Measures) in 1960 to a coherent system of units selected from metric systems. This system of units has since been adopted by ISO (International Organization for Standardization) as an international standard.

The SI system consists of seven basic units, two supplementary units, a series of derived units, and a series of approved prefixes for multiples and submultiples of the foregoing.

The basic units of the SI system are as follows:

1. Metre[1] (symbol m), unit of length, equal to 1,650,763.73 wavelengths in vacuum of the radiation corresponding to the transition between the levels $2p_{10}$ and $5d_5$ of the krypton-86 atom.
2. Kilogram (symbol kg), unit of mass, equal to that of the international prototype of the kilogram.
3. Second (symbol s), unit of time, equal to the duration of 9,192,631,770 periods of the radiation corresponding to the transition between the two hyperfine levels of the ground state of the cesium-133 atom.
4. Ampere (symbol A), unit of electric current, equal to that which, if maintained in two straight parallel conductors of infinite length, of negligible cross section, and placed one metre apart in vacuum, would produce between those conductors a force equal to 2×10^{-7} newtons per metre of length.[2]
5. Kelvin (symbol K), unit of thermodynamic temperature equal to the fraction 1/273.16 of the thermodynamic temperature of the triple point of water.
6. Candela (symbol cd), unit of luminous intensity, equal to that, in the perpendicular direction, of a surface of 1/600,000 square metres of a blackbody at the temperature of freezing platinum under a pressure of 101,325 newtons per square metre.[2]
7. Mole (symbol mol), unit of substance, equal to the amount of substance of a system which contains as many elementary entities as there are atoms in 0.012 kilogram of carbon-12. (Note that when the mole is used, the elementary entities must be specified and may be atoms, molecules, ions, electrons, other particles, or specified groups of such particles.)

The two supplementary units of the SI system are as follows:

1. Radian (symbol rad), unit of measure of a plane angle with its vertex at the center of a circle and subtended by an arc equal in length to the radius.

[1] More commonly *meter* in American usage, even though *metre* is the only correct form for SI. This spelling was officially adopted in 1971.

[2] Newton (symbol N), unit of force, equal to that which, when applied to a body having a mass of one kilogram, gives it an acceleration of one metre per second per second.

2. Steradian (symbol sr), unit of measure of a solid angle with its vertex at the center of a sphere and enclosing an area of the spherical surface equal to that of a square with sides equal in length to the radius.

For the purposes of this discussion, only those derived units which are of importance in connection with pumps will be covered, but it will be apparent from those listed that some are obvious combinations of the basic and supplementary units, while others, having distinctive names, could be derived only on the basis of accurate knowledge of the physical significance of the quantities they measure. Samples of derived units which are pertinent to subjects covered in this handbook are given in Table 1.

Prefixes for multiples and submultiples of basic and derived units, many of which are already quite familiar, are given in Table 2 for the purpose of identifying those which may not be generally known, and also to indicate the standard SI symbols for the complete range.

As with any such system, SI units will come to be used with a feeling of confidence only after they have been worked with over a period of time. This not only may necessitate the teaching of the system in our educational institutions, or the widespread use of it in industry, but may also require evolutionary changes in the system itself. That some such changes will occur is practically assured by the mere fact that the international General Conference on Weights and Measures has been meeting, since World War II, on the average of once every three years. That some changes may be desirable will be readily appreciated by considering the magnitude of some of the units encountered in the technology associated with pumps.

Consider, for example, the standard SI unit of pressure, in comparison to the units used commonly today. The pascal, equal to one newton per square metre, is indeed a miniscule value compared to the pound per square inch ($1 \text{ lb/in}^2 = 6{,}894.757 \text{ Pa}$) or to the kilogram per square centimetre ($1 \text{ kg/cm}^2 = 98{,}066.50 \text{ Pa}$). In order to eliminate the necessity for dealing with significant multiples of these already large numbers when describing pressure ratings of modern pumps, a serious effort has been made by one interested industry group to obtain acceptance of the bar ($1 \text{ bar} = 10^5 \text{ Pa}$) as an additional standard derived unit. This would indeed accomplish the purpose of reducing the size of the numbers used to describe pressures. The unit is a round multiple of a true SI unit, but it has the disadvantage of being named to infer equality to atmospheric pressure, without being exactly equivalent. In fact, it is equal to neither the standard atmosphere (101,325.0 Pa) nor the so-called metric atmosphere (98,066.50 Pa), but is close enough to be confused with both. It does not present a neat solution.

Order of magnitude is also a problem with the standard SI unit for measuring pump capacity, since in this case it is very large. The cubic metre per second is equal to 15,850.32 U.S. gallons per minute, on the basis of which all but a tiny fraction of pumps now being manufactured would carry a rated capacity value of less than 1.0. Having long recognized the undesirability of dealing with such large units of capacity, pump suppliers in metric countries have generally adopted the practice of rating all but the largest pumps in cubic metres per hour, which gets the numbers up to a more manageable range.

The examples given in the two preceding paragraphs illustrate one more point of considerable interest—the fact that adoption of SI units will require some adjustment of practices in the metric world as well as in the heretofore nonmetric countries.

There will be no country, however, where the impact of this adjustment will be more severe than in the United States. Nevertheless, and in spite of the lack of official government action, the trend here toward adoption of metric measurement, and apparently toward the SI system in particular, is accelerating. Some understanding of the system is thus becoming mandatory, and this commentary is intended to provide at least an introduction to the subject.

In addition, Table 3 will help, where it is necessary, to convert values expressed throughout the handbook in U.S. engineering units into the equivalent SI values.

TABLE 1 Typical Derived Units of the SI System

Quantity	Unit	Symbol	Formula
acceleration	metre/second^2	—	m/s^2
angular acceleration	radian/second^2	—	rad/s^2
angular velocity	radian/second	—	rad/s
area	metre^2	—	m^2
density	kilogram/metre^3	—	kg/m^3
energy	joule	J	N · m
force	newton	N	kg · m/s^2
frequency	hertz	Hz	cycle/s
power	watt	W	J/s
pressure	pascal	Pa	N/m^2
stress	newton/metre^2	—	N/m^2
velocity	metre/second	—	m/s
viscosity, dynamic	newton-second/metre^2	—	N · s/m^2
viscosity, kinematic	metre^2/second	—	m^2/s
volume	metre^3	—	m^3
work	joule	J	N · m

TABLE 2 Prefixes for SI Multiple and Submultiple Units

Prefix	SI Symbol	Multiplication Factor
tera	T	10^{12}
giga	G	10^{9}
mega	M	10^{6}
kilo	k	10^{3}
hecto	h	10^{2}
deka	da	10
deci	d	10^{-1}
centi	c	10^{-2}
milli	m	10^{-3}
micro	μ	10^{-6}
nano	n	10^{-9}
pico	p	10^{-12}
femto	f	10^{-15}
atto	a	10^{-18}

TABLE 3 Conversion of U.S. to SI Units

Multiply	By	To obtain
atmosphere (normal)	1.013250×10^{5}	pascal
barrel (oil, 42 U.S. gal)	1.589873×10^{-1}	metre^3
BTU (International Table)	1.055056×10^{3}	joule
centimetre of Hg (0 C)	1.333220×10^{3}	pascal
centipoise	1.000000×10^{-3}	newton-second/metre^2
centistoke	1.000000×10^{-6}	metre^2/second
degree (of angle)	1.745329×10^{-2}	radian
degree (Celsius)	$t_K = t_C + 273.15$	kelvin
degree (Fahrenheit)	$t_K = (t_F + 459.67)/1.8$	kelvin
degree (Rankine)	$t_K = t_R/1.8$	kelvin
dyne	1.000000×10^{-5}	newton
fluid ounce (U.S.)	2.957353×10^{-5}	metre^3
foot	3.048000×10^{-1}	metre
foot of water (39.2 F)	2.988980×10^{3}	pascal
foot of water (60 F)	2.986080×10^{3}	pascal
foot/second^2	3.048000×10^{-1}	metre/second^2
foot-pound-force	1.355818	joule
foot-pound-force/minute	2.259697×10^{-2}	watt
foot-pound-force/second	1.355818	watt
foot^2/second	9.290304×10^{-2}	metre^2/second
foot^3/second	2.831685×10^{-2}	metre^3/second
gallon (Canadian liquid)	4.546122×10^{-3}	metre^3
gallon (U.K. liquid)	4.546087×10^{-3}	metre^3
gallon (U.S. liquid)	3.785412×10^{-3}	metre^3
gallon (U.S. liquid)/minute	6.309020×10^{-5}	metre^3/second
horsepower (550 ft-lbf/s)	7.456999×10^{2}	watt
horsepower (electric)	7.460000×10^{2}	watt
inch	2.540000×10^{-2}	metre
inch of mercury (32 F)	3.386389×10^{3}	pascal
inch of mercury (60 F)	3.376850×10^{3}	pascal
kilowatt-hour	3.600000×10^{6}	joule
mil	2.540000×10^{-5}	metre
millimetre of mercury (0 C)	1.333224×10^{2}	pascal
ounce-force-inch	7.061552×10^{-3}	newton-metre
pound-force	4.448222	newton
pound-force-foot	1.355818	newton-metre
pound-force/inch^2	6.894757×10^{3}	pascal
pound-force-second/foot^2	4.788026×10	newton-second/metre^2
pound-mass	4.535924×10^{-1}	kilogram
slug	1.459390×10	kilogram
ton (long, 2,240 lb-m)	1.016047×10^{3}	kilogram
ton (short, 2,000 lb-m)	9.071847×10^{2}	kilogram
volt (international of 1948)	1.000330	volt (absolute)
watt (international of 1948)	1.000165	watt
watt-hour	3.600000×10^{3}	joule

Section **1**

Introduction and Classification of Pumps

W. C. KRUTZSCH

INTRODUCTION

Only the sail can contend with the pump for the title of the earliest invention for the conversion of natural energy to useful work, and it is doubtful that the sail takes precedence. Since the sail cannot, in any event, be classified as a machine, the pump stands essentially unchallenged as the earliest form of machine which substituted natural energy for muscular effort in the fulfillment of man's needs.

The earliest pumps of which knowledge exists are variously known, depending upon which culture recorded their description, as Persian wheels, water wheels, or norias. These devices were all undershot water wheels containing buckets which filled with water when they were submerged in a stream and which automatically emptied into a collecting trough as they were carried to their highest point by the rotating wheel. Similar water wheels have continued in existence in parts of the Orient even into the twentieth century.

The best known of the early pumps, the Archimedean screw, also persists into modern times. It is still being manufactured for low-head applications where the liquid is frequently laden with trash or other solids.

Perhaps most interesting, however, is the fact that with all the technological development which has occurred since ancient times, including the transformation from water power through other forms of energy all the way to nuclear fission, the pump remains probably the second most common machine in use, exceeded in numbers only by the electric motor.

Since pumps have existed for so long and are in such widespread use it is hardly surprising that they are produced in a seemingly endless variety of sizes and types and are applied to an apparently equally endless variety of services. While this variety has contributed to an extensive body of periodical literature, it has also tended to preclude the publication of comprehensive works. With the preparation of this handbook an effort has been made to provide just such a comprehensive treatment.

But even here it has been necessary to impose a limitation on subject matter. It has been necessary to exclude material uniquely pertinent to certain types of auxiliary pumps which lose their identity to the basic machine they serve, and where the user controls neither the specification, purchase, nor operation of the pump. Examples of such pumps would be those incorporated into automobiles or domestic appliances. Nevertheless, these pumps do fall within classifications and types covered within the handbook, and basic information on them may therefore be obtained herein once the type of pump has been identified. Only specific details of these highly proprietary applications are omitted.

Such extensive coverage has required the establishment of a systematic method of classifying pumps. Although some rare types may have been overlooked in spite of all precautions, and obsolete types that are no longer of practical importance have been deliberately omitted, principal classifications and subordinate types are covered in the following section.

CLASSIFICATION OF PUMPS

Pumps may be classified on the basis of the applications they serve, the materials from which they are constructed, the liquids they handle, and even their orientation in space. All such classifications, however, are limited in scope and tend to substantially overlap each other. A more basic system of classification, the one used

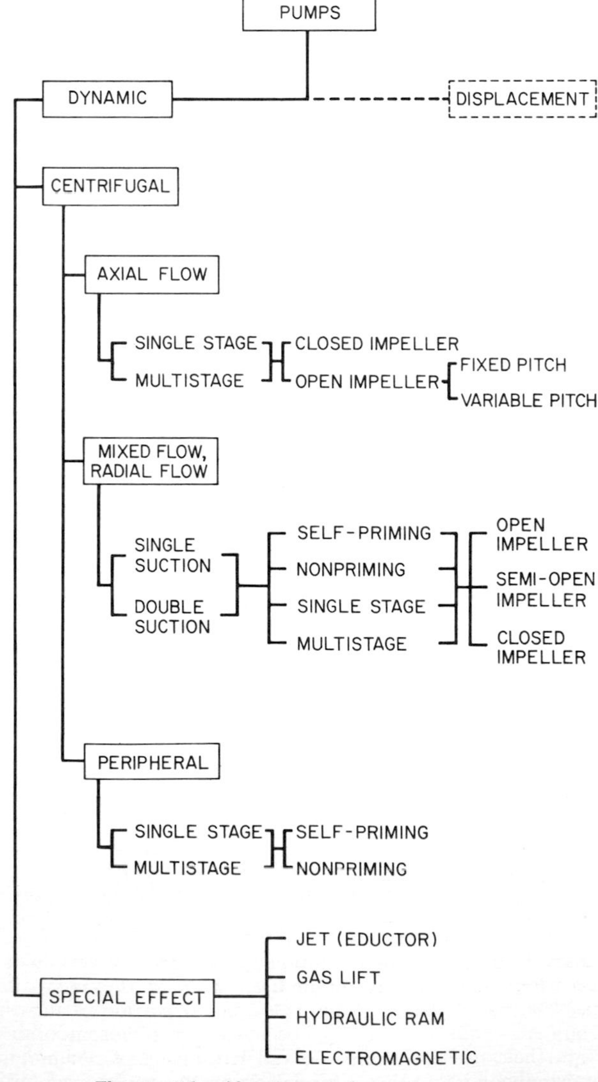

Fig. 1 Classification of dynamic pumps.

in this handbook, first defines the principle by which energy is added to the fluid, goes on to identify the means by which this principle is implemented, and finally delineates specific geometries commonly employed. This system is therefore related to the pump itself and is unrelated to any consideration external to the pump or even to the materials from which it may be constructed.

Under this system, all pumps may be divided into two major categories: (1)

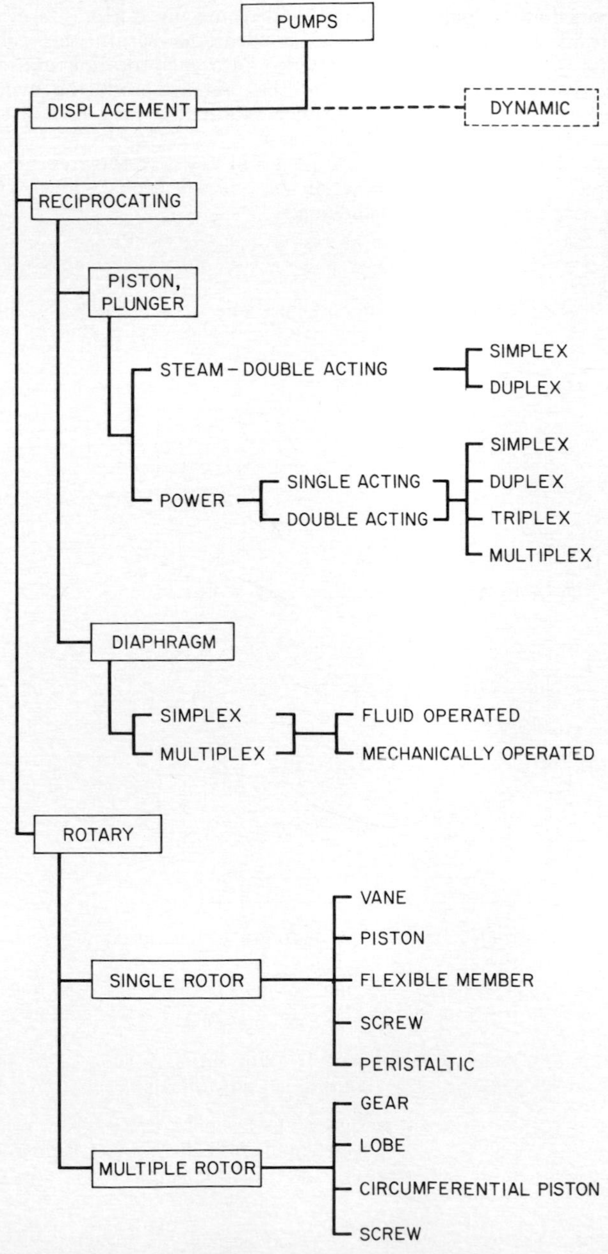

Fig. 2 Classification of displacement pumps.

dynamic, in which energy is continuously added to increase the fluid velocities within the machine to values in excess of those occurring at the discharge such that subsequent velocity reduction within or beyond the pump produces a pressure increase; (2) *displacement,* in which energy is periodically added by application of force to one or more movable boundaries of any desired number of enclosed, fluid-containing volumes, resulting in direct increase in pressure up to the value required to move the fluid through valves or ports into the discharge line.

Dynamic pumps may be further subdivided into several varieties of centrifugal and other special-effect pumps. Figure 1 presents in outline form a summary of the significant classifications and subclassifications within this category.

Displacement pumps are essentially divided into reciprocating and rotary types, depending on the nature of movement of the pressure-producing members. Each of these major classifications may be further subdivided into several specific types of commercial importance, as indicated in Fig. 2.

Definitions of the terms employed in Figs. 1 and 2, where they are not self-evident, and illustrations and further information on classifications shown are contained in the appropriate sections of the handbook.

Section **2.1**

Centrifugal Pump Theory

WALTER K. JEKAT

INTRODUCTION

The fluid motions in a centrifugal pump are complex. The velocity vectors are not parallel to the walls of the fluid passages, and appreciable secondary motions occur near the impeller discharge and in the diffusing section. These details of the true fluid motions are not well understood.

Most practical pump design is based on a one-dimensional approximation, which neglects all secondary motions and treats the main flow on the basis of available flow areas and conduit wall directions. Where this leads to obvious errors, correction factors (such as the slip factor) are introduced. While the flow areas are generally calculated under the assumption of uniform velocity over the cross section, there is sometimes a flow factor applied which recognizes the blockage effect of the boundary layers. This is particularly important for the throat area of volute or diffuser.

Because of the widespread use and considerable success of the one-dimensional analysis, the following text presents it in a generalized, though abbreviated, form. The generalization has been accomplished by the extensive use of the concept of specific speed.

Much work has been done refining the basic centrifugal pump theory by two- or three-dimensional considerations. The mathematical complexities are great and testing the detailed flow structure is difficult. Important features of the flow, such as secondary motions, are still neglected. Therefore only moderate success has been had in elucidating certain details of the flow in a centrifugal pump. No comprehensive theory, which would permit the complete hydrodynamic design, has evolved.

Nevertheless, the detail design of fluid passages can benefit from the application of two- or three-dimensional methods, approximate as they still are. Also, the "feel" of a designer for the behavior of the flow in complex passages can be improved. Therefore the subsection Advanced Methods presents some useful approaches to two-dimensional flow calculations.

The user of centrifugal pumps will not normally wish to go deeply into the design, but he can profit from familiarity with pump theory by being able to establish the rotative speed and main dimensions of an optimally designed pump. This knowledge will help him to evaluate the offers of various manufacturers. In addition, the ability to ascertain reasonably accurate efficiencies for the required pumps without the manufacturers' offers is advantageous.

NOMENCLATURE

A = sum of areas between vanes, in^2
D = diameter, in
H = total head, ft
L = flow friction loss, ft
N = rotative speed, rpm
NPSH = net positive suction head, ft
N_s = specific speed, rpm $\sqrt{\text{gpm}}/\text{ft}^{0.75}$
Q = capacity, gpm
S = suction specific speed, rpm $\sqrt{\text{gpm}}/\text{ft}^{0.75}$
T = torque, ft-lb
sp gr = specific gravity

a = area between two vanes, in^2
b = width, in
c = absolute velocity, ft/s
e = sum of widths of shrouds at outer impeller diameter, in
g = gravity constant, 32.2 ft/s^2
h = static head, ft
h = wall distance, in
m = length along streamline, starting from inlet, in
n = distance from shroud to point on orthogonal, in
p = pressure, lb/in^2
r = radius, in
s = vane thickness, in
t = clearance between impeller and volute tongue, in
u = peripheral velocity, ft/s
w = relative velocity, ft/s
y = elevation, ft
z = number of vanes

α = angle between c and u, deg
β = angle between w and u, deg
γ = density, lb/ft^3
γ = angle between streamline direction and axis of rotation, deg
η = efficiency
θ = angular distance from radial line rotating with impeller, rad
μ = slip factor
ν = kinematic viscosity, ft^2/s
ρ = radius of curvature, in
φ = central angle, deg
ψ = head coefficient
ω = angular velocity, s^{-1}

Subscripts

DF = disk friction
H = hydraulic
H = hub
I = impeller
L = leakage
M = mechanical
S = suction
S = shaft
V = volumetric
V = volute
W = water

a = at shroud
av = average
b = at hub

m = meridional
m = mean
p = at pressure side of vane
r = refers to arbitrary radius
s = at suction side of vane
th = theoretical
thr = at throat of volute or diffuser
u = denotes peripheral component

θ = denotes circumferential component
I = beginning of confined vane channel
II = end of confined vane channel

BASIC HYDRAULIC RELATIONSHIPS

Bernoulli's Equation for a Stationary Conduit Assuming no flow losses, the total head H is the same for any point along a streamline:

$$H = \frac{144p}{\gamma} + \frac{c^2}{2g} + y = \text{constant} \tag{1}$$

The individual terms in Bernoulli's equation (1) have the following physical meaning:

$144p/\gamma$ = static pressure head
$c^2/2g$ = dynamic head
y = elevation

Applying Bernouilli's equation to two stations in a conduit (Fig. 1) yields

$$\frac{144p_1}{\gamma} + \frac{c_1^2}{2g} + y_1 = \frac{144p_0}{\gamma} + \frac{c_0^2}{2g} + y_0 \tag{2}$$

When including flow friction losses, Eq. (2) reads

$$\frac{144p_1}{\gamma} + \frac{c_1^2}{2g} + y_1 = \frac{144p_0}{\gamma} + \frac{c_0^2}{2g} + y_0 - L_{0-1} \tag{3}$$

where L_{0-1} is the friction loss (measured in feet) incurred from station 0 to station 1. EXAMPLE: The static pressure at the impeller inlet (station 1), when taking suction from a large reservoir (station 0 at surface), is

$$p_1 = p_0 - \frac{1}{144}\left(\gamma y_1 - \gamma y_0 + \frac{\gamma c_1^2}{2g} + \gamma L_{0-1}\right)$$

The flow velocity c_0 at the surface of the reservoir is taken to be zero.

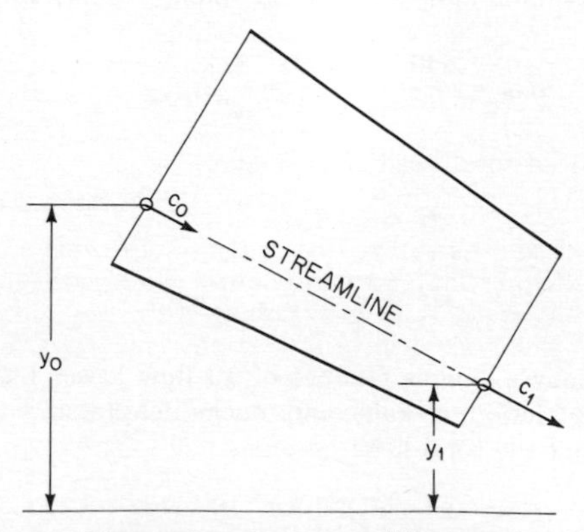

Fig. 1 Illustration for Bernoulli's equation (2).

For water of 68°F the density γ is 62.3 lb/ft³, and the above equation becomes

$$p_1 = p_0 - \frac{y_0 - y_1}{2.31} - \frac{c_1^2}{148.8} - \frac{L_{0-1}}{2.31}$$

Euler's Equation for an Impeller The torque required to drive an impeller is equal to the change of moment of momentum of the fluid passing through the impeller:

$$T = \frac{\gamma Q/449}{g} \frac{r_2 c'_{u3} - r_1 c'_{u0}}{12} \tag{4}$$

Station 0 is just in front of the vane inlets, whereas station 3 is right after the ends of the vanes (Fig. 2). The flow velocities are given a prime which

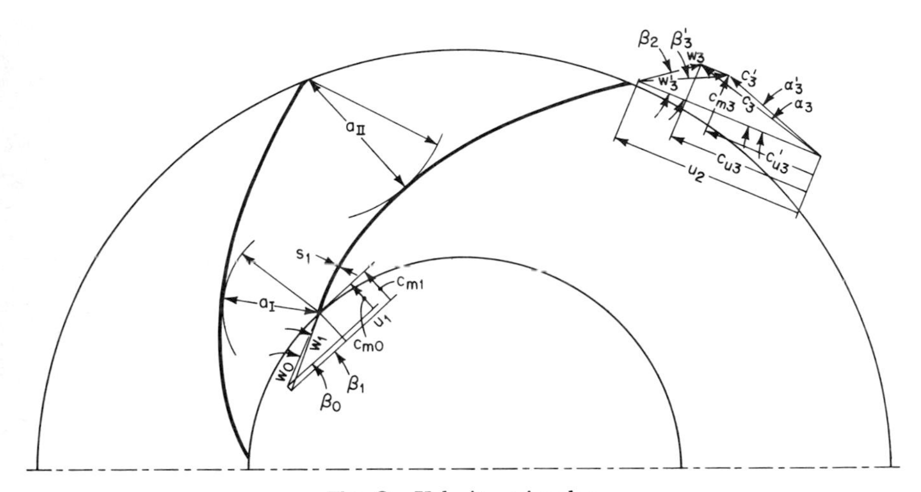

Fig. 2 Velocity triangles.

signifies that they are actual, rather than "theoretical," velocities (more about this in the subsection Velocity Triangles). Equation (4) neglects impeller disk friction, leakage losses, seal friction, and all other mechanical losses.

The power input is

$$P = \frac{\omega T}{550} = \frac{\gamma Q}{449 \times 550g} (u_2 c'_{u3} - u_1 c'_{u0}) \tag{5}$$

If there were no hydraulic losses, then the pump would produce the theoretical total head

$$H_{\text{th}} = \frac{449 \times 550P}{\gamma Q} = \frac{1}{g} (u_2 c'_{u3} - u_1 c'_{u0}) \tag{6}$$

The actually produced total head H is smaller.

$$H = \eta_H H_{\text{th}} \tag{7}$$

$$H = \frac{\eta_H}{g} (u_2 c'_{u3} - u_1 c'_{u0}) \tag{8}$$

The hydraulic efficiency η_H takes account of all flow losses inside of the pump. It does not include disk friction, leakage, and mechanical losses. Often there is no prerotation ($c'_{u0} = 0$), and the total head becomes

$$H = \frac{\eta_H}{g} u_2 c'_{u3} \tag{9}$$

Bernoulli's Equation for an Impeller The difference in static head between the two stations 0 and 3 (located on a common streamline) of an impeller can be found by subtracting from the theoretical total head as given by Euler's equation (6)

$$H_{\text{th}} = \frac{1}{g}(u_2 c'_{u3} - u_1 c'_{u0})$$

the difference in the dynamic head as given by Eq. (1) $(c'_0)^2/2g - (c'_3)^2/2g$ and the flow losses incurred from 0 to 3. The static head difference is then

$$h_3 - h_0 = \frac{1}{g}(u_2 c'_{u3} - u_1 c'_{u0}) - \frac{(c'_0)^2}{2g} + \frac{(c'_3)^2}{2g} - L_{0-3} \tag{10}$$

By observing the trigonometric relationships between the various velocities, Eq. (10) can be converted into

$$h_3 - h_0 = \frac{1}{2g}(u_2^2 - u_1^2) + \frac{(w'_0)^2}{2g} - \frac{(w'_3)^2}{2g} - L_{0-3} \tag{11}$$

which is sometimes called Bernoulli's equation for an impeller.

For no prerotation we have $(w'_0)^2 = u_1^2 + c_{m0}^2$ and obtain:

$$h_3 - h_0 = \frac{1}{2g}[u_2^2 - (w'_3)^2 + c_{m0}^2] - L_{0-3} \tag{12}$$

Instead of using the impeller loss L_{0-3}, an impeller efficiency η_I can be defined by

$$h_3 - h_0 = \frac{\eta_I}{2g}[u_2^2 - (w'_3)^2 + c_{m0}^2] \tag{13}$$

Overall Efficiency The overall efficiency (short form, efficiency) is simply the ratio of water power to shaft power:

$$\eta = \frac{P_W}{P_S} \tag{14}$$

The water power

$$P_W = QH \frac{\text{sp gr}}{3{,}960} \tag{15}$$

is the power which would be required if the desired head at the desired capacity could be produced without any losses whatsoever. The specific gravity for water of 40°F is sp gr = 1.0.

The shaft power can be added up such that

$$P_S = P_W + P_H + \left(\frac{1}{\eta_v} - 1\right)(P_W + P_H) + P_{\text{DF}} + P_M \tag{16}$$

P_H stands for the power consumed by the hydraulic losses. From Eq. (7) we have:

$$\eta_H = \frac{H}{H_{\text{th}}} = \frac{P_W}{P_W + P_H} \tag{17}$$

The third term

$$\left(\frac{1}{\eta_v} - 1\right)(P_W + P_H)$$

represents the power wasted because of the internal leakage flow Q_L (from the impeller discharge through the wearing rings or vane front clearances and thrust-equalization holes to the impeller inlet). The definition of the volumetric efficiency η_v is

$$\eta_v = \frac{Q}{Q + Q_L} \tag{18}$$

The fourth term P_{DF} is the power required to overcome the impeller-disk friction. The fifth term P_M is the sum of all mechanical power losses, such as caused by bearings and seals.

Combining Eqs. (14), (16), and (17), we arrive at

$$\eta = \frac{1}{\dfrac{1}{\eta_H \eta_v} + \dfrac{P_{DF}}{P_W} + \dfrac{P_M}{P_W}} \tag{19}$$

With this equation the efficiency can be calculated from estimates of the individual losses. Such estimates will be given under Design Procedures.

VELOCITY TRIANGLES

The velocity of a fluid element is represented by a vector. The length of the vector gives the magnitude of the velocity in feet per second, and the direction of the vector is tangential to the streamline.

The fluid velocity in a stationary conduit (inlet or discharge pipe, inlet guide vanes, diffuser vanes, volute) is measured in reference to an earthbound coordinate system. Therefore it is called *absolute velocity* and denoted c'. The prime signifies an *actual* flow velocity. Those *idealized* velocity vectors, which are calculated by assuming perfect guidance of the flow by vanes or walls, are given no prime. Failure to distinguish clearly between actual and idealized velocities produces an erroneous application of basic impeller theory and consequently a faulty design.

The fluid velocity in an impeller channel can be represented by either the absolute velocity c' or the *relative velocity* w'. The coordinate system for the relative velocity rotates with the impeller angular velocity $\omega = u/r$. Again, no prime is assigned to the idealized velocity. The idealized relative velocity w is easily calculated by dividing the flow per vane channel by the cross-sectional area of the channel.

The absolute velocity can be considered as the resultant of the relative velocity and the local impeller peripheral speed. Velocity triangles provide much information on the design under consideration and should be drawn for every calculated station. Figure 2 gives an example. The vectors of the absolute velocity and the relative velocity end at the same corner of the triangle. The peripheral velocity vector u starts at the beginning of the absolute velocity vector and goes to the beginning of the relative velocity vector.

The *meridional velocity* c_m is the component in the meridional plane of the absolute as well as the relative velocity. It forms always a right angle with u. For strictly radial flow the meridional velocity is also the radial component of the absolute and relative velocity. Likewise for strictly axial flow it is the axial component.

The peripheral components of the absolute velocity c' and the relative velocity w' are denoted c'_u and w'_u. The prime is omitted if idealized velocities are used. The angle between the absolute velocity c' and the peripheral direction is α'. If taken right after the impeller vane endings, the angle α'_3 gives the direction with which the fluid enters the diffusing system. The relative velocity w' forms with the peripheral direction the angle β'. It is only approximately equal to the vane angle β. By definition, the idealized relative velocity w forms with the peripheral direction the vane angle β.

Slip Factor Often idealized velocity triangles are drawn by assuming perfect guidance of the flow by the vanes. The vane angle β is substituted for the flow angle β'. The angle between the idealized absolute velocity c and the peripheral direction is now denoted α. The meridional velocity c_m is normally assumed to remain the same (therefore the prime was not used before). We know that the relative velocity w and the absolute velocity c are not actual velocities, but they are utilized in the design of impellers since they are much easier to calculate than actual flow velocities. However, the results must be corrected to those which would be obtained if the actual flow velocities were employed.

Deviation of the fluid from the vane direction is especially important at the impeller discharge since it reduces the peripheral component of the absolute impeller discharge velocity. This causes a proportionate reduction in head as well as power input. With the usual backward-leaning vanes ($\beta_2 < 90°$) the flow angle β_3' is smaller than the vane angle β_2. The phenomenon is often called *slip*, and is a consequence of the nonuniform velocity distribution across the impeller channels, boundary-layer accumulation, and flow separation if it occurs. Accurate prediction of slip is very difficult since the causes of slip cannot be predetermined in a practical manner. This is the greatest drawback of the existing centrifugal impeller theory. The way in which this problem is presently treated is first to define a slip factor,

$$\mu = \frac{c'_{u3}}{c_{u3}} \tag{20}$$

and then to make the assumption that the slip factor is only a function of the number of vanes, the vane discharge angle, and sometimes the impeller-radius ratio r_2/r_1. All other details of the impeller design are not taken into account. The construction of a slip-factor formula is usually based on theoretical reasoning.

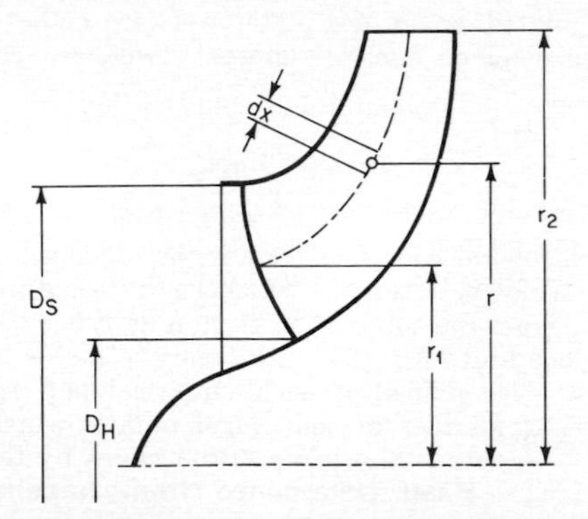

Fig. 3 Impeller with vanes extended into axial inlet.

Often empirical coefficients bring the formula into agreement with available test data. Numerous slip-factor formulas have been proposed since Stodola [1] published the first, a theoretical one:

$$\mu = 1 - \frac{\pi \sin \beta_2}{z} \tag{21}$$

It was originally derived for the case of zero flow through a centrifugal impeller but has since frequently been used for the design flow. It has enjoyed appreciable acceptance in American pump practice.

In Europe the slip factor formula by Pfleiderer [2]

$$\mu = \frac{1}{1 + a\left(1 + \dfrac{\beta_2}{60}\right)\dfrac{r_2^2}{zS}} \tag{22}$$

is in widespread use. S is the static moment of the mean streamline

$$S = \int_{r_1}^{r_2} r\,dx \tag{23}$$

which calls for a simple graphical integration in the meridional plane. Figure 3 identifies the terms in Eq. (23). For a cylindrical vane

$$S = \int_{r_1}^{r_2} r\,dr = \tfrac{1}{2}(r_2^2 - r_1^2)$$

and the slip factor

$$\mu = \frac{1}{1 + \frac{a}{z}\left(1 + \frac{\beta_2}{60}\right)\frac{2}{1 - (r_1^2/r_2^2)}} \tag{24}$$

For radius ratios r_1/r_2 below ½, experience teaches that the slip factor does not increase anymore. For such small radius ratios the slip factor for $r_1/r_2 = ½$ should be used.

Turbomachinery designers have long recognized that the slip factor is affected not only by the impeller configuration but also by the interaction of the diffusing system with the impeller. Pfleiderer [2], on the basis of a series of tests, takes this into account by keying the coefficient a in Eqs. (22) and (24) to the casing design:

Volute:	$a = 0.65$ to 0.85
Vaned diffuser:	$a = 0.6$
Vaneless diffuser:	$a = 0.85$ to 1.0

Once we have accumulated some test data, we can make better predictions of the slip factor for subsequent similar designs. First, we combine Eqs. (9) and (20) into

$$H = \mu \frac{\eta_H}{g} u_2 c_{u3} \tag{25}$$

Then we obtain the peripheral component c_{u3} of the idealized absolute velocity from the impeller-discharge velocity triangle. Furthermore, an estimate of the hydraulic efficiency (for instance, from the sub-section Hydraulic Efficiency) is made. Entering the tested total pump head into Eq. (25) yields the slip factor. A decisive advantage is gained if such empirical slip factors are applied to families of centrifugal pumps, which are similar in more detail than taken into account by the slip-factor formulas.

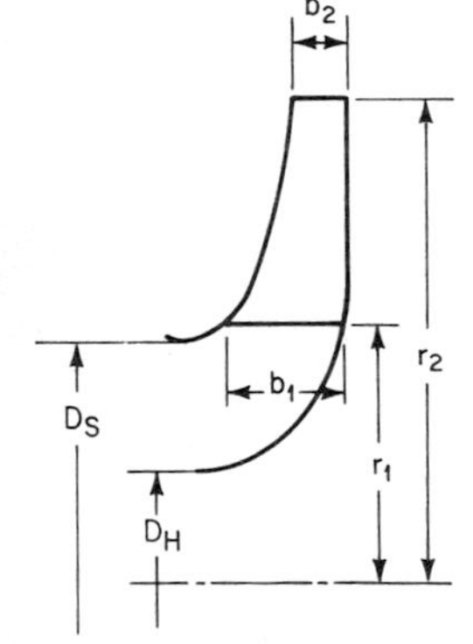

Fig. 4 Impeller with cylindrical vanes.

Head Determined from Impeller-Discharge Velocity Triangles From the impeller-discharge velocity triangle it is apparent that

$$c_{u3} = u_2 - c_{m3} \cot \beta_2 \tag{26}$$

From (20) we have

$$c'_{u3} = \mu c_{u3}$$

Entering these two expressions into Euler's equation (9)

$$H = \frac{\eta_H}{g} u_2 c'_{u3}$$

we obtain

$$H = \mu \eta_H \frac{u_2^2}{g}\left(1 - \frac{c_{m3}}{u_2} \cot \beta_2\right) \tag{27}$$

This equation enables us to ascertain the effect of the two design parameters c_m/u_2 and β_2 on the total head. If c_{m3}/u_2 is reduced, the head increases. The effect of the vane discharge angle is not so straightforward. If β_2 is increased, $\cot \beta_2$ is decreased, and the head would become larger. But the slip-factor formulas (21) and (22) indicate that larger angles β_2 decrease the slip factor μ. Therefore the effects of changes of β_2 are partly canceled out in Eq. (27). Indeed, some tests showed that varying the blade angle β_2 from 20 to 30° had only a negligible effect on the pump head. If, though, vane discharge angles as large as 90° are employed, an appreciable head increase can be expected.

The pump head as defined by Eq. (27) can be made dimensionless by dividing with $u_2^2/2g$, resulting in the head coefficient

$$\psi = \frac{H}{u_2^2/2g} = 2\mu\eta_H \left(1 - \frac{c_{m3}}{u_2} \cot \beta_2\right) \tag{28}$$

The meridional velocity c_{m3} is calculated right after the impeller discharge with

$$c_{m3} = \frac{Q/449}{2\pi r_2 b_3/144} \tag{29}$$

where b_3 is taken equal to b_2, the vane width at the impeller discharge (Fig. 4).

SIMILITUDE

A large number of centrifugal pumps, for greatly varying capacities, heads, and rotative speeds, have been built and tested. Because their efficiencies have ranged from, say, 15 to over 90 percent, it became necessary to determine whether low efficiencies were always due to poor design or whether some unfavorable conditions of service existed which precluded good performance. Conversely, it was important to find out whether certain conditions of service were conducive to high efficiencies. Furthermore it was desirable that the conditions of service could be grouped in such a way that a large number of designs could be lumped together into a single expression. Dimensional analysis suggested for this purpose the specific speed

$$\frac{N\sqrt{Q}}{(gH)^{0.75}}$$

That the specific speed is a dimensionless group can be verified by entering consistent dimensions, such as rps for the rotative speed, ft^3/s for the capacity, and ft for the head.

Normally g is dropped, and the specific speed is written

$$N_s = \frac{N\sqrt{Q}}{H^{0.75}} \tag{30}$$

In American centrifugal pump practice the following inconsistent but generally accepted dimensions are used: rpm for the rotative speed, gpm for the capacity, and ft for the head. It is in this form that the specific speed is given in this section. That the resultant value for the specific speed has now a dimension does not diminish its utility. In principle it remains nondimensional.

In 1947 Wislicenus [3] published a plot of the "approximate statistical averages" of the efficiencies of a large number of commercial centrifugal pumps versus specific speed (above $N_s = 500$). While it became obvious that a correlation existed, it was necessary to introduce an additional parameter which had something to do with the scale of the pump. At that time the capacity (from 100 to 10,000 gpm, but only 100 and 200 gpm were extended down to $N_s = 500$) was selected. This graph was quickly accepted in the United States and became a yardstick, by which designers judged the merits of their designs. Indeed, it was often referred to as "the chart," and an efficiency read from it as "chart efficiency."

An updated version of this chart, based on the evaluation of 528 new test points, is given in Fig. 5. Only those test results were used which form a cluster of points or single points of unquestionable accuracy. The chart was extended down to $N_s = 180$ and $Q = 30$ gpm over the whole range from $N_s = 180$ to 3,000. In addition, efficiencies are given for 5 and 10 gpm at specific speeds from 100 to 500.

Just as in the original Wislicenus chart, efficiencies start dropping drastically when lowering the specific speed below about 1,000. Also, smaller capacities exhibit lower efficiencies than higher capacities at all specific speeds. But the newer designs (Fig. 5) show, by comparison to the Wislicenus chart, higher efficiencies for low specific speed (especially below $N_s = 1{,}000$) and low capacities (especially below 200 gpm). Apparently, these are the regions where improvements were possible. No appreciable efficiency improvements have been achieved in the intervening 25 years for pumps of high specific speed and large capacity. For instance, at

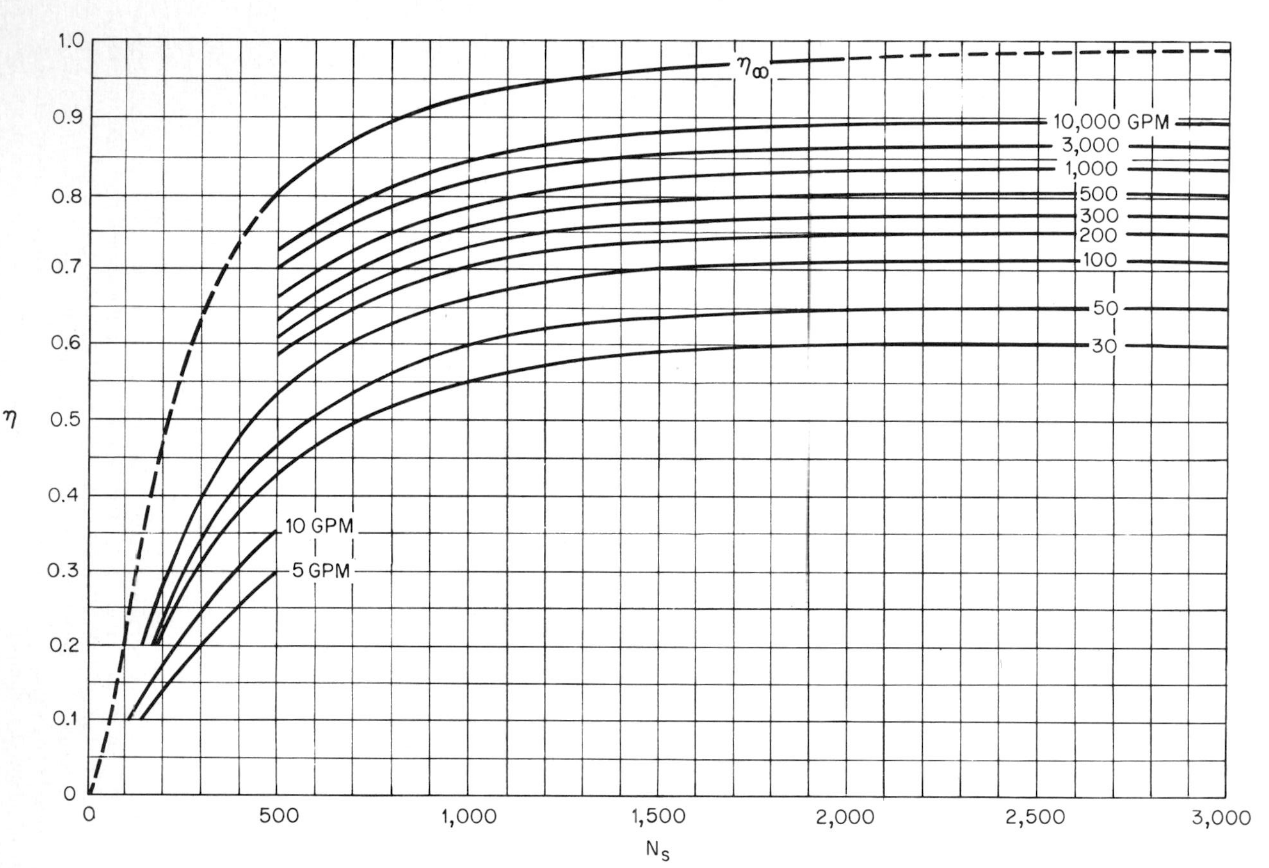

Fig. 5 Efficiency as a function of specific speed and capacity.

N_s = 2,500 and Q = 10,000 gpm we still register η = 0.89 as a representative value.

A designer used to be happy when his design exceeded chart efficiency. Our Fig. 5 will still permit this because the curves do not take into account the few pumps of exceptionally high efficiencies.

The utility of specific speed as a significant group for the prediction of efficiency rests upon the premise that geometrically similar pumps, operated at the same specific speed, have geometrically similar velocity triangles. Consequently the ratios of the flow velocities to the impeller peripheral speed are the same. Were it not for the effects of scale, their relative losses would also be the same. Scale affects losses because of its influence upon flow friction and leakage losses.

A chart such as Fig. 5 is intellectually disappointing; to represent scale, a dimensional parameter—the capacity Q—appears in an otherwise dimensionless plot (η versus N_s). It would seem more appropriate to use a dimensionless parameter, perhaps a Reynolds number. But the situation is complicated by the fact that surface roughness and clearances are not usually scaled properly. It is an empirical fact that of all known attempts to use a single parameter for scale, the use of capacity has correlated the test results best. To a large degree this is supported by another empirical finding, that the hydraulic efficiency is mainly dependent upon the capacity (see subsection Hydraulic Efficiency).

The specific speed chart (Fig. 5) is convenient to use because a designer who wants to predict efficiency quickly needs to know only the head and capacity, which must be specified by the conditions of service. The only choice he makes is that of the rotative speed. He may enter several such choices (for instance, 1,800 and 3,600 rpm) in order to ascertain the probable effect upon efficiency.

But the selection of the specific speed is found to be insufficient; nothing is yet known about the pump design which would produce the efficiency shown in Fig. 5. For this, it is necessary to have an approximation of the velocity triangles. In order to reduce the amount of information to be transmitted, only the most important—the velocity triangle after the impeller discharge—is specified (assuming that the impeller inlet and diffusing system design follow common practice; see subsections Impeller Inlet, Volute Casing, and Vaned Diffuser).

DESIGN PROCEDURES

Design Parameters Specifying Impeller-Discharge Velocity Triangle Figure 6 gives the band of commonly used values of c_{m3}/u_2. This parameter decreases with smaller specific speed, tending toward zero at zero specific speed. If higher values than those shown are attempted at low specific speeds, such small vane discharge widths would result that the usual manufacture of the impellers by casting would be difficult.

Specifying c_{m3}/u_2 ties the meridional velocity (taken right after the impeller discharge) to the impeller peripheral speed. If, in addition, one flow angle is known, then the whole impeller-discharge velocity triangle could be drawn. It is customary, instead of a flow angle, to specify the vane discharge angle, since this angle is one of the design choices which must be made. The range of commonly used values of β_2 is given in Fig. 7.

Since the vane discharge angle is not a flow angle, it is necessary to know the slip factor in order to complete the velocity triangle. It is given in an implicit manner by supplying the usual number of vanes z, which is four to eight for specific speeds up to 4,000.

The three design parameters, c_{m3}/u_2, β_2, and z, are therefore sufficient to describe the impeller discharge velocity triangle in an approximate manner. The values plotted in Figs. 6 and 7 originate from pumps designed by many engineers of several organizations. It is therefore to be expected that they do not arrange themselves neatly along lines but rather form bands.

In addition to their influence upon the design point efficiency, the three design parameters have an effect upon the shape of the head-capacity curve. It rises toward the shutoff if smaller values of β_2 and z and larger values of c_{m3}/u_2 are selected (see Sec. 2.3). When this is a consideration, one can select the extreme values within the bands in Figs. 6 and 7.

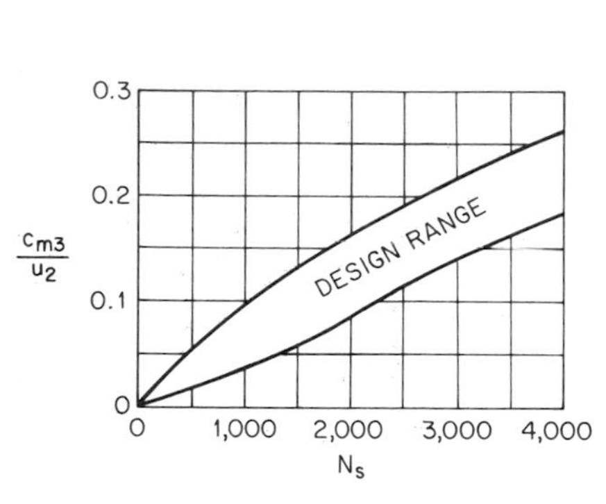

Fig. 6 c_{m3}/u_2 versus specific speed.

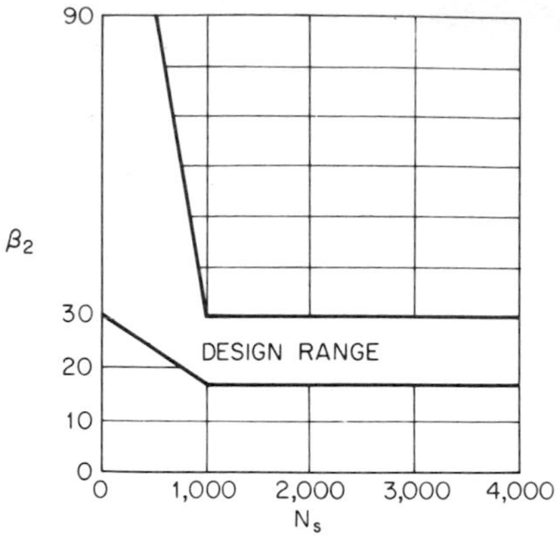

Fig. 7 Impeller discharge angle versus specific speed.

Impeller Inlet The impeller inlet design is based on the inlet flow angle β_0 (taken at the outer diameter of the inlet). The station 0 is located right in front of the vane inlet tip. The flow angle β_0 is selected between 10 and 25°, independent of specific speed. An often used flow angle is 17°, which is a compromise between efficiency and cavitation. For best impeller efficiency, the angle β_0 should be larger; for lower NPSH, it should be smaller.

It is convenient to make NPSH dimensionless by using the concept of suction specific speed (Ref. 4; see also Sec. 2.3).

$$S = \frac{N\sqrt{Q}}{\text{NPSH}^{0.75}} \tag{31}$$

A typical value is S = 9,000, which can be reached with centrifugal impellers of good manufacture having a flow angle of about 17° and approximately five to seven vanes. Many commercial pumps have lower suction specific speeds, in the area from 5,000 to 7,000. On the other hand, boiler-feed and, especially, condensate pumps require often suction specific speeds which range as high as 12,000 and 18,000. To reach such values, the flow angle is taken as low as 10° and the number of vanes is reduced to as few as four. Fewer vanes (as well as thinner vanes) are beneficial because they reduce the blockage effect. A disadvantage of low flow angles (and consequently large inlet diameters) is that the pump is more likely to run rough at part load, especially below 50 percent capacity.

Should the available NPSH be so low that the required suction specific speed is above about 18,000, then a separate axial flow impeller of special design—an inducer—is used ahead of the centrifugal impeller (Refs. 5 and 6). Its flow angle is typically between 5 and 10°. The vane angle is about 3 to 5° larger (Ref. 7). The number of vanes is often only two, and not more than four. The vane thickness is made as small as possible.

The suction diameter D_s (Figs. 3 and 4) is

$$D_S = 4.54\left(\frac{Q}{kN\tan\beta_0}\right)^{1/3} \tag{32}$$

whereby the hub ratio k is

$$k = 1 - \left(\frac{D_H}{D_S}\right)^2 \tag{33}$$

For vanes which extend into the axial portion of the inlet (Fig. 3), the flow angle β_0 varies along the leading edge. Assuming uniform meridional velocity over the radius, the flow angle is determined from

$$\tan \beta_{0(r)} = \tan \beta_0 \frac{r_1}{r} \tag{34}$$

The flow angle, and consequently the vane angle, is assumed to be constant for vanes with leading edges parallel to the axis (Fig. 4).

The vane angle β_1 is made larger than the flow angle β_0 in order to compensate for the blockage of the vanes (Ref. 2). From this approach follows (Fig. 2):

$$\tan \beta_{1(r)} = \frac{\tan \beta_{0(r)}}{1 - (zs_1/2\pi r \sin \beta_{1(r)})} \tag{35}$$

The equation is used for vanes according to Fig. 3 as well as for vanes according to Fig. 4 ($r = r_1$).

For vanes with leading edges parallel to the axis, it remains to determine the inlet width b_1 (Fig. 4). This can be done by assuming a constant and equal meridional velocity for the axial and radial portion of the inlet passage, resulting in

$$b_1 = \frac{k}{4} D_S \frac{D_S}{D_1} \tag{36}$$

whereby D_S/D_1 is between 0.8 and 1.0.

Impeller Vane Layout The selection of the impeller discharge and inlet design has so far been treated independently. Actually they should be taken together, for they greatly influence the whole vane shape and velocity distribution within the impeller.

In order to judge whether a particular vane shape is likely to lead to high impeller efficiency, it is most helpful to consider the distribution of the relative velocity along the flow path. Once the flow has entered the impeller vane system, it is not yet confined by a channel until it has reached the section a_I (Fig. 2). In order to avoid the danger of flow separation, the vane angle is increased only slightly up to section a_I. From there on to section a_{II} (Fig. 2) the flow is confined. The mean relative velocities in the two sections are:

$$w_I = \frac{Q/449}{za_I/144} \qquad \text{and} \qquad w_{II} = \frac{Q/449}{za_{II}/144}$$

The ratio of the relative velocities is

$$\frac{w_{II}}{w_I} = \frac{a_I}{a_{II}}$$

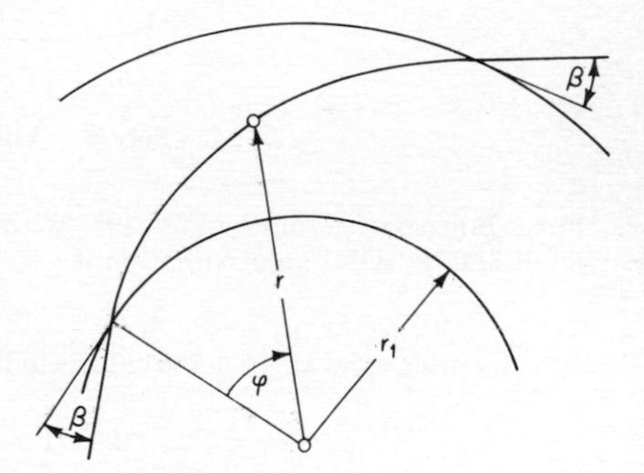

Fig. 8 Logarithmic spiral.

Normally the cross-sectional area of the channel increases gradually from a_I to a_{II}. The reduction of the relative velocity causes an increase of static head, as indicated by Eq. (11). There are limitations to the area increase if flow separation is to be avoided. For high hydraulic efficiency, the area ratio is selected as $a_{II}/a_I = 1.0$ to 1.3.

The determination of a_I and a_{II} is simple provided cylindrical vanes (Fig. 4) are used. Only the meridional section (Fig. 4) and an axial view (Fig. 2) are required. The vane layout depends mainly upon the selected angles β_1 and β_2. The recommended ranges for these two angles overlap. Therefore the outlet vane angle is occasionally equal to the inlet vane angle. In such a case, the vane takes the shape of a logarithmic spiral. Its coordinates (Fig. 8) can be calculated from

$$r = r_1 e^{(\pi\varphi/180) \tan \beta} \tag{37}$$

If β_2 is different from β_1, Eq. (37) of the logarithmic spiral can still be utilized for the vane layout by a stepwise variation of the vane angle β.

For twisted vanes (Fig. 3), areas a_I and a_{II} can be obtained only after considerable labor has been expended on the complicated vane layout. In order to reduce the number of trials, it is desirable to be able to calculate the areas between vanes at least approximately, before starting the first vane layout.

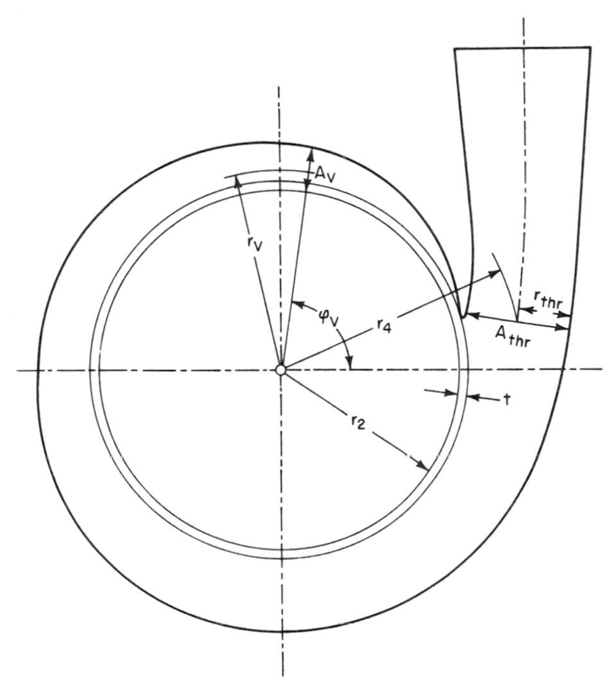

Fig. 9 Volute casing.

For Francis-type vanes (Fig. 19), with mixed flow in the vane inlet area and a small hub, a useful approximation is

$$A_I = za_I \approx \pi r_1^2 \sin \beta_{1m} \tag{37a}$$

The mean inlet vane angle β_{1m} is calculated from Eq. (35) by entering the mean radius

$$r_{1m} = \left(\frac{r_1^2 + r_H^2}{2}\right)^{1/2} \tag{38}$$

If the front shroud in the discharge region is inclined, as is usual for Francis-type impellers, the discharge area between vanes can be approximated by

$$A_{II} = za_{II} \approx b_2(2\pi r_2 \sin \beta_2 - zs_2) \tag{39}$$

If the front shroud in the discharge region is not inclined, about 25 percent should be subtracted from Eq. (39).

To make sure that the flow encounters no sudden changes of flow area and vane angle, it is recommended that their values, as determined from the vane layout, be plotted against the radius or length of flow path.

Volute Casing A large part of the dynamic head $(c_3')^2/2g$ of the fluid issuing from the impeller is converted into static head in the following diffusing system. The most popular one for centrifugal pumps is the volute.

The design of the volute starts preferably with the calculation of the area of the throat (station 4 in Fig. 9):

$$A_{thr} = \frac{Q/449}{c_{thr}/144} \tag{40}$$

The average throat velocity c_{thr} is determined from

$$\frac{c_{thr}}{c'_{u3}} = \frac{r_2}{r_4} C \tag{41}$$

for frictionless flow $C = 1$, according to the law of constant angular momentum. Experience has shown that $C = 1$ is indeed a good design value for volutes of large pumps or very smooth volutes of medium-size pumps. For commercially cast volutes of medium size and small pumps the value $C = 0.9$ gives a reasonable approximation.

The distance r_4 of the center of the throat section from the axis is (Fig. 9):

$$r_4 \approx r_2 + t + r_{thr} \tag{42}$$

whereby the throat section is assumed to be circular.

The clearance t between impeller and tongue of volute has a bearing on efficiency and pressure pulsations. If the highest possible efficiency is desired, the clearance is made as small as 1/32 in. But such a small clearance produces larger pressure pulsations and has led to impeller failure in some cases. As a remedy, the clearance is then increased to about 1/8 in. A reasonably safe value for the tongue clearance, still allowing good efficiency, is 5 to 10 percent of the impeller radius.

The calculation of the throat area can be aided by initially assuming a throat velocity c_{thr}, for example, from the ratio c_{thr}/u_2 as given by Fig. 10, which shows a band of empirical values plotted versus the specific speed. The sources are Refs. 8 and 9, in the range of N_s from 200 to 2,000 augmented by data collected by the author. With the initially assumed throat velocity c_{thr} and tongue clearance t, the throat area A_{thr} and the throat distance r_4 can be calculated. Then Eq. (41) is applied as a check, and if necessary, a correction is made for the throat area.

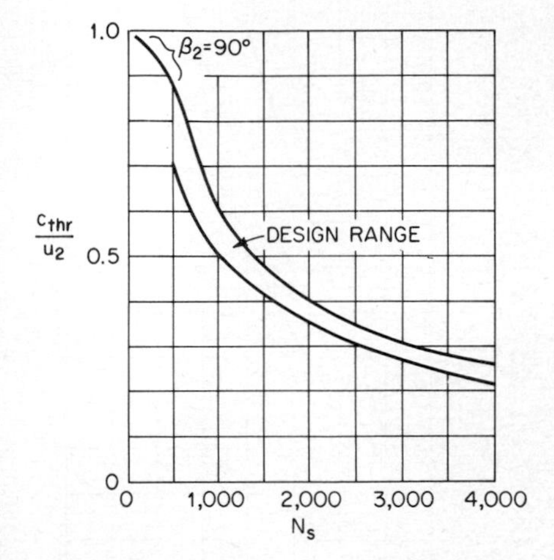

Fig. 10 c_{thr}/u_2 versus specific speed.

An alternate method for the calculation of the throat area was originally developed by Anderson [9]. It makes use of the area ratio $y = A_{II}/A_{thr}$, where $A_{II} = za_{II}$ is the total outlet area between impeller vanes (Fig. 2). Worster [10] has verified that, for the condition of constant angular momentum, a relation exists between y and N_s. Figure 11 is a plot of empirical data for $1/y = A_{thr}/A_{II}$ versus N_s. The band results from the author's data, and shows the considerable range of area ratios which are used in actual pump designs. The line in the middle of the band is from Anderson [9]. The condition of constant angular momentum is approximately fulfilled for the lower limit of A_{thr}/A_{II} in Fig. 11. High efficiency can be expected between this limit and Anderson's line.

Commercial pump lines make use of the fact that large variations of A_{thr}/A_{II} are tolerable by employing impellers of different widths in the same volute casing, or by using the same impeller in volutes of different throat areas. The intent is to arrive at new combinations of head and capacity without manufacturing additional hardware.

It is not recommended that the curves in Fig. 11 be extrapolated to lower specific speeds. The resultant impeller discharge widths would, in many cases, be too small for manufacturing. Actual successful low specific speed pumps have much wider impeller discharges.

The intermediate volute areas

$$A_V = \frac{Q/449}{c_V/144} \frac{\varphi_V}{360} \tag{43}$$

can be calculated on the basis of constant angular momentum

$$\frac{c_V}{c'_{u3}} = \frac{r_2}{r_V} C \tag{44}$$

Stepanoff [8] prefers to use a constant mean velocity for all volute sections, which results in volute areas being proportional to the central angle φ_V

$$A_V = A_{\text{thr}} \frac{\varphi_V}{360} \tag{45}$$

The resultant intermediate volute areas are larger than according to the constant angular momentum method. Still there is very little, if any, detrimental effect on the efficiency. Even a constant area collector, instead of a volute, reduces the efficiency at the design point by no more than 1 to 2 points.

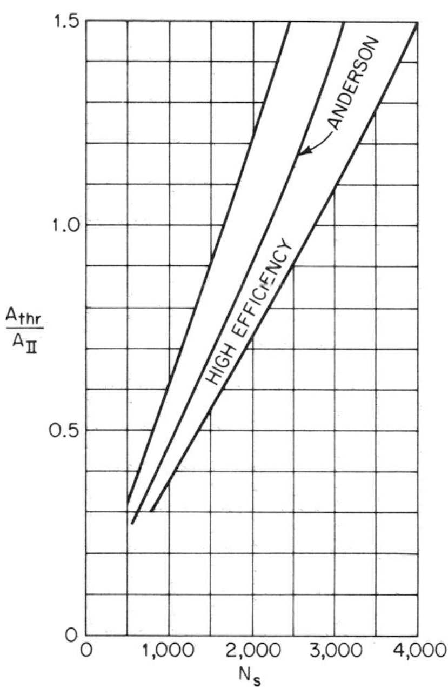

Fig. 11 $A_{\text{thr}}/A_{\text{II}}$ versus specific speed.

The volute is followed by a conical diffuser, making the transition to the discharge pipe. The total included angle of the diffuser can be taken from 6 to 10°. The volute sections may have shapes different from circles. For instance, rectangles or trapezoids are often used. These shapes make it possible for the entrance width of the volute to be appreciably greater than the impeller discharge width b_2 (the volute entrance width may be as large as two or three times the impeller discharge width). Such loose-fitting volutes are common design practice in the United States, but much less so in Europe. According to tests (Ref. 15) made in Germany, they produce slightly better efficiency, and their great advantage is that they are easier to cast (because of the better core support) and easier to clean after casting.

Volutes cause radial thrust, small at the design point, but increasing as we approach larger and, especially, smaller capacities. The reason for the radial thrust is that the pressure at the periphery of the volute entrance is no longer constant, as it should be at the design point. Twin volutes, vaned diffusers, and, to a lesser degree, concentric casings reduce the radial thrust. Agostinelli et al. [16] give numerical values based on extensive tests. This reference also describes a semiconcentric casing design, in which the collector is circular for a portion of the angular distance from the tongue before the radial extent of the casing wall is increased towards the throat section. This simple design can have radial thrust as low as the more complicated twin volute.

Vaned Diffuser Instead of a volute, a vaned diffuser may be used. It consists of a number of vanes set around the impeller. The flow from the vaned diffuser is collected in a volute or circular casing and discharged through the outlet pipe.

The design of the vaned diffuser is analogous to that of the volute. Instead of one throat section, there are now several (Fig. 12). Again, Eq. (41)

$$\frac{c_{\text{thr}}}{c'_{u3}} = \frac{r_2}{r_1} C$$

can be used to calculate the throat velocity c_{thr}. The distance r_4 of the center of the throat from the axis is now smaller than for the volute pump of equal specific speed. Therefore one could expect that the throat velocity of a diffuser would be larger. This is actually not the case because the factor C is smaller for vaned diffusers. A typical value is $C = 0.8$. The ratios c_{thr}/u_2 in Fig. 10 can also be used for vaned diffusers.

From the throat on, the area of the vane channel increases progressively, so that further, though slight, pressure increase takes place. The centerline of the vane channel after the throat may be straight or curved. The straight diffusing channel is slightly more efficient but results in a larger casing. The vane surface from the vane inlet tip to the throat can be shaped like a volute, but an approximation by a circular arc is entirely permissible. The vane entrance angle is made 3 to 5° larger than α_3'. Moderate deviations do not matter; what matters mostly is the proper choice of the throat area. The side walls of the vaned diffuser are preferably parallel, but in some designs they spread apart with increasing radius. A good choice for the number of diffuser vanes is one more than the impeller vanes. In this way, circulation around the diffuser vanes, resulting from uneven impeller channel discharge, is minimized.

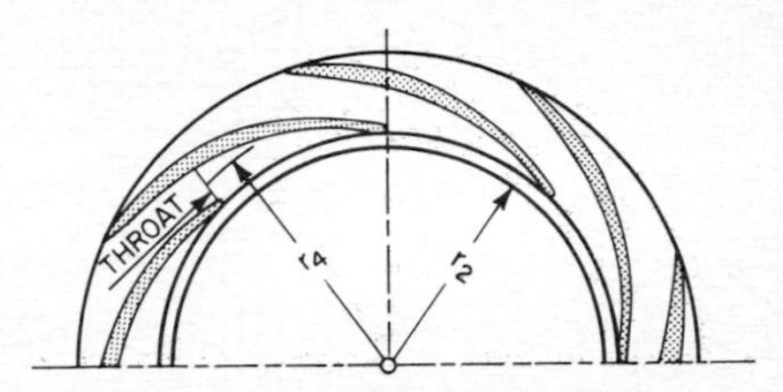

Fig. 12 Vaned diffuser.

Hydraulic Efficiency The hydraulic efficiency depends, naturally, very much upon the design and execution of the flow passages. Nevertheless, if one considers only pumps of above average performance, and plots their hydraulic efficiency (at the design point) versus the capacity, the points will arrange themselves in a narrow band. It can be approximated by

$$\eta_H \approx 1 - \frac{0.8}{Q^{0.25}} \tag{46}$$

This equation has been plotted in Fig. 13. The specific speed has only a small influence on the hydraulic efficiency. At $N_s = 500$ the hydraulic efficiency seems

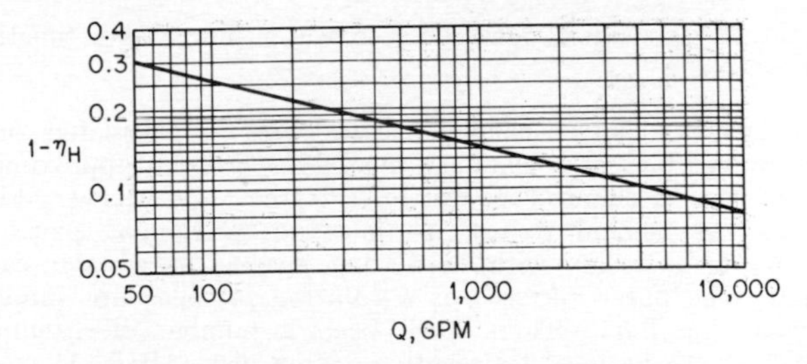

Fig. 13 Hydraulic efficiency versus capacity.

to be about 2 points lower than at $N_s = 2{,}000$. This possible variation has been neglected in Eq. (46), since it is not greater than the scatter of the utilized test data.

There are many pumps in service which have, because of mediocre design or manufacture, hydraulic efficiencies five points lower than given by Eq. (46). Even ten-point lower hydraulic efficiencies have been tolerated in some cases, especially for small capacities.

Volumetric Efficiency The definition of the volumetric efficiency has been given by Eq. (18):

$$\eta_V = \frac{Q}{Q + Q_L}$$

To calculate the leakage flow, the details of the individual pump design must be known. Standard methods can be found in Refs. 2 and 8. For an approximate prediction of η_V at the design point, Fig. 14 can be used. It shows the volumetric efficiency as a function of specific speed and flow for pumps of normal construction. For instance, for a pump handling 1,000 gpm at a specific speed of 2,500, Fig. 14 estimates $\eta_V \approx 0.98$. For 100 gpm at $N_s = 500$, the estimate is $\eta_V \approx 0.9$.

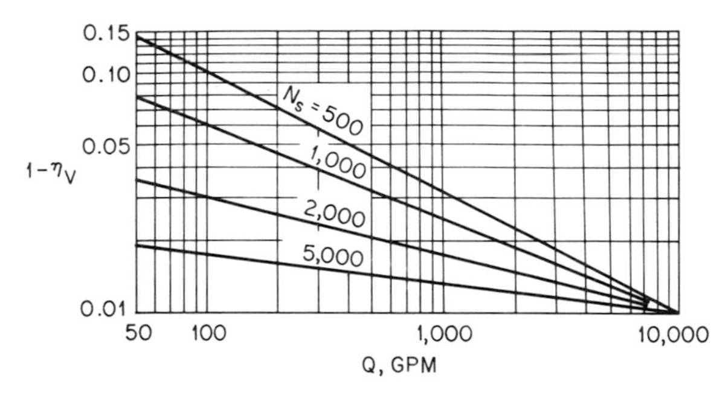

Fig. 14 Volumetric efficiency as a function of specific speed and capacity.

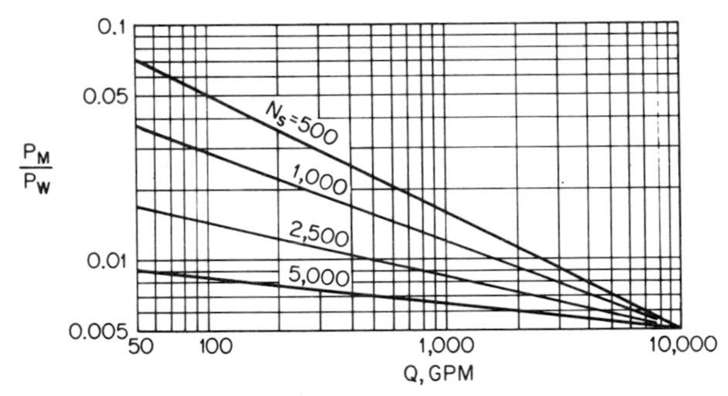

Fig. 15 Ratio of mechanical power loss to water power as a function of specific speed and capacity.

Mechanical Losses The mechanical losses can be calculated only when the design details of bearings and seals are available. For an approximate prediction of the ratio of the mechanical power loss to the water power, Fig. 15 can be used. It shows the general trends, in that the mechanical power loss relative to the water-power increases with decreasing specific speed and capacity.

Disk Friction The outer surfaces of a rotating impeller are subject to friction with the surrounding fluid. There have been a number of attempts to predict this power loss on the basis of tests with rotating disks (Refs. 11 to 13), therefore the expression *disk friction*. Pfleiderer [2] writes on the basis of tests by Schultz-Grunow [11]:

$$P_{DF} = 4.07(10^{-10})N^3D^4(D + 5e) \tag{47}$$

valid for $Re = \omega(D/2)^2/\nu = 7.0\ (10^5)$. The disk friction power increases approximately with $Re^{-0.2}$. The formula (47) aims to account not only for the friction on the two sides of the impeller, but also for the friction on the outer cylindrical surfaces of the shrouds (e is the sum of the width of two shrouds). References 8 and 14 state that Eq. (47) gives values on the high side.

The ratio of the disk friction power to the water power, as a function of the specific speed, can be derived by utilizing Eqs. 15, 28, 30, and 47:

$$\frac{P_{DF}}{P_W} = \frac{1.38(10^5)}{N_S^2\psi^{2.5}} \frac{D + 5e}{D} \tag{48}$$

In view of the smaller value obtained in the subsection Impeller-Disk Friction Deduced from Maximum Efficiency, it is recommended that only 70 to 80 percent of Eq. (48) be used.

The conditions in a centrifugal pump are more complicated than in a simple disk-friction experiment. Very probably an appreciable portion (according to Ref. 14 almost 50 percent) of the disk friction loss is recovered as contribution to the pump head, if the rotating flow induced by disk friction is allowed to freely enter the casing. The whirl velocity c'_{u3} of the flow issuing from the impeller reduces the friction on the outer cylindrical impeller surfaces (Ref. 14). The wearing-ring leakage causes a radial flow which tends to reduce disk friction.

Maximum Efficiency It is of some interest to estimate the maximum efficiency at a given specific speed. Such an estimate can be made on the basis of Fig.

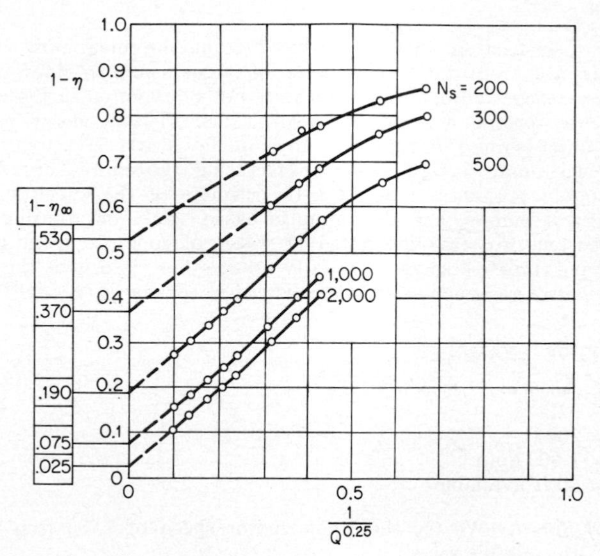

Fig. 16 Extrapolation for maximum efficiency.

5 by extrapolating the capacity to infinity. A great deal of the efficiency increase with capacity results from the rise in hydraulic efficiency. The approximate formula (46) for the hydraulic efficiency

$$\eta_H \approx 1 - \frac{0.8}{Q^{0.25}}$$

suggests therefore plotting the value $1 - \eta$ versus $1/Q^{0.25}$ (Fig. 16). The pump efficiency η is taken from the specific-speed chart, Fig. 5. Extrapolating to $1/Q^{0.25} = 0$ in Fig. 16 provides an estimate of the efficiency at infinite capacity—the maximum efficiency η_∞. Between $N_S = 500$ and 2,000, the extrapolation is reliable and can be represented by

$$\eta_\infty = \frac{1}{1 + (7{,}800/N_S{}^{5/3})} \tag{49}$$

This is the upper line denoted η_∞ in Fig. 5.

Impeller Disk Friction Deduced from Maximum Efficiency At infinite capacity, the hydraulic efficiency is unity according to Eq. (46). The volumetric efficiency reaches the same value and P_M/P_W vanishes. Equation (19) for the pump efficiency becomes simply

$$\eta_\infty = \frac{1}{1 + (P_{DF}/P_W)} \tag{50}$$

Equating this with the previous estimate of the maximum efficiency (49), the following expression for the impeller disk friction results:

$$\frac{P_{DF}}{P_W} = \frac{7{,}800}{N_S^{5/3}} \tag{51}$$

Since Eq. (49) is reliable only from $N_s = 500$ to 2,000, the same limitation applies to Eq. (51). The numerical values are approximately 70 percent of Eq. (48), which is based on Pfleiderer's disk friction formula (47).

Above specific speeds of 2,000, the disk friction power is relatively small and can be approximated by

$$\frac{P_{DF}}{P_W} \approx 0.02$$

The impeller disk friction (51), as deduced from the maximum efficiency and therefore indirectly from the performance of actual pumps, does include those effects of pump design and operation which are not represented in the usual rotating disk experiments: partial recovery of impeller disk friction energy in the casing; reduction of outer shroud friction due to impeller swirl; and wearing-ring flow. Provided typical pump design and manufacturing procedures are followed, Eq. (51) will probably provide a more realistic estimate of the effective disk friction loss of centrifugal pumps than the formulas based solely on rotating disk experiments. As written, Eq. (51) is limited to water of room temperature and those fluids which have similar kinematic viscosity. For other viscosities, the rotating-disk experiments suggest a correction by multiplying Eq. (51) by $(\nu/\nu_{water})^{0.2}$.

COMPREHENSIVE EXAMPLE

Conditions of Service

H = 200 ft
Q = 1,000 gpm
NPSH = 40 ft available

Selection of Speed. We try the 4-pole motor speed of 1,780 rpm and calculate from Eq. (30) the specific speed:

$$N_S = \frac{N\sqrt{Q}}{H^{0.75}} = \frac{1{,}780\sqrt{1{,}000}}{200^{0.75}} = 1{,}057$$

According to the specific speed chart Fig. 5, the pump efficiency to be expected is 0.785. The required suction specific speed (31) is:

$$S = \frac{N\sqrt{Q}}{\text{NPSH}^{0.75}} = \frac{1{,}780\sqrt{1{,}000}}{40^{0.75}} = 3{,}537$$

a low value for normal inlet design. Therefore we try the 2-pole motor speed of 3,560 rpm and calculate:

$$N_S = \frac{3{,}560\sqrt{1{,}000}}{200^{0.75}} = 2{,}114$$

$$S = \frac{3{,}560\sqrt{1{,}000}}{40^{0.75}} = 7{,}075$$

The expected pump efficiency is 0.83, which would reduce the input power by 5.4 percent. The suction specific speed is still below the value of 9,000, mentioned in the subchapter Impeller Inlet. Because of the appreciable efficiency gain and the reduction of pump size, we select the speed of:

$$N = 3{,}560 \text{ rpm}$$

Impeller-Discharge Velocity Triangle. Having fixed the specific speed at 2,114, we obtain $c_{m3}/u_2 = 0.12$ from Fig. 6, $\beta_2 = 25°$ from Fig. 7, and $z = 7$, to be evaluated by slip factor formulas. The slip factor according to Stodola's formula (21) is

$$\mu = 1 - \frac{\pi \sin \beta_2}{z} = 1 - \frac{\pi \sin 25°}{7} = 0.81$$

Pfleiderer's formula (24) gives

$$\mu = \frac{1}{1 + \frac{a}{z}\left(1 + \frac{\beta_2}{60}\right)\frac{2}{1 - (r_1^2/r_2^2)}} = \frac{1}{1 + (0.65/7) \times 1.42(2/0.75)} = 0.74$$

whereby we took the lower value for the volute $a = 0.65$. The radius ratio was tentatively estimated to be equal to or below $r_1/r_2 = 0.5$. We accept $\mu = 0.74$

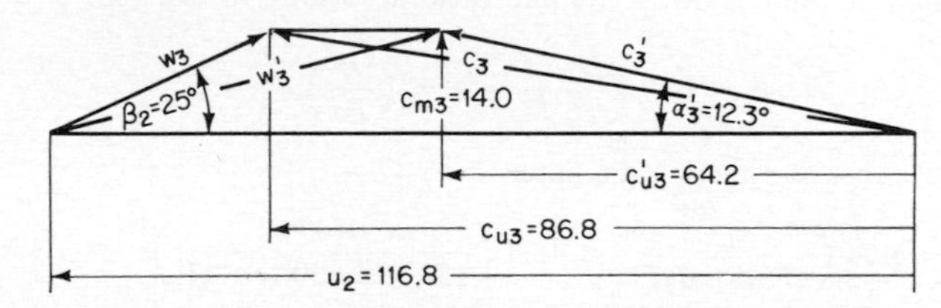

Fig. 17 Comprehensive Example: impeller-discharge velocity triangles.

since Stodola's $\mu = 0.81$ appears high from experience with similar pumps. The hydraulic efficiency is estimated from Eq. (46) or from Fig. 13:

$$\eta_H \approx 1 - \frac{0.8}{Q^{0.25}} = 1 - \frac{0.8}{1{,}000^{0.25}} = 0.86$$

The head coefficient can now be calculated from Eq. (28):

$$\psi = 2\mu\eta_H\left(1 - \frac{c_{m3}}{u_2}\cot \beta_2\right) = 2 \times 0.74 \times 0.86(1 - 0.12 \cot 25°) = 0.945$$

and the necessary impeller peripheral speed, also from Eq. (28), is

$$u_2 = \sqrt{\frac{2gH}{\psi}} = \sqrt{\frac{64.4 \times 200}{0.945}} = 116.8 \text{ ft/s}$$

With this, we calculate the meridional velocity right after the impeller discharge:

$$c_{m3} = \frac{c_{m3}}{u_2} u_2 = 0.12 \times 116.8 = 14.0 \text{ ft/s}$$

The peripheral component of the absolute impeller-discharge velocity without slip is with the geometric relationships of Fig. 2:

$$c_{u3} = u_2 - c_{m3} \cot \beta_2 = 116.8 - 14.0 \cot 25° = 86.8 \text{ ft/s}$$

With slip, we have according to Eq. (20):

$$c'_{u3} = \mu c_{u3} = 0.74 \times 86.8 = 64.2 \text{ ft/s}$$

Now all the ingredients for drawing the impeller-discharge velocity triangle in Fig. 17 are available. It reveals that the absolute flow angle α'_3, under which the flow leaves the impeller, is 12.3°. This is also roughly the angle for which the tongue of the volute is to be designed.

Impeller-Discharge Dimensions. The impeller diameter is

$$D_2 = \frac{60 \times u_2 \times 12}{\pi N} = \frac{60 \times 116.8 \times 12}{\pi \times 3{,}560} = 7.52 \text{ in}$$

and the impeller radius $r_2 = D_2/2 = 3.76$ in. The impeller discharge width is calculated from Eq. (29):

$$b_3 = \frac{Q/449}{2\pi r_2 c_{m3}/144} = \frac{1{,}000/449}{2\pi \times 3.76 \times 14.0/144} = 0.97 \text{ in}$$

Impeller Inlet Diameter. Since NPSH requirements are not critical, we select a normal flow angle of

$$\beta_0 = 17°$$

An axial inlet shall be assumed with a small hub which does not protrude far enough to block the inlet flow. The hub ratio according to Eq. (33) is then

$$k = 1 - \left(\frac{D_H}{D_S}\right)^2 = 1 - \left(\frac{0}{D_S}\right)^2 = 1.0$$

Equation (32) gives the suction diameter

$$D_S = 4.54\left(\frac{Q}{kN \tan \beta_0}\right)^{1/3} = 4.54\left(\frac{1{,}000}{1.0 \times 3{,}560 \tan 17°}\right)^{1/3} = 4.42 \text{ in}$$

and the inlet radius $r_1 = r_S = D_S/2 = 2.21$ in.

With the inlet radius and the previously calculated impeller-discharge dimensions, the impeller profile can now be drawn as in Fig. 19. The mean inlet radius, as defined by Eq. (38), is

$$r_{1m} = \left(\frac{r_1^2 + r_H^2}{2}\right)^{1/2}$$

$$= \left(\frac{2.21^2 + 0^2}{2}\right)^{1/2} = 1.56 \text{ in}$$

While Pfleiderer [2] does not specifically say so, it is proper to take r_{1m} for r_1 in his equation (24) for the slip factor if the vanes are extended into the axial inlet. The radius ratio r_1/r_2, under which no further increase of the slip factor occurs, would then become r_{1m}/r_2. In our case

$$\frac{r_{1m}}{r_2} = \frac{1.56}{3.76} = 0.41$$

Therefore the previously made assumption, that it is under 0.5, is fulfilled.

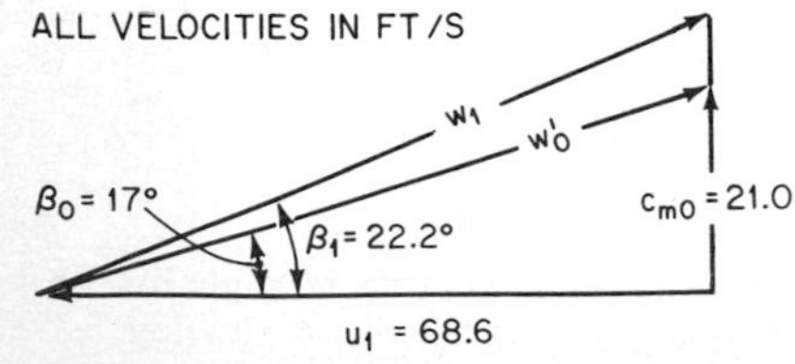

Fig. 18 Comprehensive Example: impeller-inlet velocity triangles.

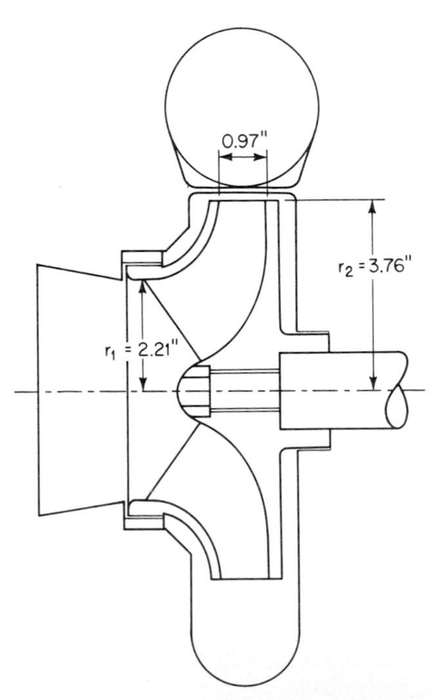

Fig. 19 Comprehensive Example: impeller profile with volute.

Inlet Vane Angles. The inlet vane thickness s_1 shall be $\frac{3}{16} = 0.1875$ in. The inlet vane angle at outer radius is calculated from (35):

$$\tan \beta_1 = \frac{\tan \beta_{0(r)}}{1 - \dfrac{zs_1}{2\pi r_1 \sin \beta_{1(r)}}} = \frac{\tan 17°}{1 - \dfrac{7 \times 0.1875}{2\pi \times 2.21 \sin \beta_1}} = 0.409$$

$$\beta_1 = 22.2°$$

Some iteration was necessary because β_1 appears on both sides of Eq. (35).

By the same procedure, one can calculate that at the mean radius $r_{1m} = 1.56$ in the mean inlet vane angle is

$$\beta_{1m} = 30.5°$$

Impeller Inlet Velocity Triangle. In order to draw the inlet velocity triangle (Fig. 18), the following calculations are performed:

The meridional velocity ahead of the vanes:

$$c_{m0} = \frac{Q/449}{\pi r_1^2/144} = \frac{1{,}000/449}{\pi \times 2.21^2/144} = 21.0 \text{ ft/s}$$

The peripheral vane velocity at the eye:

$$u_1 = \frac{\pi r_1 N}{30 \times 12} = \frac{\pi \times 2.21 \times 3{,}560}{360} = 68.6 \text{ ft/s}$$

Flow Areas between Vanes. The inlet area between vanes is approximately from Eq. (37*a*):

$$A_{\text{I}} = za_{\text{I}} \approx \pi r_1^2 \sin \beta_{1m} = \pi \times 2.21 \sin 30.5° = 7.77 \text{ in}^2$$

The discharge area between vanes, assuming a vane thickness $s_2 = \frac{3}{16}$ in $= 0.1875$ in, is approximately from Eq. (39):

$$\begin{aligned} A_{\text{II}} &= za_{\text{II}} \approx b_2(2\pi r_2 \sin \beta_2 - zs_2) \\ &= 0.97(2\pi \times 3.76 \sin 25° - 7 \times 0.1875) = 8.41 \text{ in}^2 \end{aligned}$$

With the above, the area ratio becomes

$$\frac{A_{\text{II}}}{A_{\text{I}}} = \frac{8.41}{7.77} = 1.08$$

It is well below 1.3, and therefore acceptable. The areas between vanes must be checked by the actual vane layout.

Volute Casing. Because of its simplicity, a single volute shall be used. The throat velocity c_{thr} is tentatively selected from Fig. 10 for $N_S = 2{,}114$:

$$\frac{c_{\text{thr}}}{u_2} \approx 0.35$$

$$c_{\text{thr}} = \frac{c_{\text{thr}}}{u_2} u_2 \approx 0.35 \times 116.8 = 40.9 \text{ ft/s}$$

The tentative throat area from Eq. (40) is

$$A_{\text{thr}} = \frac{Q/449}{c_{\text{thr}}/144} = \frac{1{,}000/449}{40.9/144} = 7.84 \text{ in}^2$$

Assuming a circular throat section, its radius is

$$r_{\text{thr}} = 1.58 \text{ in}$$

The tongue distance is selected to be 7 percent of the impeller radius:

$$t = 0.07 \times 3.76 = 0.26 \text{ in}$$

The distance of the throat center from the axis is about Eq. (42)

$$r_4 \approx r_2 + t + r_{thr} = 3.76 + 0.26 + 1.58 = 5.60 \text{ in}$$

Now the tentative choice of the throat velocity is checked by applying Eq. (41) in the form

$$C = \frac{c_{thr}}{c'_{u3}} \frac{r_4}{r_2} = \frac{40.9}{64.2} \frac{5.60}{3.76} = 0.95$$

According to the comments in subsection Volute Casing, this appears to be a very reasonable factor C.

A further verification is obtained by forming:

$$\frac{A_{thr}}{A_{II}} = \frac{7.84}{8.41} = 0.93$$

and entering this value into Fig. 11 at $N_S = 2{,}114$. This point falls into the region marked High Efficiency.

There is now enough assurance to make the tentative choice of the throat velocity permanent, and to proceed with the volute design. If the flow factor C had been larger than 1.0 or smaller than 0.9, then the calculation would have to be repeated with an improved choice for c_{thr}. The intermediate volute areas are first approximated by Stepanoff's equation (45):

$$A_V = A_{thr} \frac{\varphi_V}{360}$$

Assuming the intermediate volute areas to be of circular shape, their radii and distances from the axis are calculated. The results are listed in the Table 1. Some

Table 1

φ_V	A_V	r	r_V
0	0	0	4.02
45	0.98	0.56	4.58
90	1.96	0.79	4.81
135	2.94	0.97	4.99
180	3.92	1.11	5.13
225	4.90	1.25	5.27
270	5.88	1.36	5.38
315	6.86	1.48	5.50
360	7.84	1.58	5.60

designers would stop here and dimension the volute according to this table. We will go further and apply the law of constant angular momentum also to the intermediate volute areas (44). To get the calculation started, we will first use as an approximation the distances r_V from the Table 1. The flow factor shall remain constant at $C = 0.95$. Equation (44) states

$$c_V = \frac{r_2}{r_V} C c'_{u3} = \frac{3.76 \times 0.95 \times 64.2}{r_V} = \frac{229.3}{r_V}$$

and from (43)

$$A_V = \frac{Q/449}{c_V/144} \frac{\varphi_V}{360} = \frac{1{,}000/449}{c_V/144} \frac{\varphi_V}{360} = 0.891 \frac{\varphi_V}{c_V}$$

The newly calculated intermediate areas result in new center distances r_V, which are then compared with the assumed values. The results are listed in Table 2. The calculated center distances r_V are close to the assumed values. A further iteration step could be performed by entering the newly calculated center distances

Table 2

φ_V	Assumed r_V	c_V	A_V	r	Calculated r_V
0	4.02	57.04	0	0	4.02
45	4.58	50.06	0.80	0.50	4.52
90	4.81	47.67	1.68	0.73	4.75
135	4.99	45.95	2.62	0.91	4.93
180	5.13	44.70	3.59	1.07	5.09
225	5.27	43.51	4.61	1.21	5.23
270	5.38	42.62	5.64	1.34	5.36
315	5.50	41.69	6.73	1.46	5.49
360	5.60	40.90	7.84	1.58	5.60

as assumed values into the above procedure. For the present example, the resultant changes of the volute dimension would be negligible.

The volute can now be drawn using the above listed section radii r and the distances r_V of the volute centers from the axis (Figs. 20 and 19).

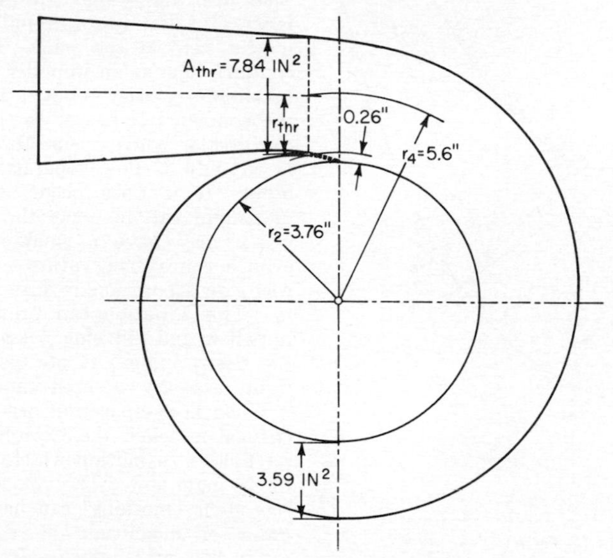

Fig. 20 Comprehensive Example: volute dimensions.

Loss Estimates, Resultant Efficiency, and Shaft Power. The following estimates can be made for the losses:

Hydraulic efficiency from Eq. (46) (or from Fig. 13):

$$\eta_H \approx 1 - \frac{0.8}{Q^{0.25}} = 1 - \frac{0.8}{1{,}000^{0.25}} = 0.86$$

Volumetric efficiency from Fig. 14 for $N_S = 2{,}114$ and $Q = 1{,}000$ gpm:

$$\eta_V \approx 0.98$$

Ratio of mechanical losses to water power from Fig. 15 for $N_S = 2{,}114$ and $Q = 1{,}000$ gpm:

$$\frac{P_M}{P_W} \approx 0.012$$

Ratio of impeller disk friction to water power from Eq. (51):

$$\frac{P_{DF}}{P_W} = \frac{7{,}800}{N_S^{5/3}} = \frac{7{,}800}{2{,}114^{5/3}} = 0.023$$

Entering the above estimates into (19) gives the pump efficiency:

$$\eta = \frac{1}{\dfrac{1}{\eta_H \eta_V} + \dfrac{P_{DF}}{P_W} + \dfrac{P_M}{P_W}} \approx \frac{1}{\dfrac{1}{0.86 \times 0.98} + 0.023 + 0.012} = 0.82$$

The pump efficiency calculated from estimated losses is close to the value of 0.83 as read from the chart Fig. 5.

With $\eta = 0.82$, the power necessary to operate the pump is [Eqs. (14) and (15)]:

$$P_S = \frac{1}{\eta} QH \frac{\text{sp gr}}{3{,}960} = \frac{1{,}000 \times 200 \times 1}{0.82 \times 3{,}960} = 61.6 \text{ hp}$$

ADVANCED METHODS

In all the preceding treatment, flow velocities, even if denoted *actual,* were always *averaged* velocities. This is a greatly simplified approach. Actually, the velocity is always nonuniformly distributed over the area. In a conduit with a straight axis, as in an axial pump inlet, the velocity is normally larger in the center than near the walls. In the boundary layer, it diminishes rapidly until it reaches zero at the walls. In a curved conduit, such as an impeller passage, the maximum velocity is not in the center of the conduit, but nearer to the wall with the greater convex curvature. But this is so only if flow separation does not occur. If it does occur, the velocity maximum can be near the other wall, which may have a smaller convex or even a concave curvature. On the wall with flow separation, reverse flow may set in. This situation can prevail in some impellers and diffusing passages, even at the design point; it always does occur if the capacity is small enough.

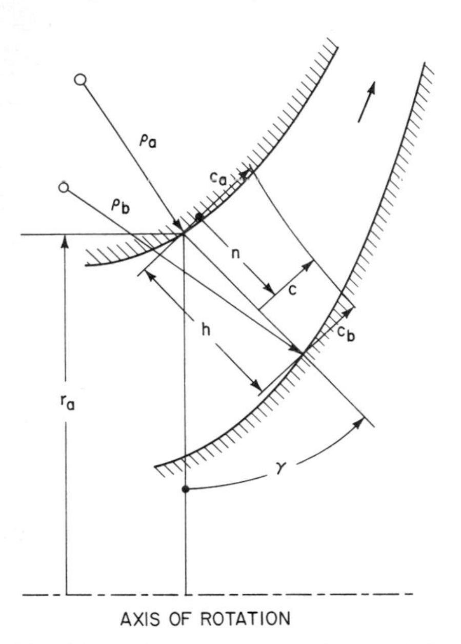

Fig. 21 Axisymmetric vaneless conduit with curved walls.

When flow separation occurs in a centrifugal impeller, the Coriolis forces can set flows into motion which go crosswise to the main flow. The velocities of these "secondary motions" can have the same order of magnitude as those of the main flow. Therefore, advanced theories, which do not include secondary motions, are not complete.

It has been the hope of many investigators that methods to calculate two- or three-dimensional velocity distributions would refine the theory of centrifugal pumps and eventually lead to improved designs.

A Method Excluding Vane Forces This method had been published by Flügel [17] as early as 1914. It is one of the first attempts to predict the velocity distribution in an axisymmetric vaneless conduit with curved walls, Fig. 21. One simplifying assumption is that the curvature $1/\rho$ of the streamlines changes linearly from wall to wall. Perfect guidance of the fluid by the walls is assumed; viscosity effects and flow separation are excluded. The method yields the ratio of the two wall velocities at the ends of an orthogonal (a curved line which intersects two opposing walls perpendicularly):

$$\frac{c_a}{c_b} = \exp\left[\frac{h}{2}\left(\frac{1}{\rho_a} + \frac{1}{\rho_b}\right)\right] \tag{52}$$

Also, the ratios of any other two velocities on the orthogonal can be obtained by inserting into Eq. (52) their distance (instead of the wall distance h) and the streamline curvature from the assumed linear curvature distribution. The absolute values of the velocities have to be found by trial and error. The resultant velocity distribution over the orthogonal has to produce the desired flow. This calls for a tedious iteration to be repeated in each case. A simplification, utilizing prepared charts, will be presented in the subchapter A Simplified Method for the Calculation of the Meridional Velocity Distribution.

But even without calculating the absolute values of the velocities, the ratio of the wall velocities, as obtained from Eq. (52), provides an indication of the uniformity or nonuniformity of the flow. For instance, if $\rho_a = \rho_b = h$, then the ratio of the wall velocities becomes

$$\frac{c_a}{c_b} = \exp\left[\frac{h}{2}\left(\frac{1}{h} + \frac{1}{h}\right)\right] = e = 2.72$$

The larger the radii of curvature with respect to the wall distance, the more uniform is the flow.

Methods Including Vane Forces A number of investigations (Refs. 18–21) have been published which aim to calculate the velocity distribution including the effects of vane forces. Reference 21 derives a linear differential equation which can be written in the form (see Fig. 22)

$$-\frac{dw}{dn} = Aw - B - C\frac{dw_\theta}{dm} \qquad (53)$$

whereby

$$A = \frac{\cos^2 \beta^*}{\rho} - \frac{\sin^2 \beta^* \cos \gamma}{r} - \frac{d\theta}{dn} \cos \beta^* \sin \beta^* \sin \gamma$$

where $\beta^* = \beta - 90°$

$$B = 2\omega \sin \beta^* \cos \gamma - r\frac{d\theta}{dn} \cos \beta^* \omega \sin \gamma$$

$$C = r\frac{d\theta}{dn} \cos \beta^*$$

$$w_\theta = w \sin \beta^*$$

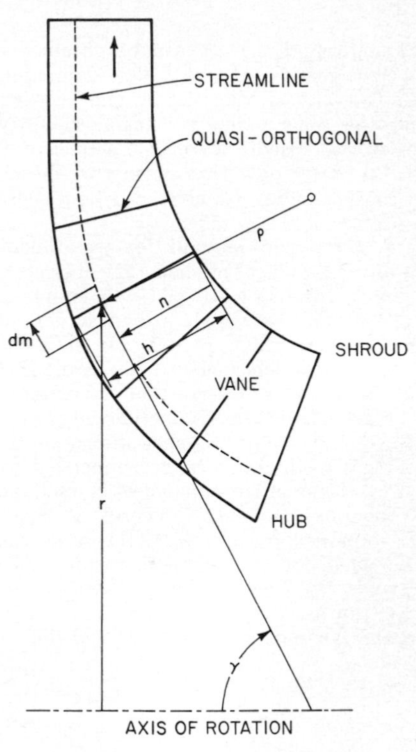

Fig. 22 Impeller configuration.

Assumptions are a perfect guidance of the flow by the vanes, no effects of viscosity, and no flow separation. Equation (53) gives the gradient of the relative velocity w along the orthogonal n in the meridional plane. Just as in the Flügel method [17], the differential equation does not give the absolute value of the velocities. To do this, it must be combined with the condition of continuity.

The velocity gradient along one orthogonal is influenced by adjacent orthogonals via the term dw_θ/dm. This imposes iteration. Equation (53) is suitable for computer programming, which is the only reasonable method of solution. Reference 21 contains instructions for programming.

A certain simplification (Ref. 22) can be introduced by approximating the orthogonals by straight lines; i.e., *quasi-orthogonals*.

The intent of this subsection is to give the reader an indication of the type of approach which is used to approximate the flow field in a centrifugal impeller. Before attempting to set up computer programs, it would be advantageous to study thoroughly the theoretical basis.

Vane-to-Vane Velocity Distribution The relative velocities obtained from Eq. (53) are only mean values for the selected point on the orthogonal. They exist somewhere in the channel between vanes. On the vane pressure side (the driving side) they are actually lower, and on the vane suction side (the trailing side) they are higher.

Reference 18 assumes a linear distribution of the relative velocity over the vane channel:

$$w_p = w - \frac{\Delta w}{2} \qquad \text{and} \qquad w_s = w + \frac{\Delta w}{2} \tag{54}$$

Reference 18 also gives

$$\Delta w = \frac{2\pi}{z} \sin \beta \, \frac{d(rc_u)}{dm} \tag{55}$$

The term $d(rc_u)/dm$ can be obtained by graphical differentiation of a plot of rc_u versus meridional length m. Computer programming is also possible.

The surface velocities w_p and w_s occur on opposite sides of the vane channel but at the same radius r. Again, viscosity effects and possible flow separation have been neglected. Normally perfect guidance is assumed. Then Eq. (55) should be used for only that portion of the channel which is confined by vanes. For the other portions, β and c_u can be calculated on the basis of assumed slip and then inserted into Eq. (55).

A Simplified Method for the Calculation of the Meridional-Velocity Distribution The Flügel method (52) neglects the effects of the vanes on the flow velocities and is therefore mathematically simple. When applying it to the meridional-velocity distribution between hub and shroud of several centrifugal impellers, it was noted that the results were not very different from the computer output obtained from methods which included the vane forces, for example, Eq. (53). The greatest deviation was observed in the region of greatest shroud curvature. There the Flügel method gives meridional velocities which are typically 15 to 20 percent lower. Near the impeller discharge there is almost no difference. Therefore the Flügel method can be recommended for the estimation of the meridional velocity field if the more sophisticated methods are not available as operating computer programs. Its only drawback is that, although the ratio of the wall velocities is obtained quickly (52), the numerical values of the meridional velocities must be calculated by tedious iteration. Therefore an approach recommends itself which avoids the time-consuming iteration by precalculating typical cases and presenting the results in graphical form.

To facilitate this, Eq. (52) and the continuity equation

$$c_{m,\text{av}} = \frac{Q/449}{2\pi h \left(r_a - \dfrac{h \cos \gamma}{2} \right)} \tag{56}$$

were combined and rearranged into

$$\frac{c_{m,a}}{c_{m,\text{av}}} = \frac{1 - \dfrac{h}{r_a} \dfrac{\cos \gamma}{2}}{\displaystyle\int_2^1 \left(1 - \frac{n}{h} \frac{h}{r_a} \cos \gamma \right) \exp \left\{ \frac{n}{h} \left[\frac{n}{h} \frac{1}{2} \left(\frac{h}{\rho_a} - \frac{h}{\rho_b} \right) - \frac{h}{\rho_a} \right] \right\} d \frac{n}{h}} \tag{57}$$

This is still a ratio of velocities, but now the easily calculated average meridional velocity (56) appears in the denominator.

If one calculates $c_{m,a}/c_{m,\text{av}}$ for h/ρ_a varying from 0 to 2, h/ρ_b from 0 to 2, and $\cos \gamma$ (h/r_a) from 0 to 1, the conditions occurring in typical meridional impeller sections are covered. The results are preserved in graphical form for use when required. Two such graphs, one for $h/\rho_b = 0$ (Fig. 23) and the other for $h/\rho_b = 0.4$ (Fig. 24), are given as samples. Figure 23 is suitable for an impeller inlet with no hub, since $h/\rho_b = 0$ can be interpreted as $\rho_b = \infty$.

To apply such graphs, one draws about eight orthogonals or quasi-orthogonals (it is sufficient to do this by eye) and obtains their lengths h, the radii of curvature of shroud ρ_a and hub ρ_b, as well as the angle γ. The average meridional velocity is calculated from (56). The meridional velocity at the shroud $c_{m,a}$ can be calculated from (57) or, as recommended, read from graphs such as Figs. 23 and 24. The meridional velocity at the hub is obtained from (52). Intermediate meridional

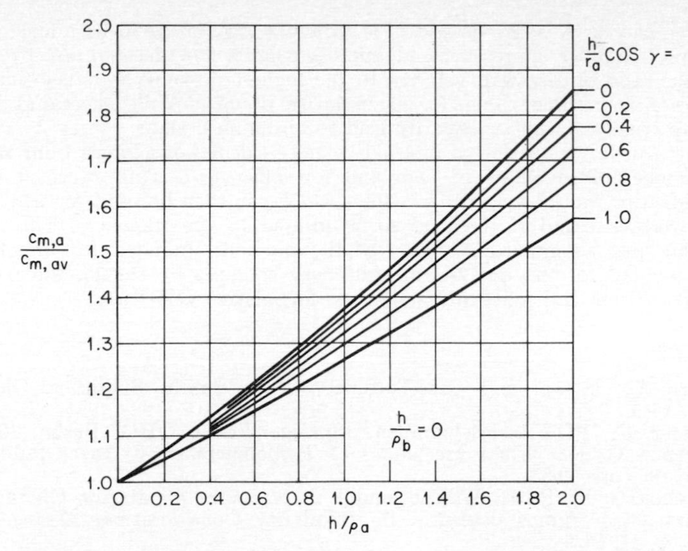

Fig. 23 Graph for determination of meridional shroud velocity.

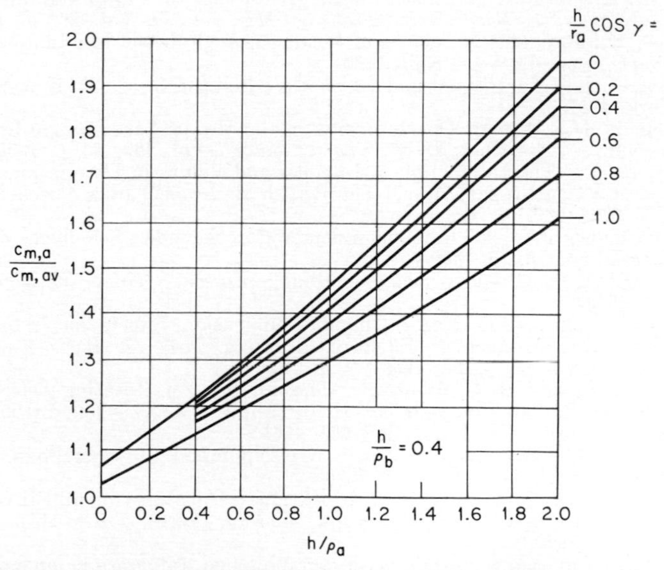

Fig. 24 Graph for determination of meridional shroud velocity.

velocities c_m can be computed from a modification of Eq. (52):

$$\frac{c_m}{c_{m,a}} = \exp\left[\frac{n^2}{2h}\left(\frac{1}{\rho_a} - \frac{1}{\rho_b}\right) - \frac{h}{\rho_a}\right] \tag{58}$$

Utilization of Advanced Methods All the referenced methods for internal-velocity calculations neglect fluid friction, boundary-layer buildup, and flow separation. Therefore their results must not be accepted at face value, but judged in view of their limitations. The calculated velocity distributions are idealized and can lead to wrong conclusions if used for new designs without reference to existing designs of known performance.

The problem is how to apply criteria to the calculated velocity schedules. One criterion, employed by some designers, is to avoid low wall velocities and certainly reverse flow. Another, more stringent approach limits flow deceleration by specifying that the ratio of any wall velocity to the highest previous wall velocity should not be lower than some value, for example 0.5 or 0.7. While successful in some cases, these criteria do not necessarily lead to optimum designs.

The most fruitful utilization of internal-velocity calculations comes from analyzing a large number of impellers of high and low efficiency. This approach is most likely to identify "good" velocity schedules which can then be used for new designs. It can nevertheless fail if the analysis is limited to the impeller. The diffusing passages are just as important. Some of the methods intended for impeller flow can be converted for the analysis of stationary conduits by setting the peripheral speed equal to zero and substituting absolute for relative velocities.

REFERENCES

1. Stodola, A.: "Steam and Gas Turbines," Peter Smith, Publisher, Gloucester, Mass., 1945.
2. Pfleiderer, C.: "Die Kreiselpumpen," Springer-Verlag OHG, Berlin, 1961.
3. Wislicenus, G. F.: "Fluid Mechanics of Turbomachinery," Dover Publications, Inc., New York, 1965.
4. Wislicenus, G. F., R. M. Watson, and I. J. Karassik: Cavitation Characteristics of Centrifugal Pumps described by Similarity Considerations, *Trans. ASME,* vol. 61, p. 17, 1939.
5. Ross, C. C., and G. Banerian: Some Aspects of High Suction Specific Speed Inducers, *Trans. ASME,* vol. 78, p. 1715, 1956.
6. Jekat, W. K.: A New Approach to the Reduction of Pump Cavitation: The Hubless Inducer, *J. Basic Eng., Trans ASME,* ser. D, vol. 89, p. 125, 1967.
7. Jekat, W. K.: Reynolds Number and Incidence-Angle Effects on Inducer Cavitation, *ASME Paper* 66-WA/FE-31, 1966.
8. Stepanoff, A. J.: "Centrifugal and Axial Flow Pumps," John Wiley & Sons, Inc., New York, 1957.
9. Anderson, H. H.: Modern Developments in the Use of Large Single-Entry Centrifugal Pumps, *Proc. Inst. Mech. Eng. (London)*, vol. 169, no. 6, 1955.
10. Worster, R. C.: "The Interaction of Impeller and Volute in Determining the Performance of a Centrifugal Pump," The British Hydromechanics Research Association, RR 679, 1960.
11. Schultz-Grunow, F.: Der Reibungswiderstand rotierender Scheiben, *Z. Angew. Math. Mech.,* vol. 15, no. 4, 1935.
12. Pantell, K.: Versuche über Scheibenreibung, *Forsch. Gebiet Ingenieurw.,* vol. 16, no. 4, 1949/50.
13. Daily, J. W., and R. E. Nece: Chamber-Dimension Effects on Induced Flow and Frictional Resistance of Enclosed Rotating Disks, *J. Basic Eng., Trans. ASME,* ser. D, vol. 82, p. 217, 1960.
14. Bennett, T. P., and R. C. Worster: "The Friction on Rotating Disks and the Effect on Net Radial Flow and Externally Applied Whirl," The British Hydromechanics Research Association, RR 691, 1961.
15. Bergen, J.: Untersuchungen an einer Kreiselpumpe mit verstellbarem Spiralgehäuse, *Konstrukt.,* vol. 22, no. 12, 1970.
16. Agostinelli, A., D. Nobles, and C. R. Mockridge: "An Experimental Investigation of Radial Thrust in Centrifugal Pumps," *J. Eng. Power, Trans. ASME,* ser. A, vol. 82, p. 120, 1960.
17. Flügel, G.: "Ein neues Verfahren der graphischen Integration, angewandt auf Strömungen," doctoral dissertation, Oldenbourg, München, 1914.
18. Stanitz, J. D., and V. D. Prian: A Rapid Approximate Method for Determining Velocity Distribution on Impeller Blades of Centrifugal Compressors, *NACA TN* 2421, 1951.
19. Wu, C-H.: A General Theory of Three-dimensional Flow in Subsonic and Supersonic Turbomachines of Axial-, Radial-, and Mixed-Flow Types, *NACA TN* 2604, 1952.
20. Smith, K. J., and J. T. Hamrick: A Rapid Approximate Method for the Design of Hub Shroud Profiles of Centrifugal Impellers of Given Blade Shape, *NACA TN* 3399, 1955.
21. Stockman, N. O., and J. L. Kramer: Method for Design of Pump Impellers Using a High-speed Computer, *NASA TN* D-1562, 1963.
22. Katsanis, T.: Use of Arbitrary Quasi-Orthogonals for Calculating Flow Distribution in a Turbomachine, *J. Eng. Power, Trans. ASME,* ser. A, 1966.

Section 2.2

Centrifugal Pump Construction

IGOR J. KARASSIK

CLASSIFICATION AND NOMENCLATURE

A centrifugal pump consists of a set of rotating vanes, enclosed within a housing or casing and used to impart energy to a fluid through centrifugal force. Thus, stripped of all refinements, a centrifugal pump has two main parts: (1) a rotating element, including an impeller and a shaft, and (2) a stationary element made up of a casing, stuffing box, and bearings.

In a centrifugal pump the liquid is forced, by atmospheric or other pressure, into a set of rotating vanes. These vanes constitute an impeller which discharges the liquid at its periphery at a higher velocity. This velocity is converted into pressure energy by means of a volute (Fig. 1) or by a set of stationary diffusion vanes (Fig. 2), surrounding the impeller periphery. Pumps with volute casings are generally called *volute pumps,* while those with diffusion vanes are called *diffuser pumps.* Diffuser pumps were once quite commonly called *turbine pumps* but this term has recently been more selectively applied to the vertical deep-well centrifugal diffuser pumps, usually referred to as *vertical turbine pumps.* Figure 1 shows the path of the liquid passing through an end-suction volute pump operating at rated capacity (capacity at which best efficiency is obtained).

Impellers are classified according to the major direction of flow in reference to axis of rotation. Thus, centrifugal pumps may have:

1. Radial-flow impellers (Figs. 29, 38, 39, etc.)
2. Axial-flow impellers (Fig. 34)
3. Mixed-flow impellers, which combine radial- and axial-flow principles (Figs. 31, 33)

Impellers are further classified as:

1. Single-suction, with a single inlet on one side (Figs. 29, 32, 37, 41)
2. Double-suction, with water flowing to the impeller symmetrically from both sides (Figs. 30, 31, 42)

The mechanical construction of the impellers gives a still further subdivision into:

1. Enclosed, with shrouds or side walls enclosing the waterways (Figs. 29, 30, 31, 32, etc.)
2. Open, with no shrouds (Figs. 34, 37, 38, etc.)
3. Semiopen, or semienclosed (Fig. 40)

If the pump is one in which the head is developed by a single impeller, it is called a *single-stage pump.* Often the total head to be developed requires

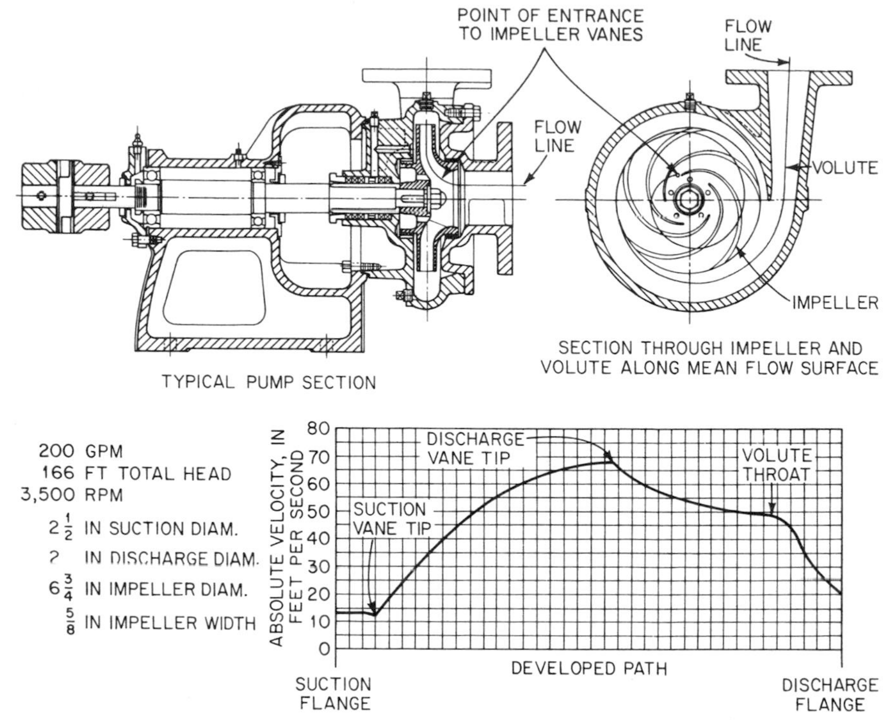

Fig. 1 Typical single-stage end-suction volute pump. (Worthington Pump, Inc.)

the use of two or more impellers operating in series, each taking its suction from the discharge of the preceding impeller. For this purpose, two or more single-stage pumps can be connected in series or all the impellers may be incorporated in a single casing. The unit is then called a *multistage pump.*

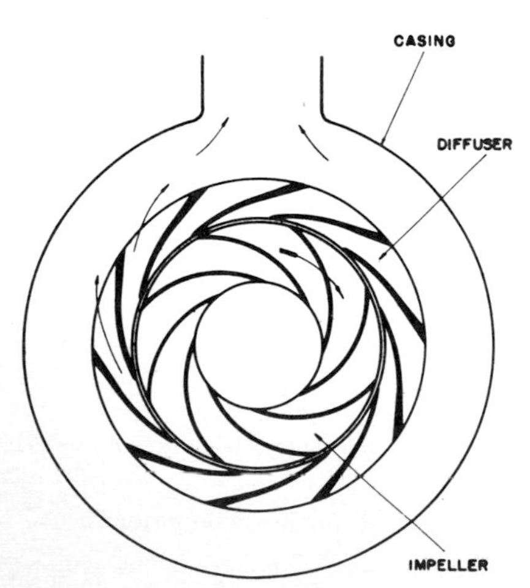

Fig. 2 Typical diffuser-type pump.

The mechanical design of the casing provides the added pump classification of *axially split* or *radially split,* while the axis of rotation determines whether it is a horizontal or vertical unit. Horizontal-shaft centrifugal pumps are classified still further according to the location of the suction nozzle:

1. End-suction (Figs. 1, 10, etc.)
2. Side-suction (Figs. 9, 12, etc.)
3. Bottom-suction (Fig. 17)
4. Top-suction (Figs. 28 and 134)

Some pumps operate in air with the liquid coming to and being conducted away from the pumps by piping. Other pumps, most often vertical types, are submerged in their suction supply. Vertical-shaft pumps are therefore called either *dry-pit* or *wet-pit* types. If the wet-pit pumps are axial-flow, mixed-flow, or vertical-turbine types, the liquid is discharged up through the supporting drop or column pipe to a discharge point above or below the supporting floor. These pumps are consequently designated as *above-ground discharge* or *below-*

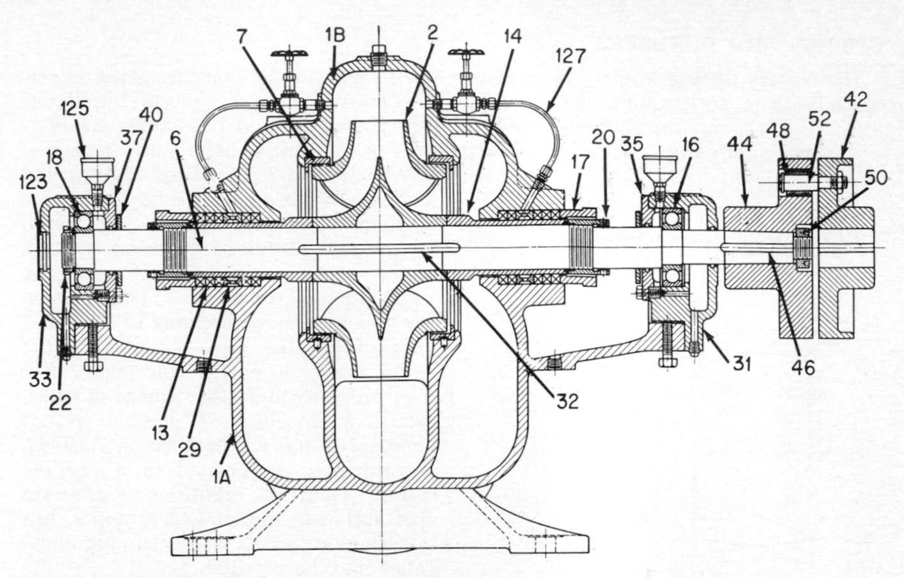

Fig. 3 Horizontal single-stage double-suction volute pump. (Numbers refer to parts listed in Table 1.) (Worthington Pump, Inc.)

ground discharge units. Figures 3, 10, and 4 show typical constructions of a horizontal double-suction volute pump, a vertical dry-pit single-suction volute pump, and the bowl section of a single-stage axial-flow propeller pump respectively. Names recommended by the Hydraulic Institute for various parts are given in Table 1.

TABLE 1 Recommended Names of Centrifugal Pump Parts

These parts are called out in Figs. 3, 4, and 10

Item No.	Name of part	Item No.	Name of part
1	Casing	33	Bearing housing (outboard)
1A	Casing (lower half)	35	Bearing cover (inboard)
1B	Casing (upper half)	36	Propeller key
2	Impeller	37	Bearing cover (outboard)
4	Propeller	39	Bearing bushing
6	Pump shaft	40	Deflector
7	Casing ring	42	Coupling (driver half)
8	Impeller ring	44	Coupling (pump half)
9	Suction cover	46	Coupling key
11	Stuffing box cover	48	Coupling bushing
13	Packing	50	Coupling lock nut
14	Shaft sleeve	52	Coupling pin
15	Discharge bowl	59	Handhole cover
16	Bearing (inboard)	68	Shaft collar
17	Gland	72	Thrust collar
18	Bearing (outboard)	78	Bearing spacer
19	Frame	85	Shaft enclosing tube
20	Shaft sleeve nut	89	Seal
22	Bearing lock nut	91	Suction bowl
24	Impeller nut	101	Column pipe
25	Suction head ring	103	Connector bearing
27	Stuffing box cover ring	123	Bearing end cover
29	Seal cage	125	Grease (oil) cup
31	Bearing housing (inboard)	127	Seal piping (tubing)
32	Impeller key		

CASINGS AND DIFFUSERS

The Volute Casing Pump This pump (Fig. 1) derives its name from the spiral-shaped casing surrounding the impeller. This casing section collects the liquid discharged by the impeller and converts velocity energy into pressure energy.

A centrifugal pump volute increases in area from its initial point until it encompasses the full 360° around the impeller and then flares out to the final discharge opening. The wall dividing the initial section and the discharge nozzle portion of the casing is called the tongue of the volute or the "cut-water." The diffusion vanes and concentric casing of a diffuser pump fulfill the same function as the volute casing in energy conversion.

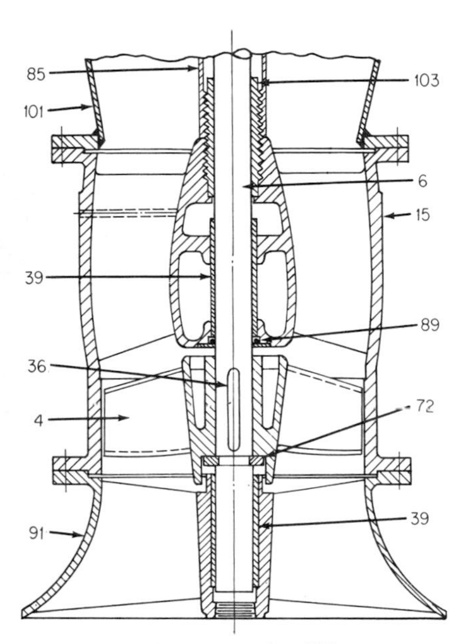

Fig. 4 Vertical wet-pit diffuser pump bowl. (Numbers refer to parts listed in Table 1.) (Worthington Pump, Inc.)

In propeller and other pumps in which axial-flow impellers are used, it is not practical to use a volute casing; instead, the impeller is enclosed in a pipelike casing. Generally diffusion vanes are used following the impeller proper, but in certain extremely low-head units these vanes may be omitted.

The diffuser is seldom applied to a single-stage radial-flow pump. Except for certain high-pressure multistage pump designs, the major application of diffusion vane pumps is in vertical turbine type pumps and in single-stage low-head propeller pumps (Fig. 4).

Radial Thrust In a *single-volute pump casing* design (Fig. 5) uniform or near uniform pressures act on the impeller when the pump is operated at design capacity (which coincides with the best efficiency). At other capacities, the pressures around the impeller are not uniform (Fig. 6), and there is a resultant radial reaction (F). A graphical representation of the typical change in this force with pump capacity is shown in Fig. 7. Note that the force is greatest at shutoff.

For any percentage of capacity, this radial reaction is a function of total head, and of the width and diameter of the impeller. Thus a high-head pump with a large impeller diameter will have a much greater radial reaction force at partial capacities than a low-head pump with a small impeller diameter. A zero radial reaction is not often realized; the minimum reaction occurs at design capacity.

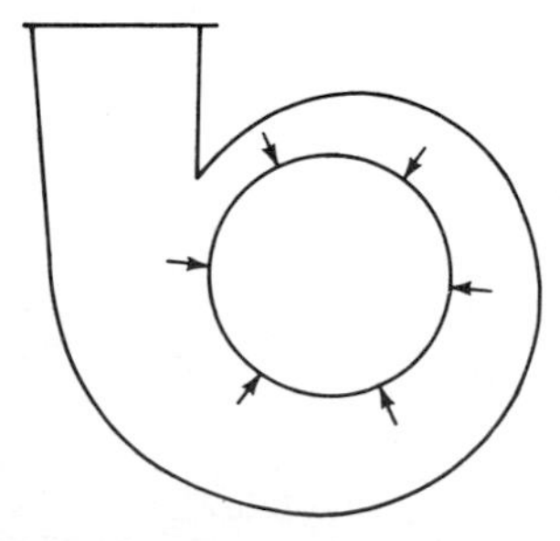

Fig. 5 Uniform casing pressures exist at design capacity resulting in zero radial reaction.

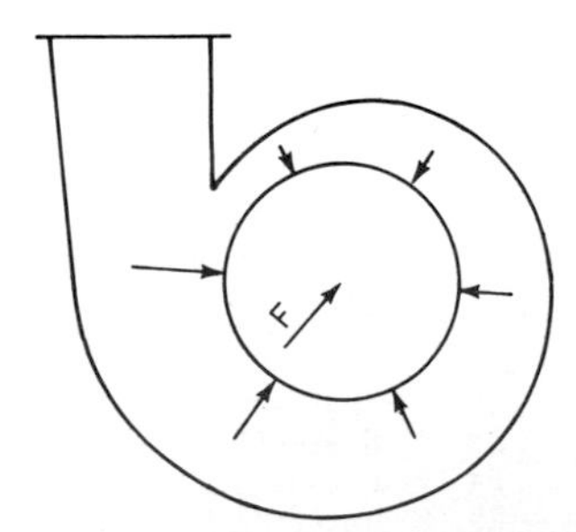

Fig. 6 At reduced capacities uniform pressures do not exist in a single-volute casing resulting in a radial reaction F.

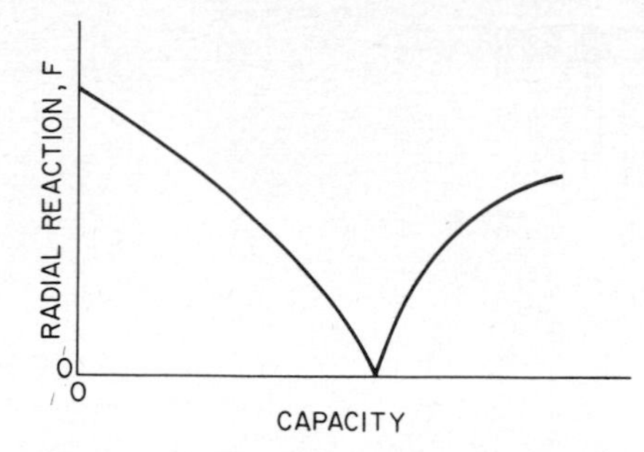

Fig. 7 Magnitude of radial reaction F decreases from shutoff to design capacity and then increases again with overcapacity in a single-volute pump. With overcapacity the reaction is roughly in opposite direction from that with part capacity.

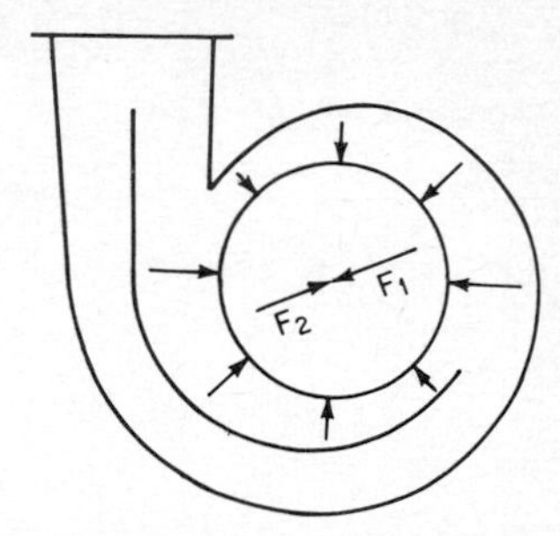

Fig. 8 While in a double-volute pump the pressures are not uniform at part-capacity operation, the resultant forces F_1 and F_2 for each 180° volute section oppose and balance each other.

While the same tendency for unbalance exists in the diffuser-type pump, the reaction is limited to a small arc repeated all around the impeller. As a result the individual reactions cancel each other.

In a centrifugal pump design, shaft diameter as well as bearing size can be affected by the allowable deflection as determined by the shaft span, impeller weight, radial reaction forces and the torque to be transmitted.

Formerly standard designs compensated for radial reaction forces of a magnitude encountered at capacities in excess of 50 percent design capacity for maximum diameter impeller for the pump. For sustained operation at lower capacities the pump manufacturer, if properly advised, would supply a heavier shaft usually at a much higher cost. Sustained operation at extremely low flows without informing the manufacturer of this at the time of purchase is a much more common practice today. The result: broken shafts, especially on high-head units.

Because of the increasing application of pumps which must operate at reduced capacities, it has become desirable to design standard units to accommodate such conditions. One solution is to use heavier shafts and bearings. Except for low-head pumps in which only a small additional load is involved, this solution is not economical. The only practical answer is a casing design that develops a much

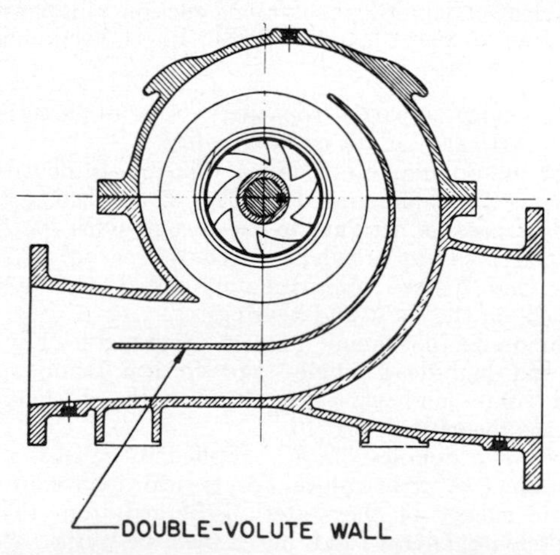

Fig. 9 Transverse view of a double-volute casing pump.

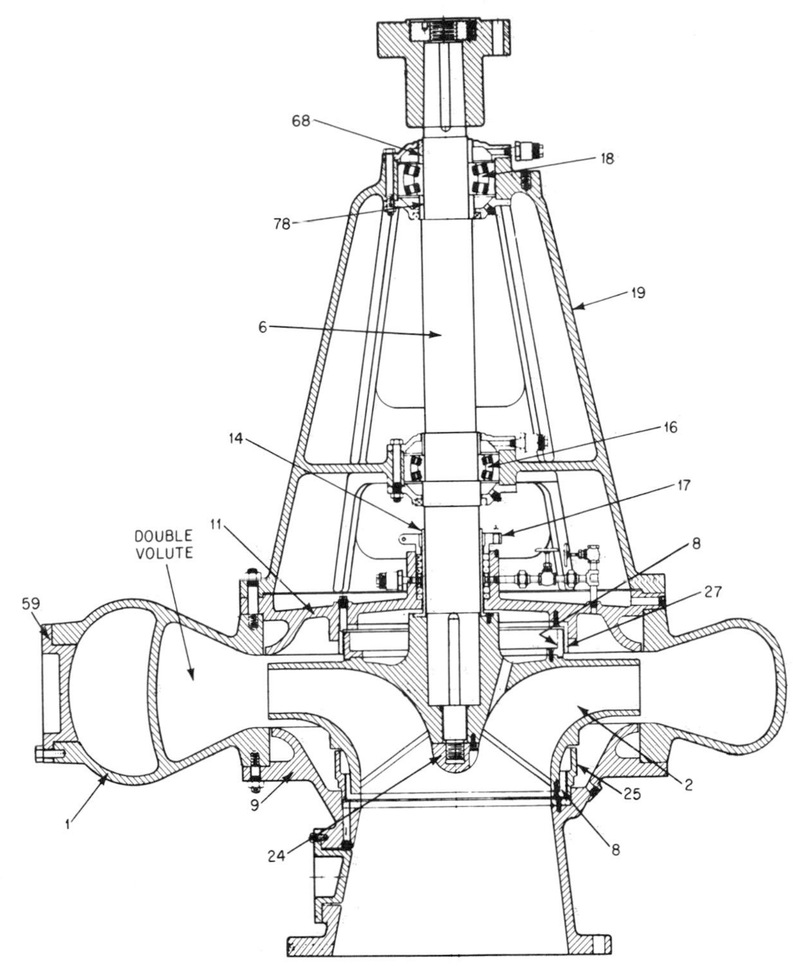

Fig. 10 Sectional view of a vertical-shaft end-suction pump with a double-volute casing. (Numbers refer to parts listed in Table 1.) (Worthington Pump, Inc.)

smaller radial reaction force at partial capacities. One of these is the *double-volute* casing design, also called *twin-volute* or *dual-volute.*

The application of the double-volute design principle to neutralize radial reaction forces at reduced capacity is illustrated in Fig. 8. Basically, this design consists of two 180° volutes; a passage external to the second joins the two into a common discharge. Although a pressure unbalance exists at partial capacity through each 180° arc, forces F_1 and F_2 are approximately equal and opposite. Thus little if any radial force acts on the shaft and bearings.

The double-volute design has many hidden advantages. For example, in large-capacity medium- and high-head single-stage vertical pump applications, the rib forming the second volute and separating it from the discharge waterway of the first volute strengthens the casing (Fig. 10).

When the principle of a double-volute is applied to individual stages of a multistage pump, it becomes a twin-volute. A typical twin-volute is illustrated in Fig. 11. The kinetic energy of the water discharged from the impeller must be transformed into pressure energy and must then be turned back 180° to enter the impeller of the next stage. The twin-volute therefore also acts as a return

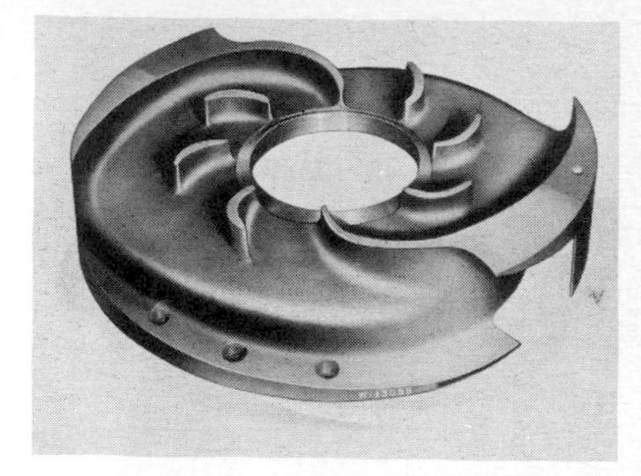

Fig. 11 Twin-volute of a multistage pump. Front view (left) and back view (right). (Worthington Pump, Inc.)

channel. The back view in Fig. 11 shows this, as well as the guide vanes used to straighten the flow into the next stage.

Solid and Split Casings *Solid casing* implies a design in which the discharge waterways leading to the discharge nozzle are all contained in one casting, or fabricated piece. It must have one side open so that the impeller may be introduced into the casing. As the sidewalls surrounding the impeller are in reality part of the casing, a solid casing, strictly speaking, cannot be used, and designs normally called solid casing are really radially split (Figs. 1, 14, and 15).

A *split casing* is made of two or more parts fastened together. The term *horizontally split* had regularly been used to describe pumps with a casing divided by a horizontal plane through the shaft centerline or axis (Fig. 12). The term *axially split* is now preferred. Since both the suction and discharge nozzles are usually in the same half of the casing, the other half may be removed for inspection of the interior without disturbing the bearings or the piping. Like its counterpart *horizontally split,* the term *vertically split* is poor terminology. It refers to a casing split in a plane perpendicular to the axis of rotation. The term *radially split* is now preferred.

Fig. 12 Axially split casing, horizontal-shaft, double-suction volute pump. (Worthington Pump, Inc.)

In some special horizontal-pump applications, it is desirable to use a casing split along a plane passing through the pump axis but inclined diagonally to the horizontal. This construction is primarily used when the need for a vertical discharge is combined with the desire for the convenience of an axially split casing (Fig. 13).

End-Suction Pumps Most end-suction single-stage pumps are made of one-piece solid casings. At least one side of the casing must be open so the impeller can be assembled in the pump, thus requiring a cover for that side. If the cover is on the suction side, it becomes the casing sidewall and contains the suction opening (Fig. 1). This is called the *suction cover* or *casing suction head.* Other designs are made with stuffing box covers (Fig. 14), and still others have both casing suction covers and stuffing box covers (Figs. 10 and 15).

For general service, the end-suction single-stage pump design is extensively used for small pumps up to 4- and 6-in discharge size both for motor-mounted and coupled types. In all these the small size makes it feasible to cast the volute and one side integrally. Whether the stuffing box side or the suction side is made integrally with the casing is usually determined by the most economical pump design. For larger pumps, especially those for special service such as sewage handling, there is a demand for pumps of both rotations. A design with separate suction and stuffing box heads permits use of the same casing for either rotation if the flanges on the two sides are made identical. There is also a demand for vertical pumps that can be disassembled by removing the rotor and bearing assembly

Fig. 13 Horizontal-shaft end-suction pump with casing split along plane through its axis, but inclined from horizontal. (Allis Chalmers)

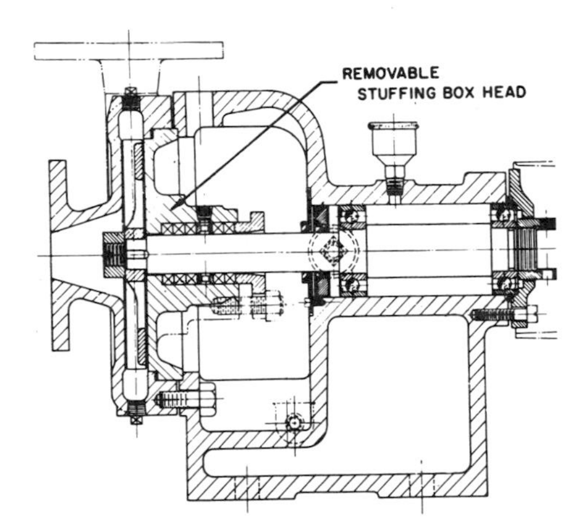

Fig. 14 End-suction pump with removable stuffing box head. (Worthington Pump, Inc.)

from the top of the casing. Many horizontal applications of the pumps of the same line, however, require partial dismantling from the suction side. Such lines are most adaptable when they have separate suction and stuffing box covers.

Casing Construction for Open-Impeller Pumps In the inexpensive open-impeller pump, the impeller rotates within close clearance of the pump casing (Fig. 14). If the intended service is more severe, a side plate is mounted within the casing to provide a renewable close-clearance guide to the liquid flowing through the open impeller. One of the advantages of using side plates is that abrasion-resistant material, such as stainless steel, may be used for the impeller and side plate while the casing itself may be of a less costly material. Although double-suction open-impeller type pumps are seldom used today, they were common in the past and were generally made with side plates.

Prerotation and Stop Pieces Improper entrance conditions and inadequate suction approach shapes may cause the liquid column in the suction pipe to spiral for some distance ahead of the actual impeller entrance. This phenomenon is called *prerotation,* and it is attributed to various operational and design factors in both vertical and horizontal pumps.

Prerotation is usually harmful to pump operation because the liquid enters between the impeller vanes at an angle other than allowed for by the designer

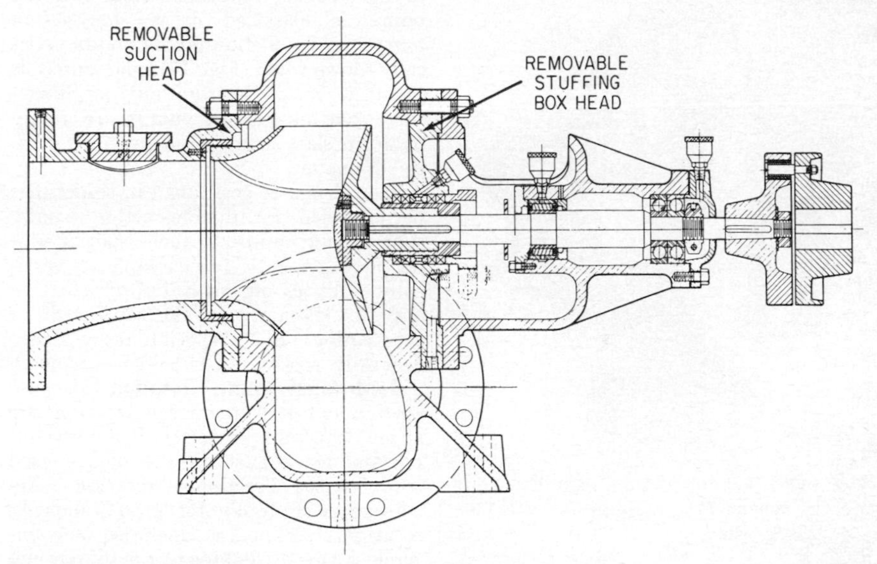

Fig. 15 End-suction pump with removable suction and stuffing box heads. (Worthington Pump, Inc.)

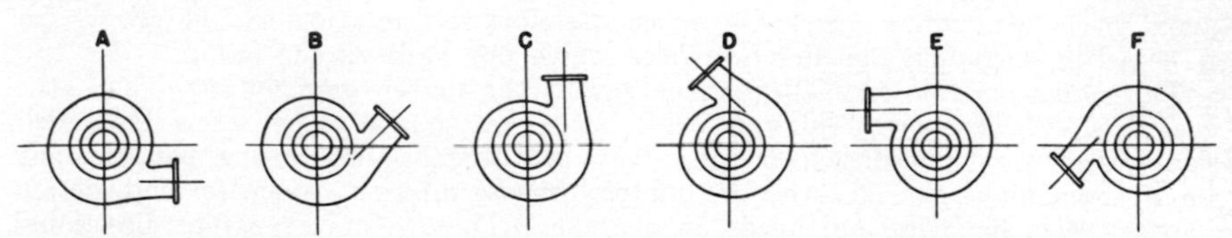

Fig. 16 Positions of discharge nozzles possible for a specific design of an end-suction solid-casing horizontal-shaft pump. Rotation illustrated is counterclockwise from suction end.

in his calculations. This frequently lowers the net effective suction head and the pump efficiency. Various means are used to avoid prerotation both in the construction of the pump and in the design of the suction approaches.

Practically all horizontal single-stage double-suction pumps and most multistage pumps have a suction volute that guides the liquid in a streamline flow to the impeller eye. The flow comes to the eye at right angles to the shaft and separates unequally on both sides of the shaft. Moving from the suction nozzle to the impeller eye, the suction waterways reduce in area, meeting in a projecting section of the sidewall dividing the two sections. This dividing projection is called a *stop piece.* To prevent prerotation in end-suction pumps, a radial-fin stop piece projecting toward the center may be cast into the suction-nozzle wall.

Nozzle Locations The discharge nozzle of end-suction single-stage horizontal pumps is usually in a top-vertical position (Figs. 1 and 14). However, other nozzle positions may be obtained, such as top-horizontal, bottom-horizontal, or bottom-vertical discharge. Figure 16 illustrates the flexibility available in discharge nozzle location. Sometimes the pump frame, bearing bracket, or baseplate may interfere with the discharge flange, prohibiting a bottom-horizontal or bottom-vertical discharge-nozzle position. In other instances, solid casings cannot be rotated for various nozzle positions because the stuffing-box seal connection would become inaccessible.

Practically all double-suction axially split casing pumps have a side-discharge nozzle and either a side- or a bottom-suction nozzle. If the suction nozzle is placed on the side of the pump casing with its axial centerline at right angles to the vertical centerline (Fig. 12), the pump is classified as a *side-suction pump.* If its suction nozzle points vertically downward (Fig. 17), the pump is called a *bottom-suction pump.* Single-stage bottom-suction pumps are rarely made in sizes below 10-in discharge-nozzle diameter.

Fig. 17 Bottom-suction, axially split casing, single-stage pump. (Worthington Pump, Inc.)

Special nozzle positions can sometimes be provided for double-suction axially split casing pumps to meet special piping arrangements, for example, a vertically split casing with bottom suction and top discharge in the same half of the casing. Such special designs are generally costly and should be avoided.

Centrifugal Pump Rotation Because suction and discharge nozzle locations are affected by pump rotation, it is very important to understand the means used to define the direction of rotation. According to Hydraulic Institute Standards, rotation is defined as clockwise or counterclockwise by looking at the driven end of a horizontal pump or looking down on a vertical pump, although some manufacturers still designate rotation of a horizontal pump from its outboard end. To avoid misunderstanding, clockwise or counterclockwise rotation should always be qualified by including the direction from which one looks at the pump.

The terms *inboard end*—the end closest to the driver—and *outboard end*—the end farthest away—are used only with horizontal pumps. The terms lose their significance with dual-driven pumps. Any centrifugal pump casing pattern may be arranged for either clockwise or counterclockwise rotation, except for end-suction pumps, which have integral heads on one side. These require separate directional patterns.

Casing Hand Holes Casing hand holes are furnished primarily on pumps handling sewage and stringy materials that may become lodged on the impeller suction vane edges or on the tongue of the volute. They permit removal of this material without dismantling the complete pump. End-suction pumps used primarily for handling liquids of this type are provided with hand holes or access to the suction side of the impellers. These are located on the suction head or in the suction elbow. Hand holes are also provided in drainage, irrigation, circulating, and supply pumps if foreign matter may become lodged in the waterways. On the very large pumps manholes provide access to the interior for both cleaning and inspection.

Mechanical Features of Casings Most single-stage centrifugal pumps are intended for service with moderate pressures and temperatures. As a result, pump manufacturers usually design a special line of pumps for high operating pressures and temperatures rather than make their standard line unduly expensive by making it suitable for too wide a range of operating conditions.

If axially split casings are subject to high pressure, they tend to "breathe" at the split joint, leading to misalignment of the rotor and, even worse, leakage. For such conditions, internal and external ribbing is applied to casings at the points subject to greatest stress. In addition, whereas most pumps are supported by feet at the bottom of the casing, high temperatures require centerline support so that, as the pump becomes heated, expansion will not cause misalignment.

Series Units For large-capacity medium-high-head service conditions that require such an arrangement, two single-stage double-suction pumps may be connected in series on one baseplate with a single driver. Such an arrangement is very common in waterworks applications for heads of 250 to 400 ft. One series arrange-

Fig. 18 Series unit (motor in middle). (Worthington Pump, Inc.)

Fig. 19 Series unit (motor at the end). (Worthington Pump, Inc.)

ment uses a double-extended shaft motor in the middle, driving two pumps connected in series by external piping (Fig. 18). In a second type, a standard motor is used with one pump having a double-extended shaft (Fig. 19). This latter arrangement may have limited applications because the shaft of the pump next to the motor must be strong enough to transmit the total pumping horsepower. If the total pressure generated by such a series unit is relatively high, the casing of the second pump may require ribbing.

MULTISTAGE PUMP CASINGS

While the majority of single-stage pumps are of the volute-casing type, both volute and diffuser casings are used in multistage pump construction. Because a volute casing gives rise to radial thrust, axially split multistage casings generally have staggered volutes so that the resultant of the individual radial thrusts is balanced out (Figs. 20 and 21). Both axially and radially split casings are used for multistage pumps. The choice between the two designs is dictated by the design pressure, with 1,600 lb/in² being the average upper limit for 3,600-rpm axially split casing pumps.

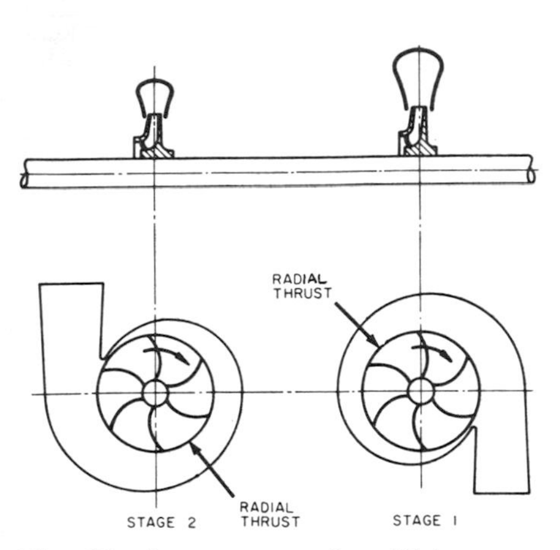

Fig. 20 Arrangement of multistage volute pump for radial-thrust balance.

Axially Split Casings Regardless of the arrangement of the stages within the casing, it is necessary to connect the successive stages of a multistage pump. In the low and medium pressure and capacity range, these interstage passages are cast integrally with the casing proper (Figs. 22 and 23). As the pressures and capacities increase, the desire to maintain as small a casing diameter as possible, coupled to the necessity of avoiding sudden changes in the velocity or the direction of flow, leads to the use of external interstage passages cast separately from the pump casing. They are formed in the shape of loops, bolted or welded to the casing proper (Fig. 24).

Interstage Construction for Axially Split Casing Pumps A multistage pump inherently has adjoining chambers subjected to different pressures, and means must

Fig. 21 Horizontal flange of an axially split casing six-stage pump. (Worthington Pump, Inc.)

be made available to isolate these chambers from one another so that the leakage from high to low pressure will take place only at the clearance joints formed between the stationary and rotating elements of the pump and so that this leakage be kept to a minimum. The isolating wall used to separate two adjacent chambers of a multistage pump is called a *stage piece, diaphragm,* or *interstage diaphragm.* The stage piece may be formed of a single piece or it may be fitted with a renewable stage-piece bushing at the clearance joint between the stationary stage piece and the part of the rotor immediately inside the former. The stage pieces, which are usually solid, are assembled on the rotor along with impellers, sleeves, bearings, and similar components. To prevent the stage pieces from rotating, a locked tongue-and-groove joint is provided in the lower half of the casing.

Clamping the upper casing half to the lower half securely holds the stage piece and prevents rotation.

The problem of seating a solid stage piece against an axially split casing is

Fig. 22 Two-stage axially split casing volute pump for small capacities and pressures up to 200 to 250 lb/in^2. (Worthington Pump, Inc.)

Fig. 23 Two-stage axially split casing volute pump for pressures up to 400 lb/in^2. (Worthington Pump, Inc.)

one which has given designers much trouble. First, there is a three-way joint; second, it is necessary that this seating be so accomplished that the joint is tight and leakproof under a pressure differential without resorting to bolting the stage piece directly to the casing.

To overcome this problem it is wise to make a pump which has a small diameter casing so that when the casing bolting is pulled tight, there is a seal fitting of the two casing halves adjacent to the stage piece. The small diameter likewise helps to eliminate the possibility of a stage piece cocking and thereby leaving a clearance on the upper half casing when it is pulled down. No matter how rigidly the stage piece may be located in the lower half casing, there must be a sliding fit between the seat face of the stage piece and that in the upper half

Fig. 24 Six-stage axially split casing volute pump for pressures up to 1,300 lb/in². (Worthington Pump, Inc.)

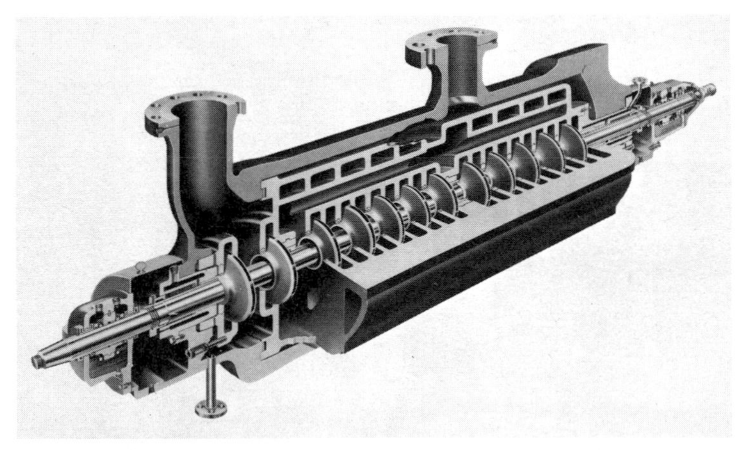

Fig. 25 Double-casing multistage pump with axially split inner casing. (Allis-Chalmers)

casing in order that the upper half casing may be pulled down. Each stage piece, furthermore, must be so arranged that the pressure differential which is developed by the pump will tend to seat it tight against the casing rather than open up the joint.

We have said that axially split casing pumps are used for working pressure up to 1,600 lb/in² gage. High-pressure piping systems, of which these pumps form a part, are inevitably made of steel because this material has the valuable property of yielding without breaking. Considerable piping strains are unavoidable, and these strains, or at least a part thereof, are transmitted to the pump casing. The

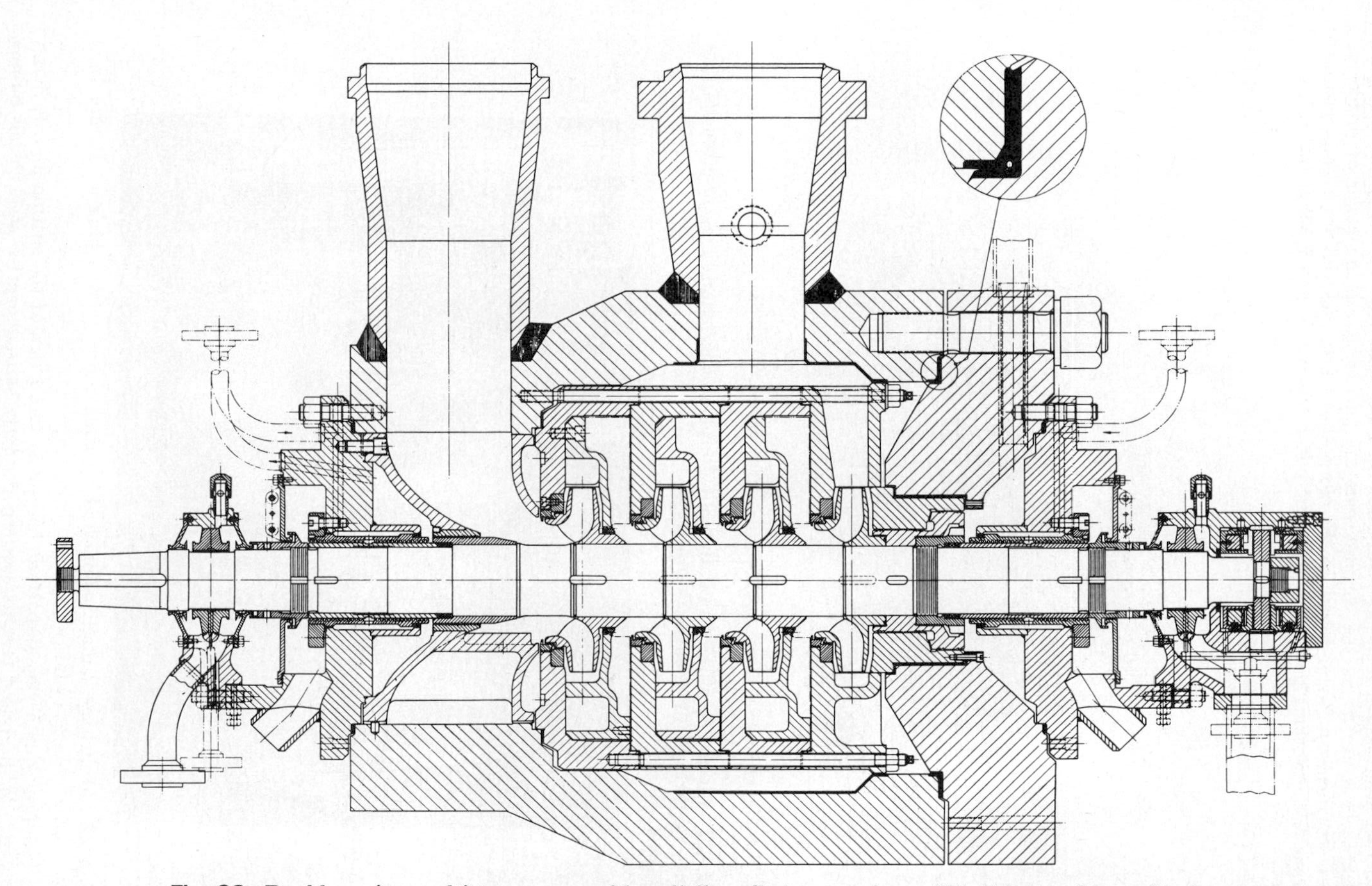

Fig. 26 Double-casing multistage pump with radially split inner casing. (Worthington Pump, Inc.)

latter consists essentially of a barrel, split axially, flanged at the split, and fitted with two necks which serve as inlet and discharge openings. When piping stresses exist, these necks, being the weakest part of the casing, are in danger of breaking off if they cannot yield. Steel is therefore the safest material for pump casings whenever the working pressures within the pump are in excess of 1,000 lb/in^2.

Fig. 27 Rotor assembly of radially split double-casing pump being inserted into its outer casing. (Worthington Pump, Inc.)

Fig. 28 Multistage radially split double-casing pump. (Worthington Pump, Inc.)

This brings up a very important feature in the design of the suction and discharge flanges. While raised face flanges are perfectly satisfactory for steel-casing pumps, their use is extremely dangerous with cast iron pumps. This danger arises from the lack of elasticity in cast iron which leads to flange breakage when the bolts are being tightened up, the fulcrum of the bending moment being located inwardly of the bolt circle. As a result, it is essential to avoid raised-face flanges with cast iron casings as well as the use of a raised-face flange pipe directly against a flat-face cast iron pump flange. Suction flanges should obviously be suitable for whatever hydrostatic test pressure is applied to the pump casing. Therefore, if this pressure is over 500 lb/in^2, a 300 lb/in^2 series suction flange would be used.

The location of the pump-casing foot support is not critical in the smaller pump sizes operating at pressures under 250 lb/in^2 and at moderate temperatures (see Fig. 22). The unit is extremely small, and very little distortion is bound to occur. However, for larger units operating at higher pressures and perhaps at higher temperatures, it is important that the casing be supported as close as possible to the horizontal centerline and immediately below the bearings (Figs. 23 and 24).

Radially Split, Double-Casing Pumps The oldest form of radially split casing multistage pump is that commonly called the *ring-casing* or the "doughnut" type. When it was found necessary to use more than one stage for the generation of higher pressures, two or more single-stage units of the prevalent radially split casing type were assembled and bolted together.

In later designs, the individual stage sections and separate suction and discharge heads were held together with large through-bolts. These pumps, still an assembly of bolted-up sections, presented serious dismantling and reassembly problems because suction and discharge connections had to be broken each time the pump was opened. The double-casing pump retained the advantages of the radially split casing design and solved the dismantling problem.

The basic principle consists in enclosing the working parts of a multistage centrifugal pump in an inner casing and in building a second casing around this inner casing. The space between the two casings is maintained at the discharge pressure of the last pump stage. The construction of the inner casing follows one of two basic principles: (1) axial splitting (Fig. 25) or (2) radial splitting (Fig. 26).

The double-casing pump with radially split inner casing is an evolution of the ring-casing type pump, with added provisions for ease of dismantling. The inner unit is generally constructed exactly as a ring-casing pump. After assembly, it is inserted and bolted inside a cylindrical casing which supports it and leaves it free to expand under temperature changes. In Fig. 27 the inner assembly of such a pump is being inserted into the outer casing. Figure 28 shows the external appearance of this type of pump. The suction and discharge nozzles form an integral part of the outer casing and the internal assembly of the pump can be withdrawn without disturbing the piping connections.

IMPELLERS

In a single-suction impeller, the liquid enters the suction eye on one side only. As a double-suction impeller is, in effect, two single-suction impellers arranged back-to-back in a single casting, the liquid pumped enters the impeller simultaneously from both sides, while the two casing suction passageways are connected to a common suction passage and a single suction nozzle.

For the general-service single-stage axially split casing design, a double-suction impeller is favored because it is theoretically in axial hydraulic balance and because the greater suction area of a double-suction impeller permits the pump to operate with less net absolute suction head. For small units, the single-suction impeller is more practical for manufacturing reasons as the waterways are not divided into two very narrow passages. It is also sometimes preferred for structural reasons. End-suction pumps with single-suction overhung impellers have both first cost and maintenance advantages not obtainable with a double-suction impeller. Most radially split casing pumps therefore use single-suction impellers. Because an overhung impeller does not require the extension of a shaft into the impeller suction eye, single-suction impellers are preferred for pumps handling suspended matters, such as sewage. In multistage pumps, single-suction impellers are almost universally used because of the design and first-cost complexity that double-suction staging introduces.

Impellers can also be classified by the shape and form of their vanes:

1. The straight-vane impeller (Figs. 29, 38, 39, 40, and 41)
2. The Francis-vane or screw-vane impeller (Figs. 30, 31, and 32)
3. The mixed-flow impeller (Fig. 33)
4. The propeller or axial-flow impeller (Fig. 34)

Fig. 29 Straight-vane single-suction closed impeller. (Worthington Pump, Inc.)

Fig. 30. Francis-vane double-suction closed impeller. (Worthington Pump, Inc.)

In a *straight-vane impeller,* the vane surfaces are generated by straight lines parallel to the axis of rotation. These are also called single-curvature vanes. The vane surfaces of a *Francis-vane impeller* have double curvature. An impeller design that has both a radial and axial flow component is called a *mixed-flow impeller.* It is generally restricted to single-suction designs with a specific speed above 4,200. Types with lower specific speeds are called Francis-vane impellers. Mixed-flow impellers with a very small radial flow component are usually referred to as *propellers.* In a true propeller or *axial-flow impeller* the flow strictly parallels the axis of rotation. In other words, it moves only axially.

Fig. 31. High-specific-speed Francis-vane double-suction closed impeller. (Worthington Pump, Inc.)

The relation of single-suction impeller profiles to specific speed is shown on Fig. 35. Classification of impellers according to their vane shape is naturally arbitrary inasmuch as there is much overlapping in the types of impellers used in the different types of pumps. For example, impellers in single- and double-suction pumps of low specific speed have vanes extending across the suction eye. This provides a mixed flow at the impeller entrance for low pickup losses at high rotative speeds, but allows the discharge portion of the impeller to use the straight-vane principle. In pumps of higher specific speed operating against low heads, impellers have double-curvature vanes extending over the full vane surface. They are, therefore, full Francis-type impel-

Fig. 32 Low-specific-speed Francis-vane single-suction closed impeller. (Worthington Pump, Inc.)

lers. The mixed-flow impeller, usually a single-suction type, is essentially one-half a double-suction high-specific-speed Francis-vane impeller.

In addition, many impellers are designed for specific applications. For instance, the conventional impeller design with sharp vane edges and restricted areas is not suitable for handling liquids containing rags, stringy materials, and solids like sew-

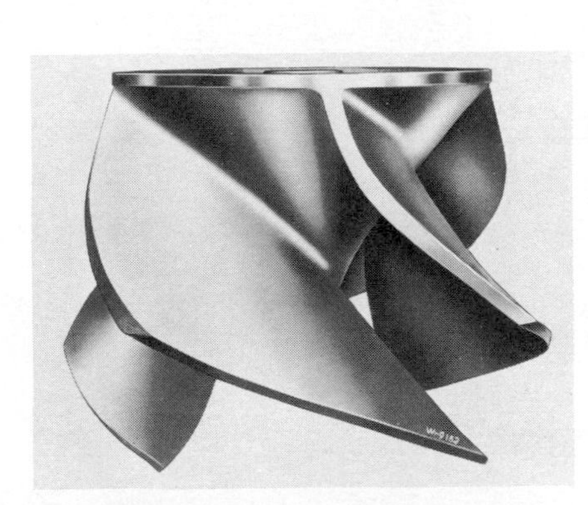

Fig. 33 Open mixed-flow impeller. (Worthington Pump, Inc.)

Fig. 34 Axial-flow impeller. (Worthington Pump, Inc.)

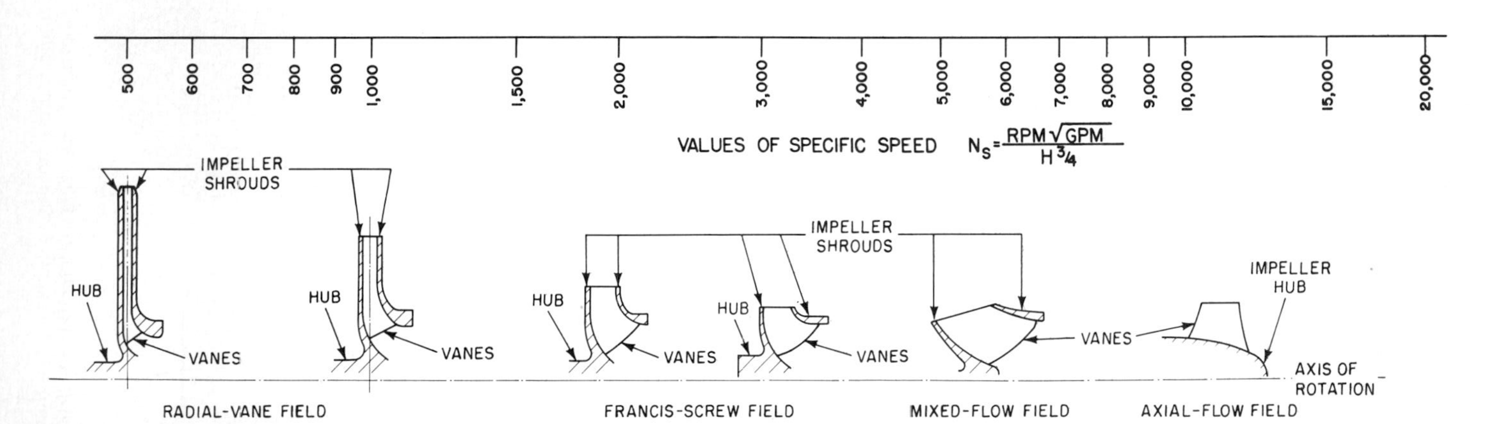

Fig. 35 Variation in impeller profiles with specific speed and approximate range of specific speed for the various types.

age because it will become clogged. Special nonclogging impellers with blunt edges and large waterways have been developed for such services (Fig. 36). For pumps up to the 12- to 16-in size, these impellers have only two vanes. Larger pump sizes normally use three or four vanes.

The impeller design used for paper pulp pumps (Fig. 37) is fully open, nonclogging, and has screw and radial streamlined vanes. The screw-conveyor end projects far into the suction nozzle, permitting the pump to handle high-consistency paper pulp stock.

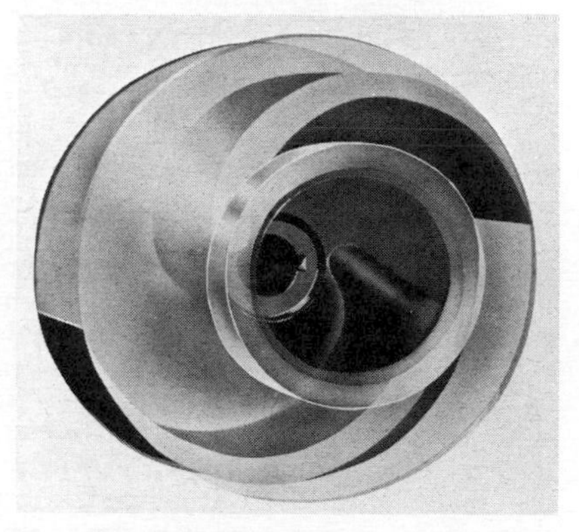

Fig. 36 Phantom view of radial-vane nonclogging impeller. (Worthington Pump, Inc.)

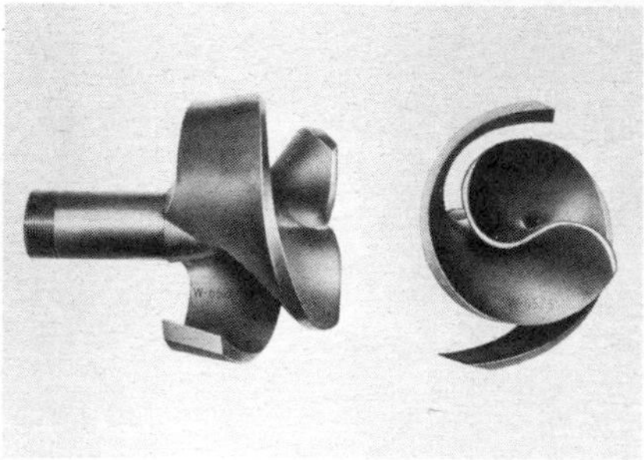

Fig. 37 Paper-pulp impeller. (Worthington Pump, Inc.)

Impeller Mechanical Types Mechanical design also determines impeller classification. Accordingly, impellers may be (1) completely open, (2) semiopen, or (3) closed.

Strictly speaking, an *open impeller* (Figs. 38 and 39) consists of nothing but

Fig. 38 Open impellers. Notice that the impellers at left and right are strengthened by a partial shroud. (Worthingon Pump, Inc.)

vanes, attached to a central hub for mounting on the shaft without any form of a sidewall or shroud. The disadvantage of this impeller is structural weakness. If the vanes are long, they must be strengthened by ribs or a partial shroud. Generally, open impellers are used in small, inexpensive pumps or pumps handling abrasive liquids. One advantage of open impellers is that they are capable of handling suspended matter with a minimum of clogging. The open impeller rotates between two side plates, between the casing walls of the volute or between the stuffing box head and the suction head. The clearance between the impeller vanes and the sidewalls allows a certain amount of water slippage. This slippage increases as wear increases. To restore the original efficiency, both the impeller and the side plate must be replaced. This, incidentally, involves a much larger expense than would be entailed in closed-impeller pumps where simple rings form the leakage joint.

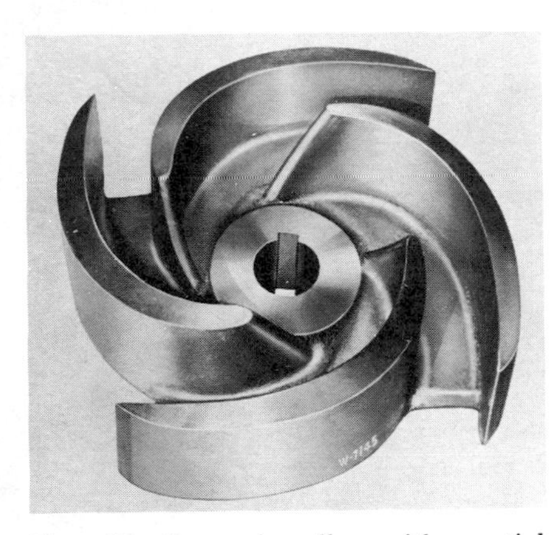

Fig. 39 Open impeller with partial shroud.

The *semiopen impeller* (Fig. 40) incorporates a shroud, or an impeller backwall. This shroud may or may not have pump-out vanes, which are vanes located at the back of the impeller shroud. Their function is to reduce the pressure at the back hub of the impeller and to prevent foreign matter from lodging back of the impeller and interfering with the proper operation of the pump and of the stuffing box.

The *closed impeller* (Figs. 29 to 32), which is almost universally used in centrifugal pumps handling clear liquids, incorporates shrouds or enclosing sidewalls that totally enclose the impeller waterways from the suction eye to the periphery. Although this design prevents the liquid slippage that occurs between an open or semiopen impeller and its side plates, a running joint must be provided between the impeller and the casing to separate the discharge and suction chambers of the pump. This running joint is usually formed by a relatively short cylindrical surface on the impeller shroud that rotates within a slightly larger stationary cylindrical surface. If one or both surfaces are made renewable, the leakage joint can be repaired when wear causes excessive leakage.

Fig. 40 Semiopen impeller. (Worthington Pump, Inc.)

Fig. 41 Front and back views of an open impeller with partial shroud and with pumpout vanes on back side. (Worthington Pump, Inc.)

If the pump shaft terminates at the impeller so that the latter is supported by bearings on one side, the impeller is called an *overhung impeller*. This type of construction is the best for end-suction pumps with single-suction impellers.

Impeller Nomenclature The inlet of an impeller just before the section at which the vanes start is called the suction eye (Fig. 42). In a closed-impeller pump, the suction eye diameter is taken as the smallest inside diameter of the shroud. In determining the area of the suction eye. the area occupied by the impeller shaft hub is deducted.

The hub is the central part of the impeller which is bored out to receive the pump shaft. The expression, however, is also frequently used for the part of the impeller which rotates within the casing fit or within the casing wearing ring. It is then referred to as the *outer impeller hub* or the *wearing ring hub of the shroud.*

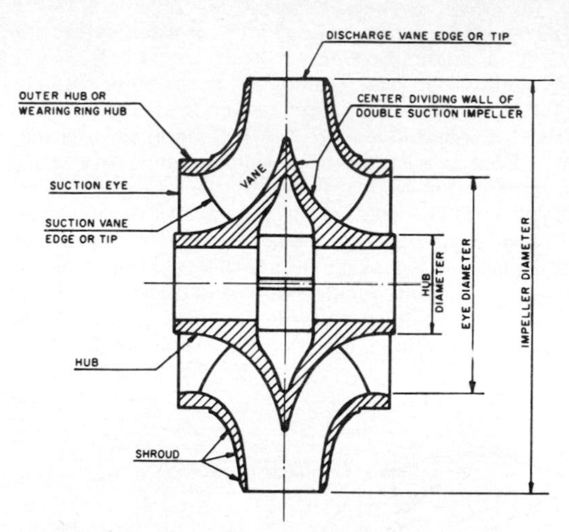

Fig. 42 Parts of a double-suction impeller.

WEARING RINGS

Wearing rings provide an easily and economically renewable leakage joint between the impeller and the casing. A leakage joint without renewable parts is illustrated in Fig. 43. To restore original clearances of such a joint after wear occurs, the

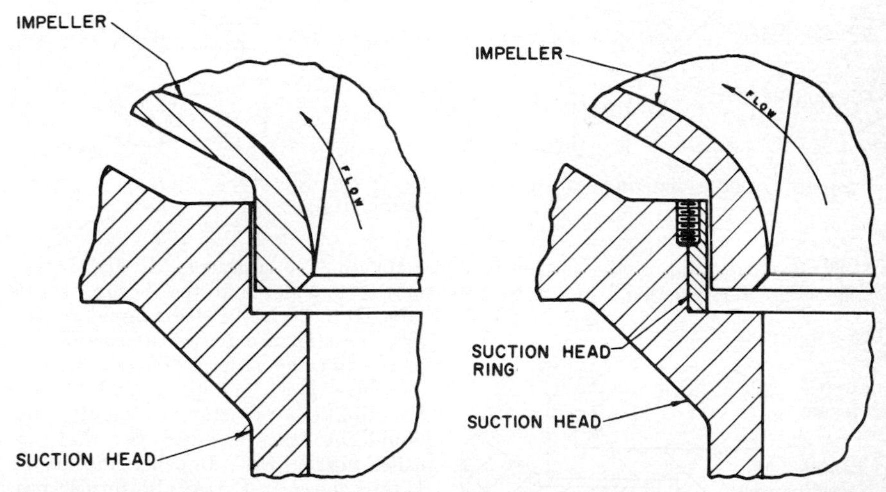

Fig. 43 Plain flat leakage joint—no rings.

Fig. 44 Single flat casing-ring construction.

user must either (1) build up the worn surfaces by welding, metal spraying, or other means and then true up the part or (2) buy new parts.

The new parts are not very costly in small pumps, especially if the stationary casing element is a simple suction cover. This is not true for larger pumps or where the stationary element of the leakage joint is part of a complicated casting. If the first cost of a pump is of prime importance, it is more economical to provide for both the stationary parts and the impeller to be remachined. Renewable casing and impeller rings can then be installed (Figs. 43, 44, and 45). Nomenclature for the casing or stationary part forming the leakage joint surface is as follows:

(1) *casing ring* (if mounted in the casing); (2) *suction-cover ring* or *suction-head ring* (if mounted in a suction cover or head); and (3) *stuffing-box-cover ring* or *head ring* (if mounted in the stuffing box cover or head). Some engineers like to identify the part further by adding the word "wearing," as, for example, *casing wearing ring*. A renewable part for the impeller wearing surface is called the *impeller ring*. Pumps with both stationary and rotating rings are said to have *double-ring* construction.

Wearing-Ring Types There are various types of wearing-ring designs, and the selection of the most desirable type depends on the liquid being handled, the pressure differential across the leakage joint, the rubbing speed, and the particular pump design. In general, centrifugal pump designers use that ring construction which they have found to be most suitable for each particular pump service. The most common ring constructions are the flat type (Figs. 44 and 45) and the L type.

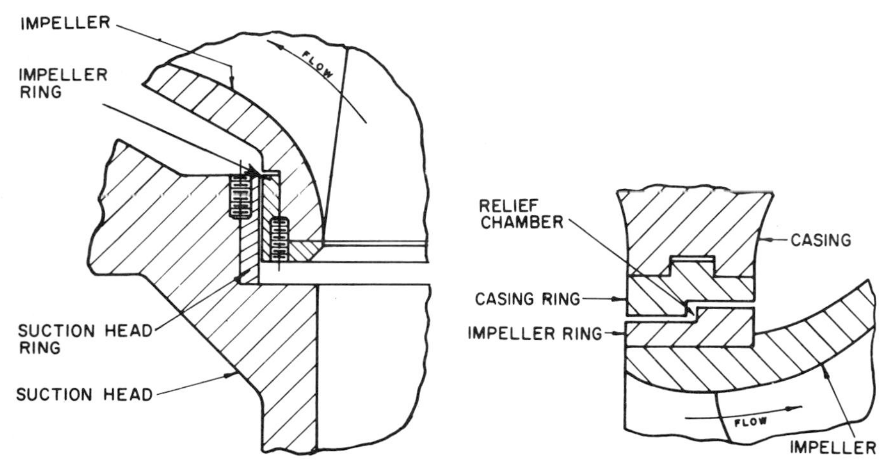

Fig. 45 Double flat-ring construction.

Fig. 46 Step-type leakage joint—double rings.

The leakage joint in the former is a straight annular clearance. In the L-type ring (Fig. 47), the axial clearance between the impeller and the casing ring is large so the velocity of the liquid flowing into the stream entering the suction eye of the impeller is low. The L-type casing rings shown in Figs. 47 and 48 have the additional function of guiding the liquid into the impeller eye; they are called *nozzle rings*. Impeller rings of the L type shown in Fig. 48 also furnish protection for the face of the impeller wearing-ring hub.

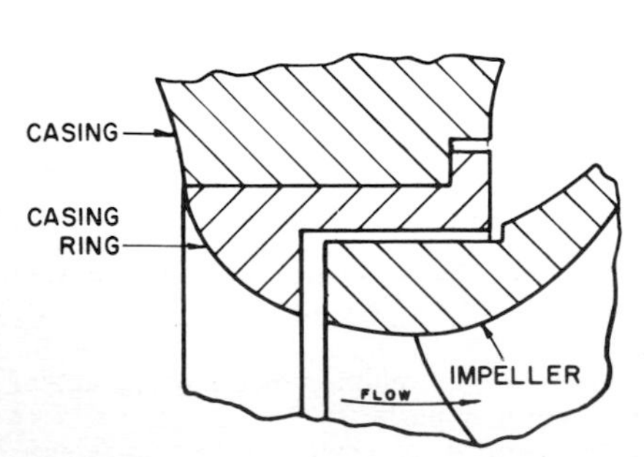

Fig. 47 Casing ring of L-nozzle type.

Some designers favor labyrinth-type rings (Figs. 49 and 50) which have two or more annular leakage joints connected by relief chambers. In leakage joints involving a single unbroken path, the flow is a function of the area and the length of the joint and of the pressure differential across the joint. If the path is broken by relief chambers (Figs. 46, 49, and 50) the velocity energy in the jet is dissipated in each relief chamber, increasing the resistance. As a result, with several relief chambers and several leakage joints for the same actual flow through the joint, the area and hence the clearance between the rings can be greater than for an unbroken, shorter leakage joint.

The single labyrinth ring with only one relief chamber (Fig. 49) is often called an *intermeshing ring.* The *step-ring type* (Fig. 46) utilizes two flat-ring elements of slightly different diameters over the total leakage-joint width with a relief

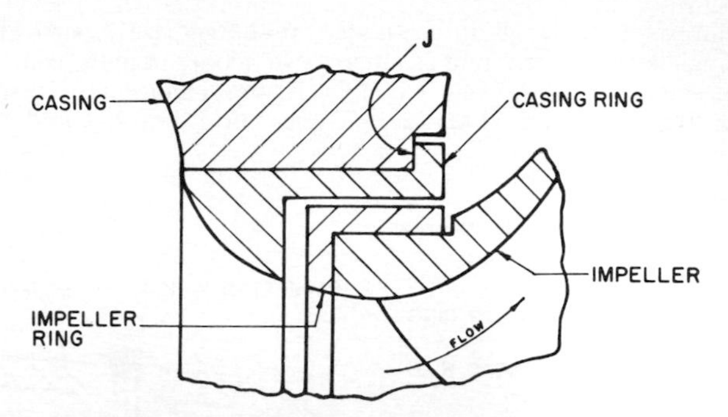

Fig. 48 Double rings—both of L type.

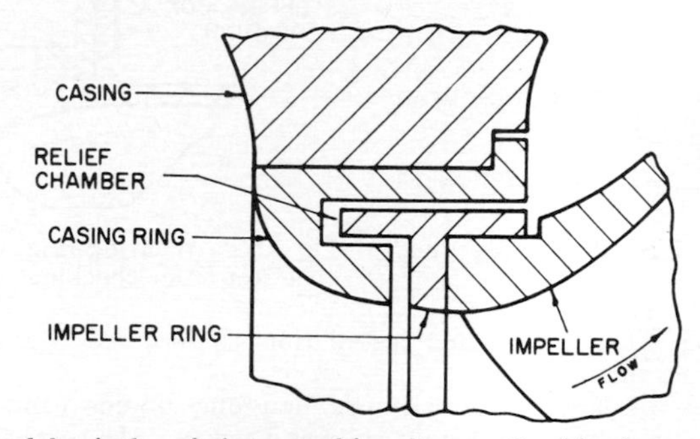

Fig. 49 Single labyrinth of intermeshing type. Double-ring construction with nozzle-type casing ring.

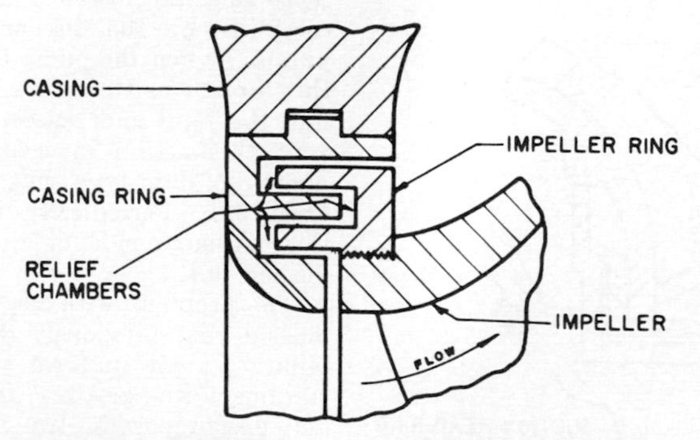

Fig. 50 Labyrinth type in double-ring construction.

chamber between the two elements. Other ring designs also use some form of relief chamber. For example, one commonly used in small pumps has a flat joint similar to that in Fig. 44, but with one surface broken by a number of grooves. These act as relief chambers to dissipate the jet velocity head, thereby increasing the resistance through the joint and decreasing the leakage.

For raw water pumps in waterworks service and for larger pumps on sewage service in which the liquid contains sand and grit, *water-flushed rings* have been used (Fig. 51). Clear water under a pressure greater than that on the discharge side of the rings is piped to the inlet and distributed by the cored passage, the holes through the stationary ring and the groove to the leakage joint. Ideally the clear water should fill the leakage joint, with some flow both to the suction and discharge sides to prevent any sand or grit from getting into the clearance space.

In large pumps (roughly 36 in or larger discharge size), particularly vertical end-suction single-stage volute pumps, mere size alone permits some refinements not found in smaller pumps. One example is the inclusion of inspection ports for measuring ring clearance (Fig. 52). These ports can be used to check the

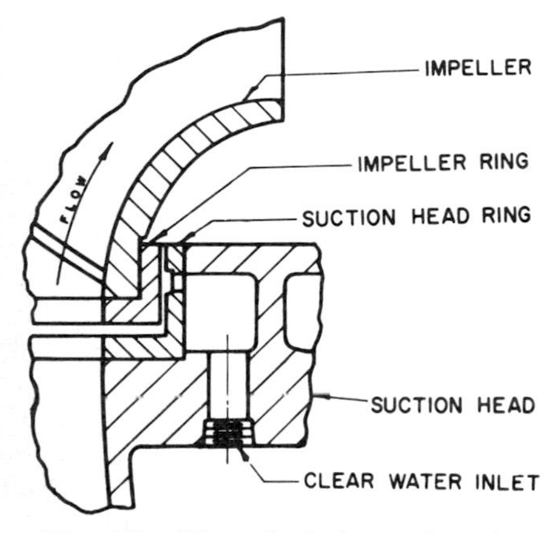

Fig. 51 Water-flushed wearing ring.

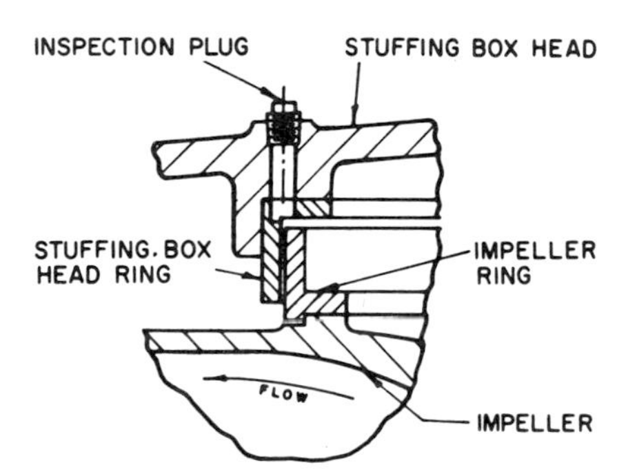

Fig. 52 Wearing-ring design with inspection-hole checking clearance.

impeller centering after the original installation as well as to observe ring wear without dismantling the pump.

The lower rings of large vertical pumps handling liquids containing sand and grit on intermittent service are highly subject to wear. During shutdown periods the grit and sand settle out and naturally accumulate in the region in which these rings are installed, as it is the lowest point on the discharge side of the pump. When the pump is started again, this foreign matter is washed all at once into the joint and causes wear. To prevent this action in medium and large pumps a dam-type ring is often used (Fig. 53). Periodically the pocket on the discharge side of the dam can be flushed out.

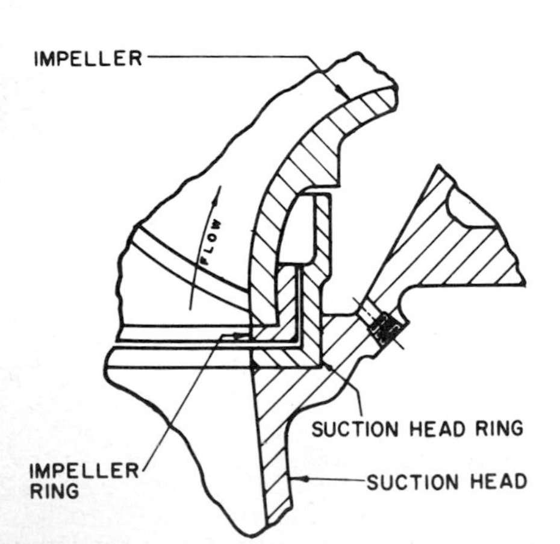

Fig. 53 Dam-type ring construction.

One problem with the simple water-flushed ring previously described is the failure to get uniform pressure in the stationary ring groove. If the pump size and design permit, two sets of wearing rings in tandem separated by a large water space (Fig. 54) provide the best solution. The large water space allows uniform distribution of the flushing water to the full 360° degrees of each leakage joint. Because ring 2 is shorter and because a greater clearance is used there than at ring 1, equal flow can be made to take place to the discharge pressure side and to the suction pressure side. This design also makes it easier to harden the surfaces with stellite or to flame-plate them with tungsten carbide.

For pumps handling gritty or sandy water, the ring construction should provide an apron on which the stream leaving the leakage joint can impinge, as sand or grit in the jet will erode any surface which it hits. Thus a form of L-type casing ring similar to that shown in Fig. 53 should be used.

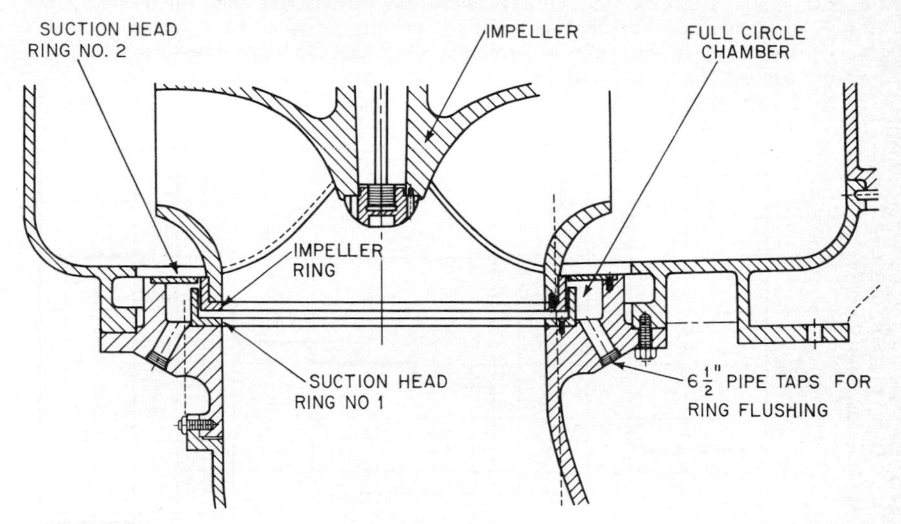

Fig. 54 Two sets of rings with space between for flushing water. (Worthington Pump, Inc.)

Wearing-Ring Location In a few designs used by one or two sewage pump manufacturers leakage is controlled by an axial clearance (Fig. 55). Generally, this design requires a means of adjustment of the shaft position for proper clearance. Then, if uniform wear occurs over the two surfaces, the original clearance can be restored by adjusting the position of the impeller. There is a limit to the amount of wear that can be compensated for since the impeller must be nearly central in the casing waterways.

Leakage joints with axial clearance are not overly popular for double-suction pumps because a very close tolerance is required in machining the fit of the rings in reference to the centerline of the volute waterways. Joints with radial clearances, however, allow some shifting of the impeller for centering. The only adverse effect is a slight inequality in the lengths of the leakage paths on the two sides.

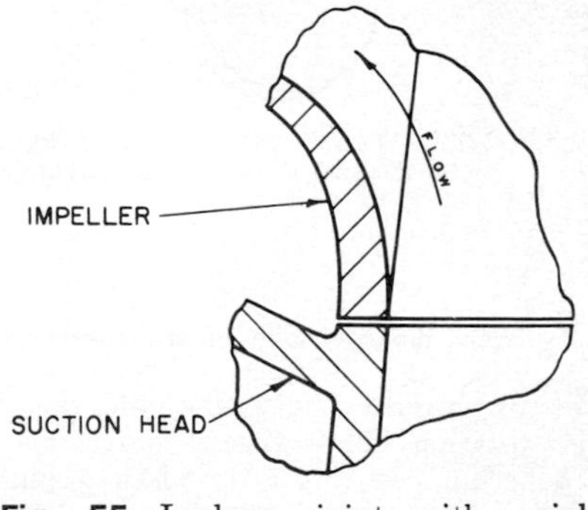

IMPELLER

FLOW

SUCTION HEAD

Fig. 55 Leakage joint with axial clearance.

So far, this discussion has treated only those leakage joints located adjacent to the impeller eye or at the smallest outside shroud diameter. There have been designs where the leakage joint has been at the periphery of the impeller. In a vertical pump this design is advantageous because the space between the joint and the suction waterways is open so sand or grit cannot collect. Because of rubbing speed and because the impeller diameters used in the same casing vary over a wide range, the design is impractical in regular pump lines.

Mounting of Stationary Wearing Rings In small single-suction pumps with suction heads, a stationary wearing ring is usually pressed into a bore in the head and may or may not be further locked by several set screws located half in the head and half in the ring (Fig. 45). Larger pumps often use an L-type ring with the flange held against a face on the head. In axially split casing pumps, the cylindrical casing bore (in which the casing ring will be mounted) should be slightly larger than the outside diameter of the ring. Unless some clearance is provided, distortion of the ring may occur when the two casing halves are assembled. However, the joint between the casing ring and the casing must be tight enough to prevent leakage. This is usually provided by a radial metal-to-metal joint (as J in Fig. 47) so arranged that the discharge pressure will press the ring against the casing surface.

As it is not desirable for the casing ring of an axially split casing pump to

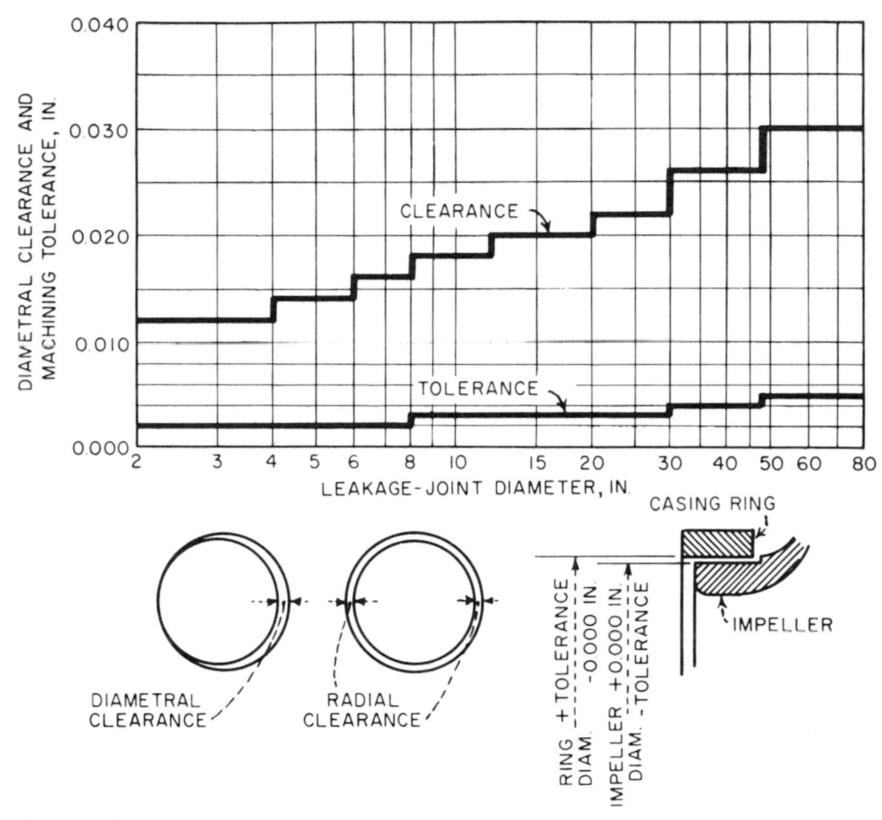

Fig. 56 Wearing-ring clearances for single-stage pumps using nongalling materials.

be pinched by the casing, the ring will not be held tightly enough to prevent its rotation unless special provisions are made to keep it in place. One means to accomplish this is to place a pin in the casing that will project into a hole bored in the ring, or, conversely, to provide a pin in the ring that will fit into a hole bored in the casing or into a recess at the casing split joint.

Another favorite method is a tongue on the casing ring extending around 180° which engages a corresponding groove in one-half of the casing. This method can be used with casing rings having a central flange by making the flange of larger diameter for 180° and cutting a deeper groove in that half of the casing.

Many methods are used for holding impeller rings on the impeller. Probably the simplest is to rely on a press fit of the ring on the impeller or, if the ring is of proper material, on a shrink fit. Designers do not usually feel that a press fit is sufficient and often add several machine screws or set screws located half in the ring and half in the impeller as in Fig. 45.

Some designers prefer to thread the impeller and ring and screw the ring onto the impeller (Fig. 50). Some even favor the use of right- and left-hand threads for the different sides so that rotation will tend to tighten the ring on the impeller. Generally some additional locking device is used rather than relying solely on the friction grip of the ring on the impeller.

In the design of impeller rings consideration has to be given to the stretch of the ring due to centrifugal force, especially if the pump is of a high-speed design for the capacity involved. For example, some boiler-feed pumps operate at speeds which would cause the rings to become loose if only a press fit were used. For such pumps shrink fits should be used or, preferably, impeller rings should be eliminated.

Wearing-Ring Clearances Typical clearance and tolerance standards for nongalling wearing joint metals in general service pumps are shown in Fig. 56. They apply to the following combinations: (1) bronze with a dissimilar bronze, (2) cast iron with bronze, (3) steel with bronze, (4) monel metal with bronze, and (5) cast iron with cast iron. If the metals gall easily (like the chrome steels), the values given should be increased by about 0.002 to 0.004 in.

In multistage pumps, the basic diameter clearance should be increased by 0.003 in for larger rings. The tolerance indicated is plus (+) for the casing ring and minus (—) for the impeller hub or impeller ring.

In a single-stage pump with a joint of nongalling components, the correct machining dimension for a casing-ring diameter of 9.000 in would be 9.000 plus 0.003 and minus 0.000 in and for the impeller hub or ring, 9.000 minus 0.018, or 8.982 plus 0.000 and minus 0.003 in. Actual diametral clearances would be between 0.018 and 0.024 in. Naturally the manufacturer's recommendation for ring clearance and tolerance should be followed.

AXIAL THRUST

Axial Thrust in Single-Stage Pumps The pressures generated by a centrifugal pump exert forces on both its stationary and rotating parts. The design of these parts balances some of these forces but separate means may be required to counterbalance others.

Axial hydraulic thrust is the summation of unbalanced forces on an impeller

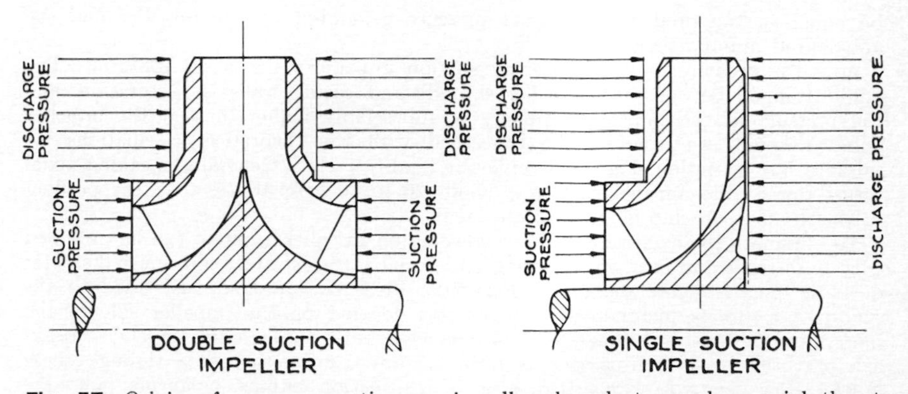

Fig. 57 Origin of pressures acting on impeller shrouds to produce axial thrust.

acting in the axial direction. As reliable large capacity thrust bearings are now readily available, axial thrust in single-stage pumps remains a problem only in larger units. Theoretically, a double-suction impeller is in hydraulic axial balance with the pressures on one side equal to and counterbalancing the pressures on the other (Fig. 57). In practice this balance may not be achieved for the following reasons:

1. The suction passages to the two suction eyes may not provide equal or uniform flows to the two sides.

2. External conditions, such as an elbow located too close to the pump suction nozzle, may cause unequal flows to the two suction eyes.
3. The two sides of the discharge casing waterways may not be symmetrical or the impeller may be located offcenter. These conditions will alter the flow characteristics between the impeller shrouds and the casing, causing unequal pressures on the shrouds.
4. Unequal leakage through the two leakage joints can upset the balance.

Combined, these factors can create axial unbalance. To compensate for this all centrifugal pumps, even those with double-suction impellers, incorporate thrust bearings.

The ordinary single-suction radial-flow impeller with the shaft passing through the impeller eye (Fig. 57) is subject to axial thrust because a portion of the front wall is exposed to suction pressure, thus exposing relatively more backwall surface to discharge pressure. If the discharge chamber pressure were uniform over the entire impeller surface, the axial force acting toward the suction would

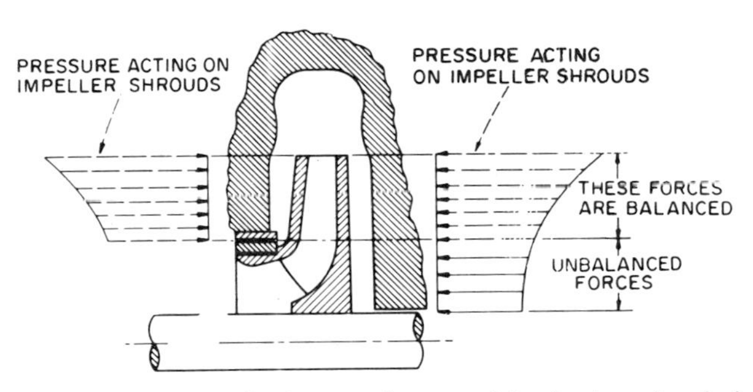

Fig. 58 Actual pressure distribution on front and back shrouds of single-suction impeller with shaft through impeller eye.

be equal to the product of the net pressure generated by the impeller and the unbalanced annular area.

Actually, pressure on the two single-suction impeller walls is not uniform. The liquid trapped between the impeller shrouds and casing walls is in rotation, and the pressure at the impeller periphery is appreciably higher than at the impeller hub. Although we need not be concerned with the theoretical calculations for this pressure variation, Fig. 58 describes it qualitatively. Generally speaking, axial thrust toward the impeller suction is about 20 to 30 percent less than the product of the net pressure and the unbalanced area.

To eliminate the axial thrust of a single-suction impeller, a pump can be provided with both front and back wearing rings; to equalize thrust areas, the inner diameter of both rings is made the same (Fig. 59). Pressure approximately equal to the suction pressure is maintained in a chamber located on the impeller side of the back wearing ring by drilling so-called balancing holes through the impeller. Leakage past the back wearing ring is returned into the suction area through these holes. However, with large single-stage single-suction pumps, balancing holes are considered undesirable because leakage back to the impeller suction opposes the main flow, creating disturbances. In such pumps, a piped connection to the pump suction replaces the balancing holes. Another way to eliminate or reduce axial thrust in single-suction impellers is by use of pump-out vanes on the back shroud. The effect of these vanes is to reduce the pressure acting on the back shroud of the impeller (Fig. 60). This design, however, is generally used only in pumps handling gritty liquids, where it keeps the clearance space between the impeller back shroud and the casing free of foreign matter.

So far, our discussion of the axial thrust has been limited to single-suction impellers with a shaft passing through the impeller eye and located in pumps with two stuffing boxes, one on either side of the impeller. In these pumps suction-pressure magnitude does not affect the resulting axial thrust.

On the other hand, axial forces acting on an overhung impeller with a single stuffing box (Fig. 61) are definitely affected by suction pressure. In addition to the unbalanced force found in a single-suction two-box design (see Fig. 58), there is an axial force equivalent to the product of the shaft area through the stuffing

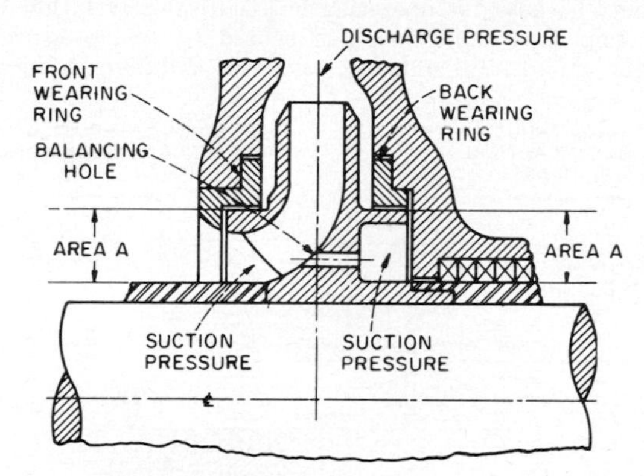

Fig. 59 Balancing axial thrust of single-suction impeller by means of wearing ring on back side and balancing holes.

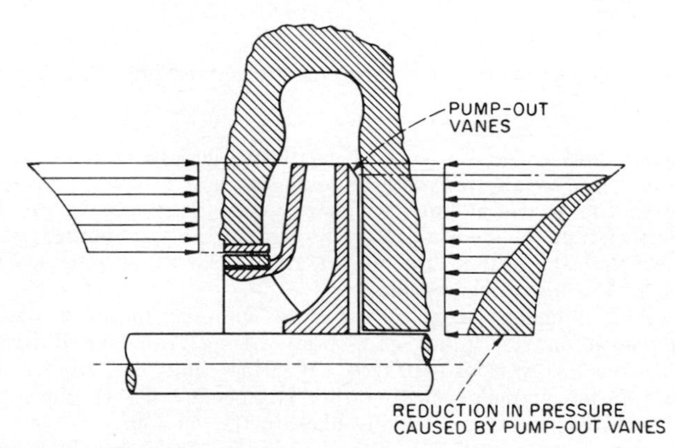

Fig. 60 Effect of pump-out vanes in a single-suction impeller to reduce axial thrust.

box and the difference between suction and atmospheric pressure. This force acts toward the impeller suction when the suction pressure is less than atmospheric or in the opposite direction when it is higher than atmospheric.

When an overhung impeller pump handles a suction lift, the additional axial force is very low. For example, if the shaft diameter through the stuffing box is 2 in (area = 3.14 in^2) and if the suction lift is 20 ft of water (absolute pressure = 6.06 lb/in^2 abs), the axial force caused by the overhung impeller and acting toward the suction will be only 27 lb. On the other hand, if the suction pressure is 100 lb/in^2, the force will be 314 lb and acts in the opposite direction. Therefore, as the same pump may be applied for many conditions of service over a wide range of suction pressures, the thrust bearing of pumps with single-suction overhung impellers must be arranged to take thrust in either direction. They must also be selected with sufficient thrust capacity to counteract forces set up under the maximum suction pressure established as a limit for that particular pump.

This extra thrust capacity may become quite significant in certain special cases, as for instance with boiler circulating pumps. These are usually of the single-suction single-stage overhung impeller type and may be exposed to suction pressures as high as 2,800 lb/in² gage. If such a pump has a shaft diameter of 6 in, the unbalanced thrust would be as much as 77,500 lb, and the thrust bearing has to be capable of counteracting this. Except for very large units and certain special applications, the maximum thrust developed by mixed-flow and axial-flow impellers is not of consequence because the operating heads are relatively low.

With axial-flow impellers, axial thrust is caused by the pressure on the vanes as they act on the liquid. In addition, there is a difference in pressure acting

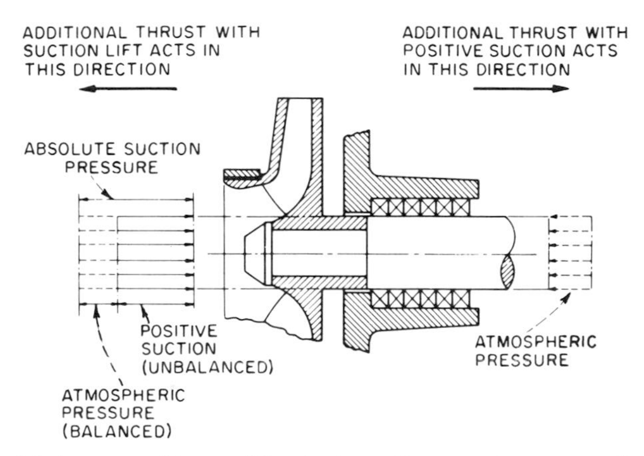

Fig. 61 Axial-thrust problem with single-suction overhung impeller and single stuffing box.

on the two shaft hub ends, one subject to discharge pressure and the other to suction pressure. Occasionally provision is made in an axial-flow pump for a leakage joint at or near the discharge-hub periphery, with balancing holes through the hub. This construction is used mainly in vertical wet-pit pumps with covered shaft designs so that the seal at the lower end of the cover pipe will be subject to suction rather than discharge pressure.

With mixed-flow impellers, axial thrust is a combination of forces caused by action of the vanes on the liquid and those arising from the difference in the pressures acting on the various surfaces. Wearing rings are often provided on the back of mixed-flow impellers, with either balancing holes through the impeller hub or an external balancing pipe leading back to the suction.

In the past, some large mixed-flow impeller designs and some high-head vertical pumps with single-suction radial-type impellers had a leakage joint on the back side of the impeller that was larger in diameter than the leakage joint on the suction side. This disparity caused the axial thrust to act upward, balancing the dead weight of the rotor. This design practice has been discarded because more reliable thrust bearings are now available.

The use of wearing rings on the back of large-capacity sewage pumps with mixed-flow impellers has not met with general approval. Therefore larger capacity thrust bearings must be used.

Axial Thrust in Multistage Pumps Most multistage pumps are built with single-suction impellers in order to simplify the design of the interstage connections. Two obvious arrangements are possible for the single-suction impellers:

1. Several single-suction impellers may be mounted on one shaft, each having its suction inlet facing in the same direction and its stages following one another in ascending order of pressure (Fig. 62). The axial thrust is then balanced by a hydraulic balancing device.

2. An even number of single-suction impellers may be used, one-half of these facing in an opposite direction to the second half. With this arrangement, axial

thrust on the first half is compensated by the thrust in the opposite direction on the other half (Fig. 63). This mounting of single-suction impellers back-to-back is frequently called *opposed impellers.*

An uneven number of single-suction impellers may be used with this arrangement,

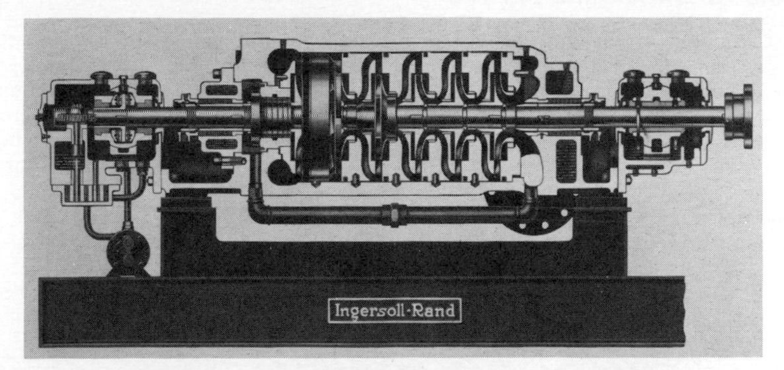

Fig. 62 Multistage pump with single-suction impellers facing in one direction and with a hydraulic balancing device. (Ingersoll-Rand)

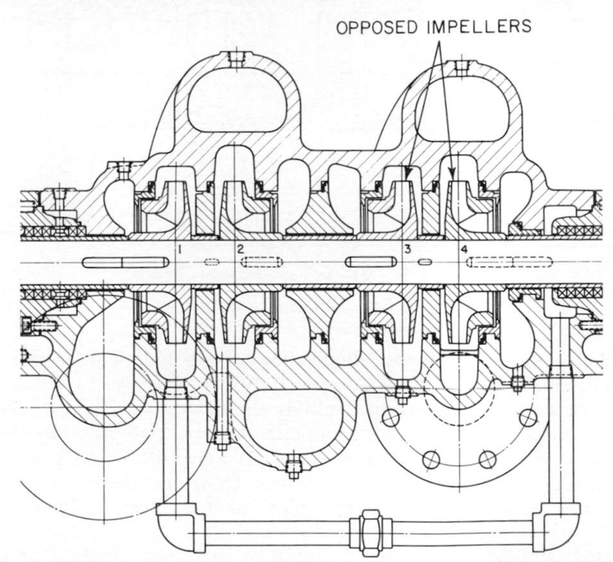

Fig. 63 Four-stage pump with "opposed impellers." (Worthington Pump, Inc.)

provided the correct shaft and interstage bushing diameters are used to give the effect of a hydraulic balancing device that will compensate for the hydraulic thrust on one of the stages.

It is important to note that the opposed-impeller arrangement completely balances axial thrust only under the following conditions:

1. The pump must be provided with two stuffing boxes.
2. The shaft must have a constant diameter.
3. The impeller hubs must not extend through the interstage portion of the casing separating adjacent stages.

Except for some special pumps that have an internal and enclosed bearing at one end, and therefore only one stuffing box, most multistage pumps fulfill the first condition. But because of structural requirements, the last two conditions are not practical. A slight residual thrust is usually present in multistage opposed-impeller pumps, and is carried on the thrust bearing.

HYDRAULIC BALANCING DEVICES

If all the single-suction impellers of a multistage pump face in the same direction, the total theoretical hydraulic axial thrust acting toward the suction end of the pump will be the sum of the individual impeller thrusts. The thrust magnitude (in pounds) will be approximately equal to the product of the net pump pressure (in pounds per square inch) and the annular unbalanced area (in square inches). Actually the axial thrust turns out to be about 70 to 80 percent of this theoretical value.

Some form of hydraulic balancing device must be used to balance this axial thrust and to reduce the pressure on the stuffing box adjacent to the last-stage

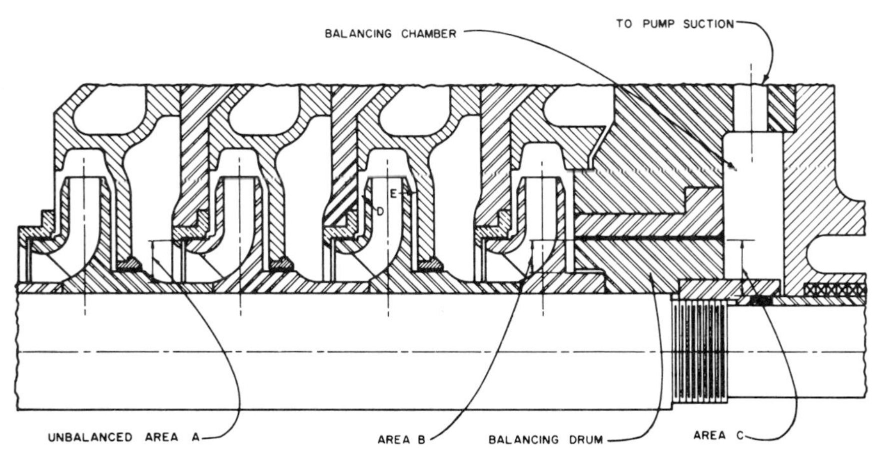

Fig. 64 Balancing drum.

impeller. This hydraulic balancing device may be a balancing drum, a balancing disk, or a combination of the two.

Balancing Drums The balancing drum is illustrated in Fig. 64. The balancing chamber at the back of the last-stage impeller is separated from the pump interior by a drum that is either keyed or screwed to and rotates with the shaft. The drum is separated by a small radial clearance from the stationary portion of the balancing device, called the *balancing-drum head,* which is fixed to the pump casing.

The balancing chamber is connected either to the pump suction or to the vessel from which the pump takes its suction. Thus the back pressure in the balancing chamber is only slightly higher than the suction pressure, the difference between the two being equal to the friction losses between this chamber and the point of return. The leakage between the drum and the drum head is, of course, a function of the differential pressure across the drum and of the clearance area.

The forces acting on the balancing drum in Fig. 64 are the following:

1. Toward the discharge end: the discharge pressure multiplied by the front balancing area (area B) of the drum.
2. Toward the suction end: the back pressure in the balancing chamber multiplied by the back balancing area (area C) of the drum.

The first force is greater than the second, thereby counterbalancing the axial thrust exerted upon the single-suction impellers. The drum diameter can be selected to balance axial thrust completely or within 90 to 95 percent, depending on the desirability of carrying any thrust-bearing loads.

It has been assumed in the preceding simplified description that the pressure acting on the impeller walls is constant over their entire surface and that the axial thrust is equal to the product of the total net pressure generated and the unbalanced area. Actually this pressure varies somewhat in the radial direction because of the centrifugal force exerted upon the water by the outer impeller shroud (see Fig. 58). Furthermore, the pressures at two corresponding points on the opposite impeller faces (*D* and *E*, Fig. 64) may not be equal because of variation in clearance between the impeller wall and the casing section separating successive stages. Finally pressure distribution over the impeller wall surface may vary with head and capacity operating conditions.

This pressure distribution and design data can be determined by test quite accurately for any one fixed operating condition, and an effective balancing drum could be designed on the basis of the forces resulting from this pressure distribution. Unfortunately varying head and capacity conditions change the pressure distribu-

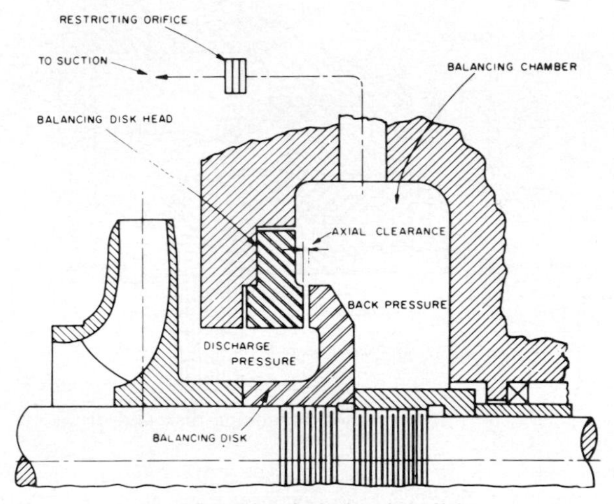

Fig. 65 Simple balancing disk.

tion, and as the area of the balancing drum is necessarily fixed, the equilibrium of the axial forces can be destroyed.

The objection to this is not primarily the amount of the thrust, but rather that the direction of the thrust cannot be predetermined because of the uncertainty about internal pressures. Still it is advisable to predetermine normal thrust direction, as this can influence external mechanical thrust-bearing design. Because 100 percent balance is unattainable in practice and the slight but predictable unbalance can be carried on a thrust bearing, the balancing drum is often designed to balance only 90 to 95 percent of total impeller thrust.

The balancing drum satisfactorily balances the axial thrust of single-suction impellers and reduces pressure on the discharge-side stuffing box. It lacks, however, the virtue of automatic compensation for any changes in axial thrust caused by varying impeller reaction characteristics. In effect, if the axial-thrust and balancing-drum forces become unequal, the rotating element will tend to move in the direction of the greater force. The thrust bearing must then prevent excessive movement of the rotating element. The balancing drum performs no restoring function until such time as the drum force again equals the axial thrust. This automatic compensation is the major feature that differentiates the balancing disk from the balancing drum.

Balancing Disks The operation of the simple balancing disk is illustrated in Fig. 65. The disk is fixed to and rotates with the shaft. It is separated from the balancing disk head which is fixed to the casing by a small axial clearance. The leakage through this clearance flows into the balancing chamber and from there either to the pump suction or to the vessel from which the pump takes its suction.

The back of the balancing disk is subject to the balancing chamber back pressure whereas the disk face experiences a range of pressures. These vary from discharge pressure at its smallest diameter to back pressure at its periphery. The inner and outer disk diameters are chosen so that the difference between the total force acting on the disk face and that acting on its back will balance the impeller axial thrust.

If the axial thrust of the impellers should exceed the thrust acting on the disk during operation, the latter is moved toward the disk head, reducing the axial clearance between the disk and the disk head. The amount of leakage through the clearance is reduced so that the friction losses in the leakage return line are also reduced, lowering the back pressure in the balancing chamber. This lower-

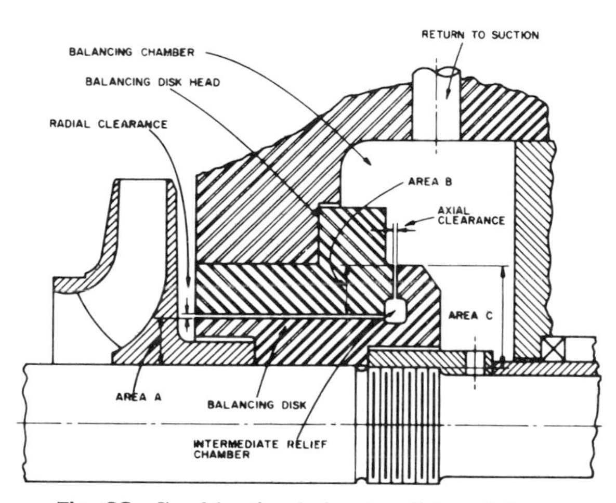

Fig. 66 Combination balancing disk and drum.

ing of pressure automatically increases the pressure difference acting on the disk and moves it away from the disk head, increasing the clearance. Now the pressure builds up in the balancing chamber, and the disk is again moved toward the disk head until an equilibrium is reached.

To assure proper balancing-disk operation, the change in back pressure in the balancing chamber must be of an appreciable magnitude. Thus, with the balancing disk wide open with respect to the disk head, the back pressure must be substantially higher than the suction pressure to give a resultant force that restores the normal disk position. This can be accomplished by introducing a restricting orifice in the leakage return line that increases back pressure when leakage past the disk increases beyond normal. The disadvantage of this arrangement is that the pressure on the stuffing box packing is variable—a condition that is injurious to the life of the packing and therefore to be avoided. The higher pressure that can occur at the packing is also undesirable.

Combination Balancing Disk and Drum For the reasons just described, the simple balancing disk is seldom used. The combination balancing disk and drum (Fig. 66) was developed to obviate the shortcomings of the disk while retaining the advantage of automatic compensation for axial thrust changes.

The rotating portion of this balancing device consists of a long cylindrical body that turns within a drum portion of the disk head. This rotating part incorporates a disk similar to the one previously described. In this design, radial clearance remains constant regardless of disk position, whereas the axial clearance varies with the pump rotor position. The following forces act on this device:

1. Toward the discharge end: the sum of the discharge pressure multiplied by area A, plus the average intermediate pressure multiplied by area B.
2. Toward the suction end: the back pressure multiplied by area C.

Whereas the position-restoring feature of the simple balancing disk required an undesirably wide variation of the back pressure, it is now possible to depend upon a variation of the intermediate pressure to achieve the same effect. Here is how it works: When the pump rotor moves toward the suction end (to the left, in Fig. 66) because of increased axial thrust, the axial clearance is reduced, and pressure builds up in the intermediate relief chamber, increasing the average value of the intermediate pressure acting on area *B*. In other words, with reduced leakage, the pressure drop across the radial clearance decreases, increasing the pressure drop across the axial clearance. The increase in intermediate pressure

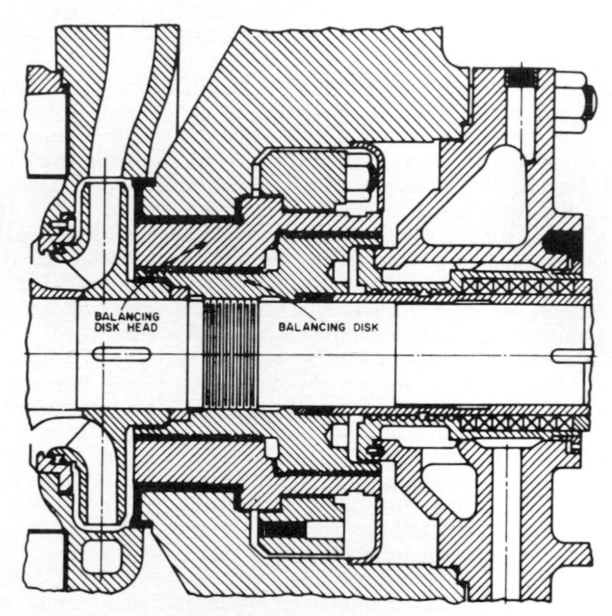

Fig. 67 Combination balancing disk and drum, with disk located in the center portion of the drum. (Worthington Pump, Inc.)

forces the balancing disk toward the discharge end until equilibrium is reached. Movement of the pump rotor toward the discharge end would have the opposite effect of increasing the axial clearance and the leakage and decreasing the intermediate pressure acting on area *B*.

There are now in use numerous hydraulic balancing device modifications. One typical design separates the drum portion of a combination device into two halves, one preceding and the second following the disk (Fig. 67). The virtue of this arrangement is a definite cushioning effect at the intermediate relief chamber, thus avoiding too positive a restoring action, which might result in the contacting and scoring of the disk faces.

SHAFTS AND SHAFT SLEEVES

The basic function of a centrifugal pump shaft is to transmit the torques encountered in starting and during operation while supporting the impeller and other rotating parts. It must do this job with a deflection less than the minimum clearance between rotating and stationary parts. The loads involved are (1) the torques, (2) the weight of the parts, and (3) both radial and axial hydraulic forces. In designing a shaft, the maximum allowable deflection, the span or overhang, and the location of the loads all have to be considered, as does the critical speed of the resulting design.

Shafts are usually proportioned to withstand the stress set up when a pump is started quickly, for example, when the driving motor is thrown directly across the line. If the pump handles hot liquids, the shaft is designed to withstand the stress set up when the unit is started cold without any preliminary warmup.

Critical Speeds Any object made of an elastic material has a natural period of vibration. When a pump rotor or shafting rotates at any speed corresponding to its natural frequency, minor unbalances will be magnified. These speeds are called the *critical speeds.*

In conventional pump designs, the rotating assembly is theoretically uniform around the shaft axis, and the center of mass should coincide with the axis of rotation. This theory will not hold for two reasons. First, there are always minor machining or casting irregularities; second, there will be variations in metal density of each part. Thus, even in vertical shaft machines having no radial deflection caused by the weight of the parts, this eccentricity of the center of mass produces centrifugal force and therefore a deflection when the assembly rotates. At the speed at which the centrifugal force exceeds the elastic restoring force, the rotor will vibrate as though it were seriously unbalanced. If it is run at that speed without restraining forces, the deflection will increase until the shaft fails.

Rigid- and Flexible-Shaft Designs The lowest critical speed is called the first critical speed; the next higher is called the second, and so forth. In centrifugal pump nomenclature, a *rigid shaft* means one with an operating speed lower than its first critical speed; a *flexible shaft* is one with an operating speed higher than its first critical speed. Once an operating speed has been selected, the designer must still determine the relative shaft dimensions. In other words, he must decide whether the pump will operate above or below the first critical speed.

Actually the shaft critical speed can be reached and passed without danger because frictional forces tend to restrain the deflection. These forces are exerted by the surrounding liquid, the stuffing box packing, and the various internal leakage joints acting as internal liquid-lubricated bearings. Once the critical speed is passed, the pump will run smoothly again up to the second speed corresponding to the natural rotor frequency, and so on to the third, fourth, and all higher critical speeds.

Designs rated for 1,750 rpm (or lower) are usually of the rigid-shaft type. On the other hand, high-head 3,600 rpm (or higher) multistage pumps, such as those in boiler feed service, are frequently of the flexible-shaft type. It is possible to operate centrifugal pumps above their critical speeds for the following two reasons: (1) very little time is required to attain full speed from rest (the time required to pass through the critical speed must therefore be extremely short), and (2) the pumped liquid in the stuffing box packing and the internal leakage joints acts as a restraining force on the vibration.

Experience has proved that, although it was usually assumed necessary to use shafts of such rigidity that the first critical speed is at least 20 percent above the operating speed, equally satisfactory results can be obtained with lighter shafts with a first critical speed of about 60 to 75 percent of the operating speed. This, it is felt, is a sufficient margin to avoid any danger caused by operation close to the critical.

Influence of Shaft Deflection To understand the effect of critical speed upon the selection of shaft size, consider the fact that the first critical speed of a shaft is linked to its static deflection. Shaft deflection depends upon the weight of the rotating element (w), the shaft span (l), and the shaft diameter (d). The basic formula is:

$$f = \frac{wl^3}{mEI}$$

where f = deflection, in
w = weight of the rotating element, lb
l = shaft span, in
m = coefficient depending on shaft-support method and load distribution
E = modulus of elasticity of shaft materials, lb/in^2
I = moment of inertia ($d^4/64$), in^4

This formula is given in its most simplified form, that is, for a shaft of constant diameter. If the shaft is of varying diameter (the usual situation), deflection calculations are much more complex. A graphical deflection analysis is then the most practical answer.

This formula solves only for static deflection, the only variable that affects critical-speed calculations. The actual shaft deflection—which must be determined to establish minimum permissible internal clearances—must take into account all transverse hydraulic reactions on the rotor, the actual weights of the rotating element, and other external loads such as belt pull.

It is not necessary to calculate exact deflection to make a relative shaft comparison. Instead, a factor can be developed that will be representative of relative shaft deflections. As a significant portion of rotor weight is in the shaft, and as methods of bearing support and modulus of elasticity are common to similar designs, deflection f can be shown as follows:

$$f = \text{function of } \frac{(ld^2)(l^3)}{d^4}$$

or

$$f = \text{function of } \frac{l^4}{d^2}$$

In other words, pump deflection varies approximately as the fourth power of shaft span and inversely as the square of the shaft diameter. Therefore the lower the l^4/d^2 factor for a given pump, the lower the unsupported shaft deflection, essentially in proportion to this factor.

For practical purposes, the first critical speed N_c can be calculated as:

$$N_c = \frac{187.7}{\sqrt{f}} \quad \text{rpm}$$

To maintain internal clearances at the wearing rings, it is usually desirable to limit shaft deflection under most adverse conditions from 0.005 to 0.006 in. It follows that a shaft design with a deflection of 0.005 to 0.006 in will have a first critical speed of 2,400 to 2,650 rpm. This is the reason for using rigid shafts for pumps that operate at 1,750 rpm or lower. Multistage pumps operating at 3,600 rpm or higher use shafts of equal stiffness (for the same purpose of avoiding wearing ring contact). However, their corresponding critical speed is about 25 to 40 percent less than their operating speed.

Shaft Sizing Shaft diameters usually have larger dimensions than are actually needed to transmit the torque. A factor that assures this conservative design is the requirement for ease of rotor assembly.

The shaft diameter must be stepped up several times from the end of the coupling to its center to facilitate impeller mounting (Fig. 68). Starting with the maximum diameter at the impeller mounting, there is a stepdown for the shaft sleeve, another for the external shaft nut, followed by several more for the bearings and the coupling. Therefore the shaft diameter at the impellers exceeds that required for torsional strength at the coupling by at least an amount sufficient to provide all intervening stepdowns.

One frequent exception to shaft oversizing at the impeller occurs in units consisting of two double-suction single-stage pumps operating in series, one of which is fitted with a double-extended shaft. As this pump must transmit the total horsepower for the entire series unit, the shaft diameter at its inboard bearing may have to be greater than the normal diameter.

Shaft design of end-suction overhung-impeller pumps presents a somewhat different problem. One method for reducing shaft deflection at the impeller and stuffing box—where concentricity of running fits is extremely important—is to considerably increase shaft diameter between the bearings.

Except in certain smaller sizes, centrifugal pump shafts are protected against wear, erosion, and corrosion by renewable shaft sleeves. In very small pumps, however, shaft sleeves present a certain disadvantage. As the sleeve cannot appreciably contribute to shaft strength, the shaft itself must be designed for the full maximum stress. Shaft diameter is then materially increased by the addition of

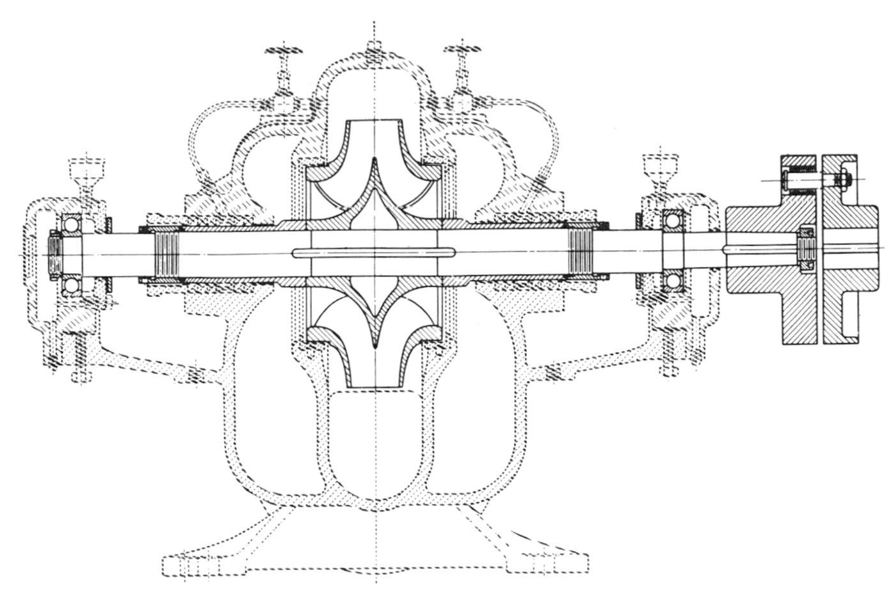

Fig. 68 Rotor assembly of a single-stage double-suction pump. (Worthington Pump Inc.)

the sleeve, as the sleeve thickness cannot be decreased beyond a certain safe minimum. The impeller suction area may therefore become dangerously reduced, and if the eye diameter is increased to maintain a constant eye area, the liquid pickup speed must be increased unfavorably. Other disadvantages accrue from greater hydraulic and stuffing box losses caused by increasing the effective shaft diameter out of proportion to the pump size.

To eliminate these shortcomings, very small pumps frequently use shafts of stainless steel or some other material that is sufficiently resistant to corrosion and wear not to need shaft sleeves. One such pump is illustrated in Fig. 69. Manu-

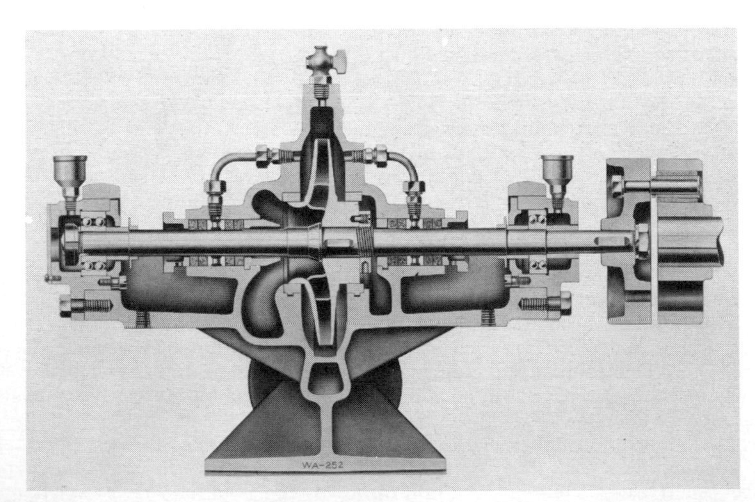

Fig. 69 Section of a small-size centrifugal pump with no shaft sleeves. (Worthington Pump, Inc.)

facturing costs, of course, are much less for this type of design, and the cost of replacing the shaft is about the same as the cost of new sleeves (including installation).

Shaft Sleeves Pump shafts are usually protected from erosion, corrosion, and wear at stuffing boxes, leakage joints, internal bearings, and in the waterways by renewable sleeves.

The most common shaft-sleeve function is that of protecting the shaft from

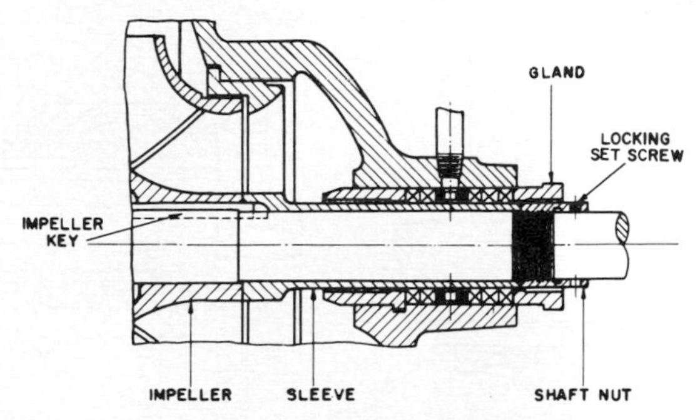

Fig. 70 Sleeve with external locknut and impeller key extending into sleeve to prevent slip.

wear at a stuffing box. Shaft sleeves serving other functions are given specific names to indicate their purpose. For example, a shaft sleeve used between two multistage pump impellers in conjunction with the interstage bushing to form an interstage leakage joint is called an *interstage* or *distance sleeve.*

In medium-size centrifugal pumps with two external bearings on opposite sides

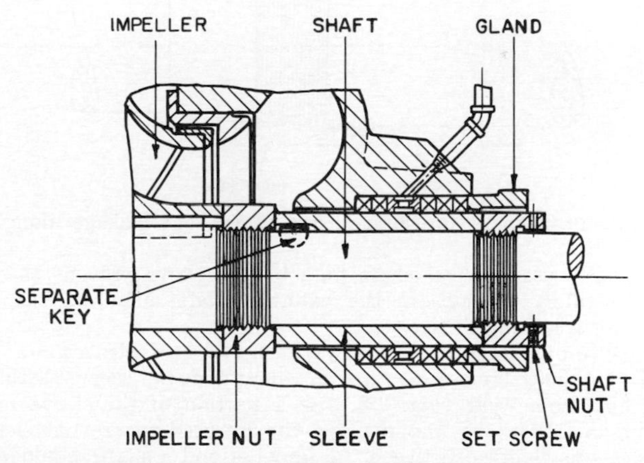

Fig. 71 Sleeve construction with internal impeller nut, external shaft-sleeve nut, and separate key for sleeve.

of the casing (the common double-suction and multistage varieties), the favored shaft-sleeve construction uses an external shaft nut to hold the sleeve in axial position against the impeller hub. Sleeve rotation is prevented by a key, usually an extension of the impeller key (Fig. 70). If the axial thrust exceeds the frictional grip of the impeller on the shaft, it is transmitted through the sleeve to the external shaft nut.

In larger high-head pumps, a high axial load on the sleeve is possible, and a design similar to that shown in Fig. 71 may be favored. This design has the commercial advantages of simplicity and low replacement cost. Some manufacturers favor the sleeve shown in Fig. 72 in which the impeller end of the sleeve is threaded and screwed to a matching thread on the shaft.

A key cannot be used with this type of sleeve, and right- and left-hand threads are substituted so that the frictional grip of the packing on the sleeve will tighten it against the impeller hub. In the sleeve designs shown in Figs. 70 and 71, right-hand

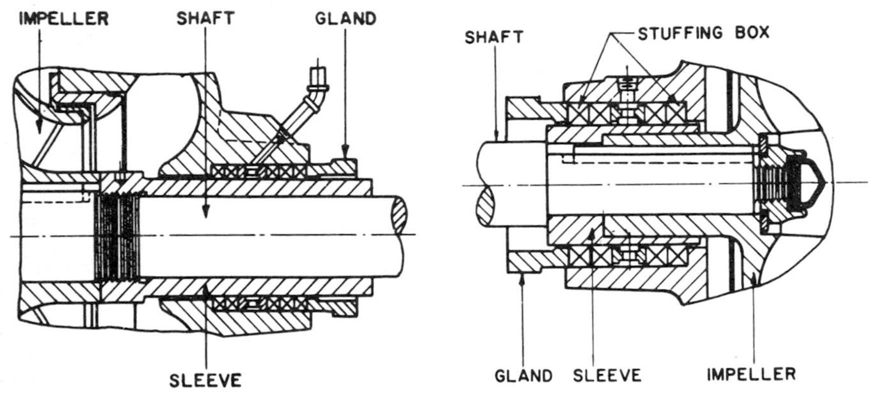

Fig. 72 Sleeve threaded onto shaft, with no external locknut.

Fig. 73 Sleeve for pumps with overhung impeller hubs extending into stuffing box.

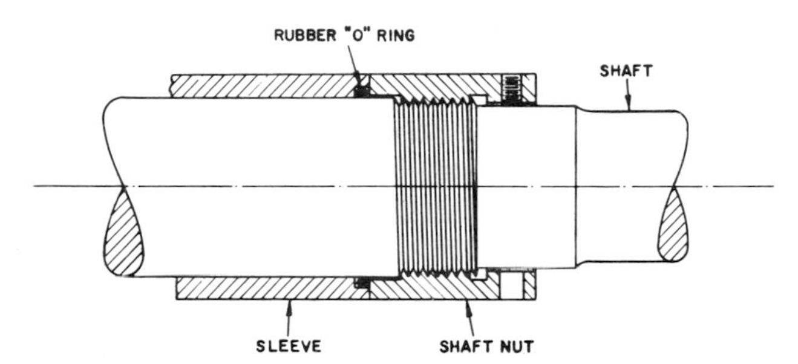

Fig. 74 Seal arrangement for shaft sleeve to prevent leakage along the shaft.

threads are usually used for all shaft nuts because keys prevent the sleeve from rotating. As a safety precaution, the external shaft nuts and the sleeve itself use set screws for a locking device.

In pumps with overhung impellers, various types of sleeves are used. Often stuffing boxes are placed close to the impeller, and the sleeve actually protects the impeller hub from wear (Fig. 73). As a portion of the sleeve in this design fits directly on the shaft, the impeller key can be used to prevent sleeve rotation. Part of the sleeve is clamped between the impeller and a shaft shoulder to maintain its axial position.

In designs with a metal-to-metal joint between the sleeve and the impeller hub (Fig. 70), operation under a positive suction head often starts liquid leakage into the clearance between the shaft and sleeve. For a pump operating under negative suction head, the various clearances may cause slight air leakage into the pump. Usually this leakage is not important; however, it occasionally causes trouble, and a sleeve design with a leakage seal may then become desirable. One possible arrangement is shown in Fig. 74. The design shown in Fig. 75 is used for high-temperature process pumps. The contact surface of the sleeve and shaft is ground at a 45° angle. That end of the sleeve is locked, but the other is free to expand with temperature changes.

Material for Stuffing Box Sleeves Stuffing box shaft sleeves are surrounded in the stuffing box by packing; the sleeve must be smooth so that it can turn without generating too much friction and heat. Thus the sleeve materials must be capable of taking a very fine finish, preferably a polish. Cast iron is therefore

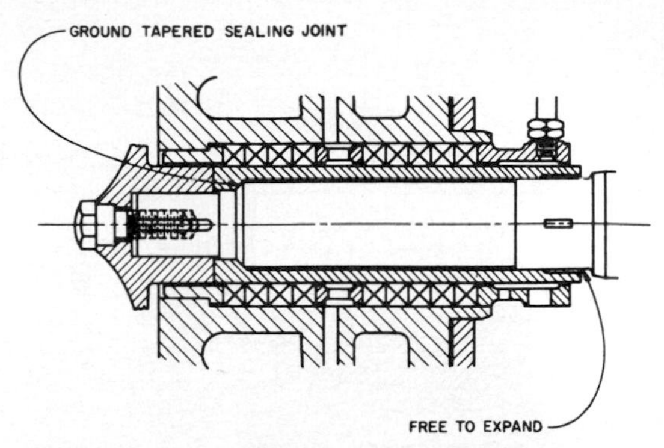

Fig. 75 Sleeve with 45° bevel contacting surface.

not suitable. A hard bronze is generally used for pumps handling clear water, but chrome or other stainless steels are sometimes preferred. For services subject to grit, hardened chrome or other stainless steels give good results. For more severe conditions, stellited sleeves are often used and occasionally sleeves that are chromium-plated at the packing area. Ceramic-coated sleeves with the coating applied by flame plating are also used for some severe services. Sleeves made entirely of a hardened chrome steel are usually the most economical and satisfactory.

STUFFING BOXES

Stuffing boxes have the primary function of protecting the pump against leakage at the point where the shaft passes out through the pump casing. If the pump handles a suction lift and the pressure at the interior stuffing box end is below atmospheric, the stuffing box function is to prevent air leakage into the pump. If this pressure is above atmospheric, the function is to prevent liquid leakage out of the pump.

For general service pumps, a stuffing box usually takes the form of a cylindrical recess that accommodates a number of rings of packing around the shaft or shaft sleeve (Figs. 76 and 77). If sealing the box is desired, a lantern ring or seal cage (Fig. 78) is used that separates the rings of packing into approximately equal sections. The packing is compressed to give the desired fit on the shaft or sleeve by a gland that can be adjusted in an axial direction. The bottom or inside end of the box may be formed by the pump casing itself (see Fig. 73), a throat bushing (Fig. 76), or a bottoming ring (Fig. 77).

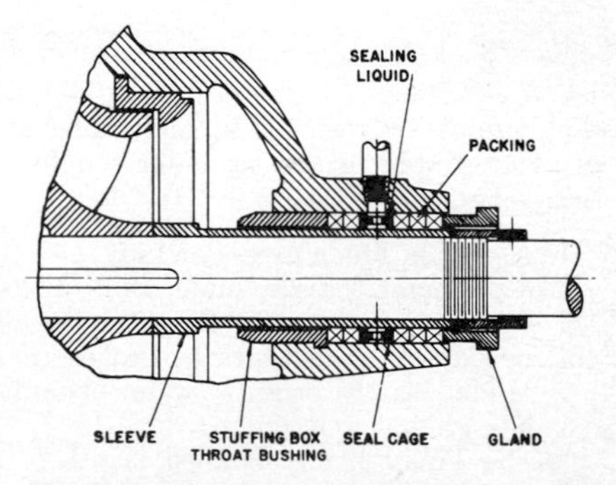

Fig. 76 Conventional stuffing box with throat bushing.

For manufacturing reasons, throat bushings are widely used on smaller pumps with axially split casings. Throat bushings are always solid rather than split. The bushing is usually held from rotation by a tongue-and-groove joint locked in the lower half of the casing.

Seal Cages When a pump operates with negative suction head, the inner end of the stuffing box is under vacuum, and air tends to leak into the pump. For this type of service, packing is usually separated into two sections by a lantern ring or seal cage (Fig. 76). Water or some other sealing fluid is introduced under

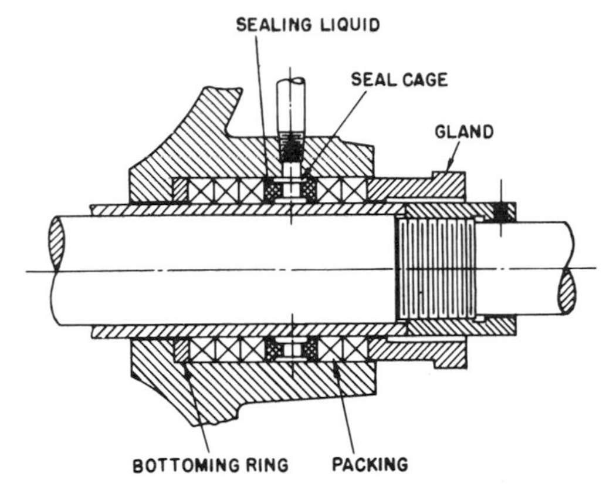

Fig. 77 Conventional stuffing box with bottoming ring.

pressure into the space, causing flow of sealing fluid in both axial directions. This construction is useful for pumps handling flammable or chemically active and dangerous liquids since it prevents outflow of the pumped liquid. Seal cages are usually axially split for ease of assembly.

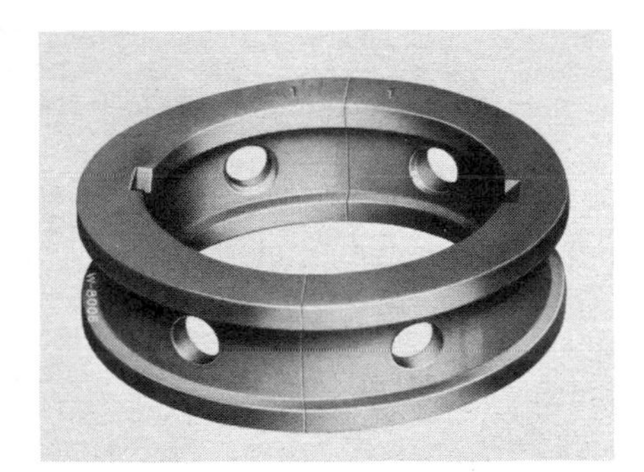

Fig. 78 Lantern gland or seal cage.

Some installations involve variable suction conditions, the pump operating part time with head on suction and part time with suction lift. When the operating pressure inside the pump exceeds atmospheric pressure, the liquid seal cage becomes inoperative (except for lubrication). However, it is maintained in service so that when the pump is primed at starting, all air can be excluded.

Sealing Liquid Arrangements When a pump handles clean, cool water, stuffing box seals are usually connected to the pump discharge, or, in multistage pumps, to an intermediate stage. An independent supply of sealing water should be provided if any of the following conditions exist:

1. A suction lift in excess of 15 ft
2. A discharge pressure under 10 lb/in^2 (or 23-ft head)
3. Hot water (over 250°F) being handled without adequate cooling (except for boiler feed pumps, in which seal cages are not used)
4. Muddy, sandy, or gritty water being handled.
5. For all hot-well pumps
6. The liquid being handled is other than water—such as acid, juice, molasses, or sticky liquids—without special provision in the stuffing box design for the nature of the liquid.

If the suction lift exceeds 15 ft, priming may be difficult unless an independent seal is provided because of excessive air infiltration through the stuffing boxes. A discharge pressure under 10 lb/in^2 may not provide sufficient sealing pressure. Hot-well (or condensate) pumps operate with as much as 28-in vacuum, and air infiltration would take place when the pumps are standing idle on standby service.

When sealing water is taken from the pump discharge, an external connection may be made through small diameter piping (Fig. 79) or internal passages. In some pumps these connections are arranged so that a sealing liquid can be introduced into the packing space through an internal drilled passage either from the pump casing or an external source (Fig. 80). When the liquid pumped is used for sealing,

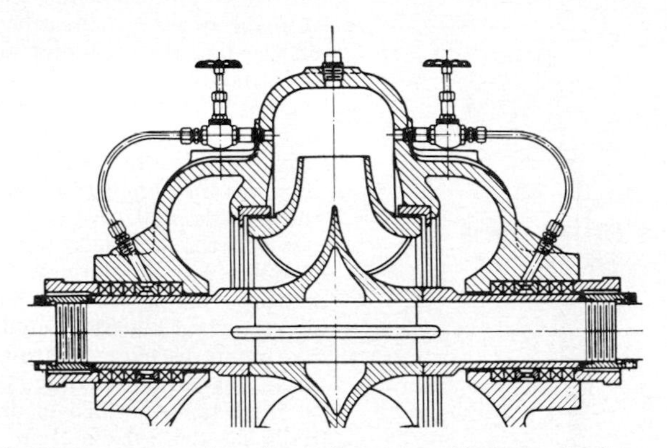

Fig. 79 Piping connections from the pump discharge to the seal cages.

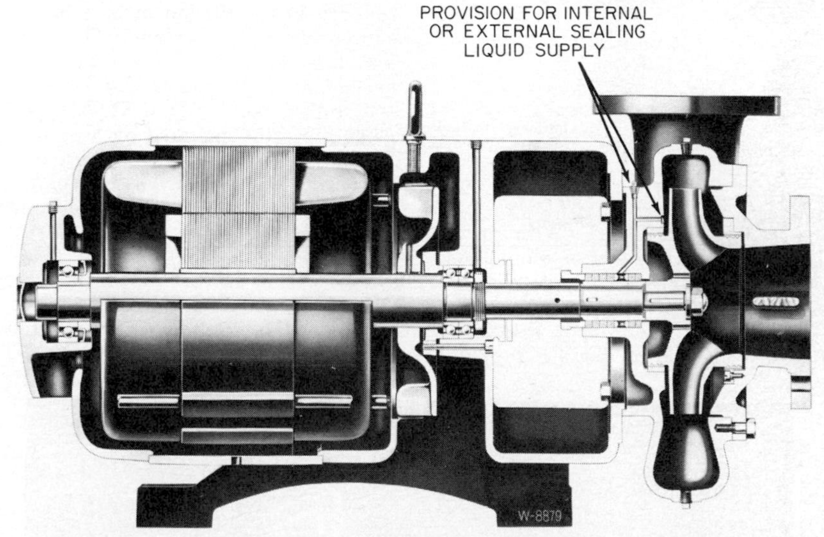

Fig. 80 End-suction pump with provision for internal or external sealing-liquid supply. (Worthington Pump, Inc.)

the external connection is plugged. If an external sealing liquid source is required, it is connected to the external pipe tap with a socket-head pipe plug inserted at the internal pipe tap.

It is sometimes desirable to locate the seal cage with more packing on one side than on the other. For example, on gritty-water service, a seal cage location closer to the inner portion of the pump would divert a greater proportion of sealing liquid into the pump, thereby keeping grit from working into the box. An arrangement with most of the packing rings between the seal cage and the inner end of the stuffing box would be applied to reduce dilution of the pumped liquid.

Some pumps handle water in which there are small, even microscopic, solids. Using water of this kind as a sealing liquid introduces the solids into the leakage path, shortening the life of the packing and sleeves. It is sometimes possible to remove these solids by installing small pressure filters in the sealing water piping from the casing to the stuffing box.

But filters ultimately get clogged unless they are frequently backwashed or otherwise cleaned out. This disadvantage can be obviated by using a cyclone (or centrifugal) separator. The operating principle of the cyclone separator is based on the fact that if liquid under pressure is introduced tangentially into a vortexing chamber, centrifugal force will make it rotate in the chamber, creating a vortex. Particles heavier than the liquid in which they are carried will tend to hug the outside wall of the vortexing chamber and the liquid in the center of the chamber will be relatively free of foreign matter. The action of such a separator is illustrated on Fig. 81. Liquid piped from the pump discharge or from an intermediate stage of a multistage pump is piped to inlet tap *A*, which is drilled tangentially to the cyclone bore. The liquid containing solids is directed downward to the apex of the cone at outlet tap *B* and is piped to the suction or to a low-pressure point in the system. The cleaned liquid is taken off at the center of the cyclone at outlet tap *C* and is piped to the stuffing box. Figure 82 shows a cyclone separator mounted on a pump.

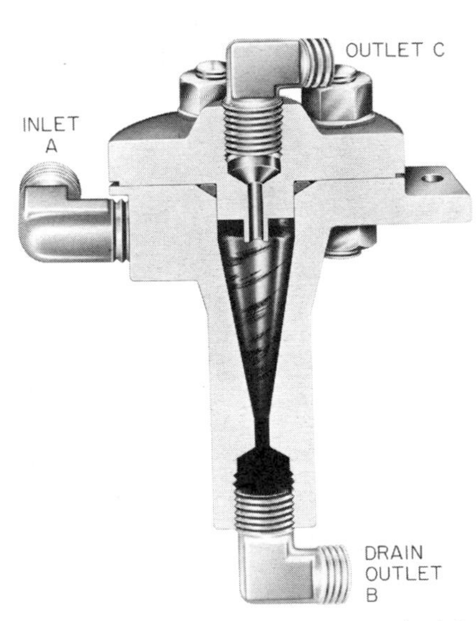

Fig. 81 Illustration of the principle of cyclone separators. (Borg-Warner Corp.)

Sand that will pass through a No. 40 sieve will be 100 percent eliminated in a cyclone separator, with supply pressures as low as 20 lb/in^2. With 100 lb/in^2 supply pressure, 95 percent of the particles of 5-micron size will be eliminated.

Most city ordinances require that some form of backflow preventer be interposed between city water supply lines and connections to equipment where backflow or siphoning could contaminate drinking water supply. This is the case, for instance, with an independent sealing supply to stuffing boxes of sewage pumps. Quite

Fig. 82 Cyclone separator mounted on a pump to supply clean sealing liquid. (Borg-Warner Corp.)

a variety of backflow preventers is available. In most cases, the device consists of two spring-loaded check valves in series and a spring-loaded, diaphragm-actuated, differential-pressure relief valve located in the zone between the check valves.

In normal operation, as long as there is a demand for sealing water, both check valves remain open. The differential-pressure relief valve remains closed because of the pressure drop past the first check valve. If the pressure downstream of the device increases, tending to reverse the direction of flow, both check valves close and prevent backflow. If the second check valve is prevented from closing tightly, the leakage past it increases the pressure between the two check valves, the relief valve opens, and water is discharged to atmosphere. Thus the relief valve operates automatically to maintain the pressure between the two check valves lower than the supply pressure.

Some local ordinances prohibit any connection between the city water line and a sewage or process liquid line. In such cases an open tank under atmospheric

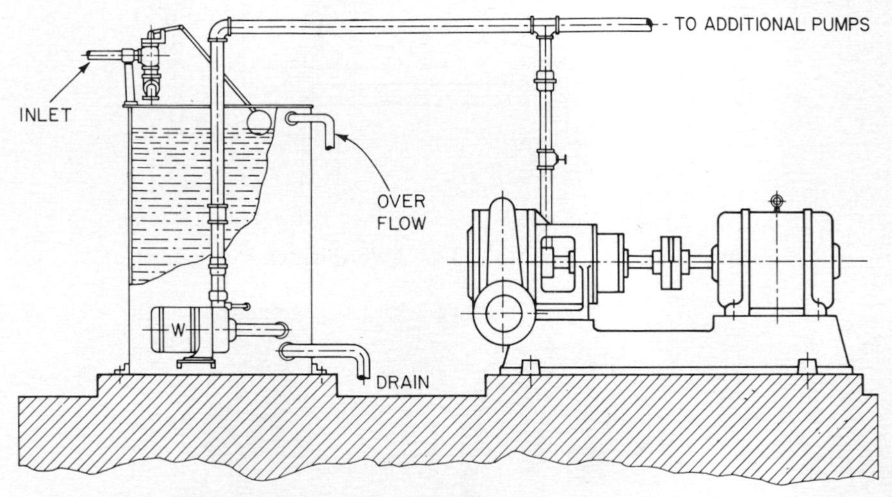

Fig. 83 Water seal unit. (Worthington Pump, Inc.)

pressure is installed into which city water can be admitted and from which a small pump can deliver the required quantity of sealing water. Such a water-sealing supply unit (see Fig. 83) can be installed in a location from which it can serve a number of pumps.

The tank is equipped with a float valve to feed and regulate the water level so that contamination of the city water supply is prevented. A small close-coupled pump is mounted directly on the tank and maintains a constant pressure of clear water at the stuffing box seals of the battery of pumps it serves. A small recirculation line is provided from the close-coupled pump discharge back to the tank to prevent operation at shutoff. The discharge pressure of the small supply pump is set by the maximum sealing pressure required at any of the pumps served. Supply at the individual stuffing boxes is then regulated by setting small control valves in each individual line.

If clean, cool water is not available (as with some drainage, irrigation, or sewage pumps), grease or oil seals are often used. Most pumps for sewage service have a single stuffing box subject to discharge pressure and operate with a flooded suction. It is therefore not necessary to seal these pumps against air leakage, but forcing grease or oil into the sealing space at the packing helps to exclude grit. Figure 84 shows a typical weighted grease sealer.

Automatic grease or oil sealers that exert pump discharge pressure in a cylinder on one side of a plunger, with light grease or oil on the other side, are available for sewage service. The oil or grease line is connected to the stuffing box seal, which is at about 80 percent of the discharge pressure. As a result, there is a slow flow of grease or oil into the pump when the unit is in operation. No flow takes place when the pump is out of service. Figure 85 shows an automatic grease sealer mounted on a vertical sewage pump.

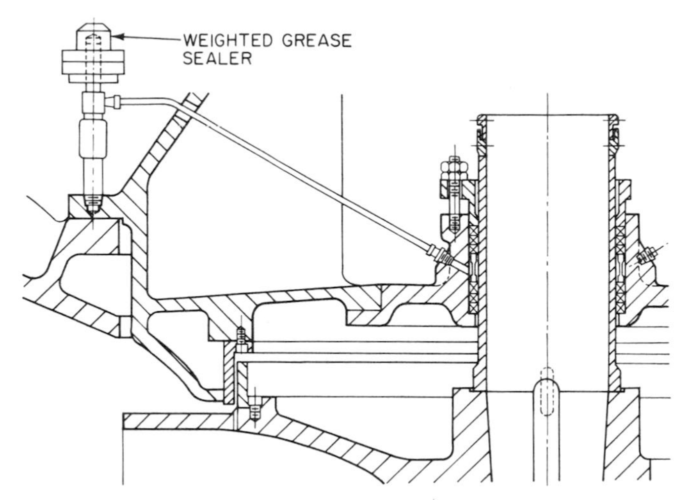

Fig. 84 Weighted grease sealer. (Worthington Pump, Inc.)

Fig. 85 Automatic grease sealer mounted on a vertical sewage pump. (Zimmer & Francescon)

Water-cooled Stuffing Boxes High temperatures or pressures complicate the problem of maintaining stuffing box packing. Pumps in these more difficult services are usually provided with jacketed, water-cooled stuffing boxes. The cooling water removes heat from the liquid leaking through the stuffing box and heat generated by friction in the box, thus improving packing service conditions. In some special cases, oil or gasoline may be used in the cooling jackets instead of water. Two water-cooled stuffing box designs are commonly used. The first (Fig. 86) provides cored passes in the casing casting. These passages, which surround the stuffing box, are arranged with in-and-out connections. The second type uses a separate

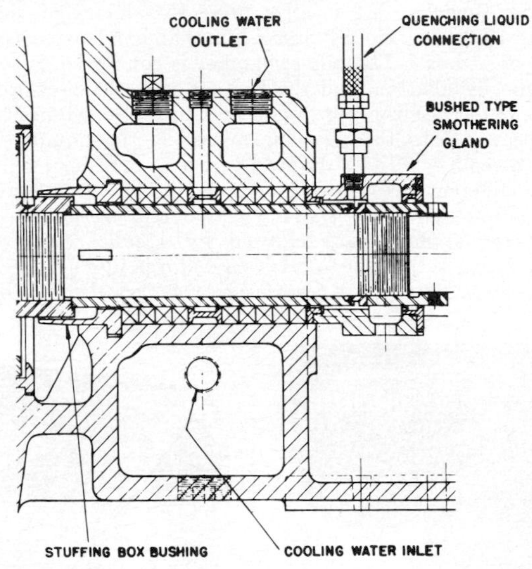

Fig. 86 Water-cooled stuffing box with cored water passage cast in casing.

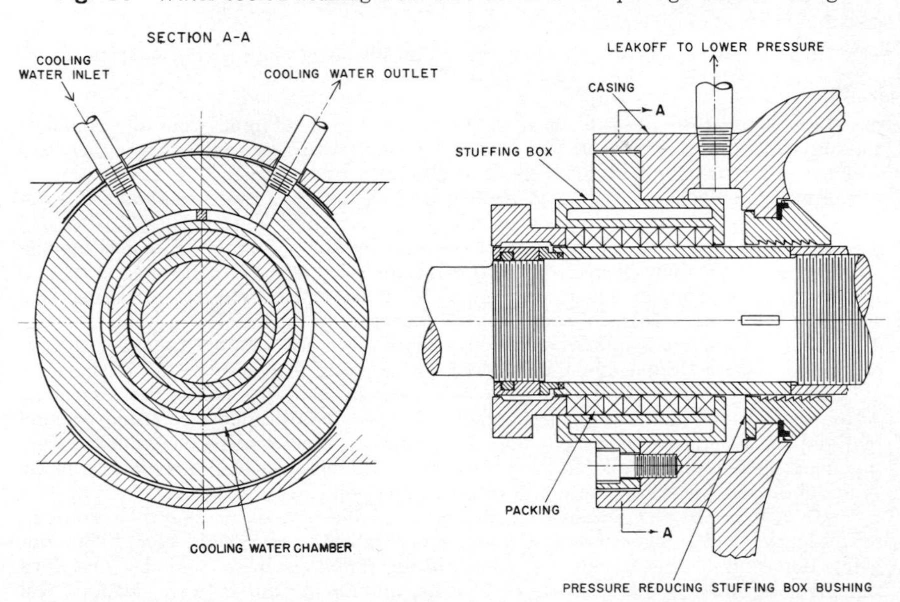

Fig. 87 Separate water-cooled stuffing box with pressure-reducing stuffing box bushing.

cooling chamber combined with the stuffing box proper, with the whole assembly inserted into and bolted to the pump casing (Fig. 87). The choice between the two is based on manufacturing preferences.

Stuffing box pressure and temperature limitations vary with the pump type, because it is generally not economical to use expensive stuffing box construction for infrequent high-temperature or high-pressure applications. Therefore, whenever the manufacturer's stuffing box limitations for a given pump are exceeded, the only solution is the application of pressure-reducing devices ahead of the stuffing box.

Pressure-reducing Devices Essentially, pressure-reducing devices consist of a bushing or meshing labyrinth, ending in a relief chamber located between the pump interior and the stuffing box. The relief chamber is connected to some suitable low-pressure point in the installation, and the leakage past the pressure-reducing device is returned to this point. If the pumped liquid must be salvaged, as with treated feedwater, it is returned into the pumping cycle. If the liquid is expendable, the relief chamber can be connected to a drain.

There are many different pressure-reducing device designs. Figure 87 illustrates a design for limited pressures. A short, serrated stuffing box bushing is inserted at the bottom of the stuffing box, followed by a relief chamber. The leakage past the serrated bushing is bled off to a low-pressure point.

With relatively high-pressure units, intermeshing labyrinths may be located follow-

Fig. 88 Graphited asbestos packing in continuous coil form. (John Crane Co.)

Fig. 89 Metallic packing in spiral form. (John Crane Co.)

ing the balancing device and ahead of the stuffing box. Piping from the chamber following pressure-reducing devices should be amply sized so that as wear increases leakage, piping friction will not increase stuffing box pressure.

Stuffing Box Packing Basically, stuffing box packing is a pressure-breakdown device. The packing must be somewhat plastic so that it can be adjusted for proper operation. It must also absorb energy without failing or damaging the rotating shaft or shaft sleeve. In a breakdown of this nature, friction energy is liberated. This generates heat that must be dissipated in the fluid leaking past the breakdown or by means of cooling-water jacketing or both.

There are numerous stuffing box packing materials, each adapted to some particular class of service. Some of the principal types are the following:

1. *Asbestos packing:* Comparatively soft and suitable for cold-water and hot-water applications in the lower temperature range. It is the most common packing material for general service under normal pressures. For pressures above 200 lb/in^2, this packing is only usable at very moderate rubbing speeds. Asbestos packing is prelubricated with either graphite or some inert oil.

2. *Metallic packing:* Composed of flexible metallic strands or foil with graphite or oil-lubricant impregnation and with either asbestos or plastic core. The impregnation makes this packing self-lubricating for its startup period. The foils are made of various metals such as babbitt, aluminum, and copper. Babbitt foil is used on water and oil service for low and medium temperatures (up to 450°F) and medium to high pressures. Copper is used for medium to high temperatures and pressures with water and low sulphur-content oils. Aluminum is used mainly on oil service and for medium to high temperatures and pressures.

Many other types of packing are regularly furnished to meet customers' special specifications, for example, hemp, cord, braided type, duck fabric, chevron type, and numerous others.

Packing is supplied either in continuous coils of square cross-section or in preformed die-molded rings. When coil-type packing (Figs. 88 and 89) is used, it is cut in lengths that make up individual rings. The ends are cut with a diagonal, or scarf joint, and with a slight clearance to provide for expansion and avoid buckling. The rings have a tendency to swell from the liquid action and the rise in temperature. The scarf joint allows the end to slide and laterally absorb expansion.

It is preferable, where possible, to use die-molded packing rings (Fig. 90) which

Fig. 90 Metallic packing in ring form. (John Crane Co.)

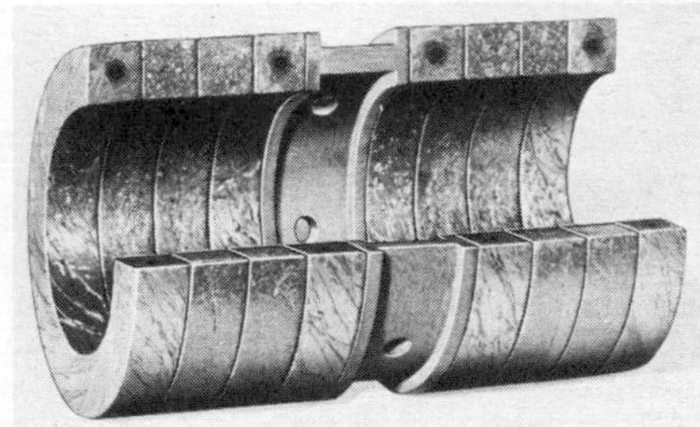

Fig. 91 Combination set of hard and soft packing. (John Crane Co.)

are available to exact size and in sets. A molded ring ensures an exact fit to the shaft or shaft sleeve and to the stuffing box bore, and also establishes equal packing density throughout the stuffing box.

Frequently more efficient packing life can be gained by a combination of two or more different kinds of packing, for example, alternating hard and soft rings (Fig. 91). Such sets are usually available in standard die-formed ring combinations from most established packing manufacturers.

For best results, the shaft or shaft sleeves should be in perfect alignment, concentric with the axis of rotation, highly polished, and should operate without vibration. The material of which they are made is also extremely important, as it directly affects the life and maintenance of the packing.

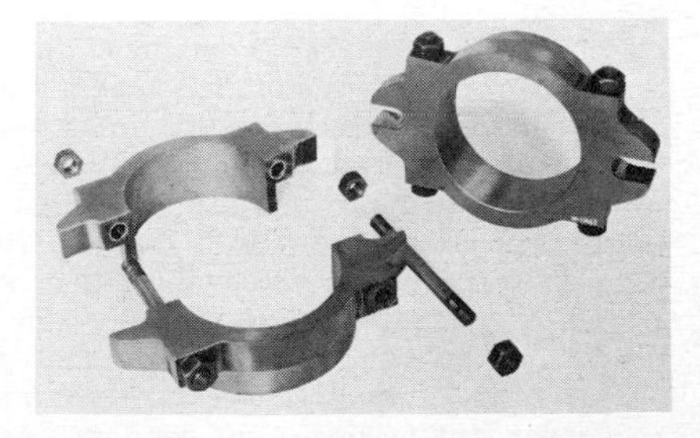

Fig. 92 Split stuffing box gland.

Stuffing Box Glands Stuffing box glands may assume several forms, but basically they can be classified into two groups: solid glands and split glands (Fig. 92). Split glands are made in halves so that they may be removed from the shaft without dismantling the pump, thus providing more working space when the stuffing boxes are being repacked. Split glands are desirable for pumps that have to be repacked frequently, especially if the space between the box and the bearing is restricted. The two halves are generally held together by bolts, although other methods are also used. Split glands are generally a construction refinement rather than a necessity, and they are rarely used in smaller pumps. They are commonly furnished for large single-stage pumps, for some multistage pumps, and for refinery pumps. Another common refinement is the use of swing bolts in stuffing box glands. Such bolts may be swung to the side, out of the way, when the stuffing box is being repacked.

Stuffing box leakage into the atmosphere might, in some services, seriously inconvenience or even endanger the operating personnel—for example, when such liquids as hydrocarbons are being pumped at vaporizing temperatures or temperatures above their flash point. As this leakage cannot always be cooled sufficiently by a water-cooled stuffing box, smothering glands are used (see Fig. 86). Provision is made in the gland itself to introduce a liquid—either water or another hydrocarbon at low temperature—that mixes intimately with the leakage, lowering its temperature, or, if the liquid is volatile, absorbing it.

Stuffing box glands are usually made of bronze, although cast iron or steel may be used for all iron-fitted pumps. Iron or steel glands are generally bushed with a nonsparking material like bronze in refinery service to prevent the ignition of flammable vapors by the glands sparking against a ferrous metal shaft or sleeve.

MECHANICAL SEALS

The conventional stuffing box design and composition packing are impractical to use as a method for sealing a rotating shaft for many conditions of service. In the ordinary stuffing box, the sealing between the moving shaft or shaft sleeve and the stationary portion of the box is accomplished by means of rings of packing forced between the two surfaces and held tightly in place by a stuffing box gland. The leakage around the shaft is controlled merely by tightening up or loosening the gland studs. The actual sealing surfaces consist of the axial rotating surfaces of the shaft, or shaft sleeve, and the stationary packing. Attempts to reduce or eliminate all leakage from a conventional stuffing box increase the gland pressure. The packing, being semiplastic in nature, forms more closely to the shaft and tends to cut down the leakage. After a certain point, however, the leakage continues no matter how tightly the gland studs are brought up. The frictional horsepower increases rapidly at this point, the heat generated cannot be properly dissipated, and the stuffing box fails to function. Even before this condition is reached the shaft sleeves may be severely worn and scored, so that it becomes impossible to pack the stuffing box satisfactorily.

These undesirable characteristics prohibit the use of packing as the sealing medium between rotating surfaces if the leakage is to be held to an absolute minimum under severe pressure. The condition, in turn, automatically eliminates use of the axial surfaces as the sealing surfaces, for a semiplastic packing is the only material that can always be made to form about the shaft and compensate for the wear. Another factor that makes stuffing boxes unsatisfactory for certain applications is the relatively small lubricating value of many liquids frequently handled by centrifugal pumps, such as propane or butane. These liquids actually act as solvents of the lubricants normally used to impregnate the packing. Seal oil must therefore be introduced into the lantern gland or a packed box to lubricate the packing and give it reasonable life. With these facts in mind, designers have attempted to produce an entirely different type of seal with wearing surfaces other than the axial surfaces of the shaft and packing.

This form of seal, called the mechanical seal, is a later development compared to regular stuffing boxes but has found general acceptance in those pumping applications in which the shortcomings of packed stuffing boxes have proved excessive. Fields in which the packed boxes gave good service, however, have shown little tendency to replace them with mechanical seals.

As both packed boxes and mechanical seals are subject to wear, neither are perfect. One or the other proves to be the better according to the application. In some fields both give good service, and choosing between them becomes a matter of personal preference or first cost.

Principles of Mechanical Seals Although they may differ in various physical respects, all mechanical seals are fundamentally the same in principle. The sealing surfaces of every kind are located in a plane perpendicular to the shaft and usually consist of two highly polished surfaces running adjacently, one surface being connected to the shaft and the other to the stationary portion of the pump.

Complete sealing is accomplished at the fixed members. The polished or lapped surfaces, which are of dissimilar materials, are held in continual contact by a spring, forming a fluidtight seal between the rotating and stationary members with very small frictional losses. When the seal is new, the leakage is negligible and can actually be considered as nonexistent. (To obtain a pressure breakdown between the internal pressure and the atmospheric pressure outside the pump, a flow of liquid past the seal faces is required. This flow may be only a drop every few minutes or even a haze of escaping vapor, if a liquid such as propane is being handled. Thus, even though leakage is negligible, technically speaking a rotating mechanical seal cannot *entirely* eliminate it.) Of course, some wear always occurs, and provision must be made for a very small amount of leakage in time.

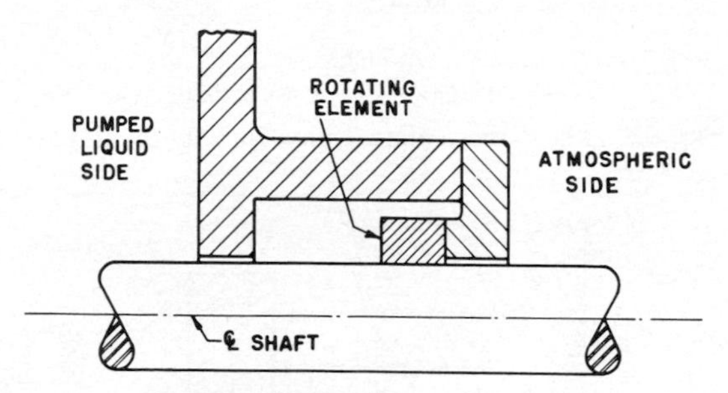

Fig. 93 Internal assembly seal.

The wide variation in seal design stems from the many methods used to provide flexibility and to mount the seals. A mechanical seal is similar to a bearing in that it involves a close running clearance with a liquid film between the faces. The lubrication and cooling provided by this film cut down the wear, as does a proper choice of seal face materials.

Seals for centrifugal pumps do not operate satisfactorily on air or gas; if run "dry," they will fail rapidly. Seals can be used in pumps handling liquids that contain solids if the solids are prevented from getting between the seal faces or interfering with the flexibility of the mounting.

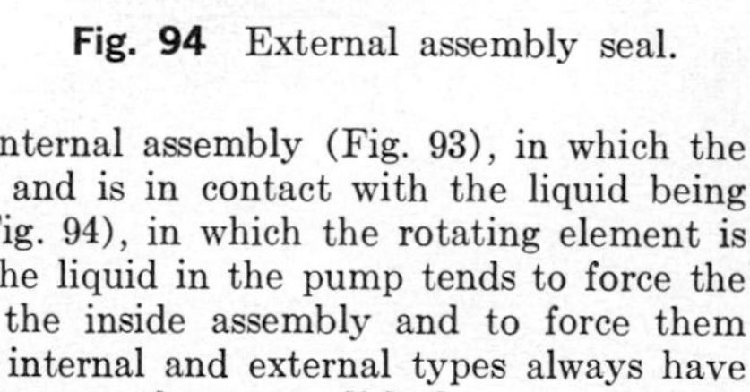

Fig. 94 External assembly seal.

Internal and External Seals There are two basic seal arrangements: (1) the internal assembly (Fig. 93), in which the rotating element is located inside the box and is in contact with the liquid being pumped, and (2) the external assembly (Fig. 94), in which the rotating element is located outside the box. The pressure of the liquid in the pump tends to force the rotating and stationary faces together in the inside assembly and to force them apart in the external assembly. But both internal and external types always have three primary points (Fig. 95) at which sealing must be accomplished:

1. Between the stationary element and the casing
2. Between the rotating element and the shaft (or the shaft sleeve, if one is used)
3. Between the mating surfaces of the rotating and stationary seal elements

To accomplish the first seal, conventional gaskets or some form of a synthetic O ring are used. Leakage between the rotating element and the shaft is stopped by means of O rings, bellows, or some form of flexible wedges. Leakage between the mating surfaces cannot be entirely stopped but can be held to an insignificant amount by maintaining a very close contact between these faces.

Balanced and Unbalanced Seals The pressure within the pump just ahead of the mechanical seal tends to keep the mating faces of the internal seal together. In the simplest design (as in Fig. 93), the entire internal pressure acts to close the faces. If the liquid handled is a good lubricant and the pressures are not excessive, this loading will not be harmful. The design is known as the unbalanced seal. A graphic description of the forces and area relationships in this seal is given

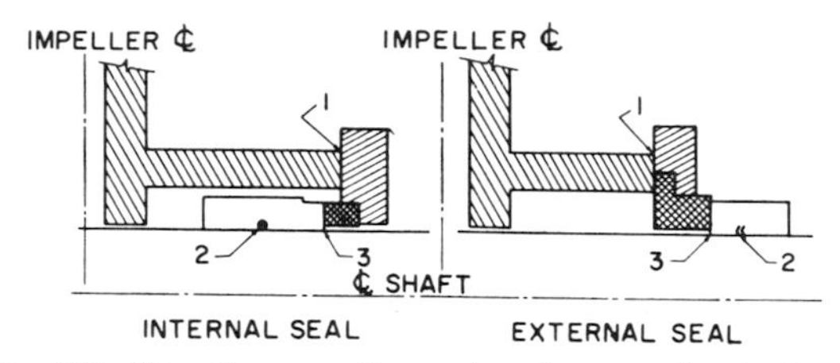

Fig. 95 The three sealing points in a mechanical seal.

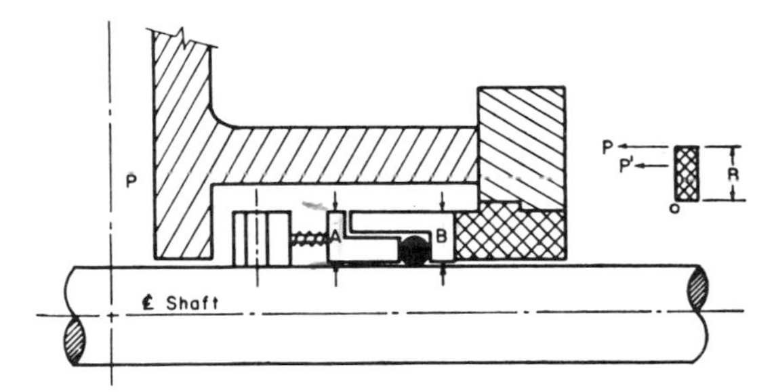

Fig. 96 Unbalanced seal construction.

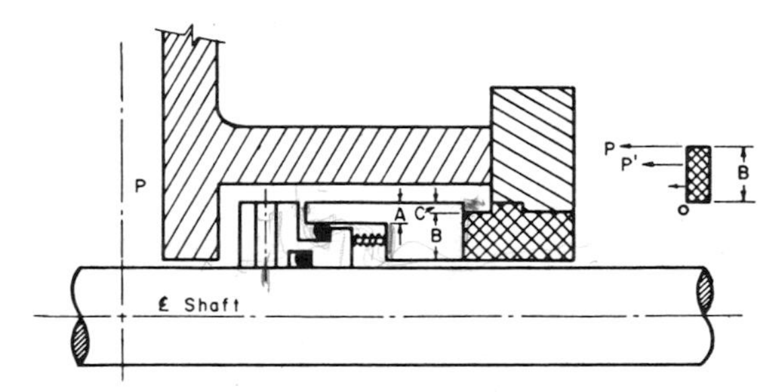

Fig. 97 Balanced seal construction.

in Fig. 96. If P = pressure of liquid in the box, and P' = average pressure across the seal faces, then

$$\text{Closing force} = (P)\ (\text{area } A) + \text{spring loading}$$
$$\text{Opening force} = (P')\ (\text{area } B)$$

Generally the application of unbalanced seals is limited to pressures lower than 100 to 150 lb/in² gage and to liquids with lubricating properties equal to or better than gasoline.

When these criteria are not met, it has been found preferable to so proportion the areas subject to pressure as to reduce the loading on the mating faces, providing what is known as a *balanced seal* (Fig. 97). Although such seals have been applied very successfully for quite high internal pressures, they are not particularly suitable for low pressures (under 50 lb/in² gage) as the sealing force is reduced to a point at which contact between the mating faces may not be sufficient to provide adequate sealing. If P and P' are the quantities described above, then

$$\text{Closing force} = (P)\ (\text{area } A - \text{area } C) + \text{spring loading}$$
$$\text{Opening force} = (P')\ (\text{area } B)$$

Double Seals Two mechanical seals may be mounted inside a stuffing box to make a *double seal* assembly, shown diagrammatically in Fig. 98. Such an arrangement is used for pumps handling toxic or highly inflammable liquids that cannot be permitted to escape into the atmosphere. It is also applicable to pumps handling corrosive or abrasive liquids at very high or very low temperatures. A clear, filtered, and generally inert sealing liquid is injected between the two seals

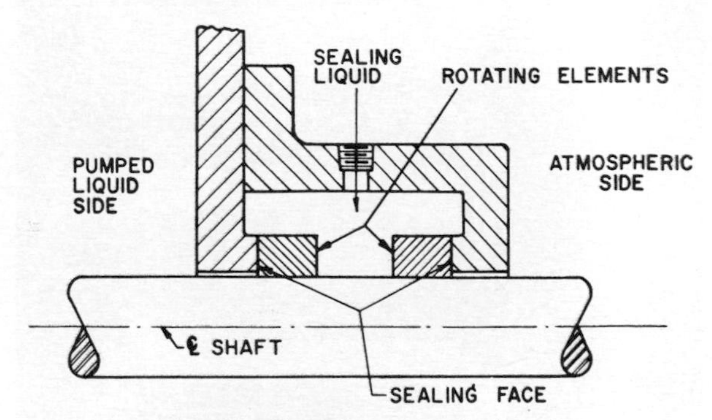

Fig. 98 Double mechanical seal.

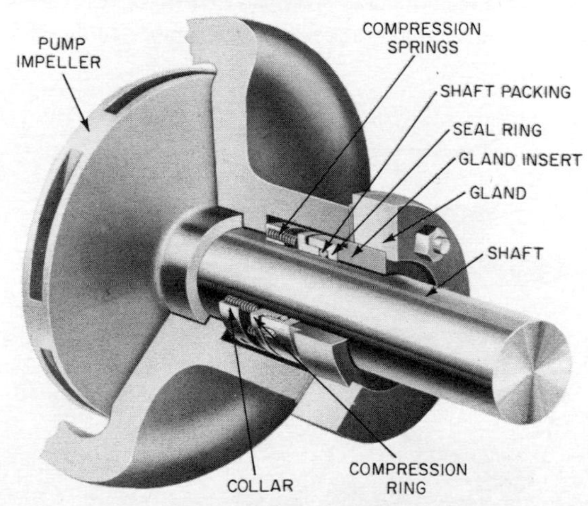

Fig. 99 Dura-seal mechanical seal. (Durametallic Co.)

at a pressure slightly in excess of the pressure in the pump ahead of the seal. This liquid prevents the pumped liquid from coming into contact with the seal parts or from escaping into the atmosphere.

Seal Designs Some manufacturers of seals have found it advisable to provide the prospective user with a very complete engineering service in order to ensure that every application is given the most thorough analysis. Such an analysis is essential to the success of a mechanical-seal appplication, and users should take full advantage of this service.

The operation of a mechanical seal can best be understood by reference to a few standard commercial units. The following discussion treats of the general characteristics of several typical mechanical seals.

One typical seal construction is illustrated in Fig. 99. The gland with its gland insert is fitted into the casing and constitutes the stationary member of the seal assembly. It provides a seal at two points: (1) between the gland and the face of the stuffing box, by means of the gaskets, and (2) between the gland insert and the seal-ring face, by contact. The mating seal ring with a hardened steel surface rotates with the shaft and is held against the stationary member by the compression ring. The latter supports a nest of springs that are connected at the opposite end to the collar, which, in turn, is fixed to the shaft. The seal

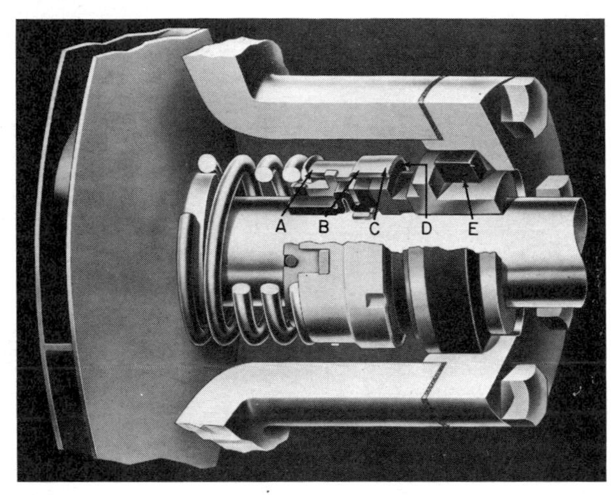

Fig. 100 Mechanical seal (rubber bellows type). (John Crane Co.)

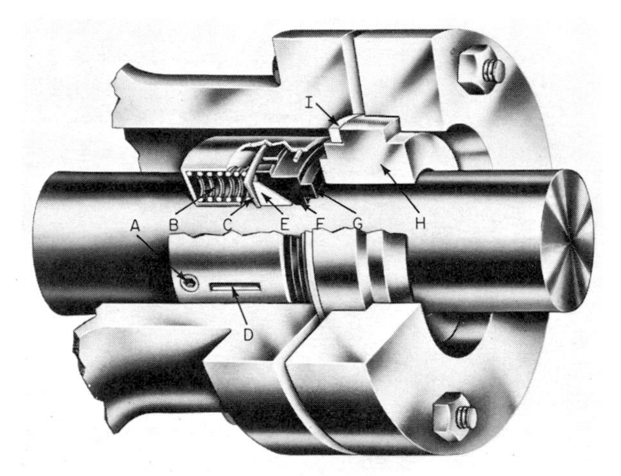

Fig. 101 Mechanical seal (Teflon type). (John Crane Co.)

ring is fitted with packing in the shape of an O ring, which prevents all leakage between the seal ring and the shaft. It is essential in such a seal that one face be flexibly mounted so as to keep the surfaces in full contact with reasonable shaft deflection. The springs and an ample clearance between the shaft and the seal ring proper accomplish this end in the unit illustrated. At the same time, the collar keeps all the other rotating members of the seal in proper position.

The gland insert in contact with the seal ring is made of antifrictional material. When necessary, it can be designed for lubrication by a liquid other than the liquid pumped.

A typical spring-loaded synthetic-rubber-bellows mechanical seal is illustrated in Fig. 100. The tail of the synthetic rubber bellows (*A*) seals against leakage between the rotating element and the shaft; the head is flexible and adjusts automatically for washer wear and shaft end play. The protective ferrule (*B*) prevents the flexing area of the bellows from sticking to the shaft. The sealing washer (*C*) has a positive drive through metal parts and seals against the stationary floating seat (*E*). The two sealing faces (*D*) are lapped at the factory and provide a seal against leakage.

For extremely high or extremely low temperature applications (for instance, the circulation of a very low temperature refrigerant), the bellows seal is not satisfactory. A Teflon seal is manufactured for this type of service (Fig. 101). The metal retainer locked to the shaft by set screws (*A*) provides a positive drive from the shaft to the carbon sealing washer (*F*) through dents (*D*), which fit into corresponding washer grooves. The seal between the shaft and the washer is insured by the Teflon wedge ring (*E*), which is preloaded by the action of multiple springs (*B*). The spring pressure is uniformly distributed by a metal disk (*C*). The lapped raised face of the rotating sealing washer mates against the precision-lapped face (*G*) of the stationary seal (*H*) to provide a positive

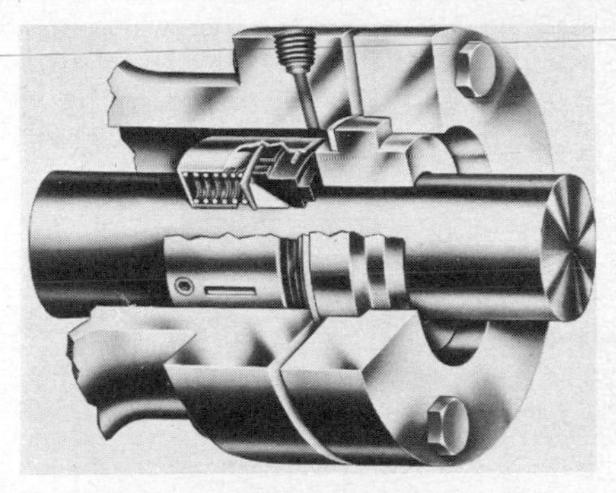

Fig. 102 Mechanical seal (Teflon type) with a connection for liquid lubrication of the seal faces. (John Crane Co.)

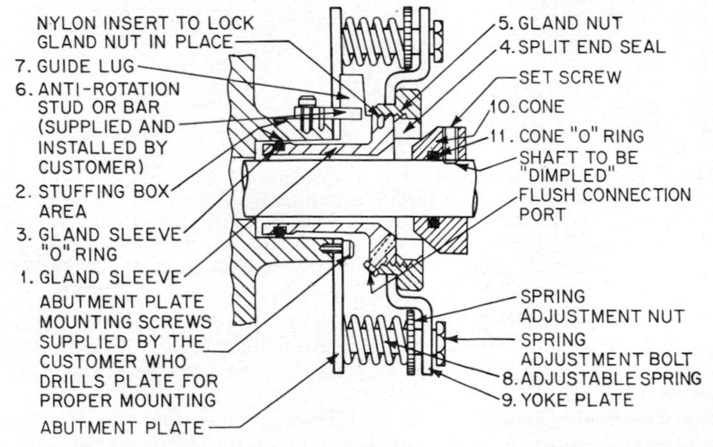

Fig. 103 External mechanical seal for slurry or sewage service.

leakproof seal with minimum running friction between the faces. The spring pressure keeps the faces in constant contact, providing automatic adjustment for wear and shaft end play. A Teflon ring (*I*) acts as a static seal between the stationary seat and the end play. When this seal is applied to pumps on vacuum service or operating with a high suction lift, a supply of lubricating liquid is provided through a connection (Fig. 102). If the liquid pumped is clear, this tap can be connected to the discharge of the pump. If the liquid pumped contains particles of foreign matter, it becomes necessary to use a cyclone separator (see Figs. 81 and 82).

Figure 103 illustrates an external seal very popular for sewage or slurry service. The entire seal assembly is outside the stuffing box, so that the split-end seal can be readily replaced without dismantling the pump or installing new shaft sleeves. The gland sleeve (1) is fitted into the stuffing box (2) and is sealed against the stuffing box bore by O ring (3). The split-end seal (4) is fitted onto the gland sleeve and held in place by the nut (5). An antirotation stud (6) is fitted to the stuffing box and passes between the guide lugs (7) on the gland sleeve. This entire assembly moves axially within the stuffing box according to the pressure exerted by the adjustable springs (8) acting on the gland nut and the abutment plate through the yoke plate (9). These springs force the end seal against the cone (10) which rotates with the shaft and is sealed by the cone O ring (11) to provide the running seal.

A number of pump manufacturers have lines of single-stage end-suction pumps that can be equipped with mechanical seals instead of conventional packing. In many cases, such seals are offered as a standard alternative construction and can be applied to the pumps without any change in machining. A few manufacturers build some sizes of standard pumps with mechanical seals only, but such practice makes these pumps less flexible as a line.

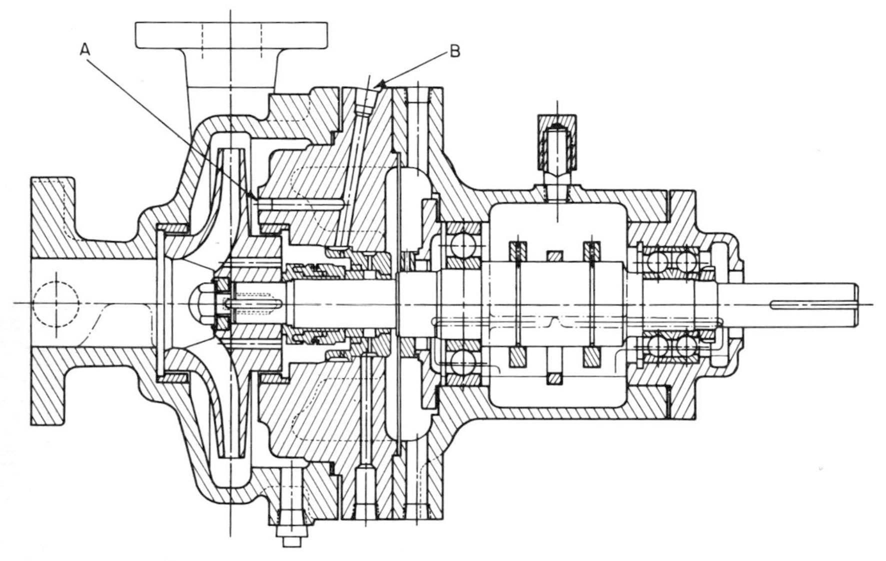

Fig. 104 Cooling circulation provisions to mechanical seal. (*A*) Use this circulation for cooling seal faces when clean liquid is pumped. For this installation, ½-in pipe must be plugged. (*B*) Provide ½-in piping from discharge for cooling seal faces when liquid contains gritty particles. A strainer should be installed for keeping these particles out of seal area. For this installation, ¼-in pipe must be plugged.

Cooling the Seal Cooling of the seal faces is important for satisfactory seal life, and a seal installed inside a pump without a suitably directed flow of liquid for cooling and flushing (to prevent solid material from settling on the springs, for instance) may have a high failure incidence.

A number of different methods are used to provide cooling and flushing. Sometimes all that may be necessary is to direct some of the pumped liquid at the sealing faces. When the pumped liquid is not suitable for this purpose, or when it must first be filtered, external circulation must be provided. Standard lines of pumps arranged for mechanical seals are therefore generally built in such a manner that either type of cooling circulation can be provided (Fig. 104).

In both methods the circulation is taken from a pressure higher than that in the stuffing box. This increased pressure provides positive circulation and prevents flashing at the seal faces caused by the heat generated by the seal when the pump handles liquids near their boiling point.

When the pumping temperatures reach 350°F, it is advisable to provide some means for cooling the chamber surrounding the seal as well. A typical arrangement for accomplishing this is shown in Fig. 105. The seal is equipped with a water-circulator ring and a heat-exchanger hookup to the seal chamber. The water-circulator ring acts as a miniature pump, causing the water to flow through the outlet piping located at the top of the seal chamber. The water then passes through the heat exchanger from which it returns directly to the seal faces through the bottom inlet at the end plate. As the water circulates back through the circulation ring, it picks up heat from the pump housing and the shaft. Because this is a closed circulating system, none of the hot pumped water enters the seal chamber.

Limitations on Seal Applications The substitution of mechanical seals for packed stuffing boxes is not always an unmixed blessing and for some services the mechani-

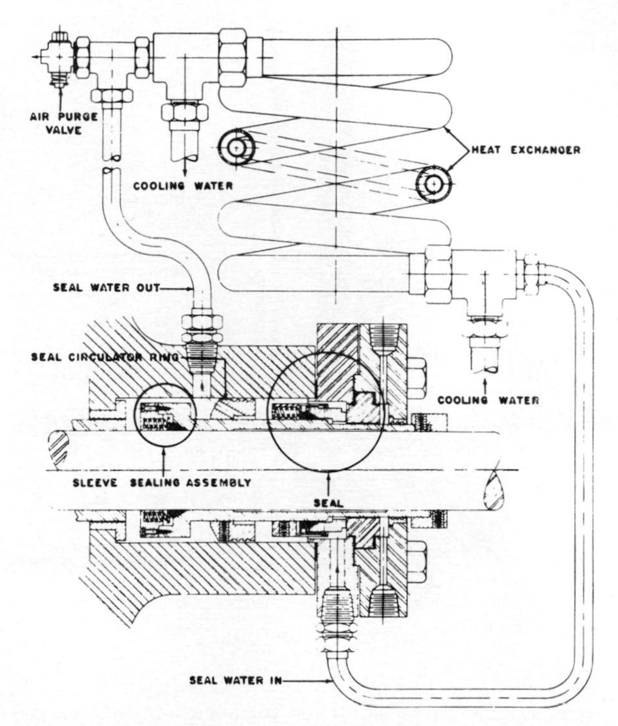

Fig. 105 Mechanical seal with external cooling arrangement. (John Crane Co.)

cal seal is not as desirable as packing, including those with conditions tending to cause the pumped liquid to form crystals after temperature changes or on settling. If mechanical seals are used for such services, it is very important to provide adequate flushing. Another condition unfavorable to mechanical seals is a pump service with long idle periods, when the pump may even be drained. The flexible materials used in the seal may harden or slight rusting may occur, thereby possibly causing the seal to stick and become damaged on restarting. Finally, mechanical seals are still subject to failure on occasion, and their failure may be more rapid than that of conventional packing. If packing fails, the pump can usually be kept running by temporary adjustments until it is convenient to shut it down. If a mechanical seal fails, the pump must be shut down at once in nearly every case.

CONDENSATE INJECTION SEALING

Condensate injection sealing (sometimes called *packless stuffing box*) has been a very successful solution of the stuffing box problem for high-speed boiler feed pumps where neither conventional packing nor mechanical seals provide a satisfactory answer. The construction of a pump with condensate injection sealing (Fig. 106) involves the substitution of a serrated breakdown bushing for the conventional packing. The pump shaft sleeve runs within this bushing with a reasonably small radial clearance. Cold condensate, at a pressure in excess of the boiler feed pump suction pressure, is introduced centrally in this breakdown bushing. A small portion of the injection water flows inwardly into the pump proper; the remainder flows out into a collecting chamber that is vented to the atmosphere. From this chamber the leakage is piped back to the condenser.

Cold condensate (at temperatures from 80 to 100°F) is available at pressures in excess of the boiler feed pump suction pressure in closed cycles as well as open

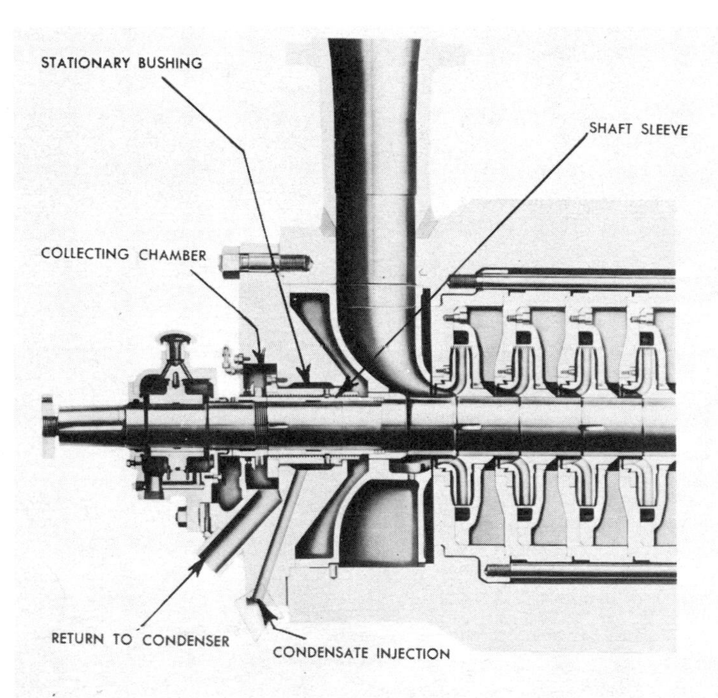

Fig. 106 Packless stuffing box construction for a modern high-pressure boiler feed pump.

cycles in which the condensate pump discharges into a deaerating heater from which the pump takes its suction. The water for the injection in both should be taken immediately from the condensate pump or condensate booster pump discharge before it has gone through any closed heaters. The amount of the injection water will depend upon (1) the diameter of the running joint between the shaft sleeve and the pressure breakdown bushing, (2) the clearance at that running joint, and (3) the injection pressure. To give some general idea of the values in question, if the sleeve diameter is 5 in and the diametral clearance 0.009 in, the amounts measured in a 3,600 rpm pump will be approximately as follows:

1. Total injection per box, 8 to 10 gpm
2. Leakage into the pump interior per box, 2 to 4 gpm
3. Return to condenser per box, 6 to 8 gpm

The injection supply must be absolutely clear and free of foreign matter. It is therefore necessary to install filters or strainers in the injection line to avoid the entrance of fine mill scale or oxide particles into the close clearances between the stationary bushings and the sleeves. Pressure gages should be installed upstream and downstream of these filters to permit the operator to follow the rate at which foreign matter clogs up the filters and to clean these when the pressure drop across them becomes excessive.

Drains from Condensate Injection Sealing Two different systems are used to dispose of the drains coming from the collecting chambers. The first utilizes traps that drain directly to the condenser. The second collects the drains in a condensate storage tank into which various other drains are also returned. As this tank is under atmospheric pressure, it must be set at a reasonable elevation below the pump centerline so that the static elevation difference will overcome friction losses in the drain piping. A pump then transfers the condensate drains from the storage tank into the condenser.

The clearances between the sleeves and the breakdown bushings will double

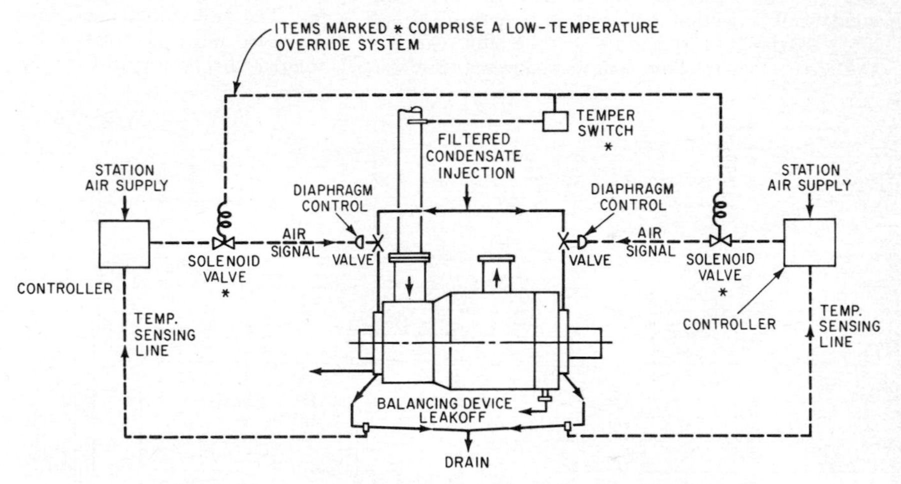

Fig. 107 Temperature control of condensate injection.

in a time approximately equal to the life of the internal wearing parts. With double clearances, the leakage will double. This factor should be considered when sizing the return-drain piping back to the condenser or to the collecting tank if friction losses are to be kept to a minimum in this piping. The collecting chamber at the pump stuffing box is vented to the atmosphere; the only head available to evacuate it is the static head between the pump and the point of return. This head must always be well in excess of the frictional losses (even after the leakage doubles), otherwise the drains will back up and run off at the collecting chamber.

Control of the Injection The variation in supply pressure to the condensate injection compared to the internal pressures at both ends of the pump makes it necessary to use a control for satisfactory seal operation. Three possible types of flow control are (1) manual, (2) pressure differential, and (3) temperature. A manual flow control requires hand-setting and readjusting a valve for each seal. This necessitates the availability of a man to recheck the setting, which varies with the load. While it is possible to use such a control it is not normally recommended.

The pressure-differential control senses the differences between injection pressure and boiler feed pump internal pressure and, by sending a signal to an automatic valve in the injection line to each seal, maintains the differential at some predetermined setting, usually 10 lb/in^2. This system tends to be unstable because a change in pressure affects valve position which changes pressure, and so on.

The temperature-sensitive control (Fig. 107) operates on signals received from a temperature-sensing probe in each seal drain line. A controller transmits an air signal to a pneumatic control valve in the injection line. In addition to providing rapid response to variations in operating conditions, this type of control will use the least amount of injection water. It will always supply just enough injection water to keep the drains at the recommended temperature range of 140–150°F. The injection valves are equipped with limit stops so that they cannot close fully regardless of the air signal.

Mechanical Modifications A number of mechanical modifications have been developed by various pump manufacturers. One of these (Fig. 108) consists of substituting a stack of individual solid rings for the serrated breakdown bushing shown in Fig. 106. Each ring is mounted in a holder, spring-loaded to produce a stationary seal face in an axial direction, and locked against rotation by a pin-and-slot arrangement. A small radial clearance is provided between the rings and the shaft sleeve. The length of each ring varies with the diameter of the condensate injection seal but is generally about ½ in. The individual seal rings are "floating" to a certain degree and can find their own position relative to the shaft. Their short length reduces the effect of angular displacement between

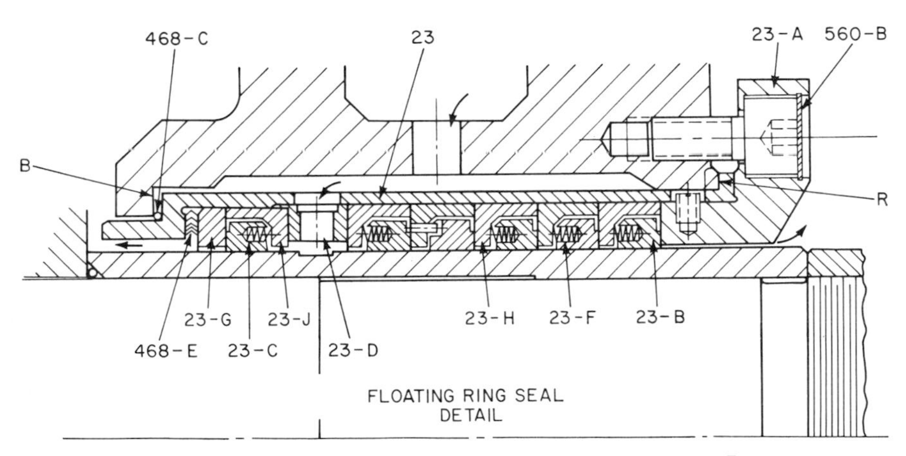

Fig. 108 Floating seal ring design. (Worthington Pump, Inc.)

the stationary and rotating components, whether this displacement arises from errors in original assembly or from distortions caused by temperature changes.

The condensate injection-sealing arrangement, which has been so successfully applied to boiler feed pump service, is of course applicable to a number of other services. For instance, it is very suitable for cold-water high-pressure pumps applied to hydraulic descaling or hydraulic press work. In such services, of course, there is no need to bring in injection supply water to the breakdowns (unless this water is not clear and free of gritty material), for the water handled by the pump is already cold.

BEARINGS

The function of bearings in centrifugal pumps is to keep the shaft or rotor in correct alignment with the stationary parts under the action of radial and transverse loads. Those that give radial positioning to the rotor are known as *line bearings,* whereas those that locate the rotor axially are called *thrust bearings.* In most applications the thrust bearings actually serve both as thrust and radial bearings.

Types of Bearings Used All types of bearings have been used in centrifugal pumps. Even the same basic design of pump is often made with two or more different bearings, required either by varying service conditions or the preference of the purchaser.

Two external bearings are usually used for the double-suction single-stage general service pump, one on either side of the casing. These were formerly of the babbitted type using oil lubrication, but in recent years most manufacturers have changed to antifriction bearings using either grease or oil lubrication.

Some of the small, inexpensive centrifugal pumps used for pumping clear liquids are provided with an internal sleeve bearing (Fig. 109). The liquid itself is used as a lubricant, although separate grease lubrication through an alemite fitting is used in some designs.

In horizontal pumps with bearings on each end, the bearings are usually designated by their location as *inboard* and *outboard*. Inboard bearings are located between the casing and the coupling. Pumps with overhung impellers have both bearings on the same side of the casing so that the bearing nearest the impeller is called inboard and the one farthest away outboard. In a pump provided with bearings at both ends, the thrust bearing is usually placed at the outboard end and the line bearing at the inboard end.

The bearings are mounted in a housing that is usually supported by brackets

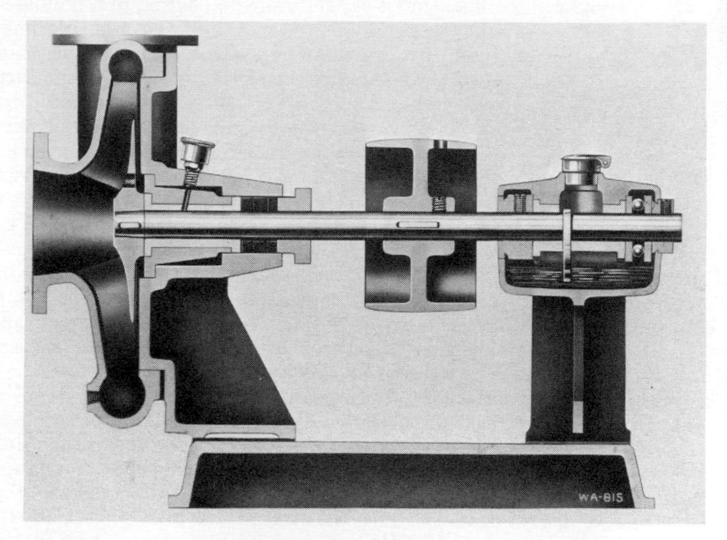

Fig. 109 Internal sleeve bearing.

attached to or integral with the pump casing. The housing also serves the function of containing the lubricant necessary for proper operation of the bearing. Occasionally the bearings of very large pumps are supported in housings that form the top of pedestals mounted on soleplates or on the pump bedplate. These are called *pedestal bearings*.

Because of the heat generated by the bearing itself or the heat in the liquid being pumped, some means other than radiation to the surrounding air must occasionally be used to keep the bearing temperature within proper limits. If the bearings have a forced-feed lubrication system, cooling is usually accomplished by circulating the oil through a separate water-to-oil cooler. Otherwise a jacket through which a cooling liquid is circulated is usually incorporated as part of the housing.

Pump bearings may be rigid or self-aligning. A self-aligning bearing will automatically adjust itself to a change in the angular position of the shaft. In babbitted or sleeve bearings, the name "self-aligning" is applied to bearings that have a spherical fit of the sleeve in the housing. In antifriction bearings it is applied to bearings the outer race of which is spherically ground or the housing of which provides a spherical fit.

The most common antifriction bearings used on centrifugal pumps are the various types of ball bearings. Roller bearings are used less often, although the spherical roller bearing (Fig. 110) is used frequently for large shaft sizes for which there is a limited choice of ball bearings. As most roller bearings are suitable only for radial loads, their use on centrifugal pumps tends to be limited to applications in which they are not required to carry a combined radial and thrust load.

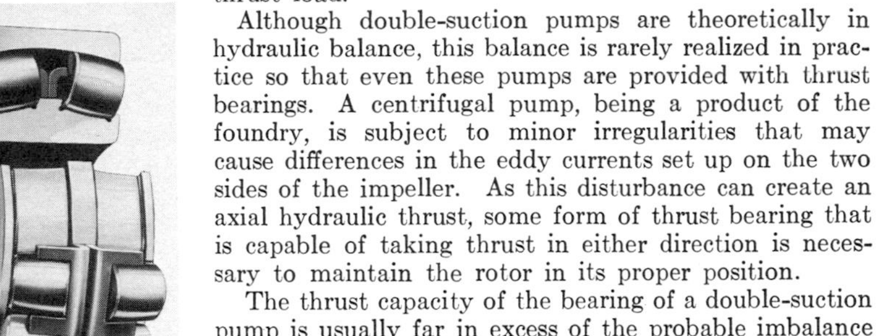

Fig. 110 Self-aligning spherical roller bearing. (SKF)

Although double-suction pumps are theoretically in hydraulic balance, this balance is rarely realized in practice so that even these pumps are provided with thrust bearings. A centrifugal pump, being a product of the foundry, is subject to minor irregularities that may cause differences in the eddy currents set up on the two sides of the impeller. As this disturbance can create an axial hydraulic thrust, some form of thrust bearing that is capable of taking thrust in either direction is necessary to maintain the rotor in its proper position.

The thrust capacity of the bearing of a double-suction pump is usually far in excess of the probable imbalance caused by irregularities. This provision is made because (1) unequal wear of the rings and other parts may cause imbalance, and (2) the flow of the liquid into the two suction eyes may be unequal and cause imbalance because of an improper suction-piping arrangement.

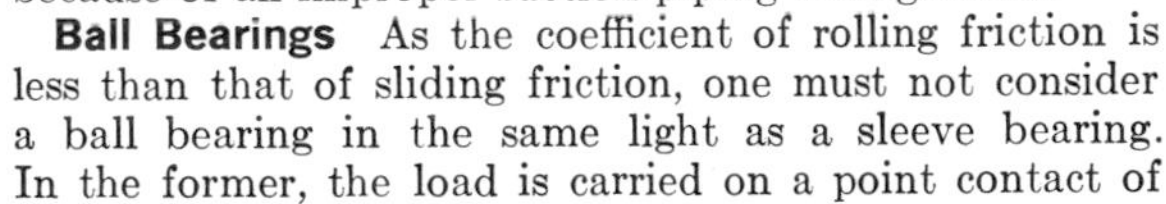

Ball Bearings As the coefficient of rolling friction is less than that of sliding friction, one must not consider a ball bearing in the same light as a sleeve bearing. In the former, the load is carried on a point contact of the ball with the race, but the point of contact does not rub or slide over the race and no appreciable heat is generated. Furthermore, the point of contact is constantly changing as the ball rolls in the race, and the operation is practically frictionless. In the sleeve bearing there is a constant rubbing of one surface over another, and the friction must be reduced by the use of a lubricant.

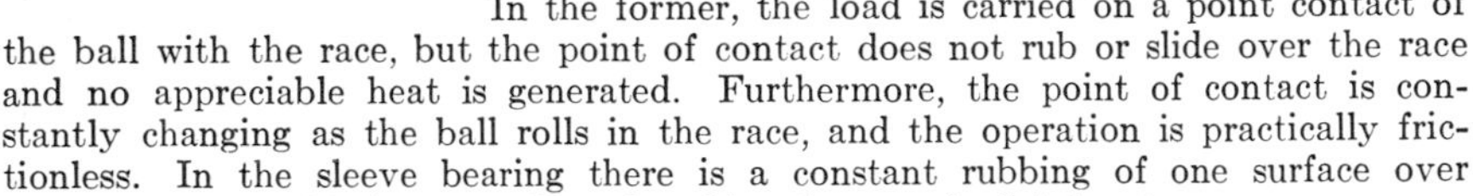

Ball bearings operated at an absolutely constant speed theoretically would require no lubricant. However, no speed can be called absolutely constant, for the conditions affecting the speed always vary slightly. For instance, a motor with a full load speed rated at 3,510 rpm might vary in speed in the course of a minute from 3,505 to 3,515 rpm. Each variation in speed has the effect of causing the balls in a ball bearing to lag or lead the race because of their inertia. Consequently, a very slight, almost immeasurable sliding action takes place. Another limiting condition is that the hardest of metals suffer minute deformation on carrying load, thus upsetting perfect point contact and adding another slight sliding action. For these reasons ball bearings must be given some lubrication.

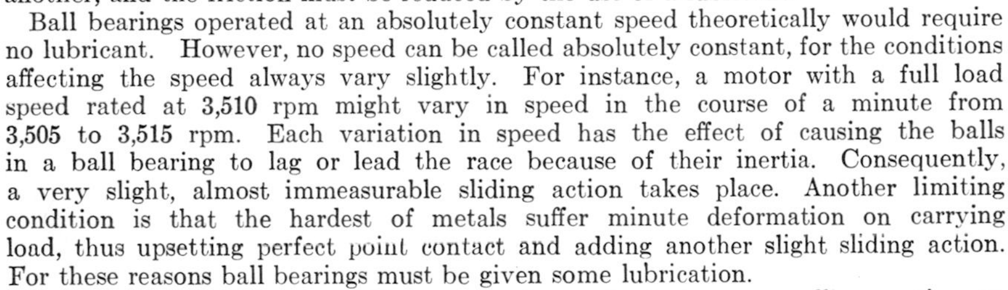

Ball thrust bearings are built to carry heavy loads by pure rolling motion on an angular contact. As thrust load is axial, it is equally distributed to all the balls around the race, and the individual load on each ball is only a very small fraction of the total thrust load. In such bearings it is essential that the balls be very equally spaced, and for this purpose a retaining cage is used between the balls and between the inner and outer race. This cage carries no load, but the contact between it and the ball produces sliding friction that generates a small amount of heat. It is for this reason that ball thrust bearings are generally water-jacketed.

Types and Applications A pump designer has a wide variety of antifriction bearings to choose from. Each type has characteristics that could make it a good or bad selection for a specific application. Although several types might sometimes be acceptable, it is best for purchasers to leave the choice to the manufacturer. For example, some purchasers specify double-row bearings whatever the size or type of pump, even though single-row bearings are often equally suitable or better.

The most common ball bearings used on centrifugal pumps are (1) single-row, deep-groove, (2) double-row, deep-groove, (3) double-row, self-aligning, and (4) angular-contact, either single- or double-row. All except the double-row, self-aligning bearings are capable of carrying thrust loads as well as radial loads.

Sealed ball bearings, adapter ball bearings, and other modifications have also found special applications. Sealed prelubricated bearings require special attention if the unit in which they are installed is not operated for a long period of time (for instance, one kept in stock or storage). The shaft should be turned over occasionally, say once every three months, to agitate the lubricant and maintain a film coating of the balls of such units.

The self-aligning ball bearing (Fig. 111) is the most serviceable bearing for heavy loads, high speeds, long-bearing spans, and no end thrust. For this reason it is ideally adapted for service as a line bearing on a centrifugal pump. Its double row of balls runs in fixed grooves in the inner, or shaft, race; its outer race is ground to a spherical seat. Any slight vibration or shaft deflection is

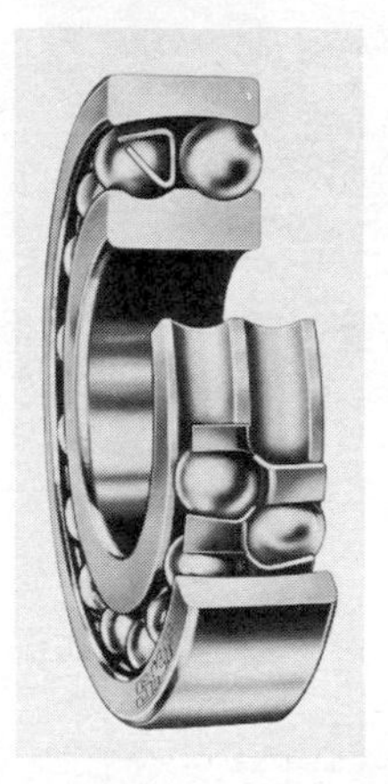

Fig. 111 Self-aligning double-row ball bearing. (SKF)

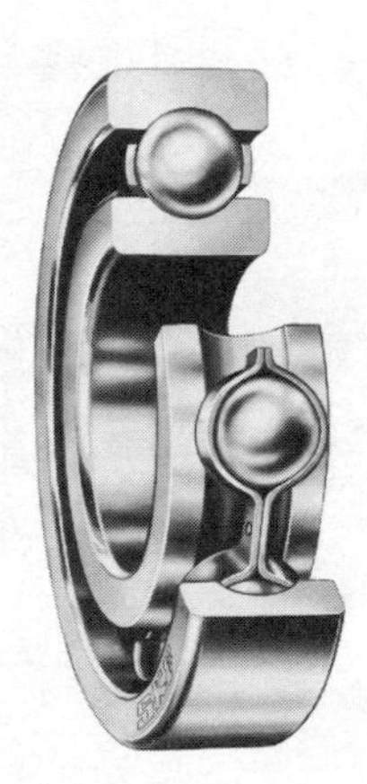

Fig. 112 Single-row deep-groove ball bearing. (SKF)

Fig. 113 Double-row deep-groove ball bearing. (SKF)

therefore taken care of by this bearing, which operates as a pivot. In lightly constructed pumps, it will also compensate for the slight misalignment caused by the "breathing" that takes place in the casing when pressure is developed.

The self-aligning ball bearing has proved very satisfactory for high speeds and has long life, even with long-bearing spans. It has very little thrust capacity, however, and is not used for combined radial and thrust loads in centrifugal pumps. For large shafts, the self-aligning spherical roller bearing (Fig. 110) is used instead, for it can carry such loads with a considerable thrust component.

The single-row deep-groove ball bearing (Fig. 112) is the most commonly used bearing on centrifugal pumps except for the larger sizes. It is good for both radial thrust and combined loads but requires careful alignment between the shaft and the housing in which the bearing is mounted. It is sometimes used with seals built into the bearing in order to exclude dirt, retain lubricant, or both.

The double-row deep-groove ball bearing—in effect, two single-row bearings placed side by side—has greater capacity both for radial and thrust loads (Fig. 113). It is used quite commonly if the loading is more than that permitted by the single-row bearing.

The angular-contact ball bearing operates on a principle that makes it good for heavy thrust loads. The single-row type (Fig. 114) is good for thrust in only one direction, whereas the double-row type (Fig. 115), which is basically two single-row bearings placed back-to-back, can carry thrust in either direction. Two single-row angular-contact bearings are frequently matched and the faces of the races ground by the manufacturer so they can be used in tandem for large, one-directional thrust loads or back-to-back (Fig. 116) for two-directional thrust loads. The two bearings are sometimes locked together by recessing the inner races and pressing them onto a short ring (Fig. 117). If two separate angular-

Fig. 114 Single-row angular-contact ball bearing. (New Departure)

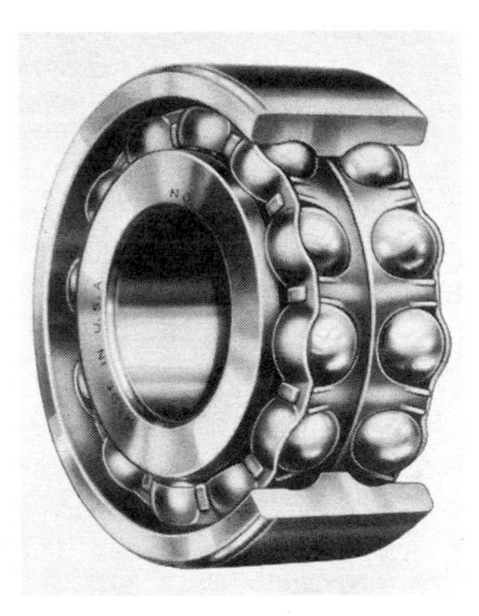

Fig. 115 Double-row angular-contact ball bearing. (New Departure)

Fig. 116 Two single-row angular-contact bearings mounted back-to-back to act as a double-row bearing. (New Departure)

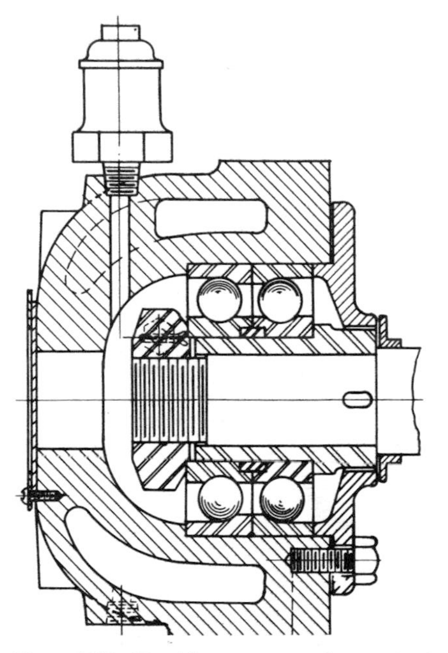

Fig. 117 Double-row angular-contact ball thrust bearing. (grease-lubricated and water-cooled).

contact bearings are used, care must be taken to mount them correctly on the shaft.

The single-row angular-contact ball bearing can be used singly on centrifugal pumps only if the thrust is always in one direction. Its field of application is thus limited primarily to vertical pumps. Another very interesting application is the use of two such bearings in an end-suction pump to take care of axial thrust in either direction. This arrangement permits a certain amount of axial adjustment of the impeller in its volute, accomplished by loosening up one bearing nut and tightening the other. Unfortunately, such adjustment is extremely delicate and requires a first-class mechanic; its commercial practicability is therefore somewhat limited.

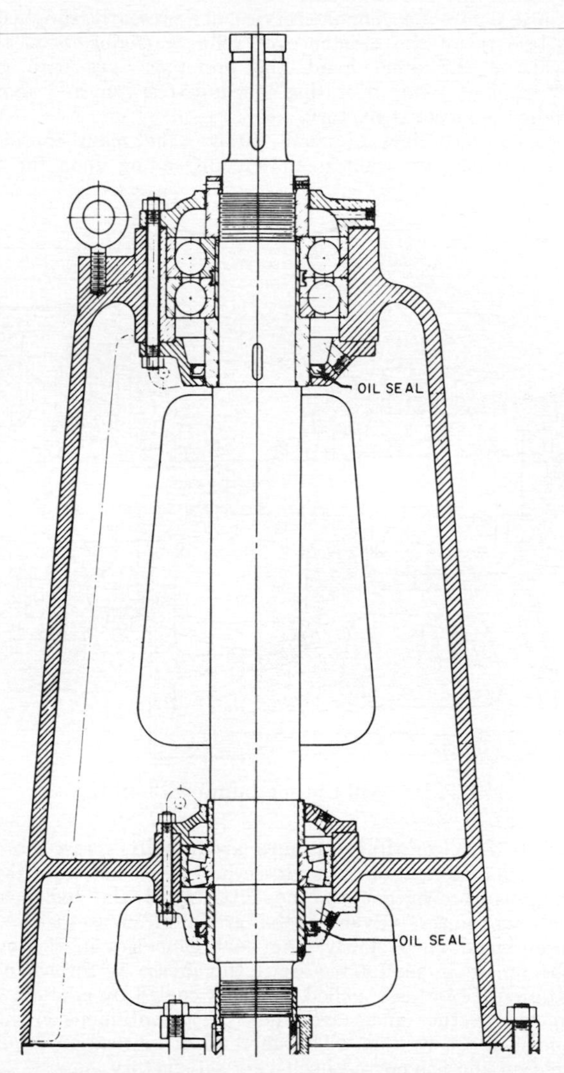

Fig. 118 Ball bearing construction with seal in vertical pump (seal guards against escape of grease).

The double-row angular-contact bearing, or its equivalent of a matched pair mounted back-to-back, has been found very satisfactory for pumps capable of a high thrust load in either direction. Some pump manufacturers standardize on this bearing for many applications.

Lubrication of Antifriction Bearings In the layout of a line of centrifugal pumps, the choice of the lubricant for the pump bearings is dictated by application requirements, by cost considerations, and sometimes by the preferences of a group of purchasers committed to the major portion of the output of that line.

For example, the application of vertical wet-pit condenser circulating pumps dictates the choice of water in preference to grease or oil lubrication. If oil or grease were used in such pumps and the lubricant leaked into the pumping system, the condenser operation might be seriously affected because the tubes would become coated with the lubricant.

Most centrifugal pumps for refinery service are presently supplied with oil-lubricated bearings because of the insistence of refinery engineers on this feature. In the marine field, on the other hand, the preference lies with grease-lubricated bearings. For very high pump operating speeds (5,000 rpm and above), oil lubrication is found to be the most satisfactory.

For highly competitive lines of small pumps, the main consideration is cost, and the most economical lubricant is chosen, depending upon the type of bearing used.

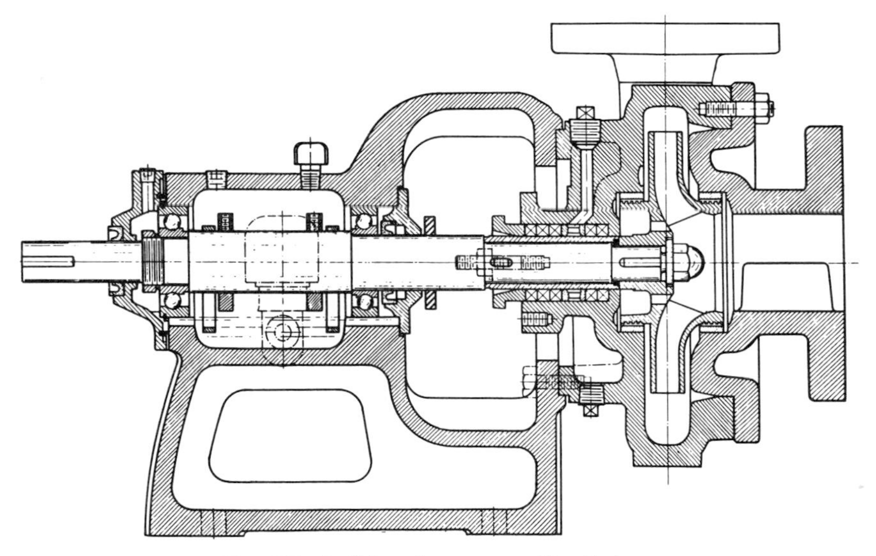

Fig. 119 Ball bearing pump with oil rings.

Ball bearings used in centrifugal pumps are usually grease-lubricated, although some services use oil lubrication. In grease-lubricated bearings, the grease packed into the bearing is thrown out by the rotation of the balls, creating a slight suction at the inner race. (Even if the grade of grease is relatively light, it is still a semisolid and flows slowly. As heat generates in the bearing, however, the flow of the grease is accelerated until the grease is thrown out at the outer race by the rotation.) As the expelled grease is cooled by contact with the housing and thus attracted to the inner race, there is a continuous circulation of grease to lubricate and cool the bearing. This method of lubrication requires a minimum amount of attention and has proved itself very satisfactory.

As housings of bearings in vertical pumps require seals to prevent the escape of the lubricant, grease is usually preferred, for it lessens the chance of leakage (Fig. 118).

A bearing fully packed with grease prevents proper grease circulation in itself and its housing. As a rough rule, therefore, it is recommended that only one-third of the void spaces in the housing be filled. An excess amount of grease will cause the bearing to heat up, and grease will flow out of the seals to relieve the situation. Unless the excess grease can escape through the seal or through the relief cock that is used on many large units, the bearing will probably fail early.

Oil-lubricated ball bearings require an adequate method for maintaining a suitable oil level in the housing. This level should be at about the center of the lowermost ball of a stationary bearing. It may be achieved by a dam and an oil slinger to maintain the level behind the dam and thereby increase the leeway in the amount of oil the operator must keep in the housing. Oil rings are sometimes used to supply oil to the bearings from the bearing housing reservoir (Fig. 119). In other designs, a constant-level oiler is used (Fig. 120).

Sleeve Bearings Although the plain cylindrical journal or sleeve bearing has

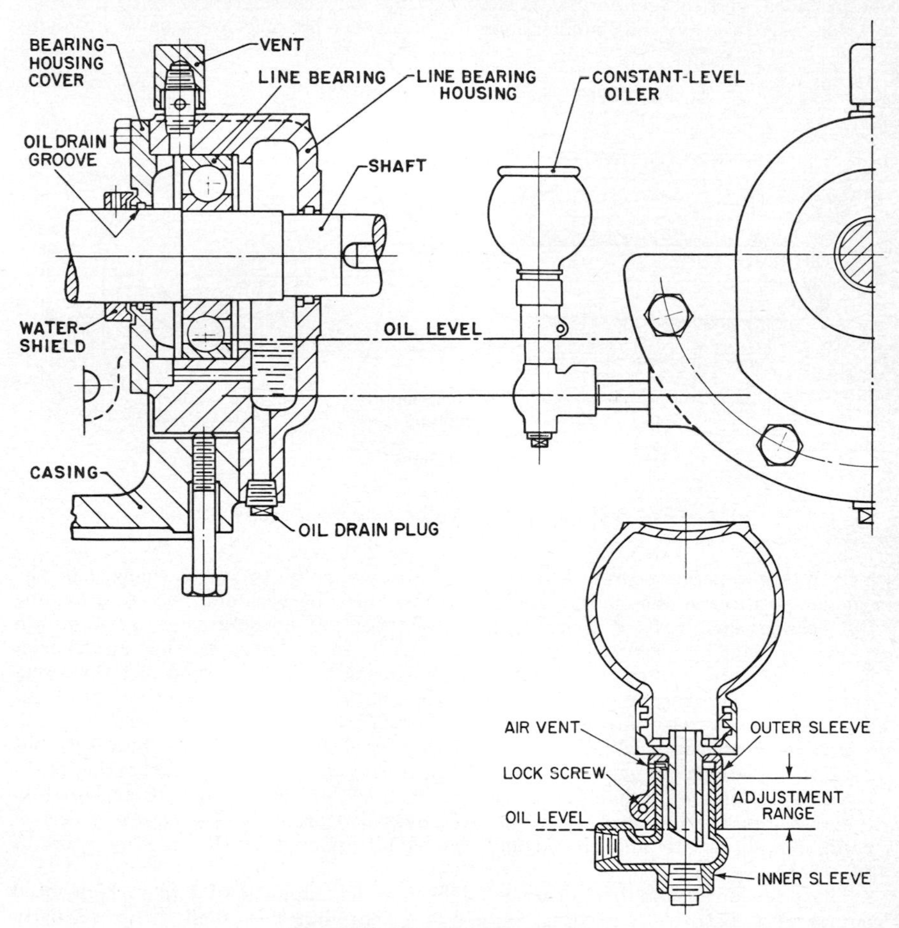

Fig. 120 Constant-level oiler.

been replaced by antifriction bearings in most designs, it still has a large field of application. At one extreme, it is used for reasons of economy of construction, for instance, in certain small pumps used strictly for pumping clear liquids. The internal sleeve bearing in these pumps (see Fig. 109) depends primarily on the liquid pumped for its lubrication (sometimes a water seal is centered in the bearing and connected to the pump discharge through which the liquid is introduced under pressure).

At the other extreme, sleeve bearings are used for large heavy-duty pumps with shaft diameters of such proportions that the necessary antifriction bearings are not commonly available. Another typical application is in high-pressure multistage pumps, like boiler feed pumps for pressures of 1,500 lb/in^2 and higher, which require sleeve bearings because of a combination of fairly large shaft diameters and high speeds (3,600 to 9,000 rpm). Still another application is in vertical submerged pumps, like vertical turbine pumps, in which the bearings are subject to a water contact, a condition that precludes the use of antifriction bearings. Moreover, a personal preference for sleeve bearings is sometimes strong enough to make a pump purchaser even pay a premium for special design modifications of a standard pump so that those bearings may be substituted for the usual ball bearing construction.

Most sleeve bearings are oil-lubricated. As a proper means of oil feed must be provided, vertical and horizontal shaft bearings vary considerably.

Especially in heavy-duty applications, the line sleeve bearings are usually self-aligning.

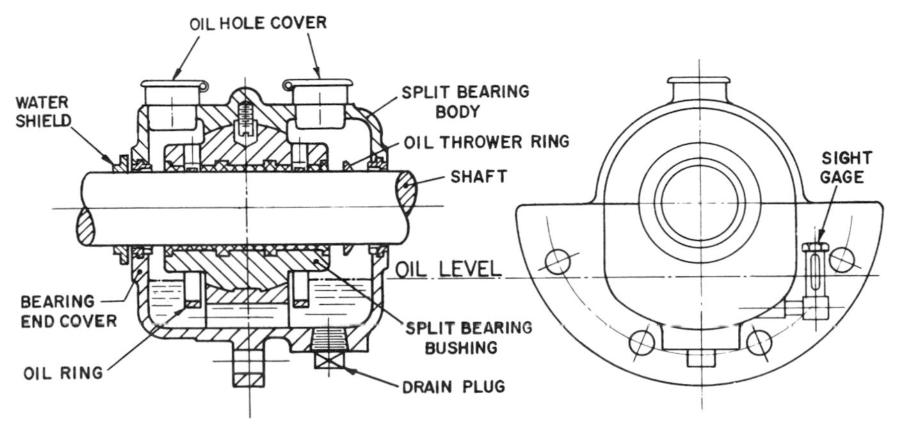

Fig. 121 Self-aligning sleeve bearing, spherically seated.

A self-aligning bearing adjusts itself automatically to small changes in the angular position of the shaft. This adjustment may be accomplished by providing the bearing shell with a spherical fit in the bearing housing (Fig. 121), or—an equally satisfactory solution and one that results in a shorter-bearing and shorter bearing span—by reducing the length of the bearing bushing engaged in the housing (Fig. 122). The second method permits the bushing to rock slightly in its seat and thus compensate for changes in the angular position of the shaft.

Babbitted Bearings Various materials are used for the bearing bushings, but babbitted bearings are generally preferred for heavy-duty sevice. The bearing bushing may consist of a babbitt lining (⅛ in thick or more) that is anchored in the cast iron bearing shell by means of dovetailed grooves. To ensure a perfect bond, the shells are first tinned and the babbitt poured at the melting point of tin.

The precision automotive bearing has been widely applied of late in high-speed pumps (Fig. 123). This bearing consists of a split thin steel shell with a similarly thin deposit of babbitt. It is available in a wide range of sizes.

To build up bearing load, the bearings of high-speed pumps are made shorter than those in a conventional unit. In addition, they may incorporate a so-called "anti-oil whip" construction. (High-speed construction results in lighter rotor weights that tend to induce oil whip if ignored.) A pocket is provided in the upper half of the bearing in which a pressure pad of oil builds up during operation. This pressure creates a downward force on the journal holding it down in the bearing and keeping the vibration to a minimum.

Sleeve bearings are seldom applied to horizontal pumps with an overhung impeller arrangement, but rather for those designed for bearings on both ends and for drive through a flexible coupling. For such pumps, provision must be made for thrust in one of the following ways:

1. A sleeve radial bearing at one end and a babbitted, combined radial and thrust bearing at the other end.

2. Two sleeve bearings with a separate thrust-bearing element of either the sleeve or antifriction type

For thrust loads heavier than those usually encountered, it is now the practice to apply a Kingsbury type thrust bearing which is also suitable for high speeds.

Kingsbury Thrust Bearings This bearing was first developed to meet the need

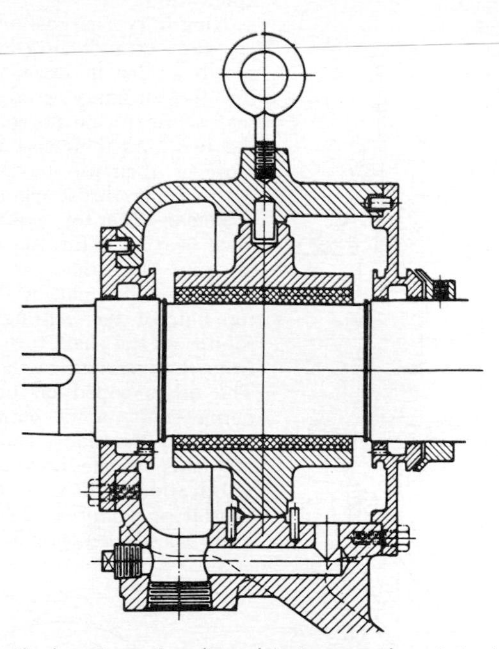

Fig. 122 Self-aligning sleeve bearing with short seating of the bearing bushing.

for a suitable pivot bearing for vertical shaft turbines and has gradually been applied to other rotary apparatus, such as the centrifugal pump. The operating principle is simple. An ordinary cylindrical or sleeve bearing has a running clearance between the bearing shell and the journal. Because of the relation of the curved surfaces and the capillary attraction of the oil particles, a "pumping" action takes place that draws a lubricating oil film into this clearance (Fig. 124). If the oil is of correct viscosity, it will resist the breakdown of the film except at excessive loads. To provide a positive and ample supply of cool oil to the bearing, a simple gravity device is ordinarily used, although operation at higher speeds resulting in maximum tendency to heat requires some form of forced-feed lubrication. In an ordinary thrust collar subjected to high pressures and high speeds, the parallel surfaces tend to squeeze out the oil film. The metal-to-metal contact that results makes this type of bearing unsuitable for heavy loads.

The principle of the Kingsbury bearing can be described as follows: Suppose that a circular collar is cut into little segments and that each block is suitably supported on its underside so that it may rock slightly on the point indicated as the suspension point and yet stay in place. When the shaft begins to rotate, the film of oil tends to be dragged in under the slightly rounded edges of the blocks. As the speed of the shaft increases, this tendency increases, the block adjusting itself slightly by tipping at a greater angle, riding up on the oil film as a sled runner rides up upon meeting the surface resistance of snow underneath (Fig. 125). The higher the speed, the greater this tendency for the block to rock forward, permitting an increased "sledding" action, and the greater the tendency to adjust itself to the increasing oil film dragged underneath it. Construction details of a typical Kingsbury bearing can be examined more closely in the sectional assembly shown in Fig. 126.

The thrust mounting of Kingsbury bearings used in horizontal pumps is arranged to take thrust in both directions. Sometimes both loads are approximately equal; at other times there may be a major thrust in one direction and an occasional minor thrust in the opposite direction. In any event, the Kingsbury bearing is provided with thrust shoes on each side to limit the axial motion of the rotor. The number of shoes on each side may or may not be equal, depending on the application.

Kingsbury bearings are capable of taking care of unit thrust loads and linear speeds so far in excess of those suitable for the ordinary straight collar thrust that there can be no comparison between the two. As their cost is relatively high, however, their use may be warranted only for extreme thrust conditions.

Sleeve Bearing Lubrication A ring-oiled bearing is furnished with a soft steel oil ring that rides on the pump shaft through a slot cut in the middle of the top half of the bearing shell. This ring rotates as the shaft turns and picks up oil from the reservoir in the bearing housing. The oil is wiped off on the top of the pump shaft, flows between the bearing bore and the shaft, and is discharged at the ends of the bearing (see Fig. 121). Lubrication by means of oil rings is fully satisfactory only at relatively low operating speeds. A provision for automatic circulation of the oil—and, if necessary, for cooling it—is an essential feature of all higher-speed sleeve bearings, especially thrust bearings.

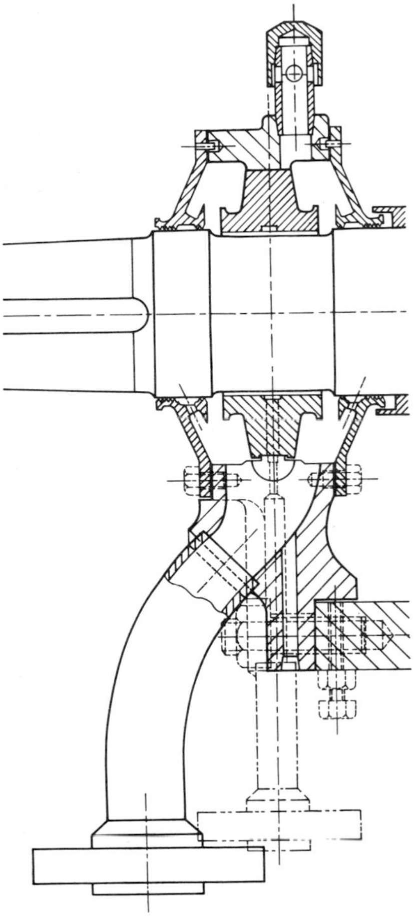

Fig. 123 "Precision" automotive-type sleeve bearing.

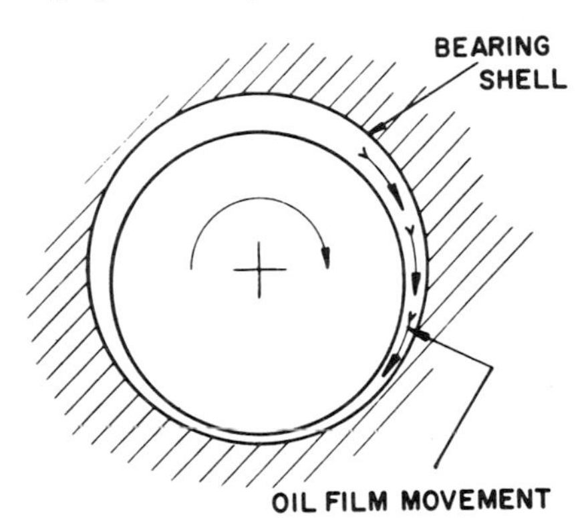

Fig. 124 Ordinary cylindrical bearing with oil film formed by the pumping action.

In some bearings, the oil circulation is effected by a rotary positive-displacement gear pump directly connected to the outboard end of the pump shaft by means of a flexible coupling (see Fig. 126). The oil pump takes the oil from a reservoir, located either in the bearing housing itself or separately on the pump baseplate, and delivers it under pressure through the oil cooler. From the cooler the oil flows in part to the outboard thrust bearing, from which it flows into the reservoir located in the lower half of the bearing housing. It then overflows by gravity from this reservoir into the main reservoir on the baseplate. This lubricating system is illustrated in Fig. 127.

General practice supplies the inboard line bearing of this system with oil under pressure through a branch line in the discharge from the oil cooler. The oil from the inboard bearing is returned by gravity through large return lines into the main

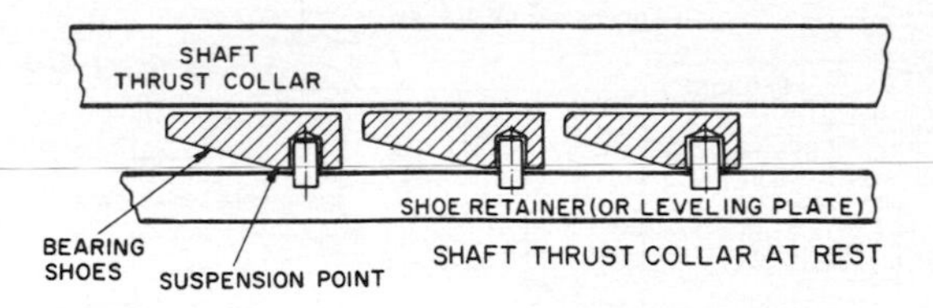

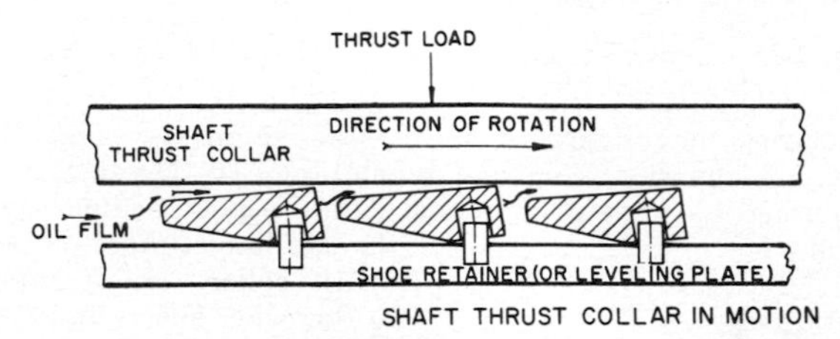

Fig. 125 Principle of Kingsbury-type thrust bearing.

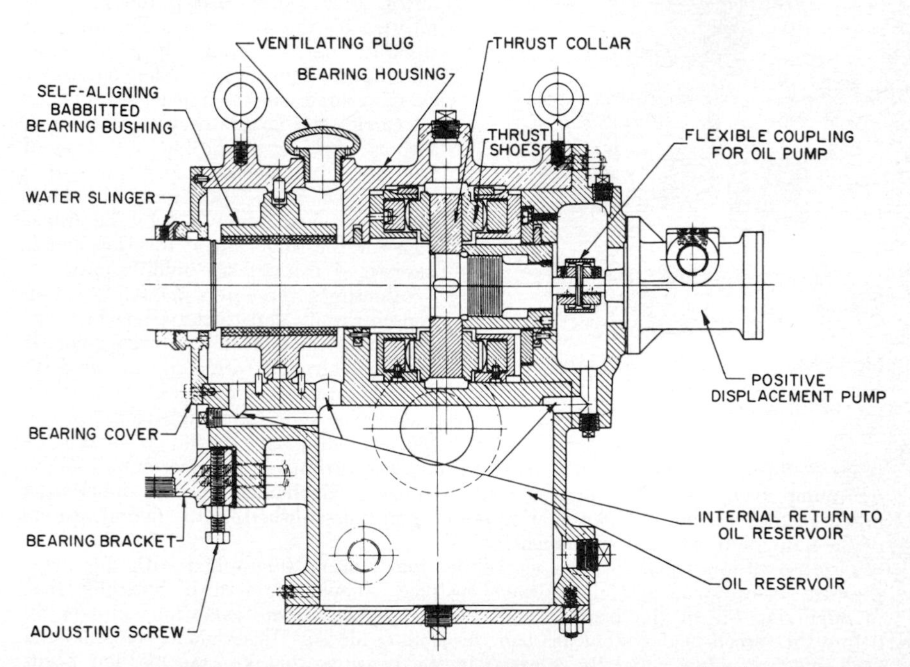

Fig. 126 Sectional assembly of Kingsbury-type thrust bearing.

reservoir. It is essential to provide an adequate pressure drop from all bearings so that the oil will not overflow because of unsatisfactory evacuation.

Numerous alternative methods exist for supplying the bearing with forced-feed lubrication. For example, some arrangements use a vertical oil pump driven from the main pump shaft by means of a worm gear (Fig. 128). Other bearings employ the Kingsbury adhesive lubrication system (Fig. 129). In this system, oil from the reservoir beneath the thrust bearing is drawn into a bronze ring called the *circulator* or *oil pumping ring,* which is around the collar. The adhesion of oil

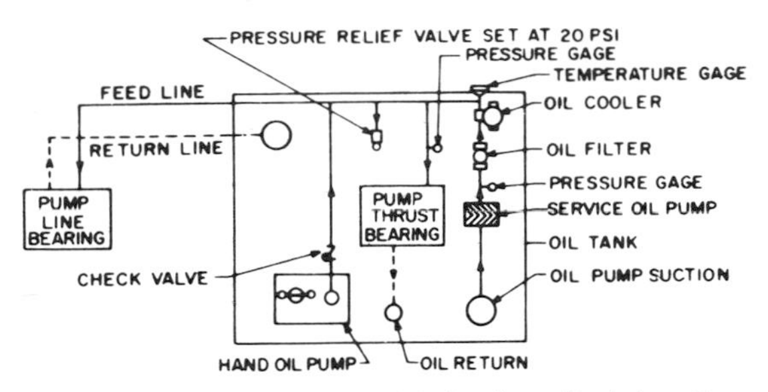

Fig. 127 Typical forced-feed lubrication oil piping diagram.

to the collar carries the oil around in the groove in the ring. The oil travels with the collar for almost a complete revolution. It then meets a dam in the groove and is pushed by the stream behind it into a port leading to spaces between the two lowest shoes on both sides of the thrust collar. Shaft rotation carries it to the other shoes, and it finally escapes, above the collar, into a passage leading down to a cooler. From the cooler it returns to the reservoir. The oil will circulate equally well with the collar running the other way. When the collar changes direction, the adhesiveness of the oil carries the circulator with it through a short angle, until the lug at the top of the circulator meets a stop. In either of the two "stop" positions, oil enters the groove in the circulator by the proper port for the direction of rotation and is discharged through the middle port.

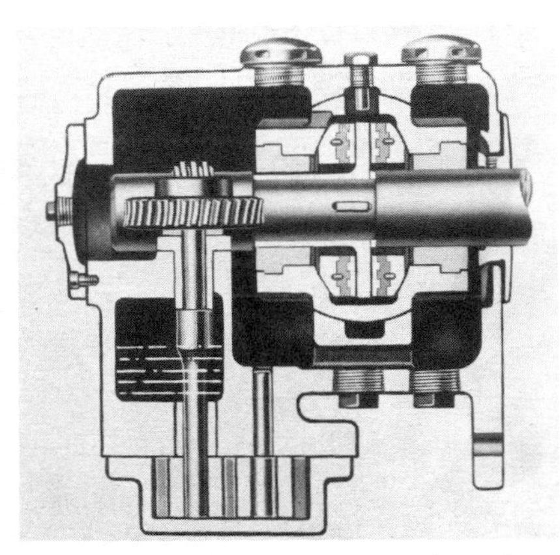

Fig. 128 Vertical oil pump driven from the main pump shaft by a worm gear. (Ingersoll-Rand)

Sometimes the forced-feed lubrication system supplies oil to the driver bearings as well, although this arrangement is usually restricted to electric motor drivers. A typical system combining pump and driver lubrication is shown in Fig. 130. If pumps are driven by steam turbines or through gears, it is customary to have the turbine or the gear supply oil to the pump bearings. Such arrangements require reconcilement of the lubricating oil characteristics and of the operating temperatures established by the manufacturers of the individual pieces of equipment.

The use of oil rings for line sleeve bearings normally supplied with oil under pressure is optional and not always justified. Their function is basically that of supplying oil to the bearing at the start of the pump operation, supposedly before the forced-feed system has had the time to do so. It should be remembered that sufficient oil is generally retained in the bearings to take care of their needs before forced-feed delivery takes place.

If the normal retention of oil in the bearing or the use of oil rings will not afford adequate protection, auxiliary oil pumps are called upon. These may be manually operated gear pumps intended for use at scheduled intervals when the pump is standing idle. Operation of this auxiliary pump at weekly or biweekly intervals is usually sufficient to keep the oil from draining out completely from the bearings or the oil piping.

More elaborate lubricating systems incorporate a motor-driven auxiliary oil pump, which is started before the main pump begins operating. The motor starter controls are interlocked in such a manner that the main motor cannot be started until the oil pressure in the system reaches a predetermined value. As soon as the oil pump driven from the main pump shaft develops sufficient pressure, the auxiliary

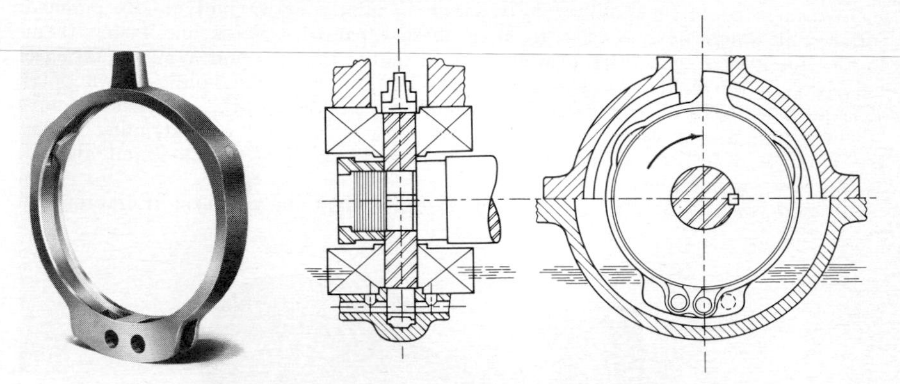

Fig. 129 Pumping ring of a Kingsbury bearing.

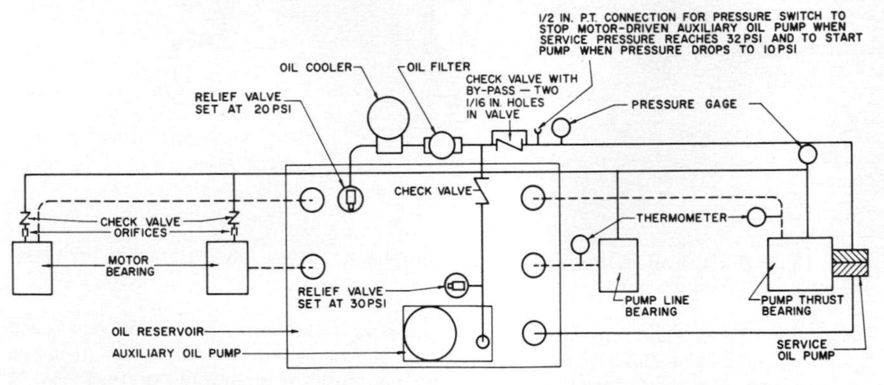

Fig. 130 Integral forced-feed lubrication system for pump and motor bearings.

pump is shut down by means of a pressure switch. A second pressure-switch setting automatically restarts the auxiliary pump on failure of the regular pump to maintain the desired pressure. This arrangement is illustrated in Fig. 130.

COUPLINGS

Centrifugal pumps are connected to their drivers through couplings of one sort or another, except for close-coupled units, in which the impeller is mounted on an extension of the shaft of the driver. Because couplings may be used with both centrifugal and positive displacement pumps, they are discussed separately in Sec. 6.3.

BEDPLATES AND OTHER PUMP SUPPORTS

For very obvious reasons, it is desirable that pumps and their drivers be removable from their mountings. Consequently, they are usually bolted and doweled to machined surfaces that in turn are firmly connected to the foundations. To simplify the installation of horizontal-shaft units, these machined surfaces are usually part of a common bedplate on which either the pump or the pump and its driver have been prealigned.

Bedplates The primary function of a pump bedplate is to furnish mounting surfaces for the pump feet that are capable of being rigidly attached to the foundation. Mounting surfaces are also necessary for the feet of the pump driver or drivers or of any independently mounted power transmission device. Although such surfaces could be provided by separate bedplates or by individually planned surfaces, it would be necessary to align these separate surfaces and fasten them to the foundation with the utmost care. Usually this method requires in-place mounting in the field as well as drilling and tapping for the holding-down bolts after all parts have been aligned. To minimize such "field work," coupled horizontal-shaft pumps are usually purchased with a continuous base extending under the pump and its driver; ordinarily, both these units are mounted and aligned at the place of manufacture.

Although such bases are designed to be quite rigid, they deflect if improperly

Fig. 131 Horizontal-shaft centrifugal pump and driver on cast iron bedplate. (Worthington Pump, Inc.)

supported. It is therefore necessary to support them on foundations that can supply the required rigidity. Furthermore, as the base can be sprung out of shape by improper handling during transit from the place of manufacture to point of installation, it is imperative that the alignment be carefully rechecked during erection and prior to starting the unit.

As the unit size increases so do the size, weight, and cost of the base required. The cost of a prealigned base for most large units would exceed the cost of the field work necessary to align individual bedplates or soleplates and to mount the component parts. Such bases are therefore used only if appearances require or if their function as a drip collector justifies the additional cost. Even in fairly small units, the height at which the feet of the pump and the other elements are located may differ considerably. A more rigid and pleasant looking installation can frequently be obtained by using individual bases or soleplates and building up the foundation to various heights under the separate portions of equipment.

Cast iron baseplates are usually provided with a raised edge or raised lip around the base to prevent dripping or draining onto the floor (Fig. 131). The base itself is suitably sloped toward one end so as to collect the drainage for further disposal. A drain pocket is provided near the bottom of the slope, usually with a mesh screen. A tapped connection in the pocket permits piping the drainage to a convenient point.

Bedplates fabricated of steel plate and structural steel shapes, now used very extensively, do not easily permit incorporation of a raised lip or drip pocket

(Figs. 132 and 133). From a utility viewpoint, however, the customary use of bearing brackets as drip pockets (to collect leakage from the stuffing boxes) now makes the use of a raised-lip bedplate unnecessary in horizontal pumps except if the pump handles a cold liquid in a moist atmosphere and thus must contend with considerable condensation on its surface.

Soleplates Soleplates are cast iron or steel pads located under the feet of the pump or its driver and embedded into the foundation. The pump or its

Fig. 132 Pump and internal combustion engine mounted on a portable steel skid base. (Worthington Pump, Inc.)

Fig. 133 Small centrifugal pump on structural steel bedplate made of a simple channel. (Worthington Pump, Inc.)

driver is doweled and bolted to them. Soleplates are customarily used for vertical dry pit pumps and also for some of the larger horizontal units to save the cost of the large bedplates otherwise required.

Centerline Support For operation at high temperatures, the pump casing must be supported as near to its horizontal centerline as possible in order to prevent excessive strains caused by temperature differences. These might seriously disturb the alignment of the unit and eventually damage it. Centerline construction is usually employed in boiler feed pumps or hot-water circulating pumps operating at temperatures around 300°F (Fig. 134).

Horizontal Units Using Flexible Pipe Connections The foregoing discussion of bedplates and supports for horizontal shaft units assumed their application to pumps with piping setups that do not impose hydraulic thrusts on the pumps themselves. If flexible pipe connections or expansion joints are desirable in the suction or discharge piping of a pump (or in both), the pump manufacturer should be so advised for several reasons. First, the pump casing will be required to withstand various stresses caused by the resultant hydraulic thrust load. Although this is rarely a limiting or dangerous factor, it is best that the manufacturer

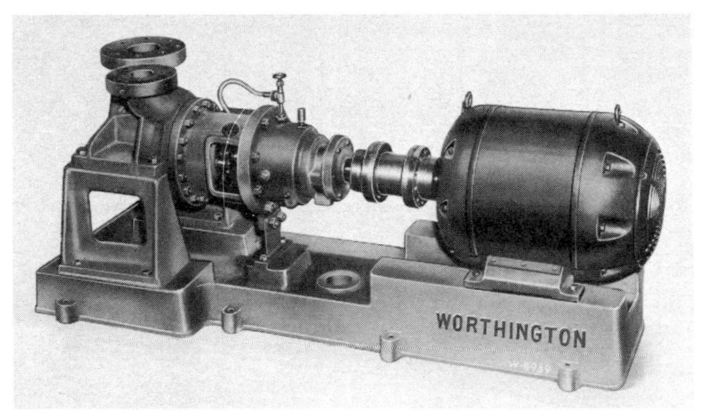

Fig. 134 Single-stage hot-water circulating pump with centerline support. (Worthington Pump, Inc.)

have the opportunity to check the strength of the pump casing. Second, the resulting hydraulic thrust has to be transmitted from the pump casing through the casing feet to the bedplate or soleplate and then to the foundation. Usually horizontal-shaft pumps are merely bolted to their bases or soleplates so that any tendency to displacement is resisted only by the frictional grip of the casing feet on the base and by relatively small dowels. If flexible pipe joints are used, this attachment may not be sufficient to withstand the hydraulic thrust. If high hydraulic thrust loads are to be encountered, therefore, the pump feet must be keyed to the base or supports. Similarly, the bedplate or supporting soleplates must be of a design that will permit transmission of the load to the foundation.

VERTICAL PUMPS

Vertical-shaft pumps fall into two separate classifications, dry pit and wet pit. Dry-pit pumps are surrounded by air, whereas the wet-pit types are either fully or partially submerged in the liquid handled.

Vertical Dry-Pit Pumps Dry-pit pumps with external bearings include most small, medium, and large vertical sewage pumps; most medium and large drainage and irrigation pumps for medium and high head; many large condenser circulating and water supply pumps; and many marine pumps. Sometimes the vertical design is preferred (especially for marine pumps) because it saves floor space. At other times it is desirable to mount a pump at a low elevation because of suction conditions, and it is then also preferable or necessary to have its driver at a high elevation. The vertical pump is normally used for very large capacity applications because it is more economical than the horizontal type, all factors considered.

Many vertical dry-pit pumps are basically horizontal designs with minor modifications (usually in the bearings) to adapt them for vertical-shaft drive. The reverse is true of small- and medium-sized sewage pumps; a purely vertical design is the most popular for that service. Most of these sewage pumps have elbow suction nozzles (Figs. 135, 136, 137) because their suction supply is usually taken from a wet well adjacent to the pit in which the pump is installed. The suction elbow usually contains a handhole with a removable cover to provide easy access to the impeller.

To dismantle one of these pumps, the stuffing box head must be unbolted from the casing after the intermediate shaft or the motor and motor stand have been removed. The rotor assembly is drawn out upward, complete with the stuffing box head, the bearing housing, and the like. This rotor assembly can then be completely dismantled at a convenient location.

Vertical-shaft installations of single-suction pumps with a suction elbow are commonly furnished with either a pedestal or a base elbow (see Fig. 135). These may be bolted to soleplates or even grouted in. The grouting arrangement is not too desirable unless there is full assurance that the pedestal or elbow will never be disturbed or that the grouted space is reasonably regular and the grout will separate from the pump without excessive difficulty.

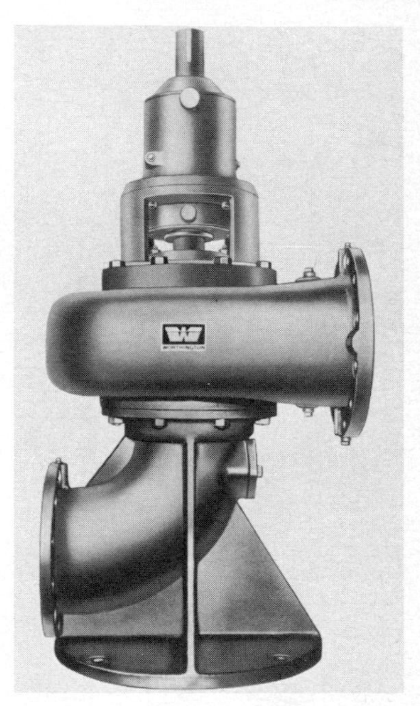

Fig. 135 Small vertical sewage pump with intermediate shafting. (Worthington Pump, Inc.)

Vertical single-suction pumps with bottom suction are commonly used for larger sewage, water supply, or condenser circulating applications. Such pumps are provided with wing feet that are bolted to soleplates grouted in concrete pedestals or piers (Fig. 138). Sometimes the wing feet may be grouted right in the pedestals. These must be suitably arranged to provide proper access to any handholes in the pump and to allow clearance for the elbow suction nozzles if these are used.

If a vertical pump is applied to condensate service or some other service for which the eye of the impeller must be vented to prevent vapor binding, a pump with a bottom single-inlet impeller is not desirable because it does not permit effective venting. Neither does a vertical pump employing a double-suction impeller (Fig. 139). The most suitable design for such applications incorporates a top single-inlet impeller (Fig. 140).

If the driver of a vertical dry-pit pump can be located immediately above the pump, it is often supported on the pump itself (see Fig. 137). The shafts of the pump and driver may be connected by a flexible coupling, which requires that each have its own thrust bearing. If the pump shaft is rigidly coupled to the driver shaft or is an extension of the driver shaft, a common thrust bearing is used, normally in the driver.

Although the driving motors are frequently mounted right on top of the pump casing, one important reason for the use of the vertical-shaft design is the possibility of locating the motors at an elevation sufficiently above the pumps to prevent their accidental flooding. The pump and its driver may be separated by an appreciable length of shafting, which may require steady bearings between the two units. Section 6.3 (Pump Couplings and Intermediate Shafting) discusses the construction and arrangement of the shafting used to connect vertical pumps to drivers located at an elevation some distance above the pump elevation.

Bearings for vertical dry-pit pumps and for intermediate guide bearings are usually antifriction grease-lubricated to simplify the problem of retaining a lubricant in a housing with a shaft projecting vertically through it. Larger units, for which antifriction bearings are not available or desirable, use self-oiling babbitt bearings with spiral oil grooves (Figs. 141 and 142). Figure 142 illustrates a vertical dry-pit pump design with a single sleeve type line bearing. The pump is connected by a rigid coupling to its motor (not shown in the illustration) which is provided with a line and a thrust bearing.

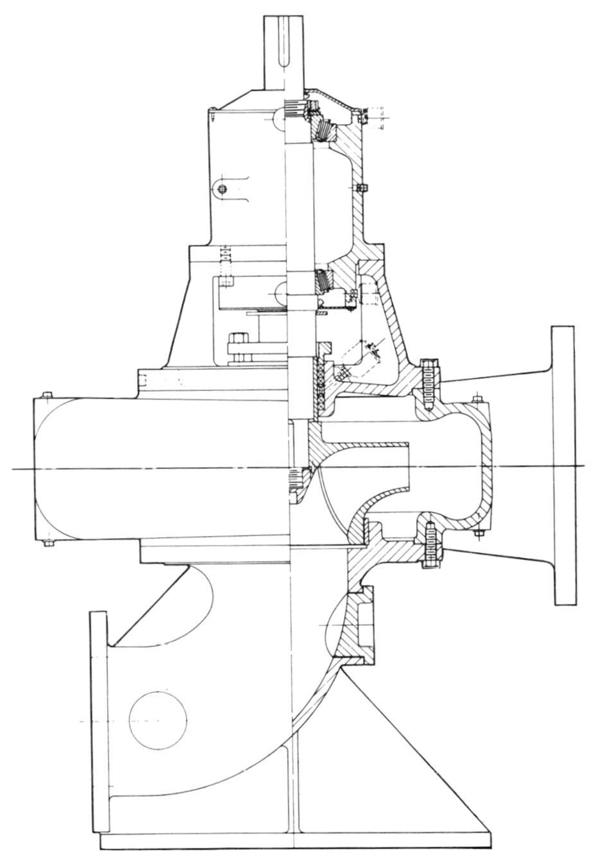

Fig. 136 Section of pump shown in Fig. 135. (Worthington Pump, Inc.)

Vertical dry-pit centrifugal pumps are structurally similar to horizontal-shaft pumps. It is to be noted, however, that many of the very large vertical single-stage single-suction (usually bottom) volute pumps that are preferred for large storm water pumpage, drainage, irrigation, sewage, and water supply projects have no comparable counterpart among horizontal-shaft units. The basic U-section casing of these pumps, which is structurally weak, often requires the use of heavy ribbing to provide sufficient rigidity. In comparable water turbine practice, a set of vanes (called a *speed ring*) is employed between the casing and runner to act as a strut. Although the speed ring does not affect the operation of a water turbine adversely, it would function basically as a diffuser in a pump because of the inherent hydraulic limitations of that construction. Some high-head pumps of this type have been made in the twin-volute design. The wall separating the two volutes acts as a strengthening rib for the casing, thus making it easier to design a casting strong enough for the pressure involved.

Bases and Supports for Vertical Pumping Equipment. Vertical-shaft pumps, like horizontal-shaft units, must be firmly supported. Depending upon the installation, the unit may be supported at one or several elevations. Vertical units are seldom supported from walls, but even that type of support is sometimes encountered.

Occasionally a nominal horizontal-shaft pump design is arranged with a vertical shaft and a wall used as the supporting foundation. Regular horizontal-shaft units can be used for this purpose without modification, except that the bedplate is attached to a wall. Careful attention must be given to the arrangement of the pump bearings to prevent the escape of the lubricant.

Installations of double-suction single-stage pumps with the shaft in the vertical position are relatively rare, except in some marine or navy applications. Hence manufacturers have very few standard pumps of this kind arranged so that a portion of the casing itself forms the support (to be mounted on soleplates). Figure 139 shows such a pump, which also has a casing extension to support the driving motor.

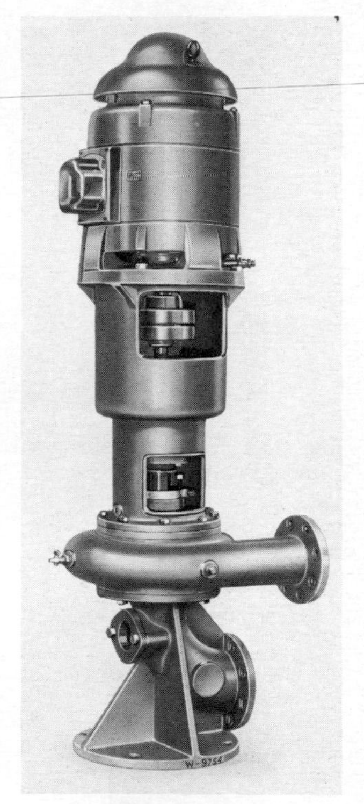

Fig. 137 Vertical sewage pump with direct mounted motor. (Worthington Pump, Inc.)

Vertical Wet-Pit Pumps Vertical pumps intended for submerged operation are manufactured in a great number of designs, depending mainly upon the service for which they are intended. Thus wet-pit centrifugal pumps can be classified in the following manner.

1. Vertical turbine pumps.
2. Propeller or modified propeller pumps
3. Sewage pumps
4. Volute pumps
5. Sump pumps

Vertical Turbine Pumps. Vertical turbine pumps were originally developed for pumping water from wells and have been called *deep-well pumps, turbine-well pumps,* and *borehole pumps*. As their application to other fields has increased, the name *vertical turbine pumps* has been generally adopted by the manufacturers. (This is not too specific a designation because the term *turbine pump* has been applied in the past to any pump employing a diffuser. There is now a tendency to designate pumps using diffusion vanes as *diffuser pumps* to distinguish them from *volute pumps.* As that designation becomes more universal, applying the term *vertical turbine pumps* to the construction formerly called *turbine-well pumps* will become more specific.)

The largest fields of application for the vertical turbine pump are pumping from wells for irrigation and other agricultural purposes, for municipal water supply, and for industrial water supplies, processing, circulating, refrigerating, and air conditioning. This type of pump has also been used for brine pumping, mine dewatering, oil field repressuring, and other purposes.

These pumps have been made for capacities as low as 10 or 15 gpm and as high as 25,000 gpm or more and for heads up to 1,000 ft. Most applications naturally involve the smaller capacities. The capacity of the pumps used for bored wells is naturally limited by the physical size of the well as well as by the rate at which water can be drawn without lowering its level to a point of insufficient pump submergence.

Vertical turbine pumps should be designed with a shaft that can be readily raised or lowered from the top to permit proper adjustment of the position of the impeller in the bowl. An adequate thrust bearing is also necessary to support

Fig. 138 Vertical bottom-suction volute pumps with piloted couplings. (Worthington Pump, Inc.)

Fig. 139 Vertical double-suction volute pump with direct mounted motor. (Worthington Pump, Inc.)

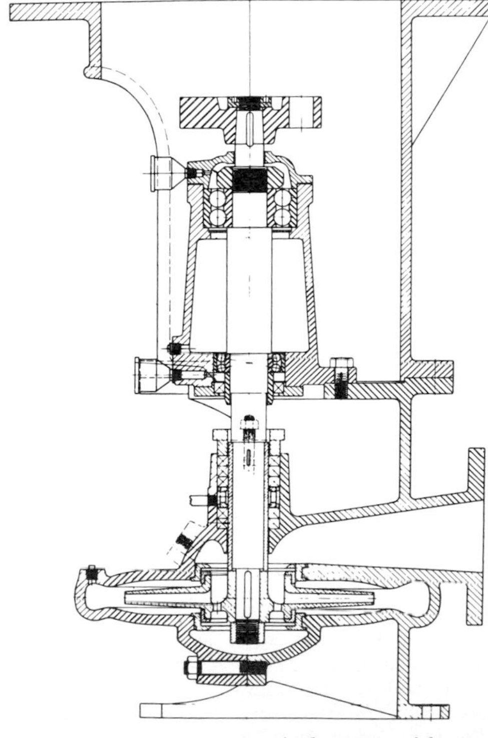

Fig. 140 Section of vertical pump with top-suction inlet impeller. (Worthington Pump, Inc.)

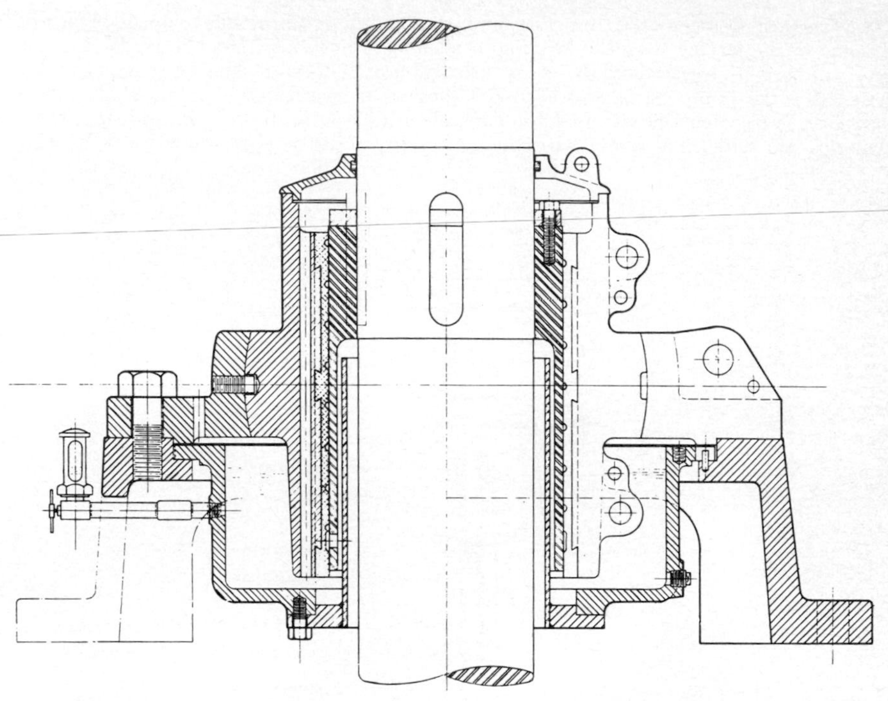

Fig. 141 Self-oiling steady bearing for large vertical shafting. (Worthington Pump Inc.)

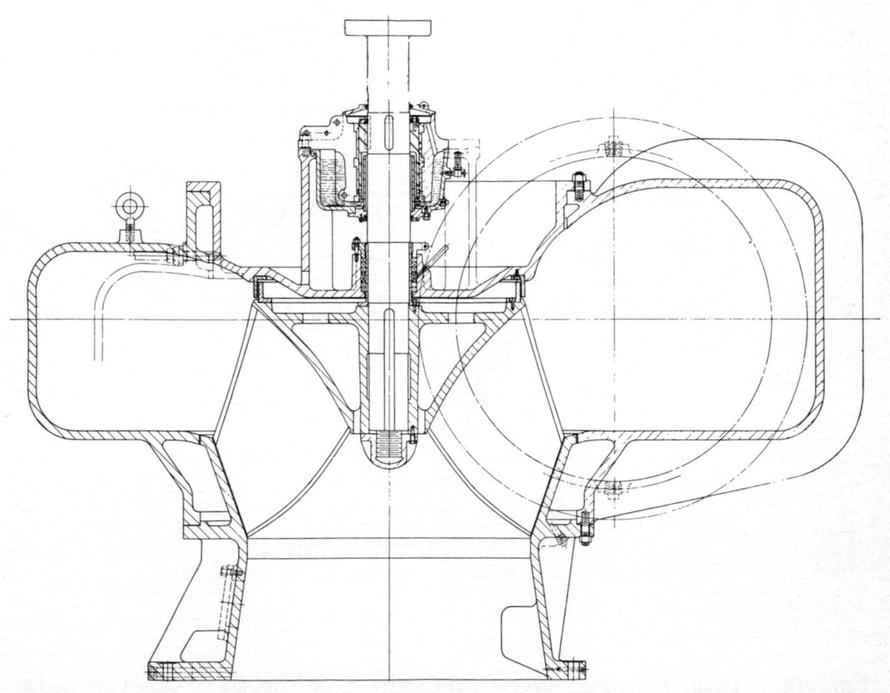

Fig. 142 Section of large vertical bottom-suction volute pump with single guide bearing. (Worthington Pump, Inc.)

the vertical shafting, the impeller, and the hydraulic thrust developed when the pump is in service. As the driving mechanism must also have a thrust bearing to support its vertical shaft, it is usually provided with one of adequate size to carry the pump parts as well. For these two reasons, the hollow-shaft motor or gear is most commonly used for vertical-turbine-pump drive. In addition, these pumps are sometimes made with their own thrust bearings to allow for belt drive

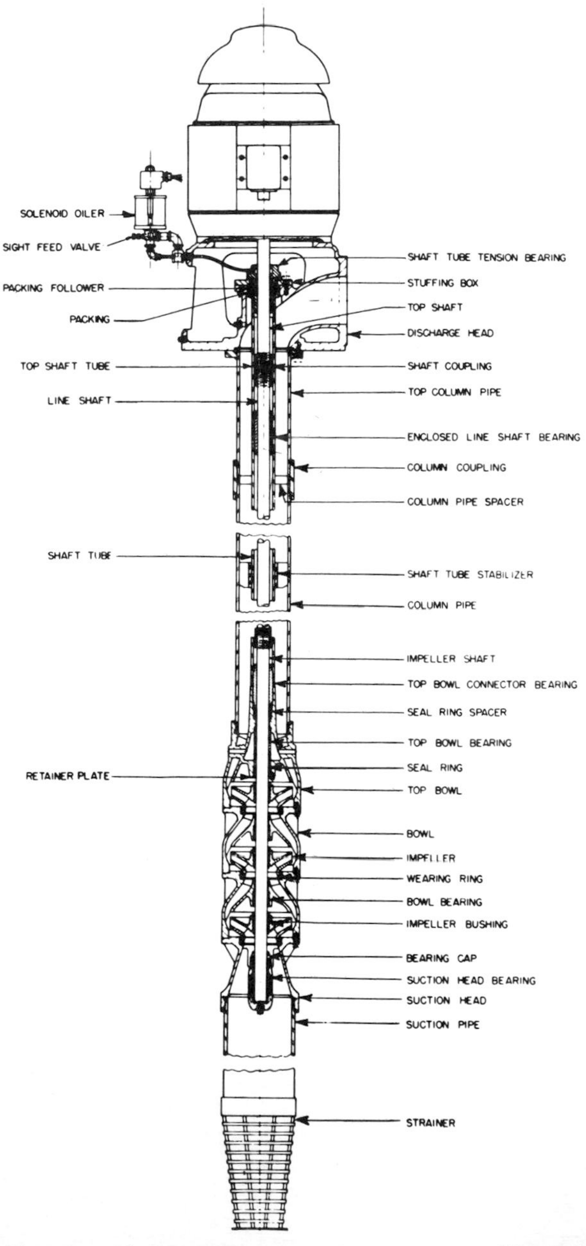

Fig. 143 Section of vertical turbine pump with closed impellers and enclosed line shafting (oil lubrication). (Worthington Pump, Inc.)

or for drive through a flexible coupling by a solid-shaft motor, gear, or turbine. Dual-driven pumps usually employ an angle gear with a vertical motor mounted on its top.

The design of vertical pumps illustrates how a centrifugal pump can be specialized

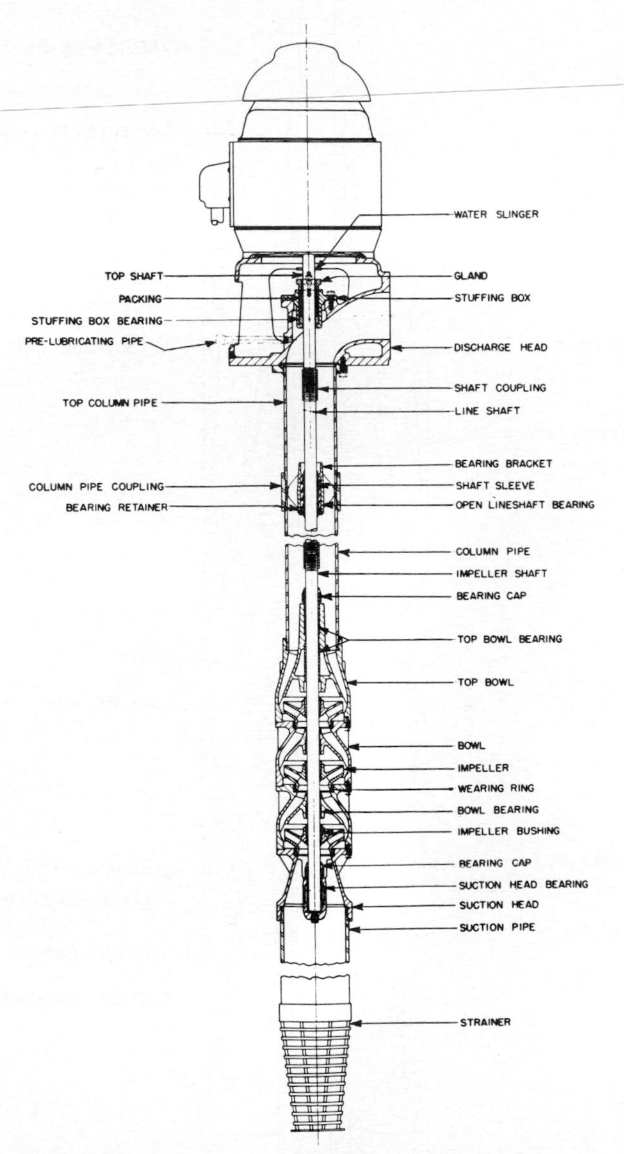

Fig. 144 Section of vertical turbine pump with closed impellers and open-line shafting (water lubrication). (Worthington Pump, Inc.)

to meet a specific application. Figure 143 illustrates a turbine design with closed impellers and enclosed-line shafting; Fig. 144 illustrates another turbine design with closed impellers and open-line shafting.

The bowl assembly or section consists of the suction case (also called suction head or inlet vane), the impeller or impellers, the discharge bowl, the intermediate

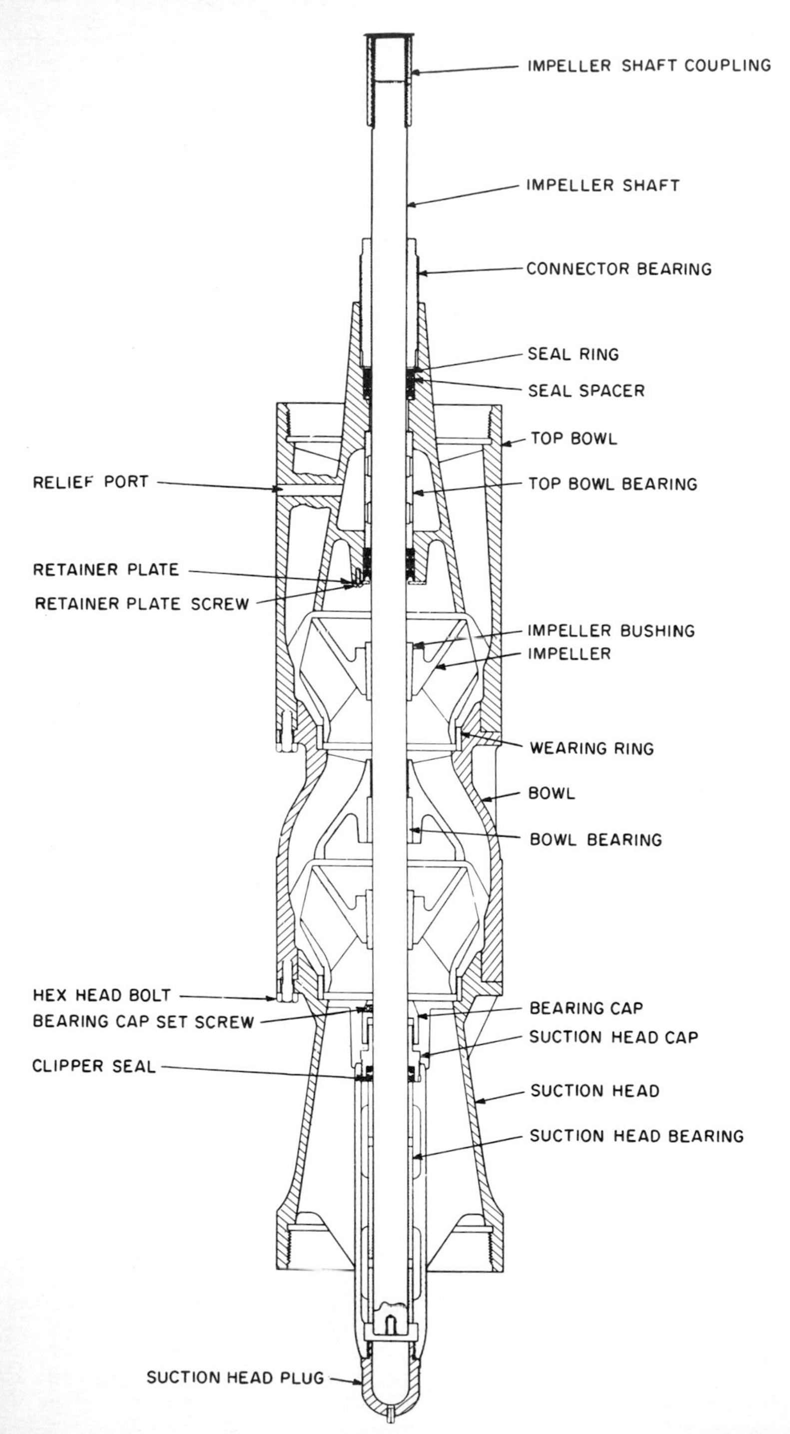

Fig. 145 Section of bowls of a vertical turbine pump with closed impellers for connection to enclosed shafting. (Worthington Pump, Inc.)

bowl or bowls (if more than one stage is involved), the discharge case, the various bearings, the shaft, and miscellaneous parts such as keys, impeller-locking devices, and the like. The column pipe assembly consists of the column pipe itself, the shafting above the bowl assembly, the shaft bearings, and the cover pipe or bearing

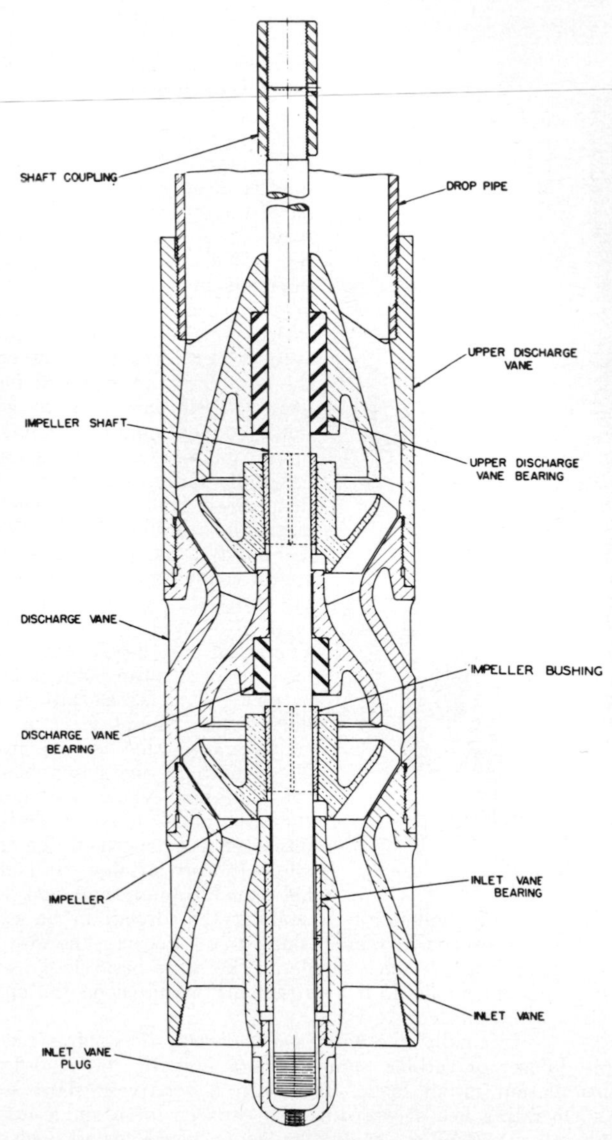

Fig. 146 Section of bowls of a vertical turbine pump with open impellers for connection to open-line shafting. (Worthington Pump, Inc.)

retainers. The pump is suspended from the driving head, which consists of the discharge elbow (for above-ground discharge), the motor or driver and support, and either the stuffing box (in open-shaft construction) or the assembly for providing tension on and the introduction of lubricant to the cover pipe. Below-ground discharge is taken from a tee in the column pipe, and the driving head functions principally as a stand for the driver and support for the column pipe.

Liquid in a vertical turbine pump is guided into the impeller by the suction case or head. This may be a tapered section (Figs. 145 and 146) for attachment of a conical strainer or suction pipe, or it may be a bellmouth.

Semiopen and enclosed impellers are both commonly used. For proper clearances in the various stages, the semiopen impeller requires more care in assembly on the impeller shaft and more accurate field adjustment of the vertical shaft position in order to obtain the best efficiency. Enclosed impellers are favored over semiopen ones, moreover, because wear on the latter reduces capacity, which cannot be restored unless new impellers are installed. Normal wear on enclosed impellers does not affect impeller vanes, and worn clearances may be restored by replacing wearing rings. The thrust produced by semiopen impellers may be as much as 150 percent greater than that by enclosed impellers.

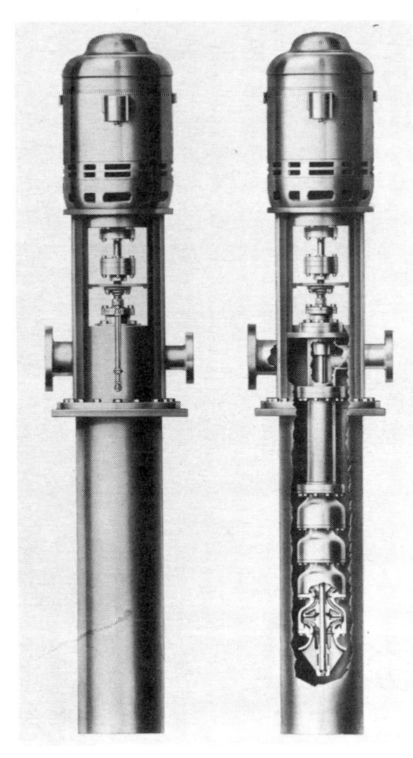

Fig. 147 Vertical turbine "can pump" for condensate service. (Worthington Pump, Inc.)

Occasionally in power plants the maximum water level that can be carried in the condenser hotwell will not give adequate NPSH (net positive suction head) for a conventional horizontal condensate pump mounted on the basement floor, especially if the unit is one that has been installed in an existing plant in a space originally allotted for a smaller pump. To build a pit for a conventional horizontal condensate pump or a vertical dry-pit pump that will provide sufficient submergence involves considerable expense. Pumps of the design shown in Fig. 147 have become quite popular in such applications. This is basically a vertical turbine pump mounted in a tank (often called a can) that is sunk into the floor. The length of the pump has to be such that sufficient NPSH will be available for the first stage impeller design, and the diameter and length of the tank have to allow for proper flow through the space between the pump and tank and then for a turn and flow into the bellmouth. Installing this design in an existing plant is naturally much less expensive than making a pit because the size of the hole necessary to install the tank is much smaller. The same basic design has also been applied to pumps handling volatile liquids that are mounted on the operating floor and not provided with sufficient NPSH.

Propeller Pumps. Originally the term *vertical propeller pump* was applied to vertical wet-pit diffuser or turbine pumps with a propeller or axial-flow impellers, usually for installation in an open sump with a relatively short setting (Figs. 148 and 149). Operating heads exceeding the capacity of a single-stage axial-flow impeller might call for a pump of two or more stages or a single-stage pump with a lower specific speed and a mixed-flow impeller. High enough operating heads might demand a pump with mixed-flow impellers and two or more stages. For lack of a more suitable name, such high-head designs have usually been classified as propeller pumps also.

Although vertical turbine pumps and vertical modified propeller pumps are basically the same mechanically and even could be of the same specific speed hydraulically, a basic turbine pump design is one that is suitable for a large number of stages, whereas a modified propeller pump is a mechanical design basically intended for a maximum of two or three stages.

Most wet-pit drainage, low-head irrigation, and storm-water installations employ conventional propeller or modified propeller pumps. These pumps have also been used for condenser circulating service, but a specialized design dominates this field. As large power plants are usually located in heavily populated areas, they frequently have to use badly contaminated water (both fresh and salt) as a cooling medium. Such water quickly shortens the life of fabricated steel. Cast iron, bronze, or an even more corrosion-resistant cast metal must therefore be used for the column pipe assembly. This requirement means a very heavy pump if large capacities are involved. To avoid the necessity of lifting this large mass for maintenance of the rotating parts, some designs (one of which is illustrated in Fig. 150) are built so that the impeller, diffuser, and shaft assembly can be removed from the top without disturbing the column pipe assembly. These designs are commonly designated as *pullout* designs.

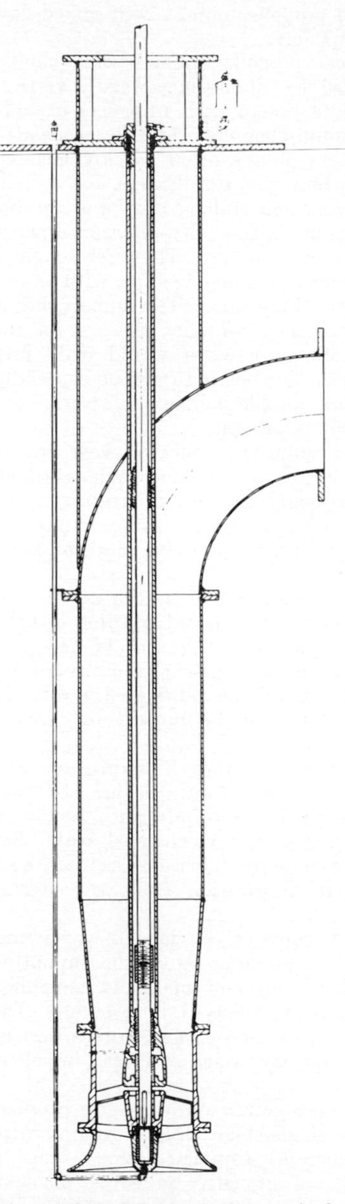

Fig. 148 Section of vertical propeller pump with below-ground discharge. (Worthington Pump, Inc.)

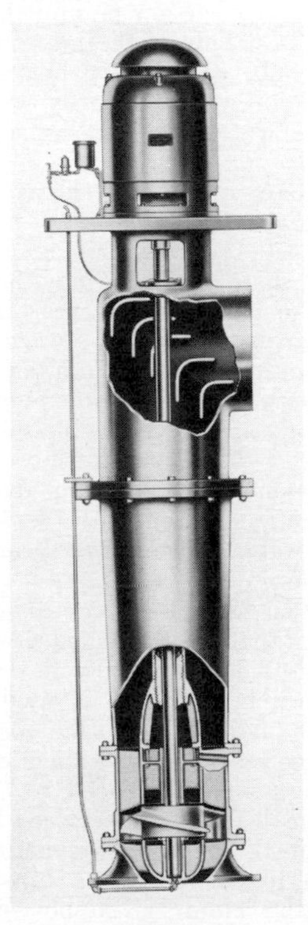

Fig. 149 Vertical propeller pump with below-ground discharge. (Peerless Pump Co.)

Like vertical turbine pumps, propeller and modified propeller pumps have been made with both open- and enclosed-line shafting. Except for condenser circulating service, enclosed shafting—using oil as a lubricant but with a grease-lubricated tail bearing below the impeller—seems to be favored. Some pumps handling condenser circulating water use enclosed shafting but with water (often from another source) as the lubricant, thus eliminating any possibility of oil getting into the circulating water and coating the condenser tubes.

Propeller pumps have open propellers. Modified propeller pumps with mixed-flow impellers are made with both open and closed impellers.

Sewage Pumps. Except for some large vertical propeller pumps that handle dilute sewage (basically storm water contaminated by domestic sewage), vertical wet-pit sewage pumps have a bottom-suction volute design with impellers capable of handling solids and stringy materials with minimum clogging. Usually suspended from a higher floor by means of a drop pipe, these pumps often employ covered or enclosed shafting like that used in vertical turbine pumps. Except for a bell-mouth suction inlet and certain differences in bearing and stuffing box construction, they usually are hydraulically and mechanically similar to their dry-pit counterparts.

Three basic constructions have been used for such pumps. The first employs impellers without back rings and a water- or grease-lubricated bearing with a seal at its lower end immediately above the impeller (Fig. 151). The upper end is vented to the suction pit to prevent any appreciable hydraulic pressure on the seal at the lower end of the shaft cover pipe; otherwise water would work into the cover pipe. The seal at the lower end of the impeller bearing must be especially effective with high pump heads; otherwise a considerable amount of water will leak through the bearing, with some cutting if grit is present.

The second construction is similar but employs pump-out vanes or wearing-ring joints on the back side of the impeller (the latter necessitates balance holes through the impeller hub) so that the bearing is subjected only to suction pressure. The third construction, used primarily with impellers having no back rings or pump-out vanes, retains a stuffing box in some form, with bearings above and separate from the box.

Although shaft seals and packing used to seal the lower end of the cover pipe or bottom bearing are intended to exclude as much water as possible, some leakage is to be expected at high-suction water levels even when the seal is new. As some of the shaft bearings may have to operate in water or a mixture of oil and water, the bearing may wear relatively faster than one lubricated positively with oil or grease. Wet-pit sewage pumps should usually be limited to services requiring operation for a very limited period of the day.

Volute Pumps. A very interesting design of wet-pit pump is shown on Fig. 152. It uses a single-stage double-suction impeller in a twin-volute casing. Because the axial thrust is balanced, the thrust bearing need carry only the weight of the rotating element. The pump requires no stuffing box or mechanical seal. The shaft is entirely enclosed and the bearings are externally lubricated, either with oil or with water. The lower bearing receives its lubrication from an external pipe connection.

Sump Pumps. The term *sump pump* ordinarily conveys the idea of a vertical wet-pit pump that is suspended from a floor plate or sump cover or supported by a foot on the bottom of a well, that is motor-driven and automatically controlled by a float switch, and that is used to remove drains collected in a sump. The term does not indicate a specific construction, for both diffuser and volute designs are used; these may be single-stage or multistage and have open or closed impellers of a wide range of specific speeds.

For very small capacities serviced by fractional horsepower motors, *cellar drainers* can be obtained. These are small and usually single-stage volute pumps with single-suction impellers (either top or bottom suction) supported by a foot on the casing; the motor is supported well above the impeller by some form of a column enclosing the shaft. These drainers are made as complete units, including float, float switch, motor, and strainers (Fig. 153).

Sump pumps of larger capacity may be vertical propeller or turbine pumps (single-stage or multistage) or vertical wet-pit sewage or volute pumps. If solids or other waste materials may be washed into the sump, the vertical-wet-pit sewage pump with a nonclogging impeller is preferred. The larger sump pumps are usually standardized but obtainable in any length, with covers of various sizes (on which a float switch may be mounted), and the like. Duplex units, that is, two pumps

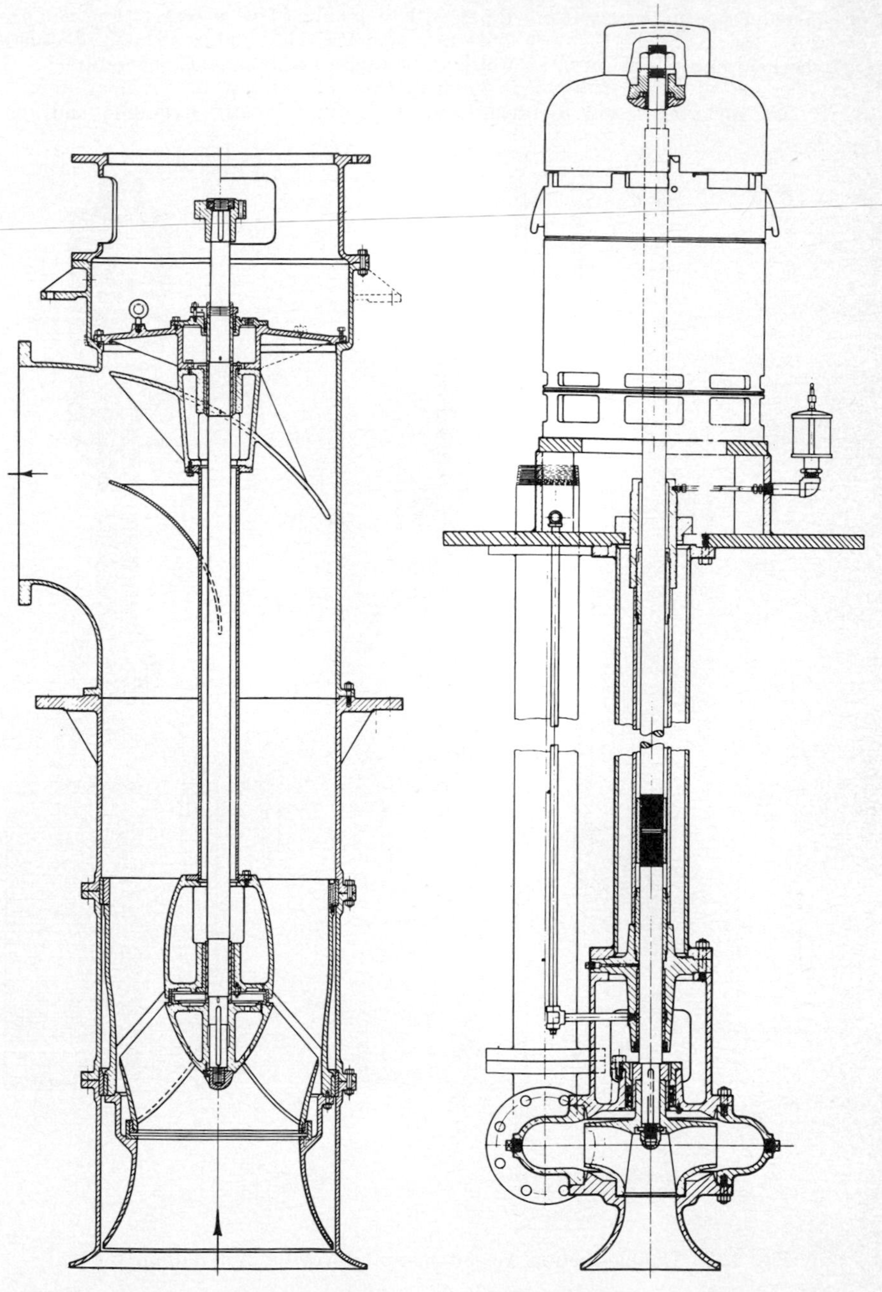

Fig. 150 Section of vertical modified propeller pump with removable bowl and shafting assembly. (Worthington Pump, Inc.)

Fig. 151 Section of a vertical wet-pit or nonclogging pump. (Worthington Pump, Inc.)

on a common sump cover (sometimes with a manhole for access to the sump), are often used (Fig. 154). Such units may operate their pumps in a fixed order, or a mechanical or electrical alternator may be used to equalize their operation.

Application of Vertical Wet Pit Pumps. Like all pumps, the vertical wet-pit pump has advantages and disadvantages, the former mostly hydraulic and the

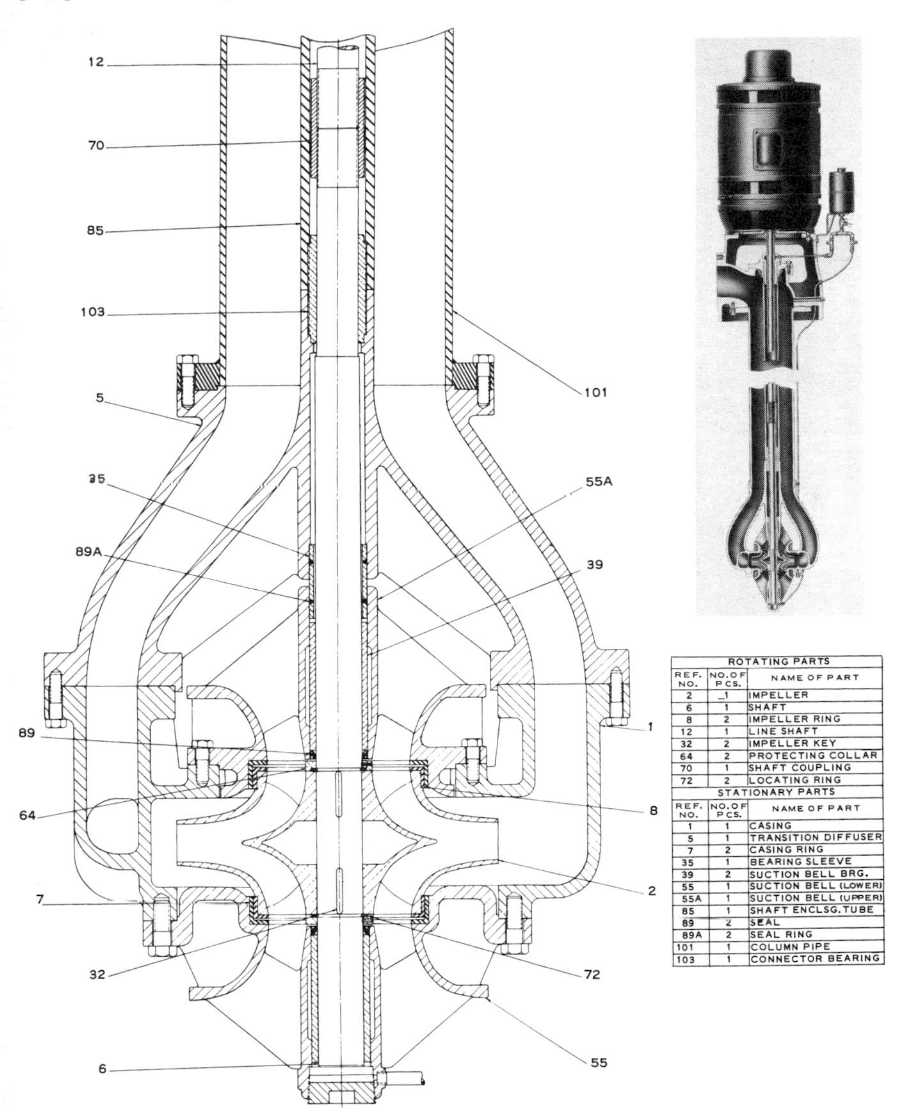

ROTATING PARTS		
REF. NO.	NO. OF PCS.	NAME OF PART
2	1	IMPELLER
6	1	SHAFT
8	2	IMPELLER RING
12	1	LINE SHAFT
32	2	IMPELLER KEY
64	2	PROTECTING COLLAR
70	1	SHAFT COUPLING
72	2	LOCATING RING
STATIONARY PARTS		
REF. NO.	NO. OF PCS.	NAME OF PART
1	1	CASING
5	1	TRANSITION DIFFUSER
7	2	CASING RING
35	1	BEARING SLEEVE
39	2	SUCTION BELL BRG.
55	1	SUCTION BELL (LOWER)
55A	1	SUCTION BELL (UPPER)
85	1	SHAFT ENCLSG. TUBE
89	2	SEAL
89A	2	SEAL RING
101	1	COLUMN PIPE
103	1	CONNECTOR BEARING

Fig. 152 Double-suction wet-pit pump. (Worthington Pump, Inc.)

latter primarily mechanical. If the impeller (first-stage impeller in multistage pumps) is submerged, there is no priming problem, and the pump can be automatically controlled without fear of its ever running dry. Moreover, the available NPSH is greater (except in closed tanks) and often permits a higher rotative speed for the same service conditions. The only mechanical advantage is that the motor or driver can be located at any desired height above any flood level. The mechanical disadvantages are the following: (1) possibility of freezing when idle, (2) possibility of damage by floating objects if the unit is installed in an open ditch or similar installation, (3) inconvenience of lifting out and dismantling for inspection and repairs, no matter how small, and (4) the relative short life of the pump bearings unless the water and bearing design are ideal. The vertical wet-pit pump is the best pump available for some applications, not ideal but

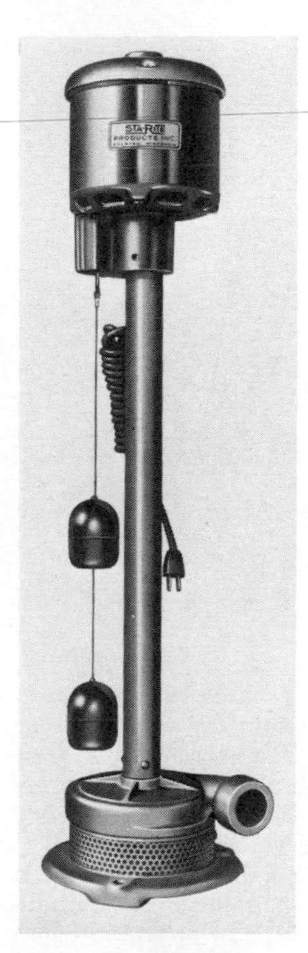

Fig. 153 Cellar-drainer sump pump. (Sta-Rite Products)

Fig. 154 Duplex sump pump. (Economy Pump Co.)

the most economical for other installations, a poor choice for some, and the least desirable for still others.

Typical Arrangements of Vertical Pumps A pump is only part of a pumping system. The hydraulic design of the system external to the pump will affect the overall economy of the installation and can easily have an adverse effect upon the performance of the pump itself. Vertical pumps are particularly susceptible because the small floor space occupied by each unit offers the temptation to reduce the size of the station by placing the units closer together. If the size is reduced, the suction arrangement may not permit the proper flow of water to the pump suction intake. Recommended arrangements for vertical pumps are discussed in detail in Sec. 11.

Section **2.3**

Centrifugal Pump Performance

C. P. KITTREDGE

DEFINITIONS

The most important operating characteristics are the *capacity* Q, the *head* H, the *power* P, and the *efficiency* η. Variables which influence these are the *speed* n and the impeller or wheel *diameter* D. It has been assumed that all other dimensions of the impeller and casing have been fixed. The *specific speed* n_s is a parameter which classifies impellers according to geometry and operating characteristics. The value at the operating conditions corresponding to best efficiency is called the specific speed of the impeller and, usually, is the value of interest.

Capacity The *pump capacity* Q is the volume of fluid per unit time delivered by the pump. In English measure it is usually expressed in gallons[1] per minute (gpm) or, for very large pumps, in cubic feet per second (ft^3/s). In metric measure the corresponding units are litres per second (ℓ/s) and cubic metres per second (m^3/s). If the capacity is measured at a location m, where the specific weight of the fluid, γ_m, is different from the specific weight at the inlet flange, γ_s, then the capacity is given by $Q = Q_m(\gamma_m/\gamma_s)$.

Head The *pump head* H represents the net work done on a unit weight of fluid in passing from the inlet or suction flange s to the discharge flange d. It is given by

$$H = \left(\frac{p}{\gamma} + \frac{V^2}{2g} + Z\right)_d - \left(\frac{p}{\gamma} + \frac{V^2}{2g} + Z\right)_s \tag{1}$$

The term p/γ, called the *pressure head* or *flow work,* represents the work required to move a unit weight of fluid across an arbitrary plane perpendicular to the velocity vector V against the pressure p. The term $V^2/2g$, called the *velocity head,* represents the kinetic energy of a unit weight of fluid moving with velocity V. The term Z, called the *elevation head* or *potential head,* represents the potential energy of a unit weight of fluid with respect to the chosen datum.

The first parenthetical term in Eq. (1) represents the *discharge head,* h_d, and the second, the *inlet* or *suction head* h_s. The difference is variously called the *pump head, pump total head,* or *total dynamic head.*

The unit of H is customarily feet of fluid pumped, in English measure, and meters of fluid pumped in metric measure. This requires that consistent units

[1] United States: 1 gal = 231 in^3; United Kingdom: 1 gal = 277.418 in^3.

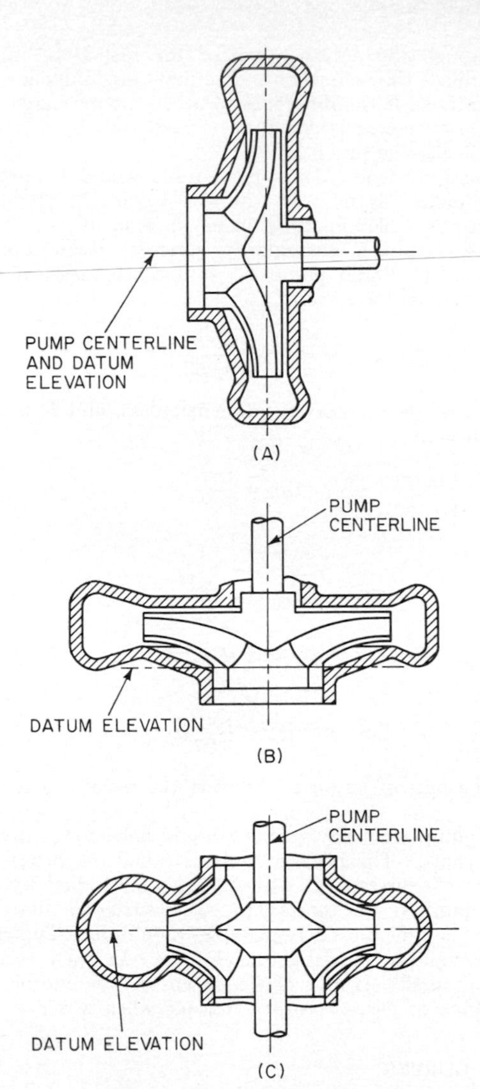

Fig. 1 Elevation datum for defining pump head. (From "Hydraulic Institute Standards," 12th ed., copyright 1969 by the Hydraulic Institute, Cleveland, Ohio)

be used for all quantities in the equation. In English measure, typical units would be:

■ Static pressure p, in pounds force per square foot (lb/ft^2)
■ Specific weight γ, in pounds force per cubic foot (lb/ft^3)
■ Average fluid velocity V, in feet per second (ft/s) where $V = Q/A$ and A = cross-sectional area of flow passage in square feet (ft^2)
■ Acceleration of gravity, g, in feet per second per second (ft/s^2) assumed to be 32.174 ft/s^2 since corrections for latitude and elevation above or below mean sea level usually are small
■ Elevation Z, in feet, above or below the datum

The standard datum for horizontal shaft pumps is a horizontal plane through the centerline of the shaft, Fig. 1A. For vertical shaft pumps the datum is a horizontal plane through the entrance eye of the first stage impeller, Fig. 1*B*, if single suction, or through the centerline of the first-stage impeller, Fig. 1*C*, if double suction. Since pump head is the difference between the discharge and suction heads, it is not necessary that the standard datum be used and any convenient datum may be selected for computing the pump head.

In the International System (SI)[1] typical units would be, for p, pascals (Pa), newtons per square metre (N/m^2), or bars (10^5 kg/m·s^2); $\gamma = \rho g$, where ρ = mass density (kg/m^3) and g = 9.8067 m/s^2; Q, in m^3/s; A, in m^2; V, in m/s; and Z, in m.

Power The pump output is customarily given as liquid horsepower or water horsepower if water is the liquid pumped. In English measure, the output power, ℓhp, in horsepower, is given by

$$\ell\text{hp} = \frac{QsH}{3{,}960} \tag{2}$$

where Q is in gpm, s is specific gravity, dimensionless, and H is in feet. If Q is in ft^3/s, the equation becomes

$$\ell\text{hp} = \frac{QsH}{8.82} \tag{3}$$

In metric measure,

$$\ell\text{hp (metric)} = \frac{QsH}{75} \tag{4}$$

where Q is in litres per second (ℓ/s), and H is in metres. If Q is in m^3/s, Eq. (4) becomes

$$\ell\text{hp (metric)} = \frac{QsH}{0.075} \tag{5}$$

In the SI system, the unit of power is the watt (kg·m^2/s^3), so that Eqs. (4) and (5) would not apply.

Efficiency The pump efficiency η is the liquid horsepower divided by the power input to the pump shaft. The latter usually is called the brake horsepower (bhp). The efficiency may be expressed as a decimal or multiplied by 100 and expressed as percent. Some pump-driver units are so constructed that the actual power input to the pump is difficult or impossible to obtain. Typical of these is the "canned" pump for volatile or dangerous liquids. In such case, only an *overall efficiency* can be obtained. If the driver is an electric motor, this is called the *wire-to-liquid efficiency* or *wire-to-water efficiency* when water is the liquid pumped.

CHARACTERISTIC CURVES

Ideal Pump and Fluid An ideal impeller contains a very large number of vanes of infinitesimal thickness. The particles of an ideal fluid move exactly parallel to such vane surfaces without friction. As shown in Sec. 2.1, an analysis of the power transmitted to an ideal fluid by such an impeller leads to

$$H_e = \frac{u_2 c_{u2}}{g} - \frac{u_1 c_{u1}}{g} \tag{6}$$

often called *Euler's pump equation.* The Euler head is the work done on a unit weight of the fluid by the impeller. In Fig. 2, u is the peripheral velocity of any point on a vane and w is the velocity of a fluid particle relative to the vane as seen by an observer attached to and moving with the vane. The absolute velocity of a fluid particle, c, is the vector sum of u and w. The subscripts 1 and 2 refer to the entrance and exit cross sections of the flow passages.

[1] SI base units include the *metre,* m, the *kilogram mass,* kg, and the *second,* s. The *newton* (N) is defined as that force which steadily applied to a mass of 1 kg produces an acceleration of 1 m/s^2.

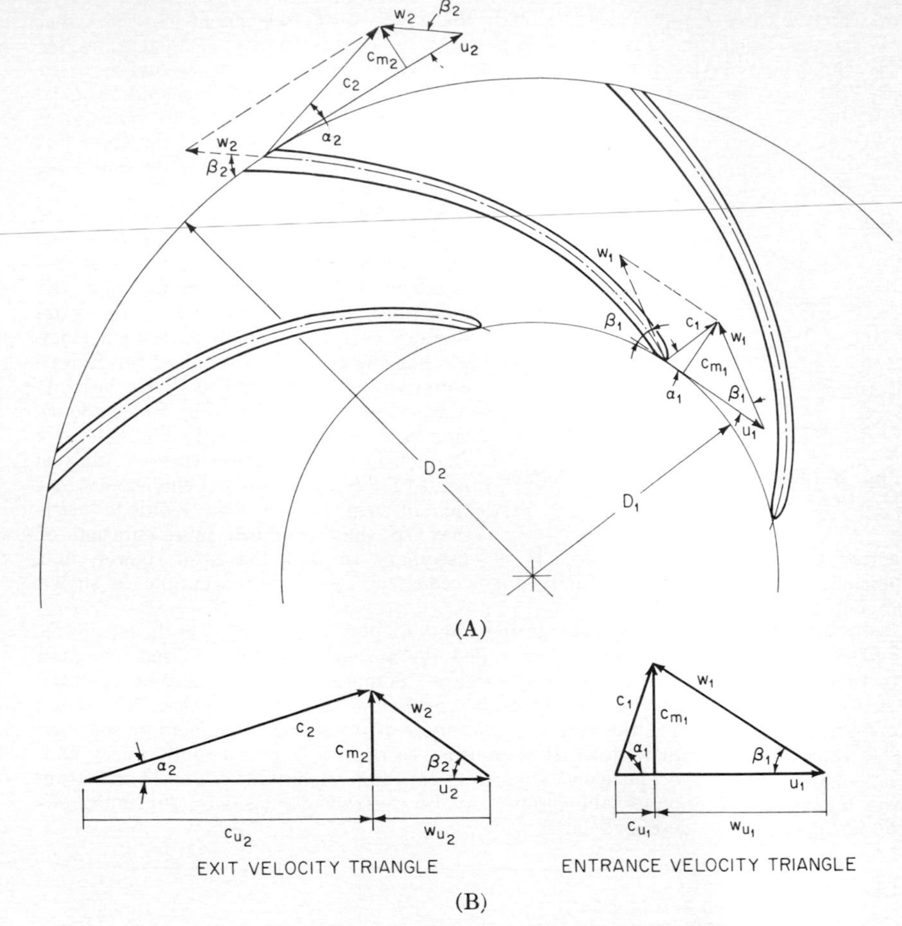

Fig. 2 Velocity diagrams for radial-flow impellers.

The second term in Euler's equation frequently is small compared to the first term and may be neglected so that Eq. (6) becomes

$$H_e \cong \frac{u_2 c_{u2}}{g} \tag{7}$$

As shown in Fig. 2*B*, the absolute velocity vector c may be resolved into the meridian or radial velocity, c_m, and the peripheral velocity c_u. From the geometry of the figure

$$c_{u2} = u_2 - \frac{c_{m2}}{\tan \beta_2} \tag{8}$$

which, substituted into Eq. (7), gives

$$H_e = \frac{u_2^2}{g} - \frac{u_2 c_{m2}}{g \tan \beta_2} \tag{9}$$

Neglecting leakage flow, the meridian velocity c_m must be proportional to the capacity Q. With the additional assumption of constant impeller speed, Eq. (9) becomes

$$H_e = k_1 - k_2 Q \tag{10}$$

in which k_1 and k_2 are constants with the value of k_2 dependent on the value of the vane angle β_2. Figure 3 shows the Euler head-capacity characteristics for the three possible conditions on the vane angle at exit β_2. The second term in Eq. (6) may be treated in like manner to the foregoing and included in Eqs. (9) and (10). The effect on Fig. 3 would be to change the value of $H_e = u_2^2/g$ at $Q = 0$ and the slopes of the lines but all head-capacity characteristics would remain straight lines.

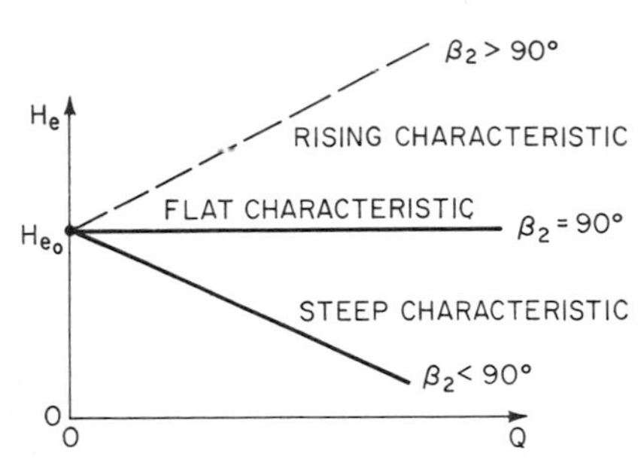

Fig. 3 Euler's head-capacity characteristics.

Real Pump and Fluid The vanes of real pump impellers have finite thickness and are relatively widely spaced. Investigations [16, 17] have shown that the fluid does not flow parallel to the vane surfaces even at the point of best efficiency so that the conditions required for Euler's equation are not fulfilled. The head is always less than predicted by Euler's equation and the head-capacity characteristic frequently is an irregular curve. Analysis has so far failed to predict the characteristics of real pumps with requisite accuracy so that graphical representation of actual tests is in common use. It is customary to plot the head, power, and efficiency as functions of capacity for a constant speed. An example is shown in Fig. 4.

Pumps are designed to operate at the point of best efficiency. The head, power, and capacity at best efficiency, often called the *normal* values, have been indicated in this section by H_n, P_n, and Q_n respectively. Sometimes a pump may be operated continuously at a capacity slightly above or below Q_n. In such case, the actual operating point is called the *rated* or *guarantee* point if the manufacturer specified this capacity in the guarantee. It is unusual to operate a pump continuously at a capacity at which the efficiency is much below the maximum value. Apart from the unfavorable economics, the pump may be severely damaged by continued off-design operation as described later.

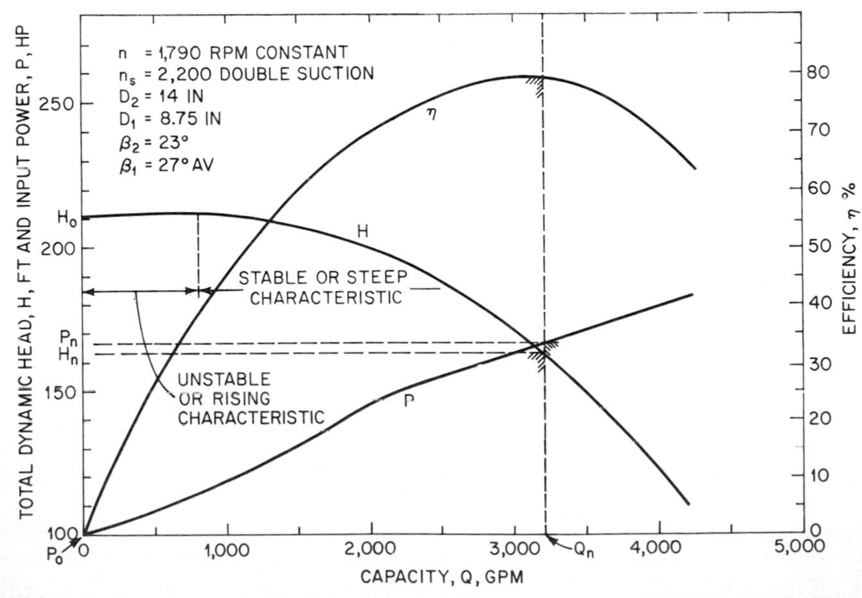

Fig. 4 Typical pump characteristics, backward-curved vanes.

Backward-curved Vanes, $\beta_2 < 90°$ Figure 4 shows the characteristics of a double-suction pump with backward-curved vanes, $\beta_2 = 23°$. The impeller discharged into a single volute casing and the specific speed was $n_s \approx 2{,}200$ at best efficiency. At shutoff, $Q = 0$, Eq. (9) predicts Euler's head H_e to be 374 ft, whereas the pump actually developed about 210 ft, and this remained nearly constant for the range $0 < Q < 1{,}000$ gpm. For $Q > 1{,}000$ gpm, the head decreased with increasing capacity but not in the linear fashion predicted by Euler's equation. At best efficiency, $Q_n = 3{,}200$ gpm, Eq. (9) predicts $H_e = 281$ ft, whereas the pump actually developed $H_n = 164$ ft.

Radial Vanes, $\beta_2 = 90°$ Large numbers of radial-vaned pumps are used as cellar drainers, cooling-water pumps for internal combustion engines, and other

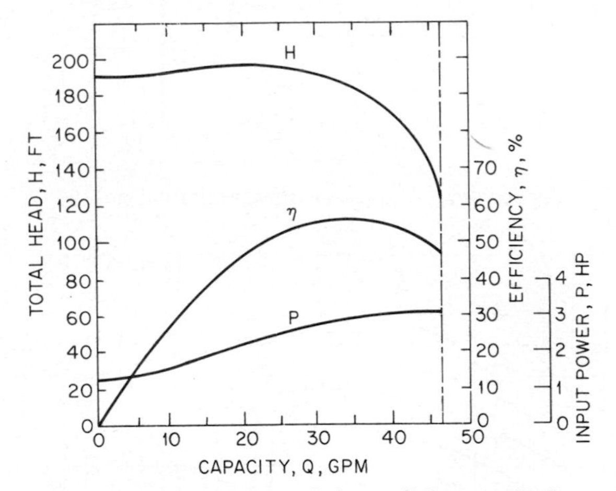

Fig. 5 Pump characteristics, radial vanes [18].

applications where low first cost is more important than high efficiency. The impellers are rarely more than 6 in in diameter but the speed range may be from a few hundred to 30,000 rpm or more. The casings usually are concentric with the impellers and have one or more discharge nozzles that act as diffusers. The impellers usually are open with from three to six flat sheet-metal vanes. The clearance between vanes and casing is relatively large for easy assembly. Such pumps exhibit a steep head-capacity curve. They develop a higher head than would be obtained from equivalent backward-curved vanes but the efficiency usually is lower.

Figure 5 shows the characteristics of a pump as reported by Rupp [18]. The impeller was fully shrouded, $D = 5.25$ in, and fitted with 30 vanes of varying length. The best efficiency, $\eta = 55$ percent, was unusually high for the specific speed $n_s = 475$, as may be seen on Fig. 7. The head-capacity curve showed a rising (unstable) characteristic for $0 < Q < 25$ gpm and a steep characteristic for $Q > 25$ gpm.

Figure 6 shows the characteristics of a pump as reported by Barske [19]. The impeller was open, $D \approx 3$ in, and fitted with six radial tapered vanes. The effective β_2 may have been slightly greater than 90° due to the taper. At 30,000 rpm the best efficiency was over 35 percent at $n_s = 355$, which is much higher than for a conventional pump of this specific speed and capacity (Fig. 7). The head-capacity curve showed a nearly flat characteristic over most of the usable range as predicted by Euler's equation, but the head was always lower than the Euler head H_e. The smooth concentric casing was fitted with a single diffusing discharge nozzle. When two or more nozzles were used, the head-capacity curves showed irregularities at low capacities and became steep characteristics at high capacities.

Manson [21] has reported performance characteristics for jet-engine fuel pumps having straight radial vanes in enclosed impellers. The head curves showed un-

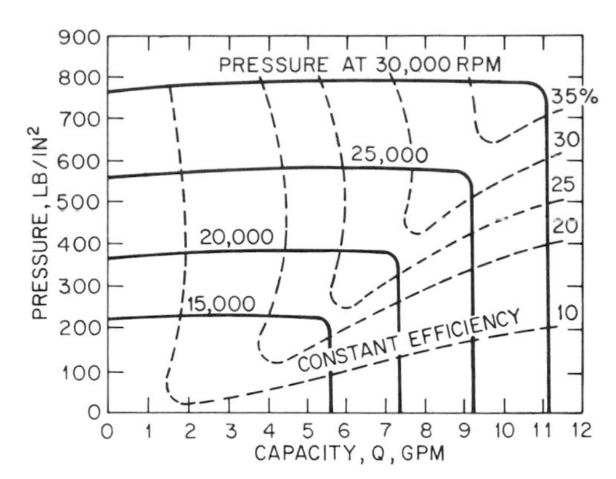

Fig. 6 Pump characteristics, radial vanes [19].

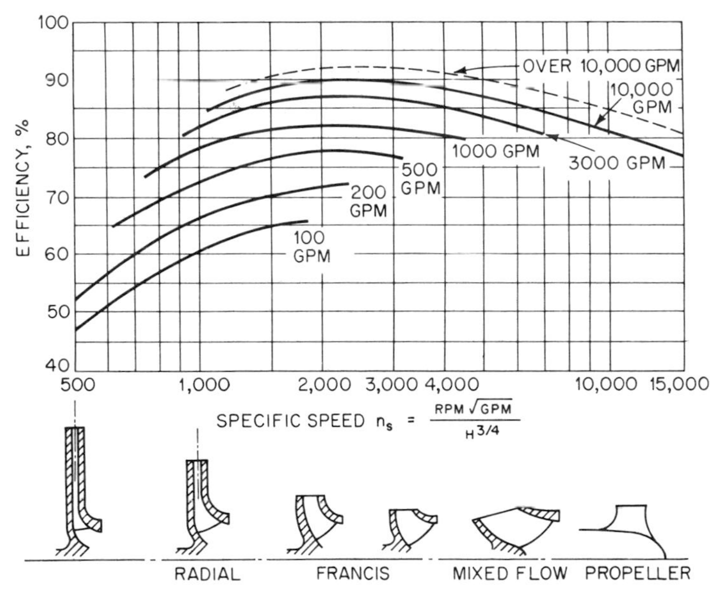

Fig. 7 Pump efficiency versus specific speed and size. (Worthington Pump International, Inc.)

stable characteristics at low capacities and steep characteristics at higher capacities. The best efficiency reported was 54.7 percent for an impeller diameter of 3.300 in and speed $n = 28{,}650$ rpm.

Forward-curved Vanes, $\beta_2 > 90°$ Pumps with forward-curved vanes have been proposed [20] but the research necessary to achieve an efficient design appears never to have been carried out. Tests have been made of conventional double-suction pumps with the impellers mounted in the reversed position but with rotation correct for the volute casing.

Table 1 shows the pertinent results for six different pumps. Both capacity and efficiency were drastically reduced and there was only a modest increase in head for five of the six pumps. The sixth pump showed a 38 percent increase in

TABLE 1 Effects of Reversed Mounting of Impeller

Number of stages	Specific speed per stage, n_s	Shutoff head in percent of normal shutoff head	Percent of normal values at best efficiency			
			Head	Capacity	Power	Efficiency
2	828	86	111	65	104	71
2	1,024	82	112	88	145	68
1	1,240	75	105	38.5	68.5	59
1	1,430	82	106	69.7	138	53.5
1	2,570	74.5	117	62	138	52.5
1	2,740	77.5	138	61.5	180	47

SOURCE: Worthington Pump International, Inc.

head over that obtained with the impeller correctly mounted. Published estimates [4, 11] of the head-capacity curves to be expected from reversed impellers predict an unstable characteristic at the low end of the capacity range and a steep characteristic at the high end of the range.

PERFORMANCE PARAMETERS

Affinity Laws Consider a series of geometrically similar pumps. It may be shown by a dimensional analysis that three important dimensionless parameters describing pump performance are $D\omega/\sqrt{gH}$, $Q/D^2\sqrt{gH}$, and $P/D^2\rho(gH)^{3/2}$, where ω is the pump speed in radians per unit time, ρ is the mass density of the fluid, and g is the acceleration of gravity, all expressed in the system of measure selected for the other variables. At a common operating point, say the best efficiency point,

$$\frac{D\omega}{\sqrt{gH}} = C_1 \qquad \frac{Q}{D^2\sqrt{gH}} = C_2 \qquad \text{and} \qquad \frac{P}{D^2\rho(gH)^{3/2}} = C_3 \tag{11}$$

for the geometrically similar series. The C's depend slightly on the efficiency which, in turn, may depend on the size of the pump, but it is customary to assume them to be constants. Since gravity is very nearly constant over the surface of the earth, it may be combined with the C's. Also it is common practice to restrict consideration to cold water and allow C_3 to absorb ρ. Equations (11) may then be written

$$\frac{Dn}{\sqrt{H}} = k_3 \qquad \frac{Q}{D^2\sqrt{H}} = k_4 \qquad \text{and} \qquad \frac{P}{D^2H^{3/2}} = k_5 \tag{12}$$

Equations (12) are known as the *affinity laws*. Being no longer dimensionless, any convenient system of units may be used but, once selected, no change in the units may be made. Usually, in the United States H = pump head in ft, Q = capacity in gpm, P = power in hp, n = pump speed in rpm, and D = impeller diameter in either ft or in. Obviously the numerical values of the k's depend upon the system of measure and units employed. k_3, k_4, and k_5 are sometimes called, respectively, the *unit speed, unit capaciy,* and *unit power*.

Specific Speed The diameter may be eliminated between the first two of Eqs. (12) to obtain the specific speed

$$n_s = k_3\sqrt{k_4} = \frac{n\sqrt{Q}}{H^{3/4}} \tag{13}$$

In the United States it is usual to express n in rpm, Q in gpm, and H in feet. Conversion to other units may be accomplished by

$$\text{rpm}\frac{\sqrt{\text{gpm}}}{(\text{ft})^{3/4}} = 21.19 \text{ rpm}\frac{\sqrt{\text{ft}^3/\text{s}}}{(\text{ft})^{3/4}} = 1.633 \text{ rpm}\frac{\sqrt{\ell/\text{s}}}{(\text{m})^{3/4}} = 51.64 \text{ rpm}\frac{\sqrt{\text{m}^3/\text{s}}}{(\text{m})^{3/4}}$$

The specific speed[1] n_s is a useful dimensional parameter for classifying the overall geometry and performance characteristics of impellers. If the impeller is double-suction, that is, two impellers in parallel, use either $Q/2$ in Eq. (13) or specify n_s *double-suction.* It has been common practice to use the term *specific speed of a pump* to mean the specific speed of the impeller. This is particularly confusing with multistage pumps for which only the specific speed of a single stage is significant. With multistage pumps, the head of a single stage is to be used in Eq. (13) and not the

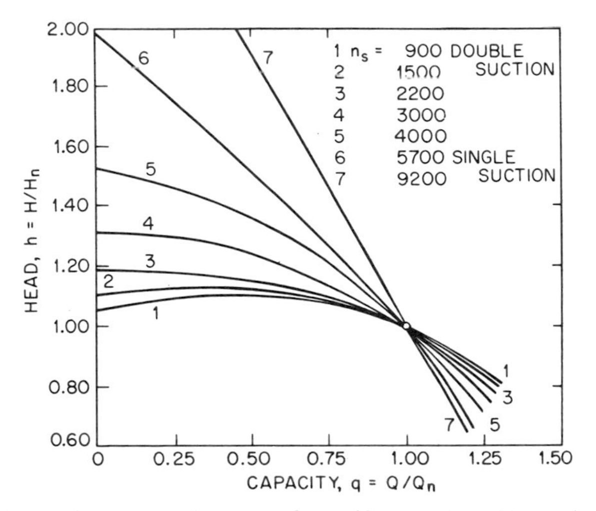

Fig. 8 Head-capacity curves for several specific speeds. (From A. J. Stepanoff, "Centrifugal and Axial Flow Pumps," 2 ed., John Wiley & Sons, Inc., New York, 1967)

total head of the pump. The head of a single stage is almost always the total pump head divided by the number of stages. Figure 7 shows efficiency as a function of specific speed and capacity for well-designed pumps. Also shown are typical impeller profiles which correspond approximately to the specific speeds indicated. For a series of *geometrically similar* impellers operated at dynamically similar conditions, that is, at identical specific speeds, the affinity laws become

$$\frac{Q_1}{Q_2} = \frac{n_1 D_1^3}{n_2 D_2^3} \tag{14a}$$

$$\frac{H_1}{H_2} = \frac{n_1^2 D_1^2}{n_2^2 D_2^2} \tag{14b}$$

$$\frac{P_1}{P_2} = \frac{n_1^3 D_1^5}{n_2^3 D_2^5} \tag{14c}$$

where subscripts 1 and 2 refer to any two noncavitating operating conditions. *See the section on diameter reduction for the affinity laws to be used when an impeller is reduced in diameter.*

Classification of Curve Shapes A useful method for comparing characteristics of pumps of different specific speed is to normalize on a selected operating condition, usually that of best efficiency. Thus

$$q = \frac{Q}{Q_n} \qquad h = \frac{H}{H_n} \qquad \text{and} \qquad p = \frac{P}{P_n} \tag{15}$$

where the subscript n designates values for the best efficiency point. Figures 8, 9, and 10 show approximate performance curves normalized on the conditions

[1] The dimensionless specific speed $\omega_s = \omega\sqrt{Q}/(gH)^{3/4}$, obtained by eliminating D between the first two of Eqs. (11), is rarely used.

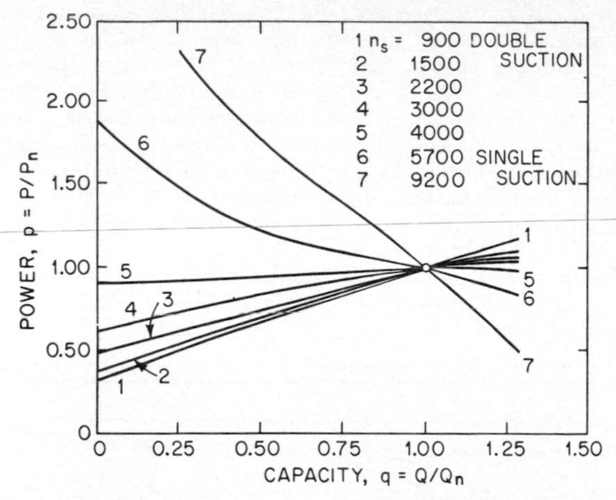

Fig. 9 Power-capacity curves for several specific speeds. (From A. J. Stepanoff, "Centrifugal and Axial Flow Pumps," 2 ed, John Wiley & Sons, Inc., New York, 1967)

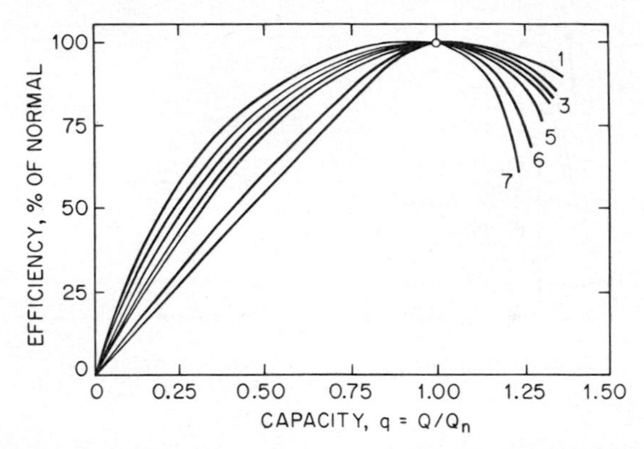

Fig. 10 Efficiency-capacity curves for several specific speeds. (From A. J. Stepanoff, "Centrifugal and Axial Flow Pumps," 2 ed., John Wiley & Sons, Inc., New York, 1967)

of best efficiency and for a wide range of specific speeds. These curves are applicable to pumps of any size because absolute magnitudes have been eliminated. In Fig. 8, curves 1 and 2 exhibit a *rising head* or *unstable* characteristic where the head increases with increasing capacity over the lower part of the capacity range. This may cause instability at heads greater than the shutoff value, particularly if two or more pumps are operated in parallel. Curve 3 exhibits an almost constant head at low capacities and is often called a *flat* characteristic. Curves 4 to 7 are typical of a *steep* or *stable-head* characteristic in which the head always decreases with increasing capacity. Although the shape of the head-capacity curve is primarily a function of the specific speed, the designer has some control through selection of the vane angle β_2, number of impeller vanes, z, and the capacity coefficient, $\phi = c_{m2}/u_2$ as described in Sec. 2.1 (see also Fig. 2). For pumps having a single-suction specific speed approximately 5,000 and higher, the power is maximum at shutoff and decreases with increasing capacity. This may require an increase in the power rating of the driving motor over that required for operation at normal capacity.

Efficiency The efficiency η is the product of three efficiencies,

$$\eta = \eta_m \eta_v \eta_h \tag{16}$$

The mechanical efficiency η_m accounts for the bearing, stuffing box, and all disk-friction losses including those in the wearing rings and balancing disks or drums if present. The volumetric efficiency η_v accounts for leakage through the wearing rings, internal labyrinths, balancing devices, and glands. The hydraulic efficiency η_h accounts for fluid friction losses in all through-flow passages, including the suction elbow or nozzle, impeller, diffusion vanes, volute casing, and the crossover passages of multistage pumps. Figure 11 shows an estimate of the losses from various sources in double-suction single-stage pumps of at least a 12-in discharge pipe diameter. Minimum losses and hence maximum efficiencies are seen to be in the vicinity of $n_s \approx 2{,}500$, which agrees with Fig. 7.

Effects of Pump Speed Increasing the impeller speed increases the efficiency of centrifugal pumps. Figure 6 shows a gain of about 15 percent for an increase

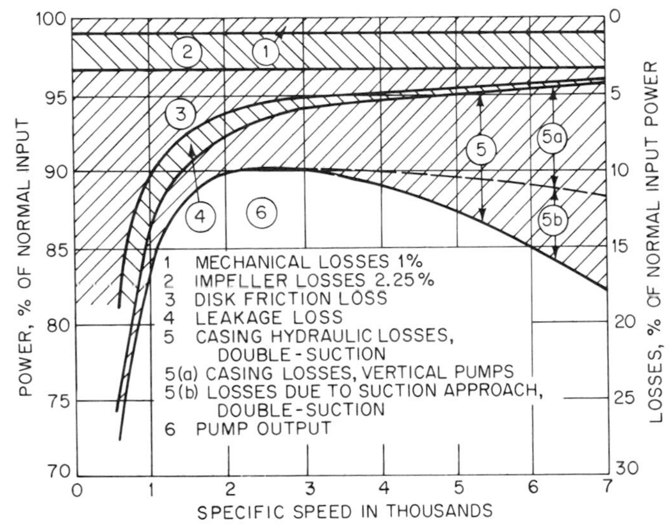

Fig. 11 Power balance for double-suction pumps at best efficiency. (From A. J. Stepanoff, "Centrifugal and Axial Flow Pumps," 2 ed., John Wiley & Sons, Inc., New York, 1967)

in speed from 15,000 to 30,000 rpm. The increases are less dramatic at lower speeds. For example, Ippen [22] reported about 1 percent increase in the efficiency of a small pump, $D = 8$ in and $n_s = 1{,}992$ at best efficiency, for an increase in speed from 1,240 to 1,880 rpm. Within limits, the cost of the pump and driver usually decrease with increasing speed. Abrasion and wear increase with increasing speed particularly if the fluid contains solid particles in suspension. The danger of cavitation damage usually increases with increasing speed unless certain suction requirements can be met as described later.

Effects of Specific Speed Figures 7 and 11 show that maximum efficiency is obtained in the range $2{,}000 < n_s < 3{,}000$, but this is not the only criterion. Pumps for high heads and small capacities occupy the range $500 < n_s < 1{,}000$. At the other extreme pumps for very low heads and large capacities may have $n_s = 15{,}000$ or higher. For given head and capacity, the pump having the highest specific speed that will meet the requirements probably will be the smallest and least expensive. However, Eq. (13) shows that it will run at the highest speed and be subject to maximum wear and cavitation damage as previously mentioned.

Effects of Clearances *Wearing-Ring Clearance.* Details of wearing-ring construction are given in Sec 2.2. Schematic outlines of two designs of rings are shown in Fig. 12. The L-shaped construction shown in Fig. 12*a* is very widely used with the close clearance between the cylindrical portions of the rings. Leakage losses increase and pump performance falls off as the rings wear. Table 2 shows

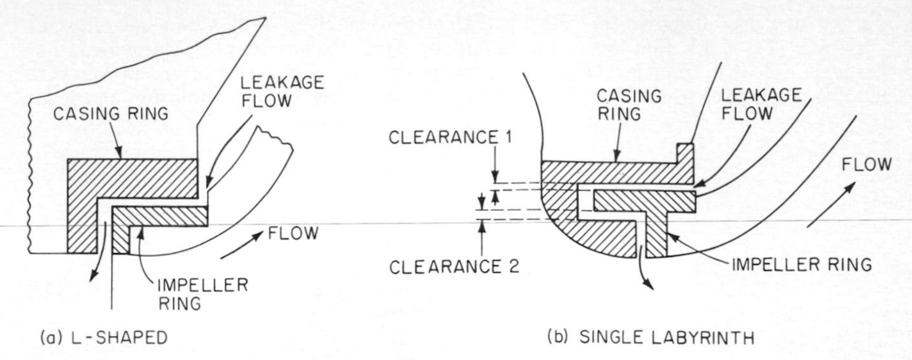

Fig. 12 Typical wearing rings.

TABLE 2 Effects of Increased Wearing-Ring Clearance on Centrifugal Pump Performance

Specific speed n_s	Design head, ft	Ring clearance, percent of normal value	Percent of values at best efficiency with normal ring clearance				Percent of values at shutoff with normal ring clearance	
			Q	H	P	η	H_0	P_0
2,100	63	178	100	98.3	98.9	99.4	97.0	100
		356	100	97.5	99.0	98.5	93.6	98.2
		688	100	96.0	98.9	97.1	91.2	94.8
		1,375	100	94.3	97.4	96.8	88.8	92.5
3,500	65	354	100	90.0	99.1	90.8	85.0	96.2
4,300	41	7,270	62	65.5	81.7	49.8	44.3	106
4,800	26	5,220	96	78.8	89.2	84.8	78.2	83.3

SOURCE: Worthington Pump International, Inc.

some of the effects of increasing the clearance of rings similar to Fig. 12*a*. The labyrinth construction shown in Fig. 12*b* has been used to increase the leakage path without increasing the axial length of the rings. If the pressure differential across these rings is high enough, the pump shaft may take on lateral vibrations with relatively large amplitude and long period, which can cause serious damage. One remedy for these vibrations is to increase clearance 2 (Fig. 12*b*), relative to clearance 1 at the expense of an increased leakage flow. High-pressure breakdown through plain rings may cause vibration (Ref. 23) but this is not usually a serious problem. It is considered good practice to replace or repair wearing rings when the nominal clearance has doubled. The presence of abrasive solids in the fluid pumped may be expected to increase wearing-ring clearances rapidly.

Vane-Tip Clearance. Many impellers are made without an outer shroud and relay on close running clearances between the vane tips and the casing to hold leakage across the vane tips to a minimum. Although this construction usually is not used with pumps having specific speeds less than about 6,000, Wood et al [24] have reported good results with semiopen impellers at $1{,}800 \leq n_s \leq 4{,}100$. It appears that both head and efficiency increase with decreasing tip clearance and are quite sensitive to rather small changes in clearance. Reducing the tip clearance from about 0.060 to about 0.010 in may increase the efficiency by as much as ten percent. Abrasive solids in the fluid pumped probably will increase tip clearances rapidly.

MODIFICATIONS TO IMPELLER AND CASING

Diameter Reduction To reduce cost, pump casings usually are designed to accommodate several different impellers. Also, a variety of operating requirements can be met by changing the outside diameter of a given radial impeller. Euler's equation (7) shows that the head should be proportional to $(nD)^2$ provided that the exit velocity triangles (Fig. 2*B*) remain similar before and after cutting with w_2 always parallel to itself as u_2 is reduced. This is the usual assumption and leads to

$$\frac{Q_1}{Q_2} = \frac{n_1 D_1}{n_2 D_2} \tag{17a}$$

$$\frac{H_1}{H_2} = \frac{n_1^2 D_1^2}{n_2^2 D_2^2} \tag{17b}$$

$$\frac{P_1}{P_2} = \frac{n_1^3 D_1^3}{n_2^3 D_2^3} \tag{17c}$$

which apply only to a given impeller with altered D and constant efficiency but *not* to a geometrically similar series of impellers. The assumptions on which Eqs. (17) were based are rarely if ever fulfilled in practice so that exact predictions by the equations should not be expected. Diameter reductions greater than from 10 to 20 percent of the original full diameter of the impeller are rarely made.

Radial Discharge Impellers. Impellers of low specific speed may be cut successfully provided the following items are kept in mind:

1. The angle β_2 may change as D is reduced but this usually can be corrected by filing the vane tips. See section on vane-tip filing.
2. Tapered vane tips will be thickened by cutting and should be filed to restore the original shape. See section on vane-tip filing.
3. Bearing and stuffing box friction remain constant but disk friction should decrease with decreasing D.
4. The length of flow path in the pump casing is increased by decreasing D.
5. Since c_{m1} is smaller at the reduced capacity, the inlet triangles no longer remain similar before and after cutting, and local flow separation may take place near the vane entrance tips.
6. The second term in Eq. (6) was neglected in arriving at Eqs. (17), but it may represent a significant decrease in head as D is reduced.
7. Some vane overlap should be maintained after cutting. Usually the initial vane overlap decreases with increasing specific speed, so that the higher the specific speed the less the allowable diameter reduction.

Most of the losses are approximately proportional to Q^2, and hence to D^2 by Eqs. (17). Since the power output decreases approximately as D^3, it is reasonable to expect the maximum efficiency to decrease as the wheel is cut and this often is the case. By Eqs. (13) and (17), the product $n_s D$ should remain constant so that the specific speed at best efficiency increases as the wheel diameter is reduced (see Table 3).

The characteristics of the pump shown in Fig. 13 may be used to illustrate reduction of diameter at constant speed. Starting with the best efficiency point and $D = 16\frac{5}{16}$ in, let it be required to reduce the head from $H = 224.4$ to $H' = 192.9$ ft and to determine the wheel diameter, capacity, and power for the new conditions. Since the speed is constant, Eqs. (17) may be written

$$H = k_H Q^2 \qquad \text{and} \qquad P = k_p Q^3 \tag{18}$$

where k_H and k_p may be obtained from the known operating conditions at $D = 16\frac{5}{16}$ in. Plot a few points for assumed capacities and draw the curve segments as shown by the solid lines in Figs. 13*B* and *C*. Then, from Eqs. (17),

$$D' = D\sqrt{H'/H} \qquad Q' = Q\sqrt{H'/H} \qquad \text{and} \qquad P' = P(H'/H)^{3/2} \tag{19a}$$

or

$$D' = D(Q'/Q) \qquad H' = H(Q'/Q)^2 \qquad \text{and} \qquad P' = P(Q'/Q)^3 \tag{19b}$$

from which $D = 15\frac{1}{8}$ in, $Q = 3{,}709$ gpm, and $P' = 215.5$ hp. In Fig. 13, the initial conditions were at points A and the computed conditions after cutting at points B. The test curve for $D = 15\frac{1}{8}$ in shows the best efficiency point a at a lower

TABLE 3 Predicted Characteristics at Different Impeller Diameters—Radial-Flow Pump

	Test values					Predicted from $D = 16^5/_{16}$			Predicted from $D = 15^1/_8$		
D, in	Q, gpm	H, ft	P, hp	n_s	n_sD	Q', gpm	H', ft	P', hp	Q', gpm	H', ft	P', hp
$16^5/_{16}$	4,000	224.4	270.4	1,953	31,860	....	......	......	3,883	227.3	272.2
Percent error in predicting best efficiency point									−2.93	1.29	0.68
$15^1/_8$	3,600	195.4	217.0	2,055	31,080	3,709	192.9	215.5			
Percent error in predicting best efficiency point						3.02	−1.28	−0.69			
14	3,200	163.6	167.2	2,214	31,000	3,433	165.3	170.9	3,332	167.4	172.1
Percent error in predicting best efficiency point						7.28	1.03	2.33	4.13	2.33	2.93

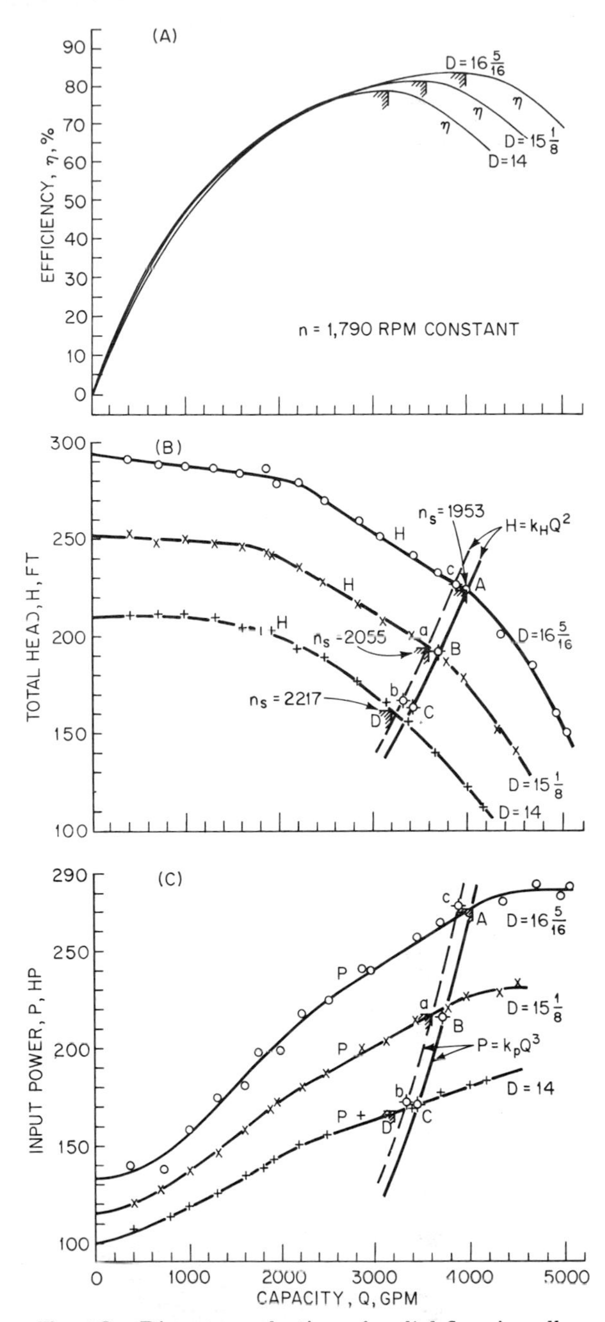

Fig. 13 Diameter reduction of radial-flow impeller.

capacity than predicted by Eqs. (19) but the head-capacity curve satisfies the predicted values very closely. The power-capacity prediction was not quite as good. Table 3 and Fig. 13 give actual and predicted performance for three impeller diameters. The error in predicting the best efficiency point was computed by (predicted value minus test value)(100)/(test value). As the wheel diameter was reduced, the best efficiency point moved to a lower capacity than predicted by Eqs. (19) and the specific speed increased, showing that dynamically similar operating conditions were not maintained.

Wheel cutting should be done in two or more steps with a test after each cut to avoid too large a reduction in diameter. Figure 14 shows an approximate correction, given by Stepanoff [12], which may be applied to the ratio of D'/D as computed by Eqs. (17) or (19). The accuracy of the correction decreases with increasing specific speed. Figure 15 shows a correction proposed by Rütschi [25] on the basis of extensive tests on low specific-speed pumps. The corrected diameter reduction, ΔD, is the diameter reduction, $D - D'$, given by Eqs. (19) and multiplied by k from Fig. 15. The shaded area in Fig. 15 indicates the range of scatter of the test points operating at or near maximum efficiency. Near shutoff the values of k were smaller, and at maximum capacity the values of k were larger than shown in Fig. 15. Table 4 shows the results of applying Figs 14 and 15 to the pump of the preceding example.

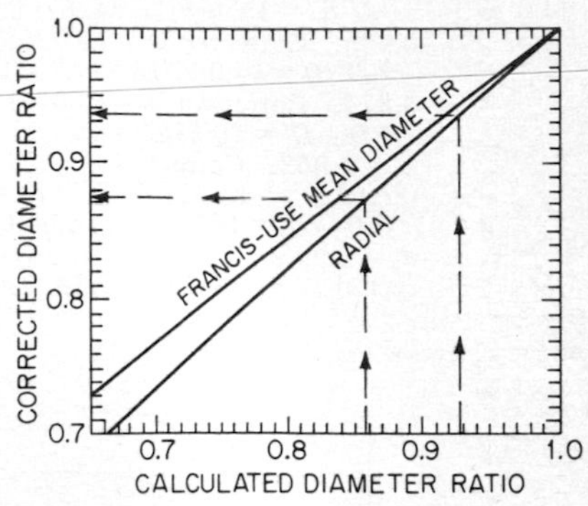

Fig. 14 Corrections for calculated impeller diameter reductions. (From A. J. Stepanoff, "Centrifugal and Axial Flow Pumps," 2 ed., John Wiley & Sons, Inc., New York, 1967)

There is no independent control of Q and H in impeller cutting although Q may be increased somewhat by underfiling the vane tips as described later. The discharge and power will automatically adjust to the values at which the pump head satisfies the system head-capacity curve.

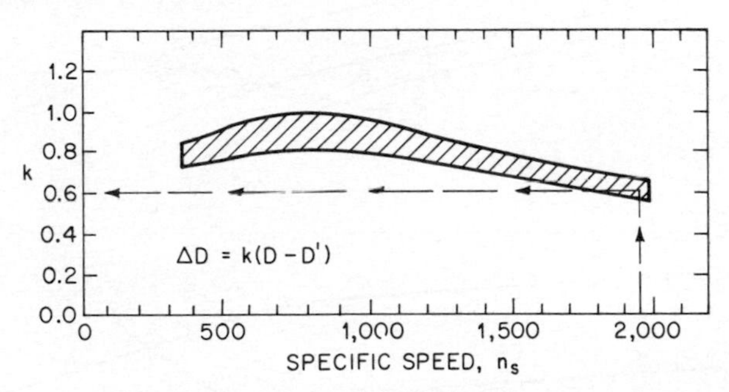

Fig. 15 Corrections for calculated impeller diameter reductions [25].

Mixed-Flow Impellers. Diameter reduction of mixed-flow impellers is usually done by cutting a maximum at the outside diameter, D_o, and little or nothing at the inside diameter, D_i, as shown in Fig. 16*D*. Stepanoff [9] recommends that the calculations be based on the average diameter, $D_{av} = (D_o + D_i)/2$ or estimated from the vane-length ratio FK/EK or GK/EK in Fig. 16*D*. Figure 16 shows a portion of the characteristics of a mixed-flow impeller on which two cuts were made as shown in Fig. 16*D*. The calculations were made by Eqs. (19) using the mean diameter $D_m = \sqrt{(D_o^2 + D_i^2)/2}$, instead of the outside diameter, in each case. The predictions and test results are shown in Fig. 16 and Table 5. It is clear that the actual change in the characteristics far exceeded the predicted values.

TABLE 4 Impeller Diameter Corrections

D before cutting	in	16.3125	16.3125
D' predicted by Eqs. (19)	in	15.125	14.000
D' corrected by Fig. 14	in	15.25[a]	14.26[b]
D' corrected by Fig. 15	in	15.60[c]	14.93[d]

[a] $D'/D = 15.125/16.3125 = 0.927$; by Fig. 14, corrected $D'/D = 0.935$. Corrected $D' = (0.935)(16.3125) = 15.25$ in.

[b] $D'/D = 14.000/16.3125 = 0.858$; by Fig. 14, corrected $D'/D = 0.874$. Corrected $D' = (0.874)(16.3125) = 14.26$ in.

[c] $D - D' = 16.3125 - 15.125 = 1.1875$; by Fig. 15, $K \cong 0.6$ at $n_s = 1{,}953$. Corrected $D - D' = (0.6)(1.1875) = 0.7125$ and corrected $D' = 16.3125 - 0.7125 = 15.60$ in.

[d] $D - D' = 16.3125 - 14.000 = 2.3125$; by Fig. 15, $K \cong 0.6$ at $n_s = 1{,}953$. Corrected $D - D' = (0.6)(2.3125) = 1.3875$ and corrected $D' = 16.3125 - 1.3875 = 14.93$ in.

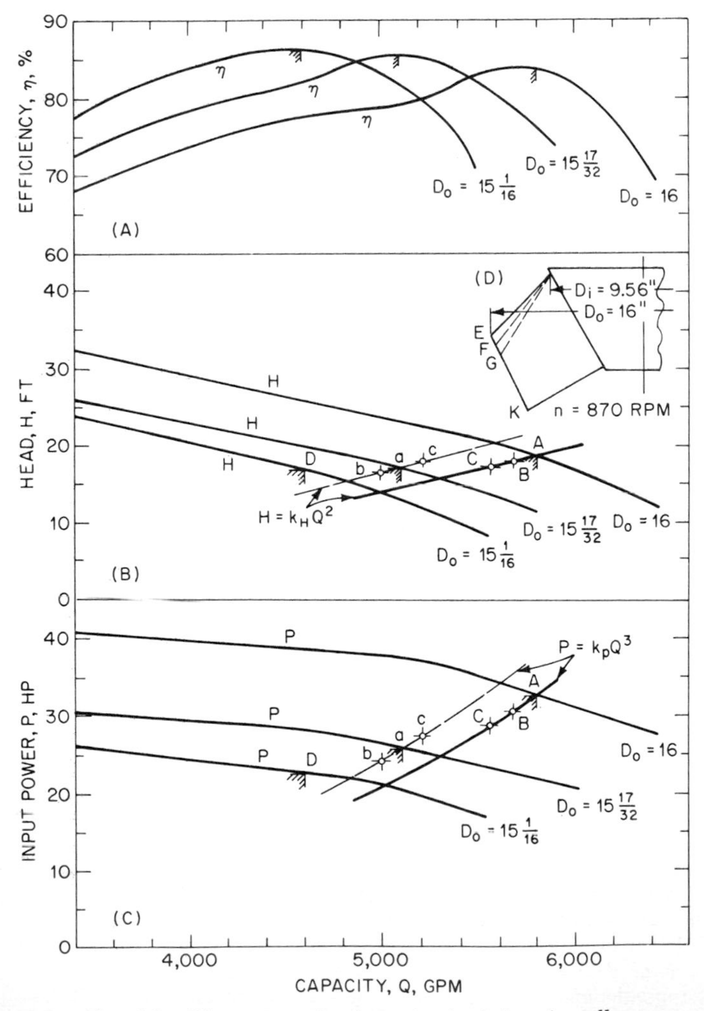

Fig. 16 Diameter reduction of mixed-flow impeller.

TABLE 5 Predicted Characteristics at Different Impeller Diameters—Mixed-Flow Pump

		Test values					Predicted from $D = 16.00, D_m = 13.17$			Predicted from $D = 15.53, D_m = 12.89$		
D, in	D_m, in	Q, gpm	H, ft	P, hp	n_s	$n_s D_m$	Q', gpm	H', ft	P', hp	Q', gpm	H', ft	P', hp
16	13.17	5,800	18.6	32.5	7,385	97,300	....	.....	....	5,210	17.9	27.4
Percent error in predicting best efficiency point										−11.3	−3.91	−18.8
$15^{17}/_{32}$	12.89	5,100	17.1	25.7	7,385	95,200	5,680	17.8	30.4			
Percent error in predicting best efficiency point							10.2	4.50	15.5			
$15^{1}/_{6}$	12.62	4,600	16.8	22.6	7,100	89,600	5,560	17.1	28.6	5,000	16.4	24.1
Percent error in predicting best efficiency point							17.3	1.79	22.4	8.00	−2.44	6.22

Except for the use of the mean diameters, the procedure was essentially the same as that described for Fig. 13 and all points and curves are similarly labeled. The corrections given in Fig. 14 would have made very little difference in the computed diameter reductions and those of Fig. 15 were not applicable to impellers having specific speeds greater than $n_s = 2{,}000$. In this case, the product $n_s D_m$ did not remain constant and the maximum efficiency increased as D_m was reduced. Although the changes in diameter were small, the area of vane removed was rather large for each cut. The second cut eliminated most of the vane overlap. The characteristics of mixed-flow impellers can be changed by cutting, but very small cuts may produce a significant effect. The impellers of propeller pumps are not usually subject to diameter reduction.

Shaping Vane Tips If the discharge tips of the impeller vanes are thick, performance usually can be improved by filing over a sufficient length of vane to produce

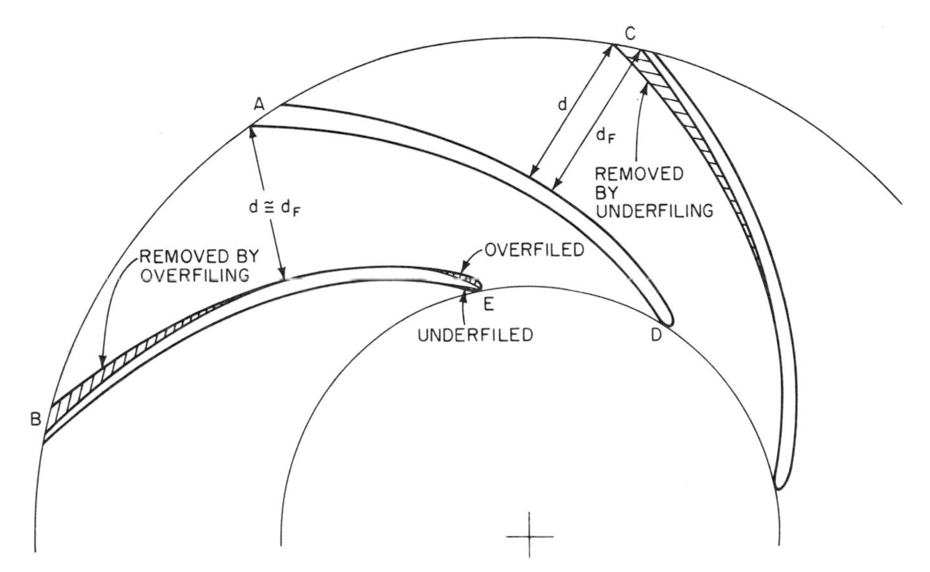

Fig. 17 Under- and overfiling vane tips.

a long gradual taper. Chamfering or rounding the discharge tips may increase the losses and should never be done. Reducing the impeller diameter frequently increases the tip thickness.

Overfiling. This is shown at *B* in Fig. 17 compared to the unfiled vane at *A*. Usually there is little or no increase in the vane spacing d before and d_F after filing so that the discharge area is practically unchanged. Experience indicates that any change in the angle β_2 due to overfiling usually produces a negligible change in performance.

Underfiling. This is shown at *C* in Fig. 17 compared to the unfiled vane at *A*. If properly done, underfiling will increase the vane spacing from d to d_F and, hence, the discharge area which lowers the average meridian velocity c_{m2} at any given capacity Q. The angle β_2 usually is increased slightly. Figure 18*A* and Eq. (7) show that the head and, consequently, the power increase at the same capacity. The maximum efficiency usually is improved and may be moved to a higher capacity. At the same head, Fig. 18*B* shows that both c_{m2} and the capacity will increase. The change in both the area and in c_{m2} may increase the capacity by as much as 10 percent. Table 6 shows the results of tests before and after underfiling the impeller vanes of nine different pumps. In general, they confirm the foregoing predictions based on changes in area and in the velocity triangles.

Inlet Vane Tips. If the inlet vane tips are blunt as shown at *D* in Fig. 17, the cavitation characteristics may be improved by sharpening them as shown at *E*. In

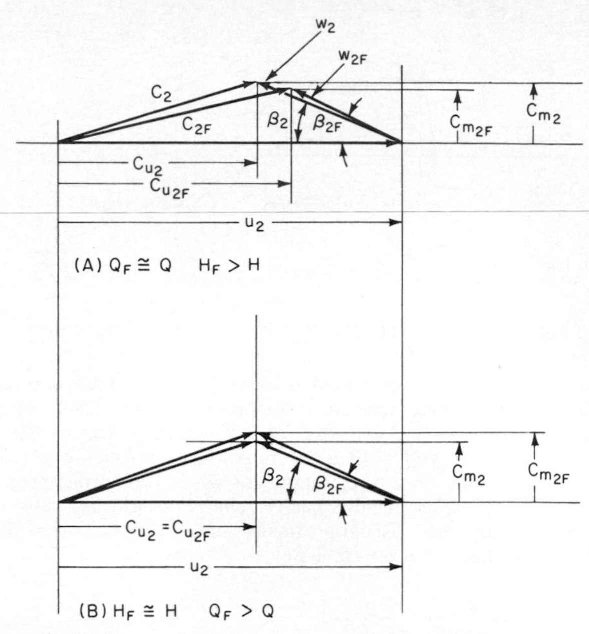

Fig. 18 Discharge velocity triangles for underfiled vanes.

TABLE 6 Changes in Performance due to Underfiling Impeller Vanes

Specific speed, n_s	No. of stages	Impeller diameter D_2, in	Change in vane spacing d_F/d	$\frac{H_F - H}{H}$ %, H_F = head after filing	Changes at the best efficiency point after filing, %				
					H_0[d]	H	Q	P	η
862	1	11½	1.13	4	1.5	4	0	9	-5
945	2	8⅜	1.05	5.5	0	3	4	-0.5	7.5[b]
1,000	4	9⅛	1.055	5.5	0	5.5	0	3	2.5
1,080	2	9 13/16	1.08	10	2.5	10	0	6.8	3
1,525	2	10½	1.035	3.2	3	0.5	4.5	3.5	1.2
1,950	1	22	1.02	1.5	1	1.5	0	1.5	0
3,300[a]	1	12	.....	7.8	2.5	7.8	0	8.5	-0.6
3,450[a]	1	30½	.....	6.5	0.5	6.5	0	7.8	-2.2
4,300	1	41¼	.....	5	...	5	0	0.5	4.5[c]

[a] Double suction.
[b] Due in part to changes in pump casing.
[c] Due in part to rounding inlet vane tips.
[d] Shutoff head.
SOURCE: Worthington Pump International, Inc.

this case overfiling increases the effective flow area which reduces c_{m1} for a given capacity. If more area is needed, it may be advantageous to cut back part of the vane and sharpen the leading edge. Overfiling tends to increase β_1, which is incompatible with a decrease in c_{m1} (see Fig. 19). The increase in β_1 increases the angle of attack of the fluid approaching the vane. In Figs. 2 and 19, w_1 is tangent to the centerline of the vane at entrance and w_0 is the velocity of the approaching fluid as seen by an observer moving with the vane. The angle δ between w_0 and w_1 is the angle of attack which increases to δ_F after overfiling. Although the increase in δ tends to increase the opportunity for the fluid to separate from the low-pressure

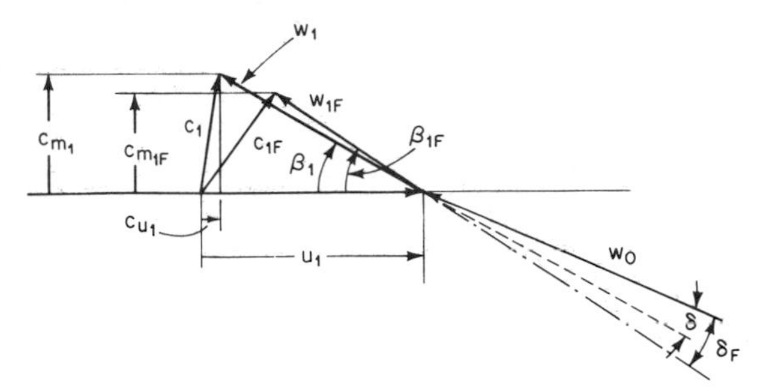

Fig. 19 Inlet velocity triangles for overfiled vanes.

face of the vane, this usually is outweighed by the improvement due to reducing c_{m1}.

Casing Tongue. The casing tongue forms part of the throat of the discharge nozzle of many volute casings. See Fig. 64. Frequently the throat area is small enough to act as a throttle and reduce the maximum capacity otherwise obtainable from the impeller. Cutting back the tongue increases the throat area and increases the maximum capacity. The head-capacity characteristic is then said to *carry out further.* Shortening the discharge nozzle may increase the diffusion losses a little and result in a slightly lower efficiency.

CAVITATION

The formation and subsequent collapse of vapor-filled cavities in a liquid due to dynamic action are called *cavitation.* The cavities may be bubbles, vapor-filled pockets, or a combination of both. The local pressure must be at or below the vapor pressure of the liquid for cavitation to begin, and the cavities must encounter a region of pressure higher than the vapor pressure in order to collapse. Dissolved gases often are liberated shortly before vaporization begins. This may be an indication of impending cavitation, but true cavitation requires vaporization of the liquid. Boiling accomplished by the addition of heat or the reduction of static pressure without dynamic action of the fluid is arbitrarily excluded from the definition of cavitation. With mixtures of liquids, such as gasoline, the light fractions tend to cavitate first.

When a fluid flows over a surface having convex curvature, the pressure near the surface is lowered and the flow tends to separate from the surface. *Separation* and *cavitation* are completely different phenomena. Without cavitation, a separated region contains turbulent eddying fluid at pressures higher than the vapor pressure. When the pressure is low enough, the separated region may contain a vapor pocket which fills from the downstream end (Ref. 26), collapses, and forms again many times each second. This causes noise and, if severe enough, vibration. Vapor-filled bubbles usually are present which collapse very rapidly in any region where the pressure is above the vapor pressure. Knapp [27] found the life cycle of a bubble to be on the order of 0.003 s.

Bubbles which collapse on a solid boundary may cause severe mechanical damage. Shutler and Mesler [28] photographed bubbles which distorted into toroidal-shaped rings during collapse and produced ring-shaped indentations in a soft metal boundary. The bubbles rebounded following the initial collapse and caused pitting of the boundary. Pressures on the order of 10^4 atm have been estimated (Ref. 29) during collapse of a bubble. All known materials can be damaged by exposure to bubble collapse for a sufficiently long time. This is properly called *cavitation erosion,* or *pitting.* Figure 20 (Ref. 30) shows extensive damage to the suction side of pump impeller vanes after about three months operation with cavitation. At two locations, the pitting has penetrated deeply into the ⅜-in thickness of stainless steel. The unfavorable inlet flow conditions, believed to have been the cause of the cavitation, were at least partly due to elbows in the approach piping. Modifications to the approach piping and the pump inlet passages reduced the cavitation to the point that impeller life was extended to several years.

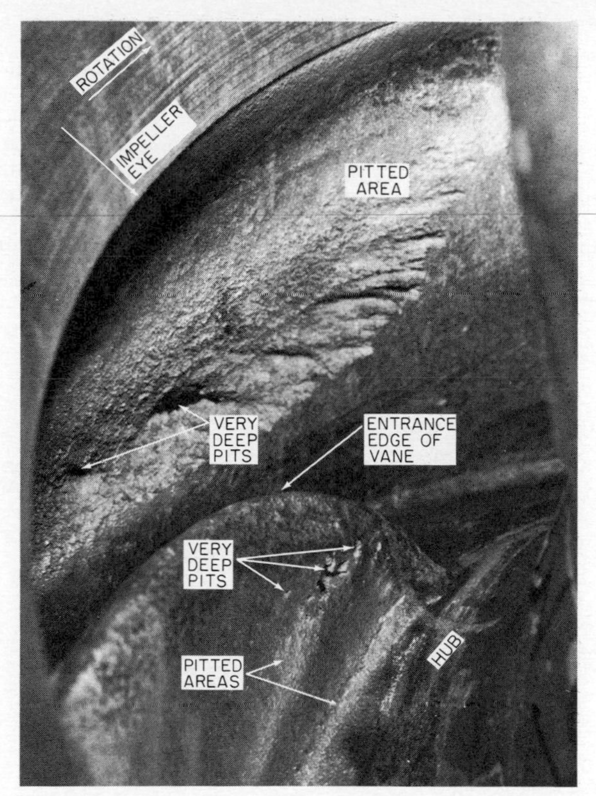

Fig. 20 Impeller damaged by cavitation [30]. (DeLaval Turbine, Inc.)

It has been postulated that high temperatures and chemical action may be present at bubble collapse but any damaging effects due to them appear to be secondary compared to the mechanical action. It seems possible that erosion by foreign materials in the liquid and cavitation pitting may augment each other. Controlled experiments [31, 32] with water indicated that the damage to metal depends on the liquid temperature and was a maximum at about 100 to 120°F. Cavitation pitting, as measured by weight of the boundary material removed per unit time, frequently increases with time. Cast iron and steel boundaries are particularly vulnerable. Controlled experiments have shown a strong dependence of cavitation pitting of metals such as aluminum, steel, and stainless steel on the velocity of the fluid in the undisturbed flow past the boundary. On the basis of tests of short duration, Knapp [26] reported that damage to annealed aluminum increased approximately as the sixth power of the velocity of the undisturbed flow past the surface. Hammitt [33] found a more complicated relationship between velocity and damage as shown in Fig. 21. It seems clear that once cavitation begins it will increase rapidly with increasing velocities. Frequently the rate of pitting accelerates with elapsed time.

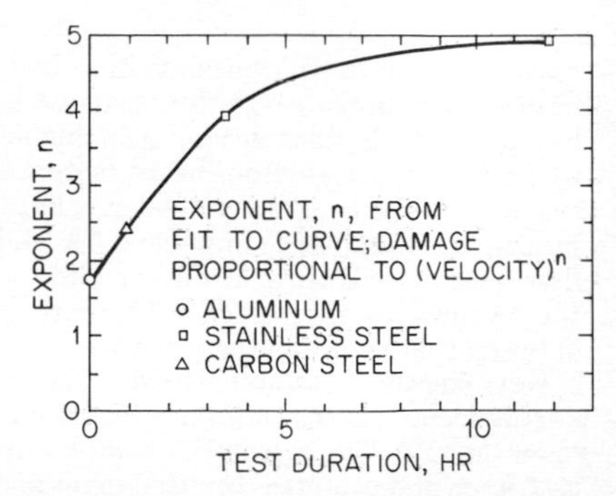

Fig. 21 Cavitation damage exponent versus test time for several materials in water [33].

Centrifugal pumps begin to cavitate when the suction head is insufficient to maintain pressures above the vapor pressure throughout the flow passages. The most sensitive areas usually are the low-pressure sides of the impeller vanes near the inlet edge and the front shroud where the curvature is greatest. Axial flow and high-specific-speed impellers without front shrouds are especially sensitive to cavitation on the low-pressure sides of the vane tips and in the close tip-clearance spaces. Sensitive areas in the pump casing include the low-pressure side of the tongue and the low-pressure sides of diffusion vanes near the inlet edges. As the suction head is reduced, all existing areas of cavitation tend to increase and additional areas may develop. Apart from the noise and vibration, cavitation damage

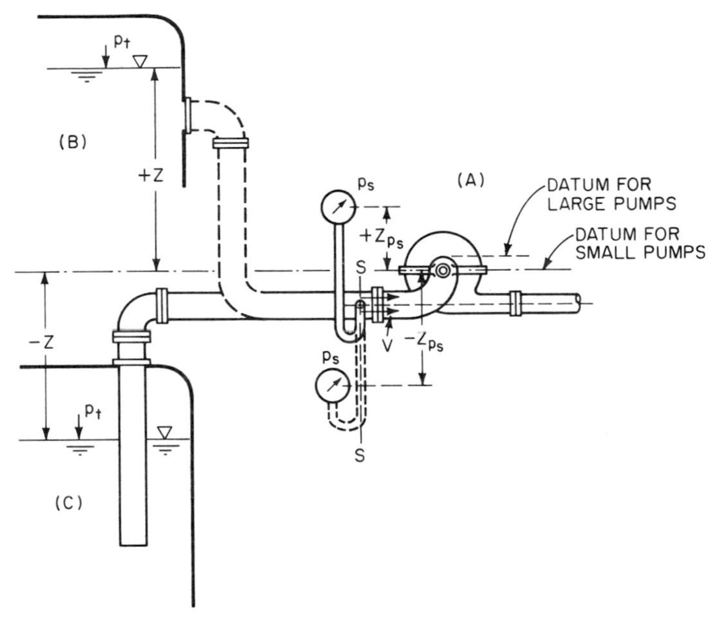

Fig. 22 Definition sketch for computing NPSH.

may render an impeller useless in as little as a few weeks of continuous operation. In multistage pumps, cavitation usually is limited to the first stage but Kovats [6] has pointed out that second and higher stages may cavitate if the flow is reduced by lowering the suction head (submergence control). Cavitation tends to lower the axial thrust of an impeller. This could impair the balancing of multistage pumps with opposed impellers. A reduction in suction pressure may cause the flow past a balancing drum or disk to cavitate where the fluid discharges from the narrow clearance space. This may produce vibration and damage due to contact between fixed and running surfaces.

Net Positive Suction Head The *net positive suction head* (NPSH), h_{sv}, is a statement of the *minimum* suction conditions required to prevent cavitation in a pump. The *required* or *minimum* NPSH must be determined by test and usually will be stated by the manufacturer. The *available* NPSH at installation must be at least equal to the required NPSH if cavitation is to be prevented. Increasing the available NPSH provides a margin of safety against the onset of cavitation. Figure 22 and the following symbols will be used to compute the NPSH:

p_a = absolute pressure in atmosphere surrounding gage, Fig. 22

p_s = gage pressure indicated by gage or manometer connected to pump suction at section *s-s*. May be positive or negative

p_t = absolute pressure on free surface of liquid in closed tank connected to pump suction

p_{vp} = vapor pressure of liquid being pumped corresponding to the temperature at section *s-s*. If the liquid is a mixture of hydrocarbons, p_{vp} must be measured by the *bubble point* method

h_f = lost head due to friction in suction line between tank and section *s-s*

V = average velocity at section *s-s*

Z, Z_{ps} = vertical distances defined by Fig. 22. May be positive or negative

γ = specific weight of liquid at pumping temperature

It is satisfactory to choose the datum for small pumps as shown in Figs. 1 and 22, but with large pumps the datum should be raised to the elevation where cavitation is most likely to start. For example, the datum for a large horizontal-shaft propeller pump should be taken at the highest elevation of the impeller-vane tips. The available NPSH is given by

$$h_{sv} = \frac{p_a - p_{vp}}{\gamma} + \frac{p_s}{\gamma} + Z_{ps} + \frac{V^2}{2g} \tag{20}$$

or

$$h_{sv} = \frac{p_t - p_{vp}}{\gamma} + Z - h_f \tag{21}$$

Consistent units must be chosen so that each term in Eqs. (20) and (21) represents feet (or metres) of the fluid pumped. Equation (20) is useful for evaluating the

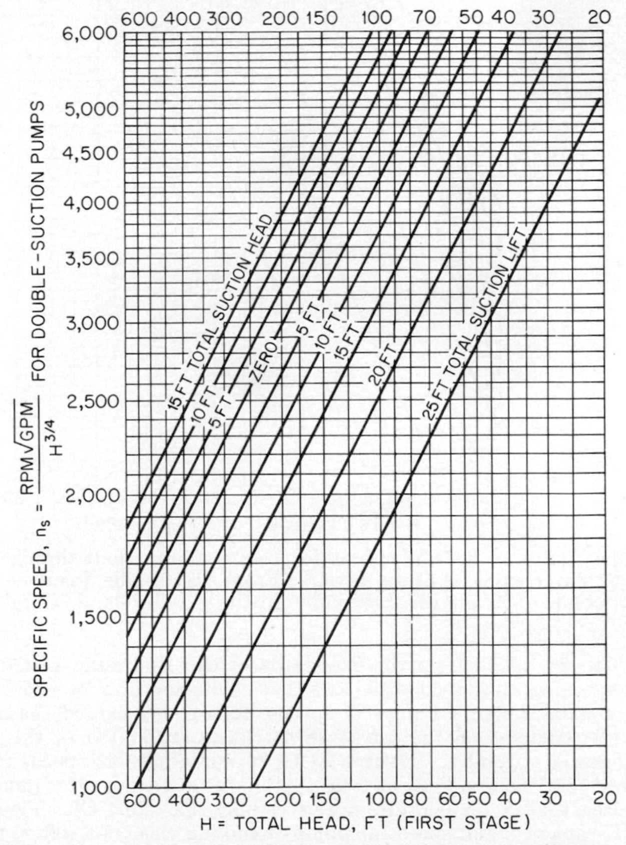

Fig. 23 Upper limits of specific speeds for double-suction pumps handling clear water at 85°F at sea level. (From "Hydraulic Institute Standards," 12th ed., copyright 1969 by the Hydraulic Institute, Cleveland, Ohio)

results of tests. Equation (21) is useful for estimating available NPSH during the design phase of an installation. In Eq. (20), the first term represents the height of a liquid barometer, h_b, containing the liquid being pumped while the sum of the remaining terms represents the suction head h_s of Eq. (1). Therefore

$$h_{sv} = h_b + h_s \tag{22}$$

Usually a positive value of h_s is called a *suction head* while a negative value of h_s is called a *suction lift.*

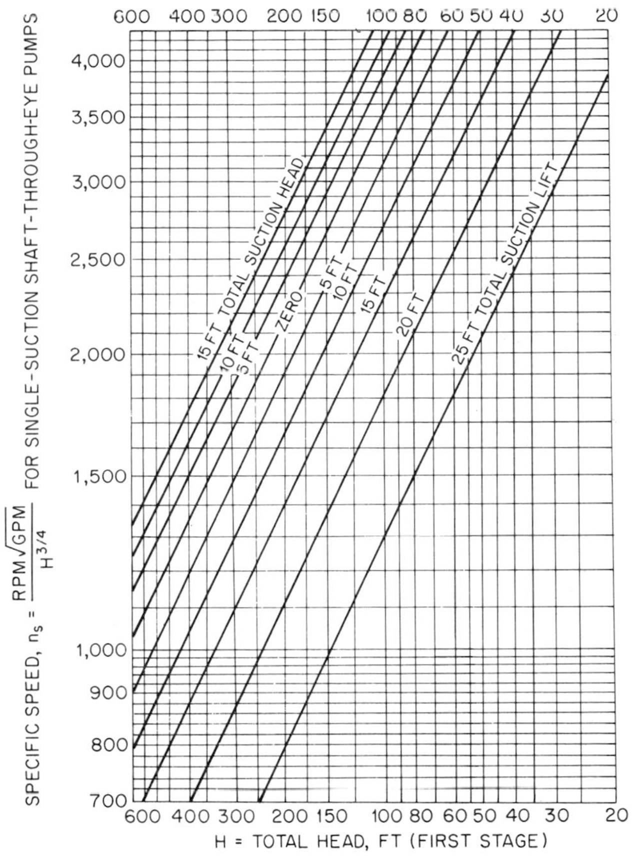

Fig. 24 Upper limits of specific speeds for single-suction shaft-through-eye pumps handling clear water at 85°F at sea level. (From "Hydraulic Institute Standards," 12th ed., copyright 1969 by the Hydraulic Institute, Cleveland, Ohio)

Figures 23 to 29, taken from the Standards of the Hydraulic Institute [13], are useful in determining suction conditions so that cavitation may be avoided in pumps of good commercial design. Pumps of special design may exceed the limits set by the charts. Restrictions on the use of these charts are stated in the legends and should be observed carefully. Figures 23 to 26 relate specific speed, total head of first stage, and total suction head or total suction lift at sea level for pumps handling clear cold water, that is, temperature not exceeding 85°F (29.4°C). Figures 27 to 28 relate NPSH, capacity, and speed for pumps handling clear hot water, that is, temperature higher than 85°F (29.4°C). These are especially useful with boiler feed pumps. Figure 29 relates NPSH, capacity, and speed for condensate pumps of not more than three stages with shaft through the eye of the first-stage impeller. Note that there are separate capacity scales for single- and double-suction impellers. This chart may be applied to single-suction overhung impellers by dividing the specified capacity by 1.2 for $Q \leqq 400$ gpm and by 1.15 for $Q > 400$ gpm. The data of Figs. 23 to 26 may be applied to other temperatures and elevations as shown in the following example:

Given a double-suction pump with $H = 200$ ft on the first stage and $n_s = 2{,}300$, Fig. 23 shows a permissible suction lift of 10 ft so that $h_{s1} = -10$ ft. Determine h_s

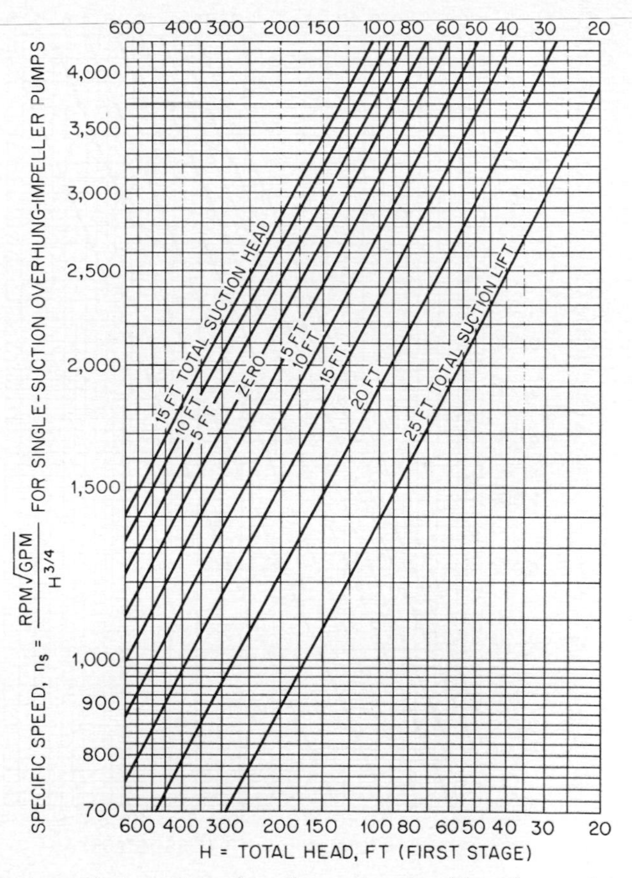

Fig. 25 Single-suction overhung-impeller pumps handling clear water at 85°F at sea level. (From "Hydraulic Institute Standards," 12th ed., copyright 1969 by the Hydraulic Institute, Cleveland, Ohio)

if this pump is installed at 1,500 ft elevation where the atmospheric pressure is $p_{a2} = 13.92$ lb/in² abs and the temperature of the water being pumped is 180°F. At sea level and $T_1 = 85$°F steam tables give $p_{a1} = 14.70$ lb/in² abs, $p_{vp1} = 0.60$ lb/in² abs, and $v_{f1} = 1/\gamma_1 = 0.01609$ ft³/lb. At 180°F, steam tables give $p_{vp2} = 7.51$ lb/in² abs and $v_{f2} = 1/\gamma_2 = 0.01651$ ft³/lb. The respective barometric heights are then $h_{b1} = (p_{a1} - p_{vp1})v_{f1} = (144)(14.70 - 0.60)(0.01609) = 32.67$ ft, and $h_{b2} = (144)(13.92 - 7.51)(0.01651) = 15.24$ ft. By Eq. (21), $h_{sv} = h_{b1} + h_{s1}, = h_{b2} + h_{s2}$, from which $h_{s2} = h_{b1} - h_{b2} + h_{s1} = 32.67 - 15.24 - 10 = 7.43$ ft, the *suction head* which will be required to prevent cavitation.

Cavitation Tests The very faint noise produced by incipient cavitation is almost always masked by machinery noise so that special facilities and procedures are required for accurate testing. Figure 22C shows a suitable arrangement in which either or both p_t and $-Z$ can be varied to control h_s and h_{sv}. A head-capacity curve is obtained with ample h_{sv} to prevent cavitation. The test is repeated at a reduced constant value of h_{sv} such that cavitation will take place at some capacity beyond which values of the head will fall below those without cavitation. Figure 30, taken from Ref. 34, shows a series of such tests on a low-specific-speed pump. The capacity at

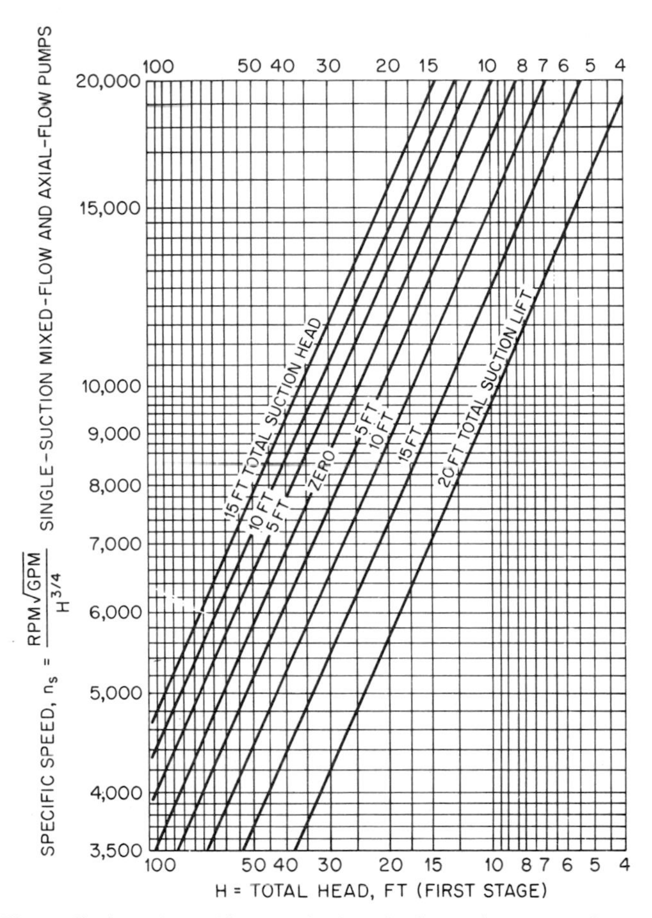

Fig. 26 Upper limits of specific speeds for single-suction and mixed- and axial-flow pumps handling clear water at 85°F at sea level. (From "Hydraulic Institute Standards," 12th ed., copyright 1969 by the Hydraulic Institute, Cleveland, Ohio)

which the head curve becomes approximately vertical is called the *cutoff capacity,* an example of which is shown at NPSH = 9 ft in Fig. 30. Cavitation is assumed to impend at a capacity above which the two head curves just begin to separate. Point C in Fig. 30 is typical of this condition. A useful cavitation test is to hold Q constant and vary h_{sv}. The results when H is plotted as a function of h_{sv} will be similar to the curve of Fig. 32 and will be discussed later. References 7 and 13 may be consulted for details of other test procedures.

Thoma Cavitation Parameter σ All the terms in Eqs. (20) to (22) may be made dimensionless by dividing each by the pump head H. The resulting parameter

$$\sigma = \frac{h_{sv}}{H} \tag{23}$$

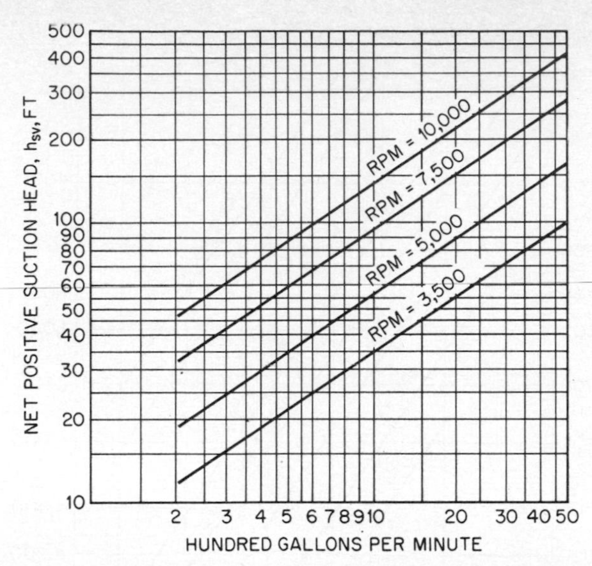

Fig. 27 NPSH for centrifugal hot-water pumps, single suction. (From "Hydraulic Institute Standards," 12th ed., copyright 1969 by the Hydraulic Institute. Cleveland, Ohio)

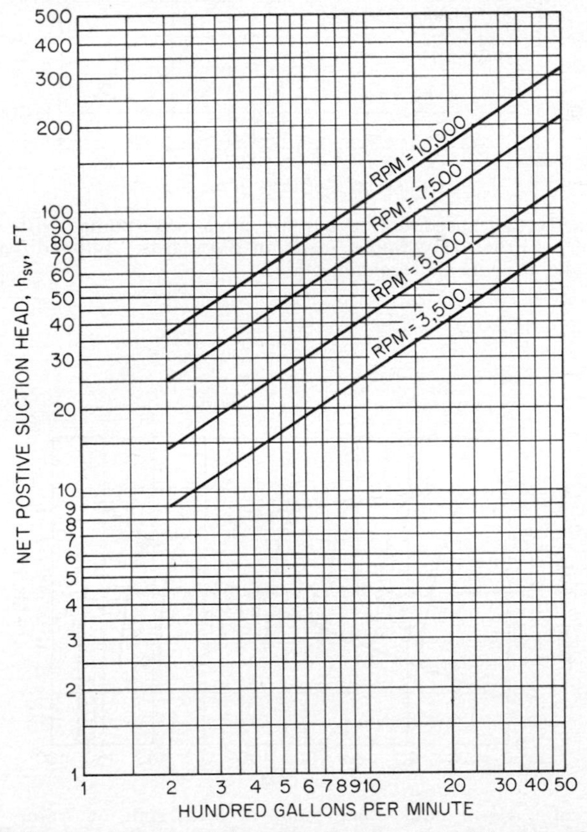

Fig. 28 NPSH for centrifugal hot-water pumps, double suction, first stage. (From "Hydraulic Institute Standards," 12th ed., copyright 1969 by the Hydraulic Institute, Cleveland, Ohio)

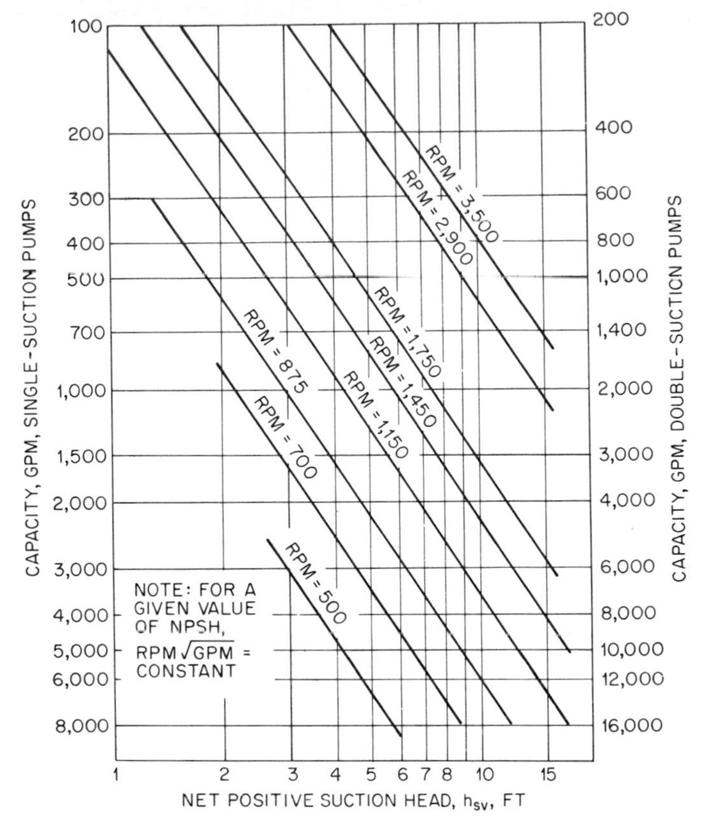

Fig. 29 Capacity and speed limitations for condensate pumps with shaft through eye of impeller. (From "Hydraulic Institute Standards," 12th ed., copyright 1969 by the Hydraulic Institute, Cleveland, Ohio)

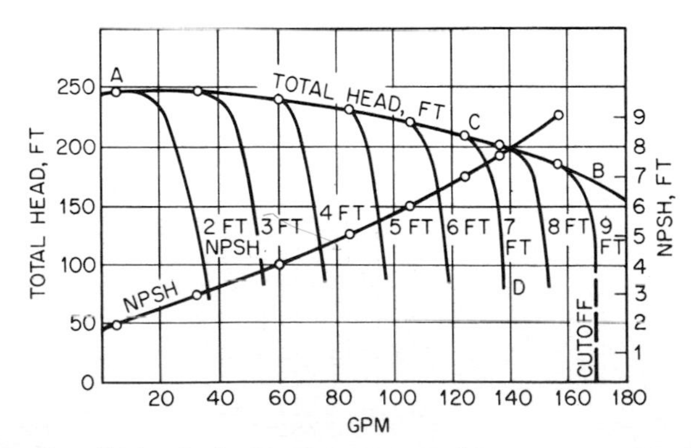

Fig. 30 Test of a 1½-in single-stage pump at 3,470 rpm, on water, 70°F. (From A. J. Stepanoff, "Centrifugal and Axial Flow Pumps," 2 ed., John Wiley & Sons, Inc., New York, 1967)

has been very useful in predicting the onset of cavitation. Equation (23) is Thoma's [35, 36] similarity law for cavitation in pumps and turbines. When the vapor pressure is reached at any point in the flow, cavitation impends and $\sigma = \sigma_C$, which is

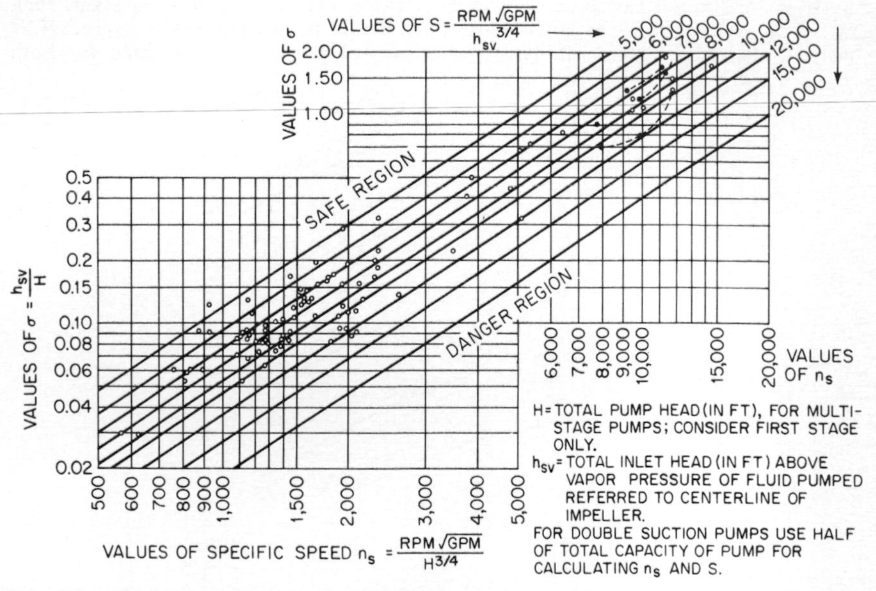

Fig. 31 Cavitation limits of centrifugal and propeller pumps. (Worthington Pump International, Inc.)

called the *critical sigma*. Experimentally determined values of $\sigma \approx \sigma_C$ are plotted against specific speed in Fig. 31. The scatter of the points is partly due to the difficulty in determining the actual onset of cavitation. For a given specific speed, the larger the value of the available σ the safer the pump will be against cavitation.

Detecting the Onset of Cavitation Geometrically similar pumps operated at the same specific speeds should have the same values of σ_C provided that disturbing influences such as vapor pockets do not destroy the dynamic similarity of the flow patterns. Figure 32 shows a curve, typical of a test to determine the onset of cavitation, in which Q is held constant and h_{sv} decreased until a drop in head indicates that cavitation has taken place. Point C is the critical point and marks the onset of cavitation. It is very difficult to determine $(h_{sv})_C$ directly and the usual practice is to use a point C_1, where the head is lower than at C by a small but measurable amount ΔH as a pseudo critical point. Then, by Eq. (23),

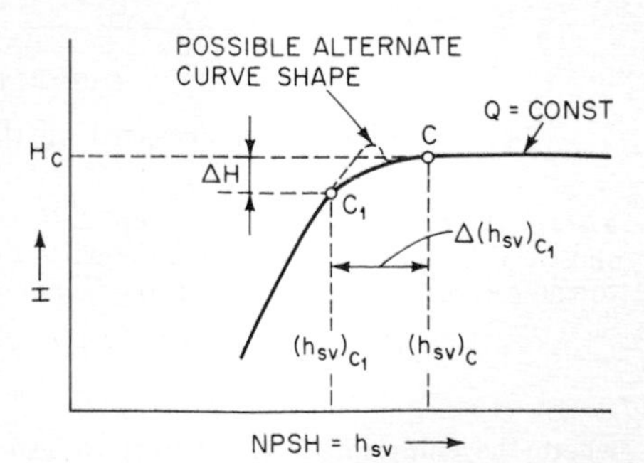

Fig. 32 Cavitation test at constant speed and constant capacity.

$$\sigma_{C1} = \frac{(h_{sv})_C - \Delta(h_{sv})_{C1}}{H} = \sigma_C - \frac{\Delta(h_{sv})_{C1}}{H} \tag{24}$$

Equation (24) may be applied to the case where $(h_{sv})_{C1} > (h_{sv})_C$ by changing the sign of $\Delta(h_{sv})_{c1}$. A common problem is the determination of σ_c or $(h_{sv})_C$ at a particular speed from tests on the same pump at a different speed or from tests of a geometrically similar pump. Two of the criteria that may be used are (1) hold $\Delta H/H$ constant for both cases or (2) hold ΔH constant regardless of the value of H. These require separate and different treatment.

Stepanoff [37], on a basis of reasonable assumptions about the volume of vapor generated by limited cavitation, has demonstrated that if $\Delta H/H$ = constant, then $\Delta h_{sv}/H$ is also constant for small values of ΔH. If the criterion is to keep $\Delta H/H$ constant, then the Thoma similarity criterion holds and σ is the same for both

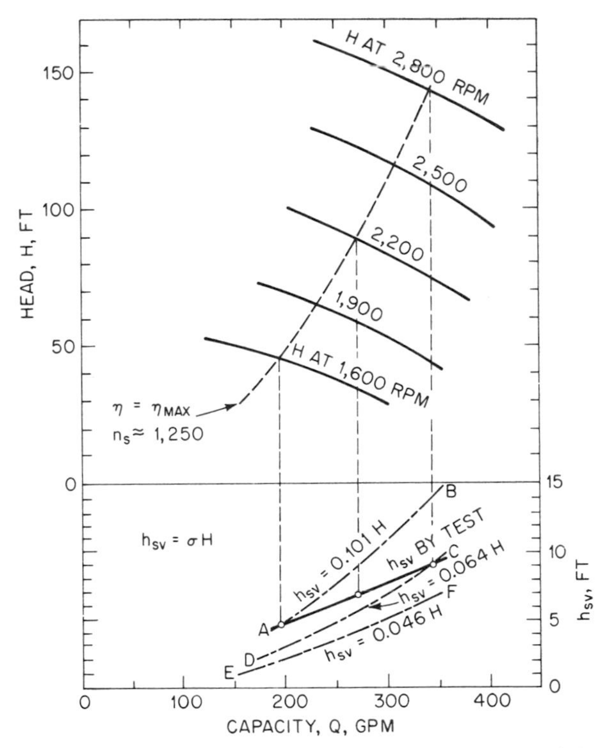

Fig. 33 Effect of pump speed on the value of σ at best efficiency [39].

cases. If the criterion is to keep ΔH constant, the Thoma criterion is violated and a different similarity relationship due to Tenot [38] is required. Applied to the case of controlled cavitation, Tenot's equation is

$$\frac{\sigma_C - \sigma_1}{\sigma_C - \sigma_2} = \frac{H_2}{H_1} \tag{25}$$

where the subscripts 1 and 2 refer to two determinations of the point C_1 in Fig. 32 with different heads H_1 and H_2. Tenot's equation may be solved for σ_c and, together with Eq. (23), gives

$$\sigma_C = \frac{\sigma_2 H_2 - \sigma_1 H_1}{H_2 - H_1} = \frac{h_{sv2} - h_{sv1}}{H_2 - H_1} \tag{26}$$

For accurately determining σ_C, $(h_{sv2} - h_{sv1})$ should be as large as possible.

Figure 33 shows the results of cavitation tests on a small pump reported by Krisam [39]. Portions of the head-capacity curves for a variety of constant speeds are shown by solid lines with the values corresponding to maximum efficiency indicated by the dashed curve. The required h_{sv} reported by Krisam is shown by curve AC. The values of σ computed for the end points of the curve differ slightly from those reported by Krisam, which were 0.093 at 1,600 rpm and 0.061 at 2,800 rpm. The two dot-dash curves show h_{sv} predicted on a basis of σ = constant. The equation of curve AB is $h_{sv} = 0.101\text{H}$ and predicts safe values of h_{sv} for all speeds. The equation of curve DC is $h_{sv} = 0.064\text{H}$. This lies well below the test curve. Assuming that the test curve represents the minimum safe values of h_{sv}, tests at the lowest head and speed should be used for extrapolation purposes to avoid the onset of cavitation at any higher head and speed. To determine σ_C,

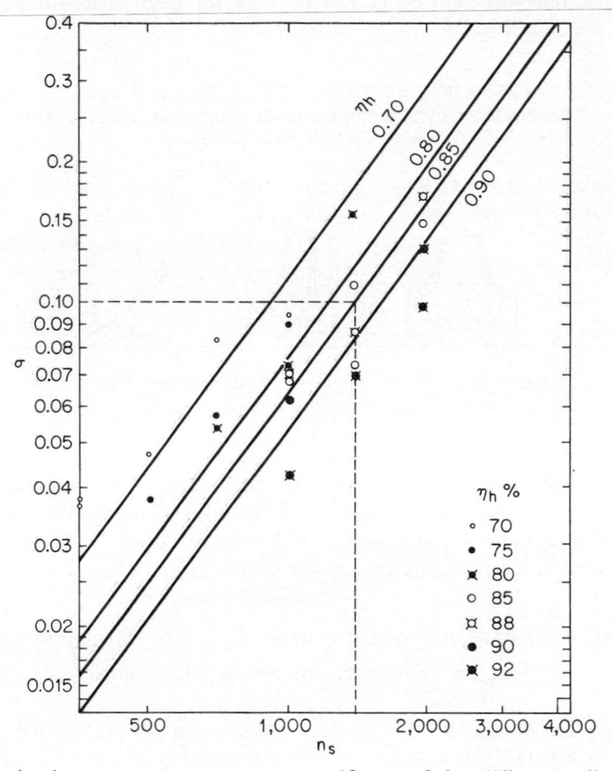

Fig. 34 Cavitation parameter σ versus specific speed for different efficiencies [25].

the following data may be read from the test curves: N = 1,600 rpm, H = 46 ft, h_{sv} = 4.63 ft; N = 2,800 rpm, H = 144 ft, h_{sv} = 9.16 ft. Then by Eq. (26),

$$\sigma_C = \frac{9.16 - 4.63}{144 - 46} = 0.046$$

The equation of curve EF in Fig. 33 is $h_{sv} = 0.046H$ and shows that curve AC has some margin of safety. Unless cavitation can be tolerated, values of $h_{sv} = \sigma_C H$ should not be used because unexpected changes in operating conditions may produce values of $\sigma < \sigma_C$ with accompanying cavitation. A further use of Eq. (26) is to compute h_{sv} for any desired speed such as 2,200 rpm for which H = 89 ft. Using the previous data

$$\sigma_C = 0.046 = \frac{9.2 - h_{sv}}{144 - 89}$$

from which h_{sv} = 6.7 ft.

Rütschi [25] has found that σ_C depends on the hydraulic efficiency as shown in Fig. 34. The hydraulic efficiency usually increases with increasing Reynolds number which may be defined by

$$\mathrm{Re} = \frac{D\sqrt{H}}{\nu} \tag{27}$$

where D = impeller diameter in ft, H = pump head in ft, and ν = kinematic viscosity of fluid pumped in ft²/s. It would be reasonable to expect that lower values of σ and h_{sv} might be used with larger pumps, for which the Reynolds numbers are larger, but this appears not to be true for large pumped-storage units as discussed in the next section.

SPECIFIC SPEED LIMIT VS. TOTAL HEAD WITH ZERO SUCTION HEAD
ASSUMING 85°F, 1.0 S.G. WATER, SEA LEVEL EQUIVALENT TO 32.6 FT NPSH
8,000 SUCTION SPECIFIC SPEED

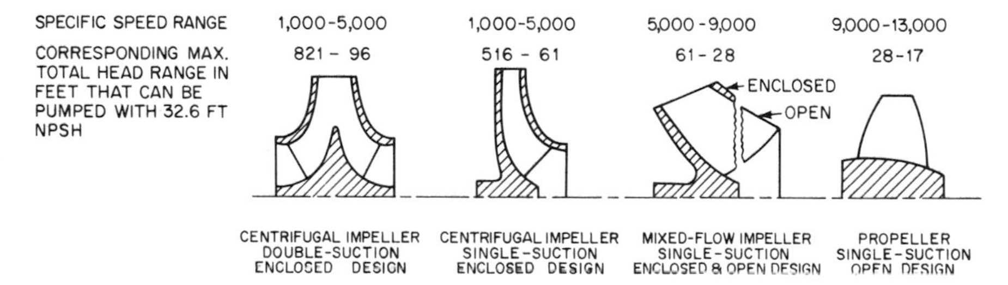

FORMULAS:

FOR SAME Q, S, h_{sv}: $n_{\text{DOUBLE SUCTION}} = \sqrt{2}\, n_{\text{SINGLE SUCTION}} = 1.414\, n_{\text{SINGLE SUCTION}}$

FOR SAME Q, H, h_{sv}: $(n_s)_{\text{DOUBLE SUCTION}} = \sqrt{2}\,(n_s)_{\text{SINGLE SUCTION}} = 1.414\,(n_s)_{\text{SINGLE SUCTION}}$

FOR SAME Q, n_s, h_{sv}: $H_{\text{DOUBLE SUCTION}} = 1.587\, H_{\text{SINGLE SUCTION}}$

FOR SAME Q, n: $(h_{sv})_{\text{DOUBLE SUCTION}} = 0.630\,(h_{sv})_{\text{SINGLE SUCTION}}$

$n_{MAX} = S(h_{sv})^{3/4}/\sqrt{Q} \cong 8000\,(h_{sv})^{3/4}/\sqrt{Q}$

Q = GPM; H = FT OF FLUID PUMPED; n = RPM; h_{sv} = FT OF FLUID PUMPED

Fig. 35 Summary for commercial pumps.

Suction Specific Speed, *S* The *suction specific speed* S [40] may be obtained by replacing H in Eq. (13) by h_{sv}

$$S = \frac{N\sqrt{Q}}{(h_{sv})^{3/4}} \tag{28}$$

Note that Q = *half the discharge* of a double-suction impeller when computing S. Equations (13), (23), and (28) may be combined to yield

$$\sigma = \left(\frac{n_s}{S}\right)^{4/3} \tag{29}$$

or

$$n_s = S(\sigma)^{3/4} \tag{30}$$

Figure 31 shows lines of constant S. Values of S computed from Figs. 23 to 26 lie within the range $7{,}500 < S < 11{,}000$. $S = 8{,}000$ is an average value frequently used for estimating purposes although much higher values may apply to special designs or service conditions such as an inducer ahead of the first-stage impeller. For a given specific speed, the lower the value of S the safer the pump will be against cavitation. Experience with large European pumped storage installations has shown that cavitation began at $S \approx 6{,}000$ and this value is recommended for large pumps. Figure 35 shows a summary of data and formulas that may be useful with commercial pumps.

German practice differs considerably from that in the United States in computing suction specific speed. Pfleiderer [5] defined a hub correction k by

$$k = 1 - \left(\frac{d_h}{D_o}\right)^2 \tag{31}$$

where d_h = the hub diameter and D_o = the diameter of the suction nozzle, in any consistent units. The suction specific speed S_G is defined by

$$S_G = \frac{(n/100)^2 Q}{k h_{sv}^{3/2}} \tag{32}$$

where n is measured in rpm, Q in cubic metres per second per impeller inlet, and h_{sv} in metres of fluid pumped. It follows that

$$S = 5{,}164\sqrt{S_G k} \tag{33}$$

NPSH for Liquids Other Than Cold Water Field experience together with carefully controlled laboratory experiments have indicated that pumps handling hot water

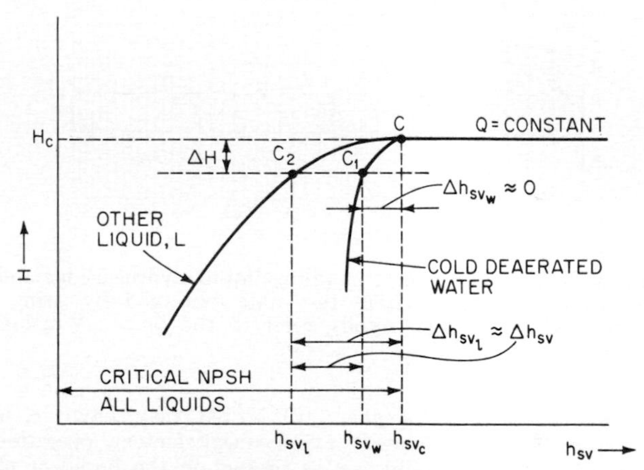

Fig. 36 Cavitation tests with different liquids at constant speed and constant capacity.

or certain liquid hydrocarbons may be operated safely with less NPSH than would normally be required for cold water. This may lower the cost of an installation appreciably, particularly in the case of refinery pumps. A theory for this has been given by Stepanoff and others [9, 34, 41, 42]. Figure 36 shows the results of cavitation tests on two liquids for constant capacity and constant pump speed. No cavitation is present at point C or at $h_{sv} > h_{svc}$. With cold deaerated water, lowering h_{sv} slightly below h_{svc} produces limited cavitation and a decrease in pump head, ΔH, to point C_1 but Δh_{svw} usually is negligible. With hot water ($T \geqq 100$°F) or with many liquid hydrocarbons, a much larger decrease in h_{sv} will be required to produce the same drop in head, ΔH, that was shown by the cold water test. The NPSH *reduction* or NPSH *adjustment* is $\Delta H_{sv} \approx \Delta h_{svl}$. In practice ΔH has been limited to $\Delta H \leqq 0.03H$ for which there is a negligible sacrifice in performance. The pumps usually are made of stainless steel or other cavitation-resistant materials and the small localized production of vapor bubbles is not enough to cause damage.

Chart for NPSH Reductions A composite chart of NPSH *reductions* for deaerated hot water and certain gas-free liquid hydrocarbons is shown in Fig. 37. The curves of vapor pressure in pounds per square inch absolute versus temperature in °F and the curves of constant NPSH reduction were based on laboratory tests with the fluids shown. Pending further experience with pumps operated at reduced NPSH, Fig. 37 should be used subject to the following limitations: (1) No NPSH reduction should exceed 50 percent of the NPSH required by the pump for cold

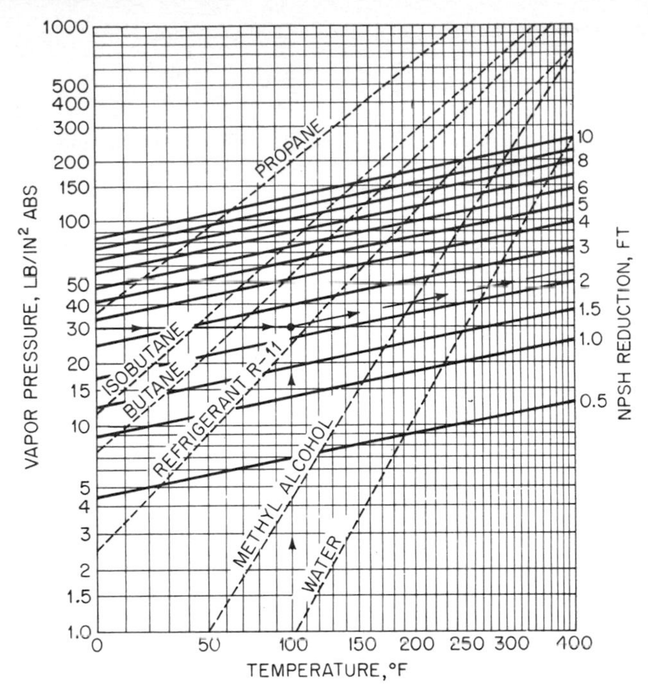

Fig. 37 NPSH reductions for pumps handling liquid hydrocarbons and hot water. This chart has been constructed from test data obtained by using the liquids shown. For applicability to other liquids, refer to the text. (Worthington Pump International, Inc.)

water or ten feet, whichever is smaller. (2) NPSH may have to be increased *above* the normal cold water value to avoid unsatisfactory operation when (*a*) *entrained* air or other noncondensable gas is present in the liquid or (*b*) *dissolved* air or other noncondensable gas is present in the liquid and the absolute suction pressure is low enough to permit release of the gas from solution. (3) The vapor pressure of hydrocarbon mixtures should be determined by the bubble point method at pumping temperature (see Ref. 7). Do not use the Reid vapor pressure or the vapor pressure of the lightest fraction. (4) If the suction system may be susceptible to transient changes in absolute pressure or temperature, a suitable margin of safety in NPSH should be provided. This is particularly important with hot water and may exceed the reduction that would otherwise apply with steady-state conditions. (5) Although experience has indicated the reliability of Fig. 37 for hot water and the liquid hydrocarbons shown, its use with other liquids is not recommended unless it is clearly understood that the results be accepted on an experimental basis.

Use of Figure 37. Given a fluid having a vapor pressure of 30 lb/in² abs at 100°F. Follow the arrows on the key shown on the chart and obtain an NPSH reduction of about 2.3 ft. Since this does not correspond to one of the fluids for which vapor pressure curves are shown on the chart, the use of this NPSH reduction should be considered as a tentative value only. Given a pump requiring 16 ft cold-water NPSH at the operating capacity; the pump is to handle propane at 55°F. Figure 37 shows the vapor pressure to be about 105 lb/in² abs and the NPSH reduction to be about 9.5 ft. Since this exceeds 8 ft, which is half the cold-water NPSH, the recommended NPSH for the pump handling propane is half the cold-water NPSH, or 8 ft. If the temperature of the propane in the previous example is reduced to 14°F, Fig. 37 shows the vapor pressure to be 50 lb/in² abs and the NPSH reduction to be about 5.7 ft which is less than half the cold-water NPSH. The NPSH required for pumping propane at 14°F is then $16 - 5.7 = 10.3 \approx 10$ ft.

Reduction of Cavitation Damage Once the pump has been built and installed,[1] there is little that can be done to reduce cavitation damage. As previously mentioned, sharpening the leading edges of the vanes by filing may be beneficial. Stepanoff [12] has suggested cutting back part of the vanes in the impeller eye together with sharpening the tips, for low specific-speed pumps, as a means of reducing the inlet velocity c_1 and thus lowering σ. Although a small amount of prerotation or prewhirl in the direction of impeller rotation may be desirable

TABLE 7 Cavitation Erosion Resistance of Metals [44]

Alloy	Magnetostriction weight loss after 2 hr, mg
Rolled stellite[a]	0.6
Welded aluminum bronze	3.2
Cast aluminum bronze	5.8
Welded stainless steel (2 layers, 17 Cr–7 Ni)	6.0
Hot rolled stainless steel (26 Cr–13 Ni)	8.0
Tempered rolled stainless steel (12 Cr)	9.0
Cast stainless steel (18 Cr–8 Ni)	13.0
Cast stainless steel (12 Cr)	20.0
Cast manganese bronze	80.0
Welded mild steel	97.0
Plate steel	98.0
Cast steel	105.0
Aluminum	124.0
Brass	156.0
Cast iron	224.0

[a] Despite the high resistance of this material to cavitation damage, it is not suitable for ordinary use because of its comparatively high cost and the difficulty encountered in machining and grinding.

(Ref. 8), excessive amounts should be avoided. This may require straightening vanes ahead of the impeller and rearranging the suction piping to avoid changes in direction or other obstructions. The cavitation damage to the impeller shown in Fig. 20 was believed to have been at least partly due to bad flow conditions produced by two 90° elbows in the suction piping. The planes of the elbows were at 90° to each other and this arrangement should be avoided.

Straightening vanes in the impeller inlet may increase the NPSH requirement at all capacities. Three or four radial ribs equally spaced around the inlet and extending inward about one-quarter of the inlet diameter are effective against excessive prerotation and may require less NPSH than full-length vanes. This is very important with axial flow pumps which are apt to have unfavorable cavitation characteristics at partial capacities. Operation near the best efficiency point usually minimizes cavitation.

The admission of a small amount of air into the pump suction tends to reduce cavitation noise (Ref. 4). This rarely is used, however, because it is difficult to inject the right amount of air under varying head and capacity conditions and, frequently, there are objections to mixing air with the fluid pumped.

If a new impeller is required because of cavitation, the design should take into account the most recent advances described in the literature. Gongwer [43] has suggested (1) the use of ample fillets where the vanes join the shrouds, (2) sharpened leading edges of vanes, (3) reduction of β_1 in the immediate vicinity

[1] Sometimes it is possible to lower the pump, and this should be considered before other alterations are made.

of the shrouds, and (4) raking the leading edges of the vanes forward out of the eye. Increasing the number of vanes for propeller pumps lowers σ for a given submergence. A change in the impeller material may be very beneficial as described below.

Resistance of Materials to Cavitation Damage Table 7 (Ref. 44) shows the relative resistance of several metals to cavitation pitting produced by magnetostriction vibration. It will be seen that cast iron, the most commonly used material for impellers, has relatively little pitting resistance compared to bronze and stainless steel which are readily cast and finished. Elastomeric coatings have been found to be highly resistant to cavitation pitting. Table 8 (Ref. 44) shows the relative

TABLE 8 Cavitation Erosion Resistance of Elastomeric Coatings [44]

Material	Subtype	Thickness of coating in rotating-disk cavitation tests, in	Cavitation test exposure period, hr	Degree of erosion after exposure period, 150 ft/s
Neoprene solvent base, brush-applied	A	0.030	24	Slight
	B	0.025	17	Slight
Neoprene, cured sheet cold-bonded	...	0.062	14	None
Neoprene, *in situ* cured and bonded	...	0.060	10½	None
Polyurethane, liquid	A	0.062	12	Slight
	B	0.018	12	Severe
	C	0.062	12	None
Polyurethane cured sheet, cold-bonded	A	0.060	14	None
	B	0.062	12	Severe
Polysulfide, liquid	...	0.062	12	Severe
Polysiloxane, liquid	...	0.062	7	Severe
Butyl, cured sheet, cold-bonded	...	0.060	2¼	Severe
Butyl, *in situ* cured and bonded	...	0.060	12	Severe
Cis-polybutadiene (98%) cured sheet, cold bonded	...	0.060	10	None
Polybutadiene (polysulfide modified) *in situ* cured and bonded	...	0.060	13	Severe
Styrene-butadiene copolymer, *in situ* cured and bonded (SBR)	...	0.060	24	None
Natural rubber, cured sheet, cold-bonded	...	0.062	10	None
Natural rubber, *in situ* cured and bonded	...	0.060	16	Severe

merits of several elastomers which were tested on a rotating disk at 150 ft/s. The best of the elastomers were even more resistant to cavitation damage than stellite 6B which leads the list of metals in Table 7. The value of such coatings has been known for a long time but only recently has it appeared possible to secure an adequate bond between the coating and the metal. Polyurethane and neoprene, which show high resistance to cavitation pitting and may be applied in liquid form, should be considered if other methods of reducing cavitation damage cannot be used.

PUMP SELECTION

Many manufacturers issue *pump selection charts* which show performance data for their commercial line of pumps. If pumping requirements can be met by one of these, there is usually a considerable saving over the cost of a custom-designed unit. A typical selection chart is shown in Fig. 38. Each numbered field covers the range of head and capacity that can be had with good efficiency from the pump bearing that particular designation. Let it be required to pump 5,000 gpm of cold water against a head of 180 ft at sea level. Figure 38, curve *A*, shows that the A1015L pump running at 1,775 rpm will meet the requirement.

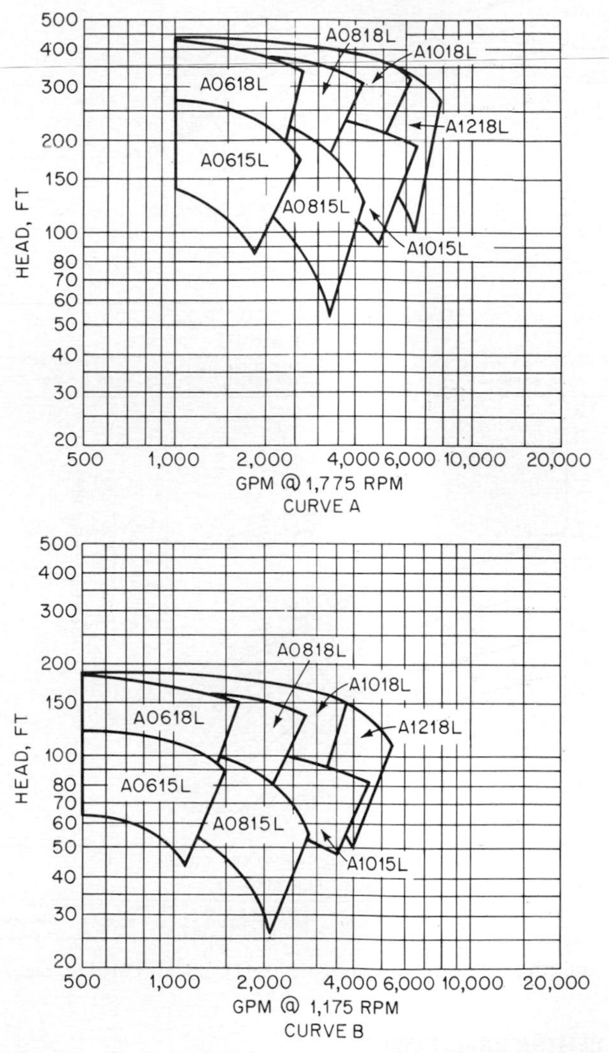

Fig. 38 Pump selection chart. (DeLaval Turbine, Inc.)

Figure 39, 1,775 rpm, which is typical of performance curves accompanying pump selection charts, shows that the impeller diameter should be about 14.6 in. The efficiency should be slightly over 89 percent and about 260 hp will be required to drive the pump. Since this impeller will be about midway between the minimum and maximum diameter, the upper curves show that about 21 ft NPSH must be provided which is equivalent to a 12 ft maximum suction lift. A reasonable number of such charts will cover a complete line of pumps and supply all the data usually needed to select a pump to meet a given set of operating conditions.

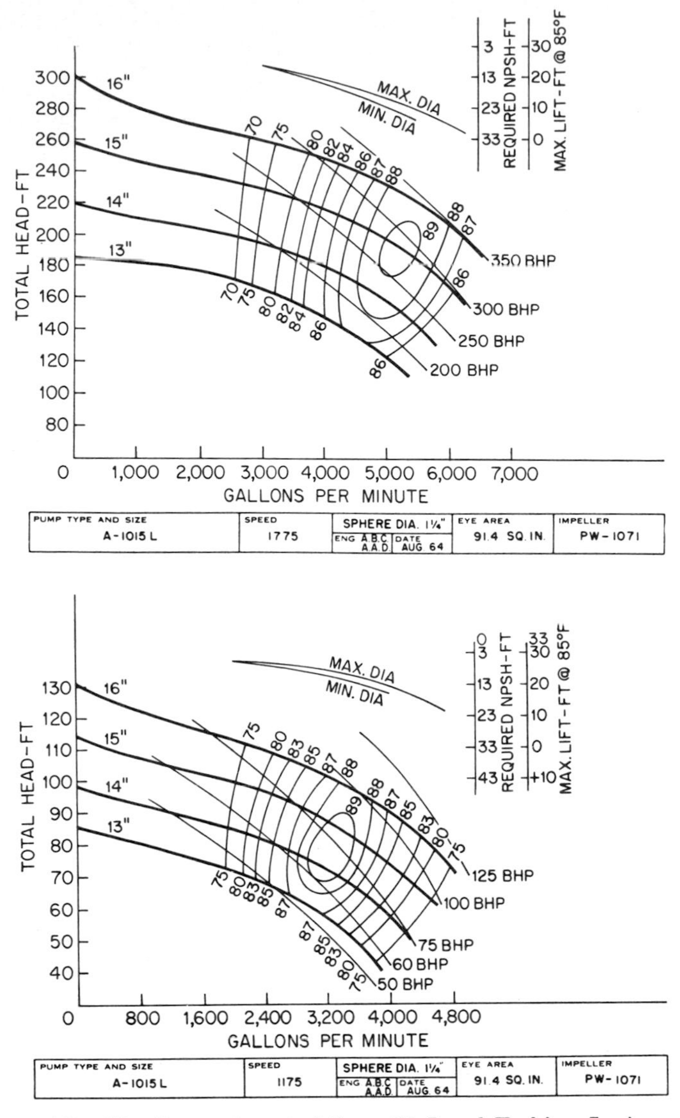

Fig. 39 Pump characteristics. (DeLaval Turbine, Inc.)

STARTING CENTRIFUGAL PUMPS

Priming Centrifugal pumps usually are completely filled with the liquid to be pumped *before starting.* When so filled with liquid the pump is said to be *primed.* Pumps have been developed to start with air in the casing and then be primed (Ref. 45). This is an unusual procedure with low-specific-speed pumps but is sometimes done with propeller pumps (Ref. 12). In many installations, the pump is at a lower elevation than the supply and remains primed at all times. This is customary for pumps of high specific speed and all pumps requiring a positive suction head to avoid cavitation. Pumps operated with a suction lift may be primed in any of several ways. A relatively inexpensive method is to install a special type of check valve, called a *foot valve,* on the inlet end of the suction pipe and prime the pump by filling the system with liquid from any available source. Foot valves cause undesirable friction loss and may leak enough to require priming before each starting of the pump. A better method is to close a valve in the discharge line and prime by evacuating air from the highest point of the pump casing. Many types of vacuum pumps are available for this service. A priming chamber is a tank which holds enough liquid to keep the pump submerged until pumping action can be initiated. Self-priming pumps usually incorporate some form of priming chamber in the pump casing. Section 2.4 and Ref. 4 may be consulted for further details.

Torque Characteristics of Drivers Centrifugal pumps of all specific speeds usually have such low starting torques (turning moments) that an analysis of the starting phase of operation seldom is required. Steam and gas turbines have high starting torques so that no special starting procedures are necessary when they are used to drive pumps. If a pump is directly connected to an internal combustion engine, the starting motor of the engine should be made adequate to start both driver and pump. If the starter does not have enough torque to handle both units, a clutch must be provided to uncouple the pump until the driver is started.

Electric motors are the most commonly used drivers for centrifugal pumps. Direct current motors and alternating current induction motors usually have ample starting torque for all pump installations provided the power supply is adequate. Many types of reduced voltage starters are available (Ref. 4) to limit the inrush current to safe values for a given power supply. Synchronous motors are often used with large pumps because of their favorable power-factor properties. They are started as induction motors and run as such up to about 95 percent of synchronous speed. At this point, dc field excitation is applied and the maximum torque the motor can then develop is called the *pull-in torque,* which must be enough to accelerate the motor and connected inertia load to synchronous speed in about 0.2 s if synchronous operation is to be achieved. Centrifugal pumps usually require maximum torque at the normal operating point and this should be considered in selecting a driver, particularly a synchronous motor to be sure that the available pull-in torque will bring the unit to synchronous speed.

Torque Requirements of Pumps The *torque* or turning moment for a pump may be estimated from the power curve by

$$M = \frac{5{,}250P}{n} \tag{34}$$

where M = pump torque, lb-ft
P = power, hp
n = speed, rpm

Equation (34) makes no allowance for accelerating the rotating elements or the fluid in the pump. If a 10 percent allowance for accelerating torque is included, the constant should be increased to 5,800. The time, Δt, required to change the pump speed by an amount $\Delta n = n_2 - n_1$ is given by

$$\Delta t = \frac{WK^2\,\Delta n}{308(M_m - M)} \tag{35}$$

where Δt = time, s
WK^2 = inertia (flywheel effect) of all rotating elements of driver, pump, and fluid, lb-ft^2
Δn = change in speed, rpm
M_m = driver torque, lb-ft
M = pump torque, lb-ft [Eq. (34)]

The inertia WK^2 of the driver and pump usually can be obtained from the manufacturers of the equipment. The largest permissible Δn for accurate calculation will depend on how rapidly M_m and M vary with speed. The quantity $M_m - M$ should be nearly constant over the interval Δn if an accurate estimate of Δt is to be obtained. Torque-speed characteristics of electric motors may be obtained from the manufacturers.

Horizontal-shaft pumps fitted with plain bearings and packed glands require a *breakaway torque* of about 15 percent of M_n, the torque at the normal operating point, to overcome the static friction. This may be reduced to about 10 percent of M_n if the pump is fitted with antifriction bearings. The breakaway torque may be assumed to decrease linearly with speed to nearly zero when the speed reaches 15 to 20 percent of normal. Construction of torque-speed curves requires a knowledge of the pump characteristics at normal speed as well as details of the entire pumping system. Some typical examples taken from Refs. 4 and 46

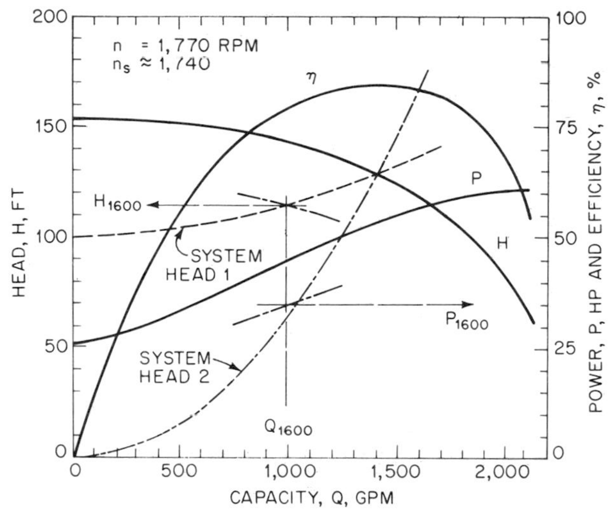

Fig. 40 Characteristics of a 6 by 8 double-suction pump at 1,770 rpm [4].

are given below. The following forms of the affinity laws, Eqs. (17), are useful in constructing the varous performance curves:

$$Q_2 = Q_1 \frac{n_2}{n_1} \tag{36}$$

$$H_2 = H_1 \left(\frac{n_2}{n_1}\right)^2 = H_1 \left(\frac{Q_2}{Q_1}\right)^2 \tag{37}$$

$$P_2 = P_1 \left(\frac{n_2}{n_1}\right)^3 \tag{38}$$

$$M_2 = M_1 \left(\frac{n_2}{n_1}\right)^2 \tag{39}$$

where Q = capacity
H = head
P = power
M = torque
n = speed

in any consistent units of measure. Once speeds n_1 and n_2 are chosen, the subscripts 1 and 2 refer to corresponding points on the characteristic curves for these speeds.

Low Specific-Speed Pumps. Figure 40 shows the constant-speed characteristics of a pump having $n_s \approx 1{,}740$ at best efficiency. This pump usually would be started with a valve in the discharge line closed. During the starting phase, the pump operates at shutoff with $P_1 = 25.8$ hp and $n_1 = 1{,}770$ rpm. Then, by Eq. (34), $M_1 = 76.6$ lb-ft. These values may be used in Eq. (39) to evaluate starting torques M_2 at as many speeds n_2 as desired and plotted in Fig. 41 as curve *BCD*. Section *AB* of the starting-torque curve is an estimate of the breakaway torque. If the discharge valve is now opened, the speed remains nearly constant but the torque increases as the capacity and power increase. If the normal operating point is $Q_n = 1{,}400$ gpm and $P_n = 53.2$ hp, the motor torque will be $M_n = 158$ lb-ft by Eq.

(34). The vertical line *DE* in Fig. 41 shows the change in torque produced by opening the discharge valve.

Instead of starting the pump with the discharge valve closed, let the pump be started with a check valve in the discharge line held closed by a static head of 100 ft. The friction head in the system may be represented by kQ^2. The value of k may be estimated from the geometry of the system or from a friction-loss measurement at any convenient flowrate Q, preferably near the normal capacity Q_n. In this example, $Q_n = 1{,}400$ gpm and $k = 14.4/10^6$. The curve "System Head 1" in Fig. 40 was computed from $H = 100 + (14.4/10^6)Q^2$ and intersects the head-capacity curve at $H_n = 128$ ft and $Q_n = 1{,}400$ gpm. The normal shutoff head is $H_1 = 153$ ft

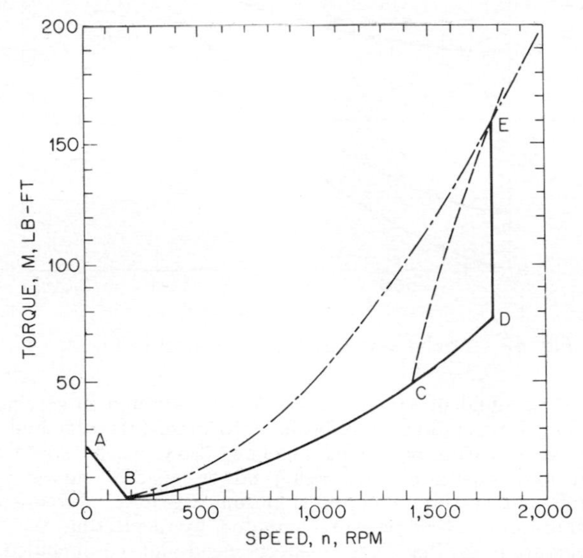

Fig. 41 Torque characteristics of 6 by 8 pump shown in Fig. 40 [4].

at $n_1 = 1{,}770$ rpm. By Eq. (37), the pump will develop a shutoff head, $H_2 = 100$ ft, at $n_2 = 1{,}430$ rpm. By Eq. (39), the torque at 1,430 rpm will be 50 lb-ft corresponding to point *C* in Fig. 41. The portion *ABC* of the starting torque curve has already been constructed. Trial and error methods must be used to obtain the portion *CE* of the starting torque curve. The auxiliary curves in Fig. 42 are useful in constructing the *CE* portion of the starting-torque curve. Select a value of n_2 intermediate between 1,427 rpm and 1,770 rpm, say $n_2 = 1{,}600$ rpm. In Fig. 40, read values of Q_1, H_1, and P_1 for speed $n_1 = 1{,}770$ rpm. By Eqs. (36) and (37), determine values of Q_2 and H_2 and plot as shown in Fig. 40 until an intersection with the system head 1 curve is obtained which provides $Q_{1,600}$ corresponding to $n = 1{,}600$ rpm. By Eq. (38), determine value of P_2 and plot as shown in Fig. 40 until an intersection is obtained with the $Q_{1,600}$ line which provides $P_{1,600}$ corresponding to $n = 1{,}600$ rpm. Equation (34) is now used to obtain $M_{1,600}$ which is one point on the desired starting-torque curve. The above process is repeated for various speeds n_2 until the curve *CE* in Fig. 41 can be drawn. The complete starting-torque curve for this example is *ABCE* in Fig. 41 with steady-state operation at point *E*.

Assume that the pump of the preceding examples is installed in a system having zero static head but a long pipeline with friction head given by $H = (65.4/10^6)Q^2$ as shown in Fig. 40 by the curve "System Head 2." The valve in the discharge line is assumed to open instantaneously when power is first applied to the pump. The procedure described to construct curve CE of the preceding example must now be used together with the system head 2 curve of Fig. 40 to obtain the curve *BE* of Fig. 41. The complete starting-torque curve for this example is *ABE* in Fig. 41 with steady-state operation at point *E*.

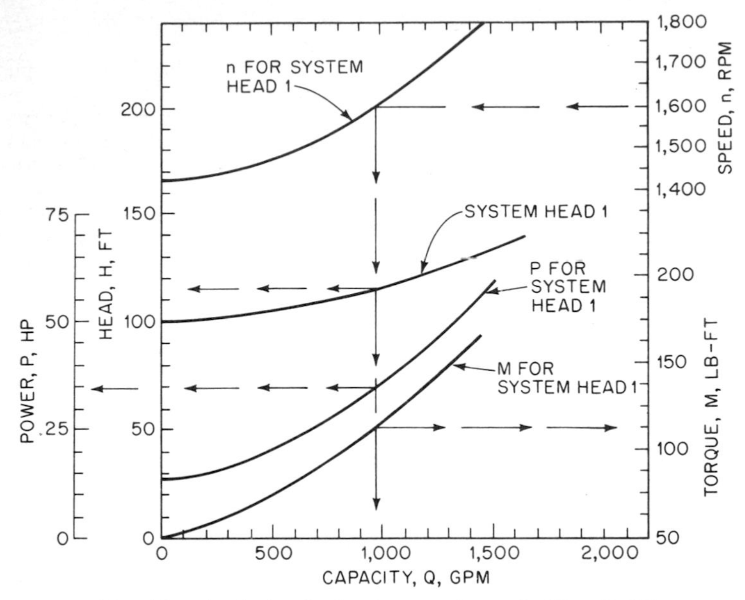

Fig. 42 Analysis of 6 by 8 pump shown in Fig. 40 [4].

The inertia of the fluid in the system has been neglected in solving the previous examples. Some of the power must be used to accelerate the fluid and this may be appreciable in the case of a long pipeline. Low specific-speed pumps, which are used with long pipelines, have rising power-capacity curves with minimum power at shutoff and maximum power at normal capacity. Experience has shown that the starting-torque-speed curves computed by neglecting the inertia of the fluid are conservative so that inertia effects need not be included. The inertia effect of the fluid does slow the starting operation. If the time required to reach any event, such as a particular speed or capacity, is required, the inertia of the fluid should be considered but including it greatly increases the difficulty of computation. References 52, 56, 57, and 58 give general methods for handling problems involving fluid transients. High specific-speed pumps have falling power-

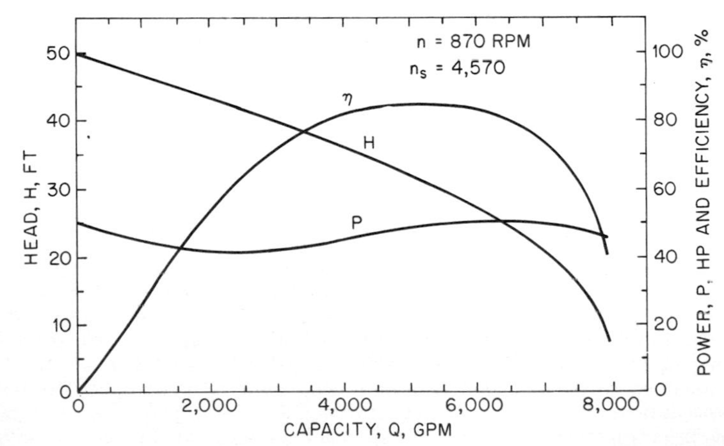

Fig. 43 Characteristics of a 16-in volute pump with mixed-flow-type impeller with flat power characteristic [4].

capacity curves with maximum power at shutoff and minimum power at normal capacity. Neglecting the inertia of the fluid probably will result in too low a value for the computed starting torque for such pumps. If fluid inertia is to be included, consult Refs. 52, 56, 57, and 58 and Sec. 9.1.

Medium and High Specific-Speed Pumps. Figure 43 shows constant-speed characteristics for a medium specific-speed pump n_s = 4,570 at best efficiency. The shutoff power is the same as the power at best efficiency and the starting-torque-speed curve is but little affected by the method of starting, as shown by Fig. 44. Figure 45 shows the constant-speed characteristics of a high specific-speed propeller pump, $n_s \approx$ 12,000 at best efficiency. Figure 46 shows the starting-torque-speed curve when the pump is started against a static head of 14 ft and a friction head of 1 ft at Q_n = 12,500 gpm. The system was assumed full of water with a closed check valve at the outlet end of the short discharge pipe. The methods of computation for Figs. 44 and 46 were the same as for Fig. 41. Sometimes propeller pumps are started with the pump submerged but with the discharge column filled with air. In such a case, the torque-time characteristic for the driver must be known and a step-by-step calculation carried out. If the discharge column is a siphon, initially filled with air, the starting torque may exceed the normal running torque during some short period of the starting operation. If the pump is driven by a synchronous motor, it is particularly important to investigate the starting torque in the range of 90 to 100 percent normal speed to make sure that the pull in torque of the motor is not exceeded. For additional information regarding starting high specific-speed pumps discharging through long and/or large diameter systems see Sec. 9.1.

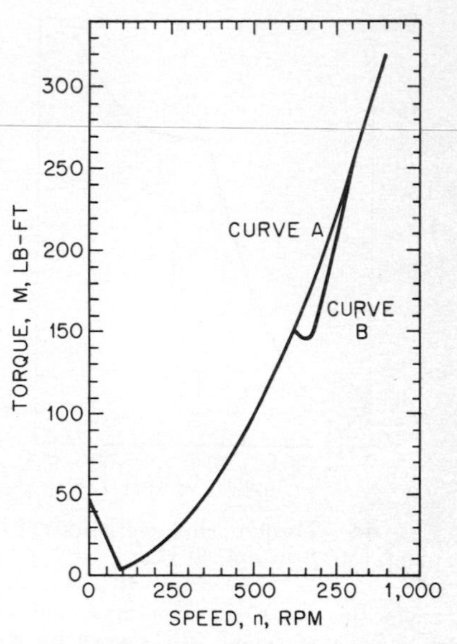

Fig. 44 Torque characteristics of pump shown in Fig. 43 [4].

Miscellaneous Requirements Pumps handling hot liquids should be warmed up to operating temperature before being started unless they have been especially

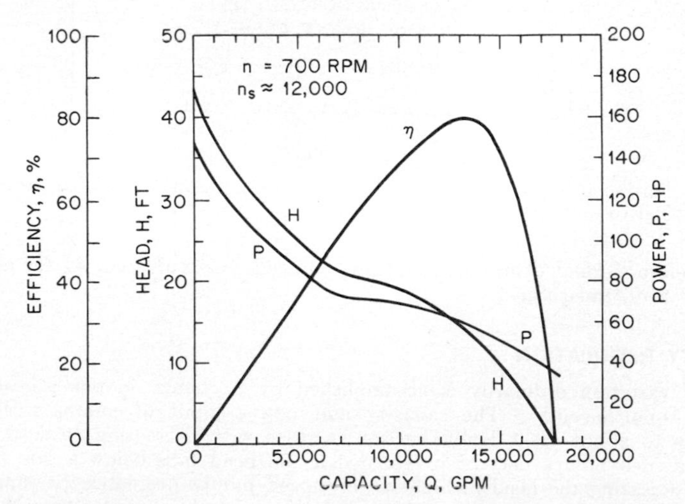

Fig. 45 Characteristics of a 30-in discharge propeller pump at 700 rpm [4].

designed for quick starting. Failure to do this may cause serious damage to wearing rings, seals, and any hydraulic balancing device that may be present. A careful check of the installation should be made before starting new pumps, pumps that have had a major overhaul, or pumps that have been standing idle for a long time. It is very important to follow the manufacturer's instructions when starting boiler feed pumps. If these are unavailable, Ref. 4 may be consulted. Ascertain that the shaft is not frozen, that the direction of rotation is correct, preferably with the coupling disengaged, and that bearing lubrication and gland cooling water meet normal requirements. Failure to do this may result in damage to the pump or driver.

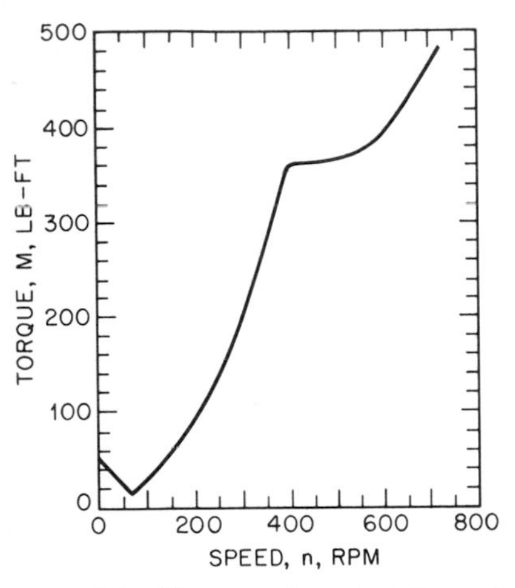

Fig. 46 Torque characteristics of pump shown in Fig. 45 [4].

A pump may run backwards at runaway speed if the discharge valve fails to close following shutdown. Any attempt to start the pump from this condition will put a prolonged overload on the motor. Figure 47 shows one example of the torque-speed transient for a pump, n_s = 1,700, started from a runaway reversed speed while normal pump head was maintained between the suction and discharge flanges. In most practical cases, water-hammer effects would make this transient even more unfavorable than Fig. 47 indicates. The duration of such a transient will always be much longer than the normal starting time so that protective devices probably would disconnect the motor from the power

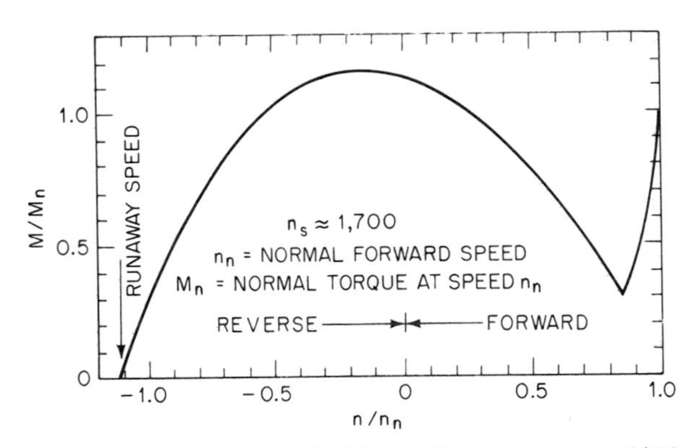

Fig. 47 Torque characteristics of a double-suction pump, $n_s \approx 1{,}700$, from reverse runway speed to normal forward speed [4].

supply before normal operation could be achieved. Consult Sec. 9.1 for additional information on this subject.

CAPACITY REGULATION

Capacity variation ordinarily is accomplished by a change in pump head, speed, or both simultaneously. The capacity and power input of pumps with specific speeds up to about 4,000 double suction increase with decreasing head so that the drivers of such pumps may be overloaded if the head falls below a safe minimum value. Increasing the head of high specific-speed pumps decreases the capacity but *increases* the power input. The drivers of these pumps should either be able to meet possible load increases or be equipped with suitable overload protection. Capacity regulation by the various methods given below may be manual or automatic. See also Refs. 4, 5, 8, 10, 12.

Discharge Throttling This is the cheapest and most common method of capacity modulation for low and medium specific-speed pumps. Usually its use is restricted to such pumps. Partial closure of any type of valve in the discharge line will increase the system head so that the system-head curve will intersect the head-capacity curve at a smaller capacity, as shown in Fig. 49. Discharge throttling moves the operating point to one of lower efficiency and power is lost at the throttle valve. This may be important in large installations where more costly methods of modulation may be economically attractive. Throttling to the point of cutoff may cause excessive heating of the fluid in the pump. This may require

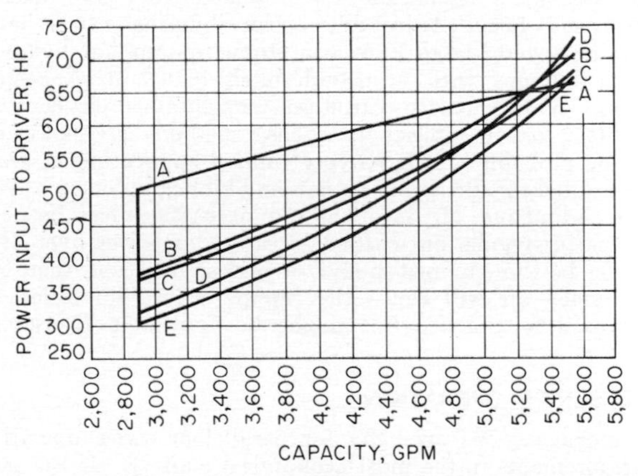

Fig. 48 Power requirements of two double-suction pumps in series operated at constant head and variable capacity. Total $H_n = 382$ ft for both pumps at 1,800 rpm [47].

Curve *AA:* constant speed with discharge throttling
Curve *BB:* synchronous motor with variable-speed hydraulic coupling on each pump
Curve *CC:* variable-speed wound-rotor induction motor
Curve *DD:* dc motor with rectifier and shunt field control
Curve *EE:* synchronous motor with variable-speed constant-efficiency mechanical speed reducer

a bypass to maintain the necessary minimum flow or use of a different method of modulation. This is particularly important with pumps handling hot water or volatile liquids as previously mentioned. Refer to Sec. 9.2 for information regarding the sizing of a pump bypass.

Suction Throttling If sufficient NPSH is available, some power can be saved by throttling in the suction line. Jet engine fuel pumps frequently are suction throttled (Ref. 21) because discharge throttling may cause overheating and vaporization of the liquid. At very low capacity, the impellers of these pumps are only partly filled with fluid so that the power input and temperature rise are about one-third the values for impellers running full with discharge throttling. The capacity of condensate pumps frequently is submergence-controlled (Ref. 4), which is equivalent to suction throttling. Special design reduces cavitation damage of these pumps to a negligible amount.

Bypass Regulation All or part of the pump capacity may be diverted from the discharge line to the pump suction or other suitable point through a bypass line. The bypass may contain one or more metering orifices and suitable control valves. Metered bypasses are commonly used with boiler feed pumps for reduced capacity operation mainly to prevent overheating. There is a considerable power saving if excess capacity of propeller pumps is bypassed instead of using discharge throttling.

Speed Regulation This can be used to minimize power requirements and eliminate overheating during capacity modulation. Steam turbines and internal combustion engines are readily adaptable to speed regulation at small extra cost. A wide variety of variable-speed mechanical, magnetic, and hydraulic drives are available as well as both ac and dc variable-speed motors. Usually variable-speed motors are so expensive that they can be justified only by an economic study of a particular case. Figure 48 shows a study, by Richardson [47], of power requirements with various drivers wherein substantial economies in power may be obtained from variable-speed drives.

Regulation by Adjustable Vanes Adjustable guide vanes ahead of the impeller have been investigated and found effective with a pump of specific speed, $n_s = 5{,}700$. The vanes produced a positive pre-whirl which reduced the head, capacity, and efficiency. Relatively little regulation was obtained from the vanes with pumps having $n_s = 3{,}920$ and 1,060. Adjustable outlet diffusion vanes have been used with good success on several large European storage pumps for hydroelectric developments. Propeller pumps with adjustable-pitch blades have been investigated with good success. Wide capacity variation was obtained at constant head and with relatively little loss in efficiency. These methods are so complicated and expensive that they probably will have very limited application in practice. Reference 8 may be consulted for further discussion and bibliography.

Air Admission Admitting air into the pump suction has been demonstrated as a means of capacity regulation with some saving in power over discharge throttling. Usually, air in the pumped liquid is undesirable and there is always the danger that too much air will cause the pump to lose its prime. The method has rarely been used in practice but might be applicable to isolated cases.

PARALLEL AND SERIES OPERATION

Two or more pumps may be arranged for parallel or series operation to meet a wide range of requirements in the most economical manner. If the pumps are close together, that is, in the same station, the analysis given below should be adequate to secure satisfactory operation. If the pumps are widely separated, as in the case of two or more pumps at widely spaced intervals along a pipeline, serious pressure transients may be generated by improper starting or stopping procedures. The analysis of such cases may be quite complicated, and Refs. 56, 57, and 58 should be consulted for methods of solution.

Parallel Operation Parallel operation of two or more pumps is a common method of meeting variable-capacity requirements. By starting only those pumps needed to meet the demand, operation near maximum efficiency can usually be obtained. The head-capacity characteristics of the pumps need not be identical, but pumps with unstable characteristics may give trouble unless operation only on the steep portion of the characteristic can be assured. Multiple pumps in a station provide spares for emergency service and for the down time needed for maintenance and repair.

The possibility of driving two pumps from a single motor should always be considered as it usually is possible to drive the smaller pumps at about 40 percent higher speed than a single pump of twice the capacity. The saving in cost of the higher-speed motor may largely offset the increased cost of two pumps and give additional flexibility of operation.

One of the first steps in planning for multiple-pump operation is to draw the system-head curve as shown in Fig. 49. The system head consists of the static head H_s and the sum. H_f, of the pipe-friction head and the head lost in the valves and fittings. See Secs. 9.1 and 9.2. The head-capacity curves of the various pumps are plotted on the same diagram and their intersections with the system-head curve show possible operating points. *Combined head-capacity curves* are drawn by adding the capacities of the various combinations of pumps for as many values of the head as necessary. The intersection of any combined H-Q curve with the system head curve is an operating point. Figure 49 shows two head-capacity curves and the combined curve. Points 1, 2, and 3 are possible operating conditions. Additional operating points may be obtained by changing the speed of the pumps or by

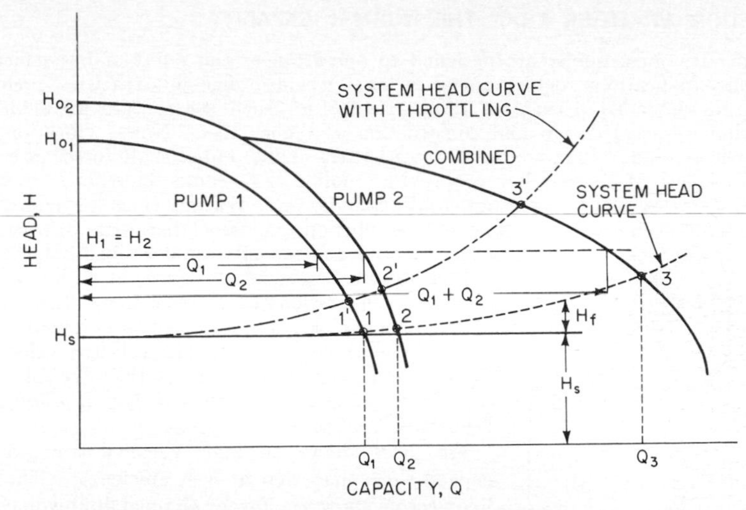

Fig. 49 Head-capacity curves of pumps operating in parallel.

increasing the system-head loss by throttling. Any number of pumps in parallel may be included on a single diagram although separate diagrams for different combinations of pumps may be preferable. The overall efficiency η of pumps in parallel is given by $\eta = (sH/3{,}960)(\Sigma Q/\Sigma P)$ where s = specific gravity of the fluid, H = head, ft. ΣQ = sum of the pump capacities, gpm, and ΣP = total power supplied to all pumps, hp.

Series Operation Pumps are frequently operated in series to supply heads greater than those of the individual pumps. The planning procedure is similar to the case of pumps in parallel. The system-head curve and the individual head-capacity curves for the pumps are plotted as shown in Fig. 50. The pump heads are added as shown to obtain the combined head-capacity curve. In this example, pump 2 operating alone will deliver no fluid because its shutoff head is less than the system static head. There are two possible operating points 1 and 2 as shown by the appropriate intersections with the system-head curve. As with parallel operation, other operating points could be obtained by throttling or by changes in the pump speeds. The overall efficiency of pumps in series is given by $\eta = (sQ/3{,}960)(\Sigma H/\Sigma P)$ wherein the symbols are the same as for parallel operation. It is important to note that the stuffing box pressure of the second pump is increased by the discharge pressure of the first pump. This may require a special packing box for the second pump with leakoff to the suction of the first pump. The higher suction pressure may increase both the first cost and maintenance costs of the second pump.

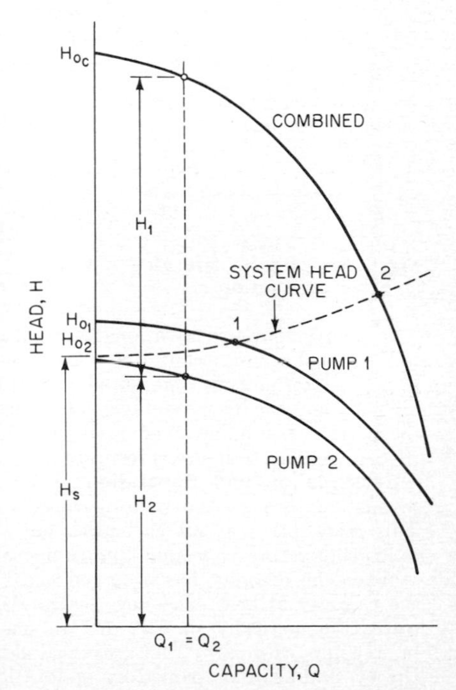

Fig. 50 Head-capacity curves of pumps operating in series.

OPERATION AT OTHER THAN THE NORMAL CAPACITY

Centrifugal pumps usually are designed to operate near the point of best efficiency but many applications require operation over a wide range of capacities, including shutoff, for extended periods of time. Pumps for such service are available but may require special design and construction at higher cost. Noise, vibration, and cavitation may be encountered at low capacities. Large radial shaft forces at shutoff as well as lack of throughflow to provide cooling may cause damage or breakage to such parts as shafts, bearings, seals, glands, and wearing rings of pumps not intended for such service. Some of the phenomena associated with operation at other than normal capacity are described below.

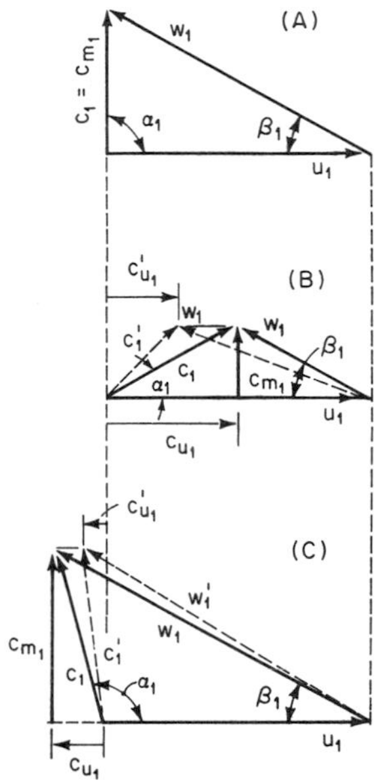

Prerotation Figure 51*A* shows an inlet velocity triangle for operation at best efficiency. The capacity is proportional to $c_{m1} = c_1$ which is perpendicular to the peripheral velocity u_1. The relative velocity w_1 is assumed parallel to the vane at the vane inlet angle β_1. There is no prerotation, and this is often called "shockless" entrance.

Fig. 51*B* shows an inlet velocity triangle for a capacity *less* than that at best efficiency. The solid-line vectors show the flow of an ideal fluid which never separates from the vanes. Assuming no guide vanes ahead of the impeller, the average real-fluid velocities will be approximately as shown by the broken-line vectors. The average absolute fluid velocity c_1' has a *prerotation* component c_{u1}' parallel to and in the direction of the peripheral velocity u_1. As throttling is increased, c_{m1} decreases and c_{u1}' increases but probably never reaches u_1 in magnitude.

Fig. 51*C* shows an inlet velocity triangle for a capacity *greater* than that at best efficiency. As before, the average real-fluid velocities will be approximately as shown by the broken-line vectors assuming no guide vanes ahead of the impeller. The *prerotation* velocity, c_{u1}', is in the opposite direction to the peripheral velocity, u_1. This prerotation is limited by the maximum capacity attainable (largest c_{m1}).

Fig. 51 Inlet triangles showing prerotation.

In pumps equipped with long, straight suction nozzles but no suction elbow, prerotation has been detected over considerable distances upstream from the impeller eye. Suction pressures measured at wall taps where prerotation is present are always higher than the true average static pressure across the measuring section. This means that the pump head as determined from wall taps is less than it would be if true average static pressures were measured.

Recirculation and Separation There is a small flow from impeller discharge to suction through the wearing rings and any hydraulic balancing device present. This takes place at all capacities but does not usually contribute to raising the fluid temperature very much unless operation is near shutoff.

When the capacity has been reduced by throttling to about one-half to one-third the capacity at best efficiency, a secondary flow called *recirculation* begins, as illustrated schematically in Fig. 52. In one case investigated by Minami et al. [48] in a pump with $n_s = 2{,}400$, reversed flow took place over the outer 10 percent of the eye radius accompanied by prerotation of about 60 to 70 percent of the impeller peripheral velocity. Some prerotation was detected in the forward flow to the impeller which took place through the core of the rotating backflow annulus. Measurements of the swirl angle just ahead of the impeller indicated that strong prerotation began at $Q \approx 0.45Q_n$, and backflow was present at all capacities less than $0.33Q_n$, where Q_n is the normal capacity at best efficiency. This is typical of flow in a suction nozzle at partial capacities. The turbulence produced when these counterflows mix usually produces noise and vibration in the suction pipe and, in some cases, severe surging of the fluid column in the suction. Velocity distributions before and after the impeller of a three-bladed propeller pump were investigated by Toyokura [49]. At less than about half the normal capacity, recirculation was detected both at the inlet and discharge sides of the impeller as shown in Fig. 53. Measurements of the velocity distribution just ahead of the impeller indicated that prerotation and backflow both commenced at $0.126 > c_m/u > 0.102$, where c_m was the average axial fluid velocity and u was the peripheral velocity of the vane tips. Prerotation

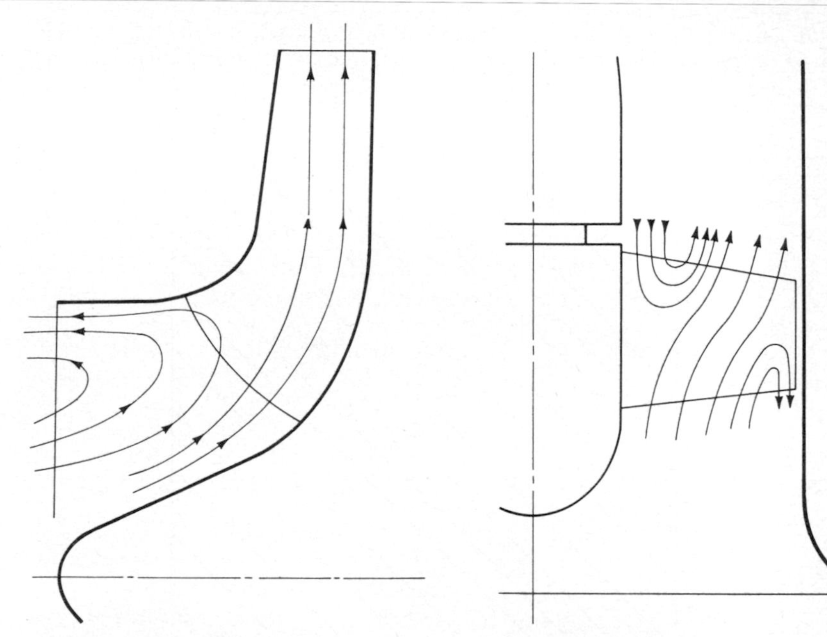

Fig. 52 Flow pattern at partial capacities [48].

Fig. 53 Flow pattern at partial capacities [49].

and recirculation at the suction side of an impeller appear to be intimately related phenomena. The presence of prerotation without recirculation suggests that it may initiate the backflow when the capacity is further reduced.

Fischer and Thoma [16] conducted a visual and photographic investigation of a radial-flow centrifugal pump, n_s = 1,400 at best efficiency. *Separation* was detected on the low-pressure faces of the vanes at all capacities, and the separated regions nearly filled the flow channels as shutoff was approached. Some separation was found on the high-pressure faces of the vanes at capacities above normal. In all cases the separated regions were filled with turbulent eddying fluid. Near shutoff, the fluid in the separated zones surged back and forth parallel to the vanes with outward flow in some channels matched by inward flow in the others. At shutoff, each impeller channel was filled with small continually changing eddies generally rotating in an opposite sense to that of the impeller. Other investigators have reported backflow along the pressure sides of the impeller vanes near the discharge tips at $Q < Q_n$.

The high turbulence produced by recirculation and separation accounts for most of the power consumed at shutoff. This may vary from about 30 percent of the normal power for pumps of very low specific speed to nearly three times the normal power for propeller pumps. Separation and, possibly, cavitation may take place on the casing tongue or diffusion vanes at very low capacities. Operation near shutoff not only causes excessive heating but also vibration and cavitation, which may cause serious mechanical damage.

Temperature Rise at Low Capacity If a pump is operated at low capacity long enough to establish equilibrium conditions, the temperature rise ΔT (in °F) will be given by

$$\Delta T = \frac{H}{778C_p}\left(\frac{1}{\eta} - 1\right) \tag{40}$$

where H = total head, ft
C_p = specific heat of the fluid, Btu/(lb) per (°F)
η = pump efficiency

Eq. (40) assumes that all the heat equivalent to the losses remains in the fluid. Figure 54 gives solutions of Eq. (40) for water, $C_p = 1$. Pumps handling hot

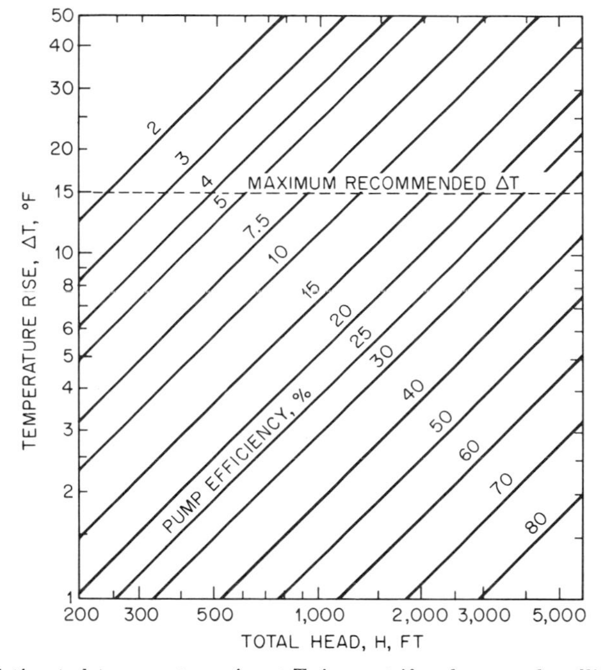

Fig. 54 Estimated temperature rise, ΔT, in centrifugal pumps handling water [4].

liquids, such as boiler feed pumps, usually are limited to ΔT = 15°F. The minimum capacity in gpm for water, Q_m, is given approximately by $Q_m = 0.3\ P_o$, where P_o is the shutoff power in hp. General service pumps handling cold liquids may be able to stand a temperature rise as great as 100°F. A rule which includes a factor of safety of approximately 20 percent is that $Q_m = 6P_o/\Delta T$, where Q_m is in gpm, P_o is in hp, and ΔT is permissible temperature rise in °F. The NPSH required at the elevated temperature may be the controlling factor if cavitation is to be avoided. Pumps handling hot liquids must be protected against operation at shutoff. This is usually done by providing a bypass line fitted with a metering orifice to maintain the minimum required flow through the pump. In the case of boiler feed pumps, the bypass flow usually is returned to one of the feed water heaters. Unless especially designed for cold starting, pumps handling hot liquids should be warmed up gradually before being put into operation.

Radial Thrust The pressure distribution at discharge from the impeller is very rarely uniform around the periphery regardless of the casing design or operating point. This leads to a radial force on the pump shaft called the *radial thrust* or *radial reaction.* The radial thrust in pounds force, F_r, may be determined by

$$F_r = 0.433 K_r s H D_2 b_2 \tag{41}$$

where K_r = an experimentally determined coefficient
s = specific gravity of the fluid pumped (equal to unity for cold water)
H = pump head, ft
D_2 = outside diameter of impeller, in
b_2 = breadth of impeller at discharge, including shrouds, in.

Values of K_r determined by Agostinelli et al. [50] for single-volute pumps are given in Fig. 55 as functions of specific speed and capacity. The magnitude and direction of F_r on the pump shaft may be estimated from Fig. 56, but Eq. (41) probably will be more accurate for determining the magnitude of the force. F_r usually is minimum near $Q = Q_n$, the capacity at best efficiency, but rarely goes completely to zero. Near shutoff, F_r usually is maximum and may be a considerable force on the shaft in high-head pumps.

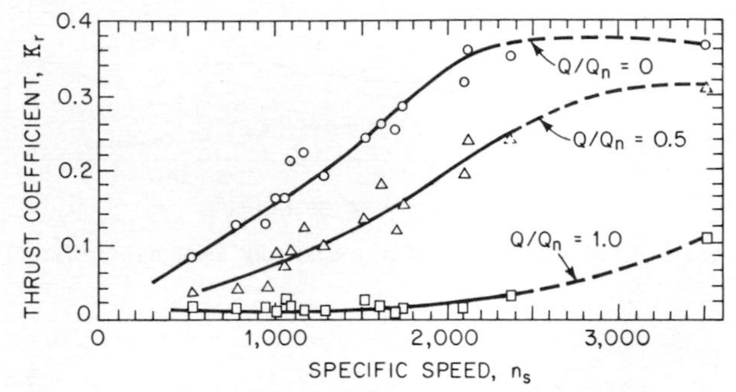

Fig. 55 K_r as a function of specific speed and capacity for single volute pumps [50].

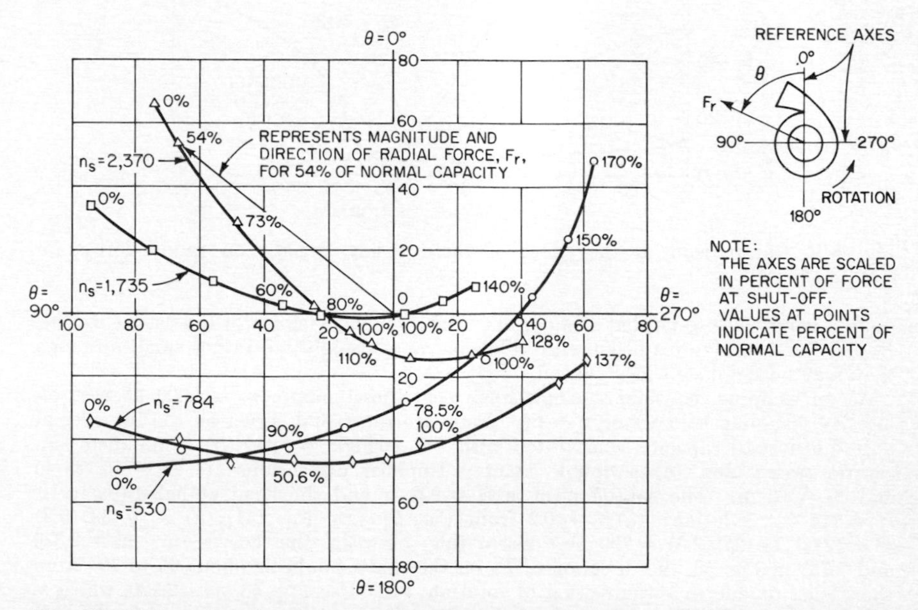

Fig. 56 Polar plot showing the direction of the resultant radial forces for single-volute pumps at various capacities and specific speeds [50].

The radial thrust can be made much smaller throughout the entire capacity range by using a double volute (twin volute) or a concentric casing, and these designs should be considered, particularly if the pump must operate at small capacities. Figures 57 to 59 compare radial forces generated by three types of casings, that is, a standard volute, a double volute, and a modified concentric casing. The last-named casing was concentric with the impeller for 270° from the tongue and then enlarged in the manner of a single volute to form the discharge nozzle. The magnitude and direction of F_r on the pump shaft for the modified concentric casing may be estimated from Fig. 60. The direction of F_r on the pump shaft with a

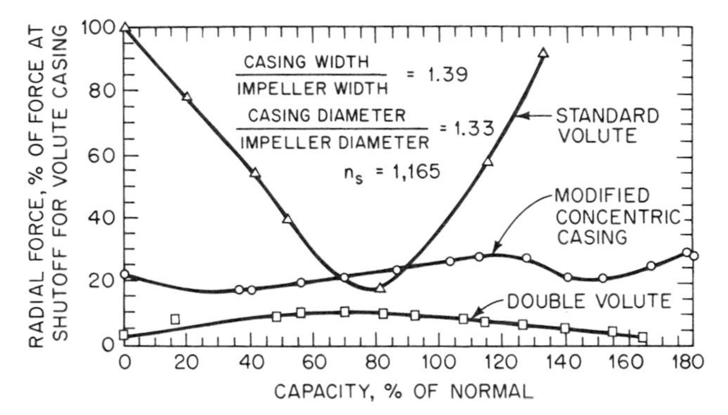

Fig. 57 Comparison of the effect of three casing designs on radial forces for $n_s = 1{,}165$ [50].

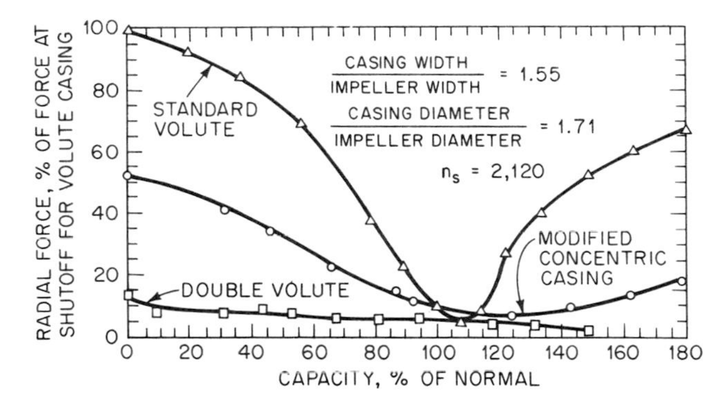

Fig. 58 Comparison of the effect of three casing designs on radial forces for $n_s = 2{,}120$ [50].

double volute was somewhat random but in the general vicinity of the casing tongue. Radial forces on pumps fitted with diffusion vanes usually are rather small although they may be significant near shutoff.

As an example, consider a single-stage centrifugal pump, $n_s = 2{,}000$ at best efficiency, handling cold water, $s = 1.0$. Estimate the radial thrust on the impeller at half the normal capacity when fitted with (*a*) a single volute, (*b*) a modified concentric casing, and (*c*) a double volute. Impeller dimensions are $D_2 = 15.125$ in and $b_2 = 2.5$ in. The shutoff head is $H = 252$ ft and the head at half-capacity is $H = 244$ ft. Solution: (*a*) $K_r = 0.2$ from Fig. 55. By Eq. (41), $F_r = (0.433)(0.2)(1.0)(244)(15.125)(2.5) = 799$ lb. Estimating between the curves for $n_s = 2{,}370$ and 1,735 in Fig. 56, the direction of F_r on the shaft should be about 65 to 70° from the casing tongue in the direction of rotation. (*b*) Use Fig. 58, $n_s = 2{,}120$, which is nearest to $n_s = 2{,}000$, and find the radial force is about 33 percent of shutoff value for a modified concentric casing. At shutoff, $F_r = (252/244)(799) = 825$ lb. At half-capacity, $F_r = (0.33)(825) = 272$ lb. From Fig. 60, the direction of F_r should

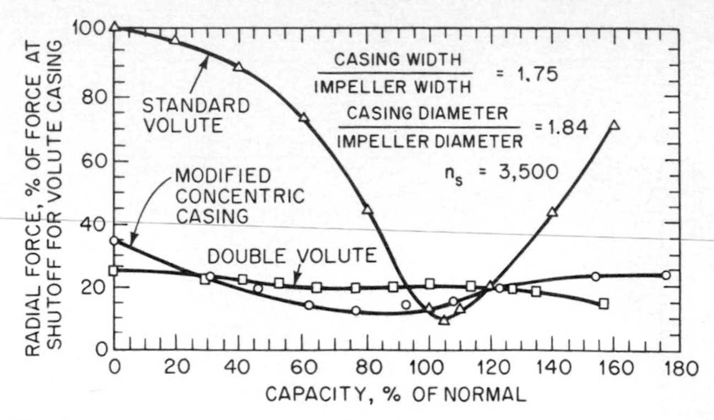

Fig. 59 Comparison of the effect of three casing designs on radial forces for $n_s = 3{,}500$ [50].

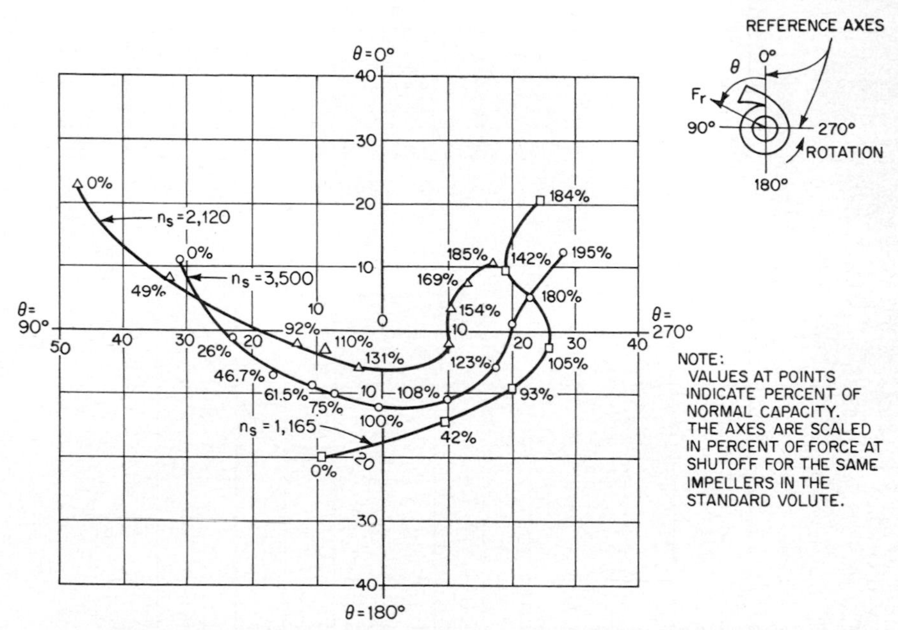

Fig. 60 Polar plot showing the direction of the resultant radial forces for the modified concentric casings at various capacities and specific speeds. The casings were concentric for 270° from the tongue [50].

be about 80 to 90° from the casing tongue in the direction of rotation. (*c*) From Fig. 58, the radial force is about 8 percent of the shutoff value so that $F_r = (0.08)(825) = 66$ lb and should be in the vicinity of 0° from the casing tongue. See also Biheller et al. [51].

ABNORMAL OPERATION

Complete Pump Characteristics Many types of abnormal operation involve reversed pump rotation, reversed flow direction, or both, and special tests are required to cover these modes of operation. Several methods of organizing the data have been proposed and each has certain advantages. The Thoma diagrams shown in Figs. 61 and 62 (Ref. 52) are easily understood and are truly *complete characteristics diagrams* because all possible modes of operation are covered. See

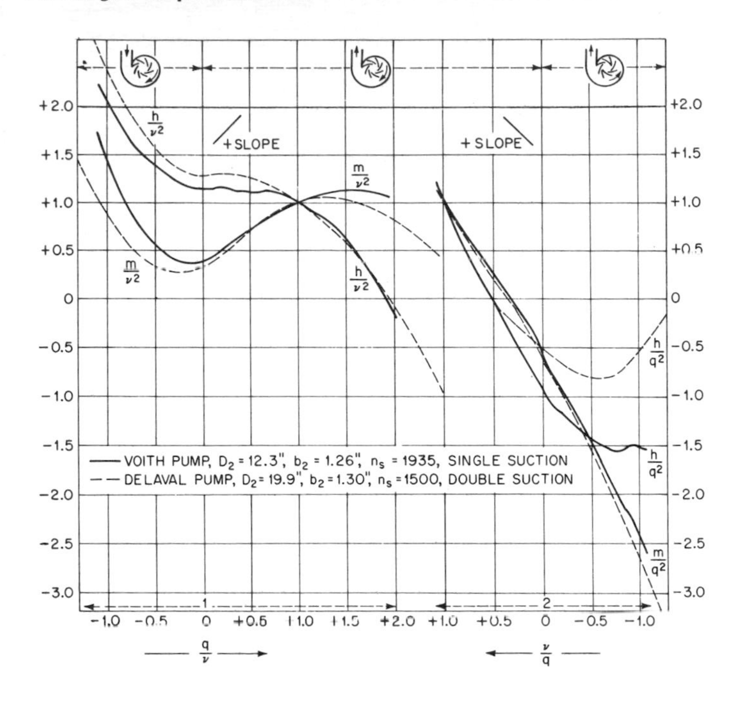

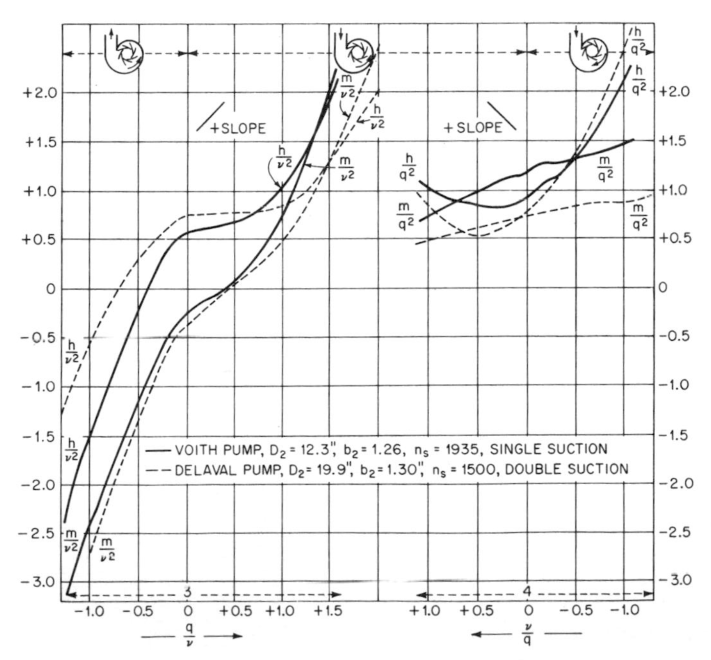

Fig. 61 Complete pump characteristics [52].

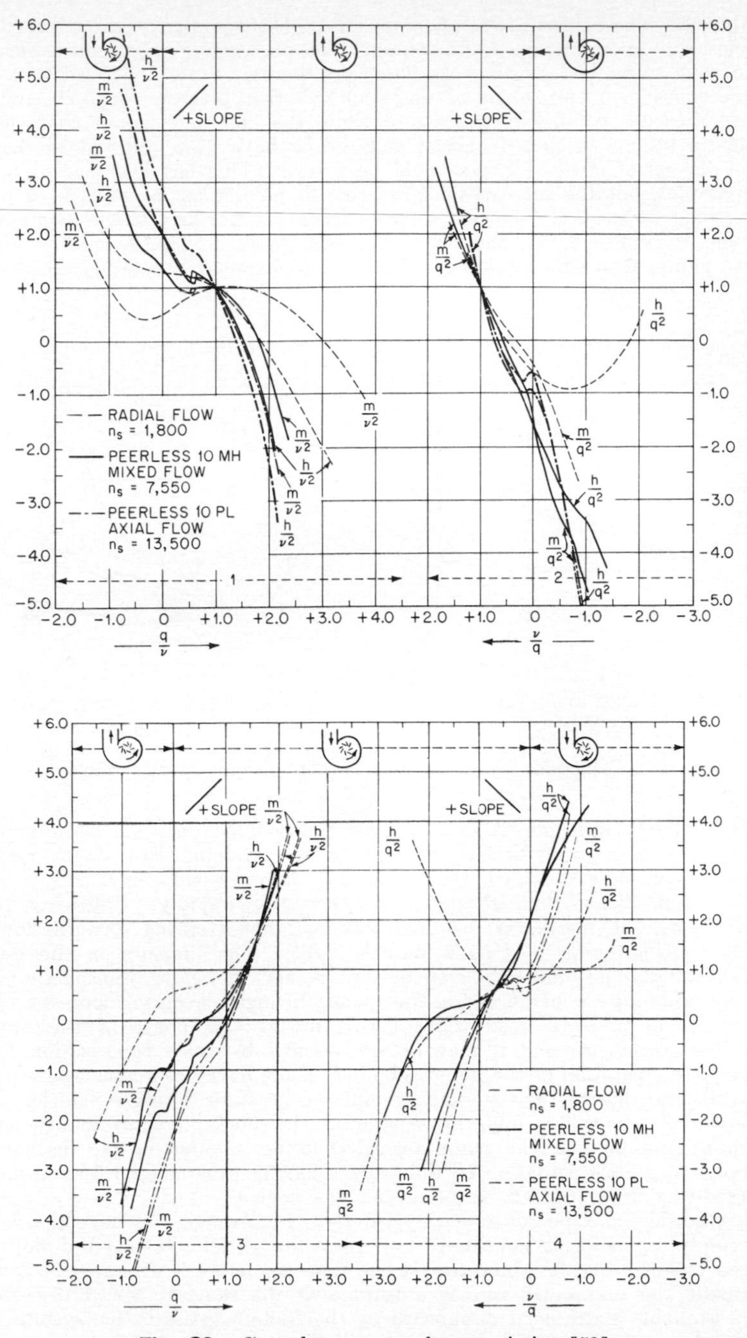

Fig. 62 Complete pump characteristics [52].

also Ref. 53. The *Karman circle diagram* (Ref. 54) attempted to show the complete characteristics as a four-quadrant contour plot of surfaces representing head and torque with speed and capacity as base coordinates. Since the head and torque tend to infinity in two zones of operation, another diagram with H and M as base coordinates would be required to show the complete pump characteristics. Frequently tests with negative head and torque have been omitted so that only half of the usual circle diagram could be shown. This has been called a *three-quadrant plot,* but the information necessary to predict an event such as possible water-column separation following a power failure is lacking. Circle diagrams may be found in Refs. 12, 54, and 55. Figure 63 shows schematic cross sections of the two pumps tested by Swanson [55] for which characteristics are given in Fig. 62.

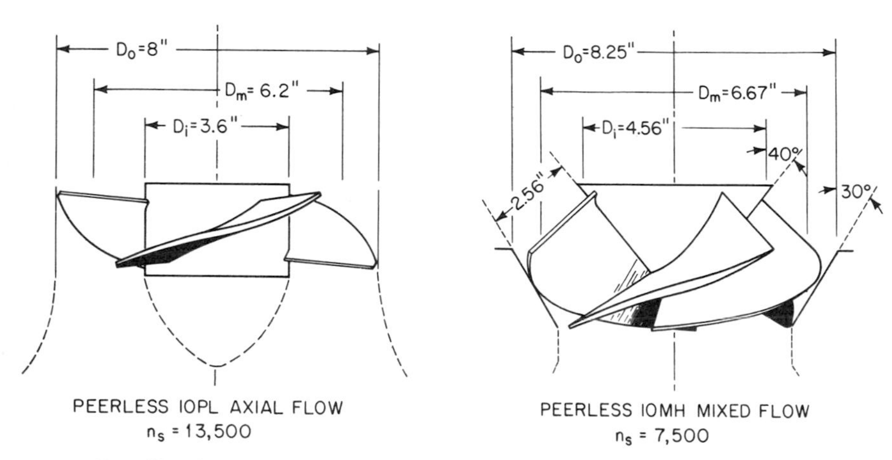

Fig. 63 Schematic cross sections of high-specific-speed pumps [52].

Power-Failure Transient A sudden power failure which leaves a pump and driver running free may cause serious damage to the system. Except for rare cases where a flywheel is provided, the pump and driver usually have a rather small moment of inertia so that the pump will slow down rapidly. Unless the pipeline is very short, the inertia of the fluid will maintain a strong forward flow while the decelerating pump acts as a throttle valve. The pressure in the discharge line falls rapidly and, under some circumstances, may go below atmospheric pressure, both at the pump discharge and at any points of high elevation along the pipeline. The minimum pressure head which occurs during this phase of the motion is called the *downsurge,* and it may be low enough to cause vaporization followed by complete separation of the liquid column. There have been cases where pipelines have collapsed due to the external atmospheric pressure during separation. When the liquid columns rejoin, following separation, the shock pressures may be sufficient to rupture the pipe or the pump casing. Closing a valve in the discharge line will only worsen the situation so that valves having programmed operation should be closed very little, if at all, before reverse flow begins.

Reversed flow may be controlled by valves or by arranging to have the discharge pipe empty while air is admitted at or near the outlet. If reversed flow is not checked, it will bring the pump to rest and then accelerate it with reversed rotation. Eventually the pump will run as a turbine at the runaway speed corresponding to the available static head diminished by the friction losses in the system. However, while reversed flow is being established, the reversed speed may reach a value considerably in excess of the steady-state runaway speed. Maximum reversed speed appears to increase with increasing efficiency and increasing specific speed of the pump. Calculations indicate maximum reversed speeds over 150 percent of normal speed for n_s = 1,935 and η = 84.1 percent [52]. This should be considered in selecting a driver particularly if it is a large electric motor.

There will be a pressure increase, called the *upsurge,* in the discharge pipe during reversed flow. The maximum upsurge usually occurs a short time before maximum reversed speed is reached and may cause a pressure as much as 60 percent or more above normal at the pump discharge. A further discussion of power-failure transients is given in Sec. 9.4.

Analysis of Transient Operation. The data of Figs. 61 and 62 have been presented in a form suitable for general application to pumps having approximately the same specific speeds as those tested. The symbols are $h = H/H_n$, $q = Q/Q_n$, $m = M/M_n$, and $\nu = n/n_n$ wherein H, Q, M, and n represent instantaneous values of head, capacity, torque, and speed respectively and the subscript n refers to the values at best efficiency for normal constant-speed pump operation. Any consistent

TABLE 9 Abnormal Operating Conditions of Several Pumps

Specific speed n_s	Downsurge: Free-running $m/q^2 = 0$		Downsurge: Locked-rotor $\nu/q = 0$		Runaway turbine: $m/\nu^2 = 0$		Runaway turbine: $h = 1$
	h/q^2	ν/q	h/q^2	m/q^2	q/ν	h/ν^2	ν
1	2	3	4	5	6	7	8
1,500[a]	−0.22	0.36	−0.55	−0.53	0.46	0.77	1.14
1,800	−0.24	0.31	−0.60	−0.44	0.56	0.75	1.16
1,935	−0.36	0.32	−0.94	−0.60	0.41	0.66	1.23
7,550	−0.23	0.56	−1.57	−1.38	0.99	0.38	1.62
13,500	−0.12	0.67	−0.96	−0.60	1.08	0.33	1.73

[a] Double-suction.

system of units may be used. According to the affinity laws, Eqs. 17, q is proportional to ν, and h and m are proportional to ν^2. Thus the affinity laws are incorporated in the scales of the diagrams. Each of Figs. 61 and 62 is divided into four sections for convenience in reading data from the curves. The curves of sections 1 and 3 extend to infinity as q/ν increases without limit in either the positive or negative direction. This difficulty is eliminated by sections 2 and 4 where the curves are plotted against ν/q which is zero when q/ν becomes infinite.

Usually any case of transient operation would begin at or near the point $q/\nu = \nu/q = 1$ which appears in both sections 1 and 2 of Figs. 61 and 62. The detailed analysis of transient behavior is beyond the scope of this treatise. An analytical solution by the rigid-column method, in which the fluid is assumed to be a rigid body, is given in Ref. 52. Friction is easily included and the results are satisfactory for many cases. The same reference includes a semigraphical solution to allow for elastic waves in the fluid but friction must be neglected. Graphical solutions including both elastic waves and friction have been discussed in Refs. 52, 56, 57, and 58. Computer solutions of a variety of transient problems have been discussed in Ref. 57. These offer considerable flexibility in the analysis once the necessary programs have been prepared.

Some extreme conditions of abnormal operation can be estimated at points where the curves of Figs. 61 and 62 cross the zero axes and are listed in Table 9. The data for columns 2 to 4 of Table 9 were read from sections 2 and the data for columns 6 and 7 were read from sections 3 of Figs. 61 and 62. Column 8 was computed from column 7 by assuming $h = 1$. Let the pump having $n_s = 1{,}500$ deliver cold water with normal head $H_n = 100$ ft and let the center of the discharge flange be 4 ft above the free surface in the supply sump. The discharge pressure head following power failure may be estimated by assuming the inertia of the rotating elements to be negligible compared to the inertia of the fluid in the pipeline. Then $q = 1$ and, from column 2 of Table 9, the downsurge pressure head is $(-0.22)(100) - 4 = -26$ ft, which is not low enough to cause separation of the water column. Actually the downsurge would be less than this because of the effects of inertia and friction which have been neglected. If this pump were stopped suddenly by a shaft seizure or by an obstruction fouling the impeller, column 4 of Table 9 shows the downsurge to be (—0.55)(100) — 4 = —59 feet which would cause water-column separation and, probably, subsequent water hammer. If, following power failure, the pump were allowed to operate as a no-load turbine under the full normal pump head, column 8 of Table 9 shows the runaway speed would be 1.14 times the normal pump speed. The steady-state runaway speed usually would be less than this because the effective head would be decreased by friction but higher speeds would be reached during the transient preceding steady-state operation. Column 8 shows that runaway speeds increase with increasing specific speed.

Incorrect Rotation Correct rotation of the driver should be verified before it is coupled to the pump (see Fig. 64). Section 3 of Figs. 61 and 62 shows that reversed rotation might produce some positive head and capacity with pumps of low specific speed but at very low efficiency. It is unlikely that positive head would be produced by reversed rotation of a high specific-speed pump.

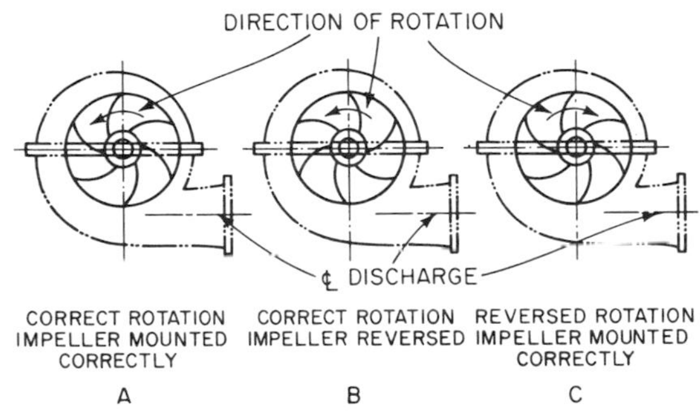

Fig. 64 Pump assembly and rotation [4].

Reversed Impeller Some double-suction impellers can be mounted reversed on the shaft. If the impeller is accidentally reversed, as at *B* in Fig. 64, the capacity and efficiency probably will be much reduced and the power consumption increased. Care should be taken to prevent this as the error might go undetected in some cases until the driver was damaged by overload. Table 1 shows performance data for six pumps with reversed impellers. At least one of these would overload the driver excessively.

Further discussion of abnormal operating conditions may be found in Refs. 4, 8, and 12.

Vibration Vibration caused by flow through wearing rings and by cavitation has been discussed in the foregoing and some remedies indicated. Vibration due to unbalance is not usually serious in horizontal units but may be of major importance in long vertical units where the discharge column is supported at only one or two points. The structural vibrations may be quite complicated and involve both natural frequencies and higher harmonics. Vibration problems in vertical units should be anticipated during the design stage. If vibration is encountered in existing units, the following steps may help to reduce it: (1) dynamically balance all rotating elements of both pump and motor; (2) increase the rigidity of the main support and of the connection between the motor and the discharge column; (3) change the stiffness of the discharge column to raise or lower natural frequencies as required. A portable vibration analyzer may be helpful in this undertaking. Kovats [59] has discussed the analysis of this problem in some detail.

PREDICTION OF EFFICIENCY FROM MODEL TESTS

Many pumps used in pumped-storage power plants and water supply projects are so large and expensive that extensive use is made of small models to determine the best design. It is often necessary to estimate the efficiency of a prototype pump, as a part of the guarantee, from the performance of a geometrically similar model. A model and prototype are said to operate under dynamically similar conditions when $Dn/\sqrt{H} = D'n'/\sqrt{H'}$, where D = impeller diameter, n = pump speed, and H = pump head in any consistent units of measure. Primed quantities refer to the model and unprimed quantities to the prototype. Dynamic similarity is a prerequisite to model-prototype testing so that losses which are proportional to the squares of fluid velocities, called *kinetic losses,* will scale directly with size and not change the efficiency. Surface friction losses are boundary-layer phenomena which depend on Reynolds number, $\mathrm{Re} = D\sqrt{H}/\nu$, where ν = kinematic viscosity of the fluid pumped. Reynolds numbers increase with increasing size and, within limits, surface friction-loss coefficients decrease with increasing Reynolds number. This leads to a gain in efficiency with increasing size. Computational difficulties have forced an empirical approach to the problem. Details of the development of a number of formulas are given in Ref. 60.

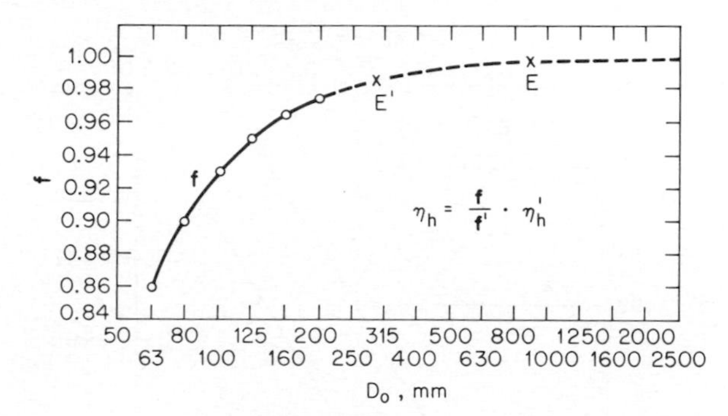

Fig. 65 The **f** function for the Rütschi formula [25,60].

Moody-Staufer Formula In 1925, L. F. Moody and F. Staufer independently developed the same formula which was later modified by Pantell to the form

$$\left(\frac{1-\eta_h}{1-\eta_h'}\right)\left(\frac{\eta_h'}{\eta_h}\right) = \left(\frac{D'}{D}\right)^n \tag{42}$$

where η_h = hydraulic efficiency, discussed previously, and D = impeller diameter. Primed quantities refer to the model and n is a constant to be determined by tests. The model must be tested with the same fluid that will be used in the prototype, that is, cold water in most practical cases. The original formula contained a correction for head which is negligible if $H' \geqq 0.8H$, and this requirement is now virtually mandatory in commercial practice. The meager information available indicates $0.2 \geqq n \geqq 0.1$ approximately, with the higher value presently favored. Improvements in construction and testing techniques very likely will move n toward the lower value in the future. The Moody-Staufer formulas have been widely used since first publication. In practice, both η_h' and η_h usually are replaced by the overall efficiencies η' and η, respectively, due to the difficulty in determining proper values for the mechanical and volumetric efficiencies [see Eq. (16)].

Rütschi Formulas. The general form of several empirical formulas due to K. Rütschi [60] and others was originally given as

$$\eta_h = \frac{\mathbf{f}}{\mathbf{f}'}\eta_h' \tag{43}$$

where η_h and η_h' are the hydraulic efficiencies of the prototype and model, respectively, and **f** and **f′** are values of an empirical **f** function for both the prototype and model. The **f** function was obtained from tests of six single-stage pumps, $n_s < 2000$, and is shown in Fig. 65 based on the *eye diameters* D_o of the pumps in *millimeters.* Thus the **f** function depends on actual size in addition to scale ratio. The extrapolated portion of the curve, shown dashed in Fig. 65, checked well with values for a model and large prototype shown by E' and E, respectively. One of several formulas which have been proposed to fit the curve in Fig. 65 is

$$\mathbf{f} = 1 - \frac{3.15}{D_o^{1.6}} \tag{44}$$

where the eye diameter D_o is in centimeters or

$$\mathbf{f} = 1 - \frac{0.0133}{D_o^{1.6}} \tag{45}$$

where D_o is in feet. Figure 66 gives a graphical solution of Eq. (45). The Society of German Engineers (VDI) has adopted a slightly modified version of the Rütschi

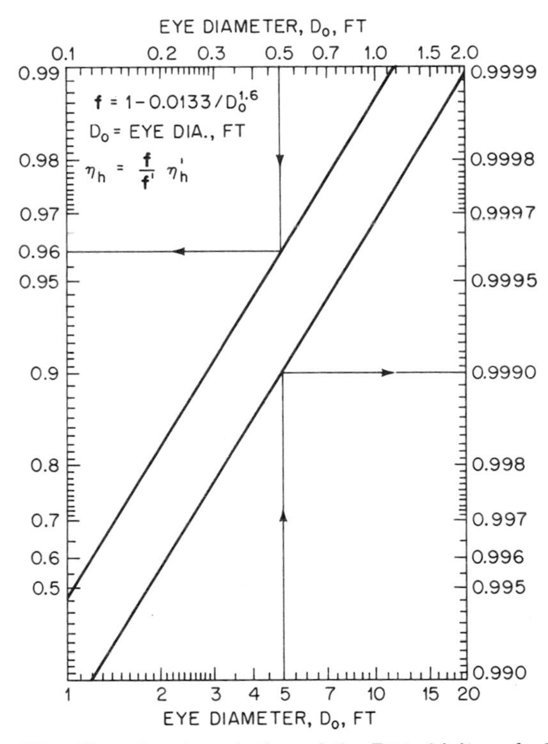

Fig. 66 Chart for the solution of the Rütschi formula [60].

formula as standard. Rütschi later recommended, in discussion of [60], that the internal efficiency, $\eta_i = \eta/\eta_m$, be used instead of the hydraulic efficiency η_h in Eq. (43). The mechanical efficiency η_m will probably be very high for both model and prototype for most cases of interest so that good results should be obtained if the overall efficiency is used in Eq. (43).

OPERATION OF PUMPS AS TURBINES

Centrifugal pumps may be used as hydraulic turbines in some cases where low first cost is paramount. Since the pump has no speed regulating mechanism, considerable speed variation must be expected unless the head and load remain very nearly constant. Some speed control could be obtained by throttling the discharge automatically, but this would increase the cost, and the power lost in the throttle valve would lower the overall efficiency.

Pump Selection Once the head, speed, and power output of the turbine have been specified, it is necessary to select a pump which, when used as a turbine, will satisfy the requirements. Assuming that performance curves for a series of pumps are available,[1] a typical set of such curves should be normalized using the head,

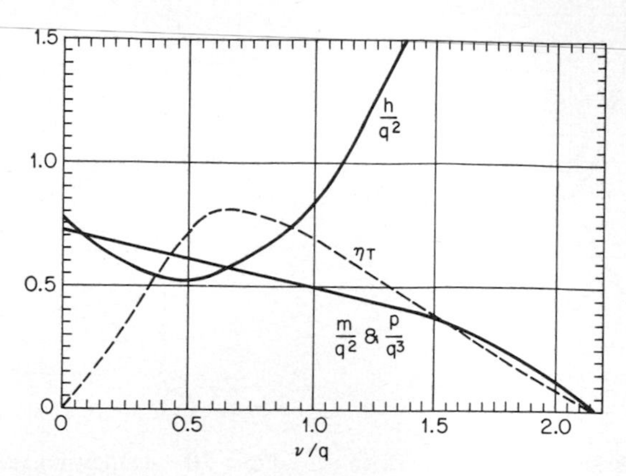

Fig. 67 Dimensionless characteristic curves for DeLaval L 10/8 pump, constant-discharge turbine operation [61].

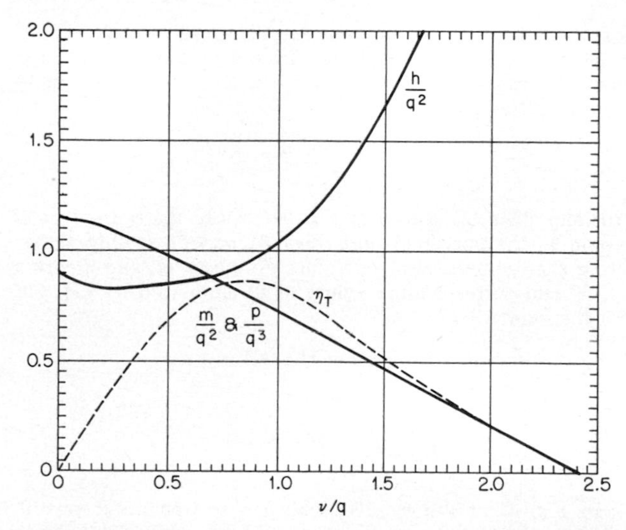

Fig. 68 Dimensionless characteristic curves for Voith pump, constant-discharge turbine operation [61].

power, and capacity of the best efficiency point as normal values. These curves will correspond to the right-hand part of section 1 of either Fig. 61 or 62. In normalizing the power P, let $p = P/P_n$ and the curve of p/ν^3 will be identical with the curve of m/ν^2 in Fig. 61 or 62. The normalized curves may be compared with the curves in sections 1 of Figs. 61 and 62 to determine which curves best represent the characteristics of the proposed pump. Once a choice has been made, the approximate turbine performance can be obtained from the corresponding figure of Figs. 67 to 70.

[1] It is assumed that turbine characteristics of the proposed pump are not available when the initial selection is made.

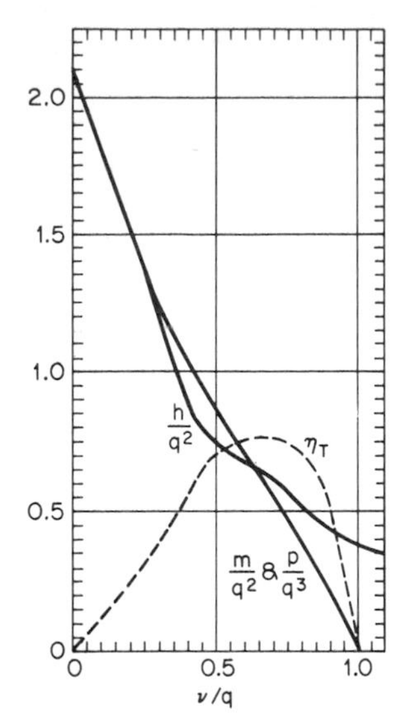

Fig. 69 Dimensionless characteristic curves for Peerless 10 MH pump, constant-discharge turbine operation [61].

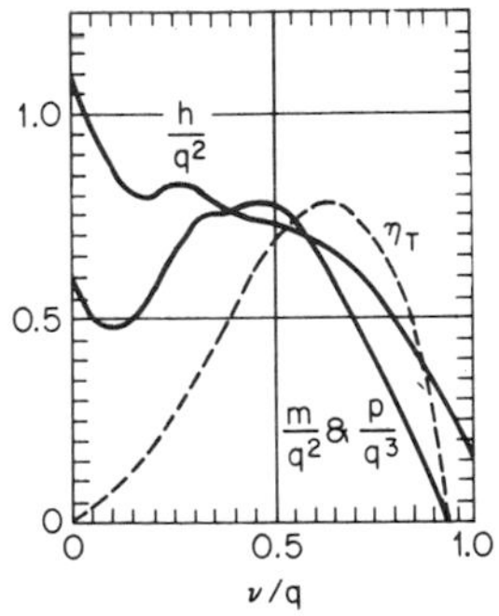

Fig. 70 Dimensionless characteristic curves for Peerless 10 PL pump, constant-discharge turbine operation [61].

Assume, for example, that the turbine specifications are $H_T = 20$ ft, $P_T = 12.75$ hp, and $n = 580$ rpm, and that the characteristics of the DeLaval L10/8 pump are representative of a series of pumps from which a selection can be made. The turbine discharge Q_T, in gpm, is given by

$$Q_T = \frac{3{,}960P_T}{H_T\eta_T} = \frac{2{,}520}{\eta_T} \tag{46}$$

where η_T is turbine efficiency shown in Fig. 67. Only the normal values Q_n, H_n, P_n, etc., are common to the curves of both Figs. 61 and 67 so that these alone can be used in selecting the required pump. Values of ν/q, h/q^2, and η_T are read from the curves of Fig. 67 and corresponding values of Q computed by Eq. (46). Values of Q_n and H_n are then given by

$$Q_n = Q(\nu/q) \tag{47}$$

and

$$H_n = \frac{H(\nu/q)^2}{h/q^2} = \frac{20(\nu/q)^2}{h/q^2} \tag{48}$$

For example, in Fig. 67 at $\nu/q = 0.700$, read $h/q^2 = 0.595$, and $\eta_T = 0.805$. By Eq. (46), $Q_T = 2{,}520/0.805 = 3{,}130$ gpm; by Eq. (47), $Q_n = (3{,}130)(0.700) = 2{,}190$ gpm; and, by Eq. (48), $H_n = (20)(0.700)^2/0.595 = 16.5$ ft. In similar manner, the locus of the best efficiency points for an infinite number of pumps, each having the same characteristics as shown in Figs. 61 and 67, is obtained and each pump would satisfy the turbine requirements. This locus of best efficiency points is plotted as curve A in Fig. 71. The head-capacity curve for a DeLaval L 16/14 pump having a 17-in-diameter impeller tested at 720 rpm is shown as curve B in Fig. 71. The best efficiency point was found to be at $Q_n = 3{,}500$ gpm and $H_n = 37.2$ ft. The locus of the best efficiency points for this pump for different speeds and impeller diameters is given by Eqs. (17) as

$$H_n = 37.2\left(\frac{Q_n}{3{,}500}\right)^2 = \frac{3.04}{10^6}Q_n^2 \tag{49}$$

and is shown by curve C in Fig. 71. Curve C intersects curve A at two points showing that the L 16/14 pump satisfies the turbine requirements. Only the intersection at the higher turbine efficiency is of interest. At this point, $Q_n = 2{,}490$ gpm and $H_n = 18.8$ ft. Since the turbine speed was specified to be 580 rpm, the required impeller diameter is given by Eqs. (17) as $D = (17)(2{,}490/3{,}500)/(580/720) = 15$ in or by $D = (17)\sqrt{18.8/37.2}/(580/720) = 15$ in. The computed head-capacity curve for the 15-in-diameter impeller at 580 rpm is shown as curve D in Fig. 71. The turbine discharge is 3,200 gpm from Eq. (46) with $\eta_T = 0.787$.

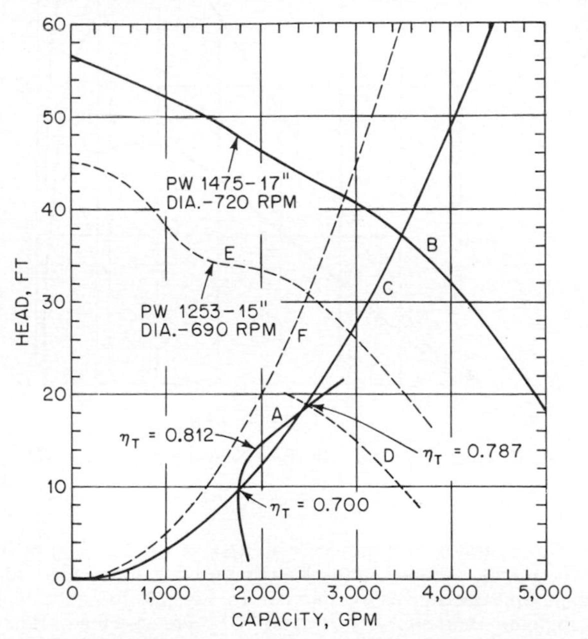

Fig. 71 Head-capacity curves for example of pump selection [61].

The optimum solution would be to have curve C tangent to curve A at the point corresponding to maximum turbine efficiency, in this case $\eta_T = 0.812$. Since this would require a smaller pump, curve E in Fig. 71 shows a head-capacity curve for a K 14/12 pump, which was the next smaller pump in the series. Curve F, the locus of the best efficiency points, does not intersect curve A showing that the smaller pump will not satisfy the turbine requirements. It is important to note that the turbine head (20 ft) specified for this example was assumed to be the net head from inlet to outlet flange of the pump when installed and operated as a turbine.

The procedure outlined above should lead to the selection of a pump large enough to provide the required power. However, it probably will be necessary to apply the affinity laws over such wide ranges of the variables that the usual degree of accuracy should not be expected. Considerable care should be exercised if it becomes necessary to interpolate between the curves of Figs. 67 to 70. The computed performance will very likely differ from the results of subsequent tests. The curves of Fig. 67 may be converted to show the constant-head characteristics of the L 16/14 pump when installed and operated as a turbine. Details of the method of computation are given in Ref. [61], and the computed characteristics are shown in Fig. 72.

A simplified method of pump selection has been proposed [62, 63]. Equating the output power of the turbine, P_T, to the input power of the pump, P_P, led to $Q_T H_T \eta_T = Q_P H_P/\eta_P$. It was then assumed that $\eta_T \cong \eta_P = \eta$, and that the capacity

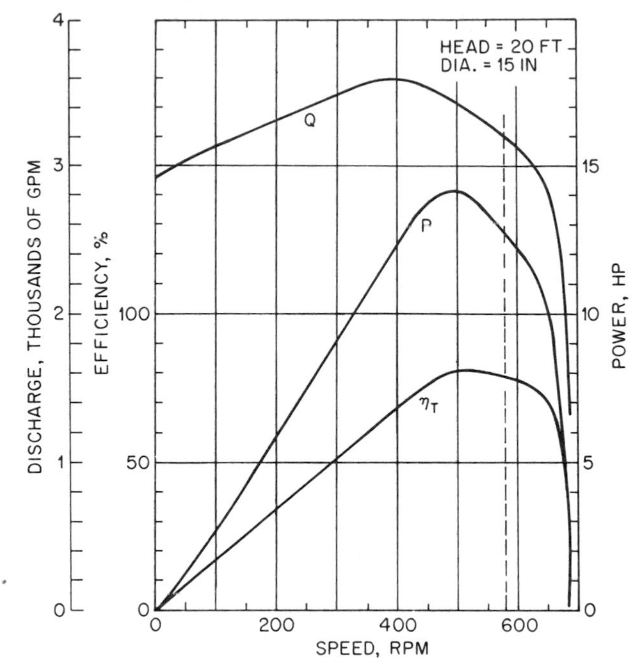

Fig. 72 Computed constant-head turbine characteristics for DeLaval L 16/14 pump [61].

and head may be separately computed from $Q_T\eta = Q_P$ and $H_T\eta = H_P$. If Q_T and H_T were known, η was assumed, and the computed values of Q_P and H_P used to enter a set of pump selection charts, similar to Fig. 39, to estimate the pump size and required impeller diameter. If P_T and H_T were specified, Eq. (46) could be used to obtain $Q_P = 3{,}960\ P_T/H_T$, where H is in ft of fluid handled, Q is in gpm, and P is in hp. A difficulty with this method is the assumption of the correct value of η. If the assumed value is too low, the pump selected may be too small to meet the requirements. If the assumed value is too high, the pump selected may be unnecessarily expensive, and it may operate at relatively poor efficiency as a turbine. Although the method described previously is more complicated, it avoids a direct assumption of η and gives an indication of the approximate efficiency at which the turbine will operate.

REFERENCES

1. Spannhake, W.: "Centrifugal Pumps, Turbines, and Propellers," The Technology Press of the Massachusetts Institute of Technology, Cambridge, Mass., 1934.
2. Church, A. H.: "Centrifugal Pumps and Blowers," John Wiley & Sons, Inc., New York, 1944.
3. Wislicenus, G. F.: "Fluid Mechanics of Turbomachinery," McGraw-Hill Book Company, New York, 1947.
4. Karassik, I. J., and R. Carter: "Centrifugal Pumps: Selection, Operation and Maintenance," McGraw-Hill Book Company, New York, 1960.
5. Pfleiderer, C.: "Die Kreiselpumpen," 5th ed., Springer-Verlag OHG, Berlin, 1961.
6. Kovats, A.: "Design and Performance of Centrifugal and Axial Flow Pumps and Compressors," The Macmillan Company, New York, 1964.
7. "Centrifugal Pumps," PTC 8.2-1965, The American Society of Mechanical Engineers, New York.
8. Łazarkiewicz, S., and A. T. Troskolański: "Impeller Pumps," Pergamon Press, New York, 1965.

9. Stepanoff, A. J.: "Pumps and Blowers—Two Phase Flow," John Wiley & Sons, Inc., New York, 1965.
10. Addison, H.: "Centrifugal and Other Rotodynamic Pumps," 3d ed., Chapman & Hall, Ltd., London, 1966.
11. Holland, F. A., and Chapman, F. S.: "Pumping of Liquids," Reinhold Publishing Corporation, New York, 1966.
12. Stepanoff, A. J.: "Centrifugal and Axial Flow Pumps," 2d ed., John Wiley & Sons, Inc., New York, 1967.
13. "Hydraulic Institute Standards for Centrifugal, Rotary & Reciprocating Pumps," 12th ed., Hydraulic Institute, Cleveland, Ohio, 1969.
14. Moody, L. F. and T. Zowski: Hydraulic Machinery, sec. 26 of "Handbook of Applied Hydraulics," Davis and Sorenson (eds.), 3d ed., McGraw-Hill Book Company, New York, 1969.
15. Gartmann, H.: "DeLaval Engineering Handbook," 3d ed., McGraw-Hill Book Company, New York, 1970.
16. Fischer, K., and D. Thoma: Investigation of the Flow Conditions in a Centrifugal Pump, *Trans ASME,* vol. 54, pp. 141–155, 1932.
17. Osborne, W. C., and D. A. Morelli: Head and Flow Observations on a High-Efficiency Free Centrifugal-Pump Impeller, *Trans ASME,* vol. 72, pp. 999–1007, 1950.
18. Rupp, W. E.: High Efficiency Low Specific Speed Centrifugal Pump, U.S. Patent No. 3,205,828, Sept. 14, 1965.
19. Barske, U. M.: Development of Some Unconventional Centrifugal Pumps, *Proc. Inst. Mech. Eng.* (*London*), vol. 174, no. 11, pp. 437–461, 1960.
20. Wislicenus, G. F.: Critical Considerations on Cavitation Limits of Centrifugal and Axial-Flow Pumps, *Trans. ASME,* vol. 78, pp. 1707–1714, November, 1956.
21. Manson, W. W.: Experience with Inlet Throttled Centrifugal Pumps, Gas Turbine Pumps, *Cavitation in Fluid Machinery,* Symposium Publication, ASME, pp. 21–27, 1972.
22. Ippen, A. T.: The Influence of Viscosity on Centrifugal Pump Performance, *Trans. ASME,* vol. 68, no. 8, pp. 823–848, November, 1946.
23. Black, H. F., and D. N. Jensen: Effects of High-Pressure Ring Seals on Pump Rotor Vibrations, *ASME Paper No.* 71-*WA*/*FE*-38, 1971.
24. Wood, G. M., H. Welna, and R. P. Lamers: Tip-Clearance Effects in Centrifugal Pumps, *Trans. ASME, J. Basic Eng.,* Ser. D, vol. 89, pp. 932–940, 1965.
25. Rütschi, K.: Untersuchungen an Spiralgehäusepumpen verschiedener Schnelläufigkeit, *Schweiz. Arch Angew. Wiss. Tech.,* vol. 17, no. 2, pp. 33–46, February, 1951.
26. Knapp, R. T.: Recent Investigations of the Mechanics of Cavitation and Cavitation Damage, *Trans. ASME.,* vol. 77, pp. 1045–1054, 1955.
27. Knapp, R. T.: Cavitation Mechanics and Its Relation to the Design of Hydraulic Equipment, James Clayton Lecture, *Proc. Inst. Mech. Eng.* (*London*), sec. A, vol. 166, pp. 150–163, 1952.
28. Shutler, N. D., and R. B. Mesler: A Photographic Study of the Dynamics and Damage Capabilities of Bubbles Collapsing near Solid Boundaries, *Trans. ASME, J. Basic Eng.,* ser. D, vol. 87, pp. 511–517, 1965.
29. Hickling, R., and M. S. Plesset: The Collapse of a Spherical Cavity in a Compressible Liquid, Division of Engineering and Applied Sciences, Report No. 85-24, California Institute of Technology, March, 1963.
30. Pilarczyk, K., and V. Rusak: Application of Air Model Testing in the Study of Inlet Flow in Pumps, *Cavitation in Fluid Machinery,* Symposium Publication, ASME, pp. 91–108, 1965.
31. Hickling, R.: Some Physical Effects of Cavity Collapse in Liquids, *Trans. ASME, J. Basic Eng.,* ser. D, vol. 88, pp. 229–235, 1966.
32. Plesset, M. S.: Temperature Effects in Cavitation Damage, *Trans. ASME, J. Basic Eng.,* ser. D, vol. 94, pp. 559–566, 1972.
33. Hammitt, F. G.: Observations on Cavitation Damage in a Flowing System, *Trans. ASME, J. Basic Eng.,* ser. D, vol. 85, pp. 347–359, 1963.
34. Stahl, H. A., and A. J. Stepanoff: Thermodynamic Aspects of Cavitation in Centrifugal Pumps, *Trans. ASME,* vol. 78, pp. 1691–1693, 1956.
35. Bischoff, A.: Untersuchungen über das Verhalten einer Kreiselpumpe bei Betrieb im Kavitationsbereich, *Mitt. Hydraul. Inst. Tech. Hochsch. Müench.,* no. 8, pp. 48–68, 1936.
36. Thoma, D.: Verhalten einer Kreiselpumpe beim Betrieb im Hohlsog—(Kavitations-) Bercich. *Z. Ver. Deut Ing.,* vol. 81, no. 33, pp. 972–973, Aug. 14, 1937.
37. Stepanoff, A. J.: Cavitation in Centrifugal Pumps with Liquids Other Than Water, *Trans. ASME, J. Eng. Power,* ser. A, vol. 83, pp. 79–90, 1961.

38. Tenot, M. A.: Phénomènes de la Cavitation, *Mem. Soc. Ing. Civils France Bull.*, pp. 377–480, May and June, 1934.
39. Krisam, F.: Neue Erkenntnisse im Kreiselpumpenbau, *Z. Ver. Deut. Ing.*, vol. 95, no. 11/12, pp. 320–326, Apr. 15, 1953.
40. Wislicenus, G. F., R. M. Watson, and I. J. Karassik: Cavitation Characteristics of Centrifugal Pumps Described by Similarity Considerations, *Trans. ASME,* vol. 61, pp. 17–24, 1939; vol. 62, pp. 155–166, 1940.
41. Salemann, V.: Cavitation and NPSH Requirements of Various Liquids, *Trans. ASME, J. Basic Eng.*, ser. D, vol. 81, pp. 167–173, 1959.
42. Stepanoff, A. J.: Cavitation Properties of Liquids, *Trans. ASME, J. Eng. Power,* ser. A, vol. 86, pp. 195–200, 1964.
43. Gongwer, C. A.: A Theory of Cavitation Flow in Centrifugal-Pump Impellers, *Trans. ASME,* vol. 63, pp. 29–40, 1941.
44. Kallas, D. H., and J. Z. Lichtman: Cavitation Erosion, in "Environmental Effects on Polymeric Materials," D. V. Rosato and R. T. Schwartz (eds.), Interscience Publishers, a division of John Wiley & Sons, Inc., New York, pp. 223–280, 1968.
45. Mechanical Engineering, vol. 93, no. 6, p. 39, June, 1971
46. Carter, R.: How much Torque Is Needed to Start Centrifugal Pumps? *Power,* vol. 94, no. 1, pp. 88–90, January, 1950.
47. Richardson, C. A.: Economics of Electric Power Pumping, *Allis-Chalmers Elec. Rev.,* vol. 9, pp. 20–24, June, 1944.
48. Minami, S., K. Kawaguchi, and T. Homma: Experimental Study on Cavitation in Centrifugal Pump Impellers, *Bull. Jap. Soc. Mech. Eng.,* vol. 3, no. 9, pp. 19–29, 1960.
49. Toyokura, T.: Studies on the Characteristics of Axial Flow Pumps (Part I), *Bull. Jap. Soc. Mech. Eng.,* vol. 4, no. 14, pp. 287–293, 1961.
50. Agostinelli, A., D. Nobles, and C. R. Mockridge: An Experimetal Investigation of Radial Thrust in Centrifugal Pumps, *Trans. ASME, J. Eng. Power,* ser. A, vol. 82, pp. 120–126, 1960.
51. Biheller, H. J.: Radial Force on the Impeller of Centrifugal Pumps With Volute, Semivolute, and Fully Concentric Casings, *Trans. ASME, J. Eng. Power,* ser. A, vol. 87, pp. 319–323, 1965.
52. Kittredge, C. P.: Hydraulic Transients in Centrifugal Pump Systems, *Trans. ASME,* vol. 78, no. 6, pp. 1307–1322, August, 1956.
53. Donksy, B.: Complete Pump Characteristics and the Effects of Specific Speeds on Hydraulic Transients, *Trans. ASME, J. Basic Eng.,* ser. D, pp. 685–699, 1961.
54. Knapp, R. T.: Complete Characteristics of Centrifugal Pumps and Their Use in the Prediction of Transient Behavior, *Trans. ASME,* vol. 59, pp. 683–689, 1937; vol. 60, pp. 676–680, 1938.
55. Swanson, W. M.: Complete Characteristic Circle Diagrams for Turbomachinery, *Trans. ASME,* vol. 75, pp. 819–826, 1953.
56. Parmakian, J.: "Waterhammer Analysis," Prentice-Hall, Inc., Englewood Cliffs, N.J. 1955.
57. Streeter, V. L., and E. B. Wylie: "Hydraulic Transients," McGraw-Hill Book Company, New York, 1967.
58. Bergeron, L.: "Waterhammer in Hydraulics and Wave Surges in Electricity," John Wiley & Sons, Inc., New York, 1961.
59. Kovats, A.: Vibration of Vertical Pumps, *Trans. ASME, J. Eng. Power,* ser. A, vol. 84, pp. 195–203, 1962.
60. Kittredge, C. P.: Estimating the Efficiency of Prototype Pumps from Model Tests, *Trans. ASME, J. Eng. Power,* ser. A, vol. 90, pp. 129–139, 301–304, 1968.
61. Kittredge, C. P.: Centrifugal Pumps Used as Hydraulic Turbines, *Trans. ASME, J. Eng. Power,* ser. A, pp. 74–78, 1961.
62. Childs, S. M.: Convert Pumps to Turbines and Recover HP, *Hydrocarbon Process. Petrol. Refiner,* vol. 41, no. 10, pp. 173–174, October, 1962.
63. Hancock, J. W.: Centrifugal Pump or Water Turbine, *Pipe Line News*, pp. 25–27, June, 1963.

Section **2.4**

Centrifugal Pump Priming

C. J. TULLO

A centrifugal pump is primed when the waterways of the pump are filled with the liquid to be pumped. The liquid replaces the air, gas, or vapor in the waterways. This may be done manually or automatically.

When first put in service, the waterways of a pump are filled with air. If the suction supply is above atmospheric pressure, this air will be trapped in the pump and compressed somewhat when the suction valve is opened. Priming is accomplished by venting the entrapped air out of the pump through a valve provided for this purpose.

If the pump takes its suction from a supply located below the pump, priming can be accomplished either by providing a foot valve in the suction line, so that the pump and suction piping can be filled with liquid, by providing a priming chamber in the suction line, or by using some form of vacuum-producing device.

FOOT VALVES

A foot valve is a form of check valve installed at the bottom or foot of a suction line (Fig. 1). When the pump stops and the ports of the foot valve close, the water cannot drain back to the suction well if the valve seats tightly. Foot valves were very commonly used in early installations of centrifugal pumps. Except for certain applications, their use is now much less common.

A foot valve does not always seat tightly, and the pump occasionally loses its prime. However, the rate of leakage is generally small, and it is possible to restore the pump to service by filling and starting it promptly. This tendency to malfunction is increased if the water contains small particles of foreign matter such as sand, and foot valves should not be used for such service. Another disadvantage of foot valves is their unusually high friction loss.

The pump can be filled through a funnel attached to the priming connection or from an overhead tank or any other source of water. If a check valve is used on the pump and the discharge line remains full of water, a small bypass around the valve permits the water in the discharge line to be used for repriming the pump when the foot valve has leaked. Provision must be made for filling all the waterways and for venting out the air.

PRIMING CHAMBERS

A priming chamber is a tank with an outlet at the bottom that is level with the pump suction nozzle and directly connected to it. An inlet at the top of the tank

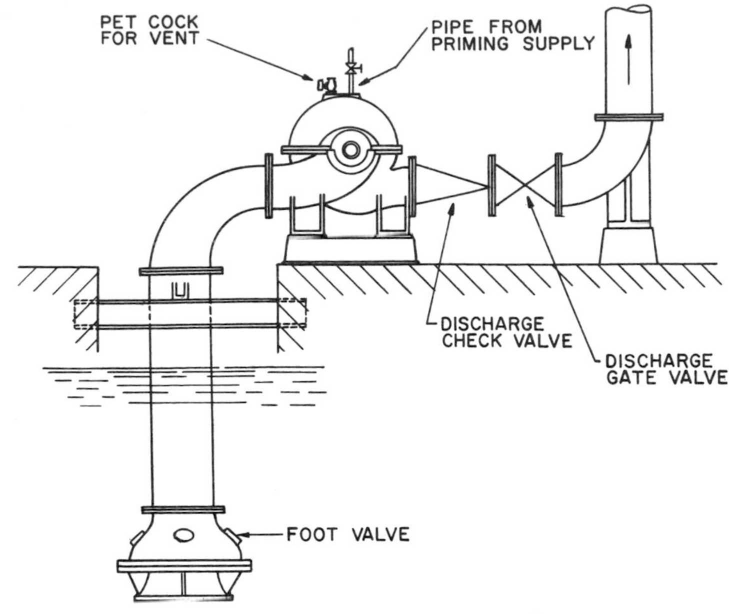

Fig. 1 Installation using foot valve.

connects with the suction line (Fig. 2). The size of the tank must be such that the volume contained between the top of the outlet and the bottom of the inlet is approximately three times the volume of the suction pipe. When the pump is shut down, the liquid in the suction line may leak out, but the liquid in the tank below the suction inlet cannot run back to the supply. When the pump is started, it will pump this entrapped liquid out of the priming chamber, creating a vacuum in the tank. The atmospheric pressure on the supply will force the liquid up the suction line into the priming chamber.

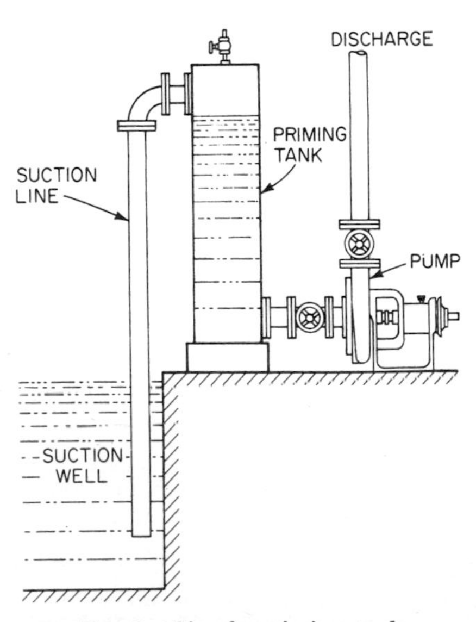

Fig. 2 Simple priming tank.

Commercial priming chambers are readily available with proper automatic vents and other features. The use of priming chambers is restricted because of their size to installations of relatively small pumps.

TYPES OF VACUUM DEVICES

Almost every commercially made vacuum-producing device can be used with systems in which pumps are primed by evacuating the air. Formerly water- or steam-jet primers had wide application, but with the increase in the use of electricity as a power source, motor-driven vacuum pumps have become popular.

Ejectors Priming ejectors work on the jet principle, using steam, compressed air, or water as the operating medium. A typical installation for priming with an ejector is shown in Fig. 3. Valve V_1 is opened to start the ejector and then valve V_2 is opened. When all the air has been exhausted from the pump, water will be drawn into and discharged from the ejector. When this occurs the pump is primed, and valves V_2 and V_1 are closed in that order.

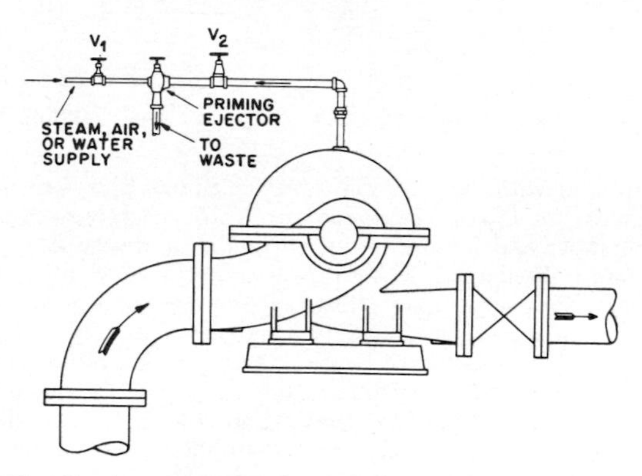

Fig. 3 Arrangement for priming with an ejector.

An ejector can be used to prime a number of pumps if it is connected to a header through which the individual pumps are vented through isolating valves.

Dry Vacuum Pumps Dry vacuum pumps, which may be of either the reciprocating or rotary type, cannot accommodate mixtures of air and water. When they are used in priming systems, some protective device must be interposed between the centrifugal pump and the dry vacuum pump to prevent water from entering the vacuum pump. The dry vacuum pump is used extensively for central priming systems.

Wet Vacuum Pumps Any rotary, rotative, or reciprocating pump that can handle air or a mixture of air and water is classified as a wet vacuum pump. The most common type used in priming systems is the Nash Hytor pump (Fig. 4). This is a centrifugal displacement type of pump consisting of a round, multiblade rotor revolving freely in an elliptical casing partially filled with liquid. The curved rotor blades project radially from the hub and, with the side shrouds, form a series of pockets and buckets around the periphery.

The rotor revolves at a speed high enough to throw the liquid out from the center by centrifugal force. This forms a solid ring of liquid revolving in the casing at the same speed as the rotor, but following the elliptical shape of the casing. It will be readily seen that this alternately forces the liquid to enter and to recede from the buckets in the rotor at high velocity.

Referring to Fig. 4 and following through a complete cycle of operation in a given chamber, we start at point *A* with the chamber (1) full of liquid. Because of the effect of the centrifugal force, the liquid follows the casing, withdraws from the rotor, and pulls air through the inlet port, which is connected to the pump inlet. At (2) the liquid has been thrown outwardly from the chamber in the rotor

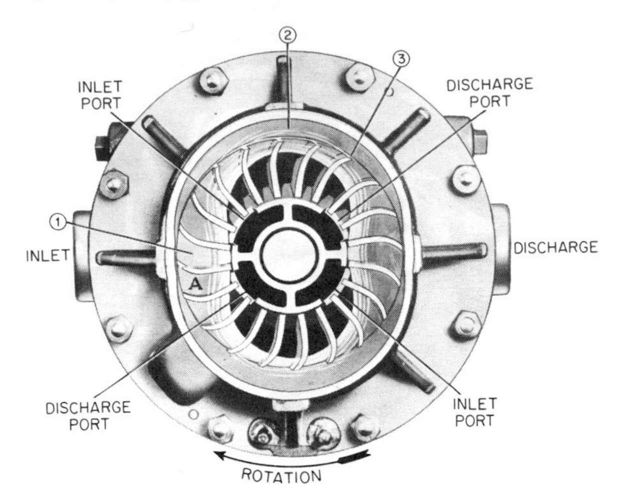

Fig. 4 Operating principle of Nash pump. (Nash Engineering Co.)

and has been replaced with air. As rotation continues, the converging wall of the casing at (3) forces the liquid back into the rotor chamber, compressing the air trapped in the chamber and forcing it out through the discharge port, which is connected with the pump discharge. The rotor chamber is now full of liquid and ready to repeat the cycle. This cycle takes place twice in each revolution.

If a solid stream of water circulates in this pump in place of air or of an air and water mixture, the pump will not be damaged, but will require more power. For this reason, in automatic priming systems using this type of vacuum pump, a separating chamber or trap is provided so that water will not reach the pump. Water needed for sealing a wet vacuum pump can be supplied from a source under pressure, with the shutoff valve operated manually or through a solenoid connected with the motor control. It is, however, preferable to provide an independent sealing water supply by mounting the vacuum pump on a base containing a reservoir. This is particularly desirable in locations where freezing may occur, as a solution of antifreeze can be used in the reservoir.

CENTRAL PRIMING SYSTEMS

If there is more than one centrifugal pump to be primed in an installation, one priming device can be made to serve all the pumps. Such an arrangement is called a central priming system (Fig. 5). If the priming device and the venting of the pumps are automatically controlled, the system is called a central automatic priming system.

Vacuum-controlled Automatic Priming System A vacuum-controlled automatic priming system consists of a vacuum pump exhausting a tank. The pump is controlled by a vacuum switch and maintains a vacuum in the tank of 2 to 6 inHg above the amount needed to prime the pumps with the greatest suction

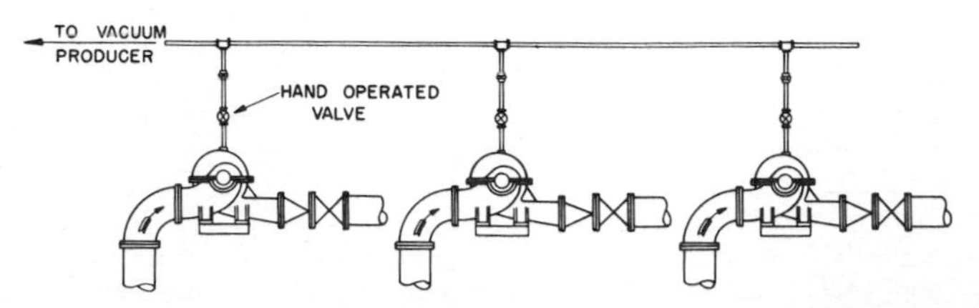

Fig. 5 Connections for a central priming system.

lift. The priming connections on each pump served by the system are connected to the vacuum tank by automatic vent valves and piping (Fig. 6). The vacuum tank is provided with a gage glass and a drain. If water is detected in the vacuum tank as the result of leakage in a vent valve, the tank can be drained. Automatic vent valves consist of a body containing a float which actuates a valve located in the upper part. The bottom of the body is connected to the space being vented. As air is vented out of the valve, water rises in the body until the float is lifted and the valve is closed.

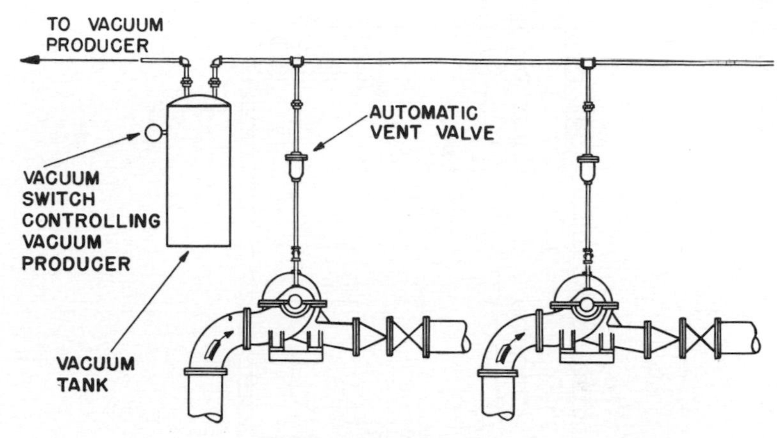

Fig. 6 Vacuum-controlled central automatic priming.

A typical vent valve designed basically for vacuum priming systems is illustrated in Fig. 7. The valve is provided with auxiliary tapped openings on the lower part of the body for connection to any auxiliary vent points on the system—for instance, when air is to be exhausted simultaneously from the high point of the discharge volute and the high point of the suction waterways. When one or more of these vent points are points of higher pressure, such as the top of the volute of the pump, an orifice is used in the vent line to limit the flow of liquid. Otherwise, a relatively high constant flow of water from the discharge back to the suction would take place, causing a constant loss. Where a unit is used more or less constantly, a separate valve should be used for each venting point.

This is the most commonly used of central automatic priming systems. It can use either wet or dry motor-driven vacuum pumps. Most central priming systems are provided with two vacuum pumps. The usual practice is to have the control of one vacuum pump switched on at some predetermined vacuum and the control of the second switched on at a slightly lower vacuum.

Automatic priming systems using ejectors are also feasible, and several such systems are commercially available.

SELF-CONTAINED UNITS

Centrifugal pumps are available with various designs of priming equipment that makes them self-contained units. Some have automatic priming devices, which are basically attachments to the pump and become inactive after the priming is accomplished. Other units, which are self-priming pumps, incorporate a hydraulic device that can function as a wet vacuum pump during the priming period (see Sec. **10.11**, Pumps for Construction Services). For stationary use, the automatically primed type is more efficient. The self-priming designs are generally more compact and are preferred for portable or semiportable use.

An automatically primed motor-driven pump uses a wet vacuum pump either directly connected to the pump or driven by a separate motor. In a direct-con-

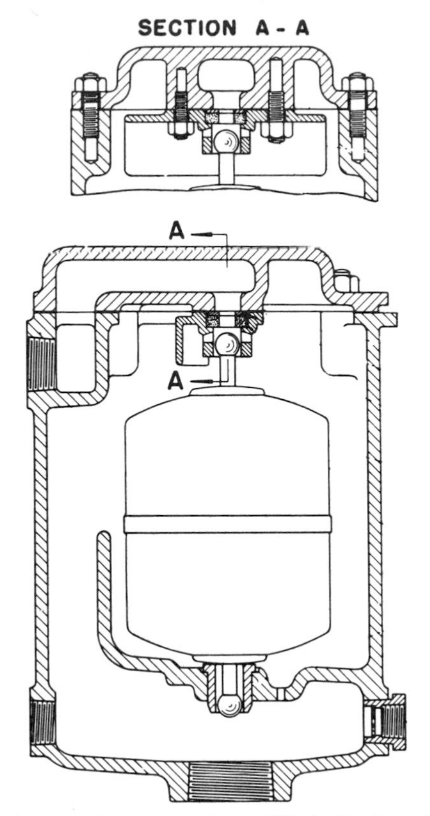

Fig. 7 Automatic vent valve. (Nash Engineering Co.)

nected unit, as soon as the centrifugal pump is primed, a pressure-operated control opens the vacuum pump suction to atmosphere so that it operates unloaded. With a separately driven vacuum pump, the controls stop the vacuum pump when the centrifugal pump is primed.

SPECIAL APPLICATIONS

Systems for Sewage Pumps A pump handling sewage or similar liquids containing stringy material can be equipped with automatic priming, but special precautions must be used to prevent carry-over of the liquid into the vacuum producing device. One approach is to use a tee on the suction line immediately adjacent to the pump suction nozzle, with a vertical riser mounted on the top outlet of the tee. This riser is blanked at the top, thus forming a small tank. The top of the tank is vented to a vacuum system through a solenoid valve. The solenoid valve in turn is controlled electrically, through electrodes located at different levels in the tank. The solenoid valve closes if the liquid reaches the top electrode and opens if the liquid level falls below the level of the lower electrode.

Another solution permits the use of an automatic priming system with a separate motor-driven vacuum pump controlled by a discharge pressure actuated pressure switch. An inverted vertical loop is incorporated in the vacuum pump suction line to the pump being primed. This prevents the sewage from being carried over into the vacuum pump because this pump shuts down before the liquid reaches the top of the loop.

Systems for Air-charged Waters Some types of water, particularly from wells, have considerable dissolved gas that is liberated when the pump handles a suction lift. In such installations, an air-separating tank (also called a *priming tank* or an *air eliminator*) should be used in the suction line. One type (Fig. 8) uses a float-operated vent valve to permit the withdrawal of air or gas. Another common arrangement uses a float valve mounted on the side of the tank to directly control the starting and stopping of the vacuum pump. Unless the air-separating tank is relatively large and the vacuum pump is not oversized, there is danger of frequent starting and stopping of the vacuum pump in such a system. When sand is present as an impurity in the water, the air-separating tank can be made to also function as a sand trap.

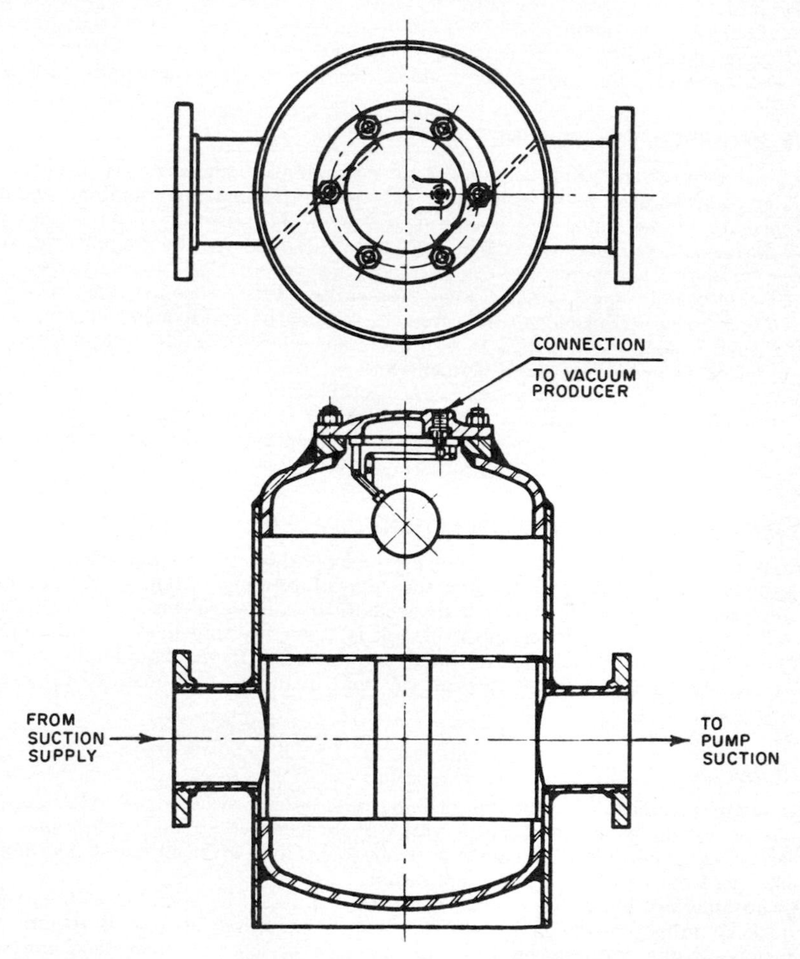

Fig. 8 Air-separating chamber.

Systems for Units Driven by Gasoline or Diesel Engines An automatic priming system using motor-driven vacuum pumps can be used for diesel engine-driven centrifugal pumps, if a reliable source of electric power is available in the station. An auxiliary gasoline engine–driven vacuum pump for emergency use might be desirable in case of electric power failure. Alternatively, a direct-connected wet vacuum pump with controls similar to those used in motor-driven automatically primed units is very satisfactory.

The choice of the priming device for a gasoline engine–driven centrifugal pump depends on the size of the pump, the required frequency of priming, and the portability of the unit. Most portable units are used for relatively low heads and small capacities, for use in pumping out excavations and ditches, for example. Self-priming pumps of various types are most satisfactory for this service and are preferable to regular centrifugal pumps.

It is possible to utilize the vacuum in the intake manifold of a gasoline engine as a means of priming or keeping the pump primed. The rate at which the air can be drawn from the pump in this manner is relatively low, so that many of these units use foot valves. They are initially primed by filling the pump manually. Provision must be made to prevent water from being drawn over into the manifold.

For larger volume low-head portable units, it is possible to use a wet-vacuum pump belted to the main shaft by a tight-and-loose pulley. The vacuum pump can then be stopped when the unit is primed. For permanent installations, a separate wet vacuum pump driven by a small gasoline engine is generally preferred.

TIME REQUIRED FOR PRIMING

The time required to prime a pump with a vacuum-producing device depends on (1) the total volume to be exhausted, (2) the initial and final vacuum, and (3) the capacity of the vacuum-producing device over the range of vacuums that will exist during the priming cycle. The actual calculations for determining the time necessary to prime a pump are complicated. To permit close approximations, jet primers are usually rated in net capacity for various lifts. It is necessary to divide the volume to be exhausted by the rating to obtain the approximate priming time. Unless such a simplified method is available, the selection of the size of a primer is best left to the vendor of the equipment.

Central automatic priming systems are usually rated for the total volume to be kept primed. The time initially required to prime each unit served by the central system is not usually considered, as the basic function of the system is to keep the pumps primed and in operating condition at all times.

PREVENTION OF UNPRIMED OPERATION

Various controls may be used to prevent the operation of a pump when it is unprimed. These controls depend upon the type of priming system used. For most installations, a form of float switch in a chamber connected with the suction line is used. If the level in the chamber or tank is above the impeller eye of the pump, the float-switch control allows the pump to operate. If the liquid falls below a safe level, the float switch acts through the control to stop the pump or to prevent its being started.

GENERAL

Since a great number of automatic priming devices and systems are available, care should be taken to use the type or variation best suited to the application. The discussion of priming in this chapter does not, of course, cover all makes and modifications available for every specific application.

An automatic priming system will often allow units to be operated with excessive air leakage into the suction lines. This is poor practice, because it requires the operation of the vacuum producer for greater periods of time than normally necessary.

REFERENCE

Karassik, I. J., and R. Carter: Chap. 21 in "Centrifugal Pumps: Selection, Operation, and Maintenance," McGraw-Hill Book Company, New York, 1960.

Section **2.5**

Special Variants of Dynamic Pumps

H. JOSEPH BIHELLER

INTRODUCTION

This section covers variations of dynamic pumps which have unusual characteristics in design or operation. Emphasis is placed on features which make them different from standard centrifugal pumps. Listings of applications are given for each type, but since more detailed application features will be found in later chapters under the particular industry in question, no descriptions of how the pump fits into the system are given.

All the pumps covered are commercially available from at least one supplier; that is, no purely experimental models or concepts have been included. Also, some variations like variable-pitch propeller pumps, jet pumps, and various positive-displacement pumps will be found elsewhere in the handbook.

Fig. 1 Reversible pump. (Gardner-Denver Co.)

REVERSIBLE CENTRIFUGAL PUMP

Characteristics. Low-capacity, low-head (20 to 50 gpm, 20 to 50 ft), specific speed approximately 900. Steep head-capacity curve (see Fig. 11).

Application. Circulating water for diesel engines, compressors, and other water-cooled machines; any service within capacity-head range where flow direction is to remain constant with reversed driver rotation.

Description. Basically a standard centrifugal pump with concentric casing having end suction and radial discharge nozzle. Impeller has straight radial vanes. See Fig. 1.

Advantages. Pump delivers the same quantity of water against same head with same efficiency with either direction of shaft rotation.

Disadvantages. Low efficiency (30 to 40 percent). At present writing made only by one manufacturer (Gardner-Denver Co.) in one small size and one impeller diameter. Different design capacities can therefore be obtained by speed change only. Maximum speed is 2,000 rpm.

HERMETICALLY SEALED MAGNETIC-DRIVE CENTRIFUGAL PUMP

Characteristics. Low-capacity, low-head (3 to 60 gpm, 10 to 50 ft), specific-speed range 1,300 to 2,300, steep head-capacity curve (see Fig. 11).

Application. Chemical industry, electroplating, process industry, any application where absolutely no leakage from stuffing box or seal can be tolerated.

Description. See Fig. 2 for schematic sectional drawing. Pump is basically a centrifugal pump with casing (usually of plastic like polypropylene) completely closed except for suction and discharge nozzles. The impeller (usually also of plastic) is made with an oversized back hub which has a circular magnet attached to its periphery. There is close clearance between the impeller magnet and the inside of the back casing. The motor drive has a circular magnet attached to its shaft, with the drive magnet located annular to the impeller magnet and having a close clearance to the outside of the back casing. The impeller is driven by magnetic interaction of the two magnets, and since there is no shaft extending through the casing, no shaft-sealing device of any kind is required.

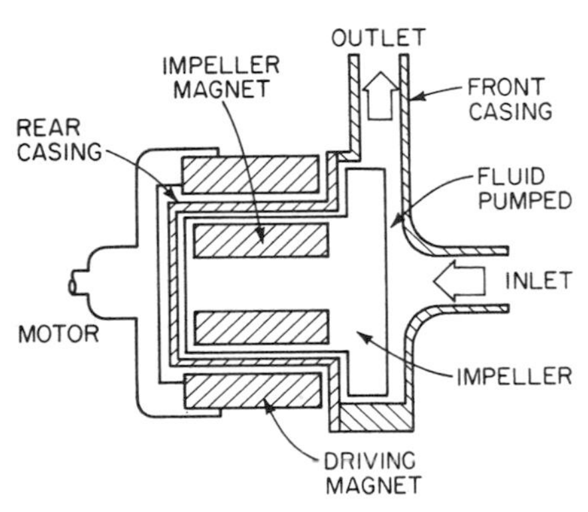

Fig. 2 Hermetically sealed magnetic drive pump. (Eastern Industries)

Advantages. No shaft seal, therefore no possible leakage.

Disadvantages. Because of limited torque which can be transmitted by the magnetic drive, this pump can be used only for relatively low-torque applications. Within head and capacity limitations listed above, specific gravity of fluid pumped is restricted to less than 1.3 and viscosity to below 150 SUS.

VORTEX (SHEAR-LIFT) PUMP

Characteristics. Medium capacities, moderate heads (to 4,000 gpm, to 100 ft), specific-speed range 1,500 to 2,800. Normal head-capacity curve (see Fig. 11).

Application. Sewage, slurries, solids handling, fluids with entrained air and gas.

Description. Pump consists of a standard concentric casing with axial suction nozzle and tangential discharge nozzle. The impeller, usually but not necessarily with straight radial blades, is axially recessed in the casing. The recess can be as much as 100 percent (that is, the impeller is completely out of the flow stream) to 50 percent (that is, impeller is halfway in the flow stream). The rotating impeller creates a vortex field in the casing, which moves the fluid from the centrally located suction to the tangentially located discharge. Since the fluid pumped does not have to flow through any vane passages, solid size is limited only by suction and discharge nozzle diameters. Also, since pumping action is by induced vortex rather than by impeller vanes, this type of pump can handle much larger percentages of air and entrained gases than a standard centrifugal pump where large percentages of air will block vane passages and choke the flow.

See Fig. 3 for one schematic section of one particular patented design.

Advantages. Handles solids and gases, stringy sewage, requires relatively low NPSH.

Disadvantages. Relatively low efficiency (35 to 55 percent).

LAMINATED ROTOR PUMP

Characteristics. Medium capacity, medium heads (to 1,000 gpm, 300 ft), specific-speed range 700 to 1,000, essentially flat head-capacity curve (see Fig. 11).

Application. Not generally used and not presently built by any major pump manufacturer. Good for relatively high viscosities (2,000 to 8,000 SUS), hence possibly applicable for viscous processes when gear-type pump cannot be used.

Description. Pump rotor consists of many flat disks, each with one or more holes in the center, stacked axially and separated by small spacers. This rotor is placed in a conventional end-suction volute casing, with the disk stack attached at one end to a shaft. Fluid enters the rotor through the holes in the disks and is energized by centrifugal force, exiting the rotor at the periphery. As the rotor has no vanes, all energy transfer between rotor and fluid is by viscous drag action between the disks and the fluid.

Advantages. Useful where flat head-capacity curve is desired, especially with relatively viscous fluids. Compared to gear pump or other positive displacement pumps, has smooth discharge flow (no pulsation). Requires low NPSH.

Disadvantages. Rotor very difficult to construct; hence, pump is expensive to produce. Cannot handle any solids entrained in the fluid because of the small clearances between the disks. Pump was invented in 1910 (by Nikola Tesla) and concept has been superseded in most applications by gear pumps; as of this writing, there is no present-day commercial manufacturer of these pumps.

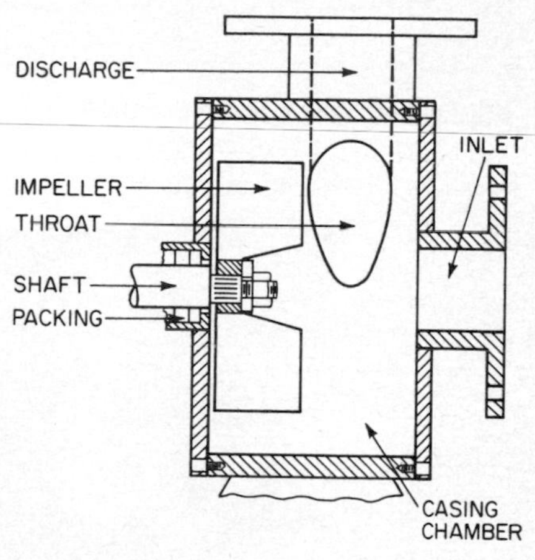

Fig. 3 Vortex pump, schematic section. (Allis Chalmers)

INCLINED ROTOR PUMP

Characteristics. Medium-capacity, low-head (to 3,000 gpm and to 75 ft, could be built for larger capacities), specific speed 2,500 to 5,000, steep head-capacity curve (see Fig. 11).

Application. Sewage, sludge, rags, fluids containing large percentages of cuttable solids.

Description. Rotor consists of a flat elliptical plate mounted at an angle of approximately 45° to a rotating shaft. Usually this plate has teeth machined all around its periphery. This rotor is placed into a concentric tubular casing having its inlet nozzle axial and in line with the shaft and a radically placed discharge nozzle. There is close clearance between the periphery of the rotor plate and the inside diameter of the casing. In the case where the rotor has teeth, the inside of the casing has matching serrations. Fluid entering the pump is energized by centrifugal action of the rotating plate and discharges through the exit nozzle. See Fig. 4 for construction.

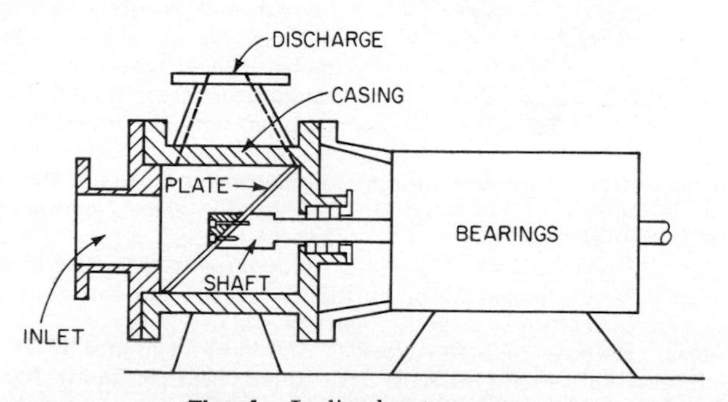

Fig. 4 Inclined rotor pump.

Advantages. Can pump raw uncomminuted sewage and liquids containing stringy materials. Low maintenance, not expensive to manufacture.

Disadvantages. Needs flooded suction (cannot pull lift). Suction head has to be removed for disassembly, hence cannot be opened without disturbing piping. Low efficiency. To writer's knowledge not manufactured in the United States, but available in Europe.

REGENERATIVE TURBINE PUMP

Characteristics. Very low-capacity, high-head (to 100 gpm and to 1,000 ft). Specific-speed range 40 to 600, very steep head-capacity curve (see Fig. 11).

Fig. 5 Regenerative turbine pump view showing impeller and inner casing. (Worthington Corp.)

Application. Laundry, drinking water, car wash, breweries, small boiler feed, chemical process, refineries, spraying systems, etc.

Description. Impeller consists of a solid disk with a large number of small vanes attached to its periphery. This impeller is contained in a radially split casing having radial suction and discharge openings next to each other but separated by a stripper with close clearance to the impeller periphery (see Fig. 5 for general view). Liquid coming in through the suction nozzle enters an impeller vane and is forced outward by centrifugal force. However, this liquid strikes the casing and is therefore turned back inward and reenters the impeller at a different vane. This cycle is repeated many times with continuous pressure buildup until the liquid completes its travel around the complete pump, reaches the discharge opening, and is forced out of the casing (through the discharge nozzel) by the close clearance stripper (see Fig. 6).

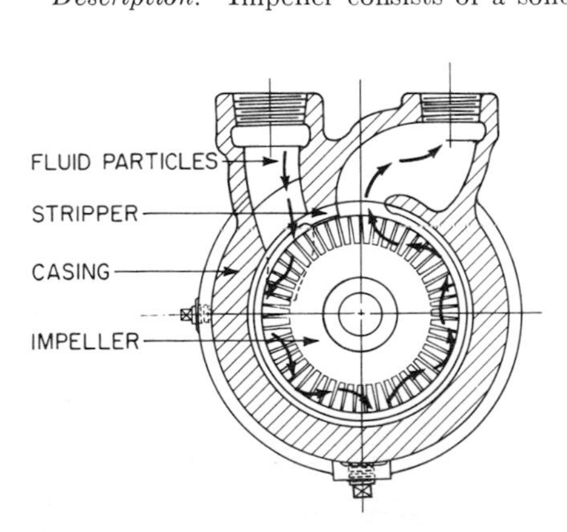

Fig. 6 Regenerative turbine pump schematic of operation. (Worthington Pump International, Inc.)

Advantages. Smaller in size and lower in cost than positive displacement or multistage centrifugal for same conditions of service. Can handle relatively large amounts of gas or vapor.

Disadvantages. Because of high velocities and close clearance in the pumps, any abrasives contained in the liquid will cause rapid wear. Cannot handle solids of any appreciable size. Maximum viscosity approximately 250 SUS, therefore good for low-viscosity clean liquids only. Requires periodic maintenance and replacement of inner casing because of erosion. Pump is slightly noisy, especially when compared to rotating casing pump.

ROTATING CASING (PITOT TUBE PUMP)

Characteristics. Low-capacity, high-head (to 250 gpm and to 2500 ft at present, could be made larger), specific speed 500 to 2,600, normal head-capacity curve (see Fig. 11).

Application. Same as for regenerative turbine pumps.

Description. (See Fig. 7.) The pump consists essentially of a rotating circular casing containing a stationary pitot tube in its center. Liquid enters the casing axially at the center and centrifugal action increases its pressure and velocity at

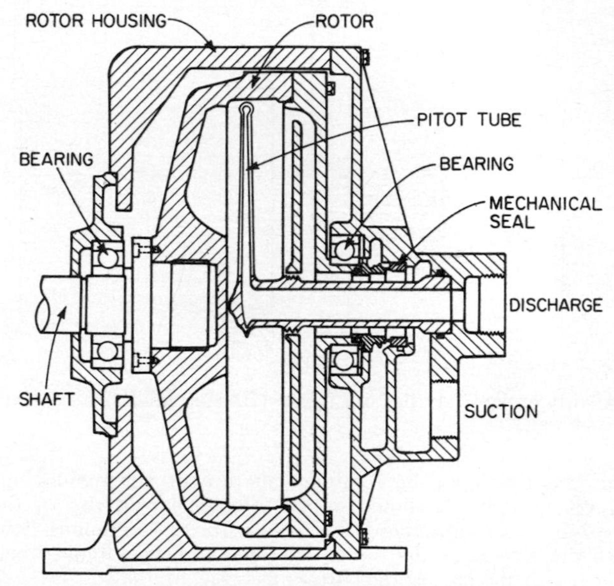

Fig. 7 "Roto Jet" rotating casing pump. (Kobe Inc.)

the casing periphery. The pitot tube collector, with opening facing opposite the direction of rotation and located near the outer wall, converts most of the kinetic energy of the liquid into static pressure. The liquid then flows through the collector arm supporting the pitot opening back down to the center of the pump and exits into the discharge pipe.

To aid in imparting velocity and pressure to the liquid, usually radial ribs are added to the inside of the casing at the side walls. Some designs have two pitot tubes located 180° from each other, thus doubling the capacity of the pump. Such an arrangement, however, owing to increased drag effect, lowers the efficiency of the pump and also results in lower head for same diameter and speed, and is therefore seldom used. For safety reasons, the rotating casing is usually enclosed in a stationary secondary casing or rotor housing.

Advantages. Smaller in size and lower in cost than positive displacement or multistage centrifugal pumps for same conditions of service, has smooth flow, no pulsations. Will not seize if run dry as it has no close-fitting parts in the fluid stream.

Disadvantages. Because of high velocities inside the casing and especially passing the pitot tube, erosion will occur if any abrasives are contained in the fluid. Although pump will not seize if run dry, its gas and vapor handling characteristics are relatively poor, as any gas or vapor will collect at the lowest pressure area of the pump, which is at the center of the casing. Since the amount of liquid pumped is small compared to the amount contained in the casing, such collections of gas or vapor will not be easily purged. Medium efficiencies, usually below 40 percent.

AIR-LIFT PUMP

Characteristics. Capacities to 2000 gpm, head to 1000 ft.

Application. Water and oil wells, especially with sandy or corrosive fluids, mine tailings.

Description. The pump basically consists of a vertical pipe submerged in the well, lake, tank of fluid, etc., and an air-supply tube enabling compressed air to be fed to the pipe at a considerable distance below the static liquid level. When air is

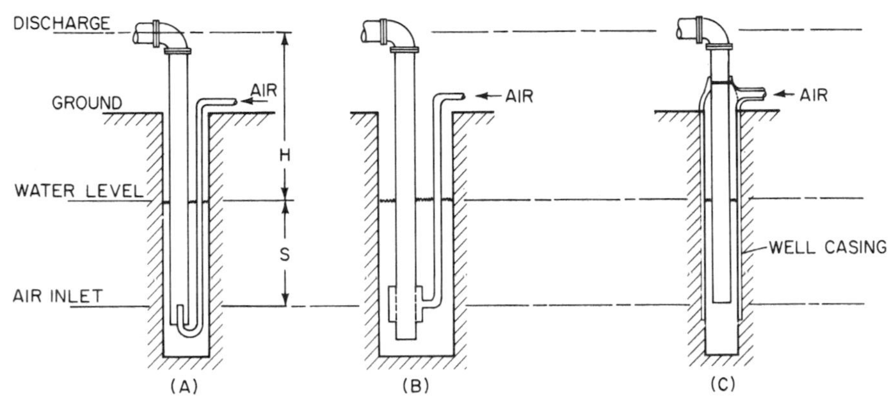

Fig. 8 Air-lift pump: (*A*) bottom inlet; (*B*) side inlet (Pohle) type; (*C*) Sauders system cased well.

introduced into the pipe, the resultant mixture of air bubbles and liquid, being lighter in weight than the liquid outside the pipe, will rise in the pipe. As air is being continuously introduced at the bottom, a continuous flow of liquid and air mixture will emerge at the top of the pipe, with new liquid from the well, open tank, etc., entering the pipe at the bottom (see Fig. 8).

As the only head-producing mechanism is the difference in specific weight of the liquid-air mixture inside the pipe and the liquid outside the pipe, the head that can be obtained from an air-lift pump depends on the distance between water level in the pit and location of air introduction. If head (H) is measured from the discharge pipe to water level and submergence (S) from water level to introduction of air, the ratio H/S is approximately 1 for most applications, reaching 3 for high heads (and low flows) and going as low as 0.4 for low heads (and high flows). The capacity of liquid pumped depends on the amount of air supplied, with capacity increasing with increased air supplied up to an optimum amount of air. However, since the effluent out of the pipe discharge is a mixture of liquid and air, introducing more air than the optimum will actually decrease the net liquid amount delivered. Figure 9 gives approximate amounts of cubic feet per minute of free air required to pump 1 gpm of water against the heads shown at relative submergence shown. The amounts of air given in the table will have to be adjusted if the specific weight of the liquid differs from that of water.

Advantages. No moving parts, can be used for corrosive and erosive fluids, gentle action (has been used to remove sand from buried undersea objects). Operates on air and can therefore be used in explosive atmospheres. Can sometimes be put into wells of irregular shape where regular deep well pumps cannot be fit.

Disadvantages. Low efficiency (less than 40 percent). Needs very large submergences compared to conventional pumps. Not available (to writer's knowledge) from

H/S					
H	3	2	1	0.67	0.4
20	....	...	...	0.22	0.15
50	....	...	...	0.3	0.2
100	....	...	...	0.4	0.3
150	....	...	0.7	0.5	
200	....	...	0.8	0.6	
300	....	2.1	1.0		
400	....	2.3	1.2		
500	3.25	2.6	1.4		
650	3.75	3.0	2.1		
800	4.2	3.5			
950	4.7	3.9			

Fig. 9 Table 1 Number of cfm free air required to pump 1 gpm of water. H = head, S = submergence (ft).

any manufacturer, but since it is simple and easy to make, usually built by user himself.

ELECTROMAGNETIC PUMP

Characteristics. Liquid metals or other highly conductive fluids only, capacities to 60,000 gpm, discharge pressures to 250 lb/in² possible, to 100 lb/in² normally. Temperatures of liquid to 1600°F.

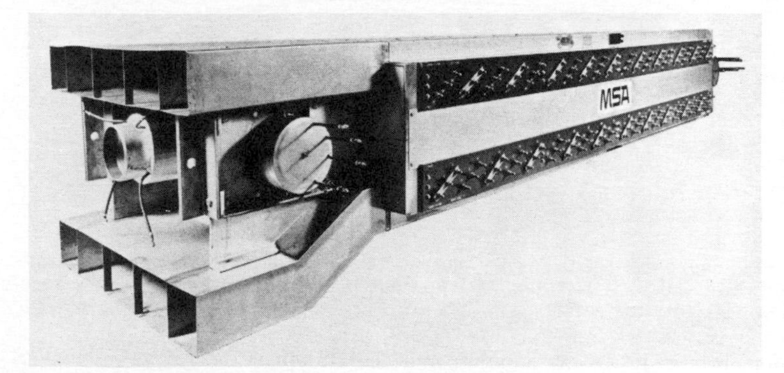

Fig. 10 Linear induction electromagnetic pump. (MSA Research Corporation)

Application. Primarily heat transfer fluids in nuclear power installations. Pumping liquid metals in other applications.

Description. Electromagnetic pumps are divided into two basic types: Conduction (ac or dc) and induction (ac only).

Conduction-type pumps consist of a tube of nonmagnetic metal containing the fluid to be pumped. Around this tube a magnet is placed so that flux lines are at right angles to the tube axis. When a current is introduced into the fluid at right angles to both the tube and the flux lines (via electrical contacts on the tube), a force is created in the liquid metal, acting at right angles to the direction of the flux and the current. The direction of the force is according to the left-hand rule of electrical theory. This force will push the liquid out one end of the tube, and if pipes or other connections are provided at the tube ends, continuous pumping action results. If the pump is powered by direct current, usually a permanent magnet is used to produce the magnetic flux, and reversal of flow direction is accomplished by reversing the polarity of the current injected into the flow tube. In the alternating current-powered type, an electric magnet is used, and reversal of direction of liquid flow is accomplished by changing the phase relationship (reversing direction) of the magnet relative to the injected current. The conduction type electromagnetic pumps are generally small and used for low capacities and low pressures. If higher pressures (and still low flows) are desired, conduction pumps can be built in two or more stages.

Induction electromagnetic pumps consist of a stator similar to the stator of an induction type three-phase electric motor. The place of the rotor of the electric motor is taken by a stationary laminated core and an annulus with stationary spiral vanes containing the liquid metal to be pumped. The annulus is made of nonmagnetic material, usually a refractory due to the high temperature of the liquid metal.

Voltage is induced in the liquid contained in the annulus; this results in a circulating field in the liquid which interacts with the magnetic field from the stator, producing a force in the liquid acting in the circumferential direction. The liquid

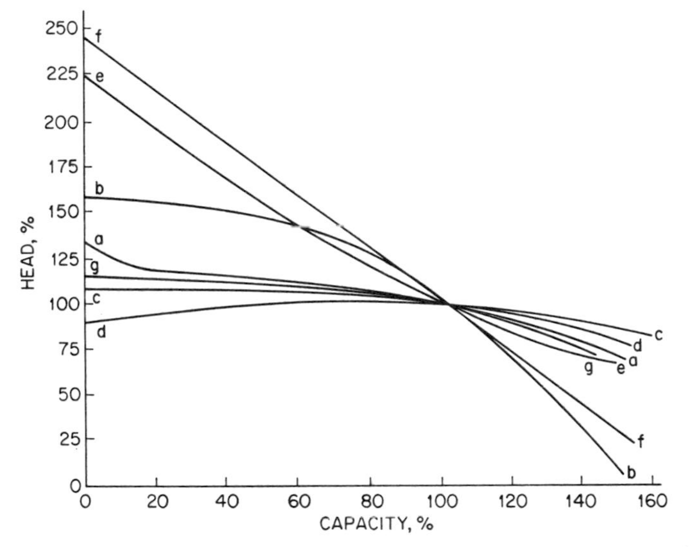

Fig. 11 Shape of head-capacity curve for various pumps: (*A*) reversible, (*B*) hermetically sealed, (*C*) vortex, (*D*) laminated, (*E*) inclined rotor, (*F*) regenerative turbine, (*G*) rotating casing.

therefore rotates inside the annulus, and the stationary helical vanes convert this rotary motion into linear motion of the liquid, pumping it out the discharge end of the tube. Flow can be only in one direction, depending only on the direction of the vanes.

A variation of the induction pump is the linear induction pump (used mostly for the larger capacities). In this construction the stator field is laid out flat and the annulus is replaced by a multitude of small tubes piped up in parallel. Since the stator windings are linear instead of circular, the induced force is in the axial direction and no turning vanes are required to produce flow in the axial direction of the tubes. Most present-day induction electromagnetic pumps are of the linear induction type, and all large capacity pumps are of this design. While conduction-type pumps are available in standard sizes, almost all linear induction type pumps are custom-designed for the particular conditions of service required.

Advantages. Can pump high-temperature liquid metals. Pump has no moving parts, seals, and power-transmission train. Direction of flow is easily reversible (for conduction and linear induction types) by electrical switching.

Disadvantages. Can only be used with liquid metals and similar fluids; low efficiency; large size in case of linear induction; high cost.

Section 3.1

Power Pumps

FRED BUSE

INTRODUCTION

A power pump is a constant-speed, constant-torque, and nearly a constant-capacity reciprocating machine, whose plungers or pistons are driven through a crank-shaft from an external source.

The pump's capacity fluctuates with the number of plungers or pistons. In general, the higher the number, the less capacity variation at a given rpm. The pump is designed for a specific speed, pressure, capacity, and horsepower. The pump can be applied to horsepower conditions less than the specific design point, but at a sacrifice of the most economical operating condition.

Pumps are built in both horizontal and vertical construction. Horizontal construction is used on plunger pumps up to 200 hp. This construction is usually below the waist level, and permits ease of assembly and maintenance. They are built with three or five plungers. Horizontal piston pumps are rated to 2,000 hp, and usually have two or three pistons, which are single- or double-acting. Vertical construction is used on plunger pumps up to 1,500 hp, with the fluid end above the power end. This construction eliminates the plunger weight on the bushings, packing, and crosshead, and has an alignment feature of the plunger to the packing. A special sealing arrangement is required to prevent the liquid of the fluid end from mixing with the oil of the power end. There can be three to nine plungers.

Plungers are applied to pumps with pressures from 1,000 to 30,000 lb/in². Maximum developed pressure with a piston is around 1,000 lb/in². The pressure developed by the pump is proportional to the power available at the crankshaft. This pressure can be greater than the rating of the discharge system or pump. When the pressure developed is greater than these ratings, a mechanical failure can result. To prevent this, a pressure relief device should be installed between the pump discharge flange and the first valve in the discharge system.

DESIGN

Brake horsepower for the pump is:

$$\text{Bhp} = \frac{Q \times Ptd}{1714 \times ME}$$

where Q = delivered capacity, U.S. gpm
Ptd = developed pressure, lb/in²
ME = mechanical efficiency, percent

Capacity The capacity (Q) is the total volume delivered per unit of time. The fluid includes liquid, entrained gases, and solids at the specified condition. Capacity is equal to the capacity displaced less slip.

$$Q = D(1 - S)$$

where D = displaced capacity
S = slip, percent

Displacement Displacement (D) is the calculated capacity of the pump with no slip losses. For single-acting plunger or piston pumps this is

$$D = \frac{A \times m \times n \times s}{231}$$

where A = cross-sectional area of plunger or piston, in²
m = number of plungers or pistons
n = rpm of pump
s = stroke of the pump, in (half the linear distance the plunger or piston moves linearly in one revolution)
231 = in³/U.S. gal

For double-acting plunger or piston pumps, this is

$$D = \frac{(2A - a)m \times n \times s}{231}$$

where a = cross-sectional area of piston rod, in²

Pressure The pressure (Pld) used to determine brake horsepower is the differential developed pressure. Because the suction pressure is usually small compared to the discharge pressure, discharge pressure is used in lieu of differential pressure. Figure 1 shows a typical performance curve for a power pump.

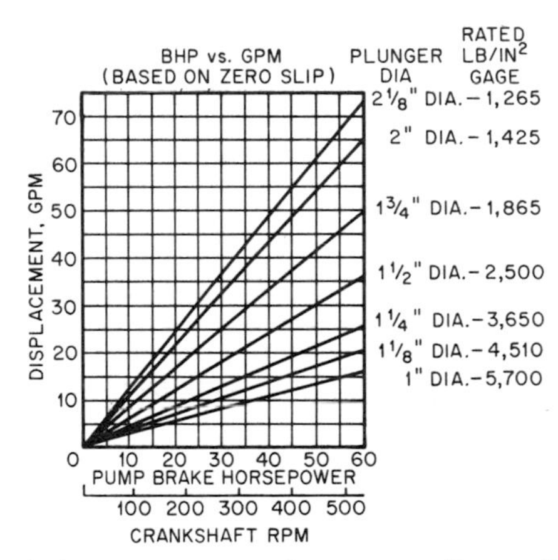

Fig. 1 Typical power-pump performance. (Ingersoll-Rand Co.)

Slip Slip (S) is the capacity loss as a percentage of the suction capacity. It consists of volumetric efficiency loss (VE_1), stuffing box loss (B_1), and valve loss (V_1).

$$S = VE_1 + B_1 + V_1$$

Volumetric Efficiency Volumetric efficiency (VE) is the ratio of the discharge volume to the suction volume expressed as a percentage. It is proportional to the ratio r and developed pressure (see Fig. 2); r is the ratio of internal volume of fluid between valves when the plunger or piston is at the top of the peak of its back stroke ($C + D$) to the plunger or piston displacement (D) (See Fig. 3). VE_1 is (1-VE) expressed as a percentage.

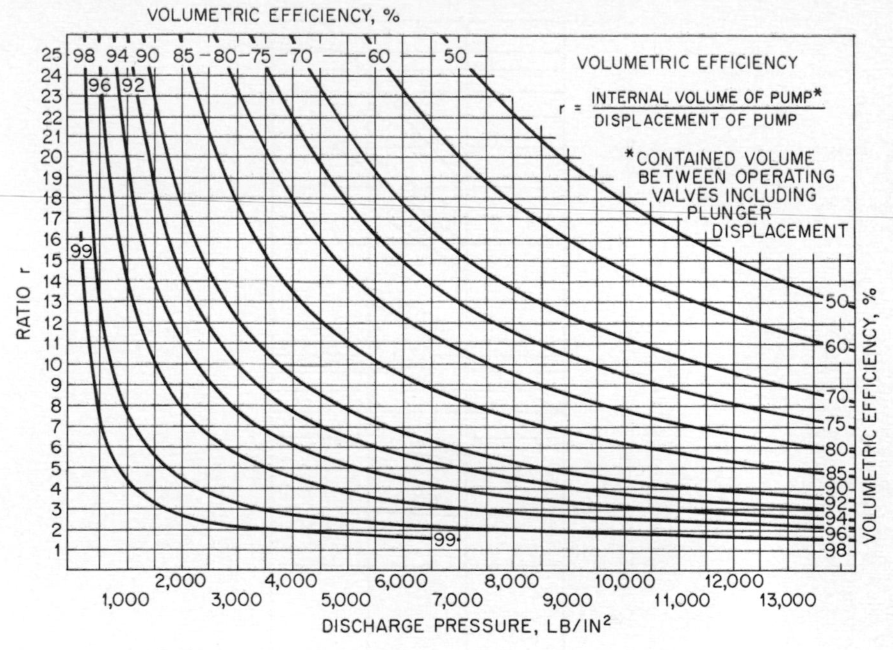

Fig. 2 Volumetric efficiency. (Ingersoll-Rand Co.)

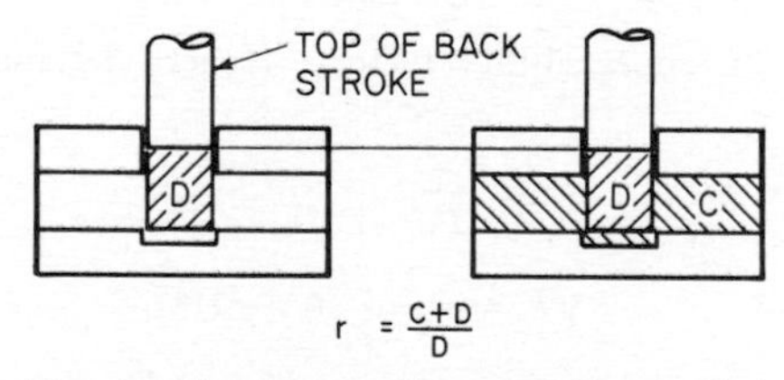

Fig. 3 Ratio r. (Ingersoll-Rand Co.)

Since discharge volume cannot be readily measured at discharge pressure, it is taken at suction pressure. Discharge volume at suction pressure will result in a higher volumetric efficiency than the calculated discharge volume at discharge pressure because of fluid compressibility. Compressibility becomes important when pumping water or other fluids over 6,000 lb/in², and it should be taken into consideration to determine the actual delivered capacity into the discharge system

Figure 2 shows the approximate *VE* for water.

VE based on expansion back to suction capacity:

$$VE = \frac{1 - Ptd \times \beta \times r}{1 - Ptd \times \beta}$$

VE based on discharge capacity:

$$VE = 1 - Ptd \times \beta \times r$$

Figure 4 shows the approximate compressibility factor β for various liquids.

When the compressibility factor is not known, but suction and discharge density can be determined in pounds per cubic foot, the following can be used for calculating *VE*.

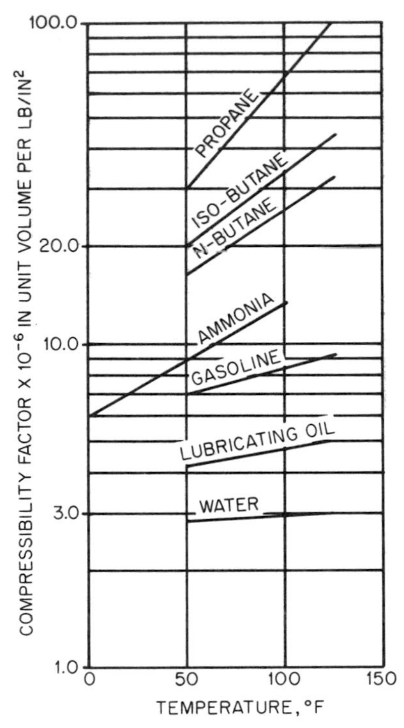

Fig. 4 Compressibility factor. (Ingersoll-Rand Co.)

Based on suction capacity:

$$VE = r - \frac{\rho_d}{\rho_s}(r - 1)$$

Based on discharge capacity:

$$VE = 1 - r\left(1 - \frac{\rho_s}{\rho_d}\right)$$

where ρ_s = suction density
ρ_d = discharge density

Stuffing Box Loss Stuffing box loss (B_1) is usually considered negligible when calculating S.

Valve Loss Valve loss (V_1) is the flow of fluid going back through the valve while closing and/or seated. This is a 2 to 10 percent loss depending on the valve design and valve condition.

Slip is affected by the viscosity of the fluid. The following is an example of a pump with a plate valve at 150 rpm:

Centistokes	*Slip, %*
100	8
1,000	8.5
2,000	9.5
6,000	20.0
10,000	41.0
12,000	61.0

Slip is also affected by speed and pressure:

psi/rpm	Slip, % 440	390	365
4,000	11	22	34
3,000	9	20	31
2,000	7	18	30
1,000	7	15	27.5

Mechanical Efficiency Mechanical efficiency (*ME*) of a power pump at full load pressure and speed is 90 to 95 percent depending on size, speed, and construction. Mechanical efficiency is affected by speed and slightly by developed pressure. When

TABLE 1 Effects of Speed and Pressure on Mechanical Efficiency

Constant Speed	
% full-load developed pressure	Mechanical efficiency
20	82
40	88
60	90.5
80	92
100	92.5
Constant Developed Pressure	
% speed	Mechanical efficiency
44	93.3
50	92.5
73	92.5
100	92.5

a single built-in-gear is part of the power frame, the pump's mechanical efficiency is 80 to 85 percent.

Speed Design speed (n) of power pumps is from 300 to 800 rpm, depending on capacity, size, and horsepower. To maintain good packing life, speed is sometimes limited to a plunger speed of 140 to 150 ft/min. Pump speed is also limited by valve life and allowable suction conditions.

Speed is the limiting factor of liquid separation from the plunger. Low speed limits with sleeve bearings must be considered to prevent a loss of the lubrication film.

Plunger Load In addition to the horsepower limit, a power pump is designed for a plunger load (*PL*) limit. This is the load in pounds that is applied to the plunger or piston and the bearing system. The industry's definition of plunger load is:

$$PL = Ptd \times A$$

where Ptd = developed pressure, lb/in^2

A = cross-sectional area of the plunger or piston, in^2

With double-acting pistons, the piston-rod area (a) is subtracted on the forward stroke.

The bearing system is rated for a specific load at design speed. Higher plunger loads will result in short bearing life. With the sleeve bearing, high plunger loads at slow speed will destroy the lubrication film. The load on the bearings from the moment of inertia of unbalanced reciprocating and unbalanced rotating parts is considered in the bearing selection, and is approximately 10 to 25 percent of the rated plunger load.

Unbalanced Reciprocating Parts Force (F_{rec}) Parts are ⅓ connecting rod, crosshead, crosshead bearing, wristpin, pony rod, and plunger. Additional parts on vertical pumps are pull rods, yoke, and plunger nut.

$$F_{rec} = \frac{W}{g} \omega^2 R \left(\cos \theta + \frac{R}{L} \cos 2\theta \right)$$

Unbalanced Rotating Parts Force (F_{rot}) Parts are ⅔ connecting rod, crank-end bearing, and crankpin.

$$F_{rot} - \frac{W}{g} \omega^2 R$$

where W = weight of all reciprocating or rotating parts, lb
g = 32.2 ft/s²
$\omega = (2\pi/60) \times n$ (n = pump rpm)
R = one-half of the stroke, ft
L = length of the connecting rod ℄ to ℄, ft
θ = crank angle; usually max force is at $\theta = 0°$, $\cos \theta = 1$

Number of Plungers or Pistons The following are the industry's terms for the number of plungers or pistons (m) on the crankshaft. They are the same for single- or double-acting pumps.

Number	*Term*
1	Simplex
2	Duplex
3	Triplex
4	Quadruplex
5	Quintuplex
6	Sextuplex
7	Septuplex
9	Nonuplex

Pulsations The pulsating characteristics of the output of power pumps are extremely important in the pump application. The magnitude of the discharge pulsation is mostly affected by the number of plungers or pistons on the crankshaft. In

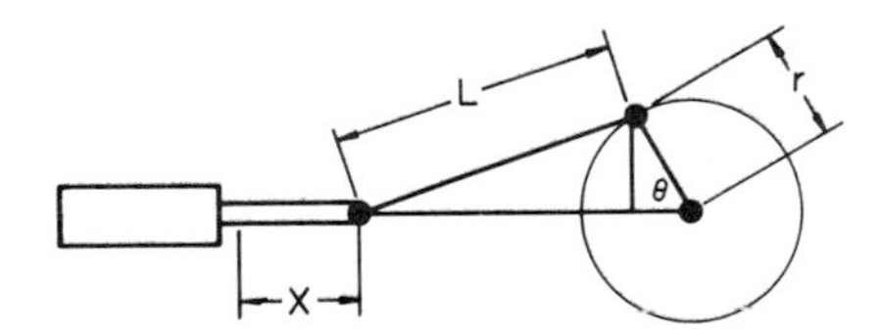

Fig. 5 Crankshaft position. (Ingersoll-Rand Co.)

Fig. 5 r = radius of crank, ft; L = length of the connecting rod, ft; $C = L/r$; and $\omega = (2\pi/60) \times$ rpm.

$$X = r \left(1 - \cos \theta + \frac{1}{2} \times \frac{1}{n} \sin 2\theta \right)$$

The approximate velocity of the plunger or piston is

$$S = \frac{dx}{dt} = r \left(\sin \theta + \frac{\sin 2\theta}{2C} \right) \omega$$

The approximate acceleration of the plunger or piston is

$$Gp = \frac{d^2x}{dt^2} = r \left(\cos \theta + \frac{\cos 2\theta}{C} \right) \omega^2$$

Momentary rate of discharge capacity is the cross-section area of the plunger or piston A times velocity.

For a single-acting plunger or piston:

$$A = 0.785 \times D^2$$

For a double-acting piston:

$$A - a = 0.785(D^2 - d^2)$$

Total discharge $Q = ALS'$ where S' = effective strokes in a given time. Q is the area of the curve. Mean height of the curve is

$$\frac{\text{Total } Q}{t} = \frac{ALS'}{t} = ALS''$$

where S'' = strokes per second and t-time in seconds.

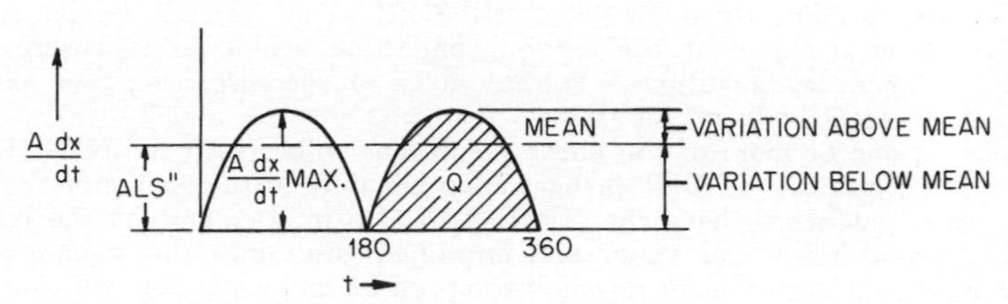

Fig. 6 Discharge rate for single double-acting pump. (Ingersoll-Rand Co.)

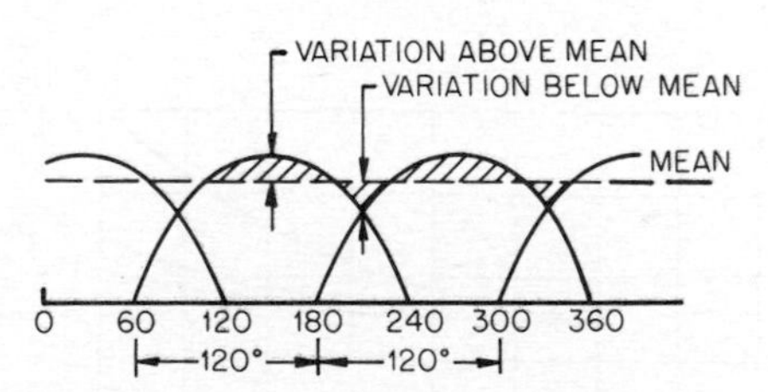

Fig. 7 Discharge rate for triplex single-acting pump. (Ingersoll-Rand Co.)

Tables 2 and 3 show how variations from the mean are related to changes in the number of plungers and changes of $C = L/r$. Table 2 shows variation with the number of plungers with a C of approximately 6:1. This table shows that

TABLE 2 Effect of the Number of Plungers in Producing Variations in Capacity from the Mean (C Approximately 6:1)

Type	Number of plungers	% above mean	% below mean	Total %	Plunger phase
Duplex (double)	2	24	22	46	180°
Triplex	3	6	17	23	120°
Quaduplex	4	11	22	33	90°
Quintaplex	5	2	5	7	72°
Sextuplex	6	5	9	14	60°
Septuplex	7	1	3	4	51.5°
Nonuplex	9	1	2	3	40°

pumps with an even number of throws have a higher variation than adjacent pumps with an odd number of throws. Table 3 shows the variations in a triplex with changes in $C = L/R$.

TABLE 3 Effect of a change in C in Producing Variations in Capacity from the Mean for a Triplex Pump

C	% above mean	% below mean	Total %
4:1	8.2	20.0	28.2
5:1	7.6	17.6	25.2
6:1	6.9	16.1	23.0
7:1	6.4	15.2	21.6

Net Positive Suction Head Required (NPSHR) The NPSHR is the required head of clean clear liquid at the suction connection centerline to ensure proper pump-suction operating conditions. For any given plunger size, rpm, gpm, and pressure, there is a specific value of NPSHR.

A change in one or more of the above conditions will change the NPSHR. For a given plunger size the NPSHR changes approximately as the square of the speed.

It is a good practice to have the NPSH 3 to 5 lb/in² greater than the NPSHR. This will prevent release of vapor and entrained gases into the suction system. Released vapors and gases under repeated compression and expansion will cause cavitation damage in the internal passages.

Figure 8 shows an NPSHR for a triplex pump vs rpm and plunger diameter.

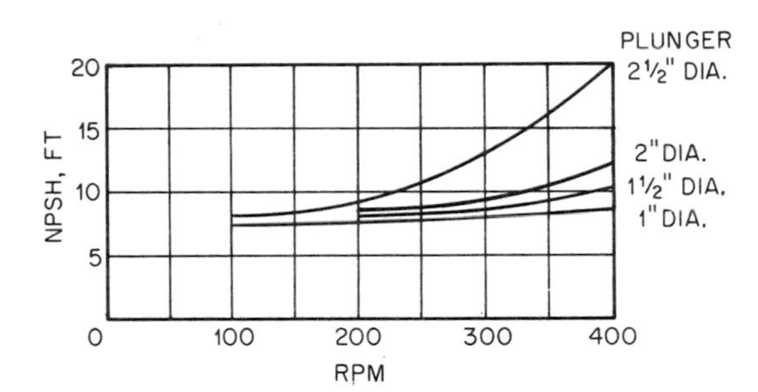

Fig. 8 NPSHR for a triplex pump. (Ingersoll-Rand Co.)

Net Positive Suction Head Available (NPSHA) NPSHA is the static head plus atmospheric head, minus lift loss, friction loss, vapor pressure, velocity head, and accleration head in feet available at the suction connection centerline.

Acceleration head can be the highest factor of NPSHA. In some cases it is ten times the total of all the other losses. Data from both the pump and the suction system are required to determine acceleration head; its value cannot be calculated until these data have been established.

Acceleration Head The flow in the suction line is always fluctuating, continuously accelerating or decelerating. The acceleration head is not a loss, because the energy is restored during deceleration. Acceleration head is defined as

$$ha = \frac{L \times V \times n \times C}{g \times K}$$

where ha = acceleration head loss, ft
L = length of the suction pipe, ft
V = mean velocity in the suction pipe, ft/s

$$V = \frac{\text{gpm} \times 0.321}{\text{area of the suction pipe, in}^2}$$

n = rpm of the pump
g = acceleration of gravity, 32.2 ft/s²
K = factors for the various fluids:

Water	1.4
Petroleum	2.5
Liquid with entrained gas	1.0

C = factor for the type of pump:

Duplex	0.115
Triplex	0.006
Quadruplex	0.08
Quintuplex	0.04
Sextuplex	0.055
Septuplex	0.028

When the suction system consists of pipes of various sizes, calculate the acceleration head for each section separately. Add the acceleration head of all sections to obtain the total.

If the calculated NPSHA, including acceleration head, is greater than the suction system can provide, the system NPSH should be increased. This can be accomplished by:

a. Increasing the static head
b. Increasing the atmospheric pressure
c. Adding a booster pump to the system
d. Adding pulsation dampener

The basic definition of acceleration head is:

$$Gs = V \times n \times C = \text{acceleration of liquid in suction line, ft/s}^2$$

$$Fs = \frac{W_s \times G_s}{g} = \text{force to produce acceleration, lb}$$

where W_s = weight of liquid in line = $L \times A_s \times$ sp gr

$$H_t = \frac{F_s \times 2.31}{A_s \times \text{sp gr}} = \text{theoretical head, ft of liquid}$$

where A_s = cross-sectional area of pipe

Substituting:

$$H_L = \frac{W_s \times V \times n \times C \times 2.31}{A_s \times \text{sp gr} \times g} = \frac{L \times V \times n \times C}{g}$$

$$Ha = \frac{Ht}{K}$$

where K = ratio of the theoretical head to the actual head

Therefore

$$Ha = \frac{L \times V \times n \times C}{g \times K}$$

Liquid Separation from the Plunger The pump speed at which water will separate from the end of the plunger may be calculated from the following:

$$\text{Rpm} = 54.5 \sqrt{\frac{(34 - h_s - h_f)A_s}{LR[1 - 1/(L/R)]A_p}}$$

where rpm = pump speed
h_s = suction head, ft
h_f = piping friction loss, ft
A_s = area of suction pipe, in²
L = length of connecting rod, ℄ to ℄, ft
R = crank radius, ft
l = length of the pipe where the resistance of the flow is to be measured in ft
A_p = area of plunger, in²

Example: h_s = 4 ft (lift)
h_f = 0.146 ft
A_s = 113 in²
L = 2.5 ft
R = 0.41 ft
l = 10 ft
L/R = 6:1
A_p = 38.48 in²

$$\text{Rpm} = 54.5 \sqrt{\frac{(34 - 4 - 0.146)113}{2.5 \times 0.41[10 - (10/6)]38.48}} = 177$$

CONSTRUCTION

The following is a description of the fluid and power-end components of a power pump.

Liquid End The liquid end consists of the cylinder, plunger or piston, valves, stuffing box, manifolds, and cylinder head.

Cylinder (Working Barrel). The cylinder is the body where the pressure is developed. It is continuously under fatigue. Cylinders on many horizontal pumps have the suction and discharge manifolds made integral with the cylinder. Vertical pumps usually have separate manifolds.

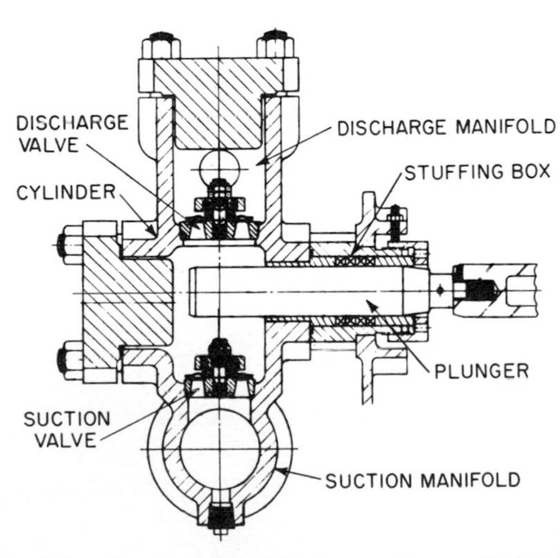

Fig. 9 Liquid end, horizontal pump. (Ingersoll-Rand Co.)

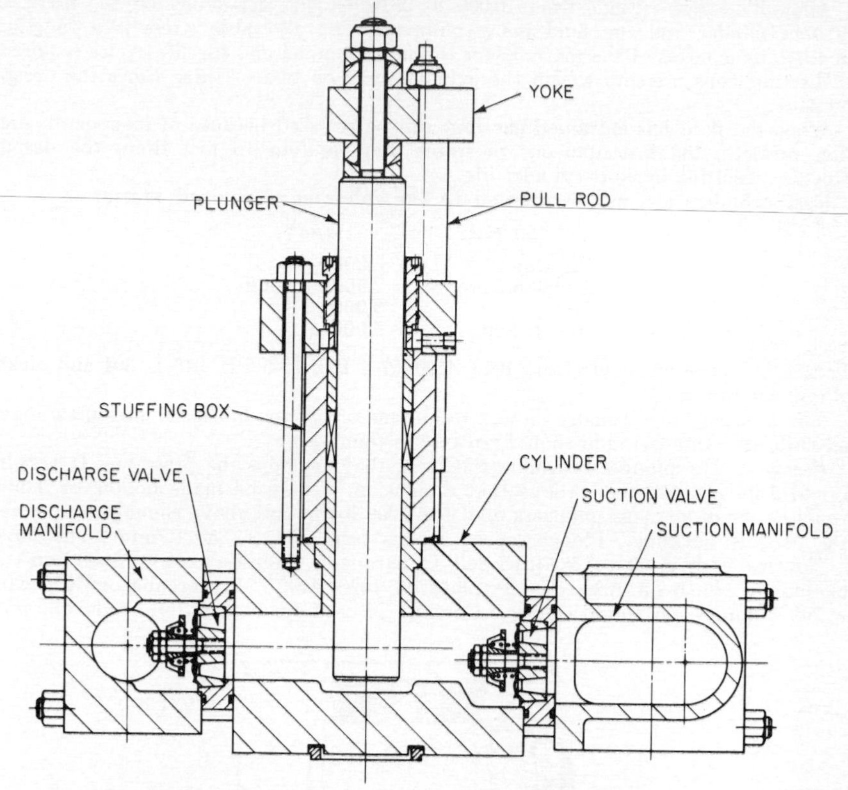

Fig. 10 Liquid end, vertical pump. (Ingersoll-Rand Co.)

When a cylinder contains the passages for more than one plunger, it is referred to as a single cylinder. When the cylinder is used for one plunger, it is called an individual cylinder. Individual cylinders are used where developed stresses are high. The forging may have a 4:1 to 6:1 forging reduction on all sides of the cylinder to obtain a homogeneous forging. Stress at the cylinder's intersecting bore can be based on a double hoop stress.

In Figs. 11 and 12 b and d are the internal diameters, a and c are the diameters of the nearest obstruction of the solid material of the cylinder, P is the developed pressure, and S is the resulting stress.

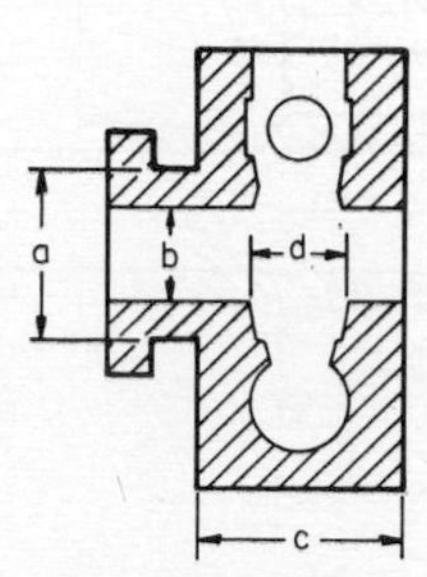

Fig. 11 Stress dimensions, horizontal liquid end. (Ingersoll-Rand Co.)

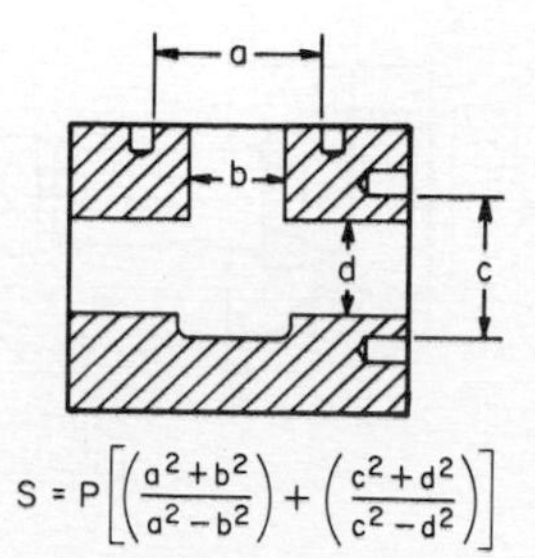

$$S = P\left[\left(\frac{a^2+b^2}{a^2-b^2}\right)+\left(\frac{c^2+d^2}{c^2-d^2}\right)\right]$$

Fig. 12 Stress dimensions, vertical liquid end. (Ingersoll-Rand Co.)

Allowable stresses range from 10,000 to 25,000 lb/in², depending on the material of the cylinder and the fluid being pumped. The allowable stress is a function of the fatigue stress of the material for the fluid pumped and the life cycles required.

Instantaneous pressure within the cylinder may be two to three times the design pressure.

When the fluid has entrained gas that can be released because of inadequate suction pressure, the instantaneous pressures may be four to five times the design pressure, resulting in short cylinder life.

Cast cylinders are usually limited to the following developed pressures:

Material	*lb/in²*
Cast iron	2,000
Aluminum bronze	2,500 to 3,000
Steel	3,000
Ductile iron	3,000

Forged cylinders are made from 1020, 4140, 17-4 PH, 15-5 PH, 316 L, 304 and nickel aluminum bronze.

The internal bores usually have a minimum of 63 rms finish for pressures above 3,000 lb/in². Internal radii should not be less than ¼ in.

Plunger. The plunger transmits the force that develops the pressure. It is solid up to 5 in. in diameter. Above that dimension, it may be made hollow to reduce weight. Small-diameter plungers used for 6,000 lb/in² and above should be reviewed for possible buckling. Plunger speed ranges from 150 to 350 ft/min. The finish is 16 rms with a 30 to 58 Rockwell C hardness. Materials of construction are Colmonoy No. 6 on 1020, chrome plate on 1020, 440C, 316, ceramic on 1020 with a 200°F limit, and solid ceramic. Ceramic is used for soft water, crude oil, mild

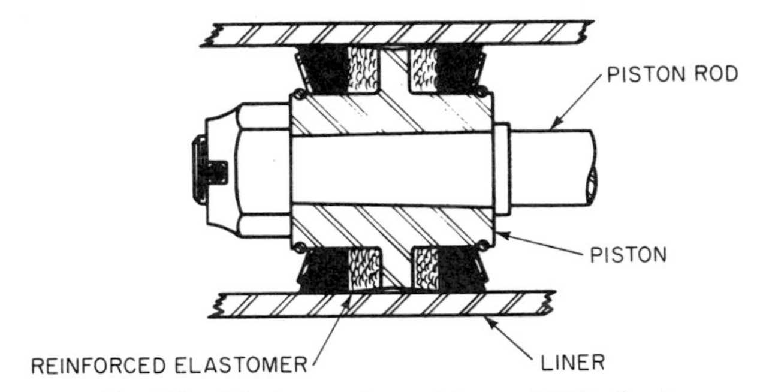

Fig. 13 Elastomer face piston. (FWI, Inc.)

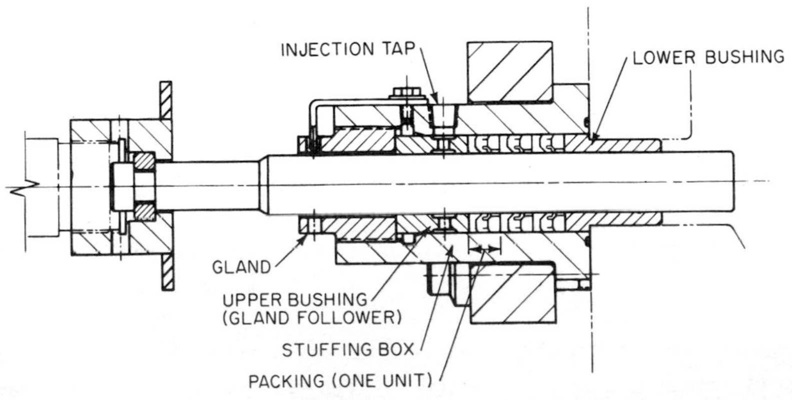

Fig. 14 Stuffing box. (Ingersoll-Rand Co.)

acids, and mild alkalies. The problem with coatings on plungers is that at high pressures the liquid gets through the pores and underneath the coating. The pressure underneath the coating will cause it to flake off the plunger.

Pistons. Pistons are used for water pressures up to 1,000 lb/in^2; for higher pressures a plunger is usually used. Pistons are cast iron, bronze, or steel with reinforced elastomer faces. They are most frequently used on duplex double-acting pumps. The latest trend is to use single-acting pistons on triplex pumps.

Stuffing Box. The stuffing box of a pump consists of the box, lower and upper bushing, packing, and gland. It should be removable for maintenance.

The stuffing box bore is machined to a 63 rms finish to ensure packing sealing and life. A single hoop stress is used to determine the stuffing thickness, with an allowance of 10,000 to 20,000 lb/in^2.

The bushings have a 63 rms finish with approximately 0.001 to 0.002 in diametrical clearance per inch of plunger diameter. The lower bushing is sometimes secured in an axial position to prevent the working of the packing. Bushings are made of bearing bronze, Ni-resist, or 316.

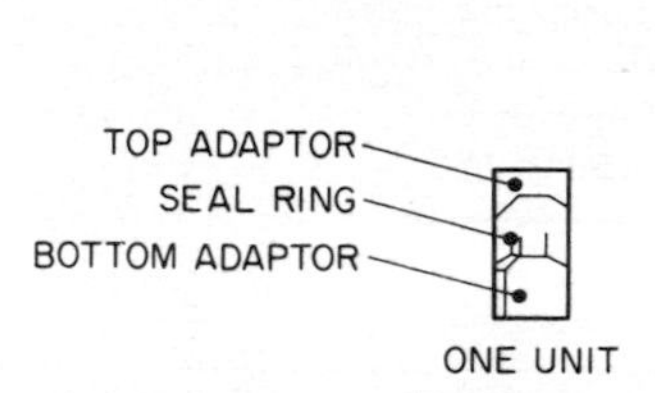

Fig. 15 Chevron packing. (Ingersoll-Rand Co.)

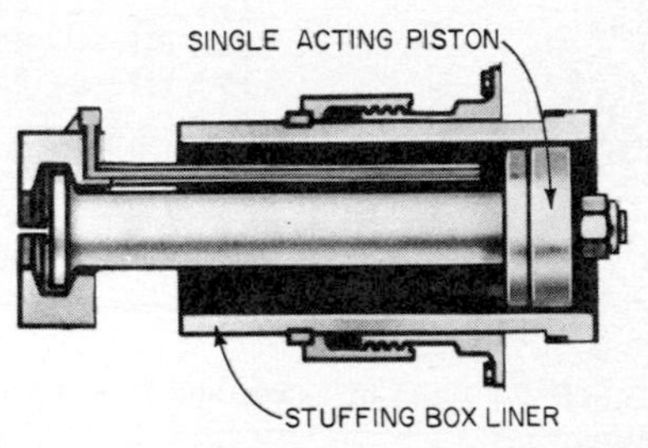

Fig. 16 Single-acting piston stuffing box. (Continental-Emsco.)

Packing is V- or chevron-shaped. Some use metal backup adapters. Unit packing consists of top and bottom adapter with a seal ring. A stuffing box can use three to five rings of packing or units, depending on the pressure and fluid. Packing or seal ring is reinforced asbestos, Teflon, duck, or neoprene.

Packing can be made self-adjusting by installing a spring between the bottom of the packing and the lower bushing or bottom of the cylinder. This arrangement eliminates overtightening and allows for uniform break-in of the packing. The packing is lubricated by injecting grease through a fitting, by gravity oil feed, or by an auxiliary lubricator driven through a take-off on the crankshaft.

For chemical or slurry service a lower injection ring is used for flushing. This prevents concentrated pumped fluid from impinging directly on the packing. This injection can be a continuous flush or can be synchronized to inject only on the suction stroke. Flush glands are employed where toxic vapor or flashing appears after the packing.

The stuffing box of a double-acting piston does not require an upper and lower bushing because the piston is guided by the cylinder liner. Single-acting pistons do not employ a stuffing box. Leakage of the piston goes into the frame extension to mix with the stuffing box continuous-circulating lubricant.

Cylinder Liner. See Figs. 13 and 16. This wear liner is usually of Ni-resist material. Its length is slightly longer than the stroke of the pump, to allow for an assembly entrance taper of the piston into the liner. On double-acting pumps the linear has packing to prevent leakage from the high pressure to the low side of the cylinder. Because of the brittleness of the liner, the construction should be such that the liner is not mechanically in compression. The finish of the liner is 16 rms.

Valves. There are many types of valves. Their use depends on the application. The main parts of the valve are the seat and the plate. The plate movement is controlled by a spring or retainer. The seat usually uses a taper where it fits into the cylinder or manifold. The taper not only gives a positive fit, but also permits easy replacement of the seat.

Some pumps use the same size suction and discharge valves for interchangeability (Fig. 10). Some use longer suction valves than discharge valves for NPSHR reasons. Others have larger discharge valves than suction valves because the discharge valve is on top of the suction valve (Fig. 9).

Because of space considerations, valves are sometimes used in clusters on each side of the plunger. This is done to obtain the required total valve area.

Table 4 shows the hardness of the seat and of the plate for some valve materials.

TABLE 4 Recommended Material Hardness for Valve Plate and Seat

	Rockwell "C"	
Material	Plate	Seat
329	30 to 35	38 to 43
440	44 to 48	52 to 56
17-4 PH	35 to 40	40 to 45
15-5 PH	35 to 40	40 to 45
	Brinell	
316	150 to 180	150 to 180

TABLE 5 Types of Valves and Their Applications

	TYPE	SKETCH	PRESSURE	APPLICATION
FIG. 17	PLATE	A = SEAT AREA B = SPILL AREA A B	5,000	CLEAN FLUID. PLATE IS METAL OR PLASTIC
FIG. 18	WING	A B	10,000	CLEAN FLUIDS. CHEMICALS
FIG. 19	BALL	A B	30,000	FLUIDS WITH PARTICLES. CLEAR, CLEAN FLUID AT HIGH PRESSURE. BALL IS CHROME PLATED
FIG. 20	PLUG	A B	6,000	CHEMICALS
FIG. 21	SLURRY	INSERT A B	2,500	MUD, SLURRY. POT DIMENSIONS TO API-12. POLYURETHANE OR BUNA-N INSERT

(Courtesy Ingersoll-Rand Co.)

316 seats and plates are chrome-plated or Colmonoy No. 6 plated to give them surface hardness.

Seats and plates have a 32 rms finish. Table 5 shows various types of valves and their applications. Spill velocity through the valves is:

$$\text{Velocity (ft/s)} = \frac{\text{gpm through valve} \times 0.642}{\text{spill area of the valve (in}^2)}$$

The quantity 0.642 is used because all the liquid passes through the valve in half the stroke.

Valve	*ft/s*
Clean-liquid suction valve	3–8
Clean-liquid discharge valve	6–20
Slurry suction and discharge valve	6–12

Manifolds. These are the chambers where liquid is dispersed or collected for distribution before or after passing through the cylinder. On horizontal pumps the suction and discharge manifold is usually made integral with the cylinder (Fig. 22). Some horizontal pumps and some vertical pumps have only the discharge manifold integral with the cylinder (Fig. 23). Most vertical pumps have the suction and discharge manifold separate from the cylinder (Fig. 10).

Suction manifolds are designed to eliminate air pockets from the flange to the valve entrance (Fig. 24). Separate suction manifolds are cast iron or fabricated steel. Discharge manifolds are steel forgings or fabricated steel. The manifolds have a minimum deflection to prevent gasket shift when subjected to the plunger load (Fig. 24).

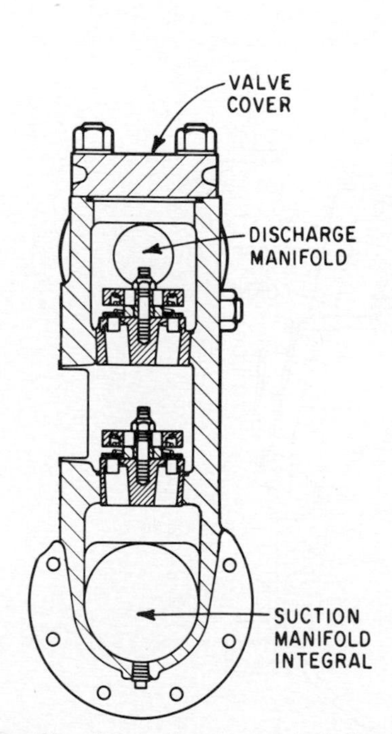

Fig. 22 Integral suction and discharge manifold. (Gardner-Denver Co.)

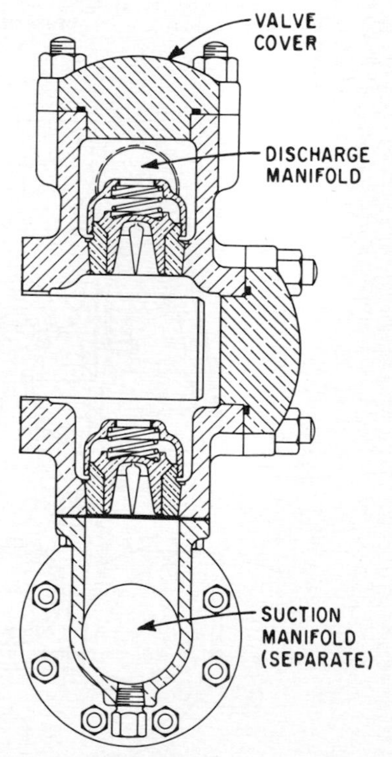

Fig. 23 Separate suction with integral discharge manifold. (Gardner-Denver Co.)

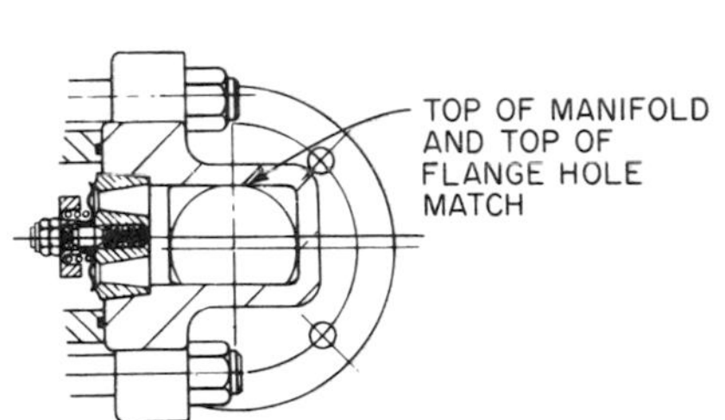

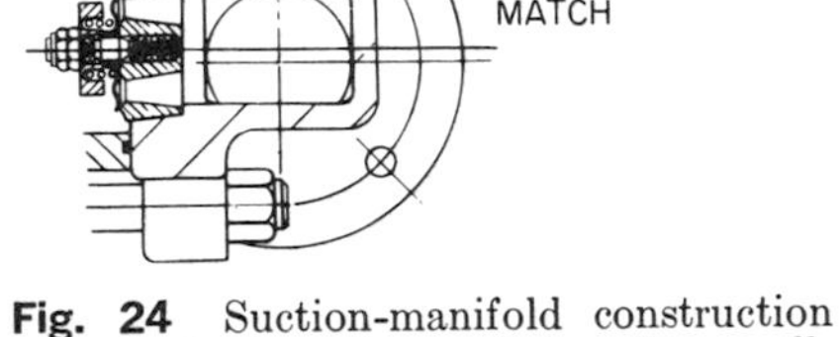

Fig. 24 Suction-manifold construction to eliminate air pockets. (Ingersoll-Rand Co.)

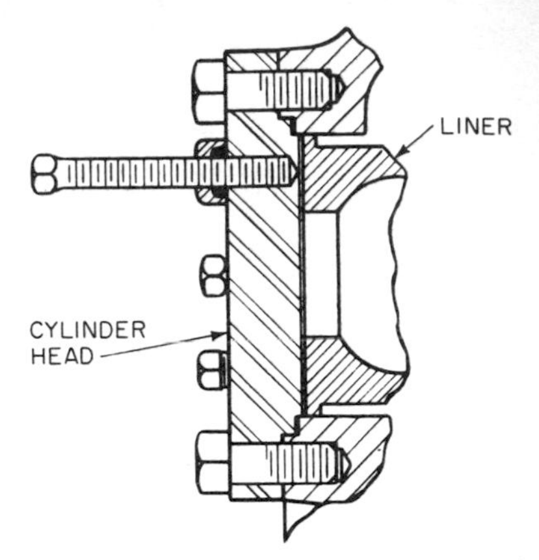

Fig. 25 Plunger cover. (FWI, Inc.)

The velocity through the manifolds of a clean liquid is 3 to 5 ft/s at the suction and 6 to 16 ft/s at the discharge. Suction and discharge manifold velocities on slurry service are 6 to 10 ft/s. Slurry service has a minimum of 6 ft/s to prevent the fallout of slurries.

$$\text{Velocity (ft/s)} = \frac{\text{gpm of pump} \times 0.321}{\text{cross-sectional area of the manifold, in}^2}$$

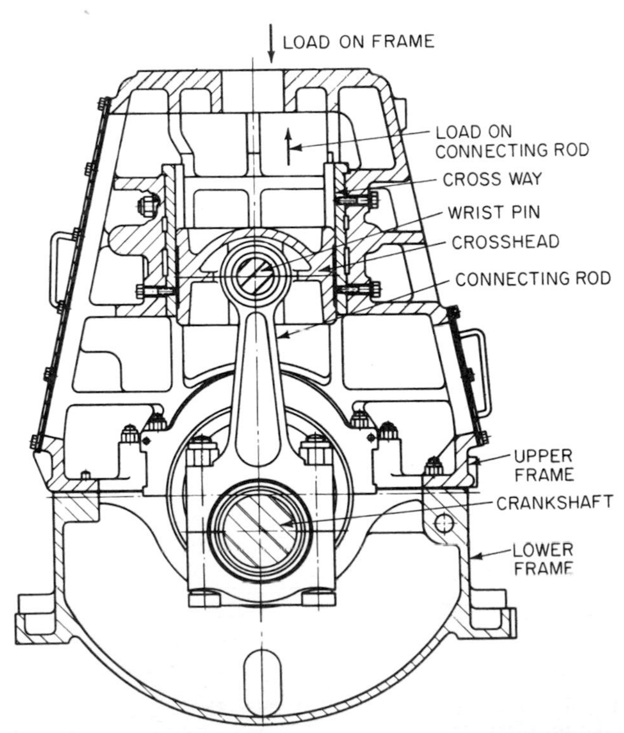

Fig. 26 Power end, vertical pump. (Ingersoll-Rand Co.)

Water hammer creates an additional pressure which is added to the rated pump pressure, as does hydraulic shock loading. The discharge manifold rating is then made equal to or greater than the sum of these pressures.

Valve Covers. Valve covers are used to provide accessibility to the valves without disturbing the cylinder or manifolds (Figs. 22 and 23).

Plunger Covers and/or Cylinder Heads. These are used on horizontal pumps to provide accessibility to the plunger, piston, and cylinder liner (Figs. 22, 23, and 25).

Power End The power end contains the crankshaft, connecting rod, crosshead, pony rod, bearings, and frame (Figs. 26 and 27). Basic designs are horizontal and vertical with sleeve or antifriction bearings.

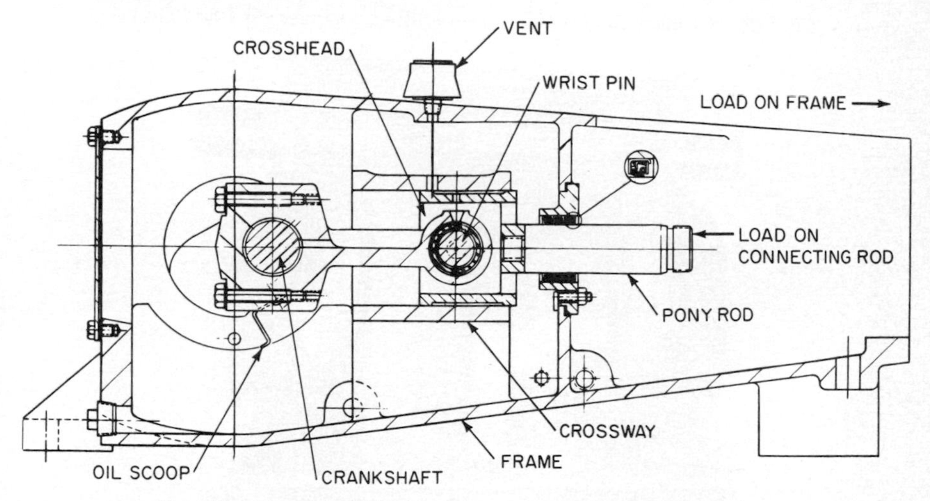

Fig. 27 Power end, horizontal pump. (Ingersoll-Rand Co.)

Frame. The frame absorbs the plunger load and torque. On vertical pumps with an outboard stuffing box (Fig. 10), the frame is in compression (Fig. 26). With horizontal single-acting pumps, the frame is in tension (Fig. 27). Frames are usually close-grain cast iron. Slurry-pump frames, designed for mobile service at various sites, are usually fabricated steel. The frame is vented to the atmosphere.

When the working atmosphere is detrimental to the working parts in the frame,

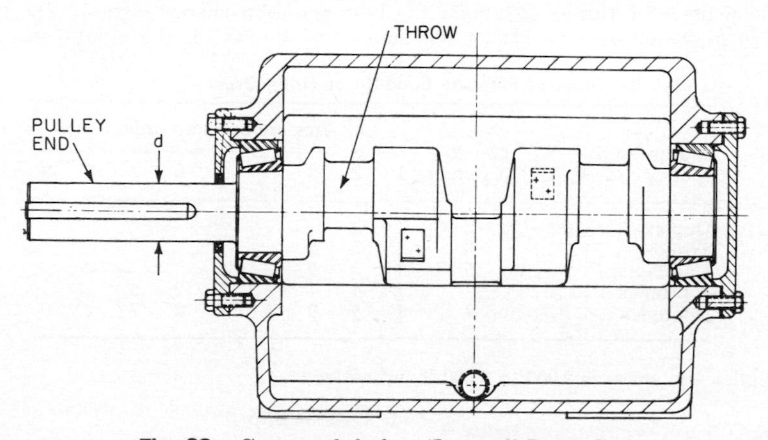

Fig. 28 Cast crankshaft. (Ingersoll-Rand Co.)

such as ammonia attack on bronze bearings, the frame may be purged continuously with nitrogen.

Crankshaft. See Figs. 28, 29, and 30. The crankshaft varies in construction depending on the design and horsepower. In horizontal pumps, the crankshafts are

Fig. 29 Crankshaft machined from a billet. (Ingersoll-Rand Co.)

Fig. 30 Cast crankshaft with integral gear. (Continental-Emsco.)

usually of nodular iron or cast steel. Vertical pumps use forged steel or machined billet crankshafts. Because the crankshafts have relatively low speeds and mass, counterweights are not used. Except for duplex pumps, the crankshaft usually has an odd number of throws to obtain the best pulsation characteristics. The firing order in a revolution depends on the number of throws on the crankshaft.

TABLE 6 Order of Pressure Build-Up or Firing Order

Throw from pulley end	No. of plungers or pistons	Pressure build-up order								
		1	2	3	4	5	6	7	8	9
Duplex	2	1	2							
Triplex	3	1	3	2						
Quintaplex.	5	1	3	5	2	4				
Septuplex	7	1	4	7	3	6	2	5		
Nonuplex	9	1	5	9	4	8	3	7	2	6

The pulley-end stress is 2,000 to 3,000 lb/in² where

$$\text{Stress} = \frac{3.21 \times 10^5 \times \text{bhp}}{d^3 \times n}$$

and d = pulley-end diameter, in and n = rpm. Bearing surfaces are ground to 16 rms finish.

Many crankshaft designs incorporate rifle drilling between throws to furnish lubrication to the crankpin bearings.

Connecting Rods and Eccentric Straps. See Figs. 31 and 32. The connecting

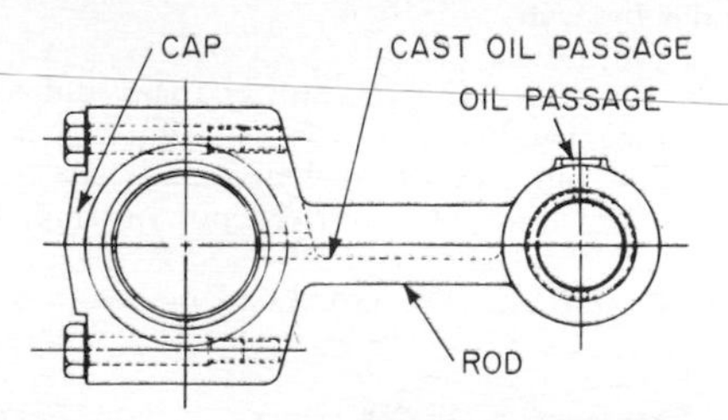

Fig. 31 Connecting rod. (Ingersoll-Rand Co.)

Fig. 32 Eccentric strap. (Gardner-Denver Co.)

rods transfer the rotating force of the crankpin to an oscillating force on the wristpin. Connecting rods are split perpendicular to their centerline at the crankpin end for assembly of the rod onto the crankshaft.

The cap and rod are aligned with a close-tolerance bushing or body-bound bolts. The rods are either rifle-drilled or have cast passages for transferring oil from the wristpin to the crankpin. A connecting rod with a tension load is made of forged steel, cast steel, or fabricated steel. Rods with a compression loading are cast nodular steel or aluminum alloy. Connecting-rod finish, where the bearings are mounted, is 32 to 63 rms.

The ratio of the distance between the centerlines of the wristpin and of the crankpin bearing to half the length of the stroke is referred to as L/R. This ratio directly affects the pressure pulsations, volumetric efficiency, size of pulsation dampener, speed of liquid separation, acceleration head, moment of inertia forces, and size of the frame. Low L/R results in high pulsations. A high L/R reduces pulsations, but may result in a large and uneconomical power frame. The common industrial L/R range is 4:1 to 6:1.

The eccentric strap has the same function as a connecting rod, except that it usually is not split. The eccentric strap is furnished with antifriction bearings whereas connecting rods are furnished with sleeve bearings. Eccentric straps are applied to mud and slurry pumps, which are started up against full load without requiring a bypass line.

Wrist Pin. Located in the crosshead, the wrist pin transforms the oscillating motion of the connecting rod to reciprocating motion.

The maximum stress in the wrist pin from deflection should not exceed 10,000 lb/in².

$$S = \frac{PL \times l}{8 \times 0.098d^3}$$

where PL = plunger load, lb
l = length under load, in
d = diameter of pin, in

Large pins are hollow to reduce the oscillating mass and assembly weight. Depending on the design, pins can have a tight straight fit, a taper fit, or a loose fit in the cross head. The pin is case-hardened and has a 16-rms finish. When needle or roller bearings are used for the wrist-pin bearing, the wrist pin is used as the inner race.

Crosshead. See Figs. 33 and 34. The crosshead moves in a reciprocating motion

Fig. 33 Crosshead. (Gardner-Denver Co.)

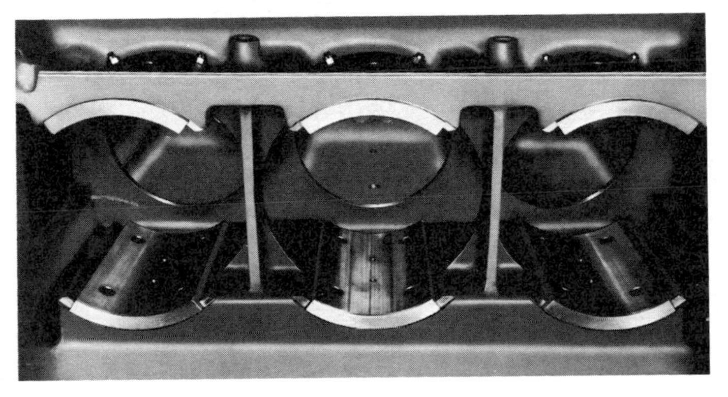

Fig. 34 Crossways. (Gardner-Denver Co.)

and transfers the plunger load to the wrist pin. The crosshead is designed to absorb the side or radial load from the plunger as it moves linearly on the crossway. The side load is approximately 25 percent of the plunger load. For cast iron crossheads the allowable bearing load is 80 to 125 lb/in². Crossheads are grooved for oil lubrication with a bearing surface of 63 rms. Crossheads are piston-type (full round) or partial-contact. The piston type should be open-end or vented to prevent air compression at the end of the stroke.

On vertical pumps the pull rods go through the crosshead so that the crosshead is under compression when the load is applied (Fig. 26).

Crossways (Crosshead Guide). The crossway is the surface on which the crosshead reciprocates. On horizontal pumps it is cast integral with the frame (Fig. 27). On large frames it is usually replaceable and is shimmed to effect proper running clearance (Figs. 26 and 34). The crossway finish is 63 rms.

Pony Rod (Intermediate Rod). The pony rod is an extension on the crosshead on the horizontal pumps (Figs. 33 and 34). It is screwed or bolted to the crosshead and extends through the frame (Fig. 27). A seal on the frame and against the pony rod prevents oil from leaking out of the frame. A baffle is fixed onto the rod to deflect leakage from coming in contact with the frame (Fig. 27).

Pull Rod (Tie Rod). On vertical pumps two pull rods go through the crosshead. The rods are secured by a shoulder and nut so that the cast iron crosshead is in compression when the load is applied (Fig. 26). The rods extend out of the top of the frame and fasten to a yoke. The plunger is attached to the middle of the yoke with an aligning feature for the plunger (Fig. 10).

Bearings Both sleeve and antifriction bearings are used in power pumps. Some frames use all sleeve, others use all antifriction, and others use a combination of both.

Sleeve Bearings. When properly installed and lubricated, sleeve bearings are considered to have infinite life. The sleeve bearing is designed to operate within a certain speed range, and too fast or too slow a speed will upset the film lubrication. Sleeve bearings cannot be operated satisfactorily below 40 rpm with the plunger fully loaded, using standard lubrication. Below this speed, film lubrication is inadequate. The finish on the sleeve bearing is 16 rms. Clearances are approximately 0.001 in/in of diameter of the bearing.

Wrist-Pin Bearings. These bearings have only oscillating motion. On single-acting pumps with low suction pressure, there is adequate reverse loading on the bearing to permit replenishment of the oil film. High suction pressure on horizontal pumps increases the reverse loading. On vertical pumps, high suction pressure can produce a condition of no reversal loading, and in this case higher oil pressure is required. Allowable projected area loading is 1,200 to 1,500 lb/in^2, with bearing bronze.

Crankpin Bearings. The crankpin bearing is a rotating split bearing and has a better oil film than the wrist-pin bearing. It is clamped between the connecting rod and cap. The bearing is bronze-backed babbitt metal or a trimetal automotive bearing. Allowable projected area load is 1,200 to 1,600 lb/in^2.

Main Bearing. The main bearing absorbs the plunger load and gear load. The total plunger load varies during the revolution of the crankshaft. The triplex main bearings receive the greatest variations in loading because the crankshaft has the greatest relative span between bearings. Sleeve bearings are flanged to lock them in an axial position, and the flange absorbs residual axial thrust. On large-stroke vertical pumps, there is a main bearing between every connecting rod. Split bearings are bronze-backed babbitt metal with an allowable projected area load of 500 lb/in^2.

Antifriction Bearings A pump with antifriction bearings can be started under full plunger load without a bypass line. Antifriction bearings allow the pump to operate continuously at low speeds with full plunger load. They are selected for a 30,000- to 50,000-hr B-10 bearing life. Slurry pipeline applications may require 100,000 hr. Eccentric straps are used in place of connecting rods.

Wrist-Pin Bearings. These are needle or roller bearings. The outer race is a tight fit into the strap, and the wrist pin is used as the inner race. This reduces the size of the bearing and strap. The wrist pin is held in the crosshead with a taper or keeper plate (Fig. 36).

Crankpin Bearings. These are roller bearings. The outer race is mounted separately from the inner race; the inner race is mounted on the crankshaft and secured axially by a shoulder on the shaft and by a keeper plate, or by keeper plates on both sides of the race. The outer race is assembled in the strap in the same way as the inner race mounting. The strap is then slipped over the shaft and assembled to the inner race (Fig. 36).

Main Bearings—Antifriction. The main bearing used in conjunction with the sleeve wrist bearing and the sleeve crankpin bearings is usually a tapered roller bearing, as in the case of a horizontal triplex plunger pump. This main bearing is ordinarily mounted directly into the frame (Fig. 35).

The main bearing used with full antifriction bearing design is a self-aligning spherical roller bearing. This bearing compensates for axial and radial movement of the crankshaft. These bearings are usually mounted in bearing holders, which in turn are mounted on the frame. This type of design is used on mud and slurry pumps. (Fig. 36).

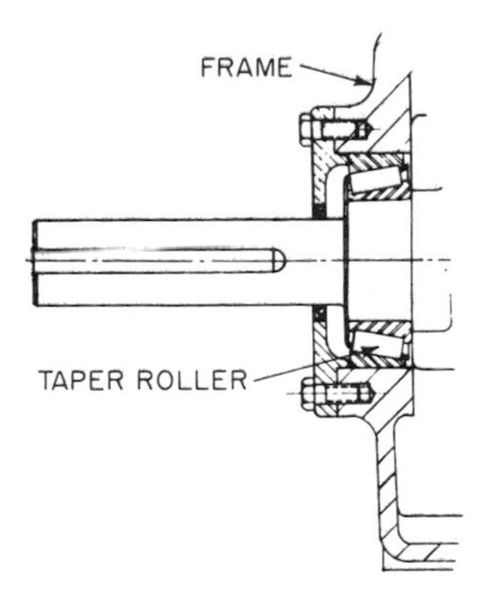

Fig. 35 Main bearing. (Ingersoll-Rand Co.)

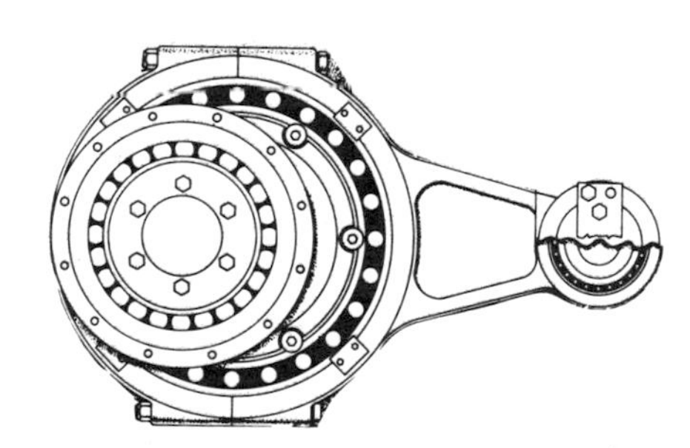

Fig. 36 Main bearing. (Gardner-Denver Co.)

Lubrication SAE 30 to 40 oil is used for bearing lubrication. With splash lubrication, the cheeks of the crankshaft or oil scoops (Fig. 27) throw oil by centrifugal force against the frame wall. It is then distributed by gravity to the crosshead, wrist pin, and crankpin. At slow speeds there is not enough centrifugal force for proper oil distribution, so that partial force feed is then employed to get proper lubrication. Force-feed lubrication requires ½ to 1 gpm per bearing with an oil pressure of 25 to 40 lb/in².

Applications Some of the applications for power pumps are:

Ammonia service
Carbamate service
Chemicals
Crude-oil pipeline
Cryogenic service
Fertilizer plants
High-pressure water cutting
Hydro forming
Hydrostatic testing
Liquid petroleum gas
Liquid pipeline
Power oil
Power press
Soft-water injection for water flood
Slurry pipeline (70 percent by weight)
Slush ash service
Steel-mill descaling
Water-blast service

The following applications should be reviewed with the pump supplier:

Cryogenic service
Highly compressible liquids
Liquids over 250°F
Liquids with high percentage of entrained gas
Low-speed operation
Slurry pipeline
Special fluid end materials
Viscosity over 250 SSU

Duty Service. In making an economical pump selection, the type of full-load service should be specified:

Continuous: 8 to 24 hr/day
Light: 3 to 8 hr/day
Intermittent: up to 3 hr/day
Cyclic: 30 s loaded out of every 3 min

Relief Valve. Relief valves should be set at the following percentage above operating pressure:

Duplex double acting: 25 percent
Triplex and above: 10 percent

Bypass Relief. Starting a power pump under full load requires a high starting torque. Also, a pump with sleeve-bearing construction when starting under full load may not have adequate bearing lubrication to reach operating speed.

The starting torque on the driver and plunger load on the bearings may be reduced by:

1. Installing a bypass line from the discharge line back to the suction or to a drain (Fig. 37). This line is located between the pump discharge and a check valve before the discharge piping system. The bypass valve is operated manually or automatically. The bypass line reduces starting torque from mechanical losses and fluid inertia of the suction and bypass systems. After the pump is up to speed, the bypass is closed and the pump goes on stream.

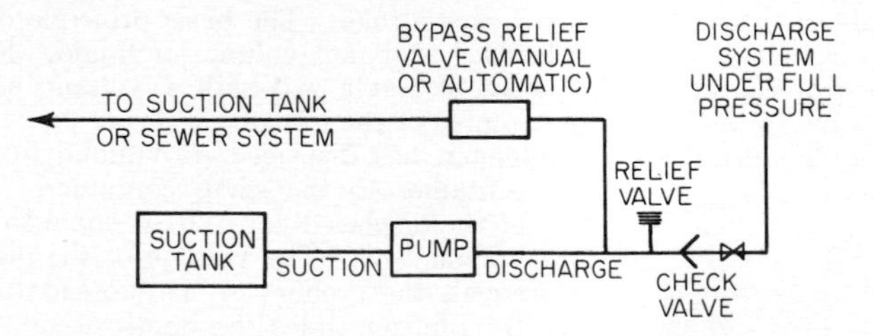

Fig. 37 Bypass relief system. (Ingersoll-Rand Co.)

2. Use of suction-valve unloaders. The unloader is a mechanism that lifts the suction-valve plates off the seats before and during start-up. This stops pumping action by the pump. There are only mechanical losses to be overcome. After the pump is up to speed, the suction valves are closed in unison. On large-flow pumps a distributor is used to close the suction valves in synchronism with the plunger-pressure build-up order.

Pulsation Dampeners. Dampeners are used to reduce suction and discharge pulsations. They are especially useful on pumps with high pressure pulsations, as in duplex and triplex pumps. A properly sized and located suction dampener can reduce the system pulsations to an equivalent pipe length of 5 to 15 pipe diameters. The dampener should be located on the same side of the manifold as the pipe, not at the dead end of the manifold.

An effective suction dampener is a vertical air chamber made from pipe with a connection at the top for recharging. The inside diameter of the chamber should be as close as possible to the inside diameter of the pipe. The height is approximately 8 to 10 pipe diameters, with a minimum of 2 ft. The air cushion should be 1 ft in height.

Dampeners on the discharge are mostly nitrogen-charged bladder bottles. Charge pressure is approximately 66 percent of the discharge system pressure.

Section **3.2**

Steam Pumps

ROBERT M. FREEBOROUGH

INTRODUCTION AND BASIC THEORY

A reciprocating positive-displacement pump is one in which a plunger or piston displaces a given volume of fluid for each stroke. The basic principle of a reciprocating pump is that a solid will displace an equal volume of liquid. For example, an ice cube dropped into a full glass of water will spill a volume of water out of the glass equal to the submerged volume of the ice cube.

In Fig. 1 a cylindrical solid, a plunger, has displaced its volume from the large container to the small container. The volume of the displaced fluid (B) is equal to the plunger volume (A). The volume of the displaced fluid equals the product of the cross-sectional area of the plunger times the depth of submergence.

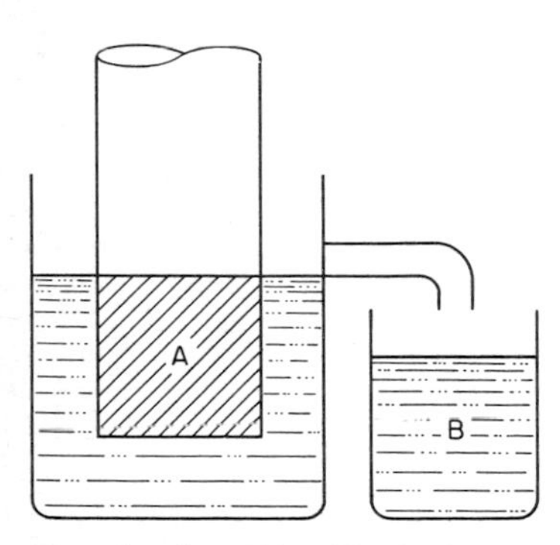

Fig. 1 A solid will displace a volume of liquid equal to its own volume.

All reciprocating pumps have a fluid-handling portion, commonly call the *liquid end,* which has:

1. A displacing solid called a *plunger* or *piston.*
2. A container to hold the liquid called the *liquid cylinder.*
3. A suction check valve to admit fluid from the suction pipe into the liquid cylinder.
4. A discharge check valve to admit flow from the liquid cylinder into the discharge pipe.
5. Packing to seal tightly the joint between the plunger and the liquid cylinder to prevent the liquid from leaking out of the cylinder and air from leaking into the cylinder.

These basic components are identified on the rudimentary liquid cylinder illustrated in Fig. 2. To pump, i.e., to move the liquid through the liquid end, the plunger must be moved. When the plunger is moved out of the liquid cylinder as shown in Fig. 2, the pressure of the fluid within the cylinder is reduced. When the pressure becomes less than that in the suction pipe, the suction check valve opens and liquid flows into the cylinder to fill the volume being vacated by withdrawal of the plunger. During this phase of operation, the discharge check valve is held closed by the higher pressure in the discharge pipe. This portion of the pumping action of a reciprocating positive-displacement pump is called the *suction stroke.*

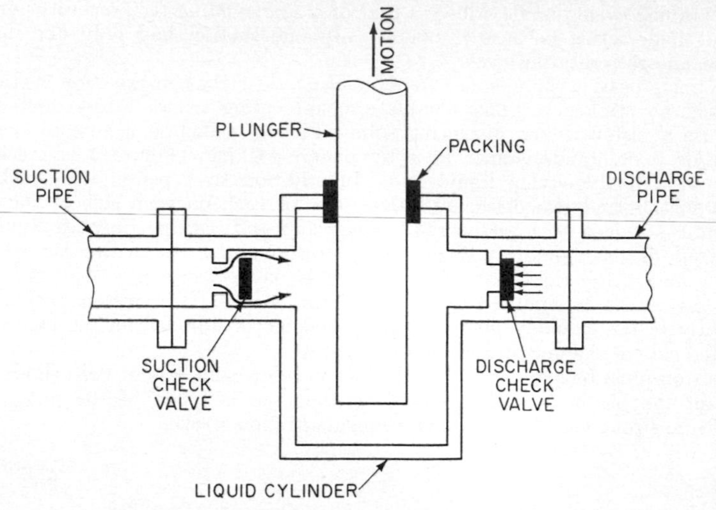

Fig. 2 Schematic of a reciprocating-pump liquid end during the suction stroke.

The withdrawal movement must be stopped before the end of the plunger gets to the packing. The plunger movement is then reversed and the *discharge stroke* portion of the pumping action is started as illustrated in Fig. 3.

Movement of the plunger into the cylinder causes an increase in pressure of the liquid contained therein. This pressure immediately becomes higher than suction-pipe pressure and causes the suction check valve to close. With further plunger movement, the liquid pressure continues to rise. When the liquid pressure in the cylinder reaches that in the discharge pipe, the discharge check valve is forced open and liquid flows into the discharge pipe. The volume forced into the discharge pipe is equal to the plunger displacement less very small losses. The plunger displacement is the product of its cross-sectional area times the length of stroke. The plunger must be stopped before it hits the bottom of the cylinder. The motion is then reversed, and the plunger again goes on suction stroke as previously described.

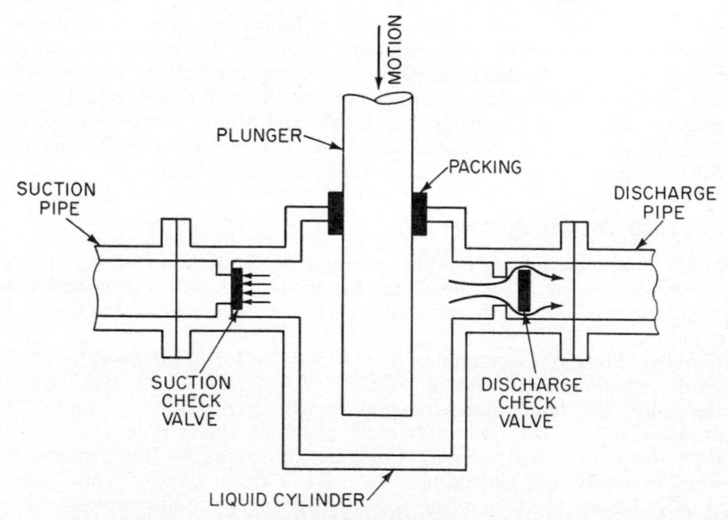

Fig. 3 Schematic of a reciprocating-pump liquid end during discharge stroke.

The pumping cycle just described is that of a *single-acting* reciprocating pump. It is called single-acting because it makes only one suction and only one discharge stroke in one reciprocating cycle.

Many reciprocating pumps are *double-acting,* i.e., they make two suction and two discharge strokes for one complete reciprocating cycle. Most double-acting pumps use a piston as the displacing solid which is sealed to a bore in the liquid cylinder or to a liquid-cylinder liner by piston packing. Figure 4 is a schematic diagram of a double-acting liquid end. In addition to a piston with packing, it has two suction and two discharge valves, one of each on each side of the piston. The piston is moved by a piston rod. The piston-rod packing prevents liquid from leaking out of the cylinder. When the piston rod and piston are moved in the direction shown, the right side of the piston is on a *discharge stroke* and the left side of the piston is simultaneously on *suction stroke.* The piston packing must seal tightly to the cylinder liner to prevent leakage of liquid from the high-pressure right-hand side of the low-pressure left-hand side.

The piston must be stopped before it hits the right-hand side of the cylinder. The motion of the piston is then reversed, so that the left side of the piston begins its discharge stroke and the right side begins its suction stroke.

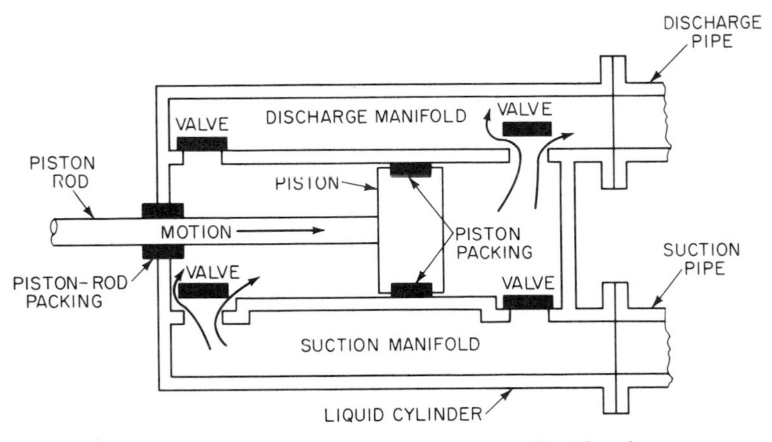

Fig. 4 Schematic of a double-acting liquid end.

A reciprocating pump is not complete with a liquid end only; it must also have a driving mechanism to provide motion and force to the plunger or piston. The two most common driving mechanisms are a reciprocating steam engine and a crank and throw device. Those pumps using the steam engine are called *direct-acting steam pumps.* Those pumps using the crank and throw device are called *power pumps.* Power pumps must be connected to an external rotating driving force such as an electric motor, steam turbine, or internal combustion engine.

DIRECT-ACTING STEAM PUMPS

Direct-acting steam pumps are mainly classified by the number of working combinations of cylinders, such as duplex (Fig. 5), which has two steam and two liquid cylinders mounted side by side, or simplex (Fig. 6), which has one steam and one liquid cylinder.

Additionally, the simplex or duplex pumps may be further defined by (1) cylinder arrangement, whether horizontal or vertical; (2) number of steam expansions in the power end; (3) liquid-end arrangement, whether piston or plunger; and (4) valve arrangement, i.e., cap and valve plate, side pot, turret type, etc.

Although this section will refer to steam as the driving medium, compressed gases such as air or natural gas can be used to drive a steam pump. These gases should have oil mist added to them prior to entering the pump to prevent wear of the steam end parts.

Fig. 5 Duplex steam pump. (Worthington Pump Corporation)

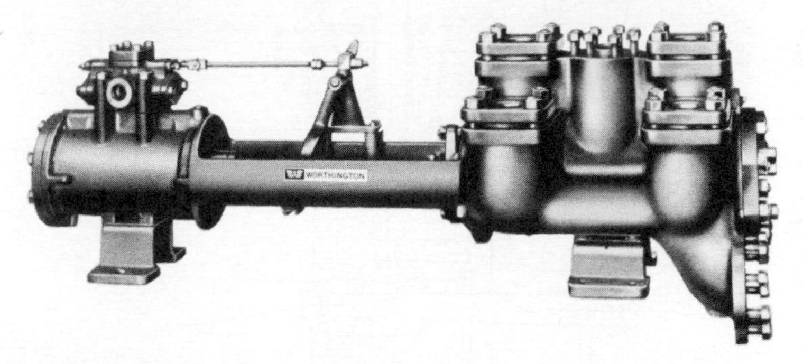

Fig. 6 Simplex steam pump. (Worthington Pump Corporation)

Steam-End Construction and Operation The driving mechanism, or *steam end,* of a direct-acting steam pump includes the following components as illustrated in Fig. 7:

1. One or more steam cylinders with suitable steam inlet and exhaust connections
2. Steam piston with rings
3. Steam-piston rods which are directly connected to the liquid-piston rods
4. Steam valves which direct the steam into and exhaust steam from the steam cylinder.
5. A steam-valve actuating mechanism which moves the steam valve in proper sequence to produce reciprocating motion.

The operation of a steam pump is quite simple. The motion of the piston is obtained by admitting steam of sufficient pressure to one side of the steam piston while simultaneously exhausting steam from the other side of the piston. There is very little expansion of the steam since it is admitted at a constant rate throughout the stroke. The moving parts, i.e., the steam piston, the liquid piston, and the piston rod or rods, are cushioned and brought to rest by exhaust steam trapped in the end of the steam cylinder at the end of each stroke. After a brief pause at the end of the stroke, steam is admitted to the opposite side of the piston and the pump strokes in the opposite direction.

Steam Valves. Since the steam valve and its actuating mechanism control the reciprocating motion, any detailed description of the construction of the direct-acting

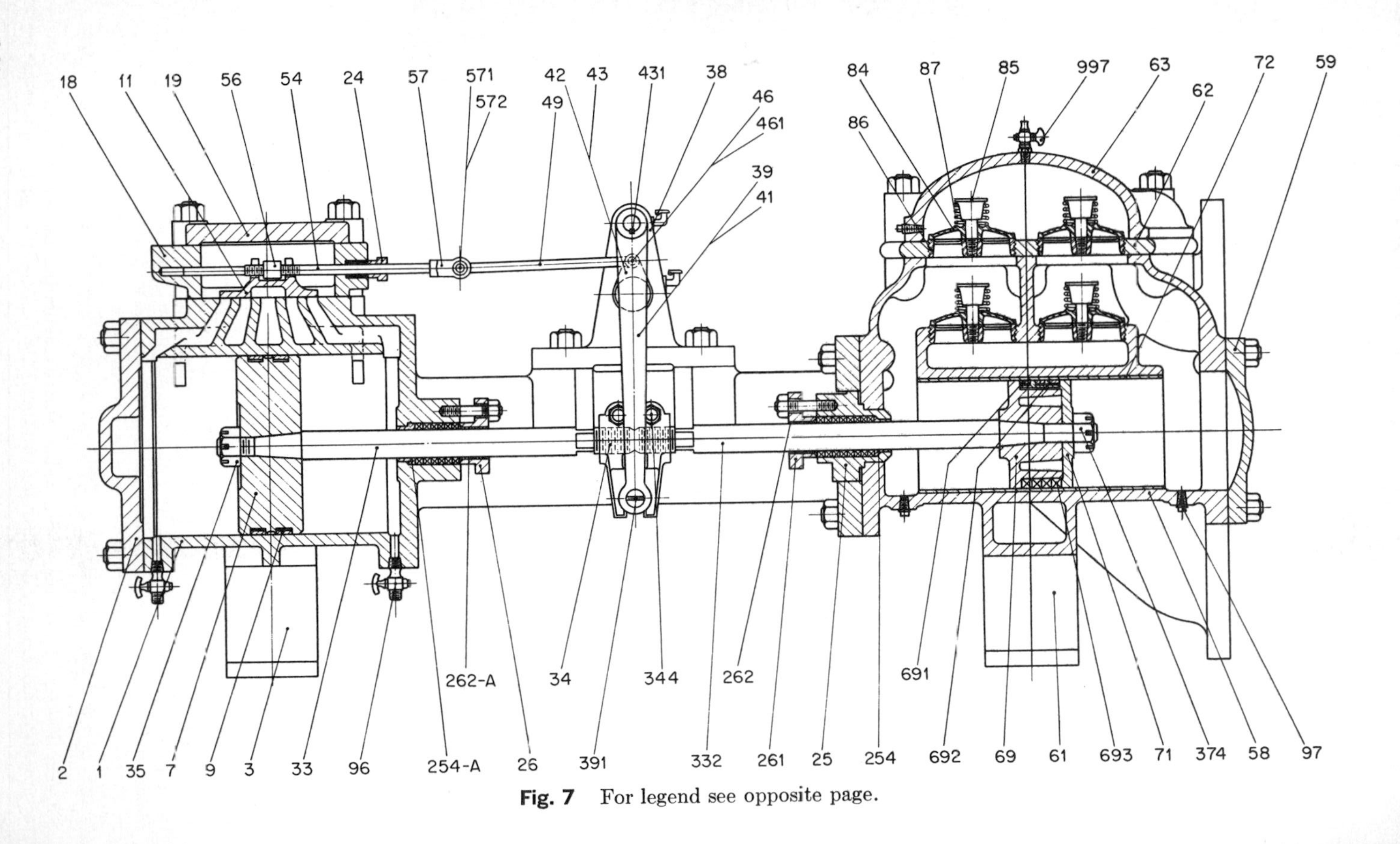

Fig. 7 For legend see opposite page.

steam pump should rightfully begin with a discussion of steam-valve types, operation, and construction.

Duplex Steam Valves. The steam valves in a duplex steam pump are less complicated than those in a simplex pump and will be described first. As previously stated, the duplex steam pump can be considered as two simplex pumps arranged side by side and combined to operate as a single unit. The piston rod of one pump in making its stroke actuates the steam valve and thereby controls the admission or exhaust of steam in the second pump. A valve gear cross-stand assembly is shown in Fig. 8. The wishbone-shaped piston-rod lever of one side is connected by a shaft to the valve-rod crank of the opposite side. The steam valve is connected to the valve-rod crank by the steam-valve rod and steam-valve link. Through this assembly, the piston rod of one side moves the steam valve of the opposite side in the same direction. When the first pump has completed its stroke, it must pause until its own steam valve is actuated by the movement of the second pump before it can make its return stroke. Since one or the other steam cylinder port is always open, there is no "dead center" condition; hence the pump is always ready to start when steam is admitted to the steam chest. The movements of both pistons are synchronized to produce a well-regulated flow of liquid free of excessive pulsations and interruptions in the flow stream.

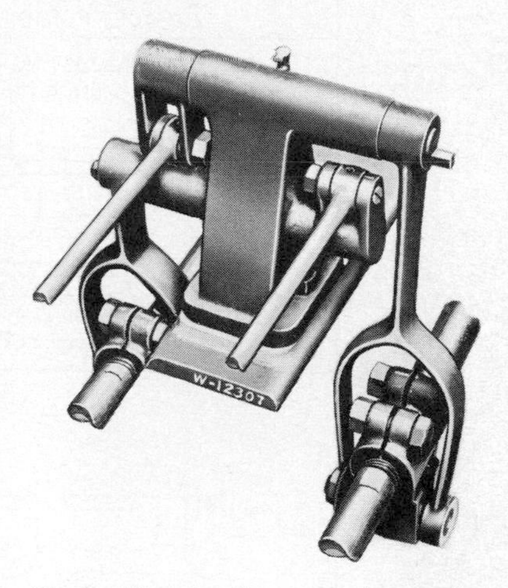

Fig. 8 Steam-valve actuating mechanism for a duplex pump. (Worthington Pump Corporation)

Flat Slide Steam Valve. Steam enters the pump from the steam pipe into the steam chest on top of the steam cylinder. Exhaust steam leaves the pump through the center port of five ports, as shown in Fig. 9. Most duplex pumps use a flat slide valve which is held against its seat by steam pressure acting upon its entire top area; this is called an *unbalanced* valve. The flat or D valve, as it is often called, is satisfactory for steam pressures up to approximately 250 lb/in^2 and has reasonable service life, particularly where steam-end lubrication is permissible. On large pumps, the force required to move an unbalanced valve is considerable, so a balanced piston valve, which will be discussed later, is used.

The slide valve shown in Fig. 9 is positioned on dead center over the five valve ports. A movement of the valve to the right uncovers the left-side steam port and the right-side exhaust port, which is connected through the slide valve to the

---

Fig. 7 Typical section of duplex steam pump. 1, steam cylinder with cradle; 2, steam-cylinder head; 3, steam-cylinder foot; 7, steam piston; 9, steam piston rings; 11, slide valve; 18, steam chest; 19, steam-chest cover; 24, valve-rod stuffing box gland; 25, piston-rod stuffing box, liquid; 26, piston-rod stuffing box gland, steam; 33, steam piston rod; 34, steam piston spool; 35, steam piston nut; 38, cross stand; 39, long lever; 41, short lever; 42, upper rock shaft, long crank; 43, lower rock shaft, short crank; 46, crankpin; 49, valve-rod link; 54, valve rod; 56, valve-rod nut; 57, valve-rod head; 58, liquid cylinder; 59, liquid-cylinder head; 61, liquid-cylinder foot; 62, valve plate; 63, force chamber; 69, liquid piston body; 71, liquid piston follower; 72, liquid cylinder lining; 84, metal valve; 85, valve guard; 86, valve seat; 87, valve spring; 96, drain valve for steam end; 97, drain plug for liquid end; 254, liquid piston-rod stuffing box bushing; 332, liquid piston rod; 344, piston-rod spool bolt; 374, liquid piston-rod nut; 391, lever pin; 431, lever key; 461, crankpin nut; 571, valve-rod head pin; 572, valve-rod head-pin nut; 691, liquid piston snap rings; 692, liquid piston bull rings; 693, liquid piston fibrous packing rings; 997, air cock; 251, liquid piston-rod stuffing box; 254A, steam piston-rod stuffing box bushing; 261, piston-rod stuffing box gland, liquid; 262, piston-rod stuffing box gland lining, liquid; 262A, piston-rod stuffing box gland lining, steam. (Worthington Pump Corporation)

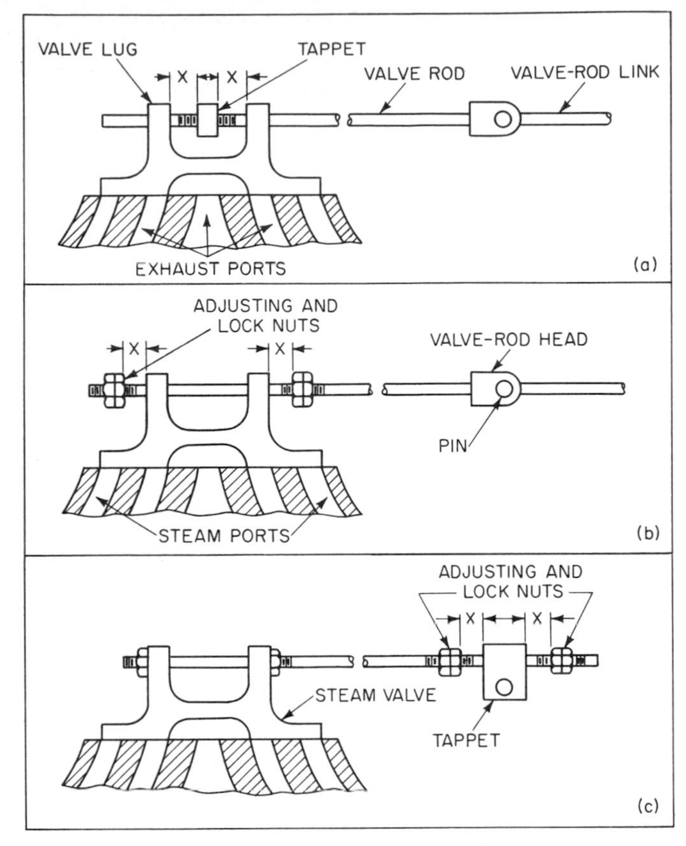

Fig. 9 Duplex-pump steam-valve lost-motion arrangements.

center exhaust port. The main steam piston would be moved from left to right by the admitted steam. Movement of the slide valve from dead center to the left would, of course, cause opposite movement of the steam piston.

The steam valves of a duplex pump are mechanically operated, and their movements are dependent upon the motion of the piston rod and the linkage of the valve gear. In order to ensure that one piston will always be in motion when the other piston is reversing at the end of its stroke, lost motion is introduced into the valve gear. Lost motion is a means by which the piston can move during a portion of its stroke without moving the steam valve. Several lost-motion arrangements are shown in Fig. 9.

If the steam valves are out of adjustment, the pump will have a tendency not to operate through its designed stroke. Increasing the lost motion lengthens the stroke; if this is excessive, the piston will strike the cylinder head. Reducing the lost motion will shorten the stroke; if this is excessive, it will cause the pump to short-stroke, with a resultant loss in capacity.

The first step in adjusting the valves is to have both steam pistons in a central position in the cylinder. To accomplish this, the piston is moved toward the steam end until the piston strikes the cylinder head. With the piston rod in this position, a mark is made on the rod flush with the steam-end stuffing box gland. Next the piston rod is moved toward the liquid end until the piston strikes, and then another mark is placed on the rod half way between the first mark and the steam-end stuffing box gland. After this the piston rod is returned toward the steam end until the second mark is flush with the stuffing box gland. The steam piston is now in central position. This procedure is repeated for the opposite piston-rod assembly.

The next step is to see that both steam valves are in a central position with equal amounts of lost motion on each side, indicated by distance X in Fig. 9.

Most small steam pumps are fitted with a fixed amount of lost motion as shown in Fig. 9*a*. With the slide valve centered over the valve ports, a properly adjusted pump will have the tappet exactly centered in the space between the valve lugs. The lost motion (X) on each side of the tappet will be equal.

Larger pumps are fitted with adjustable lost motions such as are shown in Fig. 9*b*. The amount of lost motion (X) can be changed by moving the locknuts. Manufacturers provide specific instructions concerning the setting of proper lost motion. However, one rule of thumb is to allow half the width of the steam port on each side for lost motion. A method to provide equality of lost motion is to move the valve each way until it strikes the nut and then note if both port openings are the same.

In some cases it is desirable to be able to adjust the steam valves while the pump is in motion. With the arrangements mentioned above, this cannot be done because the steam-chest head must be removed. In a pump equipped with a lost-motion mechanism as in Fig. 9*c*, all adjustments are external and can therefore be made while the pump is in operation.

Balanced Piston Steam Valve. The balanced piston steam valve previously mentioned (Fig. 10) is used on duplex steam pumps when the slide-type valve cannot

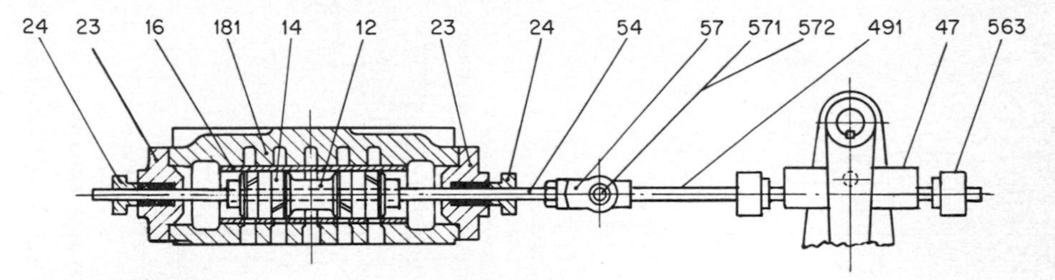

Fig. 10 Balanced piston steam valve. 12, piston valve; 181, steam chest; 23, valve-rod stuffing box; 24, valve-rod stuffing box gland; 54, valve rod, complete; 57, valve-rod head; 14, piston-valve ring; 16, piston-valve lining; 47, lost-motion block tappet; 491, valve-rod link; 563, valve-rod collar. (Worthington Pump Corporation)

be used because of size. It can also be used without lubrication at pressures over 250 lb/in^2 and temperatures over 500°F. At higher pressures, wire drawings or steam cutting can occur as the piston slowly crosses the steam ports. To prevent wear and permanent damage to the steam chest and piston, piston rings are used on the steam valve and a steam-chest liner is pressed into the steam chest to protect it.

Cushion Valves. Steam cushion valves are usually furnished on larger pumps to act as an added control to prevent the steam piston from striking the cylinder heads when the pump operates at high speeds. As previously shown, the steam end has five ports; the outside ports are for steam admission, and the inside ports are for steam exhaust. As the steam piston approaches the end of the cylinder, it covers the exhaust port, trapping a volume of steam in the end of the cylinder. This steam acts as a cushion and prevents the piston from striking the cylinder head. The cushion valve is simply a bypass valve between the steam and exhaust ports so that by opening or closing this valve the amount of cushion steam can be controlled.

If the pump is running at low speed or working under heavy load, the cushion valve should be opened as much as possible without allowing the piston to strike the cylinder head. If the pump is running at high speed or working under light load, then the cushion valve should be closed. The amount of steam cushion and, consequently, the length of stroke can be properly regulated for different operating conditions by the adjustment of these valves.

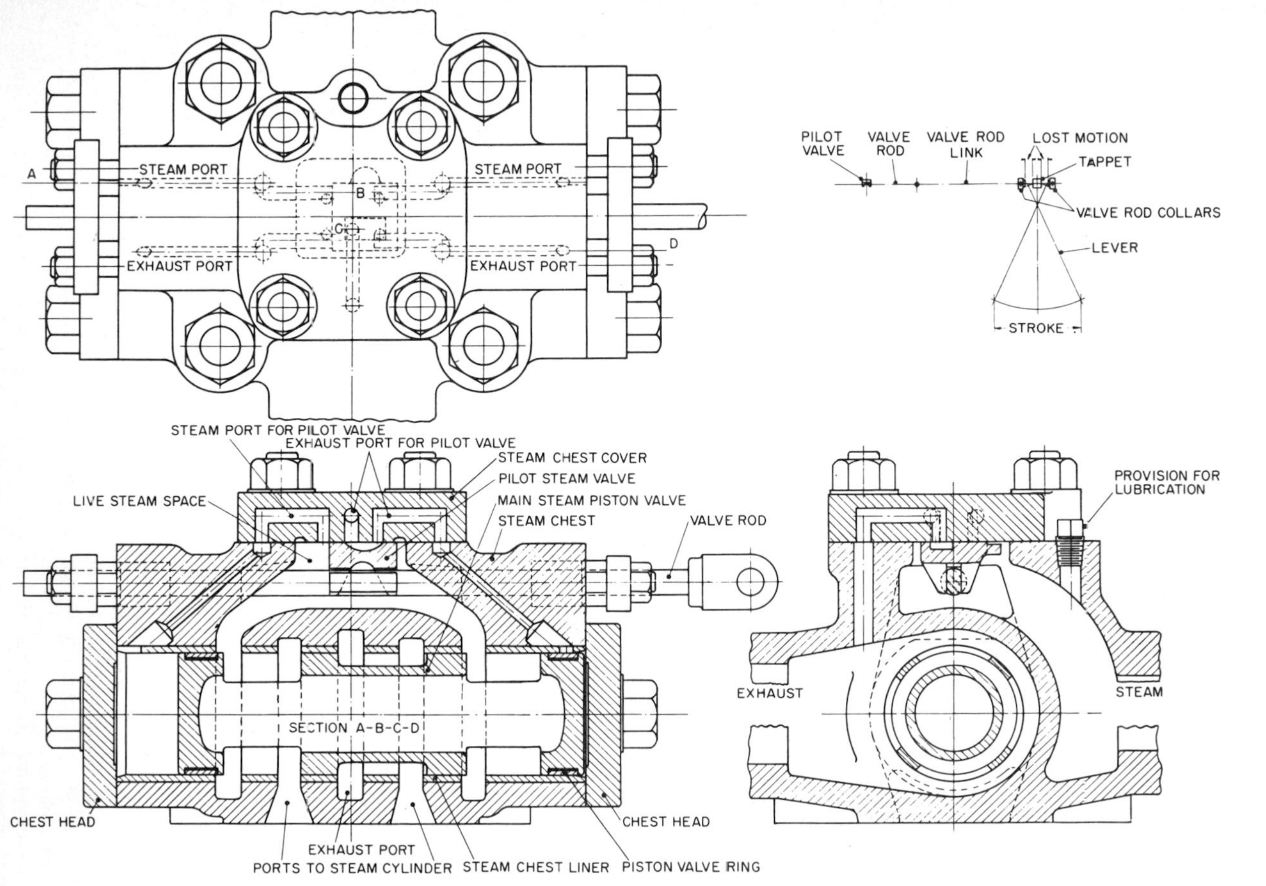

Fig. 11 Simplex type steam valve. (Worthington Pump Corporation)

Simplex Steam Valve. The simplex-pump steam valve is steam-operated, not mechanically operated like the duplex steam valves. The reason for this is that the piston-rod assembly must operate its own steam valve. Consequently the travel of the valve cannot be controlled directly by means of the piston-rod motion. Instead the piston rod operates a pilot valve by means of a linkage similar to that used with a duplex pump. This controls the flow of steam to each end of the main valve, shuttling it back and forth. The arrangement illustrated in Fig. 11 is one of the designs available to produce this motion.

With the pilot valve in the position shown in Fig. 11, steam from the "live steam space" flows through the pilot-valve steam port into the steam space at the left-hand end of the main valve (balanced piston type). Simultaneously, the D section of the pilot valve connects the steam space at the right-hand end of the main valve with the exhaust port, thereby releasing the trapped steam. The main valve has moved completely across to the right-hand end of the chest. The main valve in this position permits steam to flow from the chest to the left-hand steam-cylinder port and, at the same time, connects the right-hand steam-cylinder port with the exhaust port.

The steam piston now moves to the right and, after the lost motion is taken up in the valve gear, the pilot valve moves toward the left. In this position the same cycle described above takes place with the exception that it occurs at the opposite end of the steam chest. Since the main valve is steam-operated, the valve

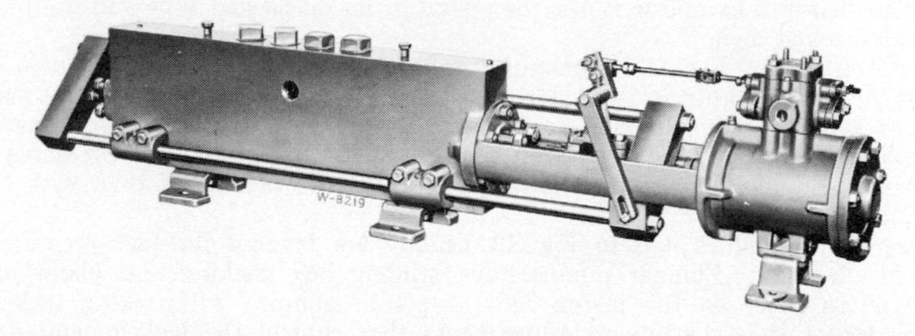

Fig. 12 Simplex-type plunger pump. (Worthington Pump Corporation)

can be in only two positions, either at the left-hand or at the right-hand end of the chest. Hence it is impossible to have it at dead center. In other words, steam can always flow either to one side or to the other of the steam piston, irrespective of the position of the steam piston.

For the valve to operate smoothly and quietly, an arrangement must be provided to create a cushioning effect on the valve travel. The steam piston, as it approaches the end of its travel, cuts off the exhaust port and traps a certain amount of steam which acts as a cushion and stops the steam piston.

All valve adjustments are outside of the steam chest, so it is possible to adjust the valve while the pump is in operation. The effect of decreasing or increasing the lost motion is the same as that described for duplex pumps. The lost-motion arrangement is the same as that shown in Fig. 9c.

Steam-End Materials. For most services cast iron is an excellent material for the steam cylinder, and as shown in Fig. 17, it is the major element of the steam end. It is readily cast in the complicated shape required to provide the steam porting. It possesses good wearing qualities, largely because of its free graphite content. This is required in the piston bores, which are continually being rubbed by the piston rings. At high steam temperatures and pressures, ductile iron or steel is used. In the latter case, however, cast iron steam-cylinder liners are frequently used for their better wear resistance.

Counter bores are provided at each end of a steam cylinder so the leading piston ring can override, for a part of its width, the end of the cylinder bore to prevent the wearing of a shoulder on the bore.

The cylinder heads and steam pistons are also usually made of cast iron. The cylinder head has a pocket cast in it to receive the piston-rod nut at the end of the stroke. Most steam pistons are made in one piece, usually with two piston-ring slots machined into the outside circumference.

The relatively wide piston rings are usually made from hammered iron. They are split so they can be expanded to fit over the piston and snapped into the grooves in the piston. They must be compressed slightly to fit into the bores in the cylinder. This ensures a tight seal with the cylinder bores even as the rings wear during operation. In services where steam-cylinder lubrication is not permissible, a combination two-piece ring of iron and bronze is used to obtain longer life than is obtained with the hammered iron rings.

The piston and valve rods are generally made from steel, but stainless steel and monel are also commonly used. Packing for the rods is usually a braided graphited asbestos.

Drain cocks and valves are always provided to permit drainage of condensation which forms in the cylinder when a pump is stopped and cools down. On each start-up these must be cracked open until all liquid is drained and only steam comes out; then they are closed.

The steam end and liquid end are joined by a cradle. On most small pumps the cradle is cast integral with the steam end. On large pumps it is a separate casting or fabricated weldment.

Liquid End Construction Steam pumps are equipped with many types of liquid ends, each being designed for a particular service condition. However, they can all be classified into two basic types, the piston or inside-packed type and the plunger or outside-packed type.

The piston pump (Fig. 7) is generally used for low and moderate pressures. Because the piston packing is located internally, the operator cannot see the leakage past it or make adjustments that could make the difference between good operation and packing failure. Generally piston pumps can be used at higher pressures with noncorrosive liquids having good lubricating properties such as oil than with corrosive liquids such as water.

Plunger pumps, illustrated in Fig. 12, usually are favored for high-pressure and heavy-duty service. Plunger pumps have stuffing box packing and gland of the same type as those on the piston rods of piston pumps. All packing leakage is external, where it is a guide to adjustments that control the leakage and extend packing and plunger life. During operation, lubrication can be supplied to the external plunger packing to extend its life. Lubrication cannot be supplied to the piston packing rings on a piston pump.

Piston-Type Liquid Ends. The most generally used piston pump is the cap and valve plate design illustrated in Fig. 7. This is usually built for low pressures and temperatures, although some designs are used up to 350 lb/in^2 discharge pressure and 350°F. The discharge-valve units are mounted on a plate separate from the cylinder and have a port leading to the discharge connection. A dome-shaped cap, subject to discharge pressure, covers the discharge-valve plate. The suction-valve units are mounted in the cylinder directly below their respective discharge valves. A passage in the liquid cylinder leads from below the suction valves down between the cylinders of a duplex pump to the suction connection.

Side-pot liquid ends are used where the operating pressures are beyond the limitations of the cap and valve plate pump. Figure 13 illustrates this design. Suction valves are placed in individual pots on the side of the cylinders and discharge valves in the pots above the cylinders. Each valve can be serviced individually by removing its own cover. The small area of the valve covers exposed to discharge pressure makes the sealing much simpler than is the case in the cap-and-valve design. Side-pot liquid ends are widely used in refinery and oil-field applications. This design is commonly employed to the maximum pressure practicable for a piston pump.

There are several specially designed piston-type liquid ends which have been developed for specific applications. One of these is the "close clearance" design illustrated in Fig. 15. This pump can handle volatile liquids such as propane or butane or a liquid which may contain entrained vapors.

The close-clearance cylinder is designed to minimize the dead space when the

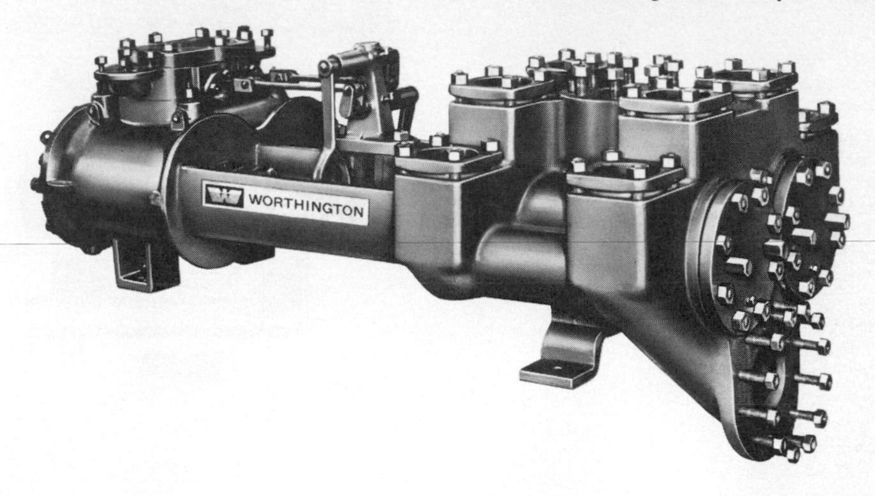

(a)

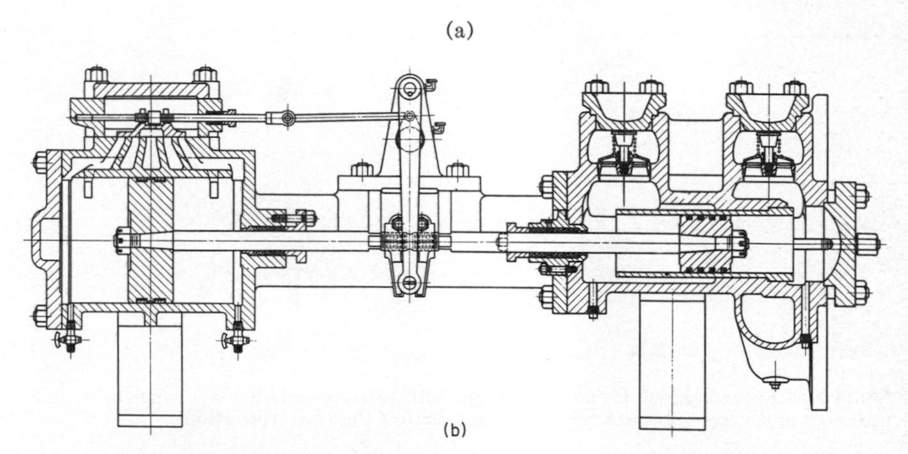

(b)

Fig. 13 Side pot-type piston pump. (Worthington Pump Corporation)

Fig. 14 Close-clearance liquid end pump. (Worthington Pump Corporation)

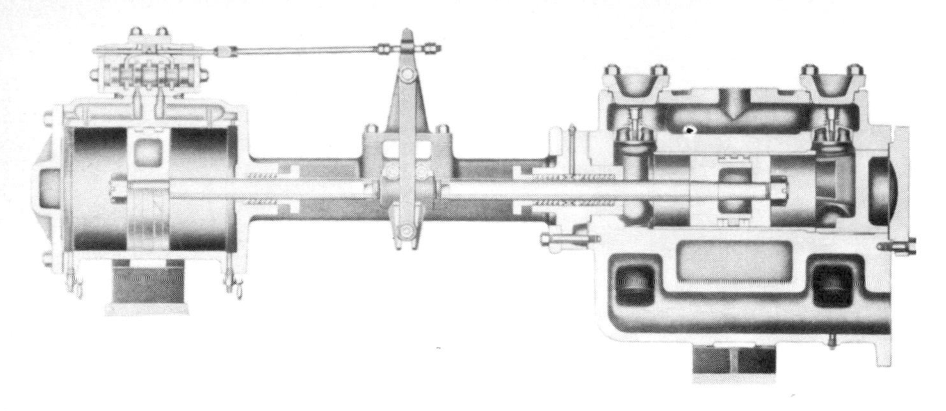

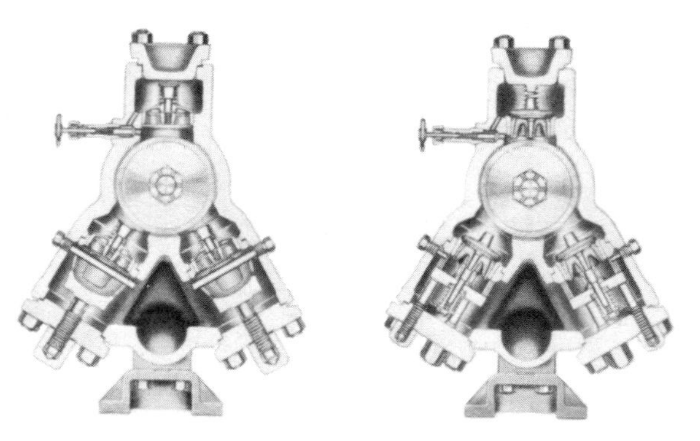

Fig. 15 Close-clearance liquid end pump, with end views below showing disk-valve assembly and wing-valve assembly. (Worthington Pump Corporation)

piston is at each end of its stroke. The liquid valves are placed as close as possible to the pump chamber to keep clearance to a minimum. The suction valves are positioned below the cylinder at the highest points in the suction manifold to ensure that all the gases are passed into the pump chamber. Although these pumps are of close-clearance design, they are not compressors and can "vapor-bind," i.e., a large amount of gas trapped below the discharge valve will compress and absorb the entire displacement of the pump. When this occurs the discharge valve will not open, and will cause a loss of flow. Hand-operated bypass or priming valves are provided to bypass the discharge valve and permit the trapped gases to escape to the discharge manifold. When the pump is free of vapors, the valves are closed.

There are a number of other special piston-pump designs for certain services in addition to the most common types just described. One of these is the wet vacuum pump which features tight-sealing rubber valves that permit the pump to handle liquid and air or noncondensable vapors. Another special design is made of hard, wear-resistant materials to pump cement grout on construction projects. Another design, Fig. 16, has no suction valves and is made for handling thick viscous products such as sugarcane pulp, soap, white lead, printer's ink, tar, etc. The liquid flows into the cylinder from above through a suction port which is cut off as the piston moves back and forth.

Piston-Pump Liquid-End Materials and Construction. The materials used for piston-pump liquid ends vary widely with the liquids handled. Most of the services

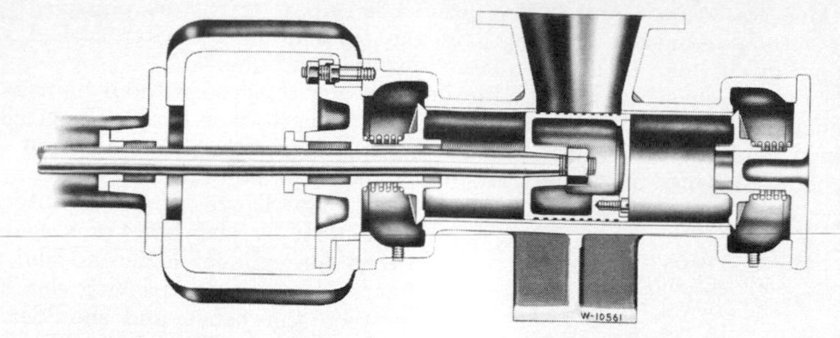

Fig. 16 Viscous liquid-handling pump. (Worthington Pump Corporation)

to which these pumps are applied use one of the common material combinations listed in Table 1.

The liquid cylinder, the largest liquid-end component, is most frequently made from cast iron or bronze. However, other materials are also used. Cast steel cylinders are used in refineries and chemical plants for high-pressure and high-temperature applications. Nickel cast steels are used for low-temperature services. Ni-Resist cast iron, chrome-alloy steels, and stainless steels are occasionally used for certain corrosive and abrasive applications, but tend to make the pump cost very high. The liquid-cylinder heads and valve covers are usually made from the same material as the liquid cylinder.

TABLE 1 Material and Service Specifications for Pump Liquid Ends.

Part	Regular fitted (RF)	Bronze fitted (BF)	Fully bronze fitted (FBF)	All iron fitted (AIF)	All bronze (AB)
Cylinders	Cast iron	Cast iron	Cast iron	Cast iron	Bronze
Cylinder liners	Bronze	Bronze	Bronze	Cast iron	Bronze
Piston	Cast iron	Cast iron	Bronze	Cast iron	Bronze
Piston packing	Fibrous	Fibrous	Fibrous	Cast iron, 3-ring	Fibrous
Stuffing boxes	Cast iron, bronze bushed	Cast iron, bronze bushed	Cast iron, bronze bushed	Cast iron	Bronze
Piston rods	Steel	Bronze	Bronze	Steel	Bronze
Valve service	Bronze	Bronze	Bronze	Steel	Bronze
Services for which most often used	Cold water; other cold liquids not corrosive to iron and bronze	Same as for RF, with reduced maintenance; continuous hot-water service	Boiler feed; intermittent hot-water service; sodium chloride brines	Oils and other hydrocarbons not corrosive to iron or steel; caustic solutions	Mild acids which would attack iron cylinders but not acid-resisting bronze

As was the case in the steam end, a liquid-cylinder liner is used to prevent wear and permanent damage to the liquid cylinder. Liners must be replaced periodically when worn by the piston packing to the point that too much fluid leaks from one side of the piston to the other. The liners may be either of a driven-in or pressed-in type or of a removable type which is bolted or clamped in position in the cylinder bore.

The pressed-in type, Fig. 7, derives its entire support from the drive fit in the cylinder bore. As a rule, such a liner is relatively thin and is commonly made from a centrifugal casting or a cold-drawn brass tube. After a driven liner is worn to the point where it must be replaced, it is usually removed by chipping a narrow groove along its entire length. This groove is cut as nearly as possible through the liner without damage to the wall of the cylinder bore. After such a groove has been cut in the liner, it can be collapsed inward and removed.

The removable type, Fig. 13, may be removed and replaced without damage. Instead of being a driven fit throughout most of its length, it is located longitudinally by a flange at the cylinder-head end and is a driven fit only at the end of the cylinder bore. A flange on the liner fits into a recess at the beginning of the cylinder bore. This flange is held in contact with a shoulder by jack bolts or a spacer between the cylinder head and end of the liner. Sometimes a packing ring is used between the flange and shoulder for a positive seal. Removable liners are heavier than pressed-in ones.

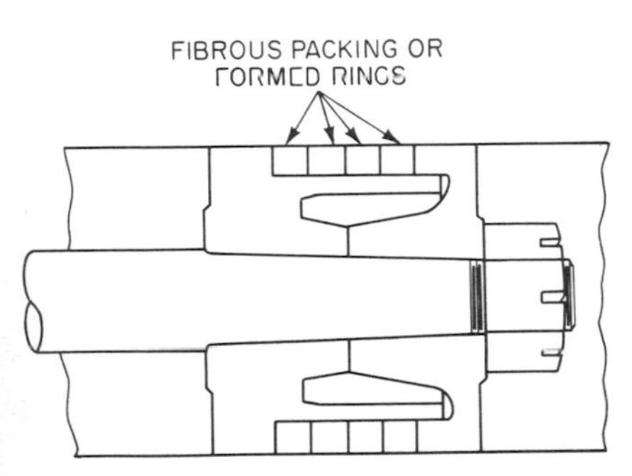

Fig. 17 Body and follower-type piston.

There are several designs of pistons and piston packings used for various applications. The three most common are as follows:

1. The body and follower-type piston with soft fibrous packing or hard-formed composition rings (Fig. 17). The packing is installed in the packing space on the piston with a clearance in both length and depth. This clearance permits fluid pressure to act on one end and the inside of the packing to hold and seal it against the other end of the packing space and the cylinder-liner bore.
2. The solid piston or, as shown in Fig. 18, a body and follower with rings of cast iron or other materials. This type is commonly used in pumps handling oil or other hydrocarbons. The metal rings are split with an angle or step-cut joint. Their natural tension keeps them in contact with the cylinder liner, assisted by fluid pressure under the ring.
3. The cup piston (Fig. 19), which consists of a body and follower-type piston with molded cups of materials such as rubber reinforced with fabric. Fluid pressure on the inside of the cup presses the lip out against the cylinder bore, forming a tight seal.

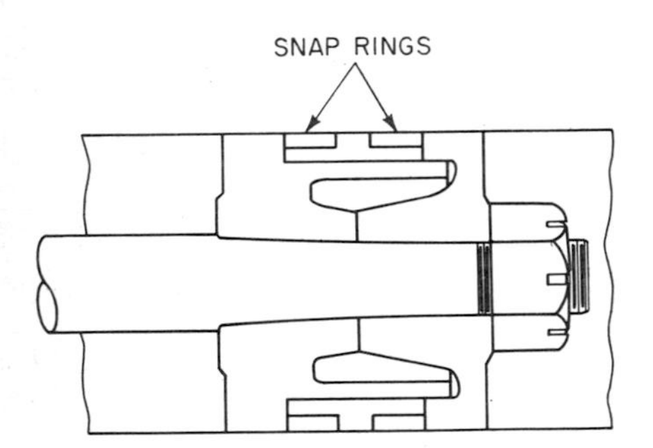

Fig. 18 Body and follower piston with snap rings.

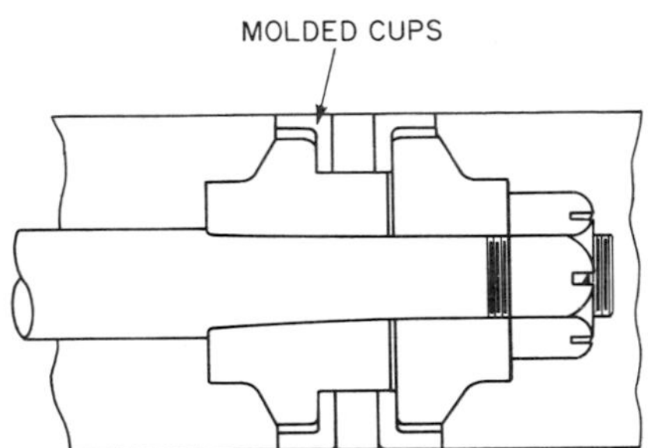

Fig. 19 Body and follower piston with molded cups.

The piston-rod stuffing boxes are usually made separate from, but of the same material as, the liquid cylinder. When handling liquids with good lubrication properties, the stuffing boxes are usually packed full with a soft, square, braided packing which is compatible with the liquid. When the liquid has poor lubricating properties, a lantern ring is installed in the center of the stuffing box with packing rings on both sides of it. A drilled hole is provided through which a lubricant, grease or oil, can be injected into the lantern ring from the outside of the stuffing box. At higher temperatures, approximately 500°F or higher, a cooling water jacket is added to the outside of the stuffing box or as a spacer between the stuffing box and the liquid cylinder. The purpose of the cooling water jacket is to extend packing life by keeping the packing cool.

The liquid-end valves of all direct-acting steam pumps are self-acting in contrast to the mechanically operated slide valves in the steam end. The liquid-end valves act like check valves; they are opened by the liquid passing through and are closed by a spring plus their weight.

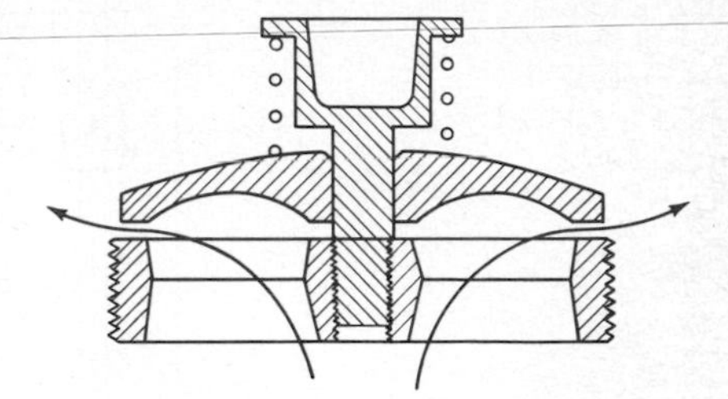

Fig. 20 Stem-guided disk valve.

Liquid-end valves are roughly divided into three types: the disk valve for general service and thin liquids, the wing-guided valve for high pressures, and the ball or semispherical valve for abrasive and viscous liquids.

The valve shown in Fig. 20 is typical of the disk type. This stem-guided design is commonly used in the cap-and-valve-plate design. For hot-water boiler feed and general service, the disk, seat, and stem are usually made of bronze, although other alloys may also be used. For lower temperatures and pressures, the valve disks may be made of rubber, which has the advantage of always making a tight seal with the valve seat.

The wing-guided valve shown in Fig. 21 is typical of the design used for high pressures. It derives its name from the wings on the bottom of the valve which guide it in its seat. The beveled seating surfaces on the valve and seat tend to form a tighter seal than the flat seating surfaces on a disk valve. There is also less danger that a solid foreign particle in the fluid will be trapped between the seat and the valve. This type of valve is commonly made from a heat-treated chrome-alloy-steel forging, although a cast hard bronze and other materials may be used.

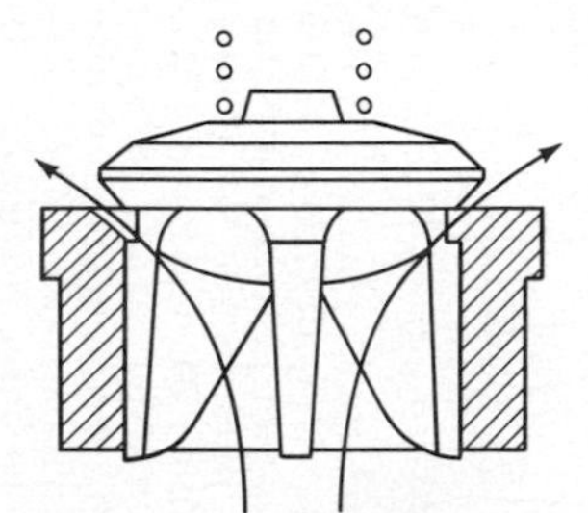

Fig. 21 Wing-guided valve.

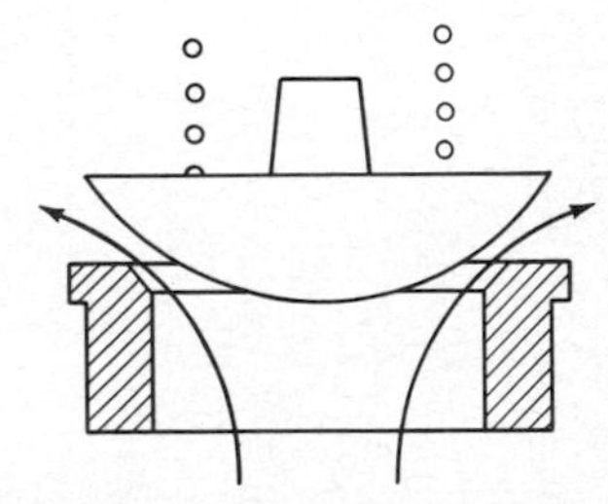

Fig. 22 Semispherical valve.

The ball-type valve, as its name suggests, is a ball which acts like a check valve. It is usually not spring-loaded, but guides and lift stops are provided as necessary to control its operation. The balls may be made of rubber, bronze, stainless steel, or other materials as service conditions require. The semispherical valve (Fig. 22) is spring-loaded and can therefore be operated at higher speeds than ball valves. Both the ball and the semispherical types have the advantage of having no obstructions to flow in the valve seat. Of the previously described types, the disk valve seat has ribs and the wing-guided valve has vanes which obstruct the flow. The one large opening in the valve seat and the smooth spherical surface of these valve types minimize the resistance to flow of thick, viscous liquids. These types are also used for liquids with suspended solids because their rolling seating action prevents trapping of the solids between the seat and the valve.

Plunger-Type Liquid Ends. As mentioned previously, plunger-type pumps are

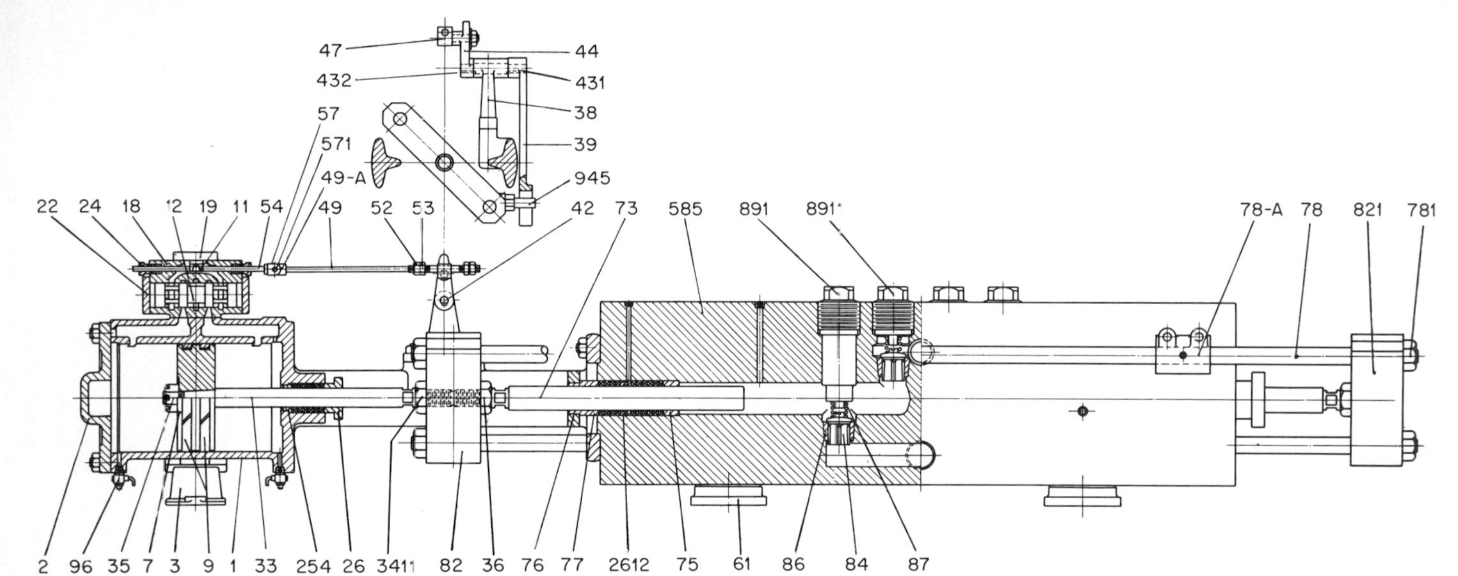

Fig. 23 Simplex plunger pump with forged steel cylinder. 1, steam cylinder with cradle; 2, steam-cylinder head; 3, steam-cylinder foot; 7, steam piston; 9, steam piston rings; 11, slide valve; 12, piston valve; 18, steam chest; 19, steam-chest cover; 22, steam-chest head; 24, valve-rod stuffing box gland; 26, steam piston-rod stuffing box gland; 33, steam piston rod; 35, steam piston-rod nut; 36, plunger nut; 38, cross stand; 39, lever; 42, fulcrum pin; 44, crank; 47, tappet; 49, valve-rod link; 49A, valve-rod link head; 52, lost-motion adjusting nut; 53, lost-motion locknut; 54, valve rod; 57, valve-rod head; 61, liquid cylinder foot; 73, plunger; 75, plunger lining; 76, plunger gland flange; 77, plunger gland lining; 78, side rod; 78A, side-rod guide; 82, plunger cross head, front; 84, metal valve; 86, valve seat; 87, valve spring; 96, steam cylinder drain valve; 254, piston-rod stuffing box bushing; 431, lever key; 432, crank key; 571, valve-rod head pin; 585, liquid cylinder; 781, side-rod nut; 821, plunger cross head, rear; 891, suction-valve plug; 945, cross-head pin; 2612, lantern gland; 8911, discharge-valve plug; 3411, steam piston-rod jam nut. (Worthington Pump Corporation)

used where dependability is of prime importance, even when the pump is operated continuously for long periods and where the pressure is very high. Cast fluid-end plunger pumps are used for low and moderate pressures. Forged liquid-end pumps (Fig. 23), which are the most common plunger types, are used for high pressures and have been built to handle pressures in excess of 10,000 lb/in^2.

Most of these designs have opposed plungers, i.e., one plunger operating into the inboard end of the liquid cylinder and one into the outboard end. The plungers are solidly secured to inboard and outboard plunger crossheads. The inboard and outboard plunger crossheads are joined by side rods positioned on each side of the cylinder. With this arrangement, each plunger is single-acting, i.e., it makes only one pressure stroke for each complete reciprocating cycle. The pump, however, is double-acting because the plungers are connected by the side rods.

Plunger-Pump Liquid-End Materials. The liquid cylinder of a forged liquid-end plunger-type pump is most commonly made from forged steel, although bronze, monel, chrome alloy, and stainless steels are also used. The stuffing boxes and valve chambers are usually integral with the cylinder (Fig. 23), which is desirable for high temperatures and pressures since high-pressure joints are minimized.

The liquid plungers may be made of a number of materials. The plungers must be as hard and smooth as possible to reduce friction and to resist wear by the plunger packing. Hardened chrome-alloy steels and steel coated with hard-metal alloys or ceramics are most commonly used.

The stuffing box packing used will vary widely depending upon service conditions. A soft square packing, cut to size, may be used. However, solid molded rings of square, V-lip, or U-lip design are commonly used at the higher pressures. A lubricant, oil or grease, is frequently injected into a lantern ring in the center of the stuffing box to reduce friction and reduce packing and plunger wear.

The liquid valves may be of any of the types or materials previously described. However, the wing-guided valve with beveled seating surfaces is the most common because it is most suitable for high pressures.

DIRECT-ACTING STEAM-PUMP PERFORMANCE

The direct-acting steam pump is a very flexible machine. It can operate at any point of pressure and flow within the limitations of the particular design. The speed of, and therefore the flow from, the pump can be controlled from stop to maximum by throttling the steam supply to it. This can be done by either a manual or an automatically operated valve in the steam supply line. The maximum speed of a particular design is primarily limited by the frequency with which the liquid valves will open and close smoothly. The pump will operate against any pressure imposed upon it by the system it is serving from zero to its maximum pressure rating. The maximum pressure rating of a particular design is determined by the strength of the liquid end. In a particular application, the maximum liquid pressure developed may be limited by the available steam pressure and the ratio of the steam-piston and liquid-piston areas.

Steam-Pump Capacity The actual flow to the discharge system is termed the pump *capacity*. The capacity, usually expressed in U.S. gallons per minute (gpm), is somewhat less than the theoretical *displacement* of the pump. The difference between displacement and capacity is called *slip*. The displacement is a function of area of the liquid piston and the speed at which the piston is moving.

The displacement of a single double-acting piston can be calculated from the formula

$$D = \frac{12 \times AS}{231} \quad \text{or} \quad 0.0408d^2S$$

where D = displacement, U.S. gpm
A = area of piston or plunger, in^2
S = piston speed, ft/min
d = diameter of the liquid piston or plunger, in.

For a duplex double-acting pump, D is multiplied by 2.

This formula neglects the area of the piston rod. For very accurate calculations it is necessary to deduct the rod area from the piston area. This is normally not done, and the resultant loss is usually considered part of the slip.

The slip also includes losses due to leakage from the stuffing boxes, leakage across the piston on packed-piston pumps, and leakage back into the cylinder from the discharge side while the discharge valves are closing. Slip for a given pump is determined by actual test. For a properly packed pump, slip is usually 3 to 5 percent. As a pump wears, slip will increase, but this can be compensated for by increasing the pump speed to maintain the desired capacity.

Piston Speed Although piston speed in feet per minute is the accepted term used to express steam-pump speed, it cannot easily be measured directly and is usually calculated by measuring the revolutions per minute of the pump and converting this to piston speed. One revolution of a steam pump is defined as one complete forward and reverse stroke of the piston. The relationship between piston speed and rpm is:

$$S = \frac{\text{rpm} \times \text{stroke}}{6}$$

where S = piston speed, ft/min
rpm = revolutions per minute
stroke = stroke of pump, ins

A steam pump must fill with liquid from the suction supply on each stroke or it will not perform properly. If the pump runs too fast, the liquid cannot flow through the suction line, pump passageways, and valves fast enough to follow the piston. On the basis of experience and hydraulic formulas, maximum piston speeds can be established which will vary with the length of stroke and the liquid handled.

Table 2 shows general averages of maximum speed ratings for pumps of specified stroke handling various liquids. Some pumps, by reason of exceptionally large valve areas or other design features, may be perfectly suitable for speeds higher than shown. From the table it should be noted that the piston speed should be reduced for viscous liquids. Unless the net positive suction head is proportionately high, viscous liquids will not follow the piston at high speeds because frictional resistance in suction lines and in the pump increases with viscosity and rate of flow. Pumps handling hot water are run more slowly to prevent boiling of the liquid as it flows into the low-pressure area behind the piston.

TABLE 2 Piston Speed

Stroke, in	3	4	5	6	7	8	10	12	15	18	24	36
	ft/min											
Cold water; oil to 250 SSU	37	47	53	60	64	68	75	80	90	95	105	120
Oil, 250 to 500 SSU	35	45	51	57	61	65	72	77	86	91	100	115
Oil, 500 to 1,000 SSU	33	42	47	53	57	61	67	71	80	85	94	107
Oil, 1,000 to 2,500 SSU	29	36	41	46	49	53	58	62	69	73	81	92
Oil, 2,500 to 5,000 SSU	24	31	35	39	42	45	49	52	57	62	68	78
Boiler feed 212° F	22	28	32	36	39	41	45	48	54	57	63	72

Size of Liquid End The size of a steam pump is always designated as follows:

Steam-piston diameter × liquid-piston diameter × stroke

For example, a 7½ × 5 × 6 steam pump has a 7½-in-diameter steam piston, a 5-in-diameter liquid piston, and a 6-in stroke.

To determine the liquid piston diameter for a specified capacity the following procedure is used: First a reasonable stroke length is assumed and the maximum piston speed for this stroke and the type of liquid pumped is selected from Table 2. Then the desired capacity is increased by 3 to 5 percent to account for slip. The result is the desired displacement. Then for either a simplex or duplex pump the liquid-piston diameter can be calculated as follows:

For simplex pumps $$d_l = 4.95\left(\frac{D}{S}\right)^{1/2}$$ *For duplex pumps* $$d_l = 3.5\left(\frac{D}{S}\right)^{1/2}$$

where d_l = liquid-piston diameter, in
D = displacement, gpm
S = piston speed, ft/min

Using the resultant liquid-piston diameter, the next larger standard piston size is selected.

Size of Steam End To calculate the size of the steam end required for a specific application, the basic principle of steam-pump operation should be considered. A simple schematic of a steam pump is shown in Fig. 24.

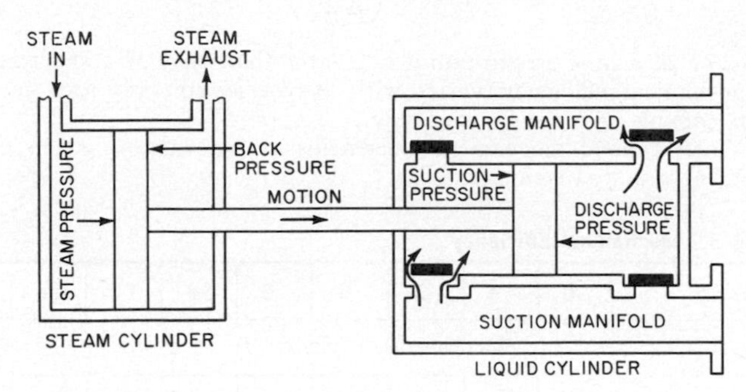

Fig. 24 Schematic showing direction of forces on pistons.

For the pump to move, the force exerted on the steam piston must exceed the force on the liquid piston which is opposing it. The force on the steam piston is the product of the net steam pressure times the steam-piston area. The *net* steam pressure is the steam inlet pressure minus the exhaust pressure. The force acting on the liquid piston is the product of the net liquid pressure times the liquid piston area. The *net* liquid pressure is the pump discharge pressure minus the suction pressure or *plus* the suction lift. This may be expressed algebraically as follows:

$$P_sA_s > P_lA_l$$

where P_s = net steam pressure
A_s = steam-piston area
P_l = net liquid pressure
A_l = liquid-piston area

Since the pistons are circular, the squares of their diameters are directly proportional to their areas, and the above formula can be rewritten as:

$$P_sd_s^2 > P_ld_l^2$$

where d_s = steam-piston diameter
d_l = liquid-piston diameter

In actual practice it is necessary for the force on the steam piston to exceed the force opposing it on the liquid piston by a considerable amount. This is because of mechanical losses. These include stuffing box friction, friction between piston rings and cylinder of both liquid and steam ends, and the operation of the valve gear.

These losses are determined by test and are accounted for in size calculations by introduction of a mechanical efficiency figure. Mechanical efficiencies are expressed as a percentage, with 100 percent being a perfect balance of forces acting on the steam and liquid pistons as expressed in the above formula. Since the efficiencies of two identical pumps may vary with stuffing box and piston-ring packing tightness, the efficiencies published by manufacturers tend to be conservative.

With the mechanical efficiency factor inserted, the formula of forces becomes:

$$P_s d_s^2 E_m = P_l d_l^2$$

where P_s = net steam pressure
d_s = steam-piston diameter
P_l = net liquid pressure
d_l = liquid-piston diameter
E_m = mechanical efficiency, expressed as a decimal

This formula is commonly used to determine the *minimum* size of steam piston required when the liquid-piston size has already been selected and the net steam and net liquid pressures are known. For this calculation, the formula is rearranged as:

$$d_s = d_l \left(\frac{P_l}{P_s E_m}\right)^{1/2}$$

The efficiency of a long stroke pump is greater than that of a short stroke pump. Although mechanical efficiency varies with stroke length, any two pumps of the same size are capable of the same efficiency.

Table 3 shows typical mechanical efficiencies in percentages which can be used to determine the required steam-end size.

TABLE 3 Mechanical Efficiency

Stroke, in	3	4	5	6	8	10	12	18	24
	Mechanical efficiency, %								
Piston pump	50	55	60	65	65	70	70	73	75
Plunger pump	47	52	57	61	61	66	66	69	71

Steam Consumption and Water Horsepower After determining the proper size of other pump types, the next concern usually is to calculate the maximum brake horsepower so that the proper size driver can be selected. With a steam pump, the next step is usually to determine the steam consumption. This must be known to ensure that the boiler generating the steam is large enough to supply the steam required by the pump as well as all that is required for all its other services.

To determine the steam consumption, it is necessary first to calculate the water horsepower as follows:

$$\text{whp} = \frac{Q \times P_l}{1{,}715}$$

where whp = water horsepower
Q = pump capacity, gpm
P_l = net liquid pressure, lb/in^2

The steam consumption chart (Fig. 25) affords a means of quickly obtaining an approximate figure for the steam rate of direct-acting steam pumps. For duplex pumps, divide the water horsepower by 2 before applying it to the curves. These curves were made up on the basis of water horsepower per cylinder; if the above procedure is not followed, inaccurate results will be obtained.

Starting with the water horsepower per cylinder:

1. Move vertically to the curve for steam cylinder size.
2. Move horizontally to the curve for 50 ft/min piston speed. This is the basic curve from which the other curves were plotted.
3. Move vertically to the actual piston speed at which the pump will run.
4. Move horizontally to the steam rate scale and read it in pounds per water horsepower-hour.

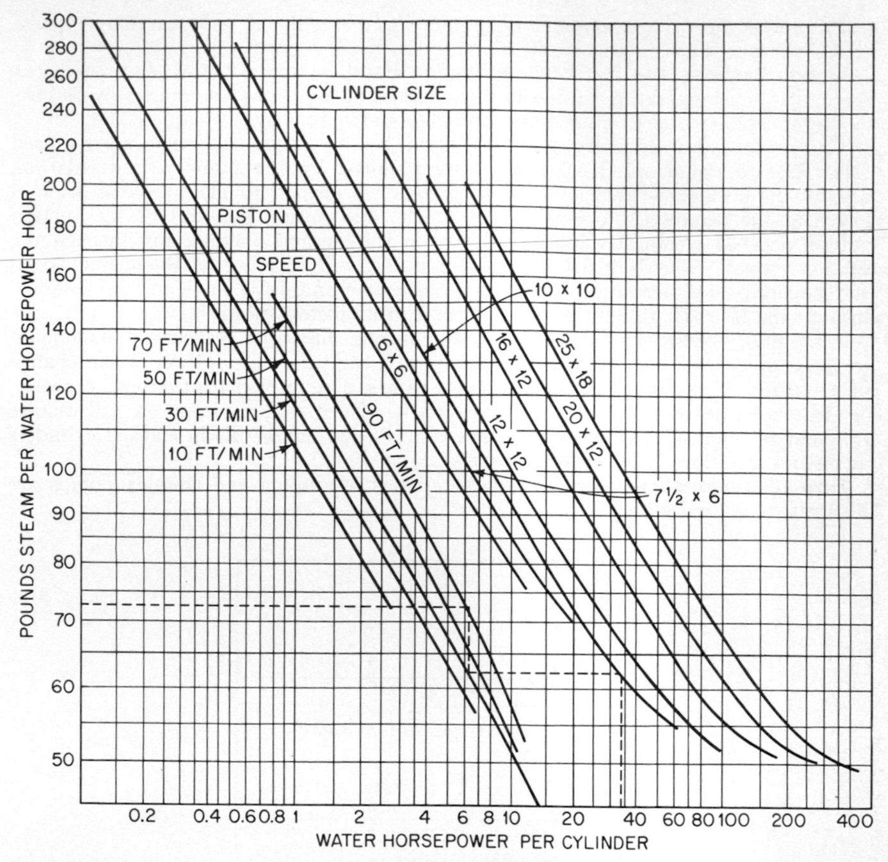

Fig. 25 Approximate steam consumption for steam pumps. (From Hydraulic Institute, "Hydraulic Institute Standards," 12th ed., 1969)

5. Multiply the above result by total water horsepower to obtain the steam rate in pounds per hour.

For steam cylinders with diameters as shown, but with longer stroke, deduct 1 percent from the steam rate for each 20 percent additional stroke. Thus, a 12×24 steam end will have a steam consumption about 5 percent less than a 12×12 steam end. For $5\frac{1}{4} \times 5$ and $4\frac{1}{2} \times 4$ steam ends, the 6×6 curve will give approximate figures. For cylinders of intermediate diameters, interpolate between the curves.

To correct for superheated steam, deduct 1 percent for each 10° of superheat. To correct for back pressure, multiply the steam rate by a correction factor equal to

$$\left(\frac{P + BP}{P}\right)^{1/2}$$

where P = net steam pressure to drive pump
BP = back pressure

Direct-acting steam pumps have inherently high steam consumption. This is not necessarily a disadvantage, however, when the exhaust steam can be used for heating the boiler feed water or for building heating or process work. Because these pumps can operate with a considerable range of back pressure, it is possible to recover nearly all the heat in the steam required to operate them. Since they do not use steam expansively, they are actually metering devices rather than heat engines, and as such consume heat from the steam only as it is lost in radiation from the steam end of the pump. These pumps act, in effect, like a reducing valve to deliver lower pressure steam that contains nearly all its initial heat.

Suction Systems and Net Positive Suction Head A majority of pump engineers agree that most operating problems with pumps of all types are caused by failure to supply adequate suction pressure to fill the pump properly.

The steam-pump industry uses the term net positive suction head required (NPSHR) to define the head or pressure required by the pump over its datum, usually the discharge-valve level. This pressure is needed to (1) overcome friction losses in the pump, (2) overcome the weight and spring loadings of the suction valves, and (3) create the desired velocity in the suction opening and through the suction valves. The NPSHR of a steam pump will increase as the piston speed and capacity are increased. The average steam pump will have valves designed to limit the NPSHR to 5 lb/in^2 or less at maximum piston speed.

If absolute pressure in the suction system minus the vapor pressure is inadequate to meet or exceed the NPSHR, the pump will cavitate. Cavitation is the change of a portion of the liquid to vapor, and it causes a reduction in delivered capacity, erratic discharge pressure, and noisy operation. Even minor cavitation will require frequent refacing of the valves, and severe cavitation can result in cracked cylinders or pistons or failure of other major parts.

Flow Characteristics The flow characteristics of duplex and simplex pumps are illustrated in Figure 26.

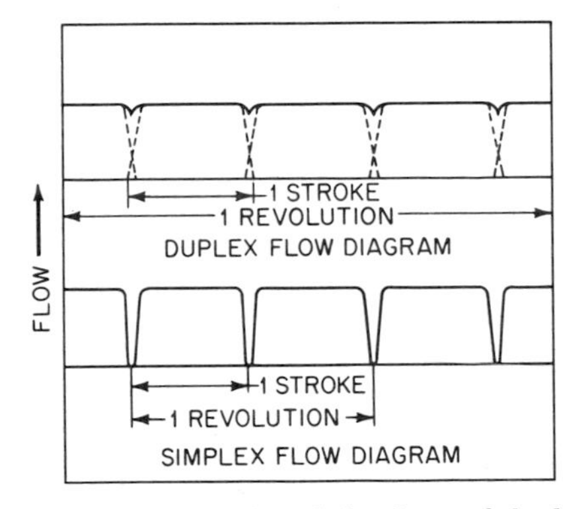

Fig. 26 Flow characteristics of simplex and duplex pumps.

The flow from a simplex is fairly constant except when the pump is at rest. However, the fact that the flow must stop for the valves to close and for the forces on both sides of the steam and liquid pistons to reverse produces an uneven pulsating flow. This can be compensated for, in part, by installing a pulsation-dampening device on the discharge side of the pump or in the discharge line.

The mechanism of a duplex pump is such that just before one piston completes its stroke, the other piston starts up and overlaps the first, eliminating the sharp capacity drop.

REFERENCES

1. Grobholz, R. K.: "Get the Right Steam Pump to Handle the Job," *Mill and Fact.*, 1955.
2. Hydraulic Institute: "Hydraulic Institute Standards," 12th ed., Hydraulic Institute, Cleveland, Ohio.
3. Schaub, D. G.: Reciprocating Steam Pumps, *South. Power and Ind.*, 1955.
4. Wright, E. F.: "Directing-acting Steam Pumps," in T. Baumeister and L. Marks, "Standard Handbook for Mechanical Engineers," 7th ed., McGraw-Hill Book Company, 1967, pp. 14-9 to 14-14.
5. Wright, E. F., in Carter, Karassik, and Wright, "Pump Questions and Answers," McGraw-Hill Book Company, 1949.

Section 3.3

Screw Pumps

G. J. CZARNECKI

J. K. LIPPINCOTT

INTRODUCTION

Screw pumps are a special type of rotary positive displacement pump in which the flow through the pumping elements is truly axial. The liquid is carried between screw threads on one or more rotors and is displaced axially as the screws rotate and mesh. In all other rotary pumps the liquid is forced to travel circumferentially, thus giving the screw pump with its unique axial flow pattern and low internal velocities a number of advantages in many application areas where liquid agitation or churning is objectionable.

The applications of screw pumps cover a diversified range of markets such as navy, marine, and utilities fuel-oil service; marine cargo; industrial oil burners; lubricating-oil service; chemical processes; petroleum and crude-oil industries; power hydraulics for navy and machine tools; and many others. The screw pump can handle liquids in a range of viscosity from molasses to gasoline, as well as synthetic liquids in a pressure range of 50 to 5000 lb/in² and flows up to 5000 gpm.

Because of the relatively low inertia of their rotating parts, screw pumps are capable of operating at higher speeds than other rotary or reciprocating pumps of comparable displacement. Some turbine-attached lubricating-oil pumps operate at 10,000 rpm and even higher. Screw pumps, like other rotary positive displacement pumps, are self-priming and have a delivery flow characteristic which is essentially independent of pressure.

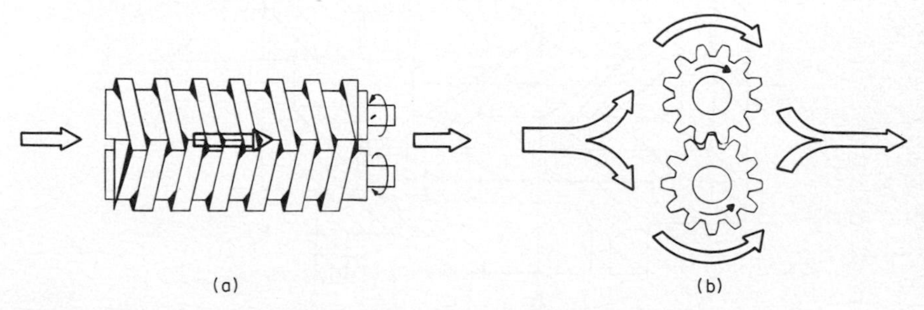

Fig. 1 Diagrams of screw and gear elements, showing (*a*) axial and (*b*) circumferential flow.

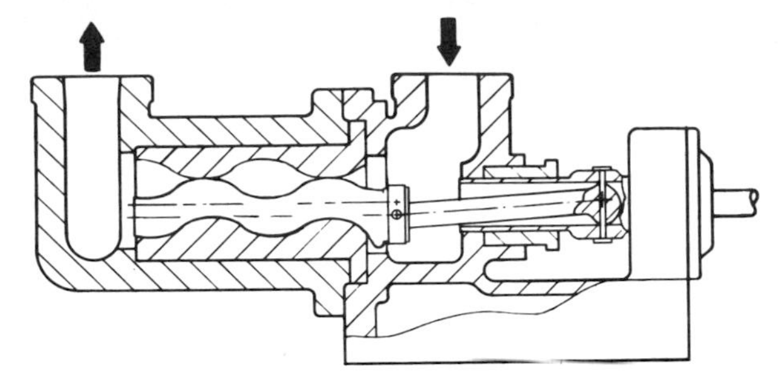

Fig. 2 Single-rotor pump.

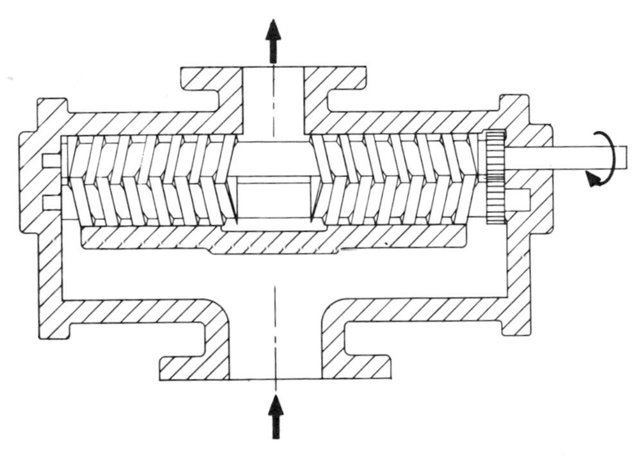

Fig. 3 Multiple-screw double-end arrangement.

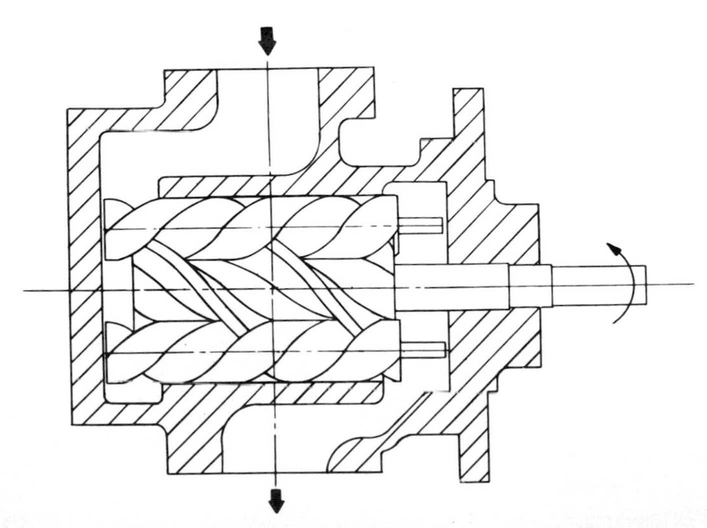

Fig. 4 Multiple-screw single-end arrangement.

According to the Hydraulic Institute Standards, screw pumps are classified into single- or multiple-rotor types. The latter are further divided into timed and untimed categories.

The single-screw pump exists only in a limited number of configurations. The rotor thread is eccentric to the axis of rotation and meshes with internal threads of the stator (rotor housing or body); alternatively the stator is made to wobble along the pump center line.

Multiple-screw pumps are available in a variety of configurations and designs. All employ one driven rotor in mesh with one or more sealing rotors. Several manufacturers have two basic configurations available—single- and double-end construction—of which the latter is the best known.

As with every pump type, there are certain advantages and disadvantages characteristic of the screw-pump design; these should be recognized in selecting the best pump for a particular application.

Advantages

1. Wide range of flows and pressures.
2. Wide range of liquids and viscosities.
3. High speed capability, allowing freedom of driver selection.
4. Low internal velocities.
5. Self-priming, with good suction characteristics.
6. High tolerance for entrained air and gases.
7. Minimum churning or foaming.
8. Low mechanical vibration, pulsation-free flow, and quiet operation.
9. Rugged, compact design—easy to install and maintain.
10. High tolerance to contamination in comparison with other rotary pumps.

Disadvantages

1. Relatively high cost because of close tolerances and running clearances.
2. Performance characteristics sensitive to viscosity change.
3. High-pressure capability requires long length of pumping elements.

THEORY

In screw pumps, it is the intermeshing of the threads on the rotors and the close fit of the surrounding housing which create one or more sets of moving seals in series between pump inlet and outlet. These sets of seals act as a labyrinth and provide the screw pump with its positive pressure capability. The successive sets of seals form fully enclosed cavities (see Fig. 5) which move continuously from inlet to outlet. These cavities trap liquid at the inlet and carry it along to the outlet, providing a smooth flow.

Delivery Because the screw pump is a positive-displacement device, it will deliver a definite quantity of liquid with every revolution of the rotors. This delivery can be defined in terms of displacement (D), which is the theoretical volume displaced per revolution of the rotors and is dependent only upon the physical dimensions of the rotors. It is generally measured in cubic inches per revolution. This delivery can also be defined in terms of theoretical capacity (Q_t), measured in U.S. gallons per minute, which is a function of displacement and speed (N):

$$Q_t = \frac{DN}{231}$$

If no internal clearances existed, the pump's actual delivered capacity or net capacity (Q) would equal the theoretical capacity. Clearances, however, do exist, with the result that whenever a pressure differential occurs, there will always be internal leakage from outlet to inlet. This leakage, commonly called *slip* (S), varies depending upon the pump type or model, the amount of clearance, the liquid viscosity at pumping conditions, and the differential pressure. For any given set of these conditions, the slip for all practical purposes is unaffected by speed. The delivered

Fig. 5 Axially moving seals and cavities. Alternate cavities filled with oil shown below. (DeLaval-IMO)

capacity or net capacity, therefore, is the theoretical capacity less slip: $Q = Q_t - S$. If the differential pressure is almost zero, the slip may be neglected and $Q = Q_t$.

The theoretical capacity of any pump can readily be calculated if all essential dimensions are known. For any particular thread configuration, assuming geometrical similarity, the size of each cavity mentioned earlier is proportional to its length and cross-sectional area. The length is defined by the thread pitch measured in terms of the same nominal diameter which is used in calculating the cross-sectional area (see Fig. 6). Therefore the volume of each cavity is proportional to the cube

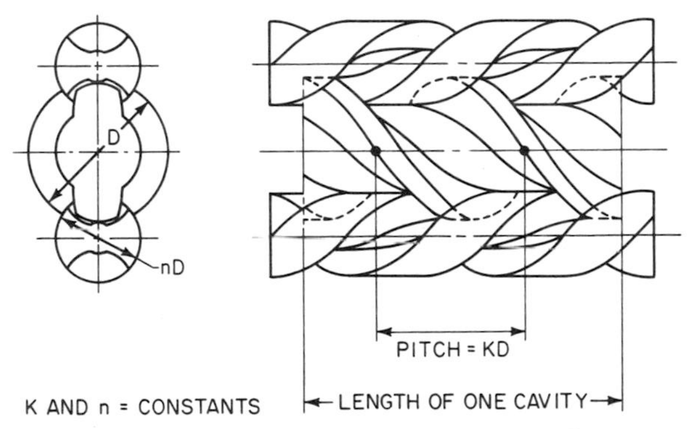

Fig. 6 Screw-thread proportions, showing lead, diameter, etc.

of this nominal diameter; and the pump theoretical capacity is also proportional to the cube of this nominal diameter and the speed of rotation, i.e.,

$$Q_t = KD^3N$$

Thus it can be seen that a relatively small increase in pump size can give a large increase in capacity.

Slip can also be calculated, but usually it depends upon empirical values developed by extensive testing. These test data are the basis of the design parameters used by every pump manufacturer. Slip generally varies approximately as the square of the nominal diameter. The net capacity therefore is

$$Q = KD^3N - S$$

Pressure Capability As mentioned earlier, screw pumps can be applied over a wide range of pressure up to 5000 lb/in² provided the proper design is selected. Internal leakage must be restricted for high-pressure applications. Close-running clearances and high accuracy of the conjugate rotor threads are requirements. In addition, an increased number of moving seals between inlet and outlet is employed, as in classic labyrinth-seal theory. The additional moving seals are obtained by a significant increase in length (see Fig. 9) of the pumping elements for a given size of rotor and pitch. Here minimum pump length is sacrificed in order to gain pressure capability.

The internal leakage within the pumping elements resulting from the differential pressure between outlet and inlet causes a pressure gradient to exist across the moving cavities. This gradient is approximately linear (see Fig. 7) when measured at

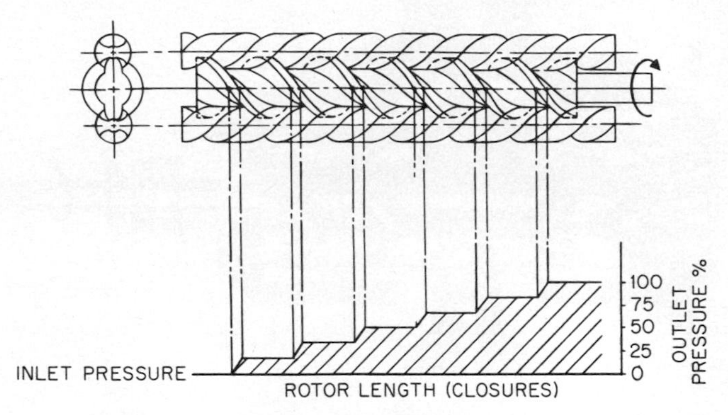

Fig. 7 Pressure gradient along a screw set.

any instant. Actually the pressure in each moving cavity builds up gradually and uniformly from inlet to outlet pressure as the cavity moves toward the outlet. In effect, the pressure capability of a screw pump is limited by the allowable pressure rise across any one set of moving seals. This pressure rise is sometimes referred to as "pressure per closure" or "pressure per stage" and generally is of the order of 125 to 150 lb/in² with normal running clearances, but it can go as high as 500 lb/in² when minimum clearances are employed.

Design Concepts The pressure gradient existing within the pump elements of all types of screw pumps produces various hydraulic reaction forces. The mechanical and hydraulic techniques employed in the different designs of screw pumps for absorbing these reaction forces are among the fundamental differences in the types produced by various manufacturers. Another fundamental difference lies in the method of engaging or meshing the rotors and maintaining the running clearances between them. Two basic design approaches are used:

1. *Timed rotors,* which rely on external means for phasing the mesh of the threads and for supporting the forces acting on the rotors (see Fig. 3). In this concept, theoretically, the threads do not come into contact with each other or with the housing bores in which they rotate.

2. *Untimed rotors* rely on the precision and accuracy of the screw forms for proper mesh and transmission of rotation. They utilize the housing bores to serve as journal bearings supporting the pumping reactions along the entire length of the rotors (see Fig. 4).

Timed screw pumps require the use of separate timing gears between the rotors and need separate support bearings at each end to absorb the reaction forces and maintain proper clearances. The untimed screw pumps do not require gears or external bearings, and thus they achieve considerable simplification of design.

CONSTRUCTION

Basic Types As indicated in the introduction, there are three major types of screw pumps:

1. Single-rotor.
2. Multiple-rotor—timed.
3. Multiple-rotor—untimed.

The second and third types are available in two basic arrangements—single- and double-ended. The double-end construction (see Fig. 8) is probably the best-known

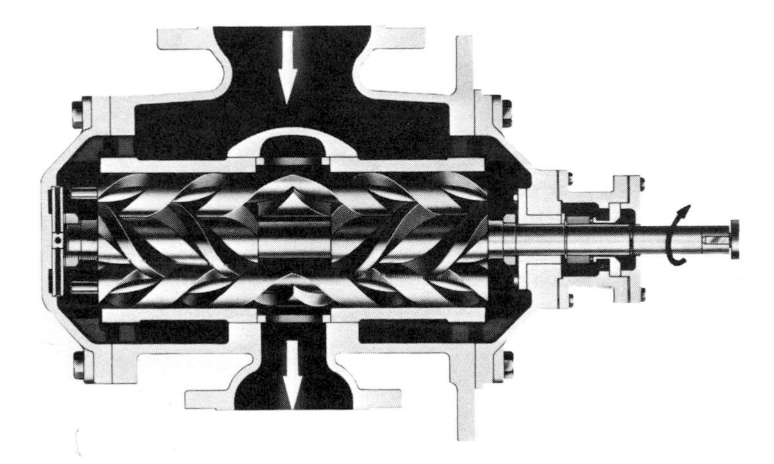

Fig. 8 Double-end pump. Flow path provides axial balance. (DeLaval-IMO)

version as it was by far the most widely used for many years because of the relative simplicity and compactness of design.

Double-end The double-end arrangement is basically two opposed single-end pumps or pump elements of the same size with a common driving rotor of opposed double-helix design within one casing. As can be seen from Fig. 8, the fluid enters a common inlet with a split flow going to the outboard ends of the two pumping elements and is discharged from the middle or center of the pump elements. The two pump elements are, in effect, pumps connected in parallel. The design can also be provided with a reversed flow for low-pressure applications. In either of these arrangements all axial loads on the rotors are balanced out as the pressure gradients in each end are equal and opposite.

The double-end screw-pump construction is usually limited to low- and medium-pressure applications, with 400 lb/in^2 a good practical limit to be used for planning purposes. However, with special design features, applications to 1,400 lb/in^2 can be handled. Double-end pumps are generally employed where large flows are required or where very viscous liquids are handled.

Single-end All three types of screw pumps are offered in the single-end construction. As pressure requirements in many application areas were raised, the single-end design came into much wider use because it provided the only practical means for obtaining the greater number of moving seals necessary for high-pressure capability. The only penalty in the use of the single-end pump is the complexity of balancing the axial loads.

The single-end construction is most often employed for handling low-viscosity fluids at medium-to-high pressures or hydraulic fluids at very high pressures.

The single-end design for high pressure is developed by literally stacking a number of medium-pressure, single-end pumping elements in series within one pump casing (see Fig. 9). The single-end construction also offers the best design arrangement for quantity manufacture.

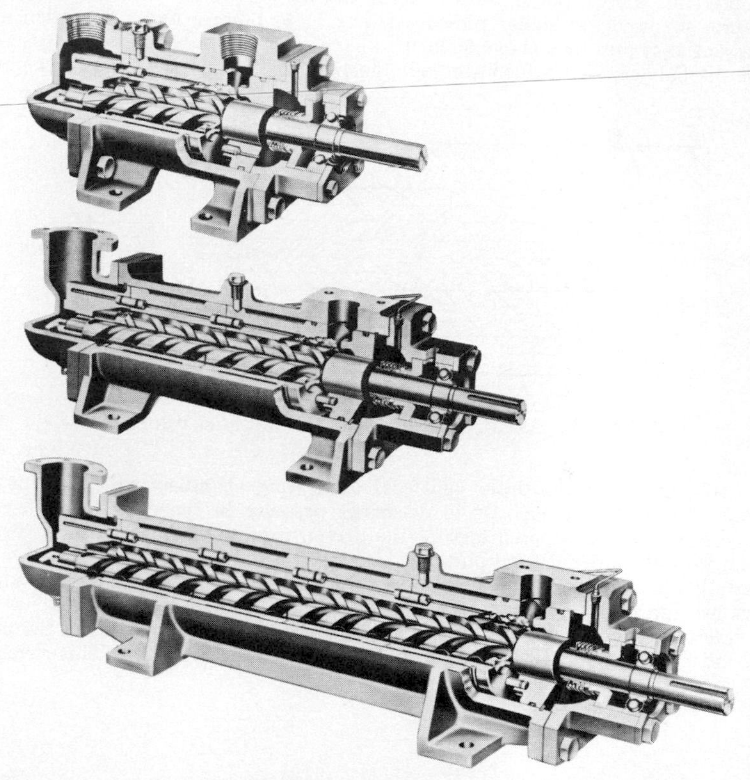

Fig. 9 Increasing pump-pressure capability by modular design. (DeLaval IMO)

Special mention must be made of the single-end, single-rotor design (see Fig. 10). The pump elements of this design consist of only a stator and one rotor. The stator has a double helical internal thread and is constructed of an elastomeric material chemically bonded to a metal tubing. Rotating inside the stator is the rotor consisting of a single external helical thread. The rotor is constructed of hard chrome stainless- or tool-steel material. One version of this design uses internally

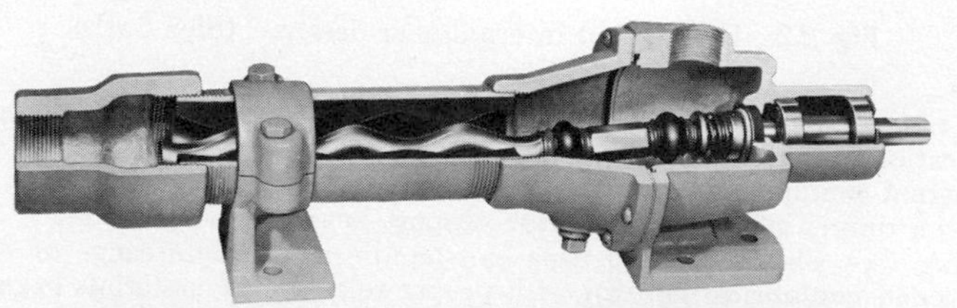

Fig. 10 Single-rotor pump. (Robbins-Meyers)

developed pressure within the pump to cause compression of the elastomeric stator on the rotor, thus maintaining minimum running clearances.

In most single-end designs, special axial balancing arrangements must be used for each of the rotors, and in this respect the design is more complicated than the double-end construction. For smaller pumps, strictly mechanical thrust bearings can be used for differential pressures up to 150 lb/in^2, while hydraulic balance arrangements are used for higher pressures. For large pumps, hydraulic balance becomes essential at pressures above 50 lb/in^2.

Hydraulic balance is provided through the balance piston (see Fig. 11) mounted

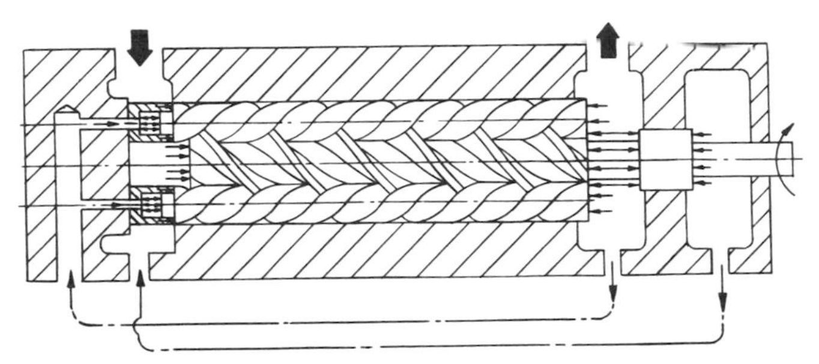

Fig. 11 Axial balancing of power and idler rotors.

on the rotors between the outlet and seal or bearing chambers which are at inlet pressure. This piston is exposed to discharge pressure in the outlet chamber and is equal in area to the exposed area of the driven rotor threads; thus the hydraulic forces on the rotor are canceled out.

Timed Design Timed-type screw pumps having timing gears and rotor support bearings are furnished in two general arrangements—internal and external.

The internal version has both the gears and the bearings located within the pumping chamber and is relatively simple and compact (see Fig. 12). This version is

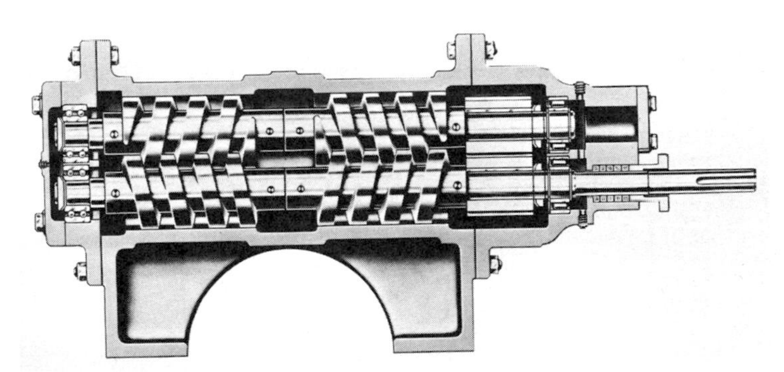

Fig. 12 Double-end internal gear design. (Sier Bath)

generally restricted to the handling of clean lubricating fluids, which serve as the only lubrication for the timing gears and bearings.

The external timing arrangement is the most popular and is extensively used. It has both the timing gears and the rotor support bearings located outside the pumping chamber (see Fig. 13). This type can handle a complete range of fluids, both lubricating and nonlubricating, and with proper selection of materials has good abrasion resistance. The timing gears and bearings are oil-bath lubricated from an ex-

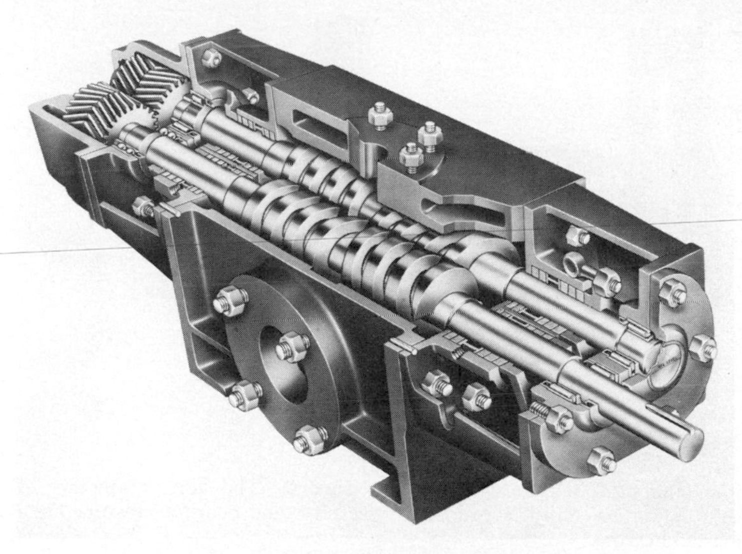

Fig. 13 Double-end external gear design. (Warren)

ternal source. This arrangement requires the use of four stuffing boxes or mechanical seals, as opposed to the internal type which employs only one shaft seal.

The various manufacturers of the timed screw pump claim that the timing gears transmit power to the rotors without the necessity of metallic contact between the screw threads, thus promoting long pump life. The gears are timed at the factory to maintain the proper clearance between the screws. They also claim that the bearings at each end of the rotating elements completely support the rotors so they do not come in contact with the housing; hence no liner is required. One set of these bearings also positions and supports the timing gears.

The timing gears are usually helical, herringbone, hardened-steel gears with tooth profiles designed for efficient, quiet, positive drive of the rotors. Cast iron timing gears are also used. Antifriction radial bearings are usually of the heavy-duty roller type, while the thrust bearings which position the rotors axially are either double-row ball-thrust or spherical-roller types.

The housing can be supplied in a variety of materials including cast iron, ductile iron, cast steel, stainless steel, and bronze. In addition, the rotor bores of the housing can be lined with industrial hard chrome for abrasion resistance.

Since the rotors are not in metallic contact with the housing or with one another, they can also be supplied in a variety of materials including cast iron, heat-treated alloy steel, stainless steel, monel, and nitralloy. The outside diameter of the rotors can also be furnished with hard coatings including tungsten carbide, chrome oxide, and ceramics.

Untimed Design The untimed type of screw pump has rotors which have generated mating-thread forms such that any necessary driving force can be transmitted smoothly and continuously between the rotors without the use of timing gears. The rotors can be compared directly to precision-made helical gears with a high helix angle. This design usually employs three rotor screws with the center or driven rotor in mesh with two close-fitting sealing or idler rotors symmetrically positioned about the central axis (see Fig. 14). A close-fitting housing provides the only transverse bearing support for both the driven and idler rotors.

The use of the rotor housing as the only means for supporting idler rotors is a unique feature of the untimed screw pump. There are no outboard support bearings required on these rotors. The idler rotors in their related housing bores are in effect partial journal bearings which generate a hydrodynamic fluid film preventing metal-to-metal contact. The load-supporting capability of this design follows

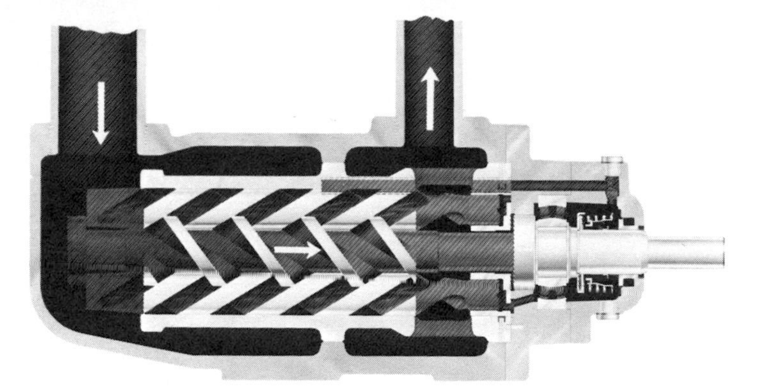

Fig. 14 Single-end design. (Roper)

closely the laws of the hydrodynamic film theory. The key parameters of rotor size, clearance, surface finish, speed, fluid viscosity, and bearing pressure are related as in a journal bearing.

Since the idler rotors are supported by the bores along their entire length, there are no bending loads applied to them. The central driven rotor is also not subjected to any bending loads because of the symmetrical positioning of the idler rotors and the use of two threads on all the rotors (see Fig. 15). This is quite different

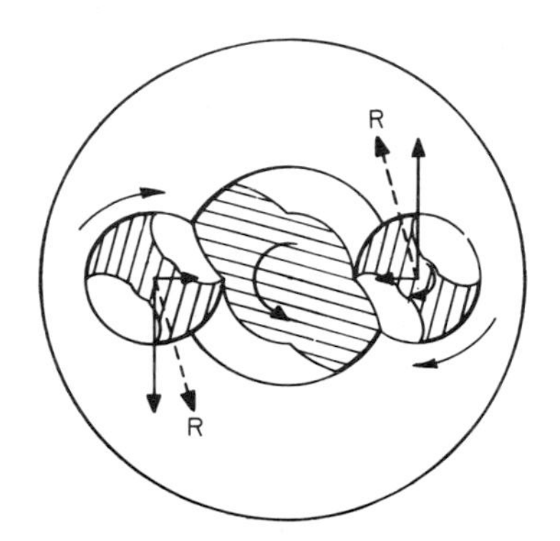

Fig. 15 Force diagram on rotor set.

from the two-rotor design common to the timed type, where the hydraulic forces generated within the pump cause significant bending loads on the intermeshing pairs of screws.

In contrast to the timed pumps, the untimed design with its absence of timing gears and bearings appears very simple, but its success depends entirely upon the accuracy and finish of the rotor threads and rotor housing bores. Special techniques and machine tools have been developed to manufacture these parts. The combination of design simplicity and manufacturing techniques has enabled this design to be used in very long rotor lengths, with a multiplicity of sealing closures, for applications up to 5000 lb/in^2.

In special applications handling highly aerated oils, a four-rotor design is sometimes employed in the untimed type. Three idler rotors are equally spaced radially at 120° around the driven rotor. This design is not a truly positive displacement pump and it falls outside the scope of this handbook.

The rotors in untimed pumps are generally made of gray or ductile iron or carbon steel. The thread surfaces are often hardened for high pressure and abrasive resistance. Flame hardening, induction hardening, and nitriding are currently used. Through-hardened tool steel or stainless steel can be used in some critical applications.

Rotor housings or liners are made of gray pearlitic iron, bronze, or aluminum alloy. In many instances the bores as well as rotors may be treated by the application of dry lubricant or toughening coatings. The pump casings are made of gray iron or ductile iron or cast steel where shock or other safety requirements demand it.

In many untimed designs an antifriction bearing is employed on the shaft end of the driven rotor (see Fig. 9). It is used to provide precise shaft positioning for mechanical seal and coupling alignment. This bearing can be either an external grease-sealed bearing or an internal type with the pumped fluid providing the lubrication. It also acts to support overhung loads with belt or gear drives.

Seals As with any rotary pump, the sealing arrangement for the shafts is very important and often is critical. Every type of rotary seal has been used in screw pumps at one time or another. All types of pumps require at least one rotary seal on the drive shaft. The timed screw pumps with external timing and bearings require additional seals at each rotor end to separate the pumped fluid from the lubricating oil necessary for the gears and bearings.

For drive shafts, rotary mechanical seals as well as stuffing boxes or packings are used, depending on the manufacturer and/or customer preference. Double back-to-back arrangements with a flushing liquid are sometimes used for very viscous or corrosive substances.

The mechanical seal is gaining wider use with the advent of new elastomers such as Viton, Butyl, Nordel, etc. The rotary components are made of carbon, bronze, cast iron, Ni-resist, carbides, or ceramics. The mechanical seal can be designed to be completely or partly independent of the fluid pressure to which it is exposed, and it has also the capability of operating at subatmospheric pressures without drawing in air. In spite of these advantages, the stuffing box is still preferred by some users. The stuffing box requires regular maintenance and tightening, which many users find objectionable, but a mechanical seal failure usually results in a major shutdown.

PERFORMANCE

Performance considerations of screw pumps are very closely related to applications, so any discussion must cover both viewpoints.

In the application of screw pumps there are certain basic factors which must be considered to ensure a successful installation. These are fundamentally the same regardless of the liquids to be handled or the pumping conditions.

The pump selection for a specific application is not difficult if all the operating parameters are known. It is often quite difficult, however, to obtain this information, particularly with reference to inlet conditions and viscosity, since it is a common feeling that, inasmuch as the screw pump is a positive displacement device, these items are unimportant.

In any screw-pump application, regardless of the design, suction lift, viscosity, and speed are inseparable. Speed of operation, therefore, is dependent upon viscosity and suction lift. If a true picture of these two items can be obtained, the problem of making a proper pump selection becomes simpler, and the selection will result in a more efficient unit.

Inlet Conditions The key to obtaining good performance from a screw pump, as with all other positive displacement pumps, lies in a complete understanding and control of inlet conditions and the closely related parameters of speed and viscosity. To ensure quiet, efficient operation, it is necessary to completely fill with liquid the moving cavities between the rotor threads as they open to the inlet, and this becomes more difficult as viscosity, speed, or suction lift increases. Basically it can be said that if the liquid can be properly introduced into the rotor elements, the pump will perform satisfactorily. The problem is getting it in!

It must be remembered that a pump does not pull or lift liquid into itself. Some external force must be present to push the liquid into the rotor threads initially. Normally atmospheric pressure is the only force present, but there are some applications where a positive inlet pressure is available.

Naturally the more viscous the liquid, the greater the resistance to flow; therefore the slower the rate of filling the moving cavities of the threads in the inlet. Conversely, light-viscosity liquids will flow quite rapidly and will quickly fill the rotor threads. It is obvious that if the rotor elements are moving too fast, the filling will be incomplete and a reduction in output will result. The rate of liquid flow into the pumping elements should always be greater than the rate of cavity travel to obtain complete filling. The following are examples of safe internal axial velocity limits found from experience by one screw-pump manufacturer for various liquids and pumping viscosities with only atmospheric pressure available at the pump inlet:

Liquid	*Viscosity, SSU*	*Velocity, ft/s*
Diesel oil	32	30
Lubricating oil	1,000	12
#6 fuel oil	7,000	7
Cellulose	60,000	½

It is thus quite apparent that pump speed must be selected to satisfy the viscosity of the liquid being pumped. The pump speed of rotation is directly related to the internal axial velocity, which in turn is a function of the screw-thread lead. The lead is the advancement made along one thread during a complete revolution of the driven rotor, as measured along the axis. In other words, it is the distance traveled by the moving cavity in one complete revolution of the driven rotor.

Fluids and Vapor Pressure In many cases the screw pumps handle a mixture of liquids and gases, and therefore the general term "fluid" is more descriptive. Most of these fluids, especially petroleum products, because of their complex nature, contain certain amounts of entrained and dissolved air or gas which is released as vapor when the fluid is subjected to reduced pressures. If the pressure drop required to overcome entrance losses is sufficient to reduce the static pressure significantly, vapors are released in the rotor cavities and cavitation results.

Vapor pressure is a key property of fluids which must always be recognized and considered. This is particularly true of volatile petroleum products such as gasoline, for example, which has a very low vapor pressure. Crude oil is an example of a volatile fluid where vapor pressure has been overlooked in the past when applying screw pumps. The vapor pressure of a liquid is the absolute pressure at which the liquid will change to vapor at a given temperature. A common example is: the vapor pressure of water at 212°F is 14.7 lb/in^2. For petroleum products, the Reid vapor pressure (absolute) is usually the only information available. This is vapor pressure determined by the ASTM Standard D-323 procedure.

In all screw pump applications, the absolute static pressure must never be allowed to drop below the vapor presure of the fluid; this will prevent boiling and the release of gases, which again will cause cavitation.

Cavitation, as has been mentioned previously, results when vaporization of fluid occurs in the pump inlet because of incomplete filling of the pump elements and a reduction of pressure. Under these conditions, vapor bubbles or voids pass through the pump and collapse as each moving cavity opens to discharge pressure. The result is noisy vibrations, the severity depending on the extent of vaporization or incomplete filling and the magnitude of the discharge pressure. There is also an attendant reduction in output. It is therefore very important to be fully aware of the characteristics of entrained and dissolved air as well as of the vapor pressure of the fluid to be handled. This is particularly true when a suction lift exists.

Suction Lift Suction lift occurs when the total available pressure at the pump inlet is below atmospheric pressure. It is normally the result of a change in elevation and pipe friction. Although screw pumps are capable of producing a high vacuum, it is not this vacuum that forces the fluid to flow. As previously explained, it is atmospheric or some other externally applied pressure that forces the fluid into the pump. Since atmospheric pressure at sea level corresponds to 14.7 lb/in^2 absolute, or 30 inHg, this is the maximum amount of pressure available for moving the fluid, and suction lift cannot exceed these figures. In practice, a lower value of pressure is available; some of it is used up in overcoming friction in the inlet lines, valves, elbows, etc. It is considered the best practice to keep suction lift just as low as possible.

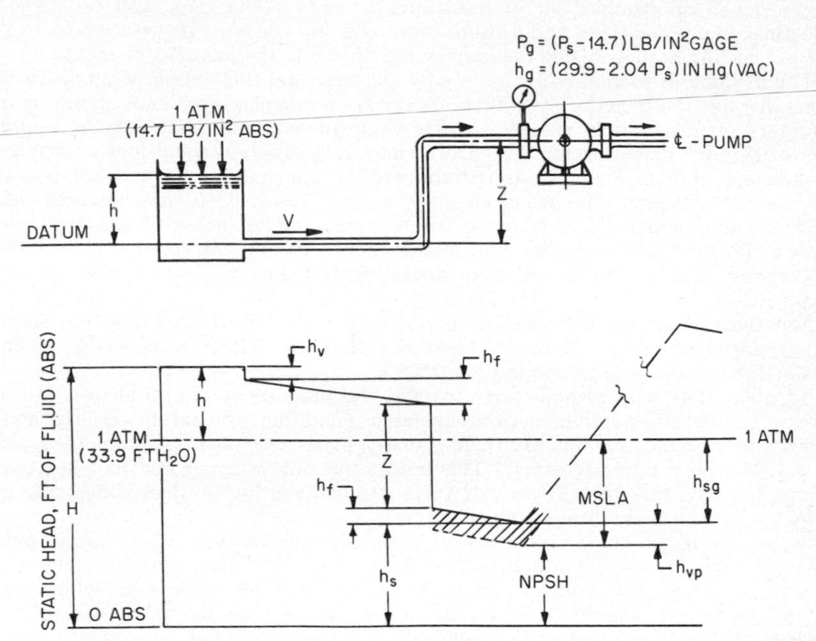

Fig. 16 Relationhip between hydraulic gradient, NPSH, and MSLA.

Total head at source = velocity head + elevation head + static head + friction head loss

$$H \text{ ft} = h + \frac{33.9}{w} = h_v + Z + h_s + \Sigma(h_f) = \frac{V^2}{2g} + Z + \frac{144P_s}{w} + \frac{144P_f}{w}$$

Static head at pump inlet = net positive suction head + liquid vapor pressure, ft abs

$$h_s = \text{NPSH} + h_{vp} \qquad \text{or} \qquad \text{NPSH} = h_s - h_{vp}$$

Maximum suction lift available = NPSH expressed in reference to atm (gage reading)

$$\text{MSLA} = 1 \text{ atm} - \text{NPSH}$$

P_g, h_g = pressure gage readings at pump inlet flange, lb/in² gage and inHg (vac)
P_s = absolute static pressure at pump inlet, lb/in² abs
h_s, h_{sg} = static head at pump inlet, ft of liquid abs or gage
Z = elevation head, ft in reference to datum
h = reservoir liquid level, ft in reference to datum
h_v, h_f = velocity head and friction head loss
P_{vp}, h_{vp} = liquid vapor pressure, lb/in² abs, or head, ft abs
P_{sv} = net positive inlet pressure, lb/in² abs
NPSH = net positive suction head, ft of liquid abs
P_{svr} = net positive inlet pressure required by pump
MSLA = maximum suction lift available from pump, ft of liquid or inHg (vac)
w = specific weight of liquid

The majority of screw pumps operate with suction lifts of approximately 5 to 15 inHg. Lifts corresponding to 24 to 25 inHg are not uncommon and there are numerous installations operating continuously and satisfactorily where the absolute suction pressure is within one-half inch of a perfect vacuum. In the latter cases, however, the pumps are usually taking the fluid from tanks under vacuum, and no entrained or dissolved air or gases are present. Great care must be taken in selecting pumps for these applications since the inlet losses can very easily exceed the net suction head available for moving the fluid into the pumping elements.

The defining of suction requirements by the user and the stating of pump suction capability by the manufacturer have always been complex problems. Some manufacturers are reluctant to publish a single value of minimum inlet pressure required for satisfactory operation for fear that it may not cover all conditions encountered in practice, such as air and gas entrainment or temperature fluctuations resulting in viscosity changes, and hence changes in line losses. One manufacturer refers to maximum suction lift available (MSLA), specified in inches of mercury, which varies with the viscosity of the fluid pumped and the pumping speed. These MSLA ratings are actually determined from operating test data under precisely controlled conditions.

For those engineers accustomed to working with NPSH (net positive suction head), it might prove helpful to point out that the NPSH required by a pump is equal to atmospheric pressure minus MSLA.

To allow the pump manufacturer to offer the most economical selection and also assure a quiet installation, accurate suction conditions should be clearly stated. Specifying a higher suction lift than actually exists results in selection of a pump at a lower speed than necessary. This means not only a larger and more expensive pump, but also a costlier driver. If the suction lift is higher than stated, the outcome could be a noisy pump installation.

There are many known instances of successful installations where screw pumps were properly selected for high suction lift. There are also, unfortunately, many other installations with equally high suction lifts which are not so satisfactory. This is because proper consideration was not given, at the time the pump was specified and selected, to the actual suction conditions at the pump inlet. Frequently suction conditions are given as "flooded" simply because the source feeding the pump is above the inlet. In many cases no consideration is given to outlet losses from the tank or to pipe friction in the inlet lines, and these can be exceptionally high in the case of viscous liquids.

Where it is desired to pump extremely viscous products such as grease, chilled shortening, cellulose preparations, and the like, care should be taken to use the largest possible size of suction piping, to eliminate all unnecessary fittings and valves, and to place the pump as close as possible to the source of supply. In addition, it may be necessary to supply the liquid to the pump under some pressure which may be supplied by elevation, air pressure, or mechanical means.

Entrained and Dissolved Air As mentioned previously, a factor which must be given careful consideration is the possibility of entrained air or gas in the liquid to be pumped. This is particularly true of installations where recirculation occurs and the fluid is exposed to air either through mechanical agitation, leaks, or improperly located drain lines.

Most liquids will dissolve air or gas, retaining it in solution, the amount being dependent upon the liquid itself and the pressure to which it is subjected. It is known, for instance, that lubricating oils under conditions of atmospheric temperature and pressure will dissolve up to 10 percent air by volume and that gasoline will dissolve up to 20 percent.

When pressures below atmospheric exist at the pump inlet, dissolved air will come out of solution; both this and entrained air will expand in proportion to the existent absolute pressure. This expanded air will accordingly take up a proportionate part of the available volume of the moving cavities, with a consequent reduction in delivered capacity. See a more detailed study at the end of this section entitled Effect of Entrained or Dissolved Gas on Performance.

One of the apparent effects of handling liquids containing entrained or dissolved air or gas is noisy pump operation. When such a condition occurs, it is usually dismissed as "cavitation" and let go at that; then too, many operators never expect anything but noisy operation from rotary pumps. This should not be the case, particularly with screw pumps. With properly designed systems and pumps, quiet vibration-free operation can be produced and should be expected. Noisy operation is inefficient; steps should be taken to make corrections until the objectionable conditions are overcome. Proper pump design and size with proper speed selection can go a long way toward overcoming the problem.

Viscosity It is not very often that a screw pump is called upon to handle liquids at a constant viscosity. Normally, because of temperature variations, a wide range of viscosity will be encountered; for example, a pump may be required to handle a viscosity range of 150 to 20,000 SSU, the higher viscosity usually resulting from cold-starting conditions. This is a perfectly satisfactory range for a screw pump, but a better and a more economical selection can be made if information can be obtained concerning such things as the amount of time the pump is required to operate at the higher viscosity, whether the motor can be overloaded temporarily, whether a multispeed motor can be used, or if the discharge pressure will be reduced during the period of high viscosity.

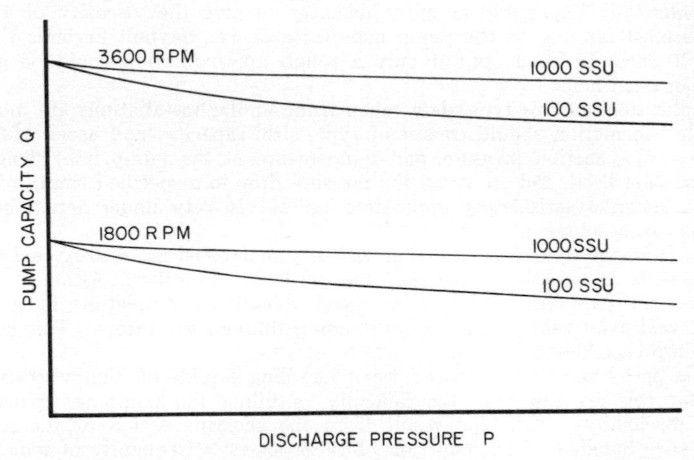

Fig. 17 Head-capacity performance curve with viscosity as parameter for two speeds.

Quite often no viscosity is specified, but only the type of liquid, and assumptions must be made for the operating range. For instance, "Bunker C or No. 6 Fuel Oil" is known to have a wide range as to viscosity and usually must be handled over a considerable temperature range. The normal procedure in a case of this type is to assume an operating viscosity range of 20 to 700 SSF. The maximum viscosity, however, might very easily exceed the higher value if extra heavy oil is used or if exceptionally low temperatures are encountered. If either should occur, the result may be improper filling of the pumping elements, noisy operation, vibration, and overloading of the motor.

Although it is the maximum viscosity and the expected suction lift that determine the size of the pump and set the speed, it is the minimum viscosity that determines the capacity. Screw pumps must always be selected to give the specified capacity when handling the expected minimum viscosity since this is the point at which maximum slip, hence minimum capacity, occurs (see Fig. 17).

It should also be noted that the minimum viscosity often determines the selection of the pump model since most manufacturers have special lower-pressure ratings for handling liquids having a viscosity of less than 100 SSU.

Non-Newtonian Liquids The viscosity of most liquids, as, for example, water and mineral oil, is unaffected by any agitation or shear to which they may be subjected as long as the temperature remains constant; these liquids are accordingly known as "true" or "Newtonian" because they follow Newton's definition of viscosity. There is another class of liquids, however, such as cellulose compounds, glues, greases, paints, starches, slurries, candy compounds, etc., which displays changes in viscosity as agitation is varied at constant temperature. The viscosity of these substances will depend upon the shear rate at which it is measured, and these fluids are termed "non-Newtonian."

If a substance is known to be non-Newtonian, the expected viscosity under actual pumping conditions should be determined, since it can vary quite widely from the viscosity under static conditions. Since non-Newtonian substances can have an unlimited number of viscosity values (as the shear rate is varied), the term "apparent viscosity" is used to describe its viscous properties. Apparent viscosity is expressed in absolute units and is a measure of the resistance to flow at a given rate of shear. It has meaning only if the shear rate used in the measurement is also given.

The grease-manufacturing industry is very familiar with the non-Newtonian properties of its products, as evidenced by the numerous curves that have been published where "apparent viscosity" is plotted against "rate of shear." The occasion is rare, however, when one is able to obtain accurate information as to viscosity when it is necessary to select a pump for handling this substance.

It is practically impossible in most instances to give the viscosity of grease in the terms most familiar to the pump manufacturer, i.e., Saybolt Seconds Universal or Saybolt Seconds Furol; but if only a rough approximation could be given, it would be of great help.

For applications of this type, data taken from similar installations are most helpful. Such information should consist of type, size, capacity, and speed of already-installed pumps, suction pressure, and temperature at the pump inlet flange, total working suction head, and above all the pressure drop in a specified length of piping. From the latter, a satisfactory approximation of viscosity under actual operating conditions can be obtained.

Speed It was previously stated that viscosity and speed are closely tied together and that it is impossible to consider one without the other. Although rotative speed is the ultimate outcome, the basic speed which the manufacturer must consider is the internal axial velocity of the liquid going through the rotors. This is a function of pump type, design, and size.

Rotative speed should be reduced when handling liquids of high viscosity. The reasons for this are not only the difficulty of filling the pumping elements, but also the mechanical losses that result from the shearing action of the rotors on the substance handled. The reduction of these losses is frequently of more importance than relatively high speeds, even though the latter might be possible because of positive inlet conditions.

Capacity The actual delivered capacity of any screw pump, as stated earlier, is theoretical capacity less internal leakage or slip when handling vaporfree liquids. For a particular speed, $Q = Q_t - S$ where the standard unit of Q and S is the U.S. gallon per minute.

The actual delivered capacity of any specific rotary pump is reduced by:

1. Decreasing speed.
2. Decreased viscosities.
3. Increased differential pressure.

The actual speed must always be known; most often it differs somewhat from the rated or nameplate specification. This is the first item to be checked and verified in analyzing any pump performance. It is surprising how often the speed is incorrectly assumed and later is found to be in error.

Because of the internal clearances between rotors and their housing, lower viscosities and higher pressures increase slip, which results in a reduced capacity for a given speed. The impact of these characteristics can vary widely for the various types of pumps encountered. The slip, however, is not measurably affected by changes in speed and thus becomes a smaller percentage of the total flow with the use of higher speeds. This is a very significant factor in the handling of light viscosities at higher pressures, particularly in the case of untimed screw pumps which favor high speed for best results and best volumetric efficiency. This will not generally be the case with pumps having support-bearing speed limits.

Pump Volumetric Efficiency (E_v) is calculated as

$$E_v = \frac{Q}{Q_t} = \frac{Q_t - S}{Q_t}$$

with Q_t varying directly with speed.

As stated previously, the theoretical capacity of a screw pump is a function which varies directly as the cube of the nominal diameter. Slip, however, varies with approximately the square of the nominal diameter. Therefore, for a constant speed

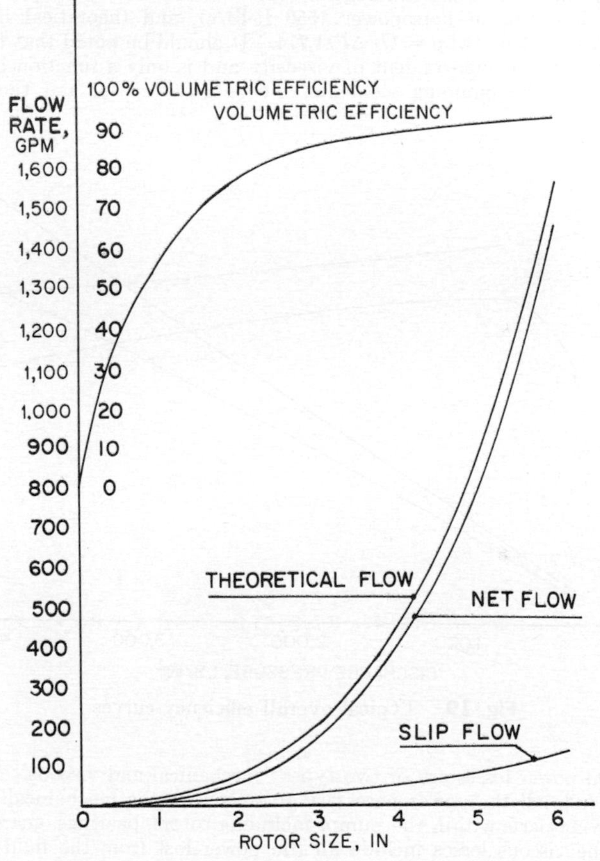

Fig. 18 Capacity and volumetric efficiency as a function of pump size.

and geometry, doubling the rotor size will result in an eightfold increase in theoretical capacity and only a fourfold increase in slip. It follows, therefore, that the volumetric efficiency improves rapidly with increase in rotor size.

On the other hand, viscosity change affects the slip inversely to some power which has been determined empirically. An acceptable approximation for the range 100 to 10,000 SSU is obtained by using the 0.5 power index. Slip varies directly with approximately the square root of differential pressure and a change from 400 SSU to 100 SSU will double the slip in the same way as will a differential pressure change of 100 to 400 lb/in^2.

$$S = K\sqrt{P/\text{viscosity}}$$

Pressure Screw pumps do not in themselves create pressure; they simply transfer a quantity of fluid from the inlet to the outlet side. The pressure developed on the outlet side is solely the result of resistance to flow in the discharge line. The slip characteristic of a particular pump type and model is one of the key factors which determine the acceptable operating range and it is generally well defined by the pump manufacturer.

Horsepower The brake horsepower (bhp) required to drive a screw pump is the sum of the theoretical liquid horsepower and the internal power losses. The theoretical liquid horsepower (twhp) is the actual work done in moving the fluid from its inlet pressure condition to the outlet at discharge pressure. Note: This work is done on all the fluid of theoretical capacity, not just delivered capacity, as slip does not exist until a pressure differential (ΔP) occurs. Screw-pump power ratings are expressed in terms of horsepower (550 ft-lb/s), and theoretical liquid horsepower can be calculated: twhp $= Q_t \, \Delta P/1{,}714$. It should be noted that the theoretical liquid horsepower is independent of viscosity and is only a function of the physical dimensions of the pumping elements, the rotative speed, and the differential pressure.

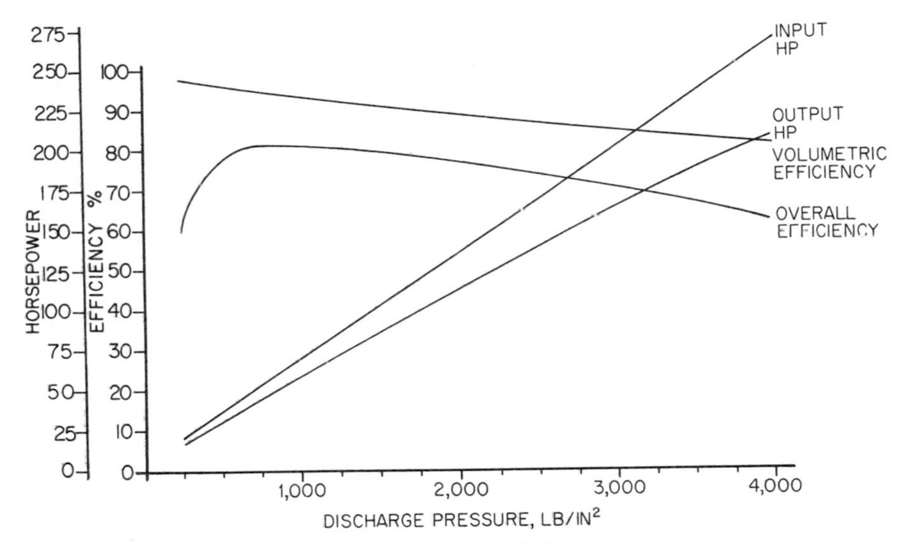

Fig. 19 Typical overall efficiency curves.

The internal power losses are of two types: mechanical and viscous. The mechanical losses include all the power necessary to overcome the mechanical friction drag of all the moving parts within the pump, including rotors, bearings, gears, mechanical seals, etc. The viscous losses include all the power lost from the fluid viscous drag effects against all the parts within the pump as well as from the shearing action of the fluid itself. It is probable that the mechanical loss is the major component when operating at low viscosities and high speeds, while the viscous loss is the larger at high viscosity and slow-speed conditions.

In general, the losses for a given type and size of pump vary with viscosity and rotative speed and may or may not be affected by pressure, depending upon the type and model of pump under consideration. These losses, however, must always be based upon the maximum viscosity to be handled since they will be highest at this point.

The actual pump power output (whp) or delivered liquid horsepower is the power imparted to the liquid by the pump at the outlet. It is computed similarly to theoretical liquid horsepower, using Q in place of Q_t; hence the value will always be less.

The pump efficiency (E_p) is the ratio of the pump power output (whp) to the brake horsepower (bhp).

INSTALLATION AND OPERATION

Rotary pump performance and life can be improved in practice by following the recommendations on installation and operation given below.

Pipe Size Resistance to flow usually consists of differences in elevation, fixed resistances of restrictions such as orifices, and pipe friction. Nothing much can be done about the first, since this is the basic reason for using a pump. Something, however, can be done about restrictions and pipe friction. Literally millions of dollars are thrown away annually because of the use of piping that is too small for the job. To be sure, all pipe friction cannot be eliminated as long as fluids must be handled in this manner, but every effort should be made to use the largest pipe that is economically feasible. Numerous tables are available from which friction losses in any combination of piping can be calculated; among the most recent are the tables in the Hydraulic Institute "Pipe Friction Manual." There are also other similar hydraulic engineering references.

Before any new installation is made, the cost of larger-size piping, which will result in lower pump pressures, should be carefully balanced against the cost of a less expensive pump, smaller motor, and a saving in horsepower over the expected life of the system. The larger piping may cost a little more in the beginning but the ultimate savings in power will often offset many times the original cost. These facts are particularly true of the handling of extremely viscous fluids. Although most engineers dealing with fluids of this type are conscious of what can be done, it is surprising how many installations are encountered where considerable savings could have been made if a little more study had been made initially.

Foundation and Alignment The pump should be placed on a smooth solid foundation readily accessible for inspection and repair. It is essential that the power shaft and drive shafts be in perfect alignment. The manufacturer's recommendation of concentricity and parallelism should always be followed and checked occasionally.

The suction pipe should be as short and straight as possible, with all joints airtight. There should be no points at which air or entrapped gases may collect. If it is not possible to have the fluid flow to the pump under gravity, a foot or check valve should be installed at the end of the suction line, or as far from the pump as possible. All piping should be independently supported to avoid strains on the pump casing.

Start-up A priming connection should be provided on the suction side, and a relief valve should be set from 5 to 10 percent above the maximum working pressure on the discharge side.

Under normal operating conditions with completely tight inlet lines and wetted pumping elements, a screw pump is self-priming. Starting the unit may involve simply opening the pump suction and discharge valves and starting the motor. It is always advisable to prime the unit on initial starting to wet the screws. On new installations, the system may be full of air which must be removed; if it is not, the performance of the unit will be erratic, and in certain cases air in the system can prevent the unit from pumping. Priming the pump should preferably consist of filling not only the pump with fluid but as much of the suction line as possible.

The discharge side of the pump should be vented on the initial starting. Venting is especially essential where the suction line is long or where the pump is discharging against system pressure upon starting.

If the pump does not discharge after being started, the unit should be shut down immediately. The pump then should be primed and tried again. If it still does not pick up fluid promptly, there may be a leak in the suction pipe, or the trouble may be traceable to excessive suction lift from an obstruction, throttled valve, or other causes. Attaching a gage to the suction pipe at the pump will help locate the trouble.

Once the screw pump is in service, it should continue to operate satisfactorily with practically no attention other than an occasional inspection of the mechanical seal or packing for excessive leakage and a periodic check to be certain alignment is maintained within reasonable limits.

Noisy Operation Should the pump develop noise after satisfactory operation, this is usually indicative of excessive suction lift due to cold liquid, air in the liquid, misalignment of the coupling, or, in the case of an old pump, excessive wear.

Shutdown Whenever the unit is shut down, if the operation of the system permits, both suction and discharge valves should be closed. This is particularly important if the shutdown is to be for an extended period because leakage in the foot valve, if the main supply is below the pump elevation, could drain the oil from the unit and necessitate repriming as in the initial starting of the system.

Abrasives There is one other point that has not as yet been discussed and that is the handling of liquid containing abrasives. Since screw pumps depend upon close clearances for proper pumping action, the handling of abrasive fluids will usually cause rapid wear. Much progress has been made in the use of harder and more abrasive-resistant materials for the pumping elements so that a good job can be done in some instances. It cannot be said, however, that performance is always satisfactory when handling liquids laden excessively with abrasive materials. On the whole, screw pumps should not be used for handling fluids of this character unless shortened pump life and an increased frequency of replacement are acceptable.

EFFECT OF ENTRAINED OR DISSOLVED GAS ON PERFORMANCE

The following data have been taken directly from "H. Gartmann, DeLaval Engineering Handbook," 3d ed., McGraw-Hill Book Company, pages 6-56 to 6-59, by permission of the publisher and DeLaval Turbine Inc., IMO Division.

A very important factor in rotary-pump applications is the amount of entrained and dissolved air or gas in the liquid handled. This is especially true if the suction pressure is below atmospheric. It is generally neglected since rotary pumps are of the displacement type and hence are self-priming. If the entrained or dissolved air and gases are a large percentage of the volume handled and if their effect is neglected, there may be noise and vibration, loss of liquid capacity, and pressure pulsations.

The amount of entrained air or gas is extremely variable depending upon the viscosity, type of liquid, and the time and manner of agitation it may have received.

There is little information available covering the solubility of air and other gases in liquids, especially all those handled by rotary pumps. Dr. C. S. Cargoe of the National Bureau of Standards developed the following formula about 1930 based on literature data available at the time to show the solubility of air at atmospheric pressure in oils, both crude and refined, and other organic liquids:

$$\log_{10} A = \frac{792}{t + 460} - 4 \log_{10} \text{sp gr} - 0.4$$

where A = dissolved air, in³/gal

t = temperature, °F

sp gr = specific gravity of the liquid

This equation is plotted as Fig. 20, taken from a paper on rotary pumps by Sweeney in the February, 1943, issue of the *Journal of the Society of Naval Engineers*. The equation and curve should be considered as approximate only, since some liquids have a higher affinity for air and gases. For example, gasoline at atmospheric pressure will dissolve as much as 20 percent of air by volume.

This actual displacement is measured in terms of volume of fluid pumped, and will be the same whether it is liquid, gas, or a mixture of both as long as the fluid can get to and fill the pump moving voids.

If the fluid contains 5 percent entrained gas by volume and no dissolved gas, and the suction pressure is atmospheric, the mixture is then 95 percent liquid and 5 percent gas. This mixture fills up the moving voids on the inlet side, but 5 percent of the space is filled with gas, the remainder with liquid. Therefore, in terms of amount of liquid handled, the output is reduced directly by the amount of gas

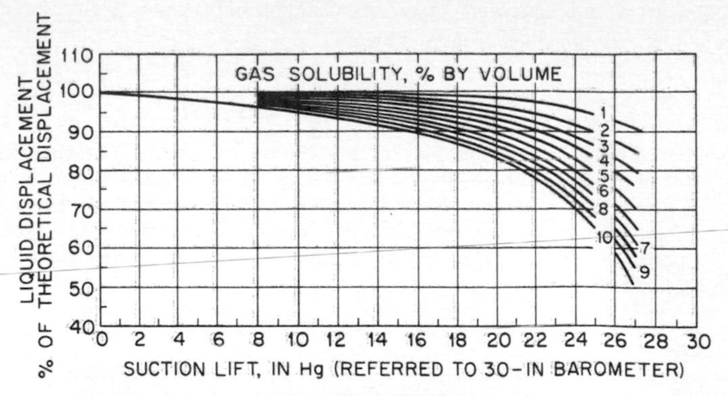

Fig. 20 Effect of dissolved gas on liquid displacement.

present, or 5 percent. The liquid displacement as a function of the theoretical displacement when the suction pressure is atmospheric then becomes

$$D' = D(1 - E)$$

where D = theoretical displacement
D' = liquid displacement
E = percent entrained gas by volume at atmospheric pressure, divided by 100

Assume the fluid handled is a liquid mixture containing 5 percent entrained gas by volume at atmospheric pressure, no dissolved gas, but with the inlet pressure at the pump p_i in psia which is below atmospheric. The entrained gas will increase in volume as it reaches the pump in direct ratio to the absolute pressures. The new mixture will have a greater percentage of gas present, and the portion of theoretical displacement available to handle liquid becomes

$$D' = \frac{D(1 - E)}{(1 - E) + Ep/p_i}$$

where p = atmospheric pressure, psia
p_i = inlet pressure, psia

Note that p_i depends upon the vapor pressure of the liquid, the static lift, and the friction and entrance losses to the pump.

In the above equation, if the atmospheric pressure is 14.7 psia, the pump inlet pressure 5 psia, and the vapor pressure is very low; the liquid displacement is 86.6 percent of the theoretical.

If dissolved gases in liquids are considered, the effect on the liquid displacement reduction is the same as that due to entrained gases, since in the latter case the dissolved gases come out of solution when the pressure is lowered. For example, assume a liquid free of entrained gas, but containing gas in solution at atmospheric pressure and the pumping temperature. So long as the inlet pressure at the pump does not go below atmospheric pressure and the temperature does not rise, gas will not come out of solution. If pressure below atmospheric does exist at the pump inlet, gas will evolve and expand to the pressure existing. This will have the same effect as entrained gas taking up available displacement capacity, and reduce the liquid displacement accordingly. The liquid displacement then will be

$$D' = \frac{D}{1 + y(p - p_i)/p_i}$$

where the symbols have the meanings given above and y is the percent of dissolved gas by volume at pressure p divided by 100. If the operating conditions are 9 percent of dissolved gas at 14.7 lb/in^2 abs with a pump inlet pressure of 5 lb/in^2 abs, the liquid displacement will be 85.2 percent of the theoretical displacement.

If both entrained and dissolved gases are considered as existing in the material to be pumped, the liquid displacement becomes

$$D' = \frac{Dp_i(1 - E)}{(1 - E)[p_i + y(p - p_i)] + Ep}$$

where the symbols have the meanings given above. For the operating conditions 5 percent entrained gas, 9 percent dissolved gas at 14.7 psia, and a pump inlet pressure of 5 lb/in² abs, the liquid displacement is 75.2 percent of the theoretical. Figure 21 shows graphically the reduction in liquid displacement as a function of pump inlet pressure, expressed in terms of suction lift, for different amounts of dissolved gas, neglecting slip.

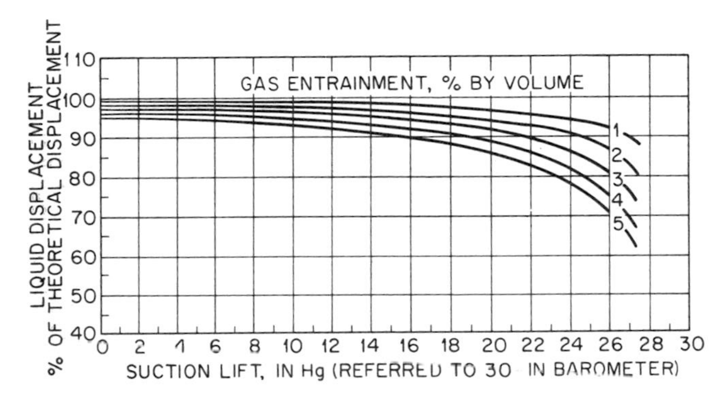

Fig. 21 Effect of entrained gas on liquid displacement.

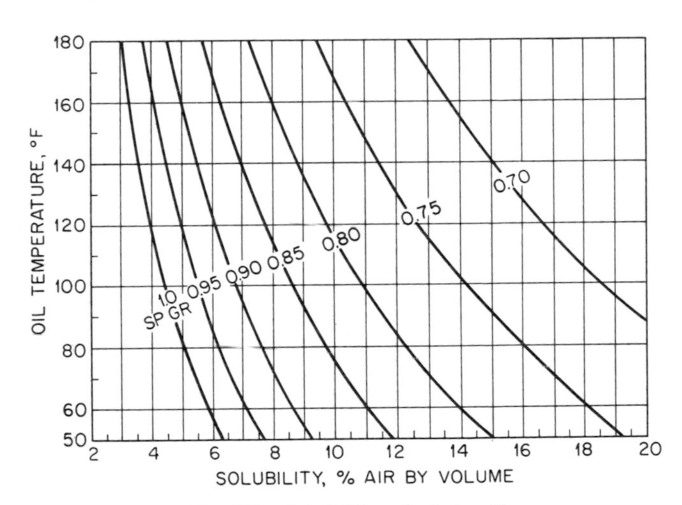

Fig. 22 Solubility of air in oil.

Figure 22 shows the reduction in liquid displacement as a function of pump inlet pressure, expressed as suction lift, for different amounts of entrained air only, neglecting slip. From this figure it may be noted that a very small air leak can cause a large reduction in liquid displacement, especially if the suction lift is high.

From these few examples and curves it would appear that the question of entrained and dissolved gases could be cared for by providing ample margins in pump capacity. Unfortunately, capacity reductions from the causes mentioned are attended by other and usually more serious difficulties.

The operation of a rotary pump is such that as rotation progresses, closures are formed which fill and discharge in succession. If the fluid pumped is compressible, such as a mixture of oil and air, the volume within each closure is reduced as it comes in contact with the discharge pressure. This produces pressure pulsations, the intensity and frequency of which depend upon the discharge pressure, the number of closures formed per revolution, and the speed of rotation. Under some conditions the pressure pulsations are of high magnitude and can cause damage to piping and fittings or even the pump, and will almost certainly be accompanied by undesirable noise.

The amount of dissolved air or gas may be reduced by lowering the suction lift. This may often be controlled by pump location, suction pipe diameter, and piping arrangement.

Many factors are associated with the amount of entrained air that can exist in a given installation. It is prevalent in systems where the liquid is handled repeatedly and during each cycle is exposed to, or mechanically agitated in, air. Unfortunately in many cases the system is such that air entrainment cannot be entirely eliminated, as in the case of the lubrication system of a reduction gear. Considerable work has been done by the various oil companies on foam dispersion, and while it has been recommended that special oils be used which are inhibited against oxidation and corrosion, all agree that the best cure is to remove or reduce the cause of foaming, namely, air entrainment.

Even though air entrainment cannot be entirely eliminated, in many cases it is possible, by adhering to the following rules, to reduce it and its ill effects on rotary-pump performance.

1. Keep liquid velocity low in the suction pipe to reduce turbulence and pressure loss. Use large and well-rounded suction bell to reduce entrance loss.
2. Keep suction lift low. If possible locate the pump to provide positive head on the inlet.
3. Locate the suction piping within a reservoir to obtain the maximum submergence.
4. Submerge all return lines, particularly from bypass and relief valves, and locate them away from the suction.
5. Keep the circulation rate low and avoid all unnecessary circulation of the fluid.
6. Do not exceed rated manifold pressures on machinery lubricating systems, as the increased flow through sprays and bearings increases the circulation rate.
7. Heat the fluid where practical to reduce viscosity and as an expedient to drive off entrained air. Fluids of high viscosity will entrain and retain more air than fluids of low viscosity.
8. Avoid all air leaks no matter how small.
9. Provide ample vents; exhauster fans to draw off air and vapors have been used with good results.
10. Centrifuging will break a foam and remove foreign matter suspended in the oil, which promotes foaming.
11. Use a variable-speed drive for the pump to permit an adjustment of pump capacity to suit the flow requirements of the machinery.

Section **3.4**

Rotary Pumps

C. W. LITTLE, Jr.

DEFINITIONS AND NOMENCLATURE

Rotary pumps are rotary positive-displacement pumps in which the main pumping action is caused by relative movement between rotating elements of the pump and stationary elements of the pump. Their rotary motion distinguishes them from reciprocating positive-displacement pumps, in which the main motion of moving elements is reciprocating. The positive-displacement nature of their pumping action distinguishes them from the general class of centrifugal pumps in which liquid displacement and pumping action depend in large part on developed liquid velocity.

It is characteristic of a rotary pump, as a positive-displacement pump, that the liquid displaced by each revolution of the pump is independent of speed. Also, it is characteristic of rotary pumps that a time-continuous liquid seal of sorts is maintained between the inlet and outlet ports of the pump by the action and position of the pumping elements and the close running clearances of the pump. Hence rotary pumps generally do not require inlet and outlet valve arrangements as reciprocating pumps do.

Certain general actions are common to all of the large number of types of rotary pumps. The following terms are useful in describing these actions. This nomenclature is consistent with that used in the Hydraulic Institute Standards.[1]

Rotary pumps are useful in handling both fluids and liquids, where *fluid* is a general term including liquids, gases, vapors, and mixtures thereof, and sometimes including solids in suspension, and where *liquid* is a more specific term that is limited to true liquids that are relatively incompressible and relatively free of gases, vapors, and solids.

Parts of a Rotary Pump The *pumping chamber* is generally defined as all the space inside the pump that may contain the pumped fluid while the pump is operating. Fluids enter the pumping chamber through one or more *inlet ports* and leave the pumping chamber through one or more *outlet ports,* all of which usually will include arrangements for liquidtight and airtight connections to external fluid systems. The *body* is that part of the pump which surrounds the boundaries of the pumping chamber and is sometimes called a *casing* or a *housing.* In a very few rotary pumps the body also may be a rotating assembly, but in most types it is stationary and is sometimes called the *stator. Endplates* are those parts of the body or those separate parts which close the ends of the body to form the pumping chamber. They are sometimes called *pump covers.*

The *rotating assembly* generally includes all the parts of the pump that rotate when the pump is operating, while the *rotor* is the specific part of the rotating assembly which rotates within the pumping chamber. Rotors may be given specific

[1] "Hydraulic Institute Standards," 13th ed., 1975, Hydraulic Institute, 2130 Keith Bldg., Cleveland, Ohio 44115.

names in specific types of rotary pumps; they may be called gears, screws, etc. Most rotary pumps have *drive shafts* which accept driving torque from a power source. The majority of rotary pumps are mechanically coupled to the driving power source with *couplings* of various types, but a few are coupled to the driving source magnetically or electromagnetically in configurations called *sealless drives*.

Pump *seals* are of two general types, static and moving. *Static seals* provide a liquidtight and airtight seal between demountable stationary parts of the pumping chamber, and *moving seals* are used at pumping-chamber boundary locations through which moving elements extend, usually shafts. Moving seals also are formed between pump rotors in some types of rotary pumps. A cavity in the pump body through which a shaft extends is called a *seal chamber,* and leakage through the seal chamber is controlled by either a *radial seal,* which seals on its ouside diameter through an interference fit with its mating bore and on the rotating shaft with a radially loaded sliding surface, or a *mechanical seal,* in which two seal faces are opposingly loaded axially and maintained close to each other at all times. Sometimes a *stuffing box* is used instead of a seal chamber. A compressible sealing material called *packing* is compressed in the stuffing box by a part called a *gland* or *gland follower,* which keeps the stuffing in intimate contact with the stationary and rotating surfaces in the stuffing box. Devices called *lantern rings* or *seal cages* are used to allow lubrication or cooling of the stuffing or to control the net pressure on the stuffing in the stuffing box.

Where drive shafts are used, the generally accepted direction of rotation is determined as clockwise or counterclockwise when viewing the pump from the driver end of the drive shaft. In multiple-rotor positive-displacement pumps, torque may be transmitted to the rotors, and the angular relationship between them maintained, by the action of *timing gears,* sometimes called *pilot gears*.

A number of auxiliary devices and arrangements may be used with rotary pumps, but two are somewhat characteristic for positive-displacement pumps. The pressure at the outlet port and in the outlet portion of the pumping chamber can become damagingly high if the pump discharge is obstructed or blocked, and *relief valves* are used to limit the pressure there by opening an auxiliary passage at a predetermined pressure. The valve may be integral with the body, or integral with an endplate, or attachable. The low fluid velocity (relative to some centrifugal pumps) of the fluid flowing through rotary pump chambers permits some control of the temperature of the pump or of the fluid in the pump by passages or jackets in or on the pump body or endplates through which an auxiliary fluid may be circulated to transfer heat to or from the pump fluid. Such pumps are called *jacketed pumps*.

Pumping Action of Rotary Pumps The pumping sequence in all rotary positive-displacement pumps includes three elementary actions. The rotating and stationary parts of the pump act to define a volume, sealed from the pump outlet and open to the pump inlet, which grows as the pump rotating element rotates. Next, the pump elements establish a seal between the pump inlet and some of this volume, and there is a time, however short, when this volume is not open to either the inlet or the outlet parts of the pump chamber. Then the seal to the outlet part of the chamber is opened, and the volume open to the outlet is constricted by the cooperative action of the moving and stationary elements of the pump. In all the many types of rotary pumps the action of the pumping volume elements must include these three conditions: *closed-to-outlet open-to-inlet, closed-to-outlet closed-to-inlet, open-to-outlet closed-to-inlet*. For a good pumping action the open-to-inlet (OTI) volume should *grow* in volume smoothly and continuously with pump rotation, the closed-to-inlet-and-outlet (CTIO) volume should *remain constant* in volume with pump rotation, and the open-to-outlet (OTO) volume should *shrink* in volume smoothly and continuously, with pump rotation. At no time should any fluid in the pumping chamber be open to both inlet and outlet ports simultaneously if the pump is truly a positive-displacement pump.

ROTARY PUMP TYPES

The major types of rotary pumps are named and described below. Although this is not an exhaustive list, the types listed illustrate all the various rotary-pump principles of operation commonly in use.

Gear Pumps Gear pumps are rotary pumps in which two or more gears mesh to provide the pumping action. It is characteristic that one of the gears be capable of driving the others. The mechanical contacts between the gears form a part of the moving fluid seal between the inlet and the outlet ports, and the outer radial tips of the gears and the sides of the gears form a part of the moving fluid seal between inlet and outlet ports. The gear contact locus moves along the tooth surfaces and then jumps *discontinuously* from tooth to tooth as the gears rotate. (These two characteristics distinguish gear pumps from lobe pumps, in which the rotors are *not* capable of driving each other, and in which fluid seal contact locus between lobes moves *continuously* across all the radial surfaces of the lobes.)

The two main types of gear pumps are *external gear pumps* and *internal gear pumps*. Generally external gear pumps are arranged so that the center of rotation of each element is external to the major diameter of an adjoining gear and all gears are of the external-tooth type. The center of rotation of at least one gear in an internal gear pump is inside the major diameter of an adjoining gear, and at least one gear is an internal-tooth type or a crown-tooth type.

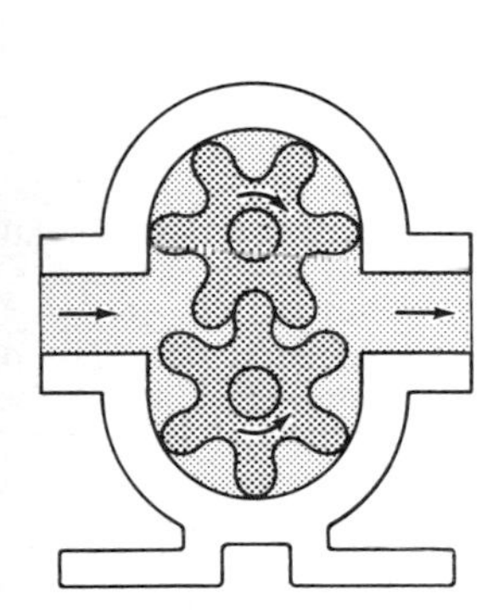

Fig. 1 External gear pump.

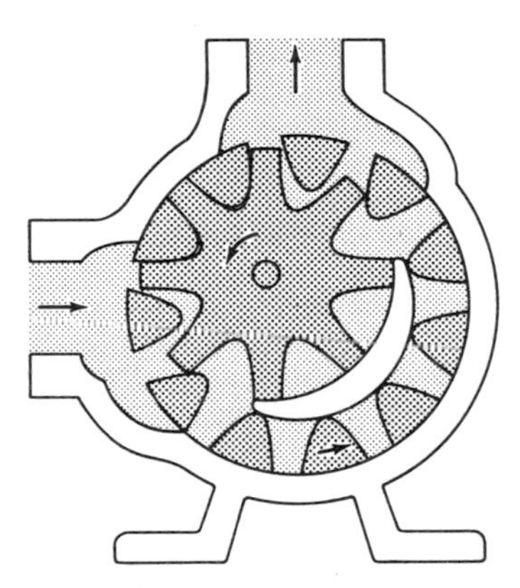

Fig. 2 Internal gear pump.

Figure 1 shows a section through an external gear pump, and Figure 2 shows a section through an internal gear pump. In these drawings, and in those of following types, the unshaded zones are parts of the body or stator, the lightly shaded zones are areas where liquid may be present in the pump chamber, and the dark zones are parts of the rotating assembly.

The OTI volume of the pump chamber in gear pumps is defined by the body walls and by the gear-tooth surfaces between the fluid seal points where the gears mesh and the fluid seal points where each gear-tooth tip meets and seals with the body walls as it leaves the OTI volume. The fluid trapped between the gear teeth and the body walls is sealed from both inlet and outlet chambers and is the CTIO volume. The OTO volume is defined by the body walls and those gear-tooth surfaces between the fluid seal points where each gear-tooth tip leaves the body wall and enters the OTO volume and the fluid seal points where the gears mesh.

A part or all of the side (or axial) surfaces of the gears runs in small-clearance contact with the axial end faces of the pumping chamber. The gear teeth run in small-clearance contact with each other where they mesh. The gear-tooth tips run in small-clearance contact with radial surfaces of the pumping chamber in their travel from the OTI volume to the OTO volume. Load-bearing contact between rotors or between rotors and the stator may exist in all three of these zones, and the apertures defined by the running clearances in these zones determine the amount of fluid leakage from the OTO volume to the OTI volume for any given pressure difference between these volumes, for any given effective fluid viscosity. (See discussion of *slip* later in this section.)

Pumping torque is shared by both rotors, and the proportional amount of the total torque felt by each rotor at any instant of time is determined by the locus of the fluid seal point between rotor gear teeth. As this fluid seal point moves *toward* the center of rotation of a rotor gear, the pumping torque on that gear *increases,* and as it moves *away* from the center the torque *decreases.* Methods of computing this torque are given later. External timing gears are not necessary in gear pumps, but they may be used to transfer torque between rotating assemblies to avoid rapid wear between rotor teeth when the pump handles nonlubricating liquids.

A special form of a gear pump is illustrated in Fig. 3. It is called a *screw and wheel pump.* The driving gear is a helical gear, and the driven gear is a special form of a spur gear. The helical gear always is the driving, or power, rotor in this type of pump, and external timing gears are not used. The pumping torque in the screw and wheel pump is felt both by the screw and by the wheel, and the amount of torque felt by each is determined by the fluid seal contact locus

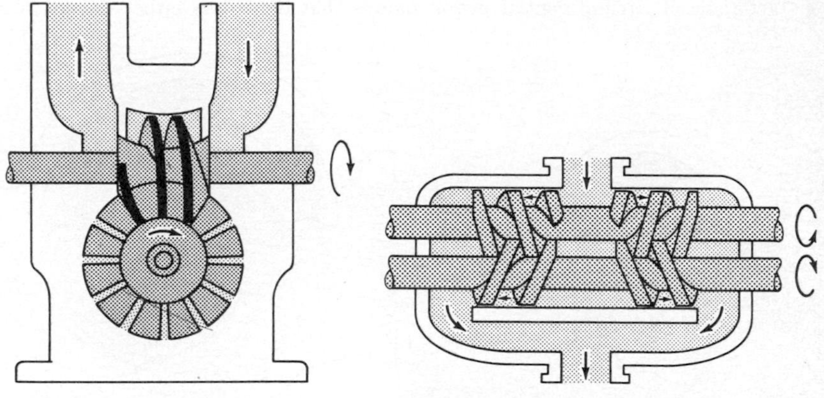

Fig. 3 Screw and wheel pump. **Fig. 4** Multiple rotor screw pump.

points between the two rotors. As in other gear pumps, the running clearances between rotors and between rotors and the body walls determine fluid leakage from the OTO volume to the OTI volume.

Multiple Rotor Screw Pumps Figure 4 shows a multiple rotor screw pump. Usually the screw-shaped rotors in this type of pump cannot drive each other, and timing gears are required. The OTI volume is defined by the body walls and by the screw surfaces between the fluid seal meshing contact between screws and the fluid seal contact between the major diameter of the screw flights and the body wall where the screws adjoin the OTI volume. The CTIO volume is trapped in the screw flights between the fluid seal points adjoining the OTI and the OTO volumes and the screw-mesh fluid seal points there. The OTO volume is defined by the body walls and by the screw surfaces between the fluid seal meshing contact between screws and the fluid seal contact between the major diameter of the screw flights and the body wall where the screws adjoin the OTO volume.

The major diameters of the screw rotors run in small-clearance contact with stator walls. The sides of the screw flights run in small-clearance contact with each other where they mesh. Load-bearing contact may exist in both these zones, and the apertures defined by the running clearances determine the amount of fluid leakage from the OTO volume to the OTI volume for any given pressure difference and effective fluid viscosity.

The nature of the contact locus between screws where they mesh determines the pumping torque felt by each rotating assembly. In most two-screw pumps the pumping torque is divided equally between the screws at all times. The axial end faces of each screw section feel thrust force generated by the pressure difference between the OTO volume and the OTI volume. Consequently, as illustrated, opposing sections are usually used in screw pumps designed for high-pressure service, to balance the large net end-thrust forces that otherwise would be imposed on thrust bearings at the rotor ends. Screw pumps are discussed in more detail in Sec. 3.3.

Circumferential Piston Pumps An external circumferential piston pump is illustrated in Fig. 5. Pistonlike rotor elements, supported from cylindrical hubs inset into the pump end plate, travel in circular paths in mating body bores. This is called "external" because the centers of rotation of the rotors are external to the major diameter of adjoining rotors. The OTI volume is defined by the body walls and by the rotor-piston element surfaces between the contact locus points where the rotors form fluid seals with stator walls as they enter (zone between rotors) or leave (top and bottom zones) the OTI volume. The rotors do *not* mesh or touch, and fluid seals exist *only* between the rotor and stator surfaces, and not between rotors. The OTO volume is defined by the surfaces of rotor-piston elements between their fluid seal contacts with stator walls where they leave (zone between rotors) or enter (top and bottom zones) the OTO volume and by body walls. It is characteristic of circumferential piston pumps that the rotors either do not mesh

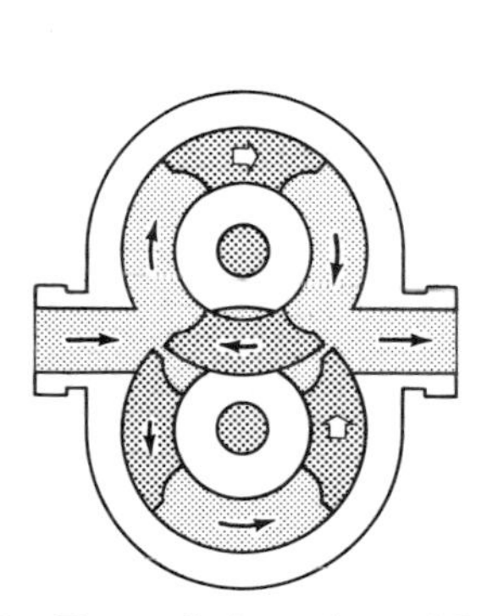

Fig. 5 External circumferential piston pump.

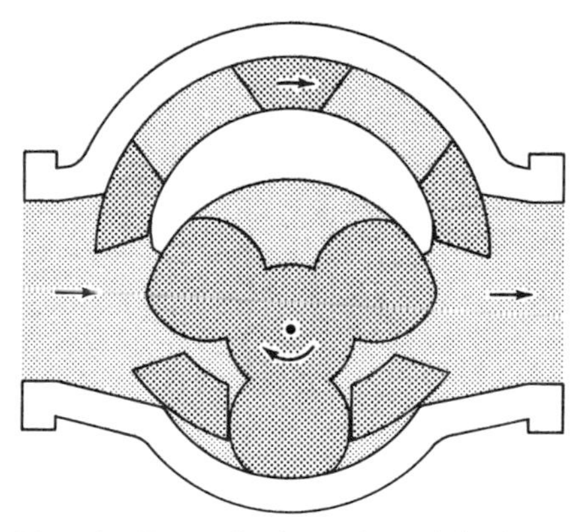

Fig. 6 Internal circumferential piston pump.

with, or contact, each other, or that they form no fluid seal if they do. This distinguishes them from lobe, gear, and screw pumps.

The radial surfaces and the axial-end surfaces of the rotor-piston elements run in close-clearance contact with body walls, and load-bearing contact may exist in these zones. The apertures defined by the running clearances there determine the amount of fluid leakage from the OTO volume to the OTI volume for a given pressure difference and a given effective fluid viscosity.

Each rotor in an external circumferential piston pump feels the full pumping torque alternately.

Figure 6 shows an internal circumferential piston pump, where the center of rotation of one of the rotors is *inside* the major diameter of the other rotor. Unlike the internal gear pump, which it physically resembles, there is no fluid seal between rotors in the internal circumferential piston pump. In this type of pump the smaller, or *idler,* rotor operates with balanced torque, and all the pumping torque is felt continuously by the larger, or power, rotor. With no pumping-torque transfer from rotor to rotor, timing gears are usually not needed, even for a pump handling nonlubricating liquids.

Lobe Pumps The lobe pump receives its name from the rounded shape of the rotor radial surfaces which permits the rotors to be continuously in contact with each other as they rotate. Figure 7 shows a single lobe pump and Figure 8 shows a multiple lobe pump.

Unlike gear pumps, neither the number of lobes nor their shapes permit one rotor to drive the other, and all true lobe pumps require timing gears. The OTI volume is defined by the body surfaces, the rotor surfaces, the contact between rotors, and the contact between rotor lobe ends and the body. The CTIO volume is defined by the contacts between lobe ends and the body wall and the adjoining body wall and lobe surfaces. The OTO volume is defined by the body walls, the rotor surfaces, the lobe-to-body-wall contacts, and the lobe-to-lobe contacts. In the two-rotor lobe pump the torque is shared by both rotors, with the proportional amount of torque dependent on the position of the rotor-to-rotor contact point on the rotor contact locus. When the contact point is at the major locus radius (maximum lobe radius

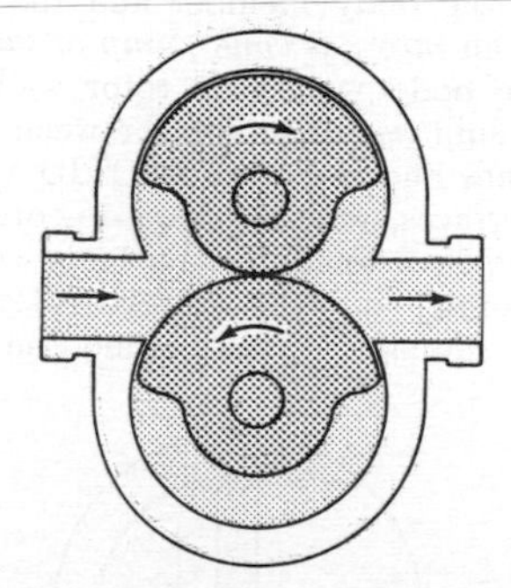

Fig. 7 Lobe pump, single lobe.

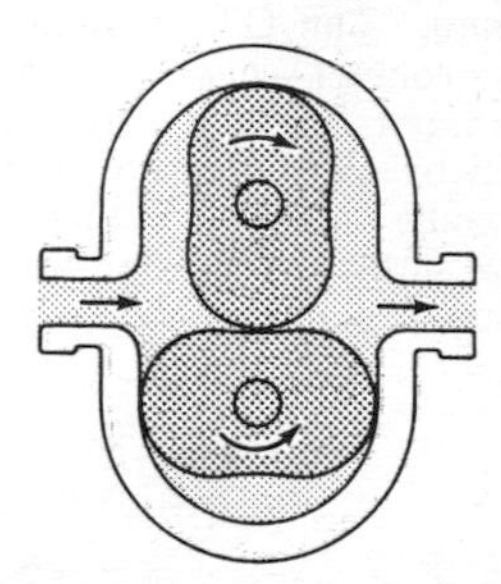

Fig. 8 Lobe pump, multiple lobes.

of one rotor in contact with minimum lobe radius of adjoining rotor), one rotor feels the full pumping torque while the other rotor sees a balanced torque. The full-pumping-torque transfer from one rotor to the other takes place as many times in each complete revolution of a rotor as there are lobes on the rotor.

The *internal lobe pump,* sometimes called an "internal gear pump without crescent," is one in which a single rotor with lobelike peripheral shape is moved in a combination of rotation and gyration of center of rotation in a body with internal lobe-shape contours in such a way that the rotor always contacts the body at two or more locations to preserve the fluid seal between OTI and OTO volumes. It is illustrated in Fig. 9.

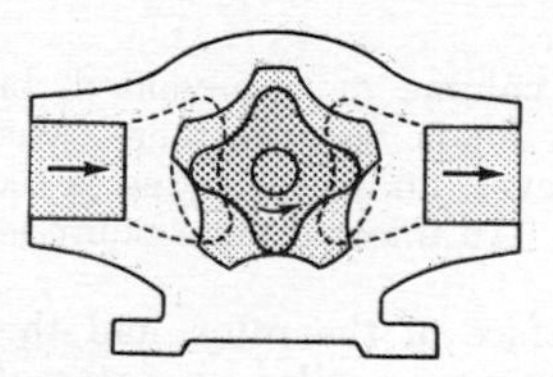

Fig. 9 "Internal gear" or "internal lobe" pump.

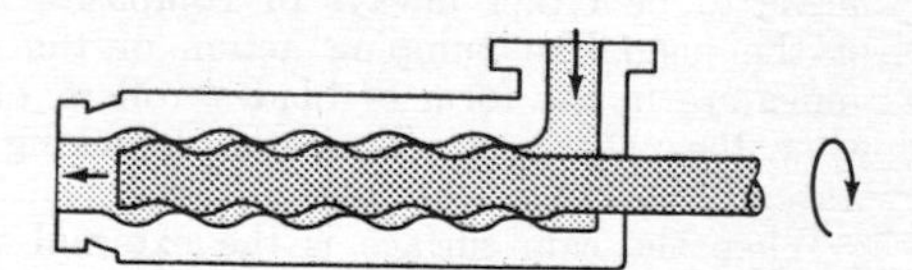

Fig. 10 Single screw pump.

The OTI volume is defined by the outer rotor surface and the inner body surface and the fluid seal points between them. The CTIO volume is defined by the outer rotor surface and the inner body surface between two adjacent fluid seal points, and the OTO volume is defined by the outer rotor surface, the inner body surface, and the rotor-to-body fluid seal points. Most pumps of this type have one less rotor lobe than internal body lobe cavity; consequently the term "progressing tooth gear pump" is sometimes used. The full pumping torque is felt by the single rotor, but the torque is cyclic, dependent on the position of the rotor and its sealing arrangement with the body, the number of torque cycles per rotor revolution being equal to the number of lobes on the rotor.

The *single screw pump,* illustrated in Fig. 10, is similar in principle of operation to the internal lobe pump described above, except that the cavity progresses *axially* along a screwlike member with lobular outer surfaces as it rotates in a body with internal matching lobular grooves. A high thrust load at high pressures is felt by the rotor in the single-ended version illustrated, but this could be balanced by the use of two opposing sections, as in the two-screw pump illustrated in Fig. 4. The full pumping torque is felt by the single rotor.

Rigid Rotor Vane Pumps In the various types of rigid rotor vane pumps, movable sealing elements in the form of rigid blades, rollers, slippers, shoes, buckets, etc., are moved, generally radially inward and outward by cam surfaces, to maintain a fluid seal or seals between the OTI and OTO volumes during the operation of the pump. When the cam surface is internal to the body member and the vanes are mounted in or on the rotor, the pump is called an *internal vane pump* or *vane-in-rotor pump*. The OTI volume is defined by the body walls, the rotor walls, the fluid seal contact between rotor and body, the fluid seal contacts between vanes and rotor, and the fluid seal contacts between vanes and body. The CTIO volume is defined by the body wall surface, the rotor surfaces, and the vane-to-rotor and vane-to-body fluid seal points. The OTO volume is defined by the body surface, the rotor surface, the vane-to-body fluid seal points, and the vane-to-rotor fluid seal points. In internal vane pumps, the volume radially inward behind the vanes

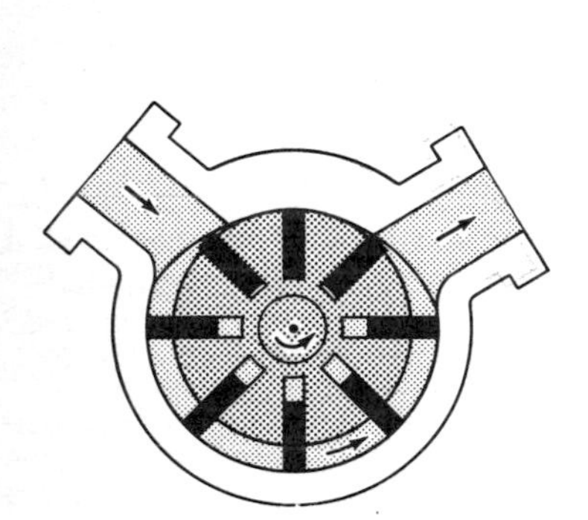

Fig. 11 Internal (vane-in-rotor) pump.

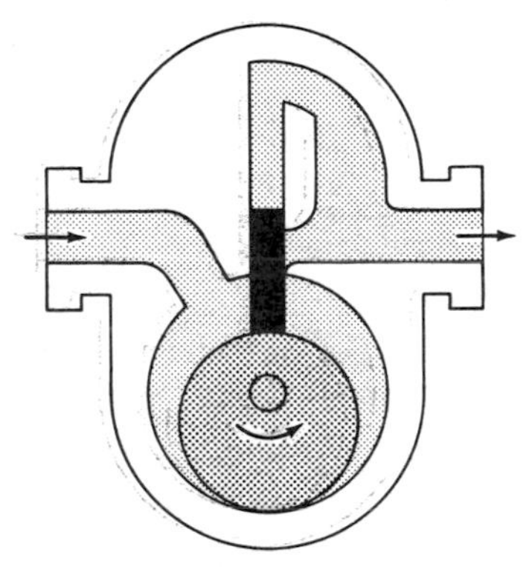

Fig. 12 External (vane-in-body) pump.

needs to be either always of composite constant volume or else vented, because of the pistonlike pumping action of the vanes on fluids trapped there, when the vanes are in the form of blades, rollers, etc. However, no such venting is required when the vanes are in the form of rocking slippers. An internal vane pump is illustrated in Fig. 11.

When the cam surface is the external radial surface of the rotor, and the vane or vanes are mounted in the body or stator, the pump is called an *external vane pump* or *vane-in-body pump*. The OTI, CTIO, and OTO volumes are defined as for internal vane pumps when multiple external vanes are used. Figure 12 shows a single-vane external vane pump. In this case the CTIO volume is defined by the rotor surface, the body surface, and the fluid seal points between them. Vane pumps are single-rotor pumps, and this rotor feels the full pumping torque.

Rotary Piston Pumps Rotary piston pumps are true rotary pumps that require no inlet or outlet valves. They get their name from the pistonlike elements that reciprocate in bores of the rotor as the pump rotor rotates. The pistonlike elements operate off camming surfaces in much the same way that blade and roller vanes do in vane pumps, but the pumping action comes directly from the reciprocal movement of the piston elements in their cavities. When the piston elements move axially as the rotating element rotates, the pump is called an *axial piston pump*. The OTI volume is defined by the inlet chamber body walls and by the part of each piston-element cylinder exposed to the inlet chamber as it passes. The CTIO volume is defined by the cylinder-to-body-wall and cylinder-to-piston-element volume when the cylinder is sealed from both inlet and outlet, and the OTO volume is defined by the outlet chamber walls and the part of each piston-element cylinder exposed by the piston when open to the outlet chamber. The camming action is provided by an inclined plate, adjustable or fixed in angle of incline, that determines the axial position of each piston element as a function of rotor angular position. Since it is a single-rotor pump, the full pumping torque is felt by the rotor.

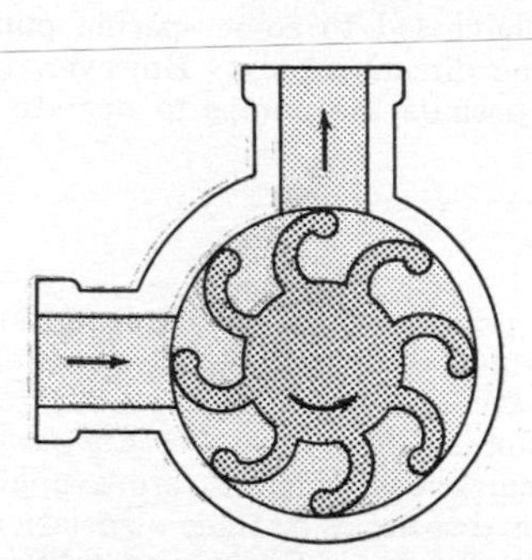

Fig. 13 Flexible vane pump.

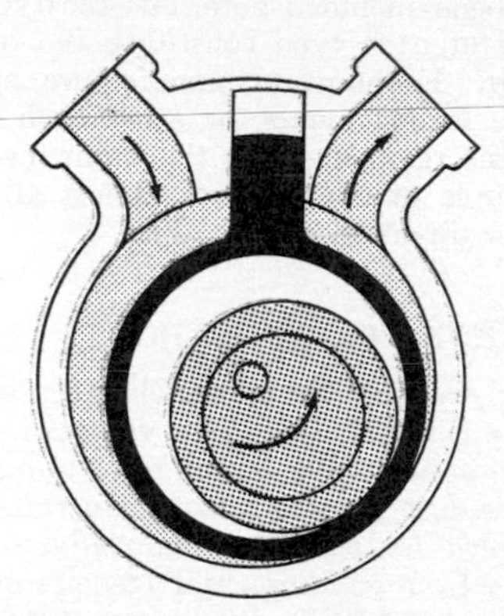

Fig. 14 Flexible liner pump.

A *radial piston pump* has the same pumping action, but the radially reciprocal piston-element movements in the cylinder and bores are generally caused by an eccentric camming surface on the shaft.

Flexible Member Pumps Several types of rotary positive-displacement pumps depend on the elasticity of flexible parts of the pump. The *flexible vane pump,* illustrated in Fig. 13, has a pumping action similar to that of an internal vane pump with the OTI, CTIO, and OTO volumes defined by the rotor surfaces and the body surfaces and the fluid seal contacts between the rotor flexible vanes and the body surfaces.

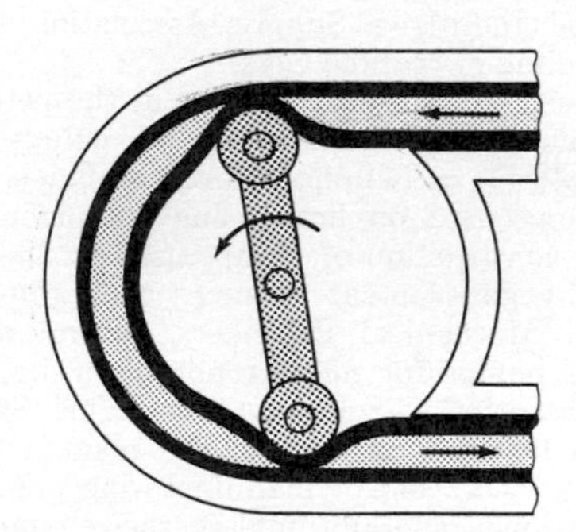

Fig. 15 Flexible tube pump.

The *flexible liner pump,* illustrated in Fig. 14, is similar in pumping action to the external vane pump, and all three chamber volumes are defined by the inner surface of the body, the outer surface of the liner, and the liquid seal contact between liner and body bore. Most flexible liner pumps, *unlike other rotary pumps,* have at least one position of the rotor in which there is no fluid seal between the OTI and OTO volumes. The pump depends only on fluid velocity and inertia to limit backflow during this position.

The *flexible tube pump,* illustrated in Fig. 15, is one in which the three volumes are bounded only by the inner surface of the flexible tube and defined by the locus of the compression points on the tube by the rollers or shoes and the body wall. The OTI volume is the tube volume between the inlet and the first compression, or "nip," point; the CTIO volume is that contained between two nip points; and the OTO volume is contained in the tube volume between the pump outlet and the adjacent nip point. Flexible member pumps, being single-rotor pumps, feel all the pumping torque on the single rotor.

Other Types There are many other possible types of rotary pumps which have not been included here, but the types listed above and the numerous design variations of each type constitute the bulk of rotary positive-displacement pumps used today. Economy of manufacture, special pressure-balancing arrangements or relief-valve arrangements, or other such considerations have led to some specific pump designs requiring that the pump be operated in one direction only. However, the principle of operation of almost all rotary pumps permits the pumps to operate in either direction equally well.

INDUSTRY CLASSIFICATIONS

Overlying the pump-type classification according to principle of operation are rotary-pump classes designed for various industrial, governmental, or military specifications. *It is important that both the manufacturer and the user know of any voluntary or regulatory specifications governing the construction and materials of rotary pumps destined for use in specific industries.* These specifications generally are available either from governmental agencies or from professional societies or trade associations.

For example, the International Association of Milk, Food and Environmental Sanitarians, the United States Public Health Service, and the Dairy Industry Committee have jointly issued specifications called "The Three-A Sanitary Standards for Pumps for Milk and Milk Products." These standards govern the materials of construction of the pump, the surface finish and shape details of fluid contact surfaces, the finish and shape of external surfaces, the method of mounting, even to the details of the legs for the pump base, and restrictions on openings, gaskets, seals and other pump auxiliary features. Pumps constructed to meet "The Three-A Sanitary Standards" are called *sanitary pumps* and can carry the Three-A seal of approval. Similar specifications have been generated by the International Association of Milk, Food and Environmental Sanitarians, the United States Public Health Service, the United States Department of Agriculture, the Institute of American Poultry Industries, and the Dairy and Food Industries Supply Association to cover design features in pumps used in the handling of cracked eggs.

Standards for pumps used in various processes in the petroleum industry are concerned with the materials and design features of pumps to prevent catastrophic failure when handling explosive or flammable or toxic fluids. Other standards govern the specifications for pumps used on firefighting equipment; military specifications govern the materials and construction of pumps used by the armed services; detailed specifications generated by government agencies and professional societies, such as the American Society of Mechanical Engineers, control manufacturing procedures and design limitations of pumps for nuclear power service, etc. In most cases, the descriptors "sanitary," "aseptic," "explosion proof," "N Stamp," etc., may not be applied casually but can be used only when the manufacturer warrants the pump to meet the specifications. The cost of manufacturing pumps to meet these various special industry specifications generally makes these pumps more expensive than those intended for general service.

Three main industry *classes* of rotary pumps are:

1. The class of rotary pumps used in commercial, industrial, governmental, or military *liquid-handling* tasks. In this class an immense variety of commercially important liquids (well over 1,500) are handled. The performance of the pump is judged on its ability to handle the specific liquids in the specific applications. In this class the hydraulic power generated by the pump is usually secondary, and the liquid-handling task is primary. Liquid handling may include liquid transfer and delivery, control of liquid amount transferred, control of liquid flow rates, control of delivery-point pressures, etc.

2. The class of rotary pumps used in *hydraulic-power* applications. In this class, the hydraulic power generated by the pump is of primary importance, and hydraulic

fluids are carefully developed and selected to complement pump performance and hydraulic power-system requirements. Rotary-pump types commonly used in this class include gear and vane pumps, with design features permitting operation at *high pressures* (up to several thousand pounds per square inch) on selected hydraulic-power fluids.

3. The class of rotary pumps specifically designed for use in a single type of application or on a single type of liquid, in the category of *retail* equipment. This is a miscellaneous class ranging from miniature pumps for home aquariums to fuel-injection pumps for automotive vehicles.

The first class has the widest variety of pump types and industry specifications.

MATERIALS OF CONSTRUCTION

General information on materials of construction is given in Section 5. The following paragraphs cover only those particular properties of materials important to rotary pumps.

Special consideration in the selection of materials for use in rotary pumps is needed in five areas: Rotary pumps constructed of rigid materials require consideration of the material's *modulus of elasticity, coefficient of friction and nongalling properties* in sliding contact, and *coefficient of thermal expansion;* rotary pumps with flexible members additionally require consideration of the material's *bulk modulus* and *time of recovery* following deformation.

The close running clearances in most rotary pumps constructed of rigid materials impose the requirement that these materials resist deformation and deflection by the various forces present when the pump is operated. Otherwise such deformation or deflection could open the clearances to lower operating efficiency dramatically, or could close the clearances to cause high mechanical loading or seizing between moving parts and stationary parts. These same considerations require the selection of materials for compatible coefficients of thermal expansion when the pump must operate at a variety of ambient and process temperatures. Even though care is taken to select materials of high rigidity and similar thermal expansion coefficients, the deflection of rotating assemblies to cause relatively high load-bearing sliding contact between rotating and stationary parts requires that these materials have good bearing characteristics and resistance to galling, up to the point of compressive yield of the mating materials. This is even more important when the pumped liquid has no lubricity. Furthermore, some materials used in centrifugal pumps to provide resistance to corrosion by the pumped fluid, where surfaces are noncontacting, may not be usable in rotary pumps where sliding contact between rotor and stator may continually wipe or wear away the passivating or protective layer on the materials.

In general, these material-selection restrictions become more severe when the pump is to handle low-viscosity fluids at high pressures, particularly those without lubricity or those that contain abrasive fines, and they become less severe when the pump is to handle lubricating, clean fluids at medium or high viscosities and at medium or low pressures, where larger operating clearances may be used. Even in those pumps where either the mechanical design or the hydraulic-pressure balancing design would permit operation of the pump without load-bearing contact between rotary and stationary parts in steady-state dynamic conditions, the high transient forces generated in startup or shutdown, or during unusual operating conditions (such as those accompanying heavy cavitation), may cause sufficient deflection to warrant the same consideration in the selection of materials as for pumps in which load-bearing contact is ordinary.

The performance of flexible-member pumps is dependent on the material chosen for the flexible member. The bulk modulus must be high enough to keep distortion of the flexible member under pressure within functional limits. In many cases, as in the flexible vane and flexible tube pumps, the ability of the flexible member to spring back to its original shape (recovery) after the flexing or compression by the pumping action is essential. If the tube in the tube pump, once flattened, were to remain flat and not spring back to its tubular shape, the pump would not operate. If the vanes in the vane pump, once deflected by a cam surface, were to remain deflected, the pump would not operate. Consequently the materials for flexible members in flexible-member rotary pumps should be chosen not only for the initial properties that produce the desired pumping effect, but also for resistance to the deterioration of these properties from fatigue, chemical attack by the fluid being pumped, and the temperatures to which they may be exposed in pumping service.

Frequently the natural properties of a flexible material may be strengthened and enhanced by lamination or by filling with other materials in various forms that may limit deflection or bulging or that may supply additional snapback or recovery after deflection. Other reinforcing materials, such as metal cores molded into flexible vane rotors, may be used to support the flexible member at locations where high stresses may exist.

Where the pump must meet limiting specifications imposed by various industry applications, additional constraints may be imposed on material selection. For example, nonsparking metals not subject to brittle fracture may be required when the pump is to be used to handle flammable or explosive fluids. Only certain materials are approved for use in sanitary pumps for food and dairy service, and the pump bodies of pumps for use in certain petroleum processing applications cannot be constructed of materials such as gray cast iron which may fail in brittle fracture. Many materials commonly used in centrifugal pumps, because of these restraints, cannot be used in rotary pumps.

Most rotary pumps are low-speed pumps. Hence the torque transmitted through the drive shaft to the rotating assembly in rotary pumps is usually much higher than that required to be transmitted to rotating impellers in centrifugal pumps for the same liquid horsepower. This is particularly true where the rotary pump is pumping a high-viscosity liquid against a high differential pressure; both the materials and the mechanical design in these high-torque locations sometimes are critical, disqualifying some materials that otherwise would be suitable for use in rotary pumps.

Because of the complexity of these many design considerations, it is not practical to outline a complete procedure for selecting materials for every given application. It should be noted, however, that a pump perfectly suitable for handling a lubricating liquid at medium or low pressures may fail miserably if used to pump a nonlubricating liquid, or a liquid containing abrasive, or a high-temperature liquid, or a high-viscosity liquid at high pressures. Thus there are some general considerations other than corrosion resistance that affect the choice of materials for each application.

OPERATING CHARACTERISTICS

The operating characteristics of rotary pumps are covered in the following paragraphs. For clarity, it is assumed first that the pumped fluid is a true liquid with Newtonian viscosity, such that the fluid is incompressible, and that the fluid's resistance to shear (shear stress) is in direct proportion to the rate of shear (shear strain). The effects of the pumped fluid on operating characteristics, including viscosity, vapor pressure, temperature, multiphase fluids, liquids with gases, liquids with solids, liquids with abrasives, shear-sensitive liquids, lubricity, etc., are discussed later. A common liquid used in evaluation tests of basic pump performance is cool water, with or without very small amounts of soluble oil for some lubricity. Hence water may be assumed to be the fluid in the following discussion of basic operating characteristics.

Pump Displacement The displacement (D) of a rotary pump is the total net fluid volume transferred from the OTI volume to the OTO volume during one complete revolution of the driving rotor. A standard unit of displacement is cubic inches per revolution. For any given pump the displacement depends only upon the physical dimensions of the pump elements and the pump geometry and is independent of other operating conditions. In those pumps designed for variable displacement, the pump usually is rated at its maximum displacement.

The displacement of any rotary pump may be computed by the general method of integrating, over one complete revolution of the driveshaft, the differential rate of net volume transfer with respect to angular displacement of the driveshaft through any complete plane segment taken through the pump chamber between the inlet and outlet ports. For any given type of pump, the coordinate system should be selected and the plane located to simplify the computation. Most pump rotors have constant radial dimensions in the axial direction in the body cavity and sweep a right circular cylinder of volume while rotating. Consequently in single-rotor pumps, or in multiple-rotor pumps in which no sealing contact exists between rotors (all dynamic seals are formed between rotor elements and body surfaces and none are formed between rotors), the volume-transfer computation may be based on polar coordinates centered on each rotor axis and the contribution to net volume transfer can be computed for each rotor independently.

In general, the axial dimension of the rotor in the body cavity can be most simply expressed if the plane segment is taken through the rotor axis, or at least parallel to the rotor axis. Also, for most types of rotary pumps, the computation is simplified if the intersections of the plane with the body cavity occur in closed-to-inlet, closed-to-outlet (CTIO) regions, usually midway between the inlet and outlet ports of the

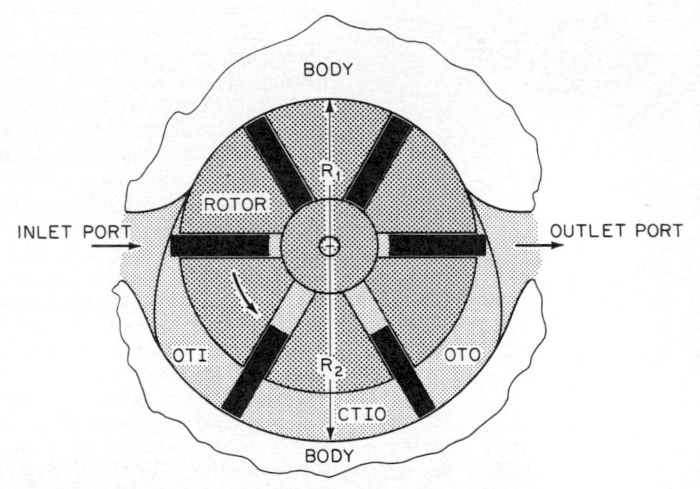

Fig. 16 Displacement calculation dimensions, internal vane pump.

pump. This is particularly true for those rotary pumps which pump equally well in either direction of rotation and are generally symmetric. In many cases the computation can be further simplified by separately stating the differential statement for volume transfer through the plane from the inlet to the outlet, and that for volume transfer through the plane from the outlet to the inlet, and expressing the result as a difference. Examples of this method of computation for some commonly used types of rotary pumps follow.

A section through a vane-in-rotor pump is shown in Fig. 16.

Let z be the axial distance toward the front end plate from the rotor end surface next to the rear end plate, and let Z be the total axial length of the rotor. Let r be the radial distance from the rotor axis. Let R_1 be the minimum radial dimension of the rotor elements at the intersection of the plane with the minor cam radius of the pump chamber in the CTIO zone, and let R_2 be the maximum radial dimension of the rotor elements at the intersection of the plane with the major cam radius of the pump chamber in the CTIO zone. Let ϕ be the angular displacement of the drive shaft (assumed to be direct-coupled to the rotor; no gear increase or decrease). Then the general equation for D is as follows.

$$D = \int_{\phi=0}^{\phi=2\pi} \int_{r=R_1}^{r=R_2} \int_{z=0}^{z=Z} kr\, d\phi\, dr\, dz = k\pi Z(R_2^2 - R_1^2) \tag{1}$$

where k is a constant used to convert D into desired units ($k = 1$, if z and r are in feet and D is in ft^3/rev).

It may be noted that the actual equation describing the transition of the major radius cam surface to the minor radius cam surface is not used or needed in the equation, because the plane segment is entirely in the CTIO zone of the pump. Also, the integration limits for r were chosen by noting that the net volume transfer for all $r < R_1$ cancels and equals zero. The same result is obtained if the integral is expressed as the difference of the positive contribution of the integration limits O to R_2 and the negative contribution of the integration limits O to R_1.

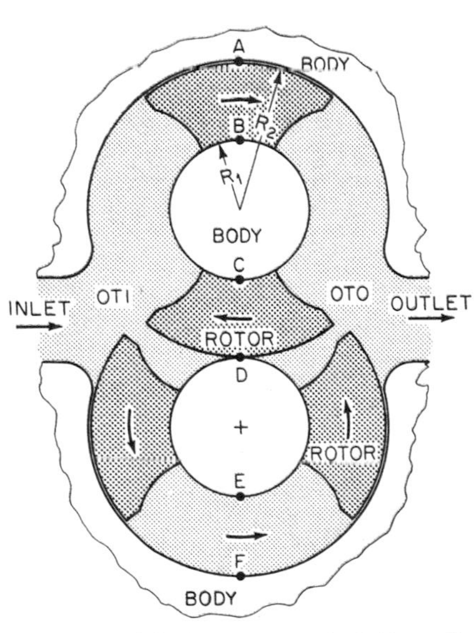

Fig. 17 Displacement calculation dimensions, external circumferential piston pump.

The same computations and formula would apply to flexible vane pumps with the vanes on the rotor, and for any vane-in-rotor pump where the camming surface creating the pumping action is formed on the interior surface of the body chamber. The equation can be used for most single-rotor positive-displacement pumps with appropriate selection of the numerical values for R_1 and R_2. For example, in the vane-in-body pump, R_1 is the minimum radius of the rotor cam and R_2 is the maximum radius of the rotor cam.

An example extending this method of computing displacement to multiple-rotor pumps that have no rotor-to-rotor sealing contact is the computation on the external circumferential piston pump shown in Fig. 17. A derivation can be made in the same manner used for the vane pump, and the individual net-volume-flow equations for each rotor can be computed and summed.

The pump is symmetrical, and the actual net contribution of each rotor occurs only during a total of half of the drive-shaft revolution. In the derivation shown in Eq. (1), this would mean a change in the integration limits of ϕ from two pi to pi for the contribution of each rotor. However, when the contribution of each of the two rotors is added, the result is the same as Eq. (1).

Another method could be used. Because of the pump symmetry, inspection would show that there are two net volume components continually transferred from the inlet to the outlet chamber by the motion of the rotors in zones A–B and E–F and one equal volume element continually transferred from the outlet to the inlet chamber by the cooperative action of the rotors in zone C–D. Consequently the net volume being transferred from outlet to inlet in zone C–D is continually canceled by one of the volumes continually being transferred from inlet to outlet at either zone A–B or zone E–F. The entire computation, then, can be made only for zone A–B or zone E–F over the entire revolution. The result would be the same; the resulting formula for displacement would be that given in Eq. (1), with R_1 being the minimum radial dimension of the piston element of the rotor and R_2 being the maximum radial dimension of the piston element of the rotor.

The direct computation of displacement for multiple-rotor positive-displacement pumps where a moving seal is formed by contact between rotors of the pump is much more complex. In such pumps the equation of motion of the locus of the contact point between rotors as a function of angular displacement of the drive shaft is needed for a rigorous solution of the displacement. The actual differential rate of volume transfer is not constant with angular displacement, but decreases as the contact locus moves toward the inlet chamber and increases as the contact locus moves toward the outlet chamber. A graph of pump displacement D versus angular displacement would not be a straight line (or constant function of angular displacement) as in the prior two examples. In effect, the motion of the contact locus superimposes a "ripple" on the steady-state component of the differential rate of volume transfer.

Figure 18 shows the locus of the contact line between lobes in a two-lobe pump. The locus is shown on a plane taken perpendicular to the plane segment used for computing displacement. The plane segment used for computing displacement is taken through the axis of both rotors, with the center of rotation of the two rotors

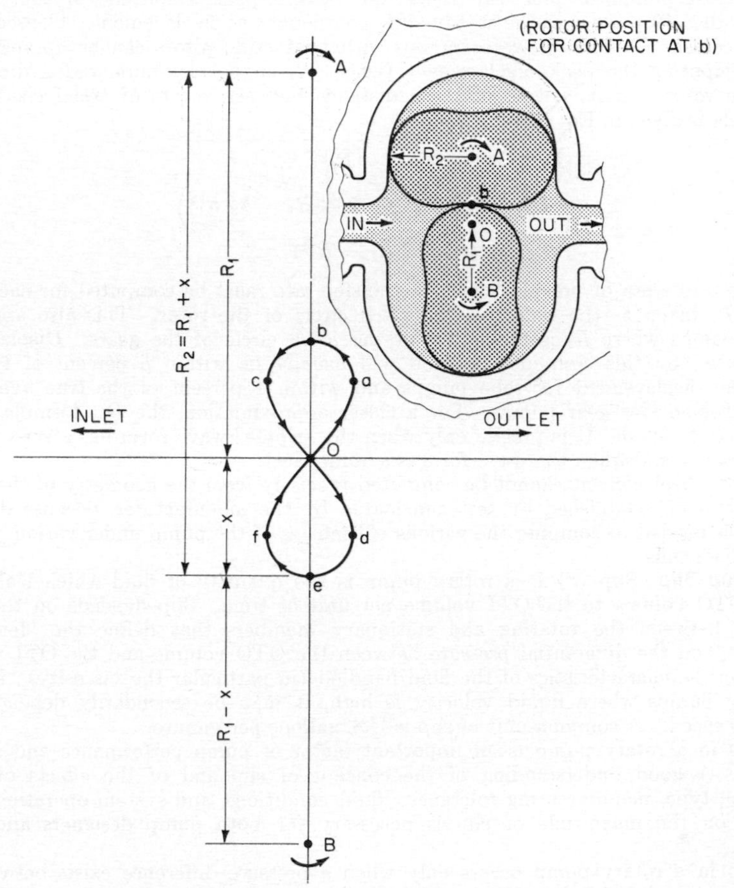

Fig. 18 Rotor contact locus of lobe pump.

being points A and B. Point O is a point midway between the two rotor centers. The small letters represent key points in the differential rate of volume transfer. Remembering that the velocity of the movement of the locus point toward or away from the outlet port determines the amount of plus or minus deviation from the average differential transfer rate, the following observations can be made. The *maximum* instantaneous differential volume-transfer rate occurs as the locus passes through point O on its way from c to d or from f to a. The *minimum* instantaneous rate of differential transfer occurs when the locus passes through point b or point f. The average differential rate of transfer occurs at points a, c, d, and f. In the symmetrical case shown, if R_1 is the distance from point O to A or B, if x is the distance from point O to b or e, and if R_2 is the maximum radial dimension of a lobe, then derivations similar to those given before will give the maximum and minimum differential rates of volume transfer shown in Eq. (2).

$$\frac{dD}{d\phi_{max}} = kZ(R_2^2 - R_1^2)$$

$$\frac{dD}{d\phi_{min}} = kZ(R_2^2 - R_1^2 - x^2) = kZ[2(R_2R_1 - R_1^2)] \qquad (2)$$

since for the lobe pump shown $R_2 = R_1 + x$

In most pumps of practical design the peak-to-peak amplitude of the "ripple" is less than 10 percent of the steady-state component of displacement. Consequently displacement for multiple-rotor pumps with contacting rotors can be *approximated* by computing the peak displacement (where R_2 is the maximum radial dimension of the rotor and R_1 is one half the distance between rotors of equal size). The formula is given in Eq. (3).

$$D_{max} = 2\pi kZ(R_2^2 - R_1^2)$$

$$D_{av} = 2\pi kZ\left(\frac{R_2^2}{2} + R_2R_1 - \tfrac{3}{2} R_1^2\right) \qquad (3)$$

$$D_{min} = 4\pi kZ(R_2R_1 - R_1^2)$$

If the rotors are of unequal size, the transfer rate must be computed for each rotor with R_1 taken as the radius of the pitch circle of the rotor. This also applies to gear pumps where R_1 is the radius of the pitch circle of the gears. Displacement computed by this simplified method will usually be within 5 percent of the true average displacement for lobe pumps and within 1 percent of the true average of displacement for gear pumps. For a closer approximation, the D_{av} formula of Eq. (3) may be used. It is precise only when the "ripple" wave form has a zero average component and when the wave form is symmetrical.

If the displacement cannot be computed precisely from the geometry of the pump, it should be established by test and stated by the manufacturer, because displacement is needed to compute the various efficiencies of the pump under various operating conditions.

Pump Slip Slip (S) in a rotary pump is the quantity of fluid which leaks from the OTO volume to the OTI volume per unit of time. Slip depends on the clearances between the rotating and stationary members that define the "leak path orifice," on the differential pressure between the OTO volume and the OTI volume, and on the characteristics of the fluid handled (in particular the viscosity); in those rotary pumps where liquid velocity is high, it may be secondarily dependent on pump speed. A common unit of slip is U.S. gallons per minute.

Slip in a rotary pump is an important factor of pump performance and applications. A good understanding of the concept of slip and of the effects of pump design, type, manufacturing tolerances, fluid conditions, and system operating conditions on the magnitude of slip is necessary for both pump designers and pump users.

Slip in a rotary pump occurs only when a pressure difference exists between the inlet and outlet chambers of the pump. This pressure difference causes the pumped fluid to flow between the outlet and inlet chambers through clearances between rotors and clearances between rotors and body members. It has the same effect as a shunt or bypass around the pump from the outlet port to the inlet port. Most rotary pumps are constructed so that the clearances in the pump generally are of the same nature as those found between two parallel flat plates, with one plate stationary and the other moving. Elements of the clearances have long, narrow, rectangular cross sections. In most pumps these clearances across the narrow dimension range from essentially zero to a few thousandths of an inch, and consequently even minor variations in manufacturing tolerances can cause considerable variations in the percentage change of the aperture volume. Also, the movement or deflection of movable elements in the pump when exposed to pressure differences can cause relatively large percentage changes in these clearances in different locations in the pump. Consequently each pump must be tested to determine slip under any given operating condition. The effect of fluid viscosity on the amount of slip is discussed later in this section. A brief discussion of the effects of pump geometry follows.

The major paths of the back flow through the pump in the presence of a positive differential pump pressure are the clearances between the end faces of the rotors and the end plates of the pump chamber and those between the outer radial surfaces of the rotors and the inner radial surfaces of the chamber. The width, length, and height of the apertures thus formed vary considerably with different positions of the rotor as the drive shaft is slowly revolved through a complete revolution. If the differential pressure across the pump remains constant during a revolution, then the instantaneous slip rate usually varies throughout the revolution. This variation in the slip is caused by the same effect that would be produced if the physical dimensions of the equivalent bypass around the pump were varied as a function of angular rotation of the drive shaft. This is one of the common causes of pulsation in flow in rotary pumps; it is particularly dominant when large amounts of slip occur while pumping low-viscosity liquids at high pressure.

The average slip for any given operating condition can be determined by measuring the actual flow rate of fluid through the outlet port (incompressible liquid) and subtracting that flow rate from the flow rate that would be produced by the total displacement of the pump at that speed, Q_d. Most slip paths are constant in width but may vary in height with run-out of the outside diameter of the rotors, or with wobble of the end faces of the rotors as they rotate. They also vary considerably in length during a revolution because of the relative positions of the mating rotary and body-sealing surfaces during different angular positions of the rotor.

In general, slip increases as the ratio of the height of the slip path to the length of the slip path increases, and decreases as this ratio decreases. During pump design the clearances or height of the slip path may be increased if the length of the slip path is increased correspondingly, without materially affecting the slip characteristic of the pump.

The effect of pressure on slip is complex. The primary effect is direct: *slip increases in direct proportion to pressure.* However there are several secondary effects. These secondary effects can be classified into three general categories.

One category is the effect of pressure differences across the pump on the actual dimensions of the slip path. This occurs because of the deflection of pump elements as a function of pressure. This effect is relatively small in most rigid-element pumps but may become extremely large in pumps like flexible vane pumps where the pressure may actually cause the vanes to flatten out and move away from the body walls. One generalization is that slip may be dramatically increased in flexible-member pumps at high pressures, but clearances may be closed by high-pressure deflection of the rotors in rigid-rotor pumps and the effective slip may be decreased.

Another category is the indirect effect of pressure on the fluid velocity through the slip paths. At any given viscosity the flow through these paths may have the characteristics of turbulent flow, or the characteristics of viscous flow, or the characteristics of slug flow, depending on fluid velocity. The majority of practical applications would require that slip be a minor percentage amount of the pump displacement. To remain so, the velocity of fluid flow through slip paths would normally be in the viscous-flow region, and slip would then be directly proportional to the pressure difference. A pressure increase could cause a change to turbulent flow and a corresponding change in slip as a function of pressure.

A third category is the indirect effect of pressure on the effective compression ratio of compressible fluids. The compression ratio reduces the amount of net volume flow through the outlet port compared to the displacement of the pump. Although not a true slip in the sense discussed up to this point, it is effective in reducing the net volume delivered through the outlet port and consequently affects the volumetric efficiency. This effect is a secondary effect in most liquids but can become a large component of "slip" in aerated or compressible liquids. An increase in compression ratio caused by an increase in pressure difference causes an increase in "slip" from this effect.

Pump Capacity The capacity (Q) of a rotary pump is the net quantity of fluid actually delivered by the pump per unit of time through its outlet port or ports under any given operating condition. When the fluid is essentially noncompressible, capacity is numerically equal to the total volume of liquid displaced by the pump per unit of times minus the slip, all expressed in the same units. The capacity of a rotary pump operating with zero slip is called the *displacement capacity* (Q_d). A common unit of capacity is U.S. gallon per minute.

$$Q = kDN - S = Q_d - S \tag{4}$$

where $k = 0.004329$, with S and Q in gpm, D in ft^3, and N in rpm.

Pump Speed The speed (N) of a rotary pump is the number of revolutions of the driving or main rotor per unit time. When no gear reduction or increase exists between the drive shaft and the main rotor, the speed may be measured or set at the drive shaft. A common unit of speed is revolutions per minute.

Pump Pressure The absolute pressure of the fluid at any location in the pump, expressed in the common unit of pounds per square inch, is the total pressure there, and is the basis for other pressure definitions associated with pump operation. Most of the pressure-associated operating characteristics of a rotary pump involve conditions where the *velocity pressure* (P_v) caused by fluid velocity is small compared to the total pressure and may be neglected, or where the fluid velocities (at pump locations used to determine pressure differences) are sufficiently alike that the velocity pressures cancel when the total pressure difference is computed. Should this not be so, velocity pressure may be computed as: $P_v = wV^2/288g$, where V is the fluid velocity, in feet per second; w is the specific weight of the fluid, in pounds per cubic foot; and g is the acceleration caused by gravity at the elevation, in feet per second; or $P_v = (0.000357w/g)(Q/a)^2$, where Q is flow in gallons per minute and a is the cross-sectional area perpendicular to flow, in square inches.

Several pressure terms are of interest. The *outlet pressure* (P_d) is the total pressure at the outlet of the pump. In pumps with multiple outlets, this pressure is usually defined at the location in the outlet manifold where the pump is mated to the external piping system. Although composed of the sum of system and velocity pressures external to the pump, the outlet pressure is most commonly expressed as the *gage pressure* at the outlet port. Gage pressure (lb/in^2 gage) is the difference between the absolute pressure and atmospheric pressure at the point of measurement.

The *inlet pressure* (P_s) is the total pressure at the inlet to the pump or, for multiple-inlet pumps, at the manifold location where the pump connects to the external piping system. In common practice the inlet pressure may be variously expressed as absolute pressure (lb/in^2 abs), as positive or negative gage pressure (lb/in^2 gage), or as vacuum (inHg).

The pump *differential pressure* (P_{td}) is the algebraic difference between the outlet pressure and the inlet pressure, with both expressed in the same units. The differential pressure is used in the determination of power input and in evaluating the slip characteristics of the pump

$$P_{td} = P_d - P_s \tag{5}$$

The *net inlet pressure* (P_{sv}) of a rotary pump is the difference between the inlet pressure expressed in absolute units and the vapor pressure of the fluid expressed in absolute units.

$$P_{sv} = P_{sa} - P_{\text{vapor}} \tag{6}$$

The *required net inlet pressure* (P_{svr}) is the minimum net inlet pressure that can exist without the creation of enough vapor in the inlet to interfere with proper operation of the pump. The inlet pressure usually is not the true minimum pressure in the OTI volume. The flow of fluid in its passage through the OTI volume causes fluid friction pressure loss, which causes the pressure to drop below the inlet pressure at some point in the OTI volume or inlet chamber. This "inlet pressure loss" increases with fluid velocity, and hence with pump speed; it increases with fluid viscosity; and it is a function of the pump geometry that determines the fluid path lengths and local velocities where the fluid changes direction in flowing around curves or corners in the pump OTI volume boundary. In pumps operating on fluids at pressures where appreciable slip occurs, the pump differential pressure may have an indirect effect on the inlet loss. Consequently the required net inlet pressure for a given pump is usually established by the pump manufacturer for specific speeds, pressures, and fluid characteristics. In practice, the pump user is warned that the net inlet pressure is near or at the required net inlet pressure by noisy and rough operation of the pump caused by incomplete filling of the CTIO volume with liquid, with an accompanying reduction in pump capacity.

Other pressure ratings for pumps are the *maximum outlet working pressure* (P_{dr}), the *maximum inlet working pressure* (P_{sr}), and the *maximum differential pressure* (P_{tdr}). The maximum outlet working pressure is the maximum gage pressure at the outlet port permitted for safe operation of the pump, and the maximum inlet working pressure is the maximum gage pressure permitted at the inlet port for safe operation. These two ratings usually are determined by the stiffness and strength of the pump body or casing and the type of seals used in the pump. The maximum differential pressure is the maximum allowable difference between the outlet pressure and the inlet pressure, measured in the same units, and is determined by the ability of the rotating assembly and its fluid seal contact zones to withstand pressure difference between the OTO and OTI volumes.

Pump Power The *total power input* to a pump (ehp) is the total power required by the pump driver or the pump prime mover for given operating conditions of the pump. Sometimes called "driver power," it is the sum of the power required to overcome losses in the pump driver or prime mover; to overcome mechanical friction, fluid friction, and slip losses in the pump; and to deliver the net power imparted by the pump to the fluid discharged from it.

The *pump power input* (php) is the net power delivered to the pump drive shaft by the pump driver at a given operating condition of the pump. It is the net power available after subtraction of the power loss of the driver and associated transmission devices from the total power input (ehp).

The *pump power output* (whp) is the power actually imparted to the fluid delivered by the pump at given operating conditions, and is frequently called "liquid horsepower." It is the power remaining after the amount of slip power loss, mechanical power loss, and fluid friction power loss in the pump is subtracted from the pump power input (php). The relationship between these power terms may be expressed as below.

$$\text{ehp} = \text{driver and transmission power loss} + \text{php} \qquad (7)$$

$$\text{php} = \text{pump power loss} + \text{whp} \qquad (8)$$

A common unit used for expressing power ratings is the *horsepower*, equal to 550 ft-lb/s. The pump power output can be computed by the formula

$$\text{whp} = \frac{QP_{td}}{1{,}714} \qquad (9)$$

The constant 1,714 gives whp in horsepower when Q is in gallons per minute and P_{td} is in pounds per square inch.

The total volume of liquid handled by the pump is larger than Q when slip is present. The amount of slip actually represents "wasted" power and affects the pump efficiency. The difference between pump power input and pump power output actually consists of three power-loss amounts: the amount of power loss represented by slip, the amount of power loss represented by mechanical friction in the pump, and the amount of power loss represented by fluid friction (a function of viscosity of the fluid and pump shear stresses on the fluid). To determine the combined mechanical and fluid friction losses, the "displacement" power output (dph) is first computed by the following formula and then subtracted from php to give the combined friction losses.

$$\text{dhp} = \frac{(Q + S)P_{td}}{1{,}714} = \frac{DNP_{td}}{395{,}934} = \frac{Q_d P_{td}}{1{,}714} \qquad (10)$$

where dhp is horsepower when Q, Q_d, and S are in gallons per minute, P_{td} is in pounds per square inch, D is in cubic inches per revolution, and N is in revolutions per minute.

Pump Efficiency Several efficiencies can be computed for a pump. The *overall unit efficiency* (E_o) is the percent of the total power input delivered as pump power output and is computed by the equation

$$E_o = \frac{\text{whp}}{\text{ehp}} \times 100\% \tag{11}$$

The *pump efficiency,* or "pump mechanical efficiency" (E_p), is the ratio of the pump power output to the pump power input. It may be computed by the equation

$$E_p = \frac{\text{whp}}{\text{php}} \times 100\% \tag{12}$$

The *volumetric efficiency* (E_v) of a pump is the percentage of pump displacement per unit time delivered as pump capacity. The equation for computing the volumetric efficiency is

$$E_v = \frac{231Q}{DN} \times 100\% = \frac{Q}{Q_d} \times 100\% \tag{13}$$

where E_v is in percent when Q and S are gallons per minute, D is in cubic feet per revolution, and N is in revolutions per minute. Equation (13) may be stated in alternative ways as follows:

$$E_v = \frac{Q}{Q + S} \times 100\% \tag{14}$$

$$E_v = \frac{kDN - S}{kDN} \times 100\% \tag{15}$$

Pump Performance Figures 19, 20, and 21 are graphs showing the change in the displacement capacity (Q_d), capacity (Q), and slip (S), as differential pressure (P_{td}) across the pump, liquid viscosity (ν), and pump speed (N) are varied.

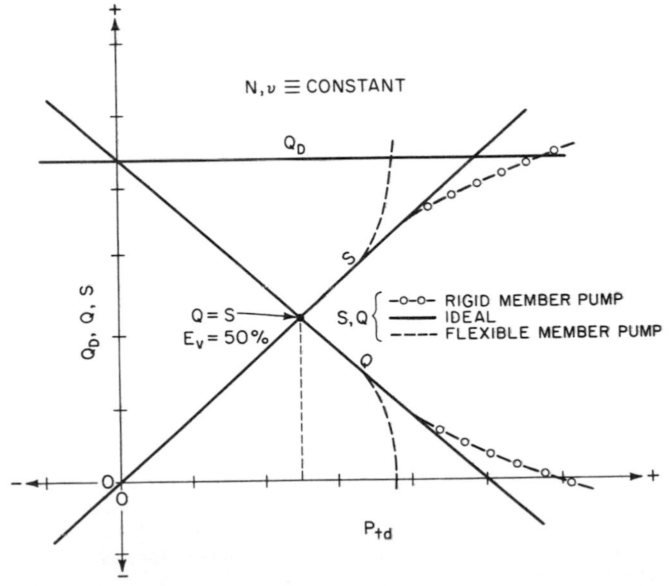

Fig. 19 Variation of Q_D, Q, and S with P_{td}, N, and ν constant.

Several assumptions are made. It is assumed that inlet conditions are satisfactory and that there is no inlet effect on pump capacity over the charted range. It is assumed that the liquid viscosity is Newtonian. It is assumed that the liquid is

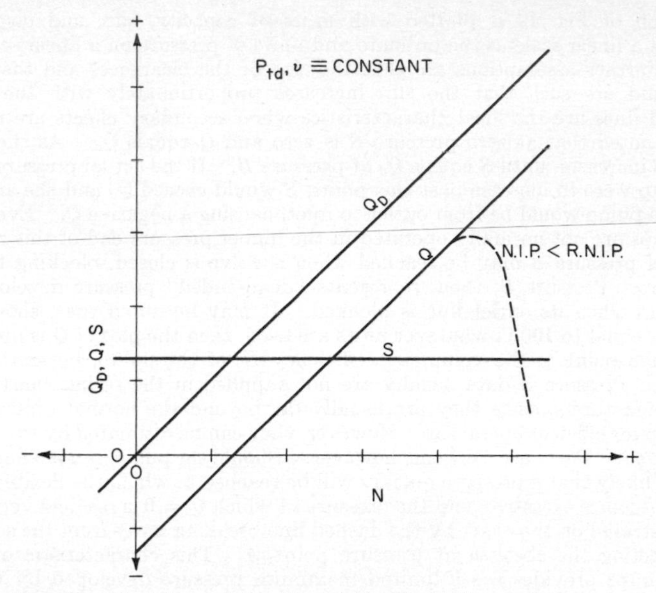

Fig. 20 Variation of Q_D, Q, and S with N, P_{td}, and ν constant.

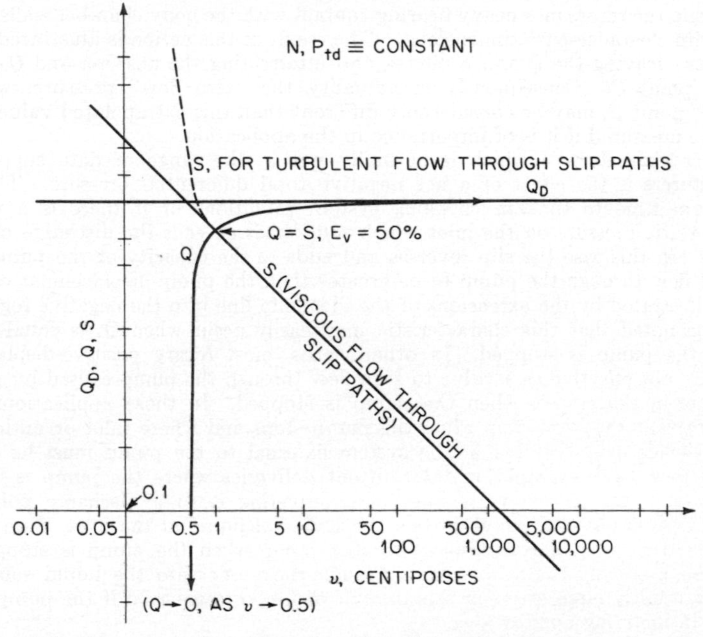

Fig. 21 Variation of Q_D, Q, and S with ν, N, and P_{td} constant.

incompressible. In Fig. 19 it is assumed that viscosity is constant at a relatively low point, approximately that of water, and that the speed is within the normal speed range of the pump. In Fig. 20 it is assumed that the viscosity is constant and relatively low and that the pressure is within the normal rated pressure range of the pump. In Fig. 21 it is assumed that both pressure and speed are within the normal ratings of the pump.

The graph of Fig. 19 is plotted with units of capacity, slip, and displacement capacity on a linear scale as the ordinate and units of pressure on a linear scale as the abscissa. Further assumptions are that the size of the clearances and the viscosity of the liquid are such that the slip increases proportionately with the pressure.

The solid lines are the ideal characteristics when secondary effects are neglected. It may be noted that at zero pressure S is zero and Q equals Q_d. As the pressure increases, S increases, until S equals Q_d at pressure B. If the actual pressure imposed on the pump were to increase past this point, S would exceed Q_d and the actual flow through the pump would be from outlet to inlet, causing a negative Q. Even though rotary pumps are not normally operated in the higher pressure end of this range, the condition of pressure B may be reached when a valve is closed, blocking the outlet of the pump. Pressure B, then, represents "dead-ended" pressure developed by a rotary pump when its outlet line is blocked. It may be noted that, should Q_d be numerically equal to 100 in whatever units are used, then the plot of Q is numerically equal at each point to the volumetric efficiency E_v of the pump, becoming zero at pressure B. Pressure B data usually are not supplied in the rating charts of most pump manufacturers, since they are usually far beyond the normal pressure rating of the pump for efficient operation. However, they can be estimated by extrapolating the data given. There are cautions, however. Should the pump be a flexible-member pump, it is likely that a pressure quickly will be reached at which the flexible-member deflection becomes excessive, and the pressure at which $Q = 0$ is reached very rapidly. This is illustrated on the chart by the dashed line breaking away from the solid Q line and intersecting the abscissa at pressure point A. This characteristic of flexible-member pumps provides a self-limited maximum pressure developed by the pump should the pump be dead-ended. In rigid rotor pumps, as the pressure increases beyond the normal operating range, deflections of the shafts and rotors within the pump bring the rotors into heavy bearing contact with the body chamber walls, reducing the slip-clearance-path dimensions. The result of this action is illustrated by the dashed line leaving the Q and S curves and intersecting the abscissa and Q_d line at pressure point C. Consequently in actuality the "zero flow" pressure, which is ideally at point B, may be considerably different than this extrapolated value, and it should be measured if it is of importance in the application.

Another characteristic which is normally not in the range of data supplied by manufacturers is the effect of a net negative total differential pressure. This may occur from time to time in changing system conditions or if there is a variable positive static pressure on the inlet which sometimes exceeds the discharge or outlet pressure. In this case the slip reverses and adds to the capacity of the pump, causing total flow through the pump to be greater than the pump-displacement capacity. This is illustrated by the extensions of the ideal slip line into the negative region. It should be noted that this characteristic may easily occur when Q_d is equal to zero because the pump is stopped. In other words, most rotary positive-displacement pumps are not effective as a valve to stop flow through the pump caused by pressure differences in the system when the pump is stopped. In those applications where it is important that flow stop when the pump stops and where inlet or outlet static pressure heads exist, valving in the system external to the pump must be used to stop the flow. For example, in intermittent deliveries where the pump is "lifting" liquid from a source below its inlet and delivering it to a discharge point with a piping system physically higher than the source of liquid at the inlet, if no valving arrangement is used in connection with the pump when the pump is stopped, the liquid will gradually drain backward through the pump into the liquid source and may create fairly large errors in the amount of liquid transferred if the pump should be used in metering applications.

Under the same assumptions given for Fig. 19, Fig. 20 shows the relative independence of slip with speed when differential pressure is constant. It may be noted that the chart of Q intersects the abscissa at the point where the speed is low enough to reduce the displacement capacity to equal slip at the pressure of operation. The speed at which Q_d equals 2 S is the speed at which S equals Q, and the volumetric efficiency E_v is 50 percent. The volumetric efficiency [see Eq. (15)] increases with speed because the ratio of $kDN - S$ to kDN increases with speed. As the speed increases, the pumping action of the shear stress in the clearances tends to reduce slip below the ideal line. However, the deterimental effects of speed increase above normal operating range for most rotary pumps usually are caused by inlet losses in the pump. This effect is discussed later in this section.

Figure 21 shows the effect of viscosity (ν) on slip and capacity of a rotary pump. The graphs are on a log-log scale. In this particular chart it is assumed that the pressure, speed, and viscosity combine to keep flow through clearances of the pump in the viscous flow region. Slip, then, is directly proportional to the total pressure difference across the pump and is inversely proportional to viscosity. This may be expressed in Eq. (16), where the constant K is a function of the pump geometry and size.

$$S = \frac{KP_{td}}{\nu} \tag{16}$$

The constant K is sometimes called the *coefficient of slip*. It may be designated as K_s. It includes all the constants needed to express slip in the desired units of flow.

As viscosity increases, slip becomes arbitrarily small and the capacity of the pump approaches the displacement capacity. As the viscosity decreases, slip very rapidly approaches the displacement capacity, and the capacity of the pump drops very rapidly to zero or a negative quantity.

For any given pump and any given pump speed (N) and pressure difference (P_{td}), there is a viscosity below which slip flow through clearances of the pump will change to turbulent flow. It is unlikely that this change occurs simultaneously through all slip paths. However, once it begins to occur, slip will increase much more rapidly with further reductions in viscosity because of the turbulent flow relationship of slip, pressure, and viscosity expressed in Eq. (17).

$$S = \frac{KP_{td}^{1/2}}{\nu^{1/x}} \tag{17}$$

where x usually is in the range of 4 to 10.

Pump volumetric efficiency (E_v) drops very rapidly with viscosity if the viscosity is lower than that required for 50 percent volumetric efficiency (represented by the crossover point of the slip and capacity curves). This crossover point occurs for almost all rotary pumps operating at rated P_{td} for viscosities between 0.1 centipoise and 10 centipoises, and for the majority of commercially available models this point usually falls in the viscosity range of 0.3 centipoise to 3 centipoises at P_{tdr}.

The effect of inlet pressure on capacity can be seen clearly if secondary effects are eliminated. It is assumed that a pump is operating within its normal pressure rating and within its normal speed rating, and that viscosity is sufficiently high to reduce slip to a negligible or zero value. A graph of capacity as a function of inlet pressure is shown in Fig. 22 with these assumptions.

There is no change in capacity as inlet pressure is lowered until the pressure reaches pressure A on the graph. If the inlet pressure were lowered further, the capacity would drop as shown. The cause of this drop is complex in detail but simple in concept. The liquid flow from the inlet port through the inlet chamber of the pump causes a pressure drop, which causes a minimum pressure point somewhere in the inlet chamber. When the pressure in the liquid at this minimum pressure point approaches the vapor pressure of the liquid, vapor begins to form there. When the amount and time persistence of this vapor cause vapor to be swept into the CTIO volume of the pump, the amount of liquid in this volume is reduced. When this volume reaches higher pressure, the vapor condenses, leaving a deficiency of liquid volume. For example, if half the actual fluid volume swept from the inlet chamber is vapor, then only half the normal-capacity liquid volume is available at the outlet chamber, and the capacity (Q) is reduced accordingly.

An increase in speed would mean an increase in capacity, all other things being equal. This would increase the pressure drop between the inlet port and the inlet chamber and correspondingly increase the absolute inlet pressure (which is measured at the inlet port) at which capacity would begin to drop (pressure A). Correspondingly, if the speed and capacity were constant and the viscosity were to increase, the pressure drop between the inlet port and the minimum pressure point in the

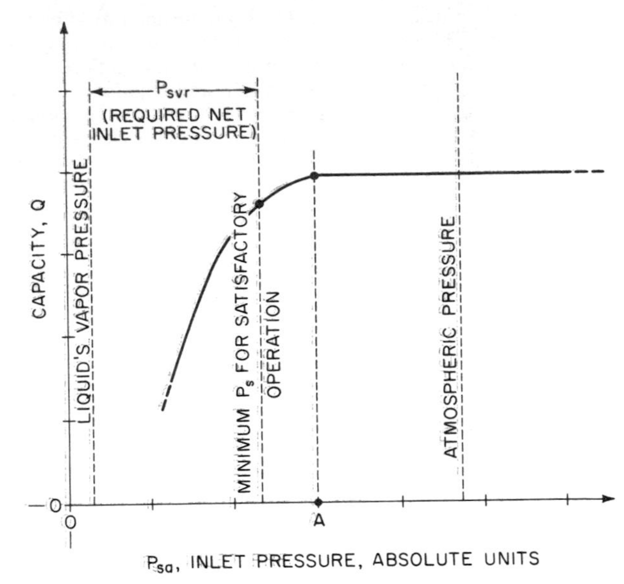

Fig. 22 Variation of Q with P_s, other operating conditions constant.

inlet chamber would increase with viscosity. This, also, would cause pressure point A to move to higher inlet absolute pressures. Operation with absolute inlet pressures below pressure A for any given speed and viscosity is usually unsatisfactory both because of the drop in capacity (and, hence, in volumetric efficiency) and because of the noisy and rough operation caused by the formation and collapse of vapor. For lower viscosity liquids, where the collapse of the vapor bubbles formed may be quite rapid, a significant amount of damage may be done to body or rotor surfaces by cavitation. It is important to understand that cavitation damage may occur even though the absolute inlet pressure is above pressure A. There may be locations in the inlet chamber, particularly where flow direction changes rapidly as around sharp corners, where cavitation (vapor formation in the form of very fine bubbles) may occur. However, these vapor cavities may be swept into higher pressure regions of the inlet chamber and collapse near body or rotor surfaces to cause cavitation damage, even though the capacity of the pump is not affected. For any given viscosity, then, there is an upper limit to the speed at which the pump may be operated.

The inlet pressure may be allowed to drop slightly below pressure A without significant deterioration of pump performance. However, there is a point at which the pressure becomes too low for satisfactory pump operation. This pressure determines the required net inlet pressure (P_{svr}) for the particular pump and particular set of operating conditions.

For any given set of operating conditions, the satisfaction of required net inlet pressure is a main limitation on pump operating speed.

The other main limit on pump operating speed is the *pump outlet pressure*. In every pump application there is some friction loss in the outlet system of the pump. Even if the pump outlet is opened to the atmosphere, a pressure drop exists between some maximum pressure point in the pump outlet chamber and the pump outlet itself. However, by far the most common situation is one in which a significant liquid friction pressure is developed in the system external to the outlet of the pump. This liquid pressure usually is a function of pump capacity and hence of pump speed. If the inlet conditions are maintained to keep the inlet pressure above pressure A as the pump increases, a speed will be reached at which the pump outlet pressure equals the outlet pressure rating of the pump. Operations at speeds higher than this would cause the pump outlet pressure to exceed the rated pressure, and this could result in possible damage to the pump.

These two limits on pump speed are illustrated in Fig. 23. Three sets of operating conditions are shown. In operating condition 1, the outlet system friction resisting liquid flow (outlet system impedance) is relatively high and the outlet pressure developed by flow is directly proportional to pump capacity (viscous flow). There is no static head in the system and the outlet pressure is developed only when the pump causes liquid to flow through the outlet system. The liquid viscosity is as-

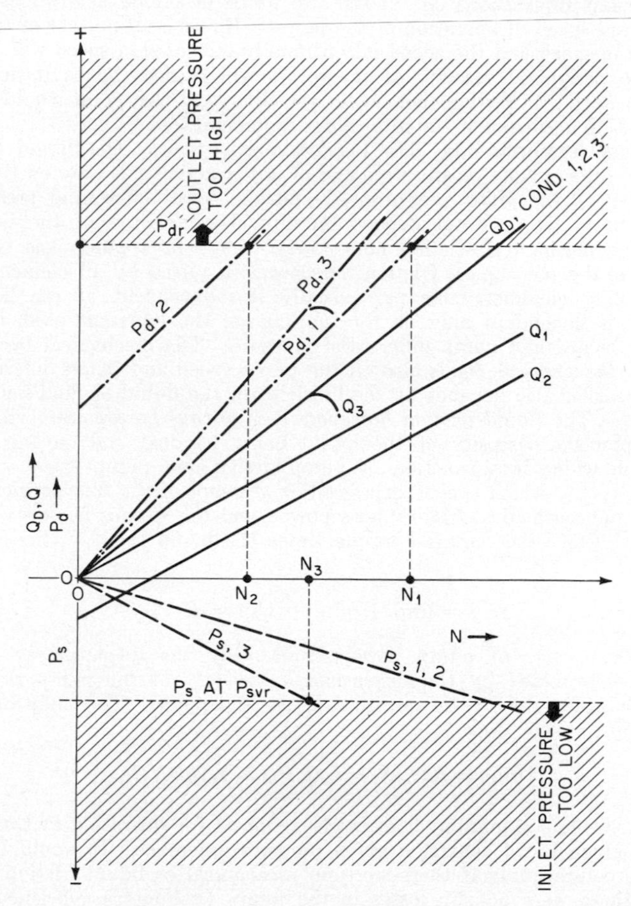

Fig. 23 P_d and P_s limits on pump speed, N.

sumed to be low enough to permit some slip in the pump. The net inlet pressure is assumed to be above pressure A over the range of operation shown. Operating condition 2 is the same as operating condition 1 except that a static "head" (static outlet pressure) exists in the outlet system. This "head" is the pressure at which the chart of pump outlet pressure in condition 2 intercepts the zero speed line. Operating condition 3 is one in which the impedance to liquid flow in the outlet system is relatively low, but the liquid viscosity is relatively high. In this condition the capacity (Q) is equal to the displacement capacity (Q_d) as speed is increased, until a speed is reached at which available net inlet pressure of the pump drops to the required net inlet pressure of the pump. If the speed were increased beyond this point the capacity of the pump would drop rapidly as an ever-increasing part of the pump fluid becomes vapor instead of liquid. The upper limit of speed for satisfactory operation of the pump is shown as N_1, N_2, and N_3 for the three conditions described. The locations of N_1, N_2, and N_3 on the speed axis are independent of each other, because they depend primarily on the operating conditions of the pump and the system in which it operates and on the conditions of the pumped fluid. For example a negative head in the outlet system could cause N_3 to move to a higher speed than N_2, and N_2 to move to a higher speed than N_1. Operation with a liquid with lower viscosity (but still sufficient to reduce slip to zero) could cause N_3 to occur at a speed higher than N_1.

In most pump applications one of the two limits described determines the maximum permitted speed of operation of the pump. However, if neither of these conditions limits the speed and the speed is continually increased, a speed will be reached at which the peripheral velocity of the rotors will exceed the cavitation velocity of the liquid itself. A further increase in speed beyond this point would be limited by the cavitation occurring at the rotor outer radial surfaces.

Pump Power Requirements The *displacement-capacity* (Q_d) *liquid horsepower* (dhp) of any rotary pump of any type and of any size depends only on the displacement capacity at the given pump speed and the total differential pressure (P_{td}) across the pump. It does not depend on any characteristics of the liquid being pumped. The mechanical friction horsepower (mhp) in a pump has two components. One is the mechanical friction horsepower required by all elements *external* to the pumping chamber; this part usually is independent of the liquid being pumped and is dependent only on the lubricity of the lubricant used, if any, and on the pump speed and pump differential pressure. The mechanical friction horsepower *inside the pump* depends also on the pump speed and pump differential pressure, but it usually also depends on the lubricity of the liquid or fluid being handled by the pump. The *liquid friction horsepower* or *viscous horsepower* (vhp) depends primarily upon the viscosity of the liquid being handled and on the shear rate in the liquid, which is a function of pump design and pump speed. The *pump power input* (php), which can be expressed as the sum of the displacement-capacity horsepower, the mechanical friction horsepower, and the viscous horsepower, is equal to a constant times the required torque times the pump speed. This is expressed in Eq. (18).

$$\text{php} = \text{dhp} + \text{mhp} + \text{vhp} = K_h T_p N \qquad (18)$$

Of the three parts of pump power input, only the displacement horsepower is uniquely determined by the displacement capacity of the pump at its elected speed of operation and by the pressure across the pump. This may be expressed as in Eq. (19).

$$\text{dhp} = K_h T_d N = k_d Q_d P_{td} = k_d(kDN)P_{td} \qquad (19)$$

This is the absolute minimum horsepower required for operation of that particular pump at a given speed against a given pressure difference. It would be the total horsepower required only if there were no mechanical or liquid friction horsepower losses and there were no slip losses in the pump (volumetric efficiency at 100%). Equation 19 can be solved in terms of torque, displacement, and pump differential pressure to give Eq. (20). The constant given is for torque in foot-pounds, displacement in cubic feet per revolution, differential pressure in pounds per square inch, and speed in revolutions per minute.

$$T_d = 0.01326 D P_{td} \qquad (20)$$

Equation (19) also may be used to generate the chart in Fig. 24.

The chart can be used to determine the minimum basic horsepower input to a pump expected to deliver a given flow rate against a given pressure differential. Equation (13) can be used to compute Q_d when the actual capacity Q is known and the pump volumetric efficiency E_v is known. In any case, if the displacement capacity is known for whatever speed is required to produce the desired capacity, the chart can be entered from the known displacement capacity and from the known

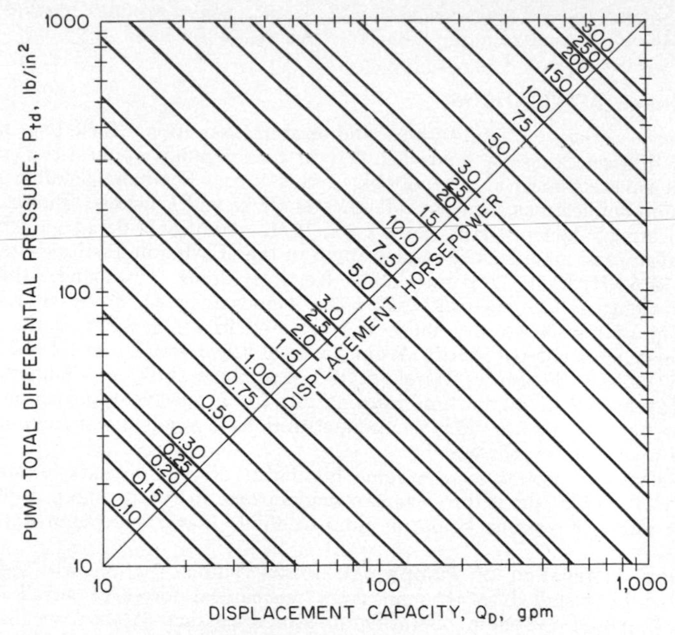

Fig. 24 Pump displacement horsepower, php, variation with Q_d and P_{td}, all rotary pumps, all liquids.

pressure difference to determine the displacement horsepower. To this displacement horsepower must be added the basic mechanical friction horsepower of the pump and the viscous horsepower of the pump. The significance of displacement horsepower is that it is a measure of all the energy required to impart energy of motion to the external capacity and internal slip flow of the pump against the specified pressure. The remaining horsepower required is consumed in the heat generated by mechanical friction and by liquid viscous friction.

The torque required to overcome the friction losses in the pump may become large when the pump handles certain kinds of liquids. When the pump handles nonlubricating liquids, and pump differential pressure is high, the mechanical friction in the pump chamber may become high and increase the pump torque requirement. When the pump handles high-viscosity liquids, the torque required by the viscous friction grows as viscosity increases, even if pump speed and pressure remain constant. This torque depends on a number of design features of the pump and usually will grow proportionally to the viscosity to the nth power, where n usually ranges between 0.3 and 1.0 for most rotary pumps. Increasing the clearances in a pump reduces the shear stress in the liquid in those pump clearances at any given speed and consequently reduces the torque to overcome the viscous friction. Too great an increase in clearances can cause slip, which will increase the total power loss for a given capacity by a drop in volumetric efficiency. If a pump is designed for a liquid with a specific viscosity, the "minimum power clearance" is that at which volumetric efficiency just reaches 100 percent. A further reduction of clearance beyond that point would not improve volumetric efficiency but would increase viscous horsepower.

Most pump manufacturers publish graphs or tables of total pump horsepower required for any set of operating conditions within the ratings of the pump. Many pump manufacturers also will identify the maximum permitted torque on the drive shaft of the pump. Once the actual horsepower requirements are determined for any specific operating conditions, the required torque may be computed from Eq. (18). In Eq. (18), K_h equals 1.903×10^{-4}, for speed in revolutions per minute and torque in foot-pounds. The torque limitation usually is of importance on low-speed high-horsepower applications, such as might occur when the pump handles a very-high-viscosity liquid at reasonably high pressures at low speeds.

ROTARY PUMP APPLICATIONS

The immense variety of applications and system conditions, including the large number of different kinds of fluids handled in rotary pump applications, precludes a comprehensive coverage in a handbook of this type. For more detailed information on pump applications, a recent reference is Hicks and Edwards, "Pump Application Engineering," McGraw-Hill, New York, 1971. Additional detail on pump testing, application, and maintenance can be found in the "Hydraulic Institute Standards," 13th ed., 1975, Hydraulic Institute, 2130 Keith Building, Cleveland, Ohio 44115, Library of Congress Catalog Card A56-4036. Good coverage of the complexity of the viscosity of non-Newtonian fluids is contained in a *Chemical Engineering* refresher course prepared by Martin Wohl and published serially in *Chemical Engineering,* in January, February, March, April, May, June, July, and August of 1968. An excellent source of application information for a specific type of pump is a manufacturer of that pump. Many manufacturers have available application, installation, and maintenance manuals.

In the remainder of this section some highlights of requirements in mechanical installation of rotary pumps, of system considerations in rotary pump applications, and of the effect of various kinds of pumped fluids in various pump applications are given.

Mechanical Installation of Pumps All rotary pumps, particularly rigid rotor pumps, must be installed so that no large mechanical forces of any kind other than those imposed by pump operating pressures can act to warp or distort the pump chamber or the pump's rotating assembly. Relatively small distortions of a few thousandths of an inch may cause interference between rotating and stationary parts of the pump and generate high wear or pump damage.

To avoid such unwanted distortions, the pump must not be installed with rigid fittings to long rigid piping systems, where either the weight of the piping system is supported by the pump body or where large mechanical forces on the pump body can be generated by thermal expansion of the piping system. The problem of distortion of operating clearances in the pump chamber is not as severe for flexible-member pumps, but large mechanical forces tending to distort the pump body may cause serious distortion of mechanical sealing arrangements and unusually high wear rates on bearing systems, etc. A cardinal rule, then, is that a *rotary pump must not be so rigidly coupled to either the fluid piping system or to the driver that it supports the weight of either, or that it is exposed to high forces caused by thermal expansion of either.* A poor mechanical installation of a pump might be the sole cause of unsatisfactory operation.

System Considerations Most rotary pumps will be damaged by continued operation without liquid present and flowing through the pump chamber. If the system provides a positive static pressure on fluid present at the inlet of the pump, the pump usually will not be damaged by running dry. If a negative inlet gage pressure, or "suction lift," is a condition at the pump inlet, the system installation should be checked to ensure airtight seals at all demountable joints in the inlet system between the pump inlet and the body of fluid to be pumped. The sealing arrangement on the pump must be monitored to prevent any air or gas leakage through the seal. If these conditions are met, liquid will enter the pump very soon after pump operation starts because of the vacuum generated by pump operation. If they are not met, a suffiicient flow of air or other gas through leaks in the inlet system or seal may satisfy the pump-capacity requirements and the pump will run dry indefinitely. An alternate way to ensure liquid at the pump inlet is to install a foot or check valve in the liquid-submerged portion of the pump inlet piping system. Once the pump is primed, the foot valve will keep liquid in the pump inlet and prevent its "running dry."

If there is a possibility that a valve in the outlet piping system of the pump may close or remain closed to block all flow from the pump, then a pressure relief (bypass) valve should be used with the pump. This valve may be externally installed or may be an integral part of the pump. A pressure relief valve usually is not necessary in flexible-member rotary pumps. Pressure relief valves also may be used to limit the pump discharge pressure to a selected predetermined maximum pressure.

The inlet system of the pump must be arranged in such a way that fluid is always present at the pump inlet at a pressure at or above the inlet pressure needed to satisfy the required net inlet pressure.

Effects of Pumped Fluid The variety of fluids handled by rotary pumps requires that each pump application be considered in terms of the effects of the pumped fluid on pump performance. General summaries of the effects of various fluid characteristics on pump performance are covered in the following paragraphs.

The *temperature* of the pumped fluid affects pump performance in three main ways. If the pump is to handle a fluid at temperatures considerably different from ambient, the materials of construction, both in the pump and in the pump seals, and the operating clearances to which the pump is manufactured must be selected to give the desired operating characteristics at the temperature of operation. The selection process becomes even more stringent when the pump is operated over a wide range of temperatures. Such a use may preclude the use of pump construction materials having high thermal coefficients of expansion. Furthermore, in the actual application it may be necessary to preheat or precool the pump to or near the temperature of operation before the pump is started, in order to avoid the thermal shock of the fluid entering the pump chamber. If this is not done, the resulting rapid heating or cooling of pump members from the inside out can cause transient conditions that may damage the pump. The preheating or precooling can be accomplished by use of a secondary medium (a jacketed pump), or, if the state of the liquid permits, the pump may be started on fluid at ambient temperature and the temperature of the fluid gradually lowered or raised to the desired operating temperature while the pump is operating.

A second effect of fluid temperature is on the vapor pressure of the fluid. For most liquids the vapor pressure increases with increased temperature; consequently the required net inlet pressure would cause a corresponding increase in the necessary inlet pressure at the pump.

A third effect of fluid temperature is on the viscosity of the pumped fluid. For the majority of commercial liquids, the viscosity tends to increase with decreasing temperature and to decrease with increasing temperature. If the pump application requires operation over a wide temperature range, the highest viscosity at this temperature range must be known to determine if the pump is operating below the upper speed limit imposed by this fluid viscosity.

Another important characteristic of the pumped fluid is its *viscosity*. Earlier parts of this section describe the general effects of viscosity upon pump performance. However, very few commercial liquids are completely Newtonian in viscosity. The actual effective viscosity of the fluid under the actual projected or actual operating conditions should be determined by test or computation. If the pump is to operate over a range of speeds and pressures, the maximum and minimum effective viscosities of the liquid over this range should be determined to provide for the additional slip or the additional horsepower that may be required, to ensure satisfactory selection of operating speeds and pump driver horsepower. A brief summary of the effects of time and shear rate on the viscosity of fluids is in "Hydraulic Institute Standards," 13th ed., page 158, and in the reference on viscosity cited earlier. The importance of a good knowledge of the viscous behavior of fluids in pump selection and application is illustrated by the following example. The shear stress in either a pseudoplastic or thixotropic fluid decreases as the shear rate increases. In such fluids a high viscosity at rest may reduce rapidly as shear rate increases. In such a case a much smaller pump than otherwise would be selected, operated at much higher speeds than otherwise would be permitted, may be the most satisfactory choice for the application.

Still another important fluid characteristic is *lubricity*. Many rotary pumps are not designed to operate on liquids with no lubricity. Either the wear rate or the pump mechanical friction may be inefficiently high if these pumps were used to pump nonlubricating fluids.

Knowledge of the *corrosiveness* of the pumped fluid on the pump materials in contact with the pumped fluid is important to satisfactory pump application. The clearances in rotary pumps are small, and corrosion rates of only a few thousandths of an inch per year may seriously affect the efficiency of the pump, particularly when the pump is handling low-viscosity fluids. Also, the general compatibility of the pumped fluid with materials of pump construction must be considered. For example, a certain solvent to be pumped may soften, dissolve, or destroy the elasticity of flexible members in the pump, including those flexible members used in the sealing arrangements. Hence the effect of the pumped fluid on the physical and chemical state of materials used in pump construction should be considered in choosing a particular pump for a particular application.

Very few commercially handled fluids are pure liquids. In actual systems most fluids have air or other gases dissolved or entrained in them, or the system may operate where liquid vapor can exist in various concentrations in the pumping chamber. In other cases, wanted or unwanted solids may be contained in the liquid. These solids may be abrasive or nonabrasive to the materials of the pump.

If appreciable amounts of gas or air are entrained or dissolved in the pumped fluid (for example, an aerated ice cream mix or various foamed plastics), it is necessary to select a type of rotary pump in which the action of the pump returns no part of the fluid in the discharge chamber to the inlet chamber during operation. One type of rotary pump usually satisfactory for operation on aerated fluids is the lobe pump. This type of rotary pump has been used as a rough vacuum pump and as a compressor, and consequently is capable of handling the gaseous phase of the fluid.

The pump geometry in applications involving nonabrasive solids, particularly those that are fibrous and tend to mat, must be such as to maintain high fluid velocities everywhere in the pump. Abrasive solids represent a formidable problem for almost any rotary pump. Again, because of the small operating clearances, the tendency of the abrasives in the pumped fluid to wear down and open clearances usually will cause a rapid decline in pump efficiency, particularly when the abrasives are carried in low-viscosity liquids. Very few rotary pumps that will handle abrasive fines in a low-viscosity liquid at medium or high pressures are commercially available. Most of the pumps used for this service are constructed of very-high-hardness (usually expensive) materials.

Many other characteristics of the pumped fluid could be of importance in the selection and application of rotary pumps. Some pumped liquids are sensitive to shear. They must be handled in pumps with low shear rates. Some liquids cannot tolerate exposure to the atmosphere either because they may explode or because they may instantly crystallize into hard crystals which would damage the pump seal arrangement or otherwise interfere with satisfactory operation of the pump. Rotary pumps for these applications should have only static seals or multiple-section rotary seals with a protective liquid or gas in the seal zones between the pumped fluid and the outside atmosphere. For example, stationary and moving seals on aseptic pumps may be designed with multiple seals allowing the use of steam under pressure as a sterile barrier between the pump chamber and the outside world. Pumps used to pump human blood, such as the tube pump, may use disposable, sterilizable pump elements. Pumps used in industrial applications where conditions may become dangerous or unsatisfactory if the pump liquid is exposed to atmosphere may use multiple sealing arrangements with an inert liquid under pressure acting as a barrier between the pumping chamber and the outside atmosphere.

For effective application of rotary pumps, then, the temperature, viscosity, lubricity, and corrosiveness of the pumped fluid must be known. Also the nonliquid content of the fluid must be known. Any characteristics of the liquid, such as shear sensitivity or atmospheric sensitivity, should be known. When these are known, the proper pump type, pump seal arrangements, pump horsepower requirements, and pump speed can be determined.

The positive-displacement nature of rotary pumps makes them suitable for many metering applications. The pump would be a perfect meter with zero error if there were no slip. It would be a good meter if the slip were low or maintained constant over the entire range of operating conditions. A useful ratio in a comparison of different types of rotary pumps for a given metering application is the ratio of the minimum capacity existing in any of the operating conditions to the maximum capacity existing in any of the operating conditions at a given speed. In variable-speed applications the ratio of minimum volumetric efficiency to maximum volumetric efficiency is substituted. The metering effectiveness is highest as this ratio approaches 1. The difference between this ratio and $1 \times 100\%$ is the percentage of change that can be expected in either the flow rate of the pump or in the total amount of fluid displaced by the pump for a given number of revolutions, under extremes of operating conditions (pressures, temperatures, viscosities, etc.).

Section 4

Jet Pumps

ALEX M. JUMPETER

JET PUMPS

Jet Pump Principle The term *jet pump,* or ejector, describes a pump having no moving parts and utilizing fluids in motion under controlled conditions. Specifically, motive power is provided by a high-pressure stream of fluid directed through a nozzle designed to produce the highest possible velocity. The resultant jet of high-velocity fluid creates a low-pressure area in the mixing chamber causing the suction fluid to flow into this chamber. Ideally, there is an exchange of momentum at this point producing a uniformly mixed stream traveling at a velocity intermediate to the motive and suction velocity. The diffuser is shaped to reduce the velocity gradually and convert the energy to pressure at the discharge with as little loss as possible. The three basic parts of any ejector are the nozzle, the diffuser, and the suction chamber or body (see Fig. 1).

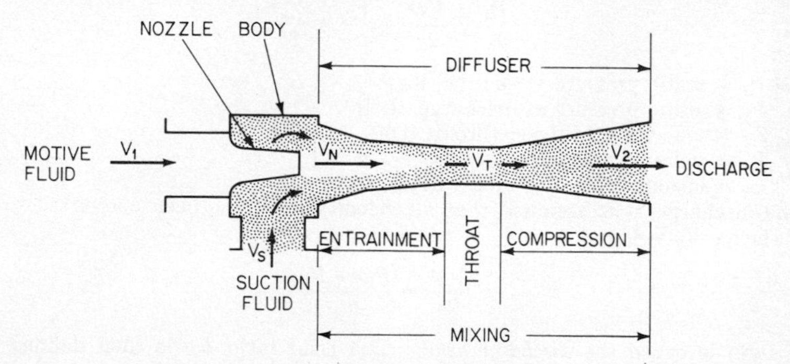

Fig. 1 Jet nozzles convert pressure energy into velocity while diffusers entrain and mix the fluids and change velocity back into pressure.

Definition of Terms A definition of standard ejector terminology is as follows:

Ejector General name used to describe all types of jet pumps which discharge at a pressure intermediate to motive and suction pressures.

Eductor A liquid jet pump using a liquid as motive fluid.

Injector A particular type of jet pump which uses a condensable gas to entrain a liquid and discharge against a pressure higher than either motive or suction pressures. Principally a boiler injector.

Jet compressor A gas jet pump used to boost pressure of gases.
Siphon A liquid jet pump utilizing a condensable vapor, normally steam, as the motive fluid.

Of concern to this text are jet pumps used to pump liquids. Eductors, being the most common, will receive the principal treatment herein. Sizing parameters for siphons will also be presented.

EDUCTORS

Theory and Design Eductor theory is developed from the Bernoulli equation. Static pressure at the entrance to the nozzle is converted to kinetic energy by permitting the fluid to flow freely through a converging-type nozzle. The resulting high-velocity stream entrains the suction fluid in the suction chamber resulting in a flow of mixed fluids at an intermediate velocity. The diffuser section then converts the velocity pressure back into static pressure at the discharge of the eductor. Writing the Bernoulli equation for the motive fluid across the nozzle of an eductor:

$$\frac{P_1}{w_1} + \frac{V_1^2}{2g} = \frac{P_s}{w_1} + \frac{V_N^2}{2g} \tag{1}$$

where P_1 = static pressure upstream, lb/ft²
P_s = static pressure at suction (nozzle tip), lb/ft²
V_1 = velocity upstream of nozzle, ft/s
V_N = velocity at nozzle orifice, ft/s
w_1 = specific weight of motive fluid, lb/ft³

Upstream of the nozzle, all the energy is considered static head so that the velocity term V_1 drops out, yielding

$$\frac{V_N^2}{2g} = \frac{P_1 - P_s}{w_1} \tag{2}$$

This term is called the *operating head.*

Across the diffuser, the same principle applies for the mixed fluid stream, except the effect is the reverse of a nozzle; hence,

$$\frac{P_s}{w_2} + \frac{V_T^2}{2g} = \frac{P_2}{w_2} + \frac{V_2^2}{2g}$$

where P_s = static pressure at suction, lb/ft²
P_2 = static pressure at discharge, lb/ft²
V_T = velocity at diffuser throat, ft/s
V_2 = velocity downstream, ft/s
w_2 = specific weight of mixed fluids, lb/ft³

At the discharge, it is assumed that all velocity head has been converted to static head; hence $V_2 = 0$ and

$$\frac{V_T^2}{2g} = \frac{P_2 - P_s}{w_2} \tag{3}$$

This term is called the *discharge head.* The head ratio R_H is then defined as the ratio of the operating head to the discharge head:

$$R_H = \frac{V_N^2/2g}{V_T^2/2g} = \frac{V_N^2}{V_T^2} = \frac{(P_1 - P_s)/w_1}{(P_2 - P_s)/w_2} = \frac{(P_1 - P_s)w_2}{(P_2 - P_s)w_1} \tag{4}$$

Since ratios are involved, it is convenient to replace specific weight with specific gravity

$$R_H = \frac{(P_1 - P_s)\ \text{sp gr}_2}{(P_2 - P_s)\ \text{sp gr}_1} \tag{5}$$

When both suction and motive fluid are the same, no gravity correction is required and Eq. (5) becomes

$$R_H = \frac{H_1 - H_s}{H_2 - H_s} \tag{6}$$

where $H_1 - H_s$ = operating head, ft
$H_2 - H_s$ = discharge head, ft

Entrainment conditions are defined by the basic momentum equation

$$M_1 V_N + M_s F_s = (M_1 + M_s) V_T$$

where M_1 = mass of motive fluid, slugs
M_s = mass of suction fluid, slugs
V_N = velocity at nozzle discharge, ft/s
V_s = velocity at suction inlet, ft/s
V_T = velocity at diffuser throat, ft/s

The velocity of approach at the suction inlet is zero; therefore rearranging yields

$$M_s = M_1 \left(\frac{V_N}{V_T} - 1 \right)$$

and the term below is defined as the *weight operating ratio:*

$$R_w = \frac{M_s}{M_1} = \frac{V_N}{V_T} - 1 \tag{7}$$

Observe that the term $V_N{}^2/V_T{}^2$ has previously been defined as the head ratio R_H; therefore

$$R_w = \sqrt{R_H} - 1 \tag{8}$$

The volume ratio R_q is then simply

$$\frac{Q_s}{Q_1} = R_w \frac{\text{sp gr}_1}{\text{sp gr}_2} \tag{9}$$

where Q_s = suction flow in volumetric units
Q_1 = motive flow in volumetric units

The maximum theoretical performance of eductors is calculated from the above relationships. In actual practice there are energy losses associated with the mixing of two fluids and frictional losses in the diffuser. These losses are accounted for by the use of an empirical factor to reduce the theoretical maximum performance. Figure 2 shows this factor plotted against NPSH (net positive suction head) for a single nozzle and annular nozzle eductor. In an annular nozzle eductor, the motive fluid is introduced around the periphery of the suction fluid, either by a ring of nozzles (Fig. 15), or by an annulus created between the inner wall of the diffuser and the outer wall of the suction nozzle (Fig. 14). The NPSH is the head available at the centerline of the eductor to move and accelerate suction fluid entering the eductor mixing chamber. NPSH is the total head in feet of fluid flowing and is defined as atmospheric pressure minus suction pressure minus vapor pressure of suction or motive fluid, whichever is higher.

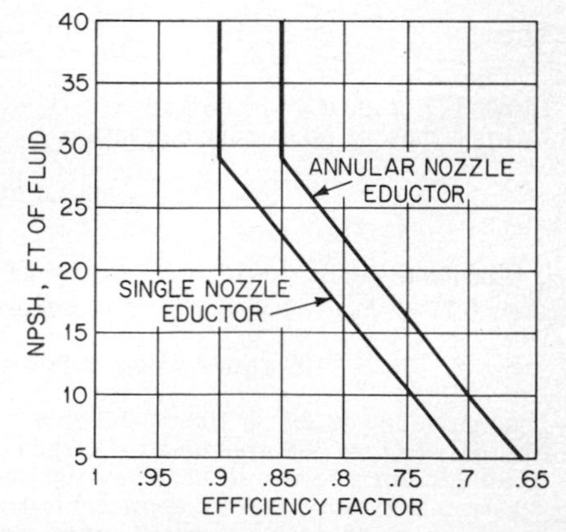

Fig. 2 NPSH versus efficiency factor. (Schutte and Koerting Co.)

The efficiency factor is introduced into the R_w equation as shown:

$$R_w = \epsilon \sqrt{R_H} - 1$$

This equation is used to calculate the motive quantity or pressure from the operating parameters. The nozzle and diffuser diameters are calculated from the equation $Q = wAV$ using suitable nozzle and diffuser entrance coefficients. The principal problems in design concern the size and proportions of the mixing chamber, the distance between the nozzle and diffuser, and the length of the diffuser. Eductor designs are based on theory and empirical constants for length and shape. The most efficient units are developed from calculated designs which are then further modified by prototype testing.

Increased viscosity of motive or suction fluid increases the friction and momentum losses and therefore reduces the efficiency factor of Fig. 2. Below 20 centipoises the effect is minimal (approximately 5 percent lowering of ϵ). Above this value, the loss of performance is more noticeable, and empirical data or pilot testing is used to determine sizing parameters.

Figure 3 shows the operating ratio R_w versus the head ratio R_H for various lift conditions. The efficiency factor has been incorporated into this curve.

The final size of the eductor is determined by the discharge line and is based on normal pipeline velocities (3 to 10 ft/s). Figures 4A and B are used for estimating eductor size.

To illustrate the use of Figs. 3 and 4 consider the following example:

EXAMPLE 1: It is desired to remove 100 gpm of water at 100°F from a pit 20 ft deep. Discharge pressure is 10 lb/in² gage. Motive water is available at 60 lb/in² gage and 80°F. The eductor is to be located above the pit. Find the eductor size and motive water quantity required.

SOLUTION. To use Fig. 3 it is necessary to determine the NPSH and the head ratio R_H. The centerline of the eductor is chosen as the datum plane and NPSH = atmospheric pressure − suction lift − vapor pressure at 100°F, or

$$\text{NPSH} = 34\text{ ft} - 20\text{ ft} - 1.933\text{ inHg}\left(\frac{13.6}{12}\right) = 11.81\text{ ft}$$

Since motive and suction are the same fluid, it is convenient to work in feet rather than pounds per square inch and

$$P_1 = 60\text{ lb/in}^2\text{ gage} = 138.6\text{ ftH}_2\text{O}$$
$$P_2 = 10\text{ lb/in}^2\text{ gage} = 23.1\text{ ftH}_2\text{O}$$
$$P_s = -20\text{ ft}$$

Then

$$R_H = \frac{138.6 - (-20)}{23.1 - (-20)} = \frac{158.6}{43.1} = 3.68$$

Enter Fig. 3 at $R_H = 3.68$ and NPSH = 11.81; read $R_w = 0.48$.

Since there is no gravity correction

$$R_w = R_q = \frac{0.48\text{ gal suction}}{\text{gal motive}}$$

The same result can be obtained by using the efficiency factor from Fig. 2. R_w is then $0.77\sqrt{R_H} - 1 = 0.48$. The required motive fluid is then

$$100\text{ gpm suction}/0.48 = 208\text{ gpm at }60\text{ lb/in}^2\text{ gage}$$

Discharge flow is 208 + 100 = 308 gpm. The size is obtained from Fig. 4. Entering Fig. 4B at $Q_2 = 308$ gpm and discharge head $(H_2 - H_s) = [23.1 - (-20)] = 43.1$ ft. Read eductor size of 4 in based on the discharge connection.

NOTE: If there were any appreciable length of run on the discharge line, it would be necessary to calculate the pressure drop in this line and recalculate the eductor after adding the line loss to the discharge head required. Friction losses on the suction side must also be included. In the example chosen, however, 100 gpm in a 4-in suction line 20 ft long will have a negligible friction loss (less than 0.25 ftH_2O) and is ignored.

Performance Characteristics Figure 5 illustrates the performance characteristics of eductors. Note the sharp break in capacity below the design point. For this reason all eductors are not designed for a peak efficiency. It is often advantageous

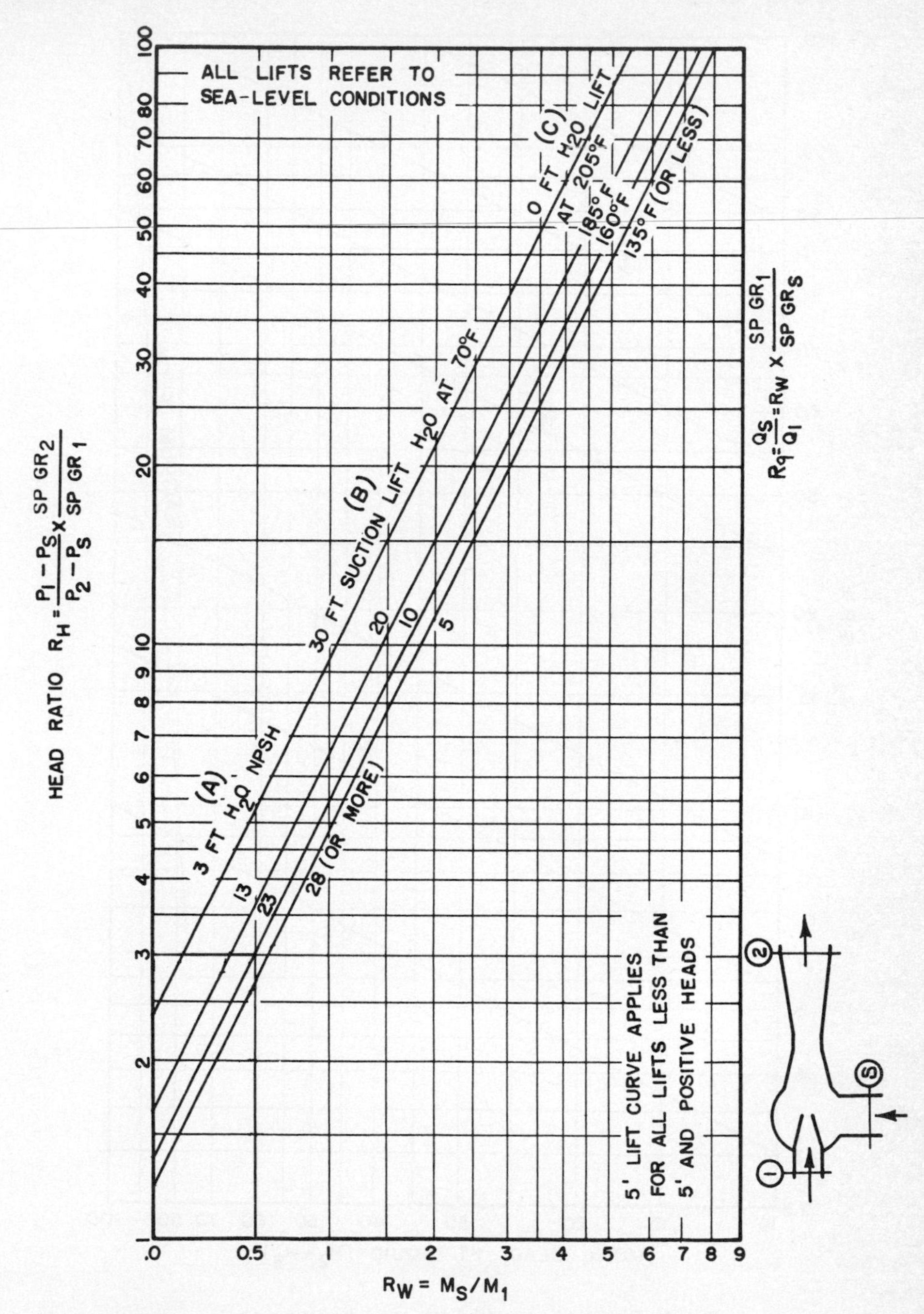

Fig. 3 Estimating operating ratios, liquid jet eductors. An eductor can be designed for only one head ratio. (Schutte and Koerting Co.)

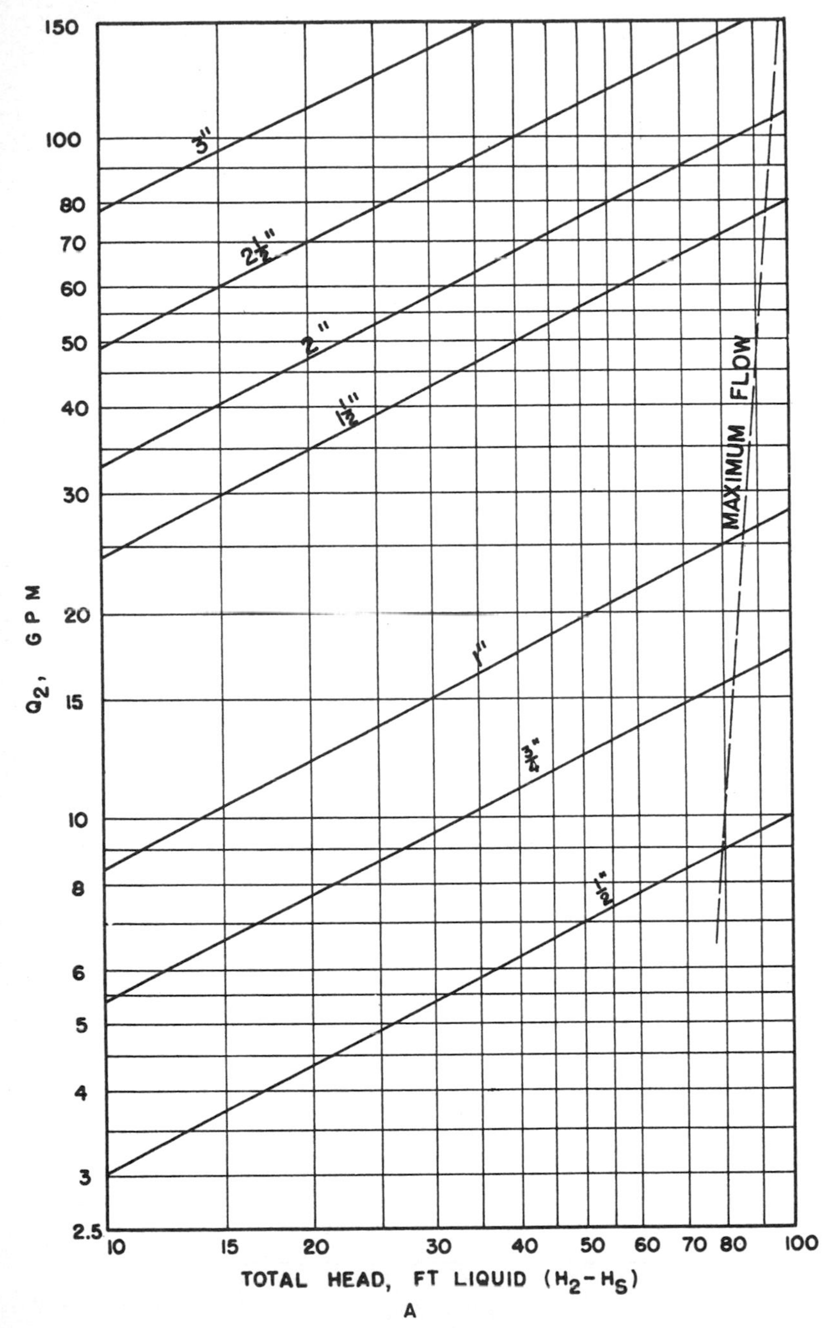

Fig. 4 Sizing curve. (Schutte and Koerting Co.)

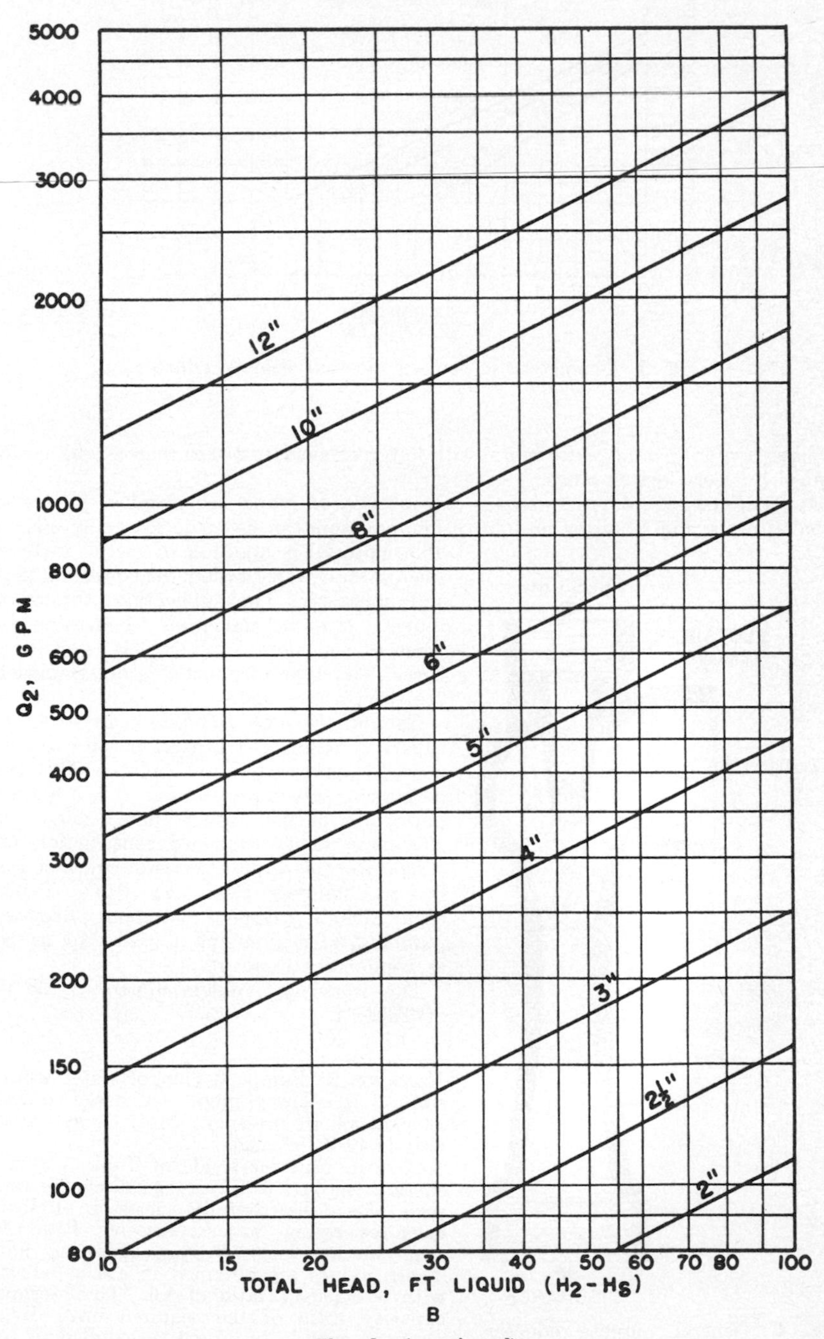

Fig. 4 *(continued)*

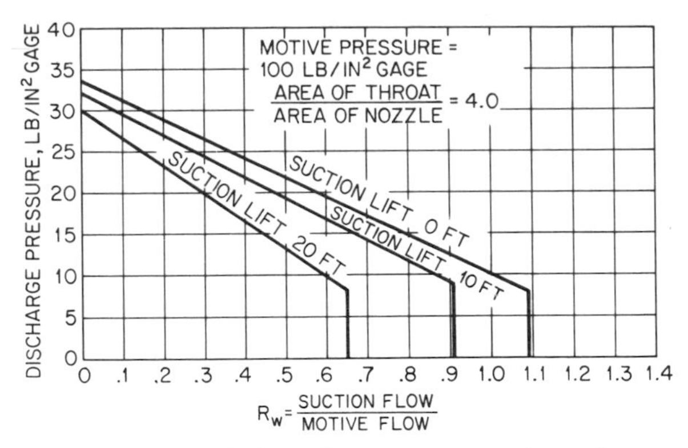

Fig. 5 Characteristic performance curve of an eductor.

to have a wide span of performance with lower efficiencies rather than a peak performance, but very limited range.

Applications Beside the obvious advantages of being self-priming, having no moving parts, and requiring no lubrication, eductors can be made from any machinable material in addition to special materials such as stoneware, Teflon, heat-resistant glass, and fiberglass. The applications throughout industry are too numerous to mention, but some of the more common will be discussed here. The type of eductor is determined by the service intended.

General Purpose Eductors. Table 1 is a capacity table for a general purpose eductor used for pumping and blending. This type of eductor illustrated in Fig. 6 has a broad performance span rather than a high peak efficiency point. Standard construction materials for this type of eductor are cast iron, bronze, stainless steel and PVC. Typical uses include cesspool pumping, deep-well pumping, bilge pumping aboard ship, as well as condensate removal.

The following problem illustrates the use of Table 1.

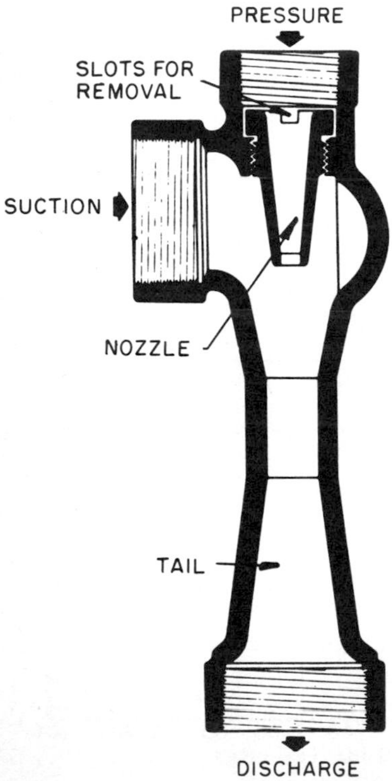

PRESSURE
SLOTS FOR REMOVAL
SUCTION
NOZZLE
TAIL
DISCHARGE

Fig. 6 General purpose eductor. (Schutte and Koerting Co.)

EXAMPLE 2: Pump 30 gpm of water from a sump 5 ft below ground. Discharge to drain at atmospheric pressure. Motive water available is 40 lb/in² gage.

SOLUTION. Enter left side of Table 1 at 5 ft suction lift and 0 lb/in² gage discharge pressure. Read horizontally across to 40 lb/in² gage operating water pressure. Read 9.6 gpm suction and 7.3 gpm operating fluid. These values are obtained in a 1-in eductor with a capacity ratio of 1.0. To determine capacity ratio of the required unit, divide required suction by quantity handled in 1-in eductor: 30/9.6 = 3.13 capacity ratio. Referring to the bottom of Table 1, a 2-in eductor with a capacity ratio of 4.0 is obtained. The required motive flow is then 4 × 7.3 = 29.2 gpm and the suction capacity is 4 × 9.6 = 38.4 gpm. A 1½-in unit can

TABLE 1 Capacity Table of Standard Eductors (Schutte and Koerting Co.)

Suction lift, ft	Discharge pressure lb/in² gage	Function	Capacity of standard 1-in water-jet eductor, gpm							
			Operating water pressure, lb/in² gage							
			10	20	30	40	50	60	80	100
0	0	Suction	5.85	8.1	9.5	10.0	12.0	12.0	12.0	12.0
		Operating	3.55	5.0	6.1	7.1	7.9	8.7	10.0	11.0
	5	Suction	...	1.4	4.1	6.0	8.0	10.0	11.0	12.0
		Operating	...	4.9	6.1	7.0	7.9	8.6	10.0	11.0
	10	Suction	...	...	0.28	2.3	4.8	6.4	8.8	11.0
		Operating	...	...	5.9	6.8	7.8	8.5	9.8	11.0
	15	Suction	...	...	...	...	1.2	3.4	5.9	8.6
		Operating	...	...	...	...	7.7	8.4	9.8	11.0
	20	Suction	...	...	...	...	...	0.3	3.5	5.9
		Operating	...	...	...	...	...	8.2	9.7	11.0
	25	Suction	...	...	...	...	...	...	0.83	3.9
		Operating	...	...	...	...	...	...	9.6	11.0
	30	Suction	...	...	...	...	...	...	...	1.7
		Operating	...	...	...	...	...	...	...	11.0
5	0	Suction	4.4	6.8	8.6	9.6	11.0	11.0	12.0	12.0
		Operating	3.9	5.3	6.4	7.3	8.1	8.8	10.0	11.0
	5	Suction	...	1.5	3.2	5.0	7.0	9.0	11.0	11.0
		Operating	...	5.2	6.3	7.2	8.0	8.7	10.0	11.0
	10	Suction	...	...	...	1.9	3.6	5.6	8.6	10.0
		Operating	...	...	...	7.1	7.9	8.6	10.0	11.0
	15	Suction	...	...	...	...	1.1	2.6	5.8	8.3
		Operating	...	...	...	...	7.8	8.6	9.9	11.0
	20	Suction	...	...	...	...	...	...	3.3	5.6
		Operating	...	...	...	...	...	...	9.8	11.0
	25	Suction	...	...	...	...	...	...	0.47	3.6
		Operating	...	...	...	...	...	...	9.8	11.0
	30	Suction	...	...	...	...	...	...	...	1.5
		Operating	...	...	...	...	...	...	...	11.0
10	0	Suction	2.0	4.6	6.7	8.3	9.0	10.0	10.0	10.0
		Operating	4.2	5.5	6.6	7.4	8.2	9.0	10.0	11.0
	5	Suction	...	...	2.0	4.3	5.9	7.7	9.9	10.0
		Operating	...	...	6.5	7.4	8.2	8.9	10.0	11.0
	10	Suction	...	...	...	1.1	3.0	4.5	8.1	9.6
		Operating	...	...	...	7.3	8.1	8.8	10.0	11.0
	15	Suction	...	...	...	...	1.1	2.1	5.6	7.3
		Operating	...	...	...	...	8.0	8.7	10.0	11.0
	20	Suction	...	...	...	...	...	...	2.8	5.3
		Operating	...	...	...	...	...	...	9.9	11.0
	25	Suction	...	...	...	...	...	...	...	2.8
		Operating	...	...	...	...	...	...	...	11.0
	30	Suction	...	...	...	...	...	...	...	1.1
		Operating	...	...	...	...	...	...	...	11.0
15	0	Suction	...	3.3	5.3	7.9	8.4	8.9	8.9	9.1
		Operating	...	5.7	6.8	7.6	8.4	9.1	10.0	12.0
	5	Suction	...	...	...	4.0	4.9	7.3	8.6	9.1
		Operating	...	...	...	7.6	8.3	9.0	10.0	11.0
	10	Suction	...	...	...	...	2.4	4.0	6.4	8.6
		Operating	...	...	...	...	8.2	9.0	10.0	11.0
	15	Suction	...	...	...	...	...	...	4.2	6.8
		Operating	...	...	...	...	...	...	10.0	11.0
	20	Suction	...	...	...	...	...	...	2.1	4.5
		Operating	...	...	...	...	...	...	10.0	11.0
	25	Suction	...	...	...	...	...	...	...	1.9
		Operating	...	...	...	...	...	...	...	11.0

TABLE 1 Capacity Table of Standard Eductors (Schutte and Koerting Co.) (Continued)

20	0	Suction	...	2.0	4.0	6.4	7.8	7.8	7.8	7.8
		Operating	...	6.0	7.0	7.8	8.6	9.3	11.0	12.0
	5	Suction	...	...	...	2.8	3.9	6.3	7.8	7.8
		Operating	...	...	...	7.7	8.5	9.2	10.0	12.0
	10	Suction	...	...	...	...	1.2	3.1	5.7	7.1
		Operating	...	...	...	...	8.3	9.1	10.0	12.0
	15	Suction	...	...	...	...	...	...	3.6	5.4
		Operating	...	...	...	...	...	...	10.0	11.0
	20	Suction	...	...	...	...	...	...	1.4	3.8
		Operating	...	...	...	...	...	...	10.0	11.0
	25	Suction	...	...	...	...	...	...	...	1.5
		Operating	...	...	...	...	...	...	...	11.0

Size educator, in	½	¾	1	1½	2	2½	3	4	6
Capacity ratio	0.36	0.64	1.00	2.89	4.00	6.25	9.00	16.00	36.00

handle 2.89 times the values in the table or 27.7 gpm suction when using 21 gpm motive water at 40 lb/in^2 gage. If the suction flow rate were not critical, some capacity could be sacrificed in order to use a smaller and therefore lower-cost eductor. If optimum performance is desired, it will be necessary to size a special eductor using Figs. 3 and 4.

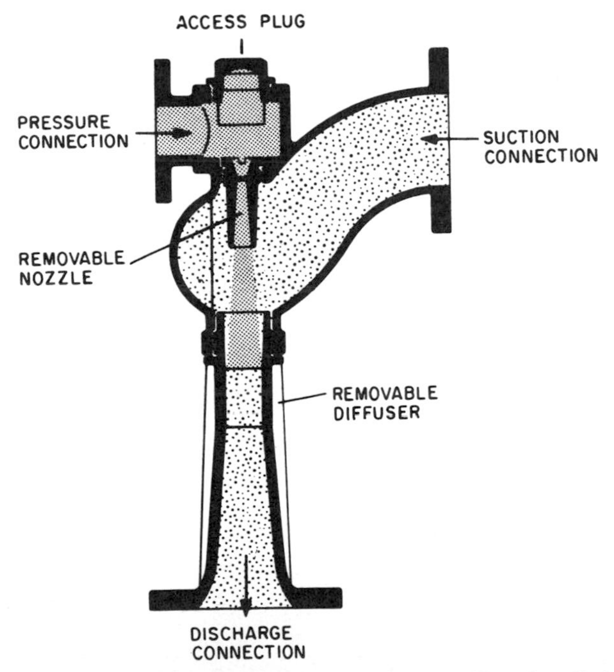

Fig. 7 Streamlined eductor. (Schutte and Koerting Co.)

Figure 7 illustrates more streamlined versions for higher suction lifts or applications involving the handling of slurries. This type of eductor is often used to remove condensate from vessels under vacuum. The advantage is that eductors require only 2 ft NPSH and, being smaller than mechanical pumps, save considerable space. Further, a partial vapor load is much less likely to vapor lock a jet pump because the venturi tube minimizes the expansion effect of flashing vapor. Sizing is done in the same manner as example 1, using Figs. 3 and 4.

Mixing Eductors. While any eductor is inherently a mixing device, some eductors are specifically designed as mixers. They are used to replace mechanical agitators and are located inside the tank containing the fluid to be agitated. Figure 8 illustrates the simplest type of eductor, the *Sparger nozzle.* These units entrain approximately 3 gal of suction fluid for each gal of motive fluid. A 20 lb/in² drop across the nozzle is recommended for proper mixing. Figure 9 shows the motive capacities for this type of eductor. Sparger nozzles are normally used for shallow tanks, whereas the tank mixer described below is preferred for deeper vessels.

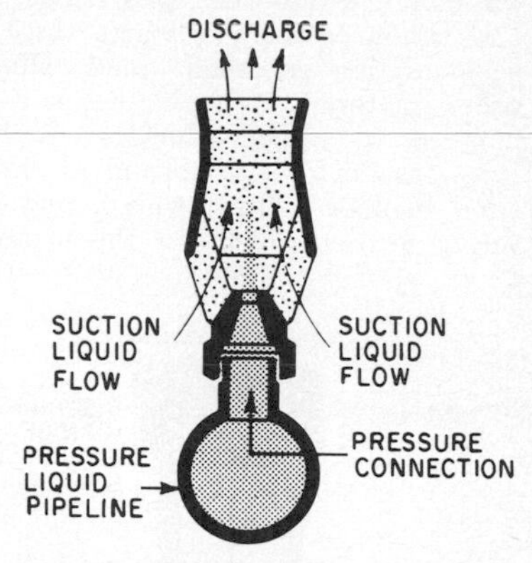

Fig. 8 Sparger nozzle. (Schutte and Koerting Co.)

Figure 10 illustrates another type of eductor called a *tank mixer.* It is installed under the tank containing the fluid to be agitated. Motive capacities are shown in Table 2. The units are usually custom designed for a specific entrainment ratio, the required unit being determined by the quantity of tank fluid, the ratio of mixture desired, and the depth of the tank being agitated.

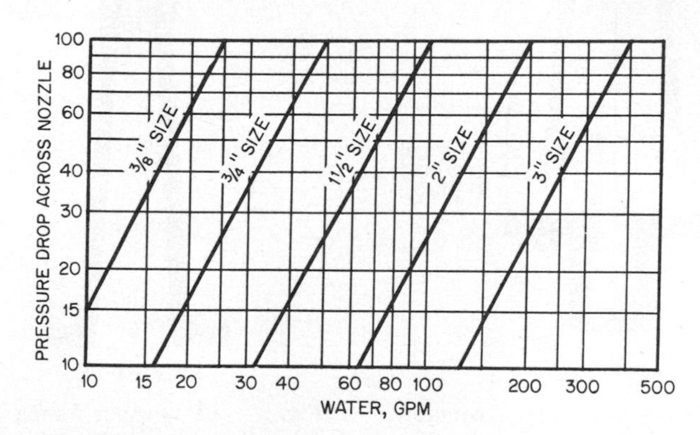

Fig. 9 Motive capacity of Sparger nozzles. (Schutte and Koerting Co.)

EXAMPLE 3: It is desired to blend recycled tank fluid into a tank 20 ft deep in the ratio of 1 gpm motive to 1.5 gpm suction. The tank contains 7,500 gal, and it is desired to turn over the tank in 30 min. The motive pump will deliver 60 lb/in² gage at the eductor nozzle. What size mixing eductor is needed?

SOLUTION: 7,500 gallons turned over in 30 min = 250 gpm. Since the motive fluid in this case is recycled from the tank itself, both motive and suction fluid contribute to the tank turnover. In the ratio of 1.5 gal suction to 1 gal of motive fluid the required motive quantity to attain a circulation rate of 250 gpm is then 100 gpm. To select the size it is necessary to obtain the differential pressure across the nozzle orifice of the eductor. Since the eductor is below the tank the net driving head is 60 lb/in² gage − 20/2.31 = 51.35 lb/in² gage across the nozzle. Enter Table 2 and interpolate between 50 and 60 lb/in² gage. A 1½-in eductor will pass only 73 gpm, while a 2 in will pass 129 gpm. The selection would then be a 2-in mixing eductor.

Spindle Proportioning Eductors. Another type of mixing eductor is illustrated in Fig. 11. Typical applications of this type include mixing hydrocarbons with caustic, oxygen or copper chloride slurries, producing emulsions, and proportioning liquids in chemical process industries. In critical applications the regulating spindle is sometimes fitted with a diaphragm operator to achieve close control. Table 3 shows capacities on several typical applications for units of this type.

Sand and Mud Eductors. Figure 12 illustrates a sand and mud eductor used for pumping out wells, pits, tanks, sumps and similar containers where there is an accumulation of sand, mud, slime, or other material of a nature not easily handled by other eductors. With this type of eductor the bottom of the pressure chamber is fitted with a ring of agitating nozzles which stir the material in

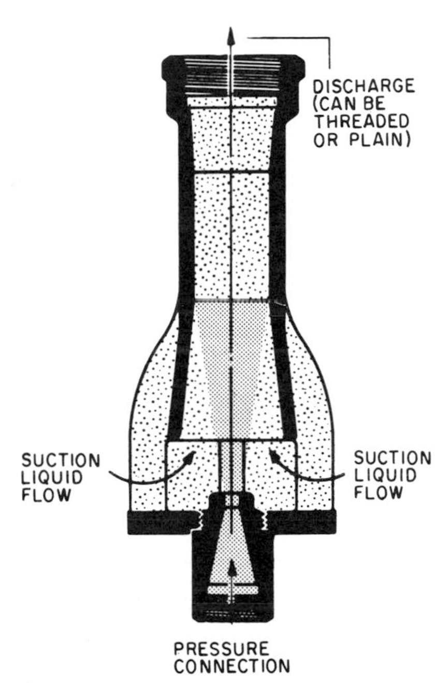

Fig. 10 Tank mixing eductor. (Schutte and Koerting Co.)

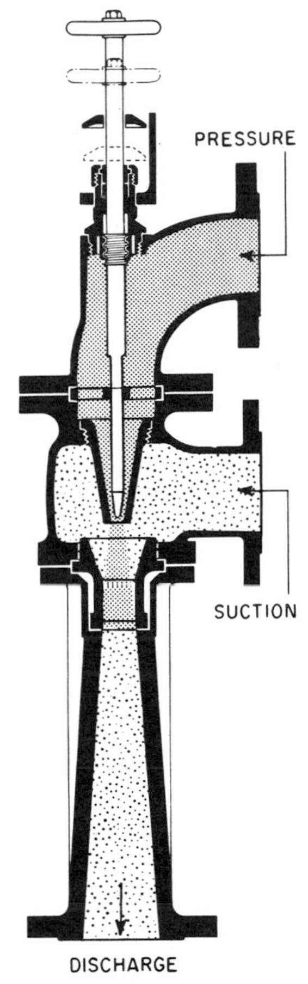

Fig. 11 Proportioning eductor. (Schutte and Koerting Co.)

which the jet is submerged to allow maximum entrainment. Capacity table for this type of eductor is shown in Table 4.

This table is used in the same manner as Table 1. The required suction flow is divided by the suction capacity selected from Table 4 under the appropriate motive pressure. This value is the capacity ratio. From the table select the eductor by choosing the next higher capacity ratio. Actual capacities are then determined by multiplying the values in the table by the capacity ratio of the eductor selected. Maximum discharge head is read from the table.

Solids-Handling Eductors. Figure 13 illustrates a specific type called a *hopper eductor* made for handling slurries or dry solids in granular form and used for ejecting sludges from tank bottoms, pumping sand from filter beds, and washing or conveying granular materials. Typical construction is cast iron with hardened steel nozzle and throat bushings. In operation, the washdown nozzles are adjusted to provide smooth flow down the hopper sides, thus preventing bridging of the material being handled and also sealing the eductor suction against excess air quantities. Without this seal, the capacities shown in Table 5 should be divided by approximately three. Table 6 shows typical materials handled by this eductor and their

TABLE 2 Motive Capacities of Tank Mixing Eductors (Schutte and Koerting Co.)

Size, in	Motive fluid water, gpm							
	Pressure-difference inlet to tank, lb/in² gage							
	10	20	30	40	50	60	80	100
½	3.5	5.0	6.0	7.0	8.0	8.5	10.0	11.0
¾	10.0	14.5	17.5	20.0	23.0	24.5	29.0	32.0
1	14.2	20.0	25.0	28.0	30.0	34.5	40.0	44.5
1¼	22.0	31.0	37.5	44.0	50.0	53.0	62.5	69.0
1½	31.5	45.0	54.0	63.0	72.0	76.5	90.0	99.0
2	56.0	80.0	96.0	112.0	128.0	136.0	160.0	176.0
3	126.0	180.0	216.0	252.0	288.0	306.0	360.0	396.0
4	224.0	320.0	384.0	448.0	512.0	544.0	640.0	704.0
5	350.0	500.0	600.0	700.0	800.0	850.0	1,000.0	1,100.0
6	494.0	720.0	864.0	1,008.0	1,152.0	1,224.0	1,440.0	1,584.0

TABLE 3 Capacities—Proportioning Eductor (Schutte and Koerting Co.)

Motive liquid	Naphtha	Hydrocarbon	Gasoline	Gasoline	Sour kerosene
Suction fluid	Copper chloride slurry	Hydrocarbon	Slurry	Water	Kerosene slurry
Pressure, lb/in² gage:					
Motive	165	295	170	75	146
Suction	40	5	75	50	60
Discharge	75	10	100	50	70
Flow, gpm:					
Motive	30	10	90	170	482
Suction	20	58	74	42	700
Discharge	50	68	164	212	1182
Eductor size, in	1½	3	4	4	6

TABLE 4 Relative Capacities of Sand and Mud Eductors (Schutte and Koerting Co.)

Suction capacity			
Operating water pressure, lb/in² gage	40.0	50.0	60.0
Total motive fluid, gpm	69.5	77.5	85.0
Net suction fluid, gpm	30.0	34.5	38.5
Maximum discharge head, ft	22.0	26.0	32.0

Size eductor, in	1½	2½	3	4	5	6
Capacity ratio	0.29	0.62	1.00	1.85	2.80	3.80

bulk density. Use of the capacity table for hopper eductors is similar to use of Tables 1 and 4, except the suction quantities required are expressed in cubic feet. Capacity ratio is determined by dividing the value in the table into the required suction flow, and the next larger size eductor is selected.

Another type of solids-handling eductor is illustrated in Fig. 14. This *annular-orifice eductor* is used where the material being handled tends to agglomerate and gum up when wetted. In this unit intimate mixing occurs in the throat and the device is virtually clogproof. Normally this unit is installed directly over the tank into which the mixture is discharged. Table 7 shows capacities for this type unit.

TABLE 5 Relative Capacities of Hopper Eductors (Schutte and Koerting Co.)

Operating water pressure, lb/in^2 gage	30	40	50	60
Suction capacity, ft^3/hr	13	36	72	90
Maximum discharge pressure, lb/in^2 gage	14	17	18	20
Motive water consumption, gpm*	35	40	45	50

* Based on using approximately 10% motive water through washdown nozzles.

Relative capacities of standard sizes					
Size in	1½	2	3	4	6
Capacity ratio	1.00	1.60	3.50	6.00	18.00

Units of this type have also been used successfully for handling and mixing hard to wet solids.

Capacity Table 7 is similar to Table 5 and the selection method is the same as discussed previously.

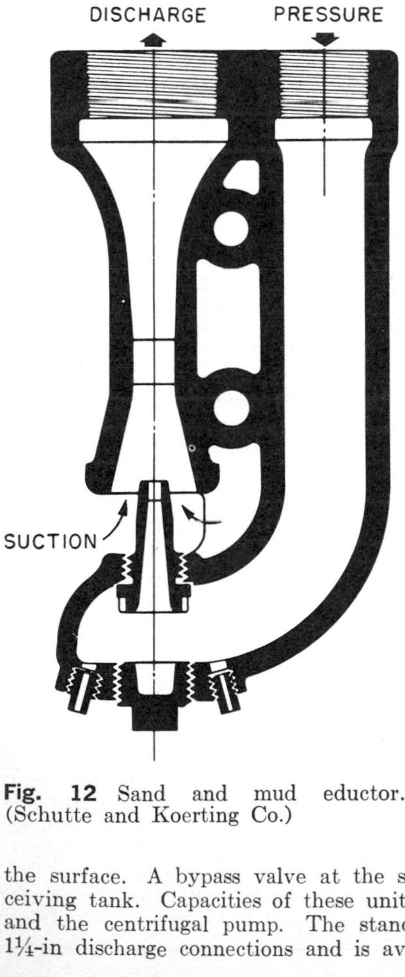

Multinozzle Eductors. Figure 15 illustrates an annular multinozzle eductor designed for special applications where the suction fluid contains solids or semisolids. It is used primarily for large flows at low discharge heads. Because these units have relatively large air-handling capacities, they are well suited for priming large pumps such as dredging pumps, where air pockets can cause these pumps to lose their prime. These eductors are designed by using the basic equations for head ratio. The appropriate efficiency factor is selected from Fig. 2 and the volumetric flow ratio is calculated. Figure 4 is used to size the eductor after discharge flow has been determined.

Deep-Well Eductors. The eductor illustrated in Fig. 16 is typical of those used in conjunction with a mechanical pump for commercial and residential water supply from a deep well. The eductor is used to lift water from a level below barometric height up to a level where the suction of the motive pump at the surface can lift the water the remaining distance. In operation, the eductor is fitted with hose connections connected to the suction and discharge of the motive pump and dropped into the well casing. An initial prime is required which is maintained by the foot valve at the suction of the eductor. When the surface pump is activated, pressure water through the eductor entrains water from the well, lifting it high enough to enable the mechanical pump to carry it to the surface. A bypass valve at the surface diverts the suction quantity to a receiving tank. Capacities of these units are dependent upon the depth of the well and the centrifugal pump. The standard commercial unit has 1-in pressure and 1¼-in discharge connections and is available with a variety of nozzle and diffuser

Fig. 12 Sand and mud eductor. (Schutte and Koerting Co.)

TABLE 6 Typical Materials Handled by Hopper Eductors (Schutte and Koerting Co.)

Material	Approx bulk density, lb/ft^3
Borax	50–55
Charcoal	18–28
Diatomaceous earth	10–20
Lime, pebble	56
Lime, powdered	32–40
Mash	60–65
Fly ash	35–40
Rosin	67
Salt, granulated	45–51
Salt, rock	70–80
Sand, damp	75–85
Sand, dry	90–100
Sawdust, dry	13
Soda ash, light	20–35
Sodium nitrate, dry	80
Sulfur, powdered	50–60
Wheat	48
Zinc oxide, powdered, dry	10–35

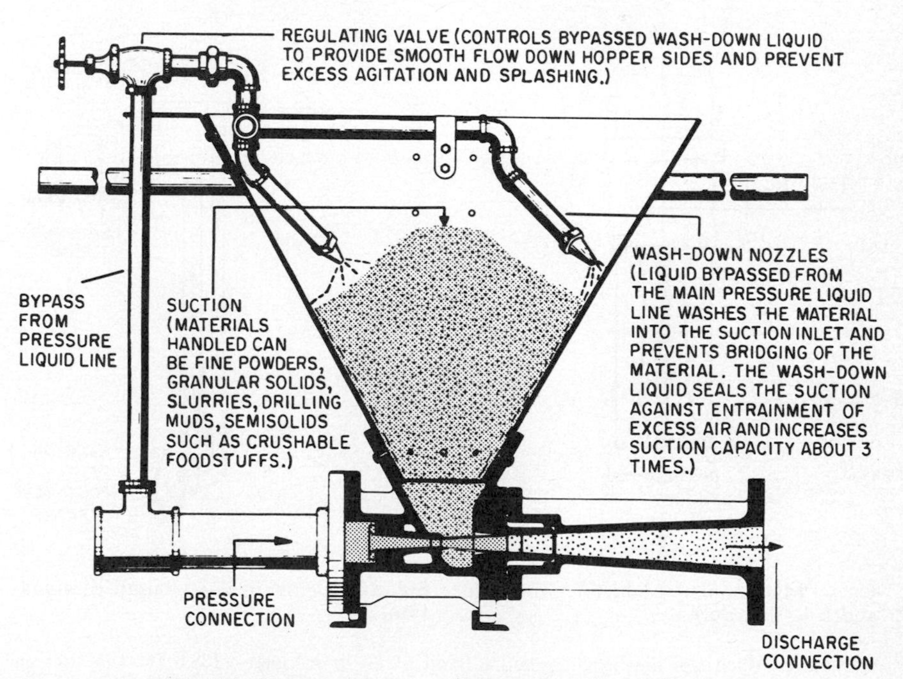

Fig. 13 Hopper eductor. (Schutte and Koerting Co.)

combinations for use with standard centrifugal pumps at varying depths. The following example illustrates how to calculate this type unit.

EXAMPLE 4: A centrifugal pump with a capacity of 100 gpm at a total discharge head of 150 ft and requiring 10 ft NPSH is available to operate an eductor to pump water at 50°F from a level 50 ft below grade. Find the quantity of water that can be delivered at 60 lb/in^2 gage (see Fig. 17).

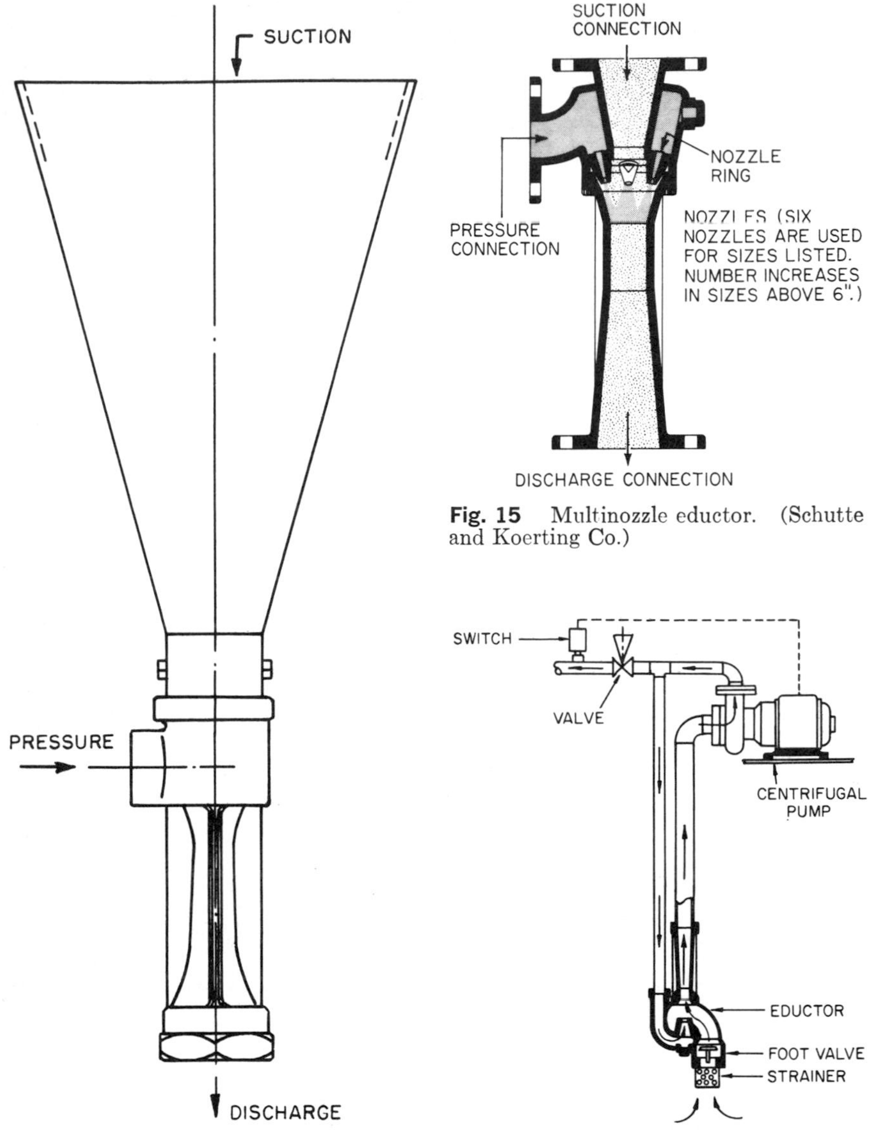

Fig. 14 Annular eductor. (Schutte and Koerting Co.)

Fig. 15 Multinozzle eductor. (Schutte and Koerting Co.)

Fig. 16 Centrifugal–jet pump combination.

SOLUTION: With the discharge pressure fixed at 60 lb/in² gage (138.6 ft) the suction lift developed by the pump is 150 − 138.6 = 11.4 ft. The eductor must then discharge to a height of 50 − 11.4 = 38.6 ft. The eductor must also overcome the friction loss in the discharge line.

Choosing the eductor suction port as the datum plane the suction lift is zero (slight positive head to balance loss through the foot valve and strainer). Since the eductor discharge flow is fixed at 100 gpm, Fig. 4 can be used to select a 3-in pipe size (100 gpm at $H_2 - H_s = 38.6$). Friction loss in 50 ft of 3-in pipe is 2.23 ftH_2O (Williams and Hazen formula). The required discharge head is then

$$38.6 + 2.23 = 40.83 \text{ ft}$$

TABLE 7 Relative Capacities of Annular Eductors (Schutte and Koerting Co.)

Size, in	1¼	1½	2	2½	3	4
Capacity ratio	0.62	1.00	1.43	2.86	4.76	8.80

Capacities for 1½-in mixing eductor, 5 lb/in² gage discharge pressure

Motive pressure, lb/in² gage	30	40	50	60	70	80	90	100
Entrainment, ft³/hr	2.6	7.1	14.4	17.9	20.3	22.0	22.9	23.8
Motive flow, gpm	12.7	14.6	16.3	17.9	19.3	20.7	21.9	23.1

The available operating head is 138.6 ft + 50 ft − friction loss. As a first assumption the friction loss is ignored and the head ratio is 188.6/40.23 = 4.69. From Fig. 2 at NPSH 33.6 ft (34 − vapor pressure at 50°F) ϵ = 0.9 and

$$R_w = R_q = 0.9\sqrt{4.69} - 1 = 0.949 \text{ [from Eqs. (8) and (9)]}$$

with Q_2 fixed at 100 gpm $Q_1 = 100/(1 + R_q) = 100/1.949 = 51.3$ gpm. The motive line size is now chosen by selecting a reasonable velocity and friction loss. Choosing a 2-in pipe size, velocity is 5.57 ft/s and friction loss is 6.1 ftH_2O. The revised operating head becomes 188.6 − 6.1 = 182.5 and $R_q = 0.9\sqrt{182.5/40.23} - 1 = 0.9$, then $Q_1 = 100/1.9 = 52.6$ gpm. (This value is close enough so that a third trial is not necessary.) The suction flow that can be delivered is then

$$100 - 52.6 = 47.4 \text{ gpm}$$

Priming Eductors—Water-Jet Exhausters. Educators are often used as priming

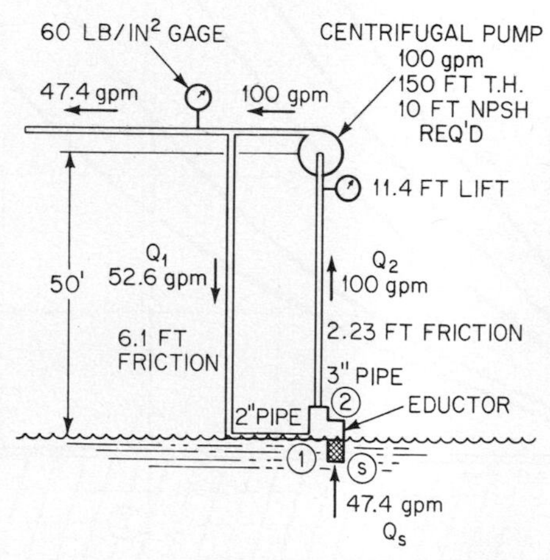

Fig. 17 Centrifugal–jet pump example 4.

devices for mechanical pumps. In this application the eductor is used to remove air rather than water. Liquid jets are not well suited for pumping noncondensables; therefore the capacities are low. However, the volume being primed is usually small, so the low capacity is not a factor. When larger volumes are involved, such as condenser water boxes, it is more feasible to use an exhauster. The water-jet eductor of Fig. 6 is converted to a water-jet exhauster by replacing the jet nozzle with a solid-cone spray nozzle. Evacuating rates and capacity tables for such a unit are shown in Fig. 18 and Table 8. Eductors have approximately one-fifth the air-handling capacities of water-jet exhausters when supplied with similar motive quantities and pressures.

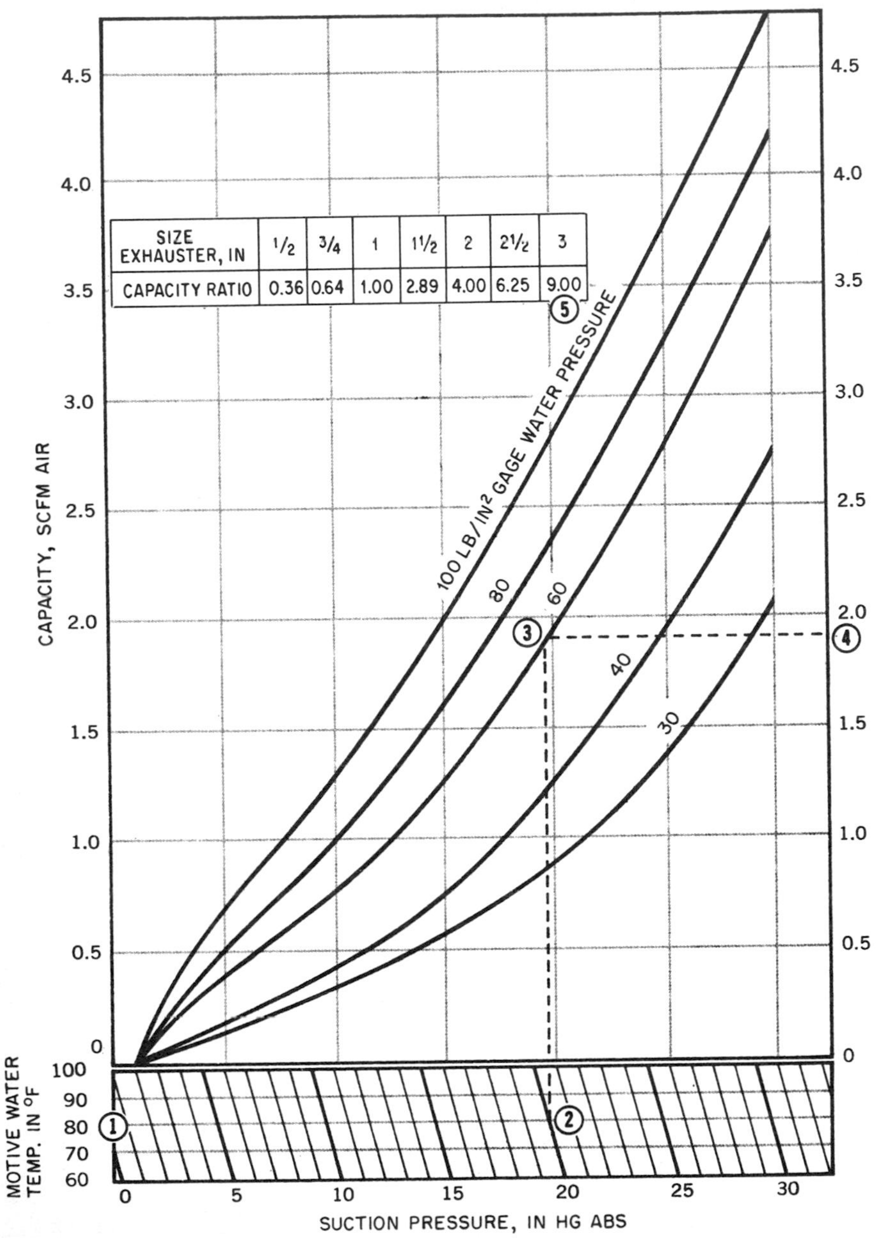

Fig. 18 Capacity curve of water-jet exhausters. (Schutte and Koerting Co.)

EXAMPLE 5: From Fig. 18 and Table 8, determine size and water consumption to exhaust 15 standard ft^3/min of air at 20 inHg abs discharging to atmosphere using 60 lb/in^2 gage motive water at 80°F.

SOLUTION: Enter Fig. 18 at 80°F (1); read horizontally to the suction pressure 20 inHg abs (2); project vertical line to 60 lb/in^2 gage motive pressure (3); project a horizontal line for the capacity of a 1-in exhauster (4); divide desired flow by the capacity of a 1-in unit (1.9 standard ft^3/min) to find capacity ratio: $15/1.9 = 7.9$.

Capacity ratio table shows that a 3-in exhauster with a capacity ratio of 9.0 is

TABLE 8 Water Consumption of Water-Jet Exhausters (Schutte and Koerting Co.)

	Approximate water consumption, gpm				
	Water pressure, lb/in² gage				
Size, in	30	40	60	80	100
½	2.6	2.9	3.4	3.8	4.2
¾	4.6	5.3	6.4	7.4	8.3
1	6.2	6.8	8.1	9	10
1½	20	23	27	30	32
2	28	31	36	41	45
2½	46	51	60	67	73
3	66	73	86	96	106

Note: All flows at 15 inHg abs.

required. The motive water quantity from Table 8 is 86 gpm. *Note:* Table 8 gives water consumption at 15 inHg abs; since flow varies as the square of pressure differential across the nozzle the exact flow is obtained as follows:

$$\text{Nozzle upstream pressure} = 60 + 14.7 = 74.7 \text{ lb/in}^2 \text{ abs}$$

$$\text{Nozzle downstream pressure} = 20 \text{ inHg abs } \frac{14.7 \text{ lb/in}^2}{30 \text{ inHg}} = 9.8 \text{ lb/in}^2 \text{ abs}$$

$$\text{Operating differential} = 74.7 - 9.8 = 64.9 \text{ lb/in}^2$$

$$\text{Table differential} = 74.7 - 15\,\frac{14.7}{30} = 67.35 \text{ lb/in}^2$$

$$\text{Actual flow} = 86\left(\frac{64.9}{67.35}\right)^{1/2} = 84.4 \text{ gpm}$$

SIPHONS

Operation As previously defined, the term siphon refers to a jet pump utilizing a condensable vapor to entrain a liquid and discharge to a pressure intermediate to motive and suction pressure. The principal motive fluid is steam.

In an eductor the high-pressure motive fluid enters through a nozzle and creates a vacuum by jet action, which causes suction fluid to enter the mixing chamber. The siphon of Fig. 19 is identical to the eductor of Fig. 6 except, unlike the eductor, the siphon motive nozzle is a converging-diverging nozzle to achieve maximum velocity at the nozzle tip. The velocity is supersonic at this point. The motive fluid is condensed into the suction fluid on contact and imparts its energy into the liquid, thus impelling it through the diffuser. The diffuser section is the same as an eductor diffuser and it converts the velocity energy into pressure at the discharge. To achieve maximum performance, the siphon nozzle must be expanded to the desired suction pressure in order to achieve the highest possible velocity. Since negligible radiation losses are encountered, the siphon is 100 percent thermally efficient in that heat in the incoming water plus heat in the operating steam must equal the heat of the mixture plus its mechanical energy. Furthermore, the momentum of the incoming water plus the momentum of the expanded steam is equal to the momentum of the discharge mixture less impact and friction losses.

It is important that the motive steam be condensed in the suction liquid prior to the throat for proper operation. If condensation does not occur full available energy is not transferred. Furthermore, energy must be expended to recompress the uncondensed steam. For this reason discharge temperature cannot exceed the boiling point at the discharge pressure. The fact that energy is required to recompress any uncondensed steam explains why air is a very poor motive fluid for liquid pumping.

Standard Siphons. Tables 9 and 10 illustrate the capacity and operating characteristics of standard siphons. These tables are similar to the eductor-capacity tables. To size a unit read the suction capacity of a 1½-in unit from Table 9 at the appropriate motive steam pressure, suction lift, and discharge head. Divide the desired suction flow by this capacity to find the capacity ratio. From Table 10 find the

TABLE 9 Relative Capacities of Steam-Jet Siphons, gpm (Schutte and Koerting Co.)

Suction lift, ft	Suction temp., °F	0-ft discharge head								20-ft discharge head							
		Operating steam pressure, lb/in² gage															
		40	50	60	80	100	120	160	240	40	50	60	80	100	120	160	240
1	70	52	51	51	49	46	43	37	30	36	47	48	48	45	41	35	29
	90	45	44	43	42	40	37	33	26	37	44	43	43	41	38	33	26
	110	40	38	36	36	35	33	28	22	38	39	37	37	36	34	30	23
	130	35	32	30	30	29	29	25	...	35	33	31	31	30	29	25	...
	150	26	25	24	24	24	24	21	...	26	25	24	24	23	22	20	...
	165	17	17	17	18	18	17	17	...	17	17	17	17	17	17	16	...
10	70	38	38	37	35	30	28	25	20	27	36	37	35	31	29	25	19
	90	34	34	33	30	27	25	21	17	26	32	32	29	26	23	20	17
	110	28	27	26	25	23	21	18	...	26	27	27	25	23	21	18	...
	130	21	21	21	20	18	16	14	...	21	22	22	21	18	16	14	...
	145	16	16	16	16	14	12	...	...	16	16	16	16	14	12	...	...
15	70	34	32	30	26	23	21	18	14	24	33	32	27	24	23	19	15
	90	29	28	26	23	20	18	16	12	23	28	27	23	20	19	16	12
	110	24	23	22	19	17	15	13	...	23	23	22	19	17	15	13	...
	130	17	17	17	15	13	11	...	...	17	17	17	14	13	...	...	...
	145	10	12	11	9	...	...	...	...	11	11	10	10	...	...	...	...
20	70	26	23	21	18	16	15	13	...	24	24	22	19	17	15	12	...
	90	22	19	17	15	14	12	11	...	19	20	18	15	14	12	11	...
	110	18	16	14	12	11	10	...	...	17	16	14	12	11	...	...	...
	125	13	12	11	...	...	...	...	...	12	11	10	...	...	...	...	...

		40-ft discharge head								50-ft discharge head							
		40	50	60	80	100	120	160	240	40	50	60	80	100	120	160	240
1	70	...	20	33	47	44	41	36	29	...	...	18	44	44	41	36	28
	90	...	18	36	43	42	39	34	27	...	...	21	42	42	39	34	27
	110	...	20	36	37	37	35	30	24	...	...	24	37	36	34	30	24
	130	...	23	31	31	30	29	25	...	...	...	26	30	30	29	25	...
	150	...	24	24	24	24	24	20	...	...	...	24	24	24	24	20	...
	165	...	17	18	18	18	18	16	...	...	...	18	18	18	18	16	...
10	70	...	...	24	34	30	27	24	18	...	...	...	35	30	27	23	18
	90	...	...	26	29	26	24	20	17	...	...	...	29	27	24	21	17
	110	...	...	27	26	23	20	18	...	...	...	...	25	23	20	18	...
	130	...	...	22	21	19	16	14	...	...	...	...	21	19	17	14	...
	145	...	...	16	16	14	12	...	...	...	...	...	16	14	13	...	...
15	70	...	...	23	28	24	22	19	15	...	...	...	27	24	21	18	14
	90	...	...	20	24	20	18	16	...	...	...	...	24	21	18	16	...
	110	...	...	21	19	17	15	...	...	...	...	...	19	17	15	...	...
	130	...	...	17	14	12	...	...	...	...	...	...	14	...	...	...	...
	145	...	...	11	...	...	...	...	...	...	...	...	...	...	...	...	...
20	70	...	...	21	19	16	15	12	...	...	...	...	...	16	15	...	...
	90	...	...	18	16	14	13	11	...	...	...	...	...	15*	13*	...	...
	110	...	...	...	11	10	...	...	...	...	...	...	...	...	...	...	...

*Suction temperature 85°.

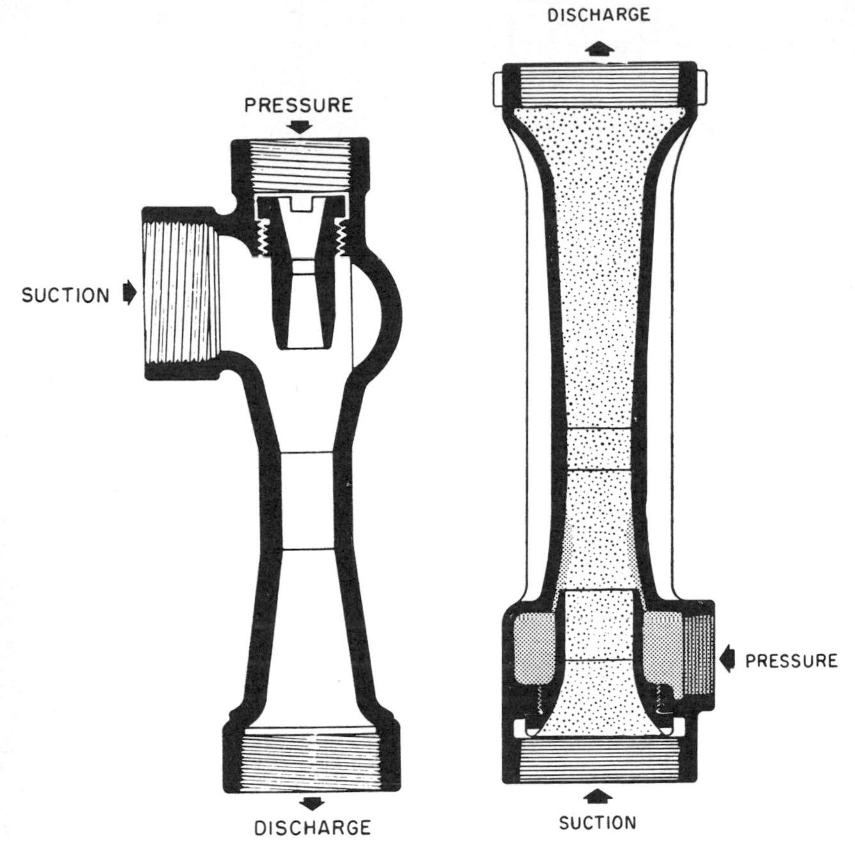

Fig. 19 Standard siphon. (Schutte and Koerting Co.)

Fig. 20 Annular siphon. (Schutte and Koerting Co.)

TABLE 10 Steam Consumption of Steam-Jet Siphons (Schutte and Koerting Co.)

Siphon size, in	Capacity ratio	Steam consumption, lb/hr — Operating steam pressure, lb/in² gage: 40	50	60	80	100	120	160	240
½	0.125	40	47	54	69	83	97	126	184
¾	0.222	70	83	96	122	147	173	222	322
1	0.346	110	130	150	190	230	270	350	510
1½	1.000	318	376	434	550	665	780	1,012	1,475
2	1.38	440	520	600	761	920	1,080	1,400	2,040
2½	2.0	635	750	865	1,100	1,329	1,558	2,020	2,940
3	3.11	990	1,170	1,350	1,710	2,065	2,425	3,145	4,590
4	5.54	1,760	2,085	2,400	3,045	3,685	4,320	5,500	8,170
6	12.45	3,960	4,680	5,400	6,850	8,280	9,710	12,600	18,360

unit with the next larger capacity ratio. Read the motive steam required under the proper motive pressure. Standard materials of construction are cast iron, bronze, stainless steel, pyrex. If desired, special capacity ratios can be achieved by using a custom-designed unit. Sizes over 6 in can be fabricated of any suitable material.

Annular Siphons. Figure 20 illustrates an annular siphon. This unit is identical to the eductor of Fig. 14 except that steam is the motive fluid. Capacity Table 11 is used in the same manner as previous examples. This type of siphon is used

TABLE 11 Relative Capacities of Annular Siphons (Schutte and Koerting Co.)

Water temperature 100° F, 0 suction lift				
Steam pressure, lb/in gage	50	75	100	125
Steam consumption, lb/hr	1,180	1,620	2,060	2,490
Max back pressure, lb/in² gage at zero flow	12	18	22	35
Suction capacity, gpm	140	130	120	110
Discharge pressure, lb/in² gage	5	8	12	30

Relative capacities of standard sizes

Size, in	1¼	1½	2	2½	3	4	6	8
Capacity ratio	0.13	0.21	0.30	0.60	1.00	1.85	4.0	7.1

when inline flow is desired or when the suction liquid contains some solids. Units are available in cast iron through an 8-in size. Special materials or fabricated designs are also available.

OTHER JET-PUMP DEVICES

Air Siphons As previously mentioned, air is a very poor motive fluid for entraining a liquid because energy must be expended in compressing the air back to the

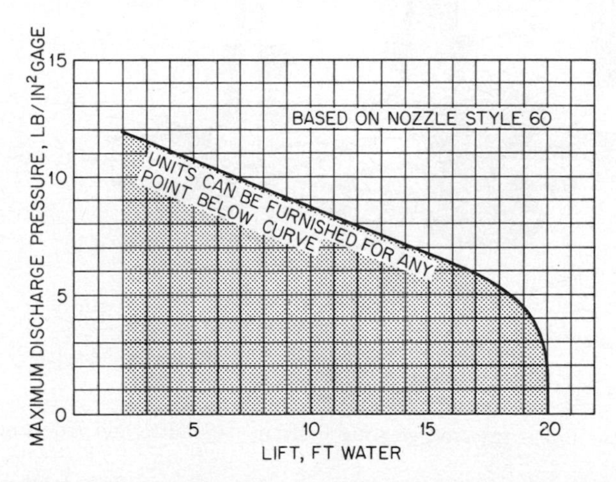

Fig. 21 Air pumping liquid. (Schutte and Koerting Co.)

discharge pressure. There are, however, applications where it is necessary to sample a liquid with no dilution.

Small units (less than 1 gpm) can be supplied for limited discharge pressures as indicated by Fig. 21. With air as the motive fluid, the suction liquid can be very close to its boiling point and only a very slight NPSH is required.

When air is used as a motive fluid, the smaller sizes operate more efficiently because the air is more intimately mixed with a suction fluid. In larger sizes the tendency is for the fluid to be discharged in slugs, since intimate mixing does not readily occur. This has a detrimental effect on the performance and especially on available discharge head.

Air-Lift Eductors Air-lift pumps are frequently used for difficult pumping operations. In operation, compressed air is forced into the bottom of a pipe submerged in the liquid to be pumped. The expanding air, as it rises up the pipe, entrains the suction fluid.

If compressed air is not available, it is possible to accomplish lifts of water greater than 34 ft with the use of an eductor-air-lift combination. Figure 22 illustrates

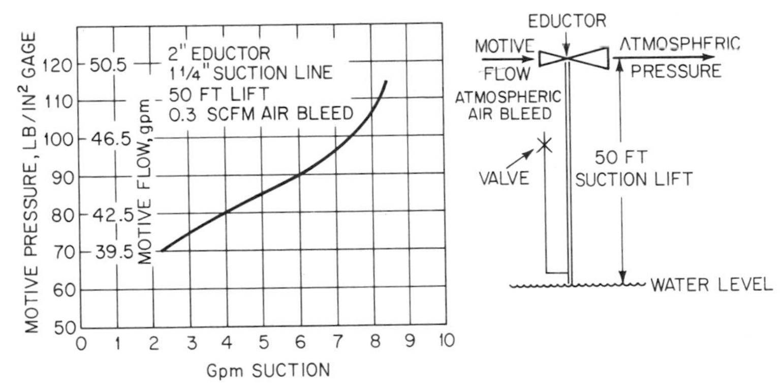

Fig. 22 Suction capacity of eductor air-lift combination.

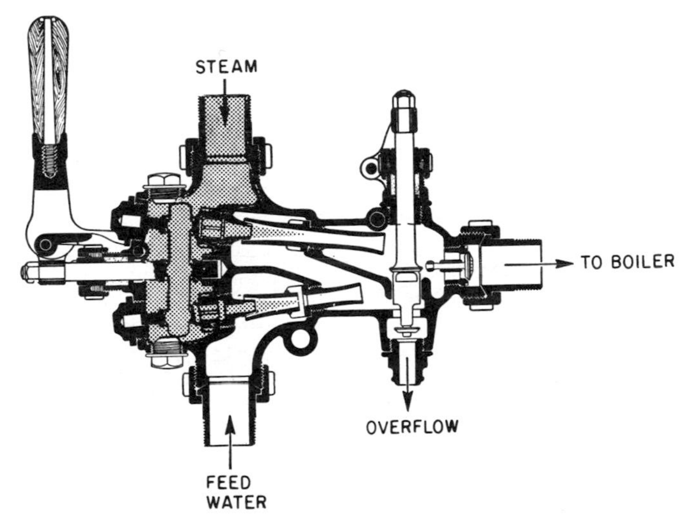

Fig. 23 Boiler injector, starting position. (Schutte and Koerting Co.)

the suction capacity of a 2-in eductor drawing water from a 50-ft depth and discharging it to the atmosphere. In operation, an air line from the atmosphere enters the suction pipe near the water level. As the eductor creates a vacuum in this line, atmospheric pressure forces air into the suction pipe. Once in the line, the rising air carries the suction fluid to the surface and both fluids are discharged to the atmosphere through the eductor.

No sizing data are presented since this type of pump is best specified according to specific conditions.

Boiler Injectors The boiler injector is a jet pump utilizing steam as a motive fluid to entrain water, and it is used as a boiler feedwater heater and pump. It

TABLE 12 Capacities of Boiler Injectors (Schutte Koerting Co.)

Size No.	Size iron pipe conn., in	Size copper pipe OD, in	Size overflow (drip funnel) pipe, in	Capacities, gph				
				Steam 50 lb	Steam 100 lb	Steam 150 lb	Steam 200 lb	Steam 250 lb
0	¼	⅜	¼	80	100	110	100	90
1	⅜	½	¼	110	140	180	140	130
2	½	⅝	⅜	170	210	230	200	190
3	¾	⅞	½	280	340	400	340	320
3½	¾	⅞	½	400	470	550	470	440
4	1	1⅛	¾	530	620	720	620	590
5	1¼	1½	1	680	800	920	800	750
6	1¼	1½	1	820	990	1,130	990	930
7	1½	1¾	1¼	1,070	1,370	1,610	1,370	1,290
8	1½	1¾	1¼	1,400	1,800	2,100	1,800	1,700
9	2	2¼	1½	1,700	2,100	2,500	2,100	2,000
10	2	2¼	1½	2,000	2,500	2,900	2,500	2,300
11	2½	2¾	2	2,500	3,000	3,500	3,000	2,800
12	2½	2¾	2	3,000	3,600	4,300	3,600	3,400
14	3	3¼	2½	3,900	4,600	5,500	4,600	4,400
16	3	3¼	2½	5,000	6,000	7,000	6,000	5,700

differs from a siphon in that the discharge pressure is higher than either motive or suction pressure. This is achieved by the double-tube design (see Fig. 23). In operation, the lower nozzle is activated by pulling the handle part way back. The lower jet creates a vacuum in the chamber causing water to be induced into the unit. When water is spilling out the overflow, the handle is drawn back all the way. This closes the overflow and simultaneously admits motive steam to the upper jet. This second jet, which is of the straight or forcing type, then picks up the discharge from the first jet and imparts a velocity to the water through the discharge tube. The energy contained is sufficient to open the check valve and discharge against the boiler pressure.

The now obsolete steam locomotives were the largest users of this type injector. Principal use at present is as a backup to a regular boiler feed pump. Capacity Table 12 illustrates the range of capacities available for the double-tube injector.

Section 5

Materials of Construction

WARREN H. FRASER

INTRODUCTION

The essential requirements for a successful pump installation are performance and life. *Performance* is the rating of the pump—head, capacity, and efficiency. *Life* is the total number of hours of operation before one or more pump components must be replaced to maintain an acceptable performance. Initial performance is the responsibility of the pump manufacturer and is inherent in the hydraulic design. Life is primarily a measure of the resistance of the materials of construction to corrosion, erosion, or a combination of both under actual operating conditions.

The selection of the most economic material for any particular service, however, requires a knowledge of not only the pump design and manufacture, but also of the erosion-corrosion properties of the materials under consideration when subjected to the velocities actually encountered in the pump. Very little corrosion data exists on the effects of the velocities encountered in centrifugal pumps for liquids other than seawater.

Despite this limitation, experience has provided the designer of pumping systems with a specialized branch of metallurgical and corrosion engineering adequate for most pumping problems. Factors that lead to a long pump life are:

1. Neutral liquids at low temperatures
2. Absence of abrasive particles
3. Continuous operation at or near the maximum efficiency capacity of the pump
4. An adequate margin of available NPSH over the NPSH required as stated on the pump manufacturer's rating curve

Any pumping installation that satisfies all these criteria will have a long performance life. A typical example would be a waterworks pump. Some waterworks pumps with bronze impellers and cast iron casings have a performance life of 50 years or more. At the other extreme would be a chemical pump handling a hot corrosive liquid with abrasive particles carried in suspension. Here the performance life might be measured in months rather than in years despite the fact that the construction has been selected based upon the most resistant materials available.

Most pumping applications lie somewhere between these two extremes. Aside from straight corrosion or erosion from abrasive particles in the fluid, the greatest single factor that reduces the performance life is operation at flows other than the maximum efficiency or rated capacity of the pump. The vane angles of the impeller are designed to match the fluid angles at the maximum efficiency capacity. At flows other than the rated capacity the fluid angles no longer match the vane angles, and separation occurs with increasing intensity as the operating point moves away from the maximum efficiency capacity. The destruction of the impeller vanes is particularly severe at the inlet to the impeller, as this is the point of lowest pressure in the pump. In addition to surface damage to the inlet vanes from separation, localized cavitation damage may occur during sustained operation at capacities less than 50 or more than 125 percent of the maximum efficiency capacity. This is not to say that many pumps do not operate under these adverse operating conditions, but their performance life is considerably less than the same pump operating at or near its maximum efficiency capacity.

TYPES OF CORROSION

Erosion Corrosion The rate of corrosion of most metals is increased when there is a flow of liquid relative to the component itself. The rate of corrosion is also dependent upon the angle of attack of the piece to the direction of flow. As a general rule the rate of corrosion increases as the angle of attack increases, that is, as the separation of the fluid from the metal surface becomes larger. The greater the separation the more intense the turbulence, and therefore the greater the rate of metal removal.

The corrosion rate of most metals and alloys in any liquid environment under static conditions depends upon the resistance of the film that forms on the surface and protects the base metal from further attack. Damage to or removal of this film from erosion exposes the unprotected base metal to the corrosive environment and metal removal continues unabated.

In centrifugal pumps the impeller is particularly susceptible to erosion corrosion. Although the casing can be damaged from erosion corrosion, the problem is usually secondary to that of the impeller. The diffuser-type casing with a multiplicity of vanes is more susceptible to erosion corrosion than is the volute type casing with only one vane—the casing tongue—as an obstruction to the line of flow.

Wearing rings are also susceptible to erosion corrosion and should receive special consideraton in material selection. The high fluid velocities through the small clearance annulus can result in a high rate of wear unless the proper material selection has been made.

When considering any material for a pump service the erosion-corrosion properties of the material must be evaluated. This can be a difficult task as many of the properties and characteristics of both the liquid to be pumped and the materials under consideration are not known in sufficient detail to arrive at an optimum solution. Most materials used in pump construction have been corrosion-tested under static conditions in a wide variety of liquids, but few have been tested in an environment of high velocity. For example, the corrosion rate of the 300 series austenitic stainless steels remains virtually unchanged over a wide range of seawater velocities, whereas the rate of corrosion for the bronzes increases rapidly over the same range. The rate of corrosion, however, is only part of the problem of evaluating erosion corrosion. The erosion resistance of a metal or of an alloy is dependent primarily on the hardness of the material. The hardness of a material, however, is not necessarily compatible with its corrosion resistance. The 300 series austenitic stainless steels are not hardenable but exhibit low rates of corrosion in most liquids. The 400 series stainless steels, on the other hand, are hardenable to 500 Brinell, but exhibit higher corrosion rates in many more liquids than do the austenitic stainless steels.

Most pump components of standard designs are limited to approximately 350 Brinell. Above 350 Brinell the standard machining operations of turning, boring, drilling, and tapping become uneconomical. Stuffing box sleeves and wearing rings of higher hardness can be provided, and the finishing operation of cylindrical components can be performed by grinding.

As a general rule, the material should be selected primarily on the basis of its velocity-corrosion resistance to the liquid being pumped provided the liquid is free of abrasive solids. If the liquid contains abrasive solids, then the material of construction should be selected primarily for abrasive-wear resistance provided the velocity-corrosion-resistance characteristics are acceptable.

Corrosion Fatigue When evaluating the selection of a material for a pump component that is subjected to a cyclic stress the endurance limit of the material must be considered. The *endurance limit* is the maximum cyclic stress that the material can be subjected to and still not fail after an infinite number of cyclic stress reversals. The endurance limit of steel, for example, is approximately 50 percent of its tensile strength. A 100,000 lb/in^2 tensile steel could be subjected to a static load in tension to produce 100,000 lb/in^2 stress, but the same steel subjected to a cyclic stress of 100,000 lb/in^2 would fail in a short period of time. If the stress were reduced to 50,000 lb/in^2, however, the endurance limit would not be exceeded, and the metal could be subjected to 50,000 lb/in^2 stress reversals and not fail.

If the same steel were subjected to a cyclic stress of 50,000 lb/in^2 in a corrosive environment, however, failure would occur quite rapidly. The failure is caused by corrosion fatigue. As a result of the cyclic stressing of the piece, minute cracks develop at the surface. In a corrosive environment the surface of the metal exposed by the cracks corrodes rapidly. The crack then penetrates deeper, corrosion develops further, and the piece will ultimately fail.

Experience has shown that the maximum combined stress in a pump shaft should not exceed 7,500 lb/in^2 for those portions exposed to the liquid pumped. Above this value the premature failure of shafts increases sharply. Since the corrosion-fatigue strength of any metal is dependent more on the corrosion resistance of the metal than on its tensile strength, the life of any pump component subjected to cyclic loading can only be estimated. A pump shaft, for example, is subjected to a complete stress reversal for each revolution, and will have some definite life based upon the corrosion-fatigue strength of the material in the particular application and the rotational speed of the pump. The best way to guard against short shaft life is to protect the shaft against exposure to the liquid by means of sleeves.

Intergranular Corrosion Intergranular corrosion is the corrosion of the grain boundaries in the body of the material itself. Unlike direct or galvanic corrosion of the metal surface itself, intergranular corrosion is insidious in that the mechanical properties of the material can be drastically reduced with little apparent surface damage. As the grain boundaries alone are affected, the material appears sound from a surface inspection. Progressive intergranular corrosion can proceed, however, to the point where the material literally disintegrates.

Intergranular corrosion of the austenitic stainless steels occurs as a result of carbides precipitating out at the grain boundaries during the slow cooling of the casting. When exposed to a corrosive environment the carbides are preferentially attacked, destroying the strongly bonded matrix of metal grains. The precipitation of carbides can be controlled by heating the casting to 2000°F and then quenching. At 2000°F the carbides are held in solution, and the rapid quench prevents their precipitation.

Susceptibility to intergranular corrosion in the austenitic stainless steels can also be reduced by controlling the carbon content of the alloy. The standard austenitic stainless steels of the 300 series have a carbon content in excess of 0.08 percent. Without proper heat treatment these steels are susceptible to intergranular corrosion. Extra low carbon steels are available in the 300 series and are identified by the suffix letter L. These steels have a carbon content less than 0.03 percent and are much less susceptible to intergranular corrosion.

The possibility of intergranular corrosion must be considered when castings of austenitic stainless steels are indicated for impellers and casings of centrifugal pumps, or for the liquid ends of reciprocating pumps. For moderate size castings, the low carbon grades are adequate when properly heat treated, provided post–heat treatment is not necessary. The use of the more costly 0.03 percent carbon steels is not justified in this case. In the case of larger more intricate castings, however, the 0.03 percent carbon steels should be considered. This is particularly true for large open or semiopen impellers used in mixed flow and propeller pumps. Here the large unshrouded vanes can be severely distorted during the quenching operation. In this case a 0.03 percent carbon steel with forced air cooling is the preferred selection.

Restoration of impeller vane tips as a result of erosion or corrosion during the life of the pump requires that the impeller casting be specified with a maximum carbon content of 0.03 percent. If not, subsequent welding will precipitate carbides at the grain boundaries, and rapid failure may occur from intergranular corrosion at the welded areas.

Cavitation Erosion Cavitation erosion is the removal of metal primarily as a result of high localized stresses produced in the metal surface from the collapse of cavitation vapor bubbles. In a corrosive environment the rate of damage is accelerated as the corrosion products are continually removed, and the corrosion proceeds unabated.

While every effort should be made in the design and application of centrifugal pumps to prevent cavitation, it is not always possible to do so at capacities less than the rated maximum efficiency capacity of the pump. It must be recognized that at low-flow operation the stated required NPSH curve is not usually sufficient to suppress all cavitation damage. The stated required NPSH is that required to produce the head, capacity, and efficiency shown on the rating curve. At low flows it must be expected that some cavitation damage will occur. It would be impractical to supply an NPSH that would suppress all cavitation at these low flows as it would be many times that required at the best efficiency point. Therefore, it should be anticipated that some cavitation will occur at low-flow operation and should be considered in any evaluation of the material for impellers.

Open-type mixed-flow impellers that produce heads in excess of 35 ft are particularly susceptible to cavitation erosion in the clearance space between the rotating vanes and the stationary housing. This is usually referred to as *vane-tip erosion* and is caused by a cavitating vortex in the clearance space between the vane and the housing. Here again it would be impractical to provide sufficient NPSH to eliminate this type of cavitation. Any evaluation of the impeller and housing material for a pump of this type should include the possibility of vane-tip cavitation.

Extensive laboratory tests of the resistance of a wide range of materials to cavitation erosion have produced data for all the materials used in centrifugal pump construction. Despite the complexity of the cavitation process itself and the mass of laboratory data available it is possible to correlate the laboratory data and field experience to develop the following tabulation of the cavitation-resistance properties of pump materials. The materials are listed with increasing cavitation resistance from cast iron through nickel-aluminum bronze.

1. Cast iron
2. Bronze
3. Cast steel
4. Manganese bronze
5. Monel
6. 400 series stainless steel
7. 300 series stainless steel
8. Nickel-aluminum bronze

Abrasive Wear Abrasive wear is the mechanical removal of metal from the cutting or abrading action of solids carried in suspension in the pumped liquid. An undulating matte-finish wear pattern can usually be identified as abrasive wear from solid particles in the liquid. The rate of wear for any given material is dependent upon the following characteristics of the suspended solids:

1. Solids concentration
2. Size and mass of the solids
3. Form of the solids, that is, spherical, angular, or sharp fractured surfaces
4. Hardness of the solids
5. Relative velocity between the suspended solids and the metal surface

The rate of wear is also dependent upon the materials selected for the rotating and stationary components of a centrifugal pump. While metal hardness is not the sole criterion of resistance to abrasive wear, hardness does provide a convenient index in the selection of ductile materials usually available for centrifugal pumps. Such an index is shown in Fig. 1, where the abrasive-wear-resistance ratio is shown as a function of Brinell hardness for various materials. It should be noted that a brittle material such as cast iron exhibits a much lower ratio than either the steels or bronzes for the same hardness. There is some evidence that while the rate of abrasive wear for a ductile material is proportional to the square of the

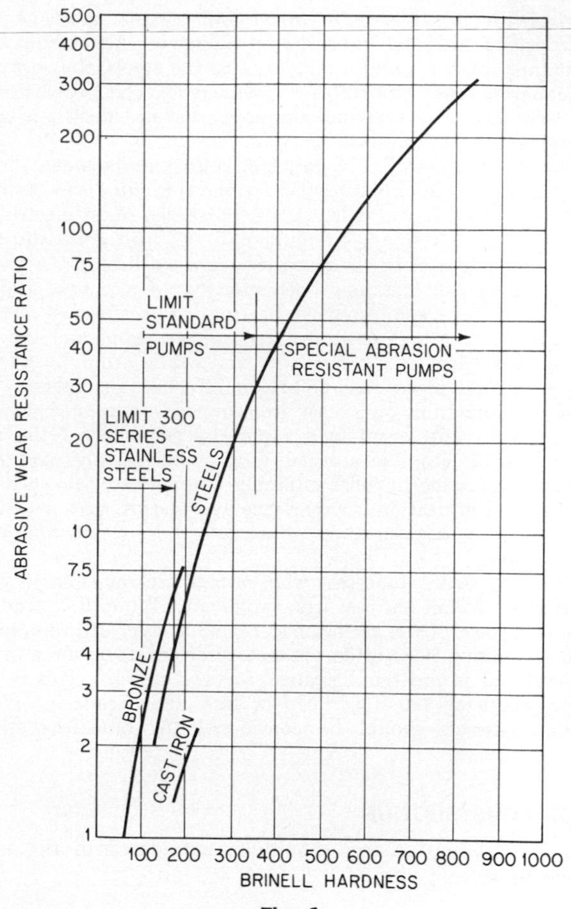

Fig. 1

velocity of the abrasive particles, the rate of abrasive wear of a brittle material may be as high as the sixth power of the particle velocity.

While it is impossible to establish any direct relation between the life of pump components and the quantity and characteristics of abrasive particles pumped, the following tabulation can be used as a guide in material selection. The materials are listed with increasing abrasive-wear resistance from cast iron through the 400 series stainless steels.

1. Cast iron
2. Bronze
3. Manganese bronze
4. Nickel-aluminum bronze
5. Cast steel

6. 300 series stainless steel
7. 400 series stainless steel

Graphitization Gray iron consists of a matrix of iron and graphite. The graphite exists as flakes and produces the characteristic gray appearance of cast iron. The presence of the graphite also provides a lubricant during machining. This property, plus the fragility of the chips, accounts for the excellent machining qualities of cast iron.

These characteristics, in addition to low foundry costs, combine to make gray iron the most widely used metal in the pump industry. Aside from the low tensile strength and ductility of cast iron as compared to the steels, the corrosion-resistance properties of cast iron must be carefully considered. The presence of graphite in the matrix of cast iron produces the unique corrosion effect known as graphitic corrosion or, more simply, graphitization.

In the presence of an electrolyte a galvanic cell exists between the iron and the graphite particles. In the combination of iron and graphite, iron becomes the anode and the graphite becomes the cathode. In the presence of an electrolyte a galvanic current flows from the iron to the graphite and the iron goes into solution. The result is a gradual depletion of the iron in the matrix until only the graphite remains. The original casting in iron has now been reduced to a porous graphite structure interspersed with the corrosion products of iron. The physical properties of the graphite structure are greatly inferior to that of cast iron, and the structure fails rapidly. In fact, the effect is so dramatic that while from outward appearances the casting appears sound, pieces may be broken off with the fingers.

The effect of graphitization on a cast iron impeller pumping seawater has been observed many times. The result is a rapid deterioration of the impeller vanes as the soft graphite structure is scoured away progressively exposing new metal to further attack. The same impeller pumping a nonelectrolyte, such as fresh water, shows no effect of graphitization. Experience has shown that a cast iron impeller should never be used on brackish water or seawater: the result is inevitably destruction by graphitization.

Experience has also shown that cast iron casings are much more resistant to destructive graphitization than are cast iron impellers. While it is true that examination of the inside surface of the casing may reveal a layer of graphite, the velocities encountered in the casing very often are not sufficient to scour away the graphite, and the base material is protected against further attack. This is true, however, only for pumps producing 100 ft of head or less. For heads in excess of 100 feet alternate casing materials should be considered for brackish-water or seawater services.

MATERIALS OF CONSTRUCTION

Impellers The following criteria should be considered in the selection of the material for the impeller:

1. Corrosion resistance
2. Abrasive-wear resistance
3. Cavitation resistance
4. Casting and machining properties
5. Cost

For most water and other noncorrosive services bronze satisfies these criteria on an evaluated basis. As a result bronze is the most widely used impeller material for these services. Bronze impellers, however, should not be used for pumping temperatures in excess of 250°F. This is a limitation imposed primarily because of the differential rate of expansion between the bronze impeller and the steel shaft. Above 250°F the differential rate of expansion between bronze and steel will produce an unacceptable clearance between the impeller and shaft. The result would be a loose impeller on the shaft.

Cast iron impellers are used to a limited extent in small low-cost pumps. As cast iron is inferior to bronze in corrosion, erosion, and cavitation resistance, low initial cost would be the only justification for a cast iron impeller on an evaluated basis.

Stainless steel impellers of the 400 series are widely used where bronze would not satisfy the requirements for corrosion, erosion, or cavitation resistance. The 400 series stainless steels are not used for seawater, however, as pitting will limit their performance life. The 400 series stainless steels should be used where the pumping temperatures exceed 250°F, as the differential expansion problem no longer exists with a steel impeller on a steel shaft.

The austenitic stainless steels of the 300 series are the next step up on the corrosion- and cavitation-resistance scale. Initial cost is a factor here that should be evaluated against the increased performance life.

Casings The following criteria should be considered in the selection of material for centrifugal pump casings:

1. Strength
2. Corrosion resistance
3. Abrasive-wear resistance
4. Casting and machining properties
5. Cost

For most pumping applications cast iron is the preferred material for pump casings when evaluated against initial cost. For single-stage pumps cast iron is usually of sufficient strength for the pressures developed. For corrosive and volatile petroleum products it may be necessary to specify cast steel or cast stainless steels of the 400 or 300 series.

Cast iron casings for multistage pumps are limited to approximately 1,000 lb/in^2 discharge pressure and 350°F. For temperatures above 350°F and pressures up to 2000 lb/in^2 discharge pressure, a cast steel is usually specified for split-casing multistage pumps. For pressures higher than 2,000 lb/in^2 a cast or forged steel barrel-type casing is required.

In any evaluation of cast iron versus steel casings, consideration should be given to the probability of casing erosion during operation. Erosion can occur from either abrasive particles in the fluid or from wire drawing across the flange of a split-case pump. While the initial cost of a steel casing is higher than that of a cast iron casing, a steel casing can often be salvaged by welding the eroded portions and then remachining. Salvaging a cast iron casing by welding is not practical, and the casing usually must be replaced.

The ductile irons are useful casing materials for pressure and temperature ratings between cast iron and the steels. While the modulus of elasticity for the ductile irons is essentially the same as that for cast iron, the tensile strength is approximately doubled. In any evaluation of the ductile irons as a substitute for the steels in the intermediate pressure and temperature range, it must be remembered that ductile iron casings cannot be effectively repair-welded in the field.

Shafts The following criteria should be considered in the selection of the material for a centrifugal pump shaft:

1. Endurance limit
2. Corrosion resistance
3. Notch sensitivity

The endurance limit is the stress below which the shaft will withstand an infinite number of stress reversals without failure. Since one stress reversal occurs for each revolution of the shaft this means that, ideally at least, the shaft will never fail if the actual maximum bending stress in the shaft is less than the endurance limit of the shaft material.

In actual practice, however, the endurance limit is substantially reduced because of corrosion and stress raisers such as threads, keyways, and shoulders on the shaft. In the evaluation of the selection of the shaft material consideration must be given to the corrosion resistance of the material in the fluid being pumped as well as the notch sensitivity. For a more detailed discussion of the effect of endurance limit on the shaft design refer to the section on Corrosion Fatigue.

Wearing Rings The following criteria should be considered in the selection of the material for the wearing rings:

1. Corrosion resistance
2. Abrasive-wear resistance
3. Galling characteristics
4. Casting and machining properties

The wearing rings of a centrifugal pump consist of the impeller ring rotating in the bore of the stationary or casing ring. As the purpose of the wearing rings is to provide a close running clearance to minimize the leakage from the discharge to the suction of the impeller, an increase in the leakage as a result of wear in the rings will have a direct effect on the head, capacity, and efficiency of the pump.

To reduce the rate of wear of the wearing rings, and thereby increase the performance life of the pump, special consideration must be given to the corrosion and abrasive wear characteristics of the ring material in any evaluation of wearing-ring selection.

Bronze is the most widely used material for wearing rings. Bronze exhibits good corrosion resistance for a wide range of water services. In addition, bronze exhibits good wear characteristics in clear liquids, but wears rapidly when abrasive particles are present. The bronzes also are low on the galling scale if metal to metal contact occurs. This is especially true for the 8 to 12 percent leaded bronzes, and the leaded bronzes should be used whenever possible. The casting and machining properties of most of the grades of bronze are excellent.

In applications where bronze is not suitable either because of corrosion or abrasive-wear limitations, or where pumping temperatures exceed 250°F, stainless steel rings would be required. Unlike bronze, the stainless steels of the 300 or 400 series exhibit a greater tendency to gall. The risk of wearing-ring seizure can be minimized by increasing the clearance between the rings, specifying a difference in hardness between the two rings of 125 to 150 Brinell, or providing serrations in one of the wearing-ring surfaces.

It is apparent that increasing the clearance between the wearing rings is the least costly procedure to reduce the risk of galling or seizure. But increasing the wearing-ring clearance will reduce the output and efficiency of the pump. In large low-head pumps the loss in efficiency is less than 1 percent, but in small high-head units the loss in efficiency can be significant. The serrated rings can be used on smaller pumps to maintain the efficiency but only at an increase in manufacturing costs.

SELECTION OF MATERIALS OF CONSTRUCTION

The selection of the materials for pumps is at best a compromise between the cost of manufacture and the anticipated maintenance costs. Many pump installations start out with a low service factor and through operating experience are gradually upgraded in materials until an acceptable and scheduled replacement program is achieved. It must be anticipated that for the more corrosive services modifications and replacement of the wetted parts will be necessary during the life of the pump. As an initial selection for the most economic materials encountered in centrifugal and reciprocating pump applications Tables 1, 2, and 3 will prove to be helpful.

When the liquid being pumped is also an electrolyte particular attention must also be directed at an evaluation of the probability of an unacceptable level of galvanic corrosion. The electrolytes most commonly encountered in pump applications are seawater, brines, and mine waters. For these services the use of incompatible materials can lead to rapid failures, and this is particularly true when a small area of the less noble metal is in contact with a larger area of a more noble metal. Bronze wearing rings, for example, should never be used with a stainless steel impeller when pumping an electrolyte such as seawater. Here the unequal area between the wearing ring and the impeller would result in a rapid preferential corrosion rate of the ring. An even more destructive combination would be the use of bronze screws or bolts in a stainless steel impeller or casing. Tables 4 and 5 list combinations of metals most commonly used in pumps that should be avoided when the application requires the pumping of an electrolyte.

TABLE 1 Material Selection Chart, Volute Casing Pumps

(1) Most economical to give acceptable service; (2) extended life at additional cost; temperatures of °F

Part	Fresh water (40–250°)	Seawater 40–80°	Seawater >80°	Boiler feed 250°	Boiler feed 350°	Boiler feed >350°	Sewage (40–90°)	Mine water (40–90°)	Condensate (90–120°)
Impeller:									
Cast iron	...	...	...	...	...	...	1		
Bronze	1	1	...	1	...	...	...	...	1
400 series stainless steel	...	...	...	2	1	1	2	...	2
300 series stainless steel	...	2	1	...	...	...	...	2	
Ni resist	...	1	...	...	...	...	...	1	
Casing:									
Cast iron	1	1	...	1	...	...	1	...	1
Bronze	...	2	2						
400 series stainless steel	...	...	...	2	1	1	...	...	2
300 series stainless steel	...	...	...	...	...	...	...	2	
Ni resist	...	2	1	...	...	...	...	1	
Wearing rings:									
Bronze	1	1	...	1	...	...	...	...	1
400 series stainless steel	2	...	...	2	1	1	1	...	2
300 series stainless steel	...	2	1	...	...	...	...	1	
Monel	...	...	2	...	...	...	...	2	
Shaft:									
Steel	1	...	...	1	...	...	1	...	1
400 series stainless steel	...	...	...	2	1	1	2	1	2
300 series stainless steel	...	1	1	...	...	...	...	2	
Monel	...	2	2						

TABLE 2 Material Selection Chart, Wet-Pit Diffuser Pumps

(1) Most economical to give acceptable service; (2) extended life at additional cost; impeller selection same as Table 1 for volute casing pumps; temperatures in °F

Part	Fresh water (40–250°)	Seawater 40–80°	Seawater >80°	Mine water (40–90°)	Condensate (90–120°)
Diffuser:					
Cast iron	1	...	...	...	1
Bronze	...	2	...	1	
400 series stainless steel	...	...	...	...	2
300 series stainless steel	...	2	2	2	
Ni resist	2	1	1	1	
Elbow and column:					
Steel	1	...	...	...	1
Cast iron	...	1			
Ni resist	...	2	1	1	
Wearing rings:					
Bronze	1	1	...	...	1
400 series stainless steel	2	...	...	...	2
300 series stainless steel	...	2	1	1	
Monel	...	...	2	2	
Shaft:					
400 series stainless steel	1	...	...	...	1
300 series stainless steel	...	1	1	1	
Monel	...	2	2	2	

TABLE 3 Material Selection Chart, Reciprocating Pumps

(1) Most economical to give acceptable service; (2) extended life at additional cost; temperatures in °F

Part	Fresh water (250°)	Seawater 40–90°	Seawater >90°	Water flood (90°)	Brines (40–90°)	Oil (250°)	Boiler feed (300°)
Valve service:							
Steel	1	...	...	1	...	1	1
Bronze	1	1	1	1	1	1	1
400 series stainless steel	2	...	...	2	...	2	2
300 series stainless steel	...	2	2	2	2		
Monel	...	2	2	2	2		
Liquid cylinder:							
Cast iron	1	1	...	1	1	1	1
Bronze	1	1	1	1	1	...	1
Steel	2	...	...	...	...	2	1
400 series stainless steel	...	...	...	2	...	...	2
300 series stainless steel	...	2	2	2	2		
Aluminum-bronze	...	2	2	2	2		
Plungers:							
Steel	...	...	...	...	...	1	
400 series stainless steel	1	...	...	1	...	1	1
300 series stainless steel	...	1	1	1	1		
Hard-faced stainless steel	2	2	2	2	2	2	2

TABLE 4 Combinations to Be Avoided When the Area of the Metal Considered Is Small as Compared with the Coupled Metal

Metal considered	Coupled metal					
	Steel	Cast iron	Bronze	400 series stainless steel	300 series stainless steel	Monel
Steel	...	X	X	X	X	X
Cast iron	...	...	X	X	X	X
Bronze	...	...	...	X	X	X
400 series stainless steel	...	...	X	...	X	X
300 series stainless steel	...	...	X	...	...	X
Monel	...	...	...	...	...	...

TABLE 5 Combinations to Be Avoided When the Area of the Metal Considered Is Equal to the Coupled Metal

Metal considered	Coupled metal					
	Steel	Cast iron	Bronze	400 series stainless steel	300 series stainless steel	Monel
Steel	...	...	X	X	X	X
Cast iron	...	...	X	X	X	X
Bronze	...	...	...	...	...	...
400 series stainless steel	...	...	X	...	X	X
300 series stainless steel	...	...	...	...	...	...
Monel	...	...	...	...	...	...

Section **6.1.1**

Electric Motors and Motor Controls

A. A. DiVONA

A. J. DOLAN

TYPES OF MOTORS

Alternating-Current Motors *Squirrel-Cage Induction Motor.* By far the most common motor used to drive pumps is the squirrel-cage induction motor. This motor consists of a conventional stator wound with a specific number of poles and phases, and a rotor which has either cast bars or brazed bars imbedded in the rotor. The squirrel-cage induction motor operates at a speed below synchronous speed by a specific slip or rpm.

The synchronous speed is defined as

$$N = \frac{f \times 60 \times 2}{p}$$

where p = number of poles
f = line-power frequency, Hz
N = velocity, rpm

The percent slip is defined as

$$\% \text{ slip} = \frac{(N - s) \times 100}{N}$$

where s = slip rpm.

When the stator winding of a squirrel-cage induction motor is connected to a suitable source of power, a magnetic flux is generated in the air gap between the stator and rotor of the motor. This flux revolves around the perimeter of the air gap and induces a voltage in the rotor bars. Since the rotor bars are short-circuited to each other at their ends (end rings), a current circulates in the rotor bars. This current and the air-gap flux interact, causing the motor to produce a torque. Figure 1 shows the cross section of a typical squirrel-cage induction motor.

The squirrel-cage induction motor exhibits a characteristic speed-torque relationship that is determined by the resistance of the rotor bars. The metal from which the rotor bars is fabricated permits a selection of this resistance at the time the motor rotor is designed and manufactured to obtain the desired speed-torque characteristics.

Figure 2 suggests several typical speed-torque characteristics which have been standardized by NEMA (National Electrical Manufacturers Association) covering motor frames 143T through 449T. Motors larger than 449T may not fulfill these same values, but have generally the same characteristic curves. Also, single-phase motors may not exhibit these charasteristics and are defined specifically by NEMA with different values.

Most pumps are driven by NEMA B characteristic motors when operated from three-phase power sources.

Wound-Rotor Induction Motor. The wound-rotor induction motor is in every respect similar to the squirrel-cage version except that the rotor is wound with insulated wire turns, and this winding is terminated at a set of slip rings on the rotor shaft. Connections are made to the slip rings through brushes, and in turn to an external resistor, which can be adjusted in ohmic value to cause the motor speed-torque characteristics to be changed.

Fig. 1 Cross section of typical squirrel-cage induction motor. (Westinghouse Electric Corporation)

Figure 3 demonstrates the speed-torque characteristics of a wound-rotor induction motor for several resistor values. It will be noticed that increasing the value of the external resistance of the control will cause the peak torque of the motor to be developed at the lower speeds until the peak torque occurs at zero speed. Increasing the resistance beyond this value will cause the motor to have a limited torque, as for example curves 4, 5, or 6. This motor can be used where torque control is required or where variable speed is necessary. In the case of the variable-speed application, the rotor resistance is adjusted to produce a motor torque that matches the load torque at the specific speed desired.

Synchronous Motor. The synchronous motor once again is similar to the squirrel-cage induction motor except that it operates at synchronous speed and its rotor is constructed with definite salient poles on which a field coil is wound and connected to a source of direct current for excitation. The most common synchronous motor is constructed with slip rings on the rotor shaft to connect the dc excitation to the field coils.

There are various means of providing the dc power to the slip rings, as follows:

1. Static excitation. The dc power to be connected to the slip-ring brushes on the motor shaft is obtained from a transformer and rectifier package external to the motor.

2. Direct-connected exciter. This arrangement has a dc generator directly connected to the synchronous motor shaft, Fig. 4. The dc power from this generator is connected to the brushes of the synchronous motor slip rings.

3. Motor-generator exciter. The dc power for exciting the synchronous motor is generated by means of a remote motor-generator set operating from normal ac power and the dc voltage generated from this motor-generator set is connected to the brushes of slip rings of the synchronous motor.

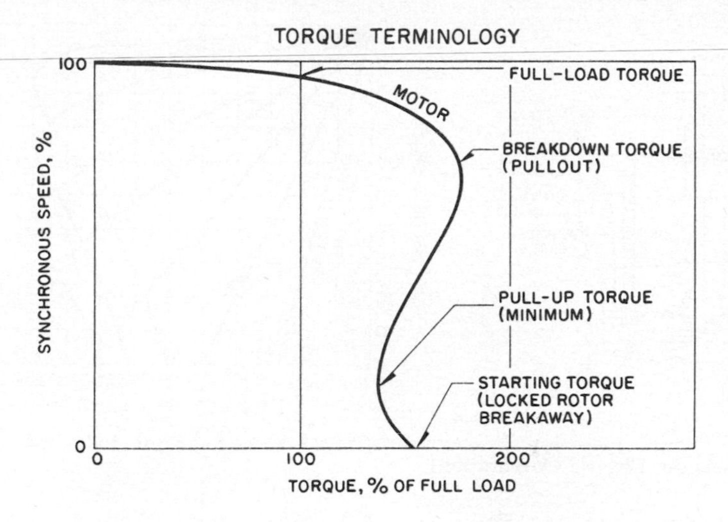

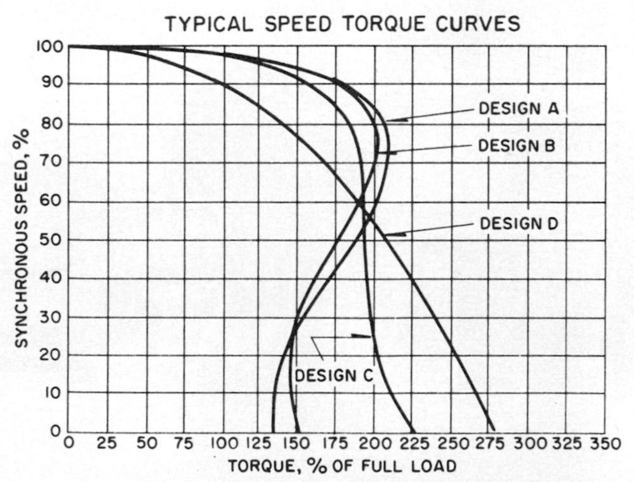

Fig. 2 Typical speed-torque characteristics of squirrel-cage induction motors standardized by NEMA. (Westinghouse Electric Corporation)

Another form of synchronous motor is known as the *brushless synchronous motor* (Fig. 5). As the name implies, this motor has its rotating field excited without the use of slip rings for connecting the external direct current to the motor field. The construction of this motor incorporates a shaft-connected ac generator. The field of the ac generator is physically stationary and connected to a source of dc voltage. The rotor of this ac generator is connected through a solid-state controlled rectifier mounted on the synchronous motor rotor and in turn connected to the synchronous motor field. This arrangement facilitates a connection between external excitation power and the rotating field of the synchronous motor through the air gap of the shaft-connected ac generator. The brushless synchronous motor has many advantages over the conventional slip-ring synchronous motor. Among these is the elimination of brushes and slip rings, which are high-maintenance items; the elimination of sparking devices, which are not permissible in certain atmospheres; and the use of static devices for field control, which are more reliable than conventional electromagnetic controls.

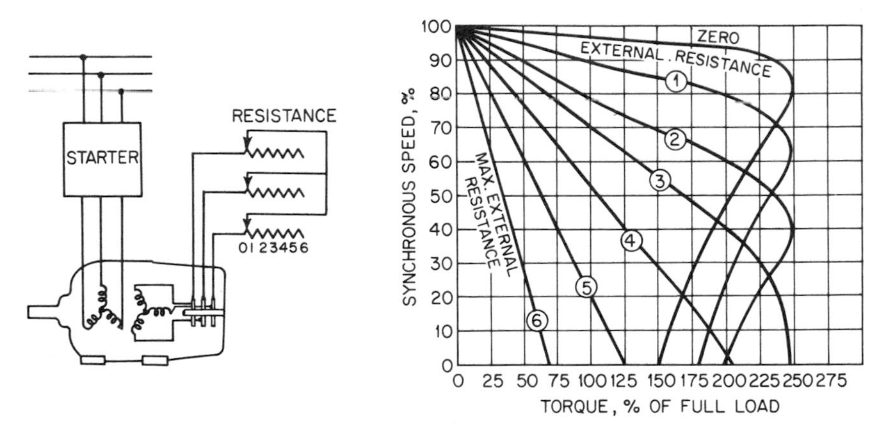

Fig. 3 Typical speed-torque characteristics of a wound-rotor induction motor. (Westinghouse Electric Corporation)

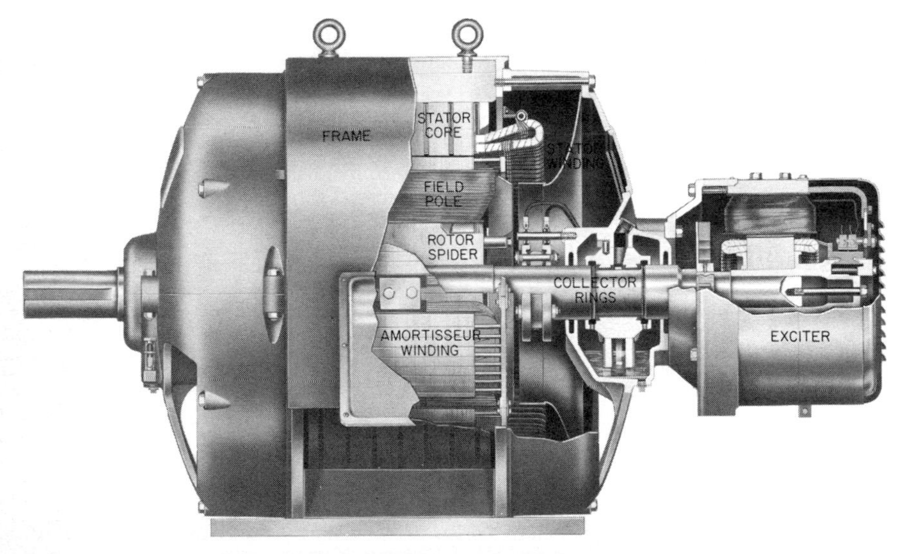

Fig. 4 Synchronous motor with direct-connected exciter. (Electric Machinery Mfg. Company)

Synchronous motors are used for pump applications requiring larger horsepower ratings at lower speed conditions, as illustrated in Fig. 6. Also, they are used on applications where it is desired to have a high power factor or a power-factor-correction capability. Of less importance is the characteristic of the synchronous motor that it will always operate at synchronous speed (does not have a slip) regardless of load. Synchronous motors are started on their damper windings (the same as squirrel-cage induction motors) and when they have accelerated to within 5 percent

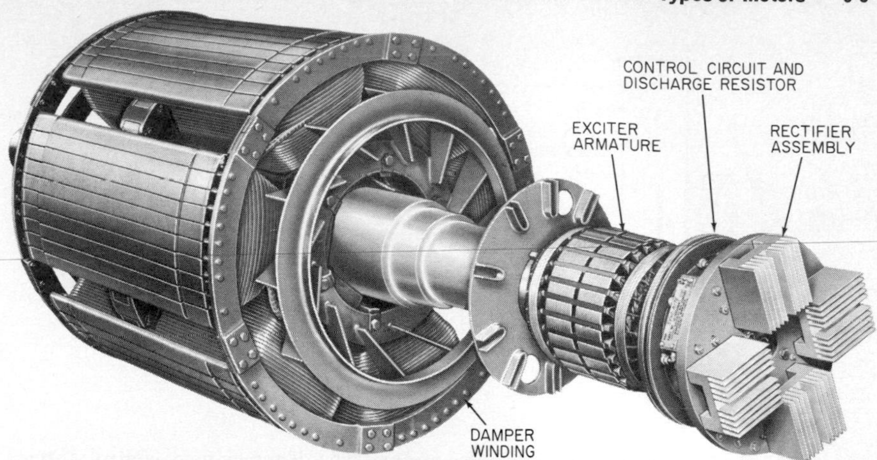

Fig. 5 Brushless synchronous motor with ac generator mounted on the shaft with rectifier and control devices. (Electric Machinery Mfg. Company)

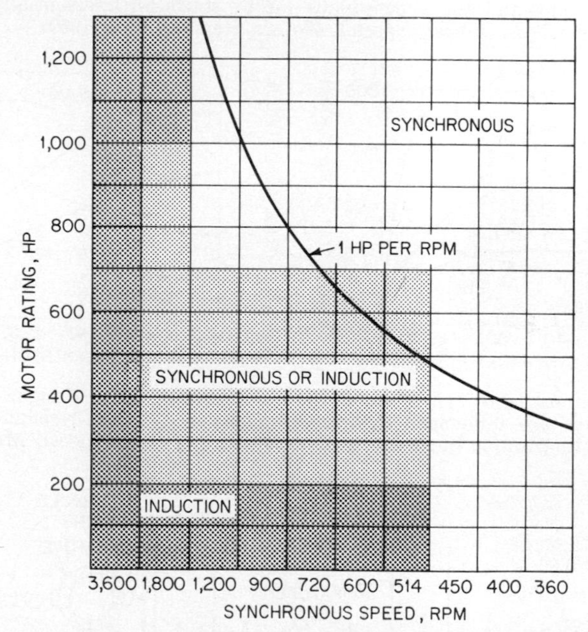

Fig. 6 Below 514 rpm, or horsepowers greater than approximately 1 per rpm, synchronous motors are a better selection than squirrel-cage induction motors because higher cost can generally be offset by higher power factor and efficiency. (*Power,* Special Motors Report, June 1969)

of synchronous speed, the field is applied and the motor will accelerate to synchronous speed (Fig. 7). Typical characteristic curves are shown in Fig. 8.

Direct-Current Motors Dc motors are only occasionally used to drive pumps. This is the case when direct current is the only power available, as in ship, railway, aircraft, and emergency-battery operation.

There are three types of dc motors available (Fig. 9): shunt, series, and compound-connected. Larger horsepower ratings of shunt-wound dc motors are frequently qualified as "stabilized shunt-wound" motors and incorporate a series field

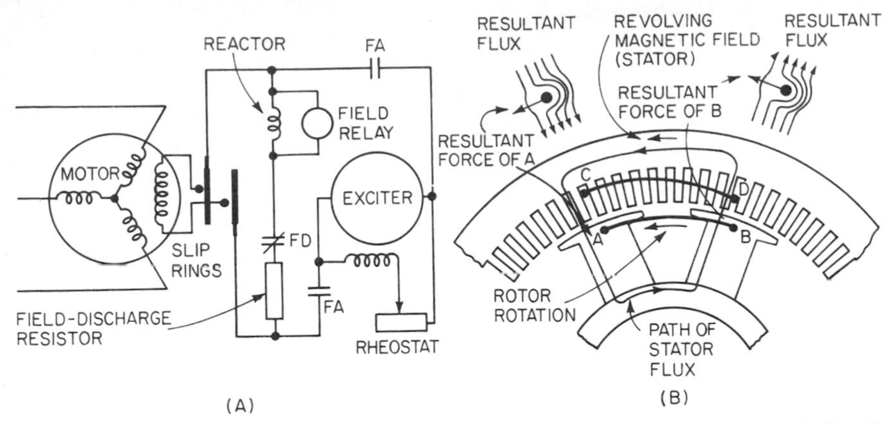

Fig. 7 Method of starting synchronous motor. (*A*) Typical field control (brush type) energizes the dc field to pull the rotor into synchronism as it comes up to speed. Field relay senses change in induced frequency as motor speeds up. (*B*) Damper winding is similar to a squirrel-cage rotor winding. It produces most of the starting torque, and the synchronous motor starts with essentially induction-motor characteristics. (*Power,* Special Motors Report, June 1969)

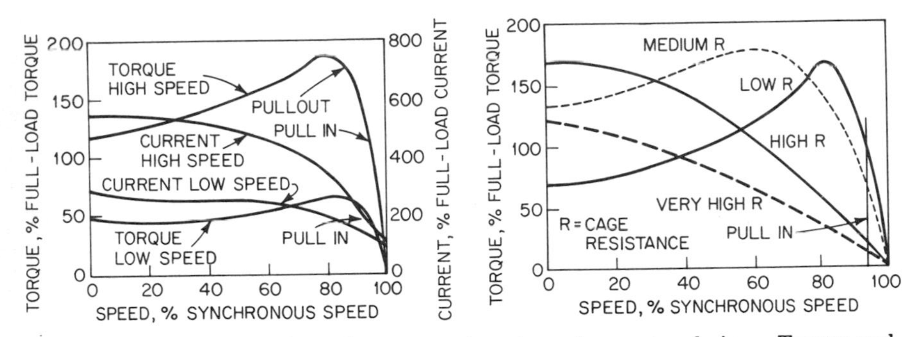

Fig. 8 Characteristics of synchronous motors depend on rotor design. Torque and current relations are influenced by synchronous speed. High-resistance cage produces high-starting torque but a low pull-in torque. (*Power,* Special Motors Report, June 1969)

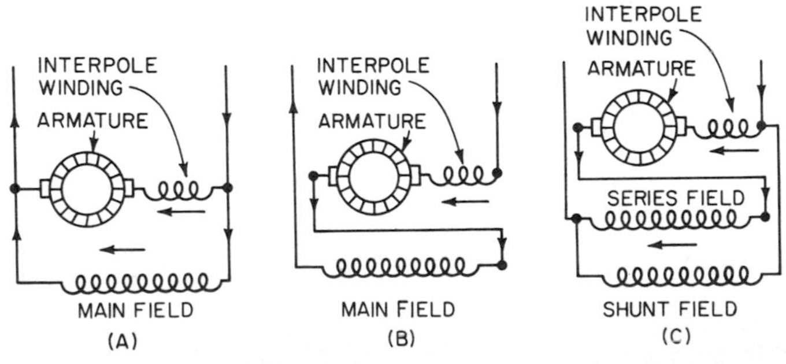

Fig. 9 Types of dc motors. (*A*) Shunt motor has field winding of many turns of fine wire connected in parallel with the armature circuit. The interpole winding aids commutation. (*B*) Series motor has field in series with the armature. Field has a few turns of heavy wire carrying full-motor current flowing in the armature. (*C*) Compound motor has both a shunt and a series field to combine characteristics of both shunt- and series-type motors in the same machine. (*Power,* Special Motors Report, June 1969)

similar to a compound wound motor. This is necessary to adjust the regulation of the shunt motor so as not to exhibit a rising speed-torque characteristic. It is important to be aware of the actual speed at which a dc motor will operate on pump applications because of pump performance guarantees.

Dc motors can be made to provide up to a 4:1 speed range with a field rheostat for special applications when operating from constant dc voltage. Many adjustable speed drives are available which utilize a dc motor and different means to produce a varying dc voltage to operate this motor. These include variable-voltage motor-generator sets, solid-state silicon-controlled rectifier packages, and variable-ratio transformers with a rectifier. Figure 10 illustrates typical characteristics of the dc motors discussed.

Motor Enclosures Electric motors are manufactured with a variety of mechanical-enclosure features to provide protection to the working parts of the motor for

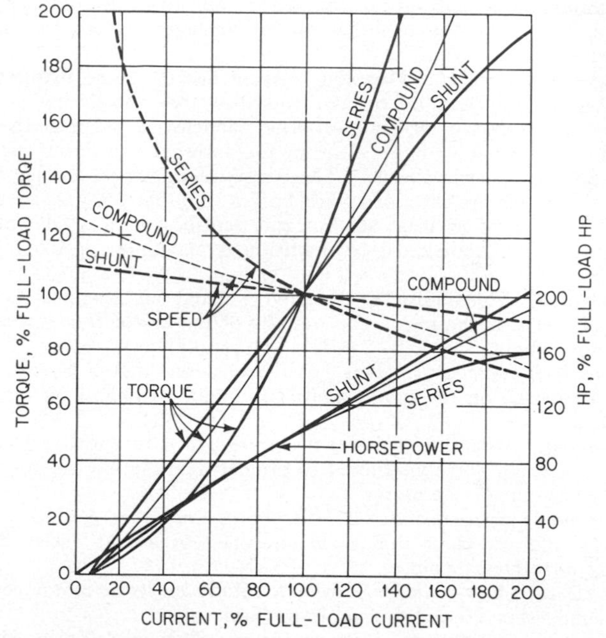

Fig. 10 Speed, torque, and horsepower characteristics of dc motors. (*Power*, Special Motors Report, June 1969)

specific environmental conditions. Although these special enclosures are classified, they are not, for reasons of economy, available for all size motors. The electric motor industry has incorporated a number of specific enclosure classifications on standard designs of small- and medium-size motors.

The specific enclosure classifications provided for electric motors are as follows:

Open. This enclosure permits passage of external cooling air over and around the windings and rotor of a motor and normally includes no special restrictions to ventilation other than that which is inherent in the mechanical parts of the motor.

Open dripproof. This is an open machine with ventilating openings so designed to permit satisfactory operation of the machine when drops of liquids or solids fall on the machine at any angle up to 15° from the vertical. This construction includes mechanical baffling to prevent materials from entering the machine within these limits.

Splashproof. This is an open machine with ventilating openings designed to permit satisfactory operation of the machine when drops of liquids or solids fall *directly* on the machine or come toward the machine in a straight line at any angle up to 100° from vertical.

Guarded enclosure. This provision limits the size of the ventilating opening to prevent accidental contact with the operating parts of the motor other than the shaft.

Semiguarded. This is an open construction where part of the ventilating openings are guarded and the remaining openings are left open.

Open, externally ventilated. This motor is ventilated by a separate motor-driven blower mounted on the motor enclosure (piggyback construction).

Open, pipe-ventilated. This motor is equipped to accommodate an air-inlet duct or pipe for accepting cooling air from a location remote from the motor. Air is circulated in the motor either by its own internal-blower parts or by an external blower, in which case the motor is said to be *forced-ventilated.*

Weather-protected type I. This is an open-construction motor with ventilating openings to minimize the entrance of rain, snow, and airborne particles to the motor electric parts, and where the openings are so arranged to prevent the passage of a ¾-in round rod.

Weather-protected type II. This construction has the same features as the type I machine, but in addition, the intake and discharge ventilating passages are so designed that high-velocity air and airborne particles blown into the machine by storms can be discharged without entering the internal ventilating passages of the motor leading to the electrical parts. The ventilating passages leading to the electrical parts of the motor are provided with baffles or other features to allow at least three abrupt changes of at least 90° for the ventilating air. The intake air path and openings are so proportioned to maintain a maximum of 600 ft/min velocity of the entering air.

Totally enclosed. This motor is designed without air openings to prevent free exchange of air between the inside and outside of the motor frame; the construction is not liquid- or airtight.

Totally enclosed, fan-cooled. This totally enclosed motor is equipped with an external fan operating on the motor shaft to circulate external air over the outside of the motor.

Explosion-proof. A totally enclosed motor designed to withstand an internal explosion of gas or vapor and constructed to prevent ignition, by the internal explosion, of gases or vapors outside the motor.

Totally enclosed, pipe-ventilated. This motor enclosure is similar to the open, pipe-ventilated motor except that it is equipped to accept outlet ducts or pipes in addition to inlet ducts or pipes.

Totally enclosed, water-cooled. This is a totally enclosed motor cooled by water passages or conductors internal to the motor frame.

Totally enclosed, water-to-air heat exchange. This is a totally enclosed motor equipped with a water-to-air heat exchanger in a closed, recirculating air loop through the motor. Air is circulated through the heat exchanger and motor by integral fans or fans separate from the rotor shaft and powered by a separate motor.

Totally enclosed air-to-air heat exchange. This motor is similar to the water-to-air heat exchanger motor, except external air is used to remove the heat from the heat exchanger instead of water.

Submersible. This totally enclosed motor is equipped with sealing features to permit operation while submerged in a specific material at a specified depth.

Bearings and Lubrication Motors are generally available with oil-lubricated sleeve bearings or antifriction bearings.

Sleeve bearings are used on smaller-sized motors of fractional horsepower ratings wherein the bearings are lubricated with an oil-impregnated wick. The wick-lubricated sleeve bearing is generally limited to motors rated approximately 1 hp. Sleeve-bearing motors larger than 1 hp are ring-lubricated. Lubricating oil is drawn up from the bearing sump to the bearing by a ring that rolls over the top of the motor shaft as the shaft rotates. Larger motors having bearing heat losses that cannot be dissipated directly may require the use of a force-lubrication system wherein oil is pumped into the bearing and allowed to recirculate through a heat-exchanger pump.

The oil delivered to each bearing is metered to provide only the required amount of lubricant. A lubrication system composed of heat exchanger, sump, and pump is normally common to a number of bearings, rather than having a single lubricating pump for each bearing. Other types of bearings must be used in place of or in addition to the sleeve-type bearing when thrust loads are present. Smaller sleeve bearings are in the form of a solid cylindrical shell and are made of bronze or babbitt metal. Larger sleeve bearings are usually split on a horizontal centerline, allowing easy assembly or disassembly. The bearing housing, which contains the bearing, is also split in the horizontal centerline and held together with bolts between the top and bottom halves.

Ball-type antifriction bearings are used on motors where thrust forces are present and where motors will not operate in a horizontal position. Ball bearings are usually grease-lubricated and can be operated for prolonged periods of time between lubrication procedures. Ball-type antifriction bearings are standard on NEMA size motors. More highly loaded ball bearing applications may require oil lubrication with a pump system similar to sleeve bearings.

Roller-type antifriction bearings are used on applications of very high radial loads. These bearings are not usually used on motors and have the inability to accept thrust loads.

Kingsbury thrust bearings are used where very high thrust loads must be accommodated by the motor. The Kingsbury bearing consists of a series of rotating segments that ride on a stationary bearing ring. This type bearing is oil-lubricated either with a sump or pump system depending on its size and rating.

Motor Insulation *Classes.* The insulation used within a motor permits the windings to be electrically insulated from the mechanical parts of the motor as well as between the turns of the coil winding. Also included in the insulation of a motor are those materials used to secure the windings and to make them rigid and impervious to ambient conditions. Consequently, motor insulation is a complex system utilizing many different materials and parts to effectively insulate the windings. The parts considered in an insulation system include slot cells, phase barriers, conductor insulation, slot wedges, end-turn supports, tie material, and winding impregnation. Motors are insulated with specific classes of insulating material.

There are four basic classes of insulating materials presently recognized by the motor industry. Each of these insulation classes differs according to its respective physical properties and can withstand a certain maximum operating temperature (frequently termed "total temperature" or "hot-spot temperature") and provide a practical and useful insulation life. The insulation classes and their maximum operating temperatures are as follows:

Class A 90°C
Class B 130°C
Class F 155°C
Class H 180°C

Those factors which contribute to the maximum operating temperature of a motor insulation system are the ambient temperature, the temperature rise within the motor winding itself caused by motor losses, and any overload allowance designed into the motor (service factor).

The present standard for motors within the range of NEMA ratings (frames 140T to 449T) requires nameplate marking for the maximum allowance ambient temperature, the horsepower rating, and the associated line current to develop this horsepower, the class of insulation used, and the service factor provided. Motors larger and smaller than NEMA frame sizes have nameplate marking including maximum allowable ambient temperature, temperature rise in degrees Celsius either by thermometer or resistance measurement, horsepower rating and the line current to develop this rating, and any service factor provided.

Environmental Factors. Motors are basically applied to a pump application with the environmental conditions of the application dictating the type of motor enclosures to be used. A brief set of rules for selecting motor enclosures follows:

1. Dripproof. For installation in nonhazardous, reasonably clean surroundings free of any abrasive or conducting dust and chemical fumes. Moderate amounts of moisture or dust and falling particles or liquids can be tolerated.

2. Lifeguard (encapsulated motor frames 364T to 449T only). For installation in nonhazardous high-humidity or chemical applications, free of clogging materials, metal dust, or chips and/or where hosing down or severe splashing is encountered.

3. Totally enclosed, nonventilated or fan-cooled. For installation in nonhazardous atmospheres containing abrasive or conducting dusts, high concentrations of chemical or oil vapors, and/or where hosing down or severe splashing is encountered.

4. Totally enclosed, explosion-proof. For installation in hazardous atmospheres containing:

a. Class I, Group D. Acetone, acrylonitrile, alcohol, ammonia, benzine, benzol, butane, dichloride, ethylene, gasoline, hexane, lacquer-solvent vapors, naphtha, natural gas, propane, propylene, styrene, vinyl acetate, vinyl chloride, or xylenes

b. Class II, Group G. Flour, starch or grain dust

c. Class II, Group F. Carbon black, coal or coke dust

d. Class II, Group E. Metal dust including magnesium and aluminum or their commercial alloys

Note. Under Class 1 only, there are two divisions which allow some latitude on motor selection:

Generally, Class 1, Division 1 locations are those in which the atmosphere is or may be hazardous under normal operating conditions. Included are the locations which can become hazardous during normal maintenance. An explosion-proof motor is mandatory, for Division 1 locations.

Class 1, Division 2 refers to locations where the atmosphere may become hazardous only under abnormal or unusual conditions (breaking of a pipe, for example). In general, motors in standard enclosures can be installed in Division 2 locations if the motor has no normally sparking parts. Thus open or standard totally enclosed squirrel-cage motors are acceptable, but motors with open slip rings or commutators (wound rotor, synchronous or dc) are not allowed unless the commutators or slip rings are in an explosion-proof enclosure.

In addition to the enclosure and the ambient temperature, there are several additional environmental conditions that must be considered when applying an electrical motor. These conditions include: chemical fumes; moisture; mechanical vibration, abrasive dust, etc.; tropical fungus conditions.

In those applications where chemical fumes or moisture are abnormal and can cause decomposition of an insulation system, standard insulation will be inadequate. These applications require a motor with a premium insulation system which will incorporate more resistant-type components and may include special impregnation techniques. Chemical fumes and moisture can also be destructive to the mechanical parts of a motor and special treatment should be provided to these parts for protection. NEMA frame size motors have a special motor for "chemical" industry applications which will have standard features to resist these environmental factors.

Applications with excessive vibration can destroy a winding and damage the mechanical parts of a motor. In such cases, it is advisable to provide a winding with extra treatment to ensure that the winding is rigid and will not vibrate and chafe the insulating materials and provide a mechanical construction that will have the strength to withstand the above.

If abrasive dust is present, the motor insulation should be protected with a resilient surface coating to withstand the impact of the abrasive particles.

Since all insulation systems employ components that can in some degree support fungus growth in tropical locations, motors applied in such areas should incorporate fungus-proofing treatment on the insulation.

Obviously applications exhibiting a combination of any or all of the environmental conditions discussed should have special protection for each condition.

Coupling Methods for Pump Applications *Direct Coupling.* Pumps are usually directly coupled to motors, and where the pump is not close-coupled, the greatest majority are coupled by means of a flexible coupling.

The use of flexible coupling permits a minor misalignment (angular and parallel) between motor and pump shafts. However, most flexible couplings tend to become rigid when they are transmitting torque and therefore can impart thrust into the motor bearing, should this thrust prevail at the time the motor and pump are started. Ball bearing motors have reasonable thrust capacity and are capable, therefore, of accepting these thrust components.

Sleeve bearings on the other hand have a negligible thrust capacity unless a thrust bearing is specifically provided in the motor (ball thrust, Kingsbury bearing, etc.). Sleeve-bearing motors must be protected from thrust components. Since a pump normally has a negligible end-float motion, it is customary to position the motor rotor halfway in its end-float capability and couple the motor and the pump with a limited end-float flexible coupling that will permit the sum of the coupling end float and pump end float to approximate 50 percent of the motor end float. This will ensure that all factors involved will never permit the motor shaft to reach either extreme shaft position and cause thrust on the motor bearings.

Larger pumps and motors can require torque values that make it necessary to use solid couplings. Extreme care must be exercised in these applications to ensure a good alignment, a stable alignment, and an alignment that establishes the motor in its mid-float position. If the motor has antifriction bearings (ball bearings), it will be necessary to have the outer races of the ball bearings free floating in their housings to prevent preloading the ball bearings.

Close-Coupled. Close-coupled pumps have become very popular. In this arrangement no coupling is provided between the pump and motor shafts and the pump housing is flange-mounted between close-tolerance fits on both the motor and the pump flanges. The pump impeller is mounted directly on the motor shaft. Care must be taken in this arrangement to ensure that the motor shaft run-out or axial movement plus machine tolerances do not cause interference between the pump housing and its rotor. The motor shaft material must be compatible with the fluid being pumped and if the pump impeller is held in place by a nut, the thread must respect the rotation of the motor. High-pressure close-coupled pumps of a nonbalanced design can cause excessive shaft thrust which may be incorporated in the motor bearing capacity. It is always good practice with close-coupled pumps to provide some form of flinger on the motor shaft to prevent liquids that leak past the pump seal from entering the motor bearing.

Flanged Motors. Flanged motors are provided to allow an easy means of aligning pump housings with motors. This construction is usually in the form of a vertical mounting in which the motor is set on top of the pump and the pump supports the motor weight. The pump and motor shaft are normally coupled, and those comments made under the subject of direct coupling methods are applicable. Also, as in the case of coupled pumps, this construction permits thrust forces that must be considered when selecting a motor if the pump does not have a thrust bearing.

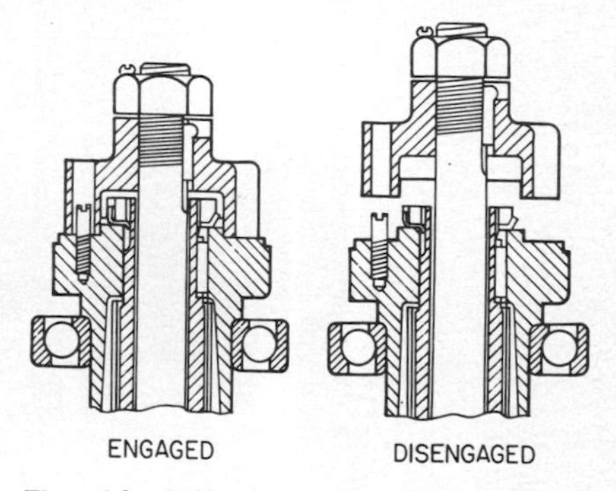

Fig. 11 Self-release coupling connecting pump-head shaft to hollow shaft of vertical motor disengages as a result of pump-shaft couplings unscrewing. (U.S. Electrical Motors, Inc.)

A further extension of flange motors includes the vertical hollow-shaft motor, and with this design a variable length of shafting connects the pump and motor. The pump shaft passes through the center of the motor bore and the motor torque is imparted to the pump shaft by a suitable coupling at the top of the motor. The weight of the shaft and the pump impeller and the force of the hydraulic thrust are assumed by the motor bearings.

The coupling on the motor can be made a *self-release coupling* (Fig. 11) to prevent the motor from delivering torque to the shaft in the event the motor is started in the wrong direction and to prevent reversed motor rotation from unscrewing the threaded joints between lengths of pump shafting.

Another modification to a coupling is a *nonreverse ratchet* (Fig. 12), which prevents the remaining head of liquid in a pump from rotating the pump in the reverse direction from normal when the pump is stopped. This prevents possible overspeeding of the pump and motor if the pump is connected to a large reservoir and most of the total pump head is static. This will also prevent a pump with a long discharge column pipe from running in reverse with no liquid in the upper portion of the pipe to lubricate the line shaft bearings. Starting a pump capable of back spinning is also prevented.

Performances Motors arc designed to produce their rated horsepower, torque, and speed at a specific line voltage, line frequency, and ambient temperature. The motor will also operate at a specific efficiency and power factor when all of these conditions are met. The normal operating conditions of a motor are stipulated

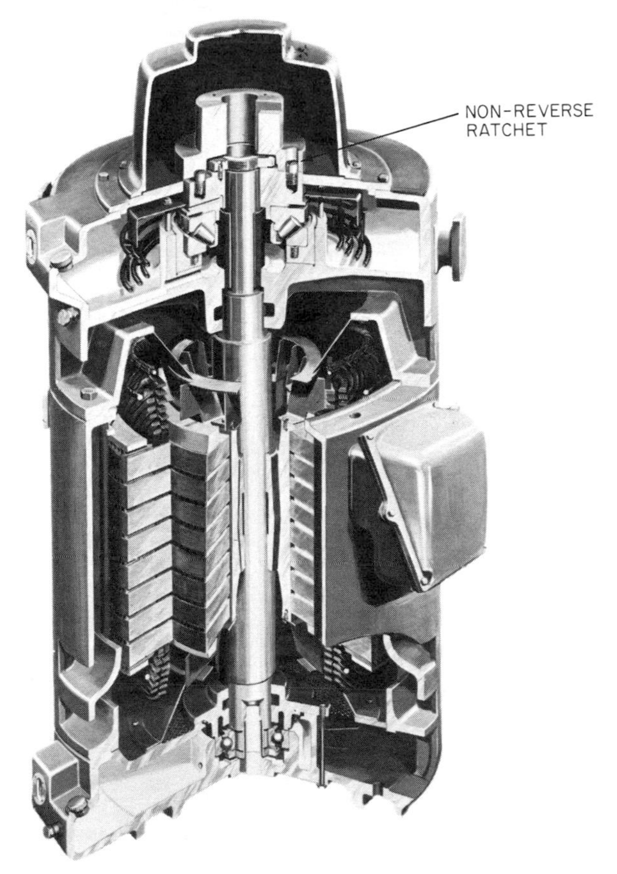

Fig. 12 Section of vertical hollow-shaft motor showing nonreverse ratchet. Spring-loaded pins ride on ratchet plate in one direction only. (General Electric Company)

on its nameplate with values for horsepower, speed, ambient temperature, and frequency. If the operating conditions are different from the motor-nameplate ratings, the motor performance will be altered.

Voltage. Ac motors are designed to operate satisfactorily at their nameplate voltage rating with a ±10 percent variation from nameplate voltage when operated at rated nameplate frequency. This dictates the motor will develop rated horsepower and speed to a pump and will operate at a safe insulation temperature over the range of voltage. The motor torque will vary directly in a ratio with the square of the applied voltage divided by the nameplate voltage. This affects the peak torque of a motor and will cause the motor speed-torque curve as shown in Fig. 2 to be altered. Within the ±10 percent voltage band, a motor can be expected to accelerate and operate a pump safely and continuously. A motor should never be expected to operate continuously beyond ±10 percent band. If variation in voltage beyond ±10 percent occurs as a transient condition, the pump and motor may not operate satisfactorily.

For example, assume the pump is operated by a NEMA design B motor (refer to Fig. 2) where this motor will produce 200 percent pull-out torque at rated voltage. If the line voltage were to fall to 70 percent of rated nameplate voltage, the motor would produce only 49 percent of its peak torque value. The pull-out torque of the motor will then become 0.49 × 200 percent rated torque, or 98 percent rated torque. It then becomes questionable whether the motor will be able to sustain the pump load and the motor can be expected to lose speed, stall, or become overloaded.

In a similar sense, a motor may be unable to accelerate a pump if low line voltage exists. In the example previously discussed, this same motor develops 150 percent of rated torque when started at zero speed and rated voltage. If the line voltage is again 70 percent of nameplate voltage, the motor will develop 0.49 × 150 percent rated torque, or 73 percent rated torque. This may be a problem with certain types of pumps, such as a constant displacement pump. It is conceivable that this would not be a problem in starting a centrifugal pump because of its square-law speed-torque characteristics. If the motor voltage never increased beyond 70 percent, the centrifugal pump would not reach normal operating speed. An exception to this rule is the commutating ac motor, for which a ±6 percent voltage variation is allowable.

Varying motor voltage from nameplate rated voltage will also affect the motor operating speed, power factor, and efficiency established for rated voltage and load. An induction motor will operate at several rpm faster than nameplate speed at a 10 percent over voltage and several rpm below nameplate speed at 10 percent under voltage. Synchronous motors on the other hand will not change speed over a ±10 percent voltage range.

Dc motors can also be operated over a ±10 percent voltage range from nameplate rated voltage. However, it should be recognized that different types of dc motors will have different speed and torque characteristics over the voltage range. This should be taken into account when meeting pump performance requirements.

Frequency. Ac motors will operate satisfactorily at rated load and voltage with a frequency variation up to ±5 percent from rated nameplate frequency. However, it will be expected that the speed of the motor will vary with the applied frequency. Synchronous motor speed will vary directly proportionate to applied frequency and induction motors will vary closely to the same proportion.

When an ac motor must operate with a varying voltage and frequency, the *combined* variation of voltage and frequency must not be more than ±10 percent from rated nameplate voltage and frequency, provided the frequency variation does not exceed ±5 percent from rated nameplate frequency.

Frequency variation from the motor nameplate frequency will cause motors to operate at a power factor and efficiency other than those established for rated frequency.

Speed and Speed Range. Synchronous and induction motors are intended basically to operate at one specific speed. Some special applications have been designed using both synchronous and induction motors with adjustable-frequency power sources to develop an adjustable-speed characteristic, but these are unique and special.

Adjustable-speed applications are usually satisfied by using one of the adjustable-speed drive packages that are available and usually incorporate a dc motor. Larger-size motors with an adjustable-speed requirement normally take the form of a wound rotor motor with a slip controller. An additional form of adjustable-speed drive incorporates an eddy-current coupling and some form of constant-speed motor, usually a squirrel-cage induction motor. The maximum range of practical speed adjustment for the drives mentioned approaches 10:1.

Pumps requiring discrete speed adjustment can be operated with motors that are designed for multiple-speed operation, and these motors may be selected from a variety of squirrel-cage induction motors. Speed ranges that are even multiples can be obtained with a squirrel-cage induction motor equipped with a reconnectable single winding. Applications requiring a speed range of an uneven ratio normally require a two-winding squirrel-cage induction motor. However, larger horsepower motors requiring speeds of an uneven ratio can be designed with a single-winding incorporated pole-amplitude modulation. This design depends upon a concept which establishes a modulation of the frequency of the flux in the motor.

Acceleration. A motor must be capable of accelerating as well as driving a pump at rated speed and horsepower. The acceleration can be analyzed by examining a typical pump-motor combination involving a centrifugal pump driven by a six-pole squirrel-cage induction motor rated 10 hp with NEMA design B torque characteristics. The curve in Fig. 13 demonstrates this combination when the pump is loaded and the motor is operating at nameplate frequency and voltage. It will be noted that the torque produced by the motor at any speed is greater than the torque

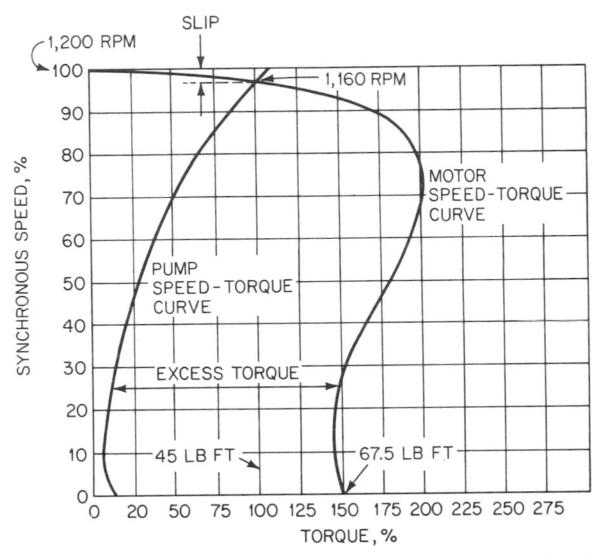

Fig. 13 Typical speed-torque characteristic curves for a centrifugal pump and a squirrel-cage induction, NEMA design B motor. Excess torque accelerates pump.

required by the pump at that speed. The excess torque at any speed is available to accelerate the mass (WK^2) of the entire motor and pump-rotating parts. The time to accelerate the pump is equal to:

$$t = \frac{WK^2 \times \Delta\text{rpm}}{308T}$$

where Δrpm = change in speed
t = time, s
WK^2 = total moment of inertia, lb-ft^2
T = torque, lb-ft

Because the difference in torque is not uniform over the speed range the curve can be analyzed by assuming discrete changes in speed and an average torque over this change in speed. A time period can be calculated for each discrete speed change and all time values can be totaled to obtain the complete acceleration time (see also Secs. 2.3 and 9.1).

Increasing the WK^2 of the pump, the operating speed of the pump and motor, or the torque required by the pump at any speed will result in a longer acceleration time, which may not be possible for the motor. Each motor has a capability to operate at a reduced speed and a torque in excess of rated torque for a given time. Beyond this time, damage can be done to the windings and/or rotor.

When an application is analyzed and found to present an acceleration-time problem, consideration should be given to:

1. Unloading the pump during acceleration
2. Reducing the WK^2 of the rotating parts
3. Applying a motor which has the capability for accelerating the WK^2

When larger pumps are started and a reduction in line voltage is realized because of the high starting current, the motor torque is reduced by the square of the voltage ratio. Naturally, the acceleration time is greater because of the reduced torque produced by the motor. This will not hamper the ability of the motor to accelerate the pump as long as the motor develops more torque than is required to drive the pump at any speed over the accelerating range.

A similar analysis can be made for synchronous motors once the accelerating-speed torque curve for the synchronous motor is known. It should be recognized that synchronous motors operate as a squirrel-cage induction motor up to the moment of synchronization. At that time the synchronous motor must have an additional capability of synchronizing torque, frequently called *pull-in torque,* to accelerate the motor from subsynchronous to synchronous speed. If the motor cannot develop this synchronous torque, it will "pull out" and usually be shut down by a control function. In the case of pump applications, an economic advantage can be realized by unloading the pump during acceleration to reduce the acceleration and synchronizing torque required by the motor.

Attention should always be given to the breakaway torque that is required to start the pump from zero speed. This is particularly important with constant displacement pumps where the pump will operate at a constant torque over the entire accelerating speed range.

Service Factor. Motors are available with service factor ratings which range from 1.0 and may go as high as 1.5. A service factor implies that a motor has a built-in thermal capacity to operate at the nameplate horsepower times the service factor stamped on the nameplate. It should be noted, however, that when the motor is operated at the service factor horsepower, the motor will operate at what is termed a *safe temperature.* This means that the motor will operate at a total temperature that is greater than the temperature for a motor designed for the same horsepower with 1.0 service factor. Consequently it is not advisable to apply a motor with a service factor larger than 1.0 where the continuous horsepower requirements will be greater than the normal horsepower. A service factor rating on a motor is to provide an increased horsepower capacity beyond nominal nameplate capacity for occasional overload conditions. Also, the speed-torque characteristics are related to the nominal horsepower rating and not the service factor horsepower.

Efficiency. Motors are designed to operate with an efficiency expressed in percent at rated voltage, frequency, and horsepower. Efficiency is delivered as:

$$\text{Efficiency } \% = \frac{\text{shaft-output power} \times 100}{\text{electrical-input power}}$$

The efficiency of a given motor design will vary slightly between duplicate motors because of manufacturing tolerance and variations in materials. For this reason guaranteed efficiencies are usually quoted at lower values than actual efficiencies. When motors are operated at reduced horsepowers, the tendency is for the efficiency to decrease.

Several other factors have an effect on motor efficiency. Increasing the applied voltage and operating a motor at its rated horsepower will increase efficiency very slightly, while decreasing applied voltage will decrease the motor efficiency noticeably. Also, increasing the frequency will cause a very slight increase in a motor efficiency while decreasing the applied frequency will cause a slight decrease in efficiency.

Power Factor. The power factor of a motor is expressed as:

$$\text{PF} = 100 \cos \Theta$$

where Θ is the angle between voltage and current at motor terminals (leading or lagging).

The power factor at which a motor operates is dependent on the design of the motor as a basic consideration and is established at rated voltage, frequency, and horsepower output.

In the case of an induction motor the power factor can never be 100 percent leading. A number of factors will influence the power factor of an induction motor:

Condition	*Effect on power factor*
Increase applied voltage	Decrease
Decrease applied voltage	Increase
Increase load	Increase
Decrease load	Decrease
Increase applied frequency	Slight increase
Decrease applied frequency	Slight increase

In the case of a synchronous motor, it is usual to use two varieties of motors, the 100 (unity) and 80 percent leading motors. The power factor of a synchronous motor operated at rated voltage and frequency is fixed by its field excitation and its horsepower output. At a given horsepower output the power factor can be adjusted over a range by adjusting the field excitation. Increasing the field excitation will cause the motor to operate at a more leading power factor and, conversely, reducing the field excitation will make the power factor lag.

Varying the horsepower output of a synchronous motor with a constant field excitation will vary the operating power factor. A decrease in horsepower output will cause a more leading power factor; conversely, increasing the horsepower output will induce operation at a lagging power factor. Consequently, to operate the motor at rated power factor with a varying output horsepower, it is necessary to adjust the field excitation. However, this is not normally done because a synchronous motor is frequently used for the purpose of improving the power factor and the more leading power factor is used to accommodate power factor improvement. When a synchronous motor is overloaded and operates at a more lagging power factor, it is not usual to increase the excitation beyond its rated excitation because of the extra heating this will develop in the motor. In this case the more lagging power factor is simply accepted.

TYPES OF CONTROLS

Alternating-Current Motor Starters *Manual Starters.* Manual motor starters are designed to provide positive overload protection and start and stop control of single-phase and polyphase motors. A single manual operating handle provides the control and indication of "on," "off," and "tripped" states.

Magnetic Starters. Magnetic motor starters are designed to control a motor by incorporating a magnetically operated contactor to apply power to the motor terminals. An overload relay is incorporated to protect the motor from overloading. Magnetic starters are available for reversing and nonreversing service and are also made as noncombination or combination types. The noncombination starter combines only the motor contactor and overload relay. The combination starter combines these parts along with either a circuit breaker or fused switch to provide short-circuit protection.

Reduced-Voltage Starters. Reduced-voltage starters are available in several types and are basically magnetic starters with additional features to provide reduced voltage, which in turn provides for reduced motor-starting current and/or torque. These starters include the following types:

Primary-resistor starters, sometimes known as *cushion-type* starters, will reduce the motor torque and starting inrush current to produce a smooth, cushioned acceleration with closed transition. Although not as efficient as other methods of reduced-voltage starting, primary resistor–type starters are ideally suited to applications where reduction of starting torque is of prime consideration. A typical diagram for this type of starter is shown in Fig. 14.

Autotransformer starters are the most widely used reduced-voltage starters because of their efficiency and flexibility. All power taken from the line, except transformer

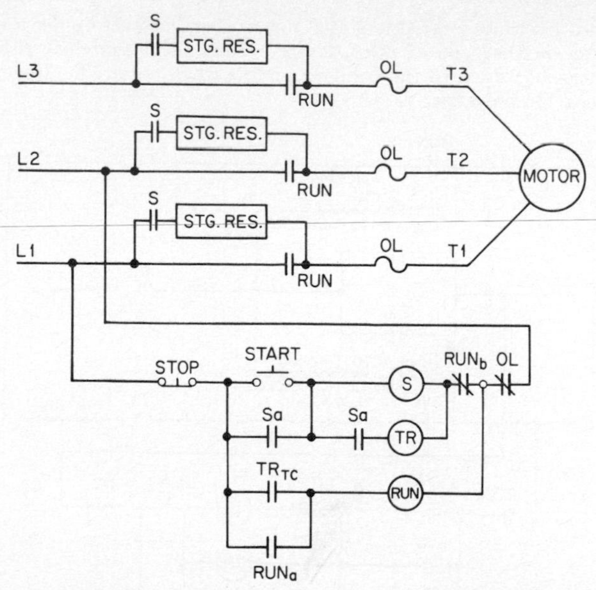

Fig. 14 Typical primary-resistor reduced-voltage starter wiring diagram. (Westinghouse Electric Corporation)

losses, is transmitted to the motor to accelerate the load. Taps on the transformer allow adjustment of the starting torque and inrush to meet the requirements of most applications. The following characteristics are produced by the three voltage taps:

Tap, %	*Starting torque, % locked torque*	*Line inrush, % locked current*
50	25	28
65	42	45
80	64	67

A typical diagram for this type starter is shown in Fig. 15.

Part-winding starting provides convenient, economical, one-step acceleration at reduced current where the power company specifies a maximum or limits the increments of current drawn from the line. These starters can be used with standard dual-voltage motors on the lower voltage and with special part-winding motors designed for any voltage. When used with standard dual-voltage motors, it should be established that the torque produced by the first half-winding will accelerate the load sufficiently so as not to produce a second undesirable inrush when the second half-winding is connected to the line. Most motors will produce a starting torque equal to between one-half to two-thirds of NEMA standard values with half the winding energized and draw about two-thirds of normal line-current inrush. A typical diagram is shown in Fig. 16.

Star-Delta Starters have been applied extensively to installations starting motors driving high inertia loads with resulting long acceleration times. They are not, however, limited to this application. When six or twelve lead delta-connected motors are started star-connected, approximately 58 percent of full-line voltage is applied to each winding and the motor develops 33 percent of full-voltage starting torque and draws 33 percent of normal locked rotor current from the line. When the motor has accelerated, it is reconnected for normal delta operation. A typical diagram is shown in Fig. 17.

Wound-Rotor Motor Starters. These magnetic motor starters are used for starting, accelerating, and controlling the speed of wound-rotor motors. The primary control includes overload protection and low-voltage protection or low-voltage release, depending on the type of pilot device. Disconnect switches, circuit breakers, and reversing can be added to the primary circuit when required. Reversing starters are not designed for plugging.

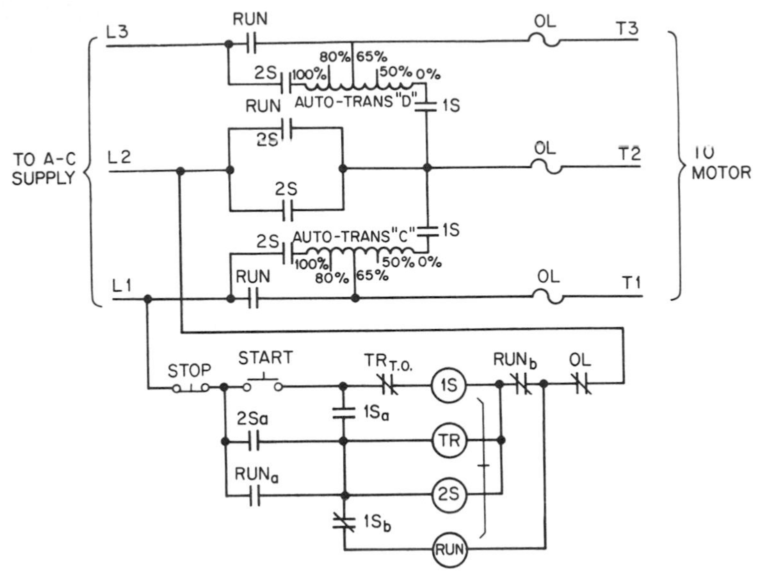

Fig. 15 Typical autotransformer reduced-voltage starter wiring diagram. (Westinghouse Electric Corporation)

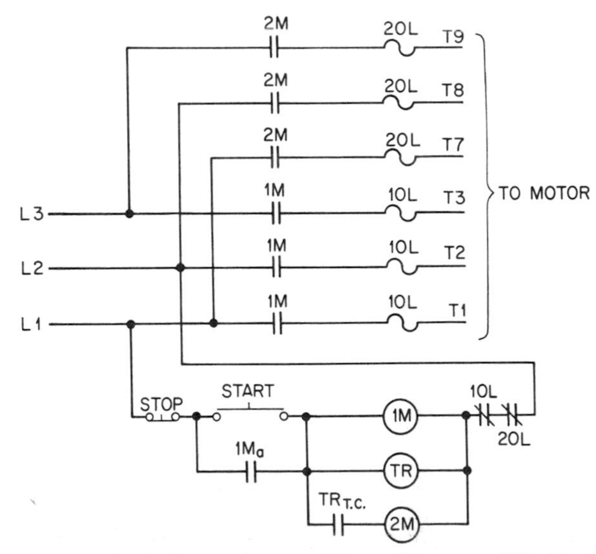

Fig. 16 Typical part-winding starter wiring diagram. (Westinghouse Electric Corporation)

The secondary circuit contains the NEMA recommended number of accelerating or running contactors and resistors to allow approximately 150 percent of motor full-load torque on first point of acceleration. Additional accelerating points can be added for high-inertia loads or exceptionally smooth starts. Adjustable timing

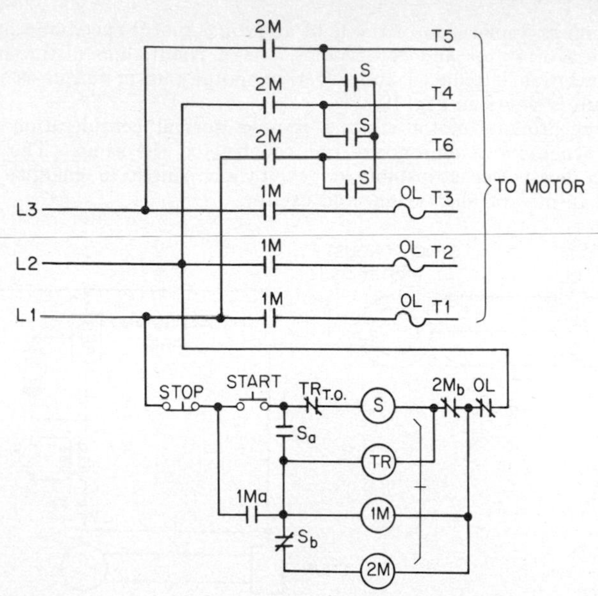

Fig. 17 Typical star-delta reduced-voltage starter-wiring diagram. (Westinghouse Electric Corporation)

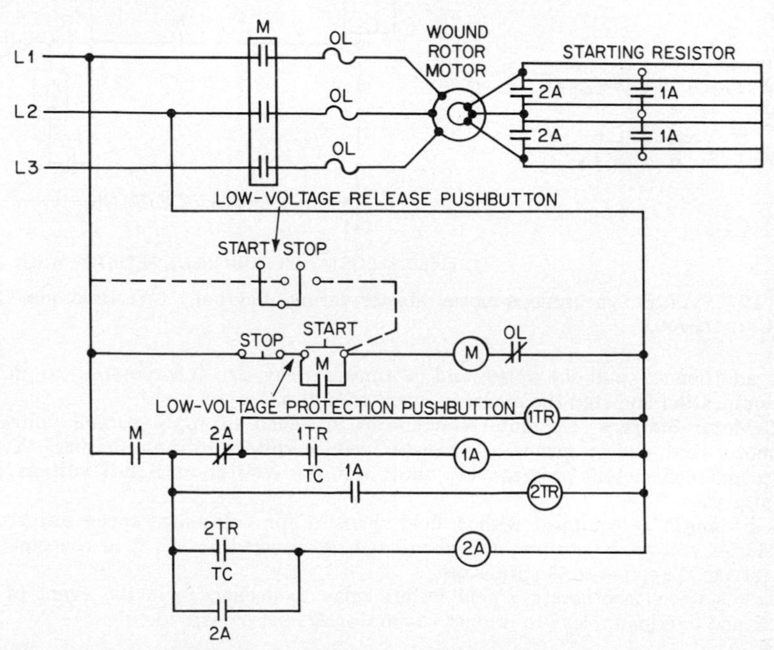

Fig. 18 Typical wound-rotor motor starter wiring diagram. (Westinghouse Electric Corporation)

relays permit field adjustment. Standard starting duty NEMA 135 resistors allow 10 s starting out of every 80 s. A typical diagram is shown in Fig. 18.

Synchronous-Motor Starter. Synchronous, magnetic, full-voltage starters provide reliable automatic starting of synchronous motors. They can be used whenever full-voltage starting is permissible. Automatic synchronization is provided by field relay which assures application of the field at proper motor speed and at a favorable angular position of stator and rotor poles. As a result, line disturbance resulting from synchronization is reduced and effective motor pull-in torque is increased. A typical diagram is shown in Fig. 19.

Brushless synchronous motor starters require special consideration inasmuch as all brushless synchronous motors are not constructed the same. The usual starter incorporates a low-power adjustable dc excitation source to energize and control the output of an integral shaft-connected exciter.

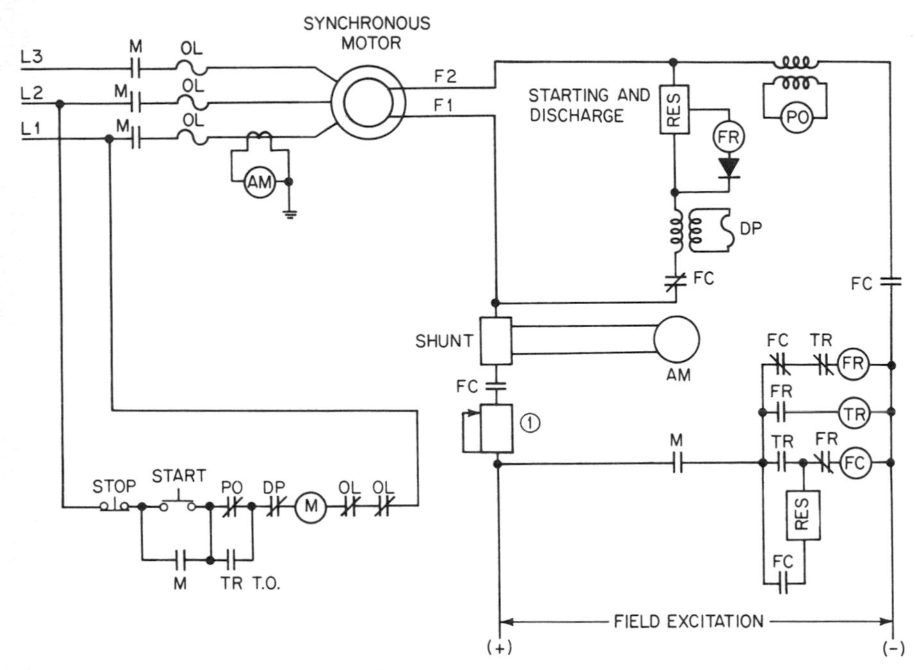

Fig. 19 Typical synchronous-motor starter-wiring diagram. (Westinghouse Electric Corporation)

In addition, a pull-out relay and a timing relay are incorporated to initiate synchronization and stop the motor in event of pull-out.

DC Motor Starters Dc motor starters are designed to apply normal voltage to the motor field, and by means of a resistor, reduce voltage to the armature. Timing relays and contractors progressively short out the resistor until full voltage is on the armature.

If the motor is equipped with a field rheostat for field-range speed adjustment, the starter will apply the preset field voltage as adjusted by the field rheostat after full voltage is applied to the armature.

These starters incorporate a field failure relay to deenergize in the event of field failure and overload relays to protect the motor against overspeed.

Section **6.1.2**

Steam Turbines

WALLACE L. BERGERON

DEFINITIONS

"A steam turbine may be defined as a form of heat engine in which the energy of the steam is transformed into kinetic energy by means of expansion through nozzles, and the kinetic energy of the resulting jet is in turn converted into force doing work on rings of blading mounted on a rotating part."[1]

This definition may be restated:

"A steam turbine is a prime mover which converts the thermal energy of steam directly into mechanical energy of rotation."[2]

REASONS FOR USING STEAM TURBINES

Steam tubines are used to drive the different types of pumps for a variety of reasons:

1. The economical generation of steam often requires boiler steam pressures and temperatures that are considerably in excess of those at which the steam is utilized for various purposes throughout the operating plant. Steam may also be used at two or more pressure levels within the same plant. The pressure reduction can be accomplished through valves, pressure-reducing stations, or use of a steam turbine.

Pressure reduction by using a steam turbine and thereby developing power to drive a pump permits lower utility costs. The incremental increase in steam flow and consequently in fuel costs for the same lower pressure steam heating load is in most instances less than the cost of purchased power for a motor-driven pump.

2. A pump driven by a steam turbine may be operated over a wide speed range utilizing the turbine governor system or a separately controlled valve in the turbine or in the steam line to the turbine. Operation at variable speeds is an inherent characteristic of steam turbines and does not require the use of special speed-changing devices as is the case with other prime movers.

The overall efficiency of the turbine and pump unit can be optimized by operating at reduced speeds and the resultant reduced hp ratings. The pump performance can be controlled by reducing the speed of the pump rather than throttling the pump. While the turbine efficiency normally declines when operating at a reduced speed, the steam flow will still be less than when throttling the pump.

[1] Edwin F. Church, "Steam Turbines," McGraw-Hill Book Company, 1950.

[2] Single-Stage Steam Turbines for Mechanical Drive Service, *Nat. Elec. Manufacturers' Ass. Publ.* SM22-1970, 1970.

Operation at reduced hp but at constant speed is also permitted by the speed governor which throttles the steam to the nozzles as the hp is reduced. Efficiency may be improved by equipping the turbine with auxiliary steam valves that are closed for reduced hp operation. Closing these valves reduces the available nozzle area and reduces the pressure drop across the governor valve.

The steam flow when operating with the auxiliary steam valve closed will approximate that for the same turbine if it were designed for the reduced rating.

3. The use of a steam turbine driver permits operation of the driven pump that is essentially independent of the electric power or distribution system. The steam turbine is not affected by electric power stoppages or interruptions. Therefore critical pumping operations may be maintained under such circumstances by utilizing a steam turbine driver.

4. A turbine may be used as a secondary driver for a pump; it may also drive an independent standby or emergency pump. The particular plant design may not afford sufficient steam for the pump to be normally driven by the steam turbine. However, in the event of an electric power failure or power system disturbance, a steam turbine may be employed as a dual drive or to drive a separate pump to assure continued operation of the plant until the electric power system is again operable.

5. The steam turbine controls—governor system and overspeed trip system—are inherently sparkproof. Consequently steam turbines can be readily applied to drive centrifugal pumps in a wide variety of hazardous atmospheres without entailing additional cost for explosionproof or sparkproof construction.

6. Steam turbines can normally be readily altered to accommodate an increase in rating for increased pump output or for new applications of the turbine-driven pump. This inherent flexibility of a steam turbine also permits it to be readily altered to accommodate changes in the initial steam pressure and temperature and the exhaust steam pressure at which the turbine operates when changes in the plant steam system so require.

7. Steam turbines have a starting or breakaway torque of approximately 150 to 180 percent of the rated torque. Additional starting torque can be readily furnished by designing the turbine for the additional required steam flow—and without reducing the efficiency at the normal operating rating by using an auxiliary steam valve. The additional starting torque can often be obtained without increasing the turbine-frame size.

8. Steam turbines can be used to drive all types of pumps.

9. Steam turbines are inherently self-limiting with respect to the power developed. Special protective devices do not have to be furnished to prevent damaging the turbine because of overload conditions. The maximum power that can be developed by a turbine is a function of the flow areas provided in the design of the nozzle ring and governor valve. Application of a load greater than that which can be developed by the turbine causes the turbine to slow down to a value at which the torque generated by the turbine matches that required by the pump.

10. When the application of the pump requires the driver to be designed with excess power or to permit operation of the pump "at the end of the curve," the steam turbine can be designed for the corresponding rating without reducing the turbine efficiency when operating at the normal rating. Closing an auxiliary steam valve furnished for operation at the normal rating permits realizing the normal efficiency because the turbine governor valve is not throttling to obtain the lower rating.

11. With respect to the operation of the various types of pump drivers and their supporting systems, steam turbines afford minimum maintenance, low vibration, and a quiet installation.

TYPES OF STEAM TURBINES

Single-Stage Turbines A single-stage steam turbine is one in which the conversion of the kinetic energy to mechanical work occurs with a single expansion of the steam in the turbine—from inlet steam pressure to exhaust steam pressure.

A single-stage turbine may have one or more rows of rotating buckets which absorb the velocity energy of the steam resulting from the single expansion of the steam.

Single-stage turbines are available in a variety of wheel diameters, that is, 9 to 28 in. The overall efficiency of a turbine for a particular operating speed and steam conditions is normally dependent on the wheel diameter. The efficiency will generally increase with an increase in wheel size, and therefore the steam rate will be less (for the more usual speeds and steam conditions).

The larger wheel diameter steam turbines can be furnished with more nozzles providing increased steam-flow capacity and consequently greater hp capabilities. The larger wheel diameter turbines are therefore furnished with larger steam connections, valves, shafts, bearings, etc. Consequently the size of the turbine will generally increase with increases in hp rating.

Multiple-Stage Turbines A multiple- (multi-) stage turbine is one in which the conversion of the energy occurs with two or more expansions of the steam within the turbine.

The number of stages (steam expansions) is a function of three basic parameters: thermodynamics, mechanical design, and cost.

The thermodynamic considerations include the available energy and speed.

The mechanical considerations include speed, steam pressure, steam temperature, etc., most of which are material limits. Cost considerations include the number, type, and size of the stages; the number of governor-controlled valves; the cost of steam; and the number of years used as a basis for the cost evaluation.

The two factors which are generally used in selecting multistage turbines are initial cost and steam rate. Since these two factors are a function of the total number of stages, the application becomes a factor of stage selection. The initial cost increases with the number of stages, but inversely the steam rate generally improves.

Multiple-stage turbines are normally used to drive pumps when the cost of steam or the available supply of steam requires turbine efficiencies which are greater than available with a single-stage turbine, or when the steam flow required to develop the desired rating exceeds the capability of single-stage turbines.

Multiple-stage turbines can be furnished with a single or multiple governor valves. A single governor valve is often of the same design whether used in a single-stage or multiple-stage turbine and generally has the same maximum steam flow, pressure, and temperature parameters. Multiple valves are used when the parameters for a single valve are exceeded, or to obtain improved efficiency, particularly at reduced power outputs.

Shaft Orientation Some steam turbines, particularly single-stage turbines, can be furnished with vertical downward shaft extensions. The application of such turbines can require considerable coordination between the pump and the turbine manufacturer to assure an adequate thrust bearing in the turbine, shaft length and details, mounting flange dimensions, and even shaft run-out.

Vertical shaft pumps are frequently driven by horizontal turbines through a right-angle speed-reduction gear unit.

Direct-connected and Geared Turbines Steam turbines can be direct-connected to the pump shaft so that the turbine operates at the pump speed or can drive the pump through a speed-reduction (and even speed increasing) gear unit, in order to permit the turbine to operate at a more efficient speed.

Turbine Stages The two types of turbine stages are impulse and reaction. The turbines discussed in this article employ impulse stages because steam turbines driving pumps normally have impulse-type stages.

In the ideal impulse stage, the steam expands only in the fixed nozzles and the kinetic energy is transferred to the rotating buckets as the steam impinges on the buckets while flowing through the passages between the buckets. The steam pressure is constant and the steam velocity relative to the bucket decreases in the bucket passages.

In a reaction stage the steam expands in both the fixed nozzles and the rotating buckets. The kinetic energy is transferred to the rotating buckets by the expansion of the steam in the passages between the buckets. The steam pressure decreases as the steam velocity relative to the bucket increases in the bucket passages.

In an impulse stage the steam can exert an axial force on the buckets as it flows through the blade passages. While this force is usually referred to as a reaction, the use of the term does not imply a reaction-type stage.

The larger buckets used in the last stages of an impulse-type multistage turbine can be of a free-vortex design—twisted and tapered. Such buckets are ideally subjected to a near pure impulse force at the root of the bucket and a near pure reaction force at the tip of the bucket, but, in reality, this bucket is a high reaction-design bucket compared to a normal impulse-stage bucket. A steam turbine stage with such a bucket design is still referred to as an impulse stage because the primary conversion of kinetic energy is by a reduction in relative steam velocity rather than by an increase in relative steam velocity.

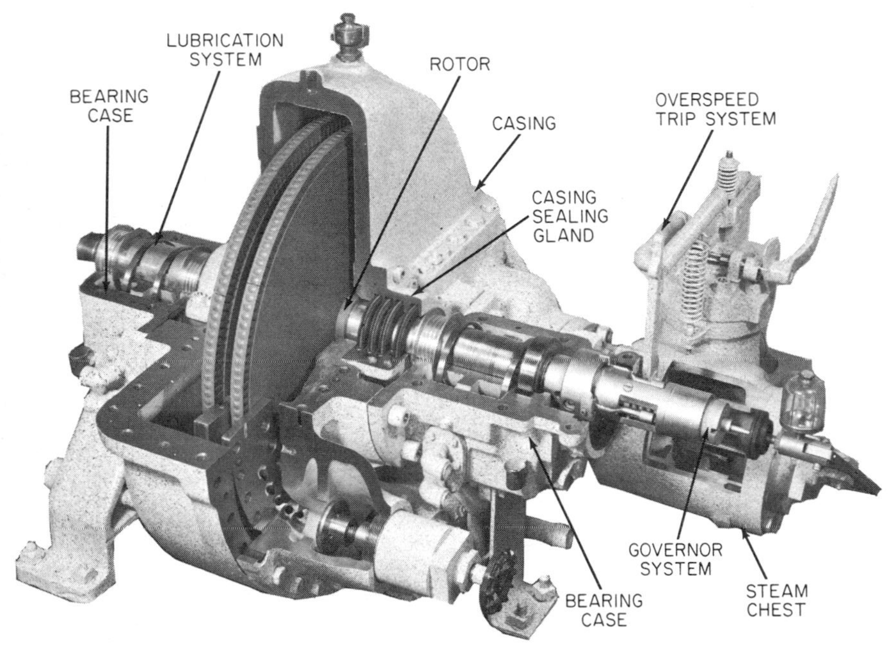

Fig. 1 Components parts of steam turbine. (Elliott Company)

A reaction turbine has more stages than an impulse turbine for the same application because of the small amount of kinetic energy absorbed per stage, and requires a larger thrust bearing and/or a balancing piston because of the pressure drop across the moving blades. The small pressure drop per stage and the pressure drop across the moving blades require that the steam-leakage losses be minimized by elaborate sealing between the tips of the nozzle blades and the rotor, and the tips of the moving blades and the casing.

The small pressure differential across the rotating blades of an impulse stage results in smaller thrust bearings and no close blade-tip clearances. Consequently impulse turbines can be started more quickly without thermal-expansion damage and their stage efficiencies remain relatively constant over the life of the turbine.

CONSTRUCTION DETAILS

Component Parts The main components of a single-stage steam turbine are listed below and shown in Fig. 1:

Casing	Casing sealing glands
Steam chest	Governor system
Rotor	Overspeed trip system
Bearing cases	Lubrication system

Function and Operation—Single Stage Turbine The *steam chest* and the *casing* contain the steam furnished to the turbine, being connected to the higher-pressure steam-supply line and the lower-pressure steam-exhaust line, respectively. The steam chest, which is in turn connected to the casing, houses the governor valve and the overspeed trip valve. The casing contains the rotor and the nozzles through which the steam is expanded and directed against the rotating buckets.

The *rotor* consists of the shaft and disk assemblies with buckets. The shaft extends beyond the casing and through the bearing cases. One end of the shaft is used for coupling to the driven pump. The other end of the shaft serves the speed-governor and the overspeed trip systems.

The *bearing cases* support the rotor and the assembled casing and steam chest. The bearing cases contain the journal bearings and the rotating oil seals which prevent outward oil leakage and the entrance of water, dust, and steam. The steam end-bearing case also contains the rotor positioning bearing and the rotating components of the overspeed trip system. An extension of the steam end-bearing housing encloses the rotating components of the speed-governor system.

The *casing sealing glands* seal the casing and the shaft with spring-backed segmental carbon rings (supplemented by a spring-backed labyrinth section for the higher exhaust-steam pressures).

The *governor system* commonly consists of spring-opposed rotating weights, a steam valve, and an interconnecting linkage or servo motor system. Changes in the turbine inlet and exhaust-steam conditions, and the power required by the pump will cause the turbine speed to change. The change in speed results in respositioning the rotating governor weights and subsequent repositioning of the governor valve.

The *overspeed trip system* usually consists of a spring-loaded pin or weight mounted in the turbine shaft or on a collar, a quick-closing valve which is separate from the governor valve and interconnecting linkage. The centrifugal force created by rotation of the pin in the turbine shaft exceeds the spring loading at a preset speed. The resultant movement of the trip pin causes knife edges in the linkage to separate and permit the spring-loaded trip valve to close.

The trip valve may be closed by disengaging the knife edges manually, by an electric or pneumatic signal, by low oil pressure, or by high turbine-exhaust steam pressure.

The two usual types of lubrication system are oil-ring and pressure lubrication systems.

The *oil-ring* lubrication system employs an oil ring(s) which rotates on the shaft with the lower portion submerged in the oil contained in the bearing case. The rotating ring(s) transfers oil from the oil reservoir to the turbine-shaft-journal bearing and rotor-locating bearing. The oil in the bearing-case reservoirs is cooled by water flowing in cooling water chambers or tubular heat exchangers.

A *pressure* lubrication system consists of an oil pump driven from the turbine shaft, an oil reservoir, a tubular oil cooler, and an oil filter with interconnecting piping. Oil is supplied to the bearing cases under pressure. The oil rings may be retained in this system to provide oil to the bearings during startup and shutdown when the operating speed and bearing design permit.

Typical sectional drawings are shown in Figs. 2, 3, and 4.

GOVERNORS AND CONTROLS

Governor systems are speed-sensitive control systems that are integral with the steam turbine. The turbine speed is controlled by varying the steam flow through the turbine by positioning the governor valve. Variations in the power required by the pump and changes in steam inlet or exhaust conditions alter the speed of the turbine, causing the governor system to respond to correct the operating speed.

Control systems, unlike governor systems, are not directly speed-sensitive but respond to changes in pump or pump-system pressures and then reposition the turbine "governor" valve to maintain the preset pressure. Consequently changes in the turbine steam conditions or the power required by the pump result in repositioning the turbine governor valve or a separate steam valve only after the pressure being sensed by the controller has changed.

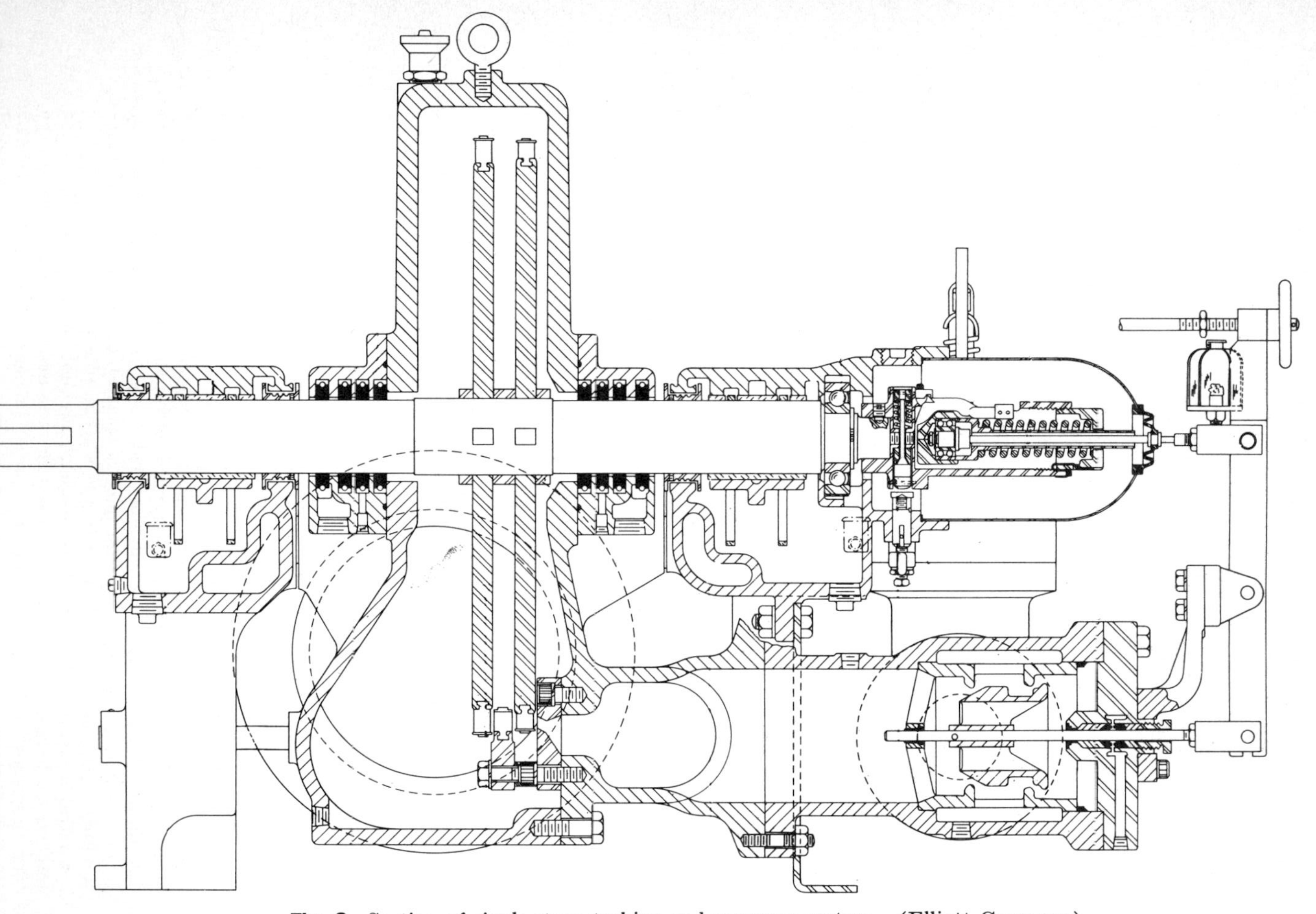

Fig. 2 Section of single-stage turbine and governor system. (Elliott Company)

Even when a control system is furnished, a speed governor is also normally furnished. The speed governor is set for a speed slightly higher than the desired operating speed in order to function as a pre-emergency governor, that is, prevent the turbine from reaching the trip speed when the controller would cause the turbine to operate at a speed above rated speed.

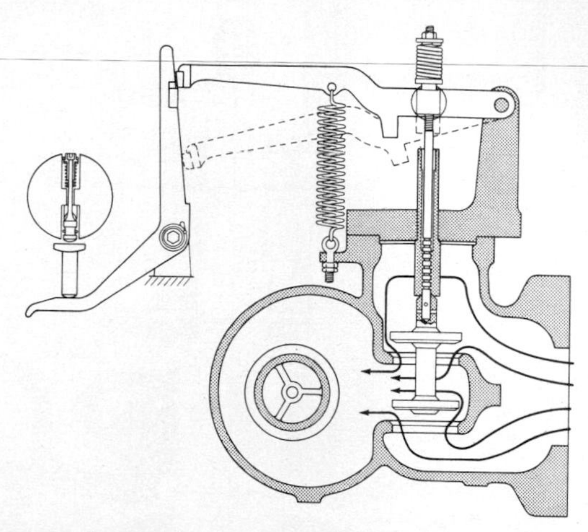

Fig. 3 Section of a turbine-overspeed trip system. (Elliott Company)

Governor systems are defined by their performance as follows:[1]

Class of governor system	*Speed range, % (as specified)*	*Maximum-speed regulation, %*	*Maximum-speed variation, %, ±*	*Maximum-speed rise, %*
A	10–65	10	0.75	13
B	10–80	6	0.50	7
C	10–80	4	0.25	7
D	10–90	0.50	0.25	7

Speed range is the percentage below rated speed for which the governor speed setting may be adjusted. For example, a turbine with 4,000 rpm rated speed and a governor system having a 30 percent range can be operated at a minimum speed of 2,800 rpm:

$$4{,}000 - \frac{30 \times 4{,}000}{100} = 2{,}800$$

If the speed range had been specified as plus 5 percent and minus 25 percent, for example, the maximum and minimum speeds would be 4,200 and 3,000 rpm, respectively.

Steady-state *speed regulation* is the change in speed required for the governor system to close the governor valve when the load is *gradually* reduced from rated load to no-load with turbine steam conditions constant. Regulation is always expressed as a percentage of rated speed and calculated as follows:

$$\frac{\text{No-load speed} - \text{rated speed}}{\text{Rated speed}} \times 100 = \%\ \text{regulation}$$

Therefore

$$\text{No-load speed} = \text{rated speed}\left(1 + \frac{\%\ \text{regulation}}{100}\right)$$

[1] Single-Stage Steam Turbines for Mechanical Drive Service, *Nat. Elec. Manufacturers' Ass. Publ.* SM21-1970, 1970.

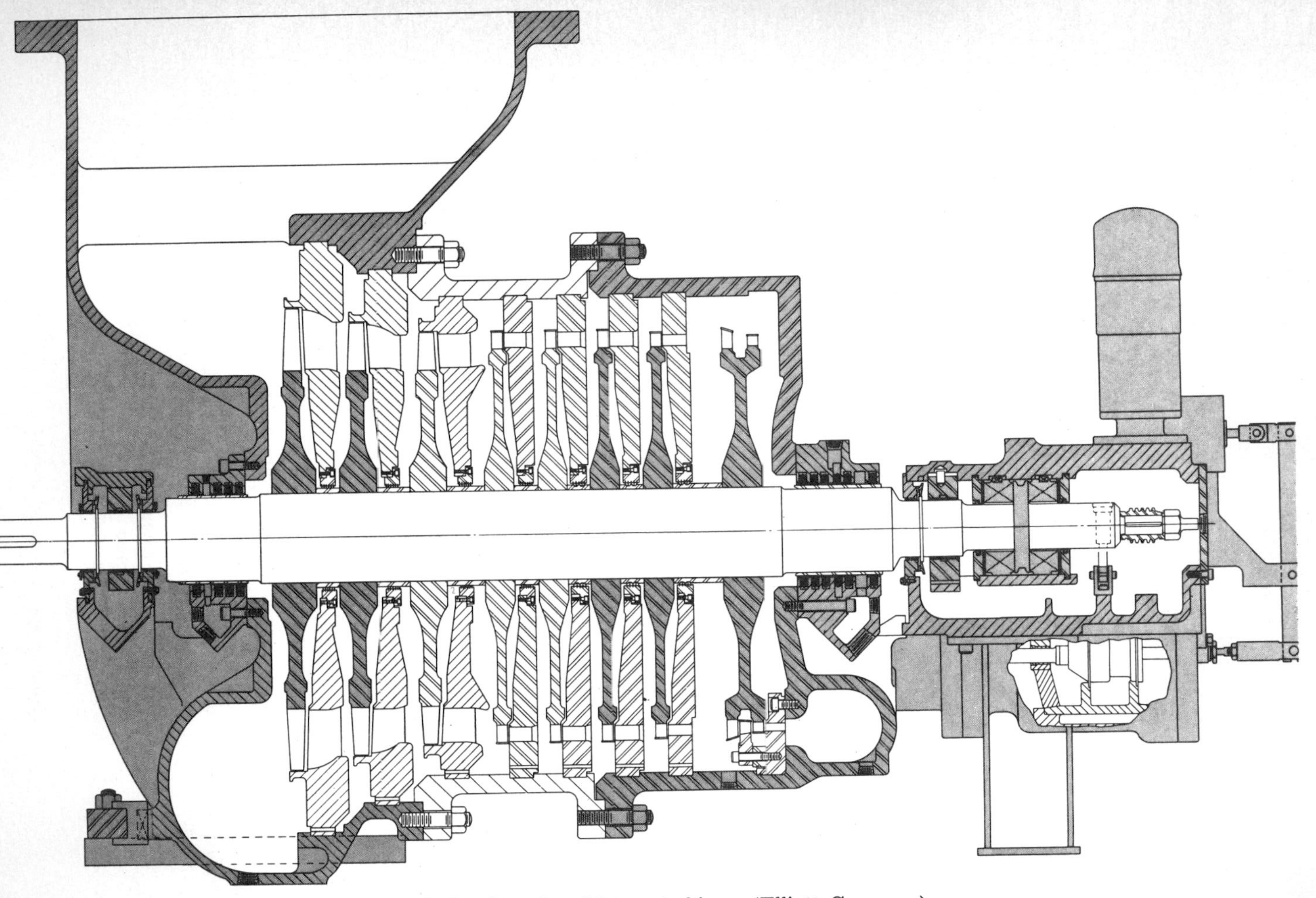

Fig. 4 Section of multistage turbine. (Elliott Company)

A turbine with 4,000 rpm rated speed and equipped with a NEMA A speed governor system would have 4,400 rpm maximum no-load speed:

$$4{,}000 \times \left(1 + \frac{10}{100}\right) = 4{,}400 \text{ rpm}$$

Consequently whenever the turbine is developing less than rated horsepower, the operating speed will be greater than rated speed as required for the governor system to reposition the governor valve.

Speed variation, expressed as a percentage, is the total magnitude of the fluctuations from the set speed permitted by the governor system when the turbine is normally operating at rated speed, power, and steam conditions. This is the "insensitivity" of the governor system.

The speed variation equation is

$$\frac{\text{Maximum speed} - \text{minimum speed}}{\text{Rated speed} \times 2} \times 100 = \pm\% \text{ speed variation}$$

The turbine with 4,000 rpm rated speed and NEMA A governor system would have ± 30 rpm maximum speed variation:

$$4{,}000 \times \frac{\pm 0.75}{100} = \pm 30 \text{ rpm}$$

Maximum-speed rise represents the momentary increase in speed when the load is suddenly reduced from rated horsepower to no-load horsepower with the pump still coupled to the turbine shaft. Shortly after the sudden loss of load, the governor system will cause the turbine speed to be reduced to the no-load speed.

Maximum-speed rise is also expressed as a percentage of rated speed and calculated as follows:

$$\frac{\text{Maximum speed} - \text{rated speed}}{\text{Rated speed}} \times 100 = \% \text{ speed rise}$$

Therefore

$$\text{Maximum speed} = \text{rated speed} + \frac{\text{rated speed} \times \% \text{ speed rise}}{100}$$

The turbine with 4,000 rpm rated speed and equipped with a NEMA A governor will have a maximum-speed rise to:

$$4{,}000 + \frac{4{,}000 \times 13}{100} = 4{,}520 \text{ rpm}$$

The setting of the *overspeed trip* is a function of the maximum-speed rise of the governor system—it must be higher than the maximum-speed rise. The recommended settings are:

Class of governor system	*Overspeed trip setting, (% of rated speed)*
A	115
B	110
C	110
D	110

The overspeed trip system for the turbine with 4,000 rpm rated speed and equipped with a NEMA A governor is set for operation at

$$4{,}000 \times \frac{115}{100} = 4{,}600 \text{ rpm}$$

The speed-sensitive portion of the speed-governor system is usually a set of spring-loaded rotating weights. Movement of the weights caused by a change in turbine speed position the governor valve through a suitable linkage.

The speed-sensitive element can also be other devices that are speed-responsive such as a positive-displacement oil pump, electric generator, or a magnetic-impulse signal generator.

The rotating-weight-type governor system is a direct-acting type and is classified as a NEMA A governor. *Direct-acting* designates a governor system in which the speed-sensitive element also provides the power for positioning the governor valve.

The NEMA B, C, and D governor systems have speed-sensitive elements which position the governor valve through a relay or servomotor system instead of actuating the valve directly. The speed-sensitive element can therefore be more precise and sensitive as required for the improved governor-system performance.

THEORY

A steam turbine develops mechanical work by converting into work the available heat energy in the steam expansion. Heat and mechanical work, being two forms of energy, can be converted from one to the other.

The heat energy is converted in two steps. The steam expands in nozzles and discharges at a high velocity, converting the available heat energy into velocity (kinetic) energy. The high-velocity steam strikes moving blades, converting the velocity energy into work. Since the total heat energy available in the steam is converted into velocity (kinetic) energy, the magnitude of the steam velocity is, therefore, dependent upon the available energy.

The mechanical work that is developed in the turbine by the high-velocity steam striking the buckets is a function of the speed of the buckets. Maximum work occurs when the bucket velocity is approximately one-half the steam-jet velocity for an impulse stage and one-fourth the steam-jet velocity for a velocity-compounded impulse stage. While the steam-jet velocity is fixed by the available heat energy, the bucket velocity is fixed by the speed of the turbine and the diameter of the turbine wheel on which the buckets are mounted. The work developed, or the efficiency of the turbine, ignoring losses in the turbine, is, therefore, determined by the size of the turbine and the turbine (pump) speed for a fixed amount of available heat energy.

The most common single-stage turbine is the velocity-compounded (Curtis) type. The complete expansion from inlet to exhaust pressure occurs in one step. The Curtis stage, with two rows of rotating buckets and two reentry-type velocity-compounded stages, is illustrated in Fig. 5.

Single-stage turbines are available with a wide range of efficiencies since they are manufactured with a variety of wheel diameters: 9 to 28 in.

Multistage turbines are manufactured with a more limited variety of wheel sizes. The efficiency of multistage turbines is varied primarily by varying the number of stages. When the total available energy of the steam results in a steam velocity which is greater than twice the bucket velocity (using convenient wheel sizes), then a multistage turbine will be more efficient. In a multistage turbine, the total steam expansion is divided among the various impulse stages to produce the desired steam velocity for each row of buckets.

A steam turbine is normally evaluated using *steam rate*—the amount of steam required by the turbine to produce the specified horsepower per hour at the specified speed—rather than *efficiency*. The steam rate is a direct function of the turbine efficiency.

The steam consumption can be expressed either as steam rate (pounds of steam per horsepower per hour) or as steam flow (pounds of steam per hour). The higher the efficiency, the lower the steam rate or steam flow, and vice versa.

The total available energy of the steam is that available from an isentropic expansion. For given initial steam pressure and temperature and exhaust pressure, the available energy (Btu per pound of steam) can be obtained from the tables or the Mollier chart (see Fig. 10) in Ref. 4.

The available energy can be converted into horsepower or kilowatt units, and expressed as the theoretical steam rate—pounds per horsepower hour or pounds per kilowatthour. The theoretical steam rate is the steam rate for a 100 percent efficient turbine, and therefore can be used more conveniently than energy, Btu per pound, for the calculation of turbine steam rates. Theoretical steam rates can be obtained directly from Theoretical Steam Rate Tables (ASME) or from the polar Mollier chart (Elliott Company).

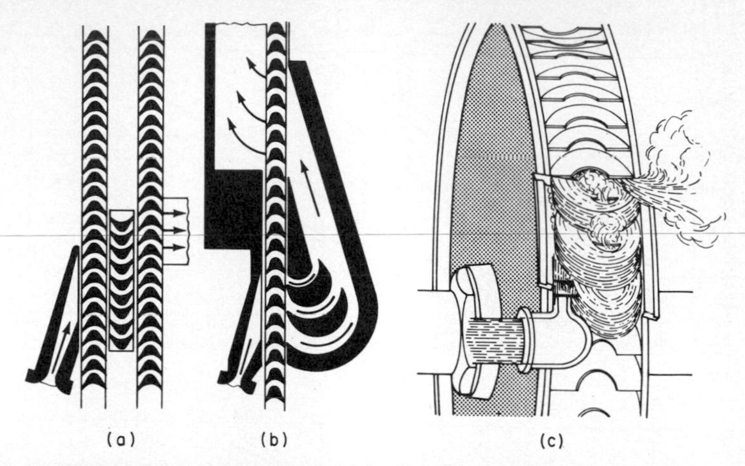

Fig. 5 Velocity-compounded stages: (*a*) Curtis stage; steam flows once through moving buckets; (*b*) reentry stage; steam flows twice through moving buckets; (*c*) reentry stage; steam flows three times through moving buckets. (Reprinted with permission from *Power,* June 1962. Copyright McGraw-Hill, Inc., 1962.)

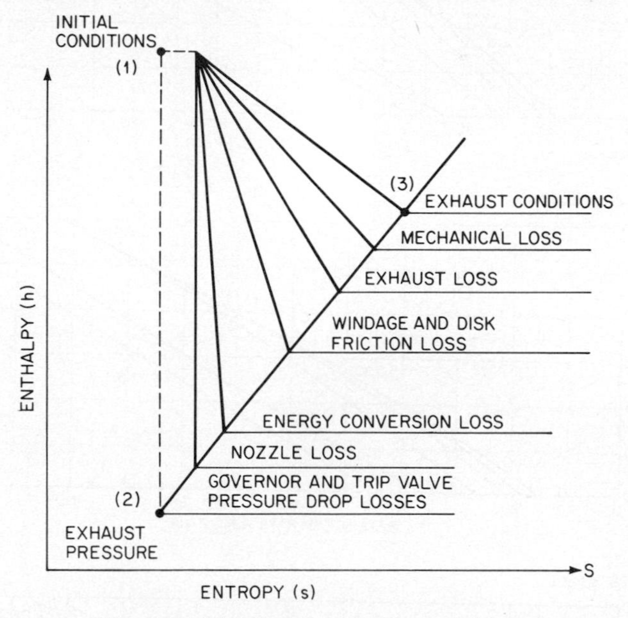

Fig. 6 Mollier diagram with energy losses.

The actual steam rate for a turbine is greater than the theoretical steam rate because of the losses that occur in the turbine when converting the available energy into mechanical work and the ratio of the steam velocity to bucket velocity. The energy remaining in the steam exhausting from the turbine is greater than that after an isentropic expansion, as illustrated in Mollier diagram shown in Fig. 6, where

1 = energy in steam at initial steam pressure and temperature
2 = energy in steam at exhaust pressure for an isentropic expansion
3 = actual energy in steam at exhaust pressure

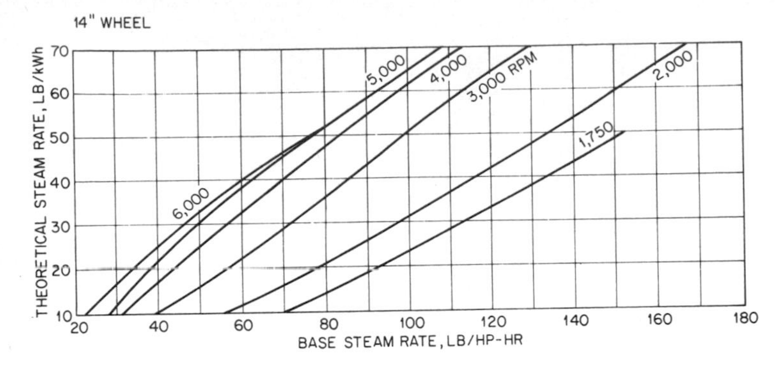

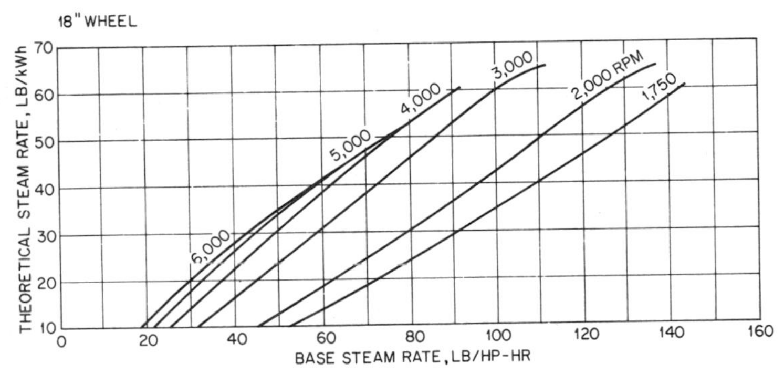

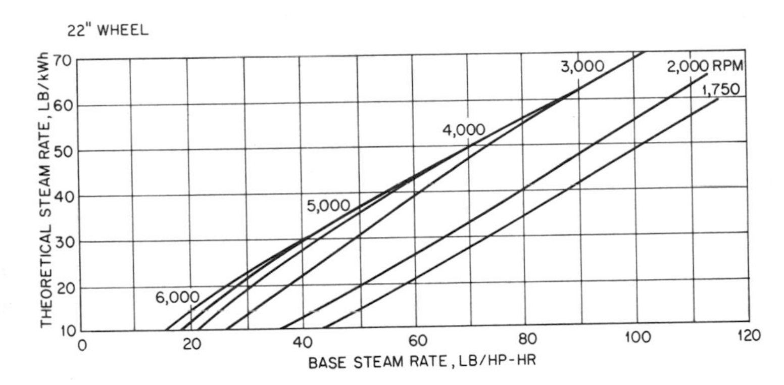

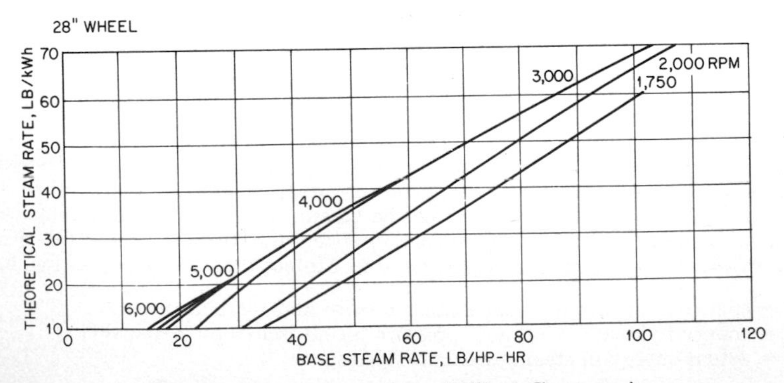

Fig. 7 Base steam rates. (Elliott Company)

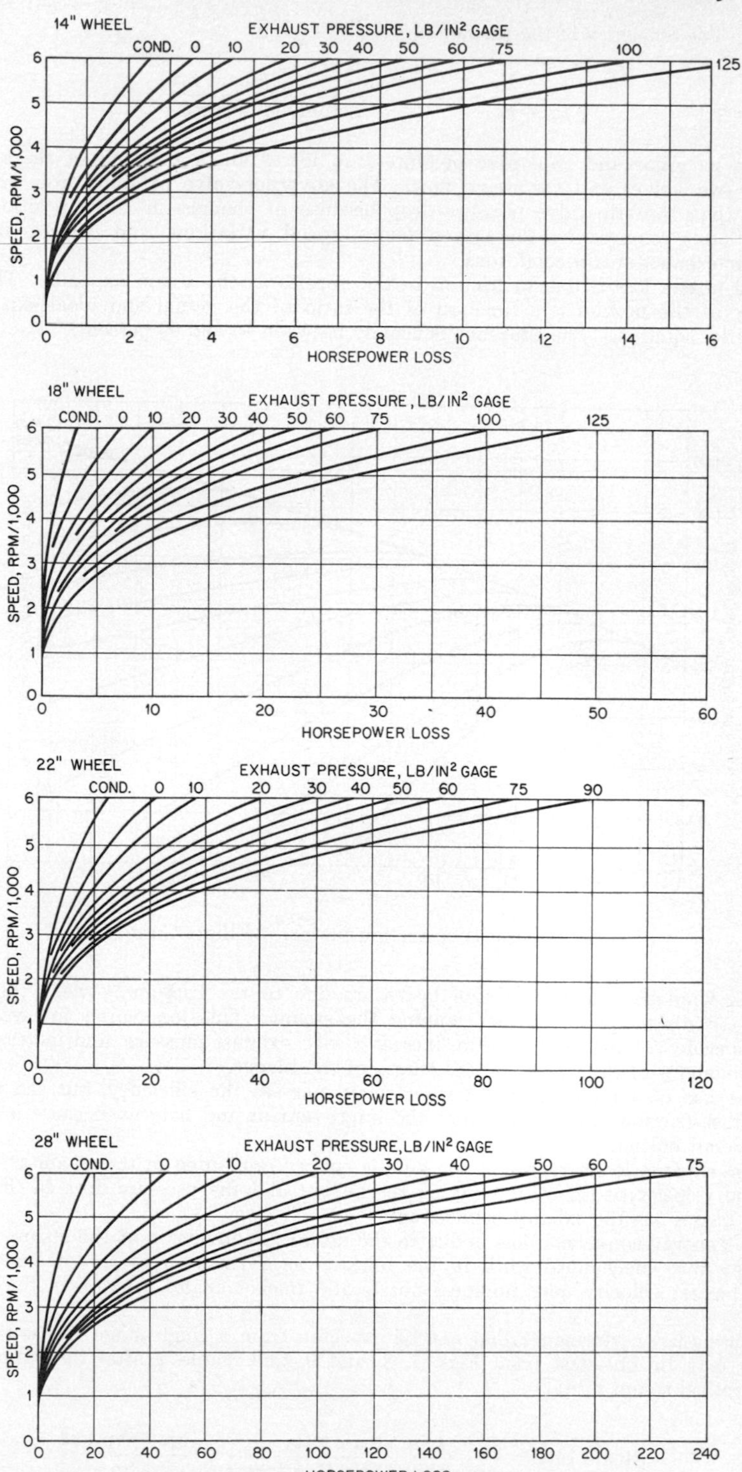

Fig. 8 Horsepower loss. (Elliott Company)

The efficiency of the turbine is

$$\frac{h_1 - h_3}{h_1 - h_2} \quad \text{or} \quad \frac{\text{theoretical steam rate}}{\text{actual steam rate}}$$

The governor and trip-valve pressure-drop losses are a function of the sizes of these two valves and the steam flow. The governor-valve pressure drop will vary more than the trip-valve pressure drop because of changes in valve position with horsepower required by the driven pump, speed variations, and changes in inlet and/or exhaust steam conditions.

The nozzle loss is due to friction in the nozzles as the steam expands. The efficiency of the nozzles is a function of the ratio of the actual and ideal exit steam velocities squared. The efficiency is usually between 95 and 99 percent.

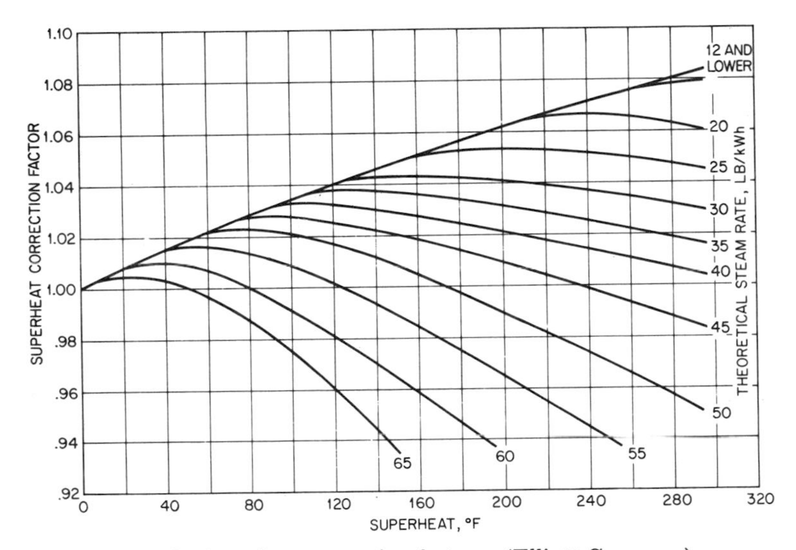

Fig. 9 Superheat correction factor. (Elliott Company)

The windage and disk friction losses are due to the friction between the steam and the disks and the blades fanning the steam. This loss varies inversely with the specific volume of the steam, increases with exhaust pressure, and increases with the diameter of the wheel and the length of the blades.

The use of a larger diameter wheel may increase the efficiency, but the windage and disk-friction losses will reduce the improvement and may even cause a net loss in overall efficiency.

The exhaust losses represent the kinetic energy remaining in the steam as a result of the velocity of the steam leaving the bucket and the pressure drop in the steam as it passes out the exhaust connection.

The energy-conversion loss is due to the nonideal conversion of the steam-velocity energy into mechanical work in the buckets as a function of the steam velocity and bucket velocity, plus nonideal nozzle and bucket angles, friction in the system, etc.

The actual performance that can be expected from a single-stage Curtis-type turbine may be obtained from Figs. 7, 8, and 9, and Table 1 after determining the theoretical steam rate:

$$\text{Steam rate} = \frac{\text{base steam rate}}{\text{superheat correction factor}} \times \frac{\text{hp} + \text{hp loss}}{\text{hp}}$$

TABLE 1 Temperature of Dry and Saturated Steam

(To obtain superheat in °F, subtract temperature given in tabulation below from total initial temperature.)

Lb/in² gage	Saturation temp., °F	Lb/in² gage	Saturation temp., °F	Lb/in² gage	Saturation temp., °F	Lb/in² gage	Saturation temp., °F
0	213	150	366	300	422	450	460
5	228	155	368	305	423	455	461
10	240	160	371	310	425	460	462
15	250	165	373	315	426	465	463
20	259	170	375	320	428	470	464
25	267	175	378	325	429	475	465
30	274	180	380	330	431	480	466
35	281	185	382	335	432	485	467
40	287	190	384	340	433	490	468
45	293	195	386	345	434	495	469
50	298	200	388	350	436	500	470
55	303	205	390	355	437	510	472
60	308	210	392	360	438	520	474
65	312	215	394	365	440	530	476
70	316	220	396	370	441	540	478
75	320	225	397	375	442	550	480
80	328	230	399	380	444	560	482
85	328	235	401	385	445	570	483
90	331	240	403	390	446	580	485
95	335	245	404	395	447	590	487
100	338	250	406	400	448	600	489
105	341	255	408	405	449	610	491
110	344	260	410	410	451	620	492
115	347	265	411	415	452	630	494
120	350	270	413	420	453	640	496
125	353	275	414	425	454	650	497
130	356	280	416	430	455	660	499
135	358	285	417	435	456	670	501
140	361	290	419	440	457	680	502
145	364	295	420	445	458	690	504

Sample calculation:

Steam conditions: 250 lb/in² gage inlet, 575°F inlet and 50 lb/in² gage exhaust
Design conditions: Turbine to develop 500 hp at 4,000 rpm

1. Theoretical steam rate = 26.07 lb/kWhr
2. Base steam rate for 28-in wheel (Fig. 7) = 36 lb/hp-hr
3. Horsepower loss for 28-in wheel (Fig. 8) = 55
4. Temperature of dry and saturated inlet steam (Table 1) = 406°F
5. Superheat (Table 1) = 575 − 406 = 169°F
6. Superheat correction factor (Fig. 9) = 1.052
7. Steam rate $= \dfrac{36}{1.052} \times \dfrac{500 + 55}{500} = 38.0$ lb/hp-hr

The ability of a particular size turbine to actually develop the required hp is determined primarily by:

1. The flow capacity of the inlet and exhaust connection from Figs. 10 and 11, where: steam flow = hp × steam rate.

2. The flow capacity of the nozzles available in a particular turbine. The number and size of the nozzles vary considerably with each design of turbine manufactured, and a meaningful plot cannot be included in this section.

EVALUATION OF COSTS

The economic selection of a turbine considers the initial cost of the turbine and the operating costs. A lower-steam-rate (more efficient) turbine generally has a higher initial cost than a turbine with a higher steam rate. The operating costs are the cost of the steam for the number of years upon which the evaluation is to be based.

Total cost = initial cost
+ (hp × steam rate × steam cost × operating hours per year × number years)

The cost of installation—foundation, steam piping, and cooling-water service (also electric service if required)—is not normally considered unless there are significant differences between the turbines: single-stage versus multistage, a direct-connected turbine versus a turbine and gear, vertical turbine versus horizontal turbine with a right-angle gear, etc.

The economic selection of a steam turbine versus an electric motor, diesel engine, gas engine, gas turbine, etc., must include an evaluation of all installation costs, costs of supporting equipment (starters, switch gear, fuel-supply system etc.), operating costs, and the initial costs of the various drivers being considered.

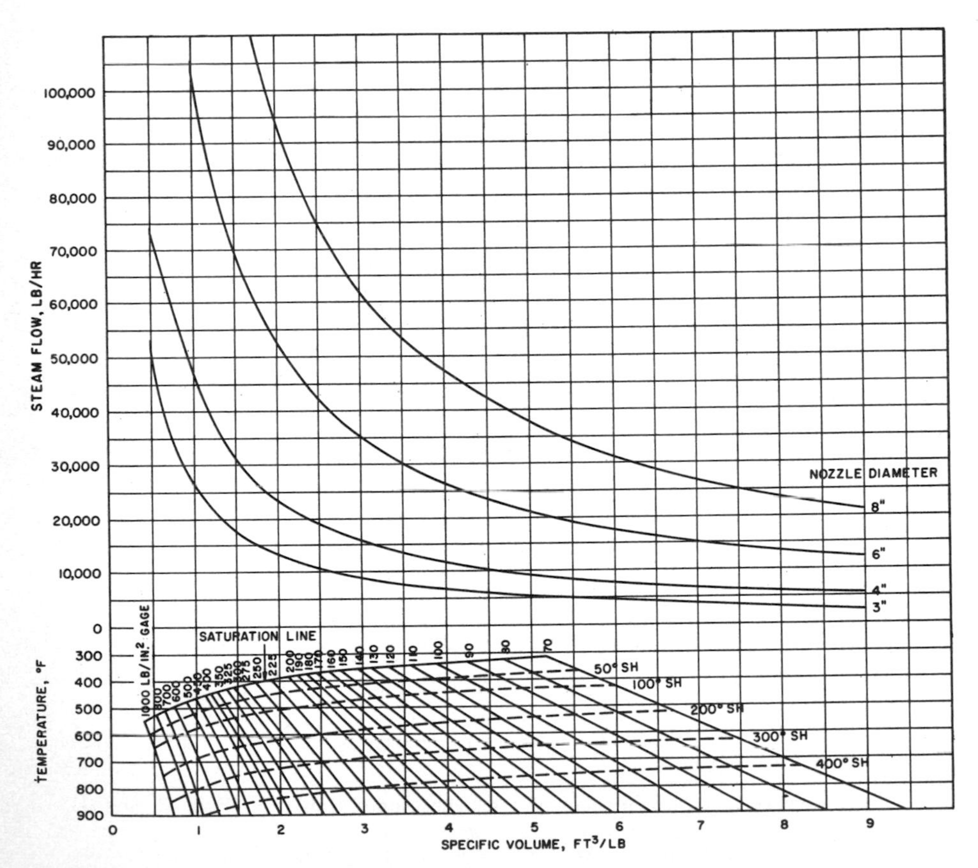

Fig. 10 Nominal inlet-flow capacity. Read inlet nozzle size required to pass maximum flow (based on 150 ft/s steam velocity).

REFERENCES

1. Edwin F. Church, "Steam Turbines," McGraw-Hill Book Company, New York, 1950.
2. Single-Stage Steam Turbines for Mechanical Drive Service, *Nat. Elec. Manufacturers' Ass. Publ.* SM22-1970, 1970.
3. Multistage Steam Turbines for Mechanical Drive Service, *Nat. Elec. Manufacturers' Ass. Publ.* SM21-1970, 1970.
4. Joseph H. Keenan and Frederick G. Keyes, "Thermodynamic Properties of Steam," John Wiley & Sons, Inc., New York, 1936.
5. "Theoretical Steam Rate Tables," ASME, New York, 1969.
6. B. G. A. Skrotzki, Steam Turbines, *Power Spec. Rept.*, 1962.
7. Wallace L. Bergeron, Governors and Controls—Each Has Specific Operation, *Sugar J.*, March, 1965.
8. Hall Steen-Johnsen, Selecting the Right Steam Turbine for Industrial Use, *Power Eng.*, April 1965.
9. Hall Steen-Johnsen, Here's What You Should Know before You Specify Control Systems for Mechanical Drive Steam Turbines, *Power Eng.*, February and March, 1966.
10. Hall Steen-Johnsen, Mechanical Drive Turbines, *Oil Gas Equip.*, March, April, and May, 1967.
11. Hall Steen-Johnsen, How to Estimate the Size and Cost of Mechanical Drive Steam Turbines, *Hydrocarbon Process.*, October, 1967.
12. Polar Mollier Chart, Elliott Company.

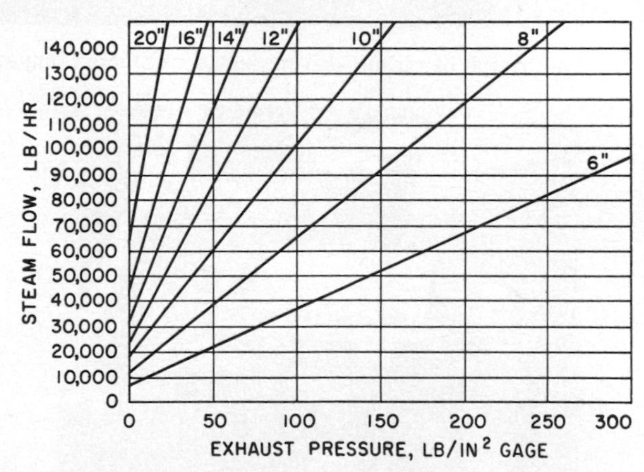

Fig. 11 Nominal exhaust-flow capacity. Noncondensing exhaust nozzles (based on 200 ft/s steam velocity).

Fig. 12 Photograph of turbine-driven pumps. (Elliott Company)

Fig. 13 Reactor feed pump, turbine, and lubricating-oil system mounted on a common bedplate. Turbine rating 11,600 hp, 5,530 rpm, 470°F, 2 inHg, double-flow top exhaust. (DeLaval Turbine Inc.)

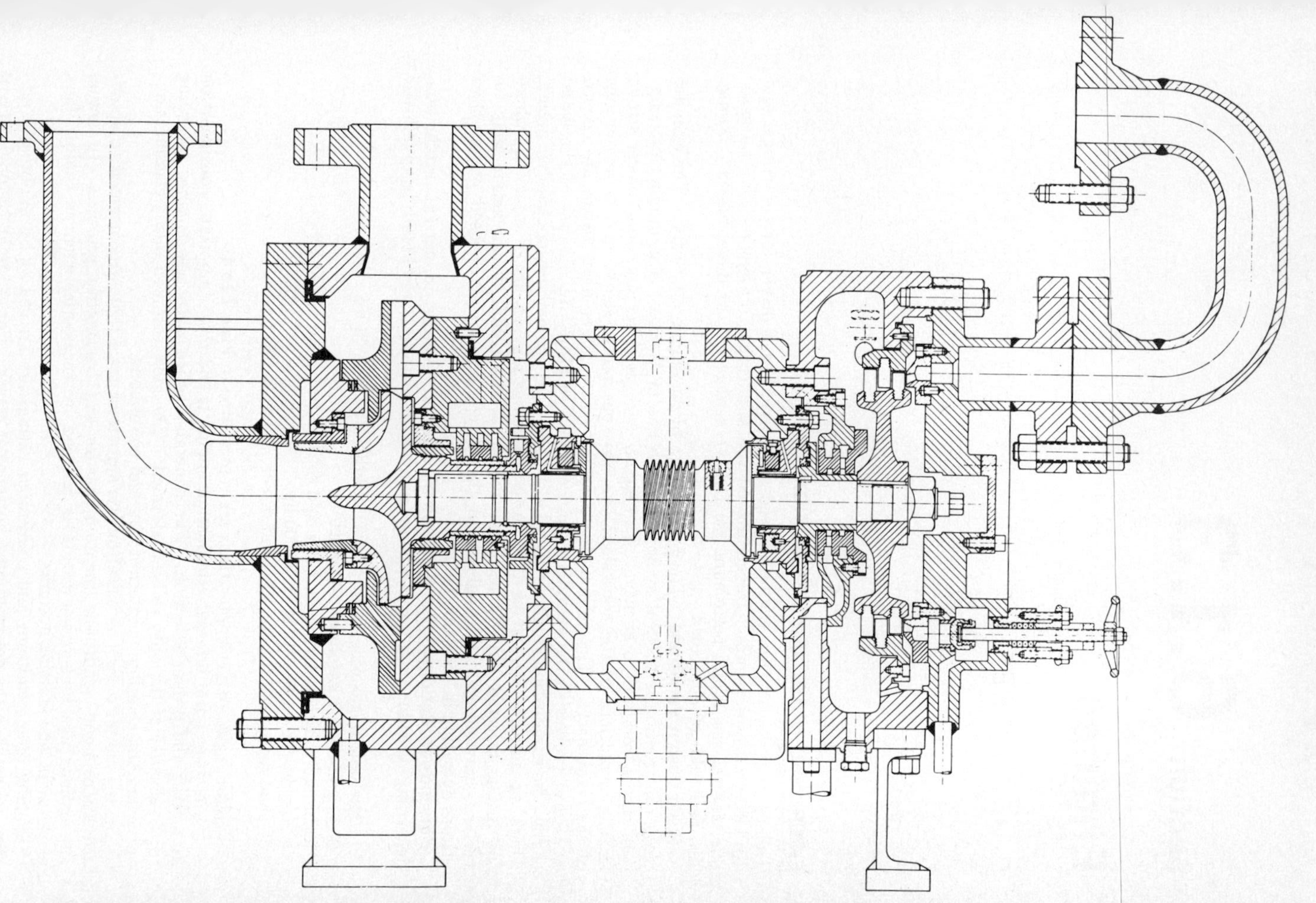

Fig. 14 High-speed centrifugal pump and single-stage steam turbine mounted on a common shaft. Package is designed for boiler feed in marine, industrial, and process steam plants. (Pacific Pumps, Division of Dresser Industries, Inc.)

Section **6.1.3**

Engines

F. J. GUNTHER

ENGINE SELECTION AND APPLICATION

The internal combustion engine is used extensively as a driver for centrifugal and displacement type pumps. Depending on the application the engine may be gasoline, natural gas, liquid petroleum gas (LPG), sewage gas, or diesel-fueled. It may also be liquid- or air-cooled.

Basic Design Variations The basic design of the engine may vary. The cylinder block construction is vertical or horizontal in-line or V-type. The number of cylinders ranges from one to twenty. The engine cycle is either four (one power stroke in four revolutions of the crankshaft) or two (one power stroke in two revolutions of the crankshaft). The combustion chamber and cylinder head design are classed as L-head (valves in the cylinder block) or valve-in head. In the case of the diesel engine, the combustion chamber may be the so-called precombustion chamber design, where an antechamber is used to initiate combustion, or direct injection, where the fuel is injected directly into the cylinder. The piston design action may be vertical, horizontal, at an angle as in the V-type engine, or opposed piston (two pistons operating in the same cylinder).

Horsepower Ratings The horsepower range of engines in current production, depending on cubic inch displacement, number of cylinders, and speed, is as follows:

1. Air-cooled—gasoline, natural gas, and diesel: 1.0 to 75 hp
2. Liquid-cooled gasoline: 10 to 300 hp
3. Liquid-cooled natural gas, LPG, and sewage gas: 10 to 15,000 hp
4. Liquid-cooled diesel: 10 to 50,000 hp
5. Dual fuel, natural gas, LPG, and diesel: 150 to 25,000 hp

Typical engine-driver applications are represented in Figs. 1 to 4.

The rating of the internal combustion engine is the most important consideration in making the proper selection. The general practice is to rate engines according to the severity of the duty to be performed. The most common rating classifications are maximum, standby or intermittent, and continuous.

The maximum output is based on dynamometer tests that are corrected to standard atmospheric conditions for temperature and barometric pressure. In actual applications this horsepower rating is reduced by accessories such as cooling fans, air cleaners, and starting systems.

Standby or intermittent and continuous ratings are arrived at by applying a percentage factor to the net maximum horsepower rating. For example, 75 to 80 percent is used for continuous, and 90 percent for intermittent.

Fig. 1 Gasoline–natural gas engine driving vertical deep-well pump for private Los Angeles water utility. (Waukesha Motor Co.)

Fig. 2 Natural gas engine driving horizontal water pump for Winnipeg, Canada, water utility. (Waukesha Motor Co.)

The *duty cycle* is a term used to describe the load pattern imposed on the engine. If the load factor (ratio of average load to maximum capabilities) is low we call the duty cycle "light," but if it is high we classify the cycle "heavy." Continuous, or heavy-duty, service is generally considered to be 24 hr/day with little variation in load or speed. Intermittent, or standby, service is classed as duty where an engine is called upon to operate in emergencies or at reduced loads at frequent intervals.

Fig. 3 Sewage gas engine driving vertical pump for Seattle Metropolitan Sewerage Commission. (Waukesha Motor Co.)

Fig. 4 Diesel engine driving pumps for flood-control station in Seattle. (Waukesha Motor Co.)

To make an analysis of power problems in the selection of a proper engine certain terms are used in the industry.

1. *Displacement.* The volume in cubic inches of the cylinders of an engine:
Displacement bore (in^2) × 0.7854 × stroke (in) × no. of cylinders
2. *Torque.* Twisting effort of engine described in pound-feet:

$$T = 5{,}252 \times \frac{\text{bhp}}{\text{rpm}}$$

3. *Engine horsepower.* Measure of theoretical characteristics of an engine. Brake horsepower is the measurable horsepower after the deduction for friction losses.

$$\text{bhp} = T \times \frac{\text{rpm}}{5{,}252}$$

4. *BMEP (brake mean effective pressure).* The average cylinder pressure to give a resultant torque at the flywheel:

$$\text{bmep (lb/in}^2) = \frac{792{,}000 \times \text{bhp}}{\text{rpm} \times \text{displacement}} \quad \text{(four-cycle)}$$

$$\frac{396{,}000 \times \text{bhp}}{\text{rpm} \times \text{displacement}} \quad \text{(two-cycle)}$$

5. *Piston speed.* Average velocity of piston at a given speed:

$$\text{Piston speed (ft/min)} = \text{stroke (in)} \times 2 \times \frac{\text{rpm}}{12}$$

The proper selection of an engine is affected by conditions as follows:

Altitude
Ambient air temperature
Rotation and speed
Bmep and piston speed
Maintenance
Type of fuel
Operating atmosphere (dust and dirt)
Vibrations and torsionals

The observed horsepower is that which an engine will produce under the existing altitude and temperature. All engine manufacturers publish horsepower ratings corrected to certain conditions; for conditions other than these it is necessary to correct by applying a percentage factor for altitude and temperature. Generally this is 3½ percent per thousand feet above sea level and 1 percent for every 10° above 60°F. In the case of a turbocharged engine there is no established standard and the engine manufacturer should be consulted.

The basic rotation of engines in current production is counterclockwise when viewed from the flywheel end of the engine although many of the larger engines are available in both counterclockwise or clockwise rotation. The speed of the engine is generally fixed by the equipment being driven. Through the use of speed-increasing or reducing gear boxes, the proper engine for a given application may be selected. A gear box may also be used to correct a rotation problem.

Speed ranges for engines generally fall into three categories:

High—above 1,500 rpm
Medium—700 to 1,500 rpm
Low—below 700 rpm

High-speed engines generally offer weight and size advantages as well as cost savings and thus are used for standby or intermittent applications. On the other hand, the medium- or slow-speed engine, although heavier and larger in size, offers a gain in service life and lower maintenance costs.

The flexibility of an engine drive as to speed plays an important role in the operation of pumps where variable quantities of liquid must be pumped. The engine speed may be changed very simply either manually or through the use of liquid or pressure controls.

Bmep is generally a measure of load, and the piston speed a measure of potential wear and maintenance. Although the introduction of the turbocharged and intercooled engine has somewhat changed the consideration given these as evaluation factors they are still important to consider in the selection of engines where long life is a factor.

Maintenance of engines has been considered by some as objectionable and more costly than electric power. A recent innovation of the engine manufacturers, in the form of a service contract for installations where trained personnel are not available or desirable for economic reasons, can eliminate these same objections and costs. The complete maintenance of the engine is done on a fixed fee basis for a designated period of time.

The remaining conditions listed above will be discussed in detail later.

FUEL SYSTEMS

Gasoline Gasoline has been used and is still used primarily with standby and intermittent pumping units. Inasmuch as the spark ignition system first introduced the internal combustion engine to power applications gasoline was used as the primary fuel. Through the use of a carburetor the mechanical mixing or blending of a volatile liquid such as gasoline with the proper proportion of air forms a combustible mixture. Commercial gasoline has an average heating value of 19,000 Btu/lb. It is easy to transport and handle and unlike gaseous fuels does not require pressure storage and regulating equipment. The starting capabilities of a gasoline engine are satisfactory providing the engine is in good operating condition. With

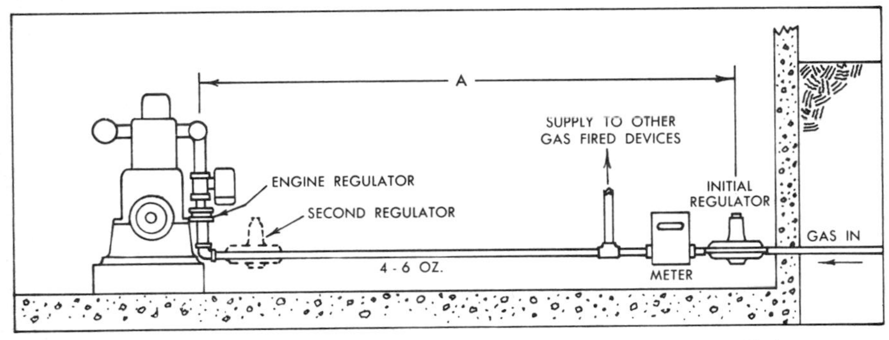

Fig. 5 Typical natural gas fuel system. (Waukesha Motor Co.)

the high-power engine of today refinery control can produce a fuel matched to the operating conditions.

Gasoline does have some disadvantages which are reducing its use in comparison with other fuels. In small volume usage it is safe and easily handled. In larger volumes it becomes expensive and hazardous. Because it is not entirely stable it will deteriorate with the introduction of gums and resins in storage over a period of time. There is also the possibility of condensation and resulting water in the fuel which is detrimental to good operation. The danger of fire is always present because of leaks in the system. Finally, the increased production of gasoline and its distribution has made it a target of increasing taxation, making it economically prohibitive in many installations.

Gas A gaseous fuel system using natural gas, LPG, or sewage gas may be a simple manually controlled system such as a gasoline engine or a carefully engineered automatic system. The basic gas carburetion system consists of a carburetor and pressure regulator mounted on the engine. A gas distribution system, like a water supply system, must be at some designated pressure and flow so that a field pressure regulator is required. The characteristics of this regulator will depend upon the gas analysis, the displacement of the engine, the speed range, and local regulations. A typical schematic of a gas fuel system is shown in Fig. 5. The location of the regulator is generally under the jurisdiction of the gas utility that supplies it. In most cases a single field regulator is all that is required, but at times this can cause problems. The installation of future gas-burning equipment used intermittently may cause gas pressure regulation not compatible with the small amount required for pilot lighting. A single regulator installed some distance from the engine could result in hard starting because the engine vacuum is not sufficient for a full gas flow. To eliminate this problem the initial regulator is set at a higher pressure in order to give a readily available supply of fuel for all devices.

Natural gas has an average heating value of 800 to 1000 Btu/ft^3. Commercial butane has a value of 2950 Btu/ft^3, and propane a valuc of 3370 Btu/ft^3. Commercial LPG fuel, which is a mixture of the two, varies in both amount and corresponding heating value.

LPG fuel is produced by mechanical and compression processes, and the methods of distribution and handling must meet regulations. Natural gas is usually supplied under moderate pressures seldom exceeding 50 lb/in^2, while LPG fuel is supplied as a liquid under pressures as high as 200 lb/in^2 in warm weather. As a result, natural gas will readily mix with air and burn, whereas LPG fuel must be changed from a liquid to a vapor with the addition of heat as shown in Fig. 6.

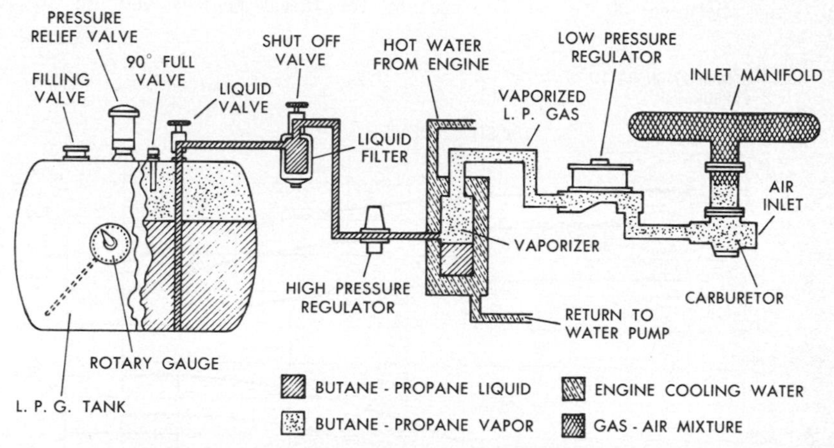

Fig. 6 Typical LPG fuel system. (Waukesha Motor Co.)

As a natural process in the modern waste-treatment plant, sewage gas may be produced in the sewage digester. This gas has the same basic qualities as natural gas, being about 65 to 70 percent methane. It has a heating value of 550 to 700 Btu/ft^3. The same basic carburetion system as natural gas is used. This gas contains inert substances, particularly hydrogen sulfide or free sulfur, which in the presence of free moisture or moisture resulting from combustion will form sulfurous acid, which is corrosive and thus damaging to the valves, pistons, and cylinder walls of an engine. An engine can tolerate from 10 to 30 g of sulfur/100 ft^3. Beyond this a filtering system to remove the sulfur and moisture is advisable.

Diesel The diesel engine over the years has been used for larger power systems. The initial cost of the system is justified to an extent by lower cost of the fuel. The better fuel economy of the diesel engine and its torque characteristics also are an important factor in the selection of this fuel for many applications. Diesel fuel has one distinct advantage in its safety from fire; it does not form the dangerous fuel vapors that other fuels do. It does require, however, a good fuel-filtering system because of contaminants in the fuel which can create problems for the precision design of the fuel-injection system.

Commercial diesel fuels are the residue that remains after the more volatile fractions of crude oil are removed. The heating value is generally about 19,000 Btu/lb. In the diesel engine the fuel is injected into the cylinder at the end of the compression stroke in an atomized form. The compression stroke results in a temperature sufficient to ignite the fuel without the use of any ignition device. Although fuel systems from different engine manufacturers vary, the basic components are the same.

The larger diesel engines may be designed to operate on a dual fuel system. The engine operates on a gaseous fuel with a pilot injection of diesel fuel for ignition. In case of a loss of the gaseous fuel supply the engine will convert to 100 percent diesel fuel.

Relative-Performance Curves Typical performance curves showing the comparison between horsepower, torque, and part-load fuel economy are shown in Fig. 7.

COOLING SYSTEMS

Cooling is essential to all internal combustion engines because only a small portion of the total heat energy of any fuel is converted into useful energy. The remainder is dissipated into the coolant, exhaust, lubricating oil, and by radiation. A hypothetical heat balance is shown in Fig. 8.

Specific data and recommendations on cooling requirements will vary with each engine manufacturer. In general the heat rejection to an engine cooling system will range between 30 to 60 Btu/hp/min for diesel engines and up to 70

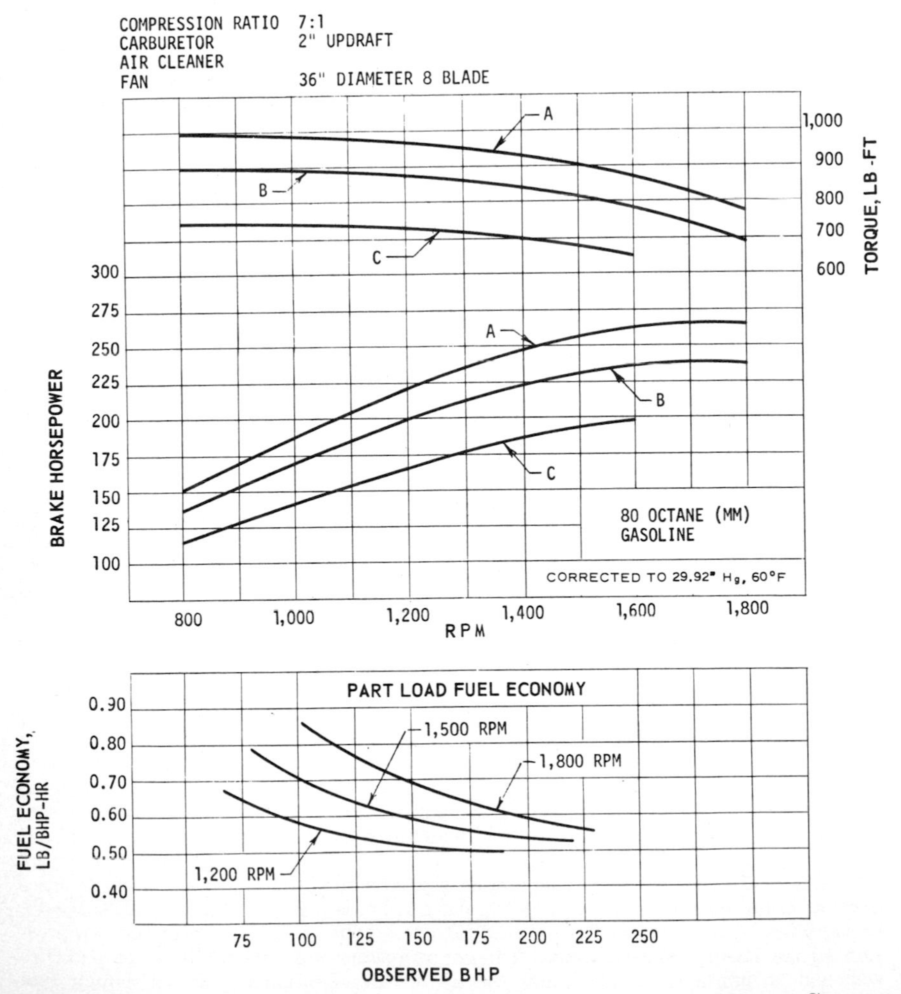

Fig. 7a Gasoline engine-performance curves: A—maximum, B—intermittent, C—continuous ratings of engine with accessories. (Waukesha Motor Co.)

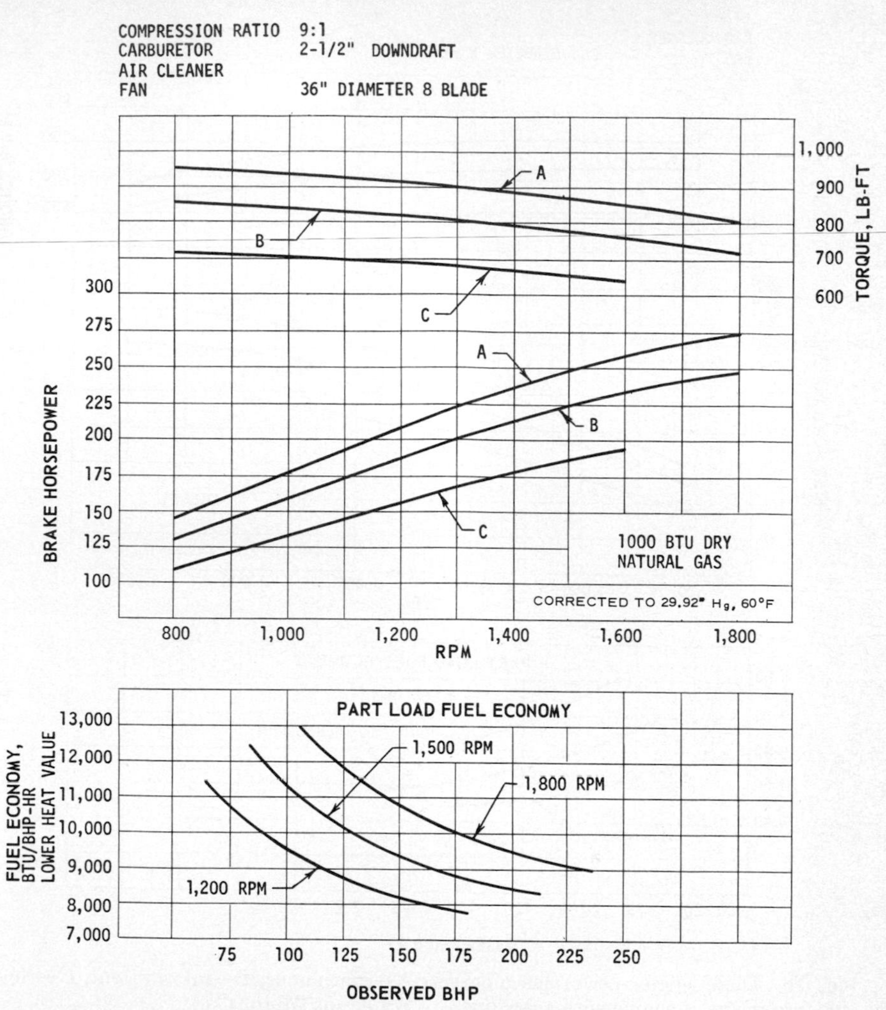

Fig. 7b Gasoline engine-performance curves: A—intermittent, B—intermittent, C—continuous ratings of engine with accessories. (Waukesha Motor Co.)

Btu/hp/min for natural gas and gasoline engines. This heat must be transferred to some form of heat-exchange medium.

In designing any cooling system certain factors must be considered:

1. Additional heat from driven equipment, such as the cooling of reducing or speed-increasing gears where the engine coolant is the medium.
2. Water-cooled exhaust manifolds on the engines or water-cooled exhaust turbochargers or after coolers.
3. High ambient temperatures or heat from nearby processing equipment.
4. A variation in the heat-exchanger coolant temperature.
5. Entrapment of substantial quantities of air in the coolant water.
6. Inability to maintain a clean cooling system.

With any cooling system one of the most important factors of design is the temperature drop across the engine. Most engine manufacturers desire a temperature differential of no more than 10 to 12°F, and closer values are desired. The jacket-water temperature across the engine of 170°F is preferred and in high-temperature

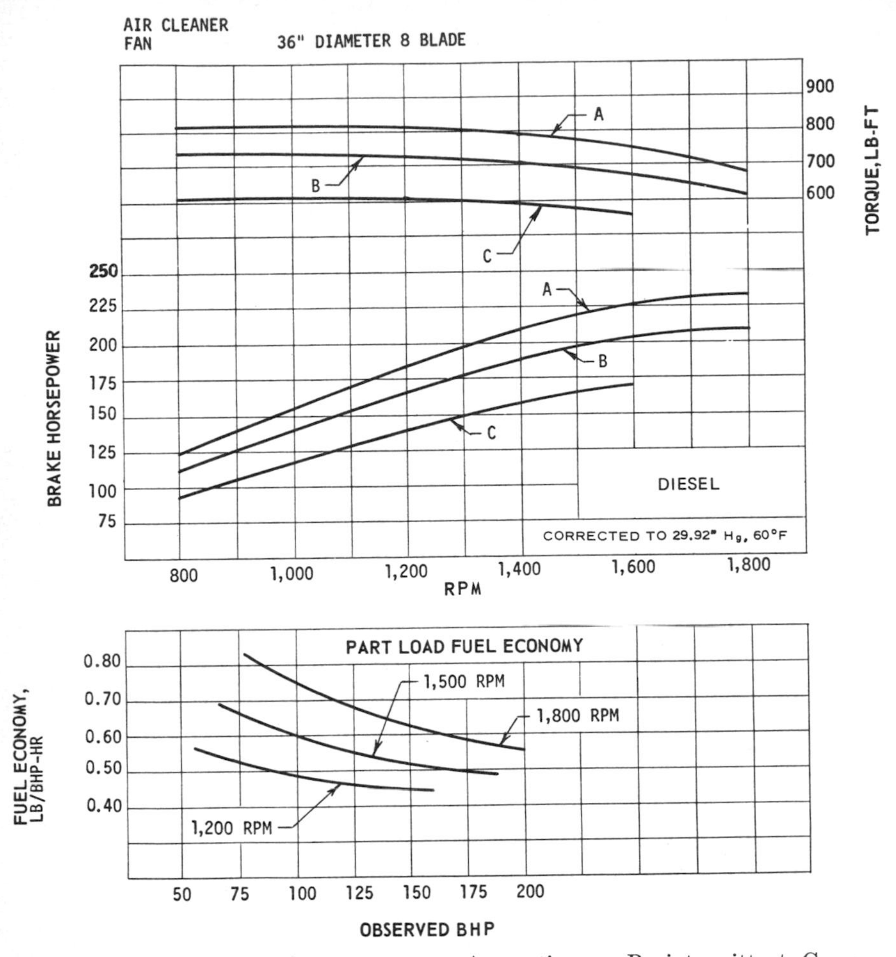

Fig. 7c Diesel engine-performance curves: A—continuous, B—intermittent, C—continuous ratings of engine with accessories. (Waukesha Motor Co.)

or waste-heat-recovery systems temperatures of 200°F or more are common and not harmful.

Radiator The radiator cooling system is perhaps the most common and best understood method of cooling. It is based on a closed system of tubes through which the jacket water passes. The heat is dissipated by a fan creating a stream of moving air passing through the tubes. The fan is either driven by the engine or by an auxiliary source of power (Fig. 9).

An engine in a fixed outside installation can be cooled without much difficulty. Certain factors must be considered, such as ambient temperature, direction of the prevailing wind, and presence of foreign airborne materials. In high temperatures (usually over 110°F) a larger radiator is required. If the prevailing winds are extremely high the unit can be located to offset normal fan flow. Screening can be used to prevent the clogging of the air passes in the radiator where the atmosphere tends to have foreign airborne material such as dust.

Radiator cooling may be used in an inside installation, but there are certain problems which unless properly anticipated limit this system. The recirculation of cooling air and the radiation of exhaust heat from the engine create a problem. As was previously pointed out, a rise of every 10° above 60°F results in a loss of

1 percent in power. When 5 to 10 percent of the total heat put into an engine is radiated, it means that some means of power ventilation must be provided to remove this heat through ducts, louvers, or forced ventilation with a separate fan. An installation of this type is illustrated in Fig. 10.

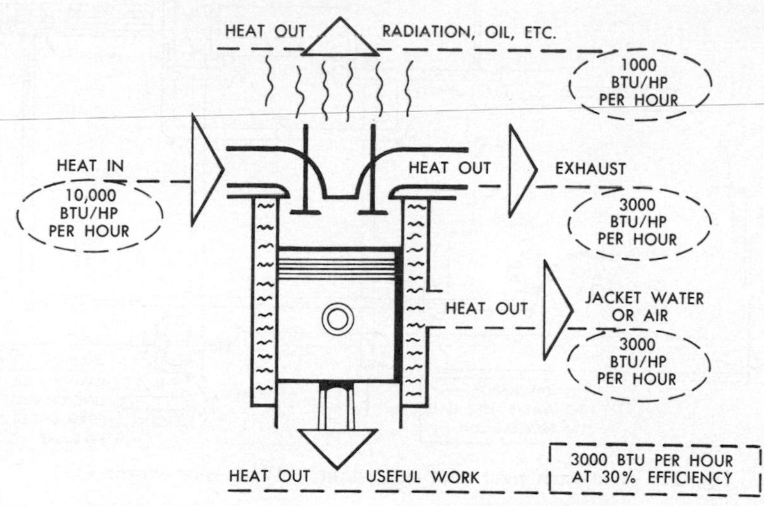

Fig. 8 Hypothetical heat balance. (Waukesha Motor Co.)

Heat Exchanger The heat-exchanger cooling system (Fig. 11) is the best system for a stationary-engine installation. Using a tube bundle in a closed shell, the cooling exchange medium is water (often called raw water). This water may be plant or process water; it may be recirculated or in standby installations allowed to pass to waste. The system has the advantage of the radiator cooling system in that it is self-contained: the quantity and quality of the water within the engine can be

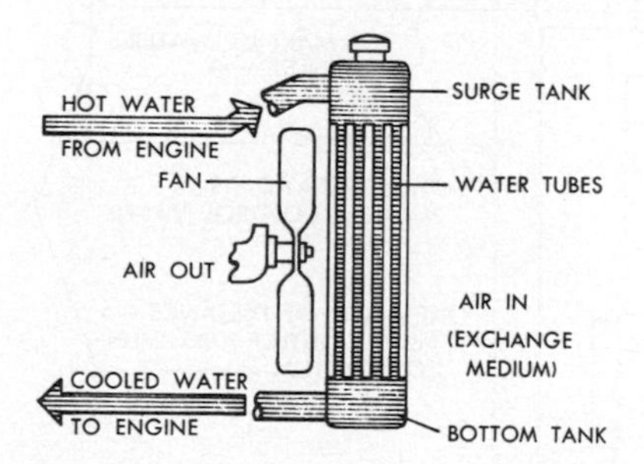

Fig. 9 Radiator cooling system. (Waukesha Motor Co.)

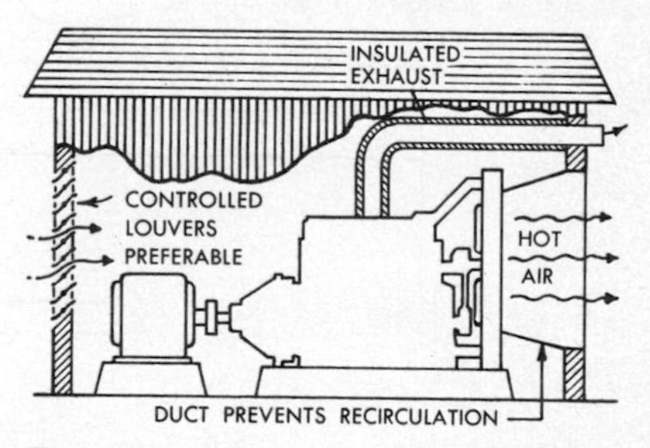

Fig. 10 Radiator cooling system for an inside installation. (Waukesha Motor Co.)

controlled. It has a further advantage in that it is not affected by the flow of heat to air movement if the heat of radiation is taken into account in the design of the system. On the other hand, the cooling medium, unless used in a plant system, is a disadvantage because it is costly. A separate pump is required to provide the necessary water for cooling unless the city water or process water is under sufficient pressure.

City Water and Standpipe City water cooling is designed to take water directly from the city main or from the pump the engine is driving. It is used on some emergency or standby installation. It is simple and inexpensive, gives unlimited cooling for moderate-size engines, is easily understood, and will operate instantly

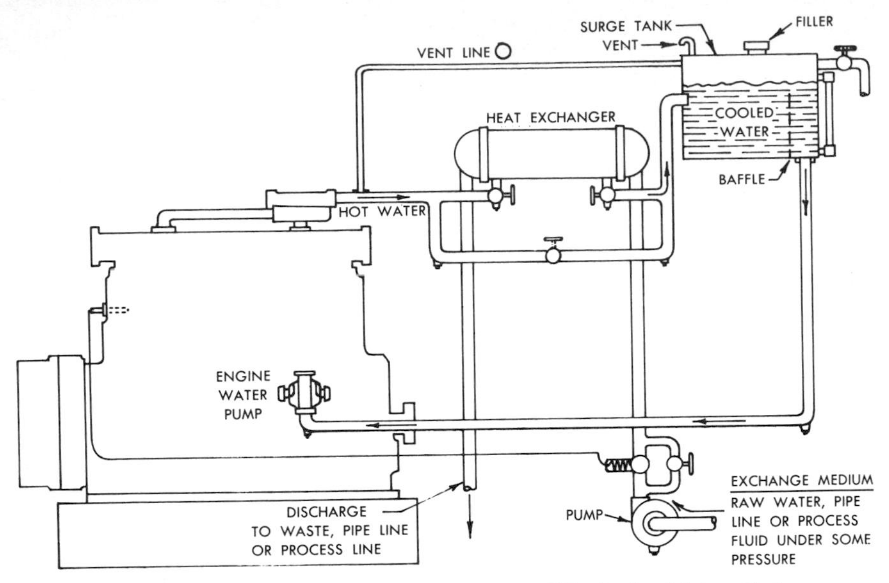

Fig. 11 Heat-exchanger cooling system. (Waukesha Motor Co.)

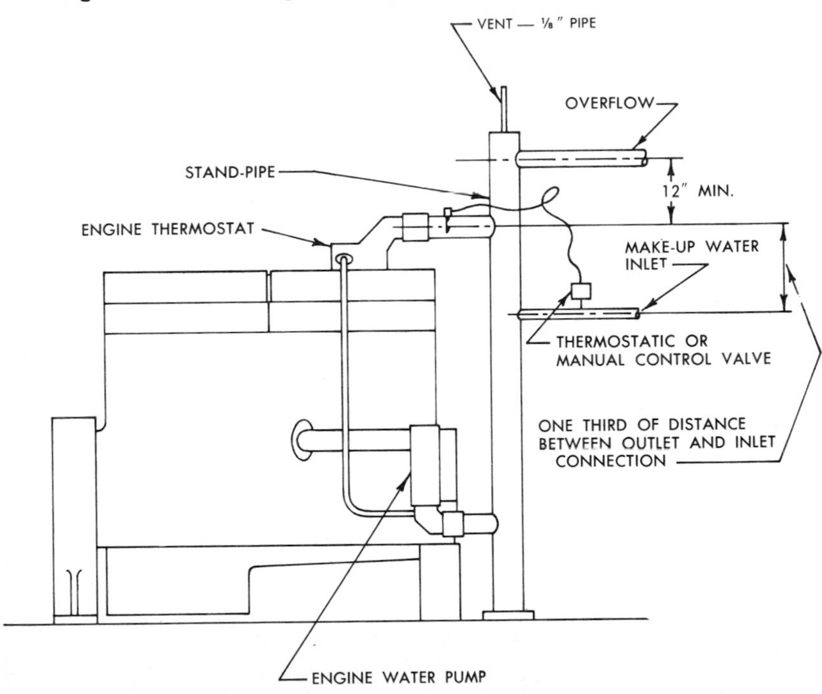

Fig. 12 Standpipe cooling system. (Waukesha Motor Co.)

under emergency conditions. On the other hand, the cooling water is wasted, corrosive elements may be introduced into the engine-jacket-water system, and it may create excessive temperature changes across the engine jacket.

Standpipe cooling as shown in Fig. 12 is basically the same as city-water cooling except that a thermostatic valve is employed to admit makeup water as required.

The vertical pipe as shown is a blending tank into which city water is introduced only in the amount necessary for makeup. It is an inexpensive and simple system to operate.

Ebullition In installations where heat is required for process equipment a method of high-temperature or ebullition cooling is being used as a very economical method, particularly with larger installations. This system has been termed *steam cooling, high-temperature,* or *Vapor-phase* (a registered trademark). In this system the coolant leaves the engine at a temperature equal to or above the atmospheric boiling point and under sufficient pressure to remain as liquid until discharged from the engine into a flash chamber, where a drop in pressure causes the formation of steam which is condensed and returned to the engine at very near the discharge temperature. A schematic of this system is shown in Fig. 13. This system has the advantages of a very small temperature differential across the engine, thus minimizing

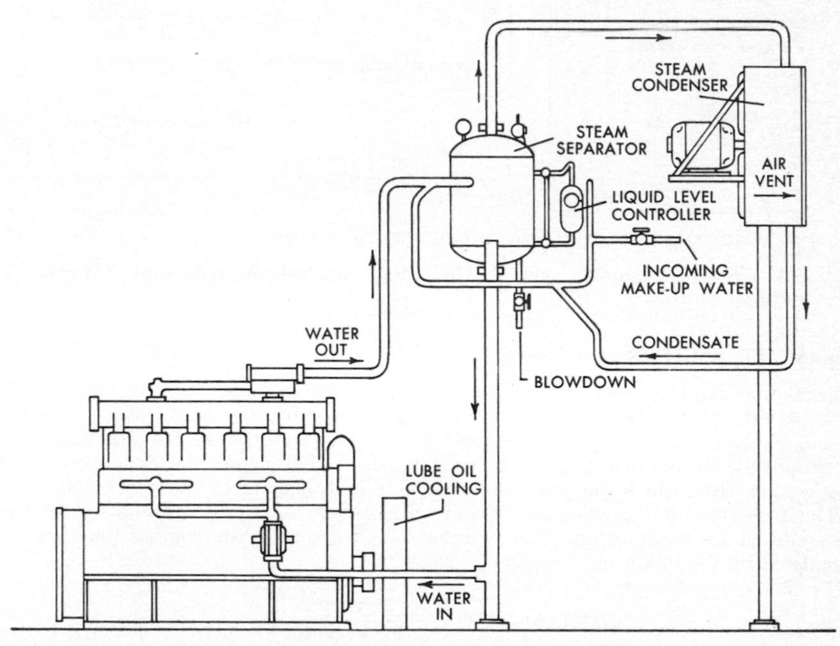

Fig. 13 High-temperature cooling system. (Waukesha Motor Co.)

distortion of all working parts, and a constant working temperature regardless of the load. Because of the higher working temperatures the combustion area and crankcase of the engine have fewer liquid byproducts of combustion and corrosive materials. Of prime importance is the waste heat that can be recovered for plant process with a very small amount of makeup water for cooling.

Cooling Tower Cooling towers are used in some large or multiple engine installations. Through the use of a current of air either produced by a natural draft or mechanical means a tower causes a sensible heat flow from the cooling water to the air. Atmospheric or natural draft towers depend upon natural wind velocities and thus can vary widely. Mechanical draft towers where the air supply can be controlled can be put in any given area, but the limit to the cooling capacity is the horsepower required to operate it. As the water volume increases, the volume of air required and the pressure the fan has to operate against increases. A point is reached where the cost of installation and operation becomes prohibitive. A typical schematic diagram of a system that combines components of the cooling systems mentioned plus the waste-heat recovery system silencer to be mentioned later is shown in Fig. 14.

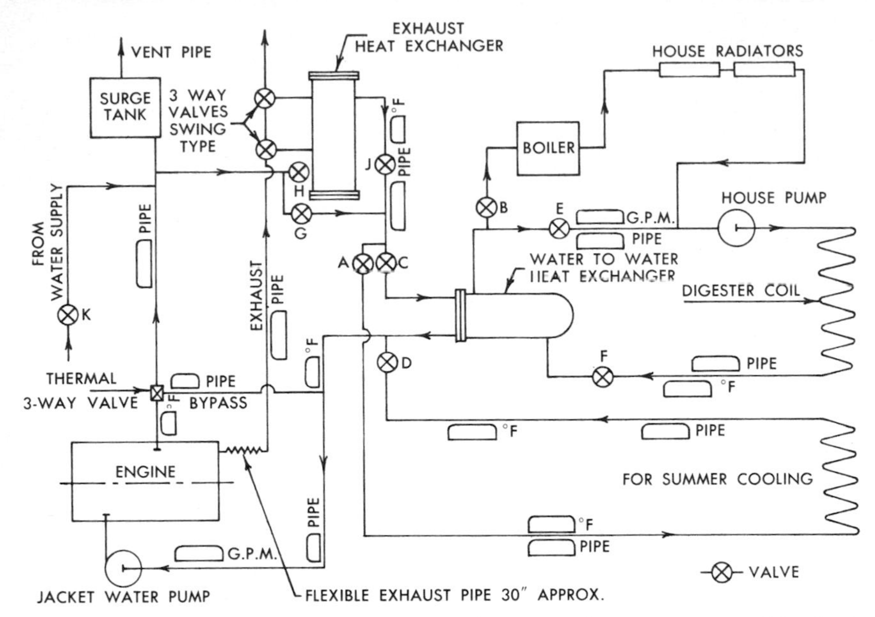

Fig. 14 Complete cooling system including waste-heat recovery. (Waukesha Motor Co.)

AIR-INTAKE SYSTEMS

A most important consideration in the application of an engine to any pump drive is its ability to breathe. As air is required for combustion the design engineers of any project must take into account the necessary provisions for this air. The environment, the service usage, the speed range of the engine, the duty cycle, and the location from which the combustion air is to be taken are of vital importance.

The first step in the selection of any intake system is a determination of the air required for combustion. The volume of air required for combustion may be calculated by the following formula:

$$\text{Required air} = \frac{B^2 \times S \times \text{rpm} \times N}{2{,}200 \times K}$$

where B = cylinder bore, in
S = piston stroke, in
rpm = engine speed, rpm
N = number of cylinders
K = constant: 2-cycle, one; 4-cycle, four

Volumetric efficiency will vary with engine design, but an average of 80 percent may be used to determine the air-cleaner size.

The two basic cleaners available are either wet or dry. The wet or oil-bath cleaner consists of either an oil wire mesh or an oil bath through which the air must pass. A dry-type cleaner uses an element of paper or cloth which will trap the dust, lint, etc., but still allow the air to pass through.

The installation of a suitable air cleaner is important. In cases where adequate air can be supplied through proper ventilation of the area surrounding the engine, it is best to mount the cleaner on the engine. If it is necessary to bring air to the engine from outside the area of the building certain design factors must be considered. The pipe connections from an outside cleaner should be tight and mechanically strong, and fabric hose should not be used unless the length is relatively short. The air to the engine should not be heated by close proximity to the engine or any other heating device because as previously mentioned there is a loss in power with air temperatures above 60°F. There should be no restrictions in the piping system caused by sharp bends in the piping system. Finally, methods should be provided with any outside air cleaner to prevent the entrance of moisture, such as rain or snow.

The introduction of supercharging or turbocharging of engines has introduced an additional problem to the air-intake system. Because of the design and operation of the turbocharger, an air-cleaner system has to be employed to remove the impurities in the air that would be detrimental to the efficiency of the turbocharger. Because of the increased air requirements a larger air-cleaner system must be used.

EXHAUST SYSTEMS

The removal of the engine exhaust is important when it is realized that an engine consumes a large volume of air for combustion and that volume must be removed after combustion. It is necessary that the exhaust back pressure be kept at a minimum while this volume is being removed. The exhaust piping should be properly sized and long sweep elbows should be employed if necessary. Unless the back pressure is kept low the following conditions can result:

1. Loss of power
2. Poor fuel economy
3. High combustion temperatures with increased maintenance
4. High jacket-water temperatures
5. Crankcase sludging with resulting corrosion and bearing wear

All internal combustion engines create noise. Depending upon the location of the engine, this noise can be a problem. Normally an unmuffled engine when measured from a distance of about 10 ft creates a decibel noise level ranging from 100 for the small- and medium-size engine to 125 for the larger engine. Thus most engine installations will incorporate some form of an exhaust silencer, or muffler. Depending upon the degree of silencing, a muffler will reduce the unmuffled decibel reading by 30 to 35 dB. Various types are manufactured to meet the required conditions and are classified as follows:

1. Standard or industrial
2. Semiresidential (high-degree)
3. Residential or hospital (supercritical)

The basic exhaust silencer is designed as either a dry type or wet type. The latter is used in installations such as sewage- or water-treatment plants where the recovery of heat for plant processes is important. Basically this type may be classified as a low-pressure boiler. Water is admitted to the silencer through tubes or coils to pick up the heat from the exhaust. As was shown in Fig. 8, approximately 30 percent of the total heat input into the engine is exhausted. The wet-type silencer is designed to regain about 60 to 70 percent of this heat. This silencer has an advantage in its ability to operate either wet or dry. Thus when heat is not required it may be operated dry and vice versa.

STARTING SYSTEMS

Direct Starting methods for engines fall into two broad categories, either direct or auxiliary. The direct system, used primarily with large engines, employs some means of exerting a rotating force on the crankshaft, such as the introduction of high-pressure air directly into the cylinders of the engine. A direct system on small air or water-cooled engines using either a rope or a hand crank has to a large extent been discarded in favor of an auxiliary method.

Auxiliary The auxiliary system employs a small gear which meshes with a larger gear (ring gear) on the engine flywheel. The ratio of the number of gear teeth on the large gear to those on the small gear is called the cranking ratio. Generally the larger the ratio the better the cranking performance will be. The auxiliary system uses several means to drive the small gear:

1. Electric motor
2. Air motor
3. Hydraulic motor
4. Auxiliary engine

Electric Motor. The electric motor may be either dc or ac. The dc motor most commonly used is available in 6, 12, 24, or 32 volts. The voltage size will depend upon the size of the engine, the ambient temperature, and the desired cranking speed of the engine. The dc system requires a source of outside power usually in the form of a storage battery. To assure prompt starting a charging system for the battery is required either in the form of a charging generator driven by the engine or an ac-powered battery trickle charger. The latter is recommended for standby installations where an engine-driven charging generator functions only when the engine is operating. During idle periods the battery will lose its charge unless maintained by a trickle charger. In recent years there has been a trend toward the use of an ac generator, or alternator, which has the advantages of small size, higher voltage and amperage, competitive price, and good charging ability under idle speed conditions.

Another type of electric motor is a line voltage starter available in 110, 220, or 440 volts ac. It has the advantages of faster and more powerful cranking, the elimination of the battery and a charging system, less maintenance, and sustained cranking through unlimited available electric power. It does have disadvantages in a higher initial cost, the requirement of high line voltage at the site, the danger to personnel due to the high voltage, and the requirement to conform to existing wiring and installation codes.

Air Motor. The air motor, which is usually of the rotary-vane type, uses high-pressure air in the range of 50 to 150 lb/in² to turn it in starting the engine. It is mounted on the engine flywheel housing to mesh with the gear on the flywheel in the same manner as the electric motor. An outside source of air from an air compressor, usually with a 250 lb/in² capacity, is required. A pressure-reducing valve is installed in the line to the engine. The high-pressure air stored in an adequate receiver is sufficient for several starting cycles. This starting system has the advantages of faster cranking, sustained cranking as long as the air supply lasts, suitability to operate in hazardous locations where an electric system might be dangerous, and ability to operate on either compressed air or high-pressure natural gas. It has certain disadvantages in a higher initial cost, the requirement of an air-compressor system, and, finally, a shutdown condition if the air supply is depleted before the engine starts.

Hydraulic Motor. The hydraulic motor system consists of the motor, an oil reservoir, an accumulator, and some means of charging the accumulator. The accumulator, which is a simple cylinder with a piston, is charged on one side with nitrogen gas. As the hydraulic fluid, usually oil, is pumped into the other side the gas is compressed to a very high pressure. When released the fluid turns the motor which in turn rotates the engine. Charging of the system can be accomplished by hand, with an engine-driven pump, or with an electric-motor-driven pump. Generally an engine-driven or electric-motor-driven pump is used in conjunction with the hand pump in case of an engine or electric failure. This system has the same basic advantages of the air motor except that there is no prolonged starting. If the engine is in good operating condition, the cranking is fast and a start is instantaneous, but, if not, it is necessary to recharge the system before another start can be made.

Auxiliary Engine. A small auxiliary air-cooled or water-cooled engine is sometimes employed for starting. It may be mounted on the engine in the same manner as the other systems, or a belt drive may be employed. Some form of speed reduction is required to reduce the higher speed of the auxiliary engine to that required for proper cranking. The principal advantage is a complete independence from outside sources of power such as batteries, air, or pumps, but this is offset by a higher initial cost and the required regular maintenance of the engine.

IGNITION SYSTEMS

The internal combustion engine requires some means to ignite the combustible charge in the cylinder at the proper time. The present high-compression gasoline and gas engines demand a system which will produce a high-tension spark across a short gap in the combustion chamber for the ignition of the charge. It is obvious that the design of the combustion chamber must be such that this combustible mixture of fuel and air be present between this discharge gap when the spark occurs or ignition will not take place.

Ignition systems for gasoline or gas engines are considered as high or low tension. The energy system for the high-tension system is either an électric generator and battery or a magneto. The generator and battery produce a direct current at 6 to 12 volt potential while a magneto produces an alternating current with higher peak voltages. With either energy source this system has a primary circuit for the low-voltage current and a secondary circuit for the high-voltage current.

The primary circuit consists of the battery, an ammeter, an ignition switch, a primary coil, and breaker points and a condenser in the distributor. When the ignition switch and the breaker points close a current flows through the circuit and builds up a magnetic field in the primary coil. Opening of the breaker points breaks this circuit causing the magnetic field to start collapsing. As the field collapses it produces the current to flow in the same direction in the primary circuit and charge the condenser plates. The condenser builds up a potential opposing flow which discharges back through the current. This results in a sudden collapse of the remaining magnetic field and the induction of a high voltage into the secondary winding of the coil. The opening and closing of the breaker points are performed by a cam which is engine driven, usually at half engine speed.

The secondary circuit consists of the secondary coil winding, the lead to the distributor rotor, the distributor, the spark plug leads or wires, and the spark plug. A magneto eliminates the battery but includes in its construction the balance of both the primary and secondary circuits. It may have either a rotating coil and stationary permanent magnets or a stationary coil and rotating magnets. The relative movement of the primary coil winding and the magnets induces an alternating current in the primary circuit, the breaking of which induces a high-voltage current in the secondary circuit.

The development of the modern engine has required many refinements in spark plug design but basically a spark plug consists of two electrodes, one grounded through the shell of the plug and the other insulated with porcelain or mica. The center electrode is exposed to the combustion process. The heat flow occurs from the center electrode to the spark plug shell through the grounded electrode.

Recent developments in ignition systems have produced the low-tension and breakerless systems. The breakerless system has eliminated most of the moving parts in the distributor system. The breaker points in the distributor system are actually a switch which opens and closes the primary circuit of an ignition coil. In the breakerless system the use of solid-state devices provides a switch with no moving parts to wear or require adjustment.

The low-tension magneto system has been developed to reduce electric stresses in the ignition circuit. The secondary coil has been removed from the magneto itself and relocated near each spark plug. The low voltage generated by the magneto is transmitted through the wiring harness to these coils. These coils then step up the voltage which is transmitted through short leads to the spark plug. These leads may be insulated to withstand the stresses imposed upon them. This results in a minimum of electric stresses with a resulting longer life for all components of the system.

As was mentioned previously under fuel systems the diesel engine uses the heat of compression for ignition; thus no auxiliary systems, such as the systems mentioned above, are required. The fuel is injected into the combustion chamber under relatively high pressures through the use of a fuel pump and injection nozzle. This system may be either an individual pump and nozzle for each cylinder, commonly called unit injection, or a multicylinder pump which maintains a high pressure in a common fuel line connected to each injection nozzle. The latter is normally called the common rail system.

ENGINE INSTALLATIONS

Foundation The correct foundation, mounting, vibration isolation, and alignment are most important to the success of any engine installation. All stationary engines require a foundation or mounting base. There are many variations but all basically serve to isolate the engine from the surrounding structures and absorb or inhibit vibrations. Such a base also provides a permanent and accurate surface upon which the engine and usually the pump may be mounted.

To meet these requirements the foundation must be suitable in size and mass, rest on an adequate bearing surface, provide an accurately finished mounting surface, and be equipped with the necessary anchor bolts.

The size and mass of the foundation will depend upon the dimensions and weight of the engine and the pump (if a common base is considered). The following minimum standards should be followed:

1. Width should exceed the equipment width and length by a minimum of 1 ft
2. The depth should be sufficient to provide a weight equal to 1.3 to 1.5 times the weight of the equipment. This depth may be determined by the following formula:

$$H = \frac{(1.3 - 1.5)W}{L \times B \times 135}$$

where H = depth of foundation, ft
L = length of foundation, ft
B = width of foundation, ft
135 = density of concrete, lb/ft^3
W = weight of equipment

The soil-bearing load should not exceed the building standard codes. It may be calculated by the following formula:

$$\text{Bearing load} = \frac{(2.3 - 2.5)W}{B \times L}$$

where W = weight of equipment
B = width of foundation, ft
L = length of foundation, ft

Foundation or anchor bolts used to hold the equipment in place should be of SAE grade No. 5 bolt material or equivalent. The diameter of course is determined by the mounting holes of the equipment. The length should be equivalent to a minimum embedded length of 30 times the diameter plus the necessary length for either a J or L hook. An additional 5 to 6 in should be provided above the top surface of the foundation for grout, sole plate, chocks, shims, equipment base washers and nuts, plus small variations in the surface level. Around the bolts it is a good practice to place a sleeve of iron pipe or plastic tubing to allow some bending of the bolts to conform with the mounting hole locations. This sleeve should be about two-thirds the length of the bolt with the top slightly above the top surface of the foundation to prevent concrete from spilling into the sleeve.

Sole plates running the length of the equipment are recommended for mounting directly to the foundation. Made of at least ¾-in hot- or cold-rolled steel and a width equivalent to the base-foot mounting of the equipment, they will provide a level means of mounting and avoid variations in the level of the concrete. These plates should be drilled for the mounting holes and drilled and tapped for leveling screws, which will permit the plates to be leveled and held during the pouring of grout.

Alignment Although the alignment will vary with the type of engine and the pumping equipment, the basic objective remains the same. The driven shaft should be concentric with the driver shaft, and the centerline of both shafts should be parallel. Rough alignment should be made through the use of chocks and shims. A dial indicator should be used to check deflection by loosening or tightening the anchor bolt nuts until there is less than a 0.005-in reading at each bolt. Shims should be added or removed to arrive at this point. A final check should be made with all the conditions "hot" as the engine and its driven equipment expand at the rate of 0.000006 in/degree above ambient hot to cold. Although the coupling or driving member between the engine and the pump are not discussed in this section, it must be considered in the final alignment.

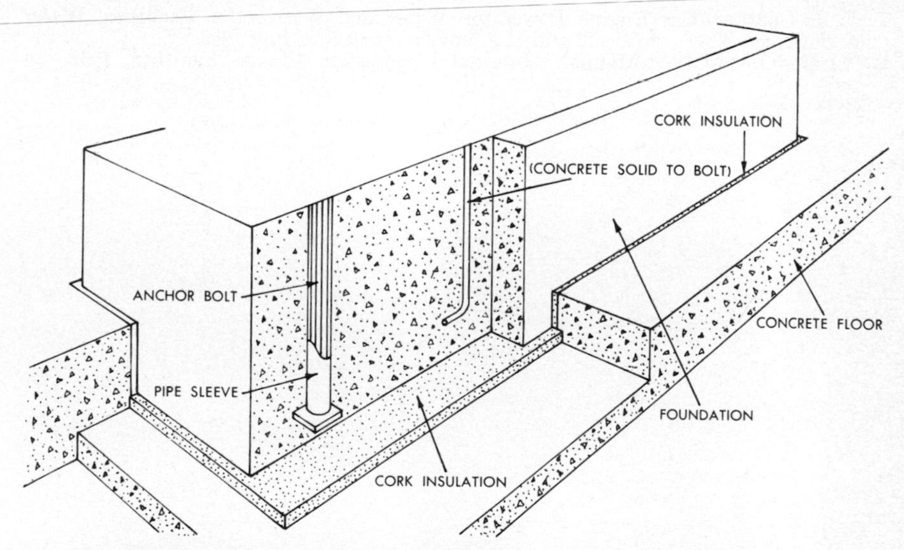

Fig. 15 Engine foundation installation. (Waukesha Motor Co.)

Vibration Isolation It is desirable to isolate the engine and at times the pumping equipment from the building structure because of vibrations. The use of cork as shown in Fig. 15 is used in the larger and heavier installations. A combination of cork and rubber pads may be used at each mounting-hole location on small or medium-size installations, and spring isolators may be used on a complete installation if flexible connections are used for fuel, water, and air connections where required. The manufacturer of the engine and the pumping equipment should be consulted in the use of any isolating material or device.

Vibrations are closely associated with the driving and driven equipment, couplings, and other connections. These linear vibrations may be caused by improper supports of the unbalanced parts. The latter produce a vibration known as torsional. An understanding of this vibration is important since the elimination is the responsibility of the engine and the driven-equipment manufacturer. It is complex and cannot be detected without the use of calculations and special instruments.

The basic concept involves an elastic element such as an engine crankshaft which will tend to twist when any firing impulses are applied. When these forces are removed, the elastic body will try to return to its original position. The driven mass and the connecting elements tend to resist these external impulses. The natural elasticity of the crankshaft and its connecting system allow a small amount of torsional deflection and will tend to reduce the deflection as the external impulses reduce in force. Other reciprocating forces within the engine and external forces in the driven equipment may excite vibrations in the entire system. When all these forces come into resonance with the natural frequency of the entire system, the torsional vibrations will occur. These vibrations may or may not be serious, but being complex cannot be solved hastily.

The engine and pump manufacturer designs and constructs the product so that critical harmonic vibrations will not be present under normal speeds and loads. However, there is no way to control the combination. An analysis of the complete system should be made. This requires a study of the mass elastic system of the combination involving the mass weight and radius of gyration of all rotating parts. This study should be made either by the engine or pump manufacturer or by a torsional-analysis specialist.

REFERENCES

1. F. J. Gunther, Gas Engine Power for Water and Wastewater Facilities, *Water & Sewerage Works,* vols. 112 and 113, November 1965 to July 1966.
2. "Engine Installation Manual," Internal Combustion Engine Institute, 1962.

Section **6.1.4**

Hydraulic Turbines

WARREN G. WHIPPEN

PUMPING APPLICATIONS

In many pumping applications the liquid remains at a high pressure after it has completed its cycle. It is often economically desirable to recover some of the energy, which will otherwise have to be dissipated when bringing the liquid back to a lower pressure. Instead of running the liquid through a pressure-reducing valve and therefore destroying the energy, a hydraulic turbine may be installed, and this turbine can therefore assist in driving the process pumps. There are many such applications in use. Among these are the use of turbine-driven pumps in gas-cleaning operations. Here the solutions from scrubber towers are passed through turbines which in turn drive pumps. Some of the solutions involved in such a process are water saturated with carbon dioxide at a temperature of 40 to 70°F, and potassium carbonate containing dissolved carbon dioxide at a specific gravity of 1.31. A liquid at the much lower specific gravity of 0.84 has been used in power-recovery turbines in glycol-ethylene hydration processes.

One of the oldest applications of turbine-driven pumps was for fire-fighting equipment in mills having a natural head of water. The water was routed through a hydraulic turbine which drove the pump to pressurize a sprinkler system or provide water to the fire-hose connections. Today hydraulic turbines are used to drive pumps that generate fire fighting foam.

At thermal power plants cooling water returning from the cooling towers has been used to drive a hydraulic turbine directly connected to a pump that provides part of the cooling water. Naturally in such an application the power from the turbine alone is not enough to maintain the pumping system and therefore an auxiliary power source is also required.

Turbines using oil pressure have been employed at thermal power plants. Large steam turbines obtain bearing oil and control oil pressure from a main feed pump, which is directly connected to the steam turbine shaft. A small amount of the oil at this high pressure is used for the relay control with the majority of the oil going to the bearings at a lower pressure. To reduce the oil pressure from that required by the relays to that needed by the bearings, the oil is passed through a turbine connected to a booster pump, which pressurizes the oil at the intake of the main pump. Since it is preferable to locate pumps and electrical equipment above flood level, some recent steam power plant installations have used the return from the condenser to the river (normally an appreciable drop in head) to drive a hydraulic turbine which in turn may assist in driving the condensate pumps.

It is often economical to use a small volume of water at a high head to move a large volume of water at a low head. Of course, the reverse application can

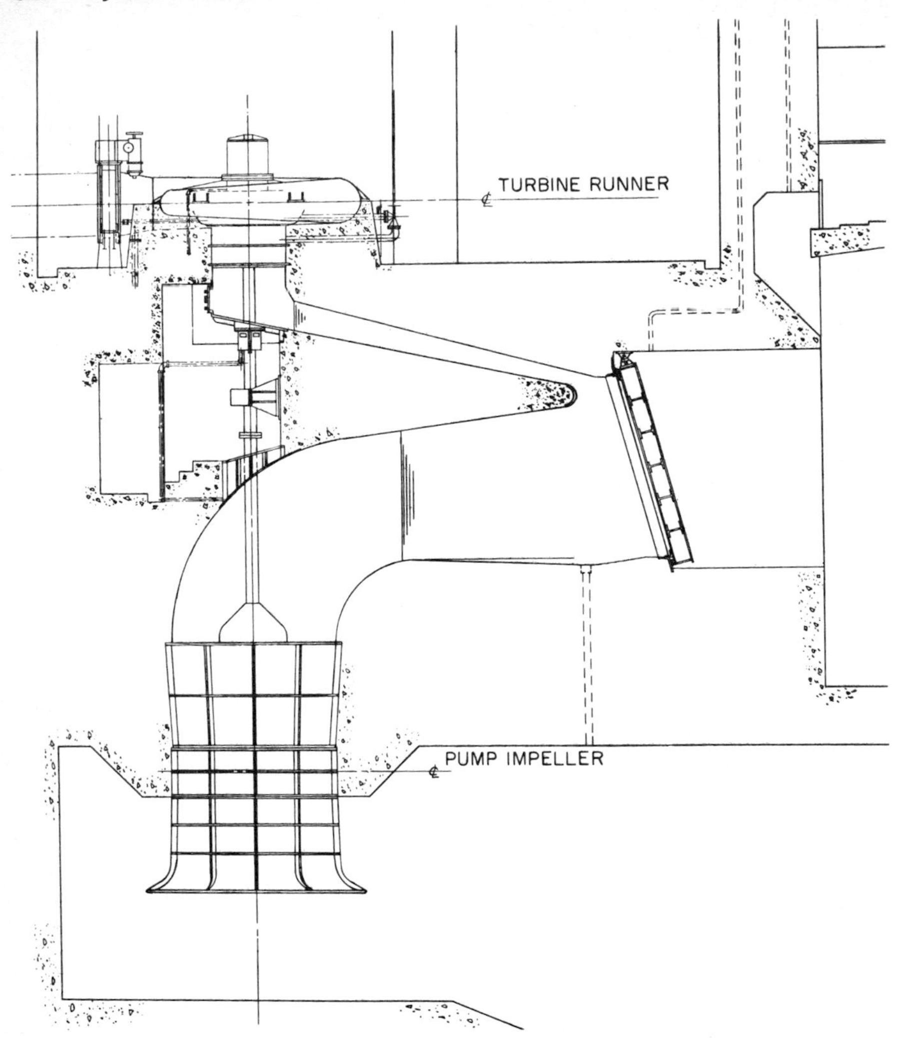

Fig. 1 Turbine-driven fishway pumps.

also be accomplished. The former procedure is employed at hydroelectric projects where it is necessary to operate fishways to enable migrating fish to continue traveling upstream over the dam. The large volume of water is used both to attract the fish to the fish flume and to transport the fish to the fish ladders. An example of such an installation can be seen in Fig. 1.

Turbines have also been used to start large pumping units when the hydraulic conditions are suitable. The impulse turbine, which develops maximum torque at zero speed, is especially useful for this application. The horsepower required on such a starting turbine would be less than the horsepower required on a starting motor to do the same job. Many electric controls can be eliminated when starting with a turbine. An illustration of this application is shown in Fig. 2.

Desalinization plants have been proposed whereby salt is removed from seawater by pumping it through a membrane at high pressure. Only part of the water goes through the membrane, with most of the water being used to carry away the salt. The excess high-pressure water is then put through a power-recovery turbine which helps to drive the high-pressure pumps.

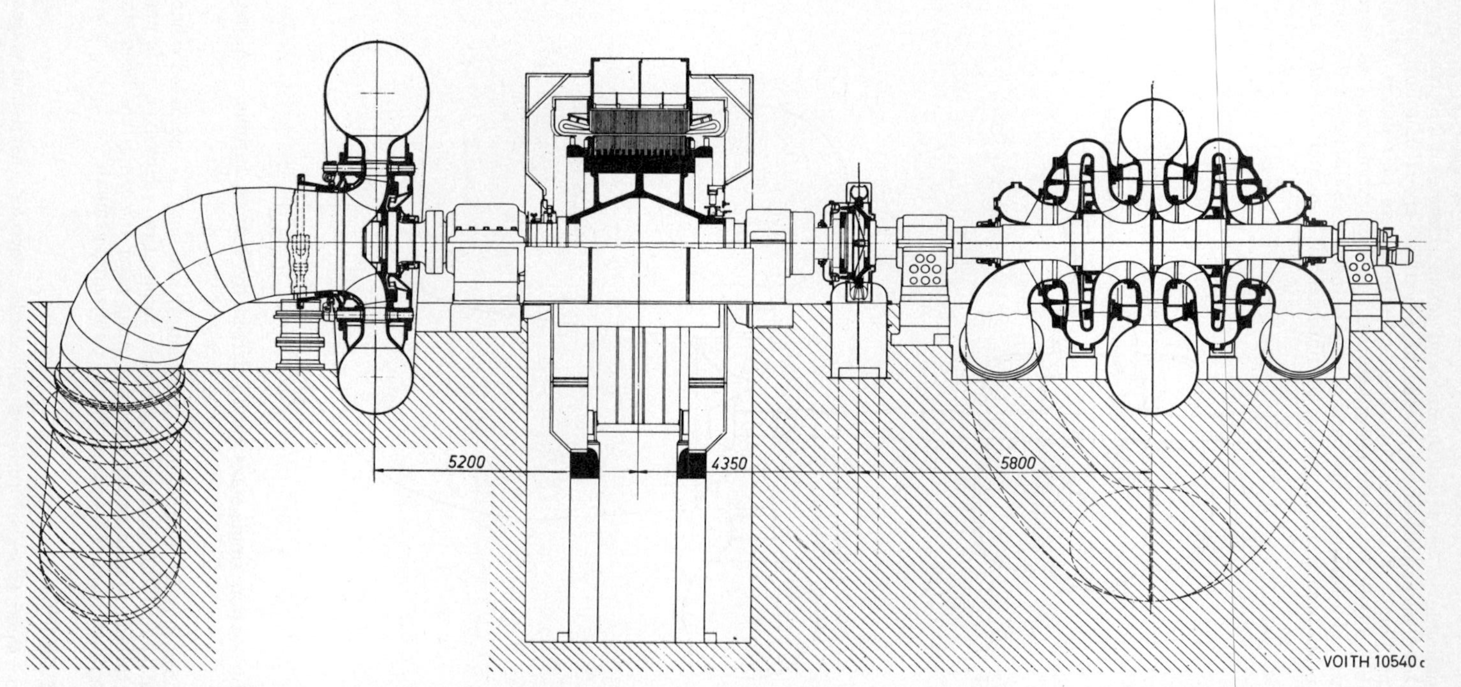

Fig. 2 Impulse runner used to start large pump. Shown from left to right: Francis turbine, generator motor, starting impulse turbine, and two-stage double-suction storage pump. (Adapted from Voith Publication 1429, J. M. Voith GmbH, Heidenheim, Germany)

It is often desirable to have the turbine and pump running at different speeds, and this has been accomplished by means of a variable-speed coupling between the turbine and pump. Excellent low-speed gear increasers or reducers of up to 35,000 hp are being built by many manufacturers. In applications such as starting a pump by means of a turbine, a mechanism may be needed to disengage the turbine after the pump has been started in order to minimize windage and friction. Disen-

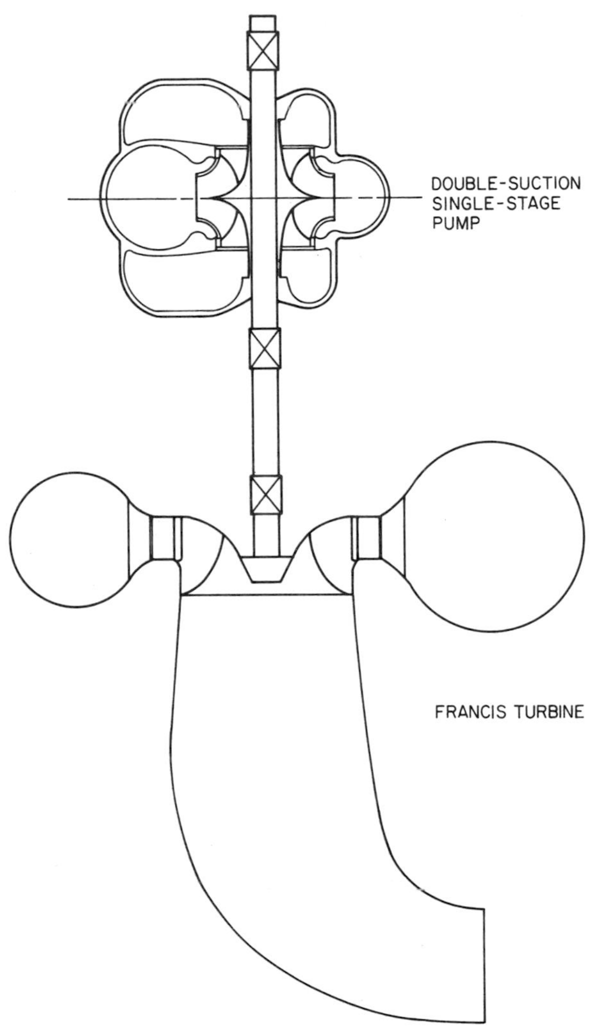

Fig. 3 Double-suction single-stage pump driven by Francis turbine. (Allis-Chalmers Corp.)

gaging couplings may be hydraulic or mechanical and may be actuated when the unit is stationary or rotating, depending on the application. However, it is possible to allow the turbine to spin in air with the pump after the pump has been started. Figure 3 is an illustration of a turbine-driven pump arrangement.

TYPES OF TURBINES

There will be three types of turbines discussed: propeller (fixed and adjustable blade), Francis, and impulse. Figures 4 to 7 are illustrations of these turbine runners. Figure 8 shows a complete adjustable-blade unit. The fixed- and adjustable-blade propellers are essentially the same type with the exception that the adjustable is suited to a much wider range of loading conditions. This versatility is reflected in a higher manufacturing cost for the adjustable blade. The Francis runner is much like a centrifugal pump running backward. The third type is the impulse

Fig. 4 Fixed-blade propeller runner.

Fig. 5 Adjustable-blade propeller runner.

Fig. 6 Francis runner. (Allis-Chalmers Corp.)

runner (or Pelton wheel). The impulse runner is for high-head applications. The water is first channeled through a nozzle which directs a jet of water into the atmosphere surrounding the runner. This jet then strikes the bowl-shaped buckets of the impulse wheel, discharging into the atmosphere. When a large back pressure is present in the system, the impulse wheel cannot discharge properly, and the performance drops off rapidly as the back pressure is increased. However, it is sometimes possible to admit air pressure into the impulse-runner housing and thereby lower the level of the liquid below the level of the runner. When the back pressure cannot be eliminated, a Francis machine will perform much better than an impulse machine. Figure 9 illustrates the relative efficiencies of the different types of runners. The fixed-blade propeller and Francis runners are designed to operate most efficiently

Fig. 7 Impulse (or Pelton) runner. (S. Morgan Smith Co.)

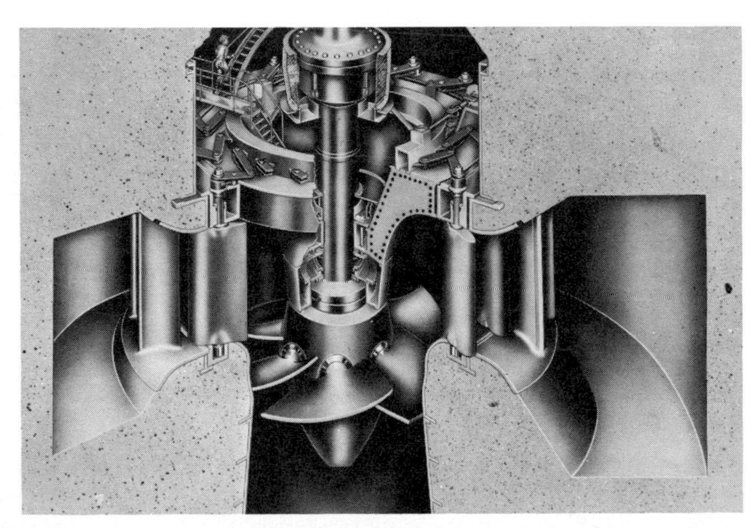

Fig. 8 Complete turbine unit with adjustable-blade propeller. (Allis-Chalmers Corp.)

at a point near full load, and no adjustment can be made when the load drops. The impulse and adjustable blade, however, are designed for high efficiencies over a large load range. The adjustable-blade propeller accomplishes this by changing the angle of its blades by means of linkage in the hub of the runner. With the impulse runner the size of the jet is controlled by the nozzle, thus enabling the runner to maintain high efficiencies at low loads.

Specific Speed (N_s) This is the speed (rpm) at which a unit will run if the runner diameter is such that running at 1 ft net head it will develop 1 hp.

$$N_s = \frac{\text{rpm} \times \text{hp}^{1/2}}{\text{head}^{5/4}}$$

where head is measured in feet. The specific speed is one of the important factors governing the selection of the type of runner best suited for a given operating range. The impulse wheels have very low specific speeds as compared to propellers

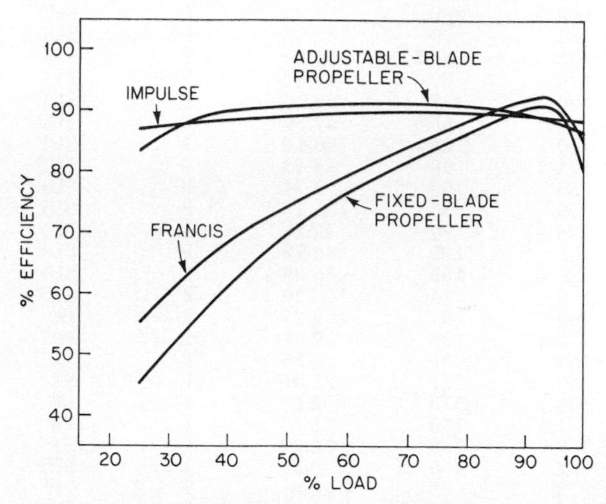

Fig. 9 Efficiency versus load for different runner types.

having high specific speeds. The Francis turbine N_s lies between the impulse and propeller. Table 1 shows some statistics on existing units.

SIZES AND RATINGS

Although power-recovery turbines are of relatively small size, there are much larger turbines in existence which could be used to drive pumps. Turbines of the following sizes do exist: propeller, 405.5-in diameter; Francis, 288-in diameter; and impulse, 176-in diameter. The extreme range of horsepower, head (feet), and rpm is shown in Table 2.

GENERAL CHARACTERISTICS

Sigma As in pumps, the problem of cavitation also exists in turbines. Sigma is defined as follows:

$$\sigma = \frac{H_a - H_s}{H}$$

where H_a = feet of atmospheric pressure minus vapor pressure

H_s = distance in feet the centerline of the blades is above tailwater for vertical units and the distance in feet the highest point of the blade is above tailwater for horizontal units

H = feet of elevation between inlet head and tailwater

TABLE 1 Turbine Statistics of Existing Units

rpm	hp	Head	N_s	Type	Use	Supplier
450	2,000	925	3.94	I	SUP	BLH
450	450	485	4.19	I	SUP	BLH
750	170	575	3.47	I	SUP	BLH
340	570	485	3.57	I	SUP	BLH
1,775	264	1,085	4.63	I	SUP	BLH
437.5	425	405	4.97	I	SUP	BLH
1,770	640	960	8.38	I	SUP	BLH
1,775	230	1,085	4.32	I	SUP	BLH
720	885	460	10.05	I	SUP	BLH
900	640	866	4.85	I	SUP	BLH
3,550	85	1,920	2.58	I	SUP	BLH
690	535	86.5	60.50	F	FW	BLH
126	670	80	13.63	F	FW	BLH
450	2,600	118	58.90	F	SUP	LEFFEL
525	1,050	83.5	67.39	F	SUP	LEFFEL
900	750	123	60.18	F	SUP	LEFFEL
1,000	375	96	64.43	F	SUP	LEFFEL
700	53	150	9.71	I	SUP	LEFFEL
750	160	30	135.13	P	SUP	LEFFEL
882	128	70	49.26	F	SUP	LEFFEL
1,750	145	135	45.78	F	SUP	LEFFEL
1,750	220	135	56.39	F	SUP	LEFFEL
280	200	14	146.20	P	SUP	A–C
550	400	35	129.21	P	IR	A–C
3,450	7.5	196	12.89	I	PP	A–C
1,775	353	1,085	5.36	I	ST	A–C
1,185	280	231	22.01	I	ST	A–C
1,775	300	1,510	3.27	I	CH	A–C
1,300	1.6	150	3.13	I	PP	A–C
700	1,760	798	6.92	I	ST	A–C
600	392	460	5.58	I	SUP	A–C
122	973	65	20.61	F	FW	A–C
108	1,200	75	16.95	F	FW	A–C

Type code: I = impulse, P = propeller, F = Francis.
Use code: SUP = supplement electric motor to drive pump, FW = drives fishway pumps, ST = scrubber-tower application, PP = turbine drives petroleum pumps, CH = power recovery in chemical plant, IR = irrigation project.
Supplier code: BLH = Baldwin-Lima-Hamilton Corp., LEFFEL = The James Leffel and Co., A–C = Allis-Chalmers Corporation.

TABLE 2 Range of Horsepower, Head, Discharge and rpm of Existing Units of One Manufacturer

	Propeller		Francis		Impulse	
	Low	High	Low	High	Low	High
Horsepower	82.5	268,000	1.2	820,000	1.6	330,000
Head (ft)	6.0	180	4.0	2,204	75.0	5,790
rpm	50	750	56.4	3,500	180	3,600

Critical sigma is defined as that point at which cavitation begins to affect the performance of the turbine. Figure 10 shows the relationship between critical sigma and specific speed. This figure also shows which type of runner is best for a given specific speed. The impulse wheel is not affected by sigma since it is a free jet action and therefore not subject to low-pressure areas.

Affinity Laws The relationships between head, discharge, speed, horsepower, and diameter can be seen in the following equations, where Q = rate of discharge, H =

head, N = speed, hp = horsepower, D = diameter, and subscripts denote two geometrically similar units with the same specific speed:

$$\frac{Q_1}{N_1D_1^3} = \frac{Q_2}{N_2D_2^3}$$

$$\frac{Q_1^2}{H_1D_1^4} = \frac{Q_2^2}{H_2D_2^4}$$

$$\frac{N_1^2D_1^2}{H_1} = \frac{N_2^2D_2^2}{H_2}$$

$$\frac{\text{hp}_1}{N_1^3D_1^5} = \frac{\text{hp}_2}{N_2^3D_2^5}$$

Most designs used are tested as exact homologous models, and performance is stepped up from the model by the normal affinity laws given above. Because of

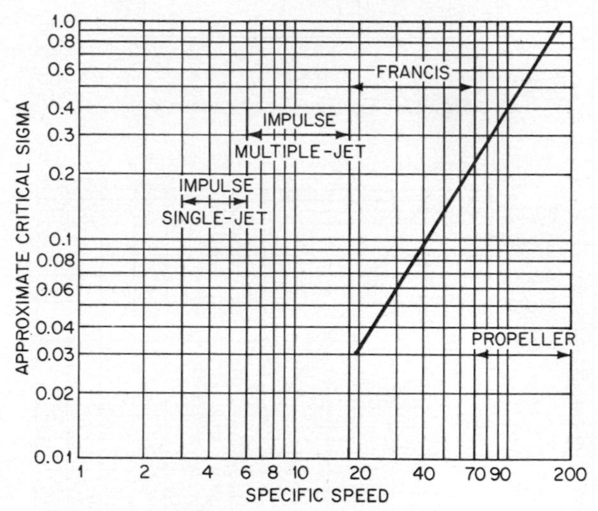

Fig. 10 Critical sigma versus specific speed.

difficulties in measuring large flows at the field installation, only approximate or relative metering is normally done.

Speed Characteristic ϕ is defined as the peripheral speed of the runner divided by the spouting velocity of the water.

$$\phi = \frac{u}{(2gH)^{1/2}}$$

where u = peripheral speed of runner (ft/s)
H = head (ft)

A plot of ϕ versus specific speed can be seen in Fig. 11; ϕ becomes constant at 0.46 after the specific speed has dropped into the impulse-runner region. Theoretically, for maximum energy conversion, ϕ should equal 0.5 for impulse runners. However, because of small losses in the runner, this value is set at approximately 0.46. The control of the hydraulic turbine can be accomplished by means of a governor or a flow or load control system.

Torque All hydraulic turbines have maximum torque at zero speed; therefore they have ideal starting characteristics. The pump can be accelerated to design speed by gradually opening the turbine gates, while keeping within the limitations of hydraulic transients.

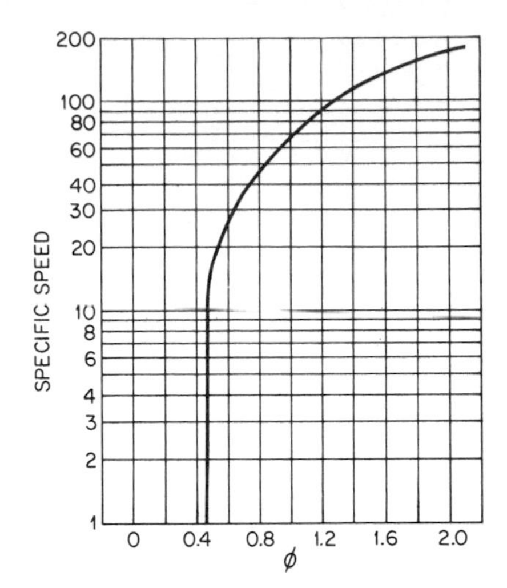

Fig. 11 Specific speed versus ϕ.

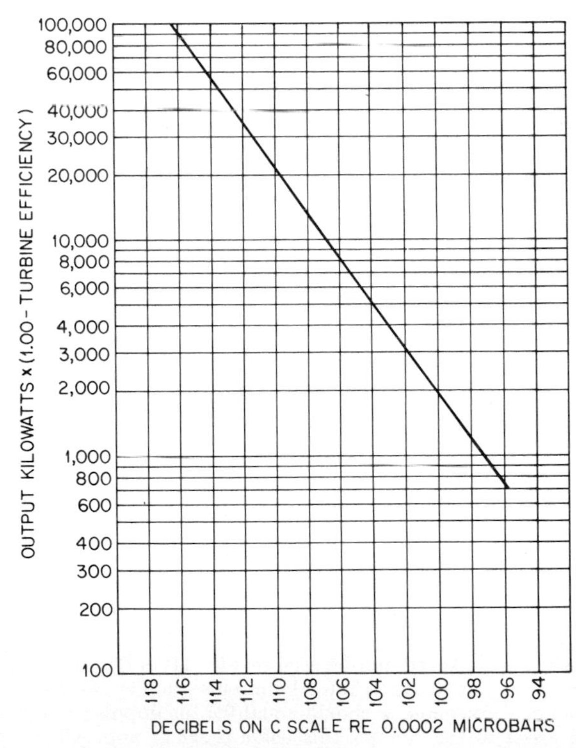

Fig. 12 Sound-level meter readings taken 3 ft from main shaft. Hydraulic losses are approximated by output kilowatts × (1.00 — turbine efficiency). (From Laurence F. Henry. Selection of Reversible Pump/Turbine Specific Speeds)

DATA REQUIRED FOR TURBINE SELECTION

The turbine supplier should have the following information in order to select the best combination of size and type of runner.

1. Head available and head range.
2. Power and speed required to drive the pump.
3. Description of fluid to be handled including chemical composition and specific gravity.
4. Possibility of adding air to the system. The turbine will operate satisfactorily without air but air may be added to system to reduce pressure fluctuations (normally one-third of rpm in frequency) at part load and possibly to smooth unit operation at full load.
5. If a corrosive fluid is to be handled a description of the materials required should be included.
6. What controls from the pumping process are going to affect the turbine operation. If speed is a controlling factor, the possibility of using a speed-varying device should be considered.
7. Back pressure on the turbine.
8. Noise-level limitations for the turbine should be considered. Figure 12 illustrates noise levels which have been recorded on some rather large turbine units.

References 4, 5, and 6 are recommended for additional information on this section.

REFERENCES

1. Henry, L. F.: Selection of Reversible Pump/Turbine Specific Speeds, presented at meeting on Pumped-Storage Development and Its Environmental Effects, University of Wisconsin, September 1971.
2. Wilson, P. N.: Turbines for Unusual Duties, *Water Power*, June 1971.
3. Roth, H. H., and T. F. Armbruster: Starting Large Pumping Units, *Allis-Chalmers Eng. Rev.*, vol. 31, no. 3, 1966.
4. Stepanoff, A. J.: "Centrifugal and Axial Flow Pumps," John Wiley & Sons, Inc., New York, 1957.
5. Daugherty, R. L., and A. C. Ingersoll: "Fluid Mechanics," McGraw-Hill Book Company, New York, 1954.
6. Schlichting, H.: "Boundary Layer Theory," McGraw-Hill Book Company, New York, 1960.

Section **6.1.5**

Gas Turbines

RICHARD G. OLSON

THERMODYNAMIC PRINCIPLE AND CLASSIFICATIONS

The gas turbine is an internal combustion engine differing in many respects from the standard reciprocating model. In the first place, the process by which it operates involves steady flow; hence the elimination of pistons and cylinders. Secondly, each part of the thermodynamic cycle is carried out in a separate piece of apparatus. The basic process involves compression of air in a compressor, introduction of the compressed air with fuel into the combustion chamber(s), and finally expansion

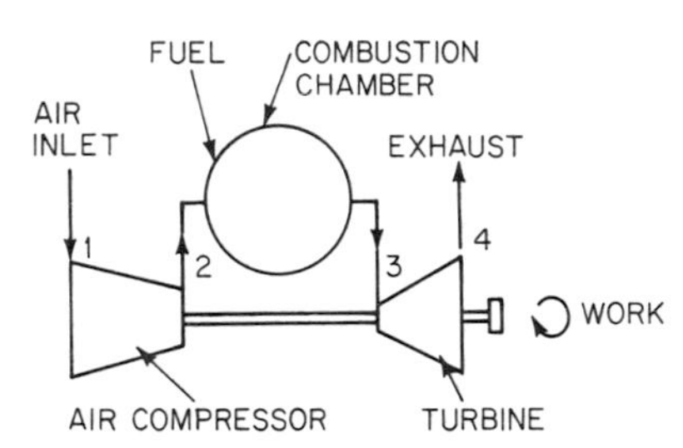

Fig. 1 Components of a simple gas turbine.

of the gaseous combustion products in a power turbine. Fig. 1 illustrates a simple gas turbine.

The Brayton or Joule Cycle Figure 2 shows an ideal Brayton or Joule cycle illustrated in p-v and t-s diagrams. This cycle is commonly used in the analysis of gas turbines.

The inlet air is compressed isentropically from point 1 to 2, then heat is added at an assumed constant pressure from point 2 to 3, followed by isentropic expansion in the power turbine from point 3 to 4, with final rejection of heat at an assumed constant pressure from 4 to 1. From Fig. 2, the compressor work is $h_2 - h_1$, the turbine work is $h_3 - h_4$, and the difference is the net work output. The heat input is $h_3 - h_2$. One can then derive an expression for thermal efficiency as follows:

$$\eta = \frac{(h_3 - h_4) - (h_2 - h_1)}{h_3 - h_2}$$

In reality the cycle is irreversible, and the efficiency of the compression, combustion, and expansion must be taken into account. However, an examination of the equation above shows the need of high compressor and turbine efficiencies in order to produce an acceptable amount of work output. The importance of h_3 is also readily apparent. In noting that h_3 is directly proportional to temperature, one can appreciate why continual emphasis is being placed on the development of materials and techniques to permit higher combustion temperatures and correspondingly higher turbine-inlet temperatures.

Classifications In an *open-cycle* gas turbine the inlet air mixes directly with the combustion products and is exhausted to the atmosphere after passing through the power turbine. The *closed-cycle* gas turbine uses a heat exchanger to transfer heat to the working fluid, which is continuously recirculated in a closed loop. A *combined cycle* uses the principles of both the open and closed cycles. In current practice the combined cycle uses an open cycle to provide shaft work while the

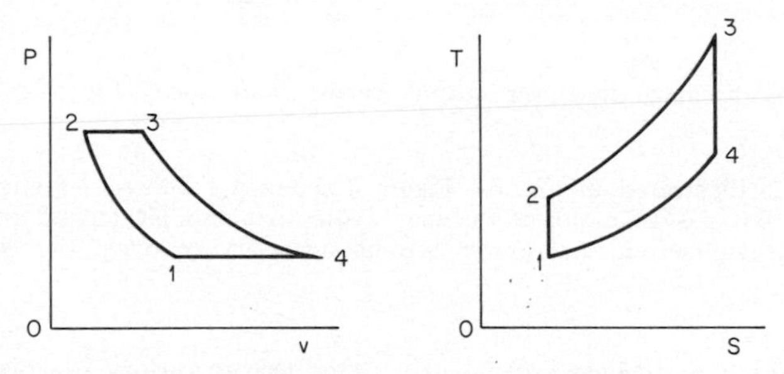

Fig. 2 *P-V* and *T-S* diagrams for an ideal Brayton or Joule cycle.

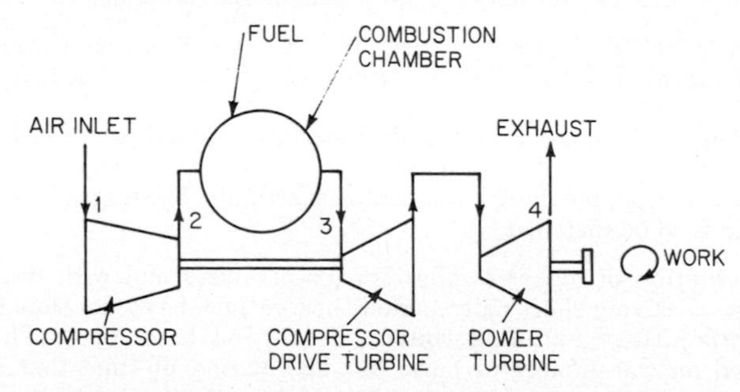

Fig. 3 Components of a split-shaft gas turbine.

heat from the exhaust is partially recovered in a waste-heat boiler. The heat recovered then proceeds through a standard steam power cycle until heat is rejected to the most readily available low-temperature reservoir.

Of the above configurations, the open-cycle gas turbine is most extensively used today for driving centrifugal pumps. This is probably due to the important consideration of minimum capital investment for each horsepower output.

Two distinct types of open-cycle gas turbines have evolved, the *single-shaft* and *split-shaft* versions. The single-shaft gas turbine was developed primarily for the electric power industry and uses a compressor and a power turbine, both integrated on a common shaft. As the unit will be used continuously at a single rotational speed, the compressor and power-turbine efficiency can be optimized.

The split-shaft gas turbine was developed primarily for mechanical drive applications where output power and speed might be expected to be variable. Figure 3 illustrates such a turbine. A typical curve of power output versus shaft speed is illustrated in Fig. 4.

Split-shaft gas turbines are available in the so-called *conventional* and the *aircraft-derivative* versions. The conventional gas turbine evolved from steam turbine tech-

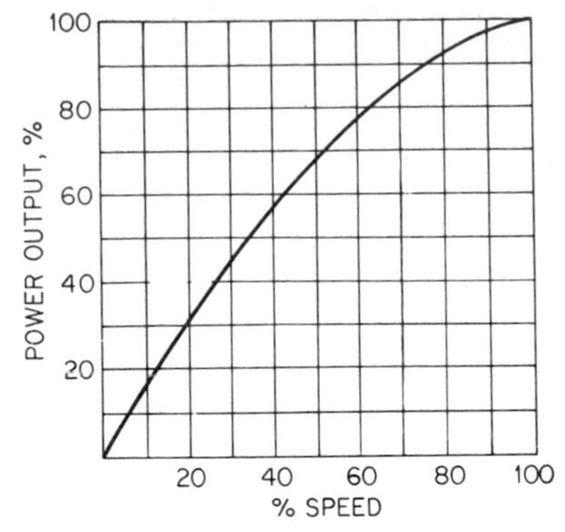

Fig. 4 Typical curve of power output versus shaft speed for a split-shaft gas turbine.

nology and is illustrated in Fig. 5. Figure 6 shows a modified jet engine used as a source of hot gas to a power turbine. Note that the jet engine combines the compression, combustion, and power turbine necessary to drive the compressor.

RATINGS

In the evolution of the gas turbine as a prime mover various organizations have put forth standard conditions of inlet temperature and elevation to allow direct comparison of various gas turbines. Four common standards exist:

ISO (International Standards Organization) Sea level and 59°F (15°C)

NEMA (National Electrical Manufacturers Association) 1,000 feet above sea level and 80°F (27°C)

CIMAC (Congrès International des Machines à Combustion) Sea level and 59°F (15°C)

Site The elevation and design temperature actually existing at the place where the gas turbine is to be installed

With the evolution of higher combustion temperatures and with the greater need for power over relatively short daily periods new ratings have developed: emergency (maximum intermittent), peaking (intermittent), and base load. These classifications are based on the number of hours per unit period of time that a gas turbine is operated and are related to the material used in the power-turbine blading. A common standard is to use materials suitable for 100,000 hr continuous operation. Higher temperatures are permitted, but at the sacrifice of the life of the material and an increase in maintenance costs.

In a pump-driving application, the cycle of operations should be considered in specifying a gas-turbine driver. A typical gas-turbine output curve as a function of inlet temperature (Fig. 7) clearly indicates the necessity of specifying an accurate design temperature. Figure 8 is a typical correction curve for altitude.

FUELS

A wide range of fuels may be burned in simple-cycle gas turbines ranging from natural gas to the bunker oils. In most cases units are available to operate on gas or liquid fuels. Some turbines are available with automatic-switchover capability while under load.

Although it would appear advantageous to choose the cheapest fuel available, there may be disadvantages in such a choice: impurities such as vanadium, sulfur, and sodium definitely result in high-temperature corrosion; solid impurities and a high ash content will lead to erosion problems. One is therefore often faced with a tradeoff between the cost of treating a fuel versus the higher original cost.

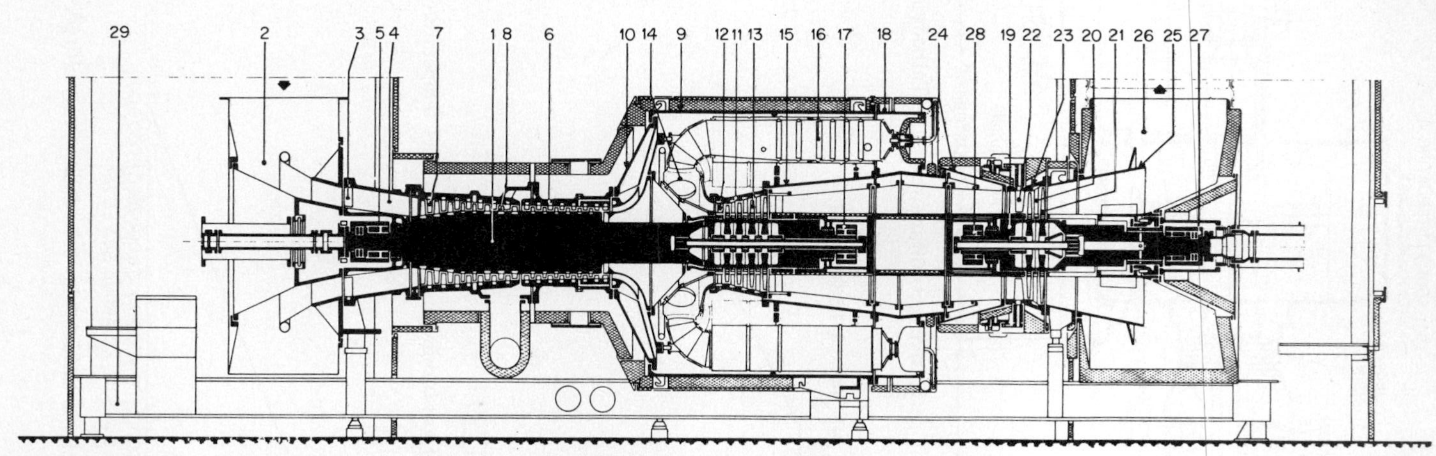

Fig. 5 Gas-turbine cross section. 1, gas-generator turbine rotor; 2, inlet air casing; 3, compressor inlet guide vanes; 4, conical inlet casing with bearing support; 5, compressor radial-axial bearing; 6, compressor stator; 7, compressor stator blades; 8, compressor rotor blades; 9, combustion-chamber main casing; 10, intermediate conical casing with diffusor; 11, gas-generator turbine stator blade carrier; 12, gas-generator turbine stator blades; 13, gas-generator turbine rotor blades; 14, hot gas casing; 15, intermediate casing; 16, nine incorporated combustion chambers; 17, compressor radial bearing; 18, central casing; 19, power turbine rotor; 20, power turbine stator blades; 21, power turbine rotor blades; 22, power turbine adjustable rotor blades; 23, power turbine stator blade carrier; 24, conical intermediate piece; 25, outlet diffusor; 26, outlet casing; 27, power turbine radial-axial bearing; 28, power turbine radial bearing; 29, metallic foundation. (Turbodyne Corp.)

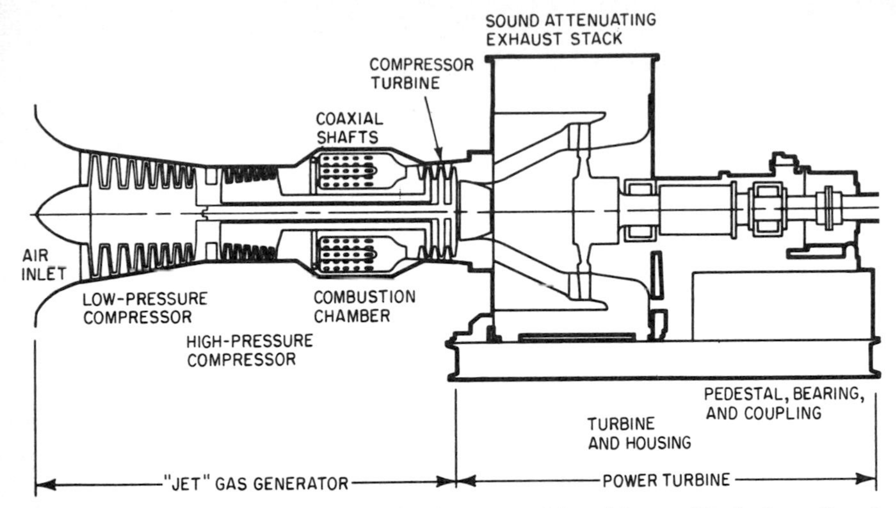

Fig. 6 Cross section of aircraft derivative gas-turbine driver. (Turbodyne Corp.)

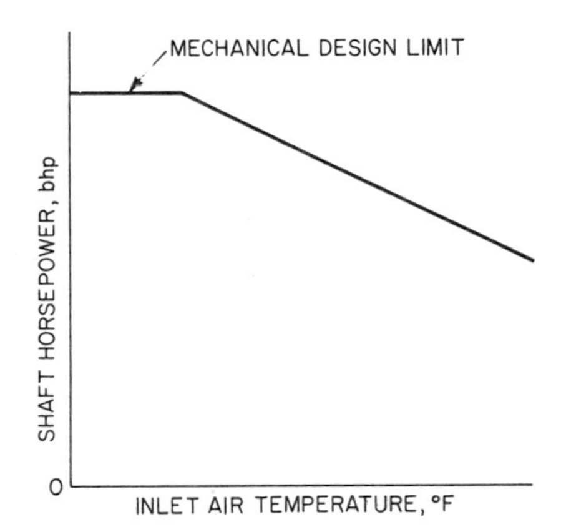

Fig. 7 Typical curve of gas-turbine output versus inlet temperature.

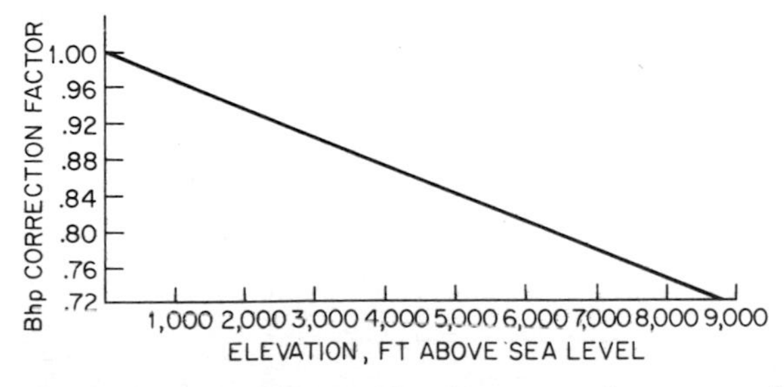

Fig. 8 Typical curve of gas-turbine bhp correction versus altitude.

With ever-increasing firing temperatures, research is continuing in the field of high-temperature corrosion with or without accompanying erosion problems. Present approaches to the problem include the use of inhibitors in the fuel as well as coatings for power-turbine blades. Another solution which has been used on some crude oil pipelines is the topping unit. This is a small still which removes a specific fraction of the crude oil for burning as gas-turbine fuel. It is good practicc to include a detailed fucl analysis as part of any request for bids.

ENVIRONMENTAL CONSIDERATIONS

Legislation at both the federal and state levels has been concerned with the protection of our environment. An engineer must now consider the effects of the project on the environment.

Noise Any piece of dynamic mechanical equipment will emit airborne sounds which, dependent on frequency and level, may be classified as noise.

The primary sources of noise in the gas turbine are the gas-turbine inlet and exhaust as well as the contribution from accessories and support systems such as fin-fan lubrication-oil coolers, auxiliary air blowers and fans, starting devices, and auxiliary lubrication-oil pumps. It is common practice for manufacturers to provide inlet and exhaust silencers as well as some form of acoustic treatment for auxiliaries.

In considering the degree of acoustic treatment necessary for each installation the practicing engineer must consider the following parameters:

1. Federal law (Walsh Healy Act) limiting the time a worker may spend in a noisy environment
2. Local codes and their interpretation
3. Plant location and existing noise level at site
4. Site topography, including any noise-reflective surfaces
5. Applicable ASME Standards; other relevant standards

Emissions The Federal Air Quality Acts require each state to develop a plan for achieving satisfactory air quality. Specifically, goals have been put forth to limit suspended particulate matter and oxides of sulfur and nitrogen. As of the time of writing, the implementation of this act has not been defined in a large number of states. Therefore the engineer must investigate existing regulations during the planning stage of a pump installation. In general, it can be stated that gas turbines have low particulate emissions. The amount of sulfur oxides exhausted to the atmosphere is in direct proportion to the content of sulfur in the fuel, and current practice calls for elimination at the source. The formation of oxides of nitrogen is a direct result of the combustion process. Manufacturers are currently committing a considerable amount of resources to the investigation and solution of this problem.

GAS-TURBINE SUPPORT SYSTEMS

Starting Systems A form of mechanical cranking is necessary to bring a gas turbine up to its self-sustaining speed. The amount of energy necessary will depend on each manufacturer's design. Available systems include electric motors, diesel engines, and gas-expander turbines.

Lubrication The gas turbine manufacturer normally provides a combined pump-turbine lubrication oil system. A main lubrication-oil pump of sufficient capacity for the combined system is necessary in addition to a standby pump in the event of failure of the main pump. A reservoir sized for at least 4 min retention time is usually specified. Filtration to 10 micron particle size should be adequate for most gas turbines and pumps.

Lubrication-oil cooling can be accomplished by various means, the selection of which depends upon local conditions. The simplest and cheapest method uses a shell and tube heat exchanger with water as the cooling medium. In arid regions a fin-fan cooler with direct air-to-water cooling is commonly used.

Inlet-Air Filtration The degree of inlet-air filtration needed is primarily a function of the size and number of particles in the atmosphere surrounding the installation. In most cases a simple tortuous-path precipitator will suffice. For dirtier atmospheres or in arid regions where sand storms occur, inertial separators followed by a rolling-media filter should be considered.

Any filtration will result in a loss in performance due to the pressure drop across the inlet filter. Conservative design practice calls for a face velocity of 500 ft/min for a rolling-media filter and 1700 ft/min for a tortuous-path precipitator.

Control Most manufacturers offer control packages which provide proper sequencing for automatic startup, operation, and shutdown. During automatic startup, the sequencer receives signals from various transmitters to ensure that auxiliaries are functioning properly, and with the aid of timers brings the unit on line through a preplanned sequence of events.

Key operating parameters are continually monitored during normal operation, for example, output speed, gas-turbine compressor speed, turbine-inlet temperature, lubrication-oil temperature and pressure. Signals from a pump discharge-pressure transmitter can be fed into a speed and fuel controller to cause the unit to respond to changes in speed and/or output.

As with the case of startup, normal and emergency shutdown are accomplished through the sequencer. Most standard control packages are easily adaptable to remote control by means of cable or microwave.

APPLICATION TO PUMP DRIVER

Gas turbines are available to drive centrifugal pumps in a wide range of speeds and sizes ranging from 40 hp to over 20,000 hp. It is not practical to list here all the available units since they are too numerous and are continually being upgraded and added to. A listing can be found in "Sawyer's Gas Turbine Catalog," which is published annually.

Pipeline Service Crude oil pipeline service has seen a tremendous growth in the use of gas-turbine drivers since the mid-fifties and this trend will continue as crude oil production becomes more and more remote from its markets. The advantages of the gas turbine for this application are as follows:

1. Installed cost is usually lower than a corresponding reciprocating engine.
2. Variable-speed operation allows maintenance of a specific discharge pressure under a wide range of operating conditions, thus achieving maximum flexibility.
3. The normal gas-turbine control system is easily adapted to unattended operation and remote control.
4. Operating experience has proved the gas turbine to have a high degree of reliability.
5. Gas turbines can be packaged into modules for ease of transportation and erection.

Water Flooding An important aspect of oil production is the use of secondary recovery methods to increase the output of crude oil reservoirs whose pressure does not allow the crude to flow freely to the surface. Flooding the reservoir with high-pressure water has been a primary technique for years. The development of high-pressure centrifugal pumps has allowed increased water flow into oil fields.

As most oil production is located in remote areas, the packaged gas-turbine driver has become increasingly popular. Figure 9 shows a typical 3,300 bhp split-shaft gas-turbine-driven centrifugal water-flood pump package. A cutaway view of a typical 1,100 bhp gas turbine is shown in Fig. 10; note the reduction gearing at the exhaust end for direct driving a pump. Supporting systems such as starter, lubrication-oil pumps, governor, and fuel-oil pumps are driven off the accessory pods located at the air-inlet end. Offshore platform installations require drivers with a minimum vibration as well as small, unbalanced inertia forces. The gas turbine fits both of these categories very well.

Cargo Loading Another interesting application of this prime mover is in the field of cargo loading, where units are presently in operation charging tankers with crude oil. Selection of a gas-turbine pumping unit with critical speeds above the

Fig. 9 3,300-bhp gas-turbine-driven waterflood package. (Solar Division of International Harvester Company)

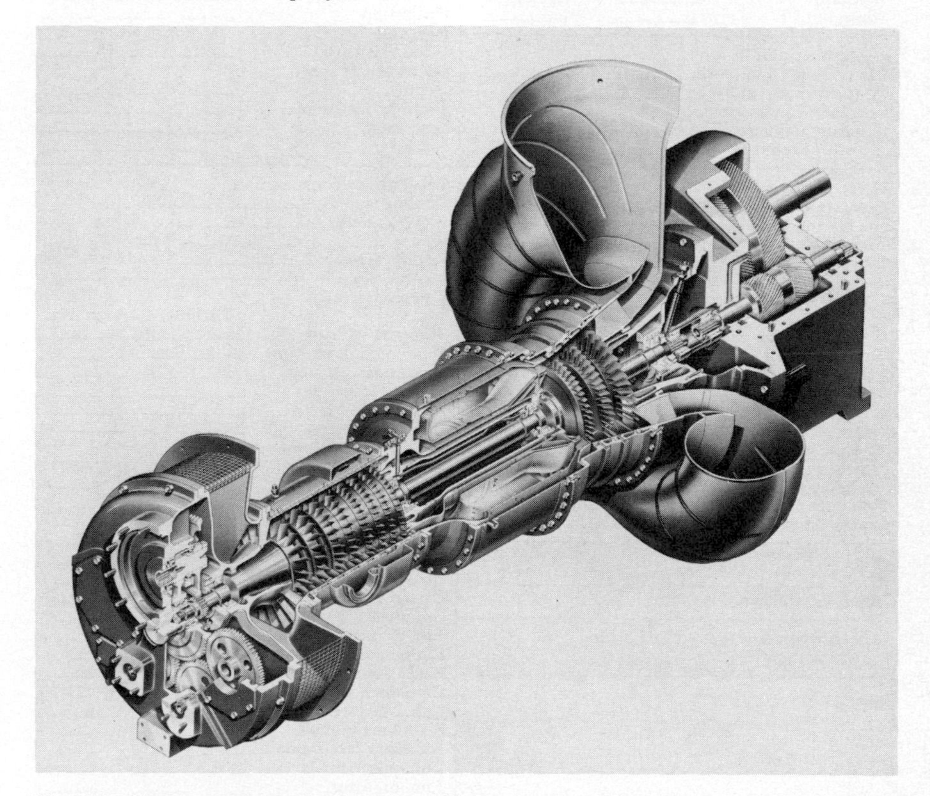

Fig. 10 Cutaway view of 1,100-bhp split-shaft gas-turbine driver. (Solar Division of International Harvester Company)

normal operating range allows great flexibility of operation, particularly during final topping operations.

Application Considerations The application of gas-turbine drivers can vary from a simple driver for one pump operating at constant flow and discharge pressure to a multiplicity of units operating at variable speed on a pipeline. For the purposes of this discussion it is assumed that the pumping system has been analyzed, a pump selection has been made, and all possible operating conditions have been analyzed resulting in known bhp and speed requirements.

TABLE 1 Typical Data Sheet, Gas-Turbine Driven Centrifugal Pump

Information by purchaser

Pumping requirements
Service ____
Liquid ____
Pumping temp. ____
Capacity (total) normal/max ____ / ____
No. pumps operating ____
Specific gravity at pump temp. ____
Viscosity at pumping temp. ____
Total head ____
NPSH available ____

Pump type, materials, and accessories

Site conditions
Elevation ____ ft
Range of site ambient temperature

	Dry bulb	Wet bulb
Design, °F	____	____
Maximum, °F	____	____
Minimum, °F	____	____

Atmospheric air
Dust below 10 microns ____ ppm
10 microns and above ____ ppm
Corrosive constituents:
Sulfur, ammonia, ammonium salts, salt or seacoast, other ____

Noise specifications
City, state, federal, other ____

Emission specifications
City, state, federal, other ____

Utilities available at site
Steam: Pressure ____ lb/in² gage
Temp., °F ____
Quantity ____ lb/hr
Electricity

	V	Phase	Cycles
Ac	____	____	____
Dc	____	____	____

Cooling water
Source ____ quality ____
Supply temp. ____ min ____ max
Supply press. ____ lb/in² gage
Max return ____ °F

Fuel
Gas ____ liquid ____
Analysis attached

Accessory items required (see list in right column)

Information by manufacturer

Pump performance
Service ____
Liquid ____
Pumping temp. ____
Capacity (total) normal/max ____ / ____
No. pumps operating ____
Specific gravity at pumping temp. ____
Viscosity at pumping temp. ____
Total head ____
NPSH required ____

Bhp required	____ normal	____ max
Speed, rpm	____ normal	____ max
Efficiency	____ normal	____ max

Pump type, materials, and accessories

Gas turbine excluding gear

	Design	Max	Min
Dry bulb temp., °F	____	____	____
Output hp	____	____	____
Heat rate, Btu/(hp)(hr)(LHV)	____	____	____
Output shaft speed, rpm	____	____	____
Turbine inlet temp., °F	____	____	____
Gear loss	____	____	____
Net hp	____	____	____

Total utility consumption
Cooling water ____ gpm
Electric power ____ kw, ac/dc
Steam ____ lb/hr
Compressed air ____ Standard ft³/min

Shipping data

	Turbine	Aux. items
Shipping wt, tons	____	____
Max erection wt, tons	____	____
Max maint. wt, tons	____	____
Length, ft-in	____	____
Width, ft-in	____	____
Height, ft-in	____	____

Accessories included (as required by purchaser)

	yes	no
Inlet-air filter	____	____
Inlet-air silencer	____	____
Exhaust silencer	____	____
Exhaust duct	____	____
Starting equipment	____	____
Load gear (if required)	____	____
Driven pump	____	____
Coupling	____	____
Fire protection system	____	____
Equipment enclosure	____	____
Baseplate or soleplates	____	____
Combined turbine-pump	____	____
Lub.-oil system	____	____
Main lube pump	____	____
Auxiliary lub.-pump	____	____
Lub.-oil reservoir	____	____
Lub.-oil filter	____	____
Lub.-oil cooler	____	____
Unit control panel	____	____
Auxiliary motor	____	____
Control center	____	____

Typical data sheet adapted from API Standard 616.

The bhp output of the gas turbine must equal or exceed that required by the pump. This output can be determined by the use of specific performance curves similar to Fig. 7 as corrected for elevation (Fig. 8). Gear losses as necessary are added to the bhp required by the pump. Intermediate bhp and speed requirements should then be checked against a gas-turbine output versus speed curve (Fig. 4).

A torsional analysis of the combined unit is made (usually by the gas-turbine manufacturer) to ensure the absence of any critical speeds in the operating range. Table 1 is a typical data sheet recommended for use when purchasing a gas turbine to drive a centrifugal pump.

REFERENCES

1. Baumeister, T., and Marks, L.: "Standard Handbook for Mechanical Engineers," 7 ed., McGraw-Hill Book Company, New York, 1967.
2. Shepherd, D. G.: "Introduction to the Gas Turbine," 2 ed., D. Van Nostrand Co., Princeton, N.J., 1960.
3. Sawyer, J. W., ed.: "Sawyer's Gas Turbine Engineering Handbook," 2 ed., Gas Turbine Publications, Inc., Stamford, Conn., 1972.
4. Dacy, J. R.: Gas Turbines Used on Arctic Pipelines, *Oil Gas J.*, vol. 200, pp. 26–32, July 8, 1973.
5. Schiefer, R. B.: The Combustion of Heavy Distillate Fuels in Heavy Duty Gas Turbines, *ASME* 71-GT-56, 1971.
6. Combustion Gas Turbines for General Refinery Services, *API Standard* 616, 1968.
7. Proposed Gas Turbine Procurement Standard, *ASME*.

Section **6.2.1**

Eddy-Current Couplings

W. J. BIRGEL

DESCRIPTION

The eddy-current coupling is an electromechanical torque-transmitting device installed between a constant-speed prime mover and a load to obtain adjustable-speed operation. Generally ac motors are the most commonly used pump drives, and they inherently operate at a fixed speed. Insertion of an eddy-current slip coupling into the drive train will allow desired adjustments of load speeds.

In most cases, the eddy-current coupling has an appearance very similar to that of its driving motor with an additional shaft extension. The shaft, which operates at constant speed, is connected to the motor while the other shaft, providing the adjustable-speed output, is connected to the load. A typical self-contained air-cooled eddy-current slip coupling is shown in Fig. 1.

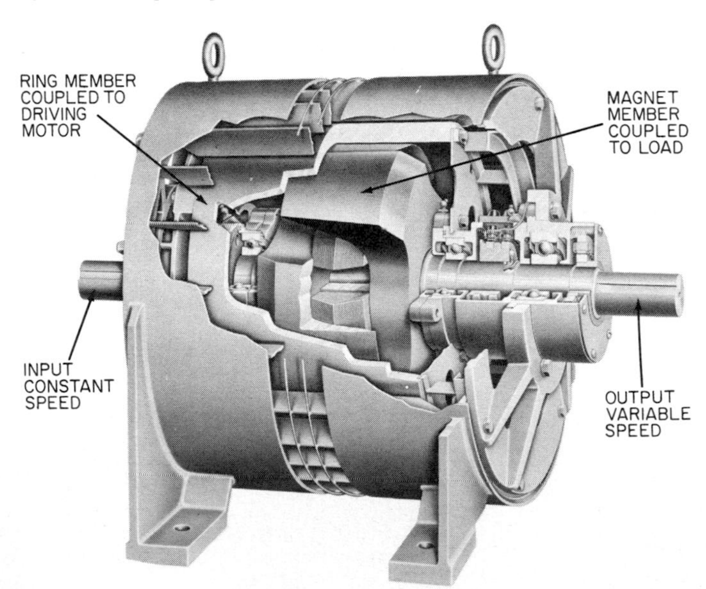

Fig. 1 Cutaway of an Ampli-Speed (eddy-current slip coupling). (Electric Machinery Mfg. Company)

The input and output members are mechanically independent, with the output magnet member revolving freely within the input ring or drum member. An air gap separates the two members, and a pair of antifriction bearings serves to maintain their proper relative position. The magnet member has a field winding which is excited by direct current, usually from a static power supply. Application of this field current to the magnet induces eddy currents in the ring. The interaction between these currents and magnetic flux develops a tangential force tending to turn the magnet in the same direction as the rotating ring. The net result is a torque available at the output shaft for driving a load. An increase or decrease in field current will change the value of torque developed, thereby allowing adjustment of the load speed. Field current, output torque, and load speed are usually not proportional. Therefore load-torque characteristics and speed range must be known for proper adjustable-speed drive size selection.

Most eddy-current couplings have load-speed control. An integral part of this system is a tachometer generator or magnetic pickup. These provide an indication of exact output speed and enable control of load speeds within relatively close tolerances regardless of reasonable variations in load-torque requirements. Slip-type adjustable-speed drives without load-speed control will experience fluctuations in load speed when load-torque requirements vary.

FUNDAMENTALS

The eddy-current coupling, like many other adjustable-speed drives, operates on the slip principle and is classified as a torque transmitter. This means that the input and output torques are essentially equal, disregarding friction and windage losses. The motor-input horsepower is equal to the sum of load horsepower and what is known as slip loss. This slip loss is the product of slip rpm, which is the difference between motor and load rpm, and the transmitted torque.

The various relationships may be expressed as follows:

$$\text{Motor hp} = \frac{\text{rpm}_1 \times T}{5{,}250}$$

$$\text{Load hp} = \frac{\text{rpm}_2 \times T}{5{,}250}$$

$$\text{Slip loss (hp)} = \frac{(\text{rpm}_1 - \text{rpm}_2) \times T}{5{,}250} = \frac{\text{rpm}_3 \times T}{5{,}250}$$

This may be further expressed as follows:

$$\text{Slip loss (hp)} = \frac{\text{rpm}_3}{\text{rpm}_2} \times \text{load hp}$$

or

$$\text{Slip loss (hp)} = \frac{\text{rpm}_3}{\text{rpm}_1} \times \text{motor hp}$$

where rpm_1 = motor rpm at designated load
rpm_2 = load rpm at designated load
rpm_3 = slip rpm at designated load
T = load torque, ft-lb

An eddy-current coupling must slip in order to transmit torque. The normal minimum value of slip for a centrifugal pump application is usually 3 percent but values from 2 to 5 percent are common. The above formulas hold true regardless of the type of load involved.

Efficiency of an eddy-current coupling can never be numerically greater than the percentage of output speed. This effectively takes into consideration only the slip losses, and a true efficiency value must also include friction and windage losses plus excitation losses. The friction and windage losses are constant for a fixed-motor speed and therefore increase in significance with speed reduction. Excitation losses, on the other hand, decrease with reduction in output speed.

The overall effect of these losses is an efficiency versus speed relationship which is somewhat linear in nature with efficiency values anywhere from 2 to 5 points less than the percentage of output speed.

LOAD CHARACTERISTICS

In the application of slip couplings for continuous pump loads, both variable- and constant-torque requirements are encountered.

Variable-Torque Loads Variable-torque loads are those where the torque increases with the speed and varies approximately as the square of the speed while the load horsepower varies approximately as the speed cubed. The centrifugal pump fits into this classification. To be specific, the above relationship exists only where

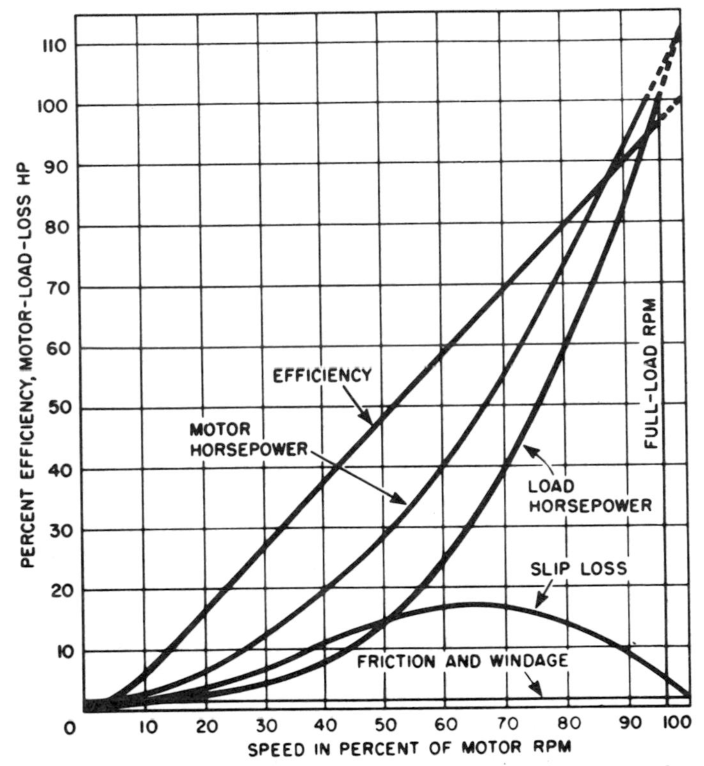

Fig. 2 Load, horsepower, and slip characteristics for a friction-only pumping system.

the friction head is the total system head, as would be the case if a centrifugal pump were pumping from and to reservoirs having the same liquid levels or in a closed loop.

The various relationships of load horsepower, motor horsepower, and slip loss applying to a typical variable-torque load are shown in Fig. 2, and again friction and windage losses have been disregarded.

Static heads, which usually exist in centrifugal pump systems, do not significantly affect the selection of a suitable slip coupling for a specific requirement. Pump efficiency, on the other hand, can be quite an important factor when it decreases significantly with speed reduction. Pumps with relatively flat efficiency curves are most desirable for adjustable-speed duty.

Centrifugal pumps usually operate against some static head. This causes the pump torque to follow a closed discharge characteristic until sufficient speed is reached to cause the resultant discharge head to equal or exceed the static head. This is equivalent to a closed discharge or normally unloaded condition. The pump load then follows a different curve to the full-load condition at minimum slip.

These characteristics are illustrated in Fig. 3. Curve OAB indicates the closed discharge horsepower. At point A the static head is overcome, and the pump load then rises along curve AC; this causes a significant difference in slip loss as shown. Under conditions as indicated, where static head is not exceeded and water does not start to flow until 80 percent speed is reached, the slip loss never exceeds 10 percent of the full-load rating. In general, this is true of a pump with high static head.

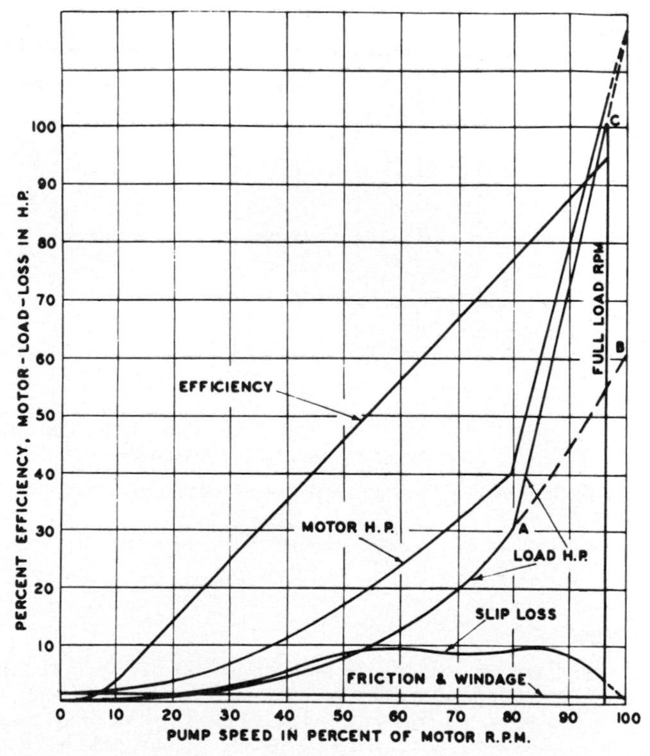

Fig. 3 Percent efficiency, motor-load loss in horsepower versus pump speed in percent of motor rpm for a static head and friction pumping system.

Constant-Torque Loads Constant-torque loads are those requiring essentially constant-torque input regardless of operating speed. Positive displacement pumps generally are of this type. The load characteristics showing the division between slip loss and load horsepower are illustrated in Fig. 4. Note that the driving-motor output does not change, regardless of load speed and horsepower.

The slip-loss characteristics make slip couplings undesirable for large horsepower loads if any appreciable speed range is required. However, in relatively small units, the simplicity and ease of speed control will frequently justify the use of slip couplings instead of more efficient but more complex speed-control systems.

The constant-torque-load capacity of a slip coupling is largely limited by slip loss and, to a lesser degree, by breakaway torque and minimum slip. Since slip loss is directly proportional to slip, the desired speed range will definitely limit the torque, and therefore the horsepower, that can be transmitted. In some cases breakaway torque is important as the static friction may be quite high. It is usually

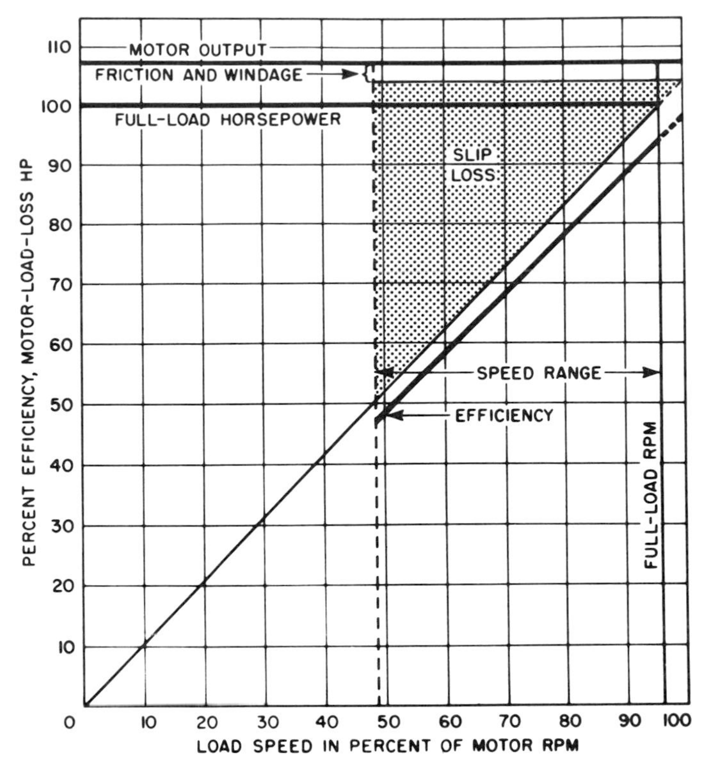

Fig. 4 Horsepower, torque, and losses; constant-torque load.

recommended that at least 150 percent starting torque be available for any constant-torque load.

CONSTRUCTION

Eddy-current couplings are available in both horizontal and vertical configurations as might be required for any specific pump mechanical arrangement. Horizontal machines, in smaller sizes, are frequently close-coupled to the drive motor in what is known as *integral construction.* Larger sizes are usually flexibly coupled to the drive motor and pump load.

Vertical motors and slip couplings are close-coupled to limit overall height and to prevent vibration problems. Pump hydraulic-thrust requirements can be accommodated, when necessary, in much the same way as in constant-speed motor applications. Because of mechanical limitations, thrust bearings are frequently located in the bottom of the adjustable-speed drive.

Enclosures available will vary depending on the type of cooling involved. Obviously water-cooled types need little if any enclosure adaptation for virtually any installation. However, caution should be exercised in outdoor use where freezing can occur. Also, vertical installations utilizing water cooling may require special considerations.

Air-cooled couplings are more universally used for centrifugal pump loads but do require attention on enclosure design. Indoor installations in clean atmospheres need only open or dripproof enclosures where heat rejection is not a problem. In some cases, intake and/or discharge covers for connection to ductwork may be necessary for environmental isolation.

Weather-protected enclosures are available for outdoor installations. A NEMA Type I rating is normally adequate because of the slip coupling's inherent mechanical design and relatively low field winding voltage levels.

CONTROLS

Precise speed control of an eddy-current coupling is obtained by the use of feedback circuitry. This consists of an output-speed sensing device which produces a voltage proportional to speed, a speed selector which establishes a command voltage, and a resultant voltage which is impressed on an amplifier circuit to control the rectifier supplying excitation to the coupling field. A simplified diagram is shown in Fig. 5.

It can be noted that modern semiconductor technology involving transistors, SCRs, and integrated circuits is used extensively in circuit design. Repositioning of the speed-control potentiometer creates an imbalance in the control circuitry. This results in an excitation change to regain balance by modification of the speed feedback signal to a suitable value at the new operation point. Speed of response of the control is extremely high. In all cases of significant speed change, the excitation is maximum or minimum resulting in a net torque differential and maximum rate of speed change.

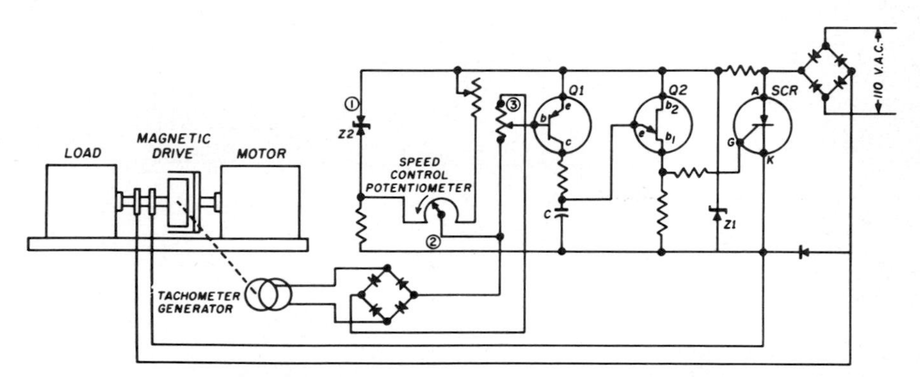

Fig. 5 Simplified diagram of an eddy-current coupling control. Semiconductor devices provide the rectifying and control elements for adjusting the speed of the magnetic drive and for maintaining output speed at the desired point. (Electric Machinery Mfg. Company)

The speed-control potentiometer may be positioned by mechanical or electric means which are responsive to liquid level, pressure, flow, or other parameters. Electric signals can, in some cases, be applied directly to the speed-control circuitry or may be conditioned for such use through amplification and isolation.

A typical two-pump liquid-level control system schematic is shown in Fig. 6. One unit is programmed for lead-pump operation with the output speed proportional to a range of liquid-level variations. Both units operate in parallel when additional pumping capacity is required. Circuitry is usually provided for lead-pump alternation, and alarms for high or low levels, for loss of air, or for other malfunctions can be readily added.

Constant-pressure system controls are also frequently encountered in eddy-current coupling applications. Pressure transmitters and two-mode controllers make up the bulk of the instrumentation package. The interface with the coupling control is otherwise similar to that in the liquid-level control.

APPLICATIONS

Slip couplings are applied to centrifugal pumps for water and waste-water pumping in municipal installations, for boiler feed pumping, for circulating water and condensate pumping in power plants, for fan and stock pumping in paper mills, and for reciprocating pumping in a multitude of applications and industries.

In the water and waste-water fields, slip couplings have developed a particularly large following for raw- and finished-water pumping, lift-station pumping, raw-sewage pumping, effluent and sludge-pumping applications. Almost any pumping problem, where cyclic constant-speed pumping or throttling or other means of flow

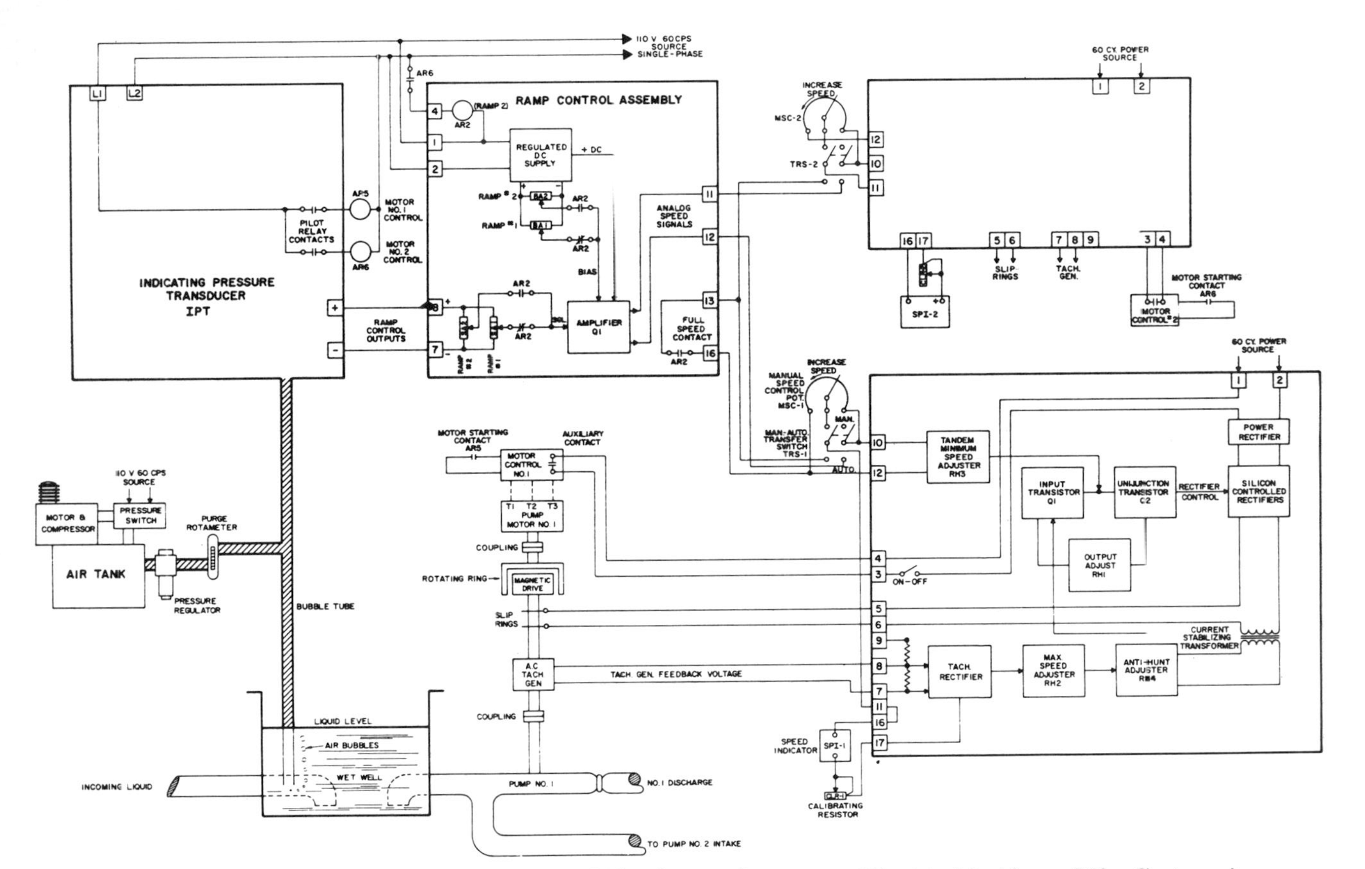

Fig. 6 Block diagram for double-ramp liquid-level control system. (Electric Machinery Mfg. Company)

TABLE 1 Eddy-Current Coupling. Selection Chart for Centrifugal Pumps

Drive-motor input speed, rpm	Unit size	Percent of motor full-load speed			
		98	97	96	95
		Horsepower output			
1750	M110	10	16	18	22
	M111	15	22	27	32
	M132	26	37	45	50
	M133	37	51	62	68
	M154	58	74	86	91
	M155	78	97	109	114
	M186	120	148	164	170
	M187	160	200	218	225
	M188	199	248	272	280
	S189	265	320	326	335
	S208	330	385	380	380
	S209	420	405	395	380
	S238	610	597	579	561
	S276	800	775	750	725
	S326	1,100	1,065	1,030	1,000
1150	M110	4	6	8	10
	M111	6	9	12	15
	M132	11	17	21	25
	M133	16	24	30	35
	M154	26	37	45	51
	M155	36	50	59	66
	M186	57	75	90	100
	M187	78	102	120	135
	M188	98	127	151	166
	S189	132	168	193	210
	S208	190	210	235	245
	S209	290	305	295	285
	S238	355	380	400	400
	S239	470	455	442	428
	S276	615	600	580	565
	S326	870	844	818	793
870	M111	3	5	7	8
	M132	6	9	12	15
	M133	9	13	17	21
	M154	15	21	27	31
	M155	21	29	36	41
	M186	33	45	56	63
	M187	45	62	75	85
	M188	58	80	97	108
	S189	75	110	128	140
	S208	120	145	160	170
	S209	170	225	240	235
	S238	241	265	280	285
	S239	336	375	364	353
	S276	355	490	480	465
	S326	718	696	675	654
700	M154	8	14	18	21
	M155	13	20	25	29
	M186	20	28	35	42
	M187	25	36	45	56
	M188	30	45	55	70
	S189		73	85	95
	S208		110	115	120
	S209		160	185	200

TABLE 1 Eddy-Current Coupling. Selection Chart for Centrifugal Pumps (Continued)

Drive-motor input speed, rpm	Unit size	Percent of motor full-load speed: 98	97	96	95
		Horsepower output			
700	S238		200	210	217
	S239		316	316	306
	S276		365	375	405
	S326		597	579	561
585	S189		49	64	75
	S208		80	90	100
	S209		110	135	150
	S238		140	160	170
	S239		223	270	272
	S276		270	280	290
	S326		520	507	492
495	S208			55	65
	S209			100	115
	S238			135	138
	S239			200	220
	S276		205	235	250
	S326		410	460	445
435	S208			55	60
	S209			80	90
	S238			115	120
	S239			167	188
	S276			190	210
	S326		330	390	405
385	S276			150	170
	S326		275	320	340

control are alternate considerations, can be conveniently solved with the use of an eddy-current slip coupling as the adjustable-speed-flow controlling device.

Potable-water treatment and distribution facilities are continually confronted with substantial fluctuations in demand through daily, weekly, and even seasonal periods. Distribution systems, which depend on direct pumping, usually must utilize total or partial adjustable-speed operation for high service and booster requirements. The quick response of eddy-current slip couplings makes them extremely well suited for this duty.

Waste-water collection systems, where inflow conditions to lift stations and treatment plants vary widely throughout the day, can realize many advantages when designs are based on adjustable speed with eddy-current slip couplings.

RATINGS AND SIZES

Eddy-current couplings are available in a wide range of ratings and sizes from fractional hp units up through 10,000 hp and beyond. The type of cooling employed is an important consideration with water and air used exclusively or together by the various manufacturers. In addition, the type of load is a factor since thermal capability will vary significantly between water- and air-cooled varieties.

Centrifugal pump variable-torque loads are usually best handled by air-cooled couplings having high-torque capabilities at low values of slip and limited heat-dissipating capabilities. The selection charts shown in Table 1 are representative of one manufacturer's line of couplings which are designed specifically for centrifugal pump loads. Where full-torque capabilities are realized at 3 percent slip below motor speed, thermal loads at two-thirds of motor speed will be 16.2 percent of rated speed-load hp. Sizes starting at 3 to 5 hp and extending up through 3,000 to 5,000 hp are generally available.

Eddy-current couplings for pump loads requiring constant-torque drives have rather limited usage. In small sizes where thermal capabilities are proportionately greater than the 6:1 ratio encountered on large units, constant-torque loads can be adequately handled by air-cooled couplings. Beyond 100 hp, air-cooled units can be impractical because of thermal and starting-torque requirements.

Water-cooled eddy-current couplings have somewhat different characteristics, making them more suitable for constant-torque and large horsepower variable-torque

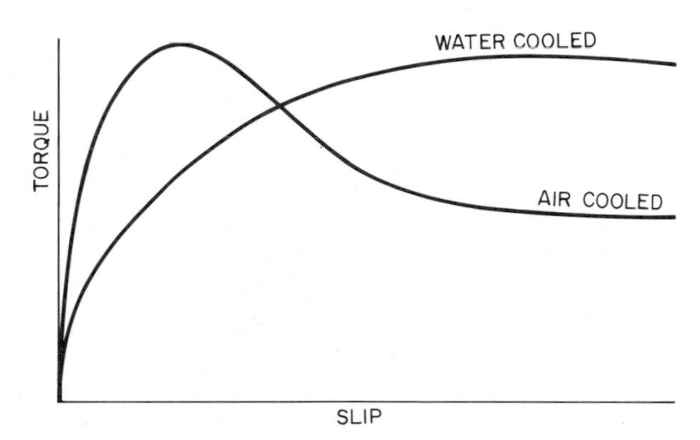

Fig. 7 Speed-torque curves for variable-torque air-cooled and constant-torque water-cooled couplings.

Fig. 8 Horizontal centrifugal pump driven by a 1,000-hp 1,780-rpm induction motor through a step-up gear and an eddy-current coupling. (Electric Machinery Mfg. Company)

loading. High starting torque, high minimum slip, and high thermal capabilities all tend to lead to those conclusions. Thermal capabilities are frequently equal to or greater than full-load horsepower ratings. The starting torque is usually the maximum torque and as a result low slip values are limited.

Figure 7 illustrates the differences in the speed-torque curves of variable-torque air-cooled and constant-torque water-cooled couplings.

THRUST CAPABILITIES: VERTICAL UNITS

Most vertical eddy-current couplings have limited external downthrust load capabilities when provided with standard bearing arrangements. Typical values are listed in Table 2 for the same sizes as listed in Table 1. Bearing life is an important factor

Fig. 9 Vertical centrifugal pumps driven by 40-hp 1,750-rpm induction motors and eddy-current couplings. (Electric Machinery Mfg. Company)

and the values shown are based on 5 years minimum life in accordance with manufacturer's standards at an average output speed of 85 percent of input speed. Bearings are angular-contact ball type with grease lubrication. Where higher thrust values are encountered, or where longer bearing life or adherence to AFBMA life standards must be met, spherical roller bearings with oil lubrication can be furnished.

TABLE 2 Typical Eddy-Current Coupling Downthrust Capabilities

	Motor speed, rpm					
	1,750	1,150	870	700	585	495
Unit size	Thrust, lb					
M110 & M111	1,665	1,900	2,150			
M132 & M133	1,555	1,790	2,040			
M154 & M155	2,615	3,065	3,465	3,825	4,005	
M186, 7 & 8	2,200	2,660	3,060	3,420	3,600	
S189	2,090	2,550	2,950	3,310	3,490	
S208 & S209	2,920	3,530	4,080	4,580	4,860	5,180
S238 & S239	2,240	2,860	3,410	3,910	4,190	4,500
S276		3,900	4,605	5,170	5,670	6,150
S326		2,545	3,260	3,810	4,290	4,780

Section **6.2.2**

Single-Unit Adjustable-Speed Electric Drives

E. O. POTTHOFF

GENERAL

Electric drives are available featuring a single rotating element and some associated control so as to fully perform the adjustable-speed driving function. The feature of using a single rotating element differentiates this type from the eddy-current coupling motor, the hydraulic-coupling-motor combinations, and the adjustable-speed belt drives, all of which are tandem drives.

Control of speed of the single-unit adjustable-speed drive results from the interaction of the control and the motor itself. For this reason one must consider these two elements as a drive and not consider the motor alone.

Section 6.1.1 described motors that will be further discussed in this section. Previous discussions of control were limited to that required for starting and protection; this section will also describe specific controls required for speed control. The discussion will cover the following types of drives:

- Ac adjustable-voltage type
- Wound-rotor induction motors with several different types of secondary controls
- Adjustable-frequency type
- Modified Kraemer type
- Dc motors with silicon-controlled-rectifier (SCR) power supplies

All have physical features, operating characteristics, or prices that make them particularly valuable in some specific segment of the pump-driver spectrum.

DRIVE DESCRIPTION

Ac Adjustable-Voltage Drive This drive consists primarily of an adjustable voltage, constant frequency control, and a motor (M) conveniently configured as shown in Fig. 1. The motor must possess high slip characteristics and other characteristics that allow it to work successfully with the associated control. The motor is designed for operation and tested with its associated control. Because the motor has high-slip characteristics, insulation is of Class F rating.

The fundamental parts of the control are a circuit protective device such as a circuit breaker, an SCR assembly, and firing circuitry identified as FC in Fig. 1.

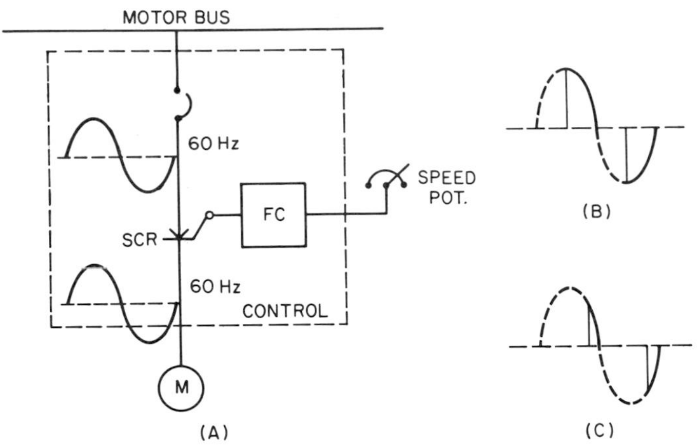

Fig. 1 Ac adjustable-voltage drive power block diagram. (General Electric Co.)

Other parts are required to complete the control but serve as auxiliaries. The main function of this control is to provide a voltage to the motor at a level that will ensure a desired motor speed. It also serves to protect the motor under abnormal operating conditions and the motor cable under short-circuit conditions, and provides a current-limit function so that the motor draws a maximum of some preselected value, such as 150 percent of normal under all operating conditions.

The SCR receives impressed voltage at 60 Hz usually. The SCR possesses the ability to be turned on (or become conducting) by means of current pulses received from the firing circuits energizing them at different points in the sinusoidal wave. Once conductance starts, it continues until voltage disappears at the end of the half-cycle. The SCR must again be pulsed in order to become conducting. This occurs while the voltage is in the negative phase of its sinusoidal generation and is performed by an SCR connected in parallel and with reverse polarity to that of the first. The negative-loop conductance continues until the voltage again returns to zero. The firing controls are designed to turn the SCR on repetitively sometime during each voltage half-cycle. Figure 1*A* shows the shape of applied voltage in the wave illustration shown between SCR and motor (M). In this case, the SCR is turned on at the beginning of each half-cycle.

By delaying the firing or turning on of the SCR until later in voltage generation, shorter intervals of applied voltage and lower levels of voltage appear across the motor. The solid line of Fig. 1*B* illustrates applied motor voltage when the voltage half-wave is half-completed. Fig. 1*C* illustrates the motor-applied voltage when the half-waves are approximately 75 percent completed. Notice in these illustrations that the frequency of the voltage applied to the motor does not vary, only the voltage magnitude.

The exact point in the half-wave when firing occurs is controlled by a low-energy electronic signal that may come from a potentiometer, as indicated in Fig. 1*A* or some process-instrument signal. By increasing the signal level, voltage applied to the motor increases; on the other hand, a decrease in signal level decreases the voltage level.

Figure 1*A* shows the SCR on the utility side of the motor as a convenience in illustration. In reality, the SCR is generally placed at the motor neutral point. This reduces voltage from SCR to ground and allows the motor impedance to provide a degree of protection for the SCR from damage due to transient voltage spikes entering from the electric utility line. However, it does require the use of six motor conductors instead of the conventional three.

Figure 2 illustrates representative motor speed-torque curves at rated and other voltages as well as a pump speed-torque curve varying as the square of the speed. The motor possesses approximately 10 percent slip under rated torque and voltage conditions. Note that motor torque does not break down at speeds lower than breakdown torque speed. These various characteristics help serve to identify it as a motor designed to operate specifically for this application. Note that by reducing applied motor voltage, motor torque decreases, causing pump operation at lower speed.

Ratings of these drives are limited essentially to low horsepowers. Figure 3 provides a guide to those available in open construction only. All ratings with maximum rated speeds existing within the solid envelope are available. Some, but very few, are available outside of this envelope. Totally enclosed motors are more restricted in supply than open motors and are limited to approximately 40 hp maximum. In addition to these restrictions, load torques must not exceed values varying as the square of the speed; thus this drive is not a candidate to drive a constant-torque load. Table 1 gives some pertinent application information.

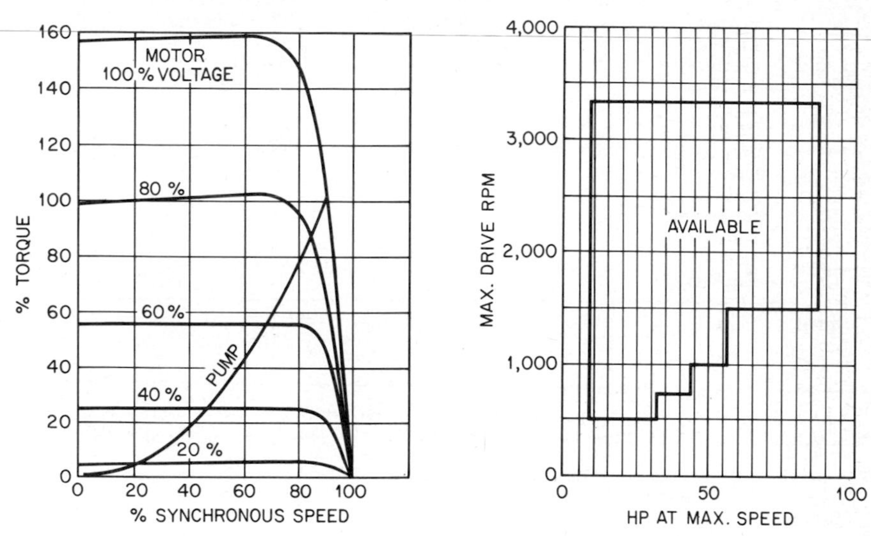

Fig. 2 Ac adjustable-voltage drive torque versus speed. (General Electric Co.)

Fig. 3 Ac adjustable-voltage drive availability.

Wound-Rotor Induction Motors with Liquid-Rheostat Controls This form of drive consists primarily of a full-voltage nonreversing (FVNR) starter, a wound-rotor induction motor, and a liquid rheostat, all of which integrate into a configuration as illustrated in Fig. 4. The FVNR starter switches power to and from the motor stator as well as provides generally accepted protection to the motor from short circuits, overloads, etc. Secondary cables interconnect motor rotor and fixed electrodes in the liquid rheostat. The fixed electrodes exist in separate cells of the liquid rheostat, one for each of the phases. Movable electrodes, one located in each cell, are suspended from a horizontal bar and the bar together with the vertical-electrode suspension bars, movable electrodes, and the electrolyte filling the cells complete the Y, or common point in the external motor rotor circuit. By moving the upper electrodes up or down, secondary circuit resistance of the drive varies and causes a change in motor speed-torque characteristic.

As already pointed out, the motor starter provides the normal protective functions for the motor. Control circuitry provides a feature of allowing motor starting with maximum secondary resistance in the rotor circuit, thus drawing minimum motor current from the power supply.

Figure 4 illustrates a motor (M3) operating a pulley to cause a change in movable-electrode position. This motor could be controlled manually by a push button, or automatically through a controller providing either a raise-lower or modulated

TABLE 1 Ac Adjustable Voltage Drive Data

Drive element	Hp ratings*	Voltage rating	Max rated speed, rpm	Speed range, %	Enclosure	Mounting	No. of speed pts.
Motor	See Fig. 3 for open frame ratings; limited availability outside these limits Totally enclosed ratings limited as shown under Enclosure	200 230 460	1,640 1,095 820 655 545 & others	100	Vertical: shielded dripproof, TEFC not generally available Horizontal: dripproof, TEFC to 40 hp approx.	Vert. or horiz.	Infinite
Control	5–150	200 230 460	. . .	. . .	NEMA 1 NEMA 12	Wall or floor	Infinite

* The drive is capable of operating with the horsepower varying as the cube of speed maximum. Specifically, these ratings are not suitable for constant torque applications.

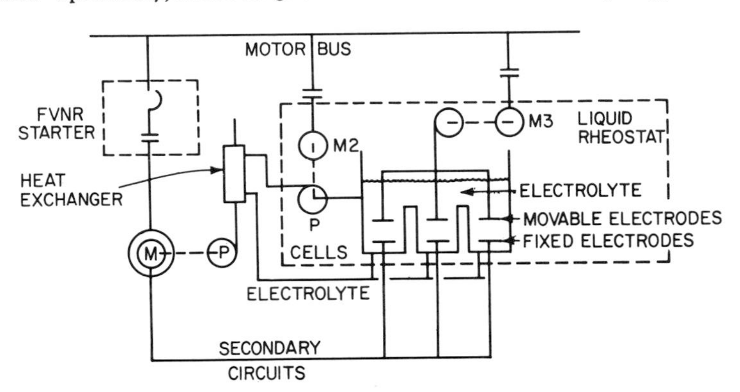

Fig. 4 Wound-rotor induction motor with liquid rheostat secondary power block diagram. (General Electric Co.)

voltage signal. If desired, a pneumatic cylinder may substitute for motor (M3) to provide power to move the electrodes.

The electrolyte receives the drive slip losses dissipated in each of the cells, and of course this heat must be removed into some heat sink capable of dissipating it. In some installations, an electrolyte pump driven by motor (M2) circulates the electrolyte through a heat exchanger before returning it to the individual cells. Returning cooled electrolyte reenters the cells under the fixed electrodes and passes through holes in the electrodes before passing vertically through the cells. The heat-exchanger primary coolant can be tap water or mill water, which provides a good means of passing drive slip losses as heat directly out of the station, or it can be the pump-discharge water itself which conveys heat directly away from the building as illustrated in Fig. 4.

The liquid rheostat is factory-assembled with the rheostat, electrolyte pump, and its motor as well as its electrode-positioning assembly packaged in a single steel enclosure suitable for control lineups except for ratings above 700 hp. The heat exchanger may be of the sleeve (or wrap-around) type, as illustrated in Fig. 4, and comes as a separate device to be attached to station effluent piping. Other forms of exchangers, such as liquid to air or shell and tube heat exchangers, are available.

Changing the movable-electrode position results in a change in motor speed-torque relations as shown in Fig. 5. Each curve shown (except the pump curve) represents a motor characteristic for a discrete secondary resistance. Notice the number associated with each curve which represents the percent of secondary resistance external to the motor rotor; 100 percent ohms provides 100 percent motor torque at zero speed. In examining individual curves note that as the percent resistance increases, slope of the motor speed-torque curve decreases, thus allowing the motor to slide down the pump speed-torque curve and assume a lower speed.

This drive utilizes resistance only in the motor secondary circuit as a means of controlling speed. Reactance is available as a substitute for resistance in other forms of secondary controllers, but its use reduces drive power factor and thereby increases motor current, reduces efficiency, and increases motor heating. As a further effect, the additional motor current may require the use of a larger motor frame to accommodate it.

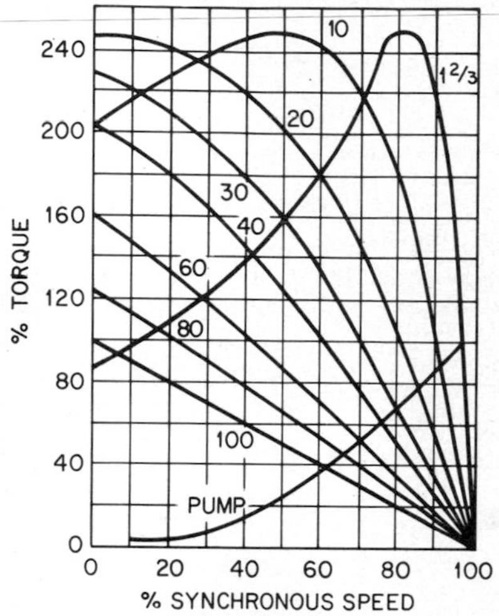

Fig. 5 Wound-rotor induction motor with variable secondary-resistance torque versus speed. (General Electric Co.)

Figures 6*a*, *b*, and *d* provide clues to the availability of enclosed vertical and horizontal wound-rotor induction motors. Figure 6*a* illustrates the point that vertical totally enclosed wound-rotor induction motors are not generally available but may be available in random horsepowers and speeds. TEWAC (totally enclosed water-to-air-cooled) construction becomes available at some minimum point as illustrated. In large ratings, shown on the right, TEWAC construction becomes very important as it provides a convenient way of capturing motor losses and expelling them in cooling water external to the building. Both curves of Fig. 6*a* terminate at 1,800 rpm synchronous speed because higher speed ratings are generally unavailable.

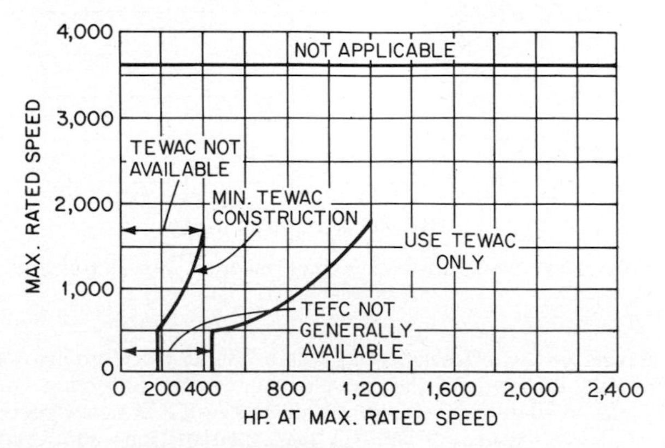

Fig. 6a Approximate availability of totally enclosed vertical wound-rotor induction motors. (General Electric Co.)

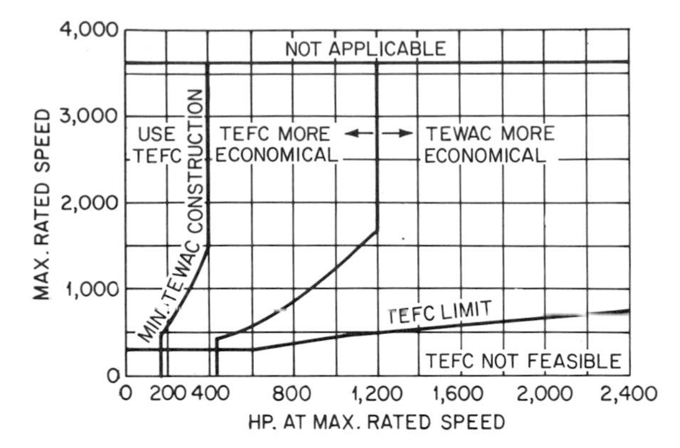

Fig. 6b Approximate availability of totally enclosed horizontal and vertical squirrel-cage induction motors and horizontal wound-rotor induction motors. Wound-rotor induction motor speeds limited to 1800 rpm maximum. (General Electric Co.)

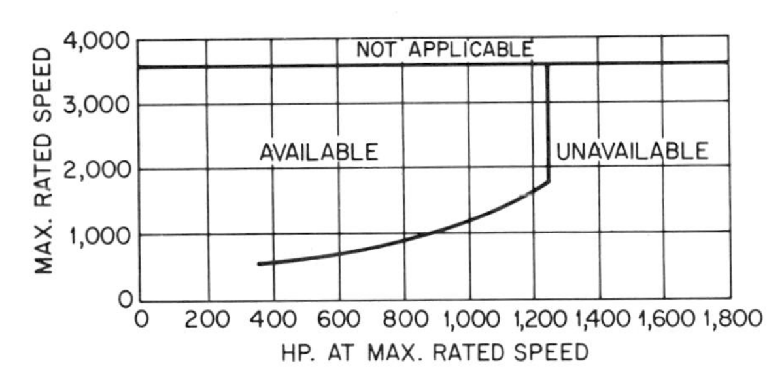

Fig. 6c Approximate availability of explosion-proof vertical squirrel-cage induction motors. (General Electric Co.)

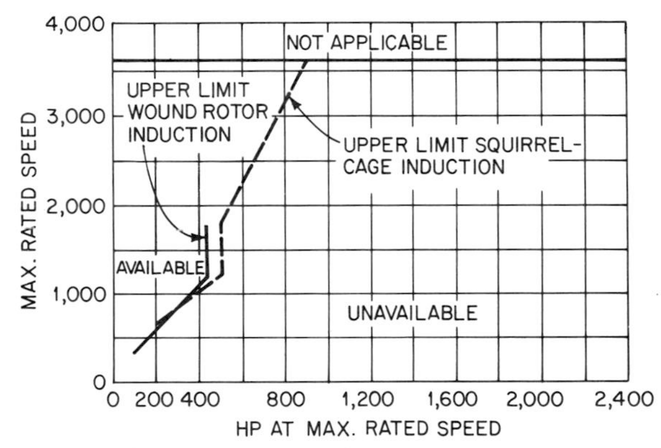

Fig. 6d Approximate availability of explosion-proof horizontal squirrel-cage and wound-rotor induction motors. (General Electric Co.)

Figure 6*b* delineates the left-hand areas where TEFC (totally enclosed fan-cooled) construction should be used for horizontal wound-rotor induction motors because it is the only one available; the middle area where TEFC construction should be used because it is less expensive than TEWAC construction; and, finally, the right-hand area where TEWAC is less expensive than TEFC construction. Again, motor speeds are limited to 1,800 rpm synchronous.

Vertical wound-rotor induction motors are not available in explosionproof construction as their omission from Fig. 6*c* implies.

Finally, the areas of availability and unavailability are delineated in Fig. 6*d* for horizontal wound-rotor induction motors in explosionproof construction. Again, the curve terminates at 1,800 rpm synchronous speed for reasons already expressed.

Delineation of motor availability is an important criterion in drive selection. Table 2 presents significant data useful in drive selection.

TABLE 2 Wound-Rotor Induction Motor with Liquid Rheostat Drive Data

Drive element	Hp ratings*	Voltage rating	Max rated speed, rpm	Speed range, %	Enclosure	Mounting	No. of speed pts.
Motor	No limits for open motors. See Fig. 6*a, b, d* for restrictions by enclosure	Any NEMA Standard	All established by number of motor poles and supply frequency	60	Vertical: shielded dripproof, TEFC generally not available, TEWAC available per Fig. 6*a* Horizontal: dripproof, TEFC, explosionproof per Fig. 6*d*	Vert. and horiz.	Infinite
Control	25 min; no stated max	For any motor secondary voltage	. . .	60	NEMA 1	Lineup or singly Floor only	Infinite

* The drive may be designed for constant torque or torque varying as the square of the speed.

Wound-Rotor Induction Motors with Static-Power-Pulse Resistance-Type Secondary Controls This drive utilizes the same wound-rotor induction motor and FVNR starter as the preceding but substitutes a static-power-pulse resistance-type controller for liquid rheostat. Figure 7*A* shows the configuration with the wound-rotor induction motor and the starter identical to that of Fig. 4.

A full-wave double bridge receives rotor slip losses at variable voltage and frequency, and rectifies them to a variable dc voltage. A filter smooths out the ac ripple, and the dc voltage is then impressed on a fixed resistor R_1. Resistor R_2 possesses very low resistance and parallels resistor R_1. The SCR in series with R_2 switches current on and off and operates under the logic of the firing circuitry, FC. The low-energy load speed potentiometer controls speed of the drive by changing reference to the firing circuit. Other forms of signal devices such as process instruments may substitute for the potentiometer.

The SCR fires and rests for a total fixed time period T during each cycle. Figure 7*B* illustrates the appearance of the secondary resistance of the controller when the SCR is on 50 percent of the time and off the remainder. Note the reappearance of the fixed time cycle T. When the SCR is off, only R_1 is in the circuit. When the SCR fires, R_1 and R_2 are in the circuit. The effective value of resistance results from the contributions of these two resistances. As the SCR fires longer and rests less, the combination R_1 and R_2 exists longer in each time period as indicated in Fig. 7*C*. The effective secondary resistance decreases from the previous case thereby increasing the speed of the motor. When the SCR remains idle continuously, only the resistor R_1 remains in the secondary circuit, and the effective secondary resistance equals the value of R_1. Under this condition the motor would operate at minimum speed.

The motor starter provides normal motor protective functions as well as short-circuit protection of the motor cables and starter. The static-power-pulse resistance-type controller monitors motor current and limits it to some preselected value such as 150 percent of normal value under all conditions.

Packaging of the secondary controller is of particular interest. Both packaging and configuration lend themselves to mounting resistors R_1 and R_2 on or near the control enclosure or removing them some distance from the control enclosure. By placing them outdoors or at some indoor location away from the operations, heat losses can be released to the environment with impunity to operators.

TABLE 3 Wound-Rotor Induction Motor with Static-Power-Pulse Resistance Controller Drive Data

Drive element	Hp ratings*	Voltage rating	Max rated speed, rpm	Speed range, %	Enclosure	Mounting	No. of speed pts.
Motor	No limits for open motors. See Fig. 6*a, b, d* for restrictions by enclosure	Any NEMA Standard	All established by number of motor poles and supply frequency	60	Vertical: shielded dripproof, TEFC generally not available, TEWAC available per Fig. 6*a* Horizontal: dripproof, TEFC, explosion-proof per Fig. 6*d*	Vert. and horiz.	Infinite
Control	7½ min 500 max	For any motor secondary voltage	...	30 or 50	NEMA 1	Lineup or singly Floor only	Infinite

* The drive may be designed for constant torque or torque varying as the square of the speed.

The speed-torque ability of this drive would be very close to that illustrated by Fig. 5. The comments contained in the text originally describing that figure apply equally well here. Table 3 presents significant data useful in drive selection.

Wound-Rotor Induction Motors with Contact-Type Secondary Controls Many pump drives utilize a very simple combination of FVNR starter, wound-rotor induction motor, and a form of contact making secondary control. Figure 8 shows the configuration with the motor and FVNR starter identical with those of Figs 4 and 7*A*.

The main difference between this drive and the two described previously lies in the physical construction of the secondary controller and the characteristics of the drive. R is a three-phase resistor connected across the slip rings of the motor. Its resistance rating is selected to provide minimum motor torque at standstill or adequate torque at minimum speed. Contacts existing in the form of magnetic contactors, drum, cam, or dial switches are closed by a logic device generally magnetic in character to short-circuit part or all of the resistor. Thus, instead of a modulated

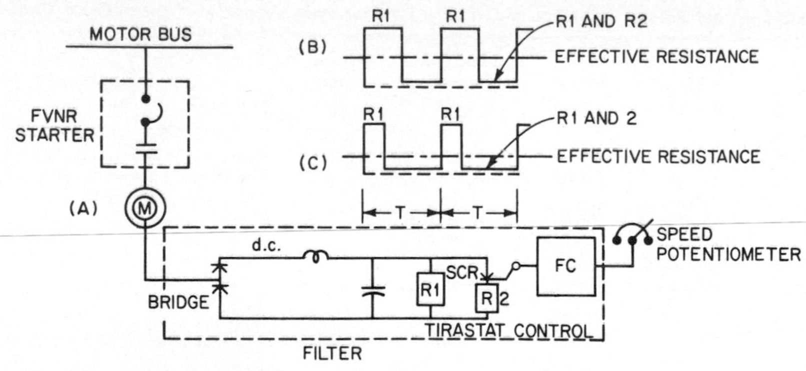

Fig. 7 Wound-rotor induction motor, Tirastat secondary-control power block diagram. (General Electric Co.)

or infinitely variable resistor as encountered with the other two drives, this drive modifies secondary resistance in discrete steps, providing discrete speed-torque curves instead of an infinite family. The number of speed-torque curves that the control generates will depend directly on the number of contacts provided in the secondary-control circuit.

A manual switch, or some automatic contact-making device actuated by the process, can be utilized to actuate the control logic.

Secondary controls are normally packaged in the factory with resistors and contact-making secondary controller integrally assembled.

The speed-torque capability of this drive would be very close to that illustrated in Fig. 5 with the exact number of curves determined by the number of contacts of the contact-making secondary controller and the shape of the curves determined by the resistance selected. Table 4 presents significant data useful in drive selection.

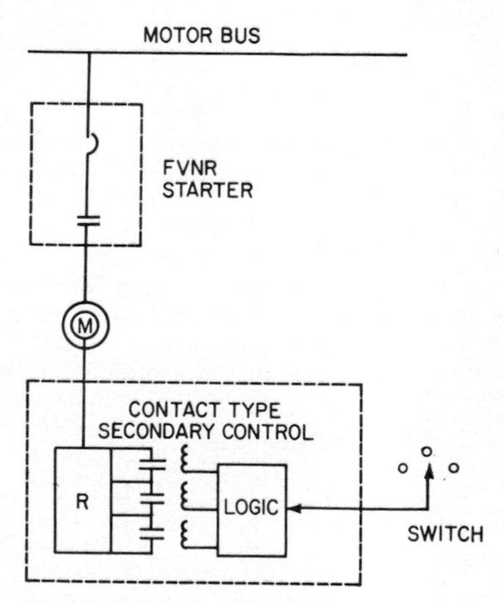

Fig. 8 Wound-rotor induction motor, magnetic secondary-control power block diagram.

Adjustable-Frequency Drive This drive consists of an adjustable-frequency rectifier-inverter and a constant-speed type of motor (M). Figure 9 shows a configuration for this drive. The control consists fundamentally of a circuit-protective device such as a circuit breaker, a diode bridge, an inverter section consists of an SCR and a firing-control section (FC) for the SCR. All are packaged in a steel enclosure. Required auxiliaries such as special power supplies, motor protective devices, and electric protective devices complete the package.

Only low voltages at 60 Hz are used to energize the motor bus. A diode bridge rectifies voltage. In turn an inverter inverts the direct current into an adjustable-frequency adjustable voltage with the voltage varying linearly with the frequency. In one design, this voltage is not sinusoidal but rather consists of a number of dc pulses of positive or negative polarity as indicated in Fig. 9 and shown between inverter and motor. The firing controls modulate the width of the pulses and the number of pulses per half-cycle to vary the apparent frequency and to maintain motor voltage at a constant volts-per-cycle value. The firing controls automatically introduce additional pulses or withdraw pulses as bandwidths reach their limits.

TABLE 4 Wound-Rotor Induction Motor with Contact-Type Secondary Control Drive Data

Drive element	Hp ratings*	Voltage rating	Max rated speed, rpm	Speed range, %	Enclosure	Mounting	No. of speed pts.
Motor	No limits for open motors. See Fig. 6*a*, *b*, *d* for restrictions by enclosure	Any NEMA Standard	All established by number of motor poles and supply frequency	60	Vertical: shielded dripproof, TEFC generally not available, TEWAC, available per Fig. 6*a* Horizontal: dripproof, TEFC, explosion-proof per Fig. 6*d*	Vert. and horiz.	
Control	No limits	For any motor secondary voltage	...	75	NEMA 1	Line-up or singly	Discrete only

* The drive may be designed for constant torque or torque varying as the square of the speed.

Different manufacturers feature different designs; however, all approach the same objective, namely varying output frequency to obtain different motor speeds.

Firing circuits composed of solid-state devices perform the triggering of SCR. They accomplish this in accordance with logic controlled by the setting adjustment of a speed potentiometer, as indicated in Fig. 9, or the signal from the process.

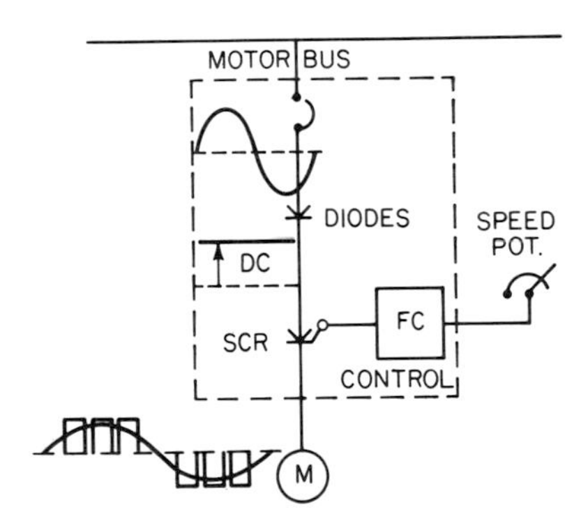

Fig. 9 Ac adjustable-frequency drive block diagram. (General Electric Co.)

The control provides normal protection of the motor as well as short-circuit protection of control, motor, and cables. In addition, the control monitors and limits current drawn by the motor under all conditions to a preadjusted value, such as 150 percent of normal.

The motor usually is a squirrel-cage induction motor. The motor need not be of special design although compatibility between motor and control unit should be confirmed by motor and control-unit suppliers.

Figure 10 displays some representative speed-torque characteristics of the drive utilizing a squirrel-cage induction motor and a representative centrifugal pump. Note that each characteristic intersects the zero-torque point at a different speed value. This characteristic differs from those illustrated for other drives illustrated in Figs. 2 and 5. The steepness of the slope of each as it rises from its zero-torque value indicates its low-speed regulation and provides a clue to its ability to maintain speed with little fluctuation as load torque varies slightly. It should be obvious that motor speed varies as a function

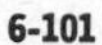

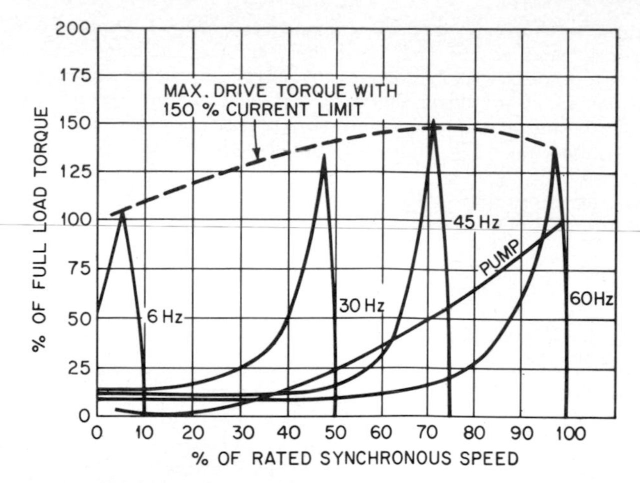

Fig. 10 Adjustable-frequency drive torque versus synchronous speed.

TABLE 5 Adjustable-Frequency Drive Data

Drive element	Hp ratings*	Voltage rating	Max rated speed, rpm	Speed range, %	Enclosure	Mounting	No. of speed pts.
Motor	All NEMA ratings for open motors. See Fig. 6*b*, *c*, *d* for restrictions by enclosures	460 or 230	All established by number of motor poles and supply frequency	100 plus limited overspeed	Vertical: shielded dripproof, TEFC per Fig. 6*b*, explosion-proof per Fig. 6*c* Horizontal: dripproof, TEFC per Fig. 6*b* explosion-proof per Fig. 6*d*	Vert. or horiz.	Infinite
Control	Up through 500 hp; higher ratings available as custom-built units	460 or 230	Capable of producing 150% of rated frequency but at reduced torque	150	NEMA 1 or NEMA 12	Floor	Infinite

* The drive may be designed for constant torque or a torque varying as the square of the speed.

of frequency adjustment; voltage adjustments occur only to accommodate changes in motor impedance. Table 5 illustrates significant application data.

Modified Kraemer Drive Kraemer drives have been used as heavy industrial drives for many years. Although functioning very successfully, each has three rotating units requiring maintenance, and each has significant losses.

In recent years the advance of semiconductors simplified the drive configuration to that of Fig. 11, which involves use of a FVNR starter, a wound-rotor induction motor, a solid state converter, and an acceleration section. The acceleration section and contactors 1 and 2 can be omitted if the converter rating is large. The FVNR starter switches power to and protects not only the motor but the converter as well. Contactor C_2, if provided, serves as a synchronizing contactor between converter and power line and is closed only when converter and line frequencies and voltages are compatible. The contactor may be placed on either side of the converter unit.

Under running conditions, rotor-slip energy of low voltage and frequency (see Fig. 11*B*) flows to the low-frequency side of the converter. The converter unit

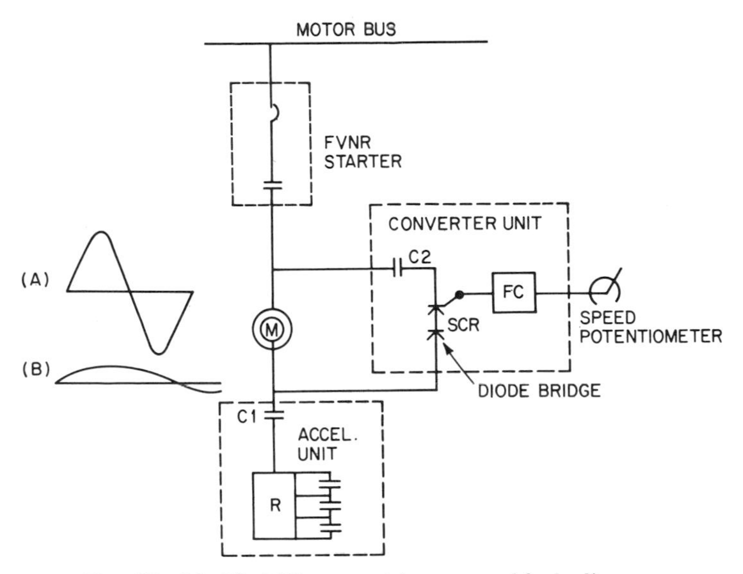

Fig. 11 Modified Kraemer drive power block diagram.

inverts this voltage to a fixed line frequency. Thus all motor rotor-slip energy except converter-unit losses are returned to the power source, thereby improving drive efficiency.

Under starting conditions, acceleration occurs with contactor C2 open and C1 closed. Accelerating contactors progressively and automatically short-circuit the resistor *R* to allow the motor to accelerate to some preselected speed. When converter output voltage and frequency match line values, contactor C2 closes and C1 opens automatically. The drive then operates as already described.

The converter generally consists of a diode bridge and SCRs with their firing circuitry for inverting. Required auxiliaries such as special power supplies complete the package. Generally all control materials are packaged in a single lineup to facilitate building and installation.

The motor starter provides normal protection of the motor as well as short-circuit protection of control, motor, cables, and converter. In addition the converter monitors and limits current drawn from the line under all conditions to a preadjusted value, such as 150 percent of normal.

Although the modified form of Kraemer drive is familiar to many, it nonetheless has been used little to drive pumps. At this writing, costs and capabilities are not sufficiently crystallized for cataloging.

Dc Motor and SCR Power Supply General industries use dc motors with SCR power supplies in large quantities. These drives perform very well in the very exacting circumstances that they normally face. Some of these drives are used to drive pumps. They are configured as shown in Fig. 12 and consist of a dc shunt-wound motor and an SCR power supply.

The dc power supply unit rectifies motor bus voltage to an adjustable dc voltage

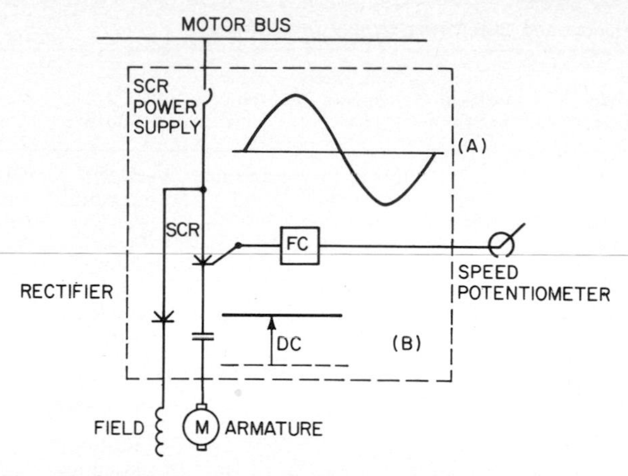

Fig. 12 Dc motor with SCR power supply power diagram.

level which in turn energizes the armature of the motor. The second function of the power supply unit is to rectify motor bus voltage and apply this constant dc voltage to the motor shunt field. By adjusting the dc voltage to the motor armature, the speed of the motor can be adjusted to a desired value.

The SCR power supply unit consists primarily of a short-circuit protective device, a switching contactor, SCR power modules, firing circuitry (FC), a speed regulator, and a shunt field rectifier. The firing circuitry responds to a low-energy-level speed potentiometer or some process variable. The SCRs in turn respond to the firing circuitry to translate those signals into the correct dc voltage to be applied to the armature of the motor. Control circuitry is included to monitor motor current and limit motor current to a preselected value, such as 150 percent of normal.

The SCR power supply unit comes packaged for easy installation in the field.

Speed-torque curves for the drive and for a representative centrifugal pump are as shown in Fig. 13. Steepness of the motor torque provides a clue to its stiffness or ability to resist speed change because of a change in load torque. Motor torques are limited to some ceiling, such as 150 percent of normal value, by the current limit circuitry of the drive. Table 6 provides significant data useful for drive selection.

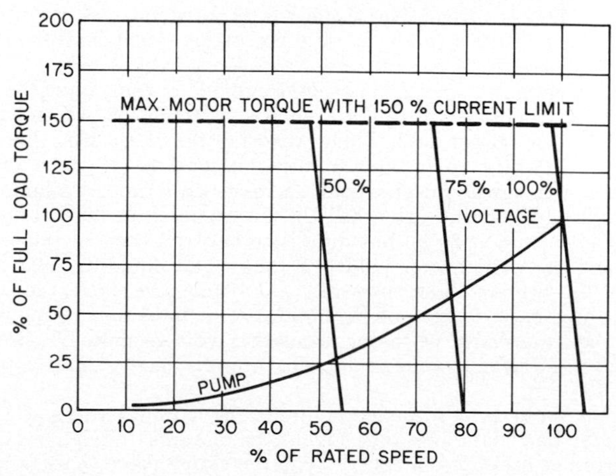

Fig. 13 Dc motor with SCR power supply torque versus speed.

TABLE 6 Dc Motor and SCR Power Supply Drive Data

Drive element	Hp ratings*	Voltage rating	Max rated speed, rpm	Speed range, %	Enclosure	Mounting	No. of speed pts.
Motor	Open horizontal machines can be built at least up to 1,500 hp; see Enclosure for other limits	240 500 550	Rating increases inversely with hp rating. Can match almost all pump speed requirements	At least 95	Vertical: open/drip-proof up through 300 hp, totally enclosed up through 200 hp Horizontal: open/drip-proof, totally enclosed up through 200 hp	Vertical; horizontal	Infinite
SCR power supply	Can be built at least up to 1,500 hp	Any NEMA ac input voltage; will match motor voltage	. . .	At least 95	NEMA 1; NEMA 12	Floor	Infinite

* Drives can power either constant-torque loads or those varying as the square of the speed.

DRIVE COMPARISONS

The remaining part of this section will deal with drive costs, features, and performances which are normally factored into criteria involved in drive selection. The relative importance of these factors seldom remain stable but change as individual circumstances change. These comparisons cover all drives discussed in this section except the modified Kraemer drive and the wound-rotor induction motor with contact-making secondary control. The former is omitted because its costs have not as yet stabilized, the latter because its costs are highly dependent upon the number of secondary-speed steps involved and motor secondary-current values.

Drive Costs Columns 1 and 2 of Table 7 provide a comparison of initial costs of the five drives discussed. Prices are shown in terms of percentages with the wound-rotor induction motor with liquid rheostat taken as 100. Percentages are used instead of monetary costs because the relationships of these drives remain fairly stable although prices increase and decrease with time. Column 1 indicates that the ac adjustable-voltage drive has the lowest price in horsepower ratings up through 50. This is true at 60 hp but difference between costs of the ac adjustable-voltage drive and the wound-rotor induction motor and liquid rheostat narrows. At 75 hp, prices of the two are about the same. At 100 hp and above, the ac adjustable-voltage drive prices exceed those of the wound-rotor induction motor with a liquid rheostat, and the availability of the ac adjustable voltage motor becomes irregular. As a result, the ac adjustable-voltage drive becomes generally unattractive at horsepowers above 75.

At 100 hp, the wound-rotor induction motor with liquid rheostat possesses the lowest initial cost and maintains this advantage upwards through maximum pump ratings of interest, as indicated in col. 2. The remaining drives possess higher prices.

TABLE 7 Comparisons of Drive Types

Type of drive	(1)* Relative initial uninstalled approx. cost (5–50 hp), %	(2)* Relative initial uninstalled approx. cost (100–500 hp), %	(3) Approx. max current during starting, %	(4) Wear points in rotating equipment	Wear points in control equipment
Ac adjustable voltage	80	....	100–150	2 motor bearings	None
Wound-rotor induction motor with liquid rheostat	100	100	100–150	2 motor bearings, 1 set collector rings, 1 set brushes	Bearings in electrolyte pump and motor electrolyte pump seals
Wound-rotor induction motor with static-power-pulse resistance-type starter	120	120	100–150	2 motor bearings, 1 set collector rings, 1 set brushes	None
Adjustable frequency	140–170	160–180	100–150	2 motor bearings	None
Dc motor with SCR power supply:					
Custom designed†	112–155	140–170	100–150	2 motor bearings, 1 commutator, 1 set brushes	None
Preengineered†	90–120	Not applicable	100–150	2 motor bearings, 1 commutator, 1 set brushes	None

* Prices cover open vertical motors only.

† Two sets of prices are provided. Use custom-designed prices for drives allowing physical amendment in factory to provide special drive sequencing and speed-control features. For applications involving manual starting and stopping and manual-speed adjustment as well as a willingness to accept factory standard ratings, use prices associated with preengineered drives.

Of course, initial cost is not the only economic factor making up total cost. Both drive efficiency and power factor may affect the total cost of the drive during its lifetime. Figure 14 is a guide to drive efficiencies for the various drives under consideration. The comparison covers a speed range of 30 percent with torque loads varying as the square of the speed. A 50-hp 1,200 rpm drive is used as a basis for comparison because it represents a probable norm for all pumps in use today. The relative spread between the various drive efficiencies shown in the figure will be maintained even with a change in horsepower or speed although the curves will be adjusted upward slightly for higher horsepowers and downward slightly for lower horsepowers.

Figure 15 illustrates representative power factors of these drives for a 50-hp drive operating through a 30 percent speed range with torque varying as the square of the speed. Power factor may be of critical importance in influencing total costs highly dependent on the terms of the rate structure applied by the electric utility furnishing power.

Operating Characteristics Operating characteristics provide another criterion for drive comparisons. In general, a comparison can be made by reviewing the follow-

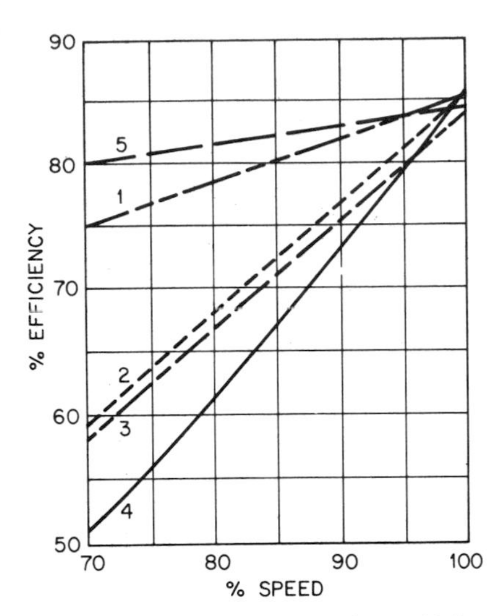

Fig. 14 Efficiency comparison: 50 hp, 1,200 rpm, synchronous speed. (1) Adjustable-frequency drive; (2) wound-rotor induction motor with liquid rheostat; (3) wound-rotor induction motor with power-pulse resistance control; (4) ac adjustable-voltage drive; (5) dc motor with SCR power supply. (General Electric Co.)

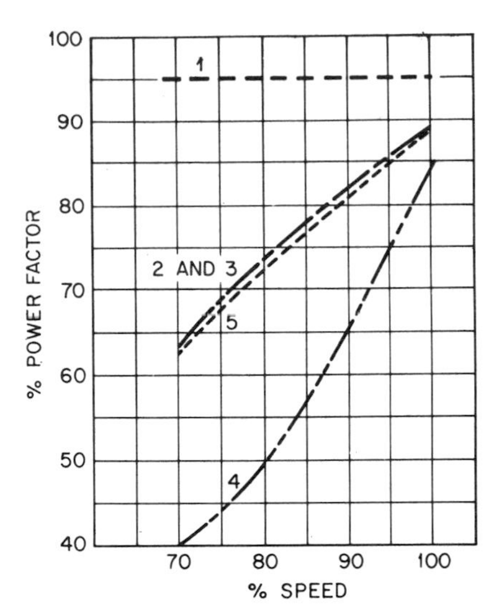

Fig. 15 Power factor comparison: 50 hp, 1,200 rpm, synchronous speed. (1) Adjustable-frequency drive; (2 and 3) wound-rotor induction motor with liquid rheostat or power-pulse resistance type control; (4) ac adjustable-voltage drive; (5) dc motor with SCR power package. (General Electric Co.)

ing characteristics of each drive:

Characteristic	*Reference*
1. Accelerating currents	Col. 3 of Table 7
2. Accelerating torques	Figs. 2, 5, 10, 13
3. Maximum rated speed	Tables 1, 2, 3, 4, 5, 6
4. Speed range	Tables 1, 2, 3, 4, 5, 6
5. Number of speed points in speed range	Tables 1, 2, 3, 4, 5, 6

All the drives examined compare favorably in each vital characteristic category and are considered generally suitable to perform functionally as pump drives.

Mechanical Features The multifaceted pump supply industry places demands on its drives requiring certain features that may serve as criteria as follows:

Mechanical feature	*Reference*
Enclosure	Tables 1, 2, 3, 4, 5, 6 and Fig. 6*a*, *b*, *c*, *d*
Mounting	Same as above
Bearing capabilities	All have generally adequate radial and thrust capabilities
Hollow shafts for verticals	All are in supply except the dc motor (available with solid shafts only)

A thorough inspection of the reference column will prove that mountings and bearing capabilities of all drives compared are generally adequate. However, the enclosure criterion shows a major variation in availability. The squirrel-cage induction motor obviously possesses all needed enclosure features for both horizontal and vertical mountings. The wound-rotor induction motor has limited availability in vertical, TEFC, enclosure, and is unavailable in vertical or horizontal explosion-proof construction. The ac adjustable voltage drive is unavailable in vertical TEFC or explosion-proof construction and is limited to approximately 40 hp maximum in horizontal TEFC or explosion-proof construction.

TEWAC (totally enclosed water to air-cooled) construction is useful as indicated in Figs. 6*a* and *b*. In usual practice, this type of construction finds use among large drives only.

Mechanical Simplicity This category is generally one of the most important in the minds of operations personnel particularly, and one of the best judgment criteria. It consists of counting the number of wear points (or parts) in both the rotating and control equipments. The rationale behind this approach is based on the concept that parts subject to wear are the eventual causes of failure in the drive: the fewer the number of parts, the less cause for failure. Column 4 of Table 7 provides the number and identity of wear points of the drives under consideration. Obviously the ac adjustable-voltage drive (utilizing a squirrel-cage induction motor) and the adjustable-frequency drive (again utilizing a squirrel-cage induction motor) have the simplest rotating and control equipments.

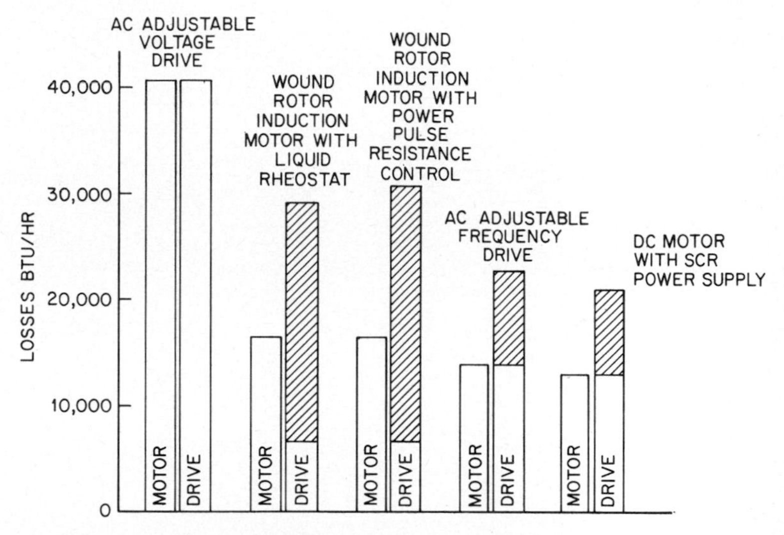

Fig. 16 Maximum loss comparison allocated to motor and drive: 50 hp, 1,200 rpm. Torque varies as the speed squared; 30 percent range. Crosshatched areas show control losses. (General Electric Co.)

Removal of Heat from Building This category has come to the foreground as a very important criterion in drive selection. Removal of heat from the pump buildings has become a steadily increasing problem as drive horsepowers have increased. Both operator comfort and successful drive operation are at stake as the area around drives becomes hotter and building temperatures increase.

This general problem is affected in a major way by the amount of heat emitted by the rotating equipment alone and the drive in general. The latter may or may not be of consequence depending on whether all or some of the drive losses are emitted in the building—and if so, where. Other factors of consequence are the relation of drive heat losses to building size and the geographic location.

Figure 16 shows maximum heat-loss distribution of 50-hp drives operating through a 30 percent speed range with torques varying as the square of the speed. Maximum motor-heat losses are shown separately to indicate the amount of heat emitted around the motor. These occur at maximum speed. The maximum drive-heat losses show losses emitted by the motor and its control at minimum or maximum speed. The crosshatched areas depict the control losses and the noncrosshatched areas show associated motor losses. It is important that the destiny of the control-heat losses be kept in mind as some of these are dissipated in the pump building. Others may be dissipated in cooling water or outdoors. Total losses dissipated in the building may range greatly, depending upon the physical handling of these losses.

Losses of drives at different ratings but operating through a 30 percent range may be approximated by extrapolation.

FINAL SELECTION OF DRIVES

In the final analysis, no simple routine can be suggested for drive selection. Each drive must be reviewed in the light of the application. By reviewing the capabilities of the various drives in each of the criteria used in this comparison and the requirements of the application, one has a starting point that should lead to successful drive selection.

Section 6.2.3

Fluid Couplings

CONRAD L. ARNOLD

TYPES OF FLUID COUPLINGS

The term *fluid coupling* can be loosely used to describe any device utilizing a fluid to transmit power. The fluid is invariably a natural or synthetic oil. This is due to the fact that oil is capable of transmitting power, is a lubricant, and is able to absorb and dissipate heat. Manufacturers have attempted to utilize water, but sealing problems (keeping water out of the bearings and oil out of the water) and corrosion have prevented its use in any standard catalog drive.

All fluid couplings may be broken down in four categories as follows:

1. Hydrokinetic
2. Hydrodynamic
3. Hydroviscous
4. Hydrostatic

HYDROKINETIC DRIVES

While all types of fluid couplings are used in starting and controlling pumps, the one most commonly used is the hydrokinetic machine (Fig. 1).

Basic Principle In the hydrokinetic drive, commonly known as a Fluid Drive or Hydraulic Coupling, oil fluid particles are accelerated in the impeller (driving member) and are then decelerated as they impinge on the blades of the runner (driven member). Thus power is delivered in accordance with the basic law of kinetic energy: $E = \frac{1}{2}M(V_1^2 - V_2^2)$. E represents energy, M is the mass of the working fluid, V_1 is the velocity of the oil particles before impingement, and V_2 is the velocity after impingement on the runner blades. These principles are used in so-called traction units and, with modification, in torque convertors. Neither of these offer controlled variable speed.

In variable-speed units, the mass of the working fluid can be changed while the machine is operating and infinitely variable output speed is achieved. Variation of oil quantity can be accomplished in four ways, which we will class as: scoop-trimming couplings, leakoff couplings, scoop-control couplings, and put-and-take couplings (Figs. 2 to 5).

Components These are common to all the above types with few exceptions:

The *housing* of the fluid drive serves four purposes—it serves as a reservoir for the nonworking oil, a support for the bearings and scoop tube, a guard to surround the moving parts, and it contains oil particles and vapors to prevent their escape

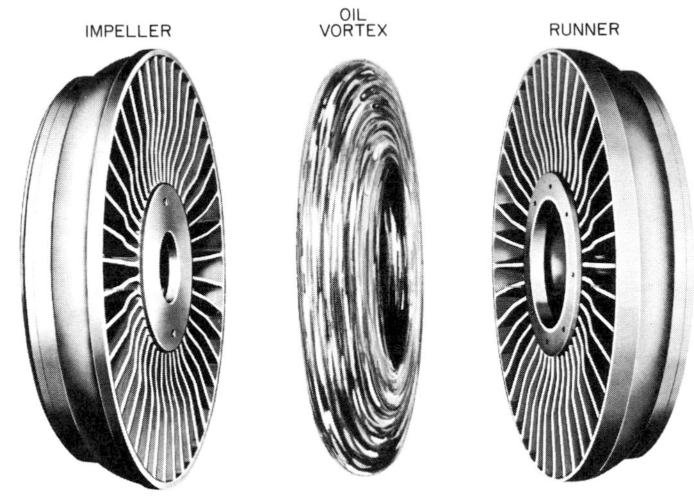

Fig. 1 Power-transmitting elements of a hydrokinetic coupling. (American Standard, Industrial Products Division)

to the atmosphere. It also provides support for the oil pump when an internal pump is used. On small units, the housing is of end-bell construction; all others are split on the horizontal centerline to facilitate inspection and maintenance.

Bearings are used to support the shafts radially and axially. In the case of smaller industrial units, ball or roller bearings are usually used; larger machines utilize babbitt radial bearings and Kingsbury-type thrust bearings. Input sleeve-bearing pillow blocks often support the internal oil pump driving and driven gears. In most cases the thrust bearings are designed to handle only the internal thrust of the fluid drive. While thrust developed by sleeve-bearing driving motors can be accepted by the hydraulic couplings, driven machines must usually have provisions to absorb their own thrust.

Shafts support the rotors and transmit driving torque to and from the rotors. In some cases they are hollow and are used to supply oil to the bearings and to the working circuit (Figs. 2 to 5).

Rotors are often compared to halves of grapefruit after the meat has been removed, and may be fabricated in three different ways. The lightest duty units are equipped with die-cast rotors of SAE 356 aluminum. Heavier duty units have rotors which are machined out of 4130 or 4340 aircraft-quality steel forgings.

Inner and outer *casings* are bowl-shaped members which bolt to the front of the impeller to contain the oil in two interconnected areas known as the working circuit (see Fig. 2). One chamber is formed by the impeller and inner casings. The other is formed between the inner and outer casings and can be called the scoop-tube chamber. Ports in the inner casing permit oil to flow between the two.

The *scoop tube,* Fig. 2, is capable of being moved radially or can be rotated inside the scoop tube chamber, and is supported by sleeve or antifriction bearings. The pickup end of the tube is between the two casings facing the direction of oil rotation. Linkages permit the tube to be moved from outside the housing, and seals prevent the leakage of oil or vapors at this penetration.

An *oil pump* is provided which may be an internally mounted gear pump driven from the input shaft or an externally mounted positive-displacement motor-driven pump. In cases where extreme reliability is required, emergency standby ac and/or dc driven pumps may be furnished. These pumps furnish light turbine oil to lubricate, transmit power, and remove heat from the fluid drive. In many cases they supply lubrication to the driver, the intermediate gear boxes, and the driven machine.

Oil coolers are required on all drives rated above 3 hp. These coolers remove heat dissipated by the fluid drive and other machines for which they furnish lubrication. Shell and tube water-to-oil exchangers are normally supplied, although finned

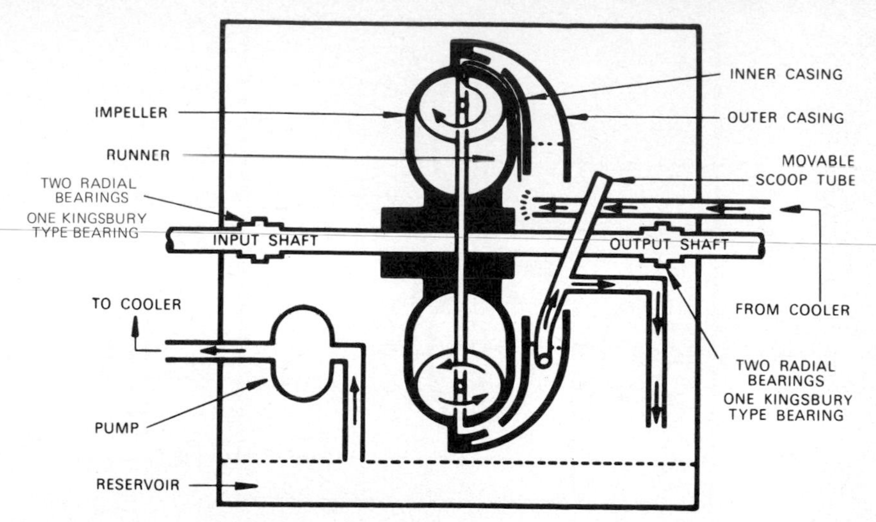

Fig. 2 Hydrokinetic coupling, scoop-trimming type. (American Standard, Industrial Products Division)

tube air-to-oil exchangers are utilized where water is not available or economical. On pipeline work, it is common to use in-line coolers, wherein the product of the pipeline is put through one side of the cooler to remove heat from the fluid-drive oil system.

Manifolds are usually used on scoop-controlled couplings in lieu of housings. They provide passages to permit oil flow to and from the working circuit, and support the scoop tube, and sometimes one bearing on the output shaft.

Operation The flow of oil in the *scoop-trimming* fluid drive is begun by the circulating pump, which is driven at constant speed by the input shaft, or external motor driver. It pumps the oil from the reservoir at the bottom of the housing to an external oil cooler (if used) and then to the rotating elements. Oil entering the rotating casing is acted upon by centrifugal force caused by the casing rotating at the input speed. This centrifugal force throws the oil outward against the side of the casing and into the impeller and runner, or working circuit, where it takes the form of an annular ring. Communication ports in the inner casing permit oil to equalize in level between the chambers.

The amount of oil in the working circuit is regulated by the scoop tube acting as a sliding weir. The scoop tube removes the oil from the casing and empties it into the oil reservoir at the bottom of the housing where it is ready to begin the cycle once more.

By using either manual or automatic control the scoop tube is moved in the casing. This, in turn, sets the level of the oil in the working circuit since the oil tends to seek the same level in the entire assembly. The scoop tube is designed to give fast response for both increase and decrease of output speed as required. In the *leakoff* type of fluid drive, Fig. 3, the scoop tube and outer casings are not used. Oil flow is initiated by a pump, usually of the viscous or centrifugal type, driven by the input shaft, through a heat exchanger (if required) and to a two-way control valve. This control valve modulates between the two extreme positions: all oil to the working circuit and all oil dumped back to the reservoir. Oil in the working circuit is thrown out through orifices called leakoff ports. Flow is created by centrifugal head which varies with the depth of oil in the coupling.

If oil is added to the coupling faster than it is thrown out of the orifices, the quantity of oil in the unit and the output speed increase. Obviously the converse is true, and if oil is put into the working circuit at *exactly* the same rate that it is "leaked off," the unit runs at constant speed. This type of unit lends itself well to closed-loop automatic control which compensates for the differential flow

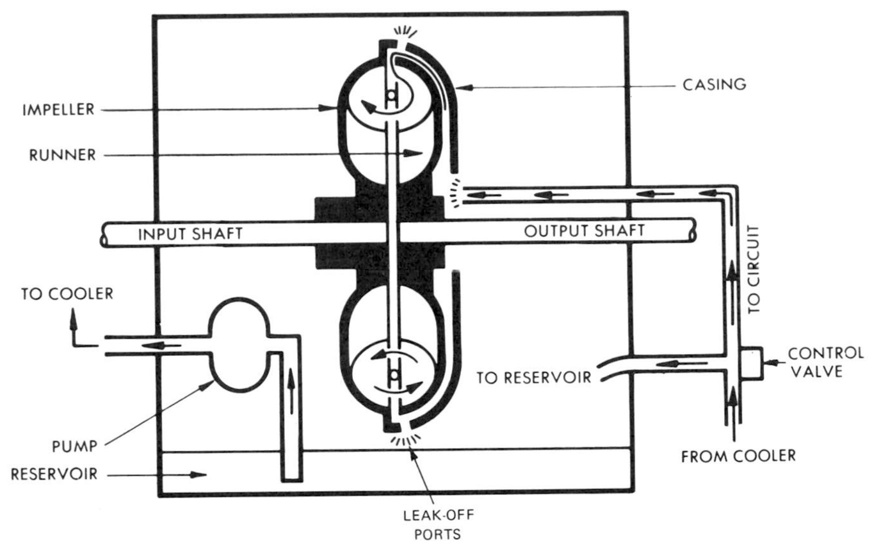

Fig. 3 Hydrokinetic coupling, leakoff type. (American Standard, Industrial Products Division)

through the leakoff ports. Manual control is questionable since oil must be added at *exactly* the rate it is discharged or output speed will drift.

In the *scoop-control* fluid drive the communication ports in the inner casing are closed to form orifices, and the scoop tube casing is sealed at the shaft. This breaks the unit into two separate chambers, the working circuit between the impeller and inner casing and the rotating reservoir between the inner and outer casings. They are connected only by the orifices, or leakoff ports. Usually the housing, three bearings, and input shaft are omitted. In this configuration, the input rotor and casings are supported by the driving motor. In some cases the mounting is accomplished through a solid hub as hown in Fig. 4, and in others through a disk which is capable of flexing to absorb slight misalignment. The runner and output shaft are supported by a pilot bearing and either an outboard bearing or by the driven machine through a piloted flexible coupling.

Oil flow is initiated by the scoop in the reservoir acting as a pump. This flow is directed out through the manifold to the oil cooler, back to the manifold, and into the working circuit.

A portion of the oil constantly flows through calibrated nozzles in the inner casing to the outer casing, where it is held in an annular ring against the outer casing by centrifugal force. The fluid drive is initially charged with just enough oil to fill the impeller and runner and the cooler circuit so that the idle oil in the outer casing is a subtraction from the working circuit. The movable scoop tube adjusts the amount of oil in the outer casing and thus regulates the oil quantity in the working circuit. The scoop tube can be fully engaged where it skims off all the oil in the casing and thus fills the working circuit, or it can be retracted completely where all the oil lies idle in the outer casing and the unit is "declutched." In-between positions regulate torque and speed of the drive.

Put-and-take couplings, Fig. 5, have not been manufactured for the past 20 years. There was, in the design of such couplings, a variation of the scoop control coupling wherein the scoop tube was located in a fixed position; thus it provided circulation only between working circuit and the cooler. The amount of oil in the coupling itself was regulated by a gear-type pump which was operated in one direction to pump oil from a reservoir into a unit, stopped to maintain constant coupling speed, or reversed to remove oil from the drive and pump it into the reservoir. This created very unwieldy control systems having very poor response characteristics with some hunting, and the design became obsolete.

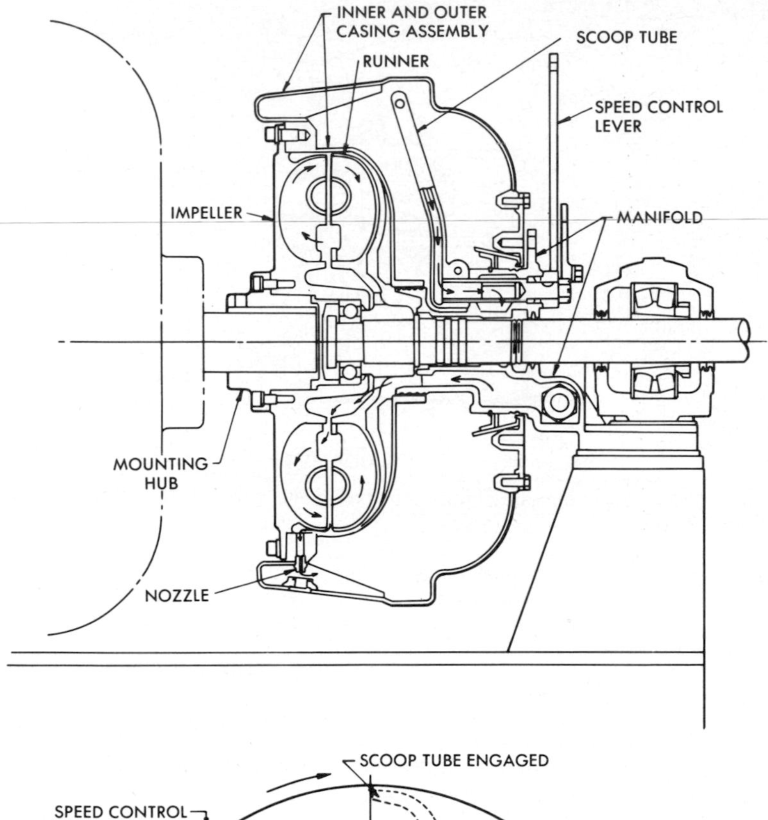

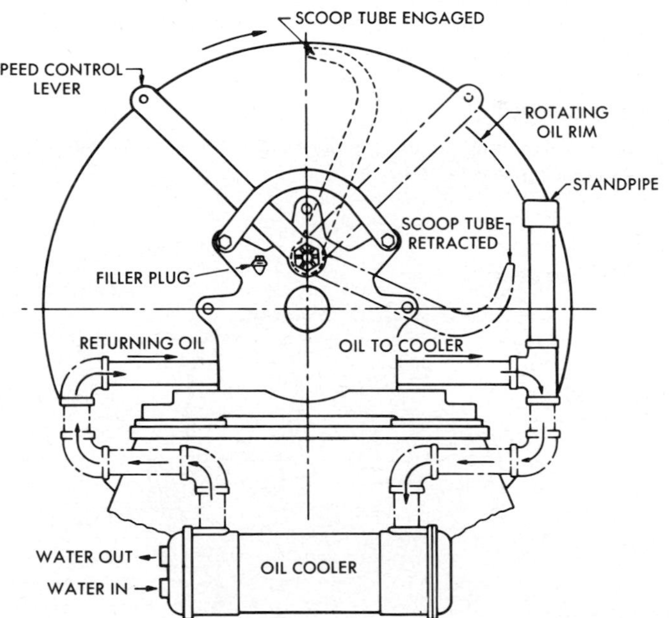

Fig. 4 Hydrokinetic coupling, scoop-control type.

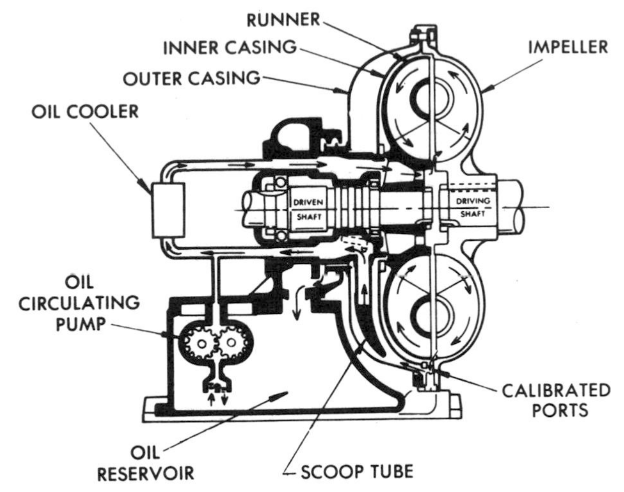

Fig. 5 Hydrokinetic coupling, put-and-take type. (American Standard, Industrial Products Division)

Reversibility can be obtained by reversing the driving motor, provided that the unit incorporates oil pumps which are not affected by input shaft rotation. In addition, units utilizing scoop tubes must have dual tips which can accept the flow of oil from either side.

In all fluid drives, the same fluid is utilized to transmit power, to remove absorbed heat, and to lubricate. Thus there is no requirement for internal seals, or slingers, and positive lubrication is assured. Since the power-transmitting medium is the heat-absorbing medium, there are no problems of heat transfer encountered in units utilizing oil pumps. This type of unit can be selected with the capability of dissipating 100 percent or more of the driving-motor rated horsepower.

HYDRODYNAMIC DRIVES

This type of fluid coupling is occasionally used to drive pumping equipment, usually in the portable pump field (Fig. 6).

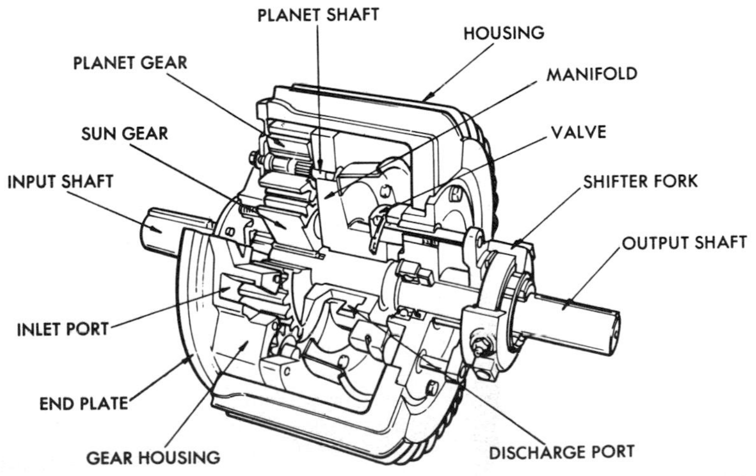

Fig. 6 Hydrodynamic coupling. (American Standard, Industrial Products Division)

Basic Principle In the most common forms of hydrodynamic drives, planetary gear trains utilize some components as oil pumps. Throttling the discharge of these pumps creates back pressure and increases drive torque.

Components The input shaft, supported by a bearing either on an independent bearing pedestal or on a packaged sub-base assembly, drives the housing, end plate, manifold, and planetary gear shafts. These planet gears are partially surrounded by the manifold, which forms a pump cavity.

The sun gear drives the output shaft. The control yolk moves the internal valve.

Operation When the driver is started, oil flow is initiated by planet gears rotating against the sun gear. With the control yolk in the low-speed position, the mixing valve is positioned to admit air mixed with oil, the pump discharge-valve ports are wide open, and the pump-developed head is approaching zero. The force on the pump gear teeth approaches zero, and the output speed is minimum. As the control yolk is moved, the pump discharge valves begin to close, less air is admitted, and the discharge pressure rises. This develops resistance to pump rotation, imparts a force on the sun gear, and the output shaft begins to rotate. If the oil discharge ports are closed, theoretically the pump pressure will rise until the pump gear is locked to the sun gear. This would rotate the output shaft at exactly input speed. In actual practice, leakage permits the pump to rotate and the output shaft turns at a slightly lesser speed than the input shaft. Reversibility can be achieved simply by reversing the driving motor.

HYDROVISCOUS DRIVES

Hydroviscous drives are relatively new in commercial use. There are several manufacturers in the United States who are marketing this type of drive for a wide range of pump applications (Fig. 7).

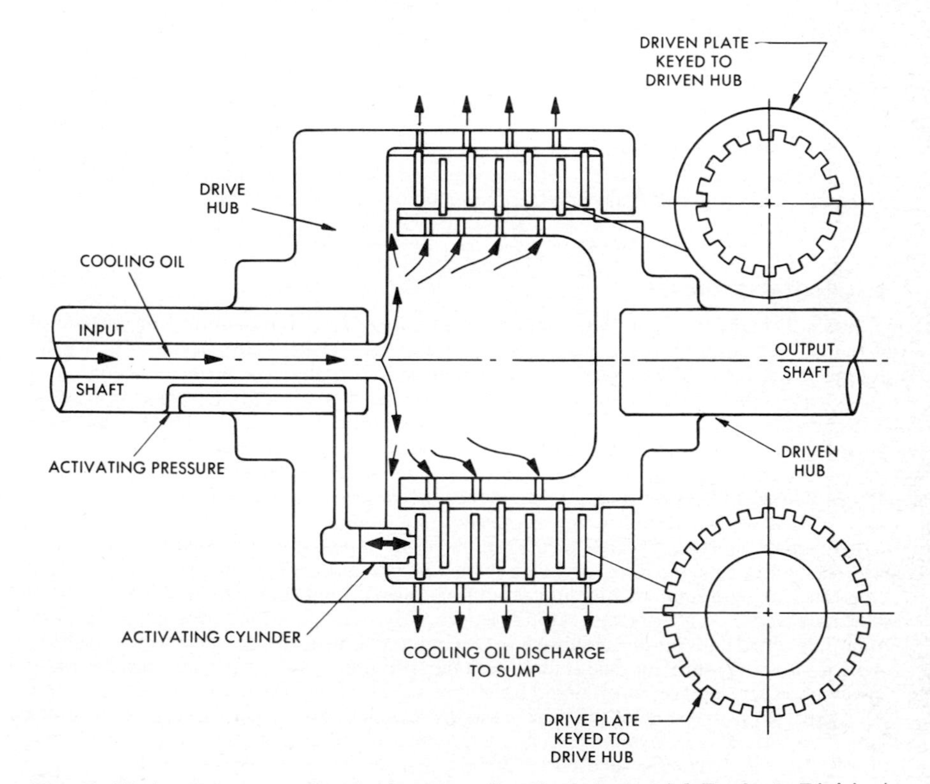

Fig. 7 Hydroviscous coupling. (American Standard, Industrial Products Division)

Basic Principle Hydroviscous drives operate on the basic principle that oil has viscosity, and energy is required to shear it. More energy is required to shear a thin film than a thick one. The hydroviscous drive varies its torque capability by varying the film thickness between driving and driven members.

Components These are common to all hydroviscous drives. The primary variations between manufacturers are in the mechanics of control, the numbers of disks, and the support of the rotors.

The *housing* serves the same purpose as in other fluid drives, supporting bearings, guarding moving parts, and containing oil and vapors. In addition, one manufacturer uses cored passages in the housing to introduce water to remove heat from the working fluid. Bearings are usually of the antifriction type in small machines, with sleeve and Kingsbury bearings available in large units. Shafts support rotors, transmit driving torque and, in most cases, are hollow to supply cooling oil and control oil.

The *rotors* have the driving hub keyed on the inside to driving disks and the driven hub keyed at its inner diameter to the driven disks.

The *disks* are made of various materials and usually grooved with some type of pattern to direct cooling-oil flow.

Pistons are hydraulic. When moved by control hydraulic oil, they force the disk stack closer together.

Oil pumps are usually motor-driven, but sometimes are driven by the input shaft of the coupling. It is not uncommon to have two separate pumping systems, one providing high-pressure control oil; the other, lower-pressure cooling oil.

Oil coolers are usually shell and tube water-to-oil heat exchangers although air-to-oil exchangers can be furnished and, as mentioned earlier, cored housings can sometimes be used.

Operation Oil flow is initiated by the oil pumps, which force cooling oil through the disk stack, draining into the sump. With the control set at minimum speed, the disks are at maximum spacing, and the coupling transmits minimum torque. As pressure is applied to the piston, it forces the disks together. This decrease in film thickness between disks increases the force transmitted between plates. At maximum piston pressure the spacing between plates is zero, and the output shaft is driven at input shaft speed. In the full-speed condition, this device is actually a lockup mechanical clutch; at reduced speeds, it is an oil shear coupling; and in a narrow band between these two points of operation it must be looked upon as an oil-cooled mechanical clutch. Reversibility can be accomplished by reversing the driving motor if oil pumps are driven by separate motors.

HYDROSTATIC DRIVES

Basic Principle There are many variations of the hydrostatic variable-speed drive, but in one form or another they invariably use positive-displacement hydraulic pumps in conjunction with positive-displacement hydraulic motors.

In some cases, varying amounts of fluid are bypassed from the pump discharge back to the pump suction. This provides a controllable, variable flow to the positive-displacement motor and, therefore, a variable output speed. This system has no particular advantages over the more common variable-speed drives. The higher than average first costs and above-average maintenance required explain why this type of hydrostatic system is seldom used.

The other method utilizes variable-flow positive-displacement pumps which may be of the sliding-vane type or axial piston type (Fig. 8). Reducing the discharge flow on the hydraulic pump reduces output speed; increasing pump flow increases output speed. This type of variable speed drive is offered in package form with pump, piping, and motor mounted in a common housing. It offers the capability of torque multiplication, maintains a relatively constant efficiency regardless of speed, has excellent control characteristics, and is widely used in the machine tool and other industries. Output shaft can be reversed by valving (without changing motor rotation). It has inherently high first cost and maintenance requirements precluding significant use as a pump driver.

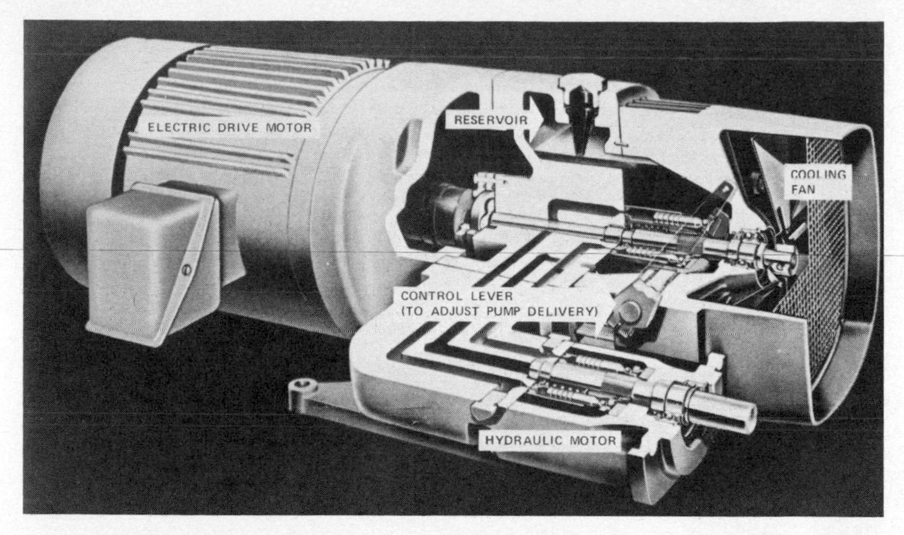

Fig. 8 Typical packaged hydrostatic drive. (Sperry Vickers)

CAPACITY

Hydrokinetic Drives Being centrifugal machines, fluid drives follow very familiar laws: horsepower varies as speeds raised to the third power, varies as diameter to the second power, and varies directly as the density of the working fluid.

Thus hydraulic capacity is governed by speed, diameter, and operating fluid; mechanical capability is governed by the structural design of housing, bearings, shafts, rotors, and casings; thermal capacity is limited by capacity of oil pumps, specific heat and thermal conductivity of the oil, and capability of the heat exchanger to dissipate heat.

It should be noted that in scoop-trimming couplings the oil pumps are usually sized solely for heat dissipation. In the leakoff coupling, the orifices are sized to permit enough oil flow to dissipate heat, and the pumps must handle this plus enough to fill the coupling in a reasonable time. The scoop-control coupling has a limited flow and pressure since both are generated by the scoop tube. This may preclude its uses on certain positive-displacement pumping applications or where installation of coolers at a remote location is required.

Hydrodynamic Drives This type of coupling varies to such a degree in design configuration that it is impossible to establish similar laws. Since any given machine has a definable torque limitation we can state that horsepower varies directly with speed.

Since standard units have no provision for removal and replacement of the working fluid, all cooling must be provided on the exterior surfaces of the rotating housing. This becomes a decided limitation if used with constant-torque loads and has limited the available sizes to some degree.

Hydroviscous Drives As would be expected, this device also follows the centrifugal laws: horsepower varies as speed raised to the third power and as diameter raised to the second. However, density of the working fluid has little or no effect; instead capacity varies directly with viscosity. Thus hydraulic capacity varies with speed, diameter, and viscosity. Mechanical capability is a relatively simple matter of structural design. However, thermal design is critical. Since power-transmitting capability varies with viscosity which, in turn, varies with temperature, disk design is most important. Free area available for cooling oil flow varies with disk spacing, output speed, and heat load.

REGULATION

The output speed of all fluid couplings (hydrostatic drives are not being considered) is affected, to some degree, by changes in load. While this may be significant in cases of single-cylinder low-speed reciprocating pumps, this effect is insignificant on multicylinder reciprocating and all centrifugal pumps. In these cases, ±1 percent speed regulation is considered normal. In special cases, regulation has been guaranteed at ±0.3 percent.

TURNDOWN

Standard cataloged hydrokinetic and hydroviscous units will offer the regulation described above over a 5 to 1 turndown on centrifugal machines and 4 to 1 when driving positive-displacement pumps on constant-pressure systems. Specially designed fluid drives have been sold which give stable control at 10 to 1 turndowns. Hydrodynamic units are limited primarily by heat dissipation capabilities and range from 100 to 1 to 1.2 to 1.

RESPONSE

It must be recognized that all fluid couplings being discussed here are slip devices. Thus any demand speed change cannot be accomplished in microseconds or milliseconds. However, the time required to change the torque applied varies from one type of unit to another.

Hydrokinetic Drives In the scoop-trimming fluid drive, speed of response is affected by many factors. The speed with which oil can be added to the working circuit (size of the oil pumps) or removed (size of the scoop tube) influences response capability.

In the *leakoff* unit, the size of the leakoff ports determines how quickly the unit will empty. However, the oil pumps must be sized to replace this oil and have additional capacity to fill the coupling in a reasonably short time.

Scoop-control units are limited by the capability of the scoop tube to pump oil from the reservoir into the working circuit and by the leakoff ports to return it to the reservoir. Some special marine couplings utilize quick-dumping valves, but these are seldom, if ever, used with pump drives.

Obviously the speed at which the scoop tube is moved is also significant. Large WK^2 of the driven equipment will increase response time.

The *scoop-trimming* coupling offers the best overall response characteristics of the hydrokinetic drives, and standard cataloged machines have "normal" fill times ranging from 10 to 15 s. They will accomplish 90 percent of a 10 percent step speed change in the 40 to 100 percent speed range in 7 to 20 s if coupled to a "normal load inertia." Special units are in operation where this change is accomplished in 2 to 6 s.

Hydrodynamic and Hydroviscous Drives Both hydrodynamic and hydroviscous couplings respond very quickly to a change in demand for torque output. Both require a mechanical motion (move or valve or change spacing between disks) followed immediately by change in pressure or change in film thickness.

In most cases the torque available for speed change and the WK^2 involved are of such a magnitude that the major portion of the response time is caused by inertial effects, rather than time required to change torque. This is particularly true in the deceleration mode of centrifugal pumps. Unless auxiliary brakes are built in, none of the hydrokinetic, hydrodynamic, or hydroviscous drives can provide dynamic braking. On a demand to decrease speed, they can at best reduce driving torque to zero. Under these circumstances, the only retarding force to slow the inertia of the driven machine is the load developed by the driven machine. In the case of centrifugal pumps on fixed systems, this load would fall off as the cube of speed, and below 40 percent speed, they have an almost insignificant braking effect.

EFFICIENCY

There are two kinds of losses present in hydrokinetic, hydrodynamic, or hydroviscous couplings. First, we will consider what are termed *circulation losses.* They are made up of bearing friction, windage, and power required to accelerate the oil within the rotor. On internal pump units, the power required to drive the oil pump is included. As an average, these losses represent approximately 1.5 percent of the unit rating and for most purposes, these losses may be considered as being constant, regardless of output speed.

Second are so-called *slip losses.* As is the case on similar slip machines such as mechanical clutches, eddy-current couplings, etc., the torque seen on the input

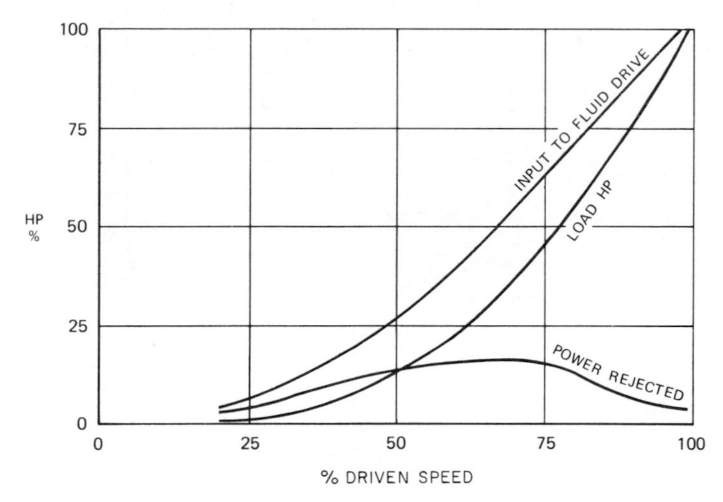

Fig. 9 Load hp varies as speed cubed. This figure represents a pump operating in a system where all the head is frictional, or where head varies as flow squared. In this system, the driven pump operates at only one point on its characteristic, and therefore the shape of the pump curve is academic. While the rapid loss of efficiency of the hydraulic coupling at reduced speeds is quite obvious, the dramatic decrease in pump-horsepower requirements results in a total system horsepower which is most acceptable. Note that the heat-rejection requirements in the fluid drive are maximum at about 18 percent.

shaft equals the torque on the output shaft. Therefore any reduction in rpm on the output shaft has a directly related power loss which occurs inside the machine. In other words, slip efficiency equals output speed/input speed times 100. Total fluid-drive losses are the sum of the two inefficiencies. The complete energy formula is as follows:

$$\text{Fluid drive input hp} = \frac{\text{output hp}}{\text{output speed/input speed}} + \text{circulation hp losses}$$

At maximum designed operating speed (which is usually about 98 percent of driven speed), the total coupling efficiency is approximately 96½ percent. 1½ percent of the losses are circulation losses and 2 percent are slip losses. The hydroviscous unit can be operated at 100 percent driven speed, but under these conditions, it is not a fluid coupling.

Since the circulation losses become relatively insignificant at reduced speeds, approximate calculations may be made using the following formula:

$$\text{Efficiency} = \frac{\text{output speed}}{\text{input speed}}$$

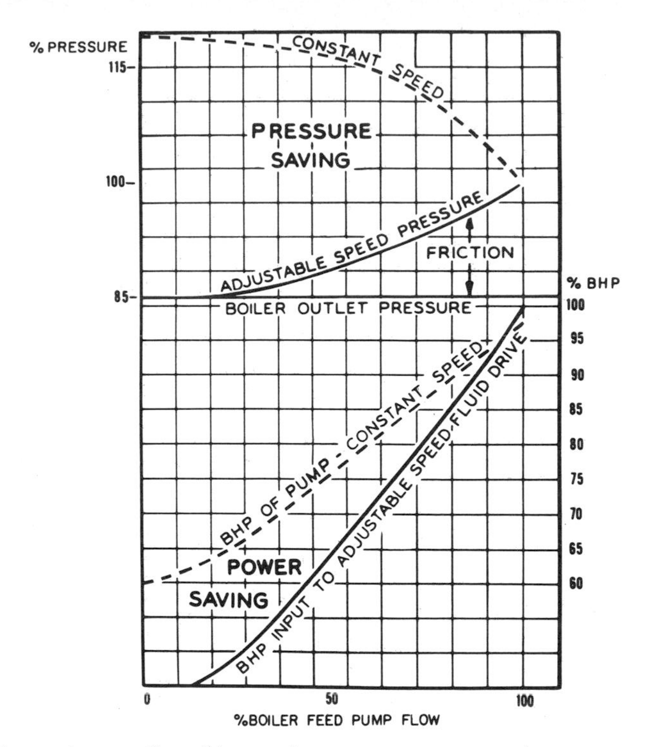

Fig. 10 Comparison: adjustable speed versus constant speed—pressure and horsepower characteristics for typical centrifugal boiler feed pump. This figure is typical of a boiler feed pump where a high percentage of the developed head is relatively constant. In this case, this fixed head is the boiler pressure. It demonstrates the savings in pressure and horsepower as compared to control with a feedwater regulating valve.

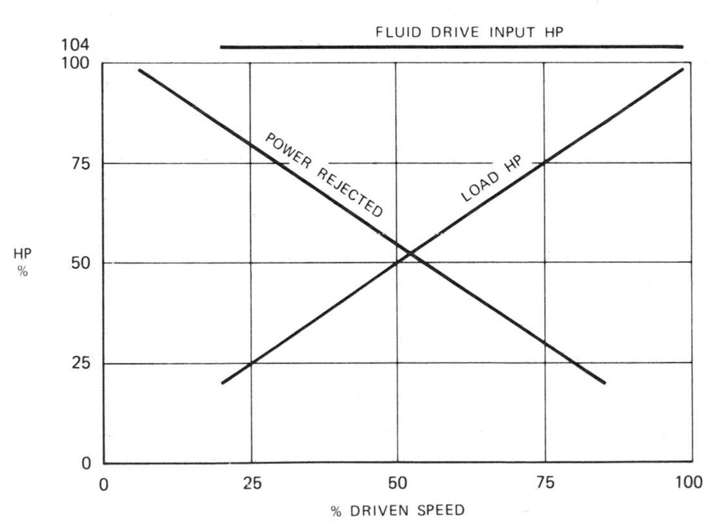

Fig. 11 Positive displacement pump with constant discharge head. This figure assumes that a positive-displacement pump is working on a system where pressure is constant. While this type of system is seldom found, it is shown to demonstrate that the rapid reduction in fluid-drive efficiency does not require overmotoring the pump. While system efficiency is much poorer than that of a bypass valve, fluid drives are used to provide no-load starting, isolation of torsional vibrations in reciprocating pumps, and elimination of the bypass valve in slurry systems where erosion is severe.

CONTROLLERS

Of the hydrokinetic devices, the scoop-trimming and scoop-control units require that a mechanical motion be imparted to the scoop tubes for control, and some devices must be furnished to provide it. This may be a hand crank on a manual control system. Simple mechanical systems are often used—a typical example is a weighted float with a rope connected to the scoop tube, controlling level in a tank. However, most installations utilize electric, electrohydraulic, hydraulic, or pneumatic actuating devices. It is not surprising that the pipeline and refinery industry use electrohydraulic actuators similar to those used on valves. The electric utility industry prefers pneumatic or electric damper operators. The only criteria for actuator selection is "compatibility with the other elements of the control system."

The leakoff type devices require a signal to the control valve. At present this is standardized as a hydraulic-pressure signal, although special transducers would permit the use of other types of signals. The hydrodynamic devices are available with manual level control (which could be adapted to actuators) or with closed-loop constant-pressure or temperature systems. All hydroviscous drives utilize oil pressure applied to a piston to "clamp" or vary the spacing of the disks. This hydraulic pressure may be varied utilizing almost any type of signal provided that the proper servos are utilized. Thus the signal may be electric, hydraulic, or pneumatic.

CAPACITIES AVAILABLE

Standard cataloged variable-speed fluid couplings are available from one or more manufacturers in the speeds and horsepowers shown in Table 1. Special designs are available for higher horsepower.

TABLE 1 Variable-Speed Coupling Capacities

	Hydrokinetic*			
Input speed, rpm	Min/max hp (H)	Max (V)	Hydrodynamic† Min/max hp	Hydroviscous† Min/max hp
720	40,000–45,000	......	....	3,000– 8,000
900	1,000–7,500	6,000	1–25	3,000–10,000
1,200	1,000–14,000	17,000	1–30	3,000–15,000
1,800	1,000–14,000	18,000	1–60	3,000–20,000
3,600	1,000–30,000	29,000	....	3,000–20,000

* Max hp (H) applies to horizontal shaft units. (V) applies to vertical machines.
† Horsepowers apply to either horizontal or vertical units on other types of drives.

DIMENSIONS

To give some idea of physical dimensions, Table 2 lists approximate dimensions for hydrokinetic drives of one U.S. manufacturer. These feature a scoop-trimming coupling and are probably the largest type unit for a given speed and horsepower.

TABLE 2 Hydrokinetic Drive Dimensions

Hp at 1,800 rpm	Length, in	Width, in	Shaft height, in
5	24	15	11½
20	39	18	12½
50	38	28	19
100	38	28	19
1,000	74	50	30
4,000	102	62	42

Fig. 12 Thirteen fluid drives are installed driving reciprocating pumps on a coal pipeline. The fluid drive absorbs a large percentage of the pulsations created by the reciprocating pumps and controls their speed to provide proper pipeline flow. (American Standard, Industrial Products Division.)

SELECTION AND PRICING

Because of variations in the design of fluid couplings from size to size and between speeds, it is virtually impossible to develop rule-of-thumb methods of estimating pricing. Examination of pricing of one major fluid drive manufacturer indicates

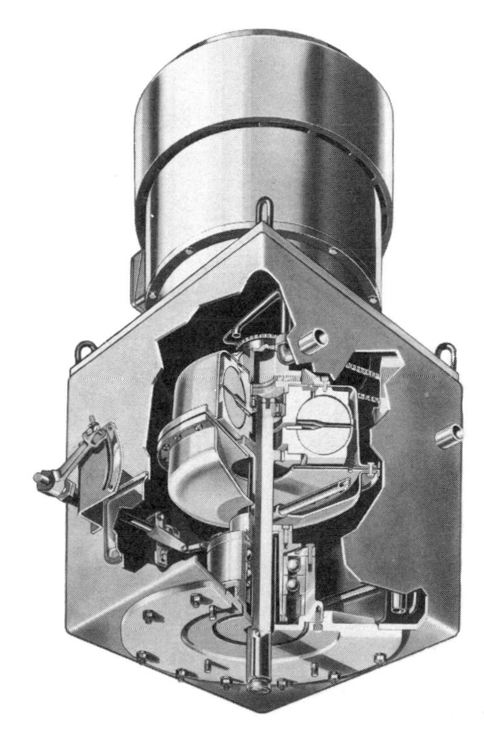

Fig. 13 Cutaway view of vertical fluid drive suitable for operation with vertical pumps. (Note NEMA P pump flange and output shaft.) (American Standard, Industrial Products Division)

that prices can range from $35.00 per horsepower for sophisticated machines down to $8.00 per horsepower for others. Pricing of fluid couplings can be increased dramatically by specification requirements for exotic controls, backup pumps and heat-exchanger equipment, and other "accessories." Because of this fact, it is recommended that the manufacturers be contacted to obtain even budget prices.

The basic information required by the manufacturer for selection and pricing is as follows:

1. Speed and type of driver
2. Horsepower required by the driven machine at, at least, one operating point
3. Character of driven machine—smooth load, pulsating load. How do torque requirements change with speed?
4. Cooling medium available, at what temperature?
5. Control type
6. Accessories
7. Special specification requirements

Reasonable budget figures can usually be obtained with only items 1 to 3.

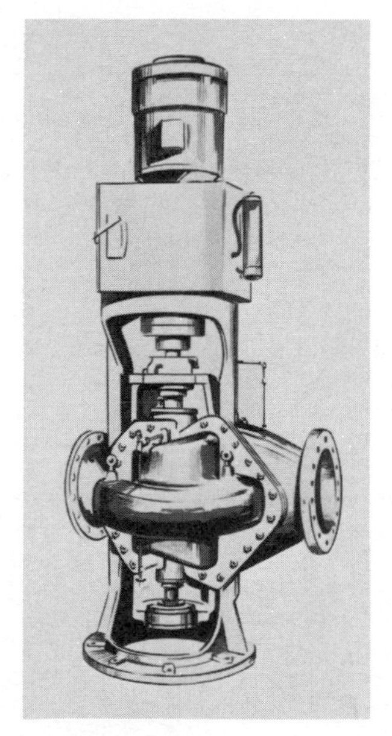

Fig. 14 Fluid drive in Fig. 13 shown mounted on a vertical shaft pump. (American Standard, Industrial Products Division)

CONCLUSION

Fluid couplings are utilized to drive pumps in virtually all pump applications requiring variable-flow or pressure. They are used primarily to improve efficiency and controllability, to permit no-load starting, and to reduce pump and system wear. They are standardized to the degree that units are available to handle most pumping applications. Most manufacturers stand ready to develop new designs as the requirements of the marketplace change.

Section 6.2.4

Gears

H. O. KRON

USE OF GEARS WITH PUMP DRIVES

The main use of gearing in pump drives is to reduce the speed of the prime mover (a motor or an engine) to the level applicable to the pump. In some cases, however, the gears are employed to step up the speed because the pump is required to operate at a speed faster than that of the prime mover.

There are other jobs that gear drives must perform with pump units. For instance, a pump may be of the vertical type, while the prime mover must operate horizontally (as in the case of a diesel engine). A right-angle gear set (Figs. 1 and 4) can then be incorporated into the drive to transmit the power "around the corner" even if the gears are of a 1:1 ratio and no speed change is involved.

At times, too, a gear drive must combine the power shafts of two prime movers, say, an electric motor and a diesel engine. This combination is desirable where there is a need for emergency power in cases of electric power failure. The motor is used ordinarily while the diesel is reserved for emergencies. In such an arrangement, the motor may be mounted vertically on top of the gear drive to drive right through the shaft, whereas the diesel is geared at right angles to the motor.

Gear drives are also used to vary the speed of a pump. Change gear arrangements, Fig. 2, may be used to facilitate varying the gear ratio. Also, numerous types of variable-speed devices, such as variable pulley belts, friction rollers, hydraulic couplings, hydrostatic and hydroviscous drives, eddy-current and electric drives, are often utilized (Sec. 6.2).

TYPES OF GEARS

The gears generally used for pump applications are parallel-shaft type helical or herringbone gears, or right-angle type spiral-bevel gears. Spur gears are used on occasion, particularly in smaller horsepower lower-speed pump drives. Straight bevel and hypoid gears are also occasionally used in right-angle drives. As in the case of spur gears, straight-bevel gears are limited in horsepower and speed. Hypoid gears are employed only infrequently for pumps because they are generally more costly than the other right-angle types. Worms and gears, too, are only employed on occasion, in cases where an overall package requires a compact gear arrangement or when a high ratio of speeds is called for. Worm gears are limited in horsepower capacity, and the efficiency of this type drive is lower than for other types (see Fig. 3 for types of gears).

Parallel-Shaft Gearing

Helical versus Spur Gears. Spur gears transmit power between parallel shafts without end thrust. They are simple and economical to manufacture and do

Fig. 1 Spiral-bevel gear, right-angle vertical pump drive.

Fig. 2 Four-speed change gear, variable-speed drive.

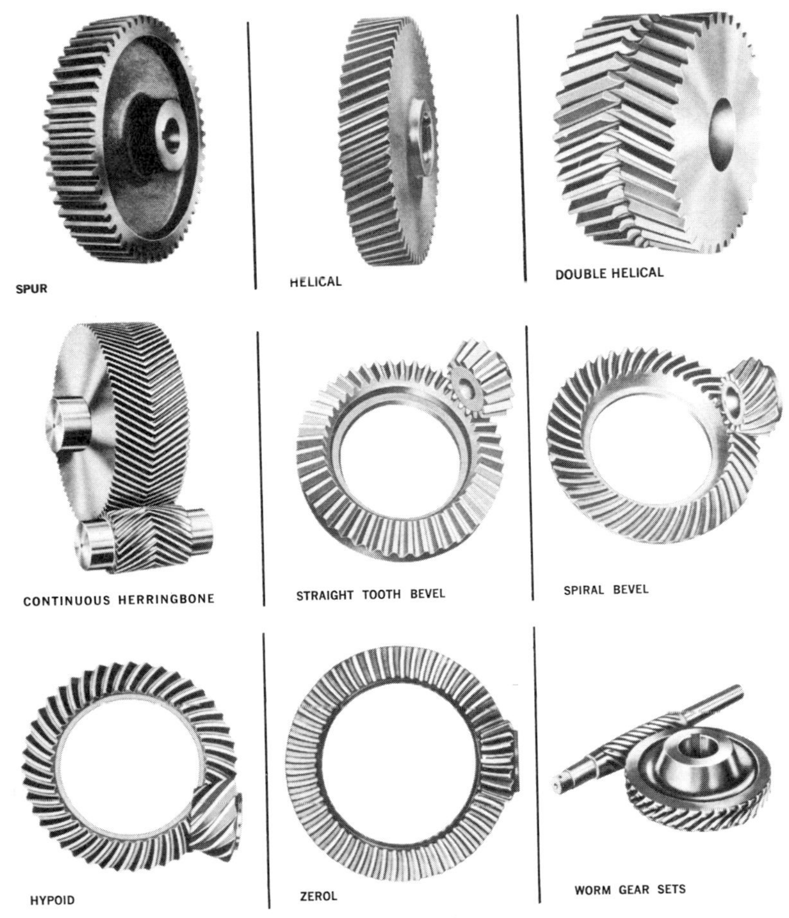

Fig. 3 Types of gears.

not require thrust bearings—but they are generally used only on moderate-speed drives.

One of the first decisions that must be made when considering a parallel-shaft gear reducer or power transmission drive is whether the gears would be of the spur or helical type, and if helical, whether they should be single or double helical gears. It has been generally acknowledged that helical gears offer better performance characteristics than do spur gears. But since helical gears, size for size, often are somewhat more expensive, some users have shied away from helical gears to keep the cost of the gear drive to a minimum. However, cost studies that compared similar size spur and helical gears found that helical gears are actually a better buy.

The geometries of the spur and helical types are both the involute tooth form. Slice a helical gear at right angles to its shaft axis and you have the typical spur-gear profile. Typical of a spur gear, however, is that when driving another spur its teeth make contact with the teeth of the mating gear along the full length of the face. The load is transferred in sequence from tooth to tooth.

A more gradual contact between mating gears is obtained by slanting the teeth in a way to form helices that make a constant angle—a helix angle with the shaft axis. Tooth contact between the teeth of mating helical gears is gradual, starting at one end and moving along the teeth so that, at any instant, the line of contact will run diagonally across the teeth. The effect of the tooth helix is to give multiple

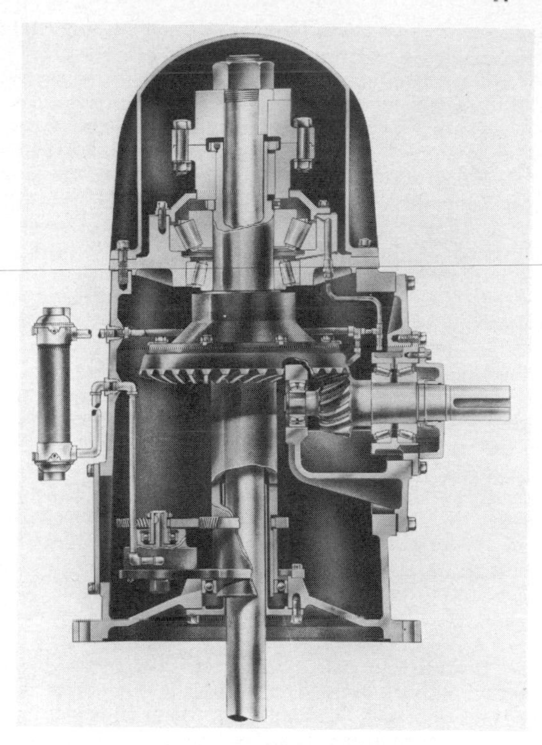

Fig. 4 Cross section of spiral-bevel vertical pump drive.

Fig. 5 High-speed parallel-shaft gear drive.

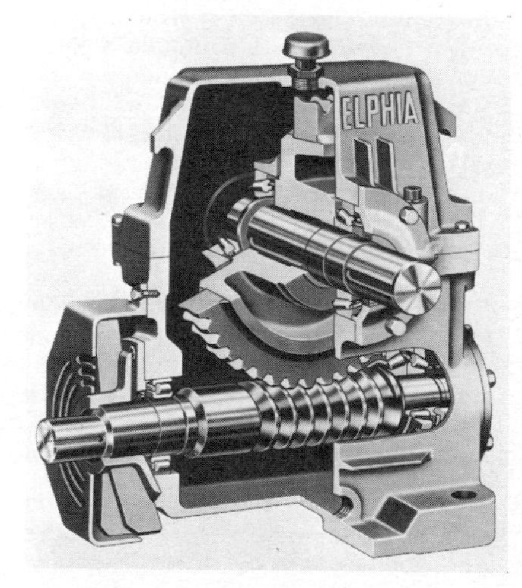

Fig. 6 Fan-cooled worm-gear drive.

teeth contact at any one time. Gear geometry can be arranged to give from two to six or more teeth contact in any time.

Because of the greater number of teeth in contact, a helical gear has a greater effective face width (up to 75 percent) than an equivalent spur gear. Also, the effect of the tooth helix on the profile geometry in the plane of rotation is to make it equivalent to a pinion with a greater number of teeth, thereby increasing its power capacity. A helical gear is capable of transmitting up to 100 percent more horsepower than an equivalent spur gear. Furthermore, a helical-gear set will give smoother, quieter operation.

The recommended upper limit of pitch-line velocity for commercial spur gears is around 1,000 ft/min. The upper limit for equivalent helical gears is about five times that, or 5,000 ft/min. Of course, as precision goes up, so do the permissible operating speeds for both spur and helical. Velocities in the 30,000 ft/min range for helical gears are not uncommon.

In addition to the normal radial loads produced by spur gears, helical gearing also produces an end thrust along the axis of rotation. The end thrust is a function of the helix angle: the larger the helix angle, the greater the thrust produced. Mounting assemblies and bearings for helical gearing must be designed to receive this thrust load.

Double Helical Gears. Gears of this type have two sets of opposed helical teeth. Each set of teeth has the same helix angle and pitch, but the helices have opposing hands of cut. Thus, the thrust loads in two sets of teeth counterbalance each other and no thrust is transmitted to shaft and bearings. Also, because end thrust is eliminated, it is possible to cut the teeth with greater helix angles than is generally used in helical gears. Tooth overlap is greater, producing a stronger and smoother tooth action.

The advantages attributed to helical gears are also applicable to double helical gears. Double helical gearing finds application in high-speed pump applications where a large helix angle must be combined with tooth sharing and elimination of end thrust for extra smooth gear action.

Single helical gears, however, have some attractive advantages over double helical gears, the most significant being that the external thrust loads do not affect gear-tooth action. With a double helical gearing, a thrust load tends to unload one of the helices and to overload the opposite one.

Furthermore, the gear face for a single helical can be made narrower than for a double helical because the need for a groove between the two helices is eliminated. This leads to the use of a narrower, stiffer pinion with less tooth deflection and torsional windup, and, generally, to a more favorable critical-speed condition.

In addition, the accuracy of a single helical gear is greater because it is not subject to apex runout. With a double helical gear, the tolerances in tooth position tend to unload cyclically one helix and induce axial vibrations in the pinion.

Continuous-Tooth Herringbone Gears. Gears of this type are double helical gears cut without a groove separating the two rows of teeth. Because of the arched construction of these gears, they are often known as "the gears with a backbone." Continuous-tooth herringbone gears are used for the transmission of heavy loads at moderate speeds where continuous service is required, where shock and vibration are present, or where a high reduction ratio is necessary in a single train. Because of the absence of a groove between opposing teeth, a herringbone gear has greater active face width than the hobbed double helical gear, and, therefore, is stronger. There is also no end thrust as the opposing helices counterbalance one another.

Much of the success of the continuous-tooth herringbone gear is due to the greater number of teeth in contact and the continuity of tooth action, which is an outgrowth of the high helix angle. These larger helix angles can be fully utilized without creating bearing thrust loads. Continuous-tooth herringbone gears normally are furnished with helix angle of 30°.

Crossed-Axis Gearing

Straight-Bevel Gears. Gears of this type transmit power between two shafts usually at right angles to each other. However, shafts other than 90° can be used.

The speed ratio between shafts can be decreased or increased by varying the number of teeth on pinion and gear. These gears are designed to operate at speeds up to 1,000 ft/min and are more economical than spiral-bevel gears for right-angle power transmission where operating conditions do not warrant the superior characteristics of spiral-bevel gearing. When shafts are at right angles and both shafts turn at the same speed, the two bevel gears can be alike and are called miter gears.

Spiral-Bevel Gears. Spiral-bevel gear teeth and straight-bevel gear teeth are both cut on cones. They differ in that the cutters for straight-bevel teeth travel in a straight line, resulting in straight teeth, whereas the cutters for spiral-bevel gear teeth travel in the arc of a circle, resulting in teeth which are curved and are called spiral.

Spiral-bevel gearing is superior to straight-bevel in that loading is always distributed over two or more teeth in any given instant. Recommended maximum pitch-line velocity for spiral bevels is about 8,000 ft/min. Spiral bevels are also smoother and more quiet in action because the teeth mesh together gradually. Because of the curved tooth, spiral-bevel pinions may be designed with fewer teeth than a straight-bevel pinion of comparable size. But thrust loads for spiral-bevel gearing are greater than for straight-tooth bevels and vary in axial direction with the direction of rotation and hand of cut of the pinion and gear. Where possible, the hand of spiral should be selected such that the pinion tends to move out of mesh.

Zerol Gears. Zerol-bevel gears are cut on conical-gear blanks and have curved teeth, similar to the spiral bevels. But the teeth are cut with a circular cutter which does not pass through the cone apex. Thus, they are spiral-bevel gears with zero-degree spiral angle (hence the name). Furthermore, the tooth bearing is localized as in spiral-bevel gearing; thus stress concentration at the tips of the gear teeth is eliminated. Zerol-bevel gears are replacing straight-bevel gearing in many installations because their operation is smoother and quieter due to their curvature and their operating life is longer. Like straight-bevel gears, zerol gears have the advantage of no inward axial thrust under any conditions. The zero-degree spiral angle produces thrust loads equivalent to straight-bevel gears.

Hypoid Gears. Hypoid-bevel gears have the general appearance of spiral-bevel gears but differ in that the shafts supporting the gears are not intersecting. The pinion shaft is offset to pass the gear shaft. The pinion and gear are cut on a hyperboloid of revolution, the name being shortened to "hypoid." Hypoid gears can be made to provide higher ratios than spiral-bevel gears. They are also stronger and operate even more smoothly and quietly. The fact that two supporting shafts can pass each other, with bearings mounted on opposite sides of the gear, provides the ultimate rigidity in mounting.

Worm Gears. In operation, the teeth on the worm of a worm-gear set slide against the gear teeth and at the same time produce a rolling action similar to that of a rack against a spur gear. Because of this screw action, worm-gear drives are quiet, vibration-free, and produce a constant output speed completely free of pulsations. Worm gearing is particularly adaptable to service where heavy shock loading is encountered.

Worm gearing is extremely compact considering load-carrying capacity. Much higher reduction ratios can be attained through a worm gear set on a given center distance than any other type of gearing. Thus the number of moving parts in a speed-reduction set is reduced to the absolute minimum. However, worm gearing is limited in horsepower capacity and has lower efficiencies than parallel-shaft and bevel-gear types (see Table 1 for generally accepted design criteria).

GEAR MATERIALS AND HEAT TREATMENT

One of the most important factors dictating the success or failure of a gear set is the choice of material and heat treatment. This is especially true for gears designed for higher horsepower. American Gear Manufacturers Association (AGMA) ratings for strength and durability of gears are dependent on the choice of material and heat treatment. It is often possible to reduce the overall size of the gear box substantially by simply changing from low hardened or medium hardened gears (about 300 Brinell) to full hardened gears (about 55 to 60 Rockwell C). Generally used methods for hardening gear sets are:

1. Through hardening
2. Nitriding
3. Induction hardening
4. Carburizing and hardening
5. Flame hardening

The use of high hardness heat-treated steels permits smaller gears for given loads. Also, hardening can increase service life up to ten times without increasing size or weight. But after hardening the gear must have at least the accuracies associated with softer gears, and for maximum service life even greater precision.

Through Hardening Suitable steels for medium to deep hardening are 4140 and 4340. These steels, as well as other alloy steels with proper hardenability characteristics and carbon content of 0.35 to 0.50, are suitable for gears requiring maximum wear resistance and high load-carrying capacity. Relatively shallow-hardening carbon steel gear materials, types 1040, 1050, 1137, and 1340, cannot be deep hardened, and are suitable for gears requiring only a moderate degree of strength and impact resistance. 4140 steel will produce a hardness of 300 to 350 Brinell. For heavy sections and applications requiring greater hardness, 4340 will provide 350 to 400 Brinell. Cutting of gears in the 380 to 400 Brinell range, although practical, is generally difficult and slow.

Nitriding Nitriding is especially valuable when distortion must be held to a minimum. It is done at a low temperature (975 to 1050°F) and without quenching—eliminating the causes of distortion common to other methods of hardening and often the necessity for finish machining after hardening. Nitrided case depths are relatively shallow so that nitriding is generally restricted to finer pitch gears (4 diameter pitch or finer). However, double-nitriding procedures have been developed for nitriding gears with as coarse a pitch as 2 diameter pitch.

Any of the steel alloys that contain nitride-forming elements, such as chromium, vanadium, or molybdenum, can be nitrided. Steels commonly nitrided are 4140, 4340, 6140, and 8740. It is possible with these steels to obtain core hardnesses of 300 to 340 Brinell and case hardnesses of 47 to 52 Rockwell C. Where harder cases are required, one of the Nitralloy steels may be used. These steels develop a case hardness of 65 to 70 Rockwell C with a core hardness of 300 to 340 Brinell. The depth of case in a 4140 or 4340 steel varies as the length of time in the nitriding furnace. A single nitride cycle will produce a case of 0.025 to 0.030 in in 72 hr. Doubling the time will produce a case depth of 0.045 to 0.050 in. For the majority of applications, the case depth obtained from a single cycle is ample.

Depth of case for Nitralloy steels is somewhat less than the depth obtainable for other alloy steels. In general, alloy steels 4140 and 4340 give up to 50 percent deeper case than Nitralloy steels for the same furnace time. These cases are tougher but less hard.

Induction Hardening Two basic types of induction hardening are used by gear manufacturers, coil and tooth-to-tooth. The coil method consists of rotating the work piece inside a coil producing high-frequency electric currents. The current causes the work piece to be heated. It is then immediately quenched in oil or water to produce the desired surface hardness. Hardnesses produced by this method range from 50 to 58 Rockwell C depending on the material. The coil method hardens the entire tooth area to below the root.

Tooth-to-tooth full-contour induction hardening is an economical and effective method for surface hardening larger size spur, helical, and herringbone gearing. In this process an inductor passes along the contour of the tooth producing a continuous hardened area from one tooth flank around the root and up the adjacent flank. The extremely high localized heat allows small sections to come to hardening temperature while the balance of the gear dissipates heat. Thus major distortions are eliminated. 4140 and 4340 alloy steels are widely used for tooth-to-tooth induction hardening. The case produced by this method ranges from 50 to 58 Rockwell C, and the flanks may be hardened to a depth of 0.160 in. These steels are air quenched in the hardening process. Plain carbon steels such as 1040 and 1045 may be used for induction hardening but these steels must be water quenched.

Carburizing Carburizing with subsequent surface hardening offers the best way to obtain the very high hardness needed for optimum gear life. It also produces the strongest gear, providing excellent bending strength and high resistance to wear, pitting, and fatigue. The residual compressive stresses inherent in the carburized case substantially improve the fatigue characteristics of this heat treated material. Normal case depths range from approximately 0.030 to 0.250 in. Case hardnesses range from 55 to 62 Rockwell C, and core hardnesses from 250 to 320 Brinell. Recommended carburizing grade steels are 4620, 4320, 3310, and 9310. The main limitation to carburizing and hardening is that it tends to distort the gear. Techniques have been developed to minimize this distortion, but generally after carburizing and hardening it is necessary that the gear be ground or lapped to maintain the required tooth tolerances.

Flame Hardening In tooth-to-tooth progressive flame hardening, an oxyacetylene flame is applied to the flanks of the gear teeth. After the surface has been heated to the proper temperature, it is air- or water-quenched. This method has some limitations; since the case does not extend into the root of the tooth, the durability is improved, but the overall strength of the gear is not necessarily improved. In fact, stresses built up at the junction of a hardened soft material may actually weaken the tooth.

Many times it is desirable to use different heat treatments for the pinion and gears. Heat-treament combinations used for pinions and gears are here listed in order of preference for optimum gear design:

1. Carburized pinion–carburized gear
2. Carburized pinion–through-hardened gear
3. Carburized pinion–nitrided gear
4. Nitrided pinion–nitrided gear
5. Nitrided pinion–through-hardened gear
6. Induction-hardened pinion–through-hardened gear
7. Carburized pinion–induction-hardened gear
8. Induction-hardened pinion–induction-hardened gear
9. Through-hardened pinion–through-hardened gear

OPTIMIZING THE GEARING

The most important factor influencing the durability of a gear set, and hence the gear size, is the hardness of gear teeth. It is often possible to reduce considerably—sometimes as much as by half—the overall dimensions of a gear set by changing from medium-hardened gears (about 300 Bhn) to full-hardened gears (about 55 to 60 Rockwell C).

There are other factors that play a role in minimizing the dimensions of a gear set, for example, the ratio between face width and pitch diameter, the selection of the proper pressure angle and pitch of teeth, etc. Thus, when it comes to deciding between selecting a set of standard catalog gears as against having the gears designed specifically to meet the requirements of the application, the question of cost versus optimizing comes to bear. In general, where there is only a limited number of units to be made, the catalog gears are much less expensive and also much more readily available. But there are many applications which call for critical power, speed, or space requirements, and it may pay in these applications to select gears that are designed for that application.

Minimizing Gear Noise Specifying or designing a gear set to produce low noise and vibration levels frequently leads to choices that are opposite from those for optimizing the gears for strength and size. Generally, parallel-shaft gearing is preferred for quiet operation rather than right-angle gearing because of greater geometric control, inherent ability to maintain tight manufacturing tolerances, and minimum friction during tooth contact. Helical gears, in particular, offer the ability to have more than one tooth in contact (helical overlap) and some experience has shown as much as a 12-dB reduction in noise using helical instead of spur gears. Double helical or herringbone type gearing creates the problem of correctly manufacturing the two helices with the exact same phase and accuracy. Therefore the optimum type of gear throughout all speed ranges becomes the single helical gear. The thrust loads and slight overturning moments of single helical contact are easily handled with modern design techniques.

For quiet, smooth operation, the gears should be designed with some or all of the following properties:

1. Select the finest pitch allowable under load considerations.
2. Employ the lowest pressure angle: 14½ or 20° are most commonly used.
3. Modify the involute profile to include tip and root relief with a crowned flank to ensure smooth sliding in and out of contact without knocking and to compensate for small misalignments.
4. Allow adequate backlash (clearance) for thermal and centrifugal expansion, but not as much extra as to prevent proper contact.
5. Specify the higher AGMA quality levels, which will reduce the total dynamic load. Generally, AGMA quality 12 or better is required for smooth, quiet operation.
6. Surface finishes of at least 20 rms should be maintained.
7. Rotor alignments and runouts must be maintained accurately.
8. Rotor unbalance should be limited to less than $3W/N$ oz-in, where W = weight, lb, and N = speed, rpm.
9. A nonintegral ratio ("hunting tooth") should be provided to prevent a tooth on the pinion from periodically contacting the same teeth on the mating gear.
10. Resonances of rotating system members (critical speeds) should be at least 30 percent away from operating speed, multiples of rotating speeds, and tooth-mesh frequencies.
11. Resonances of gear cases and other supporting members should be 20 percent away from operating speeds, multiples, and tooth-mesh frequencies.
12. Specify the highest viscosity lubricant consistent with design and application.
13. Bearing selection can influence noise levels. Rolling element bearings must be selected and applied to minimize noise generation. Generally, hydrodynamic sleeve bearings are quieter than antifriction types but are more difficult to apply.
14. Housing design is still another area where further noise and vibration reductions can be obtained. An abundance of acoustic absorbant materials are available for the housings. Moreover, the housing can be designed with built in isolation mounts to cut down any vibration attenuation.

PACKAGED GEAR DRIVES

In many cases it is preferable to select a packaged gear drive rather than a set of open gears that still must be mounted and housed.

The relative merits of a packaged drive, or "gear reducer," versus open gearing are many. The packaged drive consists essentially of gears, housing, bearings, shafts, oil seals, and a positive means of lubrication. Frequently reducers also include any or all of the following: electric motor and accessories, bedplates or motor supports, outboard bearings, a mechanical or electric device providing overload protection, a means to prevent reverse rotation, and other special features as specified.

The advantages of packaged gear drives have been well established and should be given consideration when selecting the type of gear drive for the application:

1. *Power conservation.* Because of accurate gear design, high-grade workmanship, proper bearings, and adequate lubrication, there is an assurance of minimum loss between applied and delivered power.
2. *Low maintenance.* If the correct design and power capacity for the requirements are selected and the recommended operating instructions are followed, low maintenance costs will result.
3. *Operating safety.* All gears, bearings, and shafts are enclosed in oil-tight, strongly built cast iron or steel housings.
4. *Low noise and vibration level.* Precision gearing is carefully balanced and mounted on accurate bearings. The transmitting motion is uniform and shock-free. The entire mechanism is tightly sealed in sound-damping rigid housing. Noise and vibration are reduced to a minimum.

5. *Space conservation.* Units are entirely self-contained and extremely compact; therefore, they require small space. This also enables them to be installed in out-of-the-way locations.

6. *Adverse operating conditions.* Enclosure designs have been developed to protect the mechanism from dirt, dust, soot, abrasive substances, moisture, or acid fumes.

7. *Economy.* Units permit the use of high-speed prime movers directly connected to low applied speeds.

8. *Life expectancy.* The life of a unit can be predetermined by design and made unlimited if it is correctly aligned and properly maintained.

9. *Horsepower and ratios.* Units are available in almost all desired ratios and for all practical power requirements.

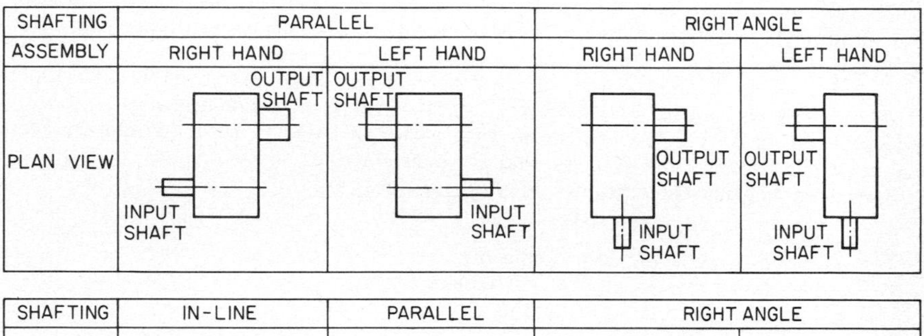

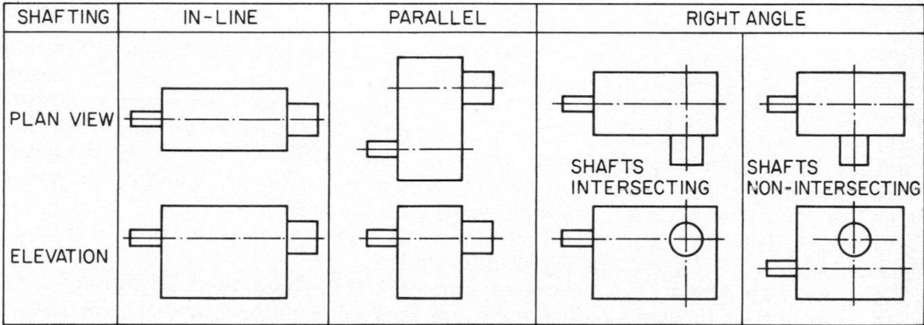

Fig. 7 Shaft arrangements for gear drives.

10. *Cooling systems.* Greater attention to sump capacity for oil and the use of fan air cooling have allowed higher horsepowers to be transmitted through smaller units without overheating.

11. *Appearance.* Housings have been streamlined for eye appeal as well as for reduction of weight and space.

12. *Rugged capabilities.* Ruggedness of steel-constructed welded housings and modern housings produces higher reliability and service life.

Types of Gear Packages Gear packages are used in multiple combinations to produce the ratio desired in the unit. Units are available in single-, double-, and triple-reduction configurations. Three stages of reduction are generally the maximum number used in standard reducers, although it is possible to use four or even more stages. Units for increasing the output speed generally have only one gear set, although at times two-stage units have been used successfully as speed increasers. Gear packages may be assembled with shaft arrangements that are right-handed or left-handed (Fig. 7).

Allowable Speeds of Gear Reducers. The maximum speed of a gear reducer is limited by the accuracy of the machined gear teeth, the balance of the rotating parts, the allowable noise and vibration, the allowable maximum speed of the bearings, the pumping and churning of the lubricating oil, the friction of the oil seals, and the heat generated in the unit.

At high speeds it is possible for inaccuracies in the gear teeth to produce failure even though no power is being transmitted. Gear reducers built in accordance with AGMA specifications are recommended to operate at speeds given in Table 1.

Horsepower Range of Gear Reducers. Horsepower-transmitting capacity of a reduction-gear unit is a function of the output torque and the speed of reducers. Some types of reducers, such as worm-gear reducers, are more satisfactory for high torques and low speeds while others, such as helical herringbone (or double helical), are suitable for high torques and also high speeds. Therefore the range of horsepowers suitable for various units is considerable. A listing of this range obtainable in standard types of reducers is given in Table 1, with a brief explanation of why the range indicated is maintained. These horsepowers are not fixed at the values given since they are continually changing. Although the values given are general, there are many special reducers available outside this range.

Ratios and Efficiencies. The ratio of a gear reducer is defined as the ratio of the input-shaft speed to the output-shaft speed. Different types of gearing allow different ratios per gear stage. Spur gears usually are used with a ratio range of 1:1 to 6:1; helical, double helical, and herringbone with ratios of 1:1 to 10:1; straight-bevel, with ratios of 1:1 to 4:1; spiral-bevel (also zerols and hypoids) with ratios of 1:1 to 9:1; and worm gears with ratios of 3¼:1 to 90:1. Planetary-gear arrangements allow ratios of 4:1 to 10:1 per gear stage.

Factors influencing the efficiency of a gear reducer are:

1. Friction loss in bearings
2. Losses due to pumping lubricating oil
3. Windage losses due to rotation of reducer parts
4. Friction losses in gear-tooth action

It is not uncommon in many types of reducers that the combined losses due to items 1 to 3, inclusive, are greater than the loss due to item 4. For this reason in some cases the horsepower lost in the reducer remains practically constant regardless of the power transmitted. Therefore it must be realized that the efficiency specified for a reducer applies only when the unit is transmitting its rated power, since when no power is being transmitted through the reducer, all the input power (small as it may be) is used in friction, and the efficiency is zero.

Ratios available in standard reducers and efficiencies to be expected when units are transmitting rated power are given in Table 1.

Epicyclic Gear Units Epicyclic gear units are sometimes used for pump drive applications. The important advantages are compact configuration, coaxial shafts, and light weight.

The most common types of epicyclic arrangements include planetary, star, and solar types, Fig. 8. The planetary configuration (with the planet carriers integral with the output shaft) is the most commonly used arrangement. It is simple and rugged and gives the maximum ratio for the size of gears. The star arrangement (where the planet carrier is fixed and the internal gear rotates integrally with the output shaft) is used for higher speed application since centrifugal loads of the planet gears are eliminated with nonrotating planet carrier. The solar arrangement (where the pinion is fixed and the planet carrier is integral with the output shaft) has the input through the internal gear. This arrangement gives epicyclic advantages but allows a low ratio, generally less than 2:1. Higher ratios (in the range of 8:1 to 60:1) can be obtained utilizing double-reduction or compound-planetary arrangements.

INSTALLATION

The basic gear unit is generally shipped from the factory completely assembled. Mating gears and pinions are carefully assembled at the factory to provide proper tooth contact. Nothing should be done to disturb this setting.

Solid Foundation The reducer foundation should be rigid enough to maintain correct alignment with connected machinery. The foundation should have a flat mounting surface in order to assure uniform support for the unit. If the unit is

TABLE 1 State-of-the-Art Guide for Gear Selection and Design (reflecting generally accepted design criteria)

	Spur gearing		Helical gearing	
	External	Internal	External	Internal
Shaft arrangement	Parallel axis	Parallel axis	Parallel axis	Parallel axis
Ratio range	1:1 to 10:1	1½:1 to 10:1	1:1 to 15:1	2:1 to 15:1 (generally feasible). Ratio depends on pinion gear tooth combination because of clearance requirements
Size availability (including maximum face widths)	Up to 150-in OD, 30-in face width. Larger segmental gears can be produced with special processing and tooling	Up to 100-in OD, 16-in face width	Up to 150-in OD, 30-in face width	Up to 100 in depending on blank configuration; 16-in maximum face width
Gear tolerances (quality requirements)	See footnote	See footnote	See footnote	See footnote
Finishing methods (singly or in combination)	Cast: rotary cut, shaped; hobbed: shaved, ground	Same as external spurs	Shaped, hobbed, shaved, ground	Shaped, hobbed, shaved, honed, lapped, ground
Horsepower range	Commercial: less than 1,000		Commercial: generally up to 50,000. However, hp only limited by maximum size capacity of design	Same as external helical gearing
Speed range, pitch-line velocity, ft/min	Commercial: normal up to 1,000; special precision up to 20,000	Commercial, standard manufacture: up to 1,000; precision manufacture: up to 20,000	To 30,000	
Gear efficiency, percent	Commercial: 95 to 98%		97 to 99%	
Quietness of operation	Commercial: quiet under 500 ft/min. Noise increases with increasing pitch-line velocity		Noise level depends on quality of gear. Higher pitch-line velocity (above 5,000 ft/min) requires higher precision gear. Gears operating 30,000 ft/min have been made with overall noise level below 90 dB. Quieter than spur gears.	
Load imposed on bearings	Radial only	Radial only	Radial and thrust	Radial and thrust

TABLE 1 State-of-the-Art Guide for Gear Selection and Design (Continued)

	External double helical gearing	Continuous-tooth herringbone gearing
Shaft arrangement	Parallel axis	Parallel axis
Ratio range	1:1 to 15:1	1:1 to 10:1
Size availability (including maximum face widths)	Up to 150-in OD, 30-in face width	Up to 150-in OD, 30-in face width
Gear tolerances (quality requirements)	See footnote	See footnote
Finishing methods (singly or in combination)	Same as helical gearing	Shaped, shaved
Horsepower range	Same as helical gearing	Up to 2,000
Speed range pitch-line velocity, ft/min	Same as helical gearing	Commercial: up to 5,000 ft/min
Gear efficiency, %	Same as helical gearing	96 to 98
Quietness of operation	Same as helical gearing	Quiet operation up to 5,000 ft/min. Not generally used at extremely high pitch-line velocity (over 20,000 ft/min)
Load imposed on bearings	Radial only	Radial only

	Straight-bevel gearing	Spiral-bevel gearing	Zerol-bevel gearing	Hypoid gearing
Shaft arrangement	Intersecting axis	Intersecting axis	Intersecting axis	Nonintersecting, nonparallel axis
Ratio range	1:1 to 6:1	1:1 to 10:1	1:1 to 10:1	1:1 to 10:1
Size availability (including maximum face widths)	Up to 102-in OD, 12-in face width	Up to 102-in OD, 12-in face width	102-in OD, 12-in face	102-in OD, 12-in face
Gear tolerances (quality requirements)	See footnote	See footnote	See footnote	See footnote
Finishing methods (singly or in combination)	Cast, generated, planed	Generated, planed, ground	Generated, planed, ground	Generated, planed, ground
Horsepower range	Up to 1,500	Up to 20,000, depending on speed	Same as straight bevel gears	Same as spiral bevel gears, with use of EP lubricants
Speed range pitch-line velocity, ft/min	Same as spur gearing	Commercial, normal: up 5,000; special precision: up to 15,000	Up to 15,000 for ground gears	6,000 to 10,000 depending on offset
Gear efficiency, %	Same as spur gearing	96 to 98 (commercial)	94 to 98	85 to 98, depending on offset
Quietness of operation	Same as spur gearing	Noise level depends on quality of gear. Higher pitch-line velocity (above 5,000 ft/min) requires higher precision gear	Quieter than straight-bevel gears	Quiet
Load imposed on bearings	Radial and thrust	Radial and thrust	Radial and thrust	Radial and thrust

	Worm gearing	
	Cylindrical	Double-enveloping cone-drive gears
Shaft arrangement	Nonintersecting, nonparallel axis	Right angle in single-reduction units
Ratio range	3½:1 to 100:1	5:1 to 70:1
Size availability (including maximum face widths)	300-in OD, 4-in circular pitch	2- to 24-in center distance. For special requirements larger and smaller sizes are available
Gear tolerances (quality requirements	Worm gear tolerances given in AGMA Standard: Inspection of Course-Pitch Cylindrical Worms and Worm Gears, Standard No. 234.01	Standard AGMA commercial tolerances. Closer tolerances available for special applications
Finishing methods (singly or in combination	Worm gears: hobbed, worm-milled, and ground	Worms: threads generated with cutter and finished by polishing. Gears: hobbed. Both members then lapped and matched together
Horsepower range	Up to 400	Fractional to 1,430, dependent on ratio, center distance and speed
Speed range pitch-line velocity, ft/min	Up to 6,000	0 to 2,400 rpm, or 2,000 rubbing speed with splash lubrication. Higher speeds permissible with special combinations
Gear efficiency, %	From 25 to 95, depending on ratio	52 to 94, depending on ratio and speed
Quietness of operation	Relatively quiet operation up to 6,000 ft/min	Smooth and quiet up to 2000 ft/min; can run quietly at higher speeds with special attention to lubrication, mounting, materials, balancing, etc.
Load imposed on bearings	Radial and thrust	Radial and thrust

Gear tolerances are dependent on method of manufacture, application, load requirements, and speeds. For spur, helical, and herringbone gears, AGMA Gear Classification Manual 390.03 lists quality numbers from 3 to 15 for coarse-pitch gears and 5 to 16 for fine-pitch gears, quality increasing as quality number increases. General range of quality gears now being manufactured is from quality numbers 5 to 14. Quality numbers relate to runout, tooth-to-tooth, spacing, profile, total composite, and lead tolerances. Bevel and hypoid gear tolerances range from 3 to 13.

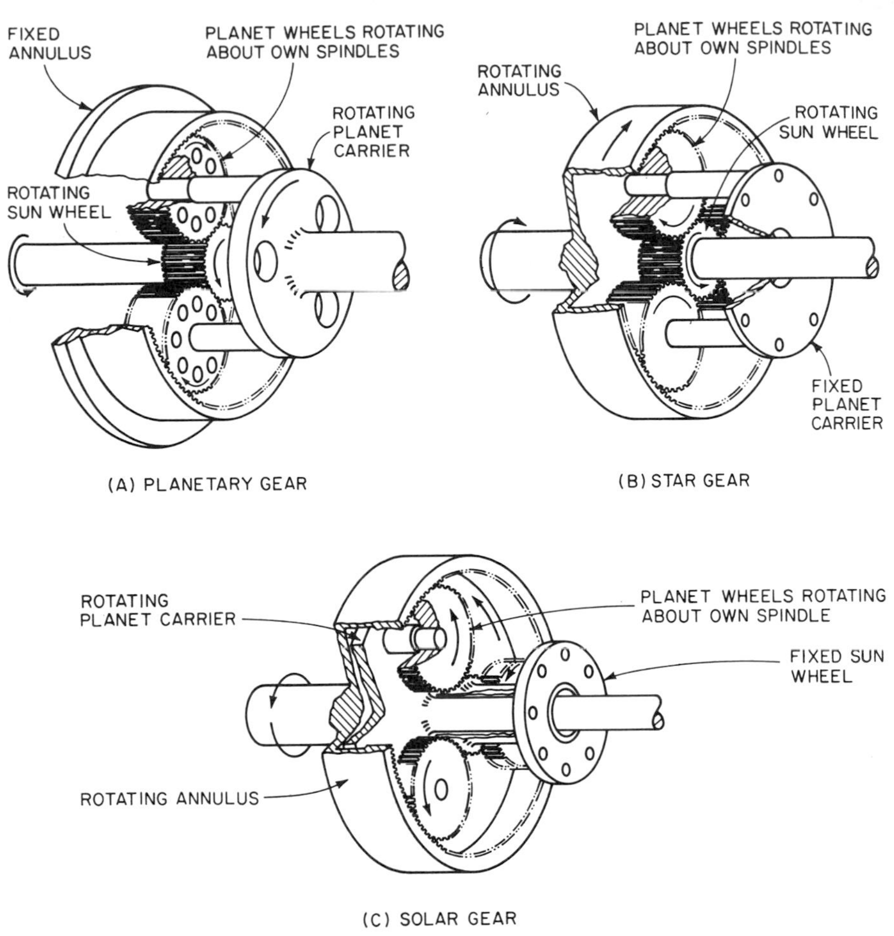

Fig. 8 Simple epicyclic gear drives.

mounted on a surface which is other than horizontal, consult factory to ensure that design provides for proper tooth contact and adequate lubrication.

Design of fabricated pedestals or baseplates for mounting speed reducers should be carefully analyzed to determine that they are sufficiently rigid to withstand operating vibrations. Vibration dampening materials may be used under the baseplate to minimize the effect of vibrations.

When mounting a drive on structural steel, the use of a rigid baseplate is strongly recommended. Bolt unit and baseplate securely to steel supports with proper shimming to ensure a level surface.

If a drive is mounted on a concrete foundation, allow the concrete to set firmly before bolting down the unit. For the best mounting, grout structural steel mounting pads into the concrete base, rather than grouting the gear unit directly into the concrete.

Leveling If shims are employed to level or align the unit, they should be distributed evenly around the base under all mounting pads to equalize the support load and to avoid distortion of the housing and highly localized stresses. All pads must be squarely supported to prevent distortion of the housing when the unit is bolted down.

Alignment If equipment is received mounted on a bedplate, it has been aligned at the factory. However, it may have become misaligned in transit. During field mounting of the complete assembly, it is always necessary to check alignment by breaking the coupling connection and shimming the bedplate under the mounting pads until the equipment is properly aligned. All bolting to the bedplate and foundation must be pulled up tight. After satisfactory alignment is obtained, close up the coupling.

Couplings Drive shafts should be connected using flexible couplings. The couplings should be aligned as closely as possible following the manufacturer's instructions.

Alignment and Bolting The gear unit, together with the prime mover and the driven machine, should be correctly aligned. After precise alignment, each member must be securely bolted and dowelled in place. Coupling alignment instructions should be carefully followed.

LUBRICATION

Types of Lubricant The recommended types of oil for use in gear units are either straight mineral oil or extreme-pressure (EP) oil. In general, the straight

Fig. 9 4,000-hp gas turbine driving through high-speed gear unit into a centrifugal pump, turbine speed 14,000 rpm, pump speed 6,000 rpm.

mineral oil should be a high grade, well-refined petroleum oil within the recommended viscosity range. It must be neutral in reaction and not corrosive to gears and ball or roller bearings. It should have good defoaming properties and good resistance to oxidation for high operating temperatures.

Gear drives that are subject to heavy shock, impact loading, or extremely heavy duty, should use an extreme-pressure lubricant. Extreme-pressure gear lubricants are petroleum-based lubricants containing special chemical additives. EP lubricants recommended are those containing either lead-naphthanate or sulfur-phosphorous additives. Sulfur-phosphorous type EP oils are generally more stable than lead-naphthanate type oils and may be used to a maximum sump temperature of 180°F. Limit lead naphthanate EP oils to a maximum sump temperature of 160°F.

In general, if units are subjected to unusually high ambient temperatures (100°F or over), extreme humidity, or atmospheric contaminants, use the straight mineral oil recommended.

TABLE 2 Troubleshooting Chart

Trouble	What to inspect	Action
Overheating	1. Unit overloaded.	Reduce the loading or replace with drive of sufficient capacity.
	2. Oil-cooler operation.	Check coolant and oil flow. Vent system of air. Oil temperatures into unit should be approximately 110°F. Check cooler internally for build up of deposits from coolant water.
	3. Has recommended oil level been exceeded or is level too low?	Check oil-level indicator to see that housing is accurately filled with lubricant to the specified level.
	4. Are bearings properly adjusted?	Bearings must not be pinched. Adjustable tapered bearings must be set at proper bearing lateral clearance. All shafts should spin freely when disconnected from load.
	5. Oil seals or stuffing box.	Oil seals should be greased on those units having grease fitting for this purpose. Otherwise, apply small quantity of oil externally at the lip until the seal is run in. Stuffing box should be gradually tightened to avoid overheating. Packing should be a self-lubricating braided-asbestos type.
	6. Breather.	Breather should be open and clean. Clean breather regularly in a solvent.
	7. Grade of oil.	Oil must be of grade specified in lubrication instructions. If it is not, clean unit and refill with correct grade.
	8. Condition of oil.	Check to see if oil is oxidized, dirty, or of high sludge content; change oil and clean filter.
	9. Forced-feed lubrication system.	Make sure oil pump is functioning. Check that oil passages are clear and permit free flow of lubricant. Inspect oil-line pressure regulators, nozzles, and filters to be sure they are free of obstructions. Make sure pump suction is not sucking air.
	10. Coupling alignment.	Disconnect couplings and check alignment. Realign as required.
	11. Coupling lateral float.	Adjust spacing between drive motor, etc., to eliminate end pressure on shafts. Replace flexible coupling with type allowing required lateral float.
	12. Speed of unit excessive?	Reduce speed or replace with drive suitable for speed.
Shaft failure	1. Type of coupling used.	Rigid couplings can cause shaft failure. Replace with coupling to provide required flexibility and lateral float.
	2. Coupling alignment.	Realign equipment as required.
	3. Is overhung load excessive?	Reduce overhung load. Use outboard bearing or replace with unit having sufficient capacity.
	4. Is unit overloaded?	Reduce the loading or replace with drive of sufficient capacity.
	5. Is unit subjected to high-energy loads or extreme repetitive shocks?	Apply couplings capable of absorbing shocks and, if necessary, replace with drive of sufficient capacity to withstand shock loads.
	6. Torsional or lateral vibration condition.	These vibrations can occur through a particular speed range. Reduce speed to at least 25% below critical speed. System mass-elastic characteristics can be adjusted to control critical-speed location. If necessary, adjust coupling weight, as well as shaft stiffness, length, and diameter. For specific recommendations contact factory.

TABLE 2 Troubleshooting Chart (Continued)

Trouble	What to inspect	Action
Shaft failure	7. Is outboard bearing properly aligned?	Realign bearing as required.
Bearing failure	1. Is unit overloaded?	See overheating (item 1). Abnormal loading results in flaking, cracks, and fractures of the bearing.
	2. Is overhung load excessive?	See shaft failures (item 3).
	3. Speed of bearing excessive?	See overheating (item 12).
	4. Coupling alignment.	See overheating (item 10).
	5. Coupling lateral float.	See overheating (item 11).
	6. Are bearings properly adjusted?	See overheating (item 4). If bearing is too free or not square with axis, erratic wear pattern will appear in bearing races.
	7. Are bearings properly lubricated?	See overheating (items 2, 3, 7, 8, 9). Improper lubrication causes excessive wear and discoloration of bearing.
	8. Rust formation due to entrance of water or humidity.	Make necessary provisions to prevent entrance of water. Use lubricant with good rust-inhibiting properties. Make sure bearings are covered with sufficient lubricant. Turn over gear unit more frequently during prolonged shutdown periods.
	9. Is bearing exposed to an abrasive substance?	Abrasive substance will cause excessive wear, evidenced by dulled balls, rollers, and raceways. Make necessary provision to prevent entrance of abrasive substance. Clean and flush drive thoroughly and add new oil.
	10. Has unit been stored improperly or damaged by prolonged shutdown?	Prolonged periods of storage in moist, ambient temperatures will cause destructive rusting of bearings and gears. When these conditions are found to have existed, the unit must be disassembled, inspected, and damaged parts either thoroughly cleaned of rust or replaced.
Oil leakage	1. Has recommended oil level been exceeded?	Check through level indicator that oil level is precisely at level indicated on housing.
	2. Is breather open?	Breather should be open and clean.
	3. Are all oil drains open?	Check that all oil drain locations are clean and permit free flow. Drains are normally drilled in the housing between bearings and bearing cap where shafts extend through caps.
	4. Oil seals.	Check oil seals and replace if worn. Check condition of shaft under seal and polish if necessary. Slight leakage normal, required to minimize friction and heat.
	5. Stuffing boxes.	Adjust or replace packing. Tighten packing gradually to break in. Check condition of shaft and polish if necessary.
	6. Force-feed lubrication to bearing excessive?	Reduce flow of lubricant to bearing by adjusting orifices. Refer to factory.
	7. Plugs at drains, levels, etc., and standard pipe fittings.	Apply pipe joint sealant, and tighten fittings.
	8. Compression-type pipe fittings.	Tighten fitting or disassemble and check that collar is properly gripping tube.
	9. Housing and caps.	Tighten cap screws or bolts. If not entirely effective, remove housing cover and caps. Clean mating surfaces and apply new sealing compound (Permatex #2 or equal). Reassemble. Check compression joints by tightening fasteners firmly.

TABLE 2 Troubleshooting Chart (Continued)

Trouble	What to inspect	Action
Gear wear	1. Backlash.	Gear set must be adjusted to give proper backlash. Refer to factory.
	2. Misalignment of gears.	Check contact pattern to be over approximately 75% of face, preferably in center area. Check condition of bearings.
	3. Housing twisted or distorted?	Check shimming and stiffness of foundation.
	4. Is unit overloaded?	See overheating (item 1).
	5. Has recommended oil level been maintaned?	See overheating (item 3).
	6. Are bearings properly adjusted?	See overheating (item 4).
	7. Grade of oil.	See overheating (item 7).
	8. Condition of oil.	See overheating (item 8).
	9. Forced-feed lubrication.	See overheating (item 9).
	10. Coupling alignment.	See overheating (item 10).
	11. Coupling lateral float.	See overheating (item 11).
	12. Excessive speeds.	See overheating (item 12).
	13. Torsional or lateral vibration condition.	See shaft failure (item 6).
	14. Rust formation due to entrance of water or humidity.	See bearing failure (item 8).
	15. Gears exposed to an abrasive substance.	See bearing failure (item 9).

Grease Lubrication Lubricants should be high-grade, nonseparating, ball bearing grease suitable for operating temperatures to +180°F. Grease should be N.L.G.I. No. 2 consistency.

The grease lubricant must be noncorrosive to ball or roller bearings and must be neutral in reaction. It should contain no grit, abrasive, or fillers; it should not precipitate sediment; it should not separate at temperatures up to 300°F.; and it should have moisture-resistant characteristics. The lubricant must also have good resistance to oxidation.

Grease Lubrication of Bearings Pressure fittings are often supplied in gear units for the application of grease to bearings that are shielded from the oil. Although a film of grease over the rollers and races of the bearing is sufficient lubrication, drives are generally designed with ample reservoirs at each grease point.

Greased bearings should be lubricated at definite intervals. Usually 1-month intervals are satisfactory unless experience indicates that regreasing should occur at shorter or longer intervals.

Oil Seals Oil seals require a small amount of lubricant to prevent frictional heat and subsequent destruction when the shaft is rotating. Normally when a single seal is utilized, sufficient lubricant is provided by spray or splash. Certain design or application requirements dictate that double seals be used at some sealing points. When this is the case a grease fitting and relief plug are located in the seal retainer to provide lubricant to the outer seal. Grease must periodically be applied between the seals by pumping through the fitting until overflow is noted at relief plug. The greases recommended for bearings may also be used for seals.

TROUBLESHOOTING TIPS

Improper lubrication causes a high percentage of gear reduction unit failures. Too frequently speed reducers are started up without any lubricant at all. Conversely, units are sometimes filled to a higher oil level than specified in the mistaken belief that better lubrication is obtained. This higher oil level usually results in more of the input power going into churning the oil, creating excessive temperatures with detrimental results to the bearings and gearing. Insufficient lubrication gives the same results.

Gear failure due to overload is a broad and varied area of misapplication. The nature of load (input torque, output torque, duration of operating cycle, shocks, speed, acceleration, etc.) determines the gear-unit sizing and other design criteria. Frequently, a gear drive must be larger than the torque output capability of the prime mover would indicate. An AGMA service factor compensates for varying severity of application conditions by providing a higher nominal horsepower which in effect increases the size of the gear unit. If there is any question in the user's mind that the actual service conditions may be more severe than originally anticipated it is recommended that this information be communicated to the gear manufacturer before startup. Often there are remedies that can be suggested before a gear unit is damaged by overload, but none are effective after severe damage.

Motors and other prime movers should be analyzed while driving the gear unit under fully loaded conditions to determine that the prime mover is not overloaded and thus putting out more than rated torque. If it is determined that overload does exist, the unit should be stopped and steps taken either to remove the overload or to contact the manufacturer to determine suitability of the gear drive under observed conditions.

REFERENCES

1. Dudley, D. W.: "Gear Handbook," McGraw-Hill Book, New York, 1962.
2. Staniar, W.: "Plant Engineering Handbook," 2d ed., McGraw-Hill Book Company, New York, 1959.
3. Gear Drives, *Design News,* Jan. 19, 1968.
4. Kron, H. O.: Optimum Design of Parallel Shaft Gearing, *Trans. ASME,* 72PTG-17, October 1972.
5. Hamilton, J. M.: Are You Paying Too Much for Gears? *Mach. Design,* Oct. 19, 1972.
6. "AGMA Standards and Technical Publications Index," AGMA 000-67, March 1974.
7. "Philadelphia Application Engineered Gearing Catalog," G-965, Philadelphia Gear Corp., King of Prussia, Pa.

Section **6.2.5**

Adjustable-Speed Belt Drives

MILTON B. SNYDER

Mechanical adjustable-speed drives for pump applications are generally of the compound adjustable-pitch-sheave and rubber-belt variety as illustrated here. This integrated drive package converts constant input speed into an output that is steplessly variable within a certain range. They are usually driven by constant-speed ac induction motors and usually contain builtin gear reducers to obtain low output speeds.

Drive packages may be mounted horizontally, vertically, or on a 45° angle and are available in standard open or totally enclosed designs. Some of the possible mounting arrangements are illustrated in Fig. 2. Speed ranges of 10:1 to 2:1 can be obtained with most units. A typical distribution of available output speeds is 4,850 to 1.4 rpm, including drives with and without reducer gearing. Increaser gearing (offered as integrally mounted packages) provides speeds to 16,000 rpm.

Fig. 1 Pump application utilizing typical mechanical adjustable-speed belt drives. (Reliance Electric Co.)

C – POWER FLOW MOUNTINGS

Vertical 45° Horizontal Trunnion

Z – POWER FLOW MOUNTINGS

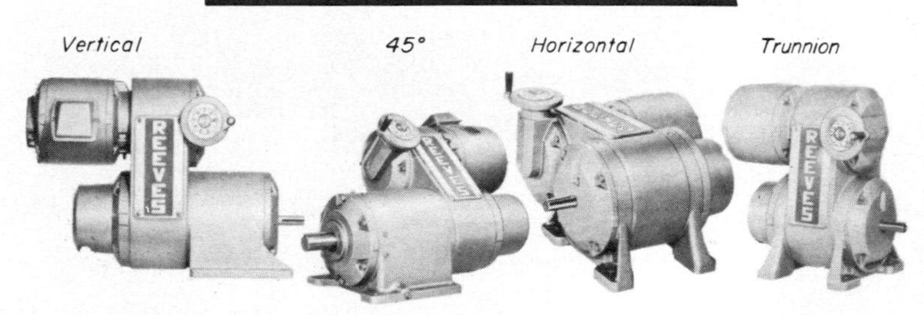

SPECIAL MOUNTINGS and CASE ENCLOSURES

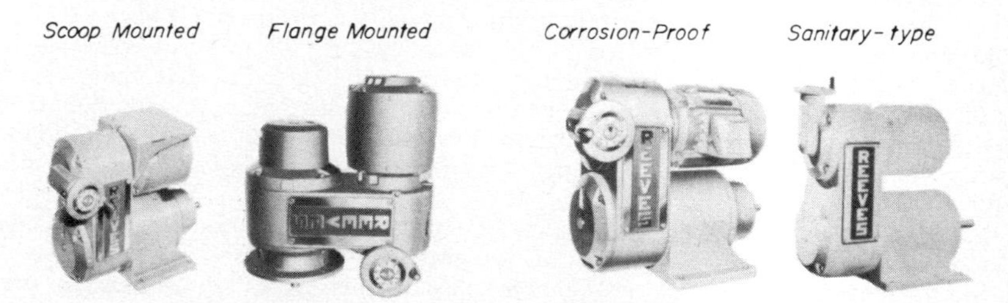

Fig. 2 Mechanical adjustable-speed drive mounting and special enclosures.

Ac induction-drive motors used with mechanical adjustable-speed drives usually operate at speeds of 1,750 or 1,160 rpm. The electric design characteristics of these motors comply with the standards of the National Electrical Manufacturers Association (NEMA). The NEMA design B motor with normal torque and normal slip characteristics is standardly furnished. NEMA design C, high-torque low starting current, and NEMA design D, high-torque high-slip motors, may be used when their specific characteristics are dictated by the application.

The mechanical mounting characteristics of the drive motors for mechanical adjustable-speed drives vary from one manufacturer to another. Motors of round body footless NEMA C-face construction are commonly used. Sometimes standard foot-mounted motors are supplied as an option in conjunction with a scoop support by some manufacturers or as standard practice by others. Still others may supply mechanically special partial motor frames with their drives.

For a given power rating, many manufacturers will supply drive motors with

Fig. 3 Cutaway of typical mechanical adjustable-speed drive. (Reliance Electric Co.)

increased service-factor horsepower. This increased horsepower of the drive motor compensates for and overcomes the inherent mechanical losses of the drive. This, in turn, causes full-rated horsepower to be developed at the output shaft of the mechanical adjustable-speed drive. Drive output horsepower ratings are more fully discussed in the later subsection Rating Basis.

The input speed of the mechanical adjustable-speed drive is typically 1,750 or 1,160 rpm as defined by the speed of the drive motor. The output speed of the internal adjustable-speed belt section goes above and below the drivemotor speed. A maximum speed of 4,200 rpm or greater is not uncommon for fractional and small-integral horsepower drives. The need for stages of output gearing is, therefore, readily apparent to obtain final output speeds that are usable for pump or other applications.

Parallel-shaft or right-angle-shaft reducer gearing, at the option of the design engineer or user, may be incorporated as an integral part of the mechanical adjustable-speed drive package. Generally speaking, gear reduction is required when drive maximum-output speed must be lower than 1,750 rpm or drive minimum output speed lower than 583 rpm.

The American Gear Manufacturers Association (AGMA) does not define standards for reducers used in adjustable-speed applications. However, most all drive manufacturers produce reducers for these drives in accordance with accepted AGMA standards for constant-speed reducers.

Where infinite or stepless adjustment over a specific finite speed range is necessary, stepless mechanical adjustable-speed drives are generally most economical for standard pump-application requirements. Initial costs are usually lower than for comparable electric or hydraulic systems, and the mechanical systems are easier to operate and maintain.

Reliability and accuracy of speed control are advantages of the mechanical adjustable-speed belt-drive package. Construction details, physical size, and mounting dimensions are not standardized and vary with the manufacturer, but all employ dual adjustable-pitch sheaves mounted on parallel shafts at a fixed center distance, and a special wide-section rubber V belt to provide a compact assembly. Most of the designs utilize spring-loaded sheaves for control of belt tension.

A high degree of flexibility in application and control is offered by these mechanical adjustable-speed belt-drive power packages. Capacities range from fractional to 100 hp, with maximum speed ratios of 10:1 in the fractional hp sizes, decreasing to 6:1 at 30 hp, and 3:1 in the largest sizes. Multiple-belt drive arrangements are usually required for capacities over 50 hp. Again, depending on capacity, speeds ranging from a maximum of 16,000 to a minimum of 1.4 rpm are possible, although most of the standard units are within the 2 to 5,000 rpm range.

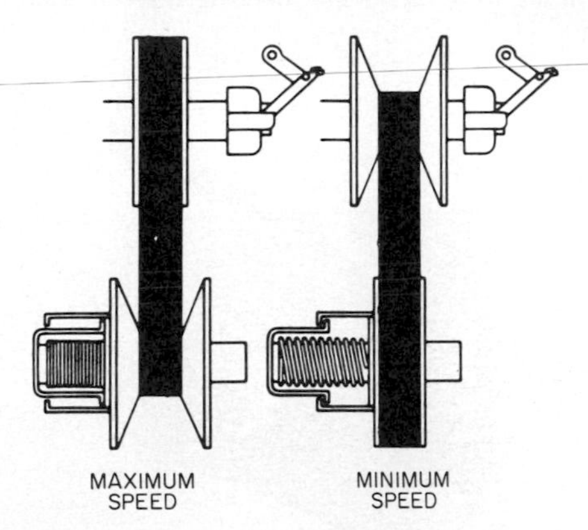

Fig. 4 Mechanical adjustable-speed drive—belt section functional diagram. (Reliance Electric Co.)

Operating Principle In Fig. 4, the upper-disk assembly is the input assembly driven directly by the shaft of the ac induction motor at a constant speed. This constant-speed disk assembly has one stationary-disk member on the *left,* and one movable or sliding-disk member on the *right.* The sliding member is mechanically attached to a positive-shifting linkage arrangement consisting of a thrust bearing, bearing housing, and shifting yoke. This linkage is actuated by a control handwheel or some other speed-changing device.

In the same view, the lower-disk assembly is the output assembly whose speed is adjustable. In this adjustable-speed disk assembly, the sliding member is next to the spring cartridge on the *left,* and the stationary member is on the *right.* Note that this is just the opposite arrangement from the upper or constant speed disk assembly. This adjustable-speed disk assembly is mounted on the adjustable-speed output shaft of the drive. A flexible wide-section rubber V belt connects the two disk assemblies.

Assuming the minimum speed-belt position as a starting point, the positive shifting linkage on the sliding member of the constant-speed disk assembly is moved to the left toward the fixed number. This positive change forces the wide-section V belt to a larger diameter in the constant-speed disk assembly. Simultaneously the belt forces the sliding member on the adjustable-speed disk assembly against its spring so that the belt assumes a smaller running or pitch diameter in this disk assembly. Speed of the output shaft increases in stepless increments while the drive-motor speed remains constant. Reversing the above procedures reduces the output-shaft speed.

When the V belt is at this maximum diameter in the constant-speed disk assembly, it is at a minimum diameter in the adjustable-speed disk assembly and the output-shaft speed is at maximum. Conversely, when the V belt is at its minimum diameter in the constant-speed disk assembly, it is at a maximum diameter in the adjustable-speed disk assembly and the output-shaft speed is at minimum.

Since one member of each disk assembly is fixed on its shaft, the V belt must move axially as well as along the inclined surface of the disk to assume different running-pitch diameters and cause speed to change. Centrifugal force makes this composite movement of the belt effortless when the drive is in operation. But the speed setting of the drive must never be changed while the drive is not operating and motionless. If speed-setting change is attempted while the drive is motionless, destructive, crushing forces are imposed on the belt. These same forces may also damage the positive shifting-linkage mechanism of the otherwise rugged mechanical adjustable-speed drive.

Other Belt Drives Other drives of the mechanical adjustable-speed type such as the single adjustable-motor-sheave or pulley and the wood block or metal-chain-belt type transmission are only mentioned here because of their rather limited application as pump drives.

Fig. 5 Typical motor-pulley belt drive—adjustable-shaft center distance. (Reliance Electric Co.)

The motor sheave or pulley illustrated in Fig. 5 is a simple, single adjustable-pitch device mounted on a motor shaft and drives by way of a standard or wide-section V belt to a companion fixed-diameter, flat-face pulley or V sheave. Driving and driven shafts are parallel but must be arranged so that their center distance is adjustable; this is generally accomplished by means of a sliding motor base. The entire drive assembly is usually operated as an open belt drive.

The wood-block or metal-chain belt transmission uses the original mechanical adjustable-speed belt drive operating principle. This drive utilizes a special wide-section wooden-block belt or laminated-metal-chain belt driven by two pairs of

Fig. 6 Adjustable-speed-transmission low-speed high-torque device. (Reliance Electric Co.)

positively controlled variable-pitch sheaves. Movement of the sheave flanges is synchronized by a positive linkage arrangement.

The transmission-type drive is generally thought of as a low-speed high-torque device that can withstand severe overload and abuse for long periods of time.

Rating Basis The mechanical adjustable-speed drive package, first discussed, is usually rated as a constant-torque variable-horsepower device. The horsepower rating is based upon the capacity of the whole unit at maximum-speed setting. When operating at any output speed below this maximum, the horsepower capacity is reduced in direct proportion.

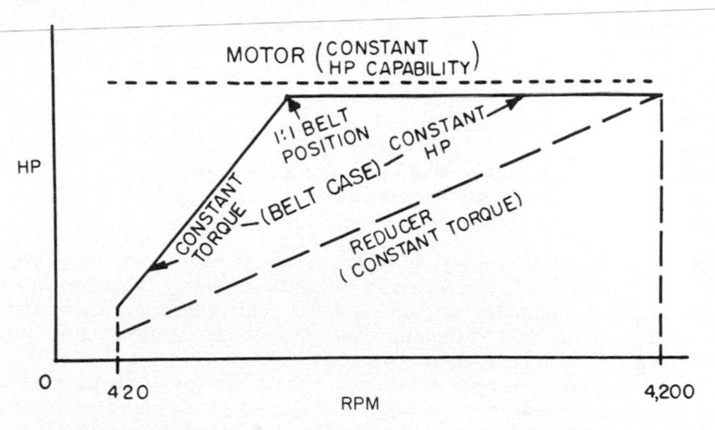

Fig. 7 Mechanical adjustable-speed belt drive; horsepower versus output-speed relationship; motor, belt section, and reducer section.

An entire drive unit is actually made up of three major component parts, each of which has its own individual torque or power-output characteristics. For example, the ac induction-drive motor or prime mover develops constant horsepower at a constant rotational speed.

The adjustable-speed belt section has an output-torque characteristic that is a mixture of both constant torque and variable torque over its speed range. When operating above a 1:1 belt position, this section has a constant-horsepower variable-torque characteristic and, when operating below a 1:1 belt position, has a constant-torque variable-horsepower characteristic. Finally, a parallel-shaft or right-angle gear reducer section has a constant-torque variable-horsepower characteristic. Figure 7 graphically illustrates the above relationships.

Since the gear reducer section has a *constant-torque* characteristic, *this section* defines the output characteristic for the entire mechanical adjustable-speed drive.

It should be noted that virtually all manufacturers rate parallel-shaft drives using output-shaft horsepower as a base, whereas right-angle output-shaft drives are rated on the basis of input horsepower to the reducer minus reducer efficiency.

Because of the relatively low efficiency of right-angle worm gear and right-angle combination worm and helical gear reducers, the power transmission industry follows the practice of rating these units in terms of horsepower at the input shaft. From the earlier subsection Operating Principle you will note that the adjustable-speed output shaft of the belt section of the drive becomes the input shaft of the reducer section of the same drive.

One or more sections of a drive, usually the belt and reducer sections, may be service factored by frame or case oversizing to permit rating these sections for constant horsepower over all or a portion of their speed range. Since the drive motor is a constant-horsepower device as already mentioned, oversizing of its frame is unnecessary.

Service Factoring Service factoring of a mechanical adjustable-speed belt drive is common practice when the normal operating requirements for steady constant-torque loads, running 8 hr/day, 5 days/week are to be exceeded.

Those drives to be used for types of service other than normal, as described above, must be selected by use of modifying factors that will provide correct service capacity. Some unusual service requirements are: moderate to heavy shock loads, 24 hr/day continuous operation, and constant-horsepower demands over a wide speed range.

Other unusual service conditions may be suggested by the following information check list. This list itemizes *Required Information* data that should be furnished the variable-speed drive manufacturer for those applications calling for unusual service:

1. Speed (rpm) and torque required for the application
2. Value and frequency of peak-load condition
3. Hours of operation per day or week
4. Frequency of starts and stops
5. Inertia (WK^2) of the load
6. Frequency of reversals of direction of rotation
7. Electric and mechanical overload protection provisions
8. Method used to connect drive-output shaft to the driven load
9. Any unusual environment or other operating conditions.

Methods of Control A variety of control systems have been developed for use with mechanical adjustable-speed drives. For the majority of pump applications, speed is controlled manually through a lever, handwheel, or knob attachment. Remote semiautomatic and automatic control methods in mechanical, pneumatic, or electric forms are also being used.

For manual operation, vernier attachments are often useful to increase accuracy of speed adjustment. Cams are occasionally employed, mounted externally or internally, to assure a prescribed pattern of output characteristics.

Remote control is usually obtained by means of a positioning motor, a fractional-horsepower motor connected to the drive-control shifting screw through reduction

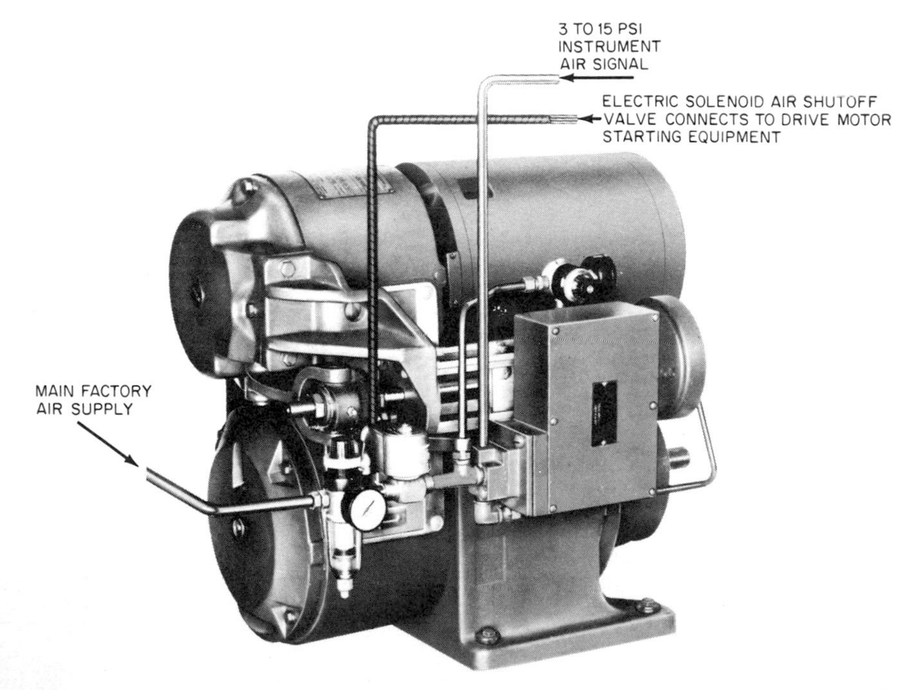

Fig. 8 Mechanical adjustable-speed drive with pneumatic actuator for automatic speed changing. (Reliance Electric Co.)

gearing. Output speed of the drive is then adjusted from a push button station at a remote location.

For semiautomatic or automatic operation, control systems usually consist of three elements: a sensing unit, a receiver, and a positioning actuator. The sensing unit detects changes in the process being controlled and transmits a signal to the receiver. At the receiver, the signal is analyzed, amplified, and transmitted to the positioning actuator which adjusts the speed of the mechanical adjustable-speed drive accordingly.

If the process and/or load requirements can be adapted to produce a signal, there should be a suitable control system that can be used for speed adjustment. The only limitation is that the load requirements must follow a specific pattern of some type, regardless of whether the pattern is based on direct, inverse, or proportional relationships.

The pneumatic actuators used for speed adjustment are usually responsive to a 3 to 15 lb/in^2 air-signal pressure. These pneumatic positioning devices are actually analog piloted servovalve positioning devices that can be used in a great variety of open- and closed-loop process control applications. This device can be used to cause a mechanical adjustable-speed drive either to follow or maintain a given process signal from variables such as liquid level, pressure, flow rate, or any other measurable value. The adjustable-speed drive thus becomes the final control element in a closed-loop process control system. Figure 8 is a typical drive equipped with a pneumatic actuator for speed changing.

Application Guidelines The following is a listing of the more common items to consider when specifying a mechanical adjustable-speed belt drive for a specific application:

1. Manufacturer's size designation of the mechanical adjustable-speed drive
2. Range of speed variation and actual output speeds
3. Motor specifications: horsepower rating, electric current (single or polyphase, frequency, and voltage), type of enclosure, and other special electric or mechanical modifications
4. Special drive-output shaft extension
5. Type of control: handwheel, electric remote, mechanical automatic, pneumatic, etc.
6. Manufacturer's assembly-configuration designation and type of mounting whether standard floor type, trunnion, ceiling, sidewall, or flange
7. Accessory equipment such as tachometer, magnetic brake, etc.
8. Horsepower rating based on constant torque and at maximum output speed
9. Type of case enclosure

Section **6.2.6**

Clutches

E. COZZARIN

INTRODUCTION

Fundamentally, a clutch is a device used to connect and disconnect a driving and a driven part of a mechanism. Clutching devices are designed to perform a variety of functions such as maintain constant speed, torque, and power, or limit torque. They are also used for automatic disconnect, quick starts and stops, gradual starts, nonreversing, and overrunning functions. Clutches currently used can be divided into three basic categories: mechanical, electric, and hydraulic or pneumatic. This section will be concerned with mechanical clutches as applied to the various types of industrial pumps.

CLUTCH TYPES: MECHANICAL

Mechanical clutches consist of two basic types: friction and positive-engagement, both of which can be subdivided into a number of discrete types. The friction types used in pump drives include centrifugal, rim or drum, multiple-disk, and overrunning or freewheeling clutches. Positive-engagement types such as jaw or tooth clutches are not generally used in pump applications; however, disconnect couplings are used extensively and will be covered in this category.

Friction Type The centrifugal, rim or drum, and multiple-disk clutches consist of three basic elements; a set of opposing friction surfaces, means for transmitting torque to and from the friction surfaces, and a mechanism for forcing the friction surfaces into contact. These units have the inherent advantage over positive-engagement clutches in that they have the ability to slip. As a result, this enables the drive to pick up and accelerate the driven load with a minimum of shock. In the event of momentary shock loads, a cushioning action is provided. Friction clutches have no limitations regarding unloading before engaging or disengaging, and both actions can be performed at virtually any speed.

In addition, special-purpose friction clutches used in pump applications are overrunning clutches. These clutches consist of spring-loaded cams or rollers that wedge between races to transmit torque in one direction of rotation only. The centrifugal, rim or drum, and multiple disk clutches perform a different function than overrunning clutches. The former are used to transmit torque in either direction as long as they are engaged automatically at some speed or at will. This feature is used to allow an engine or steam turbine to warm up or idle disengaged from a pump at part or full speed. In dual-driven pump installations, both drivers can be connected or one driver can be automatically disengaged when inoperative due to speed reduction or at will. Dual pumps can be driven by a single driver between them.

Either pump can be disconnected using a friction-type clutch which can be connected or disconnected at will. On the other hand, the latter-type overruning clutch can transmit torque in only one direction. This feature is used with dual-driven pumps also to automatically disconnect a driver which is inoperative or which has zero torque. If one race of an overrunning clutch is connected to a fixed or grounded member (torque arm), it can be used as a nonreverse ratchet (backstop). Overrunning clutches are used for backstop operation in installations where it is not desirable to have a pump or driver rotate backwards during loss of power or after a normal shutdown.

Centrifugal Clutches. Centrifugal clutches are automatic clutches used to keep the driving element unloaded until it reaches a predetermined speed. It permits the drive motor or engine to start, warm up, and accelerate to operating speed without load. These units are designed to reduce the initial starting loads on prime movers by providing gradual engagement without shock and hence smooth the pickup of load. Centrifugal clutches are particularly useful where high-inertia loads are involved or where the prime mover has an inherently low starting-torque characteristic. These clutches are also used to disengage the idle member of a dual drive. They provide overload protection since the clutch will slip when overloaded.

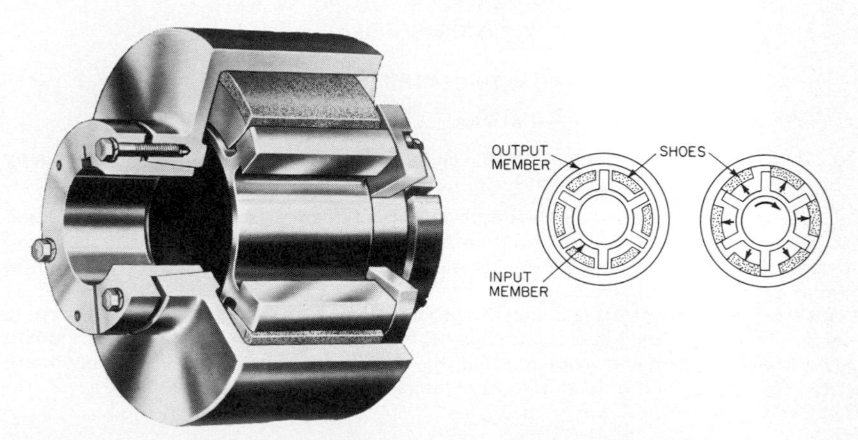

Fig. 1 Centrifugal clutch, components and operation. (Formsprag Company a subsidiary of Dana Corporation)

Essentially, centrifugal clutches are of the expanding shoe or drum type where engagement is effected by weighted friction elements which are forced outward, due to centrifugal force, against an outer drum. Shown in Fig. 1 is a representative example of a centrifugal clutch and component parts. The centrifugal clutch utilizes two fundamental forces in its operation: centrifugal force and friction force. When the inner drive member is set in motion, the shoes start to rotate and are subjected to centrifugal force. The centrifugal force developed forces the shoes against the clutch outer member (drum) creating a frictional force between the shoe and drum. As the driver's speed increases, centrifugal force increases and hence frictional force increases. When the frictional force is of sufficient magnitude to overcome the driven system's load and inertia, the shoe virtually locks against the drum to transmit load without slip.

Clutches of this type can be supplied with a variety of shoe springs to give controlled engagement characteristics (free engagement or delayed engagement). Shown in Fig. 2 are curves depicting typical motor torque and current characteristics of a NEMA B motor together with a typical centrifugal-clutch performance curve. With a centrifugal clutch, the motor is started without load because the clutch does not fully engage until peak torque speed (A) is reached. Therefore, most of the motor torque is used to accelerate the rotor only, decreasing motor-acceleration time to a fraction of a second. Since motor current is highest during acceleration, this greatly decreases the peak-current period and allows the motor to operate more efficiently (B). Thus, heat losses are much smaller so that the need for reduced-

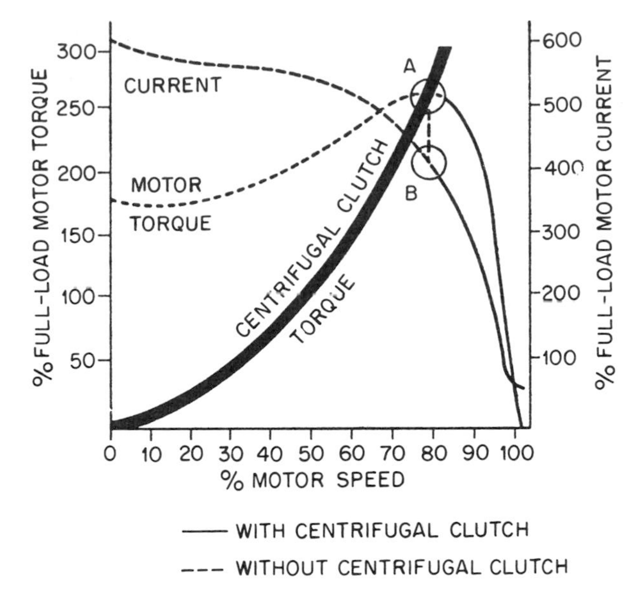

Fig. 2 Centrifugal-clutch-engagement characteristic curve. (Formsprag Company a subsidiary of Dana Corporation)

voltage starting equipment, high-torque motors, and oversized motors can be eliminated. Since the motor is up to running speed almost immediately and does not operate in the high-current low-torque range, output torque is increased and motor current and heating are decreased.

Another type of centrifugal clutch, often called a dry-fluid clutch, is shown in Fig. 3. The "fluid" used in this clutch is heat-treated steel shot. A measured amount called the flow charge is contained in the housing which is keyed to the motor shaft. When the motor is started centrifugal force throws the flow charge to the

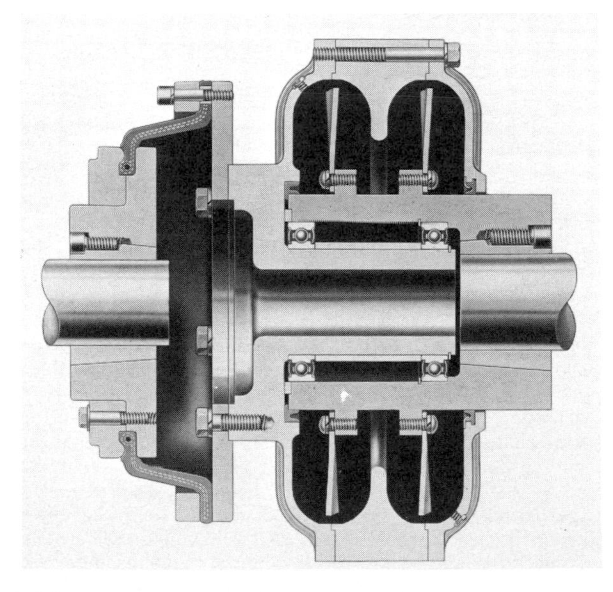

Fig. 3 Dry-fluid type clutch. (Dodge Manufacturing Div., Reliance Electric Co.)

perimeter of the housing (see Fig. 4), packing it between the housing and the rotor which transmits power to the load.

After the starting period of slippage between housing and rotor the two become locked together and achieve full-load speed, operating without slip and with 100 percent efficiency.

Rim or Drum-Type Clutches. These are of the contracting-shoe type; the friction shoes contract to engage a drum. Rim and drum clutches provide several advantages: capacity of the clutch is large for its size because of the 360° contact surface at maximum diameter; applied pressure is uniform over the friction shoes so that load is distributed evenly; constant-velocity contact assures even wear of friction surfaces; compensation for misalignment and shock absorption is provided; and no lubrication or adjustment is required. In the contracting type clutch shown in Figs. 5*A* to *C*, the air-expanded rubber tube forces the friction shoes inward. The resiliency of the tube cushions vibration, permits some parallel and angular-shaft misalignment, dampens torsional shocks, and automatically compensates for lining wear. Centrifugal force is overcome by increasing air pressure beyond that required for engagement. At the time of disengagement, centrifugal force on the friction shoes facilitates rapid release and total disengagement. During idling, with the primary clutch element rotating on the power-input side, centrifugal force ensures complete disengagement. Because torque output is directly proportional to applied air pressure, controlled output and overload protection are available. Smooth, gradual acceleration is possible by modulating the air supply.

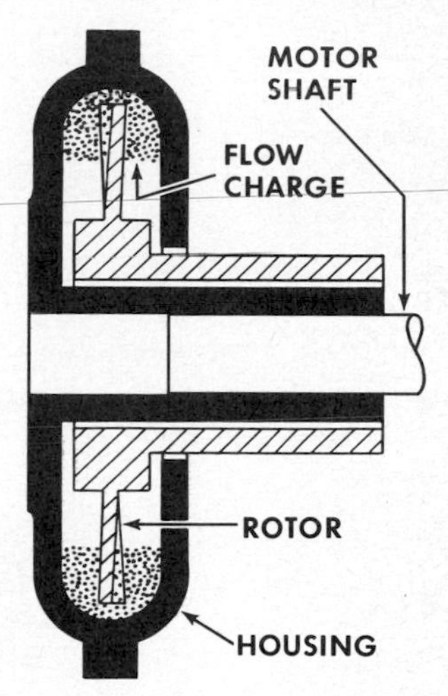

Fig. 4 Operation of dry-fluid clutch. (Dodge Manufacturing Div., Reliance Electric Co.)

The contracting-shoe clutch is well suited to cyclic applications because of its rapid response. For cyclic and continuous applications it provides constant-torque output while engaged because its capacity remains the same throughout friction-shoe life. Clutches of this design can be used in dual-pump drives.

Multiple-Disk Clutches. Clutches of the disk type are units where the clamping force is applied axially to one or more disks. In units of this design, the clamping force is completely free from effects of centrifugal force and self-energization. Also, large areas of friction surface are available for a given size and weight of clutch and the ability to dissipate heat is good.

Shown in Figs. 6*A* and *B* is a multiple-disk clutch that is frequently used in pump applications. The clutch is pneumatically actuated and utilizes the inflatable tube principle similar to that used in rim clutches. Air is supplied to the tube causing it to act axially on a pressure plate, which then moves to provide the necessary clamping force to the friction elements for engagement of the clutch. The torque developed by the clutch is directly proportional to the applied air pressure. When air pressure is released, the clutch is disengaged by means of release springs. The clutch shown is capable of handling drive-line misalignment and does not transmit thrust load to the pump thrust bearing.

Clutches of this design have been utilized in a great variety of pump applications. Compounding power (two prime movers) for various drives is commonly done by use of these clutches since they engage smoothly, are unaffected by centrifugal force, and are subject to accurate control of torque thus eliminating any possibility of two engines fighting each other.

Overruning Clutches. Overrunning or freewheeling clutches are precision devices which will lock up to transmit torque in one direction of rotation, but which will freewheel in the opposite direction of rotation. Shown in Fig. 7 is a cam-type overruning clutch which consists of an inner race, outer race, cam cage assembly, bearings, and lip-type seals.

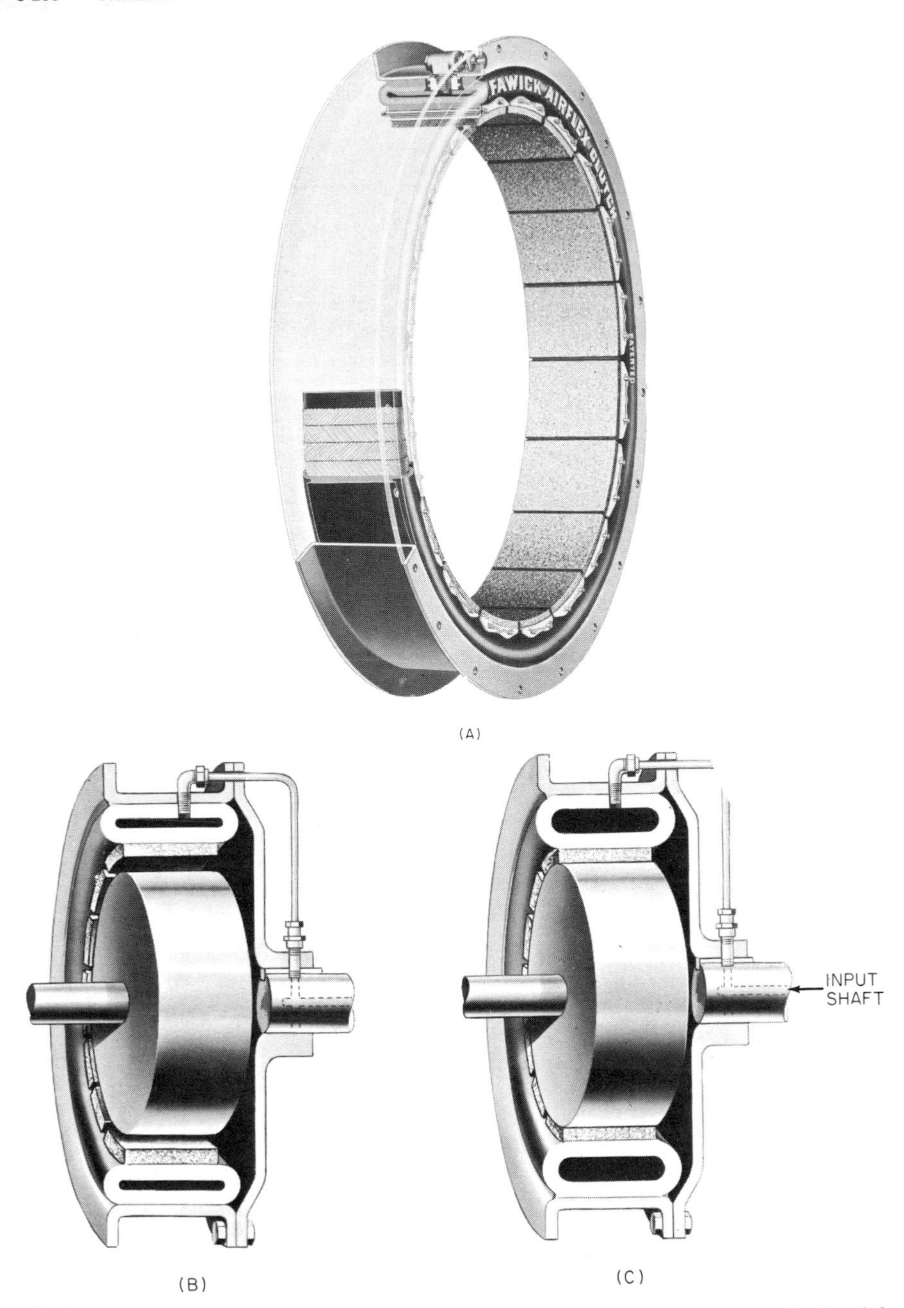

Fig. 5 Rim- or drum-type clutch: external-contracting. (Eaton Corp., Industrial Drives Div.)

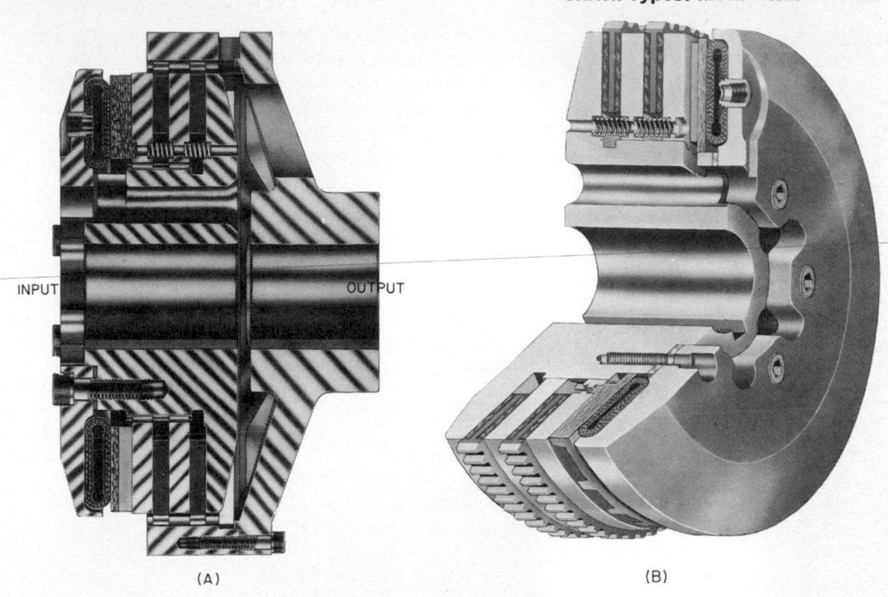

Fig. 6 Multiple-disk clutch: pneumatically operated. (Wichita Clutch Company, Inc.)

The specially designed cams operate in the annular space between the concentric inner and outer races. One race is keyed to the driver shaft and the other race to the load. Power is transmitted from one race to the other by wedging action of the cams between the races. The cams have a greater dimension (B) across one set of diametrically opposite corners than across the other (A), as shown in Fig. 8. Fig. 8*A* shows the cams in a position to allow the inner race to overrun

Fig. 7 Cam-type overrunning clutch. (Morse Chain Div., Borg-Warner Corp.)

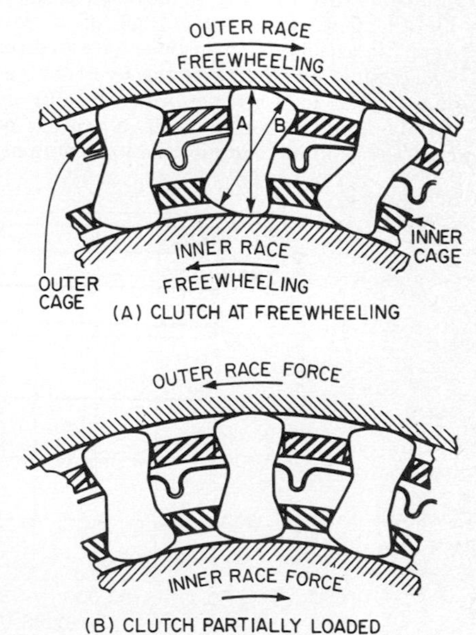

Fig. 8 Cam geometry and operation. (Morse Chain Div., Borg-Warner Corp.)

counterclockwise or the outer race to overrun in a clockwise direction. In Fig. 8*B* the inner race has rotated clockwise and has forced the cams to rotate counterclockwise partially loading the clutch. In pump applications, the clutch is positioned between the prime mover and the load. In installations of this type a clutch-coupling assembly must be used to allow for misalignment of the driving and driven shafts. Figure 9 is a cross-sectional view of a clutch-coupling assembly. The cou-

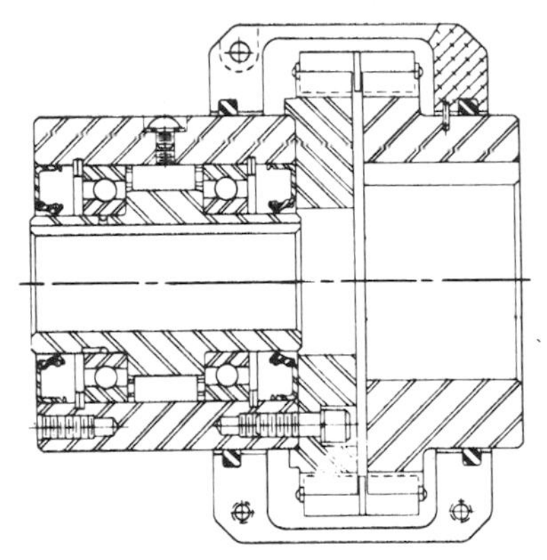

Fig. 9 Clutch coupling assembly. (Morse Chain Div., Borg-Warner Corp.)

pling depicted in this illustration is of the silent-chain type designed to accommodate angular and parallel misalignment. This coupling design also offers ample axial travel to compensate for thermal expansion of the connected equipment. Since cam clutches are available for either inner- or outer-race overrunning operation, connection of clutch inner race or coupling hub (clutch-coupling assembly) to the driver will depend on selection made. Generally, clutches of this type are used in the inner-race overrunning mode; hence, the coupling hub is connected to the driver. Figure 10 shows a typical set of overrunning wear-life curves which list clutch wear life as a function of inner-race overrunning rpm.

A distinct advantage of the overrunning clutch is that engagement and disengagement of the load are performed rapidly and automatically. Also, available are completely enclosed continuously operating overrunning clutch units as shown in Fig. 11. These units contain the overrunning clutch within a fully enclosed housing

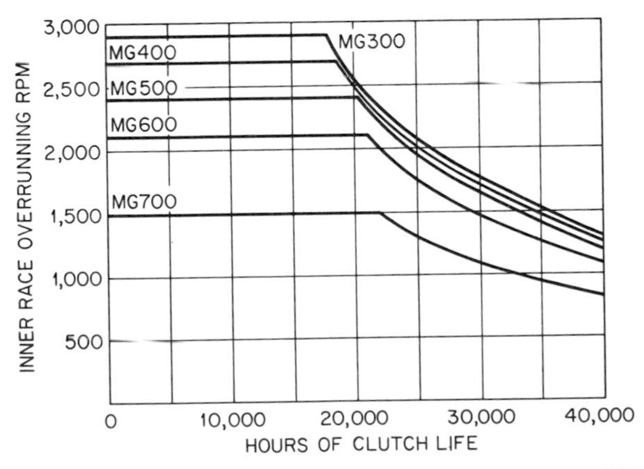

Fig. 10 Cam clutch overrunning wear life curves. (Morse Chain Div., Borg-Warner Corp.)

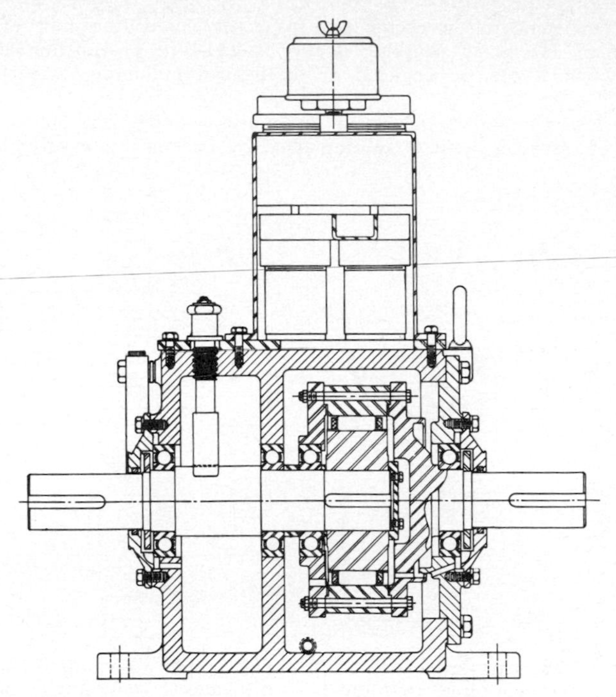

Fig. 11 Completely enclosed high-speed overrunning clutch. (Marland One-Way Clutch Co., Inc.)

which contains its own lubrication system. Units of this design are used where drive-line shaft speeds are high, where uninterrupted service (continuous operation up to one year) is required, and where atmospheric conditions are extremely severe. A typical dual-drive pump system utilizing this high-speed clutch is depicted in Fig. 12.

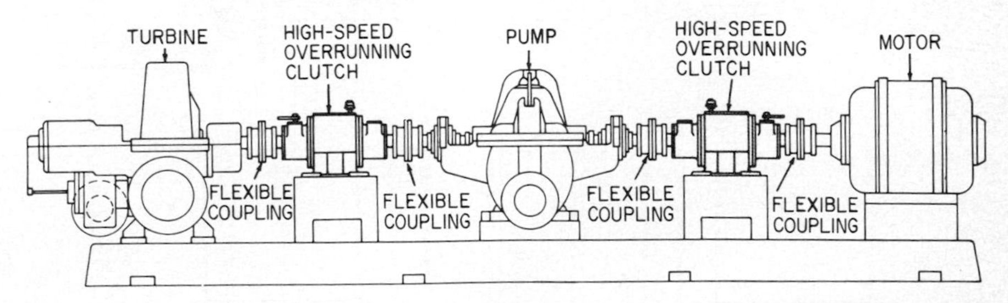

Fig. 12 Dual-drive pump system with high-speed overrunning clutches. (Marland One-Way Clutch Co., Inc.)

Positive-Engagement Type The positive-engagement type of clutch (square-jaw, spiral-jaw, multiple-tooth) is generally not used in pump applications. However, positive-engagement couplings of the disconnect type are frequently used in pump installations and are covered in this text.

Disconnect couplings are generally desirable for use in applications where quick connection or disconnection of shafts is required as in pump dual drives and emergency standby systems. A variety of disconnect couplings is available. The simplest unit is a roller-chain or silent-chain coupling which consists of a driving and driven sprocket engaged by a wrap-around chain. The drive chain can be quickly disconnected by removing a cotter pin from the connecting link of the chain. Another type of disconnect coupling is the flanged-type or continuous-sleeve-type gear coupling which can be engaged or disengaged manually or with a shifting mechanism.

Figure 13 shows the continuous-sleeve disconnect coupling without shifting collar. Engagement or disengagement is accomplished by backing outwardly the lockscrew

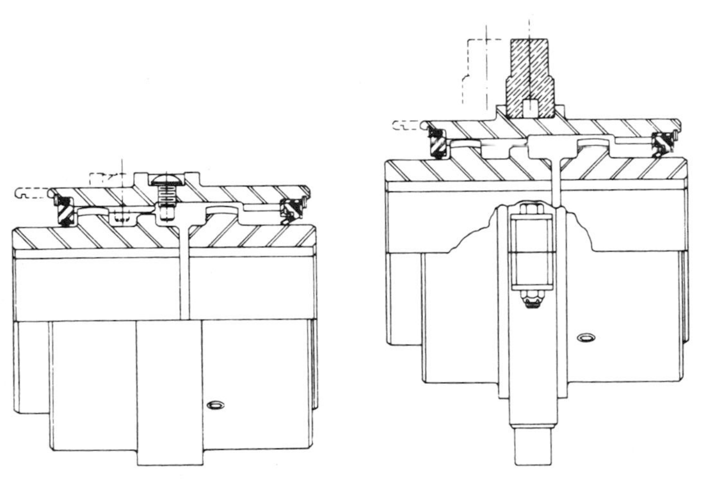

Fig. 13 Continuous-sleeve disconnect coupling (without shifting collar). (Zurn Industries, Inc., Mechanical Drives Div.)

Fig. 14 Continuous-sleeve disconnect coupling (with shifting collar). (Zurn Industries, Inc., Mechanical Drives Div.)

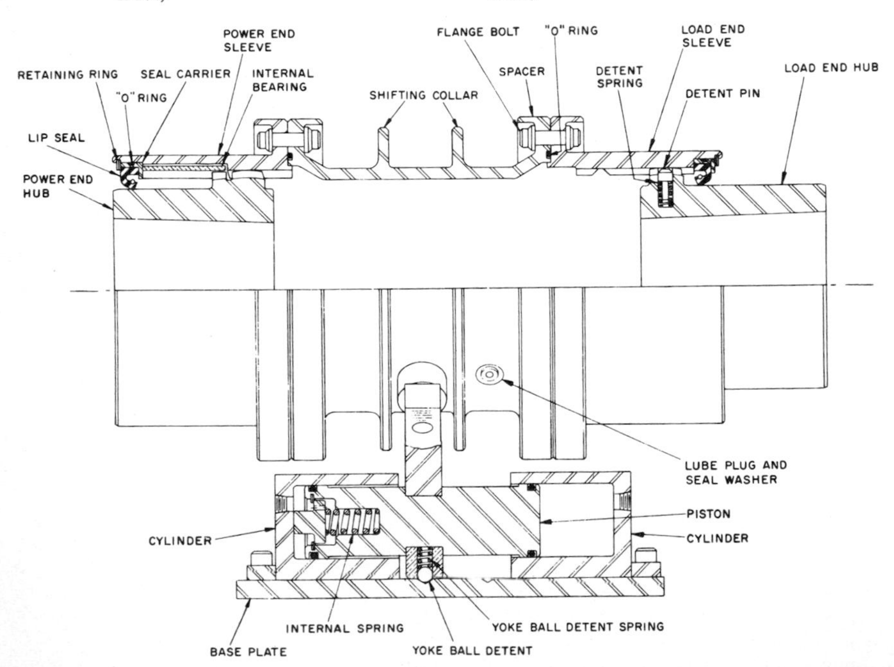

Fig. 15 High-speed disconnect coupling showing shifting-mechanism detail. (Zurn Industries, Inc., Mechanical Drives Div.)

until the sleeve is free to move axially into the desired position and the lockscrew is turned inwardly until locked. This unit must be stopped before engagement or disengagement is made. Figure 14 illustrates the continuous-sleeve disconnect coupling with shifting collar. Engagement or disengagement can be made at low differential speeds. The shifting collar is made of bearing bronze and is designed for grease lubrication. The collar is usually operated by a shifting fork and lever.

High-speed precision disconnect couplings are available for applications such as centrifugal compressors, turbines, and pumps.

A sealed-lubricant disconnect coupling is shown in Fig. 15. Engagement and disengagement are effected by axially shifting the spacer section into the engage or disengage position which is maintained by the action of the spring-loaded pin detents registering in the respective engage or disengage groove.

Axial shifting may be performed manually by way of a shifting fork and lever or with a hydraulic actuator. This high-speed gear coupling is designed to permit disconnection at speed and load conditions when rapid disconnect is required; however, the seals are capable of momentary operation at high differential speeds. Continuously lubricated disconnect couplings are a more sophisticated version of the sealed-lubricant disconnect coupling previously discussed. This general arrangement contains the coupling within an oiltight case. Engagement and disengagement are similar to the sealed-lubricated coupling. The advantage of this design is its capability of withstanding high differential speeds for extended periods.

CLUTCH APPLICATION

Unbalance, shaft misalignment, and improper maintenance are the major causes of mechanical failure. The following are required to minimize field problems and to ensure adequate service life:

1. A complete understanding of the drive system is required.
2. Pertinent application information is required in order to determine proper service factors before selecting the component. Also required are items such as life, space, speed, load (shock factor), temperature, lubrication, etc.
3. It is important that the performance capabilities of the clutch selected be understood and that the component selected be capable of fulfilling not only drive-system requirements of torque and speed but also other areas:
   a. Misalignment capability.
   b. Does clutch impart thrust load to pump bearings?
   c. Is unit subject to self-energizing?
   d. Speeds at which unit can be engaged or disengaged.
   e. Does drive require clutch that prevents reverse rotation?
   f. Must unit perform required function automatically or can mechanical, electric, or hydraulic (pneumatic) actuation be used.

Consultation with supplier or manufacturer of component selected for engineering assistance is recommended.

Section **6.3**

Pump Couplings and Intermediate Shafting

FRED K. LANDON

COUPLING TYPES USED IN PUMP DRIVE SYSTEMS

A coupling is used wherever there is a need to connect a prime mover to a piece of driven machinery. The principal purpose of a coupling is to transmit rotary motion and torque from one piece of equipment to another. Couplings may perform other secondary functions, such as accommodating misalignment between shafts, transmitting axial thrust loads between machines, permitting adjustment of shafts to compensate for wear, and maintaining precise alignment between connected shafts.

Rigid Couplings Rigid couplings are used to connect machines where it is desired

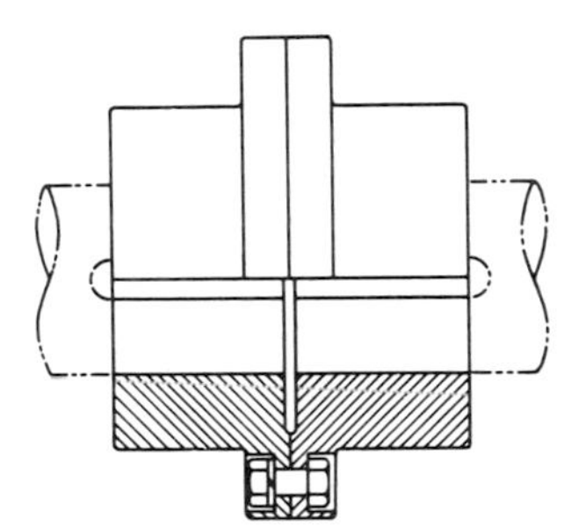

Fig. 1 Flanged rigid coupling. (Koppers Co. Inc.)

to maintain shafts in precise alignment. They are also used where the rotor of one machine is used to support and position the other rotor in a drive train. A coupling of this type cannot accommodate misalignment between shafts, so precise alignment of machinery is necessary when using a coupling of this design.

Types. There are two commonly used types of rigid couplings. One type consists of two flanged rigid members, each mounted on one of the connected shafts (Fig. 1). The flanges are provided with a number of bolt holes for the purpose of connecting the two half-couplings. Through proper design and installation of the coupling, it is possible to transmit the torque load entirely through friction between the two flanges, which assures that the flange bolts do not experience a shearing stress. This type of arrangement is especially desirable to drive systems where torque oscillations occur, as it avoids subjecting the flange bolts to a shearing stress.

A second type of rigid coupling is known as the *split rigid,* which is split along its horizontal centerline (Fig. 2). The two halves are clamped together by a series of bolts arranged axially along the coupling. The rigid coupling and machine shafts may be equipped with conventional keyways, which are in turn fitted with keys to transmit the torque load, or in certain cases the frictional clamping force may be sufficient to permit transmitting the torque by means of friction between shaft and rigid coupling. This type of coupling is commonly used to connect sections of line shafting in a drive train.

A variation to the flanged rigid coupling is known as the adjustable rigid coupling (Fig. 3). This coupling is designed along the lines of conventional rigid couplings, except that a threaded adjusting ring is placed between the two flanges. This ring engages a threaded extension on one of the shaft ends. By means of this ring, it is possible to position the pump shaft axially with respect to the driver.

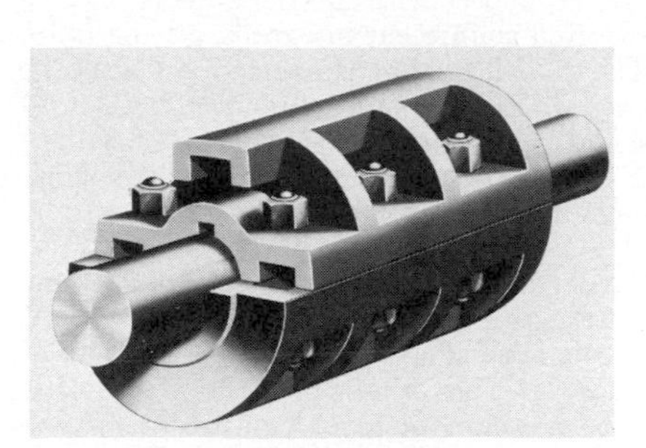

Fig. 2 Split rigid coupling. (Dodge Manufacturing Division, Reliance Electric Co.)

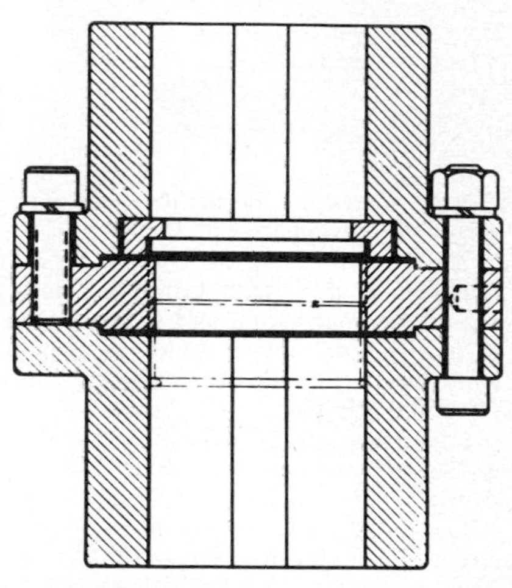

Fig. 3 Adjustable flanged rigid coupling. (Koppers Co. Inc.)

Applications. A common application for rigid couplings in the pump industry is in vertical drives, where the prime mover (generally an electric motor) is positioned above the pump. In such cases, both machines can employ a common thrust bearing, which is generally located in the motor. The coupling flange bolts must be capable of transmitting any down thrust from the pump to the motor. In applications where the thrust from the pump is *toward* the motor, it is common practice to provide shoulders on the shafts to transmit the axial force.

Many pump drive systems require a rigid coupling which is capable of providing axial adjustment to compensate for wear in the pump impeller or impellers. The adjustable rigid coupling described in the previous section is used for this purpose. A threaded adjusting ring is attached to a mating threaded extension of the pump shaft and permits vertical up-and-down positioning of the impeller or impellers. The hub which is mounted on the pump shaft is equipped with a clearance fit and feathered key to permit the hub to slide with respect to the shaft. The load capacity of a coupling of this type is generally limited by the pressure on the pump shaft key since there is no possibility of load being transmitted by interference fit.

A few words of caution should be noted when applying rigid couplings. First, precise alignment of machine bearings is absolutely necessary, since there is no flexibility within the coupling to accommodate misalignment between shafts. Secondly, accuracy of manufacture is extremely important. The coupling surfaces which interface between driving and driven shafts must be manufactured with high degrees of concentricity and squareness, to avoid the transmittal of eccentric motion from one machine to the other.

Flexible Couplings Flexible couplings accomplish the primary purpose of any coupling, that is, to transmit a driving torque between prime mover and driven machine. In addition, they perform a second important function: they accommodate unavoidable misalignment between shafts. A proliferation of designs exists for flexible couplings, which may be classified into two types: mechanically flexible, and material-flexible.

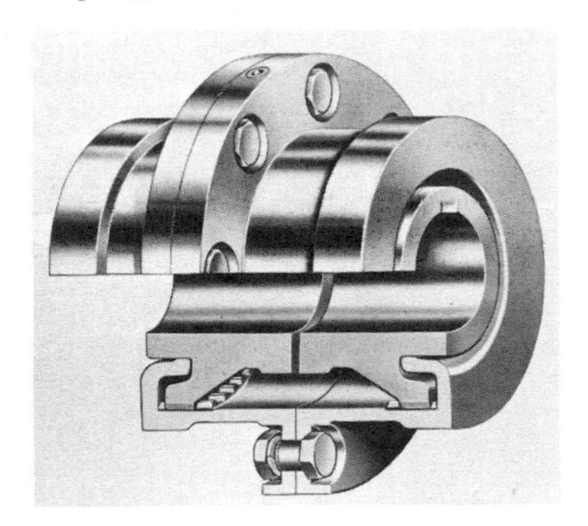

Fig. 4 Gear-type mechanically flexible coupling. (Koppers Co. Inc.)

Mechanically flexible couplings compensate for misalignment between two connected shafts by means of clearances incorporated in the design of the coupling. The most commonly used type of mechanically flexible coupling is the gear-type, or dental, coupling (Fig. 4). This coupling essentially consists of two pair of clearance-fit splines. In the most common configuration the two machine shafts are equipped with hub members having external splines cut integrally on the hubs. The two hubs are connected by a sleeve member, having mating internal gear teeth. Backlash is intentionally built into the spline connection, and it is this backlash which compensates for shaft misalignment. Sliding motion occurs in a coupling of this type, so a supply of clean lubricant (grease or oil, depending on the design) is necessary to prevent wear of the rubbing surfaces.

If interruption of operation for the purpose of relubricating couplings with oil cannot be tolerated, constantly lubricated couplings are used as shown in Fig. 5. It consists of an oiltight enclosure bolted at one end to the stationary portion of either the driving or driven piece of equipment. The other end of the enclosure has a slip fit inside a cover that is bolted to the other piece of equipment. Some form of packing is used to prevent loss of lubricant at the slip joint. Oil under

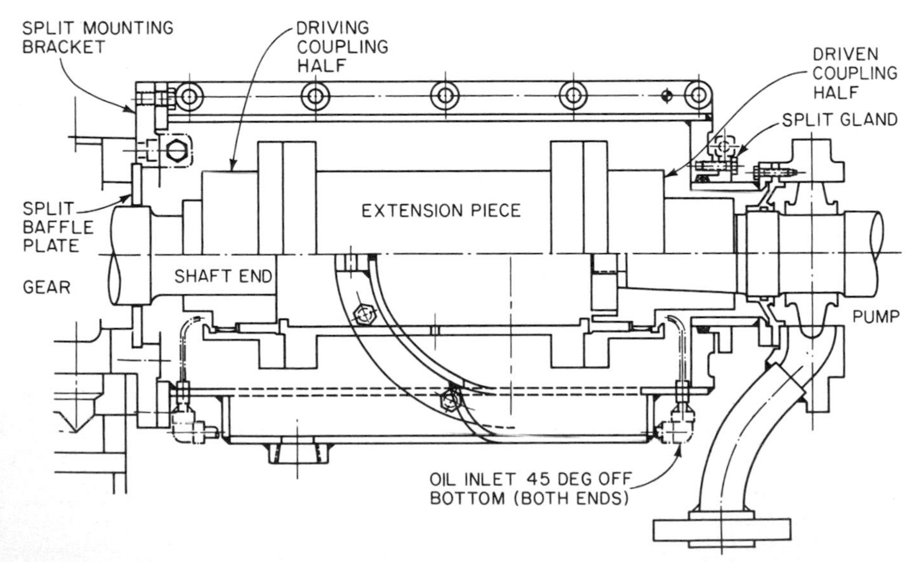

Fig. 5 Continuously lubricated coupling. (Koppers Co. Inc.)

pressure is brought through the enclosure and impinges upon the meshing gear teeth of the coupling, the excess being collected at the bottom of the enclosure and returned to the oil reservoir.

A second type of mechanically flexible coupling which sees wide usage, especially in low-cost drive systems, is known as the roller-chain flexible coupling (Fig. 6). This coupling employs two sprocket-like members which are mounted on each of the two machine shafts and are connected by an annulus of roller chain. The clearance between sprocket and roller, and, in some cases, the crowning of the rollers themselves provide mechanical flexibility for misalignment. This type of coupling is generally limited to low-speed machinery.

Fig. 6 Roller-chain mechanically flexible coupling. (Dodge Manufacturing Division, Reliance Electric Co.)

Fig. 7 Disk-type material-flexible coupling. (Formsprag Co.)

Material-flexible couplings rely on flexing of the coupling element to compensate for shaft misalignment. The flexing element may be of any suitable material (metal, elastomer, or plastic) which has sufficient resistance to fatigue failure to provide acceptable life. Some materials, such as steel, have a finite fatigue limit, and the coupling must be operated under conditions of load and misalignment which assure that the stress developed within the coupling element is within that limit. Other materials, such as elastomers, generally do not have a well-defined fatigue limit. In these cases, flexing of the coupling element develops heat within the material, which can cause failure if excessive.

One type of material-flexible coupling is the metal-disk coupling (Fig. 7). This coupling consists of two sets of thin sheet-metal disks which are attached to the driving and driven hub members by means of bolts. Each set of disks is made up of a number of thin laminations which are individually flexible, and by means of this flexibility compensate for shaft misalignment. These disks may be stacked together as required to obtain the desired torque-transmission capability. This type of coupling requires no lubrication; however, alignment of the equipment must be maintained within acceptable limits so as not to exceed the fatigue limit of the material.

Another example of an all-metal material-flexible coupling is the flexible diaphragm coupling, shown in Fig. 8. This coupling is similar in function to the metal-disk coupling in that the disk flexes to accommodate misalignment. However, the diaphragm type consists of a single element with a hyperbolic contour which is designed to produce uniform stress in the member from inner to outer diameter. By more

Fig. 8 Diaphragm-type material-flexible coupling. (Bendix Corp., Fluid Power Division)

efficiently utilizing the material the weight is reduced correspondingly, thus making this coupling suitable for high-speed applications.

Flexible-element couplings employing elastomer materials are numerous and their designs are varied. By definition, an elastomer is a material which has a high degree of elasticity and resiliency and will return to its original shape after undergoing large-amplitude deformations. One example of an elastomer coupling is the pin-and-bushing coupling (Fig. 9). This design is comprised of two flanged hub members, one mounted on each machine shaft. The flange of one hub is fitted with pins which extend axially toward the adjacent shaft. The other flange is equiped with rubber bushings, which generally have a metal sleeve at the center. The pins fit into these sleeves and provide transmission of torque through the bushings. Since the bushings are made of flexible material, they can accept slight angularity or offset conditions between the two flanges.

Fig. 9 Pin-and-bushing elastomer coupling. (Ajax Flexible Coupling Co.)

A second group of elastomer couplings employs a sleevelike element which is connected to a hub member on each shaft, and transmits torque through shearing of the flexible element. The flexible element may be attached to the machine hubs by a number of different means: it may be chemically bonded, it may be mechanically connected by means of loose-fitting splines (Fig. 10), or it may be clamped to the hubs and held in place by friction (Fig. 11). Misalignment between shafts is accommodated through flexing of the elastomer sleeve.

A third group of couplings utilizes an elastomer member which is loaded in compression to transmit load between shafts. The elastomer material may be bonded onto the machine shafts (or onto hub members attached to the shafts, Fig. 12) or may be placed loosely into cavities or pockets which are formed by members which are rigidly mounted onto the two shafts (Fig. 13). Again, the elastomer material itself deflects to compensate for shaft misalignment.

A type of elastomer coupling which is commonly used on small-horsepower drive systems is the rubber-jaw type of coupling (Fig. 14). The heart of this coupling is a "spider" member, having a plurality (usually three) of segments extending radially from a central section. The hubs, which are mounted on driving and driven

Fig. 10 Sleeve-type elastomer coupling. (T. B. Woods' Sons Inc.)

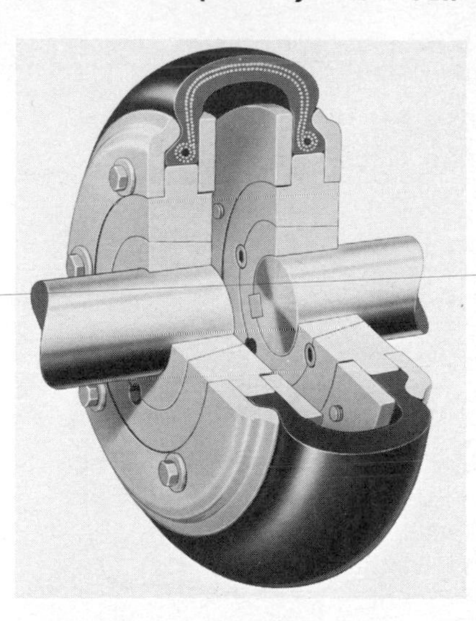

Fig. 11 Sleeve-type clamped elastomer coupling. (Dodge Manufacturing Division, Reliance Electric Co.)

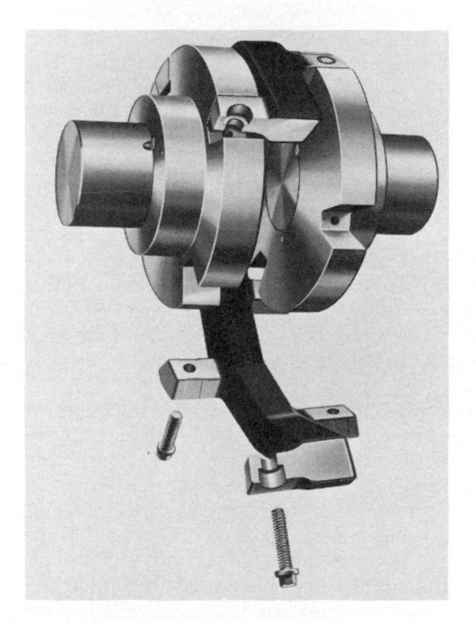

Fig. 12 Compression-loaded bonded elastomer coupling. (Koppers Co. Inc.)

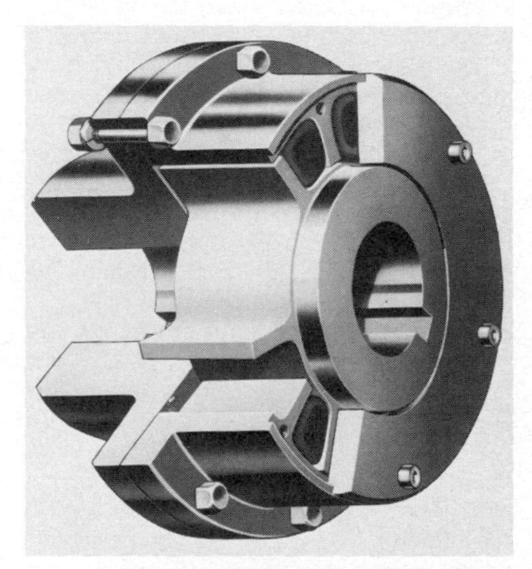

Fig. 13 Compression-loaded loosely fitted elastomer coupling. (Koppers Co. Inc.)

shafts, each have a set of jaws corresponding to the number of spiders on the flexible element. The spider fits between the two sets of jaws, and provides a flexible "cushion" between the two sets of jaws. This cushion transmits the torque load as well as compensating for misalignment.

There is a type of commercially available flexible coupling which combines the characteristics of both the mechanically flexible and material-flexible couplings. This is the spring-grid coupling (Fig. 15). This design has a hub mounted on each machine shaft. Each hub has a raised portion on which toothlike slots are cut. A spring-steel grid member is fitted, or woven, in between the slots on the two hubs. The grid element can slide within the slots to accommodate shaft misalignment and flexes like a leaf spring to transmit torque between machines. Unlike most material-flexible couplings this design requires periodic lubrication to prevent excessive wear of the grid member.

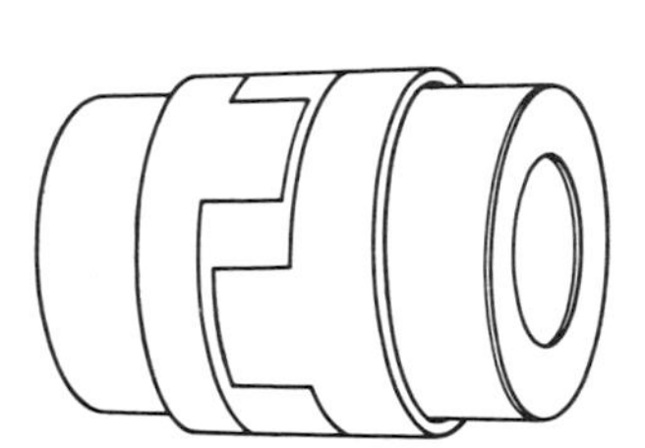

Fig. 14 Rubber-jaw-type coupling. (Lovejoy Inc.)

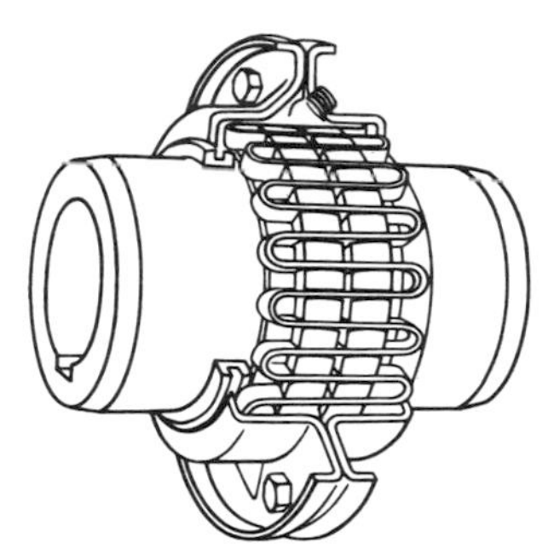

Fig. 15 Spring-grid coupling. (Falk Div. of Sundstrand)

Application of Flexible Couplings. The type of flexible coupling most suitable for a particular application depends upon a number of factors, including horsepower, speed of rotation, shaft separation, amount of misalignment, cost, and reliability. In the design of a system, it is the goal of the designer to use the least expensive coupling which will "do the job." In low-cost systems, cost alone may be the most important criterion, and the least expensive coupling which transmits the rated power and accepts some small degree of misalignment is generally the choice, albeit at some sacrifice of reliability and durability. On the other hand, high-horsepower, high-speed machinery generally represents a critical piece of equipment for a power station, sewage plant, or other vital process, and in these cases, a coupling should be selected which will not compromise the overall reliability of the system.

Low-horsepower pumps (up to about 200 hp) driven by electric motors can usually be coupled successfully by any of the types of couplings described here. Selection procedures vary from manufacturer to manufacturer, but generally the following data is required: hp, rpm, anticipated misalignment, and type of pump (reciprocating, vane-type, centrifugal, etc.).

Pumps of the same power range are very often driven by reciprocating engines (diesel, gasoline, natural gas). This is quite common in remote areas such as at pipeline pumping stations, where a source of electric power is not available. Since this type of prime mover produces a pulsating type of power, it is often necessary to perform a torsional vibration analysis of the drive system to ensure that the normal operating speed is well removed from a speed which may produce a torsional resonant vibration. Such an analysis requires that the torsional "stiffness" of the coupling be known. It is quite often possible to "tune" the drive system by selecting the proper coupling stiffness, to avoid operating at a resonant condition. The selection data required for a system of this type is the same as listed above. Most coupling manufacturers will, however, assign a higher service factor to an application involving a reciprocating prime mover, to compensate for fatigue effects due to torque fluctuations. In addition, the remote location of many engine-driven pumps indicates a special need to ensure a high degree of reliability of the system.

Another commonly employed prime mover is a steam turbine. These machines, which range from about 100 to well over 50,000 hp (for pump drives), operate very efficiently and economically, providing there is a source of steam available at the installation. Any flexible coupling employed on a steam turbine driven machine must be capable of accepting the thermal gradient at the turbine shaft, and must also accommodate the axial growth of the turbine shaft as it warms up to operating speed.

Steam turbines are generally high-speed machines (4,000 to as high as 10,000 to 15,000 rpm in some cases) and as such require a relatively high degree of system balance to avoid critical vibration. Elastomer couplings have occasionally been applied successfully to steam-turbine drives, but because of the high speeds involved, all-metal couplings are usually employed. Most common of the various metal couplings used on this type of drive is the gear-type (mechanically flexible) design. High-speed machinery requires that the weight of rotating components be minimized so as to decrease shaft deflections and hence increase the lateral critical speed of the system. The gear-type coupling is the most efficient design yet devised for transmitting large amounts of horsepower at high speeds and with minimum weight. Where coupling weight and torsional stiffness are critical values to the overall system, special designs may be created which provide the specific values required for satisfactory system operation.

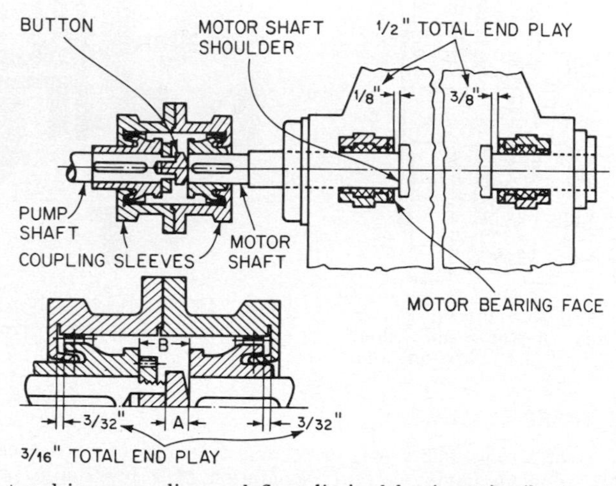

Fig. 16 Motor-driven coupling end float limited by inserting button between shaft ends. (Worthington Pump International)

Limited End Float. Many horizontal motor-driven pump systems utilize motors which are equipped with journal, or sleeve-type, bearings. These bearings are intended only to absorb the transient thrust created by the motor rotor during acceleration and deceleration. The coupling for this type of drive should be equipped with suitable provisions for limiting the axial float of the motor rotor to some fraction of its total float. This may be done by positioning the motor in the center of its axial travel, and then employing a coupling having a total float which is *less* than the float of the motor. Any motor thrust is taken by the pump bearing. Typical design practice calls for using a coupling with ⅛-in float on motors having a ¼-in total end float. Motors having ½-in (or greater) rotor float generally use couplings having $\frac{3}{16}$-in total float. Coupling total float can be limited by inserting a button between shaft ends as shown in Fig. 16. This type of coupling prevents the motor rotor from ever contacting the thrust shoulders on the shaft bearings. Certain types of elastomer and disk couplings having inherent float-restricting characteristics provide centering without any additional modifications (Fig. 17).

Vertical Operation. As previously noted, rigid couplings are commonly used on vertical-drive systems, where the system characteristics warrant such a coupling. However, many vertical-drive systems require a flexible coupling to accommodate shaft misalignment. It is generally possible to use a nonlubricated coupling, such as one of the many elastomer-element designs, in a vertical position without modification, provided that the shafts are supported in their own bearings and the coupling does not have to transmit a thrust force. Lubricated designs such as the gear and spring-grid type usually require modification of some type, to make certain that lubricant is retained in both halves of the coupling. One design which provides the advantages of a gear-type coupling, and which is commonly used on pump drives is the vertical double-engagement coupling (Fig. 18). This coupling provides two gear meshes in one half-coupling (thus permitting angular *and* offset shaft misalignment), avoiding the potential leakage problem which exists when a conventional coupling is operated vertically.

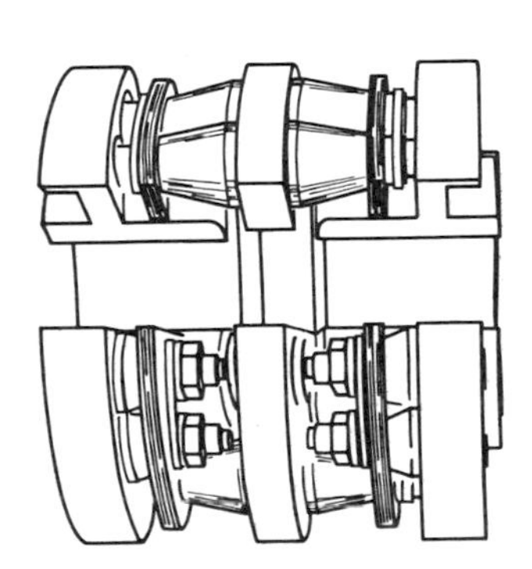

Fig. 17 Metal-disk coupling can be used to limit motor end float. (Thomas Coupling Div., Rexnord Inc.)

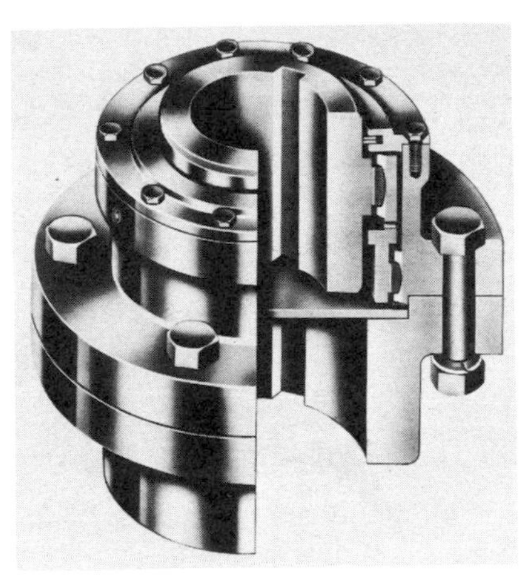

Fig. 18 Vertical double-engagement gear-type coupling. (Koppers Co. Inc.)

PUMP-DRIVE SHAFT SYSTEMS

Pump-drive system arrangements may be classified in one of two categories: those which are close-coupled, having a shaft separation of a fraction of an inch (not to be confused with integral motor pumps not having a flexible coupling), and those which, for one or more reasons, have the prime mover located a substantial distance from the pump. These latter types of systems require a modification to the basic flexible-coupling designs previously described.

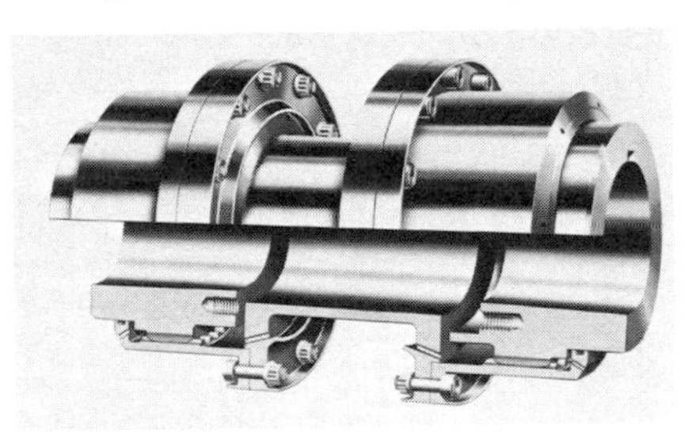

Fig. 19 Tubular spacer coupling. (Koppers Co. Inc.)

Spacers One means of accommodating shaft separation in excess of the normal amount provided in a standard coupling is to employ a flanged tubular spacer between the two coupling halves (Fig. 19). These components are lightweight and are commonly used with end-suction pumps, where it is possible to remove the pump impeller with the pump housing and piping, as well as the prime mover, still in place (Fig. 20). For high-speed drive systems, a lightweight spacer is usually the only practical means of achieving the goal. This spacer may be used to "tune" the system torsionally (as previously discussed) by varying the body diameter and wall thickness. Spacers can be manufactured in lengths up to several feet, but the cost generally restricts its usage to shorter lengths except where no other means is suitable.

Floating Shafts Floating-shaft couplings accomplish a purpose similar to that of spacer couplings, namely, to connect two widely separated shafts. The basic difference is in the construction of the component. While spacers are normally made

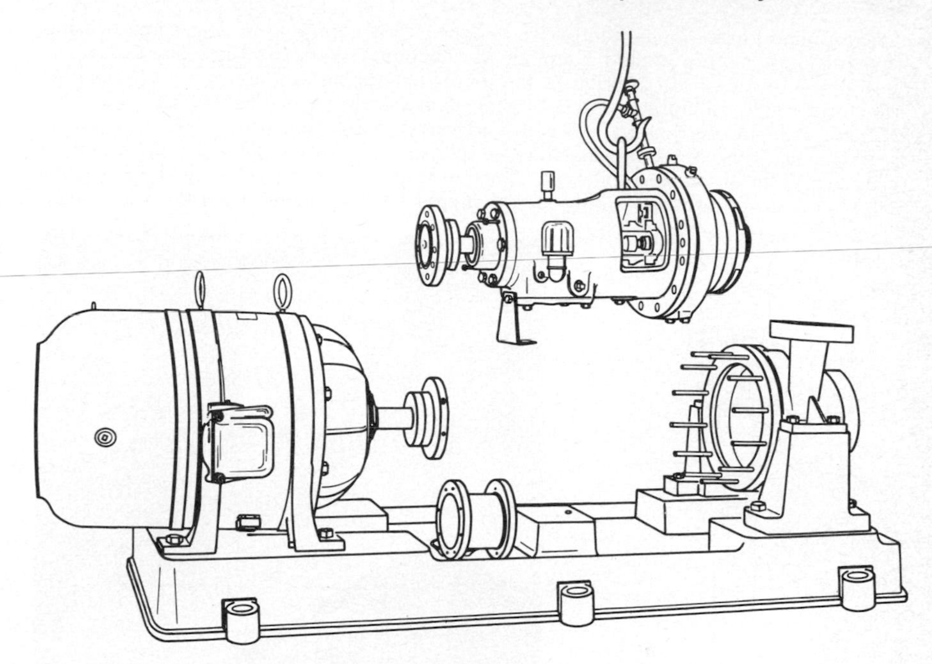

Fig. 20 Spacer coupling enables end-suction pump to be dismantled without moving piping, pump casing, or driver. (Worthington Pump International)

with the body having an integrally formed flange to connect the two coupling halves, a floating shaft usually is made by attaching a flange to a piece of solid or tubular shafting by means of mechanical keys or by welding (Fig. 21). This type of construction is generally less expensive than a one-piece spacer, especially when very large shaft separations must be spanned. Floating-shaft arrangements are widely used on horizontal pump applications of all types and are especially common on vertical pump applications, such as in water pumping and sewage treatment stations. In these latter applications, the pump may be submerged in a pit usually 30 to 50 ft below ground level, while the motor (to prevent damage during flooding) is mounted at ground level. Settings as deep as 100 ft have been constructed. Such systems require long floating shafts, which are generally made in sections and supported by line bearings at intermediate supports or floors.

Rigid Shafts Vertical centrifugal pumps can be designed to contain their own thrust and line bearings. These pumps employ a flexible coupling at the pump shaft. When vertical intermediate shafting is used, the shaft need only be designed

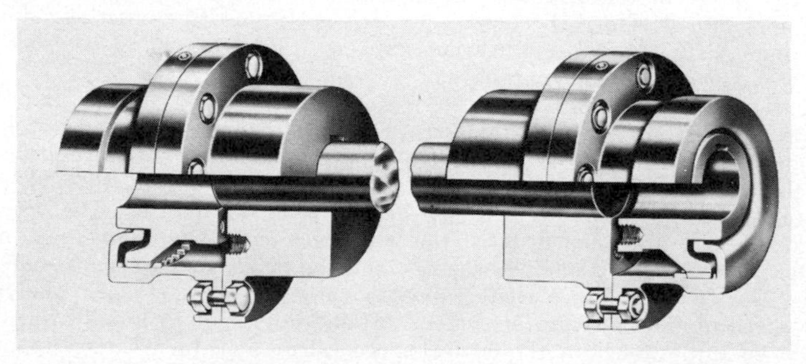

Fig. 21 Section of floating shaft and couplings. (Koppers Co. Inc.)

to transmit torque. Some pumps are designed to contain only a single line bearing. These pumps require a rigid coupling at the pump shaft so that axial thrust can be carried by the thrust bearing in the driver or gear located above. The construction of a single oil-lubricated line-bearing pump is shown in Fig. 142, Sec. 2.2, Centrifugal Pump Construction. In order to carry thrust the drive shaft, if made in sections, must use rigid intermediate couplings and the coupling at the driver or gear must also be of the rigid type. Intermediate bearings used with rigid shafting must be designed to provide only lateral support thereby assuring all the axial thrust

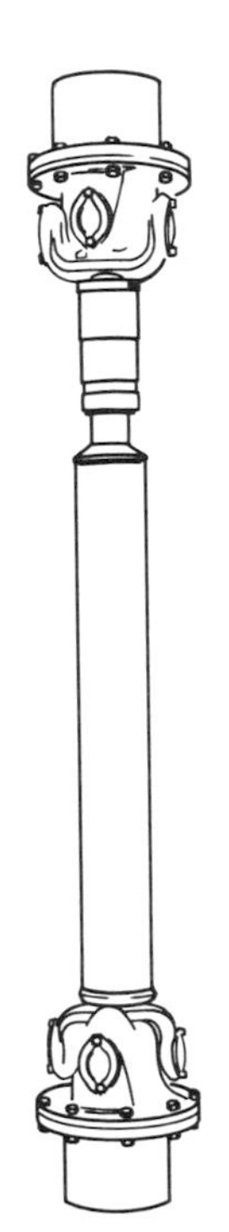

Fig. 22 Tubular intermediate shafting and universal joints. (H. S. Watson Co.)

Fig. 23 Sections of flexible shafts and intermediate guide bearings used to transmit torque from motor located above flood elevation to pump located below ground. (Worthington Pump International)

is carried by the driver or gear. An intermediate oil-lubricated sleeve-type guide bearing is shown in Fig. 141, Sec. 2.2, Centrifugal Pump Construction. Pump and intermediate shaft sleeve guide bearings can be either grease- or oil-lubricated, and they can also be antifriction if preferred.

A rigid intermediate shaft connected to a pump provided with a rigid coupling must be designed to carry the axial thrust from the pump, its own weight, and a bending load. Volute-type centrifugal pumps produce a radial reaction on the pump shaft which in turn is transmitted to the vertical shafting through the rigid pump coupling. Intermediate bearings then act as line bearings. The bearing supports must also be designed to resist this bending force. The pump shaft and bearing, intermediate shafts and line bearings, and the driver or gear shaft and bearings must be treated as a single shaft supported at each bearing when analyzing the critical speed of the entire rotor system. If the intermediate line-bearing supports are assumed to be nodes in the critical-speed calculation, then these supports must be absolutely rigid. The supplier of the shafting should, however, confirm what flexibility he will allow at these supports and give the design forces. The support for the guide bearings must also be designed not to have a natural frequency of vibration within the operating speed range of the pump.

Flexible-Drive Shafts Universal joints (Fig. 22) with tubular shafting can be substituted for flexible couplings whenever it is necessary to (*a*) eliminate the need for critical alignment, (*b*) provide wider latitude in placement of pump and driver, and (*c*) permit large amounts of relative motion between pump and driver. This type of shafting can be used horizontally as well as vertically to provide a short spacer arrangement or to provide a large separation between pump and driver as required for deep settings (Fig. 23).

Flanges are furnished to fit pump and driver shafts to the universal joints which are splined to allow movement of the shaft. An intermediate steady bearing is required at each joint, which must also support the weight of a section of shafting (Fig. 24). Pump thrust cannot be taken by the driver due to the spline and it is necessary that a combination pump thrust and line bearing be used. Because of the universal joint the intermediate bearing or bearings take no radial load from the pumps and therefore act only as a steady bearing for the shaft.

The shaft must be selected to transmit the required torque and be of a length and diameter that will have a critical speed well removed from the operating speed range of the pump. The support for the guide bearing must not have a natural frequency of vibration within the above range. Shafts of this type are often preengineered and stocked. Selection charts are available from the manufacturer to make a proper size and length selection. Standard tubular, flexible drive shafts have limited torque-carrying capacity.

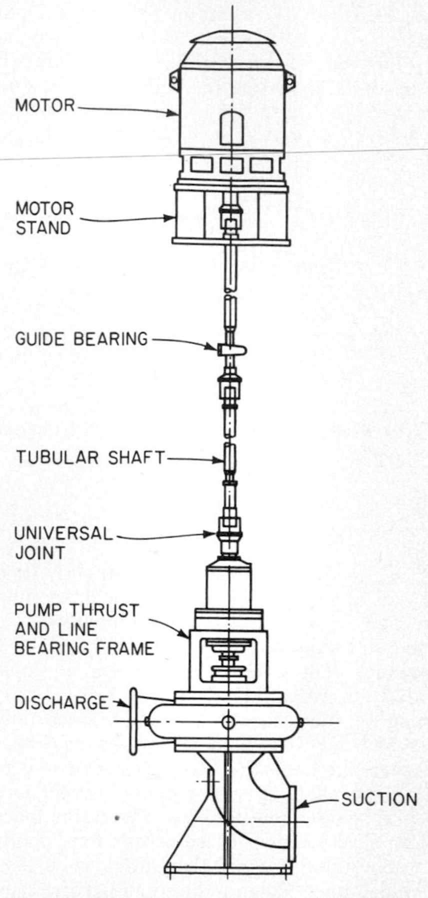

Fig. 24 Vertical pump with flexible shafting. (Worthington Pump International)

Design Criteria *Torsional Stress.* The torsional stress in the shafting may be calculated by the following equations:

$$S_s = \frac{16T}{\pi D_o^3} \quad \text{for solid shafting}$$

$$S_s = \frac{16T}{\pi D_o^3\left(1 - \dfrac{D^4}{D_o^4}\right)} \quad \text{for tubular shafting}$$

where S_s = torsional shear stress, lb/in²
T = transmitted torque, in-lb
D = shaft inner diameter (tubular shaft only), in
D_o = shaft outer diameter, in

The allowable value of shear stress depends upon the material being used and whether it is subjected to other loads, such as bending or compression. The design safety factor on the shafting should be equal to or greater than the other components in the drive train.

Critical Speed. The critical speed of a drive shaft is determined by the deflection, or "sag," of the shaft in a horizontal position under its own weight. The less the sag the higher the critical speed. In practical terms, a long, slender shaft will have a low critical speed while a short, large-diameter shaft will have a very high critical speed. Calculating the deflection of a simply supported shaft is as follows:

$$y = \frac{5wL^4}{384EI} = \text{shaft deflection}$$

noting that

$$I = \frac{\pi D_o^4}{64} \quad \text{for solid shafting}$$

$$I = \frac{\pi(D_o^4 - D^4)}{64} \quad \text{for tubular shafting}$$

where w = weight of shaft per unit length, lb/in
L = length between bearing supports, in
E = Young's modulus, lb/in^2
I = moment of inertia, in^4

Knowing the natural deflection of the shaft, it is possible to calculate the first critical speed from the equation:

$$N_{\text{crit}} = 187\sqrt{\frac{1}{y}}$$

which expresses critical speed directly in rpm of the rotating shaft.

In practice, the critical speed should be placed well away from the operating speed of the shaft. Actual operating conditions such as imbalance, clearance in bearings and couplings, etc., will tend to reduce the critical speed from its theoretical value. The designer has several options from which he may choose to attain the desired critical speed. These are (1) vary the shaft diameter, (2) use tubular shafting to increase stiffness and reduce deflection, and (3) vary the bearing span.

Varying the diameter of the shafting has a dramatic effect on shaft deflection, since the deflection decreases inversely with the cube of the diameter. The weight of the shafting will increase proportionately to the square of the diameter, and may, consequently, impose excessive loads on bearings. When this is the case, tubular shafting may be substituted for solid shafting. The ratio of I/w is much higher for a tubular shaft than for a solid shaft of the same diameter so that the shaft deflection is again reduced and critical speed is increased.

Varying the bearing span is something which may be employed when the size and weight of a single length of shafting become excessive for the application. In this case a floating shaft is broken into sections, with a single engagement coupling used to connect each section and acting as a hinge joint. A self-aligning bearing is placed on each section immediately adjacent to the coupling. Each section may then be treated as an individual shaft, which will result in a substantial reduction in shafting size. The rigidity of the support at each bearing must be taken into consideration when calculating the critical speed of the shafting.

Dynamic Balance. Drive systems which operate at high speeds and connect two shafts which are widely separated must have special consideration with regard to dynamic imbalance. Care must be taken to ensure that the long slender shaft used in such systems is not bent, either in manufacture or installation. These systems quite often require shafting which has been dynamically balanced after manufacture to compensate for these normal errors.

Section 7

Pump Controls and Valves

W. O'KEEFE

CONTROLS

Pump control in the broadest sense gives the pump user (1) the flow rate, pressure, or liquid level that he desires, (2) protection for his pump and system from damage involving the pumped liquid, and (3) administrative freedom in decisions on operations and maintenance.

Control System Types Pump control systems range in complexity from single hand-operated valves to highly advanced, automatic flow-control or pump-speed control systems. Pump type and drive type are factors in control system choice. For centrifugal pumps, either change of speed or change of valve setting can control the desired variable. For positive-displacement pumps, whether reciprocating, rotary, screw, or other types, control is by change in speed, change in setting of a bypass valve, or change in displacement. The last-mentioned method is found in metering and hydraulic-drive pumps. Although this section considers only control systems having valves as final control elements, the sensing elements discussed also serve in pump-speed control systems.

Pump control systems divide readily into on-off and modulating. The on-off system provides only two conditions: a given flow (or pressure) value, or a zero value. A valve is therefore either open or closed, and a pump driver is running or not. The modulating system, on the other hand, adjusts valve setting or speed to needs of the moment. Either type of system can be automatic or manual.

System Essentials All control systems have:

1. A sensing or measuring element
2. A means of comparing the measured value with a desired value
3. A final control element (a valve) to produce the needed change in the measured variable of the liquid
4. An actuator to move the final control element to its desired position
5. Relaying or force-building means to enable a weak sensing signal to release enough force to power the actuator

The sensing or measuring element is often physically separated from the comparison and relaying means, which are usually housed together and called the *controller*. The actuator and valve are physically connected and may be at a distance from the controller.

In a very simple control action, such as one based on an administrative decision to shut down temporarily one of several small parallel pumps in service, some of the five essentials are in the mind and body of the human operator who turns the valve handwheels and pushes the motor stop button. Nevertheless, the essentials must always be present in some form.

Effect of Rate of Change. The nature of the rate of change of the measured variable or desired value with time gives a convenient guideline in pump control. The chief types of change are:

1. Slow change (practically steady state)
2. Sudden change from one steady state to another (either a nearly instantaneous step change or a high-rate ramp change)
3. Fluctuation at varying rates and in varying amounts

Slow change involves questions of the ability of the control system to hold the desired value accurately and not lag unnecessarily during the change. Equal accuracy when approaching the new value from above or below is also desired.

Sudden change involves additional questions of how long the system will take to reach the new value, whether it will overshoot and then fluctuate back and forth, and for how long it will do this.

Fluctuation introduces serious questions of whether the system will be excited into amplification of some types of fluctuations and go totally out of control. These systems are the most difficult to design and operate. They involve such factors as inertia of the control elements, amount of liquid in the system, and dynamic behavior of each element and of the elements together.

Open-Loop Control. The simplest mode of automatic control is open-loop control, in which the pump-speed (or displacement in some pump types) or the control-valve setting is adjusted to and held at a desired value calculated or calibrated to produce the required output of flow, level, or pressure. The calculation can result in a cam for the controller or positioner or a particular characterization of a valve plug. In operation, only the deviation of the input variable from its desired value is measured, and the control system adjusts the input variable to eliminate the deviation. Because the output variable is not measured, a change in the conditions on which calculation or calibration were based will introduce output errors. Change of input variable can be done manually or by another control system. For example, a pump may be speeded up by a rheostat, or the air pressure to a valve actuator may be changed by changing a pneumatic pressure-control valve setting. Open-loop systems are also called feedforward systems, in contrast to feedback, or closed-loop systems. Open-loop systems are stable, simple, and quick in response, but they tend to error as downstream conditions change.

Closed-Loop Control. A closed-loop control system eliminates much of the error of the open-loop system. In the basic closed-loop, or feedback, system, the output variable is measured and the value compared with an arbitrary desired or set value. If the comparison reveals an error, the pump-speed or control-valve setting is changed to correct the error. Large-capacity water tanks or lag in the control system itself can introduce delays in establishment of the new output value, and the system can therefore overcorrect and oscillate back and forth unless design prevents this.

ON-OFF CONTROL. The simplest closed-loop systems operate on-off between fixed limits, such as water level or pressure. The on-off action is at extremes of a wide or narrow band that can be set at any point in the range. For example, a tank level control may work in an on-off band of 1 in or 10 in at any level in a 5-ft-deep tank.

PROPORTIONAL CONTROL. This is the basic type of closed-loop control. Within a wide or narrow band of output-variable values, the controller input, such as actuator air pressure, is proportional to the deviation from the set point, or desired value, at the band center. If the band is very narrow, say 1 in of level in a 60-in-deep tank, then the controller will apply full air pressure to the valve actuator at a ½-in deviation from the set level in one direction and minimum air pressure at a ½-in deviation in the other direction. This is close to the effect of an on-off control. If the band is wider, say 20 in of level in the 60-in tank, the air pressure will vary from minimum to full pressure over the 20-in band, and so the system will be less sensitive and apply less correction for a given small change in output variable. The lower sensitivity can make the system less likely to overshoot or hunt. Because a given controller output corresponds to every value of deviation from the set point, the simple proportional system will not come back to its set point if the output variable changes as a result of changed demand, such as for more water from the tank. The difference between set point and actual new equilibrium value of level is called *offset*. Narrowing the band will reduce the offset but may cause intolerable oscillations or hunting.

To improve response and stability and to achieve very high accuracy, however, several refinements may be needed. Addition of reset to a simple proportional controller will eliminate offset. This is the proportional-plus-reset or proportional-plus-integral system. In terms of the proportional band, reset means that the band is shifted in such a way as to produce slightly more correction and return the output variable back to what was desired. The reset feature may impair stability, however, because of the added control action.

Derivative action is an added refinement to improve stability and response. In this, the rate of change of the measured output variable is what determines the controller output. A step or sudden change in measured output variable will cause a momentary large increase in controller output which will initiate response. When the derivative action fades, the basic proportional-plus-reset action takes over to restore conditions.

The open-loop system, sensing change in input variable and therefore giving rapid response, is exploited by adding it to the closed-loop (or feedback) system. An example of a feedforward-feedback system in pump flow control is the three-element boiler-feedwater regulator.

Sensing and Measuring Elements In automatic control of pump, these elements detect values of and changes in liquid level, pressure, flow rate, chemical concentration, and temperature. The signal emitted by the element often needs amplification or conversion into another medium, which is done in a transducer. Air pressure to electric voltage or current, or rotary motion to electric voltage are common transformations.

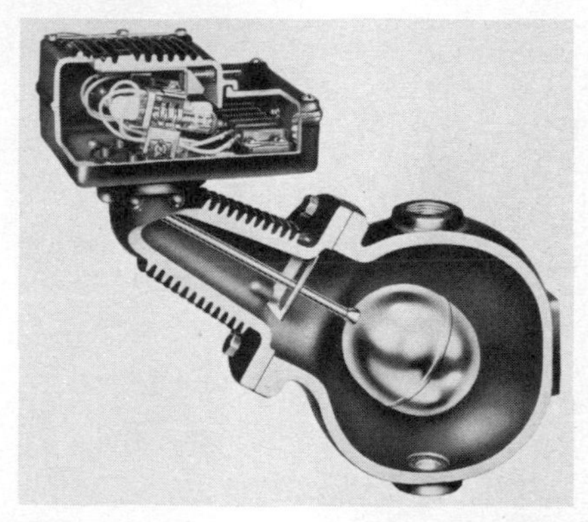

Fig. 1 Low-water cutoff and alarm are purposes of this liquid-level sensor. (McDonnell & Miller, Inc.)

Liquid-Level Sensors. The simplest of several types of sensors is the float in the main tank (or boiler drum), or in a separate float chamber connected at top and bottom to the tank or drum (Fig. 1). The float can be a pivoted type, with motion transmitted outside the chamber by a small-diameter rotating shaft or translational rod attached to the lever arm near its pivot to obtain mechanical advantage. A rod of the latter type can actuate the stem of a balanced valve to control liquid flow and thus liquid level in the supplied tank.

Floats on vertical rods can actuate switches outside and above the float chamber (Fig. 2). Depths can vary from less than 1 to over 50 ft, with rod guides often necessary at the greater depths. In some cases the floats slide on the vertical rod between adjustable stops. The floats then push upward or downward on the rod at the desired control levels and trip the switch above. In a displacer-type arrangement for open tanks, a ceramic displacer is suspended from one end of a stainless steel tape that passes over a pulley and down again to a counterweight. The counterweight compensates for part of the ceramic displacer weight, so that it floats in the liquid. The extended pulley shaft drives through a reducing gear to a shaft which carries mercury switches controlling as many as four circuits (Fig. 3). The gearing allows the displacer to travel as far as 30 ft, with level adjustment between 2 and 27 ft. A weighted overcenter mechanism in the switches gives quick make and break.

In other applications of the displacer, porcelain bodies on a cable are suspended from the armature of a magnetic head control. In one form a spring partly supports the weight of the displacers. As liquid rises to the displacers in succession, their apparent weight decreases, and the spring can move the cable and armature upward to actuate snap-action switches. The displacers can be moved up and down the cable to initiate action at the desired levels. Three displacers can be mounted on a cable for such applications as one pump actuated by the center displacer, a second pump by either top or bottom displacer, and an alarm by the third displacer. Displacers are advantageous for dirty or viscous liquids which are still. Levels covered are 1 to over 10 ft.

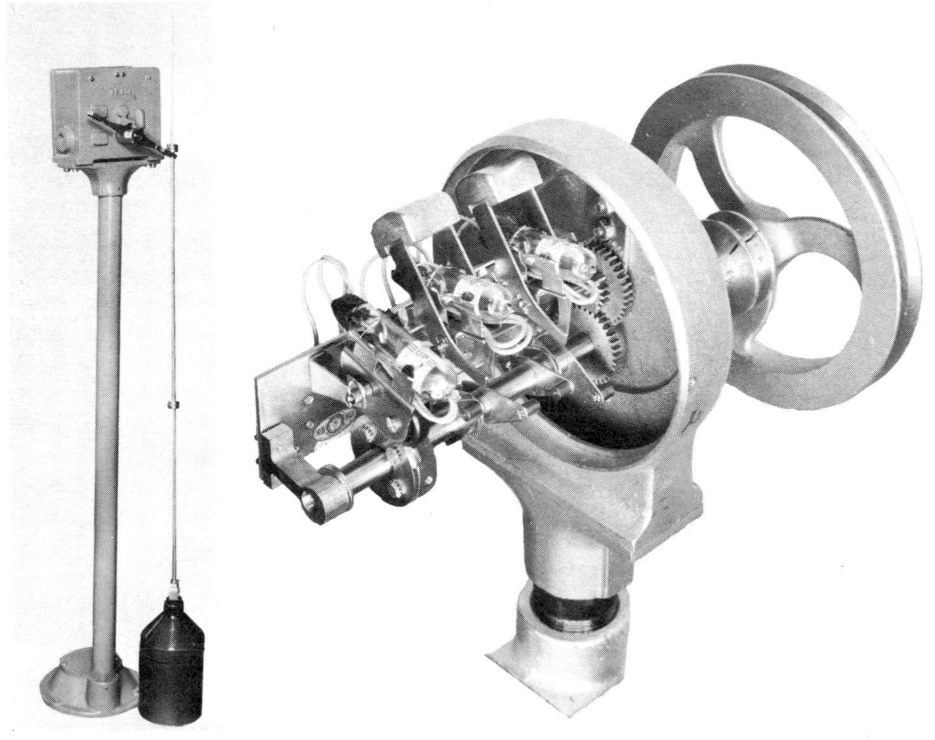

Fig. 2 Adjustable stops on rod actuate lever arm to tilt mercury switches as float moves. (Autocon Industries Inc.)

Fig. 3 Mercury switches in liquid-level sensor head are tripped by adjustable cams. (Autocon Industries Inc.)

The pulley shaft of the tape-suspended displacer can also drive a potentiometer. The potentiometer output can be applied to solid state control equipment handling recording, pump start and stop, and alarms.

Connection between float and switch need not be mechanical. An armature can be attached to the top of the float rod sliding in a tube. Outside the nonmagnetic tube is mounted the control switch, with a permanent magnet attached to it and set close to the tube. When the level rises, the armature passes the permanent magnet and attracts it, so that the switch is actuated. A spring retracts the magnet when the levels falls enough, and the switch is reactuated. A float arrangement of this kind, although limited in range, finds use as a low-water cutoff for boilers to 600 lb/in^2 gage pressure. Switches are dry-contact or mercury type. Two of these units can be placed at different levels to give both high- and low-limit control.

Liquid-level control systems without floats operate on several principles. If the liquid is at all conductive, probes can be used. The electrode probes are fixed, usually mounted in the same holder, and extend down into the tank (Fig. 4). Two or three electrodes are most common. Inductive or electronic relays are also part of the control system and actuate pumps or valves.

In a tank filled by a pump, a drop in level below the lower electrode breaks the circuit to allow a relay to start a pump or open a valve. When the liquid rises to the high-level electrode, the direct electric circuit between electrodes is established, and a relay stops the pump or closes the valve. In an induction relay (Fig. 5), the line voltage is separated from the control circuit by a primary and secondary coil arrangement. The relay design depends on the specific resistance of the liquid, which can vary from that of metallic circuits to that of demineralized water. Electronic relays have low potential and low electrode current.

Fixed probes usually do not exceed 6 ft in length, but suspension electrodes are available for deeper tanks or higher level differences. Pressuretight electrode holders capable of operating at 10,000 lb/in^2 are available. Temperatures are generally limited to 450°F. For tanks where icing is a problem, an electrode unit with a pipe sleeve in which the probes are suspended can be supplied (Fig. 4). An immersion heater near the sleeve bottom warms the water when the pump is not in operation.

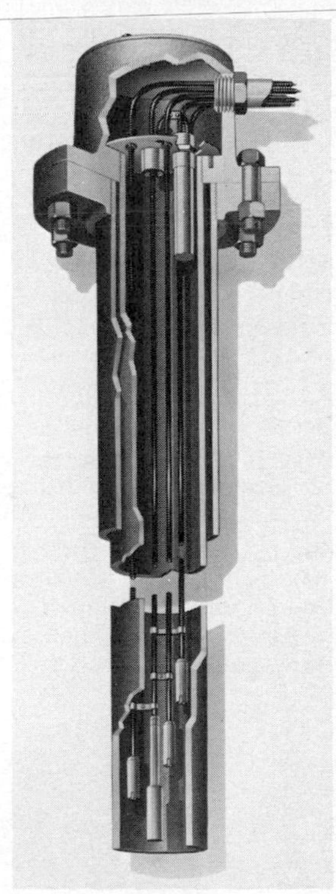

Fig. 4 Electrodes suspended on cables sense tank-water level below ice. (B/W Controls Inc.)

Bubble sensors measure liquid level by determining the air pressure required to force a small stream of air bubbles through the lower end of a tube extending to the bottom of an open tank. The air flow tends to keep the tube and tube end clear in liquids that contain solids. Floats and probes are eliminated in this method, and only the air-flow regulator and pressure switch are exposed to corrosive effects. The airstream flow rate can vary over a range without affecting air pressure. The specific gravity of the liquid must be known to allow the instrument to be calibrated. Because the measured variable is air pressure, the other instrumentation can be set at distances of 250 ft horizontally or vertically. The range of liquid level is 6 in to about 32 ft. Sewage, industrial processes, and water supply are some applications.

Notwithstanding the low air pressure involved, the differential sensitivity of pressure switches for bubbler sensors is about 0.5 percent of maximum operating range. Air consumption is about 1½ ft^3/hr when the air-flow regulator is set for 60 to 80 bubbles/min. The effects of failure of air pressure can be prevented by providing a cylinder of carbon dioxide gas; a pressure switch and solenoid valve will introduce CO_2 to the system if the compressor fails.

Pressure Sensors. Pressure controllers of the simple on-off variety may have a single-pole double-throw mercury switch actuated by a bourdon tube. A typical differential value is 2 percent of maximum scale reading. Adjustment to desired cutin (low) pressure is made by a knob on the case. Pressure ratings go to 5,000 lb/in^2 for these devices. Proportional control can be added to controllers of this type by incorporating a slidewire potentiometer. See Fig. 8.

In other types of pressure sensors, one sensor is provided for pump start and one for pump stop. Adjustable time delay prevents surging or water hammer from giving a spurious start or stop signal. Increasing liquid pressure transmitted through tubing to an air chamber acts on a bellows and, overcoming adjustable spring tension, trips a mercury switch. Differential sensitivity of the bellows-type sensors is 0.5 percent of maximum operating range. Maximum pressure is about 175 lb/in^2

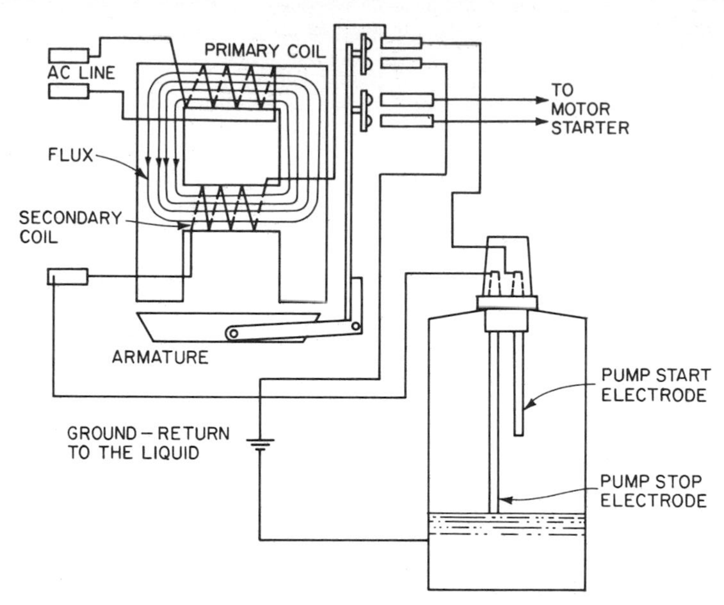

Fig. 5 In induction relay, when liquid reaches pump-start electrode, current flows in secondary coil and diverts flux to lift armature and close motor contacts. (B/W Controls Inc.)

gage, since the systems are intended for use on open tanks. Sensors, timers, and relays can be mounted in a cabinet located near tank or even near pump.

Air trapped in a bell and pressurized by rising water is the actuating mechanism for another alarm switch (Figs. 6 and 7). A synthetic rubber diaphragm in the switch body mounted above the bell and connected to it by a small pipe is caused to tilt a mercury switch and thus give the alarm. A rise in level of about 1¼ in above the bell mouth will activate the switch.

Alternators. An alternator may be installed to achieve regular use and equal

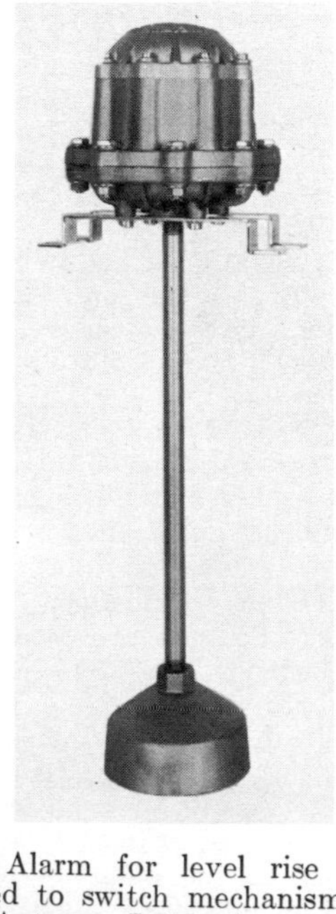

Fig. 6 Alarm for level rise has bell connected to switch mechanism by 1-ft pipe. (Autocon Industries Inc.)

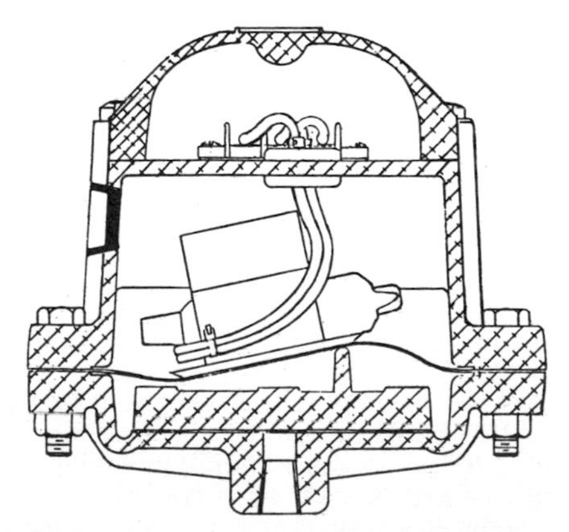

Fig. 7 Switch assembly for level-rise alarm. (Autocon Industries Inc.)

wear of each pump in multipump installations (Figs. 10, 11, and 12). The simplest versions serve on two-pump systems, but more advanced designs can rotate starting sequence of as many as twelve pumps. In one version of the two-pump alternator, a solenoid plunger picks up and causes a four-pole double-throw switch to take alternate positions, maintaining a position after the solenoid is deenergized. If one pump leads with the other coming into service only to augment the lead pump, the switching compensates accordingly.

If starting sequence is to be rotated for more than two pumps, a motor-driven rotor can be advanced a given number of degrees each time a pump motor operates. The rotor contacts are connected together in pairs to provide circuits between pairs of stator contacts. Other control variations available in this regard are a change

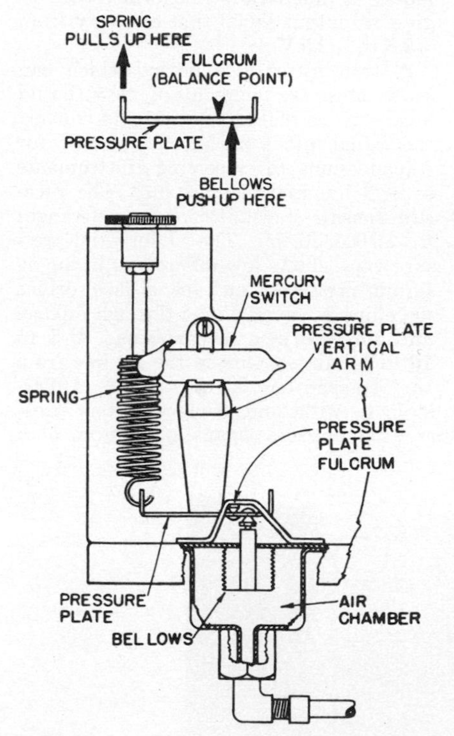

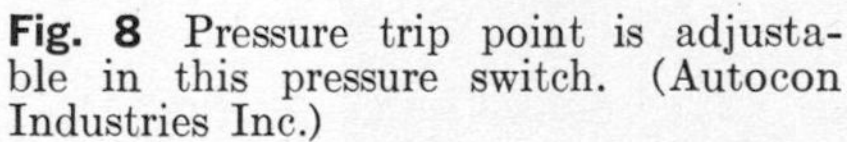
Fig. 8 Pressure trip point is adjustable in this pressure switch. (Autocon Industries Inc.)

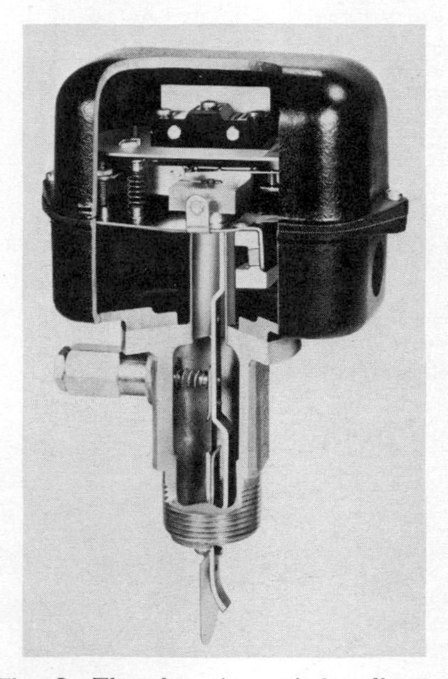
Fig. 9 Flow-detection switch relies on paddle in stream to actuate switch. (McDonnell & Miller, Inc.)

in sequence after a timed interval and an option of starting the pump that has been idle longest and stopping the pump that has run longest. Although many alternators operate on the same voltage as the loads, variants are available for operation from the low voltage and current ratings of control equipment.

Transducers and Transmitters. The variable that is most convenient or advantageous to measure is rarely the one best suited for direct use in the control system or for actuation of the final control element. A small differential pressure in a liquid-level or flow-control system can scarcely open a large valve. Conversion of measured-variable values into another signal medium is therefore necessary and is the task of transducers and transmitters. The terms are used interchangeably to some extent, although the transducer usually converts a signal into an electric current, and the output of the transmitter is usually an air pressure.

A pressure-to-voltage transducer is especially useful in pump control. In one design (Fig. 13), a bellows subjected to the pressure of the liquid transmits force to a pivoted beam that moves a core in a differential transformer or motion transducer to produce a 3 to 15 V output. The beam is balanced by a spring, and another spring allows the minimum pressure for voltage output to be set, which is equivalent to zero suppression. Zero suppression here can go as high as 95 percent.

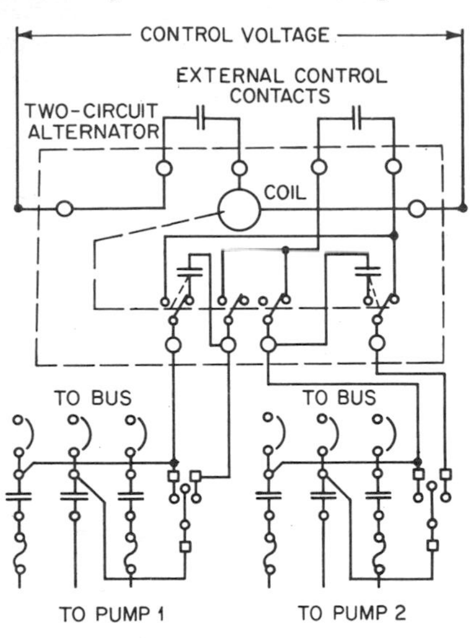

Fig. 10 Two-circuit alternator wiring. (Autocon Industries Inc.)

Differential-pressure tranducers may operate on a force-balance principle, with a very small motion of the bellows. The motion is converted into rotary motion of an arm or shaft, which then moves a differential-transformer core to give an output signal that can vary from −2.5 to +2.5 V dc.

A transmitter is a device which can sense pressure, temperature, flow, liquid level, or differential pressure and convert the signal into a pneumatic pressure for transmission to receiving instruments several hundred feet distant. The pressure-sensing transmitter can span ranges to 80,000 lb/in². The differential-pressure type allows low differences in air or liquid pressures, such as a flow orifice develops, to be amplified through linkage and force-balance mechanisms. A 3 to 15 lb/in² air pressure in the air line from the transmitter is the result. Differential-pressure transmitter designs are available to withstand primary system pressures to 6,000 lb/in² gage, whereas the differential pressures span ranges from 5 to 25 to 200 to 850 in water.

Fig. 11 Two-circuit alternator in standard enclosure. (B/W Controls Inc.)

Telemetry Systems. When pump control must be exercised over distances greater than the few hundred feet capability of most pneumatic control equipment, telemetry systems find application. In some of these systems, the sensor's output is converted to a proportionally variable 3 to 15 V dc voltage at the transmitter input. The transmitter then converts the dc voltage into a square-wave pulse with duration varying in proportion to the input signal. The resultant pulse-width-modulation

Fig. 12 Sequencer varies lead pump to equalize wear. (Cleveland Controls, Inc.)

(PWM) signal is sent directly over transmission lines or via tone carrier in microwave, VHF, or UHF systems. To transmit the bipolar PWM signal alone requires direct wiring with less than 5,000 ohms resistance. This means as much as 24 miles on direct telephone lines. The PWM signal eliminates loss of information from signal amplitude variations, and the bipolarity of pulses reduces line capacitance effects.

With tone transmission, either amplitude modulation or frequency shift is used.

The receiver converts the pulse signal back to a variable 3 to 15 V dc signal identical to the transmitter input signal and capable of use for indication or control. If transmission is not received for a certain time period, an alarm can be actuated and the pumps started or stopped.

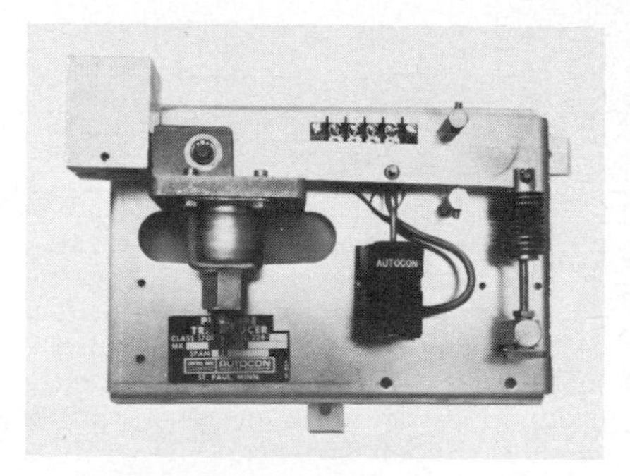

Fig. 13 In pressure-to-voltage transducer, bellows at left moves core in transformer at center. (Autocon Industries Inc.)

Constant-Speed Control Where pump-speed control is economically justified, it is a preferred method of obtaining the desired output parameter, such as flow rate, head, or liquid level. If the pump operates at constant speed, then there are four common control means:

1. On-off
2. Throttling by valve
3. Bypass by valve
4. Submergence

The on-off method with a single pump largely focuses on liquid-level or temperature-range control. Centrifugal and positive-displacement pumps can be controlled by this method. If an accumulator is installed downstream, the method can be extended to head control. With multiple pumps in parallel, flexibility is slightly greater, and a coarse control over flow rate is possible.

The simplest mechanism for on-off control of constant-speed pumps is the push button switch and starter for across-the-line start of small pumps (Fig. 14). For large pumps, reduced-voltage starting is customary. Number of starts per hour is restricted in all cases to prevent overheating. The electric impulse to cause start or stop of the motor can originate in any of the sensors or switches described above.

Throttling by valve is very common and can provide refined control under difficult conditions where rapid response and outstanding dynamic stability are sought, as in boiler feedwater control. Positive-displacement pumps cannot use this method.

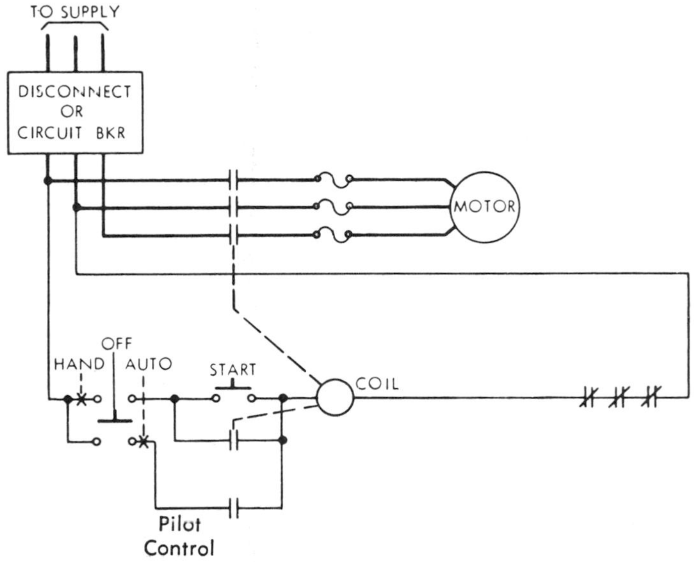

Fig. 14 Elementary wiring diagram for standard pump control. (Furnas Electric Co.)

Bypass by valve is an occasional variation of valve control based on bleedoff of discharge liquid to reduce flow rate at a downstream point or to allow a cooling flow to pass through a constant-speed pump when its discharge has been blocked. The method can serve both centrifugal and positive-displacement pumps.

Submergence control, for centrifugal pumps, relies on temporary decrease in available NPSH to reduce pump flow rate to the value at which liquid is entering the sump (Fig. 18). The method serves for condensate, and design precautions prevent rapid cavitation damage.

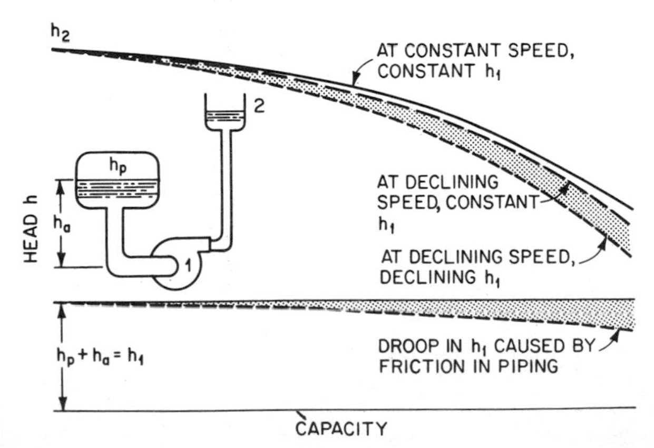

Fig. 15 Centrifugal pump characteristic changes depending on speed and head at inlet.

Valve-throttling Control. The chief elements of centrifugal-pump performance are shown in Fig. 15. At any given flow rate (capacity) a centrifugal produces a discharge head consisting of the static head on the pump inlet and the dynamic head imparted by the pump. At higher flow rates, speed usually declines slightly, lowering the actual characteristic curve as shown. In addition, the higher flow rates produce more friction head loss in the inlet piping, so that the pump senses an inlet head slightly less than the static head developed by weight of liquid column and effect of compressed gas or upstream pumps.

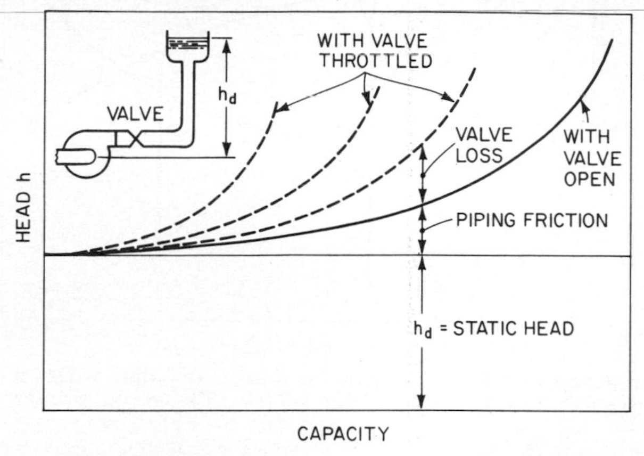

Fig. 16 Static head, piping friction, and valve loss determine the piping characteristic downstream of the pump.

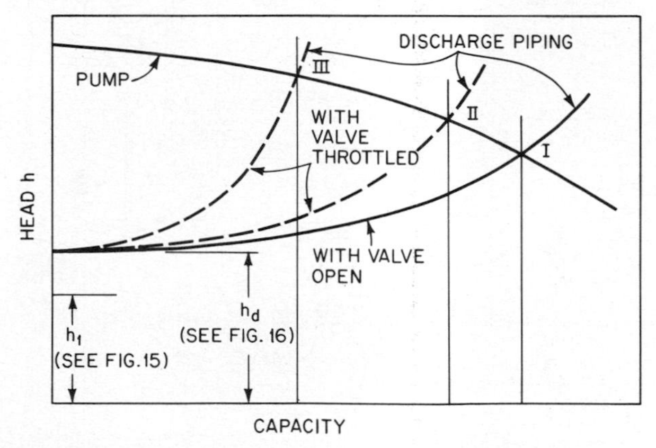

Fig. 17 Intersection of combined pump and piping characteristics is operating point (I, II, III).

The pump can deliver any flow rate along the curve. What determines the actual flow rate at any instant is the characteristic curve of the downstream piping (system curve) as shown in Fig. 16. Under zero-flow conditions, there is a gravity head of liquid and perhaps a pressure in a container, such as a boiler drum. When liquid flows, piping friction head is added. Piping friction causes the system curve to turn upward, roughly parabolically. If a downstream control valve, previously wide open, is throttled, a new and more rapidly rising system curve is established. The plot of pump curve and system curve on a single chart (Fig. 17) indicates at the intersection the actual conditions at the pump discharge.

The combined plot also shows that the flow rate or discharge head will be modified by change in other parameters besides throttle-valve setting. For example, pump-speed increase will lift the entire pump curve up and move the intersection point to higher flow rate and head. Decreased pressure in a boiler drum will lower the entire system curve and move the intersection point to higher flow rate and lower head. Throttling the inlet line to the pump will reduce inlet head and cause the pump curve to start at the same point but slope downward at a faster rate. The

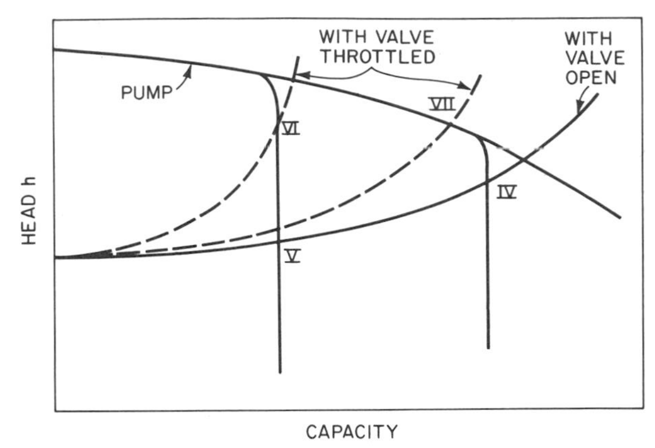

Fig. 18 In submergence control, operation can be at points where restricted-capacity curve intersects piping characteristic (IV, V, VI) or on regular pump characteristic (VII).

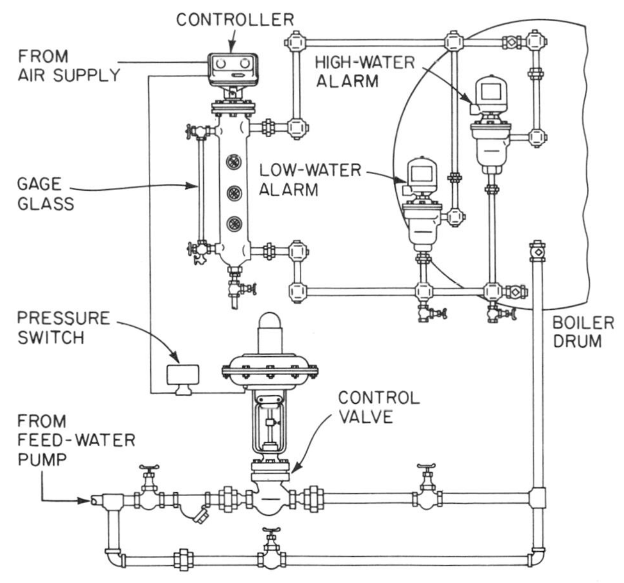

Fig. 19 Boiler-water-level control system. In the controller, an external magnet senses position of displacer. (Magnetrol, Inc.)

pump curve will then intersect the unchanged (downstream) system curve at lower head and flow rate.

Boiler Feedwater Control. In its original and simplest form, this control maintained the water level in a boiler drum. Although this is still the primary objective in many boilers, there are other applications in which balance of steam flow rate against feedwater flow rate is the primary objective, with level maintenance a secondary factor unless it exceeds preset limits. In steam generators operating above the critical pressure of 3,206 lb/in^2 abs, the feedwater turns to steam without a water level being visible, so that temperature and flow rates are the variables to be controlled.

Both on-off and modulating control are used in feedwater control systems. One classification of boiler feedwater control systems is based on whether the system is electric or pneumatic. Another classification system gives the number of variables sensed to determine control-valve position: a single-element regulator senses water level alone, while the two-element regulator also senses steam flow, and the three-element regulator adds feedwater flow sensing.

For low-pressure boilers, operating on moderate loads at no higher than 600 lb/in^2 gage and usually far below, on-off control of constant-speed pumps is sometimes used. The control can be similar to a low-water cutoff device, actuated by float, but it has two switches: the switch at the higher water level is for level control, and the one below is for low-water cutoff of fuel and for alarm. Level differences of 1 to 3 in start the pump. This type of control can have a third switch, installed to give a separate alarm for low water before the fuel cutoff. With fire-tube boilers, the third switch can give a separate high-water alarm if the pump does not stop when the pump cutout switch is actuated.

Regulators directly actuated by float are also in use (Fig. 20). Valves for these

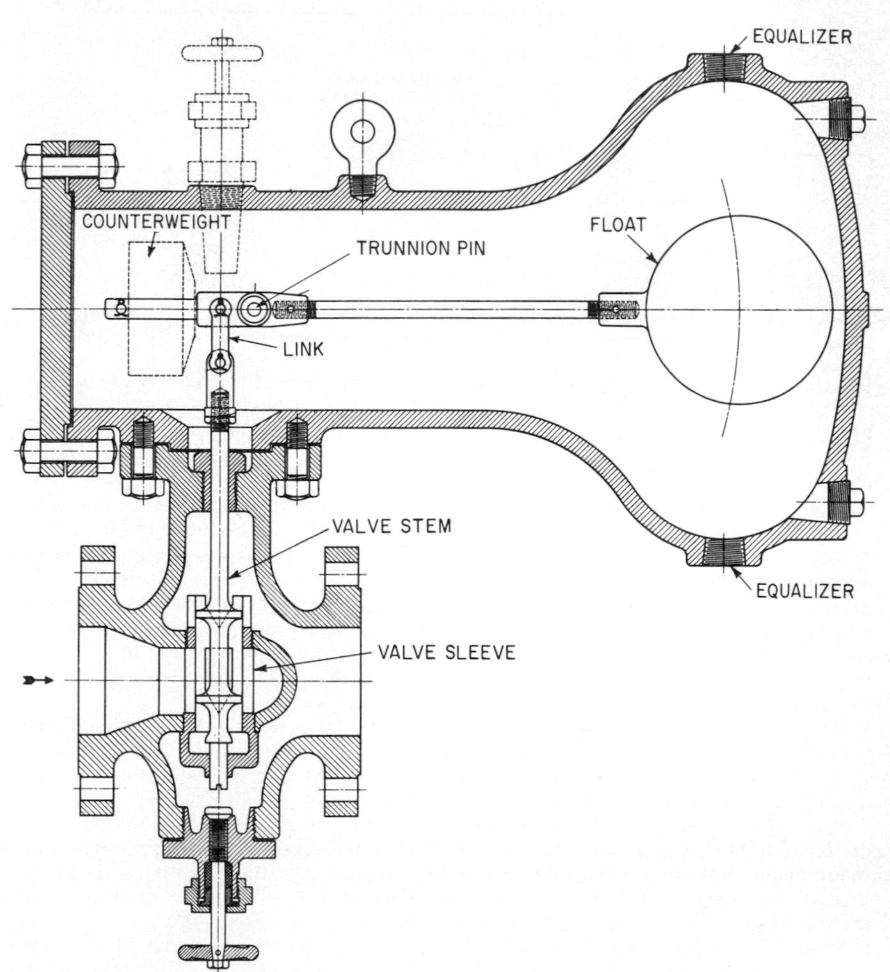

Fig. 20 Regulator with two-seated piston valve. (Stets Co., Inc.)

regulators are usually two-seated to reduce the thrust required of the float and linkage mechanism. Boiler pressures are low for this method, below 250 lb/in^2, although the regulators can be built to withstand 600 lb/in^2 gage. Capacities go to over 400,000 lb/hr at pressure drops of 100 lb/in^2 across the valve.

For more demanding service, modulating control using an amplified signal applied to a valve actuator is necessary. The simplest type of modulating control of this kind is a single-element regulator, serving for fairly constant loads and pressures (Figs. 19 and 21). In the pneumatic system, change of water level in the boiler drum provides a pneumatic output signal which is transmitted to a controller supplying air pressure to the diaphragm or piston actuator of a control valve in the discharge line from the feedwater pump. A sensing thermostat for drum water level may be designed as a proportional controller, with gain changing in proportion to the deviation from the set point. Linkage transfers the elongation of the sensing element to the pneumatic transmitter. Torque tubes or magnetic couplings (Fig. 19) may also convert the water-level sensing of a float or displacer into a pneumatic signal, at any of the standard pressure ranges. The air pressure required for the feedwater control valves must usually be at least 50 lb/in^2 gage, and up to 125

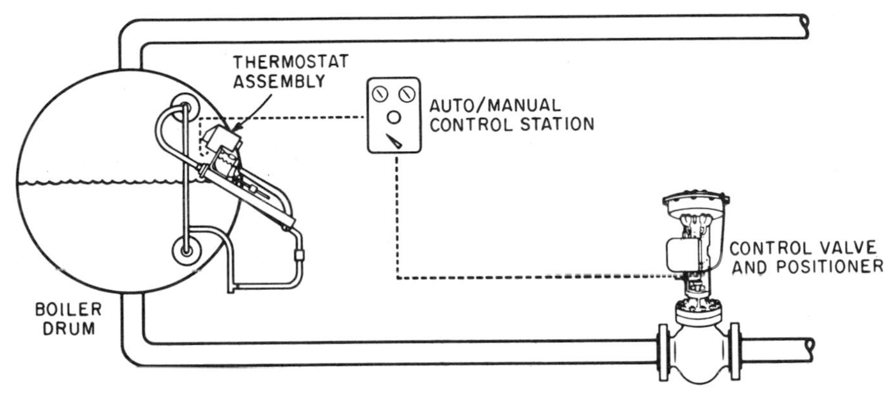

Fig. 21 Single-element boiler-feedwater regulator system. (Copes-Vulcan, Inc.)

lb/in^2 gage may be necessary. Balanced-trim valves do not require as high air pressures as do the unbalanced-plug type, and the small amount of leakage is not harmful.

In some single-element systems, the sensing element directly actuates the valve. In one system of this type, a high enough vapor pressure is produced in an enclosed tube to operate the feedwater-control valve directly. The vapor-pressure generator consists of a slanting inner tube mounted beside the boiler drum, with ends connected to top and bottom of the drum. A finned outer tube, filled with a liquid and connected only to the control-valve actuator, envelops the inner tube. A decrease in water level brings more heating steam into the inner tube to warm the liquid in the outer tube and increase its vapor pressure. The vapor pressure is transmitted to the valve actuator to open the valve.

In an electric control system, the level sensor can emit a signal modified by a slidewire potentiometer. The valve operator is an electric motor. In the larger sizes, the valve-actuating speed will be slow, so that the system cannot respond quickly to rapid change in steaming rate and water level.

Some single-element systems employ a pressure-control valve directly upstream of the main feedwater control valve. The upstream valve, called a differential valve, maintains a constant pressure on the feedwater control valve, improving its performance. Pressure-control valves of this type have been used on some more advanced systems, too.

Single-element systems are inadequate for a boiler whose steaming rate changes suddenly. The reason for this is the anomalous behavior of the water level during the change. A sudden demand for steam will reduce pressure in the drum, and steam-bubble formation will increase, temporarily raising the water level at precisely the time when a falling level is required to signal for increased feedwater flow. The anomaly is called a rising water-level characteristic. If the steaming-rate change is gradual, then the characteristic can be constant or even lowering.

The two-element regulator (Fig. 22) solves this problem by sensing steam flow through an orifice in the steam main. Flow-rate signal goes to the controller, and a sudden increase in steam flow will temporarily override the spurious water-level signal.

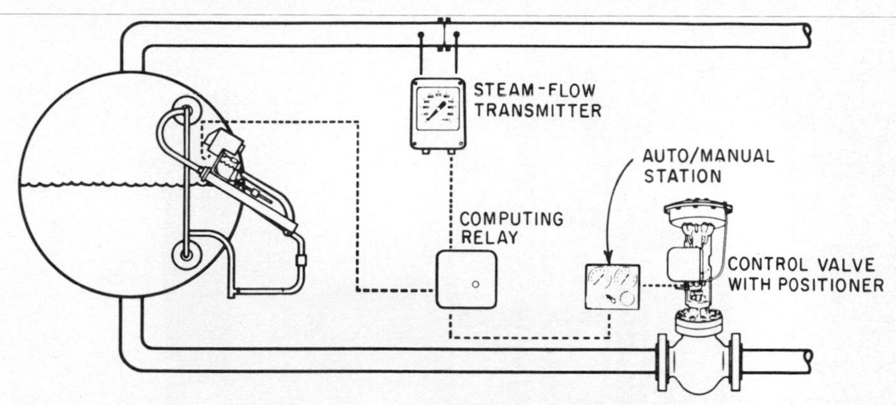

Fig. 22 Two-element regulator system, with steam-flow measurement. (Copes-Vulcan, Inc.)

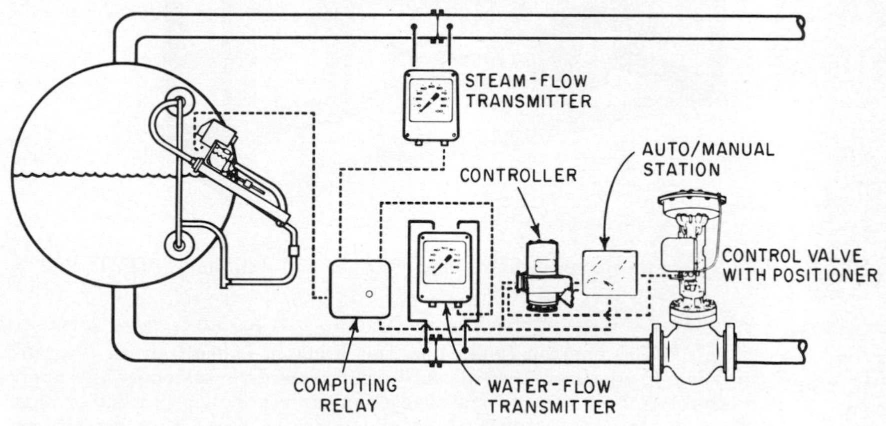

Fig. 23 Three-element regulator system action depends on water level, steam flow, and feedwater flow. (Copes-Vulcan, Inc.)

Three-element controls (Fig. 23) offer a further refinement—sensing feedwater flow rate in addition to water level and steam flow rate. During a rapid and large load swing, the feedwater flow rate can then be adjusted to the steam flow rate, while note is taken simultaneously of the water level. The feedwater flow-rate signal is converted into a linear signal for transmission to a computing relay which can be adjusted for relative influences of the three variables. A balanced signal then goes to the drum-level controller, whose air output actuates the control valve. Three-element control systems can eliminate the effects of pressure variations upstream of the regulating valve.

Building-Water Pressure Control. In tall buildings or large industrial, commercial, or housing water systems, water pressure can be maintained in several ways. If elevated or pressure tanks are not desired, a multiple-pump system may be considered. The pumps can be constant-speed or variable-speed. For constant-speed pumps, pressure sensors bring in individual pumps as required to maintain pressure (Fig. 24). If a large number of pumps is necessary, then means must be provided to prevent all pumps from starting simultaneously on restoration of power after a failure.

Hydropneumatic systems rely on air pressure in the top of a tank into which pumps deliver water intermittently. As water is drawn off, the air expands, reducing its pressure and eventually requiring another pump start. Both pressure and water

Fig. 24 Pump set to maintain head and flow in tall buildings. (ITT Bell & Gossett)

level are sensed. If the pumped liquid does not bring in enough air to the tank to make up for losses, an air compressor or the compressed-air system must supply air. Tank pressure after each level-controlled pumping cycle indicates whether more air is needed or whether air should be bled off. Float or probe type sensors can determine liquid level, and various pressure sensors are available.

VALVES

For the final control element, valves of conventional and traditional type serve for on-off control in pump systems. They also cover much of the modulation need. In recent years, modulating control in demanding services has required development of special control valves, actuators, and accessories. System dynamic characteristics, corrosion, erosion, noise, and costs have influenced the development.

Valve Elements A valve consists of:

1. A body or housing able to (*a*) withstand internal pressure (external pressure for vacuum applications); (*b*) direct the fluid in a suitable path, which is not always that of least resistance; (*c*) resist external forces that reach the valve from adjacent

piping or are caused by valve actuation; (*d*) act as support for other valve components.

2. A movable closure element (disk, ball, or plug) that can constrict the passageway through the valve, reducing or stopping flow.

3. A means of positioning the movable element; because this is usually done from outside the valve body, an opening, adequately sealed, is needed to allow the stem or rod to move without leakage.

Valve Bodies. Cylindrical (Fig. 32) or spheroidal (Fig. 92) shape, for efficient pressure resistance, is most common. Size of body in relation to the movable closure element varies widely; in butterfly valves for low pressure, the body is a thin narrow ring enclosing a large disk (Fig. 45), while in some high-pressure low-flow control valves the body may be a massive block of steel enveloping a small plug (Figs. 68, 74, and 79).

Movable Closure Elements. These differ almost infinitely. A high degree of specialization is necessary. Thin flexible flaps are one extreme (Fig. 63), and highly rigid plugs and balls are another (Figs. 41 and 68). Direction of motion toward

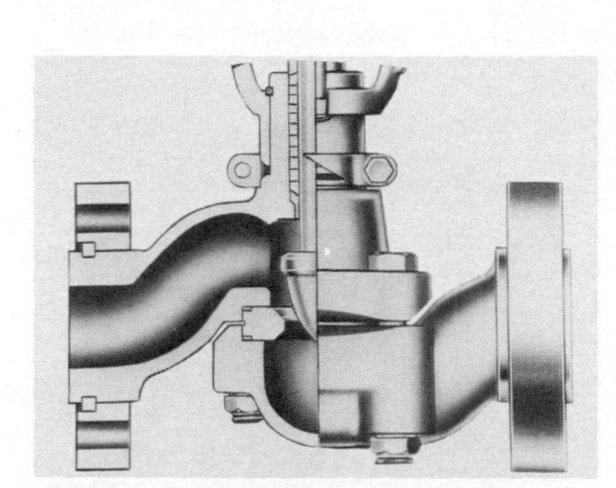

Fig. 25 Valve body can be split to allow quick trim change. (Masoneilan International, Inc.)

Fig. 26 Butterfly valve with split body. (Worcester Controls Corp.)

complete closure also differs widely. Perpendicular movement toward a seat, sliding movement across a single aperture or multiple apertures, or oblique movement toward a beveled seat are all common.

Valve Bonnet. The valve body usually has an opening covered by a bonnet. The valve bonnet has no function directly connected with the valve's primary purpose but is merely a necessary evil incorporated to facilitate machining, assembly, and maintenance, and to hold the piercement and seal for the actuating stem. The simplest bonnet might be considered the provision for bearings to hold the pivot-pin ends in some check-valve designs. The most complicated bonnets are those for large high-pressure shutoff or control valves (Figs. 32 and 77).

Small ball valves in which the ball and rings are assembled from a valve end may have a rudimentary bonnet structure (Figs. 46 and 47) with merely a counterbored hole for stem and stem seal and a platform for seal retainer and actuator support cast integral with the body.

Small gate, globe, and check valves can have a screwed bonnet (Fig. 39); the screw threads can be inside or outside the gasket or sealing surfaces. The next step is the union bonnet (Fig. 40), considered superior if the bonnet is to be removed frequently. The screw threads are outside the sealing sufaces. Larger valves tend to the bolted bonnet, especially for larger sizes and higher pressures. The bolted bonnet (Fig. 27) requires additional body metal for stud placement and reinforcement of the hole in the body top.

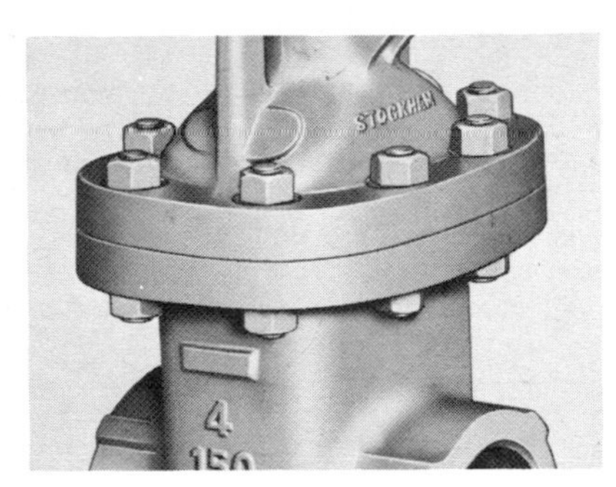

Fig. 27 Bolted bonnet allows large valve to be opened easily. (Stockham Valves & Fittings)

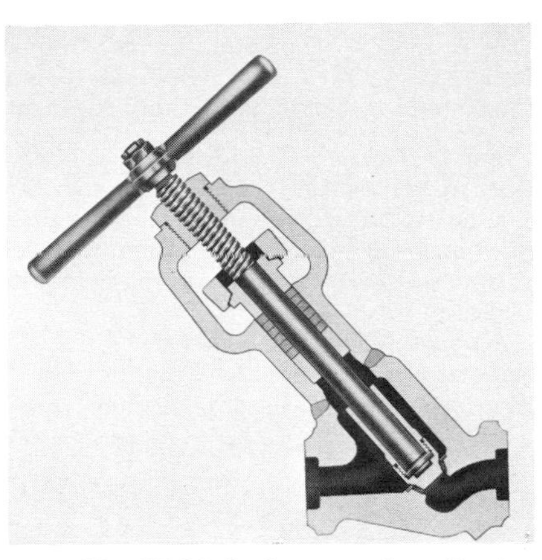

Fig. 28 Welded bonnet is effective seal on valve with small-diameter disk. (Dresser Industrial Valve & Instrument Div., Dresser Industries, Inc.)

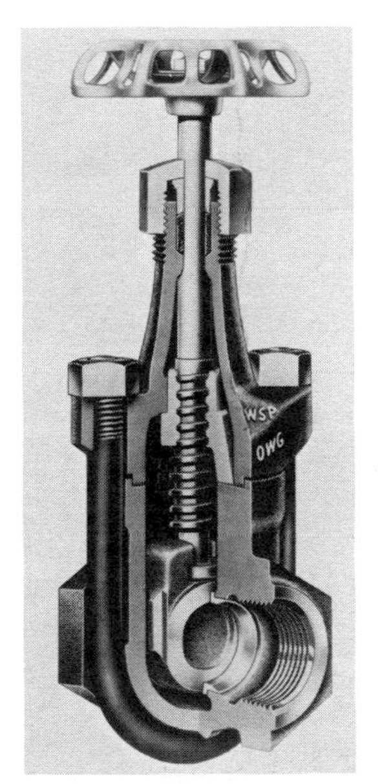

Fig. 29 U bolt holds bonnet to valve in small sizes. (Jenkins Bros.)

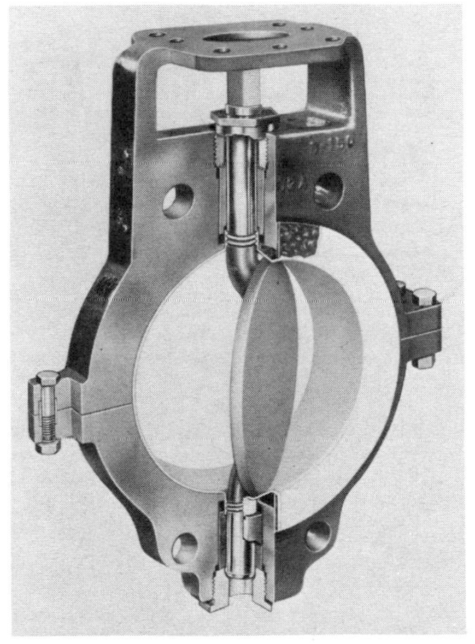

Fig. 30 Flexible liner of eccentric-butterfly valve also seals valve trunnions. (Duriron Co., Inc.)

Packing Box. A seal is necessary at the entry of the stem into the bonnet or body of the valve. It is usually a packing box designed to be packed with material such as braided asbestos or molded TFE (tetrafluoroethylene) rings, with a follower to exert sealing pressure on the packing when outside bolts or a screw thread are tightened. A spring under the packing can help keep the seal tight in some designs as initial packing compression is lost. For high-temperature service or on fluids containing abrasives, the box may be deeper, with a metal lantern ring separating the packing into two sections; a pipe connection to the lantern-ring area allows cooling or clean flushing liquid to enter under pressure. The liquid can also prevent air leakage into the system.

The stem motion influences choice and design of seal. Translational and helical motions make more demand on a seal than does the simple 90° rotation of ball, plug, and butterfly valves. The latter types often have only one or two O rings or similar molded rings for their seal (Figs. 30 and 31), and the rings require no external pressure-developing means.

A valve with a bellows seal on its stem (Fig. 33) is an attempt to eliminate the chief disadvantage of the housing piercement while retaining the mechanical link with disk or plug.

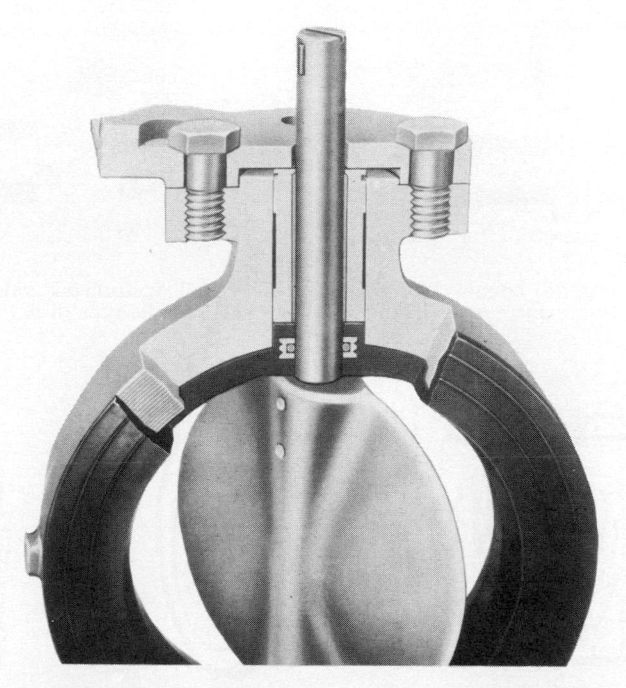

Fig. 31 Single O ring seals shaft in butterfly valve. (Garlock Inc.)

Positioning Means. These are generally translating or rotating stems passing through the housing. Some valves, including many solenoid, check and relief valves, do not require that the positioning element pierce the housing or its sealed extension. Electric forces or fluid energy actuates these valves. Many hydraulic and pneumatic valves are similar in this respect.

Other Design Factors. Manufacturing reasons, desire for flexibility in application, and considerations of maintenance have also been a major influence on valve design and construction. Heavy body walls can facilitate metal casting and improve rigidity for machining. Large openings in top or bottom of bodies allow machining and assembly and maintenance of internal parts. Separate seat rings can be replaced when worn, or rings with different-sized orifices can be inserted. The cage type of construction in control valves is another path to application versatility: cage and plug diameters and geometry of cage wall holes are variables that can be tailored to match system needs.

Protection of sensitive surfaces against erosion, corrosion, and mechanical damage is important. Location of bonnet closure threads outside machined sealing surfaces

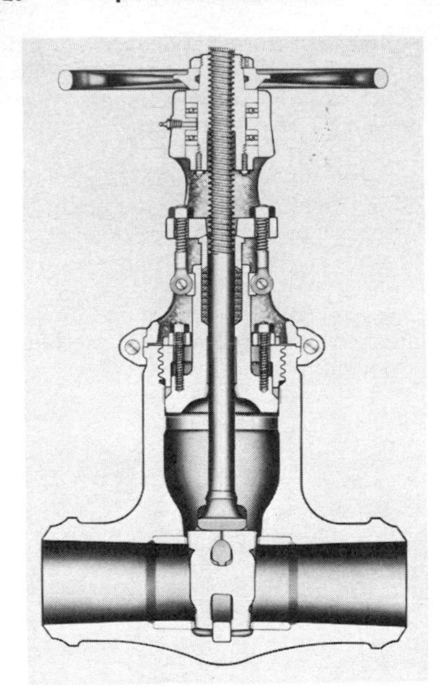

Fig. 32 High-pressure bonnet closure tightens with pressure increase. (Crane Co.)

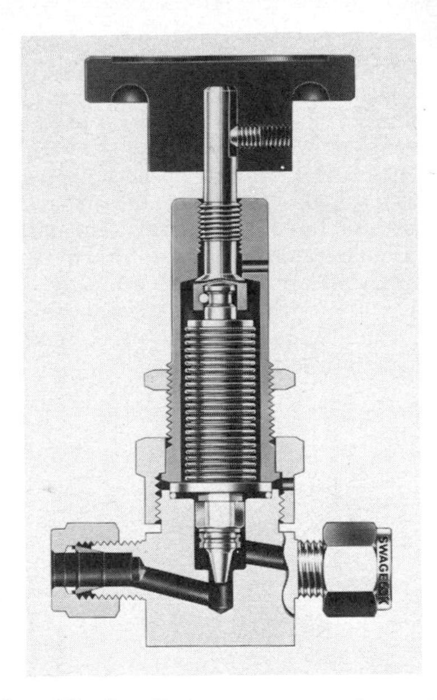

Fig. 33 Small instrument valve with bellows seal, needle-type disk. (Nupro Co.)

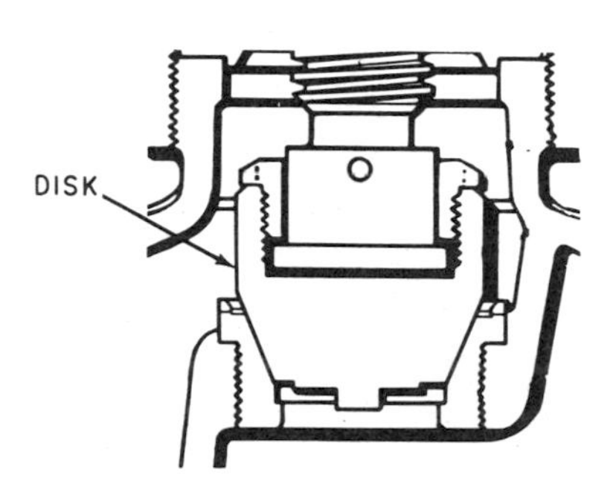

Fig. 34 Plug-type disk regulates flow closely, extends valve life. (Stockham Valves & Fittings)

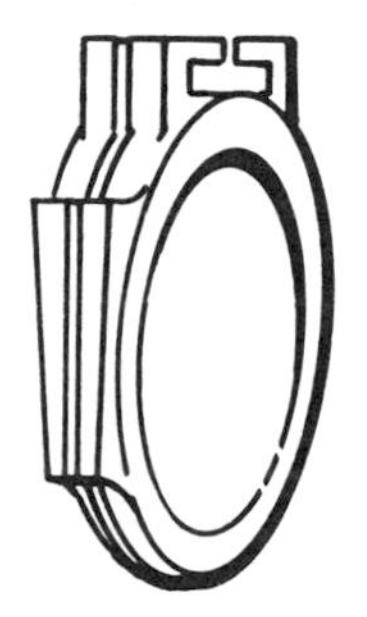

Fig. 35 Flexible disk for steel gate valve conforms to both seats. (Stockham Valves & Fittings)

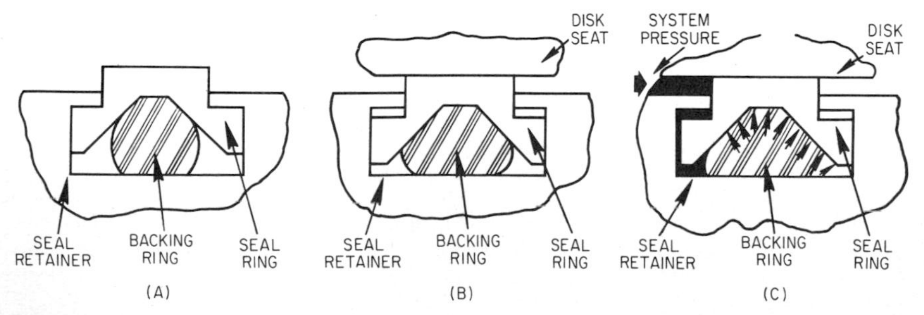

Fig. 36 Flexible backing ring in eccentric-butterfly valve helps hold seal ring against disk. (Posi-Seal International, Inc.)

is one example. Internal body erosion is countered by redesign of flow path or change of material. The disk-seat sealing area almost always receives the severest wear, and many innovations aim at prolonging life and assuring tight shutoff.

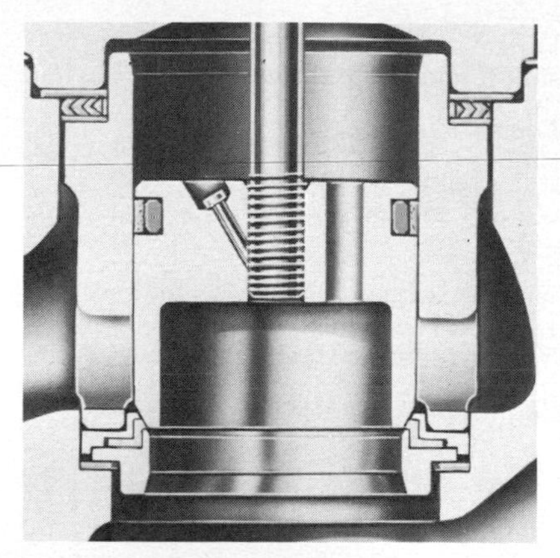

Fig. 37 Flexible element in seat of control valve promotes tight seal. (Fisher Controls Co.)

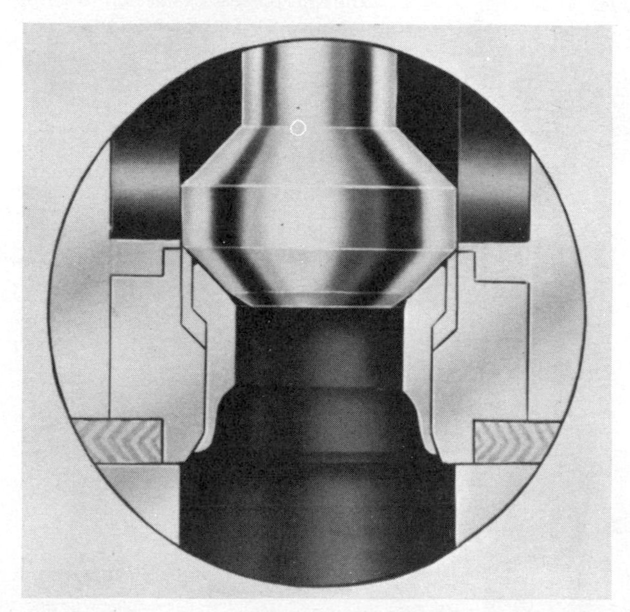

Fig. 38 Movable inside sleeve develops pressure on soft seat ring of the control valve. (Masoneilan International, Inc.)

Flexibility of individual internal parts (Figs. 35 to 38) and looseness of connection between them are two drawbacks that must be accepted to obtain better certainty in seating of disk or plug against seat. Unfortunately, a valve has moving parts that are unobservable, buffeted by liquid forces which can cause fracture or wear. Design here must be supported by experiment and feedback of field experience; operating precautions complement this.

Operation and Valve Types On-off operation of pump valves serves for

1. Isolation of a pump: protection, maintenance, removal, administrative reasons
2. Bypass or partial isolation: inlet or outlet block for protection, improved flow control, administrative reasons
3. Pressure relief: protection
4. Venting: removal of gases and vapors from the casing
5. Draining: removal of liquids from the casing

Modulating mode of operation serves for

1. Control of flow rate to pump or of pressure at inlet
2. Control of delivered flow rate or pressure
3. Control of bypass flow rate

Auxiliary flows in lines to packing boxes, seals, and sensing or measuring elements are controlled by either on-off or modulating valves.

Principal types of valves for on-off and much modulating service are:

Gate (rare for modulating)
Globe (and angle)
Butterfly
Ball
Eccentric-butterfly
Plug
Diaphragm

Check valves and relief valves, although possessing design features peculiar to their nature, make use of essential features of globe and butterfly valves.

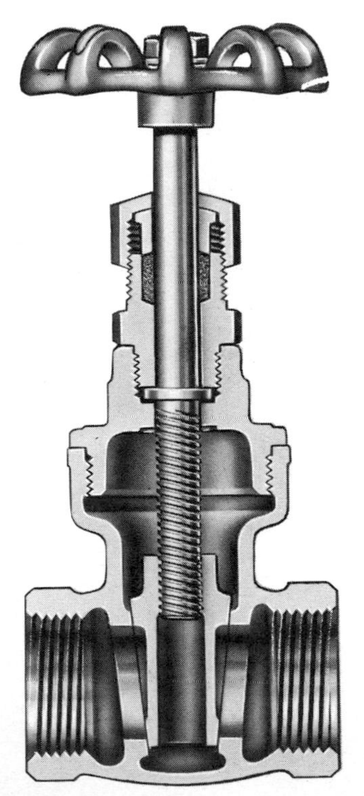

Fig. 39 Small-gate-valve stem screws into recess in disk. (Walworth Co.)

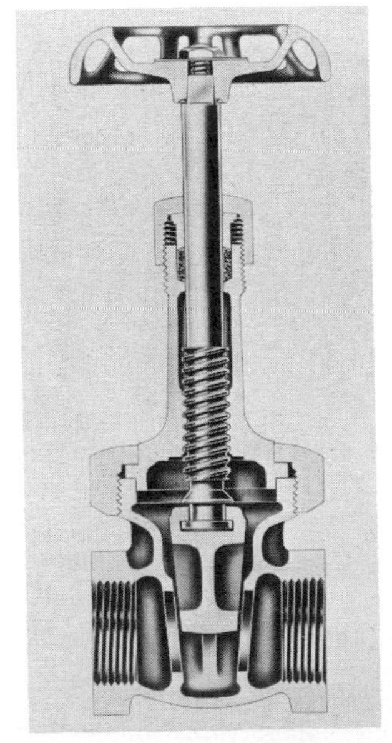

Fig. 40 Rising-stem gate valve with solid wedge. (Jenkins Bros.)

Control valves have developed as a special group for demanding services that require wide modulation range, stability, wear resistance, low noise, or a specific flow characteristic. Ingenuity and experience have combined to evolve many unusual but effective designs, each with advantages and drawbacks.

Gate Valves. These seal off flow by a disk which slides parallel or nearly so to the seat. The sealing force to hold the disk against the seat develops from a wedging action, with some aid from upstream liquid pressure on the disk. Disks may be rigid single-piece (usually called a solid wedge; see Fig. 40), two-piece (usually called a split wedge or double-disk), or single-piece with attenuated connecting material between face pieces (flexible wedge; see Fig. 35). The last-mentioned two have flexibility to allow each face to adapt to its seat.

Gate-valve disks and seats wear quickly in throttling service because of unfavorable fluid flow paths and also disk vibration caused by fluid buffeting. These valves should be either closed tightly or fully open. The space for a fully opened disk

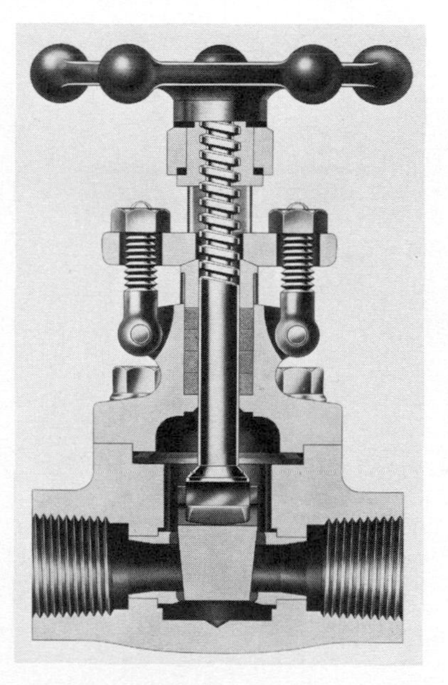

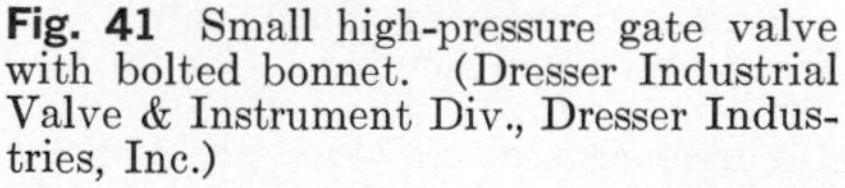

Fig. 41 Small high-pressure gate valve with bolted bonnet. (Dresser Industrial Valve & Instrument Div., Dresser Industries, Inc.)

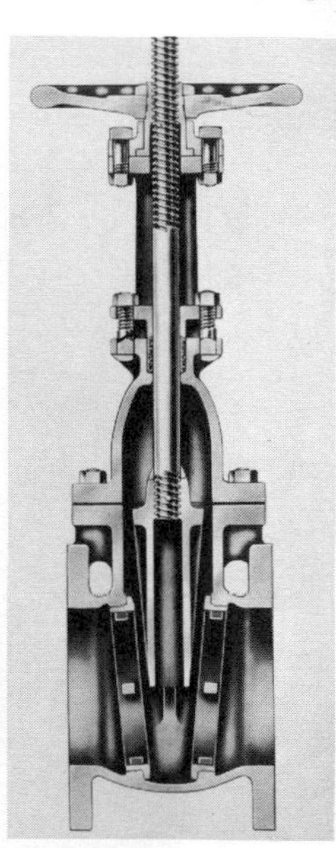

Fig. 42 Large rising-stem gate valve with screwed-in seat rings. (Jenkins Bros.)

is usually in a separate valve bonnet screwed, bolted, clamped, or even welded to the valve body. For large line sizes, this means a tall bonnet and topworks. The axial length of gate valves is relatively short, however, especially in large low-pressure valves. Gate-valve orifice size is rarely ever changed, because the valve is designed for, and should be used as a maximum-flow valve when open.

Small gate valves, in bronze, iron, or forged steel, see use in vent, drain, and instrumentation and control lines (Figs. 40 and 39). Larger gate valves serve for isolation and bypass on pump inlet and discharge lines (Fig. 42).

Globe and Angle Valves. Movement of a disk or tapered plug perpendicularly into contact with a seat, which is the machined edge of a circular orifice, closes these valves. A disk need lift only about a quarter of the orifice diameter to expose a peripheral opening equal in area to the orifice. The disk therefore does not have to lift as far as a gate-valve disk, but on the other hand the seat must be located near the valve centerline in a Z-shaped diaphragm separating upstream from downstream (Fig. 43). This location tends to increase height of bonnet and topworks, canceling part of the overall height advantage over gate valves.

A globe valve can have a separate seat ring whose orifice can vary in size depending on flow rate and purpose (Fig. 25). In addition, because the sealing action can occur along a conical bore inside the ring, as well as on the flat top of the seat ring or orifice edge, the globe valve can be provided with a tapered disk or plug (Fig. 34) that will allow efficient throttling at low flow rates. The flow patterns around the disk and seat of a partly closed globe valve are more nearly symmetric than in a partly closed gate valve, too, which makes the globe valve superior for throttling.

The globe valve is inherently a high-loss valve, with poor recovery of pressure after passage through it. The angle valve, similar in appearance to the globe valve, causes less loss, especially when flow is from above the disk and seat.

Butterfly Valves. A thin disk mounted symmetrically on journals on a diametral line through the cylindrical valve body is the closure element (Fig. 45). Sealing action occurs when the disk, rotated to block the passage, touches a sealing surface

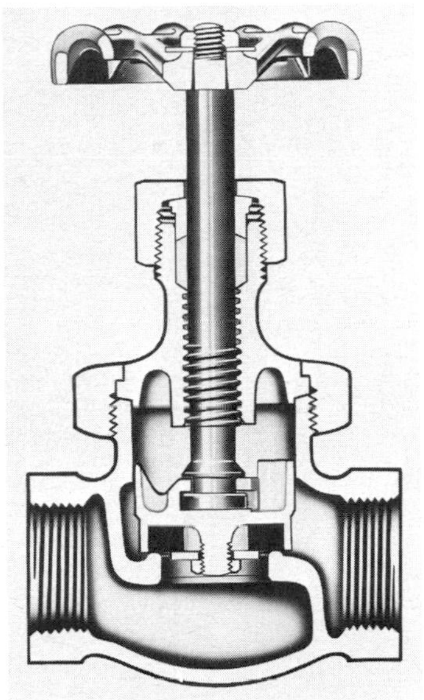

Fig. 43 Globe valve with composition disk. (Crane Co.)

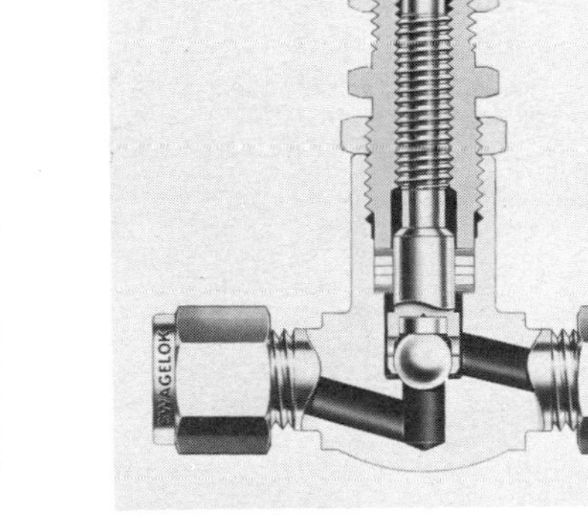

Fig. 44 Instrument globe valve with free-turning ball for disk. (Nupro Co.)

around the body perimeter. Metal-to-metal sealing surfaces are practically impossible to seal completely, but more recent designs with metal-to-plastic or metal-to-rubber seats are successful. Lack of symmetry at the piercement of journal through body wall, and the necessary discontinuity there, demand care in design.

Because the butterfly valve disk requires no more than 90° rotation between fully opened and fully closed settings, it lends itself readily to several simple types of power actuation. The fully opened disk offers some resistance to flow, but the added resistance is nearly constant within about 20° of fully open. This fact is sometimes exploited to reduce the stroke of an actuator. Butterfly disks need not be symmetric in cross section. Asymmetry or cupped wings are occasionally used in attempt to improve actuator-torque characteristics or control characteristics.

Many modern butterfly valves, dispensing with integral flanges, are retained in the line by the clamping action of line flanges on either side of the heavy body ring. The line flanges bolt to each other by long bolts outside the body-ring perimeter. In other models, bored lugs integral with the body ring permit the valve to be bolted directly to a line end.

Topworks of butterfly valves are extremely compact, being called on merely to house the extension of the stem and to support the manual or power actuation means. In the axial direction, the butterfly valve is short when closed, but the disk of large sizes requires considerable axial clearance when open. This can put the type at a disadvantage compared with the gate valve or a sliding gate.

Butterfly valves range in size from fractions of an inch to 200 in or more. Isolation, bypass, and flow control on pump inlet and discharge lines are main uses for butterfly valves in pump systems.

Ball Valves. These are basically small-size valves. Above approximately 6 in line size, the size and complication of body and cost of producing the accurately spherical ball often put this type at a disadvantage. In this valve a ball with a diametral hole through it is rotated 90° to align the hole with the pipe passage or place it at right angles for closure. The sealing action is accomplished by plastomer or elastomer rings which seat against the smooth surface of the ball. In some cases, metal or carbon-graphite rings (Fig. 51) allow ball valves to handle high-temperature liquids or seal in case of fire that would destroy softer rings. In designs with soft rings, a fire-safe feature is provided by metal seating edges that seal against the ball when the ring is melted.

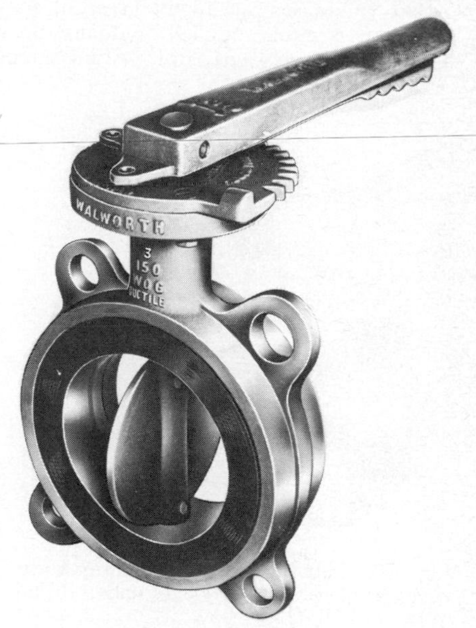

Fig. 45 Butterfly valve with lugs on body for positioning between flanges. (Walworth Co.)

In the closed setting of the ball valve, line fluids are held back by the circular contact of one or both seal rings against the ball. In the open setting, the seal rings make circular contact around the passage hole in the ball. During change from one setting to the other, the hole edges must slide past the seal rings. Assembly of ball and rings can be done from valve end openings in some designs (Figs. 46 and 47); this allows a low profile for bonnet and topworks but requires that

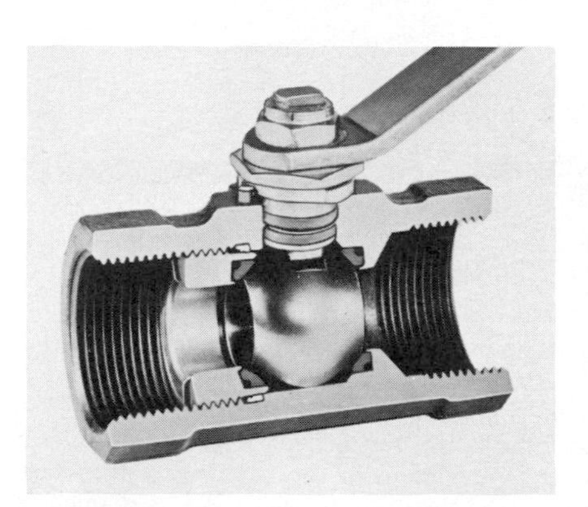

Fig. 46 Ball valve with internal screwed sleeve to retain seat rings. (Walworth Co.)

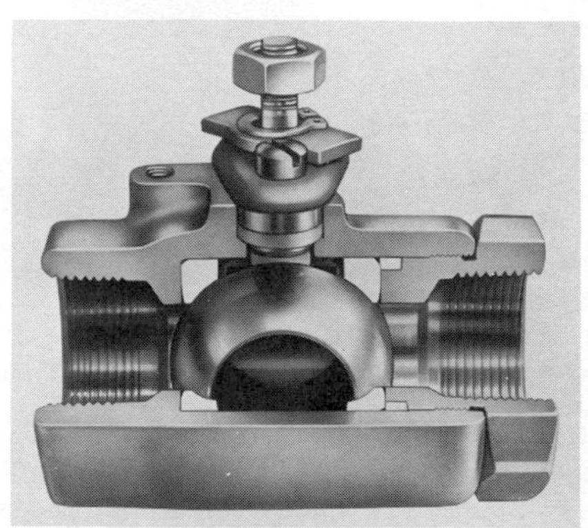

Fig. 47 Ball valve with seat-retaining ring also serving for line pipe attachment. (Jamesbury Corp.)

the fluid line be broken or that the valve body be rotated out of the line. To make possible an assembly while fluid line is intact requires a large bonnet opening (Figs. 50 and 51).

Actuation of ball valves is nearly universally a simple 90° turn, facilitating application of power. The piercement of the actuating stem through the body or bonnet requires a seal; this is often a simple TFE ring (Fig. 47), although sometimes a packed gland is used.

Ball valves see use in all types of pump piping up to their size limits. Power actuation is easily applied to many manually operated valves by simply bolting the manufacturer's actuator to the bonnet and then connecting air lines.

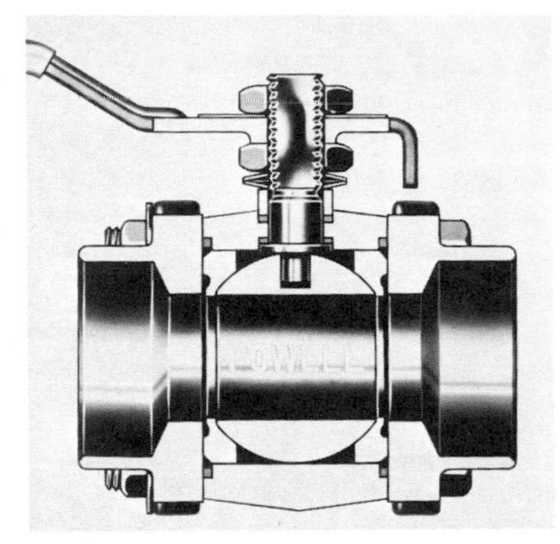

Fig. 48 Ball valve with separate body ring clamped between socket ends. (The Wm. Powell Co.)

Fig. 49 Ball valve with two-piece body. (Jenkins Bros.)

Fig. 50 Ball valve with top entry to allow maintenance without removal from the line. (Trans-Union Corp.)

Eccentric-Butterfly Valves. Also called offset-disk butterfly valves, these often are closer in action and application to the ball valve than to the true butterfly. The disk in these valves, although perpendicular to the flow line when closed, is displaced a short distance from the stem axis to give this type some important advantages over true butterfly valves. The outer surface of the disk is often machined to coincide with a sphere centered on the stem, so that the disk represents a slice out of a ball (Fig. 52). In the closed setting, sealing is against a seat ring, just as in the ball valve, but the seat ring is less rigidly supported and is often designed so that line pressure can energize the seal from each side of the disk. Higher shutoff pressures can be restrained from the smaller-diameter side of the disk, however.

Because the seat ring and its contact with the disk are uniform all around the ring, many eccentric-butterfly valves have inherently simpler shutoff seal problems than do conventional butterfly valves. Also, during opening, the disk moves away from seat contact very early in the motion, so that sliding action along the seat ring is minimal. In the open position, the eccentric-butterfly disk and stem obstruct flow more than does the conventional butterfly disk. Valve body ring is similar

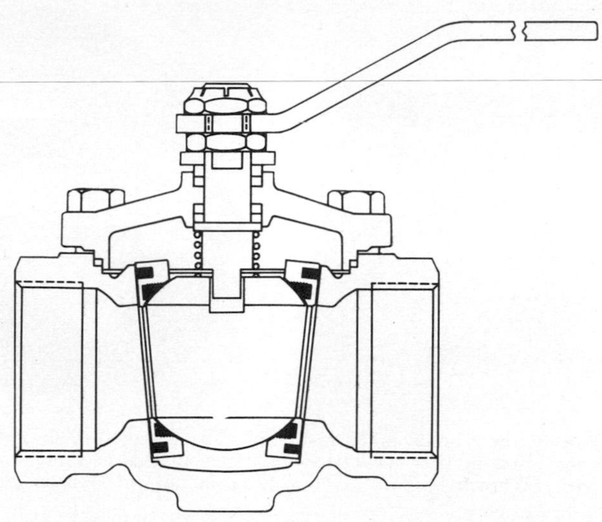

Fig. 51 Graphite and carbon rings in this top-entry ball valve can seal at high temperatures. (Hills-McCanna Co.)

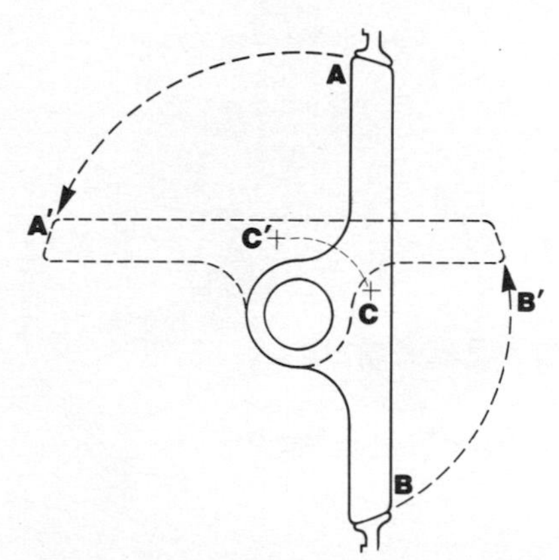

Fig. 52 Principle of operation of one eccentric-butterfly valve type. Sealing surface is part of sphere and rotates about sphere center (*A–B* to *A′–B′*). (Jamesbury Corp.)

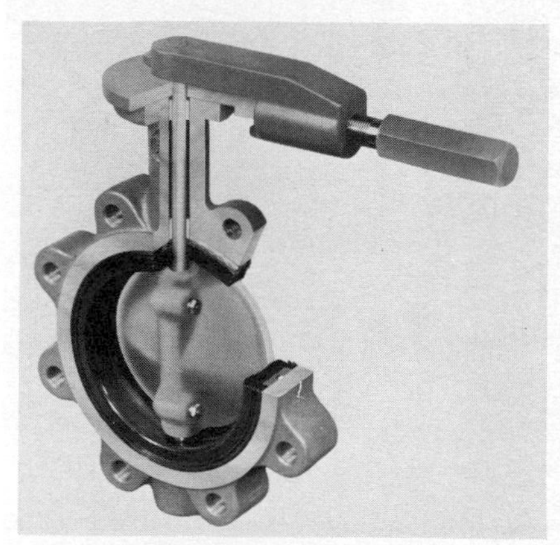

Fig. 53 Eccentric-butterfly valve with seating in two positions. (DeZurik, a unit of General Signal)

to that of the butterfly and is retained in the line in the same way. The eccentric butterfly can serve in modulating control.

Plug Valves. Seal in these is by contact of a cylindrical or taper plug against the close-fitting body wall of the valve. Sealing is improved and plug rotation eased by lubricant distributed along grooves on the plug (Figs. 57 and 58). In other plug valves the plug is jacketed in TFE or carries TFE rings in grooves on it (Figs. 59 and 60). Seat rings may also be TFE. Actuation is 90°. Plug valves are commonly used in smaller sizes in piping for pump systems.

Eccentric-plug valves resemble eccentric-butterfly valves, with a camming action seating the plug against the sealing face (Fig. 61). The general design is adaptable to modulating control.

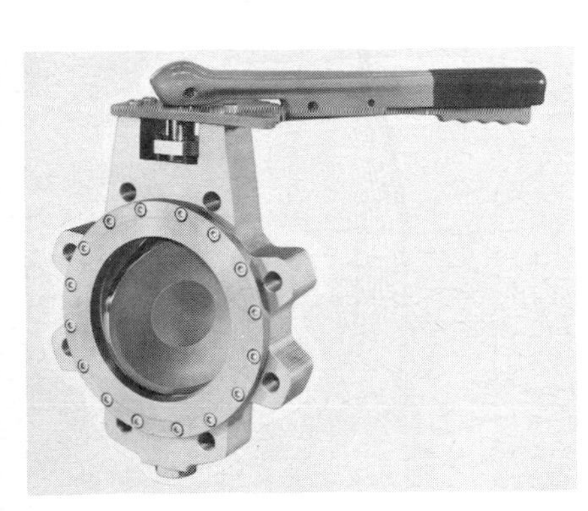

Fig. 54 Bolted retainer holds and protects flexible seat ring in this eccentric-butterfly valve. (Walworth Co.)

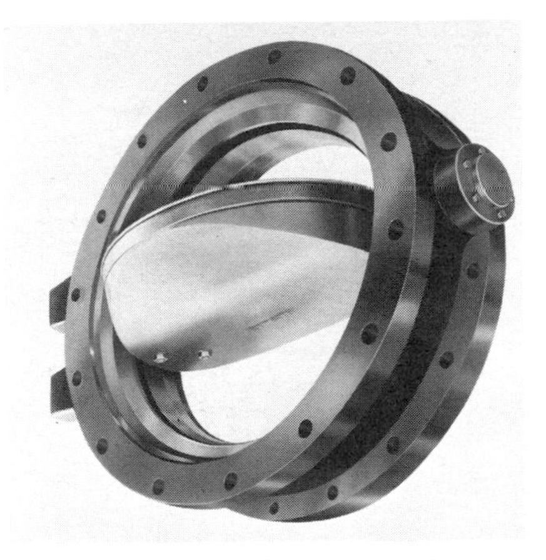

Fig. 55 Tapered disk reduces flow resistance in this eccentric-butterfly valve. (BIF, a unit of General Signal)

Fig. 56 Large eccentric-butterfly valve with flow passages through built-up disk. (Allis-Chalmers)

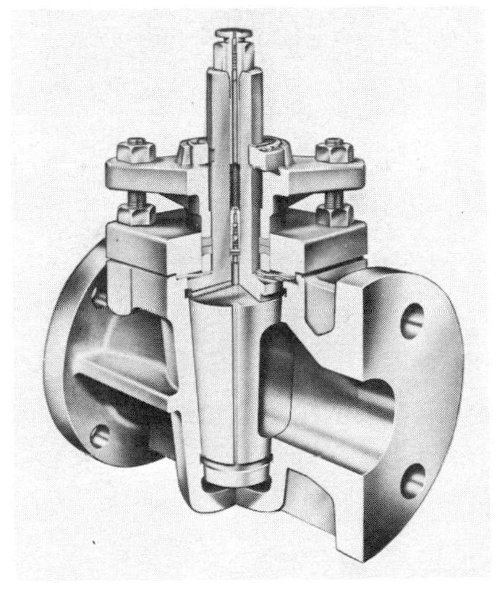

Fig. 57 Plug valve with lubricant grooves on tapered plug. (Walworth Co.)

Diaphragm Valves. Force for sealing is exerted on a flexible diaphragm to press it against a ridge or other seating surface on the opposite side of the flow passage through the valve body. A chemically resistant elastomeric diaphragm will protect the bonnet and stem and prevent leakage out of the stem piercement. Diaphragm valves find use in corrosive services at low to moderate temperature (about 400°F maximum). Liquids containing solids and fibers are also applications. Actuation is by screw action on the stem. Sizes are usually small. Modulation and power actuation are often employed.

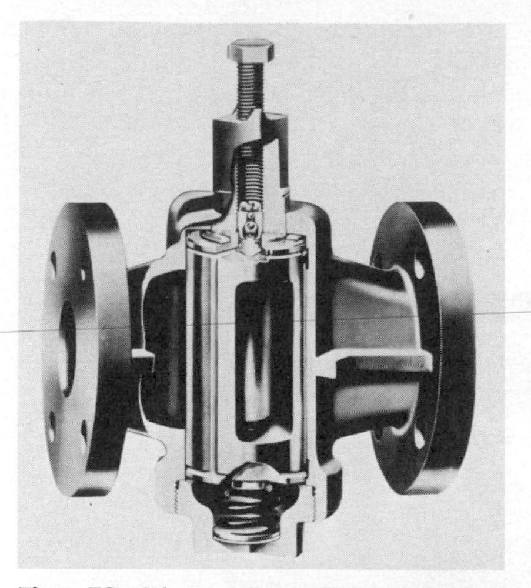

Fig. 58 Plug valve with cylindrical plug, spring-loaded, and with lubricant grooves. (Homestead Industries, Inc.)

Fig. 59 Cylindrical plug valve with TFE sleeve for sealing. (Xomox Corp.)

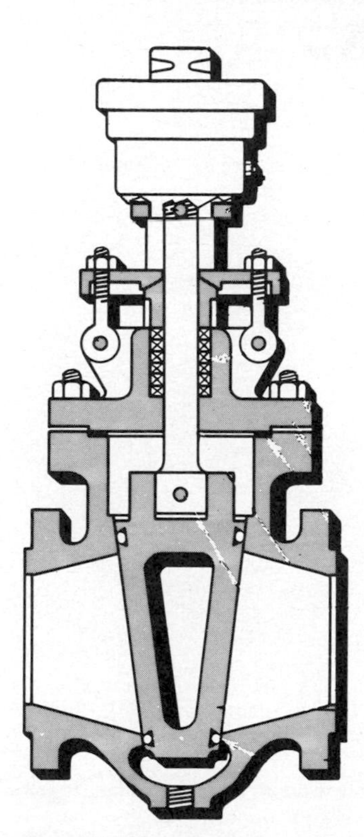

Fig. 60 Tapered plug in this valve lifts slightly, rotates, then seals on TFE O rings. (Stockham Valves & Fittings)

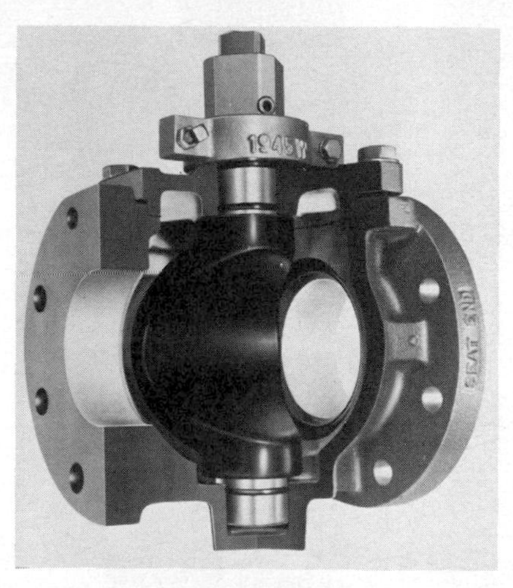

Fig. 61 Eccentric plug in this valve rotates out of fluid path. (Homestead Industries, Inc.)

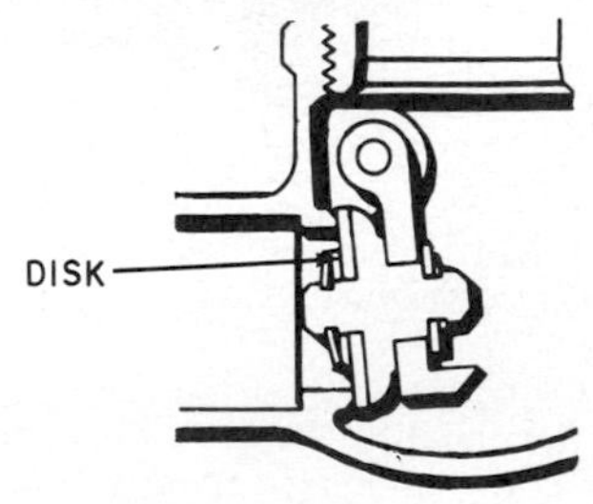

Fig. 62 Small swing-check valve with composition disk. (Stockham Valves & Fittings)

Check Valves. Preventing flow reversal in piping is the sole function of this type. The energy for opening and closing them comes entirely from the flowing liquid. The valve mechanism is out of sight inside the valve body, so that reliability is an important design factor. On the basis of closing action, the two main types of check valves are the lift check (Figs. 64 and 65) and swing check (Figs. 62

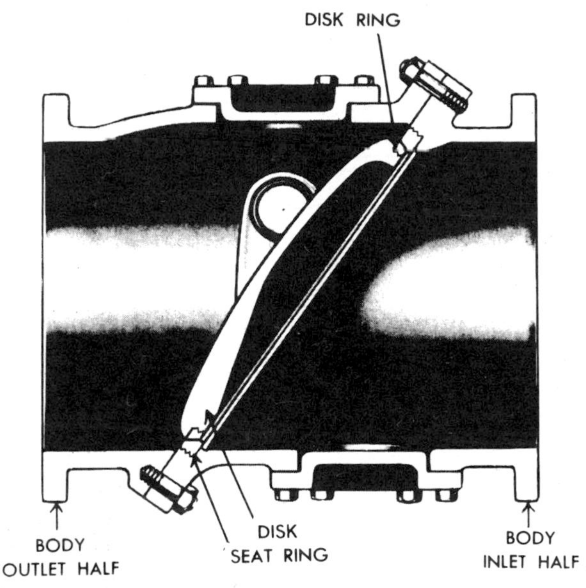

Fig. 63 Large tilting-disk check valve with metal-to-metal seat. (Crane Co.)

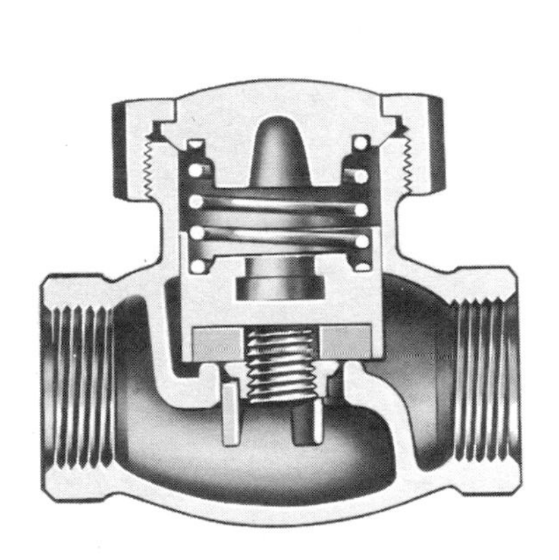

Fig. 64 Small lift-check valve with guided disk. (Walworth Co.)

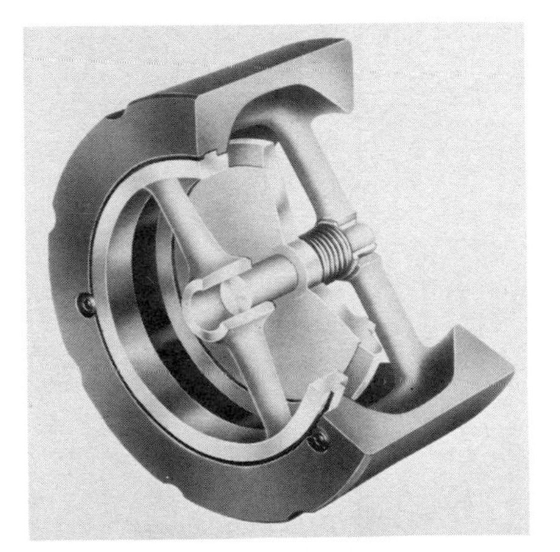

Fig. 65 Disk in this check valve is guided by a central spindle. (Val-Matic Valve & Manufacturing Corp.)

and 63). In the lift check, a disk, plug, or ball translates perpendicular to the seat and requires some form of guidance, either a sleeve on a stem or a cage in which the ball can move. In the swing check, a disk or flapper rotates around a pivot which is generally above the disk, although other locations are used.

Lift-check valves serve on steam, gases, and liquids; swing-check valves serve predominantly on liquids. If possible, a check valve should be installed in an attitude permitting gravity to help close the valve. Springs are another way of assisting closing. The lift-check valve with spring-assisted disk and stem guidance above and below the seat is often selected for vertical water lines at all pressures.

The ball type of lift check is restricted to small sizes but can operate on viscous liquids or liquids containing solids.

A stop-check valve is a special globe valve in which the plug is a lift-check plug. This valve type, often in a Y body, automatically prevents backflow and allows positive shutoff.

Dashpot effect is sometimes desirable in check valves. It is easy to obtain in the lift type by entrapment of liquid above the stem or plug. With the swing type, dashpot effect is not as easy to obtain. In large valves, the pivot shaft, which is pinned to the flap arm, is extended outside the valve body and is available for attachment of not only dashpots but also weights to assist closure or to balance the flapper.

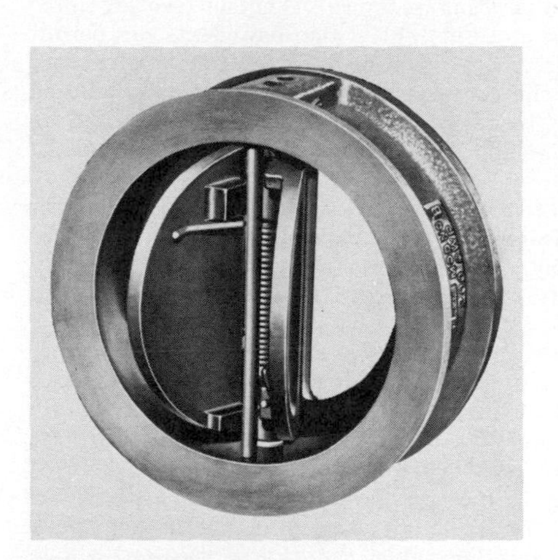

Fig. 66 Check valve with twin disks. (Mission Manufacturing Co.)

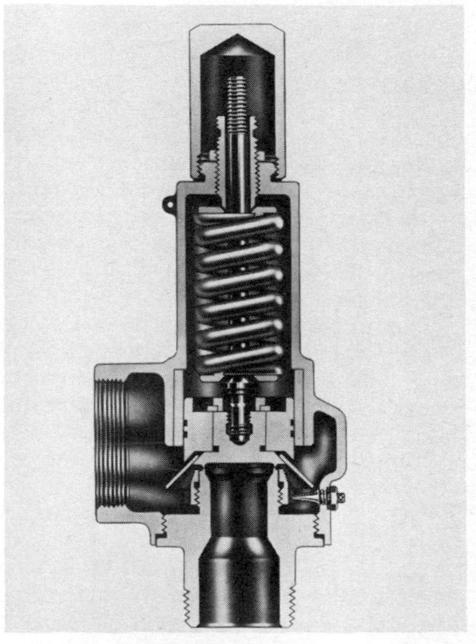

Fig. 67 Relief valve. (Dresser Industrial Valve & Instrument Div., Dresser Industries, Inc.)

Double-flap swing-check valves (Fig. 66) have a divider bar across the fluid passageway to form two D-shaped orifices. Two D-shaped flaps, pivoted on a vertical shaft downstream of the divider bar, seal off the orifices when flow reverses. A light torsion spring assists the closing action. These valves serve on horizontal and vertical lines at pressures beyond 3,000 lb/in^2 gage.

In swing checks with the disk pivoted near its center, resembling an eccentric-butterfly valve, the disk can be more nearly balanced. A beveled-edge metal disk seats on metal here (Fig. 63). The machining and fits must be accurate to reduce leakage to low values.

The smaller-size check valves usually have bonnets, such as screwed, union, flanged, or special high-pressure seal, to permit machining, assembly, inspection, and repair. Other smaller holes in the body may be desirable for pivot shaft mounting or regrinding. Some lift and double-flap swing-check valves are built in wafer form, with the valve body a short heavy-wall cylinder clamped between piping flanges (Figs. 65 and 66), much like modern butterfly valves.

Relief Valves. As shown in Fig. 67, these valves act in pump systems to prevent buildup of pressure beyond preselected limits. Positive-displacement pumps can cause the overpressure. Leakage from higher-pressure systems into lower-pressure ones, through a closed but leaking reducing valve, and thermal expansion in sections of piping isolated by check valves and closed outlet valves are the usual causes in centrifugal pump systems. The time rates of pressure rise in these cases are low enough to allow valves to function in timely fashion. Water hammer, another cause of overpressure, has a rapid rate of rise. This is also the case with internal explosions, accelerated chemical reactions, and sudden vaporization of liquids contacted accidentally by liquids at higher temperature. For rapid rates of pressure rise, rupture disks may be necessary, but their protection is not complete, either.

Vacuum-breaker valves are relief valves that protect large thin-walled components of piping systems against effects of outer atmospheric pressure when the interior is under vacuum.

In both types of valves, springs hold the disks shut against differentials up to the rated opening value, at which the disk opens and relieves pressure. If large flows of liquid or air must pass, the overpressure needed to force the flow through the orifice must be taken into account. Flow rates of relief valves are based on certain percentages of overpressure, often 10 percent but sometimes as high as 25 percent. Flashing of a hot liquid or pipe friction can add to the total overpressure required for a given capacity.

Leakage is sometimes a problem with relief valves, especially when the desired opening pressure is close to the normal operating pressure. The net seating force is then low, so that geometric irregularities and differential thermal effects can permit seepage. In applications where leakage is critical, a pilot-operated relief valve may be necessary. In a valve of this design, the pressure is on top of the disk, holding it increasingly tightly closed as pressure builds. A powered actuator, released in consequence of the opening of a small relief valve, forces the disk open against internal pressure.

Trim of relief valves can be metal or soft materials. The valve flow characteristic is usually quick-opening, although for flashing liquids, vapors, or gases it may be superquick-opening, with a slight cracking allowing pressure to build up on additional disk area and open the valve fully and suddenly. This is the familiar pop safety valve for steam work.

Control Valves Although all the above valves assist in controlling flow, they are not ordinarily called control valves. In the narrower sense, a control valve is a valve that modulates the flow through it to provide the desired downstream (or upstream) pressure, flow rate, or temperature. Although most types of valves can be partly closed and thus give a degree of control that may be acceptable for many purposes, the term control valve has come to mean a specialized type of power-actuated valve designed for good performance under steady-state or dynamic flow conditions. Many modern control valves closely resemble the types discussed above, but others are so different that they represent new types.

Before examining various control valves and their reasons for existence, some basic practical concepts must be reviewed. A control valve includes:

1. A body to contain the pressure, direct the liquid flow, and resist loads from piping and actuation
2. Variable orifice or orifices
3. A stem for positive connection of orifice elements with the actuator
4. A piercement through the body wall to allow the stem to pass
5. An actuator to adjust the orifice size

Adaptations of globe and angle valves are common. This type inherently gives tight shutoff. High-quality trim helps resist erosion and wear at low flow rates when the orifice is nearly closed. Support is often provided for the stem to prevent vibration and flutter (Fig. 68). The plug can be characterized (shaped to give certain rates of flow for given percentages of opening) as desired. The support of the stem is usually on the bonnet side of the orifice rather than opposite, to keep orifice size down. With flow upward (through the orifice, then past the plug), the actuator must overcome upstream pressure to close the valve. With flow downward (past the plug, then through the orifice), valve motion becomes unstable when the valve nears closing.

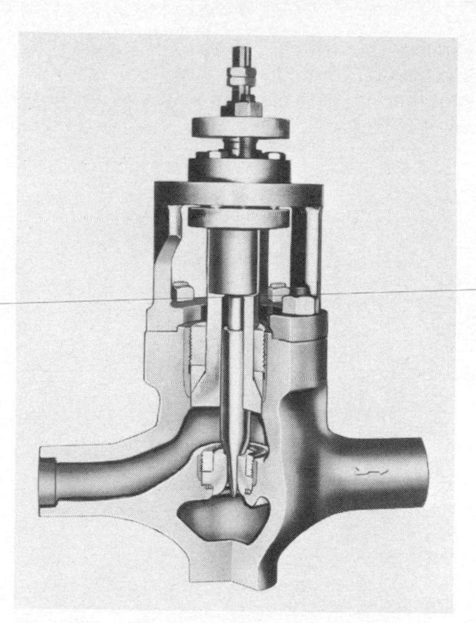

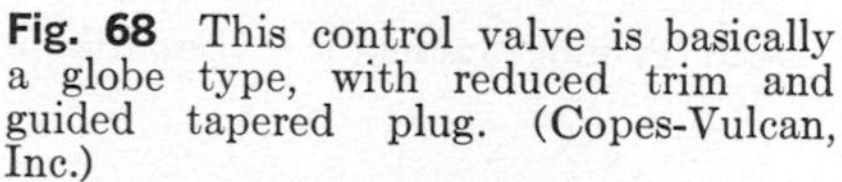

Fig. 68 This control valve is basically a globe type, with reduced trim and guided tapered plug. (Copes-Vulcan, Inc.)

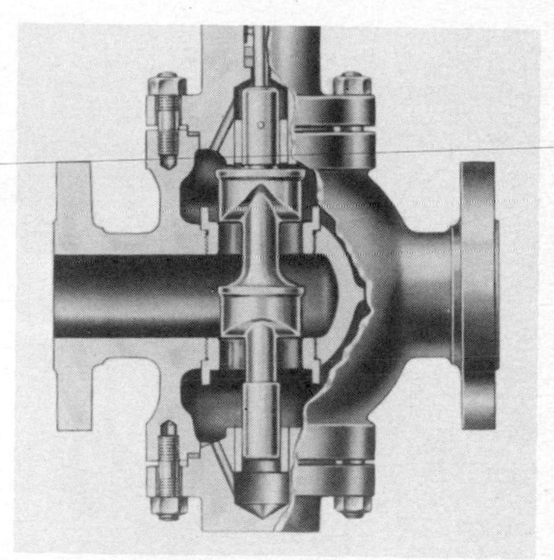

Fig. 69 Double-seated control valve with characterized plugs. (Masoneilan International, Inc.)

The basic globe valve is often modified to put two orifices and plugs on the same stem, with the upstream fluid entering the space between and passing in two opposite flows through the two orifices. This is the double-seated valve (Fig. 69). Actuator force is greatly reduced because fluid pressure tends to open one plug and close the other. The balance is not complete, however, because one orifice is usually larger than the other to permit assembly and because there is a difference in head conversion effects in the orifices at low flow rates.

Addition of an internal diaphragm and a port in the body near one orifice makes the valve a three-way type, able to divide flow between two outlet lines or, with reversed flow direction, combine two flows in desired ratio. The double-seated valve cannot seal tightly because of manufacturing tolerances and thermal and pressure effects on the valve body.

Butterfly valves can modulate flow. Special vane shapes have been introduced to improve performance. Elastomer or plastomer linings give a tight shutoff on liquids within the temperature range of the materials.

Fig. 70 Butterfly control valve with linkage connecting spindle to diaphragm actuator and positioner. (Masoneilan International, Inc.)

Ball valves as control valves may take conventional form, with an actuator and positioner atop the valve. In other designs, the ball may be merely a fraction of a spherical shell, adequate for sealing on the customary TFE seat ring but with its edge shaped to develop the required characterized flow as the shell rotates and exposes the orifice. Convex, V-notch, and parabolic edge shapes find use. The conventional ball valve has two variable orifices in series, of course, with a small chamber in between, in which some head recovery occurs as fluid momentarily slows.

Fig. 71 Eccentric-butterfly control valve, with cylinder actuator. (DeZurik, a unit of General Signal)

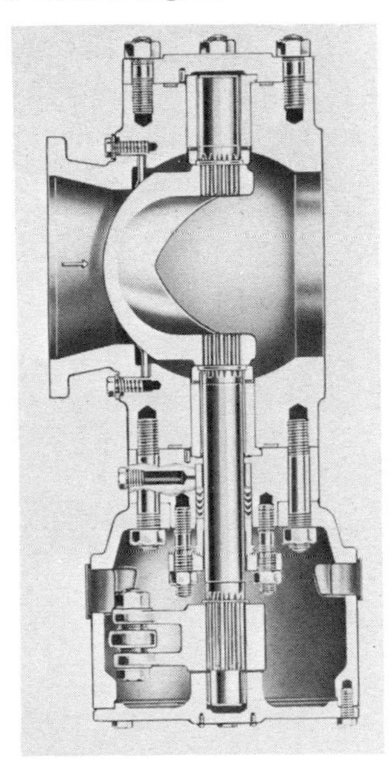

Fig. 72 Control valve with characterized plug which is part of a sphere. (Fisher Controls Co.)

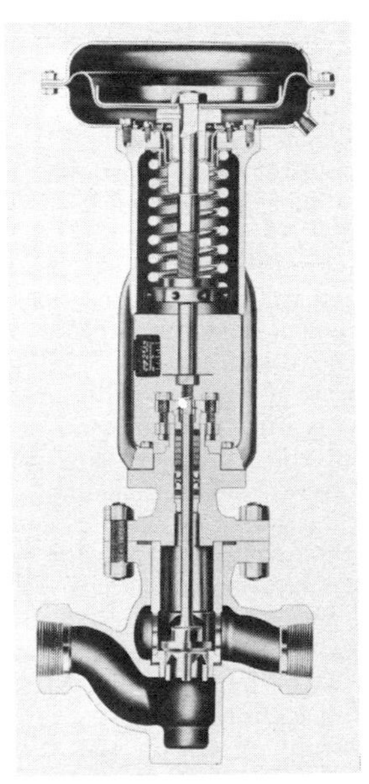

Fig. 73 A skirt guides plug of this valve in which flow is upward against disk. (Leslie Co.)

A specialized form of gate valve can serve as a control valve. This type contains a multiple-orifice plate mounted permanently as a diaphragm perpendicularly across the line of flow (Fig. 84). The "disk," a plate which also contains two or more slotted orifices, is mounted to slide vertically across the upstream side of the stationary plate. The degree of orifice coincidence determines the flow rate. Actuation is by a pin mounted on the stem and protruding through the stationary plate into a pocket on the sliding plate. Low vibration and straight-through flow are characteristic of this valve. Actuating force is low at all flow rates because of the sliding action of lapped disk and plate and because of the disk support. Stainless steel or other alloys are material of disk and plate.

Cage Valves. This type has developed into an entire group of control valves. The valve body closely resembles that of the globe valve, with a large orifice in a horizontal area of the central diaphragm. The cage is a hollow cylinder which is positioned and held between the bonnet and the edge of a hole in the diaphragm. The disk or plug, sliding up and down inside the cage, is guided by it.

In most cage valves, the lower zone of the annular cage is available for flow-control orifices, alternately exposed and covered as the plug moves up and down (Fig. 77). Tight shutoff at these orifices is impossible, of course, so the bottom of the cage or a separate seat ring is machined and finished to match a seating surface on the plug.

In other cage valves, where the plug may be much smaller than the cage inside diameter (Figs. 74 and 76), all control action is at the lower seat ring. The plug guiding is then on the stem or in the seat-ring orifice, and the holes in the cage are merely passage holes to distribute the liquid evenly around the perimeter.

The seat ring in cage valves is retained by bonnet bolting forces acting through the cage. Gaskets take up tolerances in dimensions and finish, so that the seat ring need not be pressed or screwed into the valve diaphragm.

The ordinary cage valve has several advantages. Suitable machining of cage holes can give the valve the desired characteristic. Removal and replacement of internals, such as cage and seat ring, are quick and simple. A vertical hole through the plug will make the valve nearly balanced, although considerable leakage can occur between plug and cage wall if the plug is balanced in this way. Cage-wall orifices vary not only in number, size, and cross-sectional form, but also in path and surface roughness.

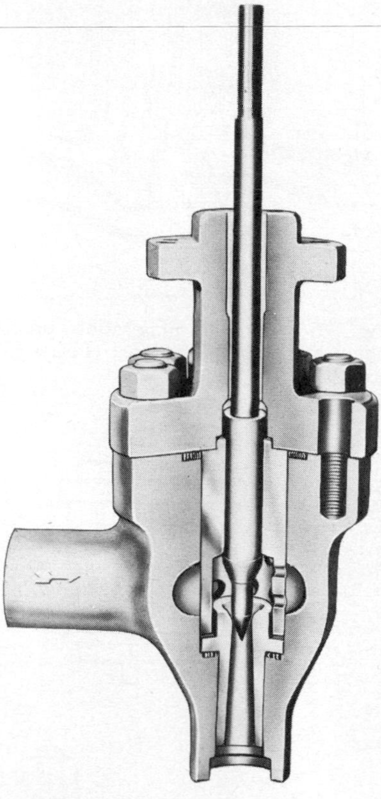

Fig. 74 Tapered outlet section improves flow pattern in this cage valve. (Copes-Vulcan, Inc.)

In one advanced variation of the cage valve, the cage wall is comparatively thick and the many pathways through it are labyrinthine, with several right-angle turns and several orifices and expansion chambers in each pathway (Fig. 82). To make the cage practical from a manufacturing standpoint, it consists of a series of thin disks each carrying a pattern of labyrinthine paths in one surface. The opposite surface is flat, so that when the disks are stacked into a cage, the paths are sealed from one another. Flow in valves of this type can be from inside or outside the cage. Characterization is possible by such means as change in number of orifices per disk at various heights in the stack. As in conventional cage valves, change of trim and characterization is quickly done after bonnet removal. The bonnet bolting, outside the liquid, holds the cage elements in place.

In other cage valves, the passage walls may be wavy, resembling screw threads. This assists in noise reduction in gas valves.

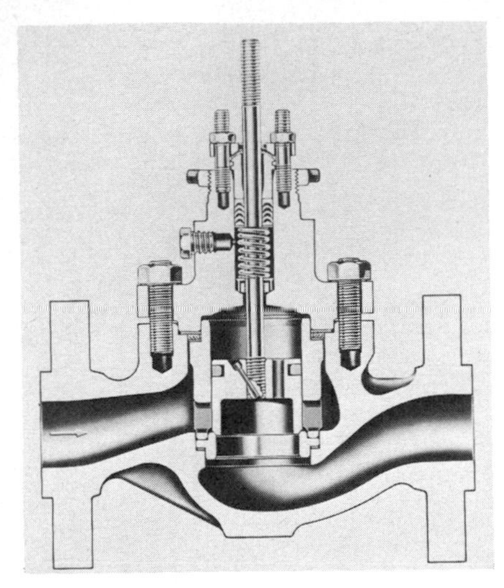

Fig. 75 Holes in plug allow pressure balance in this cage valve. (Fisher Controls Co.)

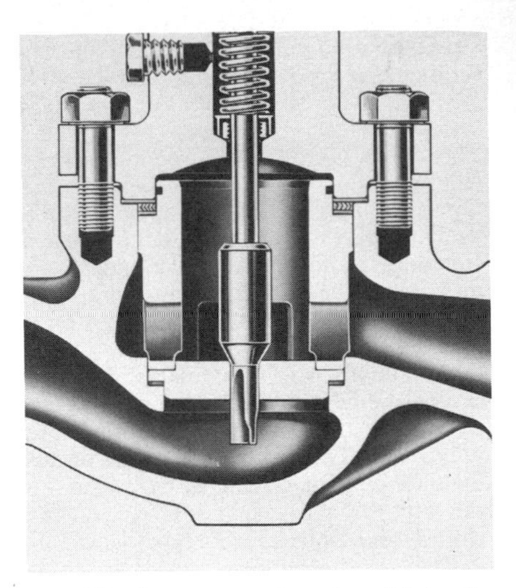

Fig. 76 An extreme in reduced trim for a cage valve. (Fisher Controls Co.)

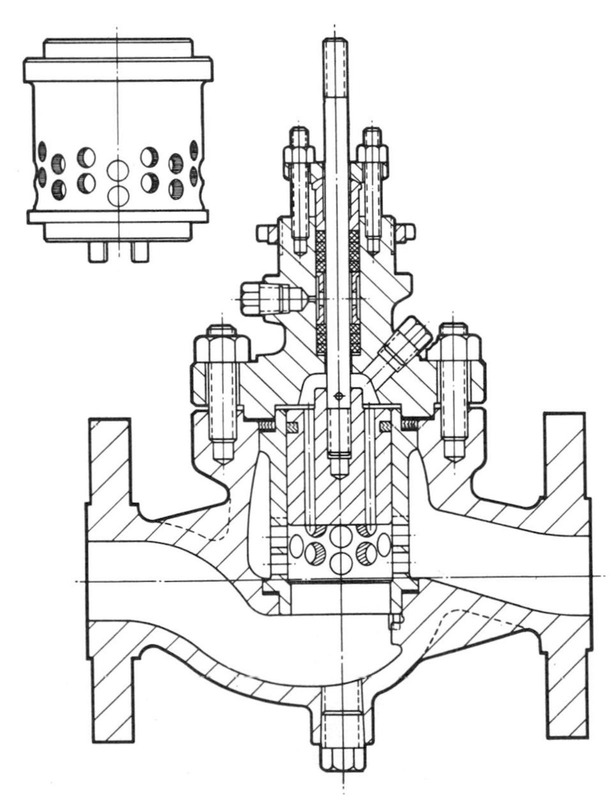

Fig. 77 Hole pattern in cage can reduce noise and cavitation. (ITT Hammel Dahl Conoflow)

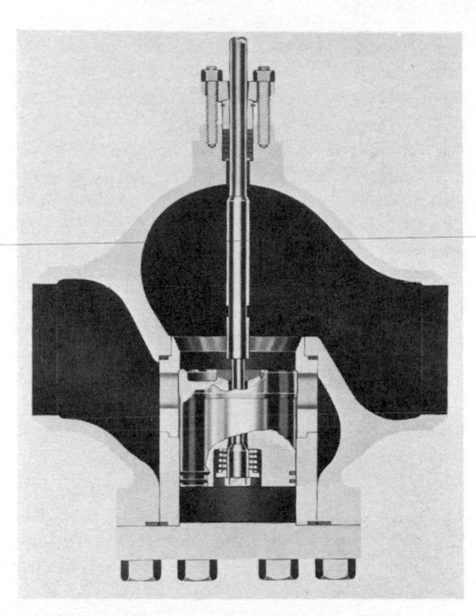

Fig. 78 Cage valve with bottom access. (Copes-Vulcan, Inc.)

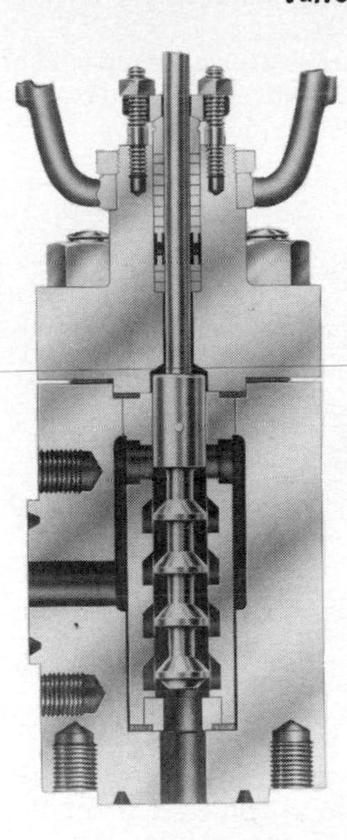

Fig. 79 Pressure breakdown occurs across several annular orifices and direction changes in this valve. (Masoneilan International, Inc.)

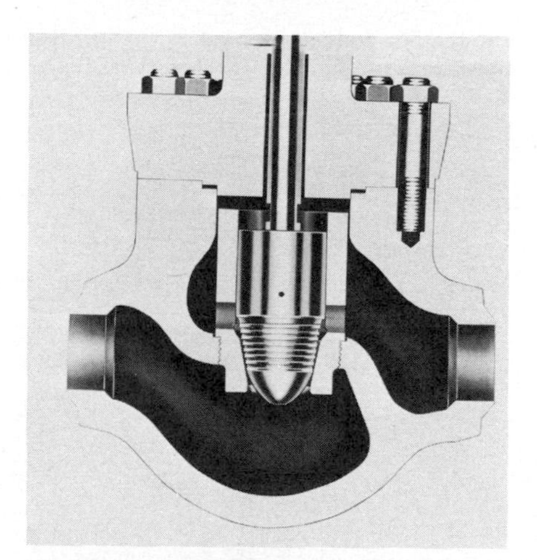

Fig. 80 Serrations on large plug give high pressure drop at low flow. (Copes-Vulcan, Inc.)

Fig. 81 Pump bypass flow at high pressure drop results when line flow at top ceases and mechanism opens valve at right. (Yarway Corp.)

Multiple Orifices in Series. Several valves have orifices in series rather than in parallel; the principle is called *cascading.* In one group, a tapered plug with a series of circumferential serrations moves in a tapered seat which may be either conical or stepped (Figs. 79 and 80). In either case the serrations or steps produce a series of small annular chambers alternating with annular restrictions that serve as orifices. Some alternating change in flow direction also occurs to create desired head loss.

In another group, annular chambers in the wall of a cylindrical cage are separated from one another by ridges that are a close fit with annular ridges on a sliding plug. The orifices narrow as the ridge sets approach one another, while the expansion chambers remain nearly constant in size. Repeated change in flow direction and speed produces head loss. In a variation of this type (Fig. 81), the chambers on the cylindrical plug are short, steeply angled helical cuts, so that the liquid takes a helical path. The purpose is to fling the liquid against the walls of the cage and displace cavitation bubbles toward the center and away from wall contact. Shutoff in these valves cannot rely on the multiple-orifice systems but instead depends on a separate conventional seat and plug surface, either upstream or downstream of the orifice system.

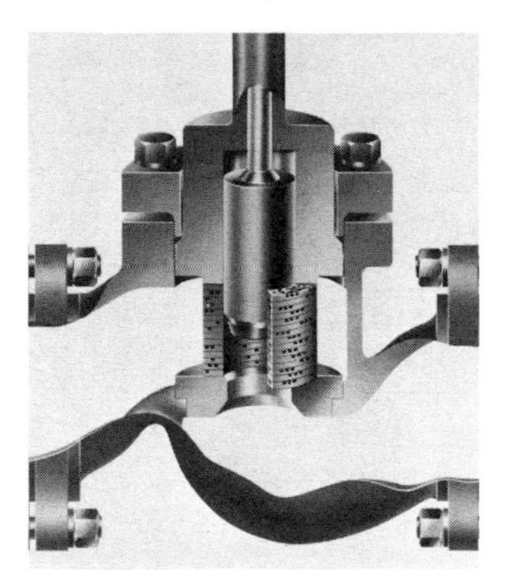

Fig. 82 Control valve with flow through labyrinthine orifices in built-up cage. (Control Components, Inc.)

Many of these special valves are very expensive because of multiplicity of complicated parts and because of the reduction in capacity caused by the advanced design measures. Larger bodies and overall sizes are required for a given flow rating. In pump control, the valves see service on high-pressure feedwater-pump minimum-flow recirculating lines, where pressures go as high as 6,000 lb/in^2 and water temperatures range to 500°F. Tight shutoff over thousands of operating cycles is the goal.

Flow Characteristics. An important parameter for valves in modulating control is the flow characteristic of the valve, often called simply the characteristic. The

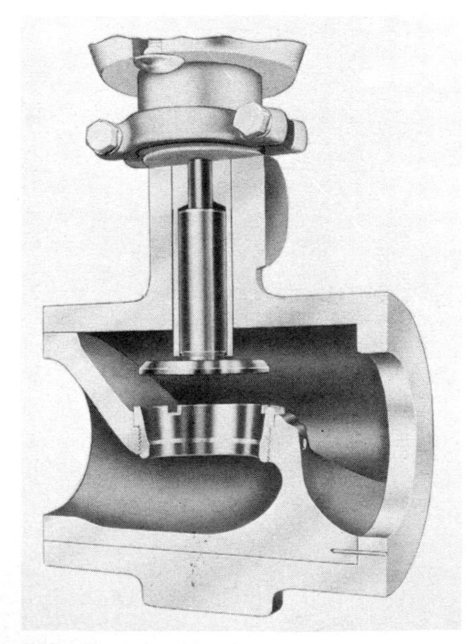

Fig. 83 Unusual control valve design with removable diaphragm and thin flat disk. Seat taper characterizes flow. (Masoneilan International, Inc.)

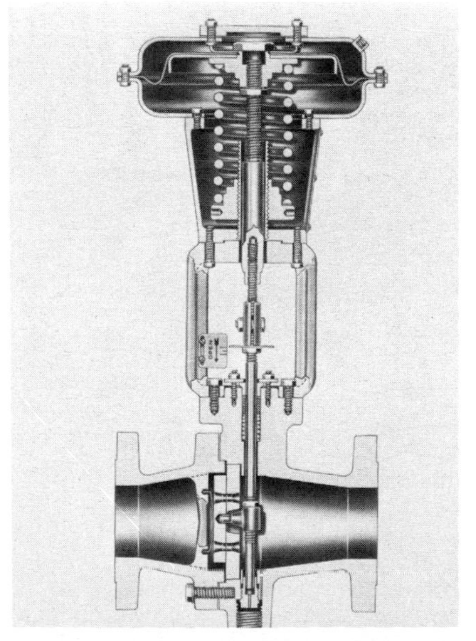

Fig. 84 Movement of one plate past another opens or closes flow orifices in this control valve. (Jordan Valve Div., Richards Industries, Inc.)

flow characteristic expresses the way in which the flow through the valve depends on percentage of valve stem travel. The latter may be translatory or rotary motion, of course. A plot of percentage of maximum flow at various percentages of stem travel is the usual quantitative way of showing a characteristic (Fig. 86). Several types of characteristic have become common, either because of inherent desirability or because familiar and traditional types of valves have them.

The linear characteristic is a straight line, with flow percentage always equal to stem-travel percentage. A quick-opening characteristic, on the other hand, produces proportionately more flow in the early stages of stem travel. An equal-percentage characteristic gives a change that, for a given percentage of lift, is a constant percentage of the flow before the change. A change of 16 percent of total stem travel will double the flow, so that at a stem travel of about 84 percent, flow will be 50 percent of maximum.

Although the linear characteristic would seem best because the rate of flow change is uniform for a given stem-travel change, incorporation of the valve into a piping system affects the decision. Since resistance to flow in a given piping system is roughly proportional to the square of the flow rate, the curve of the piping-system head loss plotted against flow rate will be a parabola, with resistance increasing at a faster rate than flow. If the piping system and valve are considered together, with the flow rate in the system plotted against percentage of valve-stem travel, the overall system characteristic will differ from the valve characteristic. The overall system characteristic is displaced upward toward the quick-opening valve characteristic but can have points of flexure. The amount of displacement depends on what part of the total system pressure drop is taken by the valve. Only with a very short outlet pipe would the valve take all the pressure drop, and then its characteristic would be that of the system.

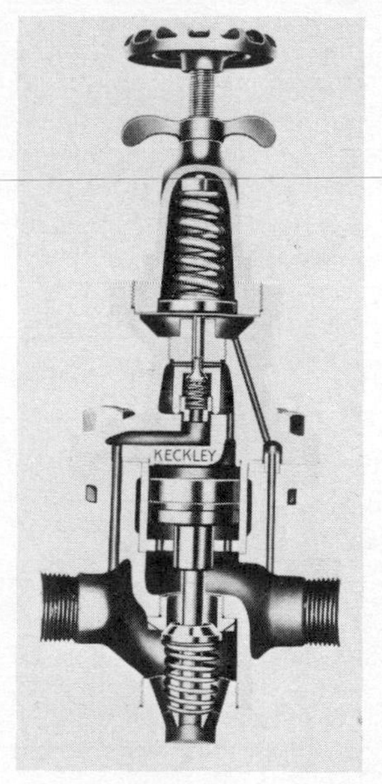

Fig. 85 Regulator, a self-contained pressure-control system with a slightly drooping curve. (O. C. Keckley Co.)

In many systems the pressure drop across the valve is designed to be from a tenth to a third of the total system drop. If the valve in such a system has an

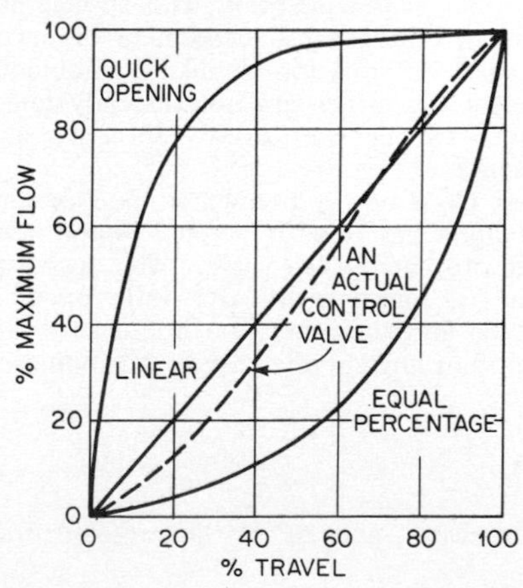

Fig. 86 Basic control valve characteristics. Actual valves are designed to approach these.

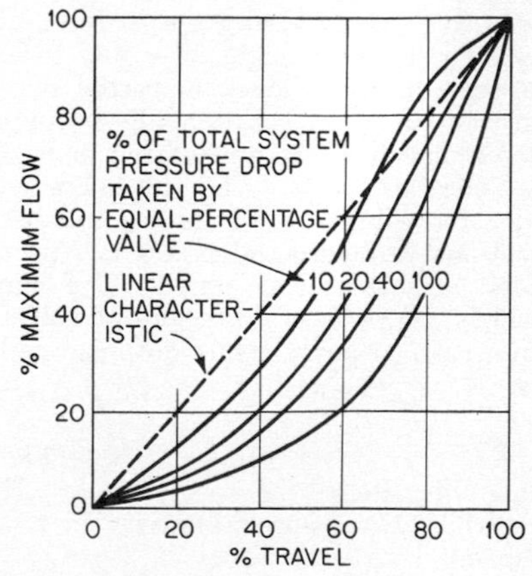

Fig. 87 With valve in a system, the overall characteristic depends on valve characteristic and pressure breakdown.

equal-percentage characteristic, then the characteristic of the overall system will be close to linear as far as the actuator of the valve is concerned (Fig. 87).

The equal-percentage characteristic is obtained by such measures as contouring the valve plug, contouring slots in plug skirt or cage, or suitably spacing holes in the cage.

Valves with characteristic between linear and equal percentage are also useful in modulating control. Ball, plug, and butterfly valves are examples. Characterized ball and plug valves are examples of modifications for control characteristic purposes.

Rangeability. Control valve rangeability can be important in some cases; it is defined as the ratio of maximum flow to the minimum flow at which the valve characteristic is still evident and control is possible. A high value means that a single valve can handle low as well as high flows, so that auxiliary valves are unnecessary. The best performance in this regard is about 100:1 for special designs under favorable circumstances; 25:1 is common for conventional valves and ordinary circumstances.

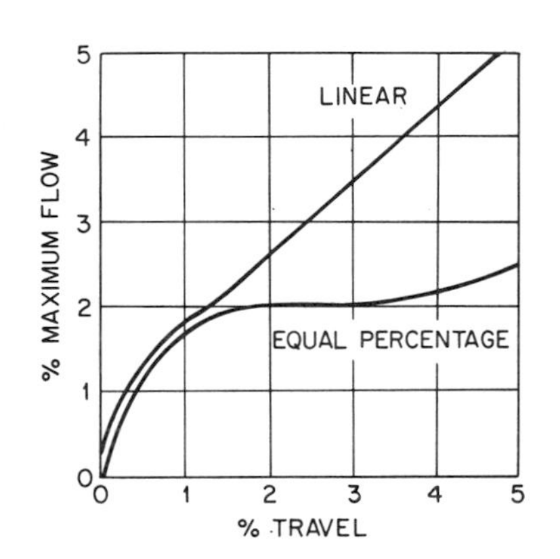

Fig. 88 Rangeability of a valve is determined by the point at which valve characteristic is still evident, between 1 and 2 percent of maximum flow for the two valve characteristics here.

Connected with rangeability is valve gain. The slope of the flow characteristic curve at any point is called the *valve gain.* In practical terms, it is the change in flow rate per unit of change in stem travel. A high gain means that a slight movement of the stem causes a large change in flow rate, so that instability occurs more readily. This sets a limit on a valve's rangeability. The quick-opening valve, with a high gain in the nearly closed position, is unsuitable for many modulating tasks.

Valves with approximately equal-percentage flow characteristic are considered most suitable for the majority of flow-control tasks; valves with linear characteristic are preferred for some applications.

"Size" of a valve is indefinite concept. In the past, a valve's size was understood to be the pipe size of the line connected to it. Venturi type valves with tapered end passages leading to reduced-diameter orifices, and valves in which orifice area is reduced for reasons such as cost cutting, characterization, and special advantages, have forced users to rate valve size in other terms. Statement of maximum orifice area might be a way to express size, but this too would be ambiguous because the head losses in partial recovery after the orifice are not the only losses in the valve. In addition, the valve may have two or more orifices in series, and the geometry of the orifice itself may affect results.

Flow Coefficient. The valve flow coefficient C_v is now a frequently used parameter for valve size. It is the number of gallons per minute of 60°F water that will flow through a valve at a 1 lb/in^2 pressure drop across the valve. The upstream test pressure is also stated. The maximum C_v, found with valve fully open, is widely accepted as a measure of valve size. To find the maximum liquid flow rate of a valve at any pressure drop and with a liquid of any specific gravity, the equation is

$$Q \text{ (in gpm)} = \frac{C_v}{(G_t/\Delta P)^{1/2}}$$

in which G_t is specific gravity in relation to water, and ΔP is the pressure drop (lb/in^2) across the valve.

Pressure Recovery and Cavitation. Drop in liquid pressure upon passing through a valve is recovered to varying extent downstream (Fig. 89). Degree of recovery depends on valve type: ball and butterfly valves have higher recovery percentages than do globe and angle valves. To avoid cavitation, which is the formation of vapor bubbles near the vena contracta of the valve, followed by sudden damaging collapse near metal, the static pressure at the vena contracta must be above the liquid vapor pressure. This is easier to do with a low-recovery valve, because the initial pressure drop need not be as high for a given downstream pressure. Several factors have been devised to indicate pressure recovery. One, C_f or critical flow factor, is the ratio of pressure recovery, varying for different valve openings. Another, K_m or valve recovery coefficient, is the ratio of pressure drop across the valve to the pressure drop between valve inlet and vena contracta at that instant when flow begins to be choked by bubble formation. Both of these factors will be higher for globe valves than for ball and butterfly valves, and the factors serve to indicate valve suitability for marginal cavitation service.

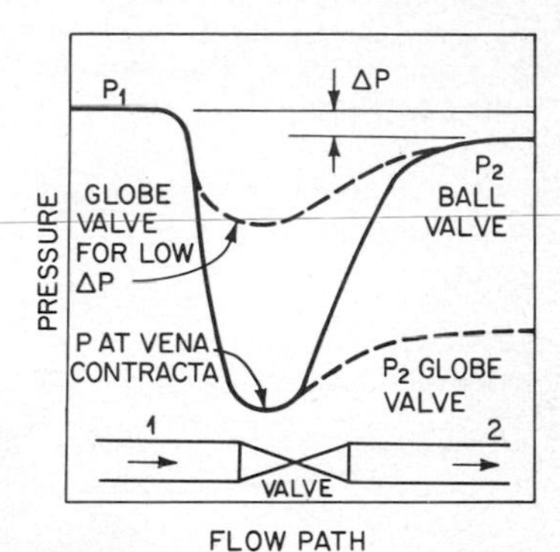

Fig. 89 For a given pressure drop ΔP across a valve, the globe will show a higher pressure at the vena contracta, making it more likely that cavitation difficulties will be avoided when the vapor pressure is high.

Actuators. The motion needed to change the valve orifice area and to close the valve tightly is produced by an actuator. The types of motion of the valve plug or disk are either linear or rotary, the latter being usually 90°, but occasionally as low as 70°. These motions can be effected in several ways. The linear translating motion can result from a cylinder or diaphragm actuator working directly or through linkage. A screw thread at stem top can convert a rotary motion into linear stem motion, or threads at stem bottom

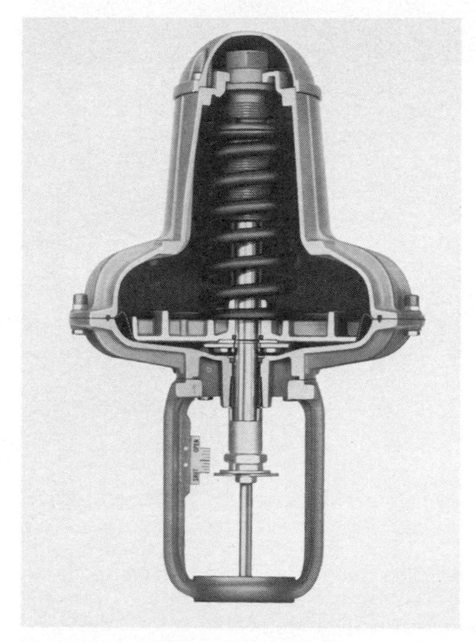

Fig. 90 Diaphragm-and-spring actuator, reversible type. (The Foxboro Co.)

Fig. 91 Linkage connects actuator and valve stem. (Masoneilan International, Inc.)

can engage threads in the valve disk so that rotation of the stem moves the disk. Geared electric motor drives, cylinders, and diaphragm-and-spring actuators are common with ball, plug, and butterfly valves. The solenoid valve (Fig. 94) relies on an electromagnetic force to move a disk directly or to initiate the piloting action that allows line fluid to open the valve.

The simplest actuator is the manually powered operator, which is a gear box. It provides enough mechanical advantage to overcome starting friction and to seal the valve tightly. Provision for an impact blow to initiate opening is found in some operators.

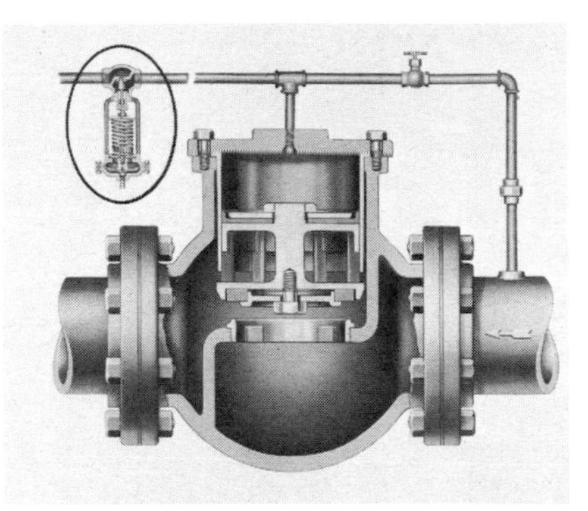

Fig. 92 Internal cylinder actuator. (O. C. Keckley Co.)

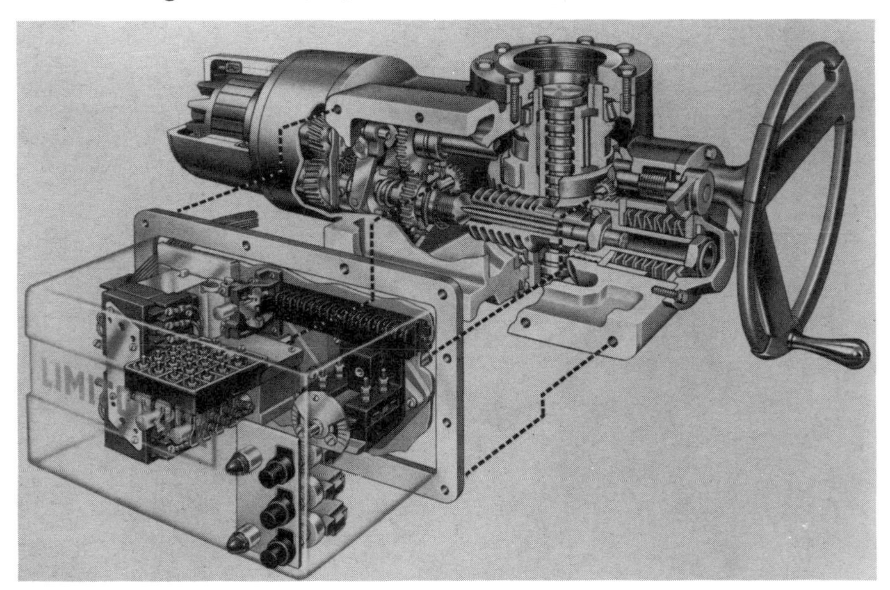

Fig. 93 Electric-motor-driven actuator with mechanism for limiting torque. (Philadelphia Gear Corp.)

The choice of actuators depends first on whether the service is on-off or modulating. For on-off service, the actuator need have only enough force to overcome breakaway force or torque and sufficient stroke to open the valve fully. Speed of operation is rarely critical, and motion limits can be designed into valve or actuator. Pneumatically and hydraulically powered actuators usually stroke rapidly but can be slowed in either direction by auxiliary valving or controls. On some pneumatic actuators, times to 5 min are possible. Electric-motor-driven actuators are slower than pneumatic or hydraulic types and require limit switches to stop motors at the end of travel.

In modulating service, where the actuator must hold a control-valve setting, demands are more severe. The speed of movement, expressed as stroking speed, is sometimes an important factor, especially in emergency shutdown or bypass. Stability of an actuator is partly its ability to hold the valve setting under fluctuating or buffeting loads from the fluid. Damping and high spring rate can help with this. The relation of natural frequency of the actuator and its adjacent elements

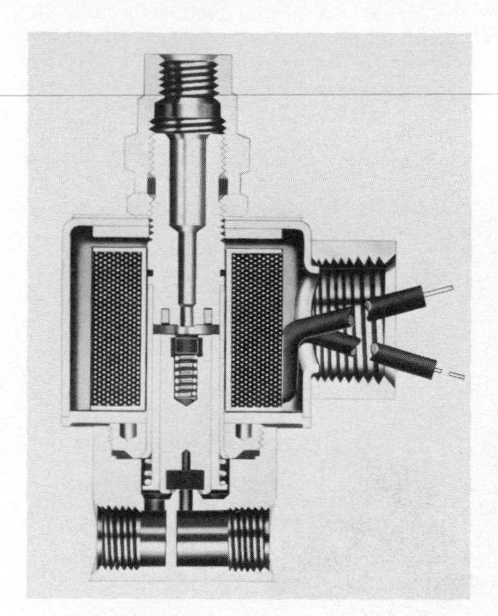

Fig. 94 Solenoid valve for three-way operation is direct-acting. (Skinner Precision Industries, Inc.)

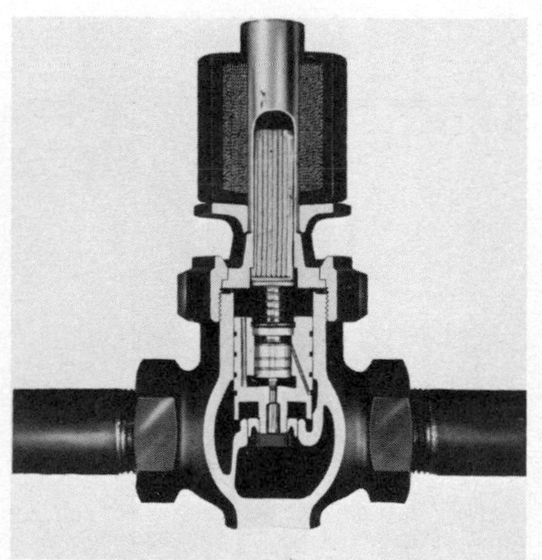

Fig. 95 Piloted solenoid valve relies on fluid pressures to open main orifice. (Magnatrol Valve Corp.)

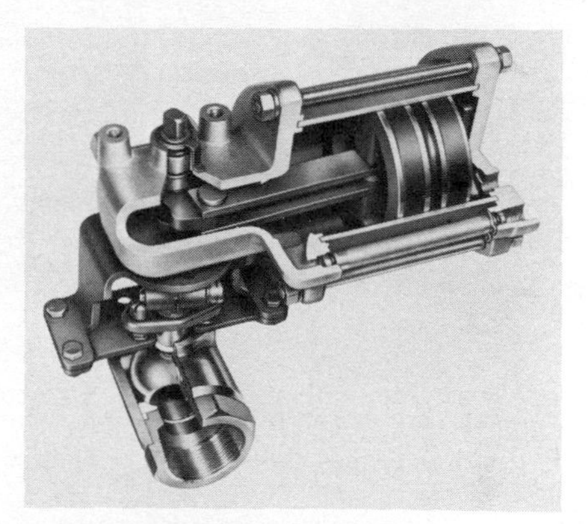

Fig. 96 Cylinder actuator and linkage for ball valve. (Jamesbury Corp.)

to the frequencies encountered in controlling the flow or experienced from fluid buffeting can also be important.

Stroke length is also a factor. Although the disk in a globe valve or similar type need lift only one-quarter of the seat diameter to give adequate area for full flow, this distance in large valves will exceed the 2-in stroke of most diaphragm actuators. If linkage with its lever advantage is needed to increase thrust, the problem becomes more acute. A cylinder or electric actuator is then necessary.

The source of power for the actuator influences choice, too. One standard may be 3 to 25 lb/in^2 instrument air, while in other cases much higher air or oil pressure is available.

The diaphragm-and-spring actuator (Fig. 90) is a very common type and gives several important advantages. The spring can be preloaded to cause the valve either to close or open fully (be fail-safe) if control air fails. The spring also opposes the force generated by the control signal, doing so in a manner giving proportional control. Of course, the spring's opposition negates much of the force available on the diaphragm, but the simplicity, low friction of diaphragm motion, ease of maintenance, and general high degree of perfection of this actuator have made it very popular. Most modern types are reversing: the fail-safe action can easily be

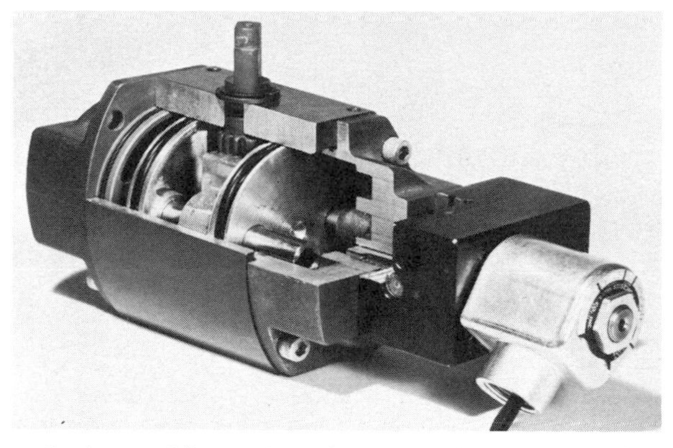

Fig. 97 Opposed pistons drive rack and gear mechanism for 90° rotation in this pneumatic actuator. (Worcester Controls Corp.)

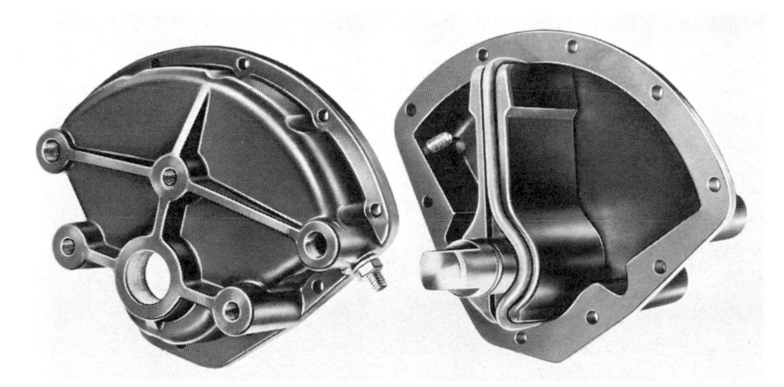

Fig. 98 Rotary actuator with sealed blade. (Xomox Corp.)

changed from open to close by turning the diaphragm enclosure upside down and reassembling.

Positioners. With actuators that lack an internal spring, a positioner is needed to adjust valve position to the desired value. A positioner is a mechanism which is actually a small feedback system receiving an input signal (usually an air pressure but sometimes an electric signal) from a controller and adjusting a valve-stem position to a prearranged corresponding value. The valve-stem position, which is the output, need not vary linearly with input pressure; cams in the mechanism can give a wide range of stem-position functions and thus apparently change the characteristic of the valve.

The positioner is a necessity for actuators in which the valve-stem position is not a function of the actuator-fluid pressure or electric current magnitude. Examples are pneumatic and hydraulic cylinders or electric motors. Even though posi-

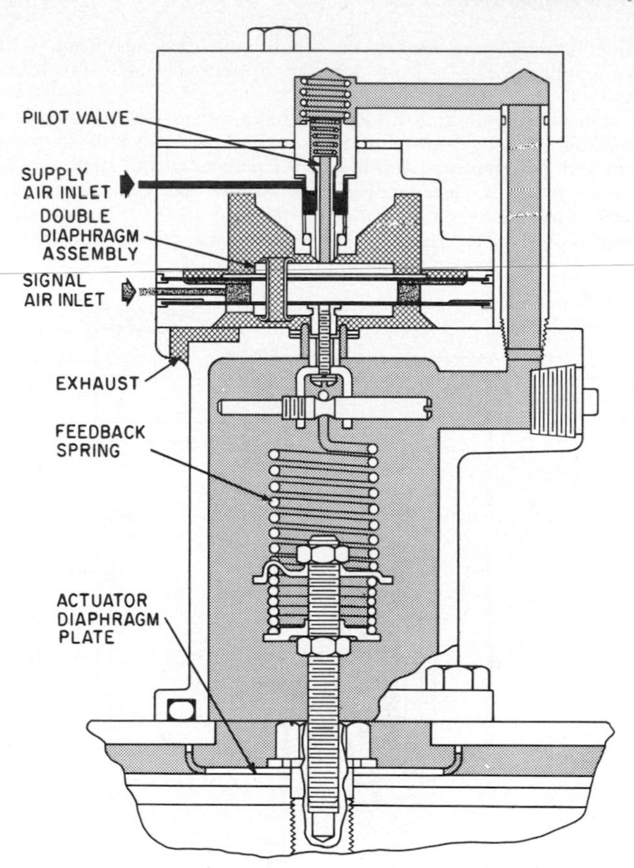

Fig. 99 Force-balance positioner with spool and sleeve pilot valve. (Masoneilan International, Inc.)

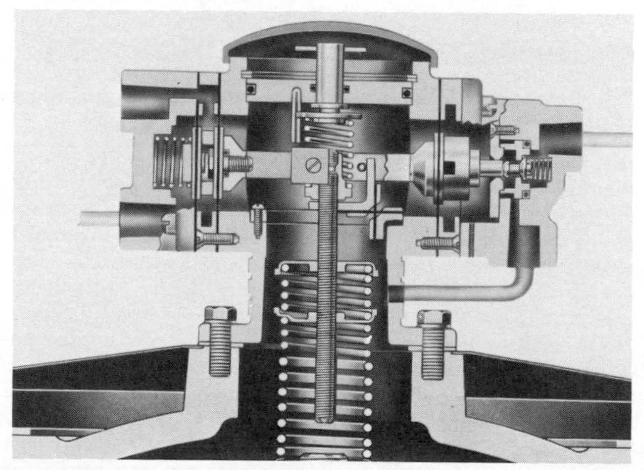

Fig. 100 Force-balance positioner with flexure linkage. (ITT Hammel Dahl Conoflow)

tioners are not inherently necessary on the diaphragm-and-spring actuator, they are sometimes applied. The reasons for the application hold for other types of actuators, too.

Friction in actuator diaphragm, in cylinder, or in valve-stem packing is one reason. The positioner can cut the dead-band from values such as 5 to 15 percent to less than ½ percent and can give repeatability of 0.1 percent of full span.

Need for more force to close a single-seat valve tightly is another reason. If loading pressure must be increased above a standard 15 lb/in² gage value, the positioner can control air at a higher pressure and thus greatly increase stem force.

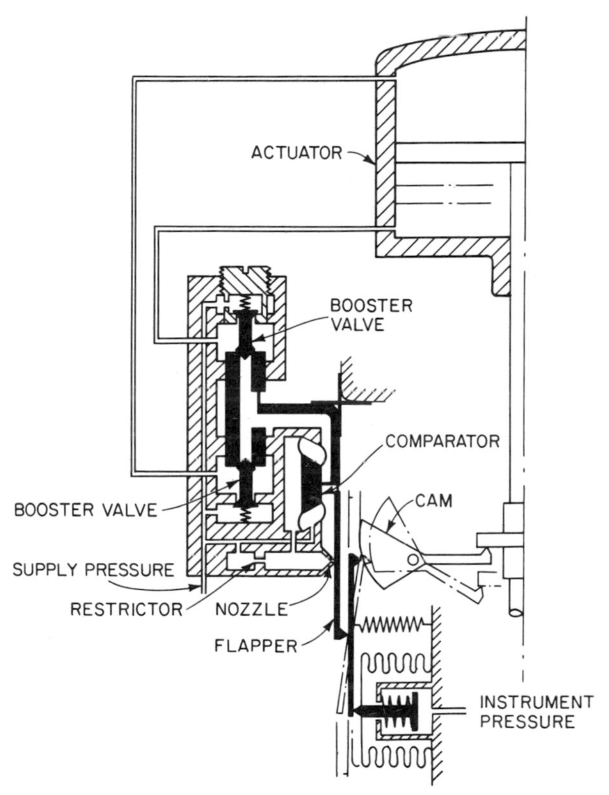

Fig. 101 Motion-balance positioner with cam and diaphragm comparator. (ITT Hammel Dahl Conoflow)

Split-range operation, in which different valves operate over different parts of the controller output pressure range, calls for positioners. Reversal of valve action, too, is easily achieved with positioners. A positioner can also speed up valve response, because the low-volume positioner will act faster than the high-volume valve actuator and can open a larger air supply than that in the controller. A pneumatic amplifier or booster is an alternative way to do this. Finally, change in control-valve characteristic, such as from linear to equal-percentage, is also possible through a positioner cam, mentioned above.

Because a positioner is another control loop added to a system, its effect under dynamic conditions may worsen overall performance. If changes or oscillations are slow, the positioner and actuator will follow them accurately and correct for them. For rapid changes, however, the effect of the positioner can be harmful. Evidence shows that if the natural frequency of the complete process control loop is more than 20 percent of the frequency at which p/a system gain is attenuated 3 dB, then the positioner will impair system performance. Liquid-level control systems are more likely to benefit from positioners than are flow or pressure control systems.

Pneumatic positioners may be classified as force-balance or motion-balance type. In the force-balance type, the force in the range spring inside the positioner is balanced against control air pressure inside a bellows or double-diaphragm assembly. In the force-balance positioner of Fig. 99, the feedback spring, which can be adjusted for range of pressure and initial actuation pressure, is attached to the actuator diaphragm plate at bottom and to a double-diaphragm assembly at top. The upper of the two diaphragms has twice the area of the lower; introduction of signal air

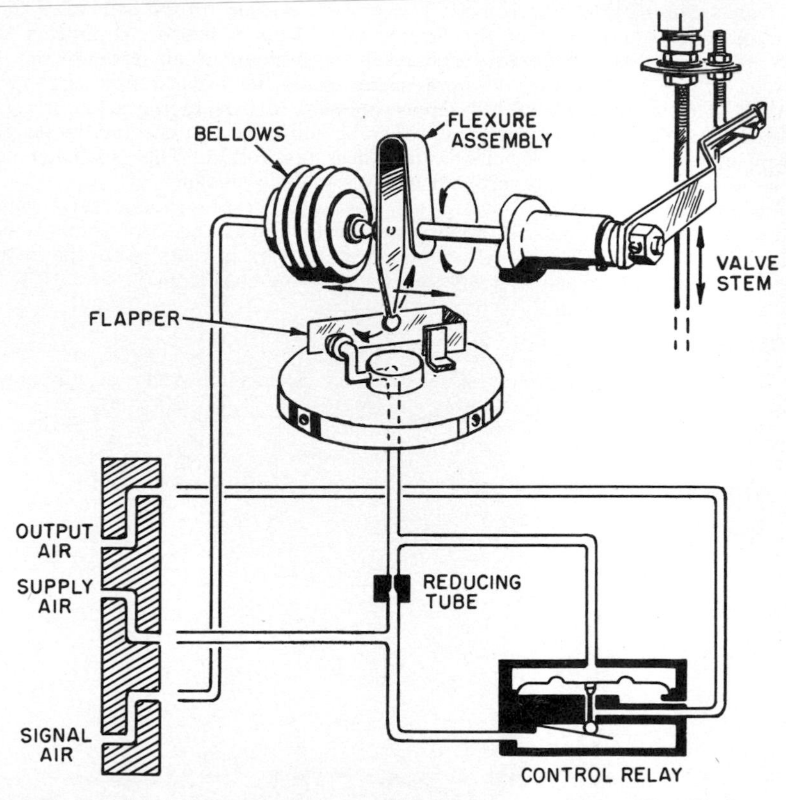

Fig. 102 Motion-balance positioner with flexure assembly. (The Foxboro Co.)

from the controller into the space between the two diaphragms forces the assembly upward very slightly but enough to lift a pilot valve at the positioner top and allow supply air pressure to flow through and downward past the feedback spring to press the actuator diaphragm down until forces balance. Reduction in signal pressure allows the double-diaphragm assembly to move downward, first closing the pilot valve and then exposing a hole through the pilot valve stem. Air then bleeds out from the actuator to atmosphere until forces are again in balance.

In the force-balance positioner of Fig. 100, flexure strips and a bell crank convert the vertical actuator motion into a horizontal motion of the double-diaphragm assembly at top left and the supply valve at right.

A motion-balance positioner showing application of a cam to impart a characteristic is shown in Fig. 101. The cam at lower right is pivoted and caused to rotate by the actuator-stem motion. Supply pressure enters the valve-assembly block at left and goes to both booster valves. It also bleeds through a restrictor and nozzle at the bottom of the valve-assembly block. The position of the flapper before the nozzle determines the pressure in the diaphragm comparator at the right of the valve-assembly block. Increase in signal air pressure to the bottom bellows in the figure moves one end of a balance beam and pushes the flapper closer to the nozzle. This builds pressure in the diaphragm comparator and moves it to the right. A linkage transforms this motion into a motion which opens the booster valve to supply air to the cylinder-actuator top and permits air to exhaust from the cylinder bottom. The actuator stem moves down until the feedback cam, aided by the comparator linkage, has repositioned the flapper in front of the nozzle.

In another motion-balance positioner (Fig. 102), signal air pressure from 3 to 15 lb/in^2 gage in a bellows opposes a flexure assembly on a shaft which is rotated by the valve-stem motion. An increase in signal air pressure to the bellows expands it and moves the lower end of the flexure away from a flapper, permitting the flapper to move toward a nozzle. The resultant buildup of air pressure on the diaphragm of the control relay at lower right closes the exhaust port and opens the supply port to allow air at full supply pressure to pass to the actuator. The valve-stem motion rotates the flexure and thereby shifts the tip touching the flapper. The flapper assumes an equilibrium position near the nozzle. This positioner produces a valve-stem position proportional to the signal air pressure.

The pneumatic amplifier or booster is a special kind of regulator valve which develops an output air pressure proportional to the input signal pressure. It can be used to boost pressure on an actuator for faster action in cases where the instrument tubing is small-bore and long and the actuator volume is large.

REFERENCE

"ISA Handbook of Control Valves," Instrument Society of America, Pittsburgh, 1971.

Section **8**

Supervisory and Monitoring Instrumentation

A. S. GRIMES

PURPOSE OF SUPERVISORY INSTRUMENTATION

The purpose of supervisory and monitoring instrumentation, as discussed below, is to monitor the routine operation of pumps, their drives, and their accessories in order to sustain a desired level of reliability and performance. Generally these instruments are not used for accurate performance tests or for automatic control, although they may share connections or functions.

Supervisory instruments should provide a sufficient guide to the hydraulic performance and mechanical condition of the pump to permit determining the need for a more detailed performance test or for a physical inspection of the equipment as well as indicating gross deficiencies or unsafe conditions. The information obtained from supervisory instruments may be useful in diagnosing failures or in indicating system problems external to the pump proper.

An approximate indication of pump performance usually can be obtained from the process that the pump serves. Is a level, pressure, or temperature being maintained? However helpful answers to these questions may be, this limited information does not tell if the margin or excess capacity of the pump is disappearing and, in the event of trouble, does not necessarily tell if the trouble is in the pump or in the process.

DETERMINING NEED FOR SUPERVISORY INSTRUMENTATION

The type and extent of the supervisory instrumentation suitable for a given pump installation are determined from the pump application, the pump design and size, experience with similar equipment, and the amount of monitoring provided by an operator. These items establish the level of reliability and safety desired, as well as the economic consequences of failure and the technical requirements of the instrumentation.

The pump application must be considered in terms of the effect of the loss of the pump on the process or system that the pump is a part of: minimal effect, partial loss of the process, total loss, and the cost or safety aspect of that effect. Loss of part, or all, of a process means the loss of production or a safety function.

A critical single-pump application, such as a single boiler feed pump for a generating unit, should be provided with a comprehensive set of supervisory instruments designed to detect changes in performance or mechanical condition at an early stage in order to try to avoid an unexpected forced shutdown of the unit. The same service but with three half-size pumps would not require such extensive instrumentation, since the risk of shutting down the entire unit is much less. Small pumps for intermittent auxiliary services require little or no instrumentation.

The fluid handled may have a pronounced effect on the rate of wear of a pump and the importance of predicting maintenance requirements. Suction-strainer cleanliness or the pumped-fluid viscosity may need to be monitored.

The pump design and type affect both the importance of monitoring and the parameters to be monitored. High-speed long-shaft span pumps may need vibration and bearing monitoring as well as hydraulic performance monitoring. Casing-metal temperature indicators are desirable as a guide to the warming of heavy-wall pump casings in high-temperature services. Pumps with hydraulic balancing devices should have a means of monitoring the balancing-device leakoff.

The users and manufacturers' previous experience with similar or identical pumps can be an important guide to applying supervisory instrumentation. Pumps with long records of troublefree service may be provided with a minimum of instrumentation while the histories of other pumps may indicate the need for careful monitoring of a specific parameter.

Pump installations with operators in attendance generally do not need monitoring of vibration or bearing temperature, as the operator can be expected to check these items. Unattended pumps in remote locations may require telemetering of key measurements or alarms. Some pumps in nuclear plant services may be relatively inaccessible during plant operation and may require more instrumentation than would a similar nonnuclear application.

Consideration should be given to providing supervisory instruments that will serve to back up automatic controllers during controller maintenance. Independent alarm devices may be needed to monitor a controller operation.

SOME APPLICATIONS OF SUPERVISORY INSTRUMENTS

Monitoring Hydraulic Performance The following recommendations are based on detecting changes from previous values of key parameters which will indicate a change in pump performance. This technique depends on the repeatability of the measurements rather than on absolute accuracy. Trends can be determined with reasonable certainty. The appropriate test procedure as described in Sec. 14 should be used if the absolute level of performance is to be measured.

Constant-Speed Centrifugal Pumps. The performance of a single constant-speed centrifugal pump can be monitored by measuring the suction and discharge pressures and the flow rate for comparison to previous values. If the suction pressure or lift is known to vary less than about 2 percent of the pump head, it need not be monitored. This is based on the pump having a head-capacity curve with the head decreasing 0.75 percent or more for 1 percent increase in flow in the monitoring range. Typically, with this condition, the monitoring instrumentation accuracy will not give closer head comparisons. The flow rate may be measured directly or determined in some other manner, such as measuring the time to change the level in a tank. In many cases, the suction and discharge conditions remain fixed; that is, there is no throttling of the flow or variable head which the pump must meet. In these cases the performance can be monitored by measuring the suction and discharge pressures only. This simplification may not be accurate enough if the pump has a relatively flat head curve.

When two or more pumps operate in parallel, the flow rate for each pump must be measured in order to monitor the performance of the individual pumps. The foregoing assumes that the pumps are handling fluids whose properties are essentially constant with time. If not, any applicable variables such as temperature, viscosity, density, or vapor pressure should be measured or otherwise accounted for in judging performances.

The pressure of seal water, if provided, should be monitored. Provision should be made for periodic checks of balancing-device leakoff or seal leakoff flows, if they exist. Continuous monitoring of such flows is not usually required to monitor hydraulic performance.

Variable-Speed Centrifugal Pumps. Monitoring the performance of variable-speed centrifugal pumps requires that speed be measured as well as the variables required for constant-speed pumps. Normally pumps with direct electric motor drives can be assumed to have constant speed for monitoring purposes. The possibility of significant speed changes should be considered for belt-driven or engine-driven pumps.

If a variable-speed pump is automatically controlled to maintain an essentially fixed-flow rate, comparison of running speeds over a period of time is a simple check of the pump-performance trend. The pump head should also be checked to assure that a process change has not caused a pump-speed change.

Positive-Displacement Pumps. Monitoring the performance of these pumps requires measuring the suction and discharge pressures, flow rate, and speed, if necessary, as in the case of centrifugal pumps. In addition, it may be necessary to measure the relief or recirculating flow used to control the discharge pressure or, for variable-displacement pumps, to monitor some index of pump displacement.

Monitoring Mechanical Condition Pumps with mechanical or labyrinth shaft seals which permit a controlled leakage should be provided with a means of measuring these flows. An increase in seal flow not only indicates wear or damage to the seal concerned, but frequently indicates a change in internal running clearances from the same cause that affected the shaft seal. For pumps with balancing disks or drums, the balancing flow is a better indication of internal-leakage flow rates.

Seals which require an injection flow either to prevent leakage of the pumped fluid or to provide smothering or quenching of hot fluids are best monitored by measuring the injection flow. In the case of pumps with multistage shaft seals, such as reactor-coolant pumps, both the injection flow and the leakoff flows should be measured. For some pumps the seal-water flow is varied to maintain a desired temperature of the seal leakoff water temperature. Indicators or alarms should be provided to monitor these leakoff temperatures.

Provision should be made for monitoring the pressure drop across pump suction strainers. An increase in pressure drop may indicate that foreign material did, or may, pass through the pump. An instrument that reads differential pressure directly, such as a manometer, is preferable to two separate pressure gages in this application.

A running-time meter is desirable, particularly for pumps that run intermittently with automatic starting and stopping. The running time for pumps in more or less continuous service possibly can be obtained readily from operating logs or other records.

Permanently installed vibration-measuring instruments have not been used extensively in pump installations. Many large turbine-driven boiler feed pumps have been so equipped. Large unattended pumps may have high-vibration alarms or trip devices. A significant change in the vibration level of a pump usually is associated with fairly extensive mechanical damage which will be indicated by changes in other supervisory instrument readings.

Bearing-temperature monitoring is desirable for large pumps equipped with sleeve bearings. The temperature element, resistance thermometer or thermocouple, should be located in the bearing so as to permit reading the bearing metal temperature. Pressure-lubricated bearings can be monitored by measuring the temperature of the return oil although this method responds less quickly to a change in bearing temperature. The supply-oil temperature and pressure should be measured.

Pumps with heavy-walled casings which handle high-temperature fluids should be provided with casing-metal temperature measurements to provide a guide to warming rates. One temperature measurement should be made near the inner surface of the casing and one or more measurement should be made on the outer surface of the casing. Alternatively, the temperature of the fluid entering the pump suction may be used in place of the inner-surface-metal temperature. Instruments are available to read the temperature difference, inner to outer, directly if desired. The pump manufacturer's recommendations for location of the measuring points should be followed.

Supervisory instrumentation diagrams are shown in Figs. 1 and 2 for a large variable-speed boiler feed pump and a reactor-coolant pump with its motor drive. These two examples include most of the supervisory instruments that are used to monitor pumps. The symbols used follow ISA-S5.1. Control functions are not shown.

In Fig. 1, instruments XR-1, XT-11, and XT12 monitor shaft vibrations and TR-2, TE-21, and TE-22 monitor bearing temperatures. Instruments TR-1 and TE-11 through TE-15 measure pump-casing temperatures to monitor pump warming. The instruments shown for the seal injection and return are for a system which controls the seal-return temperature. ST-1 and SR-1 are the speed transducer and speed recorder. The speed transducer normally is attached to the pump drive.

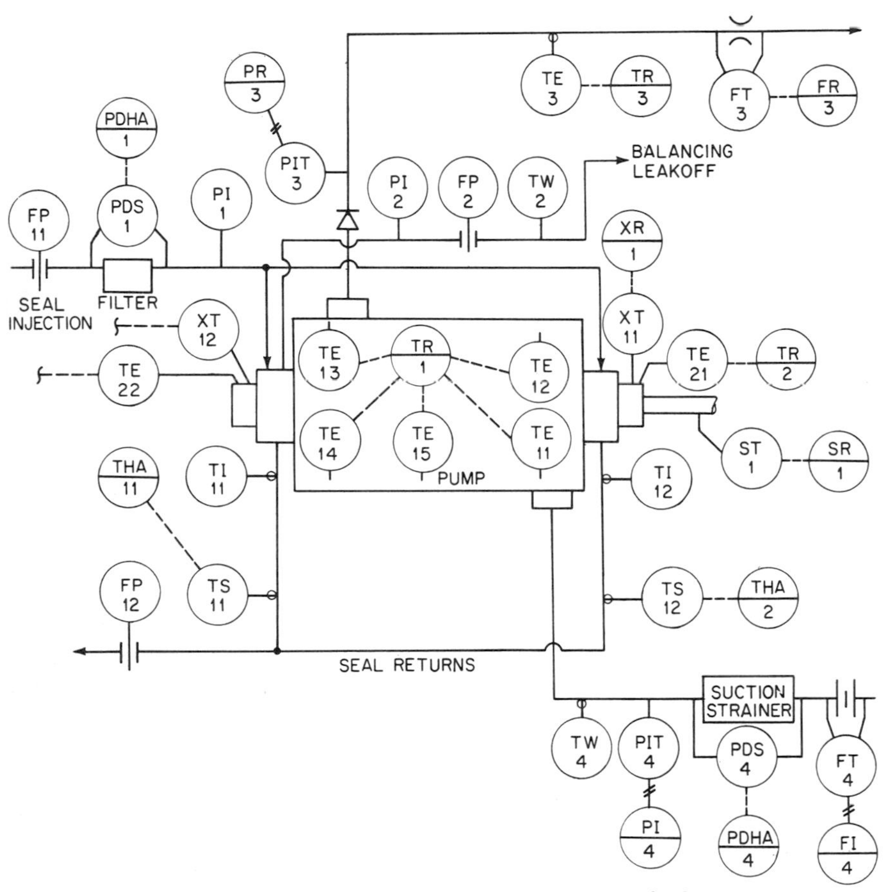

Fig. 1 Instrumentation for a variable-speed boiler feed pump.

Figure 2 shows a large constant-speed vertical pump and motor drive of a type used for reactor-coolant service. Instruments TR-10 and TE-11 through TE-14 monitor the temperature of the thrust-bearing upper and lower shoes, motor bearings, and pump bearing. Level switches LS-11 and 12 provide low-level alarms for the upper and lower oil reservoirs. Pressure gage P-12 monitors the bearing-lift oil pressure used to lift the thrust runner before starting the pump. Temperature indicator TI-5 monitors the motor-stator temperature. Usually a number of temperature detectors are installed and wired through a selector switch to an indicator. Vibration is shown on VI-1. The seal injection is regulated to maintain a pressure difference across the pump labyrinth which is monitored by PDI-32. The injection flow and the return flows are measured. In this example, the No. 2 seal return has high and low flow alarms only. This pump has a heat exchanger between the pump diffuser and the bearing and seal area. The cooling-water flow is monitored by alarms on the pressure difference across the heat exchanger and the outlet water temperature.

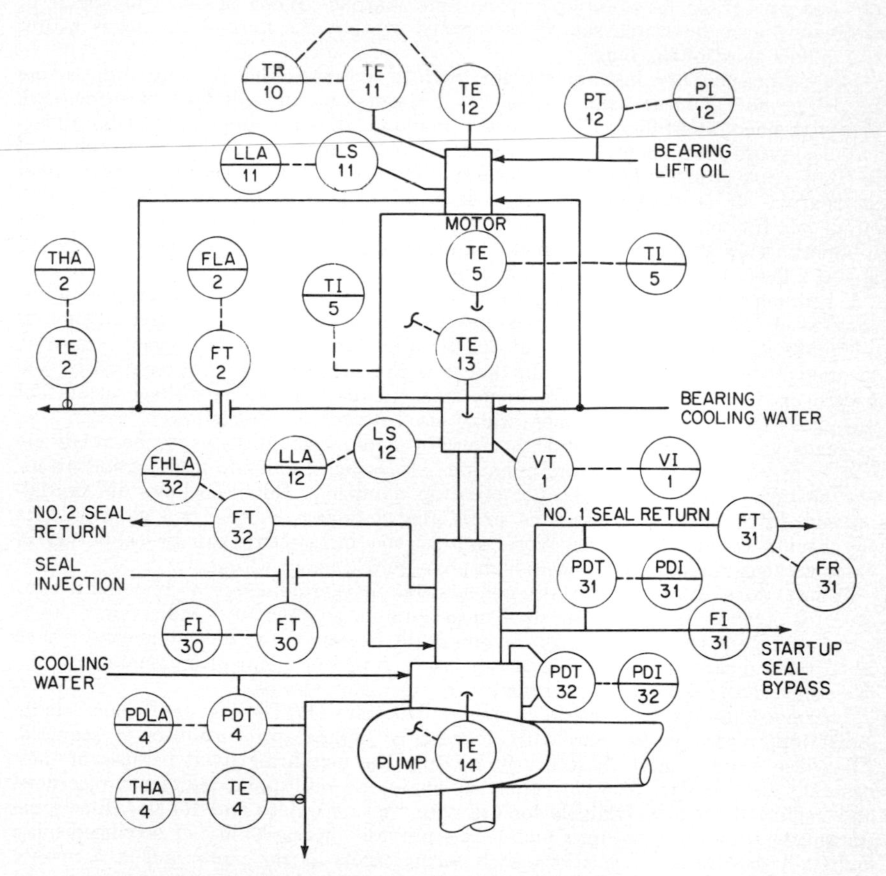

Fig. 2 Instrumentation for a reactor-coolant pump.

SUPERVISORY INSTRUMENTS SELECTION

Instruments to be used as supervisory instruments should be selected on their ability to maintain reasonable accuracy and sensitivity for extended periods of their expected operating environment. Since the purpose of most supervisory instruments is to detect changes, instruments that shift calibration or develop hysteresis quickly are of little value. This requirement does not eliminate the need for periodic calibration and preventative maintenance.

Alarm devices should be considered for operating parameters that may be subject to sudden changes indicative of impending or actual damage to the pump. Typical alarms are loss of suction head, high bearing temperature, and high vibration.

If periodic manual logging is not planned, recording or automatic logging of key variables should be considered so that normal operating values can be determined and so that changes smaller than will cause an alarm condition can be observed.

Pressure Instruments The bourdon tube gage is used commonly to indicate static pressures for pump monitoring. Metal-diaphragm measuring elements may be used for vacuum and low-pressure measurements up to about 30 lb/in^2 gage. These instruments can be equipped with low-voltage contacts to operate alarms, if desired. However, the more common practice is to use separate, nonindicating pressure switches. Pressure switches may use bourdon tubes, metal diaphragms, spring-opposed bellows, or spring-opposed pistons as the measuring elements. Mercury-in-glass and dry contact switches are available. Dry contact switches are less susceptible to vibration causing false alarms. When possible, the range of a pressure gage or switch should be selected so that the normal reading is within the middle third of the range.

Differential-pressure instruments are used to measure the pressure drop across head-type flow elements and, particularly if alarms or remote indication are needed, across strainers and filters. The mercury manometer is a simple and reliable differential-pressure instrument, but the use of mercury may be restricted or prohibited in some applications. The double-bellows differential-pressure gage is a widely used alternative. It is available as a local indicator, alarm switch, or transmitter with either electric or pneumatic output. Suitable ranges for both flow and strainer differentials are available. Pneumatic and electronic square root extractors are available if a linear output is required in flow-measuring applications.

When remote indication is required, the use of pressure transmitters should be considered, particularly if high pressures are involved. Pneumatic transmission is adequate for distances up to about 500 ft unless the transmitted signal is also to be used for a control function. In that case the speed of response required by the control must be considered. Electronic transmitters have either voltage or current outputs. The current type is more widely used for remote indication.

Gages and transmitters should be located and mounted so as to be relatively free from vibration. If pressure pulsations are present, pressure gages, transmitters, and switches should be fitted with pulsation dampers. Double-bellows differential-pressure instruments normally have an internal damping device. It is good practice to provide valving and connections so that the instruments can be calibrated in place without having to disconnect high-pressure fittings or wiring.

Temperature Instruments Armored glass-stem thermometers or bimetallic-actuated temperature indicators are used most commonly when local indication of temperature only is required. Vapor- or liquid-filled temperature indicators can be used when the indicator must be located a few feet from the point of measurement. All these devices are available with contacts for operating alarms.

Thermocouples or resistance temperature detectors (RTDs) are used when remote indication, recording, or automatic logging of temperature readings is required. They also provide more flexibility in locating the measuring point because of their generally smaller size. A wide variety of single and multipoint indicators, recorders, and logging devices is available for use with thermocouples and RTDs. Electronic transmitters for thermocouples and RTDs provide the capability of driving display devices at two or more locations, such as indicators at the pump and at a remote control point and the input to a logging device. Monitoring instruments designed for bearing-temperature monitoring are made by several manufacturers. These instruments provide continuous surveillance and alarming on high temperature. The actual temperature of a bearing is displayed on demand. Figure 3 shows a typical temperature monitor. This model monitors up to six temperatures.

Flow Measurement Head, or differential-pressure, flow elements together with suitable readout devices are the most frequently used instruments for measuring pump flows. Variable-area meters, or rotameters, may be used to advantage in measuring flows such as injection or cooling water flows. Magnetic flow meters are particularly useful when it is necessary to measure the flow of slurries, sewage, and corrosive liquids.

Square-edged orifice plates are widely used as primary flow elements. Flow nozzles, venturis, and proprietary flow tubes, although more expensive than orifice plates, have the advantage of lower unrecovered pressure losses and are less susceptible to being damaged. Special designs of pitot tubes such as the averaging impact tube can be used when very low recovery is required. Accuracy of any of these primary elements is affected by the installation. A well-developed velocity profile and freedom from swirls and vortices are required. An adequate run of straight pipe or straightening vanes should be provided. The preferred location for the primary element is in the pump suction in order to avoid swirls set up by the pump. Protruding gaskets, offset flanges, and burrs on the pressure taps must be avoided.

Magnetic flow meters can be used with conducting liquids of about 10 μmhos/cm and higher conductivity. This includes most industrial liquids except petroleum products and high-purity condensate. Magnetic flow meters require relatively short runs of straight piping for installation as these meters are relatively unaffected by swirls.

Speed Instruments The vibrating-reed tachometer is the simplest instrument for local indication of pump speed. However, it is not adaptable to remote indication or recording. Both ac and dc tachometer generators are used for remote indication and recording of speed. The ac tachometer has the advantage of not requiring

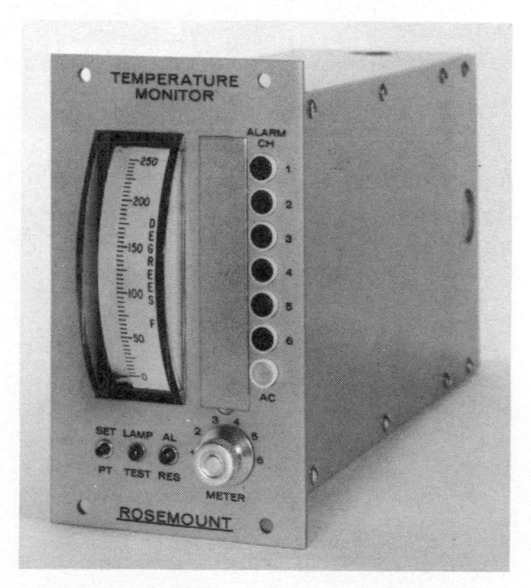

Fig. 3 Bearing temperature monitor. (Rosemount, Inc.)

brushes and a commutator. Integral gear drives provide for matching generator speed to the measured speed.

Some steam turbines for pump-drive service are furnished with electronic tachometers. A variable-reluctance pickup produces pulses at a rate proportional to turbine speed. The pulses are used to control a dc current, which drives the speed indicator or recorder.

Vibration Monitors Vibration sensors are available to measure either bearing vibration or shaft vibration directly. Direct measurement of shaft vibration is desirable for machines with stiff bearing supports where bearing-cap measurements will be only a fraction of the shaft vibration. Bearing-cap vibration pickups normally are seismic devices with an electric output proportional to the relative velocity of the mounting compared to the spring-supported weight. The monitor, or readout, usually is calibrated in terms of peak-to-peak displacement of the mounting. Shaft vibration pickups may be the seismic type with a follower riding on the shaft or the proximity type where the position of the shaft is detected electrically. Proximity-type pickups can be affected by variable magnetic properties around the circumference of a shaft such as may be caused by a varying thickness of chrome plating. Both types of pickups typically operate in the range from 600 to several thousand rpm. Low-speed applications require the use of an accelerometer-type pickup with a suitable monitor.

The monitors are available with indicators and alarm contacts or as blind instruments with alarm contacts only. Automatic gain changing features are available to provide for machines which go through a period of high vibration while coming up to speed.

Torque Measurements Although transmission dynamometers normally are used for special tests only, they can be installed as part of the pump drive for routine monitoring. Shaft-mounted strain-gage transmitters which receive their power from an external transmitter are used to avoid the problem of maintaining slip rings. Such installations are relatively expensive and, consequently, of limited application.

REFERENCES

1. Considine, D. M.: "Process Instruments and Controls Handbook," 2d ed., McGraw-Hill Book Company, New York, 1974.
2. Beckwith, T. G., and N. L. Buck: "Mechanical Measurements," Addison-Wesley Publishing Co., Inc., Reading, Mass., 1961.
3. Report of ASME Research Committee on Fluid Meters, "Fluid Meters, Their Theory and Application," 6th ed., The American Society of Mechanical Engineers, New York, 1971.
4. Standard 5.1, "Instrumentation Symbols and Identification," The Instrument Society of America, Pittsburgh, Pa., 1973.

Section **9.1**

General Characteristics of Pumping Systems and System Head Curves

J. P. MESSINA

SYSTEM CHARACTERISTICS AND PUMP HEAD

A pump is selected to deliver a specified rate of flow through a particular system. When purchasing a pump the required capacity and total head necessary to overcome the system's resistance must be specified. The total head rating of a centrifugal pump is usually measured in feet and the differential-pressure rating of a positive-displacement pump is usually measured in lb/in^2. Both express the foot-pounds of energy the pump is capable of adding to each pound of liquid pumped at the rated flow. It is the responsibility of the purchaser to calculate the resistance of the system so that the supplier can make a proper pump selection. Underestimating the total head required will result in a centrifugal pump delivering less than desired flow through the system. An underestimate of differential pressure required will result in a positive-displacement pump using more power than estimated and the design pressure limit of the pump could be exceeded. Therefore the resistance of the total system to flow which is dependent on *system characteristics* dictates the required pump head rating.

THE PUMPING SYSTEM

The piping and equipment through which the liquid flows to and from the pump comprise the pumping system. Only the length of the piping containing liquid controlled by the action of the pump is considered part of the pump system. A pump and the limit of the length of its system is shown in Fig. 1.

The pump suction and discharge piping can consist of branch lines, as shown in Fig. 2. There can be more than one pump in a pumping system. Several pumps can be piped together in series or in parallel or both, as shown in Fig. 3. When there is more than one pump, the flow through the system will be determined by the combined performance of the pumps.

The system through which the liquid is pumped offers resistance to flow for several reasons. Flow through the pipes is impeded by friction. If the liquid discharges to a higher elevation and a higher pressure, additional resistance is encountered requiring energy or pump head. The pump must therefore overcome the total sys-

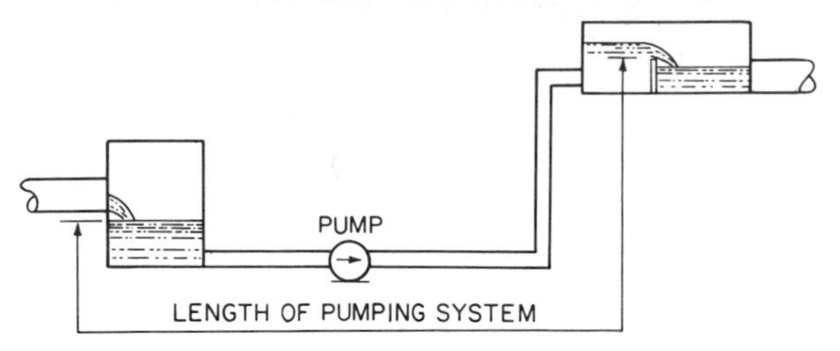

Fig. 1 Length of system controlled by pump.

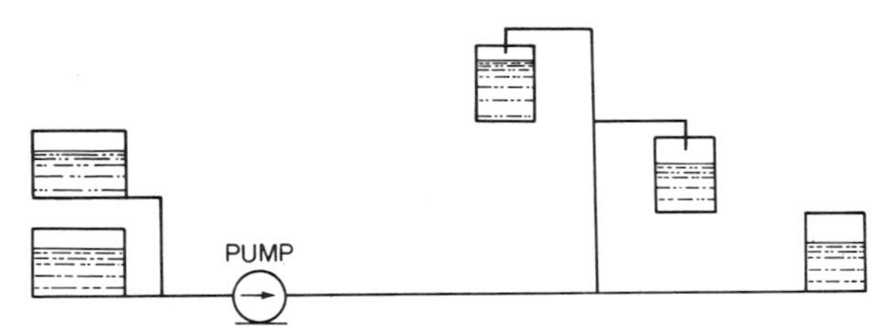

Fig. 2 Branch-line pumping system.

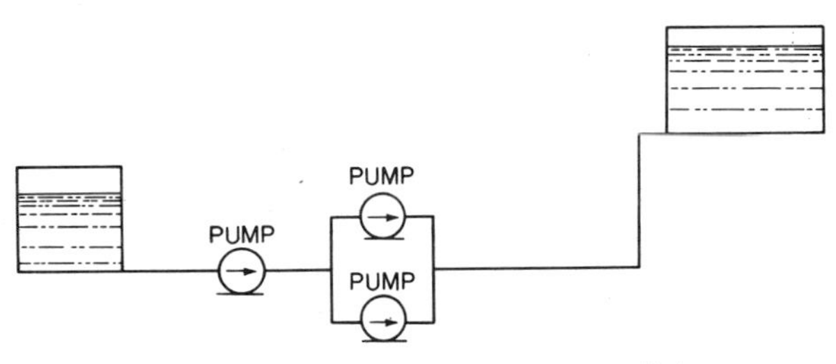

Fig. 3 Pumps in series and parallel.

tem resistance, that is, friction plus elevation and pressure heads at the desired rate of flow.

ENERGY IN AN INCOMPRESSIBLE FLUID

The total energy in feet or the differential pressure in lb/in^2 produced by a pump is a measure of the energy added to the liquid and is the energy difference between the point where it leaves the pump and the point where it enters the pump. It is also the amount of energy added to the liquid in the system. The total energy at any point in a system is a relative term and is measured above some arbitrarily selected datum plane.

An incompressible fluid has energy in the form of velocity, pressure, and elevation. Bernoulli's theorem for an incompressible fluid states that in steady flow without losses the energy at any point is the sum of the velocity head, pressure head, and elevation head, and that this sum is constant along a streamline in the conduit. Therefore the energy H in ft-lb/lb, or feet, gage or absolute, at any point in the system relative to a selected datum plane is

$$H = \frac{V^2}{2g} + \frac{144p}{w} + Z \tag{1}$$

where V = velocity, ft/s
g = acceleration of gravity, approximately 32.17 ft/s^2
p = pressure (+ or −) lb/in^2 gage or abs
w = specific weight of liquid, lb/ft^3
Z = elevation above (+) or below (−) datum, ft

Velocity Head The kinetic energy in a mass of liquid is $\frac{1}{2}mV^2$ or $\frac{1}{2}(W/g)V^2$. The kinetic energy per pound of liquid is $\frac{1}{2}WV^2/Wg$, or $V^2/2g$ measured in feet. This quantity is theoretically equal to the head of liquid required in a vessel above an opening or orifice if the discharge is to have a velocity equal to V. This is also theoretically the height a jet of liquid would rise discharging from a vertical orifice. A free-falling particle in a vacuum acquires the velocity V starting from rest after falling a distance H. Also $V = \sqrt{2gH}$. All liquid particles moving with the same velocity have the same velocity head regardless of their specific weights. The velocity of liquids in pipes and open channels invariably varies across any one section of the conduit. However, it is sufficiently accurate to use the average velocity computed by dividing the flow by the cross-sectional area of the conduit.

Pressure Head The pressure head, or flow work, in a liquid is 144 p/w, with units in feet. Liquid having pressure is capable of doing work, for example, on a piston having an area A and stroke L. The quantity of liquid required to complete one stroke is wAL. The work (force × stroke) per pound is 144 pAL/wAL, or 144 p/w. Pressure head also represents the work required per pound to give a liquid the pressure it has. The work required by a pump to produce the pressure intensity in liquids having different specific weights varies inversely with the specific weight or specific gravity of the liquid. Figure 4 illustrates liquids having specific gravities of 1.0 and 0.5. The less-dense liquid must be raised to a higher column height to produce the same pressure at the same elevation as the heavier liquid. The pressure at the bottom of each liquid column H is the weight of the liquid above the point of pressure measurement divided by the cross-sectional area A at the same point, AHw/A, which in lb/in² is simply $Hw/144$. To keep these two columns filled with equal pressures at their bases, pumps handling the same pounds per hour are required to do different amounts of work. The work required by the pump handling the lighter liquid is greater since each pound must be raised to a greater height inversely proportional to its unit weight. Note in the above that A is in ft² and L and H are in ft.

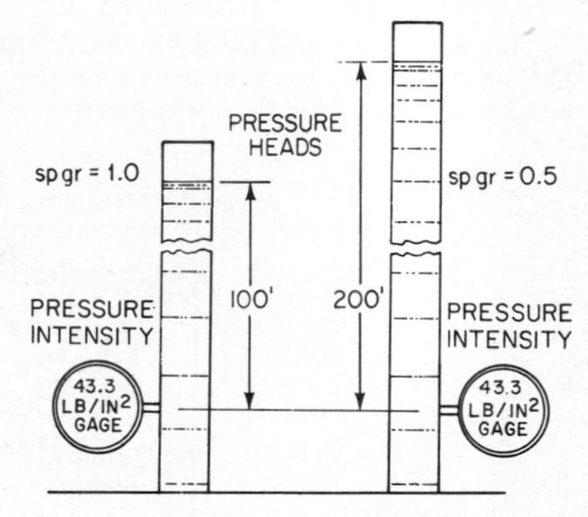

Fig. 4 Liquids of different specific weights (also specific gravities) require different column heights or *pressure heads* in ft-lb/lb to produce the same pressure intensities.

The pressure head in any liquid is also the height of the liquid column in feet above the point of pressure head measurement if the column is of constant density. Substituting pressure intensity $AHw/144A$ for p in 144 p/w, it can be seen that the pressure head H is the liquid column height. Therefore, at the base of equal columns containing different liquids (with equal surface pressures), the pressure heads in ft are the same, but the pressure intensities in lb/in² (or lb/ft²) are different. For this reason it is necessary to identify the liquids when comparing pressure *heads*.

Elevation Head The elevation energy, or potential energy, in a liquid is the distance Z ft measured vertically above or below an arbitrarily selected horizontal datum plane. Liquid above a reference datum plane has positive potential energy since it can fall a distance Z acquiring kinetic energy or vertical head equal to Z. Also, W lb of liquid requires WZ ft-lb of work to be raised above the datum plane. The work per pound is therefore WZ/W, or Z ft. In a pumping system the energy required to raise a liquid above a reference datum plane can be thought of as being provided by a pump located at the datum elevation and producing a pressure that will support the total weight of liquid in a pipe between the pump discharge and the point in the pipe where the liquid is to be raised to. This pressure is AZw/A, or simply Zw lb/ft² or $Zw/144$ lb/in². Since head is equal to pressure divided by specific weight, elevation head is Zw/w or Z ft. Liquid below the reference datum plane has negative elevation head.

Total Head Figure 5 illustrates a liquid under pressure in a pipe. To determine the total head at the pressure gage connection and relative to the datum plane, Eq. (1) may be used. Assume that the gage is at the pipe centerline.

$$\begin{aligned} H &= \frac{V^2}{2g} + \frac{144p}{w} + Z \\ &= \frac{12^2}{2 \times 32.17} + \frac{144 \times 15}{62.4} + 5 \\ &= 2.24 + 34.6 + 5 \\ &= 41.84 \text{ ft-lb/lb or ft} \end{aligned}$$

The total head may also be calculated by the addition of

$$\begin{aligned} \frac{V^2}{2g} + \text{manometer height} + Z &= 2.24 + 34.6 + 5 \\ &= 41.84 \text{ ft-lb/lb or ft} \end{aligned}$$

The total head of 41.84 ft is equivalent to 1 lb of the liquid raised 41.48 ft above the datum plane (zero velocity) or the pressure head of 1 lb of the liquid under a column height of 41.84 ft measured at the datum plane.

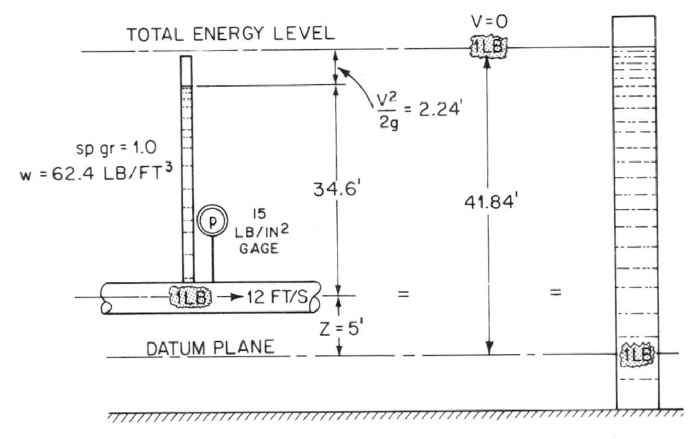

Fig. 5 The total head, or energy in ft-lb/lb, is equal to the sum of the velocity, pressure, and elevation heads relative to a datum plane. A pound of the same liquid or any liquid raised to rest at the height shown, or under a column of liquid of this same height, has the same head as the pound of liquid shown flowing in the pipe.

The gage pressure p, and consequently the pressure head, may be measured relative to atmospheric pressure. Gage pressure head can therefore be a positive or negative quantity. The pressure p may also be expressed as an absolute pressure (measured relative to a vacumn). Therefcre, when combining velocity, pressure, and elevation heads to obtain the total energy at a point, it should be clearly stated that the total head is either feet *gage* or feet *absolute,* with respect to the datum plane.

The pressure and/or velocity of a liquid may at times be expressed as a pressure head of some other liquid. In the total head, that is, the sum of the pressure, velocity, and elevation heads, the head components must be corrected to be the head of the liquid being pumped. For example, if the pressure of 60°F water at some point in a system is measured by a manometer to be 24 in of mercury absolute, the pressure head, or energy, in ft-lb/lb of water pumped is found as follows: let subscripts 1 and 2 denote different liquids, or in the above example mercury and

water, respectively:

$$h_1 = \frac{144p_1}{w_1}$$

$$p_1 = \frac{h_1 w_1}{144} = p_2$$

$$h_2 = \frac{144p_1}{w_2} = \frac{144}{w_2} \times \frac{h_1 w_1}{144}$$

$$= \frac{w_1}{w_2} h_1 \tag{2}$$

or

$$= \frac{\text{sp gr}_1}{\text{sp gr}_2} h_1 \tag{3}$$

Therefore

$$h_2 = \frac{13.6}{1} \times \frac{24}{12} = 27.2 \text{ ft abs.}$$

Also, $h_2 = 27.2 - 34 = -6.8$ ft gage if corrected to a standard barometer of 30 in of mercury ($13.6 \times 30/12 = 34$ ft of water).

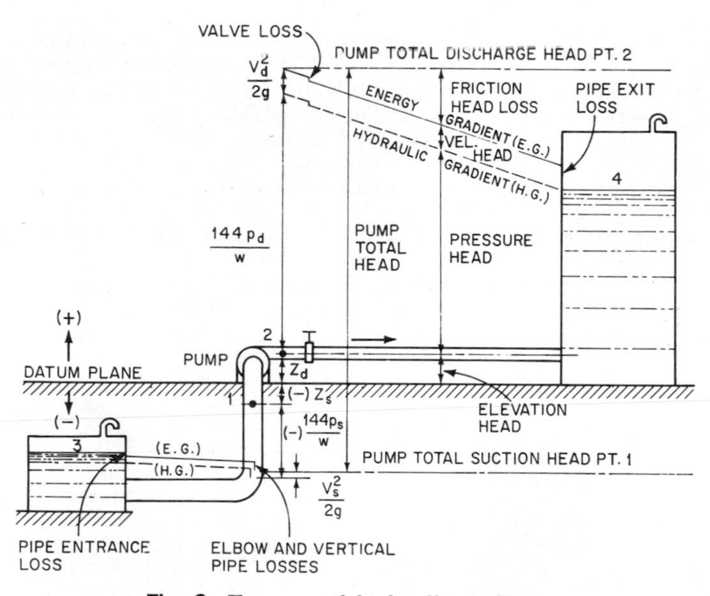

Fig. 6 Energy and hydraulic gradients.

PUMP TOTAL HEAD

The total head of a pump is the difference in energy at the pump discharge (point 2) and the pump suction (point 1) as shown in Fig. 6. Applying Bernoulli's theorem at each point, Eq. (1), the pump total head TH in feet becomes

$$TH = H_d - H_s = \left(\frac{V_d^2}{2g} + \frac{144p_d}{w_d} + Z_d\right) - \left(\frac{V_s^2}{2g} + \frac{144p_s}{w_s} + Z_s\right) \tag{4}$$

The equation for pump differential pressure P_Δ in lb/in² is

$$P_\Delta = P_d - P_s$$

$$= \left[p_d + 0.433 \text{ sp gr}_d \left(Z_d + \frac{V_d^2}{2g}\right)\right] - \left[p_s + 0.433 \text{ sp gr}_s \left(Z_s + \frac{V_s^2}{2g}\right)\right] \tag{5}$$

where for Eqs. (4) and (5) subscripts d and s denote discharge and suction, respectively, and

H = total head, (+ or −) ft gage (or ft abs)
P = total pressure, (+ or −) lb/in² gage (or lb/in² abs)
V = velocity, ft/s
p = pressure, (+ or −) lb/in² gage (or lb/in² abs)
Z = elevation above (+) or below (−) datum, ft
w = specific weight of liquid, lb/ft³
sp gr = specific gravity of liquid
g = acceleration of gravity, approx 32.17 ft/s²

Pump total head TH in ft and pump differential pressure P_Δ in lb/in² are always absolute quantities since either gage pressures or absolute pressures but not both are used at the discharge and suction connections of the pump, and a common datum plane is selected.

Pump total head in ft and pump differential pressure in lb/in² are related to each other as follows:

$$TH = \frac{144\,P_\Delta}{w} \tag{6}$$

It is very important to note that if the rated total head of a centrifugal pump is given in *feet,* this head can be imparted to *all* liquids pumped at the rated capacity and speed regardless of the specific weight of the liquid but having approximately the same viscosity. A pump handling different liquids of approximately the same viscosity will generate the same total head in ft, but will not produce the same differential pressure in lb/in², nor will the power required to drive the pump be the same. On the other hand a centrifugal pump rated in lb/in² would have to have a different pressure rating for each liquid of different specific weight. In this section, pump total head will be expressed in ft, the usual way of rating centrifugal pumps.

Pump total head can be measured by installing gages at the pump suction and discharge connections and then substituting these gage readings into Eq. (4). Pump total head may also be found by measuring the energy difference between any two points in the pumping system, one on each side of the pump, providing all losses between these points are credited to the pump and added to the energy-head difference. Therefore, between any two points in a pumping system where the energy is added only by the pump and the specific weight of the liquid does not change (for example, as a result of temperature), the following general equation for determining pump total head applies:

$$\begin{aligned} TH &= (H_2 - H_1) + \Sigma h_{f(1-2)} \\ &= \left(\frac{V_2^2}{2g} + \frac{144p_2}{w} + Z_2\right) - \left(\frac{V_1^2}{2g} + \frac{144p_1}{w} + Z_1\right) + \Sigma h_{f(1-2)} \end{aligned} \tag{7}$$

where subscripts 1 and 2 denote points in the pumping system anyplace upstream and downstream from the pump, respectively, and

H = total head, (+ or −) ft gage (or ft abs)
V = velocity, ft/s
p = pressure, (+ or −) lb/in² gage (or lb/in² abs)
Z = elevation above (+) or below (−) datum, ft
w = specific weight of liquid (assumed the same between points), lb/ft³
g = acceleration of gravity, approx. 32.17 ft/s²
Σh_f = sum of the losses between points, ft

When the specific gravity of the liquid is known, the pressure head in ft may be calculated from the following relationship:

$$\frac{144p}{w} = \frac{2.31p}{\text{sp gr}} \tag{8}$$

The velocity in ft/s in a pipe may be calculated as follows:

$$V = \frac{\text{gpm} \times 0.408}{(\text{pipe ID}'')^2} \tag{9}$$

EXAMPLE 1: Figure 7 illustrates the use of Eqs. (4) and (7) for determining pump total head.

ENERGY AND HYDRAULIC GRADIENTS

The total energy at any point in a pumping system may be calculated for a particular rate of flow using Bernoulli's equation (1). If some convenient datum plane is selected and the total energy or head at various locations along the system is plotted to scale, the resultant line drawn through these points is called the *energy gradient*. Figure 6 shows the variation in total energy H measured in ft from the suction liquid surface point 3 to the discharge liquid surface point 4. A horizontal energy gradient indicates no loss of head.

The line drawn through the sum of the pressure and elevation heads at various points represents the pressure variation in flow measured above the datum plane. It also represents the height the liquid would rise in vertical columns relative to the

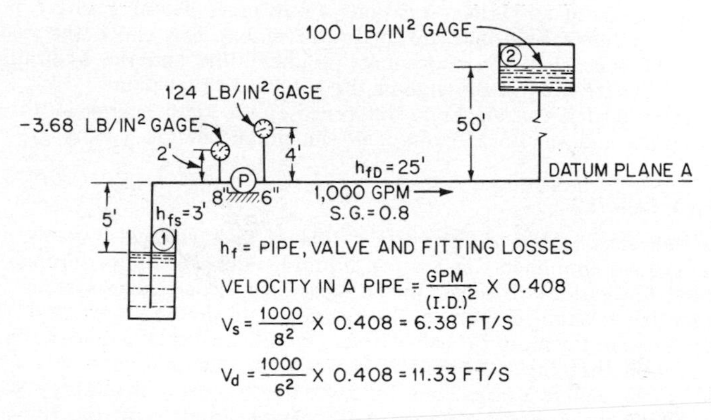

A. PUMP TOTAL HEAD BY MEASUREMENT USING GAGES AT THE PUMP AND DATUM PLANE A

FROM EQ. (4): $TH = \left(\frac{V_d^2}{2g} + \frac{144\,p_d}{w} + Z_d\right) - \left(\frac{V_s^2}{2g} + \frac{144\,p_s}{w} + Z_s\right)$

AND $\frac{144\,p}{w} = \frac{2.31\,p}{sp\,gr_1}$

$TH = \left(\frac{11.33^2}{2 \times 32.2} + \frac{2.31 \times 124}{0.8} + 4\right) - \left(\frac{6.38^2}{2 \times 32.2} - \frac{2.31(-3.68)}{0.8} + 2\right)$

$= (364) - (-8) = 372\ \frac{FT-LB}{LB}$ OR FT

B. PUMP TOTAL HEAD BY CALCULATION USING PRESSURES AT POINTS ① AND ②, ADDING FRICTION LOSSES AND USING DATUM PLANE A.

FROM EQ. (7): $TH = \left(\frac{V_2^2}{2g} + \frac{144\,p_2}{w} + Z_2\right) - \left(\frac{V_1^2}{2g} + \frac{144\,p_1}{w} + Z_1\right) + \Sigma h_{f(1-2)}$

AND $\frac{144\,p}{w} = \frac{2.31\,p}{sp\,gr}$

$TH = \left(0 + \frac{2.31 \times 100}{0.8} + 50\right) - (0 + 0 - 5) + (3 + 25)$

$= (339) - (-5) + (28) = 372\ \frac{FT-LB}{LB}$ OR FT

Fig. 7a Example 1. Pump total head by measurement and calculation.

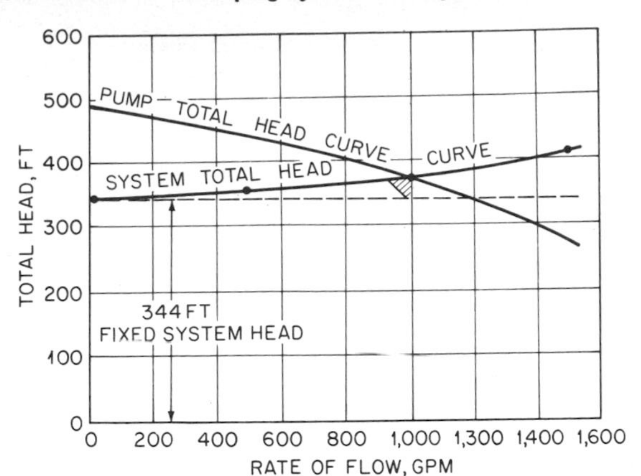

Fig. 7b System total head curve for Example 1 shown in Fig. 7*a*.

datum plane when placed at various locations along pipes having positive pressure anywhere in the system. This line shown dotted in Fig. 6 is called the *hydraulic gradient*. The difference between the energy gradient line and the hydraulic gradient line is the velocity head in the pipe or the conduit at that point.

The pump total head is the algebraic difference of the total energy at the pump discharge (point 2) and the total energy at the pump suction (point 1).

SYSTEM HEAD CURVES

A pumping system may consist of pipe, valves, fittings, open channels, vessels, nozzles, weirs, meters, process equipment, and other liquid-handling conduits through which flow is required for various reasons. When analyzing a particular system for the purpose of selecting a pump or pumps, the resistance to flow of the liquid through these various components must be calculated. It will be explained in more detail later in this section that the resistance increases with flow and at a rate approximately equal to the square of the flow through the system. In addition to overcoming flow resistance, it may be necessary to add head to raise the liquid from suction level to a higher discharge level. In some systems the pressure at the discharge liquid surface may be higher than the pressure at the suction-liquid surface requiring more pumping head. The latter two heads are fixed system heads as they do not vary with rate of flow. Fixed system heads can also be negative, as would be the case if the discharge-level elevation or the pressure above that level were lower than suction elevation or pressure. Fixed system heads are also called *static heads*.

A system head curve is a plot of total system resistance, variable plus fixed, for various flow rates, and it has many uses in centrifugal pump application. It is preferable to express system head in ft rather than lb/in² since centrifugal pumps are rated in ft as previously explained. System head curves usually show flow in gallons per minute (gpm), but when large quantities are involved, cubic feet per second (ft³/s) or million gallons per day (mgd) units are used.

When the system head is required for several flows, or when the pump flow is to be determined, a system head curve is constructed using the following procedure. Define the pumping system and its length. Calculate (or measure) the fixed system head, which is the net change in total energy from the beginning to the end of system, due to elevation and/or pressure-head difference. An increase in head in the direction of flow is a positive quantity. Next calculate the variable system total head loss through all pipe, valves, fittings, and equipment in the system for several rates of flow. As an example, see Fig. 8. The pumping system is defined as starting from point 1 and ending at point 2. The fixed system head is the net change in total energy. The total head at point 1 is $144\ p_s/w$, and at point 2 it is $144\ p_d/w + Z$. The pressure and liquid levels do not vary with flow. The variable system head is pipe friction (including valves and fittings). The fixed head and variable heads for several flow rates are added together, resulting in a total system head curve vs. flow.

The flow produced by a centrifugal pump varies with the system head, while the flow of a positive-displacement pump is independent of the system head. By superimposing the head-capacity characteristic curve of a centrifugal pump on a system head curve, as shown in Fig. 8, the unknown flow of a pump can be determined. The curves will intersect at the flow rate of the pump as this is the point at which the pump head is equal to the required system head for the same flow. When a pump is being purchased it should be specified that the pump head-capacity curve intersect the system head curve at the desired flow rate. This intersection should be the best efficiency capacity or very close to it.

The system head curve for Example 1 is shown in Fig. 7*b*. This example assumes that the suction- and discharge-liquid levels are at elevations 5 ft below and 50

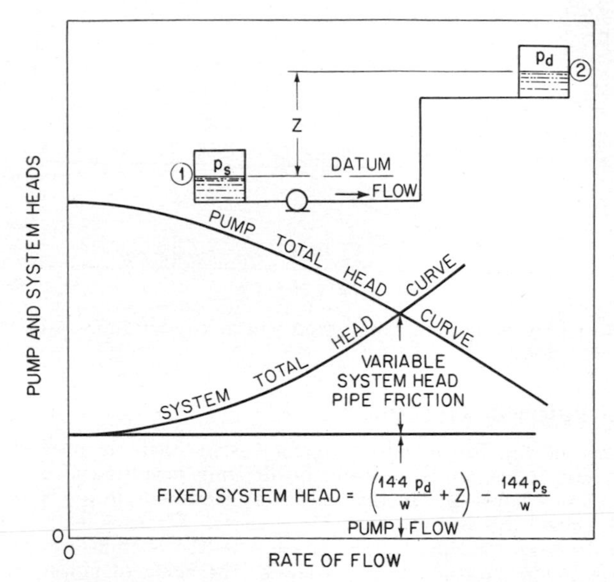

Fig. 8 Construction of system total head curve.

ft above the datum plane, respectively, and do not vary with flow. The pressure in the discharge tank is also independent of flow and is 100 lb/in² gage. These values are therefore fixed system heads. The pipe and fitting losses are assumed to vary with flow as a square function. The length of the pumping system is from point 1 to point 2. The difference in heads at these points plus the friction losses at various flow rates is the total system head and the head required by a pump for the different flows. It is only necessary to calculate the total system head for one flow rate—say, design—which in this example is 1,000 gpm. The total head at other flow conditions is the fixed system head plus the variable system head multiplied by $(\text{gpm}/1{,}000)^2$. If Example 1 is assumed to be an existing system, the total head may be calculated by using gages at the pump suction and discharge connections. The total head measured will then be the head at the intersection at the pump and system curves. as shown in Fig. 7*b*. In this example, a correctly purchased pump would produce a total head of 372 ft at the design flow of 1,000 gpm.

In systems that are open-ended and in which there is a decrease in elevation from inlet to outlet, a portion of the system head curve will be negative (Fig. 9). In this example the pump is used to increase gravity flow. Without a pump in the system, the negative resistance or static head curve is the driving head which moves the liquid through the system. Steady-state gravity flow is sustained at the flow rate corresponding to zero total system head (negative static head plus system resistance equals zero). If a flow is required at any rate greater than that which gravity can produce, a pump is required to overcome the additional system resistance.

For additional information concerning the construction of system head curves for flow in branch lines refer to Sec. 9.2, Branch-Line Pumping Systems.

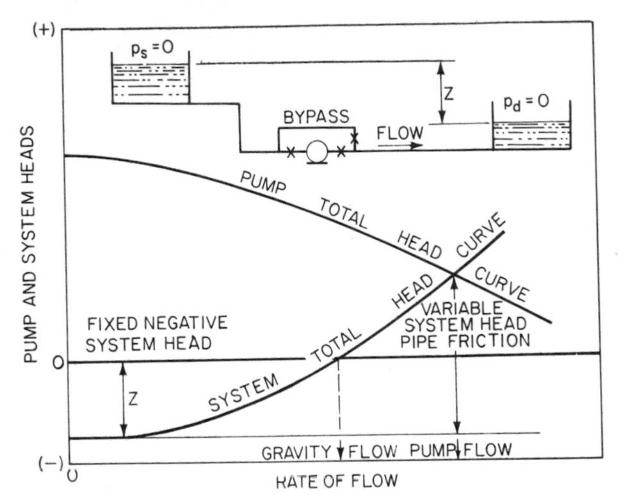

Fig. 9 Construction of system total head curve to determine gravity flow and centrifugal pump flow.

VARIANTS IN PUMPING SYSTEMS

For a fixed set of conditions in a pumping system there is just one total head for each flow rate. Consequently a centrifugal pump operating at a constant speed can deliver just one flow. In practice, however, conditions in a system vary either as a result of controllable or uncontrollable changes. Changes in the valve opening in the pump discharge or bypass line; changes in the suction- or discharge-liquid levels; changes in the pressures at these levels; the aging of pipes; changes in the size, length, or number of pipes; changes in the process; changes in the number of pumps pumping into a common header are all examples of either controllable or uncontrollable system changes. These changes in system conditions alter the shape of the system head curve and, in turn, affect the pump flow.

Methods of constructing system-head curves and determining the resultant pump flows for two of the more common of these variants in pumping systems are explained below.

Variable Static Head In a system where a pump is taking suction from one reservoir and filling another, the capacity of a centrifugal pump will decrease with an increase in static head. The system-head curve is constructed by plotting the variable system friction head vs. flow curve for the piping. To this is added the anticipated minimum and maximum static heads (difference in discharge and suction levels). The resulting two curves are the total system heads for each condition. The flow rate of the pump is the point of intersection of the pump head-capacity curve with any one of the latter two system head curves, or with any intermediate system head curve for other level conditions. A typical head vs. flow curve for a varying static-head system is shown in Fig. 10.

If it is desired to maintain a constant pump flow for different static-head conditions, the pump speed can be varied to adjust for an increase or decrease in the total system head. A typical variable-speed centrifugal pump operating in a varying static-head system can have a constant flow as shown in Fig. 11.

It is important to select a pump that will have its best efficiency within the operating range of the system and preferably at the condition the pump will operate most often.

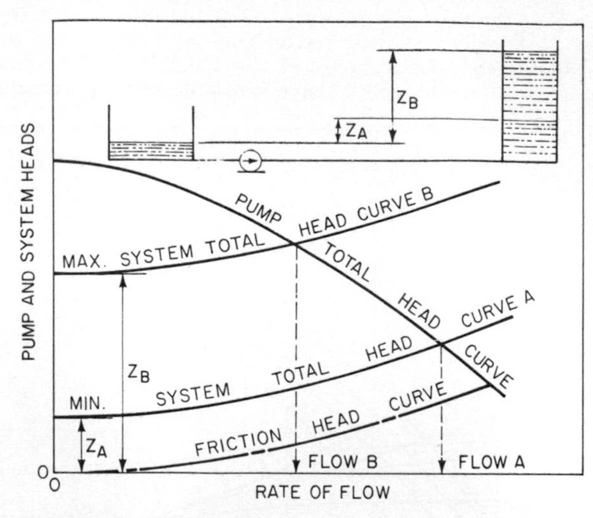

Fig. 10 Construction of system total head curves for a varying-static-head pumping system.

Variable System Resistance A valve or valves in the discharge line of a centrifugal pump alters the variable friction-head portion of the total system head curve and consequently the pump flow. Figure 12, for example, illustrates the use of a discharge valve to change the system head for the purpose of varying the pump

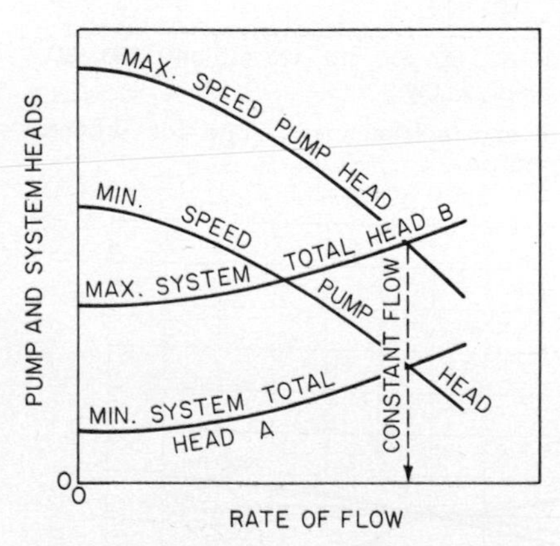

Fig. 11 Varying centrifugal pump speed to maintain constant flow for different reservoir levels shown in Fig. 10.

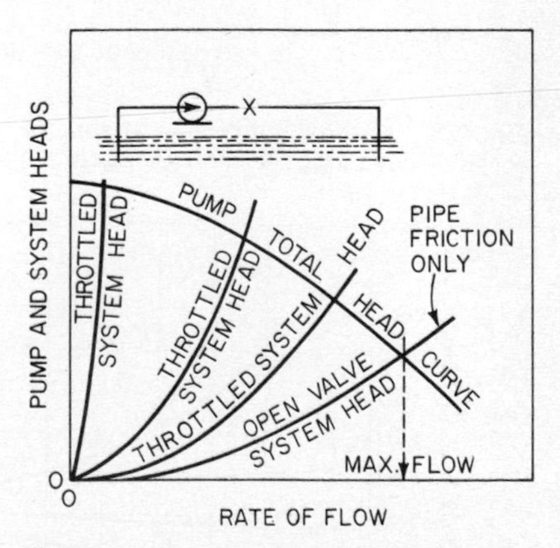

Fig. 12 Construction of system total head curves for various valve openings.

flow during a shop performance test. The maximum flow is obtained with a completely open valve and the only resistance to flow is the friction in the pipe, fittings, and flow meter. A closed valve results in the pump operating at shutoff conditions and produces maximum head. Any flow between maximum and shutoff can be obtained by proper adjustment of the valve opening.

TRANSIENTS IN SYSTEM HEADS

During the starting of a centrifugal pump and prior to the time normal flow is reached, certain transient conditions can produce or require heads and, consequently, torques much higher than design. In some cases the selection of the driver and the pump must be based on starting rather than on normal flow conditions.

Low- and medium-specific-speed pumps of the radial and mixed-flow types (less than approximately 5,000 specific speed) have favorable starting characteristics. The

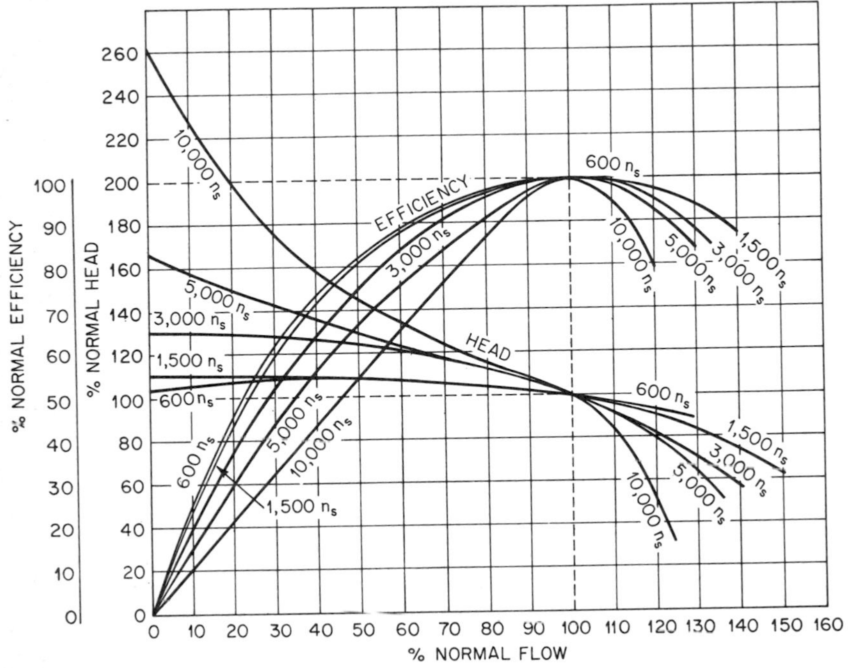

Fig. 13 Approximate comparison of head and efficiency vs. flow for different specific-speed impellers, single-stage volute pumps.

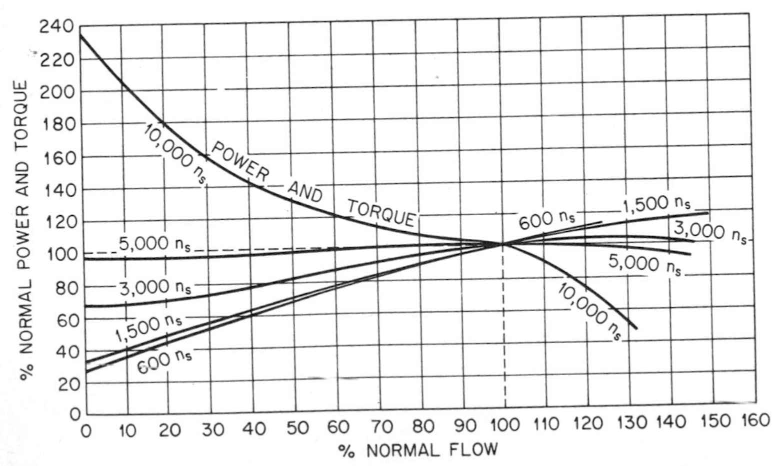

Fig. 14 Approximate comparison of power and torque vs. flow for different specific-speed impellers, single-stage volute pumps.

pump head at shutoff is not significantly higher than normal flow. The shutoff torque is less than at normal flow. High-specific-speed pumps of the mixed- and axial-flow types (greater than approximately 5,000 specific speed) develop relatively high shutoff heads. The shutoff torque is greater than at normal flow. These characteristics of high-specific-speed pumps require special attention during the starting period. Characteristics of different specific-speed pumps are shown in Figs. 13 and 14.

Starting Against a Closed Valve When any centrifugal pump is started against a closed discharge valve the pump head will be higher than normal. The shutoff head will vary with pump specific speed. As shown in Fig. 13, the higher the specific speed, the higher will be the shutoff head in percent of normal pump head. As a pump is accelerated from rest to full speed against a closed valve, the head on the pump at any speed is equal to the square of the ratio of the speed to the full speed times the shutoff head at full speed. Therefore during starting the head will vary from A to E, as shown in Fig. 15. Points B, C, and D represent intermediate

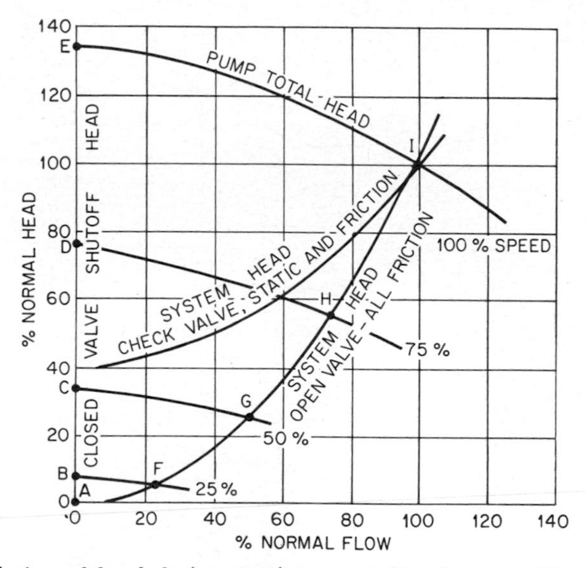

Fig. 15 Variation of head during starting a centrifugal pump with a closed valve, an open valve, and a check valve.

heads at intermediate speeds. The pump, the discharge valve, and any intermediate piping must be designed for the maximum head at point E.

The shutoff power and torque increase with specific speed as shown in Fig. 14. Pumps requiring less shutoff power and torque than at normal flow condition are usually started against a closed discharge valve. To prevent backflow from a static discharge head prior to starting, either a discharge shutoff valve, a check valve, or a broken siphon is required. When pumps are operated in parallel and are connected to a common discharge header that would permit flow from an operating pump to circulate back through an idle pump, a discharge valve or check valve must be used.

Figure 16 is a typical characteristic curve for a low-specific-speed pump. Figure 17 illustrates the variation of torque with pump speed when started against a closed discharge valve. The torque under shutoff conditions varies as the square of the ratio of speeds similar to the variation in shutoff head and is shown as curve ABC. At zero speed, the pump torque is not zero as a result of static friction in the pump bearings and stuffing box or boxes. This static friction is greater than the sum of running friction and horsepower input to the impeller at very low speeds, which explains the dip in the pump-torque curve between 0 and 10 percent speed.

Also shown in Fig. 17 is the speed-torque curve of a typical squirrel-cage induction motor. Note that the difference between motor and pump torque is the excess torque available to accelerate the pump from rest to full speed. During acceleration the pump shaft must not only be designed to transmit the pump torque, curve

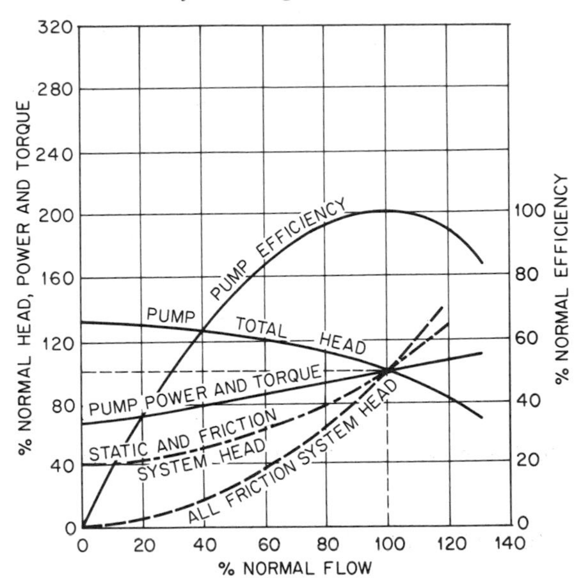

Fig. 16 Typical constant-speed characteristic curves for a low-specific-speed pump.

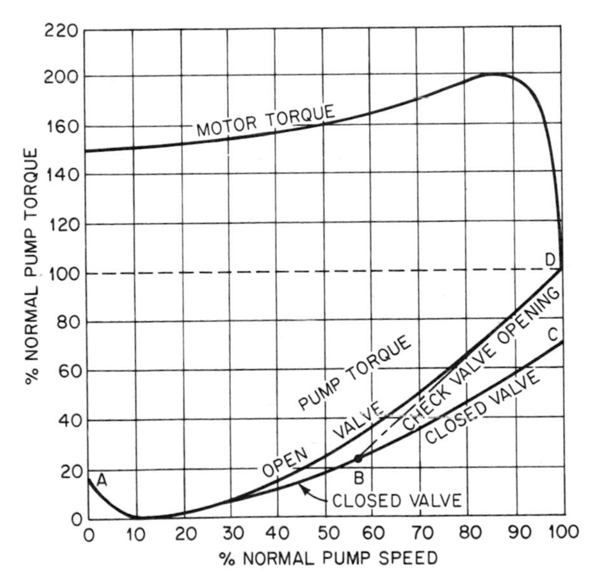

Fig. 17 Variation of torque during starting a low-specific-speed pump with a closed valve, an open valve, and a check valve. See Fig. 16 for pump characteristics.

ABC, but also the excess torque available in the motor. Therefore pump-shaft torque follows the motor speed-torque curve less the torque required to accelerate the mass inertia (WK^2) of the motor's rotor.

High-specific-speed pumps, especially propeller pumps, requiring more than normal torque at shutoff are not normally started with a closed discharge valve because larger and more expensive drivers would be required. These pumps will also produce relatively high pressures in the pump and in the system between pump and discharge valve. Figure 18 is a typical characteristic curve for a high-specific-speed pump.

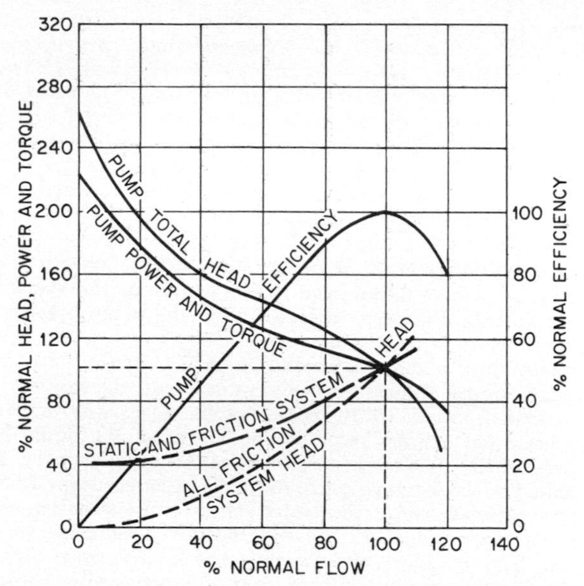

Fig. 18 Typical constant-speed characteristic curves for a high-specific-speed pump.

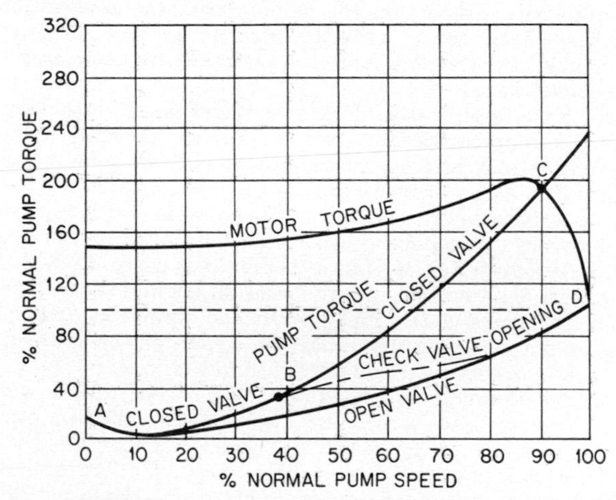

Fig. 19 Variation of torque during starting a high-specific-speed pump with a closed valve, an open valve, and a check valve. See Fig. 18 for pump characteristics.

Figure 19, curve *ABC*, illustrates the variation of torque with speed for this pump when started against a closed discharge valve. A typical speed-torque curve of a squirrel-cage induction motor sized for normal pump torque is also shown. Note that the motor has insufficient torque to accelerate to full-rated speed, which would remain overloaded at point C until the discharge valve on the pump was opened. To avoid this situation, the discharge valve should be timed to open sufficiently so as not to overload the motor when the pump reaches full speed. To accomplish this timing it may be necessary to start opening the valve in advance of energizing the motor. Care should be taken not to start opening the discharge valve too soon which would then cause excessive reverse flow through the pump and require the motor to start under adverse reverse-speed conditions. If the driver is a synchronous motor, in addition to requiring motor torque to overcome system head and accelerate the pump and driver rotors from rest, additional torque at pull-in speed is required. At the critical pull-in point sufficient torque must be available to pull the load into synchronism in the prescribed time. A synchronous motor is started on low-torque squirrel-cage windings prior to excitation of the field windings at the pull-in speed. The low starting torque, the torque required at pull-in, and possible voltage drop which will lower the motor torque (varies as the square of the voltage) must all be taken into consideration when selecting a synchronous motor to start a high-specific-speed pump against a closed valve.

If a high-specific-speed pump is to be started against a closed discharge valve, high starting torques can also be avoided by the use of a bypass valve (see Sec. 9.2) or by an adjustable-blade pump.[1]

Starting Against a Check Valve A check valve can be used to prevent reverse flow from static head and/or head from other pumps in the system. The check valve will open automatically when the head from the pump exceeds system head. When a centrifugal pump is started against a check valve, pump head and torque follow shutoff values until a speed is reached at which shutoff head exceeds system head. As the valve opens, the pump head continues to increase and, at any flow, the head will be that necessary to overcome system static head or head from other pumps, friction head, valve head loss, and the inertia of the liquid being pumped.

Figure 17, curve *ABD*, illustrates speed-torque variation starting a low-specific-speed pump against a check valve with static head and system friction as shown in Fig. 16. Figure 19, curve *ABD*, illustrates speed-torque variation starting a high-specific-speed pump against a check valve with static head and system friction as shown in Fig. 18. The use of a quick-opening check valve with high-specific-speed pumps eliminates starting against higher than full-open-valve shutoff heads and torques.

The speed-torque curves shown for the period during the acceleration of the liquid in the system have been drawn with the assumption that the head required to accelerate the liquid and overcome inertia is insignificant. Acceleration head is discussed in more detail later.

Starting Against an Open Valve If a centrifugal pump is to take suction from a reservoir and discharge to another reservoir having the same liquid elevation or the same equivalent total pressure, it can be started without a shutoff discharge valve or check valve. The system-head curve is essentially all friction plus the head required to accelerate the liquid in the system during the starting period. Neglecting liquid inertia, the pump head would not be greater than normal at any speed during the starting period, as shown in Fig. 15 as curve *AFGHI*. Pump torque would not be greater than normal at any speed during the starting period as shown in Figs. 17 and 19, curves *A-D*. Pump head and torque at any speed are equal to their values at normal condition times the square of the ratio of the speed to full speed, while the capacity varies directly with this ratio.

Starting a Pump Running in Reverse When a centrifugal pump discharges against a static head or into a common discharge header with other pumps and is then stopped, unless the discharge valve is closed or unless there is a check valve in the system or a broken siphon in a siphon system, the flow will reverse through the pump. If the pump does not have a nonreversing device, it will turn in the reverse direction. A pump that discharges against a static head through a siphon system without a valve will have a reverse flow and speed during the priming of the siphon prior to starting.

Figures 20 and 21 illustrate typical reverse-speed-torque characteristics for a low- and a high-specific-speed pump. When flow reverses through a pump and the driver offers very little or no torque resistance, the pump will reach higher than normal forward speed in the reverse direction. This runaway speed will increase with

[1] A. J. Stepanoff, "Centrifugal and Axial Flow Pumps," 2d ed., p. 366, John Wiley & Sons, Inc., New York, 1948.

specific speed and system head. Shown in Figs. 20 and 21 are speed, torque, head, and flow all expressed as a percentage of the pump design conditions for the normal forward speed. When a pump is running in reverse as a turbine under no load, the head on the pump will be a static head (or head from other pumps) minus head loss as a result of friction for the reverse flow rate.

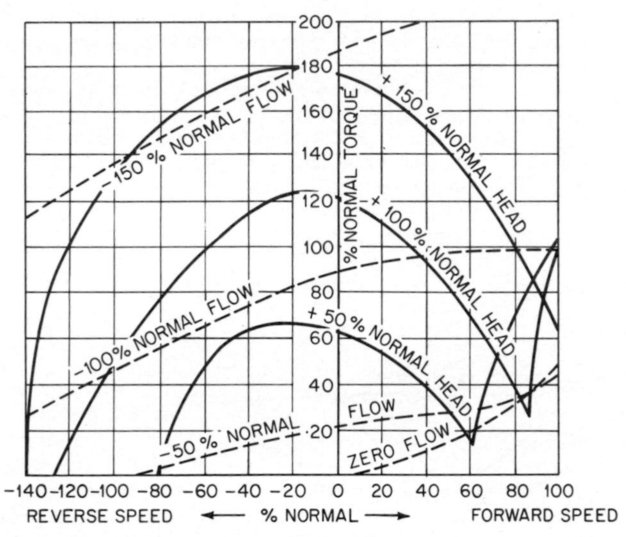

Fig. 20 Typical reverse speed-torque characteristics of a low-specific-speed radial-flow, double-suction pump.

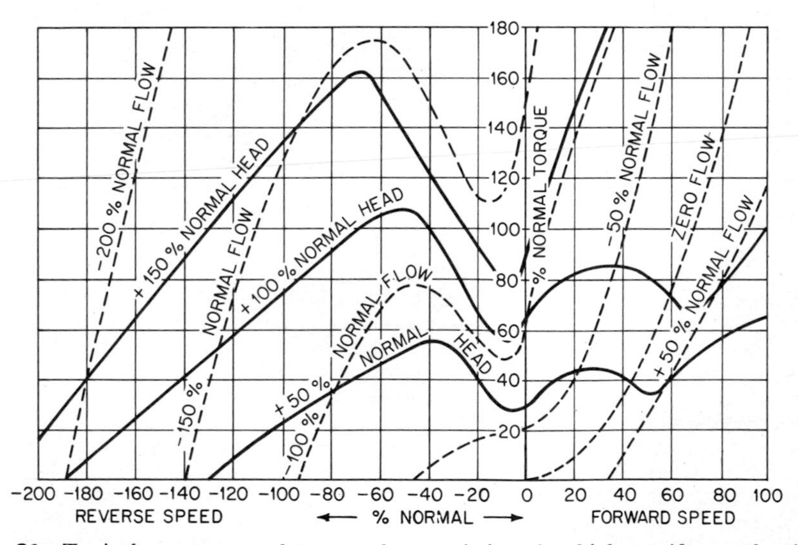

Fig. 21 Typical reverse speed-torque characteristics of a high-specific-speed axial-flow diffuser pump.

If an attempt is made to start the pump while running in reverse, an electric motor must apply positive torque to the pump while initially running in a negative direction. Figures 20 and 21 show, for the two different specific-speed pumps, the torques required to decelerate, momentarily stop, and then to accelerate the pump to normal speed. If either of these pumps were pumping into an all-static-head system, starting these pumps in reverse would require overcoming 100 percent normal head, and it can be seen that a torque in excess of normal would be required by the driver while running in reverse. In addition to overcoming positive head, the driver must add additional torque to the pump to change the direction of the liquid flowing initially in reverse. This could result in a prolonged starting time under higher than normal current demand. Characteristics of the motor, pump, and system must be analyzed together to determine actual operating conditions during this transient period. Starting torques requiring running in reverse become less severe when the system head is partly or all friction head.

Inertia Head If the system contains an appreciable amount of liquid, the inertia of the liquid mass could offer a significant resistance to any sudden change in velocity. Upon starting a primed pump and system without a valve, all the liquid in the system accelerates from rest to a final condition of steady flow. Figure 22 illustrates a typical system-head resistance that could be produced by a propeller

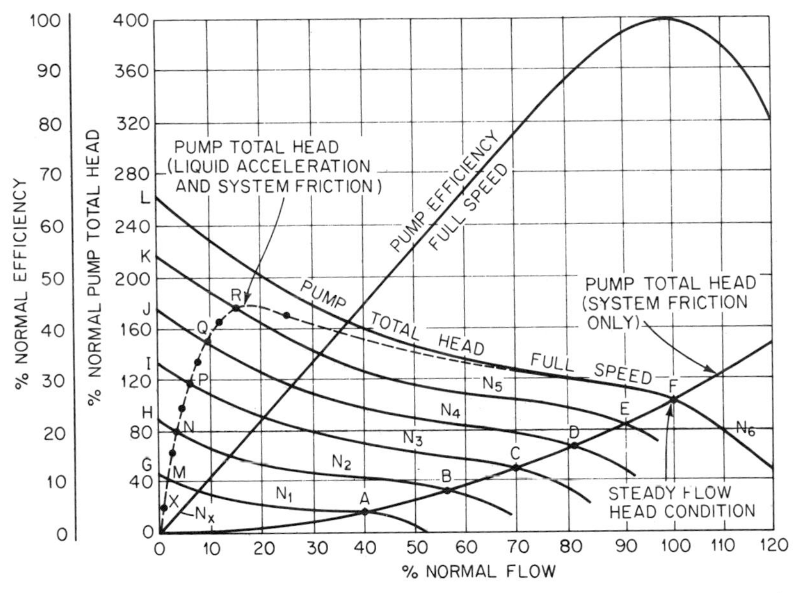

Fig. 22 Transient system head as a result of liquid acceleration and system friction when starting a propeller pump with an open valve.

pump pumping through a friction system when accelerated from rest to full speed. If the pump were accelerated very slowly it would produce an all-friction resistance with zero liquid acceleration varying with flow approaching curve *OABCDEF*. Individual points on this curve represent system resistance at various constant pump speeds. If the pump were accelerated very rapidly it would produce a system resistance approaching curve *OGHIJKL*. This is a shutoff condition which cannot be realized unless infinite driver torque was available. Individual points on this curve represent a system resistance at various constant speeds with no flow, which is the same as operating with a closed discharge valve. Curve *LF* represents the maximum total head the pump can produce as a result of both system friction and inertia at 100 percent speed. Head variation from *L* to *F* is a result of flow through the system increasing at a decreasing rate of acceleration and increasing friction to the normal operating point *F* where all the head is friction. The actual total system-head resistance curve for any flow condition will be, therefore, the sum of the frictional resistance in feet for that steady-flow rate, plus the inertial resistance also expressed in feet. The inertial resistance at any flow is dependent on the mass of liquid and the instantaneous rate of change of velocity at the flow condition. A typical total system resistance for a motor-driven propeller pump is shown as

OMNPQRF. For a particular pump and system, different driver-speed-torque characteristics will result in a family of curves in area *OLF*.

The added inertial system head produced momentarily on starting high-specific-speed pumps is important when considering the duration of high driver torques and currents, the pressure rise in the system, and the effect on the pump from operation at high heads and low flows. The acceleration head in feet required to change the velocity of a mass of liquid at a uniform rate and cross section is

$$h_a = \frac{L\,\Delta V}{g\,\Delta t} \tag{10}$$

where L = length of constant cross section conduit, ft
ΔV = velocity change, ft/s
Δt = time interval, s
g = acceleration of gravity, approx 32.17 ft/s^2

To calculate the time to accelerate a centrifugal pump from rest, or from some initial speed, to a final speed, and to estimate the pump head variation during this interim, a trial and error solution may be used. Divide the speed change into several increments of equal no flow heads such as *OGHIKL* as shown in Fig. 22. For the first incremental speed change O to N_1 estimate total system-head point M between G and A. Next estimate the total system head point X at the average speed for this first incremental speed change. These points are shown in Fig. 22. Calculate the time in seconds for this incremental speed change to take place using the equation

$$\Delta t = \frac{\Sigma WK^2\,\Delta N}{308(T_D - T_P)} \tag{11}$$

where ΣWK^2 = total pump and motor moment of inertia, lb-ft^2
ΔN = incremental speed change, rpm
T_D = driver torque at the average speed, ft-lb
T_P = pump torque at the average speed, ft-lb

Calculate the acceleration head required to change the flow in the system from point O to M using Eq. (10) and time from Eq. (11). Add acceleration head to friction head at the assumed average flow, and if it is correct, it will fall on the average pump-head-capacity curve, point X. Adjust points M and X until these assumed flows result in the total acceleration and friction heads agreeing with flow X at the average speed. Repeat this procedure for other increments of speed change adding incremental times to get total accelerating time to bring the pump up to its final speed. Plot system head vs. average flow for each incremental speed change during this transient period as shown in Fig. 22.

Figures 23 and 24 illustrate how driver and pump torques can be determined from their respective speed-torque curves. Figure 23 is a family of curves which represent the torques required to produce flow against different heads without acceleration of the liquid or pump for the various speeds selected. Pump torque for any reduced speed can be calculated from the full-speed curve using the relation that torque varies as the second power and flow varies as the first power of the speed ratio. Point X is the torque at the average speed and trial average flow during the first incremental speed change which is adjusted for different assumed conditions. Figure 24 shows a typical squirrel-cage induction motor speed-torque curve and, for the purpose of illustration, it has been selected to have the same torque rating as the pump requires at full speed (approximately 97 percent synchronous speed). Point X in this figure is the motor torque at the average speed during the first incremental speed change. The developed torque curve for the different speeds shown in Fig. 23 is redrawn as the total friction and inertia pump torque T_p in Fig. 24. For the conditions used in this example, and as shown in Fig. 24, after approximately 88 percent synchronous speed very little excess torque $(T_D - T_p)$ is available for accelerating the pump and motor WK^2. However, as long as an induction motor has adequate torque to drive the pump at the normal condition, full speed will be reached providing the time-current demand can be tolerated. A synchronous motor, on the other hand, may not have sufficient torque

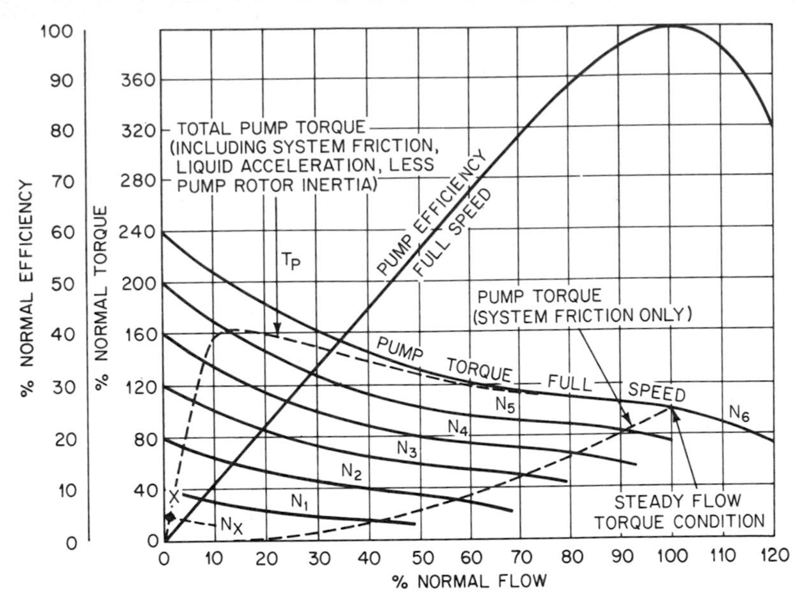

Fig. 23 Transient total pump torque as a result of liquid acceleration and system friction when starting a propeller pump with an open valve. Pump head characteristics are shown in Fig. 22.

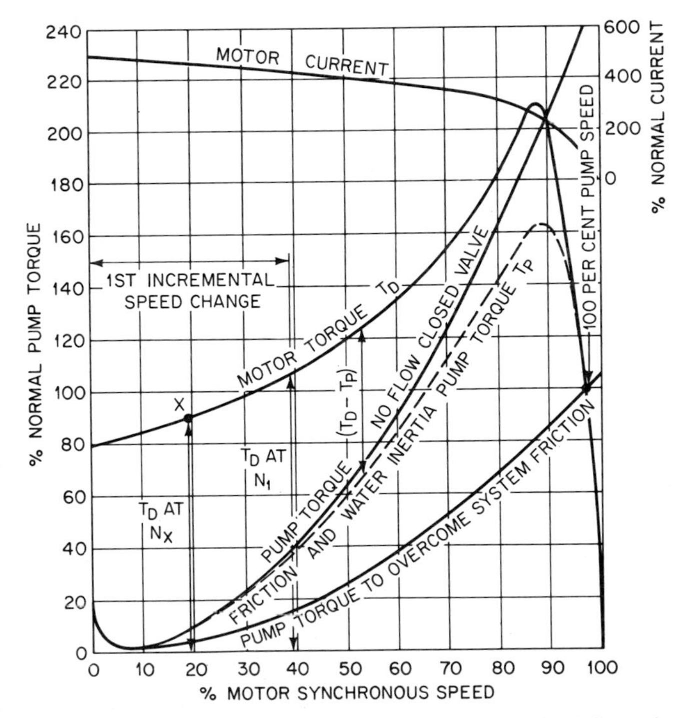

Fig. 24 Propeller pump and motor speed-torque curves showing effect of accelerating the liquid in the system during the starting period.

to pull into step. Liquid acceleration decreases as motor torque decreases adjusting to the available excess motor torque. The pump speed changes very slowly during the final period. Figure 24 also shows the motor current for the different speeds and torques.

A motor having a rating higher than normal pump torque or a motor having high-starting-torque characteristics will reduce the starting time but higher heads will be produced during the acceleration period. Note, however, heads produced cannot exceed shutoff at any speed. The total torque input to the pump shaft is equal to the sum of the torques required to overcome system friction, liquid inertia, and pump rotor inertia during acceleration. Torque at the pump shaft will therefore follow the motor-speed-torque curve less the torque required to overcome the motor's rotor inertia (WK^2). To reduce the starting-inertia pump head to an acceptable amount, if desired, other alternative starting schemes can be used. A short bypass line from the pump discharge back to the suction can be provided to divert flow from the main system. The bypass valve is closed slowly after the motor reaches full speed. A variable-speed, or a two-speed, motor will reduce the inertia head by controlling motor torque and speed thereby increasing the accelerating time.

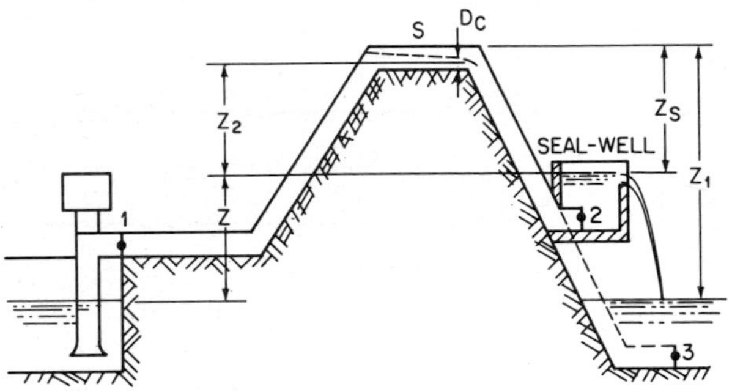

Fig. 25 Pumping system using a siphon for head recovery.

The above procedure for developing the actual system head and pump torque including liquid inertia becomes more complex if the pump must employ a discharge valve. To avoid high pump-starting heads and torques the discharge valve must be partially open on starting and then opened a sufficient amount before reaching full speed. The valve resistance must be added to the system friction and inertia curves if an exact solution is required.

Siphon Head Between any two points having the same elevation in a pumping system no head is lost because of piping elevation changes. Piping may be laid over and under obstacles without additional pumping head being required to sustain flow other than to overcome friction and minor losses. On rising the liquid pressure head is transformed to elevation head and the reverse takes place on falling. A pipe or other closed conduit that rises and falls is called a siphon, and one that falls and rises is called an inverted siphon. The siphon principle is valid provided the conduit flows full and free of liquid vapor and air, and it is this requirement that determines the limiting height of a siphon for complete recovery.

Pressure in a siphon is minimum at the summit, or just downstream from it, and Bernoulli's equation can be used to determine if the liquid's pressure is above vapor pressure. Referring to Fig. 25, observe the following: absolute pressure head H_s in feet at the top of the siphon is

$$H_S = H_B - Z_S + h_{f(S-2)} - \frac{V_S^2}{2g} \tag{12}$$

where H_B = barometric pressure, ft of liquid pumped

Z_S = siphon height to top of conduit, ft

$h_{f(S-2)}$ = friction and minor losses from S to 2 including exit velocity head loss at 2, ft

$V_S^2/2g$ = velocity head at the summit, ft

Whenever Z_1 shown in Fig. 25 is so high that it exceeds the maximum siphon capability, a seal well is necessary to increase the pressure at the top of the siphon above vapor pressure. Note $Z_1 - Z_s$ represents an unrecoverable head and increases the pumping head. Water has a vapor pressure of 0.77 ft at 68°F and theoretically a 33.23-ft-high siphon is possible with a 34-ft barometer. In practice higher water temperatures and lower barometric pressures limit the height of siphons used in condenser-cooling-water systems to 26 to 28 ft. The siphon height can be found by using Eq. (12) and letting H_S equal the vapor pressure in feet.

In addition to recovering head in systems such as condenser cooling water, thermal dilution, and levees, siphons are also used to prevent reverse flow after pumping is stopped by use of an automatic vacuum breaker located in the summit. Often siphons are used solely to eliminate the need for valves or flap gates.

In pumping systems siphons can be primed by external means of air removal. Unless the siphon is primed initially upon starting, a pump must fill the system and provide a minimum flow to induce siphon action. During this filling period and until the siphon is primed, the siphon-head curve must include this additional siphon filling head which must be provided by the pump. Pumps in siphon systems

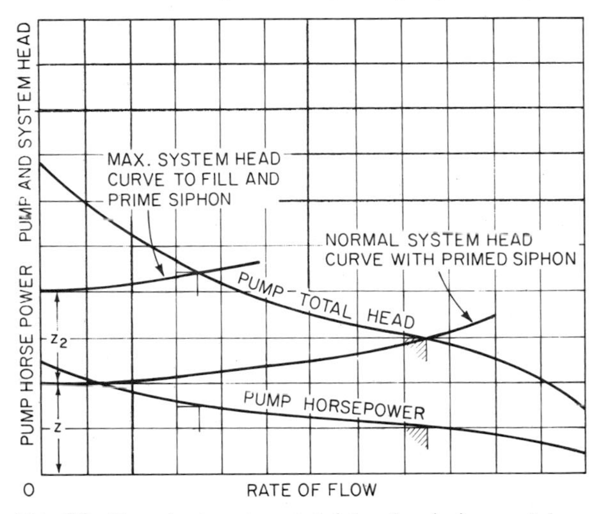

Fig. 26 Transient system total head priming a siphon.

are usually low head and they may not be capable of filling the system to the top of the siphon or of filling it with adequate flow. Low-head pumps are high-specific-speed and require more power at reduced flows than during normal pumping. Figure 26 illustrates the performance of a typical propeller pump when priming a siphon system and during normal operation.

When a pump and driver are to be selected to prime a siphon system it is necessary to estimate the pump head and the power required to produce the minimum flow needed to start the siphon. The minimum flow required increases with the length and the diameter and decreases with the slope of the down leg pipe.[1] Prior to the removal of all the air in the system, the pump is required to provide head to raise the liquid up to and over the siphon crest. Head above the crest is required to produce the minimum flow similar to flow over a broad-crested weir. This weir head may be an appreciable part of the total pump head if the pump is low-head and large-capacity. A conservative estimate of the pump head would include a full conduit above the siphon crest. In reality the down leg must flow partially empty before it can flow full, and it is accurate enough to estimate that the depth of liquid above the siphon crest is at critical depth for the cross section. Table 1 can be used to estimate critical depth in circular pipes.

[1] J. A. Moors, Criteria for the Development of Siphonic Action in Pumping Plant Discharge Lines, *Hydraul. Div. ASCE Ann. Conv. Paper,* October 1957.

Until all the air is removed, and all the piping becomes filled, the down leg is not part of the pumping system and its friction and minor losses are not to be added to the maximum system-head curve to fill and prime the siphon shown in Fig. 26. The total head in feet to be produced by the pump in Fig. 25 until the siphon is primed is

$$TH = Z + Z_2 + h_{f(1-S)} + \frac{V_C^2}{2g} \tag{13}$$

where Z = distance between suction and discharge levels, ft
Z_2 = siphon height to centerline of liquid, ft
$h_{f(1-S)}$ = friction and minor losses from 1 to S, ft
$V_C^2/2g$ = velocity head at the crest using actual liquid depth, approx critical depth, ft

Use of the above equation permits plotting the maximum system-head curve to fill and prime the siphon for different flow rates. The pump priming flow is the

TABLE 1 Critical Depth in Circular Pipes. Tabulated value = ft³/s ÷ $d^{5/2}$; d = diameter, ft; D_{crit} = critical depth, ft. (From H. W. King and E. Brater, "Handbook of Hydraulics," 5th ed. Copyright 1963 by McGraw-Hill Book Company, New York and used with their permission.)

$\frac{D_{crit}}{d}$	0.00	0.01	0.02	0.03	0.04	0.05	0.06	0.07	0.08	0.09
0.0	......	0.0006	0.0025	0.0055	0.0098	0.0153	0.0220	0.0298	0.0389	0.0491
0.1	0.0605	0.0731	0.0868	0.1016	0.1176	0.1347	0.1530	0.1724	0.1928	0.2144
0.2	0.2371	0.2609	0.2857	0.3116	0.3386	0.3666	0.3957	0.4259	0.4571	0.4893
0.3	0.523	0.557	0.592	0.628	0.666	0.704	0.743	0.784	0.825	0.867
0.4	0.910	0.955	1.000	1.046	1.093	1.141	1.190	1.240	1.291	1.343
0.5	1.396	1.449	1.504	1.560	1.616	1.674	1.733	1.792	1.853	1.915
0.6	1.977	2.041	2.106	2.172	2.239	2.307	2.376	2.446	2.518	2.591
0.7	2.666	2.741	2.819	2.898	2.978	3.061	3.145	3.231	3.320	3.411
0.8	3.505	3.602	3.702	3.806	3.914	4.023	4.147	4.272	4.406	4.549
0.9	4.70	4.87	5.06	5.27	5.52	5.81	6.18	6.67	7.41	8.83

intersection of the pump total-head curve and this system-head curve. The pump selected must have a driver with horsepower as shown in Fig. 26 to prime the system during this transient condition.

For the pumping system shown in Fig. 25, and after the system is primed, the pump total head reduces to

$$TH = Z + h_{f(1-2)} \tag{14}$$

where Z = distance between suction and discharge levels, ft
$h_{f(1-2)}$ = friction and minor losses from 1 to 2 including exit velocity head loss at 2, ft

Use of the above equation permits plotting the normal system head curve with a primed siphon shown in Fig. 26 for different flow rates. The normal pump flow is the intersection of the pump total-head curve and the normal system-head curve.

If the pump cannot provide sufficient flow to prime the siphon, or if the driver does not have adequate power, the system must then be primed externally by a vacuum or jet pump. An auxiliary priming pump can also be used to continuously vent the system as it is necessary that this be done to maintain full siphon recovery. In some systems the water pumped is saturated with air and as the liquid flows through the system the pressure is reduced and in cooling systems the temperature is increased. Both these conditions cause the release of some of the entrained air. Air will accumulate at the top of the siphon and in the upper parts of the down leg. The siphon works on the principle that an increase in elevation in the up leg produces a decrease in pressure, and an equal decrease in elevation in the down leg results in recovery of this pressure change. This cannot occur if the density of the liquid in the down leg is decreased as a result of the formation of air pockets. These air pockets also restrict the flow area. A release of entrained air and air leakage into the system through pipe joints and fittings will result in a centrifugal pump delivering less than design flow as the head will be higher than estimated. Also, high-specific-speed pumps with rising horsepower curves toward shutoff can become overloaded. In order to maintain full head recovery it is necessary to continuously vent the siphon at the top and at several points along the down leg, especially at the beginning of a change in slope.[1] These venting points can be manifolded together and connected to a single venting system.

The following examples illustrate the use of Eqs. (3), (9), (12), (13), and (14), Table 1, and Fig. 27.

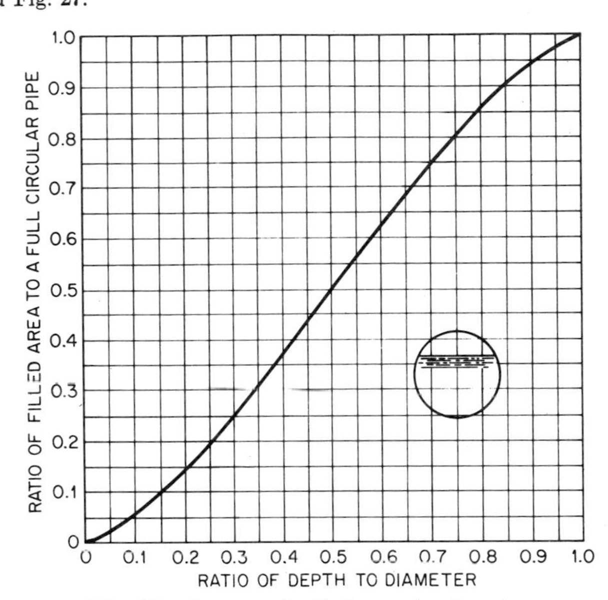

Fig. 27 Area vs. depth for a circular pipe.

EXAMPLE 2: A pump is required to produce a flow of 70,000 gpm through the system shown in Fig. 25. The conditions are:

sp gr = 0.998 for 80°F water
Barometric pressure = 29 in mercury absolute (sp gr 13.6)
Suction and discharge water levels are equal
$Z_1 = 40$ ft
$h_{f(1-S)} = 3$ ft up-leg friction head
$h_{f(S-3)} = 3.3$ ft down-leg friction head including exit loss
Pipe diameter = 48 in ID
Water-vapor pressure = 0.507 lb/in² abs at 80°F

Calculate the system total head from point 1 to point 3 (no seal well).
The maximum siphon height may be found from Eq. (12).

$$Z_S = H_B + h_{f(S-3)} - H_S - \frac{V_S^2}{2g}$$

From Eq. (3),

$$H_B = \text{barometric pressure head in feet of liquid pumped}$$
$$= \frac{\text{sp gr}_1}{\text{sp gr}_2} h_1 = \frac{13.6}{0.998} \times \frac{29}{12} = 32.9 \text{ ft abs}$$

$$H_S = \frac{144p}{w} = \frac{144 \times 0.507}{62.19} = 1.17 \text{ ft abs}$$

[1] R. T. Richards, Air Binding in Water Pipelines, Washington State University 2d Hydraulic Conference, October 1959.

From Eq. (9),

$$V_S = \frac{\text{gpm}}{(\text{Pipe ID})^2} \times 0.408 = \frac{70{,}000}{48^2} \times 0.408 = 12.4 \text{ ft/s}$$

$$Z_S = 32.9 + 3.3 - 1.17 - \frac{12.4^2}{2 \times 32.17} = 32.63 \text{ ft}$$

Since the maximum height is exceeded, $Z_1 > Z_S$, siphon recovery is not possible. The system total head is therefore found from Eq. (13).

$$TH = Z + Z_2 + h_{f(1-S)} + \frac{V_C^2}{2g}$$

The critical depth D_{crit} is found using Table 1.

$$\text{ft}^3/\text{s} = \frac{70{,}000}{7.481 \times 60} = 156$$

$$\frac{\text{ft}^3/\text{s}}{d^{5/2}} = \frac{156}{4^{5/2}} = 4.88$$

$$\frac{D_{\text{crit}}}{d} = 0.901$$

$$D_{\text{crit}} = 0.901 \times 4 = 3.6 \text{ ft}$$

From Fig. 27, ratio of filled area to area of a full pipe is 0.95 for a ratio of depth to diameter of 0.901.

$$V_C = \frac{\text{gpm}}{0.95(\text{ID})^2} \times 0.408 = \frac{70{,}000}{0.95(48)^2} \times 0.408 = 13.0 \text{ ft/s}$$

$$Z + Z_2 = 40 - 4 + \frac{3.6}{2} = 37.8 \text{ ft}$$

$$TH = 37.8 + 3 + \frac{13.0^2}{2 \times 32.17} = 43.43 \text{ ft}$$

Example 3: Calculate the minimum total system head using conditions in Example 2 and a seal well as shown in Fig. 25. Use 2.8 ft for the friction head loss $h_{f(S-2)}$.

The maximum siphon height Z_S in Example 2 was found to be 32.63 ft. Therefore from Eq. (14) the total system head after priming is

$$TH = Z + h_{f(1-2)}$$

$$h_{f(1-2)} = h_{f(1-S)} + h_{f(S-2)} = 3 + 2.8 = 5.8 \text{ ft}$$

$$Z = Z_1 - Z_S = 40 - 32.63 = 7.37 \text{ ft}$$

Note seal-well elevation is above discharge level.

$$TH = 7.37 + 5.8 = 13.17 \text{ ft}$$

Example 4: The dimensions of the down leg in Example 3 requires a minimum velocity of 5 ft/s flowing full to purge air from the system and start the siphon. Calculate the system head the pump must overcome to prime the siphon.

$$\text{gpm} = \frac{V(\text{pipe ID})^2}{0.408} = \frac{5(48)^2}{0.408} = 28{,}200$$

$$\text{ft}^3/\text{s} = 28{,}200 \div 449 = 62.8$$

The critical depth D_{crit} is found from Table 1.

$$\frac{\text{ft}^3/\text{s}}{d^{5/2}} = \frac{62.8}{4^{5/2}} = 1.97$$

$$\frac{D_{\text{crit}}}{d} = 0.6$$

$$D_{\text{crit}} = 0.6 \times 4 = 2.4 \text{ ft}$$

From Fig. 27, the ratio of filled area to area of a full pipe is 0.63 for a ratio of depth to diameter of 0.60.

$$V_C = \frac{5}{0.625} = 8.0 \text{ ft/s}$$

From Eq. (13),

$$TH = Z + Z_2 + h_{f(1-S)} + \frac{V_C^2}{2g}$$

$$Z_2 = Z_S - 4 + \frac{D_{crit}}{2}$$

$$= 32.63 - 4 + \frac{2.4}{2} = 29.83 \text{ ft}$$

$$Z = 7.37 \text{ ft} \qquad \text{(from Example 3)}$$

$$h_{f(1-S)} \text{ at } 28{,}200 \text{ gpm} = \left(\frac{28{,}200}{70{,}000}\right)^2 (3) = 0.49 \text{ ft}$$

$$TH = 7.37 + 29.83 + 0.49 + \frac{8^2}{2 \times 32.17} = 38.68 \text{ ft}$$

If the system is not externally primed, a centrifugal pump must be selected to deliver at least 28,200 gpm at 38.68 ft total head and be provided with a driver having adequate horsepower for this condition. After the system is primed, the pump must be capable of delivering at least 70,000 gpm at 13.17 ft (see Fig. 26).

HEAD LOSSES IN SYSTEM COMPONENTS

Pressure Pipes Resistance to flow through a pipe is caused by viscous shear stresses within the liquid and turbulence at the pipe walls. *Laminar* flow occurs in a pipe when the average velocity is relatively low and the energy head is lost mainly as a result of viscosity. In laminar flow, liquid particles have no motion next to the pipe walls, and flow occurs owing to particles moving in parallel lines with increasing velocity toward the center. The movement of concentric cylinders past each other causes viscous shear stresses, more commonly called *friction*. As flow increases the flow pattern changes, the average velocity becomes more uniform, and there is less viscous shear. As the laminar film decreases in thickness at the pipe walls and as the flow increases, the pipe roughness then becomes important as it causes turbulence. *Turbulent* flow occurs at relatively high average pipe velocities, and energy head is lost predominantly because of turbulence caused by the wall roughness. The average velocity at which the flow will change from laminar to turbulent is not definite, and there is a critical zone in which either laminar or turbulent flow can occur.

The dimensionless Reynolds number Re is used to describe the type of flow in a pipe flowing full and can be expressed as follows:

$$\text{Re} = \frac{VD}{\nu} \tag{15}$$

where V = average pipe velocity, ft/s
D = inside pipe diameter, ft
ν = liquid kinematic viscosity, ft²/s (=0.0000107639 × centistokes)

When the Reynolds number is 2,000 or less the flow is generally laminar, and when greater than 4,000 the flow is generally turbulent. The flow of water in pipes is usually well above 4,000 and therefore almost always turbulent.

The Darcy-Weisbach formula is most often used to calculate pipe friction. This formula recognizes that friction increases with pipe-wall roughness, with wetted surface area, with velocity to a power, and with viscosity, while friction decreases with pipe diameter to a power and with density. Specifically the friction head loss h_f in feet is

$$h_f = f \frac{L}{D} \frac{V^2}{2g} \tag{16}$$

where f = friction factor
L = pipe length, ft
D = inside pipe diameter, ft
V = average pipe velocity, ft/s
g = acceleration of gravity, approx 32.17 ft/s²

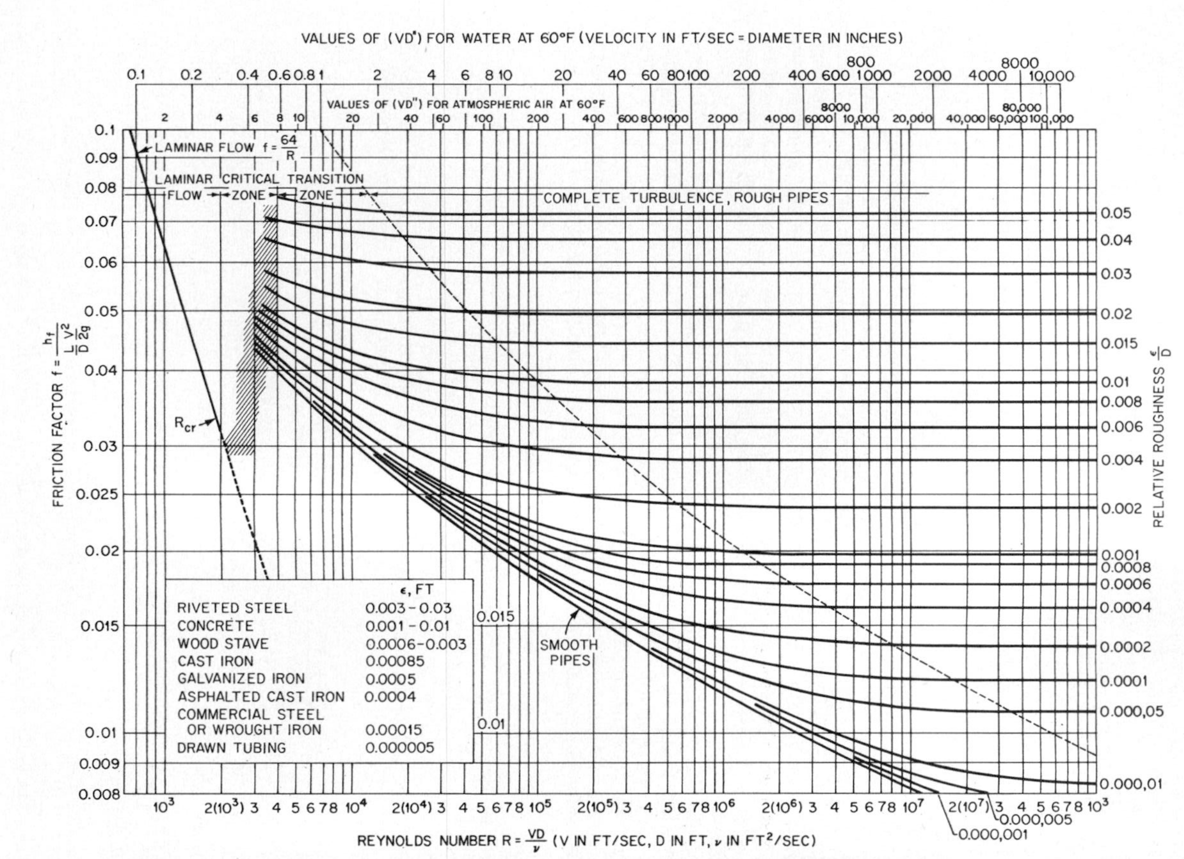

	ϵ, FT
RIVETED STEEL	0.003 - 0.03
CONCRETE	0.001 - 0.01
WOOD STAVE	0.0006-0.003
CAST IRON	0.00085
GALVANIZED IRON	0.0005
ASPHALTED CAST IRON	0.0004
COMMERCIAL STEEL OR WROUGHT IRON	0.00015
DRAWN TUBING	0.000005

Fig. 28 Moody diagram. (V. L. Streeter, "Fluid Mechanics," 5th ed. Copyright 1971 by McGraw-Hill Book Company, New York)

For laminar flow the friction factor f is equal to 64/Re and is independent of pipe-wall roughness. For turbulent flow the friction factor f for all incompressible fluids can be determined from the well-known Moody diagram shown in Fig. 28. To determine f it is required that the Reynolds number and the relative pipe roughness be known. Values of relative roughness ϵ/D (where ϵ is a measurement of pipe-wall-roughness height in feet) can be obtained from Fig. 29 for different pipe diameters and materials. Figure 29 also gives values of f for flow of 60°F water in rough pipes and with complete turbulence. Values of kinematic viscosity and Reynolds number Re for a number of different fluids at various temperatures are given in Fig. 30. The Reynolds numbers of 60°F water for various velocities and pipe diameters may be found by using the VD″ scale in Fig. 28.

There are many empirical formulas for calculating pipe friction for water flowing under turbulent conditions. The most widely used is the Hazen-Williams formula:

$$V = 1.318Cr^{0.63}S^{0.54} \tag{17}$$

where V = average pipe velocity, ft/s
C = friction factor for this formula which depends on roughness only
r = hydraulic radius (liquid area ÷ wetted perimeter) or $D/4$ for a full pipe, ft
S = hydraulic gradient or friction head loss per unit length of pipe, ft/ft

The effect of age on a pipe should be taken into consideration when estimating the friction loss. A lower C value should be used depending on the expected life of the system. Table 2 gives recommended friction factors for new and old pipes. Figure 31 is a nomogram which can be used for a solution to the Hazen-Williams formula.

The friction head loss in pressure pipes can be found using either the Darcy-Weisbach formula or the Hazen-Williams formula, Eqs. (16) and (17) respectively. The following examples illustrate how Figs. 28, 29, 30, and 31, and Table 2 may be used to solve problems of this type.

EXAMPLE 5: Calculate the Reynolds number Re for 175°F kerosene flowing through 4-in Schedule 40 (3.426 in ID) seamless steel pipe, velocity 14.6 ft/s.

$$VD'' = 14.6 \times 3.426 = 50 \text{ ft/s} \times \text{in}$$

Follow the tracer lines in Fig. 30 and read directly

$$\text{Re} = 3.5 \times 10^5$$

EXAMPLE 6: Calculate the friction head loss for 100 ft of 20-in Schedule 20 (19.350 ID) seamless steel pipe for 109°F water flowing 11,500 gpm. Use the Darcy-Weisbach formula.

$$V = \frac{\text{gpm}}{(\text{pipe ID})^2} \times 0.408 = \frac{11{,}500}{(19.35)^2} \times 0.408 = 12.53 \text{ ft/s}$$

$$VD'' = 12.53 \times 19.35 = 242 \text{ ft/s} \times \text{in}$$

From Fig. 30, $\text{Re} = 3 \times 10^6$

From Fig. 29, $\frac{\epsilon}{D} = 0.00009$

From Fig. 28, $f = 0.012$

Using Eq. (16),

$$D = \frac{19.35}{12} = 1.61 \text{ ft}$$

$$h_f = f\frac{L}{D}\frac{V^2}{2g} = 0.012\frac{100}{1.61} \times \frac{12.53^2}{2 \times 32.17} = 1.82 \text{ ft}$$

EXAMPLE 7: The flow in Example 6 is increased until complete turbulence results. Determine the friction factor f and flow.

From Fig. 28 follow the relative roughness curve $\epsilon/D = 0.00009$ to the beginning of the zone "complete turbulence, rough pipes" and read

$$f = 0.0119 \qquad \text{at Re} = 2 \times 10^7$$

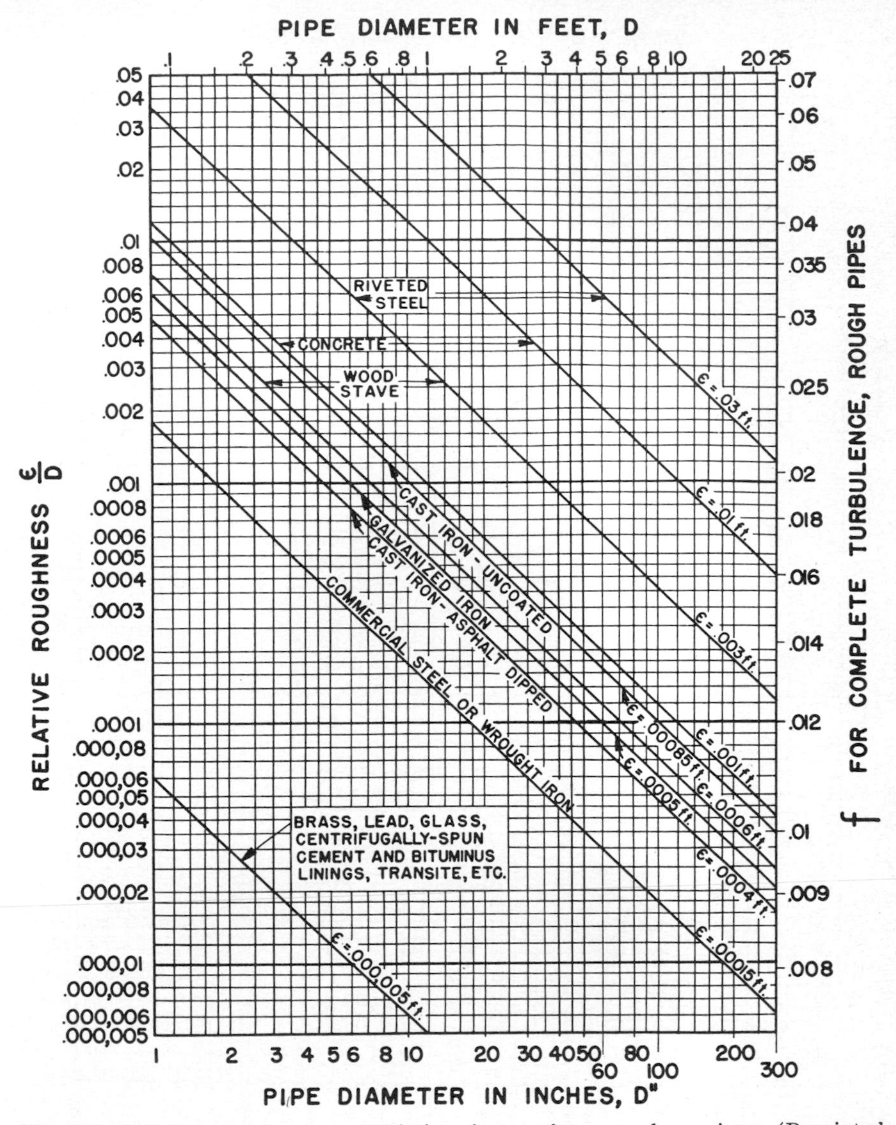

Fig. 29 Relative roughness and friction factors for new clean pipes. (Reprinted from the "Pipe Friction Manual," 3d ed. Copyright 1961 by the Hydraulic Institute, Cleveland, Ohio)

The problem may also be solved using Fig. 29. Enter relative roughness $\epsilon/D = 0.00009$; read directly across to

$$f = 0.0119$$

An increase in Re from 3×10^6 to 2×10^7 would require an increase in flow to

$$\frac{2 \times 10^7}{3 \times 10^6} \times 11{,}500 = 76{,}700 \text{ gpm}$$

EXAMPLE 8: The liquid in Example 6 is changed to water at 60°F. Determine Re, f, and the friction head loss per 100 ft of pipe.

$$VD'' = 242 \text{ ft/s} \times \text{in} \qquad \text{(as in Example 6)}$$

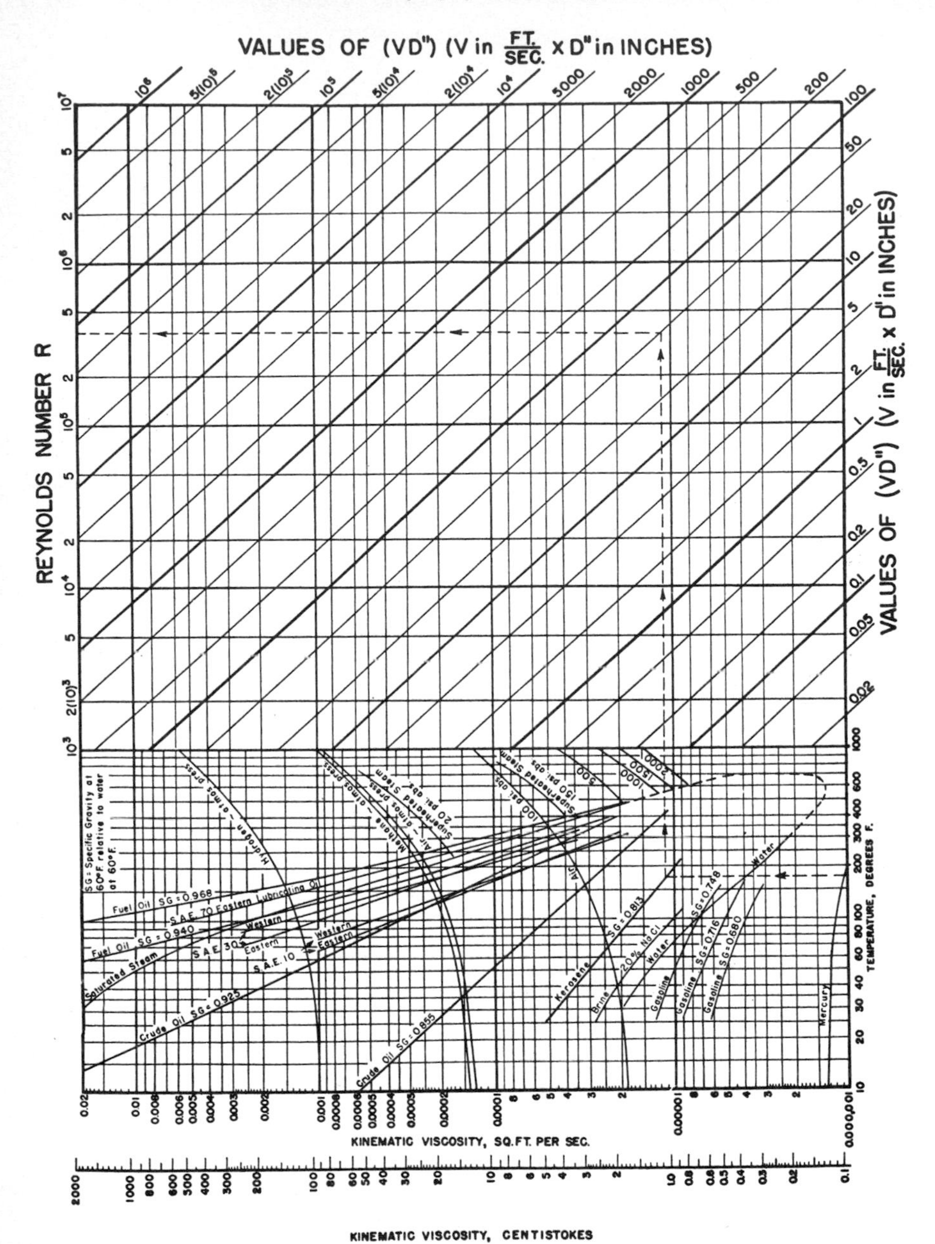

Fig. 30 Kinematic viscosity and Reynolds number. (Reprinted from the "Pipe Friction Manual," 3rd ed. Copyright 1961 by the Hydraulic Institute, Cleveland, Ohio)

Since the liquid is 60°F water, enter Fig. 28 and read directly downward from VD'' to

$$\mathrm{Re} = 1.8 \times 10^6$$

Where the line VD'' to Re crosses $\epsilon/D = 0.00009$ in Fig. 28, read

$$f = 0.013$$

60°F water is more viscous than 109°F water, and this accounts for Re decreasing and f increasing. Using Eq. (16) it can be calculated that the friction head loss increases to

$$h_f = f\frac{L}{D}\frac{V^2}{2g} = 0.013\,\frac{100}{1.61} \times \frac{12.53^2}{2 \times 32.17} = 1.97 \text{ ft}$$

EXAMPLE 9: A 102-in ID welded steel pipe is to be used to convey water at a velocity of 11.9 ft/s. Calculate the expected loss of head due to friction per 1,000 ft of pipe after 20 years. Use the empirical Hazen-Williams formula.

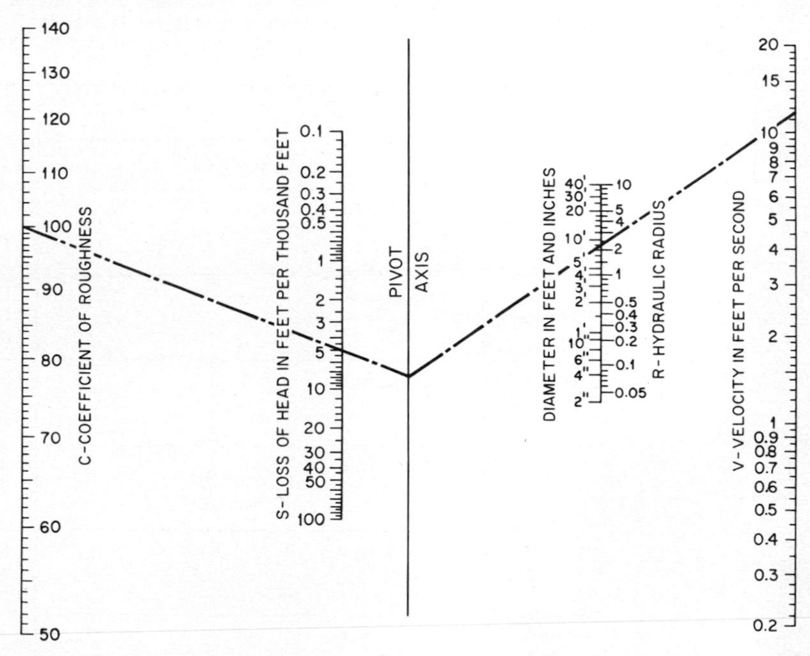

Fig. 31 Nomogram for the solution of the Hazen-Williams formula. Obtain values of C from Table 2. (C. Davis and K. Sorensen, "Handbook of Applied Hydraulics," 3d ed. Copyright 1969 by McGraw-Hill Book Company, New York, and used with their permission.)

From Table 2, $C = 100$. The hydraulic radius $r = D/4 = 102/(4 \times 12) = 2.13$ ft. Substituting in Eq. (17),

$$S^{0.54} = \frac{V}{1.318\,Cr^{0.63}} = \frac{11.9}{1.318 \times 100 \times 2.13^{0.63}} = 0.0557$$

$$S = (0.0557)^{1/0.54} = 0.0048 \text{ ft/ft}$$

$$h_f = 1{,}000 \times 0.0048 = 4.8 \text{ ft}$$

The problem may also be solved by using Fig. 31, following the trace lines:

$$h_f \approx 5 \text{ ft}$$

Partially Full Pipes and Open Channels Another popular empirical equation applicable to the flow of water in pipes flowing full or partially full or in open channels is the Manning formula

$$V = \frac{1.486}{n}\,r^{2/3}S^{1/2} \tag{18}$$

where V = average velocity, ft/s
n = friction factor for this formula which depends on roughness only
r = hydraulic radius (liquid area ÷ wetted perimeter), ft
S = hydraulic gradient or friction head loss per unit length of conduit, ft/ft

The Manning formula nomogram shown in Fig. 32 can be used to determine the flow or friction head loss in open or closed conduits. Note that the hydraulic gradient S in Fig. 32 is plotted in ft/100 ft of conduit length. Values of friction factors n are given in Table 3.

TABLE 2 Values of friction factor C to be used with the Hazen-Williams Formula in Fig. 31. (From C. V. Davis and K. E. Sorensen, "Handbook of Applied Hydraulics," 3d ed. Copyright 1969 by McGraw-Hill Book Company, New York and used with their permission.)

Type of pipe		Condition	C
Cast iron	New	All sizes, in	130
	5 years old	12 and over	120
		8	119
		4	118
	10 years old	24 and over	113
		12	111
		4	107
	20 years old	24 and over	100
		12	96
		4	89
	30 years old	30 and over	90
		16	87
		4	75
	40 years old	30 and over	83
		16	80
		4	64
		40 and over	77
		24	74
		4	55
Welded steel	Values of C the same as for cast iron pipe, 5 years older		
Riveted steel	Values of C the same as for cast iron pipe, 10 years older		
Wood-stave	Average value, regardless of age		120
Concrete or concrete-lined	Large sizes, good workmanship, steel forms		140
	Large sizes, good workmanship, wooden forms		120
	Centrifugally spun		135
Vitrified	In good condition		110

If the conduit is flowing partially full, computing the hydraulic radius is sometimes difficult. When the problem to be solved deals with a pipe which is not flowing full Fig. 33 may be used to obtain multipliers for correcting the flow and the velocity of a full pipe to the actually filled section. If the flow in a partially full pipe is known and the friction head loss is to be determined, Fig. 33 is first used to correct the flow to what it would be if the pipe were full. Then by using Eq. (18) or Fig. 32, determine the friction head loss (which is also the hydraulic gradient and the slope of the pipe). The problem is solved in reverse if the hydraulic gradient is known and the flow is to be determined.

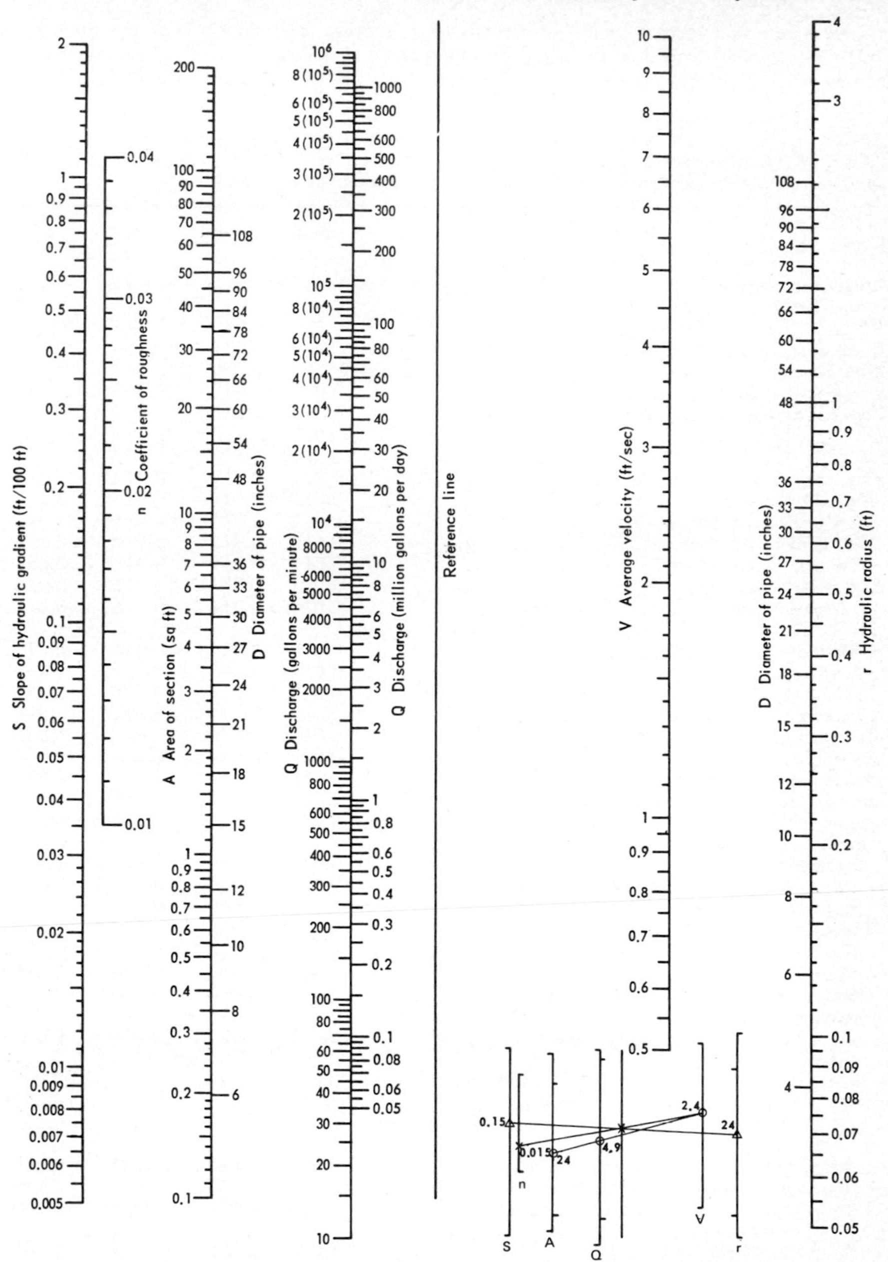

Fig. 32 Nomogram for the solution of the Manning formula. Obtain values of n from Table 3. To solve for S, align A with Q and read V, align V and n intersecting reference line, align r with reference line intersection and read S.

For conduits which are not circular in cross section and flow is full or partially full, an alternate solution to using Eq. (18) is to calculate an equivalent diameter which is equal to four times the hydraulic radius. If the conduit is extremely narrow

TABLE 3 Values of Friction Factor n to be Used with the Manning Formula in Fig. 32. (From H. W. King and E. Brater, "Handbook of Hydraulics," 5th ed. Copyright 1963 by Mc-Graw-Hill Book Company, New York and used with their permission.)

Surface	Best	Good	Fair	Bad
Uncoated cast iron pipe	0.012	0.013	0.014	0.015
Coated cast iron pipe	0.011	0.012*	0.013*	
Commercial wrought iron pipe, black	0.012	0.013	0.014	0.015
Commercial wrought iron pipe, galvanized	0.013	0.014	0.015	0.017
Smooth brass and glass pipe	0.009	0.010	0.011	0.013
Smooth lockbar and welded OD pipe	0.010	0.011*	0.013*	
Riveted and spiral steel pipe	0.013	0.015*	0.017*	
Vitrified sewer pipe	0.010 / 0.011	0.013*	0.015	0.017
Common clay drainage tile	0.011	0.012*	0.014*	0.017
Glazed brickwork	0.011	0.012	0.013*	0.015
Brick in cement mortar; brick sewers	0.012	0.013	0.015*	0.017
Neat cement surfaces	0.010	0.011	0.012	0.013
Cement mortar surfaces	0.011	0.012	0.013*	0.015
Concrete pipe	0.012	0.013	0.015*	0.016
Wood-stave pipe	0.010	0.011	0.012	0.013
Plank flumes:				
Planed	0.010	0.012*	0.013	0.014
Unplaned	0.011	0.013*	0.014	0.015
With battens	0.012	0.015*	0.016	
Concrete-lined channels	0.012	0.014*	0.016*	0.018
Cement-rubble surface	0.017	0.020	0.025	0.030
Dry-rubble surface	0.025	0.030	0.033	0.035
Dressed-ashlar surface	0.013	0.014	0.015	0.017
Semicircular metal flumes, smooth	0.011	0.012	0.013	0.015
Semicircular metal flumes, corrugated	0.0225	0.025	0.0275	0.030
Canals and ditches:				
Earth, straight and uniform	0.017	0.020	0.0225*	0.025
Rocks cuts, smooth and uniform	0.025	0.030	0.033*	0.035
Rock cuts, jagged and irregular	0.035	0.040	0.045	
Winding sluggish canals	0.0225	0.025*	0.0275	0.030
Dredged earth channels	0.025	0.0275*	0.030	0.033
Canals with rough stony beds, weeds on earth banks	0.025	0.030	0.035*	0.040
Earth bottom, rubble sides	0.028	0.030*	0.033*	0.035
Natural stream channels:				
(1) Clean, straight bank, full stage, no rifts or deep pools	0.025	0.0275	0.030	0.033
(2) Same as (1), but some weeds and stones	0.030	0.033	0.035	0.040
(3) Winding, some pools and shoals, clean	0.033	0.035	0.040	0.045
(4) Same as (3), lower stages, more ineffective slope and sections	0.040	0.045	0.050	0.055
(5) Same as (3), some weeds and stones	0.035	0.040	0.045	0.050
(6) Same as (4), stony sections	0.045	0.050	0.055	0.060
(7) Sluggish river reaches, rather weedy or with very deep pools	0.050	0.060	0.070	0.080
(8) Very weedy reaches	0.075	0.100	0.125	0.150

* Values commonly used in designing.

where width is small relative to length (annular or elongated sections) the hydraulic radius is one-half the width of the section.[1] After determining the equivalent diameter the problem may be solved by using the Darcy-Weisbach equation (16).

[1] Flow of Fluid Through Valves, Fittings, and Pipe, Crane Co., Technical Paper No. 410, 1969.

The hydraulic gradient in a uniform open channel is synonymous with friction head loss in a pressure pipe. The hydraulic gradient of an open channel or of a pipe flowing partially full is the slope of the free liquid surface. In the reach of the channel where the flow is uniform the hydraulic gradient is parallel to the slope of the channel bottom. Figure 34 shows that in a pressure pipe of uniform cross section the slope of both the energy and hydraulic gradients is a measure

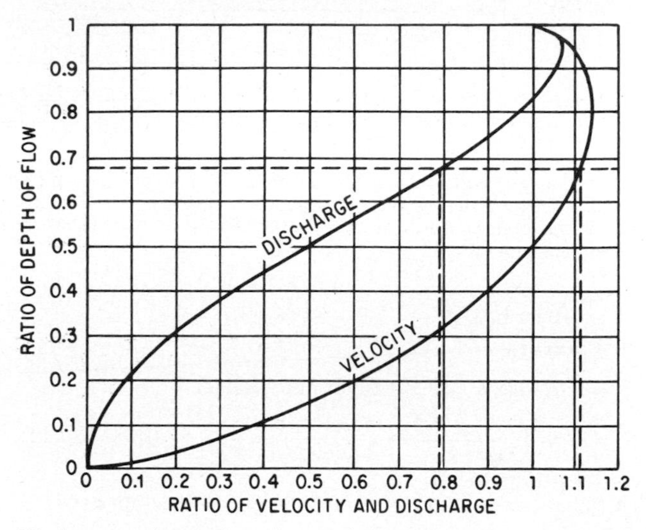

Fig. 33 Discharge and velocity of a partially full circular pipe vs. a full pipe.

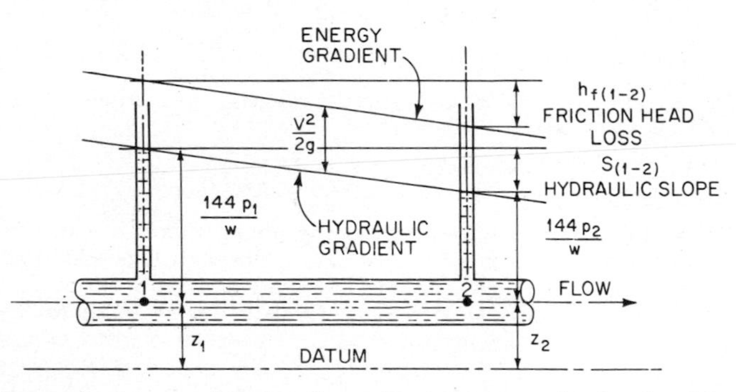

Fig. 34 Slope of energy and hydraulic gradients measure friction head loss per foot of pipe length.

of the friction head loss per foot of pipe between points 1 and 2. Figure 35 illustrates the flow in an open channel of varying slope. Between points 1 and 2 the flow is uniform and the liquid surface (hydraulic gradient) and channel bottom are both parallel and their slope is the friction head loss per foot of channel length.

A rearrangement of Eq. (18) gives

$$S = h_f = \left(\frac{Vn}{1.486r^{2/3}}\right)^2 \tag{19}$$

The following examples illustrate the application of the Manning formula, Eq. (19), Figs. 32 and 33, and Table 3 to the solution of the flow of water in open-channel problems.

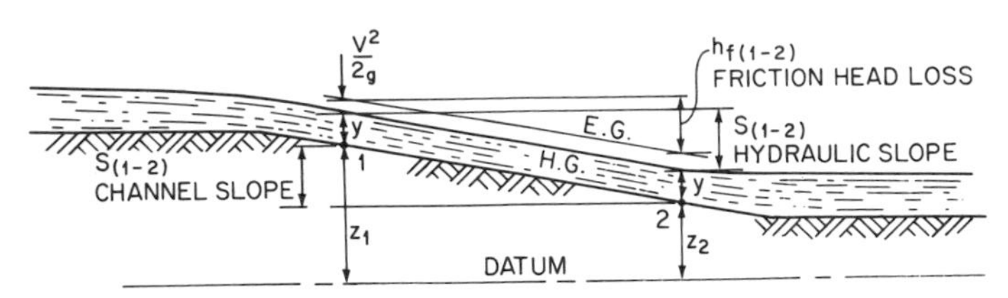

Fig. 35 In an open channel with uniform flow, slope of channel bottom, energy, and hydraulic gradients are the same as friction head loss per foot of channel length.

EXAMPLE 10: The flow through a 24-in ID commercial wrought iron pipe in fair condition is 4.9 mgd. Determine the loss of head as a result of friction in ft/100 ft of pipe and the slope required to maintain a full, uniformly flowing pipe.

The velocity in the pipe is

$$V = \frac{\text{gpm}}{(\text{pipe ID})^2} \times 0.408 = \frac{4.9 \times 10^6 \times 0.408}{24 \times 60 \times 24^2} = 2.41 \text{ ft/s}$$

From Table 3, $n = 0.015$,

$$r = D/4 = 2/4 = 0.5 \text{ ft (hydraulic radius)}$$

$$S = h_f = \left(\frac{Vn}{1.486r^{2/3}}\right)^2 = \left(\frac{2.41 \times 0.015}{1.486 \times 0.5^{2/3}}\right)^2 \qquad \text{(from Eq. 19)}$$

$$= 0.0015 \text{ ft/ft}$$

$$100 \times 0.0015 = 0.15 \text{ ft/100 ft (slope and friction head)}$$

The problem may also be solved by using Fig. 32 and following the trace lines

$$S = h_f = 0.15 \text{ ft/100 ft}$$

EXAMPLE 11: Determine what the flow and velocity would be if the pipe in Example 10 were flowing two-thirds full and were laid on the same slope.

Follow the trace lines in Fig. 33 and note the multipliers for discharge and velocity are 0.79 and 1.11, respectively. Therefore

$$\text{Flow} = 0.79 \times 4.9 = 3.87 \text{ mgd}$$
$$\text{Velocity} = 1.11 \times 2.4 = 2.66 \text{ ft/s}$$

Pipe Fittings Invariably a system containing piping will have connections which change the size and/or direction of the conduit. These fittings add friction, called *minor losses,* to the system head. Fitting losses are generally the result of changes in velocity and/or direction. A decreasing velocity results in more loss in head than an increasing velocity as the former causes energy-dissipating eddies. Experimental results have indicated that minor losses vary approximately as the square of the velocity through the fittings.

Valves and Standard Fittings. The resistance to flow through valves and fittings may be estimated by any of the following methods. The Hydraulic Institute "Pipe Friction Manual," 3d ed., lists losses through valves and fittings in terms of the average velocity head in a pipe of corresponding diameter and a *resistance coefficient* as shown in Tables 4*a* and *b*. The frictional resistance in feet is found from the equation

$$h = K\frac{V^2}{2g} \tag{20}$$

where K = resistance coefficient which depends on design and size of valve or fitting
V = average velocity in pipe of corresponding diameter, ft/s
g = acceleration of gravity, approx 32.17 ft/s²

Table 4*c* indicates the wide variation in published values of K.

A comparison of the Darcy-Weisbach equation (16) and the above equation (20) suggests that $K = f(L/D)$ to produce the same head loss in a straight pipe as in

TABLE 4a Resistance Coefficients for Valves and Fittings. (From "Pipe Friction Manual," 3d ed. Copyright 1961 by Hydraulic Institute, Cleveland, Ohio)

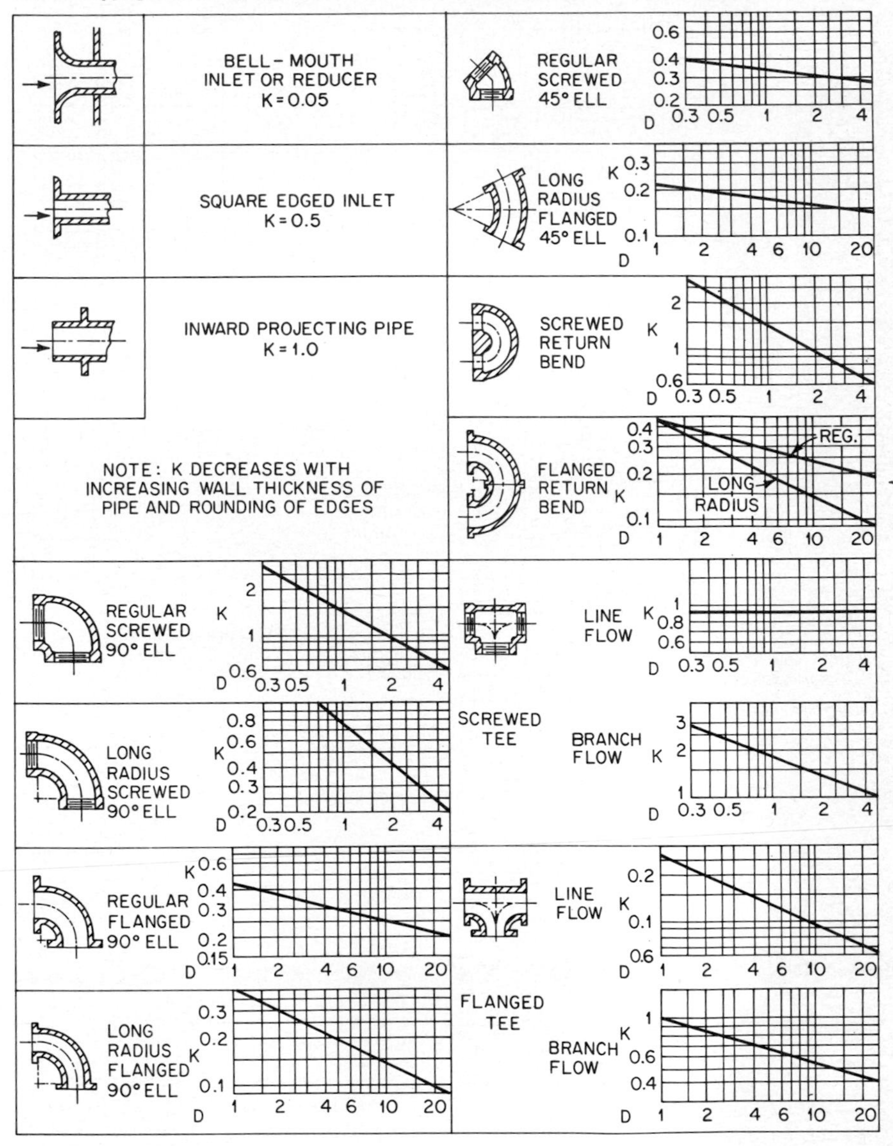

$$h = K \frac{V^2}{2g} \text{ FEET OF FLUID}$$

the valve or fitting. The ratio L/D or *equivalent length in pipe diameters* of straight pipe may then be used as another method to estimate valve and fitting losses. Tests have shown that while K decreases with size of different lines of valves and fittings, L/D is almost constant.[1] In the zone of complete turbulence, as shown in Fig. 28, K for a given size and L/D for all sizes of valves and fittings are constant. In the transition zone K increases as does the friction factor f with decreas-

[1] Flow of Fluids Through Valves, Fittings, and Pipe, Crane Co., Technical Paper 410, 1969.

TABLE 4b Resistance Coefficients for Valves and Fittings. (From "Pipe Friction Manual," 3d ed. Copyright 1961 by Hydraulic Institute, Cleveland, Ohio)

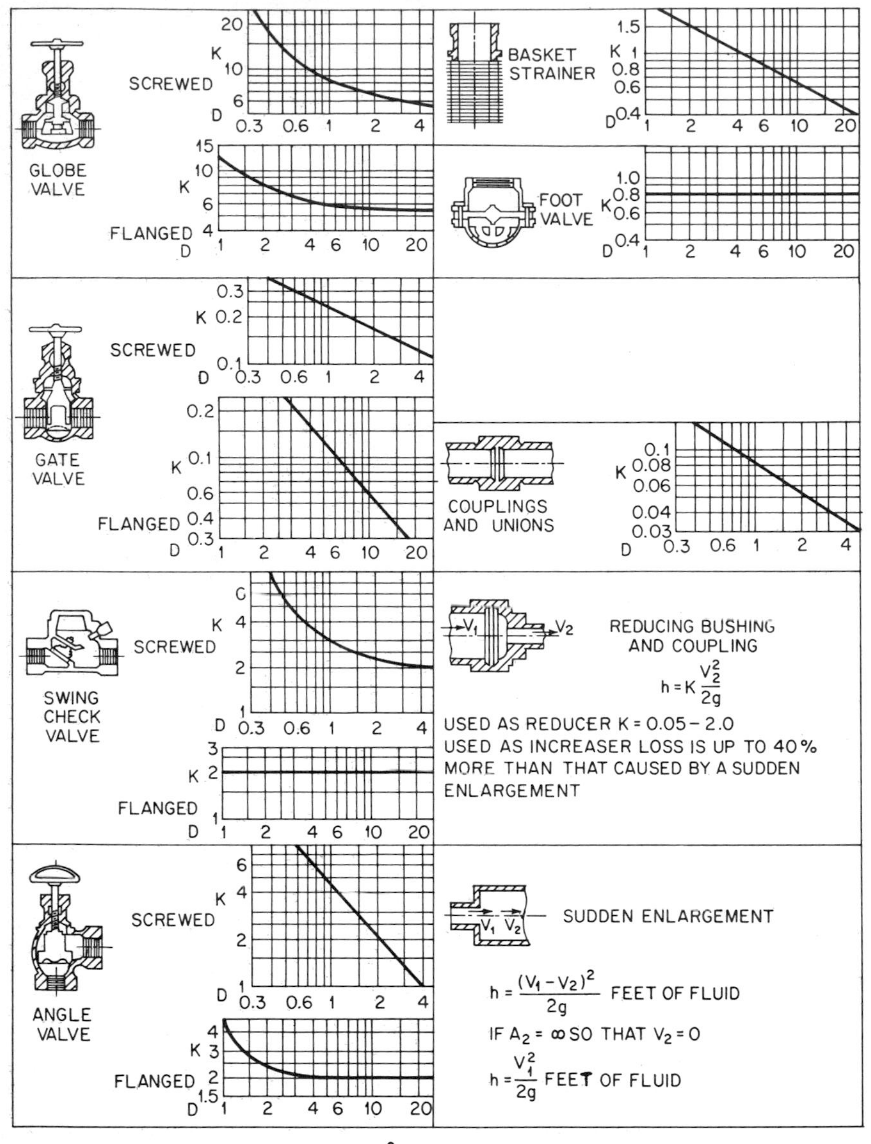

$$h = K\frac{V^2}{2g} \text{ FEET OF FLUID}$$

ing Reynolds number Re while L/D remains approximately constant.[1] Table 5*a* lists suggested values of L/D for various valves and fittings. Multiplying L/D by the inside diameter of pipe corresponding to the schedule number shown in Table 5*b* gives the equivalent length of straight pipe. When using the *equivalent-length* method, the friction head loss is determined by employing the Darcy-Weisbach equation (16). This method therefore takes into consideration the viscosity of the liquid, which in turn determines the Reynolds number and the friction factor.

[1] Flow of Fluids Through Valves, Fittings, and Pipe, Crane Co., Technical Paper 410, 1969.

TABLE 4c Approximate Variation for K Listed in Tables 4a and 4b. (Reprinted from the "Pipe Friction Manual," 3d ed. Copyright 1961 by the Hydraulic Institute, Cleveland, Ohio)

Fitting		Range of variation, %
90° elbow	Regular screwed	±20 above 2-in size
	Regular screwed	±40 below 2-in size
	Long radius, screwed	±25
	Regular flanged	±35
	Long radius, flanged	±30
45° elbow	Regular screwed	±10
	Long radius, flanged	±10
180° bend	Regular screwed	±25
	Regular flanged	±35
	Long radius, flanged	±30
T	Screwed, line or branch flow	±25
	Flanged, line or branch flow	±35
Globe valve	Screwed	±25
	Flanged	±25
Gate valve	Screwed	±25
	Flanged	±50
Check valve	Screwed	±30
	Flanged	+200 / −80
Sleeve check valve		Multiply flanged values by 0.2 to 0.5
Tilting check valve		Multiply flanged values by 0.13 to 0.19
Drainage gate check		Multiply flanged values by 0.03 to 0.07
Angle valve	Screwed	±20
	Flanged	±50
Basket strainer		±50
Foot valve		±50
Couplings		±50
Unions		±50
Reducers		±50

Notes on the Use of Tables 4a, 4b, and 4c:

1. The value of D given in the charts is nominal IPS (iron pipe size).
2. For velocities below 15 ft/s, check valves, and foot valves will be only partially open and will exhibit higher values of K than that shown in the charts.

The loss of head through valves, particularly control valves, is sometimes expressed in terms of a *flow coefficient* C_v. The flow in gpm at 60°F to produce a pressure drop of 1 lb/in² is defined as the flow coefficient for a particular valve opening. Values of C_v may be obtained from various manufacturers for their different lines of valves. The pressure loss for liquids with viscosity close to water at 60°F may be found for different flows from: lb/in² = sp gr at 60° $(\text{gpm}/C_v)^2$.

The following examples illustrate the use of the *resistance coefficient* K and the *equivalent length in pipe diameters* L/D methods for estimating losses in valves and fittings.

EXAMPLE 12: A pumping system consist of 20 ft of 2-in suction pipe and 300 ft of 1½-in discharge pipe, both Schedule 40 new steel. Also included are a bellmouth inlet, a 90° LR suction elbow, a suction-gate valve, a discharge-gate valve and a swing-check valve. The valves and fittings are screw-connected and the same size as the connecting pipe.

Determine the pipe, valve, and fitting losses when pumping 60 gpm of 60°F oil having a sp gr of 0.855. Use the resistance coefficient method.

Suction pipe:

ID = 2.067 in

From Fig. 29, $\epsilon/D = 0.00087$

From Fig. 30, $\nu = 0.0009$ ft²/s

From Eq. (9),

$$V = \frac{60}{2.067^2} \times 0.408 = 5.73 \text{ ft/s}$$

TABLE 5a Representative Equivalent Length in Pipe Diameters (L/D) of Various Valves and Fittings. (From Crane Co. Technical Paper 410, Flow of Fluids through Valves, Fittings and Pipe. Copyright 1969.)

Description of product				Equivalent length in pipe diameters, L/D
Globe valves	Stem perpendicular to run	With no obstruction in flat, bevel, or plug type seat	Fully open	340
		With wing or pin-guided disk	Fully open	450
	Y pattern	(No obstruction in flat, bevel, or plug type seat) With stem 60° from run of pipeline	Fully open	175
		With stem 45° from run of pipeline	Fully open	145
Angle valves		With no obstruction in flat, bevel, or plug type seat	Fully open	145
		With wing or pin-guided disk	Fully open	200
Gate valves	Wedge, disk, double-disk, or plug-disk		Fully open	13
			Three-quarters open	35
			One-half open	160
			One-quarter open	900
	Pulp stock		Fully open	17
			Three-quarters open	50
			One-half open	260
			One-quarter open	1200
Conduit pipeline gate, ball, and plug valves			Fully open	3*
Check valves	Conventional swing	0.5†	Fully open	135
	Clearway swing	0.5†	Fully open	50
	Globe-lift or stop; stem perpendicular to run or Y pattern	2.0†	Fully open	Same as globe
	Angle-lift or stop	2.0†	Fully open	Same as angle
	In-line ball	2.5 vertical and 0.25 horizontal†	Fully open	150
Foot valves with strainer		With poppet lift-type disk 0.3†	Fully open	420
		With leather-hinged disk 0.4†	Fully open	75
Butterfly valves (8-in and larger)			Fully open	40
Cocks	Straight-through	Rectangular plug port area equal to 100% of pipe area	Fully open	18
	Three-way	Rectangular plug port area equal to 80% of pipe area (fully open)	Flow straight through	44
			Flow through branch	140
Fittings	90° standard elbow			30
	45° standard elbow			16
	90° long radius elbow			20
	90° street elbow			50
	45° street elbow			26
	Square-corner elbow			57
	Standard T	With flow through run		20
		With flow through branch		60
	Close-pattern return bend			50

* Exact equivalent length is equal to the length between flange faces or welding ends.

† Minimum calculated pressure drop (lb/in^2) across valve to provide sufficient flow to lift disk fully.

TABLE 5b Schedule (Thickness) of Steel Pipe Used in Obtaining Resistance of Valves and Fittings Shown in Table 5a. (From Crane Co. Technical Paper 410, Flow of Fluids through Valves, Fittings and Pipe. Copyright 1969.)

Valve or fitting ANSI pressure classification			
Steam rating, lb		Cold rating, lb/in² gage	Schedule No. of pipe thickness
250 and Lower		500	S 40
300 to 600		1440	S 80
900		2160	S 120
1500		3600	S 160
2500	½ to 6″	6000	xx (double extra strong)
	8″ and larger	3600	S 160

$VD'' = 5.73 \times 2.067 = 11.8$ ft/sec × in
From Fig. 30, Re $= 1 \times 10^4$
From Fig. 28, $f = 0.032$
From Eq. (16),

$$h_{fs} = 0.032 \frac{20 \times 12}{2.067} \times \frac{5.73^2}{2 \times 32.17} = 1.9 \text{ ft}$$

Discharge pipe:
ID = 1.610 in
From Fig. 29, $\epsilon/D = 0.0011$
From Fig. 30, $\nu = 0.00009$ ft²/s
From Eq. (9),

$$V = \frac{60}{1.610^2} \times 0.408 = 9.44 \text{ ft/s}$$

$VD'' = 9.44 \times 1.610 = 15.2$ ft/sec × in
From Fig. 30, Re $= 1.5 \times 10^4$
From Fig. 28, $f = 0.030$
From Eq. (16),

$$h_{fd} = 0.030 \frac{300 \times 12}{1.610} \times \frac{9.44^2}{2 \times 32.17} = 92.9 \text{ ft}$$

Valve and fitting losses from Table 4*a*, *b*, and *c*, and Eq. (20):
2-in bellmouth, $K = 0.05$

$$h_{f1} = 0.05 \frac{5.73^2}{2 \times 32.17} = 0.026$$

2-in LR, 90° elbow, $K = 0.4 \pm 25\%$

$$h_{f2} = 0.4 \frac{5.73^2}{2 \times 32.17} = 0.202 \pm 0.051 \text{ ft}$$

2-in gate valve, $K = 0.16 \pm 25\%$

$$h_{f3} = 0.16 \frac{5.73^2}{2 \times 32.17} = 0.082 \pm 0.021 \text{ ft}$$

1½-in gate valve, $K = 0.19 \pm 25\%$

$$h_{f4} = 0.19 \frac{9.44^2}{2 \times 32.17} = 0.263 \pm 0.066 \text{ ft}$$

1½-in swing-check valve, $K = 2.4 \pm 30\%$

$$h_{f5} = 2.4 \frac{9.44^2}{2 \times 32.17} = 3.32 \pm 1.0 \text{ ft}$$

Total pipe, valve, and fitting losses:

$$\begin{aligned}\Sigma h_f &= h_{fs} + h_{fd} + h_{f1} + h_{f2} + h_{f3} + h_{f4} + h_{f5} \\ &= 1.9 + 92.9 + 0.026 + 0.202 + 0.082 + 0.263 + 3.32 \\ &= 98.69 \text{ ft}\end{aligned}$$

Total variation $= \pm(0.202 + 0.021 + 0.066 + 1.0) = \pm 1.29$ ft

EXAMPLE 13: Determine the valve and fitting losses in Example 12 by the equivalent length in pipe diameters method and compare results

From Example 12, 2-in bellmouth $h_{f1} = 0.026$ ft. From Table 5*a* and *b*:

2-in LR, 90° elbow, $L/D = 20$

$$L_2 = 20 \frac{2.067}{12} = 3.44 \text{ ft}$$

2-in gate valve, $L/D = 13$

$$L_3 = 13 \frac{2.067}{12} = 2.24 \text{ ft}$$

1½-in gate valve, $L/D = 13$

$$L_4 = 13 \frac{1.610}{12} = 1.74 \text{ ft}$$

1½-in swing-check valve, $L/D = 135$

$$L_5 = 135 \frac{1.610}{12} = 18.1 \text{ ft}$$

Using f from Example 12, total valve and fitting losses are:

$$\begin{aligned}\Sigma h_f &= h_{f1} + f_s \frac{L_2 + L_3}{D_s} \frac{V_s^2}{2g} + f_d \frac{L_4 + L_5}{D_d} \frac{V_d^2}{2g} \\ &= 0.026 + 0.032 \frac{(3.44 + 2.24)12}{2.067} \times \frac{5.73^2}{2 \times 32.17} \\ &+ 0.03 \frac{(1.74 + 18.1)12}{1.610} \times \frac{9.44^2}{2 \times 32.17} \\ &= 0.026 + 0.54 + 6.14 = 6.7 \text{ ft}\end{aligned}$$

Total valve and fitting losses from Example 12:

$$\Sigma h_f = (0.026 + 0.202 + 0.082 + 0.263 + 3.32) \pm 1.29 = 3.89 \pm 1.29 \text{ ft}$$

Increasers. The loss of head for a sudden increase in diameter with velocity changing from V_1 to V_2 in the direction of the flow can be calculated analytically. Computed results have been confirmed experimentally to be true to within ±3 percent. The head loss is expressed as shown below with K computed to be equal to unity:

$$h = K \frac{(V_1 - V_2)^2}{2g} = K \left[1 - \left(\frac{D_1}{D_2}\right)^2\right]^2 \frac{V_1^2}{2g} = K \left[\left(\frac{D_2}{D_1}\right)^2 - 1\right]^2 \frac{V_2^2}{2g} \qquad (21)$$

The value of K is also approximately equal to unity if a pipe discharges into a relatively large reservoir. This indicates that all the kinetic energy $V_1^2/2g$ is lost, and V_2 equals zero.

The loss of head for a gradual increase in diameter through a diffuser can be found from Fig. 36. The diffuser converts some of the kinetic energy into pressure. Values for the coefficient used with the above equations for calculating head loss are shown in the figure. The optimum total angle appears to be 7.5°. Angles greater than this result in shorter diffusers and less friction. but separation and turbulence occur. For angles greater than 50° it is preferable to use a sudden enlargement.

Reducers. Figure 37 gives values of the resistance coefficient to be used for sudden reducers.

Bends. Figure 38 may be used to determine the resistance coefficient for 90° pipe bends of uniform diameter. Figure 39 gives resistance coefficients for less than 90° bends and can be used for surfaces having moderate roughness such as clean steel and cast iron. Figures 38 and 39 are not recommended for elbows with R/D below 1. Table 6 gives values of resistance coefficients for miter bends.

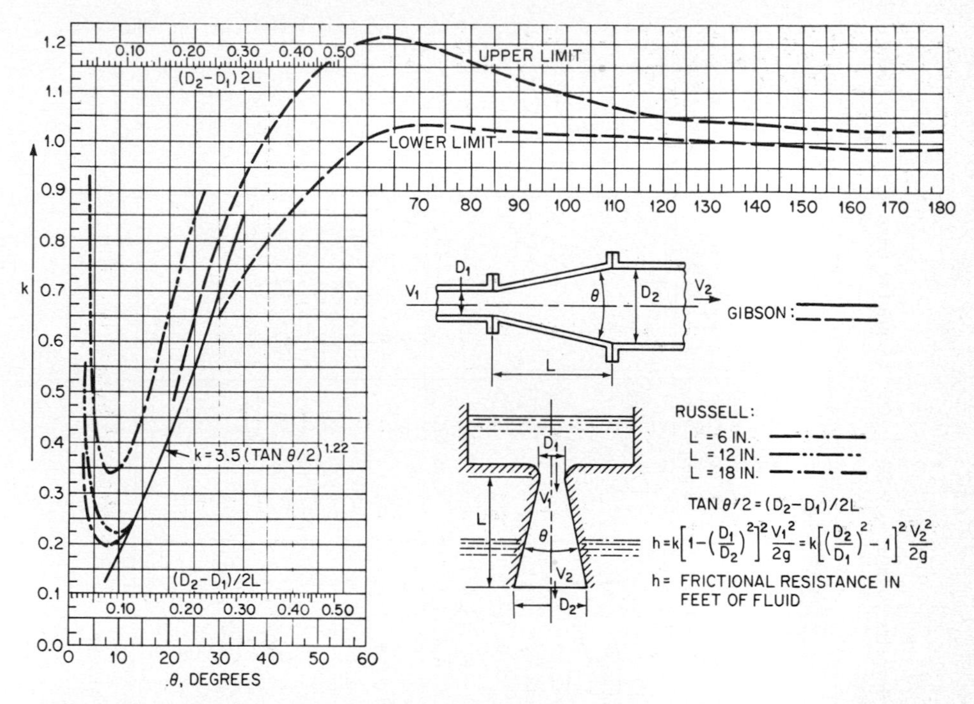

Fig. 36 Resistance coefficients for increasers and diffusers. (Reprinted from the "Pipe Friction Manual," 3d ed. Copyright 1961 by the Hydraulic Institute, Cleveland, Ohio)

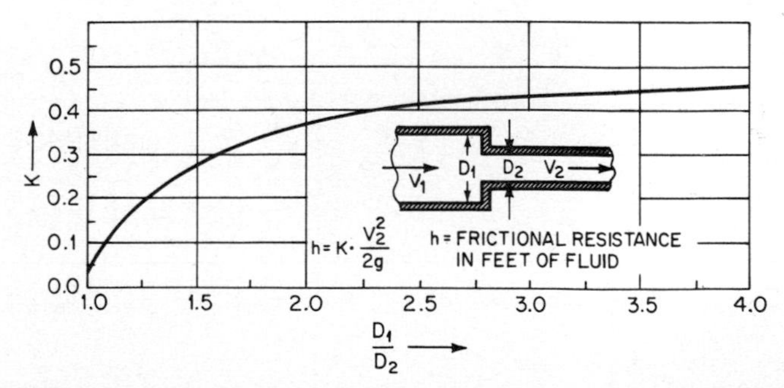

Fig. 37 Resistance coefficients for reducers. (Reprinted from the "Pipe Friction Manual," 3d ed. Copyright 1961 by the Hydraulic Institute, Cleveland, Ohio)

Pump Suction Elbows. Figures 40 and 41 illustrate two typical rectangular to round reducing suction elbows. Elbows of this configuration are sometimes used under dry pit vertical volute pumps. These elbows are formed in concrete and are designed to require a minimum height, thus permitting a higher pump setting with reduced excavation. Figure 40 shows a long-radius elbow and Fig. 41 a short-

TABLE 6 Resistance Coefficients for Miter Bends at Reynolds Number $\approx 2.25 \times 10^5$. (Reprinted from the "Pipe Friction Manual," 3d ed. Copyright 1961 by the Hydraulic Institute, Cleveland, Ohio)

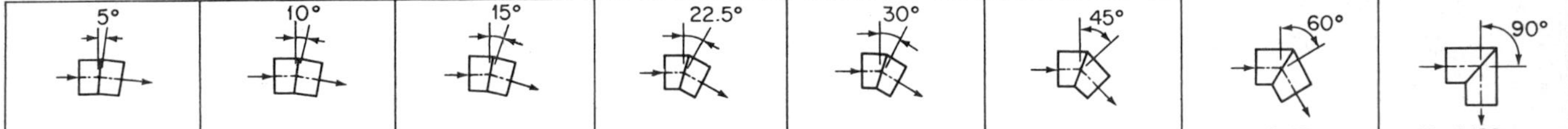

	5°	10°	15°	22.5°	30°	45°	60°	90°
K_s	0.016	0.034	0.042	0.065	0.130	0.236	0.471	1.129
K_r	0.024	0.044	0.062	0.154	0.165	0.320	0.684	1.265

	22.5°, 22.5°; 1.17 D	30°, 30°; 1.23 D	30°, 30°; 2.37 D	20°, 20°, 20°, 20°; 1.06 D	30°, 30°, 30°; 1.23 D, 2.37D	30°, 30°, 30°; 2.37 D, 1.23 D	30°, 60°; 1.44 D	60°, 30°; 1.44 D
K_s	0.112	0.150	0.143	0.108	0.188	0.202	0.400	0.400
K_r	0.204	0.268	0.227	0.236	0.320	0.323	0.534	0.601

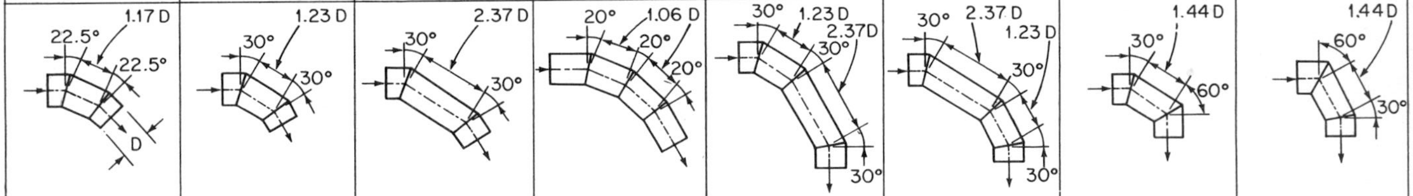

45°, 45° (spacing α, diameter D):

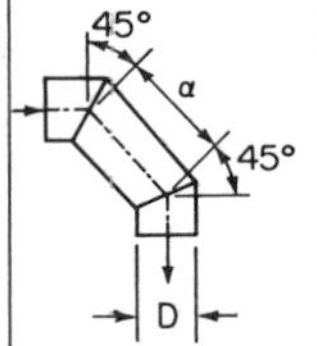

α/D	K_s	K_r
0.71	0.507	0.510
0.943	0.230	0.415
1.174	0.333	0.384
1.42	0.261	0.377
1.50*	0.280	0.376
1.88	0.269	0.390
2.58	0.338	0.429
3.14	0.346	0.426
3.72	0.356	0.490
4.89	0.389	0.455
5.59	0.392	0.444
6.29	0.399	0.444

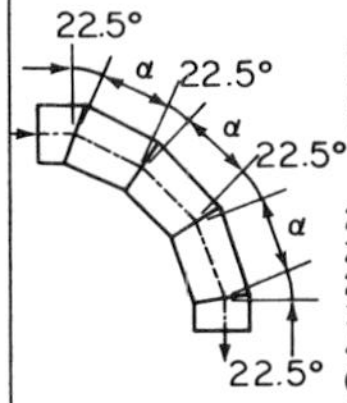

22.5°, 22.5°, 22.5°, 22.5° (spacing α):

α/D	K_s	K_r
1.186	0.120	0.294
1.40	0.125	0.252
1.50*	—	0.250
1.63	0.124	0.266
1.86	0.117	0.272
2.325	0.096	0.317
2.40*	0.095	—
2.91	0.108	0.317
3.49	0.130	0.318
4.65	0.148	0.310
6.05	0.142	0.313

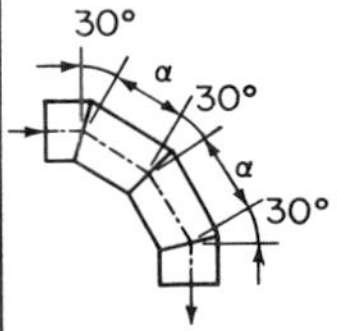

30°, 30°, 30° (spacing α):

α/D	K_s	K_r
1.23	0.195	0.347
1.44	0.196	0.320
1.67	0.150	0.300
1.70*	0.149	0.299
1.91	0.154	0.312
2.37	0.167	0.337
2.96	0.172	0.342
4.11	0.190	0.354
4.70	0.192	0.360
6.10	0.201	0.360

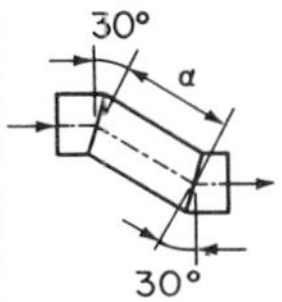

30°, 30° (spacing α):

α/D	K_s	K_r
1.23	0.157	0.300
1.67	0.156	0.378
2.37	0.143	0.264
3.77	0.160	0.242

K_s = RESISTANCE COEFFICIENT FOR SMOOTH SURFACE
K_r = RESISTANCE COEFFICIENT FOR ROUGH SURFACE, $\frac{\epsilon}{D} \cong 0.0022$

*OPTIMUM VALUE OF α INTERPOLATED

radius elbow. The resulting velocity distribution into the impeller eye and the loss of head are shown for these selected two designs.

Meters. Orifices, nozzles, and venturi meters are used to measure rate of flow. These metering devices, however, introduce additional loss of head into the pumping

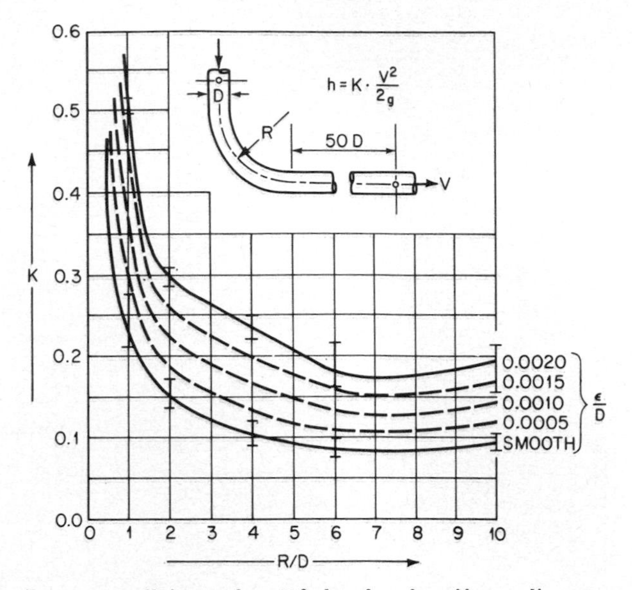

Fig. 38 Resistance coefficients for 90° bends of uniform diameter. (Reprinted from the "Pipe Friction Manual," 3d ed. Copyright 1961 by the Hydraulic Institute, Cleveland, Ohio)

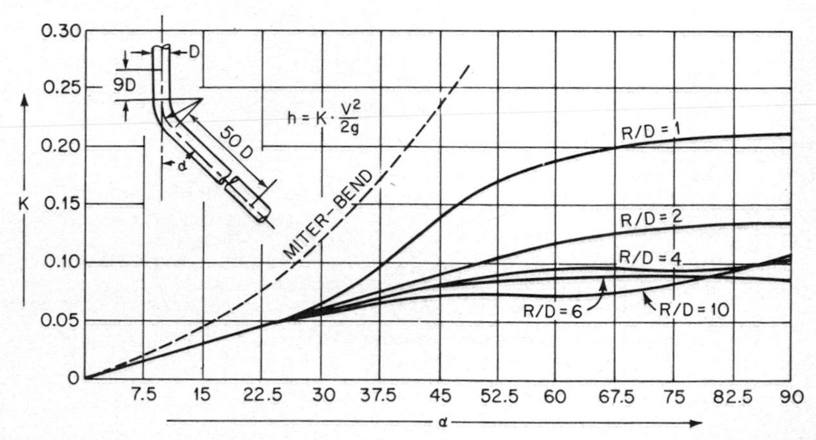

Fig. 39 Resistance coefficients for bends of uniform diameter and smooth surface at Reynolds number $\approx 2.25 \times 10^5$. (Reprinted from the "Pipe Friction Manual," 3d ed. Copyright 1961 by the Hydraulic Institute, Cleveland, Ohio)

system. Each of these meters is designed to create a pressure differential through the primary element. The magnitude of the pressure differential depends on the velocity and the density of the liquid and the design of the element. The primary element restricts the area of flow, increases the velocity, and decreases the pressure. An expanding section following the primary element provides pressure head recovery and determines the meter efficiency. The pressure differential between inlet and

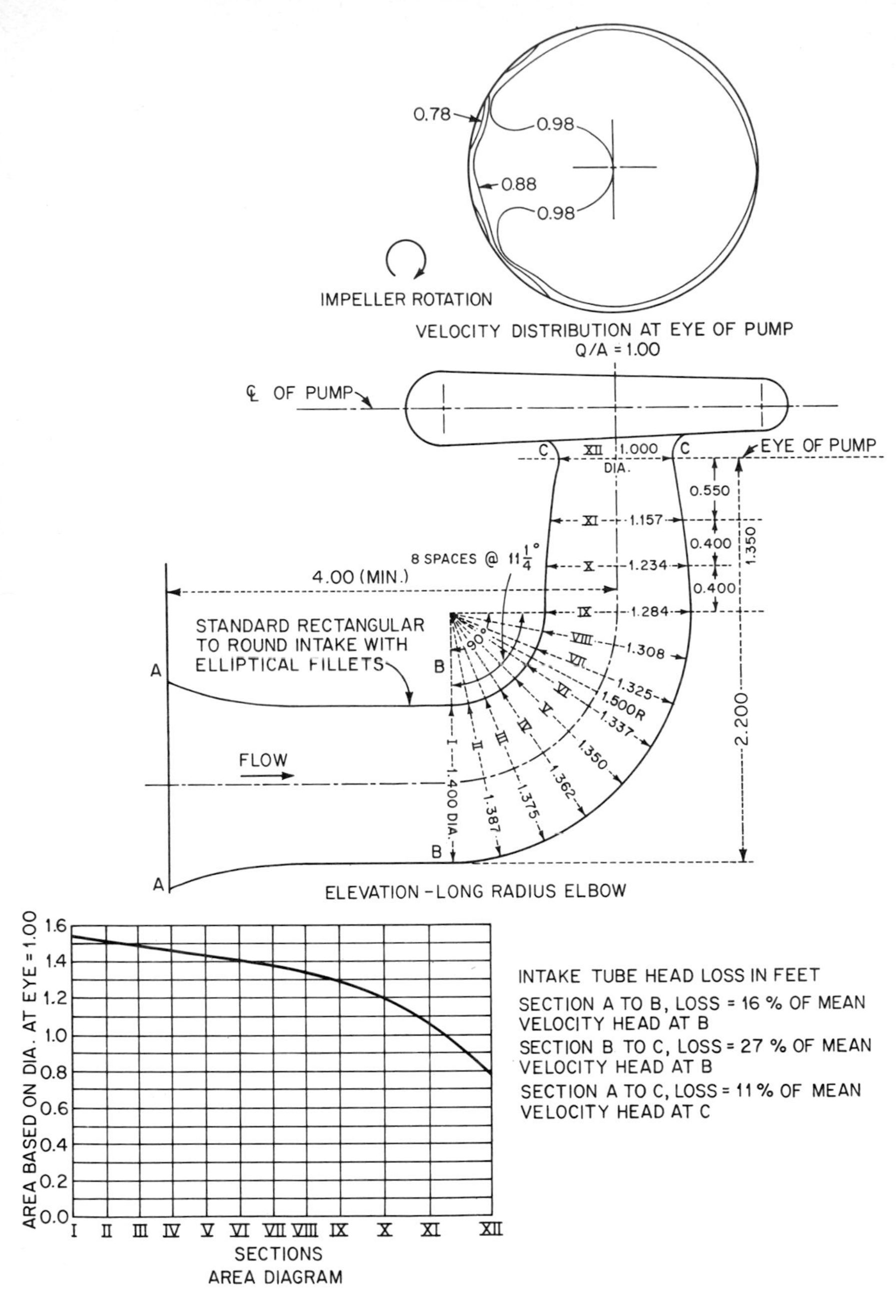

Fig. 40 Head loss in a long-radius pump suction elbow. (From U.S. Department of the Interior, Bureau of Reclamation, Design Standards No. 6, "Turbines and Pumps," 1956)

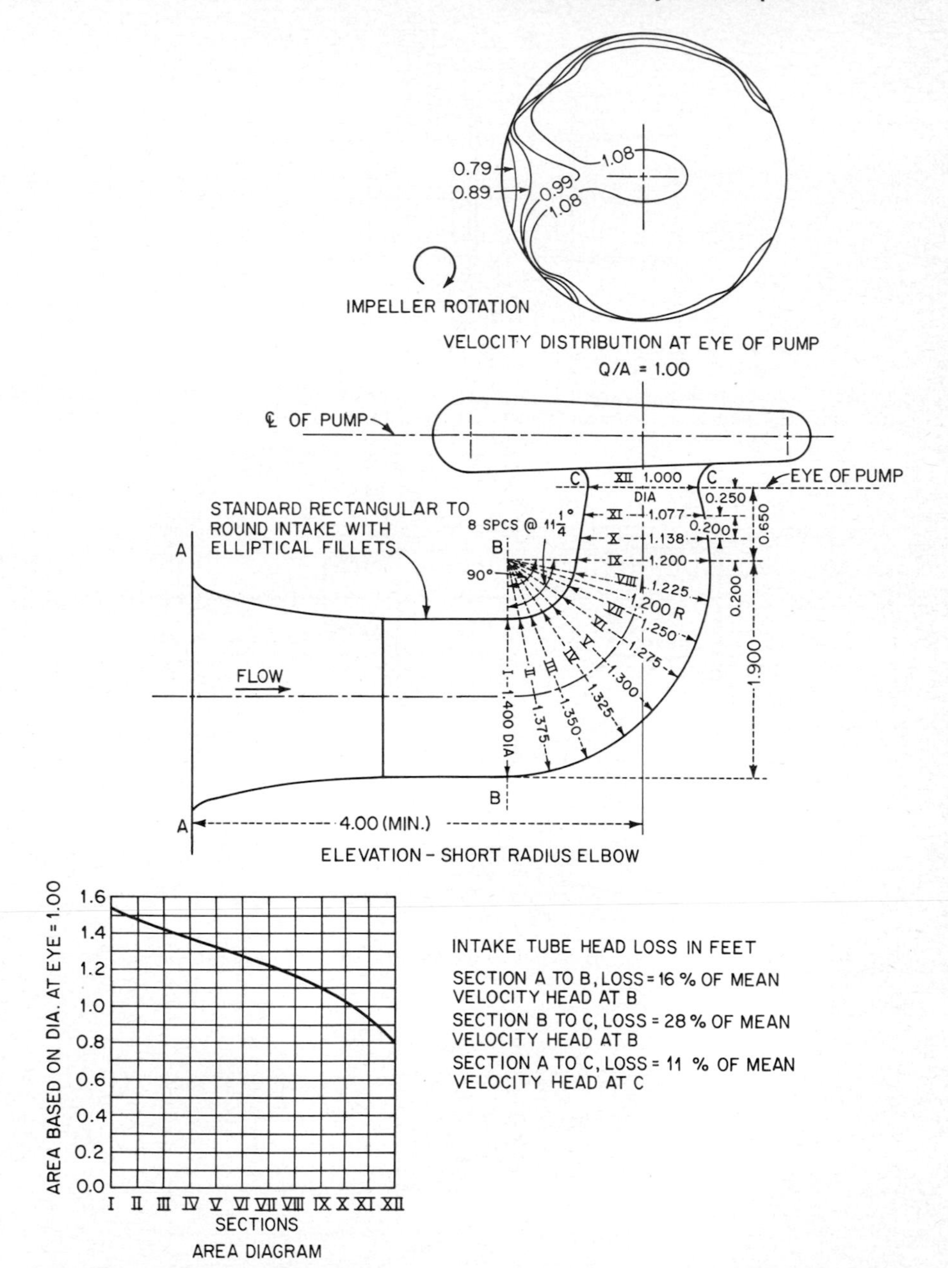

Fig. 41 Head loss in a short-radius pump suction elbow. (From U.S. Department of the Interior, Bureau of Reclamation, Design Standards No. 6, "Turbines and Pumps," 1956)

throat taps measures rate of flow; the pressure differential between inlet and outlet taps measures the meter head loss (an outlet tap is not usually provided). The meters offering the least resistance to flow are in the following decreasing order: venturi, nozzle, and orifice. Figures 42, 43, and 44 illustrate these different meter designs.

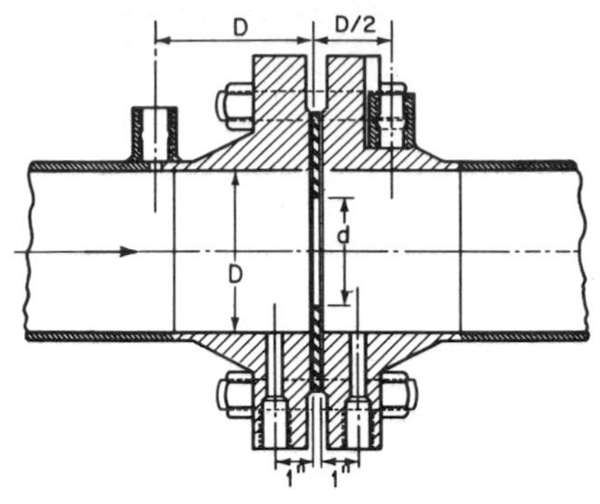

Fig. 42 Thin-plate square-edged orifice meter showing alternate locations of pressure taps. (From the American Society of Mechanical Engineers, "Fluid Meters," 6th ed. Copyright 1971, New York)

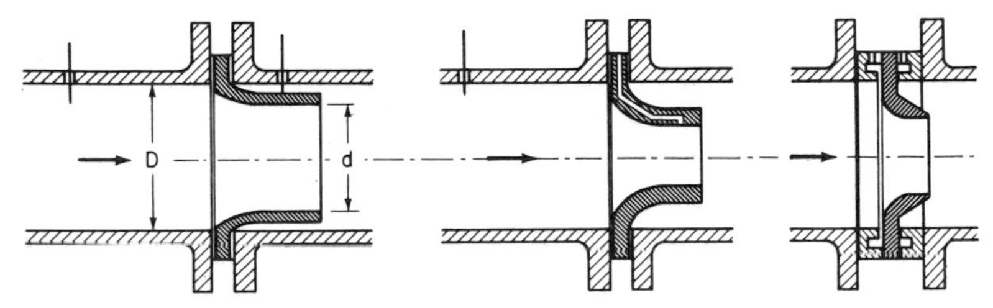

Fig. 43 Flow-nozzle-meter shapes and locations of pressure taps. (From the American Society of Mechanical Engineers, "Fluid Meters," 6th ed. Copyright 1971, New York)

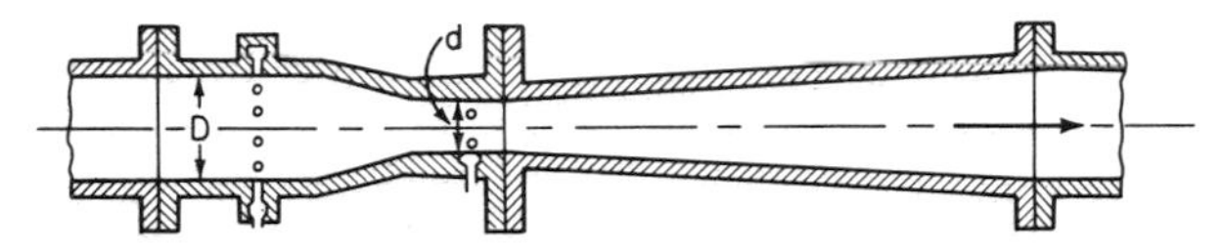

Fig. 44 Herschel type venturi meter showing locations of pressure taps. (From the American Society of Mechanical Engineers, "Fluid Meters," 6th ed. Copyright 1971, New York)

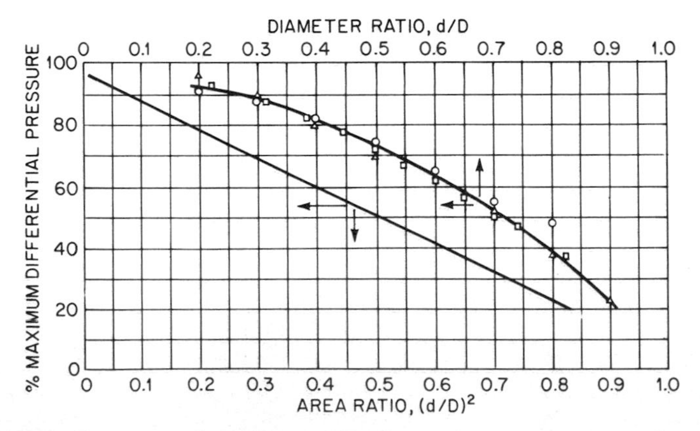

Fig. 45 Overall pressure loss across thin-plate orifices. (From the American Society of Mechanical Engineers, "Fluid Meters," 6th ed. Copyright 1971, New York)

When meters are designed and pressure taps are located as recommended (see Ref. 11), Figs. 45, 46, and 47 may be used to estimate the overall pressure loss. The loss of pressure is expressed as a percentage of the differential pressure measured at the appropriate taps and values are given for various size meters. Loss of head should be in units of feet of liquid pumped if other system losses are expressed this way. Reference 11 should be consulted for information concerning formulas and coefficients for calculating differential pressure vs. rate of flow.

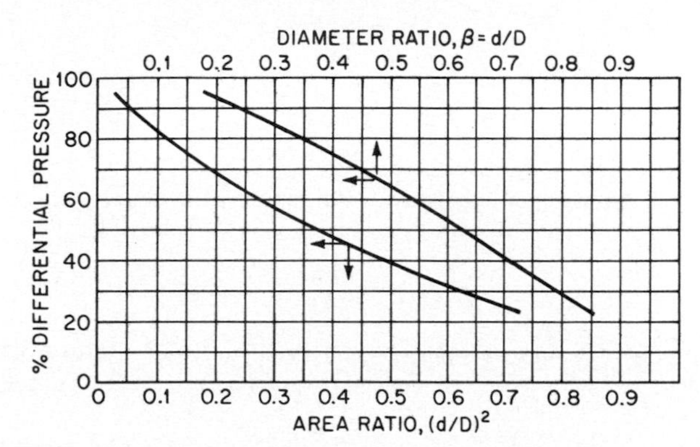

Fig. 46 Overall pressure loss across flow nozzles. (From the American Society of Mechanical Engineers, "Fluid Meters," 6th ed. Copyright 1971, New York)

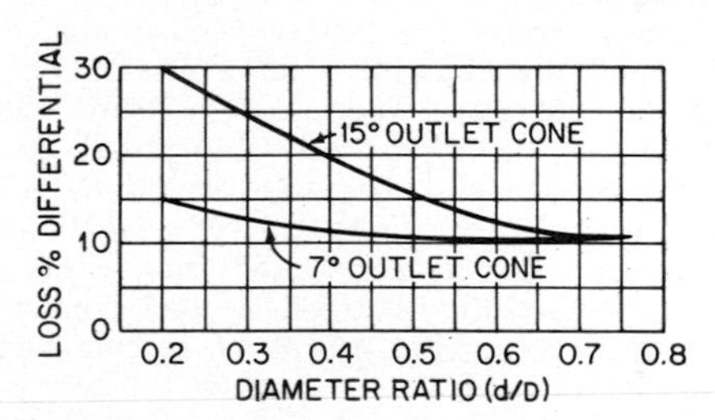

Fig. 47 Overall pressure loss across venturi tubes. (From the American Society of Mechanical Engineers, "Fluid Meters," 6th ed. Copyright 1971, New York)

TOTAL HEAD IN VARYING TEMPERATURE SYSTEMS

In a pumping system, or in a portion of a pumping system, where the weight of the liquid pumped is constant but the temperature varies, the volume flow rate will not be constant. An example would be the condensate and feedwater system in a steam generating plant. The following equation may be used to calculate flow rate in gpm using the specific gravity corresponding to the temperature of the liquid at the location in the system where the flow is required:

$$\text{gpm} = \frac{\text{lb/hr}}{500 \text{ sp gr}} \tag{22}$$

When calculating the total head required of a pump or pumps to overcome individual system-component losses, the actual flow rate and temperature through each component must be used since the loss is a function of velocity and viscosity. Figures and tables provided in this section of the handbook permit computing pipe, valve, and fitting losses in ft-lb/lb or ft of liquid passing through the component. If the pump is at a location in the system where the temperature is different than at the locations where the head losses are calculated, the total head to be produced by the pump is not the sum of the individual component heads. Because identical total heads in different density liquids do not have the same equivalent total pressures, and the same can be said for differential heads and pressures, the total head required across a pump must be the sum of the equivalent heads that will result in the pump producing the required system total differential pressure. The pump total head can be found by either correcting the total head loss in ft-lb/lb of liquid through the component to an equivalent head using the specific weight of the liquid at the pump which will produce the same differential pressure or by expressing all component losses in lb/in². From Eqs. (2) or (3) the equivalent head is

$$h_2 = \frac{w_1}{w_2} h_1$$

or

$$h_2 = \frac{\text{sp gr}_1}{\text{sp gr}_2} h_1$$

where subscript 1 denotes the component and subscript 2 denotes the pump. From Eq. (6),

$$p = \frac{hw}{144}$$

individual component head losses may be converted to lb/in². Also from Eq. (6),

$$TH = \frac{144\, P_\Delta}{w}$$

the total component pressure losses in lb/in² may be converted to an equivalent total pump head in ft-lb/lb or ft of liquid to be pumped.

In a varying temperature system the positive or negative head required to raise or lower the liquid pumped is not simply a difference in elevation. A pump must produce pressure in a pipe to raise liquid; the pressure required is proportional to the specific weight of the liquid. The head required at the pump should be found by expressing the suction and discharge elevation heads Z as pressures at the pump suction and discharge connections (corrected to the reference datum plane, if it is not at the pump centerline elevation) and using actual specific weights along the pipe. This differential pressure in lb/in² is then converted to an equivalent total elevation head using the specific weight or specific gravity at the pump in the above appropriate equations.

In designing a pumping system, there may be several locations for placing a pump to produce a specified increase in pressure. If the temperature of the liquid varies at the different locations considered, the pump total head in feet and the horsepower will be greater at the higher-temperature locations. Power is equal to the pump head in ft-lb/lb of liquid pumped times the flow rate in lb per unit of time. Both pump head in ft-lb/lb and power increase directly with the temperature at the pump location. While the total differential pressure P_Δ across the pump is the same at each location, the total head in ft-lb/lb varies with temperature or specific weight ($TH = 144\ P_\Delta/w$). Consequently pumps producing the same differential pressure must in the same period of time raise different liquids to different heights directly proportional to the temperature at the pump. Think of two piston-type pumps of equal bore producing the same differential pressure and pumping the same pounds per hour. The pump handling the less dense liquid at a higher temperature must have a larger stroke to pass a larger volume flow rate requiring proportionately more work to be done in the same period of time and therefore more power.

REFERENCES

1. Flow of Fluids through Valves, Fittings, and Pipe, Crane Co., Technical Paper 410, 1969.
2. Davis, C. V., and K. E. Sorensen: "Handbook of Applied Hydraulics," 3d ed., McGraw-Hill Book Company, New York, 1969.
3. "Pipe Friction Manual," 3d ed., Hydraulic Institute, New York, 1961.

4. Karassik, I. J., and R. Carter: "Centrifugal Pumps," McGraw-Hill Book Company, New York, 1960.
5. King, H. W., and E. Brater: "Handbook of Hydraulics," 5th ed., McGraw-Hill Book Company, New York, 1963.
6. Moors, J. A.: Criteria for the Development of Siphonic Action in Pumping Plant Discharge Lines, Hydraulic Division ASCE Annual Convention Paper, October 1957.
7. Richards, R. T.: Air Binding in Water Pipelines, Washington State University, 2d Hydraulic Conference, October 1959.
8. Stepanoff, A. J.: "Centrifugal and Axial Flow Pumps," 2d ed., John Wiley & Sons, New York, Inc.
9. Streeter, V. L.: "Fluid Mechanics," 5th ed., McGraw-Hill Book Company, New York, 1971.
10. "Turbines and Pumps," U.S. Department of the Interior, Bureau of Reclamation, Design Standards No. 6, 1956.
11. "Fluid Meters," 6th ed., The American Society of Mechanical Engineers, New York, 1971.
12. McNown, J. S.: Mechanics of Manifold Flow, *Proc. Amer. Soc. Civil Eng.,* 1953.

Section **9.2**

Branch-Line Pumping Systems

J. P. MESSINA

In some systems the liquid leaving the pump or pumps will divide into a network of pipes. If the pump is of the centrifugal type, the total pump flow is dependent on the combined system resistance. The total pump flow and flow through each branch can be determined by the following methods. (Review Pump Total Head and System-Head Curves in Sec. 9.1.)

BRANCHES IN CLOSED-LOOP SYSTEMS

Figure 1 illustrates a pump and network of piping consisting of three parallel branches in series with common supply and return headers. Junction points 1 and 2 need not be at the same elevation (provided the fluid density remains constant and the pipes flow full and free of liquid vapor) because in a closed-loop system the net change in elevation is zero. Figure 2 shows the system total head curves for each branch line and header when considered independent of each other and labeled *A*, *B*, *C*, and *D*. These curves are constructed for several flow rates by adding the frictional resistances of the pipes, fittings, and head losses through the equipment serviced from junction points 1 to 2. Curves *A*, *B*, *C*, and *D* therefore represent the variation in system resistance in feet vs. flow through each branch and header.

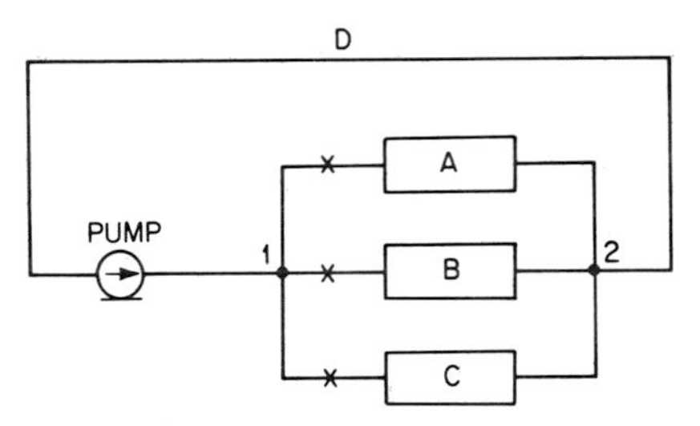

Fig. 1 Closed-loop pump system with branch lines.

If the valves are open in all branches the total system resistance, total pump flow, and individual branch flows are found by the following method. First observe

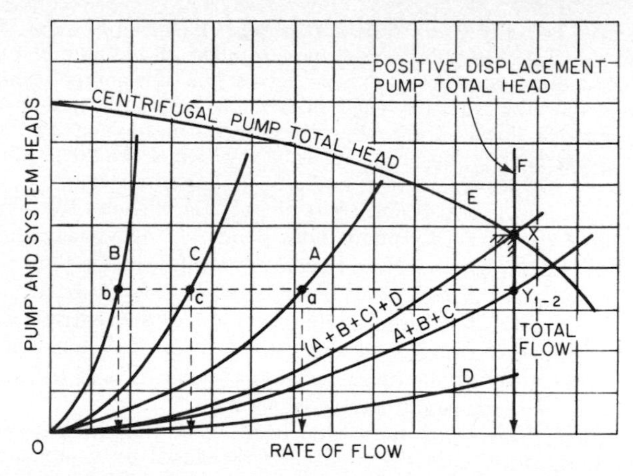

Fig. 2 System-head curves for pump and branch lines shown in Fig. 1 with all valves open.

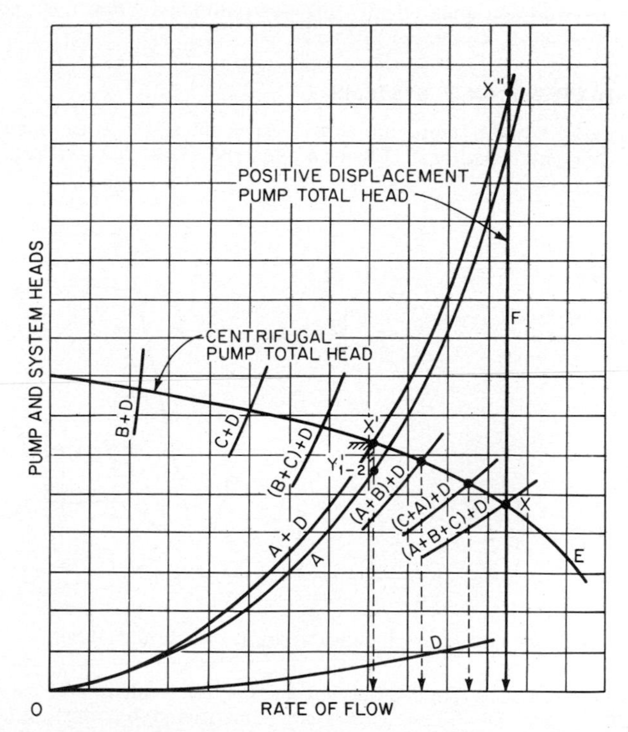

Fig. 3 System-head curves for pump and branch line shown in Fig. 1 with different combinations of open valves.

that (*a*) the total flow must be equal to the sum of the branch flows, (*b*) the head loss or pressure drop across each branch from junction 1 to junction 2 is identical, and (*c*) the flow divides to produce these identical head losses. Therefore, at several head points, add together the flow through each branch and obtain curve $A + B + C$. Header D is in series with $A + B + C$, and their system heads are added together for several flow conditions to obtain curve $(A + B + C) + D$. If curve E is the head-capacity characteristic of a centrifugal pump, point X represents the pump flow since at this point the system total head and pump total head are equal. Point Y_{1-2} represents the total head across junction points 1 and 2 and this head determines the flow through each branch; consequently points a, b, and c give individual branch flows. Curve F represents the head-capacity characteristics of a positive-displacement pump (constant capacity) which would produce the same flow conditions.

If valve A is open and valves B and C are closed, Fig. 3 shows the construction of the curves required to determine pump flow point X'. Obviously the pump flow and branch A flow are the same. Note that the total flow is less than the condition for all valves open at point X as a result of an increase in system head. If all valves were open and the total flow were obtained by a positive-displacement pump having a constant-capacity curve F, closing valves B and C would not change the flow. The system head would, however, increase to point X'' and the head would be greater than for a centrifugal pump having curve E.

Also shown in Fig. 3 are the system total head curves for different combinations of open valves A, B, and C and the resulting flow caused by a pump having characteristic curve E. For these various valve combinations the head differential across the junction points is found by subtracting the head of curve D from the system total head for the condition investigated, to obtain point Y'_{1-2}. The intersection of a horizontal line through point Y'_{1-2} and the individual branch curves gives the branch flow as previously illustrated in Fig. 2.

BRANCHES IN OPEN-ENDED SYSTEMS

Figure 4 illustrates a pump supplying three branch lines which are open-ended and terminate at different elevations. Figure 5 shows the system total head curves for

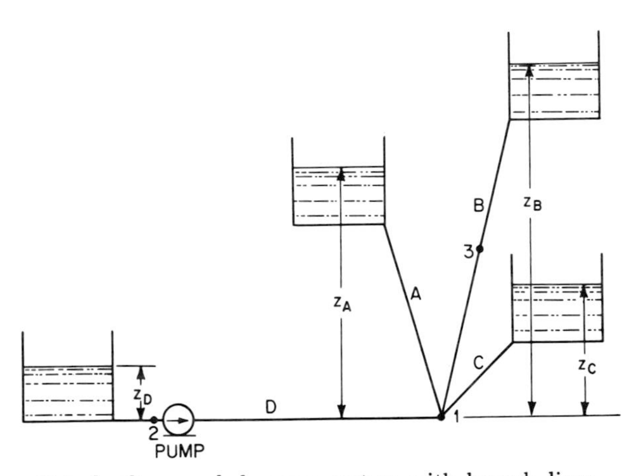

Fig. 4 Open-ended pump system with branch lines.

each branch line and main supply line considered independently of each other and labeled A, B, C, and D. These curves are constructed by starting at elevation heads Z_A, Z_B, Z_C, and Z_D at zero flow. To each of these heads is added the frictional resistances in each line for several flow rates. Friction losses from the suction tank to junction point 1 are to be included in curve D. Curves A, B, C, and D therefore represent the variation in system resistance in feet vs. flow through each branch and supply line. Note Z_D is negative because in line D there is a decrease in elevation to point 1.

The total head at the junction is the head Z_D in the suction tank measured above point 1, plus the pump total head, less the friction head loss h_{fD} in line D, and it varies with flow as illustrated by curve F.

The total system resistance, total pump flow, and individual branch flows are found by the following method. First observe that (*a*) the total flow must be equal to the sum of the branch flows, (*b*) the frictional resistance plus the elevation head measured relative to junction point 1 for each branch are identical, and (*c*) the flow divides to produce these identical total branch heads. Therefore, at several head points add together the flow of each branch to obtain curve $A + B + C$. Supply line D is in series with $A + B + C$, and their system heads are added algebraically for several flow conditions to obtain curve $(A + B + C) + D$. If curve E is the head-capacity characteristic of a centrifugal pump, point X represents the pump flow since at this point the system total head and pump total head are equal. Point Y_1 represents the total head at junction point 1 and this head determines the flow through each branch; consequently points a, b, and c give individual branch flows.

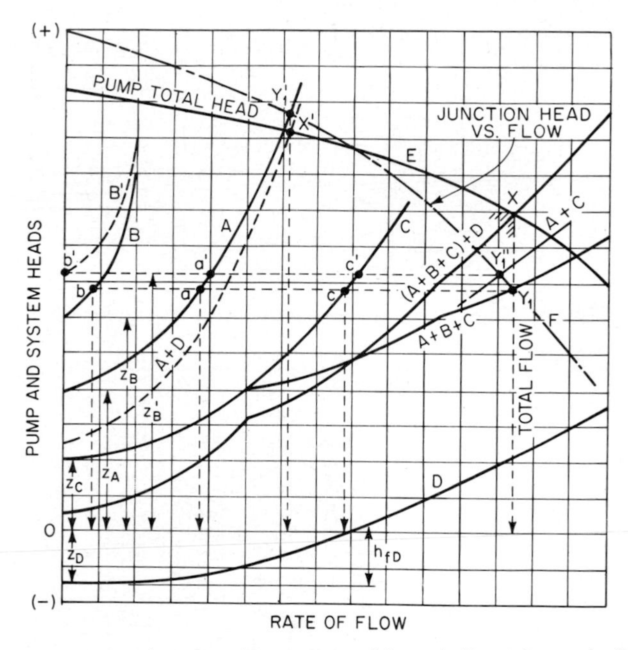

Fig. 5 System-head curves for pump and branch lines shown in Fig. 4.

In the above example the pump discharges to all tanks, but this should not be assumed. There is a limiting liquid level elevation for each tank, and, if exceeded, flow will be from the tank into the junction. Therefore it is possible for the lower level tanks to be fed by the higher-level tank and the pump. The limit for the liquid elevation in tank B is Z'_B and it is found from the intersection of curve $A + C$ with curve F, point Y''_1. The flow in branches A and C is at rates a' and c' when there is no flow in branch B. This is also a condition that would be similar to closing a valve in branch line B.

If elevation Z_B is greater than the previously found limiting height Z'_B, flow in branches A and C is determined in the following manner. Construct a curve for junction head vs. flow by adding heads and flows that result when the pump and the suction tank are in series with each other and tank B (less line losses) is in parallel with the pump and the suction tank. The intersection of this curve with curve $A + C$ will give the junction head required to determine the individual flows from the pump and tank, and the flows to tank A and tank C (not illustrated).

If flow to branch lines B and C is shut off, Fig. 5 illustrates the construction of the curves required to determine the pump flow point X' and junction head point Y'_1.

CENTRIFUGAL PUMP BYPASS

Bypass orifices around centrifugal pumps are often used to maintain a minimum flow recommended by the pump manufacturer because of one or more of the following reasons:

Limit the temperature rise
Reduce shaft and bearing loads
Prevent excessive recirculation in the impeller
Prevent overloading of the driver if pump horsepower increases with decrease in flow

Figure 6 illustrates a system which under certain conditions reduces pump flow below the recommended minimum. The pump delivers its flow to either tank A or tank

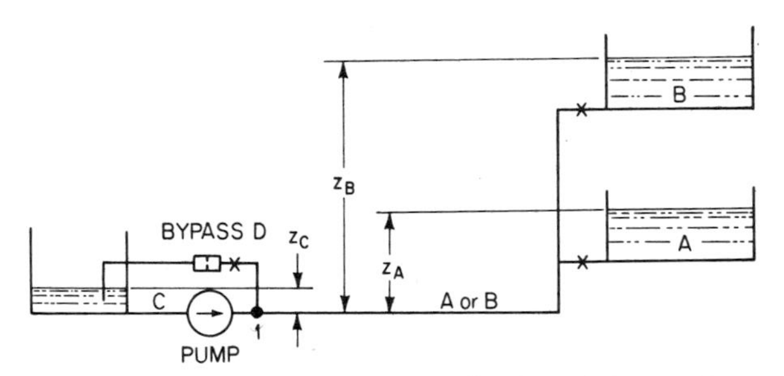

Fig. 6 Pump with bypass to maintain minimum flow.

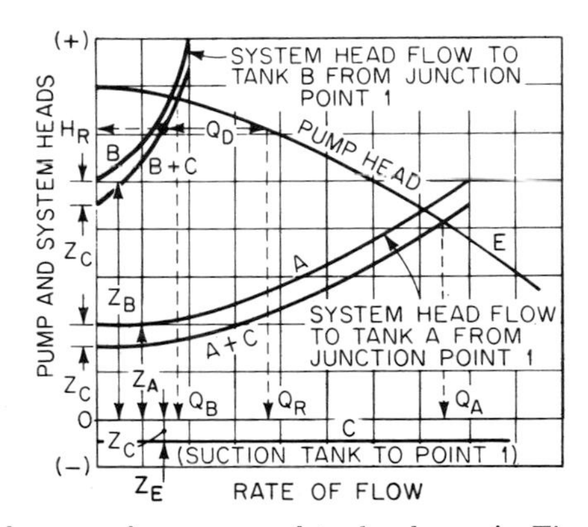

Fig. 7 System-head curves for pump and tanks shown in Fig. 6 with bypass open.

B. Figure 7 shows the separate system-head curves for flow to tank A and for flow to tank B. Curves A and B are heads required for flow from junction point 1 to tanks, and curves $A + C$ and $B + C$ are heads required for flows from suction level to tanks. Negligible losses are assumed between the suction tank and point 1. Curve E is the head-capacity characteristic of the centrifugal pump. Individual flow rates to each tank are shown as Q_A and Q_B. The recommended minimum flow is Q_R, which is greater than Q_B by the amount shown. In order to maintain the minimum flow, a bypass orifice and pipe are required to pass flow Q_D at total head H_R when pump discharges to tank B only.

Figure 8 shows the necessary construction to determine the required bypass head vs. flow characteristic of the orifice and pipe. Figure 9 illustrates the resultant pump flow with the bypass in operation. The bypass system head curve D includes elevation head Z_C from junction point 1 to suction tank surface level. Curve D is added to curves A and B using the method previously described for branch lines.

The head required to produce flow from the suction tank level to junction point 1 is negative Z_c and is shown as curve C in Fig. 7. Total system heads required for flow from the suction tank to the junction, where it divides with part flow recirculating through the bypass and part flow going to either tank, are shown in Fig. 9 as curves $C + (A + D)$ and $C + (B + D)$ for tanks A and B, respectively. These curves are constructed by series addition (addition of heads at the same flow). Points X' and X are pump flows when discharge is to tank A and B, respectively. Curve F represents junction head vs. flow and is constructed by adding Z_c to curve E. Flow through the bypass or flow to either tank can be determined by observing points Y_1 and Y_1', which are the junction heads for the two different flow conditions, and these heads produce flow through branch D and either branch A or B, shown as Q_A' and Q_B', respectively. Note that when flow is directed to tank B, pump flow is increased from

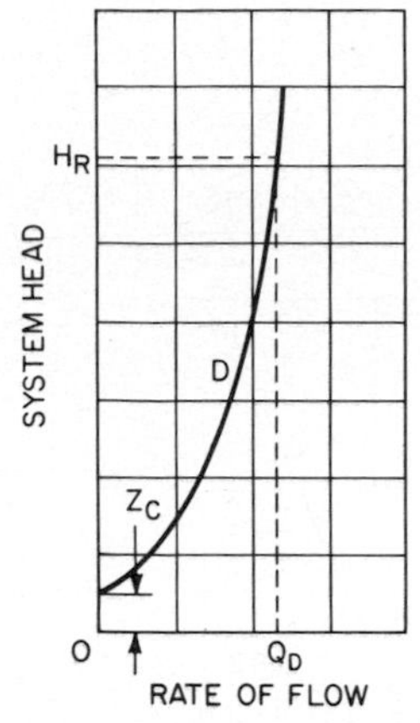

Fig. 8 Bypass orifice system requirements.

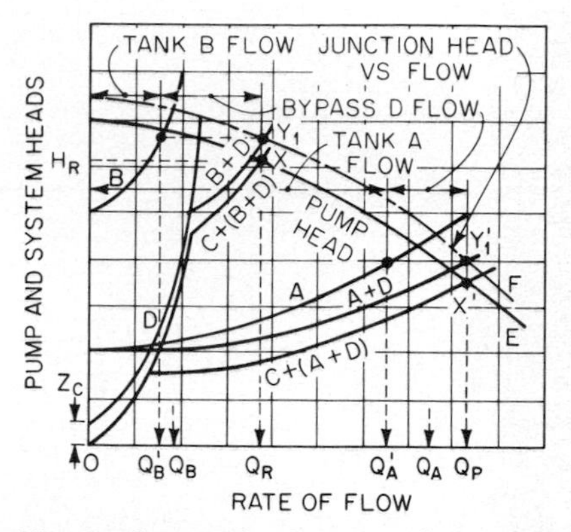

Fig. 9 System-head curves for pump and tanks shown in Fig. 6 with bypass valve open.

flow Q_B to flow Q_R. When the flow is directed to tank A, pump flow is increased from flow Q_A to flow Q_P. Also note that, while the pump flow is increased by use of the bypass, the flow to each tank has decreased. The bypass valve can be closed when discharging to tank A if it is desired to maintain maximum flow and to use minimum power. A pressure-relief valve in the bypass system could be set to open at a pressure equivalent to total head H_R in place of a manually operated valve.

FLOW CONTROL THROUGH BRANCHES

The flow through branches A, B, and C in Figs. 1 and 4 has been shown to be dependent on the individual branch characteristics. When parallel branches are connected to a pump a division of flow could result which may not satisfy the requirements of the individual lines. If it is desired that the flow to each branch meet or exceed specified individual line requirements, then it is only necessary to select a pump to provide the maximum head required by any one branch. In those branches where this head is more than required, the flow will be greater than the desired amount. A throttling valve or other flow-restricting device may be used to reduce the flow in these branches to the desired quantity. By controlling the flow in this manner the pump need only be selected to produce the minimum total-system flow at a total head required to satisfy the branch needing the highest head at junction 1 in Figs. 1 and 4.

The flow through a branch is sometimes dictated by the requirement of a component elsewhere in the system. For example, in the system shown in Fig. 10, the required flow through component D could be greater than the sum of the required flows through components A, B, and C. An additional branch line and control valve F (Fig. 10) around components A, B, and C may be used to bypass the additional flow needed by component D. Individual component throttling valves may also be used, if needed, to adjust the flow in each branch.

The following example illustrates how flow through branches may be controlled and how the pump total head is calculated.

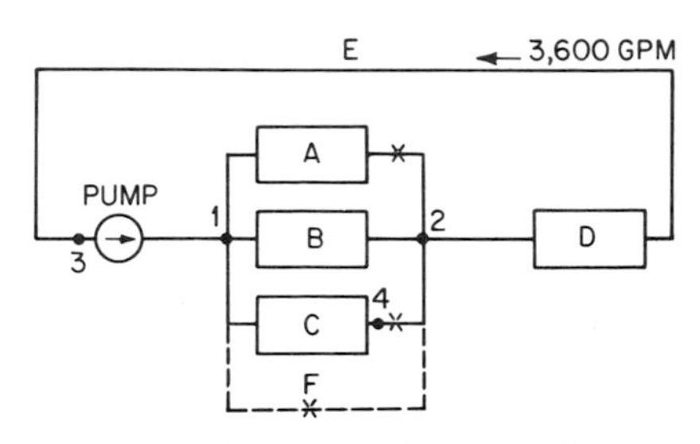

Fig. 10 Example branch-flow pumping system.

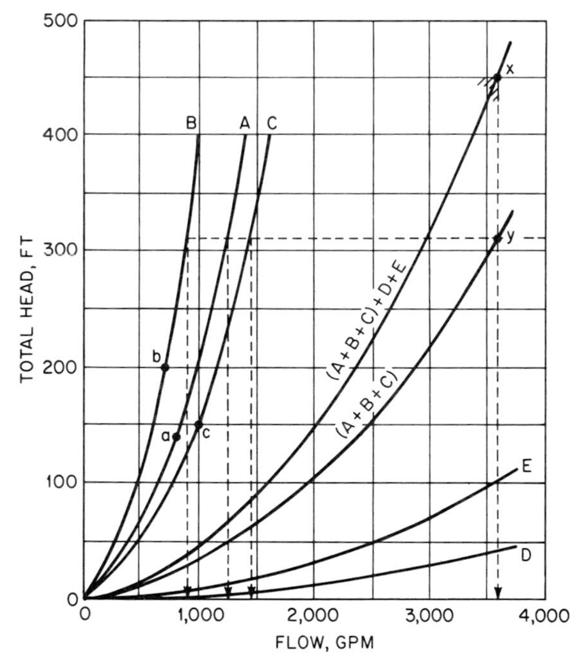

Fig. 11 System-head curves required for the solutions to the example problems.

EXAMPLE: A pump is required to circulate water at a rate of 3,600 gpm through the system shown in Fig. 10. The head vs. flow characteristics of the system components A, B, C, D, and E (system pipe and fittings) are shown in Fig. 11. The branch pipe and fitting losses from points 1 to 2 are included in the total heads for components A, B, and C.

1. Determine the pump total head required and the individual flows through components A, B, and C.
2. Determine the pump total head required if components A, B, and C need only 800, 700, and 1,000 gpm, respectively, and a bypass F is installed. Calculate the

individual throttling-valve head drops to achieve a controlled branch-flow system.
SOLUTIONS: 1. Refer to Fig. 11. Head vs. flow curves A, B, and C are added together in parallel, giving curve $(A + B + C)$. Curves $(A + B + C)$, D, and E are added together in series resulting in curve $(A + B + C) + D + E$. This latter curve indicates 450 ft total pump head is required at 3,600 gpm, point X. Curve $(A + B + C)$ crosses point Y at 3,600 gpm flow through the branches, and this condition requires 310 ft total head, which is the head across branch points 1 and 2. From each individual component curve the flow through branches A, B, and C can be read as 1,250, 900, and 1,450 gpm, respectively.

2. Refer to Fig. 11. Since the total flow through components A, B, and C need be only 2,500 gpm (800 + 700 + 1,000), the bypass should be designed to pass 1,100 gpm. Component B requires the maximum head, 200 ft differential (point b) across points 1 and 2. Throttling valves are needed in components A, C, and bypass F to increase the head in each branch to 200 ft at the required flows. Branch A (point a) requires only 140 ft total head to pass 800 gpm; therefore the throttling valve must be designed for a 60-ft head loss. Branch C (point c) requires only 150 ft total head to pass 1,000 gpm, requiring a throttling valve for 50 ft. The bypass control valve and piping should be designed to produce a 200-ft head drop at 1,100 gpm.

At 3,600 gpm the pump is now required to overcome 200 ft total head across points 1 and 2, 40 ft total head through component D, and 100 ft total head through the system pipe and fittings, component E, or a total of 340 ft. The reduction in pumping head from 450 to 340 ft, a saving of 110 ft, or 24.4 percent water horsepower, is the result of decreasing the branch head from 310 to 200 ft by bypassing the excess flow.

PUMP TOTAL HEAD IN BRANCH-LINE SYSTEMS

The total head produced is the difference in total heads measured across the suction and discharge connections of a pump. As explained in Sec. 9.1, General Characteristics of Pumping Systems and System-Head Curves, the total head is also the difference in heads between any two points in the pumping system, one on each side of the pump, plus the sum of the head losses between these two points. Confusion sometime results when the flow through the pump divides into branches in either closed-loop or open-ended systems. The points of head measurement can be in any branch line, upstream or downstream from the pump, regardless of the flow rate in these lines.

In part 2 of the above example, the pump or system total head could be measured using points 3 and 4 in Fig. 10. If the total head measured at the pump suction, point 3, were 25 ft gage, the head measured at point 4 would be 205 ft gage above the same reference datum assuming 10 ft of friction between the pump discharge and point 1. The difference in heads between points 3 and 4 is 180 ft. The loss of head due to friction and the head drop through component C is 10 + 150 or 160 ft. The pump and system total head at 3,600 gpm is therefore 180 + 160 or 340 ft.

Similarly, the pump and system total head for an open-ended system, such as Fig. 4, could be found by measuring, for example, head difference between points 2 and 3. Each head measurement would be referred to a common datum. The total head loss from 2 to 1 plus the head loss from 1 to 3 at the rate of flow in their respective lines added to the difference in heads between points 2 and 3 is the pump and system total head.

Section 9.3

Economics of Pumping Systems

G. R. KENT

INTRODUCTION

Every engineering decision must consider economics. An unprofitable venture must be guarded against to avoid the collapse of an engineering effort. It is important to consider both the capital outlay as well as the time value of money. A dollar borrowed for investment in a plant must be paid off in the future at the market value of the interest rate.

Each subsystem in a plant, such as a pumping arrangement, is subject to the same economic rules which apply to the entire project.

ECONOMIC FACTORS

There are three major costs which have to be considered in an economic evaluation: direct, indirect, and operating. Each of the basic costs is made up of items which are inherent in their meaning; these may be summarized as follows:

Direct costs (material and labor)
- Pump, drive, and driver
- Freight
- Foundations
- Piping
- Controls

Indirect costs (material and labor)
- Overhead
- Headquarters office
  - Engineering
  - Design inspection
  - Estimating
  - Purchasing
- Office Work
  - Supervision
  - Accounting
  - Field office
- Distributable cost items
  - Insurance

Temporary construction
Construction equipment
Cleanup

Operating Costs
Power: steam, electricity
Lubrication
Cooling water
Maintenance

To the total direct and indirect costs must be added local taxes and an engineering fee. The total direct and indirect costs, including local taxes and the engineering fee, represent the capital outlay for the equipment or plant, whereas the total operating costs must be evaluated on the basis of the time value of money.

In developing the economic factors covering the operating costs after determining the cost for each of the items listed under operating costs, it is necessary to establish the following factors:

Plant life expectancy, n
Annual interest rate, i
Annual fixed-charge rate, AFC
Present-worth factor, PWF
Capital-recovery factor, CRF
Annual levelized cost, ALC
Operating factor, OF

It is the annual levelized cost (ALC) which is required for addition to the total capital cost to determine the comparative investment value of alternate arrangements. For a determination of ALC we have:

$$\text{ALC} = \frac{(\text{present worth})(\text{CRF})}{\text{AFC}} \qquad (1)$$

where present worth = (hourly operating costs)(8,760)(OF)(PWF)

$$\text{CRF} = 1/\text{PWF}$$

$$\text{PWF} = [(1+i)^n - 1]/[i(1+i)^n]$$

To determine the values for PWF one may use logarithms, a desk calculator, or published tables such as in Grant's [1], where listings are given for varying interest rates at differing periods.

Plant life expectancy depends upon the industry and the total investment for the facility. An electric power generating station may vary from 30 to 50 years for plant life, depending on its being fossil- or nuclear-fueled. A petrochemical plant may vary from 15 to 20 years plant life, depending on size of facility and competitive market of the produced product.

The annual interest rate is dependent on the charge for borrowing money. Bank charge rates vary with available funds, size of loan, and period over which the payment is to be made.

Each successful industry has developed its own profit rate after taxes to defray operating costs and satisfy its shareholders. The electric power generating industry is government-regulated as to its profitability; therefore, the annual fixed charge rate AFC is set by regulation. Other industries, regulated by a competitive market, establish their profit margins by competition. Profit margins for many industries may be obtained from the Federal Trade Commission Quarterly [2], listed as profit rate after taxes which, for economic evaluation, has the same meaning as AFC.

The operating factor is dependent on a plant's design load and operating expectancy of the plant. These conditions are developed with the plant owner's cooperation so as to obtain realistic performance requirements.

PUMP AVAILABILITY

Every manufacturer designs his equipment to operate over its full life span without difficulty or mechanical problems. In spite of the manufacturer's intent, there are small items such as seals, packing, bearings, etc. which may develop an unexpected problem while operating, thus causing an unscheduled shutdown. These shutdowns for repairs are fortunately short in duration but must be allowed for in the design of a plant. For a discussion of availability and definitions of terminology, refer to the Industrial Power Systems Publication [3], which reviews the problems and presents the industries' approach to a solution. The electric utility industry [4] has developed a record keeping coding system for the purpose of obtaining equipment availability. Unfortunately there are little published data to guide one as to the availability of pumping equipment. The tabulation in Table 1 is presented as an example for evaluating unscheduled centrifugal pump failures.

TABLE 1 Typical Unscheduled Failures for Centrifugal Pumps

Type of failure	Hours per outage	Est. failures per year	Hours outage per year
Packing	4	4.0	16.0
Seals	16	1.0	16.0
Bearings	16	0.5	8.0
Unbalance	48	0.2	9.6
Total			49.6

From the tabulated estimated values, we deduce that centrifugal pumps may be expected to result in an availability of 99.3 percent. It is not enough to consider the pump alone; its driver must also be included, and since most pumps are driven by electric motors, whose availability is as good as or better than the pumps, 99.3 percent may be taken as a value for the combined assembly.

When a pump is driven by an engine, the availability of the combined assembly would be governed by the engine driver. For a guide to engine performance refer to the ASME Report [5].

The availability of the pump is considered when a decision is to be reached on a single unit or multiple units.

PUMP OPERATING RANGE

Rating vs. Operation A pump is sized to include margins for capacity and head. Figure 1 defines the pump margin as it affects the characteristics of a centrifugal pump. It is important to inform a pump supplier where the expected operation is to be so as to obtain its best efficiency within the operating range.

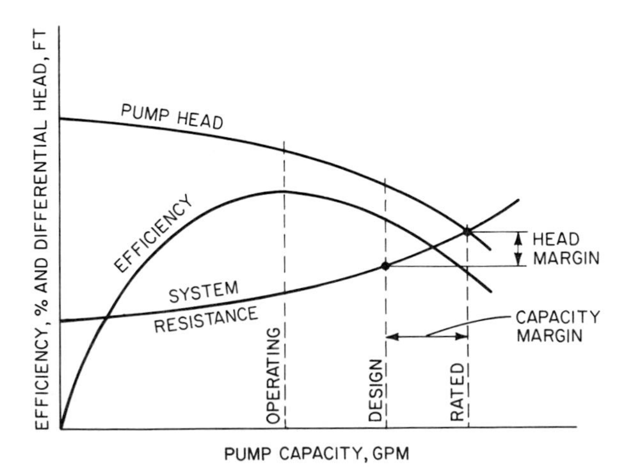

Fig. 1 Margins and characteristics for a centrifugal pump.

The capacity margin is added to ensure operability at the design value. The head margin results from the capacity margin to fit the system resistance. Other allowances for pump performance are left to the supplier because they are subject to manufacturing tolerances.

Power at Varying Loads Since a pump rarely, if ever, is called upon to operate at one condition during the life of a plant, it is necessary to define the expected operating range. Taking into account the power consumption and operating factor at differing loads allows the determination of power consumption over the plant life. Curves similar to Fig. 1 for a specific pump application are referred to for the determination of power consumption.

Multipump Selection The threat of an unscheduled shutdown with the resulting loss of productivity requires the consideration of equipment availability. By comparing the value of lost production with the additional cost of spare pump units or multiple pump selection, an economic basis is set to guide a decision. A multipump selection cuts back the shutdown loss to a fraction of what it would be for a single unit. A full spare pump results in total protection but requires investment and maintenance without affecting production except in an emergency.

For the determination of loss in production due to unscheduled shutdowns, we calculate the following for centrifugal pumps:

Net annual loss = (net product value) (annual production capacity) (0.007). The factor 0.007 results from the pump availability of 99.3 percent. The net product value is the difference between the sale price of the product and the cost of production.

The net annual loss may be totally or in part reduced by installing a spare pump, which would be placed into service during an unscheduled shutdown, or multiple parallel pumps, only one of which would be shut down while the others remained in operation.

Pump Duty Petrochemical plant pumps may be expected to operate over a cycle similar to Table 2.

TABLE 2 Process Pump Operating Cycle over a 16-Year Plant Life

Operating years	% operating factor for % plant load		
	100	50	0
1–2	90	8	2
3–12	95	3	2
13–14	90	8	2
15–16	80	10	10

A centrifugal pump service which will be evaluated over the operating cycle of Table 2 for the purpose of determining multiple pump service is presented as follows:

Service—demethanizer reflux
Plant elevation—sea level
Maximum discharge—509 lb/in^2 abs
Operating temperature— −143°F
Vapor pressure—450 lb/in^2 abs
Sp gr—0.30
Available NPSH—14 ft
Driver—electric motor (440-V, 3-phase, 60-Hz)

The demethanizer pump duty is specified on Fig. 2, which includes the duties for two 50 percent capacity units and three 33⅓ percent capacity units as well as the 100 percent capacity unit. The peak pump efficiency is required for the 100 percent capacity for each pump selection.

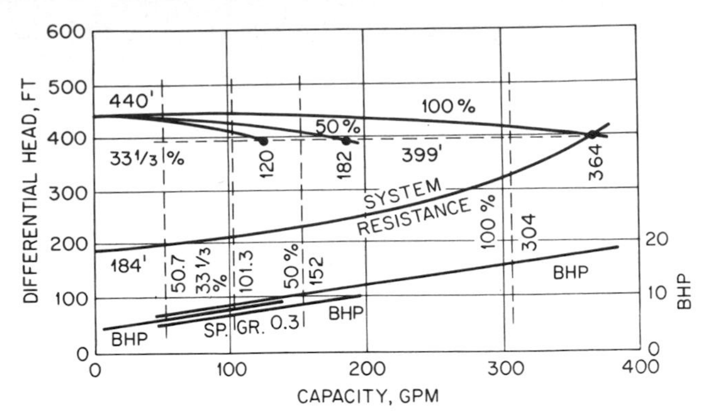

Fig. 2 Duty and service for a demethanizer pump.

Pumps in an electric power generating plant may be expected to operate over a cycle similar to Table 3.

A centrifugal pump service, which will be evaluated over the operating cycle of Table 3 for the purpose of determining economic piping size, is presented as follows:

Service—boiler feed
Plant elevation—sea level
Maximum discharge—5,520 lb/in² abs
Suction temperature—323°F
Discharge temperature—329°F
Sp gr—0.905
Available NPSH—115 ft
Driver—steam turbine (inlet 172 lb/in² abs and 677°F; exhaust 2½ in Hg)

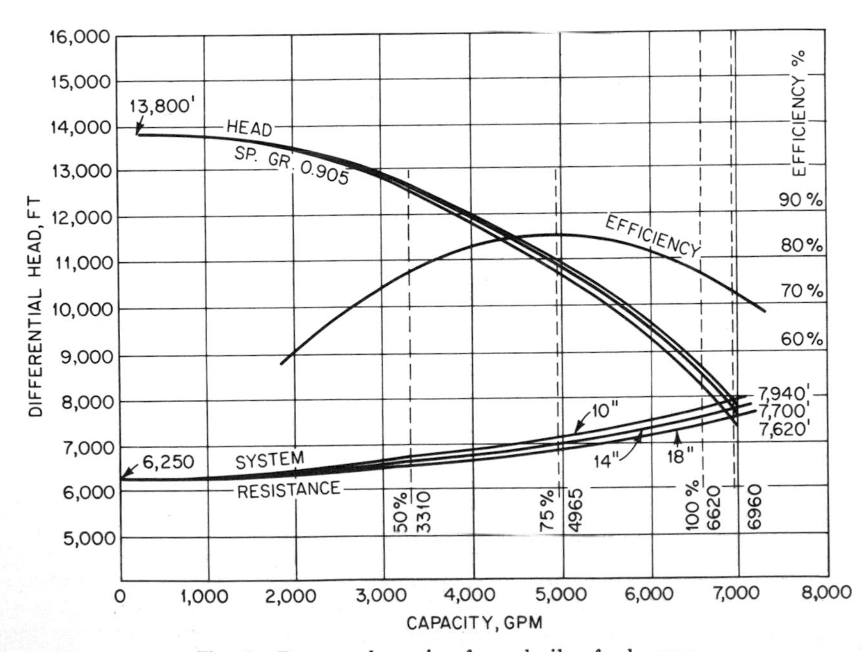

Fig. 3 Duty and service for a boiler feed pump.

TABLE 3 Boiler Feed-Pump Operating Cycle over a 35-Year Plant Life

	% operating factor for % plant load			
Years	100	75	50	0
1–5	60	0	30	10
6–10	55	0	30	15
11–15	40	5	32.5	22.5
16–20	25	10	35	30
21–25	10	15	37.5	37.5
26–35	0	0	20	80

The boiler-feed-pump duty is specified on Fig. 3, which includes the modifications imposed by changes in discharge-piping size. The peak pump efficiency is required at 75 percent capacity.

ECONOMIC EVALUATION

Multipump Selection The service conditions and operating cycle stated for a demethanizer reflux pump will be used to evaluate multipump selection. The direct and indirect costs for each pump arrangement are summarized in Table 4.

TABLE 4 Summary of Direct and Indirect Costs for Assumed Alternate Demethanizer Pump Selections

Direct and indirect costs	(1) 100% capacity	(2) 100% capacity	(2) 50% capacity	(3) 33⅓% capacity
Pump.	$ 4,900	$ 9,800	$ 9,100	$10,200
Piping.	20,210	25,920	24,870	27,910
Total. . . .	$25,110	$35,720	$33,970	$38,110

The data for evaluating operating costs are as follows:

Plant life—n = 16 yr
Annual interest rate—i = 8 percent
Profit rate after taxes—AFC = 11.9 percent
Charge for electric power—7.8¢/kWhr
Charge for lubrication—70¢/gal
Charge for cooling water—7.5¢ 1,000 gal
Charge for maintenance—$1.50/hp-yr

To evaluate loss in production resulting from unscheduled pump shutdowns, it is necessary to determine a value for the losses.

Annual production = 60×10^6 lb/yr
Unit sale price = 76.3¢/lb
Unit production cost = 18.1¢/lb
Net unit loss = 58.2¢/lb

$$\text{Net annual loss} = 0.582 \times 60 \times 10^6 \times 0.007 = \$244{,}000$$

The sum of evaluated pump cost and annual loss for each pump arrangement yields a total cost for comparison to determine the lowest sum as the economic choice.

For needed calculations, it is convenient to determine the product of operating factor (OF) and present-worth factor (PWF) over the life of the plant. Table 5 summarizes the calculations for each plant load.

TABLE 5 Adjusted Operating Factors for Process Plant Life

		100% load		50% load	
Years	PWF	OF	(OF)(PWF)	OF	(OF)(PWF)
1–2	1.783	90	160.5	8	14.28
3–12	5.753	95	547.0	3	17.30
13–14	0.708	90	63.8	8	5.66
15–16	0.607	80	48.6	10	6.07
Total			819.9		43.31

PWF at 16 yr = 8.851
CRF for plant life = 0.113
At 100% load: 819.9 × 0.113 = 92.5%
At 50% load: 43.31 × 0.113 = 4.9%

Table 6 summarizes the required power for each selected pump arrangement for the demethanizer pump service.

TABLE 6 Power Consumption Summary for Demethanizer Pump Service

		Each pump							
Pump size, %	Capacity, %	gpm	Pump head, ft	Theoretical hp	% E_p	bhp	% E_m	kWhr	Total kWhr
100	100	304	410	9.45	59.0	16.0	87.8	13.6	13.6
	50	152	435	5.0	49.6	10.1	87.1	8.7	8.7
50	100	152	408	4.7	58.7	8.0	86.8	6.9	13.8
	50	152	408	4.7	58.7	8.0	86.8	6.9	6.9
33⅓	100	101.3	405	3.1	47.8	6.5	85.9	5.7	17.1
	50	50.7	430	1.65	30.0	5.5	84.5	4.9	10.6

$$\text{Theoretical hp} = \frac{(\text{gpm})(\text{head})(\text{sp gr})}{3{,}960}$$

$$\text{bhp} = \frac{\text{theoretical hp}}{E_p}$$

$$\text{kWhr} = \frac{0.745 \text{ bhp}}{E_m}$$

Lubricating oil consumption is based on 0.02 gal/100 hp-hr for each pair of bearings. A motor driven centrifugal pump results in 0.04 gal/100 hp-hr total. Hence, gal/hr = 0.04 bhp/100.

Cooling-water requirements are based on 10°F temperature rise and 2 percent energy loss to the water for each pair of bearings. Because the demethanizer service is at −143°F it will not be necessary to furnish cooling water for the pumps; it will, however, be necessary to furnish cooling water for the motor bearings. Therefore, gpm = bhp/100.

The annual operating costs associated with each pump arrangement in demethanizer service are developed from the following:

Power—0.078 (kWhr) × 8,760
Lubricating oil—0.7 × 0.04 (bhp per 100) × 8,760
Cooling water—(0.075 per 1,000) (bhp per 100) × 60 × 8,760
Maintenance—1.5 × bhp

From the data listed in Table 6 combined with the cost values listed above, Table 7 summarizes the total annual operating costs.

TABLE 7 Operating Costs for Process Pump Arrangements

Pump arrangement	Service	100% load	50% load
1-100% capacity	Power	$ 9,300	$5,960
	Lub. oil	39	25
	Cooling water	6	4
	Maintenance	24	15
	Total	$ 9,369	$6,004
2-50% capacity	Power	$ 9,450	$4,730
	Lub. oil	39	20
	Cooling water	6	3
	Maintenance	24	12
	Total	$ 9,519	$4,765
3-33⅓ capacity	Power	$11,700	$7,260
	Lub. oil	48	30
	Cooling water	8	5
	Maintenance	29	18
	Total	$11,785	$7,313

From the total annual operating costs listed in Table 7, we determine the annual levelized costs (ALC) as follows:

$$\text{ALC} = \frac{(\text{total operating cost})(\text{OF})(\text{PWF})(\text{CRF})}{(\text{AFC})}$$

where the product (OF) (PWF) (CRF) has been determined in Table 5 for 100 and 50 percent of plant load. Table 8 summarizes all the costs associated with the alternate demethanizer pump arrangements.

TABLE 8 Total Evaluated Costs for Process Pump Arrangements

Load	100% ALC	100% ALC	50% ALC	33⅓% ALC
100%	$ 72,900	$ 72,900	$ 74,000	$ 91,400
50%.	2,480	2,480	1,965	3,010
Total (ALC).	$ 75,380	$ 75,380	$ 75,965	$ 94,410
Direct and Indirect	25,111	35,720	33,970	38,110
Evaluated costs	$100,491	$111,100	$109,935	$132,520
Annual loss	244,000	0	122,000	81,333
Comparative sum	$344,491	$111,100	$231,935	$213,853

Note that the annual loss which was calculated to be $244,000 is listed as zero for two 100 percent capacity units because one of the units is a spare to be placed in operation during an unscheduled shutdown. For the multiple parallel units, it is considered that only one of the pumps may be shut down during an unscheduled mishap; therefore, the loss is proportionately less.

Thus it may be seen that the most economic pump arrangement (the lowest comparative sum), is two 100 percent capacity units.

Optimum Piping Size The service conditions and operating cycle stated for a boiler feed pump will be used to evaluate optimum piping size. Only changes in discharge piping will be considered for study because the suction piping is subject to NPSH restrictions and, therefore, not subject to economic considerations. Table 9 summarizes the direct and indirect costs associated with the boiler feed pump service.

TABLE 9 Summary of Direct and Indirect Costs for Alternate Piping and Boiler Feed Pump

Direct and indirect costs	10 in	14 in	18 in
Pump.	$1,436,350	$1,436,350	$1,436,350
Piping.	236,784	347,170	549,310
Total.	$1,673,134	$1,783,520	$1,985,660

The data for evaluating operating costs are as follows:

Plant life—n = 35 yr
Annual interest rate—i = 7%
Annual fixed-charge rate—AFC = 13%
Charge for steam—80¢/1,000 lb
Charge for lubrication—70¢/gal
Charge for cooling water—7.5¢/1,000 gal
Charge for maintenance—$1.50/hp-yr

Since an electric generating system is made up of multiple units which permit minor unscheduled shutdowns, it is not generally necessary to consider the availability of a centrifugal pump requiring more than a one 100 percent capacity unit.

For needed calculations, it is convenient to determine the product of operating factor (OF) and present worth factor (PWF) over the life of the plant. Table 10 summarizes the calculations for each plant load.

TABLE 10 Adjusted Operating Factor for Power Plant Life

		100% load		75% load		50% load	
Years	PWF	OF	OF, PWF	OF	(OF)(PWF)	OF	(OF)(PWF)
1–5	4.100	60	246.011	0	0	30	123.006
6–10	2.923	55	160.786	0	0	30	87.701
11–15	2.084	40	83.373	5	10.422	32.5	67.741
16–20	1.486	25	37.153	10	14.861	35	52.014
21–25	1.060	10	10.596	15	15.894	37.5	39.734
26–35	1.294	0	0	0	0	20	25.882
Total.			537.919		41.177		396.078

PWF at 35 yr = 12.95
CRF for plant life = 0.07723
At 100% load: 537.918 × 0.07723 = 41.55%
At 75% load: 41.176 × 0.07723 = 3.18%
At 50% load: 396.076 × 0.07723 = 30.59%

Table 11 summarizes the steam consumption for the varying discharge piping size and the differing capacity of the boiler feed pump.

From steam tables or an H-S diagram:

Steam	*Lb/in² abs*	*°F*	*Enthalpy*
Inlet	172	677	1,363
Exhaust	2½ inHg	. . .	977
			386 Btu/lb

$$\text{Theoretical } WR = \frac{2{,}545}{386} = 6.6 \qquad WR = \frac{6.6}{E_t}$$

$$\text{Steam consumption} = (\text{bhp})(WR) \text{ lb/hr}$$

TABLE 11 Steam Consumption Summary for Boiler-Feed-Pump Service

Capacity, %	Pipe size	gpm	Pump head, ft	Theoretical hp	$\% E_p$	bhp	$\% E_t$	Steam 1,000 lb/hr
100	10	6,620	8,600	13,020	76.5	17,050	74	152.0
	14	6,620	8,500	12,850	76.5	16,800	74	150.0
	18	6,620	8,400	12,700	76.5	16,600	74	148.0
75	10	4,965	11,000	12,500	86.0	14,550	73	131.8
	14	4,965	10,900	12,380	86.0	14,400	73	130.2
	18	4,965	10,500	12,250	86.0	14,250	73	129.0
50	10	3,310	12,600	9,540	77.2	12,350	71	115.0
	14	3,310	12,580	9,500	77,2	12,300	71	114.5
	18	3,310	12,550	9,480	77.2	12,280	71	114.0

Lubricating-oil consumption, based on 0.02 gal/100 hp-hr for each pair of bearings, results in 0.04 gal/100 hp-hr for a steam-turbine-driven centrifugal pump. Hence

$$\text{gal/hr} = 0.04 \frac{\text{bhp}}{100}$$

Cooling water, based on 10°F rise and 2 percent energy loss to the water for each pair of bearings, results in:

$$\text{gpm} = \frac{\text{bhp}}{50}$$

The annual operating costs associated with the boiler feed pump and the alternate piping size are developed from the following:

Steam—0.8 (1,000 lb per hr) × 8,760
Lubricating oil—0.7 × 0.04 (bhp per 100) × 8,760
Cooling water—(0.075 per 1,000) (bhp per 50) 60 × 8760
Maintenance—1.5 × bhp

From the data listed in Table 11, combined with the cost values listed above, Table 12 summarizes the total operating costs.

TABLE 12 Operating Costs for Alternate Piping and Boiler Feed Pump

Discharge pipe size, in	Service	100%	75%	50%
10	Steam	$1,068,000	$925,000	$808,000
	Lub. oil	42,000	35,800	30,400
	Cooling water	13,500	11,500	9,750
	Maintenance	25,600	21,800	18,550
	Total	$1,149,100	$994,100	$866,700
14	Steam	$1,053,000	$915,000	$804,000
	Lub. oil	41,400	35,400	30,300
	Cooling water	13,300	11,400	9,720
	Maintenance	25,200	21,600	18,460
	Total	$1,132,900	$983,400	$862,480
18	Steam	$1,040,000	$905,000	$800,000
	Lub. oil	40,800	35,000	30,200
	Cooling water	13,100	11,280	9,700
	Maintenance	24,900	21,400	18,400
	Total	$1,118,800	$972,680	$858,300

From the total annual operating costs listed in Table 12, we determine the annual levelized costs as follows:

$$ALC = \frac{(\text{total operating cost})(OF)(PWF)(CRF)}{(AFC)}$$

where the product (OF) (PWF) (CRF) has been determined in Table 10 for 100, 75, and 50 percent load. Table 13 summarizes all the costs associated with the alternate piping sizes for the boiler feed pump.

TABLE 13 Total Evaluated Costs for Piping and Boiler Feed Pump

Load	Pipe size, in		
	10	14	18
100%	$3,670,000	$3,630,000	$3,570,000
75%	244,000	241,000	238,000
50%	2,040,000	2,030,000	2,005,000
Total ALC	$5,954,000	$5,901,000	$5,813,000
Direct and Indirect	1,673,134	1,783,520	1,985,660
Evaluated costs	$7,627,134	$7,684,520	$7,798,660

From a plot of the total evaluated costs shown on Fig. 4, one can see that the minimum piping size is 10 in. This conclusion may have been reached from the

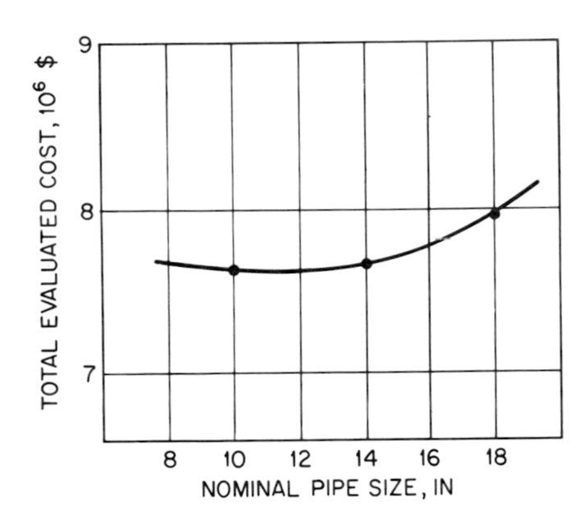

Fig. 4 Piping size vs. evaluated cost for a boiler feed pump.

tabulated data in Table 13, but it is cautioned that it is not safe to reach a decision until a plot is made.

REFERENCES

1. Grant, Eugene L., and W. Grant Ireson: "Principles of Engineering Economy," 5th ed. The Ronald Press Company, New York, 1970.
2. Federal Trade Commission, Securities and Exchange Commission, "Quarterly Financial Report for Manufacturing Corporations," Fourth Quarter 1970.
3. General Electric Company, *Industrial Power Systems,* vol. 14, No. 1, March 1971.
4. Report on Equipment Availability for The Nine Year Period 1960–1968, EEI Publication 69-33, September 1969.
5. American Society of Mechanical Engineers, "Oil and Gas Engine Power Costs," 1956.

Section 9.4

Water Hammer

JOHN PARMAKIAN

Introduction Water hammer is a very destructive force which exists in any pumping installation where the rate of flow changes abruptly for various reasons. Most engineers recognize the existence of water hammer but few realize its destructive force. Much time and expense have been spent repairing damage to pipe lines and pumps in which the cause of the failure was due to water hammer. It is thus essential for an engineer to be able to know when to expect water hammer, to estimate the possible maximum pressure rise, and if possible to provide means to reduce the maximum pressure rise to a safe limit.

The computational procedures used for the analysis of water hammer in pump discharge lines with electric-motor-driven pumps have been known for many years. It started with the basic water hammer contributions by Joukousky and Allievi over 65 years ago. This was followed by many others in later years with the application of numeric, graphic, and computer techniques. Although the theory and mechanics of computation of water hammer in pump discharge lines have advanced rapidly in recent years, there are many practical aspects of this subject which are still confusing to engineers. It is the purpose of this section to bring these to the reader's attention.

General Considerations The aim of this section is twofold: the first and major portion contains a discussion of some practical aspects of water hammer control devices used in pumping plants; the second is to indicate the source of various charts which provide ready water hammer solutions for a variety of these water hammer control devices.

Nomenclature

a = velocity of pressure wave, ft/s
D = inside diameter of conduit, ft
e = thickness of pipe wall, ft
E = Young's modulus for pipe material, lb/ft²
g = acceleration of gravity, ft/s²
H_0 = pumping head for initial steady pumping conditions, ft
H_R = rated pumping head, ft
K_1 = 91,600 $H_R Q_R / W R^2 \eta_R N_R^2$, s^{-1}
K = volume modulus of liquid, lb/ft²
L = total length of conduit, ft
$2L/a$ = round-trip wave travel time, s
N_R = rated pump speed, rpm
η_R = pump efficiency at rated speed and head, decimal form

ρ = pipe line constant = $a\bar{V}_0/gH_0$
Q_0 = initial flow through pump, ft³/s
Q_R = rated flow through pump, ft³/s
μ = Poisson's ratio of pipe material
V_0 = velocity in conduit for initial steady conditions, ft/s
w = specific weight of water, lb/ft³
WR^2 = flywheel effect of rotating parts of motor, pump, and entrained water, lb-ft²

Basic Assumptions A considerable number of assumptions were made in the derivation of the fundamental water hammer equations and in the solution of the various hydraulic transients in pumping systems. These assumptions are often overlooked and involve the physical properties of the fluid and pipeline, the kinematics of the flow, and the transient response of the pump as follows:

1. The fluid in the pipe system is elastic, of homogeneous density, and is always in the liquid state.
2. The pipe-wall material or conduit is homogeneous, isotropic, and elastic.
3. The velocities and pressures in the pipeline, which is always flowing full, are uniformly distributed over any transverse cross section of the pipe.
4. The velocity head in the pipeline is negligible when compared to the pressure changes.
5. At any time during the pump transient, when operating in the zones of pump operation, energy dissipation, and turbine operation, there is an instantaneous agreement at the pump, as defined by the steady-state complete pump characteristics, of the pump speed and torque corresponding to the transient head and flow which exist at that moment at the pump.
6. The length between the inlet and outlet of the pump is so short that water hammer waves propagate between them instantly.
7. Windage effects of the rotating elements of the pump and motor during the transients are negligible.
8. Water levels at the intake and discharge reservoirs do not change during the transient period.

High- and Low-Head Pumping Systems Water hammer is of greater significance at low-head pumping systems than at high-head systems. The normal steady water velocities in both high-head and low-head pumping systems are usually of about the same order of magnitude. However, the pressure changes are proportional to the rate of change in the velocity of the water in the line. Then, for a given rate of change in the velocity, the pressure changes in the high- and low-head pumping systems are of about the same order of magnitude. Therefore a head rise of a given amount would be a larger proportion of the pumping head at a low-head pumping system than at a high-head system.

Discharge-Line Profile The pump discharge-line profile is usually based on economic, topographic, and land right-of-way considerations. However, in selecting the alignment along which a pump discharge line is to be located, there are other considerations which often make one pipeline profile and alignment more favorable than another. For example, upon a power failure at the pump motors, the envelope of the maximum downsurge gradient along the length of the pipeline is a concave curve. Therefore it may be possible to avoid the use of expensive pressure-control devices at a pumping plant if the pipeline profile is also concave and is not located above the downsurge-gradient curve. In some cases it may even be economical to lower the profile of the discharge line at the critical locations by deeper excavation. If a surge tank at the pumping plant is definitely required, the most favorable pipeline profile is one with high ground near the pumping plant where the surge-tank structure above the natural ground line would be much shorter in height.

Rigid Water Column Theory The question is often raised as to whether the rigid water column theory is sufficiently accurate for the computation of water hammer in pump discharge lines. In the rigid water column theory the water is assumed to be incompressible and the pipe walls rigid. In the author's experience, the accuracy and limitations of the rigid water column theory are often questionable for most water hammer problems that occur in pump discharge lines.

Water Hammer Wave Velocity From a practical viewpoint a difference of 15 to 20 percent in the magnitude of the computed water hammer wave velocity usually has very little effect on the water hammer in pump discharge lines. The effect on the water hammer due to a possible error in the wave velocity can be verified by first computing the wave velocity as accurately as possible, and then recomputing the transients for the critical cases with a wave velocity which is about 20 percent higher or lower. At installations where alternative materials for the pipeline such as steel or concrete are being investigated, one water hammer wave velocity and solution for water hammer for either alternative will usually suffice regardless of the pipe material finally selected.

Pipeline Size The diameter of the pipeline is usually determined from economic considerations based on steady-state pumping conditions. However, the water hammer effects in a pump discharge line can be reduced by increasing the size of the discharge line since the velocity changes in the larger pipeline will be less. This is usually an expensive method for reducing water hammer in pump discharge lines, but there are sometimes occasions where an increase in pipe size may be justified to avoid the use of more expensive water hammer control devices.

Number of Pumps The number of pumps connected to each pump discharge line is usually determined from the operational requirements of the installation, availability of pumps, and other economic considerations. However, the number and size of pumps connected to each discharge line have some effect on the water hammer transients. For pump startup with pumps equipped with check valves, the greater the number of pumps on each discharge line the smaller the pressure rise. Moreover, if there is a malfunction at one of the pumps or check valves, a multiple pump installation on each discharge line would be preferable to a single pump installation because the flow changes in the discharge line due to such a malfunction would be less. When a simultaneous power failure occurs at all of the pump motors, the fewer the number of pumps on a discharge line the smaller the pressure changes and other hydraulic transients. For a given total flow in the discharge line, a large number of smaller pumps and motors will have considerably less total kinetic energy in the rotating parts to sustain the flow than a small number of pumps. Consequently, for the same total flow, the velocity changes and water hammer effects due to a power failure are a minimum when there is only one pump connected to each discharge line.

Flywheel Effect (WR^2) Another method for reducing the water hammer effects in pump discharge lines is to provide additional flywheel effect (WR^2) in the rotating element of the motor. As an average the motor usually provides about 90 percent of the combined flywheel effect of the rotating elements of the pump and motor. Upon a power failure at the motor, an increase in the kinetic energy of the rotating parts will reduce the rate of change in the flow of water in the discharge line. In most cases an increase of 100 percent in the WR^2 of large motors can usually be obtained at an increased cost of about 20 percent of the original cost of the motor. Ordinarily an increase in WR^2 is not an economical method for reducing water hammer, but it is possible in some marginal cases to eliminate other more expensive pressure control devices.

Specific Speed of Pumps For a given pipeline and initial steady-flow conditions, the maximum head rise which can occur in a discharge line subsequent to a power failure, where the reverse flow passes through the pump, depends first on the magnitude of the maximum reverse flow which can pass through the pump during the energy-dissipation and turbine-operation zones, and then upon the flow which can pass through the pump at runaway speed in reverse. Upon a power failure the radial-flow (low-specific-speed) pump will produce slightly more downsurge than the axial-flow (high-specific-speed) and mixed-flow pumps (see Ref. 5). The radial-flow pump will also produce the highest head rise upon a power failure if the reverse flow is permitted to pass through the pump. There is usually very little head rise at mixed-flow and axial-flow pumps when a power failure occurs and if a water column separation does not occur at some other location in the line.

During a power failure with no valves, the highest reverse speed is reached by the axial-flow pump and the lowest by the radial-flow pump. Care must therefore be taken to prevent damage to the motors with the higher specific-speed pumps because of these higher reverse speeds. Upon pump startup against an initially closed check valve, the axial-flow pump will produce the highest head rise in the discharge line since it also has the highest shutoff head. On pump startup a radial-flow pump will produce a nominal head rise but an axial-flow pump can produce a head rise of several times the static head.

Complete Pump Characteristics In order to determine the transient conditions due to a power failure at the pump motors, the water hammer wave phenomena in the pipeline, the rotating inertia of the pump and motor, and the complete pump characteristics as well as other boundary conditions and head losses must be known. In the solution of water hammer problems with computers, the complete pump characteristics are sometimes approximated by polynomial expressions in which the coefficients of the polynomial are obtained by fitting a representative curve through several points at specific locations on the pump characteristics diagram. Pump manufacturers sometimes provide limited information to determine such coefficients. However, a comparison between the polynomial values and the complete pump characteristics diagram indicates serious discrepancies in some cases especially in the zone of energy dissipation. Care must therefore be exercised in the use of an approximate polynomial expression as a substitute for the correct complete pump characteristics, to ensure that a serious error does not result in the computation of the hydraulic transients.

Complex Piping Systems As noted above in the basic assumptions, the water hammer theory is strictly applicable for a pipeline of uniform characteristics. However, for water hammer purposes a complex piping system can be reduced to a satisfactory equivalent uniform pipe system. The approximations are made by neglecting the wave-transmission effects at the junctions and points of discontinuity and by utilizing the rigid water column theory. The pertinent water hammer equations are then found to be analogous to those used in electric circuits. In practice the water hammer analysis with these approximations will usually be found to give more conservative results than those experimentally obtained from the actual pipe system (see Ref. 10).

Available Water Hammer Solutions The water hammer solutions for pumping systems with various surge control devices are given in convenient chart form in the references. These include the following:

1. Hydraulic transients at the pump and midlength of the pump discharge lines for radial-flow, mixed-flow, and axial-flow pumps with reverse flow passing through the pumps
2. Surge tanks
3. Air chambers
4. Surge suppressors
5. One-way surge tanks
6. Water column separation

The operation of these surge control devices is described below.

Power Failure at Pump Motors *Pumps with No Valves at the Pump.* When the power supply to the pump motors is suddenly cut off, the only energy that is left to drive the pump in the forward direction is the kinetic energy of the rotating elements of the pump and motor. Since this energy is usually relatively small when compared to that required to maintain the flow of water against the discharge head, the reduction in the pump speed is very rapid. As the pump speed reduces, the flow of water in the discharge line is also reduced. As a result of these rapid flow changes, water hammer waves of increasing subnormal pressure are formed in the discharge line at the pump. These subnormal pressure waves move rapidly up the discharge line to the discharge outlet where complete wave reflections occur. Soon the speed of the pump is reduced to a point where no water can be delivered against the existing head. If there is no control valve at the pump, the flow through the pump reverses, although the pump may still be rotating in the forward direction. The speed of the pump now drops more rapidly and passes through zero speed. Soon the maximum reverse flow passes through the pump. A short time later the pump, acting as a turbine, reaches runaway speed in reverse. As the pump approaches runaway speed, the reverse flow through the pump is reduced. For radial-flow pumps this rapid reduction in reverse flow produces a pressure rise at the pump and along the length of the discharge line. For a given set of radial-flow (low-specific-speed) pump characteristics, the results of a large number of water hammer solutions are given in chart form in Figs. 1 to 8 inclusive. These charts furnish a convenient method for obtaining the hydraulic transients at the pump

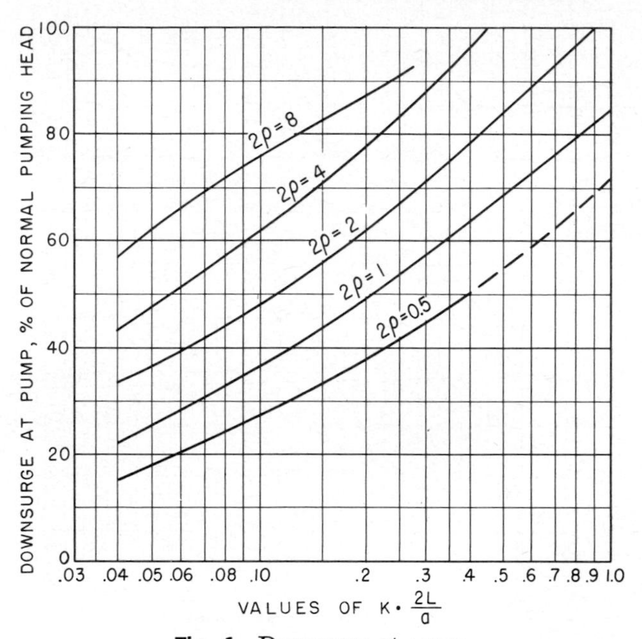

Fig. 1 Downsurge at pump.

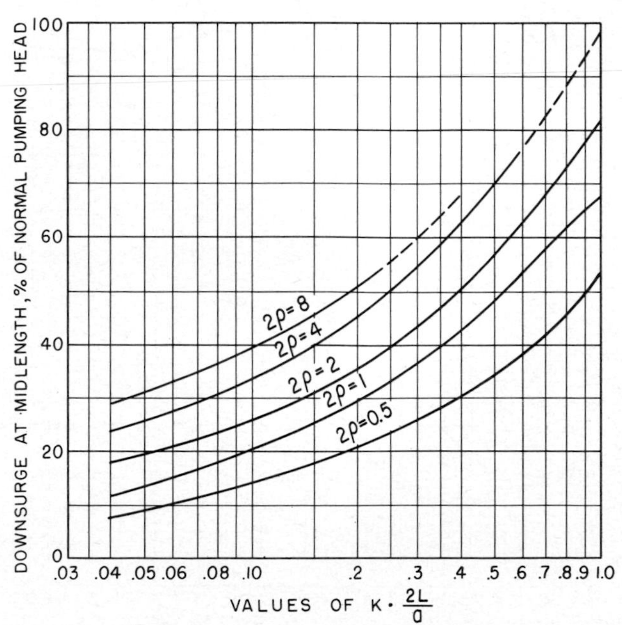

Fig. 2 Downsurge at midlength.

and midlength of the discharge line when no control valves are present at the pump. Although the charts are theoretically applicable to one particular set of radial flow pump characteristics, they are useful for estimating the water hammer effects in any pump discharge line which is equipped with radial-flow pumps. The charts are based on two independent parameters: ρ, the pipeline constant, and $K(2L/a)$,

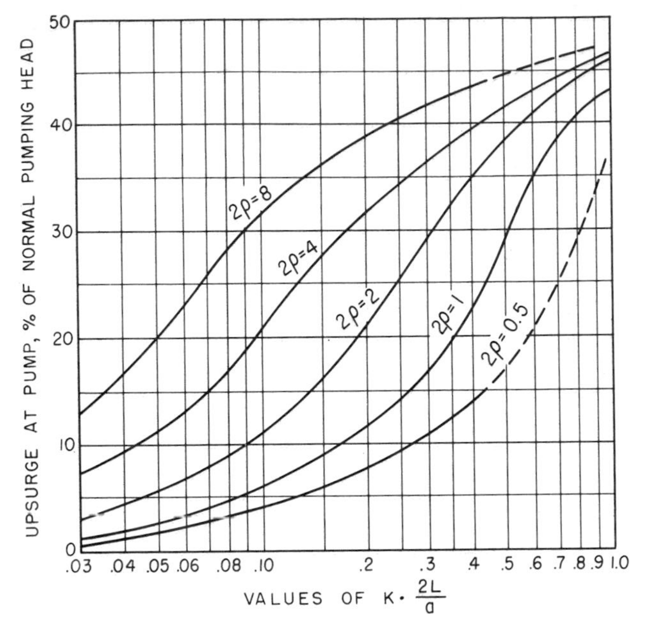

Fig. 3 Upsurge at pump.

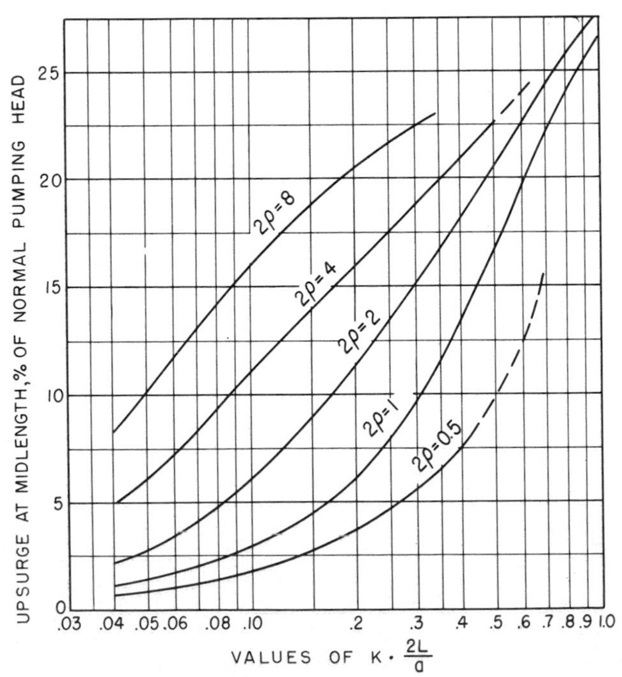

Fig. 4 Upsurge at midlength.

a constant which includes the effect of the pump and motor inertia and the water-hammer-wave travel time of the discharge line. If the friction head in the discharge line during normal operation is more than 25 percent of the total pumping head, and the water column separation does not occur at any point in the line, the maximum head at the pump with reverse flow passing through the pumps will usually not exceed the initial pumping head (see Ref. 2).

Pumps Equipped with Check Valves. There are a number of problems associated

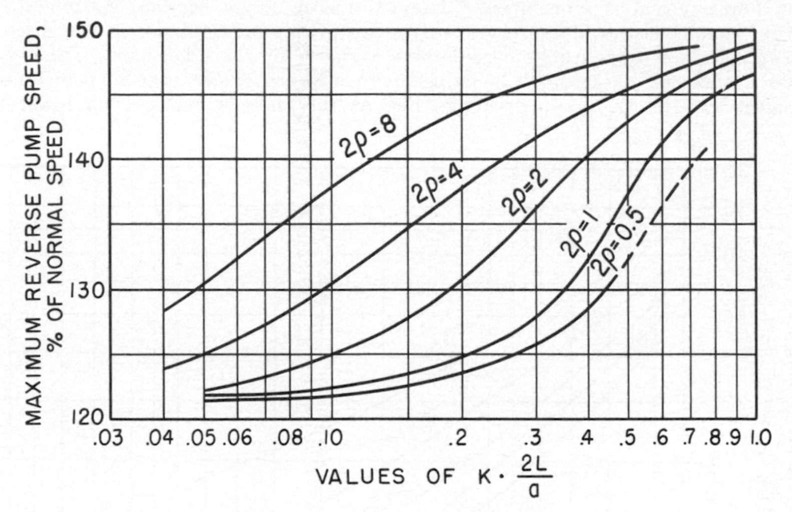

Fig. 5 Maximum reverse speed.

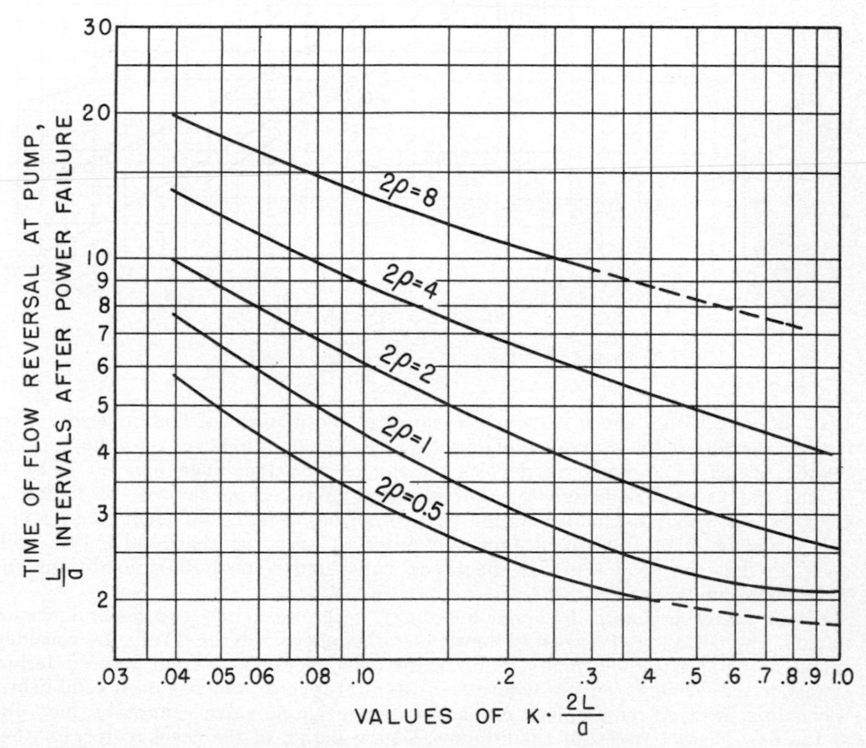

Fig. 6 Time of flow reversal at pump.

with the use of check valves in pump discharge lines. Under steady flow conditions the pump discharge keeps the check valve open. However, when the flow through the pump reverses subsequent to a power failure, the check valve closes very rapidly under the action of the reverse flow and the resulting dynamic forces on the check-valve disk. Under these conditions, neglecting pipeline friction, the head rise in the discharge line at the check valve is about equal to the head drop which existed at the moment of flow reversal. However, in the event that the check-valve closure upon flow reversal is momentarily delayed due to hinge friction, malfunction, or inertial characteristics of the check valve, the maximum head rise in the discharge line at the check valve can be considerably higher. On the other hand, if the check-valve closure can be accomplished slightly in advance of the time of flow reversal, the head rise in the pump discharge line at the check valve is even lower than

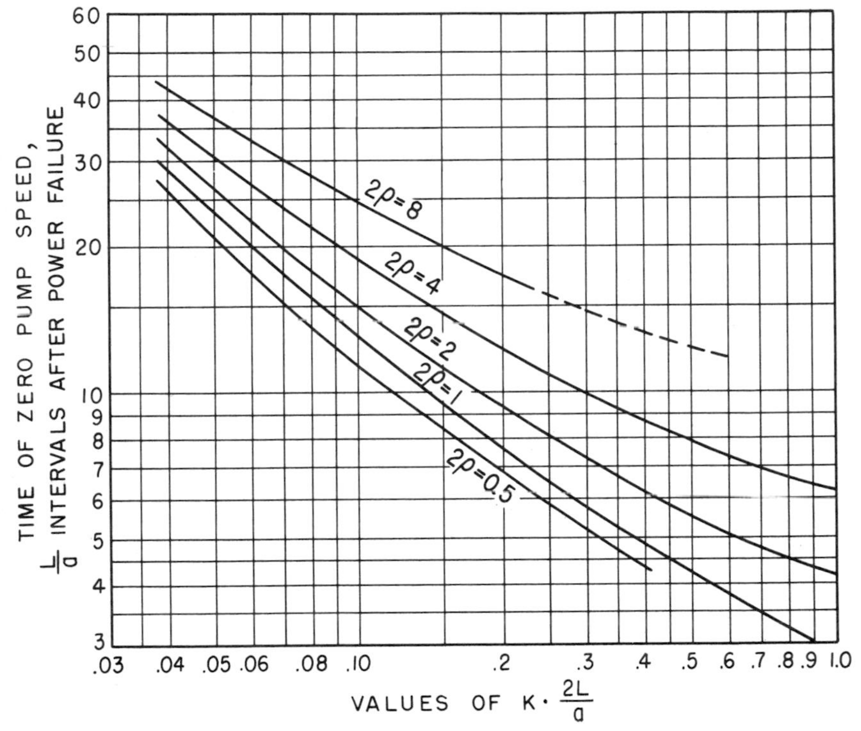

Fig. 7 Time of zero pump speed.

that obtained with a check valve which closes at the moment of flow reversal. This feature is utilized by a number of check-valve manufacturers by providing spring-loaded or lever-arm-weighted devices on the check-valve hinge pins to assist in closing the valve disk before the flow reverses. With these devices the hydraulic forces on the valve disk under normal flow conditions must be sufficient to overcome the spring or lever-arm-weight forces in order to keep the check-valve disk wide open, so that the head losses at the check valve under steady-flow conditions will be a minimum.

Check valves in pump discharge lines may be grouped into two general classes, namely, rapid-closing check valves or slow-closing check valves. From the considerations noted above, the primary requirement for a check valve upon a power failure is that it should close quickly before a substantial reverse flow has been established. When this primary requirement for a fast-closing check valve cannot be met due to the flow characteristics of the system and the design of the check valve, another alternative is to provide a device such as a dashpot which will slow down or cushion the last portion of the check-valve closure. This feature has been utilized by a number of check-valve manufacturers.

Controlled Valve Closure. At most large pumping plant installations the use of a single-speed discharge-valve closure subsequent to a power failure will usually limit the head rise in the discharge line to an acceptable value. However, it will be found that with the optimum single-speed closure some reverse rotation below the maximum runaway speed of the unit in reverse will occur. If it is desired from other considerations to prevent or to limit the reverse speed of the unit, a two-speed valve closure can be used. In such cases the discharge valve should close the major portion of its stroke very rapidly up to the moment that the flow reverses at the pump. It should then complete the remainder of its stroke at a slower

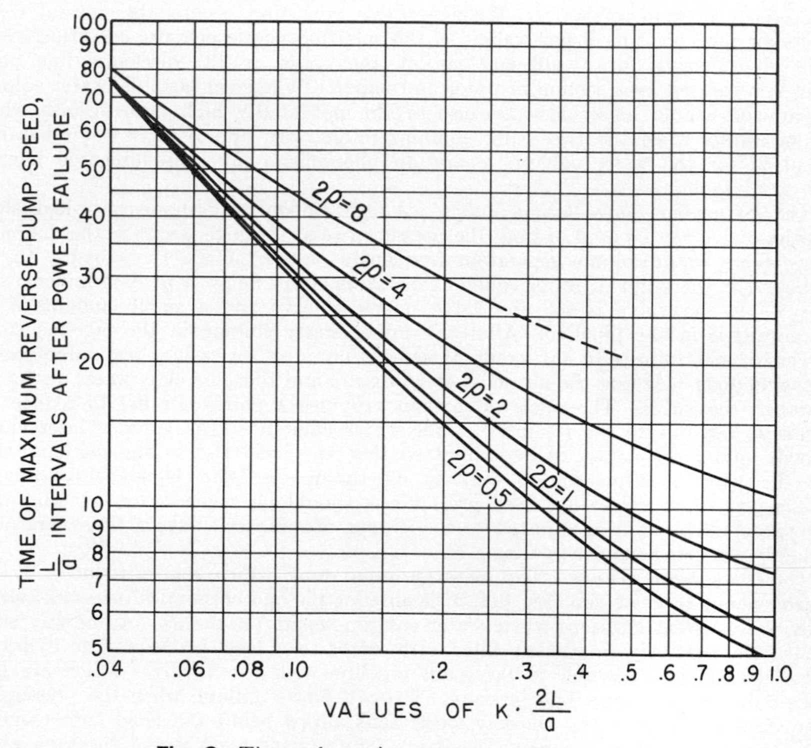

Fig. 8 Time of maximum reverse pump speed.

rate in order to limit the pressure rise in the discharge line to an acceptable valve. At pumping plants where there is more than one pump on the same discharge line, a compromise must be obtained on the optimum single speed and two-speed closure rates for the various combination of pumps which might be in operation at the time of a power failure.

Surge Suppressors. Surge suppressors are sometimes used in pumping plants to control the pressure rise that occurs in pump discharge lines subsequent to power interruptions. A typical surge suppressor consists of a pilot-operated valve which opens quickly after a power interruption through the loss of power to a solenoid, or by a sudden large pressure reduction or pressure increase at the surge suppressor. This valve provides an opening for releasing water from the pump discharge line. The valve is later closed at a slower rate by the action of a dashpot to control the pressure rise as the flow of water is shut off. A properly sized and field-adjusted surge suppressor can reduce the pressure rise in the discharge line to any desired value provided that a water column separation does not occur at other locations in the discharge line. The charts given in Ref. 3 can be used to determine the required flow capacity of the surge suppressor.

The proper field adjustment of a surge suppressor is very important. If the surge suppressor opens too rapidly subsequent to a power failure, the downsurge at the pump and along the discharge-line profile would be more than if no surge suppressor was present. As a result a water column separation may actually be produced at some locations in the discharge line by the premature opening of the surge suppressor. If the surge suppressor closes too rapidly after the maximum reverse flow has been established, a large pressure rise will occur.

Water Column Separation. Water column separation in a pump discharge line subsequent to a power failure at the pump motors occurs whenever the momentary hydraulic gradient at any location reduces the pressure in the discharge line to the vapor pressure of water. Whenever this condition occurs, the normal water hammer solution is no longer valid. If the subatmospheric pressure condition inside the pipe persists for a sufficient period, the water in the discharge line parts and is separated by a section of water and vapor. Whenever possible, water column separation should be avoided because of the potentially high pressure rise which often results when the two water columns rejoin. An approximate water hammer solution for the water column separation phenomena in pump discharge lines is given in Reference 9.

Quick-Opening Slow-Closing Valves. A quick-opening dashpot-controlled slow-closing valve can be used to limit the pressure rise at the high points in the discharge line where water column separation frequently occurs. When the pressure in the pipeline at the point of water column separation drops below a predetermined value for which the valve is set, the valve opens quickly and a small amount of air is admitted in the pipeline. After the upper water column in the pipeline stops, reverses, and returns to the point of separation near the valve, the valve should be wide open. At first the air and water mixture and then the clear water discharges through the valve. The open valve then provides a point of relief to reduce the pressure rise due to the rejoining of the water columns. The valve is later closed slowly under the action of a dashpot so that the head rise in the discharge line at the valve location due to shutting off the reverse flow is not objectionable. Whenever these valves are used precautions should be taken to ensure that they are properly sized, field-adjusted to the proper opening and closing times, and adequately protected against freezing.

One-Way Surge Tanks. The one-way surge tank which was introduced by the writer about 16 years ago (see Ref. 8) is an effective and economical pressure-control device for use at locations where water column separation occurs. A one-way surge tank is a relatively small tank filled with water to a level far below the hydraulic gradient. It is connected to the main pipeline with check valves which are held closed by the discharge-line pressure. Upon a power failure, when the pressure in the discharge line at the one-way surge tank drops below the head corresponding to the water level in the tank, the check valve opens quickly and the tank starts to drain, thus filling the void formed by the separation of the water columns. When the flow in the upper column starts to reverse, the check valves at the one-way tank close before any appreciable reverse flow is established in the discharge line. Thus the pressure rise due to the rejoining of the water columns is avoided. The initial level of water in the one-way surge tank is usually maintained automatically with float control or altitude valves. It should be noted that the one-way surge tank does not act during the starting-up cycle of the pump discharge line and that it must also be protected against freezing.

Air Chambers. An effective device for controlling the pressure surges in a long pump discharge line is a hydropneumatic tank or air chamber. The air chamber is usually located at or near the pumping plant. The air chamber can be of any desired configuration and may be placed in a vertical, horizontal, or sloping position. The lower portion of the chamber contains water, while the upper portion contains compressed air. The desired air-water levels are maintained with float level controls and an air compressor. When a power failure occurs at the pump motor, the head and flow developed by the pump decrease rapidly. The compressed air in the air chamber then expands and forces water out of the bottom of the chamber into the discharge line, thus minimizing the velocity changes and water hammer effects in the discharge line. When the pump speed is reduced to the point where it cannot deliver water against the existing head, which is usually a fraction of a second after a power failure, the check valve at the discharge side of the pump closes rapidly, and the pump then slows down to a stop. A short time later the water in the discharge line slows down to a stop, reverses, and flows back into the air chamber. As the reverse flow enters the chamber usually through a throttling orifice, the air volume in the chamber decreases and a head rise above the pumping head occurs in the discharge line. The magnitude of this head rise depends on the throttling orifice and the volume of air which was initially provided in the air chamber.

The results of a large number of graphic water hammer–air chamber solutions are given in convenient chart form in Ref. 1. Another presentation of air chamber charts using the rigid water column theory is given in Ref. 4.

Surge Tanks. A surge tank is one of the most dependable devices that can be used at a pumping plant to reduce water hammer resulting from rapid changes of flow in the discharge line subsequent to a power failure at the pump motor. It has no moving parts which can malfunction. Following a power failure the water in the surge tank provides a nearby source of potential energy which will effectively reduce the rate of change of flow and water hammer in the discharge line. The charts given in Ref. 1 provide a ready means for calculating the surges in the pipeline due to the sudden starting or stopping of a pump.

One of the disadvantages of a conventional surge tank is that, since the top of the tank must extend above the normal hydraulic gradient to avoid spilling, the tank must be quite tall and expensive at high-head pumping installations. In order to obtain the most economical surge tank design, care should be given to the proper sizing of the throttling device at the base of the tank.

Nonreverse Ratchets. Another device occasionally used for reducing water hammer in a pump discharge line upon a power failure is a nonreverse ratchet on the pump and motor shaft which prevents the reverse rotation of the pump. This device is effective for controlling water hammer upon a power failure because of the large reverse flow which can pass through the stationary impeller. The experience to date with nonreverse-ratchet mechanisms except on small pumps has been very disappointing. At a number of moderate-sized pump installations where these devices were used, the shock to the pump and motor-shaft system due to the sudden shaft stoppage created other serious mechanical difficulties.

Automatic Restart of Motors. At small unattended pumping plants it is often desirable after a power failure to automatically return the pumps to service as soon as the power is restored. However, it was found that occasionally, subsequent to a very short power outage, an induction motor could restart and come quickly up to forward speed while a reverse flow was still passing through the pump. Under these conditions the water hammer in the discharge line was very objectionable. If the pump motor has the capability of restarting under such transient conditions, a time delay or similar device should be installed at the motor controls so that the pump can be restarted only when it is safe to do so.

Normal Pump Startup with Controlled Valve Opening At some pumping plants the pump is brought up to speed against a close valve on the discharge side of the pump. The valve is then opened slowly and there is very little water hammer in the discharge line. However, it will be found that nearly all of the pump flow in the discharge line is established with only a relatively small valve opening, since the head losses across the valve decrease very rapidly during the opening stroke. For long discharge lines the head-loss and flow characteristics of the valve during the opening stroke must be considered in determining the optimum rate of opening.

Normal Pump Startup with Check Valves At pumping plants where the pipeline is held full with pump check valves, the water hammer in the discharge line due to a pump startup can be objectionable in some cases. If the motor comes up to speed very rapidly, the pump will develop a pressure rise in the discharge line as the sudden increase in flow moves into the line. As noted above and in Ref. 5, this pressure rise is lower for radial-flow (low-specific-speed) pumps than for axial-flow (high-specific-speed) pumps.

Normal Pump Startup with Casing Unwatered At pumping plants which are equipped with large pumps, normal starting of a pump is often performed with the pump casing unwatered. This is accomplished by depressing the water level below the pump impeller by means of compressed air, which is admitted into the pump casing with the pump discharge valve closed and the discharge line full. After the motor is synchronized on the line, the compressed air in the pump casing is released allowing water to reenter the pump from the suction elbow, after which the discharge valve is slowly opened. This type of operation has been satisfactory with most large pumping units, and there are normally no significant water hammer effects on the discharge line. However, there have been some difficulties with this type of operation at a few large pump units. In the latter case, when the rising water level in the suction elbow first reaches the pump impeller, a very fast pumping action occurs within a few seconds and a severe uplift of the pump and motor from the thrust bearing could occur at this moment. If the discharge valve is still closed when this fast pumping action occurs, there is no water hammer effect in the discharge line.

Normal Pump Startup with Surge Tanks or Air Chambers With a surge tank or air chamber at the pumping plant, it makes very little difference whether the increased pump flow is sudden or gradual, inasmuch as the major portion of the

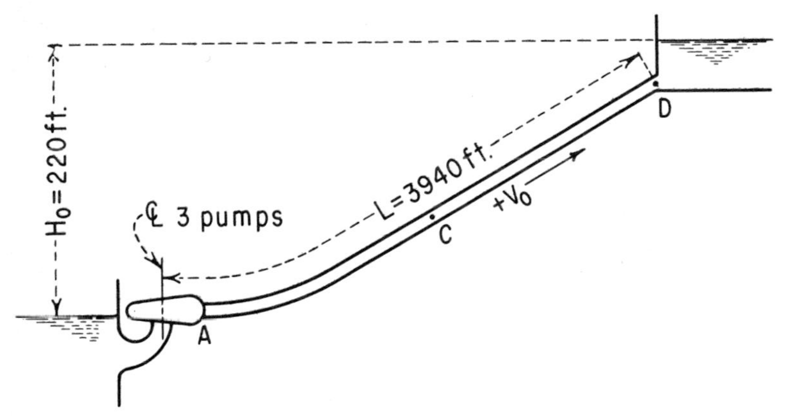

Fig. 9 Pipeline profile.

sudden increased flow will enter the surge tank or air chamber. With these devices the steep front of the pressure rise at the pump is transformed into a smaller pressure rise in the discharge line and a subsequent slow oscillating movement in the surge tank or air chamber.

Normal Pump Shutdown The pumping installation which produces the least water hammer effect in a pump discharge line during a normal pump shutdown is one in which the control valve on the discharge side of the pump is first closed and then the power to the pump motor is shut off. If only check valves are in operation on the discharge side of the pumps and the power to one of several pump motors which are connected to the same discharge line is cut off, the flow at the pump which has been shut down will reverse rapidly, and the check valve will also close rapidly. The use of antislam or slow-closing features at the check valves will reduce the water hammer effect in the discharge line.

Conclusions A variety of water hammer control devices for pumping plants are available to the designer. In most cases the experienced designer can narrow the choice of the most suitable device to a few practical alternatives. A prior knowledge of the available water hammer solutions for these devices will reduce the amount of detailed computational work which must be made to determine the critical hydraulic transient effects.

EXAMPLE: Consider a power failure at the pumping plant installation shown in Fig. 9. This installation consists of three pumps which discharge into a steel pipeline. Aside from isolation valves there are no check valves in the system. The basic data for this installation are as follows:

$D = 32$ in; $e = 3/16$ in; $Q_0 = Q_R = 33.7$ ft³/s (for three pumps)
$V_0 = 6.03$ ft/s (for three pumps); $H_0 = H_R = 220$ ft
400-hp motors at each pump
WR^2 of each pump and motor, 385 lb-ft²
$N_R = 1{,}760$ rpm; $\eta_R = 0.847$

$$\frac{D}{e} = \frac{32}{0.1875} = 171$$

$a = 3{,}000$ ft/s from Fig. 10

$$\frac{2L}{a} = \frac{2(3{,}940)}{3{,}000} = 2.63 \text{ s}$$

$$2\rho = \frac{aV_0}{gH_0} = \frac{(3{,}000)(6.03)}{(32.2)(220)} = 2.55$$

$$K = \frac{(91{,}600)(H_R Q_R)}{WR^2 \eta_R N_R^2} = \frac{(91{,}600)(220)(33.7)}{3(385)(0.847)(1760)^2} = 0.224$$

$$K\frac{2L}{a} = 0.59$$

From Figs. 1 to 8 the following results are obtained:

1. Downsurge at pump = (0.92) (220) = 202 ft
2. Downsurge at midlength = (0.64) (220) = 141 ft
3. Upsurge at pump = (0.42) (220) = 92 ft
4. Upsurge at midlength = (0.23) (220) = 51 ft
5. Maximum reverse speed = (1.45) (1760) = 2550 rpm
6. Time of flow reversal at pump = 3.5 L/a = 4.6 s
7. Time of zero pump speed = 5.8 L/a = 7.6 s
8. Time of maximum reverse speed = 10.0 L/a = 13.1 s

As noted in the previous section, Pumps Equipped with Check Valves, if there were check valves at the pumps which closed at the time of flow reversal, the upsurge or head rise at the pump above the normal head would have been about **202** ft.

PRESSURE PULSATIONS IN PUMP SYSTEMS

Introduction Motor-driven centrifugal pumps often produce objectionable pressure pulsations in pump discharge lines. The frequency of these pulsations results from the rotating and stationary parts that the pump comprises. The following pressure-pulsation frequencies have been observed at a number of major pump installations:

1. Fundamental shaft frequency, once per revolution pressure pulsation
2. Impeller-vane frequency, product of the shaft frequency and the number of impeller vanes.
3. Scroll-case frequency, product of shaft frequencies and the number of guide vanes either at the inlet or discharge side of the pump

Pressure Pulsations at Fundamental Shaft Frequency Objectionable pressure pulsations in pump systems with a frequency corresponding to the shaft frequency result primarily from a hydraulic unbalance in the pump impeller. The source of such a hydraulic unbalance is usually due to an eccentricity of the flow passage in the impeller. In some cases it may be due to the eccentricity of the outer periphery of the impeller and the pumping action exerted by this eccentricity.

Considerable care must be taken in the techniques of dynamically balancing a pump impeller to avoid pressure pulsations. In accomplishing this, balance weights are often welded to the top surface or an eccentric machine cut is taken on the outer periphery. Although such surfaces are out of sight after the unit is assembled, objectionable pressure pulsations with a frequency corresponding to the rotational speed of the unit can occur. Such sources of pressure pulsations are very difficult to correct after the unit has been installed. In order to avoid such difficulties, balance weights or metal removed at the outer periphery of the pump impeller should be done in such a manner that these surfaces remain smooth and concentric with the axis of rotation. Balance weights on surfaces normal to the axis of rotation should be applied or covered in such a manner that the surface remains flat, smooth, and normal to the axis of rotation.

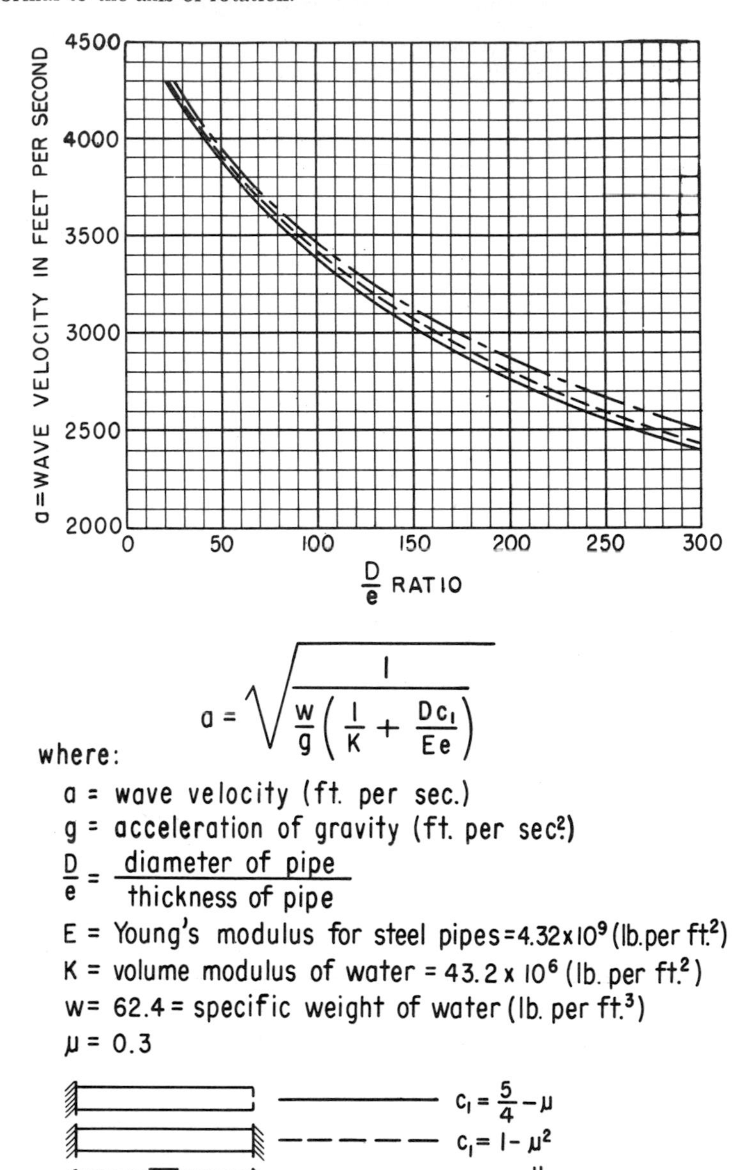

Fig. 10 Pressure wave velocity in steel pipes.

Pressure Pulsations at Impeller-Vane Frequency The tongue of the pump is the source of the pressure pulsations which are transmitted to the discharge line with a frequency equal to the product of the shaft frequency and the number of impeller vanes. An explanation of the phenomena is as follows: The velocity distri-

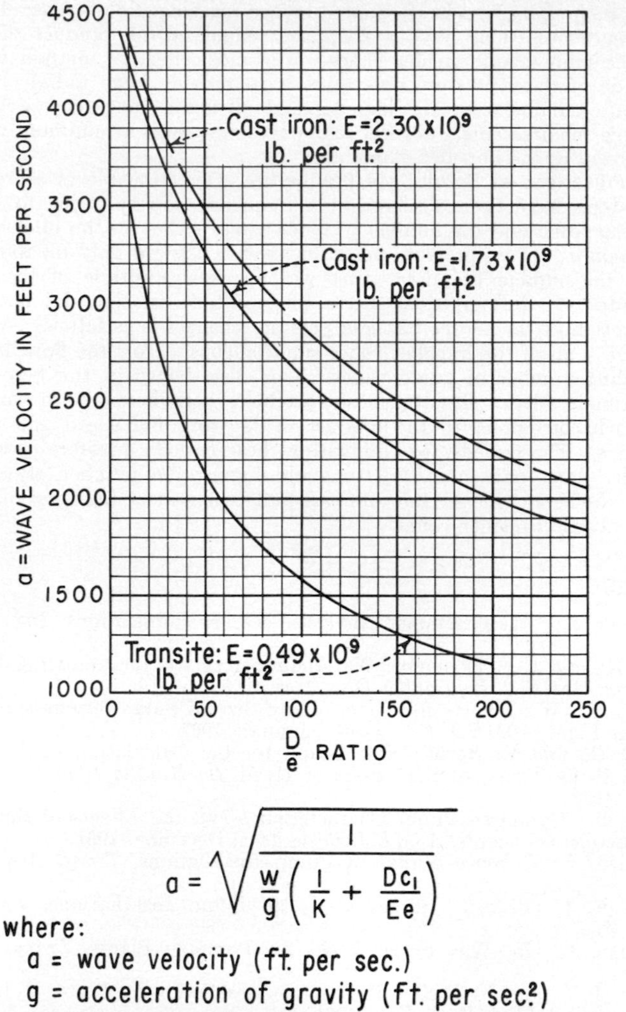

$$a = \sqrt{\frac{1}{\frac{w}{g}\left(\frac{1}{K} + \frac{Dc_1}{Ee}\right)}}$$

where:

a = wave velocity (ft. per sec.)

g = acceleration of gravity (ft. per sec.^{2})

$\frac{D}{e} = \frac{\text{diameter of pipe}}{\text{thickness of pipe}}$

E = Young's modulus for pipe material (lb. per ft.^{2})

K = volume modulus of water $= 43.2 \times 10^6$ (lb. per ft.^{2})

$w = 62.4$ = specific weight of water (lb. per ft.^{3})

$c_1 = 1 - \mu^2$

$\mu = 0.3$.

Fig. 11 Pressure wave velocity in cast iron and transite pipes.

bution at the exit of the pump impeller is not uniform. As this nonuniform flow passes the tongue of the pump casing, an abrupt change in the direction of the impeller-exit-velocity vector occurs at the proximity of the tongue. This produces a positive pressure wave at the pressure face and a negative pressure wave at the back face of the tongue. From this location the positive pressure wave travels directly up the discharge line, while the negative pressure wave travels completely around the pump casing and is attenuated before reaching the discharge line. These positive-pressure pulsations have a frequency equal to the product of the pump speed and the number of impeller vanes. The most effective method for reducing the magnitude of these pressure pulsations is to provide large radial clearance between the outer diameter of the impeller and all guide vanes consistent with the head-discharge requirements. Several manufacturers adopt a minimum radial clearance of 5 percent of the impeller diameter.

Pressure Pulsations at Scroll-Case Frequency Objectionable pressure pulsations in pump systems have been observed at a frequency corresponding to the product of the rotating speed and the number of guide vanes either on the inlet or discharge side of the pump. These pressure pulsations were observed only on those pumping units where the number of guide vanes was an exact multiple of the number of impeller blades. It did not exist where this multiple relation did not exist.

An explanation of the source of these pressure changes is as follows: As the pump impeller rotates, all of the impeller vanes simultaneously cross the flow lines between a corresponding number of guide vanes. This disturbance in the flow pattern by all of the runner blades simultaneously produces periodic pressure changes inside the unit which correspond to the product of the rotationl speed and the number of guide vanes. The remedy for eliminating these periodic pressure changes is relatively simple. It is only necessary to avoid having the number of vanes in the suction and discharge side of the pumps equal to an exact multiple of each other or of the number of impeller vanes.

REFERENCES

1. Parmakian, J.: "Waterhammer Analysis," Dover Publications, Inc., New York, 1963.
2. Kinno, H., and J. F. Kennedy: Waterhammer Charts for Centrifugal Pump Systems, *Proc. ASCE, J. Hydraulics Div.,* May 1965.
3. Lundgren, C. W.: Charts for Determining Size of Surge Suppressors for Pump-Discharge Lines, *ASME J. Eng. Power,* January 1967.
4. Combes, G., and R. Borot: New Graph for the Calculation of Air Reservoirs, Account Being Taken of the Losses of Head, *La Houille Blanche,* October–November, 1952.
5. Donsky, B.: Complete Pump Characteristics and the Effects of Specific Speeds on Hydraulic Transients, *ASME J. Basic Eng.,* December 1961.
6. Parmakian, J.: Pressure Surges in Pump Installations, *Trans. ASCE,* vol. 120, 1955.
7. Parmakian, J.: Pressure Surges at Large Pump Installations, *Trans. ASME,* August 1953.
8. Parmakian, J.: One-Way Surge Tanks for Pumping Plants, *Trans. ASME,* vol. 80, 1958.
9. Kephart, J. T., and K. Davis: Pressure Surges Following Water-Column Separation, *Trans. ASME J. Basic Eng.,* 1961.
10. Jones, S. S.: Water-Hammer in a Complex Piping System—Comparison of Theory and Experiment, *ASME* Paper 64—WA/FE-23.
11. Parmakian, J.: Unusual Aspects of Hydraulic Transients in Pumping Plants, *J. Boston Soc. Civil Eng.,* January 1968.

Section **9.5**

Vibration and Noise in Pumps

F. R. SZENASI

J. C. WACHEL

INTRODUCTION

The increased size and speed of pumps and the introduction of regulations which specify allowable noise and vibration limits in pumping systems have created new interest in the development of techniques to anticipate when vibration and noise problems will occur and methods to alleviate them. This chapter will discuss some of the more important aspects of such factors as shaft lateral and torsional vibration, fluid pulsation, cavitation, piping vibration, fluid transients, and structure-borne noise and vibration.

SHAFT VIBRATION

Pump shafts are subjected to numerous loads as torsion due to transmitted torque, bending due to static-weight considerations, the whirling caused by unbalance, and the axial loading due to axial thrusts. Of these, shaft whirling caused by unbalance is probably the most important. In the past, pumps have had stiff shafts; that is, the running speed was below the first lateral critical speed. As pumps sizes have increased, the running speed is quite often above the first critical. This means that the balance considerations are most important and extra care should be taken in calculating the critical speeds to ensure that there will be no coincidence with the running speed.

Lateral Critical-Speed Analysis A critical speed is defined as the speed at which a definite resonance is noted in the vibrations of a rotating shaft and is normally near the shaft undamped natural frequency. The critical speeds of a centrifugal pump shaft are a function of the following:

1. Length of bearing span and bearing design
2. Shaft diameter and material
3. Impeller weight and stiffness effects
4. Lubricant properties (viscosity, temperature, etc.)
5. Stiffness and damping of wear rings, seals, and packing
6. Case and foundation-support stiffness

The first critical speed of a uniform shaft with rigid bearings can be calculated from the equations for a simply supported beam.

$$f_R = \frac{\lambda}{2\pi}\sqrt{\frac{gEI}{\mu L^4}}$$

where λ = frequency factor (for rigid bearings = 9.87)
g = gravitational constant, 386.4 in/s²
E = modulus of elasticity, lb/in²
I = moment of inertia, in⁴
μ = beam weight per length, lb/in
L = length of beam, in
f_R = first critical with rigid bearings, Hz

For a uniform steel shaft with a concentrated weight (that is, impeller) at the center

$$f_w = \frac{79{,}200\ D/L^2}{\sqrt{1 + 2(W_I/\mu L)}}$$

where W_I = impeller weight, lb
f_w = frequency with impeller weights considered, Hz

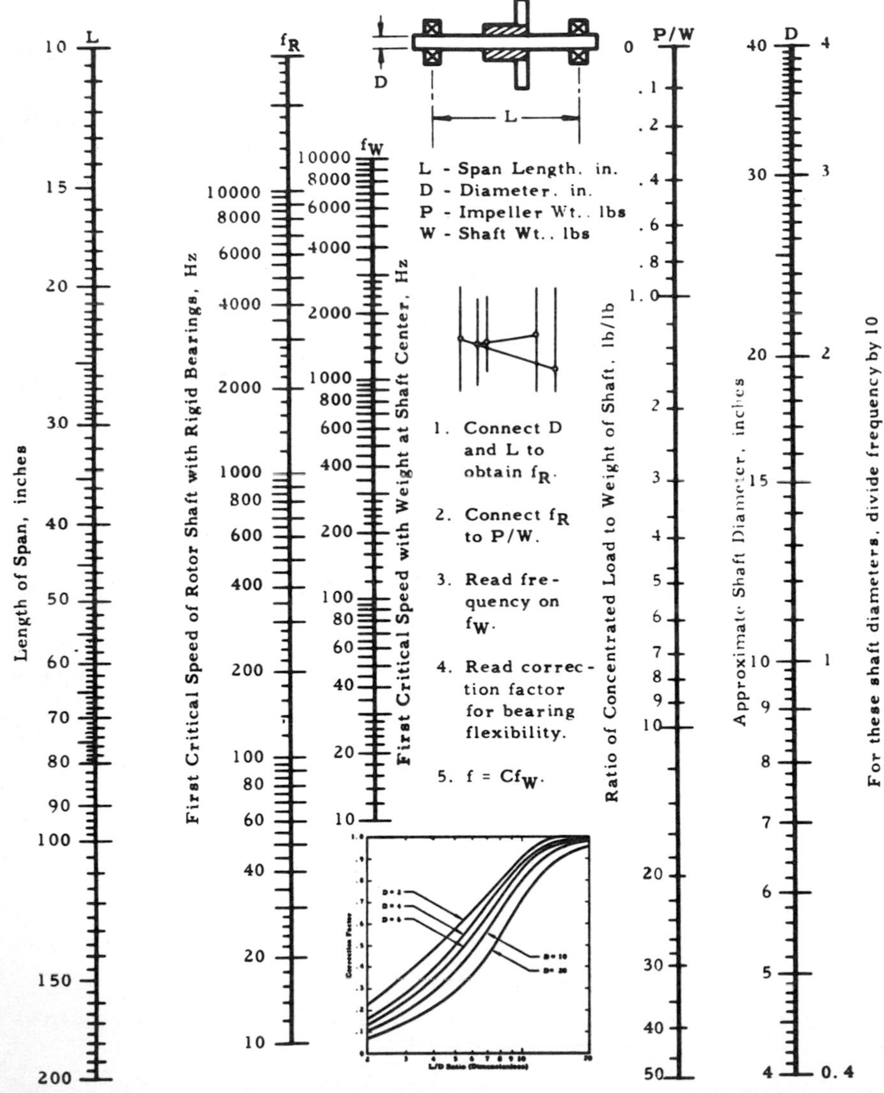

Fig. 1 Approximate first lateral critical speed of shaft including impeller weight and bearing flexibility.

The simplified solution of this equation is given in the nomogram in Fig. 1. An estimate of the effect of bearing a support flexibility upon the pump shaft critical speed can also be obtained from Fig. 1 using the accompanying graph. For example, for a pump with an average diameter of 6 in with a bearing span of 50 in and an impeller weight of 100 lb, the critical speed would be 150 Hz. If the effect of bearing flexibility is considered ($L/D = {}^{50}\!/_{6} = 8.33$), the correction factor is 0.75, making the critical speed 112 Hz (6,750 rpm).

The critical speeds of a multistage pump shaft are normally calculated on a digital computer using approaches such as the Rayleigh, Myklestad-Prohl, or the eigenvector-eigenvalue techniques. These methods utilize lumped-mass and weightless-spring considerations to develop stiffness-mass matrix equations which are solved for the critical speeds.

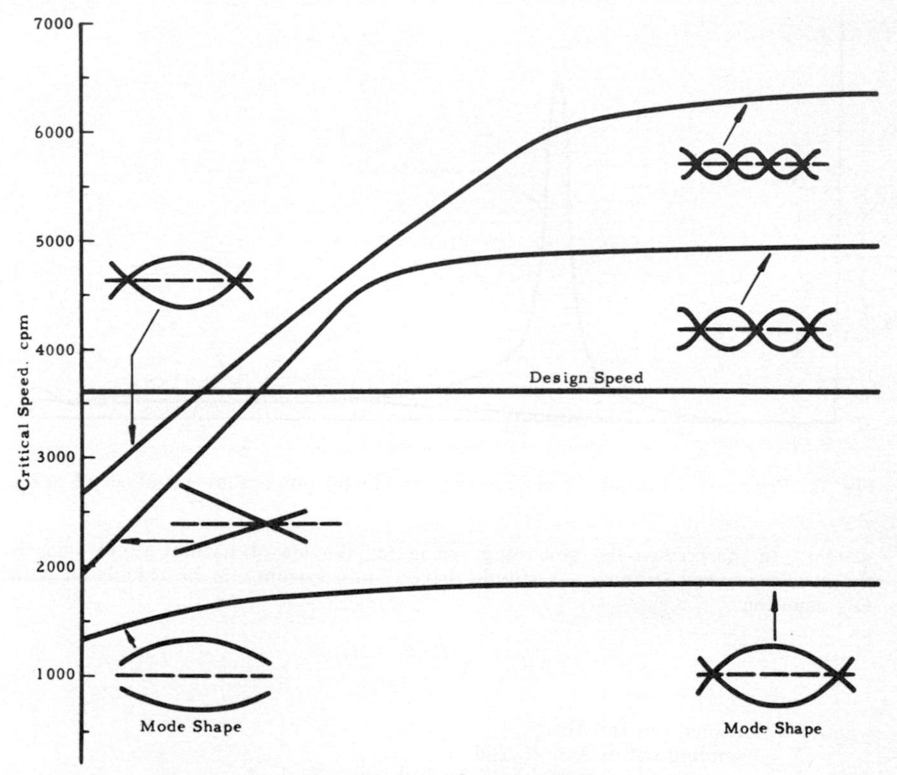

Fig. 2 Critical speed map.

A critical speed map, such as shown for a three-stage centrifugal pump in Fig. 2, conveniently describes the effect of the bearing and support flexibility upon the critical speeds and their mode shapes. Seal and wear rings can greatly affect the critical speeds of a pump shaft due to their stiffness and damping effects. The radial stiffness of seal and wear rings are a function of the fluid viscosity, pressure drop, flow rate, rotational speed, and radial clearance. Calculated stiffnesses range from approximately 10^4 to 10^5 lb/in for the typical design parameters.

One way in which the effects on vibration of the seal and wear rings can be more accurately assessed is to include their stiffness and damping characteristics in the prediction of shaft vibrations for a given unbalance. The forced-vibration response calculated on a three-stage pump for a 1 in-oz unbalance at the coupling is shown in Fig. 3. These calculations show that the stiffness and damping effects of the seals and wear rings cause the critical speed to increase and the amplitudes to be lowered. Many rotors with their critical speed in the running speed range never experience excessive vibrations because of effective seal damping.

Pump rotor problems at subharmonic frequencies such as one-half speed and the first critical speed can occur in high-speed centrifugal pumps when the running speed is greater than two times the first critical. These problems are caused by instability mechanisms within the bearing-oil film, hysteresis effects in the mechanical structure of the rotor, or aerodynamic considerations. Table 1 lists typical problems found in centrifugal pumps and their causes and can be used in troubleshooting vibration problems.

Torsional Critical Speed Analysis The torsional natural frequencies and response of a pump system are a function of the mass inertias and torsional springs in the

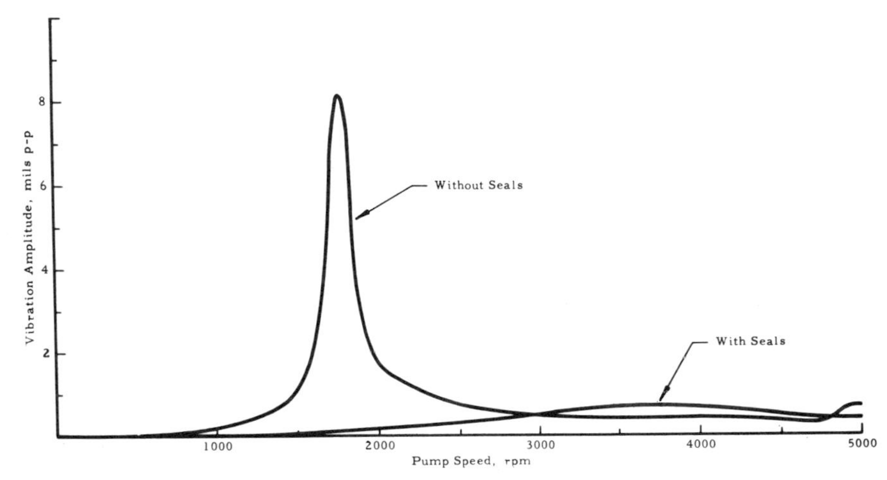

Fig. 3 Predicted vibration of three-stage centrifugal pump showing effect of seals.

system. In most cases the governing spring for the lowest natural frequencies is the coupling-spring stiffness. A simple driver-pump system can be calculated with this equation:

$$f = \frac{1}{2\pi}\sqrt{\frac{K(J_1 + J_2)}{J_1 J_2}}$$

where f = torsional critical, Hz

K = coupling stiffness, in-lb/rad

J_1, J_2 = rotary inertia (WR^2) of driver and pump, in-lb-s²

The Holzer method can be used for calculating all the system torsional criticals for multimass systems (Table 2). The Holzer tabulation method as outlined below requires the WR^2 and stiffness values of the components, which can generally be obtained or calculated from detailed drawings. Using the above equation, first obtain an estimate for ω ($\omega = 2\pi f$) and then use this value for the first trial. Following the steps in Table 2, begin with $\theta = 1$ radian for the first mass. The true solution for ω will occur when the torque summation $\Sigma J\omega^2\theta = 0$ for the last mass. If the value is not zero, make an adjustment in the natural frequency ω and reevaluate the table. Complex systems with multiple branches can be more conveniently solved by eigenvector-eigenvalue techniques.

From a practical standpoint, most torsional problems occur when calculations of natural frequencies and response are not performed in the design stage. When torsional problems occur, they can be measured with a torsiograph. Many system problems can be solved by a change of coupling stiffness; however, each case must be considered separately.

TABLE 1 Possible Causes of Problem

Vibrational frequency	Types of problems	Cause
0–40% running speed (RS)	Oil-whip resonance, friction-induced whirl, loose bearing, loose seals, bearing damage, bearing-support resonance, case distortion, poor shrink fit, torsional critical	1
40–60% RS	Half-speed whirl, oil-whip resonance, worn bearings, support resonance, coupling damage, poor shrink fit, bearing-support resonance, rotor rub (axial), seal rub, torsional critical	1
60–100% RS	Loose bearing, loose seals, poor shrink fit, torsional critical	1,2
Running speed	Unbalance, lateral critical, torsional critical, transient torsional, foundation resonance, bearing-support resonance, bent shaft, bearing damage, thrust-bearing damage, bearings eccentric, seal rub, loose impeller, loose coupling, case distortion, shaft out-of-round, case vibrations	3
2 × RS	Misalignment, loose coupling, seal rub, case distortion, bearing damage, loose coupling, support resonance, thrust-bearing damage	1,2,3
n × RS	Blade or vane frequency, pressure pulsations, misalignment, case distortion, seal rub, gear inaccuracy	3,4
Very high frequency	Shaft rub: seals, bearings, gear inaccuracy, bearing chatter, poor shrink fit	3,4
Nonsynchronous frequencies > RS	Piping vibrations, foundation resonance, case resonance, pressure pulsations, valve vibrations, noise, shaft rubs, cavitation	5

1. *Bearing-related problems*
   - Low-stability-type bearing
   - Excessive bearing clearance
   - Loose liners
   - Impurities in oil
   - Improper oil properties (viscosity, temperature)
   - Frothing of oil due to air or process fluid
   - Poor lubrication
   - Worn bearings
2. *Seal-related problems*
   - Excessive clearance
   - Loose retainers
   - Too-tight clearance
   - Worn seals
3. *Unit-design-related problems*
   - Critical speed
   - Loose coupling sleeves
   - Thermal gradients
   - Shaft not concentric
   - Inadequate support stiffness
   - Pedestal or support resonance
   - Case distortion
   - Thrust bearing or thrust balance deficiencies
   - Unbalance
   - Coupling unbalance
   - Bent shaft
   - Loose shrink fits
4. *System-related problems*
   - Torsional criticals
   - Pedestal resonances
   - Foundation resonances
   - Misalignment
   - Excessive piping loads
   - Gear tooth inaccuracies/wear
   - Piping mechanical resonances
5. *System-flow-related problems*
   - Pulsation
   - Vortex shedding
   - Piping shell resonances
   - Inadequate flow area
   - Inadequate NPSH
   - Acoustic resonance
   - Cavitation

RECIPROCATING PUMPS

Vibrations and noise in reciprocating pump systems are caused by pulsations at multiples of running speed and at the plunger frequency (speed times number of plungers) and its harmonics. Pulsations produce excessive noise and vibrations when they excite acoustic or mechanical resonances. Typical field-pulsation data obtained on a three-plunger pump are shown in Fig. 4.

TABLE 2 Holzer Tabulation Method—First Mode

$f = 11.789$ Hz $\omega = 74.077336$ $\omega^2 = 5487.4517$

Mass position	Shaft diam. d, in	Moment of inertia J, in-lb-s^2	Torque for unit deflection $J\omega^2$, in-lb/rad	Rotational deflection θ, rad 1.0*	Torque for deflection $J\omega^2\theta$, in-lb	Torque summation $\Sigma J\omega^2\theta$, in-lb	Shaft stiffness $K = 1.178 \frac{d^4}{L} \times 10^6$, in-lb/rad	Change in deflection $\frac{\Sigma J\omega^2\theta}{K}$, rad	Stress per deflection $\frac{\Sigma J\omega^2\theta}{11.25d^3}$, lb/in^2
Motor	7.75	5,662.51	31.070×10^6	1.0	31.070×10^6	31.070×10^6	34.63×10^6	0.8972	5,930
Coupling	10.0	926.79	5.085×10^6	0.1028	0.5227×10^6	31.593×10^6	52.65×10^6	0.6000	2,800
Impeller	6.0	5,778.89	31.711×10^6	−0.4972	-15.766×10^6	15.826×10^6	$10{,}000.00 \times 10^6$	0.0016	6,510
Impeller		5,778,89	31.711×10^6	−0.4988	-15.817×10^6	0.009×10^6 $\Sigma J\omega^2\theta = 0$ for last mass			

* Begin each time with $\theta = 1$ for the first mass.

Acoustic filters, surge volumes, and accumulators, when properly designed, can reduce the effect of pulsations in the piping. Accumulators tuned for narrow frequency bands are effective on constant-speed reciprocating pumps; however, in variable-speed systems, acoustic filters are more effective. A low-pass acoustic filter consisting of two volumes connected by a small-diameter choke can be designed to prevent the transmission of high-frequency pulsations into attached piping systems. The equation given in Fig. 5 can be used to calculate the cutoff frequency

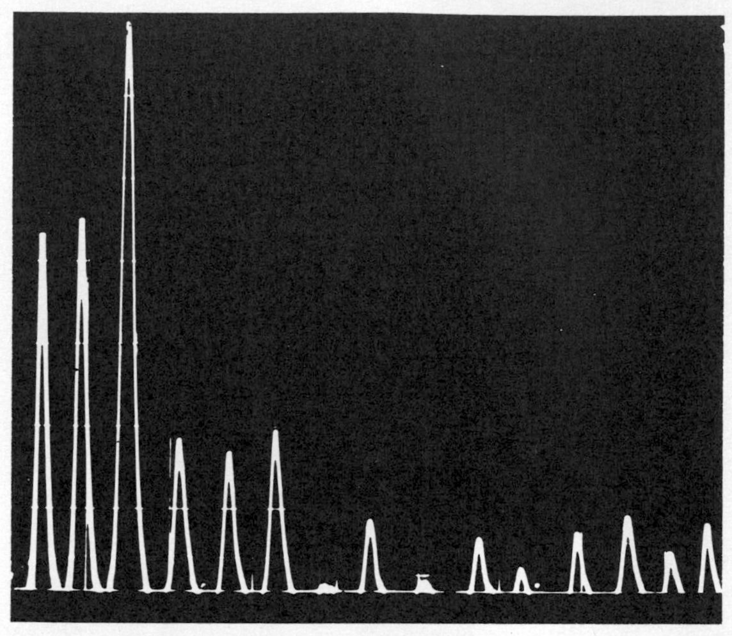

Fig. 4 Typical field data recorded on a three-plunger pump. Vertical scale: 5 lb/in^2 per division. Horizontal scale: 0 to 100 Hz.

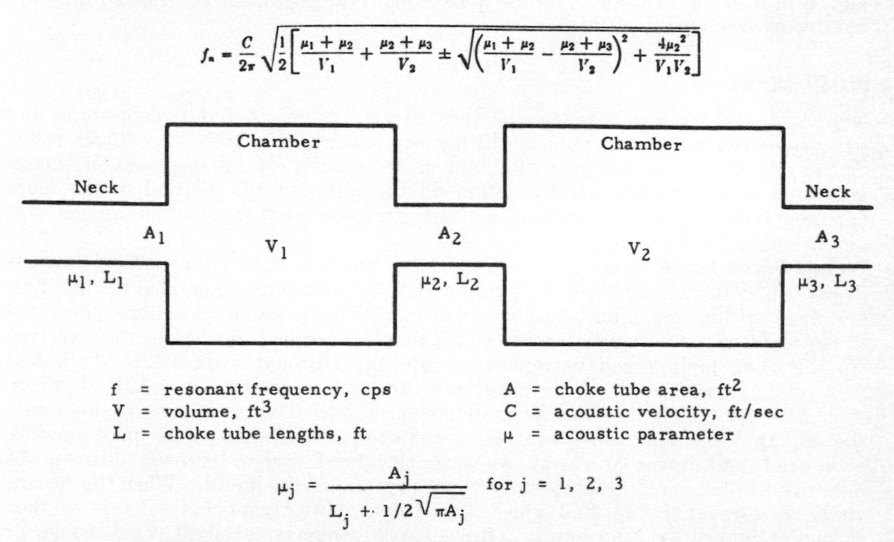

Fig. 5 Two-chamber resonator system with both ends open.

for a simple volume-choke-volume filter. For more complicated piping systems, an electroacoustic analog such as described by Nimitz [15] can be used. This design tool has become widely accepted by the industry for designing safe, trouble-free piping systems for reciprocating liquid pumps and gas-compressor units. Figure 6 shows the reduction in pulsations in a liquid pump system in a nuclear plant obtained by using an acoustic-filter system designed with the use of the electroacoustic analog.

Pulsation generated in liquid pump systems can cause excessive piping vibrations. Reduction of resonant mechanical piping vibrations can generally be accomplished by detuning the individual piping spans by the addition of clamps or braces. Generally, piping systems should be clamped at all valves and bends.

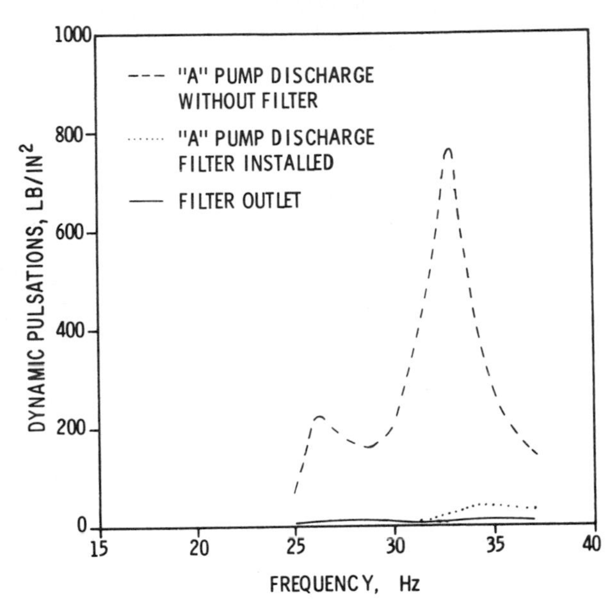

Fig. 6 Effect of acoustic filter on pulsations. (Department of Applied Physics, Southwest Research Institute)

PUMP NOISE

Noise radiated from pumps may be conveniently categorized into fluid-borne and structure-borne noise components. Either component may arise as a direct result of the rotation of the pump in the fluid or secondarily by impingement of energy generated in other parts of the system on the pump itself. Several of the more common noise problems involving primary and secondary generation mechanisms acting within the pump are described below.

Fluid-Borne Noise This may be produced in pump systems by periodic pulsations generated at integer multiples of pump speed, by turbulence generated at high flow velocities within the pump, or by cavitation. The periodic pulsations generated in the fluid contain acoustic energy which may be coupled directly to the structure (in this case, piping) and be reradiated into the adjacent work areas. Turbulent flow produces wideband energy about the Strouhal frequency ($fs = 0.2V/D$, where V is flow velocity in ft/s and D is diameter in ft). This energy may also excite the piping walls and cause acoustic reradiation. Cavitation noise may produce wide-band high-intensity energy whenever localized system pressure adjacent to a flow restriction is reduced below the vapor pressure of the liquid. When this occurs, voids are formed in the fluid which then travel downstream and collapse as they encounter higher fluid pressures. This collapse produces localized shock waves of sufficient energy to erode metallic components within the pump. As this energy

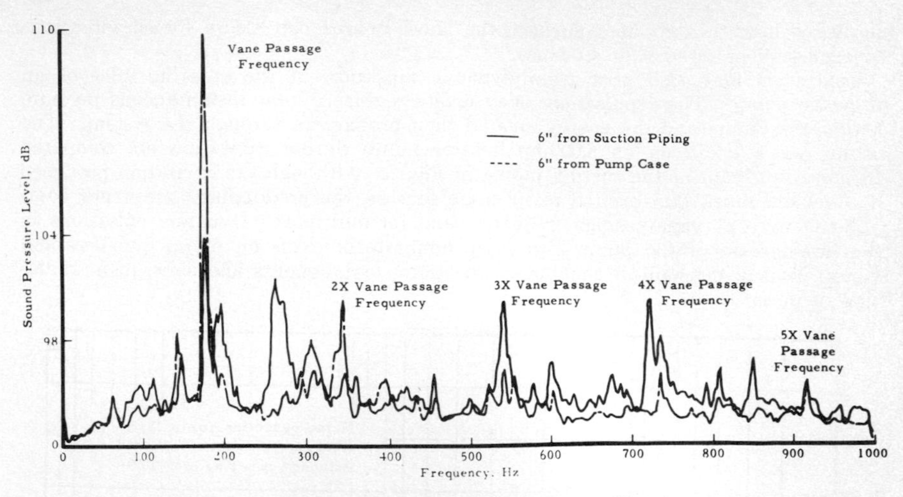

Fig. 7 Noise spectra of cavitation on centrifugal pump.

TABLE 3

Occupational safety and health act noise exposure limits

Sound level, dBA	Allowable exposure time	
	min	hr
90	480	8.00
91	420	7.00
92	360	6.00
93	320	5.33
94	280	4.67
95	240	4.00
96	210	3.50
97	180	3.00
98	160	2.67
99	140	2.33
100	120	2.00
101	105	1.75
102	90	1.50
103	80	1.33
104	70	1.16
105	60	1.00
106	54	0.90
107	48	0.80
108	42	0.70
109	36	0.60
110	30	0.50
111	27	0.45
112	24	0.40
113	21	0.35
114	18	0.30
115 (maximum allowable level)	15	0.25

impinges upon a compliant surface, the shock waves can excite forced vibrations and subsequent reradiation of noise.

Cavitation may also arise from dynamic pulsation in the entering inlet of an impeller pump. These pulsations may create regions of low instantaneous pressure within the pump, and the energy created then propagates through the system. The sound levels taken on an 8,000-hp booster pump during cavitation are compared to noise upstream of the suction piping in Fig. 7. Although the cavitation produced a wideband shock that excited many noise sources, the predominant frequency components were at vane-passage frequency and its multiples. Dynamic pulsations in the flow stream of the pump also place nonuniform loads on pump impellers and vanes, causing mechanical shaking of the pump components and subsequent radiation of noise.

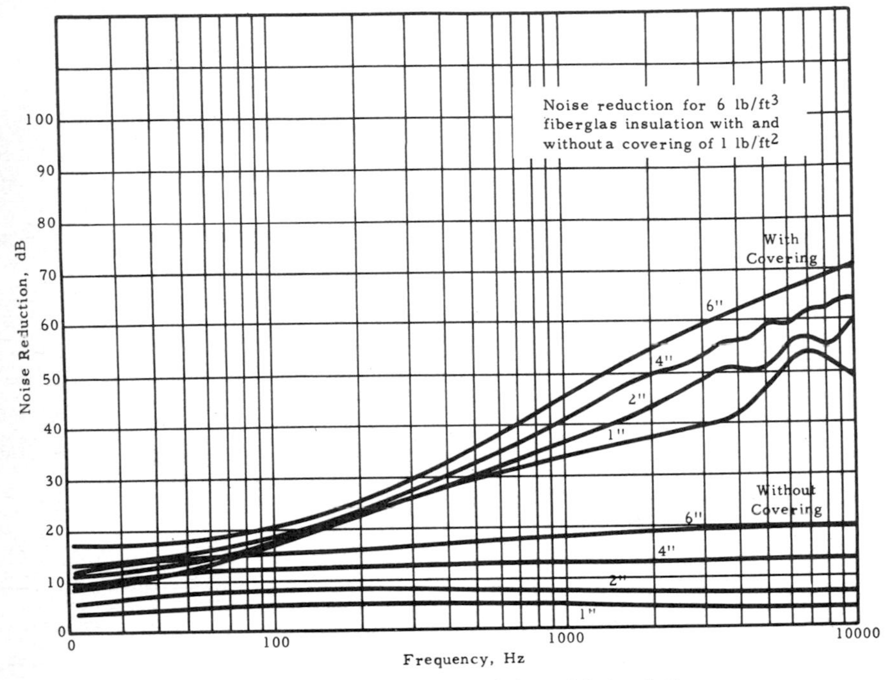

Fig. 8 Noise reduction for piping with insulation.

Structure-Borne Noise This may result from either mechanical vibration within the pump or fluid-pulsation-induced vibrations in the pump housing or piping. Whatever the source, this structure-borne energy will induce forced vibration or mechanical resonances in the piping or support structure which result in noise propagation into the surrounding environment.

Other sources of noise, such as water hammer and surge, while capable of producing severe noise problems, are not encountered in all installations and typically result from infrequent transients generated only when the pump system is operating away from its design point.

Airborne Noise When excessive noise is encountered in space, the source or sources need to be identified. This can be obtained by a detailed spectral analysis. For most cases, octave-band analysis techniques are used. Basically, an octave-band analyzer breaks the audible portion of the noise spectra into 10 adjacent bands having an upper cutoff frequency equal to two times the lower cutoff frequency.

Criteria In deciding whether treatment is necessary, one has to compare the levels to applicable criteria. The prime criterion for many pump installations will be similar to Section 50.204.10 of the Walsh-Healey Public Contracts Act as adopted under the Occupational Safety and Health Act of 1970. Table 3 presents guidelines for allowable daily noise dosages as a function of time and sound-pressure levels (in dBa). When the noise exposure for personnel is intermittent in nature, the allowable exposure may be calculated from Table 3 by computing the fractions of actual exposure time at a given noise level to the allowable exposure time at that level. If the sum of these fractions over a working day is less than one, the exposure may be considered safe.

It is important to note one underlying implication of this criterion. If a reduction of the noise levels is not economically feasible, it is possible to schedule the operator's work so that the criterion levels will not be exceeded. This rescheduling permits a wider latitude in the final solution of a plant noise problem.

Treatment Noise levels which exceeded OSHA requirements in work areas could be reduced by insulating piping, enclosing pumps, and drivers with sound barriers, or lining the building walls with a sound-absorbing material or baffles. Thermal insulation applied to piping can result in a noise reduction as well as preventing heat loss. The noise radiated by the vibrating pipe wall is partially absorbed and contained by the insulating material, resulting in a transmission loss. Increased noise reduction is achieved with an additional barrier such as a metal, fiberglass, or other rigid, high-density shell. Expected noise reduction with insulating materials of various thicknesses are plotted in Fig. 8.

REFERENCES

Books

1. Karassik, I. J., and R. Carter: "Centrifugal Pumps," McGraw-Hill Book Company, New York, 1960.
2. Advanced-Class Boiler Feed Pumps, *Proc. Inst. Mech. Eng.,* vol. 184, pt. 3N, 1969–1970.
3. Thomson, W. T.: "Mechanical Vibrations," Prentice-Hall, Inc. Englewood Cliffs, N.J., 1948.
4. Warring, R. H.: "Handbook of Noise and Vibration Control," Trade and Technical Press Ltd.
5. "Chemical Engineering Deskbook," McGraw-Hill Book Company, New York, 1971.
6. Streeter, V. L.: "Handbook of Fluid Dynamics," McGraw-Hill Book Company, New York, 1961.

Critical speeds and stability

7. Duncan, A. B.: Vibration in Boiler Feed Pumps: A Critical Review of Experimental and Service Experience, *Proc. Inst. Mech. Eng.,* vol. 181, pt. 3A, pp. 55–64, 1966–1967.
8. Fritz, R. J.: The Effects of an Annular Fluid on the Vibrations of a Long Rotor, *ASME Paper* 70-FE-30.
9. Walston, W. H., W. F. Ames, and L. G. Clark: Dynamic Stability of Rotating Shafts in Viscous Fluids, *Trans. ASME, J. Appl. Mech.,* pp. 291–299, June 1964.

Seals and labyrinths

10. Adams, M. L., and R. J. Colsher: An Analysis of Self-Energized Hydrostatic Shaft Seals, *Trans. ASME, J. of Lub. Tech.,* pp. 658–667, October 1969.
11. Black, H. F., and D. N. Jenssen: Effects of High Pressure Ring Seals on Pump Rotor Vibrations, *ASME Paper* 71-WA/FE-38.
12. Decker, O.: Dynamic Seal Technology: Trends and Developments, *Mech. Eng.,* pp. 28–33, March 1968.
13. Koenig, H. A., and W. W. Bowley: Labyrinth Seal Analysis, *ASME Paper* 72-Lub C.
14. Mallaire, F. R., L. H. Nelson, and P. S. Buckman: Evaluation of Wear Ring Seals for High-Speed, High-Pressure Turbopumps, *Trans. ASME, J. Lub. Tech.,* pp. 438–450, July 1969.

Pulsation and noise

15. Nimitz, W.: Pulsation Effects on Reciprocating Compressors, *ASME Paper* 69-PET-30.
16. Simpson, H. C., and T. A. Clark: Noise Generation in a Centrifugal Pump, *ASME Paper* 70-FE-37.
17. Streeter, V. L., and E. B. Wylie: Hydraulic Transients Caused by Reciprocating Pumps, *Trans. ASME, J. Eng. Power,* pp. 615–620, October 1967.

Section **10.1**

Water Supply

F. G. HONEYCUTT, JR.

D. E. CLOPTON

SOURCES OF WATER

Surface Water Surface water supplies are obtained from streams, rivers, natural lakes, and man-made reservoirs. The quantity of water available from a surface supply can be determined with reasonable accuracy from yield studies which take into account the local effects of rainfall conditions, runoff, evaporation and sedimentation rates, and other hydrological factors. Development of a surface supply usually requires the use of pumps to transport raw water from the source to a treatment plant, and to provide the head necessary for proper hydraulic operation of the treating facilities. Pumps utilized for this purpose are classified as low-lift pumps, because relatively low discharge heads are required.

Selection of a specific type of pump for low-lift service is dependent on the intake conditions. Since surface water supplies show significant variations in temperature, bacteria count, and turbidity at varying depths, and since the water level may be subject to considerable fluctuation, it is necessary to provide some type of intake structure which will permit withdrawal of water at several elevations. Multiple intake ports equipped with trash racks and water screens provide this capability and provide protection from fish and debris which might clog or damage the low-lift pumps. Design and location of the intake structure influence the selection of either a horizontal- or vertical-type pump for low-lift service.

Ground Water In many areas of the United States where rainfall and runoff are sparse, significant supplies of water are available from underground sources. The ground water table is formed when rainfall percolates through the soil and reaches a zone of saturation, the depth of which is governed by soil characteristics and subsurface conditions.

Ground water can be developed as a source of supply through utilization of wells or springs. Shallow wells generally utilize the water table as a source, while deep wells utilize water held in a pervious subsurface stratum (aquifer). Deep wells generally provide more constant and more prolific supplies than do shallow wells.

Artesian wells may be developed when an aquifer outcrops at a surface elevation significantly greater than the ground elevation at the well site. The aquifer is thus pressurized, and water flows from the well without pumpage. A natural artesian well may occur if a fault extends from the aquifer to the ground surface.

Springs occur at those points where the ground water table outcrops at the surface of the earth. However, the supply available from springs is seldom great enough to serve more than one or two homesteads, and is thus usually insignificant in terms of supplying communities. Surface streams may also be fed from the ground water table, and vice versa. Figure 1 depicts the various conditions described above.

Well water is usually pumped to a treatment site or into a distribution system by either vertical turbine-type line-shaft pumps or submersible pumps. Economics dictates the selection of one type over the other. In general, at depths greater than 500 ft submersible pumps become competitive economically with line-shaft pumps. For shallower depths line-shaft pumps are used almost exclusively.

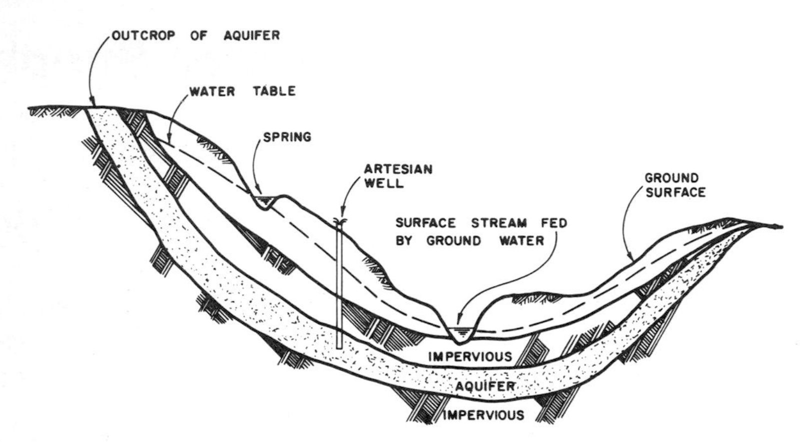

Fig. 1 Subsurface conditions for development of ground-water supplies.

Effects of Source on Water Quality Although the quality of water obtained from both ground water and surface supplies varies greatly according to local climatological, hydrological, and geological conditions, some general comparisons can be made between ground-water and surface-water supplies. Surface-water supplies generally contain greater quantities of bacteria, algae, and suspended solids than do ground water sources. This condition creates the need for specific water-treatment processes to eliminate the resulting turbidities, colors, odors, and tastes.

Ground-water supplies generally contain few bacteria and in some instances are pure enough for domestic use without chemical treatment. Frequently, however, ground-water supplies contain significant amounts of dissolved minerals. Depending upon the mineral type, the resulting supply may exhibit such characteristics as extreme hardness, toxicity, color, odor or taste. Special treatment processes such as aeration, lime softening, and disinfection may be required to remove such objectionable characteristics.

USES OF WATER

Classification of Water Usage History has shown that significant concentrations of population have always occurred at locations where a water supply was readily available. Accordingly, technology has been developed to make full use of the available supply, not only for domestic usage but also for public, commercial, industrial, and agricultural consumption.

Domestic usage consists of water used for household purposes such as washing, bathing, cooking, and watering of lawns. The amounts used for such purposes vary in accordance with the standard of living of the consumers, quality of water, use of metering devices, and other factors. Consumption for domestic purposes is generally in the range of 50 to 60 gallons per capita per day (gpcd).

Public usage includes the water used for such purposes as street cleaning, water for public parks, and supply to public buildings. Such consumption generally amounts to about 10 to 15 gpcd.

Commercial usage varies according to the number and size of shops and stores located in the area served. Attempts have been made to assign commercial consumption on the basis of gallons per day per square foot of floor space. However, the nature of business conducted in commercial installations varies widely, and accurate allocation of specific demands for commercial use must be based on a thorough investigation of the individual establishments.

Industrial usage can play a major role in design of water supply, treating, and distribution systems. In heavily industrialized areas, industrial usage can account for 30 percent or more of the total water consumption.

Agricultural usage includes the water used for irrigation purposes and for watering of livestock. In most localities irrigation water will not be a factor in system design. However, in semiarid locations where crops rely heavily on irrigation water, consideration must be given to such needs, and a thorough analysis must be made.

Water Consumption A necessary element in the design and selection of pumping equipment for water-supply projects is the determination of the amount of water required. Population forecasts and estimates of future usage serve as a basis for the design of municipal and urban systems.

The rate of consumption is usually expressed as the average annual usage in gpcd. Actual average use varies throughout the country, generally from 120 to 200 gpcd. In metropolitan areas an average value of 175 gpcd is often used for design purposes.

Demand and Daily Fluctuations Water-supply systems are subject to wide fluctuation in demand. The rate of consumption varies seasonally, monthly, daily, and hourly. For design purposes it is essential that a reasonably dependable reiationship between certain water use rates be determined from past experience records. The usage and "demand rates," which are critical in system design, are listed and described below.

1. *Average daily demand* may be expressed as gallons per capita per day (gpcd) or million gallons per day (mgd). The first represents the total gallonage of water used during a calendar year, divided by 365 times the average number of persons supplied thereby. The second represents the total year's usage, in million gallons, divided by 365. The average daily demand rate is generally used as the yardstick by which all other demand rates are measured.

2. *Maximum monthly demand* is the average daily use during the month of greatest consumption, and it is determined by dividing that month's total use, in million gallons, by the number of days in the month.

3. *Maximum daily demand* is the total use of water during the day of heaviest consumption in the year. Experience shows this demand rate to occur from 3 to 5 consecutive days during the year.

4. *Maximum hourly demand* is the rate of use during the hour of peak demand on the day of maximum demand. This demand rate normally establishes the highest rate of design for distribution systems and "peaking pumpage."

To record the hourly variations in water consumption over a 24-hr period is no simple undertaking. It involves making synchronized hourly readings of all pump discharge rates and storage levels and, from these, the computation of actual hourly rate of consumption. Such recordings of hourly consumption rates have been made in investigations for the City of Dallas Water Utilities, as shown in Fig. 2. The results shown there serve to establish system design parameters.

Water Consumption Rates as a Basis of Design The "demand rates" for proper water supply and distribution have been previously identified and described. The variations in demand, when expressed as a ratio of the average daily demand, must be considered in basic designs of water supply systems.

Figure 3 illustrates graphically the relationship that is characteristic of many water-supply demand rates. It also serves to establish percentage ratios for design purposes as follows:

$$\begin{aligned}\text{Maximum month rate} &= 155\% \times \text{average daily}\\ \text{Maximum day rate} &= 186\% \times \text{average daily}\\ \text{Maximum hour rate} &= 343\% \times \text{average daily}\end{aligned}$$

Or the design demand rates may be expressed in per capita consumption as follows:

$$\begin{aligned}\text{Average daily demand} &= 175 \text{ gpcd}\\ \text{Maximum monthly demand} &= 270 \text{ gpcd}\\ \text{Maximum daily demand} &= 325 \text{ gpcd}\\ \text{Maximum hourly demand} &= 600 \text{ gpcd}\end{aligned}$$

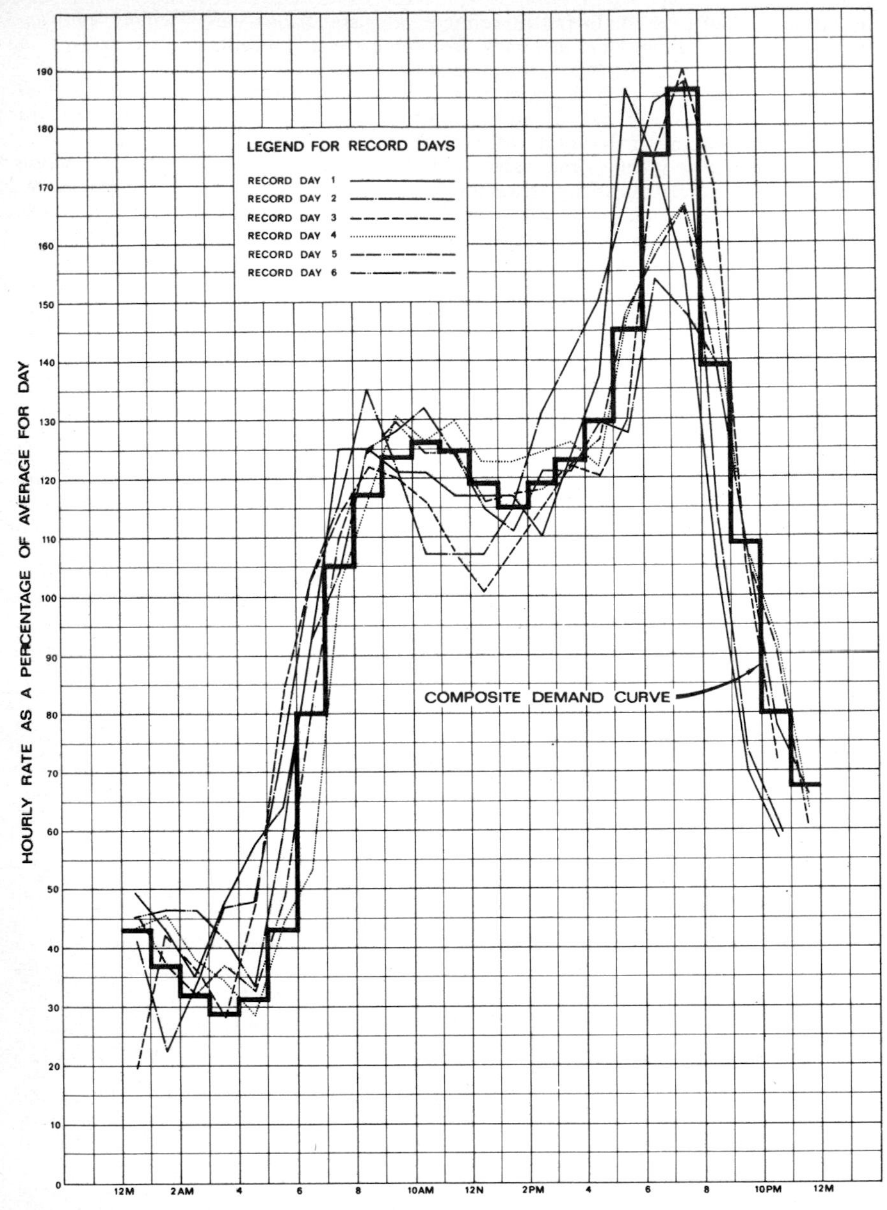

Fig. 2 Percentage rate of hourly water consumption for the city of Dallas, Texas, showing variations on days of recorded maximum day demand and composite demand curve, used as a basis of system design. (URS/Forrest and Cotton, Inc., Consulting Engineers)

The above demand rates are indicative of water requirements found in the southwest region of the United States and will vary in other regions. Nevertheless, the values shown are reasonable and will assist other designers in establishing local criteria. Above all, such values depict the broad range of pumping rates for which equipment must be chosen.

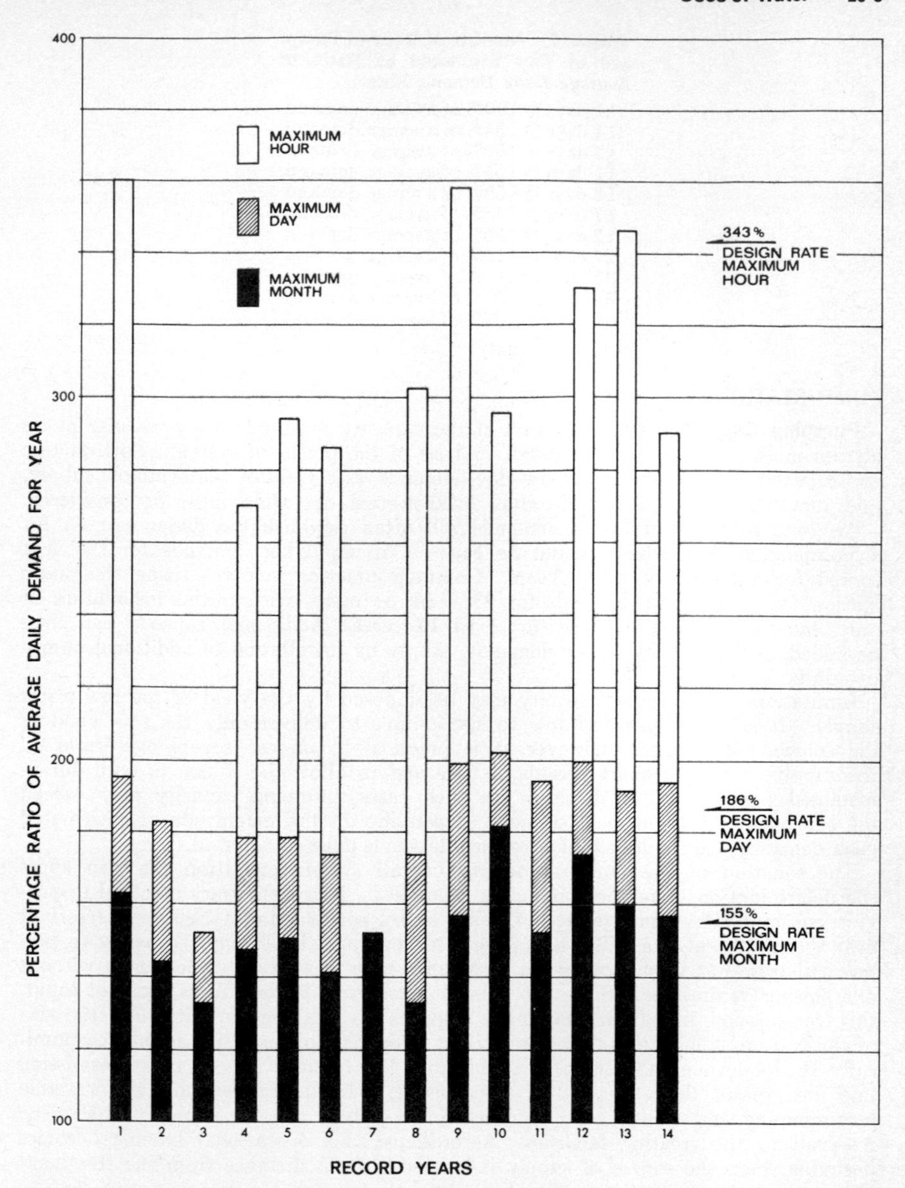

Fig. 3 A comparison diagram of maximum monthly, maximum daily, and maximum hourly demand rates, expressed as a percentage of the average daily rate for the year of their occurrence. (URS/Forrest and Cotton, Inc., Consulting Engineers)

Variations for Pumping-Time Evaluations Most water-supply systems require pumps to operate continuously throughout the year. Studies are often required for determination of operating costs and evaluation of equipment. Thus, pumping time in days, which can also be expressed as a percentage of average daily demands, is useful in various pumping design problems. Based upon studies made of municipally owned water works of various sizes in the southwestern United States, the total pumpage for the year will vary approximately as shown in Table 1.

TABLE 1 Number of Days of Pumpage in Year Expressed as Ratio of Average Daily Demand Rate

Days		Percent of average demand
11 days	@	190% of average demand
14 days	@	185% of average demand
12 days	@	180% of average demand
11 days	@	165% of average demand
13 days	@	155% of average demand
17 days	@	145% of average demand
12 days	@	135% of average demand
27 days	@	115% of average demand
95 days	@	85% of average demand
64 days	@	75% of average demand
89 days	@	65% of average demand
365 days Total		

PUMP STATIONS

Pumping Capacity Determination of the capacity required at a particular pump station must be based on a thorough analysis of the proposed system. Such factors as projected average and maximum day demands, safe yield of the available supply, and function of the pump station in total system operation must be considered.

Accurate forecasts of future demands will often establish the design criteria for a pumping station, which should be capable of supplying demands for the area served for a period of many years. Common practice involves sizing the pump station for anticipated demands for 25 years or more, with initial installation of only enough pumping capacity for 5 to 10 years. Additional capacity can then be added, as directed by future demands, simply by installation of additional pumping units.

Limitations on pumping capacity may be imposed by the yield of the raw water supply. It is sometimes desirable to size pumps to deliver only the safe yield of the source. More often, however, it is practical to impose severe overdrafts on the supply source for short periods of time and to allow the source to refill during sustained periods of low demand. In such cases, pumping capacity may exceed the safe yield by 200 percent or more, depending on the magnitude of anticipated peak demands and the length of time such demands must be satisfied.

The function of the pump station in overall system operation can also affect the determination of required pumping capacities. It is sometimes practical to provide constant-rate pumpage from the source by constructing a balancing reservoir near the treatment site. The balancing reservoir must have ample capacity to permit withdrawal at varying rates (in accordance with treated water demands) without overflowing or draining. Since the balancing reservoir is sized for a constant input, the transmission line from the source requires no peaking capacity, and the size of the transmission line is thus minimized. This type of operation is thus economically feasible when the savings realized from the reduced size of the transmission line are greater than the cost of constructing a balancing reservoir and variable rate-pumping and transmission facilities for the short distance from the balancing reservoir to the treating facilities. Accordingly, this operational scheme becomes desirable when the source of supply is located a great distance from the treatment site.

Selection of Pump Type The location and configuration of the pump station and intake structure, and the anticipated heads and capacities, are the major factors influencing the selection of a specific type of pump construction. If the pump station and intake structure are to be located within a surface reservoir, vertical turbine pumps with columns extending down into a suction well provide a logical choice.

In many cases, however, the pump station is located downstream from the dam, with connecting suction piping from the intake structure. In such instances, a horizontal centrifugal pump represents a more logical selection. Horizontal centrifugal pumps of split-case design are commonly used in the waterworks industry because the rotating element can be removed without disturbing suction and discharge piping. Selection of a bottom suction pump (in lieu of side suction) should be considered when possible, because less space is required on the station floor.

Effect of Source on Selection of Pump Materials Although many service conditions are involved in the selection of pump materials, the primary factors which can be related to source of supply are alkalinity and abrasiveness. Alkalinity (or acidity) of a raw water source is reflected by the raw water pH. In general, a pH over 8.5 or under 6.0 precludes the use of a standard bronze fitted pump (cast iron casing, steel shaft, bronze impeller, wearing rings, and shaft sleeve). The high pH values often associated with ground water sources then dictate the use of all-iron or stainless steel fitted pumps.

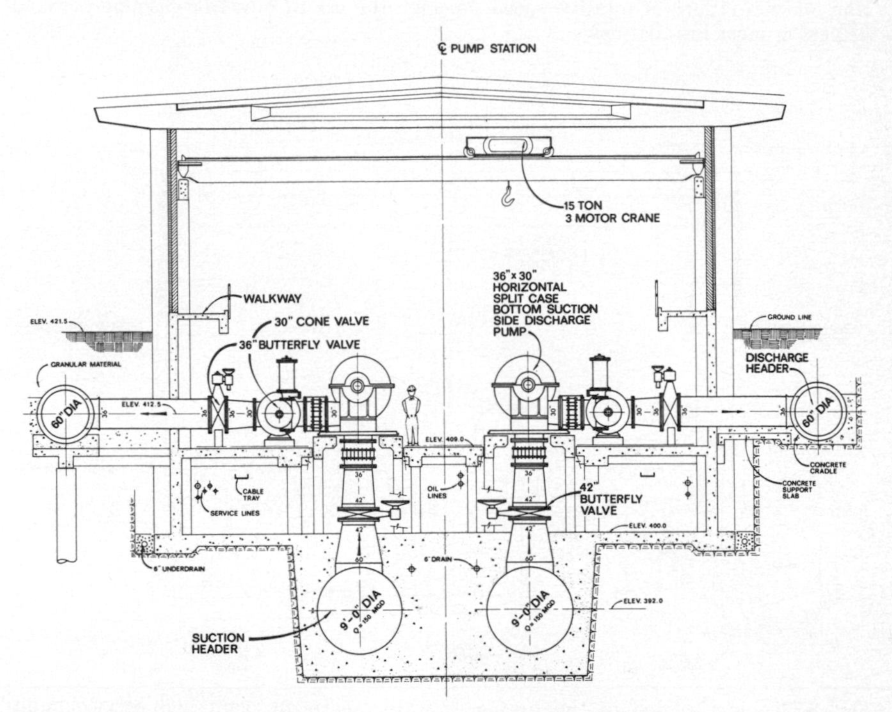

Fig. 4 Typical section of the Forney raw-water pump station in the city of Dallas showing pump suction and discharge piping. (URS/Forrest and Cotton, Inc., Consulting Engineers)

Abrasiveness, which may result from sand and other suspended matter in a surface water supply, may dictate the selection of stainless steel or nickel-cast iron casing; cast iron, nickel-cast iron, or chrome steel impellers; and stainless steel, phosphor bronze, or monel wearing rings, shafts, sleeves, and packing glands.

Suction and Discharge Piping In order to minimize head loss and turbulence, the use of long-radius bends in both suction and discharge piping is strongly recommended. American Water Works Association (AWWA) approved double-disk gate valves with outside screw and yoke or AWWA-approved butterfly valves of the proper classification are recommended for use as isolation valves in the pump suction and discharge piping. Additionally, a check valve should be provided on the discharge side of the pump to prevent backflow through the pump upon shutdown or power failure. Many types of check valves have been used satisfactorily in such applications. However, the regulated opening and closing times afforded by cone valves, combined with excellent throttling characteristics, have proven effective in minimizing surges and should be considered for pump check service if economically feasible.

Manifolding of suction and discharge headers is common practice in the design of pump stations, because parallel operation can be readily achieved with such an arrangement. Suction headers may be located directly below the pumps (Fig. 4) or along the outside wall of the pump station (Fig. 5), depending on the location of the intake structure and the configuration of the suction piping. Interconnection of discharge headers (Fig. 6) provides additional system flexibility and added protection in the event of line breakage.

Pump Drivers The waterworks industry has evolved to the point of almost exclusive use of electric motors for service as pump drivers. Diesel or gasoline engines may be used as emergency drivers or as the primary drivers at locations where reliable electric power is not available. However, the high costs of continuous operation and limitations of rotative speed preclude the use of diesel or gasoline-powered drivers in most installations.

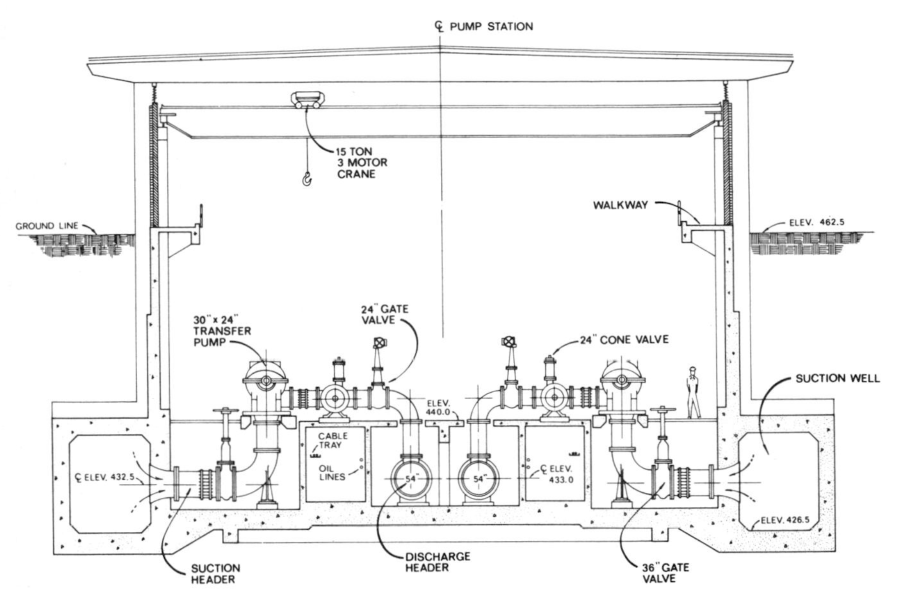

Fig. 5 Typical section of the East Side water-treatment plant high-service pump station in the city of Dallas. (URS/Forrest and Cotton, Inc., Consulting Engineers)

Although some direct current motors are used in the waterworks industry, the great majority of electric motors used as pump drivers are ac motors of the squirrel-cage-induction, wound-rotor, or synchronous type. Variable-speed devices of the hydraulic or magnetic type can be used in conjunction with the pump driver to vary pumping rates in accordance with demands.

System Head Curves The capacity of a specific pumping installation cannot be determined without an accurate determination of the head requirements of the system. Consequently, a system head curve must be derived, depicting calculated losses through the system for various pumping rates. Construction of a system head curve must be based upon a logical sequence of determinations of the various components of system losses.

A preliminary sketch or schematic should be derived, showing configuration and size of suction and discharge piping, including all pipe, valves, and fittings.

Friction Losses. Friction loss in the discharge piping can be determined for various pumping rates from the formula

$$h_f = KQ^{1.85}$$

where h_f = head loss, ft
K = a constant depending on pipe size and friction factor C
Q = flow, mgd

Fig. 6 Dallas Iron Bridge 100-mgd raw-water pump station, showing layout with bottom suction side-discharge pumping units. (URS/Forrest and Cotton, Inc., Consulting Engineers)

The above formula is a modification of the basic Hazen-Williams formula for the special case of circular pipes flowing full. Values of 10^5K for various pipe sizes and friction factors are listed in Table 2. Common practice has been to design systems on the basis of $C = 100$. However, recent investigations for the Dallas Water Utilities have indicated that a regulated cleaning program for pipelines can result in sustained values of C as high as 135 or 140. Accordingly, consideration should be given to establishment of maintenance programs to provide for periodic cleaning of pipelines so that the initial design capacity of the pumping units can be sustained.

To demonstrate the use of the above formula, assume that it is desired to compute the head losses due to pipe friction as a result of pumping at a rate of 30 mgd through 1,050 ft of 36-in-diameter pipe having a friction factor of 130.

From Table 2, $10^5K = 0.610$ or unit $K = 0.00000610$.

$$\text{Segmental } K = 0.00000610 \times 1050 = 0.006405$$
$$Q^{1.85} = 30^{1.85} = 540.3$$
$$h_f = 0.006405 \times 540.3 = 3.46 \text{ ft}$$

Entrance and exit losses and losses through valves and fittings can be calculated from the data in Sec. 9.1, "General Characteristics of Pumping Systems and System Head Curves."

Static Head and Pressure Differential. In computing total system head, consideration must be given to the effects of static head and pressure differential. Static head can be calculated simply from the difference in elevation between supply level and discharge level, and differential pressure can be calculated from the difference between terminal pressure and suction pressure.

Pump Characteristic Curves. The shape of a centrifugal pump curve must be considered in selection of a specific pump. If, for example, the water level at the source of supply or at the point of discharge is subject to wide variation, a pump

TABLE 2 Computed Values of 10^5K*

Pipe diameter, in.	Hazen-Williams *C* value								
	70	80	90	100	110	120	130	135	140
6	11,800	9,240	7,420	6,100	5,110	4,350	3,750	3,590	3,280
8	2,900	2,270	1,823	1,500	1,240	1,070	924	881	805
10	982	767	617	507	425	362	312	298	272
12	404	315	253	209	175	149	128.5	122.6	112.1
14	190	149	119.7	98.4	82.5	70.4	60.5	57.8	52.9
16	99.6	77.8	62.6	51.6	43.3	36.8	31.6	30.3	27.7
18	56.1	43.9	35.2	29.0	24.3	20.7	17.84	17.04	15.6
20	33.6	26.2	21.1	17.33	14.53	12.4	10.67	10.17	9.30
21	26.4	20.7	16.6	13.67	11.47	9.77	8.43	8.05	7.35
24	13.8	10.3	8.68	7.13	5.99	5.09	4.39	4.19	3.83
27	12.4	6.08	4.89	4.02	3.37	2.87	2.47	2.36	2.16
30	4.66	3.64	2.91	2.41	2.02	1.717	1.48	1.412	1.291
33	2.94	2.29	1.964	1.516	1.269	1.081	0.933	0.890	0.814
36	1.92	1.50	1.206	0.993	0.832	0.708	0.610	0.583	0.534
39	1.298	1.016	0.816	0.670	0.563	0.480	0.413	0.395	0.361
42	0.906	0.706	0.570	0.469	0.392	0.334	0.287	0.276	0.251
45	0.646	0.504	0.406	0.334	0.280	0.238	0.206	0.1964	0.1796
48	0.462	0.379	0.290	0.239	0.200	0.1702	0.1470	0.1401	0.1280
51	0.352	0.275	0.221	0.1816	0.1522	0.1296	0.1119	0.1067	0.0975
54	0.266	0.208	0.1673	0.1377	0.1152	0.0982	0.0848	0.0807	0.0738
57	0.204	0.160	0.1285	0.1056	0.0886	0.0753	0.0650	0.0621	0.0567
60	0.1594	0.1244	0.1000	0.0823	0.0691	0.0587	0.0507	0.0484	0.0442
63	0.1256	0.0982	0.0789	0.0650	0.0545	0.0464	0.0400	0.0382	0.0349
66	0.1000	0.0785	0.0630	0.0518	0.0434	0.0370	0.0319	0.0304	0.0278
69	0.0805	0.0630	0.0507	0.0417	0.0349	0.0298	0.0256	0.0245	0.0224
72	0.0655	0.0511	0.0412	0.0339	0.0284	0.0242	0.0209	0.01991	0.01820
75	0.0538	0.0414	0.0339	0.0278	0.0233	0.01982	0.01710	0.01632	0.01493
78	0.0444	0.0347	0.0279	0.0229	0.01924	0.01637	0.01413	0.01348	0.01231
81	0.0314	0.0289	0.0232	0.01906	0.01600	0.01360	0.01173	0.01121	0.01024
84	0.0309	0.0242	0.01942	0.01600	0.01340	0.01141	0.00984	0.00939	0.00859
90	0.0222	0.01730	0.01390	0.01143	0.00957	0.00816	0.00704	0.00672	0.00614
96	0.01614	0.01262	0.01013	0.00834	0.00700	0.00596	0.00513	0.00491	0.00448
102	0.01200	0.00939	0.00756	0.00621	0.00520	0.00444	0.00381	0.00365	0.00334
108	0.00910	0.00711	0.00572	0.00471	0.00395	0.00336	0.00289	0.00276	0.00253
120	0.00545	0.00426	0.00343	0.00283	0.00235	0.00200	0.00173	0.00165	0.00151

* For use in the formula: $h_f = KQ^{1.85}$

where h_f = head loss, ft

$K = (1{,}594/C)^{1.85}(L/d^{4.87})(1/0.446)$

C = Hazen-Williams coefficient of pipe friction

L = pipe length, ft; 10^5K values are for L = 1.0 ft and must be multiplied by true line length to determine line segmental K

d = pipe diameter, in

Q = discharge, or flow, mgd

with a steep characteristic curve near the design point should be selected to minimize the effects of head variations on pump capacity. Additionally, the pump shutoff head (or head of impending delivery) must exceed the static head to ensure that the pump will operate upon opening of the discharge valve.

Net Positive Suction Head. As a final step in selection of a specific pump for a particular application, suction conditions should be investigated to determine the net positive suction head (NPSH) available. Failure to meet NPSH requirements of the pump selected will result in cavitation of the pump impeller and very low pumping efficiency.

Total System Head. The summation of all components of system head, as calculated for various flows, results in a graphical plot of the system head curve (Fig. 7). To determine the capability of a specific centrifugal pump operating under system conditions, the pump's characteristic curve should be superimposed on the system head curve. The intersection of the two curves then represents the capacity which that specific pump can deliver.

Parallel and Series Operation The installation of multiple pumping units operating in parallel is common practice in the waterworks industry because, with proper

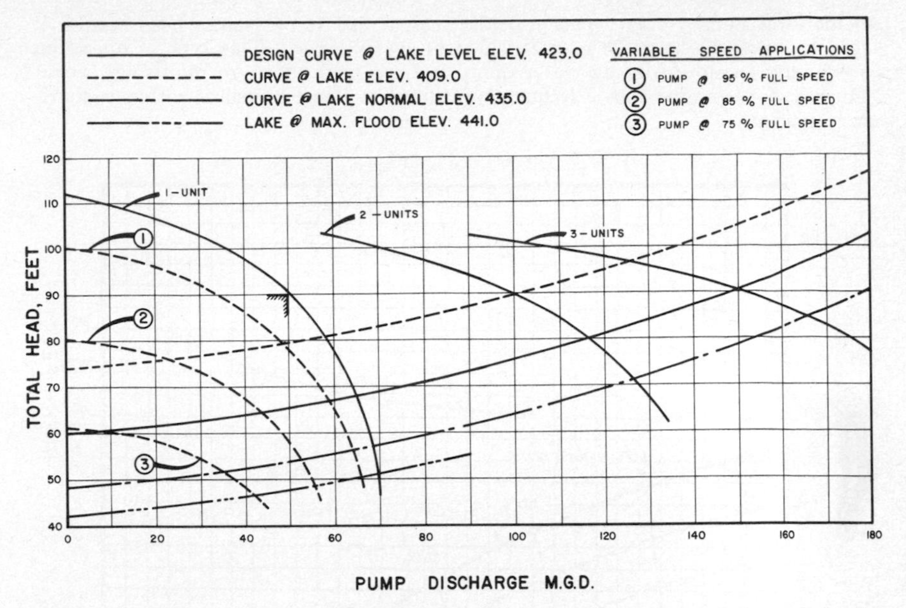

Fig. 7 System head curve for the Forney raw-water pump station of Dallas. (URS/Forrest and Cotton, Inc., Consulting Engineers)

design and regulation, it permits the most efficient use of pumping facilities and allows smooth transitions in pumping rates as demands fluctuate. This type of operation is particularly adaptable to pumpage from treating facilities into a distribution system, but is also applicable to raw-water pumpage, which is generally required to match treated-water pumpage. Special care must be taken in sizing the individual pumping units to ensure efficient operation and to prevent units from operating significantly above or below design rates during parallel operation. The principle upon which design must be based is that *total station discharge may be determined by adding the individual pump discharges associated with any particular head.* Obviously then as additional pumps are placed in operation the station flow will increase. However, since the system head curve rises with increasing flow, *the operation of two identical pumps in parallel will not produce a discharge equal to twice the capacity of one pump.* As more pumps are placed in operation, the incremental increase in pumping capacity becomes smaller.

It should be noted that, when only a portion of the pumps are in operation (total station flow is less than design capacity), the total system head is reduced, and the individual pumping units thus operate at a rate exceeding the design capacity.

If flows are increased substantially past the design point, NPSH available may become inadequate, and cavitation may occur. Additionally, the possibility of overloading the pump driver is introduced, particularly in pumps designed for high heads. In selecting a pumping unit it is therefore necessary to check conditions at runout capacity (the maximum discharge and lowest head anticipated) in order to ensure proper pump operation.

The undesirable effects of operating a pump at capacities lower than design flow are similar to those resulting from overpumping, but for different reasons. Low discharge rates result in recirculation through the pump, causing cavitation, vibration, and noise. Moreover, the radial force on the impeller increases substantially, thus increasing the stress on the shaft and bearings (to an even greater extent than would result from overpumping). Drivers of pumps designed for low heads may be subjected to overload at low capacities (and thus high heads).

In determining pumping capacities for series operation, *heads are added.* Thus two identical pumps with capacity of 30 mgd at 100 ft of system head would, if placed in series, discharge 30 mgd at 200 ft of system head. This type of operation is frequently employed in raw-water pump stations in the form of multistage, vertical-turbine-type pumps and is frequently utilized to boost pressures within a distribution system.

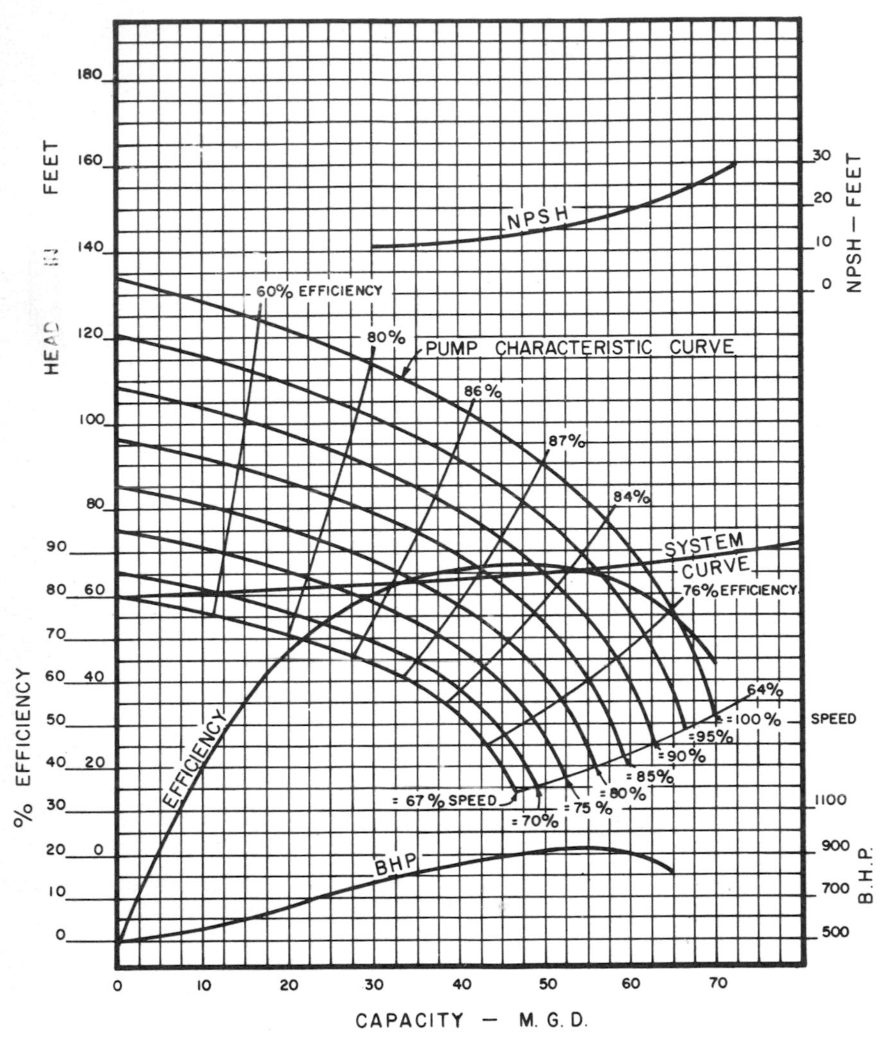

Fig. 8 System head curves for variable-speed operation for the Forney raw-water pump station in Dallas. (URS/Forrest and Cotton, Inc., Consulting Engineers)

Variable Speed Applications Installation of two or more variable-speed pumping units will allow gradual increase or decrease of station discharge as dictated by demands. When the required discharge is less than the capacity of two pumps, the variable speed units may be operated in parallel. Moreover, the installation of two variable speed units permits operation at flows in excess of the minimum allowable flow and ensures satisfactory efficiency under all conditions. Figure 8 typifies the relationships between head, capacity, and efficiency at varying speeds.

Typical operational procedure for an installation with two variable speed units is as follows:

1. If station flow is less than the capacity of one pump, operate one variable-speed unit.
2. If station flow is between one and two times the capacity of a single unit, operate both variable-speed units (in lieu of operating one unit at full speed and the other at a speed less than that required to produce minimum allowable flow).
3. If station flow is over two times the design capacity of a single unit, operate two variable-speed units and as many constant-speed units as required to ensure operation of all pumps at speeds which will produce flows exceeding minimum requirements.

USES OF PUMPS AT WATER-TREATMENT PLANTS

Variety of services The use of pumps is an integral part of the treating process at virtually every water-treatment plant in existence. Pumps serve a variety of uses, including low-lift service, coagulant feed, carbon slurry transfer and feed, fluoride feed, delivery of samples from selected points in the treating process to the chemical laboratory, plant water supply, wash water supply, surface wash supply, and high service pumpage to the distribution system (Fig. 9).

Low-Lift Pumps Low-lift pumps, located at either the source or the treatment site, are often outdoor, weatherproof-type vertical pumps (Fig. 10). They may also be horizontal centrifugal pumps housed in a separate pump station. Head and capacity requirements are as discussed previously.

Coagulant Feed Pumps Pumps used in chemical feed applications are usually not of the centrifugal type because of the kinds of materials handled. More often chemicals are fed by a jet pump or hydraulic eductor which utilizes a moving stream of water to create a vacuum at the suction port of the eductor. This type of pump has no moving parts and is thus adaptable to the feeding of coagulants such as alum or ferric sulfate in solution. A constant-level solution tank is often used to maintain a constant head on the eductor suction. The eductor capacity is then a function of water-supply pressure and friction loss in the feed lines. Eductors are normally rated according to suction capacity, which must be added to the amount of flow contributed by the motive fluid in calculating head losses through the feed system. Manufacturer's literature should be consulted to determine the water-supply requirements for various sizes of eductors.

Carbon Slurry Pumps Treatment plants which receive raw water from surface sources often require the addition of activated carbon to combat taste and odor problems caused by suspended material and microorganisms carried into the supply by floodwaters. Consequently carbon must be stored in ample quantities to supply immediate needs. The activated carbon is often transferred from storage tanks to a constant-level solution tank in the form of a slurry. Progressing cavity-type pumps equipped with variable-speed drives are suitable for this transfer service, provided the proper materials of construction are selected. The range of capacities required is simply equal to the difference between minimum and maximum carbon feed rates, and discharge head can be calculated as for any pumping system. However, selection of operating speed and horsepower requirements must be based on estimates of the abrasive and viscous characteristics of the slurry.

Fluoride Pumps Addition of hydrofluosilicic acid during the treatment process is achieving growing acceptance in the United States as an effective preventative of tooth decay. Pumps used for addition of this material should be positive-displacement, mechanical-diaphragm-type metering pumps constructed of special materials such as Kralastic, Penton, and Teflon. A relief valve should be installed on the discharge line to protect the pump in the event of a line blockage.

Diaphragm-type pumps operate on the principle of linear motion of a flexible diaphragm which pulls the solution through an intake port during the backward stroke and forces solution out the discharge port on the forward stroke. Feed rate can be varied by adjusting the stroke length or by a multistep pulley belt-connected to the motor driver. Hydrofluosilicic acid is usually supplied in slurry form, of

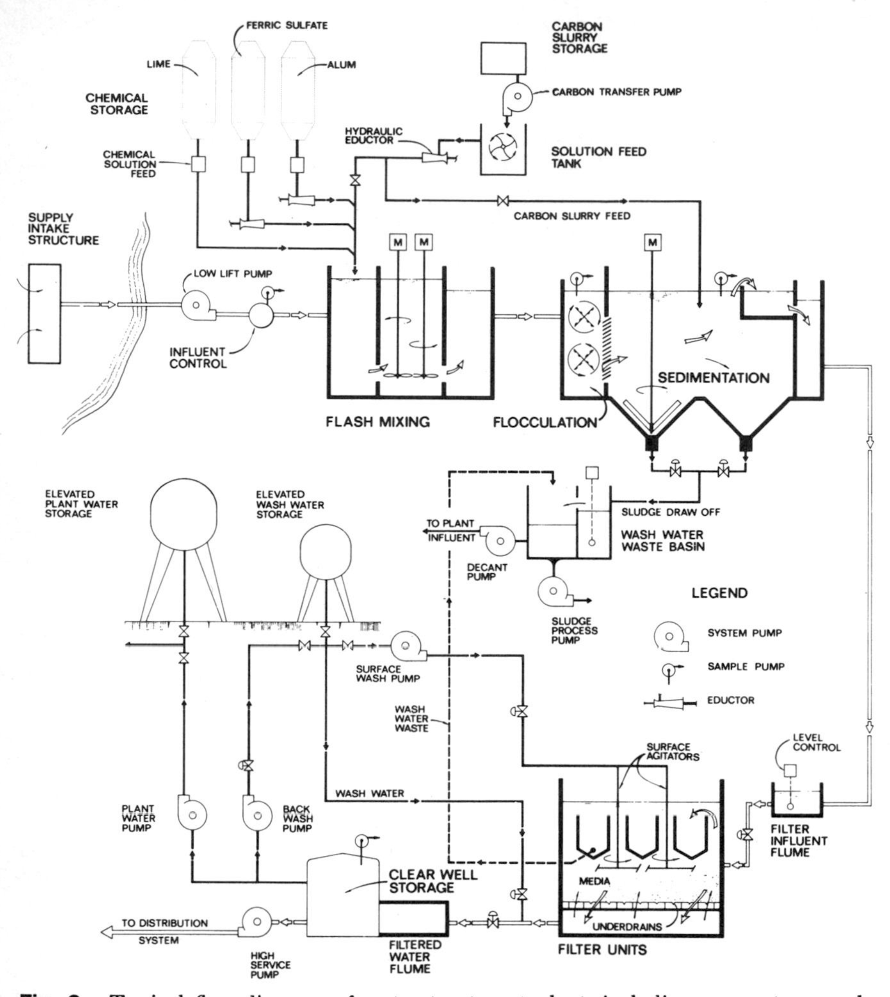

Fig. 9 Typical flow diagram of water-treatment plant, including raw-water supply.

which 20 to 30 percent is hydrofluosilicic acid. Of this acid, some 79 percent is in the form of fluoride ion. In sizing the fluoride feed pumps consideration must be given to the following factors:

1. Difference between fluoride ion content of raw water and desired content of finished water.
2. Strength of the slurry.
3. Specific gravity of the slurry.

For example, assume that the fluoride ion content of the raw water is 0.3 parts per million (ppm), the desired concentration of treated water is 0.8 ppm, slurry is 20 percent hudrofluosilicic acid with specific gravity of 1.22, and the plant raw-water inflow is 100 mgd.

Required dosage is then:

$$\frac{0.8 - 0.3}{(0.79)(0.20)(1.22)} = 0.192 \text{ ppm of slurry}$$

Quantity of slurry is then:

$$0.192 \text{ ppm} \times 100 \text{ mgd} \times 8.34 \text{ lb/gal} = 160 \text{ ppd of slurry}$$

Fig. 10 Raw-water pumping station, showing vertical turbine pumps with variable-speed drives, North Texas Municipal Water District. (Byron Jackson Pump Division Borg-Warner Corporation)

Head losses to the point of application can be calculated (with allowances for the viscosity of the slurry) and the proper feed rate accomplished by adjustment of the stroke length. The diaphragm-type pump is capable of providing a repeatable accuracy of ± 1 percent, a desirable characteristic in light of the fact that excess fluoride in drinking water can produce harmful rather than beneficial effects on the teeth.

Sampling Pumps Small-capacity centrifugal pumps are generally used for delivery of samples taken from various points in the plant to the chemical laboratory for analysis. Capacities required are generally in the vicinity of 5 to 10 gpm. Determination of the required discharge head can be made from a schematic layout of the suction and discharge piping, with provision for some 10 to 15 lb/in^2 pressure at the chemical laboratory faucet. Head losses through the system are computed as in any pumping system, and such computations are simplified somewhat by the fact that no interconnections are involved, i.e., each sample pump must have a separate discharge line to the chemical laboratory.

Plant Water Pumps Water used for various purposes throughout the treatment plant is usually taken from the treated water at the end of the process. Such water may be pumped back directly into the plant water system or to an elevated storage tank which supplies the head required for adequate pressures throughout the plant. In either case a thorough study of the plant water system is necessary to determine the amount of water required. The plant water tank can be provided with automatic start and stop controls, based on the water level in the tank, to prevent complete draining or overflowing of the tank. Allowances should be made to ensure that the plant water tank is sufficiently elevated to provide ample pressures at water-consuming devices such as hydraulic eductors and ejectors. Plant water pumps will generally be of horizontal split-case construction and must be capable of filling the elevated tank at rates determined from analysis of the plant water system.

Wash Water Pumps Selection of wash water pumps requires much the same procedure as that of plant water pumps. An elevated wash water tank should be sized large enough to permit backwashing of one filter at a time, and should be elevated as necessary to provide the required flows through the wash water piping system. Required backwash rates vary from about 15 to 22.5 gpm/ft^2 of filter surface area. In determining head losses from the elevated tank through the filters, consideration must be given not only to the head losses in piping, valves, and fittings, but also to the head losses in the filter underdrain system (which may range from 3 to 8 ft at a backwash rate of 15 gpm/ft^2, depending on the underdrain system installed) and the filter bed. The wash water tank should be sized to allow backwashing of one filter for approximately 10 min and should be equipped with controls for automatically starting and stopping the wash water pumps at predetermined levels. The required pumping capacity depends on the estimated frequency of backwashing and on the backwash rate.

Wash water pumps may also be used to supply water directly to the backwash piping system. The head required in such cases is that needed to overcome all losses in the wash water piping, underdrain system, and filter bed, and the capacity should equal the maximum backwash rate. An air release valve, check valve, and throttling valve should be provided on the discharge side of the wash water pump, and standby service is highly recommended.

Since the wash water pumps generally use treated water from clear well storage, horizontal centrifugal pumps should be used only when positive suction head is available. Otherwise a vertical pump unit may be suspended in the clear well.

Surface Wash Pumps Manufacturers of rotary agitators for surface wash systems generally specify minimum pressures for proper operation of their equipment. This pressure (generally from 40 to 100 lb/in^2) must be added to the system piping losses in determining head requirements for surface wash pumps. The required discharge may vary from 0.2 to 1.5 gpm/ft^2 of surface area depending on the supply pressure, size, and type of agitator supplied.

High-Service Pumps High-service pumps at a water treatment plant are those pumps which deliver water to the distribution system. The term "water distribution system" as used herein is defined as embodying all elements of the municipal waterworks between the facilities for the treatment of raw water and the consumer. The function of the distribution system is to provide an efficient means of delivering water under reasonable pressure in adequate volumes to meet peak consumer demands in all parts of the area served.

High-service pumps may be of either vertical or horizontal construction, depending on required capacity and on design and configuration of treated water storage facilities from which the pumps take suction. The high-service pump station often houses the plant water and wash water pumps in addition to the high-service pumps, because all such pumps utilize treated water to perform their required services (Fig. 11).

Operating conditions within the distribution system play an important role in determination of high-service pump capacities. In small municipalities, for example, it may be possible to pump from the treatment plant at a constant rate equal to the average day demand and to supply peak demands from elevated storage tanks throughout the system. However, as systems become larger, the need for variable-rate pumping from the treatment plant increases. If sufficient storage is available in the distribution system for supplying peak hour demands, it may be possible to provide variable-speed high-service pumps with total capacity equal to the maximum capacity of the treatment plant and to supply hourly peaks from storage. However, inasmuch as peak-hour demands may be two or more times as great as the peak-day demand and may be sustained for several hours, the required system storage for such operation can exceed 30 percent of the maximum day demand. In very large systems provision of such amount of storage is simply impractical. It may then become necessary to design treating facilities for capacities in excess of the maximum day demands and to supply only a portion of the peaking water from storage.

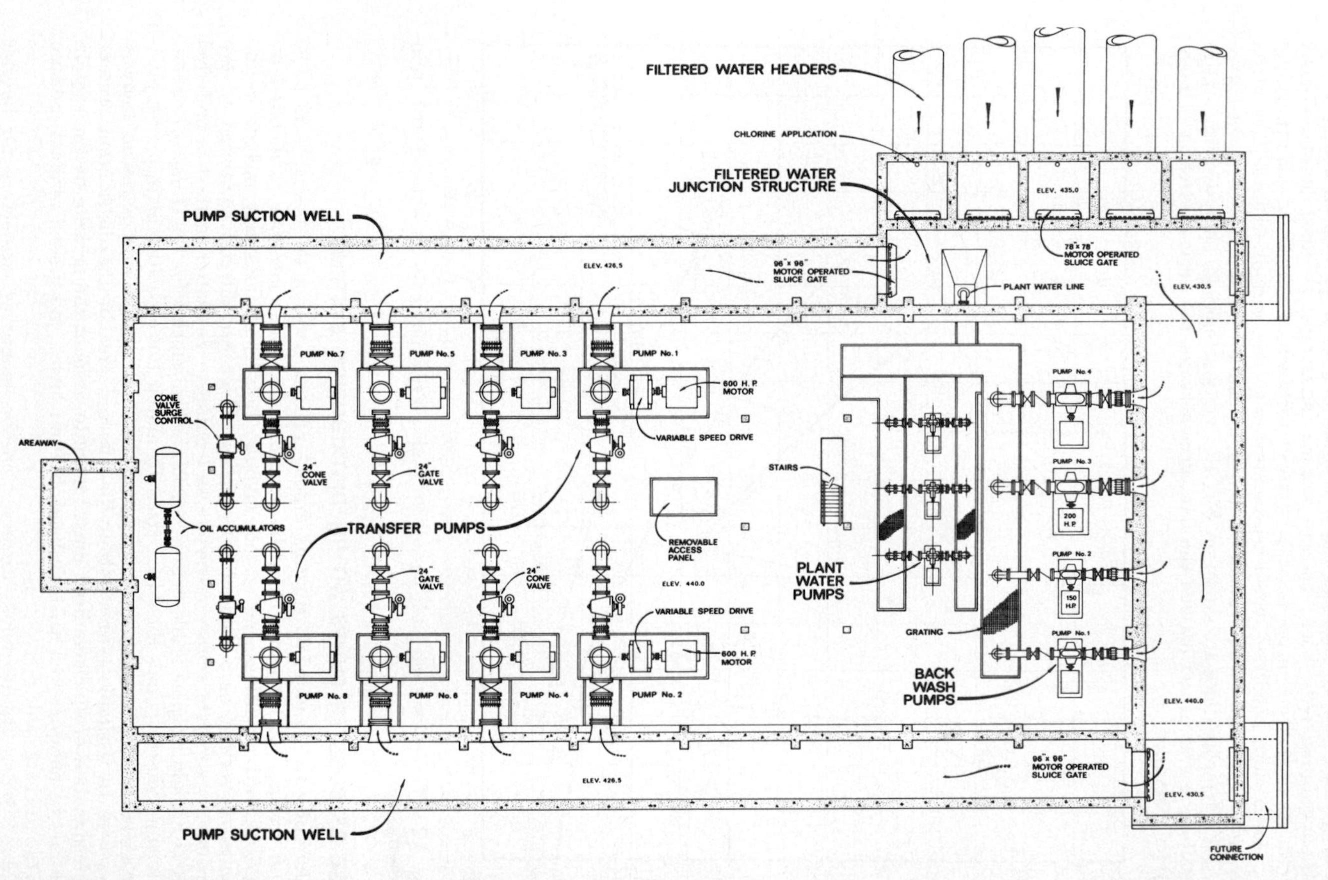

Fig. 11 Plan of the high-service pump station in the East Side water-treatment plant, Dallas, Texas. (URS/Forrest and Cotton, Inc., Consulting Engineers)

In any event, as distribution systems become larger, variable-speed pumping becomes more desirable, and development of a system head curve becomes more complex. In many cases pumpage from treatment plants during off-peak hours exceeds pumpage during peak hours, because storage tanks must be filled during off-peak hours. System head curves must be derived for each of the above conditions and also for a third condition which represents those time periods when storage tanks are full but are not being utilized to supply demands. Typical system head curves for all three conditions are shown on Fig. 12. Only a thorough analysis of the entire distribution system can provide the data necessary for the proper selection of high-service pumps.

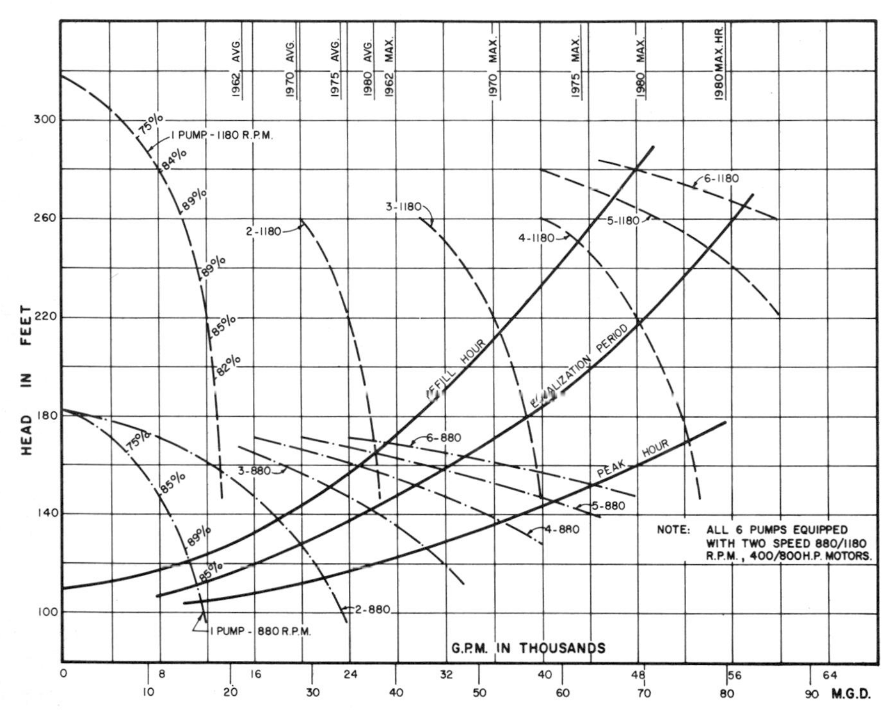

Fig. 12 Typical system head curves for high-service pumpage to water distribution system.

USE OF BOOSTER PUMPS WITHIN DISTRIBUTION SYSTEMS

In order to maintain distribution system pressures within the desirable 40 to 90 lb/in^2 range, booster pumps may be required at various locations as dictated by topographical conditions and system layout. Booster stations may include in-line pumps of vertical or horizontal construction which are directly connected to the pipeline, or separate storage reservoirs and pumps. The latter method is preferred but is not always practicable because of land requirements for the storage reservoir.

In-line booster pumps should be sized to match the capacities of incoming pipelines and should be capable of supplying the additional head required to increase pressures to desirable ranges within the affected area of the distribution system. Determination of discharge head must take into account the pressure on the suction side of the pump as well as static head, friction losses, and desired residual pressure on the discharge side.

Many water utilities refuse to accept in-line pumps as satisfactory for booster service and require construction of a pressure-equalizing storage reservoir. Booster pumps of either horizontal or vertical construction then take suction from the storage reservoir and return water at higher pressures to the mains. Pressure switches or floats provide automatic starting and stopping of the booster pumps and thus eliminate the need for continuous attendance.

GROUND-WATER WELLS

Site Selection Foremost among the factors which must be considered in selection of a site for a proposed well are local regulations governing proximity to such potential contaminants as cesspools, privies, animal pens, or abandoned wells. Also to be considered is the effect on adjacent wells, because if wells are located too close together excessive head losses may result.

Types of Pumps Used The two basic types of pump used in ground-water wells are vertical turbine, line-shaft pumps and submersible pumps. The vertical turbine pump consists of three basic components: the driving head, the column-pipe assembly, and the bowl assembly. The driving head is mounted above ground and consists of the pump discharge elbow, the motor support, and the stuffing box. The column-pipe assembly (consisting of shaft, bearings, and bearing retainers) and the bowl assembly [consisting of a suction head, impellers(s), discharge bowl and intermediate bowl(s)] are suspended from the driver head. Use of multiple bowls and impellers results in a form of series operation and permits pumpage against very high heads.

The submersible pump utilizes a waterproof electric motor located below the static level of the well to drive a series of impellers and produce a series operation similar to that of a line-shaft pump. However, the length of shafting required is greatly reduced, and thus the shaft losses and total thrust are minimized. As a result, the submersible pump becomes economically competitive with the line-shaft pump at great depths.

Air-lift pumps, which operate on the principle that a mixture of air and water will rise in a pipe surrounded by water, may be used in some cases. Such pumps are easy to maintain and operate and can be used in a crooked well or with sandy water. However, they are relatively inefficient (usually 30 to 50 percent) and allow very little system flexibility.

Reciprocating pumps are also used in some cases where small capacities are required from deep wells. Such pumps can be driven by electric motors or windmills, but they are generally noisy and are more expensive than centrifugal pumps.

Determining Pump Capacity The capacity that can be obtained from any particular well is dependent on such factors as screen size, well development, permeability of the aquifer, recharge of the ground water supply from rainfall and streams, and the head available. The basic procedure used in sizing a pump for well service involves drilling of the well and performing a test operation. The first step involves determination of the static head, or elevation of the ground water table prior to pumping. Pumpage at various rates is then conducted and the drawdown associated with each pumping rate determined. A plot of drawdown vs. pumping rate can then be derived. Pumping rate is usually measured by a weir, orifice, or pitot tube, and drawdown is determined with a detector line and gage or with an electric sounder.

From the test data and from a preliminary layout of discharge piping, a system head curve can be derived, with drawdown added to friction losses for each pumping rate. The pump characteristic curve can then be superimposed on the system head curve to determine the capacity that can be attained with a specific pump. It should be noted that pump curves for line shaft pumps are based on the results of shop tests, which do not allow for column friction or line-shaft and thrust losses. Consequently the laboratory characteristic curve for any line-shaft pump must be adjusted to actual field conditions.

Field pumping head can be determined by subtracting column friction losses from the laboratory head. Field brake horsepower is determined by adding shaft brake horsepower (which depends on shaft diameter and length, and on rotative speed) to laboratory brake horsepower; field efficiency is determined from the formula

$$\text{Field efficiency} = \frac{\text{gpm} \times \text{field head in feet}}{3960 \times \text{field brake horsepower}}$$

Since thrust loads cause additional losses in the motor bearing, it is necessary to determine the additional horsepower required to overcome thrust losses. Total thrust load is equal to the sum of the shaft weight and hydraulic thrust (which varies with laboratory head for any particular impeller), and losses due to thrust amount to approximately 0.0075 hp/100 rpm/1,000 lb of thrust. Motor efficiency is then calculated by dividing the motor's full load horsepower input (without thrust load) by the sum of full load horsepower input and loss due to thrust. Overall efficiency then equals the product of field efficiency and motor efficiency.

As a result of the efficiency losses produced by shaft weight and length in line-shaft pumps, it is usually more economical to use a submersible pump at depths over about 500 ft. Sizing a submersible pump requires calculations similar to those for the line-shaft pump. However, the submersible pump installation requires a check valve in the column pipe, which must be considered in the determination of friction losses. Moreover, the efficiency losses resulting from the motor cable (expressed as a percentage of input electric horsepower) must be considered in determining overall efficiency, which can be calculated from the formula:

$$\text{Overall efficiency} = \frac{\text{water hp} \times (\%\ \text{motor efficiency} - \%\ \text{cable loss})}{\text{shop bhp} \times 100}$$

$$\text{where water hp} = \frac{\text{gpm} \times \text{field head in feet}}{3{,}960}$$

Cable size must be selected on the basis of motor horsepower and motor input amperes, voltage, and cable length.

Well Stations A typical well station generally includes a small building for housing the pump, pump controls, metering and surge-control facilities, and chemical feed equipment. Submersible pumps do not require a pump house for protection, but if pump controls or chemical feed equipment are provided, an enclosure of some type is required. Since well-water supplies are often pumped directly into the distribution system, a differential producer is usually installed for metering purposes. Moreover, chemicals may be added at the well station to minimize corrosion, control bacteria, decrease hardness, and inject fluorides into the water supply. Surges are usually controlled by installation of a surge valve in the pump discharge line. Controls may also be installed to permit starting and stopping the pump from remote central locations and to provide for measurement and control of well drawdown.

REFERENCES

1. Fair, Geyer, and Okun, "Water and Wastewater Engineering," John Wiley & Sons, Inc., New York, 1968.
2. Karassik, Igor, and Roy Carter: "Centrifugal Pumps, Selection, Operation, and Maintenance," McGraw-Hill Book Company, New York, 1960.
3. Messina, J. P.: Operating Limits of Centrifugal Pumps in Parallel, *Water and Sewage Works,* 1969, p. R-79.
4. Singley, J. E., and A. P. Black: Water Quality and Treatment: Past, Present and Future, *Journal AWWA,* vol. 64, no. 1, p. 7.
5. Texas Water Utilities Association, "Manual of Water Utilities Operations," 5th ed., Lancaster Press, Inc., Lancaster, Penna., 1969.
6. URS/Forrest and Cotton, Inc., Consulting Engineers, "Report to the City of Dallas on Distribution System Analysis," January 1958.

Section 10.2

Sewage

H. H. BENJES, SR.

W. E. FOSTER

Sewage is defined as the spent water of a community. Although it is more than 99.9 percent pure water, it contains wastes of almost every form and description. Raw sewage, when fresh, is gray and looks something like dirty dishwater with bits of floating matter such as paper, garbage, rags, sticks, and numerous other items. If allowed to go stale, it turns black and becomes very malodorous. About 25 percent of the waste matter of normal domestic sewage is in suspension. The remainder is in solution.

Sewage contains many complex organic and mineral compounds. The organic portion of sewage is biochemically degradable and, as such, is responsible for the offensive characteristics usually associated with sewage. Sewage contains large numbers of microorganisms, most of which are bacteria. Fungi, viruses, and protozoa are also found in sewage, but to a lesser extent. Although most of the microorganisms are harmless and can be used to advantage in treating the sewage, the viruses and some of the bacteria are of the pathogenic variety which cause disease.

Sewage flow generally averages between 50 and 200 gal per capita per day (gpcd). In the absence of better information, an average figure of 100 gpcd is generally used for design purposes. The rate of flow usually varies from minimum in the early morning to maximum in the late afternoon. Minimum flow ranges from 50 to 80 percent and maximum dry-weather flow from 140 to 180 percent of average flow. The extent of variation decreases as the size of the system increases. Wet-weather flows can be 600 gpcd or more because of extraneous water entering sewers from roof drains, areaway drains, footing drains, etc.

SEWAGE SYSTEMS

In most instances, sewage systems are divided between their collection and treatment functions.

Collection Collection systems consist of a network of sewers which collect and convey sewage from individual residences, commercial establishments, and industrial plants to one or more points of disposal. Pumping stations are often needed at various points within the system to pump from one drainage area to another or to the treatment plant. The judicious location of system pumping stations enhances the economy of overall design by eliminating the need for extremely deep sewers.

Small-system pumping stations (Figs. 1 and 2) are frequently built underground and may be of the factory-built type. For larger stations, superstructures should be in keeping with surrounding development. It has been said that people smell

Fig. 1 Factory-built conventional lift station. (Ecodyne Corporation, Smith & Loveless Division)

with their eyes and their ideas as well as their noses, and for this reason above-ground structures should be attractive with landscaped grounds to overcome the popular prejudice against sewage works. Stations can be and have been designed and constructed in residential areas (Fig. 3) where the neighbors apparently are not aware or do not recognize that they are not homes.

Treatment Treatment facilities can be many and varied, with the extent and nature of the treatment determined to a large degree by the proposed use of the receiving stream and its ability to assimilate pollutants. Most conventional treatment plants being built today can be classified as either primary, biological, or advanced waste treatment. Other alternatives such as physical-chemical or chemical-biological treatment are also used on occasion, but on a lesser scale. The treatment needs of smaller communities are sometimes satisfied by package treatment plants or by waste stabilization lagoons.

Primary treatment involves removal of a substantial amount of the suspended solids but little or no colloidal or dissolved matter. Primary treatment facilities normally include screening, grit removal, and primary sedimentation. Chlorination also is often practiced along with primary treatment in order to sterilize the wastes.

Biological treatment uses bacteria and other microorganisms to break down and stabilize the organic matter. Trickling filters and the many variations of the activated sludge process are the most popular biological treatment concepts presently in use. The biological treatment facility is generally followed by final sedimentation of the solids produced by the microorganisms.

Advanced waste treatment is a very complex subject, and it can range from a limited objective such as phosphate removal to whatever additional treatment is necessary for water-reuse purposes. Advanced waste treatment usually follows conventional primary and biological treatment and can include phosphate removal,

Fig. 2 Factory-built pneumatic ejector lift station. (Ecodyne Corporation, Smith & Loveless Division)

nitrate removal, multimedia filtration, carbon absorption, and ion exchange. Where zero discharge is required, it may be necessary to follow advanced waste treatment with spray irrigation of the plant effluent or other methods of disposal.

Combined primary plus biological treatment using the activated-sludge process is perhaps the most commonly used treatment concept presently in use. A schematic drawing of a typical activated-sludge treatment plant is shown on Fig. 4. In the example, liquid treatment is accomplished by coarse screening, grit removal, fine screening or comminution, and primary settling, followed by aeration, final settling, and chlorination. Sludge processing includes thickening, dewatering, incinera-

Fig. 3 Pumping station designed for residential location.

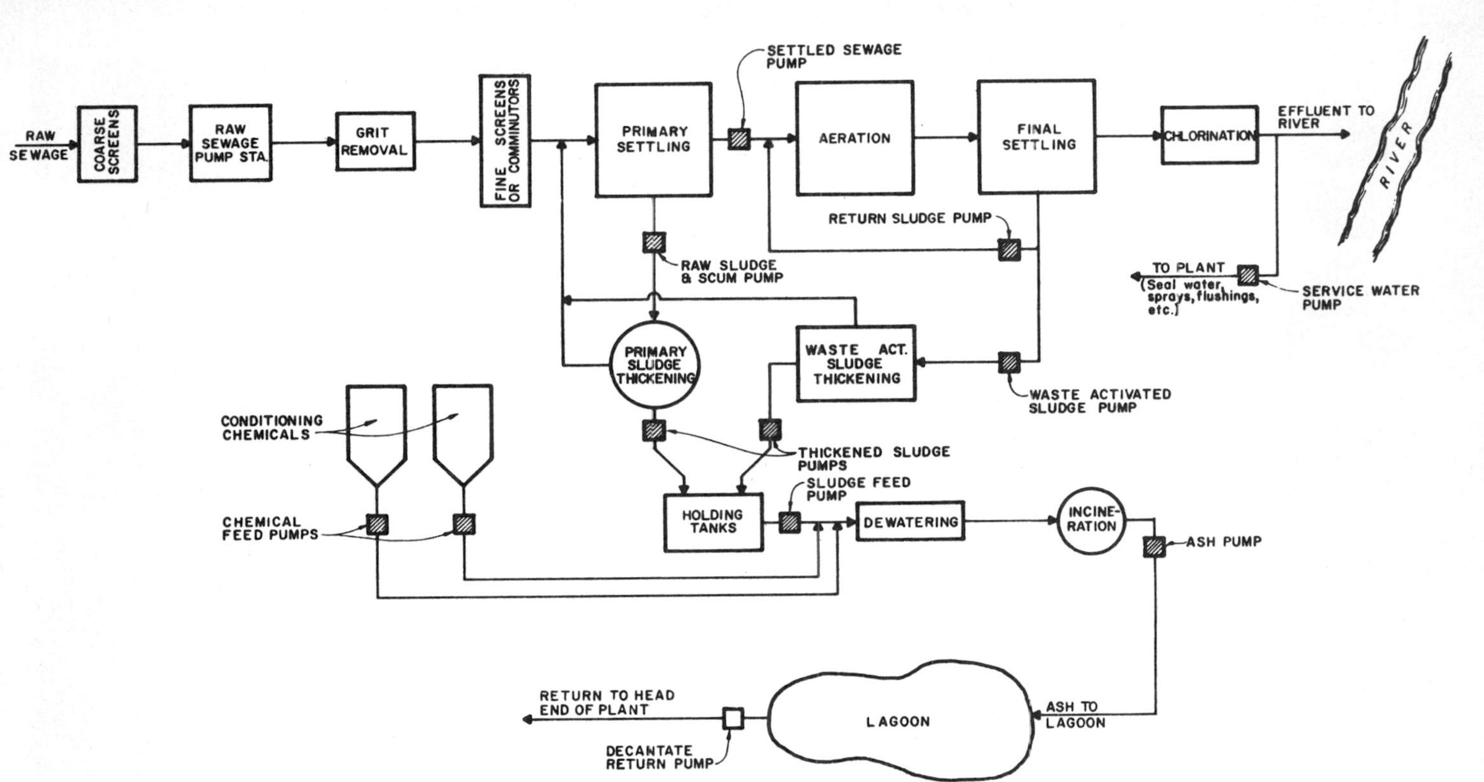

Fig. 4 Typical activated-sludge plant, with dewatering, incineration, and liquid disposal of ash.

tion, and liquid disposal of ash. There are many variations to this layout, but the one shown includes most of the pumping applications normally encountered in treatment plant design. Pumping requirements will, of course, vary from plant to plant depending on the process used, site size and topography, and the relative location of the various structures and items of equipment.

PUMP APPLICATIONS

Most of the pumping applications associated with either collection or treatment of sewage can be classified according to the general nature of the liquid to be handled. Primary classifications are (1) raw sewage, (2) settled sewage, (3) service water, and (4) sludge. There are also, however, a number of specialized applications involving the handling of abrasive materials such as grit and ash. The various types of pumps recommended for sewage associated applications are as indicated in Table 1. Although included in the table, chemical pumps are not discussed in this section. They are covered elsewhere in this handbook.

Raw Sewage Raw sewage pumps are used to lift liquid wastes from one level of the collection system to another or to the treatment plant for processing. Regardless of where the pumps are located, the basic design considerations remain the same.

Even though the sewage is normally screened at the larger installations before entering the suction wet well, it still contains a large quantity of problem material such as grit, rags, stringy trash, and miscellaneous solids small enough to pass through the coarse screens. Screens are often omitted from the smaller installations because large objects are not as much a problem because of the smaller size of the incoming sewers.

Raw sewage pumping installations are usually sized so that their firm capacity is either equal to or can be expanded to accommodate the maximum total inflow rate of the incoming sewers. Firm capacity is defined as total station capacity with one or more of the largest units out of service.

Pneumatic ejectors (Fig. 2) are sometimes used where the required capacity is less than that provided by the smallest conventional sewage pump. This type of unit, however, should not be used where more than 50 connections are expected.

Conventional sewage pumps are, by far, the most common pumps used for the handling of raw sewage. A conventional sewage pump is more specifically described as an end-suction volute-type centrifugal with an overhung impeller of either the nonclog (Fig. 5) or the radial or mixed-flow type (Figs. 6 and 7), depending on capacity and head.

The so-called "nonclog pumps" are all based on an original development by Wood at New Orleans. Actually no pump has been developed that cannot clog, either in the pump or at its appurtenances. Experience shows that rope, long stringy rags, sticks, cans, rubber and plastic goods, and grease are objects most conducive to clogging.

Nonclog impellers are used almost exclusively today for pumps smaller than 10 in in size. These pumps differ from clear-water pumps in that they are designed to pass the largest solids possible for the pump size. The conventional nonclog impeller contains two blades, although some manufacturers are now offering a single-blade ("bladeless") impeller. The two-blade impeller has thick vanes with large fillets between the vanes and the shroud at the vane entrance. The "bladeless" impeller has no vane tips to catch trash. On the other hand, it is inherently out of balance because of its lack of symmetry.

The larger raw sewage pumps are equipped with either mixed-flow or radial-flow impellers, depending on head conditions. Both have two or more vanes depending on pump size and the size of solids to be handled. The vane tips are sharper than for the nonclog impeller, resulting in a higher operating efficiency. The heavier vanes are not necessary because the vane openings can be larger than on the smaller pumps. Experience indicates that stringy trash will not clog an impeller with vane openings larger than 4 in in diameter.

The use of volute-type pumps should be limited to dry-pit applications. Units are available for wet-pit installation, but their use should be limited to temporary

TABLE 1 Types of Pumps Generally Used in Sewage Applications

Service or application	*Conventional sewage*	*Diffuser*	*Torque flow*	*Clear-water volute*	*Ash*	*Screw*	*Pneumatic ejector*	*Air lift*	*Positive displacement*	*Chemical*
Raw sewage	X	. . .	. . .	. . .	. . .	X	X			
Grit	. . .	. . .	X	. . .	. . .	. . .	. . .	X		
Primary sludge:										
Less than 2% solids	X	. . .	X							
More than 2% solids	. . .	. . .	. . .	. . .	. . .	. . .	. . .	. . .	X	
Normal primary scum	. . .	. . .	. . .	. . .	. . .	. . .	. . .	. . .	X	
Diluted scum	X									
Biological sludge	X	X	. . .	. . .	. . .	X	. . .	X		
Thickened biological sludge	. . .	. . .	. . .	. . .	. . .	. . .	. . .	. . .	X	
Digested sludge, recirculation	X	. . .	X							
Settled sewage	X	X	. . .	. . .	. . .	X				
Plant effluent	X	X	. . .	. . .	. . .	X				
Service or nonpotable water	X	X	. . .	X						
Ash sluice	. . .	. . .	. . .	. . .	X					
Decantate or supernatant liquor	X	X								
Chemical solution	. . .	. . .	. . .	. . .	. . .	. . .	. . .	. . .	. . .	X

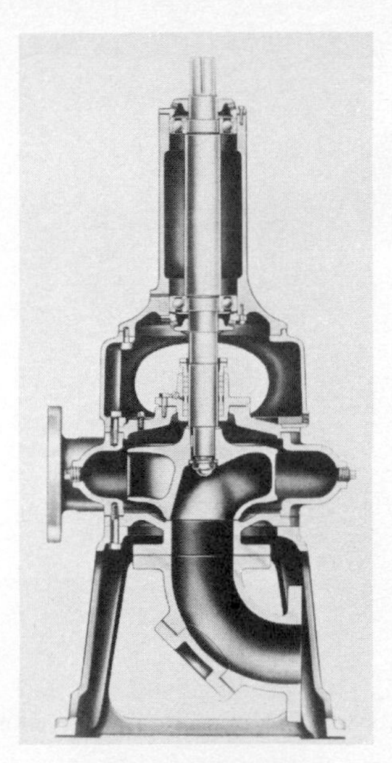

Fig. 5 Nonclog sewage pump. (Fairbanks-Morse Pump Division of Colt Industries Operating Corp.)

Fig. 6 Mixed-flow sewage pump. (Worthington Pump International)

Fig. 7 Vertical sewage pumping units, Lemay Pump Station, Metropolitan St. Louis Sewer District; 36-in, 40,000-gpm pumps close-coupled to 1250-hp motors.

Fig. 8 Self-priming nonclog sewage pump. (Fairbanks-Morse Pump Division of Colt Industries Operating Corp.)

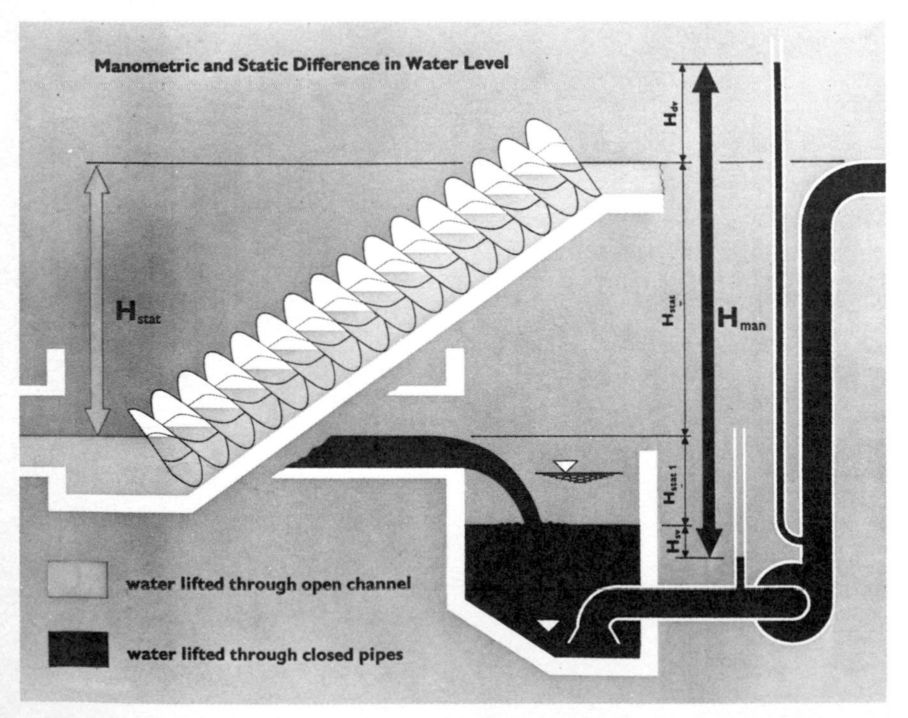

Fig. 9 Archimedean screw pump. (Beloit-Passavant Corporation)

applications with low load factors. Although nonclog pumps 8 in and smaller are available with self-priming (Fig. 8), most conventional sewage pumps are located so that the impeller is always below low water level in the suction wet well. This eliminates the need for specialized priming systems. Submersible nonclog pumps are also available, but their use generally is not recommended.

Self-priming pumps have been used successfully to pump raw unscreened sewage, particularly in the southern part of the United States. The self-priming feature eliminates the dry-pit cost and gives the centrifugal pump the gas-handling advantage of positive displacement pumps. Operating costs are higher, though, because the design efficiencies generally run about 10 to 15 percent lower than for the conventional nonclog units.

Archimedean screw pumps are occasionally used for raw sewage pumping applications. These units are advantageous in that they do not require a conventional wet well and they are self-compensating in that they automatically pump the liquid received regardless of quantity as long as it does not exceed the design capacity of the pump. This is done without benefit of special drive equipment. Also, as shown by Fig. 9, the total operating head of a screw pump installation is less than for those pumps which require conventional suction and discharge piping. Screw pumps, however, have a practical limitation as to pumping head. Generally speaking, they are not used for lifts in excess of 25 ft.

Settled Sewage Settled sewage pumps are used to lift partially or completely treated waste from one part of the plant to another or to the receiving stream. In Fig. 4, this would include the settled sewage pump, service water pump, and decantate return pump.

The liquid to be handled usually contains some solids, but grit and most of the rags and other stringy material have already been removed. Sufficient firm capacity should be provided to meet peak flow requirements. In no case should less than two units be provided.

Diffuser pumps (Fig. 10) are commonly used for the pumping of settled sewage. Depending on head conditions, they may be of either the propeller or mixed-flow design. Although normally installed in wet-pit applications, these units are sometimes mounted on suction piping and installed in a dry pit. Either type of application is acceptable, although economics usually dictates a wet-pit installation. Head and capacity conditions will determine which type of unit is applicable.

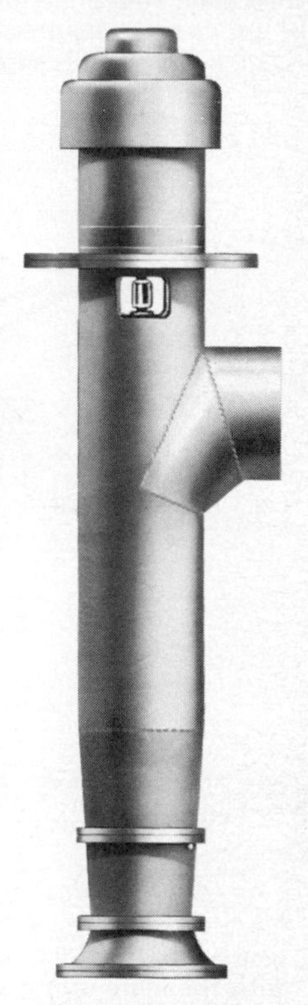

Fig. 10 Diffuser pump. (Worthington Pump International)

A conventional sewage pump may be used to pump settled sewage where a dry-pit installation is desirable. This is not usually an economical application, but it is acceptable as far as suitability of equipment is concerned. As previously indicated, wet-pit application of volute-type pumps is discouraged. This is primarily due to the high maintenance factor on the submerged bearings and the relative inaccessibility of the pumping unit for repair purposes.

Archimedean screw pumps can be used to pump settled sewage provided the lift is not excessive. As previously noted, this type of pump has certain inherent advantages.

Service Water Plant effluent water is frequently used for flushing, gland seal, foam control, sprays, chlorine injector operation, lawn sprinkling, fire protection, and various other services within a wastewater treatment plant. Except for the fact that some solids have to be contended with, this application is much the same as that found in building-water supply and small distribution systems.

Screening of solids is normally required; this can be accomplished either before or after the pumps depending upon various circumstances. Pipeline-type strainers are recommended as they are not only economical but require a minimum of space, can be automatically back-flushed, and are much easier to operate than alternative equipment.

Any type of conventional volute or diffuser clearwater pump can be used on service-water applications provided the effluent water is screened prior to entering the pump. Pumps capable of handling some solids should be used in those instances where prescreening is not practical.

Sludge and Scum This classification is divided into two separate categories, based on concentration of solids in the liquid to be handled. Specialized pumping equipment is required for the more concentrated sludges, whereas pumping of dilute sludge and scum is somewhat comparable to the handling of settled sewage.

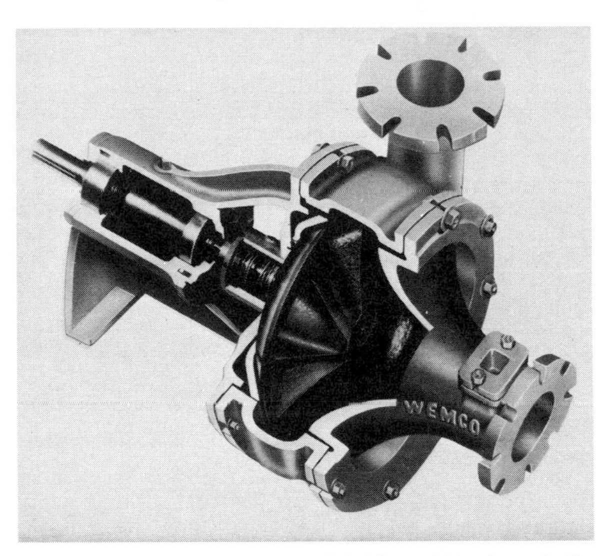

Fig. 11 Torque flow pump. (Wemco Division of Envirotech Corporation)

a. Dilute Sludge or Scum. For the purposes of this manual, dilute sludge and scum is defined as having less than 2 percent solids. An exception is digested sludge recirculation, which generally exceeds the 2 percent limit. This is included along with the more dilute sludges since the same type of pumping equipment is used.

Normally, the handling of dilute sludge is limited to the transfer of biological sludge back to the treatment process or to some other point for further concentration and/or dewatering and disposal. When digesters are used as part of the treatment facilities, sludge is often recirculated through external heat exchangers in order to maintain temperatures conducive to anaerobic bacterial action. This recirculation also helps keep the contents of the digester mixed. Occasionally primary sludge and scum are handled in diluted form.

The firm capacity of dilute sludge pumping facilities should be equal to anticipated peak loading. Biological sludge return pumps should have a capacity range from 25 to 100 percent of average design raw sewage flow to the plant. Digested sludge recirculation pumps should be sized to turn over the contents of the digester frequently enough to maintain the desired temperature. Diluted primary and waste biological sludge pumps should have sufficient capacity to handle peak sludge loading at conservative solids concentrations.

Conventional sewage pumps are suitable for handling of dilute sludge and scum. Either the nonclog or mixed-flow impeller may be used, depending upon capacity requirements. Dry-pit installations are recommended.

Diffuser pumps are particularly suitable for the handling of biological sludge provided it does not contain any appreciable amount of trash or stringy material. They are not recommended, however, for handling of diluted scum or for the recirculation of digested sludge. Depending on capacity requirements, diffuser pumps may be of either the mixed-flow or propeller design. Wet-pit applications are most common, although dry-pit installations are occasionally used.

Torque flow pumps (Fig. 11) are often used to handle dilute sludges which contain some grit. These units are particularly suitable for this type of service because their design is such that close running tolerances are not required; this allows the use of specially hardened materials such as high nickel iron which are not easily

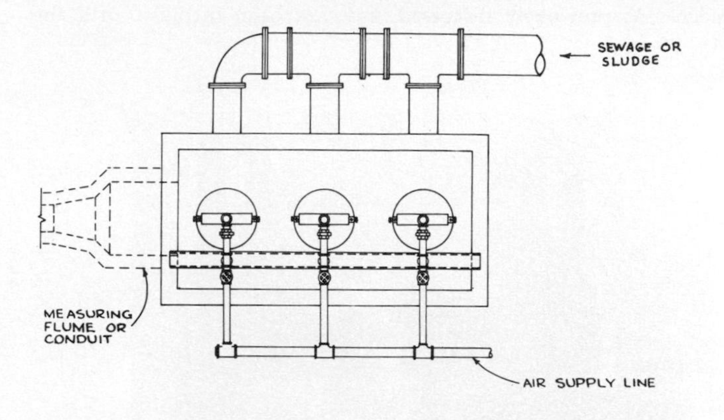

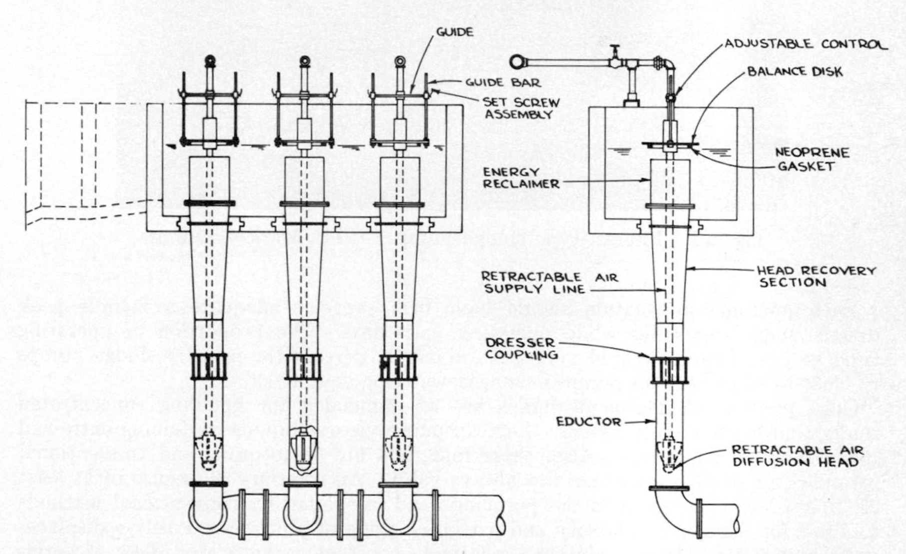

Fig. 12 Typical air-lift pump installation. (Walker Process Equipment Division of Chicago Bridge & Iron Company)

machined. The most common applications of torque flow pumps are for the pumping of nondegritted dilute primary sludge to gravity thickening and the recirculation of digested sludge.

Screw pumps can be used in certain instances for the handling of biological sludge. Use of screw pumps is generally limited to low to medium lifts and to those instances where the point of discharge is close to the sludge source.

Air-lift pumps are suitable for transfer of biological sludge where the lift is small and the point of discharge nearby. A typical air-lift pump installation is shown by Fig. 12. Total head should not exceed 4 to 5 ft. The ability of an air-lift pump to vary capacity is somewhat limited, ranging from about 60 to 100 percent of the rated amount. These pumps are inexpensive in first cost, but have an operating efficiency of only about 30 percent. They are very easy to install, and maintenance is minimal since there are no moving parts. Air-lift pumps are commonly used to transfer sludge at package treatment plants.

b. Concentrated Sludge or Scum. Concentrated sludge or scum is defined as having more than 2 percent solids. The single exception is in the case of the recirculation of digested sludge. As previously discussed, this has been included with the dilute sludge classification.

Fig. 13 Plunger-type sludge pump. (ITT Marlow Pumps)

Each pumping installation should have firm capacity adequate to handle peak design sludge quantities while operating part time. The proportion of operating time at peak loading should vary from about 25 percent for primary sludge pumps to close to 80 percent for pumps feeding dewatering equipment.

Only positive displacement pumps are recommended for handling concentrated sludge and scum, mainly because they can pump viscous liquids containing entrained gas without losing prime. Also, these materials are thixotropic, and conventional formulas for friction losses are not always valid. An arbitrary allowance of at least 25 lb/in^2 should be added to the pumping head calculated by conventional methods to allow for changes in viscosity and partial clogging of pipelines. Positive displacement pumps are able to maintain a relatively constant capacity regardless of variations in discharge head.

For most applications, positive displacement pumps may be of either the plunger (Fig. 13) or the progressing cavity design (Fig. 14). The performance of both depends upon close running clearances; consequently they have a high incidence of maintenance, especially where gritty substances are encountered. Even so, they represent the best pumping equipment presently available, and both designs have been used with success. Lobe-type gear pumps have been used for specialized applications. These are to be avoided, however, where there is any possibility that the material to be pumped will contain even a small amount of grit.

Plunger pumps should be of the heaviest design available and should be rated for capacity at about one half of full stroke. The shorter the stroke, the more stable the operation and the less maintenance required. Heads as high as 80 to

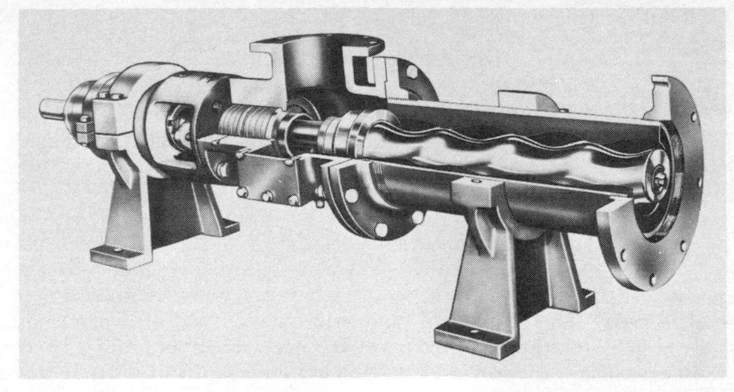

Fig. 14 Progressing cavity sludge pump. (Robbins & Myers, Moyno Pump Division)

100 lb/in^2 are available and should be specified in order to give as much flexibility as possible.

Progressing cavity pumps are available which are especially designed for the handling of sewage sludges. Wear increases along with pump speed, so excessive speed should be avoided. Ideally, the maximum speed of a progressing cavity pump should not exceed 350 rpm. These units are readily available with head capabilities up to 50 lb/in^2 and should be so specified.

Certain of the newer sludge conditioning and dewatering processes such as heat treatment and pressure filtration require use of pumps having a head capability in excess of 500 lb/in^2. This is extremely difficult service, and special care should be taken in selecting the type of equipment to be used. So far, this area of application has received very little consideration from the pump manufacturers.

Other Miscellaneous Uses Grit may be handled with reasonable success with either a torque flow or an air-lift pump. Considerable flushing water is required with a torque flow pump, and to a lesser extent with an air-lift. A special ash pump, as shown by Fig. 15, is required where it is necessary to dispose of incinerator residue in a liquid form. These units are especially designed for ash sluicing service

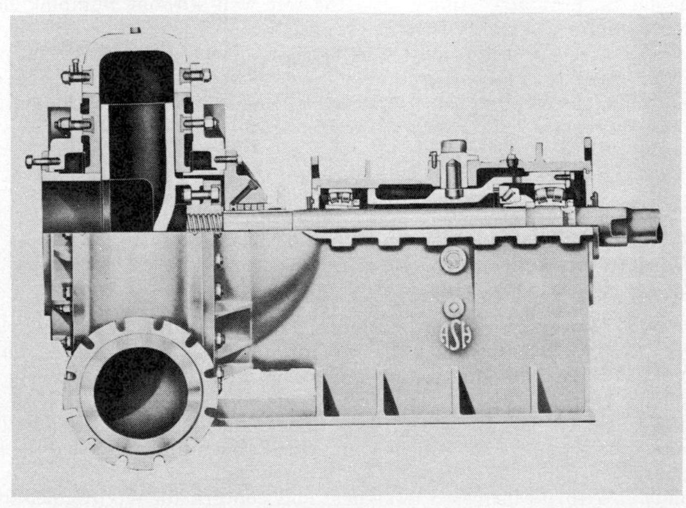

Fig. 15 Ash pump. (Allen-Sherman-Hoff Pump, Subsidiary Envirotech Corporation)

and are made of special hardened metals. No other pump should be considered for this service.

PUMP SELECTION

Various factors should be considered when selecting pumping equipment. These include the number of units to be installed, operating frequency, and station reliability requirements. Once these factors have been fully evaluated, head-capacity curves should be prepared in order to match the pumps properly with system requirements. This is necessary since the capacity of most pumps varies with the total head at which the unit operates. When a pump is referred to as having a certain capacity, this capacity applies to only one point on the characteristic curve.

Number of Pumps The number of pumps to be provided at a particular installation depends largely on the required capacity and range of flow. In considering capacity, it is customary to provide a total pumping capability equal to the maximum expected inflow with at least one of the largest pumping units out of service. A minimum of two pumps should be installed in any installation except where pneumatic ejectors are used to serve less than 50 houses. Two pumps are customarily installed where the maximum inflow is less than 1.0 mgd. At larger installations, the size and number of units should be such that the range of inflow can be met without starting and stopping pumps too frequently and without requiring excessive wet-well storage capacity. Variable-capacity pumps can be used in order to match the pumping rate with the inflow rate.

Where variable-capacity pumps are used, a minimum of two such units should be installed. In those cases where more than one variable-capacity unit is required to handle peak flow, a total of three such units should be installed. In this manner it is possible to maintain a reasonable rate of flow through each pump. Operation of a single variable-capacity pump in parallel with a constant-capacity pump requires the variable-speed unit to operate at almost no capacity whenever total inflow barely exceeds the rating of the constant-capacity unit. This is extremely difficult service and should be avoided. As a general rule, pumping rates of less than 20 percent of the rated capacity for which a pump is designed will result in excessive internal recirculation and unstable operation. Recirculation can occur in some pumps at more than 50 percent of rated capacity.

Operating Frequency Pump sizing should be coordinated with wet-well design in order to avoid frequent on-off cycling of pumps. Excessive starting will cause undue wear on the starting equipment. Also, standard motors should not be started more than six times an hour. Where more frequent starting is required, special motors should be provided. Inflow into the wet well without pumping should not exceed about 30 min if septicity is to be prevented.

Cycle time is defined as the total time between starts of an individual pump. It can be determined by comparing the volume between the on and the off levels in the wet well with the pump capacity. Cycle time is computed as follows:

$$CT = \frac{V}{D - Q} + \frac{V}{Q}$$

where CT = cycle time, min
V = wet-well volume between on and off levels, gal
D = rated pump capacity, gpm
Q = wet-well inflow, gpm

With a given wet-well volume and pumps having a uniform pumping rate, minimum cycle time will occur when the rate of inflow is equal to one half of the discharge rate of the individual pump under consideration. The formula for cycle time simplifies to $CT = 2V/Q$. An effective wet-well volume of at least 2.5 times the discharge rate of the individual pump under consideration is required in order not to exceed the six starts per hour recommended above for pumps having a uniform pumping rate.

Reliability With its increased awareness and concern for environmental matters, the public has little tolerance for bypassing of sewage equipment because of power outages, equipment failure, insufficient pumping capacity, or whatever. Reliability is of extreme importance, and the design of pumping facilities should be premised on providing continuous service. Where electric motors are used, two incoming power lines from separate sources with automatic switching from the preferred source to the standby source are the minimum required for reliability. Standby engine-driven pumps, engine-driven right-angle gear drives, or standby engine-driven generators should be provided where dual electric service cannot be obtained or where the degree of reliability provided by two feeds is not considered adequate. Raw sewage pumping installations are particularly critical. Plant pumping installations usually can be out of service as long as 4 hr without adversely affecting the treatment process provided the liquid will flow by gravity through the plant.

Speed The maximum speed at which a pump should operate is determined by the net positive suction head available at the pump, the quantity of liquid being pumped, and the total head. When specifying pumps, especially those which are to operate with a suction lift, the speed at which the pumps will operate should be checked against limiting suction requirements as set forth by the Hydraulic Institute.

In general, it is not good practice to operate sewage pumping units at speeds in excess of 1,750 rpm. This speed is applicable only to smaller units. Larger pumps should operate at slower speeds.

Preparation of Head-Capacity Curves Pump selection generally involves preparation of a system head-capacity curve showing all conditions of head and capacity under which the pumps will be required to operate. Friction losses can be expected to increase with time, materially affecting the capacity of the pumping units and their operation. For this reason, system curves should reflect extreme maximum and minimum friction losses to be expected during the lifetime of the pumping units as well as high and low wet-well levels.

Where two or more pumps discharge into a common header, it is usually advantageous to omit the head losses in individual suction and discharge lines from the system head-capacity curves. This is advisable because the pumping capacity of each unit will vary depending upon which units are in operation. In order to obtain a true picture of the output from a multiple pump installation, it is better to deduct the individual suction and discharge losses from the pump characteristic curve. This provides a modified curve which represents pump performance at the point of connection to the discharge header. Multiple pump performance can be determined by adding the capacity for points of equal head from the modified curve. Figure 16 shows a typical set of system curves, together with representative individual pump characteristic curves, modified pump curves, and combined modified curves for multiple pump operation. Intersection of the modified individual and combined pump curves with the system curves shows total discharge capacity for each of the several possible pumping combinations. A typical set of system curves consists of two curves with a Williams-Hazen coefficient of $C = 100$ (one for maximum and one for minimum static head), and two curves with a Williams-Hazen coefficient of $C = 140$ (for maximum and minimum static head). These coefficients represent the extremes normally found in sewage applications.

Pumps should be selected so that the total required capacity of the installation can be delivered with maximum water level in the wet well and a maximum friction in the discharge line. Pump efficiency should be maximum at average operating conditions. In the case of Fig. 16, assuming that the total capacity of the installation is to be obtained by operating pumps 1, 2, and 3 in parallel, the total head required at the discharge header would be approximately 51 ft. Projecting this point horizontally to the individual modified pump curves and thence vertically to the pump characteristic curves, the required head for pumps 1 and 2 should be 54 ft and for pump 3 approximately 57 ft. The difference between the head obtained from the pump characteristic curve and the modified curve is the head loss in the suction and discharge piping for the individual pumping units.

Figure 16 also shows the minimum head at which each individual pump has to operate. In the case of pumps 1 and 2, this minimum head is approximately 39 ft. It is about 42 ft for pump 3. These minimum heads are important and should be made known to the pump manufacturer since they will usually determine the maxi-

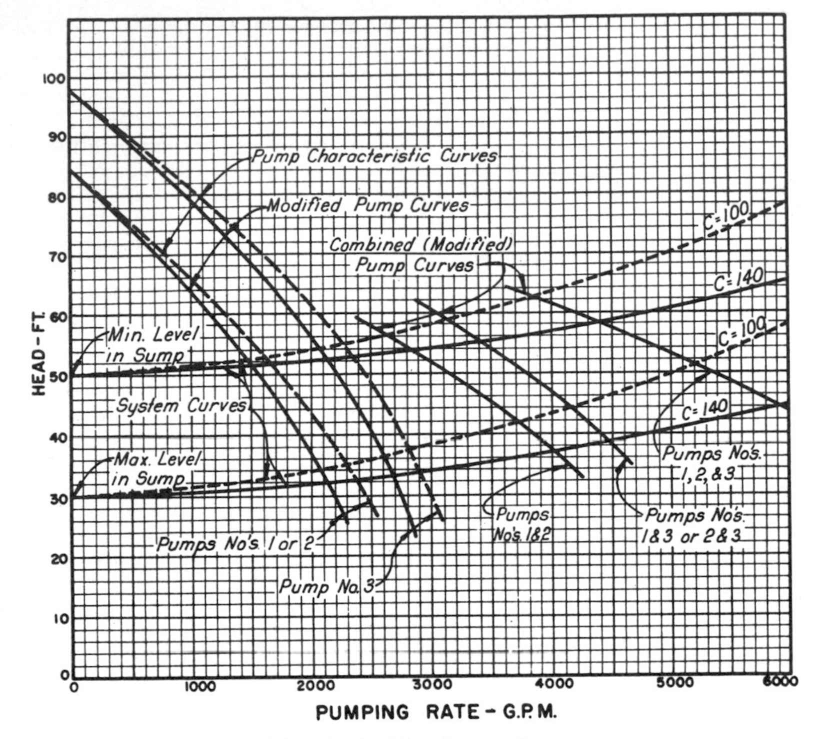

Fig. 16 Typical head-capacity curves.

mum brake horsepower required to drive the pump and the maximum speed at which the pump may operate without cavitation.

PUMP DRIVERS

In the majority of cases, pumps are driven by electric motors. Sometimes, however, they are driven by gasoline, gas, or diesel units where firm power is not available or where pumping is required only at infrequent intervals. Variable-speed drivers are used extensively in sewage applications. These units generally consist of variable-speed motors or constant-speed motors with adjustable slip couplings of either the magnetic or the hydraulic type. Selection of the type of variable-speed drive to be used is usually based on initial cost and space considerations, as there is little difference in operating efficiency.

Variable-speed drivers are particularly appropriate for raw sewage installations which discharge to a treatment plant. Use of this equipment allows the treatment facilities to operate continuously instead of intermittently surging the plant at incremental pumping rates. Variable-speed drivers are used to pump settled sewage and biological sludge where intermittent surging would adversely affect the process. Also, sludge pumps used to feed dewatering equipment are often equipped with variable-speed drives, since it is necessary to vary the rate of discharge with the dewatering characteristics of the sludge.

The choice between horizontal- and vertical-drive motors depends considerably upon station arrangement and the availability of dry-pit space. Horizontal motors are usually preferred, provided there is space and there is no potential flooding problem. Horizontal pumps are more easily maintained, and they are generally less expensive in first cost. Vertical drivers are generally used, however, for the pumping of raw sewage because of their smaller space requirements. Also, vertical units are advantageous in that the motor is located higher and is less susceptible to flooding.

Intermediate shafting is preferred with the smaller vertical dry-pit pumps. This allows the drivers to be located at ground level, out of flood range. However, intermediate shafting is not practical for large pumps because of the size of the shafting and the intermediate bearings. Where vertical-drive motors are used without intermediate shafting, the motors are set directly above the pumps and are connected by means of a flexible or rigid coupling. Separate support of the drive motors is sometimes required.

PUMP CONTROLS

Some means of controlling pump operation is required at most pumping installations. This is usually done from either wet-well level or flow.

Level Control With level control, each pump is turned on and off at specific water levels in the suction wet well; in the case of variable capacity pumps, the level control attempts to maintain a preset level, once started. Pumps turn on with a rising level, and off as the level lowers. Level control is generally used in connection with raw sewage pumping applications. In this manner it is possible to match discharge with the incoming flow.

Flow Control Flow control is used sometimes where there is no limitation on the availability of flow to the pump suction and where it is desirable to maintain a predetermined rate of discharge. Where flow control is used, a flowmeter is used as the primary instrument to measure flow and to serve as a basis for varying pump speed, which in turn controls capacity. The actual changing of speed can be accomplished either manually or automatically through closed-loop instrumentation.

Additional Level Control Low-water pump cutoff and high-level alarm are provided on most pumping installations. The low-level cutoff is required to prevent the pumps from running dry, and the high-level alarm serves to notify the operator in the event the pump should fail to operate. Annunciation of pump stop due to low-water cutoff is usually provided.

MISCELLANEOUS DESIGN CONSIDERATIONS

In addition to the matters discussed above, there are certain other items which should be given consideration in the design of pumping installations.

Piping and Valves Suction and discharge piping should normally be sized so that the maximum velocities do not exceed 5 and 8 ft/s, respectively. Higher velocities, however, may be justified by economic analysis for particular installations. Lines less than 4 in in diameter should not be used for raw sewage. Preferably sludge lines should be at least 6 in in diameter; 4-in lines are sometimes used for dilute biological sludge.

Valves should be installed on the suction and discharge sides of each pump to allow removal and maintenance of individual pumping units without disturbing the function of the remainder of the installation. It is customary to use either ball or plug valves on raw sewage and concentrated sludge applications. Either plug or butterfly valves can be used for settled sewage or for dilute sludge.

Piping should be designed with sufficient flexibility to avoid placing stress on the pump flanges. Flange-coupling adapters are sometimes used for this purpose on both the suction and discharge sides of the pump.

Surge Control Careful attention should be given to surge control wherever a pump discharges into a force main of appreciable length. Generally this is a problem only in the design of raw sewage pumping stations located within the collection system. Changes in fluid motion caused by starting or stopping of pumps or by power failure can create surge conditions.

Surges caused by normal starting and stopping of electric motor-driven pumps may be controlled (1) by selecting individual pump capacities such that the change in velocity in the system when starting or stopping a single pump will not result in excessive surges, (2) by the use of variable-speed drives to bring pumps gradually on or off line, or (3) by the use of power-operated valves which are controlled so that the pumps are started and stopped against a closed valve.

Surge control because of power failure can be effected by devices designed to open on an increase in pressure, by devices which will exhaust sewage from the system upon sudden pressure drop in anticipation of surge, or by a surge tank.

Pump Seals Most sewage and sludge pumps can be obtained with either mechanical seals or stuffing boxes. Mechanical seals have the disadvantage of requiring the pump to be dismantled so that the seal can be repaired. Often it is easier to replace the seal rather than repair it, and it is desirable to keep a spare on hand for this purpose. Water-seal stuffing boxes are recommended for most sludge pumps and for the larger sewage pumps. Grease seals are sometimes used for some of the smaller sewage pumps which do not run continuously.

Water serves multiple purposes as a sealing medium; it seals, lubricates, and flushes. Flushing is particularly important where abrasive material is involved in that it helps prevent this material from entering the seal. Grit and ash are very abrasive, and either will cut the shaft sleeves in a relatively short time. Where pumps are controlled automatically, a solenoid value interlock with the pump starting circuit should be provided in the seal water connection to each pump. A manual shutoff valve and strainer should be provided on each side of each solenoid valve, and a bypass line should be provided around it.

Mechanical seals can be lubricated by the sewage being pumped provided it is filtered. When mechanical seals are used, a connection is normally provided between the pump discharge and the seal with a 20- to 40-micron in-line strainer to prevent foreign material from entering the seal.

Pump Bearings Pump bearings must be adequate for the service and should be designed on the basis of not less than a minimum life of five years in accordance with the Anti-Friction Bearings Manufacturers Association life and thrust values. The larger sewage pumps are usually equipped with both case and impeller rings of bronze or chrome steel.

Cleanout Ports Pumps should be provided, where possible, with cleanout ports on both the suction and discharge sides of the impeller. These are desirable for inspection and maintenance purposes.

Wet-Well Design Raw sewage wet wells should not be so large that sewage is retained long enough to go septic. It is usually desirable to limit storage to a maximum of 30 min. Shorter retention time is desirable. With the variable-speed controls now available, many stations can be designed so that the pumping rate matches the inflow rate and the inherent difficulties of frequent pump cycling or long retention times in wet wells can be avoided.

REFERENCES

1. Benjes, H. H.: Design of Sewage Pumping Stations, *Public Works,* August 1960.
2. Benjes, H. H.: Sewage Pumping, *J. Sanit. Eng. Div., Proc. ASCE,* June 1958.
3. Parmakian, John: "Water Hammer Analysis," Prentice-Hall, Inc., Englewood Cliffs, N.Y., 1955.
4. Rich, George R.: "Hydraulic Transients," 1st ed., McGraw-Hill Book Company, New York, 1951.
5. "Sewage and Storm Water Pumping," Chap. 12, "Design and Construction of Sanitary and Storm Sewers," ASCE Manual of Engineering Practice No. 37, WPCF Manual of Practice No. 9, 1969.
6. "Standards for Sewage Works, Great Lakes—Upper Mississippi River Board of State Sanitary Engineers," 1968.
7. "Safety in Wastewater Works," Manual of Practice No. 1, WPCF, 1967.

Section **10.3**

Drainage and Irrigation Pumps

JOHN S. ROBERTSON

DRAINAGE PUMPS

Drainage pumps are used to control the level of water trapped within a protected area. Entrapment occurs when high lake levels, stream stages, and tides preclude the discharge of streams, storm runoff, and seepage from the protected area in the normal manner. These high-water conditions are created by floods and hurricanes or impoundment.

Floods and hurricanes occur infrequently and are relatively short-lived. However, as normal drainage is not possible at such times, all interior drainage that cannot be ponded must be pumped over the protective works if the protective works is a levee or through it if it is a concrete flood wall. Pumps for this purpose are strategically located at the edge of ponding areas and streams, in sewer systems, or within the protective works. These pumps are operated continuously or are cycled on and off as necessary to maintain the water level within the protected area below the elevation at which damage would be experienced. As their use is required only during emergency situations and usually under adverse weather conditions, reliability of operation is essential.

Where protection must be provided for valuable low-lying areas to prevent inundation due to backwater, gravity drainage from the area is generally not possible. In such situations all seepage and storm-water runoff entering the protected area must be pumped. The pumping of seepage is often a continuous operation, whereas the pumping of storm-water runoff is an infrequent type of pumping. Both can occur simultaneously because inflow due to seepage does not stop during a rainstorm. It therefore is necessary to provide pumps having the capacity to handle both seepage and storm water simultaneously.

IRRIGATION PUMPS

Irrigation pumps play an important part in making vast areas of arid and semiarid land agriculturally productive. These pumps take water from surface sources, from subsurface sources, and in ever-increasing amounts from sewage treatment facilities and pump it to the point of application.

Water from surface sources such as streams, lakes, and ponds is pumped directly into the distribution system or into a conveyance system. If pumped into a conveyance system, it flows either by gravity or under pressure to the distribution point or to a booster pumping station. Most of the pumps used in these installations are mounted permanently and are arranged to operate as necessary without constant attendance during the growing season.

Water from subsurface sources is usually pumped directly into the distribution system. In such installations a well is drilled in the vicinity of the area to be irrigated and is fitted with the proper size pump. In many instances more than one well is needed to provide the required capacity.

Effluent waters from different types of sewage treatment facilities are beginning to be used for irrigation purposes on a year-round basis in land disposal systems. In these systems, the treated effluent is temporarily stored in holding tanks or ponds and is applied to the disposal area at predetermined rates via sprinkler systems on a year-round basis. As the disposal area may be a considerable distance from the storage area and as pressure is needed for operation of the sprinkler system, pumping is required.

PUMP TYPES

Centrifugal pumps are used almost exclusively in drainage and irrigation installations. Because of the large selection of propeller, volute, turbine, and portable pumps manufactured today, there is little difficulty in finding a pump that will meet the conditions encountered in these fields. Pumps in the sizes needed to meet the requirements of the majority of installations are available as standard items. The more special pumps and the extremely large pumps are designed and built to meet the needs of the individual project.

Propeller pumps are used for low-head pumping. As most of the pumping in drainage and irrigation work is of this low-head variety, the propeller pump is the most widely used type. In general, vertical single-stage axial and mixed-flow pumps are used; however, there are instances where two-stage axial-flow pumps should be considered for economical reasons.

Horizontal axial-flow pumps are used for pumping large volumes against low heads and usually employ siphonic action when not of the submersible type. When higher heads are involved, these pumps can be arranged to operate with siphonic action until the back pressure places the hydraulic gradient above the pump.

Variable-pitch propeller pumps rotating at constant speed can be operated efficiently over a wide range of head-capacity conditions by varying the pitch of the propeller blades. These pumps are used when the head-capacity conditions cannot be met with the more economical fixed-blade pumps and where they will be operated often enough and long enough to warrant the expense. The blade-control system needed for such pumps is more sophisticated than in any of the systems usually provided on drainage and irrigation projects. As a consequence, operation and maintenance must be performed by organizations employing competent and experienced personnel.

Volute pumps are used when pumping from surface sources and in general when the total head exceeds approximately 45 ft. Such pumps are available in many types such as vertical and horizontal shaft, end suction, bottom suction, and double suction with semiopen or closed impellers, etc. These pumps are mounted in dry pits, when located below grade, and on slabs or floors when located above grade. The particular type used depends on the capacity and kind of service to be performed.

Deep-well turbine pumps are vertical-shaft single suction pumps having one or more stages. They are used in irrigation work primarily to pump water from a subsurface source into a distribution system. The head against which the pump will be required to operate will determine the number of stages that must be provided. For large capacities, more than one pump will be needed.

Submersible turbine pumps are deep-well pumps in which the motor is close-coupled to the pump and submerged in the well. This type of pump is used for high-head applications where long intermediate shafts are undesirable.

Portable pumps are used in both drainage and irrigation. In drainage work they are used as emergency equipment to control ponding elevations where mobile equipment such as tractors and trucks having power takeoffs are available to drive them. In irrigation work they are usually used to irrigate from a surface source. Such pumps are small-capacity, low-head, economical pumps that must be submerged in order to be used.

PUMP SELECTION

A pump-selection study should always be made, and its importance cannot be emphasized enough. Such studies permit selection of the type of pump and discharge system best suited to the project requirements and provide the information needed to proceed with the design of the installation.

In many cases the type of pump required will be obvious. If more than one kind of pump would be satisfactory, the specifications should permit the pump manufacturer to make the choice. This is particularly advantageous when competitive bidding is involved. The practice of the Corps of Engineers in this regard is to write a performance specification and allow the pump manufacturer to determine the type, size, and speed of the pump.

Before initiating such a study, all previous studies made to determine the total pumping requirement or station capacity, pertinent water-surface elevations, terrain, utility locations, proposed station or well locations, points of discharge, and the proposed method of operation should be reviewed. Also to be considered is the experience of the personnel that will be responsible for the operation and maintenance of the installation.

Number of Pumps First costs are generally of more concern than operating costs in drainage and irrigation work because the operating period for the majority of installations is relatively short and occurs only once a year. Costs can be minimized by using as few pumps as possible. However, one-pump installations are seldom used except in the case of wells. For reliability, a minimum of two pumps should be installed in drainage pumping stations, where the loss of even one pump during an emergency situation could result in considerable damage. Three or more pumps are preferred. Standby units are provided only in those installations where continuous operation precludes taking a pump out of service for maintenance purposes.

The number of pumps ultimately used should be consistent with the demands of the project. For instance, when the installation is located in an agricultural area or is a part of an urban sewer system, a standby pump should be provided, since these installations must always be capable of discharging project requirements during periods of blocked drainage. In this way considerable damage may be avoided. When the installation is used to pump storm water from pondage or irrigation water from a lake, the loss of a pump is not critical, so a standby pump is not needed.

If during the pump selection process it is found that the rated horsepower of the prime mover exceeds the maximum horsepower requirements of the pump by a considerable amount, the contemplated number of pumps should be increased or decreased, provided the change results in a better horsepower match without increasing the overall cost of the installation. By increasing the number of pumps and thereby reducing the required horsepower, there is a possibility that the size of the prime mover can be reduced and that the pump horsepower will approach the rated horsepower of the prime mover. On the other hand, decreasing the number of pumps will increase the horsepower requirements. This increase may be sufficient to either utilize most of the excess capacity in the prime mover or require the use of a larger one.

For installations requiring the use of large pumps, foundation conditions become important. To prevent the installation from being relocated to a less desirable site or the necessity of providing a more expensive pile foundation because the bearing pressures at the selected site exceed the allowable limit, the number of pumps should be increased, provided the loading can be reduced to an acceptable amount and the resulting installation continues to be the most economical.

Inadequate depth on the suction side of the pumps may necessitate the use of more pumps. Should the water not be deep enough to provide the submergence needed by the contemplated pump, more pumps of a smaller size may have to be used or the sump and approach channel may have to be excavated to the needed depth. The latter alternative could cause operational and maintenance problems and might be the more expensive solution.

Capacity The capacity of a pump is a function of the total pumping requirement, the number of pumps, and, in the case of wells, the capacity of the well. Whenever possible all pumps in a multiple-pump installation should be of the same capacity. This is advantageous from a cost standpoint as well as from a maintenance standpoint. In drainage installations, three pumps are generally provided in order to have the capability of pumping not less than two-thirds of the project requirement with one pump inoperative. Each pump would then have a capacity equal to one-third of the total capacity. When more than three are used, the capacity should be the total required capacity divided by the number of pumps being used. If two pumps are used, each pump should be sized to pump not less than two-thirds of the total capacity.

In irrigation installations utilizing multiple pumps, the capacity of each pump should be the same. In single-pump installations, the pump should be sized to meet project requirements. When wells are involved, the capacity of the pump will be determined by the capacity of the well. Often several wells are needed to give the capacity to satisfy the established requirements.

When a standby pump is to be provided, its capacity should be equal to that of the largest pump being furnished.

Head Total head is the algebraic difference between the total discharge head and the total suction head. In drainage and irrigation work total suction head can usually be determined, but this is not always the case for the total discharge head. The losses in the discharge systems often must be determined by hydraulic model test rather than by calculation, and therefore must be estimated for preliminary selection purposes. The head specified will have to be some head other than total head; generally pool-to-pool head is used.

Total discharge head is defined in the Hydraulic Institute Standards, and its value is to a great extent determined by the type of discharge system used. A number of the many possible discharge systems used are shown in Fig. 1. The losses in systems *A, B, C, D,* and *E* can be calculated. Thus the performance for pumps discharging into these systems can be put on a total head basis.

System A is an "over the levee" siphonic-type discharge line, the use of which can seldom be justified. However, continuous operation over long periods of time could effect a savings that would be sufficient to justify the additional cost of the installation and the taking on of the operational hazards usually encountered with such lines. The total discharge head for this system is equal to the height of the discharge pool, stream, or lake above impeller datum plus the exit loss and the discharge-line losses from the pump discharge nozzle to the line terminus. The absolute pressure at the high point of the line should be not less than 9 ft. Lower values have been used successfully but this should be the exception rather than the rule. For additional information relative to the determination of heads in a siphonic system refer to Siphon Head in Section 9.1.

Flap valves should be installed on the discharge end of all lines subjected to a cycling type of operation. The closing of these valves following pump shutdown prevents reverse flow into the protected area and permits development of the pressures needed to keep the lines full (primed) during those short periods when pumps are idle. These pressures will be less than atmospheric pressure, with the minimum pressure (absolute) being at the high point. Should there be significant leakage at the joints or in the valving, the pressures will rise and the water level at the high point will drop. If the water level drops below the invert of the line at the high point, the two legs of the line will be separated by an air space and priming will be necessary when the pump is started.

Air-release and vacuum-break valves should be installed at the high point of all siphonic-type discharge lines to provide an escape for the air being compressed by the water rising in the line during the priming phase and to prevent reverse flow into the protected area, respectively. For those lines equipped with flap valves, the vacuum-break valve should be a manually operated valve that would be used in the event the flap gate failed to seat properly or to provide a rapid means of draining the line when pumping is no longer required. Units for almost any size line can be obtained commercially from manufacturers who specialize in such equipment or they can be assembled, using swing check and angle valves.

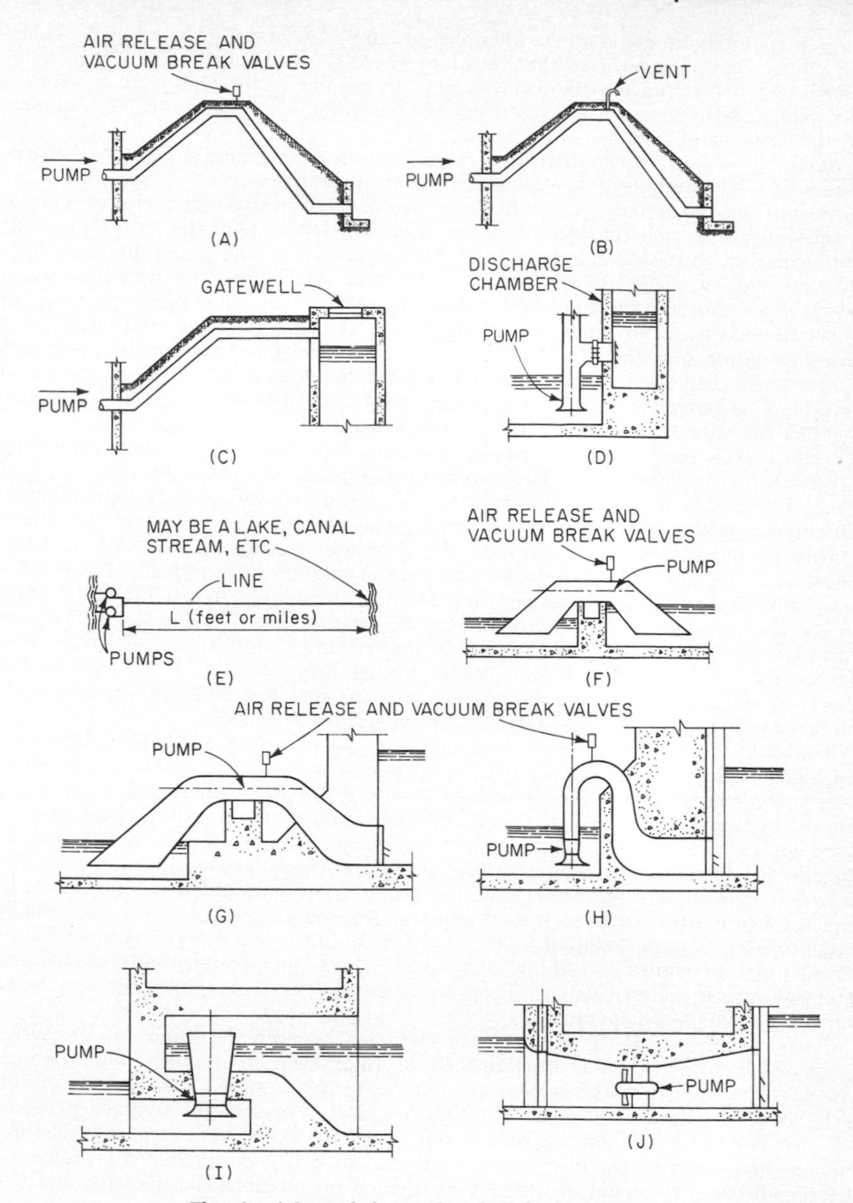

Fig. 1 Adapted from sketch 1 in Reference 4.

System B is an "over the levee" nonsiphonic-type discharge line. This line should be vented at the high point to preclude siphonic action with a vent having a diameter that is approximately one-fourth that of the line. The invert of this line at the high point should be placed at the same elevation as the top of the protective works to make possible the discharging of the pumped flow from the down leg of the line under gravity conditions without backwater effects for all discharge-pool elevations up to maximum pool elevation. Thus the total discharge head will have a constant value because it will not be affected by changes in the level of the discharge pool. To ensure adequate prime-mover capacity when using this system, it is the practice in all cases to use the top of the line at the high point in lieu of the hydraulic gradient when determining the total discharge head. Therefore total discharge head is the height of the top of the line at the high point above impeller datum plus the velocity head in the line at the high point based on a full pipe and the losses in the line from the pump discharge nozzle to the beginning of the down leg of the line.

System C is used when there is a conduit carrying the normal gravity discharge under the levee adjacent to the station that must be valved off against reverse flow into the protected area during periods of high water. The closure gate is located in a gate well constructed on the stream or lake side of the levee to prevent subjecting the gravity conduit to high-water conditions. The pump discharge lines go over the levee and are terminated in the gate well above the maximum water level. This shortens the lines and reduces the cost. The total discharge head for this system is equal to the height of the top of the line at the terminal end above impeller datum plus the exit loss and the losses in the line between the pump discharge nozzle and the terminal point. The total discharge head for this system, as in system *B,* is independent of the discharge pool and therefore is a constant value. Neither flap valves nor vents are required on these lines.

System D is used when the pumping station is constructed as an integral part of the levee or flood wall. The invert of the pump discharge line is placed at an elevation that is above the stream or lake level that will prevail approximately 70 percent of the time or as is dictated by the physical dimensions of the pump. Owing to the extreme turbulence in the discharge chamber, gates with multiple shutters which are less likely to be damaged should be used instead of flap gates on discharge lines that are larger than 36 in in diameter. When the water level in the discharge chamber is below the top of the discharge line, the total discharge head is determined in the same manner as for system *C*. For higher discharge water levels, the total discharge head is equal to the height of the water level in the discharge chamber above impeller datum plus the exit loss and the losses between the pump discharge nozzle and the chamber side of the flap valve.

System E is perhaps the most common discharge system in use today. It is used to connect one pump or several manifolded pumps with a lake, canal, stream, ditch, reservoir, or sprinkler system. For short lines and low static heads, valve and fitting losses, friction losses, and exit losses are very important, whereas in long lines or very high static head installations only friction losses are given consideration. In manifolded installations using propeller pumps, a check valve and gate valve are installed immediately downstream of the pump. The gate valve should always be opened before the pump is started because the motors provided are not usually sized to operate against shutoff head. Positive shutoff valves are placed immediately downstream of volute or turbine pumps since these pumps are usually started and stopped against a closed valve. They also prevent reverse flow into the sump when one of the pumps is inoperative.

Pool-to-pool head is the difference in elevation between the sump and discharge-pool water surfaces and is used instead of total head in drainage work because the losses in the discharge system are not easily determined. Installations of this type are exemplified by systems *F, G, H, I,* and *J*. For such installations it is best to specify the pumps on a pool-to-pool basis, to have the pump manufacturer design the pump and the discharge system, and to verify the predicted performance by model test. It should be noted that in such installations the discharge systems are usually constructed within the confines of the pumping-station structure.

System F is operated as a siphon with the pump supplying energy equivalent to the pool-to-pool head plus the system losses. The invert of the pump discharge pipe at the highest point is located above the maximum river stage, and vacuum pumps are generally used to aid in priming the pump. This system is used for pool-to-pool heads of up to approximately 6 ft and where the physical dimensions of a vertical pump would be such as to make it necessary to operate against higher heads. In estimating the losses, entrance losses which are small (approximately 0.14 ft) should be neglected. The centerline of the suction piping should be assumed to make an angle of 45° with the horizontal, and the diameter of the discharge piping as measured at the discharge flange of the discharge elbow should be such that the velocity at maximum discharge is approximately 12 ft/s or less.

System G is used for pool-to-pool heads of up to approximately 15 ft. The water passage changes in cross section from round at the mating flange to the discharge bowl to rectangular at its terminus. The width at the terminal end is the same or less than that of the suction bay, and the height is such that it is always submerged when the pump is in operation. The discharge velocity should be kept to approximately 6 ft/s. Multiple shutter gates (see Fig. 2) arranged to be raised when the pump is in operation are provided to prevent prime-mover overload when starting the pump, also preventing reverse flow when stopping the pump and reverse flow when the pump is inoperative. These pumps are usually of the larger and slower variety requiring substantial-size prime movers. Vacuum priming equipment is used in order that additional horsepower will not be required for priming purposes.

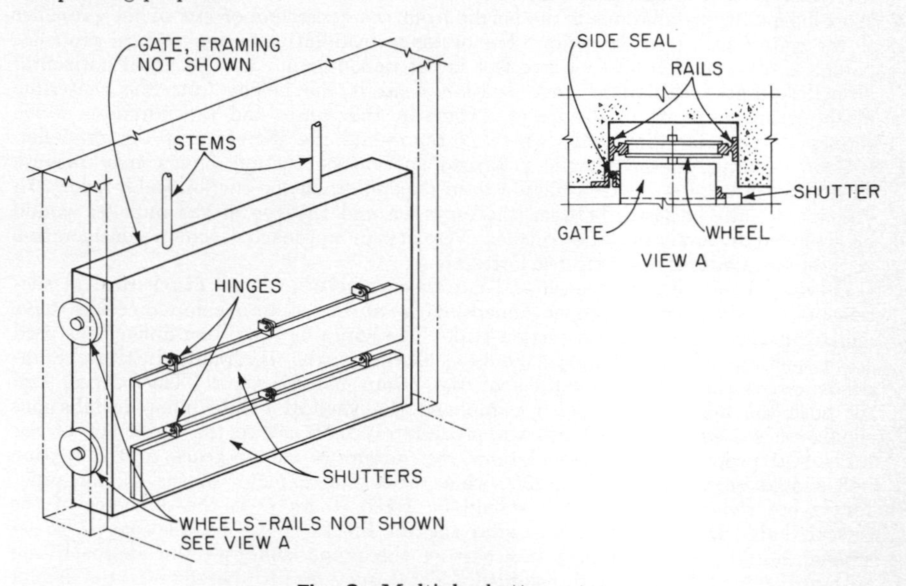

Fig. 2 Multiple-shutter gates.

System H is essentially the same as system *G*, except that it is used for pool-to-pool heads up to approximately 26 ft. Also, a splitter may be required in the discharge water passage for structural purposes as well as for keeping the multiple shutter gates to a reasonable size. The pumps used with this system are vertical and in general smaller than those used with system *G*.

System I can be constructed with the lip of the pump column either above or below the design flood elevation or the maximum surge, if it is being provided for hurricane protection. If the lip of the pump column is above and the chances of experiencing reverse flow through the pumps are extremely remote, then gating of either the pump column or the discharge water passage is unnecessary. If the lip of the pump column is below, then a decision as to whether or not to gate both the pump column opening and the discharge water passage or just the water passage must be made. In general, if the pumps will be in operation continuously during the time high water is being experienced and reverse flow through an inoperative pump will not have substantial detrimental effects, only the multiple-shutter gate at the end of the discharge water passage need be provided. Reverse flow could be experienced should the flap for some reason fail to close or be held open by debris. Like system *G*, splitter walls may be required if pumps are large. Also, the clearance between the lip of the pump column and the discharge-water-passage ceiling is critical and should be determined by model test. The shape of the pump column lip will also affect the pump efficiency and should be determined by test.

System J could have many configurations, but regardless of arrangement would have to be gated on the suction side with a positive shutoff gate such as a pressure-seating slide gate and on the discharge side with a positive shutoff gate and a multiple-shutter gate or a flap gate if the installation is small and a pipe is used in place of formed water passages. When formed water passages are used, transition sections between the pump and the gates sections will be needed and should be designed by the pump manufacturer.

Total Suction Head. The practice for propeller-pump installations is to dimension the station sump or sump bays in accordance with and to make the distance between the sump floor and the lip of the suction bowl conform to the Hydraulic Institute Standards or the recommendations of the pump manufacturer. The approach and entrance velocities resulting therefrom are small enough in magnitude to be disregarded when calculating total suction head. Total suction head in vertical propeller-pump installations is the height from the centerline or eye of the propeller to the water surface in the sump. For drainage installations using vertical propeller pumps, a total suction head of zero feet is not uncommon. In submerged horizontal propeller-pump installations, total suction head is the height from the centerline of the propeller shaft to the water surface in the sump, and the minimum value should be not less than $1.2D$, where D represents the diameter of the propeller.

Volute pumps equipped with a formed suction or suction piping may operate with either a suction lift or a suction head depending on the suction water level. In either case, all the losses between the entrance and the eye of the impeller should be included in any calculation made. Velocity of approach is not a consideration in these installations, but entrance losses are.

Total Suction Lift. Approach and entrance velocities can be disregarded in suction-lift installations by proper dimensioning of the station sump or sump bays and by proper setting of the suction bell. The sump or sump-bay dimensions used as in suction-head installations should conform to the Hydraulic Institute Standards or to the recommendations of the pump manufacturer. The suction bell for both the horizontal propeller pump and the vertical volute pump installations should be set with the lip located approximately $0.5D$ above the sump floor. For horizontal propeller pump installations the minimum submergence of the suction bell should be approximately $0.25D$, where D is the diameter of the suction pipe; for volute pump installations it should be $1.5D$, where D is the diameter of the suction bell. In determining the total suction lift for these installations, it is assumed that the suction piping is a part of the pump and that the approach and entrance velocities can be disregarded. Total suction lift therefore is the height from the water surface in the sump to the centerline of the propeller shaft or to the eye of the impeller.

Setting The setting of the pump or the locating of the centerline or eye of the propeller or impeller with respect to the water surface should be given careful consideration when selecting the pump to be used. Some installations offer few if any problems in this regard, while in others it may have considerable effect on the size and number of pumps selected.

Turbine Pumps. Drawdown is a consideration in any well installation. In setting the pump, sufficient depth over the eye of the first-stage impeller in a single or multistage installation should be maintained in order to prevent cavitation when maximum drawdown is being experienced.

Volute Pumps. Volute pumps may operate with either a suction head or a suction lift. If with a suction lift and of the horizontal type, the pump should be set above the maximum anticipated elevation of the suction-water source in order to avoid inundation. As priming will be necessary when starting, a suction lift would be considered a satisfactory arrangement when long periods of continuous operation are anticipated.

Suction-head installation is the preferred practice since priming is unnecessary; it should be used whenever conditions permit. In drainage work vertical volute pumps are used for pumping small amounts of rainfall runoff and seepage flows. These pumps are usually located in a dry sump adjacent to the storm-water pump sump with motors and valve operators located on the operating floor above. The submergence of these pumps should be such that when discharging at maximum capacity the total suction head is zero or above. In many instances water-surface fluctuations of just a few feet are experienced. When small volumes are involved, a cycling type of operation occurs. For this type of operation the pumps are usually started at the maximum water level and stopped at the minimum level.

Propeller Pumps. Sufficient water depth does not always exist or cannot always be provided to give the submergence needed to permit the smallest and perhaps the most efficient pump to be used. Excavation is one answer. However, depending on the soil type, the location of the installation, the kind of construction used, the silt-carrying characteristics of the stream, and the frequency of operation, excavation may be an operational and maintenance headache and it is impractical when sumps are to be made self-draining. Another alternative—and the one most frequently used in drainage work—is to set the centerline or eye of the propeller at or slightly below the minimum sump elevation and select a pump that will operate at this setting with little or no cavitation damage. This means that a larger pump operating at a slower speed should be used.

Prime Movers The prime movers used in drainage and irrigation installations are electric motors and diesel and gas engines. The one to be used in any particular situation must be determined before the pump can be selected.

Electric motors are the most economical installations; they should be used when a reliable source of electric power is available and when the cost of bringing it into the pumping station is not unreasonable. A reliable source of electric power is a source that historically has not suffered outages under the climatic conditions that will prevail during the time the pumps will be required to operate. Two feeders having separate origins and not subject to simultaneous outages are sometimes provided to ensure the reliability needed for drainage installations. Such an arrangement has been satisfactorily employed in urban areas but would not be a practical solution in remote areas not yet electrified. The cost of constructing and maintaining even one transmission line in such an area could be prohibitive.

The motors in all but the larger installations should be of the squirrel-cage induction type. In those installations where the motor rating is numerically larger than the speed, the type of motor used should be the one having the lowest overall first cost. It may be either a squirrel-cage-induction or a synchronous motor.

All motors should be full-voltage-starting except in those instances where the local power company indicates that reduced voltage starting is necessary. For unattended operation and for drainage stations pumping seepage or pumping from a sewer system where frequent cycling is usually necessary, control devices set to start and stop the motors automatically at predetermined sump or discharge-pool levels should be provided. In drainage pumping stations not subject to a cycling type of operation, motors are started manually by the operator and stopped automatically by a control device.

Engines are used to drive pumps when it is not feasible to use electric motors. They are more expensive than motors, but reliable if properly maintained and serviced. They are also variable-speed drives that should be operated at constant speed whenever possible. The requirements of most installations can be met with constant-speed operation. However, for those that cannot, the number of speeds used should be held to a minimum.

Engines should not be cycled on and off, but should be operated on a continuous basis. For those installations where the inflow is not sufficient to permit continuous operation, continuous operation can be obtained by returning a part of the pumped discharge back to the sump. This is accomplished by connecting the pump discharge line and the sump with a valved line.

Gas engines are seldom used, but their use should be considered when the installation is in close proximity to a natural gas main.

Right-angle reduction gears are used to transmit the power from the engine to the pump shaft of vertical propeller pumps. For horizontal pump installations where the engine shaft parallels the pump shaft but at a different elevation and off to one side, silent chain drives are used. For other horizontal installations parallel-shaft gear units may be used. A service factor of 1.50 should be used when determining the equivalent horsepower of these units. Right-angle units should be of the hollow-shaft type only if vertical adjustment of the pump impeller is required.

Adequate fuel storage in addition to the day tanks should be provided. Storage facilities should be sufficient to provide fuel at the maximum rate of consumption for a period of 48 hr or less as demanded by the anticipated pumping requirements. Larger fuel storage installations may be needed in the event that replenishment supplies are not readily available. The design of the facilities should be in accordance with the standards of the National Board of Fire Underwriters and local agencies having jurisdiction.

PUMPING STATION

Sump The sump is perhaps the most important element in the structure of the pumping station. Unless it is properly located, designed, and sized, the flow conditions within could have an adverse effect on the operation of the pump. There are many variations in sump arrangements that are acceptable; however, best results are obtained when the sumps or sump bays are oriented parallel to the line of flow. Flows approaching from an angle create dead spots and high local velocities, which result in the formation of vortices, nonuniform entrance velocities, and an increase in entrance losses. The flow to any pump should not be required to pass another pump before reaching it. When sumps or sump bays are normal to the direction of flow, such as in sewer systems, the distance between the sump or sump bay entrance and the pump must be sufficient for the flow to straighten itself out before reaching the pump. For additional information relative to sump design and sizing, refer to Section 11, Intakes and Suction Piping. If the installation is large enough to warrant it, modeling of the sump to permit the best design to be determined is advocated.

Sumps in drainage installations pumping storm water should be either located above normal water levels, in which case they would be of the self-draining type, or isolated from normal flows by gates. In small installations, motorized pressure-seating gates should be used; in large installations, roller gates raised and lowered with a crane or by some other suitable system should be used. These gates should be sized so that the velocity through them will not exceed 5 ft/s for any condition of flow. One gate should be located directly opposite each pump when all pumps are installed in a common sump or should be located at the entrance to each sump bay when pumps are separated.

Frequent cycling of pumps is encountered in installations which pump from sewer systems and which pump seepage. In such installations, the sump should be sized so that the volume stored in the sump and the ponding area or the sewer lines, as the case may be, within the limits of the operating range, will be sufficient to prevent the starting of the pumps oftener than once every 4 min.

Superstructure A superstructure is provided on practically all drainage pumping stations, but not on all irrigation pumping stations. The type of superstructure provided should be consistent with the surrounding area and should have a minimum number of windows and openings. In rural areas, corrugated sheet-metal structures which are inexpensive have been used extensively. These provide adequate protection from the elements and from vandalism. To eliminate the need for an indoor crane, hatches in the roof over each pump can be provided to permit removal of motors and pumps as a unit by use of a truck crane. For engine-driven pumps and the larger pump installations, indoor cranes of the appropriate size and type should be provided for installation, removal, and maintenance purposes.

Corrosion The corrosion of electrical and mechanical equipment in housed and unhoused stations can be controlled by painting exterior surfaces with a good paint and by installing strip heaters in all electrical enclosures. In large housed installations, heating of the operating-room area in addition to the use of strip heaters should be considered. All equipment below the operating-room floor level should be coated with a paint that is suitable for the exposure. In dry sump stations enamels should be satisfactory. In wet sumps that are kept dry during inoperative periods, cold-applied coal-tar enamel, which is easily repaired, is preferred. In installations where the equipment is continuously immersed, coal-tar epoxy paint or vinyl paint should be used. If the water is extremely corrosive, consideration should be given to mounting galvanic anodes on the pumps in addition to painting.

REFERENCES

1. Design Standard No. 6, "Turbines and Pumps," U.S. Department of the Interior, Bureau of Reclamation, Commissioner's Office, Denver, Colo., 1960.
2. Engineering Manual EM 1110-2-1410, "Interior Drainage of Leveed Urban Areas: Hydrology," Headquarters, Department of the Army, Office of the Chief of Engineers, 1965.
3. Engineering Manual EM 1110-2-3102, "General Principles of Pumping Station Design and Layout," Headquarters, Department of the Army, Office of the Chief of Engineers, 1962.
4. Engineering Manual EM 1110-2-3105, "Mechanical and Electrical Design of Pumping Stations," Headquarters, Department of the Army, Office of the Chief of Engineers, 1962.
5. Hicks, Tyler G.: "Pump Selection and Application," McGraw-Hill Book Company, Inc., New York, 1957.
6. Houk, Ivan E.: "Irrigation Engineering," John Wiley & Sons, Inc., New York, 1951.
7. "Hydraulic Institute Standards for Centrifugal, Rotary and Reciprocating Pumps," 12th ed., Hydraulic Institute, Cleveland, Ohio, 1969.
8. Israelson, Orson W.: "Irrigation Principles and Practices," 2d ed., John Wiley & Sons, Inc., New York, 1950.

Section 10.4

Fire Pumps

ROWLAND A. HARRIS

GENERAL

A fire pump is a specialized booster system meeting the design requirements set down by the National Fire Protection Association for this type of equipment. Each piece of equipment must have undergone the testing procedures of the Underwriters Laboratory or Factory Mutual Association, and be approved or listed by them. The capacity plate of each system should state the approval or listing agency, the model and serial number of the equipment, the design capacity and pressure, and the rated speed. The maximum horsepower and the shutoff pressure of the unit should also be shown.

The horizontal split-case double-suction single-stage pump is preferred for most applications. There are several other types of pumps to choose from, and close investigation is necessary to determine the best system for a particular application. The system could use a vertical turbine pump if the water source is below grade, or it could use a multistage split-case pump on high-pressure applications. The National Fire Protection Association publishes information that establishes the minimum requirements for centrifugal fire pumps.

Centrifugal fire pumps should not be purchased until conditions under which they are to be installed and operated have been carefully examined by the authority having jurisdiction. The complete system including the power supply, controlling equipment, and water supply must be approved.

The National Fire Protection Association requires that the pump, driver, and all necessary attachments be purchased under a unit contract stipulating compliance with this standard as well as satisfactory performance of the entire unit when installed. The pump manufacturer is responsible for the proper operation of the complete unit assembly as indicated by field acceptance tests.

It is the responsibility of the design engineers to submit a complete plan and detailed data describing the pump, driver, controller, power supply, fittings, suction and discharge connections, and suction conditions to the authority having jurisdiction for approval before installation. Certified shop-test characteristic curves showing the head capacity, efficiency, and horsepower should be furnished by the pump manufacturer.

Both vertical turbine and horizontal split-case pumps are sized for 500, 750, 1,000, 1,500, 2,000, and 2,500 gpm. Horizontal split-case pumps on negative suction pressure are rated at 100 lb/in^2 or more. Horizontal split-case pumps rated at 3,000, 3,500, 4,000, and 4,500 gpm have also been approved. Vertical turbine pumps are rated at 100 lb/in^2 or more. Positive-suction-pressure horizontal split-case fire pumps are rated at 40 lb/in^2 or more, and start at 250 gpm.

A customer buys a fire pump for the stated purpose of reducing his insurance rates, and in many instances he is not familiar with its operation or requirements. It is important, therefore, that adequate provisions be included in the contract for the inspection and maintenance of fire pumps and equipment, as well as for the competence of the installing contractor.

CONSTRUCTION

The recommended materials of construction for horizontal pumps are a cast iron casing and bronze fitting. Casing wearing rings are required and should be bronze. The stuffing box contains a bronze lantern packing ring, and the packing is drawn up by a split bronze gland. Horizontal pumps are equipped with grease-lubricated ball bearings. The outboard bearing should be of a duplex angular contact type. Bronze shaft sleeves with an O ring or gasket between the steel shaft and sleeve are recommended. The pump with its driver should be mounted on a sturdy steel or cast iron base to assure alignment.

For vertical pumps the materials are cast iron or bronze bowls and bronze impellers. Bronze wearing rings are required. The suction bell should be equipped with a strainer of the conical or basket type, made of nonferrous material. The strainer should have openings to restrict the passage of ½-in spheres, and it should be designed to have a free area of at least four times the suction inlet. The column should be furnished in sections not to exceed a nominal 10 ft, and should be connected by threaded sleeve-type or flange-type couplings. The ends of each section should butt to form an accurate alignment of the pump. The shafting should be interchangeable and not over 10 ft in length. The discharge head should be of the above-ground type.

PERFORMANCE REQUIREMENTS

The following performance requirements are essential for a horizontal split-case pump. The pump must be capable of not less than 150 percent of rated capacity at a total head of not less than 65 percent of total rated head. The shutoff head should not exceed 120 percent of the total rated head. Suction inlet pressure should always be specified on the basis of 150 percent of flow capacity of the pump.

The vertical turbine pump must meet the same 150 percent of capacity at not less than 65 percent of total rated head. The shutoff total head should not exceed 140 percent of total rated head. If wells are considered as a source of supply where the water level is expected to be more than 50 ft below grade, the authority having jurisdiction should be supplied with data on the drawdown characteristics of the well in addition to the pump performance to determine the available pressure at the discharge flange of the vertical pump. In any event, the point of water supply at 150 percent capacity should not be more than 200 ft below the surface. The second impeller from the bottom of the bowl assembly should be set 10 ft below the pumping water level at 150 percent of rated capacity. The minimum submergence should be increased by 1 ft for each 1,000 ft of elevation above sea level. Open-line-shaft water-lubricated columns should not be used where the distance from the discharge point from the pump head to the static water level exceeds 50 ft.

If a well is used as a source, the construction and development of the well should be as specified by the National Fire Protection Association.

DRIVERS

Fire pumps may be driven by either a motor or an engine. Steam turbines of adequate power may also be used, and all three are acceptable prime movers.

On motor-driven units the power source must be acceptable to the authority having jurisdiction. All fire-pump electric motors should be 600 volts or less, and should be rated for continuous duty at voltages not in excess of 110 percent of rated voltage. The maximum allowable drop in voltage when the pump is driven at rated output in pressure is 5 percent. Squirrel-cage induction motors for across-the-line starting, or with primary-resistance autotransformer-type starters, may also be specified. A wound-rotor-type motor with appropriate starting equipment may be substituted if it is acceptable to the authority having jurisdiction. The authority having jurisdiction may under special circumstances approve the use of 2,300 volts for motors over 75 hp and 4,000 volts for motors of 100 hp and larger. Motors used at altitudes above 3,300 ft should be derated according to NEMA standards. Motors may be open dripproof, although both splashproof and totally enclosed types may also be used when sized properly for the 150 percent capacity load.

A controller for electric-drive units of the automatic type is recommended, and must be an approved fire-pump controller bearing the label of the Underwriters Laboratory or Factory Mutual or both.

On vertical pumps either a vertical hollow-shaft electric motor or a right-angle gear drive may be specified. The right-angle gear is normally used when an internal combustion engine or a steam turbine is selected.

Internal combustion engine drives are acceptable power sources. The diesel engine is one of the most dependable sources of power. Engines should be specifically approved for fire-pump service. The engine, after being corrected for altitude and ambient temperature, should have a bare engine brake horsepower rating at least 20 percent greater than the maximum brake horsepower required to drive the fire pump at rated speed. The sea-level horsepower rating for diesel engines should be reduced by 3 percent for each 1,000 ft of elevation. A reduction in the horsepower rating of 1 percent for each 10°F above 60°F ambient temperature should also be made. When a gearhead is used between the pump and the power source, the horsepower rating should be increased by a minimum of 3 percent to cover power loss through the gears.

Dual drive units are not recommended. When a dual drive is used, the authority having jurisdiction must approve the automatic couplings as well as the complete installation.

On engine drives, the instrumentation should include an adjustable governor with a range of 10 percent. An emergency shutdown at 20 percent overspeed with a manual reset is also required, as well as a tachometer indicating engine speed and total hours of engine operation. An oil pressure gage and a gage indicating cooling-water temperature are required.

An electric starter utilizing power from storage batteries is the standard method of starting. The batteries should have sufficient capacity at 40°F to maintain the engine manufacturer's recommended cranking speed during a 6-min cycle. Lead-acid batteries are normally furnished on engine-driven units. The electrolyte should be added at the time the unit is put into service. Nickel-cadmium alkaline-type batteries may also be used in place of the lead-acid type.

Two methods of recharging the batteries are required. The engine generator is one; the other is an automatic-controlled charger taking power from an ac source. All chargers must be approved for fire-pump service. An ammeter should be furnished to indicate operation of the charger. Automatic alternation on an hourly basis from one battery to another should also be furnished. The batteries are normally located at the site of the engine to reduce power losses.

The engine cooling system should be of the closed-circuit type. The circuit consists of a circulating pump, a heat exchanger, and a reliable regulating device. The cooling water for the heat exchanger should be taken from the discharge of the fire pump before the discharge valve. The system should also include a manual shutoff valve, a strainer, a pressure-regulating valve, and an automatic solenoid valve with a second manual shutoff valve. A bypass line with a manual valve should be installed around the automatic system. The outlet from the heat exchanger must be larger than the inlet and as short as possible. The discharge line should be visible, with no valves.

The air supply should be adequate for both the engine and for ventilation of the pump room. A fuel tank meeting the capacity requirements of NFPA should be located in accordance with their regulations. Mufflers, receiving vessels, or other attachments which may accumulate unburned gases are not recommended. If they are used, they must not be located in the pump room. The exhaust should be piped to a safe point outside the pump room. A flexible connection between the engine exhaust and the exhaust pipe can be used.

An automatic-type controller listed or approved by Underwriters Laboratory or Factory Mutual must be used and must be located as close as practical to the engine. A 2½-in clearance at the rear of the enclosure should also be provided. The wiring should terminate at a suitable marked terminal to facilitate easy wiring between the terminal and the engine junction box.

PUMP FITTINGS

The following fittings become a part of each system. When the pump suction flange is a different size than the pipe size specified for the rated flow, an eccentric suction reducer is recommended. On the discharge side of the pump a concentric increaser or discharge tee must be used to match the recommended pipe size. The discharge tee is used in conjunction with a flanged elbow. The flanged elbow has mounted upon it the flanged relief valve when it is required. The relief valve is required on all adjustable-speed drives, and also when the combination of suction and discharge pressure at the shutoff condition exceeds the design pressure of the system. The relief valve and waste water pipe should be sized as follows:

Gpm	*Relief valve size, in*	*Pipe size, in*
500	3	5
750–1,000	4	6
1,500–2,000	6	8–10
3,000–4,500	8	12–14

The relief valve should be located between the pump and the pump discharge check valve. It should discharge into an open pipe in plain view near the pump, or into a cone or funnel fastened to the outlet valve. It should also be constructed in such a way that any water wasting through the relief valve can be seen. If a closed-type cone is used, provision should be made to detect movement of water through the cone. The discharge from the relief valve should be piped outside of the building, and either wasted or returned to the supply reservoir.

The pump should be equipped with a 3½-in pressure gage rated at least twice the working pressure of the pump, but not less than 200 lb/in^2. Automatic pumps should be furnished with a circulating relief valve set just below the shutoff pressure of the pump. Vertical pumps and engine-driven pumps, where cooling water is taken from the pump discharge, do not require this valve. Pumps rated at 500 to 2,500 gpm should use ¾-in valves, and pumps of 3,000 to 4,500 gpm should have at least a 1-in valve. Manually operated pumps should be equipped with umbrella cocks for air release.

A specified number of 2½-in hose valves are required in testing the pumps. The hose valves are attached to a header or manifold connected to the discharge side of the pump, preferably located outside the pump room. If danger of freezing is present, a drain valve should also be installed in the hose valve line. The following number of hose valves is usually required:

Gpm	*Number of hose valves*
500	2
750	3
1,000	4
1,500–2,000	6
2,500	8
3,000–3,500	12
4,000–4,500	16

A jockey pump is normally used to keep the pressure in a sprinkler system at its rated condition; it prevents the fire pump from operating on small pressure drops. Any pump which meets the pressure requirements can be used. The pump should be capable of starting upon a drop in pressure above the cut-in pressure of the main fire pump. It is normally set to cut out at a predetermined pressure, and may employ a minimum running timer. Since this pump is the unit that is operated most of the time, care should be taken in its selection even though it is not required to meet Underwriters or Factory Mutual standards.

TESTING

Each pump should be shop-tested with a calibrated motor or a dynamometer. The pump should also be hydrostatically tested to twice the maximum pressure developed at shutoff, but not less than 250 lb/in^2. On vertical pumps, both the discharge casting and the bowl assembly are hydrostatically tested. All gear drives are operated at the factory under full load.

Section **10.5**

Steam Power Plant Pumping Services

IGOR KARASSIK

STEAM POWER PLANT CYCLES

Power is produced in a steam power plant by supplying heat energy to the feedwater, changing it into steam under pressure, and then transforming part of this energy into mechanical energy in a heat engine to do useful work. The feedwater therefore acts merely as a conveyor of energy. The basic elements of a steam power plant include the heat engine, the boiler, and a means of getting water into the boiler. Modern power plants use steam turbines as heat engines; except for very small plants, centrifugal boiler feed pumps are used.

This basic cycle is improved by connecting a condenser to the steam turbine exhaust and by heating the feedwater with steam extracted from an intermediate stage of the main turbine. This results in an improvement of the cycle efficiency, provides deaeration of the feedwater, and eliminates the introduction of cold water into the boiler and the resulting temperature strains on the latter. The combination of the condensing and feed heating cycle (Fig. 1) requires a minimum of three

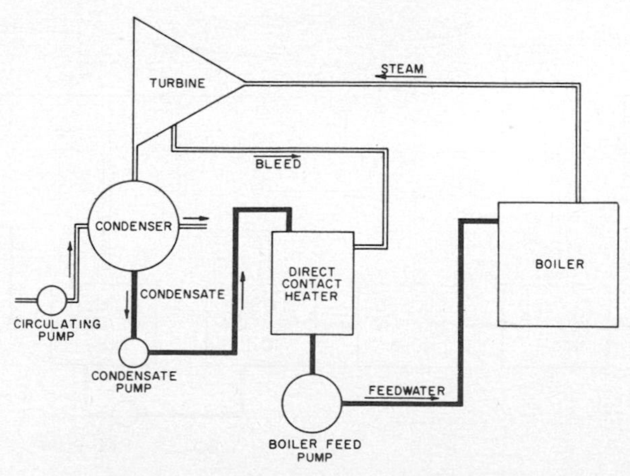

Fig. 1 Simple steam power plant cycle.

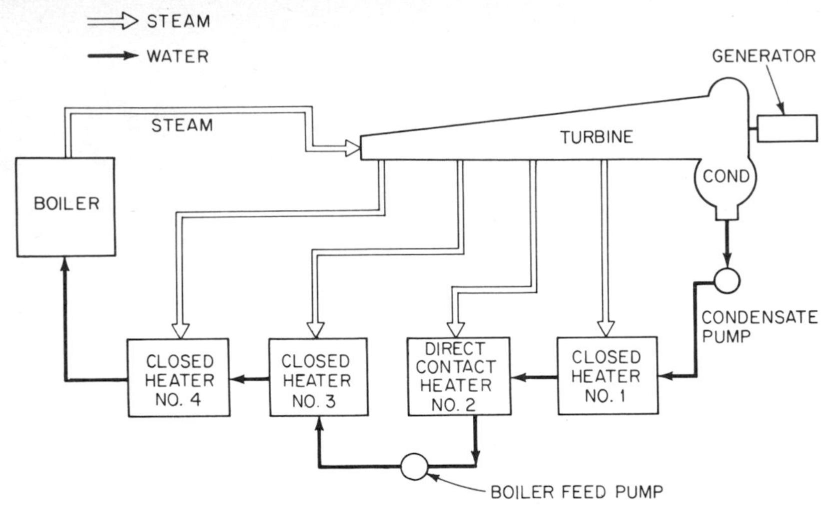

Fig. 2 Open feedwater cycle with one deaerator and several closed heaters.

pumps: the condensate pump, which transfers the condensate from the condenser hot well into the direct-contact heater; the boiler feed pump; and a circulating pump which forces cold water through the condenser tubes to condense the exhaust steam. This cycle is very common and is used in most small steam power plants. A number of auxiliary services not illustrated in Fig. 1 are normally used, such as service water pumps, cooling pumps, ash sluicing pumps, oil circulating pumps, and the like.

The desire for improvements in operating economy dictated further refinements in the steam cycle, and these have created new or altered services for power plant centrifugal pumping equipment. Some of these refinements involved a steady increase in operating pressures until 2,400 lb/in^2 steam turbines have become quite common, and many plants are operating at supercritical steam pressures of 3500 lb/in^2.

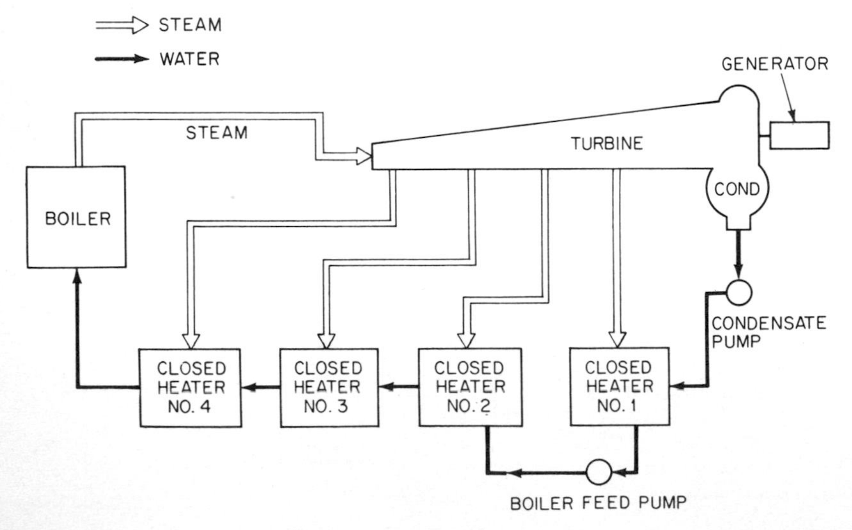

Fig. 3 Closed feedwater cycle.

Other refinements were directed toward a greater utilization of heat through increased feed heating, introducing a need for heater drain pumps—equipment with definite problems of its own. Finally, the introduction of forced or controlled circulation as opposed to natural circulation in boilers created a demand for pumping equipment of again an entirely special character.

While direct-contact heaters would have thermodynamic advantages, a separate pump would be required after each such heater. The use of a group of closed heaters permits a single boiler feed pump to discharge through these heaters and into the boiler. The average power plant is based on a compromise system: one direct contact heater is used for feedwater deaeration while several additional heaters of the closed type are located upstream as well as downstream of the direct-contact heater and of the boiler feed pump (Fig. 2). Such a cycle is termed an "open cycle." The major variation is the "closed cycle," where the deaeration is accomplished in the condenser hot well and all heaters are of the closed type (Fig. 3).

BOILER FEED PUMPS

Under the term "conditions of service" are included not only the pump capacity, discharge pressure, suction conditions, and feedwater temperature but, in addition, such data as the chemical analysis of the feedwater, the pH at pumping temperature, and other pertinent data which may reflect upon the hydraulic and mechanical design of the boiler feed pumps. Preferably a complete layout of the feedwater system and of the heat balance diagram should be supplied to the boiler feed pump manufacturer. The study of this layout will often permit the manufacturer to suggest an alternate arrangement of the equipment which would result in a more economical operation, in a lower first cost of installation, or even in longer life of the equipment so as to reduce the eventual maintenance expense.

Boiler Feed Pump Capacity The total boiler feed pump capacity is established by adding to the maximum boiler flow a margin to cover boiler swings and the eventual reduction in effective capacity from wear. This margin varies from as much as 20 percent in small plants to as little as 5 percent in the larger central stations. The total required capacity must be either handled by a single pump or subdivided between several duplicate pumps operating in parallel. Industrial power plants generally use several pumps. Central stations tend to use single full-capacity pumps to serve turbogenerators up to a rating of 100 or even 200 MW and two pumps in parallel for larger installations. There are obviously exceptions to this practice: some engineers prefer the use of multiple pumps even for small installations, while steam-turbine-driven boiler feed pumps designed for full capacity are being applied for units as large as 700 MW. A spare boiler feed pump is generally included in industrial plants. There is a trend, however, in central stations to eliminate spare pumps when two half-capacity pumps are used and, in a few cases, even if a single full-capacity boiler feed pump is installed.

Suction Conditions The net positive suction head, or NPSH, represents the net suction head at the pump suction, referred to the pump centerline, *over and above* the vapor pressure of the feedwater. If the pump takes its suction from a deaerating heater as in Fig. 2, the feedwater in the storage space is under a pressure equivalent to the vapor pressure corresponding to its temperature. Therefore the NPSH is equal to the static submergence between the water level in the storage space and the pump centerline less the friction losses in the intervening piping. Theoretically, the required NPSH is independent of the operating temperature. Practically, this temperature must be taken into account when establishing the recommended submergence from the deaerator to the boiler feed pump. A margin of safety must be added to the theoretical required NPSH to protect the boiler feed pumps against the transient conditions which follow a sudden reduction in load for the main turbogenerator. Whereas the previous discussion applies primarily to the majority of installations where the boiler feed pump takes its suction from a deaerating heater, it holds as well in the closed feed cycle (Fig. 3). The discharge pressure of the condensate pump must be carefully established so that the suction pressure of the boiler feed pump cannot fall below the sum of the vapor pressure at pumping temperature and of the required NPSH.

Discharge Pressure and Total Head The discharge pressure is the sum of the maximum boiler drum pressure and of the frictional and control losses between the boiler feed pump and the boiler drum inlet. The required discharge pressure will generally vary from 115 to 125 percent of the boiler drum pressure. The net pressure to be generated by the boiler feed pump is the difference between the required discharge pressure and the available suction pressure. This must be converted into a total head, using the formula

$$\text{Total head, in ft} = \frac{\text{net pressure, in lb/in}^2 \times 2.31}{\text{specific gravity}}$$

Slope of the Head-Capacity Curve In the range of specific speeds normally encountered in multistage centrifugal boiler feed pumps, the rise of head from the point of best efficiency will vary from 10 to 25 percent. Furthermore, the shape of the head-capacity curve for these pumps is such that the drop in head is very slow at low capacities, accelerating as the capacity is increased.

If the pump is operated at constant speed, the difference in pressure between the pump head-capacity curve and the system-head curve must be throttled by the feedwater regulator. Thus the higher the rise of head toward shutoff, the more pressure must be throttled off and, theoretically, wasted. Also, the higher the rise, the greater the pressure to which the discharge piping and the closed heaters will be subjected. However, it is not advisable to select too low a rise to shutoff because too flat a curve is not conducive to stable control; a small change in pressure corresponds to a relatively great change in capacity, and a design that gives a very low rise to shutoff may result in an unstable head-capacity curve, difficult to use for parallel operation. When several boiler feed pumps are to be operated in parallel, they must have stable curves and equal shutoff heads. Otherwise, the total flow will be divided unevenly and one of the pumps may actually be backed off the line after a change in required capacity occurs at light flows.

Driver Horsepower A boiler feed pump will generally not operate at any capacity beyond the design condition. In other words, a boiler feed pump has a very definite maximum capacity because it operates on a system-head curve made up of the boiler drum pressure plus the friction losses in the discharge. If, as it should be, the design capacity of the pump is chosen as the maximum capacity that can be expected under emergency conditions, there can be no further increase under any operating conditions since the pressure requirement corresponding to an increased capacity would exceed the design pressure of the pump. Even when the design pressure includes a safety margin, the boiler demand does not exceed the design capacity, and the feedwater regulator will impart additional artificial friction losses to increase the required pressure up to the pressure available at the pump.

When two pumps are operated in parallel, feeding a single boiler, the situation is somewhat different. If one of the pumps is taken off the line at part load, the remaining pump could easily operate at capacities in excess of its design, since its head-capacity curve would intersect the system-head curve at a head lower than the design head (Fig. 4). In such a case, it is necessary to determine the pump capacity at the intersection point; the horsepower corresponding to this capacity will be the maximum expected. It is not always necessary to select a driver of such size that it will not be overloaded at any point of the boiler feed pump operating curve. But while electric motors used on boiler feed service generally have an overload capacity of 15 percent, it is usually the practice to reserve this overload capacity as a safety margin and to select a motor that will not be overloaded at the design capacity. Exceptions occur in the case of very large motor sizes. For instance, if the pump brake horsepower is 3,100, it is logical to apply a 3,000-hp motor which will be overloaded by about 3 percent rather than a considerably more expensive 3,500-hp motor. Because steam turbines are not built in definite standard sizes but can be designed for any intermediate rating, they are generally selected with about 5 percent excess power over the maximum expected pump horsepower.

General Structural Features Boiler feed pumps designed for discharge pressures under 1,250 lb/in² are generally of the axially-split-casing type (see Sec. 2.2, Figs.

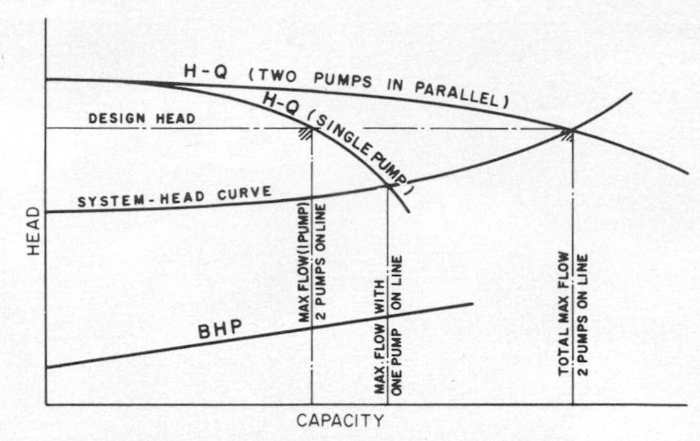

Fig. 4 Method of determining maximum pump horsepower for two boiler-feed pumps operating in parallel.

22, 23, and 24). Double-casing radially split pumps (see Sec. 2.2, Figs. 25, 26, and 28) are used for higher discharge pressures. The selection of materials for the boiler feed pump casings and internal parts has been discussed in Sec. 5.

High-Speed High-Pressure Boiler Feed Pumps As steam pressures rose from 1,250 to 1,800, then to 2,400 and even to 3,500 lb/in², the total head that had to be developed by the pump rose from somewhere around 4,000 ft to as high as 7,000 and 12,000 ft. The only means available of achieving these higher heads at 3,600 rpm was to increase the number of stages. The pumps had to have longer and longer shafts. This threatened to interfere with the long uninterrupted life between pump overhauls to which steam power plant operators were beginning to become accustomed. The logical solution was to reduce the shaft span by reducing the number of stages. Experience had indicated by 1953 that stage pressures could be increased from the 250–350 lb/in² range commonly used at 3,600 rpm to as high as 800 or 1,000 lb/in². In turn, these higher heads per stage could better be achieved by increasing the speed of rotation than by increasing impeller diameters. As a result, boiler feed pumps in large central stations today generally operate at speeds from 5,000 to 9,000 rpm.

Boiler Feed Pump Drives The majority of boiler feed pumps in small and medium size steam plants are electric-motor-driven (Fig. 5). It used to be the practice to install steam-turbine-driven standby pumps as a protection against the interruption of electric power supply, but this practice has disappeared in central steam stations and is encountered rarely even in industrial plants.

Recent years have seen a trend away from electric motor drive to steam turbine drive (Fig. 6) in large central steam stations for units in excess of 200 MW, because:

1. The use of an independent steam turbine increases plant capability by eliminating the auxiliary power required for boiler feeding.
2. Proper utilization of the exhaust steam in the feedwater heaters can improve cycle efficiency.
3. In many cases, the elimination of the boiler feed pump motors may permit a reduction in the station auxiliary voltage.
4. Driver speed can be matched ideally to the pump optimum speed.
5. A steam turbine provides variable-speed operation without an additional component such as a hydraulic coupling.

Operation of Boiler Feed Pumps at Reduced Flows Operation of centrifugal pumps at shutoff or even at certain reduced flows can lead to very undesirable results. This subject is covered in detail in Sec. 2.3 and Sec. 13, where methods for calculating minimum permissible flows and means for providing the necessary protection against operation below these flows are discussed.

Fig. 5 Battery of motor-driven double-casing boiler-feed pumps serving adjacent main T-G units. (Courtesy Worthington Pump Inc.)

Fig. 6 Steam-turbine-driven double-casing boiler-feed pump. (Courtesy Worthington Pump Inc.)

CONDENSATE PUMPS

Condensate pumps take their suction from the condenser hotwell and discharge either to the deaerating heater in open feedwater systems (Fig. 2) or to the suction of the boiler feed pumps in closed systems (Fig. 3). These pumps, therefore, operate with a very low pressure at their suction—from 1 to 3 inHg absolute. The available NPSH is obtained by the submergence between the water level in the condenser hotwell and the centerline of the condensate pump first stage impeller. Because of the desire to locate the condenser hotwell at as low an elevation in the plant as possible and to avoid the use of a condensate pump pit, the available NPSH is generally extremely low, in the order of 2 to 4 ft. The only exception to this occurs when vertical can condensate pumps are used, since these can be installed below ground elevation and higher values of submergence can be obtained. Friction losses on the suction side must be kept to an absolute minimum. The piping connection from the hotwell to the pump should therefore be as direct as possible, of ample size, and have a minimum of fittings.

Because of the low available NPSH, condensate pumps operate at relatively low speeds, ranging from 1,750 rpm in the low range of capacities to 880 rpm or even less for larger flows.

It is customary to provide a liberal excess capacity margin above the full-load steam condensing flow to take care of the heater drains that may be dumped into the condenser hotwell if the heater drain pumps are taken out of service for any reason.

Types of Condensate Pumps Both horizontal and vertical condensate pumps are used. Depending on the total head required, horizontal pumps may be either single or multistage.

Figure 7 shows a single-suction, single-stage pump with an axially-split casing used for heads up to about 100 ft. It is designed to have discharge pressure on

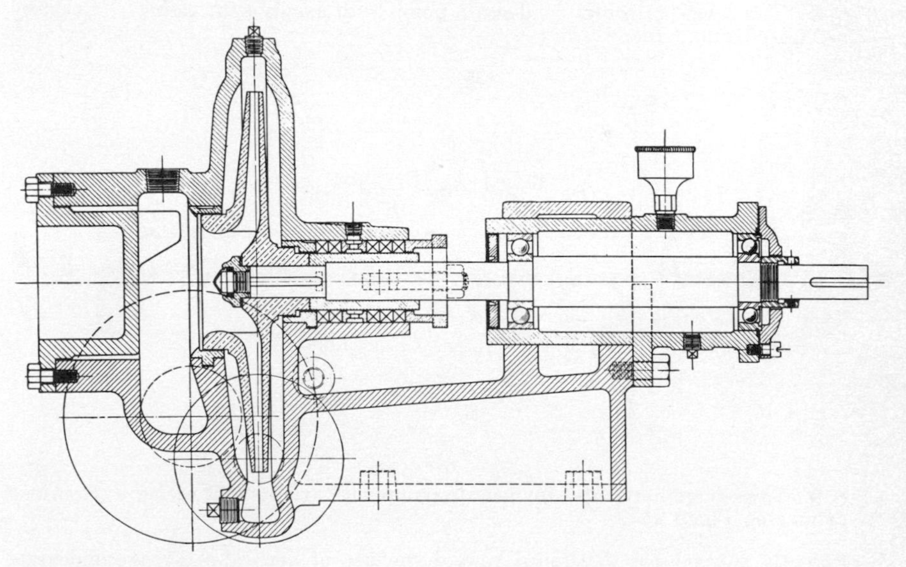

Fig. 7 Single-stage horizontal condensate pump with axially split casing. (Courtesy Worthington Pump Inc.)

the stuffing box. The suction opening in the lower half of the casing keeps the suction line at floor level. An oversize vent at the highest point of the suction chamber permits the escape of all entrained vapors that will be vented back to the condenser and removed by the air-removal apparatus.

Multistage pumps are used for higher heads. A two-stage pump is shown on Fig. 8, with the impellers facing in opposite directions for axial balance. By turning the impeller suctions toward the center, both boxes are kept under positive pressure to prevent leakage of air into the pump. For higher heads and larger capacities, a three-stage pump as in Fig. 9 may be used. The first-stage impeller is of the double-suction type and is located centrally in the pump. The remaining impellers are of the single-suction type and are also arranged so that both stuffing boxes are under pressure. Two liberal vents connecting with the suction volute on each side of the first-stage double-suction impeller permit the escape of vapor.

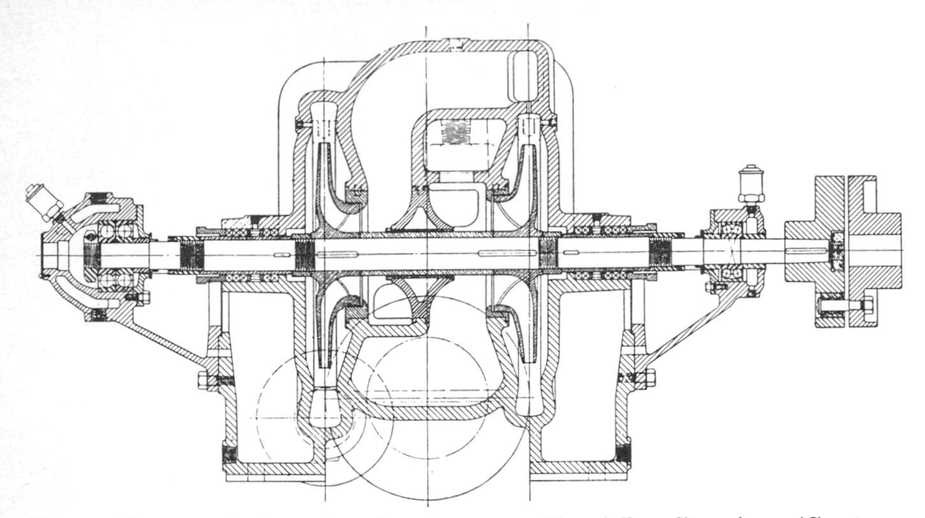

Fig. 8 Two-stage horizontal condensate pump with axially split casing. (Courtesy Worthington Pump Inc.)

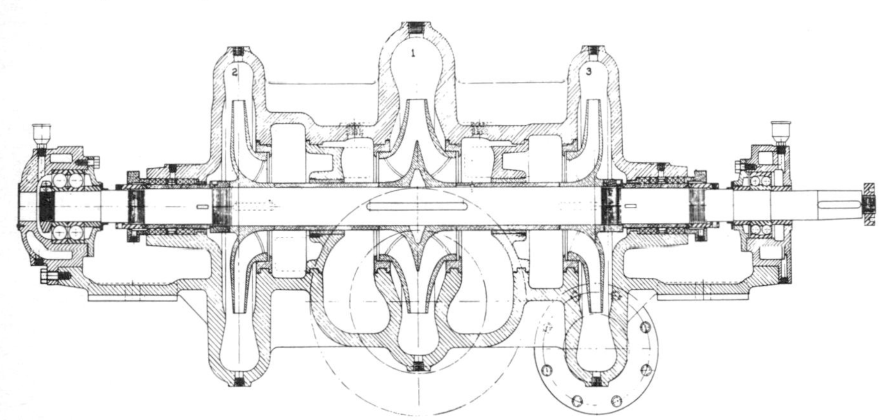

Fig. 9 Three-stage horizontal condensate pump with axially split casing. (Courtesy Worthington Pump Inc.)

Recently a trend has developed toward the use of vertical-can-type condensate pumps (Sec. 2.2, Fig. 147). The chief advantage of these pumps is that ample submergence can be provided without the necessity of building a dry pit. The first stage of this pump is located at the bottom of the pumping element, and the available NPSH is the distance between the water level in the hotwell and the centerline of the first-stage impeller.

The increasing use of full-flow demineralizers in condensate systems and, in general, the increasing discharge pressures required from the condensate pumps have resulted in the frequent need to split condensate pumping into two parts. The condensate pumps proper thus develop only a small portion of the total head required. The balance of the required head is provided by separate condensate booster pumps which have generally been of the conventional, horizontal, axially-split-casing type.

To prevent air leakage at the stuffing boxes of condensate pumps, these are always provided with seal cages. The water used for gland sealing must be taken from the condensate-pump discharge manifold beyond all the check valves.

Normally, condensate pumps have been standard fitted, with cast iron casings

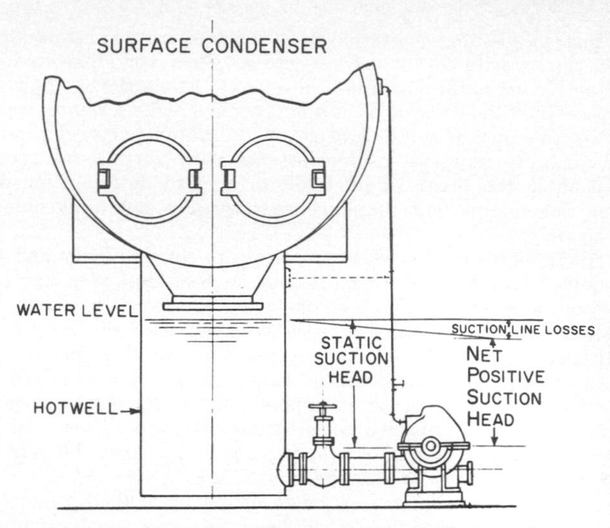

Fig. 10 Typical hookup for submergence-controlled condensate pump.

and bronze internal parts. But the emergence of once-through boilers has created the need to eliminate all copper alloys in the condensate system to avoid deposition of copper on the boiler tubes, and stainless steel fitted pumps are becoming the rule for this service.

Condensate Pump Regulation When a condensate pump operates in a closed cycle ahead of the boiler feed pump, the two pumps can be considered as a combined unit insofar as their head-capacity curve is concerned. Variation in flow is accomplished either by throttling in the boiler-feed-pump discharge or by varying the speed of the boiler feed pump.

In an open feedwater system, several means can be used to vary the condensate pump capacity with the load:

1. The condensate pump head-capacity curve can be changed by varying the pump speed. (Used very infrequently.)
2. The condensate pump head-capacity curve can be altered by allowing the pump to operate in the "break." (See Figs. 10 and 11.)
3. The system-head curve can be artificially changed by throttling the pump discharge by means of a float control.
4. The pump can operate at the intersection of its head-capacity curve and the normal system-head curve. The net discharge is controlled by bypassing all excess condensate back to the condenser hotwell.

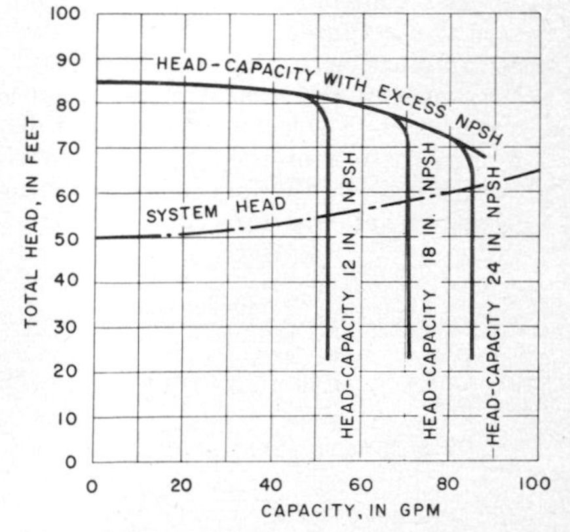

Fig. 11 Characteristics of a condensate pump operating on a submergence-controlled system.

5. Methods 3 and 4 can be combined so that the discharge is throttled back to a predetermined minimum, but if the load, and consequently the flow of condensate to the hotwell, is reduced below this minimum, the excess of condensate handled by the pump is bypassed back to the hotwell.

Operating in the "break," or "submergence control" as it has often been called, has been applied very successfully in a great many installations. Condensate pumps designed for submergence control require specialized hydraulic design, correct selection of operating speeds, and limitation of stage pressures. The pump is operating in the break (i.e., cavitates) at all capacities. However, this cavitation is not severely destructive in nature because the energy level of the fluid at the point where the vapor bubbles collapse is insufficient to create a shock wave of a high enough intensity to inflict physical damage on the pump parts. If, however, higher values of NPSH were required—as for instance with vertical-can-type condensate pumps—operation in the break would result in a rapid deterioration of the impellers. It is for this reason that submergence control is not applicable to can-type condensate pumps.

The main advantages of submergence control are its simplicity and the fact that the power required for any operating condition is less than with any other system. Disadvantages occur when the pump is operating at very light loads, however, because the system head may require as little as one half of the total head produced in the normal head-capacity curve. In this case, the first stage of a two-stage pump produces no head whatsoever, and, if the axial balance was achieved by opposing the two impellers, a definite thrust is imposed on the thrust bearing, which must be selected with sufficient capacity to withstand this condition. In addition, no control is available to provide the minimum flow that may be required through the auxiliaries such as the ejector condenser.

The condensate pump discharge can be throttled by a float control arranged to position a valve that increases the system-head curve as the level in the hot well is drawn down. It eliminates the cavitation in the condensate pump, but at the cost of a slight power increase. Furthermore, the float necessarily operates over a narrow range and the mechanism tends to be somewhat sluggish in following rapid load changes, often resulting in capacity and pressure surges.

When condensate delivery is controlled through bypassing, the hotwell float controls a valve in a bypass line connecting the pump discharge back to the hotwell. At maximum condensate flow, the float is at its upper limit with the bypass closed and all the condensate is delivered to the system. As the condensate flow to the hotwell decreases, the hotwell level falls, carrying the float down and opening the bypass. Sluggish float action can create the same problems of system instability in bypass control as in throttling control, however, and the power consumption is excessive because the pump always operates at full capacity.

A combination of the throttling and bypassing control methods eliminates the shortcoming of the excessive power consumption. The minimum flow at which bypassing begins is selected to provide sufficient flow through the ejector condenser.

A modification of the bypassing control for minimum flow is illustrated in Fig. 12, which shows a thermostatic control for condensate recirculation. With practically constant steam flow through the ejector, the rise in temperature of the condensate

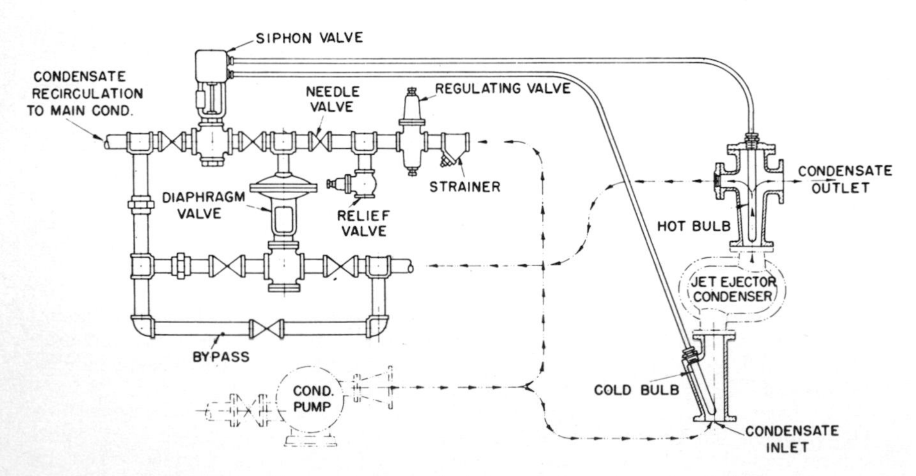

Fig. 12 Thermostatic control for condensate recirculation.

between the inlet and outlet of the ejector condenser is a close indication of the rate of flow of condensate through the ejector condenser tubes. Therefore an automatic device to regulate the rate of flow of condensate can be controlled by this temperature differential. A small pipe is connected from the condensate outlet on the ejector condenser back into the main condenser shell. An automatic valve is installed in this line and is actuated and controlled by the temperature rise of the condensate.

Whenever the temperature rises to a certain predetermined figure, indicating a low flow of condensate, the automatic valve begins to open, allowing some of the condensate to return to the condenser and then to the condensate pump, which supplies it to the ejector condenser at the increased rate. When the temperature rise through the ejector condenser is less than the limiting amount, indicating that ample condensate is flowing through the ejector condenser, the automatic valve remains closed.

HEATER DRAIN PUMPS

Service Conditions Condensate drains from closed heaters can be flashed to the steam space of a lower pressure heater or pumped into the feedwater cycle at some higher pressure point. Piping each heater drain to the next lower pressure heater is the simpler mechanical arrangement and requires no power-driven equipment. This "cascading" is accomplished by an appropriate trap in each heater drain. A series of heaters can thus be drained by cascading from heater to heater in the order of descending pressure, the lowest being drained directly to the condenser.

This arrangement, however, introduces a loss of heat since the heat content of the drains from the lowest pressure heater is dissipated in the condenser by transfer to the circulating water. It is generally the practice, therefore, to cascade only down to the lowest pressure heater and pump the drains from that heater back into the feedwater cycle, as shown on Fig. 13. Because the pressure in that heater

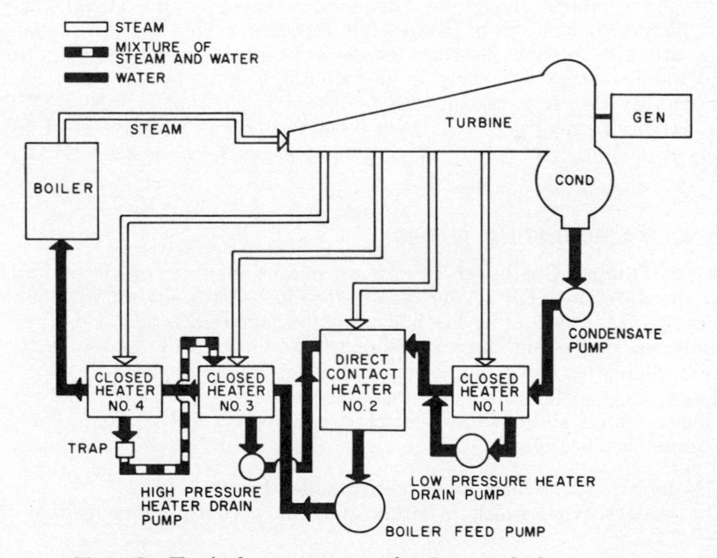

Fig. 13 Typical arrangement for heater drain pumps.

hotwell is low (frequently below atmospheric even at full load), heater drain pumps on that service are commonly described as on "low-pressure heater drain service."

In an open cycle, drains from heaters located beyond the deaerator are cascaded to the deaerator. Although the deaerator is generally located above the closed heaters, the difference in pressure is sufficient to overcome both the static and the friction losses. This difference in pressure decreases with a reduction in load, however, and at some partial main turbine load it becomes insufficient to evacuate the heater drains and they have to be switched to a lower pressure heater or even to the condenser, with a subsequent loss of heat. To avoid these complications, a "high-pressure heater drain pump" is generally used to transfer these drains to the deaerator. Actually this pump has a "reverse" system head to work against; at full load, the required total head may be negative, whereas at light loads the required head is at its maximum.

High-pressure drain pumps are subject to more severe conditions than boiler feed pumps encounter:

1. Their suction pressure and temperature are higher.
2. The available NPSH is generally extremely limited.
3. They are subject to all the transient conditions to which the feed pump is exposed during sudden load fluctuations, and these transients are more severe than those at the feed pump suction.

Types of Heater Drain Pumps In the past, heater drain pumps were always of the horizontal type, either single stage or multistage depending upon total head requirements. In the single-stage type, end-suction pumps of the heavier "process pump" type (Sec. 2.2, Fig. 134) were preferred for both low- and high-pressure service. Recently, however, just as it has invaded the field of condensate service, the vertical-can-type pump (Sec. 2.2, Fig. 147) has been frequently applied on heater drain service. As previously described, the advantages of the vertical-can pump are lower first cost and a "built-in" additional NPSH because the first-stage impeller is lowered below floor level in the can. Against these advantages, one must weigh certain shortcomings. A horizontal heater drain pump is more easily inspected than a can pump. The external grease- or oil-lubricated bearings of the horizontal pump are less vulnerable to the severe operating conditions during swinging loads than the water-lubricated internal bearings of the can pump.

Heater drain pumps should be adequately vented to the steam space of the heater. Because heater drain pumps and especially those on low-pressure service may operate with suction pressures below atmospheric, it is necessary to provide a liquid supply to the seal cages in the stuffing boxes. Low-pressure heater drain pumps can use cast iron casings and bronze fittings if no evidence of corrosion-erosion has been uncovered. On high-pressure service, stainless steel fittings are generally mandatory and 5 percent chrome stainless steel casings should preferably be used.

CONDENSER CIRCULATING PUMPS

Types of Pumps Condenser circulating pumps may be of either horizontal or vertical construction. For many years the low-speed horizontal double-suction volute centrifugal pump (Fig. 14) had been the preferred type. This type of pump incorporates a simple but rugged design which allows ready access to the interior parts for examination and rapid dismantling if repairs are required.

Although many installations of horizontal circulating pumps are still encountered, many larger central stations have switched over to vertical pumps. These fall into two separate classifications:

1. The dry-pit type, which operates surrounded by air.
2. The wet-pit type, which is either fully or partially submerged in the water pumped.

The choice between these two types is somewhat controversial; strong personal preferences exist in favor of one or the other construction. A number of factors must be considered when choosing between dry- and wet-pit pumps for condenser circulating service. Some of the factors involved lend themselves to a straight economic evaluation, their advantages and disadvantages being readily expressed in dollars. Other factors, no less important, are intangible and only experience and sound judgement can give them their proper and deserved weight.

Fig. 14 Installation of double-suction horizontal condenser circulating pumps. (Courtesy Worthington Pump Inc.)

Mechanical Considerations The most popular type of pump during the past thirty years for condenser circulating service in dry-pit installations has been the single-suction, medium-specific-speed, mixed-flow pump (Sec. 2.2, Fig. 142). This design combines the high efficiency and low maintenance of the horizontal double-suction radial-flow centrifugal with lower cost and slightly higher rotative speeds.

Because of their suction and discharge nozzle arrangements, these pumps are ideally suited for vertical mounting in a dry pit, preferably at the lowest water level, so that they are self-priming on starting. They are directly connected to solid-shaft induction or synchronous motors, either close-coupled or with intermediate shafting between the pump and the motor, which is then mounted well above the pump pit floor.

Like the horizontal double-suction pump, the vertical dry-pit mixed-flow pump is a compact and sturdy piece of equipment. Its rotor is supported by external oil-lubricated bearings of optimum design. This construction requires the least attention, for the oil level can be easily inspected by means of an oil sight glass mounted at the side of the bearing or oil reservoir.

The pump construction further facilitates maintenance and replacement, since the rotor is readily removed through the top of the casing. Therefore the pump does not have to be removed from its mounting, nor the suction and discharge connections broken to make periodic inspections or repairs.

In recent years power-plant designers have shown a preference for the wet-pit column-type condenser circulating pump, although most recently this trend has been appreciably slowed down, and several major utilities are reverting to the dry-pit type.

Unless qualified, the term "wet-pit" normally implies a diffuser-type pump. There have been instances when a conventional volute-type pump has been submerged into a pit and operated as a wet-pit pump, but this is the exception to the rule. The volute pump is essentially a dry-pit pump.

The wet-pit pump (Sec. 2.2, Figs. 148, 149, and 150) employs a long column pipe that supports the submerged pumping element. It is available with open main shaft bearings, lubricated by the water handled, or preferably with enclosed shafting and bearings, lubricated by clean, fresh, filtered water from an external source. Even in the latter case there is some danger of contamination of the lubricating water from seepage into the shaft enclosure tube during shutdowns. Even with the best design and the best care, a water-lubricated bearing is not the equal of an oil-lubricated bearing. Consequently higher maintenance costs may be expected with the use of a wet-pit pump.

Larger power plants are generally located near centers of population, and as a result, often have to use badly contaminated water—either fresh or salt—as a condenser cooling medium. With such water, a fabricated steel column pipe and elbow would give short life, so cast iron, bronze, or even a more corrosion-resistant cast metal must be used. This results in a very heavy pump when large capacities are involved. Pulling up the column in a long pump requires special and expensive facilities and, in addition, the discharge flange must be disconnected when withdrawing the pump and column from the pit. To avoid the necessity of lifting the entire pump when the internal parts require maintenance, some designs (Sec. 2.2, Fig. 150) are built so that the impeller, diffuser, and shaft assembly can be removed from the top without disturbing the column pipe assembly. (The driving motor must still be removed.) These designs are commonly designated as of the "pull-out" type.

The selection of materials for either dry- or wet-pit pumps can vary considerably, depending on the character of the circulating water. The dry-pit pump requires the least amount of changes in material, since the impeller and rings are the parts that would be primarily affected, if at all. The wet-pit pump, however, would require a careful selection of all submerged parts, particularly when sea water is to be pumped. Special alloys are usually required for the impellers and rings. Because of the danger of electrolytic attack, material changes may also often be required for such parts as the shaft, diffuser, column pipe, and shaft enclosure tube.

Performance Characteristics Condenser circulating pumps are normally required to work against low or moderate heads. Extreme care should be exercised in calculating the system friction losses, which include the condenser friction. If more total head is specified than actually required, the resulting driver size may be unnecessarily increased. For instance, an excess of 1 or 2 ft in an installation requiring only 20 ft of head represents an increase of 5 to 10 percent in excess power costs.

The range or variation of suction lift for dry-pit pumps must be determined very accurately and checked with the manufacturer to ensure that cavitation will be avoided in the installation. Priming facilities must be provided or the pump must be installed in a dry pit at such an elevation that the water in the suction channel leading to the pump will be maintained at the level recommended by the manufacturer. This presents no problem in the case of a wet-pit-type installation, because pump column can be made long enough to provide adequate submergence, even with minimum water levels in the suction well or pit. The dry-pit pump has a performance advantage over the wet-pit type. The former will generally have 3 to 4 percent higher efficiency, and therefore 3 to 4 percent lower power consumption. The two types are available for the same specific speed range. When pumping total head is 25 ft or less, an axial-flow propeller (approximately 10,000 specific speed) can be used in either type of pump.

The low-specific-speed double-suction pump has a very moderate rise in head with reducing capacities and a nonoverloading power curve with a reduction in head. The mixed-flow impeller with a higher specific speed has a steeper head-capacity curve and a reasonably flat power curve that is also nonoverloading. As the specific speed increases, the head-capacity curve increases in steepness and the curvature of the power curve reverses itself, hitting a maximum at the lowest flow. Finally, the curve of a high-specific-speed propeller pump has the highest rise in both head-capacity and power-capacity curves toward zero flow. The head range developed by the mixed-flow pump is ideal for condenser service; this pump is usually furnished with an enclosed impeller, which produces a relatively flat head-capacity curve and a flat power characteristic.

System Hydraulics The dry-pit pump is not too sensitive to the suction well design, since the inlet piping and the formed design of the suction passages into the pump normally ensure a uniform flow into the eye of the impeller. On the other hand, the higher-speed wet-pit pumps are more sensitive to departures from ideal inlet conditions than the low-speed centrifugal volute pump or the medium-speed mixed-flow pump. A discussion of the arrangements recommended for wet- and dry-pit pump installations is presented in Sec. 11.

Drivers Whether a dry-pit or a wet-pit pump is used, the axial thrust and weight of the pump rotor are normally carried by a thrust bearing in the motor, and the driver and driven shafts are connected through a rigid coupling. The higher rotative speeds of the wet-pit pumps act to reduce the cost of the electric motors somewhat. This difference may be offset considerably, however, by the fact that the thrust load of the wet-pit pump is considerably higher than that of the dry-pit pump.

BOILER CIRCULATING PUMPS

The forced-circulation or "controlled circulation" boiler requires the use of circulating pumps which take their suction from a header connected to several downcomers, which originate from the bottom of the boiler drum and discharge through the various tube circuits operating in parallel (Fig. 15). The circulating pumps, there-

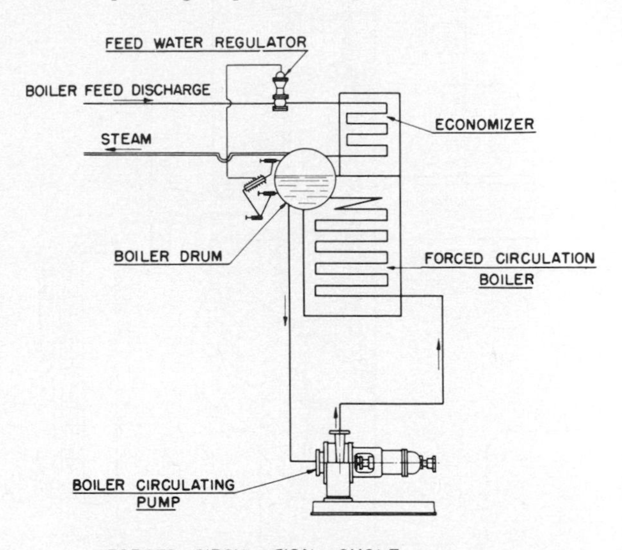

Fig. 15 Forced or "controlled circulation" cycle.

fore, must develop a pressure equivalent to the friction losses through these tube circuits. Thus, in the case of a boiler operating at 1,900 lb/in² gage, the boiler circulating pump must handle feedwater at approximately 630°F under a suction pressure of 1,900 lb/in² gage. Such a combination of high suction pressure and high water temperature at saturation imposes very severe conditions on the circulating-pump stuffing boxes, making it necessary to develop special designs for this part of the pump.

The net pressure to be developed by these pumps is relatively low, ranging from 30 to 100 or 125 lb/in². Hence these are single-stage pumps, with single-suction impellers and a single stuffing box. This imposes an extremely severe axial thrust on the pump, placing a load of several tons on the thrust bearing. In many cases a thrust relieving device must be incorporated to permit pump startup. Two general types of construction are used for this service: (1) the conventional centrifugal pump with various stuffing box modifications (Fig. 16) and (2) the submersible motor pump of either the wet- or dry-stator type (Fig. 17). In the lower boiler pressure range—up to 500 or 600 lb/in² gage—a construction as shown on Fig. 18 may be used. The pump is of the same general type as is used on high-pressure heater drain service. The packed stuffing box is preceded by a pressure-reducing bushing. Feed-water from the boiler-feed-pump discharge, at a temperature lower than in the boiler drum and at a pressure somewhat higher than pump internal pressure, is injected in the middle of this bushing. Part of this injected feedwater proceeds toward the pump interior, making a barrier against the outflow of high-

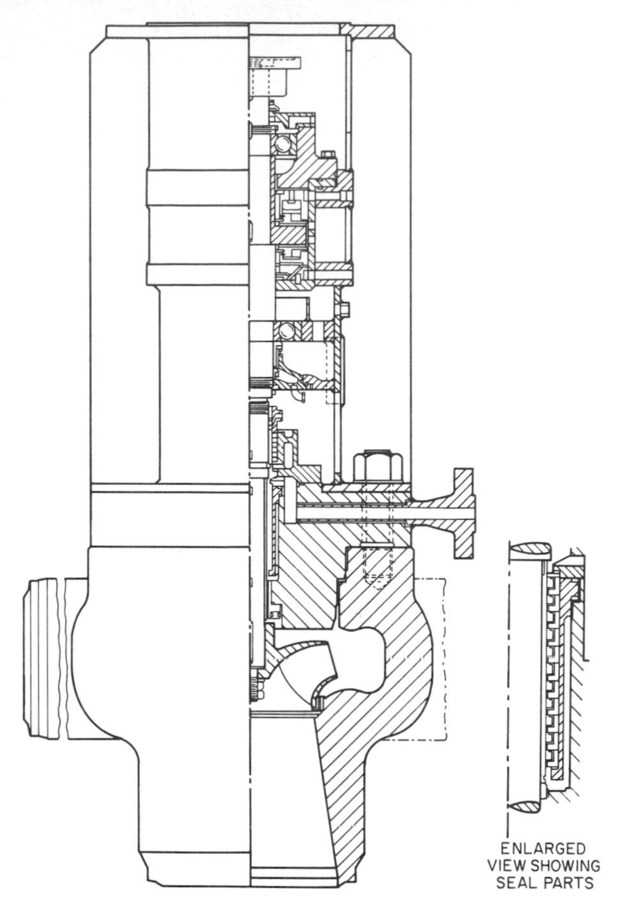

Fig. 16 Vertical injection-type boiler circulating pump for high pressures. (Courtesy Ingersoll-Rand)

temperature water. The rest proceeds outwardly to a bleed portion of the bushing, from where it is bled to a lower pressure. The packing need only withstand the lower boiler-feed pump temperature and a much lower pressure than boiler pressure.

More sophisticated designs are required for the higher pressures from 1,800 to 2,800 lb/in^2 gage (Fig. 16). The sealing of the shaft is accomplished by two floating-ring-type pressure breakdowns and a water-jacketed stuffing box. Boiler feedwater is injected at a point between the lower and upper stacks of floating-ring seals at a pressure about 50 lb/in^2 above the pump internal pressure. Here again, part of this injection leaks into the pump interior. The rest leaks past the upper stack of seals to a region of low pressure in the feed cycle. Leakage to atmosphere is controlled by the conventional stuffing box located above the upper stack.

The available NPSH may not be sufficient at startup, when the water in the boiler is cold and the pressure is low. Therefore certain installations include two-speed motors, so that a lower NPSH is required at startup. There is an added advantage to this arrangement: under normal operating conditions, the feedwater will be at boiler saturation temperature and, therefore, will have a specific gravity of as low as 0.60; at startup, however, the specific gravity will be 1.0. The power consumption on cold water would therefore be some 65 percent higher than in normal operation, if the pump operates at the same speed, and a much larger motor would

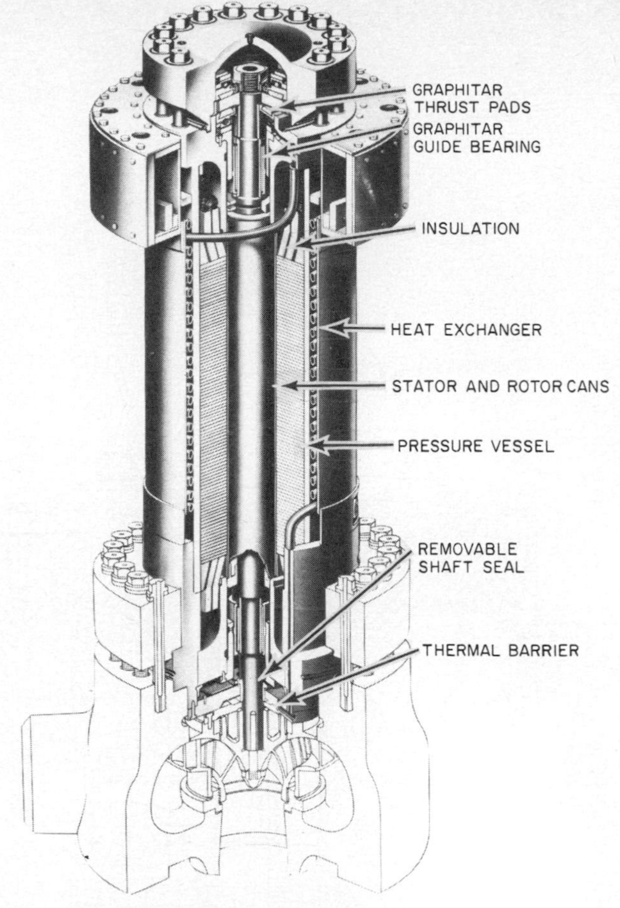

Fig. 17 Canned-rotor motor pump. (Courtesy General Electric Co.)

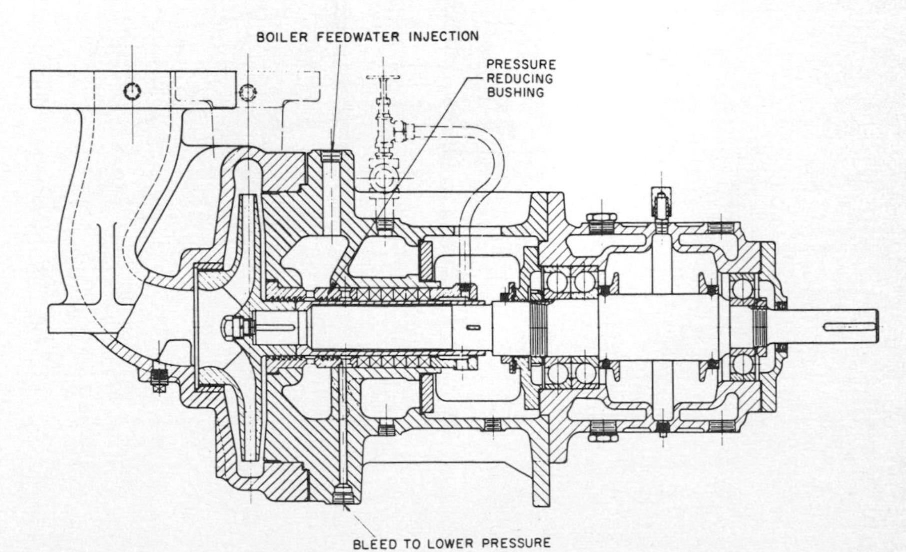

Fig. 18 End-suction boiler circulating pump for low-pressure range. (Courtesy Worthington Pump Inc.)

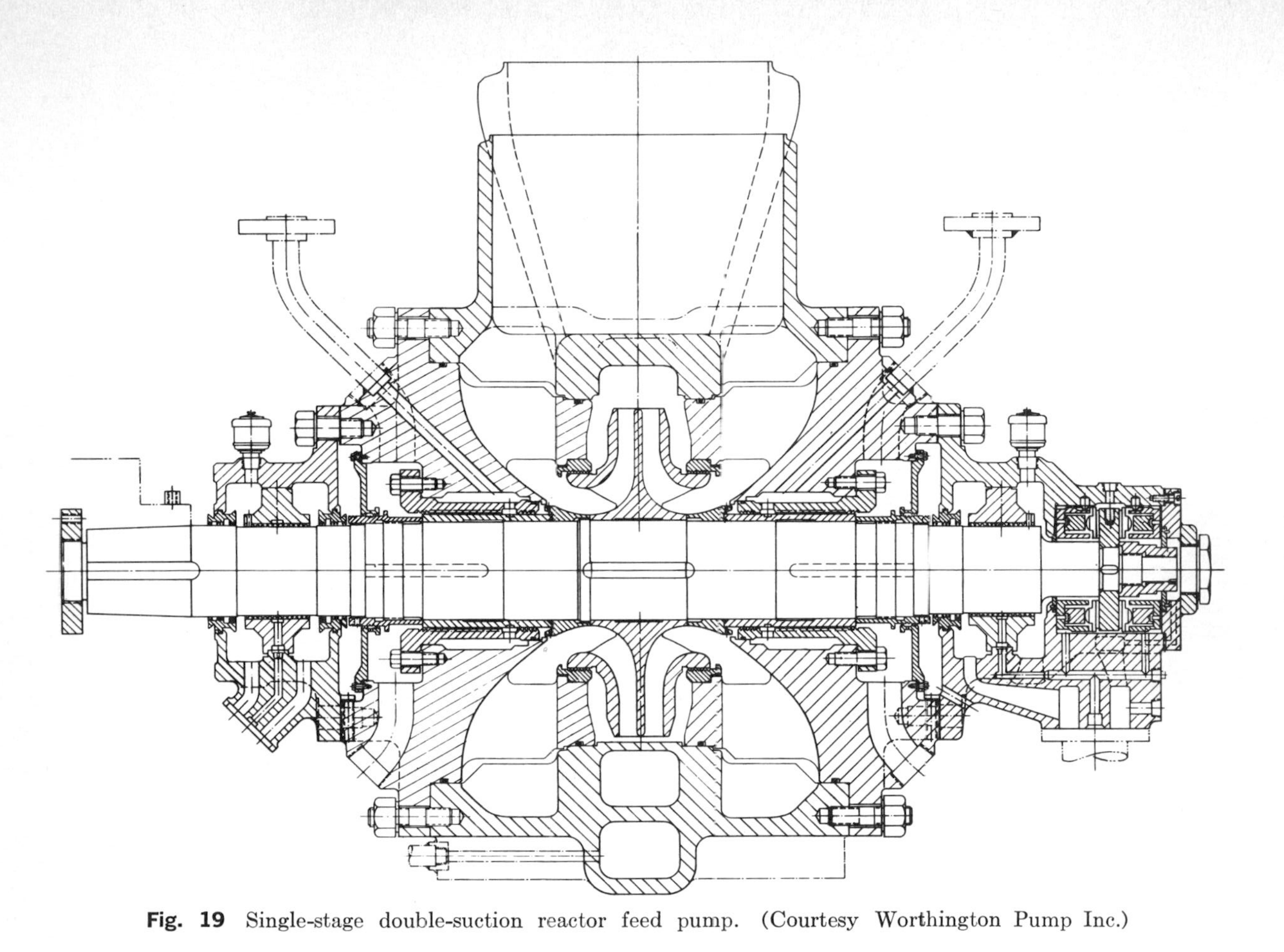

Fig. 19 Single-stage double-suction reactor feed pump. (Courtesy Worthington Pump Inc.)

be required. If a two-speed motor is used, however, the pump is operated at lower speed when the water is cold, and the motor need be no larger than the maximum horsepower under normal operating conditions.

NUCLEAR POWER PLANTS

In oversimplified form, the nuclear-energy steam power plant differs from the conventional power plant only in the fact that it uses a different fuel. Thus the so-called "secondary cycle" (consisting of turbogenerator, condenser and auxiliaries, and boiler feed pumps) is not very different from its counterpart in the conventional steam power plant. The main differences are a desire for even greater equipment reliability and a preference for absence or minimum of leakage, to avoid any possibility of contamination with radioactive material. One other difference distinguishes most nuclear power plants today from their fossil-fuel counterparts; their operating steam pressures and temperatures are much lower. Consequently, in most cases, reactor feed pumps are single-stage pumps; a typical section is shown on Fig. 19. The lower operating conditions result in higher heat rates and the flows—both of the feedwater and of the condenser circulation—are about one-third higher than for fossil-power plants of equal megawatt rating.

More detailed information on other nuclear power plant pumping services is given in Sec. 10.19.

REFERENCE

"Steam Power Plant Clinics," a series of articles in *Combustion Magazine*.

Section 10.6

Chemical Industry

JOHN R. BIRK

JAMES H. PEACOCK

SOLUTION CHARACTERISTICS

Chemical industries could be broadly defined as those which make, use, or dispose of chemicals. The pumps used in these industries are different than those used in the other industry groups of this section primarily because of the materials from which they are made. While cast iron, ductile iron, carbon steel, and aluminum- or copper-base alloys will handle a few chemical solutions, most chemical pumps are made of stainless steel, the nickel-base alloys, or the more exotic metals such as titanium and zirconium. Pumps are also available in carbon, glass, porcelain, rubber, lead, and whole families of plastics including the phenolics, epoxies, and fluorocarbons. Each of these materials has been incorporated into pump design for just one reason—to eliminate or reduce the destructive effect of the chemical liquid on the pump parts.

Since the type of corrosive liquid will determine which of these materials will be most suitable, a careful analysis of the chemical solution to be handled is the first step in selecting the proper material.

Major and Minor Constituents Foremost in importance in a study of the characteristics of any solution are the constituents of the solution. This means not only the major constituents but the minor ones as well, for in many instances the minor constituents will be the more important. They can drastically alter corrosion rates, and therefore a full and detailed analysis is most critical.

Concentration Closely allied and directly related to the constituents is the concentration of each. Merely stating "concentrated," "dilute," or "trace quantities" is basically meaningless because of the broad scope of interpretation of these factors. For instance, some interpret "concentrated" as meaning any constituent having greater than 50 percent by weight, whereas others interpret any concentration above 5 percent in a like manner. Hence it is always desirable to cite the percentage by weight of each and every constituent in a given solution. This eliminates multiple interpretation and permits a more accurate evaluation. It is also recommended that the percentage by weight of any trace quantities be cited, even if this involves only parts per million. For example, high-silicon iron might be completely suitable in a given environment in the absence of fluorides. If, however, the same environment contained even a few parts per million of fluorides, the high-silicon iron would suffer a catastrophic corrosion failure.

Temperature Generalized terms such as "hot," "cold," or even "ambient" are ambiguous in that they can be interpreted in different ways. The preferred terminology would be the maximum, minimum, and normal operating temperature in degrees, either Fahrenheit or Centigrade. Chemical reactions, in general, increase in rate of activity approximately two to three times with each increase of 18°F in temperature. Corrosion can be considered a chemical reaction, so with this in mind, the importance of temperature or temperature range is obvious. A weather-exposed pump installation is a good illustration of the ambiguity of the term "ambient." For instance, there could be as much as 150°F difference between an extremely cold climate and an extremely warm climate. If temperature cannot be cited in actual degrees, the ambient temperature should be qualified by stating the geographic location of the pump. This is particularly important for materials that are subject to thermal shock in addition to increased corrosion rate in higher temperature environments.

Acidity and Alkalinity More often than not, little consideration is given to the pH of process solutions. This may be a critical and well-controlled factor during production processing, and it can be equally revealing in evaluating solution characteristics for material selection. One reason the pH may be overlooked is that it generally is obvious whether the corrosive substance is acidic or alkaline. However, this is not always true, particularly with process solutions which may have the pH adjusted so that they will always be either alkaline or acidic. When this situation exists, the precise details should be known so that a more thorough evaluation can be made. It is also quite important to know when a solution alternates between acid and alkaline conditions because this can have a pronounced effect on materials selection. Some materials, while entirely suitable for handling a given alkaline or acid solution, may not be suitable for alternately handling the same solutions.

Solids in Suspension Erosion-corrosion, velocity, and solids in suspension are closely allied in chemical-industry pump services. Pump design is a very critical factor when solids are in the solution. It is not uncommon for a given alloy to range from satisfactory to completely unsatisfactory in a given chemical application with hydraulic design being the only variable factor. Failure to cite the presence of solids on a solution data sheet is not an uncommon occurrence. This undoubtedly is the reason for many catastrophic erosion-corrosion failures.

Aerated or Non-aerated The presence of air in a solution can be quite significant. In some instances it is the difference between success and failure in that it can conceivably render a reducing solution oxidizing and require an altogether different material. A good example of this would be a self-priming, nickel-molybdenum-alloy pump for handling commercially pure hydrochloric acid. This alloy is excellent for the commercially pure form of this acid, but any condition that can induce even slightly oxidizing tendencies renders this same alloy completely unsuitable. The very fact that the pump is a self-primer means that aeration is a factor to contend with, and extreme caution must be exercised in using an alloy that is not suitable for an oxidizing environment.

Transferring or Recirculating This item is of importance because of possible buildup of corrosion product or contaminants which can influence the service life of the pump. Such a buildup of contaminants can have a beneficial or deleterious effect, and for this reason it should be an integral part of evaluating solution characteristics.

Inhibitors or Accelerators Both inhibitors and accelerators can be intentionally or unintentionally added to the solution. Inhibitors reduce corrosivity, whereas accelerators increase corrosivity. Obviously no one would add an accelerator to increase the corrosion rate on a piece of equipment, but the addition of a minor constituent, as a necessary part of a given process, may serve as an accelerator; thus the importance of knowing of the presence of such constituents.

Purity of Product Where purity of product is of absolute importance, particular note should be made of any element that may cause contamination problems, whether it be discoloration of product or solution breakdown. In some environments, pickup of only a few parts per billion of certain elements can create severe problems. A note to this effect is particularly important in pump applications where velocity effects and the presence of solids can alter the end result, as contrasted with other types of process equipment where the velocity and/or solids may have little or no effect. When a material is basically suitable for a given environment, purity of product should not be a problem. However, this cannot be an ironclad rule, particularly with chemical pumps.

Continuous or Intermittent Duty Depending upon the solution, continuous or intermittent contact can have a bearing on service life. Intermittent duty in some environments can be more destructive than continuous duty if the pump remains half full of corrosive during periods of downtime and accelerated corrosion occurs at the liquid interface. Perhaps of equal importance is whether or not the pump is flushed and/or drained when not in service.

CORROSIVES AND MATERIALS

Metallic or Nonmetallic The selection of the proper material for a chemical-industry pump application is an extremely important item, and is dependent upon the factors mentioned in the previous section. Materials for these applications can, in general, be divided into two very broad categories; metallic and nonmetallic. The metallic category can be further subdivided into ferrous and nonferrous alloys; both of which have extensive application in the chemical industry. The nonmetallic category could be further subdivided into natural and synthetic rubbers, plastics, ceramics and glass, carbon and graphite, and wood. Of these nonmetallic categories, wood, of course, has little or no application for pump services. The other materials cited have definite application in the handling of heavy corrosives. In particular, plastics in recent years have gained widespread acclaim for their corrosion resistance, and therefore are much used in chemical environments. For a given application, a thorough evaluation of not only the solution characteristics but also the materials available should be made to ensure the most economical selection.

Source of Data To evaluate material for chemical pump services, various sources of data are available. The best source is previous practical experience within one's own organization. It is not unusual, particularly in large organizations, to have a materials group or corrosion group whose basic responsibility is to collect and compile corrosion data pertaining to process equipment in service at its various plants. These sources should be consulted whenever a materials evaluation program is being conducted. A second source of data is laboratory and pilot-plant experience. Though these sources cannot provide as valuable and detailed information as actual plant experience, they certainly can be very indicative and serve as an important guide. The experience of suppliers can be a third source of information. Though suppliers cannot hope to provide data on the specific details of a given process and the constituents involved, they normally can provide assistance and materials for test to facilitate a decision. Technical journals and periodicals are a fourth source of information. A wealth of information is contained in these publications, but if an excellent information retrieval system is not available, it can be very difficult to locate the information desired.

Reams of information have been published in the form of books, tables, charts, periodicals, bulletins, and reports pertaining to materials selection for various environments. It is not the intent of this section of the handbook to make materials recommendations. However, it is deemed advisable to provide some general comments and to point out a few applications having unusual characteristics.

Sulfuric Acid This is the most widely used chemical in industrial applications today and much time is spent in evaluating and selecting materials for applications involving sulfuric acid with and without constituents. The following are some of the applications that merit special consideration.

Dilution of Commercially Pure Sulfuric Acid. When diluting sulfuric acid with water, there is considerable evolution of heat. At times actual mixing of the acid with water does not take place in the mixing tank, but in the pump transferring the acid. This means that heat is evolved as the solution is passing through the pump. Temperatures in the magnitude of 200°F or higher are reached, depending upon the degree of dilution and the amount of mixing taking place in the pump. If, for instance, the 93 percent acid is to be diluted to 70 percent in the tank, and then transferred to another location at 100°F, a variety of materials could be used. However, if the actual mixing or diluting takes place in the pump, the heat of evolution or dilution would restrict the material selection. Very few metallics or nonmetallics are resistant to 70 percent sulfuric acid at temperatures approaching 200°F.

Sulfuric Acid Saturated with Chlorine. It is a well-known fact that any solution involving wet chlorine is one of the most corrosive solutions known. The specific weight percentage of sulfuric acid determines whether the presence of chlorine in sulfuric acid will accelerate corrosion. Because of the hygroscopic nature of concentrated sulfuric acid, it will absorb the moisture from the chlorine. This is true in the range of approximately 80 percent sulfuric acid to the completely concentrated acid. When such a condition exists, there need be little concern for the chlorine because dry chlorine is essentially noncorrosive. In such a case a material selection can be made as if sulfuric acid were the only constituent. If the solution is saturated with chlorine, but contains less than approximately 80 percent sulfuric acid, the material selection must be based on not only the sulfuric acid but also the wet chlorine. This, of course, is a very corrosive solution, and extreme caution must be exercised in selecting the material to be used.

Rayon Spin Baths. Rayon spin baths can contain up to 20 percent sulfuric acid. In addition, they contain significant quantities of zinc sulfate, sodium sulfate, carbon disulfide, and hydrogen sulfide. Temperatures up to 200°F are often encountered. In solutions of this type, it is not unusual to base a material selection strictly on the concentration of sulfuric acid and the temperature, as the other constituents may be deemed to exert a very slight influence. On the contrary, the presence of carbon disulfide and hydrogen sulfide will make these rayon spin baths strongly reducing, and can render an otherwise suitable material completely unsuitable. Even trace quantities of these sulfides exert a significant influence at any temperature and must be considered in selecting a material for rayon spin baths.

Sulfuric Acid Containing Sodium Chloride. It is quite apparent that the addition of sodium chloride to sulfuric acid will result in the formation of hydrochloric acid, and thus necessitate a material that will resist the corrosive action of hydrochloric acid also. Though this may seem obvious, it is amazing how often it is ignored. This is particularly true in 10 to 15 percent sulfuric acid pickling solutions to which sodium chloride has been added to increase the rate of pickling, with little or no consideration being given to the destructive effect of the salt addition on the process equipment handling the pickling solution.

Pigment Manufacture. A slurry of titanium dioxide in sulfuric acid is one of the processing stations in the manufacture of pigment. Again a variety of both metallics and nonmetallics would be suitable for this type of application in the absence of the titanium dioxide solids. But the presence of the solids circulating in a pump renders practically all of the normal sulfuric acid–resisting pump materials unsuitable. Special consideration must be given to evaluation of materials that will resist the severe erosion-corrosion encountered in this type of service.

Sulfuric Acid Containing Nitric Acid, Ferric Sulfate, or Cupric Sulfate. The presence of these compounds in sulfuric acid solutions will drastically alter the suitability of materials that can be used. Their presence in quantities of 1 percent or less can make a sulfuric acid solution oxidizing, whereas it would normally be reducing. Their presence, singly or in combination, could serve as a corrosion inhibitor, thus in certain instances allowing a stainless steel, such as Type 316, to be used. On the other hand, the same compounds could serve as a corrosion accelerator for a nonchromium bearing alloy such as nickel-molybdenum alloys; this would render it completely unsuitable.

Nitric Acid In the concentrations normally encountered in chemical applications, nitric acid presents less problems than the many and varied sulfuric acids services. The metallic material choices available for various nitric applications are somewhat broader than those of the nonmetallic materials. Nitric acid, being a strongly oxidizing acid, permits the use of stainless steel quite extensively; but its oxidizing characteristics restrict the application of nonmetallics in general and plastics in particular. Requiring special evaluation are such aggressive solutions as fuming nitric acid, nitric-hydrofluoric combinations, nitric-hydrochloric (some of which would fall into the aqua regia category), nitric-adipic combinations, and practically any environment consisting of nitric acid in combination with other constituents. Invariably additional constituents in nitric acid result in more aggressive corrosion; hence material selection becomes quite critical.

Hydrochloric Acid Both commercially pure and contaminated hydrochloric acids present difficult situations in selecting pump materials. The most common contaminant that creates problems with hydrochloric acid–handling equipment is ferric chloride. The presence of ferric chloride can render this otherwise reducing solution oxidizing, and thus completely change the material of construction that can be used. Addition of a very few parts per million of iron to commercially pure hydrochloric acid can result in the formation of ferric chloride to a sufficient extent to cause materials such as nickel-molybdenum, nickel-copper, and zirconium to be completely unsuitable. Conversely, the presence of ferric chloride can make titanium completely suitable. Nonmetallics find extensive application in many hydrochloric acid environments. Often the limiting factors for the nonmetallics are temperature, mechanical properties, and suitability for producing pump parts in the design desired. With the nonmetallics, the near-complete immunity from corrosion in such environments subordinates corrosion resistance to other factors.

Phosphoric Acid The increasing use and demand for all types of fertilizers have made phosphoric acid a very important commodity. In the wet process of producing phosphoric acid, the phosphate rock normally contains fluorides. In addition, at various stages of the operation the solution will also contain varying quantities of sulfuric, hydrofluoric, fluosilicic, and phosphoric acids as well as solids. In some instances the water used in these solutions may have exceptionally high chloride content, which in turn can result in the formation of hydrochloric acid, which further aggravates the corrosion problem. It is also typical for certain of these solutions to contain solids; these, of course, create an erosion-corrosion problem. Pure phosphoric and superphosphoric acids are relatively easy to cope with from a material standpoint; but when the solution contains all or some of the aforementioned constituents, a very careful material evaluation program must be conducted. Such environments are not only severely corrosive in the absence of solids, but cause severe erosion-corrosion and a drastically reduced service life when solids are present. This is particularly significant with any type of chemical pump.

Chlorine Little need be said about the corrosivity of chlorine. Wet chlorine, in addition to being extremely hazardous, is among the most corrosive environments known. Dry chlorine is not corrosive, but there are those who contend that dry chlorine does not exist. Chlorine vapor combined with the moisture in the atmosphere, for instance, can create severe corrosion problems. In any case, selecting the most suitable material for any type of chlorine environment requires very careful evaluation.

Alkaline Solutions With some exceptions, alkaline solutions such as sodium hydroxide or potassium hydroxide do not present serious corrosion problems at temperatures below 200°F. However, in certain applications, purity of product is of utmost concern, necessitating selection of a material that will have essentially a nil corrosion rate. Among the exceptions to this rule that alkaline solutions are relatively noncorrosive are bleaches, alkaline brines, and other solutions containing chlorine in some form.

Organic Acids Compared to the inorganic acids, organic acids are much less corrosive. This does not mean, however, that they can be taken lightly. For instance, organic acids such as acetic, lactic, formic, and maleic have their corrosive characteristics and must be treated accordingly when evaluating metallics and nonmetallics.

Salt Solutions Normally considered as having a neutral pH, salt solutions do not present a serious corrosion problem. In some instances, process streams have a pH adjustment to maintain the pH on the slightly alkaline side. These solutions are even less corrosive than when they are neutral. On the other hand, when a process stream has a pH adjustment to maintain a slightly acidic environment, the liquid becomes considerably more corrosive than ordinary salt solutions. This condition requires that more effort be expended in evaluating the solution before making a material selection.

Organics Most organics do not present corrosion problems of the same magnitude as inorganics. This does not mean that any material arbitrarily selected will be a suitable choice. It does mean that there will be more materials available to choose from, but each application should be considered on its own merits. Of particular concern in this area are chlorinated organics and those that will produce hydrochloric acid when moisture is present. Plastics, categorically, possess excellent corrosion resistance to inorganics within the temperature limitations for which they can be used, but they do exhibit some weaknesses in their corrosion resistance to organics.

Water Compared to most mediums encountered in chemical and allied industries, water is less corrosive. For the term "water" to be meaningful, however, it is extremely important to know the specific kind of water. Types includes demineralized, fresh, brackish, salt, boiler feed, and mine water. These waters and the various constituents in them can demand a variety of materials. The broad range of materials required for various waters is indicated by the almost complete spectrum of materials being studied and used in the various desalination programs. These programs are likely to have a very pronounced effect on our total economy, and precise material evaluation and selection are an integral part of such programs.

TYPES OF PUMP CORROSION

The types of corrosion encountered in chemical pumps may, at first, appear to be unusual compared with those found in other process equipment. Nevertheless pumps, like any other type of chemical process equipment, will experience basically only eight forms of corrosion, of which some are more predominant or characteristic in pumps than in other types of equipment. It is not the intent of this section to describe in detail these eight forms of corrosion, but it is desirable to enumerate the eight forms of corrosion and provide a brief description of each so that they can be recognized when they occur.

1. General or Uniform Corrosion This is the most common type, and it is characterized by essentially the same rate of deterioration over the entire wetted or exposed surface. General corrosion may be very slow or very rapid, but it is of less concern than the other forms of corrosion because of its predictability. However, predictability of general corrosion within a pump can be a difficult task, because of the varying velocities of the solution within the pump.

2. Concentration Cell or Crevice Corrosion This is a localized form of corrosion resulting from small quantities of stagnant solution in areas such as threads, gasket surfaces, holes, crevices, surface deposits, and under bolt and rivet heads. When concentration cell or crevice corrosion occurs, a difference in concentration of metal ions or oxygen exists in the stagnant area as compared with the main body of the liquid. This causes electrical current to flow between the two areas, resulting in severe localized attack in the stagnant area. Usually this form of corrosion does not occur in chemical pumps except in misapplications, or in designs where the factors known to contribute to concentration cell corrosion have been ignored.

3. Pitting Corrosion This is the most insidious and destructive form of corrosion, and it is very difficult to predict. It is extremely localized, is manifested by small or large holes (usually small), and the weight loss due to the pits will be only a small percentage of the total weight of the equipment. Chlorides in particular are notorious for inducing pitting, which can occur in practically all types of equipment. This form of corrosion in some instances can be closely allied to concentration cell corrosion, since pits may initiate in the same areas where concentration cell corrosion appears. Pitting is common at areas other than stagnant areas, whereas concentration cell corrosion is basically confined to areas of stagnation.

4. Stress Corrosion Cracking This is localized failure caused by a combination of tensile stresses in a specific medium. Undoubtedly more research and development have been conducted on this form of corrosion than any of the others. Nevertheless, the exact mechanism for the occurrence of stress corrosion cracking is still not well understood. Fortunately castings, because of their basic overdesign, seldom experience stress corrosion cracking. Corrosion fatigue, which can be classified as stress corrosion cracking, is of concern in chemical-pump shafts because of the repeated cyclic stressing. Failures of this type occur at stress levels below the yield point as a result of the cyclic application of the stress.

5. Intergranular Corrosion This is a selective form of corrosion at, and adjacent to, grain boundaries. It is associated primarily with stainless steels, but can also occur with other alloy systems. In stainless steels, it results from subjecting the material to heat in the 800 to 1600°F temperature range. Unless other alloy adjustments are made, this form of corrosion can be prevented only by heat treating. It is easily detectable in castings, because the actual grains are quite large by comparison with wrought material of equivalent composition. In some instances, uniform corrosion is misinterpreted as intergranular corrosion because of the etched appearance of the surfaces exposed to the environment. Even in ideally heat-treated stainless steels, very slight accelerated attack can be noticed at the grain boundaries because these areas are more reactive than the grains themselves. The untrained eye should be careful to avoid confusing general and intergranular corrosion. Stainless steel castings will never encounter intergranular corrosion if they are properly heat-treated after being exposed to temperatures in the 800 to 1600°F range.

6. Galvanic Corrosion This is a type of corrosion that occurs when dissimilar metals are in contact or are otherwise electrically connected in a corrosive medium. Corrosion of the less noble metal is accelerated, and corrosion of the more corrosion-resistant metal is decreased as compared with their behavior when not in contact. The farther apart the metals or alloys are in the electromotive series, the greater the possibility of galvanic corrosion. When it is found necessary to have two dissimilar metals in contact, caution should be exercised to make certain that the total surface area of the least resistant metal far exceeds that of the more corrosion-resistant material. This will tend to prevent premature failure by simply providing a substantially greater area of the more corrosion-prone material. This form of corrosion is not common in chemical pumps, but may be of some concern with accessory items that may be in contact wtih the pump parts, and subjected to the environment.

7. Erosion-Corrosion This type of failure is characterized by accelerated attack resulting from the combination of corrosion and mechanical wear. It may involve solids in suspension and/or high velocity. It is quite common with pumps where the erosive effects prevent the formation of a passive surface on alloys that require passivity to be corrosion resistant. The ideal material to avoid erosion-corrosion in pumps would possess the characteristics of corrosion resistance, strength, ductility, and extreme hardness. Few materials possess such a combination of properties.

8. Selective Leaching Corrosion This, in essence, involves removal of one element from a solid alloy in a corrosive medium. Specifically, it is typified by dezincification, dealuminumification, and graphitization. This form of attack is not common in chemical-pump applications because the alloys in which it occurs are not commonly used in heavy chemical applications.

TYPES OF CHEMICAL PUMPS

The second step in selecting a chemical pump is to determine which type of pump is required, based on the characteristics of the liquid and on the desired head and capacity. It should also be noted that not all types are available in every material of construction, and the final selection of pump type may depend on availability of designs in the proper material.

Centrifugal Pumps Centrifugals are used extensively in the chemical industry because of their suitability for use in practically any service. They are available in an almost unending array of corrosion-resisting materials. While not built in extremely large sizes, pumps with capacity ranges of 5,000 to 6,000 gpm are commonplace. Heads range as high as 500 to 600 ft at standard electric motor speeds. Centrifugals are normally mounted in the horizontal position, but they may also be installed vertically, suspended into a tank, or hung in a pipeline similar to a valve. They are simple, economical, dependable, and efficient. Disadvantages include reduced performance when handling liquids of more than 500 SSU viscosity and the tendency to lose prime when comparatively small amounts of air or vapor are present in the liquid.

Rotary Pumps The gear, screw, deforming-vane, sliding-vane, axial-piston, and cam types are used for high-pressure service. They are particularly adept at pumping liquids of high viscosity or low vapor pressure. Their constant displacement at a set speed makes them ideal for use in metering small quantities of liquid. Since they operate on the positive displacement principle, they are inherently self-priming. When built of materials that tend to gall or seize on rubbing contact, the clearances between mating parts must be increased, with the result of decreased efficiency. The gear, sliding-vane, and cam units are generally limited to use on clear, nonabrasive liquids.

Diaphragm Pumps These units are also classed as positive displacement pumps since the diaphragm acts as a limited displacement piston. Pumping action is obtained when the diaphragm is forced into reciprocating motion by mechanical linkage, compressed air, or oil from a pulsating external source. This type of construction eliminates any connection between the liquid being pumped and the source of energy, and thereby eliminates the possibility of leakage. This characteristic is of great importance when handling toxic or very expensive liquids. Disadvantages include a limited selection of corrosion-resisting materials, limited head and capacity range, and the necessity of using check valves in the suction and discharge nozzles. Construction details are shown in Fig. 1.

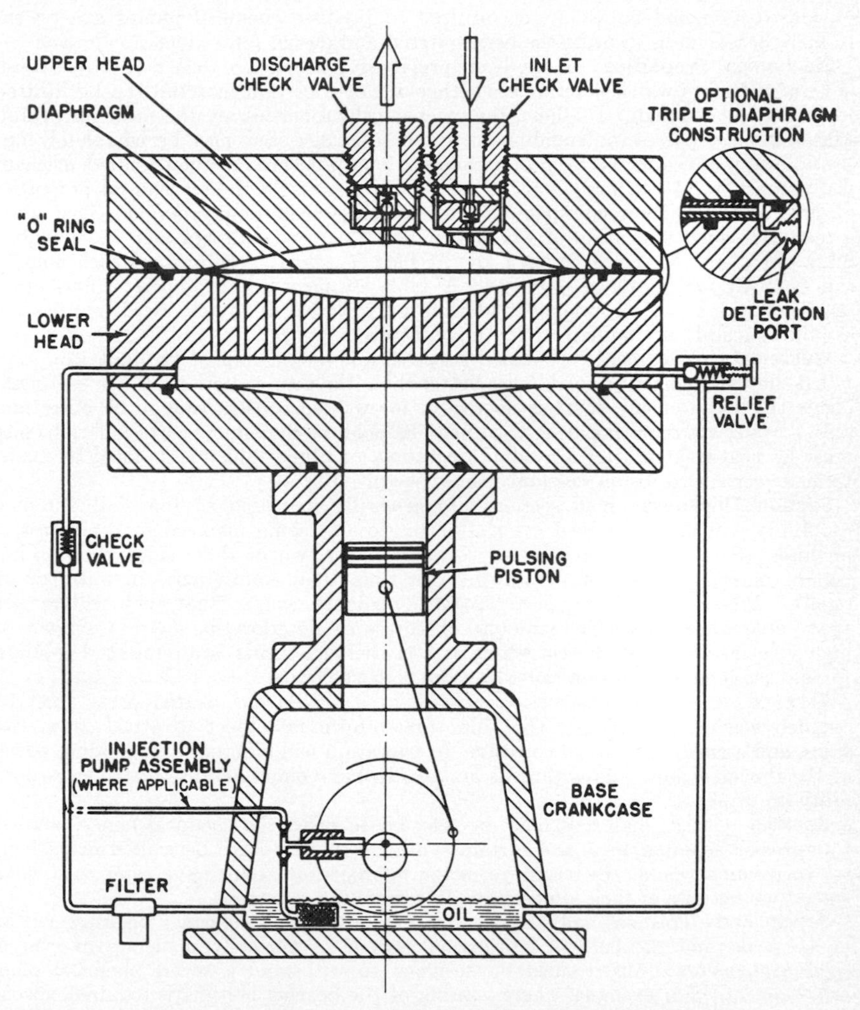

Fig. 1 Construction details of a diaphragm-type pump.

Regenerative Turbine Pumps Flow rates up to 100 gpm and heads up to 700 ft are easily handled with this type of pump. When it is used for chemical service, the internal clearances must be increased to prevent rubbing contact, which results in decreased efficiency. These pumps are generally unsuitable for solid-liquid mixtures of any concentration.

CHEMICAL-PUMP DESIGN CONSIDERATIONS

Casting Integrity Practically all the major components of chemical pumps are castings. Needless to say, it is a fruitless venture to evaluate thoroughly the detailed characteristics of the solution and the material to be used if the component castings do not satisfy the quality needed to provide good service life. This is probably of more concern in chemical-pump applications than in any other type of service application because leakage, loss of product, and downtime can be extremely costly, as well as very dangerous. Quality products in any industry are no accident, but are the result of concentrated effort. When producing pumps for chemical applications, whether these pumps be solid metallics, solid nonmetallics, or lined, there is no excuse for sacrificing quality. Being acquainted with materials of construction and how they are utilized to produce chemical pumps can be extremely beneficial in securing the best material and design for a given application.

Mechanical Properties It has been previously mentioned that there are several governing factors which determine whether or not a certain material can be utilized for a particular design. Of these, the mechanical properties are the most important. Materials may possess outstanding corrosion resistance, but may be completely impossible to produce in the form of a chemical pump because of their limited mechanical properties. Accordingly, it is advisable to be aware of the mechanical properties of any material being considered in a corrosion-evaluation program. This gives a relatively good indication of whether or not a particular design may be available. Since most materials are covered by ASTM or other specifications, such sources can be used for reference purposes. A table of mechanical properties and other characteristics and of proprietary materials not included in any standard specification should be readily available from the manufacturer.

Weldments Weldments or welded construction should impose no limitation, provided the weldment is as good as, or better than, the base material. If the weldment meets this requirement, there is no reason for welded construction to be objectionable. Materials requiring heat treatment to achieve maximum corrosion resistance must be heat-treated after a welding operation, or other adjustments must be made, to make certain corrosion resistance is not sacrificed.

Section Thickness Wall sections are generally increased so that full pumping capability will be maintained even after the loss of some material to the corrosive medium. Parts that are subject to corrosion from two or three sides, such as impellers, must be made considerably heavier than their counterparts in water or oil pumps. Pressure-containing parts are also made thicker so that they will remain serviceable after a specified amount of corrosive deterioration. Areas subject to high velocities, such as the cut water of a centrifugal casing, are reinforced to allow for the accelerated corrosion caused by the high velocities.

Threads Threaded construction of any type within the wetted parts must be avoided whenever possible. The thin thread form is subject to attack from two sides, and a small amount of corrosive deterioration will eliminate the holding power of the threaded joint. Pipe threads are also to be avoided because of their susceptibility to attack.

Gaskets Gasket materials must be selected to resist the chemical being handled. Compressed asbestos, lead, and certain synthetic rubbers have been used extensively for corrosion services. In recent years the fluorocarbon resins have come into widespread use because of their almost complete corrosion resistance.

Power End This assembly, consisting of the bearing housing, bearings, oil or grease seals, and the bearing lubrication system, is normally made up of iron or steel components; thus it must be designed to withstand a severe chemical plant environment. For example, where venting of the bearing housing is required, special means of preventing the entrance of water, chemical fumes, or dirt must be incorporated into the vent design.

The bearing which controls axial movement of the shaft is usually selected to limit shaft movement to 0.002 in or less. End play values above this limit have been found detrimental to mechanical seal operation.

Water jacketing of the bearing housing may be necessary under certain conditions to maintain bearing temperatures below 225°F, the upper limit for standard bearings.

Maintenance Maintenance of a chemical pump in a corrosive environment can be a very costly and time-consuming item. It can be divided into two categories—preventive and emergency. When evaluating materials and design factors, maintenance aspects should be high on the priority list. The ease and frequency of maintenance are two very critical items, and should be considered part of a preventive maintenance program. Such a program can be the most effective way of eliminating emergency shutdown caused by pump failure. Furthermore, the knowledge gained in a routine preventive maintenance program can be of unlimited value when a breakdown does occur, since repair personnel will have acquired a thorough knowledge of the construction details of the pump. This can be very beneficial in minimizing downtime.

PROPER STUFFING BOX DESIGN

The area around the stuffing box probably causes more chemical-pump failures than all other parts combined. The problem of establishing a seal between a rotating shaft and the stationary pump parts is one of the most intricate and vexing problems facing the pump designer.

Packings Braided asbestos, lead, fluorocarbon resins, aluminum, graphite, and many other materials or combinations of these materials have been used to establish the seal. Inconsistent as it seems, a small amount of liquid must be allowed to seep through the packing and lubricate the surface between packing and shaft. This leakage rate is hard to control; the usual result is overtightening of the packing, stopping the leak. The unfortunate result of this condition is rapid scoring of the shaft surface, making it much harder to adjust the packing to the proper tension. Recommendations as to the type of packing to be used for various chemical services should come from the packing manufacturer.

Mechanical Seals Mechanical shaft seals as described elsewhere in this manual are used extensively on chemical pumps. Once again the primary consideration is selection of the proper materials for the type of corrosive being pumped. Stainless steels, ceramics, graphite, and fluorocarbon resins are used to make most of the seal parts. Several of the large manufacturers of this type of equipment have developed very complete files on seal designs for various chemical services; they are therefore in the best position to make recommendations.

Temperature One of the most important factors affecting the stuffing box sealing medium is the operating temperature. Most packings are impregnated with a grease for lubrication purposes, but these lubricants break down at temperatures above 250°F, causing further temperature increases. The friction of the packing rubbing on the shaft can easily generate temperatures of this magnitude. One of the not-so-obvious results of this temperature increase is corrosive attack on the pump parts in the heat zone. Many materials selected for the pumping temperature will be completely unsuitable in the presence of the corrosive at elevated stuffing box temperatures. Another source of heat is, of course, the chemical solution itself. These liquids often are in the 300°F range, and some go as high as 700°F.

The best answer to the heat problem is removal of the heat by means of a water jacket around the stuffing box. While the heat conductivity is rather low for most chemical pump materials, the stuffing box area can generally be maintained in the 200° to 250°F range. This cooling is of further benefit in that it prevents the transfer of heat along the shaft to the bearing housing, thereby eliminating other problems around the bearings.

Pressure Stuffing box pressure varies with suction pressure, impeller design, and the degree of maintenance of close-fitting seal rings. Variations in impeller design would include those using vertical or horizontal seal rings in combination with balance ports, as opposed to those using back vanes or pump-out vanes. All impeller designs depend upon a close running clearance between the impeller and the stationary pump parts. This clearance must be kept as small as possible to prevent excessive recirculation of the liquid, and the resulting loss of efficiency. Unfortunately most chemical pump materials tend to seize when subjected to rubbing contact, and the running clearances must therefore be increased considerably above those used in pumps for other industries.

At pressures above 100 lb/in^2, packing is generally unsatisfactory unless the stuffing box is very deep and the operator is especially adept at maintaining the proper gland pressure on the packing. Mechanical seals incorporating a balancing feature to relieve the high face pressure are the best means of sealing at pressures above 100 lb/in^2.

Shaft The pump shaft itself can create additional stuffing box problems. Obviously a shaft that is out-of-round or bent will form a larger hole in the packing than a perfect shaft will fill, thereby allowing liquid to escape. Lack of static or hydraulic balance in the impeller produces a dynamic bend in the shaft, resulting in the same condition. Undersize shafts, or those made of materials that bend readily, will deflect from their true center in response to radial thrust on the impeller. This action produces a secondary hole in the packing and again allows the liquid to escape.

Mechanical seal operation is also impaired when the shaft is bent or deflects during operation. Since the flexible member of the seal must adjust with each revolution of the shaft, excessive deflection results in shortened seal life. If the deflection is of more than nominal value, the flexible seal member will be unable to react with sufficient speed to keep the seal faces together, allowing leakage at the mating faces.

An arbitrary limit of 0.002 in has been established as the maximum deflection or runout of the shaft at the face of the stuffing box consistent with good pump design.

Shaft Surface In the stuffing box region, the shaft surface must have corrosion resistance at least equal to and preferably better than that of the wetted parts of the pump. In addition, this surface must be hard enough to resist the tendency to wear under the packing or mechanical seal parts. Further, it must be capable of withstanding the sudden temperature changes often encountered in operation.

Since it is not economically feasible to make the entire pump shaft of stainless alloys, and physically impossible to make functional carbon, glass, or plastic shafts, chemical pumps often have carbon steel shafts with a protective coating or sleeve over the steel in the stuffing box area. Cylindrical sleeves are sometimes made so that they may be removed and replaced when they become worn. Other designs utilize sleeves that are permanently bonded to the shaft to obtain lower runout and deflection values.

Another method of obtaining a hard surface in this region is the welded overlay or spray coating of hard metals onto the base shaft. These materials are generally lacking in corrosion resistance, however, and they have not been widely accepted for severe chemical service. Ceramic materials applied by the plasma spray technique possess excellent corrosion resistance, but cannot achieve the complete density required to protect the underlying shaft.

Composite shafts utilizing carbon steel for the power end and a higher alloy for the wet end have been used extensively where the high-alloy end has acceptable corrosion resistance. The two ends are joined by various welding techniques, and the combination of metals is, therefore, limited to those that can be easily welded together. On such assemblies, the weld joint and the heat affected zone must be outside the wetted area of the shaft.

Other Designs Elimination of the stuffing box and its associated problems has been the object of several chemical-pump designs other than the diaphragm pumps previously mentioned.

Vertical submerged pumps utilize a sleeve-type bearing in the area immediately above the impeller to limit the flow of liquid up along the shaft. For chemical service, the problem of materials associated with this bearing, and its lubrication, have been major disadvantages.

Canned pumps, wherein the motor windings are hermetically sealed in stainless steel "cans," also avoid the use of a stuffing box. The pumped liquid is circulated through the motor section, lubricating the sleeve-type bearings which support the entire rotating assembly. Disadvantages have again centered around the selection of bearing materials compatible with the corrosive liquid; the lubrication of these bearings when handling nonlubricating liquids; and the probability of clogging the liquid paths through the motor section when handling solid-liquid mixtures.

DESIGNING WITH SPECIAL MATERIALS

As mentioned previously, a number of low-mechanical-strength materials have been used extensively in chemical-pump construction. While breakage problems are inherently associated with these materials, their excellent corrosion resistance has allowed them to remain competitive with higher strength alloys. Their low tensile strength and brittleness make them sensitive to tensile or bending stresses, requiring special pump designs. The parts are held together by outside clamping means, and are braced to prevent bending. The unit must also be protected from sudden temperature changes and from mechanical impact from outside sources.

High Silicon Iron (14.2 to 14.7 Percent Silicon) Although produced by very few manufacturers, high silicon iron is the most completely corrosion-resistant metallic available at an economical price. This resistance, coupled with a hardness of approximately 520 Brinell, provides an excellent material for handling abrasive chemical slurries. The same hardness, however, precludes normal machining operations, and the parts must be designed for machine grinding. It also eliminates the possibility of using drilled or tapped holes for the connection of piping to the pump parts. Special designs are therefore required for process piping, stuffing box lubrication, and drain connections.

Ceramics and Glass These materials are similar to high silicon iron in regard to hardness, brittleness, and susceptibility to thermal or mechanical shock. Pump designs must therefore incorporate the same special considerations.

Glass linings or coatings on iron or steel parts are sometimes used to eliminate some of the undesirable characteristics of solid glass. While this method provides means for connecting process piping and protects the glass from mechanical shock, the dissimilar expansion characteristics of the two materials generate small cracks in the glass, allowing the corrosive liquid to attack the armor material.

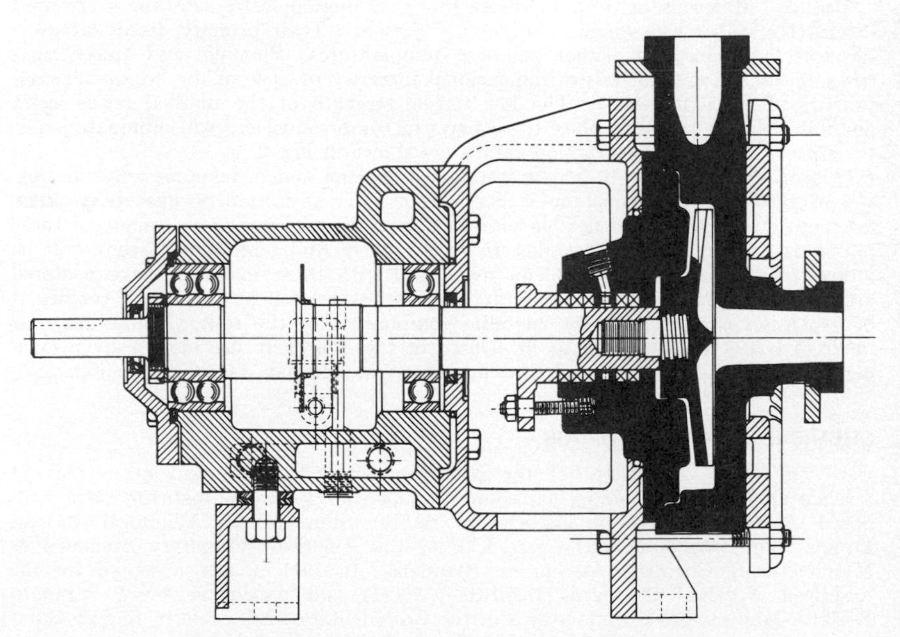

Fig. 2 Construction details of a solid plastic pump.

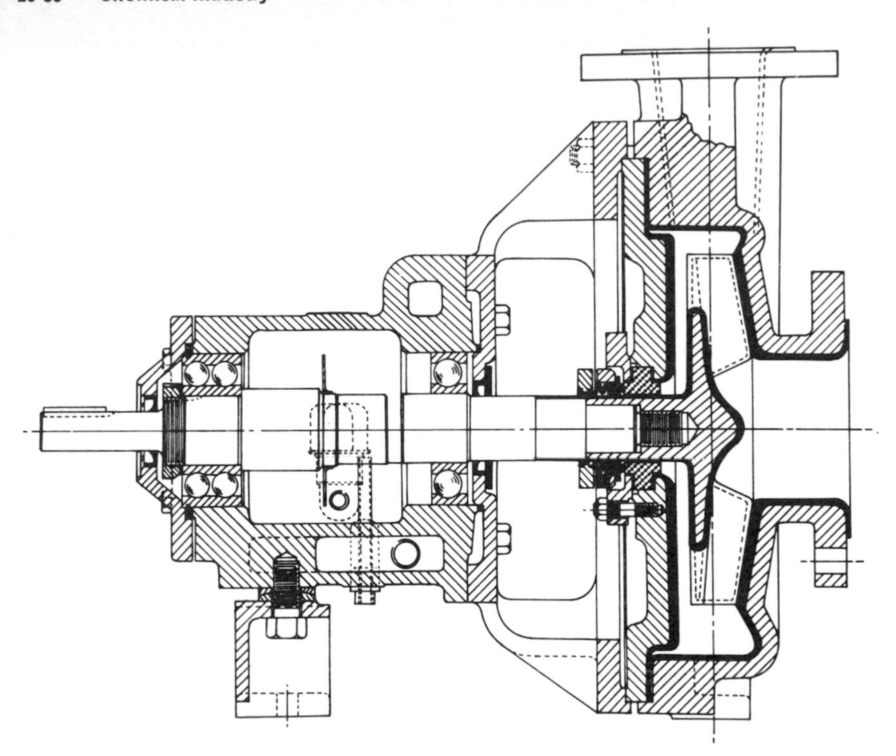

Fig. 3 Metallic pump with plastic or other nonmetallic lining.

Plastics Materials of both thermosetting and thermoplastic substances are used extensively in services where chlorides are present. Their primary disadvantage is the loss in strength at higher pumping temperatures. Phenolic and epoxy parts are subject to a gradual loss of dimensional integrity because of the "creep" characteristics of these materials. The low tensile strength of the unfilled resins again dictates a design that will place these parts in compression and will eliminate bending stresses. Typical construction details are shown in Fig. 2.

Fluorocarbon Resins Both polytetrafluoroethylene and hexafluoropropylene possess excellent corrosion resistance. These resins have been used for gaskets, packing, mechanical seal parts, and flexible piping connectors. Several pumps made of these materials have reached the market in recent years, and undoubtedly they will be followed by many more. Problems associated with these materials have centered around their tendency to cold-flow under pressure, as well as their high coefficient of expansion compared to the metallic components of the unit. Pumps may be made of heavy solid sections as illustrated in Fig. 2 or may use more conventional metallic components lined with the fluorocarbon material, as shown in Fig. 3.

CHEMICAL PUMP STANDARDS

Early in 1962 a committee of the Manufacturing Chemists Association (MCA) reached agreement with a special committee of the Hydraulic Institute on a Proposed American Standards Association (ASA) Standard for Chemical Process Pumps. This document was referred to as the American Voluntary Standard or Manufacturing Chemists Association Standard. In 1971 it was accepted by the American National Standards Institute (ANSI) and issued as ANSI Standard B123.1. Many pump manufacturers in the United States and a few in foreign countries are building pumps which meet these dimensional and design criteria.

It is the intent of this standard that pumps of similar size from all sources of supply shall be dimensionally interchangeable with respect to mounting dimensions, size and location of suction and discharge nozzles, input shafts, base plates, and foundation bolts. Table 1 lists the pump dimensions that have been standardized. A cross-section assembly of a pump meeting these criteria is shown in Fig. 4.

It is also the intent of this standard to outline certain design features which will minimize maintenance problems. The standard states, for instance, that the pump shaft should be sized so that the maximum shaft deflection, measured at the face of the stuffing box when the pump is operating under its most adverse conditions, will not exceed 0.002 in. It does not specify shaft diameter since impeller diameter, shaft length, and provision for operation with high specific gravities would determine the proper diameter.

The standard also states that the minimum bearing life, again under the most adverse operating conditions, should be not less than two years. Bearing size is not specified but is to be determined by the individual manufacturer and will be dependent upon the load to be carried.

Additional specifications in the standard include hydrostatic test pressure, shaft finish at rubbing points, packing space, and other features.

Other dimensional standards are in use in foreign countries on both horizontal and vertical pumps. In 1971 the International Organization for Standardization (ISO) reached agreement on a set of dimensional standards for horizontal, end suction, centrifugal pumps. This document, in metric units, describes a series of pumps of slightly greater capacity range than described in B123.1. It does not include design criteria as to minimum shaft deflection, minimum bearing life, or other characteristics required to reduce maintenance.

The British Standards Institution issued British Standard 4082 in 1966 to describe a series of vertical in-line type centrifugal pumps. While dimensional interchangeability was the primary reason for this standard, it also includes requirements for hydrostatic testing of the pump parts. It is made up of two sections: Part 1 cover-

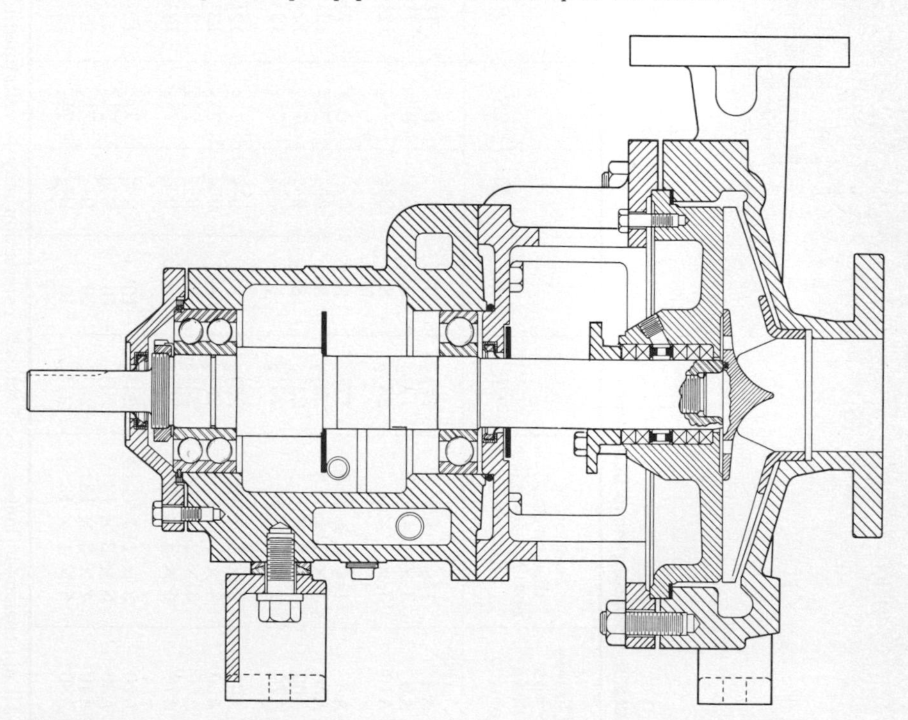

Fig. 4 Cross-section assembly of a typical pump made to the ANSI B123.1 Standard.

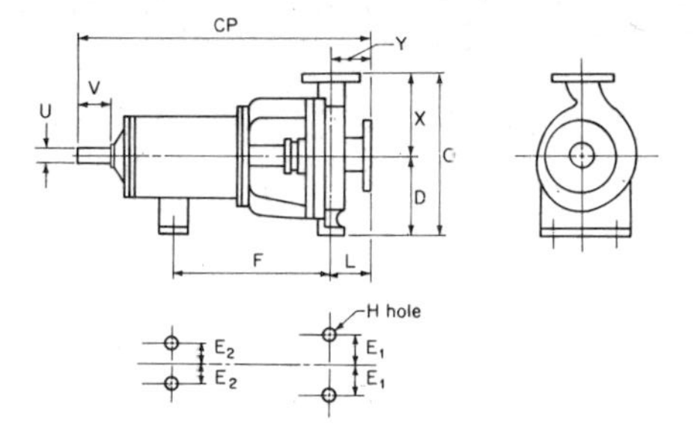

TABLE 1 Standard Pump Dimensions

Dimension designation	Suction × discharge × nominal impeller dia., in.	CP	D	$2E_1$	$2E_2$	F	H	O	U Dia.	U Keyway	V, min	X	Y
AA	1½ × 1 × 6	17½	5¼	6	0	7¼	⅝	11¾	⅞	3/16 × 3/32	2	6½	4
AB	3 × 1½ × 6	17½	5¼	6	0	7¼	⅝	11¾	⅞	3/16 × 3/32	2	6½	4
A10	3 × 2 × 6	23½	8¼	9¾	7¼	12½	⅝	16½	1⅛	¼ × ⅛	2⅝	8¼	4
AA	1½ × 1 × 8	17½	5¼	6	0	7¼	⅝	11¾	⅞	3/16 × 3/32	2	6½	4
A50	3 × 1½ × 8	23½	8¼	9¾	7¼	12½	⅝	16¾	1⅛	¼ × ⅛	2⅝	8½	4
A60	3 × 2 × 8	23½	8¼	9¾	7¼	12½	⅝	17¾	1⅛	¼ × ⅛	2⅝	9½	4
A70	4 × 3 × 8	23½	8¼	9¾	7¼	12½	⅝	19¼	1⅛	¼ × ⅛	2⅝	11	4
A05	2 × 1 × 10	23½	8¼	9¾	7¼	12½	⅝	16¾	1⅛	¼ × ⅛	2⅝	8½	4
A50	3 × 1½ × 10	23½	8¼	9¾	7¼	12½	⅝	16¾	1⅛	¼ × ⅛	2⅝	8½	4
A60	3 × 2 × 10	23½	8¼	9¾	7¼	12½	⅝	17¾	1⅛	¼ × ⅛	2⅝	9½	4
A70	4 × 3 × 10	23½	8¼	9¾	7¼	12½	⅝	19¼	1⅛	¼ × ⅛	2⅝	11	4
A20	3 × 1½ × 13	23½	10	9¾	7¼	12½	⅝	20½	1⅛	¼ × ⅛	2⅝	10½	4
A30	3 × 2 × 13	23½	10	9¾	7¼	12½	⅝	21½	1⅛	¼ × ⅛	2⅝	11½	4
A40	4 × 3 × 13	23½	10	9¾	7¼	12½	⅝	22½	1⅛	¼ × ⅛	2⅝	12½	4
A80*	6 × 4 × 13	23½	10	9¾	7¼	12½	⅝	23½	1⅛	¼ × ⅛	2⅝	13½	4

All dimensions in inches.

* Suction connection may have tapped bolt holes.

ing pumps wherein the suction and discharge nozzles are in a horizontal line (the I type), and Part 2 covering pumps wherein the nozzles are on the same side of the pump parallel to each other (the U configuration).

REFERENCES

Lee, James A.: "Materials of Construction for Chemical Process Industries," McGraw-Hill Book Company, New York, 1950.

Fontana, M. G., and N. D. Greene: "Corrosion Engineering," McGraw-Hill Book Company, New York, 1967.

"Corrosion Data Survey," National Association of Corrosion Engineers, Houston, Tex., 1967.

"Proceedings, Short Course on Process Industry Corrosion," National Association of Corrosion Engineers, Houston, Tex., 1960.

See Sec. 5 of this handbook.

Section 10.7

Petroleum Industry

A. W. ELVITSKY

USE OF PUMPS

Pumps of all types are used in every phase of petroleum production, transportation, and refining.

Production pumps include reciprocating units for mud circulation during drilling operations and sucker-rod, hydraulic rodless, or motor-driven submersible centrifugal units for lifting the crude to the surface. The largest use of centrifugal pumps in production service is for water flooding (secondary recovery, subsidence prevention, or pressure maintenance).

Transportation pumps include units for gathering, on and offshore production, pipe-lining of crude and refined products, loading and unloading of tankers, tank cars, barges, or tank trucks, and airport fueling terminals. The majority of the units are of a centrifugal type.

Refining units vary from single-stage centrifugal units to horizontal and vertical multistage barrel-type pumps handling a variety of products over a full range of temperatures and pressures. Centrifugal pumps are also used for auxiliary services such as cooling towers and cooling water.

CENTRIFUGAL PUMPS

Except for some comments about the use of displacement pumps for handling viscous liquids, this chapter is restricted to centrifugal pumps—the type most frequently used—and includes a detailed analysis of the principal types.

Refinery Pumps Major refinery processes that utilize centrifugal pumps are crude distillation, vacuum tower separation, catalytic conversion, alkylation, hydrocracking, catalytic reforming, coking, and hydrotreatment for the removal of sulfur and nitrogen. The products resulting from these processes include motor gasoline, commercial jet fuel and kerosene, distillate fuel oil, residual fuel oil, and lubricating oils. The American Petroleum Institute Standard 610, which is periodically revised, has established minimal specifications for the design features required for centrifugal pumps used for general refinery service.

Construction. Figure 1 illustrates the details of a top-suction, single-stage, double-volute, center-line mounted, overhung, radially split case refinery process pump meeting API 610 requirements. The stuffing box is equipped with a mechanical seal. Suction-nozzle location may be either end or top. By utilizing a spacer-type coupling between the pump shaft and driver shaft, the bearing bracket and cover may be removed without disconnecting the suction or discharge piping. This disassembly feature is common to all present-day refinery pumps. Use of the overhung

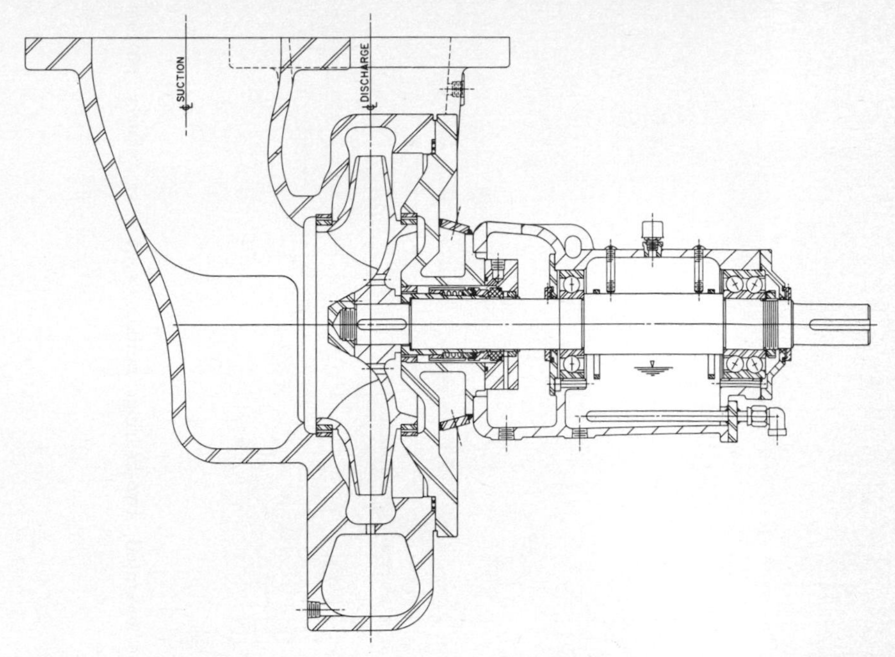

Fig. 1 Top suction, single-stage, double-volute, center-line-mounted, overhung, radially split case refinery process pump with mechanical seal. (United Centrifugal Pumps)

shaft design results in having only one stuffing box to seal from leakage to the atmosphere. The thrust bearing for this arrangement, however, is subject to a load caused by exposure of one end of the shaft to suction pressure. To meet the API Standard 610, the bearings must be designed for a minimum life of 16,000 hr.

For suction pressures in excess of 250 lb/in^2 gage, a common practice is to decrease the diameter of the back wearing rings, as compared to the suction side wearing rings, or eliminate them completely, in order to produce an equalizing thrust in the opposite direction and decrease the net load on the thrust bearing. Where casting limitations permit, pumps of double-volute construction are used to limit the radial load imposed on the impeller from uneven pressure distributions within the casing. To comply with API 610, the shaft deflection at the wearing rings for one- and two-stage pumps, under most severe dynamic conditions, must be limited to one half the minimum diametrical clearance specified. For operating temperatures above 500°F, the API recommends that consideration be given to increasing the minimum specified clearances. This is a necessary requirement if dissimilar metals are used or if a rapid warm-up procedure is to be employed. To aid in maintaining alignment at various temperatures, pump mounting feet are located on the case on the same centerline as the shaft. For pumping temperatures over 400°F, bearing cooling or stuffing-box cooling is used to limit heat transfer along the shaft and/or lower the environmental temperature of the mechanical seal or packing. For NPSH requirements lower than the capability of a single-suction design, double-suction pumps are available for capacities above 600 gpm.

Overhung shaft construction is nominally limited by most manufacturers to pumps with drivers below 500 hp. Units with greater horsepower are designed with bearings on both ends of the shaft, with the impeller or impellers in between (commonly designated as two-bearing or inboard-outboard bearing design). Ball-bearing construction in compliance with API 610 is used up to a limit of a DN factor of 300,000. The DN factor is the product of the bearing bore (D) in millimeters

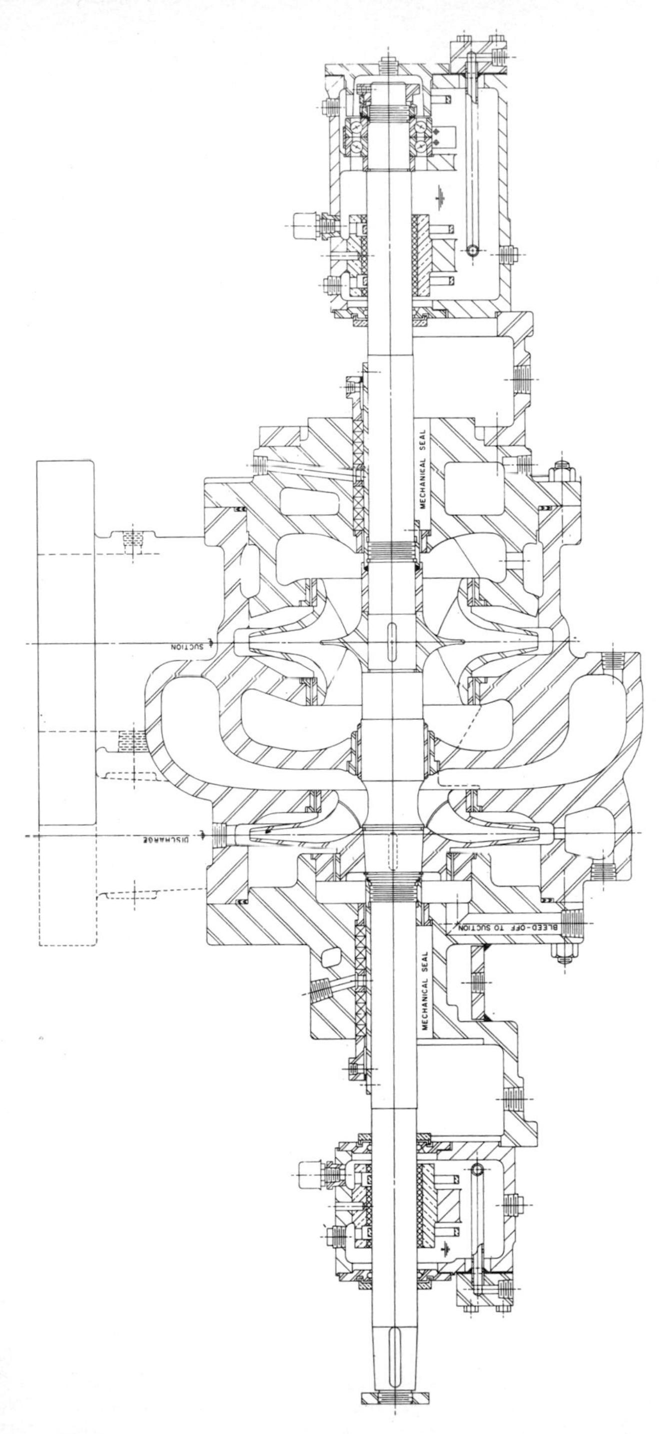

Fig. 2 Top-suction, double-suction first-stage, two-stage, center-line mounted, two-bearing, radially split case refinery process pump. (United Centrifugal Pumps)

times the operating speed (N) in rpm. For values above 300,000, double-bearing construction is used with sleeve radial bearings and ball or hydrodynamic thrust bearing. The double-suction, two-stage, two-bearing features may be combined into one design, as shown in Fig. 2. Note that two stuffing boxes are required, both under suction pressure through the use of a bleed-off arrangement from behind the second stage. Two-stage units are available in either overhung or two-bearing construction up to 500 hp and in two-bearing construction above that value. A recent innovation, which has challenged the use of the standard process unit, is the vertical in-line pump (Fig. 3). This unit has the advantages of cost savings during installa-

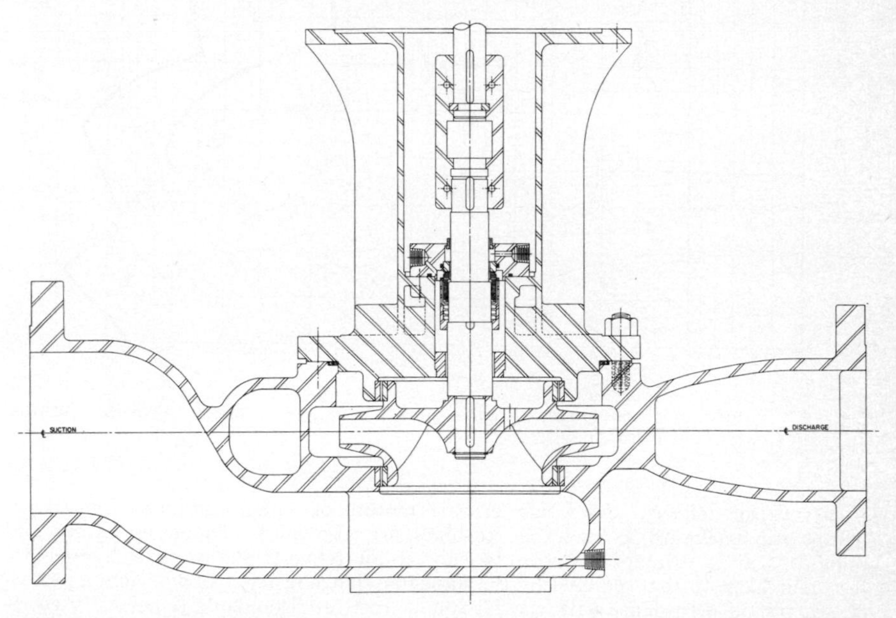

Fig. 3 Vertical, in-line, single-stage, double-volute, radially split case refinery process pump. (United Centrifugal Pumps)

tion, since it is simply mounted in the line, as a valve would be, and it is not subject to pipe strains, which are a common problem on standard units.

Performance. Figure 4 illustrates the wide range of capacity and head requirements that can be met by refinery-type pumps. The range shown varies slightly with the manufacturer. The range of in-line pumps is almost identical to that of the horizontal units.

Materials. Refinery pumps handle a variety of products with specific gravities from 0.3 to 1.3, viscosities from values below water to those as high as 15,000 SSU for centrifugal pumps and even higher for rotary pumps, over a wide range of temperatures. The product may be as inert as a lubricating oil, or extremely corrosive. Many materials are utilized to satisfy these many requirements; the most common are given in Table 1. Caution must be exercised in the selection of material for rotating parts because the physical properties of cast iron, bronze, and Ni-Resist severely limit the allowable peripheral speed for these materials. Material specifications are given in Table 2. Stuffing boxes are normally equipped with mechanical seals of a balanced-type construction. In special cases double or tandem mechanical seals are used. Material for one of the sealing rings is carbon, with the mating ring of either stellite, Ni-Resist, or tungsten carbide. Mechanical seals have proved extremely reliable and are not subject to the erratic performance sometimes encountered with packing.

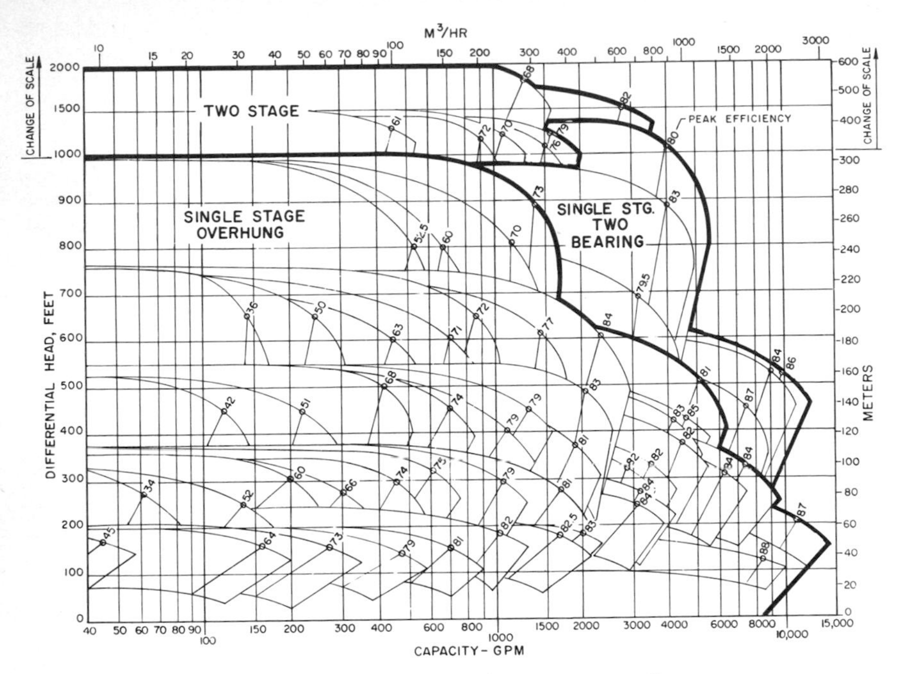

Fig. 4 Performance coverage, single- and two-stage refinery process pumps. (United Centrifugal Pumps)

Drivers for refinery pumps are electric motors or steam turbines. Centrifugal pumps run backward as hydraulic turbines are also used. To ensure safety and reliability, it is common practice to subject all refinery pumps to a hydrostatic test equivalent to that required by the discharge flange rating. In addition, a performance test in accordance with the Hydraulic Institute Standards is normally specified. If critical suction conditions warrant it, an NPSH test may also be conducted.

Pipeline Pumps Centrifugal pumps are used on every major liquid pipeline in the world to transport a variety of fluids including crude oil, motor gasoline, fuel oil, jet fuel, liquefied petroleum gases, and anhydrous ammonia. Pump efficiency is of primary importance because of the large horsepowers required to transport the liquid. Most pipeline systems install pumps in a series arrangement because the differential head is primarily made up of energy loss due to friction, and an outage of one of the units would result in only a partial loss in the through-put capacity. For pipelines where the differential head is mostly static, such as Trans-Alpine, pumps are installed in parallel. Series arrangement in a static system would be unsuitable because the differential head required could not be obtained unless all of the units were operating.

Construction. A single-stage double-suction pump, a two-stage double-suction pump, and a four-stage pipeline pump are shown in Figs. 5, 6, and 7, respectively. All the units are double-volute and axially split. The single-stage double-suction unit (Fig. 5) is typical of those used in series operation on large-diameter pipelines with capacities as high as 2 million barrels per day. One pipeline alone has installed over one hundred 5,000-hp pumps of this construction. The two-stage double suction pump (Fig. 6), is similar to the construction of the Trans-Alpine pumps. The illustrated unit is a 12,000-hp pump operating at 3,000 rpm in parallel arrangement. For lateral and smaller diameter pipeline pumping stations, multistage units (Fig. 7) are used. The number of stages chosen corresponds to the differential head that is required. Multistage units may be destaged initially to produce lower heads, with a subsequent power saving, or arranged for series or parallel operation of a

TABLE 1 Refinery Process Pump Materials

Trim part	Steel case, cast iron trim	Steel case, bronze trim	Steel case, Ni-Resist trim	Steel case, 11-13 chrome trim	11-13 Chrome case and trim	18-8 S.S. case and trim	316 S.S. case and trim
Cover	Cast steel	Cast steel	Cast steel	Cast steel	11-13 chrome	18-8 S.S.	316 S.S.
Case studs	ASTM-A193 GR B7	ASTM-A193 GR B7	ASTM-A193 GR B7	ASTM-A193 GR B7	ASTM-A193 GR B7	ASTM-A193 GR B7	ASTM-A193 GR B7
Case nuts	ASTM-A194 GR 2H	ASTM-A194 GR 2H	ASTM-A194 GR 2H	ASTM-A194 GR 2H	ASTM-A194 GR 2H	ASTM-A194 GR 2H	ASTM-A194 GR 2H
Shaft	SAE 4140 HT	SAE 4140 HT	SAE 4140 HT	SAE 4140 HT	SAE 4140 HT	18-8 S.S.	316 S.S.
Impeller	Cast iron	Bronze	Ni-Resist	Cast steel	11-13 chrome	18-8 S.S.	316 S.S.
Impeller wear ring	Cast iron	Bronze	Ni-Resist	11-13 chrome hardened	11-13 chrome hardened	18-8 S.S. hardfaced	316 S.S. hardfaced
Case wear ring	Cast iron	Bronze	Ni-Resist	11-13 chrome hardened	11-13 chrome hardened	18-8 S.S. hardfaced	316 S.S. hardfaced
Shaft sleeve, packed pump	11-13 chrome hardened	11-13 chrome hardened	11-13 chrome hardened	1020 hard-faced	11-13 chrome hardened	18-8 S.S. hardfaced	316 S.S. hardfaced
Shaft sleeve, mechanical seal	11-13 chrome	11-13 chrome	11-13 chrome	11-13 chrome	11-13 chrome	18-8 S.S.	316 S.S.
Gland	Steel	Steel	Steel	Steel	Steel	18-8 S.S.	316 S.S.
Gland studs	ASTM-A193 GR B7	ASTM-A193 GR B7	ASTM-A193 GR B7	ASTM-A193 GR B7	ASTM-A193 GR B7	ASTM-A193 GR B8	ASTM-A193 GR B8M
Lantern ring	Cast iron	Cast iron	Cast iron	Cast iron	Cast iron	18-8 S.S.	316 S.S.
Throat bushing	Cast iron	Bronze	Ni-Resist	11-13 chrome	11-13 chrome	18-8 S.S.	316 S.S.
Throttle bushing	Cast iron	Bronze	Ni-Resist	11-13 chrome hardened	11-13 chrome hardened	18-8 S.S hardfaced	316 S.S. hardfaced
Gasket, sleeve	18-8 S.S. annealed	18-8 S.S. annealed	18-8 S.S. annealed	18-8 S.S. annealed	18-8 S.S. annealed	18-8 S.S. annealed	316 S.S. annealed
Gasket, case	18-8 S.S. with asbestos	18.8 S.S. with asbestos	18-8 S.S. with asbestos	18-8 S.S. with asbestos	18-8 S.S. with asbestos	18-8 S.S. with asbestos	316 S.S. with asbestos
Impeller nut	Steel	Steel	Steel	Steel	Steel	18-8 S.S.	316 S.S.
Bearing shield	Bronze	Bronze	Bronze	Bronze	Bronze	Bronze	Bronze
Oil rings	Brass	Brass	Brass	Brass	Brass	Brass	Brass
Brearing bracket	Cast iron	Cast iron	Cast iron	Cast iron	Cast iron	Cast iron	Cast iron
Heat exchanger assembly	Steel	Steel	Steel	Steel	Steel	Steel	Steel
Coupling guard	Fab steel	Fab steel	Fab steel	Fab steel	Fab steel	Fab steel	Fab steel
Base plate	Fab steel	Fab steel	Fab steel	Fab steel	Fab steel	Fab steel	Fab steel

TABLE 2 Material Specifications (ASTM Numbers)

Material	Castings	Forgings	Bars	Studs	Nuts
Cast iron	A-48				
Ni-Resist	A-436, Types 1 and 2				
Bronze	B-143 Alloy 1A B-143 Alloy 2B B-144 Alloy 3B B-145 Alloy 4A	...	B-139 Alloy 510	B-124 Alloy 655	
Carbon steel	A-216 GR WCB	A-266, Class 1 A-266, Class 2	A-108 GR 1018 A-575 GR 1020	...	A-108 GR 1018
Alloy steel (SAE 4140)	...	...	A-434 Class BC or BD	A-193 GR B-7	A-194 GR 2H
11-13 chrome steel	A-296 GR CA-15	A-182 GR F-6 A-336 CL F-6	A-276 Type 410, 416, or 420	A-193 GR B6	A-194 GR 6
18-8 stainless steel	A-296 GR CF-8	A-182 GR F-304 A-366 CL F-8	A-276 Type 304	A-193 GR B8	A-194 GR 8
316 stainless steel	A-296 GR Cf-8M	A-182 GR F-316 A-366 CL F-8M	A-276 Type 316	A-193 GR B8M	A-194 GR 8

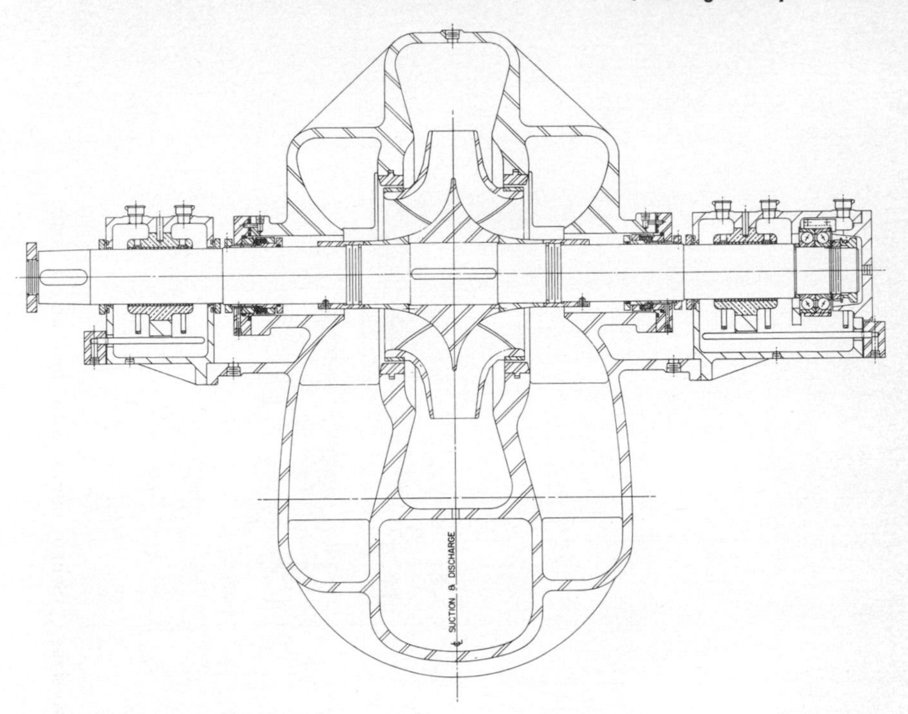

Fig. 5 Double-suction, double-volute, axially split, single-stage pipeline pump with mechanical seals. (United Centrifugal Pumps)

portion of the stages by providing proper nozzling. The exact arrangement depends on the system characteristics and initial and ultimate capacities to be pumped.

All units are equipped with sleeve radial bearings and either ball or hydrodynamic thrust bearings, if rotating speeds dictate. The single- and two-stage double-suction pumps are inherently balanced axially. The multistage units utilize opposed impeller design to obtain axial balance. All modern units are equipped with mechanical seals in the stuffing box. Since most pipeline stations are now designed for unattended outdoor service, a number of safeguard controls are used. These include warnings for low suction pressure, high discharge pressure, high bearing or case temperature, vibration monitoring, and excessive seal leakage warning. Tank farms utilize single- and double-suction in-line vertical pumps or, if the tank is to be pumped dry, vertical canned pumps.

Performance. Figures 8 and 9 indicate, typically, the present available range for single- and two-stage pipeline pumps and the range for multistage units, respectively. Pipeline units are custom-designed, and the range is being extended daily. Electric motors are normally used as drivers. With the utilization of gas turbines, the maximum horsepower employed has increased dramatically in the last few years. Present installations are 12,000-hp, with 25,000-hp units to be in operation in the near future. Another use of gas turbines on existing lines is for peak-capacity booster-station service, in which case the pumps are direct-driven, at operating speeds as high as 6,000 rpm.

Materials. Table 3 lists pump part materials commonly specified for crude and product pipeline service. Also included is a list of materials for water-flood pumps, to be discussed later.

The typical piping of a main-line pump station in which the units are arranged in series is shown in Fig. 10. As pipeline capacities have increased, one of the major problems is the pressure loss at each station. For the station shown, three discharge valves, one 16-in ball valve manifolded in parallel with two 24-in control

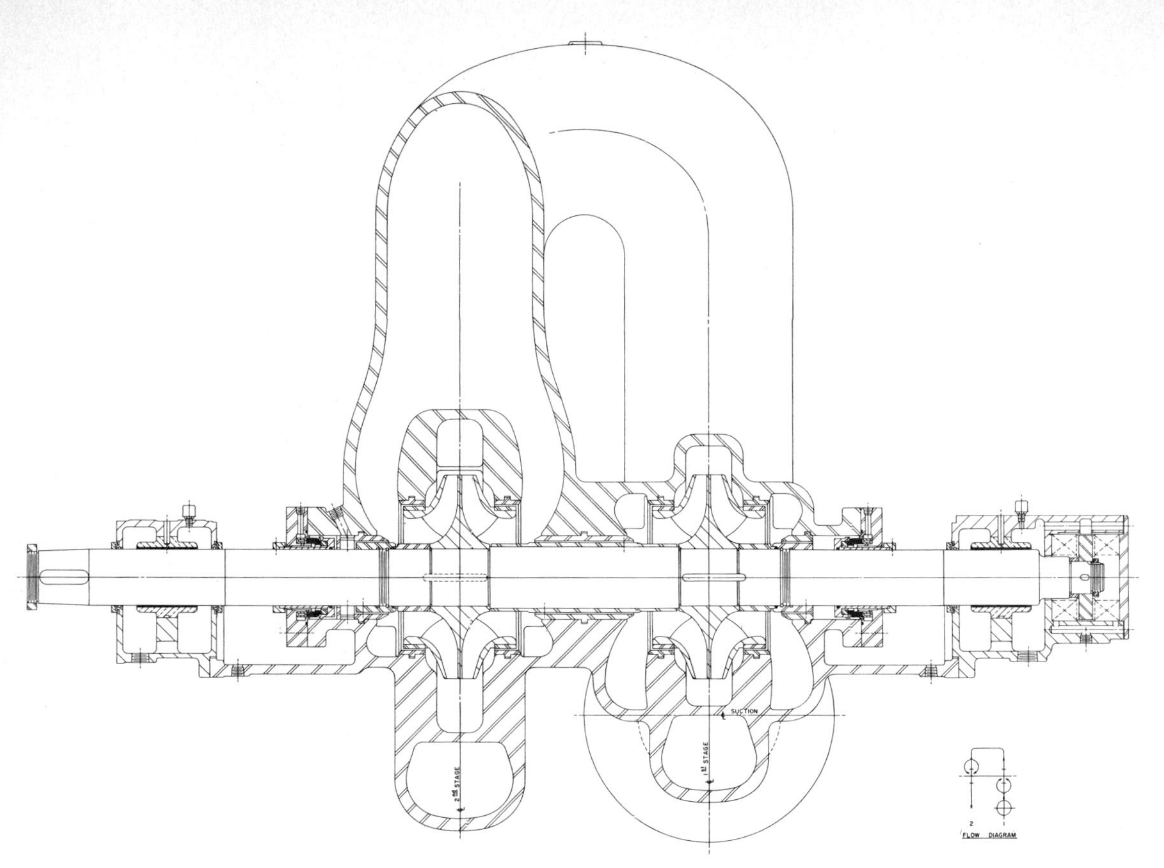

Fig. 6 Double-suction, double-volute, axially split, two-stage pipeline pump with mechanical seals. (United Centrifugal Pumps)

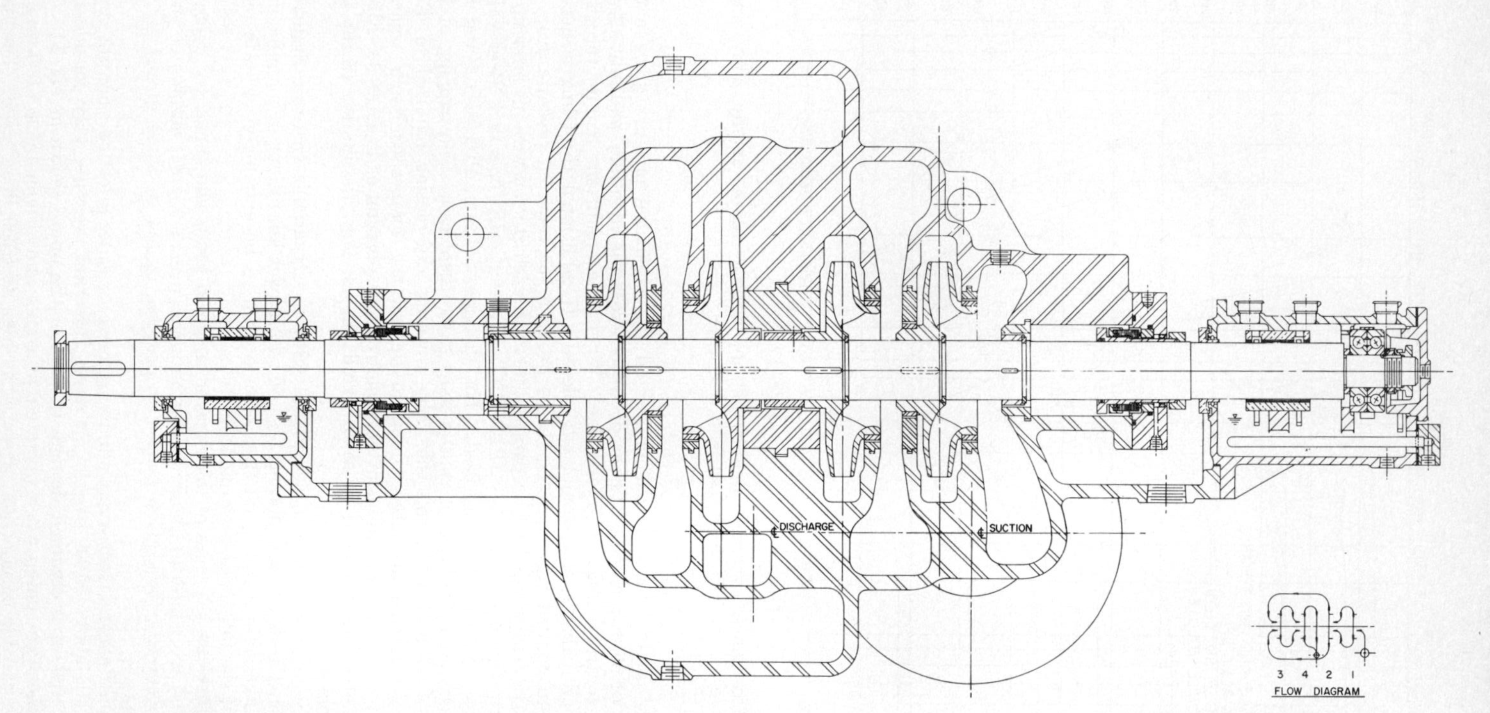

Fig. 7 Single-suction, double-volute, axially split, four-stage pipeline pump with mechanical seals. (United Centrifugal Pumps)

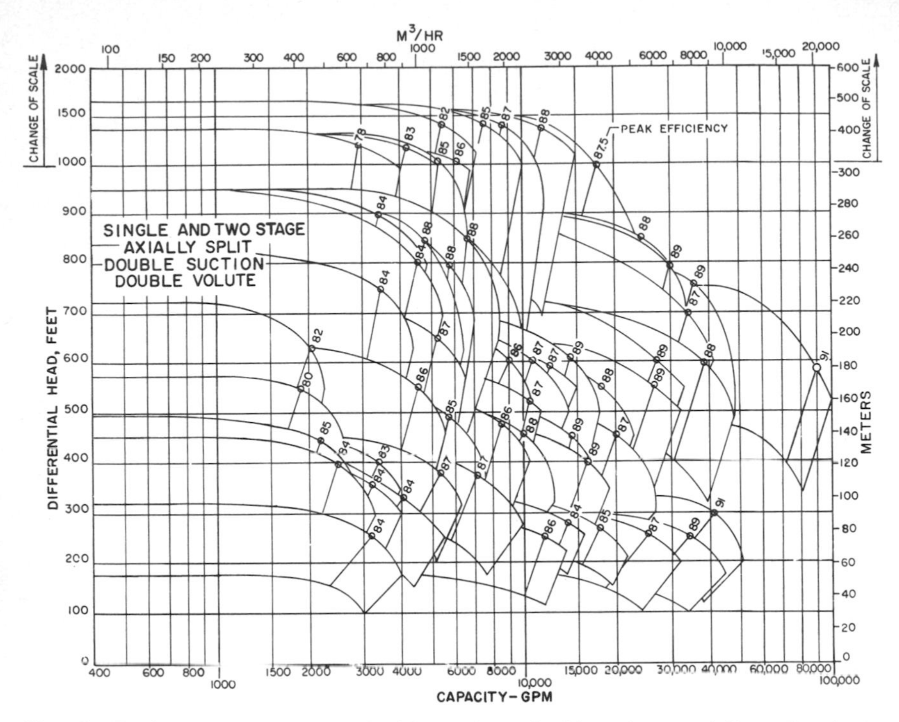

Fig. 8 Performance coverage, double-suction, double-volute, axially split single- and two-stage pumps. (United Centrifugal Pumps)

valves, were used. This arrangement limited the calculated pressure drop to that of a length of 36-in pipe equal to the distance across the manifold, and still allowed the use of control valves of proved size. When line conditions require throttling, the 16-in ball valve is first completely closed. If additional throttling is needed, the two 24-in control valves are closed to produce the required pressure drop. In order to move scrapers or batch separators through each station without interrupting flow through the pumps, signals (PIG SIG), hydraulically operated 24-in valves, and sequence control wiring are used. The distance between PIG SIG 1 and PIG SIG 2 represents the volume of station loop to be displaced. The tripping of PIG SIG 3 and PIG SIG 4 controls the valve opening and closing required to divert flow from the station discharge piping to behind the scraper or batch separator and force it to leave the station in the same relative position as it entered. Elbows at pump suction are arranged so as to avoid uneven flow distribution in the inlet of the double-suction impeller.

Pipeline pumps are tested in accordance with the Hydraulic Institute Standards. Reduced-speed tests are common because of horsepower limitations of manufacturing plants and available drivers, and they have proved to be extremely accurate representations of the full-speed performance.

Special Services Because of the extreme range in heads and capacities, operating pressures, and temperatures, the petroleum industry utilizes a great number of pumps designed specifically for a given service or process. A few of these special services are as follows.

Waterflood Pumps. Centrifugal pumps are used for water injection when the capacity required is above 10,000 barrels per day. Injection pressures vary considerably. One of the highest is obtained through the use of the pump shown in Fig. 11. Two of these nine-stage units, when operated in series and driven by a 6,000-rpm gas turbine, are capable of injection pressures as high as 8,000 lb/in^2 gage. The barrel

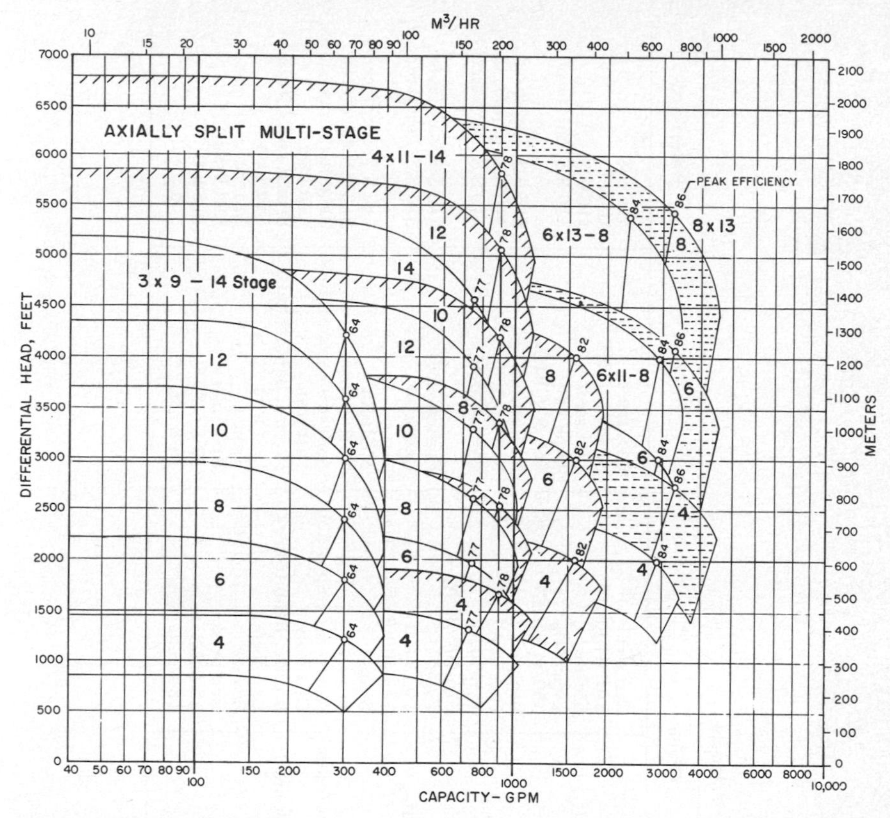

Fig. 9 Performance coverage, multistage axially split pumps. (United Centrifugal Pumps)

was constructed of forged steel with a 316 stainless steel welded overlay applied to the interior and hydrostatically tested at 12,000 lb/in² gage. Other parts were made of 18-8 chrome-nickel steel overlaid with a hard-surface material at all close-clearance parts. The shaft was of K-500 monel. Stuffing-box construction was of a limited-leakage breakdown-type bushing for increased reliability at the high operating pressures and speeds. The majority of waterflood injection pressures are such that axially split multistage pumps of the design shown in Fig. 7 may be used. As many as 14 stages in one pump are utilized to obtain the required differential head; if still greater pressures are required, the units may be piped in series. Axially split pumps have been hydrostatically tested to 6,000 lb/in² gage for a working pressure of 4,000 lb/in² gage. Materials are given in Table 3.

Reactor Feed Pumps. One of the most difficult services encountered in the petroleum industry is the high-temperature, high-pressure reactor feed or charge pump. The construction is similar to that of the waterflood pump shown in Fig. 11 but is subject to operating temperatures in the range of 600 to 700°F with discharge pressures of 2,000 lb/in² gage or more. The corrosiveness of the pumped fluid requires the use of a number of dissimilar metals, and compensation must be made for the different rates of expansion at the operating temperatures. Extreme care is required in the start-up and operation of these units.

LPG. Recently multistage axially split centrifugal pumps have been installed in pipelines transporting liquefied petroleum gases with specific gravities as low as 0.35. The low lubricity of the pump fluid requires careful selection of wearing-part materials. Stuffing boxes utilize single or double mechanical seals. At ambient tempera-

TABLE 3 Crude Pipeline, Product Pipeline, and Waterflood Pump Materials

Part	Crude		Products	Waterflood
Case	Cast steel	Cast steel	Cast steel	316 S.S.
Shaft	SAE 4140 HT	SAE 4140 HT	SAE 4140 HT	Monel
Impeller	Cast steel	Bronze	Cast steel	316 S.S.
Impeller wear ring	Type 410 350-375 BHN	Type 410 350-375 BHN	Type 410 350-375 BHN	316 S.S. hardfaced
Case wear ring	Ni-Resist	Bronze	Ni-Resist	316 S.S. hardfaced
Case separation ring	Ni-Resist	Cast iron	Ni-Resist	316 S.S. hardfaced
Shaft sleeve, high pressure	Type 410 375-400 BHN	Type 410 375-400 BHN	Carbon steel	316 S.S. hardfaced
Shaft sleeve, low pressure	Type 410	Type 410	Carbon steel	316 S.S. hardfaced
Intermediate shaft sleeve	SAE 1020 hardfaced	Type 410 375-400 BHN	Cast iron	316 S.S. hardfaced
Throttle bushing	Ni-Resist	Bronze	Ni-Resist	316 S.S. hardfaced
Throat bushing	Ni-Resist	Bronze	Ni-Resist	316 S.S. hardfaced
Gasket, case, split	Asbestos composition	Asbestos composition	Asbestos composition	Asbestos composition
Bearing bracket	Cast iron	Cast iron	Cast iron	Cast iron
Sleeve bearing	Bronze, Babbitt-lined	Bronze, Babbitt-lined	Bronze, Babbitt-lined	Bronze, Babbitt-lined
Oil-ring carrier	Cast iron	Cast iron	Cast iron	Cast iron
Bearing shield	Bronze	Bronze	Bronze	Bronze
Oil ring	Red brass	Red brass	Red brass	Red brass
Splitters	18-8 S.S.	18-8 S.S.	18-8 S.S.	316 S.S.
Impeller locating rings	18-8 S.S.	18-8 S.S.	18-8 S.S.	316 S.S.

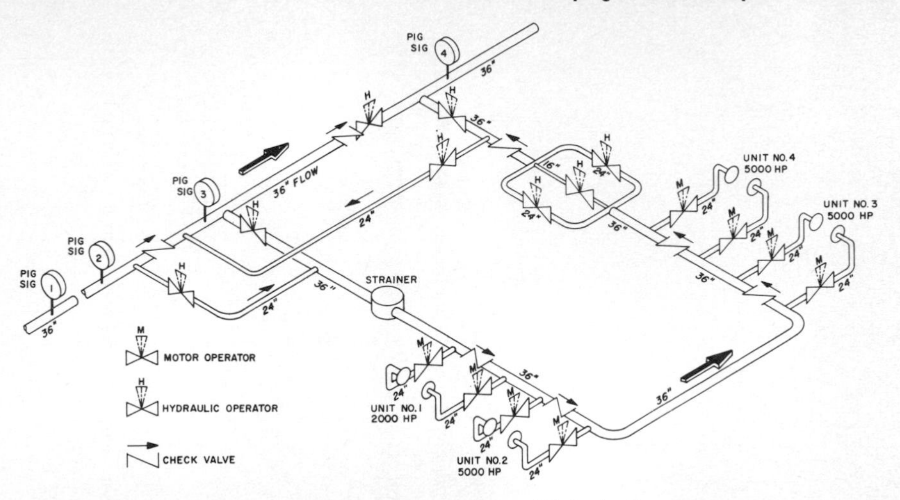

Fig. 10 Mainline pumping-station piping. (United Centrifugal Pumps)

tures, the suction pressures are as high as 1,000 lb/in^2 gage. These fluids experience a large temperature rise due to compression.

Many additional special designs could be mentioned. The wide range of operating conditions and products in the petroleum industry frequently requires pumps specifically engineered for a certain service. The pump designer is continually challenged to provide a safe, reliable, and economical design.

PUMPING OF VISCOUS LIQUIDS

In many petroleum-industry applications, an important factor in the selection of a pump is the viscosity of the liquid. As it increases, the pump is faced with a more difficult task to perform, and an understanding of the relationship between viscosity and pump performance becomes essential to proper sizing of both the pump and its driver.

Centrifugal pumps are routinely applied on services having viscosities below 3,000 SSU, and may be used up to 15,000 SSU. (For background information on viscosity, refer to Appendix.) They are sensitive, however, to changes in viscosity, and will exhibit significant reductions in capacity and head, and rather drastic reductions in efficiency, at moderate to high values of viscosity. The extent of these effects may be seen in Table 4.

Table 4 was constructed with the aid of Fig. 12, which provides a convenient means for determining the viscous performance of a centrifugal pump when its water performance is known. To use this figure, enter at the bottom with the pump capacity, proceed vertically upward to the total head (head per stage for multistage pumps), then horizontally right or left to the viscosity, and vertically upward again to the curves for correction factors. The values thus obtained for the respective correction factors are multiplied by the water-performance values for capacity, total head, and efficiency to obtain the equivalent values for viscous performance. By use of the individual correction factors for total head it is even possible to approximate the shape of the head-capacity characteristic when handling the viscous liquid, at least to 120 percent of the best efficiency point (Q_N). The total head at shutoff will remain essentially the same for viscous or nonviscous liquids.

Centrifugal pump performance is almost invariably specified by the manufacturer on the basis of handling clean cold water, even where the pump has been specifically designed for petroleum-industry applications. Pump selections, however, must necessarily be made to satisfy viscous conditions of service, and require application of these correction factors in the reverse direction. In this case, Fig. 12 provides an approximation of equivalent water performance which is probably within the

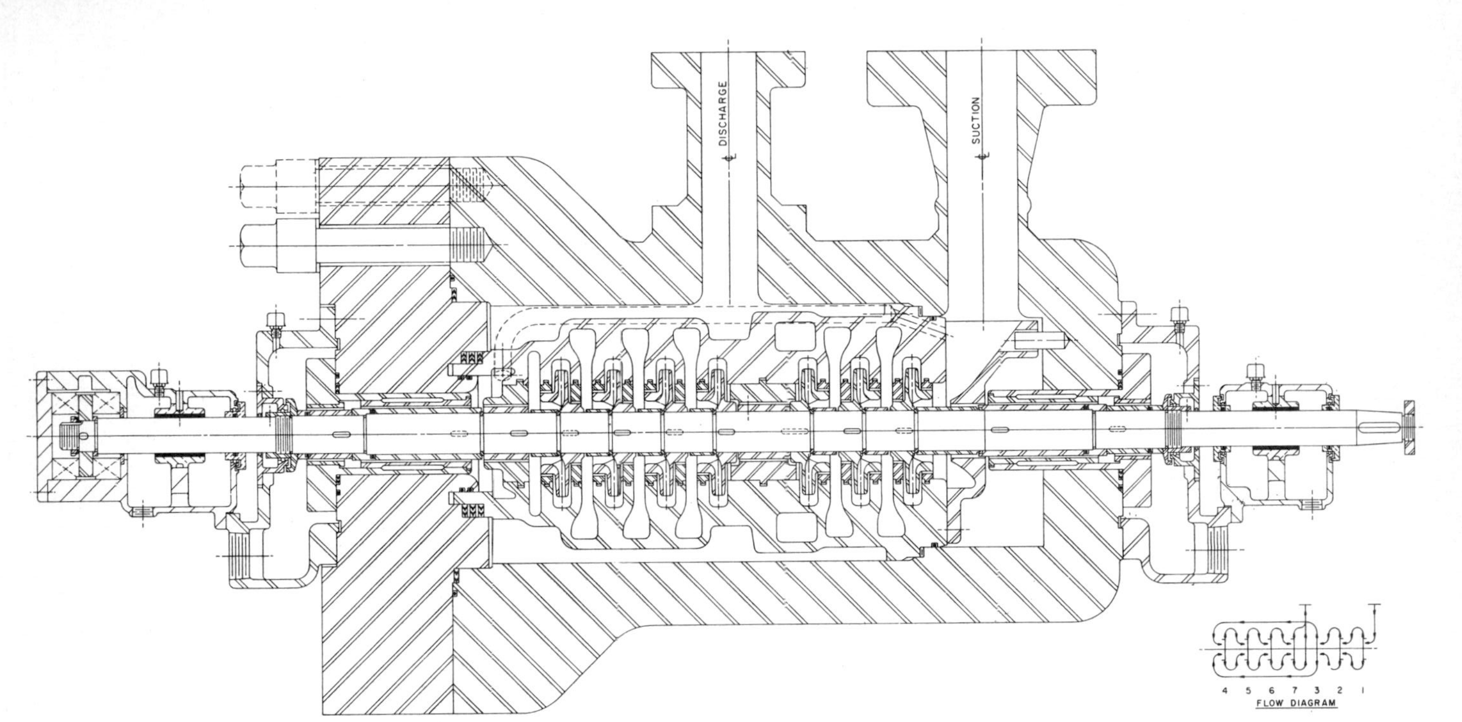

Fig. 11 Double-casing, radially split outer barrel, double-volute, axially split inner case, waterflood pump. (United Centrifugal Pumps)

TABLE 4 Effect of Viscosity on Performance of a Typical Centrifugal Pump Operating at Its Best Efficiency Point

Viscosity (SSU)	Capacity, gpm	Total head, ft	Efficiency, %	Brake horsepower*
Nil	3,000	300	85	241
500	3,000	291	71	279
2,000	2,900	279	59	312
5,000	2,670	264	43	373
10,000	2,340	243	31	417
15,000	2,100	228	23	473

* Note: All values of brake horsepower based on liquid having sp gr of 0.90.

limits of accuracy of the figure itself, for viscosities in SSU numerically equal to pump capacity in gpm. For higher viscosities the initial solution for equivalent water performance, determined as indicated in the following paragraph, may need to be adjusted and then checked back by conversion of water performance to viscous performance.

To determine approximate equivalent water performance when viscous performance is known, enter Fig. 12 at the bottom with the viscous capacity, proceed vertically upward to the desired viscous head (head per stage for multistage pumps), then horizontally right or left to the viscosity, and vertically upward to the correction-factor curves for capacity and head. In this case, divide the viscous-performance values by the correction factors to obtain the equivalent water-performance values. The pump selection can then be made on the basis of ratings established for water, and efficiency can be calculated for the viscous liquid using the efficiency correction factor applied to the pump efficiency for water.

For example, to select a pump to handle 500 gpm of 3,000 SSU liquid against a head of 150 ft, proceed as follows:

1. From Fig. 12, determine $C_Q = 0.80$ and $C_H = 0.81$.
2. Water capacity, $Q_W = 500 \div 0.80 = 625$ gpm.
3. Water head, $Q_H = 150 \div 0.81 = 185$ ft.
4. For 625 gpm, 185 ft, 3,000 SSU, Fig. 12 indicates $C_Q = 0.83$, $C_H = 0.84$, and $C_E = 0.42$. (If the values for C_Q and C_H obtained in this step are roughly within 2 percent of those in step 1, the pump may be selected on the basis of the water capacity and water head already obtained. Since, in this case, these figures differ from the first approximation by more than 3 percent, the values of water performance should be adjusted as in the following steps.)
5. Adjust water capacity:

$$Q_W \times \frac{0.80}{0.83} = 602 \text{ gpm}$$

6. Adjust water head:

$$H_W \times \frac{0.81}{0.84} = 178 \text{ ft}$$

7. Select pump for 602 gpm, 178 ft, and determine water efficiency from manufacturer's rating.
8. Assuming water efficiency in this case is 75 percent, then viscous efficiency $E_V = 0.75 \times 0.42 = 0.315$, or 31.5 percent.
9. For a specific gravity of 0.90:

$$\text{Bhp} = \frac{\text{gpm} \times \text{ft of head} \times \text{sp gr}}{3{,}960 \times \text{efficiency}} = \frac{500 \times 150 \times 0.9}{3{,}960 \times 0.315} = 54$$

The Hydraulic Institute Standards limit the use of the chart in Fig. 12 to radial-flow centrifugal pumps with open or closed impellers, handling Newtonian liquids

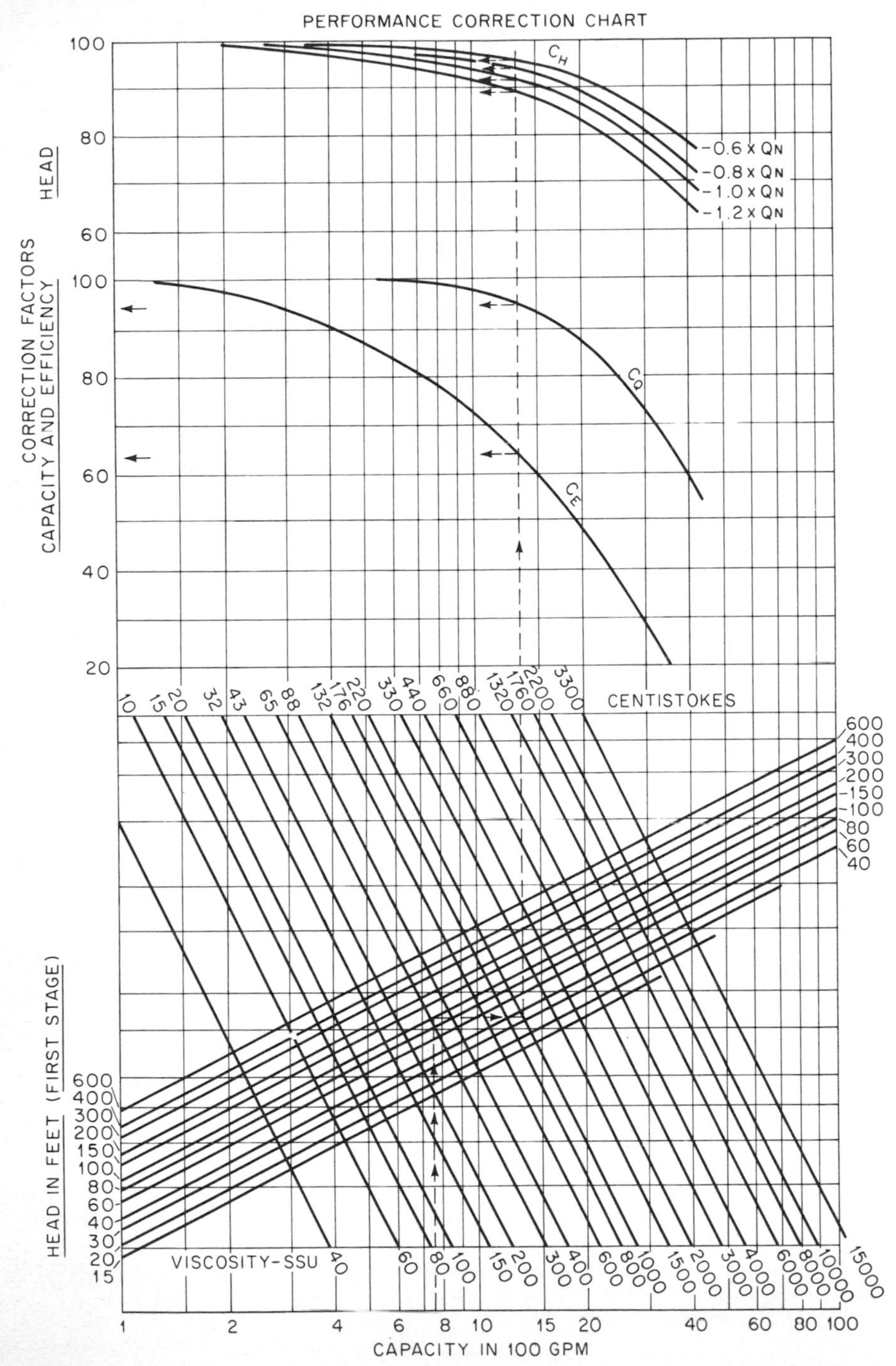

Fig. 12 Viscosity correction curves for centrifugal pumps. (Reprinted from the "Hydraulic Institute Standards," 12th ed., copyright 1969 by the Hydraulic Institute, Cleveland, Ohio)

within the pump's normal operating range, and within the capacity limits indicated on the chart. Failure to heed these limitations may result in misleading expectations of performance with viscous liquids. Tests on vertical-turbine-type pumps, for example, which normally have mixed-flow rather than radial-flow impellers, have shown that, while the viscous efficiency predicted from this chart would be quite accurate, the viscous head and viscous capacity predictions would be substantially above the demonstrable values. In other words, the performance reductions determined from the chart would not be severe enough for that type of pump. For centrifugal pumps which do not fall within the limits of applicability of Fig. 12, viscous performance can be accurately established only by test.

Displacement pumps, both rotary and reciprocating, are also frequently used for pumping viscous fluids, and some types are well suited for use where viscosities are beyond the limit that can be handled with centrifugals. In fact, many common designs are suitable only for use with liquids which are at least moderately viscous, since they depend on the viscosity to maintain the lubricating and sealing films between the various internal parts of the pumps. Most gear pumps and screw pumps are in this category.

As with centrifugal pumps, the performance of displacement pumps may be significantly affected by changes in the liquid viscosity, but the nature of these changes may be quite different. At constant speed, changes in viscosity are likely to have little or no effect on pump capacity. Total head, or differential pressure across the pump, would probably increase with increasing viscosity because of higher system resistance. Thus, brake horsepower required would increase, even though pump efficiency would not suffer nearly so drastically, if at all, as for a centrifugal pump.

The nature and extent of performance variation with changes in viscosity for displacement pumps are more significantly influenced by the pump design than is the case with centrifugal pumps. Since they are manufactured in an extremely wide variety of designs, it is not practical to attempt to provide here a general means of determining their viscous performance. This should be done only on the basis of information provided by the pump manufacturer.

Section 10.8

Pulp and Paper Mill Services

J. F. GIDDINGS

GLOSSARY OF TERMS

Additive Any material such as clay, filler, color, etc., added to the stock to contribute specific properties.

Alkali *Total alkali:* Chemical used in cooking for soda process $Na_2CO_3 + Na_2SO_4$, all expressed as Na_2. *Active alkali:* Soda-process NaOH only (as Na_2O); sulfate process $NaOH + Na_2S$ (as Na_2O). *Effective alkali:* Sulfate; $NaOH + \frac{1}{2}\ Na_2S$.

Chest Vessel for storing pulp.

Cooking The action of the chemical used to break down the lignin bond between the cellulose fiber in the wood and other organic matter.

Digester Pressure vessel used to contain the chemical action of the cooking chemical and the raw cellulose material.

Fourdrinier The continuous wire upon which the pulp or paper sheet is produced.

Freeness A measure of the degree of refinement of the stock and, hence, its ability to drain water.

Groundwood A mechanical pulp formed by simple grinding to break down the structure of the wood.

Kraft Process The method of separating cellulose fiber from the lignins by using caustic soda in the presence of the sulfur radical.

Lignin A generic term used to refer to the complex organic matter present in wood which acts as a binding agent for the cellulose fibers.

Liquor *Black liquor:* Solution of water with the residual organic matter (or lignins) in the wood after washing of raw stock. *Green liquor:* solution of the smelt from a recovery furnace when dissolved in either water or weak washed liquor. *White liquor:* Solution of caustic soda (or other alkali) sometimes in the presence of a sulfur radical. This is the liquor charged to the digester for the cooking process.

Neutral Sulfite A cooking process using a solution of about 10 percent sodium sulfite and 5 percent caustic soda mixed for cooking of wood or agricultural fiber to produce high-yield pulp.

Refining A mechanical process carried out on stock to improve the ability of the fiber to form a sheet of paper. Different techniques are employed—some designed to shorten the fiber, others to increase the amount of fibrils (or "whisker") on the fiber.

Soda Process Similar to the kraft process, but uses caustic soda without the presence of sulfur.

Stock A generic term for the suspension of cellulose fiber in water or chemicals. *Bleached stock:* The same after bleaching. *Brownstock:* The same before bleaching. *Raw Stock:* The product discharged from the digester(s) before any washing or other treatment.

Sulfidity

$$\frac{Na_2S}{NaOH + Na_2S}$$

Sulfite Process The method of separating cellulose fiber from lignin by using an acid.

Vat Semicylindrical mold or container for stock during washing or sheet information.

Yield Percentage of cellulose fiber in the form of pulp produced from a given weight of wood or other raw material. *High Yield:* The same when some of lignins are allowed to remain in the finished product.

GENERAL

Apart from the petrochemical industry there are few continuous-process plants where the pumps are so much in evidence and where reliability is so essential for successful commercial and technical operation.

Before paper leaves the machine room 100 to 200 tons of water will have been pumped to the mill for every ton of paper produced. These figures represent only the basic amount of water taken from a river and rejected as effluent. Within the process itself the amount of liquid circulated is several times greater.

Although many attempts have been made to utilize a dry process for papermaking, both the pulping (to separate crude fibers from the raw material) and the papermaking process (actual treatment of the fiber and mechanical information of the sheet) still require water as the medium to convey fiber. Throughout the mill, pumps are used to transfer or circulate fibers suspended in water (stock), chemicals or solids in solution (liquors), or residues and waste matter as slurries, as well as water supply for general services.

There are at least 150 pumps installed in a modern pulp mill, and another 50 or so in a paper mill. About 1000 kWhr is required for the production of one ton of pulp, and a further 500 kWhr for the finished paper. Of this total about 25 percent is used in pumping. This means that even in a medium-size mill the installed power required for pumps alone is approximately 10,000 hp, and often it is much higher.

PULPING FIBROUS RAW MATERIALS AND PROCESSES

Raw Materials In the past the traditional fibrous raw materials for the manufacture of paper were cloth and agricultural residues; today the vast bulk of cellulose pulp is made from wood. Although over half of this is produced from softwoods (long-fibered), an increasing amount is being now produced from hardwoods (short-fibered). Traditional raw materials such as linen, cotton waste, straw, and agricultural residues are still used in small quantities, particularly where the paper sheet requires special properties. These distinctions are important in the selection of pumps, for the liquors have different characteristics. For example, straw black liquor is much more viscous than wood black liquor, and the proper corrections must be made in calculating the pump performance and pipe friction losses.

Many grades of paper include waste paper, and specialized plants exist for the processing of waste paper before it is used for papermaking. Special care is necessary in the selection of pumps to handle waste-paper stock because of the large amount of foreign matter present—rope, string, metal, and synthetic fiber—all of which can cause problems in the process and pumping.

Groundwood Pulp This type of pulp is produced by simply grinding away wood by mechanical action. Almost all of the wood is used in the pulp, including many of the resins and other complex organic compounds. The fibers are bruised so that the pulp has inferior strength. Large amounts of water are required for cooling and for carrying away the groundwood pulp; the latter is usually acid (pH 4 to 5), so corrosion-resistant materials must be used. The pulp is used primarily for newsprint and magazines. Depending on the end use, some mechanical treatment (refining) may be required which alters the characteristics of the pulp and, in particular, the viscosity. In some cases a mild bleach process may be used to improve the color.

Semichemical or Chip Groundwood Pulp This term is applied to a process where a high yield is obtained from wood by a mild chemical treatment before subjecting the wood to mechanical treatment. The object is to break down to some extent the lignin bond, and this is achieved most efficiently when the wood is first chipped. The mechanical treatment is frequently effected by two roughened rotating disks. The chips are fed between them, and the action separates the fibers and produces a pulp. This type of pulp is often used as a filler for other products and for the manufacture of corrugating medium.

Chemical Pulping—Object of Cooking Wood is a complex nonuniform material containing about 50 percent by weight of cellulose fiber, 30 percent lignins, and 18 to 20 percent carbohydrate. The remainder is proteins, resins, and other complex organic compounds, which vary from one species to another. Cellulose resists attack from most chemicals, while the carbohydrates and other organic materials generally form compounds with the chemical cooking liquor. Some paper products can use the carbohydrate fraction to contribute bulk to the sheet, and for such papers groundwood and semichemical pulps are used. Where high strength is required, a cooking process is necessary to separate the fibers completely from the remainder of the wood.

Most cooking of wood is done in a pressure vessel at high temperature and pressure in the presence of an acid or alkali.

There is considerable tradition in chemical pulping, and a number of different processes are used. For many years the traditional method of producing pulp for high-grade papers was the acid sulfite process. This has been largely superseded in recent years by an alkaline process using sodium-based liquors in the presence of a sulfur radical; this is known as the sulfate or kraft process. The properties of the liquids pumped in the two processes are different, and the pumps require different materials in construction. The main reasons for the change to the sulfate process have been the lower corrosion rates, the ease of chemical recovery, and a stronger pulp.

Typical Sulfate Process Pulpwood logs are first chipped to about ¾ by ⅛ in and then charged into either a continuous digester (Kamyr type) or into a batch digester, each having a volume of about 5,000 ft^3. Cooking liquor (NaOH plus up to 30 percent Na_2S) is then allowed to react with the wood chips for 2 to 2½ hr at a temperature up to 350°F and a pressure in the digester of 80 to 100 lb/in^2. In many mills the heating of the chips and cooking liquor is by direct steam injection to the digester. In others some form of indirect heating is used with a closed recirculation system. In the latter case the digester circulating pumps are a critical item. These pumps must handle hot caustic solutions and entrained solid matter within a closed pressure circuit. After cooking, the contents of the digester are discharged to atmospheric pressure into a vessel called the blow tank, where the sudden expansion causes the fibers to separate from the liquid which is now known as black liquor.

From this point on the process splits into two streams—one for fiber processing, and the other for chemical recovery. The fiber is washed and screened, and then formed into a pulp or paper sheet. The back liquor is washed from the pulp and subsequently is treated for chemical recovery. Because the most troublesome liquors are to be found in the recovery process and bleach plant, the selection of these pumps is critical for the successful operation of the process.

The chemistry of the recovery process is as follows: After concentration of the black liquor in multiple-effect evaporators to about 50 percent total solids, the final concentration to 60 to 65 percent is done by direct contact with hot flue gas from the waste heat or recovery boiler. The 65 percent concentration black liquor is mixed with salt cake (Na_2SO_4) before being sprayed into the furnace under pressure generated by high-pressure pumps. The furnace atmosphere is maintained with a minimum of excess air so that the Na_2SO_4 is reduced to Na_2S, and Na_2CO_3 is formed in the process. These molten chemicals run out as a smelt and are dissolved in a tank to form green liquor. This liquor is then causticized with lime to form caustic soda (NaOH), with the Na_2S still present along with other residual chemicals, thus forming the regenerated cooking liquor known as white liquor. The calcium carbonate formed (Na_2CO_3) is burned in a lime kiln for reuse in causticizing. Various lime slurries and residues are formed during this process. The white liquor is then clarified and reused in the digesters, completing the cycle as shown in Fig. 1.

There are a variety of other pulping processes in use, but the sulfate process offers so many advantages that almost all recent installations have been of this type.

Bleaching Bleaching may be considered as an extension of the cooking process, the object being to remove the coloring matter, carbohydrate, and lignins, so that the remaining pulp has a maximum percentage of alpha cellulose present, which is the purest cellulose form and the most resistant to attack from normal chemicals.

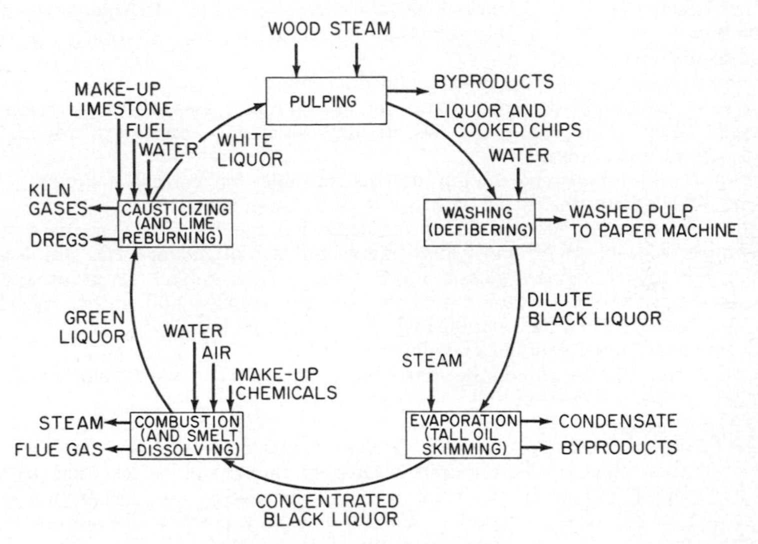

Fig. 1 The recovery cycle.

Because of this resistance, special highly reactive chemicals, such as chlorine, must be used for bleaching; these produce corrosive liquors within the bleach plant.

LIQUIDS PUMPED IN A MILL

These fall broadly into three main categories:

1. *Water* and similar fluids.
2. *Liquors* and *slurries*—mainly chemicals and solids in solution or suspension.
3. *Stock*—suspension of cellulose fibers in water.

Water Apart from the quantities involved, there are no special requirements concerning the water in pulp and paper mills, for operating conditions are well within normal limits. However, iron and carbon steel piping should not be used in bleach pulp mills because of iron pickup.

Process-water treatment is frequently used to purify process water for the mill and to remove undesirable elements such as iron. Higher quality water is required for chemical reparation in the bleach plant and for boiler feedwater where demineralizer plants are used. Rubber- or epoxy-resin-lined pumps are used for those components in contact with the demineralized water.

Pumps for Mill Water. For the majority of pumps, standard cast iron, bronze, or stainless-steel-fitted is used except as noted for demineralized water. In many mills, however, stainless-steel-fitted pumps are standard because this permits a minimum number of spares to be held in stock for other duties.

In the paper mill, water that has been removed from the sheet on the paper machine has a very low fiber content—less than 0.05 percent—and is known as white water. In some installations fiber recovery plants are used, but the fiber content does not usually cause any pumping problems. Much of this water is recirculated, and where bleached products are produced, pumps must be constructed of a low-grade stainless steel. All bronze is also suitable for service with bleached products.

Liquors and Slurries Depending on the process and the particular point in that process, the liquor characteristics may require special pumps or special materials. Although the liquor cycle is a difficult one as far as the pumps are concerned, standard designs should be used whenever possible. This reduces the number of different types of pumps in the mill, which is, of course, desirable. In some cases it may be necessary to use a higher material specification than necessary to achieve interchangeability.

Liquor and slurry pumps may be grouped as follows:

Group A. Standard designs suitable for most process uses where corrosion is not a major factor. Pumps may be 304 stainless steel with casings of Meehanite or 2 to 5 percent nickel iron.

Group B. Standard end-suction designs suitable for corrosive liquors. Pumps may be 316 stainless steel fitted with casing of 316 stainless steel.

Group C. Standard or nonstandard designs suitable for special services. Pumps are similar to group B for most applications but are of 317 or 317L stainless steel. For more corrosive services glass-reinforced epoxy resin, rubber, titanium, and stainless steels similar to Alloy 20 are used for both the impeller and casing. Mechanical seals in place of packed boxes are usually fitted to these pumps.

Recommendations for liquor and slurry pumps:

1. All liquor pumps should be classified as slurry type with open nonshrouded impellers of the end-suction and back pull-out type.
2. Where stainless-steel-fitted pumps are used, different metals or different harnesses should be used for materials in close contact. If this not feasible because of the stainless steel grades required, clearances should not be less than 0.040 in.
3. For group C pumps, in particular, it may be necessary to depart from a standard design or type of centrifugal pump. For example, if a positive displacement characteristic is required, then a screw-type pump may be used with confidence. In addition, all pump handling stock with consistency above 2 to 2½ percent must be regarded as nonstandard types.

Once the pumps are grouped as above, it becomes necessary to decide which pump may be used for specific liquors. Requirements for individual mills will differ in detail, but the following may be taken as an indication of current practice, particularly in the modern sulfate mills. In every case, manufacturers should be made aware of the liquor characteristics and the location of the pump within the process.

Cooking Liquor (White Liquor—Sulfate Process) This is essentially an alkaline solution made by causticizing green liquor. The liquor is prepared at concentrations over the range of 50 to 100 g/liter depending on the wood species, and the amount of active alkali (expressed as Na_2O) may be from 14 to 30 percent of the dry wood weight. White liquor is mainly sodium hydroxide, with a small percentage of sodium sulfide that depends on the mill sulfidity. Higher values of active chemical are used in bleached-pulp mills. The term sulfidity is used to denote the ratio of chemicals present; it is frequently calculated from the expression

$$\frac{Na_2S}{NaOH + Na_2S}$$

expressed as Na_2O.

The sulfidity value commonly used is from 20 to 30 percent; the higher values usually denote better chemical recovery. Specific gravity of the liquor will be approximately 1.2, and after clarification only small quantities of grit should be present. The liquor must be considered as an abrasive that produces a high rate of wear on pump rotating elements. White liquor has a tendency to crystallize on internal surfaces of pipes and pumps, but there are no special viscosity problems and a pump headloss allowance of about 10 percent more than water should be adequate. Group B pumps are recommended.

Cooking Liquor (Sulfite Process) This is essentially an acid solution, mainly calcium or sodium bisulfite, with an excess of sulfur dioxide present as sulfurous acid. Modern sulfite mills employ a variety of cooking liquors. A common method of preparing the cooking liquor consists of burning sulfur to form SO_2 and passing this through a packed tower of limestone or a lime solution so that calcium bisulfite is formed, along with about 1 percent of combined SO_2, the remainder being free SO_2 in amounts of about 4 to 5 percent. Some modern operations use liquors prepared with other bases such as sodium, magnesium, and ammonia; in most such cases the liquor is highly corrosive. Until the widespread use of stainless steel, corrosion was a major problem. Up to 25 percent more stainless steel is used in sulfite mills than in sulfate mills. Care should be used in material selection because 316 or 317 stainless steel is not always suitable. Group C pumps are recommended.

Blow-Tank Discharge As the liquor introduced with the chips to the digester combines with the noncellulose and hemicellulose fraction of the wood, it changes from white liquor to black liquor before reaching the blow tank. In addition, the sudden release of pressure frees the cellulose fibers from the other matter, so the blow tank contains both raw stock (pulp) and black liquor. A stock pump, therefore, is required for this duty because the stock concentration is quite high.

Black Liquor For convenience these pumps are divided into three groups:

Weak Black Liquor (Total Solids Up to 20 Percent). During the washing process hot water is used to dissolve away the surplus organic matter from the pulp, and the liquor produced is termed black liquor. This liquor is a mixture of the lignins and carbohydrates in the original wood plus the cooking chemicals; it is alkaline with a solids content of 14 to 16 percent in the case of a sulfate mill. The temperature will be about 180 to 190°F and the specific gravity about 1.08. Washing is usually carried out with a minimum of three countercurrent stages, and the solids content given above is representative of the liquor leaving the stages nearest the inlet, that is, where it is most concentrated. The quantity of recirculated liquor is quite high, and many mills have found the group A pumps with stainless trim to be satisfactory and economical. With the low solids content there are no special viscosity problems. This may be seen from Fig. 2.

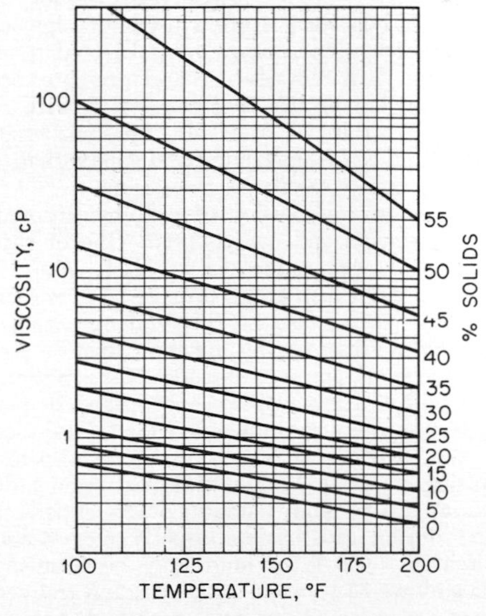

Fig. 2 Black-liquor viscosity.

With sulfite mills group B pumps should be used because of the acidity of the liquor.

Black Liquor with Total Solids of 20 to 50 Percent. This liquor is formed by the evaporation of water from weak black liquor. The concentration is accomplished in multiple-effect evaporators, which usually discharge liquor with about 50 percent total solids at close to 200°F. In some of the odor-free installations the solids concentration is much higher. Because of the nature of the evaporation, special pumps are usually required.

Liquor up to 50 percent concentration is reasonably easy to pump, but allowance must be made for viscosity effects (see Fig. 2). In noting the values in Fig. 2, it should be remembered that the plant must often start up cold, so that cold liquor with higher viscosity may have to be pumped. The liquor specific gravity rises during evaporation from about 1.1 to 1.25. Group B pumps are recommended.

Black Liquor with Total Solids of 50 to 65 Percent. This is often referred to as heavy black liquor because the specific gravity rises to 1.35. The liquor is formed by further evaporation, either in the multiple-effect units or by contact evaporators using hot flue gas.

From a pumping standpoint this liquor is probably the most difficult of all liquids to pump satisfactorily in pulp and paper mills. Continuous operation requires careful attention to pump maintenance. Steaming out at regular intervals is particularly important.

No accurate figures are available for the viscosity of solids concentrations above 55 percent as there is a wide variation between the liquor produced from the different wood species and also between the same wood of different age. Hardwood species produce a more viscous liquor, especially the eucalyptus species, as well as more liquor per ton of pulp produced. Black liquor produced from straw pulping produces even higher viscosities, and, in addition, causes the deposition of silica on the walls of pumps and piping. An approximation of the viscosity of straw-mill heavy black liquor may be determined from published figures which give viscosities up to 8,000 Redwood seconds. This is probably at least 50 percent higher than liquor from normal long-fibered softwood.

During the recovery process liquor is sprayed into the furnace for evaporation to dryness and burning. Prior to this the make-up chemical (sodium sulfate or salt cake) is added, and this chemical is reduced to Na_2S (sodium sulfide) in the reducing atmosphere of the furnace.

Little is known with certainty about heavy black liquor, but it does not seem to be very corrosive, and carbon steel is often used for pipework, although stainless steel pumps are fairly common. The pumps are subjected to severe duties—notably high heads, lumpy material, high temperature and pressure, and continuous service. Group B pumps are almost universally specified, often with casings of more wear-resistant material such as Worthite or Alloy 20. In some cases, mills making straw pulp have not found suitable centrifugal pumps and have had to resort to gear-type pumps because of the very high viscosity of the liquor.

Green Liquor Green liquor is a solution of sodium carbonate and sodium sulfide with additions of other elements and compounds. One of these compounds is iron sulfide in a colloidal form which produces a greenish color. The liquor is formed by dissolving smelt from the causticizing process. Severe erosion takes place in green-liquor pumps, primarily because of the violent action inside the dissolving tank but also because of the gritty matter always present. Green liquor also builds up on the walls of pumps and piping, causing high friction losses. The specific gravity is usually about 1.2, and an allowance of about 20 percent should be made for viscosity. Group B pumps are recommended for this service.

Lime Slurries In the causticizing process various solutions and slurries occur which, apart from causing excessive wear in standard pumps, do not cause any problems. Thus any normal slurry pump should prove satisfactory. In sulfate mills the lime mud formed during causticizing green liquor presents the most serious problem, for approximately 1,000 lb of mud may be formed for each ton of pulp produced. Solids loads above 35 percent can occur, and frequent blockages are likely to occur unless pumps are selected for minimizing downtime. Often special diaphragm-type pumps are used, but where a mild slurry duty is involved group A pumps should prove satisfactory provided the wearing parts can be readily replaced; many of the parts will have a life less than 12 months.

Bleach-Plant Liquor Most bleached-pulp mills today use at least four stages of bleaching, and often six or more. Bleaching is used to remove residual lignins or to convert them to compounds that are stable as to color and heat. The stages used include chlorination, either by hypochlorite or chlorine dioxide (usually two stages), with an alkali-extraction washing stage between. On occasions oxygen is also used. Bleach-plant chemicals are usually prepared within the mill so that solutions such as chlorine water, sulfuric acid, sodium chlorate, sodium chloride, sodium hydroxide, calcium hypochlorite, and chlorine dioxide all have to be pumped.

It cannot be emphasized too strongly that materials of construction are of vital importance in the bleach-plant chemical-preparation area.

In addition to the standard chemicals, some of the common pulp-mill bleach substances may also be included. The commonest, together with some chemical preparation systems, are as follows:

Chlorine This is usually delivered to the mill in tank cars but is always vaporized to a gas before use.

Chlorine Water (Hypochlorous and Hydrochloric Acid) Concentrations cover the range of pH 2 to 10 for bleaching pulp. In some cases the gas is mixed directly with pulp in special mixers. Group C lined pumps are essential.

Sodium Hypochlorite and Calcium Hypochlorite This is made in the mill by permitting chlorine to react with either sodium or calcium hydroxide concentrated caustic (70 percent) diluted to 5 to 6 percent before chlorination. Calcium hypochlorite is made from a 10 percent solution of slaked lime with temperatures up to 150°F, but not normally exceeding 70°F in a bleach plant. These liquors are corrosive to steel, and therefore rubber or epoxy linings are necessary when handling solutions *to* the bleach plant; *after* bleaching the filtrate may still have residual hydrochloric acid.

Chlorine Dioxide This is the most common bleach solution used because it gives an excellent brightness to the pulp and, despite corrosion problems, is usually cheaper than other bleach solutions.

After generation of the gas, during which absolute cleanliness is vital, the gas is stripped in a packed tower as an aqueous solution, and stored in plastic tanks using special resins to resist chemical attack. In modern plants an increasing use is made of glass-reinforced plastic with selected resins for piping, valves, and pump linings. This is usually a cheaper alternative than the use of exotic metals such as titanium for pumps. Pumps must be group C type, and stainless steel is not satisfactory. Solution strengths up to 8 g/per liter are used.

Sodium Peroxide and Hydrogen Peroxide These are used for bleaching ground wood pulp. Typical solutions contain sodium silicate (5 percent), sodium peroxide (2 percent), and sulfuric acid (1.5 percent). The latter is for pH control of the liquor. Concentrations of bleach liquors are up to 15 percent. Temperatures are usually less than 90°F. Group C pumps are necessary.

Wash Liquors In general the filtrate from bleach washing stages will exhibit at least some of the properties of the stage immediately before washing owing to slight excesses of chemical present. Filtrates are collected in corrosion-resistant pipes and vessels, usually made from glass-reinforced plastic, and the pumps used will be either group B or C, depending on the stage in question. The filtrate from the chlorine dioxide stages may be pumped with a 317 stainless steel case and trim pump because the filtrate is not as corrosive as the bleach solution.

Spent acid from chemical-preparation plants is also highly corrosive, and usually stainless steel is not satisfactory for use with it.

Effluent from the bleach plant, on the other hand, is usually a mixture of several liquors, and experience has shown that 317 stainless steel is a suitable material for pumps that handle it.

Chlorine Dioxide Preparation; Sodium Chlorate This chemical is used with the Mathieson process, and ClO_2 is produced by permitting sodium chlorate to react with sulfuric acid in a vessel into which sulfur dioxide is introduced in controlled quantities. Sodium chlorate solutions are usually from 43 to 46 percent, at which strength the specific gravity will be about 1.38. Stainless steel pumps may be used, but epoxy-resin-lined pumps are superior.

Foul Condensate This arises from the evaporation of water from black liquor at the multiple-effect evaporators, as these units flash vapor from the liquor in one stage and use this to evaporate the liquid in the next stage. The vapor when condensed contains some carryover from the black liquor, and thus the condensate is contaminated and corrosive. When a nickel cast iron casing and stainless trim are used, group A pumps should be satisfactory. Some liquors produce very corrosive vapors, and a stainless casing pump may prove necessary. Group B pumps are recommended.

PULP AND PAPER STOCK SYSTEMS

Nature of Stock Stock is the term applied to the suspension of cellulose fiber in water. It first appears either after grinding (in the case of mechanical pulp) or after the blow tank (in the case of chemical pulp). The quantities of stock required for a given pulp or paper production are shown in Fig. 3.

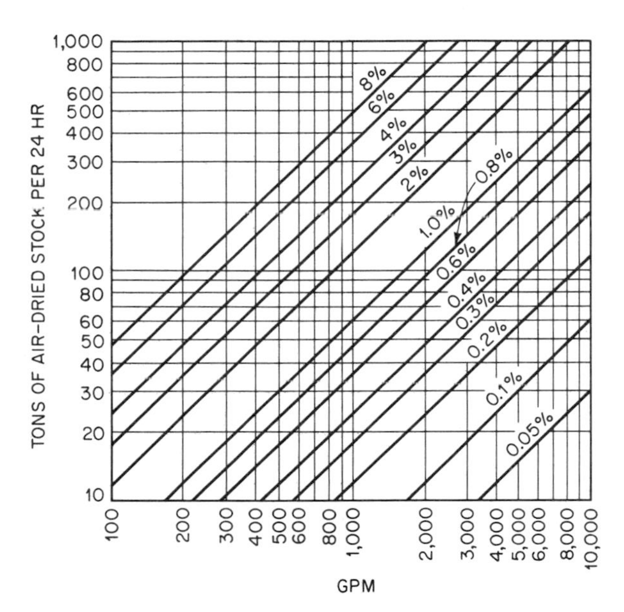

Fig. 3 Stock quantity conversion chart.

After the separation of chemicals or impurities by washing and screening, the stock is given a mechanical treatment known as either beating or refining, depending on the nature of the treatment. This enhances the sheet properties. Additives such as starch, clay fillers, alum, and size are introduced to impart special characteristics depending on the end use of the product.

Over the range of stock in normal use the specific gravity may be considered constant for all practical purposes, with a value equal to that of water at the appropriate temperature.

Cellulose fibers have a tendency to float in water, and constant agitation is required to ensure that stratification does not occur in storage. Agitation, however, can also introduce air, to the detriment of the stock.

The pH of stock varies over a wide range—from as low as 1.0 during some bleaching processes to 11.0 with others. Within the paper-machine-room area the stock will usually be within a pH range of 4 to 8. Thus from a corrosion viewpoint washed stock does not usually present special problems except when high-grade bleach products are produced. Stains will be caused by the presence of iron sulfides or oxides, and therefore stainless steel must be used—frequently 304 for washed stock, but 316 or 317 within the bleach plant before washing or where bleach liquor is likely to be present with the stock.

Unbleached-paper mills generally do not experience corrosion with washed stock, except in the case of groundwood mills where the pH is usually lower than for chemical pulp.

Fiber Characteristics Stock made from softwoods will have a predominance of fibers within the range of 2.8 to 3.5 mm long and 0.25 to 0.3 mm wide; fibers from hardwoods will be about 1.0 to 1.3 mm long and 0.1 mm wide. Straw fiber will be still shorter—0.75 mm on the average—but flax can have fibers up to 9.0 mm long. The above figures are typical, and are of interest because of their effect on pump performance.

Consistency. This is the amount of dry fiber content in the stock, expressed as a percentage by weight. Typical values will vary from about 0.1 percent for the feed to the head box of a special paper machine to 14 percent for stock between some bleaching stages or in high-density towers. The critical stock consistency in the selection of pumps is between 2 and 3 percent. Up to the 2 to 3 percent level pumps may be selected on the basis of their water performance. Above 3 percent performance of the pump decreases rapidly with increasing consistency. The magnitude of the correction will depend not only on the properties of the stock, but also on the type, size, and specific speed of the pump involved. Figure 4 illustrates the water performance of a typical stock pump and the effect of stock consistency on that performance.

Freeness. When stock is beaten or refined, it acquires an affinity for water, and the longer the stock is beaten or refined the longer the water-retention period. The retention of water by the stock increases the friction factor of the flow of stock. In pipe flow the increased friction factor is usually not significant because the pipe velocities are low. At higher velocities in a pump, however, a heavily beaten stock with a very low freeness value exhibits a very slippery characteristic, and the stock may be very difficult to pump.

Freeness is often measured by an instrument called the Canadian Standard Freeness Tester, and the range of values covers a scale from 0 to 900. This instrument measures the amount of water drained from a sample of stock under a regularly decreasing head. Its use is recommended by the Technical Association of Pulp and Paper Industries (TAPPI), and it is commonly employed in North American practice. Another measurement of freeness frequently used in Europe is the Schopper Riegler System, the numerical values of which increase in the opposite direction to that of the Canadian Standard Freeness test.

Flow Characteristics To illustrate the basic characteristics of stock, some typical curves are shown in Fig. 5.

Stock Pumps In stock pumps viscosity may be a problem, but consistency is not a major problem until a value of about 6 percent is reached. The essential requirement is to get the stock to the pump impeller, and every effort should be made to keep the piping as large and straight as possible.

Above 6 percent special pumps are required, and they are usually of the screw type. Air entrainment in the stock will reduce pump output. Air entrainment occurs from agitation in the chests, from flow over weirs, and from flow through restricted openings. The effect on pump performance of air entrained in water and in stock is shown in Fig. 6.

Piping Arrangement. Piping should be straight and as short as possible. This is particularly important on the suction side of the pump to prevent dewatering of the stock. The suction piping should be at least one pipe size larger in diameter than the pump suction size, and should project into the stock chest. The inlet end of the suction pipe should be cut at an angle, and the bottom of the pipe should be at least 1½ pipe diameters from the bottom of the chest. With the long side of the pipe on top, the probability of drawing air into the suction of the pump through vortices is reduced. Some manufacturers provide a lump-breaker device or screw feeder at the suction side of the pump for pumping stock above 4 to 5 percent consistency.

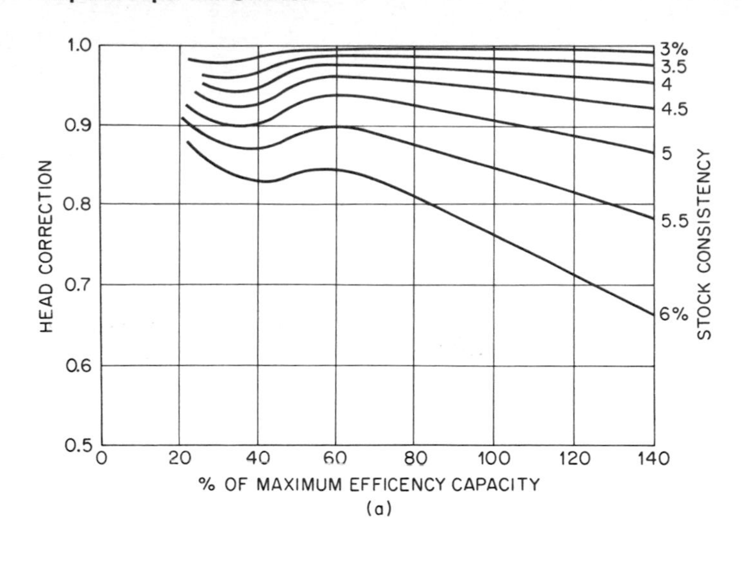

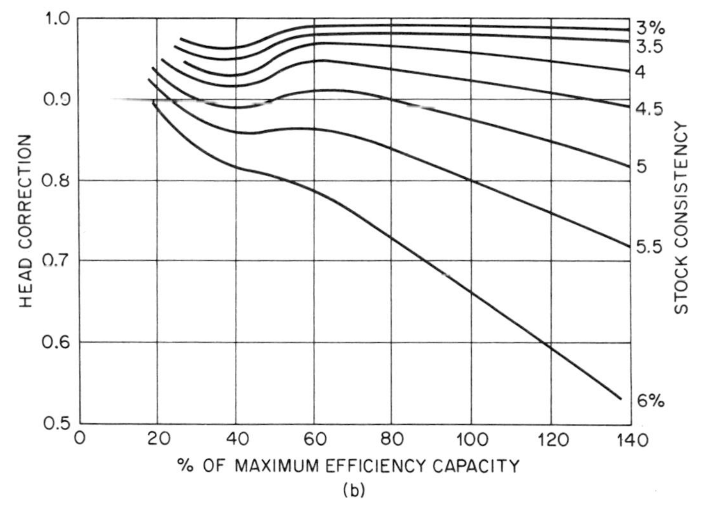

Fig. 4 Effect of stock consistency. (*a*) Groundwood stock. (*b*) Other stocks.

Size of Pumps. It is important to estimate the performance requirements of stock pumps as accurately as possible. Oversize pumps can cause an unbalanced radial thrust on the impeller, excessive wear of the sleeves and glands, and an increase in the risk of cavitation.

Economics and Pump Selection The normal economic considerations of any continuous process apply equally well to pulp and paper mills, with a few points of difference. In pump installations the improved cost figures that are possible from larger units are limited to some extent by the manufacturer's standard size units. Because the industry is capital intensive, the overriding factor in any pump installation is reliability. To achieve this it does not necessarily follow that it is better to have two pumps installed, with one as a standby. One properly designed and serviced unit may well be better than two unknown units; this is especially true

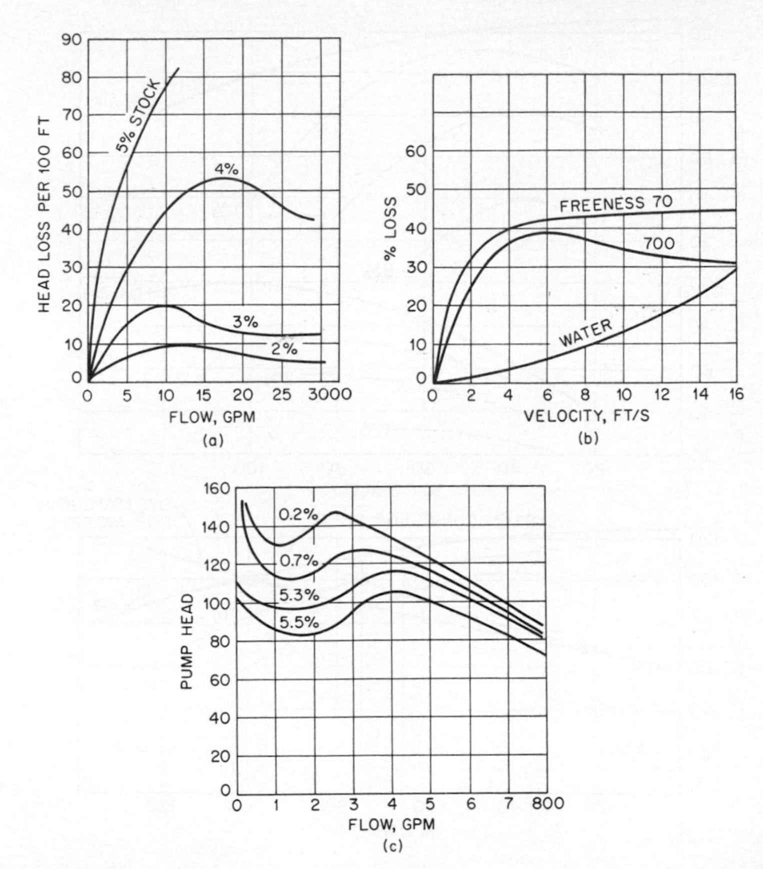

Fig 5 (*a*) Typical friction loss. (*b*) Typical freeness loss. (*c*) Typical pump-head curve with stock consistencies.

where the pump is in a portion of the process that cannot be interrupted without serious losses, either in raw materials or in quality of the finished product.

A duplication of pumps means complications in extra valves, pipework, connections for steam and viscous liquids, electric motors, cables, and starters. The result is that, in modern mills with good machinery and materials of construction, there is a strong tendency away from the duplication of pumps because of increased costs and questionable reliability. It follows that the important thing is to select the right pump and the right duty point in a particular range. In the case of stock pumps this is very important because the shape of the head-capacity curve can be changed by the consistency of the stock (see Fig. 6).

There are some 200 pumps in a modern pulp and paper mill, but the cost of these pumps is probably less than 5 percent of the total equipment cost. It is unwise, therefore, to jeopardize mill reliability by compromising pump quality. Corrosion and erosion are major factors in pump life, and even with the best materials the life of some components in severe service may be 12 months or less. Moreover, the power used by pumps is usually less than one third of the mill demand. If one remembers that the cost of total power absorbed in a mill is only around 4 percent, even a 50 percent reduction in pump power will still be less than 1 percent net.

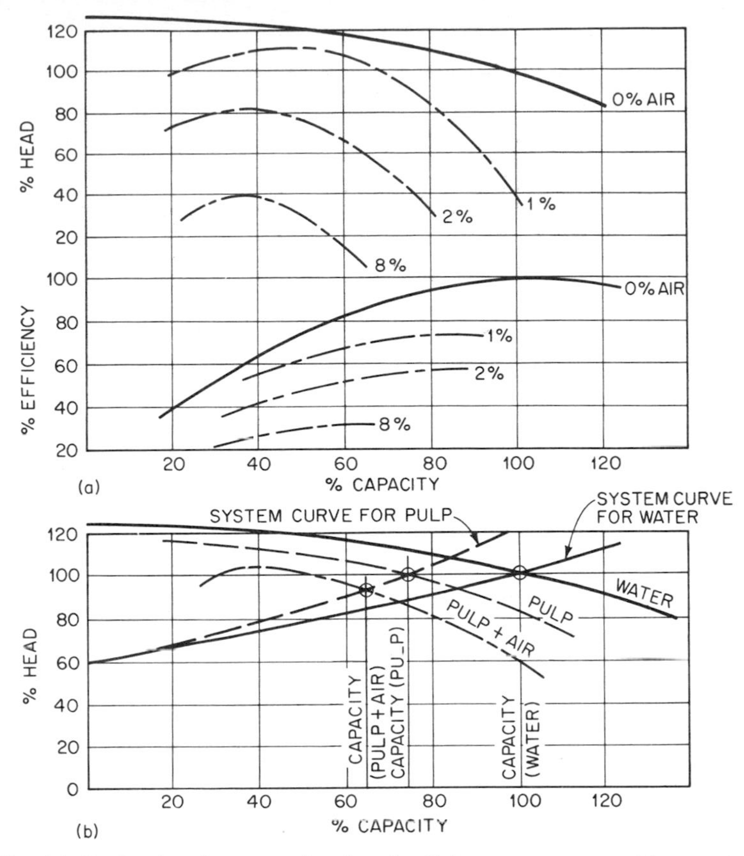

Fig. 6 (*a*) Reduction in pump head and efficiency on water with entrained air. (*b*) Reduction in pump capacity when pulp contains air.

Efficiency The best point to operate a pump is, of course, at its maximum efficiency, but this is not always possible. This is particularly true in the case of stock pumps. In any event, there will be a reduction in the water efficiency because of the viscosity effects in liquor and stock pumps. Pumps designed to produce maximum efficiency on water are not necessarily the optimum selection for the requirements in a pulp or paper mill. For example, open impellers and excess clearances reduce the efficiency, yet these factors are much more important in stock pumps than efficiency. Another important consideration is speed. Stock pumps should be chosen to run at as low a speed as possible to achieve stable operation, and this speed may not produce an efficient pump. The shape of the pump performance curve is much more important than the best efficiency quoted by a manufacturer. A flat or unstable head curve may produce surging or instability in the pump output. Good pump selection, therefore, must emphasize reliability as the first consideration and efficiency and costs as secondary considerations.

Pump Speeds Most of the pump duties in a mill can be accomplished by single-stage pumps and four-pole-motor speeds. For liquids other than water, two-pole-motor speeds should be avoided if possible. For special duties, including stock pumping, six- or eight-pole motors may be required unless some indirect or variable speed is used. While it is true that lower speeds mean larger pumps and more expensive electric motors, lower speeds are justified because of reduced maintenance and greater reliability. Some deviation from these speeds may be necessary for pumps generating heads in excess of 150 ft, but this can often be taken care of by a larger impeller rather than a higher shaft speed.

As motor power increases, the use of variable-speed pumps becomes more attractive, and stock pumps ahead of a larger paper machine may require a prime mover in excess of 500 hp. In such cases variable speeds can usually be justified, and the use of thyristors, steam turbines, and hydraulic couplings should be considered.

For pumps of lesser power, experience in several modern mills demonstrates that with conservative design V-belt drives are reliable and give excellent service.

Multistage Pumps Except with boiler feedwater, the use of multistage pumps should be avoided. This is particularly true for stock and viscous liquors. The complication of the pump design is not justified for these services.

Pipeline Systems With black liquor, green liquor, and similar high-viscosity liquors, adequate provision must be made for steaming out and subsequent liquor drainage. The pumps must also be included in this system. Although the liquor pumps should be designed to pass some solid matter, motorized strainers should be used on the pumps for cyclone evaporators and recovery boiler-fuel pumps because both pumps discharge to spray nozzles. Dead pockets and other areas where liquor can collect should be avoided, since there will build up solids from the liquor which may break away to block pipelines or pump impellers. For protection at shutdowns, even for short periods, steaming out is essential.

Positive-Displacement Pumps Apart from high-density stock pumps, the principal use of the positive-displacement pump is for consistency control of stocks above 5 percent. The normal measuring device used is quite satisfactory at low consistencies, but is less reliable at the higher values. More satisfactory control may be achieved by using a screw pump where the power is proportional to the pulp consistency at constant flow. Such pumps have been very reliable on consistency control.

Digester Circulating Pumps Digester circulating pumps are used with indirectly heated batch digesters to circulate the liquor at the digester pressure and temperature. Maintenance problems are common on these pumps because of the high heads (150 lb/in^2) and temperatures of 350°F. In addition, the circulating liquor contains some raw pulp even though screens are fitted to the digester outlets. Pumps for this service, therefore, should be of very heavy construction and have open impellers and water-cooled stuffing box assemblies. In addition, there is often considerable pipework involved, for a digester may easily be 60 ft high and pipe loads are often imposed on the pumps. Expansion joints and long radius bends are used, but it is desirable to support the pumps on springs.

Heavy-Black-Liquor Pumps Above 60 Percent Total Solids A typical pump for this service is shown in Fig. 7. An open impeller in a 316 stainless steel casing is recommended. A heavy sleeved shaft of 316 stainless steel with ample clearance between the rotating parts is also required for satisfactory operation. These pumps may be required to handle black liquor up to 8,000 Redwood seconds viscosity, and should be provided with water cooling. Steam jacketing is not always satisfactory because the liquor may tend to bake on the walls of the casing. Pump speeds should be below 1,800 rpm, if possible.

Multiple-Effect-Evaporator Pumps Pumps in this service often operate in cavitation owing to problems in regulating the flow between evaporation stages. Level control valves in the suction line to the pump can alleviate this problem, but cavitation can still be expected in the pump. A self-priming pump may increase the life of the rotating elements by comparison with the condensate-type pump usually used on this service.

Diaphragm Pumps Diaphragm pumps are used in pumping lime mud slurries of high concentration. They consist of a rubber or neoprene diaphragm with a pulsating air supply on the one side controlled by a timer and the slurry on the other.

Stock Pumps for Intermediate Consistencies Representative of pumps in this category are those provided with a breaker device and/or screw feeder at the pump inlet. The casing is often also split at 45° to permit easy access in the event of a blockage. Stock at 6 percent can be easily handled provided proper care is taken with suction piping.

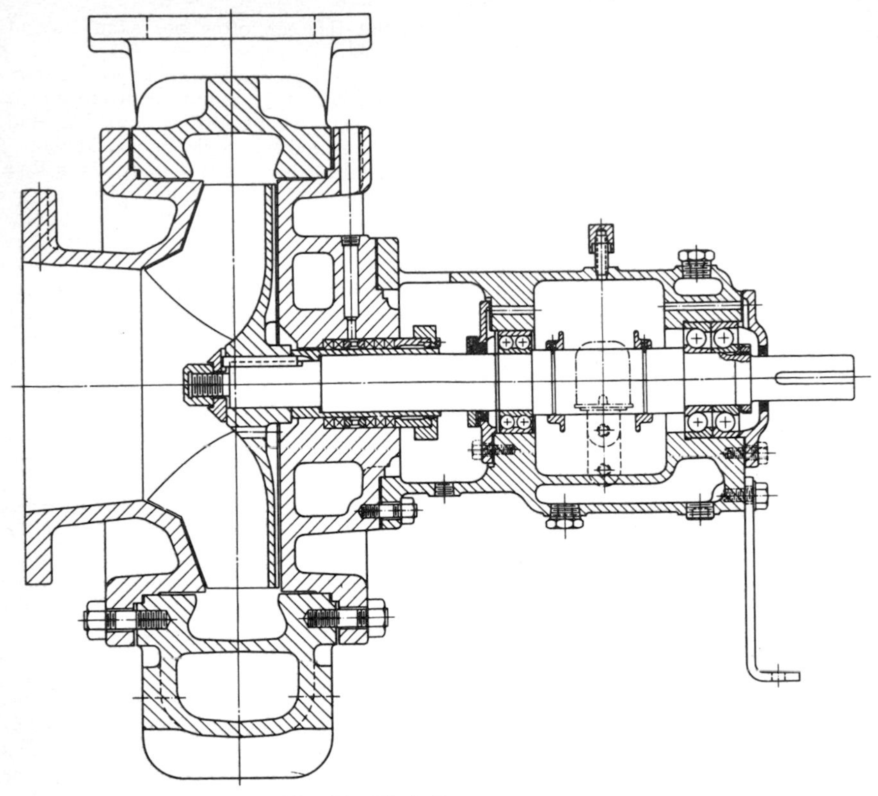

Fig. 7 Black-liquor pump.

Vacuum Pumps These are used to extract water from the sheet on the fourdinier wire and at the suction presses by means of a vacuum up to a maximum of about 25 inHg. Approximately 40,000 lb of water is extracted by this means for every ton of paper produced, and this water is removed by entrainment with the air handled by the vacuum pumps. Frequently water separators are used to remove water; their use is a matter of economics, as a reduction of up to 10 percent in power may be achieved.

Types of Vacuum Pumps Vacuum pumps are of three basic types:

1. Water ring
2. Positive displacement
3. Centrifugal or axial-flow machines

Many engineers prefer the water-ring type, probably because of its simplicity. In general, however, this type uses more power, mainly because of the heating of the circulating water, which is then discharged to a drain.

Centrifugal and axial-flow machines must be provided with water separators, but they are more efficient over all, particularly when the heat of compression is used in the machine-room ventilating system. The machines run at high speed and are usually driven by a steam turbine.

The paper-machine system requires vacuums at different levels from a few inches of mercury to the maximum possible. Often pumps are connected to a common header, and orifice plates are used to divide the flow to ensure some measure of standby capacity. This involves throttling, however, and may create flow problems unless quantities are carefully estimated. The axial-flow-type machine permits extraction at any point along the rotor within fairly close limits, and requires an accurate estimation of the quantities and specific vacuum required. A standby for axial-flow machines cannot usually be justified.

Stock and Liquor Pump Standardization Throughout the mill the duties of many pumps are similar, but different materials of construction may be used. If, at a slightly extra initial cost of the pump, the rotating elements can be standardized, this will reduce spare-parts inventories. This is also an advantage when purchasing pumps for a new mill. For example, if a standard arrangement consists of a complete rotating element including bearings, only the impeller size and material need be different. For this reason the nonstock pumps are divided into the two groups A and B. Standardization is an additional reason for recommending that all pumps be stainless fitted. Obviously, large pumps need an individual evaluation. Standardization of stock pumps is less feasible, but up to about 3 percent stock similar pumps can usually be specified.

Specification of Design Margins In many cases an excess margin on head and capacity is specified for pumps. Where margins are excessive over calculated figures, mechanical troubles and cavitation often occur. Because pump manufacturers have different standards, it is not realistic to use a fixed formula for margins, if only because the accuracy of the estimate will vary with the liquor concerned. However, experience in pump selection indicates that the following guidelines should give reasonable results in most cases.

1. Carefully estimate pump duty, using the best available data and the actual piping configuration to be used.
2. Allow in the head loss estimated for control valves the increased head loss for the partially open position.
3. Using the above figures, add 10 percent to the flows at the normal load and 10 percent to the head at the maximum duty.
4. Compare the new flow figure to the maximum flow anticipated. If it is greater, make no adjustment to the figure obtained in item 3. If the new figure is less, add 5 percent to the maximum flow anticipated.
5. To the figures derived from item 4 add a further 10 percent for flow and 5 percent for head to allow for the decrease in pump performance with time.

Notes: 1. The above margins should be applied before the motor is selected. When considering motor power, allow for the maximum impeller diameter.

2. Using the figures derived from item 5, study the manufacturer's pump curve and note the impeller diameter and range for the pump. If the duty point falls near the end of the curve and if the impeller is the largest that can be fitted in the casing, it is advisable to select the next size larger pump.

Section **10.9**

Food and Beverage Pumping

JAMES L. COSTIGAN

INDUSTRY STANDARDS

Centrifugal, rotary, and reciprocating pumps are used throughout the food and beverage industry. Unlike other industrial applications, pumps for the food and beverage industries must meet rigid sanitation codes known in the industry as the 3-A Standards. These standards have been established by the following organizations:

1. The International Association of Milk, Food, and Environment Sanitarians*
2. The U.S. Public Health Service
3. The Dairy Industry Committee composed of representatives from the following:
American Butter Institute
American Dry Milk Institute
Dairy and Food Industries Supply Association
Evaporated Milk Association
International Association of Ice Cream Manufacturers
Milk Industry Foundation
National Cheese Institute
National Creameries Association

The 3-A Standards were established originally for the dairy industry, and are rigidly enforced by the local sanitarian. The baking, bottling, and packing industries have also adopted this standard with modifications to suit their particular requirements.

The 3-A Standards specify that the wetted parts of any pump must be furnished in a type 300 series stainless steel, or in an equivalent material of the same corrosion-resistant properties. In the case of centrifugal pumps, this means that the impeller, the casing, the backplate, and the seals must be of stainless steel. Cracks or crevices in any of the wetted parts are not permitted and are a cause for rejection of the defective component. The finish and minimum radius of curvature of any wetted surface are specified by the Standards. Internal threads are not permitted except in designs where no alternative fastener is possible for the proper functioning of the pump. A further requirement of the Standards states that the pump must be designed in such a way that the wetted parts can be readily disassembled for inspection and cleaning. The purpose of these specifications is twofold. The first is to prevent the accumulation of food or beverage in isolated points and its ultimate putrefaction. The second is the ability to clean the internal components of the pump as quickly and efficiently as possible.

* Standards are available from the International Association of Milk, Food, and Environmental Sanitarians, P.O. Box 437, Shelbyville, Indiana.

PUMP TYPES

Single-stage end suction pumps are widely used throughout the food and beverage industries. They may be close-coupled or mounted on a baseplate and coupled to a separate motor. Pumps of this type were formerly used almost exclusively for transfer service, but in modern plants they are applied throughout the process cycle. The use of variable-speed motors and controls has simplified the problem of the variable-flow requirements of such process machines as separators, clarifiers, and filters. Variable-speed pumps are also used as booster pumps to the timing pumps of plate pasteurizers.

Many of the heavy viscous products in the processing of food cannot be pumped effectively by centrifugal pumps. Cottage cheese and baking doughs are typical examples. Positive-displacement pumps are required for this type of service, and the rotary pump of the lobe type with timing gears is usually selected. Internal-gear, single-screw, and flexible-vane pumps are used to a lesser extent. Sanitary lobe-type pumps are limited to 150 lb/in^2 discharge pressure, the flexible-vane and internal-gear pumps to 50 lb/in^2, and the single-screw pump to 90 lb/in^2.

The rotary lobe pump is also used where processing procedures require its use. An example of this is the timing pump used on a high-temperature short-time pasteurizer. The flow rate is critical, and the system is carefully timed by the sanitarian. The drive is then sealed to prevent an unauthorized increase in the flow rate.

MECHANICAL SEALS

Mechanical seals are used almost exclusively throughout the food and beverage industries. An acceptable seal must not only be simple in construction, but also must be readily accessible for the take-down cleaning. A sanitary version of an external balanced seal is shown in Fig. 1. An alternative design, shown in Fig. 2, can be cleaned in place, while that shown in Fig. 1 must be removed from the pump for cleaning. It is now permissible to clean pipelines and some equipment in place provided strict sanitary controls are maintained. This procedure is known in the industry as CIP (cleaned in place) as opposed to COP (cleaned out of place).

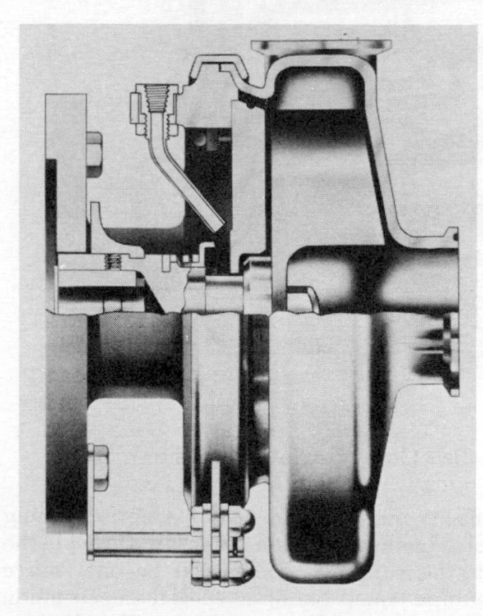

Fig. 1 A sanitary version of an external balanced seal. (Ladish Co., Tri-Clover Division)

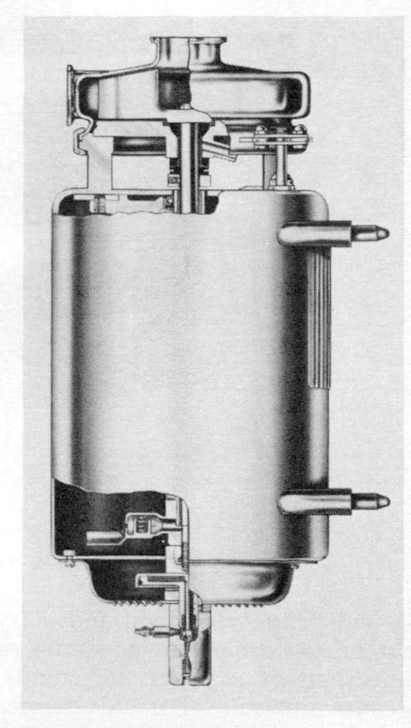

Fig. 2 An approved cleaned-in-place seal. (Ladish Co., Tri-Clover Division)

The approved mechanical seal where CIP is permissible consists of a single carbon ring between the impeller and the backplate. It is loaded by a spring coupling driving the integral impeller-shaft combination. The seal is disengaged for cleaning by a manually or pneumatically operated cylinder. The cylinder presses against a thrust bearing on the shaft and opens the seal by pushing the shaft and impeller forward. Sanitizers are circulated through the pump following the cleaning, rinse, and detergent cycles. A separate CIP supply pump is used for circulation during the cleaning and sanitizing operations. The preferred procedure is to coordinate the cleaning cycle of the pump with the cleaning cycle of the valves, the fittings, and the pipeline.

The sealing mechanism of the rotary lobe pump can be either an O-ring seal, as shown in Fig. 3, or a mechanical seal as shown in Fig. 4.

Fig. 3 An O-ring seal. (Waukesha Foundry)

Fig. 4 A mechanical seal. (Ladish Co., Tri-Clover Division)

A wide variety of food and dairy products is marketed in cans. Aseptic canning must be performed under sterile conditions. Leakage of nonsterile air through the connections or seals of the pumps used in the canning process can be one source of contamination. For this service the pumps must be of a special construction to prevent air leakage. Figures 5 and 6 show a rotary pump and a centrifugal pump seal designed for canning service. Steam or a sterile liquid is used as a sealing medium to prevent the entrance of air into the pumped fluid.

Fig. 5 Rotary pump. (Waukesha Foundry)

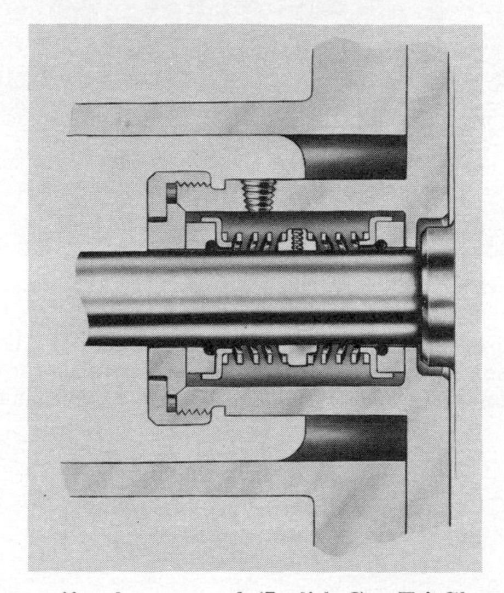

Fig. 6 A centrifugal pump seal (Ladish Co., Tri-Clover Division)

HOMOGENIZER

The homogenizer as used by the food industry incorporates a reciprocating pump—usually a triplex. On a high-temperature short-time pasteurizer the homogenizer can be used not only as a homogenizer but also as a timing pump. The homogenizer can also be used as a high-pressure pump for the jets in spray-drying operations by the simple expedient of removing the homogenizer valves.

PUMP DRIVES

Standard electric motors are used as pump drives in the food and beverage industries. The totally enclosed fan-cooled motor is preferred for cleanliness. To facilitate cleaning of the fan, the canopy should be secured with wing nuts or wing bolts. It is also desirable to specify a motor with a fan that can be readily removed from the shaft. The motor and pump of a close-coupled unit must be installed at least 2 in above the floor. The baseplate of a direct-coupled unit must be grouted

Fig. 7 A typical pump and valve installation. (Ladish Co., Tri-Clover Division)

to the floor or supported on piers of minimum height. The baseplate must be of a solid construction without pockets and sealed to the motor to facilitate cleaning.

A typical pump and valve installation in a modern food plant is shown in Fig. 7.

Section 10.10

Mining Services

W. D. HAENTJENS

INTRODUCTION

It might be questioned why pumps used in mining services receive a separate classification when the types involved do not differ in principle or in general appearance from those used in fresh-water service. The answer is the need for utmost reliability and an ability to withstand both corrosive and abrasive waters. Not all mine waters are corrosive, but almost all mine waters from active mines are abrasive because of the suspended solids from mining operations. Much can be done to limit the amount of solids handled, as noted later under Sumps, but the solution generally requires a compromise based on an economic study. In other words, removal of most coarse solids is usually justified, but removal of fine solids is generally impractical. Thus a heavy-duty type of pump has evolved which warrants the classification of mine pump.

PUMPING CONDITIONS AND TYPES USED IN MINING SERVICES

The general pumping conditions in mining services can be determined from a consideration of the types of mines and the materials being mined. For example, the broadest category would be a division between open-pit and deep mines. Open-pit mines seldom exceed 600 ft in depth, so the pumping heads generally are not much greater than this unless very long discharge lines are involved. Unless the open-pit mine is located in an arid region, the greatest pumping load is produced, not from ground water but from rainfall. Unless a certain increase in water level can be tolerated at the bottom of the pit, a study of the maximum rainfall rate and the drainage area involved will determine the required pumping capacity. These flow rates are usually large and the combination of large capacities with moderate heads generally requires a double-suction pump, either horizontal or vertical. If the capacity is large enough for good hydraulic design, a single-stage pump is preferred. Where the capacity is lower, two-stage units are common. Because of the fine solids generally present, operating speeds are usually limited to 1,800 rpm. This puts a limit on the head per stage, and thus minimizes wear.

The development of large overhung-shaft vertical pumps has been a boon to open-pit mines for these units can easily be raft-mounted. The availability of large billets of plastic foam material has simplified small raft construction and made them virtually unsinkable. The weight of the water in the pump as well as in the discharge line to the point where an off-barge float contributes to the support of the hose must be considered. The number of personnel who might be aboard at any one time must be considered, as well as the possibility that all persons might be grouped together at one side. The forces involved in either overturning or righting a barge must be determined. With the float riding level, the center of gravity and the center of buoyancy are in the same line. If someone stands on one side of the barge, the barge tilts and more of the barge becomes submerged on that side. If there is sufficient freeboard, there is increased submergence, and the barge is buoyed up by a greater force on that side because of the increased submergence. The center of gravity of the pump now acts so that its vector is displaced from the geometric center of the barge. As long as there is enough freeboard and buoyancy, the shift in the center of the buoyancy must be such as to have a righting moment. If there is insufficient freeboard, then the center-of-gravity vector may be outside the center of buoyancy, and the barge will overturn. The buoying force is dependent on the volume of displaced liquid, and once the entire section is submerged there is no increase in buoying force.

Large barges have been designed, including some for the installation of four 700-hp or three 1,500-hp pumps. Overhung-shaft pumps are available in sizes to 1,500-hp with either single or double-suction designs. Where the pumping conditions permit, a single-suction top inlet is preferred because the pump will not draw sand and mud from the bottom of the pit at minimum water level.

The selection of a single- or double-suction pump is based on the capacity-head combination at 1,800 rpm, or at the maximum operating speed. For a given set of hydraulic conditions of head and capacity, the operating speed of a double-suction pump can be increased by a factor of 1.41, from the square root of 2 times the capacity in the specific speed formula $N_s = \frac{N\sqrt{Q}}{H^{3/4}}$. Thus, in cases where a 1,200-rpm motor might be required by a single suction design, an 1,800-rpm motor, which is smaller, lighter, and cheaper, could be used for a double-suction design.

The wide variation in required capacity for open-pit mines raises the problem of parallel pumping. The head at the maximum capacity is frequently much greater than that during normal pumping, and cavitation may occur during single-pump operation. There are a number of solutions. The best solution, although not necessarily the cheapest, is to use a separate discharge line for each pump. If the discharge line is short, this may be practical; if not, groups of two pumps can be combined in a single pipeline. The alternatives should be carefully examined by a system head analysis. An alternate solution with a single discharge line is a variable-speed drive for one or more pumps. It is seldom necessary to have all pumps with variable-speed drives.

Pumps used in underground mines must produce heads above 500 ft, since that is the maximum depth of an open-pit mine. Flow rates are generally lower than in an open-pit mine because a heavy rainfall does not produce an immediate heavy flow of water. Underground mining operations also attempt to divert as much surface water as possible away from the mine. When this is done, flow rates for individual pumps are in the range from 1,000 to 5,000 gpm. Although an underground pump station might be designed for a maximum inflow of 20,000 gpm, it is unlikely that pumps larger than 5,000 gpm would be selected. There are a number of reasons for this limitation. These include the physical size of the pump and motor, as compared with the size of the mine shaft and haulage way, the maximum weight of the pump and motor that can be handled and serviced underground, the maximum water-storage capacity in the sumps, the effect of continuous as compared to timed (off-peak power) pumping, and the consequences of a power failure. The effects of a power failure require a careful analysis to determine the normal sump level and emergency storage capacity. Electrical inrush limitations frequently determine the capacity of individual pumps. For surface-mounted transformers, long power lines may limit the motor size. Although reduced-voltage motor starters may be used, the cost can rarely be justified.

The maximum size motor and total pumping capacity may still not be sufficient information to make a decision as to individual pump capacity. Obviously the pumping head must be considered. If the mine is less than 1,200 ft in total depth, there may be only one pump station, although many mines operate at higher heads in a single lift. It is necessary to consider where the greatest water inflow occurs. If it should be at the 300-ft level, it would be foolish to let all the water go to the bottom of the mine to be pumped out. The principal pump station should then be at the 300-ft level, with a smaller station at the bottom of the mine pumping to the 300-ft level or directly to the surface.

The depth of 1,200 ft is selected as a guide point since the main consideration is the suitability of 300 lb/in^2 fittings. If 600-lb/in^2 fittings must be used, the pumps and pipeline will become much more expensive. Again, this must be compared with the cost of additional pumping stations at 1,000-ft to 1,200-ft intervals. Some studies may show that 2,000-ft pumping intervals are justified.

When the pumping head exceeds 2,400 ft, it is difficult to obtain satisfactory horizontally split-case multistage pumps, and a decision must be made as to the desirability of pumping in increments or going to the barrel-type pump. Barrel pumps are available for very high pressure and were developed for high-pressure boiler-feed pump service. Since most barrel pumps for mine service operate at high heads per stage with close clearances, the mine water must contain a minimum amount of abrasive particles in suspension.

The basic decision in the design of the pump room is whether the sumps will be above or below the pumps. This will determine the type of priming system and the complexity of automatic control. The difference in available net positive suction head may also restrict the selection of pumps. A further decision is that of economic discharge-line size. Line velocity should be kept below 10 ft/s and generally in the neighborhood of 6 to 8 ft/s. This may not be possible if there is little room in the shaft. If stainless steel or lined pipe is used, higher velocities may be tolerated to reduce the cost of the piping. With higher velocities, however, an evaluation of the risk of damage from water hammer must be determined.

The benefits of a sump below the pump-room level and a positive prime of the pumps can be achieved with low head (in the range of 20- to 50-ft) vertical turbine pumps. What may appear as a complication is in effect a saving because pump priming and automatic control are simplified. Occasionally a combination of these arrangements is used for maximum safety.

The opposed-impeller horizontally split-case pump is used in most mines for underground operation. A principal reason for this is the ease of servicing and the rapidity with which a pump can be repaired. Most mines keep a completely assembled spare rotating element on hand. This includes the stationary parts (such as casing rings, partition rings, etc.) and the bearings and pump-half coupling. Removal of the top half of the pump casing permits replacement of the worn rotating element with a new or reconditioned unit. The worn unit can then be taken to the surface for repair.

Although vertically split-case pumps are employed in some foreign mines, the time required for servicing is generally greater, and except for barrel designs, they have not found wide acceptance in the United States. Another reason for this apart from the servicing time, is that the impellers all face in the same direction and produce a large hydraulic axial thrust. This thrust can be balanced by various devices such as balancing drums or pistons, but such devices are notoriously affected by abrasive solids in the water.

If the mine water cannot be economically clarified for pumping with conventional mine pumps, it is occasionally necessary to consider alternate arrangements. The use of slurry-type pumps in series at one location is a possible solution. Where there is no supply of clean water available for flushing the stuffing boxes, some mines have found it advantageous to use vertical overhung-shaft pumps in vertical open sumps at elevation intervals of 300 to 400 ft. These pumps do not have a stuffing box, so packing or seal problems are eliminated. Most underground pumps use packed stuffing boxes, although mechanical seals are available. Because of the dirty water usually present, the mechanical seals must be flushed, and some of their advantage is lost. The principal objection to mechanical seals in this service is that a seal failure requires a complete pump disassembly, whereas a stuffing box failure can be corrected simply by repacking the stuffing box.

PUMP MATERIALS

Most water pumps perform quite satisfactorily with bronze impellers, wearing rings, and shaft sleeves, but the dirty or corrosive waters in mine service require superior

	Neutral waters		Acidic waters		Basic waters	
	Moderate heads	High heads	Moderate heads	High heads	Moderate heads	High heads
Casing	Cast iron	Ductile iron/cast steel	316 S.S./Alloy 20	17-4 PH	Cast iron	Dectile iron/cast steel
				PH55A		
Impeller	28%Cr	28% Cr	PH55A/17-4 PH	PH55A/17-4 PH		
			CD4-MCu	CD4-MCu	28% Cr	28% Cr
Wear rings	28% Cr	28% Cr	Same as impeller		28% Cr	28% Cr
Shaft sleeve	28% Cr or 303 S.S. ceramic coated	28% Cr or 303 S.S. ceramic coated	316 or Alloy 20 ceramic coated, PH55A, etc.	316 or Alloy 20 ceramic coated, PH55A, etc.	28% Cr or 303 S.S. ceramic coated	28% Cr or 303 S.S. ceramic coated
Shaft	Carbon steel	Hi-Tensile alloy steel	316 S.S./Alloy 20	17-4 PH	Carbon steel	Hi-Tensile alloy steel
				17-4 PH		
	Alloy 20	21% Cr	29% Ni	2.5% Mo		
	CD4-MCu	26% Cr	5% Ni	2.0% Mo		
	28% Cr	28% Cr	—	—		
	304 S.S.	19% Cr	10% Ni	—		
	316 S.S.	19% Cr	10% Ni	2.5% Mo		
	17-4 PH	17% Cr	4% Ni	—		Hardenable
	PH55A	20% Cr	10% Ni	3.5% Mo		Hardenable

materials. Since mine waters range from neutral (pH of approximately 7), to severely acid (as low as 1.5 pH in some coal mines), and to very basic (as occurs in limestone mines, etc.), there is no universally best material. The choice obviously is the lowest priced material which gives satisfactory service life. Many times the decision cannot be made on the basis of the best available materials, but on the material that will withstand the abrasive conditions during the intended life of the project. If there is a five-year anticipated life of the mine, there is little advantage in selecting materials which will last twice as long. Conversely, a mine with a twenty-five-year projected life would require the best commercially available materials. There are also the exotic alloys, but their use is seldom justified.

Assuming that there will always be a slight amount of solids present, the metallurgy in the table on facing page should be considered.

Care should be exercised when selecting alloys because many stainless steels do not have as great a strength as carbon steel. The same applies to bolting. Thus a pump which is rated for high-pressure service with a carbon-steel or alloy-steel casing and bolting may not be suitable when made of bronze or stainless steel.

Although ceramic-coated shaft sleeves are excellent in many applications, it should be remembered that ceramic coatings are porous and that the base metal must be able to withstand the environment. Also some ceramics will not be suitable in strongly basic water. Others, however, are suitable, so a general specification for ceramic coating should never be made. A plasma-applied ceramic is generally more dense and serviceable than one which is applied by a simple flame spray.

HYDRAULIC CONDITIONS

Low-Head Pumps For low-head pumps up to approximately 150 ft, the installation must be carefully checked to prevent cavitation during periods of low-head operation. This is particularly important during one-pump operation on a parallel system. For example, if two 5,000-gpm pumps are discharging into a common discharge line against a static head of 80 ft and a friction head of 60 ft, the friction head will be only 15 ft when one pump operates alone at 5,000 gpm. The total head will now be only 95 ft, and the single pump will carry out to a much higher capacity. Unless sufficient NPSH is available for the single-pump run-out point, the pump will cavitate. Although the two pumps should be selected for 5,000 gpm at 140 ft total head, the required NPSH should be determined not at 5,000 gpm but at the capacity corresponding to the intersection of the pump and system curves.

High-Head Pumps For high-head pumps of 1,000 ft head or more, the risk of cavitation for single-pump operation on a parallel system is less than for low-head pumps. For example, if the static head is 1,000 ft and the friction head is 60 ft when two 5,000-gpm pumps are operating, the total head will decrease from 1,060 ft to only 1,015 ft when one pump operates at 5,000 gpm. This means that, for either one- or two-pump operation, the capacity of each pump will be approximately the same, and the risk of run-out cavitation is minimal.

Water Hammer and Pressure Pulsations A water hammer analysis should be made of both high- and low-pressure pumping systems. While the transient-pressure pulsations are related to the rate of change of velocity rather than the magnitude of the steady-state condition, mine experience indicates that when pipe velocities exceed 10 ft/s water hammer problems can be anticipated. In high-pressure pumping systems it is not unusual for transient-pressure pulsations to be as high as 300 lb/in^2 above or below the steady-state pressure.

Transient-pressure pulsations have been experienced in low-pressure pumping systems. The danger here is that the low-pressure portion of the cycle will fall below atmospheric pressure and the pipe will collapse.

While any pumping system for mine service should be analyzed in detail for transient-pressure pulsations, experience has shown that adequate air bottles have proved to be one of the most effective and least expensive means of surge suppression. Slow-closing valves and flywheels have been used, but they must be sized correctly. This is especially true with high-speed pumps, since they possess little rotational inertia and will decelerate very rapidly on shutdown with accompanying high-pressure surge.

SUMPS

Permanent sumps are seldom used in open-pit mines because the sump area generally moves as the mining operation progresses. This means that the pump station must be portable, and the installation of the pumps on a barge provides the most convenient arrangement. Either horizontal or vertical pumps may be used, but vertical pumps eliminate the need for priming equipment. If the vertical pumps are of the overhung-shaft design, the stuffing box may be eliminated. This is important if the water is dirty. Although small cyclones can be used to clarify the water for pumps which have stuffing boxes, care must be taken to prevent leaves and other trash from blocking the gland water line. In freezing climates, special provision must be made to prevent the suction line, pump, and even the barge itself from freezing in place.

Underground mine sumps present special problems because they function not only as sumps but also as clarifiers. A well-designed sump is a good clarifier, but all too frequently inadequate provisions are made for cleaning the sump. If it is not cleaned at regular intervals, the loss of storage capacity may be critical in the event of a power failure. Furthermore, a sump partially filled with solids does not give the proper retention time for clarification, and the solids are directed into the pump. Although few sumps can economically be made large enough for complete clarification, it is important that a large portion of the solids be removed. This is particularly true for 3,600-rpm pumps because the high speed generally produces a high head per stage. The high differential pressure between stages causes severe wear if abrasive solids are present. Some mines have installed conventional thickeners and flocculating agents in an attempt to keep a high concentration of solids from reaching the pump.

Some general rules should be considered in designing sumps for underground pump rooms:

1. Attempt to get a complete anaysis of the water (from another portion of the mine or from an adjacent mine if necessary).

2. Analyze the sample for corrosive properties to determine the proper materials of construction for the pump.

3. Analyze the sample for possible scale buildup in the pipeline and pumps. Check the velocity effect, if any, on the buildup rate.

4. Examine the sample for percentage of suspended solids, the screen analysis, and the settling rate for various fractions. Determine the sump dimensions for removal of all solids and then for progressively larger solids in order to select the most economical size.

5. Compare the sump size above with the size required for physical storage capacity for: (*a*) continuous pumping, (*b*) off-peak power pumping, (*c*) programmed pumping, and (*d*)) storage for estimated maximum length of power interruption.

6. Calculate practical sump dimensions, considering the geologic conditions.

7. Consider methods for cleaning the sump. Compare mechanical cleaning methods with cost of parallel sumps.

8. Install grit traps ahead of the sump to remove large, heavy solids. Consider methods for cleaning the grit traps.

9. Install trash screens to prevent wooden wedges, etc., from entering the sump.

10. Review sump-cleaning methods and program. The best designed sump is of no value if it is not cleaned.

11. Review the suction requirements of the pumps to be used. Because of altitude, temperature, distance from low water level to pump centerline, and suction line loss, the available NPSH may be inadequate for even an 1,800-rpm pump. If a decision as to the pump size, type, and speed has been made and an NPSH problem does exist, then a decision must be made either to use slow-speed booster pumps or to lower the pump-room level below the sump level. From a safety standpoint, the use of a booster pump is preferable, although it does add another piece of equipment.

12. Where the storage capacity is inadequate to meet possible power failures, consider either vertical pumps (possibly up to 100 ft) for the shaft bottom pumping

up to the main pump station level or sealed pump rooms that can operate over wide variations in the sump level from a 15-ft suction lift to a positive head of several hundred feet.

13. Determine the final design based on a compromise between the mine engineer (who wants maximum output), the electrical engineer (who wants small starting load), the geologist (who wants small sump dimensions), and the mechanical engineer (who wants the most reliable and easily maintained equipment).

AUTOMATIC PUMP CONTROL

Few pump stations can be operated economically with manual control. Automatic control can be a simple float switch or a pressure switch, or it can be sufficiently complex to provide reliable operation under the most critical or adverse conditions. Automatic control can provide greater reliability, and its cost can be depreciated over only a few years. Furthermore, the automatic recording of flow rate, flow totalizing, and periods of operation provides valuable data for analyzing the performance of the pumping installation as well as the possible cost savings in pumping during off-peak power periods.

Where the safety of a mine is dependent on the reliable operation of the dewatering pumps and controls, the following minimum requirements should be considered:

1. A sump-level alarm for high water. This should be both local and remote (at the surface).
2. Sump-level control should be dependable. For example, electrodes are generally unreliable in waters which leave a conducting film.
3. The control should be programmed where more than one pump is installed. However, the use of an alternator is not always desirable because all pumps are exposed to the same degree of wear. It is preferable to have one pump as a standby and programmed through a sequence selection switch so that the standby pump operates at least once per week.
4. Pump priming should be positive. Hydraulic-type devices should be combined with electric controls so that complete dependence is not on the electric control. The actual presence of water in the pump should be detected to prevent starting a dry pump.
5. A delay circuit should be provided to ensure complete priming.
6. The control should provide for at least three starting attempts (unless an overload has occurred).
7. The priming time should be limited (if under a suction lift system).
8. The control should provide for a restart in the event of a false loss of prime on startup (suction lift system).
9. Pump and motor bearings should have thermostats to stop pumps in the event of bearing failure.
10. Vibration monitoring may be important. This is particularly true for vertical pumps.
11. Pressure controls should indicate normal pressure, and fail safe in the event of a loss of pressure (broken column line, etc.)
12. Flow indication (check-valve flow switch) is needed to signal a shaft failure.
13. Remote indication (generally at the mine office) should provide at least an indication of operation and signal pump failure or high water. More detailed information may be transmitted.
14. For long distances, investigate the use of carrier-current indication schemes, together with signal multiplexing, etc.
15. Provide a method to test the control and alarm system.

DRIVERS

Open-pit mines use electrically driven mining equipment, such as drag lines and shovels, and the availability of power has permitted the use of electrically driven pumps. There are still many gasoline-driven or diesel-driven units, but the convenience of electric power, particularly for automatically controlled units, has increased the trend to electric drive. The availability of reliable high-voltage cable has made portable high-voltage equipment safe and economical. Pumps in open-pit service are seldom provided with sophisticated control or drive mechanisms. The primary requirements are reliability, portability, and wear resistance.

In locations where rainfall may be heavy and there is danger of power failure, a combination of electric drive and engine drive is used. The engine can be direct-coupled to the pump through a motor with a double-extended shaft or with a clutch between the engine and the motor. Automatic control is simple and reliable.

Although some steam-driven pumps still exist in underground service, their number is rapidly decreasing. Electric motor drive is the simplest for automatic control. Variable-speed units, however, are seldom used in underground service because the ratio of static head to total dynamic head is quite high. Thus the friction loss is not a large percentage of the total head and not much advantage is gained by variable speed. The solution is usually a multiple pump installation. This must be designed with care because it is possible to raise the friction head to a point where an additional pump produces little additional capacity. Multiple discharge lines are the answer and are frequently used for safety reasons. In normal service, both discharge lines are used in parallel, although conservative design allows each line to handle the required capacity.

As with all pumping installations, a complete set of system head curves must be prepared to analyze the power requirements under all conditions.

Motor enclosures are important in underground service. Because of the high humidity, special insulation (epoxy encapsulated, etc.) should be specified. Drip-proof enclosures are the minimum requirement, with weather-protected Type I a preferred construction. Heaters should also be provided. Screens should be installed to prevent the entrance of rats. Winding temperature detectors, bearing thermostats, and ground-fault detectors are recommended in mine service and should be incorporated in the pump-control and alarm circuits.

Although the starting torque of a centrifugal pump is low, in some cases the available torque may have to be checked. A normal torque motor should be suitable for pumps in the range of 500 to 3,000 specific speed. High-specific-speed pumps, however, have the highest horsepower at shutoff, and it will be necessary to examine the starting arrangements for such pumps.

Starting a pump against a long empty pipeline may present overload problems, and repeated starts may be necessary. In such cases the number of permissible starts per hour should be checked. Winding temperature detectors are important in such applications.

Although reduced-voltage starters may be required in some instances, most modern mines have electrical facilities designed for across-the-line starting. This is preferable because the starting equipment is cheaper. Before deciding on a reduced-voltage starter, the effect of the starting load on the transformer and line impedance should be checked. It may be that the voltage drop will eliminate the requirement for reduced-voltage starters. On the other hand, the effect on the primary side should be checked so that the voltage drop is not so large as to drop out other equipment.

Synchronous-motor drives are seldom used unless they are of large size (generally at least 1,000 hp), and then only if they are in relatively continuous service. Under such conditions they can be operated under "leading current" conditions for power-factor correction. Smaller installations frequently provide capacitors at the pump installation to provide the necessary correction for a particular installation.

Surge protection from lightning should not be overlooked. Some locations are particularly suspectible to lightning damage, especially to long surface lines. Lightning arresters should be provided at the surface, and surge arresters should be mounted at the motor location.

Section **10.11**

Pumps for Construction Services

STANLEY McFARLIN

PUMPING EQUIPMENT FOR THE CONSTRUCTION INDUSTRY

Large numbers of self-priming centrifugal, submersible, and diaphragm pumps are used on construction projects of all types. Typical applications include drainage, well pointing, jetting, and pressure grouting. Most units of this type are truck-, skid-, or wheelbarrow-mounted to allow easy movement from one site to another.

Although some are motor-driven, most centrifugal pumps for this service are engine-driven—either gasoline- or diesel-fired. Most construction jobs require the removal of soil waters, land drainage, and rainwater before and during construction. Pumping equipment in this field is called upon to perform under conditions more severe than in almost any other service.

The construction industry goes back as far as the need of man for shelter. Leather buckets and wood or stone shovels were the first tools used for dewatering work areas. Not until the discovery of electrical power and the development of the piston engine was it possible to develop any sort of improved pumping equipment to aid the construction industry.

Displacement Pumps Displacement pumps of either the piston or the diaphragm type were the first utilized by the construction industry. They were large, heavy, and difficult to maintain. Capacities were limited, so that with large volumes of water or large inflow rates dewatering could be achieved only by the installation of additional pumps.

Centrifugal Pumps The first centrifugal pump was designed and built in 1680, but not until the year 1818 was a practical pump of this kind constructed. Since centrifugal pumps are essentially high-speed machines, they were not used in the construction industry until the advent of the high-speed electric motor or steam turbine in the early twentieth century. In the 1920s the advantages of centrifugal pumps were recognized, and their smaller size, portability, and ability to pump large volumes at even flow rates and moderate heads represented a vast improvement.

The standard centrifugal pump cannot handle air or vapors unless they are located beneath its source of supply. Some means or device must be applied to maintain liquid in both the pump and its suction piping to avoid loss of prime on pump shutdown. Foot valves are used, but are a continual source of trouble because of debris in the water.

The demand naturally developed for a centrifugal pump able to handle appreciable quantities of air, and to reprime itself automatically when located above the water supply. This requirement is especially important in the construction field because pumps may be used to dewater areas into which the seepage rate is lower than the evacuation rate (snorkeling). The standard centrifugal pump will operate until the entrance to the suction pipe is uncovered or the liquid vortexes, which results in air binding and inability to reprime, even after sufficient seepage has accumulated to prevent further air infiltration.

Self-priming Centrifugal Pumps The origin of the self-priming centrifugal pump came about in the late 1920s and early 1930s. The term "self-priming" is used quite loosely for pumps in the construction industry and may be misunderstood. Only the positive displacement pumps of the reciprocating and the rotary type are truly self-priming, in that they will develop a vacuum sufficient to cause liquid to flow into the suction pipe without the addition of liquid in its pumping casing as a seal.

A self-priming pump is one that will clear its passages of air if it becomes airbound and will resume delivery of liquid without outside attention. The centrifugal self-priming pump requires that the contained liquid entrain air so that the air will be removed from the suction of the impeller. The air must be allowed to separate from the liquid once the water and air mixture has been discharged by the impeller. The separated air must be allowed to escape or be swept out through the pump discharge. The centrifugal self-priming pump, therefore, requires an air separator in the form of a large chamber or reservoir on the discharge side of the pump to accomplish this separation, and also to trap the residual liquid supply necessary to provide the liquid seal during a reprime or priming cycle. Several methods are used to make a centrifugal pump self-priming; the two most common are (1) recirculation from the discharge back into the suction channel, called "eye primer"; (2) recirculation within the discharge nozzle and the impeller itself, called "peripheral primer." These two basic methods have many variations depending upon the manufacturer and the designer.

Self-priming centrifugal pumps are built in a range of sizes from 1 in through 12 in. The most popular are the 1½-, 2-, and 3-in sizes. They are widely used by home builders and by utility and municipal maintenance crews.

In the last decade it has become apparent that the passage through the pump of stones, sticks, leaves, cans, and rags usually found on construction jobs without causing plugging would be an asset. As a result the introduction of the trash-type self-priming centrifugal pump has made a heavy impact in this field.

Self-priming Trash Pumps These were designed to handle large solids but at a lower unit efficiency. However, the overall evacuation efficiency has increased as a result of the greater resistance to plugging, and down time has been decreased because of the heavier duty parts, seals, and materials of construction and the increased accessibility of parts. For pumps of this type, the maximum solid sizes are as follows:

1½-in pump for 1-in solids
2-in pump for 1¼-in solids
3-in pump for 1½-in solids
4-in pump for 2-in solids
6-, 8-, and 10-in pumps will pass 3-in solids or larger

The most important elements of pumps for the construction industry are erosion of the wearing parts, seal life, reprime capability, ability to handle solids, and the economics of replacement parts. In the construction industry pumps of any type are usually considered a necessary evil. The pump is, therefore, probably the most misused and least understood piece of equipment that a contractor operates. From the standpoint of the contractor, he is interested in equipment that will stay "put" the longest without attention. The manufacturer is concerned with offering equipment that will provide longevity of parts, the capability of handling severely erosive liquids, the capacity for handling large solids without plugging, and easy access for construction crews to all the problem areas.

Electric Submersible Pumps With the increasing availability of electric power in remote areas in the early 1950s, electric submersible pumps were developed for the construction industry. These pumps are available in discharge sizes up to 10 in with heads and capacities from zero flow at 300 ft of head to 4,500 gpm at 10 ft of head. The submersible dewatering pump has slowly replaced the established well-pointing system for drying up excavations. Well pointing consists of a vacuum system that draws water from various suction points located in and around the construction site. The drainage from each suction point is collected in an air separator through a common header. A centrifugal pump then discharges the collected drainage in the air separator to a point of runoff.

While all these pumps and systems are still in use, most contractors recognize the economic advantage of using self-priming centrifugal and submersible pumps.

CONTRACTORS' PUMP BUREAU

Manufacturers of pumping equipment for the construction industry have established the Contractors Pump Bureau, known as the CPB. The purpose of this bureau is to establish standards for the user as well as for the manufacturer. These standards apply to quality, performance, classification, engines, and auxiliary equipment. The bureau does not test or certify pumps. The member manufacturers, however, are required to test and to certify that their products meet the CPB ratings. The bureau has published its own manual of standards and specifications. The CPB manual is required for anyone interested in this field of pumping machinery.

It is apparent that in the selection of a pump for the construction industry many aspects must be considered to select the proper pump. Listed below are the basic pump classifications used in construction installations, with their advantages and disadvantages, together with a typical pump selection table.

ADVANTAGES AND DISADVANTAGES

1. Self-priming centrifugal pump

   Advantages:

   Light weight for large capacity.
   Relatively high discharge heads.
   Large capacity for investment.
   High suction lifts.
   Simplicity of design.
   Does not require suction foot valve.
   Choice of materials.
   Multiple choice of drives.
   Pump location above the liquid.
   Can be operated without presence of attendants for repriming.
   Low maintenance costs.

   Disadvantages:

   Slightly less efficient than centrifugal pumps.
   Will not prime unless water reservoir is filled initially.
   Must be reasonably parallel to the water surface.
   Suction strainer should be used to screen out items too large to pass through the impeller (normally listed by manufacturer).

2. Trash self-priming centrifugal pump

   Advantages:

   Capable of handling large spherical solids without fear of plugging.
   Capable of pumping heavy muck and muddy water.
   Very abrasive-resistant.
   Contains easy-access cleanout for all mortality parts without disturbing the plumbing.
   Driven by heavy-duty engines and heavy-duty auxiliary equipment as required.
   Has the other advantages of standard self-primer centrifugals.

   Disadvantages:

   Less unit efficiency than standard self-priming centrifugal pumps.
   Not as portable as the self-priming centrifugal pump.
   Normally requires a suction check valve to retain maximum priming liquid.

3. Submersible pumps

   Advantages:

   Instant priming.
   No problem of NPSH requirements.
   Can run dry.

Good for mopping-up exercises.
Does not require attention.
No danger of being flooded out or frozen.
Handles solids.
No noise problems.
Easily portable.
Small dimensions to fit in well casings and coffer dams.

Disadvantages:

Electrical supply and control requirements usually more expensive than self-primer pumps of equal size.
Must be pulled out of liquid for inspection.
Will not handle solids as large as trash self-priming centrifugal pumps or diaphragm pumps.
Susceptible to damage by electrical storms and to cable damage by vehicles.

4. Diaphragm pumps—mechanically driven

Advantages:

Will handle large solids.
Will handle mucky or muddy water.
Will handle slow seepage.
Will handle air as well as liquid.
Diaphragms and valves are easily replaced.
Suitable for high lifts.
Self-priming.
Can run dry.
Multiple choice of drives.
Low maintenance costs.

Disadvantages:

Low discharge heads.
Small capacity for investment.
Pulsating flow.
Considerable vibration during pumping operation.
Must never have plugged discharge.
Capacity is determined by pump size only and not by engine speed.
Will not sustain suction heads.

5. Double diaphragm pumps—air-driven

Advantages:

High discharge heads.
High suction lifts.
Completely self-priming from dry start.
High percentage of solids handling.
Infinitely variable capacity and pressure.
Discharge can be closed and pressure is maintained with no power consumption or wear.
Can run dry indefinitely.
Low abrasive wear at high heads.
Large air-handling capability for mop-up work.
Safe for hazardous areas.

Disadvantages:

Low capacity as compared to centrifugal pump.
Less efficient at high capacity.
Compressed air required on site.

PUMP SELECTION

The Contractors Pump Bureau has established pump ratings based upon their capacity in gallons per hour at a minimal suction lift when discharging directly from the pump. There are two classification ratings set up by the CPB that identify the self-priming centrifugal pump: M-rated or MT-rated. The M-rating certification applies to the self-priming centrifugal pump and refers to size and capacity in gallons per hour. A typical CPB rating chart is shown in Table 2. The MT

TABLE 1 Pump Selection

	Self-priming centrifugal pump	Diaphragm pump	Trash pump	Submersible pump
Dewatering:				
Clear water	X			X
Slimy water	X	X	X	X
Muck water	X	X	X	X
Mud water	X	X	X	X
Silt water	X	X	X	X
Abrasive water	X	X	X	
High-solid-content water		X	X	
Slow-seepage ditch water		X		X
Fast-seepage ditch water	X		X	X
Septic tanks	X	X	X	
Manholes	X		X	X
Well points	X	X		
Cofferdams	X	X		X
Quarries	X	X		X
Deep piling dewatering				X
Supply:				
To mixer or paver	X			
Concrete curing	X			X
Water wagons	X			

trash-pump rating requirements apply to the heavy-duty self-priming trash pump and require the incorporation of a removable end plate which will provide an easy access to the interior as well as to the impeller for cleaning purposes. The MT-rated pumps can pass spherical solids in sizes as listed in Table 3. There are other rating requirements for the M and MT pumps (self-priming) such as the type of seals, engine specifications, etc. For further details, see the "Contractors' Pump Manual." However, the listings are the primary requirements. The M or MT rating seal on trash pumps constitutes the manufacturer's certification that such pumps will meet all CPB standards.

TABLE 2 Typical CPB Rating Chart

Rating	Size of pump, in	Minimum capacity, gpm
5M–MT	1½	5,000
8M–MT	2	8,000
10M–MT	2	10,000
17M–MT	3	17,000
20M–MT	3	20,000
30M–MT	4	30,000
40M–MT	4	40,000
90M–MT	6	90,000
125M–MT	8	125,000
200M–MT	10	200,000

TABLE 3 Spherical Solids Passed by MT Pumps

Pump size, in	Solid size, in
1½	1
2	1¼
3	1½
4	2
6	2½

Most of the pumping equipment advertised by the construction trade today will exceed the performance specifications outlined by the CPB. However, the customer should examine the manufacturer's published performance data to determine that his requirements will be satisfied by the pump he selects.

There are five basic facts that a manufacturer should know before recommending a piece of pumping equipment:

1. How many gallons per minute are to be pumped?
2. How high is the pump above the liquid?
3. How high must the water be lifted after it leaves the pump?
4. The total length of hose or pipe to be used.
5. The source of power required.

As in most fields of technology, the pumps in use today are the result of a long line of development improvements. Problems of wear, seal life, reprime capability, the ability to handle solids, and cleaning have been largely resolved. It should also be recognized that contractor pumps developed to resolve these difficult service requirements are finding acceptance in other industries. Thousands of contractor-type pumps are in daily use pumping raw sewage, paper stock, and industrial waste. Sanitary and paper-mill engineers have found that the nonclog capabilities of contractor pumps will often solve their pumping problems and reduce the cost of operation.

Section 10.12

Marine Pumping

G. W. SOETE

Marine pumping services can be divided into two broad categories: (1) those related to the boiler feed cycle of the power plant, and (2) those auxiliary services that support the power plant, support the hotel load, support the loading and off-loading of the ship, and, in the case of naval vessels, support the armament. Pumps of the centrifugal type and rotary displacement type predominate. Reciprocating steam pumps, fixed-stroke and variable-stroke power pumps, and regenerative pumps are used to a lesser degree.

BOILER FEED SYSTEM

Condensate The condensate pump receives water from the condenser hotwell, which is normally under a vacuum. The piping from the hotwell to the pump must be short and direct to minimize friction losses. As the hotwell is already at a very low point in the vessel, the condensate pump can be positioned only 6 to 24 in below the hotwell. The condensate pump must operate under a relatively small positive submergence (NPSH), and requires speeds of 1,750 rpm or less. The pump must operate at conditions of varying capacity to suit the power-plant load. The chief means of control is to allow the pump to operate in the cavitation break, as shown by the curves in Fig. 1. At rated load the pump operates at point R, at which time the hotwell level provides a submergence as shown. This defined submergence provides the energy required to cause the condensate to flow to the suction area of the impeller. The pump can deliver only that specific capacity. The maximum submergence available must exceed the submergence required for operation at the rated capacity R. When less flow enters the hotwell, such as at $0.9R$, the pump, operating at point R, will extract more water from the hotwell than is entering. The hotwell level will be reduced, thus lowering the submergence at the pump suction. The pump capacity will be reduced to $0.9R$, coincident with the amount entering the hotwell. At this lower flow rate the system head requirement is reduced, and the pump will develop the total head as shown at point A. If the flow is reduced further to point $0.7R$, the total head is reduced to point B on the system head curve. When flow to the condenser increases, a rise in the hotwell level follows, and the higher level of submergence furnishes the additional energy needed to increase the flow to the pump suction. The pump responds by producing the corresponding increases in head and capacity.

The operation can be very stable provided the flow to the condenser does not fluctuate violently and is not reduced below about $0.7R$. While operating in the cavitation break, the impeller is partially filled with vapor. If the pump is a multistaged type, it is possible to reduce the flow to the point where the first stage

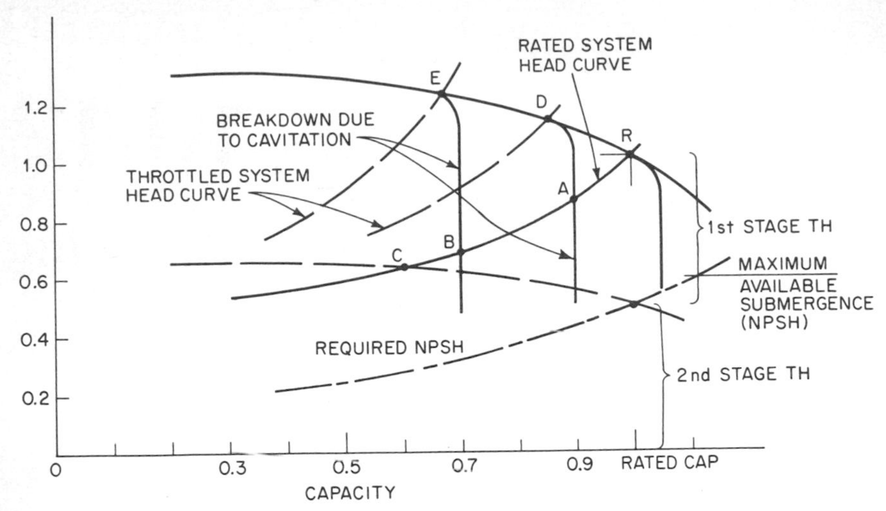

Fig. 1 Condensate-pump performance and system curves.

is completely filled with vapor, and only the remaining stage or stages develop head, as at point *C*. At lower flows, the second stage will become vapor-filled. Condensate service is severe under such conditions of fluctuating low values of submergence, and a close matching of the pump performance and system requirements is required. The term submergence is commonly used and refers to the static dimension between the liquid level and the suction nozzle of a vertical pump, or to the impeller centerline of a horizontal pump. The preferred designation is the net positive suction head (NPSH) at the impeller centerline, which is equal to the submergence less the friction losses in the suction piping.

At reduced flows it is usually necessary to open a bypass to permit a recirculation of the condensate back to the hotwell. This will reduce the noise and vibration that occur in the suction line and in the pump. The bypass may be controlled manually or automatically by the hotwell level. When the condensate is used as cooling water in the air-ejector condenser or other heat exchangers, the minimum flow is determined by those components rather than by the feedwater requirements, and the bypass may be thermostatically controlled.

It is possible to have the pump operate continuously on its head-capacity curve by throttling the discharge. This alters the system head curve to such points as *D* and *E*. A float control in the hotwell maintains a water level between a preset maximum and minimum, and throttles the pump discharge when the hotwell level falls, or opens the discharge as the level rises.

Another arrangement combines the bypass method with the throttling control. It provides throttling of the discharge over the recommended range of operation from 70 percent to the rated capacity, and also provides the additional flow at lower capacities to satisfy the cooling requirements of the heat exchangers. This combination control reduces the noise and vibration in the suction line and pump to a minimum.

The condensate is pumped from the condenser hotwell to the deaerating feedwater heater of the direct-contact type maintained at a pressure of 10 to 60 lb/in^2. The elevation of the heater is determined by the submergence required by the boiler feed pump when there is no booster pump. It is customary to pump the condensate discharge through an air-ejector condenser and gland vapor condenser. This serves the dual purpose of providing a fresh-water cooling source for these heat exchangers and of preheating the condensate before it enters the deaerating feedwater heater.

Condensate pumps may be of one, two, or three stages depending on the precise amount of total head required by a particular installation. The total head is the sum of the difference in pressure between the hotwell and the heater, the difference in their elevation, and the frictional resistance of the piping and the heat exchangers in the discharge line.

Boiler Feed The boiler feed pump receives water from the deaerating feedwater heater. The suction pressure on the boiler feed pump is equal to the sum of the pressure in the heater (usually 10 to 60 lb/in^2) and the submergence of the heater water level above the feed-pump suction, less the friction losses in the suction piping. The exact suction requirement will depend on the NPSH required by a particular pump. Boiler feed pumps customarily run at 3,500 rpm when motor-driven, and at higher speeds up to 12,000 rpm when driven by steam turbines, by the main propulsion turbine, or by the turbogenerator turbine. The relatively high speed of the pump and the higher pumping temperature require a higher NPSH as compared to the condensate pump. In addition, the feed pump cannot be allowed to operate in the cavitation break, but is required to pump at a constant or variable discharge pressure under widely fluctuating capacities.

Figure 2 shows a typical system head-capacity curve that may be required for a marine feed pump. A pump running at constant speed is shown at *a*. At all

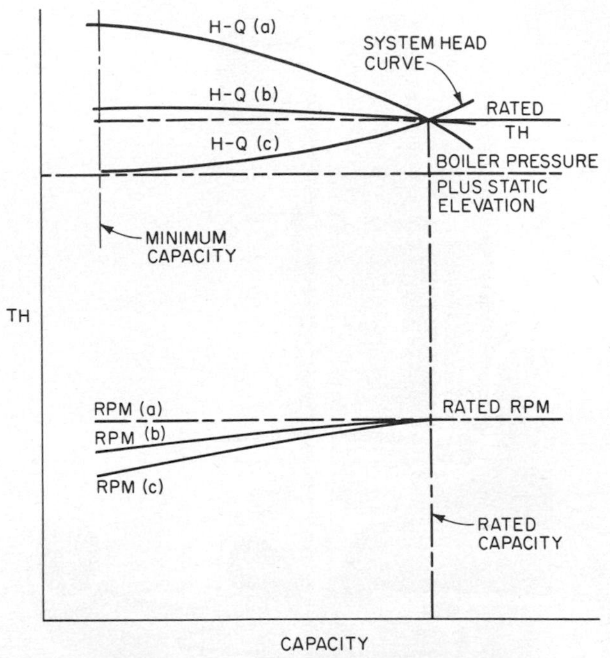

Fig. 2 Boiler-feed-pump performance and system curve.

capacities below the rated capacity there is considerable power loss because of the excess pressure produced by the pump as compared to the system head requirement. For this reason constant-speed pumps are selected only for low-capacity services such as auxiliary feed and waste heat boilers, or where the simpler motor drive offers a degree of reliability beyond that of the turbine drive. A saving in power is realized by running the feed pump at varying speed and at a constant pressure as at *b*. The variation in speed is obtained by using a constant-pressure governor which controls the turbine. The most economical arrangement permits the pump to operate at a varying speed while producing a discharge pressure coincident with the system head curve as at *c*. The driving turbine is controlled by a differential-pressure governor. Merchant vessels operating for long periods at constant load provide the special cases where the feed pump may be driven directly by the turbo-generator or the main propulsion turbine. A separate auxiliary feed pump must be installed for lower capacity or emergency conditions. A feed pump connected to the main turbine must have a special disengaging-type coupling to permit operation from an auxiliary turbine at some minimum speed when the main unit is at reduced speeds. System pressures up to 1500 lb/in^2 can be produced by one- and two-stage high-speed pumps. More stages are required for lower pump speeds and/or higher pressures.

A boiler feed pump requires a recirculation line to limit the temperature rise of the water at low flows. A common practice is to limit the temperature rise to 15°F, but limits of 20 to 25°F may be encountered. The amount of recirculation flow ranges from 5 to 15 percent of the rated capacity. It may be continuous or it may be used only to satisfy the minimum flow condition, and this fact must be considered when sizing the pump. The method of pressure governing must be considered because a pump running at constant speed will require a greater amount of recirculation than a pump running at reduced speed on the system curve. The recirculation system requires a breakdown orifice, a special valve, or friction tubing to dissipate the pressure of the recirculation flow that is returned to the deaerating feedwater heater. For automatic operation the recirculation valve is controlled by the signal from a flow transmitter connected to a flow orifice in the pump discharge line, or by a mechanical connection to a special discharge check valve.

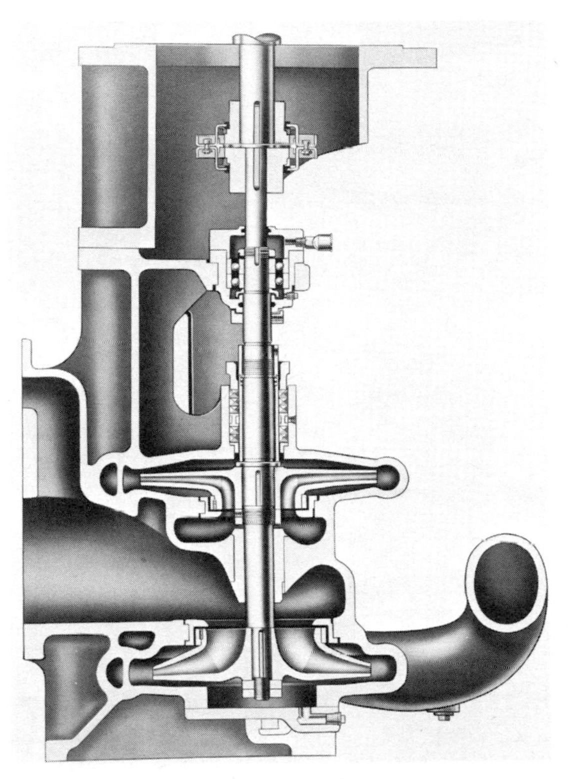

Fig. 3 Two-stage condensate pump. (Warren Pumps, Inc., subsidiary of Houdaille Industries, Inc.)

Feedwater Booster When the design requirements of the system and vessel do not permit the feedwater heater to be installed at an elevation that will provide sufficient NPSH for the feed pump, a booster pump must be installed. The booster acts essentially as the first stage of a feed pump, but at a lower speed—usually 1,750 rpm. It may have one or two stages, and its design is similar to a condensate pump. Unlike the condensate pump, however, which may operate under cavitating conditions, the booster pump forms an integral part of the boiler feed discharge system, and must furnish a stable suction pressure to the feed pump. A recirculation line must be provided to bypass a minimum flow when the boiler feed pump is not in use.

Pump Construction A typical two-stage condensate pump is shown in Fig. 3. A vertical arrangement of the rotating element is preferred to provide favorable suction conditions at the first-stage impeller located at the bottom and to permit access to the bearings, coupling, and vertical driver. Wearing rings are fitted to the casing and impellers. The internal sleeve bearing is lubricated by the pumped condensate, while the upper external combined line and thrust bearing is of the grease-lubricated ball type. Water discharged from the first stage passes to the suction of the second stage through an external crossover. A vent connection is required at the suction flange so that any vapors may pass freely back to the condenser.

Booster pumps are similar to condensate pumps in construction. Single-stage condensate or booster pumps are shown in Fig. 3 except that the second or upper stage would be omitted. Water-lubricated bearings are seldom used in booster pumps because of the higher pumping temperatures.

A typical boiler feed pump is shown in Fig. 4. The pump is driven by a separate steam turbine. A governor is required to provide either constant-pressure or differential-pressure regulation. While one- and two-stage pumps may be direct-connected to the turbine, multistage pumps up to four stages are driven by a separate steam turbine through a flexible coupling.

Fig. 4 Turbo-feed pump. (Worthington Pump International)

Positive-displacement pumps are used for temporary in-port or emergency boiler feed service. Vertical steam reciprocating pumps are simple to operate, can be operated at variable speeds, and are suitable for severe suction conditions because of their self-priming characteristics. Motor-driven fixed-stroke power pumps may be used for low-capacity boiler feed service, but their application is limited by the fact that they are essentially of constant capacity.

FRESH-WATER SERVICES

Condenser Vacuum Motor-driven vacuum pumps operating on the liquid-ring principle are frequently used in place of conventional steam-powered ejectors for extracting the saturated mixture of air and water vapor from the main and auxiliary condensers. This type of exhauster is classified as a rotary pump. Water is used as the sealing medium and serves as the liquid compressant.

A water-sealed vacuum pump is shown in Fig. 5. As can be seen from the illustration, a bladed rotor revolves within the pump body. When sealing water is introduced, the rotor carries the liquid around the eccentric casing and forms a liquid ring that revolves at nearly the same speed as the rotor. The rotating liquid almost fills and then partly empties each rotor chamber once each revolution, setting up a piston action. As liquid passes through the diverging casing sector, it draws in

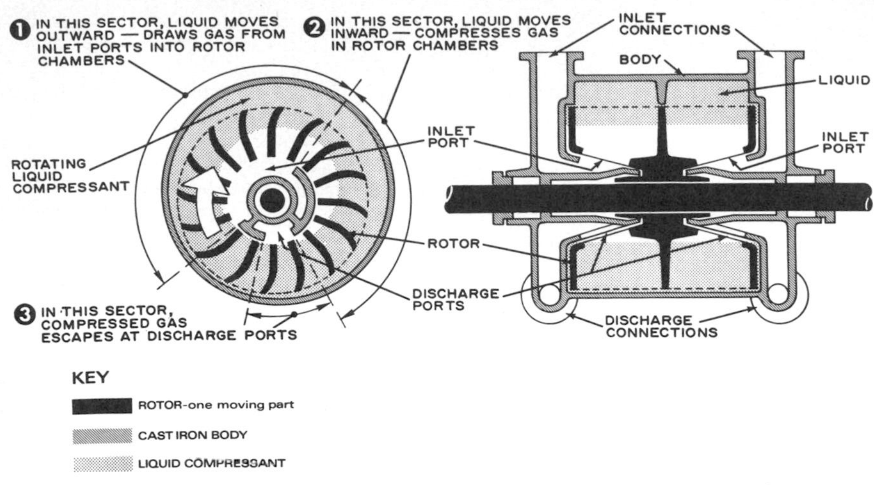

Fig. 5 Condenser exhauster. (Nash Engineering Company)

air through the inlet port near the hub. As the fluid passes through the converging sector of the pump body, the liquid moves inward, and the air and vapor are compressed and forced out through the discharge port near the hub. A portion of the liquid flows out with the compressed air and vapor and is removed in a mechanical separator. The liquid is cooled by circulation through a heat exchanger and is returned to the vacuum pump as makeup. Air and other uncondensed gases are discharged to the atmosphere.

Capacities of dry air with the water vapor required to produce saturation with 7.5°F subcooling of the mixture range from 25 standard ft^3/min at 1 inHg absolute to 65 standard ft^3/min at 3.5 inHg absolute. Speeds of 1,750 rpm and lower are common, and operation is continuous. Pumps of this type are used for main condenser vacuum as well as for auxiliary and evaporator-plant vacuum. Vacuum pumps are also applied as gland seal exhausters to draw the mixture of air, condensate, and steam from the turbine glands.

Drainage Fresh-water drainage service requires centrifugal pumps of small capacity without the capability of being self-priming. A range of capacities to 350 gpm at total heads of 180 ft and speeds of 3,500 rpm are typical. The term "fresh water" implies that the collected water has not been contaminated, and therefore may be reused in the condensate system with a minimum of treatment and deaeration. In practically all cases the pump takes suction from a collecting tank and discharges to a receiver such as the condenser hotwell or fresh-water bottom. Operation is usually on-off, controlled by a float-level switch.

Pump construction may be of several motor-driven types, as shown in Fig. 6: (*a*) the horizontal or vertical end-suction type, (*b*) the horizontal or vertical in-line type where the suction and discharge flanges provide the mounting, (*c*) the horizontal or vertical tank-mounted type where the suction flange serves as the support bracket and the pumping unit is mounted on the tank side, and (*d*) the vertical end-suction type with an extended shaft where the pumping unit is mounted on the tank top, and the extended shaft allows the pump impeller to be at an appropriate submerged level near the bottom of the tank.

Distilling-Plant Condensate Pumps for this service are similar to the general type of fresh-water drain pumps except for several added features. The condensate pump must handle water at its vapor temperature, and, therefore, requires a suitable vent at its suction near the impeller eye. In addition, the pump stuffing box must contain provision for positive sealing to ensure against loss of vacuum. A range of capacities to 350 gpm at total heads of 200 ft is typical. Speeds up to 3,500 rpm are used, but are frequently lower because of the severe suction conditions.

Construction is similar to that shown in Fig. 3 (either one or two stages) and in Fig. 6 (*a* and *c*). Operation is usually continuous, with the pump operating in the cavitation break as described previously and as shown in Fig. 1.

Fresh-Water Supply Pumps for this service are similar to the general type of fresh-water drain pump shown in Fig. 6 (*a* and *b*). A range of capacities to 350 gpm at total heads of 180 ft at 3,500 rpm is typical.

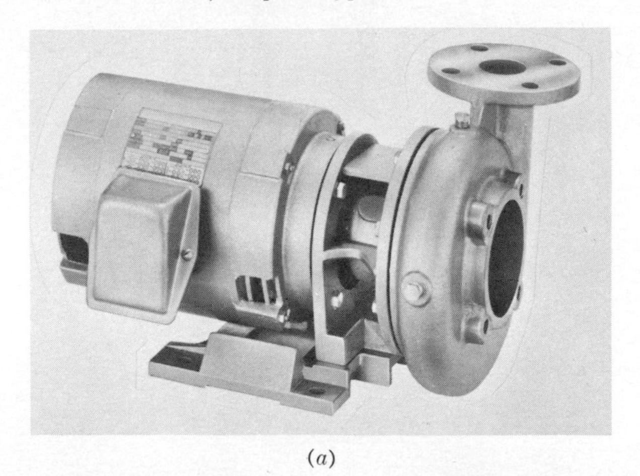

(*a*)

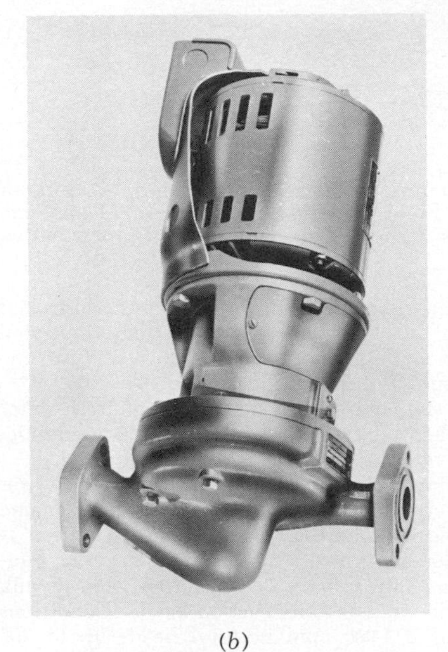

(*b*)

(*c*)

Figure 6 continued on page 10-150

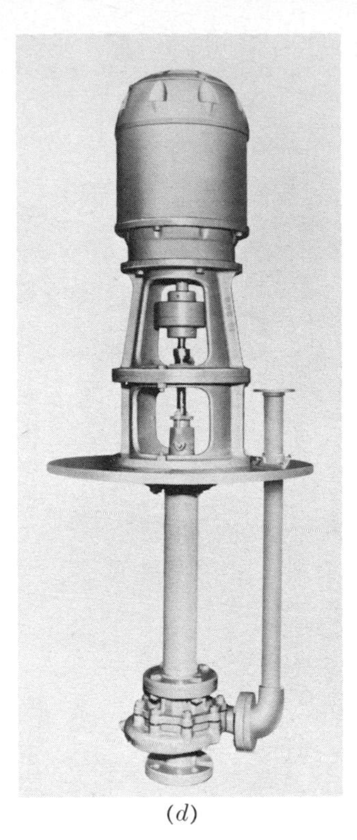

(*d*)

Fig. 6 Fresh-water pumps. (*a*) End-suction type. (Aurora Pump, a Unit of General Signal) (*b*) In-line type. (ITT Fluid Handling Division) (*c*) Tank--side mounted type. (Aurora Pump, a Unit of General Signal) (*d*) Tank-top mounted type. (Taber Pump Co., Inc.)

The operating system may use one pump running on an on-off service controlled by a pressure switch, with or without an air-charged accumulator tank connected to the discharge line. Or it may use one pump on continuous service with one or more standby pumps controlled by pressure switches.

Circulating Pumps for this service are similar to the general type of fresh-water drain pump shown in Fig. 6 (*a* and *b*). When relatively large capacities are required, double-suction types, either horizontal or vertical as shown in Fig. 7, are used.

Fresh-water circulating circuits are used for cooling purposes, namely, chilled water for air-conditioning systems, cooling of electronic components, and cooling of engine jackets, and for other auxiliaries where sea water must be excluded. Operation is usually continuous with the system capacity controlled by a thermostat or by temperature-controlled throttling valves. The pump usually takes suction from a tank or standpipe of sufficient elevation to provide an adequate suction head. Conditions of service vary widely with capacities up to 1,000 gpm at total heads up to 200 ft and speeds up to 1,750 rpm. At capacities up to 4,000 gpm the total heads are lower, ranging down to 50 ft. For higher total heads, up to 400 ft, a speed of 3,500 rpm is required, and capacities generally do not exceed 1,000 gpm. Pumping temperatures do not usually exceed 200°F.

SEA-WATER SERVICES

Condenser Circulating The main propulsion plant of a steam-powered vessel requires one or more large centrifugal pumps for this service. A range of capacities from 5,000 gpm at 90 ft total head to 25,000 gpm at 25 ft total head is typical,

Fig. 7 Vertical double-suction pump. (DeLaval Turbine, Inc.)

and can be produced by horizontal or vertical double-suction pumps. Speeds vary from 437 rpm to 1,150 rpm depending upon the pump capacity and head. A range of 14,000 gpm at 25 ft total head to 30,000 gpm at 15 ft total head is common for vertical mixed-flow pumps. Speeds usually do not exceed 870 rpm. Where the particular combination of speed, capacity, and head results in a specific speed greater than 7,500 rpm, the pump will be of the axial-flow or propeller type.

A typical circulating pump takes suction from a sea chest at the skin of the vessel and discharges to the inlet water box of the condenser. The circulated water leaving the condenser outlet is discharged overboard through another sea chest. A scoop may be fitted in a separate intake to the condenser. The scoop functions at higher ship speeds, during which time the pump may be secured or idling.

When more than one pump is installed, capacity may be varied by selecting the number of pumps operating, through speed regulation of turbine and dc motor-driven pumps, and through speed selection of multispeed ac motors. Capacity may also be varied by throttling the discharge valve, or by utilizing a bypass connecting the condenser outlet water box to the pump suction. Where the axial-flow pump is installed, it is also possible to consider the application of variable-pitch propeller blades for capacity regulation. A take-off connection is frequently fitted at the pump discharge to divert a small amount of flow to the auxiliary cooling system under emergency conditions.

For typical construction features refer to Fig. 7 for the double-suction pump and to Fig. 8 for the axial-flow pump. In the case of submersibles, the higher values of submergence require that special design considerations be given to the casing, the seals, and the thrust bearing.

Bilge and Ballast Centrifugal pumps for this service are usually of the double-suction type with a vertical shaft and with the motor or turbine mounted on top of the casing. Capacities vary up to 5,000 gpm at 90 ft total head and at speeds

Fig. 8 Condenser circulating pump. (Warren Pumps, Inc., Subsidiary of Houdaille Industries, Inc.)

up to 1,150 rpm. Higher capacities at lower heads may also be required. A typical construction is shown in Fig. 7 for medium and larger sizes, and in Fig. 6*a* for very small sizes.

Bilge service requires a priming device which may be mounted directly on the pump, or it may be a separate exhauster-type device. Vacuum pumps of the type depicted in Fig. 5 can be used for this purpose. A bilge connection may be fitted at the suction of the main condenser circulating pump and used for bilge evacuation under emergency conditions. Bilge pumps may be required to operate submerged; the driver must be mounted a suitable distance above the pump to prevent flooding or must be equipped with a protective submergence bell.

For rapid evacuation of small compartments, the bilge pump may be paired with one or more eductors. Normally the bilge pump discharges directly overboard, but for higher drainage rates it may discharge to an eductor in the same or remote compartment.

A unique type of bilge and ballast pump is shown in Fig. 9. The centrifugal pump is of the radial or mixed-flow type and is driven by a water turbine. Capacities vary from 1,000 gpm at 90 ft total head to 2,400 gpm at 50 ft total head at variable turbine speeds. The turbine is driven by water at fire-main pressures.

The steam-driven reciprocating pump may be applied to bilge and ballast service as it is inherently self-priming and the speed can be easily regulated. Where steam is not available, motor-driven horizontal or vertical power pumps are used. They are usually of the two- or three-cylinder design. Bilge and dirty ballast services can be handled by the same pump or group of pumps. For clean ballast service a separate system and pumps are required.

Fire Main For this service centrifugal pumps are of the double-suction type shown in Fig. 7 and are mounted vertically or horizontally. Pumps of the end-

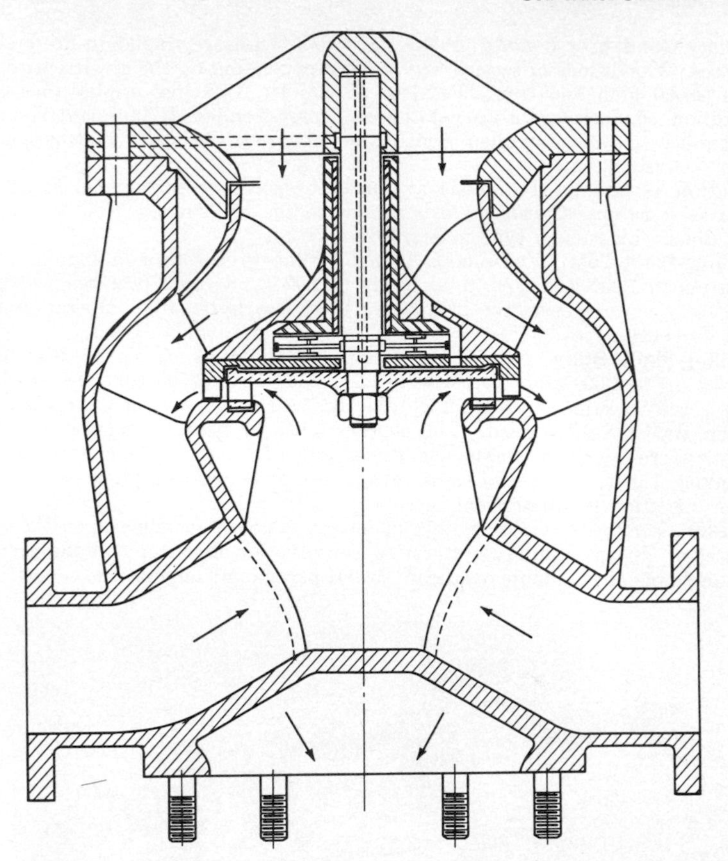

Fig. 9 Liquid turbine-driven pump. (Worthington Pump International)

suction type shown in Fig. 6*a* are also used. Capacities vary up to 1,500 gpm at total heads from 150 to 400 ft at speeds of 3,500 rpm. For higher heads a two-stage pump is used. Operation is continuous, with additional pumps placed in operation as the demand for higher volume increases. Pumps may be driven by a turbine, a diesel engine, or a gas turbine to ensure availability during a loss of electrical power. The steam-driven reciprocating pump can also be applied for fire-main service.

Flushing Pumps for this service are similar to those used for ballast and fire-main service. The capacity and total head seldom exceed 1,000 gpm and 100 ft at speeds of 3,500 rpm. Operation is usually continuous for one pump, with additional pumps activated by pressure controls when the need arises. Flushing service may be provided by the fire pump, but operating at reduced pressures through a reducing valve.

Cargo Tank Cleaning A water gun is used for cargo-tank cleaning, and capacities vary up to 4,000 gpm at heads of 100 to 400 ft, depending on the pressure required at the gun. The tank-cleaning machine may be stationary, or it may be slowly rotated by air pressure or by water pressure available from the gun supply line. When the conditions of service permit, the clean-ballast, fire-main, flushing, or auxiliary-cooling sea-water pumps may be used for tank cleaning. If not, then separate tank-cleaning pumps must be installed. Pumps of the type shown in Fig. 6*a* would be used for the smaller capacities, and pumps of the type shown in Fig. 7 would be used for the larger capacities.

Tank-cleaning operations must meet all the safety requirements in order to avoid explosions, and must be coordinated with the gas-freeing operation.

Auxiliary Condenser Cooling Pumps for this service are similar to fire and flushing pumps. Conditions of service vary greatly depending on the service, with capacities up 4,000 gpm and total heads up to 200 ft. Principal applications are for refrigeration condensers, turbogenerator condensers, and distilling-plant condensers. These services may receive their principal supply from the main condenser circulating, fire, or flushing pump.

Operation is usually continuous at constant speed, or at stepped speeds using a multispeed motor. Construction is similar to the end-suction type shown in Fig. 6*a* and the double-suction type in Fig. 7.

Distilling-Plant Feed End-suction pumps of the type shown in Fig. 6*a* with capacities up to 1,000 gpm and total heads of 200 ft at 3,500 rpm are widely used for this service. Regenerative pumps are used less frequently. Operation is continuous for both types.

Distilling-Plant Brine End-suction pumps of the type shown in Fig. 6*a* with capacities up to 1,000 gpm and total heads of 150 ft at 3,500 rpm are widely used for this service. Brine service is highly corrosive and erosive, and impellers of the semiopen type are often used. The suction must be vented, and the stuffing box must have provision for sealing against a loss of vacuum. Operation is usually continuous, although in special cases, as with submersibles, the pump is multistaged and the operation is intermittent.

Sanitary For this service a variety of pumps is used depending upon the particular function. In any sanitary system the conservation of water and the prevention of the discharge of pollutants overboard are of paramount importance.

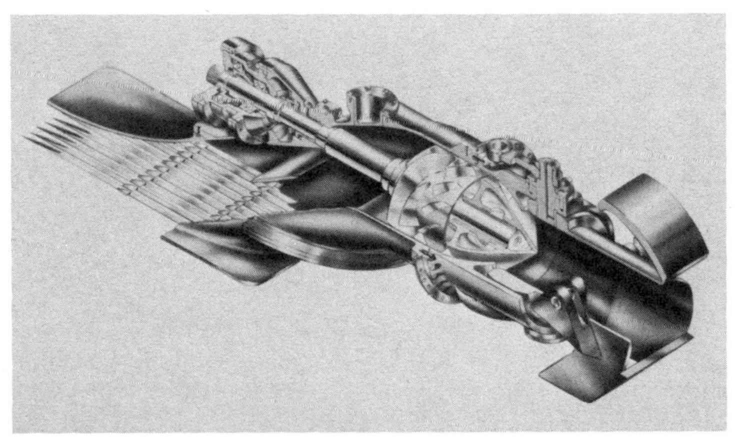

Fig. 10 Jet-propulsion pump. (Rocketdyne Division, Rocketdyne International)

The collecting circuit that operates under a vacuum requires a vacuum pump. Vacuum pumps operating on the liquid-ring principle, as shown in Fig. 5, and positive-displacement pumps of the vane or piston type are used. The vacuum is utilized to propel the liquid-solid plug to the holding tank. The containment system may involve holding only, or it may include a system for treatment. One or more pumps are used, depending on the specific system, for recirculation of the treated water for flushing purposes and for pumping overboard outside the contiguous limits or to a shoreside sewage connection. Pumps for this purpose are usually of the end-suction type, as shown in Fig. 6*a*.

Jet Propulsion The combination of pump speed, capacity, and head required for jet propulsion corresponds to a specific speed that requires a pump of the mixed-flow design, as shown in Fig. 10, or of an axial-flow design. Wide market acceptance has been established for jet propulsion as the motive power for work boats, pleasure boats, and military and commercial transportation. The pumping circuit from suction to discharge is short, but it requires careful consideration to ensure proper suction conditions at variable speeds. The overall design emphasizes weight and space saving, reliability, and proper matching to the speed of the internal combustion engine or gas turbine.

Jet thrusters may be fitted as auxiliary devices to assist in maneuvering, positioning, station keeping, docking, and undocking. Installation may be: (*a*) fixed, where the pump takes suction from the bottom of the hull and its discharge is directed to either side of the vessel; (*b*) fixed, with the pump mounted in an athwartship tunnel and discharging to either side of the vessel by reversible rotation or pitch; (*c*) retractable, where the pump direction may be fixed or turned.

VISCOUS FLUID SERVICES

Cargo Double-suction pumps are used for this service to handle the medium and high-volume rates required. Multistaged vertical mixed-flow pumps are used to a lesser extent to pump lower volume rates against higher heads. Capacities extend up to 25,000 gpm at 500 ft of total head for the single-stage double-suction pumps, and at lower capacities to 650 ft total head for the vertical multistaged pumps.

The single-stage double-suction pumps are installed in a separate pump room with two or more pumps connected through manifolds to the suction piping to the cargo tanks. The suction piping consists of double valving, cleanable suction strainers, vent connections, and stripping connections. The pumps are connected to the discharge piping leading to the vessel's deck through manifolds. From the vessel's deck connections are made to the shore-receiving facility. Double valving is often used in the discharge piping along with a relief valve, a nonreturn check valve, and an adjustable valve for throttling of the pumps.

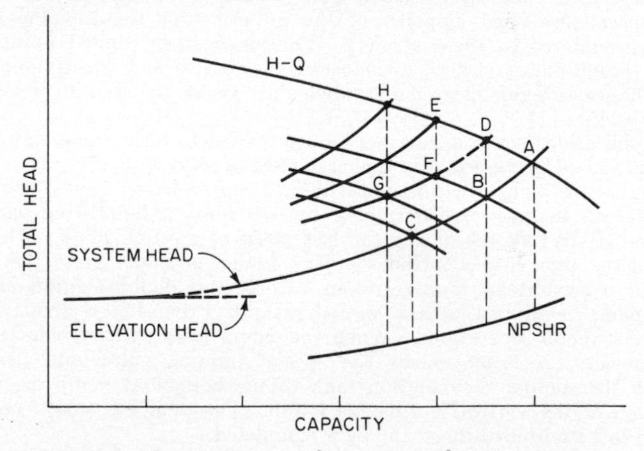

Fig. 11 Cargo pump performance and system curves.

Figure 11 shows a typical set of system conditions where point *A* is considered to be the rated condition for a particular port. The pump must be selected to produce the total head equal to the sum of the elevation and the friction head of the system piping at point *A*. In addition, the NPSH available at the pump suction must equal or exceed the NPSH required by the pump at point *A*. During off-loading of the cargo oil the level in the cargo tanks will fall until the NPSH available is inadequate at capacity *A*. At this point the speed of the pump can be reduced to produce the new condition of head and capacity at point *B*, where a lower NPSH is required. The speed may be further decreased to accommodate the continued lowering of the section level as at point *C*.

Lower capacities may also be obtained by throttling the discharge of the pump. Starting at the rating point *A* where the NPSH becomes inadequate, the system head curve is artificially changed by throttling the discharge as at point *D*. As the suction level continues to fall, further throttling will decrease the capacity at point *E*, or the pump speed may be reduced to produce the conditions at point *F*. A reduction in capacity is required to prevent cavitation noise, vibration, and possible damage to the pump and system components.

Point D represents another set of conditions that may apply to a particular port different from that at *A*. When operating at point *D* and the NPSH available is reduced to the point of cavitation, the speed of the pump must be reduced to points *F* and *G*. At constant speed the discharge of the pump must be throttled to points *E* and *H*. The degree of throttling and/or speed control must be coordinated with the driver characteristics, the system requirements for the particular port, and the time available for off-loading of the vessel. The variable-speed drive also has the advantage of providing adjustment in pump output for products of different viscosities and specific gravities.

The horizontal pump is driven by a steam turbine and reduction gear located in the adjacent machinery space, and connected by a drive shaft running in a bulkhead stuffing box. Vertical pumps are driven by vertical shafting extending to a machinery flat or deck above. The driver may be a vertical or horizontal turbine, motor, diesel engine, or gas turbine with a right-angle gear drive. The vertical multistaged pump is mounted in the cargo tank, and may be driven by a deck-mounted vertical or horizontal motor, an engine with a right-angle gear, or a hydraulic motor submerged in the cargo fluid.

The vertical multistage pump finds wide application for the pumping of crude and refined petroleum products, cryogenic fluids, and special chemical products. The canned motor pump with an isolated motor stator submerged in the fluid finds almost exclusive application for cryogenic fluids. The pump is of the end-suction type and may be mounted on the tank bottom or suspended within the discharge column. Capacities vary up to 10,000 gpm at 200 ft of total head. Both types employ a special first-stage impeller of the inducer type to meet the low values of NPSH encountered in these services. The single-stage canned pump may discharge to a second pump on deck for boosting the total head. Positive-displacement pumps of the vane, screw, gear, and sliding-shoe types are used to a lesser extent for small capacities.

For all applications some means must be provided to ensure complete unloading of the tanks. In older vessels a stripping system is used with vertical or horizontal reciprocating steam pumps or horizontal duplex motor-driven power pumps. Main cargo pumps can be made self-stripping by providing a liquid-ring exhauster for removing the air and vapors during the last phase of the unloading. The construction of modern large vessels includes a pipe tunnel near the keel; this eliminates the need for a separate stripping system. Horizontal double-suction and vertical multistage pumps equipped with a special priming valve and recirculating system can perform stripping operations. When the pump loses suction, vapors enter the impeller; the loss of prime opens the special priming valve and permits fluid to return to the suction recirculation tank of the horizontal pump or to the bell-mouth suction of the vertical multistage pump. The pumping action resumes, and the cycle repeats itself until the stripping is completed.

All modern cargo vessels have an inert-gas installation for freeing tanks and lines of explosive gases and toxic fumes. The system consists of a scrubber and gas cooler using circulating water taken from an existing flushing or general service system or from a separate double or end-suction pump.

Fuel-Oil Service Fuel pumps for marine steam generators are positive-displacement rotary pumps. Two or more are installed. Operation is continuous at constant speed when motor-driven, or at variable speed when driven by a steam turbine controlled by a constant-pressure governor. Pump discharge pressures vary from 100 to 300 lb/in^2, although higher pressures are used for special cases.

The typical service pump takes suction from either a low or a high connection at the service tank. A standby connection at a second tank is piped up for emergency use. A duplex strainer is fitted to the pump suction for protection. The discharge from the pump passes through a strainer, a heater, a meter, and control valves to the header at the boiler front. A recirculation line is provided to return fuel oil to the pump suction for control purposes, for heating oil during startup, and to allow pump operation when all burners may be temporarily secured.

The rotary fuel-oil pump for gas turbines is powered directly through auxiliary drive gears from the turbine or from a separate motor drive. The fuel-oil pump for diesel engines takes suction from a service tank and discharges through a filter to the header supplying oil to the injection pumps of the engine. The injection pumps are of the plunger type, spring loaded and cam operated, and may be combined with the injector.

Fuel-Oil Transfer Positive-displacement rotary, horizontal or vertical steam-reciprocating, and motor-driven power pumps are used for fuel-oil transfer service. Capacities vary widely depending on the precise function, and discharge pressures usually do not exceed 100 lb/in^2. Multistage mixed-flow pumps mounted within the tank may also be used for some services. Typical services include pumping to and from service tanks, to and from settling tanks, and to assist in maintaining the vessel trim.

Lubricating-Oil Service Pumps are predominantly of the positive-displacement gear type. Capacities vary widely depending on the horsepower rating of the main engine and reduction gears, or on the requirements of a particular auxiliary drive. Lubricating-oil header pressure at bearings and gears is typically 10 to 12 lb/in^2, and requires a pump discharge pressure of approximately 50 lb/in^2 for a pressurized system where the pump discharges through a strainer and cooler to the lubricating-oil headers. For a gravity system the service tank must be located 25 to 35 ft above the lubricating-oil headers to provide the required 10 to 12 lb/in^2 pressure. The pump discharges through a strainer and cooler to the overhead tank, where a constant level is maintained by an overflow. Oil temperatures are approximately 110 to 120°F.

The capacity of the service tanks varies from two times the pump capacity for small auxiliaries to five times the pump capacity for the main engine lubrication. The pump may be driven by an ac motor, by a steam turbine with a constant-pressure governor, by a dc motor, or by a direct drive from the main shaft or reduction gear. Power takeoff from the reduction gear is limited, as the capacity may be inadequate at low speeds. Separate standby and emergency pumps must also be installed.

Gas turbines and diesel engines use gear-type oil pumps driven directly from the crankshaft. Motor-driven gear or vane-type pumps must be provided for startup and cool-down as well as for emergency use. All shipboard auxiliaries that have a pressurized lubricating-oil system require one or more lubricating-oil pumps of the gear or vane type. Stern-tube lubricating systems utilize a gravity tank to supply the static head to the bearings. The operation of the service pump to the tank may be continuous or intermittent.

Other Services A wide variety of pump applications are related to the loading and off-loading of specialty cargoes. Pumps are required for slurry cargoes to furnish the high pressure at the jets that direct the solid-liquid mixture to collecting tanks. Slurry pumps off-load the mixture from the tanks to the shore facility. Additional pumps are required to flush the slurry lines.

Vacuum pumps are used for pumping fish, for cement unloading, and for vacuum cleaning of the holds of the ships and barges.

A completely independent hydraulic system fitted with one or more positive-displacement rotary pumps may be used to power several hydraulic motors. They, in turn, drive deep-well pumps for the off-loading of specialty cargoes such as lubricating oil, petrochemicals, and solvents.

Variable-displacement axial and radial piston pumps provide versatile control and instant reversibility for cargo-handling equipment and other hull machinery including the following:

1. Steering gear
2. Pipe tensioners
3. Alignment stays
4. Hydraulic rams and jacks
5. Engagement cylinders
6. Interlocks
7. Selector valves

8. Flow control gates
9. Boom and shuttle positioners
10. Elevators
11. Watertight doors
12. Hatch covers
13. Stabilizers
14. Controllable-pitch propellers

REFERENCES

1. Feck, A. W., and J. O. Sommerholder, Cargo Pumping in Modern Tankers and Bulk Carriers, *Marine Technol.*, July 1967.
2. Brandau, J. H., "Aspects of Performance Evaluation of Water-Jet Propulsion Systems and a Critical Review of the State-of-the-Art," AIAA/SNAME Paper 67-360, May 1967.
3. "Symposium on Pumping Machinery for Marine Propulsion," ed. by J. H. Brandau, ASME, 1968.
4. "Marine Engineering," ed. by R. L. Harrington, SNAME, 1971.
5. Jones, R. M., C. W. Wright, and C. S. Smith, "A Study of Large Self-Unloading Vessels," SNAME, November 1972.
6. "Guide for Inert Gas Installations on Vessels Carrying Oil in Bulk," American Bureau of Shipping, New York, May 1973.
7. Bourn, W. S., "Shipboard Solid Waste Control," ASME, July 1973.
8. "Conversion of Centrifugal Pumps to Automatic Self-Priming Operation," Ship Repair and Maintenance International, Surrey, England, October-November 1973.
9. Garber, Daniel C., An Oil-Water Separator System, *Marine Technol.*, January 1974.
10. Nielsen, R. A., and Harry H. Kendall, Stern Thruster Installation on the SS John Sherwin, *Marine Technol.*, January 1974.
11. Smith, I. W., Recent Trends in Hull Machinery, *Marine Technol.*, April 1974.
12. Fassell, W. M., and D. W. Bridges, PURETEC System for Treatment of Shipboard Waste, *Marine Technol.*, July 1974.

Section 10.13

Steel Mills

E. R. PRITCHETT

Steel is composed of iron alloyed with small quantities of carbon and other elements. The quantities of these lesser constituents are modified to obtain varying metal properties. Because steel is a rugged and tough material and one used in sizable quantities, the equipment that is used to produce it must also be rugged and tough and large. Steel is supplied in many forms, such as plates, bars, strip, structural shapes, etc., requiring many different operations. It is worked both hot, as in a slabbing mill, and cold, as in a cold reducing mill. In any case, from raw materials to finished product, the steel is processed through many operations requiring considerable amounts of energy in varying forms. For example, before a ton of steel is shipped in a finished condition, its production could require up to 50,000 gal of water, 30 million Btu of gaseous and liquid fuels, 500 kWh of electricity, and 5,000 standard ft^3 of compressed air.

Pumps play an indispensable role in moving the sources of energy, such as a liquid fuel, around a mill. Pumps are equally important in providing water at required locations to absorb waste energy produced by normal inefficiencies in rotating equipment and to maintain required equipment temperatures to allow safe operation. High-pressure water is also required for blasting scale off semifinished material before hot-working. Pumps are also used as energy converters. In the hydraulic system of a mill, valves, cylinders, and sensors may be operated by high-pressure oil. Pumps, either steam-turbine or motor-driven, convert the thermodynamic or electrical energy to energy in the fluid pumped. Another important role for pumps is in the area of pollution abatement and treatment. In treating contaminated water, the water must be moved through various pieces of equipment, such as filters and cooling towers, and then recycled to service if possible. Pumps are also of vital importance in air-pollution control when cooling of fan bearings, gas cooling, and washing of dust-laden gases are considered.

There are tremendous sums of capital invested in a typical plant. One blast furnace, for example, can cost more than 70 million dollars, and this investment only provides pig iron to a steelmaker! For this reason down time is extremely costly, and reliability of equipment is mandatory. On pumping systems where reliability is a major factor, spare capacity, in whole or in part, is installed as a part of the initial installation. In critical applications, the spare may be programmed to start automatically in the event of failure of an operating unit.

The following sections explore the use of pumps in modern integrated (raw material to finished product) mills.

MATERIAL HANDLING AND COKE OVENS

Process Coke is carbon in a concentrated form and is made from coal. Various grades of coal are blended in desired proportions and charged into tall refractory ovens. Here the coal is "baked" (away from the presence of air) to remove volatiles and moisture. After sufficient "baking time," the coke is pushed from the oven into a quencher car, quenched in water to prevent destruction by burning, and transferred to the blast-furnace area for use. Several by-products are generated from this process. Some are coke-oven gas, tars, light oils (if removed from the gas), and ammonium sulfate. The former items are used in other mill processes as energy sources, and the last is sold as a fertilizer.

The first necessity in this process is to unload coal. A device used for this purpose is a car dumper which "grabs" a railroad car of coal, inverts it to dump the contents, and returns it to its upright position on the tracks. The clamps and holding devices which support the car in lifting and turning are hydraulically operated.

Pump Applications The energy in a hydraulic system can be supplied either electrically or mechanically, and pumps are used as the energy converters. A car-dumper hydraulic system will utilize several levels of pressure, say, 400, 800, and 2,600 lb/in.^{2} Operating pressures for given pieces of equipment are largely determined by force requirements and practical cylinder or prime-mover sizes. To meet the flow and pressure needs of the two lower pressure systems, multiple variable-volume vane pumps could be utilized, each being equipped with a dual pressure control. These positive-displacement pumps achieve a variable volume by balancing a manually set spring pressure against an existing system pressure (Fig. 1). Until the two are equal, an eccentricity exists between the pump rotor containing the floating vanes and the vane-containing ring which allows fluid to be pumped. A maximum volume is set by limiting the amount of eccentricity permissible. The dual pressure

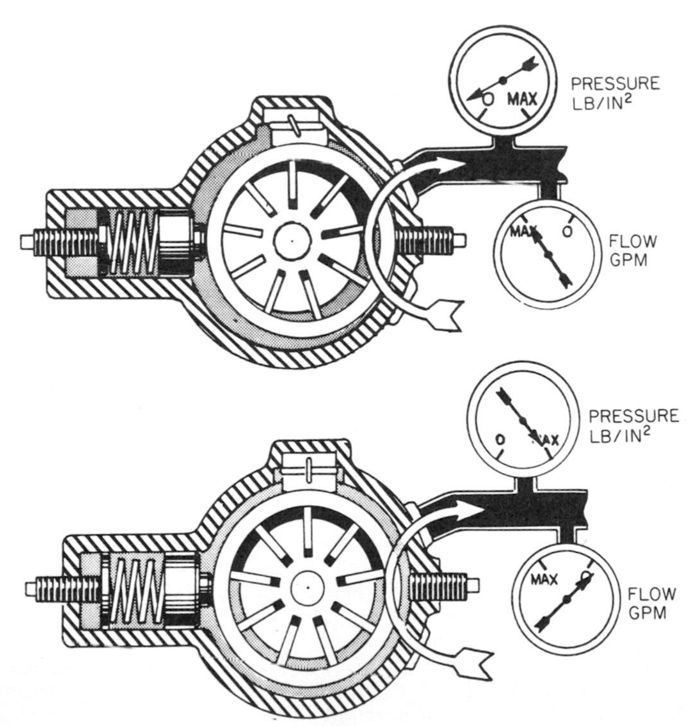

Fig. 1 Sectional representations of variable-volume pump operation. Top, full flow position. Bottom, minimum flow position. (Racine Hydraulics, Inc.)

control works by adding the system pressure to the manually set spring force, giving a higher control pressure (Fig. 2). The addition is made by the use of solenoid valves either exhausting or directing system pressure to the control piston. Solenoid valves direct the fluid to the proper system. The high-pressure (2,600 lb/in^2) system could be supplied by a radial piston pump or group of pumps as required. As the shaft of this pump rotates, an eccentric on the shaft depresses the pistons sequentially, providing a positive displacement of fluid (Fig. 3). The pump runs continuously, loading and unloading by external valve positions as directed by the system accumulator level. It would have to be equipped with high-pressure shaft seals if suction is taken from a pressurized system (say 50 lb/in^2).

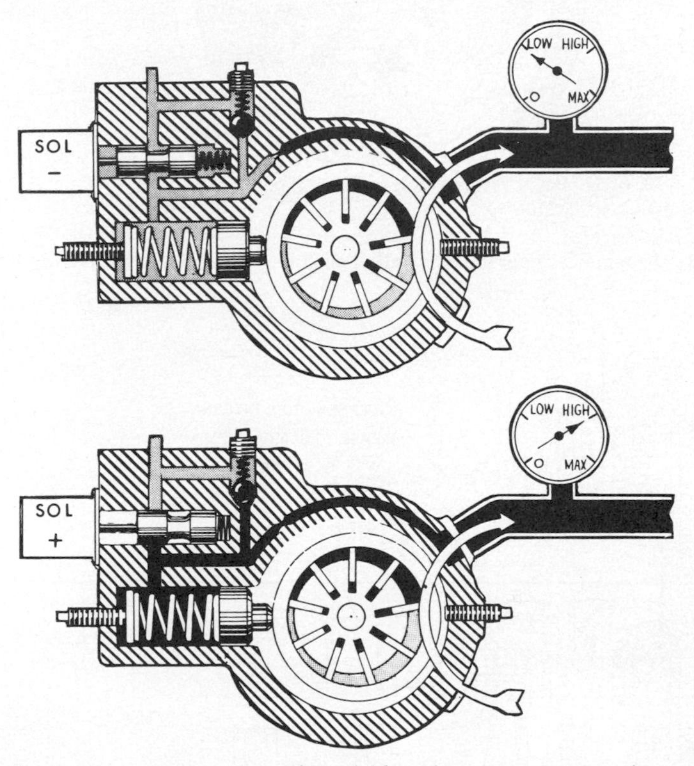

Fig. 2 Sectional representations of variable-volume pump operation with dual pressure control. Top, low-pressure contol. Bottom, high-pressure control. (Racine Hydraulics, Inc.)

Coke, after it is pushed from an oven into a car, is quenched with water from an overhead storage tank. Some of this water is vaporized and passes off into the atmosphere as steam. The balance of the water at 150 to 175°F is collected in a pit under the quencher car and pumped back up to the storage tank. These pumps can be either horizontal or vertical depending on system layout. A typical pump would be a heavy-duty, horizontal, single-stage, end-suction centrifugal (Fig. 4). To provide abrasion resistance the wearing parts would be frabricated of an alloy such as a 28 percent chrome material with a hardness of 550 to 650 BHN. To prevent abrasive material from wearing the shaft and sleeve, a lantern and liquid seal connection should be provided at the stuffing box. Either a clean source of water or filtered process water can be used as a seal water supply, possibly in conjunction with a booster pump.

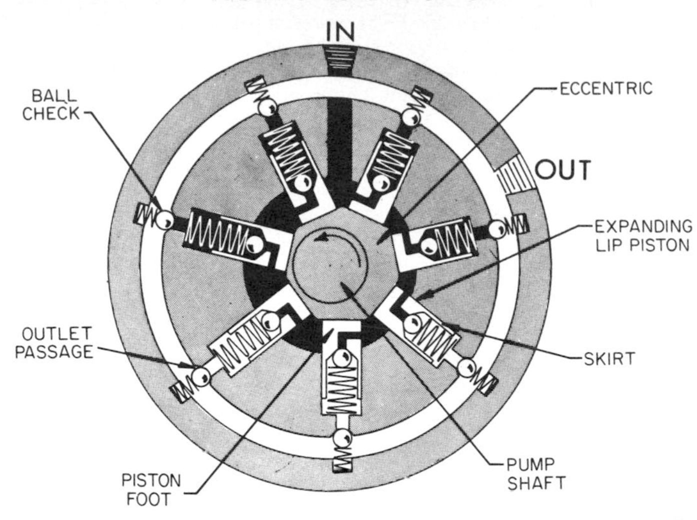

Fig. 3 Schematic diagram of a radial piston pump. (Racine Hydraulics, Inc.)

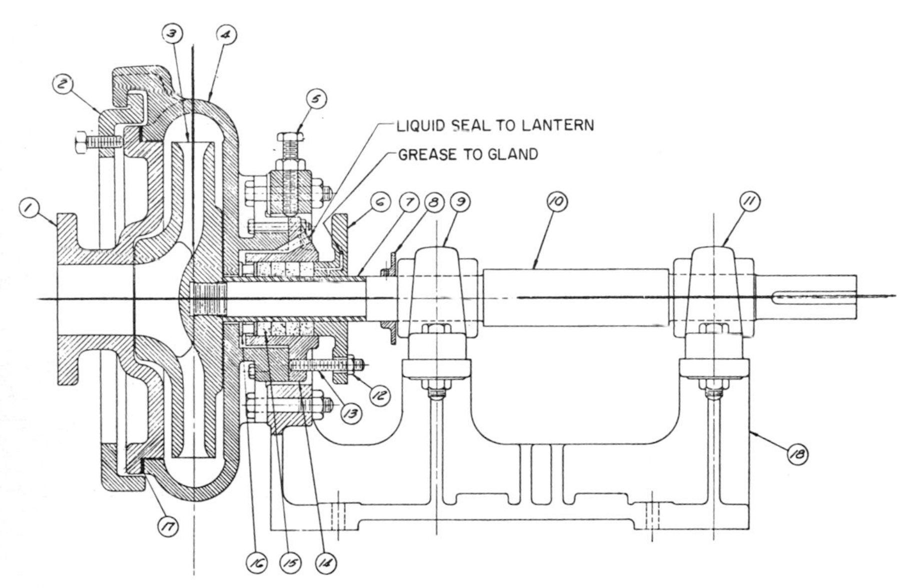

Fig. 4 Typical pump used for recycling coke quench water. 1, plate; 2, ring; 3, impeller; 4, casing; 5, arbor bolt; 6, gland follower; 7, shaft sleeve; 8, flinger; 9, thrust bearing; 10, shaft; 11, radial bearing; 12, gland nut; 13, gland stud; 14, stuffing box; 15, packing; 16, lantern; 17, gaskets; 18, bearing stand. (Nagle Pumps, Inc.)

BLAST FURNACES

Process Blast furnaces are large smelters receiving as a raw-material charge a mixture of iron-bearing materials, limestone or another fluxing agent, and coke. The most frequently used iron-bearing materials are ores, sinter (agglomerated ore fines), and iron-ore pellets. These materials may be used singly or in combinations. The raw materials are continuously charged into the top of the furnace (which is like a cylinder sitting on end) while hot air is blown into the furnace around its circumference near the bottom. The resulting combustion provides a source of heat to melt the iron. Also, the carbon in the coke unites with oxygen carried with the iron-bearing material in the form of Fe_2O_3. The limestone acts as a flux to produce a basic slag, which has an affinity for sulfur and other impurities. Molten iron and slag accumulate at the bottom of the furnace, and every few hours a "cast" is made, i.e., the iron is tapped into insulated vessels on railroad cars, and the slag is tapped into slag pots or pits. Thus the raw iron-bearing materials are purified, and the end result is a molten iron suitable for further processing into steel.

The air which is blown into the furnace passes up through the furnace's near-cylindrical shape, uniting with carbon and producing a mixture of gases known as blast-furnace gas. At the furnace top, this dirty gas, which has some heating value, is collected and cleaned and utilized in heating applications in the blast furnace and in other areas.

Pump Applications Gas cleaning is accomplished by means of a mechanical dust catcher and wet scrubbers, with recycling of water being a main feature of the system (Fig. 5). The gas is first washed with water in venturi scrubbers and then

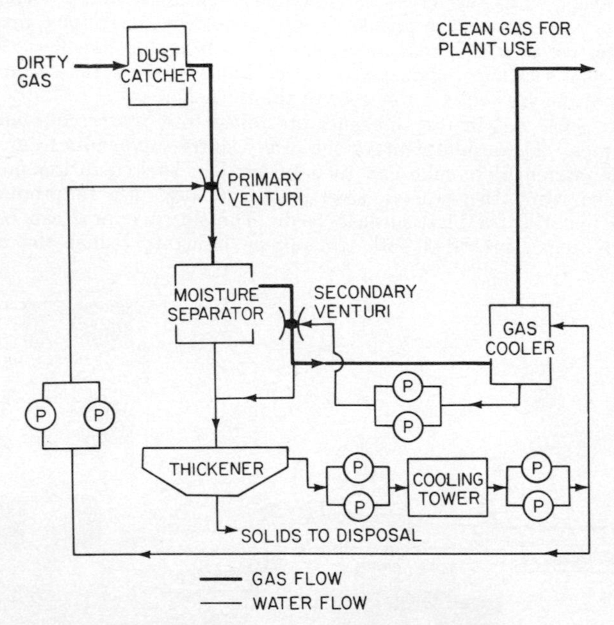

Fig. 5 Typical gas-cleaning system on a blast furnace showing application of pumps for recycling of water.

cooled before sending it to the plant for use. Dirty water from the venturi scrubbers is sent to a thickener, where solids are allowed to settle. The reasonably clean water effluent (normally 50 ppm suspended solids—100 ppm peak) from the thickener is pumped through a cooling tower to allow reuse. The tower hot-well pumps are heavy industrial abrasive-handling vertical-turbine types direct-motor-driven (see Section 2.2). A standby pump is considered to be necessary. Cast-iron bowls with 4340 cast-steel impellers and 410 stainless steel bowl and impeller wear rings are recommended materials. They are selected to provide greater abrasion resistance and thereby give longer pump life. No special control on the pumps is required as the point of operation is fixed by the pump characteristic curve and system resistance.

After passing through the cooling tower and being chemically treated, the water is combined with required make-up and pumped back to the gas cleaning system. Pumping is again accomplished by a set of vertical pumps (with installed spare) in the cold well of the tower. These pumps are identical in capacity and can be of the same type as those in the hot well except for the possible addition of a second stage which may be required because of an increased head. Control is as described for the hot-well pumps. Vertical pumps are recommended for both the hot- and cold-well applications because of their adaptability to a system design and layout featuring minimum space and high lift requirements.

Most of this cooled water goes to the gas cooler, with flow being controlled by exit gas temperature. The balance of the water is boosted in head and sent to the primary venturi scrubber. Booster pumps recommended are horizontal split-case, double-suction centrifugal pumps, as illustrated in Sec. 2.2. An 11–13 percent chrome steel can be used for the casing and impeller wearing rings, while a hardened similar material can be used for the shaft sleeves. These special features are added to an all-iron pump to provide longer life for the wearing parts. Pump drives can be split between direct-motor and turbine to provide a more reliable gas-cleaning capability. Venturi water flow is controlled by manually setting a throttling valve opening to produce a given flow as indicated by a flowmeter.

More than half the water that is actually utilized in the gas cooler is boosted in pressure and used as supply to the secondary venturi scrubber. These secondary pumps can be the same as the primary venturi boosters except for a material change which calls for use of a 2 percent nickel cast iron casing and impeller. This provides greater abrasion resistance against the dirtier water used on the secondary venturi. The balance of the gas cooler water goes to the thickener.

Solids that settle out in the thickener are raked to a center cone and under-flow to slurry pumps. These pumps move the heavy slurry to vacuum filters for dewatering. The dewatered filter cake can then be used in agglomeration facilities, where it is combined with other wastes, steel scale, and ore fines to produce a suitable iron-bearing material for blast-furnace feed. The slurry pumps can be heavy-duty centrifugals (Fig. 6) provided with a neoprene lining to reduce the abrasive wear

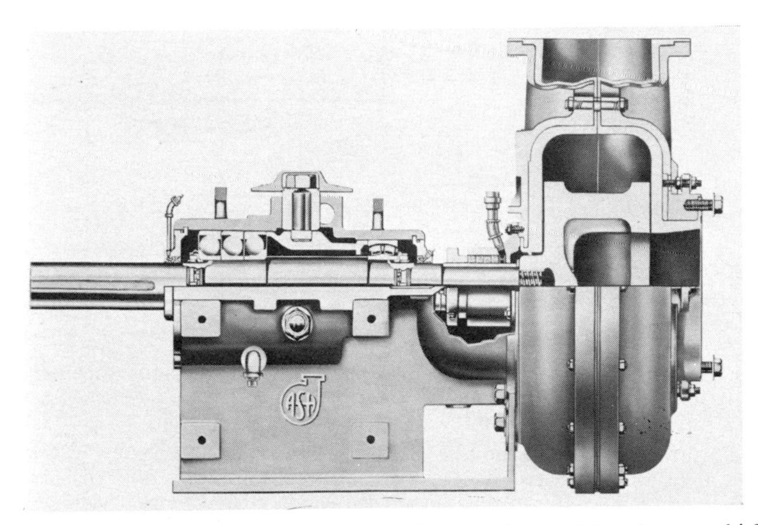

Fig. 6 Lined materials--handling pump used for transferring blast-furnace thickener underflow slurry. (A.S.H. Pump, Division of Envirotech Corp.)

caused by water that contains approximately 50 percent solids by weight. A water jet exhauster maintains a vacuum between the pump casing and lining, thereby holding the lining against the casing under suction lift conditions. The pumps may be driven by electric motors through belts that allow a speed reduction (say from 1800 to 1200 rpm) to minimize wear. An auxiliary pumping system can be used for seal water in which a turbine-type pump provides clean water at the required head for the seals. This type of slurry pump can also be used in transferring ore slurries in mining operations.

STEELMAKING

Process There are three basic methods of making steel presently in use today. The newest and most widely used is that of injecting oxygen directly into molten pig iron and scrap in what is known as a basic oxygen furnace, or BOF. An older method, for which the tonnage capacity is gradually declining because of the BOF, is the open hearth process. These two techniques involve refining pig iron and obtaining precise chemical compositions in the steel produced. The third method is the electric furnace. This is not so much a refiner as a very precise tool for melting scrap and/or pellets and maintaining close tolerances of the steel's constituents. Because the BOF is presently the highest tonnage steelmaker, the following description and examination of pump services will be confined to that process.

Three basic materials are charged into the BOF. They are molten pig iron, scrap, and fluxes. After charging, a lance is lowered into the furnace (which resembles an egg with an end lopped off) and oxygen injection begins. The combination of carbon (from the pig iron), other oxidizable elements, and oxygen produces heat that melts the scrap. Some iron ore is charged to absorb excess heat produced, resulting in precise end temperature control. The oxygen is blown until the carbon content of the molten material in the furnace is at the desired level. Other alloys required are added as the steel is tapped into a ladle.

Pump Applications One of the primary requirements for pumps is meeting cooling-water requirements for the oxygen lance. This lance, which is approximately 10 in in diameter and perhaps 65 ft long, carries oxygen in a pipe in the lance center. The oxygen pipe is surrounded by cooling water, approximately 800 gpm being required per lance at system pressures of 170 lb/in.^{2} A pumping system would consist of several pumps connected in parallel (with an installed spare), each of which is direct-driven by an electric motor. Double-suction centrifugal pumps (see Sec. 2.2) with cast iron casings and bronze impellers equipped with standard grease-lubricated line-and-thrust ball bearings can be used for this service. The required pump pressure is the sum of water-elevation differences, friction losses, etc., plus an allowance to maintain a lance full of water under all conditions. A fixed orifice or head tank is placed in the water line downstream of each lance. Pump control is "on-off." When the pumps are on, they operate along the system curve, with no further control being required. This system can be "once-through" or recirculating, utilizing heat exchangers to maintain a stable system temperature.

Another pump requirement in the BOF consists of water-cooling gases generated during the oxygen blow. When the blow occurs, large volumes of dirty gas are generated at temperatures approaching 3000°F. A water-tube boiler is used to absorb some of the heat, and the balance is cooled by water sprays in a quencher. Water is sprayed into the gas stream and a portion is evaporated until the gas temperature is adiabatically cooled to 170°F. and the gas is ready for cleaning.

Gas cleaning can be accomplished by several means, the most frequently used being electrostatic precipitators and wet scrubbers. A system using a wet scrubber, a thickener, a cooling tower, and recirculating water for reuse would be similar to that discussed in the blast furnace section.

Of special interest is the use of pumps for moving underflow slurry to centrifuges or vacuum filters for drying. The pumps are on at all times as the heavy slurry is directed to recirculate back through the thickener or to drying processes by an operator visually monitoring the solids level in the thickener. A typical specification for such pumps would call for volute-type centrifugals

Fig. 7 Pump used for recirculating and transferring BOF thickener underflow. (Galigher Pumps, Inc.)

(Fig. 7) rated at 40 gpm and a total head of 70 ft. Pumps would be motor-driven by a sheave and belt arrangement (to allow for future capacity changes and also for lower speed operation to cut down abrasion). Typical service conditions in the thickener would be: temperature at 150°F.; specific gravity of 1.6, and water containing 25 to 60 percent solids by weight of less than 270-mesh iron oxide. Because of the abrasive properties of the material, a pump utilizing a rubber (neoprene) liner and impeller would be selected.

SLABBING MILLS

Process Steel, as produced in a BOF, an open hearth, or an electric furnace, is tapped into ladles where final alloy materials are added. The steel can then follow one of two paths. The first and traditional path is that of teeming into molds for solidification as ingots, reheating the ingots in pit-type furnaces, and hot-rolling the ingots into slabs on a slabbing mill (or blooms on a blooming mill). The second possible route is that of teeming the steel into a continuous casting machine which forms a slab on a continuous basis. The first process will be described here.

After the steel in the molds has solidified, the molds are stripped off and the ingots are charged into soaking pits for heating to rolling temperature. When they are at the desired temperature, the ingots are pulled from the pits and laid on a driven roll-mill approach table which feeds the slabbing mill. Passing the ingot back and forth in the mill gradually gives it the shape of a slab (for example, 12 by 80 in in cross section), and rolling continues until final size is achieved. The slab then passes down a roller line to a scarfing machine, which consists of many oxygen-natural gas torches surrounding the entire perimeter of the slab. As the slab moves through the scarfer, the external surface up to a 3/16-in. depth can be burned away, removing any rolling defects and producing a clean surface. The slab then moves to a shear, where it is cut to length. It then goes to a storage yard for cooling, where it awaits processing into plates or other flat-rolled products.

Essentially the same procedure is followed in the production of blooms from ingots. Blooms, which are smaller than slabs (for example, 12 in square), are used in the production of billets (for example, 6 in square), which in turn are processed into bars and rods. Blooms are also processed directly into structural shapes.

Pump Applications As an ingot approaches the mill and is rolled, scale that was formed during cooling and reheating falls through the rolls of the mill approach and delivery tables into a flume. This flume is continually flushed with water to move scale to the mill scale pit. There the scale settles to the bottom, where it is cleaned out with a clam bucket on an overhead crane. The water in the pit overflows a weir into a clear well and is pumped directly back to the flume as the flushing-water supply. Pumping from the clear well minimizes scale loadings, but specially hardened materials (nickel alloys) are required for the pump casing, the impeller, and the wear plates to ensure good pump life. Two end-suction volute-type centrifugal pumps (Fig. 8) can be used for this service, one operating and one stand-by. The pumps are controlled from level switches in the scale pit in conjunction with a make-up water supply.

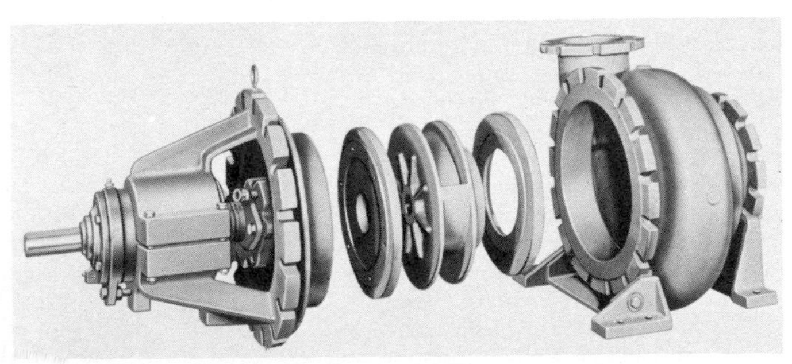

Fig. 8 Typical flume-flushing pump at slabbing-mill primary scale pit. (Allis Chalmers Corp.)

Another mill water recirculating system services the scarfer scale pit and provides scarfer flume flushing and foundation spray cooling. Pumping from the scarfer scale pit to the spray cooling, etc., can be by horizontal, single-stage, double-suction, split-case, centrifugal direct-motor driven pumps (see Sec. 2.2), with one operating and one stand-by. A double-suction pump is recommended to reduce impeller thrust and resulting bearing maintenance. A hardened impeller material is used for abrasion resistance, and for the same reason stainless steel hardened impeller and casing rings and shaft sleeves are recommended. Control is again from pit level.

Many of the devices on the mill itself (slab manipulators, clamps, etc.) are moved and controlled by hydraulic cylinders. A typical hydraulic system includes an oil reservoir, pumps, accumulators, supply and return lines, etc. (Fig. 9). As long as

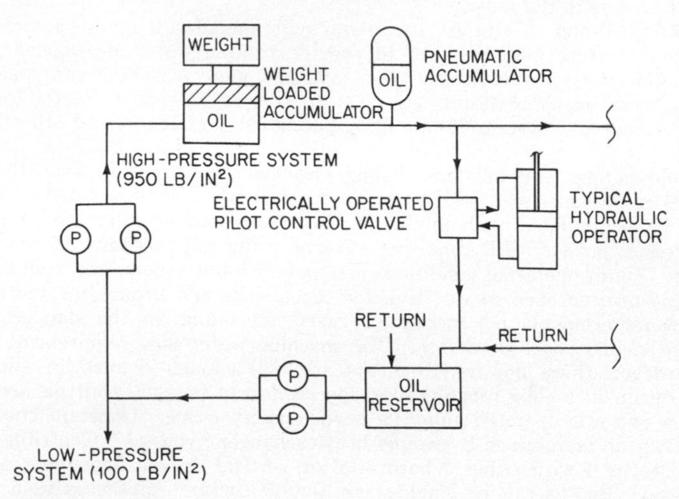

Fig. 9 Schematic drawing of a typical mill hydraulic system.

fluid remains in the weight-loaded accumulator, it establishes system pressure as weight divided by piston area. A pneumatic accumulator acts as a surge suppressor. Electrically controlled pilot valves can be used to direct hydraulic fluid to and from the cylinders when they are required to operate. Hydraulic fluid is exhausted from the cylinders to a reservoir at atmospheric pressure where make-up fluid is added as required. Horizontal end-suction centrifugal pumps, as illustrated in Sec. 2.2, are recommended for providing hydraulic fluid to low-pressure cylinders and controls. Usually pump impellers are bronze, and casings are cast iron. These pumps are connected in parallel, with one operating and one stand-by, and are controlled by their characteristic curves and system resistance. Centrifugals are selected for meeting demands of rapidly changing loads. Relief valves throughout the system discharge to the reservoir and protect against blockages and resulting overpressure. These centrifugals can also be used to supply pumps that maintain fluid level in the high-pressure system. These high-pressure pumps can also be arranged in a one-operating-and-one-stand-by setup. At low accumulator level, the pump will deliver fluid to the high-pressure system; at high level, it will continue to run, but in an unloaded condition. These pumps would be positive-displacement reciprocating vertical triplex plunger type (see Sec. 3), selected by virtue of their efficiency with a design requirement of low capacity and high pressure (typically 80 gpm and 975 lb/in^2). The pumps are made with 400-series steel plungers, valves and glands, and forged steel cylinders.

CONTINUOUS CASTING

Process A modern alternative to the process described above of converting liquid steel to ingots and then to slabs in a slabbing mill is casting steel in slab (or bloom) form direct from the steelmaking ladle. In this operation, steel is continuously poured into a tundish, where the level is maintained to control liquid steel flow to the casting machine. From the bottom of the tundish a stream of steel pours into a water-cooled mold of the desired slab (or bloom) size. A thin layer of the steel surface is solidified immediately. The slab with its molten core then progresses through a series of spray-cooling zones until the entire slab is solidified. During this cooling the moving slab is in contact on both sides with a series of water-cooled rollers. The continuous slab is then torch- or shear-cut to provide slabs of the length desired for further processing. Each such mold, cooling zone, and cutter is called a *strand*. Because of large heat sizes, machines installed today have multiple strands to permit casting more steel per unit time, thereby preventing excessive heat loss in the ladle.

Pump Applications There are two major requirements for pumps associated with a continuous casting machine: (1) to provide cooling water at sufficient pressure for the mold, the spray-cooling system, and the many machine components such as rollers, torch, machine frame, etc.; (2) to convert electrical energy to pressure energy in a hydraulic system for use in operating roll positioners and other hydraulic equipment.

The mold-cooling and machine-cooling systems are similar in that the cooling water does not touch steel, and hence does not become contaminated. Water can therefore be recirculated with minimum blow-down and chemical treatment. The mold system is most critical, requiring softened water not exceeding 20 ppm calcium carbonate. Pump operating conditions are therefore not severe, and standard materials can be utilized such as cast iron for the casings and bronze for the impellers. Mold flow requirements per strand will vary depending on the slab section cast and the particular machine design. The machine water flow requirement will also vary. However, these flow requirements are well enough defined to allow pump selection within such close range that no special flow or pressure controls are required and thus a pump is operated along its head-capacity curve. Constant cooling temperatures can be maintained by simple heat-exchanger bypasses. Centrifugal pumps are therefore used with either a horizontal or vertical design, depending on system configuration. Pumps can be single-stage, double-suction, split case with cast iron casings and bronze impellers. A vertical pump may be quite desirable in the well of a cooling tower or scale pit, but horizontal pumps would have a lower first cost for pumping from a tank above pump elevation. Motor or steam drives can be used depending upon an individual plant's projected availability and the cost of each power source. Pumps should be installed in a parallel group for maximum reliability. It may be desirable to provide two operating pumps, each sized for approximately 60 percent of maximum flow, plus an installed spare of the same size

The spray water-cooling system is more complicated in that the water comes in contact with the slab and becomes mixed with scale (see Fig. 10). After water is sprayed on the slab, it falls to the bottom of the spray chamber and into a flume which runs to a scale pit. Scale falls from the slab and is washed to the scale pit by the flushing action of the water in the flume. Vertical centrifugal pumps are utilized to move water from the pit through a cooling tower and from there through filters and back to the spray cooling system. The filters are placed last to assure the best possible water quality.

Flow requirements can vary drastically depending upon slab size, machine speed, and metallurgical steel grade. A flow of 2,000 gpm is reasonable for strip mill slabs 10 in thick by 70 in wide. Pressure required will be between 100 and 200 lb/in^2, depending on piping, elevation differences, and spray nozzle design. Depending on system design, either horizontal or vertical centrifugal pumps could be used. Pump drivers and the number of pumps for reliable operation would be selected on the same basis as for the mold and machine water systems. Individual zone water flows are varied to obtain cooling rates required, which depend upon casting speed, slab cross section, and metallurgical grade. Throttling valves in each zone water-supply line are positioned to provide this flow control. The pumps will operate along the system head-capacity curve as this control action takes place.

The hydraulic system on a casting machine is functionally the same as that on the slabbing mill or the car dumper.

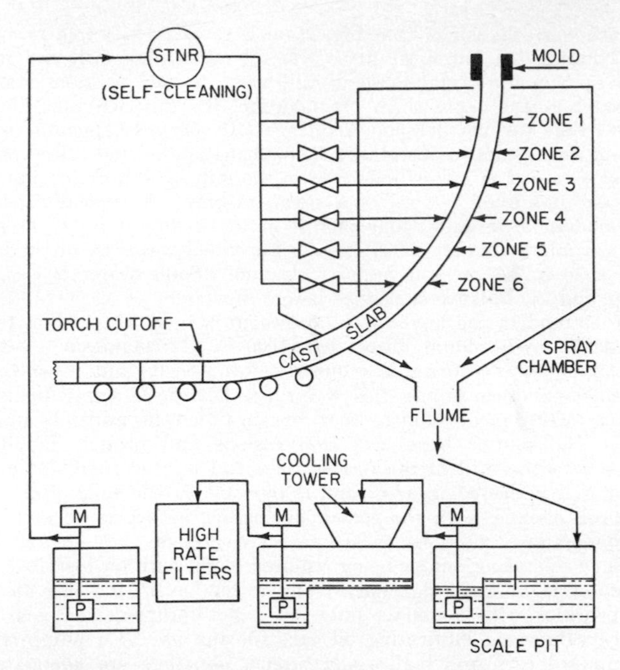

Fig. 10 Slab-casting-machine spray cooling system.

HOT-ROLLING MILLS

Process Semifinished steel, i.e., slabs and blooms, can be processed in several ways to produce a finished steel-mill product. Slabs are rolled into plates or into strip in coil form. Blooms are rolled into large structural shapes or into billets, which are subsequently rolled into bars, rods, or small structural units. This is accomplished on various types of hot-rolling mills, plate mills, hot strip mills, structural mills, and billet mills.

Functionally the processes are similar. The piece to be rolled is heated to rolling temperature in a furnace. As the piece moves to the mill, it is descaled, normally by high-pressure jets of water. This descaling is important in making a high-quality product because rolled-in scale can produce surface defects practically impossible or uneconomical to remove. As the piece is processed in the mill or mills to its final shape, descaling is continued. The process of rolling is one of "squeezing" to final shape a piece of steel made relatively plastic by its high temperature. For example, an 8-in-thick slab may be rolled to a ½-in-thick plate, or a 10-in slab may be rolled to a ⅛-in-thick strip. After the piece has been final-rolled, it must be leveled if it is a plate or straightened if it is a structural member. Strip is coiled after rolling, as are small bars and rods. Large bars (say above 1½ in), plates, and structural pieces must be cut to proper length by torches or shears. Hot-rolled strip is still not a finished product, and other operations are necessary for completion.

Pump Applications As noted previously, high-pressure water descaling is important to product quality. Providing this water is one of the more important pumping functions of a hot-rolling mill. Another pump requirement is providing water for cooling rolls used in shaping the product. Scale handling, flume flushing, and water recirculation are similar to the same operations in a slabbing mill, with the same requirements for the pumping systems. A need also exists for hydraulic system pumps, lubricating-oil pumps, sump pumps, etc.

High-pressure descaling water is supplied by multistage horizontal centrifugal pumps. These are "barrel" type pumps using 13 percent chrome steel impellers and forged carbon steel casings (see Fig. 26 of Sec. 2.2). Multiple pumps are provided for maximum demand flexibility. Not all mill stands will call for water at the same time; therefore total demand will vary as one or more stands happen to call for water at the same time. In addition, the mill will not always run at the same pace, thus varying demand frequency. In any case, pneumatic accumulators are used in the system to absorb shocks of changing demand. Pumps are motor-driven because a loss of power will shut down the mill as well as the pumps; overall reliability is not improved by use of a steam turbine. A typical capacity of one pump on a modern strip mill (four pumps in total) would be 1,500 gpm with a head increase of approximately 2,200 lb/in². Expended water is directed to the mill scale pit for scale collection and primary cleaning. Some water is recirculated for flume flushing, and the balance is cleaned before discharge.

The steel rolls used in the hot-rolling process are subject to product temperatures as high as 2350°F and seldom lower than 1600°F. To maintain a "cool" roll of the proper shape, water cooling is required continuously and must be thorough. Generally pumps used to supply this water are horizontal centrifugals similar to those used for BOF lance cooling; they are installed in multiple units and are motor-driven. The pumps have cast iron casings and bronze impellers. Water initially comes into the mill in the furnace area and is used there for indirect cooling. This water is collected in a pit and is pumped to the mill proper, where one of its uses is roll cooling, with the centrifugal pumps acting as boosters. A typical capacity of one pump would be 5,000 gpm at a pressure rise of 150 lb/in², with as many as three operating pumps being required along with an installed stand-by.

Lubricating oil is in great demand in any rolling mill. To provide enough oil at the mill bearings with adequate pressure, direct motor-driven gear pumps are used (Fig. 11). Pumping lubricating oil calls for the use of a pump with internal bearings. Standard cast iron bodies and bronze impellers are adequate materials.

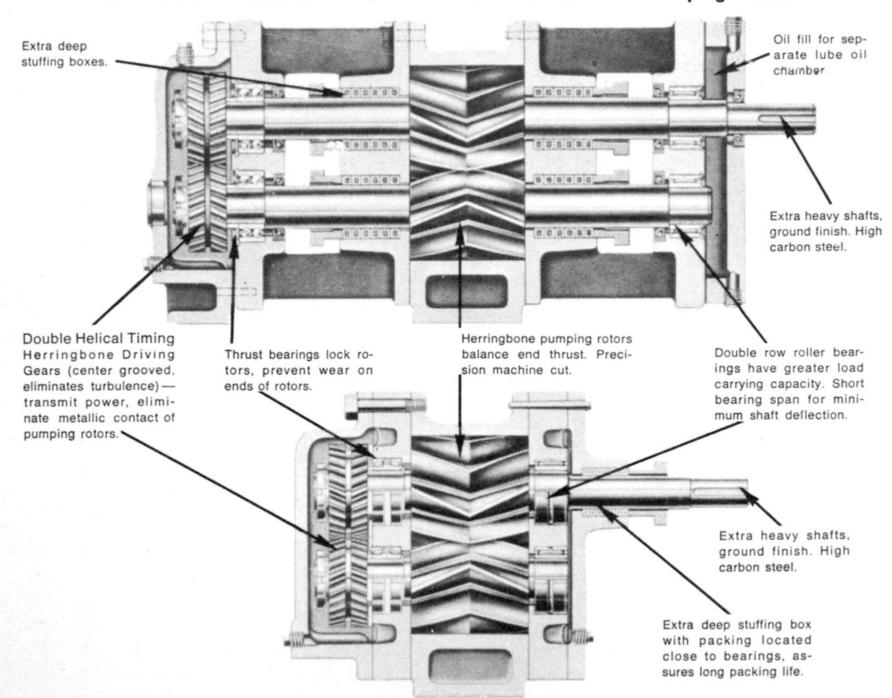

Fig. 11 Gear pumps. The internal-bearing type is used for circulation and distribution of mill lubricating oil. (Sier-Bath Pump, Division of Gilbarco, Inc.)

Mechanical seals would be preferred for long trouble-free operation. These positive-displacement pumps are discharge-pressure-controlled by bleeding excess fluid back to the oil reservoir. Holding a constant discharge pressure helps ensure even oil distribution to lubricated elements.

COLD-ROLLING MILLS

Process Coils of hot-rolled strip are sent to a complex of mills called a *cold mill.* The first operation in this complex is pickling, which involves acid-cleaning the strip of scale that was formed during the hot-rolling process. The cleaned material is then reduced in thickness in a series of mill stands called a *cold-reducing mill.* As in the case of hot-rolling, the strip is squeezed to a smaller thickness and made longer, but in this case the strip is rolled cold and the percentage of reduction is, of course, less. The cold-reduced strip, now called *sheet,* is hard and to enable forming and drawing it must be made soft. This is done by the annealing process, which involves heating the coils, holding the material at annealing temperatures allowing grain growth, and thus softening it. After annealing, the sheet coils are temper-rolled to provide flatness and a hard surface. The sheet can then be shipped in coil or cut-length form.

Another possible routing is that of galvanizing after cold-reducing. This involves running the sheet through molten zinc to provide a weather-resistant coating.

Still another process is that of tin or chrome coating in a continuous process. The sheet, after annealing, is sent to a tin mill, where it may be double-reduced (cold-reduced a second time) and sent to a tinning line where caustic cleaning, acid cleaning, rinsing, drying, and electrolytic tin plating are accomplished. Chrome coating is similar to tinning in that the chrome is also deposited on the moving sheet electrolytically.

Pump Applications Most of the above operations utilize hydraulically operated equipment, and the hydraulic system may serve more than one facility. A major pump application is providing pressure to the hydraulic systems. Pumps utilized in this service are similar to those previously discussed under slabbing mills.

Pumps are required for circulating caustic cleaning solutions, acids, hot water, and chemical treatment solutions from storage and make-up tanks to operating tanks and processes. They are also needed for moving waste solutions to pollution-abatement facilities. Pumping of lubricating and rolling oils is another major application.

A typical pump recommended for circulating fluids and for moving caustic cleaning solutions, plating solutions, acid wastes, and chemical treatment solutions from storage tanks to line tanks is a horizontal centrifugal with end suction (Fig. 12). Materials of construction would be varied to meet individual requirements. A 29 percent Ni–20 percent Cr alloy has been used successfully on the casing, impeller, cover plate, and gland of circulating pumps handling tin-line-plating solution. The same construction could be used with a caustic solution or with a 4 to 13 percent solution of sulfuric acid encountered in a pickling operation. Pumps such as these that encounter abrasive or corrosive materials should be equipped with a double seal to allow the use of a clean external water supply for seal flushing. Transfer or circulating pumps such as those on a finishing mill like a tin line will normally be installed in pairs, with one operating and one on a stand-by basis. Motor drives are satisfactory since a power outage will stop the line as well as the pump. Circulation normally involves pumping from a low storage tank up to the line tank with a gravity overflow return; therefore no special pump controls are required.

When sheet is rolled on a cold-reducing mill, some lubrication and pressure distribution are required between the rolls and the sheet. A mixture of a rolling oil and water is used for this purpose, the water acting as a carrier and making up 92 to 97 percent of the mix. Two systems of providing this mixture are used: (1) a batch of solution is made up and recirculated until its properties deteriorate, at which time a new batch is introduced; (2) a once-through system is utilized, with continuous mixing in proportions and quantities as required by the mill rollers. The latter system could use two triplex plunger pumps (Fig. 13) ganged together to a single direct motor drive to give a double triplex arrangement. One pump would be used for providing water and the other for oil. Stroke length on each pump can be adjusted to provide a changeable flow relationship. The single-drive

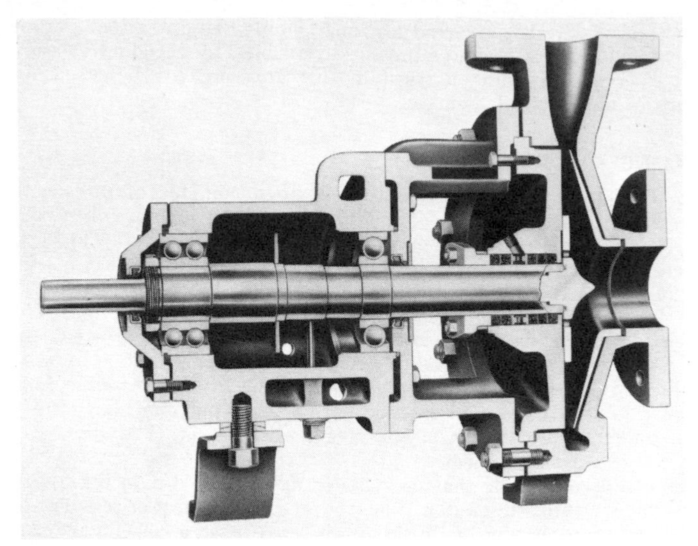

Fig. 12 Typical pump used in circulating and transferring chemical solutions on cold mill processing lines. (Duriron Company, Inc.)

unit can be equipped with a variable-speed capability to alter total flow variation as required.

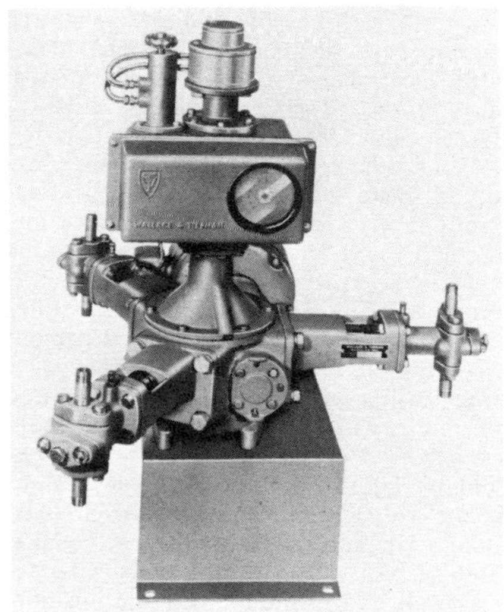

Fig. 13 Single triplex plunger pump with positioner for remote adjustment of stroke length. (Wallace and Tiernan Division, Pennwalt Corp.)

UTILITIES

Steel-mill operations must be supported by reliable utilities—electricity, steam, water, air, and fuels. The importance of the reliability of the supply stations must be emphasized strongly, since a shutdown of any one of them can bring all production to a standstill. Therefore great efforts have been made to improve reliability. Generally there are two basic methods: (1) a centralized supply station with installed spares, automatic equipment transfers upon failures, etc.; (2) a decentralized system, which because of lack of connection between stations (except for pipe, etc.) provides odds against a total failure. The trend lately has been a compromise wherein two stations are installed rather than one or many, and both stations are provided with equipment to minimize local failures. Another reliability assist is in using more than one source of power for driving equipment. Where reliable steam is available, equipment drives for air compressors or fuel pumps, etc., can be split into electric and steam units, thus minimizing the effects of driver-energy failure.

There are many pump requirements in utility supply stations. A water-supply station is primarily a pumping station. One of the many pump applications in a steam station is feed-water pumping. Most types of liquid fuel stations will require pumps for providing fuel flows at burner pressures in addition to transfers between stations.

Section 10.14

Hydraulic Presses

A. B. ZEITLIN

TYPES OF PRESSES

Hydraulic presses are used in many industrial technologies. Among them are:

Vertical Presses 1. Forging presses with flat dies are used for hot work to break down ingots and to shape them into rolls, pressure vessels (mandrel forgings), forged bars, rods, plates, etc.

2. Forging presses with closed dies are used to process preheated billets into various shapes. For example, bulkheads of aircraft structures, engine supports, main fuselage and wing beams, etc., are forged on hydraulic presses.

3. Upsetting presses are used for the production of items with elongated shafts—long hollow bushings, pipes, vessels, etc.

Horizontal Presses Presses are used primarily for conventional hot extrusion. Lately cold hydrostatic extrusion is beginning to find application in producing various end products and bulk items like very thin wire in long strands.

ACCUMULATORS

Until about 1932 all power systems for hydraulic presses were operated with water. With the advent of reliable, fast, high-pressure oil pumps, there developed a trend to utilize the significantly less expensive oil pumps.

Installations requiring a relatively uniform and more or less constant supply of hydraulic power are preferably designed with drives drawing the pressurized liquid direct from the pumps. In installations with high peaks of power (and liquid) demand of short duration, it is in most cases advantageous to arrange for a constant averaged flow of pressurized liquid from the pumps while providing equipment in which pressurized liquid can be stored in times of low demand and from which additional quantities of liquid can be drawn during demand peaks. These storage facilities are called accumulators and are illustrated in Fig. 1. A modern accumulator is buffered by a large pressurized gas (mostly air) cushion.

Until about 1960, the direct contact surface between gas and liquid virtually precluded the use of oil in accumulator installations for larger presses. Gas diffuses into oil under pressure. When oil is discharged from the press at the end of the pressing cycle, the gas begins to bubble out of the liquid; the liquid foams, and, because of its large volume, this foam is rather difficult to handle. This condition exists whether the pneumatic cushion is air or nitrogen. In addition, if air is used as the pneumatic cushion, the oxygen diffused into the oil oxidizes it, producing

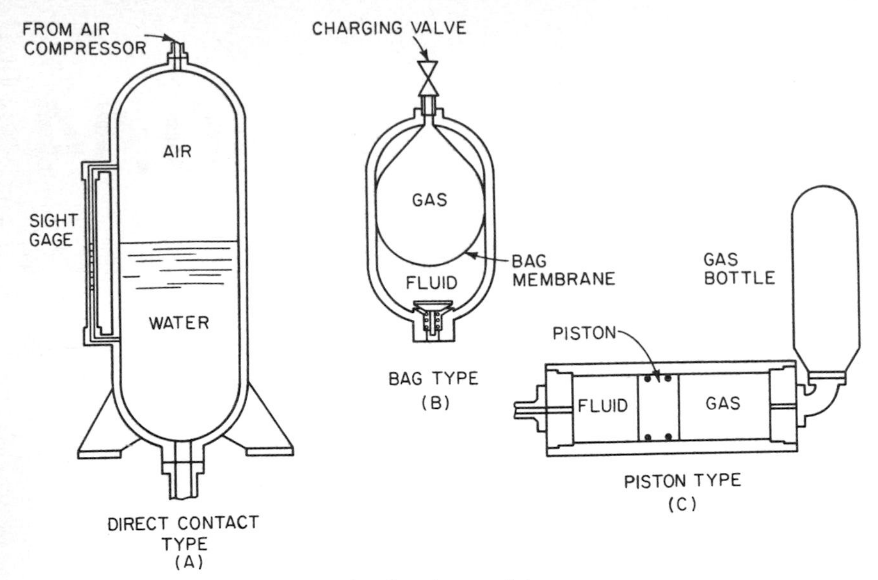

Fig. 1 Accumulators.

sludge and gum. (Spring and weight-loaded accumulator installations have other serious drawbacks and have almost completely disappeared.)

A modification of the accumulator station introduced a few years ago uses a floating piston separating oil from the pneumatic cushion (see Fig. 1C). This design eliminates foaming and oxidizing, thus allowing the inclusion of accumulators into oil systems.

Still, power plants operated with water are used in special cases, for instance, where extreme precaution must be taken against fires caused by leaking oil. Often hydraulic systems filled with noninflammable liquid, instead of oil, are used.

CENTRIFUGAL VS. RECIPROCATING TYPE PUMPS

The competition between centrifugal and reciprocating pumps in hydraulic plants for presses has been decisively won by reciprocating pumps. Under full load, centrifugal pumps have higher efficiency than reciprocating pumps (Fig. 2). However, in press installations, the power plants idle a significant portion of the time. The

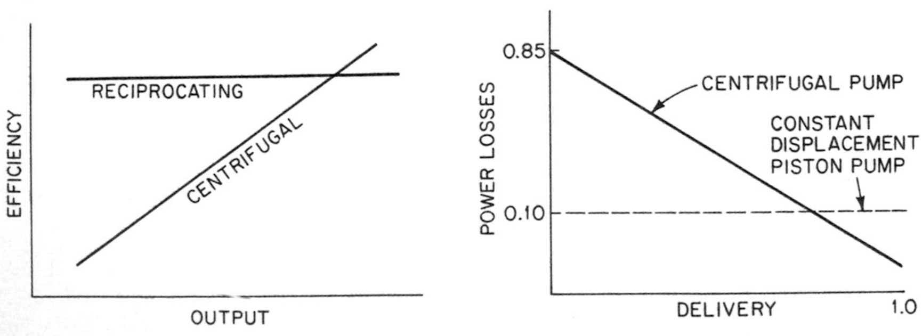

Fig. 2 Comparison of pump efficiency.

Fig. 3 Comparison of pump power losses.

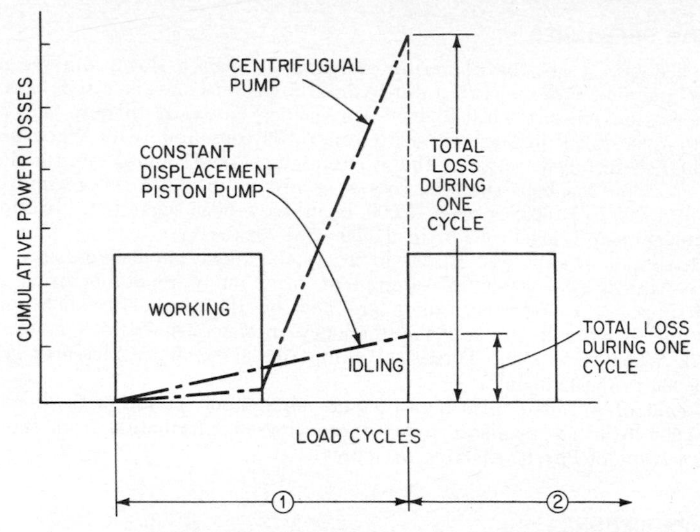

Fig. 4 Comparison of load cycling losses.

losses during idling are only about 10 percent of full load for reciprocating pumps and well over 60 to 70 percent for centrifugal units. The total combined losses in installations with reciprocating pumps are less than half the losses in centrifugal pump plants, as shown in Figs. 3 and 4.

Reciprocating pumps for hydraulic presses which use water as the hydraulic medium are generally of the single-acting multiplunger type. Although there are large installations in operation using double-acting pumps, the faster vertical single-acting pump, which requires less floor space, offers a considerable saving in capital expenditure. They have proved themselves dependable in service. In most cases they are used in conjunction with an accumulator, which allows the averaging of demand, thus reducing the required pump capacity. There are, however, installations in which these pumps drive hydraulic cylinders directly.

Idling is controlled by remotely operated bypass valves or pump suction-valve lifters which take command from the operator, accumulator level control, or some other sensor-monitoring press action.

Oil pumps most commonly used for power strokes of presses are:

1. Vane-type, rotary, constant-delivery pumps, generally for pressures not in excess of 2,500 lb/in^2.
2. Piston-type, rotary, constant-delivery pumps.
3. Piston-type, rotary, variable-delivery pumps.

The constant-delivery pumps require a bypass system for idling. This is generally accomplished by an arrangement in directional control valves or by relief bypass valves (see above).

The variable-delivery pump has the advantage of providing efficient press speed control and idling by an adjustment of piston stroke. In cases of multiple pump application, constant- and variable-delivery pumps are often used jointly, their selection depending on speed ranges required. It is possible to obtain these pumps designed for use with nonflammable fluid as a hydraulic medium.

Generally, both water and oil pumps, in press applications, are driven directly by electric motors, the motor rpm matching the pump speed. Sometimes, on large water pumps, a geared speed reducer is installed between motor and pump.

OPERATING PRESSURES

Over the last fifty years, the operating pressures have been slowly moving upward. The most popular pressure for water-hydraulic installations is rated 5,000 lb/in^2 (actual operating pressure 4,500 lb/in^2). Intensifiers are used to raise this pressure of 6,750 to 7,500 lb/in^2 in large presses (from 7,000 tons upward). When used, the intensifiers are installed between the accumulators (and pump) on the low side and the press on the high side. A bypass is always provided. For giant presses, pressures of 9,000 to 12,000, or even 15,000, lb/in^2 have been suggested. In the Soviet Union, designers are using 15,000 lb/in^2 (1000 atm) regularly.

In oil-hydraulic installations, 3,000 lb/in^2 is the most popular pressure because of the availability of 3,000-lb/in^2 equipment from many reputable firms (pumps, valves, fittings, etc.). However, there are now on the market excellent units for pressures up to 6,600 lb/in^2 and the tendency in new installations is to provide 5,000 lb/in^2 instead of 3,000. Today oil pumps for even higher pressures are being offered (9,000 to 15,000 lb/in^2).

In the field of so-called ultra-high-pressure equipment, pumps with direct action up to 225,000 lb/in^2 are available today (according to information from the Soviet-Russian Institute of Physics at High Pressures).

DESIGN PROCEDURE

Considerations and calculations required to dimension the hydraulic power plant for a hydraulic press are given in Table 1. A standardized procedure for the determination of the basic design features of the power plant for a hydraulic press is offered in the left column. The right column gives an example of how the standard procedure should be used. The press selected for the example is an open die press for forging ingots into billets or bars through gradual reduction of cross section on narrow, flat dies which, at each stroke, penetrate several inches into the hot metal (cogging). The ingot is advanced and rotated and, when it is close to desired size, the surface is made smooth by applying shallow penetration (1/8 to 1/2 in) fast strokes (planishing) of the press (Fig. 5).

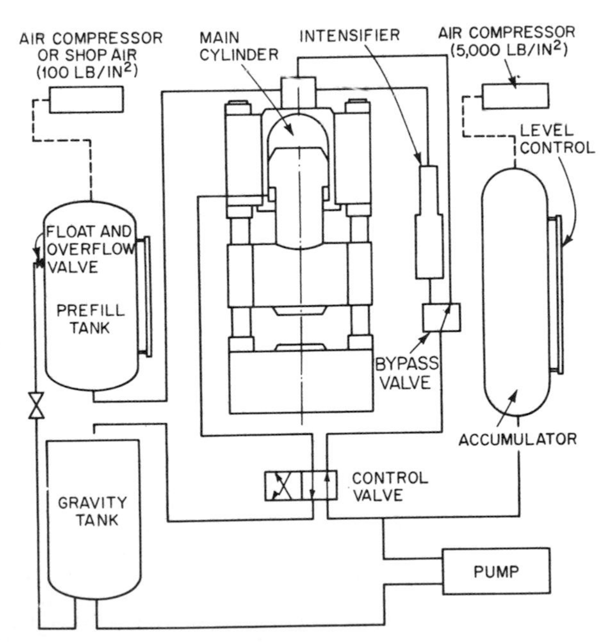

Fig. 5 Vertical forging press with double-acting cylinder.

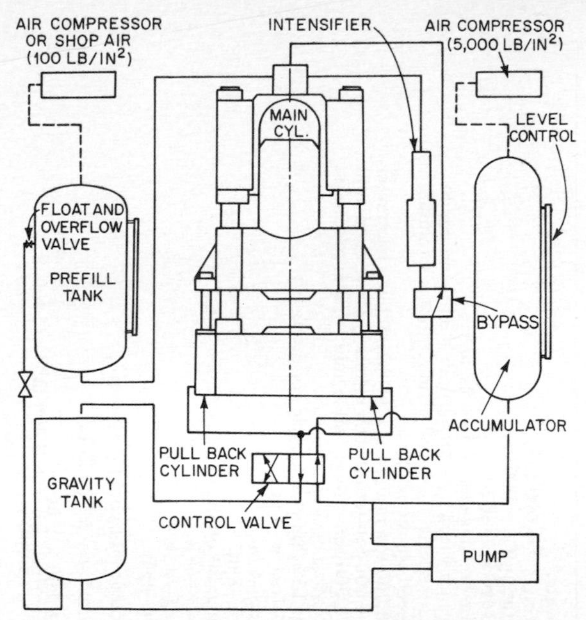

Fig. 6 Vertical forging press with pull-back cylinders.

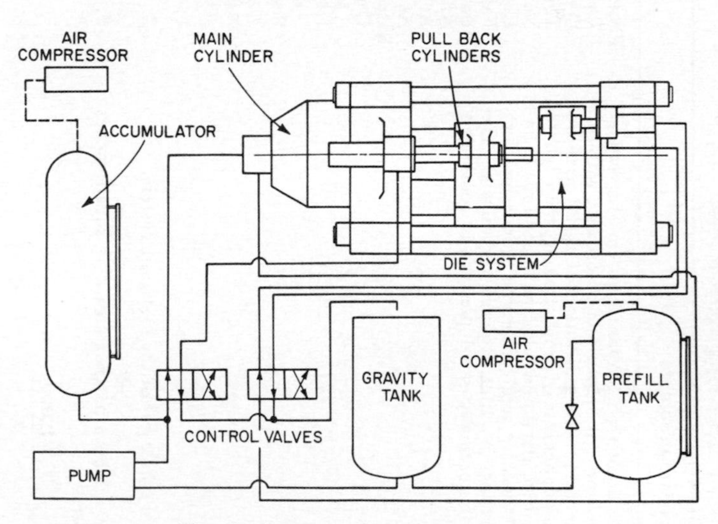

Fig. 7 Horizontal extrusion press.

TABLE 1 Example of Power-Plant Design for a Hydraulic Press

Standard procedure	Example
1. Water or oil can be used as liquids in a hydraulic system of a vertical or horizontal press. The decision can be made in accordance with the general discussion in this section. The generation of a full pressure during planishing is not required; because of rapid stroking, there would not be enough time to pressurize the liquid fully and expand the cylinder fully. The pullback cylinders are kept continuously under pressure during planishing, thus reducing the time required for valving. The generation of about 10% of nominally rated pressure is considered sufficient for planishing.	Requirements: Press rating $T = 6{,}000$ tons Full-back rating $T_R = 600$ tons Total stroke $S_t = 48$ in Maximum pressing stroke $S_p = 12$ in Fast advance speed $v_a = 360$ in/min = 6 in/s minimum Fast return speed $v_r = 360$ in/min = 6 in/s minimum Pressing speed V_p required: At full 6,000-ton rating: 120 in/min = 2 in/s maximum At $^2/_3$ = 4,000-ton rating: 180 in/min = 3 in/s maximum At $^1/_3$ = 2,000-ton rating: 360 in/min = 6 in/s maximum Rated operating cycles: Cogging: 15 c/min, each cycle consisting of: Fast Advance—4 in Pressing—4 in Return—8 in Planishing: 80 c/min, each cycle consisting of: Advance—¼ in Pressing—¼ in Return—½ in 1. Selected: oil hydraulic system
2. Select operating pressure p in lb/in² of the hydraulic system.	2. Selected: $p = 5{,}000$ lb/in²
3. Determine the required effective pressing area A in square inches of the main plunger of the press $A = \frac{T \times 2{,}000}{p}$ where T is the rated power in tons	3. $A = \frac{6{,}000 \times 2{,}000}{5{,}000} = 2{,}400$ in²

4. Select the number N_c of cylinders comprising the main system of the press and their individual ratings $T_1, T_2, T_3, \ldots$

4. Select 3 cylinders:
Center cylinder $T_2 = 4{,}000$ tons
Two side cylinders $T_1 = T_3 = 1{,}000$ tons
Operating all three cylinders: $T = 6{,}000$ tons
Operating the center cylinder alone: 4,000 tons
Operating the side cylinders alone: 2,000 tons

5. Subdivide A in accordance with selected ratings and determine the individual diameters

$$D = \sqrt{\frac{A}{0.785}}$$

If any piston rods detract from the effective plunger area, their area should be added to A before determining D.

5. $A_1 = A_3 = \frac{1}{6} \times 2{,}400 = 400\ \text{in}^2$
$A_2 = \frac{4}{6} \times 2{,}400 = 1{,}600\ \text{in}^2$

$$D_1 = D_3 = \sqrt{\frac{400}{0.785}} = 23\ \text{in}^*$$

Adjusted area $A'_1 = A'_3 = 415\ \text{in}^2$

$$D_2 = \sqrt{\frac{1{,}600}{0.785}} = 45\ \text{in}^*$$

Adjusted area $A'_2 = 1590\ \text{in}^2$

Total adjusted area $A' = 2 \times 415 + 1{,}590 = 2{,}420\ \text{in}^2$

6. Using the total stroke S_t and the rated pressing stroke S_p, determine the geometric volumes V_t and V_p corresponding to these strokes.

6. $V_t = S_t \times A' = 48 \times 2{,}420 = 116{,}160\ \text{in}^3$
For cogging V_{pc}: $4 \times 2{,}420 = 9{,}680\ \text{in}^3$/per cogging stroke
For planishing V_{pp}: $\frac{1}{4} \times 2{,}420 = 605\ \text{in}^3$/per planishing stroke

7. Using a general compressibility curve, determine the compressibility of liquid c between atmospheric pressure and rated operating pressure.

7. For 5,000 lb/in² the oil compressibility may be assumed with $c_o = 1.5\%$.

8. Assume a reasonable approximate level of hoop stresses s_h along the inner surface of the cylinder barrel. The cylinder diameter expansion will be

$$d_e = \frac{s_h}{E} \times D$$

where E = modulus of elasticity assumed 30×10^6 psi

8. Assume $s_h = 30{,}000\ \text{lb/in}^2$

$$\text{Main cylinder: dem} = \frac{30{,}000}{30 \times 10^6} \times 45 = 0.045\ \text{in}$$

$$\text{Side cylinders: des} = \frac{23}{1{,}000} = 0.023\ \text{in}$$

where dem = main cylinder diameter expansion
des = side cylinder diameter expansion

---

* Note: It is customary to select diameters of large plungers in full or ½-in sizes; this requires adjustment of the area.

TABLE 1 Example of Power-Plant Design for a Hydraulic Press (Continued)

Standard procedure	Example
9. On the basis of c_o and s_h determine the volume of liquid required to compensate for compression of liquid V_{c1} as well as for expansion of cylinder V_{ec}. During the planishing operation the pressures rise to only about 10% of the rated; therefore the compressibility of liquid and expansion of the cylinder may be neglected.	9. The compression of the liquid takes place along the entire length of the cylinder barrel. Assuming the length of the barrel to be approximately $S + \frac{1}{2}S$, or 48 in + 24 in = 72 in: $V_{c1} = 72 \times 2{,}420 \times 0.015 = 2{,}614 \text{ in}^3$ Omitting quantities small in the second order, $V_{ec} = 72 \frac{\pi}{4} \left(2D \times \frac{30{,}000}{30 \times 10^6} \times D\right)$ $= 72 \times 1.57 \times 0.001 \times D^2 = 0.113 \times D^2$ For the center cylinder: $V_{ec2} = 0.113 \times 45^2 = 229 \text{ in}^3$ For the side cylinders: $V_{ec1} = V_{ec3} = 0.113 \times 23^2 = 60 \text{ in}^3$ $V_{ec} = 229 + (2 \times 60) = 349 \text{ in}^3$ Total additional volume of liquid required: $2{,}614 + 349 = 2{,}963 \text{ in}^3$ per stroke
10. Determine the total amount of liquid V required per cogging stroke.	10. $V = 9{,}680 + 2{,}963 \text{ in}^3 = 12{,}643 \text{ in}^3$ This figure shows what additional burden can be imposed on the power plant by unnecessarily generous dimensioning of the total stroke.
11. Check whether specified speeds are compatible with the required number of cogging strokes. If not, increase the specified speeds (or reduce the number of strokes).	11. Fast advance time: $4/6 = 0.67$ s Pressing time: $4/2 = 2$ s Fast return time: $8/6 = 1.33$ s Valving time (3 switches) = 0.45 s Total cycle time = 4.45 s This time is too long to allow 15 cogging strokes per minute. Increase fast advance and fast return to 480 in/min = 8 in/s and pressing to 180 in/min = 3 in/s. Fast advance time: $4/8 = 0.5$ s Pressing time: $4/3 = 1.33$ s Fast return time: $8/8 = 1$ s Valving time = 0.45 s

12. Determine the liquid requirements for the pull-back stroke (Figs. 6 and 7). In general, the pull-back cylinders have an area equal to 10% of the area of main cylinders and twice as long a stroke compared with the pressing stroke; in general, the required volume is, therefore, with sufficient accuracy:

$$V \times 0.1 \times 2 = 0.2 \times V$$

13. Determine the gpm requirements for the pump station with or without an accumulator. Without an accumulator:

$$\text{gpm} = \frac{V \times 60}{231 \times t_p}$$

where t_p = pressing time in seconds and V = volume required for one pressing stroke.

With accumulator:

$$\text{gpm (acc)} = \frac{V_{tc} \times 60}{231 \times t_c}$$

where t_c = cycle time and V_{tc} = total volume of pressurized liquid required during one cycle

14. Determine size of the accumulator required to store the accumulated volume. Stored volume V_s of liquid:

$$V_s = \text{gpm (acc)} \times (t_c - t_p) - V_R$$

where V_R is the volume required for return stroke in gallons

Air volume required, based on 10% pressure fluctuation for isothermic condition, is $10V_s$. For adiabatic or more often polytropic condition and 10% pressure fluctuation, the air volume is

$$V_{\text{air}} = \frac{V_s}{1.11^{1/n} - 1}$$

Although $n = 1.4$ is the exponent for adiabatic compression, $n = 1.3$ is considered satisfactory, even for severe and demanding conditions, which require full utilization of press stroke and a fast cycle. Based on $n = 1.3$,

$$V_{\text{air}} = 12\ V_s$$

12. $12{,}643\ \text{in}^3 \times 0.2 = 2{,}528\ \text{in}^3$

13. Without an accumulator:

$$\text{gpm} = \frac{12{,}643 \times 60}{231 \times 1.33} = 2{,}469\ \text{gpm}$$

With accumulator:

$$\text{gpm (acc)} = \frac{(12{,}643 + 2{,}528) \times 60}{231 \times 4} = 985\ \text{gpm}$$

It is evident that the use of an accumulator will result in significant savings and in a substantial reduction of the peak load.

14.

$$V_s = 985 \times \frac{(4 - 1.33)}{60} - \frac{2{,}528}{231} = 32.9\ \text{gal}$$

$$= \frac{32.9}{7.48} = 4.4\ \text{ft}^3$$

$$V_{\text{air}}\ \text{(isothermic)} = 10 \times 32.9 = 329\ \text{gal}$$
$$= 10 \times 4.4 = 44\ \text{ft}^3$$

$$V_{\text{air}}\ \text{(polytropic)} = 12 \times 32.9 = 394.8\ \text{gal}$$
$$= \frac{394.8}{7.48} = 53\ \text{ft}^3$$

Total accumulator volume = $44 + 4.4 = 48.4\ \text{ft}^3$, respectively;
= $53 + 4.4 = 57.4\ \text{ft}^3$

TABLE 1 Example of Power-Plant Design for a Hydraulic Press (Continued)

Standard procedure	Example
15. Check planishing conditions.	15. Timing: Fast advance: $\frac{1/4}{8} = 0.032$ s Pressing: $\frac{1/4}{3} = 0.084$ s Fast return: $\frac{1/2}{8} = 0.063$ s Valving (2 switches) = 0.3 s Total = 0.479 s There will be more than enough time for 80 planishing strokes per minute. Only two valve switches are required for planishing because, during planishing, the pull-back cylinders are permanently connected to the pressure source. During the planishing operation, the pressure reaches only about 10% of the rated pressure; therefore compressibility and expansion require only 10% of the previously calculated amount. The pull-back system is constantly pressurized and does not require any liquid for compression of liquid and expansion of cylinder. The total requirements are thus: For the pressing stroke: $605 + 296 = 901$ in^3 For the pull-back stroke: $605 \times 0.2 = 121$ in^3 Total $= 1{,}022$ in^3 Gpm required $= \frac{1{,}022}{231} \times 80 = 354$ gpm The selected accumulator and pump power plant will be more than sufficient for planishing. As a matter of fact, 100 or even 110 planishing strokes per minute will be feasible.

Section 10.15

Hydraulic Servosystems

J. N. CORONEOS

INTRODUCTION

Since the concept of electrohydraulic servo systems was introduced, most design and research have been in the development of the electrohydraulic servo valve. Since 1950 electrohydraulic servo controls for variable-displacement piston pumps have also been designed and developed. The combination of this type of pump and the concept of manipulating it by servo control has made, in many ways, an ideal servo systems component. In many cases, the most economical fluid power source for the servo valve system will be the fixed-displacement gear, piston, or vane pump which delivers a constant fluid volume.

A typical servo valve system for velocity control of feed rolls is shown in Fig. 1. In this example the selections of the servo valve, hydraulic motor, and command and feedback devices are the main system design considerations. The choice of the hydraulic pump type would usually be based on the designer's preference; it must be selected with the capability to maintain the maximum system pressure and flow requirements on a continuous basis. The amount of fluid which passes the system relief valve and is converted to heat losses is inversely proportional to the commanded velocity of the driven rolls. Consideration must therefore be given to ensuring the proper sizing of the heat exchanger and the fluid circulation system to operate within the allowable system operating temperature.

PUMP TYPES

To increase the overall efficiency, many hydraulic servo valve systems are designed with the variable-displacement pressure-compensated pump for the fluid power source. A wide variety of such units is available; they may be of the vane or piston-type design, with the capability to vary the delivery rate at a set operating pressure to the demand of the servo system. Figure 2 illustrates a variable-displacement vane pump having a pressure-compensator control to maintain the set operating pressure from zero to maximum system flow requirements. Care must be used, however, to ensure that the response time of the pressure-compensator control is sufficient to follow the servo system. Accumulators may be used to compensate for pump response time, but the natural frequency of the pump and compensator control must still be considerably above the system operating frequency to ensure against uncontrolled pump oscillation. When using servo-controlled pumps rather than servo valves, these will normally consist of a variable-displacement pumping mechanism, usually of the axial or radial-piston type, capable of reversible flow. Functionally the unit consists of a continuously variable source of oil flow in either

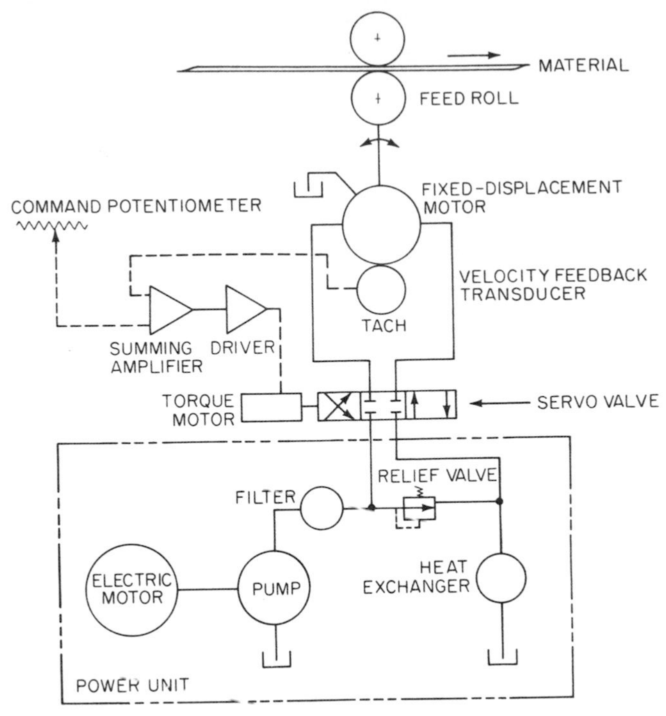

Fig. 1 A typical servo value system for velocity control of feed rolls. (Racine & Vickers—Armstrongs, Inc.)

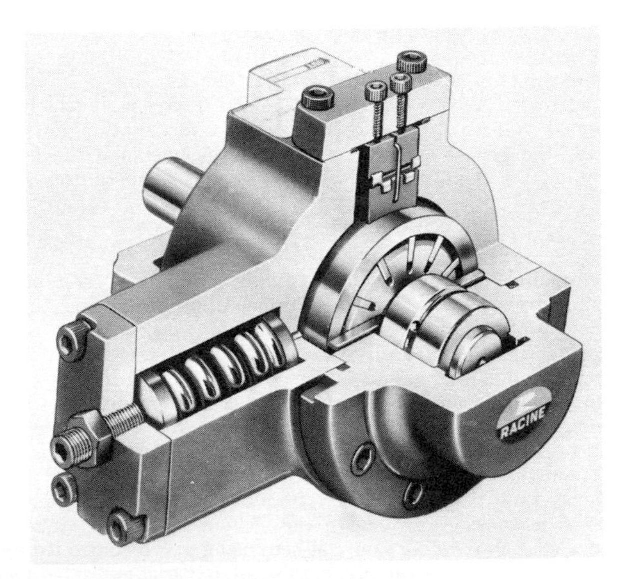

Fig. 2 A variable-displacement vane pump having a pressure-compensator control. (Racine & Vickers—Armstrongs Inc.)

direction from the pump. A basic servo-controlled pump is shown in Fig. 3. The pumping mechanism, consisting of the cylinder barrel, socket ring, and pump pistons, rotates with the main shaft. The cylinder barrel contains the piston bores, which terminate in ports that rotate against the valve plate. Kidney-shaped ports in the valve plate alternately connect the two pump ports to the pumping-piston bores. The tilt box, mounted on bearings, carries the socket ring, one end of the pumping pistons, and one end of the control pistons. The control pistons produce a controlled movement of the tilt box. At the neutral position, rotation of the main shaft pro-

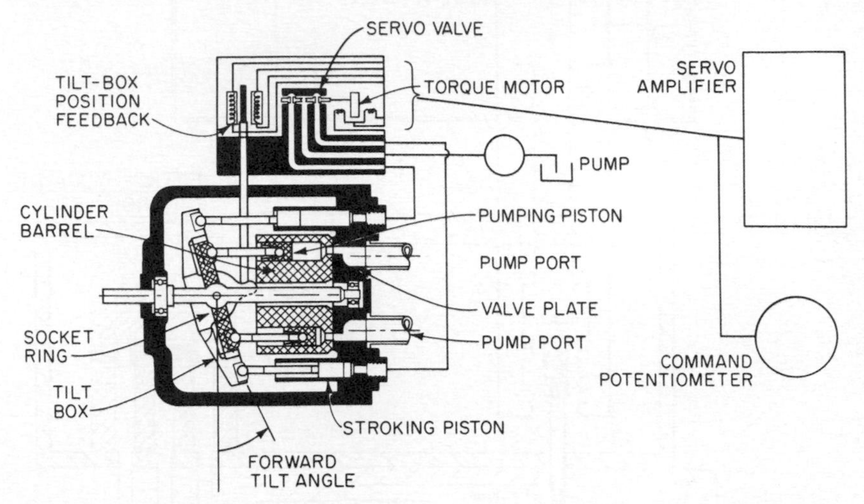

Fig. 3 A basic servo-controlled pump. (Racine & Vickers—Armstrongs, Inc.)

duces no axial motion of the pumping pistons with respect to the cylinder barrel, with no oil flow in or out of either pump port. The unit delivers flow in the direction of and in direct proportion to the tilt-box angle from neutral.

SERVO CONTROL SYSTEMS

The pump servo control is a basic servo valve system, with the valve being the prime electrohydraulic control element. A fixed-displacement gear or vane pump is the fluid source for the valve. The servo valve output ports are connected to the pump-control pistons. Fluid is supplied at a rate and in a direction dependent upon the voltage supplied to the torque motor which controls the servo valve. A feedback transducer is mechanically controlled by the tilt box or stroking mechanism of the pump; it supplies a feedback voltage proportional to the pump-stroke position. One of the main advantages of a control of this type is that power requirements are normally under 1 hp, with a capability of controlling a pump of many hundred horsepower.

A typical electrohydraulic servo control for the electrohydraulic servo pump is shown in Fig. 4. The feedback transducer signal from the linear variable differential transformer (LVDT) is applied to a summing junction on an amplifier where it is summed with the command signals. Any error signal which develops is amplified and demodulated to drive the electrohydraulic servo valve, which changes the tilt-box angle to eliminate the error signal. Since the feedback voltage is constantly compared to the command voltage, the tilt box and, therefore, the pump delivery can be infinitely and accurately positioned over its entire range.

To describe the performance characteristics of electrohydraulic servo-controlled pumps, a comparison will be made with the characteristics of a servo valve.

Resolution and sensitivity are generally interrelated and refer to the smallest change in pump flow which can be consistently and repeatably made. The resolution of a servo-controlled pump can be as high as one part in 100,000. This is at least two orders higher in magnitude than that of most servo valves.

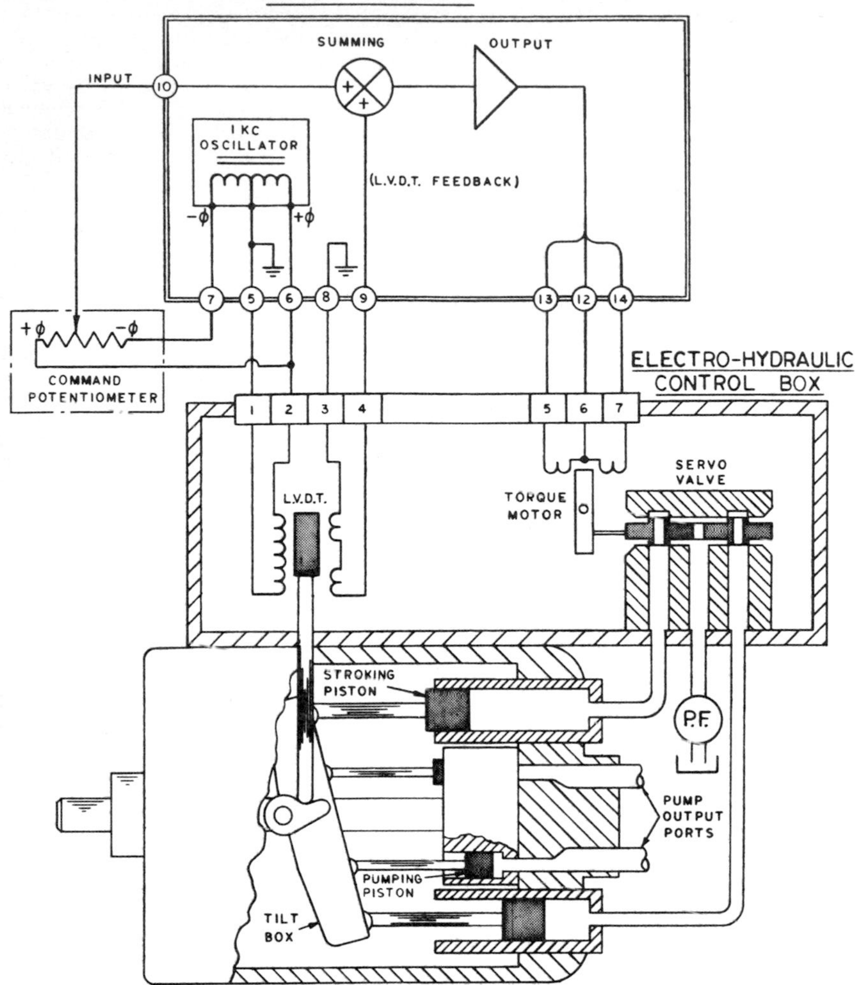

Fig. 4 Variable-volume hydraulic pump. (Racine & Vickers—Armstrongs, Inc.)

Hysteresis is the ability of the pump to supply the same output when a signal is received calling for a change from full delivery in one direction to full delivery in the opposite direction and back to full delivery in the original direction. The hysteresis of a well-designed electrohydraulic servo pump is often less than 0.01 percent. This is far better than most servo valves. The advantage lies in the inherent integrating action of the tilt box, which is similar to the second-stage feedback of a two-stage valve.

The speed of response can be expressed as follows:

1. *Slewing or saturation speed* is the time required to change from zero to full output of the pump. Normal slewing speed can be in the order of 25 ms. Servo valves generally will have somewhat higher slewing speeds, and therefore faster speed of response.

2. *Frequency response* is usually expressed in the form of a Bode plot giving phase and amplitude changes with frequency. A high-performance servo-controlled pump

may have a flat amplitude response up to 80 to 100 Hz with a 90° phase shift from 40 to 50 Hz. These figures may be exceeded by some valves. A frequency-response test applied to a 6-in^3/rev servo pump driven at 1,200 rpm with 50 ms slewing time produces a Bode plot as shown in Fig. 5. The applied amplitude was approximately 2 percent of full.

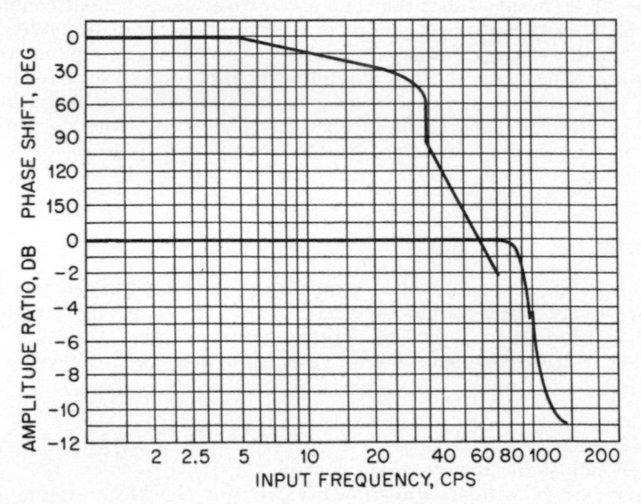

Fig. 5 Frequency response of electronically controlled pumps. (Racine & Vickers—Armstrongs, Inc.)

Pressure gain is an important characteristic of both servo pumps and servo valves, and is directly related to the sensitivity and accuracy of the overall servo system. A pressure gain curve shows the output pressure change with respect to input signal changes. A comparison of the pressure gain of a servo valve and a servo-controlled pump is shown in Fig. 6.

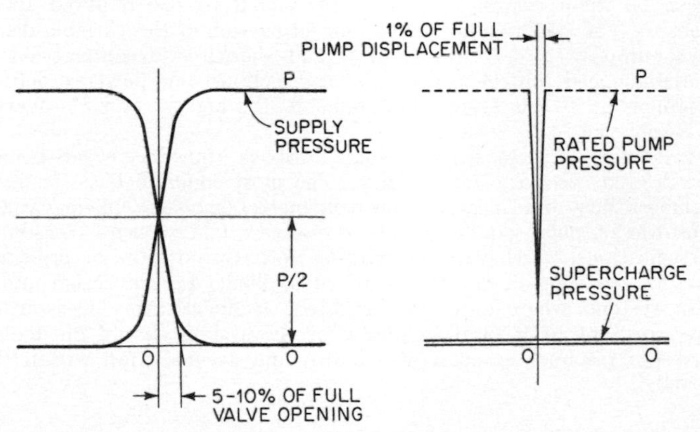

Fig. 6 Comparison of the pressure gain of servo valves (left) and electronically controlled pumps (right). (Racine & Vickers—Armstrongs, Inc.)

The pressure gain of the servo pump is normally much higher than that for the servo valve. A high-performance pump may require a signal of only 1 percent of that required for full pump output to produce rated pressure. A servo valve normally requires a 5 to 10 percent rated signal to produce a differential pressure equal to the supply pressure.

Probably the most important difference between the servo-valve and servo-pump pressure gain is the change with respect to the input signal. A servo-controlled pump will maintain approximately the same pressure gain independently of the set operating flow. As a servo valve is operated at increasing flow, the pressure gain will be far lower than for zero flow. At high flows the pressure gain of a servo pump can easily be twenty times that of a servo valve under similar conditions.

The use of a servo-controlled pump in a servo system will normally greatly increase the overall system efficiency. This is because the servo pump controls the rate of flow according to the system requirements, while the servo valve is essentially a throttling device. In a servo-valve system the pressure must be set high enough to control peak loads. This results in relatively high continuous losses as compared to the servo pump, which can often operate close to zero pressure for long periods of time.

APPLICATIONS

The electrohydraulic servo-controlled pumps have solved a wide range of industrial control applications ranging from the relatively simple to highly complex systems. When used in an open-loop system with no feedback other than the basis servo control, the servo-controlled pump can provide the following desirable characteristics:

1. High overall system efficiency
2. Infinitely adjustable remote speed control
3. Reversing drives without directional valves for shockfree system operation
4. Controlled acceleration and deceleration
5. Accurate stopping and drive reversal
6. Good speed regulation versus load change

Press applications of the servo-controlled pump are relatively simple but effective examples of a typical open-loop system. Power requirements normally begin at about 20 hp and may increase to several hundred. Remote speed-setting potentiometers provide any required number of press platen speeds. Limit or pressure switches can be incorporated to furnish the signal to the required speed-setting potentiometer. The system allows maximum utilization of the variable-displacement cross-center pump with valveless reversal and shockfree decompression using infinitely variable speed control in both platen directions. As power is delivered only in direct proportion to the system requirements, the highest possible overall system efficiency is achieved.

As system demands increase, outer-loop feedback transducers can be applied to supply the desired system characteristics. The most common basic feedback transducers include rotary and linear motion tachometers, synchros, linear variable differential transformers, pulse generators, phase resolvers, and pressure transducers. The system classifications, which will normally be velocity, position, or pressure control, will determine the class of transducer to be applied. In the design and selection of all servo systems where outer or closed-loop feedback transducers are used, the spring-mass constant or natural frequency of the system should be determined to ensure that the response speed, power limits, and accuracy fall within the known requirements.

DESIGN CALCULATIONS

While it is possible to make a complicated mathematical analysis of servo systems, simple and effective formulas can be used to determine the basic results that can be expected from the majority of systems. Because of the compressibility of the fluid, the cylinder or hydraulic motor-mass combination is equivalent to the spring-mass system. The natural frequency of the cylinder-mass or hydraulic motor-mass assembly can be expressed as follows:

Hydraulic cylinder-mass system:

$$Wn = \frac{A}{\sqrt{BVM}}$$

where Wn = natural frequency, rad/s
A = cylinder area, ft^2
V = volume of fluid in one side of piping with the cylinder centered, ft^3
B = bulk modules of the fluid (oil = $2.8 \times 10^{-8}\ ft^2/lb$)
M = Mass being driven, $lb\text{-}s^2/ft$

Hydraulic motor:

$$Wn = \frac{d}{\sqrt{BVI}}$$

where Wn = natural frequency, rad/s
d = hydraulic motor displacement, ft^3/rad
V = volume of fluid in one side of piping, ft^3
I = inertia reflected to hydraulic motor shaft and including inertia of the motor, $lb\text{-}ft\text{-}s^2$
B = bulk modulus of the fluid (oil = $2.8 \times 10^{-8}\ ft^2/lb$)

It has been found that, to have a simple positioning system stable, no element in the system should have a natural frequency less than three times greater than the frequency at which the system is to operate. Since in most cases the motor-mass or cylinder-mass frequency will be the lowest of all frequencies of the system, the most common optimum operating frequency will be expressed by $Wn/3$ rad/s. The response speed of a servo system is expressed by the system time constant and relates directly to the time required for the correction of any magnitude step-response signal.

One time constant is equal to the reciprocal of the system operating frequency, and is expressed in the terms TC = $1/(Wn/3)$ seconds. The response of the servo system to a step input signal is:

Time constant (TC)	% of correction made
1	63
2	87
3	95
4	98
5	99

Thus in 5 time constants the servo system would be within 1 percent of the step input position signal. This assumes that the step response signal is within the response rate of the servo pump. When saturation or maximum slew rate of the pump occurs, the correction time of the system will be increased. Another fundamental constant of the servo positioning system is the velocity error constant (VEC), which is the measurement of the system velocity change as the object being positioned approaches its final point. In this case error represents the object distance from final position. The system operating frequency relates directly to the VEC. For example, a system with an operating frequency of 20 rad/s driving a linear positioning device to some predetermined position would have a VEC of 20 in/s/in error. Assuming a maximum system velocity of 100 in/s, and dividing by the VEC of 20 in/s/in error, a 5-in error is obtained. This is the error lag of the positioned device to the commanded position at a velocity of 100 in/s. It also represents the required stopping distance to the final position of the system. To determine the final position accuracy of the system, it is helpful to know the slip-stick friction characteristics and component accuracies used in the system. If an error signal of 1 percent is required to raise the required torque or force to overcome the system friction, and other system components have accuracies within this figure, the expected overall accuracy of the system would be within ±1 percent of the velocity error, or 0.05 in. Many systems can, however, be provided with gain compensators to improve the accuracy when required. The method described was made with the simplest mathematics, but it will provide an effective guide for determining the accuracy and response of most servo systems.

The following examples are systems using servo-controlled pumps with velocity, position, and pressure control of the driven-machine element. The requirement to match velocity of driven rolls over considerable speed and load changes is frequent in the paper and metal-processing industry. By the use of a command tachometer driven by the element of the machine whose velocity is to be matched, together with a feedback tachometer attached to the driven roll, velocity matching of the two elements can be maintained over a infinitely variable speed range. The velocity of the driven member with respect to the command can normally be held within ±0.25 percent with a 100 percent load change. Therefore a command with a maximum velocity of 1,000 rpm can be matched with ±2.5 rpm over the entire speed range with a 100 percent load change. Many variations can be effected with a velocity feedback drive. By replacing the driven command tachometer with an adjustable command potentiometer, the drive becomes an excellent speed regulator

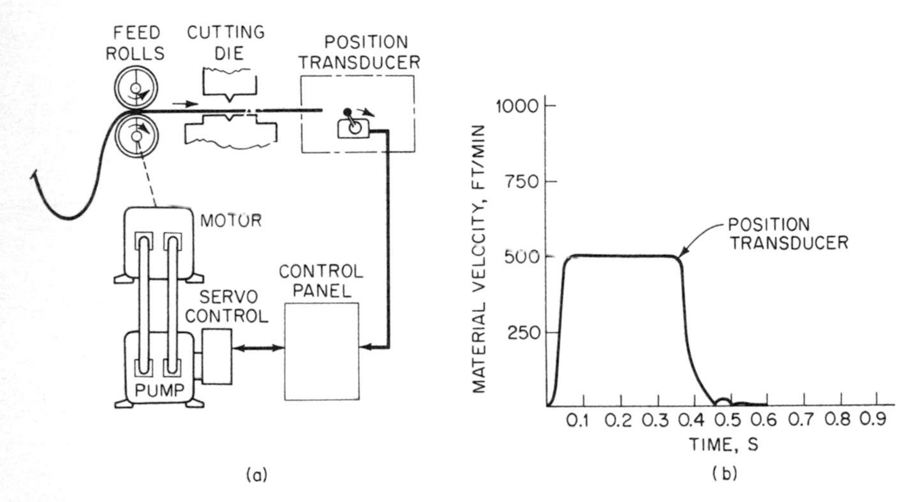

Fig. 7 (*a*) Transducer position control. (*b*) Typical material speed curve for 3-ft feed length. (Racine & Vickers—Armstrongs, Inc.)

for spindle, metal slitter, and main printing press drives where operator control to the overall machine speed is required. By the addition of other electronic devices, the acceleration and deceleration of the drives can be infinitely adjustable over a wide range.

Another common use of the closed-loop servo pump drive includes a positioning system for cutting material such as sheet or tubing to accurate lengths. The material is normally pulled off the parent roll and fed into a storage loop, and feed rolls then pull the material out of the loop. The requirements of the feed-roll device are high rates of acceleration and deceleration and accurate stopping of the drive at the desired position. Since electrohydraulic servo-controlled transmissions provide the required power levels without adding high inertial effects in themselves, they are ideal for this type of system. Figure 7 illustrates one of the simplest types of feed-roll drive with transducer position control. An electrical signal applied to the pump servo system, usually as the cutting die leaves the material, commands the servo pump to full output flow, thereby producing full motor speed. The material is then driven by the feed rolls until contact is made with the transducer. As the transducer is deflected, the position feedback signal provides a stepless deceleration signal that reduces the motor speed and stops the transmission at the required length. Systems of this type operate at speeds in excess of 1,200 ft/min with accuracies of ±0.030 in. Stopping time of $\frac{1}{10}$ s or less is possible.

The addition of a digital position control provides versatility for the system (see Fig. 8). By replacing the fixed transducer with a rotary type such as a resolver or pulse generator, the amount of material passing throughout the feed rolls can be continuously measured. The rotary transducer driven by the material is normally selected to provide an electrical signal that is a function of the shaft rotation or position. The command control is designed electronically to produce command signals equivalent to the feedback positioning device, and the hydraulic drive is made to follow this command. The versatility of the system is that any desired feed length can be digitally set at the control. When cycling is initiated, the control begins to provide the equivalent transducer signals at a rate equal to the desired rotational speed of the feedback transducer signal. The electronically produced command transducer automatically stops upon reaching the desired position. As the velocity-error lag is reduced, the resultant position-error signal decreases, thereby

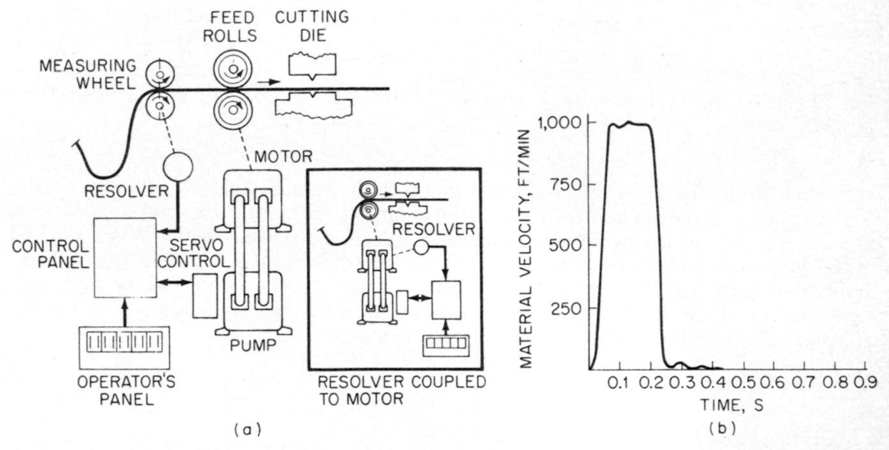

Fig. 8 (*a*) Digital position control. (*b*) Typical material speed curve for 3-ft feed length. (Racine & Vickers—Armstrongs, Inc.)

reducing the pump output and roll speed until the drive stops at the commanded material feed length. In accuracy and speed, this type of control will be equal to or will surpass the fixed position type transducer control.

A rewind or unwind system in a metal or paper process is an excellent application for the servo-controlled pump (see Fig. 9). In many of the systems the selected drive must control both the speed and torque applied to the driven core. Core buildup in many applications may be 6:1 or higher, and may require inputs that vary between constant torque and constant horsepower. A drive of this type serves to illustrate any features which can be used on a hydraulic servo pump drive. In this example, three basic transducers are used. A tachometer driven by the machine provides the base material speed and is used for a feed forward command. A position-type transducer driven by a rider arm on the material roll provides roll-diameter information. A pressure transducer in the hydraulic pressure line provides a signal in proportion to the system pressure for torque control feedback. In this system the operator can set the control for the desired initial line pull applied to the material with an override command to set a desired tapering of the line pull with roll buildup. The command signals are summed with the tachometer, the pressure transducer, and the diameter-measuring transducer in the totalizer section of the electronic control to produce the desired torque output of the servo-controlled transmission. The high response of many servo-controlled transmissions makes it possible to maintain the desired material tension with large and rapid changes in the line speed.

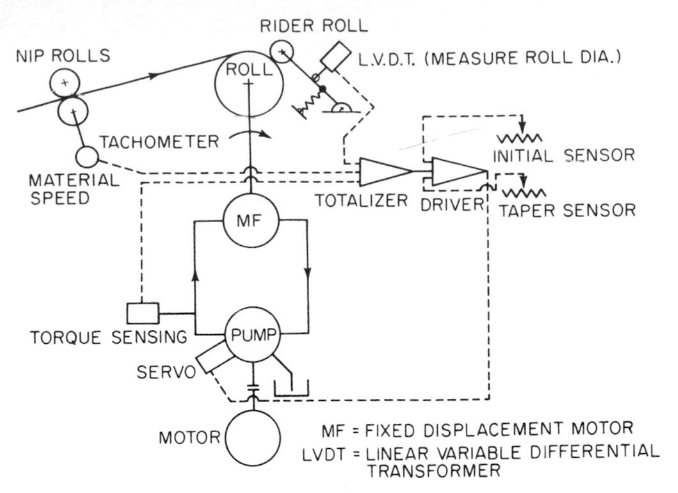

Fig. 9 A rewind system with a servo-controlled pump. (Racine & Vickers—Armstrongs, Inc.)

The electrohydraulic servo-controlled pump with its high linearity, pressure gain, and frequency response is ideally suited for a large variety of industrial applications. Only a few of the many possible applications have been described. When applied with the proper electronic control and feedback transducer systems, many difficult industrial drive and control problems can be solved. As the versatility and high performance of the servo-controlled pump become more widely known, it is certain that its use throughout industry will greatly increase.

Section **10.16**

Pumps for Machine-Tool Services

J. ARTHUR LORD

WILLIAM E. O'NEEL

INTRODUCTION

A machine tool is defined as a nonportable power-driven machine partly or wholly automatic in action and capable of shaping or forming metal or other components. The many functions of machine tools require movement of components in horizontal, vertical, or rotational planes with sufficient force to accomplish work. Force or power is applied to components through the use of mechanical or hydraulic devices such as gears, cams, pistons, screws, or other mechanisms. Pumps are vital to the circulation of the various fluids required for lubrication, hydraulic force, or cooling.

LUBRICATION

Spindles Spindles may be designed to operate in antifriction or sleeve-type bearings, or a combination of both. Some of the most accurate sleeve-type bearings used in machine tools incorporate the hydrostatic principle of bearing design. It is necessary to maintain a film of oil to support the spindle in either the sleeve-type or hydrostatic bearings. The oil is delivered to the bearing with sufficient pressure to maintain the film, and the volume of oil required is based on the size and clearance of the bearing. High-speed spindles require low-viscosity oil to reduce friction and operating temperatures; slow-speed spindles use a high-viscosity oil for greater film strength.

The lubrication of hydrostatic bearings is more critical than most other components of machine tools; it usually requires a separate pump and filter, together with an accumulator to minimize pressure pulsations. A separate reservoir is not required when the oil used for the lubrication of the machine tool is also suitable for the lubrication of the spindle bearings. Antifriction bearings require much less lubrication—only a few drops per minute—as compared to sleeve bearings, but in either case too much oil will cause overheating.

The best pumps for spindle lubrication are either gear- or vane-type positive-displacement pumps. Figure 1 shows a conservative method of determining the maximum pump speed for a given fluid viscosity. It is important to recognize that pump speed must be reduced with an increase in viscosity.

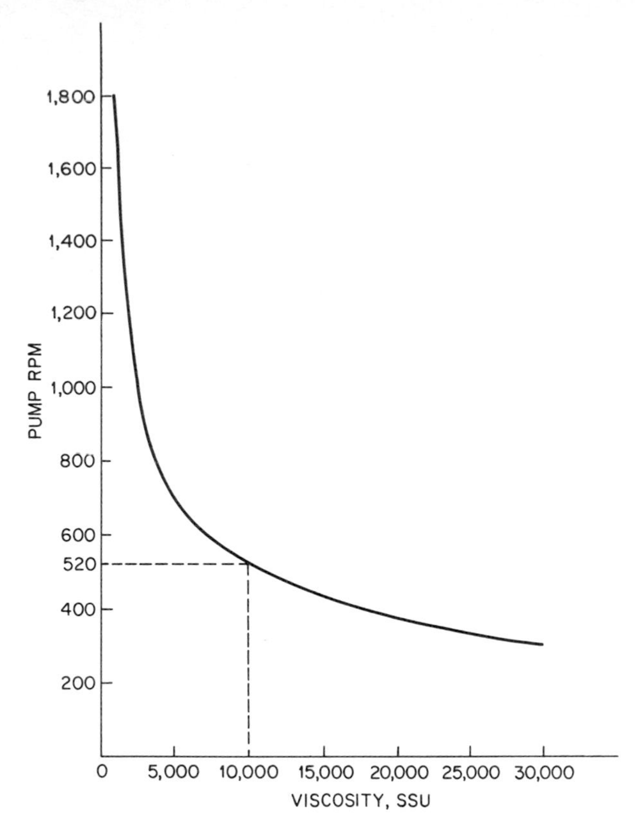

Fig. 1 Recommended maximum speed, in revolutions per minute, vs. maximum viscosity. Example indicated by dotted lines. Viscosity 10,000 SSU; rpm 520 maximum, lower rpm allowable; higher rpm creates cavitation.

Gear Cases The size and complexity of a gear case vary with the size and complexity of the machine tool, and a positive method of lubrication is used in most designs. This does not preclude the use of splash lubrication for large slow-speed units.

Very small gear cases may contain a large number of fine-pitch gears, which can be lubricated with low-viscosity oils; larger gear cases with fewer coarse-pitch gears require high-viscosity oils—1,000 SSU or more—to carry the greater tooth and bearing loads.

Gear- or vane-type pumps are used for the lubrication of gear cases, and the recommended viscosities and pump speeds are the same as that for spindles, as shown in Fig. 1.

Drive Mechanisms Drive mechanisms include power-transfer devices such as cams, chain drives, levers, racks, and pinions, which require lubrication of the wearing surfaces. Recovery of the oil for recirculation is not practical for many of these parts since they are not enclosed. The oil is lost, but metering devices are used to regulate the flow of oil to each location. The loss of oil is limited by reducing the flow to a few drops per minute. Where additional lubrication is required, a recovery and recirculation system must be installed. Gear- or vane-type pumps are used for this service, and the choice of the lubricant will depend on the area and loads on the surface to be lubricated.

Ways Cross slides and tables of machine tools are supported on ways that require lubrication to reduce friction and metal-to-metal contact. The amount and viscosity of the lubricant are determined from the weight of the sliding piece and the forces imposed on it during machining operations. The gear- or vane-type pump is used for this service.

HYDRAULIC POWER

Significant improvements in the design and operation of machine tools have been possible by the development of hydraulic power devices to replace mechanical linkages and gear trains. Because hydraulic fluid must be retained under high pressure, the application of hydraulic power to machine tools was possible only after the perfection of the O-ring seal.

Cylinders Longitudinal movement in any plane or direction is produced by one or more hydraulic cylinders. Stepless speed control is possible through an adjustment of the rate of flow of the hydraulic fluid in or out of the cylinder. The pressure required to actuate the cylinder is dependent not only on the force required to move the weight of the machine component and the loads imposed on it but also to the size of the cylinder. The following example shows a typical calculation to determine the pressure required to move a machine component:

Load on machine component = 500 lb
Cylinder diameter = 2 in
Cylinder area = 3.142 in^2

$$\text{Pressure to actuate cylinder} = \frac{500}{3.142} = 159\ \text{lb/in}^2$$

(The area of the piston rod is not considered in this example.)

Gear and vane pumps are used for pressures up to 2,000 lb/in^2, and piston or plunger pumps are used for pressures over 2,000 lb/in^2.

Servo Drives Fixed or variable positive-displacement pumps are used for servo drive applications. The major factors to be considered in an evaluation of the two types are pump response time and power consumption. A fixed-displacement pump will deliver the maximum flow rate continuously at maximum pressure. When the pump operates against a closed servo valve, however, the power input is converted into heat, and no useful work is done. An unloading or bypass valve in conjunction with a check valve and an accumulator will reduce the heat load. The check valve and accumulator, however, increase the response time to the servo valve demands and reduce sensitivity to flow fluctuations.

The variable-displacement pump incorporates an integral pressure compensator that controls the output flow as required by the servo valve. If no flow is required, the pump dead-heads at the system pressure. Since the pump is delivering only enough fluid to maintain the system pressure, the input horsepower and heat rise are kept at a minimum. Because the variable-displacement pump is similar in performance to the fixed-displacement pump with an unloading valve, a check valve and an accumulator are usually added to increase the response time and to reduce the flow sensitivity.

The system pressure in a servo drive application is determined by the pressure rating of the servo valve. For example, if the servo valve is rated at 10 gpm at a 1,000-lb/in^2 pressure drop and the moving component of the machine requires a maximum of 10 gpm and 500 lb/in^2 at a given speed, then the pump must be selected for 10 gpm at 1,500 lb/in^2 discharge pressure.

The successful operation of a servo power system depends on the quick response of the system for fluid without a drop in pressure. The volume of the system should be as small as possible, and the hydraulic fluid should be filtered constantly before entering the servo valve.

Hydraulic Motors Hydraulic motors are similar in design to the gear- or vane-type pump, although modifications in construction and tolerances are required to achieve optimum performance. The hydraulic motor is a high-torque machine, and it has the advantage of a stepless speed output. The selection of the pump to operate the hydraulic motor is the same as that described for hydraulic cylinders.

COOLANT SYSTEMS

The coolant system is used to reduce the temperature of the tool and the workpiece as well as to flush away the chips in turning, milling, drilling, grinding, and other machine operations. Chip removal is always a problem in machining, and a strict chip-removal program is required to prevent blockage of the system. Chips recirculated in the flow of coolant to the workpiece may cause scratches or blemishes in the surface finish.

Centrifugal pumps are used to provide the pressure required at the nozzle of a coolant system. Unlike the positive-displacement pump, the large internal clearances of the centrifugal pump will permit chips in the coolant to pass through with minimum wear or damage to the pump. Because the mechanical seal of a conventional centrifugal pump can be damaged by chips in the coolant, special pumps without seals have been developed for coolant service. These pumps are vertically motor-driven units with the pumping element submerged in the coolant reservoir; they have proved to be more reliable than any other design.

Machine tools equipped with elaborate hydraulic systems may have an auxiliary coolant system to recirculate the hydraulic fluid through filters and heat exchangers. Machines that produce pieces with fine surface finishes, as in gun drilling or deep-hole boring, are usually equipped with an auxiliary coolant system. The primary system usually requires a gear or vane pump to produce the pressure required at the nozzle.

PUMP SELECTION

Capacity The rate of flow of the coolant system will depend upon the size of the machine tool, the number of cutters, and the velocity at the nozzle required to wash away the chips. The determination of the flow rate is largely empirical, based upon the experience of the machine-tool manufacturers. While the pump must be selected with a reserve capacity, an oversize pump will increase the operating costs of the coolant system without an advantage in the operation of the machine tool.

The rate of flow for the lubrication system should be based on a tabulation of the quantity of oil required at each station of the machine tool that requires lubrication. Antifriction bearings, for example, may require only a few drops of oil per minute, while sleeve bearings of a large machine may require flows of a few gallons per minute.

The capacity of a pump for a hydraulic power system can be calculated with greater accuracy because it pumps into a system consisting primarily of a cylinder or hydraulic motor of fixed dimensions. A small margin is added for valve leakage and slip loss in the hydraulic motor. Once the flow rate has been established, an evaluation should be made between a fixed or variable positive-displacement pump.

When the hydraulic system is so designed that a piston in a cylinder is required to hold at a given position, usually under high pressure, a variable positive-displacement pump is an excellent choice. The variable positive-displacement pump has the capability of adjusting to the flow rate automatically or manually to meet the demands of the system. The flow rate can be high during the piston travel, while the pressure is relatively low. Upon completion of the travel cycle and for the duration of the high-pressure holding cycle, the pump flow is reduced to the desired amount. This reduces the input horsepower and the heat generated from bypassing flow through a relief valve. When the hydraulic system does not require a high-pressure holding cycle, then a fixed positive-displacement pump will meet the requirements at a lower cost.

Pressure The total system pressure is the sum of the static, friction, and velocity pressure heads. The friction head losses should be based on the maximum viscosity of the lubricating oil or hydraulic fluid. The following examples show typical losses and total pressure calculations for each of the three machine-tool services:

Lubrication system	Lb/in^2
Static pressure required at the bearings	25
Static, friction, and velocity head pressure loss through pipes, fittings, and filter	22
Static, friction, and velocity head pressure loss through suction line, fittings, and strainer	3
Total system pressure	50

Hydraulic system	
Static pressure at the cylinder	325
Static, friction, and velocity head pressure loss through valves, pipes, fittings, and filter	32
Static, friction, and velocity head pressure loss through suction line, fittings, and strainer	5
Total system pressure	362

Coolant system	
Static pressure at the nozzle	10
Static, friction, and velocity head pressure loss through pipe and fittings	5
Total system pressure	15

TEMPERATURE AND VISCOSITY

The temperature of the oil in the lubrication system must be controlled to prevent a reduction in the viscosity of the oil to the point where the film strength and lubricating properties are impaired and bearing failures occur. Although some machine tools are designed with low-viscosity lubrication, most operate with oil viscosities of 100 SSU or higher. Antifriction bearings can operate with lower viscosity oils than sleeve bearings. Since many pumps use the oil being circulated for their own lubrication, it is advisable to select a pump with antifriction bearings when circulating low-viscosity or low-lubricity oils.

Precision machining requires control of the temperatures of the coolant, of the hydraulic fluid for power systems, and of the lubricating oils. The accuracy of the machine tool is affected by the rate of expansion of the metal parts in contact with these fluids. It is essential, therefore, that the temperature be controlled and maintained within fixed limits.

The power requirements of pump motors on the coolant, hydraulic power, and lubricating systems increase with an increase in the viscosity of the pumped fluid. This is a common problem where cold starting conditions may exist, as in northern climates or when machines have been idle during a shutdown at low ambient temperature. The majority of machines are designed to operate with hydraulic-fluid and lubricating-oil temperatures between 100° and 150°F. A warmup period is usually recommended for a machine tool before any work is performed.

Some of the high-performance hydraulic pumps, especially the high-pressure piston type, are manufactured with close clearances; they will perform within their ratings only if the viscosity of the oil is kept within a specified range.

SELECTION OF OILS AND HYDRAULIC FLUIDS

Petroleum-base fluids are preferred for the lubrication and hydraulic power systems because of the broad range of viscosities available to meet the requirements of every size of machine tool. These fluids are also compatible with the materials of construction for most pumps and machine tools. This is particularly true for such critical components as the synthetic rubber O rings and mechanical seal parts.

Fire-resistant fluids are usually used in high-pressure hydraulic systems. This is particularly true for high-temperature or hot-metal machines. Fire-resistant fluids have been developed to prevent the serious fires that can occur when petroleum-base fluids leak from the system and ignite upon contact with hot metal surfaces. Manufacturers of the fire-resistant fluids, as well as pump manufacturers, should be consulted for pump and seal recommendations.

Water-soluble oil is used extensively for coolant service because it is inexpensive and effective over a broad range of working conditions. Centrifugal pumps are used to circulate the coolant. Cast iron casings and brass impellers are the preferred materials of construction.

PUMP DRIVES

Pump drives may be direct-coupled to a power takeoff shaft of the machine tool or to a separate motor. Flexible couplings should be used for separate motor or power takeoff shaft drives.

Belt, chain, or gear drives are often unavoidable. An outboard bearing on the drive is usually required to reduce the side load on the pump shaft. These drives should be limited to low-pressure applications.

STARTUP AND MAINTENANCE

Every pump should be primed before startup. Priming fills the internal clearances of positive-displacement pumps with fluid and aids in developing the vacuum necessary to lift the fluid from the reservoir to the pump. Priming also prelubricates internal bearings and close running clearances. Lubricating oils should be continuously filtered to protect not only the pump but also the machine tool.

Section **10.17**

Refrigeration, Heating, and Air Conditioning

MELVIN A. RAMSEY

HEATING

Heat is usually generated at a central point and transferred to one or more points of use. The transfer may be by means of a liquid (usually water), which has its temperature increased at the source and which gives up its heat at the point of use by reduction of its temperature. It may also be transferred by means of a vapor (usually steam), which changes from a liquid to a vapor at the source, giving up its heat at the point of use by condensation. Pumps may be required in both of these methods.

Hot-Water Circulating A centrifugal pump best meets the requirements of this service. Water is usually used in a closed circuit so that there is no static head. The only resistance to flow is that from friction in the piping and fittings, the heater, the heating coils or radiators, and the control valves. In selecting the pump the total flow resistance at the required flow rate should be calculated as accurately as possible, with some thought as to how much variation there might be as a result of inaccuracy of calculations or changes in the circuit because of installation conditions. It is not good practice to select a pump for a head or capacity considerably higher than that required, as this is likely to result in a higher noise level as well as increased power.

When hot water is used for radiation in a single circuit through several radiators, the water temperature variation is usually only about 20°F at the time of maximum requirements, so that there is not too great a difference in heat output between the first and last radiator in the circuit. With the flow rate based on water at 180° to 200°F to heat the air to about 75°F, a 10 percent reduction in the flow would have little effect, as the actual difference would increase to only 22°F, and the reduction in the heat output of the radiator with 178°F water would be only about 2 percent. Reference to Sec. 9.1, on the selection of pumps and the prediction of performance from the head-capacity pump curves and system head-flow rate curves, will show that a rather large percentage undercalculation of head loss of the circuit would be necessary to produce a flow rate 10 percent less than desired.

Greater temperature differences are frequently used for other radiation circuits, and a reduced flow rate may have a greater temperature differential than in the single circuit. Whatever the condition, the pump should be selected only after full consideration of all the factors, and not by use of so-called "safety factors," which are likely instead to be "trouble factors."

Air in the Circuit Initially the entire circuit will be full of air which must be displaced by the water. Arrangements should be provided to vent most of the air before the pump is operated. Even if all the air is eliminated at the start, more will be separated from the water when it is heated. Any water added later to replace the loss will result in additional trapped air when the water is heated. Means must be provided for continuous air separation, but this cannot be accomplished by vents at high points in the piping because the flow is usually turbulent, and the air is not separated at the top of the pipe.

A separator installed before the pump intake will remove the air circulating in the system. In a heating system an air separation device is often provided at the point where the water leaves the boiler or other heating source. If the pump intake is immediately after this point, this is the point of the lowest pressure and the highest temperature in the system, and therefore it is the point where separation of air from the water can be most effectively achieved (Ref. 1).

If there are places in the system where the flow is not turbulent, air may accumulate and remain at these points and interfere with heat transfer. Automatic air vents should seldom or never be used. If they are used, it is important that they be located only where the pressure of the water is always above that of the surrounding air, whether the pump is operating or idle. Otherwise the air vent becomes an air intake.

Several important factors influence the choice of a pump for a hot-water system with a number of separate heating coils, each having a separate control. Many systems in the past used three-way valves to change the flow from the coil to the bypass.

When two-way valves are used, a low flow operation may occur for a large portion of the operating time. For this type of operation, therefore, the pump selected should have a flat performance curve so that the head rise is limited at reduced flows. A very high head rise can cause problems when many of the valves are closed. Excessive flow rates through the coils and greater pressure difference across the control valves are some of the problems that can be avoided with a flat pump curve. A centrifugal pump should not operate very long with zero flow, for it would overheat. This condition is controlled by using one or two three-way valves, a relief bypass, or a continuous small bleed between the supply and the return line. Whichever means is used to control minimum flow, the circuit must be such that it can dispose of the heat corresponding to the pump power at that operating condition, without reaching a temperature detrimental to the pump.

Types of Pumps Many pumps for hot-water circulation are for flow rates and heads in the range of in-line centrifugal pumps that are supported by the pipeline in which they are installed. Such pumps are available up to at least 5 hp and operate with good efficiency. More important than the type of pump is the performance and efficiency.

For greater flow rates and heads (and even for the smaller ones), the standard end-suction pump can be used. In the intermediate range the use of an in-line or end-suction pump is not a question of one being better than the other, but whether one or the other is better suited to the overall design and arrangement. Practically all the in-line or end-suction pumps for this service use seals instead of packing.

If the hot-water system is of the medium- or high-temperature type, above 250°F, the pump must be carefully selected for the pressure and temperature at which it will operate.

Type of Water Circuit There are several types of water circuits. Those shown in Figs. 1 and 2 are suitable for the smaller systems, and can be used for larger systems by having several of these circuits in parallel. The one shown in Fig. 3 is suitable for small or very large circuits, but the reverse return would add considerably to the cost if the circuit extended in one direction instead of in a practically closed loop as shown. For the extended circuit, a simple two-pipe circuit, with proper design for balancing, would be used.

There are a number of reasons for using other circuits, particularly primary-secondary pumping where the system is more extended or complicated, such as continuous circulation branches with controlled temperature. When a coil heats air,

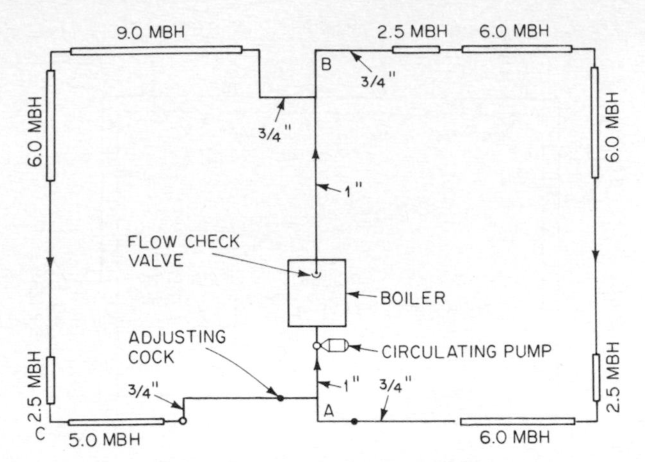

Fig. 1 A series-loop system.

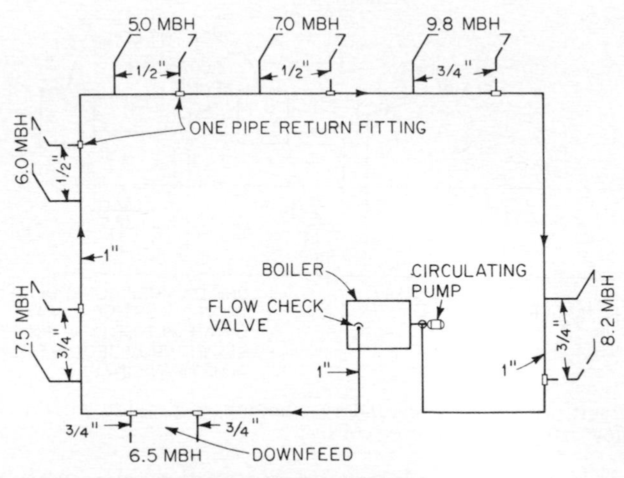

Fig. 2 A one-pipe system.

part or all of which may be below freezing, the velocity of the water in the tubes and its temperature at any point in the coil must be such that the temperature of the inside surface of the tube is not below freezing (Ref. 2). The circuit shown in Fig. 4 makes this possible.

Primary-secondary pumping permits flow rates and temperatures in branch circuits different from those in the main circuit without the flow and pressure differences in the mains or branches having a significant effect on each other. There are many possible primary-secondary circuits to meet different requirements.

Steam-Heating Systems No pumping is required with the smallest and simplest steam systems if there is sufficient level difference between the boiler and condensers (radiators, heating coils, etc.) to provide the required flow. When insufficient head exists between the level of the condensate in the condenser and the boiler to produce the required flow to the boiler, then a pump must be introduced to provide the required head. Since the condensate in the hot well will be at or near its saturation temperature and pressure, the only NPSH available to the pump will be the submergence less the losses in the piping between the hot well and the pump. A pump must be selected that will operate on these low values of NPSH without destructive cavitation.

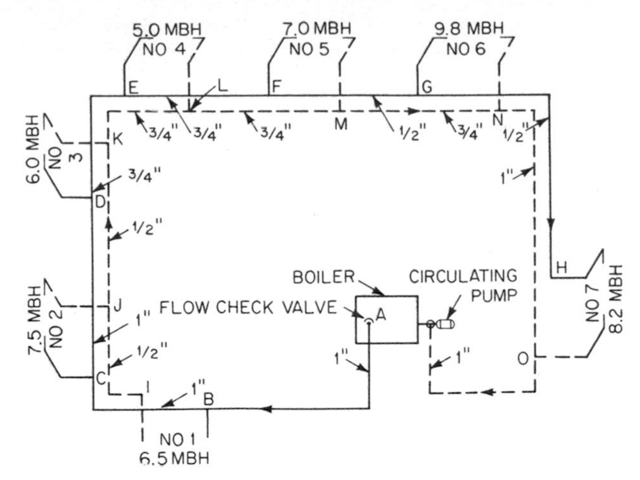

Fig. 3 A two-pipe reverse-return system.

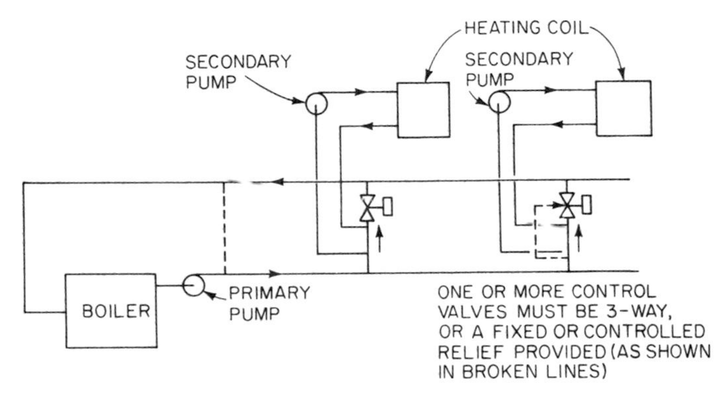

Fig. 4 A circuit, with primary-secondary pumping, to provide variable temperature at constant flow rate for two or more coils.

In many cases, particularly for very large systems, vacuum pumps are used to remove both the condensate and air from the condensers. This permits smaller piping for the return of condensate and air, more positive removal of condensate from condensers, and, when rather high vacuums (above 20 in) are possible, some control of the temperature at which the steam condenses. The use of vacuum return, particularly with higher vacuums, helps reduce the possibility of freezing heating coils exposed to outside air, or to stratified outside and recirculated air. Vacuum return pumps are available for handling air and water. Vacuum, condensate, and boiler-feed pumps with condensate tanks are all available in package form.

Most condensate pumps are centrifugal. Vacuum pumps may be rotary, including a rotary type with a water seal and displacement arrangement.

Fuel Oil When oil burners are fairly far from the oil storage tank or when there are a number of burners at different locations in a building, a fuel-oil circulating system is required. The flow rate is relatively low (1 gpm would provide over 8,000,000 BTUH), and a small gear pump is usually used.

AIR CONDITIONING

Many air-conditioning systems produce chilled water at a central location and distribute it to air-cooling coils in various locations throughout the building or group of buildings. Centrifugal pumps are particularly well suited for this service.

The type of circuit and the number of pumps used require an evaluation of several factors:

1. The cooling requirements usually vary over a wide range.

2. Flow rate through a chiller must be kept above the low point where freezing would be possible and below the point where tube damage would result. Some methods of chiller capacity control require a constant flow rate through the chiller.

3. The temperature of the surface cooling the air must be low enough to control the relative humidity (Ref. 3). This limits the use of parallel circuits through chillers when one circuit may not be in operation and permits unchilled water to mix with water of the operating chiller. Under these conditions it is now difficult or impossible to attain a sufficiently low mixture temperature. It also limits the control of flow rate and water temperature in the air-cooling coils (Ref. 4).

4. Below-freezing air may sometimes pass over all or part of a coil. This condition would require a flow rate and water temperature adequate to keep the temperature of all water side surfaces of tubes above freezing. Many water circuits are available to achieve the desired results. For control of the relative humidity, the air-flow circuit must also be considered. Figure 5 shows a circuit for cooling coils with a variable air-flow rate at constant air-leaving temperature and with two chillers in series. In addition, the two chillers are shown in parallel with a third chiller. The arrangement permits continuous flow through the coils to reduce the possibility of freezing when the average temperature of the air entering the coil is above freezing, but the usual stratification results in a below-freezing temperature for some of the air entering the coil. The word "reduce" is used because full prevention requires appropriate air-flow patterns, water velocities, and temperatures to assure that the water side of the surface will not be below freezing at any point in the coil. One of the coils is also arranged to add heat when the temperature of the air leaving the coil must sometimes be above that of the average air-entering temperature. Some circuits attempt to obtain the desired results from the circuit in Fig. 5 with fewer pumps. However, the use of fewer pumps, although it would

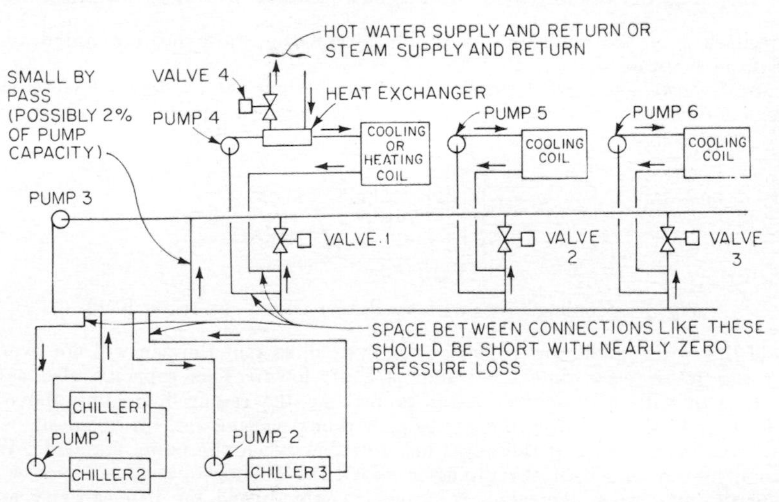

Fig. 5 Pump 3 does not operate unless pump 1 or pump 2 operates. Pump 1 operates only if chiller 1 or 2 is required and operating. Pump 2 operates only if chiller 3 is required and operating. Pump 4 operates when air circulates over the coil to which it is connected, if cooling or heating is required or if any air enters this coil below about 35°F. Pumps 5 and 6 operate in the same manner for the coil to which each is connected. Operation of pumps 4, 5, or 6 helps to equalize the temperature of air streams which enter the coil at different temperatures, and thus it may be desirable to operate these pumps continuously when air circulates over the respective coil. Valves 1 and 4 will be interlocked so the one must be closed before the other can open. Also valve 1 should be prevented from opening if the water in the pump 4 circuit is over about 90°F.

reduce the cost for pumps slightly, would also require three-way instead of two-way valves, would make control somewhat more complicated, and would almost certainly result in greater power consumption. The circuit shown permits pump heads to match the requirements exactly. It also permits stopping an individual pump when flow is not required in one of the circuits; the two-way valves 1, 2, and 3 will reduce pump circulation and the power of pump 3 at partial cooling load.

Air Separation and Removal The methods for handling air with chilled water are about the same as those for hot water except that there is not usually a rise in temperature above that of the make-up water to produce additional separation of air. An expansion tank is required, but the reduced temperature difference, as compared to that of hot water, requires a much smaller tank size.

Condenser-Water Circulation Condenser water may be recirculated and cooled by passing through a cooling tower, or it may be pumped from a source such as a lake, the ocean, or a well.

Cooling-Tower Water Centrifugal pumps are used for the circulation of cooling-tower water. The circuit is open at the tower where the water falls or is sprayed through the air and transfers heat to the air before the water falls to the pan at the base of the tower. A pump then circulates the water through the condenser as shown in Fig. 6. In this case the pump must operate against a head equal to

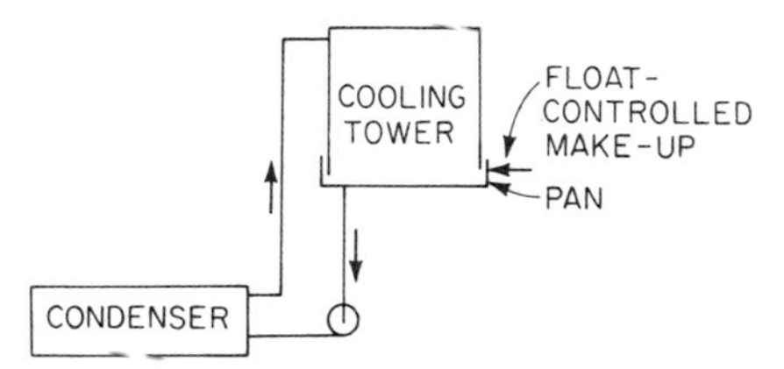

Fig. 6 Cooling tower with condenser below pan water level.

the resistance of the condenser and piping plus the static head required to the tower from the water level in the pan.

Figure 7 shows a somewhat similar circuit except that the condenser level is above the pan water level. The size of the pan of a standard cooling tower is sufficient

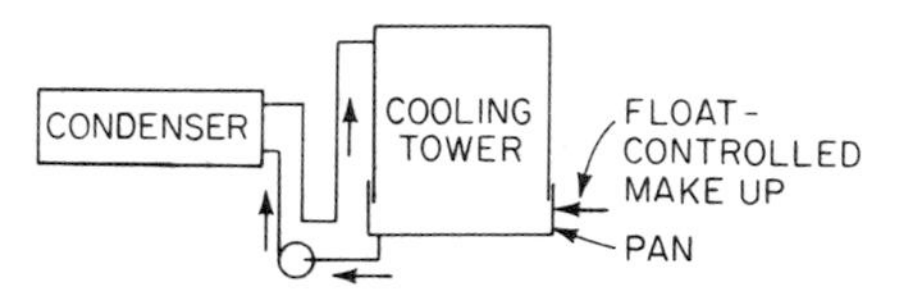

Fig. 7 Cooling tower with condenser above pan water level.

to hold the water in the tower distribution system, so that the pan will not overflow and waste water each time the pump is shut down. This capacity also assures that the pan will have enough water to provide the required amount above the pan level immediately after the start-up, without waiting for the make-up which would have been needed if there was any overflow when the pump stopped. When the condenser or much of the piping is above the cooling-tower-pan overflow, the amount draining when the pump is stopped would exceed the pan capacity unless means are provided to keep the condenser and lines from draining. In Fig. 7 it will be noted that the line from the condenser drops below the pan level before rising at the tower. This keeps the condenser from draining by making it impossible for air to enter the system. This is effective for levels of a few feet, but it is useless if the level difference approaches the barometric value. Such large level differences should be avoided, if possible, since they require special arrangements and controls.

When a cooling tower is to be used at low outside temperatures, it is necessary to avoid the circulation of any water outside unless the water temperature is well above freezing. The arrangement shown in Fig. 8 provides this protection.

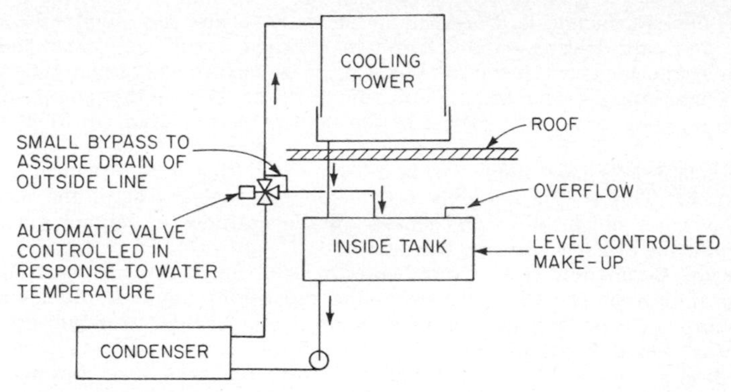

Fig. 8 Cooling tower with inside tank to permit operation at outside wet-bulb temperatures below freezing.

The inside tank must now provide the volume previously supplied by the pan in addition to the volume of the piping from the tower to the level of the inside tank. Condensers or piping above the new overflow level must be treated as already described and illustrated in Fig. 7, or additional tank volume must be provided.

The only portion of the inside tank that will be available for the water that drains down after the pump stops is that above the operating level. This operating level is fixed by the height of liquid required to avoid cavitation at the inlet to the pump. The suction piping to the pump must remove only water from the tank without air entrainment (Ref. 5). The size of this pipe at the tank outlet should not be determined by pressure loss, but by the velocity that can be attained from the available head. Exact data on this are not available, but the required velocity at the vena contracta (about 0.6 of the pipe cross-section area) can be calculated from $V = \sqrt{2gh}$, where h is the height of the operating level above the vena contracta. The outlet from the tank should be at least as large as that from the cooling tower.

Well, Lake, or Sea Water Centrifugal pumps are used for all of these services. The level from which the water is pumped is a critical factor. The level of the water in a well will be considerably lower during pumping than when the pump does not operate. In the case of pumping from a lake or from the ocean, the drawdown is usually not significant. When pumping from a pit where the water flows by gravity, there will be a drawdown that will depend on rate of pumping. With a sea-water supply there will be tidal variations. A lake supply may have seasonal level differences.

All these factors must be taken into consideration in selecting the level for mounting the pump to assure that it will be filled with water during startup. Check or foot valves may be used for this purpose. Also the head of the water entering the pump at the time of highest flow rates must not be so low that the required NPSH is not available.

To assure proper pump-operating conditions, the pump is frequently mounted below the lowest level expected during zero flow conditions, as well as below the lowest level expected at the greatest flow rate. These conditions may require a vertical turbine-type pump. The motor should be above the highest water level with a vertical shaft between the motor and the pump bowls, or the motor can be of the submerged type connected directly to the pump bowls.

REFRIGERATION

For refrigeration systems with temperatures near or below freezing, pumps are often required for brine or refrigerant circulation. The transfer of lubrication oil also frequently requires pumps.

Brine Circulation The word "brine," as used in refrigeration, applies to any liquid which does not freeze at the temperatures at which it will be used, and which transfers heat by a change solely in its temperature without a change in its physical state. As far as pumping is concerned, brine systems are very similar to systems for circulating chilled water or any liquid in a closed circuit. A centrifugal pump is the preferred choice for this service, but it must be constructed of materials suitable for the temperatures encountered. For some of the brines, the pump materials must be compatible with other metals in the system to avoid damage from galvanic corrosion.

Tightness is usually more important in a brine circulating system than in a chilled-water system. This is true not only because of the higher cost of the brine, but also because of problems that are caused by the entrance of minute amounts of moisture into the brine at very low temperatures.

Refrigerant Circulation For a number of reasons—including pressure and level as well as improvement of heat transfer—the refrigerant liquid is often circulated with a pump. The centrifugal pump is usually preferred for this purpose.

The actual liquid being pumped as a refrigerant may be the same one which is pumped as a brine. Whereas the material is all in liquid form throughout the brine circuit, some portion of it is in vapor form during its circulation as a refrigerant. In a refrigerant circulating system most of the heat transfer is by evaporation or condensation or both.

As there are changes from liquid to vapor, the liquid to be pumped must be in a saturated condition in some portion or portions of the circuit. Sufficient NPSH for the pump must be provided by the level of saturated liquid which is maintained in the tank where the liquid is collected. The level difference required for the NPSH must provide adequate margin to compensate for any temperature rise between the tank and the pump. This is an important consideration because the liquid temperature will usually be considerably lower than that of the air surrounding the pump intake pipe (Ref. 5).

When the pump is not operating, it may be warm and may contain much refrigerant in vapor form. It is usually necessary to provide a valved bypass from the pump discharge back to the tank so that gravity circulation can cool the pump and establish the prime.

Pumps for this service may require a double seal with the space between the seals containing circulated refrigerant oil at an appropriate pressure. This will reduce the possibility of the loss of relatively expensive refrigerant, and eliminate the entrance of any air or water vapor at pressures below atmospheric. A hermetic motor may also be used for this service and thus avoid the use of seals.

Lubricating-Oil Transfer Because the flow rates for lubricating-oil transfer are rather low, the gear pump is usually preferred. The NPSH requirement is also critical here because, although the oil itself is well below the saturation temperature at the existing pressure, it contains liquid refrigerant in equilibrium with the refrigerant gas. Any temperature rise or pressure reduction will result in the separation of refrigerant vapor. It is important, therefore, to design the path for oil flow from the level in the tank where it is saturated with the same safeguards as is necessary for refrigerant pumping.

To reduce the oil pumping problem, the oil can be heated to a temperature above that of the ambient air and vented to a low pressure in the refrigerant circuit. This eliminates temperature rise in the pump as well as in the suction with the corresponding reduction of available NPSH.

Usually the oil flow is intermittent, and the best results are obtained by continuous pump operation discharging to a three-way solenoid valve. The discharge would be bypassed back to the tank whenever transfer from the tank is not required. This assures even temperature conditions and a pump free of vapor.

REFERENCES

1. "Tested Solutions to Design Problems in Air Conditioning and Refrigeration," Industrial Press. New York, Section 3.
2. *Ibid.*, Section 10
3. *Ibid.*, Section 1, pp. 19–38; Section 4, pp. 63–65.
4. *Ibid.*, Section 9, pp. 119–125
5. *Ibid.*, Section 2, pp. 44–47.

Section **10.18**

Pumped Storage

GEORGE R. RICH

SIZE OF INSTALLATION

In the typical steam-based utility, it is the function of pumped storage to (1) furnish peaking capacity on the weekly load curve (Fig. 1) and (2) generate full pumped storage in an emergency for from 10 to 15 hr, as required by the particular system. On the basis of comparative cost estimates, that size of installation is selected which is the most economical.

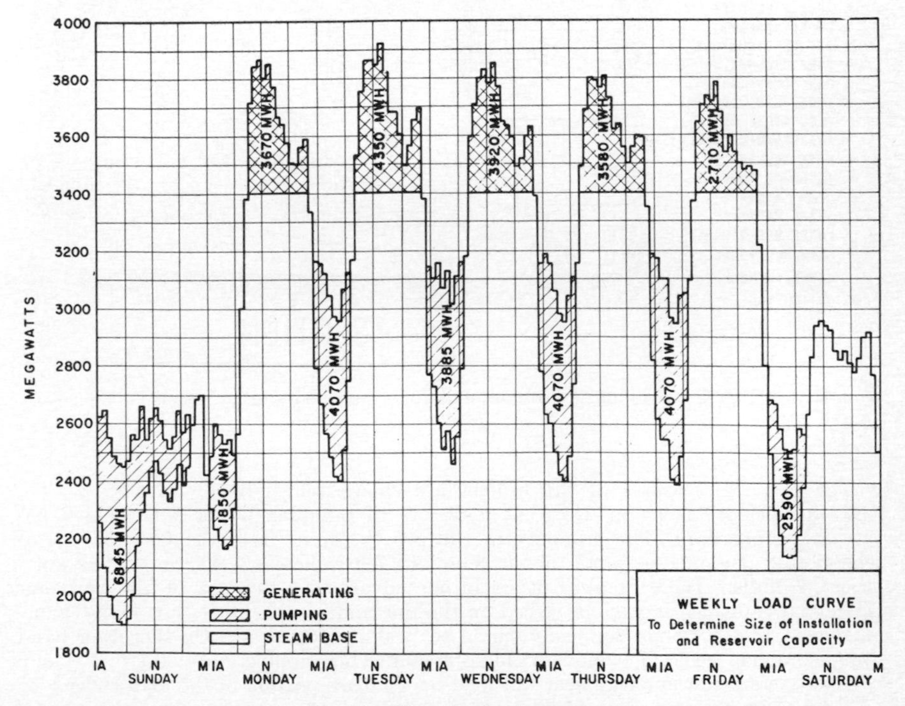

Fig. 1 Weekly load curve.

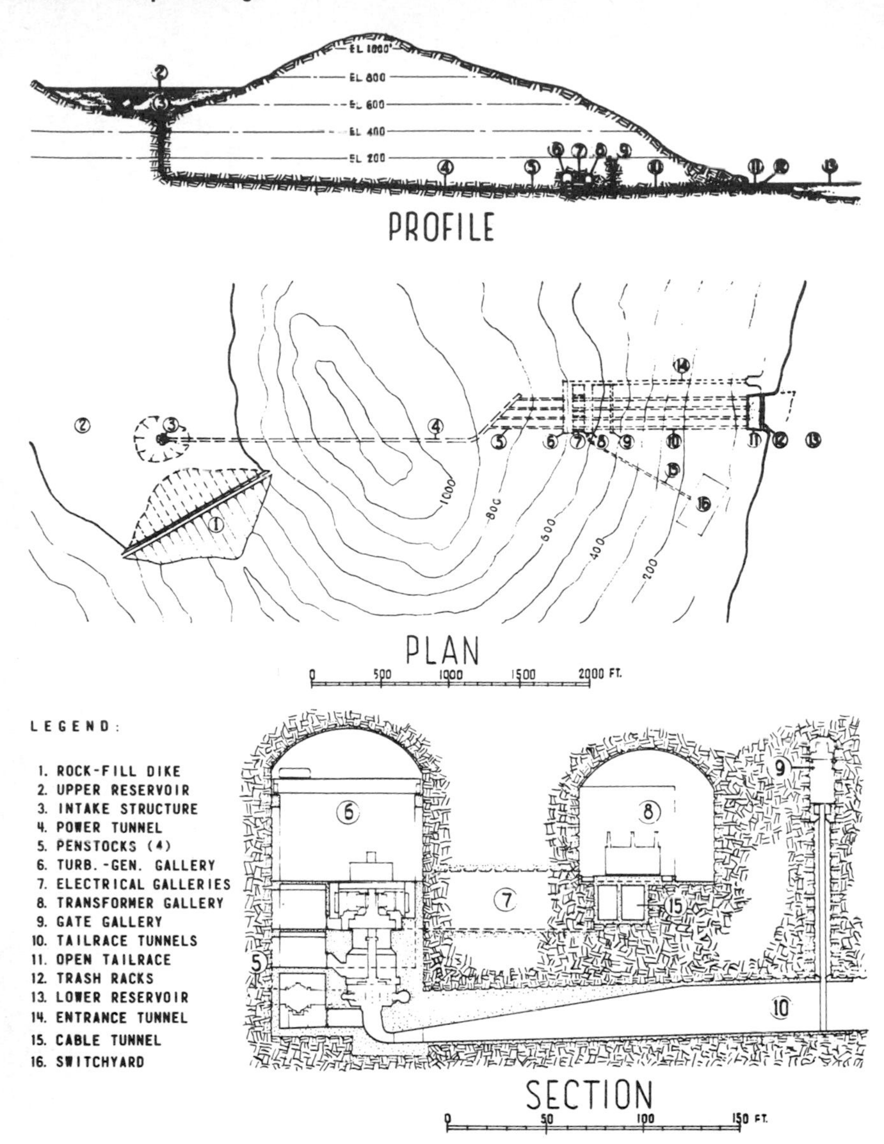

Fig. 2 Schematic arrangement for a pumped-storage project.

The principal features of the project are shown schematically in Fig. 2. The overall efficiency is about 2:3, that is, 3 kW of pumping power will yield 2 kW of peak generation. The economy of the process stems from the fact that dump energy for pumping is worth about 3 mills/kWhr whereas peak energy is worth about 7 mills. There is also an operating advantage. Because of the ease and rapidity with which it may be placed on the line and because of its low maintenance charges, pumped storage is ideally suited to peaking operation. On the other hand, for maximum economy, modern high-pressure, high-temperature thermal plants should operate continuously near full load on the base portion of the load curve.

Table 1 shows a typical calculation to determine the reservoir capacity needed

TABLE 1 Determination of Reservoir Capacity

Pertinent data:
Generating = rated net head = 900 ft
Generating capacity at rated head for 3 units = 525 MW
Pumping = rated net head = 930 ft
Pumping power at rated head for 3 units = 555 MW
Pumping-to-generating ratio = 3 to 2

Day		(1) Generating, MW	(2) Pumping, MWH	(3) Equivalent generation (col. 2 × $^2/_3$), MWH	(4) Net daily change in reservoir, MWH	(5) Cumulative change in reservoir, MWH	Remarks
Monday	P.M.	3,670			−3,670	−3,670	
Tuesday	A.M.		4,070	2,710			
	P.M.	4,350			−1,640	−5,310	
Wednesday	A.M.		3,885	2,585			
	P.M.	3,920			−1,335	−6,645	
Thursday	A.M.		4,070	2,710			
	P.M.	3,580			−870	−7,515	Reservoir empty
Friday	A.M.		4,070	2,710			
	P.M.	2,710			0	−7,515	
Saturday			2,590	1,725	+1,725	−5,790	
Sunday			6,845	4,560	+4,560	−1,230	
Monday	A.M.		1,850	1,230	+1,230	0	Reservoir full

Required reservoir capacity to sustain weekly load curve

$$= \frac{7{,}515 \times 1{,}000 \times 550 \times 3{,}600}{62.4 \times 43{,}560 \times 0.746 \times 900 \times 0.85} = 9{,}600 \text{ acre-ft}$$

Hours of capacity at full load and 900 ft head

$$= \frac{9{,}600 \times 62.4 \times 43{,}560 \times 0.746 \times 900 \times 0.85}{525 \times 1{,}000 \times 550 \times 3{,}600} = \pm 14.3 \text{ hr}$$

for sustaining the weekly load curve. The reservoir capacity to carry full load for a 10- to 15-hr emergency is obtained simply by equating the electrical energy (kWhr) in the load to the potential hydraulic energy stored in the upper reservoir.

SELECTION OF UNITS

At this stage of the basic engineering, it is necessary to make (in collaboration with the equipment manufacturers) at least a tentative selection of capacity, diameter, speed, and submergence for the turbo-machine. This will be required for refinement of the calculations in the preceding subsection and also for the calculation of hydraulic transients in the following subsection. In the head range most attractive for overall project economy, 500 to 1,500 ft, the manufacturers are prepared to offer a single turbo-machine capable of operating as a pump, and in the opposite direction of rotation as a turbine. The electrical manufacturers offer a similar machine capable of operating as a synchronous motor for pumping, and in the opposite direction of rotation as a generator. These machines are designated as pump turbines and generator motors.

For best economy the speed of the unit should be as high as practicable without involving an objectionable degree of cavitation of the impeller under the assumed submergence below minimum tail water. This speed is established (1) by model tests for cavitation at the hydraulic laboratories of the manufacturers and (2) by evaluating experience with similar prototype installations.

Figure 3 is an "experience" chart showing specific speed $N_s = \text{rpm}\sqrt{\text{hp}}/H^{5/4}$ versus head for the machine acting as a turbine, and Fig. 4 shows the specific speed $N_s = \text{rpm}\sqrt{Q}/H^{3/4}$ for the machine acting as a pump. These charts presuppose mod-

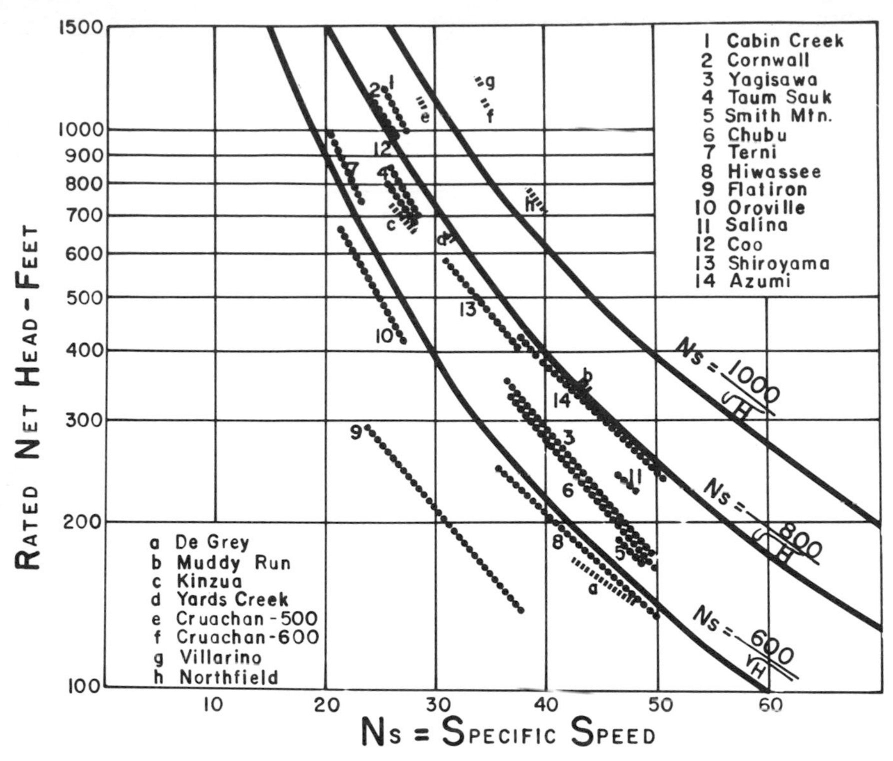

Fig. 3 Rated net head vs. specific speed for a turbine.

erate values of submergence, since unusually deep settings are uneconomical from the structural standpoint. Three curves are "fitted" to the installations shown, the equation of the curves being $N_s = K/\sqrt{H}$. The depth of submergence may be verified by checking against the value of the cavitation constant $\sigma = (H_a - H_{vp} - H_s)/H$ given by the manufacturer's cavitation model-test curves. Here H = total head, H_{vp} = vapor pressure, and H_a = atmospheric head. A typical curve of the family is shown in Fig. 5.

When the particular unit has been selected, the manufacturers will furnish (in advance of bid invitations) prototype performance curves similar to Figs. 6, 7, 8*a*, and 8*b*. Figures 8*a* and 8*b* are designated as "four-quadrant" synoptic charts and are required for the calculation of hydraulic transients in the next subsection.

Figure 8*b* is for a 5.59-in gate opening. This is the largest gate opening at which the unit will be operated in the pumping cycle.

Figure 8*a*, for an 8.94-in gate opening, is for operation on the turbine cycle only.

HYDRAULIC TRANSIENTS

In preparing purchase specifications for the generator motor, it is necessary to establish the value of the maximum transient speed and the moment of inertia of the rotor, WR^2. Similarly for the pump turbine and penstock, it is necessary to determine the maximum water hammer. This primary calculation, as summarized in Fig. 9 and Table 2, is made by the trial-and-error method of arithmetic integration, using various trial values of WR^2 for the condition of full-load rejection on all units during the generating mode with the turbine gates assumed "stuck" in the full-gate position.

Upon instantaneous loss of load, the unit builds up overspeed. The increase in speed above normal causes a reduction in turbine discharge, which causes water

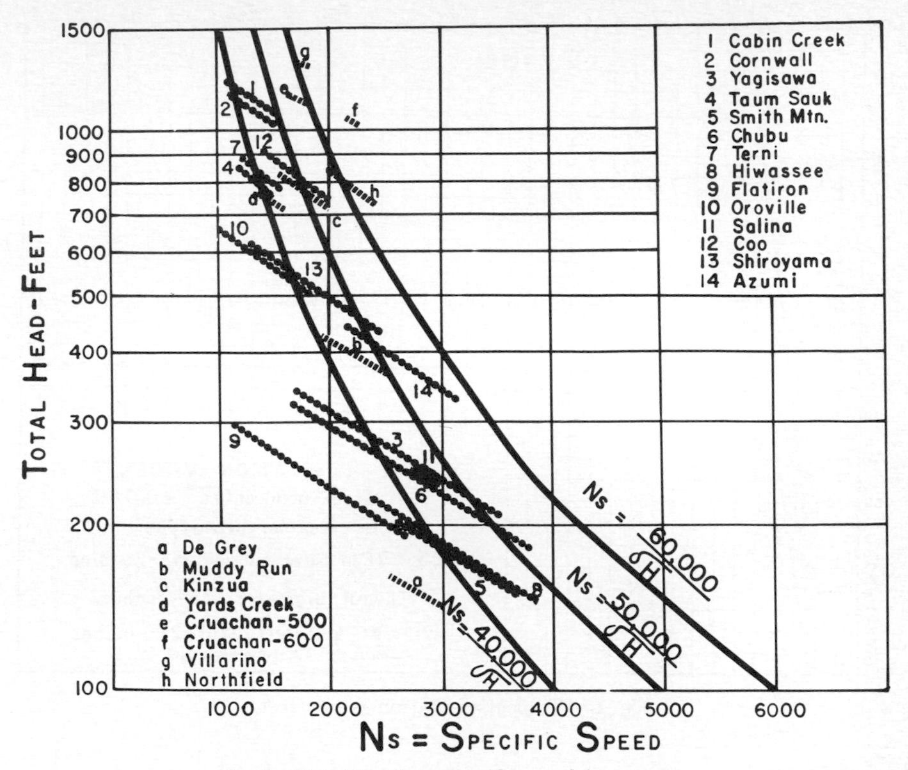

Fig. 4 Total head vs. specific speed for a pump.

hammer. The water hammer still further increases the power delivered to the rotor. This pyramiding continues until the arrival of negative reflected water hammer from the upper reservoir. The head then decreases. The unit is then so much overspeed that it begins to act as a brake, as shown by the four-quadrant synoptic chart. As shown by Fig. 9 and Table 2, the process gradually damps down to the steady-state runaway speed and head. Many additional and more refined calculations are made later in the course of the design to establish the optimum governor time and rate of turbine-gate closure, as given in detail in the references at the end of this section.

STARTING THE UNIT; SPINNING RESERVE

The procedure for starting the units is an essential feature of the design. There are three cases to be considered: (1) the pumping mode, (2) the conventional generating mode, and (3) rotating spinning reserve in the generating mode.

1. If we attempted to start the units as pumps, from rest and with the scroll case and draft tube filled, the power and inrush current would be excessive. Accordingly, the standard type of compressed-air system is provided to depress the tailwater elevation below the bottom of the impeller, with the wicket gates closed. The load to be overcome at starting then consists of accelerating the rotating masses to synchronous speed and overcoming the windage. Owing to inevitable leakage past the wicket gates, this windage is substantially greater than that due to dry air. The main penstock valves must also be closed during starting, or the leakage and "wet" windage would be still further increased at the much higher head.

For the larger units generally employed, a separate starting motor, of the induction type with wound rotor, is mounted directly above the main generator motor. It

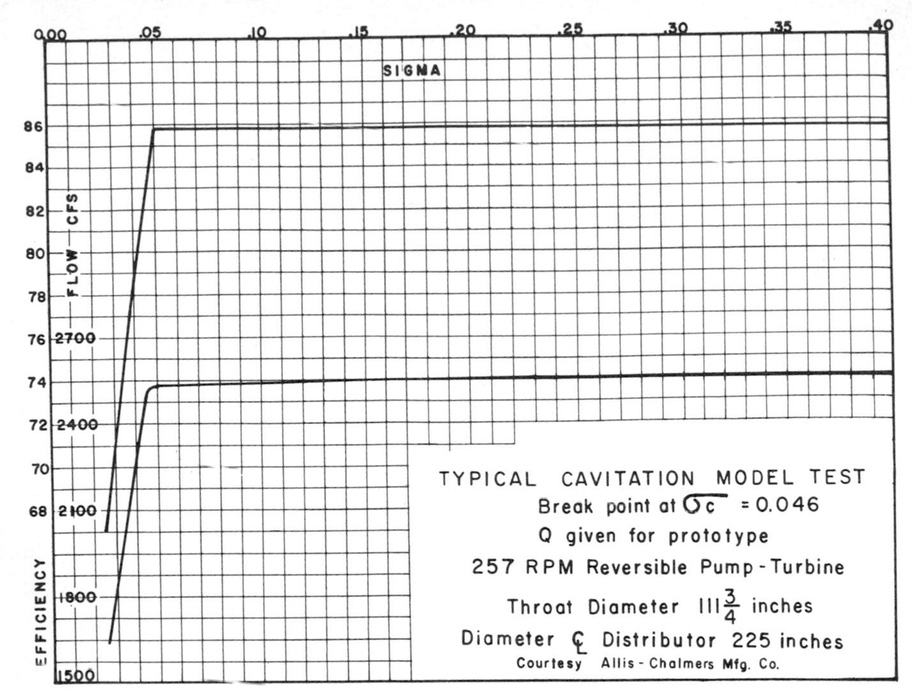

Fig. 5 Typical cavitation model test.

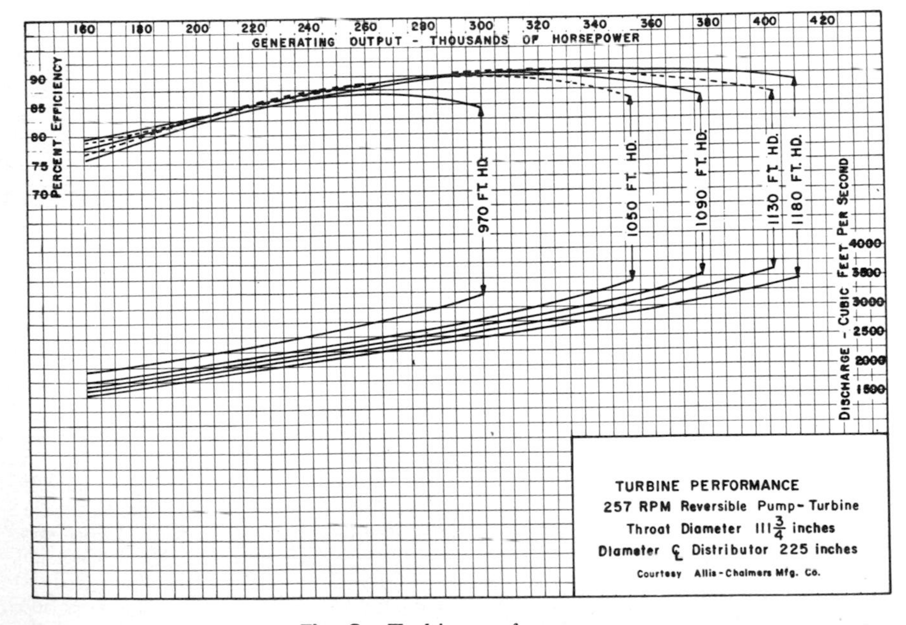

Fig. 6 Turbine performance.

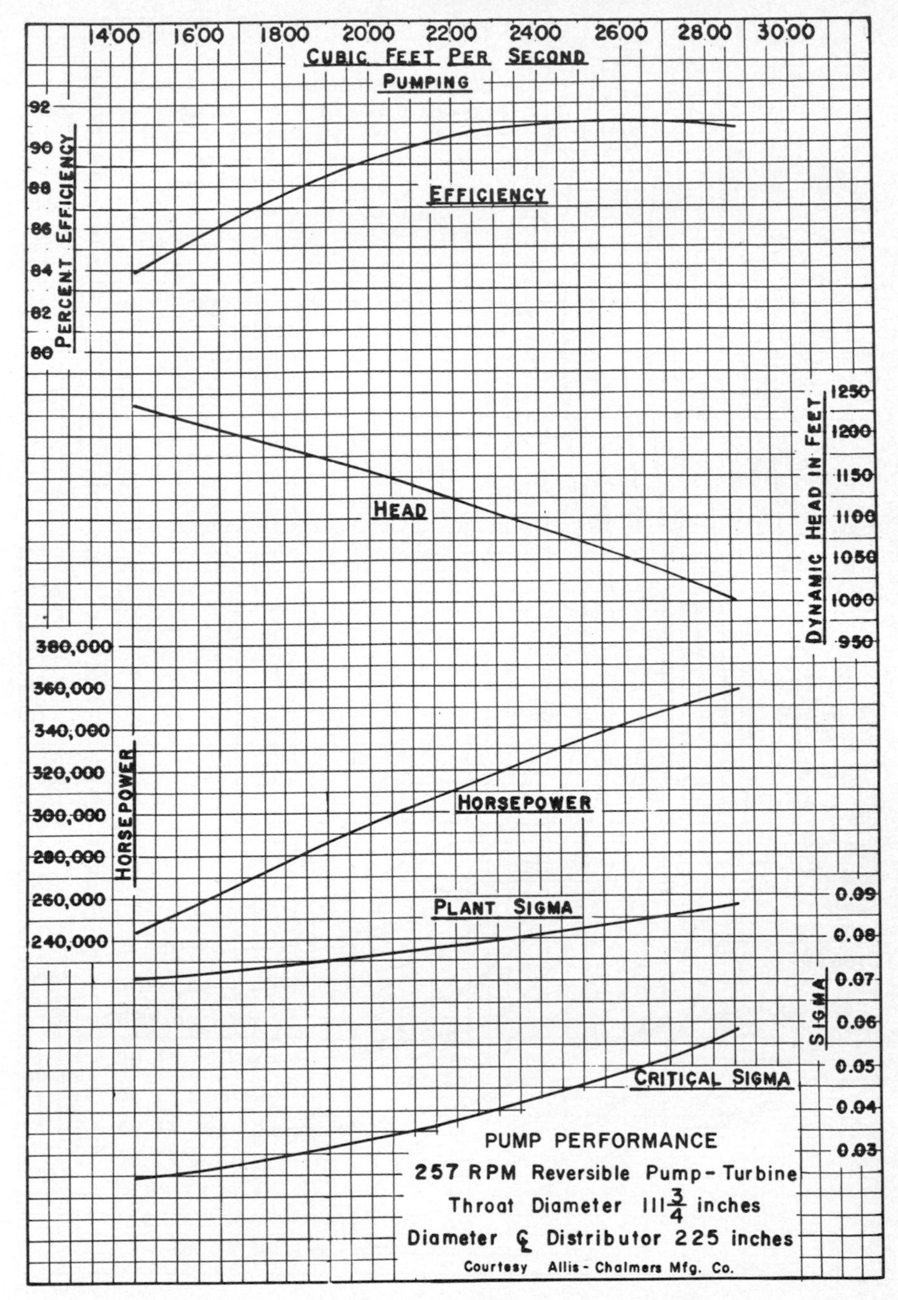

Fig. 7 Pump performance.

has not been found feasible, in the motor space available, to design an amortisseur winding capable of sustaining the heat due to the inrush current resulting from across-the-line starting of the main generator motor, even at reduced voltage. For the smaller units, however, this may be accomplished. In rare instances, where a main unit is always available, this spare may be electrically coupled to the pump-

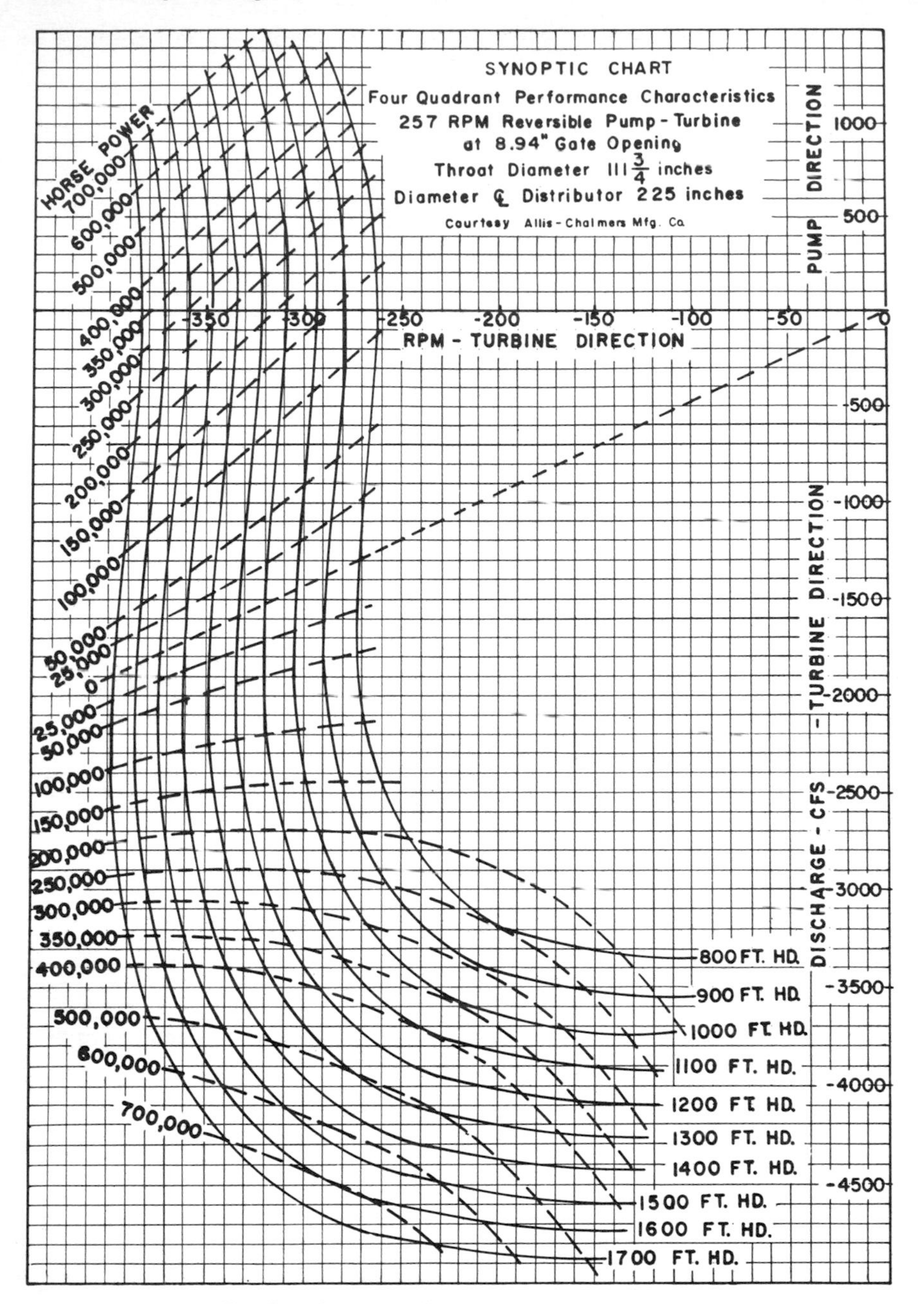

Fig. 8a Synoptic chart for an 8.94-in opening.

ing unit and the two started from rest in back-to-back synchronism. The separate starting motor is usually sized to bring the main unit up to synchronous speed in about 10 min, as shown by Fig. 10.

When the unit has attained full speed, it is synchronized to the line, the compressed air is cut off, the tail water rises to fill the draft tube, the wicket gates

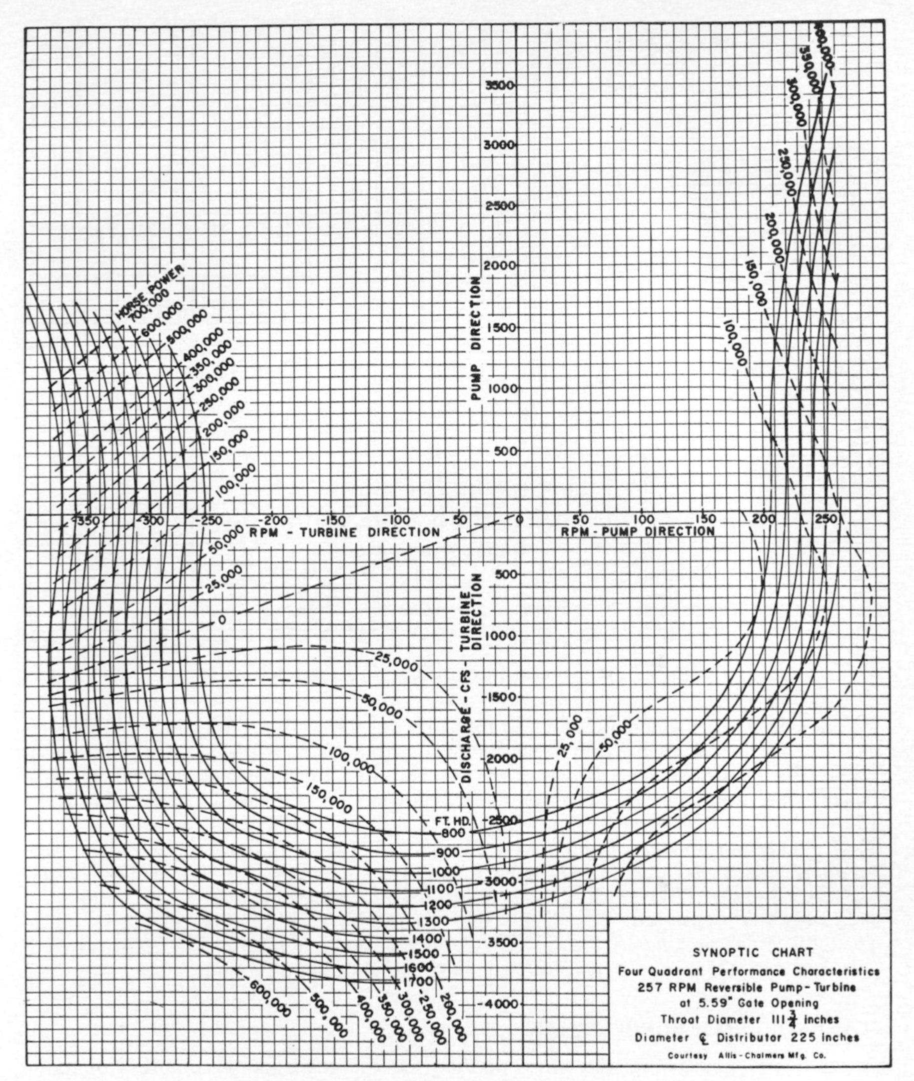

Fig. 8b Synoptic chart for a 5.59-in opening.

and main penstock valve are opened gradually to prevent shock, and pumping to the upper reservoir begins.

The maximum transient load on the generator-motor thrust bearing has been found to occur just as pumping begins. Prior to the advent of pumped storage, thrust bearings were designed to carry the weight of the rotating parts plus the hydraulic thrust at the steady-state condition. It is now found that a greatly increased thrust of short duration must also be accommodated. Because of the short duration of this transient excess load, it may usually be carried safely by the bearing as designed for the steady-state requirement, depending on the detailed design of the bearing. It is now standard practice to provide a high-pressure oil pumping system to ensure that there will be a film of oil between the bearing surfaces prior to starting rotation of the unit.

2. For generation in the conventional manner, the unit may be started from rest under its own power without assistance from the starting motor.

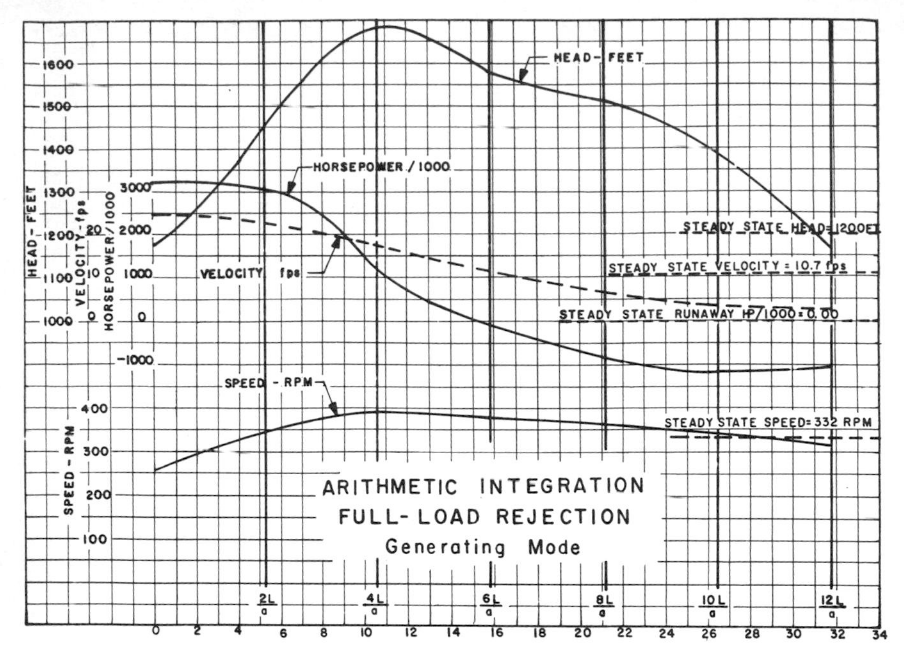

Fig. 9 Arithmetic integration, full load rejection.

3. In considering the requirements for starting the unit for rotating spinning reserve, it will be assumed that the utility is a participant in a grid system of interconnection by means of extra-high-voltage transmission lines of high capability. This means that, immediately upon loss of generation by the local utility, the EHV interconnection will carry the necessary load for the short time required (± 30 s) for the local pumped-storage units to absorb full load. In readiness for just such an emergency, these local pump turbines will be motoring on the line in the generating direction, with the wicket gates and main penstock valves closed, and with the tail water depressed. The loads and procedure for bringing the units up to speed for synchronizing to the line for rotating spinning reserve are identical with those given in (1) except that rotation is in the generating direction.

GOVERNOR TIME; SURGE TANKS

In pumped-storage plants the time specified for the governor servomotors to move the wicket gates through a complete stroke is generally not less than 30 s. The reasons for this are: (1) about 30 s is the minimum practicable time for penstock valve operation without undue complication of the valve operating machinery by the incorporation of dashpots and accessories; (2) one of the primary purposes of massive EHV interconnection is to carry loads from emergency outages for 30 to 60 s or more until the local pumped-storage units take over. In the lengths of waterways that are permissible economically, penstock and tunnel velocities may usually be accelerated to full load in a much shorter time without excessive positive or negative waterhammer, so that surge tanks are not necessary. However, in the exceptional case of a long tailrace tunnel flowing as a closed conduit, under the relatively low head of tail water, even a 30-s closure could be sufficient to produce negative water hammer great enough to cause separation of the water column and damage to the unit. For such cases a surge tank[1] at the downstream face of the power station is required. In a dual-purpose project for municipal water supply and by-product power, the length of the tunnel is dictated by water-supply eco-

[1] See Ref. 2 in bibliography for details of surge-tank calculation.

TABLE 2 Arithmetic Integration at Full-Load Rejection (Generating Mode; Wicket Gates Stuck at 92% Opening)

Interval $\left(\frac{2L}{a}\right)$	Time, s	Trial V (ΔV), ft/s	ΔH, ft	$\Sigma\Delta H$, ft	H_f, ft	$H_O + H_f + \Sigma\Delta H =$ total H, ft	Trial rpm	Q (8 units), ft³/s*	Check (V) ft/s	Hp, 8 units (000)	Average hp, 8 units (000)	N^2 rpm²	$N_2^2 - N_1^2$	Check (T), s
0	0	24.5	. . .	. . .	−36.8	1,173.2	257	28,100 (3,510)	24.7	3,200	. . . . .	66,000		
1	5.3	−1.9 22.6	274	274	−28.8	1,455.2	342	25,800 (3,220)	22.6	3,040	3,120	117,000	51,000	5.30
2	10.6	−5.3 17.3	766	492	−16.9	1,685.1	390	19,700 (2,460)	17.3	1,180	2,110	152,000	35,000	5.30
3	15.9	−6.0 11.3	867	375	−7.2	1,577.8	378	12,900 (1,610)	11.3	−104	538	142,900	9,100	5.30
4	21.2	−4.7 6.6	680	305	−2.5	1,512.5	368	7,520 (940)	6.6	−784	−444	135,500	−7,400	5.30
5	26.5	−3.25 3.35	470	165	−0.6	1,374.4	345	3,819 (477)	3.35	−1,200	−992	119,000	−16,500	5.30
6	31.8	−0.85 2.50	123	−42	−0.4	1,167.0	316	2,850 (256)	2.5	−1,080	−1,140	100,000	−19,000	5.30

$L = 12{,}322$ ft $A = 1{,}140$ ft² $a = 4{,}650$ ft/s $H_f = 0.0565V^2$ $WR^2 = 1{,}023 \times 10^6$ (8 units)

$2L/a = 5.3$ s $H_O = 1{,}210$ ft $\Delta H = a\,\Delta V/g = (4{,}650 \times \Delta V)/32.2 = 144.5\,\Delta V$

$$\text{Check } \Delta T = \frac{4\pi^2 WR^2(N_2^2 - N_1^2)}{2g \times \text{av. hp} \times 550 \times 3{,}600} = \frac{4\pi^2 \times 1{,}023 \times 10^6(N_2^2 - N_1^2)}{64.4 \text{ av. hp} \times 550 \times 3{,}600} = \frac{320(N_2^2 - N_1^2)}{\text{av. hp}}$$

* Numbers on second lines indicate quantity per unit.

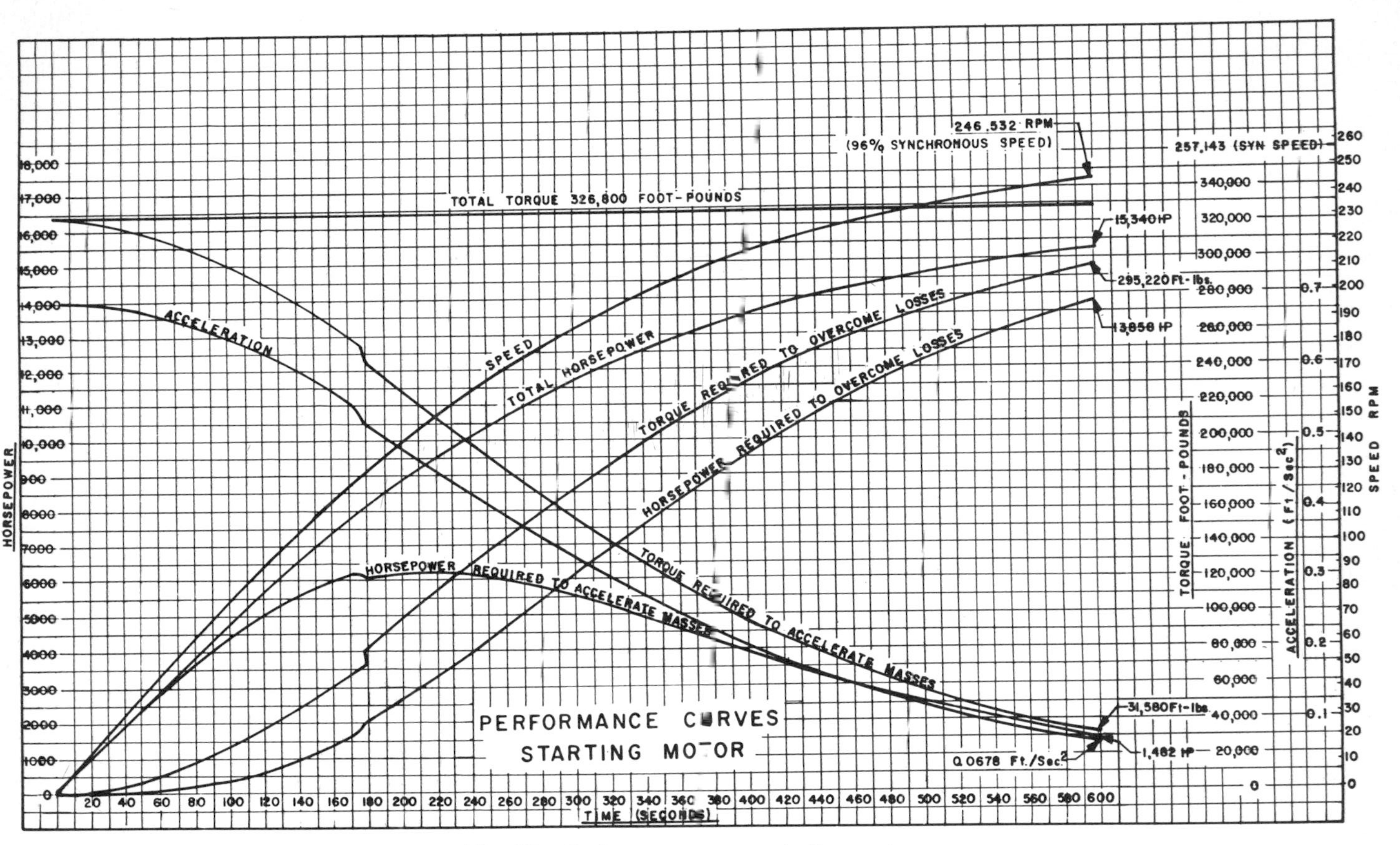

Fig. 10 Performance curves, starting motor.

TABLE 3 Economic Evaluation of Hypothetical Pumped-Storage Plant vs. Conventional Steam Reheat Plant

	Reheat plant		Hypothetical plant	
Installed capacity	3,600 MW		3,600 MW	
Number of units	4		16	
Plant investment (000 omitted)				
Generation	$450,000		$260,000	
Transmission	......		47,100	
Total	$450,000		$307,100	
Total per kW	$ 125		$ 85	
Annual generation (10^6 kWhr)				
Peak load	1,300		1,300	
Base load	20,000			
Total	21,300		1,300	
Annual capacity factor	67.5%		4.1%	
Annual fixed charge rates				
Generating plant	13.35%		10.90%	
Transmission plant	......		11.35%	
Annual costs	Total (000)	Per kW-yr	Total (000)	Per kW-yr
Fixed charges:				
Generating plant	$ 60,076	$16.69	$28,340	$ 7.87
Transmission plant	......	....	5,346	1.49
Total	$ 60,076	$16.69	$33,686	$ 9.36
Fuel	59,044	16.40	5,018*	1.39
Operation and maintenance	9,900	2.75	3,492	.97
Total	$129,020	$35.84	$42,196	$11.72
Credit to reheat unit:				
Replacement for base				
Load generation	68,480	19.02		
Net cost of 3,600 MW capacity and peak-load generation	$ 60,540	$16.82	$42,196	$11.72
Differential annual cost in favor of hypothetical plant			$ 9,172	$ 5.10

* Fuel Cost for Pumping Energy.

nomics, and may be as great as 40 to 50 miles. In such cases a surge tank on the upstream side of the power station may be indicated.

ECONOMIC DESIGN

A proposed pumped-storage project for peaking service must be able to show a liberal margin of economic superiority over competing thermal types. This requires that the undertaking be designed in accordance with the so-called "lean project concept" in which all elements are closely tailored to the specific purpose of pumped storage and many features considered standard in the conventional hydroelectric plant are found superfluous.

Selection of the proper site is of paramount importance. The rock should be strong and tight, so that no more than local grouting is needed to inhibit leakage and no drainage system for the concrete lining of the tunnels is required. The upper storage reservoir should preferably be located in a natural basin in elevated highlands so as to provide the requisite storage capacity by use of low rimdikes. It is a great advantage if the lower reservoir can be located on tidewater. The head, probably the most important natural feature, should be high, in the range from 500 to 1,500 ft. High head means smaller water quantities and consequently smaller

TABLE 4 Hypothetical Pumped-Storage Plant Cost Estimate—Summary

Account no.	Item	Cost
330	Land and land rights	$ 4,311,000
331	Structures and improvements	
−1	Powerhouse substructure	20,791,000
−2	Service bay substructure	2,023,000
−3	Powerhouse superstructure	553,200
−4	Cofferdam	600,000
332	Reservoirs, dams, and waterways	
−2	Reservoir clearing	123,000
−4	Dikes and embankments	25,162,000
−8	Intakes	1,203,000
−12	Power tunnels and waterways	76,526,500
	Tailrace	4,940,000
333	Waterwheels, turbines, and generators	
−2	Fire protection system	86,000
−5	Motor generators	31,776,800
−8	Spherical valves	7,680,000
−9	Turbines and accessories	23,915,520
−11	Piezometer system	24,000
−13	Spiral case and draft tube unwatering systems	136,000
	Air depression system	208,000
	Draft tube racks, piers, and guides	2,071,600
334	Accessory electrical equipment	6,496,000
335	Miscellaneous power plant equipment	2,046,000
335	Roads, railroads, and bridges	2,393,600
352	Transmission plant—structures and improvements	2,226,500
353	Station equipment	12,119,400
		$227,412,120
	Contingencies, overhead, and engineering	32,587,880
	Total estimated cost of project	$260,000,000

TABLE 5 Operating Range of Hypothetical Project

Reservoir elevation, ft	1,000	1,060	1,120	1,160
Tail-water elevation, ft	0	0	0	0
Gross head, ft	1,000	1,060	1,120	1,160
Generating cycle:				
Friction head loss, ft	30	30	30	30
Net head, ft	970	1,030	1,090	1,130
Best gate turbine output, hp (000)	—	310		
Full-gate turbine output, hp (000)	310	—	380	400
Turbine output blocked at, hp (000)	—	—	350	350
Turbine efficiency, %	85	88.3	87.5	88.5
Turbine discharge, ft³/s	3,300	3,000	3,200	3,100
Generator output at 98% efficiency, MW	227	227	256	256
Generator MVA (0.90 PF)	252	252	284	284
Pumping cycle:				
Friction head loss, ft	20	20	20	20
Net head, ft	1,020	1,080	1,140	1,180
Maximum possible discharge, ft³/s	2,700	2,350	2,000	1,730
Pump power, hp (000)	350	324	298	277
Pump efficiency, %	89.5	89	87	84
Motor power, at 98% efficiency, MW	262	242	223	206
Motor MVA (0.90 PF)	297	275	252	234

physical sizes for a given power output. High head also permits higher turbine speeds, lower torque, and a smaller generator. The intake need be only a bell mouth at the end of the supply tunnel; consequently headgates, cranes, and accessories can be eliminated. The individual units should be of large capacity, to ensure minimum equipment costs.

Table 3 shows a typical form for the economic comparison of pumped storage with competitive thermal peaking, and Table 4 is a typical form for the project cost estimate with its overheads.

Table 5 shows the range of machine requirements for a deep upper reservoir in which the head variation due to daily drawdown is relatively large. It will be noted that the maximum electrical motor load to develop the full hydraulic capability of the pump, occurring at the minimum head of 970 ft, is 297 MVA. However, this loading is of comparatively short duration and may be sustained at 80°C temperature rise, which corresponds to 115 percent of normal rating. The normal rating at 60°C temperature rise would then be 297/1.15, or 258 MVA, which is shown by the tabulation to be adequate to carry the pumping and generating loadings of more protracted duration. This utilization of overload rating for Class B insulation affords substantial economies in electrical machine cost.

BIBLIOGRAPHY

1. Parmakian, John: "Waterhammer Analysis," Dover Publications, Inc., New York, 1963.
2. Rich, George R.: "Hydraulic Transients," Dover Publications, Inc., New York, 1963.
3. "Waterhammer in Pumped Storage Projects," The American Society of Mechanical Engineers, 1965.
4. Rich, George R., and Walter B. Fisk: "The Lean Project Concept in the Economic Design of Pumped-Storage Hydroelectric Plants," World Power Conference Paper 60, September 1964.
5. Rudulph, E. A.: Taum Sauk Pumped Storage Power Project, *Civ. Eng.*, January 1963.
6. McCormack, Webster J.: Taum Sauk Pumped-Storage Project as a Peaking Plant, *Water Power,* June 1962.
7. Chapin, W. S.: The Niagara Power Project, *Civ. Eng.*, vol. 31, no. 4, pp. 36–39, April 1961.

Section **10.19**

Nuclear Services

W. M. WEPFER

APPLICATION

General If there is a single item which distinguishes pumps for nuclear service from conventional pumps, that item would be attention to safety. Cleanup from spills of radioactive fluid may be costly and require protection for personnel. Moreover, functioning of some pumps is vital to certain emergency conditions. Most of the rules and regulatory standards which have been prepared for nuclear pumps are dedicated to the promotion of public safety. It is necessary for a pump builder to be fully knowledgeable of these procedures before he can offer his pumps in the nuclear market.

All nuclear power plants built for public-utility service involve the generation of steam. It is usual for the plant to contain one major system, or loop, and a number of supporting systems. Since virtually every system requires some form of pumping, it is conventional to designate the pump by the name of the system with which it is identified. Principal methods of steam production are shown below. For each the main and auxiliary systems are listed, with a brief description of the pumping needs.

PWR Plants Pressurized water reactor (PWR) plants employ two separate main systems in the generation of steam. In the primary system the water is circulated by reactor coolant pumps through the nuclear core and large steam generators. The secondary side of the steam generator provides nonradioactive steam to the steam turbines. It is usual to have two to four reactor coolant pumps and primary loops. Overpressure is provided to prevent vapor formation. Typical primary water conditions are 2,250 lb/in^2 gage and 550°F.

Table 1 lists the principal pumps which are provided in systems serving a PWR-type plant. The *reactor coolant pumps* continuously circulate water from the reactor to the steam generator. *Spent-fuel pit pumps* provide the necessary cooling of the fuel elements which have been removed from the reactor; *component cooling-water pumps* circulate clean water at low system pressures for the purpose of cooling (1) primary water which is continuously bled for purification, (2) main and auxiliary pump bearings and seals, (3) primary-pump thermal barriers, (4) large motors, and (5) the containment vessel where the nuclear components are housed.

Since a reactor, once it has been critical, continues to generate heat even when shut down, *residual heat-removal pumps* are provided to circulate reactor water through coolers at any time the reactor is inoperable, even during refueling. In the event of a failure of the primary pressure boundary, the containment vessel will fill with steam; at such a time *containment spray pumps* function to condense the steam in order to lower the temperature and pressure of the environment within

TABLE 1 PWR Plants—Typical Nuclear-Pump Parameters

Pump name	Number per plant	Flow, gpm	Head, ft	Design pressure, lb/in²	Design tempera-ture, °F	Driver, hp	Horizontal or vertical	Length or height including driver, in	Speed (nominal), rpm	Notes
Reactor coolant	4	93,000	305	2,500	550	8,000	Vertical	340	1,200	
Component cooling water	2	4,800	250	250	180	500	Horizontal	92	1,800	
Residual heat removal	2	3,000	270	600	400	400	Vertical	104	1,800	
Containment spray	2	2,600	450	300	250	400	Horizontal	115	1,800	
Spent-fuel pit	2	2,300	125	250	180	100	Horizontal	66	1,800	
Charging (centrifugal)	2	650	3,000	2,800	300	900	Horizontal	225	4,850	Gear drive
Safety injection	2	425	2,500	2,500	300	400	Horizontal	160	3,600	
Chilled water	2	400	150	150	200	40	Horizontal	58	3,600	
Spent-resin sluicing	1	140	250	150	200	25	Horizontal	28	3,600	Canned
Spent-fuel pit skimmer	2	100	50	150	180	5	Horizontal	35	3,600	
Reactor-coolant drain tank	2	100	300	150	200	20	Horizontal	28	3,600	Canned
Charging (positive displacement)	1	98	5,800	3,200	300	200	Horizontal	170	50/250	Variable speed
Boric acid transfer	4	75	235	150	200	15	Horizontal	27	3,600	Canned
Floor drain tank	1	35	250	150	200	15	Horizontal	52	3,600	
Recycle evaporator feed	2	30	320	150	200	20	Horizontal	28	3,600	Canned
Boron injection recirculation	2	20	100	150	250	3	Horizontal	16	3,600	Canned

Data courtesy of the Westinghouse Electric Corporation.

the containment vessel. *Charging pumps,* both centrifugal and positive displacement, supply high-pressure make-up water to the primary system, and function as feeders to the main and auxiliary pump seals. Under other emergency conditions involving breach of primary pressure boundary, *safety injection pumps* rapidly supply cold water to the primary system. Resin beds, which are a part of the water-purificaton system, are flushed to a storage tank by *spent-resin sluicing pumps.* An evaporator package, partly for the purpose of removing boron from the primary water, is supplied by *recycle evaporator feed pumps.* Similarly, other pumps, some not listed in Table 1, support auxiliary subsystems.

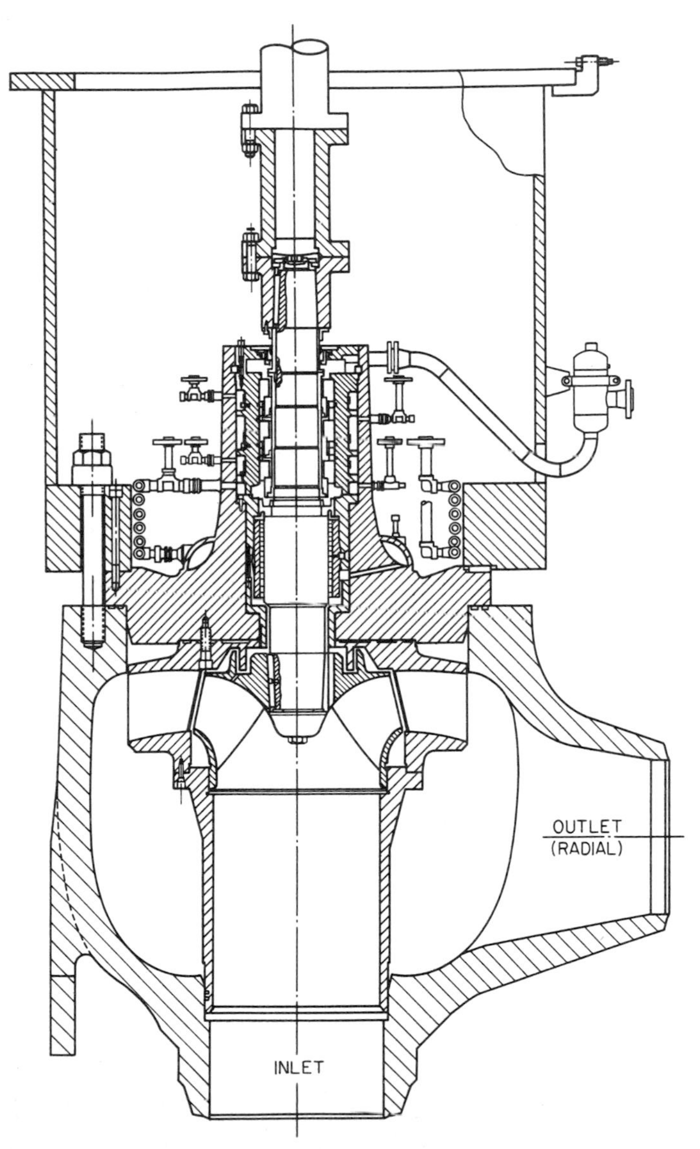

Fig. 1 Reactor-coolant pump type RDV. (Bingham-Willamette Company)

Figure 1 shows a large reactor coolant pump for a PWR-type reactor. This pump has a rated head of 400 ft at 105,000 gpm, supplying radioactive water to steam generators at 2,200 lb/in^2 and 598°F. Its speed is 1,190 rpm and it requires a driving motor of 10,000 hp. Two, three, or four of these radial-discharge pumps would be required in a nuclear plant, depending upon the plant power rating.

Figure 2 shows a pump for PWR charging/safety injection service. It is rated at 150 gpm at 5,800 ft head, at up to 300°F water. Maximum design pressure is 2,800 lb/in^2.

BWR Plants In the boiling water reactor (BWR) plants, water is boiled directly in the nuclear core and piped to the steam turbine. To assure a constant and high reactor flow rate, and to avoid local areas of core overheating, large-volume low-head *recirculation pumps* provide forced flow through the reactor. In an auxiliary system which maintains chemical cleanliness of the primary coolant, *reactor water cleanup pumps* are used to circulate a portion of the primary fluid through filters and several heat exchangers. *Closed cooling-water pumps* provide closed loop circulation of cooling water to a wide variety of components. *Residual heat-removal-service water pumps* provide heat removal to the ultimate heat-sink facility. *Core spray pumps* are used to supply cooling water directly to the core in the event of a major pipe break or component pressure boundary failure.

The *reactor-coolant isolation system pumps* are of the steam-driven turbine type and operate automatically in time and with sufficient flow to maintain adequate water level in the reactor vessel in the event this vessel becomes isolated. *Standby liquid-control-system pumps* provide water for a redundant system which functions to bring the nuclear fission reaction to subcriticality and to maintain subcriticality as the reactor cools. Control rods are operated hydraulically, with fluid pressurization supplied by *control-rod drive-system pumps.*

Typical reactor-water conditions are 1,000 lb/in^2 gage and 550°F. Other auxiliary systems are employed as in the PWR plants discussed above, and most of these systems will have special-purpose pumps identified with them.

Table 2 shows the principal pumps used in BWR plants, together with significant characteristic data.

Fast Breeder (Liquid Metal) Plants Many fast breeder plants, currently in design or use, make use of liquid sodium as the primary coolant. Often a primary and a secondary system, both employing liquid sodium, are used, with a sodium-water steam generator in the secondary circuit to generate steam for the turbine. Circulating pumps in these systems will operate at temperatures of approximately 1250°F at relatively low system pressures. In addition, auxiliary liquid-metal pumps may be used to supply necessary subsystems.

PRINCIPAL TYPES OF PUMPS

Most of the pumps in nuclear service are one- or two-stage centrifugal motor-driven pumps. Both vertical and horizontal types are used. Charging, safety-injection, feedwater, and other high-head pumps are usually multistage centrifugals, motor-driven. Some requiring high horsepower are turbine-driven. Double-suction designs are frequently used for the residual-heat-removal pumps, where service requires operation under low available NPSH conditions. Reciprocating pumps find limited service in nuclear plants for make-up flow, seal injection flow, or chemical mixing service. Canned-type pumps are frequently used in subsystems where their zero-leakage capabilities can be exploited.

MATERIALS OF CONSTRUCTION

Material Limitations When pumps built to ASME standards are required, materials for the pressure-retaining parts must be selected from a list approved by the ASME. Section III (Nuclear Components) of the ASME Boiler and Pressure Vessel Code lists these materials, their allowable stresses, and the examination requirements which must be applied to ensure their suitability.

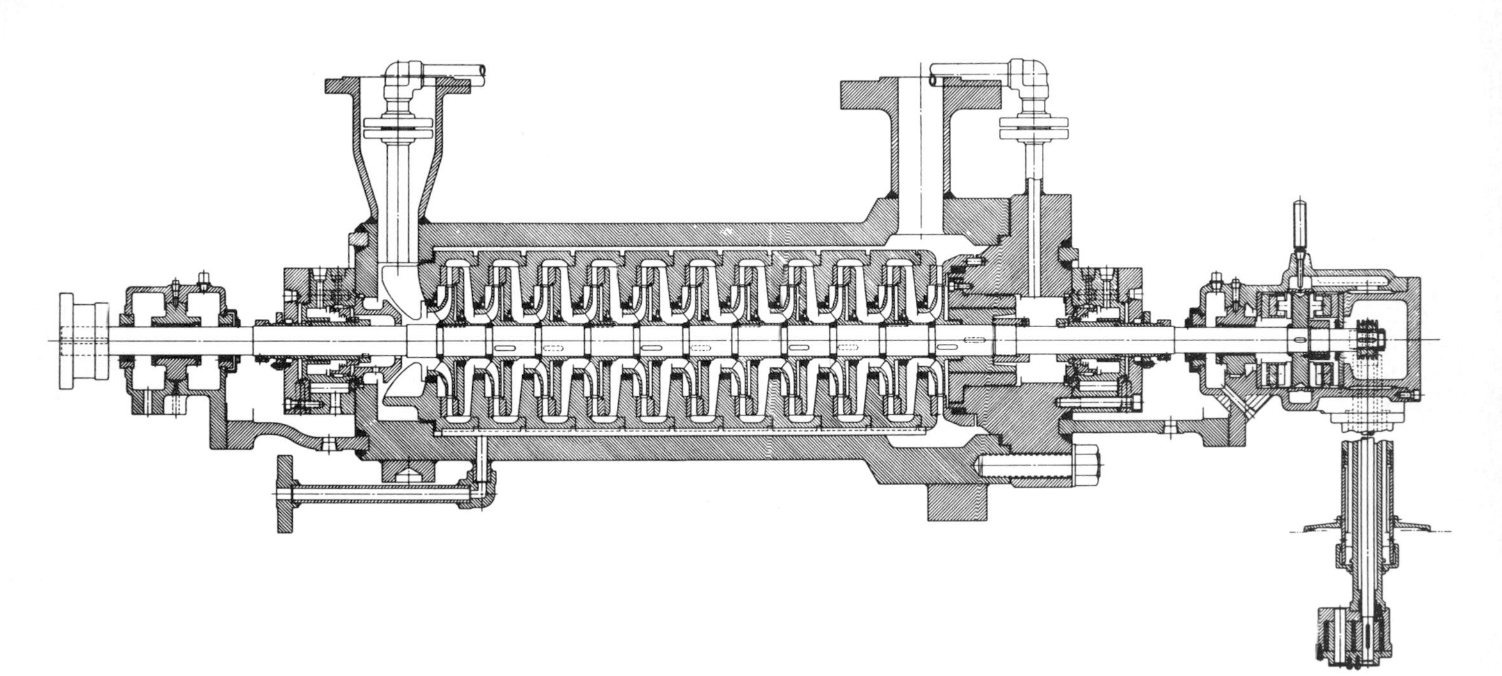

Fig. 2 Pump for PWR charging/safety injection service. (Pacific Pumps Division, Dresser Industries)

TABLE 2 BWR Plants—Typical Nuclear Pump Parameters

Pump name	Number per plant	Flow, gpm	Head, ft	Design pressure, lb/in²	Design tempera-ture, °F	Driver, hp	Horizontal or vertical	Length or height including driver, in	Speed (nominal), rpm	Notes
Recirculating coolant	2	35,400	865	1,675	575	7,200	Vertical	240	1,800	
RHR service water	4	7,300	115	150	150	300	Vertical	200	900	
Residual heat removal	3	7,100	275	450	360	900	Vertical	350	1,800	
High-pressure core spray	1	6,110	900	1,600	212	3,000	Vertical	500	1,800	
Low-pressure core spray	1	6,110	730	550	212	1,750	Vertical	400	1,800	
Closed cooling water	3	1,700	110	250	150	60	Horizontal	100	1,800	
Reactor-core isolation cooling	1	725	2,800	1,500	140	800	Horizontal	123	4,000	Turbine-driven-variable speed
Fuel-pool cooling	2	600	300	150	150	75	Horizontal	95	1,800	
Reactor-water cleanup	2	150	500	1,400	560	50	Horizontal	78	3,600	
Control rod drive hydraulic system	2	95	3,500	1,750	150	300	Horizontal	168	1,800	
Standby liquid control	2	40	2,800	1,400	150	40	Horizontal	60	1,800	Reciprocating pumps

Data courtesy of the General Electric Company.

Typical Materials Examples of acceptable materials for pressure-retaining boundary parts are shown below. They are suitable for all three ASME Code classes; however, other acceptable materials may be restricted to use for a particular class.

Carbon Steel:

Castings	SA-216, Gr WCA, WCB, WCC
Forgings	SA-105, Gr I, II
Plate	SA-515, Gr 55, 60, 65, 70
	SA-516, Gr 55, 60, 65, 70
Bolting	SA-193, Gr B6, B7, B8, B16

Stainless Steel:

Castings	SA-351, Gr CF8 (304), CF8M (316)
Forgings	SA-182, Gr 304, 316, 321, 347
Plate	SA-240, Gr 304, 316, 321, 347

Nonferrous:
A limited number of nonferrous materials are permitted.

Hazardous and porous materials are generally avoided, as are materials such as cobalt which, though normally harmless, may become hazardous from radioactive considerations. See Considerations of Radioactivity, below. Cobalt content is often limited for large stainless steel parts, but is usually permitted in concentrated form in small areas, for example, where hard facing is required.

Materials of construction should not be affected by the usual decontamination chemicals.

Nonpressure boundary parts may be made of conventional materials but will usually require the buyer's approval. Certain elastomers such as ethylene propylene, which has good radiation stability, are excellent for seal parts. Many fine grades of carbon-graphite are available for water-lubricated bearings and for mechanical seal facings.

SPECIAL REQUIREMENTS FOR NUCLEAR SERVICE PUMPS

General It is in the area of special requirements that nuclear service pumps differ most widely from commercial products. These special requirements, described in greater detail below, far exceed the requirements of the general industrial field, and illustrate the strong emphasis placed upon pressure integrity and pump operability.

Nuclear grade pumps are designed, built, inspected, tested, and installed to rigid standards of the U.S. Nuclear Regulatory Commission, the American Society of Mechanical Engineers, the American National Standards Institute, and other regulatory agencies such as state and local jurisdictional bodies. Established rules can be divided into two distinct categories: (1) those controlling integrity of the pressure retaining boundary and (2) functional considerations.

Hydraulic design of nuclear pumps is the same as that of pumps in conventional service. For recirculation and reactor coolant pumps, a radial discharge is preferred by some users because it tends to simplify certain aspects of the plant design.

Vibration characteristics of pumps in nuclear service are especially important because of the relative inaccessibility of the equipment for checking and servicing and because safety requirements permit only limited outage of pumps; otherwise, the plant must come to standby condition.

An analytical or experimental determination of lateral and torsional natural frequencies is routine for most pump-driver combinations, and occasionally a transient analysis may be required.

Design Under ASME Code Rules Under the rules established by the ASME Boiler and Pressure Vessel Code, Section III (Nuclear Components), the "owner," such as a public utility, either directly or through his agent prepares a Design Specification for a specific pump, which in Code terms is a "component."

The "manufacturer" builds the pump in accordance with the Design Specification. Verification is provided through the combined efforts of the material supplier, manufacturer, insurance inspector, and state and local enforcement agencies.

The Design Specification requires compliance with the rules of the Code with regard to the design of the pressure boundary and, in addition, includes supplementary requirements prescribed by the owner. Other standards are invoked as applicable to meet the safety and environmental requirements of the U.S. Nuclear Regulatory Commission. Functional needs may be included, and the ASME Code class to which the pump is to be built must be identified.

ASME Code classes for pumps are three in number. For the most critical service, a Class 1 pump is specified. Class 2 represents a pump serving a less critical system, and Class 3 is the lowest class pump for nuclear service, except for a few pumps which are classified as nonnuclear class (NNC) and are permitted to be used in a nuclear power plant. It is the owner's responsibility to establish the pump class, with guidance provided by the Nuclear Regulatory Commission and the manufacturers.

For the pressure boundary evaluation, a Class 1 pump, by Code rules, receives a detailed analysis by modern design techniques supported, if necessary, by experimental stress analysis, and a stress report is prepared to document adherence to rules. Fatigue analyses of critical portions of the pump may be required, and behavior under all plant conditions, including accident, must be investigated. Class 2 pumps require less analysis and no stress report, but a data report certifying compliance with the Code is required. For Class 3 pumps, wider latitude of design is permitted; the certified data report is also required. All three classes of pumps may be given the ASME N-stamp by a qualified pump builder upon completion of design, manufacture, inspection, and test.

Additionally, nuclear service pumps are usually examined in design for thermal steady state and transient conditions, behavior under seismic disturbances, conditions of nozzle loadings imposed by system piping, means of support, and accessibility for service, inservice inspection, and replacement.

Rules for the control of quality during material procurement, manufacture, and test are also contained in the Code. Nondestructive examination and document control are detailed, and a quality-assurance program requiring a formal procedure manual must be prepared and implemented by the manufacturer. Periodic surveys by the ASME verify adherence to Code rules. Local jurisdictional authorities provide day-to-day inspection services as required by the manufacturer.

Provision for "Inservice Inspection" in accordance with Section XI of the ASME Code is also a requirement for nuclear pumps. This document requires reinspection of certain critical portions of pumps in the field at specified intervals, usually on a ten-year overall cycle. In addition, periodic testing to verify operability and performance is required.

Design of Noncoded Parts Parts of the pump which are not classified as pressure boundary items or attachments to the pressure boundary are not covered by the ASME Code. The owner's Design Specification may describe applicable requirements for these nonpressure parts, referencing other documents or excluding use of certain objectionable materials. If the pump is in critical service or is needed in emergency situations, certain ANSI standards may be invoked. Accompanying these rules will be extensive quality assurance and documentation to demonstrate compliance.

Considerations of Radioactivity Radioactivity may become a serious consideration in the design of nuclear pumps because of the need for servicing the equipment. The water used in the primary system becomes contaminated with metallic elements through solubility, corrosion, and erosion. The metallic elements become radioactive because of interaction with neutrons when circulated through the core region. These radioactive contaminants may, if soluble, remain in solution in the water or, if unsoluble, may plate out on metal surfaces or become lodged in "crud traps," such as fit interfaces, screw threads, and certain types of weld configurations such as socket welds, base metal porosity, extreme surface roughness, and cracks. In an actual case, for instance, a pump impeller returned for overhaul defied attempts at decontamination until it was discovered that a repair had been made to a presumably integral wear ring by undercutting and shrinking on a new ring. The interface was barely perceptible, but once it had been found and the ring had been removed, the impeller was readily decontaminated.

Soluble contaminants are most easily removed by providing the pump with complete drainage features, that is, leaving no internal pockets which are not naturally drainable. Ease and speed of parts replacement are, therefore, also important items of design since they reduce the time the serviceman is exposed to radiation.

The Nuclear Regulatory Commission has specified that no person shall be exposed to more than 100 millirems of radiation in seven consecutive days. A pump producing radioactivity to this extent could be brought into a conventional machine shop and handled in a conventional manner. After prolonged service in an atomic plant, however, a pump may emit, say, 2 to 50 rems/hr of radioactivity. Hence provision for decontamination also becomes an important design characteristic of the nuclear service pump. It should also be mentioned that not all nuclear pumps operate in highly radioactive environments; for these pumps, radioactivity levels may be low, but the equipment will usually require some degree of decontamination before it can be freely handled.

Seals Seals, both static and dynamic, are very similar to those used in nonnuclear pumps. Occasionally a material substitution is made to avoid poor radiation properties; otherwise standard designs are usually acceptable. For pumps serving critical systems, some form of a backup seal is usually a requirement of the Design Specification. On large-size pumps employing hydrostatic seals or multiple mechanical seals, it is usual to supply filtered injection water to the seal system.

Testing Performance of nuclear pumps is verified by procedures common to nonnuclear pumps. In addition, however, it may be necessary to demonstrate pump performance under some emergency condition such as loss of cooling water or seal-injection water. Also, a test under simulated total-loss-of-power condition may be required, in which the reactor coolant pump must coast safely to a stop and withstand loss of both cooling and seal-injection water for a finite time.

Hydrostatic testing in accordance with ASME Code rules is conventional except that specific documentation is required.

Section **10.20**

Metering

ROBERT H. GLANVILLE

METERING OR PROPORTIONING

These are special adaptations of the conventional reciprocating pumps which are designed primarily to transfer liquids. The principal differences are means to vary the pumping rate and to predict what that rate will be. This makes these pumps into metering devices which can be used as final control elements in continuous-flow processes. They are often employed where two or more liquids must be proportioned or where mixture ratios must be controlled. These effects are achieved by changing the displacement per stroke or by changing the stroking speed. In the first case this is accomplished by moving the crankpin by special linkages or by partial stroking. Capacity changes by regulating pump speed are achieved through the use of variable-speed transmissions or electric motors. Three basic types of positive-displacement reciprocating pumps and several variations are used for this service.

Packed Plunger The packed-plunger pump is the most commonly used type because of its relatively simple design and wide range of pressure capability. It

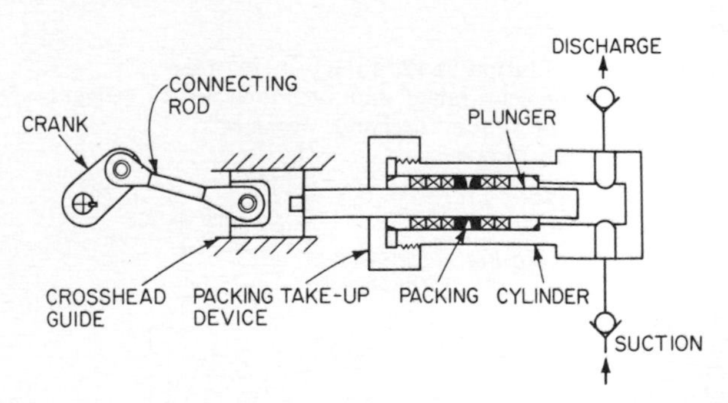

Fig. 1 Packed plunger.

is an adaptation of the conventional reciprocating transfer pump (see Fig. 1). Its advantages are:

1. Relatively low cost
2. Pressure capability to 50,000 lb/in^2 gage
3. Mechanical simplicity

4. Capacities ranging from a few cubic centimeters per hour to 20 gpm
5. Accuracy better than 1 percent over a 15:1 range
6. Least effect from changes in discharge pressure

Its disadvantages are:

1. Packing leakage making it unsuitable for corrosive or dangerous chemicals
2. Packing and plunger wear and the resulting need for gland adjustment
3. Inability to pump abrasive slurries or chemicals which crystallize

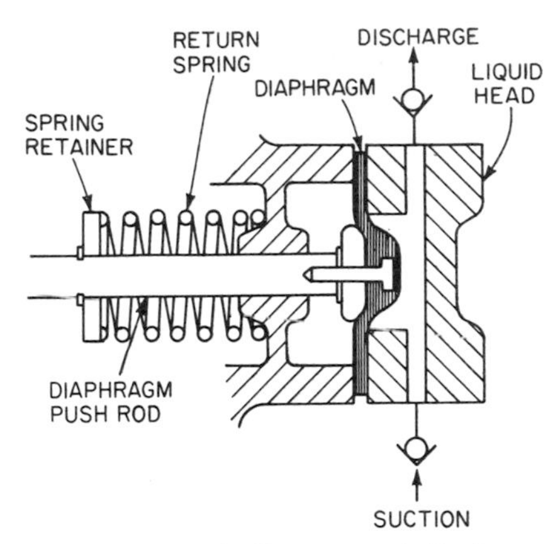

Fig. 2 Mechanically actuated diaphragm.

Mechanically Actuated Diaphragm The mechanically actuated diaphragm pump is commonly used for low-pressure service where freedom from leakage is important. This pump utilizes an unsupported diaphragm which is moved in the discharge direction by a cam and returned by a spring (see Fig. 2). Its advantages are:

1. Relatively low cost
2. Minimum maintenance at six- to twelve-month intervals
3. Zero chemical leakage
4. Ability to pump slurries and corrosive chemicals

Its disadvantages are:

1. Discharge pressure limitation of 125 to 150 lb/in^2 gage
2. Accuracy in the 5 percent range and as much as 10 percent zero shift with change from minimum to maximum discharge pressure
3. Capacity limit of 12 to 15 gph

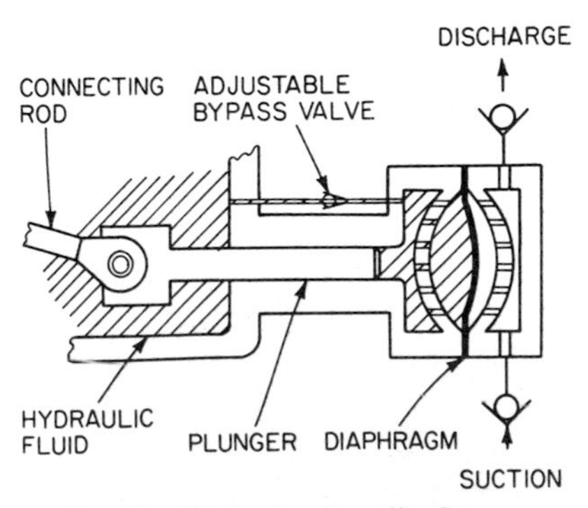

Fig. 3 Flat circular diaphragm.

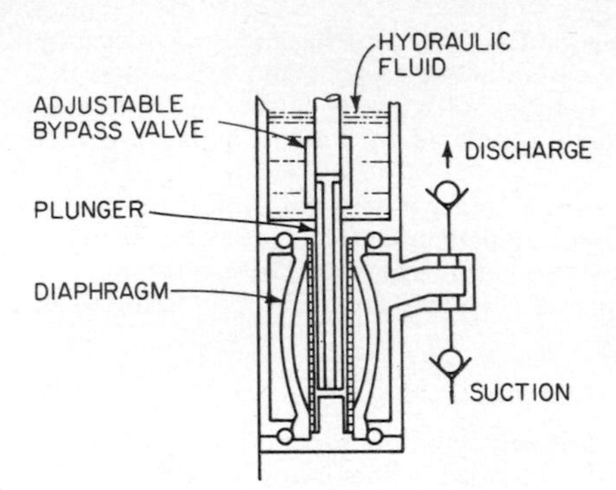

Fig. 4 Tubular diaphragm.

Hydraulically Actuated Diaphragm The hydraulically actuated diaphragm pump is a hybrid design which provides the principal advantages of the other two types. A packed plunger is used to pulse hydraulic oil against the back side of the diaphragm. The reciprocating action thus imparted to the diaphragm causes it to pump in the normal manner without being subjected to high pressure differences. A flat diaphragm design is shown in Fig. 3, and a tubular diaphragm design in Fig. 4. Its advantages are:

1. Pressure capability to 5000 lb/in^2 gage
2. Capacities to 20 gpm
3. Minimum maintenance
4. Zero chemical leakage
5. Ability to pump slurries and corrosive chemicals
6. Accuracy around 1 percent over 10:1 range

Its disadvantages are:

1. It is subject to a predictable zero shift of 3 to 5 percent per 1000 lb/in^2 gage
2. Higher cost

CAPACITY CONTROLS

There are five or more methods commonly used to adjust the capacity of metering pumps. The choice of these is determined by the application for which the pump is intended.

Manually Adjustable While Stopped This control is generally found on packed-plunger pumps of conventional design. Capacity changes are effected by moving the crankpin in or out of the crank arm while the pump is not in motion. This is the least expensive method and is used where frequent changes in pump displacement are not required.

Manually Adjustable While Running This feature is most frequently found on mechanically actuated diaphragm pumps, where it is accomplished by limiting the return stroke of the diaphragm with a micrometer screw. On packed-plunger pumps, stroke adjustment while running is relatively complicated. Stroke length is set by some type of adjustable pivot, compound linkage, or tilting plate which is manually positioned by turning a calibrated screw. On hydraulically actuated diaphragm pumps, relatively simple control is provided by manually adjustable valving which changes the amount of the intermediate liquid bypassed at each stroke.

Pneumatic Where metering pumps are utilized in continuous processes, it is necessary that they be controlled automatically. In pneumatic systems the standard 3 to 15 lb/in^2 gage air signal is utilized to actuate air cylinders or diaphragms directly connected to the stroke-adjusting mechanism.

Electric On electric-control systems, stroke adjustment is through electric servos which actuate the mechanical stroke-adjusting mechanism. These accept standard electronic control signals.

Variable Speed This method of adjusting capacity is achieved by driving a reciprocating pump with a variable-speed prime mover. Since it is necessary to reduce stroke rate to reduce delivery, discharge pulses are widely spaced when the pump is turning slowly. Surge chambers or holdup tanks are used when this factor is objectionable.

In control situations involving pH and chlorinization, two variables exist at once, e.g., flow rate and chemical demand. This is easily handled by a metering pump driven by a variable-speed prime mover. Flow rate can be adjusted by changing the speed of the pump, and chemical demand by changing its displacement.

SERVICE AND MAINTENANCE

Proportioning pumps utilizing manual capacity controls can be installed, serviced, and operated by plant personnel. Pneumatic capacity controls are more sophisticated, but once their construction and function are understood, maintenance and service should become routine. Electric capacity controls require a basic understanding of electric circuits for installation, operation, and routine service. Modular construction allows repair service by replacement of component groups, thus diminishing the task of troubleshooting.

All proportioning pumps utilize suction and discharge check valves. These require regular maintenance and service since they experience high-frequency operation, encounter corrosion, and must be kept in good working order to ensure accurate pump delivery. Service periods are greatly dependent on liquids pumped, pump operating speed, and daily running time. Usual service periods run from 30 days to 6 months or longer. Check valves are designed to facilitate service with a minimum of downtime, thus allowing replacement of wearing parts at minimum expense. Good design allows this service to be accomplished without breaking pipe connections to the pump.

Packed-plunger pumps require periodic adjustments of the packing take-up device to compensate for packing and plunger wear. Packing has to be replaced periodically, and regular lubrication of bearings and wear points is required.

Speed-reduction gearing, either separate integral units or built-in type, requires changing of the lubricating oil at six-month intervals.

Diaphragm pumps usually require replacement of the diaphragm as part of routine service at six-month intervals. Hydraulically actuated diaphragm pumps also require changing of the hydraulic fluid.

Pneumatic and electric controls generally present no special maintenance problems. The frequency of routine cleaning and lubrication is dependent on environmental conditions and should be consistent with general plant maintenance procedures.

INSTALLATION

Proper installation of metering pumps is very important if reliable pump operation is to be obtained.

NPSH must be kept as high as possible, and manufacturers' recommendations as to pipe size and length, strainers, relief valves, and bypasses must be observed.

MATERIALS OF CONSTRUCTION

With the tremendous variety of liquid chemicals used in industry today, an all-inclusive guide to suitability of materials for pump construction is virtually impossible. For the great majority of common chemicals, pump manufacturers publish data on materials of construction. In general, packed-plunger and hydraulically actuated diaphragm pumps are available as standard construction in mild steel, cast or ductile iron, stainless steel, and plastic. Mechanically actuated diaphragm pumps are usually available as standard construction in plastic and stainless steel. Almost any combination of materials can be furnished on special order.

Diaphragms are available as standard construction in Teflon, chemically resistant elastomers, and stainless steel from various manufacturers.

Section **10.21.1**

Hydraulic Transport of Solids

ELIE CONDOLIOS

EDMOND E. CHAPUS

NOTATION

τ Shear stress, dyn/cm^2
μ Dynamic viscosity, poises ($dyn\text{-}s/cm^2$)
ν Kinematic viscosity, stokes (cm^2/s)
$\frac{du}{dy}$ Velocity gradient, 1/s
τ_y Yield value or limiting shear stress, dyn/cm^2
τ_p Shear stress at wall, dyn/cm^2
η Rigidity coefficient or plasticity of the slurry, poises ($dyn\text{-}s/cm^2$)
r Radius of a cylinder, cm
R Radius of a pipe, cm
V Flow velocity, cm/s
D Pipe diameter, cm
L Length of pipeline, cm
P Pressure, dyn/cm^2
$\frac{\Delta P}{L}$ Pressure gradient, dyn/cm^3
H Head of liquid, cm
$\frac{\Delta H}{L}$ Pressure drop gradient expressed in head of liquid, dimensionless
λ Roughness coefficient of pipe, dimensionless
g Acceleration due to gravity, cm/s^2
V_c Critical velocity, cm/s
V_L Critical deposit velocity, cm/s
ρ Specific gravity of water or liquid medium, dimensionless
ρ' Specific gravity of the solid particle, dimensionless
W Settling velocity of a solid particle, cm/s
k Coefficient, dimensionless
d Mean particle size or diameter of a spherical particle, cm
d_n Nominal particle size or diameter of a sphere of same volume, cm
v Volume of a particle, cm^3
S Particle's projected area on a plane perpendicular to the movement, cm^2
C_x Drag coefficient, dimensionless

$\sqrt{C'_x}$ Apparent drag coefficient, dimensionless
Ψ Particle form coefficient: ratio between the cross section of a sphere of the same nominal diameter and the area of the largest cross section, dimensionless
Re Reynolds number, dimensionless
ϕ Pressure-drop increase coefficient to allow for presence of solids, dimensionless
C_t Volume concentration of solids, percent
Q_s Discharge of solids, g/s
Q_f Discharge of liquid before overflow, cm^3/s
Q Total liquid to be discharged, cm^3
q Overflow or additional liquid, cm^3/s

INTRODUCTION

The purpose of this section is to set forth the fundamental concepts, definitions, and main hydraulic laws of solids pipelining. Operating conditions for solids-pumping installations are also discussed in detail.

The first part of this section deals with the main laws for solids in pipelines as established from many years of intensive laboratory studies and applied with success to a number of industrial installations. Also in this first part, the economics and limitations of this transportation process are discussed, and some typical figures are given.

The second part focuses on operating conditions for hydraulic transportation systems including concepts of pipeline installations and operational reliability of pipeline installations.

Section 10.21.2, which follows, discusses in detail construction of solids-handling pumps.

LAWS OF SOLID-LIQUID FLOWS IN PIPELINES

The main laws formulated in the course of large-scale general studies are briefly reviewed in this section, without going into the details of the mechanics of solid-liquid flow in pipes, which have been dealt with elsewhere (see references).

A number of years ago, it was shown that slurries, or solid-liquid mixtures, can be broadly divided into two categories, according to the size of the solid particles: solids of less than approximately 270 mesh form *homogeneous slurries,* while coarser solids form *heterogeneous slurries.* However, more recent investigations have shown that this now-conventional division of homogeneous and heterogeneous slurries or mixtures is incomplete insofar as materials in the 270-65 mesh range are concerned. Slurries made up of these materials do not present any clearly defined plastic characteristics, nor do they behave like those more granular materials for which the flow laws have been firmly established.

Although there is a whole range of slurries, all differing in their behavior according to the particle size and the concentration of the solids transported, only the two main types will be reviewed, considering both their fundamental characteristics and the economics of transporting them.

Homogeneous Slurries When very fine materials (i.e., less than 50 microns, or 270 mesh) are placed in suspension in water, they generally form, above a minimum concentration, a *plastic slurry,* the flow properties of which can be deduced from Bingham's theories. The limiting dimension associated with the appearance of plastic properties is somewhat variable. This appearance is linked not only to the particle size and to the specific surface area of the particles in the mixture but also to the concentration and size distribution of the material, its specific gravity, its chemical nature, and the pH of the conveying water.

The technical characteristics of a given transport system must be based on the rheological properties of the mixture, in addition to other engineering considerations.

Although certain products such as cement slurry (ground limestone) do not present any fundamental difficulties, many others, such as the residual sludge produced from bauxite processing, are more complex. Such mixtures may behave quite differently as a result of vigorous turbulence or modification of the electrical properties of the suspension due, for example, to even a slight degree of contamination. Whatever the case, a very thorough rheological study is necessary.

In a true fluid, devoid of all rigidity, the shear stress produced in a laminar flow is related to the viscosity by Newton's relation:

$$\tau = \mu \frac{du}{dy} \tag{1}$$

For liquids with plastic properties (Fig. 1), i.e., having a certain rigidity, Bingham proposed a new relationship

$$\tau = \tau_y + \eta \frac{du}{dy} \tag{2}$$

where η is the plasticity of the slurry, also called *rigidity coefficient.*

Head Losses for Homogeneous Mixtures. LAMINAR FLOW CONDITIONS According to Bingham, who devised a formula based on the tangential load along a cylinder of radius r,

$$\tau = \tau_p \frac{r}{R} \tag{3}$$

With a Newtonian fluid, the Poiseuille formula is derived:

$$\frac{\Delta P}{L} = 32\mu \frac{V}{D^2}$$

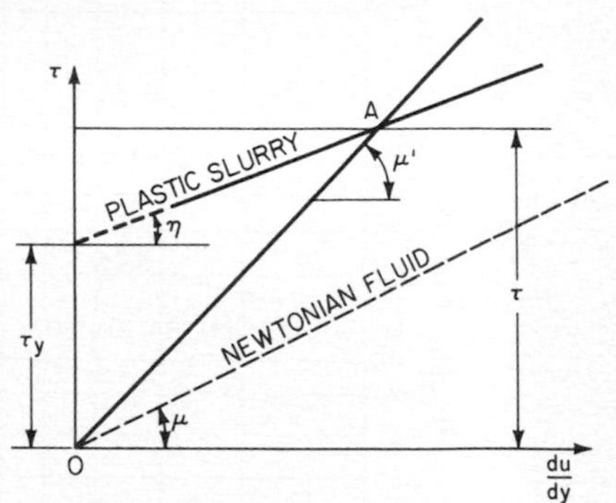

Fig. 1 Shear stress vs. velocity gradient for a plastic slurry and a Newtonian fluid.

For a liquid with a certain rigidity, Bingham found that there is a certain central section in the pipe in which the plastic fluid moves as a single solid mass, provided τ is less than the limiting shear stress τ_y.

In this case, the calculations show that, to within a close approximation

$$\frac{\Delta P}{L} = 32 \left(\frac{\tau_y}{6D} + \frac{\eta V}{D^2} \right) \tag{4}$$

TURBULENT FLOW CONDITIONS Above a certain velocity, the head losses become proportional to the second power of the flow velocity and are expressed, as a height of mixture, by the formula

$$\frac{\Delta H}{L} = \frac{\lambda}{D} \frac{V^2}{2g} \tag{5}$$

This is an experimental result from which it can be seen that the pressure losses expressed as a *height of mixture* are the same as with clear water.

Plotted logarithmically, Fig. 2 shows the head loss in terms of height of mixture as a function of the velocity for homogeneous mixtures of water and finely ground materials in concentrations of 47, 60, and 63 percent (concentration expressed in weight of solids per weight of slurry) in a 2-in pipe. The plastic properties of the 60 percent mixture (specific gravity of solids 2.55) are as follows:

$$\tau_y = 43.3 \text{ dyn/cm}^2$$
$$\eta = 0.24 \text{ poise}$$

The clear-water head losses appear as a straight line at a slope of 2:1 approximately.

Under laminar flow conditions, the head losses are noticeably greater than those with clear water and depend to only a small extent on the flow velocity, especially when the yield value of the slurry is high.

Under turbulent flow conditions, the head losses are the same as with clear water. This agrees well with the commonly used practical "mixture density" rule.

Critical Velocity. The critical velocity is the velocity at the transition between laminar and turbulent flow. As in the case of clear water, this critical velocity

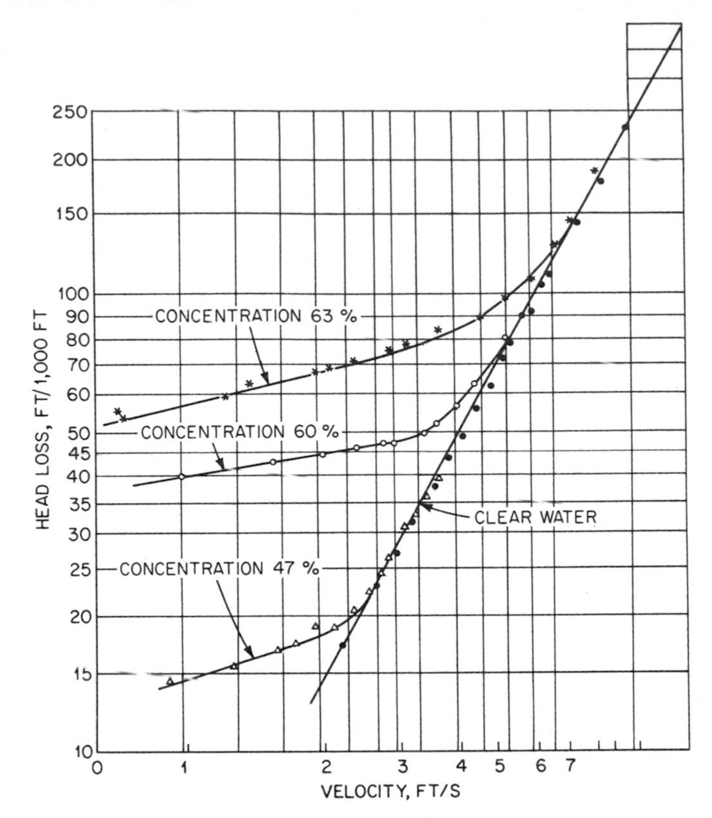

Fig. 2 Hydraulic transportation of a typical homogeneous mixture; head loss vs. velocity curves (expressed in height of mixture). (SOGREAH Co.)

occurs at a Reynolds number of about 2,000 provided, however, that the apparent viscosity deduced from the Bingham relationship is used, and not the water viscosity, which would give very different results.

The critical velocity V_c is thus arrived at from the expression

$$V_c = \frac{1{,}000\eta + 1{,}000\sqrt{\eta^2 + \dfrac{D^2\tau_y\rho}{3{,}000}}}{\rho D} \tag{6}$$

Correct operation of a line with homogeneous slurry requires that the solids be transported under turbulent flow conditions to avoid any settling out in the pipe. Although the risks are small over short distances, they can become much greater in pipes extending over long distances of, say, several miles. Consequently it is very important to determine the critical velocity V_c.

Economic Operation Conditions. For a given material, the economic transport conditions are associated with the concentration of the material in the mixture, which affects the plastic properties of the slurry as well as the rate at which the solids are transported.

In practice, the admissible concentration can be chosen only from within a narrow range of concentrations, for conflicting reasons.

If the concentration is low, the plastic properties will evanesce, the viscosity of the mixture will be close to that of water, and the transport velocity will have to be high to prevent deposition. In addition, the line would carry too much water. Conversely, for high concentrations, the plastic properties and more especially the critical shear stress of the mixture (*yield value*) will be such that high velocities will be required to produce a turbulent flow, needed for the prevention of solids deposition. Consequently the head losses would be prohibitively large. In either case, the system would be operated beyond its economical capacity.

For a given homogeneous slurry, Fig. 3 shows the variation of the minimum power required to transport 1 lb of material per second over 1 mile in terms of concentration.

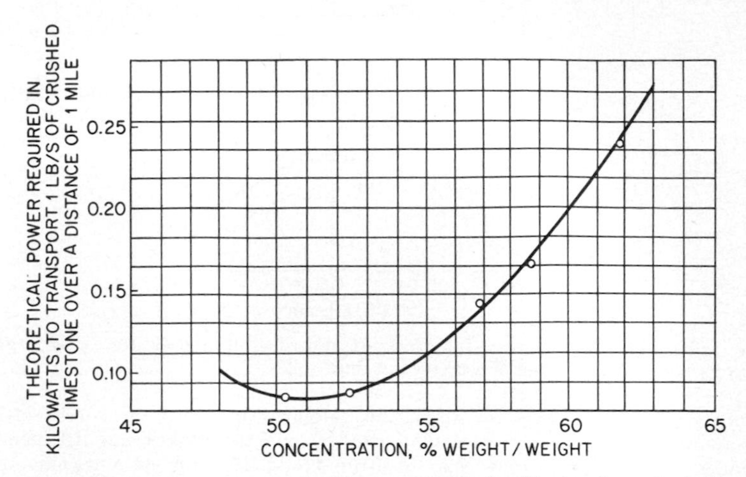

Fig. 3 Theoretical power required to transport 1 lb (dry) of crushed limestone over a distance of 1 mile through a 6-in-diameter pipe as a function of the concentration. (SOGREAH Co.)

It can be seen that the power requirement reaches a minimum for a concentration of 51 or 52 percent and varies considerably with the concentration. For example, the power requirement doubles when the concentration increases from 52 to 58 percent. This example confirms that it is always advisable to study closely the economic feasibility of a hydraulic transportation project before proceeding to its detailed technical design.

The economic conditions are also affected by the size distribution of the material to be transported. In industrial applications, where the product to be conveyed is obtained by crushing and where the granulometric curve of the material is of only minor importance for the end use of the product, the crushing process should be designed to give a size distribution that leads to an optimal concentration of the slurry together with an acceptable yield value.

In other cases, where the size distribution is fixed, it is sometimes worth while to add certain dispersing agents, such as quebracho or hexametaphosphate, in order to reduce the yield value of the slurry and, hence, the power required for transport.

Heterogeneous Mixtures Mixtures of solid particles of more than, say 270 mesh form a system in which the carrying liquid retains its own individuality and viscosity. In other words, the liquid and the solid particles behave independently.

The particles move with the flow by means of two different processes: *in suspension* if the particles are small and the flow velocity is high; *by saltation* (that is, moving along in a series of short intermittent jumps) if the particles are large or the flow velocity is relatively low.

Materials forming heterogeneous mixtures can be divided into the following three categories (Fig. 4).

First Category. Fine materials settling out in accordance with Stokes' law ($W = kd^2$), where W is the settling velocity, d the mean particle size, and k a constant. For fine sands, which have a specific gravity of about 2.65, the upper size limit of this category is about 200 microns (65 mesh). In practice, such materials are always transported in suspension.

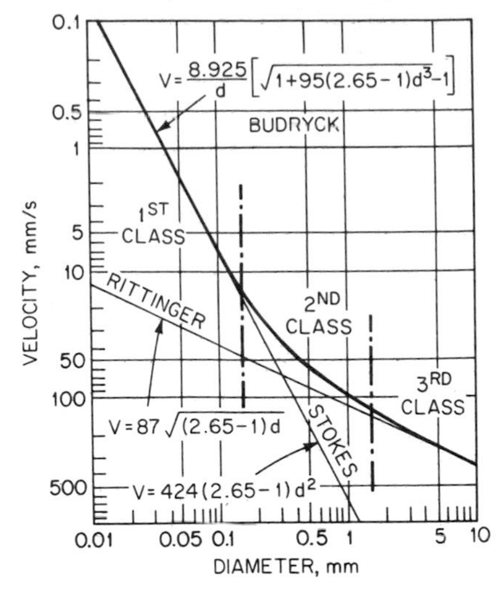

Fig. 4 Settling velocity as a function of nominal diameter for quartz grains (Richard's tests) of 2.65 specific gravity. (Budryck)

Second Category. This is the transition category, in which the solids settle out in accordance with transition laws that are between the Stokes and Rittinger laws. The category includes particle sizes of from 65 to 10 mesh at a specific gravity of 2.65, which includes the majority of medium or coarse natural sands and mine tailings. Such materials are transported either in suspension or by saltation, depending on the flow velocity.

Third Category. These materials settle out in accordance with Rittinger's law $W = k(d)^{1/2}$. At a specific gravity of 2.65, their particle size is anything over 10 mesh. They are always transported by saltation.

The two transport processes—suspension and saltation—are superimposed on the hydraulic regime in the pipeline, which may be of two types:

1. A *nondeposit regime,* which is a forced-flow condition. None of the particles can become stationary on the pipe, and they all move along with the flow, with a certain lag that depends on their size.

2. A *deposit regime,* which corresponds to an equilibrium condition between the liquid flow and the solids. The mean velocity in the pipe is low enough for the particles to settle out. The solids form a uniform layer on the bottom of the pipe. This builds up until the mean flow velocity in the unobstructed section reaches a limiting value, called the *deposit velocity.* The materials themselves create the equilibrium conditions required for their transport, the effective pipe size being automatically regulated by alteration of the flow cross section.

This latter transport condition may be unstable if the design of the pumping station is not properly suited to the particular application, with the possibility of a total blockage of the pipe.

It is very important to know the critical deposit velocity that determines the transition from one regime to the other, because this critical velocity is a determining factor in the economic operating conditions that, in turn, determine the correct pipe diameter and the characteristics of the pumping station(s).

Settling Velocity and Drag Coefficient. These are two essential parameters for the transport of heterogeneous mixtures.

When a particle falls through a liquid at rest, its maximum settling velocity is reached when the apparent weight of the particle equals the resistance opposing it, called "drag." The equation for the balanced force condition is

$$v(\rho' - \rho)g = C_x \rho S \frac{W^2}{2} \tag{7}$$

where v = particle volume, ρ = specific gravity of the water or fluid, ρ' = specific gravity of the material, g = acceleration due to gravity, S = projected area of the particle on a plane perpendicular to the movement, W = settling velocity, and C_x = drag coefficient, which is a nondimensional parameter depending on the Reynolds number associated with the particle movements.

If the Reynolds number is less than 1:

$$C_x = \frac{\text{Re}}{24} \tag{8}$$

in which Re = Wd/ν, where ν = kinematic viscosity.

This is Stokes' law, already mentioned, which covers the laminar conditions where the viscosity of the fluid is an important factor:

$$W = Kd^2\left(\frac{\rho' - \rho}{\rho}\right) \tag{9}$$

The materials placed in the first category follow this law.

If the Reynolds number exceeds 1,000, however, the drag coefficient C_x becomes constant and the field of turbulent drag is entered, where viscosity has no effect. Under these conditions:

$$\frac{W}{\sqrt{gd}} = \sqrt{\frac{4}{3}}\sqrt{\frac{\rho' - \rho}{\rho}}\frac{1}{\sqrt{C_x}} \tag{10}$$

The settling velocity becomes $W = K\sqrt{d}$.

Materials in the third category, i.e., with particles that are over 10 mesh (for a specific gravity of 2.65), follow this law:

For nonspherical and irregular particles that, in falling, present their largest cross section to the fluid, the above equation becomes

$$\frac{W}{\sqrt{gd_n}} = \sqrt{\frac{4}{3}}\sqrt{\frac{\rho' - \rho}{\rho}}\sqrt{\frac{\Psi}{C_x}} \tag{11}$$

where d_n = nominal particle size, or the diameter of a sphere of the same volume; Ψ = particle form coefficient, always less than 1, representing the cross section of a sphere of the same nominal diameter and the area of the largest cross section.

From extensive tests conducted in the laboratory, it was found that the true criterion characterizing materials being hydraulically transported was, in fact, the ratio C_x/Ψ. This leads to the definition of a new parameter C'_x:

$$\sqrt{C'_x} = \sqrt{\frac{gd_n}{W}}\sqrt{\frac{3}{4}\frac{\rho}{\rho' - \rho}\frac{C_x}{\Psi}} \tag{12}$$

$\sqrt{C'_x}$, called the apparent drag coefficient, is easily determined because it is a function of the readily measured parameters W and d_n. This nondimensional coefficient, in fact, characterizes a Froude number of the decantation.

Pressure Drop Gradient. For material above 65 mesh, the pressure drop gradient is given by:

$$\frac{\Delta H}{L} = \frac{\lambda}{D}\frac{V^2}{2g}(1 + \phi C_t) \tag{13}$$

For materials of specific gravity 2.65, the value of ϕ is found from

$$\phi = 180\left(\frac{V^2}{gD}\sqrt{C'_x}\right)^{-3/2} \tag{14}$$

In this case, a similitude criterion comes into play, in the form of the Froude number. Recent studies would seem to reveal the existence of an intermediate class of materials within a range of 270 and 65 mesh. For this class, a complex similitude criterion applies which includes not only the Froude number for the flow but also the Reynolds numbers for the flow and for the particles.

Economics and Limitations of Pipelining Solids. Tables 1 and 2 summarize the main characteristics of installations of solids pipelines for both homogeneous and heterogeneous slurries and for various materials.

TABLE 1 Typical Characteristics of Solids-Pipeline Installations for Homogeneous Mixtures

Material handled	4-in inside dia. of pipe	8-in inside dia. of pipe	12-in inside dia. of pipe
Crushed limestone uranium ore (sp gr 2.65)	40 tph*	170 tph*	400 tph*
Fine iron ore	60 tph*	250 tph*	550 tph*

* Average capacity, tons/hr
Length of line: Up to several hundred miles
Transport velocities: From 3 to 8 ft/s
Power requirement for horizontal transportation (capacity of 100 tph approx): From 0.08 to 0.13 kW/ton-mile/hr; in some especially difficult cases (pure clay, etc.) up to 1 kW/ton-mile/hr
Type of suitable pump unit:
For short distances: Diaphragm, reciprocating, or centrifugal pumps
For long distances: Reciprocating or centrifugal pumps in series

TABLE 2 Typical Characteristics of Solids-Pipeline Installations for Heterogeneous Mixtures

Material handled and average dimension of particles	4-in inside dia. of pipe	8-in inside dia. of pipe	12-in inside dia. of pipe
Coal (sp gr 1.6):			
d_n = 35 mesh	80 tph*	450 tph*	1,200 tph*
d_n = 9 mesh	70 tph*	400 tph*	1,000 tph*
Tailings (sp gr 2.65):			
d_n = 35 mesh	85 tph*	480 tph*	1,350 tph*
d_n = 9 mesh	75 tph*	420 tph*	1,150 tph*

* Average capacity, tons/hr
Length of line: Up to several miles, depending upon geographical conditions
Transport velocities: From 6 to 15 ft/s
Power requirement for horizontal transportation (capacity of 100 tph approx.):
From 1.9 to 2.5 kW/ton-mile/hr for materials with an average grain size of 35 mesh
From 4 to 7 kW/ton-mile/hr for materials with an average grain size of 9 mesh
Type of suitable pump units:
Materials with a maximum grain size of 6 mesh: centrifugal pumps (steel, special cast iron, or rubber lining)
Materials with grain size greater than 6 mesh: centrifugal pumps (Ni-Hard and special alloys); all types of clear-water pumps associated with a lock-chamber system

It should be noted that ϕ is valid for saltation conditions only and varies with the grain size of the solids transported. In particular, it increases with grain size up to a point where the drag coefficient, and thus ϕ, becomes constant. This occurs at a grain diameter in the region of 10 mesh. For materials of grain size over 10 mesh, the pressure drops will remain constant at a given concentration.

A set of pressure-drop curves is given in Fig. 5 for sand of a mean diameter of 32 mesh. These curves, which were plotted for regimes with and without deposi-

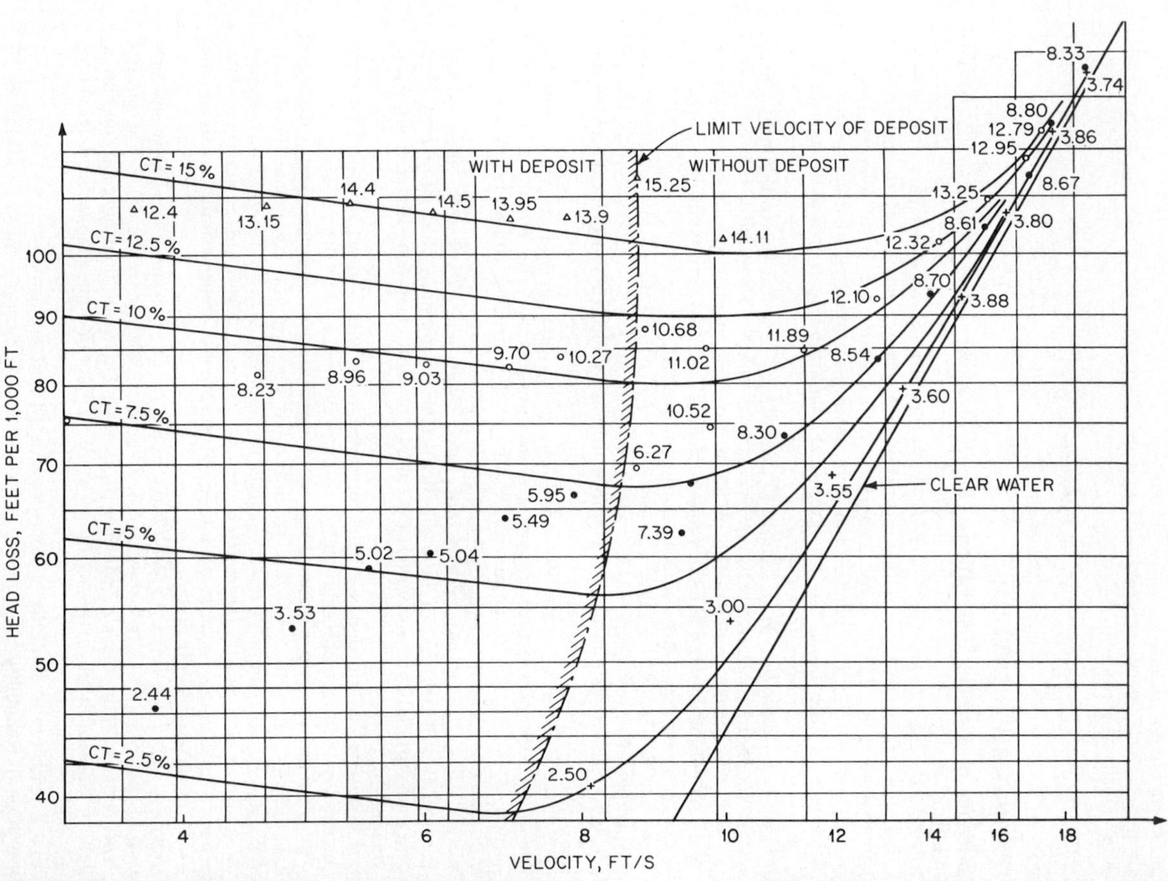

Fig. 5 Hydraulic transportation of a typical heterogeneous mixture; head loss vs. velocity curves (sand, mean size 35 mesh). (SOGREAH Co.)

tion for series of constant concentrations, show, in particular, the effect of the concentration on the pressure-drop values.

The pressure-drop concepts defined in Eqs. (13) and (14) should be complemented by that of the *critical deposit velocity,* which is the velocity below which the solids remain in permanent contact with the bottom of the pipe. In the case of an installation operating at a constant concentration, this velocity corresponds to the economic optimum.

For materials of specific gravity 2.65 and of grain size over 65 mesh, the critical settling velocity is given by

$$V_L = 1.46C_t^{0.15}\sqrt{gD} \tag{15}$$

OPERATING–SOLIDS PIPELINES

Concepts of Pipeline Installations Figure 6 shows various concepts in solids pipelines.

Gravity Designs (Types A). Gravity-plant operations costs are very low, often less than for any other type of transportation. A gravity plant may operate either under a constant head A.1 or at a constant rate of flow A.2, with delivery ranges depending on the head available and the physical properties of the material to be conveyed. Under favorable topographical conditions such installations can be made to deliver over long distances.

Designs with Solids-Handling Pumps (Types B). These are the commonest types of system for hydraulic transportation over any distance. Frequently they are integrated with the other circuits in the plant or mine, connecting one or more stations along a production line and satisfactorily replacing conventional handling equipment, particularly when circuits are elaborate and complicated.

When constant-speed centrifugal pumps are used, the rate of flow is controlled by a constant-level tank (type B.1).

In other types of installation with pumps driven by electric motor or diesel engine (Fig. 7), flow control is effected by varying the pump speed.

Long-distance lines, i.e. over several dozen miles, may feature several pumping stations, with one or more pumps to a station, which may be of either constant- or variable-speed type. Economic conditions need very close consideration for this type of line, especially if it is to convey coarse material or waste.

Designs with Clear Water Pump and Lock Chambers (Types C). For special applications, solids can be fed into the line after the pump. The pump may then be of a conventional type for water, e.g., single or multistage centrifugal, reciprocating or membrane type.

If the material is fine and is to be conveyed over only a short distance, ejectors can be employed to feed the solids into the line (types C.1 and C.2).

If the material has larger dimensions, or is very abrasive, and if transport distance is long or vertical, it is advisable to introduce the material by means of a lock system. This system may be of the single or multiple type, depending on whether solids transport rate can be intermittent or not (types C.3 and C.4). Such a system is in use in coal mines to hoist coal (Fig. 8) and in the nuclear industry to handle radioactive containers (Fig. 9).

Dredging and Back-Filling Installations (Types D, E, and F). These involve suction of the solids from rivers, sea, dam reservoirs, settling basins, etc. The pumping station is usually mobile, whether it is installed on a floating pontoon, on a dredge, or on land-bound rigs, as on a truck or on rails, on an overhead traveling crane, etc.

Type F permits the dredging of deposits located in deep water.

Operational Reliability A number of operating conditions must be fulfilled to assure the correct functioning of solids pipelines. For example, two major faults can initiate transport interruptions: mechanical blockage of the pipe and hydraulic instability of the flow regime.

Mechanical Blockage of the Pipeline. To have absolute safety of operation, it is necessary that the pipe diameter be equal to, or greater than, three times the diameter of the largest solid particles. If the percentage of large particles is small,

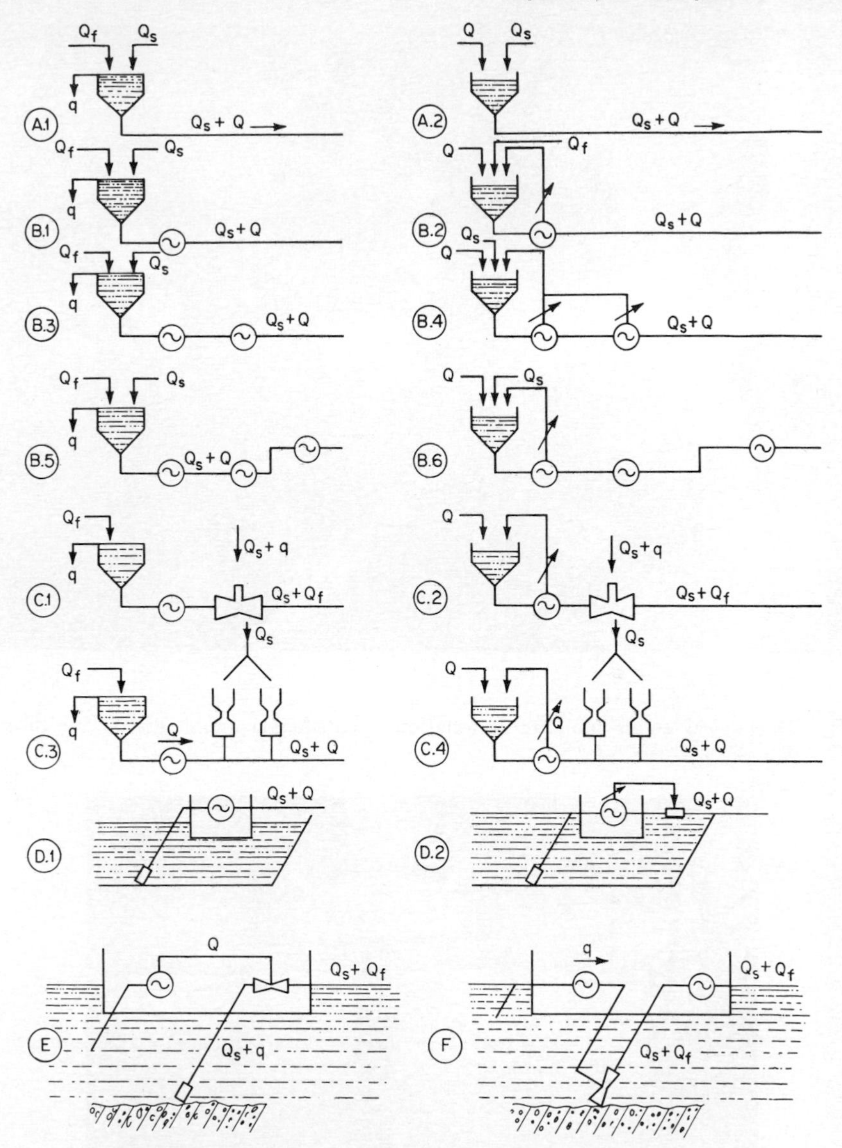

Fig. 6 Different concepts of solids-pipeline installations.

a minimum pipe diameter of twice the diameter of these particles might be sufficient. On the other hand, if the materials are moved by a procedure that requires their passing through a pump, the maximum allowable particle diameter would be one-half the diameter of the smallest canal part in the pump.

In certain applications, a suitably chosen intake screen must be provided to eliminate the risk of abnormally large particles blocking up either the pipeline or the pump.

Hydraulic Stability of Operation. Stability limits of a system are given by a comparison of the characteristic head-discharge curves of the pumping station and the discharge-head loss curves of the pipeline.

For any operating point, the system is stable if, for decreasing discharges, the characteristic curve of the pumping station is such that the pressure furnished is

Fig. 7 Typical solids-pumping installation; centrifugal pumps driven by diesel engines. (SOGREAH Co.)

Fig. 8 Example of lock-feed system for handling solids. (SOGREAH Co.)

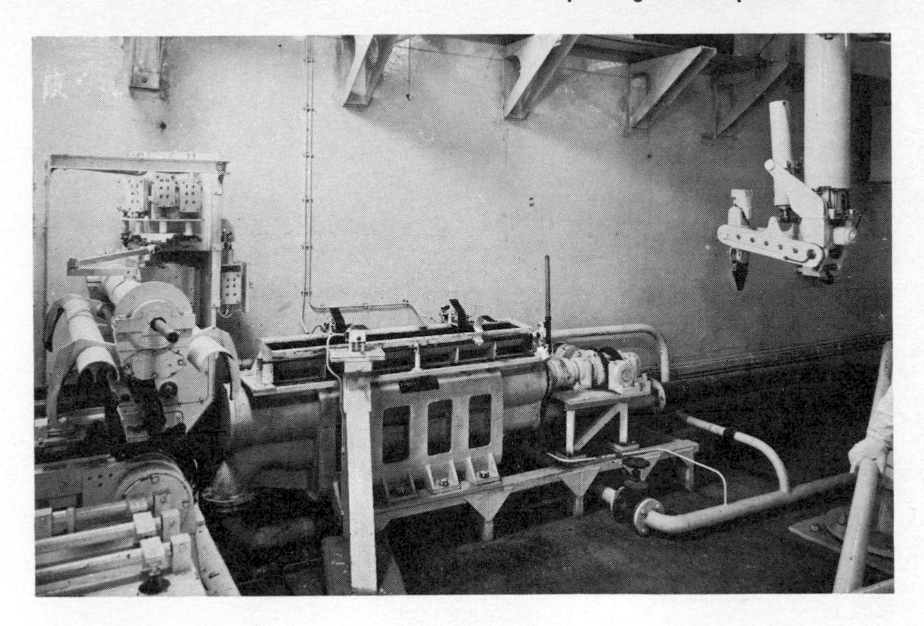

Fig. 9 Example of lock-feed system for container pipeline. (La Photothèque and SOGREAH Co.)

always greater than that required for the pipeline. The larger the algebraic angle between the two characteristic curves, the greater will be the stability around this operating point. It will, therefore, be possible to handle safely accidental variations of operating conditions—for example, an accidental increase in concentration.

Figure 10 shows the two extreme operating curves corresponding to solids discharges Q_{s1} and Q_{s2}, for a typical horizontal pipeline transporting coarse particles with a solids discharge which fluctuates during operation. The cross-hatched area between the two curves corresponds to the possible range of operating conditions for the pipeline. The shaded line V_L indicates the appearance of the deposit regime. If the operating point is to be to the left of this curve, transport occurs with a deposit in the pipeline; if it is to the right, the transport regime has no deposit. The curve J_e shows the pipeline head losses for clear water.

On the same graph the operating curves for three different solids-transport cases have been superimposed:

Curve I shows the head-discharge curve of a gravity installation delivering from a constant-head reservoir.

Curve II shows the characteristics of a pumping station with centrifugal pumps driven by constant-speed electric motors.

Curve III shows the characteristics of a pumping station equipped with either piston pumps or centrifugal pumps, driven at variable speeds to give a constant discharge.

Curve I crosses J_e and Q_{s1} at the points G_e and G_1 but does not cross curve Q_{s2}. For the minimum solids discharge Q_{s1} the transport will be stable. However, if the solids discharge approaches the maximum, Q_{s2}, the pipeline, which originally was functioning without deposit, would begin to deposit material and become blocked along its entire length. As a result of the shape of the operating curves for the pipeline, this type of gravity installation cannot work with a deposit regime.

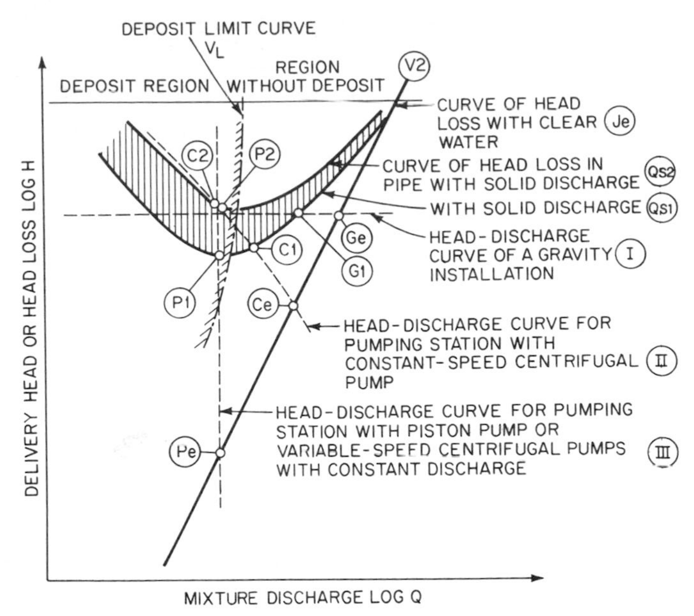

Fig. 10 Stability conditions for various types of solids-pipelining installations. (SOGREAH Co.)

Fig. 11 Frictionless automatic pressure-relief valve protecting ash-disposal pipeline against water-hammer damage; valve at left. (NEYRPIC, INC.)

In the case of an installation using constant-speed centrifugal pumps, curve II crosses curves J_e and Q_{s1} at points C_e and C_1; it crosses curve Q_{s2} at point C_2 as well. The operating points C_e and C_1 are very stable, since the algebraic angles between curves Q_{s1} or J_e and II are large. Operating point C_2 is barely stable, since the algebraic angle between the curves II and Q_{s2} is small; therefore, the least increase in solids discharge over Q_{s2} would initiate the total obstruction of the pipeline. Nevertheless, this pumping station allows operation either in regime without deposit (point C_1) or with a deposit in the pipeline (point C_2).

Finally, in the case of curve III, the curve cleanly cuts each of the curves J_e, Q_{s1}, and Q_{s2}; consequently safety of operation is complete. Depending on the delivery discharge chosen, the pipeline can operate either with or without a deposit.

Water Hammer A plant featuring a single pump delivering to a horizontal or near-horizontal pipe is fairly easy to start up and shut down. Both operations should, however, be carried out with clear or almost clear water in the system and by increasing or decreasing the pump speed very steadily to avoid pressure surges. A plant with pumps in series is trickier to start up, and this should be done in a predetermined order.

A plant with vertical or steeply inclined pipes cannot be stopped while there is still material in the pipe, as this is likely to lead to blockage. Suitable emptying arrangements have to be provided for use in a sudden shutdown.

A careful evaluation of the water-hammer risks in the installation must be prepared by a competent hydraulic engineer. If necessary, the installations should be protected with anti-water-hammer devices, which must function without fail even in the presence of the solids materials. Figure 11 shows an ash-disposal installation protected with *frictionless pressure relief valves,* available commercially.

REFERENCES

1. Durand, R., and E. Condolios: Données techniques sur le refoulement hydraulique des matériaux solides en conduite, *Rev. ind. Minér.*, June 1966.
2. Condolios, E., and E. E. Chapus: Transporting Solid Materials in Pipelines, *Chem. Eng.*, June 24, 1963; July 8, 1963; and July 22, 1963.
3. Condolios, E., E. E. Chapus, and J. A. Constans: New Trends in Solid Pipelines, *Chem. Eng.*, May 8, 1967.

Section **10.21.2**

Construction of Solids-Handling Pumps

G. WILSON

ABRASION WEAR

A number of conflicting requirements have to be considered in selecting materials for pumps to resist abrasive wear. The material may have to withstand not only abrasion but also high, moderate, or low impact, fatigue stresses, shock loads, and corrosion. The predominant factor that will cause wear must be understood and recognized in order that the most suitable materials are selected for the construction of the pump.

Abrasive wear in pumps is generalized into three types:

Gouging abrasion, which occurs when coarse particles impinge with such force that high-impact stresses are imposed, resulting in the tearing of sizable pieces from the wearing surfaces.

Grinding abrasion, which results from the crushing action on the particles between two rubbing surfaces.

Erosion abrasion, which occurs from the impingement of free-moving particles (sometimes parallel to the surface) at high or low velocities on the wearing surface.

ABRASIVENESS OF SOLID LIQUID MIXTURES

The rate of wear is related to the mixture being pumped and the materials of construction.

Wear increases with increasing particle size.

Wear increases rapidly when the particle hardness exceeds that of the metal surface being abraded; there is little to be gained in increasing the hardness of the metal unless it can be made to exceed that of the particles. The effective abrasion resistance of any metal will depend on its position on the Mohs or Knoop hardness scale. Figure 1 shows an approximate hardness-value relationship of various common ore minerals and metals.

Angular shaped particles cause greater wear than smooth rounded particles. Wear increases with increasing particle concentration.

The particle velocity, its kinetic energy, and its angle of impact are prime considerations in the selection of the material. Metals with high elastic limit are required to resist direct impact; metals with high hardness are used when there is relatively low impact, i.e., when the flow is almost parallel to the surface.

A simple expression, frequently quoted, is that the rate of wear $\propto$ (particle velocity)^m, where m will vary between 2.2 and 3.

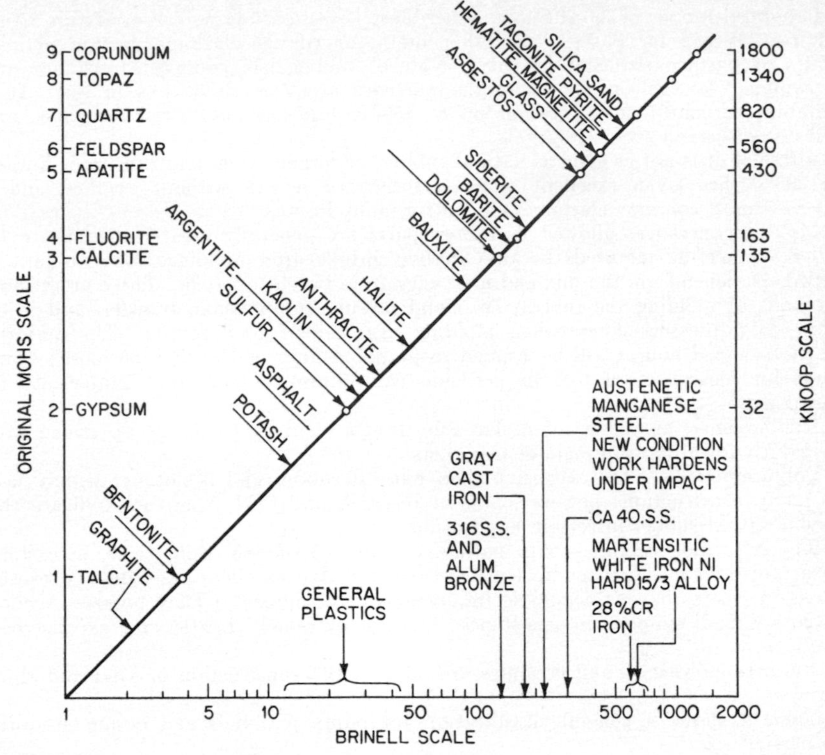

Fig. 1 Approximate comparison of hardness values of various common ore minerals and metals.

It is essential to keep the kinetic energy of the particles in the mixture as low as possible in order to give good life; this will be achieved by reducing the total head requirements of the system or by staging the pumps in series.

CENTRIFUGAL PUMP MATERIALS

All the wetted parts within the pump are subjected to varying degrees of wear. In order of decreasing severity these are impeller, suction side of casing adjacent to impeller, casing volute (particularly at the tongue and suction nozzle), and gland side of casing adjacent to impeller.

Austenetic manganese steel of 220 Bhn (12 to 14 percent manganese) is used for high-impact gouging abrasion. It has high toughness, it work-hardens under impact, it is free from high residual stresses in castings, and it can be machined by orthodox methods and welded. This material is used for the construction of dredge and heavy-duty gravel pumps.

Martensitic white irons such as Ni-Hard (4 percent nickel, 2 percent chrome) and 15/3 alloy (15 percent chrome, 3 percent molybdenum) are used for grinding and erosion abrasion where the particles are very coarse and the impact is moderately low; these irons are cast in the 550 to 650 Bhn range. In special cases 730 Bhn can be obtained by heat treatment. The harder the material, the more brittle it will be. For alkaline mixtures 28 percent chrome alloy iron of 550 Bhn may be preferred because of its superior corrosion resistance. All these hard irons have to be machine-ground and are used in the construction of sand and gravel pumps as well as slurry pumps.

If natural rubber of 38–44 Shore A hardness is compatible with the mixture being pumped, it will far outlast any other metal or rubber elastomer within definite limits of particle size and velocity. Natural rubber has poor cut resistance and is confined to applications where the particles are 7 mesh or less in size. It is suitable for continuous operation up to 150°F, but has poor swell resistance and will not withstand attack from oils.

Although it is not as good as natural rubber, neoprene, when temperature is limited to 200°F, has given excellent abrasion resistance in oils without swelling and is the next most common elastomer chosen for pump linings.

Molding tolerances allowed in rubber parts are generally greater than those for other engineering materials because rubber shrinks after molding. The amount of shrinkage depends on the mix and may vary from batch to batch. There are various methods of molding the rubber, i.e., hand lay-up, compression, transfer, and injection. High-pressure compression molding has yielded good results. The majority of rubber-lined pumps will be applied to pumping fines or less than 60 mesh; therefore side clearances of 0.06 in per side will minimize wear and compensate for shrinkage.

Rubber liners should be of replaceable design, clamped to the pump casing walls to reduce downtime and maintenance costs.

Polyurethane is more expensive than natural rubber and is not yet widely used in pump construction; however, present research and field experience indicate that it will outlast rubber in certain applications.

Silicon carbide ceramics, with hardness about 9.5 on the Mohs scale, in castable form are used in slurry pump construction and give excellent resistance to erosion abrasion where there is negligible impingement and impact. They possess excellent corrosion and temperature resistance, but are extremely brittle and expensive to produce.

Rubbers, urethanes, and ceramics are used in the construction of sand and slurry pumps.

Table 1 shows a general classification for pump materials and design according to particle size.

CENTRIFUGAL PUMP DESIGNS

Elimination of wear is impossible; however, the life of the parts can be appreciably prolonged and the cost of maintenance reduced by a sound design approach. This means:

Construct the pump with good abrasion-resistant materials.
Provide generous wear allowances on all parts subjected to abrasion.
Adopt hydraulic designs which minimize wear.
Adopt mechanical designs which are suitable for the materials of construction and allow ready access to the wearing parts for renewal.

The casing of a centrifugal pump has a profound effect on performance. If designed incorrectly, it can waste a large part of the energy given to the flow by the impeller, resulting in low efficiency, high hydraulic forces, and excessive wear on both the casing and the impeller.

In a conventional spiral volute, the static pressure around the impeller is uniform only when there is a free-vortex velocity distribution in the volute, i.e., when the spiral flow from the impeller coincides with that of the volute spiral. This occurs only at the best efficiency point (BEP). Departure from the free-vortex velocity distribution in the volute at partial capacity or overcapacities produces large circumferential pressure gradients at the impeller periphery, which are directly proportional to the specific gravity of the mixture being pumped. At partial capacity or overcapacities the flow will not tolerate the abrupt deflection at the tongue, so severe energy losses and wear occur at the vicinity of the casing throat and on the impeller.

Concentric and semiconcentric volute designs produce more uniform static pressures at the impeller periphery and make the pump much less sensitive to "off" BEP operation; consequently the hydraulic forces and wear are considerably re-

TABLE 1 Classification of Pumps According to Solid Particle Size

Tyler standard sieve series			
Aperture			
In	Mm	Mesh	Grade
3			
2			
1.5			
1.050	26.67		
0.883	22.43		
0.742	18.85		Scree shingle gravel
0.624	15.85		
0.525	13.33		
0.441	11.20		
0.371	9.423		
0.321	7.925	2.5	
0.263	6.68	3	
0.221	5.613	3.5	
0.185	4.699	4	
0.156	3.962	5	
0.131	3.327	6	
0.110	2.794	7	
0.093	2.362	8	Very coarse sand
0.078	1.981	9	
0.065	1.651	10	
0.055	1.397	12	
0.046	1.168	14	Coarse sand
0.039	0.991	16	
0.0328	0.833	20	
0.0276	0.701	24	
0.0232	0.589	28	Medium sand
0.0195	0.495	32	
0.0164	0.417	35	
0.0138	0.351	42	
0.0116	0.295	48	
0.0097	0.248	60	Fine sand
0.0082	0.204	65	
0.0069	0.175	80	
0.0058	0.147	100	
0.0049	0.124	115	
0.0041	0.104	150	
0.0035	0.089	170	Silt
0.0029	0.074	200	
0.0024	0.061	250	
0.0021	0.053	270	
0.0017	0.043	325	Pulverized
0.0015	0.038	400	
	0.025	*500	
	0.020	*625	
	0.010	*1250	
	0.005	*2500	
	0.001	*12500	
			Mud clay

* Theoretical value.

Rubber-lined pumps, closed impellers; particles must be round in shape.
Rubber-lined pumps, closed impeller
Rubber-lined pumps, open impeller
Hard iron pumps
Austenitic manganese steel pumps

General pump classification
Dredge pump
Sand and gravel pump
Sand pump
Slurry pump

duced. The elimination of the tongue, in casings of these designs, prevents abrupt deflection of flow, minimizing turbulence, and the velocity in the casing volute and throat is reduced.

For these reasons concentric and semiconcentric casing designs are adopted for solids-handling pumps; however, the degree of concentricity must be reconciled with efficiency, hydraulic radial load, and wear for each specific speed.

Up to 1,200N_s (specific speed) the castings can be fully concentric without too much sacrifice in efficiency.

Above 1,200N_s the casings can be semiconcentric; the angle of concentricity will progressively reduce as the specific speed increases. Semiconcentric designs can still be adopted in the 2,000N_s range.

Support feet are not provided on the casings since these are frequently replaced and realignment would be necessary. The absence of feet on the hard metal casing makes the casing easier to cast and eliminates a difficult grinding operation. The ability to rotate the casing to give alternative choices of discharge nozzle positions has advantages when pumping in series.

Although they are more expensive to produce, closed impeller designs are preferred to open impellers because the passages between the inlet and the outlet are usually relatively evenly worn. They are more robust and are not sensitive to falloff in performance when wear causes the clearance between the suction side shroud and the casing wall to increase.

The requirement for extra thick impeller vanes causes severe restrictions at the impeller eye, and the number of vanes has to be reconciled with the impeller geometry, NPSH, and wear. Three or four vanes are usual, depending on the specific speed.

Impeller nuts are not recommended since they are subject to wear; it is good practice to screw the impeller onto the shaft in opposite direction to the rotation of the pump. Heavy-duty thread forms are used to withstand the torsional stresses, shock, and wear and tear of maintenance. Aluminum, nodular iron, or steel can be used for the rubber impeller skeleton; however, steel is the best choice since it will not fracture under impact and can be repaired by welding.

Sealing Sealing of abrasive mixtures under pressure has long been a problem in centrifugal slurry pumps. Excessive gland leakage, costly maintenance, and even failure can occur if the seal is not effective. It is essential to keep abrasive particles in the mixture being pumped away from the seal, or at least to minimize the pressure to such an extent that excessive wear is prevented.

The rotational flow between a conventional impeller shroud (without sealing rings) and the casing wall is a forced vortex. In this case the angular velocity of the liquid in this vicinity will be approximately one-half the angular velocity of the impeller. Consequently high pressures are obtained at the seal, combined with extreme axial hydraulic unbalance, both of which are directly proportional to the specific gravity of the mixture being pumped. Steps are taken to minimize these effects by the addition of hydrodynamic pumping vanes on the back shrouds of the impeller. Properly designed, vanes can effectively increase the angular velocity of the liquid to 95 percent of the impeller angular velocity.

Pumping vanes on the impeller back shroud will therefore effectively reduce the pressure at the stuffing box and will greatly reduce the axial hydraulic unbalance.

Unless precautions are taken to prevent, or at least minimize, the ingress of solids into the stuffing box area, rapid wear will result and the gland will require constant attention. There are many methods proposed for the sealing of the gland, all of which may work under certain limitations.

In slurry pumps the shaft is always protected at the stuffing box by means of a sleeve. A hook-type design is preferred since this has the effect of reducing the stress values on the shaft and compensates for differential expansion when pumping high-temperature mixtures. The choice of sleeve material must be based on operating experience and can range from centrifugally spun cast iron to vacuum-hardened stainless steel, stellite faced and ceramic faced.

Mechanical Seals Face-type seals (even with tungsten carbide faces) that depend on an interface film for lubrication are not suitable for rough abrasive service. They require precision adjustment, they are sensitive to impact and shaft deflection, and they usually fail catastrophically. If failure occurs, the pump has to be stripped down in order to replace the seal. Split seals are complex and expensive.

Double seals with a clean barrier liquid have proved more successful, but again these are complex and expensive.

Packed Gland Seals A combination of hydrodynamic sealing vanes to minimize the pressure at the shaft and an orthodox packed-box arrangement using split rings of graphited asbestos or Teflon-filled asbestos are recommended for abrasive slurries.

If dilution of the product being pumped can be accepted, it is strongly recommended that a clean liquid service be provided under pressure at the bottom of the stuffing box. The abrasives will be prevented from entering the box and causing wear to the packing and the sleeve.

Within certain limitations of suction pressure, it will be possible to operate with a dry box. The sealing vanes will effectively prevent abrasives from reaching the box area, although a seal is still required to prevent leakage in the static condition.

If a dry box is adopted and the suction pressure is high, fines will soon penetrate into the box and become embedded in the soft packing; eventually the packing will become saturated with the fines, and leakage due to wear on the sleeve will occur.

The seal options should therefore be:

1. A packed box with clean fluid for flushing connected to the lantern ring located at the bottom of the box. This causes dilution of the product being pumped (unsuitable for cement slurries).
2. A packed box with clean fluid for flushing connected to the lantern ring located two rings from the bottom of the box. This causes less dilution, but greater wear will take place.
3. A packed box with a grease seal connected to the lantern ring located two rings from the bottom of the box. This dry-box arrangement causes no dilution of the product.

Wear Plates An examination of Figure 2 shows that impeller radial wearing rings have been eliminated from the design. This simplifies the whole of the liquid end construction and makes the pump amenable to external end clearance adjustment to compensate for wear. The worst wear occurs on the suction side, and all good slurry pump designs should incorporate an easily replaceable suction side wear plate or liner. A wear plate or liner is not always provided on the gland side adjacent to the impeller. However, this area is also subjected to wear, and despite the increased initial cost it is advisable to have one fitted.

Bearing Cartridge Assembly A calculated bearing life of over 50,000 hr is normal to meet the arduous conditions of service. This is appreciably higher than the specified life of conventional pump bearings. Referring to Fig. 2, it can be seen

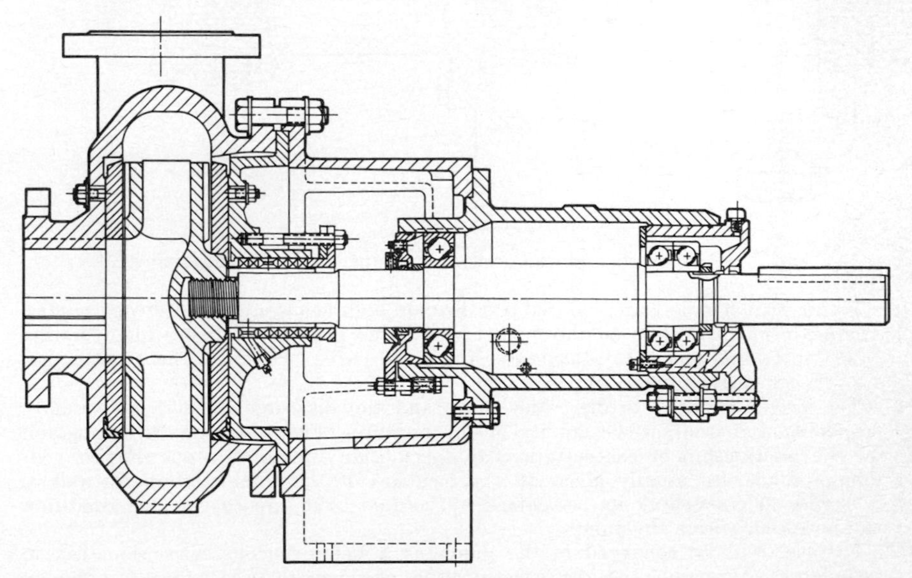

Fig. 2 Hard iron abrasive solids-handling pump for slurries, sand, and gravel.

that the bearing cartridge assembly is one complete element, which can be easily interchanged for another sealed element at short notice. Oil lubrication is preferred to grease because less attention is required. Grease lubrication should be provided as an option.

Slurry pumps are usually situated in a dust-laden atmosphere, and adequate precautions should be taken to prevent the ingress of liquid or dust into the housing. A labyrinth-type deflector is almost mandatory at the line-bearing cover to expel gland leakage from the bearing.

PUMP DRIVE

Hard metal or rubber impellers cannot be reduced in diameter to meet a given condition of service. For this reason, the pumps are usually belt-driven. The required pump performance can be achieved by selecting the desired sheave ratio that will give the pump speed required.

The motors are often fitted directly above the pump; this saves floor space and minimizes the risk of flooding. The largest motor frame that can be accommodated above the pump is determined by the bearing centers in the cartridge assembly and the shaft overhang at the drive end, as well as by the stability of the motor-supporting structure.

Figure 2 shows a typical hard-iron slurry, sand, and gravel pump.

Figure 3 shows a typical replaceable lined rubber pump for slurries and sand.

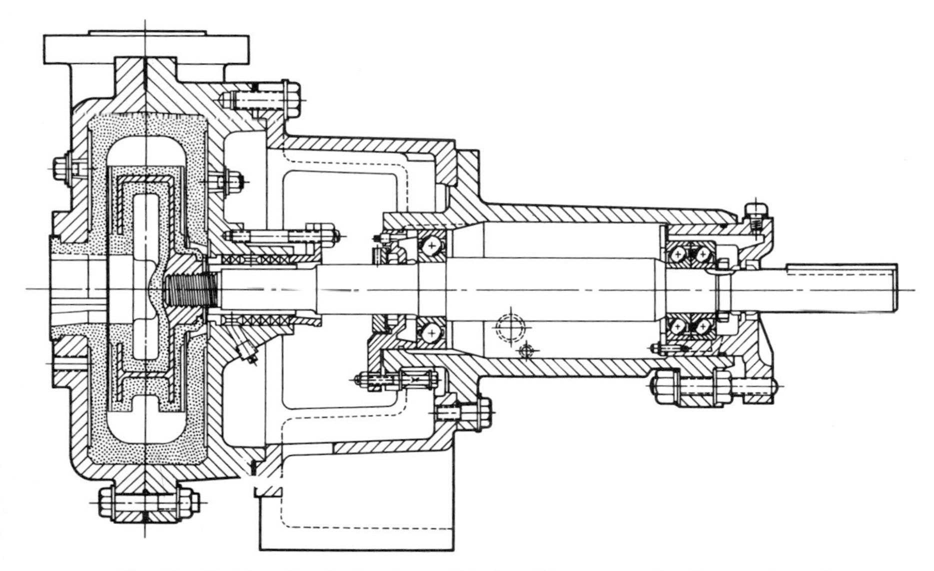

Fig. 3 Rubber-lined abrasive solids-handling pump for fines and sand.

Pumps that handle large material and develop high heads are called *dredge pumps*. There are many designs on the market with large impeller diameter for operating at low rotational speeds to prolong the life of the parts. Pumps of this construction are very expensive to produce.

The power required to drive the pump and the discharge pressure are directly proportional to the specific gravity of the mixture. Figure 4 shows a nomograph for the relationship of concentrations to specific gravities of aqueous slurries. Although solids are usually given as a percentage of the total mixture by weight, it is wise to cross-check its percentage by volume to ensure that a plug condition will not occur within the pump.

Particles will be conveyed in the pump as a heterogeneous suspension, i.e., in turbulence. Depending on the concentration and particle sizes, there can be an apparent viscosity effect, due to the failure of the solid particles to follow the liquid path. Solids cannot possess or transmit pressure energy, but only kinetic energy

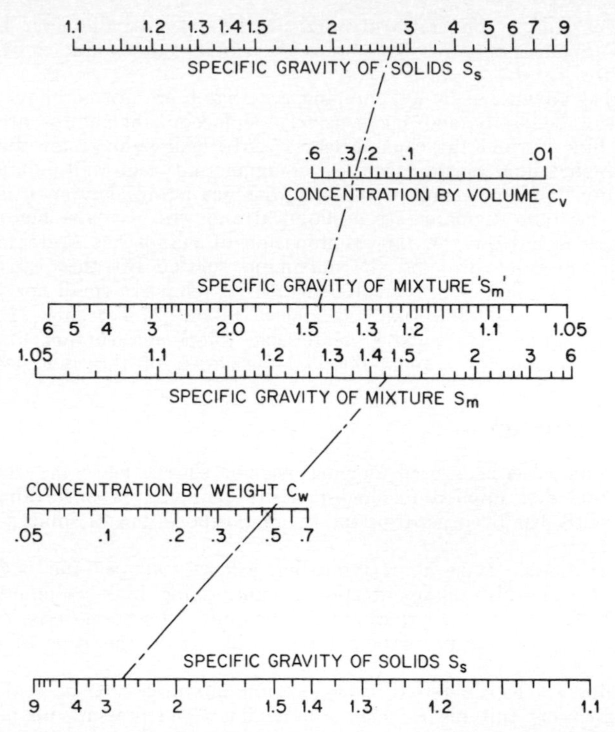

Fig. 4 Nomograph for the relationship of concentrations to specific gravities in aqueous slurries.

which is unrecoverable; therefore the total head and efficiency are bound to be lower. The predictions of these losses should be based on experience or calculated by the pump supplier. All performance curves for solids-handling pumps are for clear water only. A standard procedure often adopted is to rate the pump at 10 percent lower head than when pumping clear water and to select the motor for at least 30 percent greater than calculated bhp.

JET PUMP

A jet pump is a device which is operated by a pump producing a driving jet of clean liquid through a nozzle which will entrain vapor, liquid, or solid/liquid mixtures. A mixing tube is necessary to effect the momentum change between the driving jet and the entrained stream. The high kinetic energy of the mixed stream is converted into pressure energy through a diffuser. Figure 5 shows a diagrammatic

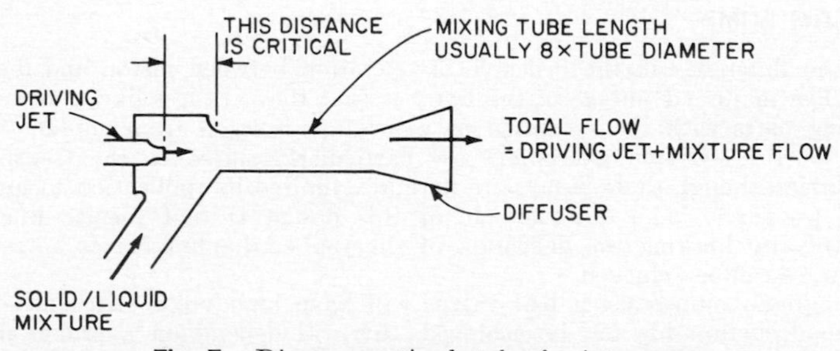

Fig. 5 Diagrammatic sketch of a jet pump.

sketch of a jet pump. The ratio of nozzle to mixing-tube diameter has a critical effect on performance; therefore jet pumps are usually tailor-made to suit the conditions of service.

The principal advantage in jet pumping are that there are no moving parts and worn parts can be easily and inexpensively replaced. Jet pumps are capable of high suction lifts, beyond the capabilities of centrifugal pumps; for this reason the device has applications in the deep-sea dredging and deep-well pumping field.

High-pressure, low-flow pumps combined with jet pumps having a large ratio of nozzle to mixing-tube diameter are economical and will give the highest efficiency (about 40 percent). However, this combination of pumps has a greater restriction on the maximum solid size and the maximum suction lift that can be handled.

Low-pressure, high-flow pumps combined with jet pumps of small nozzle to mixing-tube diameter ratio will give lower efficiencies (about 25 percent). However, this combination of pumps can handle greater-size solids and suction lifts. Refer to Section 4, Jet Pumps, for a more detailed discussion of this kind of pump.

RECIPROCATING PUMP

This type of pump has been used for pumping fine slurries since the early twentieth century. It was first applied for high-pressure pumping in oil-drilling operations and more recently has been used for the hydrotransportation of solids through long-distance pipelines.

There are two basic types of reciprocating pumps—*plunger* and *piston*—both of which have the desirable characteristic of maintaining high volumetric efficiency at any desired flow rate, which gives greater flexibility in system design. The degree of abrasiveness of the slurry to be pumped will decide the type of pump to be selected.

Piston pumps are considered for low-abrasion mixtures such as coal slurries and mud; they are more suitable for high-volume flows and pressure up to 2,000 lb/in^2 gage with solid sizes limited to 8 mesh or lower.

Plunger pumps are considered for high-abrasion fine mixtures such as magnatite, silica, and sand slurries. This type is usually considered for pressures above 2,000 lb/in^2 gage.

Valves are the most important element in reciprocating pumps, for no matter what configuration of pump is adopted, whether conventional, diaphragm, or surge leg, the valves cannot be isolated from the slurry. Consequently the pumps are limited in application because of valve wear and the ability of the valves to pass large solids. Pistons, rods, cylinders, and packing are other liquid end parts that are susceptible to severe wear.

The choice of reciprocating versus centrifugal pump will depend on the application and economics. Centrifugal pumps in abrasive services are limited to 100 lb/in^2 gage per unit, and the practical limit of the number of pumps that would be situated in series in one location would be about six, giving a total pressure of 600 lb/in^2 gage. Reciprocating pumps, however, can develop pressures in the region of 6,000 lb/in^2 gage.

DIAPHRAGM PUMP

There is no difference in the principle of operation between piston and diaphragm pumps. The major advantage of the latter is that they are less likely to wear since all working parts with the exception of valves are isolated from contact with the abrading slurries. Diaphragm pumps are particularly suited to the transportation of fine slurries, though these pumps are normally limited in application to low capacities and pressures. The disadvantage of this design is that service life will be assured only by limiting the deflection of the rubber diaphragm; as a result only low flow rates can be achieved.

All diaphragm pumps using ball valves will have high volumetric efficiency; although good suction lift can be achieved, this will depend on a proper fitting of the valves.

REFERENCES

Lipson, C.: "Wear Considerations in Design," Prentice-Hall, Inc., Englewood Cliffs, N.J., 1967.

Tsybaev, N.: "Wear-resistant Rubber Linings for Pumps Delivering Abrasive Mixtures." *Tsvetnye Metally,* no. 2, February, 1965.

Bitter, J. G. A.: A Study of Erosion Phenomena, Part I, *Wear,* vol. 6, no. 3, 1964.

Lindley, P. B.: "Engineering Design with Natural Rubber," Natural Rubber Producers Research Association.

Welte, A.: "Wear Phenomena in Centrifugal Dredging Pumps," *V.D.I. Berichte,* no. 75, 1964, pp. 111–127.

Agostinelli, A.: An Experimental Investigation of Radial Thrust in Centrifugal Pumps, ASME Paper 59.

Worster, R. C.: The Flow in Pump Volutes and Its Effect on Centrifugal Pump Performance. *Ind. Mech. Eng.,* 1963.

Wilicenus, G. F.: "Fluid Mechanics of Turbomachinery," Dover Publications, Inc., New York, 1965.

Stepanoff, A. J.: "Pumps and Blowers; Two-Phase Flow," John Wiley & Sons, Inc., New York, 1965.

Aude, T. C., et al.: Slurry Piping Systems: Trends, Design Methods, Guidelines. *Chem. Eng.,* June 28, 1971.

Section **11**

Intakes and Suction Piping

WILSON L. DORNAUS

IMPORTANCE OF SUCTION INLETS

The most critical part of a system involving pumps is the suction inlet, whether in the form of piping or open pit. A centrifugal pump that lacks proper pressure or flow patterns at its inlet will not respond properly or perform to its maximum capability. Uniformity of flow and flow control up to the point of impeller contact is most important. Part of this may be controlled by proper pump design, but the pit designer and suction-piping designer have definite responsibilities to satisfy good pump operation. In open-suction pit (wet-well) design the flow must be as uniform as possible right up to contact with the pump suction bell or suction pipe, preferably without a change in direction or velocity.

Examples of dry pit and wet pit centrifugal pumps connected to open-suction pits are shown in Fig. 1.

In dry-pit pumping, the suction pipe leading to the pump flange should not include elbows close to the pump in any plane, or other fittings which change flow direction and velocity and which may impart a spinning effect to the flow. Centrifugal pumps not designed for prerotation, either dry pit or wet pit, will suffer loss of efficiency and an increase in noise. Rotation with the impeller can result in a decrease in pump head. Rotation against the impeller can result in an increase in pump head and power, and possible driver overload. If the total system is to operate efficiently at minimum cost of maintenance, close attention must be paid to the suction environment of the pumps.

INTAKES

Purpose Water-circulating systems must have either a continuously renewable source, such as an ocean, lake or river, or they must recirculate the same water as from cooling ponds or cooling towers. Whichever type of pump is selected (wet pit or dry pit) the suction water itself will come from an open pit of some sort, or from a pressure pipeline.

Types *Once-Through: Ocean-, Lake-, or River-Source Wet Pit.* This type calls for an intake structure, usually of concrete, to gather the water into a localized spot for pickup and to support the pumps. The optimum design will bring relatively clear water directly into the pump suction area at a low velocity. To accomplish this the design procedure is as follows:

1. Establish minimum water depth required at pump at maximum flow rate (usually given to bottom of suction bell).
2. Add distance to bottom of pit from bell.

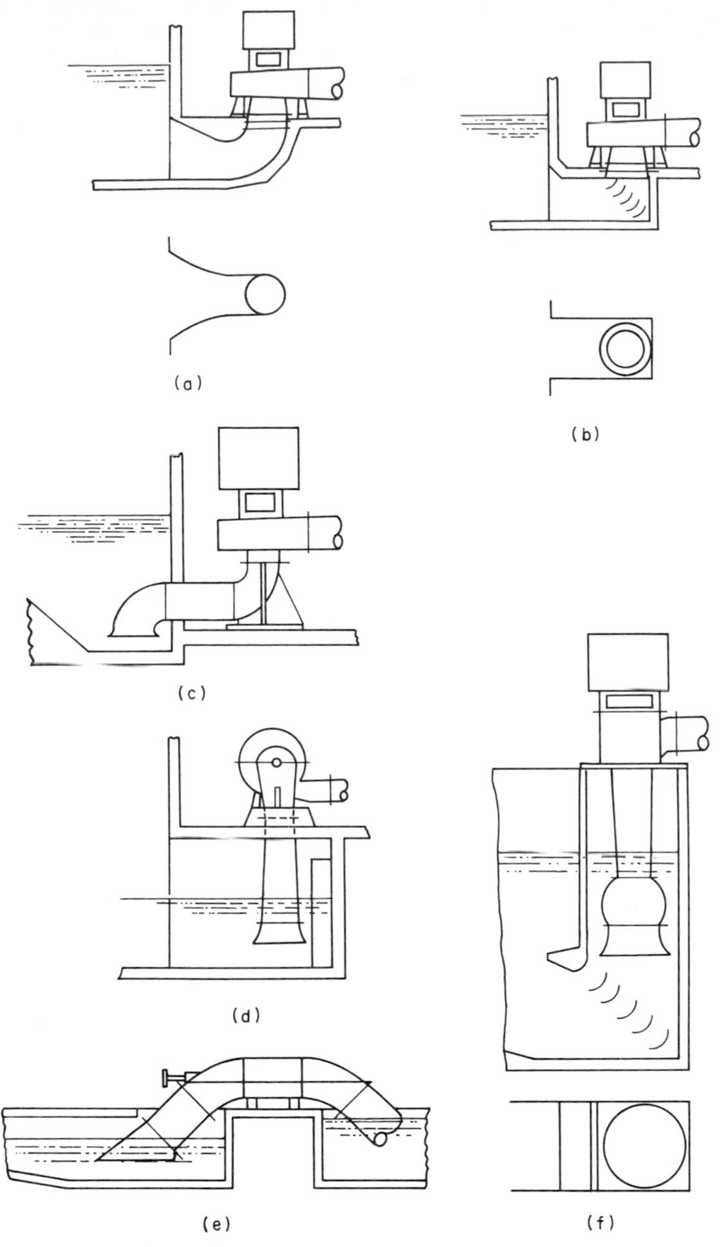

Fig. 1 (*a*) Vertical dry-pit pump suction to minimize excavation: reducing elbow. (*b*) Vertical dry-pit pump suction to minimize pit width: turning vanes. (*c*) Vertical dry-pit pump suction for solids handling: commercial piping with flared suction elbow turned down for maximum submergence; floor sloped to prevent solids deposition. (*d*) Horizontal dry-pit pump suction pipe with suction-bell inlet submerged like wet-pit pump. (*e*) Horizontal dry-pit pump suction specially shaped to provide large volumes of water at very low heads with minimum submergence and excavation. (*f*) Vertical wet-pit pump suction through turning vanes to allow higher velocity and narrower pit.

3. Calculate width of pump basin required per pump to give average velocity of approach equal to 1 ft/s.

If the maximum flow rate and/or minimum water level occurs less than 10 percent of the total operating time, the velocity based on these limits can be increased to 1½ ft/s.

4. Recess the pump location a minimum of two pump-suction-bell diameters from traveling screens or bar racks.
5. Space bar racks (if used) one pit width in front of traveling screens.
6. Project wing walls (walls between pumps if required structurally) into source area far enough to provide parallel uniform flow at pit velocity for a distance of one pit width prior to the trash racks.

The above design assumes the source area to be a lake, or a river with maximum velocity of 2 ft/s. For higher river velocities use correspondingly greater distances from source wing-wall contact to trash rack or screen. (The choice of providing trash rack or screens, or both, is based on the type and amount of debris likely to be encountered at the inlet.)

An ocean inlet has additional requirements because of tidal action and variable-direction currents which may exist. For this reason it may be necessary to create a forebay inlet basin independent of the pump pit, and fed by a submerged inlet tube extending out into the ocean for some distance (Fig. 2). This tube may utilize normal pipe velocities but the inlet must turn upward, and the opening should be protected by a horizontal cap of such size and shape that the inlet water travels horizontally at a velocity high enough to scare fish away. The discharge into the forebay can be pipe velocity, since the turbulence will be dissipated by low (2 ft/s) outlet velocities through trash racks and traveling screens. From here on to the pump chamber the velocity should be low and constant.

Any necessary change in inlet-channel dimensions should be made gradually. Tapering walls should not diverge at more than 14° including angle. If it is necessary to slope the floor, a maximum of 7° is recommended and, if possible, it should level off before reaching the pump area, as far back as possible. No sharp drops (waterfall effect) should be permitted.

If the inlet channel is a closed pipe with full-wetted perimeter, pipe velocities can be used up to within such a distance of the pump chamber that the tapered-wall rule can be applied to the increase in pipe size as it discharges into the pump chamber.

Unless the inlet velocity into the pump chamber is kept below 2 ft/s, extensive baffling and additional space will be necessary which can only be determined by model pit testing.

Recirculating System—Cooling-Tower or Pond Wet Pit. For cooling-tower systems the pump pit is normally adjacent to the tower basin. This probably means a shallow-water source. To get proper submergence a deeper pit will be called for unless a more conservative pump is selected (lower speed than normal). If a sharp shoulder drop into the pump area cannot be avoided, a long approach channel or a forebay is recommended. The pump pit depth may be dictated by a preference for a certain total screen width and velocity.

Sometimes the topography is such that the cooling towers are on a hill, so that a substantial drop in elevation provides pressure available nearer the flow area. If dry pit pumps are used this pressure can be utilized to reduce pump requirements. When this is done, the pump suction becomes a pipe, and the latter part of this section should be consulted.

The intake structure in a cooling pond should be located as far as possible from the inlet pipe to the pond to generate the maximum cooling effect. If spray equipment is used it should be so arranged in relation to the intake building that a minimum surface disturbance is encountered. Prevailing winds should be such that the building is on the lee side of the pond, and if side and bottom areas could be easily disturbed to include silt in the flow, riprap should be applied to approach slopes as well as bottom mats well beyond inlet wing walls.

The handling of silt is usually not desirable in a pumping system. A high velocity through the pump will accelerate wear. At low velocities at other points in the system silt will settle out and produce higher velocities and wear factors as a result of area blockage.

If space is available a silt-settling basin can be constructed ahead of the inlet basin. Cross baffles should be provided to impede and slow the inlet flow to a velocity of less than 1 ft/s. Most stream debris will settle out at this velocity

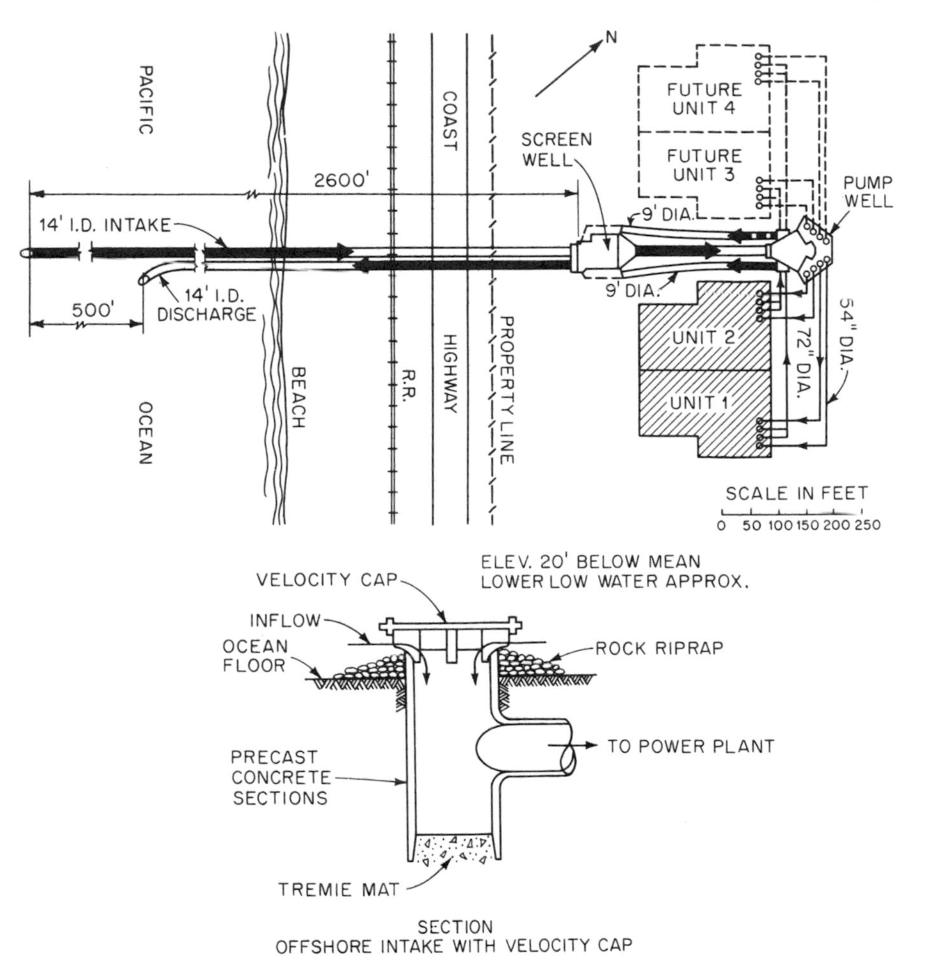

Fig. 2 Ocean intake with velocity cap to minimize fish pickup.

and the flow into the pump suction pit will be relatively clean, preventing the deposit of additional silt around the pump suction bell.

If space is not available for silt beds in large-capacity installations, the main channel can be furnished with a weir across the flow path. The height of the weir should be selected to give an overflow depth above the weir of not over one-third to one-fourth the water depth just preceding the weir. The velocity over the crest should not exceed the intake channel velocity. Weirs of this type are particularly effective when the intake channel is at right angles to the supply mainstream.

Another scheme to be considered when silt is a problem is a manifold of round suction-pipe screens. If the area is sufficient to reduce the inlet velocity to below 0.5 ft/s, the screens will not plug up too quickly.

In areas where river levels vary considerably throughout the year problems arise not only from silt and debris accumulation but from the structures required to prevent damage to motors and electric switch gear. If the required flow is moderate a system known as the Ranney well (Fig. 3) can be constructed. A Ranney well is a concrete silo 13 ft in diameter, which becomes the collecting basin and pump well. It may be situated in or near a river, and can be partly below and partly above water level with settings up to 100 ft. Small perforated pipes radiate hori-

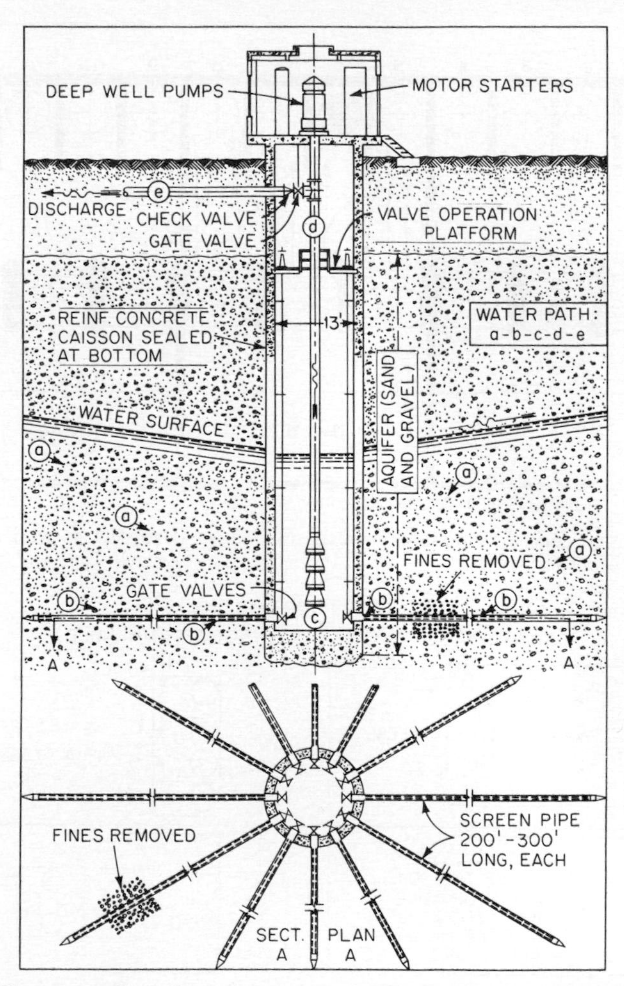

Fig. 3 A Ranney well collects into a pool from underground strata. (Ranney Method Division of Pentron Industries)

zontally from the base of the well and tap into porous strata, bringing small flows of water into the main well from which it can be pumped with a deep-well type wet pit vertical pump.

Multiple Pumps. Some pump requirements are easily met with 100 percent capacity single pumps, but more reliability usually requires two 50 or three 33 percent pumps, or if the service is critical enough, three 50 or four 33 percent pumps, etc. There are practical limits of size for various pump types, so that large flow demands will undoubtedly call for a multiple pump arrangement.

The problem arising from multiple pumps arranged in a common pit is from the probable nonuse of some of the pumps while others are operating. This causes variables in flow patterns which may lead to eddying and vortexing (Fig. 4). Installation of separating walls in the common pit may introduce additional problems, since the ends of the separating walls create eddy corners for the dead-end pocket at the unused pump. A back vent in the dividing wall will relieve this situation providing it vents at the water surface (see Fig. 5, lower left). If walls are extended past the screens and trash racks to a forebay, this problem will not occur, but the design has then become single-pump-basin design.

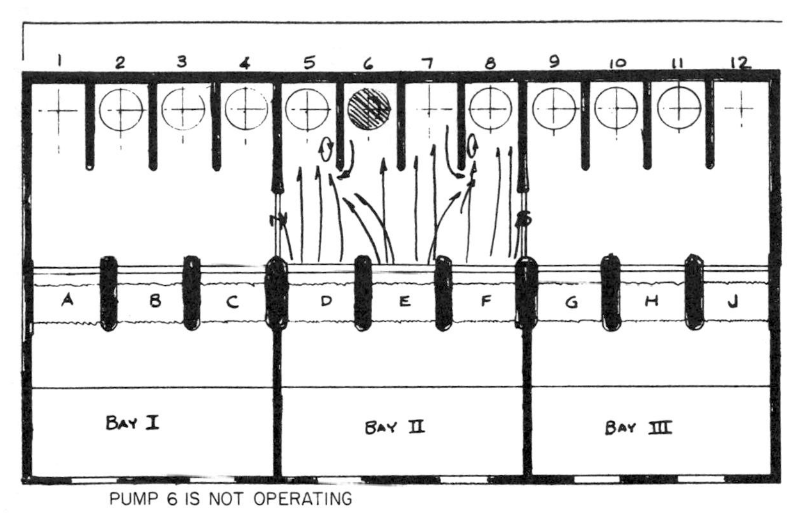

Fig. 4 Empty spaces or nonoperating pumps cause vortexing at wall ends because of reversal of flow.

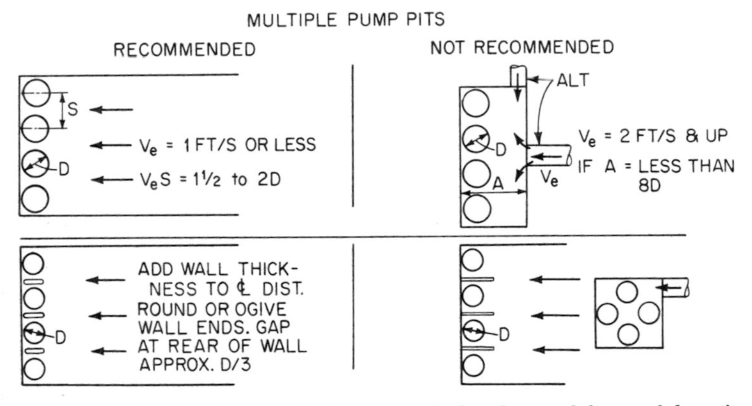

Fig. 5 Using basics of good pit design precludes the need for model testing.

The same velocity rules apply to multiple arrangements as in single-pump-basin design. Odd arrangements should be avoided even when they look invitingly symmetric—fan-shaped, round, radial, peripheral; all have directional problems that are not easily overcome.

A basic pit design consisting of a number of equally sized pumps in a common pit with flow entering parallel and straight-in at 1 ft/s or less would not need to be model-tested to assure reliability (Fig. 5). If separating walls are required for structural support and they are properly shaped and vented, no model will be required. The pumps should be located to the extreme rear of the pit so that the whole approach assumes the characteristics of a suction pipe. Individual pump manufacturers may vary the location of the pump as to the relation to the pit bottom, velocity of inlet spacing, etc. It has been found that some of these variations require additional splitters or baffles below the pump, up the back wall behind the pump, or centered in the flow ahead of the pump. If so, a model test should be run, and the additional pit cost weighed against other alternatives (changing pit shape, size, pump location, pump size and speed).

A dry-pit pump installation will either have the pumps in a dry well at or below wet-well water level (Fig. 6*a* and *b*), or located directly above the wet well using a suction lift (Fig. 1*d*), which calls for priming equipment. The additional cost of this equipment (vacuum pumps, etc.) may be partially offset by the additional space and valve requirement for the first option. In either case, the suction piping in the wet well should be treated in the same fashion as the pump suction bells in wet-pit installations as far as spacing, direction, and velocity of flow are concerned.

When the pump is installed at an elevation that may be below suction pit water level, a valve must be installed at the suction of the pump. The temptation to reduce the inlet size to get a smaller valve should be avoided. Pipe size at a pump inlet may decrease into the pump down to the pump suction size, but it should not be reduced below that size so that it must increase again as it enters the pump. The pipe in the wet well should preferably have a bell end and project downward. The minimum water level above the top edge of the pipe or the lip of the bell should be at least 5 ft for a recommended entrance velocity of 5 ft/s.

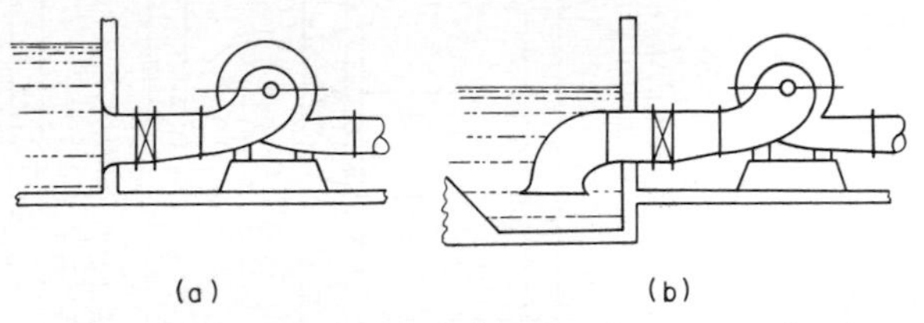

Fig. 6 Horizontal dry-pit pumps set below suction water level are self-priming.

The bell mouth should project downward to assure uniform inlet flow and to attain maximum submergence.

A wet-pit intake style that closely approximates a suction pipe arrangement and uses what is essentially a dry-pit pump has been expanded by the U.S. Department of the Interior, Bureau of Reclamation, to an elbow-type suction-tube design. This incorporates a formed-concrete suction inlet with a swing of 135° in the vertical plane and a gradual decrease in area to the suction eye of the pump (Fig. 7). The resulting design saves considerable excavation in the wet-well area, reduces losses into the pump, and allows a smaller higher-speed pump to be used. With higher velocities debris dropout is reduced so that silt buildup will not occur as readily and a smaller trash-rack area will remain effective longer. These inlets are independently self-sufficient and may be grouped into multiples as long as the forebay is designed to adhere to the basic design rules for wet-pit approach channels.

Screens and Trash Racks While economically it may not be practical to eliminate all refuse from a pumping system, it probably will be necessary to limit the size and amount of debris or sediment carried into the system. Depending on the probable source of debris, such as a river subject to flooding with considerable flotsam in the runoff and with a very loose bottom, or a lake at constant level without disturbing inlet flows near the structure and with a solid bottom, the protection needed may include only a bar trashrack to avoid large floating objects, or a raked trashrack plus rotating flushed fine-mesh screens, and if sediment deposit is likely, a settling basin may be required.

The designer must note that if this equipment is useful it will pick up debris and gradually increase the velocity through the openings as the net area decreases with blockage. When this occurs at the trash rack, the water-level differential will build up causing a waterfall with increased velocity and turbulence on the pump side of the rack. In addition the increase in velocity may pull more debris through the bars than can be tolerated. It is best to rake these racks frequently enough to keep the differential head across the rack below 6 in. The spacing of the bars should be such that objects which cannot be pumped will be excluded from passing through. This, in general, will call for the bar spacing to be in proportion to the size of the pump. A pump manufacturer can determine the maximum size sphere his pump will handle, and the bar spacing should be limited to 50 percent of that value. The size of the bar, the distance between supports laterally, and the pier spacing will influence the rate of debris accumulation and the allowable design differential head.

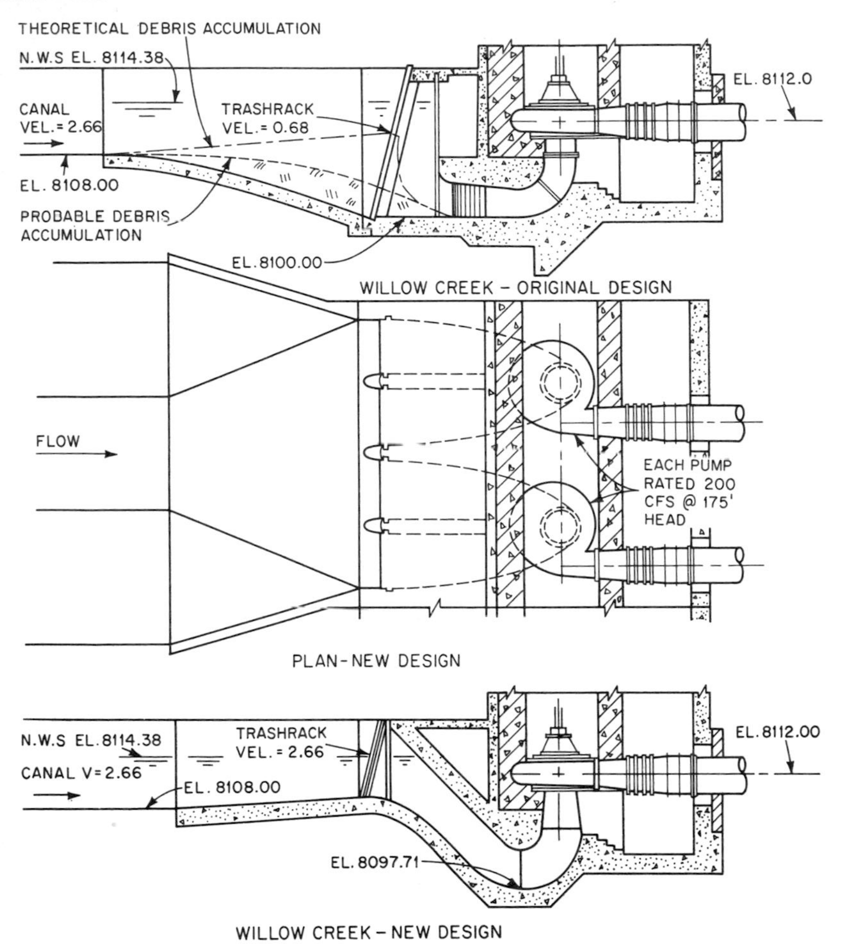

Fig. 7 Bureau of Reclamation elbow-type suction-tube inlet: improved 135° design. (U.S. Department of the Interior; Bureau of Reclamation)

Rotating screens will remove trash to a much smaller size since the accumulation is continually removed and the open area is kept fairly uniform. Finer screening than that required by the pump may be necessary in installations where the liquid pumped must pass through small openings in equipment serviced, such as condenser tubes, spray nozzles, etc. Screens are usually installed in conjunction with trashracks so that large, heavy pieces will not have to be handled by the screens. Since velocity through the screen is limited to 2 ft/s, the pit cross section may be determined by screen requirements. If flow is such that a maximum-width screen available would be too long (deep) for practical or economic reasons, two screens may be employed with a center pier. In this case the distance to the pump should be increased 50 percent over single-screen distance. Piers should be rounded (radiused) on the upstream side and ogived (tapered to a small radius) on the downstream side. Any corners at the side walls should be "faired" at small angles to the opening and wall to prevent pockets where eddies can form.

Flow from the forebay through the screens and into the pump area should be continuous as to direction. Avoid right-angle screens through which flow must change direction at least once or possibly twice. If screens must be at an angle with flow into pumps, increase the screen to pump distance by 100 percent of normal. Environmental considerations may increase the possibility of problems in this area.

Environmental Considerations Suction pit requirements will vary according to whether hydraulic or structural standpoints are being considered. Both of these may also be in conflict with environmental considerations.

A design to accommodate fish limitations was mentioned briefly in a previous paragraph. Fish react to a horizontal velocity but are not aware of a pull in a vertical direction. Thus, to keep them from entering the inlet, a horizontal flow must be established at a velocity low enough that will permit them to escape.

Intakes which take their flow directly from a river may have a high velocity that would trap fish. Even if the velocities are lowered to reasonable screen levels (2 ft/s), fish may still be drawn into the screen area and carried up to trash disposal.

When a water supply system is to derive its source from a body of water containing fish, positive steps must be taken to prevent undue disturbance and destruction of the fish.

A site survey should determine:

1. The intake location furthest removed from a natural feeding area and as remote as possible from attractive or "trap" areas.
2. The number of species involved.
3. The size range of each species—length and weight.
4. The population of each species, and whether they are anadromous or settled.

Next the plant needs should be determined as to total flow, probable intake size, velocities at inlet, through screens and at trashracks. Variations in flow throughout the year and temperature ranges from winter to summer should be available.

Sites for intakes should not be selected near feeding areas for large schools of fish, kelp beds, coral reefs, and similar attractive spots. Sheltered spots most suitable for intake flows may also be most attractive to fish.

The best source of information about the local fish is a marine biologist who has made a study of the local areas. He may not only have information on fish habits, feeding patterns, population, etc., but may also have test information about the fish swimming ability. If he does not already have this information, he can probably run a survey to develop this data.

The most difficult problem to overcome in design is related to the small fish. Screen openings must be held to a minimum, and under velocity conditions they have much less swimming "sustaining"' ability than larger fish, both in velocity and duration time. In a given stream flow (such as is generated by pumps with inlet water going through screens) a fish must have the ability to sustain a given speed against this flow for a certain length of time. When he weakens he will fall into the current flow, and in the case of pump inlets he will be impaled against the screen and destroyed. If he senses the velocity early enough and has an alternate route he can use "darting" speed to escape, or can follow another attraction (cross velocity flow into a separate chamber, or a light attraction to the chamber) and be removed on an elevator or pumped out to a safe channel. Migrating fish need a continuation channel to restore their interrupted journey.

In designing an intake it is necessary to:

1. Keep the velocity below 0.5 ft/s to avoid drawing fish into the screens. (Note that for a tube inlet away from shore a horizontal *velocity cap* should be placed over the inlet to prevent fish from being subject to a vertical velocity, which they cannot sense, and to maintain a horizontal velocity that will attract their attention

and direct them away from the inlet. Or,

2. Create a cross-flow which will propel or attract the fish to one side of the inlet area where they can be directed into a bypass pool and lifted back to their own living area or sent around the plant to a downstream location, or, if an anadromous fish, sent along to an upstream rendezvous; and,

3. Keep piers and screens flush across the inlet face to prevent attractive pockets in which fish can hide and be drawn into the screens when they weaken.

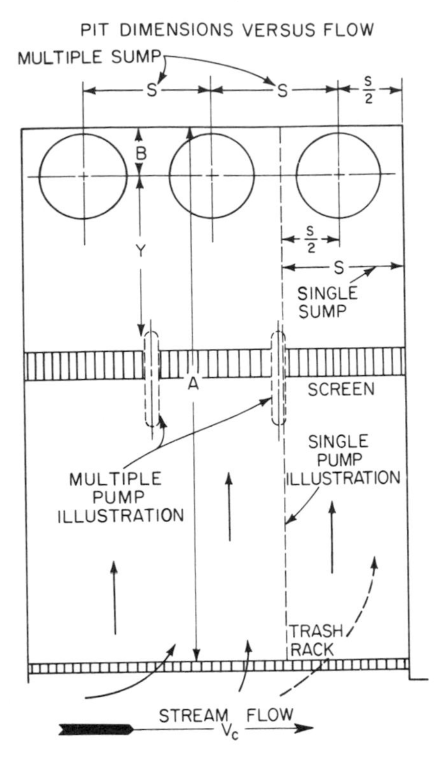

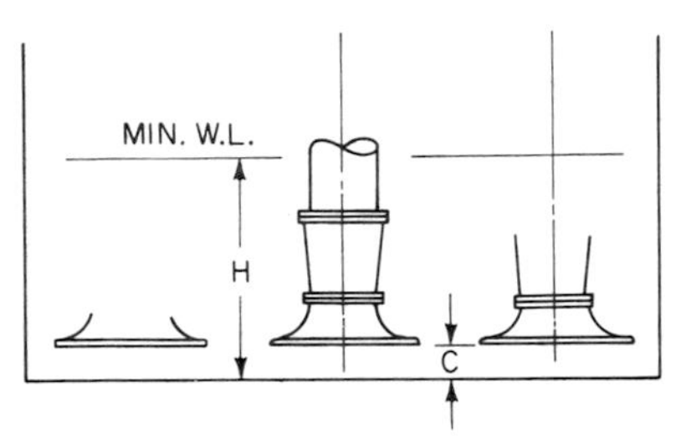

Fig. 8 Proportions for standard pump pit. See Fig. 9. (Reprinted from Hydraulic Institute Standards, 12th ed. Copyright 1969 by the Hydraulic Institute, 122 East 42d Street, New York)

Inlet screen areas should be in small sections rather than one long face, so that a fish will not be trapped in the center and find it too far to swim to safety after it realizes its predicament. The maximum deterrent flow is about 10 ft/s, but this may be too much disturbance for the flow to the pumps. Smaller areas allow short-term duration limits for enticing fish away from the inlet, and survival will be much higher.

Fish congregating at an inlet or in a forebay pool can be crowded or herded to an outlet point by using vertical nets or horizontal screens. In a direct channel, horizontal moving screens can route fish past a sloping (relative to stream flow, say 35°) moving screen which directs fish to a narrow outlet at one side leading to an outlet channel away from the main inlet.

Testing *Model Pit Design.* When the basic rules for good pump suction pit design are adhered to, no model test will be required to ensure proper operation of the pumps and pumping system. The substance of these rules is to keep a straight-in approach flow at a constant low velocity from the water source to the pump chamber. The Hydraulic Institute has established dimensions and charts to satisfy these criteria for the average pump in general application. Figure 8 shows the basic layout and Fig. 9 gives the average dimensions in relation to the flow required per pump.

Site layout problems may make the ideal solution impossible. Structural requirements may outweigh hydraulic requirements in some instances. When the ideal pit may not be possible or economically feasible a model test should be considered.

It should be noted, in regard to the ideal dimensions, that they are a composite, covering not only a range of specific speeds but also a complex of pump design philosophy. Some variation from the ideal dimensions should be expected from individual pump manufacturers.

Since pump manufacturers are not in a position to guarantee the pump pit design, differences of opinion between the structural and hydraulic pit design engineers and the pump design engineers may best be resolved by resorting to a model pit test (Fig. 10).

Model Tests. Since the gravity factor cannot be modeled, some parameters must be devised to make the model-prototype relationship meaningful. Surface-flow phenomena can be modeled by considering gravity forces. A dimensionless parameter called the Froude number will give relationships for model work and prototype to determine pit design free from vortexing and other surface disturbances. The

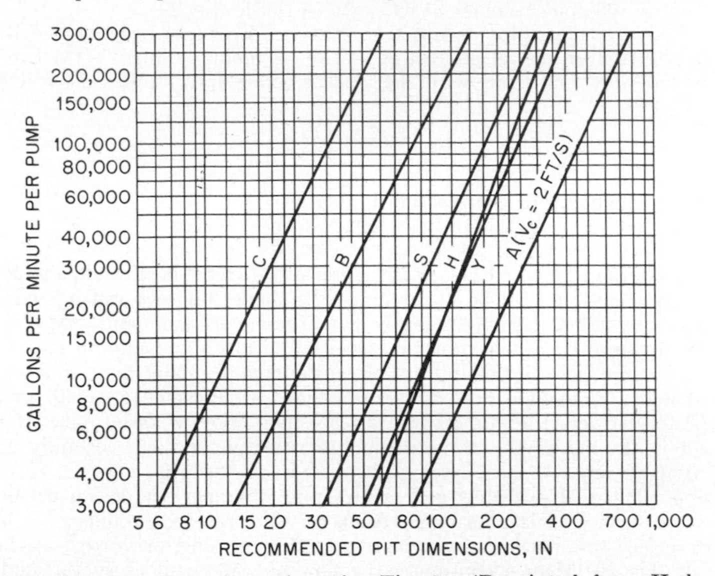

Fig. 9 Recommended pit dimensions for Fig. 8. (Reprinted from Hydraulic Institute Standards, 12th ed. Copyright 1969 by the Hydraulic Institute, 122 East 42d Street, New York)

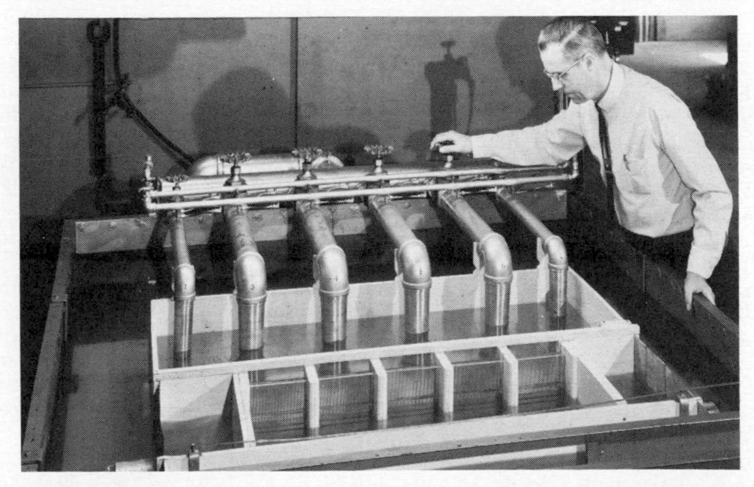

Fig. 10 A model pit test for a multiple-pump station. (Peerless Pump, Hydrodynamic Division, FMC Corp.)

Froude number is defined as

$$\mathrm{Fr} = \frac{V^2}{gh}$$

where V = flow velocity, ft/s
g = acceleration of gravity, ft/s^2
h = pump head, or any linear dimension, ft

If a model to prototype ratio of 1:10 is used, the prototype velocity of 1.0 ft/s becomes a model velocity of $(1 \div 10)^{1/2} = 0.316$ ft/s. All physical dimensions of the pump pit and related components including the physical makeup of the simulated pump are a direct ratio of prototype-model size: 1:10, or 0.10. Submergence also follows this rule. Flow is equal to area times velocity, $(0.1)^2(0.1)^{1/2} = 0.00316$, and a model flow of 100 gpm becomes 31,645 gpm in the prototype.

As the Froude relationships do not take friction forces into account the Reynolds number is also used in model pit testing. The Reynolds number is used in closed-conduit or pipe flow and deals with the viscous forces in a liquid. It is defined as

$$\mathrm{Re} = \frac{lV}{\nu}$$

where l = pipe diameter or channel depth, ft
V = velocity of flow, ft/s
ν = kinematic viscosity, ft²/s

Since the values of flow and velocity as determined from the Froude and Reynolds numbers vary widely, it will be apparent that a model pit test and a model pump test cannot be conducted simultaneously with the same equipment. Model pumps are usually tested at or near prototype velocity and prototype suction and discharge heads. To be sure that all possibilities are considered, a model pit tested at Froude number should also include flow rates throughout a sufficient range up to and including prototype velocities to observe all possible velocity variations. If no vortexes occur in the model pit at the maximum velocity they will certainly not form in the prototype pit.

Vortexing. The real problems generating from improper pit design occur largely on the water surface in the form of vortexes. Vortexes are produced by localized eddies on surface water, which can progress to form a cone, or vortex, at the water surface. If this disturbance continues, the flow of water will carry the underwater part of the vortex down toward the pump suction bell and ultimately into the pump itself. This introduces air into the impeller and will affect mechanical radial balance of the impeller due to the interruption of the normal solid-liquid flow pattern. This type of disturbance will produce hydraulic pulsations in the pump flow and mechanical overloading of the sleeve bearings and impeller guides.

Underwater vortexing sometimes occurs in round pits, or in pits where the pump suction bell is at some distance from the rear wall. Flow past the suction bell strikes the rear wall and rolls back toward the bell, forming an eddy which disturbs the normal flow into the pump. In a round pit a cross baffle below the pump bell will reduce this effect. Where the pump is some distance from the back wall a wall can be installed near the pump, or a horizontal baffle at suction-bell level behind the pump will also reduce the disturbance. In all cases the suction bell should not be more than one-half the suction-bell diameter above the bottom of the pit, and preferably one-third the diameter.

The use of the suction-bell diameter as a basis for spacing should be carefully evaluated. It will be seen that there is nothing magic in this relationship, especially when several pump manufacturers all use different bell diameters. The real criterion is the allowable velocity at the suction bell. It has been found that very low-head pumps are much more sensitive to bell velocity over 6 ft/s. For example, the velocity head loss at the bell inlet with a high velocity may be such a large percentage of total head that efficiency could drop as much as 10 percent. A good design rule for safe operation can be related to the pump head. For pumps up to a 15-ft head, the suction-bell velocity should be held to 2½ ft/s; up to a 50-ft head, 4 ft/s; and above a 50-ft head, 5½ ft/s. These values should be used for any substantial amount of pumping, but for occasional short-term pumping these values can be exceeded without destroying the pump.

Vortexes may be broken up and effectually nullified by arrangements of baffles and vanes, or they may be prevented from occurring initially by a proper pit design. The only way to determine what baffling should be used, and its effectiveness, is by model testing. Methods to try in the elimination of vortexing are shown in Fig. 11.

Vortexes are usually generated by a change in direction of flow of the liquid to be pumped, or by a high velocity past an obstruction such as a gate-inlet corner, screen pier, dividing wall, etc. In combination, these two causes invariably generate vortexes. For this reason the pump suction pit should be immediately preceded by a straight channel at velocities ranging up to not over 1.5 ft/s. Satisfactorily operating pump pits with higher velocities are rare, and should not be put into operation without the assurance of a model study. An additional condition likely to generate vortexes is in a multiple pump pit with individual cells in which only a portion of the pumps will operate simultaneously. The "dead" space behind the nonoperating pump will have flowing water tending to reverse direction and form

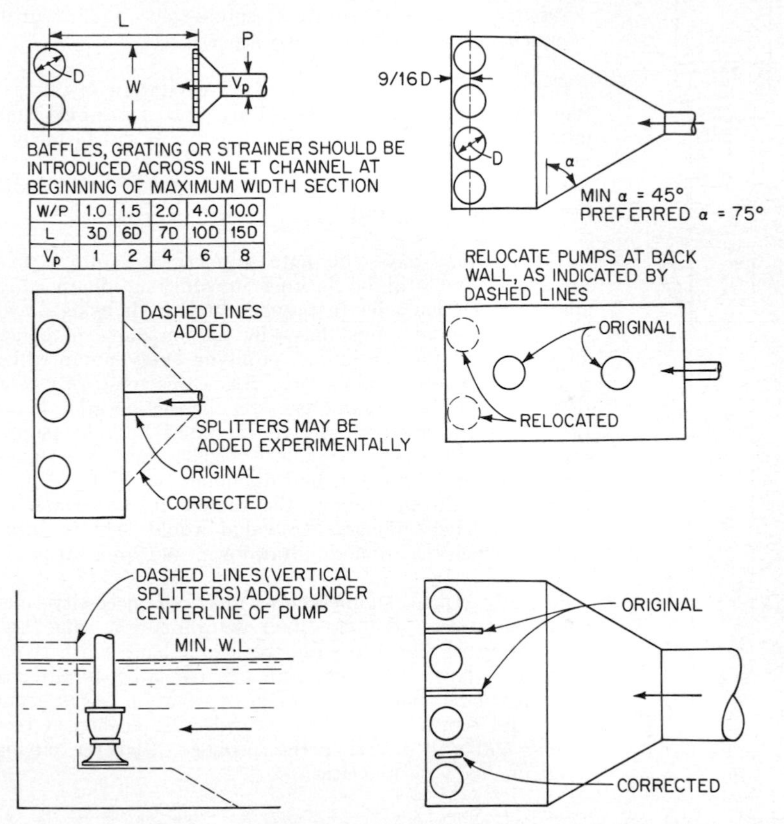

Fig. 11 Methods to correct problem design pits. (Reprinted from Hydraulic Institute Standards, 12th ed. Copyright 1969 by the Hydraulic Institute, 122 East 42d Street, New York)

eddies. Eddies and vortexes can be avoided by eliminating the walls or venting them at the rear, and by positioning the pumps at the extreme rear of the pit, as well as the expensive modifications to the pit such as splitter walls and baffles (see Fig. 4).

Round pits tend to generate vortexes, especially when the pump is centered in the pit. These vortexes are usually centered around the pump column and are generated because of the eccentric inlet flow. Special cases of the round pit are tolerable under the following controlled conditions: (1) Ranney wells (Fig. 3) and (2) booster pumps in pipelines (Fig. 12). In the Ranney well, the ratio of pump size (and flow) to pit size (and capacity) is such that a very low velocity exists, as in a lake inlet. Water comes into the Ranney well all around the periphery.

These conditions of direct flow and low velocity prevent vortexing. Booster pumps installed in a circular can (suction tank) must be centered in the can, and all velocities of inlet to the can and flow in the can and into the pump must be uniform and high enough to provide fluid control. This velocity will vary from 4 to 6.5 ft/s.

Vortexes are not generated by a pump or pump impeller and so do not fall into clockwise or counterclockwise rotation because of the pump rotation. Also, in a pump pit vortexes do not have a directional rotation induced by the rotation of the earth and therefore are not of opposite rotations above and below the equator, as is the case with "bathtub vortexing," which occurs only in tanks being drained without pumps.

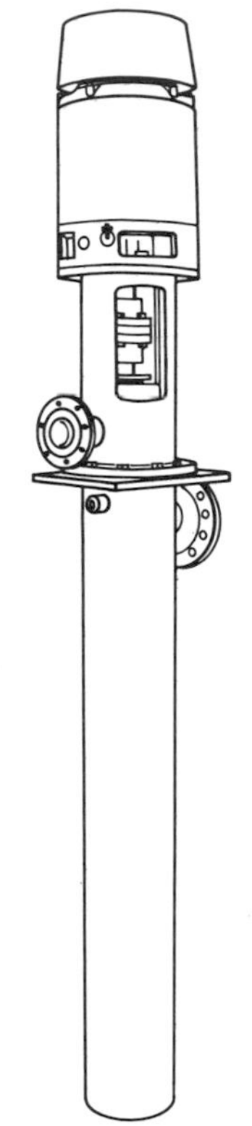

Fig. 12 Booster pumps suspended in steel wells utilize minute space for suction pit. (Byron Jackson Pumps)

Submergence. The required submergence for centrifugal pumps in intake sumps has two aspects:

1. A pressure sufficient to prevent cavitation in pump first-stage impellers referred to as NPSH. (It is assumed that the proper pump has been selected to perform satisfactorily with available NPSH.)
2. Vortex prevention and associated pit-flow problems detrimental to pump operation.

A pump may have adequate submergence from a pressure standpoint and still be lacking in sufficient depth of cover above the suction inlet to prevent surface air being drawn in. Any wet-pit pump must have its suction inlet submerged at all times, and for continuous pumping every pump will have a fixed minimum requirement. Since this relates to velocity, there are two basic parameters: (1) the suction inlet diameter; (2) the depth of water above the inlet lip. As pump size (and flow) increases the inlet velocity may stay constant along with an increase in bell diameter, but at the same time the impeller distance above the suction inlet becomes larger, so that a fixed submergence value would lead to increased surface velocity, peripheral drawdown, and an increase in air intake.

Basically, then, submergence must of necessity increase with pump size. For the final determination some balance must be struck between submergence and pit width to satisfy an average flow velocity of 1 to 1½ ft/s and maintain reasonable economic balance between excavation costs, concrete costs, and screen costs without neglecting ecological requirements, and still fulfilling the primary need for circulating water in adequate quantities.

SUCTION PIPING

Single Pumps Piping to the suction of a dry-pit centrifugal pump (Fig. 13) must be carefully worked out to provide a reasonable uniform velocity, straight-line flow, and adequate pressure and sealing against leakage, in or out. Air pockets just prior to entrance to the pump should be avoided, as well as downflow lines subject to sudden pressure changes. Air pockets can be prevented by proper elevations (Fig. 13). Pressure surges can be controlled by gas bags, air tubes, etc., which may require a system pulsation study to determine possible need and solution.

Flow to the suction nozzle of centrifugal pump should be parallel and of uniform velocity to avoid impeller disturbance. This suggests the need for a straight run of suction pipe at least eight diameters immediately prior to the pump suction flange. Suction pipe should be at least as large as the pump suction nozzle. If the pipe is larger, a reducing section may be used, taking care to avoid air pockets and uneven flow patterns.

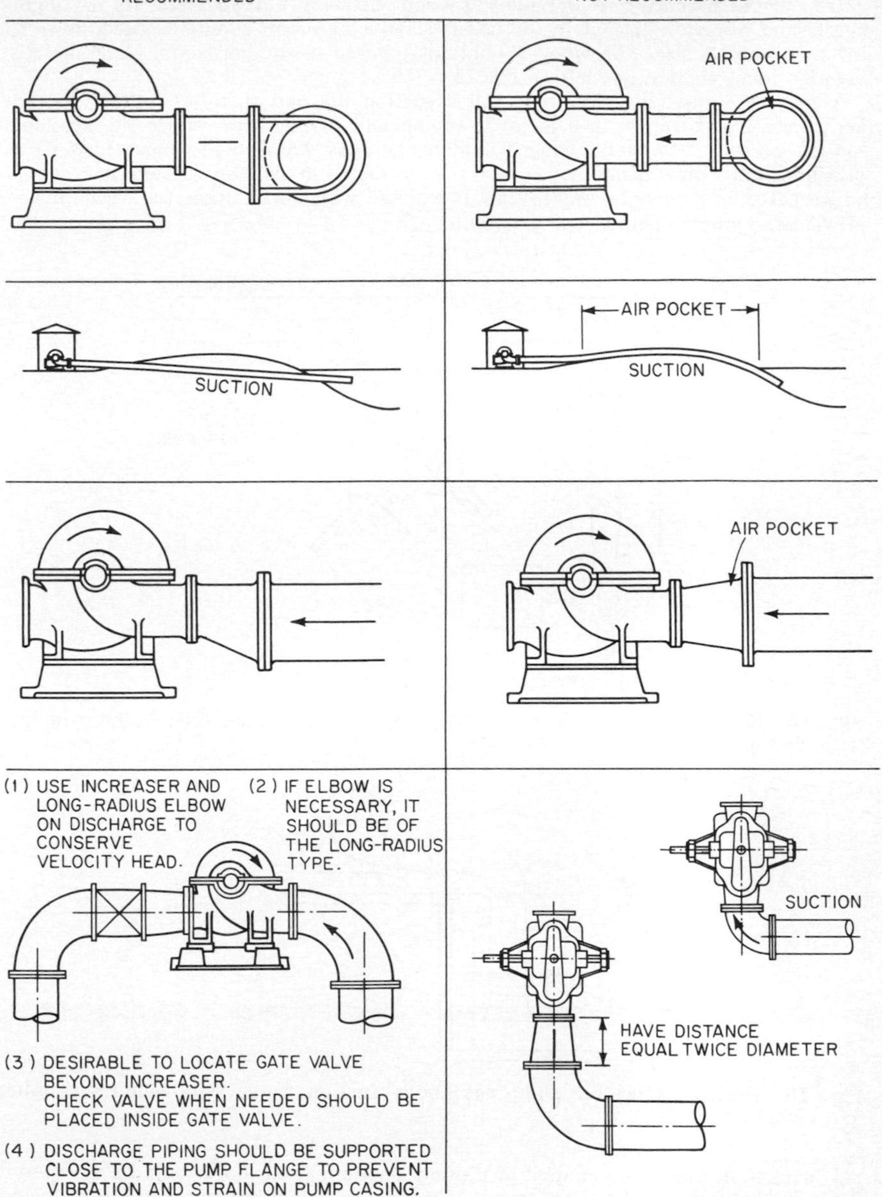

Fig. 13 Methods of connecting suction pipe to pumps to avoid air binding, surges, and prerotation. (Reprinted from Hydraulic Institute Standards, 12th ed. Copyright 1969 by the Hydraulic Institute, 122 East 42nd Street, New York)

Suction inlet velocities should be limited to 5 ft/s if flow is uniform from a basin area. In a suction manifold the main-line flow should be not over 3 ft/s. Branch outlets should be at 30 to 45° relative to main-line flow rather than 90° right-angle outlets, and the velocity can increase to 5 ft/s through a reducer (Fig. 14). With such velocities branch outlets can be spaced to suit pump dimensions to avoid crowding. Also, if the angled manifold outlet is used, pumps can be set as close to the manifold as the elbow, valve, and tapered reducer will allow.

If it seems necessary to provide a suction inlet pipe at an angle to the pump nozzle and directly adjacent to the nozzle a reducing elbow should be used, down to the pump-nozzle size. An elbow should not be used in the horizontal plane directly before a pump suction nozzle (see Fig. 13).

A dry-pit pump (Fig. 15) may pull a suction lift and therefore will be located above the liquid source. All losses in piping and fittings will reduce the available suction pressure. Suction piping should be kept as simple and straightforward as possible. Any pipe flange joint or thread connection on the suction line should be gasketed or sealed to prevent air in-leakage which will upset the vacuum and prevent the proper operation of the pump.

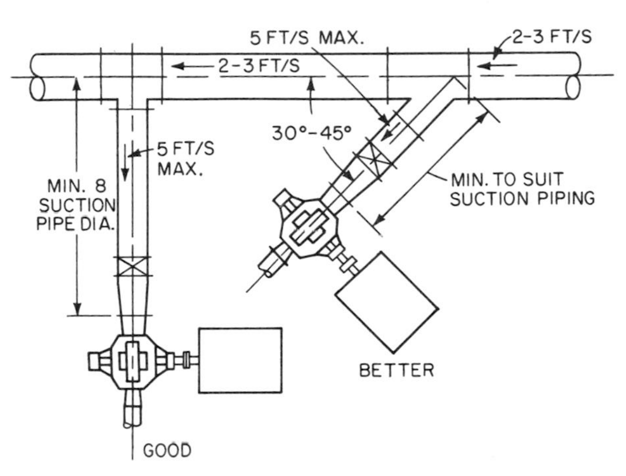

Fig. 14 Flow patterns are better even at low velocity when inlet is more in line with flow path.

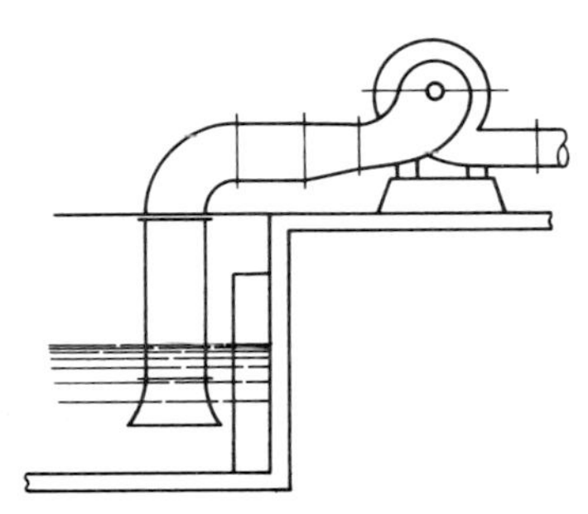

Fig. 15 Horizontal dry-pit pump set above suction water level requires priming and airtight piping.

If expansion joints are required at a pump suction, an anchor should be interposed between the nozzle and the expansion joint to prevent additional forces from being transmitted to the pump case and upsetting rotating clearances. The same requirement applies to a sleeve coupling used to facilitate installation alignment.

Reciprocating pumps must have additional consideration due to the pulsating nature of their flow. Suction piping should be as short as possible and have as few turns as possible. Elbows should be long radius. Pipe should be large enough to keep velocity between 1 and 2 ft/s. Generally this will provide pipe one to two sizes larger than the pump nozzle.

High points that may collect vapor are to be avoided. Eccentric reducers should have the flat side up. A pulsation dampener or suction bottle should be installed next to the pump inlet.

NPSH available should be sufficient to cover not only pump requirements and friction losses but also acceleration head (see Surge and Vibration).

Manifold Systems All comments relative to single pumps apply to manifold-pump systems, and additional points need to be noted.

Right-angle outlet connections to individual pump suction nozzles should be avoided to reduce shock load and losses difficult to overcome. Outlets between a 30 and 45° angle with the manifold will provide the smoothest flow transition and allow optimum pump and piping layout.

Manifold sections beyond each branch takeoff should be reduced to such a size that the velocity will remain constant. One exception to this scheme is a tunnel suction (Fig. 16). The flow through the tunnel may operate independently from the pumps which are suspended in boreholes drilled into the roof of the tunnel. Boreholes at least one-third tunnel diameter or larger should be horizontally spaced at least 12 borehole diameters apart. Smaller ratios can be closer to a minimum of six diameters. Pump suction bells should be at least two borehole diameters above tunnel roof. Velocities in the tunnel should be kept below 8 ft/s for best pump performance.

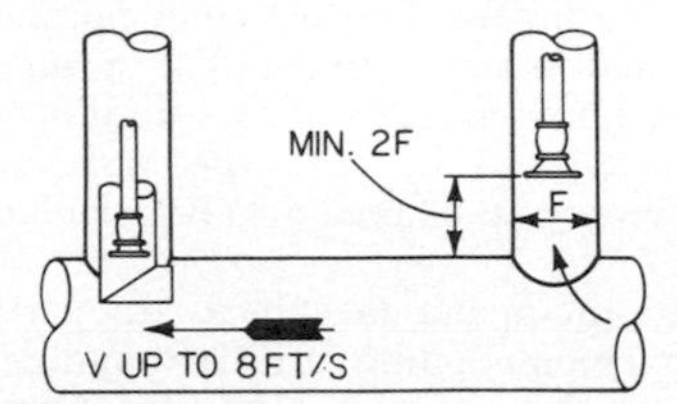

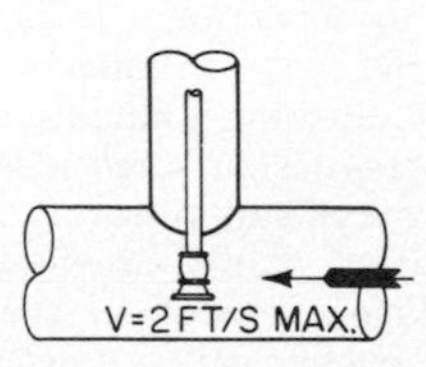

Fig. 16 Higher tunnel velocities require isolation of pump from direct flow to prevent distortion of close clearance parts and shaft. (Hydraulic Institute Standards, 12th ed. Copyright 1969 by the Hydraulic Institute, 122 East 42d Street, New York)

NPSH The net positive suction head so essential to correct and trouble-free pump operation is always reduced by losses in suction piping. An economic balance must be obtained between the pump size and speed, the NPSH required, pipe size, and suction velocity. If the suction source is a tank, such as a deaerator in a power plant, it is quite expensive to elevate the tank. Therefore the available NPSH is low, the pipe size from the tank to the pump should be large for a low velocity, and the length should be short for minimum losses. A cooling tower on a hill supplying water to a pump station many feet below could have a much higher pipe velocity and a longer length without forcing an extremely low NPSH requirement on the pump.

A dry-pit pump should be as close to the suction source as possible. When NPSH required indicates a suction lift is possible the most advantageous solution is to reduce the suction losses by increasing the pipe size, flaring the inlet bell, and maintaining a close proximity of the pump suction eye of the first-stage impeller to the minimum water level.

High-Pressure Inlets Pumps in series build up pressure so that the second and following pumps will have a high-pressure suction pipe connection. This emphasizes the need for tight joints and flanges, and careful welding. Expansion joints should not be used since the hydraulic force on the pump would be large, difficult to restrain, and perhaps impossible for the pump to handle without distortion. As NPSH will not be a problem the pipe size can be kept small to minimize the design problem and valve cost.

Effect on Efficiency of Pump Other than mechanical operation, the greatest effect of flow disturbance at the pump suction is on the pump efficiency. The higher the pumping head the less this effect becomes. For very high head pumps a high velocity may be ignored completely unless there is an extremely large horsepower evaluation factor. Greater attention should be given to the suction pipe design for pumps producing 100 ft of head or less than for those with high head, as efficiency may be worth many dollars in power cost over the life of the plant and equipment.

As a part of a total system, the pump efficiency relates to the system efficiency, which may make it well worth while to invest more money in a larger suction pipe.

Surge and Vibration One of the possibilities arising from a power failure at a pump station is the reversal of the pump if a valve fails to close, and its subsequent operation as a turbine. Under rated head a pump will run from 20 to 60 percent above rated speed in the reverse direction. In its transition to that phase the forward motion of the water is interrupted and gradually reversed. At the time of the power failure the velocity of flow in the suction pipe may not decelerate at a rate sufficiently low to prevent a surge in the direction of the pump. The suction piping should be designed to withstand the resulting pressure rise since absolute integrity of valving and power supply will be too costly as a design parameter (see Sec. 9 for additional information on this subject).

Rotating machinery has a design vibration frequency which should be removed from a system frequency far enough to avoid sympathetic activity. Any combination of vibration frequencies near enough to each other to react will do so when a prime source such as a pump excites them. When a piping system has been designed to allow only very low stresses transmitted to the pump nozzle it is quite vulnerable to vibration. It is usually good practice to analyze a suction system including pipe, valves, hangers, restraints, pump nozzle loads, pump speed and impeller configuration, foundation, and anchors to be sure the system is not "in tune." At the design stage it is relatively easy to change a valve or elbow location, or to add a surge suppressor.

Reciprocating pumps are basically surge-producing machines. In particular, they require sufficient energy at suction to overcome pump NPSHR and pipe friction, and also an energy called *acceleration head.* This energy must overcome the acceleration-deceleration pulsation flow in the suction end which could lead to liquid flashing with pump noise and vibration. Surges of sufficient magnitude to rupture the pump cylinders may also be produced.

The total NPSH required by a reciprocating pump must include NPSHR (pump) plus friction loss (suction pipe) plus acceleration head. Acceleration head is

$$Ha = \frac{LVNC}{32.2K}$$

where L = actual (not equivalent) length of suction pipe, ft
V = velocity of flow in suction line, ft/s
N = pump rpm
C = pump constant, decreasing with number of cylinders from 0.2 to 0.04
K = liquid constant: 2.5 for compressible fluids, 1.5 for water

The acceleration head can be reduced by installing a pulsation dampener in the suction line near the pump.

REFERENCES

1. Hydraulic Institute Standards for Centrifugal, Rotary and Reciprocating Pumps, 12th ed., New York, 1969.
2. Proportions of Elbow-Type Suction Tubes for Large Pumps, *Bur. Reclamation Rept.* HM-2, June 1964.
3. Stepanoff, A. J.: "Centrifugal and Axial Flow Pumps," 2d ed., John Wiley & Sons, Inc., New York, 1957.
4. Baker, O.: Multiphase Flow in Pipelines, *Oil and Gas J.*, vol. 56, no. 45, November 10, 1958.
5. Acrivos, A.: Flow Distribution in Manifolds, *Chem. Eng. Science,* vol. 9, no. 4, February 1959.
6. Patterson, I. S.: Pump Intake Design Investigations, *MSJ Mech. Eng., Proc.,* vol. 31, no. 1, London, January 1967.
7. Whistler, A. M.: New Look at Plunger Pump Suction Requirements, *ASME Publ.* 69-PET-35.
8. Langewis, C., Jr., and C. W. Gleeson: Practical Hydraulics of Positive Displacement Pumps for High-Pressure Waterflood Installations, *J. Petrol. Technol.,* vol. 23, February 1971.

9. Hugley, D., et al.: Pulsation Primer, 7 pts., *Petrol. Equip. Serv.*, May/June 1967 to May/June 1968.
10. Weltmer, W. W.: Proper Suction Intakes Vital for Vertical Circulating Pumps, *Power Eng.*, June 1950.
11. Gegan, A. J., and J. T. Fong: Upside-down Pump Solves Design Puzzle, *Power*, February 1955.
12. Scarola, J. A.: American Power Plant Cooling Towers, *ASCE Power Div. Proc.*, October 1959.
13. Mariner, L. T., and W. A. Hunsucker: Ocean Cooling Water Systems for Two Thermal Plants, *ASCE Power Div. Proc.*, August 1959.
14. Messina, J. P.: "Sump and Pit Design Data for Vertical Wet Pit Pumping Installations," Worthington Corporation, Harrison, N.J.
15. Brkich, A.: "Vertical Pump Model Tests and Intake Design," Ingersoll-Rand Co., February 1953.
16. Dornaus, W. L.: Stop Pump Problems Before They Begin with Proper Pump Pit Design, *Power Eng.*, February 1960.
17. Glass, L. M.: "Unusual Raw Water Intake Design," Matz, Childs and Assoc., Inc., Baltimore.
18. Iversen, H. W.: Air Entrainment in Pump Suction Open Sumps, *Test Lab. Report 25-S*, University of California, Berkeley.
19. Dornaus, W. L.: Flow Characteristics of a Multiple-Cell Pump Basin, *Trans. ASME*, July 1958.
20. Fraser, W. H.: Hydraulic Problems Encountered in Intake Structures of Vertical Wet Pit Pumps and Methods Leading to Their Solution, *Trans. ASME*, May 1973.
21. Messina, J. P.: Periodic Noise in Circulating Water Pumps Traced to Underwater Vortices at Inlet, *Power*, September 1971.
22. Dicmas, J. L.: Development of an Optimum Sump Design for Propeller and Mixed-Flow Pumps, *ASME Paper* 67-FE-26.
23. Bechtel Associates: "Survey and Performance Review of Fish Screening Systems," Job 8693, December 1970 report, San Francisco.

Section 12

Selecting and Purchasing Pumps

S. I. HEISLER

The entire series of operations, following the initial decision that pumping equipment is required for a system and culminating with the act of purchasing the equipment, can be divided into the following general steps:

Engineering of system requirements
Selection of pump and driver
Specifying the pump
Bidding and negotiation
Evaluation of bids
Purchasing of the selected pump

In the process of specifying pumping equipment the engineer is required to determine system requirements, select the pump type, write the pump specification, and develop all information and data necessary to define the required equipment for the supplier.

Having completed this phase of work, the engineer is then ready to take the necessary steps leading to purchase of the equipment. These steps include issuing the specifications for bids or negotiations, evaluation of pump bids, analysis of purchasing conditions, selection of supplier, and release of all necessary data for purchase order issuance.

ENGINEERING OF SYSTEM REQUIREMENTS

The first decision the engineer must make is to determine the various requirements and conditions under which the equipment will operate.

Fluid Type One of the initial steps in the defining of the pumping equipment is the development of physical and chemical data on the fluid handled, such as viscosity, density, corrosiveness, lubricating properties, chemical stability, volatility, amount of suspended particles, etc. Depending upon the process and the system, some or all of these properties may have an important influence on pump and system design; for example, the degree of corrosiveness of the fluid will influence the engineer's choice of materials of construction, while if the fluid contains solids in suspension, suitable types of pump seal designs and abrasion-resistant pump construction materials may have to be considered.

Fluid physical and chemical data of interest to the engineer should cover the entire expected operating range of the pumping equipment. The influence of such parameters as temperature, pressure, time, etc., upon the fluid properties should also be considered with regards to its possible effects on pump operation.

System-Head Curves The engineer should have a clear concept of the system wherein the pumping equipment being specified is expected to operate. A preliminary design of the system should be made, which should include an equipment layout and a piping and instrumentation diagram (or other suitable type of diagram) showing the various flow paths, their preliminary size and length, elevation of system components and including valves, equipment, piping specialties, etc., which establish the system-head losses.

The engineer should then determine the flow paths, flow rates, pressures, and temperatures for various system operating conditions, after which he can proceed to determine line or conduit sizes and estimate pipe or conduit runs.

With the above information the engineer can proceed to develop system-head curves. These curves show the graphic relationship between flow and hydraulic losses in a given pipe system. In calculating the hydraulic losses, the engineer must include adequate allowances for corrosion and scale deposits, etc., in the system over the life of the plant.

Since hydraulic losses are a function of flow-rate piping sizes and layout, each flow path in a system will have its own characteristic curve. Care must be taken when specifying pump characteristics to take into account the characteristic curve of each possible flow path served by the equipment. For use in specifying pump equipment, it is convenient to add the effect of static pressure and elevation differences to the system-head curve to form a combined system-head curve. The resultant curve would then show the total head required of the pumping equipment in order to overcome system resistance. The pump head must be at or above the combined system-curve at all expected operating points and for all flow paths the pumping equipment is expected to serve.

A sample calculation which establishes one point on a system-head curve (the rated point) is given below. To establish other points accurately the calculation should be repeated for different values of flow. For rough estimates it can be assumed that frictional head loss is proportional to the square of the flow. (For additional information see Sec. 9.)

KNOWN AND ASSUMED DATA:

1. NaOH solution is kept at 60°F
2. Required flow = 1,100 gpm
3. Assume the NaOH solution has the properties of pure water, i.e., dilute solution. (Specific weight = 62.37 lb/ft^3)
4. Assume pipe is clean and new
5. System arrangement Fig. 1
6. Pump and system-head curves Fig. 2 determined as follows:

A. Head requirement at 1,100 gpm

$$TH = \frac{144}{w}(p_D - p_S) + Z + h_f$$

1. Pressure head = (144/62.37)(75 − 0) = 173 ft
2. Static head (elevation) = 327 − 306 = 21 ft
3. Friction losses:
   a. Entrance loss at tank = 0.4 ft
   b. Losses in 8-in suction valves and 2 LR elbows = 0.82 ft
   c. Loss in 116 ft of 8-in suction pipe = 2.2 ft
   d. Losses in 6-in discharge valves and 7 LR elbows = 10.95 ft
   e. Loss in 263 ft of 6-in discharge pipe = 19.93 ft
   f. Exit loss at injection line = 2.3 ft

$$TH = 173 + 21 + (0.4 + 0.82 + 2.2 + 10.95 + 19.93 + 2.3) = 230.6 \text{ ft}$$

Specify rated head = 235 ft

B. Flow requirement = 1,100 gpm
Add 5% allowance for wear
1,100 gpm (1.05) = 1,155 gpm
Specify rated flow = 1,160 gpm

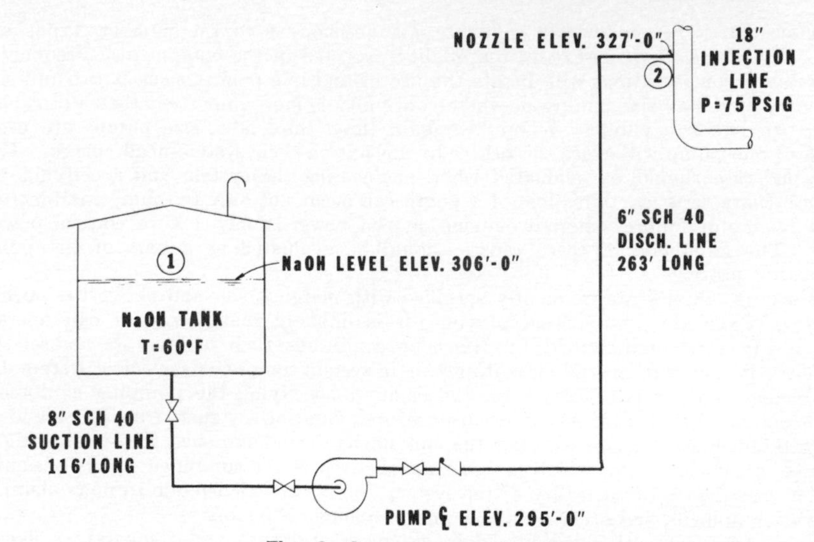

Fig. 1 System arrangement.

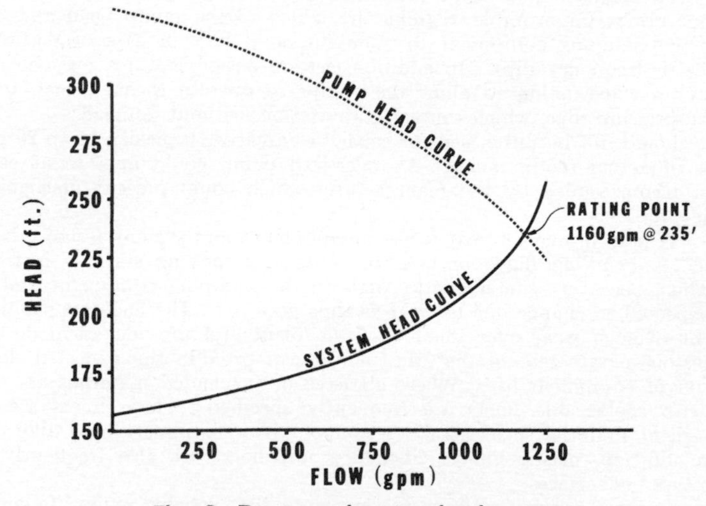

Fig. 2 Pump- and system-head curves.

Since elevation and pressure can be measured or read reasonably accurately, these values tend to be much more accurate than calculations of friction loss in piping systems. Thus, for systems with a large portion of the head loss found in fluid friction, there is a larger sensitivity to error and greater margins should be employed.

Alternate Modes of Operation The various modes of operation of the system are important considerations when specifying pumping equipment. Is operation of the pumping equipment continuous or intermittent? Is the flow or head fluctuating or constant? Is there a great difference in flow and head requirements for different flow paths? These and other questions arising from different modes of operation greatly influence such decisions as to the number of pumps, their capacities, and whether booster pumps are needed in some flow paths.

The engineer should also consider the continuity of service expected of his pumping system. This factor will influence his decision as to the number, type, and capacity of installed spares and the quality expected of the equipment. Frequently reliability considerations will dictate the use of multiple pumps, such as two full-size pumps, three half-size pumps, or where continuity is more important than full capacity, two half-size pumps. Where two half, three third, etc., size pumps are used, loss of one pump will cause the others to run out on their system-head curves. This run-out case should be evaluated when engineering the system and specifying the pump characteristics. This loss of a pump can occur not only by pump malfunction but by motor failure, external damage, loss of power supply, loss of control power, etc. The likelihood of these causes should be evaluated as a part of the pump selection process.

Margins Pumps are frequently specified with margins over and above the normal rating. Over any long period of time, it is unlikely that a system can operate at steady state undisturbed by transient conditions such as may be caused by changes in modes of operation, malfunction in system components, electric system disturbances, etc. It is necessary for the engineer specifying the pumping equipment to examine the probability of occurrence and duration of such transients, and to specify adequate margins to allow the equipment to undergo such transients without damaging effects. This involves also an evaluation of the combined effects of equipment cost, degree of criticality of the system, the inconvenience due to unavailability of the equipment, and other economic and technological factors.

Some transients often considered in design are pressure and temperature fluctuations, electric voltage and frequency dips, loss of cooling water, etc. If the maintenance of continuous flow is important, then adequate margins must be allowed in the pump rating; for example, margins are added to the pump head and capacity allowing the pumping equipment to maintain rated flow in case of occurrence of small electric-frequency dips. In addition, certain design features may be included in the pump construction to allow the pumps to operate through such transients as suction-pressure dips, which can cause cavitation, without damage.

Pumps should not be purchased for capacities greatly (typically 15 to 20 percent) in excess of actual requirements. An oversized pump could operate at capacities less than recommended by the manufacturer which could present mechanical and hydraulic problems.

Wear Wear is an ever-present factor in equipment and system design. No material which is handling fluids or used in contacting moving surfaces is free from wear. Thus, operating characteristics of both the pumping equipment and system can be expected to change due to wear as time goes by. The engineer should assess the extent of such wear over the life of the plant and provide adequate margins in his system parameters so that the pumps can provide the expected flow even at the end of equipment life. Where abrasive or suspended materials are handled, pumps with replaceable liners are frequently specified. These liners are usually either resilient material such as rubber compounds or extremely hard alloys of cast iron. In addition, plastic linings (including impellers) are also frequently chosen for these types of services.

In some applications, especially in power plants, the expected pump life is specified as the same as plant life. However, the design life of a pump is a decision based on evaluation of economic factors. The wear margin to be added is a function of such factors as mode of operation (continuous or intermittent), fluid properties (abrasiveness, corrosiveness), etc.

Future System Changes A final factor to be considered in the engineering system requirements for pumping equipment is the possibility of providing for future system changes. If the system changes can be predicted with any degree of certainty, then the system can be designed to enable the changes to be effected with minimum disturbance to operation. Thus it is important to review the possibilities and effects of such future system changes as well as provide pumping equipment to satisfy the immediate system requirements. The engineer should attempt to present future requirements in the system capacity and system parameters based on projection of available data, and then evaluate the possibility and desirability of designing his pumping equipment to allow for the future system changes (such as providing extra flow or head margins, or specifying a pump with impeller less than the maximum for a given casing size, etc.) vs. modification of his system, including the pumps, when the system changes are made in the future.

In any event, it must always be kept in mind that the equipment must operate satisfactorily in the *present* system for which the engineer is specifying the pumps and this should be a factor in whatever evaluation he is making.

SELECTION OF PUMP AND DRIVER

The selection of the pump class and type for a particular application is influenced by such factors as system requirements, system layout, fluid characteristics, intended life, energy cost, code requirements, materials of construction, etc.

Basically, a pump is expected to fulfill the following functions: (1) to pump a given capacity in a given length of time and (2) to overcome the resistance in the form of head or pressure imposed by the system, while providing the required capacity.

The behavior of the system has a very important bearing on the choice of the basic types of pumps. What are the required heads and capacities at different loads? Does the required head increase or decrease with changes in capacity? Does the required head remain substantially constant? These are some of the questions for which the engineer must seek to find answers.

Pump Characteristics Characteristically, constant-speed reciprocating pumps are suitable for applications where the required capacity is expected to be constant over a wide range of system-head variations. This type of pump is available in a wide range of design pressures from low to the highest produced; however, the capacity is relatively small for the size of the equipment required.

The fact that the output from a reciprocating pump will be pulsating is a factor to be considered. Where this is objectionable, the use of rotary pumps may be required. However, the application of rotary pumps is limited to low-to-medium pressure ranges.

Centrifugal type pumps are often used in variable-head variable-capacity applications. Straight centrifugal pumps are generally used in low-to-medium-capacity and medium-to-high-pressure applications, while a low-head high-flow condition suggests that an axial-flow type may be more suitable. Mixed-flow impellers are used in intermediate situations.

It should be noted that while reciprocating and rotary pumps are self-priming, centrifugal pumps, unless specifically designed as such, are not. This may be an important consideration in certain applications.

In some cases the system layout can influence the decision on the choice of pump type. In general, centrifugal pumps will require less floor space for installation than reciprocating pumps, and vertical pumps less floor space than horizontal pumps. However, more head room may be required for handling the vertical pumps for maintenance and installation.

Where the available NPSH is limited, such as when handling a saturated liquid, and the application calls for the use of a centrifugal pump, the engineer may have to investigate the use of a vertical canned-suction-type centrifugal pump to gain adequate NPSH. In other cases the design may call for the installation of a pump immersed in the liquid handled and here, the use of a vertical turbine-type pump may be advantageous.

Code Requirements The construction, ratings, and testing of most pumps normally used in industry are governed by codes such as the A.P.I. or the Standards of the Hydraulic Institute. However, other codes of regulatory bodies may impose additional requirements which can affect both pump rating and construction. For example, the ASME Boiler Code requires feed pumps to be capable of feeding the boiler when the highest set safety valves are discharging. Again, in critical components for nuclear power plants, the AEC sometimes has imposed much more stringent quality-assurance requirements than specified in normal industrial codes. Such additional requirements may affect the engineer's choice.

Fluid Characteristics Fluid characteristics such as viscosity, density, volatility, chemical stability, solids content in the fluid, etc., are also important factors for consideration. Sometimes exceptionally severe service may rule out some classes of pumps at once. For example, the handling of fluids having high solids content will exclude the use of reciprocating type piston pumps, or pumps with close clearances. Rotary-type pumps are suitable for use with viscous fluids such as oil or grease, while centrifugal pumps can be used for both clean and clear fluids or fluids with high solids content. On the other hand, if it is undesirable for the process liquid to come into contact with the moving parts, diaphragm type pumps may have to be used.

Pump Materials Materials are affected both by the pumped fluid and the environment. Resistance to corrosion and wear are two of the more important material properties in this regard and the engineer should evaluate the choice of material to determine which is most suitable and economical for the purpose intended. Often this becomes an evaluation of the desirability of specifying expensive long-life materials vs. specifying relatively cheaper materials which must be frequently replaced.

Operating factors such as type of service (continuous or intermittent, critical or noncritical), running speed preferences (high or low speed), and intended life are also factors which will influence the engineer's judgment. For example, equipment used in continuous and/or critical service will generally demand heavier duty design and construction than would equipment for intermittent and/or noncritical service. High-speed operation, if allowed, will permit the use of smaller, usually less expensive equipment. The life of equipment cannot be predicted with certainty. For a given life expectancy, the engineer must evaluate the effects of materials of construction, design of parts, severity of service, etc., before making his choice.

Driver Type The choice of the driver type for the pumping equipment is as important as choosing the pump, for frequently the driver can cost more than the pump itself. Depending on the available energy sources, pumps may be driven by electric motors, steam turbines, steam engines, gas turbines, internal combustion engines, etc. Also, pumps may be driven at constant speed or at variable speed. For centrifugal pumps, the variable speed can enable the pumps to operate along the system characteristic curve and thus save on expenditures for power for part-load operations.

Electric motor drives are usually used in constant-speed service unless a hydraulic coupling or other speed-varying device is introduced in the system for varying the speed. Internal combustion engine drives are usually chosen because of location (no electric power available), mobility (portability), or redundancy (loss of power back up) requirements. They can operate either as constant-speed or variable-speed drivers. Steam turbines, eddy-current couplings, adjustable-speed motors, fluid couplings, gears and belts are frequently used where variable-speed operation is required.

In large complex installations where the equipment is to be operated continuously, the decision as to the choice of the type of driver and variability of the pump speed should be based on the comparison of the total operating and capital costs for the pump system over the intended plant life for the several alternatives. Variable-speed operation would usually result in lower operating costs; however, the total first cost of the driving equipment to accomplish this would frequently be higher than for constant-speed equipment. The first cost should include cost of equipment, building space, etc., while the operating cost should include such factors as energy costs, maintenance, and replacement costs. This comparison would usually result in choosing a pumping unit providing the lowest cost per gallon pumped over the useful life of the plant.

The above discussion covers some of the factors to be considered when choosing the class and type of pump which would be most suitable for a given application. It should be recognized that these are general guidelines and that there may be other overriding nontechnical and noneconomic factors, such as prior satisfactory experience, excellent technical or service assistance, etc., which may dictate the final choice of pumping equipment.

SPECIFYING THE PUMP

Specification Types When selecting and purchasing a pump, the first decision to be made is whether or not the procurement will be based upon a formal specification or whether some abbreviated form of requirements will be suitable.

For relatively simple or inexpensive pumps, or for replacement purposes, where duplication is desired, a specification is frequently not used. For inexpensive pumps the time and cost required to write a specification and obtain and analyze competitive bids frequently exceed the potential cost savings. For this case, and where the pump supplier is already established (replacement/duplication) a direct quotation is frequently requested from the supplier. It is important, when requesting this quotation to have the principal requirements well defined and known to the supplier so that he can properly include them in his technical and priced offering. Thus, while a formal specification may not be appropriate, the purchaser should have his requirements well established. A pump data sheet (see Figs. 3 and 4) is helpful in establishing these and can be used in lieu of a complete specification.

Service	Req'n. No.	Item No.
Plant	Spec. No.	No. Req'd.
Location	P. O. No.	Cost Code
Draw'g. Ref.	(Layout) (Process)	Mfr.

LINE NO.		
1	Liquid Pumped	
2	Viscosity (SSU) / Vapor Pressure (PSIA)	
3	Temperature (°F): Max. / Min. // Specific Gravity @ °F.	
4	Flow: Rating / Min. / Max. (GPM) (Lbs./Hr.)	
5	Suction Pressure @ (Flange) (Water Level) (PSIA)	
6	Discharge Pressure @ Flange (PSIA)	
7	Diff. Press.: Rating / Shutoff (Feet) (PSI)	
8	NPSH or Submergence: Available / Req'd @.................... Elev. (Ft.)	
9	Type of Pump / Model / No. Stgs.	
10	RPM / Rotation (View from Motor Facing Pump)	
11	Efficiency / BHP at Rating / BHP Max. @..................GPM	
12	Impeller Diameter: Bid / Max. / Min.	
13	Imoeller: Eye Area / Periph. Vel.	
14	Working Press. Max./Hydrotest PSIG	
15	Clearance: Wear Ring / Bearing / Impeller	
16	Hydraulic Thrust: Rating / Max. / Up	
17	WK² / Speed-Torque: Rating / Speed-Torque Max. (Lb.Ft.²)/(Lb.Ft.)	
18	Suction: Size / Rating / Facing / Position	
19	Discharge: Size / Rating / Facing / Position	
20	Base Plate / Sole Plate	
21	Coupling: Type / Mfr. / Furn. By	
22	Suction: Strainer Mtl. / Splitter Mtl.	
23	Bearing Lube: Type / GPM / Pressure / Max. Micron	
24	Shaft Seal: Type / Sealing Conn. / Cooling (GPM)	
25	Coupling Guard Type (Horiz. only)	
26	Material: Case or Bowl	
27	(& Size) Barrel	
28	Shaft: Case or Bowl / (Dia.)	
29	Shaft Sleeve: Brg. / Stuff. Box	
30	Wear Ring: Case / Impeller	
31	Impellers / Liners	
32	VERT. ONLY: Disch. Head / Column (Dia. x Thickness)	
33	VERT. ONLY: Shaft Encl. Tube / (Dia. x Thickness)	
34	VERT. ONLY: Lineshaft / (Dia.) / (Brg. Spacing)	
35	VERT. ONLY: Sleeve Bearing: Bottom / Bowl / Lineshaft	
36	Driver: Type (Motor–Turbine) (Solid–Hollow Sh.) / RPM / HP	
37	Furn. By / Weight / Dwg. Ref. / Mfr.	
38	Bearing Description / Thrust Rating	
39	Lubrication: Thrust / Radial / Cooling	
40	Drawing No.: Outline / Sectional / Performance Curve	
41	Net Weight: Pump / Removable / Rotating Elem.	
42	Inspection: Std. / Nuc. Class I, II or III / ASME III or VIII	
43	Testing: Ultrasonic / Eddy Cur. / Mag. Part. / Liq. Pen. / Radio.	
44	Hydrostatic / Witness // NPSH / Witness	
45	Performance / Witness / Field	
46	Quality Assurance: Mfr. Std. / Documented	
47	Seismic Design Req's: Class I / Class II	

FILL IN ALL BLANKS: IF NOT APPLICABLE MARK "NA"

NO.	DATE	REVISIONS	NOTES
△			
△			
△			

ORIGIN	CENTRIFUGAL PUMP DATA SHEET	JOB NO.	
		DATA SHEET NUMBER	REV.

Fig. 3 Centrifugal-pump data sheet.

LINE NO.				
	Service	Req'n. No.		Item No.
	Plant	Spec. No.		No. Req'd.
	Location	P.O. No.		Cost Code
	Draw'g. Ref.	(Layout)	(Process)	Mfr.
1	Liquid Pumped / Newtonian or Non-N.			
2	Viscosity (SSU) (CS): Norm. Temp. / Min. Temp. / Max. Temp.			
3	Vapor Press. (PSIA): Norm. Temp. / Min. Temp. / Max. Temp.			
4	Temperature (°F): Norm. / Min. / Max.			
5	Specific Gravity / Pour Point / Flash Point @..............°F			
6	Solids: Type / % by vol. / Abrasive / Shape / Distr. / Density			
7	Gases: Entr. / Dis'd. / Solubility (% by Vol. at 30" Hg. abs.)			
8	Flow: Rating / Min. / Max. (GPM)			
9	Pressure: Outlet / Inlet / Diff. Max ΔP (PSIG)			
10	NPSH: Available / Req'd. @ ℄ Suction (PSIG)			
11	Type of Pump			
12	RPM / Rotation Facing Coupling			
13	Efficiency / BHP Rating / BHP Max			
14	Relief Valve: Internal - Setting / External - Setting (PSIG)			
15	Suction Conn.: Size / Rating / Facing / Position			
16	Discharge Conn.: Size / Rating / Facing / Position			
17	Mfr. Brg. No.: Radial / Thrust			
18	Base Plate: Pump (& Gear) Only / Comb. W. Driver (& Gear)			
19	Coupling: Mfr. / Model Size / Guard / Dr. Half Mtd.			
20	Seal: Packing Size, No., Type / Mech. Seal Mfr., No.			
21	Lubrication: Main, Type / Secondary, Type / Max. Micron			
22	Cooling: Water GPM., Pr. / Quench / Steam Jacket			
23	Material: Case			
24	Rotors			
25	Shaft / Shaft Sleeve			
26	Throat Bushing			
27	Lantern Ring / Gland			
28	Relief Valve			
29	Base Plate			
30	Cooling Piping / Steam Piping			
31	Seal Piping / Lubrication Piping			
32	Driver: Type (Motor - Turbine – Gear) / RPM / HP			
33	Furn. By / Weight / Dwg. / Mfr.			
34	Bearing Description: Radial, No. / Thrust, No.			
35	Lubrication: Radial / Thrust / Cooling			
36	High Speed Cplg.: Mfr. / Model / Furn. By			
37	Drawings: Outline / Sectional / Perf. Curve			
38	Net Weight			
39	Inspection: Std. / QA			
40	Testing: Ultrasonic / Eddy Cur. / Mag. Part. / Liq. Pen. / Radiogr.			
41	Hydrostatic / Witness // NPSH / Witness			
42	Performance / Witness / Disassembly			
43	Quality Assurance: Mfr. Std. / Documented			
44				
45				
46				
47				

FILL IN ALL BLANKS: IF NOT APPLICABLE MARK "NA"

NO.	DATE	REVISIONS	NOTES
△			
△			
△			

ORIGIN		ROTARY PUMP DATA SHEET	JOB NO.	
			DATA SHEET NUMBER	REV.

Fig. 4 Rotary-pump data sheet.

Frequently the supplier can assist in developing and clarifying these requirements, although the engineer should recognize the fact that, naturally, each supplier would tend to favor his own equipment.

For those procurements where a formal specification is indicated, the type of specification to be written is of fundamental importance. In most cases the specification will be of the performance type rather than the construction type. The performance specification basically establishes the performance which the pump must achieve and does not attempt to dictate pump design or construction methods, although certain details of construction are frequently established, particularly where choices may exist. For example where either leak-off or mechanical shaft seals may be offered, the performance specification would usually state a preference. The performance specification, however, basically establishes "what," not "how."

The construction basis specification establishes in some detail the type of design, construction, and methods to be employed in designing the pump and certain other features, which, if the performance specification method is utilized, are left to the manufacturer's discretion. From the standpoint of legal responsibility, if a construction basis specification is used, manufacturers may respond and advise that since the purchaser has established certain of the design features of the pump the manufacturer cannot be responsible for the performance of the pump.

It is therefore important that care be taken when writing the construction type specification not to relieve the manufacturer of his responsibilities for applicability, suitability, and performance of the pump, and that care also be exercised by the purchaser to avoid the unnecessary assumption of responsibility for the proper performance of the pump.

In short, unless there are unusual circumstances, it is far more appropriate to specify the pump on the basis of the performance required rather than the construction, unless the purchaser has a high degree of assurance that the requirements which are called out in the specification can be met and will not relieve the pump supplier of responsibility.

Codes and Standards When specifying a pump the codes and standards which apply are of major importance. Standards which relate to quality of materials should be referenced. ANSI, ASTM, MIL, or other industrial standards which establish such factors as metallurgy, dimensions, tolerances, flange facing and drilling, etc., should be referenced where appropriate. Similarly, if a pump is to meet certain critical service requirements there are in many cases industrial codes which apply to the design, construction, and application of the pump. An example of this is the ASME Boiler and Pressure Vessel Code Section III. For nuclear service pumps, which are pressure-retaining parts, this section establishes very specific requirements for control of the materials used in the pump construction as well as specifying certain of the pump design features and quality of manufacturing control. These codes in some cases are extremely detailed in specifying the construction of the pump and will serve in part to provide a rather well-defined specification within themselves. It is, of course, essential to establish the dimensional standards which apply, such as metric or English, the codes which may apply to the construction and fittings of the pump and of course the industrial codes which apply to the application of the pump for the service intended.

In all cases, however, the engineer must review each reference to ensure that it does not introduce conflict. Some codes and standards include alternate choices of material or inspection methods, requiring selection by the specifying engineer. Others may include cross-references to additional codes which the engineer may wish to exclude.

Alternates It is extremely difficult for a specification to cover all the possible pumps which may be offered by the various manufacturers. Coupled with that is the problem faced by a potential user in remaining up to date with the changing state of the art and the development work being performed by the manufacturers.

It is a good practice to allow the offering of alternates by the manufacturers. This provides the pump suppliers with an opportunity to present their best offer and also provides the buyer with the advantage of obtaining from the several manufacturers, or bidders, their recommended modification to the specification in the form of potentially attractive alternate offerings. The specification should encourage alternate offerings, recognizing, however, that the choice of the alternates is fully up to the purchaser, who may choose to reject any and all bids including alternates.

Bidding Documents The bidding documents for pumps will normally consist of two major parts.

Technical specification
Commercial terms

The technical specification establishes the performance requirements, materials of construction, and major technical features. The commercial terms include the contract language and cover such items as the location of the work, requirements for guarantees/warranties, shipping method, time of delivery, method of payment, normal inspection, and expediting requirements, etc.

Frequently the commercial terms and conditions are relegated to second place especially when purchasing standard low-dollar-value pumps, but in many cases, the commercial terms can assume more significance than many of the performance requirements.

Technical Specification The technical specification should consist of a series of sections which are rather carefully defined and distinct. The more complete and specific the specification, the more competitive will be the bid prices offered. A typical specification might have the following sections and content:

1. Scope of Work. Pump, baseplate, driver (if included), interconnecting piping, lubricating-oil pump and piping, spare parts, instrumentation (pump-mounted), erection supervision, etc.

2. Work Not Included. Foundations, installation labor, anchor bolts, external piping, external wiring, motor starter, etc.

3. Rating and Service Conditions. Fluid pumped, chemical composition, temperature, flow, head, speed-range preference, load conditions, overpressure, runout, off-standard operating requirements, transients, etc.

4. Design and Construction. (Care should be taken to provide latitude in this section as this borders on dictating construction requirements.) Codes, standards, materials, type of casing, stage arrangement, balancing, nozzle orientation, special requirements for nozzle forces and moments (if known), weld-end standards, supports, vents and drains, bearing type, shaft seals, baseplates, interconnecting piping, resistance temperature detectors, instruments, insulation, appearance jacket, etc.

5. Lubricating-Oil System (If Applicable). System type, components, piping, mode of operation, interlocks, instrumentation.

6. Driver. Motor voltage standards, power supply and regulation, local panel requirements, wiring standards, terminal boxes, electric devices. For internal combustion drivers, fuel type preferred (or required), number of cylinders, cooling system, speed governing, self or manual starting, couplings or clutches, exhaust muffler, etc.

7. Cleaning. Cleaning, painting and preparation for shipment, allowable primers and finish coats, flange and nozzle protection, integral piping protection, storage requirements.

8. Performance Testing. Satisfactory for the service, smooth-running, free of cavitation and vibration, shop tests (Standards of the Hydraulic Institute) for pump and spare rotating elements, overspeed tests, hydrostatic tests, test curves, field testing.

9. Drawings and Data. Drawings and data to be furnished, outline, speed vs. torque curves, WK^2 data, instruction manuals, completed data sheets, recommended spare parts.

10. Tools. One set of any special tools including wheeled carriage for rotor if needed for servicing and maintenance.

11. Evaluation Basis. Horsepower, efficiency, proven design, etc. Supplementing these may be technical specifications relating to other requirements of the order such as:

Electric motor specifications or specifications for the steam turbine or other type of driver.

A specification on marking for shipment

A specification on painting.

Requirements for any supplementary quality control testing. (These are particularly important for pumps for nuclear services where quality control and quality assurance are critical requirements of the Atomic Energy Commission.)

In addition, within the body of the technical specification, it is important that any unusual requirements be carefully listed so that the manufacturers are aware of them. Examples of these are:

Special requirements for repair of defects in pump castings.

For wet-pit applications, a sketch of the intake arrangement.

Special requirements regarding unique testing, for example, metallurgical testings which may be required during the manufacture of the pump apart from the testing for performance.

It is helpful to the pump supplier to provide system-head curves, sketches of the piping system (dimensioned, if this is significant), listings of piping and accessories required, etc.

Pump data sheets are extremely useful in providing in a concise fashion a summary of information to the bidder and also allowing the ready comparison of bids by various manufacturers. Samples of two typical pump data sheets are shown (Figs. 3 and 4). As can be noted by inspecting the pump data sheets, some of the items are filled in by the purchaser and the balance by the bidder to provide a complete summary of the characteristics of the pump, the materials to be furnished, accessories, weight, etc. The data sheets should be included with the technical specification.

Commercial Terms The commercial terms which are included with the bidding documents should cover the following information:

General, including information such as the name of the buyer, the place to which the proposals are to be sent, information on ownership of the documents, time allowed to bid, the governing laws and regulations.

Location of plant site. This establishes the geographic area in which the equipment is to perform and in a broad way the scope of the work. It should also state maximum temperatures, humidity, storage provisions (indoor or outdoor), altitude (so that the motor drivers can be selected for the proper cooling), etc.

Definitions establish the buyer, the agent, engineer, seller, whether "or equal" is intended to be exclusive or not, etc.

Proposal, which establishes the format of the proposal, number of copies, the owner's right to accept or reject any bids, the status of alternates.

Schedule, including the requirements for all drawings and design data submittals, manufacturing schedule, and equipment delivery.

Acceptable terms of payment, retentions, etc.

Transportation to and from the point of use (or installation) is frequently a consideration since with very large equipment it may not be possible to ship by truck and it may be necessary to either barge- or rail-ship the material. In addition, it is important to establish the method of shipment which may be used so that the bidder can include the proper allowance for freight as well as to establish responsibility for the risk of loss. Thus, a request for bid could include either a freight allowance; that is, a bid (*a*) FOB manufacturer's plant WFA (with freight allowed) to the point of use, or (*b*) that the equipment is bid FOB the point of use, in which case freight is included. In either case, the risk of loss remaining either with the seller or assumed by the purchaser should be clearly stated.

These items represent the principal considerations that must be known to each bidder.

SPECIAL CONSIDERATIONS

Performance Testing An important part of any specification is the requirement for testing. Normally small commercial pumps which are routinely produced by a manufacturer up to about 6 in in size would be tested on a sample-selection quality-control basis and from that, standardized curves of pump performance would be available. Thus for pumps in this range of size it is not necessary to require certified tests unless they are used in critical service such as fire protection, boiler feed, etc.

For larger pumps or pumps with more critical service requirements, however, a certified performance test should be required. This test requires that the manufacturer test the pump at several points on its performance curve to establish its exact head curve. Since it is necessary to assure that the pump driver is of the proper size a horsepower curve must be furnished with the head curve. Occasionally pumps for special services or pumps of extremely large size cannot be tested in the manufacturer's shop. Examples of this are very large low-head pumps used for circulating-water service, low-lift irrigation purposes, liquid-metal service, etc. A usual method for testing of these pumps is to extrapolate the performance tests of a geometrically similar model (usually smaller) to the pump size under consideration and use that information for the development and engineering of the pump. The actual performance testing in this case takes place following installation of the pump. It is important that the purchaser and the supplier agree upon a proper (field) test method in some detail. This method should include the number of points at which the head curve will be determined, the applicable code (Hydraulic Institute Standards, etc.), the specific method of traversing the pump discharge characteristics across the cross section of the discharge pipe, the manner in which the head will be varied, etc. Care should be taken in establishing this procedure to set forth the characteristics of the fluid and other variables which can affect the performance test of the pump. The specification should establish the performance testing requirements for the pump and whether or not it is necessary that the testing of the pump be witnessed. Witnessing of the pump and the furnishing of certified test data (including the actual test work done), are frequently priced separately and if not specified can be a source of dispute between the purchaser and supplier.

Pump Drivers Pump drivers (motors, turbines, engines, etc.) can be purchased either with the pumps or separately. As a general rule with pump drivers in the range of 50 hp and larger, consideration should be given to purchasing the driver separately since this can frequently result in cost savings. Exceptions to this are small pumps, pumps using "monobloc" construction (where the pump is mounted on and supported by the motor), and pumps built to special codes such as the Underwriters engine-powered fire pumps.

Where the driver is excluded from the pump scope of supply, the specification should require the pump supplier to determine the proper characteristics of the driver. This includes establishing the proper motor speed, sizing the driver for both the accelerating and running loads, assuring end-float compatibility and/or thrust-bearing loadings including direction, and selection and fitting of the couplings. If the driver is purchased separately and can be economically and conveniently shipped to the pump supplier's plant, the pump supplier should be required to mount the motor half of the coupling as well as align and mount the motor (for common baseplate installations). For very large motors or where it is costly or impractical to ship the motor to the pump supplier, it may be necessary to perform this work at the point of installation. To assure compatibility with the other motors in the plant it is important to specify the motor enclosure type, insulation standards, special features required, such as motor heaters, oversize junction boxes, etc.

For steam turbine drivers, speed range, throttle pressure, steam quality, exhaust pressure and control method should, in addition, be specified.

When reciprocating pumps or drivers are specified the bidders should be asked to furnish the magnitude and direction of the shaking forces, as well as weight, size, etc., to permit preliminary evaluation of the foundation requirements.

NEMA (National Electrical Manufacturers Association) standards for certain type of motors provide a horsepower overload allowance (service factor) of 15 percent based upon internal motor temperatures. It is good practice to conserve this margin for short-term overloads. Since it is based on the internal heating characteristics, short-term encroachment (starting surges, etc.) is frequently used to avoid increasing the motor to the next larger frame size. For large motors 200 hp and over, where heat dissipation is more difficult, the motor nameplate usually carries starting-frequency limitations. For very large motors, typically 1,000 hp and larger, the purchaser should state his starting-frequency requirements.

Typical limitations for these very large motors will be 3 to 5 starts per hour but may vary due to the WK^2 of the load, the load torque, voltage applied, etc. The motor manufacturers should be consulted where there is any consideration for repetitive starting.

Synchronous motors are frequently used as drivers on large (generally slow-speed) pumps. A rule of thumb often used is that synchronous motors should be considered for the application where the horsepower is equal to or greater than the pump speed. Synchronous motors require more complex control equipment but can be tailored to the specific starting, pull-in (to synchronism) and pull-out (of synchronism) torques of the load. Since these motors operate at synchronous speed, pump-driver "slip" can be disregarded. In addition these motors can be operated at unity power factor to avoid power factor changes in energy bills or if desired can be operated with leading power factors to improve the average power factor of the other equipment at the installation.

Intake Vertical wet-pit pumps are sensitive to the proper geometry of their suction pit. Factors to consider include: clearance beneath the bottom of the suction bell and the floor of the pit, spacing between pumps or between the pump and

DATE | APPROVALS | MATL. | SUPV. | CHK. | DR. | ENG | DESCRIPTION | REV. | 0

This schedule of drawing and data requirements is to be fulfilled before rendering final invoices. See below for drawings required and dates due. Failure of Seller to comply with drawing and data requirements may result in order cancellation in the case of initial drawings, or final payment being withheld in the case of final drawings. Drawings are to be forwarded to:

Attention: ....

IN ADDITION, FORWARD WITH SHIPMENT, ONE SET OF ANY DRAWINGS NECESSARY FOR FIELD INSTALLATION. FORWARD COPY OF LETTER OF TRANSMITTAL TO:

	TYPE OF DRAWINGS AND OTHER REQUIREMENTS	APPROVAL BEFORE FAB (YES/NO)	KIND OF COPIES	NUMBER REQUIRED	
				INITIAL	FINAL
A	OUTLINE DIMENSIONS AND FOUNDATION REQUIREMENTS		TRANSPARENCY		
			PRINTS		
B	CROSS SECTION WITH PARTS LISTS WITH PRICES		TRANSPARENCY		
			PRINTS		
C	SHOP DETAIL DRAWINGS		TRANSPARENCY		
			PRINTS		
D	CERTIFIED PERFORMANCE DATA AND TEST REPORTS		TRANSPARENCY		
			PRINTS		
E	WIRING DIAGRAMS		TRANSPARENCY		
			PRINTS		
F	CONTROL LOGIC DIAGRAMS		TRANSPARENCY		
			PRINTS		
G	WELDING PROCEDURES		TRANSPARENCY		
			PRINTS		
H	CODE CERTIFICATES, INSPECTION AND TEST REPORTS		ORIGINAL COPIES		
J	INSTRUCTIONS FOR ERECTION OR INSTALLATION, OPERATION AND MAINTENANCE		MANUALS OF EACH TYPE		
K	LIST OF RECOMMENDED SPARE PARTS FOR ONE YEAR'S OPERATION, WITH PRICES		LISTS		
L	COMPLETED DATA SHEETS		TRANSPARENCY		
M	MATERIAL CERTIFICATIONS				
N	MANUFACTURERS QUALITY CONTROL, INSPECTION AND TEST PROCEDURES AND REPORTS				

Seller's drawings will be reviewed and approved only as to arrangement and conformance to the specifications and related drawings, and approval shall not be construed to relieve or mitigate the Seller's responsibility for accuracy or adequacy and suitability of materials and/or equipment represented thereon.

Final drawings must be certified and must show adjacent to the title block, Buyer's equipment title and number, manufacturer's serial number and purchase order number. Initial transparencies must be made from faultless masters. Final transparencies shall be on wash-off Mylar. Additional drawing requirements will be specified in the Technical Specification.

Initial drawings required within..............days of receipt of firm order. Final drawings required within..............days of receipt of initial drawings, or within..............days of receipt of firm order if no initial drawings are requested. The finalized drawing transmittal requirement dates will be specified in the purchase order and will take precedence over the above.

		DRAWINGS AND DATA REQUIREMENTS	JOB NO.	
			ATTACHMENT TO REQUISITION NUMBER	REV.

Fig. 5 Drawings and data requirements.

the pit walls (both side and rear), approach angle of the floor of the pit, including surging and surcharge, submergence, and lack of uniform approach flow.

The Standards of the Hydraulic Institute include recommendations on the geometry of intakes. These as well as the recommendations of the manufacturer of the pumping equipment should be carefully reviewed. Suction piping, where complex or unusual, should be treated in a similar manner.

When specifying vertical wet pit pumps a layout of the installation should be furnished to the bidders for their information and comment. In many cases, if the geometry of the installation is not fixed, bidders can recommend small changes to improve pump performance. Where the geometry is fixed it may be necessary to add antivortexing baffles, surge walls, or flow-directing vanes (or walls) to avoid pump operating problems.

For moderate or large installations where any design question exists, model testing should be considered. Several pump manufacturers offer this as a service, as do a number of universities and commercial testing laboratories. Responsibility for proper pump performance will rarely be assumed by the bidder where the intake pit is of nonoptimum size or shape. The use of model testing is usually resorted to in these cases also. (For a more complete discussion see Sec. 11.)

Special Control Requirements Special control requirements is an area where care must be exercised to assure that pumps are not damaged by misapplication. Pumps that do not require more horsepower at shutoff, that is, centrifugal and mixed-flow pumps up to approximately 5,000 specific speed, can be started with a closed discharge valve and will not overload their drivers.

Propeller pumps require very high horsepower and produce very high pressures (relative to design) at shutoff, and they are started with valves that are timed to be sufficiently open when the driver comes up to full speed. In systems with a large mass of water to accelerate, opening the pump discharge valve is of no value since the inertia of the water creates a shutoff condition or close to it. For these applications a bypass or means of slowing the acceleration of the driver must be considered (see also Sec. 9).

In addition to the question of starting control, consideration must be given to pump shutdown and closing of discharge valves. For systems with large diameter piping, typically 24 in and larger, motor-operated butterfly valves are located in many cases in the pump discharge in lieu of check valves. These butterfly valves act as isolation valves and by proper control can ensure that reverse flow does not occur when not wanted. A careful analysis must be made of the pump starting requirement including the motor accelerating capability and the butterfly-valve characteristics, that is, flow vs. opening and opening angle vs. port area.

Drawing and Data Requirements This form indicates the type of drawings and data required both to define requirements for preliminary design purposes and for final information, that is, the "as built" dimensions and the certifications which demonstrate the pump meets the specified requirements. The schedule for information submittals is required to provide information for structural, electrical, or piping design necessary to incorporate the pump into the plant installation. A typical form is shown in Fig. 5.

BIDDING AND NEGOTIATION

Public and Private Sector Depending on whether the enterprise is in the public or private sector, certain requirements are placed upon the enterprise relating to the manner of bidding or negotiating for the award of major equipment contracts. Public bodies are required by law, in most instances, to take competitive bids from a broad range of bidders and to have public bid openings of their offerings. The bid openings are usually well documented and the bids recorded for public inspection. For work in the private sector there are rarely such *legal* requirements, although in the case of certain regulated industries such as public utilities it is a normal procedural requirement that bids are taken. Similarly, many firms in the private sector require as a matter of policy that bids be taken for all purchases of a substantial nature.

Where bids are to be taken, a specification must be written in accordance with the guidelines previously presented. Where there is an option, it may be preferable to preselect a bidder and negotiate the price for the recommended equipment with him. This is particularly true where the equipment is relatively low in cost or may be desired as a replacement or duplicate for an existing piece of equipment. If negotiation is used, it is still important that the specification requirements be developed, that they be used as the basis for negotiations to ensure that major considerations and characteristics are not overlooked, and that a clear basis for understanding between the supplier and the purchaser be established. Thus the only real difference between a negotiation and competitive bidding lies in the case of the formalization of the bid documents.

Bid List The development of a bid list is extremely important for both the owner and the potential supplier. In the case of the owner the bid list provides a checklist of suppliers who are able to furnish pumps of the type and quality required on schedule and are in the position to offer the proper degree of support service necessary both during design of the pump installation and over the life of the equipment. This support service includes the ability to provide adequate engineering data on the pump to suit the schedules for the process engineering, piping, electrical, and foundation design together with requirements and data for auxiliary systems required with the equipment. In the case of very large equipment, installation supervision service may be part of the contract. In addition, the supplier may be called on to supervise the startup of the installation, and over the life of the equipment, provide field service and spare parts.

Firms in the private sector have the ability to limit the bid list to selected bidders, either on the basis of experience, investigation, or reputation. Prior experience indicating satisfactory performance or reputation is frequently used as a guide to reducing the bid list to manageable size. Where complex or sophisticated equipment is involved bidders should be prequalified to assure their ability to design and produce equipment of a satisfactory nature. This typically includes inspection of the manufacturer's shops (size, tooling, capacity vs. backlog, etc.), detail reviews of other installations and pumps in service, the design, manufacturing, and quality-assurance organization, financial strength etc.

It is convenient to try to limit the number of bids received to about five since this provides an adequate number of bids for effective price competition without making the bid evaluation burdensome. To assure five bids it is prudent to solicit inquiries from up to seven suppliers if there is expectation that some "no bids" will be received. This limitation on the bid presents a problem since typically there are many more qualified suppliers. The manner of selecting the bidders varies widely but where qualified suppliers have indicated a genuine interest they should be allowed to bid even if the bid list is enlarged somewhat.

Bid Time The preparation of bids for pumping equipment will take different amounts of time depending upon the complexity of the equipment and whether or not prime movers are included in the proposal requirement. As a guideline the list below represents the typical time that vendors would like to have to prepare their proposal (from time of receipt of the bid request). This presumes that a request for proposal calls for a pump application that is relatively straightforward and does not require exhaustive analysis or research. A heavy work backlog within a firm, or the entire industry, may also extend the bid preparation time.

Type of application	*Time required*
Preengineered and conventional pumps 6-in discharge and smaller	3 weeks
Pumps 6 to 48-in discharge; pumps with other than electric motor drivers	5 weeks
Large pumps (circulating-water, large sewage pump, etc.)	6 to 12 weeks
Special pumps (liquid-metal, special case construction, ultra-high-speed, etc.)	Consult bidders

Bid extensions if they are reasonable and require only modest changes in the bidding time should be granted to a qualified supplier. Occasionally pump suppliers have numerous bid requests falling due on the same date and they may require an extension to meet the bid due dates. If the bid due date is arbitrarily maintained it is possible to lose the advantage of a potentially favorable bid. If a bid extension is granted, all bidders should be notified so that they may have the advantage of the additional time to improve their proposals.

Since bid openings for public bodies are usually advertised long in advance, bid extensions are generally not granted for these cases.

Bidding of Alternates The bidding of alternates should be allowed for the reasons cited earlier in this section. Unless there are compelling reasons, alternates should be permitted although the purchaser should maintain the right to reject any and all bids including alternates.

Legal Restrictions. When working in the public sector particularly, there may be restrictions sometimes related to financing upon the use of materials or equipment fabricated outside the United States. Frequently where non–United States materials are allowed, penalties are assessed against the bidder for the use of these materials. In addition, in certain unique situations a bidder will be given an evaluation credit. One example of this has been the credit given to bidders on public works projects whose plants are located in depressed labor areas.

In the case of certain large regulated utilities it is sometimes preferred that awards for major equipment contracts be placed with firms in the service territory of the utility. There is usually no cost penalty in placing the order within the utilities service territory; however, care must be taken to assure proper supplier cooperation if the plant design work is being performed elsewhere. This is particularly true for the timely furnishing of design information, coordination of motor or driver selection, and other interface areas.

EVALUATION OF BIDS

Perhaps the most important consideration after having specified the pumps is the manner of evaluating the bids. Evaluation factors that are used should consider not only the first cost but the pump performance, as well as guarantees, economic advantage of alternates, delivery, maintenance, installation service, etc. The sum of these several factors serves to provide the purchaser with a broader base for his purchase decision and will assure more satisfactory performance over the life of the equipment.

It is convenient to separate the evaluation of bids into two categories: those that relate to price, and those that relate to technical features. In this way the technical features which are important to performance but are difficult to quantify are treated separately and can be compared to the cost differences from the priced evaluation.

Cost First cost is a major consideration in the evaluation of bids. In many cases where offerings may be almost identical because of a commonality among equipment supplied by various bidders, the only real difference may be cost itself. Where cost alone is considered, or where cost is of paramount concern, it is prudent to consider not only the basic price offered by the supplier but the difference in price between the first, second, and third bidder, and the percentage of the bid that these differences represent. In many cases it will be found that the difference in bid price may be tenths of percents and thus the absolute values of the differences when viewed in this perspective become less significant. Where this is the case, more detailed attention should be given to the technical factors.

More usually, however, and this is particularly true with large and complex equipment, where relatively small details of performance can over the life of the equipment yield fairly high capitalized values, the technical factors require detailed consideration.

Efficiency Evaluations should include an estimate of the effect of the different efficiencies quoted by the suppliers. This evaluation is normally made at the warranted point (usually full load) while efficiencies at other points are treated qualitatively unless extended periods of part-load operations are contemplated. The efficiency should be rated from *base,* the highest efficiency quoted, with an increasing evaluation penalty against pumps having decreasing efficiency. The actual evaluation or quantification of the effect of the efficiency can be done in several ways.

A direct charge in dollars per year for one or more years for the difference in pump horsepower required. For example: a 70-hp difference, measured at the motor terminals and thus including the combined pump and driver efficiency effect, would have an evaluation value of $6,580, based upon the following assumed factors:

1. Electric motor driver
2. 15 mills/kWhr power cost
3. No demand charge
4. Full-load operation 50 weeks/year, 7 days/week, 8 hr/day
5. 3-year amortization (or payout) period

$$70 \times 0.746 \times 15/1{,}000 \times 50 \times 7 \times 8 \times 3 = \$6{,}580$$

The charge for the differential energy based upon the purchased cost of energy evaluated over the life of the pump installation using the present-worth technique. For this type of analysis, the required rate of interest on the investment is predetermined as well as the estimated economic life of the facility. From these values a present-worth factor can be determined from the standard tables of present worth given in any of the principal texts on engineering economics. For example: A 70-hp difference would have an evaluation (present worth) value of $23,200 based upon the following factors:

1. Electric motor driver
2. 15 mills kWhr power cost
3. No demand charge
4. Full-load operation 50 weeks/year, 7 days/week, 8 hr/day
5. 20-year economic life
6. 7 percent cost of money (interest rate)

The present worth of 7 percent for 20 years is 10.594. The annual difference is multiplied by this to determine the present-worth value.

$$70 \times 0.746 \times 15/1{,}000 \times 50 \times 7 \times 8 \times 10.6 = \$23{,}200$$

A charge including the cost of providing the capacity. For those cases such as public utilities or large industrial plants which generate their own power the calculation is similar to that above excepting that the efficiency cost will be made up of two elements: the cost of the energy, and the cost to provide the equipment to generate the power required. This latter cost is usually called the *capacity charge*. For example: With an energy cost of 9 mills/kWhr and a $200/kW capacity charge and the previous values, the 70-hp difference has an evaluation value, on a present-worth basis, of $24,400.

$$70 \times 0.746 \times \left(\frac{9 \times 50 \times 7 \times 8 \times 10.6}{1{,}000} + 200\right) = \$24{,}400$$

For some very large companies, the method of calculating these values, the factors to be included, etc. are standardized and frequently more detailed than the foregoing examples (for additional information, see Sec. 9.3).

Economic Life The economic life of the equipment is a consideration in evaluation of any of the proposals but it is an extremely difficult item to measure. To some extent the weight of the equipment (discussed later in the chapter) is an indicator of the ruggedness and the durability of the equipment, all other things being equal. Another measure of the economic life would be the speed of the equipment. Thus, if one pump operates at 1,800 rpm while an alternate pump operates at 3,600 rpm, assuming other construction details are roughly equivalent, it is likely that the 1,800 rpm pump, because of its lower speed, will have a longer economic life and will be less likely to require premature replacement. The concept of economic life is extremely difficult to quantify and this type of item is best left to the technical spread sheet where it is dealt with qualitatively.

Alternates For reasons cited earlier in this chapter the offering of alternates by the manufacturers should be encouraged. In the evaluation of these alternates, however, considerable care must be exercised since they may depart in many cases from the basic concept of the specification and thus will require careful review to assure compatibility. Since the application engineering by the purchaser will not have been done on these during preparation of the specification, it will be necessary to perform this work as a part of the evaluation to ensure proper consideration of the effect of the alternate.

Shipping Cost Shipping cost may be a factor in the evaluation particularly if the pumps are quoted FOB manufacturer's plant with freight paid by the purchaser. The cost for shipping should be determined for each of the alternates, assuming they are quoted other than FOB point of installation or point of usage, particularly if the pumps are large and require special handling. The shipping cost also becomes a factor from the standpoint of the assumption of responsibility for any damage to the pump prior to arrival at the point of use.

Delivery Time An important element of the quotation is the delivery time required for the pump. The amount of time required to design and manufacture the pump, ready for delivery, after approval to manufacture, will vary between bidders but will be generally in accordance with the following list.

Type of application	*Time required*
Preengineered and conventional pumps 6-in discharge and smaller	Off the shelf to 16 weeks
Pumps 6 to 48-in discharge with other than electric motor drivers	12 to 48 weeks
Large pumps (circulating-water, sewage, etc.)	26 to 78 weeks
Special pumps (liquid-metal, special case construction, ultra-high-speed, etc.)	Consult manufacturer

On large and complex projects the scheduling of delivery of the pumps and their prime movers is often extremely critical and it is necessary in many cases to stagger delivery of large pumps or perhaps components in multiple pump installations and provide adequate time for certain fitup and field assembly. Delivery for very large equipment may therefore take place in a staged fashion.

In general, deliveries that are earlier than scheduled are preferred; however, with very large pumping systems if materials are delivered too early (and invoices paid upon shipment), money is expended too early. The extra interest charges on that money should be considered. Early deliveries can also be troublesome if storage space at the point of use is not available. This can cause both storage and rehandling problems, with the storage problems causing extra costs while rehandling not only causes extra cost but affords a chance of damage to the equipment.

Spare Parts It is important to evaluate the cost of spare parts for each of the pumps offered by the bidders. Since the manufacturer is usually in a better position to evaluate the spare parts requirements, it is usual that the specification will call for the manufacturer to provide a priced list of the spare parts which are recommended for one year's maintenance and operation of the equipment. The cost of the spare parts should be reviewed, and if it appears that a bidder's spare parts requirements are either extremely high or low, supplementary information or clarification should be obtained. Either a prebid or pre-award conference with the bidders can be helpful to evaluate these requirements.

Maintenance Costs Maintenance costs are extremely difficult to evaluate in a quantified fashion. It is possible, however, using the estimated spare parts to evaluate in a very rough way what the spare parts costs would be for over a year and to make an estimate of the number of hours required to install at least some of the principal spares. For instance: If the design quoted had two packed glands it might be reasonable to assume that on a yearly basis the two glands should be removed and the packing replaced. If an alternate for a mechanical seal is offered it would be reasonable to expect that the mechanical seal would have a life of perhaps three to five years in clear-water service, and the cost, both parts and labor, evaluated on this basis.

Of more importance is the complexity of construction of the pump, arrangement of case, diffuser, bearings, etc., and an evaluation of which pump would require more disassembly time and hence more labor for the same kind of maintenance work to be performed.

Maintenance also may require the addition of lifting gear or special provisions in building structures, for example, hatches over the pumps, etc. Thus, it may be necessary to install monorails or bridge cranes over the pumps and motors or perhaps pad eyes for chain falls or other lifting devices. In this case, if the pumps that are offered are significantly different, the fixed or permanently installed handling equipment required should be considered. Care should be taken however to also consider the possibility of temporary equipment, such as, rubber-tired mobile cranes being used for certain infrequent operations as opposed to committing a large amount of capital to permanent installations which may be infrequently used.

Space and Auxiliary Requirements For horizontal and vertical (other than wet-pit pumps) space requirements may differ markedly among several pumps offered because of the manufacturer's design or the type of prime mover offered. Thus, if the specification allows bids of pumps at various speeds one manufacturer could end up with a much smaller pump than another if for example he has bid a pump with a 3,600 rpm motor rather than one with an 1,800 rpm motor. For this case the space requirement for the 3,600 rpm unit would be markedly smaller. Where pumps are offered having different physical features which affect the space required, a design study to compare the installed space requirements of the pumps should be performed. This study should be conducted prior to bidding and should include consideration of the equipment foundations, the volumetric requirements of the pumps which affect heights of building, its excavation, etc. It should also include the requirements for weight-handling and servicing equipment discussed earlier, for access of personnel, and for platforms, handrails, and stairways to properly service the motor or driver. Space for motor control centers and control panels should also be considered.

The requirements for a careful analysis of space become extremely important when alternates are offered. For instance, because of economics and space studies which were prepared during the engineering analysis which preceded the specification it may be written around a horizontal centrifugal pump, whereas a manufacturer may choose to offer, as an alternate, a vertical wet-pit pump for this same application. Usually these alternates are offered by manufacturers because they allow an attractive bid price. In this case it will be necessary to perform a rather careful engineering study of the installation of the vertical pump including the problems of the cost associated with the additional foundations, underfloor piping, special forming of underfloor water conduits (if required), etc. An example of this is shown in Fig. 6.

As a part of the considerations of the space requirement, it is necessary to develop an estimate of the *differential* cost for the bids and the alternates offered based upon an evaluation of the following items:

Typical items	*Typical values*
Foundations including embedments, anchor bolts, etc.	\$70.00–200.00/yd^3
Excavation:	
Common	\$1.50–3.00/yd^3
Rock	\$5.00–10.00/yd^3
Building volume	\$1.00–3.00/yd^3
Additional or special motor control	Cost + 50–100%
Special heating and ventilating	Cost + 50–100%
Piping, valves, piping specialties	Cost + 100–150%
Handrails, stairways, ladders and other maintenance and access devices	\$600.00–\$1,500/ton
Weight-handling equipment	Cost + 50%
Instrumentation (cost is equipment cost)	Cost + 100–150%

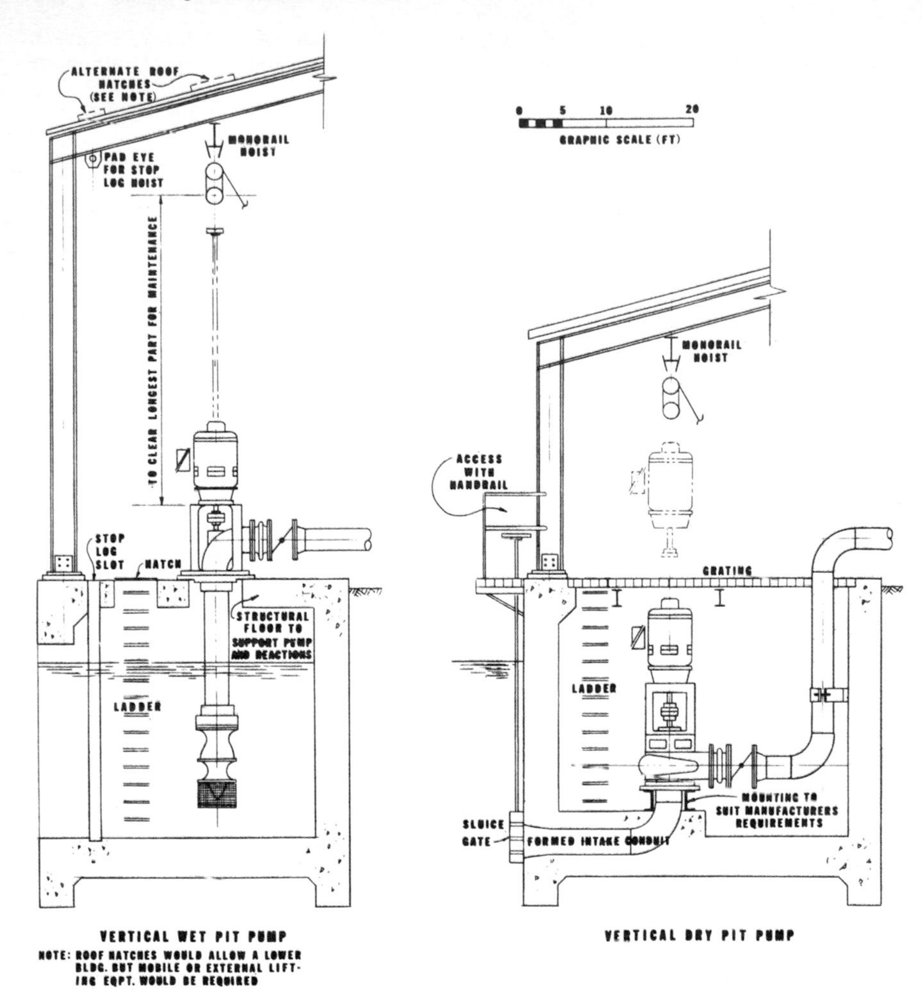

Fig. 6 Typical arrangements of wet- and dry-pit pumps.

Apart from the question of space and physical requirements, there is frequently a difference in the auxiliary requirements for the various pumps offered. Very large pumps will usually have separate self-contained lubrication-oil pump and reservoir systems. These may be driven by their own motors and include heaters, oil coolers, and other auxiliaries. For a proper evaluation it is important that these additional power or utility service requirements (that is, heating, cooling, etc.) be clearly defined so that a proper comparison can be made. This is particularly significant in instances where manufacturers will offer pumps with all this contained within the pump package, whereas others offer it as external to the pump.

Frequently controls will be included with the pump package. Typically, the control voltage will be 120-V ac single-phase, whereas motor voltage will be 480-V three-phase. For this case it is likely that *all* bidders would not offer a stepdown control transformer and its associated wiring even if the specification calls for it. This should be carefully reviewed and considered in the bid evaluation.

Similarly, with large pump packages there may be wiring and piping internal to the package which varies between the bidders. It is usually to the advantage of the purchaser to have fewer (and preferably single) points of connection of these services.

Weight of Equipment Weight of equipment, while not a critical factor, will tend to provide an indication of the ruggedness and potential life of the equipment, as discussed earlier. In evaluating bids, it is important to consider the weight of the largest piece to be handled for both installation and maintenance. This is of great importance as equipment sizes increase. For example, in engineering a pump installation for the top floor of a high-rise building with pumps that weigh 9,000 lb, the lifting ability of the tower cranes which the contractors expect to use must be considered; insufficient lifting ability could require rental of a special crane or additional disassembly plus reassembly. For this case a pump weighing 5,000 lb offered by another manufacturer should be given an evaluation advantage. There are numerous instances of these kinds of items which are significant not only to the cost but also the time required for installation.

Installation Service Installation service should always be requested with a proposal when large or unusual pumps are purchased. The installation service request should be specific about requiring the bidder to state the number of days of installation service time included, as well as the requirements for special tools. If the installation period is expected to be an extended one, arrangements for housing, transportation, and living expenses for the installation service representative should be clearly defined. Manufacturers will usually provide with their bid installation service of a specified and limited amount. Service over and above this, or extensions of service caused by delays (not attributable to the supplier) are usually quoted on a per diem basis for the additional cost for housing, living expenses, etc. Frequently, these per diem rates are requested in case these services are needed. All the same case (margins for additional head), prior relationship with a potential these items should be established prior to placement of the purchase order although they are usually evaluated as third- or fourth-order rank and rarely cause a change in evaluation results. The differences in the installation services offered by various bidders are usually not major and if the installation service time offered by the lower bidders are reasonably equivalent, in general the owner need have no major concern.

Lowest Evaluated Bid Lowest evaluated bid reflects the effect on cost of all of the various items previously referred to and represents the evaluation of bids to include the effect of efficiency, alternates, shipping cost, delivery, erection service, etc. In addition, to evaluate the lowest evaluated bid there are other considerations which cannot be priced but which must be evaluated on a qualifications basis and include items such as extended guarantees, ability to accept larger impellers within supplier, etc. It is important that these also be considered since in some cases the difference between the lowest evaluated bids is small and these factors can assume increased significance. Evaluation of nonquantifiable factors is sensitive, however, as the concept of competitive bidding precludes buying other than the lowest evaluated price that meets the specification.

In any case, these nonquantifiable factors should be reviewed and listed to assure that an item of significance is not overlooked.

Commercial Terms The commercial terms offered by the suppliers can be more significant than the evaluated effect of technical differences. When evaluating the commercial terms, the following items should be considered.

Has the period for acceptance of the proposal expired? With evaluations which have lengthy technical questions to resolve, some of the proposals may expire. Extensions of the acceptance period for the proposals should be obtained from the bidders. Bidders in general are willing to extend their proposals a reasonable amount under these circumstances.

What is the price basis required and what was offered by the bidder? Purchasers prefer firm price bids while bidders particularly on pumps requiring long delivery or development work prefer to quote on a nonfirm basis such as "price in effect at present date plus escalation" or "time and material," etc.

Are the payment terms offered by the bidder acceptable? A typical preference by purchasers is for terms which pay 90 percent of the equipment price on delivery with 10 percent retained for either a year after shipment or until the satisfactory completion of acceptance testing. (For complex equipment, partial or progress payments may additionally be allowed.) Frequently, bidders will wish smaller or

REV.	DESCRIPTION	ENG	DR	CHK	SUPV	MATL	APPROVALS	DATE
0	ISSUED FOR CLIENT APPROVAL	ABC		DEF	JK		M. NDP ST.	XXX

NOTES: ENCIRCLED ITEMS ARE UNDESIRABLE.
A CHECK ✓ INDICATES COMPLIANCE WITH SPECS

	1	2	3	4	5	6	SPEC
MANUFACTURER	MTM	PROSPER	FLJ	EUREKA			
NAME OF BIDDER	"	LUXOR ASSOC.	"	"			
PROPOSAL NO.	N-36473	LA 624	P4372-5F				
DATE OF PROPOSAL	11-14-71	11-9-71	11-16-71				
MFR'S. MODEL NO.	4XL39H	16-32NR	32-11CP9				
TYPE	7 STG.	9 STG.	6 STG.				
MOTOR MANUFACTURER	ELEC. PROD.						
ADDITIONS							
ONE SET OF ACCESSORIES	NOT REQ'D	INCL.	NOT REQ'D				REQ'D
ONE SET SPARE PARTS	INCL	+$1,550.-	INCL.				"
ONE SET SPECIAL TOOLS	NOT REQ'D	+180 (1)	NOT REQ'D	NO			"
ERECTION SUPERVISION	+$2,540	+1,130 (1)	+2,540 (3)				"
FEATURES (PER UNIT)	(10 DAYS)	(5 DAYS) (4)	(10 DAYS)	BID			
SAVINGS TO ELIMINATE UP THRUST	<3,440>	REQUIRED	<3,620>				ALT. QUOTE
SAVINGS WITH SHORTER BARREL	<2,330>	—	<2,840> (3)				
ADD FOR HYD. TESTING	INCL	+200	INCL				
" " INSTALL. COUPLING	✓	✓	+320				
" " TEMP. SUCTION STRAINER	✓	+350 (2)	INCL				
" " 11-13 Cr, 1ST STG. IMP.	✓	+1500 (2)	INCL				
" " 11-13 Cr, SUCTION BELL	+3,140	DECLINED TO BID	+2,650				
ENERGY PENALTY @ $127/HP (PER UNIT)	+12,650	BASE	+4,427				
TOTAL ADDITIONS & PENALTY (TWO UNITS)	+33,100	+8,090	+18,374				
PRICE FOB SHIPPING POINT (TWO)	$209,325	$239,760	$214,782				
ADDITIONS AS ITEMIZED ABOVE "	33,100	8,090	18,374				
COMPARATIVE COST AT SOURCE "	242,425	247,850	233,156				
TRANSPORTATION COST (ESTIMATED) "	5,430	6,580	7,240				BIDDER TO EST.
COMPARATIVE COST AT DESTINATION "	247,855	254,430	240,396				
TOTAL COST (2 UNITS)	247,855	254,430	240,396				
SHIPPING POINT	SERRA	UMPQA	LAKEVIEW				
SHIPPING WEIGHT (TOTAL INCL. DRIVERS)	137,000#	142,000#	131,000#				
DELIVERY (MO. AFTER PO)	12	14	12				12
ESCALATION	NONE	NONE	NONE				FIRM

RECOMMEND: FLJ
REASON: LOWEST EVALUATED PRICE, MEETS SPEC., DELIVERY SATISFACTORY
(1) LETTER OF DEC. 3; (2) LETTER OF DEC. 8; (3) LETTER OF NOV. 29 (4) DOUBLED FOR EVALUATION COMPARISON

SUMMARY OF BIDS (PRICED)
CONDENSATE PUMPS
XYZ POWER COMPANY

JOB No XXXX
P.O. NO.
M-23
SHT. 1 OF 4
REV. 0

Fig. 7A Condensate-pump bid evaluation (sheet 1 of 4).

shorter retentions to improve their cash flow position. In addition, purchasers will sometimes ask for deferred payment terms wherein the supplier receives no payments until completion of acceptance tests one year after installation, etc.

Are the shipment terms offered satisfactory? If the equipment is bid FOB factory with full freight allowed (W/FFA) to point of use, will supplier prepare and handle any shipping claims for damage? Will purchaser handle traffic arrangements for shipment to point of use? Who is responsible for risk of loss?

Does the schedule of shipments offered match the requirements of the project? Usually, payment terms are tied to shipments. If the cause for delay is the supplier's, it usually is paid by him; if the delay was caused by the purchaser, these extra costs are normally added to his account.

	MTM	PROSPER	FLJ	SPEC
PERFORMANCE				
RATED FLOW GPM	✓	✓	✓	6900 GPM
EFFICIENCY @ 6900 GPM	76.5%	83.0%	73.0%	—
DESIGN FLOW GPM	✓	✓	✓	8400 GPM
DESIGN TEMP. °F	✓	✓	✓	112°F
SPEC. GRAVITY @ 112°F	0.991	✓	0.991	—
DIFF. HEAD @ 8400 GPM	✓	✓	✓	1080' TDH
PRESS. @ MIN. FLOW PSIA	710	670 ②	610	675 PSIA
MIN. CONT. FLOW GPM	800	1700	800	—
FLOW @ RUN OUT GPM	10,000	9,400	11,500	9400 GPM
NPSH REQ. @ 9400 GPM	15'	21.4'	15'	—
NPSH REQ @ DESIGN (8400 GPM)	14.5'	17.6'	14.5'	—
BHP @ 6900 GPM HP	2700	2600 ②	2650	—
BHP @ 9400 GPM HP	2900	2880	3100	—
SPEED RPM	1180	1180	1180	900/1200
1ST CRITICAL SPEED RPM	NS	>9500	NS	—
Nº OF STAGES	7	9	6	—
WEIGHT ROT. PART LBS	NS	3700	3100	—
REVERSE FLOW RPM @ 100% HEAD	<125%	1250 RPM	<125%	—
DESIGN				
SUCTION: SIZE/FLANGE RATING	24"/150#	30"/150#	24"/150#	24" PREF.
DISCHARGE: SIZE/FLANGE RATING	18"/300#	18"/300#	18"/300#	16" PREF.
DIAM (O.D.) OF SHELL	42"	44"	42"	—
LENGTH OF SHELL	18'-6" ①	22'-5½"	20'-8"	—
WEIGHT OF PUMP (EACH) LBS	43,000	46,500	40,000	—
WEIGHT OF MOTOR LBS	26,000	23,000	26,000	—
SUCTION: DESIGN PRESS PSI	1.5×MAX. SUCT.	50	1.5×MAX. SUCT.	-
DISCHARGE: DESIGN PRESS. PSI	1.5×SHUT-OFF	650	1.5×SHUT-OFF	-
SUCTION: TEST PRESS. PSI	1.5×MAX. SUCT.	75	1.5×MAX. SUCT.	-
DISCHARGE: TEST PRESS. PSI	1.5×SHUT-OFF	975	1.5×SHUT-OFF	1.5×MAX. SHUT-OFF

① AS PER LETTER, MARCH 9, 1970; INCREASE IN AVAILABLE SUCTION HEAD BY 8 FT. ENABLES MTM TO SHORTEN LENGHT OF BARREL FROM 21'-6" TO 18'-6".

② ADJUSTED TO PROSPER LETTER MARCH 6, 1970

NOTES: ENCIRCLED ITEMS ARE UNDESIRABLE
A CHECK ✓ INDICATES COMPLIANCE WITH SPECS

REV.	DESCRIPTION	ENG	DR	CHK	SUPV	MATL	APPROVALS	DATE
0	ISSUED FOR CLIENT APPROVAL	abc		gef	gk		[illegible]	XXX

SUMMARY OF BIDS
CONDENSATE PUMPS
XYZ POWER COMPANY

JOB No. XXXX
P.O. NO. M-23
SHT. 2 OF 4
REV. 0

Fig. 7B Condensate-pump bid evaluation (sheet 2 of 4).

Is force majeure properly defined in the commercial terms or proposal? There are frequently differences between purchasers and suppliers as to exactly what is included: these should be carefully reviewed.

Has the bidder agreed to accept the other commercial terms included with the bidding documents? The other commercial terms may relate to rights of access to the fabrication shop for inspection or expediting and review of supplier's quality

	MTM	PROSPER	FLJ	SPEC
CASE: MAT. OF SUCTION BELL	11-13Cr ST	C.I. ①	11-13 Cr. ST	11-13 Cr. ST
MAT. OF WEAR RINGS	BRZ	C.I.	BRZ	—
MATERIAL OF CASE	A-48C.I.	A-48C.I.	A-48 C.I.	—
STD. WALL THICKNESS / MIN.	1"/7/8"	1"/13/16"	1"/7/8"	—
VANES & WATERWAYS	STD ✓	AS CAST	STD ✓	—
WEAR RINGS SUCT / DISCH.	– / ✓	✓ / –	– / ✓	—
CLEARANCE	0.02-0.024"	① NS	0.02-0.024"	—
IMPELLER: MAT. 1st STG. / OTHERS	11-13Cr ST/BRZ	11-13Cr ST/BRZ	11-13Cr ST/BRZ	11-13Cr ST/BRZ
TYPE 1st STG. / OTHERS	ENCL./SEMI-E	ENCL.	ENCL/SEMI-E	—
DIAM. 1st STG. DESIGN / MAX.	19 1/2"/20 5/8"	20 5/8"/20 5/8"	19 3/8"/20 11/16"	—
OTHER STG. DESIGN/MAX./MIN.	19 1/8"/20 7/16"	20"/20 5/8"/18 5/8"	19 3/8"/20 11/16"	
BALANCE-HYDR. / STAT. / DYN.	✓ / – / ✓	– / ✓ / ✓	✓ / – / ✓	REQ'D.
WEAR RINGS SUCT. / DISCH.	– / ✓	✓ / –	– / ✓	—
WR^2 LB FT^2	1250	650	1250	—
HYDR. THRUST MAX. DOWN	40,000	74700 ②	40,000	—
HYDR. THRUST MAX. UP	NONE	22400 ②	NONE	—
VELOCITY-EYE DESIGN/MAX.	15 1/2 / 23	17/19 FPS	16/21	—
				—
BEARING: MATERIAL	BRZ	BRZ	BRZ	BRZ
BOTTOM LUBE / L/D	SELF/2.4	SELF/1.5 ③	SELF/2.4	SELF
BOTTOM GROOVED / TYPE	✓/SLEEVE	✓/JOURN.	✓/SLEEVE	—
SERIES LUBE / L/D	SELF/2	SELF/1.0 ③	SELF/2	SELF
SERIES GROOVED / TYPE	✓/SLEEVE	✓/JOURN.	✓/SLEEVE	—
TOP LUBE / L/D	SELF/2	SELF/1.0 ③	SELF/2	SELF
TOP GROOVED / TYPE	✓/SLEEVE	✓/JOURN.	✓/SLEEVE	—
THROTTLE LUBE/L/D/GROOV.	SELF/2/✓	SELF/1.0 ③/✓	SELF/2/✓	SELF
SHAFT: MAT. PUMP ELEMENT	SS410	SS416	SS410	SS400 SERIES
MAT. COLUMN	SS410	C. ST.	SS410	—
DIAM. / LONG BEARING SPAN	5"/48"	5 1/2"/45"	5"/48"	—
COMBINED STRESS MAX.	10,000	14,500PSI TORS. ONLY	10,000	—
SURFACE FINISH & BEAR'G.	32	32 RMS	32	GRD
SURFACE @ STUFF. BOX	32	125 RMS	32	GRD
SLEEVE @ STUFF. BOX / THRU GLAND	✓/✓	✓/✓	✓/✓	REQ'D/REQ'D.
PRESS @ STUFF. BOX	15PSI > SUCT.	15-20PSI > SUCT.	15PSI > SUCT.	—
SLEEVE MAT.	SS-420	ALLOY-ST.	SS-420	—

① AS PER PROSPER LETTER MARCH 4, 1970. (SPEC. REQUESTED 11-13 Cr ST SUCTION BELL & 1st ST'G. IMPELLER MATERIAL)
② HIGH DOWN THRUST.
③ SPEC. RECOMMENDS A MIN. RATIO OF 2. PROSPER DECLINES TO UPGRADE SUBJECT RATIO.

NOTES: ENCIRCLED ITEMS ARE UNDESIRABLE
A CHECK ✓ INDICATES COMPLIANCE WITH SPECS

REV.	DESCRIPTION	ENG	DR	CHK	SUPV	MATL	APPROVALS	DATE
0	ISSUED FOR CLIENT APPROVAL							XXX

SUMMARY OF BIDS
CONDENSATE PUMPS
XYZ POWER COMPANY

JOB No XXXX
P.O. NO. M-23
SHT. 3 OF 4
REV. 0

Fig. 7C Condensate-pump bid evaluation (sheet 3 of 4).

assurance. Where significant exceptions to the commercial terms are taken by the bidders, review by legal counsel should be obtained.

Guarantee/Warranty The layman, that is the nonlawyer, must realize that the terms "guarantee" and "warranty" have lost some of the preciseness they once had. In the following comments, the word "guarantee" is used to mean "guarantee/warranty.")

	MTM	PROSPER	FLJ	SPEC
COUPLINGS: INTERNAL SPAN/TYPE	–/KEYED	45°/THREAD ①	–/KEYED	NOT SPEC'D.
DRIVE: SOLID/FLANGED	✓/✓	✓/✓	✓/✓	–
ADJUST	✓	✓	✓	–
MAKE/SIZE	MTM/#5	PROSPER/15½	FLJ/#5	–
SUCT. BARREL: WALL/WANTED	3/8"/✓	1/4"/✓	3/8"/✓	–
INT. VELOCITY DESIGN/MAX.	–/6	3½/4	–/6 ②	–
BASE PART FLANGE	–	57×57×2¼	–	–
GASKET MAT./TYPE	–/ROUND	–/FLAT	–/ROUND	–
MATERIAL	FAB. ST.	C. ST.	FAB. ST.	–
NOZZLE HEAD: NET AREA MOT. BARREL	–	52"	–	–
MAX. DIAM. BARREL	–	46"	–	–
BASE	–	52 x 1½	–	–
FASTENINGS: MAT. INTERNAL	ST	–	ST	–
EXTERNAL	ST	A-193-B7	ST	–
PERFORMANCE TEST IN SHOP	✓	NO	✓	REQ'D
FACSIMILE TEST/SAME PUMP TYPE	NOT REQ'D	✓/✓	NOT REQ'D	–
MECH. TEST IN SHOP, FULLY ASSEM.	✓	③	✓	MECH. ROT.
WELDING REQUIREMENTS TO ASME	✓	④	✓	
COMPL. OF MATERIAL TO ASTM	✓	⑤	✓ ⑥	ASTM

① KEY DRIVEN COUPLINGS ARE RECOMMENDED
② RECOMMENDED RANGE 4-6½ FPS
③ BOWL ASSEM. ONLY, ROTATED BY HAND ONLY
④ WELDERS NOT CODE QUALIFIED
⑤ PHYS. & CHEM. PROPERTY ONLY
⑥ COMPLIES WITH ASTM PROPERTIES BUT DOES NOT SUPPLY MILL TEST REPORTS.

NOTES: ENCIRCLED ITEMS ARE UNDESIRABLE
A CHECK ✓ INDICATES COMPLIANCE WITH SPECS

REV.	DESCRIPTION	DATE	ENG	DR	CHK	SUPV	MATL	APPROVALS
0	ISSUED FOR CLIENT APPROVAL	XXX	ABC		DEF. JK.			JW. NOP. ST.

SUMMARY OF BIDS
CONDENSATE PUMPS
XYZ POWER COMPANY

JOB No XXXX
P.O. NO. M-23
SHT. 4 OF 4
REV. 0

Fig. 7D Condensate-pump bid evaluation (sheet 4 of 4).

An important consideration in evaluating the pump bids is the question of the type of guarantee/warranty provided by the manufacturer. It is suggested that the purchaser, working with appropriate legal counsel, establish an acceptable clause on this subject which he desires to apply to the purchase of pumping equipment. It should be tailored to suit his business philosophies and the law applicable to his business. This clause should then be included in the bidding documents, along with the other suitable commercial terms of purchase, with the request that the bidder base his proposal on these terms.

Typically, these provide assurance of freedom from defects in material and workmanship for one year following installation of the pumps. They usually provide at no charge the necessary replacement parts. For larger more complex pumps longer warranties can usually be obtained, particularly if a pump is not put in service immediately after it is installed. Manufacturers are reluctant to provide extended warranties, however, since these increase the amount of financial risk by extending the period of their exposure for replacement. Therefore it is good practice to obtain the bidder's standard terms and to ask the bidder, as a bid alternate, for the price differential for warranties longer than his standard.

When equipment fails to perform in accordance with specified requirements, suppliers will typically offer new parts, field service as necessary, etc., to upgrade actual performance to meet specified requirements. In those cases where specified performance cannot be achieved, a remedy (usually financial) is stated. This may be a penalty in dollars per percent efficiency short of specified, dollars per additional horsepower required, etc.

Preaward Conference For complex or costly orders, conferences with the evaluated low bidder are frequently held prior to placement of the order. The purpose of this meeting is to resolve open areas before award and thus ensure a clear understanding by both parties. In the event there are major differences which cannot be resolved, it is then possible to award to the second low bidder without invoking cancellation actions and thus avoiding embarrassment to both parties.

Sample Bid Evaluation Figures 7 to 10 show a sample bid evaluation, including the items previously discussed.

REFERENCES

1. "Hydraulic Institute Standards for Centrifugal, Rotary and Reciprocating Pumps," 12th ed., Hydraulic Institute, Cleveland, Ohio, 1969.
2. ASME Boiler and Pressure Vessel Code, Sec. III, 1971.
3. Grant, E. L., and W. G. Ireson: "Engineering Economy," 4th ed., The Ronald Press Co., New York, 1964.
4. Karassik, I. J.: "Engineers Guide to Centrifugal Pumps," McGraw-Hill Book Company, New York, 1964.
5. Martinez, S.: "Equipment Buying Decisions," *Chem. Eng.,* April 1971.
6. Underwriters' Standards.
7. National Electrical Manufacturers Association (NEMA) Standard MG-1-12.47 (a), MG-1-14.30 to 1-14.42, and MG-1-20.43.
8. Flow of Fluids Through Valves, Fittings and Pipe, Crane Co., Technical Paper 410, 1969.

Section 13

Installation, Operation, and Maintenance

IGOR J. KARASSIK

The information contained in this section is general, and should be supplemented by the specific instructions prepared by the manufacturer for the particular type of pump in question.

INSTALLATION

Instruction Books Instruction books are intended to help keep the pumps in an efficient and reliable condition at all times. It is necessary, therefore, that instruction books be available to all the personnel who are involved in this function.

Preparation for Shipment After a pump has been assembled in the manufacturer's shop, all flanges and exposed machined metal surfaces are cleaned of foreign matter and treated with an anticorrosion compound such as grease, Vaseline, or heavy oil. For protection during shipment and erection, all pipe flanges, pipe openings, and nozzles are protected by wooden flange covers or by screwed-in plugs.

Usually the driver is delivered to the pump manufacturer, where it is assembled and aligned with the pump on a common baseplate. The baseplate is drilled for driver mounting, but the final doweling is performed in the field after final alignment. When size and weight permit, the unit is shipped assembled with pump and driver on the baseplate. If drivers are shipped directly to be mounted in the field, the baseplate should be drilled at the job site.

Care of Equipment in the Field If the pumping equipment is received before it can be used, it should be stored in a dry location. The protective flange covers and coatings should remain on the pumps. The bearings and couplings must be carefully protected against sand, grit, and other foreign matter.

More thorough precautions are required if a pump must be stored for an extended period of time. It should be carefully dried internally with hot air or by a vacuum-producing device to avoid rusting of internal parts. Once free of moisture, the pump internals should be coated with a protective liquid such as light oil, kerosene, or antifreeze. Preferably all accessible parts, such as bearings and couplings, should be dismantled, dried, and coated with Vaseline or acidfree heavy oil, and then should be properly tagged and stored.

If rust preventive has been used on stored parts, it should be removed completely before final installation, and the bearings should be relubricated.

Pump Location Working space must be checked to assure adequate accessibility for maintenance. Axially split-casing horizontal pumps require sufficient headroom to lift the upper half of the casing free of the rotor. The inner assembly of radially split multistage centrifugal pumps is removed axially (see Sec. 2.2, Fig. 27). Space must be provided so that the assembly can be pulled out without canting it. For large pumps with heavy casings and rotors, a traveling crane or other facility for attaching a hoist should be provided over the pump location.

Pumps should be located as close as practicable to the source of liquid supply. Whenever possible, the pump centerline should be placed below the level of the liquid in the suction reservoir.

Foundations Foundations may consist of any structure heavy enough to afford permanent rigid support to the full area of the baseplate and to absorb any normal strains or shocks. Concrete foundations built up from solid ground are the most satisfactory. Although most pumping units are mounted on baseplates, very large equipment may be mounted directly on the foundations by using soleplates under the pump and driver feet. Misalignment is corrected with shims.

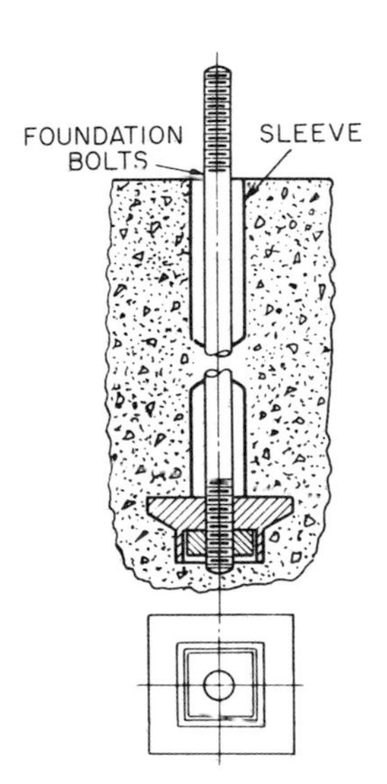

Fig. 1 Foundation bolt.

The space required by the pumping unit and the location of the foundation bolts are determined from the drawings supplied by the manufacturer. Each foundation bolt (Fig. 1) should be surrounded by a pipe sleeve, three or four diameters larger than the bolt. After the concrete foundations are poured, the pipe is held solidly in place, but the bolt may be moved to conform to the corresponding hole in the baseplate.

When a unit is mounted on steelwork or other structure, it should be placed directly over or as near as possible to the main members, beams, or walls, and should be supported so that the baseplate cannot be distorted or the alignment disturbed by any yielding or springing of the structure or of the baseplate.

Mounting of Vertical Wet-Pit Pumps A curb ring or soleplate must be used as a bearing surface for the support flange of a vertical wet-pit pump. The mounting face must be machined because the curb ring or soleplate will be used in aligning the pump.

If the discharge pipe is located below the support flange of the pump (below-ground discharge), the curb ring or soleplate must be large enough to pass the discharge elbow during assembly. A rectangular ring should be used (Fig. 2). If the discharge pipe is located above the support flange (above-ground discharge), a round curb ring or soleplate should be provided with clearance on its inner diameter to pass all sections of the pump below the support flange (Fig. 3). A typical method of arranging a grouted soleplate for vertical pumps is shown in Fig. 4.

If the discharge is below ground and an expansion joint is used, it is necessary to determine the moment that may be imposed on the structure. The pump casing should be attached securely to some rigid structural members with tie rods. If vertical wet-pit pumps are very long, some steadying device is required irrespective of the location of the discharge or of the type of pipe connection. Tie rods can be used to connect the unit to a wall, or a small clearance around a flange can be used to prevent excessive displacement of the pump in the horizontal plane.

Alignment When a complete unit is assembled at the factory the baseplate is placed on a flat, even surface, the pump and driver are mounted on the baseplate, and the coupling halves are accurately aligned, using shims under the pump and driver-mounting surfaces where necessary. The pump is usually doweled to the baseplate at the factory, but the driver is left to be doweled after installation at the site.

The unit should be supported over the foundation by short strips of steel plate or shim stock close to the foundation bolts, allowing a space of ¾ to 2 in between

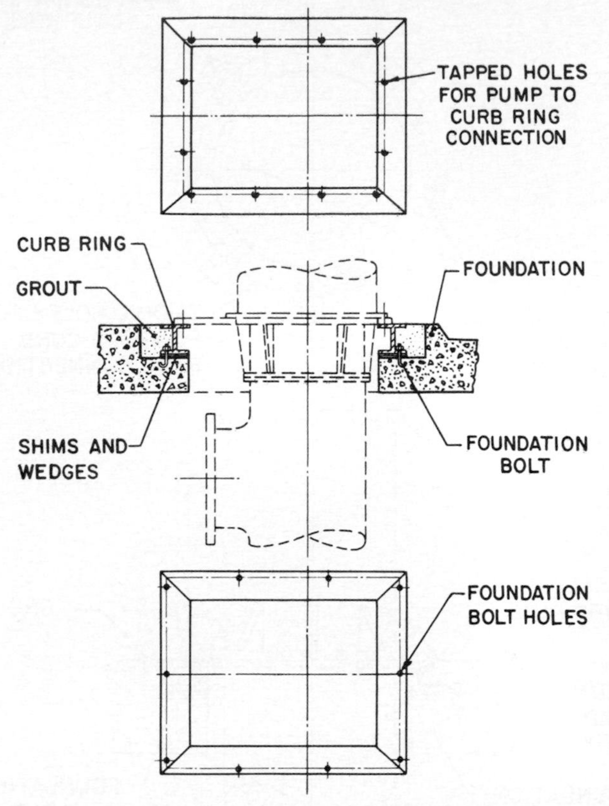

Fig. 2 Rectangular-type curb ring for below-ground discharge vertical pump.

the bottom of the baseplate and the top of the foundation for grouting. The shim stock should extend fully across the supporting edge of the baseplate. The coupling bolts should be removed before the unit is leveled and the coupling halves are aligned. Where possible, it is preferable to place the level on some exposed part of the pump shaft, sleeve, or planed surface of the pump casing. The steel supporting strips or shim stock under the baseplate should be adjusted until the pump shaft is level, the suction and discharge flanges are vertical or horizontal as required, and the pump is at the specified height and location. When the baseplate has been leveled, the nuts on the foundation bolts should be made handtight.

During this leveling operation, accurate alignment of the unbolted coupling halves must be maintained. A straightedge should be placed across the top and sides of the coupling, and at the same time the faces of the coupling halves should be checked with a tapered thickness gage or with feeler gages (Fig. 5) to see that they are parallel. For all alignment checks including parallelism of coupling faces, both shafts should be pressed hard over to one side when taking readings.

When the peripheries of the coupling halves are true circles of equal diameter and the faces are flat and perpendicular to the shaft axes, exact alignment exists when the distance between the faces is the same at all points and when a straightedge will lie squarely across the rims at any point. If the faces are not parallel, the thickness gage or feelers will show a variation at different points. If one coupling is higher than the other, the amount may be determined by the straightedge and feeler gages.

Sometimes coupling halves are not true circles or are not of identical diameter because of manufacturing tolerances. To check the trueness of either coupling half,

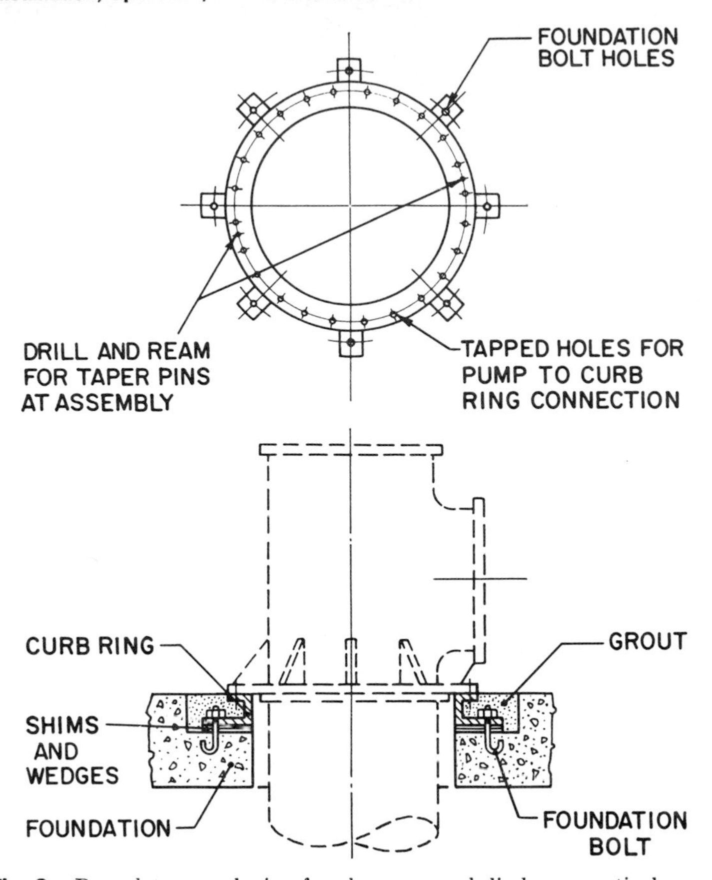

Fig. 3 Round type curb ring for above-ground discharge vertical pump.

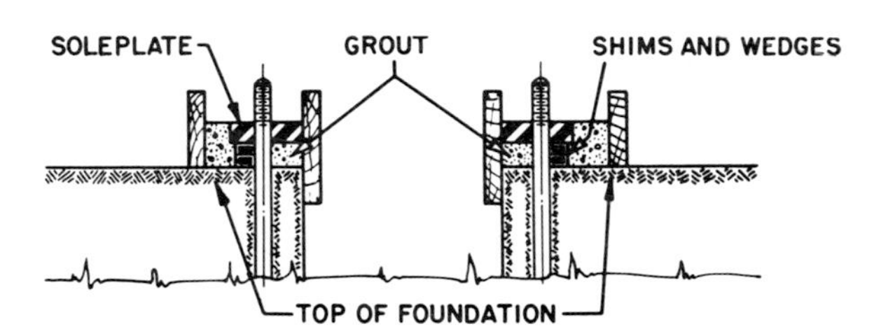

Fig. 4 Grouting form for vertical pump soleplate.

rotate it while holding the other coupling half stationary, and check the alignment at each quarter turn of the half being rotated. Then the half previously held stationary should be revolved and the alignment checked. A variation within manufacturing limits may be found in either of the half-couplings, and proper allowance for this must be made when aligning the unit.

A dial indicator bolted to the pump half of the coupling may be used to check both radial and axial alignment instead of a straightedge and feeler gage (Fig. 6). With the button resting on the periphery of the other coupling half, the dial should be set at zero and a mark chalked on the coupling half at the point where the button rests. For any check (top, bottom, or sides) both shafts should be rotated

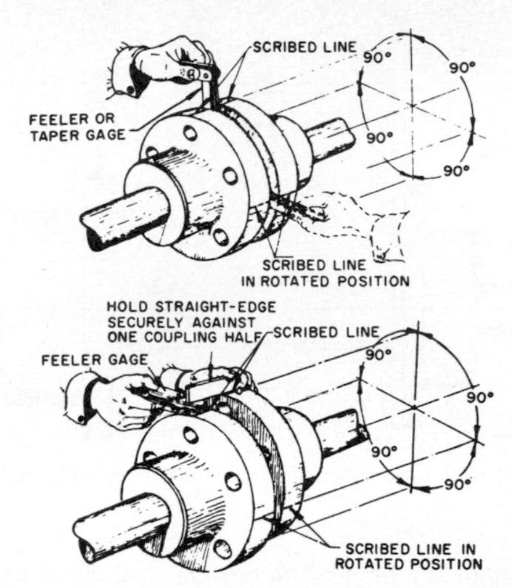

Fig. 5 Coupling alignment, using feeler gages.

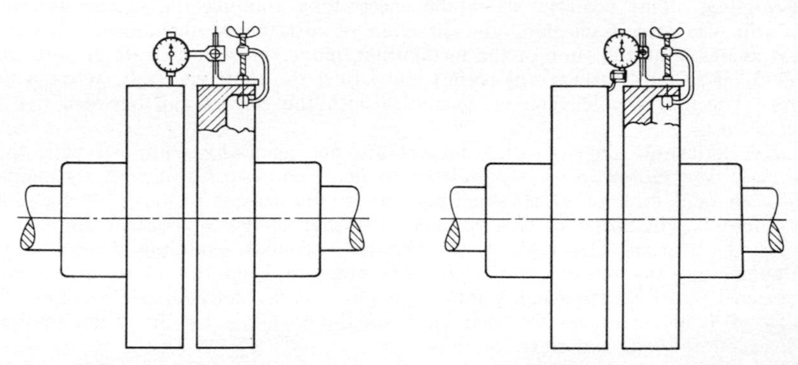

Fig. 6 Use of dial indicator for alignment of standard coupling.

the same amount, that is, all readings on the dial should be made with its button on the chalk mark. The dial readings will indicate whether the driver must be raised, lowered, or moved to either side. After any movement, it is necessary to check that the coupling faces remain parallel to one another.

For example, if the dial reading at the starting point is set to zero and the diametrically opposite reading at the bottom or sides shows ± 0.020 in, the driver must be raised or lowered by shimming or moved to one side or the other by half of this reading. The same procedure is used to align gear-type couplings, but the coupling covers must first be moved back out of the way and all measurements should be made on the coupling hubs.

When an extension coupling connects the pump to its driver, a dial indicator should be used to check the alignment (Fig. 7). The extension piece between the coupling halves should be removed to expose the coupling hubs. The coupling nut on the end of the shaft should be used to clamp a suitable extension arm or bracket long enough to extend across the space between the coupling hubs. The dial indicator is mounted on this arm, and alignment is checked both for concentricity of the hub diameters and parallelism of the hub faces. Changing the arm from one hub to the other provides an additional check.

The clearance between the faces of the coupling hubs and the ends of the shafts should be such that they cannot touch, rub, or exert a pull on either pump or driver. The amount of this clearance may vary with the size and type of coupling used. Sufficient clearance will allow unhampered endwise movement of the shaft

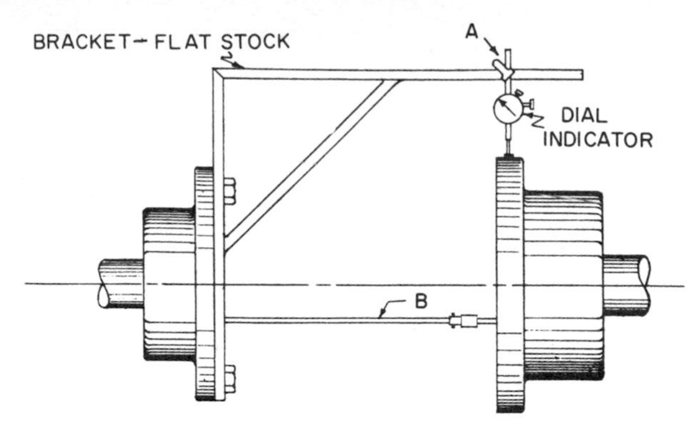

Fig. 7 Use of dial indicator for alignment of extension-type coupling.

of the driving element to the limit of its bearing clearance. On motor-driven units, the magnetic center of the motor will determine the running position of the motor half-coupling. This position should be checked by running the motor uncoupled. This will also permit checking the direction of rotation of the motor. If current is not available at the time of the installation, move the motor shaft in both directions as far as the bearings will permit and adjust the shaft centrally between these limits. The unit should then be assembled with the correct gap between the coupling halves.

Large horizontal sleeve-bearing motors are not generally equipped with thrust bearings. The motor rotor is permitted to float, and as it will seek its magnetic center, an axial force of rather small magnitude can cause it to move off this center. Sometimes it will move enough to cause the shaft collar to contact and possibly damage the bearing. To avoid this, a "limited-end float" coupling is used between the pump and the motor on all large-size units to keep the motor rotor within a restricted position (see Sec. 6.3, Pump Couplings and Intermediate Shafting). The setting of axial clearances for such units should be given by the manufacturer in his instruction books and elevation drawings.

When the pump handles a liquid at other than ambient temperature or when it is driven by a steam turbine, the expansion of the pump or turbine at operating temperature will alter the vertical alignment. Alignment should be made at ambient temperature, making suitable allowances for the changes in pump and driver centerlines after expansion takes place. The final alignment must be made with the pump and driver at their normal temperatures, and adjusted as required before placing the pump into permanent service.

For large installations, particularly with steam-turbine-driven pumps, more sophisticated alignment methods are sometimes employed, using proximity probes and optical instruments. Such procedures permit checking the effect of temperature changes and machine strains caused by piping stresses while the unit is in operation. When such procedures are recommended, they are included with the manufacturer's instructions.

When the unit has been accurately leveled and aligned, the hold-down bolts should be gently and evenly tightened before grouting. The alignment must be rechecked after the suction and discharge piping have been bolted to the pump to test the effect of piping strains. This can be done by loosening the bolts and reading the movement of the pump, if any, with dial indicators.

The pump and driver alignment should be occasionally rechecked, for misalignment may develop from piping strains after a unit has been operating for some time. This is especially true when the pump handles hot liquids, as there may be a growth or change in the shape of the piping. Pipe flanges at the pump should be disconnected after a period of operation to check the effect of the expansion of the piping, and adjustments should be made to compensate for this.

Grouting Ordinarily, the baseplate is grouted before the piping connections are made and before the alignment of the coupling halves is finally rechecked. The purpose of grouting is to prevent lateral shifting of the baseplate, to increase its mass to reduce vibration, and to fill in irregularities in the foundation.

The usual mixture for grouting a pump baseplate is composed of one part pure portland cement and two parts building sand, with sufficient water to cause the mixture to flow freely under the base (heavy-cream consistency). To reduce settling, it is best to mix the grout and let it stand for a short period, then remix it thoroughly before use without adding any more water.

The top of the rough concrete foundation should be well saturated with water before grouting. A wooden form is built around the outside of the baseplate to retain the grout. Grout is added until the entire space under the baseplate is filled to the top of the underside (Fig. 8). The grout holes in the baseplate serve as vents to allow the air to escape. A stiff wire should be used through the grout holes to work the grout and release any air pockets.

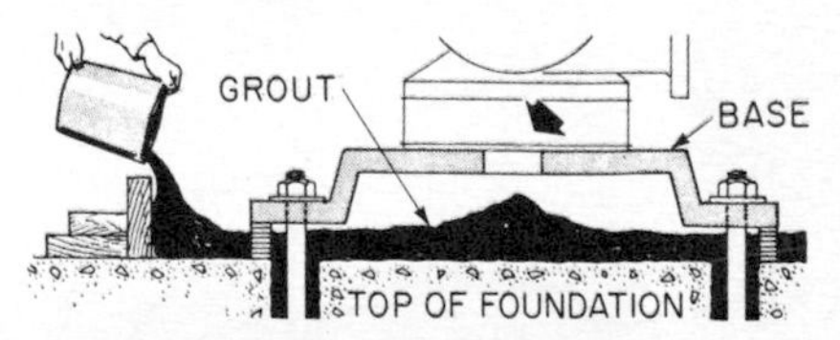

Fig. 8 Application of grouting. Grout is added until the entire space under the base is filled. Holes in the base (see arrow) allow air to escape and permit working of the grout to release air pockets.

The exposed surfaces of the grout should be covered with wet burlap to prevent cracking from drying too rapidly. When the grout is sufficiently set so that the forms can be removed, the exposed surfaces of the grout and foundations are finished smooth. When the grout is hard (72 hr or more), the hold-down bolts should be finally tightened and the coupling halves should be rechecked for alignment.

There is considerable controversy over whether the leveling strips or wedges should be removed after grouting. The best practice is to remove these in all cases for reciprocating machinery because pounding action or vibration will ultimately loosen the unit from the foundation. The space formerly occupied by shims or wedges must be regrouted. There is less danger in not removing the strips or wedges with rotating machinery, provided care is used in mixing the grout material and there is no shrinkage or drying. The strips or wedges can also be removed; erectors can follow their own preference in this matter.

The pump and driver alignment must be rechecked thoroughly after the grout has hardened permanently, and at reasonable intervals thereafter.

Doweling of Pump and Driver When the pump handles hot liquids, doweling of both the pump and its driver should be delayed until the unit has been operated. A final recheck of alignment with the coupling bolts removed and with the pump and driver at operating temperature is advisable before doweling.

Large pumps handling hot liquids are usually doweled near the coupling end, allowing the pump to expand from that end out. Sometimes the other end is provided with a key and a keyway in the casing foot and the baseplate.

PIPING

Suction Piping The suction piping should be as direct and short as possible. If a long suction line is required, the pipe size should be increased to reduce friction losses. (The exception to this recommendation is in the case of boiler feed pumps, where difficulties may arise during transient conditions of load change if the suction piping volume is excessive. This is a special and complex subject, and the manufacturer should be consulted.) Where the pump must lift the liquid from a lower level, the suction piping should be laid out with a continual rise toward the pump, avoiding high spots in the line so as to prevent the formation of air pockets. Where a static suction head will exist, the pump suction piping should slope continuously downward to the pump.

Generally the size of the suction line is larger than the pump suction nozzle, and eccentric reducers should be used. If the source of supply is located below the pump centerline, the reducer should be installed straight side up. If the source of supply is above the pump, the straight side of the reducer should be at the bottom (Fig. 9).

Elbows and other fittings next to the pump suction should be carefully arranged or the flow into the pump impeller will be disturbed. Long-radius elbows are preferred for suction lines because they create less friction and provide a more uniform flow distribution than standard elbows.

Pump installations with a static suction lift preferably should have the inlet of the vertical suction piping submerged in the liquid to four times the piping diameter. If a tendency appears for a vortex to form at the surface of the liquid supply, a floating vortex breaker (raft) or similar structure should be provided around the suction pipe. (For more details, see Sec. 11, Intakes and Suction Piping.)

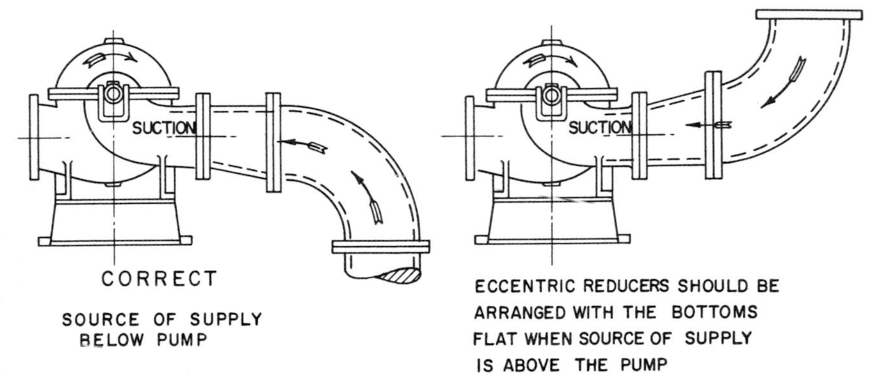

Fig. 9 Recommended installation of reducers at pump suction.

Discharge Piping Generally both a check valve and a gate valve are installed in the discharge line. The check valve is placed between the pump and the gate valve and protects the pump from reverse flow in the event of unexpected driver failure or from reverse flow from another operating pump. The gate valve is used when priming the pump or when shutting it down for inspection and repairs. Manually operated valves that are difficult to reach should be fitted with a sprocket rim wheel and chain. In some cases discharge gate valves are motorized and can be operated by remote control.

Piping Strains Cast iron pumps are never provided with raised face flanges. If steel suction or discharge piping is used, the pipe flanges should also be of the flat-face and not of the raised-face type. Full-face gaskets must be used with cast iron pumps.

Piping should not impose excessive forces and moments on the pump to which it is connected, since these might spring the pump or pull it out of position. Piping flanges must be brought squarely together before the bolts are tightened. The suction and discharge piping and all associated valves, strainers, etc., should be supported and anchored near to but independent of the pump, so that no strain will be transmitted to the pump casing.

When large-size pumps are involved or when major temperature changes are expected, the pump manufacturer generally indicates to the user the maximum moments and forces that can be imposed on the pump by the piping. A typical diagram is illustrated in Fig. 10 for a radially split double-casing multistage pump with top suction and discharge.

Expansion Joints Expansion joints are sometimes used in the discharge and suction piping to avoid transmitting any piping strains caused by expansion when handling hot liquids or by misalignment. On occasion, expansion joints are formed by looping the pipe. More often they are of the slip-joint or corrugated-diaphragm type. However, they transmit to the pump a force equal to the area of the expansion joint times the pressure in the pipe. These forces can be of very significant

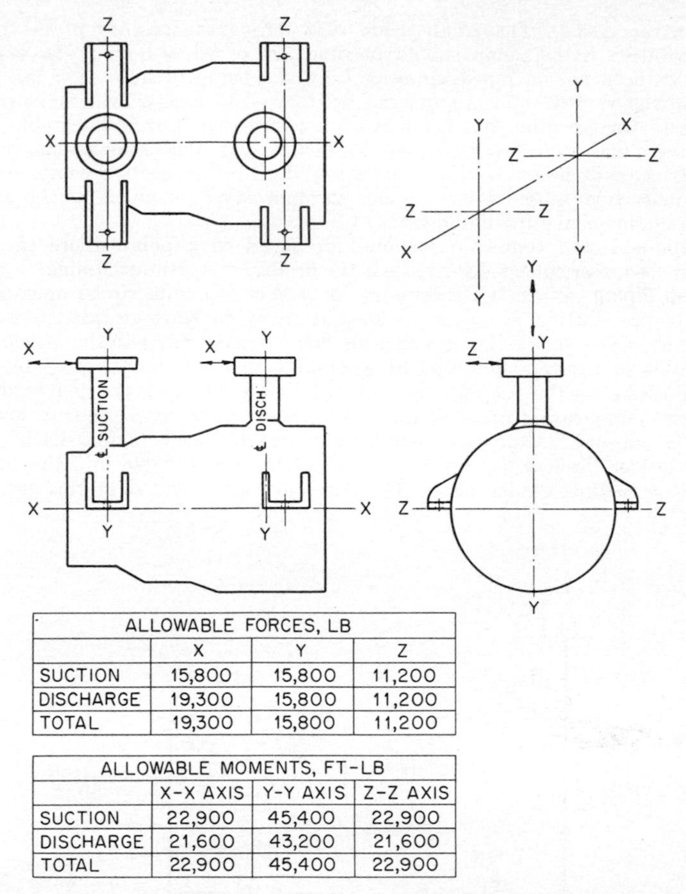

ALLOWABLE FORCES, LB			
	X	Y	Z
SUCTION	15,800	15,800	11,200
DISCHARGE	19,300	15,800	11,200
TOTAL	19,300	15,800	11,200

ALLOWABLE MOMENTS, FT-LB			
	X-X AXIS	Y-Y AXIS	Z-Z AXIS
SUCTION	22,900	45,400	22,900
DISCHARGE	21,600	43,200	21,600
TOTAL	22,900	45,400	22,900

Fig. 10 Diagram of permissible pipe stresses and moments for a radially split double-casing multistage pump with top suction and discharge.

magnitude, and it is impractical to design the pump casings, baseplates, etc., to withstand such forces. Consequently, when expansion joints are used, a suitable pipe anchor must be installed between it and the pump proper. Alternately, tie rods can be used to prevent the forces from being transmitted to the pump.

Suction Strainers Except for certain special designs, pumps are not intended to handle liquid containing foreign matter. If the particles are sufficiently large, such foreign matter can clog the pump, reduce its capacity, or even render it altogether incapable of pumping. Small particles of foreign matter may cause damage by lodging between close running clearances. Therefore proper suction strainers may be required in the suction lines of pumps not specially designed to handle foreign matter.

In such an installation the piping must first be thoroughly cleaned and flushed. The recommended practice is to flush all piping to waste before connecting it to the pump. Then a temporary strainer of appropriate size should be installed in the suction line as close to the pump as possible. This temporary strainer may have a finer mesh than the permanent strainer installed after the piping has been thoroughly cleaned of all possible mill scale or other foreign matter. The size of the mesh is generally recommended by the pump manufacturer.

Venting and Draining Vent valves are generally installed at one or more high points of the pump-casing waterways to provide a means of escape for air or vapor trapped in the casing. These valves are used during the priming of the pump or during operation if the pump should become air- or vapor-bound. In most cases these valves need not be piped up away from the pump because their use is infrequent, and the vented air or vapors can be allowed to escape into the surrounding atmosphere. On the other hand, vents from pumps handling inflammable or toxic or corrosive fluids must be connected in such a way that they endanger neither the operating personnel nor the installation itself. The suction vents of pumps taking liquids from closed vessels under vacuum must be piped to the source of the pump suction above the liquid level.

All drain and drip connections should be piped to a point where the leakage can be disposed of or collected for reuse if the drains are worth reclaiming.

Warm-up Piping When it is necessary for a pump to come up to operating temperature before starting it up, or to keep it ready to start at rated temperature, provision should be made for a warm-up flow to pass through the pump. There are a number of arrangements used to accomplish this. If the pump operates under positive pressure on the suction, the pumped liquid can be permitted to drain out through the pump-casing drain connection to some point at a pressure lower than the suction pressure. Alternately, some liquid can be made to flow back from the discharge header through a jumper line around the check valve into the pump and out into the suction header (Fig. 11). An orifice is provided in this jumper line

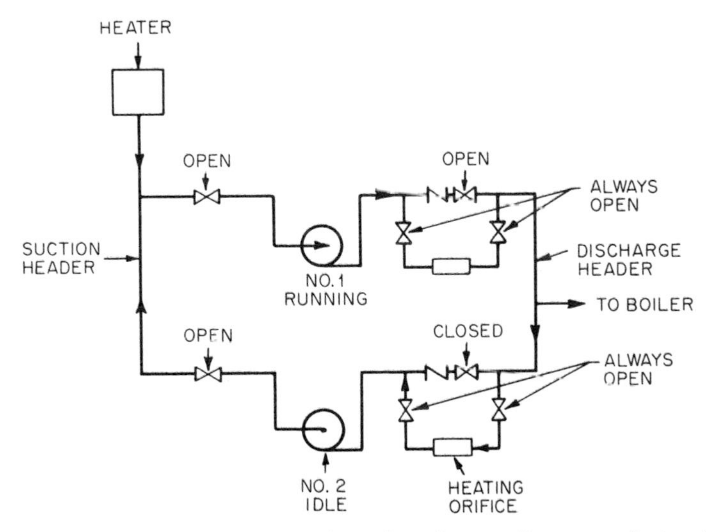

Fig. 11 Arrangement for warm-up through a jumper line around the discharge check valve.

to regulate the amount of warm-up flow. The pump manufacturer's recommendations should be sought in all cases as to the best means to provide an adequate warm-up procedure.

Relief Valves Positive-displacement pumps such as rotary and reciprocating pumps can develop discharge pressures much in excess of their maximum design pressures. To protect these pumps against excessive pressures when the discharge is throttled or shut off, a pressure relief valve must be used. Some pumps are provided with internal integral relief valves. But unless operation against a closed discharge is both infrequent and of very short duration, a relief valve with an external return connection must be used, and the liquid from the relief valve must be piped back to the source of supply.

Surge Chambers Generally centrifugal pumps do not require surge chambers in their suction or discharge piping. Reciprocating pumps may have a suction and discharge piping layout that does not require compensation for variations in the flow velocity in the piping system. But in many cases reciprocating-pump installations require surge chambers when the suction or discharge lines are of considerable length, when there is an appreciable static head on the discharge, when the liquid pumped is hot, or when it is desirable to smooth out variations in the discharge flow. The type, size, and arrangement of the surge chamber should be chosen on the basis of the manufacturer's recommendations.

Instrumentation There are a number of instruments which are essential to maintaining a close check on the performance and condition of a pump. A compound pressure gage should be connected to the suction of the pump, and a pressure gage should be connected to its discharge at the pressure taps which may be provided in the suction and discharge flanges. The gages should be mounted in a convenient location so that they can be easily observed.

In addition, it is advisable to provide a flow-metering device. Depending upon the importance of the installation, indicating meters may be supplemented by recording attachments.

Whenever pumps incorporate various leak-off arrangements, such as a balancing device or pressure-reducing labyrinths, a check should be maintained on the quantity of these leak-offs by installing measuring orifices and differential gages in the leak-off lines.

Pumps operating in important or complex services or operating completely unattended by remote control may have additional instrumentation, such as speed indicators, vibration monitors, bearing or casing temperature indicators, etc.

For more detail, see Sec. 8, Supervisory and Monitoring Instrumentation.

OPERATION

Pumps are generally selected for a given capacity and total head when operating at rated speed. These characteristics are referred to as "rated conditions of service" and, with few exceptions, represent those conditions at or near which the pump will operate the greatest part of the time. Positive-displacement pumps cannot operate at any greater flows than rated except by increasing their speed; nor can they operate at lower flows except by reducing the operating speed or by-passing some of the flow back to the source of supply.

On the other hand, centrifugal pumps can operate over a wide range of capacities, from near zero flow to well beyond the rated capacity. Because a centrifugal pump will always operate at the intersection of its head-capacity curve with the system-head curve, the pump operating capacity may be altered either by throttling the pump discharge (hence altering the system-head curve) or by varying the pump speed (changing the pump head-capacity curve). This makes the centrifugal pump very flexible in a wide range of services and applications which require the pump to operate at capacities and heads differing considerably from the rated conditions. There are, however, some limitations imposed upon such operation by hydraulic, mechanical, or thermodynamic considerations (see Sec. 2.3, Centrifugal Pump Performance).

Operation of Centrifugal Pumps at Reduced Flows. The subject of radial thrust in volute pumps has been discussed in Secs. 2.2 and 2.3. It was indicated there that not even dual-volute pumps are always suitable for operation at all flows down to zero. Therefore it is imperative to adhere to the limitations on the minimum recommended flow for sustained operation given by the pump manufacturer.

The thermodynamic problem that arises from the operation of a centrifugal pump at extremely reduced flows is caused by the heating up of the liquid handled by the pump. The difference between the brake horsepower consumed and the water horsepower developed represents the power losses within the pump itself, except for a small amount lost in the pump bearings. These power losses are converted into heat and transferred to the liquid passing through the pump.

If the pump were to operate against a completely closed valve, the power losses would be equal to the shut-off brake horsepower and, since there would be no flow through the pump, all this power would go into heating the small quantity of liquid contained within the pump casing. The pump casing itself would heat up, and a certain amount of heat would be dissipated by radiation and convection to the surrounding atmosphere. However, because the temperature rise in the liquid pumped could be quite rapid, it is generally safer to ignore the dissipation of heat through radiation and the absorption of heat by the casing. On this basis, the temperature rise in the liquid pumped can be determined from the formula:

$$T_{rm} = \frac{42.4 \times P_{so}}{W_w \times C_w}$$

where T_{rm} = temperature rise, °F/min
P_{so} = brake horsepower at shut-off
42.4 = conversion from bhp to Btu/min
W_w = net weight of liquid in pump, lb
C_w = specific heat of liquid (1.0 if liquid is water)

For example, if the pump handles water (C_w = 1.0) and contains 100 lb of liquid and if the bhp at shut-off is 100, the water temperature will increase at the rate of 42.4°F/min. Operation at shut-off under these conditions would be highly dangerous. But with a low-head, high-capacity pump that contains 5,000 lb of water and which takes the same 100 bhp at shut-off, the rate of temperature increase would be only 0.85°F/min—hardly serious if the operation against a closed discharge valve were not prolonged.

If flow is taking place through the pump, conditions become stabilized and it is possible to calculate the temperature rise through the pump for any given flow. Assuming that the liquid is water, the following formula can be used:

$$T = \frac{(\text{bhp} - \text{whp}) \times 2{,}545}{\text{capacity in lb/hr}}$$

where T = temperature rise, °F/min.
2,545 = Btu equivalent of 1 hp-hr

A more convenient formula relates the temperature rise to the total head and to the pump efficiency:

$$T = \frac{H}{778}\left(\frac{1}{e} - 1\right)$$

where H = total head, ft
e = pump efficiency at a given capacity

(Note that these formulas neglect the effect of the compressibility of the water. For a more exact calculation of temperature rise, especially when dealing with very high pressures, more precise thermodynamic calculations are indicated.)

Figure 12 gives a graphical solution of the formula for temperature rise and the minimum permissible capacity, once the maximum permissible temperature rise has been selected. When liquids other than water are handled by the pump, it is necessary to correct the answer for the difference in specific heats of the liquids.

The chart can be used to plot a temperature-rise curve directly on the performance curve of a centrifugal pump, as shown in Fig. 13. It will be noted that the temperature rise increases very rapidly with a reduction in flow. This is caused by the fact that the losses at low capacities are greater when the quantity of liquid which must absorb the heat generated in the pump is low. For example, it will be seen that for a capacity of 50 gpm, the temperature rise is 17°F while at the full capacity of 550 gpm it is less than 1°F.

The maximum permissible temperature rise varies over a wide range, depending on the type of installation. For hot water pumps—as on boiler feed service—it is generally advisable to limit the temperature rise to about 15°F. As a general rule, the minimum permissible flow to hold the temperature rise in boiler feed pumps to 15°F is 30 gpm for each 100 bhp at shut-off.

When the pump handles cold water, the temperature rise may be permitted to reach 50 or even 100°F. The minimum capacity based on thermodynamic considerations is then established as that capacity at which the temperature rise corresponds to the maximum permitted.

There are also hydraulic considerations which may affect the minimum flow at

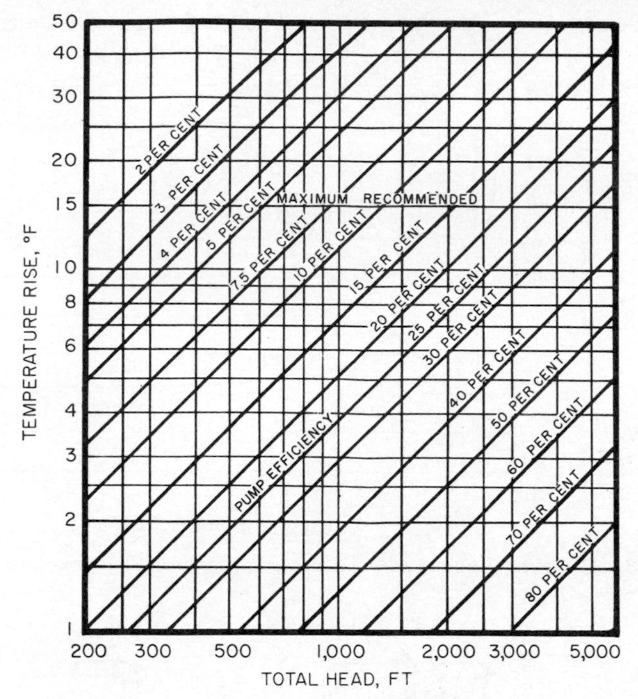

Fig. 12 Temperature rise in centrifugal pumps in terms of total head and pump efficiency.

which a centrifugal pump may be permitted to operate. Within recent years a correlation has been developed between operation at low flows and the appearance of hydraulic pulsations both in the suction and in the discharge of centrifugal impellers. It has been proved that these pulsations are caused by the development of an internal recirculation at the inlet and discharge of an impeller at flows below the capacity of best efficiency. This subject has been treated in Sec. 2.3. Recommendations of the pump manufacturer about minimum flows, dictated by hydraulic considerations, must be followed.

If the service is such that a pump may be required to operate at shut-off or at extremely low flows, means must be provided to prevent pump operation below minimum permissible flows, regardless of whether the discharge valve or check valve is closed. This is accomplished by installing a bypass in the discharge line from the pump side of the check and gate valves, leading to some lower pressure point in the pumping system where the excess heat may be dissipated. Under no circumstances should this bypass lead directly back to the pump suction.

A typical bypass arrangement is shown in Fig. 14. Since the pressure in the bypass line is equal to the pump discharge pressure, it is necessary to locate an orifice in this line to limit the bypass flow to the desired value. When the differential pressure to be broken down by the orifice is relatively low, a single drilled orifice in a 3- to 6-in stainless steel rod can be used. Figure 15 shows the flows through orifices of various diameters and with various head differentials. (For calculations of orifices, see Sec. 9.2.)

An elbow should never be located too close after an orifice. A recommended arrangement for piping up an orifice where the piping must make a turn is shown on Figs. 14 and 16. The pipe plug at the end of the coupling should be made of stainless steel. It can ultimately be replaced if the impingement of high-velocity water causes it to wear.

When higher pressures than indicated on Fig. 15 are encountered, or when it is

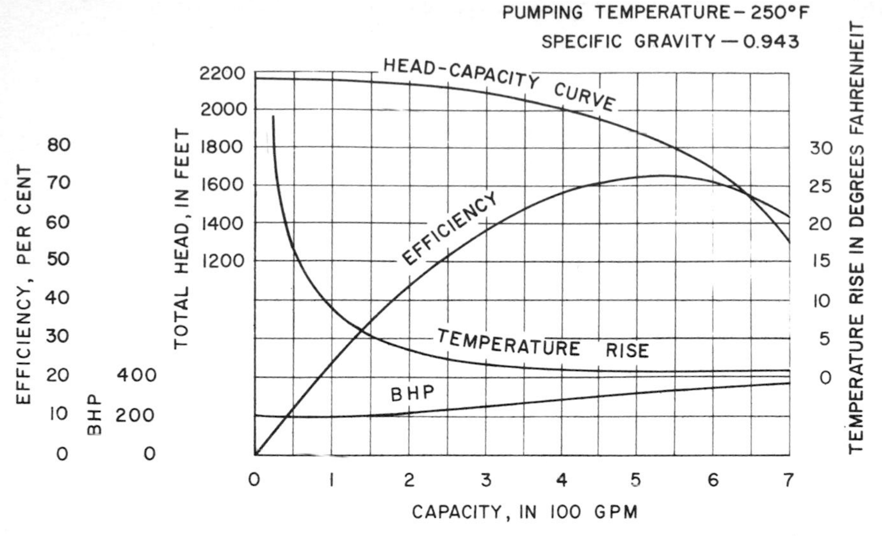

Fig. 13 Centrifugal pump performance curve indicating temperature rise.

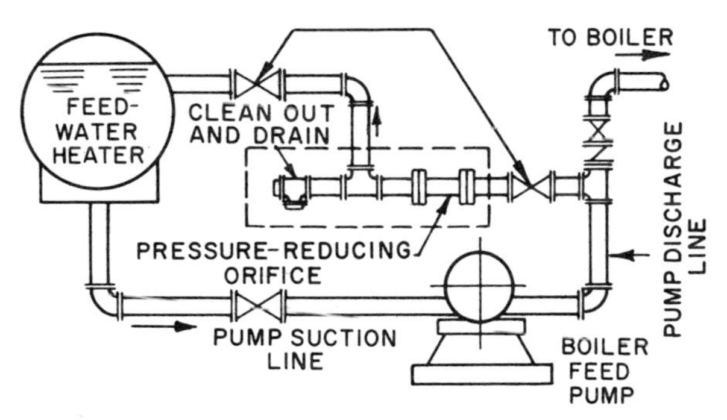

Fig. 14 Arrangement of bypass recirculation for boiler feed service. These valves must be good for full pump discharge pressure and must always be fully open or fully closed.

desired to reduce the noise incidental to the breaking down of the discharge pressure through a single orifice, special multiple-pressure reducing orifices can be used.

Figure 14 shows a manual control of the bypass. Two valves are used, one on each side of the orifice. One of these remains locked in the open position and is closed only to isolate the orifice for inspection or replacement. The second is the control valve and remains either open or closed. The control of this valve is frequently made automatic and responsive to the metering of the flow through the pump.

Priming With some very few exceptions, no centrifugal pump should ever be started until it is fully primed, that is, until it has been filled with the liquid pumped and all the air contained in the pump has been allowed to escape. The exceptions involve self-priming pumps and some special large-capacity, low-head, and low-speed installations where it is not practical to prime the pump prior to starting; the priming takes place almost simultaneously with the starting in these cases.

Reciprocating pumps of the piston or plunger type are in principle self-priming. However, if quick starting is required, priming connections should be piped to a supply above the pump.

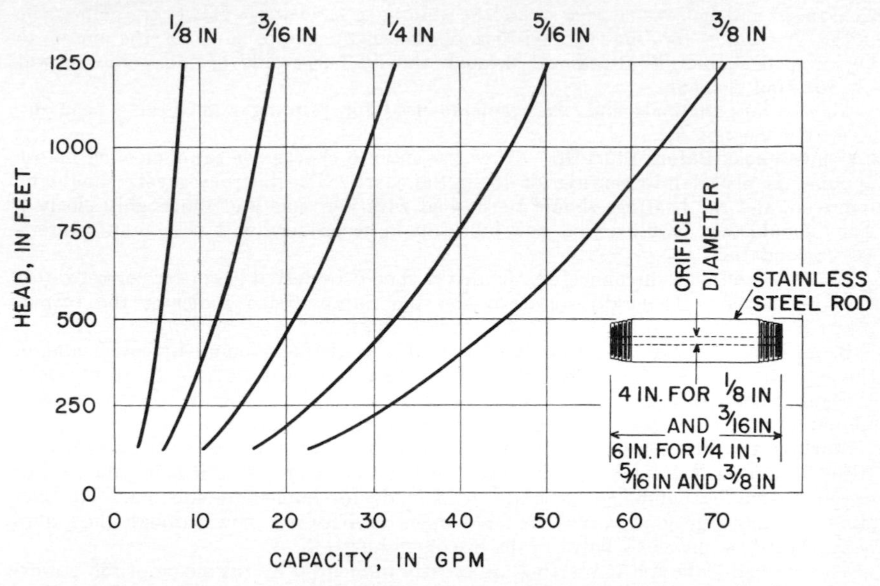

Fig. 15 Flows through single orifices drilled in a stainless steel rod.

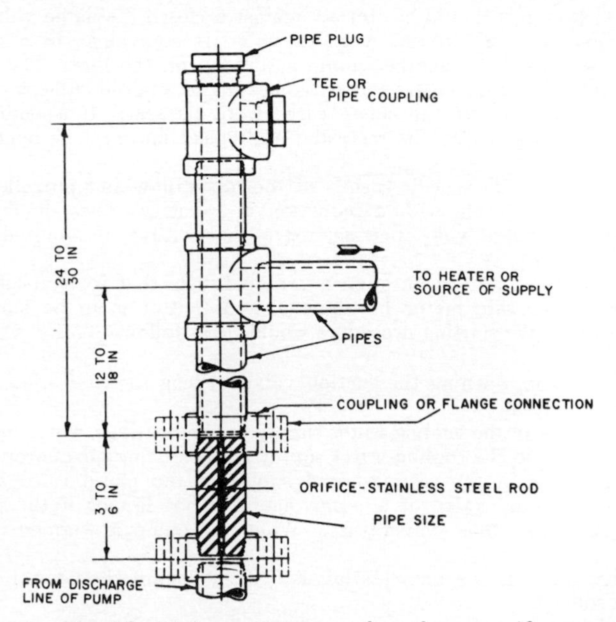

Fig. 16 Piping arrangement for a bypass orifice.

Positive-displacement pumps of the rotating type, such as rotary or screw pumps, have clearances so that the liquid in the pump will drain back to the suction. When pumping low-viscosity liquids, the pump may completely dry out when it is idle. In such cases a foot valve may be used to help keep the pump primed. Alternately, a vacuum device may be used to prime the pump. When handling liquids of higher viscosity, foot valves are usually not required because liquid is retained within the clearances and acts as a seal when the pump is restarted. However, before the initial start of a rotating-type positive-displacement pump, some of the liquid to be pumped should be introduced through the discharge side of the pump to wet the rotating element.

The various methods and arrangements used for priming pumps have been described in Section 2.4.

Final Checks Before Start-Up A few last-minute checks are recommended before a pump is placed into service for its initial start. The bearing covers should be removed, and the bearings should be flushed with kerosene and thoroughly cleaned. They should then be filled with new lubricant in accordance with the manufacturer's recommendations.

With the coupling disconnected, the driver should be tested again for correct direction of rotation. Generally an arrow on the pump casing indicates the correct rotation.

It must be possible to rotate the rotor of a centrifugal pump by hand, and in the case of a pump handling hot liquids, the rotor must be free to rotate with the pump cold or hot. If the rotor is bound or even drags slightly, do not operate the pump until the cause of the trouble is determined and corrected.

Starting and Stopping Procedures The necessary steps in the starting of a centrifugal pump will depend upon its type and the service on which it is installed. For example, standby pumps are generally held ready for immediate starting. The suction and discharge gate valves are held open, and reverse flow through the pump is prevented by the check valve in the discharge line.

The methods followed in starting are greatly influenced by the shape of the power-capacity curve of the pump. High- and medium-head pumps (low and medium specific speeds) have power curves that rise from zero flow to the normal capacity condition. Such pumps should be started against a closed discharge valve to reduce the starting load on the driver. A check valve is equivalent to a closed valve for this purpose, as long as another pump is already on the line. The check valve will not lift until the pump being started comes up to a speed sufficient to generate a head high enough to lift the check valve from its seat. If a pump is started with a closed discharge valve, the recirculation bypass line must be open to prevent overheating.

Low-head pumps (high specific speed) of the mixed-flow and propeller type have power curves rising sharply with a reduction in capacity; they should be started with the discharge valve wide open against a check valve, if required, to prevent backflow.

Assuming that the pump in question is motor-driven, that its shut-off horsepower does not exceed the safe motor horsepower, and that it is to be started against a closed gate valve, the starting procedure would be as follows:

1. Prime the pump, opening the suction valve, closing the drains, etc., to prepare the pump for operation.
2. Open the valve in the cooling-water supply to the bearings, where applicable.
3. Open the valve in the cooling-water supply if the stuffing boxes are water-cooled.
4. Open the valve in the sealing-liquid supply if the pump is so fitted.
5. Open the warm-up valve of a pump handling hot liquids if the pump is not normally kept at operating temperature. When the pump is warmed up, close the valve.
6. Open the valve in the recirculating line if the pump should not be operated against dead shut-off.
7. Start the motor.
8. Open the discharge valve slowly.
9. Observe the leakage from the stuffing boxes and adjust the sealing-liquid valve for proper flow to ensure the lubrication of the packing. If the packing is new, do not tighten up on the gland immediately, but let the packing run in before reducing the leakage through the stuffing boxes.
10. Check the general mechanical operation of the pump and motor.
11. Close the valve in the recirculating line once there is sufficient flow through the pump to prevent overheating.

If the pump is to be started against a closed check valve with the discharge gate valve open, the steps would be the same, except that the disharge gate valve would be opened some time before the motor is started.

In certain cases the cooling water to the bearings and the sealing water to the seal cages are provided by the pump itself. This, of course, eliminates the need for the steps listed for the cooling and sealing supply.

Just as in starting a pump, the stopping procedure depends upon the type and service of the pump. Generally the steps followed to stop a pump which can operate against a closed gate valve would be:

1. Open the valve in the recirculating line.
2. Close the gate valve.
3. Stop the motor.
4. Open the warm-up valve if the pump is to be kept up to operating temperature.
5. Close the valve in the cooling-water supply to the bearings and to water-cooled stuffing boxes.
6. If the sealing liquid supply is not required while the pump is idle, close the valve in this supply line.
7. Close the suction valve, open the drain valves, etc., as required by the particular installation or if the pump is to be opened up for inspection.

If the pump is of a type which does not permit operation against a closed gate valve, steps 2 and 3 are reversed.

In general, the starting and stopping of steam-turbine-driven pumps require the same steps and sequence prescribed for a motor-driven pump. As a rule, steam turbines have various drains and seals which must be opened or closed before and after operation. Similarly, many turbines require warming up before starting. Finally some turbines require turning-gear operation if they are kept on the line ready to start up. The operator should therefore follow the steps outlined by the turbine manufacturer in starting and stopping the turbine.

Most of the steps listed for starting and stopping centrifugal pumps are equally applicable to positive-displacement pumps. There are, however, a few notable exceptions:

1. Never operate a positive-displacement pump against a closed discharge. If the gate valve on the discharge must be closed, always start a positive-displacement pump with the recirculation bypass valve open.
2. Always open the steam-cylinder drain cocks of a steam reciprocating pump before starting, to allow condensate to escape and to prevent damage to the cylinder heads.

Auxiliary Services on Standby Pumps Standby pumps are frequently started up from a remote location, and several methods of operation are available for the auxiliary services, such as the cooling-water supply to the bearings or to water-cooled stuffing boxes:

1. A constant flow may be kept through the bearing jackets or oil coolers and through the stuffing box lantern rings, whether the pump is running or on standby service.
2. The service connections may be opened automatically whenever the pump is started up.
3. The service connections may be kept closed while the pump is idle, and the operator may be instructed to open them within a short interval after the pump has been put on the line automatically.

The choice between these methods must be dictated by the specific circumstances surrounding each case. There are, however, certain cases where sealing-liquid supply to the pump stuffing boxes must be maintained whether the pump is running or not. This is the case when the pump handles a liquid which is corrosive to the packing or which may crystallize and deposit on the shaft sleeves. It is also the case when the sealing supply is used to prevent air infiltration into a pump when it is operating under a vacuum.

Restarting Motor-driven Pumps after Power Failure Assuming that power failure will not cause the pump to go into reverse rotation, that is, that a check valve will protect the pump against reverse flow, there is generally no reason why the pump should not be permitted to restart once current has been reestablished. Whether the pump will start again automatically when power is restored will depend on the type of motor control used. (Sections 2.3 and 9.1 give reasons why some pumps should not be started in reverse.)

Because pumps operating on a suction lift may lose their prime during the time that power is off, it is preferable to use starters with low load protection for such installations to prevent an automatic restart. This does not apply, of course, if the pumps are automatically primed, or if some protection device is incorporated so that the pump cannot run unless it is primed.

MAINTENANCE

Because of the wide variation in pump types, ranges in sizes, differences in design, and materials of construction, these comments on maintenance are restricted to those types of pumps most commonly encountered. The manufacturer's instruction books must be carefully studied before any attempt is made to service any particular pump.

Daily Observation of Pump Operation When operators are on constant duty, hourly and daily inspections should be made, and any irregularities in the operation of a pump should be reported immediately. This applies particularly to changes in the sound of a running pump, abrupt changes in bearing temperatures, and stuffing box leakage. A check of the pressure gages and of the flowmeter, if installed, should be made hourly. If recording instruments are provided, a daily check should be made to determine whether the capacity, pressure, or power consumption indicate that further inspection is required.

Semiannual Inspection The free movement of stuffing box glands should be checked semiannually, gland bolts should be cleaned and oiled, and the packing should be inspected to determine whether it requires replacement. The pump and driver alignment should be checked and corrected if necessary. Oil-lubricated bearings should be drained and refilled with fresh oil. Grease-lubricated bearings should be checked to see that they contain the correct amount of grease, and that it is still of suitable consistency.

Annual Inspection A very thorough inspection should be made once a year. In addition to the semiannual procedure, bearings should be removed, cleaned, and examined for flaws. The bearings housings should be carefully cleaned. Antifriction bearings should be examined for scratches and wear. Immediately after cleaning and inspection, antifriction bearings should be coated with oil or grease.

The packing should be removed and the shaft sleeves—or shaft, if no sleeves are used—should be examined for wear.

When the coupling halves are disconnected for the alignment check, the vertical shaft movement of a pump with sleeve bearings should be checked at both ends with the packing removed. Any vertical movement exceeding 150 percent of the original clearance requires an investigation to determine the cause. The end play allowed by the bearings should also be checked. If it exceeds that recommended by the manufacturer, the cause should be determined and corrected.

All auxiliary piping, such as drains, sealing-water piping, and cooling-water piping should be checked and flushed. Auxiliary coolers should be flushed and cleaned.

The pump stuffing boxes should be repacked, and the pump and driver should be realigned and reconnected.

All instruments and flow-metering devices should be recalibrated, and the pump should be tested to determine whether proper performance is being obtained. If internal repairs are made, the pump should again be tested after completion of the repairs.

Complete Overhaul It is difficult to make general rules about the frequency of complete pump overhauls, as it depends on the pump service, the pump construction and materials, the liquid handled, and the economic evaluation of overhaul costs versus the cost of power losses resulting from increased clearances or of un-

TABLE 1 Check Chart for Centrifugal-Pump Troubles

Symptoms	*Possible cause of trouble (Each number is defined in the list below)*
Pump does not deliver water:	1, 2, 3, 4, 6, 11, 14, 16, 17, 22, 23
Insufficient capacity delivered:	2, 3, 4, 5, 6, 7, 8, 9, 10, 11, 14, 17, 20, 22, 23, 29, 30, 31
Insufficient pressure developed:	5, 14, 16, 17, 20, 22, 29, 30, 31
Pump loses prime after starting:	2, 3, 5, 6, 7, 8, 11, 12, 13
Pump requires excessive power:	15, 16, 17, 18, 19, 20, 23, 24, 26, 27, 29, 33, 34, 37
Stuffing box leaks excessively:	13, 24, 26, 32, 33, 34, 35, 36, 38, 39, 40
Packing has short life:	12, 13, 24, 26, 28, 32, 33, 34, 35, 36, 37, 38, 39, 40
Pump vibrates or is noisy:	2, 3, 4, 9, 10, 11, 21, 23, 24, 25, 26, 27, 28, 30, 35, 36, 41, 42, 43, 44, 45, 46, 47
Bearings have short life:	24, 26, 27, 28, 35, 36, 41, 42, 43, 44, 45, 46, 47
Pump overheats and seizes:	1, 4, 21, 22, 24, 27, 28, 35, 36, 41

Suction troubles:

1. Pump not primed
2. Pump or suction pipe not completely filled with liquid
3. Suction lift too high
4. Insufficient margin between suction pressure and vapor pressure
5. Excessive amount of air or gas in liquid
6. Air pocket in suction line
7. Air leaks into suction line
8. Air leaks into pump through stuffing boxes
9. Foot valve too small
10. Foot valve partially clogged
11. Inlet of suction pipe insufficiently submerged
12. Water-seal pipe plugged
13. Seal cage improperly located in stuffing box, preventing sealing fluid from entering space to form the seal

System troubles:

14. Speed too low
15. Speed too high
16. Wrong direction of rotation
17. Total head of system higher than design head of pump
18. Total head of system lower than pump design head
19. Specific gravity of liquid different from design
20. Viscosity of liquid different from that for which designed
21. Operation at very low capacity
22. Parallel operation of pumps unsuitable for such operation

Mechanical troubles:

23. Foreign matter in impeller
24. Misalignment
25. Foundations not rigid
26. Shaft bent
27. Rotating part rubbing on stationary part
28. Bearings worn
29. Wearing rings worn
30. Impeller damaged
31. Casing gasket defective, permitting internal leakage
32. Shaft or shaft sleeves worn or scored at the packing
33. Packing improperly installed
34. Incorrect type of packing for operating conditions
35. Shaft running off center because of worn bearings or misalignment
36. Rotor out of balance, causing vibration
37. Gland too tight, resulting in no flow of liquid to lubricate packing
38. Failure to provide cooling liquid to water-cooled stuffing boxes
39. Excessive clearance at bottom of stuffing box between shaft and casing, causing packing to be forced into pump interior
40. Dirt or grit in sealing liquid, leading to scoring of shaft or shaft sleeve

TABLE 1 Check Chart for Centrifugal-Pump Troubles (Continued)

41. Excessive thrust caused by a mechanical failure inside the pump or by the failure of the hydraulic balancing device, if any
42. Excessive grease or oil in antifriction-bearing housing or lack of cooling, causing excessive bearing temperature
43. Lack of lubrication
44. Improper installation of antifriction bearings (damage during assembly, incorrect assembly of stacked bearings, use of unmatched bearings as a pair, etc.)
45. Dirt in bearings
46. Rusting of bearings from water in housing
47. Excessive cooling of water-cooled bearing, resulting in condensation of moisture from the atmosphere in the bearing housing

TABLE 2 Check Chart for Rotary-Pump Troubles

Symptoms	*Possible cause of trouble (Each number is defined in the list below)*
Pump fails to discharge:	1, 2, 3, 4, 5, 6, 8, 9, 16
Pump is noisy:	6, 10, 11, 17, 18, 19
Pump wears rapidly:	11, 12, 13, 20, 24
Pump not up to capacity:	3, 5, 6, 7, 9, 16, 21, 22
Pump starts, then loses suction:	1, 2, 6, 7, 10
Pump takes excessive power:	14, 15, 17, 20, 23

Suction troubles:

1. Not properly primed
2. Suction pipe not submerged
3. Strainer clogged
4. Leaking foot valve
5. Suction lift too high
6. Air leaks in suction
7. Suction pipe too small

System problems:

8. Wrong direction of rotation
9. Low speed
10. Insufficient liquid supply
11. Excessive pressure
12. Grit or dirt in liquid
13. Pump runs dry
14. Viscosity higher than specified
15. Obstruction in discharge line

Mechanical troubles:

16. Pump worn
17. Bent drive shaft
18. Coupling out of balance or alignment
19. Relief valve chatter
20. Pipe strain on pump casing
21. Air leak at packing
22. Relief valve improperly seated
23. Packing too tight
24. Corrosion

scheduled downtime. Some pumps on very severe service may need a complete overhaul monthly, while other applications require overhauls only every two to four years, or even less frequently.

A pump should not be opened for inspection unless either factual or circumstantial evidence indicates that overhaul is necessary. Factual evidence implies that the pump performance has fallen off significantly, or that noise or driver overload indicate trouble. Circumstantial evidence refers to past experience with the pump in question or with similar equipment on similar service.

TABLE 3 Check Chart for Reciprocating-Pump Troubles

Symptoms	*Possible cause of trouble (Each number is defined in the list below)*
Liquid end noise:	1, 2, 7, 8, 9, 10, 14, 15, 16
Power end noise:	17, 18, 19, 20
Overheated power end:	10, 19, 21, 22, 23, 24
Water in crankcase:	25
Oil leak from crankcase:	26, 27
Rapid packing or plunger wear:	11, 12, 28, 29
Pitted valves or seats:	3, 11, 30
Valves hanging up:	31, 32
Leak at cylinder-valve hole plugs:	10, 13, 33, 34
Loss of prime:	1, 4, 5, 6

Suction troubles:

1. Insufficient suction pressure
2. Partial loss of prime
3. Cavitation
4. Lift too high
5. Leaking suction at foot valve
6. Acceleration head requirement too high

System problems:

7. System shocks
8. Poorly supported piping, abrupt turns in piping, pipe size too small, piping misaligned
9. Air in liquid
10. Overpressure or overspeed
11. Dirty liquid
12. Dirty environment
13. Water hammer

Mechanical troubles:

14. Broken or badly worn valves
15. Packing worn
16. Obstruction under valve
17. Loose main bearings
18. Worn bearings
19. Low oil level
20. Plunger loose
21. Tight main bearings
22. Inadequate ventilation
23. Belts too tight
24. Driver misaligned
25. Condensation
26. Worn seals
27. Oil level too high
28. Pump not set level and rigid
29. Loose packing
30. Corrosion
31. Valve binding
32. Broken valve spring
33. Loose cylinder plug
34. Damaged O-ring seal

In order to ensure rapid restoration to service in the event of an unexpected overhaul, an adequate store of spare parts should be maintained at all times.

The relative complexity of the repairs, the facilities available at the site, and many other factors enter into the decision whether the necessary repairs will be carried out at the installation or at the pump manufacturer's plant.

Spare and Repair Parts The severity of the service in which a pump is used will determine, to a great extent, the minimum number of spare parts which should be carried in stock at the site of an installation. Unless prior experience is available, the pump manufacturer should be consulted on this subject. As an insurance against

TABLE 4 Check Chart for Steam-Pump Troubles

Symptoms	*Possible cause of trouble (Each number is defined in the list below)*
Pump does not develop rated pressure:	4, 5, 7, 8
Pump loses capacity after starting:	1, 2, 6
Pump vibrates:	9, 10, 11, 14
Pump has short strokes:	12, 13, 14
Pump operation is erratic:	1, 2, 3, 6

Suction troubles:

1. Suction line leaks
2. Suction lift too high
3. Cavitation

System problems:

4. Low steam pressure
5. High exhaust pressure
6. Entrained air or vapors in liquid

Mechanical troubles:

7. Worn piston rings in steam end
8. Binding piston rings in liquid end
9. Misalignment
10. Foundation not rigid
11. Piping not supported
12. Excessive steam cushioning
13. Steam valves out of adjustment
14. Liquid piston packing too tight

delays, spare parts should be purchased at the time the order for the complete unit is placed. Depending upon the contemplated methods of overhaul, certain replacement parts may have to be supplied either oversized or undersized instead of the same size used in the original unit.

When ordering spare parts after a pump has been in service, the manufacturer should always be given the pump serial number and size as stamped on the nameplate. This information is essential in identifying the pump exactly and in furnishing repair parts of correct size and material.

Record of Inspections and Repairs The working schedule of the semiannual and annual inspection programs should be entered on individual pump maintenance cards, which contain a complete record of all the items requiring attention. These cards should also contain space for comments and observations on the conditions of the parts to be repaired or replaced, on the rate and appearance of the wear, and on the repair methods followed. In many cases it is advisable to take photographs of badly worn parts before they are repaired.

In all cases complete records of the cost of maintenance and repairs should be kept for each individual pump, together with a record of its operating hours. A study of these records will generally reveal whether a change in materials or even a minor change in construction may not be the most economical course of action.

Diagnosis of Pump Troubles Pump operating troubles may be either of a hydraulic or of a mechanical nature. In the first category, a pump may fail to deliver liquid, it may deliver an insufficient capacity or develop insufficient pressure, or it may lose its prime after starting. In the second category, it may consume excessive power, or symptoms of mechanical difficulties may develop at the stuffing boxes or at the bearings, or vibration, noise, or breakage of some pump parts may occur.

There is a definite interdependence between some difficulties of both categories. For example, increased wear at the running clearances must be classified as a mechanical trouble, but it will result in a reduction of the net pump capacity—a hydraulic symptom—without necessarily causing a mechanical breakdown or even excessive vibration. As a result, it is most useful to classify symptoms and causes separately, and to list for each symptom a schedule of potential contributory causes. Such a diagnostic analysis is presented in Tables 1, 2, 3, and 4 for centrifugal, rotary, reciprocating, and steam pumps, respectively.

REFERENCES

1. "Hydraulic Institute Standards," Hydraulic Institute, Cleveland, Ohio.
2. Karassik, Igor J., and Roy Carter, "Centrifugal Pumps," McGraw-Hill Book Company, New York, 1960.
3. Karassik, Igor J., "Engineers' Guide to Centrifugal Pumps," McGraw-Hill Book Company (out of print but available in reprint edition by Hoepli, Milan).

Section 14

Pump Testing

WESLEY W. BECK

INTRODUCTION

14.1 Object Down through the years since man has first used a device to pump or lift water, pump testing of a sort has been used. Maybe the Egyptian slave owner rewarded or punished the slave operators of his shadoof around 1500 B.C. by index tests. Man, being eager to reduce his work, has constantly improved the pumping device. Each improvement was accepted only after being tested, which was the proof of its worthiness. As pumping equipment has become more refined, so has the art of pump testing, both in the shop or laboratory and in the field. For very large pumps model testing is being used to develop the optimum refinement in prototype design.

Every pump regardless of size or classification should have at least some kind of test before final acceptance by the purchaser. If not, the customer or user does not have any way of knowing that his requirements have been fulfilled. Tests of pumps and the test methods will depend on the ultimate purpose of the tests. Tests normally fall into one of two purposes or objectives:

1. Improvement in design, or actual operation, thus enabling any effect on performance by a change or modification in design to be evaluated.
2. To determine if contractual commitments have been met, thus making possible the comparison of specified, predicted, and actual performance.

In most cases the manufacturer supplies a test report or log and certifies the characteristics of the pump being furnished. Even these can be given a cursory check by the customer from time to time to give a record of performance or to give an indication of the need for replacement or overhaul. If at all possible, the pump should be tested as actually installed, with repeated tests from time to time to give indication of the operation of the pump.

The main object of this section on pump testing is to present a set of procedures and rules for conducting, computing, and reporting on tests of pumping units and for obtaining the head, capacity, power, efficiency, and suction requirements of a given pump.

14.2 Classification of Tests Pump tests should be classified as follows:

1. Shop tests are also called laboratory, manufacturer's, or factory acceptance tests. These are conducted in the pump manufacturer's plant under geometrically similar, ideal, and controlled conditions, and are usually assumed to be the most accurate of testing methods.

2. Field tests are made with the pumping unit installed in its exact environment and operating under actual existing field or ultimate conditions. The accuracy and reliability of field testing depend on the instrumentations used, installation, and advance planning during the design stages of the installation. By mutual agreement field tests can be used as acceptance tests.

3. Index tests are a form of field testing usually made to serve as a standard of comparison to indicate wear, changing conditions, or overhaul evaluation. Index tests should be made by the same procedures, instruments, and personnel where possible, and a very accurate record and log of events should be kept to give as complete and comparable a history of the results as possible.

4. Model tests precede the design of the prototype and are usually quite accurately conducted. They supplement or complement field tests of the prototype for which the model was made. The role of the model test must be clearly established as early in the design as possible, preferably in the specification or invitation to bid. Model tests may be used when very large units are involved, when the comparative performance of several models may be involved for evaluation, and when an advance indication of prototype design is required.

DEFINITIONS, SYMBOLS, AND UNITS

14.3 Definitions and Description of Terms The English gravitational system of units is to be used in this section. Where the user desires he should use standard conversion methods to convert to another system. For a detailed discussion of letter symbols, definitions, description of terms, and table of letter symbols in general use, the user is referred to ASME Power Test Code [2] (see also Sec. 14.16).

14.4 Standard Units Used in Pump Testing The following definitions and quantities from "Hydraulic Institute Standards," 12th ed., 1969 [1] are used as standard throughout the industry in pump testing.

1. *Volume.* The standard unit of volume is the U.S. gallon or the cubic foot. The standard U.S. gallon contains 231.0 in³. 1 ft³ = 7.4805 gal. The rate of flow is expressed in gallons per minute (gpm), cubic feet per second (ft³/s), or million gallons per twenty-four-hour day (mgd). The specific weight (w) of pure water at a temperature of 70°F, at sea level, and 40° latitude is taken as 62.1932 lb/ft³. For other temperatures or locations, proper specific-weight corrections should be made. See Table 1 and appropriate ASME Power Test Codes.

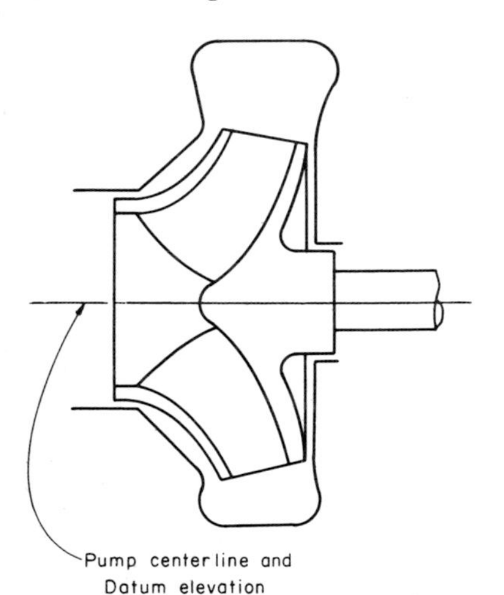

Fig. 1

2. *Head.* The unit for measuring head is the foot. The relation between a pressure expressed in pounds per square inch (lb/in²) and that expressed in feet of head is:

$$\text{Head, ft} = \text{lb/in}^2 \times \frac{144}{w}$$

where w = specific weight in lb/ft³ of liquid being pumped under pumping conditions. All pressure readings must be converted into feet of the liquid being pumped, referenced to a datum elevation. The datum elevation is defined as follows: For horizontal shaft unit it is the centerline of the pump shaft, Fig. 1. For vertical-shaft single-suction pumps it is the entrance eye to the first-stage impeller, Fig. 2. For vertical double-suction pumps it is the impeller-discharge horizontal centerline, Fig. 3.

3. *Velocity Head* (h_v). The velocity head is computed from the average velocity V obtained by dividing the flow in cubic feet per second (ft³/s) by the actual area of pipe cross section in square feet and determined at the point of the gage connec-

TABLE 1 Specific Weight of Water in Air, lb/ft³

Latitude	Temperature, °F 32	40	50	60	70	80	90	100
At sea level								
0°	62.1741	62.1823	62.1654	62.1227	62.0578	61.9729	61.8701	61.7514
10	62.1840	62.1921	62.1753	62.1325	62.0677	61.9828	61.8800	61.7612
20	62.2125	62.2206	62.2038	62.1610	62.0961	62.0112	61.9083	61.7895
30	62.2562	62.2643	62.2475	62.2046	62.1397	62.0547	61.9518	61.8328
40	62.3098	62.3179	62.3011	62.2582	62.1932	62.1082	62.0051	61.8861
50	62.3670	62.3751	62.3582	62.3153	62.2503	62.1652	62.0620	61.9429
60	62.4208	62.4289	62.4120	62.3691	62.3040	62.2188	62.1156	61.9963
70	62.4647	62.4728	62.4559	62.4130	62.3478	62.2626	62.1593	62.0399
At 2,000 ft								
0°	62.1665	62.1747	62.1578	62.1151	62.0502	61.9654	61.8626	61.7438
10	62.1764	62.1845	62.1677	62.1250	62.0601	61.9752	61.8724	61.7537
20	62.2049	62.2130	62.1962	62.1534	62.0885	62.0036	61.9008	61.7819
30	62.2486	62.2567	62.2399	62.1970	62.1321	62.0471	61.9442	61.8253
40	62.3022	62.3103	62.2935	62.2506	62.1856	62.1006	61.9976	61.8786
50	62.3594	62.3675	62.3506	62.3078	62.2427	62.1576	62.0545	61.9354
60	62.4132	62.4213	62.4044	62.3615	62.2964	62.2112	62.1080	61.9888
70	62.4571	62.4652	62.4484	62.4054	62.3402	62.2550	62.1517	62.0324
At 4,000 ft								
0°	62.1588	62.1669	62.1501	62.1073	62.0424	61.9576	61.8549	61.7361
10	62.1686	62.1767	62.1599	62.1172	62.0523	61.9675	61.8647	61.7459
20	62.1971	62.2052	62.1884	62.1456	62.0807	61.9959	61.8930	61.7742
30	62.2408	62.2489	62.2321	62.1893	62.1243	62.0394	61.9365	61.8176
40	62.2944	62.3025	62.2857	62.2428	62.1779	62.0929	61.9899	61.8709
50	62.3516	62.3597	62.3429	62.3000	62.2349	62.1498	62.0467	61.9277
60	62.4054	62.4135	62.3967	62.3537	62.2886	62.2035	62.1003	61.9811
70	62.4493	62.4574	62.4406	62.3976	62.3325	62.2472	62.1440	62.0247
At 6,000 ft								
0°	62.1508	62.1589	62.1421	62.0993	62.0345	61.9497	61.8469	61.7282
10	62.1607	62.1688	62.1520	62.1092	62.0444	61.9595	61.8568	61.7380
20	62.1891	62.1972	62.1804	62.1377	62.0728	61.9879	61.8851	61.7663
30	62.2328	62.2409	62.2241	62.1813	62.1164	62.0315	61.9286	61.8097
40	62.2864	62.2946	62.2777	62.2349	62.1699	62.0849	61.9819	61.8630
50	62.3436	62.3517	62.3349	62.2920	62.2270	62.1419	62.0388	61.9198
60	62.3974	62.4055	62.3887	62.3458	62.2807	62.1955	62.0924	61.9732
70	62.4413	62.4495	62.4326	62.3896	62.3245	62.2393	62.1361	62.0168
At 8,000 ft								
0°	62.1426	62.1507	62.1339	62.0912	62.0264	61.9416	61.8388	61.7201
10	62.1525	62.1606	62.1438	62.1011	62.0362	61.9514	61.8487	61.7300
20	62.1810	62.1891	62.1723	62.1295	62.0646	61.9798	61.8770	61.7582
30	62.2246	62.2328	62.2159	62.1731	62.1082	62.0233	61.9205	61.8016
40	62.2783	62.2864	62.2696	62.2267	62.1618	62.0768	61.9738	61.8549
50	62.3354	62.3436	62.3267	62.2838	62.2188	62.1338	62.0307	61.9117
60	62.3893	62.3974	62.3805	62.3376	62.2725	62.1874	62.0843	61.9651
70	62.4332	62.4413	62.4244	62.3815	62.3164	62.2312	62.1280	62.0087

TABLE 1 (Continued)

Latitude	32	40	50	60	70	80	90	100
	Temperature, °F							
	At 10,000 ft							
0°	62.1343	62.1424	62.1256	62.0828	62.0180	61.9333	61.8305	61.7119
10	62.1442	62.1523	62.1355	62.0927	62.0279	61.9431	61.8404	61.7217
20	62.1726	62.1807	62.1639	62.1212	62.0563	61.9715	61.8687	61.7500
30	62.2163	62.2244	62.2076	62.1648	62.0999	62.0150	61.9122	61.7934
40	62.2699	62.2781	62.2612	62.2184	62.1534	62.0685	61.9656	61.8466
50	62.3271	62.3352	62.3184	62.2755	62.2105	62.1255	62.0224	61.9034
60	62.3809	62.3890	62.3722	62.3293	62.2642	62.1791	62.0760	61.9568
70	62.4248	62.4330	62.4161	62.3732	62.3080	62.2229	62.1197	62.0005
	At 12,000 ft							
0°	62.1258	62.1339	62.1171	62.0743	62.0095	61.9248	61.8221	61.7035
10	62.1357	62.1438	62.1270	62.0842	62.0194	61.9347	61.8319	61.7133
20	62.1641	62.1722	62.1554	62.1127	62.0478	61.9630	61.8603	61.7416
30	62.2078	62.2159	62.1991	62.1563	62.0914	62.0066	61.9037	61.7849
40	62.2614	62.2695	62.2527	62.2099	62.1450	62.0600	61.9571	61.8382
50	62.3186	62.3267	62.3099	62.2670	62.2020	62.1170	62,0140	61.8950
60	62.3724	62.3805	62.3637	62.3208	62.2557	62.1706	62,0675	61.9484
70	62.4163	62.4245	62.4076	62.3647	62.2996	62.2144	62.1112	61.9920

This table was compiled using the following:
References:
1. "Smithsonian Physical Tables," 9 Rev. Ed.
2. National Advisory Committee for Aeronautics, TN 3182
3. American Society of Mechanical Engineers, PTC 2–1971
4. "Smithsonian Meteorological Tables," 6 Rev. Ed.

Conversion factors (from page 60 of Ref. 1 and page 8 of Ref. 3):
1 lb = 453.59237 g
1 ml = 1.000028 cm
1 in = 2.54 cm
1 ft = 0.3048 m

Density of water from page 296 of Ref. 1
Gravity formula from page 488 of Ref. 4:
$G_\phi = 980.616\,(1 - 0.0026373 \cos 2\phi + 0.0000059 \cos^2 2\phi)$
Altitude correction = 0.0003086 × altitude in meters
Standard gravity = 980.665 cm/s/s
Density of air from Ref. 2:
Density = $0.001225\,(1 - 0.0065H/288.16)^{4.2561}$
Where H is in meters

tion. Velocity head is expressed by the formula:

$$h_v = \frac{V^2}{2g}$$

where g = the acceleration due to gravity and is 32.17 ft/s² at sea level and approximately 45° latitude. See Table 2.
V = velocity in the pipe, ft/s

4. *Flooded Suction.* Flooded suction implies that the liquid must flow from an atmospheric vented source to the pump without the average or minimum pressure at the pump datum dropping below atmospheric pressure with the pump operating at specified capacity.

5. *Total Suction Lift* (h_s). Suction lift exists where the total suction head is below atmospheric pressure. Total suction lift, as determined on test, is the reading of a liquid manometer or pressure gage at the suction nozzle of the pump, converted to feet of liquid, and referred to datum minus the velocity head at the point of gage attachment.

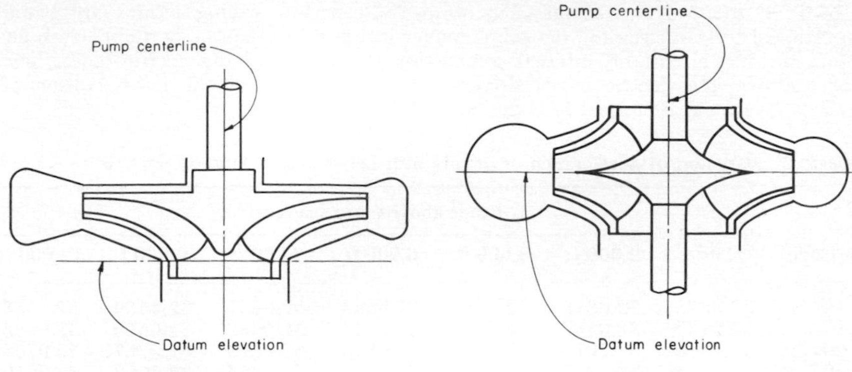

Fig. 2 **Fig. 3**

6. *Total Suction Head* (h_s). Suction head exists when the total suction head is above atmospheric pressure. Total suction head, as determined on test, is the reading of a gage at the suction of the pump converted to feet of liquid and referred to datum plus the velocity head at the point of gage attachment.

7. *Total Discharge Head* (h_d). Total discharge head is the reading of a pressure gage at the discharge of the pump, converted to feet of liquid and referred to datum plus the velocity head at the point of gage attachment.

8. *Total Head* (H). Total head is the measure of the work increase per pound of the liquid, imparted to the liquid by the pump, and is therefore the algebraic difference between the total discharge head and the total suction head. Total head, as determined on test where suction lift exists, is the sum of the total discharge head and total suction lift. Where positive suction head exists, the total head is the total discharge head minus the total suction head.

9. *Net Positive Suction Head, NPSH* (h_{sv}). The net positive suction head is the total suction head in feet of liquid absolute determined at the suction nozzle and referred to datum less the vapor pressure of the liquid in feet absolute.

10. *Driver Input* (ehp). The driver input is the input to the driver expressed in horsepower. Usually this is electric input horsepower.

11. *Pump Input* (bhp). Pump input is the horsepower delivered to the pump shaft and is designated as brake horsepower.

12. *Liquid* or *Water Horsepower* (whp). This is the useful work delivered by the pump and is usually expressed by the formula

$$\text{whp} = \frac{\text{sp gr} \times \text{gpm} \times H}{3{,}960}$$

where sp gr = specific gravity of liquid refined to water at 68°F
gpm = gallons per minute
H = total head, ft

Note: See Sec. 14.19 for derivation of formula.

13. *Efficiency.* Pump efficiency (E_p) is the ratio of the power delivered by the pump to the power supplied to the pump shaft, that is, the ratio of the liquid horsepower (also known as water horsepower, see Sec. 14.19) to the brake horsepower expressed in percent:

$$E_p = \frac{\text{whp}}{\text{bhp}} \times 100$$

Overall efficiency (E_o) is the ratio of the power delivered by the pump to the power supplied to the input side of the pump driver, that is, the ratio of the output horsepower to the input horsepower to the driver:

$$E_o = \frac{\text{whp}}{\text{ehp}} \times 100$$

14.5 Prime Mover Ratings The prime movers for driving pumps are based on established standards for the prime mover in question, such as hydraulic turbine, steam turbine, air motor, internal combustion engine, windmill, electric motor, etc. For example, for electric motor drivers see the standards of the latest edition of NEMA (National Electrical Manufacturers Association).

TABLE 2 Variation of Acceleration of Gravity with Latitude and Altitude, Ft/S/S

	Altitude above mean sea level						
Latitude	0	2,000 ft	4,000 ft	6,000 ft	8,000 ft	10,000 ft	12,000 ft
0°	32.0878	32.0816	32.0754	32.0693	32.0631	32.0569	32.0508
10	32.0929	32.0867	32.0805	32.0744	32.0682	32.0620	32.0558
20	32.1076	32.1014	32.0952	32.0890	32.0829	32.0767	32.0705
30	32.1301	32.1239	32.1177	32.1115	32.1054	32.0992	32.0930
40	32.1577	32.1515	32.1454	32.1392	32.1330	32.1269	32.1207
50	32.1872	32.1810	32.1748	32.1687	32.1625	32.1563	32.1501
60	32.2149	32.2087	32.2026	32.1964	32.1902	32.1841	32.1779
70	32.2375	32.2314	32.2252	32.2190	32.2129	32.2067	32.2005

References:

1. "Smithsonian Physical Tables," 9 Rev. Ed.
2. American Society of Mechanical Engineers, PTC 2–1971

Gravity = 980.616 (1 − 0.0026373 cos 2Ø + 0.0000059 $\cos^2$ 2Ø) (1.0/30.48)

Correction for altitude = −0.003086 ft/s^2/1,000 ft

The international standard value of gravity adopted by the International Commission on Weights and Measures is 980.665 cm/s^2 (32.17405 ft/s^2) corresponding to sea level and approximately latitude 45°.

Sample Computation:

Given:
Altitude = 12,000 ft
Latitude = 70°
Water temperature = 40°F

Altitude in meters:
Altitude = 0.30480 × 12,000 = 3657.60 m

Altitude correction for gravity:
Correction = 0.0003086 × 3,657.60 = 1.128735 cm/s/s

Gravity corrected for latitude and altitude:
Gravity = 980.616 (1 − 0.0026373 cos 140° − 0.0000059 $\cos^2$ 140°) − 1.128735 = 981.471784 cm/s/s

Density in g/cm^3 corrected for gravity:
Density = (0.9999983 × 981.471784/980.6650 × 1.000028) = 1.0007930 g/cm^3

Correcting for buoyancy of air:
Density = 1.0007930 − 0.0008491 = 0.9999438 g/cm^3

Density in lb/ft^3:
Density = 0.9999438 × 62.4279606 = 62.4244543 lb/ft^3

Rounding to four decimals:
Density = 62.4245 lb/ft^3

ACCURACY AND TOLERANCES

14.6 Accuracy The accuracy to which tests can be made depends on the instruments used, their proper installation, the skill of the test engineer, and the shop tests for the simulation of actual field conditions. The test engineer must have sufficient knowledge of the characteristics and limitations of his test instruments to obtain maximum accuracy when using them, along with a thorough understanding of the pumps, prime movers, controls, and installation peculiarities to interpret the results obtained. For shop testing the acceptable deviations and fluctuations of the instrumented test readings are given in Table 2 in ASME Power Test Code [2].

These deviations are not to be misconstrued as tolerances which must be spelled out in the specifications. The limits of accuracy of pump test measuring devices for use in field testing of pumps are shown in Fig. 4.

Quantity to be measured	Type of measuring device	Calibrated limit of accuracy plus or minus, %
Capacity	Venturi meter	¾
	Nozzle	1
	Pitot tube	1½
	Orifice	1¼
	Disk	2
	Piston	¼
	Volume or weight—Tank	1
	Propeller meter	4
Head	Electric sounding line	¼
	Air line	½
	Liquid manometer (3–5 in deflections)	¾
	Liquid manometer (over 5-in deflections)	½
	Bourdon gage—5-in min dial:	
	¼–½ full scale	1
	½–¾ full scale	¾
	Over ¾ scale	½
Power input	Watthour meter and stopwatch	1½
	Portable recording wattmeter	1½
	Test-type precision wattmeter:	
	¼–½ scale	¾
	½–¾ scale	½
	Over ¾ scale	¼
	Clamp on ammeter	4
Speed	Revolution counter and stopwatch	1¼
	Handheld Tachometer	1¼
	Stroboscope	1½
	Automatic counter and stopwatch	½
Voltage	Test meter: ¼–½	1
	Test meter: ½–¾	¾
	Test meter: ¾–full	½
	Rectifier voltmeter	5

Source: Reprinted from ANSI B-58.1 (AWWAE 101-61).

Fig. 4 Limits of accuracy of pump-test measuring devices in field use.

Using these accuracy limits, the combined accuracy of the efficiency is square root of the quantity of the square of the head accuracy plus the square of the capacity accuracy plus the square of the power input accuracy.

$$A_c = \sqrt{(\pm H^2) + (\pm Q^2) + (\pm \overline{\text{ehp}}^2)} = A_c \qquad \text{(percent)}$$

Pump speed and voltage are not required for efficiency computations, so the values for these are not included in the above formula.

14.7 Instrumentation All instruments should be calibrated before the tests, and all calibration and correction data or curves should be prepared in advance. Where required, a certified calibration curve showing the calibration of the instrument, including any procedures for establishing a coefficient, shall be furnished before actual testing begins. The specifications should be explicit in regard to a waiving of these calibration requirements. After testing all instruments should be recalibrated. Any difference discovered between the before and after calibration must be resolved either by retest or spelled out in the specifications in regard to acceptable variations.

14.8 Tolerances of Pump Performance The tolerances in regard to pump performance permitted are usually given or referred to in the specifications. The user can and should make his requirements known before his firm order for pumping apparatus is placed. The test tolerances permitted by the Hydraulic Institute are quite common; they state that no minus tolerance or margin shall be allowed with respect to capacity, total head, or efficiency at the rated condition. Also a plus tolerance of not more than 10 percent with respect to rated capacity shall be allowed at the rated head and speed.

The tolerances are quite easy to meet; they protect the user from getting pumps that are too small to do the job and also from getting oversized pumps and drivers that would increase building, installation, and operating costs. These tolerances also give the manufacturer liberal leeway when impeller trim is required to meet the specified conditions.

TEST REQUISITES

14.9 Operating Conditions The primary factors affecting the operation of a pump are the inlet (suction), outlet (discharge or total head), and speed. The secondary factors, not necessarily in this order, are the physical and climatic, such as temperature of liquid, viscosity, elevation above sea level, specific weight, and turbidity (silt and/or solids). In some installations it is impossible to measure accurately discharge or even head. In these instances good shop tests are essential. It follows then that in order for the shop test to predict the field performance of a pump the field operating, installation, and suction conditions should be simulated.

The inlet, or intake, passages are critical, and the actual sump, where used, on the suction lift must be carefully repeated or duplicated as near as possible. During the shop tests no total suction head less than specified shall be permitted, nor should

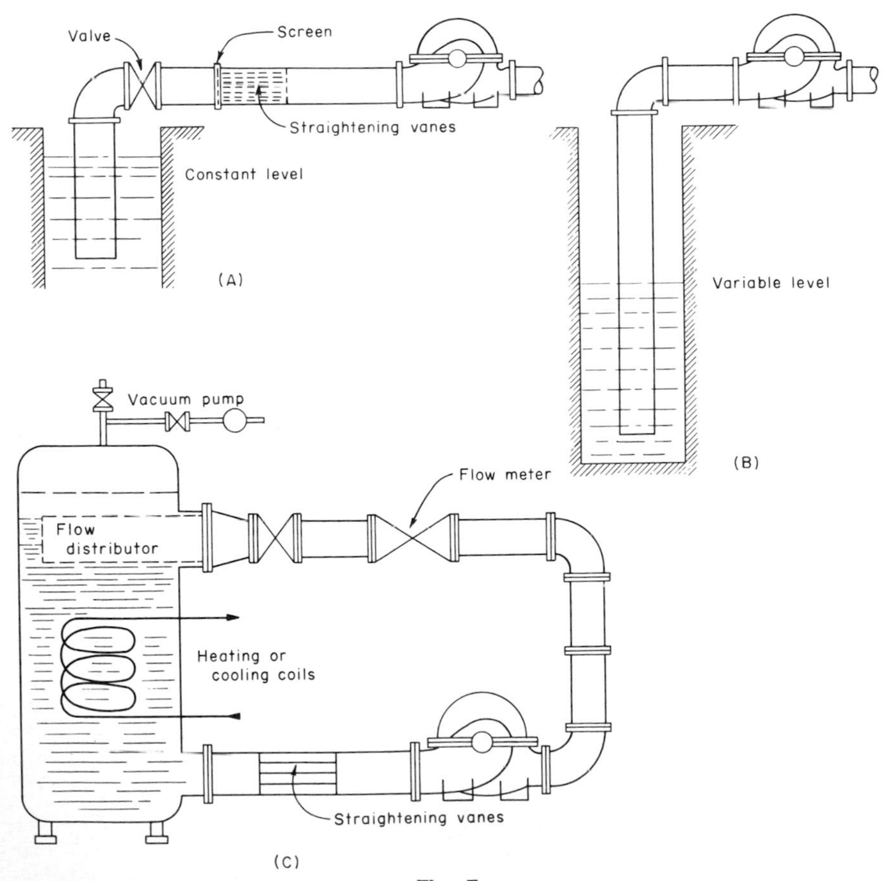

Fig. 5

the suction head exceed the specified amount in cases where cavitation or possibly operating "in the break" could occur.

For field installations above sea level the difference in elevation between the shop test site and actual field installation must be taken into account by reducing to the actual barometric pressures at the specified elevation. This is especially true if a suction lift or negative suction head is involved. Standard tables of barometric pressures are available for use in computing the data, and the tables to be used should be mutually acceptable to the interested parties.

14.10 Cavitation Tests Cavitation tests should be run if required by the specifications (provided such tests are needed and have not been previously conducted on similar pumps and certified by the manufacturer), or if such tests are needed to assume a successful pump installation.

The suction requirements that must be met by the pump are usually defined by the cavitation coefficient σ. Plant σ is defined as the NPSHA (net positive suction head available) divided by the total pump head per stage.

$$\sigma = \frac{\text{NPSHA}}{H}$$

Three typical, or standard, arrangements for determining the cavitation characteristics of pumps (from "Hydraulic Institute Standards") are illustrated in Figs. 5A, B, and C. In Fig. 5A the suction is taken from a sump with a constant-level surface. The liquid is drawn first through a valve (throttle) and then through a section of pipe containing screens and straightening devices such as vanes and baffles. This setup will dissipate the tubulence created by the suction valve and will also straighten the flow so that the pump suction flow will be relatively free from undue turbulence. In Fig. 5B the suction is taken from a relatively deep sump or well in which the water surface can be varied over a fairly large range to provide the designed variation in suction lift. In Figure 5C the suction is taken from a closed vessel in a closed loop in which the pressure level can be adjusted or varied by a gas pressure over the liquid, by temperature of the liquid, or by a combination of these variables.

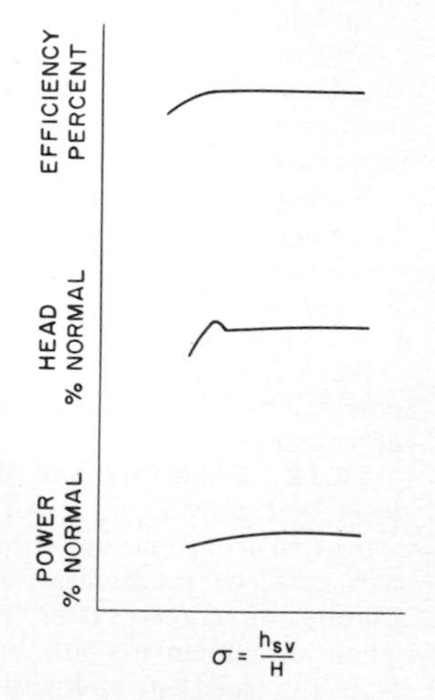

Fig. 6 Functions of sigma (σ) at constant capacity and speed; suction pressure varied.

By using one of the cavitation test arrangements above, the critical value of sigma, at which cavitation will begin, can be found by one of the following two methods or test procedures:

1. Constant speed and capacity vary the suction lift. Run the pump at constant speed and capacity with the suction lift varied to produce unstable or cavitation conditions. Plots of the head, efficiency, and power input against σ as shown in Fig. 6.

When the values of σ are held high, the values of head, efficiency, and power should remain relatively constant. As σ is reduced, a point is reached when the curves break from the normal, indicating an unstable condition. This unstable, or breakaway, condition may and usually does impair the operation of the pump. How much will depend on the actual pump in regard to size, specific speed, service, and characteristics of the pumped fluid. A variation of this method is to plot results using capacities both greater and less than normal as shown in Fig. 7.

2. Constant speed and suction lift vary the capacity. Run the pump at constant speed and suction lift or pressure and vary the capacity. For a given suction lift, the pumping head is plotted against capacity. A group or series of such tests will result in a family of curves as shown in Fig. 8.

Where the plotted curve for any suction condition breaks away from the normal, cavitation has occurred. σ may be calculated at the breakaway points by dividing the NPSHA by the total head H at the point under consideration.

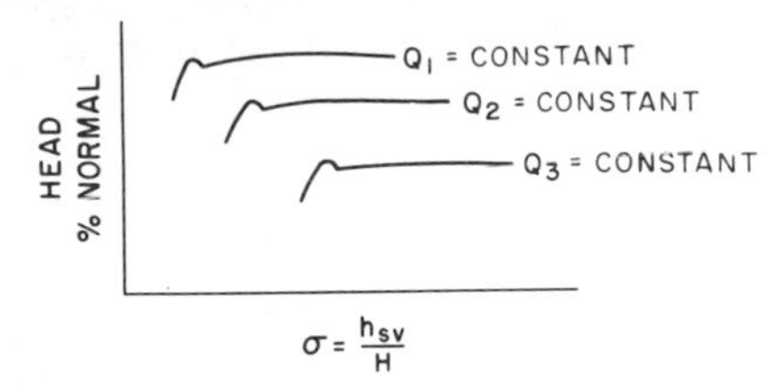

Fig. 7 Sigma (σ) capacities above and below normal; suction pressure varied.

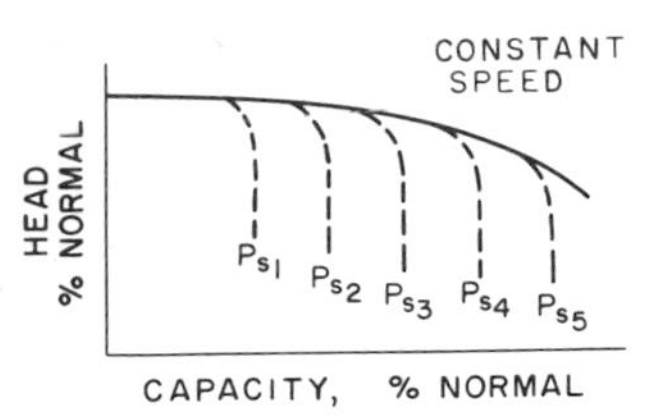

Fig. 8 Typical cavitation curves at constant speed and suction pressure.

TEST PROCEDURE

14.11 Agreements The specifications and contract should be very clear on any special points that must be covered by the pump testing. Both parties shall be represented and given equal rights in regard to test date, setup, conditions, instrumentation, calibration of instruments, examination of pump and test setup, and accuracy of results and computations. Any controversial points or methods not otherwise provided for in the specifications should be resolved to the satisfaction of all interested parties before actual testing is begun.

In some special instances and by agreement between parties an independent test expert or test engineer may be engaged to take over full responsibility of the test stand and apparatus. He would make all decisions after due consultation with the interested parties. A test expert should be used only where a complete impasse is reached or declared.

Normally the manufacturer will establish the time and date for the pump tests. In some cases the specifications or contract will cover such items as duration, limiting date, and notification time. Reasonable notice must be given to all official witnesses or representatives. A 30-day notice is preferred, and 1 week should be considered a bare minimum.

Any time limitation regarding mechanical defects or malfunction of equipment corrections that arise during the actual testing must be resolved by mutual agreement.

14.12 Observers and Witnesses Representatives from each party to the contract shall have equal opportunity to attend the testing. Where more than one representative from one of the parties is present then their function as observers, official witnesses, or representatives must be made clear before starting of testing. The number of representatives present from any one party shall not be a deciding factor when disagreements are being resolved. Any comments or constructive criticism from the observers and witnesses should be duly considered.

14.13 Inspection and Preliminary Operation All interested parties shall make as complete an inspection as possible before, during, and after the test to determine compliance with specifications requirements and correct connection of all instrumentation. The following items should be inspected before or during the test:

Impeller and casing passages
Pump and driver alignment
Piezometer openings
Electric connections
Lubricating devices and system
Wearing ring and other clearances
Stuffing box or mechanical seal adjustment and leakage

The above is not intended to be a complete list of items but rather to serve as a guide for the interested parties.

Instruments installed on the pump to obtain the necessary test information shall not affect the pump operation or performance. If a question arises as to the effect an instrument has on the operation, it should be resolved by both parties. Where necessary comparative preliminary tests can be conducted with the disputable equipment removed and then installed. The actual dimensions at the piezometer connections both on the suction and discharge sides must be accurately determined to permit accurate determination of the velocity head correction.

On satisfactory completion of the before or initial inspection the pump may be started. The pump and all instrumentation should then be checked for proper operation, within scale readings, or any evidence of malfunction. When all equipment and apparatus are functioning properly, a preliminary test run should be made. If possible, this run should be made at or near the rated condition. The correct procedures for observations and recording of data should be established during this run. Also the duration or time it takes to obtain steady test conditions is determined for use in conducting the actual pump-test runs. The acceptable deviations and fluctuations for test readings are given in Sec. 14.6 under Accuracy.

14.14 Suggested Test-Procedure Outline

1. *Time and Date.* Once tests have been decided upon, it is to the best interests of all parties to conduct them with the least delay. The contract instrument normally will not give a time or date for the actual test, but will state or specify a completion date for submission or approval of a final type of test report.

2. *Personnel.* Pump testing, regardless of classification, should be made by personnel specially trained and skilled in the operation of the test equipment used. Representatives from each party to the contract shall be given equal opportunity to attend and witness the test or tests and shall also have equal voice in commenting on the conduct of the tests or compliance with specifications or code requirements where applicable.

3. *Schedule.* A schedule or program of the sequence of events should be agreed upon by all parties in advance of the actual test. In addition to a schedule of events, it should also be as complete a program as possible, and give some particulars on the range of test heads, discharge rates, and speed to be used. This schedule should be flexible and subject to change especially after the preliminary runs have been made.

4. *Inspection.* The pump and test setup should be thoroughly inspected both before and after the tests. Special attention should be given to the hydraulic passages and pressure taps near the suction and discharge sections. Also the discharge measuring device should be inspected.

5. *Calibration of Instruments.* While performing the above inspection, all measuring devices and instruments should be calibrated and adjusted as explained in Sec. 14.7.

6. *Preliminary Tests.* After determining that the test setup complies with the installation and specification requirements and that the instrumentation is properly installed, the pump is started. A sufficient number of preliminary test runs should be made to check for the correct functioning of the test stand and all control and measuring devices. These preliminary tests also give the test personnel and representatives an opportunity to check and correct the proper function of the entire setup and serve as a basis for any agreements on accuracy and compliance before conducting the actual tests. Each test point is held until satisfactory stable conditions exist. The acceptable fluctuations in test readings are given in Sec. 14.6. It is suggested that preliminary computations be made, plotted, and analyzed prior to actual test runs. (*Note:* Most pump manufacturers will conduct a complete preliminary shop test to enable them to be sure of specifications and contractual compliance before inviting the purchaser's representatives to witness the official test of record.)

7. *Official Test Runs.* The actual test points for the official test runs must be sufficient in number to establish the head-discharge curve over the specified range, in addition to any other data needed to compute or plot the information required by the specifications, such as efficiency, horsepower, etc. It is suggested that one test run be as near the rated condition as possible and that at least three runs be in the specified operating range of the pump.

8. *Logging of Events.* In most instances the official pump test is conducted by only two, three, or four test engineers and no record other than test data is needed. In more complicated or important tests, it is sometimes very desirable to assign

someone the task of recording and logging the happenings and events as they chronologically occur. This is most important if reruns are indicated or required. Complete records, including any notes or comments on inspection and calibration, shall be kept of all data, readings, observations, and information relevant to the test. A suggested form for shop and field tests is shown in Sec. 14.22.

9. *Preliminary Computations.* Sufficient preliminary computations should be made to determine that all specification requirements have been met and whether reruns will be necessary.

10. *Reruns.* When reruns are necessary as indicated by the preliminary computations, they should be run immediately or as soon as possible after the official test runs and with the same personnel, instruments, and devices. Sometimes mechanical or electric faults will occur which will necessitate a rerun. If after correction of these faults several reruns or repeat tests indicate a change, a complete retesting may be required. Any official representative shall have the right to ask for a rerun or be shown to his satisfaction that a rerun is not required.

11. *Computations.* The results of the pump tests are carried out as provided in Secs. 14.19 to 14.21.

12. *Plotting.* See Sec. 14.23.

13. *Reports.* The final report should be written as provided in Sec. 14.24.

TEST MEASUREMENTS

14.15 Discharge Measurements *General.* The application or use of a specific method of discharge measurement to a particular situation or test should be made by agreement between all parties concerned. Some test codes and procedures in regular use permit or even recommend certain methods when conducting model or shop testing, but restrict their use in field or index testing. Some methods are more readily adaptable to the actual site conditions than others, so the test engineers and interested parties should be completely familiar with each of the several methods possible or applicable before settling on the method to be used.

The most commonly used methods of discharge measurement for testing pumps are the quantity and rate-of-flow meters. These two general types of discharge measurements are usually classified as liquid meters. Listed below are liquid meters according to their functions:

Quantity meters:

1. *Weighing meters*
   - Weighing tank
   - Tilting trap
2. *Volumetric meters*
   - Tank
   - Reciprocating piston
   - Rotary piston
   - Nutating disk

Rate of flow meters:

1. *Head (kinetic) meters*
   - Venturi
   - Nozzle
   - Orifice plate
   - Pitot tube
2. *Head-area meters*
   - Weir
   - Flume
3. *Current meters*
4. *Special methods*
   - Salt velocity
   - Acoustic flow meter
   - Magnetic flow meter
   - Radioisotope
   - Field approximating

Quantity Meter. GENERAL. The term *quantity* is here used to designate those meters through the primary element of which the fluid passes in successive and more or less completely isolated quantities, either weights or volumes, by alternately filling and emptying containers of known or fixed capacities. The secondary element of a quantity meter consists of a counter with suitably graduated dials for registering the total quantity that has passed through the meter. The quantity meters are classified into two groups, the weighing meters and the volumetric meters.

WEIGHING METERS The weighing meters are made up of the weighing tank and the tilting-trap types. These meters can be described as those meters in which the equilibrium of a container is upset by a rise of the center of gravity as the container is filled, or to those which employ a container suspended from a counter-balanced scale beam. The weighing tank and the tilting trap are affected slightly by the temperature of the liquid but not enough to cause concern in normal testing.

VOLUMETRIC METERS The volumetric meters measure volumes instead of weights and are made up of the tank, reciprocating piston, rotary piston, and nutating disk types. Volumetric meters are described as follows:

1. Tanks. This is a very elementary form of meter of limited commercial importance. As the name implies, these meters consist of one or more tanks which are alternately filled and emptied. The height to which they are filled can be regulated manually or automatically. In some cases, the rise of the liquid operates a float which controls the inflow and outflow; in others, it may start a siphon. Occasionally some tank meters have been erroneously classified as weighers.

2. Reciprocating Piston. These meters use one or more members having a reciprocating motion, which operate in one or more fixed chambers. Adjustment of the quantity per cycle can be effected either by varying the magnitude of movement of one or more of the reciprocating members or by varying the relation between the primary and secondary elements.

3. Rotary (or Oscillating) Piston. Meters of this group have one or more vanes which serve as pistons or movable partitions for separating the fluid segments. These vanes may be either flat or cylindrical and rotate within a cylindrical metering chamber. The axis of rotation or annular movement of the vanes may or may not coincide with that of the chamber. The portion of the chamber in which the fluid is measured usually includes about 270°. In the remaining 90°, the vanes are returned to the starting position for closing off another segment of fluid. This may be accomplished by the use of an idle rotor or gear, a cam, or a radial partition. The vanes must make almost a wiping contact with the walls of the measuring chamber. The rotation of the vanes operates the secondary element or counter.

4. Nutating Disk. Meters of this type have the disk mounted in a circular chamber with a conical roof and either a flat or conical floor. When in operation, the motion of the disk is such that the shaft on which it is mounted generates a cone with the apex down. However, the disk does not rotate about its own axis; this is prevented by a radial slot which fits about a radial partition extending in from the chamber sidewall nearly to the center. The peculiar motion of the disk is called *nutating*. The inlet and outlet openings are in the sidewall of the chamber on either side of the partition. Adjustment of these meters is usually effected by changing the relation between the primary and secondary elements.

Rate-of-flow Meters. GENERAL The term *rate,* or *rate of flow,* is applied to all meters through which the fluid does not pass in isolated (separately counted) quantities but in a continuous stream. The movement of this fluid stream flowing through the primary element is directly or indirectly utilized to actuate the secondary element. The quantity of flow per unit of time is derived from the interactions of the stream and the primary element by known physical laws supplemented by empirical relations.

In rate-of-flow meters, the functioning of the primary element depends upon some property of the fluid other than, or in addition to, volume or mass. Such a property may be kinetic energy (head meters), inertia (gate meters), specific heat (thermal meters), or the like. The secondary element is designed to utilize a change in the property, or properties, concerned for obtaining an indication of the rate of flow and usually embodies some device which draws the necessary inferences automatically, so that the observer can read the result from a dial or chart. In some cases, the secondary element indicates or records pressures, such as static and differential, from which the rate of flow and time-quantity flow must be obtained by computation. In others, the secondary element not only indicates the rate of flow, but also integrates it with respect to time and records the total quantity that has passed through the meter. In some cases, the indications and recordings of the secondary element are transmitted to a point some distance from the primary element.

DIFFERENTIAL-PRESSURE METERS With this group of meters the stream of fluid creates a pressure difference as it flows through the primary element. The magnitude of this pressure difference depends upon the speed and density of the fluid and features of the primary element (see Ref. 3).

MEASUREMENTS IN PRESSURE CONDUITS Measurements of flows in a pipeline or closed pressure conduit can be accomplished by a wide variety of methods, and the choice of a method for a particular installation will depend upon prevailing conditions. The accuracy of flow measurements in pressure conduits by means of properly selected, installed, and maintained measuring equipment, such as the venturi meters, flow nozzles, orifice meters, and pitot tubes, can be very high. A discussion of some of the measuring equipment and methods is contained in the following paragraphs:

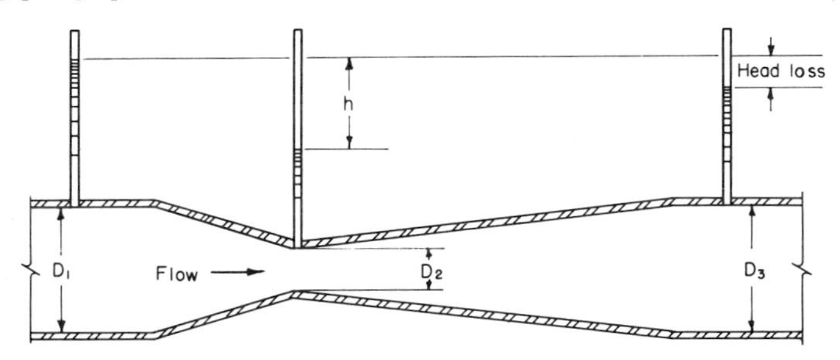

Fig. 9 Diagram of venturi meter.

1. Venturi Meters. The venturi meter (see Fig. 9) is perhaps the most accurate type of flow measuring device that can be used in a water supply system. They contain no moving parts, require very little maintenance, and cause very little head loss. Venturi meters operate upon the principle that flow in a given closed-conduit system moves more rapidly through areas of small cross section (D_2) than through areas of large cross section (D_1). The total energy in the flow, consisting primarily of velocity head and pressure head, is essentially the same at D_1 and D_2 within the meter. Thus the pressure must decrease in the constricted throat D_2 where the velocity is higher, and conversely the pressure must increase at D_1 upstream from the throat where the velocity is lower. This reduction in pressure from the meter entrance to the meter throat is directly related to the rate of flow passing through the meter, and is the measurement used to determine flow rate.

The coefficient of discharge for the venturi will range from an approximate value of 0.935 for small-throat velocities and diameters to 0.988 for relatively large-throat velocities and diameters.

Equations for the venturi meter are

$$Q = \frac{CA_2\sqrt{2gh}}{\sqrt{1 - R^4}}$$

$$Q' = 3.118\frac{CA'_2\sqrt{2gh}}{\sqrt{1 - R^4}}$$

where Q = rate of flow, ft³/s

Q' = rate of flow, gpm

C = coefficient of discharge for meter

A_2 = area of throat section, ft^2
A'_2 = area of nozzle throat, in^2
R = ratio of throat to inlet diameter (D_2/D_1)
g = acceleration of gravity, 32.17 ft/s^2
h = differential head between meter inlet and throat in feet of liquid being measured

2. Flow Nozzles. Flow nozzles operate upon the same basic principle as venturi meters. In effect, the flow nozzle is a venturi meter that has been simplified and shortened by omitting the long diffuser on the outlet side (Fig. 10). The streamlined entrance of the nozzle provides a straight cylindrical jet without contraction,

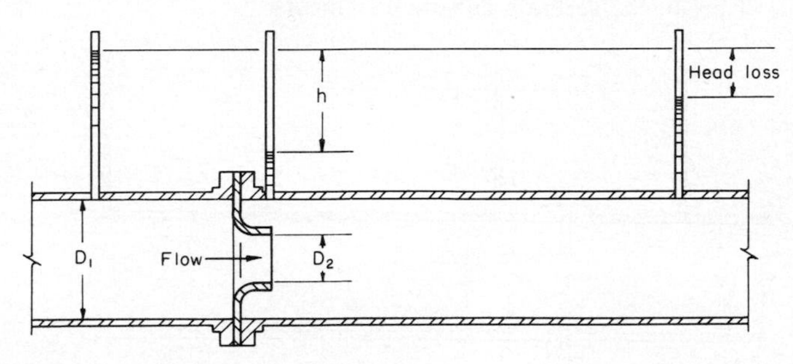

Fig. 10 Diagram of flow nozzle.

so that the coefficient is almost the same as that for the venturi meter. In the flow nozzle the jet is allowed to expand of its own accord, and the high degree of turbulence created downstream from the nozzle causes a greater loss of head than occurs in the venturi meter where the diffuser suppresses turbulence. The relationship of rate of flow to the head and the dimensions of the flow nozzle is:

$$Q = \frac{CA_2\sqrt{2gh}}{\sqrt{1 - R^4}}$$

$$A' = \frac{3.118CA'_2\sqrt{2gh}}{\sqrt{1 - R^4}}$$

where Q = rate of flow, ft^3/s
Q' = rate of flow, gpm
C = coefficient of discharge for nozzle
A_2 = area of nozzle throat, ft^2
A'_2 = area of nozzle throat, in^2
R = ratio of throat to inlet diameter (D_2/D_1)
g = acceleration of gravity, 32.17 ft/s^2
h = head at or across the nozzle, ft of liquid being measured

3. Orifice Meters. A thin-plate orifice inserted across a pipeline can be used for measuring flow in much the same manner as a flow nozzle (Fig. 11). The upstream pressure connection is often located at a distance of about one pipe diameter upstream from the orifice plate. The pressure of the jet ranges from a minimum at the vena contracta (the smallest cross section of the jet), to a maximum at about four or five conduit diameters downstream from the orifice plate. The downstream pressure connection (the center connection shown on Fig. 11) is usually made at the vena contracta to obtain a large pressure differential across the orifice.

The pressure tap openings should be free from burrs and flush with interior surfaces of pipe. Equations for the orifice plate are:

$$Q = CA_2\sqrt{2gh}$$

$$Q' = 3.118CA'_2\sqrt{2gh}$$

where Q = rate of flow, ft^3/s
Q' = rate of flow, gpm
C = coefficient of discharge for orifice plate
A_2 = area of orifice, ft^2
A_2' = area of orifice, in^2
R = ratio of throat to inlet diameter (D_2/D_1)
g = acceleration of gravity, 32.17 ft/s^2
h = head across the orifice plate, ft of liquid being measured

The principal disadvantage of orifice meters, as compared to venturi meters or flow nozzles, is their greater loss of head. On the other hand, they are inexpensive and are capable of producing accurate flow measurements.

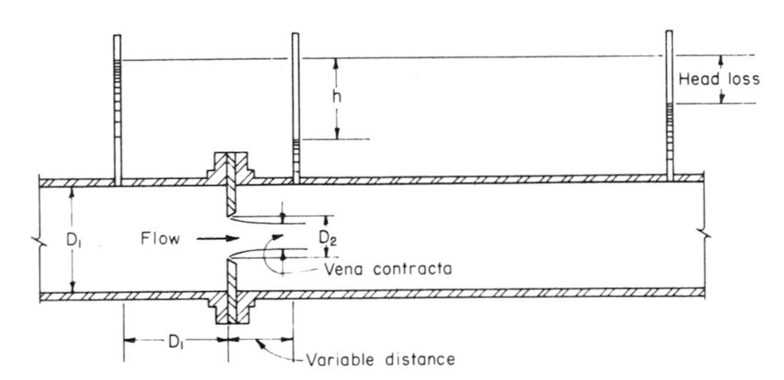

Fig. 11 Diagram of orifice meter.

It should be noted that the relationship of rate of flow to the head and dimensions of the metering section is identical for the venturi meter, flow nozzle, and orifice meter except that the coefficient c of the respective meter varies.

4. Pitot Tubes. Where it is impossible to employ one of the methods described above, the pitot tube is often used. A pitot tube in its simplest form is an instrument consisting of a tube with a right-angle bend which, when partly immersed with the bent part underwater and pointed directly into the flow, indicates velocity of flow by the distance that water rises in the vertical stem. The pitot or impact tube makes use of the difference between the static and total pressures at a single point.

The height of rise, h, of the water column above the water surface, expressed in feet and tenths of feet, equals the velocity head, $v^2/2g$. The velocity of flow, v, in ft/s, may thus be determined from the relation $v = \sqrt{2gh}$.

In a more complete form known as the pitot-static tube, the instrument consists of two separate, essentially parallel parts, one for indicating the sum of the pressure and velocity head (total head) and the other for indicating only the pressure head. Manometers are commonly used to measure these heads, and the velocity head is obtained by subtracting the static head from the total head. A pressure transducer may also be used instead of the manometer for measuring the differential head. Oscillograph or digital recording of the electric signal from the transducer provides a continuous record of the changes in head.

The simple form of the pitot tube has little practical value for measuring discharges in open channels handling low velocity flows because the distance the water in the manometer tube rises above the flowing water surface is difficult to measure. This limitation is overcome to a large extent by using a pressure transducer for the measurement and precise electronic equipment for the data readings. The pitot-static tube, on the other hand, works very well for this purpose if the tube is used with a differential manometer of the suction-lift type shown in Fig. 12.

In this manometer the two legs are joined at the top by a T that connects to a third line in which a partial vacuum can be created. After bleeding the pitot tube to remove all air, water flows up through the pitot tube into the manometer

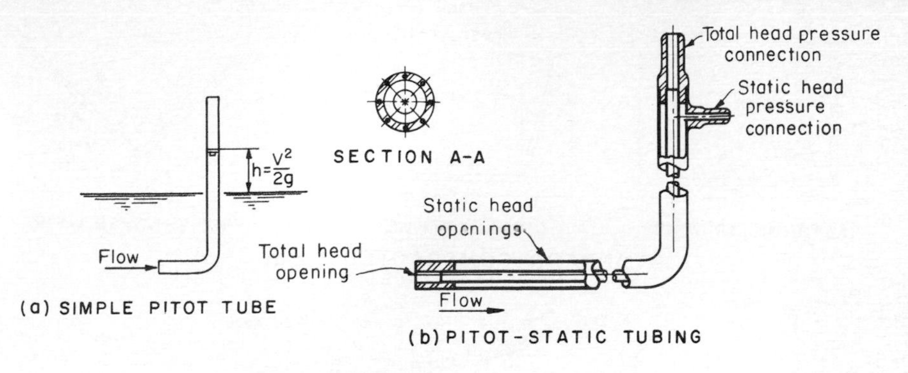

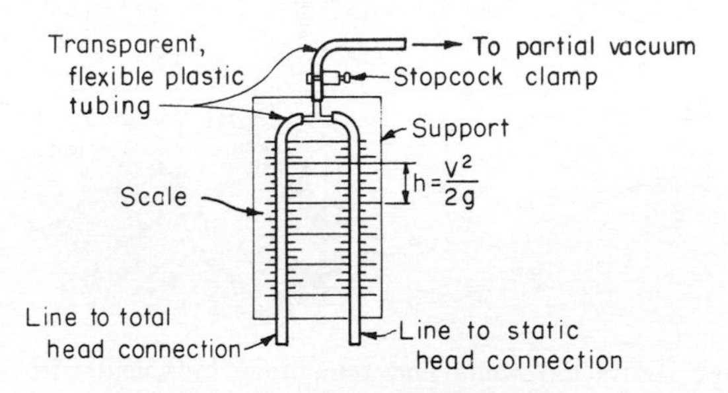

Fig. 12 Pitot tubes and manometer.

to the height desired for easy reading. Then the stopcock or clamp on the vacuum line is closed. The partial vacuum acts equally on the two legs and does not change the differential head. The velocity head h will then be the difference between the total head reading and the static head reading. If desired, a pressure transducer could also be used for the head measurement.

Pitot tubes can be used to measure relatively high velocities in canals, and it is often possible to make satisfactory discharge measurements at drops, chutes, overfall crests, or other stations where the water flows rapidly and fairly large velocity heads (h) occur. At low velocities, values of h become quite small. The pitot tube head for a low velocity will lead to a much larger inaccuracy in discharge computation than the same error when the velocity is high. The velocity traverse with pitot tube may be made in the same manner as with the current meter.

Measurements in open conduits. Measurements of flow in an open channel are normally classified as head-area meters, the most common of which are the weir and the flume.

WEIRS 1. General. A weir is an overflow structure built across an open channel. Weirs are one of the oldest, simplest, and most reliable structures that can be used to measure the flow of water in canals and ditches. These structures can be easily inspected and any improper operations can be detected and quickly corrected.

The discharge rates are determined by measuring the vertical distance from the crest of the overflow portion of the weir to the water surface in the pool upstream from the crest and referring to computations of tables which apply to size and shape of the weir. For standard tables to apply, the weir must have a regular shape, definite dimensions, and be placed in a bulkhead and pool of adequate size so the system performs in a standard manner.

Weirs may be termed rectangular, trapezoidal, or triangular, depending upon the shape of the opening. In the case of rectangular or trapezoidal weirs, the bottom edge of the opening is the crest, and the side edges are called *sides* or *weir ends* (Figs. 13 and 14). The sheet of water leaving the weir crest is called the *nappe*. In

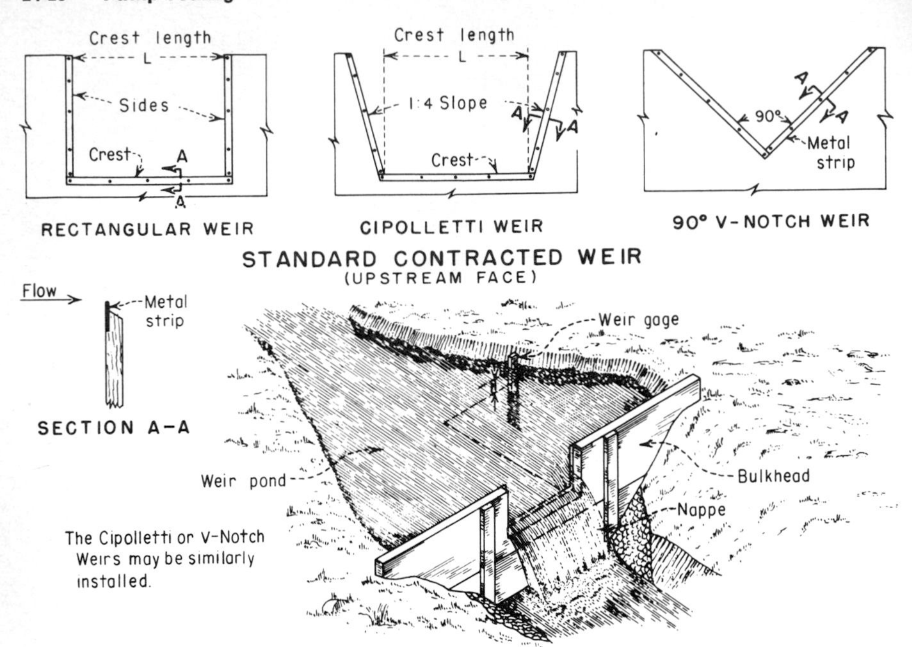

Fig. 13 Standard contracted weirs and temporary bulkhead with contracted rectangular weir discharging at free flow.

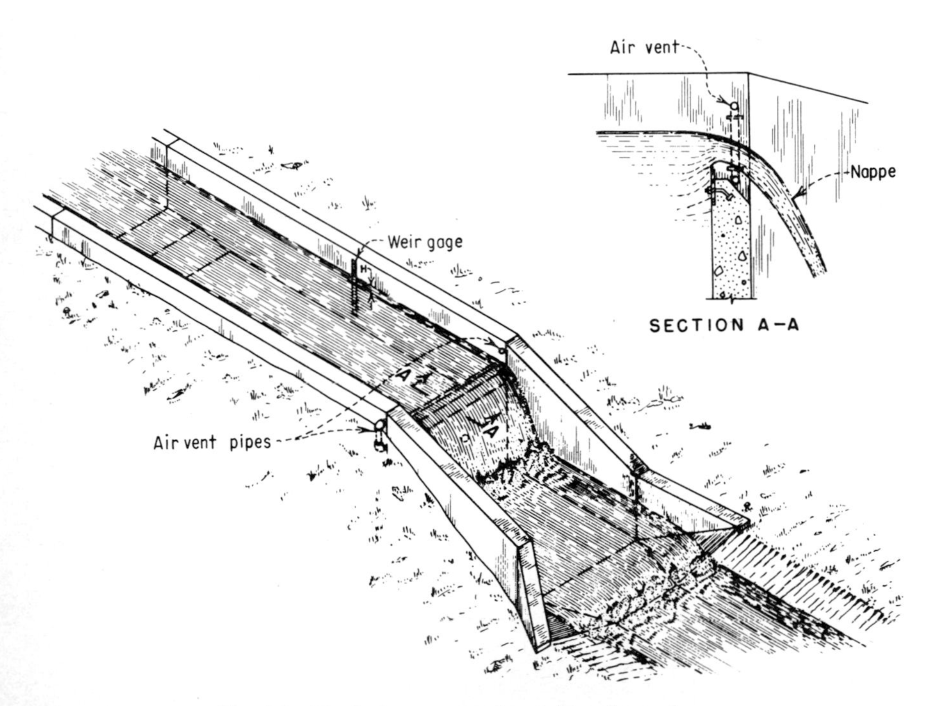

Fig. 14 Typical suppressed weir in a flume drop.

certain submerged conditions, the under-nappe airspace must be ventilated to maintain near-atmospheric pressure.

2. Types of Standard Weirs. The types of weirs most commonly used to measure water are:

Sharp-crested and sharp-sided cipolletti weirs
Sharp-sided 90° V-notch weirs
Sharp-crested contracted rectangular weirs
Sharp-crested suppressed rectangular weirs

For measuring water the type of weirs used has characteristics that fit it for a particular operating condition. In general, for best accuracy, a rectangular suppressed weir or a 90° V-notch weir should be used.

The discharge in second-feet over the crest of a contracted rectangular weir, a suppressed rectangular weir, or a cipolletti weir is determined by the head H in feet and by the crest length L in feet. The discharge of the standard 90° V-notch weir is determined directly by the head on the bottom of the V notch.

As the stream passes over the weir, the top surface curves downward. This curved surface, or drawdown, extends upstream a short distance from the weir notch. The head H must be measured at a point on the water surface in the weir pond beyond the effect of the drawdown. This distance should be at least four times the maximum head on the weir, and the same gage point should be used for lesser discharges. A staff gage (Fig. 15) having a graduated scale with the zero placed at the same elevation as the weir crest is usually provided for the head measurements.

3. Formulas for Standard Contracted Rectangular Weirs. Two widely used formulas for computing the discharges over standard contracted rectangular weirs are those of Hamilton Smith, Jr. [6] and J. B. Francis [7]. The formulas proposed by Smith require the use of coefficients of discharge varying with the head of water on the weir, and also with the length of the weir. Consequently the Smith formulas are somewhat inconvenient to use, although they are accurate for the ranges of coefficients usually given. The Francis formula for this type of weir, operating under favorable conditions as prescribed in preceding paragraphs and neglecting velocity of approach, is

$$Q = 3.33H^{3/2}(L - 0.2H) \tag{1}$$

and including velocity of approach:

$$Q' = 3.33[(H + h)^{3/2} - h^{3/2}](L - 0.2H) \tag{2}$$

where Q = discharge in s-ft neglecting velocity of approach
Q' = discharge in s-ft considering velocity of approach
L = the length of weir, ft
H = head on the weir, ft
h = head in ft due to the velocity of approach = $v^2/2g$

It will be noted that Francis' formulas contain constant discharge coefficients which facilitate computation without the use of tables.

A table of $\frac{3}{2}$ powers of numbers in the reference section in the back of the handbook provides values of $H^{3/2}$, $h^{3/2}$, and $(H = h)^{3/2}$ for convenience in computing discharge with the formulas given in this section.

4. Formulas for Sharp-Crested Suppressed Rectangular Weirs. The two principal formulas used for computing the discharge of the standard suppressed rectangular weir were also proposed by Smith and Francis. In the Smith formulas for suppressed weirs, as for contracted weirs, coefficients of discharge vary with the head on the weir and with the length of the weir; therefore, these formulas are not convenient for use in computations without tables of coefficients.

The Francis formula for the standard suppressed rectangular weir, neglecting velocity of approach, is

$$Q = 3.33LH^{3/2} \tag{3}$$

and including velocity of approach,

$$Q' = 3.33L[(H + h)^{3/2} - h^{3/2}] \tag{4}$$

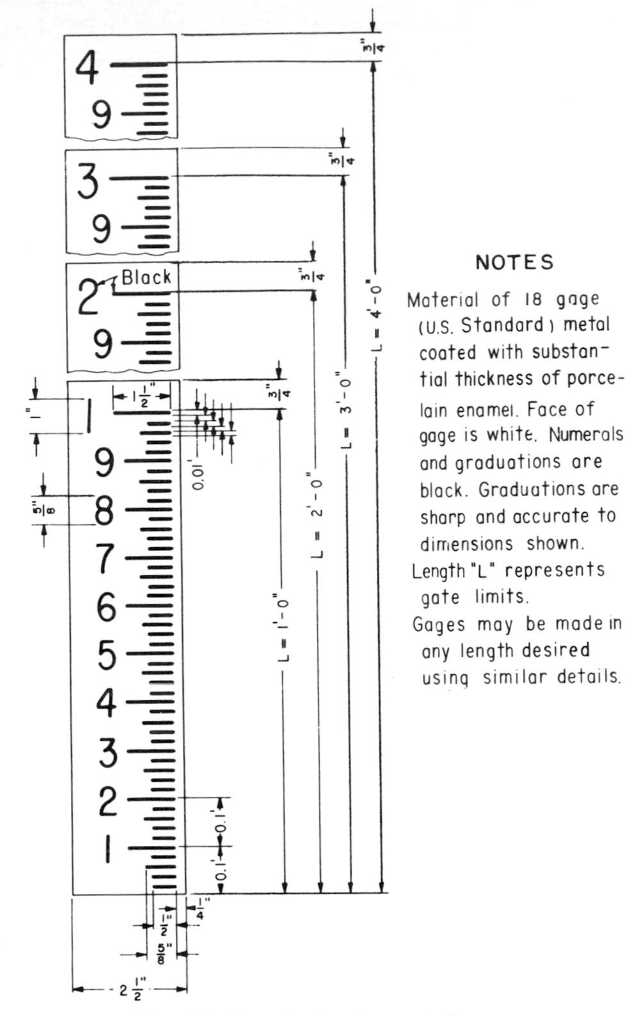

Fig. 15 Standard weir or staff gage.

In these formulas the letters have the same significance as in the formulas for contracted rectangular weirs. The coefficient of discharge was obtained by Francis from the same general set of experiments as those stated for the contracted rectangular weir. No extensive tests have been made to determine the applicability of these formulas to weirs less than 4 ft in length.

5. Formulas for Sharp-Crested Cipolletti Weirs. The Cipolletti weir is by definition a contracted weir and must be installed as such to obtain reasonably correct and consistent discharge measurements. However, Cipolletti has allowed in his formula for the reducing effect in the discharge due to end contractions by sloping the sides of the weir sufficiently to overcome the effect of contraction.

The Cipolletti formula, in which the Francis coefficient is increased by about 1 percent, and neglecting velocity of approach, is

$$Q = 3.367LH^{3/2} \tag{5}$$

The discharge for this type of weir including velocity of approach may be obtained from the formula

$$Q' = 3.367L(H + 1.5h)^{3/2} \tag{6}$$

where Q' = discharge considering velocity of approach, ft^3/s
L = length of weir crest, ft
H = head on weir crest, ft
h = head in ft due to velocity of approach = $v^2/2g$

The accuracy of measurements obtained by use of Cipolletti weirs and the above formula is inherently not as great as that obtainable with suppressed rectangular or V-notch weirs (Ref. 8). It is, however, acceptable where no great precision is required.

6. Formulas for Standard 90° Contracted V-Notch Weirs. There are several well-known formulas used to compute the discharge over 90° V-notch weirs. The most commonly used in the field of irrigation are the Cone formula and the Thomson formula. The Cone formula is considered by authorities to be the most reliable for small weirs and for conditions generally encountered in measuring water for open channels. The formula is

$$Q = 2.49H^{2.48} \tag{7}$$

where Q = discharge over weir, s-ft
H = head on the weir, ft

Ordinarily V-notch weirs are not appreciably affected by velocity of approach. If the weir is installed with complete contraction, the velocity of approach will be low.

MEASURING FLUMES 1. General. The measuring flumes have a measuring section which is produced by contraction of the channel sidewalls, raising the bottom to form a hump or by both of these. The Parshall flume is the most common and best known of the measuring flumes especially in irrigation canals.

2. Parshall Flume. A Parshall flume [9] is a specially shaped open-channel flow section which may be installed in a canal, lateral, or ditch, to measure the rate of flow of water. The flume has four significant advantages: (1) it can operate with relatively small head loss, (2) it is relatively insensitive to velocity of approach, (3) it has the capability of making good measurements with no submergence, moderate submergence, or even with considerable submergence downstream, and (4) its velocity of flow is sufficiently high to virtually eliminate sediment deposition within the structure during operation.

Discharge through a Parshall flume can occur for two conditions of flow. The first, free flow, occurs when there is insufficient backwater depth to reduce the discharge rate. The second, submerged flow, occurs when the water surface downstream from the flume is far enough above the elevation of the flume crest to reduce the discharge. For free flow, only the flume head H_a at the upstream gage location is needed to determine the discharge from a standard table. The free-flow range includes some of the range which might ordinarily be considered submerged flow because Parshall flumes tolerate 50 to 80 percent submergence before the free-flow rate is measurably reduced. For submerged flows (when submergence is greater than 50 to 80 percent, depending upon flume size), both the upstream and downstream heads H_a and H_b are needed to determine the discharge (see Fig. 16 for location of the gages).

A distinct advantage of the Parshall flume is its ability to function as a flow meter over a wide operating range with minimum loss of head while requiring but a single head measurement for each discharge. The head loss is only about one-fourth of that needed to operate a weir having the same crest length. Another advantage is that the velocity of approach is automatically controlled if the correct size of flume is chosen and the flume is used as it should be, that is, as an "in-line" structure. The flumes are widely used because there is no easy way to alter the dimensions of flumes that have been constructed or change the device or channel to obtain an unfair proportion of water.

The main disadvantages of Parshall flumes are: (1) they cannot be used in close-

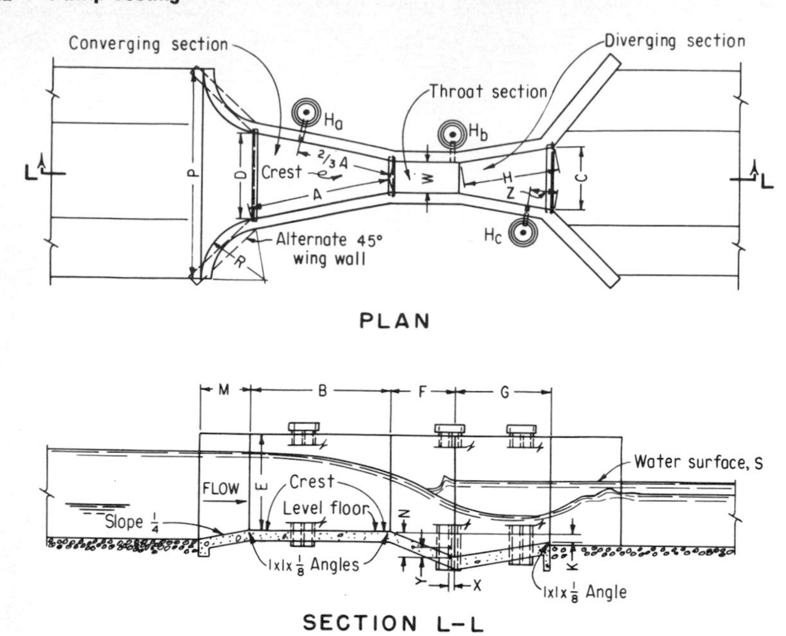

Fig. 16a Parshall flume dimensions (sheet 1 of 2). (U.S. Soil Conservation Service.)

coupled combination structures consisting of turnout, control, and measuring device, (2) they are usually more expensive than weirs or submerged orifices; (3) they require a solid, watertight foundation; and (4) they require accurate workmanship for satisfactory construction and performance.

Parshall flume sizes are designated by the throat width *W*, and dimensions are available for flumes from the 1-in size for discharges as small as 0.01 s-ft (Ref. 10) up to the 50-ft size for discharges as large as 3,000 s-ft. The flumes may be built of wood, concrete, galvanized sheet metal, or other desired materials. Large flumes are usually constructed on the site, but smaller flumes may be purchased as prefabricated structures to be installed in one piece. Some flumes are available as lightweight shells, which are made rigid and immobile by placing concrete outside of the walls and beneath the bottom. The larger flumes are used in rivers and large canals and streams; the smaller ones are used for measuring farm deliveries or for row requirements in the farmer's field.

3. Principles of Flume Operation. (*a*) Free Flow: In free flow the discharge depends solely upon the width of the throat, *W*, and the depth of water, H_a, at the gaging point in the converging section (Figs. 16 and 17). Free-flow conditions in the flume are similar to those that occur at a weir or spillway crest in that water passing over the crest is not impeded or slowed by downstream conditions.

(*b*) Submerged Flow: In most installations, when the discharge is increased above a critical value the resistance to flow in the downstream channel becomes sufficient to reduce the velocity, increase the flow depth, and cause a backwater effect at the Parshall flume. It might be expected that the discharge would begin to be reduced as soon as the backwater level H_b exceeds the elevation of the flume crest; however, this is not the case.

Calibration tests show that the discharge is not reduced until the submergence ratio H_b/H_a expressed in percent, exceeds the following values:

50 percent for flumes 1, 2, and 3 in wide
60 percent for flumes 6 and 9 in wide
70 percent for flumes 1 to 8 ft wide
80 percent for flumes 8 to 50 ft wide

	W		A		2/3 A		B		C		D		E		F		G		H		K		M		N		P		R		X		Y		Z		FREE-FLOW CAPACITY	
																																					MINIMUM	MAXIMUM
	FT.	IN.	FT.	IN.	FT.	IN.	FT.	IN.	FT.	IN.	FT.	IN.	FT.	IN.	FT.	IN.	FT.	IN.	FT.	IN.	FT.	IN.	FT.	IN.	FT.	IN.	FT.	IN.	FT.	IN.	FT.	IN.	FT.	IN.	FT.	IN.	SEC - FT.	SEC. - FT.
	0	1 [1]	1	2 9/32	0	9 17/32	1	2	0	3 21/32	0	6 19/32	0	6 TO 9	0	3	0	8	0	8 1/8	0	3/4	–	–	0	1 1/8		–	–		0	5/16	0	1/2	0	1/8	.01	0.19
[2]		2 [1]	1	4 5/16		10 7/8	1	4		5 5/16		8 13/32		6 TO 10		4 1/2		10		10 1/8		7/8		–		1 11/16		–	–			5/8		1		1/4	.02	.47
		3 [1]	1	6 3/8	1	1/4	1	6		7		10 3/16	1 TO 1 1/2			6	1	0	1	5/32		1	–			2 1/4		–	–			1		1 1/2		1/2	.03	1.13
	0	6	2	7/16	1	4 5/16	2	0	1	3 1/2	1	3 5/8	2	0	1	0	2	0	–		0	3	1	0	0	4 1/2	2	11 1/2	1	4	0	2	0	3	–		.05	3.9
		9	2	10 5/8	1	11 1/8	2	10	1	3	1	10 5/8	2	6	1	0	1	6	–			3	1	0		4 1/2	3	6 1/2	1	4		2		3	–		.09	8.9
	1	0	4	6	3	0	4	4 7/8	2	0	2	9 1/4	3	0	2	0	3	0	–			3	1	3		9	4	10 3/4	1	8		2		3	–		.11	16.1
	1	6	4	9	3	2	4	7 7/8	2	6	3	4 3/8	3	0	2	0	3	0	–			3	1	3		9	5	6	1	8		2		3	–		.15	24.6
	2	0	5	0	3	4	4	10 7/8	3	0	3	11 1/2	3	0	2	0	3	0	–			3	1	3		9	6	1	1	8		2		3	–		.42	33.1
[3]	3	0	5	6	3	8	5	4 3/4	4	0	5	1 7/8	3	0	2	0	3	0	–			3	1	3		9	7	3 1/2	1	8		2		3	–		.61	50.4
	4	0	6	0	4	0	5	10 5/8	5	0	6	4 1/4	3	0	2	0	3	0	–			3	1	6		9	8	10 3/4	2	0		2		3	–		1.3	67.9
	5	0	6	6	4	4	6	4 1/2	6	0	7	6 5/8	3	0	2	0	3	0	–			3	1	6		9	10	1 1/4	2	0		2		3	–		1.6	85.6
	6	0	7	0	4	8	6	10 3/8	7	0	8	9	3	0	2	0	3	0	–			3	1	6		9	11	3 1/2	2	0		2		3	–		2.6	103.5
	7	0	7	6	5	0	7	4 1/4	8	0	9	11 3/8	3	0	2	0	3	0	–			3	1	6		9	12	6	2	0		2		3	–		3.0	121.4
	8	0	8	0	5	4	7	10 1/8	9	0	11	1 3/4	3	0	2	0	3	0	–			3	1	6		9	13	8 1/4	2	0		2		3	–		3.5	139.5
	10	0	–		6	0	14	0	12	0	15	7 1/4	4	0	3	0	6	0	–		0	6	–		1	1 1/2	–		–		0	9	1	0	–		6	200
	12	0	–		6	8	16	0	14	8	18	4 3/4	5	0	3	0	8	0	–			6	–		1	1 1/2	–		–			9	1	0	–		8	350
	15	0	–		7	8	25	0	18	4	25	0	6	0	4	0	10	0	–			9	–		1	6	–		–			9	1	0	–		8	600
	20	0	–		9	4	25	0	24	0	30	0	7	0	6	0	12	0	–		1	0	–		2	3	–		–			9	1	0	–		10	1000
[4]	25	0	–		11	0	25	0	29	4	35	0	7	0	6	0	13	0	–		1	0	–		2	3	–		–			9	1	0	–		15	1200
	30	0	–		12	8	26	0	34	8	40	4 3/4	7	0	6	0	14	0	–		1	0	–		2	3	–		–			9	1	0	–		15	1500
	40	0	–		16	0	27	0	45	4	50	9 1/2	7	0	6	0	16	0	–		1	0	–		2	3	–		–			9	1	0	–		20	2000
	50	0	–		19	4	27	0	56	8	60	9 1/2	7	0	6	0	20	0	–		1	0	–		2	3	–		–			9	1	0	–		25	3000

[1] Tolerance on throat width (w) ± $\frac{1}{64}$ inch; tolerance on other dimensions ± $\frac{1}{32}$ inch. Sidewalls of throat must be parallel and vertical.

[2] From Colorado State University Technical Bulletin No. 61.

[3] From U.S. Department of Agriculture Soil Conservation Circular No. 843.

[4] From Colorado State University Bulletin No. 426-A.

Fig. 16b Parshall flume dimensions (sheet 2 of 2).

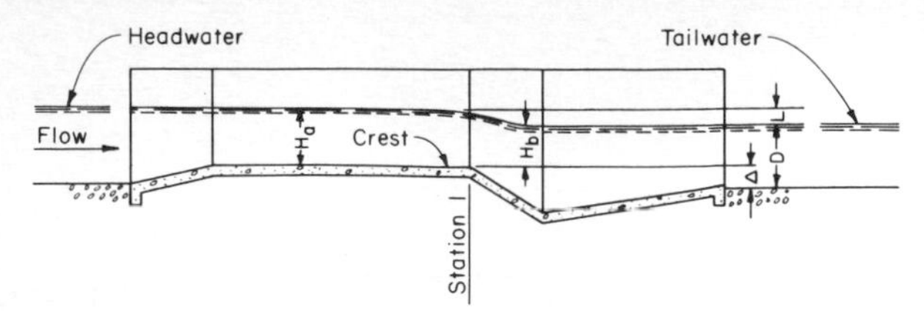

Fig. 17 Relationships of flow depths to the flume-crest elevation. (U.S. Soil Conservation Service.)

4. Discharge Equations for Free Flow over Flumes. The equation which expresses the relationship between upstream head H_a and discharge Q for widths W from 1 to 8 ft is

$$Q = 4WH_a^{1.522W^{0.026}}$$

If this equation is used to compute discharges through the larger flumes ranging from 10 to 50 ft wide, the computed discharges are always larger than actual discharges. Therefore a more accurate equation was developed for the large flumes. The equation is

$$Q = (3.6875W + 2.5)H_a^{1.6}$$

The difference in computed discharges obtained by using the two above equations for an 8-ft flume is normally less than 1 percent; however, the difference becomes greater as the flume size increases. Because of the difficulties in regularly using these equations, discharge tables have been prepared for use with the flumes 1 to 50 ft wide.

CURRENT METERS The essential features of a conventional current meter are a wheel which rotates when immersed in flowing water and a device for determining the number of revolutions of the wheel. For open-channel flow measurement a type generally used is the Price meter with five or six conical cups. The relationship between the velocity of the water and the number of revolutions of the wheel per unit of time for various velocities is determined experimentally for each instrument. Also, considerable skill is required by the operator to obtain consistent satisfactory results.

A detailed explanation of the use of current meters is contained in chapter 5 of the "Water Measurement Manual." [5]

OTHER METHODS The discharge measurement methods given above are the ones in common use; however, a number of other methods, some newer and more sophisticated, are recognized and well established. The use of these special methods is acceptable provided their limitations are recognized and all parties to the testing program are in agreement to their use. A few of these methods, not in any particular order, are as follows:

Salt-velocity
Salt-dilution
Color-velocity
Radioisotope
Acoustic flow meters
Slope-area
Deflection meters
Propeller meters
Float movement
Gates and sluices
Color dilution

Field Approximating. FOR HORIZONTAL DISCHARGE PIPE Often a field approximation of water flow from a pump discharge becomes necessary, especially if no other methods are practical or readily available. One of the accepted methods is by trajectory. The discharge from the pipe may be either vertical or horizontal, the principal difficulty being in measuring the coordinates of the flowing stream accurately. The pipes must be flowing full and the accuracy of this method varies from 85 to 100 percent. Figure 18 illustrates the approximation from a horizontal pipe.

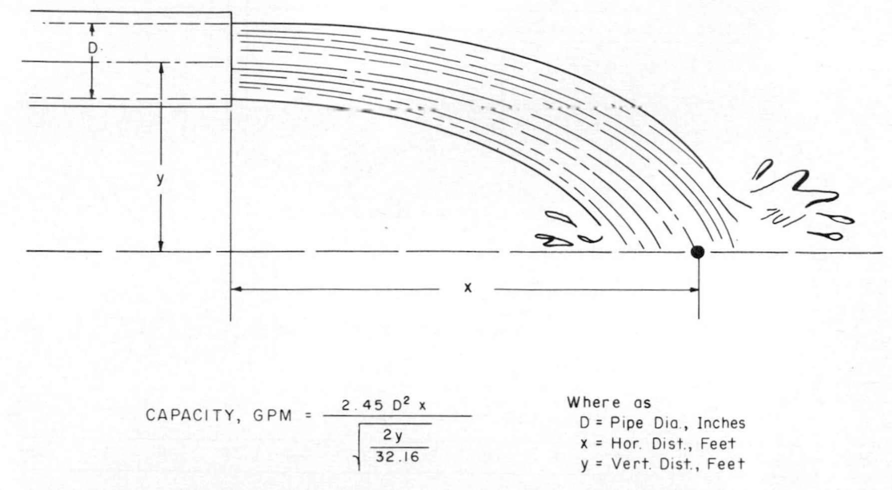

Fig. 18 Approximating flow from horizontal pipe.

This method can be further simplified by measuring to the top of the flowing stream and always measuring so that y will equal 12 in and measuring the horizontal distance X in inches as illustrated in Fig. 19.

FOR VERTICAL DISCHARGE PIPE Figure 20 illustrates a method of measuring for vertical discharge from a vertical pipe.

14.16 Head Measurements *General.* Head is a quantity used to express a form (or combination of forms) for the energy content of the liquid per unit weight of the liquid referred to any arbitrary datum. In terms of foot-pounds of energy per pound of liquid, all head quantities have the dimensions of feet of liquid. The unit for measuring head is the foot. The relation between a pressure expressed in pounds per square inch (lb/in²), and that expressed in feet of head is

$$\text{Head, ft} = \text{lb/in}^2 \times \frac{2.31 \times 62.3}{W} = \text{lb/in}^2 \times \frac{2.31}{\text{sp gr}}$$

where W = specific weight, lb/ft³
sp gr = specific gravity of the liquid

The following excerpt from the "Hydraulic Institute Standards" is used by the Bureau of Reclamation throughout their test program:

It is important that steady flow conditions exist at the point of instrument connection. For this reason, it is necessary that pressure or head measurement be taken on a section of pipe where the cross-section is constant and straight. Five to ten diameters of straight pipe of unvarying cross-section following any elbow or curved member, valve, or other obstruction, are necessary to insure steady flow conditions.

The following precautions shall be taken in forming orifices for pressure measuring instruments and for making connections: The orifice in the pipe should be flush with and normal to the wall of the water passage. The wall of the water passage should be smooth and of unvarying cross section. For a distance of at least 12 in preceding the orifice, all tubercles and roughness should be removed with a file or emery cloth, if necessary. The orifice should be of a diameter from ⅛ to ¼ in and of a length equal to twice the diameter.

The edges of the orifice should be provided with a suitable radius tangential to the wall of the water passage and shall be free from burrs or irregularities. Two

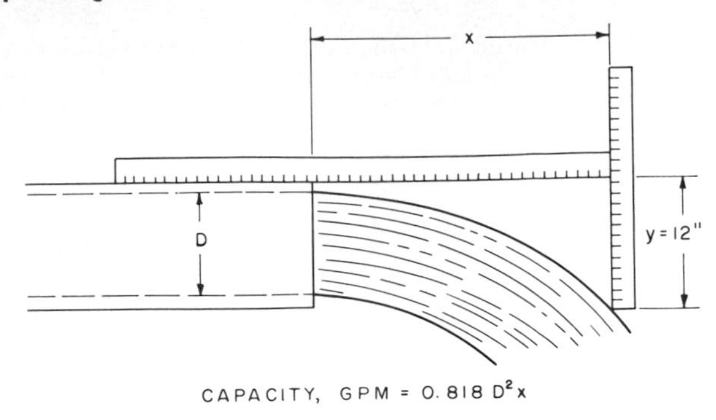

CAPACITY, GPM = $0.818\ D^2 x$

APPROXIMATE CAPACITY, GPM,

FOR FULL FLOWING HORIZONTAL PIPES

STD. WT. STEEL PIPE, INSIDE DIA., IN.		DISTANCE X, IN., WHEN Y = 12"										
NOMINAL	ACTUAL	12	14	16	18	20	22	24	26	28	30	32
2	2.067	42	49	56	63	70	77	84	91	98	105	112
$2\frac{1}{2}$	2.469	60	70	80	90	100	110	120	130	140	150	160
3	3.068	93	108	123	139	154	169	185	200	216	231	246
4	4.026	159	186	212	239	266	292	318	345	372	398	425
5	5.047	250	292	334	376	417	459	501	543	585	627	668
6	6.065	362	422	482	542	602	662	722	782	842	902	962
8	7.981	627	732	837	942	1047	1150	1255	1360	1465	1570	1675
10	10.020	980	1145	1310	1475	1635	1800	1965	2130	2290	2455	2620
12	12.000	1415	1650	1890	2125	2360	2595	2830	3065	3300	3540	3775

Fig. 19 Approximating flow from horizontal pipe.

pressure-tap arrangements shown on Fig. 27 indicate taps or orifices in conformity with the above. Where more than one tap or orifice is required at a given measuring section, separate connections, properly valved, should be made. As an alternative, separate instruments should be provided.

Multiple orifices should not be connected to an instrument except on those metering devices such as venturi meters where proper calibrations have been made on an instrument of this form.

All connections or leads from the orifice tap should be tight. These leads should be as short and direct as possible. For the dry-tube type of leads, suitable drain pots should be provided and a loop should be formed of sufficient height to keep the pumped liquid from entering the leads. For the wet-tube type of leads, vent cocks for flushing should be provided at any high point or loop crest to assure that tubes do not become air-bound.

All instrument hose, piping, and fittings should be checked under pressure prior to test to assure that there are no leaks. Suitable damping devices may be used in the leads.

If the conditions specified above cannot be satisfied at the point of measurement, it is recommended that four separate pressure taps be installed, equally spaced about the pipe, and that the pressure or head at that section be taken as the average of the four separate values of head. If the separate readings show a difference of static pressure such as might well affect the head beyond the contract tolerances, the installation shall be corrected or an acceptable tolerance determined.

Figures 21 to 27 show the suitable arrangements for various types of instruments and formulas for transforming instrument readings into feet of the liquid pumped,

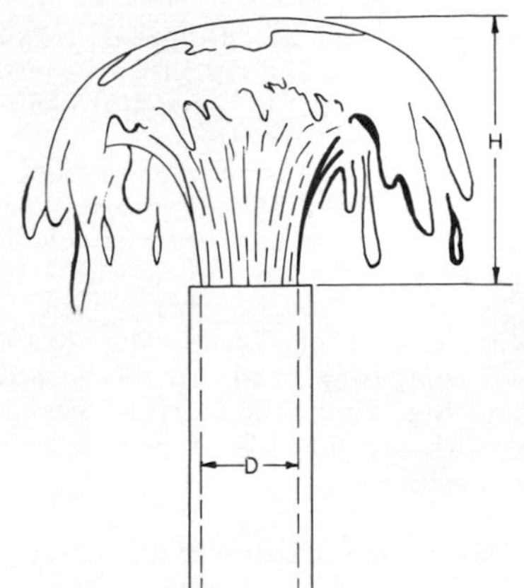

APPROXIMATE CAPACITY, GPM,
FOR FLOW FROM VERTICAL PIPES

NOMINAL	VERTICAL HEIGHT, H, OF WATER JET, IN.										
I.D. PIPE, IN.	3	3.5	4	4.5	5	5.5	6	7	8	10	12
2	38	41	44	47	50	53	56	61	65	74	82
3	81	89	96	103	109	114	120	132	141	160	177
4	137	151	163	174	185	195	205	222	240	269	299
6	318	349	378	405	430	455	480	520	560	635	700
8	567	623	684	730	776	821	868	945	1020	1150	1270
10	950	1055	1115	1200	1280	1350	1415	1530	1640	1840	2010

Fig. 20 Approximating flow from vertical pipe.

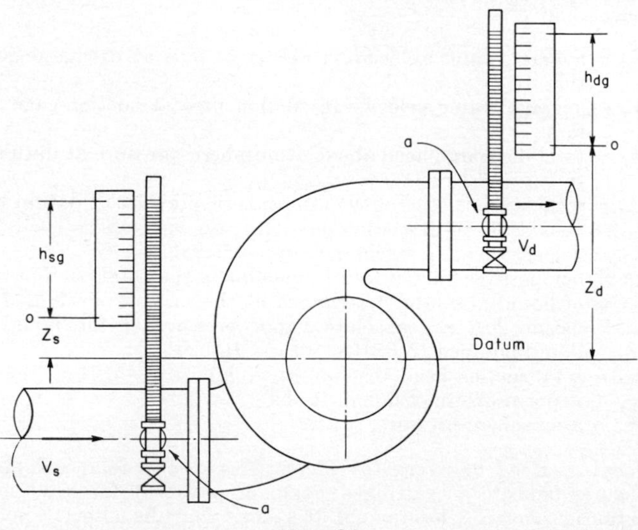

Fig. 21

for expressing instrument head as elevation over a common datum, and for correcting these formulas for the velocity head existing in the suction and discharge pipes.

The datum is taken as the centerline of the pump for horizontal shaft pumps, and as the entrance eye of the impeller for vertical shaft pumps.

The instruments, when practicable, are water columns or manometers, and for high pressures are mercury manometers, bourdon gages, electric pressure transducers, or dead-weight gage testers. When water columns are used, care should be taken to avoid errors due to the difference between the temperature of the water in the gage and that of the water in the pump.

Definitions and Symbols Used for Measurement of Head. The symbols used throughout this section for expressing and computing head are those used by the ASME Power Test Code [2] for pumps and the "Hydraulic Institute Standards" [1] (also see par. 14-3). The symbols below with explanations or definitions apply to Figs. 21 to 27 where temperature effects are negligible:

H = total head or dynamic head in feet is the measure of the energy increase per pound of liquid imparted to the liquid by the pump and is therefore the algebraic difference between the total discharge head and the total suction head ($H = h_d - h_s$). The quantities of h_d and h_s are negative if the corresponding values at the datum elevation are below atmospheric pressure.

h_{gd} = discharge gage reading, ftH_2O

h_{gs} = suction gage reading, ftH_2O

Both the above can be direct-reading water manometers, or a conversion from mercury manometers or calibrated bourdon pressure gages.

Z_d = elevation of discharge gage, zero above datum elevation, ft

Z_s = elevation of suction gage, zero above datum elevation, ft

The quantities Z_d and Z_s are negative if the gage zero is below the datum elevation.

Y_d = elevation of discharge-gage connection to discharge pipe above datum elevation, ft

Y_s = elevation of suction-gage connection to suction pipe above datum elevation, ft

The quantities Y_d and Y_s are negative if the gage connection to the pipe lies below the datum elevation.

V_d = average water velocity in discharge pipe at discharge-gage connection, ft/s

V_s = average water velocity in suction pipe at suction-gage connection, ft/s

h_d = total discharge head above atmospheric pressure at datum elevation, ft

h_s = total suction head above atmospheric pressure at datum elevation, ft

h_{vs} = velocity head in suction pipe ($Vs^2/2g$)

h_{vd} = velocity head in discharge pipe ($Vd^2/2g$)

NPSHA = net positive suction head available is the total suction head in feet of liquid absolute, determined at the suction nozzle and referred to datum less the absolute vapor pressure of the liquid in feet of liquid pumped (NPSHA $= h_a - H_{vpa} + h_s$)

h_{sa} = total suction head absolute ($h_a + h_s$)

H_{vpa} = vapor pressure of liquid, ft abs

h_a = atmosphere pressure, ft abs

Measurement of Head by Means of Water Gages. The following examples and associated figures indicate a centrifugal pump arrangement for head measurement under conditions where the location of the gages may be either above or below atmospheric pressure as indicated.

Example: The pressure at the gage connection a is above atmospheric pressure with the line between discharge or suction pipe and the corresponding gage is filled completely with water as shown in Fig. 21. The following equations apply:

$$h_d = +h_{gd} + Z_d + \frac{V_d^2}{2g}$$

$$h_s = +h_{gs} + Z_s + \frac{V_s^2}{2g}$$

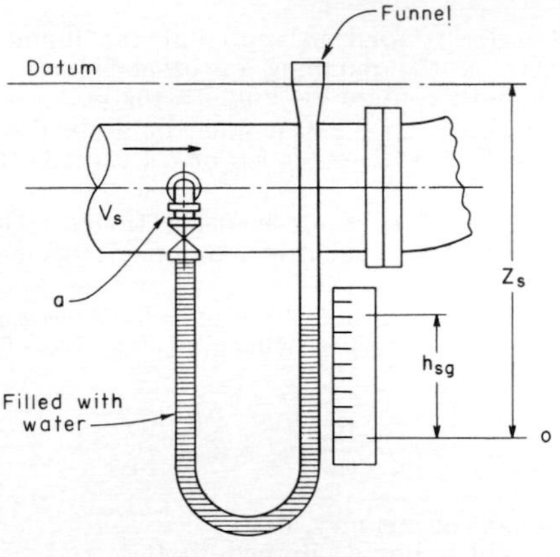

Fig. 22

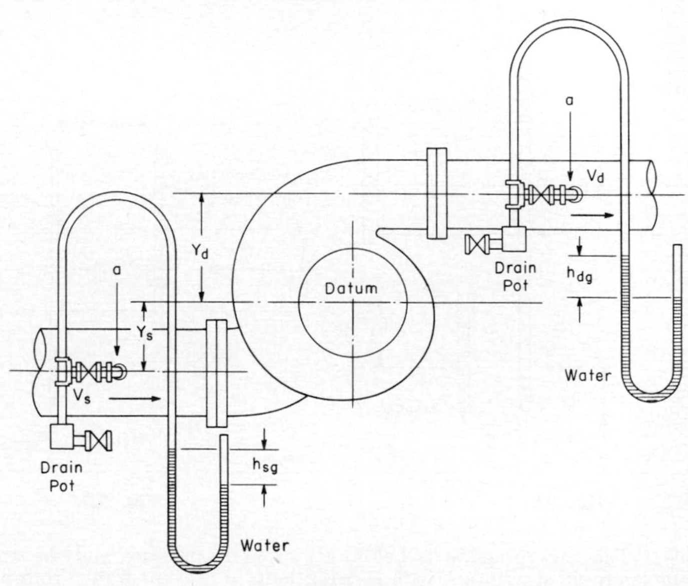

Fig. 23

Example: The pressure at gage connection a is below atmospheric pressure as shown in Fig. 22. The following equation applies:

$$h_s = h_{gs} - Z_s + \frac{V_s^2}{2g}$$

The negative sign of Z_s indicates that the gage zero is located below the datum.

Example: The pressure at the gage connection a is below atmospheric pressure pipe and the corresponding gage is filled completely with air as shown in Fig. 23. The following equations apply:

$$h_d = -h_{gd} + Y_d + \frac{V_d^2}{2_g}$$

$$h_s = -h_{gs} - Y_s + \frac{V_s^2}{2_g}$$

Note 1: The word *water* is used to represent the liquid being pumped. The provisions are applicable to the pumping of other liquids, such as oil, provided the gages and connecting lines contain the liquid being pumped.

Note 2: If connecting pipe is air-filled, it must be drained before reading is made. Water cannot be used in U tube if either h_{dg} or h_{ds} exceeds the height of the rising loop.

Measurement of Head by Means of Mercury Gages. The following examples and associated figures indicate a centrifugal pump arrangement for head measurement using mercury gages.

Example: The gage pressure is above the atmospheric pressure and the connecting line is completely filled with water, as shown in Fig. 24. The following equation applies:

$$h \doteq \frac{W_m}{W} h_g + Z + \frac{V^2}{2g}$$

where W_m = specific weight of mercury, lb/ft³
W = specific weight of liquid pumped, lb/ft³
h_g = suction- or discharge-gage reading, ftHg

The quantities h, Z, Y, and V without subscripts apply equally to suction- and discharge-head measurements.

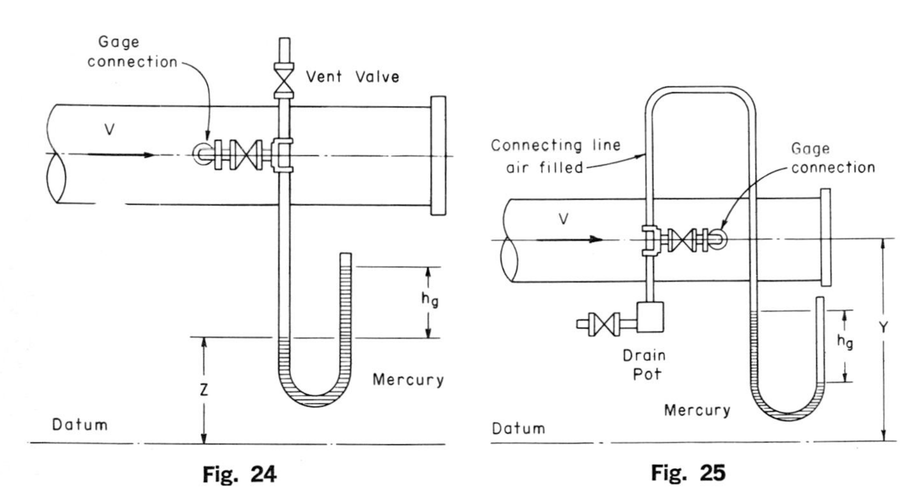

Fig. 24 **Fig. 25**

Example: The gage pressure is below atmospheric pressure and the connecting line is completely filled with air, with a rising loop to prevent water from passing to the mercury column, as shown in Fig. 25. The following equation applies

$$h = \frac{W_m}{W} h_g = Y + \frac{V^2}{2g}$$

Measurement of Head by Means of Differential Mercury Gages. The following example and Fig. 26 indicate a centrifugal pump arrangement for head measurement using a differential-type mercury gage.

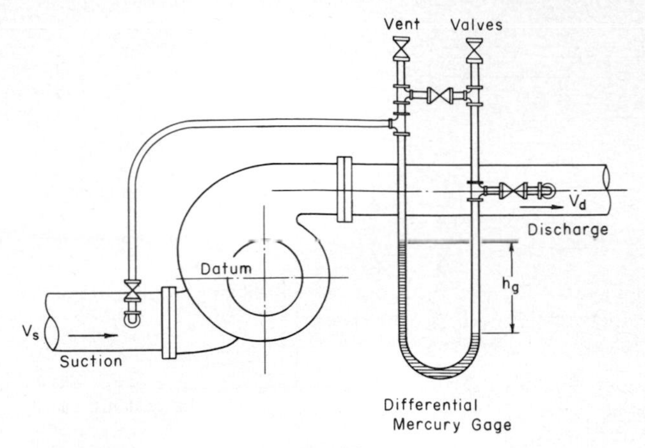

Fig. 26

When reading a differential mercury gage in feet of mercury (h_g) and with the connecting lines completely filled with water, then

$$H = \left[\frac{W_m}{W} - 1\right] h_g + \frac{V_d^2}{2g} - \frac{V_s^2}{2g}$$

In addition to the differential gage, use a separate suction gage as shown in Figs. 22 and 25, where

$$h_s = \frac{W_m}{W} h_{gs} - Z + \frac{V_s^2}{2g}$$

Measurement of Head by Means of Calibrated Bourdon Gages. An example of a centrifugal pump arrangement using calibrated bourdon gages for head measurement is shown in Example 3 of Fig. 27 with the gage pressure above atmospheric pressure. The distances Z_s or Z_d are measured to the center of the gage and are negative if the center of the gage lies below the datum line.

Measurement of Head on Vertical Suction Pumps in Sumps and Channels. Installations of vertical shaft pumps drawing water from large open sumps and having short inlet passages of a length not exceeding about three diameters of the inlet opening, such inlet pieces having been furnished as part of the pump, the total head should be the reading of the discharge connection in feet plus the vertical distance from the gage centerline to the free water level in the sump in feet (see Example 2 of Fig. 27).

14.17 Power Measurement *General.* The pump input power (horsepower) may be determined by one of the following methods:

Calibrated motor
Transmission dynamometer
Torsion dynamometer

The "Hydraulic Institute Standards" are generally used as the basis for most power measurement procedures.

Calibrated Motors. When pump input horsepower is to be determined by the use of a calibrated motor, measurements of power input should be made at the terminals of the motor to exclude any line losses that may occur between the switchboard and the driver itself. Certified calibration curves of the motor must be obtained. The calibration should be conducted on the specific motor in question, and not on a similar machine. Such calibrations must indicate the true input-output value of motor efficiency and not some conventional method of determining an

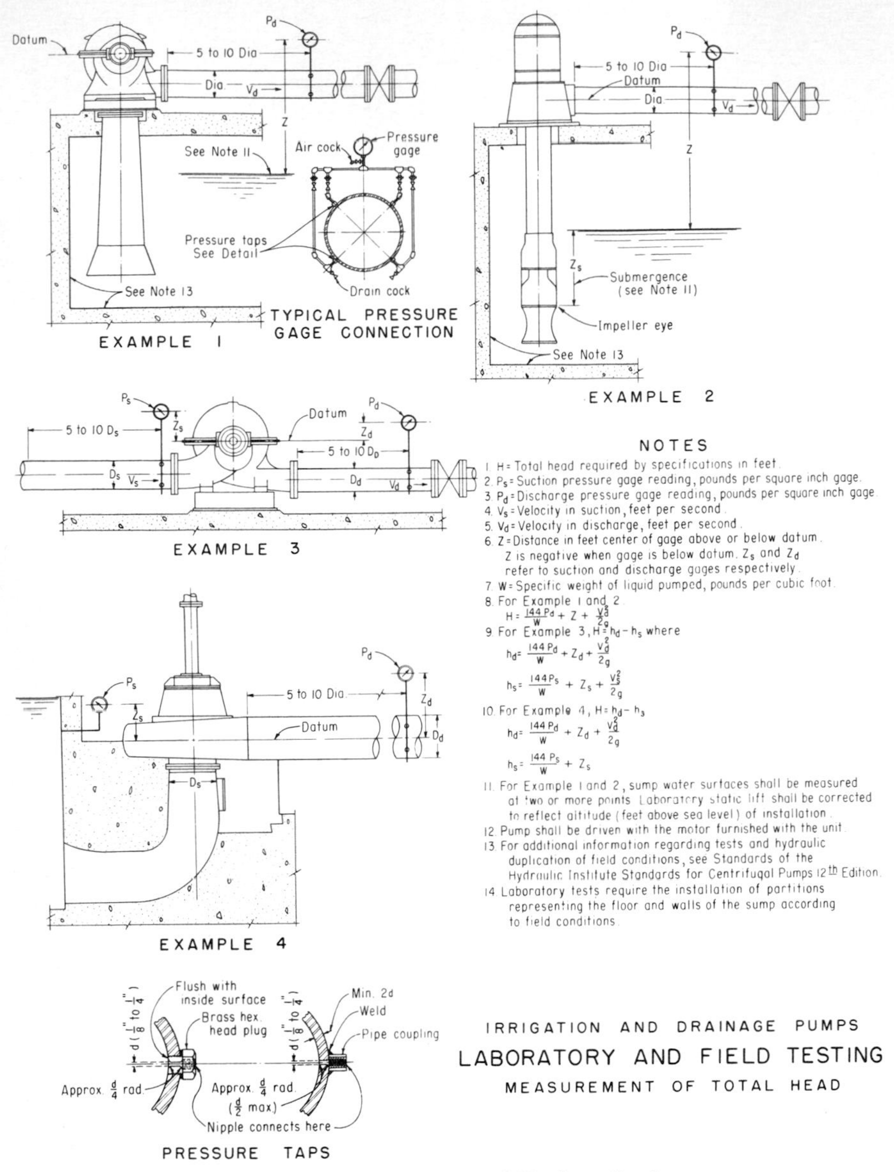

Fig. 27 (U.S. Bureau of Reclamation.)

arbitrary efficiency. Calibrated laboratory-type electric meters and transformers should be used to measure power input to all motors.

Transmission Dynamometers. The transmission or torque-reaction dynamometer consists of cradled electric motor with its frame and field windings on one set of bearings and the rotating element on another set so that the frame is free to rotate but is restrained by means of some weighting or measuring device.

When pump input horsepower is to be determined by transmission dynamometers, the unloaded and unlocked dynamometer must be properly balanced, prior to the test, at the same speed at which the test is to be run. The scales should be checked against standard weights. After the test the balance must be rechecked to assure that no change has taken place. In the event of an appreciable change, the test should be rerun. An accurate measurement of speed is essential and should not vary from the pump rated speed by more than 1 percent. Power input is calculated as shown in Sec. 14.19.

Torsion Dynamometers. The torsion dynamometer consists of a length of shafting whose torsional strain when rotating at a given speed and delivering a given torque is measured by some standard method. When pump input horsepower is to be determined by torsion dynamometers, the unloaded dynamometer should be statically calibrated prior to the test by measuring the angular deflection for a given torque.

Immediately before and after the test the torsion dynamometer must be calibrated dynamically at the rated speed. The best and simplest method to accomplish this is to use the actual job driver to supply power and use a suitable method of loading the driver over the entire range of the pump to obtain the necessary calibrations. The calibration of the torsion dynamometer after the pump tests should be within ½ of 1 percent of the original calibration. During the actual test runs the speed shall not vary from the pump rated speed by more than 1 percent. The temperature of the torsion dynamometer during the test runs shall be within 10°F of the temperature when the dynamic calibrations were made. All torsion dynamometer calibrations should be witnessed and approved by both parties to the test. In the event that a variation greater than allowed above exists, then a rerun of the test must be made. Power-input calculations are shown in Sec. 14.19.

14.18 Speed Measurement The speed of the pump under test is determined by one of the following methods:

Revolution counter (manual or automatic)
Tachometer
Stroboscopic device
Electronic counter

In all cases the instruments used must be carefully calibrated before the tests to demonstrate that they will produce the required speed readout to within the desired accuracy. Accepted accuracy is usually ±0.1 percent. Should cyclic speed change result in power fluctuations, then at least five equally spaced timed readings should be taken to give a satisfactory mean speed for the test point.

COMPUTATIONS

14.19 Pump Power Calculation *Power Output.* The water horsepower (whp) or useful work done by the pump is found by the formula:

$$\text{whp} = \frac{\text{lb of liquid pumped/min} \times \text{total head in ft of liquid}}{33{,}000}$$

If the liquid has a specific gravity of 1 and the specific weight of the liquid is 62.3 lb/ft^3 at a temperature of 68°F, the formula is

$$\text{whp} = \frac{\text{gpm} \times \text{head in feet}}{3{,}960}$$

Power Input. The brake horsepower required to drive the pump is found by the formula:

$$\text{bhp} = \frac{\text{gpm} \times \text{total head in ft}}{3{,}960 \times \text{pump efficiency}}$$

where pump efficiency is obtained by the formula

$$\text{Pump efficiency} = \frac{\text{output}}{\text{input}} = \frac{\text{whp}}{\text{bhp}}$$

The electrical horsepower input to the motor is

$$\text{ehp} = \frac{\text{bhp}}{\text{motor efficiency}}$$

$$= \frac{\text{gpm} \times \text{head in ft}}{3{,}960 \times \text{pump efficiency} \times \text{motor cfficiency}}$$

The kilowatt input to the motor

$$\text{kW input} = \frac{\text{bhp} \times 0.746}{\text{motor efficiency}}$$

$$= \frac{\text{gpm} \times \text{head} \times 0.746}{3{,}960 \times \text{pump efficiency} \times \text{motor efficiency}}$$

14.20 Pump efficiency The pump efficiency is found by:

$$\text{Pump efficiency} = \frac{\text{output}}{\text{input}} = \frac{\text{whp}}{\text{bhp}}$$

For an electric-motor-driven pumping unit the overall efficiency is found by

$$\text{Overall efficiency} = \text{pump efficiency} \times \text{motor efficiency}$$

In many specifications it is required that the actual job motor be used to drive its respective pump during shop or field testing. Using this test setup the overall efficiency then becomes what is commonly called "wire-to-water" efficiency, which is expressed by the formula

$$\text{Overall efficiency} = \frac{\text{water horsepower}}{\text{electrical horsepower input}} = \frac{\text{whp}}{\text{ehp}}$$

14.21 Speed Adjustments or Corrections The best and standard practice is to use the actual job motor to drive the pump under test. However, for purposes of plotting test results it becomes necessary to correct the test values at test speed to a rated pump speed. The rated pump speed should always be less than the actual test speed since even a small increase in speed beyond the actual test speed may result in going into an unstable zone of the pump. Also it is recommended that the speed change from test speed to rated or specified speed not be greater than 3 percent.

To adjust the pump capacity, head, NPSH, and power from that actually recorded during test to another or specified speed, the following formula from the "Hydraulic Institute Standards" [1] should be used:

Capacity:
$$Q_2 = \frac{N_2}{N_1} = Q_1$$

where Q_1 = capacity at test speed, gpm
Q_2 = capacity at rated speed, gpm
N_1 = test speed, rpm
N_2 = rated speed, rpm

Head:
$$H_2 = \left(\frac{N_2}{N_1}\right)^2 \times H_1$$

where H_1 = head at test speed, ft
H_2 = head at rated speed, ft

Horsepower:
$$\text{hp}_2 = \left(\frac{N_2}{N_1}\right)^3 \times \text{hp}_1$$

where hp_1 = horsepower at test speed
hp_2 = horsepower at rated speed

Net positive suction head (NPSH): $\text{NPSH}_2 = \left(\frac{N_2}{N_1}\right)^2 \times \text{NPSH}_1$

where NPSH_1 = net positive suction head at test speed, ft
NPSH_2 = net positive suction head at rated speed, ft

Shop or field testing at either reduced or increased speed should be permitted only when absolutely no alternates are available. It is recommended that if reduced or increased speed tests are used as official performance or acceptance tests, the specifications or agreement must state the test head and speed, and that the performance warranties be based on the specified head and speed condition.

If reduced or increased speed tests are considered, then certain affinity laws must be observed to maintain hydraulic similarity between the actual and test condition. These affinity-law relationships between the several variables can be expressed by

$$\frac{Q_1}{Q} = \frac{N_1}{N} = \left(\frac{H_1}{H}\right)^{1/2}$$

where: test = Q_1 = capacity and H_1 = head at H_1 = rpm
actual = Q = capacity and H = head at N = rpm

RECORDS

14.22 Data There probably are as many test data forms as there are test laboratories. Each may or may not have an advantage for its particular application. A pumping unit test data form for recording the pump performance data is shown in Fig. 28.

The manufacturer's serial number, type, and size or other means of identification of each pump and driver involved in the test should be carefully recorded in order that mistakes in identity may be avoided. The dimensions and physical conditions, not only of the machine tested but of all associated parts of the plant which have any important bearing on the outcome of a test, should be determined.

Normal practice suggests that one test run be at or as near the rated condition as possible and that at least three runs be within the specified range of heads.

14.23 Plotting Test Results In plotting curves from the test results it should be kept in mind that any one point may be in error but that all the points should establish a trend.

Unless some external factor is introduced to cause an abrupt change, a smooth curve can be drawn for the points plotted, not necessarily through each and every one. Figure 29 is a graphic representation showing the determination of pump performance with the total head, power input (hp), and efficiency in percent, all plotted on the same graph with the capacity as the abscissa of the curves.

14.24 Reports In some instances a preliminary report may be issued as part of a contract agreement. However, normally a final or official report is all that is required.

On shop tests of relatively small pumps the test log and plotted results constitute the entire report. The reports get progressively more involved as the pumps become larger. In the case where model testing is used, the final report covers a complete record of the agreements, inspections, personnel, calibration data, sample computations, tabulations, descriptions, discussions, etc.

MODEL TESTING

14.25 Purpose Models are used and tested for one or more of the following purposes:

1. To develop new ideas and new designs
2. To give the manufacturer a range of warranties in regard to performance and efficiency
3. To give the customer or buyer some assurance that his requirements are being met

4. To replace or supplement the field testing of the prototype
5. To compare the performance of several models.

Model testing in advance of final design and installation of a large unit not only provides advance assurance of satisfactory performance but allows for alterations in time for incorporation in the prototype.

14.26 Testing Procedures Early editions of the "Hydraulic Institute Standards" [1] state that in the comparison of model to the prototype head, the minimum acceptable is 80 percent. There are differences of opinion concerning this require-

RECORD OF PUMP PERFORMANCE TEST

DATE OF TEST____ TEST NO.____ CUSTOMER____ CUST. ORDER NO.____
MANUFACTURER'S ORDER NO.____ PLANT____ UNIT NO.____
PROJECT____

RATED CONDITIONS:
CAPACITY (G.P.M.)____ TOTAL HEAD (FEET)____ R.P.M.____
OVERALL EFFICIENCY PERCENT____ RANGE OF HEAD____

DRIVER:
TYPE____ HORSEPOWER____
MANUFACTURER____ SERIAL NO.____ TEST VOLTAGE____

TEST EQUIPMENT:
DISCHARGE MEASUREMENT METHOD____ CONVERSION FACTOR____
DISCHARGE GAGE____ CORRECTION____ SUCTION GAGE____ CORRECTION____
DIFFERENTIAL BETWEEN GAGES____ INSIDE DIAMETER SUCTION____ INSIDE DIAMETER DISCHARGE____

PUMP DATA:
TYPE OF PUMP____ SIZE____ NO. STAGES____
MANUFACTURER____ SERIAL NO.____ SUCTION SIZE____ DISCHARGE SIZE____

	RUN NO.	1	2	3	4	5	6	7	8	9	10
HEAD	PRESSURE, P.S.I.										
	HEAD, FEET										
	GAGE ℄ TO WATER LEVEL, FEET										
	VELOCITY HEAD, FEET										
	TOTAL HEAD, FEET										
CAPACITY	READING										
	CONVERSION										
	FLOW, G.P.M.										
POWER DATA	MOTOR VOLTAGE										
	AMPERES										
	KILOWATTS										
	TOTAL HORSEPOWER INPUT										
	MOTOR EFFICIENCY, %										
	SPEED, R.P.M.										
	DYNAMOMETER										
	BRAKE HORSEPOWER										
	WATER HORSEPOWER										
	PUMP EFFICIENCY, %										
OVERALL EFFICIENCY, %											

TESTED BY____ WITNESSED BY____
TYPE OF TEST____
(FIELD OR SHOP)
REMARKS:

IRRIGATION AND DRAINAGE PUMPS
LABORATORY AND FIELD TESTING
PUMPING UNIT TEST DATA

Fig. 28 (U.S. Bureau of Reclamation.)

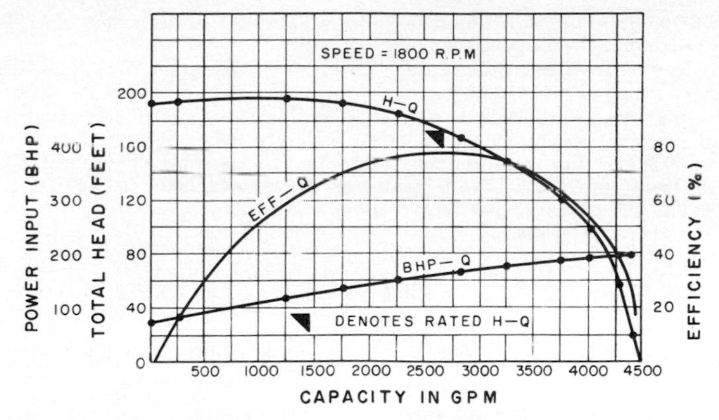

Fig. 29 Plotted pump performance.

ment. This percentage of prototype head has not been clearly substantiated either by theoretical considerations or by comparison of performance data from tests of model and their full-sized prototypes, and at present this ratio is in a transitional state.

The model should have complete geometric similarity to the prototype in all wetted parts between the intake and discharge sections of the pump. The model should be tested in the same horizontal or vertical shaft position as that of the prototype. The speed of the model should be such that at the test head the specific speed for each test run is the same as that of the installed unit or prototype. Unless otherwise specified the suction head or suction lift should be such as to give the same σ value.

If corresponding diameters of model and prototype are D_1 and D, respectively, then the model speed N and capacity Q_1 under the test head H_1 must agree with the relations

$$\frac{N_1}{N} = \frac{D}{D_1}\sqrt{\frac{H_1}{H}} \qquad \text{and} \qquad \frac{Q_1}{Q} = \left(\frac{D_1}{D}\right)^2\sqrt{\frac{H_1}{H}}$$

In testing a model of reduced size with the above conditions being observed, complete hydraulic similarity will not be secured unless the relative roughness of the impeller and pump casing surfaces are the same. With the same surface texture in model and prototype, the model efficiency will be lower than that of the larger unit, and greater relative clearances and shaft friction in the model will also reduce its efficiency.

The efficiency of a pump model can conveniently be stepped up to the prototype by applying a formula of the same general form as the Moody formula used for hydraulic turbines

$$\frac{1 - e_1}{1 - e} = \left(\frac{D}{D_1}\right)^n$$

The exponent n should be determined for a given laboratory and given type of pump on the basis of an adequate number of comparisons of the efficiencies of models and prototypes, with consistent surface finish of the models and prototypes. The "Hydraulic Institute Standards" [1] states that the values for the exponent n have been found to vary from zero when the surface roughness and clearances of the model and of the prototype are proportional to their size, to 0.26 when the absolute roughness is the same in both model and prototype.

When model tests are to serve as acceptance tests, it is generally recommended that the efficiency guarantees be stated in terms of model performance, rather than in terms of calculated prototype performance. In the absence of such provision, the efficiency stepup formula and the numerical value of its exponent should be clearly specified, or agreed upon in advance of tests.

The "Hydraulic Institute Standards" [1] gives an example of model testing as follows:

A single-stage pump to deliver 200 ft³/s against a head of 400 ft at 450 rpm and with a positive suction head, including velocity head, of 10 ft has an impeller diameter of 6.8 ft. The pump being too large for a shop or laboratory test, a model with 18-in impeller is to be tested at a reduced head at 320 ft. At what speed, capacity, and suction head should the test be run?

Applying the above relations:

$$N_1 = N\frac{D}{D_1}\sqrt{\frac{H_1}{H}} = 450\left(\frac{6.8\text{ ft}}{1.5\text{ ft}}\right)\left(\frac{320}{400}\right) = 1{,}825\text{ rpm}$$

$$Q_1 = Q\left(\frac{D_1}{D}\right)^2\sqrt{\frac{H_1}{H}} = 200\left(\frac{1.5}{6.8}\right)^2\sqrt{\frac{320}{400}} = 8.73\text{ ft}^3/\text{s, or }3{,}290\text{ gpm}$$

To check these results, the specific speed of the prototype is:

$$N_s = N\frac{\sqrt{Q}}{H^{3/4}} = 450\frac{\sqrt{200}}{400^{3/4}} = 71.2\text{ in the ft}^3/\text{s system}$$

and that of the model is

$$N_{s1} = 1{,}825\frac{\sqrt{8.73}}{320^{3/4}} = 71.2 \qquad \text{(or 1,510 in the gpm system)}$$

The cavitation factor σ for the field installation, assuming a water temperature of 80°F as a maximum probable value and $H_b = 32.8$ ft as in the first example, will be

$$\sigma = \frac{H_b - H_s}{H} = \frac{32.8 - 10}{400} = 0.107$$

where $H_b = h_{sa} - H_{vpa}$ (absolute atmospheric pressure minus absolute vapor pressure)
H_s = distance from datum to suction level

which should be the same in the test. With the water temperature approximately the same,

$$\sigma = \frac{H_b - H_{s1}}{H_1}$$

and

$$H_{s1} = H_b - \sigma H = 32.8 - (0.107)(320) = 1.45\text{ ft}$$

Hence the model should be tested with a positive suction head of 1.45 ft, to reproduce the field conditions.

Normally one of the requirements when using model tests as acceptance tests is to make sure true geometric similarity actually exists between the model and the installed prototype. True values of all required and specified dimensions should be determined. The actual parts, areas, shape, clearances, and positions should be clearly understood or spelled out between all of the parties. Also the amount of permissible geometric deviation between prototype and model should be agreed to, in writing, before starting of testing program.

OTHER OBSERVATIONS

14.27 Testing Other Than Centrifugal Pumps The next largest class of pumps other than centrifugal is the displacement pump. This classification includes reciprocating, rotary, screw, and other miscellaneous displacement-type pumps. Testing of these closely parallels the centrifugal procedures. Normally the capacities are smaller and the heads are higher but the objectives, methods, measurements are all about the same. The "Hydraulic Institute Standards" [1] quite thoroughly covers the testing of rotary and displacement-type pumps.

The testing of pumps not falling into the two broad classifications of centrifugal and displacement is usually very special and each case must be treated on that basis. The test procedures are normally spelled out in the specifications or an agreement between all parties must be made before testing is started. The testing of eduction or jet-type pumps falls under this special category of testing for which no normal or standard test procedures are set up.

14.28 Other Test Phenomena When testing pumps, other special phenomena of interest should also be checked, observed, and noted on the test record. The two phenomena normally reported on are vibration and noise. The acceptable limits of these plus instrumentation for their measuring are special and are normally covered in the contract documents. If the pumps are to be installed in a special environment, this should also be taken into consideration during testing.

REFERENCES

1. "Hydraulic Institute Standards" (Centrifugal, Rotary, and Reciprocating Pumps), 12th ed., 1969.
2. "Power Test Code, Centrifugal Pumps," PTC 8.2 1965, American Society of Mechanical Engineers.
3. "Fluid Meters, Their Theory and Application," Report of ASME Research Committee on Fluid Meters, 6th ed., 1971.
4. American Standards for Vertical Turbine Pumps, ANSI B-58.1 (AWWA E101-61) including appendix, American Water Works Association.
5. "Water Measurement Manual," 2d ed., Bureau of Reclamation, United States Department of Interior, 1967.
6. Smith, H., Jr.: "Hydraulics," John Wiley & Sons, Inc. New York, 1884.
7. Francis, J. B.: "Lowell Hydraulic Experiments," D. Van Nostrand, New York, 1883.
8. Shen, J.: A Preliminary Report on the Discharge Characteristics of Trapezoidal-Notch Thin-Plate Weirs, *U.S. Gel. Surv.*, 1959.
9. Parshall, R. L.: Improving the Distribution of Water to Farmers by Use of the Parshall Measuring Flume, *Soil Conserv. Sev., Bull* 488, U.S. Department of Agriculture, 1945.
10. Parshall, R. L.: Measuring Water in Irrigation Channels with Parshall Flumes and Weirs, *Soil Conserv. Serv., Bull.* 843, U.S. Department of Agriculture, 1950.

Appendix

Technical Data

Appendix

Technical Data

J. P. MESSINA

IGOR J. KARASSIK

This section contains technical data useful in pump application engineering. Considerably more tabulated data often referred to in the design and application of pumps may be found in other sections of the handbook. A list of the tables and figures in this section is given below; where necessary, an explanation of their use has been added.

TABLE 1 SELECTED LIST OF USEFUL REFERENCE TABLES AND FIGURES IN OTHER SECTIONS OF THE HANDBOOK

TABLES 2 AND 3 AND FIGURE 1 HEAD LOSS OLD PIPES

TABLE 2 Velocity and Friction Head Loss in Old Piping. This table gives pipe friction losses for 3-in to 24-in inside-diameter pipes based on the Hazen-Williams formula with a C of 100. Head loss is in feet of water per 100 ft of pipe.

TABLE 3 Values of Friction Factor C to Be Used with the Hazen-Williams Formula. This table is used to estimate values of C for various pipe materials and conditions.

FIGURE 1 Change in Hazen-Williams Coefficient C with Years of Service for Cast Iron Pipes Handling Soft, Clear Unfiltered Water. This figure gives conversion factors for changing $C = 100$ in Table 2 to other values. For example, with a flow of 700 gpm through a 6-in pipe, the friction head loss is 6.23 ft per 100 ft of pipe with $C = 100$. For $C = 130$, the conversion factor is 0.613; therefore the friction head loss will be 6.23×0.613, or 3.82 ft per 100 ft of pipe. See also Sec. 9.1.

TABLE 4 VELOCITY AND FRICTION HEAD LOSS IN NEW PIPING
This table gives the friction head loss in feet for flow of water through 100 ft of new steel (Schedule 40, standard weight) or wrought-iron pipe using the Darcy-Weisbach formula. For pipes of different relative roughnesses, refer to Sec. 9.1. The correct friction factor f is found for different Reynolds numbers, using the Moody diagram. From the Darcy-Weisbach formula given in Sec. 9.1, calculate the fiction head loss for actual conditions.

TABLES 5 AND 6. PRESSURE LOSS, VISCOUS LIQUIDS

Viscosity The property of a liquid which offers resistance to shearing forces in flow is called viscosity. When the flow is *laminar* (relatively low velocity), viscosity is the major cause of flow resistance and loss of head. For this flow condition viscous forces are high because particles move with increasing velocity toward the center of the pipe and have no motion next to the walls. When the flow is *turbulent* (relatively high and more uniform velocity), viscous shear is less and flow resistance is caused predominantly due to wall roughness and turbulence.

Viscosity can be visualized as follows: If the space between two plane surfaces is filled with a liquid, a force will be required to move one surface at a constant velocity relative to the other. The velocity of the liquid will vary linearly between the surfaces. The ratio of the force per unit area, called *shear stress,* to the velocity per unit distance between surfaces, called *shear rate,* is a measure of a liquid's *dynamic* or *absolute viscosity.* Liquids which exhibit shear stresses proportional to shear rates have constant viscosities for a particular temperature and pressure and are called *Newtonian* or *true liquids.* For these liquids viscosity and resistance to flow increase with decrease in temperature.

In the normal pumping range, however, viscosity may be considered independent of pressure. In the English system of measure, the unit is pound-seconds per square foot or its numerical equivalent, slugs per foot-second. The metric unit is dyne-second per square centimetre, called *poise,* which is numerically the same as the gram per centimetre-second. The common measure of dynamic viscosity is the *centipoise* (1/100 poise).

The viscous property of a liquid is also sometimes expressed as *kinematic viscosity.* This is the dynamic viscosity divided by the mass density (specific weight/g). In the English system the unit is square feet per second. The unit in the metric system is square centimeters per second or *stoke.* The common measure of kinematic viscosity is the *centistoke* (1/100 stoke).

The conversions from metric to English units are:

$$\frac{\text{lb-s}}{\text{ft}^2} = 0.0000208855 \text{ centipoise}$$

$$\frac{\text{ft}^2}{\text{s}} = 0.107639 \text{ centistoke}$$

It is usual in the United States to measure viscosity in Seconds Saybolt Universal (SSU) for liquids of medium viscosity and Seconds Saybolt Furol (SSF) for liquids of high viscosity. These viscosities are determined by using an orifice-type instrument which measures the time of discharge of a standard volume of the sample. Different measures of viscosity for some common Newtonian liquids and conversions from one system to another are given in tables in this section.

TABLE 5 Friction Loss for Viscous Liquids. This table gives the friction loss for viscous liquids flowing in new Schedule 40 steel pipe. Loss is in pounds per square inch per 100 ft of pipe and is based on a liquid specific gravity of 1.0. For liquids of other densities multiply the values in the table by the actual specific gravity. It is recommended that 15 percent be added for commercial installations. No allowance for aging has been included. Values of pressure loss are for laminar flow. The first two pages also give values for turbulent flow, which are shown to the left of the heavy line.

For laminar flow, the pressure loss is directly proportional to the viscosity and the velocity of flow, and inversely proportional to the pipe diameter to the fourth power. Therefore, for intermediate values of viscosity and flow, obtain the pressure loss by direct interpolation. For pipe sizes not shown, multiply the fourth power of the ratio of any tabulated diameter to the actual pipe diameter wanted by the tabulated loss shown. The gpm and viscosity must be the same for both diameters.

For turbulent flow, and for rates of flow and pipe sizes not tabulated, the following procedures may be followed: For the viscosity and pipe size required, an intermediate flow loss is found by selecting the pressure loss for the next lower flow and multiplying by the square of the ratio of actual to tabulated flow rates. For the viscosity and flow required, an intermediate pipe-diameter flow loss is found by selecting the pressure loss for the next smaller diameter and multiplying by the fifth power of the ratio of tabulated to actual inside diameters.

TABLE 6 Representative Equivalent Length in Pipe Diameters (L/D) of Various Valves and Fittings. For turbulent flow, the friction head loss in various valves and fittings is found by using the loss in an equivalent length of straight pipe as given in Table 6 and as explained in Sec. 9.1.

For laminar flow, the friction head loss in various valves and fittings can only be approximated. For liquids of relatively low viscosity, where the flow is close to being turbulent, the values of the equivalent length of straight pipe given in Table 6 can be used. For liquids of relatively high viscosity, i.e., having Reynolds numbers Re less than 1,000, the values of equivalent lengths given in Table 6 can be reduced. *Crane Technical Paper* 410, Flow of Fluids Through Valves, Fittings, and Pipe, suggests that L/D in Table 6 be corrected by multiplying by Re/1,000.

TABLE 7 VISCOSITY OF COMMON LIQUIDS

TABLE 8 VISCOSITY CONVERSION

TABLE 9 PROPERTIES OF WATER AT VARIOUS TEMPERATURES FROM 40 TO 705.4 F.

TABLE 10 THEORETICAL DISCHARGE OF NOZZLES, IN U.S. GALLONS PER MINUTE

TABLE 11 CAST IRON PIPE DIMENSIONS

TABLE 12 DIMENSIONS OF WELDED AND SEAMLESS STEEL PIPE

TABLE 13 CAPACITY EQUIVALENTS

FIGURE 2 CAPACITY CONVERSIONS

TABLE 14 PRESSURES AND HEAD EQUIVALENTS

FIGURE 3 PRESSURE HEAD AND HEAD CONVERSION CHART

FIGURE 4 ATMOSPHERIC PRESSURE FOR ALTITUDES UP TO 12,000 FT

TABLE 15 CONVERSIONS, CONSTANTS, AND FORMULAS

TABLE 16 MEASUREMENT CONVERSIONS

TABLE 1 Selected Lst of Useful Reference Tables and Figures in the Handbook

Head Loss in Pipe, Valves, and Fittings	Sec. & Table No.
Values of friction factor C to be used with Hazen-Williams formula	9-2
Values of friction factor n to be used with Manning formula	9-3
Resistance coefficients for valves and fittings	9-4a, b
Representative equivalent length in pipe diameters of various valves and fittings	9-5a
Resistance coefficients for miter bends at Reynolds number $\approx 2.25 \times 10^5$	9-6

	Sec. & Fig. No.
Moody diagram	9-28
Relative roughness and friction factors for new clean pipes	9-29
Kinematic viscosity and Reynolds number	9-30
Nomogram for the solution of the Hazen-Williams formula	9-31
Nomogram for the solution of the Manning formula	9-32
Discharge and velocity of a partially full circular conduit vs. a full conduit	9-33
Resistance coefficients for increasers and diffusers	9-36
Resistance coefficients for reducers	9-37
Resistance coefficients for 90° bends of uniform diameter	9-38
Resistance coefficients for bends of uniform diameter and smooth surface at Reynolds number $\approx 2.25 \times 10^5$	9-39
Head loss in a long-radius pump suction elbow	9-40
Head loss in a short-radius pump suction elbow	9-41
Overall pressure loss across thin-plate orifices	9-45
Overall pressure loss across flow nozzles	9-46
Overall pressure loss across venturi tubes	9-47

Pump Pit Dimensions	Sec. & Fig. No.
Proportions for standard pump pit	11-8
Recommended pit dimensions	11-9

SI Units	Sec.	Table No.
Typical derived units of the SI system	xvi	1
Prefixes for SI multiple and submultiple units	xvii	2
Conversion of U.S. to SI units	xvii	3

Miscellaneous Data	Sec. & Table No.
Temperature of dry and saturated steam	6-1
Critical depth in circular pipes	9-1

	Sec. & Fig. No.
Area vs. depth for a circular pipe	9-27

TABLE 2 Velocity and Friction Head Loss in Old Piping. (Friction values apply to cast-iron pipes after fifteen years service handling average water. Based on Hazen-Williams formula with $C = 100$)

gpm	v[1]	f[2]	v	f	v	f	v	f	v	f	v	f	v	f	v	f	v	f	gpm
							3-in. ID pipe		4-in. ID pipe										
30							1.36	.534	.77	.131	5-in. ID pipe								30
40							1.81	.910	1.02	.224									40
50							2.27	1.38	1.28	.338	.82	.114	6-in. ID pipe						50
60							2.72	1.92	1.53	.475	.98	.160							60
70							3.18	2.56	1.79	.631	1.14	.213	.79	.088					70
80							3.63	3.28	2.04	.808	1.31	.273	.91	.112					80
90							4.08	4.08	2.30	1.01	1.47	.339	1.02	.139					90
100							4.54	4.96	2.55	1.22	1.63	.412	1.14	.170	8-in. ID pipe				100
125							5.68	7.50	3.19	1.85	2.04	.623	1.42	.256					125
150							6.81	10.5	3.83	2.59	2.47	.874	1.70	.360	.96	.089			150
175							7.95	14.0	4.47	3.44	2.86	1.16	1.99	.478	1.12	.118			175
200							9.08	17.9	5.10	4.41	3.27	1.49	2.27	.613	1.28	.151	10-in. ID pipe		200
225							10.2	22.3	5.74	5.48	3.68	1.85	2.55	.762	1.44	.188			225
250							11.3	27.1	6.38	6.67	4.08	2.25	2.84	.926	1.60	.228	1.02	.077	250
275							12.5	32.3	7.02	7.96	4.50	2.68	3.12	1.11	1.76	.272	1.12	.092	275
	12-in. ID pipe																		
300							13.6	37.9	7.65	9.34	4.90	3.13	3.41	1.30	1.91	.320	1.23	.108	300
350	.99	.059	14-in. ID pipe				15.9	50.4	8.93	12.4	5.72	4.20	3.97	1.73	2.23	.425	1.43	.144	350
400	1.13	.076					18.2	64.6	10.2	15.9	6.54	5.38	4.54	2.21	2.55	.545	1.63	.184	400
450	1.28	.094	.94	.044					11.5	19.8	7.36	6.68	5.10	2.75	2.87	.678	1.84	.228	450
500	1.42	.114	1.04	.054	16-in. ID pipe				12.8	24.1	8.18	8.12	5.68	3.34	3.19	.823	2.04	.278	500
550	1.56	.136	1.15	.064					14.0	28.7	8.99	9.69	6.24	3.99	3.51	.982	2.24	.331	550
600	1.70	.160	1.25	.076	.96	.039			15.3	33.7	9.81	11.4	6.81	4.68	3.82	1.15	2.45	.389	600
650	1.84	.186	1.36	.088	1.04	.046	18-in. ID pipe		16.6	39.1	10.6	13.2	7.38	5.43	4.15	1.34	2.65	.452	650
700	1.99	.214	1.46	.100	1.12	.052			17.9	44.9	11.4	15.1	7.94	6.23	4.47	1.53	2.86	.518	700
750	2.13	.242	1.56	.114	1.20	.060	.95	.034			12.3	17.2	8.51	7.08	4.78	1.74	3.06	.589	750
800	2.27	.273	1.67	.129	1.28	.067	1.01	.038	20-in. ID pipe		13.1	19.4	9.08	7.98	5.10	1.97	3.26	.666	800
900	2.55	.339	1.88	.160	1.44	.084	1.13	.047			14.7	24.1	10.2	9.92	5.74	2.44	3.67	.825	900
1,000	2.83	.412	2.08	.195	1.60	.102	1.26	.057	1.02	.034	16.3	29.3	11.4	12.1	6.38	2.97	4.08	1.00	1,000
1,100	3.12	.492	2.29	.232	1.76	.121	1.39	.068	1.12	.041	18.0	35.0	12.5	14.4	7.02	3.55	4.50	1.20	1,100
1,200	3.40	.578	2.50	.273	1.92	.143	1.51	.080	1.23	.048			13.6	16.9	7.66	4.17	4.90	1.41	1,200

	12-in. ID pipe		14-in. ID pipe		16-in. ID pipe		18-in. ID pipe		20-in. ID pipe		24-in. ID pipe		6-in. ID pipe		8-in. ID pipe		10-in. ID pipe		
1,300	3.69	.671	2.71	.316	2.08	.165	1.64	.093	1.33	.056			14.8	19.6	8.30	4.83	5.31	1.63	1,300
1,400	3.97	.770	2.92	.363	2.24	.190	1.76	.107	1.43	.064			15.9	22.5	8.93	5.54	5.72	1.87	1,400
1,500	4.25	.875	3.12	.413	2.40	.215	1.89	.121	1.53	.073	1.06	.030	17.0	25.5	9.55	6.30	6.12	2.13	1,500
1,600	4.54	.985	3.33	.465	2.55	.243	2.02	.137	1.63	.082	1.13	.034	18.2	28.8	10.2	7.10	6.53	2.39	1,600
1,800	5.11	1.22	3.75	.578	2.87	.302	2.27	.170	1.84	.102	1.28	.042			11.5	8.83	7.35	2.98	1,800
2,000	5.67	1.49	4.17	.703	3.19	.367	2.52	.207	2.04	.124	1.42	.051			12.8	10.7	8.17	3.62	2,000
2,500	7.09	2.25	5.21	1.06	3.99	.555	3.15	.312	2.55	.187	1.77	.077			16.0	16.2	10.2	5.48	2,500
3,000	8.51	3.16	6.25	1.49	4.78	.778	3.78	.438	3.06	.262	2.13	.108			19.1	22.8	12.3	7.67	3,000
3,500	9.93	4.20	7.29	1.98	5.59	1.04	4.41	.583	3.57	.349	2.48	.143					14.3	10.2	3,500
4,000	11.3	5.38	8.33	2.54	6.39	1.33	5.04	.746	4.08	.447	2.83	.184					16.3	13.1	4,000
4,500	12.8	6.68	9.38	3.15	7.18	1.65	5.67	.928	4.59	.555	3.19	.228					18.4	16.3	4,500
5,000	14.2	8.13	10.4	3.83	7.98	2.00	6.30	1.13	5.10	.675	3.54	.278							5,000
5,500	15.6	9.70	11.5	4.58	8.78	2.39	6.93	1.35	5.61	.806	3.90	.332							5,500
6,000	17.0	11.4	12.5	5.38	9.68	2.81	7.56	1.58	6.12	.947	4.25	.390							6,000
6,500	18.4	13.2	13.6	6.24	10.4	3.26	8.19	1.83	6.73	1.10	4.61	.452							6,500
7,000	19.9	15.2	14.6	7.16	11.2	3.74	8.82	2.11	7.15	1.26	4.96	.518							7,000
7,500			15.6	8.13	12.0	4.24	9.45	2.39	7.66	1.43	5.32	.589							7,500
8,000			16.7	9.16	12.8	4.79	10.1	2.69	8.17	1.61	5.66	.664							8,000
9,000			18.8	11.4	14.4	5.95	11.3	3.39	9.18	2.01	6.38	.825							9,000
10,000					16.0	7.24	12.6	4.07	10.2	2.44	7.09	1.00							10,000
11,000					17.6	8.63	13.9	4.86	11.2	2.91	7.80	1.20							11,000
12,000					19.2	10.1	15.1	5.71	12.3	3.42	8.51	1.41							12,000
13,000							16.4	6.62	13.3	3.96	9.12	1.63							13,000
14,000							17.6	7.59	14.3	4.54	9.93	1.87							14,000
15,000							18.9	8.63	15.3	5.27	10.6	2.13							15,000
16,000									16.3	5.82	11.3	2.40							16,000
18,000									18.4	7.24	12.8	2.98							18,000
20,000											14.2	3.62							20,000
25,000											17.7	5.48							25,000

[1] Velocity, in feet per second.
[2] Friction head loss, in feet of water per 100 feet of pipe.

TABLE 3 Values of Friction Factor C to Be Used with the Hazen-Williams Formula. (From C. V. Davis and K. E. Sorensen, "Handbook of Applied Hydraulics," 3d ed. Copyright 1969 by McGraw-Hill Book Company, New York and used with their permission.)

Type of pipe		Condition	C
Cast iron	New	All sizes, in	130
	5 years old	12 and over	120
		8	119
		4	118
	10 years old	24 and over	113
		12	111
		4	107
	20 years old	24 and over	100
		12	96
		4	89
	30 years old	30 and over	90
		16	87
		4	75
	40 years old	30 and over	83
		16	80
		4	64
		40 and over	77
		24	74
		4	55
Welded steel	Values of C the same as for cast iron pipe, 5 years older		
Riveted steel	Values of C the same as for cast iron pipe, 10 years older		
Wood-stave	Average value, regardless of age		120
Concrete or concrete-lined	Large sizes, good workmanship, steel forms		140
	Large sizes, good workmanship, wooden forms		120
	Centrifugally spun		135
Vitrified	In good condition		110

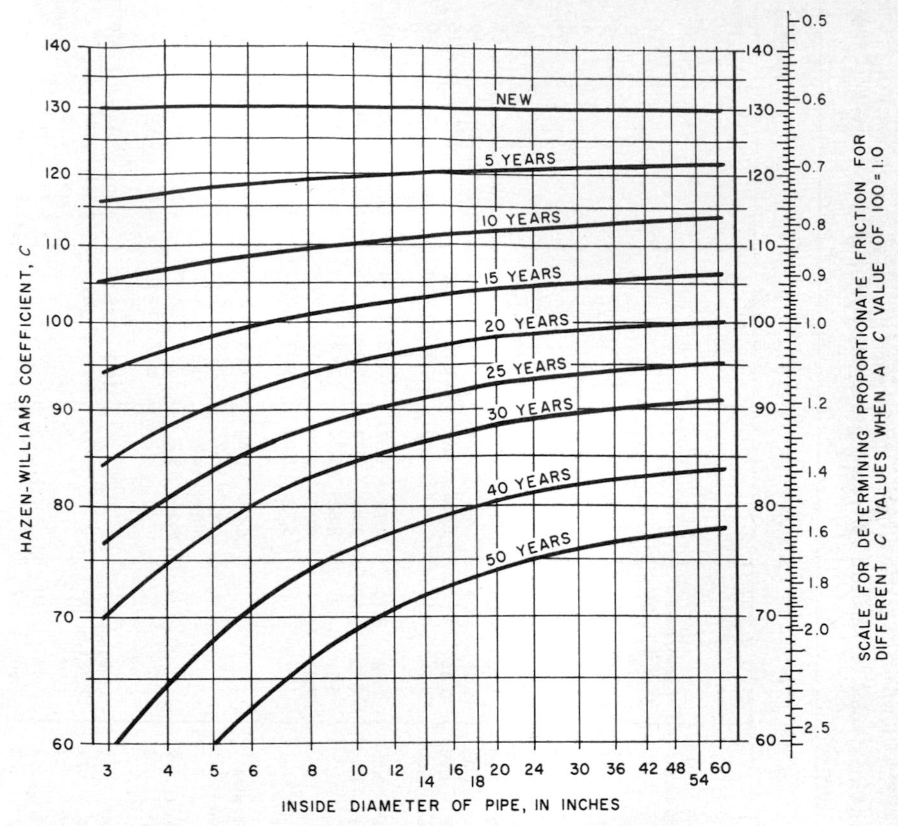

Fig. 1 Change in Hazen-Williams coefficient C with years of service, for cast iron pipes handling soft, clear unfiltered water.

TABLE 4 Velocity and Friction Head Loss in New Piping (Friction values apply to Schedule 40 (standard weight) steel pipe carrying water; based on Darcy-Weisbach formula)

gpm	v[1]	f[2]	v	f	v	f	v	f	v	f	v	f	v	f	v	f	gpm
					1-in. pipe (1.049-in. ID)		1¼-in. pipe (1.380-in. ID)		1½-in. pipe (1.610-in. ID)								
1					.37	.11											1
2					.74	.39	.43	.10									2
3					1.11	.82	.64	.21	.47	.10							3
4					1.49	1.37	.86	.36	.63	.17							4
5					1.86	2.08	1.07	.54	.79	.26							5
											2-in. pipe (2.067-in. ID)		2½-in. pipe (2.469-in. ID)				
6					2.23	2.83	1.28	.76	.95	.35	.57	.10					6
8					2.97	4.88	1.72	1.29	1.26	.61	.76	.17					8
10					3.71	7.12	2.14	1.95	1.57	.90	.96	.26	.67	.11			10
15					5.56	15.0	3.21	4.06	2.36	1.87	1.43	.54	1.00	.23			15
															3-in. pipe (3.068-in. ID)		
20					7.41	25.6	4.28	6.80	3.15	3.12	1.91	.92	1.34	.38	.87	.13	20
	3½-in. pipe (3.548-in. ID)																
25	.81	.10					5.35	10.3	3.94	4.70	2.38	1.39	1.67	.58	1.08	.20	25
30	.97	.13					6.43	14.4	4.72	6.60	2.86	1.92	2.00	.81	1.30	.28	30
			4-in. pipe (4.026-in. ID)														
40	1.30	.23	1.01	.12					6.30	11.2	3.82	3.35	2.68	1.36	1.73	.47	40
50	1.62	.34	1.26	.18					7.87	16.6	4.77	5.00	3.34	2.06	2.16	.72	50
60	1.94	.48	1.51	.25							5.72	7.00	4.02	2.85	2.60	.99	60
					5-in. pipe (5.047-in. ID)		6-in. pipe (6.065-in. ID)										
70	2.27	.63	1.76	.34	1.12	.11					6.68	9.40	4.68	3.80	3.03	1.33	70
80	2.59	.82	2.01	.43	1.28	.15					7.62	11.9	5.35	4.95	3.46	1.72	80
90	2.91	1.00	2.27	.54	1.44	.18					8.60	14.7	6.02	6.05	3.89	2.13	90
100	3.24	1.24	2.52	.65	1.60	.22	1.11	.09			9.56	18.7	6.70	7.47	4.83	2.58	100
125	4.05	1.82	3.15	1.00	2.00	.33	1.39	.13					8.37	11.1	5.41	3.90	125
									8-in. pipe (7.981-in. ID)								
150	4.86	2.55	3.78	1.39	2.40	.47	1.66	.18					10.0	15.4	6.50	5.44	150
175	5.66	3.40	4.40	1.90	2.80	.62	1.94	.24					11.7	20.8	7.58	7.30	175
200	6.48	4.35	5.04	2.40	3.20	.80	2.22	.31							8.66	9.18	200
225	7.30	5.44	5.66	2.98	3.60	.97	2.50	.38	1.44	.10					9.75	11.6	225
250	8.10	6.59	6.29	3.68	4.00	1.19	2.77	.47	1.60	.12					10.8	14.0	250

	3½-in. pipe (3.548-in. ID)		4-in. pipe (4.026-in. ID)		5-in. pipe (5.047-in. ID)		6-in. pipe (6.065-in. ID)		8-in. pipe (7.981-in. ID)		10-in. pipe (10.020-in. ID)		3-in. pipe (3.068-in. ID)		
275	8.91	7.90	6.92	4.35	4.40	1.43	3.05	.56	1.76	.15			11.9	16.9	275
300	9.72	9.30	7.55	5.04	4.80	1.65	3.32	.66	1.92	.17			13.0	19.6	300
350	11.3	12.2	8.80	6.85	5.60	2.21	3.88	.88	2.24	.23					350
400	13.0	15.9	10.1	8.67	6.40	2.89	4.44	1.12	2.56	.29	1.62	.10			400
450	14.6	20.0	11.3	10.9	7.20	3.56	4.99	1.40	2.88	.37	1.82	.12			450
500			12.6	13.3	8.00	4.36	5.54	1.72	3.20	.45	2.03	.15			500
550			13.9	16.0	8.80	5.17	6.10	2.06	3.52	.55	2.23	.18			550
600			15.1	19.1	9.60	6.16	6.65	2.42	3.84	.63	2.44	.21			600
650					10.4	7.22	7.20	2.78	4.16	.73	2.64	.24			650
700					11.2	8.29	7.75	3.25	4.47	.85	2.84	.28			700
750					12.0	9.40	8.31	3.63	4.80	.97	3.04	.31			750
800					12.8	10.3	8.87	4.11	5.11	1.11	3.25	.35			800
900					14.4	13.0	9.96	5.12	5.75	1.33	3.65	.44			900
1,000					16.0	15.8	11.1	6.17	6.40	1.64	4.06	.55			1,000
1,100					17.6	19.0	12.2	7.45	7.04	1.98	4.46	.64			1,100
1,200							13.3	8.73	7.67	2.36	4.87	.75			1,200
1,300							14.4	10.2	8.31	2.71	5.27	.88			1,300
1,400							15.5	11.9	8.95	3.10	5.68	1.02			1,400
1,500							16.7	13.2	9.60	3.49	6.09	1.18			1,500
1,600							17.8	15.0	10.2	3.92	6.49	1.31			1,600
1,800							20.0	18.5	11.5	4.99	7.30	1.60			1,800
2,000									12.8	5.96	8.11	1.97			2,000
2,500									16.0	9.00	10.2	2.95			2,500
3,000									19.2	12.5	12.2	4.15			3,000
3,500									22.4	16.6	14.2	5.60			3,500
4,000											16.2	6.90			4,000
4,500											18.3	8.80			4,500
5,000											20.3	10.8			5,000
5,500											22.3	13.0			5,500
6,000											24.4	15.3			6,000

[1] Velocity, in feet per second.

[2] Friction head loss, in feet of water per 100 feet of pipe.

TABLE 5 Friction Loss for Viscous Liquids

Loss in pounds per square inch per 100 feet of new Schedule 40 steel pipe based on specific gravity of 1.00. (For a liquid having a specific gravity other than 1.00, multiply the value from the table by the specific gravity of that liquid. For commercial installations, it is recommended that 15 per cent be added to the values in this table. No allowance for aging of pipe is included.)

GPM	Pipe Size	VISCOSITY—SAYBOLT SECONDS UNIVERSAL																	
		100	200	300	400	500	1000	1500	2000	2500	3000	4000	5000	6000	7000	8000	9000	10,000	15,000
3	½	11.2	23.6	35.3	47.1	59	118	177	236	294	353	471	589	706	824	942	......	......	......
	¾	3.7	7.6	11.5	15.3	19.1	38.2	57	76	96	115	153	191	229	268	306	344	382	573
	1	1.4	2.9	4.4	5.8	7.3	14.5	21.8	29.1	36.3	43.6	58	73	87	101	116	131	145	218
5	¾	6.1	12.7	19.1	25.5	31.9	64	96	127	159	191	255	319	382	446	510	573	637	956
	1	2.3	4.9	7.3	9.7	12.1	24.2	36.3	48.5	61	73	97	121	145	170	194	218	242	363
	1¼	0.77	1.6	2.4	3.3	4.1	8.1	12.2	16.2	20.3	24.3	32.5	40.6	48.7	57	65	73	81	122
7	¾	8.5	17.9	26.8	35.7	44.6	89	134	178	223	268	357	446	535	624	713	803	892	......
	1	3.2	6.8	10.2	13.6	17	33.9	51	68	85	102	136	170	203	237	271	305	339	509
	1¼	1.1	2.3	3.4	4.5	5.7	11.4	17	22.7	28.4	34.1	45.4	57	68	80	91	102	114	170
10	1	4.9	9.7	14.5	19.4	24.2	48.5	73	97	121	145	194	242	291	339	388	436	485	727
	1¼	1.6	3.3	4.9	6.5	8.1	16.2	24.3	32.5	40.6	48.7	65	81	97	114	130	146	162	243
	1½	0.84	1.8	2.6	3.5	4.4	8.8	13.1	17.5	21.9	26.3	35	43.8	53	61	70	79	88	131
15	1	11	14.5	21.8	29.1	36.3	73	109	145	182	218	291	363	436	509	581	654	727	......
	1¼	2.8	4.9	7.3	9.7	12.2	24.3	36.5	48.7	61	73	97	122	146	170	195	219	243	365
	1½	1.3	2.6	3.9	5.3	6.6	13.1	197	26.3	32.8	39.4	53	66	79	92	105	118	131	197
20	1	18	18	29.1	38.8	48.5	97	145	194	242	291	388	485	581	678	775	872	......	......
	1¼	4.9	6.4	9.7	13	16.2	32.5	48.7	65	81	97	130	162	195	227	260	292	325	487
	1½	2.3	3.5	5.3	7	8.8	17.5	26.3	35	43.8	53	70	88	105	123	140	158	175	263
	2	0.64	1.3	1.9	2.6	3.2	6.4	9.6	12.9	16.1	19.3	25.7	32.1	38.5	45	51	58	64	96
25	1½	3.5	4.4	6.6	8.8	11	21.9	32.8	43.8	55	66	88	110	131	153	176	197	219	328
	2	1	1.6	2.4	3.2	4	8	12.1	16.1	20.1	24.1	32.1	40.2	48.2	56	64	72	80	121
	2½	0.4	0.79	1.2	1.6	2	4	5.9	7.9	9.9	11.8	15.8	19.7	23.7	27.6	31.6	35.5	39.5	59
30	1½	5	5.3	7.9	10.5	13.1	26.3	39.4	53	66	79	105	131	158	184	210	237	263	394
	2	1.4	1.9	2.9	3.9	4.8	9.6	14.5	19.3	24.1	28.9	38.5	48.2	58	67	77	87	96	145
	2½	0.6	0.95	1.4	1.9	2.4	4.7	7.1	9.5	11.8	14.2	19	23.7	28.4	33.2	37.9	42.6	47.4	71
40	1½	8.5	9	10.5	14	17.5	35	53	70	88	105	140	175	210	245	280	315	350	526
	2	2.5	2.5	3.9	5.1	6.4	12.9	19.3	25.7	32.1	38.5	51	64	77	90	103	116	129	193
	2½	1.1	1.3	1.9	2.5	3.2	6.3	9.5	12.6	15.8	19	25.3	31.6	37.9	44.2	51	57	63	95
50	1½	12.5	14	14	17.5	21.9	43.8	66	88	110	131	175	219	263	307	350	394	438	657
	2	3.7	4	4.8	6.4	8	16.1	24.1	32.1	40.2	48.2	64	80	96	112	129	145	161	241
	2½	1.6	1.7	2.4	3.2	4	7.9	11.8	15.8	19.7	23.7	31.6	39.5	47.4	55	63	71	79	118
60	2	5	5.8	5.8	7.7	9.6	19.3	28.9	38.5	48.2	58	77	96	116	135	154	173	193	289
	2½	2.2	2.4	2.8	3.8	4.7	9.5	14.2	19	23.7	28.4	37.9	47.4	57	66	76	85	95	142
	3	0.8	0.8	1.2	1.6	2	4	6	8	9.9	11.9	15.9	19.9	23.9	27.9	31.8	35.8	39.8	60
70	2½	2.8	3.2	3.4	4.4	5.5	11.1	16.6	22.1	27.6	33.2	44.2	55	66	77	88	100	111	166
	3	1	1.1	1.4	1.9	2.3	4.6	7	9.3	11.6	13.9	18.6	23.2	27.8	32.5	37.1	41.7	46.4	70
	4	0.27	0.31	0.47	0.63	0.78	1.6	2.4	3.1	3.9	4.7	6.3	7.8	9.4	11	12.5	14.1	15.6	23.5
80	2½	3.6	4.2	4.2	5.1	6.3	12.6	19	25.3	31.6	37.9	51	63	76	88	101	114	126	190
	3	1.3	1.4	1.6	2.1	2.7	5.3	8	10.6	13.3	15.9	21.2	26.5	31.8	37.1	42.4	47.7	53	80
	4	0.36	0.36	0.54	0.72	0.89	1.8	2.7	3.6	4.5	5.4	7.2	8.9	10.7	12.5	14.3	16.1	17.9	26.8
100	2½	5.3	6.1	6.4	6.4	8	15.8	23.7	31.6	39.5	47.4	63	79	95	111	127	142	158	237
	3	1.9	2.2	2.2	2.7	3.3	6.6	9.9	13.3	16.6	19.9	26.5	33.1	39.8	46.4	53	60	66	99
	4	0.52	0.57	0.67	0.89	1.1	2.2	3.4	4.5	5.6	6.7	8.9	11.2	13.4	15.6	17.9	20.1	22.3	33.5

←TURBULENT FLOW→ ←LAMINAR FLOW→

TABLE 5 Friction Loss for Viscous Liquids (Continued)

Loss in pounds per square inch per 100 feet of new Schedule 40 steel pipe based on specific gravity of 1.00. (For a liquid having a specific gravity other than 1.00, multiply the value from the table by the specific gravity of that liquid. For commercial installations, it is recommended that 15 per cent be added to the values in this table. No allowance for aging of pipe is included.)

GPM	Pipe Size	VISCOSITY—SAYBOLT SECONDS UNIVERSAL																	
		100	200	300	400	500	1000	1500	2000	2500	3000	4000	5000	6000	7000	8000	9000	10,000	15,000
120	3	2.7	3.1	3.2	3.2	4	8	11.9	15.9	19.9	23.9	31.8	39.8	47.7	56	64	72	80	119
	4	0.73	0.81	0.81	1.1	1.3	2.7	4	5.4	6.7	8	10.7	13.4	16.1	18.8	21.4	24.1	26.8	40.2
	6	.098	0.11	0.16	0.21	0.26	0.52	0.78	1.0	1.3	1.6	2.1	2.6	3.1	3.6	4.2	4.7	5.2	7.8
140	3	3.4	4	4.3	4.3	4.6	9.3	13.9	18.6	23.2	27.8	37.1	46.4	56	65	74	84	93	139
	4	0.95	1.1	1.1	1.3	1.6	3.1	4.7	6.3	7.8	9.4	12.5	15.6	18.8	21.9	25	28.2	31.3	46.9
	6	0.13	0.15	0.18	0.24	0.30	0.61	0.91	1.2	1.5	1.8	2.4	3.0	3.6	4.2	4.9	5.5	6.1	9.1
160	3	4.4	5	5.7	5.7	5.7	10.6	15.9	21.2	26.5	31.8	42.4	53	64	74	85	95	106	159
	4	1.2	1.4	1.4	1.4	1.8	3.6	5.4	7.2	8.9	10.7	14.3	17.9	21.5	25	28.6	32.2	35.7	54
	6	0.17	0.18	0.21	0.28	0.35	0.69	1.0	1.4	1.7	2.1	2.8	3.5	4.2	4.9	5.5	6.2	6.9	10.4
180	3	5.3	6.3	7	7	7	11.9	17.9	23.9	29.8	35.8	47.7	60	72	84	95	107	119	179
	4	1.5	1.8	1.8	1.8	2	4	6	8	10.1	12.1	16.1	20.1	24.1	28.1	32.2	36.2	40.2	60
	6	0.2	0.24	0.24	0.31	0.39	0.78	1.2	1.6	2	2.3	3.1	3.9	4.7	5.5	6.2	7	7.8	11.7
200	3	6.5	7.7	8.8	8.8	8.8	13.3	19.9	26.5	33.1	39.8	53	66	80	93	106	119	133	199
	4	1.8	2.2	2.2	2.2	2.2	4.5	6.7	8.9	11.2	13.4	17.9	22.3	26.8	31.3	35.7	40.2	44.7	67
	6	0.25	0.3	0.3	0.35	0.43	0.87	1.3	1.7	2.2	2.6	3.5	4.3	5.2	6.1	6.9	7.8	8.7	13
250	4	2.6	3.2	3.5	3.5	3.5	5.6	8.4	11.2	14	16.8	22.3	27.9	33.5	39.1	44.7	50	56	84
	6	0.36	0.43	0.45	0.45	0.54	1.1	1.6	2.2	2.7	3.3	4.3	5.4	6.5	7.6	8.7	9.8	10.8	16.3
	8	.095	0.12	0.12	0.15	0.18	0.36	0.54	0.72	0.9	1.1	1.5	1.8	2.2	2.5	2.9	3.3	3.6	5.4
300	4	3.7	4.3	5	5	5	6.7	10.1	13.4	16.8	20.1	26.8	33.5	40.2	47	54	60	67	101
	6	0.5	0.6	0.65	0.65	0.65	1.3	2	2.6	3.3	3.9	5.2	6.5	7.8	9.1	10.4	11.7	13	19.5
	8	0.13	0.17	0.17	0.18	0.22	0.43	0.65	0.87	1.1	1.3	1.7	2.2	2.6	3	3.5	3.9	4.3	6.5
400	6	0.82	1	1.1	1.2	1.2	1.7	2.6	3.5	4.3	5.2	6.9	8.7	10.4	12.1	13.9	15.6	17.3	26
	8	0.23	0.27	0.29	0.29	0.29	0.58	0.87	1.2	1.5	1.7	2.3	2.9	3.5	4.1	4.6	5.2	5.8	8.7
	10	0.08	0.09	0.1	0.1	0.12	0.23	0.35	0.47	0.58	0.7	0.93	1.2	1.4	1.6	1.9	2.1	2.3	3.5
500	6	1.2	1.5	1.6	1.8	1.8	2.2	3.2	4.3	5.4	6.5	8.7	10.8	13	15.2	17.3	19.5	21.7	32.5
	8	0.33	0.39	0.44	0.47	0.47	0.72	1.1	1.5	1.8	2.2	2.9	3.6	4.3	5.1	5.8	6.5	7.2	10.8
	10	0.11	0.14	0.15	0.15	0.15	0.29	0.44	0.58	0.73	0.87	1.2	1.5	1.8	2	2.3	2.6	2.9	4.4
600	6	1.8	2.2	2.3	2.4	2.6	2.7	3.9	5.2	6.5	7.8	10.4	13	16	18.2	20.8	23.4	26	39
	8	0.47	0.57	0.62	0.67	0.67	0.87	1.3	1.7	2.2	2.6	3.5	4.3	5.2	6.1	6.9	7.8	8.7	13
	10	0.16	0.18	0.2	0.22	0.22	0.35	0.52	0.7	0.87	1.1	1.4	1.8	2.1	2.4	2.8	3.3	3.5	5.2
700	6	2.3	2.7	3	3.2	3.5	3.6	4.6	6.1	7.6	9.1	12.1	15.2	18.4	21.2	24.3	27.3	30.3	45.5
	8	0.6	0.74	0.82	0.89	0.93	1	1.5	2	2.5	3	4.1	5.1	6.1	7.1	8.1	9.1	10.1	15.2
	10	0.2	0.25	0.27	0.3	0.3	0.41	0.61	0.82	1	1.2	1.6	2	2.4	2.9	3.3	3.7	4.1	6.1
800	6	2.8	3.5	3.7	4	4.2	4.8	5.2	6.9	8.7	10.4	13.9	17.3	20.8	24.3	27.7	31.2	34.7	52
	8	0.78	0.94	1	1.1	1.2	1.2	1.7	2.3	2.9	3.5	4.6	5.8	6.9	8.1	9.3	10.4	11.6	17.3
	10	0.26	0.3	0.34	0.38	0.4	0.47	0.7	0.92	1.2	1.4	1.9	2.3	2.8	3.3	3.7	4.2	4.7	7
900	6	3.5	4.3	4.6	5.0	5.2	6	6	7.8	9.8	11.7	15.6	19.5	23.4	27.3	31.2	35.1	39	58.5
	8	0.95	1.1	1.3	1.4	1.5	1.5	2	2.6	3.3	3.9	5.2	6.5	7.8	9.1	10.4	11.7	13	19.5
	10	0.32	0.37	0.43	0.46	0.5	0.52	0.79	1.1	1.3	1.6	2.1	2.6	3.1	3.7	4.2	4.7	5.2	7.9
1000	8	1.1	1.4	1.5	1.6	1.8	1.9	2.2	2.9	3.6	4.3	5.8	7.2	8.7	10.1	11.6	13	14.5	21.7
	10	0.38	0.45	0.5	0.55	0.6	0.6	0.87	1.2	1.5	1.8	2.3	2.9	3.5	4.1	4.7	5.2	5.8	8.7
	12	0.17	0.2	0.22	0.24	0.25	0.29	0.43	0.58	0.72	0.87	1.2	1.5	1.7	2	2.3	2.6	2.9	4.3

← TURBULENT FLOW → ← LAMINAR FLOW →

TABLE 5 Friction Loss for Viscous Liquids (Continued)

Loss in pounds per square inch per 100 feet of new Schedule 40 steel pipe based on specific gravity of 1.00. (For a liquid having a specific gravity other than 1.00, multiply the value from the table by the specific gravity of that liquid. For commercial installations, it is recommended that 15 per cent be added to the values in this table. No allowance for aging of pipe is included.)

GPM	Pipe Size	VISCOSITY—SAYBOLT SECONDS UNIVERSAL														
		20,000	25,000	30,000	40,000	50,000	60,000	70,000	80,000	90,000	100,000	125,000	150,000	175,000	200,000	500,000
3	2	19.3	24.1	28.9	38.5	48.2	58	67	77	87	96	120	145	169	193	482
	2½	9.5	11.8	14.2	19	23.7	28.4	332	37.9	42.6	47.4	59	71	83	95	237
	3	4	5	6	8	9.9	11.9	13.9	15.9	17.9	19.9	24.9	29.8	34.8	39.8	99
5	2	32	40	48.2	64	80	96	112	129	145	161	201	241	281	321	803
	2½	15.8	19.7	23.7	31.6	39.5	47.4	55	63	71	79	99	118	138	158	395
	3	6.6	8.3	9.9	13.3	16.6	9.9	23.2	26.5	29.8	33	41.4	49.7	58	66	166
7	2	45	56	67	90	112	135	157	180	202	225	281	337	393	450	......
	2½	22.1	27.6	33.2	44.2	55	66	77	88	100	111	138	166	194	221	553
	3	9.3	11.6	13.9	18.6	23.2	27.8	32.5	37.1	41.7	46.4	58	70	81	93	232
10	2½	31.6	39.5	47.4	63	79	95	111	126	142	158	197	237	276	316	790
	3	13.3	16.6	19.9	26.5	33.1	39.8	46.4	53	60	66	83	99	116	133	331
	4	4.5	5.6	6.7	8.9	11.2	13.4	15.6	17.9	20.1	22.3	27.9	33.5	39.1	44.7	112
15	2½	47.4	59	71	95	118	142	166	190	213	237	296	355	415	474	....
	3	19.9	24.9	29.8	39.8	49.7	60	70	80	89	99	124	149	174	199	497
	4	6.7	8.4	10.1	13.4	16.8	20.1	23.5	26.8	30.2	33.5	41.9	50	59	67	168
20	3	26.5	33.1	39.8	53	66	80	93	106	119	133	166	199	232	265	663
	4	8.9	11.2	13.4	17.9	22.3	26.8	31.3	35.7	40.2	44.7	56	67	78	89	223
	6	1.7	2.2	2.6	3.5	4.3	5.2	6.1	6.9	7.8	8.7	10.8	13	15.2	17.3	43.3
25	3	33.1	41.4	49.7	66	83	99	116	133	149	166	207	249	290	331	828
	4	11.2	14	16.8	22.3	27.9	33.5	39.1	44.7	50	56	70	84	98	112	279
	6	2.2	2.7	3.3	4.3	5.4	6.5	7.6	8.7	9.8	10.8	13.5	16.3	19	21.7	54
30	3	39.8	49.7	60	80	99	119	139	159	179	199	249	298	348	398	....
	4	13.4	16.8	20.1	26.8	33.5	40.2	46.9	54	60	67	84	101	117	134	335
	6	2.6	3.3	3.9	5.2	6.5	7.8	9.1	10.4	11.7	13	16.3	19.5	22.7	26	65
40	3	53	66	80	106	133	160	186	212	239	265	331	398	464	532	....
	4	17.9	22.3	26.8	35.7	44.7	54	63	72	80	89	112	134	156	179	447
	6	3.5	4.3	5.2	6.9	8.7	10.4	12.1	13.9	15.6	17.3	21.7	26	30.3	34.7	87
50	4	22.3	27.9	33.5	44.7	56	67	78	89	101	112	140	168	196	223	559
	6	4.3	5.4	6.5	8.7	10.8	13	15.2	17.3	19.5	21.7	27.1	32.5	37.9	43.3	108
	8	1.5	1.8	2.7	2.9	3.6	4.3	5.1	5.8	6.5	7.2	9	10.8	12.6	14.5	36.1
60	4	26.8	33.5	40.2	54	67	80	94	107	121	134	168	201	235	268	670
	6	5.2	6.5	7.8	10.4	13	16	18.2	20.8	23.4	26	32.5	39	45.5	52	130
	8	1.7	2.2	2.6	3.5	4.3	5.2	6.1	6.9	7.8	8.7	10.8	13	15.2	17.3	43.4
70	4	31.3	39.1	46.9	63	78	94	110	125	141	156	196	235	274	313	782
	6	6.1	7.6	9.1	12.1	15.2	18.4	21.2	24.3	27.3	30.3	37.9	45.5	53	61	152
	8	2	2.5	3	4.1	5.1	6.1	7.1	8.1	9.1	10.1	12.6	15.2	17.7	20.2	51
80	6	6.9	8.7	10.4	13.9	17.3	20.8	24.3	27.7	31.2	34.7	43.3	52	61	69	173
	8	2.3	2.9	3.5	4.6	5.8	6.9	8.1	9.3	10.4	11.6	14.5	17.3	20.2	23.1	58
	10	0.93	1.2	1.4	1.9	2.3	2.8	3.3	3.7	4.2	4.7	5.8	7	8.2	9.3	23.3
90	6	7.8	9.8	11.7	15.6	19.5	23.4	27.3	31.2	35.1	39	48.7	59	68	78	195
	8	2.6	3.3	3.9	5.2	6.5	7.8	9.1	10.4	11.7	13	16.3	19.5	22.8	26	65
	10	1.1	1.3	1.6	2.1	2.6	3.1	3.7	4.2	4.7	5.2	6.6	7.9	9.2	10.5	26.2
100	6	8.7	10.8	13	17.3	21.7	26	30.3	34.7	39	43.3	54	65	76	87	217
	8	2.9	3.6	4.3	5.8	7.2	8.7	10.1	11.6	13	14.5	18.1	21.7	25.3	28.9	72
	10	1.2	1.5	1.8	2.3	2.9	3.5	4.2	4.7	5.2	5.8	7.3	8.7	10.2	11.6	29.1

← LAMINAR FLOW →

TABLE 5 Friction Loss for Viscous Liquids (Continued)

Loss in pounds per square inch per 100 feet of new Schedule 40 steel pipe based on specific gravity of 1.00. (For a liquid having a specific gravity other than 1.00, multiply the value from the table by the specific gravity of that liquid. For commercial installations, it is recommended that 15 per cent be added to the values in this table. No allowance for aging of pipe is included.)

GPM	Pipe Size	VISCOSITY—SAYBOLT SECONDS UNIVERSAL														
		20,000	25,000	30,000	40,000	50,000	60,000	70,000	80,000	90,000	100,000	125,000	150,000	175,000	200,000	500,000
	6	10.4	13	15.6	20.8	26	31.2	36.4	41.6	46.8	52	65	78	91	104	260
120	8	3.5	4.3	5.2	6.9	8.7	10.4	12.1	13.9	15.6	17.3	21.7	26	30.4	34.7	87
	10	1.4	1.8	2.1	2.8	3.5	4.2	4.9	5.6	6.3	7	8.7	10.5	12.2	14	34.9
	6	12.1	15.2	18.2	24.3	30.3	36.4	42.5	48.5	55	61	76	91	106	121	303
140	8	4	5.1	6.1	8.1	10.1	12.1	14.2	16.2	18.2	20.2	25.3	30.4	35.4	40.5	101
	10	1.7	2	2.4	3.3	4.1	4.9	5.7	6.5	7.3	8.1	10.2	12.2	14.3	16.3	40.7
	6	13.9	17.3	20.8	27.7	34.7	41.6	48.5	56	62	69	87	104	121	139	347
160	8	4.6	5.8	6.9	9.3	11.6	13.8	16.2	18.5	20.8	23.1	28.9	34.7	40.5	46.2	116
	10	1.9	2.3	2.8	3.7	4.7	5.6	6.5	7.5	8.4	9.3	11.6	14	16.3	18.6	46.6
	6	15.6	19.5	23.4	31.2	39	46.8	55	62	70	78	98	117	137	156	390
180	8	5.2	6.5	7.8	10.4	13	15.6	18.2	20.8	23.4	26	32.5	39	45.5	52	130
	10	2.1	2.6	3.1	4.2	5.2	6.3	7.3	8.4	9.4	10.5	13.1	15.7	18.3	21	52
	8	5.8	7.2	8.7	11.6	14.5	17.3	20.2	23.1	26	28.9	36.1	43.4	51	58	145
200	10	2.3	2.9	3.5	4.7	5.8	7	8.2	9.3	10.5	11.6	14.6	17.5	20.4	23.3	58
	12	1.2	1.5	1.7	2.3	2.9	3.5	4.1	4.6	5.2	5.8	7.2	8.7	10.1	11.6	28.9
	8	7.2	9	10.8	14.5	18.1	21.7	25.3	28.9	32.5	36.1	45.2	54	63	72	181
250	10	2.9	3.6	4.4	5.8	7.3	8.7	10.2	11.6	13.1	14.6	18.2	21.8	25.5	29.1	73
	12	1.5	1.8	2.2	2.9	3.6	4.3	5.1	5.8	6.5	7.2	9	10.9	12.7	14.5	36.2
	8	8.7	10.8	13	17.3	21.7	26	30.4	34.7	39	43.4	54	65	76	87	217
300	10	3.5	4.4	5.2	7	8.7	10.5	12.2	14	15.7	17.5	21.8	26.2	30.6	34.9	87
	12	1.7	2.2	2.6	3.5	4.3	5.2	6.1	7	7.8	8.7	10.9	13	15.2	17.4	43.4
	8	11.6	14.5	17.3	23	28.9	34.7	40.5	46.2	52	58	72	87	101	116	289
400	10	4.7	5.8	7	9.3	11.6	14	16.3	18.6	21	23.3	29.6	34.9	40.7	46.6	116
	12	2.3	2.9	3.5	4.6	5.8	7	8.1	9.3	10.4	11.6	14.5	17.4	20.3	23.2	58
	8	14.5	18.1	21.7	28.9	36.1	43.4	51	58	65	72	90	108	126	145	361
500	10	5.8	7.3	8.7	11.6	14.6	17.5	20.4	23.3	26.2	29.1	36.4	43.7	51	58	146
	12	2.9	3.6	4.3	5.8	7.2	8.7	10.1	11.6	13	14.5	18.1	21.7	25.3	28.9	72
	8	17.3	21.7	26	34.7	43.4	52	61	69	78	87	108	130	152	173	434
600	10	7	8.7	10.5	14	17.5	21	24.4	27.9	31.4	34.9	43.7	52	61	70	175
	12	3.5	4.3	5.2	7	8.7	10.4	12.2	13.9	15.6	17.4	21.7	26.1	30.4	34.7	87
	8	20.2	25.3	30.3	40.5	51	61	71	81	91	101	126	152	177	202	506
700	10	8.2	10.2	12.2	16.3	20.4	24.4	28.5	32.6	36.7	40.7	51	61	71	82	204
	12	4.1	5.1	6.1	8.1	10.1	12.2	14.2	16.2	18.2	20.3	25.3	30.4	35.5	40.5	101
	8	23.1	28.9	34.7	46.2	58	69	81	93	104	116	145	173	202	231	578
800	10	9.3	11.6	14	18.6	23.3	27.9	32.6	37.3	41.9	46.6	58	70	82	93	233
	12	4.6	5.8	7	9.3	11.6	13.9	16.2	18.5	20.8	23.1	28.9	34.7	40.5	46.3	116
	8	26	32.5	39	52	65	78	91	104	117	130	163	195	228	260	650
900	10	10.5	13.1	15.7	21	26.2	31.4	36.7	41.9	47.1	52	66	79	92	105	262
	12	5.2	6.5	7.8	10.4	13	15.6	18.2	20.8	23.4	26.1	32.6	39.1	45.6	52	130
	8	28.9	36.1	43.4	58	72	87	101	116	130	145	181	217	253	289	723
1000	10	11.6	14.6	17.5	23.3	29.1	34.9	40.7	46.6	52	58	73	87	102	116	291
	12	5.8	7.2	8.7	11.6	14.5	17.4	20.3	23.2	26.1	28.9	36.2	43.4	51	58	145

←——— LAMINAR FLOW ———→

TABLE 6 Representative Equivalent Length in Pipe Diameters (L/D) of Various Valves and Fittings. (From Crane Co. Technical Paper 410, Flow of Fluids through Valves, Fittings and Pipe. Copyright 1969.)

Description of product					Equivalent length in pipe diameters, L/D
Globe valves	Stem perpendicular to run	With no obstruction in flat, bevel, or plug type seat		Fully open	340
		With wing or pin-guided disk		Fully open	450
	Y pattern	(No obstruction in flat, bevel, or plug type seat) With stem 60° from run of pipeline		Fully open	175
		With stem 45° from run of pipeline		Fully open	145
Angle valves		With no obstruction in flat, bevel, or plug type seat		Fully open	145
		With wing or pin-guided disk		Fully open	200
Gate valves	Wedge, disk, double-disk, or plug-disk			Fully open	13
				Three-quarters open	35
				One-half open	160
				One-quarter open	900
	Pulp stock			Fully open	17
				Three-quarters open	50
				One-half open	260
				One-quarter open	1200
Conduit pipeline gate, ball, and plug valves				Fully open	3*
Check valves	Conventional swing		0.5†	Fully open	135
	Clearway swing		0.5†	Fully open	50
	Globe-lift or stop; stem perpendicular to run or Y pattern		2.0†	Fully open	Same as globe
	Angle-lift or stop		2.0†	Fully open	Same as angle
	In-line ball		2.5 vertical and 0.25 horizontal†	Fully open	150
Foot valves with strainer		With poppet lift-type disk	0.3†	Fully open	420
		With leather-hinged disk	0.4†	Fully open	75
Butterfly valves (8-in and larger)				Fully open	40
Cocks	Straight-through	Rectangular plug port area equal to 100% of pipe area		Fully open	18
	Three-way	Rectangular plug port area equal to 80% of pipe area (fully open)		Flow straight through	44
				Flow through branch	140
Fittings	90° standard elbow				30
	45° standard elbow				16
	90° long radius elbow				20
	90° street elbow				50
	45° street elbow				26
	Square-corner elbow				57
	Standard T	With flow through run			20
		With flow through branch			60
	Close-pattern return bend				50

* Exact equivalent length is equal to the length between flange faces or welding ends.

† Minimum calculated pressure drop (lb/in^2) across valve to provide sufficient flow to lift disk fully.

TABLE 7a Viscosity of Common Liquids

Liquid	*Sp Gr at 60 F	VISCOSITY		At F
		SSU	Centistokes	
Freon	1.37 to 1.49 @ 70 F		.27–.32	70
Glycerine (100%)	1.26 @ 68 F	2,950	648	68.6
		813	176	100
Glycol:				
Propylene	1.038 @ 68 F	240.6	52	70
Triethylene	1.125 @ 68 F	185.7	40	70
Diethylene	1.12	149.7	32	70
Ethylene	1.125	88.4	17.8	70
Hydrochloric Acid (31.5%)	1.05 @ 68 F		1.9	68
Mercury	13.6		.118	70
			.11	100
Phenol (Carbolic Acid)	.95 to 1.08	65	11.7	65
Silicate of Soda	40 Baumé	365	79	100
	42 Baumé	637.6	138	100
Sulfuric Acid (100%)	1.83	75.7	14.6	68
FISH AND ANIMAL OILS:				
Bone Oil	.918	220	47.5	130
		65	11.6	212
Cod Oil	.928	150	32.1	100
		95	19.4	130
Lard	.96	287	62.1	100
		160	34.3	130
Lard Oil	.912 to .925	190 to 220	41 to 47.5	100
		112 to 128	23.4 to 27.1	130
Menhadden Oil	.933	140	29.8	100
		90	18.2	130
Neatsfoot Oil	.917	230	49.7	100
		130	27.5	130
Sperm Oil	.883	110	23.0	100
		78	15.2	130
Whale Oil	.925	163 to 184	35 to 39.6	100
		97 to 112	19.9 to 23.4	130
MINERAL OILS:				
Automobile Crankcase Oils (Average Midcontinent Paraffin Base):				
SAE 10	**.880 to .935	165 to 240	35.4 to 51.9	100
		90 to 120	18.2 to 25.3	130
SAE 20	**.880 to .935	240 to 400	51.9 to 86.6	100
		120 to 185	25.3 to 39.9	130
SAE 30	**.880 to .935	400 to 580	86.6 to 125.5	100
		185 to 255	39.9 to 55.1	130

*Unless otherwise noted.
**Depends on origin or percent and type of solvent.

Reprinted from the "Pipe Friction Manual," 3d ed., copyright 1961 by the Hydraulic Institute, Cleveland, Ohio.

TABLE 7b Viscosity of Common Liquids

Liquid	*Sp Gr at 60 F	VISCOSITY		At F
		SSU	Centistokes	
SAE 40	**.880 to .935	580 to 950 255 to 80	125.5 to 205.6 55.1 to 15.6	100 130 210
SAE 50	**.880 to .935	950 to 1,600 80 to 105	205.6 to 352 15.6 to 21.6	100 210
SAE 60	**.880 to .935	1,600 to 2,300 105 to 125	352 to 507 21.6 to 26.2	100 210
SAE 70	**.880 to .935	2,300 to 3,100 125 to 150	507 to 682 26.2 to 31.8	100 210
SAE 10W	**.880 to .935	5,000 to 10,000	1,100 to 2,200	0
SAE 20W	**.880 to .935	10,000 to 40,000	2,200 to 8,800	0
Automobile Transmission Lubricants: SAE 80	**.880 to .935	100,000 max	22,000 max	0
SAE 90	**.880 to .935	800 to 1,500 300 to 500	173.2 to 324.7 64.5 to 108.2	100 130
SAE 140	**.880 to .935	950 to 2,300 120 to 200	205.6 to 507 25.1 to 42.9	130 210
SAE 250	**.880 to .935	Over 2,300 Over 200	Over 507 Over 42.9	130 210
Crude Oils: Texas, Oklahoma	.81 to .916	40 to 783 34.2 to 210	4.28 to 169.5 2.45 to 45.3	60 100
Wyoming, Montana	.86 to .88	74 to 1,215 46 to 320	14.1 to 263 6.16 to 69.3	60 100
California	.78 to .92	40 to 4,840 34 to 700	4.28 to 1,063 2.4 to 151.5	60 100
Pennsylvania	.8 to .85	46 to 216 38 to 86	6.16 to 46.7 3.64 to 17.2	60 100
Diesel Engine Lubricating Oils (Based on Average Midcontinent Paraffin Base): Federal Specification No. 9110	**.880 to .935	165 to 240 90 to 120	35.4 to 51.9 18.2 to 25.3	100 130
Federal Specification No. 9170	**.880 to .935	300 to 410 140 to 180	64.5 to 88.8 29.8 to 38.8	100 130
Federal Specification No. 9250	**.880 to .935	470 to 590 200 to 255	101.8 to 127.8 43.2 to 55.1	100 130
Federal Specification No. 9370	**.880 to .935	800 to 1,100 320 to 430	173.2 to 238.1 69.3 to 93.1	100 130
Federal Specification No. 9500	**.880 to .935	490 to 600 92 to 105	106.1 to 129.9 18.54 to 21.6	130 210

*Unless otherwise noted.
**Depends on origin or percent and type of solvent.

TABLE 7c Viscosity of Common Liquids

Liquid	*Sp Gr at 60 F	VISCOSITY SSU	VISCOSITY Centistokes	At F
Diesel Fuel Oils:				
No. 2 D	**.82 to .95	32.6 to 45.5	2 to 6	100
		39	4 to 3.97	130
No. 3 D	**.82 to .95	**45.5** to **65**	6 to 11.75	100
		39 to 48	3.97 to 6.78	130
No. 4 D	**.82 to .95	**140 max**	29.8 max	100
		70 max	13.1 max	130
No. 5 D	**.82 to .95	**400 max**	86.6 max	122
		165 max	35.2 max	160
Fuel Oils:				
No. 1	**.82 to .95	34 to 40	2.39 to 4.28	70
		32 to **35**	2.69	100
No. 2	**.82 to .95	36 to 50	3.0 to 7.4	70
		33 to **40**	2.11 to 4.28	100
No. 3	**.82 to .95	**35** to **45**	2.69 to .584	100
		32.8 to 39	2.06 to 3.97	130
No. 5A	**.82 to .95	**50** to **125**	7.4 to 26.4	100
		42 to 72	4.91 to 13.73	130
No. 5B	**.82 to .95	125 to	26.4 to	100
		400	86.6	122
		72 to 310	13.63 to 67.1	130
No. 6	**.82 to .95	**450** to **3,000**	97.4 to 660	122
		175 to 780	37.5 to 172	160
Fuel Oil — Navy Specification	**.989 max	**110** to **225**	23 to 48.6	122
		63 to 115	11.08 to 23.9	160
Fuel Oil — Navy II	1.0 max	**1,500 max**	324.7 max	122
		480 max	104 max	160
Gasoline	.68 to .74		.46 to .88	60
			.40 to .71	100
Gasoline (Natural)	76.5 degrees API		.41	68
Gas Oil	28 degrees API	73	13.9	70
		50	7.4	100
Insulating Oil: Transformer, switches and circuit breakers		115 max	24.1 max	70
		65 max	11.75 max	100
Kerosene	.78 to .82	35	2.69	68
		32.6	2	100
Machine Lubricating Oil (Average Pennsylvania Paraffin Base): Federal Specification No. 8	**.880 to .935	112 to 160	23.4 to 34.3	100
		70 to **90**	13.1 to 18.2	130

*Unless otherwise noted.

**Depends on origin or percent and type of solvent.

Reprinted from the "Pipe Friction Manual," 3d ed., copyright 1961 by the Hydraulic Institute, Cleveland, Ohio.

TABLE 7d Viscosity of Common Liquids

Liquid	*Sp Gr at 60 F	VISCOSITY		At F
		SSU	Centistokes	
Federal Specification No. 10	**.880 to .935	160 to 235 90 to 120	34.3 to 50.8 18.2 to 25.3	100 130
Federal Specification No. 20	**.880 to .935	235 to 385 120 to 185	50.8 to 83.4 25.3 to 39.9	100 130
Federal Specification No. 30	**.880 to .935	385 to 550 185 to 255	83.4 to 119 39.9 to 55.1	100 130
Mineral Lard Cutting Oil: Federal Specification Grade 1		140 to 190 86 to 110	29.8 to 41 17.22 to 23	100 130
Federal Specification Grade 2		190 to 220 110 to 125	41 to 47.5 23 to 26.4	100 130
Petrolatum	.825	100 77	20.6 14.8	130 160
Turbine Lubricating Oil: Federal Specification (Penn Base)	.91 Average	400 to 440 185 to 205	86.6 to 95.2 39.9 to 44.3	100 130
VEGETABLE OILS: Castor Oil	.96 @ 68 F	1,200 to 1,500 450 to 600	259.8 to 324.7 97.4 to 129.9	100 130
China Wood Oil	.943	1,425 580	308.5 125.5	69 100
Cocoanut Oil	.925	140 to 148 76 to 80	29.8 to 31.6 14.69 to 15.7	100 130
Corn Oil	.924	135 54	28.7 8.59	130 212
Cotton Seed Oil	.88 to .925	176 100	37.9 20.6	100 130
Linseed Oil, Raw	.925 to .939	143 93	30.5 18.94	100 130
Olive Oil	.912 to .918	200 115	43.2 24.1	100 130
Palm Oil	.924	221 125	47.8 26.4	100 130
Peanut Oil	.920	195 112	42 23.4	100 130
Rape Seed Oil	.919	250 145	54.1 31	100 130
Rosin Oil	.980	1,500 600	324.7 129.9	100 130

*Unless otherwise noted.
**Depends on origin or percent and type of solvent.

Reprinted from the "Pipe Friction Manual," 3d ed., copyright 1961 by the Hydraulic Institute, Cleveland, Ohio.

TABLE 7e Viscosity of Common Liquids

Liquid	*Sp Gr at 60 F	VISCOSITY SSU	VISCOSITY Centistokes	At F
Rosin (Wood)	1.09 Avg.	500 to 20,000 1,000 to 50,000	108.2 to 4,400 216.4 to 11,000	200 190
Sesame Oil	.923	184 110	39.6 23	100 130
Soja Bean Oil	.927 to .98	165 96	35.4 19.64	100 130
Turpentine	.86 to .87	33 32.6	2.11 2.0	60 100
SUGAR, SYRUPS, MOLASSES, ETC.				
Corn Syrups	1.4 to 1.47	5,000 to 500,000 1,500 to 60,000	1,100 to 110,000 324.7 to 13,200	100 130
Glucose	1.35 to 1.44	35,000 to 100,000 4,000 to 11,000	7,700 to 22,000 880 to 2,420	100 150
Honey (Raw)		340	73.6	100
Molasses "A" (First)	1.40 to 1.46	1,300 to 23,000 700 to 8,000	281.1 to 5,070 151.5 to 1,760	100 130
Molasses "B" (Second)	1.43 to 1.48	6,400 to 60,000 3,000 to 15,000	1,410 to 13,200 660 to 3,300	100 130
Molasses "C" (Blackstrap or final)	1.46 to 1.49	17,000 to 250,000 6,000 to 75,000	2,630 to 5,500 1,320 to 16,500	100 130
Sucrose Solutions (Sugar Syrups):				
60 Brix	1.29	230 92	49.7 18.7	70 100
62 Brix	1.30	310 111	67.1 23.2	70 100
64 Brix	1.31	440 148	95.2 31.6	70 100
66 Brix	1.326	650 195	140.7 42.0	70 100
68 Brix	1.338	1,000 275	216.4 59.5	70 100
70 Brix	1.35	1,650 400	364 86.6	70 100
72 Brix	1.36	2,700 640	595 138.6	70 100
74 Brix	1.376	5,500 1,100	1,210 238	70 100
76 Brix	1.39	10,000 2,000	2,200 440	70 100

*Unless otherwise noted.

TABLE 7f Viscosity of Common Liquids

Liquid	*Sp Gr at 60 F	VISCOSITY		At F
		SSU	Centistokes	
TARS:				
Tar-Coke Oven	1.12+	3,000 to 8,000 650 to 1,400	600 to 1,760 140.7 to 308	71 100
Tar-Gas House	1.16 to 1.30	15,000 to 300,000 2,000 to 20,000	3,300 to 66,000 440 to 4,400	70 100
Road Tar:				
Grade RT-2	1.07+	200 to 300 55 to 60	43.2 to 64.9 8.77 to 10.22	122 212
Grade RT-4	1.08+	400 to 700 65 to 75	86.6 to 154 11.63 to 14.28	122 212
Grade RT-6	1.09+	1,000 to 2,000 85 to 125	216.4 to 440 16.83 to 26.2	122 212
Grade RT-8	1.13+	3,000 to 8,000 150 to 225	660 to 1,760 31.8 to 48.3	122 212
Grade RT-10	1.14+	20,000 to 60,000 250 to 400	4,400 to 13,200 53.7 to 86.6	122 212
Grade RT-12	1.15+	114,000 to 456,000 500 to 800	25,000 to 75,000 108.2 to 173.2	122 212
Pine Tar	1.06	2,500 500	559 108.2	100 132
MISCELLANEOUS				
Corn Starch Solutions:				
22 Baumé	1.18	150 130	32.1 27.5	70 100
24 Baumé	1.20	600 440	129.8 95.2	70 100
25 Baumé	1.21	1400 800	303 173.2	70 100
Ink—Printers	1.00 to 1.38	2,500 to 10,000 1,100 to 3,000	550 to 2,200 238.1 to 660	100 130
Tallow	.918 Avg.	56	9.07	212
Milk	1.02 to 1.05		1.13	68
Varnish — Spar	.9	1425 650	313 143	68 100
Water — Fresh	1.0		1.13 .55	60 130

*Unless otherwise noted.

Reprinted from the "Pipe Friction Manual," 3d ed., copyright 1961 by the Hydraulic Institute, Cleveland, Ohio.

TABLE 8a Viscosity Conversion Table

The following table will give an approximate comparison of various viscosity ratings so that if the viscosity is given in terms other than Saybolt Universal, it can be translated quickly by following horizontally to the Saybolt Universal column.

Seconds Saybolt Universal ssu	Kinematic Viscosity Centistokes*	Approx. Seconds Mac Michael	Approx. Gardner Holt Bubble	Seconds Zahn Cup #1	Seconds Zahn Cup #2	Seconds Zahn Cup #3	Seconds Zahn Cup #4	Seconds Zahn Cup #5	Seconds Demmler Cup #1	Seconds Demmler Cup #10	Approx. Seconds Stormer 100 gm Load	Seconds Pratt and Lambert "F"
31	1.00	–	–	–	–	–	–	–	–	–	–	–
35	2.56	–	–	–	–	–	–	–	–	–	–	–
40	4.30	–	–	–	–	–	–	–	1.3	–	–	–
50	7.40	–	–	–	–	–	–	–	2.3	–	2.6	–
60	10.3	–	–	–	–	–	–	–	3.2	–	3.6	–
70	13.1	–	–	–	–	–	–	–	4.1	–	4.6	–
80	15.7	–	–	–	–	–	–	–	4.9	–	5.5	–
90	18.2	–	–	–	–	–	–	–	5.7	–	6.4	–
100	20.6	125	–	38	18	–	–	–	6.5	–	7.3	–
150	32.1	145	–	47	20	–	–	–	10.0	1.0	11.3	–
200	43.2	165	A	54	23	–	–	–	13.5	1.4	15.2	–
250	54.0	198	A	62	26	–	–	–	16.9	1.7	19	–
300	65.0	225	B	73	29	–	–	–	20.4	2.0	23	–
400	87.0	270	C	90	37	–	–	–	27.4	2.7	31	7
500	110.0	320	D	–	46	–	–	–	34.5	3.5	39	8
600	132	370	F	–	55	–	–	–	41	4.1	46	9
700	154	420	G	–	63	22.5	–	–	48	4.8	54	9.5
800	176	470	–	–	72	24.5	–	–	55	5.5	62	10.8
900	198	515	H	–	80	27	18	–	62	6.2	70	11.9
1000	220	570	I	–	88	29	20	13	69	6.9	77	12.4
1500	330	805	M	–	–	40	28	18	103	10.3	116	16.8
2000	440	1070	Q	–	–	51	34	24	137	13.7	154	22
2500	550	1325	T	–	–	63	41	29	172	17.2	193	27.6
3000	660	1690	U	–	–	75	48	33	206	20.6	232	33.7
4000	880	2110	V	–	–	–	63	43	275	27.5	308	45
5000	1100	2635	W	–	–	–	77	50	344	34.4	385	55.8
6000	1320	3145	X	–	–	–	–	65	413	41.3	462	65.5
7000	1540	3670	–	–	–	–	–	75	481	48	540	77
8000	1760	4170	Y	–	–	–	–	86	550	55	618	89
9000	1980	4700	–	–	–	–	–	96	620	62	695	102
10000	2200	5220	Z	–	–	–	–	–	690	69	770	113
15000	3300	7720	Z2	–	–	–	–	–	1030	103	1160	172
20000	4400	10500	Z3	–	–	–	–	–	1370	137	1540	234

* Kinematic Viscosity (in centistokes)

$$= \frac{\text{Absolute Viscosity (in centipoises)}}{\text{Density}}$$

When the Metric System terms centistokes and centipoises are used, the density is numerically equal to the specific gravity. Therefore, the following expression can be used which will be sufficiently accurate for most calculations:

Kinematic Viscosity (in centistokes)

$$= \frac{\text{Absolute Viscosity (in centipoises)}}{\text{Specific Gravity}}$$

When the English System units are used, the density must be used rather than the specific gravity.

For values of 70 centistokes and above, use the following conversion:

$$\text{SSU} = \text{centistokes} \times 4.635$$

Above the range of this table and within the range of the viscosimeter, multiply the particular value by the following approximate factors to convert to SSU:

Viscosimeter	Factor
Mac Michael	1.92 (approx.)
Demmler #1	14.6
Demmler #10	146.
Stormer	13. (approx.)

Reprinted from the "Pipe Friction Manual," 3d ed., copyright 1961 by the Hydraulic Institute, Cleveland, Ohio.

TABLE 8b Viscosity Conversion Table

The following table will give an approximate comparison of various viscosity ratings so that if the viscosity is given in terms other than Saybolt Universal, it can be translated quickly by following horizontally to the Saybolt Universal column.

Seconds Saybolt Universal ssu	Kinematic Viscosity Centistokes *	Seconds Saybolt Furol ssf	Seconds Redwood 1 (Standard)	Seconds Redwood 2 (Admiralty)	Degrees Engler	Degrees Barbey	Seconds Parlin Cup #7	Seconds Parlin Cup #10	Seconds Parlin Cup #15	Seconds Parlin Cup #20	Seconds Ford Cup #3	Seconds Ford Cup #4
31	1.00	-	29	-	1.00	6200	-	-	-	-	-	-
35	2.56	-	32.1	-	1.16	2420	-	-	-	-	-	-
40	4.30	-	36.2	5.10	1.31	1440	-	-	-	-	-	-
50	7.40	-	44.3	5.83	1.58	838	-	-	-	-	-	-
60	10.3	-	52.3	6.77	1.88	618	-	-	-	-	-	-
70	13.1	12.95	60.9	7.60	2.17	483	-	-	-	-	-	-
80	15.7	13.70	69.2	8.44	2.45	404	-	-	-	-	-	-
90	18.2	14.44	77.6	9.30	2.73	348	-	-	-	-	-	-
100	20.6	15.24	85.6	10.12	3.02	307	-	-	-	-	-	-
150	32.1	19.30	128	14.48	4.48	195	-	-	-	-	-	-
200	43.2	23.5	170	18.90	5.92	144	40	-	-	-	-	-
250	54.0	28.0	212	23.45	7.35	114	46	-	-	-	-	-
300	65.0	32.5	254	28.0	8.79	95	52.5	15	6.0	3.0	30	20
400	87.60	41.9	338	37.1	11.70	70.8	66	21	7.2	3.2	42	28
500	110.0	51.6	423	46.2	14.60	56.4	79	25	7.8	3.4	50	34
600	132	61.4	508	55.4	17.50	47.0	92	30	8.5	3.6	58	40
700	154	71.1	592	64.6	20.45	40.3	106	35	9.0	3.9	67	45
800	176	81.0	677	73.8	23.35	35.2	120	39	9.8	4.1	74	50
900	198	91.0	762	83.0	26.30	31.3	135	41	10.7	4.3	82	57
1000	220	100.7	896	92.1	29.20	28.2	149	43	11.5	4.5	90	62
1500	330	150	1270	138.2	43.80	18.7	-	65	15.2	6.3	132	90
2000	440	200	1690	184.2	58.40	14.1	-	86	19.5	7.5	172	118
2500	550	250	2120	230	73.0	11.3	-	108	24	9	218	147
3000	660	300	2540	276	87.60	9.4	-	129	28.5	11	258	172
4000	880	400	3380	368	117.0	7.05	-	172	37	14	337	230
5000	1100	500	4230	461	146	5.64	-	215	47	18	425	290
6000	1320	600	5080	553	175	4.70	-	258	57	22	520	350
7000	1540	700	5920	645	204.5	4.03	-	300	67	25	600	410
8000	1760	800	6770	737	233.5	3.52	-	344	76	29	680	465
9000	1980	900	7620	829	263	3.13	-	387	86	32	780	520
10000	2200	1000	8460	921	292	2.82	-	430	96	35	850	575
15000	3300	1500	13700	-	438	2.50	-	650	147	53	1280	860
20000	4400	2000	18400	-	584	1.40	-	860	203	70	1715	1150

* Kinematic Viscosity (in centistokes)

$$= \frac{\text{Absolute Viscosity (in centipoises)}}{\text{Density}}$$

When the Metric System terms centistokes and centipoises are used, the density is numerically equal to the specific gravity. Therefore, the following expression can be used which will be sufficiently accurate for most calculations:

Kinematic Viscosity (in centistokes)

$$= \frac{\text{Absolute Viscosity (in centipoises)}}{\text{Specific Gravity}}$$

When the English System units are used, the density must be used rather than the specific gravity.

For values of 70 centistokes and above, use the following conversion:

$$\text{SSU} = \text{centistokes} \times 4.635$$

Above the range of this table and within the range of the viscosimeter, multiply the particular value by the following approximate factors to convert to SSU:

Viscosimeter	Factor	Viscosimeter	Factor
Saybolt Furol	10.	Parlin cup #15	98.2
Redwood Standard	1.095	Parlin cup #20	187.0
Redwood Admiralty	10.87	Ford cup # 4	17.4
Engler - Degrees	34.5		

TABLE 9 Properties of Water at Various Temperatures from 40 to 705.4 F

Temp. F	Temp. C	Specific Volume Cu Ft/Lb	SPECIFIC GRAVITY 39.2 F Reference	60 F Reference	70 F Reference	Wt in Lb/Cu Ft	Vapor Pressure Psi Abs
40	4.4	.01602	1.000	1.001	1.002	62.42	0.1217
50	10.0	.01603	.999	1.001	1.002	62.38	0.1781
60	15.6	.01604	.999	1.000	1.001	62.34	0.2563
70	21.1	.01606	.998	.999	1.000	62.27	0.3631
80	26.7	.01608	.996	.998	.999	62.19	0.5069
90	32.2	.01610	.995	.996	.997	62.11	0.6982
100	37.8	.01613	.993	.994	.995	62.00	0.9492
120	48.9	.01620	.989	.990	.991	61.73	1.692
140	60.0	.01629	.983	.985	.986	61.39	2.889
160	71.1	.01639	.977	.979	.979	61.01	4.741
180	82.2	.01651	.970	.972	.973	60.57	7.510
200	93.3	.01663	.963	.964	.966	60.13	11.526
212	100.0	.01672	.958	.959	.960	59.81	14.696
220	104.4	.01677	.955	.956	.957	59.63	17.186
240	115.6	.01692	.947	.948	.949	59.10	24.97
260	126.7	.01709	.938	.939	.940	58.51	35.43
280	137.8	.01726	.928	.929	.930	58.00	49.20
300	148.9	.01745	.918	.919	.920	57.31	67.01
320	160.0	.01765	.908	.909	.910	56.66	89.66
340	171.1	.01787	.896	.898	.899	55.96	118.01
360	182.2	.01811	.885	.886	.887	55.22	153.04
380	193.3	.01836	.873	.874	.875	54.47	195.77
400	204.4	.01864	.859	.860	.862	53.65	247.31
420	215.6	.01894	.846	.847	.848	52.80	308.83
440	226.7	.01926	.832	.833	.834	51.92	381.59
460	237.8	.0196	.817	.818	.819	51.02	466.9
480	248.9	.0200	.801	.802	.803	50.00	566.1
500	260.0	.0204	.785	.786	.787	49.02	680.8
520	271.1	.0209	.765	.766	.767	47.85	812.4
540	282.2	.0215	.746	.747	.748	46.51	962.5
560	293.3	.0221	.726	.727	.728	45.3	1133.1
580	304.4	.0228	.703	.704	.704	43.9	1325.8
600	315.6	.0236	.678	.679	.680	42.3	1542.9
620	326.7	.0247	.649	.650	.650	40.5	1786.6
640	337.8	.0260	.617	.618	.618	38.5	2059.7
660	348.9	.0278	.577	.577	.578	36.0	2365.4
680	360.0	.0305	.525	.526	.527	32.8	2708.1
700	371.1	.0369	.434	.435	.435	27.1	3093.7
705.4	374.1	.0503	.319	.319	.320	19.9	3206.2

Computed from Keenan & Keyes' Steam Table.

TABLE 10 Theoretical Discharge of Nozzles, in U.S. Gallons per Minute

Head		Velocity of discharge, feet per second	DIAMETER OF NOZZLE IN INCHES																									
Lbs.	Feet		1/16	1/8	3/16	1/4	3/8	1/2	5/8	3/4	7/8	1	1 1/8	1 1/4	1 3/8	1 1/2	1 3/4	2	2 1/4	2 1/2	2 3/4	3	3 1/2	4	4 1/2	5	5 1/2	6
10	23.1	38.58	0.37	1.48	3.30	5.90	13.2	23.6	36.8	53.2	72.2	94.4	119	148	178	212	289	378	478	590	715	850	1160	1512	1912	2360	2880	3400
15	34.7	47.25	0.45	1.81	4.02	7.23	16.2	28.7	45.0	65.1	88.4	116	146	181	218	260	354	463	586	723	880	1040	1420	1852	2344	2800	3520	4160
20	46.2	54.55	0.52	2.09	4.66	8.35	18.7	33.4	52.0	75.3	102	134	169	209	252	300	409	534	676	835	1018	1200	1640	2136	2700	3340	4072	4800
25	57.8	60.99	0.58	2.33	5.23	9.33	20.9	37.2	58.2	84.1	114	149	189	233	282	336	457	597	756	933	1138	1350	1828	2390	3024	3730	4582	5400
30	69.3	66.82	0.64	2.56	5.71	10.2	22.8	40.9	63.7	92.2	125	164	207	256	309	368	501	654	828	1022	1240	1480	2000	2616	3312	4038	4960	5920
35	80.9	72.16	0.69	2.76	6.16	11.0	24.7	44.2	68.8	99.6	135	177	223	276	334	397	541	707	895	1104	1340	1590	2160	2828	3580	4416	5360	6360
40	92.4	77.14	0.74	2.95	6.60	11.8	26.4	47.2	73.6	106	144	189	239	295	357	425	578	755	956	1180	1430	1700	2320	3020	3824	4720	5720	6800
45	104.0	81.83	0.78	3.13	6.99	12.5	28.0	50.2	78.1	113	153	200	253	313	378	450	613	801	1014	1252	1520	1800	2450	3200	4056	5000	6080	7200
50	115.5	86.26	0.82	3.30	7.37	13.2	29.5	52.8	82.3	119	161	211	267	330	399	475	646	845	1069	1320	1600	1900	2580	3380	4276	5280	6400	7600
55	127.1	90.46	0.86	3.46	7.73	13.8	30.9	55.4	86.3	125	169	221	280	346	418	498	678	886	1122	1385	1680	2000	2720	3544	4488	5540	6720	8000
60	138.6	94.49	0.90	3.62	8.08	14.5	32.3	57.8	90.1	130	177	231	293	362	437	520	708	925	1171	1446	1755	2080	2840	3700	4684	5784	7020	8320
65	150.2	98.35	0.94	3.77	8.40	15.1	33.6	60.2	93.8	136	184	241	305	377	455	542	737	963	1219	1506	1830	2165	2950	3850	4876	6024	7320	8660
70	161.7	102.06	0.97	3.91	8.73	15.6	34.9	62.5	97.4	141	191	250	316	391	472	562	765	999	1265	1561	1895	2250	3060	4000	5060	6244	7580	9000
75	173.3	105.65	1.01	4.04	9.03	16.2	36.1	64.6	101	146	198	259	327	404	488	582	792	1034	1309	1616	1960	2330	3170	4136	5236	6464	7840	9320
80	184.8	109.11	1.04	4.18	9.33	16.7	37.8	66.6	104	150	204	267	338	418	504	601	818	1068	1352	1669	2030	2404	3280	4272	5400	6676	8120	9616
85	196.4	112.46	1.07	4.31	9.62	17.2	38.5	68.8	107	155	210	275	348	431	520	620	843	1101	1394	1720	2080	2480	3380	4400	5576	6880	8320	9920
90	207.9	115.72	1.10	4.43	9.89	17.7	39.6	70.8	110	160	217	283	358	443	535	637	867	1133	1434	1770	2150	2540	3470	4532	5736	7080	8600	10160
95	219.5	118.89	1.13	4.55	10.2	18.2	40.7	72.8	113	164	223	291	368	455	550	655	891	1164	1474	1820	2200	2620	3560	4656	5896	7280	8800	10480
100	231.1	121.98	1.16	4.67	10.4	18.7	41.7	74.6	116	168	228	299	378	467	564	672	914	1194	1512	1866	2260	2700	3650	4776	6048	7464	9040	10800
105	242.6	125.00	1.19	4.78	10.7	19.1	42.8	76.5	119	172	234	306	387	478	578	688	937	1224	1549	1912	2320	2760	3750	4896	6200	7648	9280	11040
110	254.2	127.94	1.22	4.90	10.9	19.6	43.8	78.3	122	177	239	313	396	490	591	705	959	1253	1586	1957	2380	2820	3840	5012	6344	7828	9520	11280
115	265.7	130.82	1.25	5.01	11.2	20.0	44.8	80.1	125	181	245	320	405	501	605	720	980	1281	1621	2002	2430	2880	3920	5124	6484	8008	9720	11520
120	277.3	133.63	1.27	5.12	11.4	20.4	45.7	81.8	127	184	250	327	414	512	618	736	1001	1308	1656	2044	2480	2950	4004	5232	6624	8176	9920	11800
125	288.8	136.38	1.30	5.22	11.7	20.9	46.7	83.5	130	188	255	334	422	522	630	751	1022	1335	1691	2086	2540	3000	4100	5340	6764	8344	10160	12000
130	300.4	139.08	1.33	5.32	11.9	21.3	47.6	85.1	133	192	260	341	431	532	643	766	1042	1362	1724	2128	2580	3070	4160	5448	6896	8512	10320	12280

The actual quantity discharged by a nozzle will be less than above table. A well tapered smooth nozzle may be assumed to give above 94 per cent of the values in the tables.

TABLE 11 Cast Iron Pipe Dimensions

Nominal Diameter	CLASS A 100 Foot Head 43 Pounds Pressure			CLASS B 200 Foot Head 86 Pounds Pressure			CLASS C 300 Foot Head 130 Pounds Pressure			CLASS D 400 Foot Head 173 Pounds Pressure		
	Outisde Dia-meter	Wall Thick-ness	Inside Dia-meter	Outside Dia-meter	Wall Thick-ness	Inside Dia-meter	Outside Dia-meter	Wall Thick-ness	Inside Dia-meter	Outside Dia-meter	Wall Thick-ness	Inside Dia-meter
Inches	Inches	Inches	Inches	Inches	Inches	Inches	Inches	Inches	Inches	Inches	Inches	Inches
3	3.80	0.39	3.02	3.96	0.42	3.12	3.96	0.45	3.06	3.96	0.48	3.00
4	4.80	0.42	3.96	5.00	0.45	4.10	5.00	0.48	4.04	5.00	0.52	3.96
6	6.90	0.44	6.02	7.10	0.48	6.14	7.10	0.51	6.08	7.10	0.55	6.00
8	9.05	0.46	8.13	9.05	0.51	8.03	9.30	0.56	8.18	9.30	0.60	8.10
10	11.10	0.50	10.10	11.10	0.57	9.96	11.40	0.62	10.16	11.40	0.68	10.04
12	13.20	0.54	12.12	13.20	0.62	11.96	13.50	0.68	12.14	13.50	0.75	12.00
14	15.30	0.57	14.16	15.30	0.66	13.98	15.65	0.74	14.17	15.65	0.82	14.01
16	17.40	0.60	16.20	17.40	0.70	16.00	17.80	0.80	16.20	17.80	0.89	16.02
18	19.50	0.64	18.22	19.50	0.75	18.00	19.92	0.87	18.18	19.92	0.96	18.00
20	21.60	0.67	20.26	21.60	0.80	20.00	22.06	0.92	20.22	22.06	1.03	20.00
24	25.80	0.76	24.28	25.80	0.89	24.02	26.32	1.04	24.22	26.32	1.16	24.00
30	31.74	0.88	29.98	32.00	1.03	29.94	32.40	1.20	30.00	32.74	1.37	30.00
36	37.96	0.99	35.98	38.30	1.15	36.00	38.70	1.36	39.98	39.16	1.58	36.00
42	44.20	1.10	42.00	44.50	1.28	41.94	45.10	1.54	42.02	45.58	1.78	42.02
48	50.50	1.26	47.98	50.80	1.42	47.96	51.40	1.71	47.98	51.98	1.96	48.06
54	56.66	1.35	53.96	57.10	1.55	54.00	57.80	1.90	54.00	58.40	2.23	53.94
60	62.80	1.39	60.02	63.40	1.67	60.06	64.20	2.00	60.20	64.82	2.38	60.06
72	75.34	1.62	72.10	76.00	1.95	72.10	76.88	2.39	72.10			
84	87.54	1.72	84.10	88.54	2.22	84.10						

TABLE 11 Cast Iron Pipe Dimensions (Continued)

Nominal Diameter	CLASS E 500 Foot Head 217 Pounds Pressure			CLASS F 600 Foot Head 260 Pounds Pressure			CLASS G 700 Foot Head 304 Pounds Pressure			CLASS H 800 Foot Head 347 Pounds Pressure		
	Outisde Dia-meter	Wall Thick-ness	Inside Dia-meter	Outside Dia-meter	Wall Thick-ness	Inside Dia-meter	Outside Dia-meter	Wall Thick-ness	Inside Dia-meter	Outside Dia-meter	Wall Thick-ness	Inside Dia-meter
Inches	Inches	Inches	Inches	Inches	Inches	Inches	Inches	Inches	Inches	Inches	Inches	Inches
6	7.22	0.58	6.06	7.22	0.61	6.00	7.38	0.65	6.08	7.38	0.69	6.00
8	9.42	0.66	8.10	9.42	0.71	8.00	9.60	0.75	8.10	9.60	0.80	8.00
10	11.60	0.74	10.12	11.60	0.80	10.00	11.84	0.86	10.12	11.84	0.92	10.00
12	13.78	0.82	12.14	13.78	0.89	12.00	14.08	0.97	12.14	14.08	1.04	12.00
14	15.98	0.90	14.18	15.98	0.99	14.00	16.32	1.07	14.18	16.32	1.16	14.00
16	18.16	0.98	16.20	18.16	1.08	16.00	18.54	1.18	16.18	18.54	1.27	16.00
18	20.34	1.07	18.20	20.34	1.17	18.00	20.78	1.28	18.22	20.78	1.39	18.00
20	22.54	1.15	20.24	22.54	1.27	20.00	23.02	1.39	20.24	23.02	1.51	20.00
24	26.90	1.31	24.28	26.90	1.45	24.00	27.76	1.75	24.26	27.76	1.88	24.00
30	33.10	1.55	30.00	33.46	1.73	30.00						
36	39.60	1.80	36.00	40.04	2.02	36.00						

Data taken from "Handbook of Cast Iron Pipe," Cast Iron Pipe Research Association, Chicago, Illinois, 1927. For a more complete description of cast iron pipe specifications and dimensions, refer to the "Handbook of Cast Iron Pipe," Latest Edition.

The A.W.W.A. Standard Specifications, Section 3 states: "For pipes whose standard thickness is less than 1 inch, the thickness of metal in the body of the pipe shall not be more than 0.08 of an inch less than the standard thickness, and for pipes whose standard thickness is 1 inch or more, the variation shall not exceed 0.10 of an inch, except that for spaces not exceeding 8 inches in length in any direction, variations from the standard thickness of 0.02 of an inch in excess of the allowance above given shall be permitted."

TABLE 12a Dimensions of Welded and Seamless Steel Pipe

Nominal Diameter	Schedule	Outside Diameter	Wall Thickness	Internal Diameter	Internal Area	Internal Diameter	Internal Area	ϵ/D $\epsilon = 0.00015$ ft
Inches		Inches	Inches	Inches	Square Inches	Feet	Square Feet	
⅛	40 (S)	0.405	0.068	0.269	0.0568	0.0224	0.00039	0.00669
	80 (X)		0.095	0.215	0.0363	0.0179	0.00025	0.00837
¼	40 (S)	0.540	0.088	0.364	0.1041	0.0303	0.00072	0.00495
	80 (X)		0.119	0.302	0.0716	0.0252	0.00050	0.00596
⅜	40 (S)	0.675	0.091	0.493	0.1909	0.0411	0.00133	0.00365
	80 (X)		0.126	0.423	0.1405	0.0353	0.00098	0.00426
½	40 (S)	0.840	0.109	0.622	0.3039	0.0518	0.00211	0.00289
	80 (X)		0.147	0.546	0.2341	0.0455	0.00163	0.00330
	160		0.187	0.466	0.1706	0.0388	0.00118	0.00386
	(XX)		0.294	0.252	0.0499	0.0210	0.00035	0.00714
¾	40 (S)	1.050	0.113	0.824	0.5333	0.0687	0.00370	0.00218
	80 (X)		0.154	0.742	0.4324	0.0618	0.00300	0.00243
	160		0.219	0.612	0.2942	0.0510	0.00204	0.00293
	(XX)		0.308	0.434	0.1479	0.0362	0.00103	0.00415
1	40 (S)	1.315	0.133	1.049	0.8643	0.0874	0.00600	0.00172
	80 (X)		0.179	0.957	0.7193	0.0798	0.00500	0.00188
	160		0.250	0.815	0.5217	0.0679	0.00362	0.00221
	(XX)		0.358	0.599	0.2818	0.0499	0.00196	0.00301
1¼	40 (S)	1.660	0.140	1.380	1.496	0.1150	0.01039	0.00130
	80 (X)		0.191	1.278	1.283	0.1065	0.00890	0.00141
	160		0.250	1.160	1.057	0.0967	0.00734	0.00155
	(XX)		0.382	0.896	0.6305	0.0747	0.00438	0.00201
1½	40 (S)	1.900	0.145	1.610	2.036	0.1342	0.01414	0.00112
	80 (X)		0.200	1.500	1.767	0.1250	0.01227	0.00120
	160		0.281	1.338	1.406	0.1115	0.00976	0.00135
	(XX)		0.400	1.100	0.9503	0.0917	0.00660	0.00164
2	40 (S)	2.375	0.154	2.067	3.356	0.1723	0.02330	0.00087
	80 (X)		0.218	1.939	2.953	0.1616	0.02051	0.00093
	160		0.344	1.687	2.235	0.1406	0.01552	0.00107
	(XX)		0.436	1.503	1.774	0.1253	0.01232	0.00120
2½	40 (S)	2.875	0.203	2.469	4.788	0.2058	0.03325	0.000729
	80 (X)		0.276	2.323	4.238	0.1936	0.02943	0.000775
	160		0.375	2.125	3.547	0.1771	0.02463	0.000847
	(XX)		0.552	1.771	2.464	0.1476	0.01711	0.00102
3	40 (S)	3.500	0.216	3.068	7.393	0.2557	0.05134	0.000587
	80 (X)		0.300	2.900	6.605	0.2417	0.04587	0.000621
	160		0.438	2.624	5.408	0.2187	0.03755	0.000685
	(XX)		0.600	2.300	4.155	0.1917	0.02885	0.000783
3½	40 (S)	4.000	0.226	3.548	9.887	0.2957	0.06866	0.000507
	80 (X)		0.318	3.364	8.888	0.2803	0.06172	0.000535
4	40 (S)	4.500	0.237	4.026	12.73	0.3355	0.08841	0.000447
	80 (X)		0.337	3.826	11.50	0.3188	0.07984	0.000470
	120		0.438	3.624	10.32	0.3020	0.07163	0.000496
	160		0.531	3.438	9.283	0.2865	0.06447	0.000524
	(XX)		0.674	3.152	7.803	0.2627	0.05419	0.000571

S = Wall thickness formerly designated, "standard weight."
X = Wall thickness formerly designated, "extra strong."
XX = Wall thickness formerly designated, "double extra strong."

Note: Extracted from American Standard Wrought Steel and Wrought Iron Pipe. (ASA B36.10-1959), with the permission of the publisher, The American Society of Mechanical Engineers, 345 East 47th Street, New York, N. Y.
The decimal wall thicknesses for welded wrought iron pipe may vary slightly from those shown in the Tables. For a complete listing of all schedules, refer to ASA B36.10-1959.

TABLE 12b Dimensions of Welded and Seamless Steel Pipe (Continued)

Nominal Diameter	Schedule	Outside Diameter	Wall Thickness	Internal Diameter	Internal Area	Internal Diameter	Internal Area	ϵ/D $\epsilon=0.00015$ ft
Inches		Inches	Inches	Inches	Square Inches	Feet	Square Feet	
5	40 (S)	5.563	0.258	5.047	20.01	0.4206	0.1389	0.000357
	80 (X)		0.375	4.813	18.19	0.4011	0.1263	0.000374
	120		0.500	4.563	16.35	0.3803	0.1136	0.000394
	160		0.625	4.313	14.61	0.3594	0.1015	0.000417
	(XX)		0.750	4.063	12.97	0.3386	0.09004	0.000443
6	40 (S)	6.625	0.280	6.065	28.89	0.5054	0.2006	0.000293
	80 (X)		0.432	5.761	26.07	0.4801	0.1810	0.000312
	120		0.562	5.501	23.77	0.4584	0.1650	0.000327
	160		0.719	5.187	21.13	0.4323	0.1467	0.000347
	(XX)		0.864	4.897	18.83	0.4081	0.1308	0.000368
8	20	8.625	0.250	8.125	51.85	0.6771	0.3601	0.000222
	30		0.277	8.071	51.16	0.6726	0.3553	0.000223
	40 (S)		0.322	7.981	50.03	0.6651	0.3474	0.000226
	60		0.406	7.813	47.94	0.6511	0.3329	0.000230
	80 (X)		0.500	7.625	45.66	0.6354	0.3171	0.000236
	100		0.594	7.437	43.44	0.6198	0.3017	0.000242
	120		0.719	7.187	40.57	0.5989	0.2817	0.000250
	140		0.812	7.001	38.50	0.5834	0.2673	0.000257
	(XX)		0.875	6.875	37.12	0.5729	0.2578	0.000262
	160		0.906	6.813	36.46	0.5678	0.2532	0.000264
10	20	10.75	0.250	10.250	82.52	0.85417	0.5730	0.000176
	30		0.307	10.136	80.69	0.84467	0.5604	0.000178
	40 (S)		0.365	10.020	78.85	0.83500	0.5476	0.000180
	60 (X)		0.500	9.750	74.66	0.8125	0.5185	0.000185
	80		0.594	9.562	71.81	0.7968	0.4987	0.000188
	100		0.719	9.312	68.11	0.7760	0.4730	0.000193
	120		0.844	9.062	64.50	0.7552	0.4479	0.000199
	140 (XX)		1.000	8.750	60.13	0.7292	0.4176	0.000206
	160		1.125	8.500	56.75	0.7083	0.3941	0.000212
12	20	12.75	0.250	12.250	117.86	1.0208	0.8185	0.000147
	30		0.330	12.090	114.80	1.0075	0.7972	0.000149
	(S)		0.375	12.000	113.10	1.0000	0.7854	0.000150
	40		0.406	11.938	111.93	0.99483	0.7773	0.000151
	(X)		0.500	11.750	108.43	0.97917	0.7530	0.000153
	60		0.562	11.626	106.16	0.96883	0.7372	0.000155
	80		0.688	11.374	101.61	0.94783	0.7056	0.000158
	100		0.844	11.062	96.11	0.92183	0.6674	0.000163
	120 (XX)		1.000	10.750	90.76	0.89583	0.6303	0.000167
	140		1.125	10.500	86.59	0.87500	0.6013	0.000171
	160		1.312	10.126	80.53	0.84383	0.5592	0.000178
14 OD	10	14.00	0.250	13.500	143.14	1.1250	0.9940	0.000133
	20		0.312	13.376	140.52	1.1147	0.9758	0.000135
	30 (S)		0.375	13.250	137.89	1.1042	0.9575	0.000136
	40		0.438	13.124	135.28	1.0937	0.9394	0.000137
	(X)		0.500	13.000	132.67	1.0833	0.9213	0.000138
	60		0.594	12.812	128.92	1.0677	0.8953	0.000140
	80		0.750	12.500	122.72	1.0417	0.8522	0.000144
	100		0.938	12.124	115.45	1.0104	0.8017	0.000148
	120		1.094	11.812	109.58	0.98433	0.7610	0.000152
	140		1.250	11.500	103.87	0.95833	0.7213	0.000157
	160		1.406	11.188	98.31	0.93233	0.6827	0.000161

S = Wall thickness formerly designated, "standard weight."
X = Wall thickness formerly designated, "extra strong."
XX = Wall thickness formerly designated, "double extra strong."

Note: Extracted from American Standard Wrought Steel and Wrought Iron Pipe, (ASA B36.10-1959), with the permission of the publisher, The American Society of Mechanical Engineers, 345 East 47th Street, New York, N. Y.

The decimal wall thicknesses for welded wrought iron pipe may vary slightly from those shown in the Tables. For a complete listing of all schedules, refer to ASA B36.10-1959.

TABLE 12c Dimensions of Welded and Seamless Steel Pipe (Continued)

Nominal Diameter	Schedule	Outside Diameter	Wall Thickness	Internal Diameter	Internal Area	Internal Diameter	Internal Area	ε/D ε = 0.00015 ft
Inches		Inches	Inches	Inches	Square Inches	Feet	Square Feet	
16 OD	10	16.00	0.250	15.500	188.69	1.2917	1.3104	0.000116
	20		0.312	15.376	185.69	1.2813	1.2895	0.000117
	30 (S)		0.375	15.250	182.65	1.2708	1.2684	0.000118
	40 (X)		0.500	15.000	176.72	1.2500	1.2272	0.000120
	60		0.656	14.688	169.44	1.2240	1.1767	0.000121
	80		0.844	14.312	160.88	1.1927	1.1172	0.000126
	100		1.031	13.938	152.58	1.1615	1.0596	0.000129
	120		1.219	13.562	144.46	1.1302	1.0032	0.000133
	140		1.438	13.124	135.28	1.0937	0.9394	0.000137
	160		1.594	12.812	128.92	1.0677	0.8953	0.000140
18 OD	10	18.00	0.250	17.500	240.53	1.4583	1.6703	0.000103
	20		0.312	17.376	237.13	1.4480	1.6467	0.000104
	(S)		0.375	17.250	233.71	1.4375	1.6230	0.000104
	30		0.438	17.124	230.00	1.4270	1.5993	0.000105
	(X)		0.500	17.000	226.98	1,4167	1.5762	0.000106
	40		0.562	16.876	223.68	1.4063	1.5533	0.000107
	60		0.750	16.500	213.83	1.3750	1.4849	0.000109
	80		0.938	16.124	204.19	1.3437	1.4180	0.000112
	100		1.156	15.688	193.30	1.3073	1.3423	0.000115
	120		1.375	15.250	182.65	1.2708	1.2684	0.000118
	140		1.562	14.876	173.81	1.2397	1.2070	0.000121
	160		1.781	14.438	163.72	1.2032	1.1370	0.000124
20 OD	10	20.00	0.250	19.500	298.65	1.6250	2.0739	0.0000923
	20 (S)		0.375	19.250	291.04	1.6042	2.0211	0.0000935
	30 (X)		0.500	19.000	283.53	1.5833	1.9689	0.0000947
	40		0.594	18.812	277.95	1.5677	1.9302	0.0000957
	60		0.812	18.376	265.21	1.5313	1.8417	0.0000980
	80		1.031	17.938	252.72	1.4948	1.7550	0.000100
	100		1.281	17.438	238.83	1.4532	1.6585	0.000103
	120		1.500	17.000	226.98	1.4167	1.5762	0.000106
	140		1.750	16.500	213.83	1.3750	1.4849	0.000109
	160		1.969	16.062	202.62	1.3385	1.4071	0.000112
24 OD	10	24.00	0.250	23.500	433.74	1.9583	3.0121	0.0000766
	20 (S)		0.375	23.250	424.56	1.9375	2.9483	0.0000774
	(X)		0.500	23.000	415.48	1.9167	2.8852	0.0000783
	30		0.562	22.876	411.01	1.9063	2.8542	0.0000787
	40		0.688	22.624	402.00	1.8853	2.7917	0.0000796
	60		0.969	22.062	382.28	1.8385	2.6547	0.0000814
	80		1.219	21.562	365.15	1.7802	2.5358	0.0000835
	100		1.531	20.938	344.32	1.7448	2.3911	0.0000857
	120		1.812	20.376	326.92	1.6980	2.2645	0.0000878
	140		2.062	19.876	310.28	1.6563	2.1547	0.0000906
	160		2.344	19.312	292.92	1.6093	2.0342	0.0000929
30 OD	10	30.00	0.312	29.376	677.76	2.4480	4.7067	0.0000613
	(S)		0.375	29.250	671.62	2.4375	4.6640	0.0000615
	20 (X)		0.500	29.000	660.52	2.4167	4.5869	0.0000621
	30		0.625	28.750	649.18	2,3958	4.5082	0.0000626

S = Wall thickness formerly designated, "standard weight."
X = Wall thickness formerly designated, "extra strong."
XX = Wall thickness formerly designated, "double extra strong."

Note: Extracted from American Standard Wrought Steel and Wrought Iron Pipe, (ASA B36.10-1959), with the permission of the publisher, The American Society of Mechanical Engineers, 345 East 47th Street, New York, N. Y.
The decimal wall thicknesses for welded wrought iron pipe may vary slightly from those shown in the Tables. For a complete listing of all schedules, refer to ASA B36.10-1959.

TABLE 13 Capacity Equivalents (For conversion charts, see Fig. 2)

Various units	U.S. gpm
1 second-foot or cubic foot per second (cfs)	448.8
1,000,000 U.S. gallons per day (mgd)	694.4
1 imperial gallon per minute	1.201
1,000,000 imperial gallons per day	834.0
1 barrel (42 U.S. gal) per day (bbl/day)	0.0292
1 barrel per hour (bbl/hr)	0.700
1 acre-foot per day	226.3
1,000 pounds per hour (lb/hr)	2.00[1]
1 cubic meter per hour (m^3/hr)	4.403
1 liter per second (l/s)	15.851
1 metric ton per hour	4.403[1]
1,000,000 liters per day = 1,000 cubic meters per day	183.5

[1] These equivalents are based on a specific gravity of 1 for water at 62°F for English units and a specific gravity of 1 for water at 15°C for metric units. They can be used with little error for cold water of any temperature between 32°F and 80°F. For specific gravity of water at various temperatures, see Table 9.

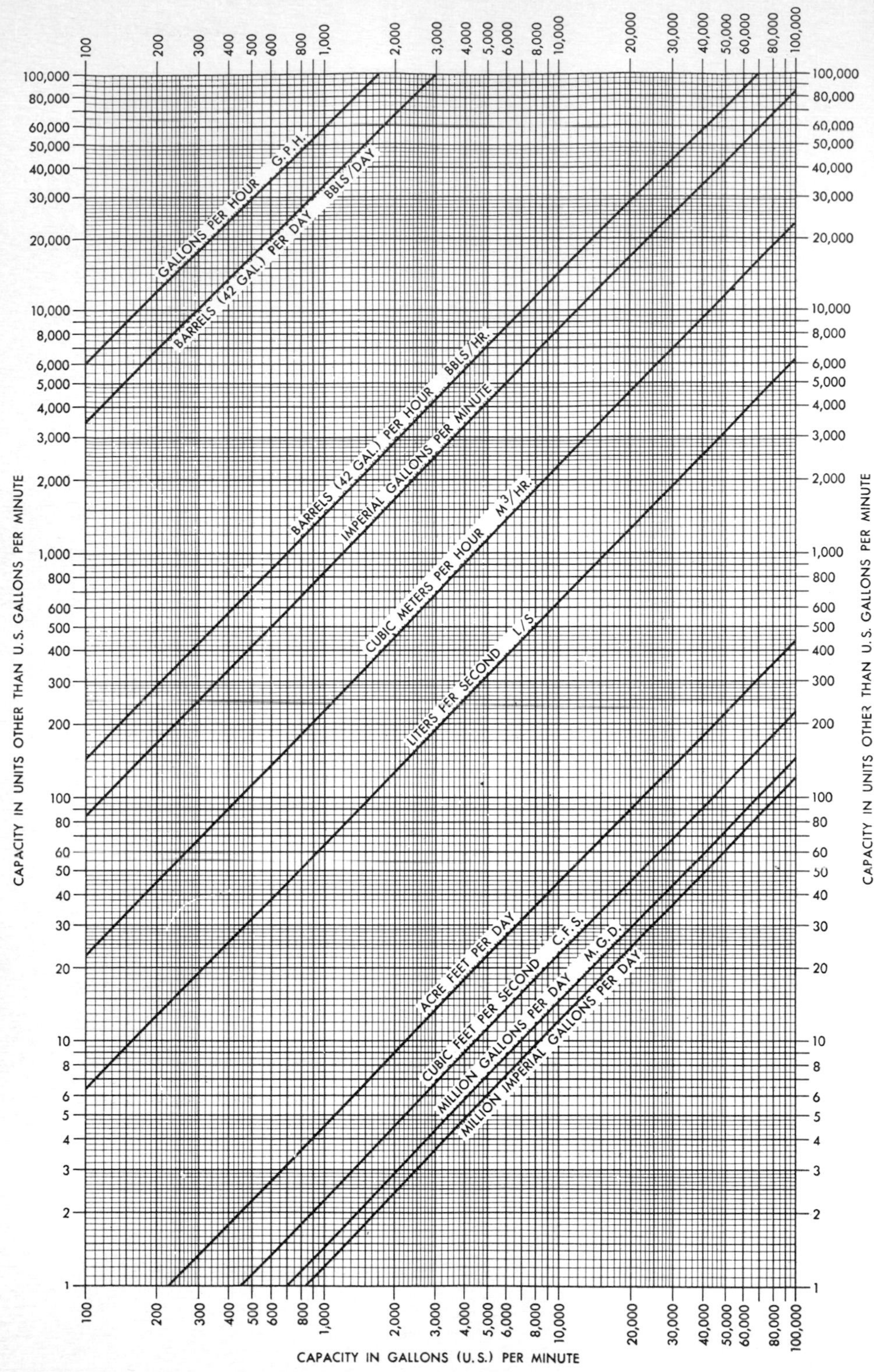

Fig. 2 Capacity conversions. For more accurate values, calculate from the equivalents shown in Table 13.

TABLE 14 Pressure and Head Equivalents (For conversion chart, see Fig. 3)

1 lb/sq in.	=	$\dfrac{2.310}{\text{specific gravity}}$[1] ft of liquid	= 2.310 ft of 62°F water
1 in. mercury (32°F)	=	$\dfrac{1.134}{\text{specific gravity}}$[1] ft of liquid	= 1.134 ft of 62°F water
1 atmosphere[2]	=	$\dfrac{33.95}{\text{specific gravity}}$[1] ft of liquid	= 33.95 ft of 62°F water
1 kilogram/sq cm	=	1 metric atmosphere	
	=	$\dfrac{32.85}{\text{specific gravity}}$[1] ft of liquid	= 32.85 ft of 62°F water
	=	$\dfrac{10.01}{\text{specific gravity}}$[1] m of liquid	= 10.01 m of 15°C water
1 meter			= 3.281 ft

[1] These equivalents are based on a specific gravity of 1 for water at 62°F for English units and a specific gravity of 1 for water at 15°C for metric units. They can be used, with little error, for cold water of any temperature between 32°F and 80°F. For the actual specific gravity of water for temperatures to 705.4°F see Table 9.

[2] Not used in conjunction with pumps.

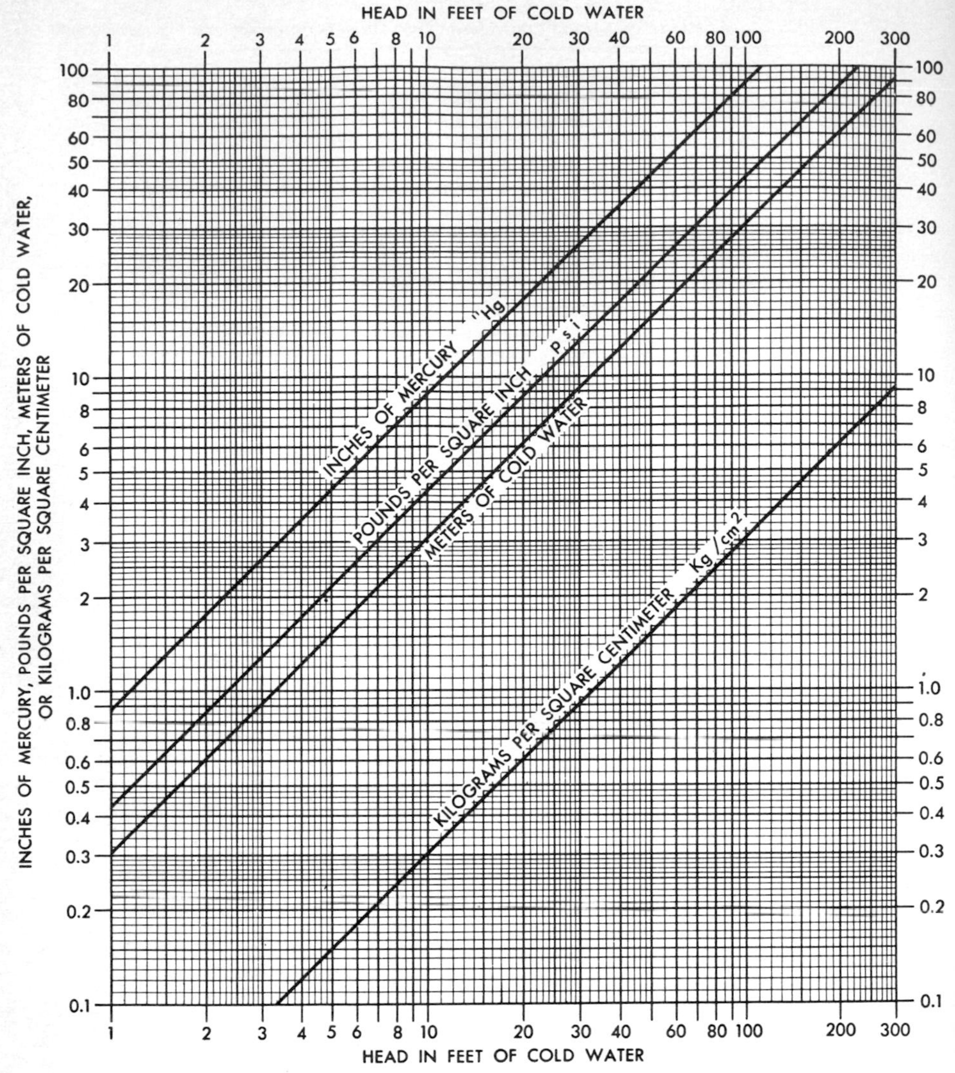

Fig. 3 Pressure and head conversion chart. Values are plotted for 62°F (18.7°C) but can be used for water between 32° and 80°F. For liquids other than cold water, divide the head by the specific gravity (62°F water = 1.0) of the liquid at the pumping temperature to get the head in feet. For more accurate values, calculate heads from the head equivalents in Table 14.

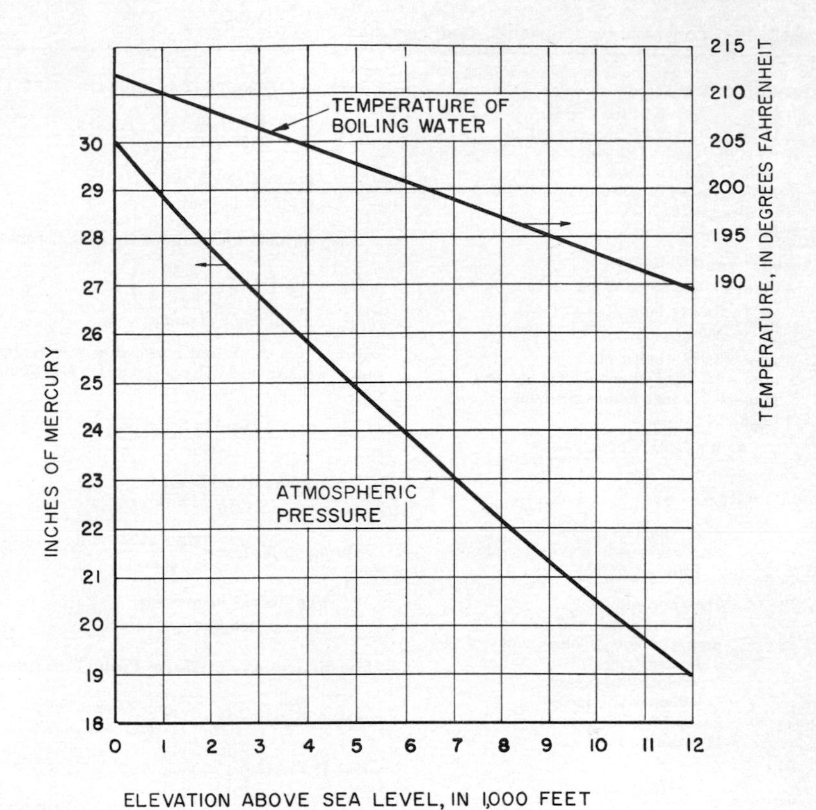

Fig. 4 Atmospheric pressures for altitudes up to 12,000 ft.

TABLE 15 Conversions, Constants, and Formulas

Volume and Weight

1 U. S. gallon = 8.34 lbs × Sp Gr
1 U. S. gallon = 0.84 Imperial gallon
1 cu ft of liquid = 7.48 gal
1 cu ft of liquid = 62.32 lbs × Sp Gr
Specific gravity of sea water = 1.025 to 1.03
1 cu meter = 264.5 gal
1 barrel (oil) = 42 gal

Capacity and Velocity

1 gpm = 0.002228 cu ft per sec

$$\text{gpm} = \frac{\text{lbs per hour}}{500 \times \text{Sp. Gr.}}$$

gpm = 0.069 × boiler Hp
gpm = 0.7 × bbl/hour = 0.0292 bbl/day
gpm = 0.227 metric tons per hour

1 mgd = 694.5 gpm

$$V = \frac{\text{gpm} \times 0.321}{\text{area in sq in}} = \frac{\text{gpm} \times 0.409}{D^2}$$

$$V = \sqrt{2gH}$$

gpm = gallons per minute
Sp Gr = specific gravity based on water at 62°F

Hp = horsepower
bbl = barrel (oil) = 42 gal
mgd = million gallons per day of 24 hours
V = velocity in ft/sec
D = diameter in inches
g = 32.16 ft/sec/sec
H = head in feet

Head

$$\text{Head in feet} = \frac{\text{Head in psi} \times 2.31}{\text{Sp Gr}}$$

1 foot water (cold, fresh) = 0.8859 inches of mercury
1 psi = 0.0703 kilograms per sq centimeter
1 psi = 0.068 atmosphere

$$H = \frac{V^2}{2g}$$

psi = pounds per square inch

Power and Torque

1 horsepower = 550 ft-lb per sec
= 33,000 ft-lb per min
= 2545 btu per hr
= 745.7 watts
= 0.7457 kilowatts

Power and Torque—(*Continued*)

$$\text{bhp} = \frac{\text{gpm} \times \text{Head in feet} \times \text{Sp Gr}}{3960 \times \text{efficiency}}$$

$$\text{bhp} = \frac{\text{gpm} \times \text{Head in psi}}{1714 \times \text{efficiency}}$$

Navy formula to determine Hp rating of motor:

$$\text{Hp} = \frac{\text{ohp}\left(1.05 + \dfrac{1.35}{\text{ohp} + 3}\right)}{\text{efficiency of pump}}$$

where ohp is the output horsepower or water horsepower work done by the pump which is determined by:

$$\text{ohp} = \frac{\text{gpm} \times \text{Head in feet} \times \text{Sp Gr}}{3960}$$

$$\text{or ohp} = \frac{\text{gpm} \times \text{Head in psi}}{1714}$$

$$\text{Torque in lbs feet} = \frac{\text{Hp} \times 5252}{\text{rpm}}$$

bhp = brake horsepower
rpm = revolutions per minute

Miscellaneous Centrifugal Pump Formulas

$$\text{Specific speed} = N_s = \frac{\sqrt{\text{gpm}} \times \text{rpm}}{H^{3/4}}$$

where H = head per stage in feet

$$\text{Diameter of impeller in inches} = d = \frac{1840\ \text{Ku}\sqrt{H}}{\text{rpm}}$$

where Ku is a constant varying with impeller type and design. Use H at shut-off (zero capacity) and Ku is approx. 1.0

$$\text{At constant speed: } \frac{d_1}{d_2} = \frac{\text{gpm}_1}{\text{gpm}_2} = \frac{\sqrt{H_1}}{\sqrt{H_2}} = \frac{\sqrt[3]{\text{bhp}_1}}{\sqrt[3]{\text{bhp}_2}}$$

At constant impeller diameter

$$\frac{\text{rpm}_1}{\text{rpm}_2} = \frac{\text{gpm}_1}{\text{gpm}_2} = \frac{\sqrt{H_1}}{\sqrt{H_2}} = \frac{\sqrt[3]{\text{Bhp}_1}}{\sqrt[3]{\text{Bhp}_2}}$$

TABLE 16 Measurement Conversions

To convert	*Multiply by*	*To obtain*
	A	
acres	4.35×10^{4}	sq. ft.
acres	4.047×10^{3}	sq. meters
acre-feet	4.356×10^{4}	cu. feet
acre-feet	3.259×10^{5}	gallons
atmospheres	2.992×10^{1}	in. of mercury (at 0°C.)
atmospheres	1.0333	kgs./sq. cm.
atmospheres	1.0333×10^{4}	kgs./sq. meter
atmospheres	1.47×10^{1}	pounds/sq. in.
	B	
barrels (u.s., liquid)	3.15×10^{1}	gallons
barrels (oil)	4.2×10^{1}	gallons (oil)
bars	9.869×10^{-1}	atmospheres
btu	7.7816×10^{2}	foot-pounds
btu	3.927×10^{-4}	horsepower-hours
btu	2.52×10^{-1}	kilogram-calories
btu	2.928×10^{-4}	kilowatt-hours
btu/hr.	2.162×10^{-1}	ft. pounds/sec.
btu/hr.	3.929×10^{-4}	horsepower
btu/hr.	2.931×10^{-1}	watts
btu/min.	1.296×10^{1}	ft.-pounds/sec.
btu/min.	1.757×10^{-2}	kilowatts
	C	
centigrade (degrees)	$(°C \times 9/5) + 32$	fahrenheit (degrees)
centigrade (degrees)	$°C + 273.18$	kelvin (degrees)
centigrams	$1. \times 10^{-2}$	grams
centimeters	3.281×10^{-2}	feet
centimeters	3.937×10^{-1}	inches
centimeters	$1. \times 10^{-5}$	kilometers
centimeters	$1. \times 10^{-2}$	meters
centimeters	$1. \times 10^{1}$	millimeters
centimeters	3.937×10^{2}	mils
centimeters of mercury	1.316×10^{-2}	atmospheres
centimeters of mercury	4.461×10^{-1}	ft. of water
centimeters of mercury	1.934×10^{-1}	pounds/sq. in.
centimeters/sec.	1.969	feet/min.
centimeters/sec.	3.281×10^{-2}	feet/sec.
centimeters/sec.	6.0×10^{-1}	meters/min.
centimeters/sec./sec.	3.281×10^{-2}	ft./sec./sec.
cubic centimeters	3.531×10^{-5}	cubic ft.
cubic centimeters	6.102×10^{-2}	cubic in.
cubic centimeters	1.0×10^{-6}	cubic meters
cubic centimeters	2.642×10^{-4}	gallons (u.s. liquid)
cubic centimeters	2.113×10^{-3}	pints (u.s. liquid)
cubic centimeters	1.057×10^{-3}	quarts (u.s. liquid)
cubic feet	2.8320×10^{4}	cu. cms.
cubic feet	1.728×10^{3}	cu. inches
cubic feet	2.832×10^{-2}	cu. meters
cubic feet	7.48052	gallons (u.s. liquid)
cubic feet	5.984×10^{1}	pints (u.s. liquid)
cubic feet	2.992×10^{1}	quarts (u.s. liquid)
cubic feet/min.	4.72×10^{2}	cu. cms./sec.
cubic feet/min.	1.247×10^{-1}	gallons/sec.
cubic feet/min.	4.720×10^{-1}	liters/sec.
cubic feet/min.	6.243×10^{1}	pounds water/min.
cubic feet/sec.	6.46317×10^{-1}	million gals./day
cubic feet/sec.	4.48831×10^{2}	gallons/min.
cubic inches	5.787×10^{-4}	cu. ft.
cubic inches	1.639×10^{-5}	cu. meters

TABLE 16 Measurement Conversions (Continued)

To convert	*Multiply by*	*To obtain*
cubic inches	2.143×10^{-5}	cu. yards
cubic inches	4.329×10^{-3}	gallons
	D	
degrees (angle)	1.745×10^{-2}	radians
degrees (angle)	3.6×10^{3}	seconds
degrees/sec.	2.778×10^{-3}	revolutions/sec.
dynes/sq. cm.	4.015×10^{-4}	in. of water (at 4°C.)
dynes	1.020×10^{-6}	kilograms
dynes	2.248×10^{-6}	pounds
	F	
fathoms	1.8288	meters
fathoms	6.0	feet
feet	3.048×10^{1}	centimeters
feet	3.048×10^{-1}	meters
feet of water	2.95×10^{-2}	atmospheres
feet of water	3.048×10^{-2}	kgs./sq. cm.
feet of water	6.243×10^{1}	pounds/sq. ft.
feet/min	5.080×10^{-1}	cms./sec.
feet/min.	1.667×10^{-2}	feet/sec.
feet/min.	3.048×10^{-1}	meters/min.
feet/min.	1.136×10^{-2}	miles/hr.
feet/sec.	1.829×10^{1}	meters/min.
feet/100 feet	1.0	per cent grade
foot-pounds	1.286×10^{-3}	btu
foot-pounds	1.356×10^{7}	ergs
foot-pounds	3.766×10^{-7}	kilowatt-hrs.
foot-pounds/min.	1.286×10^{-3}	btu/min.
foot-pounds/min.	3.030×10^{-5}	horsepower
foot-pounds/min.	3.241×10^{-4}	kg.-calories/min.
foot-pounds/sec.	4.6263	btu/hr.
foot-pounds/sec.	7.717×10^{-2}	btu/min.
foot-pounds/sec.	1.818×10^{-3}	horsepower
foot-pouhds/sec.	1.356×10^{-3}	kilowatts
furlongs	1.25×10^{-1}	miles (u.s.)
	G	
gallons	3.785×10^{3}	cu. cms.
gallons	1.337×10^{-1}	cu. feet
gallons	2.31×10^{2}	cu. inches
gallons	3.785×10^{-3}	cu. meters
gallons	4.951×10^{-3}	cu. yards
gallons	3.785	liters
gallons (liq. br. imp.)	1.20095	gallons (u.s. liquid)
gallons (u.s.)	8.3267×10^{-1}	gallons (imp.)
gallons of water	8.337	pounds of water
gallons/min.	2.228×10^{-3}	cu. feet/sec.
gallons/min.	6.308×10^{-2}	liters/sec.
gallons/min.	8.0208	cu. feet/hr.
	H	
horsepower	4.244×10^{1}	btu/min.
horsepower	3.3×10^{4}	foot-lbs./min.
horsepower	5.50×10^{2}	foot-lbs./sec.
horsepower (metric)	9.863×10^{-1}	horsepower
horsepower	1.014	horsepower (metric)
horsepower	7.457×10^{-1}	kilowatts
horsepower	7.457×10^{2}	watts
horsepower (boiler)	3.352×10^{4}	btu/hr.
horsepower (boiler)	9.803	kilowatts
horsepower-hours	2.547×10^{3}	btu
horsepower-hours	1.98×10^{6}	foot-lbs.

TABLE 16 Measurement Conversions (Continued)

To convert	*Multiply by*	*To obtain*
horsepower-hours	6.4119×10^{5}	gram-calories
hours	5.952×10^{-3}	weeks
	I	
inches	2.540	centimeters
inches	2.540×10^{-2}	meters
inches	1.578×10^{-5}	miles
inches	2.54×10^{1}	millimeters
inches	1.0×10^{3}	mils
inches	2.778×10^{-2}	yards
inches of mercury	3.342×10^{-2}	atmospheres
inches of mercury	1.133	feet of water
inches of mercury	3.453×10^{-2}	kgs./sq. cm.
inches of mercury	3.453×10^{2}	kgs./sq. meter
inches of mercury	7.073×10^{1}	pounds/sq. ft.
inches of mercury	4.912×10^{-1}	pounds/sq. in.
in. of water (at 4°C.)	7.355×10^{-2}	inches of mercury
in. of water (at 4°C.)	2.54×10^{-3}	kgs./sq. cm.
in. of water (at 4°C.)	5.204	pounds/sq. ft.
in. of water (at 4°C.)	3.613×10^{-2}	pounds/sq. in.
	J	
joules	9.486×10^{-4}	btu
joules/cm.	1.0×10^{7}	dynes
joules/cm.	1.0×10^{2}	joules/meter (newtons)
joules/cm.	2.248×10^{1}	pounds
	K	
kilograms	9.80665×10^{5}	dynes
kilograms	1.0×10^{3}	grams
kilograms	2.2046	pounds
kilograms	9.842×10^{-4}	tons (long)
kilograms	1.102×10^{-3}	tons (short)
kilograms/sq. cm.	9.678×10^{-1}	atmospheres
kilograms/sq. cm.	3.281×10^{1}	feet of water
kilograms/sq. cm.	2.896×10^{1}	inches of mercury
kilograms/sq. cm.	1.422×10^{1}	pounds/sq. in.
kilometers	1.0×10^{5}	centimeters
kilometers	3.281×10^{3}	feet
kilometers	3.937×10^{4}	inches
kilometers	1.0×10^{3}	meters
kilometers	6.214×10^{-1}	miles (statute)
kilometers	5.396×10^{-1}	miles (nautical)
kilometers	1.0×10^{6}	millimeters
kilowatts	5.692×10^{1}	btu/min.
kilowatts	4.426×10^{4}	foot-lbs./min.
kilowatts	7.376×10^{2}	foot-lbs./sec.
kilowatts	1.341	horsepower
kilowatts	1.434×10^{1}	kg.-calories/min.
kilowatts	1.0×10^{3}	watts
kilowatt-hrs.	3.413×10^{3}	btu
kilowatt-hrs.	2.655×10^{6}	foot-lbs.
kilowatt-hrs.	8.5985×10^{3}	gram calories
kilowatt-hrs.	1.341	horsepower-hours
kilowatt-hrs.	3.6×10^{6}	joules
kilowatt-hrs.	8.605×10^{2}	kg.-calories
kilowatt-hrs.	8.5985×10^{3}	kg.-meters
kilowatt-hrs.	2.275×10^{1}	pounds of water raised from 62° to 212°F.

TABLE 16 Measurement Conversions (Continued)

To convert	*Multiply by*	*To obtain*
	L	
links (engineers)	1.2×10^1	inches
links (surveyors)	7.92	inches
liters	1.0×10^3	cu. cm.
liters	6.102×10^1	cu. inches
liters	1.0×10^{-3}	cu. meters
liters	2.642×10^{-1}	gallons (u.s. liquid)
liters	2.113	pints (u.s. liquid)
liters	1.057	quarts (u.s. liquid)
	M	
meters	1.0×10^2	centimeters
meters	3.281	feet
meters	3.937×10^1	inches
meters	1.0×10^{-3}	kilometers
meters	5.396×10^{-4}	miles (nautical)
meters	6.214×10^{-4}	miles (statute)
meters	1.0×10^3	millimeters
meters/min.	1.667	cms./sec.
meters/min.	3.281	feet/min.
meters/min.	5.468×10^{-2}	feet/sec.
meters/min.	6.0×10^{-2}	kms./hr.
meters/min.	3.238×10^{-2}	knots
meters/min.	3.728×10^{-2}	miles/hr.
meters/sec.	1.968×10^2	feet/min.
meters/sec.	3.281	feet/sec.
meters/sec.	3.6	kilometers/hr.
meters/sec.	6.0×10^{-2}	kilometers/min.
meters/sec.	2.237	miles/hr.
meters/sec.	3.728×10^{-2}	miles/min.
miles (nautical)	6.076×10^3	feet
miles (statute)	5.280×10^3	feet
miles/hr.	8.8×10^1	ft./min.
millimeters	1.0×10^{-1}	centimeters
millimeters	3.281×10^{-3}	feet
millimeters	3.937×10^{-2}	inches
millimeters	1.0×10^{-1}	meters
minutes (time)	9.9206×10^{-5}	weeks
	O	
ounces	2.8349×10^1	grams
ounces	6.25×10^{-2}	pounds
ounces (fluid)	1.805	cu. inches
ounces (fluid)	2.957×10^{-2}	liters
	P	
parts/million	5.84×10^{-2}	grains/u.s. gal.
parts/million	7.016×10^{-2}	grains/imp. gal.
parts/million	8.345	pounds/million gal.
pints (liquid)	4.732×10^2	cubic cms.
pints (liquid)	1.671×10^{-2}	cubic ft.
pints (liquid)	2.887×10^1	cubic inches
pints (liquid)	4.732×10^{-4}	cubic meters
pints (liquid)	1.25×10^{-1}	gallons
pints (liquid)	4.732×10^{-1}	liters
pints (liquid)	5.0×10^{-1}	quarts (liquid)
pounds	2.56×10^2	drams
pounds	4.448×10^5	dynes
pounds	7.0×10^1	grains
pounds	4.5359×10^2	grams
pounds	4.536×10^{-1}	kilograms
pounds	1.6×10^1	ounces

TABLE 16 Measurement Conversions (Continued)

To convert	*Multiply by*	*To obtain*
pounds	3.217×10^{1}	poundals
pounds	1.21528	pounds (troy)
pounds of water	1.602×10^{-2}	cu. ft.
pounds of water	2.768×10^{1}	cu. inches
pounds of water	1.198×10^{-1}	gallons
pounds of water/min.	2.670×10^{-4}	cu. ft./sec.
pound-feet	1.356×10^{7}	cm.-dynes
pound-feet	1.3825×10^{4}	cm.-grams
pound-feet	1.383×10^{-1}	meter-kgs.
pounds/cu. ft.	1.602×10^{-2}	grams/cu. cm.
pounds/cu. ft.	5.787×10^{-4}	pounds/cu. inches
pounds/sq. in.	6.804×10^{-2}	atmospheres
pounds/sq. in.	2.307	feet of water
pounds/sq. in.	2.036	inches of mercury
pounds/sq. in.	7.031×10^{2}	kgs./sq. meter
pounds/sq. in.	1.44×10^{2}	pounds/sq. ft.
	Q	
quarts (dry)	6.72×10^{1}	cu. inches
quarts (liquid)	9.464×10^{2}	cu. cms.
quarts (liquid)	3.342×10^{-2}	cu. ft.
quarts (liquid)	5.775×10^{1}	cu. inches
quarts (liquid)	2.5×10^{-1}	gallons
	R	
revolutions	3.60×10^{2}	degrees
revolutions	4.0	quadrants
rods (surveyors' meas.)	5.5	yards
rods	1.65×10^{1}	feet
rods	1.98×10^{2}	inches
rods	3.125×10^{-3}	miles
	S	
slugs	3.217×10^{1}	pounds
square centimeters	1.076×10^{-3}	sq. feet
square centimeters	1.550×10^{-1}	sq. inches
square centimeters	1.0×10^{-4}	sq. meters
square centimeters	1.0×10^{2}	sq. millimeters
square feet	2.296×10^{-5}	acres
square feet	9.29×10^{2}	sq. cms.
square feet	1.44×10^{2}	sq. inches
square feet	9.29×10^{-2}	sq. meters
square feet	3.587×10^{-8}	sq. miles
square inches	6.944×10^{-3}	sq. ft.
square inches	6.452×10^{2}	sq. millimeters
square miles	6.40×10^{2}	acres
square miles	2.788×10^{7}	sq. ft.
square yards	2.066×10^{-4}	acres
square yards	8.361×10^{3}	sq. cms.
square yards	9.0	sq. ft.
square yards	1.296×10^{3}	sq. inches
	T	
temperature (°C.) +273	1.0	absolute temperature (°K.)
temperature (°C.) +17.78	1.8	temperature (°F.)
temperature (°F.) +460	1.0	absolute temperature (°R.)
temperature (°F.) −32	5/9	temperature (°C.)
tons (long)	2.24×10^{3}	pounds

TABLE 16 Measurement Conversions (Continued)

To convert	*Multiply by*	*To obtain*
tons (long)	1.12	tons (short)
tons (metric)	2.205×10^{3}	pounds
tons (short)	2.0×10^{3}	pounds
	W	
watts	3.4129	btu/hr.
watts	5.688×10^{-2}	btu/min.
watts	4.427×10^{1}	ft.-lbs./min.
watts	7.378×10^{-1}	ft.-lbs./sec
watts	1.341×10^{-3}	horsepower
watts	1.36×10^{-3}	horsepower (metric)
watts	1.0×10^{-3}	kilowatts
watt-hours	3.413	btu
watt-hours	2.656×10^{3}	foot-lbs.
watt-hours	1.341×10^{-3}	horsepower-hours
watt (international)	1.000165	watt (absolute)
weeks	1.68×10^{2}	hours
weeks	1.008×10^{4}	minutes
weeks	6.048×10^{5}	seconds

Index